Wolf Prize in Agriculture

Wolf Prize in Agriculture

Edited by

Ilan Chet

The Hebrew University of Jerusalem, Israel

World Scientific

NEW JERSEY • LONDON • SINGAPORE • BEIJING • SHANGHAI • HONG KONG • TAIPEI • CHENNAI

Published by

World Scientific Publishing Co. Pte. Ltd.

5 Toh Tuck Link, Singapore 596224

USA office: 27 Warren Street, Suite 401-402, Hackensack, NJ 07601

UK office: 57 Shelton Street, Covent Garden, London WC2H 9HE

British Library Cataloguing-in-Publication Data
A catalogue record for this book is available from the British Library.

The editors and publisher would like to thank the following organisations and publishers of the various journals for their assistance and permission to reproduce the selected reprints found in this volume:

ACU Today magazine
AAAS American Association for the Advancement of Science
American Society of Plant Biologists
Atlanta Magazine
Avian Pathology
BioScientifica Ltd
Blackwell Ltd
Elsevier Ltd
FASS
Journal of Animal Science
Nature Publishing Group
PNAS
Society for the Study of Reproduction, Inc.
Springer
The Endocrine Society
The Entomological Society of America

WOLF PRIZE IN AGRICULTURE

ISBN-13 978-981-283-584-0
ISBN-10 981-283-584-9

Disclaimer
Most of the material from this volume was extracted from old archives of papers containing almost illegible text. Hence, they have been reproduced to the best of the Editor's and publisher's ability.

Printed in Singapore by B & JO Enterprise

CONTENTS

PREFACE

The Wolf Foundation began its activities in 1976, with an initial endowment fund donated by the Wolf family. The Foundation's founders and major donors were Dr. Ricardo Subirana y Lobo Wolf and his wife Francisca. The annual income from investments is used to award prizes and scholarships, as well as fund the operating expenses of the Foundation.

One of the aims of the Wolf Foundation, as stated by the law is "to award prizes to outstanding scientists and artists — irrespective of nationality, race, color, religion, sex or political views — for achievements in the interest of mankind and friendly relations among peoples" in the fields of agriculture, chemistry, mathematics, medicine and the arts. The official prize award ceremony takes place at the Knesset building (Israel's Parliament) in Jerusalem and the President of the State of Israel hands the awards to the winners. Through the years, the Wolf Prize has become one of the most prestigious prizes. In agriculture, it is probably the most highly esteemed prize in the world.

The list of laureates in agricultural sciences from 1978 to 2008 contains 41 scientists in fields such as genetics, bio-control, ecology, plant sciences, animal breeding, soil chemistry and physics, and plant biochemistry. Because of the depletion in natural resources, agricultural development and human well-being became one of the major concerns in the world. The Wolf Prize winners in the field of agriculture are awarded for their remarkable innovative and pioneering discoveries, development of new technologies and/or extraordinary contribution to agricultural research, ecological conservation and food produce.

Now, at the end of 2008, there is an extremely high interest in agriculture, as the concern of the world's population is not only food supply but also alternative sources of energy and neutraceuticals.

In this volume, we documented the Wolf Prize winners' bibliography, curriculum vitae (CV), autobiographical accounts and/or reports on their work and achievements by others, important papers, lectures and other relevant information. We did our best to include the data of prize winners who are no longer with us.

The diverse fields of interest of the laureates provide a unique overview of the work done by exceptional scientists in different institutes around the world. I wish to share with the readers some of the deliberations and excitement involved in discovering novel ideas and approaches, and reassessing the old ones. In rare disclosures, updated findings alongside their views and speculations on various subjects in agriculture are offered in this volume. I sincerely hope that the readers, be they researchers, students, biotechnologists, or extension specialists, find the following pages stimulating and exciting as I did.

I would like to thank my talented secretary Mrs. Nili Ben-Yehezkel for putting this volume together with all of her enthusiasm and devotion, and to the prize winners and colleagues who helped in collecting the material for the volume.

I would also like to thank my publisher, World Scientific Publishing Co., especially Ms. Joy Quek, for her cooperation and assistance.

Ilan Chet
Rehovot, Israel

John C. Walker

University of Wisconsin
Madison, Wisconsin, USA

1893–1994

1978 — for his research in plant pathology, developing of disease-resistant varieties of major food plants.

ABOUT DR. JOHN CHARLES WALKER

Plant Pathologist
University of Wisconsin

This is a story of a local industry and a Racine man. The industry is one of the prime industries of Racine and Kenosha counties; the man has contributed more basic knowledge to that industry here and throughout the world than any other living person. This is the story of a man who attended Beebe grade school, Racine High School, and graduated from the University of Wisconsin. Today he is world-renowned in his field, a member of the National Academy of Science (membership in this society is the highest recognition possible in the United States for scientific research) and an honored member in many international scientific societies.

John Charles Walker, because of his inherent modesty, is hardly known in his home town. He, who has contributed so very much toward the economic and physical welfare of ourselves and the world, has never been formally recognized by his fellow townspeople.

In the early 1800's a few rugged individualists from Northern Europe settled near Racine. These individuals founded the type of specialized agriculture for which Racine and the surrounding area has long been noted, namely the production of truck crops — cabbage — onions — potatoes — etc.

About the time of the Civil War, Racine had become the major source of supply for cabbage in the Midwest. In the middle 1800's, Ben Bones operated a farm just east of Lathrop Road, and south of Chicory Road. He found, in this particular year, that he had a surplus of cabbage, more than could be marketed through the local trade channels. Learning that the beer industry of St. Louis had started using a refrigerated freight car to ship beer, Mr. Bones conceived the idea of loading one of these cars with cabbage and shipping this cabbage under refrigeration to the German settlement of St. Louis. It is believed that this was the first time that perishable produce was shipped under refrigeration. Two carloads were shipped this first year which proved so profitable that Mr. Bones increased his acreage the following year. Thus Wisconsin's Cabbage Industry was born.

A disease (cabbage "yellows") was noted in the area during the late 1800's. It was serious enough that the growers called a special meeting at the Berryville School in 1890. Professor H. L. Russell, later Dean of the College of Agriculture, was present at this meeting. The disease problem was discussed and reviewed but nothing constructive was accomplished. By 1910, the disease was epidemic, resulting in a complete crop failure. The Durand farm (now Case South Works) operated by the Gunthers was replanted twice, and each time the crop died as a result of this new disease. Those who were fortunate enough to have a partial crop found yields reduced to one or two tons per acre. Today, a yield of 30 tons per acre is not exceptional.

Dr. L. R. Jones, a native of Wisconsin and a graduate of Ripon College, had been teaching botany at Vermont. In 1910 the University of Wisconsin enticed L. R. Jones to join its staff, and at the second cabbage meeting held at the Berryville school, 1910, Dr. Jones was present. This time definite plans were made to learn more about the disease, to get it under control if possible. All growers offered full cooperation with Dr. Jones and the University. This disease had to be controlled, or farmers would have to change to different crops, and the kraut packers would have to move into an area in which the disease was not a factor.

A "cabbage sick" field of 1913. This field is located just north of Chickory Road and east of the North Shore tracks.

The few cabbage plants that could survive the disease were interesting to these men. Healthy plants from the most severely infected field were to be selected for seed production — stored over winter and replanted the following spring. Three plants were selected from a field of over 25,000, they were carefully stored and replanted in the same infected field of Matt Broesch. One night Mr. Broesch's cow got loose and partially destroyed two of the plants! The cabbage industry was saved, however, by that single plant that was spared.

John Charles Walker was born on July 6, 1893 in a house on Lathrop Road that still stands just north of 21st Street. He entered Beebe grade school, attended Racine High School where he graduated with the highest scholastic record of any boy in his class. In 1910 he entered the University of Wisconsin. In the spring of 1912 he went to Dr. L. R. Jones and stated that he thought he would like to major in plant diseases, the new science recently named Plant Pathology. Dr. Jones' answer to this was, "That is fine, Charley. You have quite a problem in your own backyard, a disease in cabbage that we know very little about. You can start working on that this summer during your vacation" — and for more than 45 years J. C. Walker has been working with cabbage.

RESEARCH CONTRIBUTIONS TO WISCONSIN

Wisconsin is a giant among states in the production of vegetable crops for fresh market and processing. It leads all states in production for processing of peas, sweet corn, beets, and carrots. It is second in production of cabbage for sauerkraut, and cucumbers for pickles. It is a leader in production of onions (both sets and bulbs), potatoes, beans, and lima beans.

ACCOMPLISHMENTS

Cabbage

Seed of the first yellows-resistant variety of cabbage, Wisconsin Hollander No. 8, was being increased locally as rapidly as possible. Slightly less than 100 pounds were produced in 1916, hardly enough to meet the needs of the local growers. Seed production in 1917 exceeded 800 pounds. (In 1957, forty years later, total production of yellows resistant cabbage seed in this country exceeded 135,000 pounds!) At this critical time another disease, Blackleg hit Wisconsin. This was a disease that was transmitted from one crop to another through seed, unlike cabbage yellows which was soilborne. The success of the new yellows-resistant program was doomed unless blackleg could be controlled. J. C. Walker was handed this assignment and in short order he had the disease under control. He showed that if cabbage seed was produced in the Pacific Northwest instead of in Wisconsin it was disease-free. This not only saved the day for the cabbage industry in Wisconsin but was a prime factor in the shift of the cabbage seed industry from Northern Europe to the United States.

Walker next turned his attention to the yellows disease which had been brought under control by the development of the Wisconsin Hollander No. 8 variety. This variety was only partially resistant and was not suitable for all uses. Walker discovered a more desirable type of resistance and has given to the growers some 12 or 15 different varieties of resistant cabbage.

The Clubroot disease of cabbage is as ancient as the crop and control of it has defied man through the centuries. Walker turned his plant breeding skill to this problem and has developed resistance to it. He took resistance from curly kale and transferred it to cabbage. Control of this most serious disease is now certain.

1914. Same field, the first yellows resistant strain of cabbage. In the background, right to left, the student J. C. Walker, Dr. L. R. Jones and W. J. Honsche.

In recent years kraut manufacturers in Wisconsin have been deeply concerned about a disease called Tip-burn. This disease occurs in lettuce also and results in shreds of dead tissue in both crops. Walker is now working on this problem and is on the track of a way to control it.

Onion

Walker did some of the early work showing the effect of artificial heat on the drying of onions in storage and the relationship of their preservation to disease control. This work was done about 40 years ago. With the recent change in onion production, mechanical harvesting and storage in bulk, with large amounts of air being forced through these bulk piles of onions, every grower today is referring to Dr. Walker's paper and is using heat to cure and control the diseases of onion in storage. He made the original technical descriptions of two of the neckrot diseases.

Smut is Wisconsin's worst disease of onions. Our basic knowledge on the effect of soil temperature on the occurrence of this disease was made by Walker. He took the idea of dripping formaldehyde solution on the onion seed as they are planted and worked out a practical application that saved the onion industry in Wisconsin.

Onion smudge is a disease of white but not of yellow or red onions. Walker showed that the pigments that form the yellow color in onions acted as a chemical substance forming resistance to smudge in onions. White onions, without this chemical substance, are susceptible to smudge. This is the most classical piece of research on the nature of resistance due to chemicals that has ever been worked out.

Walker was instrumental in starting a national program for the development of onions resistant to pink root, a disease of increasing importance in Wisconsin.

Peas

The story of peas differs little from that of cabbage. Pea wilt was ravaging the fields of southern Wisconsin. Walker tested some 250 varieties of peas and found some to be naturally resistant. With Professor E. J. Deiwiche, he developed resistant varieties suitable to Wisconsin's needs. Today, Wisconsin's 150,000 acres of peas are planted to these resistant varieties.

Beets

As Wisconsin's beet canning industry developed, it was faced with a serious disease known as black spot or heart rot. This disease was shown to be due not to a parasite but to a lack of boron in the soil. Walker showed that the disease could be controlled by adding boron in minute amounts to the soil or by spraying it on the leaves.

A few years ago Wisconsin's great pickle industry was threatened with extinction by epidemic occurrences of the scab disease and by severe losses from the mosaic disease. Walker crossed a scab-resistant slicing variety with a mosaic-resistant pickle variety and developed varieties of the pickle-type resistant to both diseases. These varieties are saving Wisconsin's agriculture over one million dollars each year.

This picture shows the difference between the death and survival af an annual 2½ million dollar industry which is directly attributable to the work of Dr. John Charles Walker.

Bean

The growers and canners of beans have also been on the receiving end of J. C. Walker's service. In the 1920's common bean mosaic was causing great losses to the canning industry. In 1930 Walker and one of his students developed 2 varieties of bean resistant to mosaic as well as making available to bean breeders everywhere, information on the inheritance of the resistance.

ACADEMIC RECOGNITION

Few scientists have received greater acclaim for their contributions than J. C. Walker. He is a member of The Wisconsin Academy of Science, The American Association for the Advancement of Science, The American Institute of Biological Sciences, The American Phytopathological Society (president in 1943), the Genetics Society, The Botanical Society of America, and The American Society of Naturalists.

He has earned the coveted honor of election to the National Academy of Sciences consisting of one hundred of the leading scientists of this country. Dr. Walker was also elected to the College of Electors Hall of Fame of New York University and was named "Man of the Year" by the National Vegetable Growers Association in 1953.

Walker has been invited to present papers at several International Congresses. He was a guest lecturer in 1953 at the Agricultural Institute of Sao Paulo, Brazil. He has published well over 300 technical papers and has guided some 60 students through their doctorate in Plant Pathology.

National recognition services for J. C. Walker have been held by the American Seed Trade Association, The National Kraut Packers Association, The Vegetable Growers of America, and the Manufacturers of Processing Equipment.

DR. J. C. WALKER — THE MAN

Any review of the accomplishments of Dr. J. C. Walker, scientist, would be incomplete without a description of "Charlie" Walker, the warm, wonderful human personality loved by scores. His intimate friends, his colleagues, his many acquaintances within the fields of science and industry echo the praises for a man whose singular feats have won admiration because of his dogged determination, his faith, and his ability to inspire others to similar virtues by his good example. But those who know him best say his devotion to duty and his vision are exceeded many times by his ready wit, his love of good fun and good fellowship. His career is truly an inspiration; his personality is a real pleasure to share.

The income to Wisconsin in 1956 from vegetable crops, the production of which would have been impossible without the work of Dr. J. C. Walker, was over 17 million dollars.

Annual value of crops — 1956

Cabbage	$ 1,540,000.00 in Racine and Kenosha counties
	10,957,000.00 in Wisconsin
Beets	1,302,000.00 in Wisconsin
	2,562,000.00 in Wisconsin
	1,296,000.00 in Wisconsin

Much of this saving has been recurring annually for many years. The people of Wisconsin shall, indeed, be forever indebted to John Charles Walker.

List of Publications

1. Walker, J. C. Control of potato diseases in Wisconsin. Wis. Country Magazine 7: 34. 1913.
2. Walker, J. C. Studies upon the anthracnose of onion. Phytopathology 7: 59. 1917.
3. Walker, J. C. Control of neck rot and anthracnose of onion sets. Phytopathology 8: 70. 1918.
4. Walker, J. C. Notes on the resistance of onions to anthracnose. Phytopathology 8: 70-71. 1918.
5. Walker, J. C. Onion diseases and their control. U.S.D.A. Farm. Bul. 1060, 23 pp. 1919.
6. Vaughan, R. E., and J. C. Walker. Onion smut. Wis. Agr. Expt. Sta. Circ. 114, 4 pp. 1919.
7. Walker, J. C. Occurrence and control of black leg of cabbage. Phytopathology 10: 64. 1920.
8. Walker, J. C. Experiments upon formaldehyde-drip control of onion smut. Phytopathology 10: 323-327. 1920.
9. Walker, J. C., and W. B. Tisdale. Observations on seed transmission of the cabbage black rot organism. Phytopathology 10: 175-177. 1920.
10. Jones, L. R., J. C. Walker, and W. B. Tisdale. Fifth progress report on Fusarium-resistant cabbage. Phytopathology 10: 64. 1920.
11. Jones, L. R., J. C. Walker, and W. B. Tisdale. Fusarium resistant cabbage. Wis. Agr. Expt. Sta. Res. Bul. 148, 314 pp. 1920.
12. Vaughan, R. E., and J. C. Walker. Onion smut. Wis. Horticulture 10: 133-145. 1920.
13. Walker, J. C. A Macrosporium rot of onion. Phytopathology 11: 53. 1921.
14. Walker, J. C. The occurrence of dodder on onions. Phytopathology 11: 53. 1921.
15. Walker, J. C. Onion smudge. Jour. Agr. Res. 20: 685-722. 1921.
16. Walker, J. C. Rust of onion followed by a secondary parasite. Phytopathology 11: 87-90. 1921.

17. Walker, J. C., and L. R. Jones. The relation of soil temperature and other factors to onion smut and infection. Phytopathology 11: 52-53. 1921.
18. Walker, J. C., and L. R. Jones. Relation of soil temperature and other factors to onion smut infection. Jour. Agr. Res. 22: 235-262. 1921.
19. Walker, J. C. Onion diseases and their control. U.S.D.A. Farm. Bul. 1060 (1st rev.), 28 pp. 1922.
20. Walker, J. C. Seed treatment and rainfall in relation to the control of cabbage black leg. U.S.D.A. Bul. 1029, 27 pp. 1922.
21. Walker, J. C., and W. B. Tisdale. Further notes on the occurrence of cabbage black leg. Eiiytppathology 12: 143. 1922.
22. Walker, J. C. The hot water treatment of cabbage seed, Phytopathology 13: 251-253. 1923.
23. Walker, J. C. Disease resistance to onion smudge. Jour. Agr. Res. 214: 1019-1040. 1923.
24. Jones, L. R., and J. C. Walker. Yellows-resistant cabbage varieties. Seed World (Feb. 2, 1923): 20-21. 1923.
25. Jones, L. R., J. C. Walker, and E. C. Tims. Work upon Fusarium-resistant cabbage in 1922. Phytopathology 13: 57. 1923.
26. Harter, L. L., and L. R. Jones. Cabbage diseases. (Revised by J. C. Walker.) U.S.D.A. Farm. Bul. 1351, 29 pp. 1923.
27. Walker, J. C. Occurrence of white rot of Alliuin (Sclerotium opivorum Berk.) in Europe and America. Phytopathology 114: 26. 19214.
28. Tims, E. C., and J. C. Walker. A Fusarium bulb rot of onion. Phytopathology 114: 26-27. 1924.
29. Walker, J. C., and F. L. Wellman. Temperature relations of Urocystis cepulae (Frost). Phytopathology 114: 26. 1924.
30. Walker, J. C. White rot of Allium in Europe and America. Phytopathology 114: 315-322. 1924.
31. Walker, J. C. Cabbage—seed treatment. U.S.D.A. Circ. 311, 4 pp. 1924.
32. Walker, J. C. Resistant varieties and disease-free seed save cabbage industry. Wis. Horticulture 114: 117-118; 122. 1924.
33. Walker, J. C. Observations on the cultivation and diseases of cabbage arid onions in Europe, 1922. Plant Disease Reporter, Suppl. 32, 314 pp. 1924.
34. Walker, J. C. On the nature of disease resistance in plants. Trans. Wis. Acad. Science, Arts, and Letters 21: 225-2147, 19214.
35. Walker, J. C., and C. C. Lindegren. Further studies on the relation of onion scale pigmentation to disease resistance. Jour. Agr. Research 29: 507-514. 1924.
36. Walker, J. C., and E. C. Tims. A Fusariun bulb rot of onion and the relation of environment to its development. Jour. Agr. Research 28: 683-694. 1924.
37. Walker, J. C. An investigation into the nature of disease resistance in plants. Science 59: 448. 1924.

38. Walker, J. C. Onion diseases and their control. U.S.D.A. Farm. Bul. 1066 (2nd rev.), 18 pp. 1925.
39. Walker, J. C. Two undescribed species of Botrytis associated with the neck rot diseases of onion bulbs. Phytopathology 15: 708-713. 1925.
40. Walker, J. C. Control of mycelial neck rot of onion by artificial curing. Jour. Agr. Research 30: 365-373. 1925.
41. Walker, J. C. Studies on disease resistance in the onion. Proc. Nat. Acad. Sci. 11: 183-189. 1925.
42. Walker, J. C., C. C. Lindegren, and Freda N. Bachmann. Further studies on the toxicity of juice extracted from succulent onion scales. Jour. Agr. Research 30: 175-187. 1925.
43. Jones, L. R., J. C. Walker, and J. Monteith, Jr. Fusarium resistant cabbage: progress with second early varieties. Jour. Agr. Research 30: 1027-1034. 1925.
44. Walker, J. C. Studies upon the inheritance of Fusarium-resistance in cabbage. Phytopathology 16: 87. 1926.
45. Walker, J. C. The influence of soil temperature and soil moisture upon white rot of Allium. Phytopathology 16: 697-710. 1926.
46. Walker, J. C. Botrytis neck rots of onions. Jour. Agr. Research 33: 893-928. 1926.
47. Walker, J. C. Three new yellows-resistant mid-season cabbage varieties. Seed World (Dec. 31, 1926.). 126.
48. Walker, J. C., and F. L. Wellman. Relation of temperature to spore germination and growth of Urocystis cepulae. Jour. Agr. Research 32: 133-146. 1926.
49. Walker, J. C., John Monteith, Jr., and F. L. Tellman. A new Fusarium resistant cabbage. Phytopathology 16: 72-73. 1926.
50. Walker, J. C. Newer aspects of breeding yellows-resistant cabbage. Nkt. Grow. Jour. 38: 412-414. Also in The Canner 62 (10-2): 158. 1926.
51. Walker, J. C. Diseases of cabbage and related plants. U.S.D.A. Farm. Bul. 1439, 30 pp. 1927.
52. Walker, J. C. Onion curing to prevent decay while in storage. U.S.D.A. Yearbook 1926: 558-559. 1927.
53. Walker, J. C., John Monteith, Jr., and F. L. Wellman. Development of three midseason varieties of cabbage resistant to yellows (Fusarium conglutinans (Woil.). Jour. Agr. Research 35: 785-809. 1927.
54. Walker, J. C. Yellows-resistant midseason cabbage varieties now available. Wis. Horticulture 17: 113-114. 1927.
55. Walker, J. C. An onion disease under successful control. Wis. Horticulture 17: 155-156. 1927.
56. Walker, J. C. Spraying and dusting calendar for North Central States (Minnesota, Wisconsin, and Michigan). Amer. Produce Grower, April, 1927.

57. Walker, J. C. New yellows-resistant cabbage varieties. Nkt. Grow. Jour. 40: 273-281. 1927.
58. Walker, J. C., and F. L. Weilman. A survey of the resistance of subspecies of Brassica oleracea to yellows (Fusarium conglutinans). Jour. Agr. Research 37: 233-241. 1928.
59. Walker, J. C. Onion diseases and their control. U.S.D.A. Farm. Bul. 1060 (3rd rev.), 18 pp. 1928.
60. Walker, J. C., and F. L. Weilman. A Fusarium resistant cabbage of Jersey Wakefield type. Phytopathology 18: 142. 1928.
61. Walker, J. C. Yellows resistant cabbage developed by plant breeding. U.S.D.A. Yearbk 1928: 635-636. 1929.
62. Walker, J. C., K. P. Link, and H. R. Angell. Chemical aspects of disease resistance in the onion. Science 69: 555. 1929.
63. Walker, J. C., K. P. Link, and H. R. Angell. Chemical aspects of disease resistance in the onion. Proc. Nat. Acad. Sci. 15: 845-850. 1929.
64. Walker, J. C. Some remarks on the physiological aspects of parasitism. Proc. Int. Cong. Plant Sciences 2: 1263-1270. 1929.
65. Link, K. P., H. R. Angell, and J. C. Walker. The isolation of protocatechuic acid from pigmented onion scales and its significance in relation to disease resistance in onions. Jour. Biol. Chem. 81: 369-375. 1929.
66. Link, K. P., A. D. Dickson, and J. C. Walker. Further observations on the occurrence of protocatechuic acid in pigmented onion scales and its relation to disease resistance in the onion. Jour. Biol. Chem. 84: 719-725. 1929.
67. Walker, J. C. The nature of yellows resistance in cabbage. Phytopathology 20: 114-115. 1930.
68. Walker, J. C. Inheritance of Fusarium resistance in cabbage. Jour. Agr. Research 140: 721-745. 1930.
69. Walker, J. C., and Rose Smith. Effect of environmental factors upon the resistance of cabbage to yellows. Jour. Agr. Research 141: 1-15. 1930.
70. Smith, Rose, and J. C. Walker. A cytological study of cabbage plants in strains susceptible or resistant to yellows. Jour. Agr. Research 141: 17-35. 1930.
71. Angell, H. R., J. C. Walker, and K. P. Link. The relation of protocatechuic acid to disease resistance in the onion. Phytopathology 20: 1431-1438. 1930.
72. Walker, J. C. Progress in development of yellows-resistant cabbage. The Canner 70 (10-2): 88. 1930.
73. Walker, J. C. Onion diseases and their control. U.S.D.A. Farm. Bul. 1060 (14th rev.), 214 pp. 1931.
74. Walker, J. C. Resistance to Fusarium wilt in garden, canning, and field peas. Wis. Agr. Expt. Sta. Res. Bul. 107, 15 pp. 1931.
75. Walker, J. C. Propagation of yellows-resistant cabbage seed. The Canner 72 (10-2): 71. 1931.

76. Walker, J. C. Pea disease situation in Wisconsin. The Canner 72 (10-2): 123-124. 1931.
77. Walker, J. C. Cabbage variety Jersey Queen adds early strain resistant to yellows. U.S.D.A. Yearbook 1932: 119-121. 1932.
78. Walker, J. C. Present status of cabbage improvement with regard to kraut type and disease control. The Canner 74 (11-2): 51. 1932.
79. Walker, J. C. Comparative studies of peas resistant and susceptible to Fusarium wilt. Phytopathology 23: 36. 1933.
80. Walker, J. C. Relation of soil fertility to incidence of Aphanomyces root rot of pea. Phytopathology 23: 36. 1933.
81. Walker, J. C. Comparison of yellows resistant kraut cabbage varieties in 1932. The Canner 76 (11-2): 41-42. 1933.
82. Walker, J. C., and W. C. Snyder. Pea wilt and root rots. Wis. Agr. Expt. Sta. Bul. 424, 16 pp. 1933.
83. Link, K. P., and J. C. Walker. The isolation of catechol from pigmented onion scales and its significance in relation to disease resistance in onions. Jour. Biol. Chem. 100: 379-383. 1933.
84. Walker, J. C. The Jersey Queen cabbage. Ann. Rpt. Md. Agr. Soc. 1932 (Rpt. Md. Veg. Growers' Assoc., p. 24-26). 1933.
85. Walker, J. C. Yellows-resistant lines of Jersey Wakefield cabbage. Jour. Agr. Research 46: 639-648. 1933.
86. Blank, L. N., and J. C. Walker. Inheritance of Fusarium resistance in Brussels sprouts and kohlrabi. Jour. Agr. Research 46: 1015-1022. 1933.
87. Pierce, W. H., and J. C. Walker. The development of mosaic resistant Refugee beans. Canner 77 (26): 7-9. 1936.
88. Walker, J. C. Production of cabbage seed free from Phoma lingam and Bacterium campestre. Phytopathology 24: 158-160. 1934.
89. Walker, J. C. Diseases of cabbage and related plants. U.S.D.A. Farm. Bul. 1439 (1st rev.), 28 pp. 1934.
90. Walker, J. C., and Albert Murphy. Onion-bulb decay caused by Aspergillus alliaceus. Phytopathology 24: 289-291. 1934.
91. Boswell, V. H., W. C. Edmundson, O. H. Pearson, J. E. Knott, C. E. Myers, R. A. McGinty, W. H. Friend, H. H. Zimmerley, and J. C. Walker. Descriptions of types of principal American varieties of cabbage. U.S.D.A. Misc. Publ. 169, 22 pp. 1934.
92. Larson, R. H., and J. C. Walker. Soil treatment in relation to clubroot of cabbage. Jour. Agr. Research 48: 749-759. 1934.
93. Walker, J. C., and L. N. Blank. Fusarium-resistant Danish Ballhead cabbage. Jour. Agr. Research 49: 983-989. 1934.
94. Snyder, W. C., and J. C. Walker. Fusarium near-wilt of pea. Cent. f. Bakt. II. 91: 355-378. 1935.

95. Walker, J. C., J. F. Adams, and W. H. Pierce. Results of planting tests with Wisconsin Refugee and Idaho Refugee beans. The Canner 80 (12): 7-8: 22. 1935.
96. Walker, J. C. Diseases of vegetable crops. 65 pp. Ann Arbor. 1935.
97. Walker, J. C., and K. P. Link. Toxicity of phenolic compounds to certain onion bulb parasites. Bot. Gaz. 96: 468-484. 1935.
98. Anderson, N. E., and J. C. Walker. Histological studies of Wisconsin Hollander and Wisconsin Ballhead cabbage in relation to resistance to yellows. Jour. Agr. Research 50: 823-826. 1935.
99. Walker, J. C., and R. H. Larson. Calcium cyanamide in relation to control of clubroot of cabbage. Jour. Agr. Research 51: 183-189. 1935.
100. Walker, J. C. Types of disease resistance. Proc. Sixth int. Bot. Cong. (Amsterdam) 2: 206-208. 1935.
101. Walker, J. C. A study of resistance to Fusarium wilt in Alaska peas. Amer. Jour. Botany 22: 849-857. 1935.
102. Walker, J. C. Resistance to clubroot in Brassica. Phytopathology 26: 112. 1936.
103. Walker, J. C. Discussion of disease-free and disease-resistant beans. The Canning Age 17: 171-172; 178. 1936.
104. Walker, J. C., and Otis C. Whipple. Tomato-seed certification. In Reports and observations of some 1936 tomato growers etc. Crop Service Dept., Campbell Soup Co., Chicago. Jan., 1937.
105. Walker, J. C. The fungicidal value of mustard oils. Phytopathology 27: 142. 1937.
106. Walker, J. C., and R. H. Larson. The increasing importance of cabbage mosaic. Phytopathology 27: 142. 1937.
107. Walker, J. C. Varieties of beans for canning. Canning Trade 59 (30): 20-21. 1937.
108. Walker, J. C., Sam Morell, and H. H. Foster. Toxicity of mustard oils and related sulfur compounds to certain fungi. Amer. Jour. Botany 214: 536-541. 1937.
109. Walker, J. C. Injury to -cabbage by lightning. Phytopathology 27: 858-861. 1937.
110. Walker, J. C. Onion diseases and their control. U.S.D.A. Farm. Bul. 1060 (5th rev.), 24 pp. 1937.
111. Larson, R. H., and J. C. Walker. Properties and host range of a cabbage mosaic virus. Phytopathology 28: 13. 1938.
112. Walker, J. C., and R. H. Larson. Soil temperature in relation to potato yellow dwarf. Phytopathology 28: 21. 1938.
113. Whipple, O. C., and J. C. Walker. Two strains of cucumber virus on pea and bean. Phytopathology 28: 22. 1938.

114. Walker, J. C. Diseases of cabbage and related plants. U.S.D.A. Farm. Bul. 1439 (2nd rev.), 35 pp. 1938.
115. Walker, J. C., R. H. Larson, and A. Ft. Albert. Studies of resistance to potato scab in Wisconsin. Amer. Potato Jour. 15: 246-252. 1938.
116. Walker, J. C. Cabbage disease controlled in South. Southern Seedsman. Nov. 1938, pp. 8, 20. 1938.
117. Larson, R. H., A. R. Albert, and J. C. Walker. Soil reaction in relation to potato scab. Amer. Potato Jour. 15: 325-330. 1938.
118. Walker, J. C., J. P. Jolivette, and J. G. McLean. Internal black spot of canning beets and its control. Canning Age 19: 489-491; 508. 1938.
119. Albert, A. R., R. H. Larson, and J. C. Walker. The comparative productiveness of seed potatoes grown oh sandy and on peat soils in central Wisconsin. Amer. Potato Jour. 16: 16-23. 1939.
120. Walker, J. C. Diseases of vegetable crops. 2nd rev., 67 pp. Ann Arbor. 1939.
121. Walker, J. C. Disease resistant pea varieties. Canning Trade 61 (27): 10. 1939.
122. Walker, J. C. Use of boron to control internal black spot of beets. Canning Trade 61 (27): 16. 1939.
123. Walker, J. C. Internal black spot of garden beet. Phytopathology 29: 120-128. 1939.
124. Walker, J. C. Freezing injury to canning peas in Wisconsin. Phytopathology 29: 188-194. 1939.
125. Delwiche, E. J., F. L. Musbach, W. B. Sarles, Emil Truog, J. C. Walker, and H. F. Wilson. Canning peas in Wisconsin. Wis. Agr. Expt. Sta. Bul. 444, 24 pp. 1939.
126. Pryor, Dean E., and J. C. Walker. A method for testing the toxicity of volatile compounds. Phytopathology 29: 641-643. 1939.
127. Walker, J. C., and R. H. Larson. Yellow dwarf of potato in Wisconsin. Jour. Agr, Research 59: 259-280. 1939.
128. Larson, R. H., and J. C. Walker. A mosaic disease of cabbage. Jour. Agr. Research 59: 367-392. 1939.
129. Walker, J. C., and F. L. Musbach. Effect of moisture, fertility and fertilizer placement on root rot of canning peas in Wisconsin. Jour. Agr. Research 59: 579-590. 1939.
130. Virgin, W. J., and J. C. Walker. Relation of temperature and moisture to near-wilt of pea. Jour, Agr. Research 59: 591-600. 1939.
131. Walker, J. C. Resistance to clubroot in varieties of turnip and rutabaga. Jour. Agr. Research 59: 815-828. 1939.
132. Larson, R. H., and J. C. Walker. A necrotic virosis of cabbage. Phytopathology 30: 15. 1940.

133. Walker, J. C., M. A. Stahmann, and D. E. Pryor. A method of reducing club-root infection at transplanting. Phytopathology 30: 26-27. 1940.
134. Pryor, D. E., J. C. Walker, and M. A. Stahmann. Toxicity of allyl isothiocyanate vapor to certain fungi. Amer. Jour. Bot. 27: 30-38. 1940.
135. Virgin, W. J., and J. C. Walker. Relation of the near-wilt fungus to the pea plant. Jour. Agr. Research 60: 241-248. 1940.
136. Walker, J. C. Yield and canning quality of mosaic resistant Refugee beans for 1937, 1938, and 1939. Canning Trade 62 (42): 13. 1940.
137. Flory, W. S., Jr., and J. C. Walker. Effect of different environments on head shape in Marion Market cabbage. Proc. Amer. Soc. Hort. Science 37: 778-782. 1940.
138. Walker, J. C. The use of borax for the control of internal black spot of beets. Canning Trade 67 (49): 6. 1940.
139. Walker, J. C. Tissue reactions: plant kingdom. In Third Intern. Cong. Microbiology Proc. (Sept, 2-9, 1939). Mew York. 1940.
140. Walker, J. C., E. J. Delwiche, and P. G. Smith. Pea seed treatment with chemical dusts. Jour. Agr, Research 61: 143-148. 1940.
141. Walker, J. C. Cabbage, turnip and rutabaga. Nat. Seedsman 7 (6): 22-25. 1940.
142. Whipple, O. C., and J. C. Walker. Strains of cucumber mosaic virus pathogenic on bean and pea. Jour. Agr. Research 62: 27-60. 1941.
143. Larson, R. H., J. C. Walker, and S. O. Fogelberg. Bacterial ring rot in relation to the tomato. Phythpathology 31: 14-15. 1941.
144. Walker, J. C., and J. P. Jolivette. Studies on boron deficiency in garden beet (Beta vulgaris). Phytopathology 31: 23. 1941.
145. Walker, J. C. Disease resistance in crucifers. In Proc. 7th Genetics Cong. (Edinburgh). Jour. Gen. Suppl. Vol. (1941): 312-313. 1941.
146. Walker, J. C. Origin of cabbage black rot epidemics. Plant Disease Reporter 25: 91-94. 1941.
147. Larson, R. H., and J. C. Walker. Ring necrosis of cabbage. Jour. Agr. Research 62: 475-491. 1941.
148. Walker, J. C., J. G. McLean, and J. P. Jolivette. The boron deficiency disease in cabbage. Jour. Agr. Research 62: 573-587. 1941.
149. Smith, P. G., and J. C. Walker. Certain environal and nutritional factors affecting Aphanomyces root rot of garden pea. Jour. Agr. Research 63: 1-20. 1941.
150. Walker, J. C. Disease resistance in the vegetable crops. Bot. Rev. 7: 458-506. 1941.
151. McLean, J. G., and J. C. Walker. A comparison of Fusarium avenaceuin, F. oxysporum, and F. solani var. eumartii in relation to potato wilt in Wisconsin. Jour. Agr. Research 63: 495-525. 1941.

152. Walker, J. C. Physiologic races of Plasmodiophora brassicae. Phytopathology 32: 18. 1942.
153. Walker, J. C. Systemic invasion of cabbage by Plasmodiophora brassicae. Phytopathology 32: 18. 1942.
154. Walker, J. C., and W. C. Snyder. Pea wilts and root rots. Wis. Agr. Expt. Sta. Bul. 424 (1st rev.), 16 pp. 1942.
155. Walker, J. C. Otis Chester Whipple. Phytopathology 32: 355-356. 1942.
156. Schroeder, W. T., and J. C. Walker. Influence of controlled environment and nutrition on the resistance of garden pea to Fusarium wilt. Jour. Agr. Research 65: 221-248. 1942.
157. Walker, J. C., and W. W. Hare. Survey of pea diseases in Wisconsin in 1942. Phytopathology 33: 15. 1943.
158. Walker, J. C., and W. J. Hooker, The relation of host nutrition to the development of cabbage yellows. Phytopathology 33: 15. 1943.
159. Walker, J. C., and W. W. Hare. Pea diseases in Wisconsin in 1942. Wis. Agr. Expt. Sta. Res. Bull. 145, 32 pp. 1943.
160. Walker, J. C., J. P. Jolivette, and J. C. McLean. Boron deficiency in garden and sugar beet. Jour. Agr. Research 66: 97-123. 1943.
161. Jolivette, J. P., and J. C. Walker. Effect of boron deficiency on the histology of garden beet and cabbage. Jour. Agr. Research 66: 167-182. 1943.
162. Stahmann, M. A., K. P. Link, and J. C. Walker. Mustard oils in cucumbers and their relation to resistance to clubroot. Jour. Agr. Research 67: 49-63. 1943.
163. Walker, J. C., and J. P. Jolivette. Productivity of mosaic-resistant Refugee beans. Phytopathology 33: 778-788. 1943.
164. Hooker, W. J., J. C. Walker, and F. G. Smith. Toxicity of beta-phenethyl isothiocyanate to certain fungi. Amer. Jour. Bot. 30: 632-637. 1943.
165. Walker, J. C. Histologic-pathologic effects of boron deficiency. Soil Science 57: 51-54. 1944.
166. Walker, J. C., N. A. Stahmann, and D. E. Pryor. Efficacy of fungicidal transplanting liquids for control of clubroot of cabbage. Phytopathology 34: 185-195. 1944.
167. Walker, J. C. Diseases of cabbage and related plants. U.S.D.A. Farm. Bul. 1439 (3rd rev.), 38 pp. 1944.
168. Walker, J. C., W. C. Edmundson, and H. A. Jones. Onion-set production. U.S.D.A. Farm. Bul. 1955, 21 pp. 1944.
169. Walker, J. C., W. T. Schroeder, and J. E. Kuntz. Borax sprayed on beets controls black spot. Better Crops 28 (14): 19-21: 49-50. 1944.
170. Walker, J. C. Methods of controlling diseases. In Thompson, R. C. Cauliflower and broccoli varieties and culture. U.S.D.A. Farm. Bul. 1957: 16-17. 1944.
171. Walker, J. C. Onion diseases and their control. U.S.D.A. Farm Bul. 1060 (6th rev.), 25 pp. 1944.

172. Walker, J. C., H. A. Jones, and A. E. Clarke. Smut resistance in an Allium species hybrid. Jour. Agr. Research 69: 1-8. 1944.
173. Hare, W. W., and J. C. Walker. Ascochyta diseases of canning pea. Wis. Agr. Expt. Sta. Res. Bul. 150, 31 pp. 1944.
174. Walker, J. C., O. C. Whippie, J. P. Jolivette, and W. J. Hooker. The limitations of spraying tomatoes in Wisconsin. Wis. Agr. Expt. Sta. Res. Bul. J52, 23 pp. 1944.
175. Walker, J. C. Progress in combination of yellows and mosaic resistance with high ascorbic content in cabbage. Phytopathology 34: 1012-1013. 1944.
176. Walker, J. C., E. J. Delwiche, and W. W. Hare. A major gene for resistance to near-wilt in pea. Phytopathology 34: 1013. 1944.
177. Walker, J. C., and W. W. Hare. Lima bean seed treatments. Plant Disease Reporter Suppl. 145: 19-21. 1944.
178. Walker, J. C., and W. W. Hare. Snap bean seed treatments. Plant Disease Reporter Suppl. 45: 56-57. 1944.
179. Hooker, W. J., J. C. Walker, and K. P. Link. Effects of two mustard oils on Plasmodiophora brassicae and their relation to resistance to c1ubroot. Jour. Agr. Research 70: 63-78. 1945.
180. LeBeau, F. J., and J. C. Walker. Turnip mosaic viruses. Jour. Agr. Research 70: 347-364. 1945.
181. Walker, J. C., F. J. LeBeau, and G. S. Pound. Viruses associated with cabbage mosaic. Jour. Agr. Research 70: 379-404. 1945.
182. Walker, J. C., and W. J. Hooker. Plant nutrition in relation to disease development. I. Cabbage yellows. Amer. Jour. Bot. 32: 314-320. 1945.
183. Walker, J. C., J. P. Jolivette, and W. W. Hare. Varietal susceptibility in garden beet to boron deficiency. Soil Science 59: 461-464. 1945.
184. Ou, S. H., and J. C. Walker. Anthracnose of garden pea. Phytopathology 35: 565-570. 1945.
185. Pound, Glenn S., and J. C. Walker. Differentiation of certain crucifer viruses by the use of temperature and host immunity reactions. Jour. Agr. Research 71: 255-278. 1945.
186. Walker, J. C., and W. J. Hooker. Plant nutrition in relation to disease development. II. Cabbage clubroot. Amer. Jour. Bot. 32: 487-490. 1945.
187. Pound, G. S., and J. C. Walker. Effect of air temperature on the concentration of certain viruses in cabbage. Jour. Agr. Research 71: 471-485. 1945.
188. Walker, J. C., et al. Vegetable Disease Control War Emergency Committee. Upper Mississippi Valley Division, Amer. Phytopathological Society. Report No. 2. 1942.
189. Walker, J. C., et al. Idem. Report No. 10. 1943.
190. Walker, J. C., et al. Idem. Report No. 25. 1944.

191. Walker, J. C., and J. P. Jolivette. Relation of boron to sugar beet production in Wisconsin. Amer. Soc. Sugar Beet Tech. East U.S. and Canada Proc. 2nd Region Meet. 1941: 30-31. 1941.
192. Walker, J. C. Soil management and plant nutrition in relation to disease development. Soil Science 61: 47-54. 1946.
193. Smith, F. O., and J. C. Walker. Relation of environmental and hereditary factors to ascorbic acid in cabbage. Amer. Jour. Bot. 33: 120-129. 1946.
194. Felton, H. W., and J. C. Walker. Environal factors affecting downy mildew of cabbage. Jour. Agr. Res. 72: 69-81. 1946.
195. Jones, H. A., J. C. Walker, T. H. Little, and R. H. Larson. Relation of color-inhibiting factor to smudge resistance in onion. Jour. Agr. Res. 72: 259-264. 1946.
196. Walker, J. C., and R. E. Foster. Plant nutrition in relation to disease development. III. Fusarium wilt of tomato. Amer. Jour. Bot. 33: 259-264. 1946.
197. Smith, F. G., J. C. Walker, and W. J. Hooker. Effect of hydrogen-ion concentration on the toxicity to Colletotrichum circinans (Berk.) Vogl. of some carboxylic acids, phenols, and crucifer extracts. Amer. Jour. Bot. 33: 351-356. 1946.
198. Kuntz, J. E., and J. C. Walker. Virus inhibitors in spinach extract. Phytopath. 36: 404. 1946.
199. Foster, R. E., and J. C. Walker. Improvement of ascorbic acid content in yellows-resistant cabbage. Phytopath. 36: 398. 1946.
200. Walker, J. C. James Peter Jolivette. Phytopath. 36: 415-417. 1946.
201. Walker, J. C., W. C. Edmundson, and H. A. Jones. Onion-set production. U. S. Dept. Agr. Farmers' Bul. 1955. 21 p. Rev. 1946.
202. Walker, J. C., and R. E. Foster. The inheritance of ascorbic acid content in cabbage. Amer. Jour. Bot. 33: 758-76l. 1946.
203. Gorenz, A. H., and J. C. Walker. Influence of method sticker on the effectiveness of Arasan for onion-smut control. Phytopath. 37: 7-8. 1947.
204. Walker, J. C., and Glenn S. Pound. Improvement of cabbage for disease resistance. Phytopath. 37: 23. 1947.
205. Foster, R. E., and J. C. Walker. Predisposition of tomato to Fusarium wilt. Jour. Agr. Res. 74: 165-185. 1947.
206. Smith, F. G., K. P. Link, and J. C. Walker. Acidic and phenolic fractions of crucifer roots in relation to clubroot. Jour. Agr. Res. 74: 193-204. 1947.
207. Kuntz, J. E., and J. C. Walker. Virus inhibition by extracts of spinach. Phytopath. 37: 561-579. 1947.
208. Walker, J. C. Onion diseases and their control. U.S. Dept. Agr. 1060. 26 p. 7th Rev. 1947.
209. Walker, J. C. Seed-borne pea diseases and their control. Can. Seed Growers' Assoc. Ann. Rpt. 1946-1947: 12-13. 1947.

210. Pound, G. S., and J. C. Walker. Strains of cucumber mosaic virus pathogenic on crucifers. Phytopath. 38: 21-22. 1948.
211. Walker, J. C., and R. H. Larson. Development of clubroot resistance in cabbage. Ibid. 38: 28. 1948.
212. Kendrick, J. B., Jr., and J. C. Walker. Anthracnose of tomato. Phytopath. 38: 247-260. 1948.
213. Walker, J. C., and J. B. Kendrick, Jr. Plant nutrition in relation to disease development. IV. Bacterial canker of tomato. Amer. Jour. Bot. 35: 186-192. 1948.
214. Walker, J. C., and J. P. Jolivette. Yellows-resistant varieties of cabbage in the early and midseason roundhead groups. U. S. Dept. Agr. Circ. 775. 20 p. 1948.
215. Walker, J. C., R. H. Larson, R. E. Foster, and J. E. Kuntz. Yellows- resistant cabbage varieties in the Danish Ballhead-Hollander group. U.S. Dept. Agr. Circ. 776. 12 p. 1948.
216. Grogan, R. G., and J. C. Walker. Interrelation of bean virus 1 and bean virus 2 as shown by cross-protection tests, Phytopath. 38: 1489-1493. 1948.
217. Pound, O. S., and J. C. Walker. Strains of cucumber mosaic virus pathogenic on crucifers. Jour. Agr. Res. 77: 1-12. 1948.
218. Hatfield, W. C., J. C. Walker, and J. H. Owen. Antibiotic substances in onion in relation to disease resistance. Jour. Agr. Res. 77: 115-135. 1948.
219. Kendrick, J. B., Jr., and J. C. Walker. Predisposition of tomato to bacterial canker. Jour. Agr. Res. 77: 169-186. 1948.
220. Gorenz, A. N., J. C. Walker, and R. H. Larson. Morphology and taxonomy of the onion pink-root fungus. Phytopath. 38: 831-840. 1948.
221. Walker, J. C. Diseases of cabbage and related plants. U. S. Dept. Agr. Farmers' Bul. 1439. 38 p. 4th Rev. 1948.
222. Walker, J. C., G. S. Pound, and J. E. Kuntz. Development of Wisconsin 55 tomato. Wis. Agr. Expt. Sta. Bul. 478, 20 p. 1948.
223. Walker, J. C. Vegetable seed treatment. Bot. Rev. 14: 588-601. 1948.
224. Grogan, R. G., and J. C. Walker. A pod-distorting strain of the yellow mosaic virus of bean. Jour. Agr. Res. 77: 301-314. 1948.
225. Grogan, R. G., and J. C. Walker. The relation of common mosaic to black root of bean. Jour. Agr. Res. 77: 315-331. 1948.
226. Gorenz, A. N., R. H. Larson, and J. C. Walker. Factors affecting pathogenicity of pink root fungus of onions. Jour. Agr. Res. 78: 1-18. 1949.
227. Hagedorn, D. J., and J. C. Walker. Cross protection tests between Wisconsin pea streak virus and several strains of bean virus 2 from pea. Phytopath. 39: 11. 1949.
228. Walker, J. C. Resistance in cucumber to scab. Phytopath. 39: 27. 1949.
229. Chiu, W. F., and J. C. Walker. Morphology and variability of the cucurbit black rot fungus. Jour. Agr. Res. 78: 81-102. 1949.

230. Hare, W. W., J. C. Walker, and E. J. Delwiche. Inheritance of a gene for near-wilt resistance in the garden pea. Jour. Agr. Res. 78: 239-250. 1949.

231. Walker, J. C., G. S. Pound, and J. E. Kuntz. Wisconsin 55 tomato. Seed World 64(10): 8-10; 15. 1949.

232. Calrert, O. H., G. S. Pound, J. C. Wa1ker, M. A. Stahmann, and J. F. Stauffer. Induced variability in Phoma lingam. Jour. Agr. Res. 78: 571-588. 1949.

233. Chiu, N. F., and J. C. Walker. Physiology and pathogenicity of the cucurbit black-rot fungus. Jour. Agr. Res. 51: 589-615. 1949.

234. Hagedorn, D. J., and J. C. Walker. Wisconsin pea stunt, a newly described disease. Jour. Agr. Res. 78: 617-626. 1949.

235. Wells, D. G., N. W. Hare, and J. C. Walker. Evaluation of resistance and susceptibility in garden pea to near-wilt in the greenhouse. Phytopath. 39: 771-779. 1949.

236. Hagedorn, D. J., and J. C. Walker. Wisconsin pea streak. Phytopath. 39: 837-847. 1949.

237. Gallegly, M. E., Jr., and J. C. Walker. Plant nutrition in relation to disease development. V. Bacterial wilt of tomato. Amer. Jour. Bot. 36: 613-623. 1949.

238. Wells, D. G., J. C. Walker, and W. W. Hare. A study of linkage between factors for resistance t5 wilt and near-wilt in garden peas. Phytopath. 39: 907-912. 1949.

239. Gallegly, M. E., Jr., and J. C. Walker. Relation of environmental factors to bacterial wilt of tomato. Phytopath. 39: 936-946. 1949.

240. Ladeburg, R. C., R. H. Larson, and J. C. Walker. The ringspot type of potato virus X. Amer. Potato Jour. 26: 432-435. 1949.

241. Ladeburg, R. C., R. H. Larson, and J. C. Walker. Origin interrelation and properties of ringspot strains of virus X in American potato varieties. Wis. Agr. Expt. Sta. Res. Bul. 165. 1950.

242. Walker, J. C. The mode of seed infection by the cabbage blackrot organism. Phytopath. 40: 30. 1950.

243. Walker, J. C., J. H. Owen, and M. A. Stahmann. Relative importance of phenols and volatile sulfides in disease resistance in the onion, Phytopath. 40: 30. 1950.

244. Walker, J. C. The breeding and development of pea varieties for processing. Canner 110(7): 50, 52. 1950.

245. Owen, J. H., J. C. Walker, and M. A. Stahmann. Pungency, color, and moisture supply in relation to disease resistance in the onion. Phytopath. 40: 292-297. 1950.

246. Hagedorn, D. J., and J. C. Walker. The relation of bean virus 2 to pea mosaic in Wisconsin. Phytopath. 40: 684-698. 1950.

247. Owen, J. H., J. C. Walker, and M. A. Stahmann. Variability in onion neck- rot fungi. Phytopath. 40: 749-768. 1950.

248. Walker, J. C. Can a staff member do research and teach at the same time? Yes. P1. Dis. Rptr. 34: 281-282. 1950.
249. Larson, R. H., R. E. F. Mathews, and J. C. Walker. Relationships between certain viruses affecting the genus Brassiea. Phytopath. 40: 955-962. 1950.
250. Stahmann, M. A., and J. C. Walker. Synthesized chemical can stop virus action. Wis. Agr. Expt. Sta. Bul. 491: 60-62. 1950.
251. Walker, J. C. History of plant disease control. Ann. Rpt. Veg. Growers. Assoc. Amer. 1950: 209-212. 1950.
252. Walker, J. C. Environment and host resistance in relation to cucumber scab. Phytopath. 40: 1094-1102. 1950.
253. Walker, J. C., and R. H. Larson. Progress in the development of clubroot-resistant cabbage. Phytopath. 41: 37. 1951. (Abst,) Jan.
254. Walker, J. C., and A. B. Wiles. Development of a scab-resistant pickling cucumber for Wisconsin. Phytopath. 41: 37. 1951. Jan.
255. Walker, J. C. Plant pathology. 699 p. McGraw-Hill, New York, 1950.
256. Walker, J. C. Genetics and plant pathology. In Genetics in the Twentieth Century, pp. 527-554. McMillan Co., NewYork, 1951. Mar.
257. Wellman, F. L., J. C. Walker, Allyn Cook, and M. E. Gallegly, Jr. Effects of temperature on vegetative growth of five coffee disease fungi. In: Primera Asamblea Latinoamericana de Fitoparasitologia. Misc. Folleto Mex. Dept. of Agr, No. 4, pp. 251-257. 1951.
258. Webb, R. E., R. H. Larson, and J. C. Walker. Naturally occurring strains of the potato leaf roll virus. Amer. Potato Jour. 28: 667-671. 1951. July.
259. Darby, J. F., R. H. Larson, and J. C. Walker. Variation in virulence and properties of potato virus I strains. Wis. Agr. Expt. Sta. Res. Bul. 177. 1951. Aug.
260. Bridgmon, G. H., and J. C. Walker. The relation of southern bean mosaic to black root. Phytopath. 41: 865-871. 1951. Oct.
261. Walker, J. C., and M. E. Gallegly, Jr. Plant nutrition in relation to disease development. VI. Black rot of' cabbage and ring rot of tomato. Amer. Jour. Bot. 38: 663-665. 1951. Oct.
262. Wiles, A. B., and J. C. Walker. The relation of Pseudomonas lachrymans to cucumber fruits and seeds. Phytopathology 41: 1059-1064. 1951.
263. Walker, J. C. Potato diseases: past, present, and future. In Ann. Rpt. Veg. Growers Assn. America. 1951: 36-40. 1951.
264. Pound, G. S., and J. C. Walker. Mosaic resistance in cabbage. Phytopath. 41: 1083-1090. 1951.
265. Stahmann, M. A., L. H. Graf, E. L. Patterson, J. C. Walker, and D. W. Watson. The inhibition of tobacco mosaic virus by synthetic lysine polypeptides. Jour, Biol. Chem. 189: 45-52. 1951.
266. Sill, W. H., Jr., W. C. Burger, M. A. Stahmann, and J. C. Walker. Electron microscopy of cucumber virus 1. Phytopathology 42: 19-20. 1952.

267. Larson, R. H., M. A. Stahmann, J. C. Walker, and R. W. Colburn. Mutations of potato virus X induced by nitrogen mustard. Phytopathology 42: 25. 1952.

268. Bridgmon, O. H., and J. C. Walker. Gladiolus as a virus resarvoir. Phytopathology 42: 65-70. 1952.

269. Wiles, A. B., and J. C. Walker. Epidemiology and control of angular leaf spot of cucumber. Phytopathology 112: 105-108. 1952.

270. Cooks A. A., J. C. Walker, and R. H. Larson. Studies on the disease cycle of black rot of crucifers. Phytopathology 112: 162-167. 1952.

271. Wei, C. T., J. C. Walker, and R. P. Scheffer. Plant nutrition in relation to disease development. VII. Cucurbit wilts. Amer. Jour. Bot. 39: 245-249. 1952.

272. Gasiorkiewicz, E. C., R. H. Larson, J. C. Walker, and M. A. Stahmann. Induced variability in Pyrenochaeta terrestris by nitrogen mustard. Phytopathology 112: 183-192. 1952.

273. Wa1ker, J. C. Diseases of vegetable crops. 529 p. McGraw-Hill, New York. 1952.

274. Cook, A. A., R. H. Larson, and J. C. Walker. Relation of the black rot pathogen to cabbage seed. Phytopathology 112: 316-320. 1952.

275. Sill, W. H., Jr., and. J. C. Walker. Cowpea as an assay host for cucumber virus 1. Phytopathology 142: 328-330. 1952.

276. Sill, W. H., Jr., and J. C. Walker. A virus inhibitor in cucumber in relation to mosaic resistance. Phytopathology 142: 349-352. 1952.

277. Sill, W. H., Jr., W. C. Burger, M. A. Stahmann, and J. C. Walker. Electron microscopy of cucumber virus 1. Phytopathology 42: 420-422. 1952.

278. Sill, W. H., Jr., and J. C. Walker. A strain of cowpea susceptible to systemic infection by various strains of common cucumber mosaic virus. Phytopathology 42: 442. 1952.

279. Larson, R. H., and J. C. Walker. Arasan for smut control in set-onion plantings. Phytopathology 42: 469. 1952.

280. Scheffer, R. P., and J. C. Walker. Distribution of Fusarium resistance in the tomato plant. Phytopathology 42: 474. 1952.

281. Webb, R. E., R. H. Larson, and J. C. Walker. Relationships of potato leaf roll virus strains. Wis. Agr. Expt. Sta. Res. Bul. 178. 1952.

282. Walker, J. C., R. H. Larson, and G. S. Pound. New cabbage meets needs of consumers. Wis. Agr. Expt. Sta. Bul. 498: 5-6. 1952.

283. Walker, J. C., and R. H. Larson. New cabbage lines resist clubroot. Wis. Agr. Expt. Sta. Bul. 498: 8-9. 1952.

284. Walker, J. C. New bean developed for mechanized picking. Wis. Agr. Expt. Sta. Bul. 498: 13-14. 1952.

285. Larson, R. H., M. A. Stahmann, J. C. Walker, and H. W. Colburn. Nitrogen mustard gas causes virus mutation. Wis. Agr. Expt. Sta. Bul. 498: 62-63. 1952.

286. Scheffer, H. P., and J. C. Walker. The physiology of Fusarium wilt of tomato. Phytopathology 43: 116-125. 1953.
287. Walker, J. C., C. F. Pierson, and A. B. Wiles. Two new scab-resistant cucumber varieties. Phytopathology 43: 215-217. 1953.
288. Scheffer, R. P., S. S. Gothoskar, M. A. Stahmann, and J. C. Walker. Tomato wilt caused by fungus enzymes. Wis. Agr. Expt. Sta. Bul. 500: 11. 1953.
289. MacLachlan, D. S., R. H. Larson, and J. C. Walker. Strain interrelationships in potato virus A. Wis. Agr. Expt. Sta. Res. Bul. 180. 1953.
290. Larson, R. H., and J. C. Walker. Arasan controls onion smut. Wis. Agr. Expt. Sta. Bul. 509: 34. 1953.
291. Walker, J. C. Wisconsin SR6 cucumber resists scab disease. Wis. Agr. Expt. Sta. Bul. 509: 36. 1953.
292. Pound, G. S., and. J. C. Walker. Autogenous necrosis of cabbage. Phytopathology 43: 415-418. 1953.
293. Winstead, N. N., and J. C. Walker. Metabolites of vascular Fusaria of cabbage and cotton in relation to resistance. Phytopathology 43: 489-490. 1953.
294. Winstead, N. N., and J. C. Walker. Production of vascular browning by metabolites from several pathogens. Phytopathology 43: 490. 1953.
295. Gothoskar, S. S., R. P. Scheffer, J. C. Walker, and M. A. Stahmann. The role of pectic enzymes in Fusarium wilt of tomato. Phytopathology 43: 535-536. 1953.
296. Walker, J. C. Cauliflower, cabbage, and others. U. S. Dept. Agr. Yearbook 1953: 425-430. 1953.
297. Walker, J. C. Hazards to onions in many areas. U. S. Dept. Agr. Yearbook 1953: 431-435. 1953.
298. Larson, R. H., and J. C. Walker. Thiram for smut control in onion set plantings. Phytopathology 43: 596-597. 1953.
299. Walker, J. C. Disease resistance in the vegetable crops. II. Bot. Rev. 19: 606-614. 1953.
300. Walker, J. C., R. H. Larson, and G. S. Pound. Badger Market, a new disease resistant cabbage. Phytopathology 43: 649-650. 1953.
301. Walker, J. C., and E. K. Wade. The potato rot nematode. Special Leaflet Dept. Plant Pathology, University of Wisconsin. 1954.
302. Walker, J. C. A few recollections and observations from thirty years of battling vegetable diseases. Ann. Rept. Veg. Growers Assn. of America 1953: 75-78. 1954.
303. Scheffer, R. P., and J. C. Walker. 1954. Distribution and nature of Fusarium resistance in the tomato plant. Phytopathology 44: 94-101.
304. Winstead, N. N., and J. C. Walker. 1954. Production of vascular browning by metabolites from several pathogens. Phytopathology 44: 153-159.

305. Winstead, N. N., and J. C. Walker. 1954. Toxic metabolites of the athogen in relation to Fusarium resistance. Phytopathology 44: 159-166.
306. Hagedorn, D. J., and J. C. Walker. Virus diseases of canning peas in isconsin. Wis. Agr. Expt. Sta. Res. Bul. 185. 1954.
307. MacLachlan, D. S., R. H. Larson, and J. C. Walker. Potato virus A. Amer. Potato Jour. 31: 67-72. 195)-i.
308. MacLachlan, D. S., R. H. Larson, and J. C. Walker. Interveinal mosaic of potato. Amer. Potato Jour. 31: 101-105. 1954.
309. Walker, J. C., and C. F. Pierson. SMR12 cucumber combines scab, mosaic resistance. In What's New in Farm Science. Wis. Agr. Expt. Sta. Bul. 511. 1954.
310. Pierson, C. F., and J. C. Walker. Relation of Cladosporium cucumerinum to susceptible and resistant cucumber tissue. Phytopathology 44: 459-465. 1954.
311. Walker, J. C. Methods of controlling diseases of cauliflower and broccoli. In Cauliflower and broccoli, varieties and culture. U.S.D.A. Farm. Bul. 1957.
312. Rice, R. V., M. A. Stahmann, G. D. Lindberg, and J. C. Walker. Some physical biochemical characteristics of squash mosaic virus. Phytopathology 144: 503. 1954.
313. Sinclair, J. B., and J. C. Walker. Inheritance of resistance to common cucumber mosaic virus in cowpea. Phytopathology 144: 506. 19514.
314. Walker, J. C., M. E. Gallegly, Jr., J. R. Bloom, and R. P. Scheffer. Plant nutrition in relation to disease development. VIII. Verticillium wilt of tomato. Amer. Jour. Bot. 141: 760-762. 1954.
315. Walker, J. C. 19514. Better cabbage through breeding. Vegetable Growers' Messenger. 6 (3): 30.
316. Walker, J. C. 1955. Competition, cooperation, and evolution. Second Annual Forty Miner Service Award Presentation.
317. Rice, R. V., G. D. Lindberg, P. Kaesberg, J. C. Walker, and M. A. Stahmann. The three components of squash mosaic virus. Phytopathology 45: 145-148. 1955.
318. Gothoskar, S. S., R. P. Scheffer, M. A. Stahmann, and J. C. Walker. Further studies on the nature of Fusarium resistance in tomato. Phytopathology 45: 303-307. 1955.
319. Walker, J. C., and M. A. Stahmann. Chemical nature of disease resistance in plants. Ann. Rev. Plant Physiology 6: 351-366. 1955.
320. Gothoskar, S. S., R. P. Scheffer, J. C. Walker, and M. A. Stahmann. The role of enzymes in the development of Fusarium wilt of tomato. Phytopathology 145: 381-387. 1955.
321. Bloom, J. R., and J. C. Walker. Effect of nutrient sprays on Fusarium wilt of tomato. Phytopathology 45: 443-444. 1955.
322. Walker, J. C., and C. F. Pierson, Two new cucumber varieties resistant to scab and mosaic. Phytopathology 145: 1451-1453. 1955.

323. Pierson, C. F., S. S. Gothoskar, J. C. Walker, and M. A. Stahmann. Histological studies on the role of pectic enzymes in the development of Fusarium wilt symptoms in tomato. Phytopathology 145: 524-527. 1955.
324. Sinclair, J. B., and J. C. Walker. 1955. Inheritance of resistance to cucumber mosaic in cowpea. Phytopathology 145: 563-5614.
325. Larson, R. H., M. A. Stahmann, and J. C. Walker. The induction of mutants in potato virus X by nitrogen mustard. Proc. 2nd Conf. on Potato Virus Diseases (Wageningen): 93-99. 1955.
326. Walker, J. C. Crop rotation advised for limas. In "What's New in Farm Science". Wis. Agr. Expt. Sta. Bul. 517: 20-21. 1955.
327. Walker, J. C. 1955. New disease (Angular leaf spot of bean). In "What's New in Farm Science". Wis. Agr. Expt. Sta. Bul. 517: 21.
328. Walker, J. C. Breeding for disease resistance. Report XIVth Internat. Hort. Cong., The Netherlands, 1955. 1: 51-63. 1955.
329. Busch, L. V., and J. C. Walker. Resistance to anthracnose in cucumber. Phytopathology 146: 8. 1956.
330. Lindberg, G. D., D. H. Hall, and J. C. Walker. A study of squash and melon mosaic viruses. Phytopathology 146: 18. 1956.
331. Sinclair, J. B., and J. C. Walker. A survey of ring spot on cucumber in Wisconsin. Plant Disease Reporter 140: 19-20. 1956.
332. Scheffer, H. P., S. S. Gothoskar, C. F. Pierson, and R. P. Collins. Physiological aspects of Verticillium Wilt. Phytopathology 146: 83-87. 1956.
333. Sinclair, J. B., and J. C. Walker. Extent of' cross protection among strains of cucumber mosaic virus in cucumber and cowpea. Phytopathology 146: 367-371. 1956.
334. Lindberg, G. D., D. H. Hail, and J. C. Walker. A study of melon and squash mosaic viruses. Phytopathology 46: 489-495. 1956.
335. Sinclair, J. B., and J. C. Walker. Assays for resistance to cucumber mosaic in the pickling cucumber. Phytopathology 46: 519-522. 1956.
336. Bagnall, R. H., R. H. Larson, and J. C. Walker. 1956. Three virus components of "interveinal mosaic—Irish Cobbler potato". Amer. Potato Jour. 33: 271-273.
337. Bagnall, R. H., R. H. Larson, and J. C. Walker. Potato viruses M, S, and X in relation to interveinal mosaic of the Irish Cobbler variety. Wis. Agr. Expt. Sta. Res. Bul. 198. 1956.
338. Cardona-Alvarez, C., and J. C. Walker. Angular leaf spot of bean. Phytopathology 46: 610-615. 1956.
339. Larson, R. H., J. C. Walker, and G. S. Pound. YR Charleston Wakefield, a new ye1lows resistant cabbage variety. Phytopathology 46: 623-624. 1956.
340. Edgington, L. V., arid J. C. Walker. The influence of soil and air temperature on Verticillium wilt of tomato. Phytopathology 47: 8. 1957.

341. Van Gundy, S. D., and J. C. Walker, The amino-nitrogen nutrition of Pseudomonas lachrymans. Phytopathology 47: 35. 1957.
342. Walker, J. C. Plant Pathology. 2nd Edit. 707 p. McGraw-Hill. 1957.
343. Van Gundy, S. D., and J. C. Walker. Seed transmission, overwintering, and host range of the cucurbit-angular-leaf-spot pathogen. Plant Dis. Reptr. 41: 137-140. 1957.
344. Walker, J. C., R. H. Larson, and O. S. Pound. Badger Balihead, a new cabbage resistant to yellows and mosaic. Phytopathology 47: 269-271. 1957.
345. Echandi, E., and J. C. Walker. Pectolytic enzymes produced by Scierotinia sclerotiorum. Phytopathology 47: 303-306. 1957.
346. Sinclair, J. B., P. H. Geil, and P. Kaesberg. Biophysical studies on wild cucumber mosaic virus. Phytopathology 47: 372-377. 1957.
347. Walker, J. C., Benjamin Minge Duggar, 1872-1596. Phytopathology 47: 379-380. 1957.
348. Thaung, M. M., and J. C. Walker. Studies on bacterial blight of lima bean. Phytopathology 47: 413-417. 1957.
349. Darling, H. N., J. C. Walker, G. Thorne, et al, Soil fumigation beats potato rot nematode. Wis. Agr, Expt. Sta. Bul. 527: 66. 1957.
350. Bagnall, R. H., R. H. Larson, and J. C. Walker. Potato interveinal mosaic involves virus S, M. X. Wis. Agr. Expt. Sta. Bul. 527: 68. 1957.
351. Walker, J. C., and L. V. Edgington. Studies of internal tipburn of cabbage. Phytopathology 47: 537. 1957.
352. Echandi, E., S. D. Van Gundy, and J. C. Walker, Pectolytic enzymes secreted by soft-rot bacteria. Phytopathology 47: 549-552. 1957.
353. Edgington, L. V., and J. C. Walker. Influence of soil and air temperature on Verticillium wilt of tomato. Phytopathology 47: 594-598. 1957.
354. Van Gundy, S. D., and J. C. Walker. Relation of temperature and host nutrition to angular leaf spot of cucumber. Phytopathology 147: 615-619. 1957.
355. Robinson, Dean, R. H. Larson, and J. C. Walker. Verticillium wilt of potato. Wis. Agr. Expt. Sta. Res. Bul. 202. 1957.
356. Robinson, D. B., R. H. Larson, and J. C. Walker. Verticillium wilt of potato. Amer. Potato Jour. 35: 444-445. 1958.
357. Walker, J. C., and A. J. Hiker. 1958. Lewis Ralph Jones. December 5, 1864-April 1, 1945. Nat. Acad. 3rd U. S. Biog. Memoirs 31: 156-179. 1958.
358. Busch, L. V., and J. C. Walker. 1958. Studies of cucumber anthracnose. Phytopathology 48: 302-304.
359. Edgington, L. V., and J. C. Walker. Influence of calcium and boron nutrition on development of Fusarium wilt of tomato. Phytopathology 48: 324-326. 1958.
360. Rohringer, R., M. A. Stahmann, and J. C. Walker. Effect of Fusarium oxysporum f. lycopersici and its metabolites on leaf constituents of susceptible and resistant tomatoes. Jour. Agr. and Food Chem. 6: 838-843. 1958.

361. Walker, J. C. Two new pickling varieties resistant to scab and mosaic. Plant Dis. Reptr. 42: 1337-1338. 1958.
362. Walker, J. C. Benjamin Minge Duggar. Nat. Acad. Sci. U. S. Biog, Memoirs 32: 113-131. 1958.
363. Walker, J. C., R. H. Larson, and A. L. Taylor. Diseases of cabbage and related plants. U. S. Dept. Agr. Handbook 144. 1958.
364. Tomlinson, J. A., R. J. Shepherd, and J. C. Walker. Purification and serology of cucumber mosaic virus. Nature 182: 1616. 1958.
365. Walker, J. C. Investigation of virus diseases of Brassica crops, by L. Broadbent, Cambridge Univ. Press, New York, 94 p. 1957. A review. Quarterly Rev. of Biology 33: 280. 1958.
366. Walker, J. C. Plant Pathology 2nd Edit. Review by Hassebrauk in Angewandte Botanik No. 31. 1957.
367. Walker, J. C. Plant Pathology 2nd Edit. Review by S. D. Garrett in The New Phytologist Vol 57, No. 1, March, 1958.
368. Srivastava, D. N., E. Echandi, and J. C. Walker. Pectolytic and cellulytic enzymes produced by Rhizopus stolonifer. Phytopathology 49: 145-148. 1959.
369. Walker, J. C. Enfermedades de las hortizalizas. Translation by Antonia Arnal Verderol. Salvat Editores, Barcelona. 624 pp. 1959.
370. Tomlinson, J., R. J. Shepherd, and J. C. Walker. Purification, properties, and serology of cucumber mosaic virus. Phytopathology 49: 293-299. 1959.
371. Srivastava, D. N., and J. C. Walker. Mechanisms of infection of sweet potato roots by Rhizopus stolonifer. Phytopathology 49: 400-406. 1959.
372. Walker, J. C. Disease resistance in crucifers. Proc. IXth Int. Botanical Congress (Montreal) Vol. II, p. 420. 1959.
373. Walker, J. C. Progress and problems in controlling plant diseases by host resistance. Plant Pathology. Problems and Progress 1908-1958: 32-41. Univ. of Wis. Press, 1959. 1959.
374. Walker, J. C. Disease resistance. Sigma Xi lecture, Univ. of Nebraska, Nov. 18, 1959.
375. Walker, J. C., and R. H. Larson. 1959. Badger shipper cabbage resists clubroot and yellows. Wis. Agr. Expt. Sta. Bul. 538: 100.
376. Heitefuss, R., D. J. Buchanan-Davidson, M. A. Stahmann, and J. C. Walker. Electrophoretic and immunochemical studies of proteins in cabbage infected with Fusarium oxysporuza f. conglutinans. Phytopathology 50: 198-205. 1960.
377. Nayudu, M. V., and J. C. Walker, 1960. Bacterial spot of tomato as influenced by temperature and by age and nutrition of the host. Phytopathology 50: 360-364.
378. Heitefuss, R., M. A. Stahmann, and J. C. Walker. Production of pectolytic enzymes and fusaric acid by Fusarium oxysporum f. conglutinans in relation to cabbage yellows. Phytopathology 50: 367-370. 1960.

379. Heitefuss, H., M. A. Stahmann, and J. C. Walker. Oxidative enzymes in cabbage infected by Fusariurn oxysporum f. conglutinaris. Phytopathology 50: 370-375. 1960.
380. Remarks by Dean Hanaw in presenting the honorary Doctorate of Science to J. C. Walker at Gottingen University on June 16, 1960.
381. Walker, J. C. Soil-plant relationships by C. A. Black. John Wiley and Son, 1957. A review. Quart. Rev. of Biol. 35: 143-144. 1960.
382. Walker, J. C. Plant pathology. An advanced treatise. Vol. I. The Diseased Plant, Ed. Horsfall and Dimond. Academic Press, 1959. A review, Quart. Rev. of Biol. 35: 149. 1960.
383. Nayudu, M. V., and J. C. Walker. Physiology of bacterial spot of tomato. Phytopathology 50: 648. 1960.
384. Walker, J. C. Seed borne disease spreads fast in beans. Wis. Agr. Expt. Sta. Bul. 547: 11. 1960.
385. Walker, J. C., and R. H. Larson. Development of the first c1ubroot-resistant cabbage variety. Wis. Agr. Expt. Sta. Bul. 547: 12-16. l960.
386. Walker, J. C. Plant diseases: past, present and future. Ann. Rpt. Veg. Growers Amer. 1960: 82-87. 1960.
387. Nayudu, N. V., arid J. C. Walker. In vitro nutrition of the tomato bacterial spot organism. Phytopathology 51: 32-34. 1961.
388. Walker, J. C., L. V. Edgington, and N. V. Nayudu. Cabbage varieties in relation to tipburn. Pl. Dis. Rptr. 45: 29. 1961.
389. Seaman, W. L., R. H. Larson, and J. C. Walker. Tinsel-type structure of flagella of zoospores associated with cabbage clubroot tissue. Nature 190: 186-187. 1961.
390. Walker, J. C. Plant pathology. An advanced treatise. Vol. 1. The diseased plant. Vol. 2: The pathogen. Ed. by J. G. Horsfall and A. E. Dimond. Academic Press, 1959 1960. A review. Arch. of Biochem. and Biophys. 93: 466. 1961.
391. Nayudu, N. V., and J. C. Walker. Tomato leaf composition in relation to bacterial spot. Phytopathology 51: 368-372. 1961.
392. Wasuwat, S. L., and J. C. Walker. Inheritance of resistance in cucumber to cucumber mosaic virus. Phytopathology 51: 423-428. 1961.
393. Walker, J. C., L. V. Edgington, and N. V. Nayudu. Tipburn of cabbage. Nature and control. Wis. Agr. Expt. Sta. Res. Bul. 230. 1961.
394. Walker, J. C. Plant disease control. Lecture on "Survey of Agriculture" course. Oct. 31, 1961.
395. Walker, J. C., and R. H. Larson. Onion diseases and their control. U. S. Dept. Agr. Res. Serv. Agr. Handbook 208. 27 p. Oct. 1961.
396. Seaman, W. L., R. H. Larson, and J. C. Walker, Resistance to clubroot in cabbage and related crops. Proc. Canadian Phytopathological Society 1961: 14-15. 1961.

397. Wasuwat, S. L., and J. C. Walker. 1961. Relative concentration of cucumber mosaic virus in a resistant and a susceptible cucumber variety. Phytopathology 51: 614-616.
398. Walker, J. C. Plant pathology. An advanced treatise. Vol. II. The pathogen. Vol. III. The diseased population. Epidemics and control. Ed. by J. G. Horsfall and A. E. Dimond. Academic Press, 1959-60. A review. Quart. Rev. Biology 36: 221-222. 1961.
399. Walker, J. C. laboratory manual. Second edition. By J. P. Pulton, D. A. Slack, N. D. Fulton, et al., Burgess Publ. Co., Mpls. 1960. A review, Quar. Rev. Biology 36: 222. 1961.
400. Walker, J. C. Disease resistance in crucifers. In Recent Advances in Botany. (Proc. IX int. Botanical Cong., Montreal, 1959.) Vol. I. 1961.
401. Walker, J. C., L. V. Edgington, and M. V. Nayudu. Possible answer to cabbage tipburn found. Wis. Agr. Expt. Sta. Bul. 553: 2-5. 1961.
402. Barker, K. R., and J. C. Walker. Factors affecting the pathogenicity of Pellicularia filaanentosa. Phytopathology 52: 2. 1962.
403. Patel, P. N., and J. C. Walker. Physiology of halo blight in resistant and susceptible beans. Phytopathology 52: 24. 1962.
404. Walker, J. C., L. V. Edgington, and M. V. Nayudu. Cabbage tipburrn. U. W. Coll Agr. Press Release, Marsh 2, 1962.
405. Stahmann, M. A., and J. C. Walker. Research digs deeper into plant disease resistance. Wis. Agr. Expt. Sta. Bull. 554: 57-58. 1962.
406. Anderson, J. L., and J. C. Walker. Histology of watermelon anthracnose. Phytopathology 52: 650-653. 1962.
407. Barker, K. R., and J. C. Walker. Relationship of pectolytic and cellulytic enzyme production by strains of Pellicularia filamentose to their pathogenicity. Phytopathology 52: 1119-1225. 1962.
408. Walker, J. C. Russell Harold Larson, 1904-1961. Phytopathology 52: 1233. 1962.
409. Walker, J. C., J. N. Chand, and E. K. Wade. Relation of seed- and soil-borne inoculum to epidemiology of angular leaf spot of cucumber in Wisconsin. Plant Dis. Reptr. 47: 15. 1963.
410. Chand, J. N., K. K. Wade, and J. C. Walker. Sprays for control of angular leaf spot of cucumber in Wisconsin, Plant Dis. Reptr. 74: 94-95. 1963.
411. Walker, J. C., and R. H. Larsen. Development of the first clubroot-resistant cabbage variety. Vee-Gee Messenger 15(1): 10. 1963.
412. Patel, P. N., and J. C. Walker. Relation of air temperature and age and nutrition of the host to the development of halo arid common bacterial blights of bean. Phytopathology 53: 407-411. 1963.
413. Fatel P. N., and J. C. Walker. Changes in free amino acid and amide content of resistant and susceptible beans after infection with the halo blight organism. Phytopathology 53: 522-528. 1963.

414. Walker, J. C. Control of plant diseases through breeding for host resistance, lecture delivered at Yale Forestry School, May 14, 1963.
415. Deverall, B. J., and J. C. Walker. A physiological difference between bean leaves (Phaseolus vulgaris), resistant and susceptible to halo blight, caused by Pseudomonas phaseolicola. Ann. Appl. Biology 52: 105-115. 1963.
416. Williams, P. H., and J. C. Walker. Races of clubroot in North America. Plant Dis. Reptr. 47: 608-611. 1963.
417. Menke, G. H. Physiology of Rhizopus infection on carrot. Phytopathology 53: 882. 1963.
418. Patel, P. N., and J. C. Walker. Free amino acid and amide content of tobacco and oats infected by wildfire and halo blight bacteria. Phytopathology 53: 885. 1963.
419. Menke, G. H., and J. C. Walker. Metabolism of resistant and susceptible cucumber varieties infected with cucumber mosaic virus. Phytopathology 53: 1349-1355. 1963.
420. Walker, J. C. Bacterial plant pathogen. By C. Stapp: translated by A. Schoenfeld. Oxford Univ. Press. 1961. A Review. Quar. Rev. Biol. 38: 265-266. 1963.
421. Walker, J. C. Diseases of sorghum, sudan grass and broom corn. By S. A. J. Tazrr. Commonwealth Myco. Inst. 1962. A review. Quart. Rev. Biology 38: 266. 1963.
422. Seaman, W. L., J. C. Walker, and R. H. Larson. A new race of Plasmodiophora brassicae affecting Badger Shipper cabbage. Phytopathology 53: 1426-1429. 1963.
423. Walker, J. C. The future of plant pathology. Ann. Rev. of Phytopathology 1: 1-4. 1963.
424. Walker, J. C. The physiology of disease resistance. W. Va. Agr. Exp. Sta. Bull. 488T: 1-25. 1963.
425. Walker, J. C., and P. N. Patel. Changed bean disease may cause trouble. Wis. Agr. Expt. Sta. Bull. 566: 44. 1963.
426. Walker, J. C., G. S. Pound, and P. H. Williams. TBR Globe arid Globelle cabbages which resist tipburn. Wis. Agr. Expt. Sta. Bull. 566: 46. 1963.
427. Chand, J. N., and J. C. Walker. Relation of age of leaf and varietal resistance to bacterial multiplication in cucumber inoculated with Pseudomonas lachrymans. Phytopathology 54: 49-50. 1964.
428. Chand, J. N., and J. C. Walker. Inheritance of resistancc to angular leafspot of cucumber. Phytopathology 54: 51-53. 1964.
429. Walker, J. C., and P. N. Patel. Splash dispersal and wind as factors in epidemiology of halo blight of bean. Phytopathology 54: 140-141. 1964.
430. Menke, G. H., P. N. Patel, and J. C. Walker. Physiology of Rhizopus stolonifer infection on carrot. Z. f. Pflanzenkrankheiten und Pflanzenschutz 71: 128-140. 1964.

431. Walker, J. C. Interview on disease resistance in plants. U. of Wis. Coll. of Agr. Newsrelease May 11, 1964.
432. Patel, P. N., J. C. Walker, D. J. Hagedorn, C. DeLeon-G, and N. Teliz-Ortiz. Bacterial brown spot of bean in central Wisconsin. Plant Dis. Reptr. 48: 335-337. 1964.
433. Walker, J. C., and P. N. Patel. Inheritance of resistance to halo blight in bean. Phytopathology 54: 952-954. 1964.
434. Patel, P. N., and J. C. Walker. Reactions of Phaseolus spp. to two races of Pseudomonas phaseolicola and a brown-spot organism. Phytopathology 54: 903. 1964.
435. Bickard, S. F., and J. C. Walker. Mode of inoculation and host nutrition in relation to bacterial wilt of bean. Phytopathology 55: 174-178. 1965.
436. Walker, J. C., and P. H. Williams. The inheritance of powdery mildew resistance in cabbage. Plant Dis. Reptr. 49: 198-201. 1965.
437. Walker, J. C. Black speck of cabbage. Report to the Central Florida Experiment Station on observations made January-March, 1965, on the disease in the vicinity of Sanford, Florida.
438. Walker, J. C. Host resistance as it relates to root pathogens and soil microorganisms. In Ecology of soil-borne plant pathogens. K. F. Baker and W. C. Snyder Ed. Univ. of California Press. pp 314-320. 1965.
439. Walker, J. C., P. H. Williams, and. G. S. Pound. Internal tipburn of cabbage. Its control through breeding. Wis. Agr. Expt. Sta. Res. Bul1. 258. 11 p. 1965.
440. Pound, G. S., P. H. Williams, and. J. C. Walker. Mosaic and. Yellows resistant inbred cabbage varieties. Wis. Agr. Expt. Sta. Res. Bull. 259. 30 p. 1965.
441. Rickard, S. F., and J. C. Walker. Carbon nutrition of the bean wilt bacterium. Phytopath. Z. 52: 19-28. 1965.
442. Rickard, S. F., and J. C. Walker. Nitrogen nutrition of the bean wilt bacterium. Phytopath. Z. 52: 131-144. 1965.
443. Nichols, C. G., W. H. Gabelman, R. H. Larson, and J. C. Walker. The expression and inheritance of resistance to pink root in onion seedlings. Phytopathology 55: 752-756. 1965.
444. Patel, P. N., and J. C. Walker. Resistance in Phaseolus to halo blight. Phytopathology 55: 889-894. 1965.
445. Walker, J. C. Patologia Vegetal. Translation into Spanish of Plant Pathology. 2nd edit. by Antonio Aguirre Azpeita. 818 p. Barcelona, 1965.
446. Walker, J. C. Disease resistance in the vegetable crops. III. Bot. Rev. 31: 331-380. 1965.
447. Patel, P. N., and J. C. Walker. Inheritance of tolerance to race 1 and race 2 of Pseudomonas phaseolicola. Phytopathology 55: 1070. 1965.
448. Walker, J. C. Use of environmental factors in screening for disease resistance. Ann. Rev. Phytopath. 3: 197-208. 1965.

449. Williams, P. H., and. J. C. Walker. Inheritance of Rhizoctonia bottom rot resistance in cabbage. Phytopathology 56: 367-368. 1966.
450. Darby, J. F., and J. C. Walker. Progress report of cabbage variety trials. 1965-1966. Central Fla. Experiment Sta. 15 pp. 1966.
451. Patel, P. N., and. J. C. Walker. Inheritance of tolerance to halo blight in bean. Phytopathology 56: 681-682. 1966.
452. Annon. Fellows of the American Phytopathological Society. Phytopathology 56: 380- 388. 1966.
453. Walker, J. C. Translation of Plant Pathology, 2nd ed., to Arabic. El Alwan Press, Cairo. 1966.
454. Walker, J. C. The role of pest resistance in new varieties. In Plant Breeding, a symposium held at Iowa State University (1965). K. J. Frey Ed. pp 219-242. 1966.
455. Dodman, R. L., K. R. Barker, and. J. C. Walker. Auxin production by Rhizoctonia solani. Phytopathology 56: 875. 1966.
456. Dodman, R. L., and J. C. Walker. Modes of penetration of different isolates of Rhizoctonia solani. Phytopathology 56: 875. 1966.
457. Walker, J. C., and J. F. Derby. Black speck and bacterial diseases. Cabbage variety trials 1966-1967. Central Fla. Agr. Exp. Sta. Mimeo Report CFES 67-1. 10 p. 1967.
458. Keen, N. T., P. H. Williams, and J. C. Walker. Characterization of a protease produced Pseudomonas lachrymans. Phytopathology 57: 257-262. 1967.
459. Keen, N. T., P. H. Williams, and J. C. Walker. Protease of Pseudomonas lachrymans in relation to cucumber angular leaf spot. Phytopathology 57: 263-271. 1967.
460. Struckmeyer, B. E., and J. C. Walker. The anatomy of internal tipburn of cabbage. Amer. J. Bot. 54: 228-231. 1967.
461. Patel, P. N., and J. C. Walker. Changes in free amino acid and amide content of bean, tobacco and oat leaves after infection with Pseudomonas phaseolicola, P. tabaci, and. P. coronafaciens, respectively, and in leaves treated with toxins. Intl. Symposium on Plant Pathology, New Delhi, India, Dec. 27, 1966-Jan. 1, 1967.
462. Walker, J. C. Concluding remarks. Intl. Symposium on virus diseases of rice. Apr. 28, 1967. Intl. Rice Research Institute, Los Banos, P.1. In The Virus Diseases of the Rice Plant. John Hopkins Press, 1969.
463. Dodman, R. L., K. R. Barker, and J. C. Walker. Modes of penetration by different isolates of Rhizoctonia solani. Phytopathology 58: 31-33. 1968.
464. Williams, P. H., J. C. Walker, and G. S. Pound. Hybelle and Sanibel multiple disease-resistant F1 hybrid cabbage. Phytopathology 58: 791-796. 1968.
465. Dodman, R. L., K. R. Barker, and J. C. Walker. A detailed study of the different modes of penetration by Rhizoctonia solani. Phytopathology 58: 1271-1276. 1968.

466. Walker, J. C. Plant Pathology. 3rd. edition. McGraw-Hill Co., New York. 1969. Not bound herein.
467. Chand, J. N., and J. C. Walker. Relation of free amino acids in cucumber leaves to the development of angular leaf spot. Phyt. Ztschr. 64: 94-97. 1969.
468. Walker, J. C. Plant Pathology, its background and importance. BSA McGraw-Hill Tape. 1969.
469. Walker, J. C. The story of disease resistance in cabbage. Tape I. 1969.
470. Walker, J. C. The story of disease resistance in cabbage. Tape II. 1969.
471. Walker, J. C. Six decades with cabbage clubroot. Tape 1969.
472. Halloin, J. M., J. C. Walker, G. A. de Zoeten, and G. Gaard. Effects of tentoxin on plastids of cucumber and cabbage. Phytopathology 59: 1028-1029. 1969.
473. Strandberg, J. O., J. F. Darby, J. C. Walker, and P. H. Williams. Black speck, a nonparasitic disease of cabbage. Phytopathology 59: 1879-1883. 1969.
474. Halloin, J. M., G. A. de Zoeten, G. Gaard, and J. C. Walker. The effects of tentoxin on chlorophyll synthesis and plastid structure in cuctumber and cabbage. Plant Physiol. 45: 310-314. 1970.
475. Yoshii, K., and J. C. Walker. Production of indolacetic acid in culture by Pythium debaryanum. Phytopathology 60: 1321. 1970.
476. Main, Charles E., and J. C. Walker. Physiological responses of susceptible and resistant cucumber to Erwinia tracheiphila. Phytopathology 61: 518-522. 1971.
477. Walker, J. C. Fusarium wilt of tomato. Amer. Phytopathol. Soc. Monograph No. 6. 1971.

George F. Sprague

University of Illinois
Urbana, Illinois, USA

1902–1998

1978 — for his research in plant pathology, developing of disease-resistant varieties of major food plants.

Professor Sprague (a geneticist – plant breeder) has achieved world eminence in the development of improved crop plants to the extent of greatly stabilizing the culture, increasing yields, and improving the quality of a number of major vegetable crops, and of maize – one of the world's dominant grain crops. He has developed novel and efficient methodology of breeding techniques and procedures. He was a towering figure in contributing to both applied and basic research. He had most enviable records of having trained scientific manpower for agriculture's needs. His students are to be found throughout the world, many of them having established themselves as distinguished researchers, educators and/or administrators. He served well the organizations of his respective science and agricultural industry, as well as the governments of the world.

In the annals of world agriculture, hybrid maize is the most spectacular example of the exploitation of a genetic phenomenon heterosis for increasing food production. Unquestionably Professor Sprague's name ranks foremost in the 40-year-long history of this great achievement of agricultural science. Moreover, many of the concepts, much of the theory and, any of the breeding models, which he developed for maize, have influenced the improvement of other crop plants.

Throughout his long and distinguished scientific career, Professor Sprague has worked untiringly to link theoretical quantitative genetic theory to practical plant breeding. Among his most fruitful basic studies was the development of a mathematical genetic model for selection that led to the development of an improved gene pool of maize germ plasm. Professor Sprague's genetic research laid the ground work, for improvement in nutritional quality in maize. A fact, which holds great promise to maize-eating nations. He conducted investigations, which demonstrated that protein quality of maize was genetically modifiable.

In summary, few people in the history of agriculture have had such a profound impact on the improvement of a major crop as has Professor Sprague.

CURRICULUM VITAE

Personal data:

Born: September 3, 1902; Crete, Nebraska

Degrees

B.S.	University of Nebraska	1924
M.S.	University of Nebraska	1926
Ph.D.	Cornell University (Genetics)	1930

Membership in Honorary Academic Societies

Alpha Zeta
Gamma Sigma Delta
Sigma Xi

Honors and Awards

a. Fellow, American Society of Agronomy (1947)
b. Crop Science Award, American Society of Agronomy (1957)
c. Faculty Citation, Iowa State Alumni Association (1958)
d. Honorary Doctor of Science, University of Nebraska (1958)
e. Gamma Sigma Delta Superior Teaching Award, Iowa State College (1958)
f. Distinguished and Superior Service Awards, USDA (1959 and 1965)
g. Corresponding Academician, Academia di Agricoltura de Bologna (1960)
h. Vice President and President, American Society of Agronomy (1960)
1. President, Crop Science Society of America (1961)
j. Elected, National Academy of Sciences (1968)
k. National Council of Commercial Plant Breeders Award (1972)

Professional Positions held

1924-29	Jr. Agronomist, I
1929-34	Asst. Agronomist,

1934-39 Assoc. Agronomist, USDA, Columbia MO
1939-58 Agronomist, USDA, Ames IA
1958-72 Leader, Corn and Sorghum Investigations, USDA, Washington, DC
1972 Present Professor, Plant Breeding and Genetics, University of Illinois, Urbana IL

SIGNIFICANCE OF CONTRIBUTIONS

Dr. Sprague's accomplishments have been primarily in three categories: development of new breeding methodology and the consequent production of exceedingly important inbreds widely used in commercial hybrid production, the training of graduate students, and the elaboration of genetic models dealing with such diverse traits in maize as endosperm color, leaf and scutellum conformation and chemical characteristics as oil content and protein quality.

We particularly cite Dr. Sprague's untiring effort to join theoretical quantitative genetic theory and practical plant breeding. Throughout his career he has had, by his own wish, responsibility both for basic research and applied plant breeding Unquestionably he is preeminent in his success in simultaneously conducting basic studies on the nature of heterosis and in developing hybrids for the farmer. Basic studies on the mathematics of selection led to the development of a gene pooi, "Super Stiff Stalk Synthetic", from which he isolated two lines, perhaps more widely grown than any others, B14 and B37. The nominators believe that 70% of the corn acres in the central Corn Belt in 1973 were planted to hybrids carrying either B14 or B37, or both.

Sprague's interest in nutritional quality led him to pioneer in the development of waxy (straight chain starch) hybrids which were used during World War II to produce a substitute for tapioca. In the 1940's, he and his students began a series of investigations which demonstrated that protein quality in corn was genetically modifiable. Unfortunately, the group failed to analyze the opaque-2 mutant and it remained for Purdue scientists to discover high lysine corn 20 years later. Oil content of the kernel also came under his study and he developed breeding schemes which rapidly increased it. Most of the research conducted by Sprague has been of long-range interest and fundamental to the successes of corn agriculture.

Most important and consequential to the development of modern plant breeding have been the following principles for which he was primarily responsible:

1. Early testing can be effective in the identification and isolation of superior germ plasm.
2. Specific and general combining ability are meaningful characteristics of the genetic variability of populations, can be described mathematically, and can provide a meaningful basis for the selection of appropriate testers.

3. Components of genetic variances and their interaction with different environments can be related to breeding methodology as well as provide information on the nature of gene action in populations.
4. The relative superiority of recurrent selection as an effective breeding procedure is dependent on the nature of gene action affecting the trait under investigation.
5. It is possible, and feasible, to alter composition of corn grain to make it of superior nutritional value.
6. Basic genetic information can he derived from properly designed breeding programs and breeding methods can become more effective by application of basic genetical and statistical principles.

Even though Sprague was deeply involved with research, he devoted a great deal of time and interest to graduate teaching. He taught a widely acclaimed course in corn breeding during his 18-year stay at Iowa State. He attracted students from around the world and served as major advisor for more than 50 M.S. and Ph.D. candidates. His excellence as a teacher was recognized by the Gamma Sigma Delta Superior Teaching Award, an unusual award for teachers of graduate courses.

Sprague's influence has not been confined to the U. S. He has long been a consultant with the Rockefeller Foundation and has traveled widely in that role. He also was involved with the Marshall Plan (ECA) after the war and was instrumental in the rapid spread of hybrid corn in Europe. Since 1963 he has played a central role in the U. S. Department of Agriculture activity in improvement of cereal production in Africa. He has represented agriculture ably in the National Academy of Science as chairman of the Section of Applied Biology.

Sprague's monumental contributions to theoretical plant breeding and to the improvement of maize are now made most evident by the wide use of his lines and of the basic breeding pools that arose under his direction. Breeding methodologies that he pioneered are widely used by commercial corn breeders in the U. S. and around the world. The success of corn improvement programs in East Africa, Latin America and Europe trace largely to the procedures he worked out and the carefully planned selection experiments which generated data to support his basic theories. Few people in the history of agriculture have had such a profound impact on the improvement of a major crop.

We particularly wish to emphasize the impact Sprague's research has had on practical agriculture, particularly that of the US Corn Belt. For example, inbred B37 was developed by Sprague from Iowa Stiff Stalk Synthetic, a variety developed by recurrent selection particulary to serve as a source of superior inbred. Because of the superiority of B37, hybrids carrying it in their pedigree have dominated the commercial market because of their superiority in yield, standability and early maturity. In 1970, it was the most extremely important, but other lines developed by Sprague himself and by others using his synthetics have been almost as consequential and continue to dominate corn agriculture.

LIST OF PUBLICATIONS

Sprague, G.F. 1927. Heritable characters of maize XXVII colored scutellum. J. Hered. XVIII: 41-44.

Sprague, G.F. 1929. Heterofertilization in maize. Science LXIX: 526-527.

Richey, F.D., and G.F. Sprague. 1931. Experiments on hybrid vigor and convergent improvement in corn. USDA Tech. Bull. 267. 22 p.

Sprague, G.F. 1932. The inheritance of colored scutellum in maize. USDA Tech. Bull. 292. 43 p.

Sprague, G.F., and M.N. Pope. 1932. A case of two simultaneous mutations for virescent seedlings in maize. Amer. Nat. LXVI: 284-85.

Sprague, G.F. 1932. The nature and extent of hetero-fertilization in maize. Genetics 17: 58-68.

Richey, F.D., and G.F. Sprague. 1932. Some factors affecting the reversal of sex expression in the tassels of maize. Amer. Nat. LXVI: 433-443.

Sprague, G.F. 1933. Pollen tube establishment and the deficiency of waxy seeds in certain maize crosses. Proc. Nat. Acad. Sci. 19: 838-841.

Sprague, G.F. 1934. Experiments on iarovising corn. J. Agr. Res. 48:1113-1120.

Richey, F.D., G.H. Stringfield, and G.F. Sprague. 1934. The loss in yield that may be expected from planting second generation double-crossed seed corn. J. Amer. Soc. Agron. 26: 196-199.

Sprague, G.F. 1935. Random sampling and the distribution of phenotypes on ears of backcrossed maize. J. Agr. Res. 51: 751-758.

Sprague, G.F. 1936. The relation of moisture content and time of harvest to germination of immature corn. J. Amer. Soc. Agron. 28: 472-478.

Stadler, L.J., and G.F. Sprague. 1936. Genetic effects of ultra-violet radiation in maize. I. Unfiltered radiation. Proc. Nat. Acad. Sci. 22: 572-578.

Stadler, L.J., and G.F. Sprague. 1936. Genetic effects of ultra-violet radiation in maize. II. Filtered radiation. Proc. Nat. Acad. Sci. 22: 579-583.

Stadler, L.J., and G.F. Spraque. 1936. Genetic effects of ultra-violet monochromatic 2537, and comparison-of effects of X-ray and ultra-violet treatment. Proc. Nat. Acad. Sci. 22: 584-591.

Sprague, G.F. 1936. Hybrid vigor and growth rates in a maize cross and its reciprocal. J. Agr. Res. 53: 819-830.

Perry, H.S., and G.F. Sprague. 1936. A second chromosome gene, Y_3, producing yellow endosperm color in maize. J. Amer. Soc. Agron. 28: 990-996.

Stadler, L.J., and G.F. Sprague. 1937. Contrasts in the genetic effects of ultra-violet radiation and X-rays. Science 85: 57-58.

Sprague, G.F. 1939. Corn hybrids for Missouri. Mo. Agr. Exp. Sta. Circ. 201. 27 p.

Sprague, G.F. 1939. An estimation of the number of top-crosses plants required for adequate representation of a corn variety. J. Amer. Soc. Agron. 31: 11-16.

Sprague, G.F. 1939. Heritable characters in maize 50 vestigal glume. J. Hered. 30: 143-145.

Bryan, A.A., R.C. Eckhardt, and G.F. Sprague. 1940. Spacing experiments with corn. J. Amer. Soc. Agron. 32:707-714.

Millang, Amy, and G.F. Sprague. 1940. The use of punched card equipment in predicting the performance of corn double crosses. J. Amer. Soc. Agron. 32: 815-816.

Sprague, G.F. 1941. The location of dominant favorable genes in maize by means of an inversion. Genetics 26: 170.

Sprague, G.F., and A.A. Bryan. 1941. The segregation of genes affecting yield prepotency, lodging and disease resistance in F_3 and- F_4 lines-of corn. J. Amer. Soc. Agron.- 33: 207-214.

Sprague, G.F. 1941. Transmission tests of maize mutants induced by ultra-violet radiation. Ia. Agr. Exp. Sta. Res. Bull. 292. p. 389-407.

Sprague, G.F. Building new corn hybrids. Ia. Farm Science Reporter 2: 7-9.

Sprague, G.F. 1942. Production of hybrid corn. Ia. Agr. Exp. Sta. Bull. P48. p. 556-582.

Sprague, G.F. 1942. Trends in corn breeding. 33rd Ann. Report; Nebr. Crop Imp. Assc. p. 50-57.

Sprague, G F., and L.A. Tatum. General vs. combining ability in single crosses of corn. J. Amer. Soc. Agron. 34: 923-932.

Hixon, R.M., and G.F. Sprague. 1942. Waxy starch of maize and other cereals. Ind. Eng. Chem. 34: 959-962.

Sprague, G.F., and R.M. Hixon. 1942. Waxy corn, a new crop. Ia. Farm Science Reporter 3: 10-11.

Sprague, G.F., and M.T. Jenkins. 1943. A comparison of synthetic varieties, multiple crosses and double crosses in corn. J. Amer. Soc. Agron. 35: 137-147.

Sprague, G.F., B. Brimhall, and R.M. Hixon. 1943. Some effects of the waxy gene in corn on properties of the endosperm starch. J. Amer. Soc. Agron. 35: 817-822.

Sprague, G.F. 1943. The problem of heterosis. Chronica Botanica 7: 418-419.

Hixon, R.M., G.S. Shepherd, and G.F. Sprague. 1943. Technological problems as related to war demands on corn. 8th Ann. Ia. Corn Res. Inst. Rpt. p. 7-10.

Sprague, G.F. 1944. The value of hybrid corn to the Iowa farmer. 9th Ann. Ia. Corn Res. Inst. Rpt. p. 7-11.

Kinman, M.L., and G.F. Sprague. 1945. Relation between number of parental lines and theoretical performance of synthetic varieties of corn. J. Amer. Soc. Agron. 37: 341-351.

Brimhall, B., G.F. Sprague, and J.E. Sass. 1945. A new waxy allele in corn and its effect on the properties of the endosperm starch. J. Amer. Soc. Agron. 37: 937-944.

Sprague, G.F. 1946. Early testing of inbred lines of corn. J. Amer. Soc. Agron. 38: 108-117.

Sprague, G.F. 1946. The experimental basis for hybrid maize. Biol. Review 21: 101-120.

Hansen, D.W., Brimhall, B., and G.F. Sprague. 1946. Relationship of zein to the total protein of corn. Cereal Chem. 23: 329-335.

Sprague, G.F., and J.C. Cunningham. 1946. Growing the bumper corn crop. Ch. 3 in A century of farming in Iowa — 1846–1946. p. 32–44.

Sprague, G.F., and J.C. Cunningham. 1946. One hundred years of corn growing in Iowa. Ia. Yearbook of Agr. (1945) 46: 258-274.

Federer, W.T., and G.F. Sprague. 1947. A comparison of variance components in corn yield trials: 1. Error, tester x line and line components in top cross experiments. J. Amer. Soc. Agron. 39: 453-463.

Sprague, G.F., and M.T. Jemkins. 1947. The development of waxy corn for industrial use. Ia. State Colil. Jour. Sci. 22: 205-213.

Sprague, G.F. Breeding for improved nutritional and industrial use. p. 42-49. In Growth and development of the corn plant. Published by Amer. Seed Trade Assn.

Groszmann, A., and G.F. Sprague. 1948. Comparative growth rates in a reciprocal maize cross. The kernel and its component parts. J. Amer. Soc. Agron. 40: 88-98.

Sprague, G.F. 1948. Les Bases Experirnentales de la production de Maiz Hybrid. p. 77-127. In L'Ameliorat.ion du Mais. Rabat, Morocco.

Sprague, G.F. 1948. Better protein, more oil in your corn. Iowa Farm Science 2: 3-4.

Sprague, G.F., and M.T. Jenkins. 1949. O Desenvolvimento do Metho Ceraceo para uso Industrial. Ceres 8: 112-123.

Frey, K.J., B. Brimhall, and G.F. Sprague. 1949. The effect of selection upon protein quality in the corn kernel. Agron. J. 41: 399-403.

Sass, J.E., and G.F. Sprague. 1949. Histological development of "accessory blade" and associated abnormalities in maize. Ia. State Coll. J. Sci. 23: 301-309.

Rossman, E.C., and G.F. Sprague. 1949. Effect of 2,4-D on yields of maize in the succeeding. generation after treatment. Plant Physiol. 24: 770-773.

Sprague, G.F., and B. Brimhall. 1949. Quantitative inheritance of oil in the corn kernel. Agron. J. 41: 30-33.

Sprague, G.F. 1949. Feeding quality of corn improved by breeding. Crops and Soils.

Sass, J.E., and G.F. Sprague. 1950. The embryology of "germiess" maize. Ia. State Coll. J. Sci. 24: 209-218.

Sprague, G.F., and B. Brimhall. 1950. Relative effectiveness of two systems of selection for oil content of the corn kernel. Agron. J. 42: 83-88.

Sprague, G.F., and P.A. Miller. 1950. A suggestion for evaluating current concepts of the genetic mechanism of heterosis in corn. Agron. J. 42: 161-162.

Sprague, G.F., J.H. Lilly, and D.D. Rubis. 1950. Can planting dates beat borers. Ia. Farm Sci. 4: 150-151.

Brimhall, B., and G.F. Sprague. 1951. Unsaturates of corn oil-inheritance and maturity studies. Cereal Chem. 28: 225-231.

Sprague, G.F., and W.T. Federer. 1951. A comparison of variance components in corn yield trials. II. Error, year x variety, location x variety and variety components. Agron. J. 43: 535-541.

Sprague, G.F., and P.A. Miller. 1951. New corn hybrids for Iowa. Ia. Farm Sci. 5: 137-138.

Sprague, G.F., and A.L. Lang. 1951. Further progress in hybrid maize production in European countries. O.E.E.C.

Sprague, G.F., P.A. Miller, and B. Brimhall. 1952. Additional studies of the relative effectiveness of two systems of selection for oil content of the corn kernel. Agron. J. 44: 329-331.

Sprague, G.F. 1952. Early testing and recurrent selection. Chap. 26, p. 400-417. In Heterosis. Ia. State Coil. Press.

Sprague, G.F., and P.A. Miller. 1952. The influence of visual selection during inbreeding on combining ability in corn. Agron. J. 44: 258-261.

Rojas, B.A., and G.F. Sprague. 1952. A comparison of variance components in corn yield trials. III. General and specific combining ability and their interaction with locations and years. Agron. J. 44: 462-466.

Sprague, G.F. 1953. Heterosis. Chapt. 7. In Growth and differentiation in plants. p. 113-136.

Sprague, G.F. 1954. Breeding for resistance to stalk rot. Ninth Ann. Hybrid Corn Industry Research Conf., Proc. p. 38-43.

Sprague, G.F. 1955. Corn breeding. Chapt. 5. In Corn., and corn improvement. p. 221-292. Academic Press. New York.

Sprague, G.F. 1955. Industrial utilization. Chapt. 14. In Corn and corn improvement. p. 613-636. Academic Press. New York.

Sprague, G.F. 1955. The world production of corn. Chap. 16. In Corn and corn improvement. p. 679-688. Academic Press. New York.

Sprague, G.F. 1954. The use of variance components as a guide in designing an adequate testing program. 2nd Pan Amer. Agron. Congress.

Sprague, G.F. 1954. Factors affecting choice of tester for the evaluation of inbred lines of corn. 2nd Pan Amer. Agron. Congress.

Sprague, G.F. 1954. A comparison of theoretical genetic implications of continuous self-fertilization and recurrent selection as methods of corn breeding. 2nd Pan Amer. Agron Congress.

LeRoux, P M., J.G. Dickson, and G.F. Sprague. 1954. A genetic basis for rust resistance in corn. Phytopathology 44: 496.

Hooker, A.L., G.F. Sprague, and W.A. Russell. 1954. Correlation of resistance to several stalk rots, ear rots and seedling blights and root necrosis in corn. Agron. Abstracts. p. 70.

Hooker, A.L., G.F. Sprague, and W.A. Russell. 1955. Resistance to rust Puccinia sorghi in corn. Agron. J. 47: 388.

Schuler, J.F., and G.F. Sprague. 1955. Natural mutations in inbred lines of maize and their heterotic effect. II. Comparisons of mother line vs. mutant when outcrosses to unrelated inbrecls. Genetics 41: 281-291.

Clem, Mary A., C.C. Mosier, and G.F. Sprague. 1956. Simplified punched card procedures for predicting double cross performance. Agron. J. 48: 319-320.

Sprague, G.F. 1955. Problems in the estimation and utilization of genetic variability. Cold Spring Harbor Symposia on Quantitative Biology, XX. p. 87-92.

Sprague, G.F., and A. Tavcar. 1956. Nais (Zea mays). General considerations and American breeding work. Handbuch der Pflanzenzuchtung II Band 2: 103-143.

Sprague, G.F. 1955. Replications vs. location. 10th Ann. Hybrid Corn Industiry-Research Conf., Proc. p. 10-13.

Pesek, J., G.F. Sprague, John Hanway, and F.A. Loeffel. 1955. The uptake of nitrogen, phosphorus and potassium by different corn single crosses as affected by population and nitrogen levels. Agron. Abst. 47: 32.

Sprague, G.F., and W.A. Russell. 1956. Some evidence on type of gene action involved in yield heterosis in maize. Int. Genenetics Symposia, Proc. (Tokyo). p. 522-526.

Sprague, G.F., W.A. Russell, and L.H. Penny. 1959. Further studies on convergent improvement in corn. Genetics 44: 341-346.

Sprague, G.F. 1969. Germplasm manipulations of the future. In Physiological aspects of crop yield. p. 375-396.

Sprague, G.F. 1970. World research needs for increasing corn production. Agr. Eng. 51: 71, 91 and 108.

Sprague, G.F. 1970. High lysine corn. Span. 13: 1-4.

Sprague, G.F. 1970. A method of breeding for resistance. 6th Indian Phytopathological Cof. Vol 23, June.

Russell, W.A., L.H. Peny, G.F. Sprague, W.D. Guthrie, and F.F. Dicke. 1970. Registration of maize (PL1-13) parental lines. Crop Sci. 11: 143.

Sprague, G.F., and H.H. McKinney. 1971. Further evidence of the behavior of AR in maize. Genetics 67: 533-543.

Sprague, G.F., and R.G. Dahms. 1971. Development of crop resistance to insects. J. Envir. Quality 1: 28-34.

Sprague, G.F. 1971. Genetic vulnerability in corn and sorghum. 26th Ann. Corn and Sorghum Res. Conf., Proc. p. 96-104.

Sprague, G.F. 1972. The genetics of corn breeding. Stadler Symposium 4: 69-81.

Eberhart, S.A., and G.F. Sprague. 1973. A major cereals project to improve maize, sorghum and millet in Africa. Agron. J. 65: 365-373.

Sprague, G.F. 1973. Increasing crop yields: technical measures for increasing productivity. In Man, food and nutrition. CRC Press. M. Rechcigl, Jr. (ed.). p. 91-107.

Sprague, G.F. 1975. Opportunities and limiations in improving corn quality through genetic changes. In Lowell D. Hill (ed.) Corn quality in world markets. Interstate Printers and Publishers, Inc. Danville. p. 77-86.

Sprague, G.F. 1975. The development of hybrid corn technology in the United States and selected countries. Agency for International Development, Technical Assistance Bureau. Tech. Series Bul. 16. p. 1-80.

Sir Kenneth Blaxter

Rowett Research Institute
Aberdeen, Scotland, United Kingdom

1919–1991

1979 — for his fundamental contributions to the science and practice of ruminant nutrition and livestock production.

BIOGRAPHICAL SKETCH

Sir Kenneth was trained in agricultural science at the University of Reading and in 1939 he joined the staff of the National Institute for Research in Dairying to work on aspects of the nutrition of cattle. After a period in the Armed Services he returned to Reading and later moved to the Central Veterinary Laboratory at Weybridge. Prom there, after a period spent working with the late Dr H.H. Mitchell at the University of Illinois, he was appointed Head of the Nutrition Department of the Hannah Research Institute, Ayr, where he remained for 17 years until, in 1995, he was appointed Director of the Rowett Research Institute, Aberdeen.

Sir Kenneth has made outstanding contributions to knowledge of nutrition, particularly ruminant nutrition. In his earlier days he did important work on the nutritive value of feeds for dairy cattle, on protein metabolism and on the protein requirements of cattle. In the fields of endocrinology he tackled the then controversial subject of iodinated proteins. Some of the work on protein requirements formed part of an extensive series of experiments on the metabolism and nutritional needs of the very young calf and its relation to maternal feeding which Sir Kenneth began soon after going to the Hannah Research Institute in 1948. This work is so outstanding in its originality and elegance that it is now regarded as classic; it provided a basis for the development of systems of artificial rearing of calves with specially formulated diets which have revolutionized commercial calf production.

These researches on calf nutrition led to an important series of original investigations on muscular dystrophy in calves and its relation to dietary vitamin E. The fundamentals of the subject were studied by Sir Kenneth at the Hannah Institute, and in collaboration with a Veterinary Investigation Officer in the North of Scotland he initiated large-scale field trials which showed the value of vitamin E, and selenium, in preventing the occurrence of muscular dystrophy, which at that time was accounting for a high proportion of calf mortality, in some northern areas of the country. During this period Sir Kenneth also studied magnesium metabolism in cattle and was the first to produce an experimental magnesium deficiency and to demonstrate that the disease in the field is a deficiency state.

The main work on which Sir Kenneth has latterly been engaged is the energy metabolism of sheep and cattle. He has made an extensive study of the energy requirements of these animals for growth and maintenance and for the production of milk, meat and wool, and of the ability of different feeds and different environments to meet these requirements. This subject is of the greatest importance, but because of the many difficulties involved in its study it had been neglected for many years. It had become evident that the subject should be thoroughly reinvestigated with modern equipment and in the light of modern knowledge, and. this task Sir Kenneth undertook. He designed and built a series of very efficient respiration chambers for animals of various sizes, from goats and. sheep up to adult cows and steers, his largest chamber involving a completely new conception in calorimetric technique.

Sir Kenneth's work on the energy metabolism and requirements of farm animals has culminated in a new system for expressing the energy requirements of ruminants and for the evaluation of ruminant feeds, in which the energy requirements of the animal are expressed in terms of units of "metabolizable energy" (ME). This system has the great merit of expressing nutritive requirements in terms of a unit by which feeds can be readily assayed and it has been adopted as the basis of feeding standards in the UK and in many other countries. That this work is well-known is due in part to Sir Kenneth's support of international vehicles of communication, particularly the conferences of the European and World Associations of Animal Production.

Besides being the foremost authority in the United Kingdom, and probably in the world, on the energy metabolism and energy requirements of ruminants, Sir Kenneth now occupies a pre-eminent international position in nutritional science; he is a former Chairman of the British National Committee for Nutritional Sciences of the Royal Society, and was a member of Commission VI of the International Union of Nutritional Sciences.

More specifically in the field of human nutrition Sir Kenneth has been a member of the UK Joint Agricultural Research Council/Medical Research Council Committee on Food and Nutrition and Chairman of its Working Party on Energy Foods for Man. He was also a member of the UN Committee on Proteins.

His total concern with the effectiveness of animal production as a source of human food is reflected in a number of very significant papers Relative efficiencies of farm animals in using crops and by-products in production of foods, World Conference on Animal Production, Maryland 1969; Approaches to the problem of augmentation of animal protein food sources in Asia, Asian Congress of Nutrition, Hyderabad 1972; Nutrients required for animal production as related to the world food supply, Centennial Symposium, Guelph 1975; which demonstrate his determination that the scientific resources of his chosen field shall be directed to the betterment of mankind and the enhancement of the brotherhood of the peoples of the world.

Through each Annual Report of Sir Kenneth's Institute attention is drawn to the part played in the training of visiting research workers, many of them from overseas, and bringing them together into a scientific and social fraternity. The Institute also houses the Commonwealth Bureau of Nutrition of which Sir Kenneth is Chairman and Consultant Director. The Bureau, which provides an information service for research workers in human and animal nutrition, is organized by nations of the British Commonwealth and it is of great value not only to the Commonwealth, but throughout the world.

CURRICULUM VITAE

Academic Training:

1930-1935 - City of Norwich School
1936-1939 - University of Reading

Academic Ttraining:

1939 - B.Sc. (Agric.) University of Reading
1939 - N.D.A. (Hons.)
1944 - Ph.D. University University of Reading
1952 - D.Sc. University of Reading
1974 - D.Sc. (honoris causa) Queen's University, Belfast
1975 - D.Sc. Agr. (honoris causa) University, Norway
1977 - D.Sc. Agr. (honoris causa) University of Leeds

Professional Qualifications:

1962 - Fellowship of Royal Society of Arts (F.R.S.A.)
1963 - Fellowship of Institute of Biology (F. Inst. Biol.)
1964 - Fellowship of Royal Agricultural Society of England (F.R.Agr.E.)
1965 - Fellowship of Royal Society of Edinburgh (F.R.S.E.)
1967 - Fellowship of Royal Society (F.R.S.)
1970 - Foreign Member of the Lenin Academy of Agricultural Sciences (USSR)

CAREER:

1939-1940 - Research Assistant at The National Institute for Research in Dairying
1941-1944 - "
1940-1941 - War Service (Royal Artillery)
1944-1946 - Research Officer, Ministry of Agriculture, Veterinary Laboratory, Weybridge
1947-1948 - "
1946-1947 - Commonwealth Fund Fellow at the University of Illinois, Division of Nutrition under Professor H. H. Mitchell
1948-1965 - PSO then SPSO, Head, Nutrition Department, Hannah Dairy Research Institute, Kirkhill, Ayr, Scotland, DCSO, Special Merit
Since 1965 - CSO, Director, Rowett Research Institute, Bucksburn, Aberdeen, and Consultant
1975 - Director, Commonwealth Bureau of Nutrition, Under Secretary, Special Merit

HONORARY APPOINTMENTS:

1958-1961 - Olive Belirens Lecturer in Faculty of Agriculture, University of Leeds
1962 - Visiting Lecturer, Berlin Academy, D.D.R.
1964-1965 - Visitiiag Lecturer, The National Research Council of Canada, to Universities of British Columbia
1965 - Manitoba, Saskatchewan. and Alberta, Corresponding member of the French Academy of Sciences

PROFESSIONAL ACTIVITIES:

1970-1971 - President, British Society of Animal Production
1974-1977 - President, The Nutrition Society
1965 - Editor, Journal of Agricultural Science

PROFESSIONAL HONORS AND PRIZES:

1939 - Fream Memorial Prize
1960 - Thomas Baxter Prize and Gold Medal (for research), on the nutrition of cattle)
1964 - Gold Medal of the Royal Agricultural Society of England (for research on the nutrition of farm animals)
1969 - Silver Medal of the Czechoslovakian Academy of Agriculture
1971 - Gold Award "Uovo D'Oro" of the Italian Society of Animal Production (for research in animal production)
1973 - Wooldridge Memorial. Medal of the British Veterinary Association (for research in animal science)

1976 - Messel Medal of Society of Chemical Industry (for contribution to agricultural chemistry)
1976 - Gustaf de Laval Medal of the Royal Swedish Academy of Engineering Sciences, Sweden
1977 - Keith Prize, Royal Society of Edinburgh (for services to the Society)
1977 - Knight Bachelor, H.M. The Queen, Jubilee Birthday Honours List
1977 - Massey-Ferguson National Award. (for services to United Kingdom agriculture)
1978 - Honorary Associateship of the Royal College of Veterinary Surgeons

NOMINATED LECTURES:

1962 - Fernhurst Lecture, Royal Society of Arts
1964 - Samuel Brody Memorial Lectures, University of Missouri, U.S.A.
1965 - Scott Robertson Memorial Lecture, Queen's University, Belfast
1972 - Blackman Lecture, University of Oxford
1973 - Wooldridge Memorial Lecture (British Veterinary Association)
1974 - Jubilee Lecture of the Agricultural Research Council of Norway
1976 - Hammond Lecture, British Society of Animal Production
1976 - Jubilee Lec1ture of the University of Reading
1977 - Wi1liam Dick Memorial Lecture, Edinburgh
1978 - Storer Lectures, University of California, Davis, California, U.S.A.

LIST OF PUBLICATIONS

1. Straw pulp: Recent experiments. K. L. Blaxter, S. Bartlett. Journal of the Ministry of Agriculture, 50, 224-226 (1943).
2. The normal variation in the heart rate of dairy cows. K. L. Blaxter. British Veterinary Journal, 99, 2-4 (1943).
3. Stimulation of the milk production of dairy cows by feeding thyroid-active iodinated proteins. K. L. Blaxter. Nature, London, 152, 751-752 (1943).
4. Experiments on the use of homegrown foods for milk production. 1. The effect of war-time can change food supply on the nutrient intake and milk production of dairy cows. K. L. Blaxter. Journal of Agricultural Science, 34, 22-26 (1944).
5. Experiments on the use of homegrown foods for mill: production. 2. The effect of feeding concentrated and bulky foods prior to calving on subsequent milk production. K. L. Blaxter. Journal of Agricultural Science, 34, 37-44 (1944).
6. Experiments on the use of homegrown foods for milk production. 3. The effect of over and under feeding in mid lactation. K. L. Blaxter. Journal of Agricultural Science, 34, 213-216 (1944).

7. Experiments on the use of homegrown foods for milk production. 4. Methods of feeding balky foods. K. L. Blaxter, T. H. French. Journal of Agricultural Science, 34, 217-222 (1944).
8. Food preference and habits of dairy cows. K. L. Blaxter. Proceedings of the British Society of Animal Production, pp. 85-94 (1945).
9. Variation in some physiological activities of dairy cows. K. L. Blaxter, H. Price. British Veterinary Journal, 101, 18-21 & 39-48 (1945).
10. The preparation and biological effects of iodinated protein. 1. The effects of preparations of low activity and of iodinated protein. K. L. Blaxter. Journal of Endocrinology, 4, 266-269 (1945).
11. The preparation and biological effects of iodinated protein. 2. The effects of iodinated casein. K. L. Blaxter. Journal of Endocrinology, 4, 237-265 (1945).
12. Experiments with iodinated casein on farms in England and Wales. K. L. Blaxter. Journal of Agricultural Science, 36, 117-150 (1946).
13. Experiments on the use of homegrown foods for milk production. 5. The protein requirements of growing dairy heifers. K. L. Blaxter, H. Price. Journal of Agricultural Science, 36, 301-309 (1946).
14. Excretion of lead in the bile. K. L. Blaxter, A. T. Cowie. Nature, London, 151, 588 (1946).
15. The effect of feeding shark liver oil to cows on the yield, composition, vitamin A and carotene content of the milk. K. L. Blaxter, S. L Kon, S. Y. Thompson. Journal of Dairy Research, 14, 225-230 (1946).
16. The value of urea as a substitute for protein in the ration of dairy cattle: field trials with dairy cows. K. L. Blaxter, S. Bartlett. Journal of Agricultural Science, 37, 32-34 (1947).
17. The factorization pf the protein requirements of ruminants and of the protein value of feeds with particular reference to the metabolic faecal nitrogen. K. L. Blaxter, H. H. Mitchell. Journal of Animal Science, 7, 351-372 (1948).
18. Severe experimental hyperthyroidism in the ruminant. 1. Metabolic effects. K. L. Blaxter. Journal of Agricultural Science, 38, 1-19 (1948).
19. Severe experimental hyperthyroidism in the ruminant. 2. Physiological effects. K. L. Blaxter. Journal of Agricultural Science, 38, 20-27 (1948).
20. The effect of iodinated. casein on the basal metabolism of sheep. K. L. Blaxter. Journal of Agricultural. Science, 38, 207-215 (1948).
21. The evaluation of the nutritive value of feeding stuffs. K. L. Blaxter. Analyst, 73, 11-15 (1948).
22. The role of thyroidal materials and of synthetic goitrogens in animal production and an appraisal of their practical use. K. L. Blaxter, E. P. Reineke, E. W. Crampton, W. B. Petersen. Journal of Animal Science, 8, 307-352 (1949).

23. The protein and energy nutrition of the young calf. K. L. BLaxter. Agricultural Progress, 25, 1-9 (1949).
24. Iodinated, protein and lactation. K. L. Blaxter. Journal of the Ministry of Agriculture, 57, 413-417 (1950).
25. Lead as a nutritional hazard to farm animals. 1. The determination of lead in biological material. K. L. Blaxter, R. Allcroft. Journal of comparative Pathology, 60, 133-139 (1950).
26. Lead as a nutritional hazard to farm livestock. 2. The absorption and excretion of lead by sheep and rabbits. K. L. Blaxter. Journal of Comparative Pathology (1950).
27. Lead as a nutritional hazard to farm livestock. 3. Factors affecting the distribution of lead in the tissues. K. L. Blaxter. Journal of Comparative Pathology, 60, 177-189 (1950).
28. Lead as a nutritional hazard to farm livestock. 5. The toxicity of lead to cattle and sheep and an evaluation of the lead hazard under farm conditions. R. Allcroft, K. L. Blaxter. Journal of Comparative Pathology, 60, 209-218 (1950).
29. Energy feeding standards for dairy cattle. Nutrition Abstracts & Reviews, 20, 1-21 (1950).
30. The nutrition of the young Ayrshire calf. I. The endogenous nitrogen and basal energy metabolism in the calf. K. L. Blaxter, W. A. Wood. British Journal of Nutrition, 5, 11-25 (1951).
31. The nutrition of the young Ayrshire calf. II. A spirometer for the determination of the respiratory exchange of the calf. K. L. Blaxtar, A. Howells. British Journal of Nutrition, 5, 25-29 (1951).
32. The nutrition of the young Ayrshire calf. III. The metabolism of the calf during starvation and. subsequent realimentation. K. L. Blaxter, W. A. Wood. British Journal of Nutrition, 5, 29-55 (1951).
33. The nutrition of the young Ayrshire calf. IV. Some factors affecting the biological value of protein determined by nitrogen-balance methods. K. L. Blaxter, W. A. Wood. British Journal of Nutrition, 5, 55-67 (1951).
34. Separation of the tocopherols by paper chromatography. P. Brown, K. L. Blaxter. Chemistry & Industry, pp. 633-634 (1951).
35. The importance of fat-soluble vitamins for calves. K. L. Blaxter. Journal of the Royal Agricultural Society of England, 112, 25-35 (1951).
36. Conversion factors for vegetable and animal foods for human consumption. K. L. Blaxter. British Journal of Nutrition, 5, 250-255 (1951).
37. The nutrition of the young Ayrshire calf. V. The nutritive value of whole cows' milk. K. L. Blaxter, W. A. Wood. British Journal of Nutrition, 6, 1-11 (1952).

38. The nutrition of the young Ayrshire calf. VI. The utilization of the energy of whole milk. K. L. Blaxter. British Journal of Nutrition, 6, 12-18 (1952).
39. The nutrition of the young Ayrshire calf. VII. The biological value of gelatin and casein when given as the sole source of protein. K. L. Blaxter, W. A. Wood. British Journal of Nutrition, 6, 56-70 (1952).
40. The nutrition of the young Ayrshire calf. VIII. Muscular dystrophy in the growing calf. K. L. Blaxter, P. S. Watts, W. A. Wood. British Journal of Nutrition, 6, 125-143 (1952).
41. The nutrition of the young Ayrshire calf. IX. The composition of the tissues of normal and dystrophic calves. K. .L. Blaxter, W. A. Wood. British Journal of Nutrition, 6, 144-163 (1952).
42. The nutrition of the young Ayrshire calf. X. The histopathology of muscular dystrophy and its relation to muscle chemistry. A. M. MacDonald, K. L. Blaxter, P. S. Watts, W. A. Wood. British Journal of Nutrition, 6, 164-169 (1952).
43. Some effects of thyroxine and iodinated casein on dairy cows and their practical significance. K. L. Blaxter. "Vitamins & Hormones", 10, 218-250 (1952).
44. Vitamin B in the nutrition of farm animals. K. L. Blaxter, P. Brown. Nutrition Abstracts & Reviews, 22, 1-21 (1952).
45. The utilization of the minerals, vitamins and other constituents of grass. K. L. Blaxter. British Journal of Nutrition, 6, 110-117 (1952).
46. The influence of diet on the development of the alimentary tract of the calf. K. L. Blaxter, M. K. Hutcheson, J. N. Robertson, A. L. Wilson. British Journal of Nutrition, 6, i-ii (1952).
47. The effect of Mg deficiency on the energy exchange of calves. K. L. Blaxter, J. A. F. Rook. Journal of Physiology, 121, 48-49 (1953).
48. The nutrition of the young Ayrshire calf. XI. The toxicity of cod-liver oil. K. L. Blaxter, W. A. Wood, A. N. MacDonald. British Journal of Nutrition, 7, 34-50 (1953).
49. The nutrition of the young Ayrshire calf. XII. Factors affecting the tocopherol reserves, muscle composition and muscle histology of 4-day old calves. K. L. Blaxter, P. Brown, A. M. MacDonald. British Journal of Nutrition 7, 105-123 (1953).
50. The nutrition of the young Ayrshire calf. XIII. The toxicity of the unsaturated acids of codliver oil. K. L. Blaxter, P. Brown, A. M. MacDonald. British Journal of Nutrition, 7, 287-298 (1953).
51. The nutrition of the young Ayrshire calf. XIV. Some effects of natural and synthetic antioxidants on the incidence of muscular dystrophy induced by codliver oil. K. L. Blaxter, F. Brown, W. A. Wood, A. M. MacDonald. British Journal of Nutrition, 7, 337-349 (1953).

52. Tocopherol acetate for muscular dystrophy in calves. K. L. Blaxter. Veterinary Record, 65, 496 (1953).
53. Muscular dystrophy. K. L. Blaxter. Veterinary Record, 65, 835-838 (1953).
54. Diet and muscular disease in calves. K. L. Blaxter. Proceedings of the British Society of Animal Production, pp. 94-101 (1953).
55. Prevention and cure of enzootic muscular dystrophy. K. L. Blaxter, G. A. M. Sharman. Nature, London, 172, 1006 (1953).
56. The heat of combustion of the tissues of cattle in relation to their chemical composition. K. L. Blaxter, J. A. P. Rook. British Journal of Nutrition, 7, 83-91 (1953).
57. Indirect determination of water retention in cattle as a means of estimating energy exchange. K. L. Blaxter, J. A. P. Rook. Nature, London, 171, 609-610 (1953).
58. Respiration calorimetry with farm animals. K. L. Blaxter, N. McC. Graham, J. A. F. Rook. Journal of Physiology, 121, 39-40 (1953).
59. Some observations on the biochemical and physiological events associated with diarrhoea in calves. K. L. Blaxter. Veterinary Record, 65, 889-892 (1953).
60. Experimental magnesium deficiency in calves. I. Chemical and pathological observations. K. L. Blaxter, J. A. F. Rook, A. M. MacDonald. Journal of Comparative Pathology, 64, 157-175 (1954).
61. Experimental magnesium deficiency in calves. II. The metabolism of calcium, magnesium and nitrogen and magnesium requirements. K. L. Blaxter, J. A. F. Rook. Journal of Comparative Pathology, 64, 176-186 (1954).
62. The origin of tetany in magnesium deficiency. K. L. Blaxter, J. A. F. Rook, A. M. MacDonald. Proceedings of the Nutrition Society, 13, ii-iii (1954).
63. Problems associated with the feeding of whole milk. K. L. Blaxter. Proceedings of the British Veterinary Conference, Publication No. 24, 38-43 (1954).
64. Apparatus for the determination of the energy exchange of calves and of sheep. K. L. Blaxter, N. McC. Graham, J. A. F. Rook. Journal of Agricultural science, 45, 10-18 (1954).
65. Plane of nutrition and energy utilization by sheep. K. L. Blaxter, N. McC. Graham. Proceedings of the Nutrition Society, 13, vii-viii (1954).
66. Calf rearing on dairy farms. K. L. Blaxter. Farming Leader, No. 26, 3-4 and 15 (1954).
67. Ruminant nutrition. Chapter in 'Progress in the Physiology of Farm Animals'. K. L. Blaxter. (J. Hammond, Editor) Butterworth's Press, London. Vol. 1, 3-39 (1954).
68. Hypomagnesaemic tetany in beef cattle. K. L. Blaxter, G. M. Sharman. Veterinary Record, 67, 108-115 (1955).
69. Energy and carbohydrate metabolism in magnesium-deficient calves. K. L. Blaxter, J. A. F. Rook. British Journal of Nutrition, 9, 121-132 (1955).

70. Vitamin E and experimental muscular degeneration in calves. K. L. Blaxter. Proceedings of the IIIrd International Vitamin E Congress, Venice, I, p. 79, no. 174 (1955).
71. Enzootic muscular dystrophy in cattle in Scotland. K. L Blaxter. Proceedings of the IIIrd International Vitamin B Congress, Venice, I, p. 79, no. 175 (1955).
72. The myopathies of cattle, sheep, horses and pigs and the role of vitamin B in their prevention. K. L. Blaxter. Proceedings of the IIIrd International Vitamin E Congress, Venice, II, pp. 622-630 (1954).
73. Muscular dystrophy. K. L. Blaxter, Rosemary P. McGill. Veterinary Reviews and Annotations, 1, 91-114 (1955).
74. Triorthocresyl phosphate and vitamin B metabolism in calves. B. Cowlishaw, K. L. Blaxter. Proceedings of the Nutrition Society, 14, p. iii (1955).
75. Plane of nutrition and starch equivalents. K. L. Blaxter, N. McC. Graham. Journal of Agricultural Science, 46, 292-306 (1955).
76. Methods of assessing the energy value of foods for ruminant animals. K. L. Blaxter, N. McC. Graham. Proceedings of the Nutrition Society, 14, 131-139 (1955).
77. Inter-relationships between passage of food through the digestive tract and its digestibility. K. L. Blaxter, N. McC. Graham, F. W. Wainman. Proceedings of the Nutrition Society, 14, p. iv (1955).
78. Calorimetric measurements of the nutritive value for sheep of dried grass prepared in different ways. K. L. Blaxter, N. McC. Graham. Proceedings of the Nutrition Society, 14, p. xv (1955).
79. Magnesium content of bone in hypomagnesaemic disorders of livestock. K. L. Blaxter. 'Bone Structure and Metabolism', pp. 117-131 (1956) (Churchill, London).
80. Magnesium metabolism in cattle. K. L. Blaxter, Rosemary P. McGill. Veterinary Reviews & Annotations, 2, 35-55 (1956).
81. Muscular dystrophy of beef cattle and unsaturated fats. G. A. Garton, W. R. H. Duncan, K. L. Blaxter, Rosemary P. McGill, G. A. M. Sharman, M. K. Hutcheson. Nature, London, 177, 792-793 (1956).
82. Myopathy in sheep. K. L. Blaxter, A. M. MacDonald. Lancet, 271, 895 (1956).
83. A note on the occurrence of muscular dystrophy in hoggs in the north of Scotland. T. Marr, G. A. M. Sharman, K. L. Blaxter. Veterinary Record, 68, 408-410 (1956).
84. The effect of the grinding and cubing process on the utilization of the energy of dried grass. K. L. Blaxter, N. McC. Graham. Journal of Agricultural Science, 47, 207-217 (1956).
85. The heat increment of a mixture of steam volatile fatty acids and of glucose in the fasting sheep. D. G. Armstrong, K. L. Blaxter. Proceedings of the Nutrition Society, 15, p. i (1956).

86. Frequency of feeding and energy utilization by sheep. K. L. Blaxter, N. McC. Graham, F. W. Wainman. Proceedings of the Nutrition Society, 15, p. ii (1956).
87. The nutritive value of feeds as sources of energy. K. L. Blaxter. Journal of Dairy Science, 39, 1396-1424 (1956).
88. Heat increments of feeding in ruminants. D. G. Armstrong, K. L. Blaxter. Nature, London, 177, 1183-1184 (1956).
89. The indirect determination of energy retention in farm - animals. I. The development of method. K. L. Blaxter, J. A. F. Rook. Journal of Agricultural Science, 48, 194-209 (1956).
90. The indirect determination of energy retention in farm animals. II. Studies on the validity of the method. K. L. Blaxter, J. A. F. Rook. Journal of Agricultural Science, 48, 210-219 (1956).
91. Starch equivalents, ration standards and milk production. K. L. Blaxter. Proceedings of the British Society of Animal Production, pp. 3-31 (1956).
92. Some observations on the digestibility of food by sheep and on related problems. K. L. Blaxter, N. McC. Graham, F. W. Wainman. British Journal of Nutrition, 10, 69-91 (1956).
93. Myopathic conditions in animals. K. L. Blaxter. Proceedings of the 75th Annual Congress of the British Veterinary Association, pp. 3-8 (1957).
94. Utilization of the end. products of ruminant digestion. D. G. Armstrong, K. L. Blaxter, N. McC. Graham. Proceedings of the British Society of Animal Production, pp. 3-15 (1957).
95. The heat increment of steam volatile fatty acids in fasting sheep. D. O. Armstrong, K. L. Blaxter. British Journal of Nutrition, 11, 247-272 (1957).
96. The heat increments of mixtures of steam volatile fatty acids in fasting sheep. D. O. Armstrong, K. L. Blaxter, N. McC. Graham. British Journal of Nutrition, 11, 392-408 (1957).
97. The utilization of acetic, propionic and butyric acids by fattening sheep. D. O. Armstrong, K. L. Blaxter. British Journal of Nutrition, 11, 413-425 (1957).
98. The energetic efficiency of fat synthesis from mixtures of steam-volatile fatty acids in sheep. D. G. Armstrong, K. L. Blaxter, N. McC. Graham, F. W. Weinman. Proceedings of the Nutrition Society, 16, p. xvii (1957).
99. Energy retention of sheep in relation to environmental temperature. K. L. Blaxter, N. McC. Graham, F. W. Wainman. Proceedings of the Nutrition Society, 16, pp. iii-iv (1957).
100. The effects of defective nutrition during pregnancy in farm livestock. K. L. Blaxter. Proceedings of the Nutrition Society, 16, 52-58 (1957).
101. Iron-deficiency anaemia in calves. K. L. Blaxter, G. A. M. Sharman, A. M. MacDonald. British Journal of Nutrition, 11, 234-246 (1957).

102. Anaemia in calves. K. L. Blaxter, M. K. Hutcheson, G. A. M. Shaman, A. M. MacDonald. Proceedings of the Nutrition Society, 16, p. i (1957).
103. Vitamin D_3 toxicity in calves. P. S. Blackburn, K. L. Blaxter, Elizabeth J. Castle. Proceedings of the Nutrition Society, 16, p. xvi (1957).
104. The fa-soluble vitamins. K. L. Blaxter. Annual Review of Biochemistry, 26, 275-306 (1957).
105. The utilization of the energy of two mixtures of steam-volatile fatty acids by fattening sheep. D. G. Armstrong, K. L. Blaxter, N. McC. Graham, F. W. Wainman. British Journal of Nutrition, 12, 177-188 (1957).
106. The heat increment of acetic-acid partially with sodium hydroxide, in fasting sheep. D. G. Armstrong, K. L. Blaxter, N. McC. Graham. Proceedings of the Nutrition Society, 17, xxxiii (1957).
107. Closed circuit respiration apparatus for the cow and steer. F. W. Wainman, K. L. Blaxter. Proceedings of the 1st International Symposium on Energy Metabolism, Copenhagen, Publication no. 8 from the European Association for Animal Production, pp. 80-84 (1958).
108. Developments in the design of closed-circuit respiration apparatus for animals weighing up to 70 kg. F. W. Waimnan, K. L. Blaxter. Proceedings of the 1st International Symposium on Energy Metabolism, Copenhagen, Publication no. 8 from the European Association for Annual Production, pp. 85-95 (1958).
109. 1958 109 Analytical and. other techniques used. in respiration calorimetry and their errors. N. MMc. Graham, K. L. Blaxter, D. G. Armstrong. Proceedings of the 1st International Symposium on Energy Metabolism, Copenhagen, Publication no. 8, Publication no. 8, pp. 157-163 (1958).
110. Nutrition and climatic stress in farm animals. K. L. Blaxter. Proceedings of the Nutrition Society, 17, 191-197 (1958).
111. The effect of environmental temperature and level of feeding on food utilization by sheep. D. G. Armstrong, K. L. Blaxter, N. McC. Graham, F. W. Wainman. Proceedings of the Nutrition Society, 17, p. xxxii (1958).
112. The nutrition of dairy cows. K. L. Blaxter. The Ayrshire Cattle Society's Journal, 30, 10-12 and 14-17 (1958).
113. The nutritional requirements of dairy cows. K. L. Blaxter. Proceedings of the Conference on the Scientific Principles of Feeding Farm Live stock, Brighton, 21-36 (1958).
114. Prevention of enzymatic muscular dystrophy by selenium administration. G. A. M. Shaman, K. L. Blaxter, R. S. Wildon. Veterinary Record, 71, 536 (1959).
115. Environmental temperature, energy metabolism and heat regulation in sheep. I. Energy metabolism in closely clipped sheep. N. McC. Graham, F. W. Wainman, K. L. Blaxter, D. G. Armstrong. Journal of Agricultural Science, 52, 13-24 (1959).

116. Environmental temperature, energy metabolism and heat regulation in sheep. II. The partition of heat losses in closely clipped sheep. K. L. Blaxter, N. McC. Graham, F. W. Wainman, D. G. Armstrong. Journal of Agricultural Science, 52, 25-40 (1959).
117. Environmental temperature, energy metabolism and heat regulation in sheep. III. The metabolism and thermal exchanges of sheep with fleeces. K. L. Blaxter, N. McC. Graham, F. W. Wainman. Journal of Agricultural Science, 52, 41-49 (1958).
118. The effect of environmental conditions on food. Utilization by sheep. D. G. Armstrong, K. L. Blaxter, N. McC. Graham, F. W. Wainman. Animal Production, 1, 1-12 (1959).
119. Feeding standards for livestock. K. L. Blaxter. Journal of the Farmers' Club, part 2, 14-26 (1959).
120. Minerals in relation to disease. K. L. Blaxter. Veterinary Annual, pp. 184-193 (1959).
121. Potassium and hypomagnesaemic tetany in calves. K. L. Blaxter, B. Cowlishaw, J. A. F. Rook. Animal Production, 2, 1-10 (1959).
122. The utilization of' the energy of grassland products. K. L. Blaxter. Proceedings of the 8th International Grassland Congress, Reading, pp. 479-484 (1960).
123. The effect of' muscular work on the energy metabolism of sheep. J. L. Clapperton, D. G. Armstrong, K. L. Blaxter. Proceedings of the Nutrition Society, 19, p. xvii (1960).
124. Fat synthesis from glucose by sheep. D. G. Armstrong, K. L. Blaxter, N. McC. Graham. Proceedings of the Nutrition Society, 19, pp. xxi-xxxii (1960).
125. Energy metabolism. K. L. Blaxter. State Veterinary Journal, 15, 114-120 (1960).
126. Energy utilisation in the ruminant. K. L. Blaxtor. Proceedings of the 7th Easter School, University of Nottingham. Edited D Lewis, pp. 183-197 (1960).
127. The effect of environmental temperature on the energy metabolism of steers. K. L. Blaxter, F. W. Wainman. Proceedings of the Nutrition Society, 19, pp. xvi-xvii (1960).
128. Heat production and heat emission of two breeds of sheep. D. G. Armstrong, K. L. Blaxter, J. L. Clapperton, N. McC. Graham, F. W. Wainman. Journal of Agricultural Science, 55, 395-401 (1960).
129. The voluntary food intake of sheep. K. L. Blaxter, F. W. Wainman, R. S. Wilson. Proceedings of the Nutrition Society, 19, pp. xxx-xxxi (1960).
130. Minerals in relation to disease. K. L. Blaxter. Veterinary Annual, pp. 168-175 (1960).
131. Prevention of enzootic muscular dystrophy by selenium administration. K. L. Blaxter, E. S. R. McCallum, R. S. Wilson, G. A. M. Sharman, L. O. Donald. Proceedings of the Nutrition Society, 20, pp. vi-vii (1961).

Jay L. Lush

Iowa State University
Ames, Iowa, USA

1979 — for his outstanding and pioneering contributions to the application of genetics to livestock improvement.

Jay L. Lush has one more than any agricultural scientist both to investigate the genetic basis of traits important for animal production, to assess their value and to improve them by initiating breeding programmes of proven practical effectiveness. He has been responsible for inspiring a great band of graduate students and others who have disseminated his ideas and practical suggestions throughout the world and so led to the improvement of poultry, pigs, beef and dairy cattle. His contributions are of several kinds. First, he pioneered the value to the scientific breeder of adequate measurement of the performance of individuals, their progeny and their pedigree in order to assess the genetic worth of an animal. Secondly, he pioneered the assessment of the 'heritability' of a trait, i.e. the degree to which its expression reflected a genetic basis rather than the effects of a favorable or unfavorable environment. Thirdly, having demonstrated how the inherited potentialities could be measured and distinguished from those due to non-heritable causes, he synthesized these ideas into practical and workable animal breeding plans to promote an inherited increase in the yield of meat, milk or eggs. He has exploited his ideas as a University teacher, at a diversity of national and international gatherings and enshrined much of it in his major and most influential text 'Animal Breeding Plans'. He may truly be described as the 'father' of scientific animal breeding in the twentieth century.

CURRICULUM VITA

Date and place of Birth: January 3, 1896, Shambaugh, Iowa

QUALIFICATIONS OF NOMINEE:

a) Degrees received

Kansas State College	B.S.	1916
Kansas State College	M.S.	1918
University of Wisconsin	Ph.D.	1921

b) Membership in Honorary Academic Societies

Phi Kappa Phi
Sigma Xi
Gamma Sigma Delta
Foreign Member of:
Royal Swedish Academy of Agriculture
Norwegian Academy of Sciences
Genetics Society of South Africa
National Academy of Sciences
National Medal of Sciences

c) Honors and Awards Received

1946 - Morrison Award, American Society of Animal Production
1956 - Portrait painted and hung, Saddle and Sirloin Club, Chicago, Illinois
1956 - Iowa State University, Outstanding Faculty Citation
1957 - American Society of Animal Production, Honor Guest
1957 - Royal Agricultural College, Sweden, Honorary Doctor of Agriculture
1957 - University of Giessen, Germany, Honorary Ph.D.
1957 - Iowa State University, Curtiss Distinguished Professor of Agriculture
1958 - Medal for Mathematics and Biology, Oslo, Norway
1958 - American Society of Animal Science, Honorary Fellow
1958 - Royal Danish Veterinary and Agricultural College, Copenhagen, Denmark, Honorary Doctor of Agriculture
1958 - American Dairy Science Association, Borden Award
1960 - German Society of Animal Breeding, von Nathusius Medal
1964 - Michigan State University, Honorary L. L. D.
1965 - Mendel Memorial Symposium, Brno, Czechoslovakia, Mendel Medal
1965 - Order of Merit, Republic of Italy
1965 - American Society of Animal Science, Genetics Award
1966 - Distinguished Service Award, American Dairy Science Association
1966 - Italian Republic, Official Knight of the Order of Merit

1967 - Member, National Academy of Sciences
1968 - National Medal of Science Award
1969 - University of Illinois, Honorary Doctorate
1969 - Portrait hung in Dairy Shrine Club
1969 - Citation of Excellence, Poultry Breeders of America
1970 - University of Wisconsin, Honorary Doctorate
1971 - Swiss Federal Institute of Technology, Honorary Doctorate
1973 - Beef Improvement Federation, Beef Research Pioneer
1975 - Agricultural University of Norway, Honorary Doctorate

d) Professional Positions Held

1917-18	Kansas State College	Fellow
1919-21	University of Wisconsin	Research Assistant
1921-29	Texas A & N College	Animal Husbandman
1930-57	Iowa State University	Professor
1957-	Iowa State University	Distinguished Professor

e) See Bibliography Attached

f) Education Contributions Other than Publications

Dr. Jay L. Lush was a pioneer in the application of genetics and statistics to animal breeding problems. He established a graduate program in animal breeding at Iowa State University beginning in 1930 which attracted students from many countries. His graduates have become leaders in education, in research, in industry, and in university administration. His book, "Animal Breeding Plans", first published in 1937, was the standard collegiate text for many years and remains the classic text in the field. His mimeographed notes, "The Genetics of Population" have been widely used for graduate teaching all over the world since 1948.

Educational contributions made in short time commitments away from Iowa State University are as follows:

1941 - Brazil. Lecturing on population genetics and biometry at Vicosa.
1951 - Colombia, Chile, Peru. Survey of teaching and research in animal production for Rockefeller Foundation.
1954 - India. Lecturing at New Delhi on biometrical techniques useful in agriculture.
1966 - Argentina. Teaching and advising agricultural research workers.

g) Development or Improvement of Programs, Practices, and Products for the Improvement of Food Sources

Dr. Lush can properly be called the father of modern animal breeding. Dr. Lush made a primary contribution in clarifying the breeding systems and practices which permit maximum rates of genetic improvement in farm animals. His clear insight into the roles of heredity, gene action and interaction, and the effects of environment permitted him to develop methods of measuring the relative importance of these sources. These results have been applied to poultry, swine, beef cattle and dairy cattle in ways most suitable to the particular species and to the economically important traits of each species. For example, family selection has been of considerable use in improving egg production, progeny testing, and artificial insemination in milk production, individual selection in growth rate for beef and swine.

Dr. Lush has worked closely with breed organizations and public institutions in developing plans for improved animal production. Some of the foreign assignments to which he has devoted time and effort are as follows:

1934 - Denmark. Study and Analysis of the Danish Pig Progeny Testing System.
1948 - Great Britain. Survey of research in genetics and animal husbandry for U.S.D.A.
1948 - Australia. Survey of research in animal production for C.S.I.R.O.
1951 - Paraguay. Developed cattle breeding plans in Paraguay for A.I.D.
1961 - Argentina. Survey and plans for more efficient beef industry in Argentina.

h) Other Professional Contributions

Dr. Lush has served in an advisory capacity to U.S.D.A., to breed organizations, to artificial insemination centers, and to many privately owned seed stock organizations concerning problems of genetic improvement.

He has served as advisor to the Poultry Breeders Roundtable since its foundation in 1952. This is an organization of commercial poultry seedstock producers which meets annually to discuss genetic problems.

He became advisor to the American Hereford Association when dwarfism became a problem in the U.S. beef industry in the early 1950's. His suggestions and recommendations were instrumental in bringing the problem under control without greatly upsetting beef production.

His studies of the Danish Pig Testing System were important in establishing pig testing systems which operated efficiently in the U.S., for example, averages from the Iowa Swine Testing Station in operation at Ames since 1958 indicate that ham and loin as a percent of carcass weight has increased about three quarters of one percent per year.

i) Service to Organized Groups Related to Food and Agriculture

Dr. Lush has served in official capacity with many national and private organizations in positions involving responsibility to agriculture. These include U.S.D.A., Rockefeller Foundation, C.S.I.R.O. in Australia, A.I.D. foreign agriculture, United Nations, F.A.O., C.A.F.A.D.E. in Argentina, Poultry Breeders Roundtable, American Hereford Association, Hoistein-Friesian Record Association, and Ayshire Cattle Club.

List of Publications

1. Inheritance in Swine. J. Hered. 12: 57-71 (1921).
2. An Hereditary Notch in the Ears of Jersey Cattle. J. Hered. 13: 8-13 (1922).
3. The Influence of Age and Individuality upon the Yield of Wool. Proc. Amer. Soc. An. Prod. 1922, 105-109 (1922).
4. The Practicability of the Milking Machine. Texas Sta. Circular 30. Inheritance in Swine. (with E. N. Wentworth). J. Agric. Res. 23: 557-582 (1923).
5. I. Fattening Steers on Cottonseed Meal and Hulls with and without Corn. II. The Influence of Age on Fattening Steers. (with J. M. Jones and J. H. Jones). Texas Sta. Bull. 309 (1923).
6. The Influence of Individuality, Age and Season Upon the Weight of Fleeces Produced by Range Sheep. (with J. H. Jones). Texas Sta. Bull. 31 (1923).
7. Twinning In Brahma Cattle. J. Hered. 15: 93-96 (1924).
8. "Double Ears" in Brahma Cattle. J. Hered. 15: 93-96 (1924).
9. The Influence of Individuality, Age and Season Upon the Weights of Fleeces Produced by Angora Goats Under Range Conditions. (with J. M. Jones). Texas Sta. Bull. 320 (1924).
10. Nature's Score Card for Feeder Steers. Proc. Amer. Soc. An. Prod. 1924, 98-101 (1924).
11. The Possibility of Sex Control by Artificial Insemination with Centrifuged Spermatozoa. J. Agric. Res. 30: 893-913 (1925).
12. Methods of Selecting Wool Samples in Shrinkage Studies. (with J. M. Jones). Proc. Amer. Soc. An. Prod. 1925, 115-117 (1925).
13. Practical Methods of Estimating the Proportions of Fat and Bone in Cattle Slaughtered in Commercial Packing Plants. J. Agric. Res. 32: 727-755 (1926).
14. Inheritance of Horns, Wattles, and Color in Grade Toggenburg Goats. J. Hered. 17: 72-91 (1926).
15. How Much Accuracy is Gained by Weighing Cattle Three Days Instead of One at the Beginning and End of Feeding Experiments. (with W. H. Black). Proc. Amer. Soc. An. Prod. 1926, 206-210 (1926).
16. Rice Bran as a Feed for Dairy Cows. (with Fred Hale). Texas Sta. Bull. 352 (1927).
17. "Percentage of Blood" and Mendelism. J. Hered. 18: 351-367 (1927).
18. The Production of Clean Milk. (with Fred Hale). Texas Sta. Circular 48 (1927).
19. Practices and Problems Involved in Crossbreeding Cattle in the Coastal Plain of Texas. Proc. Amer. Soc. An. Prod. 1927, 58-61 (1927).
20. A Statistical Interpretation of Some Texas Lamb Feeding Data. (with J. M. Jones). Proc. Amer. Soc. An. Prod. 1927, 167-170 (1927).
21. Changes in Body Measurements of Steers During Intensive Fattening. Texas Sta. Bull. 385 (1928).

22. The Accuracy of Cattle Weights. (With F. W. Christensen, C. V. Wilson and W. H. Black). J. Agric. Res. 36: 551-580 (1928).
23. The Use of Dressed Beef Appraisals in Measuring the Market Desirability of Beef Cattle. (with W. H. Black and A. T. Semple). J. Agric. Res. 39: 147-162 (1929).
24. Atavism in Jersey Cattle. J. Hered. 20: 381-383 (1929).
25. Twins in Jersey Cattle. J. Hered. 20: 510-513 (1929).
26. The Inheritance of Cryptorchidism. (with J. M. Jones). Proc. Amer. Soc. An. Prod. 1929, 57-61 (1929).
27. Interpreting Figures from Feeding Experiments. Proc. Assoc. Southern Agric. Workers. 1929, 177-186 (1929).
28. "Duck-legged" Cattle on Texas Ranches. J. Hered. 21: 84-90 (1930).
29. Karakul Sheep. (with J. M. Jones and R. E. Dickson). Texas Sta. Bull. 405 (1930).
30. The Inheritance of Cryptorchidism in Goats. (with J. M. Jones and W. H. Dameron). Texas Sta. Bull. 407 (1930).
31. Normal Growth of Range Cattle. (with J. M. Jones, W. H. Dameron and O. L. Carpenter). Texas Sta. Bull. 409 (1930).
32. A Study of the Accuracy of Measurements of Dairy Cattle. (with O. C. Copeland). J. Agric. Res. 41: 37-49 (1930).
33. Earlessness in Karakul Sheep. J. Hered. 21: 107-112 (1930).
34. "Nervous" Goats. J. Hered. 21: 242-247 (1930).
35. How Farm Animals Inherit. (A review of Christian Wriedt's "Heredity in Livestock"), J. Hered. 21: 306-308 (1930).
36. Interpreting the Results of Group Feeding Experiments. Proc. Amer. Soc. An. Prod. 1930, 44-55 (1931).
37. The Number of Daughters Necessary to Prove a Sire. J. Dairy Sd. 14: 209-220 (1931).
38. Predicting Gains in Feeder Cattle and Pigs. J. Agric. Res. 42: 853-881 (1931).
39. Genetic Aspects of the Record of Performance Work with Swine. Proc. Amer. Soc. An. Prod. 1931, 51-62 (1932).
40. The Ages of Breeding Cattle and the Possibilities of Using Proven Sires. (with N. D. Lacy). Iowa Sta. Bull. 290 (1932).
41. How Old are your Bulls and Cows and What Difference Does it Make? (abridged edition of Iowa Sta. Bull. 290). (with M. D. Lacy). Iowa Sta. Bull. 290A (1932).
42. An Empirical Test of the Approximate Method of Calculating Coefficients of Inbreeding and Relationship from Livestock Pedigrees. J. Agric. Res. 45: 565-569 (1932).
43. Twinning, Sex-ratios, and Genetic Variability in Birth Weight in Sheep. (with A. B. Chapman). J. Hered. 23: 473-478 (1932).

44. The Relation of Body Shape of Feeder Steers to Rate of Gain, to Dressing Percent, and to Value of Dressed Carcass. Texas Sta. Bull. 471 (1932).
45. The Amount and Kind of Inbreeding which has Occurred in the Development of Breeds of Livestock. Proc. Sixth International Congress Genetics. 2: 123-126 (1932).
46. Mutton, and How it Gets that Way. (A review of John Hammond's "Growth and the Development of Mutton Qualities in the Sheep"). J. Hered. 23: 312-314 (1932).
47. Inbreeding and the Genetics History of the Rambouillet Sheep in America. (with W. F. Dickson). J. Hered. 24: 19-33 (1933).
48. The Use of Statistical Methods in Animal Husbandry. Proc. Amer. Soc. An. Prod. 1932, 15-19 (1933).
49. Linebreeding. Iowa Agric. Exp. Sta. Bull. 301 (1933).
50. A Linebreeding Program for Horse Breeding. (with P. B. Pearson). J. Hered. 24: 185-191 (11. Paper No. B62) (1933).
51. The Bull Index Problem in the Light of Modern Genetics. J. Dairy Sci. 16: 501-522 (1933).
52. The Reliability of Some Measures of Productiveness in Brood Sows. (with A. L. Anderson, C. C. Culbertson and W. E. Hammond). Proc. Amer. Soc. An. Prod. 1933, 282-287 (1934).
53. Freshening Ages of Purebred Cows in Iowa Cow Testing Associations. (with Mogens Plum). J. Dairy Sci. 17: 625-636 (1934).
54. Progress Report on Comparison of Lactation and Yearly Records. (with C. N. Harris and E. N. Schultz). J. Dairy Sci. 17: 737-742 (1934).
55. Beef Production arid Quality as Influenced by Crossing Brahman with Hereford and Shorthorn Cattle. (with W. H. Black and A. T. Semple) U.S.D.A. Tech. Bull. 417 (1934).
56. A Herd of Cattle Bred for Twenty Years without New Blood. J. Hered. 25: 209-216. (J. Paper No. J-133) (1934).
57. Factors Affecting Birth Weights of Swine. (with H. O. Hetzer and C. C. Culbertson). Genetics 19: 329-343 (1934).
58. Progeny Test and Individual Performance as Indicators of an Animal's Breeding Value. J. Dairy Sci. 18: 1-19 (1935).
59. The Inheritance of Productivity in Farm Livestock. V. Discussion of Preceding Contributions. Empire J. Exp. Agric. 3: 25-30 (1935).
60. Genetics History of the Hoistein-Friesian Cattle in the United States. (with J. C. Holbert and 0. S. Wiliharn). 3. Hered. 27: 61-72 (1936).
61. Genetics and Animal Breeding. (A review of C. Kronacher's "Genetik und Tierzuchtung"). J. Hered. 27: 201-203 (1936).
62. Genetic Aspects of the Danish System of Progeny-testing Swine. Iowa Agric. Exp. Sta. Res. Bull. 204 (1936).

63. Variations in the Softness of Lard Produced in the Record of Performance Testing. (with B. H. Thomas, C. C. Culbertson and F. J. Beard). Proc. Amer. Soc. An. Prod. 1936, 258-259 (1936).
64. A Genetic History of the Brown Swiss Cattle in the United States. (with Dorsa N. Yoder). J. Hered. 28: 154-160 (1937).
65. Identical Twins in Cattle. (a review of C. Kronacher's "Neue Ergebnisse Der Zwillingsforschung Bèim Rind"). J. Hered. 28: 415-418 (1937).
66. Animal Breeding Plans. 366 P1. Collegiate Press, Ames, Iowa (1937).
67. The Degree to which Litter Size is a Constant Characteristic of Sows. (with A. E. Molln). Proc. Amer. Soc. An. Prod. 1937, 133-137 (1937).
68. Pedigree Promise and Progeny Test Among Sires Proved in Iowa Cow Testing Associations. (with Earl N. Schultz). J. Dairy Sci. 21: 421-432 (1938).
69. Teaching Animal Breeding. Proc. Amer. Soc. An. Prod. pp. 175-180 (1938).
70. Herd Averages Computed by the Cow-Year Method versus Herd Averages based only on Cows on Test at Least 10 Months. (with Floyd Johnston). 3. Dairy Sci. 21: 127-128 (1938).
71. Crossbreeding Hogs for Pork Production. (with P. S. Shearer and C. C. Culbertson). Iowa Agric. Exp. Sta. Bull. 380 (1939).
72. High Records Contrasted with Unselected Records and with Average Records as a Basis for Selecting Cows. (with J. C. Berry). J. Dairy Sd. 22: 607-617 (1939).
73. A Genetic History of Poland-China Swine. (with A. L. Anderson). J. Hered. 30: 149-156 and 219-224 (1939).
74. President's Address. Proc. Amer. Soc. An. Prod. pp. 11-18.
75. Intra-sire Correlations or Regressions of Offspring on Dam as a Method of Estimating Heritability of Characteristics. Proc. Amer. Soc. An. Prod. pp. 293-301 (1940).
76. "Nicking" in Dairy Cattle. (with D. M. Sheath). J. Dairy Sci. 23: 103-113 (1940).
77. Effects which Selection of Dams may have on Sire Indexes. (with H. W. Norton, III and Floyd Arnold). J. Dairy Sci. 24: 695-721 (1941).
78. Crossbreeding hogs? (with P. S. Shearer and C. C. Culbertson), Farm Science Reporter for January, pp. 8-11 (1941).
79. Applications of Genetics to Animal Breeding. Proc. Iowa Acad. Sci. 48: 65-72 (A Portuguese translation of this appeared in 1944 in "Ceres" 6: 44-51. Viçosa, Minas Gerais, Brazil) (1941).
80. Facts of Life for the Farmyard. J. Hered. 32(4): 122 (1941).
81. Repeatability of Type Ratings in Dairy Cattle. J. Dairy Sci. 25(1): 45-56. (with Arthur E. Johnston) (1941).
82. The Efficiency of Three Methods of Selection. (with L. N. Hazel). J. Hered. 33: 393-399 (1942).

83. The Heritability of Butterfat Production in Dairy Cattle. (with F. S. Straus). J. Dairy Sci. 25: 975-982 (1942).
84. Litter Size and Weight as Permanent Characteristics of Sows. (with A. E. Molln). U.S. Dept. Agric. Tech. Bull. 836 (1942).
85. Heritability of Conformation in Poland-China Swine as Evaluated by scoring. (with H. H. Stonaker). J. An. Sci. 1: 99-105 (1942).
86. Repeatability of Type Ratings in Dairy Cattle. (with Leslie E. Johnson). J. Dairy Sci. 25: 45-56 (1942).
87. Constructing Selection Indexes to Make Maximum Progress. (with L. N. Hazel). J. An. Sci. 1(1): 69-70 (1942).
88. Animal Breeding Plans. 2nd Ed. 437 p. The Collegiate Press. Ames, Iowa (1943).
89. The Quest for a Common Denominator. J. Hered. 34(8): 235-238. (Review of "Lekiels of Integration in Biological and Social Systems", edited by Robert Redfield) (1943).
90. The Optimum Emphasis on Dam's Records when Proving Dairy Sires. J. Dairy Sci. 27: 937-951 (1944).
91. Are Better Hogs Coming? Farm Science Report for April, pp. 3-6 (1944).
92. Animal Breeding Plans. 3rd Ed. 443 p. The Collegiate Press. Ames, Iowa (1945).
93. Chance as a Cause of Changes in Gene Frequency within Pure Breeds of Livestock. Amer. Naturalist. 80: 318-342 (1946).
94. Hybrid Vigor in Single Crosses between Inbred Lines of Poland China Swine. (with G. E. Dickerson and C. C. Culbertson). J. An. Sci. 5 (1946).
95. Factors Affecting Rate of Gain and their Relation to Allotment of pigs for Feeding Trials. (with Roberto M. Miranda and C. C. Culbertson). J. An. Sci. 5: 243-250 (1946).
96. Out on First Record? Iowa Farm Science, Jan. 1946, pp. 6-7 (1946).
97. The Genetics of Cattle. (with Robert R. Shrode). Advances in Genetics. 1: 209-261 (1947).
98. Heritability of Amount of Spotting in Holstein-Friesian Cattle. (with Raul Brequet, Jr.). J. Hered. 38: 98-105 (1947).
99. Family Merit and Individual Merit as Bases for Selection. Am. Naturalist. 81: 241-261 and 362-379 (1947).
100. "Genetics for the Millions" (with R. C. Cook). (Review of Dunn and Dobquhansky's "Heredity, Race and Society"). J. Hered. 38: 299-305 (1947).
101. Variations in Yield of Milk under the Pen Keeping System in Brazil. (with G. G. Carneiro). J. Dairy Sci. 31: 203-211 (1948).
102. The Heritability of Resistance to Death in the Fowl. (with Lamoreux and Hazel). Poultry Sci. 27(4): 375-388 (1948).
103. Heritability of Quantitative Characters in Farm Animals. Proc. Eighth International Genetics Congress. pp. 356-375 (1949).

104. Ernest W. Lindstrom. 1891-1948. J. Hered. 40(2): 44-46 (1949).
105. The Algebra of Genetics. (Review of C. C. Li's, "An Introduction to Population Genetics"). J. Hered. 40: 156 (1949).
106. The Accuracy of Linear Body Measurements of Dairy Cattle. (with R. W. Touchberry). J. Dairy Sci. 33(1): 72-80 (1950).
107. Inheritance of Susceptibility to Mastitis. J. Dairy Sci. 33(2): 121-125 (1950).
108. The Effects of Mild Inbreeding on a Herd of Hoistein-Friesian Cattle. J. Dairy Sci. 33: 186-193. (with R. H. Nelson) (1950).
109. Changes in Milk Production with Age and Milking Frequency. J. Dairy Sci. 33(5): 338-357. (with R. R. Shrode) (1950).
110. Review of R. A. Fisher's, "The Theory of Inbreeding". Amer. J. of Human Genetics. 2(1): 97-100 (1950).
111. Mating Systems in Honey Bees. (with Martin Polhemus and Walter C. Rothenbuhier). J. Hered. 41(6): 151-155 (1950).
112. Computing Inbreeding and Relationship Coefficients from Punched Cards. (with L. N. Hazel). J. Hered. 41: 301-306 (1950).
113. Summary. (Symposium on selection at Chicago in November, 1949). J. An. Sci. 10: 18-21 (1951).
114. The Impact of Genetics on Animal Breeding. A General Invitation Review and Forecast. J. An. Sci. 10(2): 311-321 (1951).
115. Inbreeding and Outbreeding as Practiced with Poultry. pp. 11-18. Numbers of Sires, Dams and Progeny Required for Reliable Progeny Testing. pp. 68-74. "Proceedings Fourth Pacific Northwest Chicken and Turkey Breeders Roundtable" (1951).
116. Genetic Correlation between Type and Production in Jersey Cattle. J. Dairy Sci. 35(3): 199-2l3. (with Water R. Harvey) (1952).
117. How Dominance and Gene Reaction Modify the Effectiveness of Breeding Plans. pp. 15-25. First Poultry Breeder's Roundtable (1952).
118. Results of Swine Breeding Research. U.S.D.A. Circular No. 916. (W. A. Craft, sole author, with acknowledgements to 15 or 20 others) (1953).
119. Breeding Structure of Populations. I. General Considerations (Chapter 42, page 537—542 in) Statistics and Mathematics in a Selection Index for Fat Production in Dairy Cattle: Utilizing the Fat Yields of the Cow and Her Close Relatives. (with J. E. Legates). J. Dairy Sci. 37(6): 744-753 (1954).
120. Reproductive Rates and Growth of Purebred Brown Swiss Cattle in Brazil. (with G. G. Carneiro). J. Dairy Sci. 37: 1145-1157.
121. Rates of Genetic Changes in Populations of Farm Animals. Proc. Ninth International Congress of Genetics. 1: 589-599 (1954).
122. Estimates of Heritability in Breeding Problems. (A chapter in the book; "Breeding Beef Cattle Adapted to Unfavorable Environments", published by the University of Texas Press-King Ranch Centennial.) pp. 113-126 (1955).

123. Proving Sires and Dams. (with Lon D. McGilliard). J. Dairy Sci. 38: 163-180 (1955).
124. Relations Between Parts of Lactations and Producing Ability of Holstein Cows. (with D. E. Madden and L. D. McGilliard). J. Dairy Sci. 38(11): 1264-71 (1955).
125. Statistics in Investigations in Animal Production. J. of the Indian Society of Agricultural Statistics. 7: 7-22 (1955).
126. Dairy Cattle Genetics. J. Dairy Sci. 39(6): 693-694 (1956).
127. Changes in Type Classifications of Dairy Cattle. J. Dairy Sci. 39(7): 1015-1026 (1956).
128. Theoretical Consequences of Breeding for the Heterozygote in Intra and Interbreeding Populations. p. 3-26 in Fifth Poultry Breeders Roundtable (1956).
129. Studies on Bovine Ocular Squamous Carcinoma ("Cancer Eye"). (with David E. Anderson and Doyle Chambers) (1957).
130. II. Relationship Between Eyelid Pigmentation and Occurrence of Cancer Eye Lesions. J. An. Sci. 16: 739-746 (1957).
131. III. Inheritance of Eyelid Pigmentation. J. An. Sci. 16: 1007-1016 (1957).
132. Twinning in Dairy Cattle and its Relation to Production. J. Dairy Sci. 40: 1430-1436. (with C. E. Meadows) (1957).
133. Effect of Inbreeding on Production in Holsteins. J. Dairy Sci. 41: 105-113. (with C. N. von Krosigk) (1958).
134. Practical Application of Performance Testing. The Shorthorn World. 43(11): 44 and 318-321 (1958).
135. Genetics in Plant and Animal Breeding. (Translated Title). Tolvmandsbladet 30(10): 403-409. Copenhagen (1958).
136. Genetic Relations between Body Measurements at Three ages in Holsteins. J. Dairy Sci. 41: 1045-1049. (with D. W. Blackmore and L. D. McGilliard) (1958).
137. Relationships between Body Measurements, Meat Conformation, and Milk Production. J. Dairy Sci. 41: 1050-1056. (with D. W. Blackmore and L. D. McGilliard) (1958).
138. Genetic and Environmental Portions of the Variation among Herds in Butterfat Production. J. Dairy Sci. 42: 115-122. (with F. Pirchner) (1959).
139. Making Use of New Knowledge about Basic Principles. pp. 141-155, Eighth Poultry Breeders' Roundtable (1959).
140. Controls for Selection Experiments. pp. 46-55 in Proc. NC-1 Technical Committee Meeting at Ames, July 23-24. Mimeo (no reprints) (1959).
141. Improving Dairy Cattle by Breeding. I. Current Status and Outlook. J. Dairy Sci. 43: 702-706 (1960).
142. Accuracy of Partial Trapnest Records. Poultry Science. (with John D. Wheat) (1960).

143. Practical Application of Performance Testing. Gulf Coast Cattleman. 26(7): 6-8 (1960).
144. How Selection May Change a Mendelian Population. Quart Review of Biology. 35(3): 221-223 (1960).
145. Breeding (Animal). pp. 326-327 in: The McGraw-Hill Encyclopedia of Science and Technology, 15 Volumes (1960).
146. Genetic Environmental Interactions. pp. 104-107 in Proc. of 1960 Tech. Comm. Mtg., NC-1. July 25-27. Oklahoma. Mimeo (1960).
147. Scientific Problems of Recording Systems and Breeding Plans for Domestic Animals. Lectures of II International Summer School, Sept. 12-27 at Nariensee, Schriftenrejhe des Nax-Planck-Instituts fur Tierzucht und Tierernahrung (1961).
148. Selection Indexes for Dairy Cattle, Sonderdruch aus zertschrift fur Tierzuchtung und zuchtungsbiologie. 75: 249-261 (1961).
149. Animal Improvement and Adaptation. United Nations Conference on the Application of Science and Technology for the Benefit of the Less Developed Areas, Nov. 1962. (with Jorge de Alba) (1962).
150. Effects of Inbreeding on Size and Type in Hoistein-Friesian Cattle. J. Dairy Sci. XLV 390-395. (with T. N. Sutherland) (1962).
151. Improving Agriculture Through Genetics. Proc. Annu. Mtg. Agric. Res. Inst. Oct. 14-15, 1963. pp. 159-172 (1963).
152. Population Genetics, The Voice of America Forum Lectures, Agricultural Series 4, pp. 1-11 (1963).
153. The Importance of Quantitative Genetics for Animal Breeding. Tijdschrift voor Diergeneeskunde. 22: 1592-1598 (1964).
154. Genetic and Environmental Factors in Dairy Sire Evaluation. III. Influence of Environmental and Other Extraneous Correlations among the Daughters. J. Dairy Sci. XLVIII: 356-360. (with B. Bereskin) (1965).
155. Ways of Utilizing Heterosis in Animal Breeding. Crencia E Cultura. 19: 221-222 (1967).
156. Importance of Family Structure in the Dairy Cattle Population. Symposium on Dairy Cattle Breeding. J. Dairy Sci. 51: 296-306 (1967).
157. Genetic Progress Through Selection in a Closed Herd of Indian Cattle. J. Dairy Sci. 51: 1059-1064. (with R. M. Acharya) (1967).
158. Importance of Bull x Herd-Year-Season Interaction in Milk Production. J. Dairy Sci. 50: 1703-1707. (with D. J. Kelleher and A. E. Freeman) (1967).
159. Genetic and Biometric Foundations of Populations. J. Hered. 59: 357-359 (1968).
160. Genetic Unknowns and Animal Breeding a Century After Mendel. Transactions of the Kansas Academy of Science. 71: 309-314 (1968).

161. Animal Breeding in the Age of A.I. "Animal Breeding and Application, present, and future." pp. 8-25. (North America). Mimeo. (Sponsored jointly with Univ. Wisconsin) (1968).
162. Research in Animal Production, its Past Accomplishments and Present Prospects. pp. 45-65. (In Bull. 160. 38th Jahrgang 38. The centenary of the Abteilung fur Landwirtschaft an der Eidegenossischen Technischen Hochschule, Zurich. June 25-26, 1971.
163. Early Statistics at Iowa State University. Chap. 13 (p. 221-226 in the book, "Statistical Papers in Honor of George W. Snedecor." ISU press (1972).
164. Teaching Animal Breeding and Teaching Graduate Students. An. Breeding and Genetics Symposium. pp. 78-88. (By the American Society of Animal Science and the American Dairy Science Assoc.) (1972).
165. Response to the Dedication of the Jay L. Lush Auditorium. Iowa State Journal of Science. 48(4): 281-284 (1974).

Karl Maramorosch

Rutgers University
New Brunswick, N.J., USA

1980 — for his pioneering and wide-ranging studies on interactions between insects and disease agents in plants.

BIOGRAPHICAL INFORMATION

Born on January 16, 1915, Vienna Austria. Diploma in Agricultural Engineering (M.Sc. Equivalent in Entomology) from Agricultural University (SGGW), Warsaw Poland 1938.

Came to the United States in 1947. Naturalized US citizen in 1952.

Ph.D. in Botany and Plant Pathology, Columbia University, New York, 1949.

Positions

Rockefeller University, New York City, from Assistant to Associate Professor, 1949-1961.

Boyce Thompson Institute for Plant Research, Yonkers, New York from Senior Entomologist to program Director on Insect Physiology and Virology, 1961-1974.

Rutgers- The State University of New Jersey, Professor II (Distinguished Professor)-1974 - until now; Named Professor (Robert L. Starkey Professor of Microbiology) 1983 - until now; Adjunct Professor of Entomology 1985 - until now.

Visiting Professorship

1) Wageningen Agric. University, Netherlands 1953.
2) Cornell University 1957.
3) Agric. University, Bucharest, Romania 1964.
4) Agric. University, Skierniewice and Warsaw (SGGW), Poland 1964.
5) Rutgers University 1968.
6) Fordham University, NY 1972.
7) Fulbright Senior Prof. Yugoslavia 1972 and 1977.
8) Agric. University, Bangalore, India 1979.
9) Dist Senior Prof. Hokkaido University, Sapporo, Japan 1980.
10) Fudan University, Shanghai, China 1982.
11) Justus Liebig University, Giessen, Germany 1983.

Professional Affiliations

1) American Phytopathological Society (Fellow).
2) Entomological Society of America (Fellow and Honorary Member).
3) Indian Virological Society (Honorary Fellow).
4) American Association for the Advancement of Science, AAAS (Fellow).
5) New York Academy of Sciences (Fellow).
6) New York Academy of Sciences (Recording Secretary, 1960-1962).
7) New York Academy of Sciences (Vice-President, 1961-1963).
8) National Academy of Sciences, India (Honorary Fellow).
9) Leopold ma Academy, Germany.
10) Microscopy Society of America.
11) International Organization for Mycoplasmology.
12) Society for Invertebrate Pathology.
13) Society of In Vitro Biology.
14) American Council on Science and Health (Board of Science and Policy Advisors).
15) International Association for Research on Medicinal Forest Plants (President).
16) Phi Beta Kappa Award in Science Committee.
17) Tropical Medicine and Parasitology USPH-NIH Panel (1971-1976).
18) American Institute of Biological Sciences
19) American Society of Virology.

Honors and Awards

1) Magna cum Laude Diploma, Agric University (SGGW) Warsaw, Poland 1938.
2) Predoctoral Fellow, American Cancer Society, Columbia University 1947-1949.
3) Cressy Morrison Prize, New York Academy of Science 1951.

4) New York Academy of Science, Recording Secretary 1961.
5) New York Academy of Science, Vice President 1962-1963.
6) Lalor Foundation Senior Research Award 1955.
7) American Association for the Advancement of Science, AAAS Prize and Campbell Medal 1958.
8) Honorary Member, Mendel Society 1963.
9) Member, Leopoldina Academy 1971.
10) Honorary Fellow, National Academy of Science, India 1979.
11) Member of Executive Committee of Virus Classification 1966-1968.
12) President of Executive Committee of Virus Classification, First (1969), Second (1973), Third (1977) and Fourth (1982) International Comparative Virology Conferences.
13) Senior Fuibright-Hays Professorship, Yugoslavia 1972 and 1977.
14) Award of Merit, American Phytopathological Society (APS) 1954.
15) Fellow of the American Phytopathological Society (APS) 1972.
16) Fellow of the American Association for the Advancement of Science, AAAS 1955.
17) President of the Tissue Culture Association North Eastern Branch 1980.
18) Entomological Society of America (ESA)—Ciba-Geigy National Award in Agriculture 1976.
19) Wolf Prize in Agriculture 1980.
20) Jurzykowski Prize in Biology 1981.
21) American Institute of Biological Sciences, Distinguished Service Award 1983.
22) NIH Fogarty International Center-Biomedical Exchange Program with Poland Award 1985.
23) Japan Society for Promotion of Science Distinguished Professorship, Hokkaido University, Japan 1980.
24) American Society for Microbiology Honorary Lectureship Award 1978.
25) Award of Asian Indians in America Award 1981.
26) Indian Virological Society Honorary Fellow, 1987.
27) Santokhba Durlabhji Award, Jaipur, India 1993.
28) Waksman Award and Medal, Theobald Smith Society 1978.
29) Named Professorship at Rutgers University 1983.
30) President of International Association of Medicinal Forest Plants 1991.
31) Founders Lecturer in the Society for Invertebrate Pathology, Adelaide, Australia 1990.
32) Founders Honoree, Sapporo 1998.
33) International Organization of Mycoplasmologists Recognition Award, Sydney, Australia 1998.
34) Distinguished Lifetime Achievement Award from the Society of In Vitro Biology 2001.
35) L.O. Howard Distinguished Achievement Award, Entomol. Soc. America, 2006.

Visiting Professorships

During several visits to the International Center for Insect Physiology (ICIPE) in Nairobi and Mbita Point, Kenya, Professor Maramorosch taught about novel insect control measures using insect cell culture.

In the former Soviet Union, he lectured and consulted in the Republics of Russia, Armenia and Uzbekistan.

As an invited Guest Professor at Fudan University in Shanghai, China, he lectured on applications of invertebrate cell culture.

Maramorosch's fluency in seven languages was very helpful in his international activities, When the Justus Leibig University in Giessen, Germany invited him, he lectured there in German. In Romania, as a guest of their Academy of Sciences, he made use of knowledge of Romanian, acquired during his years as civilian internee during World War II in that country. In St. Petersburg, Moscow, Yerevan and Tashkent, he lectured in Russian, and in Poland in Polish.

Maramorosch's creativity has been demonstrated throughout his career. His lecturing commitments often took him outside the U.S. borders. He has tremendous energy to meet people from around the world, to learn new cultures, and to pass on his knowledge wherever he steps. His passion for science and for learning is his inspiration. Numerous postgraduate associates from around the globe owe their own enthusiasm to having been fortunate enough to spend time in his laboratory.

Professor Karl Maramorosch has served the discipline of entomology and plant pathology with uninterrupted distinction across a span of seven decades. His hundreds of scholarly publications document the remarkable scientific journey of one of the giants and icons of entomology. He has gained true world renown in insect pathology, plant-insect vectors, insect cell culture, and virology. To have contributed to such a breadth of disciplines is notable but to have had substantial impact in each is extraordinary.

Research Interests and Expertise

The study of interactions between pathogens, vectors, and hosts, and the emergence of "new" infectious diseases during the past decades has seen a large expansion of information on viral genetics, molecular biology, and pathogen evolution. Polymerase chain reaction has provided a sensitive tool for the detection of pathogen components. Prevention of vector-borne diseases by botanical compounds such as derivatives of neem (*Azadirachta indica*) and biological control of pests and vectors were studied. Invertebrate cell culture became an area of special interest. Contributions were made to comparative virology, spiroplasma and phytoplasma pathogens, biotechnology and environmental safety, and emerging infectious diseases.

RECENT PUBLICATIONS (SINCE 1996):

- Maramorosch, K. 1996. The cadang-cadang disease of palms. In Forest Trees and Palms: Diseases and Control. S.P. Raychaudhuri and K. Maramorosch, eds. Science Publishers, New Hampshire: 309-318.
- Maramorosch, K. 1996. Lethal yellowing disease of palms. In Forest Trees and Palms: Diseases and Control. S.P. Raychaudhuri and K. Maramorosch, eds. Science Publishers, New Hampshire: 253-265.
- Maramorosch, K. 1997. Mites (Acarina: Eriophyoidea) as vectors of plant pathogens. In: Ecology and Evolution of Plant-Feeding Insects in Natural and Man-Made Environments. Anantharayanan Raman, ed. Backhuis Publishers, Leiden, The Netherlands: 197-201.
- Maramorosch, K. 1998. Punta morada de Ia papa (Potato purple top wilt) Memoirs Potato Symposium, Toluca, Mexico, Conferencias magistrales. 12 pp.
- Maramorosch, K. and Adlakha, K. 1998. Natural products from medicinal forest plants: Prospects for mutual agreements between industrialized and developing countries. In: Prospects of Medicinal Plants. P.L. Guatham et al., eds. Indian Society of Plant Genetic Resources, New Delhi: 52-58.
- Maramorosch, K. 1999. Conservation of threatened medicinal forest plants. Technology 6: 269-273.
- Maramorosch, K. 1999. Ecologically disturbed forests linked to new emerging human diseases. In: Biotechnology and Plant Protection in Forestry Science. S.P. Raychaudhuri & K. Maramarosch, eds. Science Publishers, Enfield, NH: 253-258.
- Maramorosch, K. 1999. Suggestions for lethal yellowing and cadang-cadang disease prevention. Acta Horticulturae 486: 141-148.
- Maramorosch, K. 1999. The discovery of 2-4D. International J. Tropical Plant Diseases 15: 7-15.
- Maramorosch, K. 1999. Neem for reforestation. In: Proceedings Neem Conference, Gaton College, Australia, R.P. Singh and R.C. Saxena, eds. Science Publishers, Enfield, NH: 11-20.
- Maramorosch, K. 1999. Leafhopper and plant hopper rearing. In: Maintenance of human, animal, and plant pathogen vectors. K. Maramorosch & F. Mahmood, eds. Science Publishers, Enfield, NH: 194-211.
- Maramorosch, K. 1999. The early history of invertebrate cell culture. In Vitro Report 33: 1-6.
- Maramorosch, K. & Hillman, B. 2000. Plant Disease Resistance: Natural mechanisms and engineered resistance. In: Encyclopedia of Microbiology, Second Edition. Joshua Lederberg, ed. Academic Press, Vol. III: 662-675.
- Maramorosch, K. 2000. The end of polio?. The Sciences, Sept. & Oct. 2000: 48.

- Maramorosch, K. 2002. The neem tree — source of unique natural products for integrated pest management, medicine, and other purposes. H. Schmutterer, ed. 893 pp. Neem Foundation, Vitnagar, Mumbal, India. Foreword.
- Maramorosch, K. 2002. Diseases and pathogens of Mollusca. McGraw-Hill Encyclopedia of Science & Technology, 9th Ed.: 393-397.
- Maramorosch, K. 2004. The phantom vector of an emerging viroid. In: Life Sciences for the 21st Century. Ehud Keinan et al., eds. VCH- Viley: 297-303 Science at the Turn of the Century. Ehud Keinan, ed.
- Perez, B.A., Barreto, D. and Maramorosch, K. 2004. Hop and olive diseases. In: Crop Management and Postharvest Handling of Horticultural Products. Vol. IV- Diseases and disorders of fruits and vegetables. Ramdane Dris et al., eds. Science Publishers, Enfield, NH, USA: 289-303.
- Granados, R.R. and Maramorosch, K. 2004. Molecular engineering and biology of invertebrate cell culture: a tribute to Dr. Thomaas Grace and Prof. Shangyin Gao. SIP Newsletter 37: 3.
- Maramorosch, K. and Perez, B.A. 2005. Cuscuta: possible role in emerging virus diseases. Abstracts, IX Plant Virus Epidemiology Symposium, Lima, Peru.
- Maramorosch, K. 2005. Women pioneers of invertebrate cell culture. SIP Newsletter 38: 39-40.
- Maramorosch, K. 2006. Peeping into the past: Pioneers in Plant Protection. Leslie C. Coleman (1878-1954). Indian J. Plant Protection 34: 145-146.
- Maramorosch, K. 2006. Lindsay M. Black (1907-1997). Indian J. Plant Protection 34: 146-147.
- Raychaudhuri, S.P. and Maramorosch, K. 2006. William Conway Proce (1906-1991). Indian J. Plant Protection 34: 150-151.
- Maramorosch, K. 2006. Radcliffe Nathan Salaman (1874-1955) Indian J. Plant Protection 34: 148.
- Maramorosch, K. 2006. Kenneth M. Smith (1892-1981). Indian J. Plant Protection 34: 148-149.
- Maramorosch, K. 2006. Helen Purdy Beale: Pioneer of Plant Virus Serology. (1893-1976). Indian J. Plant Protection 34: 149-150.
- Maramorosch, K. 2007. Viruses, vectors, and vegetation: an autobiography. Advances in Virus Research 70: 1-31.
- Maramorosch, K. 2007. Emergence of new phytoplasma and virus diseases. Abstracts, 10th Plant Virus Epidemiology Symposium: Controlling Epidemics of Emerging and Established Plant Virus Diseases — The Way Forward. ICRISAT, Hyderabad, India: 16.
- Maramorosch, K. 2007. Foreword. International Phytoplasmologist Working Group, Bologna, Italy. Assunta. Bertaccini and Stefano Maini, eds. Bulletin of Insectology 60 (2), December 2007: 99.

Books from 1962-2006:

1. Biological Transmission of Disease Agents. Maramorosch, K., ed. Academic Press. 192 pp. 1962.
2. A Survey of Coconut Diseases of Uncertain Etiology. Maramorosch, K., FAO, Rome 1964.
3. Insect Viruses. Maramoroach, K., ed. Springer. 192 pp. 1968.
4. Viruses, Vectors and Vegetation. Maramorosch, K., ed. Wiley- Interscience. 666 pp. 1969.
5. Methods in Virology, 8 Volumes. Maramorosch, K. and Koprowski, H., eds. Academic Press. 1967-1984.
6. Comparative Virology. Maramorosch, K. and Kurstak, E., eds. Academic Press. 584 pp. 1971.
7. Mycoplasma and Mycoplasma-Juice Agents of Human, Animal and Plant Diseases. Maramorosch, K., ed. New York Academy of Sciences. Annal # 225, 532 pp. 1973.
8. Viruses, Evolution and Cancer. Kurstak, E. and Maramorosch, K., eds. Academic Press. 813 pp. 1974.
9. Invertebrate Immunity. Maramorosch, K. and Shope, R.E., eds. Academic Press. 365 pp. 1975.
10. Tropical Diseases of Legumes. Bird, J. and Maramorosch, K., eds. Academic Press. 171 pp. 1975.
11. Advances in Virus Research. 12 volumes. Lauffer, A.H., Smith, K.M., Bang, F.B. and Maramorosch, K., eds. Academic Press. 1973-1982.
12. Invertebrate Tissue Culture: Research Applications. Maramorosch, K., ed. Academic Press. 393 pp. 1976.
13. Invertebrate Tissue Culture: Applications in Medicine, Biology and Agriculture. Kurstak, E. and Maramorosch, K., eds. Academic Press. 398 pp. 1976.
14. The Atlas of Insect and Plant Viruses. Maramorosch, K., ed. Academic Press. 478 pp. 1977.
15. Aphids as Virus Vectors. Harris, K.F. and Maramorosch, K., eds. Academic Press. 559 pp. 1977.
16. Viruses and environment. Kurstak, E. and Maramorosch, K., eds. Academic Press. 677 pp. 1978.
17. Practical Tissue Culture Applications. Maramorosch, K. and Hirumi, H., eds. Academic Press. 426 pp. 1979.
18. Leafhopper Vectors and Plant Disease Agents. Maramorosch, K. and Harris, K.F., eds. Academic Press. 654 pp. 1979.
19. Invertebrate Systems *in vitro*. Kurstak, E., Maramorosch, K. and Dubendorfer, A., eds. Elsevier-North Holland Biomedical Press. 598 pp. 1980.
20. Vectors of Plant Pathogens. Harris, K.F. and Maramorosch, K., eds. Academic press. 467 pp. 1980.

21. Vectors of Disease Agents: Interactions with Plants, Animals and Man. McKelvey, J.J., Eldridge, B.F. and Maramorosch, K., eds. Praeger. 243 pp. 1980.
22. Plant Diseases and Vectors: Ecology and Epidemiology. Maramorosch, K. and Harris, K.F., eds. Academic Press. 368 pp. 1981.
23. Mycoplasma Diseases of Trees and Shrubs. Maramorosch, K. and Raychaudhuri, S.P., eds. Academic Press. 362 pp. 1981.
24. Pathogens, Vectors, and Plant Diseases: Approaches to Control. Harris, K.F. and Maramorosch, K., eds. Academic Press. 310 pp. 1982.
25. Advances in Cell Culture. 5 volumes, (I–V) Maramorosch, K., ed. Academic Press. 1981-1987.
26. Advances in Cell Culture. Vols. VI and VII. Maramorosch, K. and Sato, G.H., eds. Academic Press. 1988-1989.
27. Mycoplasma and Allied Pathogens of Plants, Animals and Human Beings. Govindu, H.C., Maramorosch, K., Raychaudhuri, S.P. and Munlyappa, V., eds. Univ. Agric. Sciences, Bangalore, India. 118 pp. 1981.
28. Invertebrate Cell Culture Applications. Maramorosch, K. and Mitsuhashi, J., eds. Academic Press. 245 pp. 1982.
29. Advances in Virus Research. Two volumes (Nos. 28 and 29). Lauffer, A.M. and Maramorosch, K., eds. Academic Press. 1983 and 1984.
30. Subviral Pathogens of Plants and Animals: Viroids and Prions. Maramorosch, K. and McKelvey, J.J., eds. Academic Press. 550 pp. 1985.
31. Viral insecticides for biological control. Maramorosch, K. and Sherman, K.E., eds. Academic Press. 809 pp. 1985.
32. Biotechnology in Insect Pathology and Cell Culture. Maramorosch, K., ed. Academic Press. 511 pp. 1987.
33. Mycoplasma Diseases of Crops: Basic and Applied Aspects. Maramorosch, K. and Raychaudhuri, S.P., eds. Springer Verlag. 456 pp. 1987.
34. Invertebrate and Fish Tissue Culture. Kurstak, E. Kuroda, Y. and Maramorosch, K., eds. Springer 1988.
35. Biotechnology for Biological Control of Pests and Vectors. Maramorosch, K. ed. CRC Press. 278 pp. 1991.
36. Viroids and Satellites: Molecular Parasites at the Frontier of Life. Maramorosch, K., ed. CRC Press. 174 pp. 1991.
37. Plant Diseases of Viral, Viroid, Mycoplasma and Uncertain Etiology. Maramorosch, K., ed. Oxford and IBH. 190 pp. 1992.
38. Arthropod Cell Culture Systems. Maramorosch, K. and McIntosh, A.H., eds. CRC Press. 236 pp. 1994.
39. Insect Cell Biotechnology. Maramorosch, K. and McIntosh, A.H., eds. CRC Press. 190 pp. 1994.
40. Forest Trees and Palms: Disease and Control. Raychaudhuri, S.P. and Maramorosch, K., eds. Science Publishers, New Hampshire. 334 pp. 1996.

41. Advances in Virus Research. Vol. 46 and 47. Maramorosch, K., Murphy, F.A. and Shatkin, A.J., eds. 1996.
42. Advances in Virus Research. Vol. 48. Maramorosch, K., Murphy, F.A. and Shatkin, A.J., eds. 1997.
43. Invertebrate Cell Culture: Novel Directions and Biotechnology Applications. Maramorosch, K. and Mitsuhashi, J., eds. Science Publishers, New Hampshire. 296 pp. 1997.
44. Invertebrate Cell Culture: Looking Toward The Twenty First Century. Proceedings of the IX International Conference on Invertebrate Cell Culture. Maramorosch, K. and Loeb, M.J., eds. Society for In Vitro Biology, Columbia, MD. 1997.
45. Advances in Virus Research. Vols. 49. Maramorosch, K., Murphy, F.A. and Shatkin, A.J., eds. Academic Press, New York. 1998.
46. Advances in Virus Research. Vols. 50 and 51. Maramorosch, K., Murphy, F.A. and Shatkin, A.J., eds. Academic Press, New York. 1998.
47. Advances in Virus Research. Vols. 52, 53, and 54. Maramorosch, K., Murphy, F.A. and Shatkin, A.J., eds. Academic Press, New York. 1999.
48. Maintenance of Human, Animal, and Plant Pathogen Vectors. Maramorosch, K. and Mahmood, F., eds. Science Publishers, Enfield, NH. 1999.
49. Biotechnology and Plant Protection in Forestry Science. Raychaudhuri, S.P. and Maramorosch, K., eds. Science Publishers, Enfield, NH. 1999.
50. Advances in Virus Research. Forty volumes, Nos. 30-70. Maramorosch, K., Murphy, F.A. and Shatkin, A.J., eds. Academic Press/Elsevier. 1987-2007.

CHAPTER 1

Viruses, Vectors, and Vegetation: An Autobiography

Karl Maramorosch

Contents

I. FROM CHILDHOOD IN EUROPE TO AMERICA

When the suggestion was made to write my biographical chapter for *Advances in Virus Research*, I did not realize how difficult a task this would be—where to start, what to say, and what to omit? I decided to start with my childhood and describe events in my life that inspired me to become a virologist and that were responsible for my scientific career.

In the summer of 1914, shortly after World War I started and the Tsarist army approached the family farm located in the village of Soroki in the eastern part of Austria, my parents escaped to Vienna, the capital of

Department of Entomology, Rutgers-The State University of New Jersey, New Brunswick, New Jersey 08901 (maramors@rci.rutgers.edu)

Advances in Virus Research, Volume 70
ISSN 0065-3527, DOI: 10.1016/S0065-3527(07)70001-5

This article was published in *Advances in Virus Research* Vol. 70, Maramorosch, K., viruses, vectors, and vegetation: an autobiography, pp. 1-31, Copyright Elsevier Inc. (2007).

the Austro-Hungarian Monarchy. There I was born in 1915. The farm did not move, but the borders moved many times. The family estate found itself under no less than seven different regimes: Austria, Poland, Petlura's Ukraine, Romania, again Poland, USSR, Nazi Germany, USSR, and currently Ukrainian Republic.

My father, a graduate of the Vienna Agricultural University, started Ph.D. studies in Halle/Saal, Germany in 1898 but after 1 year returned home to manage a 4000 acres estate, Kamionki Wielkie near Kolomyja, owned by my grandfather. Around 1900, the estate was sold and the smaller farm, Soroki, was purchased. My father considered himself a Pole of Jewish creed. My mother, born in Zagreb, Croatia, was an accomplished pianist and a linguist, fluent in German, English, French, Italian, and Serbo-Croatian. My siblings, Alfred, 6 years older, and Karla Bronia, 5 years older, spoke only Polish with my father and only German with my mother. I grew up into this system, not realizing that it was not usual for everyone to speak only Polish to one's father and only German to one's mother. I grew up bilingual and only realized this clearly when I left home and started writing letters to my parents—my thoughts were in Polish when I addressed my father, and German toward my mother, and I had to write not one, but two letters during my studies in Warsaw. I was often asked how my parents spoke to each other. They spoke German because, despite the great language skills of my mother, she could not speak Polish without an accent, and it was, unfortunately, customary in Poland to make fun of everybody who mispronounced Polish words. My mother used Polish only when she went shopping or when she spoke with people who helped at home, but never with friends or visitors.

My third language was Ukrainian, which was spoken by all peasants in the village where our farm was located. In high school I had 4 years of Ukrainian and learned the Cyrillic script and some Ukrainian poetry by Taras Shevchenko and Ivan Franko.

When I was 14 years old, my brother came home for his winter vacation from Lwow (Lviv), where he was studying medicine. He told me how his biology professor, Rudolf Weigl, invented a vaccine during World War I that protected against exanthematous typhus. I was completely fascinated, hearing how Professor Weigl was giving enemas to individual body lice. Weigl infected the lice with *Rickettsia prowazekii*, inserting glass micropipettes into their anal openings. Afterward he maintained the inoculated lice on human volunteers for several days. Subsequently, he removed the intestines from batches of 140 inoculated lice, crushed the intestines in a tiny glass mortar with a few drops of formalin, and obtained a single doze of his vaccine. Later I found out that this was in Europe the only available vaccine against trench fever until the end of World War II (Szybalski, 1999). The information about the

report of gibberellic acid treatment of "virus-diseased plants." A few years later it became apparent that, while wound tumor was a virus disease, the two other diseases were not viral, but phyto- and spiroplasma diseases.

III. VIRUS NOMENCLATURE AND CLASSIFICATION

For several years I was actively involved in virus nomenclature and classification (Maramorosch, 1974). My interest stemmed from the finding that several leafhopper-borne viruses that were inducing plant diseases were multiplying not only in plants but also in their invertebrate animal vectors. The finding that little or no harm was observed in the virus-carrying insects could suggest that these viruses originated as insect viruses and over long periods of evolution became harmless to their animal hosts, while their newer plant hosts were severely affected and often killed. Should these viruses be considered as plant, or as insect, viruses? Plant pathogenic viruses may exert a beneficial, or a harmful, effect on their specific insect vectors (Maramorosch, 1968, 1969, 1970; Maramorosch and Jensen, 1963). My popular article in *Scientific American* (Maramorosch, 1953) also focused on these aspects. Incidentally, the honorarium received for this article provided funds for my first movie camera and my new hobby, that later changed to video photography.

I thought that the affinity of vector-borne viruses to certain plant or animal hosts should not be used as a classifying criterion. The naming of viruses was for a long time highly controversial, particularly the naming of plant viruses. European colleagues opposed the use of Latin names for many years and plant virologists had little, if any, contact with animal virologists who created their separate system of virus nomenclature and classification. Already in 1947, shortly before I came to the United States, at the International Microbiological Congress in Copenhagen, it had been approved that the bacterial code in its Latin form applies to viruses and bacteria. This was also stated at the 1953 International Microbiology Congress in Rome, which I attended. Yet, even in 1966, papers were still being published in which disease organisms were described as belonging "in between viruses and bacteria," which Dr. Andre Lwoff pointedly called complete nonsense, since an organism defined as a virus is entirely different from one that is a bacteria. The International Committee of Microbiological Societies appointed a provisional committee to deal with the nomenclature of viruses and since then the nomenclature was in the hand of a powerless committee, which could make recommendations but these were not binding to anyone concerned. In 1963, Sir Christopher Andrews as Chairman of the provisional committee dissolved it and the International Committee for Virus Nomenclature (ICVN) was created for the first time. This ICVN consisted of members nominated by all the National Microbiological Societies that belong to the International Association of Microbiological Societies. For each of the 10 member countries, 5 delegates were nominated. The United States representatives were Drs. Harold S. Ginsberg, Chair, Jordi Casals, Karl Maramorosch, Joseph L. Melnick, and William R. Romig. I was happy when Dr. Riley D. Housewright, President of the American Society for Microbiology, informed me of my election.

In 1966, a symposium was held in Moscow and two papers dealing with plant viruses, Dr. B. D. Harrison's and mine, met with a very lively discussion. There were 600 virologists seated in the auditorium and another 200 were listening by shortwave receivers outside the hall. When the Executive Committee was elected, only one plant virologist, Dr. A. J. Gibbs, was included. He was in violent opposition to all proposals that were not in conformity with his own postulates. Following the symposium, I decided to devote my time to my laboratory research and field work and I lost interest in the controversial fights between plant and animal virologists.

My luck continued at Rockefeller University when one day at lunch Dr. Rene Dubos asked whether I knew of a virus that would be beneficial. I replied that during the past centuries the smallpox virus was probably beneficial, by marking afflicted women and making them homely and less likely to fall pray to invading and raping enemy solders. This was not what Dr. Dubos was interested in at that time. He told me that he was working on an article describing how once variegated tulips became the vogue in the Netherlands and how the "tulipomania" rewarded a few families that knew how to transmit the variegation virus to healthy tulip bulbs. A few days later, also in the Rockefeller lunchroom, Dr. Dubos told me that he was requested to give a popular course on viruses at the New School in Manhattan, but that he was too busy and suggested me instead of him as a lecturer. A day later I was called by phone and offered the teaching position. It involved 14 weekly 2-h lectures and the remuneration was of considerable help. The students in the New School had a variety of backgrounds. One was a microbiology professor at a medical school, another was a nurse in a local hospital, but at the other end there was a cashier at an A&P store and a barber who had no high school education. The course was my first teaching experience and I learned how to avoid technical terms when explaining viruses to a lay audience.

IV. COLD SPRING HARBOR

In 1951 Dr. Keith Porter, a leading cytologist, suggested that I apply to Dr. Milislav Demerec, Director of the Cold Spring Harbor Laboratories, for accommodations during the summer. Dr. George Palade, also a distinguished cell biologist at Rockefeller and later a Nobel laureate, was driving with Dr. Porter to take part in the June symposium and they took me along, to see the beautiful location and to apply to Dr. Demerec personally. With no written application and no formalities, the permission was granted and this stroke of good luck had a profound influence on my career.

During the 1950s, I spent eight summers at the Cold Spring Harbor Laboratories on Long Island, New York. Dr. Barbara McClintock permitted me to use her greenhouses while she was working outdoors with corn (*Zea mays*). Each year at the end of August, Dr. Alfred Hershey organized a bacteriophage symposium for invited bacterial virologists. Although I did not work on bacteriophages, I was permitted to attend these meetings, where as yet unpublished findings were presented by the virologists. Throughout the summers, Drs. Max Delbruck, Salvator Luria, and Ernst Mayr were working in Cold Spring Harbor, lecturing, and socializing with the small group of scientists. Dr. Luria was working on his textbook on viruses and I was greatly impressed watching him dictate into a tape recorder each morning, and then mailing the tape to his secretary in Urbana for typing. When the typed version came back to Cold Spring Harbor, Dr. Luria made small corrections and each chapter was ready for publication. Few times he asked me for comments and when the book was published, he donated a copy to me and I found that he acknowledged my reviewing of a couple of chapters in his book.

One day Dr. Luria suggested that I should invite Japanese postdoctoral scientists to my laboratory and he added: "Get a good Japanese postdoc, but never more than one. You will rapidly advance with your work, but if you get more than one Japanese associate, you will no longer have any time with your daughter and your wife, because you will try to keep up with your Japanese postdocs and spend 7 days a week in the lab." I remembered the first part of Luria's suggestion and followed it when I left Rockefeller University in 1961 and moved to the Boyce Thompson Institute. But I did not follow the advice concerning the limitation of invited Japanese postdocs. I soon found out how correct Luria was when he told me never to get more than one Japanese coworker at a time. When I got three Japanese associates at the same time, my own working habit changed as I felt compelled to keep up with my Japanese coworkers.

Thanks to Dr. McClintock's generosity in Cold Spring Harbor where she permitted me to use her greenhouses during the summer, I could carry out an experiment in which I injected antibiotics into abdomens of leafhoppers, exposed to presumptive viruses of aster yellows and corn stunt. I used penicillin, streptomycin, and tetracycline, convinced that the causative agents of the two plant diseases were viruses. Penicillin and streptomycin injections did not prevent transmissions, but tetracycline-injected leafhoppers failed to infect the exposed seedlings. Convinced that tetracycline has no effect on viruses, I did not believe the results of the tests, and assumed that the failed transmission was due to the heat in the greenhouses. I did not repeat the experiment after I returned to the Rockefeller greenhouses and I published the results and my wrong conclusion in the *Transactions of the New York Academy* (Maramorosch, 1954). Had I repeated the tests, I would have found that not the summer heat but tetracycline interfered with the presumptive viruses. Ten years later, my Japanese colleagues in Tokyo discovered the phytoplasma nature of the aster yellows disease, but I missed the boat.

V. THE CADANG-CADANG DISEASE

A Food and Agriculture Organization assignment in the Philippines in 1960 was an eye opener to a very important, different, and most interesting world. I was expected to find the vector of the presumptive virus that had already killed 30 million coconut palms on Luzon and nearby islands (Maramorosch, 1961). While trying to find an insect vector, I learned also first hand about people in the Philippines. I made the decision to combine future laboratory basic research with applied field work to increase food and fiber production in developing countries.

In the Philippines, I was not able to find a vector of the palm disease but I became well acquainted with numerous owners of larger and smaller coconut plantations. One observation which I made, but did not dare to publish in my final FAO report, had to wait 14 years before it made some sense. I found that palm owners, who spoke the local Bicolano language, were losing their palms to the cadang-cadang disease, while owners who spoke Tagalog, the official Philippine language, had healthy palms. The spread of the disease seemed halted at the provinces where Tagalog was the predominant language, sparing completely areas close to Manila and Los Banos. Of course, I did not dare to mention that the virus, or its vector, seemed to distinguish whether the palm owners used one or another language, but the consistent correlation was striking and puzzling.

Fourteen years passed before the viroid cause of the cadang-cadang disease was established by the Australian virologist Dr. J. W. Randles (1975). Almost all viroids require humans to spread from plant to plant and only Avocado blotch is transmitted by pollen. The transmission of cadang-cadang viroids seemed linked to the Bicolano-speaking plantation workers and the contamination of their tools used to make incisions in the palms and the flowers of the trees. Bicolano plantation owners were hiring Bicolano-speaking workers, while plantation owners who spoke Tagalog employed "their own" Tagalong workers. Apparently, one group carried viroid-contaminated "bolos" (machetes), while the other did not. It would seem easy to stop the spread of the cadang-cadang viroid by dipping the cutting tools of plantation workers in a calcium chloride solution (Maramorosch, 1985, 1993). As far as I know, this has not been implemented and more than 50 million palms have been destroyed by cadang-cadang in the Philippines (Maramorosch, 2004). Losses are partially alleviated by replanting with susceptible, but earlier maturing, coconut palms.

VI. DARK CLOUDS ON THE HORIZON

Not everything was rosy during my last year at Rockefeller University. When I asked President Detlev Bronk about my future at the University, he first congratulated me to my AAAS Prize but then said that the study of virology was declining all over the world and that it reached its peak in 1935, when Stanley crystallized TMV. "It no longer is an important science," he said. Therefore in the US National Academy, where Dr. Bronk was the President, botany, zoology, chemistry, physics, history, mathematics, and so on were represented, but there was no virology. Although my work was interesting and important, it did not fit into his university and he, as university president, decided to remove both plant and animal virology from Rockefeller University. Dr. Igor Tamm was heading animal virology at that time and Dr. Bronk mentioned him as well as the group in which I and Dr. F. O. Holmes were working with plant viruses. When I got up to leave the President's office, I could barely walk. In the corridor, the newly appointed vice president, Dr. Douglas M. Whittaker, met me and noticed that I looked sick. He put his arm around my shoulder, took me to his office, and asked whether I was ill. I repeated, almost verbatim, what I was told by Dr. Bronk. Dr. Whittaker assured me that my position was secure and he tried to console me. Just that week I received an invitation to go to the Philippines for 6 months, to work on the devastating cadang-cadang coconut palm disease that was believed to be caused by a virus. Dr. Whittaker told me that there was a recent precedent of a leave of absence request and that he, and not Dr. Bronk, could therefore give me permission to take a paid leave of absence for the UNDP consultancy in the Orient. The precedent was a leave granted by President Bronk to Professor Paul Weiss, to teach for 4 months at Harvard University.

VII. INSECT VIRUSES AND CELL CULTURE

When I realized that my tenure at Rockefeller University could be ended by President Bronk, who considered virology an unfit subject for university studies, I approached Dr. Richard Shope to assist me in searching for a different position elsewhere. Dr. Shope left the Rockefeller branch at Princeton in 1945, when the decision to close the Princeton Laboratory was announced. He went to Merck Laboratories in Rahway, New Jersey, but returned to the Rockefeller University a few years later. When I told him about my predicament, he called Merck's President, Dr. Max Tishler, and the latter contacted Dr. Maurice R. Hilleman, Director of Merck's Virus & Cell Biology Research Institute for Therapeutic Research at West Point, PA. Dr. Hilleman invited me to West Point and offered me a research position at a salary that was 50% higher than my Rockefeller salary. My European prejudice against industrial research prevented me from accepting the very tempting offer. When I discussed this with my wife, she reminded me that I have never been unhappy with any kind of work and that, when I was making a living at the refugee camps in Romania as a cobbler, or as a piano teacher of a young singer, I seemed quite happy. She did not think that I would miss the glamour of Rockefeller University if I would accept Dr. Hilleman's job offer at West Point. Yet, I was unable to decide and mentioned this to Dr. Shope. A few days later he told me that another, temporary position would be offered by Merck. The pharmaceutical company decided to investigate the feasibility of producing insect viruses for biological control of pests. I was hired as consultant for 6 months to organize a conference on insect viruses. For the following half year, I was reading the available literature on a subject that was completely new to me—baculoviruses. I gave the names of all prominent insect pathologists to Merck in Bradenton, where their animal farm was located. A 3-day conference was arranged and I met Professor Edward Steinhaus and a score of prominent US and Canadian insect virologists. After the conference I prepared a report and my final recommendation was that it was too early to start commercial production of baculoviruses because the subject was still in its early stage of university investigation. The 6 months during which I prepared the baculovirus conference got me very interested in insect viruses. I thought that their growth in tissue culture, rather than in living insects, could eventually lead to large-scale commercial production. This did not yet materialize because *in vitro* production of viral pesticides remained more costly than production *in vivo* (Maramorosch, 1979a,b, 1991).

During the following years, at the Boyce Thompson Institute and later at the Waksman Institute of Microbiology, I worked with invertebrate cell culture and baculoviruses. I was joined by excellent postdoctoral associates from Argentina, Australia, Canada, Chile, China, France, Germany, India, Israel, Japan, Korea, the Netherlands, Philippines, Poland, Romania, Slovakia, Turkey, the United States, and Yugoslavia. I shall mention but a few. Among the first was Dr. Robert R. Granados, an insect virologist, who came to my laboratory from Madison, Wisconsin. In subsequent years, he became Program Director for biological control. His research interests focused on insect pathology, the ultrastructure of insect virus replication and insect cell culture methods for virus studies, mechanisms of infection, and pathogenesis of insect viruses. Dr. Granados provided the first evidence for the accumulation of the *Wound tumor virus* in various organs of an inefficient vector (Granados *et al.*, 1967) and for insect viremia, caused by the virus invasion of vector hemocytes (Granados *et al.*, 1968). From Japan I was joined by Professor Jun Mitsuhashi who spent 2 years with me at Boyce Thompson Institute. He developed the widely used M&M insect cell culture medium and aseptically grew plant virus vectors (Mitsuhashi and Maramorosch, 1963). Using these vectors, he inoculated plant tissue cultures with the aster yellows "virus" (Mitsuhashi and Maramorosch, 1964). After returning to Japan, Professor Mitsuhashi became known for his work on mosquito cell lines and the development of new invertebrate cell culture media for virus propagation. We also published several books jointly. Among my Japanese associates Dr. Hiroyuki Hirumi worked for 10 years with me. He distinguished himself studying the aster yellows "virus" in various organs of an insect vector (Hirumi and Maramorosch, 1963). His extensive work with Hemiptera cell culture (Hirumi and Maramorosch, 1971) included the *in vitro* cultivation of embryonic leafhopper tissues (Hirumi and Maramorosch, 1964) and the localization of the *Wound tumor virus* in embryonic nonvector cells (Hirumi and Maramorosch, 1968). Dr. Hirumi also studied the Friend murine leukemia virus in mosquitoes (Hirumi *et al.*, 1971), Marek's herpes virus, and Type C virus (Hirumi *et al.*, 1974), and the growth of mouse trophoblastic cells stimulated in culture by polyoma virus (Koren *et al.*, 1971). After I moved from Boyce Thompson Institute to Rutgers University, Dr. Hirumi joined the International Laboratory for Animal Diseases (ILRAD) in Nairobi, Kenya. During the following years, he made significant contributions to the study of the tsetse fly borne Nagana disease of cattle.

Attempts were made to maintain aphid cells *in vitro*. My daughter, Lydia, assisted during her summer high school vacation and learned how to remove unborn aphids from adult insects by cesarean section. Dr. Takashi Tokumitsu was able to maintain surviving aphid cells for limited period *in vitro* (Tokumitsu and Maramorosch, 1966). He also studied cytoplasmic protrusions that formed in cultured leafhopper cells during mitosis *in vitro* (Tokumitsu and Maramorosch, 1967). From Tubingen, Germany came Dr. Gert Streissle, who worked with me for 7 years. He was the first who compared immunologically animal reo viruses with the plant pathogenic *Wound tumor virus*. Subsequently, the latter was classified as a plant reo virus. I was alerted to the striking morphological resemblance of the plant and the animal reo viruses by Drs. Albert Sabin and Andre Lwoff. Both suggested that we try to compare them immunologically. In our article in *Science* (Streissle and Maramorosch, 1963), we acknowledged their suggestions. Dr. Streissle returned to Germany to head the antiviral research at Bayer in Wuppertal. My international connections became a most gratifying experience. I would like also to brag about my former graduate students and I shall mention but two. Dr. Dennis M. Schmatz from Merck's Research Laboratories is currently a Vice President, heading research at the Merck Research Building in Tsukuba, Japan. Professor Kenneth E. Sherman, Ph.D., M.D., leads a large group of virologists working with hepatitis virus in Cincinnati, OH.

The etiology of whitefly-borne pathogens remained an enigma for many years. Repeated attempts made in my and in other laboratories to find viruses in extracts of diseased plants or by electron microscopy in thin sections of plant tissues were fruitless. Finally, in 1975, Professor Robert M. Goodman, at the Plant Pathology Department, Illinois University in Urbana, IL, made the brilliant discovery of the whitefly-borne Gemini viruses and of their single-stranded DNA genome (Goodman, 1977). Professor Goodman left Urbana for the University of Wisconsin in Madison and others continued his seminal work on Gemini viruses. I was very happy when he came to Rutgers University 2 years ago, becoming my Executive Dean. In 1975, he generously provided some of his excellent electron micrographs of Gemini viruses for our book on tropical diseases of legumes (Bird and Maramorosch, 1975; Maramorosch, 1975).

After I joined the faculty at Rutgers' Waksman Institute, I was joined by Dr. Arthur H. McIntosh, who for 7 years worked with me on baculoviruses. At Rutgers University, Dr. McIntosh studied the retention of insect virus infectivity in mammalian cell cultures (McIntosh and Maramorosch, 1973) and the localization of a baculovirus in a vertebrate cell line (McIntosh *et al.*, 1979). He continued his baculovirus studies after joining the Biological Control Laboratory of the US Department of Agriculture in Columbia, MO. Among my later postdoctoral Japanese associates was Dr. Ken-ichi Yamada, who carried out studies on *Heliothis zea* nuclear polyhedrosis virus (Yamada and Maramorosch, 1980, 1981; Yamada *et al.*, 1981). He continued his research in Japan at Tokyo's National Institutes of Health.

Over the years, I organized several national and international conferences dealing with this subject. I edited a number of books, published by Academic Press and by others (Section IX).

In later years, I improved and modified the leafhopper injection technique. In 1958, I saw in Warsaw the last remaining insect holder used by Professor Weigl, years earlier. A similar holder for 20 leafhoppers was constructed for me by the head of the Rockefeller University Instrument Shop, Mr. Niels Jernberg (Maramorosch and Jernberg, 1970). Using carbon dioxide and this modified device, it was easy to inject 20 leafhoppers in 1 min and perform statistically significant tests with several plant viruses, phytoplasmas, and spiroplasmas.

During the past three decades, invertebrate cell culture became widely used in biotechnology and basic research in virology. Use of baculoviruses in insect cell cultures is gaining popularity for the production of recombinant proteins, viral insecticides, and the production of vaccines. *In vitro* techniques are indispensable for studies of insect virus expression systems. Application of invertebrate cell culture and molecular biology is leading to significant progress in the understanding of cellular and molecular interactions between insect cells and viruses. Often unexpected results are obtained as was the case with our M&M medium, developed for leafhopper cell culture, and later found best suitable for mosquito cell cultivation and the study of arboviruses in mosquito cells (Maramorosch, 1979b).

VIII. ELECTRON MICROSCOPY

In 1957, I took a course in electron microscopy, offered to Rockefeller faculty members by Drs. Keith Porter and George Palade. The course gave me a good knowledge of the fixation procedures, the use of the Porter-Bloom ultramicrotome, the glass, and diamond knives, and so on. When I came to the Boyce Thompson Institute and obtained outside funding from the National Science Foundation and NIH of the US Public Health, I decided to apply for a supplement to my NIH grant to purchase an electron microscope. I consulted Dr. Palade who advised me to specify that I am planning to use the expensive instrument not only for the current grant-funded project but also for long-term studies of vector-borne viruses. Dr. Palade also suggested that I should get a Siemens Elmiscope, and not an RCI electron microscope. I followed both suggestions. My application was approved and the large supplement funded. The advice of Dr. Luria to get a Japanese associate proved excellent. I was fortunate to get Professor Eishiro Shikata from Hokkaido University in Sapporo for 2 years. Dr. Shikata worked as an assistant professor with Professor Teikichi Fukushi, studying the fine structure of leafhopper vectors transmitting the rice stunt virus. He was the first to visualize the rice plant virus not only in diseased plants but also in the invertebrate animal vectors (Shikata *et al.*, 1964). During his 2 years of our collaboration, Dr. Shikata worked 7 days a week, taking no vacation or holidays. Every day he entered the EM room early morning, leaving it late in the afternoon, and taking to the darkened room a sandwich for lunch. He told me that it would have taken him at least 20 min to get his eyes adjusted to the darkened room if he were to leave the room for lunch, and he did not want to lose so much time. During his 2 years at Boyce Thompson, Shikata, as senior author, published a series of articles in *Virology*, *Journal of Virology*, *Nature*, and the *Journal of the National Cancer Institute* (Shikata and Maramorosch, 1965a,b, 1966a,b, 1967a,b, 1969; Shikata *et al.*, 1964, 1966). A few years after returning to Japan, he succeeded Professor Fukushi as head of the Botany and Plant Pathology Departments at Hokkaido University in Sapporo. Later he was elected to Japan's National Science Academy as the only plant pathologist in this Academy.

IX. BOOKS

In 1960 at the AAAS Annual Meeting, I stopped at the book exhibit of Academic Press where I met Vice President, Kurt Jacoby. We spoke for quite a while about his former work in Germany and the creation of Academic Press in New York. I asked Mr. Jacoby whether symposium papers on biological transmission of animal and plant disease agents could be published by Academic Press. I was organizing a 2-day symposium on this subject, to be held at the Annual Meeting of the Entomological Society of America in Atlantic City, NY. Mr. Jacoby agreed and my first book, of 192 pages, "Biological Transmission of Disease Agents," was published in 1962. As agreed, I received no royalties. Years later, I was told that all 1800 copies were sold when the book went out of print. The idea of publishing the presentations came only after the conference participants had agreed to be symposium speakers. I had considerable difficulty in persuading some authors to submit manuscripts for publication. Foreign participants, Dr. W. C. Willett from Kaduna, Nigeria, and Dr. D. Blascovic from Bratislava, Slovakia, were among the first to send their contributions. The Rockefeller Foundation arranged the travel of these eminent participants through a grant to the Entomological Society of America.

The success of my first book prompted me to again try Academic Press for the publication of a more voluminous volume of 666 pages. The treatise was based on a United States–Japan seminar, which I organized in Tokyo together with Dr. Paul Oman. Mr. Jacoby was not interested this time because as he explained, symposia were not selling well. Wiley Interscience agreed to publish the book when I added several additional authors who did not participate in the Tokyo meeting. I also used the title of this second book, "Viruses, Vectors, and Vegetation" (1969) for the title of my current autobiographical chapter.

During the following years several volumes on viruses, edited by me alone or jointly with other virologists, were mainly published by Academic Press. In 1968, Springer published "Insect Viruses" (192 pp.). In 1971, "Comparative Virology," edited by me and E. Kurstak, (Academic Press, 584 pp.) was followed by "Viruses, Evolution, and Cancer" (813 pp., 1974). In 1975, with R. E. Shope, we edited "Invertebrate Immunity" (Academic Press, 365 pp.)

Viruses and virus diseases were included in "Tropical Diseases of Legumes," edited by Julio Bird and me in 1975. In 1977, I edited the "Atlas of Insect and Plant Viruses" as Vol. 8 of Academic Press' "Ultrastructure in Biological Systems." "Aphids as Virus Vectors," edited by K. F. Harris and me in 1977, "Leafhopper Vectors and Plant Disease Agents," by me and Harris, 1979, "Vectors of Plant Pathogens," by Harris and Maramorosch, 1980, "Plant Diseases and Vectors: Ecology and Epidemiology," by Maramorosch and Harris, 1981, and "Pathogens, Vectors, and Plant Diseases: Approaches to Control," by Harris and Maramorosch, 1982, as well as "Viruses and Environment," by Kurstak and Maramorosch, 1978, were all published by Academic Press. "Vectors of Disease Agents: Interactions with Plants, Animals and Man," edited by J. J. McKelvey, Jr. *et al.* was published by Praeger in 1980. The voluminous "Viral Insecticides for Biological Control," by Maramorosch and K. E. Sherman, and "Subviral Pathogens of Plants and Animals: Viroids and Prions," by Maramorosch and McKelvey, were published by Academic Press in 1985. In 1965, Ms. Lore Henlein of Academic Press suggested that I should start a series dealing with "Methods in Virology." Together with Hilary Koprowski, eight volumes of "Methods in Virology" were published by Academic Press between 1967 and 1984. Maintenance of "Animal/Human and Plant Pathogen Vectors," by Maramorosch and F. Mahmood, was published by Science Publishers in 1999.

At Rockefeller University Professor William Trager and Dr. Maria Rudzinska gave me excellent suggestions how to attempt the cultivation of leafhopper tissues and cells for virus studies. I was able to maintain leafhopper tissues *in vitro* but not cells or cell layers (Maramorosch, 1956). I continued my attempts and organized several conferences nationally and internationally. Proceedings of the conferences were published by Academic Press in the following volumes: "Invertebrate Tissue Culture: Research Applications" (Maramorosch, 1976), "Invertebrate Tissue Culture: Applications in Medicine Biology and Agriculture" (E. Kurstak and Maramorosch, 1976), "Invertebrate Cell Culture Applications" (Maramorosch and Mitsuhashi, 1982), and "Biotechnology in Insect Pathology and Cell Culture" (Maramorosch, 1987). Springer Verlag published "Invertebrate and Fish Tissue Culture," edited by E. Kurstak *et al.* (1988); CRC Press published "Biotechnology for Biological Control of Pests and Vectors" (Maramorosch, 1991); "Arthropod Cell Culture Systems" (Maramorosch and McIntosh, 1994); "Insect Cell Biotechnology" (Maramorosch and McIntosh, 1994). In 1997, "Invertebrate Cell Culture: Novel Directions and Biotechnology Applications," by Maramorosch and Mitsuhashi, was published by Science Publishers. "Invertebrate Cell Culture: Looking Toward the XXI Century," by Maramorosch and M. J. Loeb, was published by SIVB, Columbia, MD (1997).

In 1976, I started the new Academic Press series "Advances in Cell Culture." Volumes I–V appeared between 1981 and 1987. Dr. Gordon Sato joined me as an editor and Volumes VI and VII were published in 1988 and 1989.

In 1970, Dr. Kenneth M. Smith, whose influence on my decision to become a virologist I described earlier, suggested to Academic Press that I should join him, Drs. Max A. Lauffer, and Frederik B. Bang as an editor of "Advances in Virus Research (AVR)." Starting with Vol. 18 till Vol. 27, all four editors worked jointly, but unfortunately in 1981, both Kenneth Smith and Frederik Bang passed away. Dr. Lauffer and I continued editing AVR and after Dr. Lauffer retired, I was able to persuade Drs. Aaron Shatkin and Frederick A. Murphy to join me as AVR editors. Starting with Vol. 30 in 1985, all three of us still continue to edit AVR.

In addition to plant and insect viruses, I worked shortly with other viruses, including Friend murine leukemia virus in experimentally infected mosquitoes, Marek's herpes disease virus, and yellow fever virus.

X. INTERNATIONAL CONNECTIONS

In 1953, shortly after I became a naturalized US citizen, I was invited by Professor H. Thung to come for 3 months to his virology laboratory in Wageningen, the Netherlands. At the same time, Dr. Kenneth M. Smith invited me to a symposium organized by him at the VI International Virology Congress in Rome, Italy. The trip to Europe was only partially reimbursed by Professor Thung and Rockefeller University. The bulk of the expenses required a personal loan from my bank, which I repaid in 24 monthly installments. I never regretted the personal expenses, realizing how important the trips abroad were in making contacts with numerous virologists from different countries. During subsequent years my urge to travel did not subside and my visits to research institutes in Europe, Asia, Australia, and Africa became a constant feature. Contacts with virologists, entomologists, and plant pathologists were made during 28 visits to India. several trips to Japan, China, and Southeast Asia, to West, East, and South Africa, South and Central America, Australia, and several European countries. My knowledge of 7 languages was an important asset during these trips abroad.

In 1962, the New York Academy of Science elected me Vice President and Recording Secretary. In 1970, I was elected to the Leopoldina, oldest European Science Academy. At first, I hesitated to accept the membership in the Academy, located in Germany, but when I found out that among former members were Bohr, Curie Sklodowska, Liebig, Linne, Pavlov, Planck, and Rutherford, I accepted the election and the invitation to give an inaugural lecture in Halle, Germany. A few years later, in 1979, the Indian National Academy of Science elected me an Honorary Fellow, followed by the Indian Virological Society in 1987.

I became a Rockefeller Foundation grantee in 1955 to work on virus diseases of corn with Mexican agricultural scientists. In 1964, the Romanian Academy of Sciences invited me as guest lecturer for 4 weeks to Bucharest. This was followed by the USSR Academy of Medicine, then by the Indian Academy of Sciences, and the Polish Academy. In addition to virology, my special interest became the promoting of international scientific cooperation. As a consultant of the United Nations Development Program, I worked with Indian plant virologists and entomologists at the University of Agricultural Sciences at Hebbal-Bangalore, India in 1978–79, studying virus and phytoplasma diseases of food and fiber plants.

Various honors and awards were received during the past years but I will mention just one, the Wolf Prize in Agriculture, received in 1980 and often called the Nobel Prize in Agriculture. It was received for my studies on interactions between insect vectors, viruses, and plants.

I have listed several lucky events that promoted my scientific career, but the most important and luckiest was—you guessed it—my wife Irene. Without her unwavering support and devotion, I would not have been able to follow my chosen profession. In 1957, when I was invited to succeed Professor Edward Steinhaus as chair of the Entomology Department at the University of California in Berkeley, she was willing to leave her beloved work at the New York Public Library to follow me to California. It took me 4 months before I declined the very tempting offer and we remained in the East. When I accepted the position as Distinguished Professor at Waksman Institute, Irene noticed how the 200 km of daily driving to and from Rutgers University was taken a toll on my energy and health. After 30 years, she sacrificed her own career, resigned, took early retirement, and moved with me to New Jersey (Fig. 6).

I would like to end this biographical sketch by citing my acceptance remarks made in Jerusalem when I received the Wolf Prize:

Mr. President, Members of the Knesset, members of the Wolf Foundation, colleagues and friends. It is difficult for me to find the proper words to express my feelings and emotions on this solemn occasion and this beautiful ceremony. I feel humble and proud of having been selected the recipient of the coveted Wolf Prize in Agriculture and I would like to express my deep appreciation to the Wolf Foundation and to its Founder for their vision and foresight in recognizing agriculture as one of the important fields of modern science. Over the past 30 years numerous associates have contributed significantly to projects carried out in my laboratory and several national and international organizations, and foundations have sponsored my research. Many conferences, surveys, consultancies, and visits to developing and developed nations have been made possible through the excellent international collaboration of scores of scientists and institutions. All of them

FIGURE 6 At home with my wife. Photo by the author.

have contributed to the success of my career and to the honor bestowed on me today.

My only regret is that neither my nor my wife's parents, nor my brother, nor my wife's sister can be with us today. Unfortunately, they perished during the holocaust, together with more than 150 of our closest relatives. Only their names are left here in Jerusalem at Yad Vashem. Let us hope that the ideals that are so aptly expressed by the Wolf Foundation, the fostering of international understanding among scientists throughout the world, will prevent future wars and assure lasting peace on earth. Science recognizes no political, religious, ethnic, or geographic borders and we, scientists, speak only one language—the language of science. I sincerely hope that real peace can be achieved through the efforts of scientists collaborating with each other, irrespective of background and political beliefs. I address my heartiest thanks to the Wolf Foundation for fostering international understanding."

ACKNOWLEDGMENTS

I want to express my special thanks to all my past graduate students and postdoctoral associates who worked with me in the Unites States and abroad during the past years. To them I express my sincere thanks and best wishes for their continuous successful research and happy life.

REFERENCES

Bird, J., and Maramorosch, K. (1975). Viruses and virus diseases associated with whiteflies. *Adv. Virus Res.* **22:**55–110.

Black, L. M. (1953). Transmission of plant viruses by Cicadellids. *Adv. Virus Res.* **1:**69–89.

Brakke, M. K., Maramorosch, K., and Black, L. M. (1953). Properties of the wound-tumor virus. *Phytopathology* **43:**387–390.

Corner, G. W. (1964). "A History of the Rockefeller Institute 1901–1953." The Rockefeller Institute Press, New York City.

Fukushi, T. (1935). Multiplication of virus in its insect vector. *Proc. Imp. Acad. Japan* **11:**301–303.

Ginsberg, H. S. (1999). The life and times of adenoviruses. *Adv. Virus Res.* **54:**1–13.

Goodman, R. M. (1977). Single-stranded DNA genome in a whitefly-transmitted virus. *Virology* **83:**171–179.

Granados, R. R., Hirumi, H., and Maramorosch, K. (1967). Electron microscopic evidence for wound-tumor virus accumulation in various organs of an inefficient leafhopper vector, *Agalliopsis novella. J. Invertebr. Pathol.* **9:**147–159.

Granados, R. R., Ward, L., and Maramorosch, K. (1968). Insect viremia caused by a plant-pathogenic virus: Electron microscopy of vector hemocytes. *Virology* **34:**790–796.

Hirumi, H., and Maramorosch, K. (1963). Recovery of aster yellows virus from various organs of the insect vector, *Macrosteles fascifrons. Contrib. Boyce Thompson Inst.* **22:**165–173.

Hirumi, H., and Maramorosch, K. (1964). The *in vitro* cultivation of embryonic leafhopper tissues. *Exp. Cell Res.* **36:**625–631.

Hirumi, H., and Maramorosch, K. (1968). Electron microscopy of wound-tumor virus in cultured embryonic cells of the leafhopper *Macrosteles fascifrons. In* "Second International Colloquium on Invertebrate Tissue Culture" (C. Barigozzi, ed.), pp. 203–217. Lombardo Acad. Sci. Lettere, Milano.

Hirumi, H., and Maramorosch, K. (1971). Cell culture of Hemiptera. *In* "Invertebrate Tissue Culture" (C. Vago, ed.), Vol. 1, pp. 307–340. Academic Press, New York.

Hirumi, H., Burton, G. J., and Maramorosch, K. (1971). Preliminary studies on electron microscopy of Friend murine leukemia virus in the midgut of experimentally infected mosquitoes. *J. Virol.* **8:**801–804.

Hirumi, H., Frankel, J. W., Prickett, C. O., and Maramorosch, K. (1974). Coexistance of particles resembling herpes virus and Type-C virus in chicken feather follicle epithelium. *J. Natl. Cancer Inst.* **52:**303–306.

Koren, Z., Van Damme, O., Hirumi, H., and Maramorosch, K. (1971). The growth and stimulation of mouse trophoblastic cells in culture by polyoma virus. *Am. J. Obstet. Gynecol.* **111:**846–850.

Maramorosch, K. (1950). Influence of temperature on incubation and transmission of the wound-tumor virus. *Phytopathology* **40:**1071–1093.

Maramorosch, K. (1951a). Handy insect-vector cage. *J. NY Entomol. Soc.* **59:**49–50.

Maramorosch, K. (1951b). A simple needle for micro-injections. *Nature.* **167:**734.

Maramorosch, K. (1952a). Multiplication of aster-yellows virus in its vector. *Nature* **169:**194–195.

Maramorosch, K. (1952b). Toxicity of cellulose acetate sheets to plants and fish. *Science* **115:**236.

Maramorosch, K. (1953). A versatile virus. *Sci. Am.* **188:**78–86.

Maramorosch, K. (1954). Biological transmission of plant viruses by animal vectors. *Trans. NY Acad. Sci.* **16:**189–195.

Maramorosch, K. (1955). Multiplication of plant viruses in insect vectors. *Adv. Virus Res.* **3:**221–249.

Maramorosch, K. (1956). Multiplication of aster yellows virus in *in vitro* preparations of insect tissues. *Virology* **2:**369–376.
Maramorosch, K. (1957). Reversal of virus-caused stunting in plants by gibberellic acid. *Science* **126:**126–651.
Maramorosch, K. (1958). Beneficial effect of virus diseased plants on non-vector insects. *Tijdschr. Plziekt.* **63:**383–391.
Maramorosch, K. (1959). Reversal of virus-caused stunting by gibberellic acid in plants infected with corn stunt, aster yellows and wound tumor. *Proc. 6th Int. Congress of Crop Protection, Hamburg* **I:**271–272.
Maramorosch, K. (1961). Report to the Government of the Philippines on the cadang-cadang disease of coconut. *FAO ETAP Bull.* 1333. Rome.
Maramorosch, K. (1968). Plant pathogenic viruses in insect vectors. *Curr. Top. Microbiol.* **42:**94–107.
Maramorosch, K. (1969). Effects of rice-pathogenic viruses on their insect vectors. *In* "IRRI Symposium on Rice Virus Diseases," pp. 179–203. Johns Hopkins University Press, Baltimore, MD.
Maramorosch, K. (1970). Insect infection caused by a plant tumor virus. *World Rev. Pest Control* **9:**29–41.
Maramorosch, K. (1974). Introduction to the systematics of viruses: Viruses of plants, viruses of invertebrates. *In* "Handbook of Microbiology" (Laskin and Lechevalier, eds.), pp. 579–594. CRC Press, Boca Raton, FL.
Maramorosch, K. (1975). Etiology of whitefly-borne diseases. *In* "Tropical Diseases of Legumes" (J. Bird and K. Maramorosch, eds.), pp. 71–89. Academic Press, New York.
Maramorosch, K. (1979a). Biological control of insect pests with viruses. *In* "Practical Tissue Culture Applications" (K. Maramorosch and H. Hirumi, eds.), pp. 387–398. Academic Press, New York.
Maramorosch, K. (1979b). Leafhopper tissue culture. *In* "Leafhopper Vectors and Plant Disease Agents" (K. Maramorosch and K. F. Harris, eds.), pp. 485–514. Academic Press, New York.
Maramorosch, K. (1985). Control of viroid diseases. *In* "Subviral Pathogens of Plants and Animals: Viroids and Prions" (K. Maramorosch and J. J. McKelvey, eds.), pp. 151–162. Academic Press, Orlando, New York, London.
Maramorosch, K. (1991). Schadlingsbekampfung mit Mikroorganismen und Viren zur Schonung der Umwelt. *Zeitschrift Pflanzenkrankheiten und Pflanzenschutz* **98:**344–350.
Maramorosch, K. (1993) The threat of cadang-cadang disease. *Principes* **37:**187–196.
Maramorosch, K. (2004). The phantom vector of an emerging viroid. *In* "Life Sciences for the 21st Century" (E. Keinan, I. Schechter, and M. Sela, eds.), pp. 297–303. VCH-Wiley, New York.
Maramorosch, K., and Jensen, D. D. (1963). Harmful and beneficial effects of plant viruses in insects. *Annu. Rev. Microbiol.* **17:**495–530.
Maramorosch, K., and Jernberg, N. (1970). An adjustable multiple insect holder for microinjection. *J. Econ. Entomol.* **63:**1216–1218.
Maramorosch, K., Brakke, M. K., and Black, L. M. (1949). Mechanical transmission of a plant tumor virus to an insect vector. *Science* **110:**162–163.
McIntosh, A. H., and Maramorosch, K. (1973). Retention of insect virus infectivity in mammalian cell cultures. *J. NY Entomol. Soc.* **81:**175–182.
McIntosh, A. H., Maramorosch, K., and Riscoe, R. (1979). *Autographa californica* nuclear polyhedrosis virus (NPV) in a vertebrate cell line: Localization by electron microscopy. *J. NY Entomol. Soc.* **87:**55–58.
Mitsuhashi, J., and Maramorosch, K. (1963). Aseptic cultivation of four virus transmitting species of leafhopper (Cicadellidae). *Contrib. Boyce Thompson Inst.* **22:**165–173.

Mitsuhashi, J., and Maramorosch, K. (1964). Inoculation of plant tissue cultures with aster yellows virus. *Virology* **23:**277–279.

Randles, J. W. (1975). Association of two ribonucleic acid species with cadang-cadang disease of coconut palm. *Phytopathology* **65:**163–167.

Shikata, E., and Maramorosch, K. (1965a). Plant tumor virus in arthropod host: Microcrystal formation. *Nature* **288:**507–508.

Shikata, E., and Maramorosch, K. (1965b). Electron microscopic evidence for the systemic invasion of an insect host by a plant pathogenic virus. *Virology* **27:**461–475.

Shikata, E., and Maramorosch, K. (1966a). An electron microscope study of plant neoplasia induced by wound tumor virus. *J. Natl. Cancer Inst.* **36:**97–116.

Shikata, E., and Maramorosch, K. (1966b). Electron microscopy of a pea enation mosaic virus in plant cell nuclei. *Virology* **30:**439–454.

Shikata, E., and Maramorosch, K. (1967a). Electron microscopy of wound tumor virus assembly sites in insect vectors and plants. *Virology* **32:**363–377.

Shikata, E., and Maramorosch, K. (1967b). Electron microscopy of the formation of wound tumor virus in abdominally inoculated insect vectors. *J. Virol.* **1:**1052–1073.

Shikata, E., and Maramorosch, K. (1969). Electron microscopy of insect-borne viruses *in situ*. *In* "Viruses, Vectors, and Vegetation" (K. Maramorosch, ed.), pp. 393–415. Interscience, John Wiley & Sons, New York.

Shikata, E., Maramorosch, K., and Granados, R. R. (1966). Electron microscopy of pea enation mosaic virus in plants and aphid vectors. *Virology* **29:**426–436.

Shikata, E., Orenski, S. W., Hirumi, H., Mitsuhashi, J., and Maramorosch, K. (1964). Electron micrographs of wound-tumor virus in an animal host and in a plant tumor. *Virology* **23:**441–444.

Storey, H. H. (1933). Investigations of the mechanism of the transmission of plant viruses by insect vectors. I. *Proc. R. Soc. Lond. Ser. B* **113:**463–485.

Streissle, G., and Maramorosch, K. (1963). Reo virus and wound-tumor virus: Serological cross reactivity. *Science* **140:**996–997.

Szybalski, W. (1999). Maintenance of human-fed live lice in the laboratory and production of Weigl's exanthematous typhus vaccine. *In* "Maintenance of Human, Animal, and Plant Pathogen Vectors" (K. Maramorosch and F. Mahmood, eds.), pp. 161–179. Science Publishers, Enfield, NH, USA.

Tokumitsu, T., and Maramorosch, K. (1966). Survival of aphid cells *in vitro*. *Exp. Cell Res.* **44:**652–655.

Tokumitsu, T., and Maramorosch, K. (1967). Cytoplasmic protrusions in insect cells during mitosis *in vitro*. *J. Cell Biol.* **34:**677–683.

Yamada, K., and Maramorosch, K. (1980). Propagation of *Heliothis zea* single embedded baculovirus in a *H. zea* cell line. Abstracts 16th International Entomology Congress Tokyo, Japan, p. 193.

Yamada, K., and Maramorosch, K. (1981). Plaque assay of *Heliothis zea* baculovirus employing a mixed agarose overlay. *Arch. Virol* **67:**187–189.

Yamada, K., Sherman, K. E., and Maramorosch, K. (1981). *In vivo* infectivity of early and late passaged *Heliothis zea* polyhedra produced in tissue culture. *Appl. Entomol. Zool.* **16:**504–505.

BOOKS

K. Maramorosch (ed.) (1962). Biological Transmission of Disease Agents. Academic Press, New York and London.

K. Maramorosch and H. Koprowski (eds.) (1967–1984). Methods in Virology, Vols. 1–8. Academic Press, New York and London.

K. Maramorosch (ed.) (1968). Insect Viruses, Springer Verlag, New York, Berlin.
K. Maramorosch (ed.) (1969). Viruses, Vectors, and Vegetation. Wiley-Interscience, New York, London, Toronto.
K. Maramorosch and E. Kurstak (eds.) (1971). Comparative Virology. Academic Press, New York.
M. A. Lauffer, F. B. Bang, K. Maramorosch, and K. M. Smith (eds.) (1973–1981). Advances in Virus Research, Vols. 22–26. Academic Press, New York.
E. Kurstak and K. Maramorosch (eds.) (1974). Viruses, Evolution, and Cancer. Academic Press, New York, London.
K. Maramorosch and R. E. Shope (eds.) (1975). Invertebrate Immunity. Academic Press, New York and London.
J. Bird and K. Maramorosch (eds.) (1975). Tropical Diseases of Legumes. Academic Press, New York and London.
K. Maramorosch (ed.) (1976). Invertebrate Tissue Culture: Research Applications. Academic Press, New York, London.
E. Kurstak and K. Maramorosch (eds.) (1976). Invertebrate Tissue Culture: Applications in Medicine, Biology, and Agriculture. Academic Press, New York and London.
K. Maramorosch (ed.) (1977). The Atlas of Insect and Plant Viruses. Academic Press, New York and London.
K. Maramorosch and H. Hirumi (eds.) (1979). Practical Tissue Culture Applications. Academic Press, New York.
E. Kurstak, K. Maramorosch, and A. Dubendorfer (eds.) (1977). Invertebrate Systems *In Vitro*. Elsevier-North Holland Biomedical Press.
K. F. Harris and K. Maramorosch (eds.) (1977). Aphids as Virus Vectors. Academic Press, New York and London.
K. Maramorosch and K. F. Harris (eds.) (1979). Leafhopper Vectors and Plant Disease Agents. Academic Press, New York.
K. F. Harris and K. Maramorosch (eds.) (1980). Vectors of Plant Pathogens. Academic Press, New York.
M. A. Lauffer and K. Maramorosch (eds.) (1981–1984). Advances in Virus Research, Vols. 27 and 28. Academic Press, New York.
K. Maramorosch and K. F. Harris (eds.) (1981). Plant Diseases and Vectors: Ecology and Epidemiology. Academic Press, New York.
K. F. Harris and K. Maramorosch (eds.) (1982). Pathogens, Vectors, and Plant Diseases: Approaches to Control. Academic Press, New York.
J. J. McKelvey, Jr., B. F. Eldridge, and K. Maramorosch (eds.) (1980). Vectors of Disease Agents: Interactions with Plants, Animals and Man. Praeger, New York.
E. Kurstak and K. Maramorosch (eds.) (1978). Viruses and Environment. Academic Press, New York.
K. Maramorosch (ed.) (1981–1987). Advances in Cell Culture, Vols. 1–5. Academic Press, New York.
K. Maramorosch and G. H. Sato (eds.) (1988–1989). Advances in Cell Culture, Vols. 6 and 7. Academic Press, New York.
K. Maramorosch and J. Mitsuhashi (eds.) (1982). Invertebrate Cell Culture Applications. Academic Press, New York.
K. Maramorosch and J. J. McKelvey, Jr. (eds.) (1985). Subviral Pathogens of Plants and Animals: Viroids and Prions. Academic Press, New York.
K. Maramorosch, F. A. Murphy, and A. J. Shatkin (eds.) (1985–2007). Advances in Virus Research Vols. 29–70. Academic Press, Elsevier, London, San Diego, Amsterdam etc.
K. Maramorosch and K. E. Sherman (eds.) (1985). Viral Insecticides for Biological Control. Academic Press, Orlando, New York, London, Tokyo.

K. Maramorosch (ed.) (1987). Biotechnology in Insect Pathology and Cell Culture. Academic Press, San Diego, CA.

E. Kurstak, Y. Kuroda, and K. Maramorosch (eds.) (1988). Invertebrate and Fish Tissue Culture. Springer, Berlin and Tokyo.

K. Maramorosch (ed.) (1991). Viroids and Satellites: Molecular Parasites at the Frontier of Life. CRC Press, Boca Raton, FL, USA.

K. Maramorosch (ed.) (1992). Plant Diseases of Viral, Viroid, Mycoplasma, and Uncertain Etiology. Oxford and IBH, New Delhi, INDIA.

K. Maramorosch and A. H. McIntosh (eds.) (1994). Arthropod Cell Culture Systems. CRC Press, Boca Raton, Ann Arbor, London, Tokyo.

K. Maramorosch and A. H. McIntosh (eds.) (1994). Insect Cell Biotechnology. CRC Press, Boca Raton, FL.

K. Maramorosch and J. Mitsuhashi (eds.) (1997). Invertebrate Cell Culture: Novel Directions and Biotechnology Applications. Science Publishers, Enfield, NH, USA.

K. Maramorosch and M. J. Loeb (eds.) (1997). Invertebrate Cell Culture: Looking Toward the XXI Century. SIVB, Columbia, MD.

K. Maramorosch and F. Mahmood (eds.) (1999). Maintenance of Animal/Human and Plant Pathogen Vectors. Science Publishers, Enfield, NH, USA.

John O. Almquist

Pennsylvania State University
University Park, Pennsylvania, USA

1981 — for his significant contributions to the application of artificial insemination to livestock improvement.

Professor Almquist's early work on the addition of antibiotics to bull semen resulted in substantial increases in breeding efficiency. This, coupled with his remarkable achievements in developing methods for processing, freezing, and thawing of frozen semen, significantly enhanced the practical utilization of artificial insemination in the livestock industry. Many techniques developed by him for cattle have been applied to other species, including the human male.

CURRICULUM VITAE

Date and Place of Birth - February 10, 1921, Hoidrege, Nebraska

Academic Degrees

Cornell University, B.S., 1942
Purdue University, M.S., 1944
The Pennsylvania State University Ph.D., 1947

Membership in Professional or Honorary Academic Societies

American Dairy Science Association
American Society of Animal Science

Fellow, American Association for the Advancement of Science
Society for the Study of Fertility (Great Britain)
Society for the Study of Reproduction (USA)
International Fertility Association
Alpha Zeta
Phi Kappa Phi
Sigma Xi
Gamma Sigma Delta

Honors and Awards Received; Special Programs and Activities

1959 - Glycerine Producers' Association Research Award
1963 - Borden Award and Gold Medal for Research in Dairy Production (American Dairy Science Association)
1971 - Outstanding Teacher Award, Gamma Sigma Delta Agricultural Honorary Society
1972 - Medal from Italian Society for Progress in Animal Production
1974 - Animal Physiology and Endocrinology Award (American Society of Animal Science)
1976 - National Association of Animal Breeders Research Award (American Dairy Science Association)
1977 - Professor I.I. Ivanov AI Pioneer Commemorative Medal from Ministry of Agriculture, Soviet Union
1952 - Member of Standing Committee for International Congresses on Animal Reproduction and Artificial Insemination
1956 - Plenary Speaker, III International Congress on Animal Reproduction and. Artificial Insemination, Cambridge, England
1961 - Speaker, IV International Congress on Animal Reproduction and Artificial Insemination, The Hague, The Netherlands
1972 - Moderator and Summary Speaker, VII International Congress on Animal Reproduction and Artificial Insemination, Munich, Germany
1976 - Chairman and Summary Speaker, VIII International Congress on Animal Reproduction and Artificial Insemination, Cracow, Poland
1976 - Speaker, Bilateral Reproduction Symposium, Moscow, USSR
1976 - Invited member of eight-man, U.S. Scientific Exchange team on Farm Animal Reproduction to visit Soviet Union

Consultant to nearly every artificial insemination organization in USA, as well as in Puerto Rico and in Ontario, Alberta and British Columbia, Canada

Editorial Board, Journal of Dairy Science for 10 years

Consultant on infertility problems in large dairy herds, University of Hawaii

Essential Bibliography

Author or coauthor of 160 scientific publications, including 78 papers in refereed journals, 3 bulletins, 3 book chapters and over 70 papers given at scientific meetings.

DESCRIPTION OF SCIENTIFIC ACTIVITIES

1. Evidence of Originality, Vision, Creativity:

Dr. John O. Almquist is one of the recognized leaders and pioneer researchers in artificial insemination and reproductive physiology of farm livestock. It was largely through his efforts that the "Dairy Breeding Research Center" was established on the Campus of The Pennsylvania State University in 1949 to assist in the development of artificial insemination (AI) in dairy cattle. As the AI industry progressed, research activities were expanded to include the broader field of animal reproduction with special emphasis on the comparative aspects of male reproductive physiology, bull sexual behavior, bull nutrition, spermatogenesis, sperm metabolism, sperm ultra-structure and female infertility.

As a result of the nominee's early recognition of the importance of the male in animal reproduction and in livestock improvement through artificial insemination, the Research Center has gained international recognition for basic and applied contributions to male reproductive physiology. Many of the techniques developed by him and his graduate students for bulls have been applied to other species, including the human male. Thus, Dr. F.P. Amann (first as a graduate student and then as a coworker with Dr. Almquist since 1955) has procured substantial National Institutes of Health grants and contracts for basic research on the physiology of the testis and epididymis. This research not only adds new knowledge on sperm production and maturation mechanisms in animals, but may lead to development of a male contraceptive for humans.

Because of the quality and practicality of his research program, Almquist has received substantial long-term financial support from the Artificial Breeding Cooperatives of Pennsylvania. During the past 33 years, these organizations have contributed about $500,000 to this program and have cooperated in conducting numerous field fertility trials of new procedures developed by Dr. Almquist. Further evidence of the overall quality of the nominee's research is the excellent industry support that has been received from the Pennsylvania Department of Agriculture, the National Association of Animal Breeders, purebred livestock associations and individual breeders.

Twenty-eight graduate students have been trained as an integral part of the program conducted by Dr. Almquist, and. many of his former students hold positions of leadership in the fields of reproductive physiology, artificial insemination, biology and veterinary medicine.

The nominee's research efforts have contributed greatly to the success of artificial insemination and, therefore, to genetic improvement of dairy cattle. He has been able to identify the most important problems and design critical experiments to obtain the answers. His earliest research enabled AI organizations to identify semen of different breeds by the addition of harmless colors to the semen extender. His pioneer work on the addition of antibiotics to bull semen resulted in a dramatic increase of 11 percentage units in breeding efficiency. This work is still one of the most significant contributions to the successful application of AI to breeding better livestock. Much of his research has been aimed toward the constant improvement of semen extenders, including early development of heated cows' milk in various forms. Of equal importance was his finding that addition of glycerol to milk extenders doubled the fertile life of sperm in unfrozen semen.

Believing that the male reproductive potential was seriously underrated, he initiated a series of investigations on the factors associated with the quantity and quality of semen produced by both young and sexually mature bulls, with special emphasis on nutrition, frequency of semen collection and management of sexual behavior. In addition, he and his colleagues have contributed much to our understanding of costly infertility problems in heifers and cows. A 1967 bulletin on genital abnormalities in sub-fertile dairy heifers by Dr. T. Y. Tanabe and Dr. Almquist has gained international recognition for its thoroughness, usefulness and the excellent quality of the 24 color plates.

Dr. Almquist recognized early in his career the tremendous potential of artificial insemination in beef cattle and other animals, as well as dairy cattle. He suggested over 20 years ago that widespread application of frozen semen, ovulation control and improved feeding and management practices could result in more rapid expansion of AI in beef cattle and improve worldwide animal agriculture. He stressed that maximum genetic gain could not be attained unless beef bulls were sampled at the earliest possible age. Thus, much of his research was designed to obtain the information necessary to make AI in beef cattle more effective? He was the first to investigate in detail the capacity of beef bulls for producing, storing and ejaculating large numbers of fertile spermatozoa. Many people assumed that beef and dairy bulls were similar in reproductive ability. Dr. Almquist expressed caution about such generalizations and his recent findings showed that important reproductive traits vary widely, not only between beef and dairy bulls, but also among different beef breeds.

After participating in an international congress in Italy in 1961, he became enthusiastic about the potential use of plastic strays for packaging frozen semen. With the financial aid of AI organizations in Pennsylvania, he equipped a laboratory to study the use of different kinds of straws for frozen semen. This research resulted in improved methods for processing, freezing and thawing frozen semen in straws which since have been adopted for field use.

Another example of his foresight and creativity is the pioneering research in the physiology of ejaculation. Dr. Almquist and graduate students developed a new electronic differential recording technique for measuring vas deferens and vesicular gland contractions during semen collections from rabbits. Silicone tubing sensors were implanted into the male sex organs and then attached to catheter-transducer systems.

2. **Soundness of Experimental Methodology and Adequacy Interpretation:**

Dr. Almquist is a meticulous investigator whose experiments have been well-designed and thorough in their husbandry and management techniques. In studies on beef bulls, for example, representatives of three major breeds — Angus, Hereford and Charolais — were used. They were obtained as weanlings, penned individually and fed a carefully controlled and measured nutrient intake because it was realized that nutrition might influence some of the traits being studied, such as age at onset of puberty and growth rate of the testes. Since initiation of different frequencies of ejaculation at puberty were part of the design, the bulls were adjusted to weekly collection procedures starting at 24 weeks of age — well before time of puberty was anticipated. Puberty was carefully defined at the outset as the age at which an ejaculate containing a minimum of 50 million sperm with at least 10% showing progressive motility was first collected. Detailed and repeated observations were made on carefully selected indices of growth, reproductive development and sexual behavior.

Obviously, fertility data are of prime importance in the research conducted by Dr. Almquist. The confidence in his research designs is well-illustrated by the fact that commercial breeding organizations have been willing to assist with numerous field trials. More recently, out-of-state breeding studs have been added to gain additional fertility data. In a recent study, data from over 29,000 first insemination cows were collected.

3. **Principles Discovered or Established by the Research:**

Many of the techniques and procedures presently used by AI organizations are based upon the sound principles established through research conducted by Dr. Almquist and his colleagues at The Pennsylvania State University.

a. Dr. Almquist has received international recognition for his early research which first demonstrated the value of adding antibiotics (penicillin in 1946; streptomycin in 1947; and penicillin plus streptomycin in 1949) to diluted bull semen to control bacterial growth, reduce early embryonal mortality and increase fertility. His classic series of papers on these antibiotics included their effects on the livability, metabolism, bacterial content and fertility of bull semen, as well as their stability during semen storage. The AT industry adopted the use of these antibiotics almost universally.

b. Nearly equal in importance to the AI industry was his research in 1951 which pioneered use of milk as a practical medium to extend the life of bull sperm. He collaborated with biochemists to identify the toxic factor to spermatozoa in unboiled milk which was destroyed by heating. In 1956, he and his graduate students showed that addition, of glycerol to milk extender doubled the fertile life of hull sperm in liquid semen from 2 to 4 days. Combined with concurrent research which showed that semen could be collected six times weekly without harm to the bull or his fertility, breeding organizations were then able to offer dairymen "every sire every day" of the week at low cost using liquid semen. Techniques were developed in 1954 for freezing bull spermatozoa in glass ampules using homogenized milk and skimmilk extenders.

c. Dr. Almquist's recent research concerns the reproductive development and capacity of Angus, Hereford and Charolais bulls, With emphasis on the application of artificial insemination to beef cattle as a means of increasing meat production by using sires of outstanding genetic merit. Techniques developed over the past 20 years to measure quantitatively the sexual behavior and semen production of dairy bulls were applied to assess the potentials and limitations of the three breeds of beef bulls. It was found that there are a number of major differences in reproductive capacity among beef and dairy breeds. They also demonstrated conclusively that many beef bulls can be used for breeding at an earlier age and more intensively than previously had been thought possible. Highlights of the findings from three 10-year-long experiments with (a) Holstein, (b) Angus and Hereford, and, most recently, (c) Charolais bulls include:

   1. Starting at puberty, bulls of each breed can routinely be ejaculated four to six times per week, provided frequent changes in the stimulus situation are made.
   2. Starting at puberty, collection of semen from dairy and beef bulls at a high frequency of six ejaculates per week continuously for up to 1 year does not have a deleterious effect on body growth, testis development, quantity and quality of spermatozoal production, freezability of spermatozoa and fertility.
   3. Proper bull management before semen collection (e.g., false mounting and restraint) greatly increases the number of sperm obtained per ejaculation. Holstein bulls require three false mounts prior to each ejaculation to maximize the number of sperm obtained in both first and second ejaculates. However, use of three false mounts with Angus, Hereford and Charolais bulls is effective only for the first of two successive ejaculates.

4. Most young dairy and beef bulls can ejaculate sufficient sperm for progeny testing by 14 months of age.
5. The testes of mature bulls produce prodigious numbers of sperm — about 7 billion per day in Holsteins.
6. Postpuberal increases in weekly sperm output are much more rapid in Charolais and Holstein than in Angus and Hereford bulls.
7. Charolais bulls attain puberty at a significantly earlier age (41 wks) than Angus and Hereford bulls (45 wks). This finding contradicts the commonly held viewpoint that Charolais bulls mature rather slowly.
8. On a within herd basis, much valuable growth data can be gathered for Charolais if the top 60% of bulls at 365 days of age are continued on gain test to 550 days of age.
9. Bull calves should be fed *ad libitum* for achieving puberty at the earliest possible age.

d. From a 10-year study of 18 Holstein bulls, he and his collaborator (Dr. R.G. Thompson, Dept. of Pathology, University of Guelph, Ontario, Canada) demonstrated that there was no difference in spinal bone lesions between bulls ejaculated continuously six times per week from 1 to 9.5 years of age and the controls ejaculated only once weekly to the same age. This report helped to allay fears in the AI industry that high frequency of ejaculation involving three false mounts preceding each ejaculation shortens the useful reproductive life of a bull by increasing the number and severity of vertebral body osteophytes.

e. Important findings of recent laboratory and fertility studies include the following:
   1. Developed new antibiotic combinations, lincomycin-spectinomycin and penicillin-neomycin, which were shown to be satisfactory substitutes if penicillin-streptomycin is not available. Each combination effectively controlled bacterial growth in semen without reducing fertility (1973–74).
   2. Explored an area of male reproduction — physiology of ejaculation — which had not been studied extensively despite its obvious importance in animal and human reproduction. Three recent publications described a new technique for making the first quantitative in viva pressure measurements of contractions of the vas deferens and vesicular glands in both anesthetized and conscious rabbits by implanting silicone tubing sensors.
   3. Published in 1973 the first two papers on new methods for freezing and thawing semen packaged in Continental US plastic straws. Dr. Almquist's research showed that fast thawing was much mare important than rate

of freezing. Freezing rate over a rather wide range did not affect sperm survival in US straws provided they were thawed rapidly (7 to 15 seconds). He and his graduate students also found that both acrosomal maintenance and motility of spermatozoa were affected more by thawing rate than by time of exposure to glycerol, a cryoprotective agent. Optimum glycerol level for freezing semen in straws was 11%, or the same as that reported earlier for freezing semen in glass ampules using skimmilk extender.

4. Showed that semen in straws could be placed onto canes before freezing and that loads of up to 7500 straws could be frozen in an automatic forced vapor freezer. Results were equally as good as those with the procedure presently used by the AI industry in which about 350 straws are individually cooled and then placed into goblets and onto canes while at 196°C. This automatic method provides a 20-fold increase in the number of straws which can be frozen at one time, reduces labor and improves product quality.
5. Conventionally, semen is warmed to 5°C and then used for insemination. From studies with thawing baths ranging from 5°C to 135°C, he showed that warming semen rapidly in straws directly to body temperature while thawing was more desirable. By placing straws of frozen semen into a 35°C water bath for 30 seconds, seminal temperature was raised rapidly to near body temperature and fertility involving over 21,000 experimental inseminations was significantly higher than for semen warmed to only 5°C.

f. Other important early research findings include the following:

1. Developed a safe, practical method of coloring semen to distinguish semen from bulls of different breeds. Reported in 1946 and still used by AI studies in most countries.
2. Established relationships between numbers and types of bacteria in bull semen and fertility.
3. Developed a commonly used method of staining bull sperm to differentiate living and dead cells.
4. Collaborated with Dr. R.J. Flipse to show that large quantities of corn and grass silage could be fed safely to dairy breeding bulls and that dairy bull calves mature later on low-energy rations.
5. Collaborated with Dr. B.B. Hale to conduct the first quantitative studies on sexual behavior and semen production by dairy bulls.
6. Cooperated with Dr. T.Y. Tanabe to show that pregnancy interruption in AI of dairy cattle is reduced by addition of antibiotics to bull semen and deposition of semen in the mid-cervical region of the uterus.

4. Practical Significance and Value of Research by John O. Almquist for Improving Food Production Worldwide

Mankind has one thing in common, an interest in how to get more food, preferably food that is lower in cost and higher in quality. Just before World War II, while working as a student assistant in Cornell University's newly established laboratory for research on artificial insemination (AI) of cattle, John Almquist recognized that AI had tremendous potential in the science of animal breeding for increasing the milk supply. In the past 30 years, average milk yield in the United States has increased from about 2,350 kg to 5,100 kg per cow largely as a result of AI and improved management. By breeding better cows through increased use of outstanding bulls in AI, farmers produce more milk from fewer cows resulting in an abundant supply of milk and milk products for the consumer. How did AI make its major contribution to this achievement?

For exploitation of AI for herd improvement, the question to be answered was: how to find superior bulls and how to make the most use of them through AI when they were found. Early in his career at The Pennsylvania State University, Dr. Almquist decided to investigate this question. With State government and AI industry support, a research facility was completed in 1949 which is unique in that it includes housing for bulls from calfhood to maturity. The pioneering long-term bull studies completed by Dr. Almquist and coworkers have greatly enhanced the world's supply of milk and meat through wider use of the best bulls in AI. By more liberal feeding and not being hesitant to initiate collection of semen for use in AI at puberty, they (1) hastened the onset of puberty by several months, thereby enabling young bulls to be sampled for proving at an earlier age than previously believed, (2) demonstrated that postpuberal fertility is satisfactory so that genetic sampling of most young bulls can be completed by 14 months rather than 18 to 24 months of age, (3) showed that mature bulls can be mated by AI to from 50,000 to 100,000 cows per year compared to the 1950 USA average of 1,250 cows serviced annually per AI sire 30 years earlier. This was accomplished by experiments showing that more spermatozoa could be harvested for use in AI by collecting bulls more frequently than previously believed possible and by developing practical procedures for expeditious removal of billions of high-quality spermatozoa during the semen collection procedure.

Examples of the practical impact of this research on the reproductive capacity of dairy bulls include:

1) The best breeding bulls are being identified by 4.5 to 5 years of age, rather than 7 to 8 years of age, thus greatly extending their useful life for breeding purposes.

2) The top breeding AI bulls sire many more offspring. Faster genetic improvement has been achieved by identifying and using the best bulls more efficiently. Fewer bulls are needed to breed the same number of females. One Holstein-Friesian bull in the USA sired about 250,000 offspring in his lifetime. Based on actual production records for his 27,754 tested daughters, extensive use of this superior sire resulted in over 17.5 million dollars more income to the owners of these daughters. The estimated value of the extra milk produced by the approximately 100,000 tested and untested daughters of this bull amounted to over 600 million dollars. Breeding potential for top bulls is tremendous; recently two outstanding bulls in a Pennsylvania AI stud using Dr. Almquist's recommendations were mated to 120,000 and 125,000 cows in one year. Such figures seemed impossible a short while ago.

Such potent "bull power" was unknown until the prodigious production of spermatozoa by bulls (6 to 9 billion per day) and means of obtaining them were revealed through original research done by Almquist and coworkers starting in 1951. They also were the first to provide quantitative data on the sexual behavior of bulls. The new techniques developed in the 1950's for measuring the numbers of sperm which bulls are capable of producing, storing and ejaculating have been adopted by reproductive researchers for determining these parameters in many species, including man. This pioneering work on male reproductive physiology has received international recognition and has aided the surge in knowledge about the role of the male in the reproductive process and in overcoming infertility in both man and domestic animals.

Dr. Almquist became convinced at an early age that the usefulness of AI should be expanded beyond dairy cattle to increase production in other species of economic importance for providing food and fiber to man. With characteristic foresight, he initiated research on the reproductive capacity of beef bulls in 1962. Much to the surprise of many animal breeders, he demonstrated that many beef bulls possess great reproductive potential for use in AI to increase meat production. Prior to his research, it was commonly believed that beef bulls should not be used for breeding before 18 months of age. He showed that semen collected, frozen and used from beef bulls under 1 year of age yielded economically satisfactory AI fertility rates, and thus, beef bulls could be sampled in a progeny testing program at a much earlier age than previously believed possible.

The original research by Dr. Almquist showing considerably higher AI conception rates with antibiotic-treated semen provided a major impetus for widespread acceptance and more rapid growth of cattle AI in Pennsylvania and the USA. Antibiotics in semen aid in controlling spread of reproductive diseases and significantly reduce the costly losses due to early death of embryos in infected cows.

The 50-year commemorative issue of the Journal of Dairy Science stated that the increase in fertility of Al in the USA from 61% in 1948 to 72% in 1949 can be credited mainly to the addition of antibiotics to bull semen. Attention was drawn to this important advantage of AI in a recent article (R.G. Saacke, Hoard's Dairyman, January 10, 1975, p. 20). This major advancement for improved fertility by 11 percentage points is of large economic importance to the producer and consumer. Based on extensive farm cost, account data (Durward Olds, Hoard's Dairyman, September 10, 1979, p. 118), over 31 million dollars per year currently is saved in the USA by adding antibiotics to bull semen. Since use of antibiotics in bull semen was adopted virtually worldwide, this practice saves about 580 million dollars annually (based on Al of 150 million cows; cited from T. Bonadonna and G. Succi, VIII International Congress on Animal Reproduction and AT, Cracow, 1976, Vol. IV, p. 769). These huge savings attest that the pioneering research by Dr. Almquist on antibiotics represents a substantial contribution to successful use of AI for cattle improvement.

Research completed by Almquist and coworkers has contributed greatly to the feasibility of distributing germ plasm from the best sires worldwide. This permits rapid increases in milk and meat production in countries lacking genetically superior bulls. Exportation of several million doses of frozen semen annually by the USA.; would have been impossible without the knowledge that semen can be collected at very high ejaculation frequencies and not harm the bull or lower his fertility. These new techniques made possible the procurement of sufficient semen from top bulls for use in both domestic and export markets. Exportation of bull semen also was aided greatly by Almquist's research showing that fertility of semen was not affected when the prefreeze numbers of living sperm per inseminating dose were reduced from the commonly used 20 million to 10 million per straw. This finding enables twice as many straws to be produced from each ejaculate of bull semen. His research on milk as an improved medium for preserving semen and more efficient freezing and thawing methods have provided the AI industry with practical, inexpensive means for processing, handling and shipping frozen semen.

Overcoming infertility continues to be one of the most costly and perplexing problems in the livestock industry. Almquist is among the few who have made important contributions to solving male infertility. Combinations of penicillin and streptoinycin were used for treating the semen of bulls of low fertility and dramatic increases in. fertility resulted. Average fertility in one study increased by 22 percent; fertility of some bulls nearly doubled.

Dr. Almquist has published extensively on improved media for preserving the life of bull spermatozoa and. extending the number of cows which can be serviced with semen from a single ejaculation. He developed an effective semen extender by heating cows' milk. This research struck a major blow to lowering the cost of existing extenders and subsequent widespread use in the USA and excellent levels of fertility.

Before frozen semen became practical, he combined two of his research findings to enable dairymen for the first time to make selected matings to the bull of their choice every day of the week. Using his milk-glycerol extender and collecting semen more frequently than previously believed possible, he demonstrated to the AI industry that semen from superior bulls could be available every day.

Recognizing that use of AI was limited or nonexistent in many areas of the world where refrigeration to preserve spermatozoa was not readily available or was very costly, in 1962 he originated an extender which enables semen to be stored at ambient room temperatures. The essential ingredient was skimmilk, together with glucose, sodium citrate and citric acid.

With his keen desire to develop practical procedures for the entire AT industry and his awareness of the need to cut production costs, in 1978 Dr. Almquist introduced procedures for processing bull semen for freezing which make it possible for laboratory technicians to do their work faster and more comfortably at room temperature rather than in a 5°C walk — in cold room. Using This egg-yolk-glycerol extender, freshly collected semen is extended completely and packaged in plastic straws at room temperature rather than at 50°C before being frozen. This new method has been tested successfully and used under routine field conditions in Pennsylvania for 3 years.

In 1979, Dr. Almquist reported a practical method for thawing frozen bull semen rapidly which significantly improved fertility by about 2 percentage points. While this might appear to be a small increase, nevertheless, use of the new procedure could save annually over 5 million dollars in the USA over 100 million dollars worldwide. Several AI organizations already have adopted this new thawing method.

Conclusions:

The founding fathers of artificial insemination probably did not envision that it would evolve into an extensive, daily agricultural practice affecting the lives of people throughout the world. Ever increasing world population has and will continue to mandate that animal agriculture must become more efficient and productive. Hunger and malnutrition know no race or creed and no human is far removed from its overtones.

It is the responsibility of scientists such as Dr. Almquist to utilize their mental and physical resources to develop programs that will undergird the food production potential of the world. The contribution that artificial insemination has made to both quality and, quantity of available food may be one of the greatest advances ever offered through animal agriculture.

It may have been in a laboratory in France or a group of scientists from Italy or Russia in Dr. Almquist's laboratory at Penn State; teaching in a classroom or at a farm; alone or working in close cooperation with another scientist to perfect a technique; regardless, it has been John Almquist at the forefront of idea development, investigation and perfection. He is a respected leader, a highly proficient teacher and a meticulous researcher.

The pioneering efforts of Dr. John Almqnist, through his lifetime of dedicated service making AI a more effective everyday practice, place him in the unique position of being a major contributor to the world's food basket and to the healthful nutrition of his fellow man.

Henry A. Lardy

University of Wisconsin
Madison, Wisconsin, USA

1981 — for his pioneering research on storage and preservation of spermatozoa thus enabling artificial insemination to become a universal practice.

Professor Lardy, with the late Professor Paul H. Phillips, made the basic discovery in the early forties that a nutrient medium of egg yolk dispersed in a phosphate buffer forms an ideal environment for the storage of spermatozoa of bull, ram, stallion, and turkey. This breakthrough paved the way to the rapid expansion of the practical use of artificial insemination in livestock breeding. Further, the ability to store semen made it possible to initiate fundamental studies on the metabolism of spermatozoa. The results of this outstanding research work enabled improvements to semen storage technique leading to more efficient and widespread utilization of semen.

CURRICULUM VITAE

Birthdate: August 19, 1917

EDUCATION:

B.S., Chemistry, South Dakota State University, Brookings, South Dakota, 1939.
M.S., Biochemistry, University of Wisconsin, Madison, Wisconsin, 1941.
Ph.D., Biochemistry, University of Wisconsin, Madison, Wisconsin, 1943.

PROFESSIONAL EXPERIENCE:

1944-45	- N.R.C. Post-Doctoral Fellow, University of Toronto.
1945-47	- Assistant Professor, Department of Biochemistry, University of Wisconsin.
1947-50	- Associate Professor, Department of Biochemistry, University of Wisconsin.
1950-88	- Professor, Department of Biochemistry, University of Wisconsin.
1950-present	- Chairman, Section II, Institute for Enzyme Research, University of Wisconsin.
1966-88	- Vilas Professor of Biological Sciences, University of Wisconsin.
1988	- Professor Emeritus.

MEMBERSHIP IN SOCIETIES:

American Chemical Society (Chairman, Biological Division, 1958)
American Society of Biological Chemists (President, 1964)
National Academy of Sciences
American Academy of Arts and Sciences
American Philosophical Society
Harvey Society
Biochemical Society (Great Britain)
American Diabetes Association
The Society for the Study of Reproduction

HONORS AND AWARDS:

1949	Paul Lewis Laboratories Award in Enzyme Chemistry, American Chemical Society.
1956	Carl Neuberg Award, American Society of European Chemists.
1966	Distinguished Alumnus Award, South Dakota State University.
1978	Honorary Member, Japanese Biochemical Society.
1979	D.Sc., South Dakota State University.
1981	Wolf Foundation Award in Agriculture.
1982	National Award for Agricultural Excellence.
1984	Carl Hartman Award, Society for the Study of Reproduction.
1984	Amory Prize, American Academy of Arts and Sciences.
1985	Henry Lardy Annual Lectureship, Established at Department of Chemistry, South Dakota State University.
1986	Distinguished Service Award, University of Wisconsin.
1986	Elected Fellow, Wisconsin Academy of Arts, Sciences and Letters.
1988	William C. Rose Award, American Society of Biochemical and Molecular Biology.
1988	Hilldale Award (Physical Sciences), University of Wisconsin.

Scientific Advisory Councils:

American Vitamin Association
Life Insurance Medical Research Fund
National Science Foundation Panel on Molecular Biology
Oregon Regional Primate Laboratory
National Heart Institute of the National Institutes of Health
The American Cancer Society
Howard Hughes Medical Institute

Other Positions Held in Scientific Societies:

1957-1960 - Member of the Council of the American Society of Biological Chemists.
1971-1975 - Member, Publications Committee, American Society of Biological Chemists.
About 1954 - Member of the Executive Committee, Biological Division of the American Chemical Society, three-year term.
1957 - Vice-Chairman, Biochemical Division, American Chemical Society.

Editorial Boards:

Archives of Biochemistry and Biophysics - 1957-1960.
Journal of Biological Chemistry - 1958-1964 and 1980-1985.
Biochemical Preparations - 1952-1968.
Methods of Biochemical Analysis - 1954-1971.
Biochemistry - 1962-1973 and 1975-1981.
FEBS Letters - 1970-1975.

PUBLICATIONS:

1. The Vitamin D content of some South Dakota roughages. Wallis, G.C. and Lardy, H.A. (1938). *Proc. South Dakota Academy of Science*, 18, 46.
2. Manganese content of some South Dakota feeds. Moxon, A.L. and Lardy, H.A. (1939). *Proc. South Dakota Academy of Science*, 19, 57.
3. The effect of selenium and arsenic at various ratios on the fermentation of glucose by baker's yeast. Lardy, H.A., and Moxon, A.L. (1939). *Proc. South Dakota Academy of Science*, 19, 109.
4. Preservation of spermatozoa. Lardy, H.A., and Phillips, P.H. (1939). *Am. Soc. Animal Production, 32nd Annual Proceedings of American Society of Animal Production*, p. 219.
5. Increasing the rate of excretion of selenium from selenized animals by the administration of bromobenzene. Moxon, A.L., Schaefer, A.E., Lardy, H.A., DuBois, K.P., and Olson, O.E. (1940). *J. Biol. Chem.* 132, 785.

6. A yolk-buffer pablum for the preservation of bull semen. Phillips, P.H. and Lardy, H.A. (1940). *J. Dairy Sci.*, **23**, 399.
7. H ion concentration of various fluids of the genital tract of the cow. Lardy, H.A., Pounden, W.D., and Phillips, P.H. (1940). *Proc. Soc. Exp. Biol. Med.*, **44**, 517.
8. Sperm stimulation in the bull through the subcutaneous administration of ascorbic acid. Phillips, P.H. and Lardy, H.A. (1940). *J. Dairy Sci.*, **23**, 873.
9. The relationship of ascorbic acid to reproduction in the cow. Phillips, P.H., Lardy, H.A., Boyer, P.D., and Werner, G.M. (1941). *J. Dairy Sci.*, **24**, 153.
10. The effect of certain inhibitors and activators on sperm metabolism. Lardy, H.A. and Phillips, P.H. (1941). *J. Biol. Chem.*, **138**, 195.
11. The interrelation of oxidative and glycolytic processes as sources of energy for bull spermatozoa. Lardy, H.A., and Phillips, P.H. (1941). *Am. J. Physiol.*, **133**, 602.
12. Phospholipids as a source of energy for motility of bull spermatozoa. Lardy, H.A. and Phillips, P.H. (1941). *Am. J. Physiol.*, **134**, 542.
13. The effect of colored ions on the photo-inactivation of invertase. Lardy, H.A. and Anderson, T.F. (1942). *Science* **95**, 330.
14. The ascorbic acid content of the livers of selenized rats and chicks. Lardy, H.A. and Moxon, A.L. (1942). *Proc. South Dakota Acad. Sci.*, **22**, 39.
15. The role of potassium in muscle phosphorylations. Boyer, P.D., Lardy, H.A., and Phillips, P.H. (1942). *J. Biol. Chem.*, **146**, 673.
16. Effect of pH and certain electrolytes on the metabolism of ejaculated spermatozoa. Lardy, H.A. and Phillips, P.H. (1943). *Am. J. Physiol.*, **138**, 741.
17. Inhibition of sperm respiration and reversibility of the effects of metabolic inhibitors. Lardy, H.A. and Phillips, P.H. (1943). *J. Biol. Chem.*, **148**, 333.
18. Inhibition of sperm glycolysis and reversibility of the effects of metabolic inhibitors. Lardy, H.A. and Phillips, P.H. (1943). *J. Biol. Chem.*, **148**, 343.
19. The effect of thyroxine and dinitrophenol on sperm metabolism. Lardy, H.A. and Phillips, P.H. (1943). *J. Biol. Chem.*, **149**, 177.
20. Further studies on the role of potassium and other ions in the phosphorylation of the adenylic system. Boyer, P.D., Lardy, H.A., and Phillips, P.H. (1943). *J. Biol. Chem.*, **149**, 529.
21. Acetate utilization for maintenance of motility of bull spermatozoa. Lardy, H.A. and Phillips, P.H. (1944). *Nature* **153**, 168.
22. Ineffectiveness of sodium succinate in control of duration of barbiturate anesthesia. Lardy, H.A., Hansen, R.G., and Phillips, P.H. (1944). *Proc. Soc. Exp. Biol. Med.*, **55**, 277.
23. Determination of citric acid in fermentation media and biological materials. Perlman, D., Lardy, H.A., and Johnson, M.J. (1944). *Ind. Eng. Chem.*, **16**, 515.

24. Oxalacetate and the dephosphorylation of adenosinetriphosphate. Lardy, H.A., Boyer, P.D., and Phillips, P.H. (1944). *Arch. Biochem.*, **5**, 295.
25. The effect of various gonadotropic hormone preparations of the blood plasma ascorbic acid of sheep and rabbits. Lardy, H.A., Casida, L.E., and Phillips, P.H. (1944). *Endocrinology* **35**, 363.
26. The respiratory metabolism of ram spermatozoa. Lardy, H.A., Winchester, B., and Phillips, P.H. (1945). *Arch. Biochem.* **6**, 33.
27. The metabolism of bovine epididymal spermatozoa. Lardy, H.A., Hansen, R.G., and Phillips, P.H. (1945). *Arch. Biochem.* **6**, 41.
28. Studies of fat and carbohydrate oxidation in mammalian spermatozoa. Lardy, H.A., and Phillips, P.H. (1945). *Arch. Biochem.* **6**, 53.
29. Biological oxidations and reductions. Lardy, H.A., and Elvehjem, C.A. (1945). *Ann. Rev. Biochem.* **XIV**, 1.
30. The enzymatic synthesis of phosphopyruvate from pyruvate. Lardy, H.A., and Ziegler, J.A. (1945). *J. Biol. Chem.* **159**, 343.
31. Phosphoric esters of biological importance. I. The synthesis of glucose-6-phosphate. Lardy, H.A., and Fischer, H.O.L. (1946). *J. Biol. Chem.* **164**, 513.
32. Feeding experiments with DDT-treated pea vine silage with special reference to dairy cows, sheep and laboratory animals. Wilson, H.F., Allen, N.H., Bohstedt, G., Betheil, J., and Lardy, H.A. (1946). *J. Econ. Entomol.* **39**, 801.
33. DDT residues on pea vines and canned peas from fields treated with DDT dusts. Wilson, H.F., Srivastava, A.S., Hull, W.B., Betheil, J., and Lardy, H.A. (1946). *J. Econ. Entomol.* **39**, 806.
34. 1,2,3,4-tetracetyl-α-D-glucopyranose. Lardy, H.A. (1947). *J. Am. Chem. Soc.* **69**, 518.
35. The metabolism of *meso*-inositol in the rat. Wiebelhaus, V.D., Betheil, J.J., and Lardy, H.A. *Arch. Biochem.* **13**, 379.
36. The role of biotin in bicarbonate utilization by bacteria. Lardy, H.A., Potter, R.L., and Elvehjem, C.A. (1947). *J. Biol. Chem.* **169**, 451.
37. Studies of thyroid toxicity. I. A nutritional factor which alleviates the toxicity of ingested thyroid substance. Betheil, J.J., Wiebelhaus, V.D., and Lardy, H.A. (1947). *J. Nutrition* **34**, 431.
38. The synthesis of D-fructomethylose by biochemical oxidation. Anderson, L., and Lardy, H.A. (1948). *J. Am. Chem. Soc.* **70**, 594.
39. Experiments with peas and sweet corn treated with DDT insecticides. Lardy, H.A. (1948). *Ind. Eng. Chem.* **40**, 710.
40. Vitamins and carbohydrate metabolism. Lardy, H.A. (1948). *J. Chem. Ed.* **25**, 262.
41. Studies of thyroid toxicity. II. The effects of desiccated thyroid and anti-thyroid agents upon plasma and tissue ascorbic acid of rabbits. Johnson, R.B., Hansen, R.G., and Lardy, H.A. (1948). *Arch. Biochem.* **19**, 246.

42. Circular bath and shaking mechanism for manometric microapparatus. Lardy, H.A., Gilson, W.E., Hipple, J., and Burris, R.H. (1948). *Anal. Chem.* **20**, 1100.
43. A metabolic regulator in mammalian spermatozoa. Lardy, H.A., Ghosh, D., and Plaut, G.W.E. (1949). *Science* **109**, 365.
44. Phosphorylation of hexoses by brain hexokinase. Wiebelhaus, V.D., and Lardy, H.A. (1949). *Arch. Biochem.* **21**, 321.
45. Comparative effectiveness of vitamin B_{12}, whole liver substance and extracts high in APA activity, as growth promoting materials for hyperthyroid animals. Betheil, J.J. and Lardy, H.A. (1949). *J. Nutrition* **37**, 495.
46. The net utilization of ammonium nitrogen by the growing rat. Lardy, H.A. and Feldott, G. (1949). *J. Biol. Chem.* **179**, 509.
47. A study of the metabolic activity of bull semen and spermatozoa in relation to their fertilizing ability. Ghosh, D., Casida, L.E., and Lardy, H.A. (1949). *J. Animal Sci.* **8**, 265.
48. Metabolic functions of biotin. I. The role of biotin in bicarbonate utilization by *lactobacillus arabinosus* studied with C^{14}. Lardy, H.A., Potter, R.L., and Burris, R.H. (1949). *J. Biol. Chem.* **179**, 721.
49. Metabolic functions of biotin. II. The fixation of carbon dioxide by normal and biotin-deficient rats. MacLeod, P.R., and Lardy, H.A. (1949). *J. Biol. Chem.* **179**, 733.
50. The oxalacetate decarboxylase of azotobacter vinelandii. Plaut, G.W.E. and Lardy, H.A. (1949). *J. Biol. Chem.* **180**, 13.
51. The synthesis of D-tagatose by biochemical oxidation and by an improved chemical method. Totton, E.L., and Lardy, H.A. (1949). *J. Am. Chem. Soc.* **71**, 3076.
52. The effect of the administration of various fatty acids on the blood ketone levels of ruminants. Schultz, L.H., Smith, V.R., and Lardy, H.A. (1949). *J. Dairy Sci.* **XXXII**, 817.
53. Metabolic functions of biotin. III. The synthesis of citrulline from ornithine in liver tissue from normal and biotin-deficient rats. MacLeod, P.R., Grisolia, S., Cohen, P.P., and Lardy, H.A. (1949). *J. Biol. Chem.* **180**, 1003.
54. Phosphoric esters of biological importance. II. The synthesis of glucose-6-phosphate from 1,2-isopropylidene-5,6-anhydro-D-glucofuranose. Lampson, G.P., and Lardy, H.A. (1949). *J. Biol. Chem.* **181**, 693.
55. Phosphoric esters of biological importance. III. The synthesis of propanediol phosphate. Lampson, G.P., and Lardy, H.A. (1949). *J. Biol. Chem.* **181**, 697.
56. Phosphoric ester of biological importance. IV. The synthesis and biological activity of D-tagatose-6-phosphate. Totton, E.L., and Lardy, H.A. (1949). *J. Biol. Chem.* **181**, 701.
57. Relationship of vitamin B_6 to the metabolism of D-amino acids. Armstrong, K.L., Feldott, G., and Lardy, H.A. (1950). *Proc. Soc. Exp. Biol. Med.* **73**, 159.

58. Orthophosphate uptake during the oxidation of fatty acids. Johnson, R.B. and Lardy, H.A. (1950). *J. Biol. Chem.* **184**, 235.
59. Systems for the separation of phosphoric esters by solvent distribution. Plaut, G.W.E., Kuby, S.A., and Lardy, H.A. (1950). *J. Biol. Chem.* **184**, 243.
60. 1,2-isopropylidene-D-glucitol, 1,2:5,6-diisopropylidene-D-glucitol and a new synthesis of D,L-glyceraldehyde dimer. Pressman, B.C., Anderson, L., and Lardy, H.A. (1950). *J. Am. Chem. Soc.* **72**, 2404.
61. The relationship of folic acid to formate metabolism in the rat. Plaut, G.W.E., Betheil, J.J., and Lardy, H.A. (1950). *J. Biol. Chem.* **184**, 795.
62. Stereochemical studies in the amino-desoxyinositol series. *Meso*-inosamine-2 and *scyllo*-inosamine. Anderson, L. and Lardy, H.A. (1950). *J. Am. Chem. Soc.* **72**, 3141.
63. The net utilization of ammonium nitrogen by the growing rat. Lardy, H.A. and Feldott, G. (1950). *J. Biol. Chem.* **186**, 85.
64. Metabolism of ejaculated bull spermatozoa with particular reference to the glycolysis of maltose. Plaut, G.W.E., and Lardy, H.A. (1950). *Am. J. Physiol.* **162**, 598.
65. Incorporation of the carbons of acetone, formate, and carbonate into acetoacetate. Plaut, G.W.E. and Lardy, H.A. (1950). *J. Biol. Chem.* **186**, 705.
66. The mode of action of the antibiotic, usnic acid. Johnson, R.B., Feldott, G., and Lardy, H.A. (1950). *Arch. Biochem.* **28**, 317.
67. The mechanism by which glyceraldehyde inhibits glycolysis. Lardy, H.A., Wiebelhaus, V.D., and Mann, K.M. (1950). *J. Biol. Chem.* **187**, 325.
68. Phosphoric esters of biological importance. V. The synthesis of L-sorbose-1-phosphate and L-sorbose-6-phosphate. Mann, K.M. and Lardy, H.A. (1950). *J. Biol. Chem.* **187**, 339.
69. Metabolic functions of the micro-catalytic B vitamins. Lardy, H.A. (1951). *Rec. Chem. Progress, Winter*, p. 9.
70. The excretion of the citrovorum factor by hyperthyroid rats. Drysdale, G.R., Betheil, J.J., Lardy, H.A., and Baumann, C.A. (1951). *Arch. Biochem. Biophys.* **33**, 1.
71. The nomenclature of the cyclohexitols and their derivatives. Fletcher, H.G., Jr., Anderson, L., and Lardy, H.A. (1951). *J. Organ. Chem.* **16**, 1238.
72. Enzymatic incorporation of C^{14}-bicarbonate into acetoacetate in the presence of various substrates. Plaut, G.W.E., and Lardy, H.A. (1951). *J. Biol. Chem.* **192**, 435.
73. Metabolic functions of biotin. IV. The role of carbamyl-L-glutamic acid in the synthesis of citrulline by normal and biotin-deficient rats. Feldott, G., and Lardy, H.A. (1951). *J. Biol. Chem.* **192**, 447.
74. 1,6-dibenzyl-3,4-isopropylidene-D-mannitol. Anderson, L., Lueptow, A.J., and Lardy, H.A. (1951). *J. Am. Chem. Soc.* **73**, 5002.

75. The influence of inorganic ions on phosphorylation reactions. Phosphorus Metabolism. Lardy, H.A. (1951). (McElroy, W.D. & Glass, B., eds.), p. 477, The Johns Hopkins Press, Baltimore.
76. The relationship of folic acid to formate metabolism in the rat: Formate incorporation into purines. Drysdale, G.R., Plaut, G.W.E., and Lardy, H.A. (1951). *J. Biol. Chem.* **193**, 533.
77. Metabolic effects of thyroxine *in vitro*. Lardy, H.A. and Feldott, G. (1951). *Ann. N.Y. Acad. Sci.* **54**, 636.
78. Synthesis of the optically active enantiomorphic 2,3-butanediols. Rubin, L.J., Lardy, H.A., and Fischer, H.O.L. (1952). *J. Am. Chem. Soc.* **74**, 425.
79. Inhibition of the Pasteur effect in yeast by tumor extracts and differences in lability of sulfur in normal and tumor tissues. Ghosh, D. and Lardy, H.A. (1952). *Cancer Res.* **12**, 232.
80. Oxidative phosphorylations: Role of inorganic phosphate and acceptor systems in control of metabolic rates. Lardy, H.A. and Wellman, H. (1952). *J. Biol. Chem.* **195**, 215.
81. Oxidative phosphorylations: Pathways and yield in mitochondrial preparations. Copenhaver, J.H., Jr., and Lardy, H.A. (1952). *J. Biol. Chem.* **195**, 225.
82. The metabolic regulator in mammalian spermatozoa. Lardy, H.A. (1952). In *Studies on Testis and Ovary, Eggs and Sperm* (Engle, E.T., ed.), pp. 111-122, Charles C. Thomas, Publisher, Springfield, Il.
83. The nature of a yeast-stimulating agent in mammalian spermatozoa. Ghosh, D. and Lardy, H.A. (1952). *J. Animal Science*, **11**, 545.
84. Influence of potassium and other alkali cations on respiration of mitochondria. Pressman, B.C. and Lardy, H.A. (1952). *J. Biol. Chem.* **197**, 547.
85. Studies on the particulate α-glycerophosphate dehydrogenase of muscle. Tung, Ta-Cheng, Anderson, L., and Lardy, H.A. (1952). *Arch. Biochem. Biophys.* **40**, 194.
86. Comparative metabolic behavior of epididymal and ejaculated mammalian spermatozoa. Lardy, H.A. and Ghosh, D. (1952). *Ann. N.Y. Acad. Sci.* **55**, 594.
87. Fatty acid oxidation by a soluble enzyme system from mitochondria. *Phosphorus metabolism, II.* Drysdale, G.R., and Lardy, H.A. (1952). (McElroy, W.D. & Glass, B., eds), p. 281, The Johns Hopkins Press, Baltimore.
88. A theory concerning the mechanism of fatty acid oxidation and synthesis, and of carbon dioxide fixation. Lardy, H.A. (1952). *Proc. Nat. Acad. Sci.*, **38**, 1003.
89. Factors controlling rates of metabolism in mammallian spermatozoa. Lardy, H.A. (1953). In *Mammalian Germ Cells, A Ciba Foundation Symposium*, pp. 59-65, J. & A. Churchill Ltd., London, England.
90. Stereochemical studies in the aminodesoxy-inositol series. II. D,L-*myo*-inosamine-4, D,L-*epi*-inosamine-2, and streptamine. Straube-Rieke, H., Lardy, H.A., and Anderson, L. (1953). *J. Am. Chem. Soc.* **75**, 694.

91. Purification and kinetics of β-D-glactosidase from *Escherichia coli*, strain K-12. Kuby, S.A. and Lardy, H.A. (1953). *J. Am. Chem. Soc.* **75**, 890.
92. Properties of thioesters: Kinetics of hydrolysis in dilute aqueous media. Noda, L., Kuby, S.A., and Lardy, H.A. (1953). *J. Am. Chem. Soc.* **75**, 913.
93. Lactose synthesis in the mammary gland perfused with 1-^{14}C-glucose. Dimant, E., Smith, V.R., and Lardy, H.A. (1953). *J. Biol. Chem.* **201**, 85.
94. The catalytic effect of 2,4-dinitrophenol on adenosinetriphosphate hydrolysis by cell particles and soluble enzymes. Lardy, H.A. and Wellman, H. (1953). *J. Biol. Chem.* **201**, 357.
95. Fatty acid oxidation by a soluble enzyme system from mitochondria. Drysdale, G.R. and Lardy, H.A. (1953). *J. Biol. Chem.* **202**, 119.
96. Metabolic effects of thyroid hormones *in vitro*. II. Influence of thyroxine and triiodothyronine on oxidative phosphorylation. Maley, G.F. and Lardy, H.A. (1953). *J. Biol. Chem.* **204**, 435.
97. Metabolic functions of biotin. Lardy, H.A. and Peanasky, R. (1953). *Physiol. Rev.* **33**, 560.
98. Growth inhibition of a biotin-less mutant of *Escherichia coli* by purines and pyrimidines. Berger, M. and Lardy, H.A. (1954). *Proc. Soc. Exp. Biol. Med.* **85**, 89.
99. Uridine- and inosine-triphosphates as phosphate donors for phosphohexokinase. Ling, K.H. and Lardy, H.A. (1954). *J. Am. Chem. Soc.* **76**, 2842.
100. Metabolic effects of thyroid hormones *in vitro*. Recent progress in hormone research. Lardy, H.A. and Maley, G.F. (1954). *Proc. Laurentian Hormone Conf.*, X, 129.
101. Efficiency of oxidative phosphorylation. Lardy, H.A. and Copenhaver, J.H., Jr. (1954). *Nature*, **174**, 231.
102. Adenosinetriphosphate-creatine transphosphorylase. I. Isolation of the crystalline enzyme from rabbit muscle. Kuby, S.A., Noda, L., and Lardy, H.A. (1954). *J. Biol. Chem.* **209**, 203.
103. Adenosinetriphosphate-creatine transphosphorylase. II. Homogeneity and physicochemical properties. Noda, L., Kuby, S.A., and Lardy, H.A. (1954). *J. Biol. Chem.* **209**, 203.
104. Substrate specificity of muscle aldolase. Tung, T.C., Ling, K.H., Byrne, W.L., and Lardy, H.A. (1954). *Biochim Biophys. Acta* **14**, 488.
105. Pentose phosphates formed by muscle aldolase. Byrne, W.L. and Lardy, H.A. (1954). *Biochim. Biophys. Acta* **14**, 495.
106. Adenosinetriphosphate-creatine transphosphorylase. III. Kinetic studies. Kuby, S.A., Noda, L., and Lardy, H.A. (1954). *J. Biol. Chem.* **210**, 65.
107. Adenosinetriphosphate-creatine transphosphorylase. IV. Equilibrium studies. Noda, L., Kuby, S.A., and Lardy, H.A. (1954). *J. Biol. Chem.* **210**, 83.
108. Inositol, discussion of terminology, biochemical systems. Lardy, H.A. (1954). In *The Vitamins*, p. 323, Ch. 8, Academic Press.

109. Phosphorylation coupled with the oxidation of reduced cytochrome c. Maley, G.F. and Lardy, H.A. (1954). *J. Biol. Chem.* **210**, 903.
110. Effect of thyroid hormones on enzyme systems. Lardy, H.A. (1955). In *The Thyroid*, p. 90, Brookhaven Symposia in Biology, No. 7.
111. Water-soluble vitamins, Part II. Fried, R., and Lardy, H.A. (1955). *Ann. Rev. Biochem.* **24**, 393.
112. 111A. Phosphohexokinase. Ling, K.H., Byrne, W.L., and Lardy, H.A. (1955). In *Methods in Enzymology*, p. 306, Vol. I.
113. Synthesis of N-substituted glucosamines and their effect on hexokinase. Maley, F. and Lardy, H.A. (1955). *J. Biol. Chem.* **214**, 765.
114. Efficiency of phosphorylation in selected oxidations by mitochondria from normal and thyrotoxic rat liver. Maley, G.F. and Lardy, H.A. (1955). *J. Biol. Chem.* **215**, 377.
115. The sulfhydryl groups of crystalline proteins. I. Some albumins, enzymes, and hemoglobins. Benesch, R.E., Lardy, H.A., and Benesch, R. (1955). *J. Biol. Chem.* **216**, 663.
116. The biological activity of O-methyl thyroxine. Lardy, H.A. (1955). *Endocrinology* **57**, 566.
117. Further studies on the potassium requirements of mitochondria. Pressman, B.C. and Lardy, H.A. (1955). *Biochim. Biophys. Acta* **18**, 482.
118. Energetic coupling and the regulation of metabolic rates. Lardy, H.A. (1955). *Proc. Third International Congress of Biochemistry*, Brussels (Liebecq, C., ed.), p. 287, Academic Press.
119. Synthesis and biological activity of some triiodinated analogues of thyroxine. Tomita, K. and Lardy, H.A. (1956). *J. Biol. Chem.* **219**, 595.
120. Synthesis of succinate from propionate and bicarbonate by soluble enzymes from liver mitochondria. Lardy, H.A. and Adler, J. (1956). *J. Biol. Chem.* **219**, 933.
121. The carboxylation of propionic acid by liver mitochondria. Friedberg, F., Adler, J., and Lardy, H.A. (1956). *J. Biol. Chem.* **219**, 943.
122. Phosphoric esters of biological importance. VI. The synthesis of D-glucosamine 6-phosphate and N-acetyl-D-glucosamine 6-phosphate. Maley, F. and Lardy, H.A. (1956). *J. Am. Chem. Soc.* **78**, 1393.
123. Influence of ATP concentration on rates of some phosphorylation reactions: Added comment. Lardy, H.A., and Parks, R.E., Jr. (1956). In *Enzymes: Units of Biological Structure and Function*, Henry Ford Hospital International Symposium, p. 584, Academic Press, Inc.
124. Enzymatic conversion of thyroxine and triiodothyronine to the corresponding acetic acid analogues. Albright, E.C., Larson, F.C., Tomita, K., and Lardy, H.A. (1956). *Endocrinology* **59**, 252.
125. Effect of surface active agents on the latent ATPase of mitochondria. Pressman, B.C. and Lardy, H.A. (1956). *Biochim. Bophys. Acta* **21**, 458.

126. The synthesis of α-D-glucosamine-1-phosphate and N-acetyl-α-D-glucosamine-1-phosphate. Enzymatic formation of uridine diphosphoglucosamine. Maley, F., Maley, G.F., and Lardy, H.A. (1956). *J. Am. Chem. Soc.* **78**, 5303.
127. Formation of ADPGla and related compounds by the soluble fraction of liver. Maley, F. and Lardy, H.A. (1956). *Science* **124**, 1207.
128. Enzymatic conversion of thyroxine to tetraiodothyroacetic acid and of triiodothyronine to triiodothyroacetic acid. Tomita, K., Lardy, H.A., Larson, F.C., and Albright, E.C. (1957). *J. Biol. Chem.* **224**, 387.
129. The metabolism of thyroid hormones by kidney and the biological activity of the products. Lardy, H.A., Tomita, K., Larson, F.C., and Albright, E.C. (1957). *Ciba Foundation Colloquia on Endocrinology* **10**, 156.
130. Liver fructokinase. Parks, R.E., Jr., Ben-Gershom, E., and Lardy, H.A. (1957). *J. Biol. Chem.* **227**, 231.
131. Effect of certain steroids on metabolic rate of hyperthyroid rats. Doisy, R.J. and Lardy, H.A. (1957). *Am. J. Physiol.* **190**, 142.
132. Hormonal regulation of enzymatic activity. Lardy, H.A. In *Hormonal Regulation of Energy Metabolism* (Kinsell, L.W., ed.), p. 45, Charles C. Thomas.
133. A role for coenzyme A in phosphorylations associated with electron transport. McMurray, W.C. and Lardy, H.A. (1957). *J. Am. Chem. Soc.* **79**, 6563.
134. The metabolism of itaconic acid by liver mitochondria. Adler, J., Wang, S.F., and Lardy, H.A. (1957). *J. Biol. Chem.* **229**, 865.
135. Oxidative phosphorylation by sonic extracts of mitochondria. McMurray, W.C., Maley, G.F., and Lardy, H.A. (1958). *J. Biol. Chem.* **230**, 219.
136. Substrate-selective inhibition of mitochondrial oxidations by enhanced tonicity. Johnson, D. andLardy, H.A. (1958). *Nature* **181**, 701.
137. Bovine liver aldolase. I. Isolation, crystallization, and some general properties. Peanasky, R.J. and Lardy, H.A. (1958). *J. Biol. Chem.* **233**, 365.
138. Bovine liver aldolase. II. Physical and chemical measurements on the crystalline enzyme. (1958). *J. Biol. Chem.* **233**, 371.
139. Purification and properties of pigeon liver malic enzyme. Rutter, W.J. and Lardy, H.A. (1958). *J. Biol. Chem.* **233**, 374.
140. A requirement for coenzyme A in oxidative phosphorylation. McMurray, W.C. and Lardy, H.A. (1958). *J. Biol. Chem.* **233**, 754.
141. 3'-hydroxy-3,5-diiodo-L-thyronine. Doskotch, R. and Lardy, H.A. (1958). *J. Am. Chem. Soc.* **80**, 6230.
142. Antibiotics as tools for metabolic studies. I. A survey of toxic antibiotics in respiratory, phosphorylative, and glycolytic systems. Lardy, H.A., Johnson, D., and McMurray, W.C. (1958). *Arch. Biochem. Biophys.* **78**, 587.
143. Monoamine oxidase activity in liver of thyroid-fed rats. Zile, M. and Lardy, H.A. (1959). *Arch. Biochem. Biophys.* **82**, 411.

144. Enhanced oxidation of α-glycerophosphate by mitochondria of thyroid-fed rats. Lee, Y-P, Takemori, A.E., and Lardy, H.A. (1959). *J. Biol. Chem.* **234**, 3051.
145. A method for the separation of orthophosphate from other phosphate compounds. Hagihara, B. and Lardy, H.A. (1960). *J. Biol. Chem.* **235**, 889.
146. Enzyme responses to thyroid hormones. Lardy, H.A., Lee, Y-P, and Takemori, A. (1960). *Ann. N.Y. Acad. Sci.* **86**, 506.
147. Enzymic conversion of iodinated thyronines to iodinated thyroacetic acids. Tomita, K. and Lardy, H.A. (1960). *J. Biol. Chem.* **235**, 3292.
148. The pathway of itaconate metabolism by liver mitochondria. Wang, S-F, Adler, J., and Lardy, H.A. (1961). *J. Biol. Chem.* **236**, 26.
149. Synthesis and some biochemical properties of phosphohydroxypyruvic aldehyde and of 3-phosphoglyceryl glutathione thiol ester. Weaver, R.H. and Lardy, H.A. (1961). *J. Biol. Chem.* **236**, 313.
150. Reactions involved in oxidative phosphorylation as disclosed by studies with antibiotics. Lardy, H.A. (1961). *Biological Structure and Function: Proceedings of the First IUB/IUBS International Symposium*, held in Stockholm. (Goodwin, T.W. & Lindberg, O., eds.), Vol. 11, Academic Press, New York.
151. Sub-cellular distribution of rat liver inorganic pyrophosphatase activity. Nordlie, R.C. and Lardy, H.A. (1961). *Biochim. Biophys. Acta* **50**, 189.
152. Adenine nucleotide inhibition of mitochondrial inorganic pyrophosphatase activity. Nordlie, R.C. and Lardy, H.A. (1961). *Biochim. Biophys. Acta* **53**, 309.
153. Control of some reactions in glycolysis and its reversal. Lardy, H.A. (1961). Reprinted from *Proceedings of the Robert A. Welch Foundation Conference on Chemistry Research*, p. 267, Vol. 5.
154. Synthesis and Biological Activity of O-methyl derivatives of thyroid hormones. Tomita, K., Lardy, H.A., Johnson, D., and Kent, A. (1961). *J. Biol. Chem.* **236**, 2981.
155. Phosphoryl group transfer (survey). Nordlie, R. and Lardy, H.A. (1962). I., in *The Enzymes, VI*, p. 3, Academic Press, New York.
156. Phosphohexokinases. Lardy, H.A. (1962). III, in *The Enzymes, VI*, p. 67, Academic Press, New York.
157. The conjugation of cysteine during its oxidation by 2,6-dichloroindophenol. Hadler, H.I., Erwin, M.J. and Lardy, H.A. (1963). *J. Am. Chem. Soc.* **85**, 458.
158. Mammalian liver phosphoenolpyruvate carboxykinase activities. Nordlie, R.C. and Lardy, H.A. (1963). *J. Biol. Chem.* **238**, 2259.
159. The synthesis of phosphoenolpyruvate in liver mitochondria. Nordlie, R.C. and Lardy, H.A. (1963). *Biochem. Zeitschrift* **338**, 356.
160. Metabolic and hormonal control of phosphoenolpyruvate carboxykinase and malic enzyme in rat liver. Shrago, E., Lardy, H.A., Nordlie, R.C., and Foster, D.O. (1963). *J. Biol. Chem.* **238**, 3188.

161. Nucleoside triphosphate-nucleoside diphosphate transphosphorylase (nucleoside diphosphokinase). I. Isolation of the crystalline enzyme from brewers' yeast. Ratliff, R.L., Weaver, R.H., Lardy, H.A., and Kuby, S.A. (1964). *J. Biol. Chem.* 239, 301.
162. Metabolic and hormonal regulation of phosphopyruvate synthesis. Lardy, H.A., Foster, D.O., Shrago, E., and Ray, P.D. (1964). In *Advances in Enzyme Regulation*, Vol. 2, p. 39.
163. Enzymic O-methylation of iodinated phenols and thyroid hormones. Tomita, K., Mo-Cha, C-J., and Lardy, H.A. (1964). *J. Biol. Chem.* 239, 1202.
164. Metabolic effects of thyroid hormones *in vitro*. Lardy, H.A. and Kent, A.B. (1964). In *Biochemical Aspects of Hormone Action*.
165. Effect of antimycin A on oxidative phosphorylation with ferricyanide as electron acceptor. Walter, P. and Lardy, H.A. (1964). *Biochemistry* 3, 812.
166. The effect of adenosine triphosphate and substrate on orthophosphate-induced mitochondrial swelling at acid pH. Connelly, J.L. and Lardy, H.A. (1964). *J. Biol. Chem.* 239, 3065.
167. Mode of action of glucocorticoids. I. Stimulation of gluconeogenesis independent of synthesis *de Novo* of enzymes. Ray, P.D., Foster, D.O., and Lardy, H.A. (1964). *J. Biol. Chem.* 239, 3396.
168. Metabolic control of enzymes involved in lipogenesis and gluconeogenesis. Young, J.W., Shrago, E., and Lardy, H.A. (1964). *Biochemistry* 3, 1687.
169. Antibiotics as tools for metabolic studies. II. Inhibition of phosphoryl transfer in mitochondria by oligomycin and aurovertin. Lardy, H.A., Connelly, J.L., and Johnson, D. (1964). *Biochemistry* 3, 1961.
170. 169. Antibiotics as tools for metabolic studies. III. Effects of oligomycin and aurovertin on the swelling and contraction processes of mitochondria. Connelly, J.L. and Lardy, H.A. (1964). *Biochemistry* 3, 1969.
171. Antibiotics as tools for metabolic studies. IV. Comparative effectiveness of oligomycins A,B,C, and rutamycin as inhibitors of phosphoryl transfer reactions in mitochondria. Lardy, H.A., Witonsky, P., and Johnson, D. (1965). *Biochemistry* 4, 552.
172. Influence of thyroid hormones on L-α-glycerophosphate dehydrogenases and other dehydrogenases in various organs of the rat. Lee, Y-P and Lardy, H.A. (1965). *J. Biol. Chem.* 240, 1427.
173. Control of choline oxidation in liver mitochondria by adenine nucleotides. Kagawa, T., Wilken, D.R., and Lardy, H.A. (1965). *J. Biol. Chem.* 240, 1836.
174. The role of adenine nucleotides in control of choline oxidation by mitochondria. Wilken, D.R., Kagawa, T., and Lardy, H.A. (1965). *J. Biol. Chem.* 240, 1843.
175. Purification and some properties of rabbit skeletal muscle phosphofructokinase. Ling, K-H., Marcus, F., and Lardy, H.A. (1965). *J. Biol. Chem.* 240, 1893.

176. Paths of carbon in gluconeogenesis and lipogenesis: The role of mitochondria in supplying precursors of phosphoenolpyruvate. Lardy, H.A., Paetkau, V., and Walter, P. (1965). *Proc. Nat. Acad. Sci. USA* **53**, 1410.
177. Hormonal control of enzymes participating in gluconeogenesis and lipogenesis. Lardy, H.A., Foster, D.O., Young, J.W., Shrago, E., and Ray, P.D. (1965). *J. Cell and Comp. Physiol.* **66**, 39.
178. Phosphofructokinase from rabbit muscle. Lardy, H.A. (1965). In *Control of Energy Metabolism*, p. 69, Academic Press, New York.
179. On the direction of pyridine nucleotide oxidation-reduction reactions in gluconeogenesis and lipogenesis. Lardy, H.A. (1965). In *Control of Energy Metabolism*, p. 245, Academic Press, New York.
180. Studies on the mechanisms underlying adaptive changes in rat liver phosphoenolpyruvate carboxykinase. Foster, D.O., Ray, P.D., and Lardy, H.A. (1966). *Biochemistry* **5**, 555.
181. A paradoxical *in Vivo* effect of L-tryptophan on the phosphoenolpyruvate carboxykinase of rat liver. Foster, D.O., Ray, P.D., and Lardy, H.A. (1966). *Biochemistry* **5**, 563.
182. Paths of carbon in gluconeogenesis and lipogenesis. II. Conversion of precursors to phosphoenolpyruvate in liver cytosol. Shrago, E. and Lardy, H.A. (1966). *J. Biol. Chem.* **241**, 663.
183. Gluconeogenesis: Pathways and hormonal regulation. Lardy, H.A. (1966). In *The Harvey Lectures, Series 60*, p. 261, Academic Press, New York.
184. Antibiotics as tools for metabolic studies. V. Effect of nonactin, monactin, dinactin, and trinactin on oxidative phosphorylation and adenosine triphosphatase induction. Graven, S.N., Lardy, H.A., Johnson, D., and Rutter, A. (1966). *Biochemistry* **5**, 1729.
185. Antibiotics as tools for metabolic studies. VI. Damped oscillatory swelling of mitochondria induced by nonactin, monactin, dinactin, and trinactin. Graven, S.N., Lardy, H.A., and Rutter, A. (1966). *Biochemistry* **5**, 1735.
186. Paths of carbon in gluconeogenesis and lipogenesis. III. The role and regulation of mitochondrial processes involved in supplying precursors of phosphoenolpyruvate. Walter, P., Paetkau, V., and Lardy, H.A. (1966). *J. Biol. Chem.* **241**, 2523.
187. Alkali metal cation release and respiratory inhibition induced by nigericin in rat liver mitochondria. Graven, S.N., Estrada-O., S., and Lardy, H.A. (1966). *Proc. Nat. Acad. Sci. USA* **56**, 654.
188. Paths of carbon in gluconeogenesis and lipogenesis. IV. Inhibition by L-tryptophan of hepatic gluconeogenesis at the level of phosphoenolpyruvate formation. Ray, P.D., Foster, D.O., and Lardy, H.A. (1966). *J. Biol. Chem.* **241**, 3904.

189. Regulation of energy-yielding processes in muscle. Lardy, H.A. (1966). In *The Physiology and Biochemistry of Muscle as a Food* (Briskey, Cassens & Trautman, eds.), p. 31, The University of Wisconsin Press.
190. Phosphofructokinase. I. Skeletal muscle. Ling, K.H., Paetkau, V., Marcus, F., and Lardy, H.A. (1966). In *Methods in Enzymology*, IX, p. 425, Academic Press, New York.
191. Cellular oxidative phosphorylation. I. Measurement in intact spermatozoa and other cells. Morton, B.E. and Lardy, H.A. (1967). *Biochemistry* **6**, 43.
192. Cellular oxidative phosphorylation. II. Measurement in physically modified spermatozoa. Morton, B.E. and Lardy, H.A. (1967). *Biochemistry* **6**, 50.
193. Cellular oxidative phosphorylation. III. Measurement in chemically modified cells. Morton, B.E. and Lardy, H.A. (1967). *Biochemistry* **6**, 57.
194. Antibiotics as tools for metabolic studies. VIII. Effect of nonactin homologs on alkali metal cation transport and rate of respiration in mitochondria. Graven, S.N., Lardy, H.A., and Estrada-O, S. (1967). *Biochemistry* **6**, 365.
195. Pigeon liver malic enzyme. II. Isolation, crystallization, and some properties. Hsu, R.Y. and Lardy, H.A. (1967). *J. Biol. Chem.* **242**, 520.
196. Pigeon liver malic enzyme. III. Fluorescence studies of coenzyme binding. Hsu, R.Y. and Lardy, H.A. (1967). *J. Biol. Chem.* **242**, 527.
197. The relation of protein synthesis to the function of adrenal cortical hormones. Lardy, H.A., Foster, D.O., and Ray, P.D. (1967). In *Memoirs of the Society for Endocrinology*, No. 15, *Endocrine Genetics* (Spickett, S.G., ed.), p. 25, Cambridge University Press, London.
198. Phosphofructokinase: Correlation of physical and enzymatic properties. Paetkau, V. and Lardy, H.A. (1967). *J. Biol. Chem.* **242**, 2035.
199. Pigeon liver malic enzyme. IV. Pyruvate reductase activity of the crystalline enzyme. Hsu, R.Y. and Lardy, H.A. (1967). *Acta Biochim. Polonica*, **XIV**, 183.
200. Potassium ion-dependent hydrolysis of adenosine triphosphate induced by nigericin in mitochondria. Estrada-O, S., Graven, S.N., and Lardy, H.A. (1967). *J. Biol. Chem.* **242**, 2925.
201. Alteration of rat liver phosphoenolpyruvate carboxykinase activity by L-tryptophan *in Vivo* and metals *in Vitro*. Foster, D.O., Lardy, H.A., Ray, P.D., and Johnston, J.B. (1967). *Biochemistry* **6**, 2120.
202. Influence of L-tryptophan and its metabolites on gluconeogenesis in the isolated, perfused liver. Veneziale, C.M., Walter, P., Kneer, N., and Lardy, H.A. (1967). *Biochemistry* **6**, 2129.
203. The removal of the cytosol and the reinitation of glagellation of bull sperm. Morton, B.E. and Lardy, H.A. (1967). *Exp. Cell Res.* **46**, 511.
204. Isolation of liver or kidney mitochondria. Johnson, D. and Lardy, H.A. (1967). In *Methods in Enzymology*, X, p. 15, Academic Press, New York.

205. Specific induction and inhibition of cation and anion transport in mitochondria. Lardy, H.A., Graven, S.N., and Estrada-O, S. (1967). *Federation Proc.* **26**, 1355.
206. Paths of carbon in gluconeogenesis and lipogenesis. VII. The synthesis of precursors for gluconeogenesis from pyruvate and bicarbonate by rat kidney mitochondria. Mehlman, M.A., Walter, P., and Lardy, H.A. (1967). *J. Biol. Chem.* **242**, 4594.
207. Antibiotics as tools for metabolic studies. Walter, P., Lardy, H.A., and Johnson, D. (1967). *J. Biol. Chem.* **242**, 5014.
208. Pigeon liver malic enzyme. V. Kinetic studies. Hsu, R.Y., Lardy, H.A., and Cleland, W.W. (1967). *J. Biol. Chem.* **242**, 5315-5322.
209. Carbohydrate supply as a regulator of rat liver phosphoenolpyruvate carboxykinase activity. Shrago, E., Young, J.W., and Lardy, H.A. (1967). *Science* **158**, 1572.
210. Rabbit muscle phosphofructokinase: Reactivity and function of thiol groups. Younathan, E.S., Paetkau, V., and Lardy, H.A. (1968). *J. Biol. Chem.* **243**, 1603.
211. Phosphofructokinase: Studies on the subunit structure. Paetkau, V.H., Younathan, E.S., and Lardy, H.A. (1968). *J. Mol. Biol.* **33**, 721.
212. Antibiotics as tools for metabolic studies. XI. Specific inhibition of ion transport in mitochondria by the monensins. Estrada-O, S., Rightmire, B., and Lardy, H.A. (1968). In *Antimicrobial Agents and Chemotherapy-1967*, p. 279, Am. Soc. for Microbiology.
213. Influence of antibiotics and cyclic polyethers on ion transport in mitochondria. Lardy, H.A. (1968). *Federation Proc.* **27**, 1278.
214. Specificity of ion transport induced by beauvericin. Dorschner, E. and Lardy, H.A. (1969). In *Antimicrobial Agents and Chemotherapy-1968*, p. 11, Am. Soc. for Microbiology.
215. Oxidative phosphorylation in mitochondria. Lardy, H.A. and Ferguson, S.M.F. (1969). *Ann. Rev. Biochemistry* **38**, 991.
216. Mode of action of hypoglycemic agents. I. Inhibition of gluconeogenesis by quinaldic acid and 5-methoxyindole-2-carboxylic acid. Hanson, R.L., Ray, P.D., Walter, P., and Lardy, H.A. (1969). *J. Biol. Chem.* **244**, 4351.
217. The oxidation of thioethers by bromine. A model system for oxidative phosphorylation. Lambeth, D.O and Lardy, H.A. (1969). *Biochemistry* **8**, 3395.
218. Malic enzyme. Hsu, R.Y. and Lardy, H.A. (1969). In *Methods in Enzymology, XIII*, p. 230.
219. The metabolism of itaconate and mesaconate in mammalian liver. Lardy, H.A. (1969). In *Methods in Enzymology, XIII*, p. 314.
220. Inhibition of mitochondrial oxidative phosphorylation by aurovertin. Lardy, H.A. and Lin, C-H. (1969). In *Colloq. der Gesellschaft fur Biol. Chem.*, p. 279, Mosbach-Baden, Germany, April 14-16.

221. Paths of carbon in gluconeogenesis. Lardy, H.A., Veneziale, C., and Gabrielli, F. (1969). *FEBS Symposium* **19**, 55.
222. Site-specific inhibitors of gluconeogenesis. Lardy, H.A. In *Colloq. der Gesellschaft fur Biol. Chem.*, p. 374, Mosbach-Baden, Germany, April 14-16.
223. Inhibition by hydrazine of gluconeogenesis in the rat. Ray, P.D., Hanson, R.L., and Lardy, H.A. (1970). *J. Biol. Chem.* **245**, 690.
224. Bongkrekic acid: An inhibitor of the adenine nucleotide translocase of mitochondria. Henderson, P.J.F. and Lardy, H.A. (1970). *J. Biol. Chem.* **245**, 1319.
225. Inhibition of phosphofructokinase by quinone methide and α-methylene lactone tumor inhibitors. Hanson, R.L., Lardy, H.A., and Kupchan, S.M. (1970). *Science* **168**, 378.
226. Antibiotic inhibition of mitochondrial energy-transfer reactions. Henderson, P.J.F. and Lardy, H.A. (1970). In *Antimicrobial Agents and Chemotherapy-1969*, p. 18, Am. Soc. for Microbiology.
227. Gluconeogenesis. Lardy, H.A. (1970). In *Nobel Symposium 13* (Cerasi, E. & Luft, R., eds.), p. 199, Wiley and Sons.
228. Factors affecting the inhibition of adenine nucleotide translocase by bongkrekic acid. Henderson, P.J.F., Lardy, H.A., and Dorschner, E. (1970). *Biochemistry* **9**, 3453.
229. Gluconeogenesis from pyruvate in isolated perfused rat liver. Veneziale, C.M., Gabrielli, F., and Lardy, H.A. (1970). *Biochemistry* **9**, 3960.
230. Mode of action of hypoglycemic agents. III. Studies on 5-methoxyindole-2-carboxylic acid and quinaldic acid. Reed, J. and Lardy, H.A. (1970). *J. Biol. Chem.* **245**, 5297.
231. Effects of phosphodiesterase inhibitors and cyclic nucleotides on sperm respiration and motility. Garbers, D.L., Lust, W.D., First, N.L., and Lardy, H.A. (1971). *Biochemistry* **10**, 1825.
232. Rat liver pyruvate carboxylase. I. Preparation, properties, and cation specificity. McClure, W.R., Lardy, H.A., and Kneifel, H.P. (1971). *J. Biol. Chem.* **246**, 3569.
233. Rat liver pyruvate carboxylase. II. Kinetic studies of the forward reaction. McClure, W.R., Lardy, H.A., Wagner, M., and Cleland, W.W. (1971). *J. Biol. Chem.* **246**, 3579.
234. Rat liver pyruvate carboxylase. III. Isotopic exchange studies of the first partial reaction. McClure, W.R., Lardy, H.A., and Cleland, W.W. (1971). *J. Biol. Chem.* **246**, 3584.
235. Rat liver pyruvate carboxylase. IV. Factors affecting the regulation *in vivo*. McClure, W.R., and Lardy, H.A. (1971). *J. Biol. Chem.* **246**, 3591.
236. Contaminant of malic acid: A malic enzyme inhibitor. Anderson, D.B., Ferguson, S.M.F., and Lardy, H.A. (1971). *FEBS Letters* **14**, 283.

237. The role of tryptophan metabolites in regulating gluconeogenesis. Lardy, H.A. (1971). *Amer. J. Clin. Nutr.* 24, 764.
238. Potassium-specific uncoupling by nigericin. Ferguson, S.M.F., Estrada-O, S., and Lardy, H.A. (1971). *J. Biol. Chem.* 246, 5645.
239. Purification and properties of rat liver mitochondrial adenosine triphosphatase. Lambeth, D.O. and Lardy, H.A. (1971). *Eur. J. Biochem.* 22, 355.
240. Reassessment of the molecular weight of mitochondrial ATPase from beef heart. Lambeth, D.O., Lardy, H.A., Senior, A.E., and Brooks, J.C. (1971). *FEBS Letters* 17, 330.
241. Effect of a new monocarboxylic acid antibiotic, A204, on the monovalent cation permeability of rat liver mitochondria. Wong, D.T., Horng, J-S, Hamill, R.L., and Lardy, H.A. (1971). *Biochem. Pharmacol.* 20, 3169.
242. Response of phosphopyruvate carboxylase to tryptophan metabolites and metal ions. Snoke, R.E., Johnston, J.B., and Lardy, H.A. (1971). *Eur. J. Biochem.* 24, 342.
243. Stimulation and maintenance of ejaculated bovine spermatozoan respiration and motility by caffeine. Garbers, D.L., First, N.L., Sullivan, J.J., and Lardy, H.A. (1971). *Biol. Reprod.* 5, 336.
244. Pyruvate uptake in rat liver mitochondria: Transport or adsorption? Zahlten, R.N., Hochberg, A.A., Stratman, F.W., and Lardy, H.A. (1972). *FEBS Letters* 21, 11.
245. Glucagon-stimulated phosphorylation of mitochondrial and lysosomal membranes of rat liver *in Vivo.* Zahlten, R.N., Hochberg, A.A., Stratman, F.W., and Lardy, H.A. (1972). *Proc. Nat. Acad. Sci. USA* 69, 800.
246. Pathogenesis of Leigh's encephalomyelopathy. Tang, T.T., Good, T.A., Dyken, P.R., Johnsen, S.D., McCreadie, S.R., Sy, S.T., Lardy, H.A., and Rudolph, F.B. (1972). *J. Pediatrics* 81, 189.
247. Properties of mitochondrial ATPase. Lardy, H.A. and Lambeth, D. (1972). In *Energy Metabolism and the Regulation of Metabolic Processes in Mitochondria*, p. 287, Academic Press, New York.
248. Incorporation of L-[methyl-^{14}C]- and [^{35}S] methionine into mitochondrial proteins. Hochberg, A.A., Zahlten, R.N., Stratman, F.W., and Lardy, H.A. (1972). *Biochemistry* 11, 3143.
249. Incorporation of L-[1-ethyl-^{14}C]- and L-[^{35}S] ethionine into mitochondrial proteins. Hochberg, A.A., Zahlten, R.N., Stratman, F.W., and Lardy, H.A. (1972). *Biochemistry* 11, 3149.
250. Synthesis, transfer and specific binding of purified L-[^{35}S] methionine-labeled rat liver mitochondrial adenosine triphosphatase and its subunits to mitochondrial inner membrane. Stratman, F.W., Zahlten, R.N., Hochberg, A.A., and Lardy, H.A. (1972). *Biochemistry* 11, 3154.

251. Binding of rat liver and hepatoma polyribosomes to stripped rough endoplasmic reticulum *in vitro.* Hochberg, A.A., Stratman, F.W., Zahlten, R.N., Morris, H.P., and Lardy, H.A. (1972). *Biochem J.* **130**, 19.
252. A23187: A divalent cation ionophore. Reed, P.W. and Lardy, H.A. (1972). *J. Biol. Chem.* **247**, 6970.
253. Artifacts in protein synthesis by mitochondria *in vitro.* Hochberg, A.A., Stratman, F.W., Zahlten, R.N., and Lardy, H.A. (1972). *FEBS Letters* **25**, 1.
254. Antibiotic A23187 as a probe for the study of calcium and magnesium function in biological systems. Reed, P.W. and Lardy, H.A. (1972). In *The Role of Membranes in Metabolic Regulation,* p. 111, Academic Press, New York.
255. Properties of adenosine 3',5'-monophosphate-dependent protein kinases isolated from bovine epididymal spermatozoa. Garbers, D.L., First, N.L., and Lardy, H.A. (1973). *J. Biol. Chem.* **248**, 875.
256. Inactivation of phosphofructokinase by 6-mercapto-9-β-D-ribofuranosylpurine 5'-triphosphate. Bloxham, D.P., Clark, M.G., Holland, P.C., and Lardy, H.A. (1973). *Biochemistry* **12**, 1596.
257. Phosphofructokinase. Bloxham, D.P. and Lardy, H.A. (1973). In *The Enzymes,* Vol. 8, p. 239.
258. Influence of glucagon on the metabolism of xylitol and dihydroxyacetone in the isolated perfused rat liver. Blair, J.B., Cook, D.E., and Lardy, H.A. (1973). *J. Biol. Chem.* **248**, 3601.
259. Interaction of propionate and lactate in the perfused rat liver. Blair, J.B., Cook, D.E., and Lardy, H.A. (1973). *J. Biol. Chem.* **248**, 3608.
260. Biosynthesis of glutamate dehydrogenase in rat liver. Demonstration of its microsomal localization and hypothetical mechanism of transfer to mitochondria. Godinot, C. and Lardy, H.A. (1973). *Biochemistry* **12**, 2051.
261. The stimulation of bovine epididymal sperm metabolism by cyclic nucleotide phosphodiesterase inhibitors. Garbers, D., First, N.L., and Lardy, H.A. (1973). *Biology of Reproduction* **8**, 589.
262. The effects of cyclic nucleotide phosphodiesterase inhibitors on ejaculated porcine spermatozoan metabolism. Garbers, D.L., First, N.L., Gorman, S.K., and Lardy, H.A. (1973). *Biology of Reproduction* **8**, 599.
263. A model study of the fructose diphosphatase-phosphofructokinase substrate cycle. Bloxham, D.P., Clark, M.G., Holland, P.C., and Lardy, H.A. (1973). *Biochem. J.* **134**, 581.
264. The theoretical estimation of substrate cycling *in Vivo.* Bloxham, D.P., Clark, M.G., Goldberg, D.M., Holland, P.C., and Lardy, H.A. (1973). *Biochem. J.* **134**, 586.
265. Estimation of the fructose diphosphatase-phosphofructokinase substrate cycle in the flight muscle of *Bombus affinis.* Clark, M.G., Bloxham, D.P., Holland, P.C., and Lardy, H.A. (1973). *Biochem. J.* **134**, 589.

266. Mode of action of hypoglycemic agents. IV. Control of the hypoglycemic activity of phenethylbiguanide in rats and guinea pigs. Cook, D.E., Blair, J.B., Gilfillan, C., and Lardy, H.A. (1973). *Biochem. Pharmacol.* **22**, 2121.

267. Mode of action of hypoglycemic agents. V. Studies with phenethylbiguanide in isolated perfused rat liver. Cook, D.E., Blair, J.B., and Lardy, H.A. (1973). *J. Biol. Chem.* **248**, 5272.

268. Mechanism of action of the hypoglycemic agent diphenyleneiodonium. Holland, P.C., Clark, M.G., Bloxham, D.P., and Lardy, H.A. (1973). *J. Biol. Chem.* **248**, 6050.

269. Accelerated substrate cycling of fructose-6-phosphate in the muscle of malignant hyperthermic pigs. Clark, M.G., Williams, C.H., Pfeiffer, W.F., Bloxham, D.P., Holland, P.C., Taylor, C.A., and Lardy, H.A. (1973). *Nature* **245**, 99-101.

270. Regulation of glucose synthesis in hormone-sensitive isolated rat hepatocytes. Zahlten, R.N., Stratman, F.W., and Lardy, H.A. (1973). *Proc. Nat. Acad. Sci. USA* **70**, 3213-3218.

271. Reduction of pyridine nucleotides induced by adenosine diphosphate in kidney mitochondria. Weiner, M.W., and Lardy, H.A. (1973). *J. Biol. Chem.* **248**, 7682-7687.

272. Rabbit muscle phosphofructokinase: The kinetic mechanism of action and the equilibrium constant. Hanson, R.L., Rudolph, F.B., and Lardy, H.A. (1973). *J. Biol. Chem.* **248**, 7852-7859.

273. Estimation of the fructose 1,6-diphosphatase-phosphofructokinase substrate cycle and its relationship to gluconeogenesis in rat liver *in Vivo*. Clark, M.G., Bloxham, D.P., Holland, P.C., and Lardy, H.A. (1974). *J. Biol. Chem.* **249**, 279-290.

274. The influence of ammonium and calcium ions on gluconeogenesis in isolated rat hepatocytes and their response to glucagon and epinephrine. Zahlten, R.N., Kneer, N.M., Stratman, F.W., and Lardy, H.A. (1974). *Arch. Biochem. Biophys.* **161**, 528-535.

275. Effects of respiratory chain inhibitors on mitochondrial ATPase activity. Weiner, M.W. and Lardy, H.A. (1974). *Arch. Biochem. Biophys.* **162**, 568-577.

276. Ultraviolet and fluorescent spectral properties of the divalent cation ionophore A23187 and its metal ion complexes. Pfeiffer, D.R., Reed, P.W., and Lardy, H.A. (1974). *Biochemistry* **13**, 4007-4014.

277. The fructose 1,6-diphosphatase-phosphofructokinase substrate cycle. A site of regulation of hepatic gluconeogenesis by glucagon. Clark, M.G., Kneer, N.M., Bosch, A.L., and Lardy, H.A. (1974). *J. Biol. Chem.* **249**, 5695-5703.

278. Regulation of hepatic gluconeogenesis by glucagon. Lardy, H.A., Zahlten, R.N., Stratman, F.W., and Cook, D.E. (1974). In *Regulation of Hepatic Metabolism* (Lundquist, F. & Tygstrup, N., eds.), (Alfred Benzon Symposium VI), p. 19-31, Academic Press, New York.
279. Differential inhibition of the exchange reactions associated with oxidative phosphorylation. Holland, P.C., LaBelle, W.C., and Lardy, H.A. (1974). *Biochemistry* 13, 4549-4553.
280. Glucose inhibition of epinephrine stimulation of hepatic gluconeogenesis by blockade of the α-receptor function. Kneer, N.M., Bosch, A.L., Clark, M.G., and Lardy, H.A. (1974). *Proc. Nat. Acad. Sci. USA* 71, 4523-4527.
281. Divalent cation transport by antibiotic A23187. Reed, P.W. and Lardy, H.A. (1974). In *Proc. of the 2nd Annual New England Bioengineering Conference*, March 29-30 (Peura, R.A., et al., eds), pp. 73-80.
282. Stimulation of rat liver mitochondrial adenosine triphosphatase by anions. Ebel, R.E. and Lardy, H.A. *J. Biol. Chem.* 250, 191-196.
283. Rat liver pyruvate carboxylase. V. Reversible dissocation by chloride salts of monovalent cations. Nakashima, K., Rudolph, F.B., Wakabayashi, T., and Lardy, H.A. (1975). *J. Biol. Chem.* 250, 331-336.
284. Antibiotics as tools for metabolic studies. XVIII. Inhibition of sodium- and potassium-dependent adenosine triphosphatase. Susa, J.B. and Lardy, H.A. (1975). *Molecular Pharmacology* 11, 166-173.
285. Uncoupling and specific inhibition of phosphoryl transfer reactions in mitochondria by antibiotic A20668. Reed, P.W. and Lardy, H.A. (1975). *J. Biol. Chem.* 250, 3704-3708.
286. Antibiotic inhibitors of mitochondrial ATP synthesis. Lardy, H.A., Reed, P., and Lin, C-H. (1975). *Fed Proc.* 34, 1707-1710.
287. Influence of aurovertin on mitochondrial ATPase activity. Ebel, R.E. and Lardy, H.A. (1975). *J. Biol. Chem.* 250, 4992-4995.
288. Transport mechanism for succinate and phosphate localized in the plasma membrane of bovine spermatozoa. Babcock, D.F., First, N.L., and Lardy, H.A. (1975). *J. Biol. Chem.* 250, 6488-6495.
289. Exploring sites on mitochondrial ATPase for catalysis, regulation, and inhibition. Lardy, H.A., Schuster, S.M., and Ebel, R.E. (1975). *Journal of Supramolecular Structure* 3, 214-221.
290. Kinetic studies on rat liver and beef heart mitochondrial ATPase. Evidence for nucleotide binding at separate regulatory and catalytic sites. Schuster, S.M., Ebel, R.E., and Lardy, H.A. (1975). *J. Biol. Chem.* 250, 7848-7853.
291. Kinetic studies on rat liver and beef heart mitochondrial adenosine triphosphatase: The effects of the chromium complexes of adenosine triphosphate and adenosine diphosphate on the kinetic properties. Schuster,

S.M., Ebel, R.E., and Lardy, H.A. (1975). *Arch. Biochem. Biophys.* **171**, 656-661.

292. Regulation of intermediary carbohydrate metabolism. Clark, M.G. and Lardy, H.A. (1975). In *Biochemistry of Carbohydrates* (Whelan, W.J., ed), pp. 223-266, University Park Press, Baltimore, MD.

293. Ionophore A23187: The effect of H^+ concentration on complex formation with divalent and monovalent cations and the demonstration of K^+ transport in mitochondria mediated by A23187. Pfeiffer, D.R., and Lardy, H.A. (1976). *Biochemistry* **15**, 935-943.

294. Essential arginyl residues in mitochondrial adenosine triphosphatase. Marcus, F., Schuster, S.M., and Lardy, H.A. (1976). *J. Biol. Chem.* **251**, 1775-1780.

295. Effect of epinephrine on the concentration of cGMP in rat liver cells. Bosch, A., Kneer, N., and Lardy, H.A. (1976). In *Use of Isolated Liver Cells and Kidney Tubules in Metabolic Studies* (Tager, J.M., Söling, H.D. & Williamson, J.R., eds.), pp. 285-290, North-Holland Publ. Co., Amsterdam, The Netherlands.

296. Interaction of anions and divalent metal ions with phosphoenolpyruvate carboxykinase. Bentle, L.A. and Lardy, H.A. (1976). *J. Biol. Chem.* **251**, 2916-2921.

297. A protein factor required for activation of phosphoenolpyruvate carboxykinase by ferrous ions. Bentle, L.A., Snoke, R.E., and Lardy, H.A. (1976). *J. Biol. Chem.* **251**, 2922-2928.

298. Some effects of ionophore A23187 on energy utilization and the distribution of cations and anions in mitochondria. Pfeiffer, D.R., Hutson, S.M., Kauffman, R.F., and Lardy, H.A. (1976). *Biochemistry* **15**, 2690-2697.

299. Action of ionophore A23187 at the cellular level. Separation of effects at the plasma and mitochondrial membranes. Babcock, D.F., First, N.L., and Lardy, H.A. (1976). *J. Biol. Chem.* **251**, 3881-3886.

300. Effect of cations and anions on the steady state kinetics of energy-dependent Ca^{2+} transport in rat liver mitochondria. Hutson, S.M., Pfeiffer, D.R., and Lardy, H.A. (1976). *J. Biol. Chem.* **251**, 5251-5258.

301. Activation of bovine epididymal sperm respiration by caffeine. Its transient nature and relationship to the utilization of acetyl carnitine. Milkowski, A.L., Babcock, D.F., and Lardy, H.A. (1976). *Arch. Biochem. Biophys.* **176**, 250-256.

302. Effect of inosine 5'-(β,γ-imido) triphosphate and other nucleotides on beef heart mitochondrial ATPase. Schuster, S.M., Gertschen, R.J., and Lardy, H.A. (1976). *J. Biol. Chem.* **251**, 6705-6710.

303. Studies on the kinetic mechanism of oxidative phosphorylation. Schuster, S.M., Reinhart, G.D., and Lardy, H.A. (1977). *J. Biol. Chem.* **252**, 427-432.

304. Pyruvate metabolism in bovine epididymal spermatozoa. Van Dop, C., Hutson, S.M., and Lardy, H.A. (1977). *J. Biol. Chem.* 252, 1303-1308.
305. Mitochondrial metabolism of pyruvate in bovine spermatozoa. Hutson, S.M., Van Dop, C., and Lardy, H.A. (1977). *J. Biol. Chem.* 252, 1309-1315.
306. P-enolpyruvate carboxykinase ferroactivator. Purification and some properties. Bentle, L.A. and Lardy, H.A. (1977). *J. Biol. Chem.* 252, 1431-1440.
307. Factors affecting the redox state of bovine epididymal spermatozoa. Milkowski, A.L. and Lardy, H.A. (1977). *Arch. Biochem. Biophys.* 181, 270-277.
308. Metabolism of pyruvate and carnitine esters in bovine epididymal sperm mitochondria. Hutson, S.M., Van Dop, C., and Lardy, H.A. (1977). *Arch. Biochem. Biophys.* 181, 345-352.
309. Regulation of phosphopyruvate synthesis in normal and pathological states. Regulatory mechanisms of carbohydrate metabolism. Lardy, H.A., MacDonald, M.J., Huang, M-T, and Bentle, L.A. (1977). *Federation of European Biochemical Societies, 11th Meeting, Copenhagen* 42 (Symposium A1), (Esmann, V., ed.), p. 93-103, Pergamon Press.
310. P-enolpyruvate carboxykinase ferroactivator; Distribution, and the influence of diabetes and starvation. MacDonald, M.J., Bentle, L.A., and Lardy, H.A. (1978). *J. Biol. Chem.* 253, 116-124.
311. Copper as an essential element for iron utilization. Lardy, H.A. (1978). *TIBS* 3, p. 93-94.
312. 311.3-aminopicolinate: A synthetic metal-complexing activator of phosphoenolpyruvate carboxykinase. MacDonald, M.J. and Lardy, H.A. (1978). *J. Biol. Chem.* 253, 2300-2307.
313. Selective inhibition of pyruvate and lactate metabolism in bovine epididymal spermatozoa by dinitrophenol and α-cyano-hydroxycinnamate. Van Dop, C., Hutson, S.M., and Lardy, H.A. (1978). *Arch. Biochem. Biophys.* 187, 235-242.
314. Ionophore A23187: Cation binding and transport properties. Pfeiffer, D.R., Taylor, R.W., and Lardy, H.A. (1978). *Ann. N.Y. Acad. Sci.* 307, 402-423.
315. Norepinephrine, vasopressin, glucagon, and A23187 induce efflux of calcium from an exchangeable pool in isolated rat hepatocytes. Chen, J-L., Babcock, D.F., and Lardy, H.A. (1978). *Proc. Natl. Acad. Sci. USA* 75, 2234-2238.
316. Increased calcium-ion influx is a component of capacitation of spermatozoa. Singh, J.P., Babcock, D.F., and Lardy, H.A. (1978). *Biochem. J.* 172, 549-556.
317. Structural requirements for pyridine carboxylate effects on phosphoenolpyruvate carboxykinase. MacDonald, M.J., and Lardy, H.A. (1978). *Bioorg. Chem.* 7, 251-262.
318. Effects of *N*-ethylmaleimide on the limited uptake of Ca^{2+}, Mn^{2+}, and Sr^{2+} by rat liver mitochondria. Pfeiffer, D.R., Kauffman, R.F., and Lardy, H.A. (1978). *J. Biol. Chem.* 253, 4165-4171.

345. The role of calcium in the stimulation of gluconeogenesis by catecholamines. Yip, B.P. and Lardy, H.A. (1981). *Arch. Biochem. Biophys.* 212, 370-377.
346. Ca^{2+} stimulation of rat liver mitochondrial glycerophosphate dehydrogenase. Wernette, M.E., Ochs, R.S., and Lardy, H.A. (1981). *J. Biol. Chem.*, 12767-12771.
347. Inhibition of phosphoenolpyruvate carboxykinase by streptoningrin. Merryfield, M.L. and Lardy, H.A. (1982). *Biochemical Pharmacology* 31, 1123-1129.
348. Ca^{2+}-mediated activation of phosphoenolpyruvate carboxykinase occurs via release of Fe^{2+} from rat liver mitochondria. Merryfield, M.L., and Lardy, H.A. (1982). *J. Biol. Chem.* 257, 3628-3635.
349. Purification and characterization of a rat liver ferroactivator with catalase activity. Merryfield, M.L., Kramp, D., and Lardy, H.A. (1982). *J. Biol. Chem.* 257, 4646-4652.
350. Purification and properties of oxidized and reduced forms of hepatic fructose 1,6-bisphosphatase. Moser, U.K., Althaus-Salzmann, M., Van Dop, C., and Lardy, H.A. (1982). *J. Biol. Chem.* 257, 4552-4556.
351. Purification and characterization of a calcium transport inhibitor protein from bovine seminal plasma. Rufo, G.A. Jr., Singh, J.P., Babcock, D.F., and Lardy, H.A. (1982). *J. Biol. Chem.* 257, 4627-4632.
352. Alterations in membrane permeability to Ca^{2+} and their consequences during maturation of mammalian spermatozoa. Babcock, D.F. and Lardy, H.A. (1982). In *Membranes in Transport*, Vol. 1 (Martonosi, A.N., ed.), pp. 671-676, Plenum Pub. Corp.
353. The development and application of a radioimmunoassay for rat phosphoenolpyruvate carboxykinase. Moore, R.E., Hansen, J.B., Lardy, H.A., and Veneziale, C.M. (1982). *J. Biol. Chem.* 257, 12546-12552.
354. Motility activation, respiratory stimulation, and alteration of Ca^{2+} transport in bovine sperm treated with amine local anesthetics and calcium transport antagonists. Singh, J.P., Babcock, D.F., and Lardy, H.A. (1983). *Arch. Biochem. Biophys.* 221, 291-303.
355. Potassium-dependent increases in cytosolic pH stimulate metabolism and motility of mammalian sperm. Babcock, D.F., Rufo, G.A., Jr., and Lardy, H.A. (1983). *Proc. Natl. Acad. Sci. USA* 80, 1327-1331.
356. Regulation of gluconeogenesis by catecholamines, vasopressin and angiotensin. Lardy, H.A., Kneer, N., and Wernette, M.E. (1983). In *Isolation, Characterization, and Use of Hepatocytes* (Harris, R.A. & Cornell, N.W., eds.), pp. 445-454, Elsevier Science Publishing Co., Inc.
357. The effect of substrate cycling on the ATP yield of sperm glycolysis. Hammerstedt, R.H. and Lardy, H.A. (1983). *J. Biol. Chem.* 258, 8759-8768.

358. Regulation of gluconeogenesis by norepinephrine, vasopressin, and angiotensin II: A comparative study in the absence and presence of extracellular Ca^{2+}. Kneer, N.M., and Lardy, H.A. (1983). *Arch. Biochem. Biophys.* **255**, 187-195.
359. Regulation of carbohydrate metabolism by 2,5-anhydro-D-mannitol. Riquelme, P.T., Wernette-Hammond, M.E., Kneer, N.M., and Lardy, H.A. (1983). *Proc. Natl. Acad. Sci. USA* **80**, 4301-4305.
360. Catecholamine stimulation of hepatic gluconeogenesis at the site between pyruvate and phosphoenolpyruvate. Ochs, R.S. and Lardy, H.A. (1983). *J. Biol. Chem.* **258**, 9956-9962.
361. Regulation of calcium content in bovine spermatozoa. Rufo, G.A. Jr., Schoff, P.K., and Lardy, H.A. (1984). *J. Biol. Chem.* **259**, 2547-2552.
362. Effects of fluoracetate on hepatic gluconeogenesis. Bobyleva-Guarriero, V., Hughes, P.E., and Lardy, H.A. (1984). *Fluoride* **17**, 94-104.
363. Mechanism of action of 2,5-anhydro-D-mannitol in hepatocytes. Riquelme, P.T., Wernette-Hammond, M.E., Kneer, N.M., and Lardy, H.A. (1984). *J. Biol. Chem.* **259**, 5115-5123.
364. Site of insulin inhibition of cAMP-stimulated glycogenolysis. Gabbay, R.A. and Lardy, H.A. (1984). *J. Biol. Chem.* **259**, 6052-6055.
365. 3-aminopicolinate inhibits phosphoenolpyruvate carboxykinase in hepatocytes and increases release of gluconeogenic precursors from peripheral tissues. Chen, K.S. and Lardy, H.A. (1984). *J. Biol. Chem.* **259**, 6920-6924.
366. The effect of nutritional state and hormones on the number of circulating red cells in gallus domesticus. Hughes, P.E., Bobyleva-Guarriero, V., and Lardy, H.A. (1984). *Comp. Biochem. Physiol.* **78B**, 621-625.
367. The influence of fasting on chicken liver metabolites, enzymes and mitochondrial respiration. Bobyleva-Guarriero, V., Hughes, P.E., Ronchetti-Pasquali, I., and Lardy, H.A. (1984). *Comp. Biochem. Physiol.* **78B**, 627-632.
368. Regulation of gluconeogenesis at phosphoenolpyruvate carboxykinase. Lardy, H.A. and Hughes, P.E. (1984). In *Current Topics in Cellular Regulation*, p. 171-179, Academic Press, Inc.
369. The antilipolytic effect of insulin does not require adenylate cyclase or phosphodiesterase action. Gabbay, R.A. and Lardy, H.A. (1985). *FEBS Lett.* **179**, 7-11.
370. Inhibition by 2,5-anhydromannitol of glycolysis in isolated rat hepatocytes and in Ehrlich ascites cells. Riquelme, P.T., Kneer, N.M., Wernette-Hammond, M.E. and Lardy, H.A. (1985). *Proc. Natl. Acad. Sci. USA* **82**, 78-82.
371. Regulation of gluconeogenesis in glycogen-depleted hepatocytes from alloxan-diabetic rats. Wernette-Hammond, M.E., and Lardy, H.A. (1985). *Diabetes* **34**, 767-773.

372. The structure of caltrin, the calcium transport structure of caltrin, the calcium transport inhibitor of bovine inhibitor of bovine seminal plasma. Lewis, R.V., San Agustin, J., Kruggel, W., and Lardy, H.A. (1985). *Proc. Natl. Acad. Sci. USA* **82**, 6490-6491.
373. Catecholamine and vasopressin stimulation of gluconeogenesis from dihydroxyacetone in the presence of atractyloside. Wernette-Hammond, M.E., and Lardy, H.A. (1985). *J. Biol. Chem.* **260**, 12647-12652.
374. Multiple requirements for glycogen synthesis by hepatocytes isolated from fasted rats. Chen, K. and Lardy, H.A. (1985). *J. Biol. Chem.* **260**, 14683-14688.
375. A ^{31}P NMR study of the epididym is and epididymal sperm of the bull and hamster. Smith, M.B., Babcock, D.F., and Lardy, H.A. (1985). *Biol. of Reprod.* **33**, 1029-1041.
376. A half century biochemistry. Lardy, H.A. (1985). In *Comprehensive Biochemistry. Personal Recollections*, II, Vol. 36 (Semenza, G., ed), p. 297-325, Elsevier.
377. The effect of different types of physical exercise on glucose and citrulline synthesis in isolated rat liver parenchymal cells. Bobyleva-Guarriero, V., and Lardy, H.A. (1986). *FEBS Letters* **194**, 56-59.
378. The role of malate in exercise-induced enhancement of mitochondrial respiration. Bobyleva-Guarriero, V. and Lardy, H.A. (1986). *Arch. Biochem. Biophys.* **245**, 470-476.
379. The role of malate in hormone-induced enhancement of mitochondrial respiration. Bobyleva-Guarriero, V., Wehbie, R.S., and Lardy, H.A. (1986). *Arch. Biochem. Biophys.* **245**, 477-482.
380. Methylisobutylxanthine blocks insulin antagonism of cAMP-stimulated glycogenolysis at a site distinct from phosphodiesterase. Evidence favoring an insulin-insensitive calcium release mechanism. Gabbay, R.A. and Lardy, H.A. (1986). *J. Biol. Chem.* **261**, 4002-4007.
381. Enzymatic adaptation to physical training under β-blockade in the rat: Evidence of a β_2-adrenergic mechanism in skeletal muscle. Ji, L.L., Lennon, D.L.F., Kochan, R.G., Nagle, F.J., and Lardy, H.A. (1986). *J. Clin. Invest.* **78**, 771-778.
382. Stereochemical course of thiophosphoryl transfer catalyzed by cytosolic phosphoenolpyruvate carboxykinase. Konopka, J.M., Lardy, H.A., and Frey, P.A. (1986). *Biochemistry* **25**, 5571-5575.
383. Caltrin: A regulator of calcium transport by bovine spermatozoa, Development, Growth andDifferentiation. Supplement, 1986. Lardy, H.A., San Agustin, J., Lewis, R., and Hughes, P. Abst. 48.
384. Chronic exercise training alters kinetic properties of rat skeletal muscle and myocardial lactate dehydrogenase. Ji, L.L., Stratman, F.W., and Lardy, H.A. (1986). *FEBS* **208**, 297-300.

385. Edwin Broun Fred (1887-1981). Lardy, H.A. (1986). In *American Philosophical Year Book—1985*, p. 122-125.
386. Insulin inhibition of hepatic cAMP-dependent protein kinase: Decreased affinity of protein kinase for cAMP and possible differential regulation of intrachain sites 1 and 2. Gabbay, R.A. and Lardy, H.A. (1987). *Proc. Natl. Acad. Sci. USA* **84**, 2218-2222.
387. Effect of verapamil and sulfhydryl reagents on calcium transport in bovine spermatozoa. Breitbart, H. and Lardy, H.A. (1987). *Biol. of Reprod.* **36**, 658-663.
388. The effects of β-adrenergic blockade on body composition in free-fed and diet-restricted rats. Ji, L.L., Doan, T.D., Lennon, D.L.F., Nagle, F.J., and Lardy, H.A. (1987). *J. Am. Coll. Nutr.* **6**, 175-177.
389. Phosphoenolpyruvate carboxykinase ferroactivator 1: Mechanism of action and identity with glutathione peroxidase. Punekar, N.S. and Lardy, H.A. (1987). *J. Biol. Chem.* **262**, 6714-6719.
390. γ-butyrobetaine hydroxylase and the protective role of glutathione peroxidase. Punekar, N.S., Wehbie, R.S., and Lardy, H.A. (1987). *J. Biol. Chem.* **262**, 6720-6724.
391. The antibiotic W341C, its ion transport properties and inhibitory effects on mitochondrial substrate oxidation. Wehbie, R.S., Cai, R., and Lardy, H.A. (1987). *J. Antibiotics* **40**, 887-893.
392. Properties and function of caltrin, the calcium-transport inhibitor of bull seminal plasma. San Agustin, J.T., Hughes, P., and Lardy, H.A. (1987). *FASEB J.* **1**, 60-66.
393. Carnitine supplementation and depletion: Tissue carnitines and enzymes in fatty acid oxidation. Negrao, C.E., Ji, L.L., Schauer, J.E., Nagle, F.J., and Lardy, H.A. (1987). *J. Appl. Physiol.* **63**, 315-321.
394. Amino acid metabolism during exercise in trained rats: The potential role of carnitine in the metabolic fate of branched-chain amino acids. Ji, L.L., Miller, R.H., Nagle, F.J., Lardy, H.A., and Stratman, F.W. (1987). *Metabolism* **36**, 748-752.
395. Effects if β_1 and β_1 + β_2—antagonists on training-induced myocardial hypertrophy and enzyme adaptation. Ji, L.L, Stratman, F.W., and Lardy, H.A. (1987). *Biochem. Pharmacol.* **36**, 3411-3417.
396. Effects of fluoride and caffeine on the metabolism and motility of ejaculated bovine spermatozoa. Schoff, P.K., and Lardy, H.A. (1987). *Biol. Reprod.* **37**, 1037-1046.
397. Characterization of Ca^{2+} uptake by guinea pig epididymal spermatozoa. Coronel, C.E. and Lardy, H.A. (1987). *Biol. Reprod.* **37**, 1097-1107.
398. Rat liver γ-butyrobetaine hydroxylase catalyzed reaction: Influence of potassium, substrates, and substrate analogues on hydroxylation and

decarboxylation. Wehbie, R.S., Punekar, N.S., and Lardy, H.A. (1988). *Biochemistry* 27, 2222-2228.

399. Identification and partial characterization of caltrin-like proteins in the reproductive tract of the guinea pig. Coronel, C.E., San Agustin, J., and Lardy, H.A. (1988). *Biol. Reprod.* 38, 713-722.
400. Enzymatic down regulation with exercise in rat skeletal muscle. Ji, L.L., Stratman, F.W., and Lardy, H.A. (1988). *Arch. Biochem. Biophys.* **263**, 137-149.
401. Antioxidant enzyme systems in rat liver and skeletal muscle: Influences of selenium deficiency, chronic training, and acute exercise. Ji, L.L., Stratman, F.W., and Lardy, H.A. (1988). *Arch. Biochem. Biophys.* **263**, 150-160.
402. Caltrin: A versatile regulator of calcium transport in spermatozoa. Lardy, H.A., San Agustin, J., and Coronel, C. (1988). In *The Roots of Modern Biochemistry* (Kleinkauf, von Dohren, Jaenicke, eds.), pp. 759-763, Walter de Gruyter and Co., Berlin-New York.
403. Pathway of gluconeogenesis from D- and L-glycerate in rat hepatocytes. Chen, K.S. and Lardy, H.A. (1988). *Arch. Biochem. Biophys.* **265**, 433-440.
404. Dehydroepiandrosterone induces enzymes that permit thermogenesis and decrease metabolic efficiency. Lardy, H.A., Su, C-Y, Kneer, N., and Wielgus, S. (1988). In *Hormones, Thermogenesis, and Obesity* (Lardy, H.A. & Stratman, F., eds.), pp. 415-426, Elsevier Science Publishing Co., Inc.
405. Effects of acute and chronic exercise in rats on energy transduction under the influence of dehydroepiandrosterone and a β-agonist. Schauer, J.E., Rohrer, M.J., Hanson, P., Lardy, H.A., and Stratman, F.W. (1988). In *Hormones, Thermogenesis, and Obesity* (Lardy, H.A. & Stratman, F., eds.), pp. 485-501, Elsevier Science Publishing Co., Inc.
406. Caltrin and calcium regulation of sperm activity. Lardy, H.A. and San Agustin, J. (1989). In *The Cell Biology of Fertilization* (Schatten, H. & Schatten, G., eds.), pp. 29-39, Academic Press, Inc.
407. Adenylate kinase activity in ejaculated bovine sperm flagella. Schoff, P.K., Cheetham, J., and Lardy, H.A. (1989). *J. Biol. Chem.* **264**, 6086-6091.
408. Sources of intramitochondrial malate. Bobyleva-Guarriero, V., Battelli, D., Bellei, M., and Lardy, H.A. (1989). *FASEB J.* 3, 2208-2211.
409. Glutathione peroxidase and iron-thiol dependent lipid peroxidation. Punekar, N.S. and Lardy, H.A. (1989). *Biochem. Internat.* **19**, 1381-1386.
410. Inhibition by caltrin of calcium transport into spermatozoa, liver and heart mitochondria. Breitbart, H., Wehbie, R.S., San Agustin, J., and Lardy, H.A. (1990). *Biochim. Biophys. Acta* **1022**, 27-32.
411. Purification and structure of caltrin-like proteins from seminal vesicle of the guinea pig. Coronel, C.E., San Agustin, J., and Lardy, H.A. (1990). *J. Biol. Chem.* **265**, 6854-6859.

412. Bovine seminal plasma constituents modulate the activity of caltrin, the calcium-transport regulating protein of bovine spermatozoa. San Agustin, J.T. and Lardy, H.A. (1990). *J. Biol. Chem.* **265**, 6860-6867.
413. Biochemical aspects of obesity. Lardy, H.A. and Shrago, E. (1990). *Annu. Rev. Biochem.***59**, 689-710.
414. Calcium transport in sperm mitochondria: Effect of substrates and phosphate. Breitbart, H., Wehbie, R., and Lardy, H.A. (1990). *Biochim. Biophys. Acta* **1026**, 57-63.
415. Regulation of calcium in bovine spermatozoa. Breitbart, H., Wehbie, R., and Lardy, H.A. (1990). *Biochim. Biophys. Acta* **1027**, 72-78.
416. Stimulation of Ca^{2+} uptake into epididymal bull spermatozoa by analogues of amiloride. Breitbart, H., Cragoe, E.J. Jr., and Lardy, H.A. (1990). *Eur. J. Biochem.* **192**, 529-535.
417. Role of malate in the enhancement of mitochondrial respiration *in vitro*. Bobyleva-Guarriero, V., Wehbie, R.S., Muscatello, U., and Lardy, H.A. (1991). *Biokhimiya* **56**, 542-551.
418. Induction of hepatic mitochondrial glycerophosphate dehydrogenase in rats by dehydroepiandrosterone. Su, C-Y. and Lardy, H.A. (1991). *J. Biochemistry* **110**, 207-213.
419. Antioxidant enzyme response to selenium deficiency in rat myocardium. Ji, L.L., Stratman, F.W., and Lardy, H.A. (1992). *J. Am. College of Nutrition* **11**, 79-86.
420. Bovine spermatozoa incorporate $^{32}P_i$ into ADP by an unknown pathway. Cheetham, J. and Lardy, H.A. (1992). *J. Biol. Chem.* **267**, 5186-5192.
421. Changes in liver structure and function after short-term and long-term treatment of rats with dehydroepiandrosterone. Bellei, M., Battelli, D., Fornieri, C., Mori, G., Muscatello, U., Lardy, H.A., and Bobyleva-Guarriero, V. (1992). *J. Nutrition* **122**, 967-976.
422. Functional properties of caltrin proteins from seminal vesicle of the guinea pig. Coronel, C. and Lardy, H.A. (1992). *Molecular Reprod. and Development* **33**, 74-80.
423. Purification, structure and characterization of caltrin proteins from seminal vesicle of the rat and mouse. Coronel, C.E., Winnica, D.E., Novella, M.L., and Lardy, H.A. (1992). *J. Biol. Chem.* **267**, 20909-20915.
424. Isolation and characterization of a 54-kilodalton precursor of caltrin, the calcium transport inhibitor protein from seminal vesicles of the rat. Coronel, C.E., Novella, M.L., Winnica, D.E., and Lardy, H.A. (1993). *Biol. Repro.* **48**, 1326-1333.
425. Comparative studies of effects of dehydroepiandrosterone on rat and chicken liver. Bobyleva, V., Kneer, N., Bellei, M., Battelli, D., Muscatello, U., and Lardy, H.A. (1993). *Comp. Biochem. Physiol.* **105B**, 643-647.

426. Concerning the mechanism of increased thermogenesis in rats treated with dehydroepiandrosterone. Bobyleva, V., Kneer, N., Bellei, M., Battelli, D., and Lardy, H.A. (1993). *J. Bioener. Biomem.* **25**, 313-321.
427. Lysogenic activity of enhancer caltrin and the influence of phospholipids on its expression. San Agustin, J.T., and Lardy, H.A. (1993). *Biol. Repro.* **49**, 723-729.
428. A study of possible involvement of glycerol 3-phosphate and malate shuttles in the regulation of thermogenesis. Bobyleva, V., Bellei, M., Battelli, D., and Lardy, H.A. (1994). In *Advances in Free Radicals in Disease* (Tomasi, A., ed.), pp. 281-288.
429. Effects of dehydroepiandrosterone and carnitine treatment on rat liver. Battelli, D., Bellei, M., Kneer, N., Contri, M., Ronchetti, I., Bobyleva, V., and Lardy, H.A. (1994). *Biochem. and Molec. Biol. International* **33**, 1063-1071.
430. Electron microscopic immunolocalization of caltrin proteins in guinea pig seminal vesicles. Coronel, C.E., Maldonado, C., Aoki, A., and Lardy, H.A. (1995). *Archives of Andrology*, **35**, 233-246.
431. Ergosteroids: Induction of thermogenic enzymes in livers of rats treated with steroids derived from dehydroepiandrosterone. Lardy, H.A., Partridge, B., Kneer, N., and Wei, Y. (1995). *Proc. Natl. Acad. Sci.* **92**, 6617-6619.
432. Induction of thermogenic enzymes by DHEA and its metabolites. Lardy, H.A., Kneer, N., Bellei, M., and Bobyleva, V. (1995). *Ann. N.Y. Acad. Sci.* **774**, 171-179.
433. Steroidal allylic fluorination using diethylaminosulfur trifluoride: A convenient method for the synthesis of 3β-acetoxy-7α- and 7β-fluoroandrost-5-en-17-one. Marwah, P., Thoden, J.B., Powell, D.R., and Lardy, H.A. (1996). *Steroids* **61**, 453-460.
434. Effects of the ergosteroid 7-oxo-dehydroepiandrosterone on mitochondrial membrane potential: Possible relationship to thermogenesis. Bobyleva, V., Bellei, M., Kneer, N., and Lardy, H.A. (1997). *Arch. Biochem. Biophys.* **341**, 122-128.
435. Ergosteroids II: Biologically active metabolites and synthetic derivatives of dehydroepiandrosterone. Lardy, H.A., Kneer, N., Wei, Y., Partridge, B., and Marwah, P. (1998). *Steroids* **63**, 158-165.
436. Δ^5-Androstenediol is a natural hormone with androgenic activity in human prostate cancer cells. Miyamoto, H., Yeh, S., Lardy, H.A., Messing, E., and Chang, C. (1998). *Proc. Natl. Acad. Sci. USA* **95**, 11083-11088.
437. Ergosteroids III. Syntheses and biological activity of seco-steroids related to dehydroepiandrosterone. Reich, I.L., Lardy, H.A., Wei, Y., Marwah, P., Kneer, N., Powell, D.R., and Reich, H.J. (1998). *Steroids* **63**, 542-553.

438. Determinants of disease in the simian immunodeficiency virus-infected rhesus macaque: Characterizing animals with low antibody responses and rapid progression. Dykhuizen, M., Mitchen, J.L., Montefiori, D.C., Thomson, J., Acker, L., Lardy, H.A., and Pauza, C.D. (1998). *Journal of General Virology* **79**, 2461-2467.
439. Acute oral gavage study of 3β-Acetoxyandrostene-7,17-dione (7-oxo-DHEA-acetate) in rats. Lardy, H.A., Henwood, S.M., and Weeks, C.E. (1999). *Biochem. Biophys. Res. Commun.* **254**, 120-123.
440. An escalating dose oral gavage study of 3β-Acetoxyandrost-5-ene-7,17-dione (7-oxo-DHEA-acetate) in rhesus monkeys. Henwood, S.M., Weeks, C.E., and Lardy, H.A. (1999). *Biochem. Biophys. Res. Commun.* **254**, 124-126.
441. Development and validation of a high-performance liquid chromatography assay for the quantitative determination of 7-oxo-dehydroepiandrosterone-3β-sulfate in human plasma. Marwah, A., Marwah, P., and Lardy, H.A. (1999). *Journal of Chromatography B* **721**, 197-205.
442. Flux of the L-Serine metabolism in rat liver: The predominant contribution of serine dehydratase. Xue, H.-H., Fujie, M., Sakaguchi, T., et al. (1999). *J. Biol. Chem.* **274**, 16020-16027.
443. Dehydroepiandrosterone and ergosteroids affect energy expenditure. Lardy, H.A. (1999). From: *Health Promotion and Aging* (R.R. Watson, Ed.), Harwood, Amsterdam.
444. Suppression of Δ5-androstenediol-induced androgen receptor transactivation by selective steroids in human prostate cancer cells. Chang, H.-D., Miyamoto, H., Marwah, P., Lardy, H.A., et al. (1999). *Proc. Natl. Acad. Sci.*, **96**, 11173-11177.
445. The possible role of pyrroloquinoline quinone in redox reactions in rat liver. Bobyleva, V., DiCarlo, G.A., Kneery, N., and Lardy, H.A., pp. 133-145.
446. The effect of 7-oxo-DHEA acetate on memory in young and old C57BL/6 mice. Shi, J., Schulze, S., and Lardy, H.A. (2000). *Steroids* **65**, 124-129.
447. Thyroid hormone and dehydroepiandrosterone permit gluconeogenic hormone responses in hepatocytes. Kneer, N. and Lardy, H.A. (2000). *Arch. Biochem. Biophys.* **375**, 145-153.
448. Short-term hypothermia activates hepatic mitochondrial sn-glycerol-3-phosphate dehydrogenase and thermogenic systems. Bobyleva, V., Pazienza, L., Muscatello, U., Kneer, N., and Lardy, H.A. (2000). *Arch. Biochem. Biophys.* **380**, 367-372.
449. Safety and pharmacokinetic study with escalating doses of 3-acetyl-7-oxo-dehydroepiandrosterone in healthy male volunteers. Davidson, M., Marwah, A., Sawchuk, R.J., et al. (2000). *Clin. Invest. Med.* **23**, 300-310.

450. Liquid chromatography—electrospray ionization mass spectrometric analysis of corticosterone in rat plasma using selected ion monitoring. Marwah, A., Marwah, P., and Lardy, H.A. (2001). *a. J. Chromatog. B*, **757**, 333-342.
451. Ergosteroids IV: Synthesis and biological activity of steroid glucuronosides, ethers, and alkylcarbonates. Marwah, P., Marwah, A., Kneer, N., and Lardy, H.A. (2001). *Steroids* **66**, 581-595.
452. Ketone bodies, potential therapeutic uses. Veech, R.L., Chance, B., Kashiwaya, Y., Lardy, H.A., and Cahill, G.F. (2001). *IUBMB Life* **51**, 241-247.
453. High-performance liquid chromatographic analysis of dehydroepiandrosterone. Marwah, A., Marwah, P., and Lardy, H.A. (2001). *J. Chromatog. A.* **935**, 279-296.
454. Ergosteroids V: Preparation and biological activity of various D-ring derivatives in the 7-oxo-dehydroepiandrosterone series. Reich, I.L., Reich, H.J., Kneer, N., and Lardy, H.A. (2002). *Steroids* **67**, 221-233.
455. Ergosteroids VI: Metabolism of dehydroepiandrosterone by rat liver in vitro: A liquid chromatographic—mass spectrometric study. Marwah, A., Marwah, P., and Lardy, H.A. (2002). *J. Chromatog. B*, **767**, 285-299.
456. Molecular differences caused by differentiation of 3T3-L1 preadipocytes in the presence of either dehydroepiandrosterone (DHEA) or 7-oxo-DHEA. Gomez, F.E., Miyazaki, M., Kim, Y.-C., et al. (2002). *Biochemistry* **41**, 5473-5482.
457. Ergosteroids VII: Perchloric acid-induced transformations of 7-oxygenated steroids and their bio-analytical applications—A liquid chromatographic—mass spectrometric study. Marwah, A., Marwah, P., and Lardy, H.A. (2002). *Bioorg. Chem.* **30**, 233-248.
458. Analysis of ergosteroids VIII: Enhancement of signal response of neutral steroidal compounds in liquid chromatographic—electrospray ionization mass spectrometric analysis by mobile phase additives. Marwah, A., Marwah, P., and Lardy, H.A. (2002). *J. Chromatog. A.*, **964**, 137-151.
459. Generation and characterization of androgen receptor knockout (ARKO) mice: An in vivo model for the study of androgen functions in selective tissues. Yeh, S., Tsai, M.-Y., Xu, Q., et al. (2002). *Proc. Natl. Acad. Sci.*, **99**, 13498-13503.
460. Transformations of DHEA and its metabolites by rat liver. Lardy, H.A., Marwah, A., and Marwah, P. (2002). In: *Lipids*, Vol. 37, pp. 1187-1191, AOCS Press.
461. Happily at Work. Lardy, H.A. (2003). Reflections: Paper in a series commissioned to celebrate the centenary of the J. Biol. Chem. in 2005. *J. Biol. Chem., Centennial, 1905-2005, 100 Years of Biochemistry and Molecular Biology.* **278**, 3499-3509.
462. How I became a biochemist. Lardy, H.A. (2003). *Recollections, IUBMB Life*, Taylor & Francis Health Sciences.

463. Microwave induced selective enolization of steroidal ketones and efficient acetylation of sterols in semisolid state. Marwah, P., Marwah, A., and Lardy, H.A. (2002). *Tetrahedron* **59**, 2273-2287, Elsevier Science Ltd.

464. 3β-Acetoxyandrost-1,5-diene-17-ethylene ketal functions as a potent antiandrogen with marginal agonist activity. Miyamoto, H., Marwah, P., Marwah, A., Lardy, H.A., and Chang, C. (2003). *Proc. Natl. Acad. Sci.* **100**, 4440-4444.

465. Inactivation of androgen receptor coregulator ARA55 inhibits androgen receptor activity and agonist effect of antiandrogens in prostate cancer cells. Rahman, M.M., Miyamoto, H., Lardy, H.A., and Chang, C. (2003). *Proc. Natl. Acad. Sci.* **100**. 5124-5129.

466. Infertility with defective spermatogenesis and hypotestosteronemia in male mice lacking the androgen receptor in Sertoli cells. Chang, C., Chen, Y.-T., Yeh, S.-D., et al. (2004). *Proc. Natl. Acad. Sci.* **101**, 6876-6881.

467. Mitotic and neurogenic effects of dehydroepiandrosterone (DHEA) on human neural stem cell cultures derived from the fetal cortex. Suzuki, M., Wright, L.S., Marwah, P., Lardy, H.A., and Svendsen, C.N. (2004). *Proc. Natl. Acad. Sci.* **101**, 3202-3207.

468. An economical and green approach for the oxidation of olefins to enones. Marwah, P., Marwah, A., and Lardy, H.A. (2004). *Green Chem.* **6**, 570-577.

469. C_{19}-5-ene steroids in nature. Lardy, H.A., Marwah, A., and Marwah, P. (2005). In: *Vitamins and Hormones*, Vol. **71**, pp. 263-299, Elsevier Inc.

470. Development and validation of a high performance liquid chromatography assay for 17a-methyltestosterone in fish feed. Marwah, A., Marwah, P., and Lardy, H.A. (2005). *J. Chromatog. B*, **824**, 107-115.

471. Identification of steroid derivatives that function as potent antiandrogens. Miyamoto, H., Marwah, P., Marwah, A., et al. (2005). *Int. J. Cancer* **117**, 866-872, Wiley-Liss, Inc.

472. A postulated role for 16α-hydroxydehydroepiandrosterone in the prevention of respiratory distress syndrome. Lardy, H.A. (2006). *Medical Hypotheses* **66**, 107-109.

473. C19 steroids as androgen receptor modulators: Design, discovery, and structure-activity relationship of new steroidal androgen receptor antagonists. Marwah, P., Marwah, A., Lardy, H.A., et al. (2006). *Bioorg. Med. Chem.* **14**, 5933-5947.

474. Oligozoospermia with normal fertility in male mice lacking the androgen receptor in testis peritubular myoid cells. Zhang, C., Yeh, S., Chen, Y.-T., et al. (2006). *Proc. Natl. Acad. Sci.* **103**, 17718-17723.

475. Hep27, a member of the short-chain dehydrogenase/reductase family, is an NADPH-dependent dicarbonyl reductase expressed in vascular endothelial tissue. Shafqat, N., Shafqat, J., Eissner, G., et al. (2006). *Cell. Mol. Life Sci.* **63**, 1205-1213.

476. Redox reactions of dehydroepiandrosterone and its metabolites in differentiating 3T3-L1 adipocytes: A liquid chromatographic—mass spectrometric study. Marwah, A., Gomez, F.E., Marwah, P., et al. (2006). *Arch. Biochem. Biophys.* **456**, 1-7.

The Main Scientific Achievements:

A. Discovered that bromobenzene is an effective treatment for selinium poisoning by forming Se-containing mercapturic acid analogs that are excreted in the urine.
B. Egg yolk in buffered solution permits spermatozoa to be stored for more than a week, or to be frozen for decades, with retention of motility and ferility.
C. Found that L-glyceraldehyde inhibits glycolysis by forming L-sorbose-1-phosphate — an inhibitor of hexokinase. This disproved Needham's "non-phosphorylating glycolysis" in embryonic and tumor tissue.
D. Found that both Thyroid hormone as well as dehydroepiandrosterone induce the formation of mitochondrial glycero-phosphate dehydrogenaase and cytosolic malic enzyme. These enzymes catalyze electron transport by a path that generates heat rather than ATP. This explains their anti-obesity action.
E. We found a protein (Caltrin) in semen that regulates calcium function in spermatozoa and thus also regulates fertility. The structure of caltrins from five different species were determined.
F. We discovered that dehydroepiandrosterone (DHEA), the most abundant steroid in humans is converted metabolically to the 7-alpha derivative which in turn is converted to 7-ketoDHEA and the latter forms 7-beta-hydroxy DHEA. The biological specific activity of these steroids increases as they progress along this pathway and we now have two additional metabolites with interesting activities. Since retirement, studies of steroid structure and activity occupies our small research group. In collaboration with Professor Chawnshang Chang we found a new androgen — Androstenediol. It is produced in the adrenals and is responsible for the terminal, fatal, stage of prostate cancer because it is not inhibited by the usual antitestosterone drugs.

Glenn Wade Salisbury
University of Illinois at Urbana-Champaign

1910–1994

1981 — for his outstanding achievements in basic and applied research on artificial insemination.

While working at Cornell University and the University of Illinois, Salisbury did extensive research on semen extenders, number of sperm per insemination, and insemination techniques. His contributions have been pivotal in the resolution of problems limiting successful reproduction following artificial insemination. He is known for his ability to stimulate thinking and interest in his field and for his bold scientific approach to the problems facing the animal industry.

BRIEF DESCRIPTION OF SCIENTIFIC CONTRIBUTIONS:

G. W. Salisbury has been an outstanding leader in basic and applied research in reproductive physiology and artificial insemination for over 40 years. His vision, foresight, and originality in scientific development were dominant factors in the development and growth of the artificial insemination industry. His pioneering research on semen extenders, number of sperm per insemination, use of antibacterial agents, and insemination techniques made major contributions to the successful use of artificial insemination. This practice has made possible the maximum use of genetically superior sires which has markedly increased the farm animal productivity. It has been estimated that increased dairy cattle productivity has resulted, in the United States, in a savings of one billion dollars per year. He has been highly instrumental in leading and developing groups that have made significant contributions to agricultural industry. His service to agriculture cannot be measured by just his awn accomplishments but must include the influence of all who have been associated with him.

ESSENTIAL BIOGRAPHY:

Glenn Wade Salisbury was born at Sheffield, Ohio, in 1910. He received the B. S. degree in agriculture from Ohio State University in 1931 and the Ph.D. degree from Cornell University in 1934. He served as graduate assistant, instructor, assistant professor, associate professor, and professor at Cornell University from 1931 to 1947. He served as Professor and Head, Department of Dairy Science, University of Illinois, fran 1947 to 1969 and then became Associate Dean, and Director of the Illinois Agricultural Experiment Station until his retirement in 1978. He is now Professor Emeritus. He has had an outstanding career in research, teaching and public service. He has been a prolific writer with over 260 scientific and popular articles published.

CITATION

To Professor Glenn W. Salisbury

In appreciation for 22 years of leadership as Head of the Department of Dairy Science at the University of Illinois Urbana-Champaign Campus.

His research accomplishments and foreign travels have given him an international reputation as an extraordinary scientist, scholar, and teacher. His abilities and dedication to serve have resulted in his appointment to numerous local, national, and international committees, as well as to his new position. These services have brought recognition to the department and to the university. His initiative and drive, and his broad experience and understanding have served as a stimulus for conducting and coordinating basic and applied research, and extension activities of the department. His example has been an inspiration to his colleagues. He is warm and considerate on a professional and personal basis. We wish to acknowledge his contributions to the department and consider it a privilege to have had this association with him.

PROFESSORIAL STAFF
DEPARTMENT OF DAIRY SCIENCE
UNIVERSITY OF ILLINOIS
JUNE 11, 1969

Wendell L. Roelofs

New York State Agricultural Experiment Station, College of Agriculture and Life Sciences, Cornell University, Geneva, New York, USA

1982 — for his fundamental chemical and biological research on pheromones and their practical use in insect control.

Wendell L. Roelofs, is the Liberty Hyde Bailey Professor of Insect Biochemistry in the Department of Entomology at the New York State Agricultural Experiment Station in Geneva. Roelofs also has served as chair of the department since 1991 to July 1, 2007.

Roelofs has been instrumental in establishing Cornell as a leader in the field of chemical ecology and in maintaining Cornell's preeminence in the field to this day. Roelofs and those who work in his laboratory have contributed greatly to our understanding and practical use of chemical insect communication systems over the past four decades. He and his co-workers have been key in developing our understanding of biochemical pathways for the synthesis of insect pheromones, male behavioral responses to female-produced pheromones, and the evolution of chemical communication systems. In a very specific context, Roelofs and his team identified over 50 new species-specific pheromones that have been used to construct insect lures that in turn have changed the way in which farmers practice pest management. These lures have been responsible for greatly reducing the use of chemical pesticide in several cropping systems.

Roelofs has made prodigious contributions in the fields of chemical ecology, insect chemical communication systems, insect behavior and physiology, and insect pest management. He is considered the patriarch of the nexus of these subject domains. Roelofs has made important contributions in extension by advocating for and exemplifying linkages among fundamental science, applied research, and extension. He has played important roles in fostering an environment that has led to the department being recognized as one of the top in the field. During his tenure as chair, the department has become a vibrant and successful unit with several new faculty and a clear sense of mission and purpose.

Roelofs has authored nearly 400 publications. His research has been widely acclaimed by his peers, resulting in him being awarded the Wolf Prize for International Agriculture in 1982, and the National Medal of Science in 1983. He is also a member of the National Academy of Sciences.

CURRICULUM VITAE

Education

1964 - Ph.D., Indiana University
1960 - B.S., Central College

Academic Ranks (year achieved)

1976-present - Professor
1969-1976 - Associate Professor
1965-1969 - Assistant Professor

Primary Departmental Program Area

Entomology - Geneva: Fundamental research with applied mission

Areas of Expertise (key words)

Sex pheromones, biosynthesis, behavior, genes, evolution, mating disruption

Professional Experience

1962-1964 - NIH Predoctoral Fellow
1964 - NIH Postdoctoral Fellow, Massachusetts Institute of Technology

Sabbaticals and Study leaves (year, project, location)

1972 TNO, Delft, The Netherlands
1983 DSIR, Auckland, New Zealand (Senior Fellowship from National Research Advisory Council)

Honors and Awards (past and current year)

1973 J. Everett Bussart Memorial Award (Entomological Society of America)
1977 Alexander von Humboldt Award
1978 Outstanding Alumni Award, Central College
1980 ESA Founder's Memorial Award Lecturer
1982 Wolf Prize (Agriculture)
1983 Distinguished Achievement Award, Eastern Branch, ESA
1983 Distinguished Alumnus Award, Indiana University
1983 National Medal of Science
1985 Distinguished Scientist Luncheon, White House
1985 Fellow, American Association for the Advancement of Science
1985 Elected to National Academy of Sciences
1986 USDA Distinguished Service Award
1986 Elected to American Academy of Arts and Sciences
1987 Fellow, Entomological Society of America
1990 Silver Medal, International Society of Chemical Ecology
1994 Sterling Hendricks Award, ACS, Agrochemicals Division
2001 Spencer Award, American Chemical Society
2007 President, International Society of Chemical Ecology
2007 CALS Outstanding Faculty Award

Academic Responsibilities

Research Responsibilities

- *Current Postdoctoral Associates (list names)*
  Paul Robbins
- *Past Postdoctoral Associates*
  Bingye Xue 2003-2006
  Grace Hao, 2000-2002
  Satoshi Nojima, 2000-2003
  Bruce Morris, North Dakota St. 2000-2001
  Aijun Zhang, USDA, Beltsville, MD 1997-2000
  Weitian Liu, Univ. South Dakato Medical School 1998-2002
  Nancy Murray, Hobart and William Smith Colleges 2000-2001
  Prior to 5 years ago from 1969:
  Thomas Baker, Jan Kochansky, Jim Miller, Ring Cardé, Ashok Tamhankar
  Lou Bjostad, Henry Arn, Jim Tette, Ada Hill, Walter Wolf, Ritzuo Nishida,
  Boris Kovalev, Jia-Wei Du, Stephen Foster, Ralph Charlton, Stuart Krasnoff,
  Russell Jurenka, Peter Ma,
- *Other Current Research Professionals Supervised (list names)*
  Charles Linn, Senior Research Associate

Graduate Field Membership

Field of Entomology
Field of Neurobiology and Behavior

Other Current Professional Activities

Professional Societies

American Chemical Society
Entomological Society of America
Sigma Xi
American Association for the Advancement of Science
National Academy of Sciences
American Academy for Arts and Sciences
International Society for Chemical Ecology
Asia-Pacific Association for Chemical Ecology

Professional Honoraries

1985	Honorary Doctor of Science, Central College
1988	Honorary Doctor of Science, Hobart & William Smith Colleges, Geneva, NY
1988	Honorary Doctor of Science, Indiana University
1989	Honorary Doctor of Philosophy, University of Lund, Sweden
1989	Honorary Doctor of Science, Free University of Brussels, Belgium

Editorial Boards

Journal of Insect Physiology
Insect Biochemistry and Molecular Biology
Journal of Chemical Ecology

Committee Assignments

- *International/National:*

2000-03	Secretary, Class VI, National Academy of Sciences
1999	Chairman, Section B, National Entomology Society of America
1998	Vice Chairman, National Entomology Society of America
1997	Secretary, National Entomology Society of America

- *University:*

2005	Chairman, CALS Research and Extension Awards Committee
2003	Geneva Director Search Committee
2000-present	Chairman, Facility Use Committee at Geneva

Other Current Professional Contributions

Assignments and Panels:

1976	Delegation on Insect Control to People's Republic of China
1976	Committee on Biology of Pest Species, National Research Council
1977	AIBS-EPA Pheromones Task Group
1977-80	Steering Committee for Research Grants to University of New Brunswick for Research and Development of Spruce Budworm Pheromone
1977	Delegation to Japan on Insect Pheromones and their applications
1978	EPA Ad Hoc Committee for Registration of Non-Biocidals
1978	US-USSR Scientific Exchange Delegation, Tbilisi, USSR
1980	US-USSR Scientific Exchange Conference, Tashkent, USSR
1987	US-New Zealand Exchange Program (Auckland, NZ, 1 month)
1989-92	Center for Insect Science, University of Arizona, Advisory Committee
1989-92	Boyce Thompson Institute Research Oversight Committee
1989	NSF Cellular Biochemistry Review Panel
1990-	Cornell Biotechnology Scientific Administrative Board
1991-	Chairman, Department of Entomology, Cornell University - Geneva
1992	University of California - Berkeley, Department of Natural Resources, External Review Committee
1993	Center of the Environment, Cornell University, Board of Advisers
1994	Dean's Search Committee, Cornell University
1995	Class Membership Committee, National Academy of Sciences
1996	NRI Review Panel - biological control
2000	NRI Panel Manager – Entomology/Nematology
2003	External Review for RAMP grant in Western States

LIST OF PUBLICATIONS

1. Campaigne, E., D. R. Maulding, and W. L. Roelofs. 1964. Ring closure of ylidenemalononitriles. III. Formation of six-membered rings and related chemistry. J. Org. Chem. 29: 1543-1549.
2. Campaigne, E. and W. L. Roelofs. 1965. Ring closure of ylidenemalononitriles. IV. Attempted cyclizations of saturated malononitrile derivatives. J. Org. Chem. 30: 396-400.
3. Campaigne, E. and W. L. Roelofs. 1965. Reductions of 2-carboxamidotetrahydroacenaphthenone derivatives. J. Org. Chem. 30: 2610-2614.
4. Campaigne, E., W. L. Roelofs, and R. F. Weddleton. 1966. Succinimido [3,4-b]-3a,4,5,6-tetrahydroacenaphthen-10-ones. U. S. Pat. 3, 227,729, Jan. 4.

5. House, H. O., W. L. Roelofs, and B. M. Trost. 1966. The chemistry of carbanions. XI. Michael reactions with 2-methylcyclopentanone and 2-methylcyclohexanone. J. Org. Chem. 31: 646-655.
6. Lloyd, W. G. and W. L. Roelofs. 1966. Hydroperoxybenzylpyridinium salts. U. S. Pat. 3,234,230, Feb. 8.
7. Campaigne, E., W. L. Roelofs, and R. F. Weddleton. 1966. Some bridgehead-substituted tetrahydroacenaphthenones. J. Med. Chem. 9: 779-781.
8. Lloyd, W. G. and W. L. Roelofs. 1966. Cationic hydroperoxides. U. S. Pat. 3,262,010, Nov. 1.
9. Roelofs, W. L. 1966. Sex attractants used to combat insects. Farm Research 32(2): 2-3.
10. Roelofs, W. L., D. W. Pulver, K. C. Feng, and F. L. Gambrell. 1967. Attractancy tests with chromatographed European chafer extracts. J. Econ. Entomol. 60: 869-870.
11. Roelofs, W. L. 1967. Possibilities of sex attractants in controlling fruit pests. Proc. NYS Hort. Soc. 112: 163-166.
12. Roelofs, W. L. 1967. Agarless medium for mass rearing red-banded leafroller. J. Econ. Entomol. 60: 1477-1478.
13. Roelofs, W. L. 1967. Sex attractants for insect control. Health News 44: 4-9.
14. Roelofs, W. L. and K. C. Feng. 1967. Isolation and bioassay of the sex pheromone of the red-banded leaf roller, *Argyrotaenia velutinana* (Lepidoptera: Tortricidae). Ann. Entomol. Soc. Amer. 60: 1199-1203.
15. Roelofs, W. L. and W. G. Lloyd. 1967. Halomethyl dimethylbenzyl hydroperoxides. U. S. Pat. 3,358,038, Dec. 12.
16. Roelofs, W. L. and K. C. Feng. 1968. Sex pheromone specificity tests in the Tortricidae - an introductory report. Ann. Entomol. Soc. Amer. 61: 312-316.
17. Campaigne, E., W. L. Roelofs, and R. F. Weddleton. 1968. 3a,4,5,6 tetra-hydrosuccinimido13,4-blacenaphthen-10-one. Potent anticonvulsant. J. Med. Chem. 11: 395-396.
18. Roelofs, W. L. and H. Arn. 1968. Red-banded leafroller sex attractant characterized. New York's Food and Life Sciences 1(1): 13.
19. Bronskill, J. F., W. L. Roelofs, P. J. Chapman, and S. E. Lienk. 1968. The sex pheromone as a taxonomic principle. Inst. Ent. Soc. abstr., Moscow.
20. Roelofs, W. L. and H. Arn. 1968. Sex attractant of the red-banded leafroller moth. Nature 219 (5153): 513.
21. Campaigne, E. E., W. L. Roelofs, and R. F. Weddleton. 1968. Dialkylaminoalkyl 1-oxo-2a,3,4,5-tetrahydroacenaphthen-2a-carboxylates. U. S. Pat. 3,391,178, July 2.

22. Roelofs, W. L. and A. Comeau. 1968. Sex pheromone perception. Nature 220: 600-601.
23. Roelofs, W. L. and A. Comeau. 1969. Sex pheromone specificity - taxonomic and evolutionary. Science 165: 398-400.
24. Roelofs, W. L., A. Comeau, and R. Selle. 1969. Sex pheromone of the Oriental fruit moth. Nature 224: 723.
25. Roelofs, W. L. and A. Comeau. 1970. Lepidopterous sex attractants discovered by field screening tests. J. Econ. Entomol. 63: 969-974.
26. Roelofs, W. L. and J. Tette. 1970. Sex pheromone of the oblique-banded leafroller. Nature 226: 1172.
27. Roelofs, W. L., E. H. Glass, J. Tette, and A. Comeau. 1970. Sex pheromone trapping for red-banded leafroller control: Theoretical and actual. J. Econ. Entomol. 63: 1162-1167.
28. Dean, R. and W. L. Roelofs. 1970. Synthetic sex pheromone of the red-banded leafroller as a survey tool. J. Econ. Entomol. 63 (2): 584-586.
29. Glass, E. H., W. L. Roelofs, H. Arn, and A. Comeau. 1970. Sex pheromone trapping red-banded leafroller moths and development of a long-lasting polyethylene wick. J. Econ. Entomol. 63(2): 370-373.
30. Roelofs, W. L., E. H. Glass, H. Arn, and A. Comeau. 1970. Modulating sex attractant reception for insect control. NY Food and Life Sciences Quarterly 3(2): 19-21.
31. Taschenberg, E. F. and W. L. Roelofs. 1970. Large-scale rearing of cecropia (Lepidoptera: Saturniidae). Ann. Entomol. Soc. 63: 107-111.
32. Roelofs, W. L. and R. Cardé. 1971. Hydrocarbon sex pheromone in tiger moths (Arctiidae). Science 171: 684-686.
33. Roelofs, W. L. and A. Comeau. 1971. Sex pheromone perception: Synergists and inhibitors for the red-banded leafroller attractant. J. Insect Physiol. 17: 435-449.
34. Roelofs, W. L., A. Comeau, A. Hill, and G. Milicevic. 1971. Sex attractant of the codling moth: Characterization with electroantennogram technique. Science 174: 297-299.
35. Roelofs, W. L. and A. Comeau. 1971. Sex pheromone perception: Electroantennogram responses of the red-banded leafroller moth. J. Insect Physiol. 17: 1969-1982.
36. Weatherston, J., W. L. Roelofs, A. Comeau, and C. J. Sanders. 1971. Studies of physiologically active arthropod secretions. X. Sex pheromone of the eastern spruce budworm, *Choristoneura fumiferana* (Lepidoptera: Tortricidae). Can. Entomol. 103: 1741-1747.
37. Roelofs, W. L. and A. Comeau. 1971. Sex attractants in Lepidoptera. In: "Chemical Releasers in Insects," III, A. S. Tahori [ed.] Gordon and Breach, New York, NY: 91-114.

38. Roelofs, W. L., J. P. Tette, E. F. Taschenberg, and A. Comeau. 1971. Sex pheromone of the grape berry moth: Identification by classical and electroantennogram methods and field tests. J. Insect Physiol. 17: 2235-2243.
39. Roelofs, W. L. 1971. Rapid identification of insect sex attractants. Food and Life Sciences 4: 7-8.
40. Roelofs, W. L., R. Cardé, G. Benz, and G. von Salis. 1971. Sex attractant of the larch bud moth found by electroantennogram method. Experientia 27: 1438.
41. Roelofs, W. L., R. T. Cardé, R. J. Bartell, and P. G. Tierney. 1972. Sex attractant trapping of the European corn borer in New York. Environ. Entomol. 1(5): 606-608.
42. Roelofs, W. L., R. J. Bartell, A. S. Hill, R. T. Cardé, and L. H. Waters. 1972. Codling moth sex attractant — field trials with geometrical isomers. J. Econ. Entomol. 65: 1276-1277.
43. Kuhr, R. J., A. Comeau, and W. L. Roelofs. 1972. Measuring release rates of pheromone analogues and synergists from polyethylene caps. Environ. Entomol. 1(5): 625-627.
44. Roelofs, W. L. 1972. Insect sex attractants. Plants & Gardens 27: 28-30.
45. Sanders, C. J., R. J. Bartell, and W. L. Roelofs. 1972. Field trials for synergism and inhibition of *trans*-11-tetradecenyl, the sex pheromone of the eastern spruce budworm. Can. Dept. Environ. Bimon. Res. Notes 28: 10.
46. Hamstead, E. O., R. E. Dolphin, and W. L. Roelofs. 1972. Field test of virgin females and synthetic sex-lure traps for male redbanded leafrollers. Environ. Entomol. 1: 488-489.
47. Cardé, R. T., W. L. Roelofs, and C. C. Doane. 1972. Natural inhibitor of the gypsy moth sex pheromone. Nature 241: 474-475.
48. Bartell, R. J. and W. L. Roelofs. 1973. Inhibition of sexual response in males of *Argyrotaenia velutinana* (Lepidoptera) by brief exposure to synthetic pheromone or its geometrical isomer. J. Insect Physiol. 19: 655-663.
49. Roelofs, W. L., R. T. Cardé, and J. Tette. 1973. Oriental fruit moth attractant synergists. Environ. Entomol. 2: 252-254.
50. Comeau, A. and W. L. Roelofs. 1973. Sex attraction specificity in the Tortricidae. Ent. Exp. Appl. 16: 191-200.
51. Bartell, R. J. and W. L. Roelofs. 1973. Evidence for natural secondary compounds which modify the response of male redbanded leafroller (*Argyrotaenia velutinana*) to female sex pheromone. Ann. Entomol. Soc. Amer. 66: 481-483.
52. Roelofs, W. L., A. Hill, and A. Comeau. 1973. Dodeca-8, 10-dien-l-ols. British Pat. 1,299,691, April 9.

53. Minks, A. K., W. L. Roelofs, F. J. Ritter, and C. J. Persoons. 1973. Reproductive isolation of two tortricid moth species by different ratios of a two-component sex attractant. Science 180: 1073-1074.
54. Roelofs, W. L., J. Tette, E. F. Taschenberg, and A. Comeau. 1973. Method of attracting and trapping male moths of the species *Paralobesia viteana.* U. S. Pat. 3,743,718, July 3.
55. Roelofs, W. L., J. Kochansky, R. Cardé, H. Arn, and S. Rauscher. 1973. Sex attractant of the grape vine moth, *Lobesia botrana.* Bull. Soc. Entomol. Suisse 46: 71-73.
56. Cardé, R. T. and W. L. Roelofs. 1973. Temperature modification of male sex pheromone response and factors affecting female calling in Holomelina immaculata (Lepidoptera: Arctiidae). Can Entomol. 105: 1505-1512.
57. Roelofs, W. L., J. Kochansky, and R. Cardé. 1974. *Trans*-7-*cis*-9-dodecadien-1-yl acetate. U. S. Pat. 3,845,108, Oct. 29.
58. Roelofs, W. L., A. Comeau, and A. Hill. 1974. *Trans*-8-*trans*-10-dodecadien-1-ol as an attractant. U. S. Pat. 3,852,419, Dec. 3.
59. Roelofs, W. L., A. S. Hill, T. C. Baker, and R. Cardé. 1975. Novel attractant components for males of the tobacco budworm moth. U. S. Pat. 3,917,711, Nov. 4.
60. Roelofs, W. L., J. Kochansky, and R. Cardé. 1974. *Trans*-7-*cis*-9-dodecadien-1-yl acetate. U. S. Pat. 3,834,108, Oct. 29.
61. Taschenberg. E. F., R. T. Cardé, A. Hill, J. P. Tette, and W. L. Roelofs. 1974. Sex pheromone trapping of the grape berry moth. Environ. Entomol. 3: 192-194.56. Cardé, R. T. and W. L. Roelofs. 1973. Temperature modification of male sex pheromone response and factors affecting female calling in *Holomelina immaculata* (Lepidoptera: Arctiidae). Can Entomol. 105: 1505-1512.
62. Smith, R. G., G. E. Daterman, G. D. Daves, Jr., K. D. McMurtrey, and W. L. Roelofs. 1974. Sex pheromone of the European pine shootmoth: Chemical identification and field tests. J. Insect Physiol. 20: 661-668.
63. Roelofs, W., A. Comeau, and A. Hill. 1974. *Trans*-8-*trans*-10-dodecadien-1-ol as an attractant. U. S. Pat. 3,852,419, Dec. 3.
64. Taschenberg, E. F., R. T. Cardé, and W. L. Roelofs. 1974. Sex pheromone mass trapping and mating distruption for control of redbanded leafroller and grape berry moths in vineyards. Environ. Entomol. 3: 239-242.
65. Trammel, K., W. L. Roelofs, and E. H. Glass. 1974. Sex pheromone trapping of males for control of redbanded leafroller in apple orchards. J. Econ. Entomol. 67: 159-164.
66. Hill, A., R. Cardé, A. Comeau, W. Bode, and W. L. Roelofs. 1974. Sex pheromones of the tufted apple bud moth (*Platynota ideausalis*). Environ. Entomol. 3: 249-252.

67. Roelofs, W. L. and R. Cardé. 1974. Sex pheromones in the reproductive isolation of Lepidopterous species. In: "Pheromones," M. C. Birch [ed.] Elsevier Pub. Co., pp. 96-114.
68. Roelofs, W. L., A. S. Hill, R. T. Cardé, and T. C. Baker. 1974. Two sex pheromone components of the tobacco budworm moth, *Heliothis virescens*. Life Sciences 14: 1555-1562.
69. Roelofs, W. L. and R. T. Cardé. 1974. Oriental fruit moth and lesser appleworm attractant mixtures refined. Environ. Entomol. 3: 586-588.
70. Cardé, R. T., C. Doane, and W. L. Roelofs. 1974. Diel rhythms of male sex pheromone response and female attractiveness in the gypsy moth. Can. Entomol. 106: 479-484.
71. Roelofs, W. L., A. Hill, R. T. Cardé, J. Tette, H. Madsen, and J. Vakenti. 1974. Sex pheromones of the fruittree leafroller moth, *Archips argyrospilus*. Environ. Entomol. 3: 747-751.
72. Persoons, C. J., A. K. Minks, S. Voerman, W. L. Roelofs, and F. J. Ritter. 1974. Sex pheromones of the moth, *Archips podana*: Isolation, identification and field evaluation of two synergistic geometrical isomers. J. Insect Physiol. 20: 1181-1188.
73. Minks, A. K., W. L. Roelofs, E. Schuurmans-van Dijk, C. J. Persoons, and F. J. Ritter. 1974. Electroantennogram responses of two tortricid moths using two-component sex pheromones. J. Insect Physiol. 20: 1659-1665.
74. Hill, A. S., R. T. Cardé, H. Kido, and W. L. Roelofs. 1975. Sex pheromones of the orange tortrix moth, *Argyrotaenia citrana*. J. Chem. Ecol. 1(2): 215-224.
75. Roelofs, W. L., A. S. Hill, and R. T. Cardé. 1975. Sex pheromone components of the redbanded leafroller moth, *Argyrotaenia velutinana*. J. Chem. Ecol. 1: 83-89.
76. Kochansky, J., R. T. Cardé, J. Liebherr, and W. L. Roelofs. 1975. Sex pheromones of the European corn borer in New York. J. Chem. Ecol. 1(2): 225-231.
77. Baker, J. L., A. S. Hill, R. T. Cardé, A. Kurokawa, and W. L. Roelofs. 1975. Sex pheromone field trapping of the omnivorous leafroller, *Platynota stultana*. Environ. Entomol. 4: 90-92.
78. Cardé, R. T., A. Comeau, T. C. Baker, and W. L. Roelofs. 1975. Moth mating periodicity: Temperature regulates the circadian gate. Experientia 31: 46-48.
79. Cardé, R. T., J. Kochansky, J. F. Stimmel, A. G. Wheeler, Jr., and W. L. Roelofs. 1975. Sex pheromones of the European corn borer: *cis* and *trans* responding males in Pennsylvania. Environ. Entomol. 4: 413-414.
80. Cardé, R. T., C. C. Doane, J. L. Granett, and W. L. Roelofs. 1975. Disruption of pheromone communication in the gypsy moth: Behavioral effects of the sex pheromone and its inhibitor. Environ. Entomol. 4(5): 793-796.

81. Cardé, R. T., T. C. Baker, and W. L. Roelofs. 1975. Behavioral role of individual components of a multichemical attractant system in the oriental fruit moth. Nature 253: 348-349.
82. Roelofs, W. L. 1975. Manipulating sex pheromones for insect suppression. Environ. Lett. 8: 41-59.
83. Liebherr, J. and W. L. Roelofs. 1975. Laboratory hybridization and mating period studies using two pheromone strains of *Ostrinia nubilalis*. Ann. Entomol. Soc. Amer. 68: 305-309.
84. Hill, A. and W. L. Roelofs. 1975. Sex pheromone components of the omnivorous leafroller moth, *Platynota stultana*. J. Chem. Ecol. 1: 91-99.
85. Roelofs, W. L. 1975. Insect communication — chemical. In: "Insects, Science and Society," D. Pimentel [ed.] Academic Press, NY: 79-99.
86. Cardé, R. T., K. Trammel, and W. L. Roelofs. 1975. Disruption of sex attraction of the redbanded leafroller (*Argyrotaenia velutinana*) with microencapsulated pheromone components. Environ. Entomol. 4: 448-450.
87. Roelofs, W. L., J. P. Kochansky, R. T. Cardé, G. G. Kennedy, C. A. Henrick, J. N. Labovitz, and V. L. Corbin. 1975. Sex pheromone of the potato tuberworm moth, *Phthorimaea operculella*. Life Sciences 17: 699-706.
88. Roelofs, W. L., J. Kochansky, E. Anthon, R. Rice, and R. Cardé. 1975. Sex pheromone of the peach twig borer moth (*Anarsia lineatella*). Environ. Entomol. 4(6): 580-582.
89. Kochansky, J., J. Tette, E. F. Taschenberg, R. T. Cardé, K. E. Kaissling, and W. L. Roelofs. 1975. Sex pheromone of the moth, *Antheraea polyphemus*. J. Insect Physiol. 21: 1977-1983.
90. Cardé, R. T., T. C. Baker, and W. L. Roelofs. 1975. Ethological function of components of a sex-attractant system for oriental fruit moth males, *Grapholitha molesta* (Lepidoptera: Tortricidae). J. Chem. Ecol. 1: 475-491.
91. Roelofs, W. L., A. S. Hill, T. C. Baker, and R. T. Cardé. 1975. Novel attractant components for males of the tobacco budworm moth. U. S. Pat. 3,917,711, Nov. 4.
92. Roelofs, W. L., J. R. Miller, and T. C. Baker. 1976. Pheromones of lepidopterous insects. In: "Perspectives of Forest Pest Management," Anderson and Kaya [eds.] Academic Press, NY.
93. Roelofs, W. L., A. S. Hill, and T. C. Baker. 1976. Novel attractant components for males of the tobacco budworm moth. U. S. Pat. 3,952,093, Apr. 20.
94. Roelofs, W. L., R. Cardé, E. F. Taschenberg, and R. W. Weires, Jr. 1976. Pheromone research for control of lepidopterous species in New York. In: Beroza, M. ed., ACS Symp. Ser. 23: 75-87.
95. Miller, J. R., T. C. Baker, R. T. Cardé, and W. L. Roelofs. 1976. Reinvestigation of oak leafroller sex pheromone components and the hypothesis that they vary with diet. Science 192: 140-143.

96. Roelofs, W. L., A. Hill, A. Cardé, R. Cardé, H. Madsen, and J. Vakenti. 1976. Sex pheromone of the European leafroller, *Archips rosanus*. Environ. Entomol. 5: 362-364.
97. Roelofs, W. L. 1976. The importance of defining lepidopteran pheromone blends. NYS Agr. Exp. Sta., Geneva. Search Agriculture 6: 1-4.
98. Roelofs, W. L. 1976. Pheromones. In: "The Future for Insecticides: Needs and Prospects," R. L. Metcalf and J. J. McKelvey, Jr. [eds.] Wiley & Sons, Inc., NY, pp. 445-466.
99. Roelofs, W. L., A. Cardé, A. Hill, and R. Cardé. 1976. Sex pheromone of the threelined leafroller. Environ. Entomol. 5: 649-652.
100. Taschenberg, E. F. and W. L. Roelofs. 1976. Pheromone communication disruption of the grape berry moth with microencapsulated and hollow fiber systems. Environ. Entomol. 5: 688-691.
101. Baker, T. C., R. T. Cardé, and W. L. Roelofs. 1976. Behavioral responses of male *Argyrotaenia velutinana* (Lepidoptera: Tortricidae) to components of its sex pheromone. J. Chem. Ecol. 2: 333-352.
102. Comeau, A., R. T. Cardé, and W. L. Roelofs. 1976. Relationship of ambient temperature to diel periodicities of sex attraction in six species of Lepidoptera. Can. Entomol. 108: 415-418.
103. Roelofs, W. L. 1976. Pheromone chemistry: Overview and discussion. In: "Chemical Control of Insect Behavior: Theory and Application," J. J. McKelvey, Jr. and H. Shorey [eds.] Wiley & Sons, NY: 287-297.
104. Baker, T. C. and W. L. Roelofs. 1976. Electroantennogram responses of the male moth *Argyrotaenia velutinana* to mixtures of sex pheromone components of the female. J. Insect Physiol. 22: 1357-1364.
105. Roelofs, W. L. and R. T. Cardé. 1976. Responses of Lepidoptera to synthetic sex pheromone chemicals and their analogs. Ann. Rev. Entomol. 22: 377-399.
106. Roelofs, W. L. 1976. Sex pheromones, hormones and trapping. In: Nat'l Acad. Sci. Report. Insect Control in China.
107. Arn, H., S. Rauscher, H-R. Buser, and W. L. Roelofs. 1976. Sex pheromone of *Eupoecilia ambiguella*: *cis*-9-dodecenyl acetate as a major component. Z. Naturforsch. 31: 499-503.
108. Cardé, R. T., T. C. Baker, and W. L. Roelofs. 1976. Sex attractant responses of male oriental fruit moths to a range of component ratios-pheromone polymorphism? Experientia 32: 1406-1407.
109. Doolittle, R. E., W. L. Roelofs, J. D. Solomon, R. T. Cardé, and M. Beroza. 1976. (*Z,E*)-3,5-tetradecadien-1-ol acetate sex attractant for the carpenterworm moth, *Prionoxystus robiniae* (Peck) (Lepidoptera: Cossidae). J. Chem. Ecol. 2: 399-410.

110. Roelofs, W. L. 1976. Communication disruption by pheromone components. Proc. Symp. "Insect Pheromones and their Applications," Nagaoka and Tokyo, Japan, Dec. 8-11, pp. 123-134.
111. Roelofs, W. L., A. S. Hill, and T. C. Baker. 1976. Novel attractant components for males of the tobacco budworm moth. U. S. Pat. 3,952,093, Apr. 20.
112. Hill, A. S., R. T. Cardé, W. M. Bode, and W. L. Roelofs. 1977. Sex pheromone components of the variegated leafroller moth, *Platynota flavedana*. J. Chem. Ecol. 3: 369-376.
113. Hill, A. S., C. W. Berisford, N. E. Brady, and W. L. Roelofs. 1977. Sex pheromone of the pitch pine tip moth, *Rhyacionia rigidana*. Environ. Entomol. 6: 959-960.
114. Miller, J. R., and W. L. Roelofs. 1977. Sex pheromone titer correlated with pheromone gland development and age in the redbanded leafroller moths, *Argyrotaenia velutinana*. Ann. Entomol. Soc. Amer. 70: 136-139.
115. Miller, J. R., K. Mori, and W. L. Roelofs. 1977. Gypsy moth field trapping and electroantennogram studies with pheromone enantiomers. J. Insect Physiol. 23: 1447-1453.
116. Hill, A. S., W. L. Roelofs, R. W. Rings, and S. R. Swier. 1977. Sex pheromone of the black cutworm, *Agrotis ipsilon* (Hufnagel) (Lepidoptera: Noctuidae). JNY Entomol. Soc. 85: 179-180.
117. Roelofs, W. L. 1977. The scope and limitations of the electroantennogram technique in identifying phermone components. In: "Crop Protection Agents - Their Biological Evaluation," N. R. McFarlane [ed.] Academic Press, NY, pp. 147-165.
118. Cardé, R. T., A. M. Cardé, A. S. Hill, and W. L. Roelofs. 1977. Sex attractant specificity as a reproductive isolating mechanism among the sibling species *Archips argyrospilus* and *mortuanus* and other sympatric tortricine moths (Lepidoptera: Tortricidae). J. Chem. Ecol. 3: 71-84.
119. Taschenberg, E. F., and W. L. Roelofs. 1977. Mating disruption of the grape berry moth, *Paralobesia viteana*, with pheromone released from hollow fibers. Environ. Entomol. 6: 761-763.
120. Feng, K. C., and W. L. Roelofs. 1977. Sex pheromone gland development in redbanded leafroller moth, *Argyrotaenia velutinana*, pupae and adults. Ann. Entomol. Soc. Amer. 70: 721-732.
121. Roelofs, W. L., W. H. Reissig, and R. W. Weires. 1977. Sex attractant for the spotted tentiform leaf miner moth, *Lithocolletis blancardella*. Environ. Entomol. 6: 373-374.
122. Roelofs, W. L., M. J. Gieselmann, A. M. Cardé, H. Tashiro, D. S. Moreno, C. A. Henrick, and R. J. Anderson. 1977. Sex pheromone of the California red scale, *Aonidiella aurantii*. Nature 267: 698.

123. Cardé, R. T. and W. L. Roelofs. 1977. Attraction of redbanded leafroller moths, *Argyrotaenia velutinana*, to blends of (Z)-and (E)-11-tridecenyl acetates. J. Chem. Ecol. 3: 143-149.
124. Cardé, R. T., C. C. Doane, J. Granett, A. S. Hill, J. Kochansky, and W. L. Roelofs. 1977. Attractancy of racemic disparlure and certain analogues to male gypsy moths and the effect of trap placement. Environ. Entomol. 6: 765-767.
125. Baker, J. L., A. S. Hill, and W. L. Roelofs. 1978. Seasonal variations in pheromone trap catches of male omnivorous leafroller moths, *Platynota stultana*. Environ. Entomol. 7: 399-401.
126. Roelofs, W. L. 1978. Threshold hypothesis for pheromone perception. J. Chem. Ecol. 4: 685-699.
127. Cardé, R. T., W. L. Roelofs, R. G. Harrison, A. T. Vawter, P. F. Bussard, A. Mutura, and E. Monroe. 1978. European corn borer: Pheromone polymorphism or sibling species? Science 199: 555-556.
128. Taschenberg, E. F. and W. L. Roelofs. 1978. Male redbanded leafroller moth orientation disruption in vineyards. Environ. Entomol. 7: 103-106.
129. Miller, J. R. and W. L. Roelofs. 1978. Gypsy moth responses to pheromone enantiomers as evaluated in a sustained-flight tunnel. Environ. Entomol. 7: 42-44.
130. Miller, J. R. and W. L. Roelofs. 1978. Sustained-flight tunnel for measuring insect responses to wind-borne sex pheromones. J. Chem. Ecol. 4: 187-198.
131. Roelofs, W. L., M. Gieselmann, A. Cardé, H. Tashiro, D. Moreno, C. Henrick, and R. Anderson. 1978. Identification of the California red scale sex pheromone. J. Chem. Ecol. 4: 211-224.
132. Kochansky, J. P., W. L. Roelofs, and P. Sivapalan. 1978. Sex pheromone of the tea tortrix moth (*Homona coffearia* Neitner). J. Chem. Ecol. 4: 623-631.
133. Roelofs, W. L. 1978. Chemical Control of Insects by Pheromones. In: "Biochemistry of Insects," M. Rockstein [ed.] Academic Press, NY, pp. 419-464.
134. Weires, R. W., W. L. Roelofs, and R. E. Robinson. 1978. Pheromone disruption of leafrollers in the Hudson Valley, 1976-1977. NYS Hort. Soc. Proc. 123: 130-135.
135. Reissig, W. H., M. Novak, and W. L. Roelofs. 1978. Orientation disruption of *Argyrotaenia velutinana* and *Choristoneura rosaceana* male moths. Environ. Entomol. 7: 631-632.
136. Hill, A. S. and W. L. Roelofs. 1978. Two sex attractants for male speckled green fruit worm moths, *Orthonia hibisci* Guenee (Lepidoptera: Noctuidae). J.N.Y. Entomol. Soc. 4: 296.
137. Novak, M. A., W. H. Reissig, and W. L. Roelofs. 1978. Orientation disruption of *Argyrotaenia velutinana* and *Choristoneura rosaceana* (Lepidoptera: Tortricidae) male moths. J.N.Y. Entomol. Soc. 4: 311-312.

138. Fein, B. L., W. H. Reissig, and W. L. Roelofs. 1978. Attraction of apple maggot *Rhagoletis pomonella* (Walsh) (Diptera: Tephritidae) females to apple volatiles in wind tunnel bioassays. J.N.Y. Entomol. Soc. 4: 286.
139. Hill, A. S. and W. L. Roelofs. 1979. Sex pheromone components of the obliquebanded leafroller, *Choristoneura rosaceana*. J. Chem. Ecol. 5: 3-11.
140. Gieselmann, M., D. J. Moreno, J. Fargerlund, H. Tashiro, and W. L. Roelofs. 1979. Identification of the sex pheromone of the yellow scale. J. Chem. Ecol. 5: 25-31.
141. Tashiro, H., M. J. Gieselmann, and W. L. Roelofs. 1979. Residual activity of a California red scale synthetic pheromone component. Environ. Entomol. 8: 931-934.
142. Hill, A. S., R. W. Rings, S. R. Swier, and W. L. Roelofs. 1979. Sex pheromone of the black cutworm moth, Agrotis ipsilon. J. Chem. Ecol. 5: 439-457.
143. Gieselmann, M. J., R. E. Rice, R. A. Jones, and W. L. Roelofs. 1979. Sex pheromone of the San Jose scale. J. Chem. Ecol. 5: 891-900.
144. Roelofs, W. L. 1979. Pheromone perception in Lepidoptera. In: "Neurotoxicology of Insecticides and Pheromones," T. Narahashi, [ed.], Plenum Press, NY, pp. 5-25.
145. Roelofs, W. L. 1979. Production and perception of lepidopterous pheromone blends. In: "Chemical Ecology: Odour Communication in Animals," F. J. Ritter [ed.] Elsevier Press, NY, Pages 159-168, 427 pp.
146. Roelofs, W. L., A. S. Hill, and C. W. Berisford. 1979. Sex pheromone of the subtropical pine tip moth, *Rhyacionia subtropica*. Environ. Entomol. 8: 894-895.
147. Roelofs, W. L. 1979. Electroantennograms. Chemtech. 9: 222-227.
148. Roelofs, W. L. 1979. Establishing efficacy of sex attractants and disruptants for insect control. Entomol. Soc. Amer., 97 pp.
149. Roelofs, W. L. 1979. Pheromones and other chemical attractants: pheromone trap specificity and potency. In: "Movement of Highly Mobile Insects: Concepts and Methodology in Research," R. L. Rabb and G. G. Kennedy [eds.] University Graphica, North Carolina State University, Raleigh, NC, pp. 272-285.
150. Gieselmann, M. J., C. A. Henrick, R. J. Anderson, D. S. Moreno, and W. L. Roelofs. 1980. Responses of male California red scale to sex pheromone isomers. J. Insect Physiol. 26: 179-183.
151. Svirskaya, P. I., C. C. Leznoff, and W. L. Roelofs. 1980. A stereoselective synthesis of a *cis,cis* conjugated dienol, a candidate pheromone. Syn. Comm. 10: 391-397.
152. Roelofs, W. L., A. J. Tamhankar, A. Comeau, A. S. Hill, and E. F. Taschenberg. 1980. Moth activity periods for uglynest caterpillar, *Archips cerasivoranus* and identification of its sex pheromone. Ann. Entomol. Soc. Amer. 73: 631-634.

153. Hill, A. S. and W. L. Roelofs. 1980. A female-produced sex pheromone component and attractant for males in the common armyworm moth, *Pseudoletia unipuncta*. Environ. Entomol. 9: 408-411.
154. Bjostad, L. B., E. F. Taschenberg, and W. L. Roelofs. 1980. Sex pheromone of the woodbine leafroller moth, *Sparganothis* spp. J. Chem. Ecol. 6: 797-804.
155. Bjostad, L. B., E. F. Taschenberg, and W. L. Roelofs. 1980. Sex pheromone of the chokecherry leafroller moth, *Sparganothis directana*. J. Chem. Ecol. 6: 487-498.
156. Bjostad, L. B. and W. L. Roelofs. 1980. An inexpensive electronic device for measuring electroantennogram responses to sex pheromone components with a voltmeter. Physiol. Entomol. 5: 309-314.
157. Miller, J. R. and W. L. Roelofs. 1980. Individual variation in sex pheromone component ratios in two populations of the redbanded leafroller moth, *Argyrotaenia velutinana*. Environ. Entomol. 9: 359-363.
158. Roelofs, W. L. 1980. Pheromones and Their Chemistry. In: "Insect Biology in the Future," M. Locke and D. S. Smith [eds.] Academic Press, Inc., NY, pp. 583-602.
159. Roelofs, W. L. and M. A. Novak. 1980. Small-plot disorientation tests for screening potential mating disruptants. In: "Management of Insect Pests with Semiochemicals: Concepts and Practice," E. R. Mitchell [ed.] Plenum Pub. Co., NY.
160. Roelofs, W. L. 1980. Attractive and aggregating pheromones. In: "Semiochemicals," D. A. Nordlund [ed.] John Wiley & Sons, NY.
161. Roelofs, W. L. 1980. Developing the potential of lepidopterous pheromones in insect control. In: "Environmental Protection and Biological Forms of Control of Pest Organisms," B. Lundholm and M. Stackerud [eds.] Ecol. Bull. (Stockholm).
162. Hill, A. S., C. W. Berisford, U. E. Brady, and W. L. Roelofs. 1981. The Nantucket pine tip moth, *Rhyacionia frustrana*: Identification of two sex pheromone components. J. Chem. Ecol. 7: 517-527.
163. Bjostad, L. B., W. A. Wolf, and W. L. Roelofs. 1981. Total lipid analysis of the sex pheromone gland of the redbanded leafroller moth, *Argyrotaenia velutinana*, with reference to pheromone biosynthesis. Insect Biochem. 11: 73-79.
164. Wolf, W. A., L. B. Bjostad, and W. L. Roelofs. 1981. Biosynthesis of insect sex pheromones: Fatty acid content of pheromone glands of several lepidoptera species. Environ. Entomol. 10: 943-946.
165. Baker, T. C. and W. L. Roelofs. 1981. Initiation and termination of Oriental fruit moth male response to pheromone concentrations in the field. Environ. Entomol. 10: 211-218.

166. Baker, T. C., W. Meyer, and W. L. Roelofs. 1981. Sex pheromone dosage and blend discrimination by Oriental fruit moth males. Ent. exp. appl. 30: 269-279.

167. Linn, C. E. and W. L. Roelofs. 1981. Modification of sex pheromone blend discrimination in male Oriental fruit moths by pre-exposure to (E)-8-dodecenyl acetate. Physiol. Entomol. 6: 421-429.

168. Suguro, T., W. L. Roelofs, and K. Mori. 1981. Stereoselective synthesis of (+)-(E)-3,9-dimethyl-1-isopropyl-5,8-decadienyl acetate, the racemate of the yellow scale pheromone, and its (Z)-isomer. Agric. Biol. Chem. 45: 2509-2514.

169. Bjostad, L. B., E. Gurevitz, S. Gothilf, and W. L. Roelofs. 1981. Sex attractant for the honeydew moth, *Cryptoblabes gridiella*. Phytoparasi. 9: 95-99.

170. Roelofs, W. L. 1981. ESA Founders' Memorial Award Lecture: Pheromones, Plateaus and Platitudes. Bull. Entomol. Soc. Amer. 27: 3-7.

171. Hill, A. S. and W. L. Roelofs. 1981. Sex pheromone of the saltmarsh caterpillar moth, *Estigmene acrea*. J. Chem. Ecol. 7: 655-668.

172. Clement, S. L., A. S. Hill, E. Levine, and W. L. Roelofs. 1981. Trap catches of male *Agrotis ipsilon* with synthetic sex pheromone emitted for different dispensers. Environ. Entomol. 10: 521-523.

173. Bjostad, L. B. and W. L. Roelofs. 1981. Sex pheromone biosynthesis from radiolabeled fatty acids in the redbanded leafroller moth. J. Biol. Chem. 256: 7936-7940.

174. Anderson, R. J., M. J. Gieselmann, H. R. Chinn, K. G. Adams, C. A. Henrick, R. E. Rice, and W. L. Roelofs. 1981. Synthesis of identification of a third component of the San Jose scale sex pheromone. J. Chem. Ecol. 7: 695-706.

175. Baker, T. C., R. Nishida, and W. L. Roelofs. 1981. Close-range attraction of female Oriental fruit moths to herbal scent of male hairpencils. Science 214: 1359-1361.

176. Hill, A. S., B. G. Kovalev, L. N. Nikolaeva, and W. L. Roelofs. 1982. Sex pheromone of the fall webworm, *Hyphantria cunea*. J. Chem. Ecol. 8: 383-396.

177. Nishida, R., T. C. Baker, and W. L. Roelofs. 1982. Hairpencil pheromone components of male Oriental fruit moths, *Grapholitha molesta*. J. Chem. Ecol. 8: 947-959.

178. Fein, B. L., W. H. Reissig, and W. L. Roelofs. 1982. Identification of apple volatiles attractive to apple maggot flies, Rhagoletis pomonella. J. Chem. Ecol. 8: 1473-1487.

179. Meyer, W. L., G. L. DeBarr, C. W. Berisford, L. R. Barber, and W. L. Roelofs. 1982. Identification of the sex pheromone of the webbing coneworm moth, *Dioryctria disclusa* (Lepidoptera: Pyralidae). Environ. Entomol. 11: 986-988.

180. Roelofs, W. L., M. J. Gieselmann, K. Mori, and D. Moreno. 1982. Sex pheromone chirality comparison between sibling species - California red scale and yellow scale. Naturwissenschaften 69: 348.
181. Roelofs, W. L., A. S. Hill, C. E. Linn, Jr., J. Meinwald, S. C. Jain, H. J. Herbert, and R. F. Smith. 1982. Sex pheromone of the winter moth, a geometrid with unusually low temperature precopulatory responses. Science 217: 657-659.
182. Roelofs, W. L. and R. L. Brown. 198d2. Pheromones and evolutionary relationships of tortricidae. Ann. Rev. Ecol. Syst. 13: 395-422.
183. Reissig, W. H., B. L. Fein, and W. L. Roelofs. 1982. Field tests of synthetic apple volatiles as apple maggot (Diptera: Tephritidae) attractants. Environ. Entomol. 11: 1294-1298.
184. Turgeon, J. J., J. N. McNeil, and W. L. Roelofs. 1983. Responsiveness of *Pseudaletia unipuncta* males to the female sex pheromone. Physiol. Entomol. 8: 339-344.
185. Turgeon, J. J., J. N. McNeil, and W. L. Roelofs. 1983. Field testing of various parameters for the development of a pheromone-based monitoring system for the armyworm *Pseudaletia unipuncta* (Haw.) (Lepidoptera: Noctuidae). Environ. Entomol. 12: 891-894.
186. Bjostad, L. B. and W. L. Roelofs. 1983. Sex pheromone biosynthesis in *Trichoplusia ni*: key steps involve delta-11 desaturation and chain shortening. Science 220: 1387-1389.
187. Jain, S. C., W. L. Roelofs, and J. Meinwald. 1983. Synthesis of a sex attractant pheromone from a geometrid moth, *Operophtera brumata* (the winter moth). J. Org. Chem. 48: 2274-2275.
188. Linn, C. E. and W. L. Roelofs. 1983. Effect of varying proportions of the alcohol component on sex pheromone blend discrimination in male Oriental fruit moths. Physiol. Entomol. 8: 291-306.
189. Wolf, W. A. and W. L. Roelofs. 1983. A chain-shortening reaction in orange tortrix sex pheromone biosynthesis. Insect Biochem. 13: 375-379.
190. Roelofs, W. L. 1984. Electroantennogram Assays: Rapid and convenient screening procedures for pheromones. In: "Techniques in Pheromone Research," H. E. Hummel [ed.], in series "Experimental Entomology," T. A. Miller [Ed.], Springer, N.Y. pp. 131-159.
191. Linn, C. E. and W. L. Roelofs. 1984. Sublethal effects of neuroactive compounds on pheromone response thresholds in male Oriental fruit moths. Arch. Insect Biochem. Physiol. 1: 331-344.
192. Linn, C. E., Jr., L. B. Bjostad, J.-W. Du, and W. L. Roelofs. 1984. Redundancy in a chemical signal: Behavioral responses of male *Trichoplusia ni* to a 6-component sex pheromone blend. J. Chem. Ecol. 10: 1635-1658.

193. Bjostad, L. B. and W. L. Roelofs. 1984. Biosynthesis of sex pheromone components and glycerolipid precursors from sodium [1-^{14}C] acetate in redbanded leafroller moth. J. Chem. Ecol. 10: 681-691.
194. Bjostad, L. B. and W. L. Roelofs. 1984. Sex pheromone biosynthetic precursors in *Bombyx mori.* Insect Biochem. 14: 275-278.
195. Roelofs, W. L. and L. B. Bjostad. 1984. Biosynthesis of Lepidopteran pheromones. Bioorg. Chem. 12: 279-298.
196. Meyer, W. L., R. S. Cameron, A. Tamhankar, G. DeBarr, C. W. Berisford, and W. L. Roelofs. 1984. Sex pheromone of the blister coneworm moth, *Dioryctria clarioralis* (Lepidoptera: Pyralidae). Environ. Entomol. 13: 854-858.
197. Bjostad, L. B., C. E. Linn, Jr., J.-W. Du, and W. L. Roelofs. 1984. Identification of new sex pheromone components in *Trichoplusia ni,* predicted from biosynthetic precursors. J. Chem. Ecol. 10: 1309-1323.
198. Reissig, W. H., R. W. Weires, C. G. Forshey, W. L. Roelofs, R. C. Lamb, H. S. Aldwinckle, and S. R. Alm. 1984. Management of the apple maggot, *Rhagoletis pomonella* (Walsh) (Diptera: Tephritidae) in disease-resistant dwarf and semi-dwarf apple trees. Environ. Entomol. 13: 684-690.
199. Du, J.-W., C. E. Linn, Jr., and W. L. Roelofs. 1984. Artificial selection for new pheromone strains of redbanded leafroller moths, *Argyrotaenia velutinana.* Contr. Shanghai Inst. Entomol. 4: 21-30.
200. Bjostad, L. B., C. E. Linn, Jr., W. L. Roelofs, and J.-W. Du. 1985. Identification of new sex pheromones in *Trichoplusia ni* and *Argyrotaenia velutinana,* predicted from biosynthetic precursors. In: "Semiochemistry: Flavors and Pheromones," T. E. Acree and D. M. Soderlund [Eds.] Walter de Gruyter and Co., Berlin, New York, pp. 223-237.
201. Roelofs, W. L., J.-W. Du, X.-H. Tang, P. S. Robbins, and C. J. Eckenrode. 1985. Three European corn borer populations in New York based on sex pheromones and voltinism. J. Chem. Ecol. 11: 829-836.
202. Novak, M. A. and W. L. Roelofs. 1985. Behavior of male redbanded leafroller moth (*Argyrotaenia velutinana*) in small disruption plots. Environ. Entomol. 14: 12-16.
203. Silk, P. J., L. P. S. Kuenen, S. H. Tan, W. L. Roelofs, C. J. Sanders and A. R. Alford. 1985. Identification of sex pheromone components of jack pine budworm, *Choristoneura pinus pinus* Freeman. J. Chem. Ecol. 11: 159-167.
204. Linn, C. E., Jr. and W. L. Roelofs. 1985. Multiple effects of octopamine and chlordimeform on pheromone response thresholds in the cabbage looper moth, *Trichoplusia ni.* Pestic. Sci. 16: 445-446.
205. Linn, C. E., Jr. and W. L. Roelofs. 1985. Multiple effects of octopamine and chlordimeform on pheromone response thresholds in the cabbage looper moth. [Abstract] Neurotox '85, Neuropharmacology and Pesticide Action. Soc. of Chem. Ind., London.

206. Linn, C. E., Jr. and W. L. Roelofs. 1985. Response specificity of male pink bollworm moths to different blends and dosages of sex pheromone. J. Chem. Ecol. 11: 1583-1590.
207. Reissig, W. H., B. H. Stanley, W. L. Roelofs, and M. R. Schwarz. 1985. Tests of synthetic apple volatiles in traps as attractants for apple maggot flies (Diptera: Tephritidae) in commercial apple orchards. Environ. Entomol. 14: 55-59.
208. Löfstedt, C. and W. L. Roelofs. 1985. Sex pheromone precursors in two primitive New Zealand tortricid moth species. Insect Biochem. 15: 729-734.
209. Glass, E. H. and W. L. Roelofs. 1985. *Argyrotaenia velutinana.* In: "Handbook of Insect Rearing," Vol. 2, Elsevier Science Publishers, BV Amsterdam. pp. 197-205.
210. Linn, C. E., Jr. and W. L. Roelofs. 1986. Neuropharmacological effects on olfactory perception. In: "Mechanisms of Insect Olfaction." T. L. Payne, M. C. Birch, and C. E. J. Kennedy [Eds.] Oxford Univ. Press, pp. 209-215.
211. Linn, C. E., Jr., M. G. Campbell and W. L. Roelofs. 1986. Male moth sensitivity to multicomponent pheromones: The critical role of the female released blend in determining the functional role of components and the active space of the pheromone. J. Chem. Ecol. 12: 659-668.
212. Foster, S. P., J. R. Clearwater, S. J. Muggleston and W. L. Roelofs. 1986. Probable sibling species complexes within two described New Zealand leafroller moths. Naturwissenschaften 73: 156-158.
213. Wolf, W. A. and W. L. Roelofs. 1986. Properties of the "11-desaturase enzyme used in cabbage looper moth sex pheromone biosynthesis. Arch. Insect Biochem. & Physiol. 3: 45-52.
214. Bjostad, L. B. and W. L. Roelofs. 1986. Sex pheromone biosynthesis in the redbanded leafroller moth, studied by mass-labeling with stable isotopes and analysis with mass spectrometry. J. Chem. Ecol. 12(2): 431-450.
215. Linn, C. E., Jr. and W. L. Roelofs. 1986. Modulatory effects of octopamine and serotonin on thresholds for male sensitivity and periodicity of response to sex pheromone in the cabbage looper, *Trichoplusia ni.* Arch. Insect Biochem. Physiol. 3: 161-172.
216. Meyer, W. L., G. L. DeBarr, J. L. Hanula, B. Kovalev, R. S. Cameron, C. W. Berisford, and W. L. Roelofs. 1986. Z-11-Hexadecenyl acetate, a sex pheromone component for the southern pine coneworm, *Dioryctria amatella* (Lepidoptera: Pyralidae). Environ. Entomol. 15: 316-320.
217. Roelofs, W. L., J. W. Du, C. E. Linn, Jr., T. J. Glover, and L. B. Bjostad. 1987. The potential for genetic manipulation of the redbanded leafroller moth sex pheromone gland. In: "Evolutionary Genetics of Invertebrate Behavior," M. D. Huettel [Ed.] Plenum Pub. Corp., New York, pp. 263-272.
218. Glover, T. J., X.-H. Tang, and W. L. Roelofs. 1987. Sex pheromone blend discrimination by male moths from E and Z strains of European corn borer. J. Chem. Ecol. 13: 143-151.

219. Foster, S. P. and W. L. Roelofs. 1987. Sex pheromone differences in populations of the brownheaded leafroller, *Ctenopseustis obliquana.* J. Chem. Ecol. 13: 623-629.
220. Foster, S. P., J. R. Clearwater and W. L. Roelofs. 1987. Sex pheromone identification of a *Planotortrix* species found on mangrove. J. Chem. Ecol. 13: 631-637.
221. Carle, S. A., A. L. Averill, G. S. Rule, W. H. Reissig, and W. L. Roelofs. 1987. Variation in host fruit volatiles attractive to apple maggot fly, *Rhagoletis pomonella.* J. Chem. Ecol. 13: 795-805.
222. Krasnoff, S. B., L. B. Bjostad, and W. L. Roelofs. 1987. Quantitative and qualitative variation in male pheromones of *Phragmatobia fulinginosa* and *Pyrrharctia isabella* (Lepidoptera: Arctiidae). J. Chem. Ecol. 13: 807-822.
223. Wolf, W. A. and W. L. Roelofs. 1987. Reinvestigation confirms action of "11-desaturases in spruce budworm moth sex pheromone biosynthesis. J. Chem. Ecol. 13: 1019-1027.
224. Linn, C. E., Jr., J. Du, A. Hammond, and W. L. Roelofs. 1987. Identification of unique pheromone components for soybean looper moth, *Pseudoplusia includens.* J. Chem. Ecol. 13: 1351-1360.
225. Stanley, B. H., W. H. Reissig, W. L. Roelofs, M. R. Schwarz, and C. A. Shoemaker. 1987. Timing treatments for apple maggot (Diptera: Tephritidae) control using traps baited with synthetic apple volatiles. J. Econ. Entomol. 80: 1057-1063.
226. Linn, C. E., Jr., M. G. Campbell, and W. L. Roelofs. 1987. Pheromone components and active spaces: What do male moths smell and where do they smell it? Science 237: 650-652.
227. Foster, S. P. and W. L. Roelofs. 1987. Comparative sex pheromone biosynthesis in the obliquebanded leafroller, *Choristoneura rosaceana*, and the redbanded leafroller, *Argyrotaenia velutinana*, moths. Z. Naturforsch. 42c: 961-964.
228. Hansson, B. S., C. Löfstedt, and W. L. Roelofs. 1987. Mendelian inheritance of olfactory response to sex pheromone components in male European corn borer, *Ostrinia nubilalis.* Naturwissenschaffen 74: 497-499.
229. Roelofs, W. L., T. J. Glover, X.-H. Tang, I. Sreng, P. Robbins, C. J. Eckenrode, C. Löfstedt, B. S. Hansson, and B. O. Bengtsson. 1987. Sex pheromone production and perception in European corn borer moths requires both autosomal and sex-linked genes. Proc. Natl. Acad. Sci. 84: 7585-7589.
230. Roelofs, W. L. and C. E. Linn, Jr. 1987. Neuroactive substances as probes of central thresholds in sex pheromone perception. J. Cell Biochem. Supp. 10: 54.
231. Linn, C. E., Jr. and W. L. Roelofs. 1987. Effects of octopaminergic agents on the sex pheromone mediated behavior of adult Lepidoptera. In: "Sites of Action for Neurotoxic Pesticides," R. M. Hollingworth and M. B. Green [Eds.] American Chemical Society Symposium Series, pp. 162-173.

232. Roelofs, W. L. and C. E. Linn. 1987. Neuroactive substances as probes of central thresholds in sex pheromone perception. In: "Molecular Entomology." J. Law [Ed.], UCLA Symposium on Molecular and Cellular Biology, New Series, Alan R. Liss, Inc., New York, NY, Vol. 49, pp. 67-77.
233. Bjostad, L. B., W. A. Wolf, and W. L. Roelofs. 1987. Pheromone biosynthesis in Lepidoptera: Desaturation and chain shortening. In: "Pheromone Biochemistry," G. D. Prestwich and G. J. Blomquist [Eds.]. Academic Press, New York, pp. 77-120.
234. Linn, C. E., Jr., A. Hammond, J. Du, and W. L. Roelofs. 1988. Specificity of male response to multicomponent pheromones in the noctuid moths Trichoplusia ni and Pseudoplusia includens. J. Chem. Ecol. 14: 47-57.
235. Linn, C. E., Jr., M. G. Campbell, and W. L. Roelofs. 1988. Temperature modulation of behavioral thresholds controlling male moth sex pheromone response specificity. Physiol. Entomol. 13: 59-67.
236. Krasnoff, S. B. and W. L. Roelofs. 1988. Sex pheromone released as an aerosol by the moth *Pyrrharctia isabella*. Nature 333: 263-265.
237. Foster, S. P. and W. L. Roelofs. 1988. Pink bollworm sex pheromone biosynthesis from oleic acid. Insect Biochem. 18(3): 281-286.
238. Foster, S. P. and W. L. Roelofs. 1988. Sex pheromone biosynthesis in the leafroller moth, *Planotortrix excessana* by $\Delta 10$ desaturation. Arch. Insect Biochem. Physiol. 8: 1-9.
239. Averill, A. L., W. H. Reissig, and W. L. Roelofs. 1988. Specificity of olfactory responses in the tephritid fruit fly, *Rhagoletis pomonella*. Entomol. Exp. Appl. 47: 211-222.
240. Roelofs, W. L. and W. A. Wolf. 1988. Pheromone biosynthesis in the Lepidoptera. J. Chem. Ecol. 14: 2019-2031.
241. Glover, T. J. and W. L. Roelofs. 1988. Genetics of Lepidopteran sex pheromone systems. ISI Atlas of Sci.: Animal & Plant Sci. 1: 279-282.
242. Glover, T. J., N. Perez, and W. L. Roelofs. 1989. Comparative analysis of sex-pheromone-response antagonists in three races of European corn borer. J. Chem. Ecol. 15: 863-873.
243. Tang, J. D., R. E. Charlton, R. A. Jurenka, W. A. Wolf, P. L. Phelan, L. Sreng, and W. L. Roelofs. 1989. Regulation of pheromone biosynthesis by a brain hormone in two moth species. Proc. Natl. Acad. Sci. USA. 86: 1806-1810.
244. Krasnoff, S. B. and W. L. Roelofs. 1989. Quantitative and qualitative effects of larval diet on male scent secretions of *Estigmene acrea, Phragmatobia fuliginosa* and *Pyrrharctia isabella* (Lepidoptera: Arctiidae). J. Chem. Ecol. 15: 1077-1093.
245. Linn, C. E., Jr. and W. L. Roelofs. 1989. Response specificity of male moths to multicomponent pheromones. Chem. Senses. 14(3): 421-437.

246. Sreng, I., T. J. Glover, and W. L. Roelofs. 1989. Canalization of the redbanded leafroller moth sex pheromone blend. Arch. Insect. Biochem. Physiol. 10: 73-82.
247. Wolf, W. A. and W. L. Roelofs. 1989. Enzymes involved in the biosynthesis of sex pheromones in moths. In: "Biocatalysis in Agricultural Biotechnology," J. R. Whitaker and P. E. Sonnet [Eds.], American Chemical Society Symposium Series, pp. 323-331.
248. Jurenka, R. A. and W. L. Roelofs. 1989. Characterization of the acetyltransferase used in pheromone biosynthesis in moths: specificity for the Z isomer in Tortricidae. Insect Biochem. 19(7): 639-644.
249. Löfstedt, C., N. J. Vickers, W. L. Roelofs, and T. C. Baker. 1989. Diet related courtship success in the Oriental fruit moth, *Grapholita molesta* (Tortricidae). Oikas 55: 402-408.
250. Rule, G. S. and W. L. Roelofs. 1989. Biosynthesis of sex pheromone components from linolenic acid in arctiid moths. Arch. Insect Biochem. Physiol. 12: 89-97.
251. Löfstedt, C., B. S. Hansson, W. L. Roelofs, and B. O. Bengtsson. 1989. The linkage relationship between factors controlling female pheromone production and male pheromone response in the European corn borer, *Ostrinia nubilalis*. Genetics 123: 553-556.
252. Foster, S. P. and W. L. Roelofs. 1990. Biosynthesis of a monene and a conjugated diene sex pheromone component of the light brown apple moth by E11-desaturation. Experientia 46: 269-273.
253. Glover, T. J., M. G. Campbell, P. S. Robbins, and W. L. Roelofs. 1990. Sex-linked control of sex pheromone behavioral responses in European corn borer moths (*Ostrinia nubilalis*) confirmed with TPI marker gene. Arch. Insect Biochem. Physiol. 15: 67-77.
254. Krasnoff, S. B. and W. L. Roelofs. 1990. Evolutionary trends in the male pheromone systems of arctiid moths: Evidence from studies of courtship in *Phragmatobia fuliginosa* and *Pyrrharctia isabella* (Lepidoptera: Arctiidae). Zool. J. Linn Soc. 99: 319-338.
255. Dennehy, T. J., W. L. Roelofs, E. F. Taschenberg, and T. N. Taft. 1990. Mating disruption for control of grape berry moth in New York vineyards. In: "Behavior-Modifying Chemicals for Insect Management," R. Ridgeway, R. M. Silverstein, and M. Insco (Eds.), Marcel Dekker, Inc., New York. pp 223-240.
256. Tang, J. D., W. L. Roelofs, and D. C. Knipple. 1990. Construction of a cDNA library from cabbage looper moth sex pheromone glands. In: "Molecular Insect Science," H. Hagedorn, J. Hildebrand, M. Kidwell, and J. Law, [Eds.], Plenum, New York. [Abstract]
257. Jurenka, R. A., G. Fabrias, and W. L. Roelofs. 1991. Hormonal control of female sex pheromone biosynthesis in the redbanded leafroller moth, *Argyrotaenia velutinana*. Insect Biochem. 19: 639-644.

258. Charlton, R. E., J. A. Wyman, J. R. McLaughlin, J.-W. Du, and W. L. Roelofs. 1991. Identification of the sex pheromone of the tomato pinworm, *Keiferia lycopersicella* (WALS.). J. Chem. Ecol. 17: 175-183.
259. Roelofs, W. L. and T. J. Glover. 1991. Genetics of a moth pheromone system. In: "Chemical Senses" Vol. 3 of "Genetics of Perception and Communication," C. J. Wysocki and M. R. Kare [Eds.], Marcel Dekker, Inc., New York. pp. 109-124.
260. Tang, J. D., W. A. Wolf, W. L. Roelofs and D. C. Knipple. 1991. Development of functionally competent cabbage looper moth sex pheromone glands. Insect Biochem. 21: 573-581.
261. Jurenka, R. A., E. Jacquin and W. L. Roelofs. 1991. Control of pheromone biosynthetic pathways in *Helicoverpa zea* by the pheromone biosynthesis activating neuropeptide. Arch. Insect Biochem. Physiol. 17: 81-91.
262. Charlton, R. E. and W. L. Roelofs. 1991. Biosynthesis of a volatile, methyl-branched hydrocarbon sex pheromone from leucine by arctiid moths (*Holomelina* spp.). Arch. Insect Biochem. Physiol. 18: 81-97.
263. Linn, C. E., Jr., M. G. Campbell, and W. L. Roelofs. 1991. The effects of different blend ratios and temperature on the active space of the Oriental fruit moth sex pheromone. Physiol. Entomol. 16: 211-222.
264. Roelofs, W. L. 1991. Insect Communication. In: "Entomology Serving Society: Emerging Technologies and Challenges," S. B. Vinson and R. Metcalf [Eds.], Entomological Society of America, Lanham, MD, pp. 179-206.
265. Jurenka, R. A., E. Jacquin, and W. L. Roelofs. 1991. Stimulation of pheromone biosynthesis in the moth *Helicoverpa zea*: Action of a brain hormone on pheromone glands involves Ca^{2+} and cAMP as second messengers. Proc. Natl. Acad. Sci. USA 88: 8621-8625.
266. Glover, T. J., M. G. Campbell, C. E. Linn, Jr., and W. L. Roelofs. 1991. Unique sex chromosome mediated behavioral response specificity of hybrid male European corn borer moths. Experientia 47: 980-984.
267. Glover, T. J., J. J. Knodel, P. S. Robbins, C. J. Eckenrode, and W. L. Roelofs. 1991. Gene flow among three races of European corn borers (Lepidoptera: Pyralidae) in New York State. Environ Entomol. 20: 1356-1362.
268. Roelofs, W. L. 1991. Feromonaj sexuales de las pollillas: aspectos basicos. Phytoma : 44-55.
269. Linn, C. E., Jr., W. L. Roelofs, W. Sun, and G. Prestwich. 1992. Activity of perfluorobutyl-containing pheromone components in the pheromone blend of the cabbage looper moth, *Trichoplusia ni.* J. Chem. Ecol. 18: 737-748.
270. Glover, T. J., P. S. Robbins, C. J. Eckenrode, and W. L. Roelofs. 1992. Genetic control of voltinism characteristics in European corn borer races assessed with a marker gene. Arch. Insect. Biochem Physiol. 20: 107-117.

271. Fabrias, G., R. A. Jurenka, and W. L. Roelofs. 1992. Stimulation of sex pheromone production by proteinaceous extracts of the bursa capulatrix in the redbanded leafroller moth. Arch. Insect Biochem. Physiol. 20: 75-86.
272. Linn, C. E., M. G. Campbell and W. L. Roelofs. 1992. Photoperiod cues and the modulatory action of octopamine and 5-hydroxytryptamine on locomotor and pheromone response in male gypsy moths, *Lymantria dispar.* Arch. Insect Biochem. Physiol. 20: 265-284.
273. Linn, C. E., Jr. and W. L. Roelofs. 1992. Role of photoperiod cues in regulating the modulatory action of octopamine on pheromone-response thresholds in the cabbage looper moth. Arch. Insect Biochem. Physiol. 20: 285-302.
274. Linn, C. E., Jr. and W. L. Roelofs. 1993. Levels of biogenic amines and peptides in individual corn earworm moths, Helicoverpa zea, using high performance liquid chromoatography with electrochemical detection. Insect Biochem. Molec. Biol. 23(3): 367-373.
275. Jurenka, R. A. and W. L. Roelofs. 1993. Biosynthesis and endocrine regulation of fatty acid derived sex pheromones in moths. In: "Insect Lipids: Chemistry, Biochemistry, & Biology", D. W. Stanley-Samuelson and D. R. Nelson [Eds.], University of Nebraska Press, Lincoln, NE, pp. 353-388.
276. Charlton, R. E., F. X. Webster, A. Zhang, C. Schal, D. Liang, I. Sreng, and W. L. Roelofs. 1993. Sex pheromone for the brownbanded cockroach is a dialkyl-substituted α-pyrone. Proc. Natl. Acad. Sci. USA 90: 10202-10205.
277. Jurenka, R. A., G. Fabrias, S. Ramaswamy, and W. L. Roelofs. 1993. Control of pheromone biosynthesis in mated redbanded leafroller moths. Arch. Insect Biochem. Physiol. 24: 129-137.
278. Jacquin, E., R. A. Jurenka, H. Ljungberg, P. Nagnan, C. Löfsted, C. Descoins, and W. L. Roelofs. 1994. Control of sex pheromone biosynthesis in the moth, *Mamestra brassicae* by the pheromone biosynthesis activating neuropeptide. Insect Biochem. Mol. Biol. 24: 203-211.
279. Roelofs, W. L. 1993. Book review of "Plant Kairomones in Insect Ecology and Control", J. Nat. Res. and Life Sci. Ed. 22 (2): 184-185.
280. Biddinger, D. J., W. Roelofs, and A. J. Howitt. 1994. The development of a sex pheromone lure for the American plum borer (Lepidoptera: Pyralidae), a major pest of cherry orchards in Michigan. Great Lakes Entom. J. 26: 311-317.
281. Jurenka, R. A., G. Fabrias, L. DeVoe and W. L. Roelofs. 1994. Action of PBAN and related peptides on pheromone biosynthesis in isolated pheromone glands of the redbanded leafroller moth, *Argyrotaenia velutinana.* Comp. Biochem. Physiol. 108C: 153-160.
282. Jurenka, R. S., K. R. Haynes, R. O. Adlof, M. Bengtsson, and W. L. Roelofs. 1994. Sex pheromone component ratio in the cabbage looper moth altered by a mutation affecting the fatty acid chain-shortening reactions in the pheromone biosynthetic pathway. Insect Biochem. Molec. Biol. 24: 373-381.

283. Foster, S. P. and W. L. Roelofs. 1994. Regulation of pheromone production in virgin and mated females of two tortricid moths. Arch. Insect Biochem. Physiol. 25: 271-285.

284. Roelofs, W. L., C. Schal, F. X. Webster, R. E. Charlton, A. Zhang. 1994. Cockroach attractant. U. S. Pat. 5, 296, 220, Mar. 22.

285. Linn, C. E., Jr., K. R. Poole and W. L. Roelofs. 1994. Studies on biogenic amines and their metabolites in nervous tissues and hemolymph of adult male cabbage looper moths. I. Quantitation of photoperiod changes. Comp. Biochem. Physiol. 108C: 73-85.

286. Linn, C. E., Jr., M. G. Campbell, K. R. Poole and W. L. Roelofs. 1994. Studies on biogenic amines and their metabolites in nervous tissues and hemolymph of adult male cabbage looper moths. Comp. Biochem. Physiol. 108C: 87-98.

287. Ma, P. W. K., D. C. Knipple, and W. L. Roelofs. 1994. Structural organization of the *Helicoverpa zea* gene encoding the precursor protein for pheromone biosynthesis-activating neuropeptide and other neuropeptides. Proc. Natl. Acad. Sci. USA. 91: 6506-6510.

288. Facundo, H. T., A. Zhang, P. S. Robbins, S. R. Alm, C. E. Linn, Jr., M. G. Villani, and W. L. Roelofs. 1994. Sex pheromone responses of the Oriental beetle (Coleoptera: Scarabaeidae). Environ. Entomol. 23: 1508-1515.

289. Zhang, A., H. T. Facundo, P. S. Robbins, C. E. Linn, Jr., J. L. Hanula, M. G. Villani, and W. L. Roelofs. 1994. Identification and synthesis of the female sex pheromone of the Oriental beetle, *Anomala orientalis* (Coleoptera: Scarabaeidae). J. Chem. Ecol. 20: 2415-2427.

290. Ma, P. W. K. and W. L. Roelofs. 1995. Sites of synthesis and release of a PBAN-like factor in female European corn borer, *Ostrinia nubilalis*. J. Insect Physiol. 41: 339-350.

291. Ma, P. W. K. and W. L. Roelofs. 1995. Calcium involvement in the stimulation of sex pheromone production in the European corn borer, *Ostrinia nubilalis* (Lepidoptera: Pyralidae). Insect Biochem. Molec. Biol. 25: 467-473.

292. Linn, C. E., Jr. and W. L. Roelofs. 1995. Pheromone communication in moths and its role in the speciation process. In: "Speciation and the Recognition Concept", D. M. Lambert and H. G. Spencer [Eds.]. John Hopkins University Press, Baltimore and London, pp. 263-300.

293. Roelofs, W. L. 1995. Chemistry of sex attraction. Proc. Natl. Acad. Sci. USA 92: 44-49.

294. Ramaswamy, S. B., R. A. Jurenka, C. E. Linn, Jr., W. L. Roelofs. 1995. Evidence for the presence of a pheromonotropic factor in hemolymph and regulation of sex pheromone production in *Helicoverpa zea*. J. Insect Physiol. 41: 501-508.

295. Ma. P. W. K. and W. L. Roelofs. 1995. Anatomy of the neurosecretory cells in the cerebral and subesophageal ganglia of the female European corn borer moth, *Ostrinia nubilalis* (Hübner)(Lepidoptera: Pyralidae). Int. J. Insect Morphol. Embryol. 24: 343-359.
296. Linn, C. E., K. R. Poole, W.-Q. Wu, W. L. Roelofs. 1995. Circadian changes in melatonin in the nervous system and hemolymph of the cabbage looper moth, *Trichoplusia ni.* J. Comp. Physiol. A 176: 761-771.
297. Cossé, A. A., Campbell, M. G., Glover, T. J., Linn, C. E., Todd, J. L., Baker, T. C. and Roelofs, W. L. 1995. Pheromone behavioral responses in unusual male European corn borer hybrid progeny not correlated to electrophysiological phenotypes of their pheromone-specific antennal neurons. Experientia 51: 809-816.
298. Roelofs, W. L. 1995. The chemistry of sex attraction. In: "Chemical Ecology: The Chemistry of biotic Interaction", T. Eisner and J. Meinwald [Eds.]. National Academy of Sciences, Wash. D. C., pp. 103-117.
299. Ma, P. W. K., Roelofs, W. L. and Jurenka, R. A. 1996. Characterization of PBAN and PBAN-encoding gene neuropeptides in the central nervous system of the corn earworm moth, *Helicoverpa zea.* J. Insect Physiol. 42: 257-266.
300. Kochansky, J., Hill, A., Neal, J. W., Bentz, J-A., Roelofs, W. 1996. Pheromone of the eastern tent caterpillar, *Malacosoma americanum* (F.) and an isomeric barrier in the reproductive systems of sympatric *Malacosoma* species (Lepidoptera: Lasiocampidae). J. Chem. Ecol. 22: 2251-2261.
301. Rasmussen, L. E. L., Lee, T. D., Roelofs, W. L., Zhang, A., Daves, Jr., G. D. 1996. Insect pheromone in elephants. Nature 379: 684.
302. Roelofs, W. L.,C. Schal, F. X. Webster, R. E. Charlton, A. Zhang. 1996. Cockroach attractant. U. S. Pat. 5,505,951 April 9.
303. Roelofs, W. L. and Jurenka, R. A. 1996. Biosynthetic enzymes regulating ratios of sex pheromone components in female redbanded leafroller moths. Bioorg. & Med. Chem. 4: 461-466.
304. Linn, C. E., Jr., Campbell, M. G., Poole, K. R., Wen-Q, W., Roelofs, W. L. 1996. Effects of photoperiod on the circadian timing of pheromone response in male *Trichoplusia ni*: Relationship to the modulatory action of octopamine. J. Insect Physiol. 42: 881-891.
305. Foster, S. P. and Roelofs, W. L. 1996. Sex pheromone biosynthesis in the tortricid moth, *Ctenopseustis herana* (Felder & Rogenhofer). Arch. Insect Biochem. Physiol. 32: 135-147.
306. Lawson, D. S., Reissig, W. H., Agnello, A. M., Nyrop, J. P., and Roelofs, W. L. 1996. Interference with the mate-finding communication system of the obliquebanded leafroller (Lepidoptera: Tortricidae) using synthetic sex pheromones. Environ. Entomol. 25: 895-905.

307. Roelofs, W. L. and Jurenka, R. A. 1997. Interaction of PBAN with biosynthetic enzymes. In: "Pheromone Research: New Directions", R. T. Cardé and A. K. Minks [Eds.]. Chapman and Hall, NY, pp. 42-45.
308. Roelofs, W. L., Jurenka, R. A., and Ma, P. 1997. Involvement of the nervous system with PBAN. In: "Pheromone Research: New Directions", R. T. Cardé and A. K. Minks [Eds.]. Chapman and Hall, New York, pp. 96-100.
309. Zhang, A., Robbins, P. S., Leal, W. S., Linn, C. E., Jr., Villani, M. G., and Roelofs, W. L. 1997. Essential amino acid methyl esters: major sex pheromone components of the cranberry white grub, *Phyllophaga anxia* (Coleoptera: Scarabaeidae). J. Chem. Ecol. 23: 231-245.
310. Elliott, J. T., Jurenka, R. A., Prestwich, G. D., and Roelofs, W. L. 1997. Identification of soluble binding proteins for an insect neuropeptide. Biochem. Biophys. Res. Commun. 238: 925-930.
311. Linn, C.E. Jr., Young, M.S., Gendle, M., Glover, T.J., and W.L. Roelofs. 1997. Sex pheromone blend discrimination in two races and hybrids of the European corn borer moth, *Ostrinia nubilalis*. Physiol. Entomol. 22: 212-223.
312. Rasmussen, L. E. L., Lee, T. D., Zhang, A., Roelofs, W. L., and Daves, Jr., G. D. 1997. Purification, identification, concentration and bioactivity of (Z)-7-dodecen-1-yl acetate: Sex pheromone of the female Asian elephant, *Elephas maximus*. Chem. Senses 22: 417-437.
313. Ma, P. W. K., Knipple, D. C. and Roelofs, W. L. 1998. Expression of a gene that encodes multiple neuropeptides in the central nervous system of corn earworm, *Helicoverpa zea*. Insect Biochem. Molec. Biol. 28: 373-385.
314. Knipple, D. C., Miller, S. J., Liu, W., Tang, J., Ma, P. W. K., and Roelofs, W. L. 1998. Cloning and characterization of a cDNA encoding a pheromone gland-specific acyl-CoA "11-desaturase of the cabbage looper moth, *Trichoplusia ni* . Proc. Natl. Acad. Sci. 95: 15287-15292.
315. Ma, P. W. K., Davis, T. R., Wood, H. A., Knipple, D. C., and Roelofs, W. L. 1998. Baculovirus expression of an insect gene that encodes multiple neuropeptides. Insect Biochem. Molec. Biol. 28: 239-249.
316. Einhorn, J., Guerrero, A., Ducrot, P.-H., Boyer, F.-D., Gieselmann, M., Roelofs. W. 1998. Pheromone of the oleander scale, *Aspidiotus nerii*: Structural characterization and absolute configuration of an unusual functionalized cyclobutane. Proc. Natl. Acad. Sci. 95: 9867-9872.
317. Facundo, H. T., Villani, M. G., Linn, Jr., C. E., and Roelofs, W. L. 1999. Temporal and spatial distribution of the oriental beetle, *Exomala orientalis* (Coleoptera: Scarabaeidae), in a golf course environment. Environ. Entomol. 28: 14-21.
318. Knipple, D. C., Roelofs, W. L., and Miller, S. J. 1999. Pheromone Desaturases. U. S. Pat. 5,876,994 March 2.

319. Facundo, H. T., Linn Jr., C.. E., Villani, M. G., and Roelofs, W. L. 1999. Emergence, mating, and postmating behaviors of the Oriental beetle (Coleoptera: Scarabaeidae). J. Insect Behav. 12: 175-192.
320. Liu, W., Ma, P. W. K., Marsella-Herrick, P., Rosenfield, C.-L., Knipple, D. C., Roelofs, W. 1999. Cloning and functional expression of a cDNA encoding a metabolic acyl-CoA "9-desaturase of the cabbage looper moth, *Trichoplusia ni.* Insect Biochem. Molec. Biol. 29: 435-443.
321. Linn, Jr., C. E., Poole, K., Zhang, A., and Roelofs, W. 1999. Pheromone-blend discrimination by European corn borer moths with inter-race and inter-sex antennal transplants. J. Comp. Physiol. A 184: 273-278.
322. Zhang, A., Linn, Jr., C., Wright, S., Prokopy, R., Reissig, W., Roelofs, W. 1999. Identification of a new blend of apple volatiles attractive to the apple maggot *Rhagoletis pomonella.* J. Chem. Ecol. 25: 1221-1232.
323. Alm, Steven R., Villani, M. G., and Roelofs, W. 1999. Oriental beetles (Coleoptera: Scarabaeidae): Current distribution in the United States and optimization of monitoring traps. J. Econ. Entomol. 92: 931-935.
324. Ma, P. W. K., Garden, R., Niermann, J. T., O'Connor, M., Sweedler, J. V., and Roelofs, W. L. 2000. Characterizing the Hez-PBAN gene products in neuronal clusters with immunocytochemistry and MALDI MS. J. Insect Physiol. 46: 221-230.
325. Dalleerac, R., Labeur, C., Jallon, J.-M., Knipple D. C., Roelofs, W., Wicker-Thomas, C. 2000. A Δ9 desaturase gene with a different substrate specificity is responsible for the cuticular diene hydrocarbon polymorphism in *Drosophila melanogaster.* Proc. Natl. Acad. Sci. USA 97: 9449-9454.
326. Rosenfield, C-L., You, K-M., Marsella-Herrick, P., Roelofs, W. L., and Knipple, D. C. 2001. Structural and functional conservation and divergence among acyl-CoA desaturase-encoding genes of two noctuid species, the corn earworm, *Helicoverpa zea,* and the cabbage looper, *Trichoplusia ni.* Insect Biochem. Molec. Biol. 31: 949-964.
327. Liu, W., Jiao, H., O'Connor, M. and Roelofs, W. L. 2002. Cloning and expression of a moth desaturase that specifically produces a mixture of Z/E11-tetradecenoic acids. Biochemistry 32: 1489-1495.
328. Liu, W., Jiao, H., Murray, N., O'Connor, M., and Roelofs, W. L. 2002. Gene characterized for membrane desaturase that produces (E)-11 isomers of mono- and diunsaturated fatty acids. Proc. Natl. Acad. Sci. USA 99: 620-624.
329. Hao, G., Liu, W., O'Connor, M., and Roelofs, W. L. 2002. Characterization and functional expression of cDNAs encoding acyl-CoA Z9- and Z10-desaturase of *Planotortrix octo.* Insect Biochem. Molec. Biol. 32: 961-966.
330. Ochieng, S. A., Robbins, P. S., Roelofs, W. L., and Baker, T. C. 2002. Pheromone reception in the scarab beetle, *Phyllophaga anxia* (Coleoptera: Scarabaeidae). Ann. Entomol. Soc. Amer. 95: 97-102.

331. Ma, W. K. P. and Roelofs, W. L. 2002. Sex pheromone gland of the female European corn borer moth, *Ostrinia nubilalis* (Lepidoptera, Pyralidae): ultrastructural and biochemical evidences. Zool. Sci. 19: 501-511.

332. Roelofs, W. L., Liu, W., Hao, G., Jiao, H., Rooney, A. P., and Linn, Jr., C. E. 2002. Evolution of moth sex pheromones via ancestral genes. Proc. Natl. Acad. Sci. USA 99: 13621-13626.

333. Hao, G., O'Connor, M., Liu, W., Roelofs, W. L. 2002. Characterization of Z/E11- and Z9-desaturases from the obliquebanded leafroller moth, *Choristoneura rosaceana* J. Insect Sci. 2:26 online: insectscience.org/2.26.

334. Ochieng, S. A., Poole, K., Linn, Jr., C., Vickers, N., Roelofs, W. L., Baker, T. C. 2002. Unusual pheromone receptor neuron responses in heliothine moth antennae derived from inter-species imaginal disc transplantation. J. Comp. Physiol. A 189: 19-28.

335. El-Sayed, A.M., Delisle, J., De Lury, N., Gut, L. J., Judd, G. J. R., Legrand, S., Reissig, W. H., Roelofs, W. L., Unelius, C. R., and Trimble, R. M. 2003. Geographic variation in pheromone chemistry, antennal electrophysiology, and pheromone-mediated trap catch of North American populations of the obliquebanded leafroller. Environ. Entomol. 32; 470-476.

336. Nojima, S., Sakata, T., Yoshimura, K., Robbins, P. S., Morris, B. D., and Roelofs, W. L. 2003. Male-specific EAD active compounds produced by female European chafer, *Rizotrogus majalis* (Razoumowshy) (Coleoptera: Scarabaeidae, Melolonthinae). J. Chem. Ecol. 29: 503-507.

337. Nojima, S., Linn, Jr., C., Morris, B. D., Zhang, A. and Roelofs, W. L. 2003. Identification of host fruit volatiles from hawthorn (*Crataegus* spp.) attractive to hawthorn-orign *Rhagoletis pomonella* flies. J. Chem. Ecol. 29: 321-336.

338. Linn, Jr., C., O'Connor, M., Roelofs, W. 2003. Silent genes and rare males: A fresh look at pheromone blend response specificity in the European corn borer moth, *Ostrinia nubilalis*. Journal of Insect Science, 3:15, Available online: insectscience.org/3.15.

339. Zhang, A., Robbins, P. S., Averill, A. L., Weber, D. C., Linn, C. E., Jr., Roelofs, W. L., Villani, M. G. 2003. Identification of the female-produced sex pheromone of the scarab beetle, Hoplia equina. J. Chem. Ecol. 29: 1635-1642.

340. Roelofs, W. L. and Rooney, A. P. 2003. Molecular genetics and evolution of pheromone biosynthesis in Lepidoptera. Proc. Natl. Acad. Sci. USA 100: 9179-9184.

341. Linn, Jr., C., Feder, J. L., Nojima, S., Dambroski, H. R., Berlocher, S. H., Roelofs, W. L. 2003. Fruit odor discrimination and sympatric host race formation in *Rhagoletis*. Proc. Natl. Acad. Sci. USA 100: 11490-11493.

342. Robbins, P. S., Crocker, R. L., Nojima, S., Morris, B. D., Roelofs, W. L., Villani, M. G. 2003. Methyl 2-(methylthio)benzoate: the unique sulfur-containing sex pheromone of *Phyllophaga crinita*. Naturwissenschaften 90: 517-520.
343. Nojima, S., Robbins, P. S., Salsbury, G. A., Morris, B. D. Roelofs, W. L., Villani, M. G. 2003. L-Leucine methyl ester: the female-produced sex pheromone of the scarab beetle, *Phyllophaga lanceolata*. J. Chem. Ecol. 29: 2439-2446.
344. Knipple, D. C. and W. L. Roelofs. 2003. Molecular biochemical investigations of pheromone desaturases. In: "Insect Pheromone Biochemistry and Molecular Biology," G. J. Blomquist and R. G. Vogt [Eds.]. Elsevier Academic Press, New York, pp. 81-106.
345. Nojima, S., C. Linn Jr., and W. Roelofs. 2003. Identification of host fruit volatiles from flowering dogwood (*Cornus florida*) attractive to dogwood-origin *Rhagoletis pomonella* flies. *J. Chem. Ecol.* 29: 2347-2357.
346. Linn, Jr., C. E., Dambroski, H. R., Feder, J. L., Berlocher, S. H., Nojima, S., Roelofs, W. L. 2004. Postzygotic isolating factor in sympatric speciation in Rhagoletis flies: Reduced response of hybrids to parental host-fruit odors. Proc. Natl. Aca. Sci. USA 10: 17753-17758.
347. Nojima, S., Kiemle, D. J., Webster, F. X., Roelofs, W. L. 2004. Submicro scale NMR sample preparation for volatile chemicals. J. Chem. Ecol. 30: 2153-2161.
348. Rodriguez, S., Hao, G., Liu, W., Pina, B. Rooney, A., Camps, F., Roelofs, W., Fabrias, G. 2004. Expression and evolution of delta-9 and delta-11 desaturase genes in the moth Spodooptera littoralis. Insect Biochem. Molec. Biol. 34: 1315-1328.
349. Liu, W., Rooney, A. P., Xue, B., Roelofs, W. L. 2004. Desaturases form the spotted fireworm moth (Choristoneura parallela) shed light on the evolutionary origins of novel moth sex pheromone desaturases. Gene 342: 303-311.
350. Nojima, S., Kiemle, D. J., Webster, F. X., Roelofs, W. L. 2004. Submicro scale NMR sample preparation for volatile chemicals. J. Chem. Soc. 30: 2147-2155.
351. Linn, C., Jr., Nojima, S., Roelofs, W. 2005. Antagonist effects of non-host fruit volatiles on discrimination of host fruit by Rhagoletis flies infesting apple (Malus pumila), hawthorn (Crataegus spp.), and flowering dogwood (Cornus florida). Entomol. Exp. Appl. 114: 97-105.
352. Nojima, S., Schal, C., Webster, F. X., Santangelo, R. G., Roelofs, W. L. 2005. Identification of the sex pheromone of the German cockroach, *Blattella germanica*. Science 307: 1104-1106.

353. Burand JP, Tan W, Kim W, Nojima S, Roelofs W. 2005. Infection with the insect virus Hz-2v alters mating behavior and pheromone production in female *Helicoverpa zea* moths. 6pp. Journal of Insect Science 5:6, Available online: insectscience.org/5.6

354. Heath, Jeremy J. , Zhang, Aijun, Roelofs, Wendell L., Smith, Robert F. 2005. Flight activity and further evidence for a female-produced sex pheromone of the apple leaf midge, *Dasineura mali* , in Nova Scotia. Northeastern Naturalist 12 (1), 93-102.

355. Linn, Jr., C. E. Dambroski, Hattie, Nojima, Satoshi, Feder, Jeffrey L., Berlocher, Stewart H., Roelofs, Wendell L. 2005. Variability in response specificity of apple, hawthorn, and flowering dogwood-infesting *Rhagoletis* flies to host fruit volatile blends: Implications for sympatric host shifts. Entomol. Exp. Appl. 116: 55-64.

356. Dambroski, H. R., Linn, Jr., C. E., Berlocher, S. H., Forbes, A., Roelofs, W., Feder, J. L. 2005. The genetic basis for fruit odor discrimination in *Rhagoletic* flies and its significance for sympatric host shifts. Evolution 59: 1953-1964.

357. Olsson, S. B., Linn, Jr., C. E., Roelofs, W. L. 2006. The chemosensory basis for behavioral divergence involved in sympatric host shifts II: olfactory receptor neuron sensitivity and temporal firing pattern to individual key host volatiles. J. Comp. Physiol. A. 192: 289-300.

358. Olsson, S. B., Linn, Jr., C. E., Roelofs, W. L. 2006. The chemosensory basis for behavioral divergence involved in sympatric host shifts. I. Characterizing olfactory receptor neuron classes responding to key host volatiles. J. Comp. Physiol. A. 192: 279-288.

359. Robbins, P. S., Zhang, A., Averill, A. L., Linn, Jr., C. E., Roelofs, W. L., Sylvia, M. M., Villani M. G. 2006. Sex pheromone of the cranberry root grub *Lichnanthe vulpine*. J. Chem. Ecol. 32: 1663-1672.

360. Robbins, P.S. and 44 collaborators. 2006. Trapping Phyllophaga spp. (Coleoptera: Scarabaeidae: Melolonthinae) in the United States and Canada using sex attractants. http://www.insectscience.org/6.39/

361. Olsson, S. B., Linn, Jr., C. E., Michel, A., Dambroski, H. R., Berlocher, S. H. Feder, F. L., Roelofs, W. L. 2006. Receptor expression and sympatric speciation: unique olfactory receptor neuron responses in F1 hybrid *Rhagoletis* populations. J. Exp. Biol. 209: 3729-3741.

362. Linn, C., Jr., Musto, C., Roelofs, W. 2007. More Rare Males in Ostrinia: Response of Asian Corn Borer Moths to the Sex Pheromone of the European Corn Borer. J. Chem. Ecol. 33: 199-212.

363. Domingue M. J., Roelofs, W. L., Linn Jr., C. E., Baker, T.C. 2006. Effects of egg-to-adult development time and adult age on olfactory neuron response to semiochemicals in European corn borers. J. Insect. Physiol. 52: 975-983.

364. Xue, B., Rooney, A. P., Kajikawa, M., Okada, N., Roelofs, W. L. 2007. Novel sex pheromone desaturases in the genomes of corn borers generated through gene duplication and retroposon fusion. Proc. Natl. Aca. Sci. USA 104: 4467-4472.

365. Linn, C. E., Jr., Domingue, M. J., Musto, C., Baker, T. C., Roelofs, W L. 2007. Support for (Z)-11-hexadecanal as a pheromone antagonist in Ostrinia nubilalis: flight tunnel and single sensillum studies with a New York population. J. Chem. Ecol. 33: 909-921.

366. Domingue, M. J., Musto, C., Linn, C. E., Jr., Roelofs, W L., Baker, T. C. 2007. Evidence of olfactory antagonistic imposition as a facilitator of evolutionary shifts in pheromone blend usage in Ostrinia spp. (Lepidoptera: Crambidae). J. Insect Physiol. 53: 488-49.

Cornelis T. De Wit

Agricultural University
Wageningen, Netherlands

1924–1993

1983/4 — for their innovative contributions to the quantitative understanding of soil-water and other environmental interactions influencing crop growth and yield.

IN MEMORIAM

On October 10, 2008 it is the 40th anniversary of Kees de Wit's inaugural address (**Theory and model**, in Dutch) at Wageningen Agricultural University, as Professor of Theoretical Production Ecology. Although it is 15 years since he passed away, much regretted, his mark on Wageningen University is still present, through the C.T. De Wit Graduate School for Production Ecology and Resource Conservation and through the 'PhD children and grandchildren' that serve the University in various positions. This year therefore seems an appropriate moment to reflect on **40 years Theory and Model** at Wageningen University and to take stock of the developments in the last 15 years.

Kees (Cornelis) De Wit grew up in a rural village in the eastern part of the Netherlands, and spent most of World War II as a farm laborer. This aroused his interest in the complexities of farming and in farmers and was to guide his professional career. He completed his studies after the war with a Ph.D. thesis ('A physical theory on placement of fertilizers', 1953). His subsequent employment at the Ministry of National Planning of the Union of Birma laid the foundation for his strong commitment to agriculture in developing countries.

After his return in 1956, De Wit was employed at the Institute for Biological and Chemical Research on Field Crops and Herbage (IBS) and its successor, CABO (Centre for Agrobiological Research), and in the next decade produced some of his most influential papers: 'Transpiration and crop yields' (1958), a major re-interpretation of crop water use data, based on a physical analysis of canopy processes; 'On competition' (1960) describing in physical and mathematical terms the interactions between plants of different species; 'Ionic balance and growth of plants' (1963), elaborating on the chemical composition of plants and the crucial role of anion/cation balances; 'Photosynthesis of leaf canopies' (1965) introducing the brute force of the computer in crop physiology; and 'A dynamic model of the vegetative growth of crops' (1968), the first publication on crop growth simulation. These publications are still widely quoted and constitute significant steps in the progress in agricultural science.

De Wit was appointed professor at Wageningen Agricultural University in 1968 to create the Department of Theoretical Production Ecology. Through its research and teaching (De Wit (co-)supervised 32 Ph.D. theses), under his strong leadership, systems analysis and modeling gained a firm footing in the agricultural research community. 'No simulation without experimentation', and later also the reverse, were among his most vivid expressions, always presented with characteristic conviction. He was the initiator of the series 'Simulation Monographs' (Publisher PUDOC, Wageningen), that in the '70s and '80s formed the major outlet for publication of models of agricultural production systems.

In the eighties, as member of the Dutch Scientific Council for Government Policy (WRR) and the Technical Advisory Committee (TAC) of the CGIAR, De Wit either initiated or took a very active part in discussions on environment, sustainability and development. For him, these appointments were challenges to combine biophysical and social sciences at a scientific and at operational level. His continuing innovative efforts yielded a critical review of the consequences of the Common Agricultural Policy of the European Union, a new approach to multi-stakeholder planning of land use and agricultural development (1988), an unconventional view on 'Resource use efficiency in agriculture' (1990) and an innovative transparent methodology for priority setting in international research (1992).

De Wit's scientific qualities, his keen interest in human beings, deep feelings of justice and equality, his informality and his very original, sharp and systematic mind made him welcome, unavoidable and outstanding in any meeting. He received the Wolf price of the State of Israel, the 'Nobel Price for Agriculture', in 1983/4.

His approach to science has formed the starting point for many of the recent developments in the international science community, both within and outside agriculture.

List of Publications

Wit, C.T. de, 1953. A physical theory on placement of fertilizers. Agricultural Research Reports 59.4, 71 pp.

Wit, C.T. de, 1958. Transpiration and crop yields. Agricultural Research Reports 64.6, Pudoc, Wageningen, 88 pp.
Wit, C.T. de, 1960. On competition. Agricultural Research Reports 66.8, Pudoc, Wageningen, 82 pp.
Wit, C.T. de, 1965. Photosynthesis of leaf canopies. Agricultural Research Reports 633, Pudoc, Wageningen, 57 pp.

List of Publications since 1968 (Start Dept of Theorical Production Ecology, Wageningen Agr. University)

Wit, C.T. de, 1968. Theorie en model. Inaugurele rede Landbouwhogeschool. 10 oktober 1968, 13 pp.
Wit, C.T. de, 1968. Simulatie van continue processen. Landbouwkundig Tijdschrift 80, 215-218.
Wit, C.T. de, 1968. Plant production. Misc. Papers Landbouwhogeschool, Wageningen, no. 3, 25-50.
Wit, C.T. de, 1968. Tussen schaarste en overvloed. Internationale Spectator 23, 392-406.
Wit, C.T. de, 1969. Een bodemvruchtbaarheidstheorie uit de 1[e] helft van de 19[e] eeuw. Landbouwkundig Tijdschrift 81, 245-251.
Brennan, R.D., C.T. de Wit, W.A. Williams & E.V. Quattrin, 1970. The utility of a digital simulation language for ecological modeling. Oecologia (Berlin) 4, 113-132.
Idso, S.B. & C.T. de Wit, 1970. Light relations in plant canopies. Applied Optics 9, 177-184.
Wieringa, P.J. & C.T. de Wit, 1970. Simulation of heat transfer in soils. Soil Science Society of America Proceedings, Vol. 34(6), 845-848.
Wit, C.T. de, 1970. Crop production yesterday, today, tomorrow. Prizoda 11, pp. 20-25 (Russisch).
Wit, C.T. de, 1970. Dynamic concepts in biology. In: Prediction and measurement of photosynthetic productivity. Proceedings of the IBP/PP Technical Meeting, Trebon, 1969. Ed. I. Setlík, Pudoc, Wageningen, pp. 17-23.
Wit, C.T. de, R. Brouwer & F.W.T. Penning de Vries, 1970. The simulation of photosynthetic systems. In: Prediction and measurement of photosynthetic productivity. Proceedings of the IBP/PP Technical Meeting, Trebon, 1969. Ed. I. Setlík, Pudoc, Wageningen, pp. 47-50.
Wit, C.T. de, 1971. Voedselproduktie: verleden, heden en toekomst. Stikstof 69, 396-408.
Wit, C.T. de, R. Brouwer & F.W.T. Penning de Vries, 1971. A dynamic model of plant and crop growth. In: Potential Crop Production. Eds P.F. Wareing & J.P. Cooper, Heinemann, London, pp. 117-142.

Wit, C.T. de, 1972. Food production: past, present and future. Stikstof 15, 65-80.

Wit, C.T. de, 1972. De moderne landbouw in het westen. Landbouwkundig Tijdschrift 84(9), 310-312.

Wit, C.T. de & H. van Keulen, 1972. Simulation of transport processes in soils. Simulation Monographs, Pudoc, Wageningen, 109 pp.

Goudriaan, J. & C.T. de Wit, 1973. A re-interpretation of Gause's population experiments by means of simulation. Journal of Animal Ecology 42, 521-530.

Wit, C.T. de & A.Th.J. Nooij, 1973. Over eten en over leven. Diësrede LH. Intermediair 9, nr 11, 24 pp.

Dekkers, W.A., J.M. Lange & C.T. de Wit, 1974. Energy production and use in Dutch agriculture. Netherlands Journal of Agricultural Science 22, 107-118.

Wit, C.T. de, 1974. Early theoretical concepts in soil fertility. Netherlands Journal of Agricultural Science 22, 319-324.

Wit, C.T. de & J. Goudriaan, 1974. Simulation of ecological processes. Simulation Monographs, Pudoc, Wageningen, 159 pp.

Sinclair, T.R. & C.T. de Wit, 1975. Photosynthesis and nitrogen requirements for seed production by various crops. Science 189, 565-567 (15 August).

Wit, C.T. de, 1975. Energetische aspecten van de plantaardige produktie. Nederlands Tijdschrift. voor Natuurkunde 41(12), 166-168.

Wit, C.T. de, 1975. Agriculture's uncertain claim on world energy resources. SPAN 18(1), 2-4.

Wit, C.T. de, 1975. Landbouw in Nederland. In: Nederland en de grenzen aan de groei. Symposium Tussentijds Bestek. Aula 548, Utrecht/Antwerpen, pp. 29-34.

Wit, C.T. de, 1975. Modellen als brug tussen proces en systeem. In: Produktiviteit in biologische systemen. Pudoc, Wageningen, pp. 157-170.

Wit, C.T. de, 1975. Substitution of labour and energy in agriculture and options for growth. Netherlands Journal of Agricultural Science 23, 145-162.

Wit, C.T. de, 1975. Over de bruikbaarheid van simulatiemodellen van ecologische systemen. In: Verhandeling van de Faculteit Landbouwwetenschappen te Gent, nr 13, XXIVe Post-Universitaire Onderwijsdag, 5 pp.

Arnold, G.W. & C.T. de Wit (Eds), 1976. Critical evaluation of systems analysis in ecosystems research and management. Simulation Monographs, Pudoc, Wageningen, 114 pp.

Keulen, H. van, C.T. de Wit & H. Lof, 1976. The use of simulation models for productivity studies in arid regions. In: Ecological Studies. Analysis and Synthesis, Vol. 19. Eds O.L. Lange, L. Kappen & E.D. Schulze. Springer Verlag, Heidelberg, pp. 408-420.

Sinclair, T.R. & C.T. de Wit, 1976. Analysis of the carbon and nitrogen limitations to soybean yield. Agronomy Journal 68, 319-324.

Wit, C.T. de, 1976. On the usefulness of simulation models of ecological systems (some thoughts for further discussion). In: Simulation of continuous systems. Proceedings of a seminar. August 25, 26, 27. SAHR/NRMG, pp. 126-130.

Wit, C.T. de & H.D.J. van Heemst, 1976. Aspects of agricultural resources. In: Chemical Engineering in a changing world. Ed. W.T. Koetsier. Proceedings of the Plenary Sessions of the First World Congress on Chemical Engineering. Amsterdam, June 28 - July 1, 1976. Elsevier Scientific Publishing Company, Amsterdam, pp. 125-145.

Laar, H.H. van, D. Kremer & C.T. de Wit, 1977. Maize. In: Crop photosynthesis: methods and compilation of data obtained with a mobile field equipment. Eds Th. Alberda et al. Agricultural Research Reports no. 865. Pudoc, Wageningen, pp. 12-21.

Sinclair, T.R., J. Goudriaan & C.T. de Wit, 1977. Mesophyll resistance and CO_2 compensation concentration in leaf photosynthesis models. Photosynthetica 11(1), 56-65.

Wit, C.T. de, 1977. Model studies on actual and potential herbage production in arid regions. Proceedings of the Seminar: Evaluation and mapping of tropical African rangelands. Bamako, Mali, 3-8 March 1975. ILCA, Addis Ababa, pp. 329-331.

Wit, C.T. de, 1977. Selectieve ontwikkeling van de landbouw in een kader van verantwoord energie- en omgevingsbeleid. In: Verslag van het Symposium Landbouwkundig Onderzoek in Nederland. Pudoc, Wageningen, pp. 93-108.

Wit, C.T. de, 1977. Problemen van de voedselproduktie. De Ingenieur 89(17), 334-338.

Wit, C.T. de, 1977. Modelle der Ertragsbildung als Brücke zwischen Prozess und System. In: Bio-physikalische Analyse pflanzlicher Systeme. Ed. K. Unger. VEB Gustav Fischer Verlag, Jena, pp. 19-30.

Wit, C.T. de, 1977. Concluding remarks. In: Environmental Effects on Crop Physiology. 5th Long Ashton Symp., 13-16 April 1975. Eds J.J. Landsberg & C.V. Cutting, Academic Press, London, pp. 365-369.

Wit, C.T. de, 1978. Simulatie van levende systemen. Landbouwkundig Tijdschrift/pt 90-8a, 237-240.

Wit, C.T. de, 1978. Een elementair model van gewasgroei. Landbouwkundig Tijdschrift/pt 90-8a, 275-281.

Wit, C.T. de, 1978. Summative address: One. In: Plant Relations in pastures. Ed. John R. Wilson, C.S.I.R.O., Australia, pp. 405-410.

Wit, C.T. de & J. Goudriaan, 1978. Simulation of ecological processes. Revised version. Simulation Monographs, Simulation Monographs, Pudoc, Wageningen, 183 pp.

Wit, C.T. de et al., 1978. Simulation of assimilation, respiration and transpiration of crops. Simulation Monographs, Pudoc, Wageningen, 148 pp.

Don Kirkham

Iowa State University
Ames, Iowa, USA

1908–1998

1983/4 — for their innovative contributions to the quantitative understanding of soil-water and other environmental interactions influencing crop growth and yield.

Professor Don Kirkham's pioneering mathematical and physical analyses of the movement of water in the soil has provided solutions to a number of important practical problems. An outstanding example is the establishment, on a rigorous scientific basis, of efficient and economic designs for land drainage systems that have had a significant and worldwide impact on the maintenance of soil fertility and land productivity over a range of irrigated and rain-fed agricultural situations. Many of the standard techniques of measurement, as well as of analysis and design, now used in the study of soil-water relations, were developed initially by Kirkham. His combination of rigorous mathematical-physical analysis, carefully designed laboratory experimentation and extensive field trials and measurements has had a seminal influence on soil physics through his training of a large number of students, many of whom later attained distinction in this branch of science, as well as through his extensive folio of scientific publications, textbooks and manuals, and his lectures.

Kirkham's long and productive career in soil physics, by its demonstration of the applicability of the fundamental laws of physics to soil and environmental problems and the advantages of mathematical analysis for their description and solution, heralded a new and productive era in soil science.

EXPANDED CURRICULUM VITAE

Title: Charles F. Curtiss Distinguished Professor Agriculture;
Professor of Soils and Physics, Professor Emeritus, Iowa
State University, Ames, Iowa, U. S. A.

Don Kirkham was born February 11, 1908 in Provo, Utah. He attended the public schools in Salt Lake City and Berkeley, California. During high school and his first year at the University of Utah he also studied music, graduating as a clarinetist from the McCune School of Music, Salt Lake City, in 1926. As a youth he worked summers on farmland owned by his father. He spent two and a half years in Germany as a missionary for his church in 1927-30, and then continued his education at Columbia University in New York City.

Columbia University awarded him the A. B. degree (with Honors in Physics) in 1933, the A. M. degree in 1934, and the Ph.D. degree, under S. L. Quimby, in 1938. His Ph.D. degree was received in Physics, his dissertation title being "The variation of the initial susceptibility with temperature and the variation of the magnetostriction and reversible susceptibility with temperature and magnetization in nickel."

From 1938 to 1940 he was instructor and assistant professor in mathematics and physics at Utah State University. Here, through his colleague, Professor Willard Gardner, he became interested in soil physics research and published several articles.

During the years of World War II (1940–46) he served as civilian scientist with the United States Navy, working on the research physics of anti-mine warfare. His naval work included designing, setting up, and operating the Navy's anti-magnetic mine program for all vessels using New York harbor. In 1946 he headed a group of Navy physicists at the Bikini Atom Bomb Tests.

During the war years, in addition to his Navy work, he continued study in theoretical soil physics problems and publication of articles. In 1946 he joined the staff of Iowa State University as Associate Professor of Agronomy and Physics, and was made Professor in 1949. In 1959 he was appointed Curtiss Distinguished Professor of Agriculture, the Distinguished Professor Title being this institution's highest recognition for faculty excellence. He served as Director of the Iowa State Water Resources Research Institute from 1964–73. Since 1973 he has continued to carry on a full program of teaching and research. He was named Professor Emeritus in 1978.

He was married and the father of three children.

Scientific Societies

International Water Resources Association
Netherlands Society of Agricultural Science

Iowa Academy of Science
Council for Agricultural Science and Technology

Honorary Societies

Sigma Xi, national honorary scientific research society
Gamma Sigma Delta, national honorary agricultural society
Phi Kappa Phi, national honorary scholastic society

DESCRIPTION OF SCIENTIFIC CONTRIBUTION

The Food and Agricultural Organization, Rome, Italy, states that 42 million hectares of land need to be reclaimed for agriculture. Most of this, in particular irrigated land, will need drainage. According to the same FAO report, the number of hectares of land under irrigation needing drainage increased from 194 million to 226 million in the years 1964 to 1974. (Annual Report 1978, International Institute for Land Reclamation and Improvement, Wageningen, the Netherlands, p. 7, p. 15.)

Kirkham's early experiments on land drainage pertained to land overlying an artesian aquifer under pressure. His results showed that spacing of horizontally laid drain tiles over such an area would not be economical. Vertical pumped drainage wells should be used. He made sand tank models showing how the artesian drainage occurred and demonstrated that the seepage patterns depended on the boundary conditions rather than on whether the soil was a clay or a sand or a more permeable soil (3)*, (4).

He next considered tile drainage of land over an impervious layer. He obtained equations for the streamlines, or water flow paths, and matched them against the sand tank model flow paths obtained with dyes. A striking discovery of the theory was that by placing a drain line deeper and closer to ait impervious layer, the flow would be less (rather than more) when compared with the flow from a less deep drain line in the soil. Figure 6 of a 1940 paper (7) shows this phenomenon, along with his theoretical streamlines matched against sand tank model streamlines. Details and amplified theory of this problem were published after World War II, in a paper (15) which, although one in agriculture, received the 1948 mathematical award of the Iowa section of the American Mathematical Society. The next year his paper (23) on the seepage of water into drainage ditches was designated the best paper of the Iowa Academy of Science for 1949. In this paper hederived theoretically the amount of water that would flow into the drainage ditches. He gave the result in an explicit formula; other researchers in the past had failed to find an explicit formula.

In an earlier paper (9), Kirkham had attacked the problem of simultaneous upward and downward seepage flow into tile drains with excellent agreement of theoretical streamlines with experimental streamlines. But this paper, like others up to then, was for homogeneous soil. So Kirkham now analyzed the problem of non-homogeneous soil and with a resulting two-part paper (31), (32), he received the American Geophysical Union's citation for the best paper in hydrology of the year 1951.

These earlier papers were for ponded water, that is, completely saturated soil and represented thus an extreme condition. He next considered the more common condition of a curved water table resulting from successive recharges of water as from irrigation, or as from successive rains. By first neglecting the hydraulic friction in the soil in the water table arch region and then later correcting for this neglect, he derived an explicit relation for the height of water table to be found between drain lines at certain depths and spacings. He compared his theoretical results with field data and found good agreement (64). In a paper with a graduate student as senior author, the solution to this drainage problem was put in a graphical form so that the various physical factors that entered into the problem could be considered in the graph and the solution obtained graphically for the depth and spacing of drains (98), (99), (99a). This same student in working for his Ph.D. degree considered stratified soil. Two other papers resulted in which the ratio of hydraulic conductivities of the layered soil could be taken into account. This long two-part paper was published in the Irrigation and Drainage Division of the American Society of Civil Engineers (173), (174).

All the foregoing work was for a simplified steady state condition of either ponded water or steady recharge to the drain tubes. Kirkham next analyzed the problem of a falling water table. He found equations which predicted where the water table arch would be after a certain time when the steady recharge was discontinued. He compared his theoretical results with actual field data that he had obtained with other scientists in the Netherlands many years before and with other field data. There was excellent agreement (130).

In all the above work involving tile drains, it was assumed that the water entered the drain tube uniformly over its length. This would be true if there were an envelope of coarse material around the drain tube. Actually drain pipes are about 30 cm to 60 cm long and have about 1 to 6 mm spacing between each individual drain pipe. At least this has been the condition up until recent times when plastic tubes, perforated to let the water enter, are pulled into the soil. To see just how much the cracks of individual tile lengths or the perforations in tubes would control drainage flow if there was not a pervious envelope around them, Kirkham authored or co-authored several articles — (28), (29), (197), (200). A first one, in two parts, (28) and (29), was co-authored by one of his joint soil physics-agricultural engineering Ph.D. candidates, who conceived the idea of using plastic flexible tubing instead of clay pipe in soil. The two-part paper on the influence of perforations on drainage flow to drain tubes received a citation of merit from the American Society of Agricultural Engineers. Perforated flexible plastic tubing is now used almost universally in drainage. Kirkham's theoretical work on the effect of perforations, cracks, and other openings in drain tubes continues to be of great importance. According to pp. 13-14 of the Annual Report 1978, International Institute of Land Reclamation and Drainage, cited earlier, the study of envelope materials around perforated drain tubes is a number one problem.

As was pointed out in the first of Kirkham's studies, drainage of land overlying an artesian aquifer is not economical when done with horizontal drains. He concluded it is better to use vertical drains. These vertical wells are pumped and if the water is of high quality the drainage procedure lowers the water table while developing irrigation water. If the water is salty it can be mixed with fresh water and still be used. Papers written in 1971 analyzed the vertical well drainage problem (177), (178).

In an artesian basin in Utah, it had been observed that tile drains would work in certain areas where it might be predicted that such drainage would be uneconomical. Kirkham reasoned that the drainage could be economical, provided there existed an upper layer of soil near the surface, highly permeable compared with an existing stratum of soil overlying the artesian gravel. With a student and a colleague, he analyzed this problem in a paper recently published in the Irrigation and Drainage Division of the American Society of Civil Engineers (210). Just how much more permeable the upper soil layer must be, compared with that overlying the aquifer, is pointed out in the paper, which contains numerous graphs of a number of situations to be used in practical application.

All the above papers on water movement into horizontal or vertical drainage pipes depend on knowledge of a soil coefficient called the hydraulic conductivity. This coefficient gives the ease with which the soil conducts water under an energy gradient. Early, Kirkham recognized that the use of soil cores would not give the hydraulic conductivity accurately because of too small samples and because of their disturbed condition (10). He reasoned that, instead of measuring the hydraulic conductivity of soil in a core removed from the soil, measuring the conductivity of the soil around the location where the core was removed or its equivalent would be a better procedure. A practical way to proceed would be to auger a hole into the soil, while simultaneously driving a casing into the hole to seal off the walls of the hole against leakage, except for a cavity at its bottom. Flow would be allowed to come into the cavity, and, by rating this flow, the conductivity of the soil around the cavity could be determined. This method, called the piezometer method, is used for measuring the conductivity of field soil from layer to layer in the soil (21).

If a casing is not driven into a soil hole while it is being augered out, one obtains simply a cylindrical auger hole in the soil and it can be used to measure

49. Rollins, R.L., M.G. Spangler, and Don Kirkham. 1954. Movement of soil moisture under a thermal gradient. Highway Research Board Proc. 33: 492-508.
50. J-2605. Kirkham, Don, and W.V. Bartholomew. 1955. Equations for following nutrient transformations in soil, utilizing tracer data: II. Soil Sci. Soc. Amer. Proc. 19: 189-192.
51. J-2628. Maasland, Marinus, and Don Kirkham. 1955. Theory and measurement of anisotropic air permeability in soil. Soil Sci. Soc. Amer. Proc. 19: 395-400.
52. J-2641. Stone, J.F., Don Kirkham, and A.A. Reed. 1955. Soil moisture determination by a portable neutron scattering moisture meter. Soil Sci. Soc. Amer. Proc. 19: 419-423.
53. van Schilfgaarde, Jan, Don Kirkham, and R.K. Frevert. 1956. Physical and mathematical theories of tile and ditch drainage and their usefulness in design. Agricultural Experiment Station, Iowa State College, Research Bulletin No. 436: 667-706.
54. J-2773. Swartzendruber, Dale, and Don Kirkham. 1956. Capillary fringe and flow of water in soil. Soil Sci. 81: 473-484.
55. J-2832. Swartzendruber, Dale and Don Kirkham. 1956. Capillary fringe and flow of water in soil. II. Experimental results. Soil Sci. 82: 81-95.
56. Kirkham, Don. 1956. Book review of L.D. Baver,s "Soil Physics". Soil Sci. Soc. Amer. Proc. 20: 296; also in Agronomy Journal, 48:195. 1956.
57. J-2913. Kirkham, Don. 1956. Mathematical aspects of soil nitrogen studies. Synposium on use of isotopes in agriculture sponsored by Argonne National Laboratory. Government Printing Office, Washington 25, D.C. January, 1956.
58. Kirkham, Don. 1957. Theory of seepage of ponded water into drainage facilities, p.139-181. In J.N. Luthin (ed.) Drainage of Agricultural Lands. Agron. Monogr. 7, ASA and SSSA, Madison, Wisconsin.
59. J-2998. Kirkham, Don. 1958. Graphs and formulas for zeros of cross product Bessel functions. Journal of Mathematics and Physics 36: 371-377.
60. J-3033. Schwab, G.O., Don Kirkham and H.P. Johnson. 1957. Effect of tile spacing on crop yield and water table level in a planosol soil. Soil Sci.Soc. Amer. Proc. 21: 448-452.
61. J-3052. Willis, W.O., W.E. Larson and Don Kirkham. 1957. Corn growth as affected by soil temperature and mulch. Agronomy Journal 49: 323-328.
62. J-3073. Adams, J.E., Don Kirkham and D.R. Nielsen. 1957. A portable rainfall simulator, infiltrometer and physical measurements of soil in place. Soil Sci. Soc. Amer. Proc. 21: 473-477.
63. J-3106. Kirkham., Don. 1957. Potential and capacity of concentric coaxial capped cylinders. Journal of Applied Physics 28: 724-731.
64. J-3179. Kirkham, Don. 1958. Seepage of steady rainfall through soil into drains. Trans. Amer. Geophys. Union 39: 892-908.

65. J-3247. Jackson, Ray D. and Don Kirkham. 1958. Method of measurement of the real thermal diffusivity of moist soil. Soil Sci. Soc. Amer. Proc. 22: 479-482.
66. J-3254. Kirkham., Don. 1958. Theory of seepage into an auger hole above an impermeable layer. Soil Sci. Soc. Amer. Proc. 22: 204-208.
67. Nestroy, O. 1958. Report of two lectures delivered in German by Don Kirkham. Wasserhaushalt und Pflanzenwachstum. Oesterreichische Wasserwirtschaft. Hochschule fur Bodenkultur, Vienna, Austria, March 17 and 18.
68. Kirkham., Don. 1958. Advanced Soil Physics. A set of twenty lectures. Soil Science Institute, Ghent, Belgium (mimeographed), 147 pages.
69. J-3255. Burrows, W.C., and Don Kirkham. 1958. Measurement of field capacity with a neutron meter. Soil Sci. Soc. Amer. Proc. 22: 103-105.
70. J-3329. Adams, J.E., Don Kirkham and W.H. Scholtes. 1958. Soil erodibility and other physical properties of some Iowa soils. Iowa Journal of Science 32: 485-540.
71. J-3331. Scholtes, W.H. and Don Kirkham. 1959. The use of radiocarbon for dating soil strata. (Belgian Journal of) Pedologie 7: 316-323.
72. Nielsen, D.R., Don Kirkham and R.E. Phillips. 1959. Synthetic soil conditioners: Soil aggregates and corn yields. Iowa Farm Science 13: 8-10.
73. J-3483. Nielsen, D.R., Don Kirkham and W.C. Burrows. 1958. Solids-water-air space relations of some Iowa soils. Iowa State College Journal of Science 33: 111-116.
74. J-3566. Kirkham, Don, M.F. DeBoodt and L. De Leenheer. 1959. Air permeability at the field capacity as related to soil structure and yields. Mededelingen van de Landbouwhogeschool van de Staat te Gent 24: 377-391.
75. J-3567. Kirkham, Don, M.F. DeBoodt and L. De Leenheer. 1959. Modulus of rupture determination on cylindrical core samples. Mededelingen van de Landbouwhogeschool van de Staat te Gent 24: 369-376.
76. J-3568. Kirkham, Don, L. De Leenheer and M.F. DeBoodt. 1959. Physical measurements and yields on some loam and clay soils in Belgium. Mededelingen van de Landbouwhogeschool van de Staat te Gent 24: 324-334.
77. J-3598. Maasland, Marinus and Don Kirkham. 1959. Measurement of the permeability of tri-axially anisotropic soil. Proc. Amer. Soc. Civil Engineers, Soil Mechanics and Foundations Division, No. 2063: 25-34.
78. Kirkham, Don, M.F. DeBoodt and L. De Leenheer. 1959. Modulus of rupture determination on undisturbed soil core samples. Soil Sci. 87: 141-144.
79. Kirkham, Don. 1959. Review of Drainage Investigations and Design, Adana Plain Irrigation and Drainage Plan. Republic of Turkey Ministry of Public Works and Tippetts-Abbett-McCarthy-Stratton Engineers, Adana, Turkey and New York City, (mimeographed), 109 pages.

80. Kirkham, Don. 1959. Lorentz, Einstein and the discovery of relativity, a letter to the Editor. American Scientist 47: 232A, 236A.
81. Kirkham, Don. 1959. Abstract: Method of solving a class of mixed boundary value problems. The Am. Math. Monthly 66: 627.
82. J-3610. Nielsen, D.R., Don Kirkham and W.R. van Wijk. 1959. Measuring water stored temporarily about the field moisture capacity. Soil Sci. Soc. Amer. Proc. 23: 408-412.
83. J-3644. Phillips, R.E., C.R. Jensen and Don Kirkham. 1960. Use of radiation equipment for plow-layer density and moisture. Soil Science 89: 2-7.
84. J-3670. Kirkham, Don. 1959. Exact theory of flow into a partially penetrating well. Jour. Geophys. Res. 64: 1317-1327.
85. J-3713. Nielsen, D.R., Don Kirkham and E. R. Perrier. 1960. Soil capillary conductivity: comparison of measured and calculated values. Soil Sci. Soc. Amer. Proc. 24: 157-160.
86. J-3795. Stone, J.F., R.H. Shaw and Don Kirkham. 1960. Statistical parameters and reproducibility of the neutron method of measuring soil moisture. Soil Sci. Soc. Amer. Proc.24: 435-438.
87. J-3796. Kirkham, Don. 1960. Seepage into ditches from a plane water table overlying a gravel substratum. J. Geophys. Res. 65: 1267-1272.
88. J-3803. DeBoodt, M.F., L. De Leenheer and Don Kirkham. 1961. Soil aggregate stability indexes and crop yields. Soil Science 91: 138-146.
89. J-3830. Kunze, R.J. and Don Kirkham. 1961. Deuterium and the self-diffusion coefficient of soil moisture. Soil Sci. Soc. Amer. Proc. 25: 9-12.
90. J-3874. Grover., Ben L. and Don Kirkham. 1961. A glassbead-glycerol model for non-steady-state tile drainage. Soil Sci. Soc. Amer. Proc. 25: 91-94.
91. J-3878. Grover, Ben L., J.T. Ligon and Don Kirkham. 1960. Operational characteristics of the laterals near the edge of a tile drainage system. J. Geophys. Res. 65: 3733-3738.
92. Kirkham, Don. 1960. Book review of Dr. H. Franz's treatise "Feldbodenkunde. Soil Sci. Soc. Amer. Proc. 24.
93. J-3902. Nielsen, D.R., Don Kirkham and W.R. van Wijk. 1961. Diffusion equation calculations of field soil water infiltration profiles. Soil Sci. Soc. Amer. Proc. 24: 165-168.
94. J-3904. Kirkham, Don. 1961. An upper limit for the height of the water table in drainage design formulas. Proc. Int. Soil Sci. Soc., Seventh Congress, Madison, Wisconsin, August, 1960. Commission VI, Vol 1: 486-492.
95. J-3919. Bartholomew, W.V. and Don Kirkham. 1961. Mathematical descriptions and interpretations of culture induced soil nitrogen changes. Proc. Int. Soil Sci. Soc., Seventh Congress, Madison, Wisconsin, August, 1960. Commission III. Vol. 2: 471-477.

96. Kirkham, Don. 1961. "Soil Physical Properties", article in Agricultural Engineer,s Handbook edited by C. B. Richey, Paul Jacobson and Carl W. Hall. McGraw Hill Book Co. Inc., New York.
97. J-3978. Phillips, R.E. and Don Kirkham. 1962. Soil compaction in the field - and corn growth. Agronomy Journal 54: 29-34.
98. J-3983. Toksöz, S. and Don Kirkham. 1961. Graphical solution and interpretation of a new drain-spacing formula. J. Geophys. Res. 66: 509-515.
99. J-3983a. Toksöz, Sadik and Don Kirkham. 1961. Yeni kir dren araligi - fornulunum grafik cozumu ve tefsiri. Tamin Bakanligi Topraksu 9: 72-78.
100. J-4082. Phillips, R.E. and Don Kirkham. 1962. Mechanical impedance and corn seedling root growth. Soil Sci. Soc. Amer. Proc. 26: 319-322.
101. J-4142. Rogowski, A.S. and Don Kirkham. 1962. Moisture, pressure, and formation of water-stable soil aggregates. Soil Sci. Soc. Amer. Proc. 26: 213-216.
102. J-4152. Kirkham, Don. 1961. Soil Physics 1936-1961 and a look ahead. Soil Sci. Soc. Amer. Proc. 25: 423-427.
103. J-4169. Kirkham, Don. 1961. Evaluation of (cot x sin 2mx ln(sin x/sin a)dx, etc. Iowa Academy of Science 68: 416-426.
104. Kirkham, Don. 1964. Chapter on Soil Physics. Handbook of Hydrology. McGraw Hill, New York.
105. Kirkham, Don. 1961. Lectures on agricultural drainage. Institute of Land Reclamation, College of Agriculture, Alexandria University, Alexandria, Egypt. 152 pages.

<u>Patents (Assigned to Iowa State University Research Foundation)</u>

106. C.H.C. van Bavel - student of Don Kirkham's. Soil Stabilizer and Soil Treated Thereby. USA patent no. 2, 793. May 28, 1957.
107. Don Kirkham and John E. Adams. Apparatus for measuring diffusion coefficients of soil in place. USA patent no. 2, 913. Nov. 4, 1959.
108. Don Kirkham and John E. Adams. Apparatus for measuring erodibility, runoff, infiltration and other physical properties of soil in place. USA patent no. 2, 923, 148. Feb. 29, 1960.
109. Don Kirkham and Ben L. Grover. Apparatus for measuring fluid permeability of porous materials USA patent no. 2, 949, 766. August 23, 1960.
110. Don Kirkham and John F. Stone. Apparatus for measuring the water content of soil. USA patent no. 2, 999, 160. Sept. 5, 1961.
111. Kirkham, Don and Gizbili Muzaffer. 1961. Tarla Drenaji. Devlet su isleri. Ankara, Turkey. 85 pages. (Drainage Lectures in Turkish, State Hydraulic Works).

219. J-8723. Kirkham, Don, and Lyle Prunty. 1977. Upslope recharge, downslope waterlogging and interceptor drains. The Third National Drainage Symposium Proceedings, American Society of Agricultural Engineers. P.O. Box 410, St. Joseph, Michigan 49085.
220. J-8824. Mousavi, S.F., and Don Kirkham. 1978. Porous media tests of groundwater mounds. Soil Science 125: 160-164.
221. J-8835. Kirkham, Don. 1976. Soil Physics and Plant Growth. In Anuario del Centro de Edafologia y Biologia Aplicada de Salamanca.
222. Kirkham, Don. 1978. Water, our prime resource. Commencement Address, Iowa State University, February 25, 1978. (Unpublished).
223. J-8853. Cushman, John, and Don Kirkham. 1978. A two-dimensional linearized view of one dimensional unsaturated-saturated flow. Water Resour. Res. 14: 319-323.
224. J-8865. Kirkham, Don, and M. Olga Sotres. 1978. Casing depths and solute travel times to wells. Water Resour. Res. 14: 237-243.
225. Kirkham, Don. 1979. Some concepts and misconcepts of drainage theory. Virgil Overholt Drainage Hall of Fame Award talk, at Ohio State University, March 9, 1979, Columbia, Ohio. (Unpublished).
226. J-8881. Battikhi, Anwar, and Don Kirkham. 1979. Casing and leak depths, and solute travel times to wells. Water Resources Bulletin 15(4): 1004-1015.
227. J-9111. Kirkham, Don, and Olga Sotres. 1979. Influence of cover slab diameter on solute travel time to wells. Water Resour. Res.15(4): 941-948.
228. J-9126. Prunty, Lyle, and Don Kirkham. 1979. A drainage model for a reclaimed surface mine. Soil Sci. Soc. Amer. J. 43: 28-34.
229. J-9339. Cushman, John H., Don Kirkham, and Roy F. Keller. 1979. A Galerkin in time, linearized finite element model of 2-dimensional unsaturated porous media drainage. Soil Sci. Soc. Amer. J. 43: 638-641.
230. J-9369. Prunty, Lyle, and Don Kirkham. 1980. Seepage versus terrace density in reclaimed mineland soil. Journal of Environ. Qual. 9: 273-278.
231. J-9518. Henning, Stanley J., Don Kirkham, and Stephen B. Affleck. 1979. Proceedings of the 8th Conference of the International Soil Tillage Research Organization, ISTRO, Bundesrepublik Deutschland. Stuttgart, Hohenheim, Germany. Sept. 8-15, 1979, Vol. 1: 103-108.
232. No J-no.. Kirkham, Don. 1980. Soil Physics. In Encyclopedia of Physics, pp. 927-928. Addison-Wesley Publishing Co., Reading, Mass.
233. J-9924. Kanwar, R.S., J.L. Baker, H.P. Johnson, and Don Kirkham. 1980. Nitrate movement with zero-order denitrification in a soil profile. Soil Sci. Soc. Amer. J. 44: 898-902.
234. J-10055. Kanwar, R.S., H.P. Johnson, and Don Kirkham. 1982. Transport of nitrate and gaseous denitrification in soil columns during leaching. J. Hydrol. 55: 171-184.

235. J-10221. Troeh, Frederick R., Jalal D. Jabro, and Don Kirkham. 1982. Gaseous diffusion equations for porous materials. Geoderma 27: 239-243.
236. Brandyk, Tomasz, and Don Kirkham. 1982. Examination of diffusivity theory for muck and sands. Annals of Warsaw Agricultural University, Land Reclamation 19: 3-8.
237. J-10676. Kirkham, Don. 1982. Free surface potential theory for a gravity well. In Papers International Symposium. Polders of the World, Vol. II, pp. 13-23. International Institute for Land Reclamation and Improvement, Wageningen, The Netherlands.
238. Mousavi, S.F., and Don Kirkham. 1982. Mineland seepage and its control., Iran Agricultural Research 1 (no. 2): 164-179.
239. Powers, W.L., and Don Kirkham. 1984. Advanced Soil Physics, revised edition, 539 pp., Robert E. Krieger Publishing Company, Inc., Malabar, Florida 32950.
240. J-11127. Modaihsh, Abdullah Saad, Robert Horton, and Don Kirkham. 1985. Soil water evaporation suppression by sand mulches. Soil Science 137: 357-361.
241. On my lecture tour of the People's Republic of China 1985, I was presented two volumes of translations into Chinese of sections from Kirkham and Powers Advanced Soil Physics, Vol. I, October, 1980, dealt with Unsaturated Water Flow; Vol. II, 1982, dealt with Saturated Water Flow. The publisher was Tsinghua University, Department of Hydraulics, Water Resources, Irrigationand Drainage Division. Beijing, People's Republic of China. I did not know until 1985 about these published translations.
242. J-12554. Kirkham, Don, and Robert Horton. 1986. Subirrigation with drainage for drought-prone areas. Proceedings, International Conference of Water Resources Needs and Planning in Drought-Prone Areas, Khartoum, Sudan, Dec. 6-12. Published by Khartoum University Press. Designated a "distinguished" paper, pp. 957-979.
243. J-12603. Walczak, R.T., R.R. van der Ploeg, and Don Kirkham. 1988. An algorithm for the calculation of drain spacings for layered soils. Soil Sci. Soc. Amer. J. 52: 336-340.
244. J-13651. Kirkham, Don, and Robert Horton. 1989. Groundwater management by a dual-pipe subirrigation system. Proceedings of the Benidorm (Spain) Symposium Oct. 1989 of the International Association of Hydrological Sciences (IAHS) Publ. 188, pp. 53-56.
245. Kirkham, Don. 1991. Article: Soil Physics. P. 1124-1126. In Rita G. Lerner and George L. Trigg (eds.) Encyclopedia of Physics. VCH Publishers, Inc., New York.
246. J-13907. Kirkham, Don, and Robert Horton. 1992. The stream function of potential theory for a dual-pipe subirrigation-drainage system. Water Resour. Res. 28(2): 373-387.

247. J-13655. Kirkham, Don, and Robert Horton. 1993. Modeling water flow from subirrigation with drainage. Soil Sci. Soc. Amer. J. 57: 1451-1457.
248. Kirkham, Don, R.R. van der Ploeg, and R. Horton. 1997. Potential theory for dual-depth subsurface drainage of ponded land. Water Resour. Res. 33: 1643-1654.
249. van der Ploeg, R.R., Maria Marquardt, and Don Kirkham. 1997. On the history of the ellipse equation for soil drainage. Soil Sci. Soc. Am. J. 61: 1604-1606.
250. van der Ploeg, R. R., R. Horton and Don Kirkham. 1999. Steady flow to drains and wells. p.213-263. In: R.W. Skaggs and J. van Schilfgaarde (eds.) Agricultural Drainage. Agronomy Monograph No. 38, ASA/CSSA/SSSA, Madison (WI), USA

Robert H. Burris

University of Wisconsin
Madison, Wisconsin, USA

1984–5 — for his pioneering fundamental research on the mechanisms of biological nitrogen fixation and its application in crop production.

SCIENTIFIC BACKGROUND

In the air above every hectare of land there are about 78 tons of elemental nitrogen. The latter is chemically inert and, under ordinary conditions, does not react with other elements. Hence most crop plants are likely to be starving in this sea of nitrogen. Only when the nitrogen in the air is combined with other elements it is possible for plants to use this nitrogen in their growth processes.

The mere fact that some bacteria are able to use the elemental form of nitrogen that exists in the atmosphere has been known for nearly 100 years since its discovery by the Russian biochemist S. N. Vinogradsky. Thus, these nitrogen-fixing bacteria can grow in the absence of combined nitrogen and, at the same time, produce nitrogenous substances in the soil that may be used later by crop plants.

However, during the first fifty years since the discovery of biological nitrogen fixation by soil-inhabiting bacteria, very little information was available as regards the biochemical mechanisms involved. This was quite surprising in view of the paramount importance of this process in the nitrogen budget of our globe.

The state of the knowledge regarding this wonder of nature has changed dramatically in the late 1930's and early 1940's when Professor Burris investigations on the basic chemistry and physiology of the microorganisms that fix nitrogen, actually opened up an entire new vista of basic research on biological nitrogen fixation. He was the first to use a radioactive isotope of nitrogen (15N) in this study, which led to the discovery of ammonia to be the key intermediate in this biochemical process.

For over 45 years, following these pioneering efforts, Professor Burris and his colleagues have purified and studied in great detail and depth the complex of enzymes involved, and the metabolic pathways followed in this natural phenomenon. At the same time they developed novel techniques for reliable, quantitative assay of the nitrogen fixing activity of microorganisms.

The scope of his research in this particular area has not been limited to the well recognized nitrogen-fixing bacteria that live in symbiosis with leguminous, plants in nodules formed on their roots. His studies also embraced the free-living microorganisms that are able to fix atmospheric nitrogen. Mention should be made in this connection of Prof. Burris pioneering work on *Azospirillum* — a free-living nitrogen-fixing genus of bacteria associated with plant roots, in particular cereal plants.

One may sum up Burris' contributions to fundamental biochemical research on nitrogen fixation by concluding that his very careful, thorough investigations have been in the heart of the efforts to elucidate the mechanism of this life-supporting process. He has thus rightly earned the title of 'world doyen of nitrogen fixation'.

Because of the pressing food needs of the world on one hand, and the rising cost of chemical nitrogen fixation for use in mineral fertilizers on the other, the subject of biological nitrogen fixation is currently receiving greatly increased attention throughout the world.

The lifelong fundamental studies of Prof. Burris provided the scientific background for the now worldwide practice of inoculating legume crops with industrially manufactured pre-cultured strains of nitrogen-fixing bacteria. The latter are capable of supplying essentially all the nitrogen requirements of the plan. Furthermore, the highly sensitive techniques developed by Prof. Burris for measuring this biological activity, proved most instrumental in the selection of more effective strains of bacteria for this purpose.

More recently, a new science-based biotechnology industry has been established for the production of cultures of the aforementioned *Azospirillum.* These are intended for inoculation of graminaceous crops in order to replace mineral nitrogen fertilizers. Enhanced yields of these important staple food crops of the world can thus be achieved at a remarkably reduced cost.

CURRICULUM VITAE

Date and Place of Birth: April 13, 1914, Brookings, South Dakota

EDUCATION:

South Dakota State College, B.S. 1936 (Chemistry)
University of Wisconsin - Madison, M.S. 1938 (Bacteriology)
University of Wisconsin - Madison, Ph.D. 1940 (Bacteriology)

POSITIONS:

National Research Council Postdoctoral Fellow, Columbia University, 1940-41
Postdoctoral Appointment in ~Bacteriology, University of Wisconsin, 1941-44
Assistant Professor of Biochemistry, University of Wisconsin, 1944-46
Associate Professor of Biochemistry, University of Wisconsin, 1946-51
Professor of Biochemistry, University of Wisconsin, 1951-date
Guggenheim Fellow, University of Helsinki (Finland) and Cambridge University (England), 1954
Chairman, Department of Biochemistry, University of Wisconsin, 1958-70
W. H. Peterson Professor of Biochemistry, 1976-date

PROFESSIONAL SOCIETIES:

National Academy of Sciences (Chairman, Section of Botany 1971-74; Executive Committee, Assembly of Life Sciences 1973-78; Chairman, Division of Biological Sciences 1977-78)
American Academy of Arts and Sciences
American Association of the Advancement of Science (Fellow)
American Society of Biological Chemists
American Chemical Society
American Society of Plant Physiologists (President, 1960)
American Society of Microbiologists
Biochemical Society (England)
American Philosophical Society
Japanese Society of Plant Physiologists
Sigma Xi
Gamma Alpha
Phi Sigma

HONORS AND AWARDS:

National Research Council Postdoctoral Fellowship, 1940
Guggenheim Fellowship, 1954
Elected to National Academy of Sciences, 1961
Honorary Doctor of Science, South Dakota State University, 1966
Merit Award, Botanical Society of America, 1966
Stephen Hales Award, American Society of Plant Physiologists, 1968

Prather Lecturer, Harvard University, 1973
Elected to American Academy of Arts and Sciences, 1975
W. H. Peterson Professorship in Biochemistry (University of Wisconsin), 1976
Charles Thom Award, Society of Industrial Microbiologists, 1977
Charles Reid Barnes Life Membership Award, American Society of Plant Physiologists, 1977
Edward W. Browning Award, American Society of Agronomy, 1978
Elected to American Philosophical Society, 1979
National Medal of Science presented by the President of the United States, January 14, 1980
Honorary Member, Wisconsin Academy of Sciences, Arts & Letters, 1980
Hilidale Professorship, University of Wisconsin, July, 1982.
Service on National Committees and Advisory Boards:
Study Section, National Science Foundation, 1950-54
President, American Society of Plant Physiologists, 1960
U.S. State Department Representative to Evaluate Biochemistry in India, 1960
Study Section, NIH Biochemistry Training Grants (Chairman of panel for two years), 1962-67
National Biotron Committee, 1968-76
Advisory Committee, NIH General Support, 1969-71
National Academy of Sciences:
Chairman, Section of Botany, 1971-74
Executive Committee, Assembly of Life Sciences, 1973-79
Chairman, Division of Biological Sciences, 1977-79
Liaison Representative to U.S.D.A. Competitive Grants Program, 1977
National Research Council Commission on Human Resources (Glaser Committee) -Subcommittee, 1974-75, Central Committee, 1976-77.
Board on Human Resource Data Analysis, Commission of Human Resources, NRC, 1974-77
U.S. State Department Representative to Evaluate Biological Nitrogen Fixation Research in Brazil, 1976-77
Advisory Committee to "Program for International Cooperation in Training and Research on Nitrogen Fixation in the Tropics," 1974-78
Advisory Committee to Biological Division, Brookhaven National Laboratory 1976-80; Chairman, 1979-80
U.S. National Committee for the International Union of Biological Sciences, 1978-81
Committee on Planetary Biology and Chemical Evolution (Space Science Board, NRC), 1980-81
Committee on Nitrogen Fixation and Photosynthesis Research in the People's Republic of China, Chairman, 1979-81 (Committee in Scholarly Communication with the People's Republic of China, National Acad. Sci.)

Service on Editorial Boards:

Burgess Publishing Co., Life Science Series, 1960-68
Plant Physiology, 1962-66; 1972-date
Biological Handbook Series of FASEB, 1966-72
Physiological Reviews, 1966-72
Chemistry-Biology Interface Series, 1968-72
American Society of Plant Physiologists, Monograph Series, 1972-74
Analytical Biochemistry, 1975-80
Study Section, NIH Biochemistry Training Grants (Chairman of panel for two years), 1962-67
National Biotron Committee, 1968-76
Advisory Committee, NIH General Support, 1969-71
National Academy of Sciences:
  Chairman, Section of Botany, 1971-74
  Executive Committee, Assembly of Life Sciences, 1973-79
  Chairman, Division of Biological Sciences, 1977-79
  Liaison Representative to U.S.D.A. Competitive Grants Program, 1977-date
National Research Council Commission on Human Resources (Glaser Committee) - Subcommittee, 1974-75, Central Committee, 1976-77.
Board on Human Resource Data Analysis, Commission of Human Resources, NRC, 1974-77
U.S. State Department Representative to Evaluate Biological Nitrogen Fixation Research in Brazil, 1976-77
Advisory Committee to "Program for International Cooperation in Training and Research on Nitrogen Fixation in the Tropics," 1974-78
Advisory Committee to Biological Division, Brookhaven National Laboratory 1976-80; Chairman, 1979-80
U.S. National Committee for the International Union of Biological Sciences, 1978-81
Committee on Planetary Biology and Chemical Evolution (Space Science Board, NRC), 1980-81
Committee on Nitrogen Fixation and Photosynthesis Research in the People's Republic of China, Chairman, 1979-81 (Committee in Scholarly Communication with the People's Republic of China, National Acad. Sci.)

Service on Editorial Boards:

Burgess Publishing Co., Life Science Series, 1960-68
Plant Physiology, 1962-66; 1972-date
Biological Handbook Series of FASEB, 1966-72
Physiological Reviews, 1966-72

Chemistry-Biology Interface Series, 1968-72
American Society of Plant Physiologists, Monograph Series, 1972-74
Analytical Biochemistry, 1975-80

LIST OF MOST SIGNIFICANT PUBLICATIONS

W.W Umbreit and RH Burris. Composition of Soybean Nodules. Soil Sci. 45:111-126 (1938).

R.H. Burris and P.W. Wilson. Respiratory Enzyme Systems in Symbiotic Nitrogen Fixation. Cold Spring Harbor Symposia on Quant. Biol. 7:349-361 (1939).

Robert H. Burris and Charles E. Miller. Applications of N^{15} to the Study of Biological Nitrogen Fixation. Science 93:114-115 (1941).

P.W. Wilson, R.H. Burris and W.B. Coffee. Hydorgenase and Symbiotic Nitrogen Fixation. J. Biol. Chem. 147:475-481 (1943).

R.H. Burris, F.J. Eppling, H.B. Wahlin and P.W. Wilson. Detection of Nitrogen Fixation with Isotopic Nitrogen. J. Biol. Chem. 148:349-357 (1943).

R.H. Burris and P.W. Wilson. Ammonia as an intermediate in Nitrogen Fixation by Azotobacter. J. Bacteniol. 52:505-512 (1946).

N.E. Tolbert and R.H. Burris. Light Activation of the Plant Enzyme which Oxidizes Glycolic Acid. J. Biol. Chem. 186:791-804 (1950).

I. Zelitch, E.D. Rosenblum, R.H. Burris and P.W. Wilson. Isolation of the Key Intermediate in Biological Nitrogen Fixation by Clostnidium. J. Biol. Chem. 191:295-298 (1951).

R. Schollhorn and R.H. Burris. Acetylene as a Competitive Inhibitor of N_2 Fixation. Proc. Natl. Acad. Sci. 58:213 (1976).

Harry C. Winter and R.H, Burris. Stoichiometry of the Adenosine Tniphosphate Requirement for N_2 Fixation and H_2 Evolution by a Partially Purified Preparation of *Clostridium pasteunianum*. J. Biol. Chem. 243:940-944 (1968).

BRIEF DESCRIPTION OF SCIENTIFIC CONTRIBUTION:

Prof. Burris discovered the role of hydrogenase in N_2 fixation. He was the first to use isotopic nitrogen to study the intermediate and mechanism of nitrogen fixation. He and his students discovered the ATP requirement for nitrogen reduction and has studied its stoichiometry. He and his students discovered glycolic oxidase of plants which was demonstrated to be the key reaction in photorespiration.

A former student in Australia, and Burris himself independently discovered acetylene reduction by nitrogenase and Burris has used this reaction for quantitative studies of eutrophication in lakes and rivers.

His accomplishments have been widely recognized (see list of Honors) and he was a highly worthy recipient of the Wolf Foundation Prize in Agriculture.

Sir Ralph Riley, F.R.S.

Agricultural and Food Research Council
London, United Kingdom

1924–1999

1986 — for fundamental research in cytogenetics of wheat, providing the basis for genetic improvement of cereal grains.

Sir Ralph Riley led a research group, which increased precision of plant breeding by experimentally describing the chromosomal architecture of wheat and related species, and by comparing genetic activities of corresponding chromosomes. During investigations which Dr. Riley started in 1958, he and his colleagues described the genetic systems by which pairing of wheat chromosomes at meiosis is limited to those which are fully homologous, and by which pairing between distantly related chromosomes is precluded. This basic knowledge allowed Dr. Riley to pair and recombine chromosomes in a way that is normally illegitimate.

CURRICULUM VITAE

Born 23 October 1924;
married 1949 Joan Elizabeth Norrington.

Education:

Audenshaw Grammar School; University of Sheffield.

Services:

1943-47 - Army; 1944, commissioned.
Served in 6th King's Own Scottish Borders (15th Division) and 1st South Lancashire Regiment (1st Division).
Final rank - Captain.

University:

1947-50 - University of Sheffield, B.Sc. 1st Class Honours (Botany)

Degrees:

1955 - Ph.D., University of Sheffield
1964 - D.Sc., University of Sheffield
1967 - M.A., University of Cambridge (Statute B, 111, 6).

Career:

1950-51 - Research student, Department of Botany, University of Sheffield, on DSIR Studentship.
1951-52 - Demonstrator, Department of Botany, University of Sheffield
1952-78 - Staff of the Plant Breeding Institute, Cambridge:
from 1955-71 - Head of Cytogenetics Department
from 1971-78 - Director of the Institute
Since 1978 - Secretary to the Agricultural Research Council, London.

Distinctions:

1965 - Fellow of the Institute of Biology,
1967 - Fellow of The Royal Society
1967 - Fellow of Wolfson College, Cambridge
1969 - William Bate Hardy Prize of the Cambridge Philosophical Society
1973 - Sir Henry Tizard Memorial Lecturer
1973-75 - Member of the Council of The Royal Society
1973-75 - President of the Genetical Society
1975 - Foreign Fellow Indian National Science Academy
1976 - Woodhull Lecturer, Royal Institutions
1977 - Nilsson-Ehle Lecturer, Swedish Mendelian Society

Social Appointments:

1970-78 - Special Professor of Botany in the University of Nottingham.

Activities in International Science:

1958-68 - Member of the International Organizing Committee of the International Wheat Genetics Symposia
1964 - Recorder of the Section for Cytology and Genetics of the Tenth International Botanical Congress.
1966 - Nuffield Foundation/National Research Council of Canada Lecturer in Canadian Universities.
1973-78 - Secretary General of the International Genetics Federation.
1973-78 - Member of the Board of Trustees of the International Rice Research Institute, The Philippines.
1976 - Member of the Committee on Genetic Experimentation (COGENE) of the International Council of Scientific Unions (responsible for the international co-ordination of work on the genetic manipulation).

Editorial Activities:

Geneitical Research
Caryologia
Wheat Information Service
Theoretical and Applied Genetics
Current Advances in Plant Science
Family Circumstances: Married with two daughters

BRIEF DESCRIPTION OF SCIENTIFIC CONTRIBUTION

Dr. Riley is a scientist of international distinction in the field of cytogenetics and plant breeding. Under his leadership, the department of cytogerietics at the Plant Breeding Institute, Cambridge, developed the most fruitful and dominant part of the Institute's programme of fundamental research.

Dr Riley's major scientific contributions have been in the genome analysis of the common wheat, *Triticum aestivum*. He determined the activity of its 5B chromosome which prevents homeologous pairing at meiosis, so enabling this hexaploid plant to function as a diploid. He devised procedures for modifying the 5B system which permitted alien chromosome addition and substitution. This enabled the transfer into wheat of such qualities as disease resistance, frost hardiness and drought resistance from neighbouring genera of wild and cultivated plants (Agropyron, Aegilops and Secale). This was of key and basic importance in the production of the outstandingly successful and commercially important new varieties of wheat that have come from the Plant Breeding Institute. Although great practical benefits have accrued from Dr Riley's work, it was originally based upon fundamental research of the highest quality.

ESSENTIAL BIOGRAPHY

Appointments:

Plant Breeding Institute, Cambridge, from 1952 to 1978, head of cytogenetics department 1954-72, director 1971-78; special professor of botany, University of Nottingham, 1970-78; secretary, Agricultural Research Council since 1978. Fellow of Wolfson College, Cambridge.

Lectures, awards, etc:

William Bate Hardy Prize of Cambridge Philosophical Society, 1969; Sir Henry Tizard Memorial Lecture, 1973; Holden Lecture, University of Nottingham, 1975; Woodhull Lecture of Royal Institution, 1976.

Offices held:

President, Genetical Society, 1973-75; Secretary, International Genetics Federation, 1973-78.

Relevant publications

R. Riley and V. Chapman, Genetic control of the cytologically diploid behaviour of hexaploid wheat. Nature Lond. 182: 713-715, 1958.

R. Riley, Diploidization of polyploid wheat. Heredity, 15: 407-429, 1960.

R. Riley, V. Chapman and R. Johnson, The incorporation of alien disease resistance in wheat by genetic intergerence with the regulation of meiotic chromosome synapsis. Genet. Res., Camb. 12: 199-219, 1968.

SCIENTIFIC CONTRIBUTIONS

A principal achievement of Dr. R. Riley has been to device and use a form of genetic engineering by which useful genetic variation from related species of wild grass can be incorporated in cultivated wheat. Relevant findings arose from cytogenetic investigations which commenced in 1952 at the Plant Breeding Institute, Cambridge, which is supported financially by the Agricultural Research Council, London. The research of Riley's group had the general purpose of increasing the precision of plant breeding and improving its rationale. Components of the work included the experimental description of the chromosomal architecture of wheat and related species and the comparison of the genetic activities of corresponding chromosomes of wheat and of its relatives.

During these investigations, starting in 1958 Riley and his colleagues discovered and described the genetic systems by which the pairing of chromosomes at meiosis in wheat is limited to those which are fully homologous and by which pairing between more distantly related chromosomes is precluded. They quickly recognised that experimental intervention in this genetic system could create conditions in the meiocyte under which pairing and recombination would take place in a way that is normally illegitimate, and that this would enable foreign genes to be incorporated in homologous locations in wheat chromosomes. The first description of the use of modification of the genetic control of meiosis was published in 1968 (Riley, Chapman and Johnson, Nature, 1968 and Genetical Research 1968) and involved the transference of a gene determining rust resistance from a wild grass (*Aegilops comosa*) into a wheat chromosome.

The potentialities of manipulation of its genetic control provided new experimental access to meiosis and gave further knowledge of this crucial biological process which is still inadequately understood. Riley and his colleagues as well as developing practical procedures for use in plant breeding also studied the functional basis of meiosis using the method of genetics, cytology and cell biology. In addition work was undertaken on the cytogenetics of other important crop species particularly oats.

Dr. Riley's contribution can, therefore, be described as the contribution of new understanding of the meiotic process which have led to the development of new means of plant breeding particularly by the introduction of useful foreign genes into wheat.

Ernest R. Sears
University of Missouri
Columbia, Missouri, USA

1910–1999

1986 — for his fundamental research in cytogenetics of wheat, providing the basis for genetic improvement of cereal grains.

Ernest R. Sears synthesized for the first time in 1946 a hexaploid wheat. In 1950, he completed work, which established the monosomic series in wheat, and later created a mullisomic series, a trisomic series, and a telesomic series of chromosomes in wheat. These have greatly enhanced development and breeding of modern wheat varieties.

CURRICULUM VITAE

Education:

B.S.	Oregon State College	1932
Ph.D.	Harvard University	1936

U.S.D.A. Geneticist:
1936- Professor of Genetics, University of Missouri, Columbia

Major Honors and Awards:

Fellow, American Society of Agronomy	1956
Hoblitzelle Award	1958
U.S.D.A. Superior Service Award	1958
Gamma Sigma Delta National Award	1958
Member, National Academy of Sciences, U.S.	1964

Fellow, Indian Society of Genetics and Plant Breeding	1966
Honorary Doctor, University of Götingen, Germany	1970
Distinguished Service Citation, University of Oregon	1973
Honorary Member, Genetics Society of Japan	1975
Professor E.R. Sears Day in Missouri by Declaration of the Governor April 10, 1976, on the occasion of his 40 years' service at the Missouri Agricultural Experiment Station	1976

E.R. Sears led the pioneering effort in transferring rust resistance from aegilops to common wheat by a unique procedure of hybridization and X-ray-induced translocations. According to the estimates of the Texas Research Foundation of Renner, Texas, the resistance factor that E R Sears introduced into the hexaploid wheat can save from destruction 100 million bushels of wheat annually in the United States alone. For this achievement the Texas awarded Dr. Sears with the $10,000 Hoblitzelle National Award in the Agricultural Sciences in 1958.

All of the scientific recognitions of E. R. Sears cannot be listed here; only the major ones are mentioned:

Fellow, American Society of Agronomy	1956
Hoblitzelle Award	1958
U.S.D.A. Superior Service Award	1958
Gamma Sigma Delta National Award	1958
Member, National Academy of Sciences, U.S.	1965
Fellow, Indian Society of Genetics and Plant Breeding	1966
Honorary Doctor, University of Götingen, Germany	1970
Distinguished Service Citation, University of Oregon	1973
Honorary Member, Genetics Society of Japan	1975
Professor E. R. Sears Day in Missouri by Declaration of the Governor, April 10, 1976, on the occasion of his 40 years of service at the Missouri Agricultural Experiment Station	1976
Award of Excellence, Genetics Society of Canada	1977

There can be no question in the mind of anyone who is familiar with the recent status of agriculture and genetics that Dr. E. R. Sears has made one of the most serious impacts on the cultural and economic history of mankind.

E. R. Sears has excelled, not only as a brilliant scientist and as an effective teacher, but also as a great humanitarian. He is one of the most modest and compassionate of men with the highest ethical standards.

CHROMOSOMAL ENGINEERING OF WHEAT FOR HIGHER AGRONIMIC PERFORMANCE

Bread wheat is by far the most important crop plant in the world. According to 1970 FAO data, the production figures of the major crop plants are as follows:

Crop	Area (million hectares)	Production (million tons)
wheat	210.3	311.6
rice	135.5	306.8
corn	110.8	266.8
millet and sorghum	112.3	92.5

A very quick comparison indicates that yield per area units is variable among these species. These figures also indicate where the greatest possibilities for genetic improvement exist. In order to realize the biological potentials of plant breeding, scientifically well-founded and practical methods are required.

E. R. Sears, working for over forty years as a U.S.D.A. geneticist at the Missouri Agricultural Experiment Station, has made far greater contributions to the scientific development of wheat breeding than any other basic scientist, past or present.

In 1946 E. R. Sears (with E. S. McFadden) synthesized for the first time a hexaploid wheat, thus laying down the foundations of experimental evolutionary work in the species.

In the 1950's Dr. Sears completed the monosomic series of wheat. This material, available to research workers around the world, has remained the most important tool of basic studies on this plant. Thus Dr. Sears is truly the founder of genetic analysis at the chromosomal level in the allopolyploids.

E. R. Sears' basic work on monosomics led him to the solution of the problem of chromosome substitution. In the last few years the chromosome substitution method has become the most powerful method of breeding hexaploid wheat, as well as tobacco, cotton, oats, etc. It is now possible to introduce single isolated chromosomes into an agronomically useful variety from other valuable strains. Actually, practical techniques are available for the transfer of scientifically selected desirable gene complexes, rather than entire chromosomes, from one variety to another without losing any of the useful traits of the recipient. Never before in the history of plant breeding, has it been possible to synthesize a favorable genetic constitution in such a predictable manner.

This synopsis does not permit an adequate expositions of the theory or practice of the development of substitution lines. It may suffice to mention that in 1971 the European Wheat Aneuploid Cooperation was formed with the participation of practically all European nations (including West and East) for the sole purpose of implementing these scientific discoveries based upon E.R. Sears' genetic research on wheat.

In the 1950's E.R. Sears pioneered in transferring agronomically useful chromosomes (addition) or chromosomal segments (transfer) from wild relatives of wheat (alien species).

During this period E. R. Sears, with his graduate student M. Okamoto, discovered the mechanism of control of chromosome pairing in hexaploid wheat. Again, this has been a major milestone in exploiting the genes of alien species useful in breeding better wheat varieties. This method, as well as other methods based upon E. R. Sears' work, led to the development of several commercial wheat varieties in the United States, Germany, and other countries.

Two very recent contributions of E. R. Sears [Genetics 80 (3) Suppl. 74 (1975) and Ann. Rev. Genet. 10:31 (1976)] outlined a corollary to this sequence of systematic engineering. Dr. Sears has isolated a mutant which lacks the gene that normally prevents chromosome pairing between wheat chromosomes and their homoeologous counterparts in alien species. The facilitation of pairing opens up a simple and gentle procedure for introducing practically any gene from the species with which wheat can be crossed. The wild relatives of wheat contain an immense — so far unexploited-reservoir of genes of agronomic interest. In the past, the utilization of these genes was not possible because there were no acceptable means by which they could be transferred into the cultivated varieties without disrupting the harmony of agronomically needed, co-adapted gene complexes.

This latest discovery, I believe, is one of the greatest potential value for securing practical breakthroughs in wheat breeding. Since wheat is the most important food source of the world, this series of discoveries will most likely shape the cultural and economic history of mankind in the forthcoming century. It is not farfetched to say that the research of E. R. Sears has had or will have a direct or indirect impact on the life of every man of the world' s population.

In 1986 Dr. Ernest R. Sears will have completed fifty years of work as a USDA research geneticist and as a professor of genetics at the University of Missouri. During that period he has made contributions to the scientific development of wheat.

In 1946 E. R. Sears (with E. F. McFadden) synthesized for the first time a hexaploid wheat. This has served as the foundation for experimental evolutionary work in the species. By 1950, Dr. Sears had completed the establishment of the monosomic series in wheat and later created a nullisomic, a trisomic series, a tetrasomic series and a telosomic series of chromosomes in wheat.

In the late 1950's E. R. Sears and one of his students, M. Okamoto discovered that the pairing of the wheat chromosome is controlled primarily by the Ph. gene present in the long arm of chromosome SB. The recognition of this basic biological principle opened up a most rewarding procedure for introducing useful genes into wheat from the immense pool of alien species. Sears transferred rust resistance from a wild species (Aegilops) to common wheat. According to estimates, the resistance factor that Sears introduced into wheat resulted in an increase of approximately 3 million tons of wheat annually in the United States alone.

Dr. Sears has meticulously documented all of the genetic stocks he produced and his data are always adequate in amount and quantity before he will report or speak about his discoveries. Dr. Sears is also known for his tremendous generosity in the distribution of the genetic stocks that he developed. This legacy of Ernest Sears will continue to enrich plant science into the future.

A Brief Appraisal of the Scientific Contributions

E. R. Sears, working for over forty years as a U.S.D.A. geneticist at the Missouri Agricultural Experiment Station and as Professor of Genetics at the University of Missouri, has made far greater contributions to the scientific development of wheat breeding than any other basic scientist, past or present. His contributions to the genetics of polyploid species of plantsare unmatched in depth in the field. Many of the discoveries of E. R. Sears concerning chromosomal mechanisms in aneuploidy serve as models for human clinical cytology.

In 1946 E. R. Sears (with E. S. McFadden) synthesized for the first time a hexaploid wheat, thus laying down the foundations of experimental evolutionary work in the species.

By 1950 Sears had completed the establishment of the monosomic series in wheat. The monosomic plants, which have 41 chromosomes rather than 42, usually produce egg cells with 20 chromosomes. When a 20-chromosome egg cell fuses with the 21-chromosome sperm of any variety, the genetic constitution of this 21st chromosome can be assessed.

If the 21 chromosomes that make up the complete set are so tested, the gene content of each of them can be determined. Such a procedure is also used for chromosome substitution, which is one of the most powerful tools of scientific plant breeding of the second half of the 20th century. The European Wheat Aneuploid Cooperative was formed by a unique international cooperation for the main purpose of exploiting these discoveries of E. R. Sears for the production of more and better-quality bread.

After the completion of the nullisomic series of wheat, E. R. Sears synthesized trisomic and tetrasomic lines to evaluate dosage effects of genes. The genetic engineering of wheat was greatly refined when Sears synthesized a series of telosomics. Because the telosomics carry only one arm of a chromosome, they can be used to locate any gene to any of the 42 chromosome arms.

All of these methods are now being applied to the precise analysis of the chromosomes of the major relatives of wheat. The techniques used to do this, again worked out by Sears, have become known as alien addition, alien substitution and alien transfer.

In the late 1950's E. R. Sears and one of his students, M. Okamoto, discovered that the pairing of the wheat chromosomes is controlled primarily by the Ph gene of the long arm of chromosome 5B. The recognition of this basic biological principle opened up a most rewarding procedure for introducing useful genes into wheat from the immense pool of wild relatives of wheat. Within a few years after this discovery numerous disease-resistant varieties of wheat were developed and introduced into commercial production in the United States as well as in several other countries. These first success stories are just the beginning of a long and comprehensive reshaping of the present varieties of wheat.

Primarily through the initiative and analytical endeavors of Aaron Aaronsohn, the late Director of the Jewish Agricultural Experiment Station of Haifa, an invaluable treasure chest of wild wheats exists now containing genes for large kernel size, high amount and better quality of protein, disease resistance, etc. All these can now be selectively and systematically incorporated into the best commercial varieties of the world by the application of the genetic principles worked out by E. R. Sears.

In such a brief summary it is not possible to point out — and even more difficult to explain — the majority of the numerous contributions of Sears which are detailed in his nearly one hundred scholarly papers and book chapters.

Theodor O. Diener

Plant Protection Institute
USDA, Beltsville, Maryland, USA

1987 — for his discovery and pioneering fundamental research on viroids and his applied work on viroid detection in crops.

Dr. Diener's fundamental and applied studies have firmly established the existence of viroids, a new group of subviral pathogens.

Dr. Diener discovered that the pathogen causing potato spindle tuber disease is not a virus, as previously believed, but a much smaller, free RNA molecule, which he named viroid. The discovery by Diener and his colleagues of other viroids affecting various cultivated plants such as chrysanthemum stunt, coconut cadang-cadang and planta macho viroid of tomato soon followed; so that today viroids are considered to be an important group of disease agents affecting potatoes, tomatoes, citrus, avocado, etc. His pioneering studies on isolation and purification showed inequivocally that viroids have a unique molecular structure, different from any other pathogen. Dr. Diener found that viroids replicate without helper virus; that viroid-complementary DNA, produced by insertion into bacteria, are themselves infectious and that the RNA transcribed in potato cells is identical with the viroid. Cloned viroid-specific cDNAs were also developed by him for novel diagnostic tests.

Dr. Diener's work has direct application to the control of viroid diseases of many crops, as well as to the elucidation of the unique biological and molecular properties of these small disease agents.

CURRICULUM VITAE

Date and place of birth: 28 February 1921, Zurich, Switzerland

EDUCATION AND POSITIONS HELD:

1942-46	- Swiss Federal Institute of Technology, Dipl. sc. nat. ETH, 1946.
1946-48	- Research Assistant, Department of Botany, Swiss Federal Institute of Technology, Dr. sc. nat. ETH, 1948.
1948-49	- Plant Pathologist, Swiss Federal Horticultural Experiment Station, Waedenswil.
1950	- Assistant Professor of Plant Pathology, Rhode Island State University, Kingston.
1950-59	- Assistant to Associate Plant Pathologist, Washington State University, Prosser, Washington.
Since 1959	- Research Plant Pathologist, U.S. Department of Agriculture, Beltsville, Maryland.

HONORS AND AWARDS:

1968	- American Institute of Biological Sciences – Campbell Award
1973	- Appointed Fellow, American Phytopathological Society
1975	- Alexander von Humboldt Award
1976	- Ruth Allen Award, American Phytopathological Society
1977	- Elected Member, National Academy of Sciences USA
1977	- Distinguished Service Award, U.S. Department of Agriculture
1978	- Elected Fellow, American Academy of Arts and Sciences
1979-81	- Elected Andrew D. White Professor-at-Large, Cornell University
1980	- Appointed Member, Leopoldina, German Academy of Natural Scientists
1981	- James Law Distinguished Lecturer, New York State College of Veterinary Medicine
1984	- Dean's Distinguished Lecturer "Frontiers in Plant Science," University of California, Davis

LIST OF SIGNIFICANT PUBLICATIONS

Diener, T. O. and Raymer, W. B. Potato spindle tuber virus: a plant virus with properties of a free nucleic acid. Science 158:378-381. 1967.

Diener, T. O. Potato spindle tuber virus. III. Subcellular location of PSTV-RNA and the question of whether virions exist in extracts or in situ. Virology 43:75-89. 1971.

Diener, T. O. Potato spindle tuber "virus". IV. A replicating, low molecular weight RNA. Virology 45:411-429. 1971.

Diener, T. O. Is the scrapie agent a viroid? Nature New Biology 235:218-219. 1972.

Diener, T. O. Potato spindle tuber viroid. VIII. Correlation of infectivity with a UV-absorbing component and thermal denaturation properties of the RNA. Virology 50:606-609. 1972.

Sogo, J. H., Koller, T. and Diener, T. O. Potato spindle tuber viroid. X. Visualization and size determination by electron microscopy. Virology 55:70-80. 1973.

Davies, J. W., Kaesberg P. and Diener, T. O. Potato spindle tuber viroid, XII. An investigation of viroid RNA as a messenger for protein Synthesis. Virology 61:281-286. 1974.

Owens, R. A., Erbe, E., Hadidi, A., Steere, R. L. and Diener, T. O. Separation and infectivity of circular and linear forms of potato spindle tuber viroid, Proc. Nat. Acad. Sci. U.S.A. 74:3859-3863. 1978.

Diener, T. O. Viroids and Viroid Diseases. John Wiley & Sons, New York. 252 pp. 1979.

Diener, T. O. Are viroids escaped introns? Proc. Natl. Acad. Sci., U.S.A. 78:5014-5015. 1981,

Owens, R. A. and T. O. Diener. Sensitive and rapid diagnosis of potato spindle tuber viroid disease by nucleic acid hybridization. Science 213:670-672. 1981.

Owens, R. A., and T. O. Diener. RNA intermediates in potato spindle tuber viroid replication Proc. Natl. Acad. Sci., U.S.A. 79:113-117. 1982.

Diener, T. O., H. P. McKinley, and S. B. Prusiner. Viroids and prions. Proc. Natl. Acad. Sci., U.S.A. 79:5220-5224. 1982.

Kiefer, M. C., R. A. Ovens, and T. 0, Diener. Structural similarities between viroids and transposable genetic elements. Proc. Natl. Acad. Sci., U.S.A. 80:6234-6238. 1983.

Galindo, A. J., O. B. Smith and T. O. Diener. A disease-associated host protein in viroid-infected tomato. Physiol. Plant Pathol. 24:257-275. 1984.

Owens, R. A. and Diener, T, O. Sensitive and rapid diagnosis of viroid diseases and viruses. U.S. Patent No. 4,480,040 (1984).

DESCRIPTION OF SCIENTIFIC CONTRIBUTION

Dr. Diener's discovery and characterization of the viroid is an outstanding contribution to plant pathology, virology, and molecular biology. His discovery has been compared to the recognition of bacteria as pathogens in the late 1800's and to the discovery of viruses at the turn of this century.

Viroid Discovery and Characterization

In 1963, when Dr. Diener began intensive studies on a transmissible disease of potato, spindle tuber, all infectious diseases of plants and animals were believed to be caused either by microorganisms or by viruses; and viruses were universally accepted as the smallest possible agents of such diseases. By 1971, however, Dr. Diener's work had revealed that the potato spindle tuber agent (PSTV) is neither a microorganism nor a virus, but instead the first representative of a previously unknown type of pathogen, far smaller than any virus, for which Dr. Diener proposed the term viroid.

His investigations showed that no viral nucleoprotein particles exist in infected tissue and that the causal agent consists solely of a free RNA molecule that is several times smaller than the smallest viral nucleic acids known. Dr. Diener's studies revealed that the viroid, despite its small size, could replicate autonomously (i.e. without a helper virus) in susceptible cells and that PSTV consists of one molecular species only. These findings seemed incompatible with tenets widely held at the time by virologists and molecular biologists; and the demonstration, in 1974, by Dr. Diener and collaborators, that PSTV does not act as a messenger RNA in Cell-free protein-synthesizing systems further deepened the viroid enigma.

Even today, no viroid-coded proteins have been detected by Dr. Diener or by other investigators. It follows that viroids must trigger their detrimental effects, not via pathogen-specific proteins, as is believed to be the case with most viruses, but by direct interaction of the viroid with host constituents. This is one of the reasons why viroids are so intriguing to the molecular biologist. As suggested by Dr. Diener in 1971, viroids may attach to regulatory sequences on host DNA and thus interfere with gene regulation. If so, study of molecular aspects of viroid infection may help solve the problem of gene regulation in higher organisms — one of the major remaining problems of modern biology.

Viroid Purification and Molecular Structure

Although the infectious RNA was known to constitute only an exceedingly small portion of total cellular RNA, application by Dr. Diener of sophisticated nucleic acid separation and purification procedures led to its isolation and purification in 1972. In collaboration with colleagues, Dr. Diener succeeded in visualizing the viroid, for the first time, by electron microscopy and to confirm its low molecular weight by direct length measurements. Determination of the viroid's thermal denaturation properties confirmed that the RNA has a unique, previously unknown, molecular stucture.

Viroid Replication and Pathogenesis

With this background firmly established, Dr. Diener initiated studies designed to answer several puzzling questions posed by the existence of the viroid. Foremost among these are the following:

1) How can viroids replicate autonomously in susceptible cells even though they contain only very limited genetic information and are not translated into viroid-specific proteins?
2) What are the molecular mechanisms by which virolds cause diseases?
3) What did viroids originate from?

Dr. Diener, sometimes in collaboration with colleagues, greatly contributed toward answers to these and other questions. His demonstration that viroid replication is sensitive to actinoinycin D and, more recently, the finding in Dr. Diener's laboratory of replication intermediates much larger than the viroid itself are important contributions toward the elucidation of viroid replication and have suggested a model involving a rolling-circle type mechanism.

Molecular Probes

Dr. Diener's laboratory was first in applying modern methods of recombinant DNA technology to the viroid problem. Synthesis of viroid-complementary DNAs (cDNAs) has yielded valuable probes with which to study viroid-specific molecular events in infected cells. By insertion into bacteria, such DNAs can be produced in any desired amounts. Of great importance is the recent demonstration, first made in Dr. Diener's laboratory and now confirmed by other investigators, that artificially produced cDNAS representing the complete structure of PSTV are infectious (exhibit #10). Because the DNA mirrors the structure of PSTV, the RNA transcribed in susceptible cells from the DNA is identical with PSTV. Thus, plants into which the DNA has been introduced, produce PSTV and display the characteristic symptoms of PSTV infection. Recombinant DNA technology makes it now possible to modify the recombinant DNA in various ways, to introduce precisely known nucleotide exchanges, insertions, or deletions, or to produce chimeric viroids. Because each DNA mutation is faithfully and predictably reflected in the transcribed RNA, these experiments are apt to identify specific regions of the viroid molecule involved in symptom formation, replication efficiency, or host specificity. Today, Dr. Diener's laboratory is in the forefront of these important investigations which, undoubtedly, will permit us to correlate viroid structure with biological function and result in the elucidation of the molecular mechanism of viroid pathogenesis. Most types of symptoms of viroid infection have their counterpart in plants infected by conventional viruses, and vice versa, indicating that viroids and viruses may, in many cases, affect the same or similar plant metabolic systems. Thus, it is likely that results obtained with viroids will explain mechanisms of virus, as well as viroid, pathogenesis.

Viroid Origin

With the discovery of split genes and RNA splicing, it has been suggested (by Dr. Diener and others) that viroids might have originated by circularization of intervening sequences (introns) that are spliced out of precursor RNAs. The identification, by Dr. Diener, of a region in the nucleotide sequence of PSTV that is homologous with the 5' end of small nuclear RNA Ui (which is involved in the splicing of nuclear mRNA precursors) lent credence to the possibility that viroids represent "escaped introns". Recently, again in Dr. Diener's laboratory, striking sequence similarities have been discovered between viroids and class I introns, further supporting a possible intron origin of viroids.

Control of Viroid Diseases

Although Dr. Diener's laboratory is dedicated primarily to fundamental research, application of newly acquired knowledge for the diagnosis and control of viroid diseases has not been neglected.

Thus, when viroid-specific, cloned cDNAs became available in his laboratory, Dr. Diener realized their potential usefulness as diagnostic tools for viroid detection in crop plants. He was aware of the urgent need for greatly improved methods of viroid detection, particularly for the large-scale screening for PSTV in seed potato and foundation stocks. In collaboration with Dr. H. A. Owens of his laboratory, Dr. Diener has developed a reliable, extremely sensitive, and specific diagnostic procedure that is based on molecular hybridization of radioactive PSTV-specific cDNA to PSTV immobilized on nitrocellulose membranes, followed by autoradiographic detection of the hybrids formed (exhibit #12). Ease of sample preparation (no extensive purification is required) and rapidity of the procedure make the test admirably suited for the assay of large numbers of individual samples. Indeed, the test (which has been awarded a U. S. Patent) has revolutionized PSTV diagnosis worldwide; and commercial kits for the detection of PSTV by variations of Dr. Diener's method are today available from at least two biotechnology companies. Furthermore, "dot blot" nucleic acid hybridization tests are now widely used for the detection of viroids other than PSTV and, in preference over available immunological procedures, even for the detection of certain conventional plant viruses.

Although prevention rather than cure of plant diseases still is the rule, results from Dr. Diener's laboratory suggest entirely novel approaches to the problem, including the possibility of permanently protecting plants from the damaging effects of viroid infection.

Epilogue

Dr. Diener has now been active in plant pathological research for 35 years and has published some 180 scientific papers, including many book chapters and one book. He has long been a recognized authority in plant pathology and particularly in plant virology. With his discovery of the viroid, Dr. Diener's international recognition and stature have been extended to other areas of science, such as animal virology, molecular biology, veterinary and human medicine, and microbiology. He has received numerous invitations to lecture at prominent research centers, medical and veterinary schools, and institutes of molecular biology, in the U.S.A. and abroad. He has received many awards, and has been elected as a member of several honorary societies, such as the U. S. National Academy of Sciences, the American Academy of Arts and Sciences, and the German Academy of Natural Scientists, Leopoldina.

Dr. Diener was actively engaged in research with his small research unit, to remain in the forefront of fundamental research on viroids and viroid diseases.

Ernest John Christopher Polge

Biotechnology Cambridge Ltd., Cambridge
United Kingdom

1926–2006

1988 — for pioneering work in reproductive physiology including cell preservation, fertilization processes, egg biology and embryo manipulations for domestic animal improvement.

Dr. Ernest John Christopher Polge is renowned for his pioneering work on the cryopreservation of living cells, such as sperm and in embryos. His work in this area promoted the development of the science of cryobiology. During the last 30 years he has also contributed significantly to our understanding of embryonic development, in vitro fertilization and embryo transplants. In addition, he led a group to international recognition for pioneering techniques in cloning livestock animals by embryo division and nuclear transfer, and initiated transgenic research by insertion of gene constructs into pronuclei of recently fertilized eggs.

Throughout his career he has been concerned with the transfer and application of science for the improvement of livestock production.

IN MEMORIAN[1]

Christopher Polge, who died on August 17 aged 80, attained scientific eminence at a remarkably early age: his discovery, whilst still a doctoral student, of how to preserve living cells and tissues at very low temperatures solved a long-standing and intractable problem in biology.

This breakthrough not only formed the basis for the new science of cryobiology but has also had profound and continuing practical implications for agriculture and medicine.

[1] *Telegraph, UK — 11/09/2006.*

The key to cell survival during the freezing process was the discovery, by Polge and his colleague Dr Audrey Smith, of a class of chemicals now known as cryoprotective agents. Although the potential importance of cell preservation by freezing had long been recognized, experiments in freezing living cells or tissues had invariably resulted in their death.

It rapidly became clear from the work of Polge and Smith that many of the cell types in the body had highly specific requirements for freezing and thawing. Polge chose to focus on the preservation of sperm and eggs in mammals, and in 1950 produced the first chicks from eggs fertilized with frozen sperm.

Although these chicks were the first vertebrates to be produced in this way, much greater acclaim followed Polge's report two years later of high pregnancy rates in cattle using sperm that had been frozen for periods in excess of a year. These reports had far-reaching consequences for the future of artificial insemination and genetic improvement in livestock.

Within 10 years virtually every cattle-breeding centre in the world had converted to the use of frozen semen; Polge gave assistance and advice to these centres, travelling throughout Britain and across the world, particularly in North and South America.

In the early 1970s pig semen was successfully frozen in Polge's laboratory, and he finally felt that he had fulfilled his objective of developing methods for the cryopreservation of semen from all the major species of livestock.

Ernest John Christopher Polge was born on August 16, 1926, the son of a Buckinghamshire farmer. He was educated at Bootham School, York, then read Agriculture at the University of Reading. After working briefly as an agricultural economist he decided to follow his vocation as a research scientist.

Polge studied for his doctorate at the Division of Experimental Biology at the National Institute for Medical Research at Mill Hill, London, before moving to the Animal Research Station at Cambridge under Sir John Hammond.

Hammond's method was to carry out rigorous fundamental science with a view to embracing a practical outcome. This approach appealed to Polge, and was to guide his research throughout his lifelong association with the Research Station, which he was to lead for the last eight years of his career.

After his initial successes in the cryobiological field, Polge realised that the freezing of embryos for use in embryo transfer programs offered the next practical means of contributing to the improvement of livestock. He recruited Ian Wilmut (later to lead the team which cloned Dolly the Sheep) to work specifically on the problem of the low temperature preservation of embryos.

A series of advances by this team resulted in the birth, in 1973, of Frosty, the first calf from a cryopreserved embryo. At that time it was not foreseen that embryo freezing would find its greatest use in assisted human reproduction; rather the focus was on the genetic improvement of livestock.

As problems in the cryopreservation of gametes were solved, Polge concentrated on other methods of improving livestock production. He became an international authority on reproductive biology and was in the forefront of new developments in artificial insemination, control of the reproductive cycle, embryo transplantation and in vitro fertilization in the pig.

In his final years as head of the Animal Research Station he fostered and encouraged projects aimed at the production of chimaeras and identical offspring in domestic animals. Success in these programs provided the foundation for subsequent progress in cloning and in the production of transgenic animals.

Polge published more than 160 papers. In addition to his research, he taught at Cambridge University, becoming an honorary professor and a fellow of Wolfson College. He was a member of many national and international bodies.

He was generous in sharing research ideas with his colleagues, his many students and his visitors from overseas. Even when he became head of the Animal Research Station he never allowed himself to become tied down by administration, instead spending as much time as possible in his laboratory or with his experimental animals.

After his retirement from the Animal Research Station in 1986 Polge co-founded Animal Biotechnology Cambridge, a commercial enterprise set up to bridge the gap between research discoveries and their conversion into profitable agricultural products and practices.

Polge was elected to the Royal Society in 1983 and was appointed CBE in 1992.

He was elected to the United States Academy of Sciences as a foreign associate.

BIOGRAPHICAL DATA

Date and place of birth: 16 August 1926 - Jordans, Buckinghamshire

Education and positions held:

1940-1944 - Bootham School, York
1944-1946 - University of Reading, B.Sc. (Agriculture)
1947-1948 - Staff member, Department of Agricultural Economy, University of Bristol
1948-1954 - Staff member, MRC National Institute for Medical Research, London
1953 - Ph.D., University of London
1954 - Animal Research Station, Cambridge
1972 - Senior Principal, Scientific Officer (special merit), ARS, Cambridge
1979-1986 - Officer-in-Charge, ARS, Cambridge
Since 1986 - Scientific Director, Animal Biotechnology Cambridge Limited

Honors and Awards:

1969 - John Scott Award, City of Philadelphia
1971 - Sir John Hammond
Memorial Prize, British Society of Animal Production
1983 - Fellow of The Royal Society
1984 - Fellow of the Royal Agricultural Society of England
1986 - Honorary Associate of the Royal College of Veterinary Surgeons
1987 - Pioneer Award, International Embryo Transfer Society

BRIEF DESCRIPTION OF SCIENTIFIC CONTRIBUTION:

When Dr. Polge was elected a Fellow of the Royal Society in 1983 the citation read - *'distinguished for his pioneering work on cryopreservation of semen for artificial insemination of farm animals and later of embryos for implantation by synchronising oestrus. His work has revolutionised animal breeding throughout the world'.*

Certainly the discovery in the early 1950's by Dr. Polge and his colleagues of a method for the preservation of living cells at very low temperatures has led to major developments. It has promoted the growth of what has become a new science of cryobiology with practical application in various spheres in medicine and agriculture. There is no doubt that the largest application has been in the deep-freezing of semen for artificial insemination, particularly in cattle in which the impact on breeding and livestock improvement has been very great. Dr. Polge was concerned not only with the development of the basic principles of semen preservation, but also with its application in Britain and elsewhere through active collaboration with such organizations as the Milk Marketing Board and the Meat and Livestock Commission.

During the last 30 years Dr. Polge's work on reproductive physiology in farm animals has also made important contributions in a number of other areas. This is particularly true in relation to pig breeding. He was responsible for the control of reproductive cycles and carried out experiments on fertilization and early embryonic development. More recently he has been concerned with embryo transplantation and fertilization in vitro. The development of techniques for the preservation of embryos at low temperatures is now being applied in a number of farm animal species.

For seven years Dr. Polge was Officer-in-Charge of the world famous Animal Research Station, Cambridge, which became part of the present AFRC Institute of Animal Physiology and Genetics. Under his leadership the Animal Research Station pioneered techniques for cloning livestock animals by embryo division and nuclear transfer and initiated transgenic research by insertion of gene constructs into pronuclei of recently fertilized eggs. Since his retirement from the Institute in

1986 he has continued at the Animal Research Station site as Research Director of a new Biotechnology Company — Animal Biotechnology Cambridge Ltd — where his expertise is being utilized to develop commercial applications of much of the breeding technology developed under his earlier guidance.

Throughout his career Dr Polge has been concerned with the application of science to the practice of animal breeding. He does not live in an 'ivory tower', but is as much concerned with the practical problems of farming as with basic science. This philosophy has led to close liaisons with the industry in matters concerned with animal breeding and has often enabled applications derived from basic research to be applied quickly and effectively as is evidenced by new developments in artifical insemination and embryo transplantation. He has undertaken numerous lecture tours in the United Kingdom and abroad. He is the United Kingdom representative on the EEC committee concerned with the physiology of reproduction in farm animals. He is a consultant to the Food and Agriculture Organization on the conservation of genetic resources, a member of the Genome Specialist Group of the International Union for the Conservation of Nature and advises the Technical Committee of the Rare Breeds Survival Trust. He has previously been Secretary and Chairman of the Society for the Study of Fertility, Chairman of the Society for Low Temperature Biology and is presently Chairman of the Journals of Reproduction and Fertility Ltd. He is also a member of the Society for Endocrinology, Society for the Study of Animal Breeding, Society for Cryobiology, Research Defence Society and the Royal Institution.

LIST OF SIGNIFICANT PUBLICATIONS

Storage of bull spermatozoa at low tamperatures. Vet. Rec. 62, 115-116 (1950). (With A.U. Smith).

The immobilization of spermatozoa by freezing and thawing and the protective action of glycerol. Biochem. J. 58, 618-622 (1951). (With J.E. Lovelock).

Low temperature storage of mammalian spermatozoa. Proc. R. Soc. Lond. B 147, 498-508 (1957).

Effective synchronisation of oestrus in pigs after treatmant with ICI compound 33282. Vet. Rec. 77, 232-236 (1965).

Synchronization of ovulation and artificial insemination in pigs. Vet. Rec. 83, 136-142 (1968). (With B.N. Day & T.W. Groves).

Cytogenetic analysis. of pig oocytes matured in vitro. J. Exp. Zool. 176, 383-396 (1971). (With R.W. McGaughey).

Recent progress in techniques for increasing reproductive potential in farm animals. In Proceedings of the 3rd World Conference in Animal Production (Melbourne) 3, 6(a) 1-14 (1973). (With L.E.A. Rowson).

The viability of deep frozen cow embryos. J. Repro. Fert. 52, 391-393 (1978). (With S.M. Willadsen & L.E.A. Rowson.).

Fertilization in the pig and horse. J. Repro. Fert. 514, 461-470 (1978).
Embryo transplantation in the large domestic species: applications and perspectives in the light of recent experiments with eggs and embryos. J. R. Agric. Soc. 141, 115-126 (1980). (With S.M. Willadsen).
Embryo transfer and preservation. In Control of Pig Reproduction (ed. D.J.A. Cole & G.E. Foxoroft), pp. 277-291. London: Butterworths (1982).
Analysis of slow-warming injury of mouse embryos by cryomicroscopical and physiochemical methods. Cryobiology 21, 106-121 (1984). (With W.F. Rall & D.S. Reid).
How does embryo manipulation fit into present and future pig reproduction. J. Repro. Fert. S33, 93-100 (1985).

Charles Thibault

Universite de Paris VI, Paris, France

1919–2003

1988 — for pioneering work in reproductive physiology including cell preservation, fertilization processes, egg biology and embryo manipulations for domestic animal improvement.

Professor Emeritus Charles Thibault is renowned for his pioneering research on the maturation of the egg and its fertilization. In this regard he had a major role in characterizing the control of meiosis in the egg and providing the first cytological descriptions of fertilization events leading up to the first cleavage. His work progressed from these studies to in vitro fertilization and the conditions for culturing eggs, the study of sperm capacitation and gamete ageing in a number of different species. The results of his work improved our understanding of embryonic mortality differences among species, elevating conception rates and factors involved in the successful use of in vitro fertilization. Throughout his career he has been recognized for his leadership in reproductive biology.

BIOGRAPHICAL DATA

Date and place of birth: July 14, 1919, Paris.

Education and positions held:

1949 - Doctorat-es-Science, Paris
1944-1949 - Scientific assistant, C.N.R.S.
1948-1949 - Assistant, Assistant-Professor, University of Paris
1950-1967 - Senior Scientist, I.N.R.A.
1962-1976 - Head of the Department of animal Physiology, INRA
1968-1987 - Professor of reproductive Physiology, University of Paris

Professor Emeritus

1950-1978 - Member of the CNRS Scientific Committees (Biology, Physiology)
1954-1958 - Dean of the National Center of Zootechnical Research, INRA
1960-1964 - Adviser of the Prime Minister for Biology
1976-1979 - President of Ph.D. Grant Committee (Biology, Agriculture, Medicine)
1979-1981 - President of the CNRS
Since 1986 - President of the French Society for the Study of Fertility

Honors and Awards:

Academy of Sciences: Prix FOULON (Zoology), 1950
Prix SERRES (Agriculture), 1955
Prix FOULON (Agricultural Economy), 1962
Prix A. & J. D'ALSACE (Gynecology), 1972
John HAMMOND lecture, London, 1977
Marechall Medal, Cambridge, 1982
Prix de la Ville de Paris (biology), 1982
Prix de l'Assocjation Etude Endocrjnologie et Fertilité, Paris, 1987
Officier Legion d'Honneur; Commandeur Méite agicole, Palmes academiques, né le 14 juillet 1919
Docteur es Sciences, Paris 1949

Fonction de recherches et d'enseignment

1944-1946 Attaché do Recherches. CNRS.
1947-1949 Assistant puis Chef de Travaux Université de Paris.
1950-1967 Maître puis Directeur de Recherches. INRA.
1962-1975 Chef du Département de Physiologie animale. INRA.
1967-1987 Professeur de Physiologie de la reproduction. Université Paris VI. Professeur Emerite. Université de Paris VI.

Fonctions d'adminsitration de la recherche

1954-1958 Administrateur du Centre de Recherches de Jouy-en-Josas. INRA.
1960-1964 Membre du Comité consultatif de la recherche (Premier Ministre).
1951-1963 Membre scientifique du Conseil économique et social.
1950-1979 Membre des Commissions de Biologie puis Physiologie. CNRS.
1976-1979 Président de Is Commission des Bourses de Theses (Biologie).
1979-1981 Président du CNRS.
1986 Président de Is Société Française pour l'Etude de Is Fertilité.

Prix scientifiques

1950 Prix Foulon (Zoologie).
1955 Prix Serres (Agriculture).
1962 Prix Foulon (Economie rurale).
1972 Prix do la Société française do Gynécologie (Annie. Jean d'ALSACE).
1980 Marschall Medal.
1983 Prix do Biologie do la Ville do Paris.
1987 Prix do l'Association pour l'Etude do. l'Endocrinologie et do Is Fertilité.

Decorations

1975 Officier de l'Ordre du Mérite (1967) et do la Legion d'Honneur.
1976 Commandeur des Palmes Académiques (1974) et du Mérite Agricole.

DESCRIPTION OF SCIENTIFIC ACHIEVEMENTS

Charles Thibault has been trained for research in a marine laboratory in Marseille and, after the war, in a Paris university laboratory.

First he studied mechanisms of color adaptation in fish (1,2,6,7) and then he used the effect of the angle of a beam light on fish equilibration to explore the color sensitivity of fish retina. His conclusions were very similar to those obtained in other Vertebrates by electrophysiological approaches: the retina is more sensitive in the wavelength ranges of 650 mμ (red) 530 mμ (green) and 450 mμ (bleu). The overall sensitivity is greater in central retina (8, 11, 12, 37). He has always remained interested in fish physiology and created in 1966 a research group on freshwater fish reproduction which has later become a department of our Institute. However the major part of his scientific activity has been focused on the Biology and Endocrinology of domestic mammals. As soon as 1947, his first experiments singled out the possibility to increase ovulation rate in sheep of whatever breed by gonadotropins and to initiate a second pregnancy per year by combining steroids and gonadotropins (15, 19, 20, 42, 44).

Studying with farmers the limiting factors in animal production he concluded that: increase of birth rates; control of birth period; extension of artificial Insemination to sheep, goat and pig; possibility of a balance between milk yields and meat production by Inducing twin or triplet pregnancies and improvement of growth would be the keys to an industrialization of breeding at low cost. However around the fifties the scarcity of the knowledge of domestic mammal physiology did not allow his conclusions to be applied straightaway. Therefore he initiated basic researches on gametogenesis and, estrous cycles and their neuroendocrine regulations, on fertilization and embryonic development both in vivo and in vitro, on mammary gland and lactation, on muscle and fat tissue differentiation, on blood groups and finally on feeding and sexual behavior.

He has regularly reminded us that fundamental research is a prerequisite to application provided that solutions to practical problems should remain the leading strand (see ref. 57 and 131).

Results of research on domestic mammals proved to be very useful for the understanding or treatment of human pathology, especially in reproduction. So he has been solicitude many times to give lectures In gynecological meetings (91, 92, 95, 100,109, 115, 118, 126, 128, 129, 130, 137) and at present he is the President of the French Fertility Society.

His list of publications is limited to his personal contribution yet he has prompted research works in many fields. His main work on applied research concerns estrous synchronization, super ovulation, induction of a double sexual season in sheep and production culture and preservation of gametes and embryos, (sheep, goat, bovine and rabbit) (13, 19, 20, 25, 26, 2?, 31, 32, 35, 39, 42, 43, 48, 51, 55, 71, 73, 74, 80, 89, 90, 102, 127).

His personal contribution to basic knowledge may be summarized as

-***Egg maturation.*** He discovered that:

1) granulosa cells are responsible for the maintenance of the oocyte meiotic arrest in the graafian follicle up to ovulatory gonadotropin discharge at estrus (86, 94, 99, 134);
2) meiosis resumption at estrus is only one aspect of the final oocyte maturation since isolatedly cultured oocytes are unable to assume sperm nucleus decondensation, although meiosis has normally resumed (87, 88, 105, 108). Granulosa and cumulus cells are necessary for what he has called cytoplasmic and membrane maturation of the oocyte. This conclusion has been confirmed many times in bovine, sheep, pig, rabbit and human oocytes. In vitro production of normal oocytes is possible either in the follicle (94, 97, 105, 106, 107, 111, 113, 114, 116, 117, 121, 134) or in the presence of granulosa cells (see STAIGMILLER and MOOR, 1984; CROZET et al., 1987).
3) After ovulation the fertilizable life of oocytes depends on the species; in rabbit, pig and probably man ageing of the cocyte occurs very rapidly in few hours but, in sheep and bovine, oocytes remain fertilizable during at least one day. In the first group of species when aged oocytes are fertilized, polyspermy and digyny are very frequent. Consequently there is an increase of embryonic mortality and artificial insemination must be performed in these species in such a way that sperm can be present in the ampullary portion of the fallopian tube at the time of ovulation.

-***Fertilization process.*** He gave the first cytological descriptions of natural fertilization in rabbit, pig, sheep and bovine with the chronology of the successive events up to the first cleavage (39, 47, 59, 76), What is more, he obtained, with his group, the first true in vitro fertilization in 1954 (45, 47, 53). The birth of young rabbits in 1959 in USA (CHANG) and in Jouy en Josas in 1961 (61) has testified to the "normality" of the experimental procedure. Later on, in vitro fertilization has been extended to domestic mammals in England, in USA, in Canada and in our Institute. Trials on practical application are now in progres in France.

He used the in vitro fertilization technique as a tool for the study of sperm capacitation, gamete ageing and the consequence of triploidy on subsequent development (65, 66, 67, 76, 84, 85, 91, 99, 102). Triploid embryos (rabbit) or parthenogenetically activated rabbit or sheep eggs are able to cleave but in no case develop beyond implantation (10, 14, 23, 34, 36).

He studied the conditions of culture of rabbit, sheep, goat and bovine eggs fertilized either in vivo or in vitro and was the first to observe that sheep, goat and bovine egg development stops in culture after few cleavages whatever the medium (43, 74) showing that fallopian tubes play a role in egg development in these species but, the nature of the factors involved so far remains unknown.

-*Other topics.* He published papers on sperm survival in vitro (46, 48) but, above all has been interested in sperm ascent in the female tract in relation to insemination practice. He showed that insemination stress is able to postpone sperm ascent in the ewe and that ocytocin stimulates sperm transport in unstressed ewes (75); as ocytocin is released by cervical stimulation this hormone probably plays a physiological role in female tract sperm transport after natural as well as artificial insemination. He demonstrated that in the cow, after mating, spermatozoa are present in the ampullary portion of the oviduct only 8 hours after natural mating and that in lower isthmus and uterotubal junction polynuclear cells and inacrophages are normally absent. This region, and not the cervix as generally assessed, is the reservoir from which the fertilizing sperm moves to the ampulla (95, 98, 101, 104, 135). This conclusion fits well with the more recent observations in rabbit, sheep, hamster and bats.

We must also mention his enlightening participation in the studies of the role of the photoperiodism on sexual activity and mainly on spermatogenesis (50, 52, 54, 69, 73). He gave a general survey on the importance of photoperiodism in the reproduction of domestic and wild mammals, mainly based on the results of the Department of Physiology in Jouy en Josas (73) and this paper has remained a reference.

During the past twenty years he has been Professor of Reproductive Physiology in Paris VI University. He has trained many hundred students, Biologists, Veterinarians, Agronomists and Physicians. Nany of them are now in applied or basic research teams in medicine, agriculture and biology. Five of them are university professors. He has been a jury member at more than 300 PhD. He has published three books and contributed to many others.

LIST OF SIGNIFICANT PUBLICATIONS

1947. - Superfecondité experimentale chez la Brebis. (+M. LAPLAUD), C.R. Acad. Sci. 224, 1786-1788.

1948. - Possibilité d'une gestation supplémentaire chez la Brebis (+M. LAPLAUD et R. ORTAVANT), C.R. Acad. Agric., 34, 1-5.

- L'électroéjaculation chez le Taureau. Techniques et résultats (+M. LAPLAUD et R. ORTAVANT), C.R. Acad. Sci., 226, 2006-2008.

1949. - L'oeuf des Nammifères. Son développement parthénogènétique (first Thesis), Ann. Sci. Nat. Zool. 11, 1436-219.

- Action de la lumière cur la posture des poissons Tèlèostéens. Son utilisation pour 1'étude de la vision. (2d Thesis) Archiv. Physiol. 3, 101-124.

1953. - Recherches expérimentales sur le role de la progesterone dans la cycle sexual de la Brebis et de la Chèvre (+L. DAUZIER, S. WINTENBERGER). Ann. Endocrinol., 14, 553-559.

1954. - La fècondation in vitro de l'oeuf de la lapine (+L. DAUZIER at S. WINTENBERGER), C.R. Acad. Sci. 238, 644-845.

1962. - Endocrine factors in the survival of spermatozoa in the female reproductive tract (+ R. NOYES), Fert. Steril., 13, 346-365.

1966. - Regulation of breeding season and estrous cycles by light and external stimuli (+ N. COUROT et al). J. Anim. Sci., 25, suppl. 119-139.

1969. - In vitro fertilization of the mammalian egg , In "Fertilization", Vol. 2, NETZ C.B., MONROY A. Eds, Acad. Press, N. Y. 405-435.

- Recherches expérimentales sur la maturation in vitro des ovocytes de truie et de veau (+D. FOOTE), Ann. Biol. Anim. Bioch. Biophys. 9, 329-349.

1972. - Final steps of mammalian oocyte maturation and fertilization of rabbit and cattle oocytes. In: "Oogenesis". BIGGERS, J., SCHUETZ, A. W. Eds. Univ. Park Press, Baltimore, 397-411.

1973. - In vitro maturation and fertilization of rabbit and cattle oocytes. In "The regulation of mammalian reproduction" SEGAL S.J. et al, Eds, 231-46.

- Cytoplasmic and nuclear maturation of rabbit oocytes in vitro. Ann. Biol. Anim. Bioch. Biophys., 13 (suppl.), 145-156.

- Sperm transport and storage in Vertebrates. J. Repr. Fert.18, 39-53.

1975. - Transport and survival of spermatozoa in cattle (+M. GERARD at Y. HEYMAN). In "The biology of the spermatozoa" HAFEZ E. et al, Eds, 156-165.

1977. - Are follicular maturation and oocyte maturation independent processes? J. Reprod. Fertil. 51, 1-15.

1978. - Permeability of ovarian follicle; corona cell-oocyte relationship in mammals (+SZOLLOSI D., GERARD M., MENEZO Y.) Ann. Biol. Biochem. Biophys. 511-521.

1985. - Le corps jaune (+LEVASSEUR M.C.) Gynecologie, 36, 265-272.

1987. - Catecholamines within the rabbit oviduct at fertilization time. (+C. KATCHADOURIAN, KENEZO Y., GERARD M.), Human Reprod. 2, 1-5.

- Fécondation in vitro d'ovocytes bovine maturés in vivo ou in vitro. (+B. LE GUIENNE) Coll. Soc. fr. Etude Fertil., 25, 145-150.
- Nammalian oocyte maturation (+SZÔLLÖSI D.,GERARD M.), in press.

Books: Bases physiologiques de la contraception (+M.C. LEVASSEUR), 1967, Doin, Paris. - La fonction ovarienne (+M. C. L.), 1979, Masson, Paris. - De la puberté à la sénescence (+M.C.L.), 1980, Masson, Paris.

Chapters: 3 in "Traité de Zoologie" GRASSE P. Ed. 1969, Masson, Paris; 2 in "Reproduction in Farm Animals" HAFEZ E., Ed., 1974, 80, Lea Febiger, Philadelphia, and some others.

Peter M. Biggs

AFRC Institute for Animal Health, Huntingdon
United Kingdom

1989 — for distinguished contributions to basic science and its successful translation into practice in the fields of animal health and crop protection.

Narrative *curriculum vitae* for Professor P.M.Biggs CBE, DSc, Ph.D, CBiol, FIBiol, FRCPath, FRCVS, FMed Sci, FRS.

After schooling in England and the USA Professor Biggs joined the Royal Air Force in 1944 and spent the first six months of his service at the Queen's University, Belfast reading physics and engineering. On demobilisation in 1948 he entered the Royal Veterinary College. London and graduated in veterinary science and as a veterinary surgeon in 1953. He then took a PhD on the subject of lymphoid tissues in the domestic chicken at Bristol University where he subsequently joined the staff of the Veterinary School in 1955 as a lecturer in Clinical Pathology during which time he started studies on the avian leucosis complex. In 1959 he moved to the Houghton Poultry Research Station to form and Head a unit to study the economically important lymphoid tumour conditions of the domestic fowl. He built and led a group that established that there were two important lymphoid tumour conditions of the fowl. One that had been shown in the USA to be caused by what is now known as a retrovirus, now called lymphoid leucosis, and a second, which he termed Marek's disease after the Hungarian veterinary pathologist who first described the disease, which was shown by the group to be caused by an highly cell associated herpesvirus. Subsequently they developed an effective vaccine against this serious tumour condition. This was the first vaccine effective against a tumour in any species. Professor Biggs also contributed with his colleagues to knowledge of the genetic control of resistance and susceptibility to infection with retroviral tumour viruses of the fowl. He also described a novel disease in turkeys which he named lymphoproliferative disease which he and colleagues showed was caused by a retrovirus. Professor Biggs has published over 130 papers mainly on viruses and poultry disease.

In 1971 Professor Biggs was appointed Deputy Director of Houghton Poultry Research Station and its Director in 1974. During his Directorship he encouraged the use of molecular biological techniques for the studies of avian disease pathogens and appointed a number of molecular biologists to the staff of the Station. His described a novel lymphoproliferative disease in turkeys and he and his research group defined its retroviral cause.

On the re-organisation of the AFRC Institutes in 1986 he was appointed AFRC Director of Animal Health charged with the remit of forming an Institute for Animal Health comprising the former Animal Virus Research Institute, Pribright, Houghton Poultry Research Station, Institute for Research on Animal Diseases, Compton and the Neuropathogenesis unit in Edinburgh. He retired from this post in 1988.

Between 1988 and 1994 he was an Andrew D. White Professor-at-Large at Cornell University, Ithaca, N.Y. USA. and has been a Visiting Professor in Veterinary Microbiology at the Royal Veterinary College, London University since 1982.

Professor Biggs is a Fellow of the Royal College of Veterinary Surgeons, founding Fellow of the Academy of Medical Science, Fellow of the Royal College of Pathologists and has been a Fellow of the Institute of Biologists since 1973 and was its President from 1990 to 1992. He has been President of the British Veterinary Poultry Association, the World Veterinary Poultry Association, The International Association for Comparative Research on Leukaemia and Related Diseases and a Vice President of the British Veterinary Association.

He has been a member of a number of Government and other Committees and Working Groups including the Ashby Working Party on Genetic Manipulation in Micro-organisms, the AFRC/ADAS Working Party on Animal Disease Research in the United Kingdom, the Virus Working Group of the Advisory Committee on Genetic Manipulation, Advisory Committee on Dangerous Pathogens chairing its Working Group on the handling of Transmissible Spongiform Encephalopathies, the Veterinary Products Committee, the Royal College of veterinary Surgeons' Committee of Enquiry into Veterinary Research. He also chaired Visiting Groups to the Ministry of Agriculture, Fisheries and Food Poultry Unit at the Experimental Husbandry Farm, Gleadthorpe, the Pig Unit at the Experimental Husbandry Farm, Terrington and the Central Veterinary Laboratory, Weybridge.

Professor Biggs has received an Honorary Doctorate of Veterinary Medicine from the Ludwig-Maximilian University, Munich, Germany, Honorary Doctorate of the University of Liège, Belgium, and the Joszef Marek Memorial Medal from the Veterinary University of Budapest, Hungary. He was elected a Fellow of the Royal Society in 1976 and an Honorary Fellow of the Royal Agricultural Society of England 1986. He was created a Commander of the Order of the British Empire (CBE) in 1987. He has been presented with the Tom Newman Memorial Award for Poultry Husbandry Research by the Poultry Association of Great Britain, the Chalmers Watson Turkey Award of the British Turkey Federation, the Poultry Stock Association Distinguished Service to the Poultry Industry Award, the J.T. Edwards Memorial Medal of the Royal College of Veterinary Surgeons for outstanding work in the fields of pathology and virology, the Dalrymple-Champneys Cup and Medal and the Chiron Award of the British Veterinary Association, the Bledisloe Veterinary Award of the Royal Agricultural Society of England, the Wooldridge Memorial Medal of the British Veterinary Association and the Gordon Memorial Medal and in 1989 he was awarded The International Wolf Foundation Prize in Agriculture.

Until 2007 he was Chairman of The Scientific Advisory Committee of the Animal Health Trust and Vice Chairman of its Executive Committee. He is currently a Trustee and Chairman of the Gordon Memorial Trust, a Trustee and Chairman of the British Egg Marketing Board Research and Education Trust, Chairman of the Committee of Management of the journal Avian Pathology and the Houghton Trust, Vice Chairman of the Council of the Royal Veterinary College and a member of the Poultry Sector Board of Assured Chicken Production and Chairman of its Technical Advisory Committee.

LIST OF PUBLICATIONS

Scientific papers

KING, A.S. and BIGGS, P.M. (1955) Experimental observations on the respiration through the humerus of *Gallus domesticus.* J. Anat. (Lond.), **89**: 567-568.

BIGGS, P.M. (1956) Lymphoid haemopoietic tissue. Vet. Rec., **68**: 525-526.

BIGGS, P.M. (1957) The association of lymphoid tissue with the lymph vessels in the domestic chicken (*Gallus* domesticus). Acta Anat., **29**: 36-47.

KING, A.S. and BIGGS, P.M. (1957) General anaesthesia in *Gallus domesticus* for non-survival laboratory experiments. Poult. Sci., **36**: 490-495.

BIGGS, P.M. and KING, A.S. (1957) A new experimental approach to the problem of the air pathway within the avian lung. J. Physiol., **138**: 282-299.

BIGGS, P.M. and PAYNE, L.N. (1959) Cytological identification of proliferating donor cells in chick embryos injected with adult chlcken blood. Nature, (Lond.), **184**: 1594.

BIGGS, P.M. and PAYNE, L.N. (1961) Pathological changes following the inoculation of chick embryos with adult cells. 1. Spleen cells. Immunology, **4**: 24-37.

BIGGS, P.M. and PAYNE, L.N. (1961) Pathological changes following the inoculation of chick embryos with adult cells. 2. Blood cells. Immunology, **4**: 38-48.

BIGGS, P.M. (1962) Some observations on the properties of cells from the lesions of Marek's disease and lymphoid leukosis. Proceedings of the 13th Symposium of the Colston Research Society on Animal Health and Production, pp. 83-99. Butterworth, London.

BIGGS, P.M. and PAYNE, L.N. (1963) Transmission experiments with Marek's disease (fowl paralysis). Vet. Rec., **75**: 177-179.

BIGGS, P.M. and PAYNE, L.N. (1963) Some properties of the agent of Marek's disease (fowl paralysis, neuro-lymphomatosis). Proceedings of the International Symposium on Comparative Leukemia, IV, 2. Hannover.

PAYNE, L.N. and BIGGS, P.M. (1964) Differences between highly inbred lines of chickens in the response to Rous sarcoma virus of the chorio-allantoic membrane and of embryonic cells in tissue culture. Virology, **24**: 610-616.

PAYNE, L.N. and BIGGS, P.M. (1964) A difference in susceptibility to lymphoid leukosis virus and Rous sarcoma virus between cells from two inbred lines. Nature (Lond.), **203**: 1306-1307.

PAYNE, L.N. and BIGGS, P.M. (1964) Transmission experiments with Marek's disease (fowl paralysis) and lymphoid leukosis. World's Poultry Science Journal, **20**: 284.

Biggs, P.M. and Payne, L.N. (1964) The relationship of Marek's disease (neuro-lymphomatosis) to lymphoid leucosis. National Cancer Institute Monograph, 17: 83-98

BIGGS, P.M., PURCHASE, H.G., BEE, B.R. and DALTON, P.J. (1965) Preliminary report on acute Marek's disease (fowl paralysis) in Great Britain. Vet. Rec., 77: 1339-1340.

PAYNE, L.N. and BIGGS, P.M. (1965) Effect of lymphoid leucosis virus on the response of the chorio-allantoic membrane to Rous sarcoma virus. Virology, 27. 621-622

PAYNE, L.N., BIGGS, P.M., CHUBB, R.C. and BOWDEN R.S.T.(1966) Contamination of egg-adapted canine distemper vaccine by avian leucosis virus. Vet. Rec., 78: 45-49

HARRIS, R.J.C., DOUGHERTY, R.M., BIGGS, P.M., PAYNE, L.N., GOFFE, A.P., CHURCHILL, A.E. and MORTIMER, R. (1966) Contaminant viruses in two live virus vaccines produced in chick cells. J. Hyg. Camb., 64; 1-7.

PAYNE, L.N. and BIGGS, P.M. (1966) Genetic basis of cellular susceptibility to the Schmidt-Ruppin and Harris strains of Rous sarcoma virus. Virology, 29: 190-198.

OWEN, J.J.T., MOORE, M.A.S. and BIGGS, P.M. (1966) Chromosome studies in Marek's disease. J. natn. Cancer Inst., 37: 199-209.

KENZY, S.G. and BIGGS, P.M. (1967) Excretion of the Marek's disease agent by infected chickens. Vet. Rec., 80: 565-568.

BIGGS, P.M. and PAYNE, L.N. (1967) Studies on Marek's disease. I. Experimental transmission. J. natn. Cancer Inst., 39: 267-280.

PAYNE, L.N. and BIGGS, P.M. (1967) Studies on Marek's disease. II. Pathogenesis. J. natn. Cancer Inst., **39**: 281-302.

CHURCHILL, A.E. and BIGGS, P.M. (1967) Agent of Marek's: disease in tissue culture. Nature (Lond), **215**: 528-530.

PURCHASE. H.G. and BIGGS, P.M. (1967) Characterization of five isolates of Marek's disease. Res. vet. Sci., **8**: 440-449.

BIGGS, P.M., THORPE, R.J. and PAYNE, L.N. (1968) Studies on genetic resistance to Marek's disease in the domestic chicken. Brit. Poult. Sci., **9**: 37-52.

PURCHASE, H.G., CHUBB, R.C. and BIGGS, P.M. (1968) Effect of lymphoid leucosis and Marek's disease on the immunological responsiveness of the chicken. J. natn. Cancer Inst., **40**: 583-592.

CHUBB. R.C. and BIGGS. P.M. (1968) Neutralization of Rous sarcoma virus. J. gen. Virol., **3**: 87-96.

LONG, P.L., KENZY, S.G. and BIGGS, P.M. (1968) Relationship between Marek's disease and coccidiosis. I. Attempted transmission of Marek's disease by avian coccidia. Vet. Rec., 83: 260-262.

BIGGS, P.M., LONG, P.L., KENZY, S.G. and ROOTES, D.G. (1968) Relationship between Marek's disease and coccidiosis. II. The effect of Marek's disease on the susceptibility of chickens to coccidial infection. Vet. Rec., **83**: 284-289.

EPSTEIN, M.A., ACHONG, B.G., CHURCHILL, A.E. and BIGGS, P.M. (1968) Structure and development of a herpes-type virus of Marek's disease. J. natn Cancer Inst., **41**: 805-820.

CHURCHILL, A.E. and BIGGS, P.M. (1968) A herpes-type virus isolated in cell culture from the tumours of chickens with Mareik's disease. II. Studies *in vivo*. J. natn. Cancer Inst., **41**: 951-956.

BIGGS, P.M., CHURCHILL, A.E., ROOTES, D.G. and CHUBB, R.C. (1968) The etiology of Marek's disease - an oncogenic herpes-type virus. In: Perspectives in Virology, vol. **6**, pp. 211-230. Edit. Pollard, M., Academic Press, New York and London.

BIGGS, P.M., LONG, P.L., KENZY, S.G. and ROOTES. D.G. (1969) Investigations into the association between Marek's disease and coccidiosis. Proceedings of International Symposium on Coccidiosis. Brno. Czechoslovakia. Acta. veterinaria (Brno). **38**: 65-75.

KENZY, S.G., LONG. P.L. and BIGGS, P.M. (1970) Relationship between Marek's disease and coccidiosis. III. Effects of infection with *Eimeria mivati* on Marek's disease in the chicken. Vet. Rec., **86**: 100-104.

PAYNE, L.N. and BIGGS, P.M. (1970) Genetic resistance of fowl to MH2 reticuloendothelioma virus. J. gen.Virol., **7**: 177-185.

BIGGS, P.M. PAYNE, L.N. MILNE, B.S. CHURCHI.LL. A.E. CHUBB. R.C. POWELL, D.G. and HARRIS. A.H. (1970) Field trials with an attenuated cell associated vaccine for Marek's disease. Vet. Rec., **87**: 704-709.

BIGGS, P.M. and MILNE, B.S. (1971) Use of the embryonating egg in studies on Marek's disease. Am, J. vet. Res., **32**: 1795-1809.

B1GGS, P.M. and MILNE, B.S. (1972) Biological properties of a number of Marek's disease virus isolates. In: Oncogenesis and Herpesviruses, pp. 88-94. Edit. Biggs, P.M., deThé, G. and PAYNE, L.N.. IARC scientific publication No.2. International Agency for Research on Cancer, Lyon.

FRAZIER, J. A. and BIGGS, P.M. (1972) Marek's disease herpesvirus particles in tissues from chickens free of precipitating antibodies. J. natn. Cancer Inst., **48**: 1519-1523.

BIGGS, P.M., POWELL, D.G., CHURCHILL, A.E. and CHUBB, R.C. (1972) The epizootiology of Marek's disease. **1**. Incidence of antibody, viraemia and Marek's disease in six flocks. Avian Path., **1**: 5-25.

PHILLIPS., P.A. and BIGGS, P.M. (1972) Course of infection in tissues of susceptible chickens after exposure to strains of Marek's disease virus and turkey herpesvirus. J. natn. Cancer Inst., **49**: 1367-1373.

WEISS, R.A. and BIGGS, P.M. (1972) Leukosis and Marek's disease viruses of feral red jungle fowl and domestic fowl in Malaya. J. natn. Cancer Inst., **49**: 1713-1725.

ROSS, L.J.N., FRAZIER, J.A. and BIGGS, P.M. (1972) An antigen common to some avian and mammalian herpesviruses. Oncogenesis and Herpesviruses. pp. 480-484. Edit. Biggs, P.M., de Thé, G. and Payne, L.N., IARC scientific publication No.2. International Agency for Research on Cancer, Lyon.

BIGGS, P.M., JACKSON, C.A.W., BELL, R.A., LANCASTER, F.M. and MILNE, B.S. (1972) A vaccination study with attenuated Marek's disease virus. In: Qncogenesis and Herpesviruses, pp. 139-46. Edit. Biggs, P.M., de Thé, G. and Payne, L.N., IARC scientific publication No.2. International Agency for Research on Cancer, Lyon.

JACKSON, C.A.W., BIGGS, P.M., BELL, R.A., LANCASTER., F.M. and MILNE, B.S. (1972) Vaccination against Marek's disease with attenuated Marek's disease virus in a commercial egg laying flock. In: Proceedings 1972 Australasian Poultry Science Convention, Auckland, New Zealand, pp. 229-240.

BIGGS, P.M., MILNE, B.S., GRAF, T. and BAUER, H. (1973) Oncogenicity of non-transforming mutants of avian sarcoma viruses. J. gen. Virol., **18**: 399-403.

ROSS, L.J.N., BIGGS, P.M. and NEWTON, ALISON A. (1973) Purification and properties of the 'A' antigen associated with Marek's disease virus infections. J. gen. Virol., **18**: 291-304.

PANI, P.K. and BIGGS, P.M. (1973) Genetic control of susceptibility to an A subgroup sarcoma virus in commercial chickens. Avian Path., **2**: 27-41.

BIGGS, P.M., JACKSON, C. A. W. and POWELL, D.G. (1973) The epizootiology of Marek's disease. II. The effect of supply flock, rearing house, and production house on the incidence of Marek's disease. Avian Path., 2: 127-134.

PANI, P. K. and BIGGS, P.M. (1974) Heritability and number of genes controlling susceptibility of chick embro chorioallantoic membranes (CAM) to RSV. Avian Path., 3: 115-122.

BIGGS, P.M. MILNE, B.S., FRAZIER, J.A., McDOUGALL, J.S. and STUART, J.C. (1974) Lymphoproliferative disease in turkeys. Proceedings XVth World's Poultry Congress, New Orleans, WPSA, Washington, pp. 55-57.

JACKSON, C.A.W., BIGGS, P.M., BELL, R.A., LANCASTER, F.M. and MILNE, B.S. (1974) A study of vaccination against Marek's disease with an attenuated Marek's disease virus. Avian Path., 3: 123-144.

BULOW von V, and BIGGS. P.M. (1975) Differentiation between strains of Marek's disease virus and turkey herpesvirus by immunofluorescence assays. Avian Path. 4: 133-146.

BULOW von, V. and BIGGS, P.M. (1975) Precipitating antigens associated with Marek's disease viruses and herpesvirus of turkeys. Avian Path., 4: 147-162.

BULOW von, V., BIGGS, P.M. and FRAZIER, J.A. (1975) Characterization of a new sero-type of Marek's disease herpesvirus. In: Oncogenesis and Herpesviruses II, pp. 329-336. Edit. de Thé, G., Epstein, M.A. and zur Hausen, H. IARC Scientific Publication No. 11. International Agency for Research on Cancer, Lyon.

JACKSON, C.A.W., BIGGS, P.M., BELL, R.A., LANCASTER, F.M. and MILNE, B.S. (1976) The epizootiology of Marek's disease. 3. The inter-relationship of virus pathogenicity, antibody and the incidence of Marek's disease. Avian Path., 5: 105-123.

RIDDELL, C., MILNE, B.S. and BIGGS, P.M. (1978) Herpes Virus of turkey vaccine: Viraemias in field flocks and in experimental chickens. Vet. Rec., 102: 123-126.

BIGGS, P.M., McDOUGALL, J.S., FRAZIER, J.A. and MILN'E, B.S. (1978) Lymphoproliferative Disease of Turkeys. I. Clinical Aspects. Avian Path., 7: 131-139.

McDOUGALL, J.S., BIGGS, P.M.., SHTILLETO, R.W. and MILNE. B.S. (1978) Lymphoproliferative Disease of Turkeys. II. Experimental Transmission and Aetiology. Avian Path., 7: 141-155.

McDOUGALL, J.S., BIGGS, P.M. and SHILLETO, R.W. (1978) A leukosis in turkeys associated with infection with reticuloendotheliosis virus. Avian Path., 7: 557-568.

BIGGS, P.M., SHILLETO, R.W. and MILNE, B.S. (1980) A quantitative study of viraemia and neutralising antibody in a commercial flock vaccinated with HVT. In: Resistance and Immunity to Marek disease, pp. 483-497. Edit. Biggs, P.M., C.E.C. Luxemburg.

BIGGS, P.M. (1975) Marek's disease virus, a model for oncogenic herpesviruses - chemical and antigenic properties. Proceedings of Xlth International Cancer Congress, Florence. Excerpta Medica, Amsterdam. Vo. 1: 159-165.

BIGGS, P.M. (1976) Criteria for the differential diagnosis of avian lymphoid leukosis and Marek's disease. In: Differential diagnosis of avian lymphoid Leukosis and Marek's disease, pp. 67-85, Edit. Payne, L.N. CEC Publication, EURm 5494e, Luxembourg, 1976.

BIGGS, P.M. (1977) Marek's disease and leukosis. In: Poultry Health, pp. 65-81. Edit. Gordon, R.F. Bailliere Tindall, London.

BIGGS. P.M. (1978) The veterinary profession and an intensive animal industry. (Wooldridge Memorial Lecture). Vet. Rec., **103:** 251-255.

BIGGS, P.M. (1979) Marek's disease now and then. In: Advances in Comparative Leukemia Research, pp. 353-358. Proceedings of the IXth International Symposium on Comparative Research on Leukemia and Related Diseases, Pitsunda, U.S.S.R. Edit. Yohn, D.S., Lapin, B.A. and Blakeslee, J.R. New York, Elsevier, 1980.

BIGGS, P.M. (1980) The biology of avian leukosis. In: Leukemias, Lymphomas, and Pappillomas: Comparative Aspects, pp. 233-246. Edit. Bachmann, P.A. London, Taylor and Francis.

BIGGS, P.M. (1982) The world of poultry disease. Avian Path. **11:** 281-300.

ALLAN, W.H., ALEXANDER, D.J., BIGGS, P.M., GORDON, R.F., JORDAN, F.T.W. and McFERRAN, J.B. (1982) Viral diseases. In: Poultry Diseases, 2nd ed., pp. 76-159. Edit. Gordon, R.F. and Jordan, F.T.W. London, Baillière Tindall.

BIGGS, P.M. (1982) Vaccination against Marek's disease in layers and breeders. In: Marek's vaccination. Present situation and prospects for the future, pp.39-52. Proceedings of the 3rd scientific meeting, IVAZ Laboratori Biologici, Padova, Italy.

BIGGS, P.M. (1982) The epizootiology of avian herpesviruses in veterinary medicine. Devl. Biol. Stand., **52:** 3-11.

BIGGS, P.M. (1983) Disease prevention and control in poultry production. In: Disease Prevention and Control in poultry Production. pp. 313-319. Sydney Post-graduate Committee in Veterinary Science.

ALEXANDER. D., ALLAN. W.H., BIGGS. P.M.. BRACEWELL. C.D., DARBYSHI.RE, J.R., DAWSON, P.S., HARRIS. A.H., JORDAN, F.T.W., MACPHERSON, I., McFERRAN, J.B., RANDALL, C.J., STUART, J.C., SWARBRICK, O. and WILDING. G.P. (1983) A standard technique for haemagglutination inhibition tests for antibodies to avian infectious bronchitis virus. Vet. Rec., **113:** 64.

BIGGS, P.M. (1983) Vaccination. In: The Control of Infectious Diseases in Farm Animals, pp. 21-27. BVA Trust Project: "Future of Animal Health Control". London, British Veterinary Association".

BIGGS, P.M. (1984) Contagious diseases: an economic hazard to the poultry industry. siipikarja, **66:** 246-252.

BIGGS, P.M. (1984) Lympoproliferative disease of turkeys. In: Diseases of Poultry, 8th ed., pp. 418-419. Edit. Hofstad, M.S., Barnes, H.J., Calnek, B.W., Reid, W.M. and Yoder, H.W. Iowa State University Press. Ames, Iowa, U.S.A.

BIGGS, P.M. (1985) Summing up of the 2nd International Symposium Marek's disease. Proceedings of the Internationa1 Symposium on Marek's Disease, 1984, pp. 590-598. Cornell University, New York, U.S.A. Edit. Calnek, B.W. and Spencer, J.L. Pennsylvania, American Association of Avian Pathologists, Inc.

BIGGS, P.M. (1985) Infectious animal disease and its control. In: Philosophical Transactions of the Royal Society, **30:** 259-274.

BIGGS, P.M. (1985) Spread of Marek's disease. In: Marek's disease. Scientific Basis and Methods of Control, pp. 329-340. Edit. Payne, L.N. Boston/Dordrecht/ Lancaster, Martinus Nijhoff Publishing.

ROSS, L.J.N. and BIGGS. P.M. (1986) Vaccination against Marek's disease. In: Leukaemia and Lymphoma Research. 3. Vaccine intervention against virus induced tumours. Eds. Goldman. J.M. and Epstein, M.A. Macmillan Press.

BIGGS, P.M. (1988) Disease and its control in poultry production. First International Poultry and Poultry Diseases Symposium, Manisa, Turkey.

BIGGS, P.M. (1988) Summing up of the 3rd International Symposium on Marek's disease. In: Advances in Marek's disease research, 1998. Edits. Kato, S., Horuichi, T., Mikami, T. and Hirai, K. Japnese Association on Marek's disease.

BIGGS, P.M. (1988) The place of molecular biology in veterinary research and teaching. Vet. Rec., **123:** 176-179.

BIGGS, P.M. (1989) La place de la biologie moléculaire dans la recherche et l'enseignment vétérinaiares. Ann. Méd. Vét., **133:** 185-189.

BIGGS, P.M. (1990) Use and desired properties of poultry vaccines. In: Genetic engineering of animals. Edit. Hansel, H. and Weir, Barbara J. J. Reprod. Fert. Suppl, **41,** pp 149-152.

BIGGS, P.M. (1990) Gordon Memorial Lecture. Vaccines and vaccination – past, present, and future. Brit. Poult. Sci. **31:** 3-32.

BIGGS, P.M. (1990) Why biological control? In: Biological control of pests and diseases. Edit. McCrachen, A.R. and Mercer, P.C.. Agriculture and Food Science Centre, Belfast. 1990.

BIGGS, P.M. (1991) Lymphoprolifcrative disease of turkeys. In: Diseases of Poultry, 9th Edn. Edit. Calnek, B.W. pp. 456-459. Iowa State University Press, Ames, Iowa, U.S.A.

BIGGS, P.M. (1992) Society's changing view of science and the effect on scientific research. Biologist, **39:** 197-200.

BIGGS, P.M. (1993) Factors affecting Livestock production in the 21st century. In: Animal Health and production for the 21st century. Edit. Beh, K.J. pp. 148-154. CSIRO, Australia.

BIGGS, P.M. (1993) Developments in Vaccines. In: The advancement of veterinary science. Vol. 4. Veterinary Science – Growth points and comparative medicine. Edit. Michell, A.R. pp. 13-127. C.A.B. International.

BIGGS, P.M. (1997) Lympoproliferative disease of turkeys. In: Diseases of poultry, 10th Edn. Edit. Calnek, B.W. pp.485-489. Iowa State University Press. Ames, Iowa, U.S.A.

BIGGS, P.M. (1997) The Leeuwenhoek Lecture 1997. Marek's disease herpesvirus: oncogenesis and prevention. Phil. Trans, R. Soc. London B 352: 1951-1962.

BIGGS, P.M. (2001) The history and biology of Marek's disease virus. In: Marek's disease. Edit.Hirai, K. Springer-Verlag, Berlin.

BIGGS, P.M. (2002) Response to Whitby, Millett and Dando. Medicine, Conflict and Survival. 18: 157-160.

BIGGS, P.M. (2004) Marek's disease: long and difficult beginnings. In: Marek' disease – an evolving problem. Edit. Davison, F. and Nair, V. Elsevier Academic Press, London.

BIGGS, P.M. (2006) 50 years of Poultry Research. In 50th Anniversary of the British Veterinary Poultry Association. British Veterinary Poultry Association.

BIGGS, P.M. (2006) The World Veterinary Poultry Association: the Beginnings and first 25 years. Avian Path., 35: 417-428.

Books edited

Resistance and immunity to Marek's disease. Edit. Biggs, P.M., C.E.C. Luxemburg.
Oncogenesis and Herpesviruses, pp. 88-94. Edit. Biggs, P.M., deThé, G. and PAYNE, L.N.. IARC scientific publication No.2. International Agency for Research on Cancer, Lyon.

Book reviews and editorials.

BIGGS, P.M. (1994) Review of "Virus infections of birds". Edit. McFerran, J.B. and McNulty, M.S. Elsevier Science Publications. In: Avian Path. 23: 189-190.

BIGGS, P.M. (1989) Review of "Improving genetic disease resistance in Farm animals". Edit. Van der Zijp and Sybesma, W. Current Topics in Veterinary Medicine and Animal Science, vol. 52. Kluwer Academic Publishers, Dordrecht, Boston, London. In Vet. Rec.

BIGGS, P.M. (2000) Guest editorial, Whither poultry disease? Vet J. 160: 90-91.

Abstracts

BIGGS, P.M. (1966) Avian Leukosis. Paper presented at Symposium on "Tumour viruses and other non-cytocidal agents". Virus group of the Society for General Microbiology, London. J. gen. Microbiol., 42: No.3, p. xv.

BIGGS, P.M. and ROOTES, D.G. (1967) Some properties of the infectious agent of Marek's disease (neurolymphomotosis) (Abs) J. Am. Vet. med. Ass., **150**: 1341.

PURCHASE, H.G., CHUBB, R.C. and BIGGS, P.M. (1967) Immunologic disturbances in birds infected with lymphoid leukosis and Marek's disease. (Abs.) J. Am. vet. med. Ass., **150**: 1341-1342.

BIGGS, P.M. (1975) Epidemiology and control of Marek's disease. (Abs) 3rd International Congress for Virology, Madrid, p. 168.

BIGGS, P.M. (1977) Vertical transmission of avian RNA tumour Viruses and its significance in oncogenesis. Proceedings of the Society for General Microbiology, **4**: 62.

BIGGS, P.M., McDOUGALL, J.S., MILNE, B.S. and SHILLETO, R.W. (1977) Leukosis in turkeys, p. 13 (abstract). VIth International Congress of the World Veterinary Poultry Association, Atlanta, Georgia.

Popular Articles

BIGGS, P.M. (1966) Avian leucosis and Marek's disease. Poultry Digest, **25**: 68-72.

BIGGS, P.M. (1977) Animal Health: the next ten years. 4. Poultry Span, Vol. 20, No. 3: 106-107.

BIGGS, P.M. (1988) Perspectives. Turkeys, Vol. 36, No. 4, p. 1.

Avian Pathology, 15: 319–322, 1986

PETER M. BIGGS – AN APPRECIATION

Professor Peter M. Biggs, on the occasion of his 60th birthday, deserves appreciative recognition by his many friends and colleagues for the varied, substantial and highly influential contributions he has made to his chosen field of avian medicine. It is especially appropriate that his central role in the establishment, nurturing and continuing stewardship of *Avian Pathology* be recognised. Additionally, the international stature he has attained as a scientist, skilled administrator and as a person bears noting.

An international flavour to Peter Biggs' career may have been shaped in part during his early education which took place both in Petersfield, Hampshire, England, where he was born, and in the United States near Boston, Massachusetts. He has been internationally prominent since then through his research, through membership of committees, and especially through leadership roles with international organisations.

After service in the Royal Air Force, he attended the Royal Veterinary College, London, from where he received, in 1953, the Bachelor of Science degree and was appointed a Member of the Royal College of Veterinary Surgeons. His commitment to research was immediately apparent through his first postgraduate appointment as a Research Assistant in the Department of Veterinary Anatomy at the University of Bristol. In 1955, he became a Lecturer in Veterinary Clinical Pathology at the same institution.

His destiny in the field of avian neoplasia was shaped by the topic of his doctoral thesis research: *Lymphoid Tissue in the Endocrine Glands of the Domestic Chicken: its Significance in Health and Disease.* The importance and eventual impact of this start most certainly could not have been imagined at that time but it earned a PhD degree in 1958 and led to a most significant appointment in the following year when he joined the Houghton Poultry Research Station as Head of the new Leukosis Experimental Unit. There, along with his friend and colleague from Bristol, Dr L.N. ('Jim') Payne, he initiated a series of studies which were to have a marked impact on the poultry industry and on avian disease research. Research efforts for which he is best known, of course, are those dealing with Marek's disease in chickens. Indeed, in 1960 at a meeting of the World Veterinary Poultry Association at Utrecht, he successfully helped to promote the adoption of the name 'Marek's disease' to distinguish that group of apparently aetiologically related neoplasms from others now referred to as the avian leukoses.

The period during which he served as Head of the Leukosis Experimental Unit (1959-73) was one of enormous fecundity. Work which he carried out personally, or in collaboration with colleagues he assembled as a team, was instrumental in establishing methods for the regular transmission of Marek's disease, characterisation of the aetiologic agent, isolation and identification of the causative cell-

This article is reprinted with permission from *Avian Pathology* (published by Taylor & Francis, Informa Ltd., London, UK).

320

associated herpesvirus and, as a crowning achievement, development of the first effective vaccine against a neoplastic disease of any type.

The true significance of these achievements can be appreciated by all members of the scientific community, but especially by workers active in the field of avian medicine during that time. The technical problems were great in view of the cell-associated nature of the Marek's disease agent, and the pressure from the industry for a solution to this most serious poultry disease was enormous. The remarkable output of his Unit during that period is a tribute to his leadership-by-example, to his ability to select bright and productive scientific and support staff and continually justify the funding base needed, and to his determined and exceptionally well organised approach to solving a difficult problem. In a highly competitive field, as Marek's disease research was at that time, it can be said that Peter Biggs and his colleagues commanded the great respect of the entire world-community of avian disease researchers. This has been reflected by the large number of scientists who have chosen to work in his laboratories while on leave from their own institutions and to the warm welcome he receives when he visits other laboratories around the world.

Although Marek's disease research occupied much of his time, it was not the only problem to which he directed efforts. A review of the more than 100 scientific publications which he authored or coauthored reveals an interest in immunologic and genetic aspects of avian diseases. A substantial number of studies were on other tumour viruses including lymphoid leukosis, avian sarcomas, and reticuloendotheliosis. In 1974, he described an entirely new disease entity in turkeys, called lymphoproliferative disease.

That so much could be accomplished in the face of heavy administrative duties is in itself a real tribute to Peter Biggs' dedication to scientific endeavour. In 1974, after heading the Leukosis Experimental Unit for nearly fifteen years, and having served for three years as Deputy Director of the Station, he was appointed Director of the Houghton Poultry Research Station. In that capacity he accepted responsibility for approximately 220 staff members. He created a 'Director's Department' in order to continue his own research, albeit at a somewhat less personal level. More important, he took the opportunity to strengthen existing programmes and develop new initiatives capitalising on molecular biological approaches to the study and control of poultry diseases. Houghton has been essentially unique in its broadly based, multidisciplinary approach to the study of diseases of poultry. Under the effective leadership of Peter Biggs, it has not rested on its laurels.

In June of this year, Peter Biggs was appointed Director of the Agricultural and Food Research Council's newly constituted Institute of Animal Disease Research, of which Houghton will form a part.

A long series of honours attests to the respect he has commanded. He is a Fellow of the Institute of Biology, the Royal College of Pathologists, the Royal College of Veterinary Surgeons, and the very prestigious Royal Society. He was given the Doctor of Science degree from the University of London in 1975, and in 1976 he was awarded an Honorary Doctorate of Veterinary Medicine from the Ludwig-Maximilians University, Munich. In 1982, he was appointed Visiting Professor of Veterinary Microbiology at the Royal Veterinary College, London.

Numerous awards have followed Professor Biggs' many accomplishments. He shared, with Dr L.N. Payne, the 1964 Tom Newman Memorial Award for Poultry Husbandry Research, presented by the Poultry Association of Great Britain, for their work on Marek's disease. Other awards which recognised his contributions to veterinary medicine and poultry science include the BOCM Poultry Science Award, the J.T. Edwards Memorial Medal of the Royal College of Veterinary Surgeons, the Brian Hanson Memorial Award of the British Veterinary Poultry Association, the Dalrymple-Champneys Cup and Medal and the Wooldridge Memorial Medal of the British Veterinary Association, the Bledisloe Veterinary Award of the Royal Agricultural Society of England, and the Central Veterinary Society Victory Medal. An award he especially prized was the Joszef Marek Memorial Medal presented to him in 1979 by the Veterinary University of Budapest.

It would require more than a page simply to list the various committee memberships and other offices which have been (and continue to be in many cases) held by Peter Biggs. In addition to his editorial responsibilities with *Avian Pathology*, he has served on the editorial boards of *Excerpta Medica, Antiviral Research* and the *Journal of Biological Standardization*. Also he has been Editor or Co-editor of numerous Proceedings Books resulting from national and international symposia he has helped organise. He has been, and is, a willing and contributing member of more than 20 domestic and international committees dealing with avian medicine, virus nomenclature, tumour viruses, comparative leukaemia, genetic manipulation of micro-organisms, poultry science, etc. He has served as Chairman of several of these. He served as President of the British Veterinary Poultry Association during 1974-75. As one of the founding members of the International Association for Comparative Research on Leukaemia and Related Diseases, he served on the Committee for 11 years before becoming Vice-President (1979) and then President (1981) of the Association. This exemplifies his interest in international affairs which are broader than just the field of avian oncology, poultry diseases or even veterinary medicine.

The offices he has held which are most apparent to the readers of *Avian Pathology* are those within the parent organisation of the journal, the World Veterinary Poultry Association. As in most associations of this type, the success of the organisation often depends on the chief coordinator of affairs, the Secretary/Treasurer. Peter Biggs served in this very influential post for the period of 1971 to 1981, when he was elected President, and in 1985 he was made an Honorary Life President of the Association.

Of all of the activities he was responsible for overseeing and coordinating, clearly the most important and visible was the establishment of the journal *Avian Pathology* in 1973. He was a founding Editor, and soon became the Editor-in-Chief, a responsibility he still carries. The difficulties involved in the initiation of a totally new venture, without a strong financial base and without any guarantees of survival, not to mention achieving success in terms of credibility as a scientific journal, were enormous. However, to his everlasting credit, Peter Biggs accomplished the publishing objectives of the WVPA in all possible respects. Not only is the journal financially stable and successful, but it has the respect of the scientific community, and contributions for publication come from all parts of the world. Avian medicine in particular, and the biological sciences in general, are deeply indebted to Professor Biggs for his many hours of hard work as well as for his dedication, understanding

322

and resourcefulness as editor. Due to his skilful nurturing and highly successful stewardship, *Avian Pathology* has 'made its mark'. So, indeed has Peter Biggs!

Finally, it should be noted that Peter Biggs is well known, well liked and well respected for his good humour, his infectious enthusiasm, his humaneness and his moral integrity. His friendship is valued by many. Together, we salute and congratulate him on the occasion of his 60th year.

B.W. Calnek

Michael Elliott

AFRC Institute of Arable Crops Research
Rothamsted, United Kingdom

1924–2007

1989 — for distinguished contributions to basic science and its successful translation into practice in the fields of animal health and crop protection.

Dr. Michael Elliot has demonstrated scientific vision in the field of crop protection against insects. In the 1940's he recognized that the naturally occurring insecticidal compounds of plant in the Composite (Sunflower family) *Chrysanthemum cinevariaefolium* known as pyrethroids had important crop protective potential but suffered from serious disadvantages of lack of potency, stability and selectivity. He, therefore, embarked on an imaginative program for the development of synthetic analogues of the naturally occurring compounds and produced two, resmethrin (1962) and bioresmethrin (1985) which had greatly enhanced potency and insect specificity, combined with low mammalian toxicity. These compounds were particularly effective for use in domestic and stored products, but were photodegradable and therefore relatively ineffective under the field conditions. By 1973 Dr. Elliott had produced further analogues which combined crop surface stability and insecticidal potency with exceptionally low mammalian toxicity; they also had the advantage over the organochlorine insecticides of being readily biodegradable in the soil without the formation of toxic residues. These compounds have proved to be of enormous economic Importance in the protection of a very wide variety of herbaceous and tree crops. At the fundamental level Dr. Elliott's work has made possible the establishment of relationships between chemical structure and insecticidal activity in pyrethroids; improved the understanding of factors which determine photostability selectivity and mammalian toxicity and provided a series of steroisomers with contrasting insecticidal activities which are proving to be powerful tools for neurophysiological "site of action" studies.

CURRICULUM VITAE

Brief details of career

1935-1942	The Skinners Company's School, Tunbridge Wells, Kent
1942-1945	University College, Southampton
1945-1948	Postgraduate research on the chemistry of the pyrethrins leading to an external PhD of the University of London; 1945-6 at University College, Southampton and 1946-8 at King's College, London under Professor S H Harper
1948	Appointed to Insecticides and Fungicides Department, Rothamsted Experimental Station as Scientific Officer to study relationships between molecular structure and biological activity
1953	Promoted to Senior Scientific Officer
1961	Promoted to Principal Scientific Officer
1970	Promoted to Senior Principal Scientific Officer (Individual Merit)
1969 & 1974	Visiting Lecturer, Division of Entomology, University of California, Berkeley, USA
1978	Visiting Professor, Imperial College of Science & Technology, London
1979	Promoted to Deputy Chief Scientific Officer (Individual Merit). Appointed Head of Department of Insecticides and Fungicides and Deputy Director, Rothamsted Experimental Station
1983	Resumed full-time research in the Department of Insecticides and Fungicides
1984	Retired from staff of Rothamsted Experimental Station (30.9.1984)
1984-1985	Consultant on chemistry of insecticides
1986-1988	Visiting Research Scientist Pesticide Chemistry and Toxicology Laboratory, Division of Entomology, University of California at Berkeley, CA 94720
1989-present	Consultant on chemistry of insecticides (UK); Lawes Trust Senior Fellow, Rothamsted Experimental Station

Academic, professional and technical qualifications and awards

1945	BSc (Special) Chemistry (External, University of London)
1952	PhD (External, University of London)

1971	DSc University of London
1975	Burdick and Jackson International Award for Pesticide Research (American Chemical Society)
1977, 1980	ARC Awards to Inventors
1978	Second Holroyd Memorial Lecturer and Medallist, Society of Chemical Industry, London
1978/9	John Jeyes Medallist and Lecturer of the Chemical Society, London
1979	Elected to Fellowship of the Royal Society of London
1982	Appointed Commander of the Order of the British Empire
1982	Royal Society Mullard Medallist*
1983	Awarded La Grande Medaille de la Societe Francaise de Phytiatrie et de Phytopharmacie
1984	Elected Fellow of King's College, University of London
1984	Elected to Fellowship of the Royal Society of Chemistry
1984	Received the first Royal Society of Chemistry Fine Chemicals and Medicinals Group Award
1985	Awarded Honorary Doctorate of Science by the University of Southampton
1986	British Crop Protection Council Medal for outstanding service to British Crop Protection
1988, 1992	BTG Awards to Inventors
1989	Awarded Wolf Foundation Prize in Agriculture
1989	Awarded Prix de la Fondation de la Chimie, Paris
1993	Awarded Environment Medal of the Society of Chemical Industry, London
1996	Elected a Foreign Associate of the National Academy of Sciences, USA

*The Royal Society Mullard Award is awarded in recognition of outstanding contributions to the advancement of science or technology leading directly to national prosperity in the United Kingdom.

Other relevant information

1976	Rothamsted Experimental Station received a Queen's Award for Technological Achievement for work on first generation Pyrethroid Insecticides (NRDC 104, 107)
1978	Research team at Rothamsted Experimental Station received the UNESCO Science Prize, a biennial award, for developing pyrethroid insecticides
1980	Rothamsted Experimental Station received a second Queen's Award for Technological Achievement for work on second generation Pyrethroid Insecticides (NRDC 143, 149, 161)

Over 120 patents and applications in the field of insecticides.

BRIEF DESCRIPTION OF SCIENTIFIC ACHIEVEMENTS

Protecting Food Production Safely

All forms of food are subject to loss through competing organisms. These organisms have to be controlled and, against a wide range of crop pests and animal disease vectors, the pyrethroids represent the most intrinsically benign and effective of the available control agents. They also embody substantially enhanced compatibility with integrated pest management systems over earlier pesticides. The pyrethroids were developed principally through the scientific commitment of Michael Elliott in transforming the low mammalian toxicity, but high insecticidal activity, of the naturally-occurring pyrethrin insecticides into a new generation of field-stable insecticides.

After attaining his B.Sc. in 1945, Elliott chose for his Ph.D. to study the natural pyrethrins, a group of highly potent insecticides found in a daisy-like plant, *Chrysanthemum (Tanacetum) cinerariaefolium*, but which lacked sufficient chemical stability for field use. During this work, he became convinced of the value of these compounds and of the possibility of synthesising analogues with even better properties, particularly stability in the field. It was therefore entirely appropriate that he was appointed, in 1948, to the Department of Insecticides and Fungicides (now the Department of Biological Chemistry) at Rothamsted Experimental Station (now Rothamsted Research) to establish relationships between molecular structure and biological activity. Over a period of nearly 15 years, Dr. Elliott collaborated with biologists in developing and applying a suitable insect bioassay procedure to his compounds and, with outstanding dedication, pursued his initially defined goal. At this time, he was the only chemist involved in the project, and the disadvantages of the organochlorines and other types of insecticides had not then become apparent. Nonetheless, he had sufficient confidence in his own judgement not to allow the research to be diverted. Eventually, in 1962, promising results were obtained. He then took the extremely far-sighted decision to approach the National Research Development Corporation (now the British Technology Group) with his proposals; financial support was granted and has continued thereafter.

The team of chemists led by Elliott (never more than six in all) has made a series of discoveries of enormous importance to world agriculture. Essential factors for this success were Elliott's constantly cheerful enthusiasm, his ability to get the best from each individual, his good judgement and his insistence on the very best equipment and apparatus for the group.

Resmethrin and bioresmethrin, with greatly enhanced potency against some insect species and outstandingly low mammalian toxicity, were discovered in 1965. In 1972/3, the NRDC-supported project led to a further series of compounds including permethrin, cypermethrin and deltamethrin, which proved to be markedly more stable in light and air, rendering them suitable for use in a wide range of agricultural situations. In addition, they were highly active (deltamethrin is one of the most potent insecticides known) and had the low mammalian toxicity generally associated with pyrethroids. The first indications of all these favourable properties were shown in experiments at Rothamsted (insecticidal activity and photostability) and at the MRC unit at Carshalton and the University of California at Berkeley (mammalian toxicity and metabolism), all initiated by Dr. Elliott at exactly the time they were needed.

The natural pyrethrins are a mixture of six complex molecules, each containing over fifty atoms arranged in structures with three chiral centres, and three opportunities for additional isomerism. Despite this daunting complexity, Elliott was able to establish which parts of the molecule could be replaced, and to determine the stereochemical features that control the shape of active analogues. The culmination of this research was deltamethrin, a single crystalline isomer (of the eight possible), in which well over half of the structural features of the natural pyrethrin molecule have been replaced by new groups. These pyrethroids, particularly deltamethrin, have substantially enhanced compatibility with integrated pest management (IPM) over the earlier organochlorines, organophosphates and carbamates, and include safety to pollinators under prescribed conditions. The advances are summarised in the following table:

	Representative LD_{50} (mg/kg)			
Compound	**Insects**	**Mammals**	**Safety Factor**	**Persistence**
Aldicarb (carbamate)	10	10	1	weeks
Dimethoate (organophosphate)	15	550	37	weeks
Pyrethrins	0.2	900	4,500	hours
Permethrin	~0.2	~2,000	~10,000	weeks
Cypermethrin	~0.06	~500	~8,000	weeks
Deltamethrin	~0.008	~100	~13,000	weeks

The research led by Dr. Elliott and the discovery of exceptionally effective synthetic pyrethroids has been described as "a classic example of applied science". His modesty made him reluctant to accept another phrase frequently applied to him: "The father of modern pyrethroids". The international significance of the work has been recognised in several important ways, the most prestigious being the award to Rothamsted of the UNESCO Science Prize for 1978.

Elliott eagerly accepted the responsibilities that went with his success — to respond to requests for information, to write reviews and to present lectures.

Many can confirm his willingness to help, either by telephone or by encouraging visits to him at Rothamsted. His publication list and curriculum vitae indicate the scope and number of reviews he has written, in addition to original research papers, and his enthusiasm to travel to talk about his work.

Statement of Impact

Of a million insect species, approximately 10,000 can be regarded as pests, and about 30% of all the food we grow feeds insects rather than people.[1] Despite the success of alternative control methods, the protection of world food production continues to rely very heavily on insecticides. Consequently, much effort has been expended towards developing safer and more effective insecticides. Insects, being animals, have many physiological processes in common with ourselves, and it is therefore technically extremely demanding to produce insecticides that are also benign to human beings and other warm-blooded animals. The immense impact of the pyrethroids was assured by their being so highly insecticidal and yet rapidly detoxified by mammals. With the discovery of the field-stable pyrethroids, e.g. permethrin, and subsequently those optically resolved such as deltamethrin, commercial development followed rapidly. The compounds have proved to be suitable for lepidopterous pests worldwide and also for a multiplicity of other uses in agriculture, animal husbandry and horticulture, generally at extremely low dose rates, down to 10 g/ha. They were swiftly accepted because, unlike DDT and other organochlorines, they are readily metabolised to harmless products in biotic media such as soil. The research provided a new group of insecticides which immediately rivalled the organophosphates as leading products for commercial crop protection. This was achieved at a time when new insecticides were urgently needed to replace compounds that had properties no longer considered suitable, or to which insects had become resistant.

Once the discoveries had been announced, they were eagerly taken up by industry, initially by eight separate firms. Even today, much of Elliott's chemistry remains the basis for the modern pyrethroid industry. For example, deltamethrin is produced, stereochemically pure, on an enormous scale (hundreds of tons) essentially by the originally published route (Hoechst Roussel), and Shell (now Cyanamid) have responded to Elliott's constant stress on the importance of stereochemical purity to reduce environmental loading by developing a two-isomer form of cypermethrin, again using similar chemistry. The immense impact of Elliott's work lies, therefore, not only in providing safer food protectants but also in opening up the opportunity for discovering new and more acceptable insecticides in a novel chemical class.

The new class of agricultural insecticides that resulted from Elliott's work remains important more than 20 years after their introduction. Pyrethroids currently (1997) represent 19% of the total world sales of insecticide ($1,530M of $8070M).[2]

Of the sales of pyrethroids, about 70% are used on food crops. Although generic products have subsequently been developed, the Elliott compounds still comprise about half of the pyrethroid market and continue to be used extensively in developing agricultural regions.*

Rarely in the field of agricultural pest control can one contribution have had such a significant and lasting influence on food production. Dr. Elliott's research, combined with that of John E. Casida, undoubtedly constitutes a major contribution to the efficiency and safety of food production worldwide, whilst minimising undesirable effects on the environment. The award of the Wolf Foundation Prize in Agriculture is a fitting tribute to the outstanding scientific integrity he has shown so clearly throughout the whole of his career and his single-minded determination to help in the never-ending quest for the increased availability of healthy food.

References

1. Elliott, M. (1990). The Contribution of Pyrethroid Insecticides to Human Welfare. *L'actualité chimique*, 57-70.
2. Wood Mackenzie report, 1998, 52-53. **(Confidential - Not for general dissemination).**

LIST OF PUBLICATIONS

CROMBIE, L., ELLIOTT, M and HARPER, S.H. 1948. Total synthesis of some pyrethrins. *Nature* **162**, 222.

CROMBIE, L., ELLIOTT, M. and HARPER, S.H. 1950. Experiments on the synthesis of the pyrethrins. III - Synthesis of dihydrocinerin-I and tetrahydropyrethrin-I; a study of the action of *N*-bromosuccinimide on 3-methyl-1-*n*-alkyl (and alkenyl)- cyclopent-3-en-1-ones. *Journal of the Chemical Society*, 971.

ELLIOTT, M., NEEDHAM, P.H. and POTTER, C. 1950. The insecticidal activity of substances related to the pyrethrins. I - The toxicities of two synthetic pyrethrin-like esters relative to that of the natural pyrethrins and the significance of the results in the bioassay of closely related compounds. *The Annals of Applied Biology* 37, 490.

ELLIOTT, M. 1951. The insecticidal activity of the pyrethrins and related compounds. *Pyrethrum Post* 2(3), 18.

ELLIOTT, M. 1954. Allethrin. *Journal of the Science of Food and Agriculture* **11**, 505.

ELLIOTT, M. 1956. The pyrethrins and related compounds. I - The preparation of cyclopentenones from the products of Stobbe condensations with aliphatic ketones. *Journal of the Chemical Society*, 2231.

POTTER, C., ELLIOTT, M and WARD, J. 1956. The pyrethrins - a British viewpoint. *Pyrethrum Post* 4(1), 15.

ELLIOTT, M. 1958. Isolation and purification of (+)-pyrethrolone from pyrethrum extract: reconstitution of pyrethrins I and II. *Chemistry and Industry*, 685.

ELLIOTT, M., OLEJNICZAK, J.S. and GARNER, J.J. 1959. Laboratory scale molecular distillation of the pyrethrins. *Pyrethrum Post* 5(2), 8.

ELLIOTT, M. 1960. The structures of the enols of pyrethrolone. *Proceedings of the Chemical Society* 406.

ELLIOTT, M. 1960. Pyrethrolone and related compounds. *Chemistry and Industry*, 1142.

ELLIOTT, M. 1961. Pyrethrolone. *Japanese Journal of Pharmacy and Chemistry* 33(12), 22.

CROMBIE, L. and ELLIOTT, M. 1961. Chemistry of the natural pyrethrins. *Fortschritte der Chemie organischer Naturstoffe* **19**, 120.

ELLIOTT, M. and JEFFS, K.A. 1961. A synthesis of calythrone. *Proceedings of the Chemical Society*, 374.

ELLIOTT, M. 1961. The pyrethrins and related compounds. II - Infra-red spectra of the pyrethrins and of other constituents of pyrethrum extract. *Journal of Applied Chemistry* **II**, 19.

ELLIOTT, M. 1961. La distillation moleculaire des pyrethrines. *Feuilles d'information du Cofap* 5, 6.

SAWICKI, R.M., ELLIOTT, M., GOWER, J.C., SNAREY, M. and THAIN, E.M. 1962. Insecticidal activity of pyrethrum extract and its four insecticidal constituents against house flies. I - Preparation and relative toxicity of the pure constituents; statistical analysis of the action of mixtures of these components. *Journal of the Science of Food and Agriculture* **13**, 172.

ELLIOTT, M. 1964. The pyrethrins and related compounds.III - Thermal isomerization of *cis*-pyrethrolone and its derivatives. *Journal of the Chemical Society*, 888.

ELLIOTT, M. 1964. The pyrethrins and related compounds. V - Purification of (+)-pyrethrolone as the monohydrate, and the nature of 'Pyrethrolone-C'. *Journal of the Chemical Society*, 5225.

ELLIOTT, M. 1965. The pyrethrins and related compounds. VI - The structures of the 'enols' of pyrethrolone. *Journal of the Chemical Society*, 3097.

SAWICKI, R.M. and ELLIOTT, M. 1965. Insecticidal activity of pyrethrum extract and its four insecticidal constituents against house flies. VI - Relative toxicity of pyrethrin I and pyrethrin II against four strains of house flies. *Journal of the Science of Food and Agriculture* **16**, 85.

ELLIOTT, M., JANES, N.F., JEFFS, K.A. NEEDHAM, P.H. and SAWICKI, R.M. 1965. New pyrethrin-like esters with high insecticidal activity. *Nature* **207**, 938.

ELLIOTT, M., JANES, N.F., JEFFS, K.A., NEEDHAM, P.H., SAWICKI, R.M. and STEVENSON, J.H. 1965. New insecticidal esters of chrysanthemic acid. *Proceedings of the 3rd British Insecticide and Fungicide Conference* 432.

CORRAL, C. and ELLIOTT, M. 1965. The pyrethrins and related compounds. VII - New pyrethrin-like compounds with ester and ketonic groups in the alcoholic side chain. *Journal of the Science of Food and Agriculture* **16**, 514.

ELLIOTT, M., FARNHAM, A.W., JANES, N.F., NEEDHAM, P.H., PEARSON, B.C.and STEVENSON, J.H. 1967. New synthetic insecticidal compounds related to the pyrethrins. *Proceedings of the 4th British Insecticide and Fungicide Conference* 437.

ELLIOTT, M. 1967. Synthesizing pyrethrin-like insecticides. *Science Journal* (3 March).

ELLIOTT, M., JANES, N.F. and JEFFS, K.A. 1967. 2,4-Dimethylphenylacetic acid from cyanodihydrocarvone. *Chemistry and Industry*, 1175.

ELLIOTT, M., FARNHAM, A.W., JANES, N.F., NEEDHAM, P.H. and PEARSON, B.C. 1967. 5-Benzyl-3-furylmethyl chrysanthemate: a new potent insecticide. *Nature* **213**, 493.

ELLIOTT, M., HARPER, S.H. and KAZI, M.A. 1967. Experiments on the synthesis of the pyrethrins XIV -Rethrins and the cyclopentadienone related to 3-methylcyclopent-2-enone. (The pyrethrins and related compounds. VIII-) *Journal of the Science of Food and Agriculture* **18**, 167.

ELLIOTT, M., JANES, N.F. and PEARSON, B.C. 1967. The pyrethrins and related compounds. IX - Alkenylbenzyl and benylbenzyl chrysanthemates. *Journal of the Science of Food and Agriculture* **18**, 325.

ELLIOTT, M., and JANES, N.F. 1967. Synthesis of 4-allyl- and 4-benzyl-2,6-dimethylbromobenzene from 2,6-xylidine. *Journal of the Chemical Society* (C), 1780.

ELLIOTT, M., 1968. The search for safer insecticides. Article No. 0911/8, No.7 in *Features Newsletter* No.35, 1968.

ELLIOTT, M. and JANES, NF. 1969. Pyrethrin II and related esters obtained by reconstitution. *Chemistry and Industry*, 270

BRAMWELL, A.F., CROMBIE, L., HEMESLEY, P., PATTENDEN, G., ELLIOTT, M. and JANES, N.F. 1969. Nuclear magnetic resonance spectra of the natural pyrethrins and related compounds. *Tetrahedron* **26**, 1727.

ELLIOTT, M., JANES, N.F., JEFFS, K.A., CROMBIE, L., GOULD, R. and PATTENDEN, G. 1969, Oxidative dimerisation of 2-alkenylcyclopent-2-ene-1, 4-diones by manganese dioxide. *Tetrahedron Letters*, 373.

ELLIOTT, M., NEEDHAM, P.H. and POTTER, C. 1969. Insecticidal activity of the pyrethrins and related compounds. II - Relative toxicity of esters from optical and geometrical isomers of chrysanthemic, pyrethric and related acids and optical isomers of cinerolone and allethrolone. *Journal of the Science of Food and Agriculture* **20**, 561.

BURT, P.E., ELLIOTT, M., FARNHAM, A.W., JANES, N.F., NEEDHAM, P.H. and PULMAN, D.A. 1974. The pyrethrins and related compounds. XIX - Geometrical and optical isomers of 2,2-dimethyl-3(2,2-dichlorovinyl)-cyclopropanecarboxylic acid and insecticidal esters with 5-benzyl-furylmethyl and 3-phenoxybenzyl alcohols. *Pesticide Science* 5, 791-799.

JANES, N.F. and ELLIOTT, M. 1975. Utilisation de la RMN en chimie des pyrethrinoids. *Feuille d'information du CILDA, No. 6*, 5-14.

ELLIOTT, M and JANES N.F. 1075. Quelques aspects du developpement des pyrethrinoides de synthese. *Feuille d'information du CILDA, No.6*, 15-26.

ELLIOTT, M., JANES, N.F. and GRAHAM-BRYCE, I.J. 1975. Retrospect on the discovery of a new insecticide. *Proceedings of 8th British Insecticide and Fungicide Conference*, 373-379.

ELLIOTT, M., FARNHAM, A.W., JANES, N.F., NEEDHAM, P.H and PULMAN, D.A. 1975. Insecticidal activity of the pyrethrins and related compounds. VII - Insecticidal dihalovinyl analogues of *cis* and *trans* chrysanthemates. *Pesticide Science* 6, 537-542.

ELLIOTT, M., JANES, N.F., PULMAN, D.A., GAUGHAN, L.C., UNAI, T. and CASIDA, J.E. 1976. Radiosynthesis and metabolism in rats of the 1*R* isomers of the insecticide permethrin. *Journal of Agricultural and Food Chemistry* 24, 270-276.

ELLIOTT, M., FARNHAM, A.W., JANES, N.F., NEEDHAM, P.H. and PULMAN, D.A. 1976. Insecticidal activity of the pyrethrins and related compounds. IX - 5-Benzyl-3-furylmethyl-2,2-dimethylcyclopropane carboxylate with non-ethylenic substituents at position 3 on the cyclopropane ring. *Pesticide Science* 7, 492-498.

ELLIOTT, M., FARNHAM, A.W., JANES, N.F., NEEDHAM, P.H. and PULMAN, D.A. 1976. Insecticidal activity of the pyrethrins and related compounds. X - 5-Benzyl-3-furylmethyl-2,2-dimethylcyclopropanecarboxylate with ethylenic substituents at position 3 on the cyclopropane ring. *Pesticide Science* 7, 499-502.

ELLIOTT, M. 1976. Chemistry, biochemistry and insecticidal action of synthetic pyrethroids. *Pesticide Science* 7, 223-224.

BRIGGS, G.G., ELLIOTT, M., FARNHAM, A.W., JANES, N.F., NEEDHAM, P.H., PULMAN, D.A. and YOUNG, S.R. 1976. Insecticidal activity of the pyrethrins and related compounds. VII - Relation of polarity with activity in pyrethroids. *Pesticide Science* 7, 236-240.

ELLIOTT, M. 1976. Properties and application of pyrethroids. *Environmental Health Perspectives* 14, 3-13.

BURT, P.E., ELLIOTT, M., FARNHAM, A.W., JANES, N.F., NEEDHAM, P.H and STEVENSON, J.H. 1977. Evaluation of pyrethroids for insect control in *'Crop Protection Agents'*, Academic Press, 393-411.

ELLIOTT, M. 1977. Synthetic pyrethroids, in *'Synthetic Pyrethroids'. ACS Symposium Series No.42*, 1-28.

ELLIOTT, M. and JANES. N.F. 1977. Preferred conformations of pyrethroids, in 'Synthetic Pyrethroids'. *ACS Symposium Series No.42*, 29-36.

NARAHASHI, T., NISHIMURA, K., PARMENTIER, J.L., TAKENO, K. and ELLIOTT, M. 1977. Neurophysiological study of the structure-activity relation of pyrethroids, in *'Synthetic Pyrethroids'. ACS Symposium Series No.42*, 85-97.

ELLIOTT, M., JANES, N.F., PULMAN, D.A. and SODERLUND, D.M. 1978. The pyrethrins and related compounds. Part XXII - Preparation of isomeric cyano-substituted 3-phenoxybenzyl esters. *Pesticide Science* 9, 105-111.

ELLIOTT, M., FARNHAM, A.W., JANES, N.F. and SODERLUND, D.M. 1978. Insecticidal activity of the pyrethrins and related compounds. Part XI - Relative potencies of isomeric cyano-substituted 3-phenoxybenzyl esters. *Pesticide Science* 9, 112-116.

ELLIOTT, M. JANES, N.F. and POTTER C. 1978. The future of pyrethroids in insect control. *Annual Reviews of Entomology* 13, 443-469.

ELLIOTT, M. and JANES, N.F. 1978. Synthetic pyrethroids — a new class of insecticides. *Chemical Society Reviews* 7, 473-505.

ELLIOTT, M. and JANES, N.F. 1978. Recent structure-activity correlations in synthetic pyrethroids. *Advances in Pesticide Science.* (Ed. H. Geissbuhler) Pergamon, Oxford. 2, 166-173.

ELLIOTT, M. 1979. Progress in the design of insecticides. *Chemistry and Industry*, 757-768.

ELLIOTT, M. 1980. Recent advances in the chemistry of pyrethoids. *Pesticide Science* 11, 101.

ELLIOTT, M. 1980. Established pyrethroid insecticides. *Pesticide Science* 11, 119-128.

ELLIOTT, M., JANES, N.F. and KHAMBAY, B.P.S. 1980. The pyrethrins and related compounds. XXIII -Kinetic control in the formation of pyrethroidal esters. *Pesticide Science* 11, 219-223.

ELLIOTT, M., FARNHAM, A.W., JANES, N.F., JOHNSON, D.M. and PULMAN, D.A.1980. The pyrethrins and related compounds. Part XXIV - Synthesis, ^{13}C-nuclear magnetic resonance spectra and insecticidal activity of cycloalkyl analogues of fenvalerate. *Pesticide Science* 11, 513-525.

ELLIOTT, M. 1980. The future for Insecticides. *Insect Biology in the Future.* Academic Press, London, 879-903.

ELLIOTT, M. 1980. Development of synthetic pyrethroids. *Table Ronde Roussel Uclaf.* Pyrethroid Insecticides: Chemistry and Action, 1-4.

ELLIOTT, M., FARNHAM, A.W., JANES, N.F. and KHAMBAY, B.P.S. 1981. The pyrethrins and related compounds. Part XX V - Synthesis and insecticidal activity of conformationally restrained compounds related to 3-phenoxybenzyl esters. *Pesticide Science* 12, 503-508.

ELLIOTT, M., FARNHAM, A.W., JANES, N.F. and KHAMBAY, B.P.S. 1982. Insecticidal activity of the pyrethrins and related compounds. Part XIII - α-Substituted 3-phenoxybenzyl esters. *Pesticide Science* **13**, 407-414.

ELLIOTT, M. 1982. University research and industrial development. *Proceedings of a symposium, Research and Industry: a community?* Paris: l'Ecole Nationale Superieure Chimie de Paris.

ELLIOTT, M., FARNHAM, A.W., JANES, N.F., KHAMBAY, B.P.S. and PULMAN, D.A. 1983. The pyrethrins and related compounds. Part XXVI - Analogues of fenvalerate from acyclic substituted 3-methylbutyric and related acids. *Pesticide Science* **14**, 1-8.

ELLIOTT, M., JANES, N.F. and KHAMBAY, B.P.S. 1983. The pyrethrins and related compounds. Part XXVII - Esters of pulegenic acid and related acids. *Pesticide Science* **14**, 9-16.

ELLIOTT, M., JANES, N.F. and KHAMBAY, B.P.S. 1983. Insecticidal activity of the pyrethrins and related compounds. Part XIII - Comparison of the effects of a-substituents in different esters. *Pesticide Science* **14**, 182-190.

BRIGGS, G.G., ELLIOTT, M. and JANES, N.F. 1983. Present status and future prospects for synthetic pyrethroids. IUPAC Pesticide Chemistry. Human Welfare and the Environment. Ed. J Miyamoto *et al.*, Pergamon Press, Oxford and New York, 157164.

ELLIOTT, M. 1983. Developments in the chemistry and action of pyrethroids. Natural products for innovative pest management. Ed. D.L. Whitehead and W.S. Bowers. Pergamon Press, Oxford and New York, 127-150.

ELLIOTT, M., JANES, N.F., STEVENSON, J.H. and WALTERS, J.H.H. 1983. Insecticidal activity of the pyrethrins and related compounds. Part XIV - Selectivity of pyrethroid insecticides between *Ephestia kuehniella* and its parasite *Venturia canescens. Pesticide Science* **14**, 423-426.

ELLIOTT, M and JANES, N.F. 1983. Evolution of pyrethroid insecticides for crop protection. *Proceedings of the 10th International Congress of Plant Protection Brighton, 1983. Plant Protection for Human Welfare.* Volume 1, 216-223.

LAUFER, J., ROCHE, M., PELHATE, M., ELLIOTT, M., JANES, N.F. and SATELLE, D.B. 1984. Pyrethroid insecticides: action of deltamethrin and related compounds on insect axonal sodium channels. *Journal of Insect Physiology*, Volume 30, No. 5, 341-349.

ELLIOTT, M., JANES, N.F. and KHAMBAY, B.P.S. 1984. Relative insecticidal activity of various 3-polyhaloethyl- 2,2-dimethylcyclopropane carboxylates. *Proceedings of the 1984 British Crop Protection Conference*, Volume 3, 849-852.

ELLIOTT, M. 1985. A survey of the development and environmental chemistry of pyrethroids, *Pesticide Science* **16**, 192-193.

ELLIOTT, M. 1985. Lipophilic insect control agents. In: *Recent Advances in the Chemistry of Insect Control, R.S.C. Special Publication No. 53*; Ed. JANES, N.F., 73-102.

ELLIOTT, M. FARNHAM, A.W., JANES, N.F., JOHNSON, D.M., PULMAN, D.A. and SAWICKI R.M. 1986. Insecticidal amides with selective potency against a resistant (*super-kdr*) strain of houseflies (*Musca domestica L.*). *Agricultural and Biological Chemistry* 50(5), 1347-1349.

BAYDAR, A.E., ELLIOTT, M., JANES, N.F. and KHAMBAY, B.P.S. 1985. Structure-activity relationships in non-ester pyrethroids. *190th American Chemical Society National Meeting, Chicago, 1985*, Abstract No.18.

BRIGGS, G.G., ELLIOTT, M., JANES, N.F. and KHAMBAY, B.P.S. 1985. The influence of fluorine substituents on the biological activity of pyrethroids. *190th American Chemical Society National Meeting, Chicago, 1985*, Abstract No.90.

ELLIOTT, M. 1985. A survey of the development and environmental chemistry of pyrethroids. *Pesticide Science* 16, 192-193.

ELLIOTT, M. and JANES, N.F. 1984. L'evoluzione degli insetticidi piretroidi in agricoltura. *Informatore Fitopatologica* (9), 11-14.

ELLIOTT, M., ELLIOTT, R.L., JANES, N.F., KHAMBAY, B.P.S. and PULMAN, D.A. 1986. The pyrethrins and related compounds. Part XXVIII. Alkenyl- and Alkynyl-substituted Benzyl Esters. *Pesticide Science* 7, 691-700.

ELLIOTT, M., ELLIOTT, R.L., JANES, N.F., KHAMBAY, B.P.S. and PULMAN, D.A. 1986. The pyrethrins and related compounds. Part XXIX. Haloallylbenzyl Esters. *Pesticide Science* 17, 701-707.

ELLIOTT, M., ELLIOTT, R.L., JANES, N.F., KHAMBAY, B.P.S. and PULMAN, D.A. 1986. The pyrethrins and related compounds. Part XXX. Esters from Acids with Mono-Halovinyl Side Chains. *Pesticide Science* 17, 708-714.

ELLIOTT, M., FARNHAM, A.W., JANES, N.F., JOHNSON, D.M. KHAMBAY, B.P.S. and SAWICKI R.M. 1987. Selectivity and resistance to non-ester pyrethroids and *N*-akylamides in houseflies (*Musca domestica* L.) In: *Combating resistance to xenobiotics. Biological and chemical approaches.* Eds. M.G. Ford, D.W. Hollomon, B.P.S. Khambay and R.M. Sawicki. Chichester: Ellis Horwood, pp. 306-313.

ELLIOTT, M., FARNHAM, A.W., JANES, N.F., JOHNSON, D.M. and PULMAN, D.A. 1987. Synthesis and insecticidal activity of lipophilic amides. Part 1: Introductory survey, and discovery of an active synthetic compound. *Pesticide Science* 18, 191-201.

ELLIOTT, M., FARNHAM, A.W., JANES, N.F., JOHNSON, D.M. and PULMAN, D.A. 1987. Synthesis and insecticidal activity of lipophilic amides. Part 2: Effect of heteroatom replacements and of introducing methyl groups in the pentadiene system. *Pesticide Science* 18, 203-209.

ELLIOTT, M., FARNHAM, A.W., JANES, N.F., JOHNSON, D.M. and PULMAN, D.A. 1987. Synthesis and insecticidal activity of lipophilic amides. Part 3: Influence of chain length and terminal group in *N*-(2- methylpropyl)-2,4-dienamides. *Pesticide Science* **18**, 211-221.

ELLIOTT, M., FARNHAM, A.W., JANES, N.F., JOHNSON, D.M and PULMAN D.A. 1987. Synthesis and insecticidal activity of lipophilic amides. Part 4: The effect of substituents on the phenyl group of 6-phenylhexa-2,4-dienamides. *Pesticide Science* **18**, 223-228.

ELLIOTT, M., FARNHAM, A.W., JANES, N.F., JOHNSON, D.M. and PULMAN. D.A. 1987. Synthesis and insecticidal activity of lipophilic amides. Part 5: Influence on activity of varying the substituent on nitrogen. *Pesticide Science* **18**, 229-238.

ELLIOTT, M., FARNHAM, A.W., JANES, N.F., JOHNSON, D.M. and PULMAN, D.A. 1987. Synthesis and insecticidal activity of lipophilic amides. Part 6: 6-(Disubstituted-phenyl)hexa-2,4-dienamides. *Pesticide Science* **18**, 239-244.

ELLIOTT, M., FARNHAM, A.W., JANES, N.F. and KHAMBAY, B.P.S. 1988. The pyrethrins and related compounds. Part XXXI: Alkoxyimino- substituted esters. *Pesticide Science* 22, 231-249.

ELLIOTT, M., FARNHAM, A.W., JANES, N.F. and KHAMBAY, B.P.S. 1988. The pyrethrins and related compounds. Part XXXIII: Effective combinations in the gemdimethyl-central group region of non-ester pyrethroids. *Pesticide Science* **23**, 231-246.

BAYDAR, A.E., ELLIOTT, M., FARNHAM, A.W., JANES, N.F. and KHAMBAY, B.P.S. 1988. The pyrethrins and related compounds. Part XXXIV: Optimisation of insecticidal activity in non-esters. *Pesticide Science* **23**, 247-257.

ELLIOTT, M., FARNHAM, A.W., JANES, N.F., JOHNSON, D.M. and PULMAN, D.A. 1989. Synthesis and insecticidal activity of lipophilic amides. Part 7: Alternative aromatic groups for phenyl in 6-phenylhexa-2,4- dienamides. *Pesticide Science* **26**, 199-208.

ELLIOTT, M. 1989. The pyrethroids: Early Discovery, Recent Advances and the Future. *Pesticide Science* **17**, 337-351.

ELLIOTT, M. 1990. The Contribution of Pyrethroid Insecticides to Human Welfare. *L'actualité chimique*, 57-70.

ELLIOTT, M. 1990. Safer Crop Protection - A Citation Classic Commentary. Current Contents (Agriculture, Biology and Environmental Sciences) **21**(36), 20.

ELLIOTT, M., PULMAN, D.A., LARKIN, J.P. and CASIDA, J.E. 1992. Insecticidal 1,3-dithianes. *J. Agricultural Food Chemistry* **40**, 147-151.

ELLIOTT, M. 1992. Pyrethroids at Rothamsted: A Case History. *Proceedings of the Conference: Technology Transfer and Implementation*, London July 1992.

ELLIOTT, M. 1995. Chemicals in Insect Control. in *'Pyrethrum Flowers: Production, Chemistry, Toxicology and Uses.* Eds. J.E. Casida and G.B. Quistad, Oxford University Press, New York and Oxford, pp. 3-26.

DENHOLM, I. and ELLIOTT, M. 1995. Roman Mieczyslaw Sawicki. *Biographical Memoirs of Fellows of the Royal Society* **41**, 395-418.

ELLIOTT, M. 1996. Synthetic Insecticides Related to the Natural Pyrethrins. In: *Crop Protection Agents from Nature: Natural Products and Analogues.* Critical Reports on Applied Chemistry, Volume 35. Edited by L.G. Copping, pp. 254-300.

Jozef Stefaan Schell

Max-Planck-Institute for Plant Breeding
Cologne, Germany

1935–2003

1990 — for his pioneering work in genetic transformation of plants, thereby opening up new horizons in basic plant science and breeding.

Professor Jozef Stefaan Schell played a key role in the development of Agrobacterium and its Ti plasmid as a vector for inserting genes into plants.

His subsequent work has moved to the use of transgenic plants to bring a new dimension into the study of classical problems in plant physiology, particularly in relation to hormone receptors and the mechanisms of hormone action.

Another particularly notable feature of Professor Schell's work has been his orchestration of collaborative efforts to spread the knowledge and use of these techniques around the world.

The impact of his pioneering contribution and leadership is now beginning to appear as companies are developing transgenic plants for pest and disease resistance and for effecting a range of compositional changes of potential industrial importance.

Ti-plasmid-derived genetic constructions are the only proven vehicles for the introduction of foreign DNA into plants. The newly acquired DNA integrates into the plant's genome and becomes a stable constituent of the genetic make-up of the transformed plant.

Professor Schell's studies and research work of over many years led to the elucidation of the molecular basis of crown gall formation on plants by *Agrobacterium tumefaciens* and the mechanism of DNA transfer to higher plants, and formed the basis for the development of sophisticated vectors and methods for the introduction of novel or modified genes into plants. These developments now allow the stable introduction of desirable genetic traits into plants, also those of agricultural interest. Furthermore, the techniques and vector systems originating from this work are being used to study gene expression in higher plants, in response to hormones, heat shock, pathogens, light and symbiotically nitrogen fixing bacteria and in a tissue specific manner. This led to the identification of tissue specific promoters which can be used to direct the expression of newly introduced genes in the desired tissue (organ) of the plant.

The work of Schell and his coworkers, therefore, represents a major contribution not only to a scientific (molecular) understanding of gene transfer to and gene expression in higher plants, but also offers major new tools for modern plant breeding.

CURRICULUM VITAE

Personal data

Place of birth	Antwerp (Belgium)
Date of birth	July 20th, 1935
Citizenship	Belgian
Marital status	Married
Children	2

Education

October 1953-July 1957 - State University Gent (Belgium), Licentiate Zoology

September 1957-November 1960 - State University Gent (Belgium), Laboratory of Microbiology (Dir.: Prof. J. De Ley). Field: Ph.D. in Microbiology (Comparative Biochemistry)

April 1959-September 1959 - State University Utrecht (The Netherlands), Laboratory of Microbiology (Dir.: Prof. K.C. Winkler). Field: Part of Ph.D. training in Microbiology

September 1962-July 1963 - Hammersmith Hospital, London (U.K.), Microbial Genetics Research Unit of the Medicinal Research Council (Dir.: Prof. W. Hayes). Field: Post-doctoral fellow in Microbial Genetics

September 1967-October 1967 - National Institutes of Health (Dir.: Dr. A. Weissbach), Bethesda, Maryland (USA). Field: Isolation of circular DNA species from bacteria

June 1967-August 1967 - Albert Einstein College of Medicine, Department of Developmental Biology (Dir.: Dr. J. Hurwitz). Field: Methylating enzymes

June 1968-September 1968 - University of Toronto (Canada), Department of Cancer Research (Dir.: Prof. L. Siminovitch). Field: Virus host relationships between phage lambda and *E. coli*

Positions held

1957-1960	Bursar of the "Instituut tot Aanmoediging van het Wetenschappelijk Onderzoek in Nijverheid en Landbouw" (I.W.O.N.L. - Belgium)
1960-1961	"Aspirant" of the "Nationaal Fonds voor Wettenschappelijk Onderzoek" (N.F.W.O. - Belgium)
1961-1967	"Werkleider" of the Laboratory of Microbiology State University of Gent (Belgium)
1967-1970	Associate Professor (Docent) and Director of the Laboratory of General Genetics, State University of Gent (Belgium)
1970-1978	Full Professor (Gewoon Hoogleraar) and Director of the Laboratory of General Genetics, State University Gent (Belgium)
1972-1978	Extraordinary Professor (Buitengewoon Hoogleraar), Free University Brussels (Belgium)
1978-1995	Extraordinary Professor (Buitengewoon Hoogleraar) and Director (until 1988) of the Laboratory of General Genetics, State University Gent (Belgium)
1978-2000	Director of the Max-Planck-Institut für Züchtungs-forschung, Köln (Germany), (Emeritus since 01.08.2000)
1980-	Honorary Professor at the University of Cologne (Germany)
10/1994-09/1998	Professeur de biologie moléculaire des plantes, Collège de France, Paris (France)
10/1998-	Professeur honoraire, Collège de France, Paris (France)

Scientific and Academic Distinctions

1959 - Scholarship of the Belgian-Dutch Cultural Agreement

1961 - Travel Scholarship of the Belgian National Education Ministry

1962 - Stipend of the "Van't Hoff Foundation" of the Royal Dutch Academy of Sciences

1964 - Award of the Prize Questions Competition of the Royal Flemish Academy of Sciences, Letters and Fine Arts of Belgium

1967 - "Advanced Fellow" of the Belgian-American Educational Foundation

1976 - Appointment as Guest Professor at the Australian National University (Department of Genetics, Research School of Biological Sciences), Canberra (Australia)

1977 - Chairman and Organizer of the "Sessions on Crown gall" of the 3rd Gordon Research Conference on Plant Cell and Tissue Culture"
- Offer to become a scientific member of the Max-Planck-Gesellschaft"

1978 - Offer to become Full Professor of Biology at Harvard University (Cambridge, USA)
- Appointment as Guest Professor at the Aarhus University (Institut for Molekyaer Biologi og Plantenfysiologi), Aarhus (Denmark)
- Nomination as Editorial Board Member of Molecular and General Genetics (Springer Verlag)
  Director of the Max-Planck-Institut für Züchtungsforschung, Köln (Germany) (emeritus since 01.08.2000)

1979 - Award of the Francquiprize (Belgium)
- Election as Active Member of The New York Academy of Sciences

1980 - Nomination as Editorial Board Member of Journal of Molecular and Applied Genetics (Raven Press)
- Nomination as Editorial Board Member of Plant Cell Reports (Springer Verlag) until June 1991
- Co-organizer of EMBL Symposium on "Molecular Biology Looks at Green Plants"
- Honorary Professor at the University of Köln (FRG)

1981 - Appointment as Guest Professor at the University of California, Riverside (Department of Biochemistry), Riverside, CA (USA)
- Elected member of the "Deutsche Akademie der Naturforscher Leopoldina", Halle (GDR)
- Nomination as Editorial Board Member of The EMBO Journal (IRL Press)

1982 - ONR Lectureship awarded by the "Society for Industrial Microbiology"
- Chairman of the EMBO Fund Committee (until 1984)

1983 - Member of the "Conseil Scientifique de l'ICP" (International Institute of Cellular and Molecular Pathology), Brussels-Leuven (Belgium)

1984 - Holder of the Belgian Francqui chair at the Katholieke Universiteit Leuven (K.U.L.), Belgium
- Member of the Scientific Board of the "Gesellschaft Deutscher Naturforscher und Ärzte"
- Member of "Gesellschaft für Biologische Chemie", Wuppertal (Germany)
- Chairman of the "Scientific Advisory Board" of the Otto Warburg Center, Rehovot, Israel (until 2001)

1985 - Award of the "Mendel-Medaille" of the "Deutsche Akademie der Naturforscher Leopoldina"
- Award of the "Otto Bayer-Preis" of the Otto Bayer-Stiftung
- Award of the "Prix Alexandre de Humboldt" of the Ministère de la Recherche et de la Technologie and Alexander von Humboldt-Stiftung

- Elected Foreign Associate of the National Academy of Sciences (USA)
- Member of the Scientific Advisory Board of the Department of Molecular Biology, Massachusetts General Hospital, Boston, USA (until 1989)
- Appointment as member of the "Scientific Advisory Panel" of the Ciba Foundation

1986 - Member of the "Societé Royale des Sciences de Liège", Belgium

1987 - Award of the "Rank-Prize for Nutrition" of the Rank Prize Funds, London, U.K.
- Elected member of the "Nordrhein-Westfälische Akademie der Wissenschaften"
- Nomination as Associate Editor of Cell (MIT Press) until 12/2001
- Member of the Scientific Committee of "Deutsches Krebsforschungszentrum Heidelberg" (until 1990)
- Fellow of the Institute for Biotechnological Studies (IBS), Canterbury, U.K.
- Member of "Vereinigung Deutscher Wissenschaftler E.V." (until 1996)

1988 - Award of the "IBM Europe Science and Technology Prize 1987"
- Elected as Foreign Fellow of the "Indian National Science Academy"
- Member of the Scientific Advisory Board of the "European Institute of Technology" (EIT), Paris, France
- Nomination as Editorial Board Member of "Cell Differentiation and Development"
- Nomination as Editorial Board Member of the "International Journal of Plant Genetic Manipulation"
- Nomination as Member of the Advisory Board for the "International Review of Cytology" (Academic Press)
- Member of the Editorial Board of "Current Methods in Molecular Biology" (John Wiley & Sons)

1989 - Elected Member of the "Royal Swedish Academy", Stockholm
- Member of the Advisory Group "Cell and Developmental Biology" of the Academia Europaea, Cambridge, England
- Member of the Scientific Council of the Institute of Biotechnology, University of Helsinki (until 1994)
- Member of "Fondation Pacis Magna", Brussels, Belgium
- Member of "The American Society of Plant Physiologists" (ASPP), Maryland, USA
- Member of the Scientific Advisory Committee for the IMBB (Institute of Molecular Biology & Biotechnology), Heraklion, Crete (until 1997)

1990 - Award of the "Wolf Prize in Agriculture 1990" by the Wolf Foundation, Israel
- Award of the "Australia Prize 1990" of the Australian Academy of Science

- Award of the "Prix Charles Leopold Mayer" of the Academie des Sciences, Paris, France
- Member of the "Koninklijke Academie voor Wetenschappen, Letteren en Schone Kunsten van Belgie", Brussels, Belgium
- Member of the John Innes Council, John Innes Institute, Norwich, G.B. (until 06/1998)
- Member of the Conseil Scientifique of ORSTOM (Institut Francais de Recherche Scientifique pour le Développement en Coopération), Paris, France
- Member of the Peer Advisory Group for Biotechnology Research at the International Rice Research Institute, Manila, Philippines
- Foreign Member of Biological Research Center, Hungarian Academy of Sciences, Szeged
- Member of "Arbeitsgruppe Biowissenschaften und Medizin", Wissenschaftsrat
- Member of the Editorial Board of "Mechanisms of Development" (MOD), Elsevier Science Publishers B.V.
- Member of the Editorial Advisory Board of "Critical Reviews in Plant Sciences"
- Senior Editor of "The Plant Journal" (Blackwell Scientific Publications), until 10/1998
- Elected Member of the Board of Governors of the Weizmann Institute of Science, Rehovot, Israel (until 2000)
- Chairman of the EMBO Council (until 1995)
- Chairman AMICA Science Board until 12/1997; since 01/1998 member

1991 - Award of the "Hansen Gold Medal 1991" of the Emil Christian Hansen Foundation, Denmark
- Member of the International Board of the Shasha Institute for International Seminars, Jerusalem
- Full member of the Board of Governors of the Hebrew University of Jerusalem, Israel
- Member of the Scientific Advisory Council of "Alfried Krupp von Bohlen und Halbach Stiftung", Essen, Germany
- Chairman of the Founding Committee of the Institute of Plant Biochemistry, Halle

1992 - Award of the "Feodor Lynen Lecture 1992" medal (Miami Winter Symposium)
- Member of the Scientific Committee on the Application of Science to Agriculture, Forestry and Aquaculture (CASAFA)
- Guest Professor at the Collège de France in Paris, from March 18 to April 8, 1992

- Chairman Evaluation Committee for "Biotechnology in Plant Production and Processing" of the "Dutch Landbouwkundig Onderzoek" of the Dutch University of Agriculture, Environment and Fisheries
- "Max-Planck-Forschungspreis 1992" (Alexander von Humboldt-Stiftung/ Max-Planck-Gesellschaft)
- "Docteur Honoris Causa", Université Louis Pasteur, Strasbourg (France)

1993 - Nomination as full Board Member of the Biotechnology Advisory Commission (BAC) of the Stockholm Environment Institute (SEI) (Sweden)
- Member of the Board of Reviewing Editors of Science (USA), until 12/1999
- Elected Foreign Honorary Member of the Academy of Arts and Sciences, Cambridge (USA)
- Elected Honorary Member of the Hungarian Academy of Sciences (Budapest, Hungary)
- Charter Member of the International Society of Molecular Evolution

1994 - Chairman of the Scientific Advisory Board, IPB Halle (Germany), until 12/1997; 01/1998-12/1999 Member
- Member of the Advisory Board of "Regio Köln/Bonn und Nachbarn e.V." and "Vereinigung von Wirtschaftskammern zur Förderung der Region Köln/ Bonn" (Germany)
  Member of "Landesinitiative Bio- und Gentechnik NRW" (Germany)
- The John M. Chemerda Lecturer in Science (April 25-27), Pennsylvania State University, University Park (USA)
- "Doctor Philosophiae Honoris Causa", Hebrew University, Jerusalem (Israel)
- Member of the Board of Trustees "The Darwin Trust of Edinburgh", Edinburgh (Great Britain)
- Member of European Science and Technology Assembly (ESTA), Brussels (Belgium) (until 10/1997)
- Member of the Advisory Board of KOWI (Koordinierungsstelle EG der Wissenschaftsorgani-sationen), Bonn (Germany), until 2000
- Elected member of the Scientific and Academic Advisory Committee (SAAC) and ad hoc Committee for Technology Transfer, Weizmann Institute of Science, Rehovot (Israel), until 2000
- Professor of Plant Molecular Biology, Collège de France, Paris (France), until 09/1998

1995 - Chairman AMICA Science EEIG (until 07/2000)
- Elected member of "CAESAR-Fachausschuß der Region" (Center of Advanced European Studies and Research), Bonn (Germany)
- Award of the "Wilhelm-Exner-Medaille", Vienna (Austria)

1996 - Nominated Honorary Member of "Gesellschaft für Pflanzenzüchtung e.V.", Göttingen (Germany)

16. Schreier, P.H., Reif, H.-J., and Schell, J. Gentechnik. In: *"Pflanzenproduktion im Wandel"*, G. Haug, G. Schuhmann, G. Fischbeck (Eds.), VCH Verlagsgesellschaft, Weinheim, p. 27-39 (1990).
17. Staiger, D., Kaulen, H., and Schell, J. A nuclear factor recognizing a positive regulatory upstream element of the *Antirrhinum majus* chalcone synthase promoter. *Plant Physiol.* 93, 1347-1353 (1990).
18. Steinbiss, H.-H., Töpfer, R., Pröls, M., and Schell, J. Transient gene expression in tobacco protoplasts and seed derived embryos of wheat. In: *"Proceedings of the International Congress of Plant Physiology"*, New Delhi, India, February 15-20, 1988, S.K. Sinha, P.V. Sane, S.C. Bhargava, P.K. Agrawal (Eds.) Society for Plant Physiology and Biochemistry, New Delhi, India, Vol. 1, pp. 658-659 (1990).
19. Töpfer, R., Gronenborn, B., Schaefer, S., Schell, J., and Steinbiß, H.-H. Expression of engineered wheat dwarf virus in seed-derived embryos. *Physiol. Plantarum* 79, 158-162 (1990).
20. Töpfer, R., and Schell, J. Ansätze zur Isolation von Genen der *de novo* Fettsäurebiosynthese. In: *"Pflanzliche Öle im chemisch-technischen Sektor"*, Schriftenreihe des Bundesministers für Ernährung, Landwirtschaft und Forsten, Reihe A: Angewandte Wissenschaft, Heft 391, Landwirtschaft-sverlag GmbH., Münster-Hiltrup, pp. 221-225 (1990).
21. Walden, R., Koncz, C., and Schell, J. The use of gene vectors in plant molecular biology. *Meth. Mol. Cell. Biol.* 1, 175-194 (1990).
22. Walden, R., and Schell, J. Techniques in plant molecular biology - progress and problems. *Eur. J. Biochem.* 192, 563-576 (1990).
23. Wingender, R., Röhrig, H., Höricke, C., and Schell, J. *cis*-regulatory elements involved in ultraviolet light regulation and plant defense. *The Plant Cell* 2, 1019-1026 (1990).
24. Barbier-Brygoo, H., Ephrotikhine, G., Klämbt, D., Maurel, C., Palme, K., Schell, J. and GUERN, J. Perception of the auxin signal at the plasma membrane of tobacco mesophyll protoplasts. *Plant J.* 1, 83-93 (1991).
25. Campos, N., Feldwisch, J., Zettl, R., Boland, W., Schell, J., and Palme, K. Identification of auxin-binding proteins using an improved assay for photoaffinity labeling with 5-N_3-(7-^{3}H)-indole-3-acetic acid. *Technique* 3, 69-75 (1991).
26. Davidson, A., Kendiek, J., Peerenbaum, E., Schell, J., and Steinbiss, H.-H. Genome analysis of Barley Yellow Mosaic Virus RNA1 and RNA2. In: *"Barley Genetics VI"* Vol. I, Short Papers, Munck, L., Kirkegaard, K., and Jensen, B. (Eds.), (Proceedings of the Sixth International Barley Genetics Symposium, July 22-27, 1991, Helsingborg, Sweden), Munksgaard International Publishers Ltd., Copenhagen, pp. 608-609 (1991).

27. Davidson, A.D., Pröls, M., Schell, J., and Steinbiss, H.-H. The nucleotide sequence of RNA 2 of barley yellow mosaic virus. *J. Gen. Virol.* 72, 989-993 (1991).
28. Estruch, J.J., Chriqui, D., Grossman, K., Schell, J., and Spena, A. The plant oncogene *rolC* is responsible for the release of cytokinins from glucoside conjugates. *EMBO J.* 10, 2889-2895 (1991).
29. Estruch, J.J., Prinsen, E., van Onckelen, H., Schell, J., and Spena, A. Viviparous leaves produced by somatic activation of an inactive cytokinin-synthesizing gene. *Science* 254, 1364-1367 (1991).
30. Estruch, J.J., Schell, J., and Spena, A. The protein encoded by the *rolB* plant oncogene hydrolyses indole glucosides. *EMBO J.* 10, 3125-3128 (1991).
31. Fraley, R., and Schell, J. Plant biotechnology - Editorial overview. *Current Opinion in Biotechnol.* 2, 145-146 (1991).
32. Fritz, C.C., Herget, T., Wolter, F.P., Schell, J., and Schreier, P.H. Reduced steady-state levels of *rbcS* mRNA in plants kept in the dark are due to differential degradation. *Proc. Natl. Acad. Sci. USA* 88, 4458-4462 (1991).
33. Fritze, K., Staiger, D., Czaja, I., Walden, R., Schell, J., and Wing, D. Developmental regulation and UV light regulation of the snapdragon chalcone synthase promoter. *The Plant Cell* 3, 893-905 (1991).
34. Kammann, M., Matzeit, V., Schmidt, B., Schell, J., Walden, R., and Gronenborn, B. Geminivirusbased shuttle vectors capable of replication in *Escherichia coli* and monocotyledonous plant cells. *Gene* 104, 247-252 (1991).
35. Kammann, M., Schalck, H.-J., Matzeit, V., Schaefer, S., Schell, J., and Gronenborn, B. DNA replication of wheat dwarf virus, a geminivirus, requires two *cis*-acting signals. *Virology* 184, 786-790 (1991).
36. Koncz, C., Schmülling, T., Spena, A., and Schell, J. T-DNA gene-functions. In: *"Plant Molecular Biology 2"* (Proceedings on Elmau-Meeting, May 13-23, 1990) R.G. Herrmann, and B. Larkins (Eds.) Plenum Press, New York, pp. 205-209 (1991).
37. Kondorosi, A., Kondorosi, E., John, M., Schmidt, J., and Schell, J. The role of nodulation genes in bacterium-plant communication. In: *"Genetic Engineering"*, Vol. 13, Setlow, J.K. (Ed.), Plenum Press, New York and London, pp 115-136 (1991).
38. Kondorosi, E., Pierre, M., Cren, M., Haumann, U., Buire, M., Hoffmann, B., Schell, J., and Kondorosi, A. Identification of NolR, a negative transacting factor controlling the *nod* regulon in *Rhizobium meliloti*. *J. Mol. Biol.* 222, 885-896 (1991).
39. Körber, H., Strizhov, N., Staiger, D., Feldwisch, J., Olsson, O., Sandberg, G., Palme, K., Schell, J., and Koncz, C. T-DNA gene 5 of *Agrobacterium* modulates auxin response by autoregulated synthesis of a growth hormone antagonist in plants. *EMBO J.* 10, 3983-3991 (1991).

40. Leung, J., Fukuda, H., Wing, D., Schell, J., and Masterson, R. Functional analysis of *cis*-elements, auxinresponse and early developmental profiles of the mannopine synthase bidirectional promoter. *Mol. Gen. Genet.* 230, 463-474 (1991).
41. Matzeit, V., Schaefer, S., Kammann, M., Schalck, H.-J., Schell, J. and Gronenborn, B. Wheat dwarf virus vectors replicate and express foreign genes in cells of monocotyledonous plants. *The Plant Cell* 3, 247-258 (1991).
42. Mayerhoffer, R., Koncz-Kalman, Z., Nawrath, C., Bakkeren, G., Crameri, A., Angelis, K., Redei, G.P., Schell, J., Hohn, B., and Koncz, C. T-DNA integration: a mode of illegitimate recombination in plants. *EMBO J.* 10, 697-704 (1991).
43. Mukhopadhyay, A., Töpfer, R., Pradhan, A.K., Sodhi, Y.S., Steinbiß, H.-H., Schell, J., and Pental, D. Efficient regeneration of *Brassica oleracea* hypocotyl protoplasts and high frequency genetic transformation by direct DNA uptake. *Plant Cell Rep.* 10, 375-379 (1991).
44. Palme, K., Hesse, T., Moore, I., Campos, N., Feldwisch, J., Garbers, C., Hesse, F., and Schell, J. Hormonal modulation of plant growth: the role of auxin perception. *Mech. Dev.* 33, 97-106 (1991).
45. Palme, K., and Schell, J. Auxin receptors take shape. *Current Biology* 1, 228-230 (1991).
46. Pechan, P.M., Bartels, D., Brown, D.C.W., and Schell, J. Messenger-RNA and protein changes associated with induction of *Brassica* microspore emryogenesis. *Planta* 184, 161-165 (1991).
47. Prinsen, E., Chauvaux, N., Schmidt, J., John, M., Wieneke, U., DE Greef, J., Schell, J., and van Onckelen, H. Stimulation of indole-3-acetic acid production in *Rhizobium* by flavonoids. *FEBS Letters* 282, 53-55 (1991).
48. Schmidt, J., John, M., Kondorosi, E., Kondorosi, A., and Schell, J. Recent progress in the elucidation of the function of *nodAB* and *C* genes of *Rhizobium meliloti*. *Israel J. Bot.* 40, 165-169 (1991).
49. Schmidt, J., John, M., Wieneke, U., Stacey, G., Röhrig, H., and Schell, J. Studies on the function of *Rhizobium meliloti* nodulation genes. In: *"Molecular Genetics of Plant-Microbe Interactions"*, Hennecke, H. and Verma, D.P.S. (Eds.), Kluwer Academic Publishers, Dordrecht, The Netherlands, pp. 150-155 (1991).
50. Staiger, D., Becker, F., Schell, J., Koncz, C., and Palme, K. Purification of tobacco nuclear proteins binding to a CACGTG motif of the chalcone synthase promoter by DNA affinity chromatography. *Eur. J. Biochem.* 199, 519-527 (1991).
51. Stieger, M., Neuhaus, G., Momma, T., Schell, J., and Kreuzaler, F. Self assembly of immunoglobulins in the cytoplasm of the alga *Acetabularia mediterranea*. *Plant Science* 73, 181-190 (1991).

52. Walden, R., Hayashi, H., and Schell, J. T-DNA as a gene tag. *Plant J.* 1, 281-288 (1991).
53. Walden, R., and Schell, J. Tissue culture and the use of transgenic plants to study plant development. *In Vitro Cell. Dev. Biol.* 27P, 1-10 (1991).
54. Zettl, R., Campos, N., Feldwisch, J., Schell, J., Boland, W., and Palme, K. Synthesis and application of 5'-Azido-[3,6-$^3H_2$]- naphthylphthalamic acid, a photo-activatable probe for auxin efflux carrier proteins. *Technique* 3, 151-158 (1991).
55. Becker, D., Kemper, E., Schell, J., and Masterson, R. New plant binary vectors with selectable markers located proximal to the left T-DNA border. *Plant Mol. Biol.* 20, 1195-1197 (1992).
56. Campos, N., Bako, L., Feldwisch, J., Schell, J., and Palme, K. A protein from maize labeled with azido-IAA has novel ß-glucosidase activity. *Plant J.* 2, 675-684 (1992).
57. Feldwisch, J., Zettl, R., Hesse, F., Schell, J., and Palme, K. An auxin-binding protein is localized to the plasma membrane of maize coleoptile cells: Identification by photoaffinity labeling and purification of a 23-kDa polypeptide. *Proc. Natl. Acad. Sci. USA* 89, 475-479 (1992).
58. Fraley, R., and Schell, J. Plant biotechnology - Editorial overview. *Current Opinion in Biotechnol.* 3, 139-140 (1992).
59. Grevelding, C., Becker, D., Kunze, R., von Menges, A., Fantes, V., Schell, J., and Masterson, R. High rates of *Ac/Ds* germinal transposition in *Arabidopsis* suitable for gene isolation by insertional mutagenesis. *Proc. Natl. Acad. Sci. USA* 89, 6085-6089 (1992).
60. Kemper, E., Grevelding, C., Schell, J., and Masterson, R. Improved method for the transformation of *Arabidopsis thaliana* with chimeric dihydrofolate reductase constructs which confer methotrexate resistance. *Plant Cell Rep.* 11, 118-121 (1992).
61. Klein, B., Pawlowski, K., Höricke-Grandpierre, C., Schell, J., and Töpfer, R. Isolation and characterization of a cDNA from *Cuphea lanceolata* encoding a ß-ketoacyl-ACP reductase. *Mol. Gen. Genet.* 233, 122-128 (1992).
62. Koncz, C., Nemeth, K., Redei, G.P., and Schell, J. T-DNA insertional mutagenesis in *Arabidopsis. Plant Mol. Biol.* 20, 963-976 (1992).
63. Koncz, C., and Schell, J. Exploitation of *Agrobacterium tumefaciens.* In: *"Development The Molecular Genetic Approach"*, V.E.A. Russo, S. Brody, D. Cove, and S. Ottolenghi (Eds.) Springer Verlag, Berlin Heidelberg, pp. 217-224 (1992).
64. Koncz, C., Schell, J., and Redei, G.P. T-DNA transformation and insertion mutagenesis. In: *"Methods in Arabidopsis Research"*, C. Koncz, N.-H. Chua, and J. Schell (Eds.), World Scientific, Singapore, pp. 224-273 (1992).

65. Logemann, J., Jach, G., Tommerup, H., Mundy, J., and Schell, J. Expression of a barley ribosome-inactivating protein leads to increased fungal protection in transgenic tobacco plants. *Bio/Technol.* 10, 305-308 (1992).
66. Logemann, J., and Schell, J. Das Dilemma der Landwirtschaft. In: *"Edition Zeitthema"*, 1/92 Gentechnologie I, Hargitay & List Verlag, Wien, pp. 53-55 (1992).
67. Logemann, J., and Schell, J. Die Landwirtschaft zwischen Produktivitätssteigerung und zunehmender Umweltbelastung: Kann das Dilemma durch Gentechnologie gelöst werden? GRUR, Heft 4, VCH Verlagsgesellschaft mbH., Weinheim, pp. 248-252 (1992).
68. Maas, C., Reichel, C., Schultze, S.C., Matzeit, V., Schell, J., and Steinbiss, H.-H. Modular structure of monocot expression vectors: A novel stimulatory element combined with *sh 1* intron 1. In: *"Technological Impacts and Limiting Factors of Plant Genetic Transformation"* (Biotechnologies'92) Le Conseil Régional de Picardie (Ed.), pp. 78-88 (1992).
69. Maas, C., Schell, J., and Steinbiss, H.-H. Applications of an optimized monocot expression vector in studying transient gene expression and stable transformation of barley. *Physiol. Plant.* 85, 367-373 (1992).
70. Moore, I., Feldwisch, J., Campos, N., Zettl, R., Brzobohaty, B., Bako, L., Schell, J., and Palme, K. Auxin-binding proteins of *Zea mays* identified by photoaffinity labelling. Proceedings of "Molecular Biology of Plant Hormones and their Receptors", Edinburgh, Sept. 45, 1991. *Biochem. Soc. Trans.* 20, 70-73 (1992).
71. Nitschke, K., Fleig, U., Schell, J., and Palme, K. Complementation of the *cs dis 2-11* cell cycle mutant of *Schizosaccharomyces pombe* by a protein phosphatase from *Arabidopsis thaliana. EMBO J.* 11, 1327-1333 (1992).
72. Palme, K., Dieffenthal, T., Vingron, M., Sander, C., and Schell, J. Molecular cloning and structural analysis of genes from *Zea mays* (L.) coding for members of the *ras*-related *ypt* gene family. *Proc. Natl. Acad. Sci. USA* 89, 787-791 (1992).
73. Palme, K., Feldwisch, J., Peters, W.S., Schell, J., Zettl, R., Campos, N., and Felle, H. Auxin binding proteins are located in the ER and in the plasma membrane: Identification by photoaffinity labeling and characterization. In: *"Progress in Plant Growth Regulation"*, C.M. Karssen, L.C. Van Loon, and D. Vreugdenhill (Eds.), Kluwer Academic Publishers, Dordrecht, pp. 73-81 (1992).
74. Palme, K., Hesse, T., Campos, N., Garbers, C., Yanofsky, M.F., and Schell, J. Molecular analysis of an auxin-binding protein gene located on chromosome 4 of *Arabidopsis. The Plant Cell*, 4, 193-201 (1992).
75. Peerenbaum, E., Pröls, M., Schell, J., Steinbiss, H.-H., and Davidson, A.D. The complete nucleotide sequence of RNA 1 of a German isolate of barley yellow

mosaic virus and its comparison with a Japanese isolate. *J. Gen. Virol.* 73, 1303-1308 (1992).

76. Roberts, D.M., Besl, L., Oh, S.-H., Masterson, R.V., Schell, J., and Stacey, G. Expression of a calmodulin methylation mutant affects the growth and development of transgenic tobacco plants. *Proc. Natl. Acad. Sci. USA* 89, 8394-8398 (1992).
77. Schell, J. Plant biotechnology: A powerful tool to use plant resources and to improve the environmental impact of agriculture. In: *"Natural Resources and Human Health - Plants of medicinal and nutritional value"* Proceedings of the 1st WHO Symposium on Plants and Health for All: Scientific Advancement, Kobe, Aug. 26-28, 1991, S. Baba, O. Akerele, and Y. Kawaguchi (Eds.), Elsevier Science Publishers B.V., Amsterdam, pp. 49-60 (1992).
78. Schell, J. Isolation and characterization of genes involved in phytohormone activity. In: *"Proceedings of the VII International Symposium in Conjunction with the Awarding of the International Prize for Biology - Cellular Basis of Growth and Development in Plants, Osaka, Nov. 26-28, 1991"*, H. Shibaoka (Ed.), Osaka University, Osaka, pp. 113-118 (1992).
79. Schell, J. *Agrobacterium*-Mediated Gene Transfer to Plants: Its Impact on Plant Breeding, Studies of Gene Expression and Plant Developmental Biology. In: *"Molecular Biology of the Cell"* (Final Report of the Sonderforschungsbereich "Molekularbiologie der Zelle" 1970-1988, DFG Deutsche Forschungsgemein-schaft), W. Doerfler (Ed.), VCH Verlagsgesellschaft mbH, Weinheim, pp. 125-143 (1992).
80. Schell, J. Plant biotechnology: A powerful tool to use plant resources and to improve the environmental impact of agriculture. In: *"Basic Life Sciences and Human Society"*, H. Pfeiffer and T. Schmidt-Dörr (Eds.), Proceedings of the 10th Discoveries Symposium 1991, Bonn. Alexander von Humboldt Foundation and Honda Foundation. Druckpartner Moser GmbH., Rheinbach, pp. 141-152 (1992).
81. Schell, J., Koncz, C., Spena, A., Walden, R., John, M., and Schmidt, J. *Agrobacterium* and *Rhizobium* and the genetic control of plant development. In: *"Advances in Gene Technology: Feeding the World in the 21st Century"* W.J. Whelan, F. Ahmad, H. Bialy, S. Black, M.L. King, M.B. Rabin, L.P. Solomonson, and I.K. Vasil (Eds.), Proceedings of the 1992 Miami Bio/Technology Winter Symposium. Miami Short Reports Vol. 2, IRL Press at Oxford University Press, appendix (1992).
82. Schell, J., Koncz, C., Spena, A., Walden, R., and Palme, K. Characterization of genes affecting plant growth and development. In: *"Harnessing Biotechnology for the 21st Century"* M.R. Ladisch, and A. Bose (Eds.), Proceedings of the Ninth International Biotechnology Symposium and Exposition, Crystal City, Aug. 16-21, 1992, American Chemical Society, Washington, pp. 496-499 (1992).

83. Schell, J., and Logemann, J. Pflanzliche Gentechnologie - Perspektiven für die praktische Anwendung in der Landwirtschaft. *Schering Lecture*, Heft 6, Schering Berlin (1992).
84. Schell, J., Schmidt, J., John, M., and Röhrig, H. Potential and limitations of developing new plant-microbe interactions. In: *"Nodulation and Nitrogen Fixation in Rice - Potential and Prospects"*, G.S. Khush and J. Bennett (Eds.), (Proceedings of the Workshop on Nodulation and Nitrogen Fixation in Rice, Manila, Febr. 24-27, 1992), IRRI, Manila, pp. 103-105 (1992).
85. Schmülling, T., Schell, J., and Spena, A. Construction of cytokinin overproducing mutants in higher plants. In: *"Physiology and Biochemistry of Cytokinins in Plants"* M. Kaminek, D.W.S. Mok, and E. Zazimalová (Eds.), Proceedings of the International Symposium on Physiology and Biochemistry of Cytokinins in Plants, held in Liblice, Czechoslovakia, Sept. 10-14, 1990, Academic Publishing bv, The Hague, The Netherlands, pp. 87-94 (1992).
86. Spena, A., Estruch, J.J., Aalen, R.D., Prinsen, E., Parets-Soler, A., Nacken, W., Sommer, H., Chriqui, D., Grossmann, K., van Onckelen, H., and Schell, J. Transgenic plants and transgenic plant mosaics for the expression of pathogen derived genes able to affect phytohormone activity. In: *"Progress in Plant Growth Regulation"* C.M. Karssen, L.C. Van Loon, and D. Vreugdenhil (Eds.), Proceedings of the 14th International Conference on Plant Growth Substances, Amsterdam, July 21-26, 1991, Kluwer Academic Publishers, Dordrecht, pp. 724-730 (1992).
87. Spena, A., Estruch, J.J., Hansen, G., Langenkemper, K., Berger, S., and Schell, J. The *Rhizogenes* tale: modification of plant growth and physiology by an enzymatic system of hydrolysis of phytohormone conjugates. In: *"Advances in Molecular Genetics of Plant-Microbe Interactions"* Vol. 2, E.W. Nester and D.P.S. Verma (Eds.), Proceedings of the Sixth International Symposium on Molecular Plant-Microbe Interactions, July 11-16, 1992, Seattle. Kluwer Academic Publishers, Dordrecht, pp. 109-124 (1992).
88. Spena, A., Estruch, J.J., and Schell, J. On microbes and plants: new insights in phytohormonal research. *Current Opinion in Biotechnol.* 3, 159-163 (1992).
89. Tamas, I.A., Langbridge, W.H.R., Abel, S.D., Crawford, S.W., Randall, J.D., Schell, J., and Szalay, A.A. Hormonal control of apical dominance. Studies in tobacco transformed with bacterial luciferase and *Agrobacterium rol* genes. In: *"Progress in Plant Growth Regulation"* Proceedings of the 14th International Conference on Plant Growth Substances, Amsterdam, 21-26 July, 1991, C.M. Karssen, L.C. Van Loon, and D. Vreugdenhil (Eds.), Kluwer Academic Publishers, Dordrecht, pp. 418-430 (1992).
90. Zettl, R., Feldwisch, J., Boland, W., Schell, J., and Palme, K. Azido-(3,6-$^3H_2$)-N-1-naphthylphthalamic acid, a photoactivatable probe for naphthylphthalamic acid receptor proteins from higher plants: Identification of a 23-kDa protein

from maize coleoptile plasma membranes. *Proc. Natl. Acad. Sci. USA* 89, 480-484 (1992).

91. Bachmair, A., Becker, F., and Schell, J. Use of a reporter transgene to generate *Arabidopsis* mutants in ubiquitin-dependent protein degradation. *Proc. Natl. Acad. Sci. USA* 90, 418-421 (1993).
92. Becker, F., Buschfeld, E., Schell, J., and Bachmair, A. Altered response to viral infection by tobacco plants perturbed in ubiquitin system. *Plant J.* 3, 875-881 (1993).
93. Kendiek, J., Davidson, A.D., Schultze, S.C., Schell, J., and Steinbiss, H.-H. Identification and classification of a resistance breaking strain of barley yellow mosaic virus. *Ann. Appl. Biol.* 122, 481-491 (1993).
94. Brzobohaty, B., Moore, I., Kristoffersen, P., Bako, L., Campos, N., Schell, J., and Palme, K. Release of active cytokinin by a ß-glucosidase localized to the maize root meristem. *Science* 262, 1051-1054 (1993).
95. de Bruijn, F.J., and Schell, J. Regulation of plant genes specifically induced in developing and mature nitrogen-fixing nodules: *Cis*-acting elements and *trans*-acting factors. In: *"Control of Plant Gene Expression"*, D.P.S. Verma (Ed.), CPC Press, Boca Raton, pp. 241-257 (1993).
96. Dehio, C., Grossmann, K., Schell, J., and Schmülling, T. Phenotype and hormonal status of transgenic tobacco plants overexpressing the *rolA* gene of *Agrobacterium rhizogenes* T-DNA. *Plant Mol. Biol.* 23, 1199-1210 (1993).
97. Dehio, C., and Schell, J. Stable expression of a single-copy *rolA* gene in transgenic *Arabidopsis thaliana* plants allows an exhaustive mutagenic analysis of the transgene-associated phenotype. *Mol. Gen. Genet.* 241, 359-366 (1993).
98. Dehio, C., Schell, J., and Koncz, C. *Agrobacterium* und *Rhizobium* - Bodenbakterien als Pflanzeningenieure. In: *"Extremophile"* Mikro-organismen in ausgefallenen Lebensräumen, K. Hausmann, B.P. Kremer (Eds.), VCH Verlagsgesellschaft, Weinheim, pp. 257-277 (1993).
99. Fraley, R., and Schell, J. Plant biotechnology - Editorial overview. *Current Opinion in Biotechnol.* 4, 133-134 (1993).
100. Grevelding, C., Fantes, V., Kemper, E., Schell, J., and Masterson, R. Single-copy T-DNA insertions in *Arabidopsis* are the predominant form of integration in root-derived transgenics, whereas multiple insertions are found in leaf-discs. *Plant Mol. Biol.* 23, 847-860 (1993).
101. Grosskopf, E., Cam Ha, D.T., Wingender, R., Röhrig, H., Szecsi, J., Kondorosi, E., Schell, J., and Kondorosi, A. Enhanced levels of chalcone synthase in alfalfa nodules induced by a Fix$^-$ mutant of *Rhizobium meliloti. Molec. Plant-Microbe Interact.* 6, 173-181 (1993).
102. Hesse, T., Garbers, C., Brzobohaty, B., Kreimer, G., Söll, D., Melkonian, M., Schell, J., and Palme, K. Two members of the *ERabp* gene family are expressed

differentially in reproductive organs but to similar levels in the coleoptile of maize. *Plant Mol. Biol.* 23, 57-66 (1993).

103. Heyraud, F., Matzeit, V., Kammann, M., Schaefer, S., Schell, J. and Gronenborn, B. Identification of the initiation sequence for viral-strand DNA synthesis of wheat dwarf virus. *EMBO J.* 12, 4445-4452 (1993).
104. Heyraud, F., Matzeit, V., Schaefer, S., Schell, J., and Gronenborn, B. The conserved nonanucleotide motif of the geminivirus stem-loop sequence promotes replicational release of virus molecules from redundant copies. *Biochimie* 75, 605-615 (1993).
105. John, M., Röhrig, H., Schmidt, J., Wieneke, U., and Schell, J. *Rhizobium* NodB protein involved in nodulation signal synthesis is a chitooligosaccharide deacetylase. *Proc. Natl. Acad. Sci. USA* 90, 625-629 (1993).
106. Logemann, J., Jach, G., Logemann, S., Leah, R., Wolf, G., Mundy, J., Oppenheim, A., Chet, I., and Schell, J. Expression of a ribosome inhibiting protein (RIP) or a bacterial chitinase leads to fungal resistance in transgenic plants. In: *"Mechanisms of Plant Defense Responses"* B. Fritig and M. Legrand (Eds.), Kluwer Academic Publishers, Dordrecht, pp. 446-448 (1993).
107. Logemann, J., and Schell, J. The impact of biotechnology, or how to combine increases in agricultural productivity with an improved protection of the environment. In: *"Biotechnology in Plant Disease Control"*, Chet, I. (Ed.), Wiley-Liss, Inc., New York, pp. 1-14 (1993).
108. Murphy, P.J., Trenz, S.P., Grzemski, W., de Bruijn, F.J., and Schell, J. The *Rhizobium meliloti* Rhizopine *mos* locus is a mosaic structure facilitating its symbiotic regulation. *J. Bacteriol.* 175, 5193-5204 (1993).
109. Palme, K., and Schell, J. On plant growth regulators and their metabolites: a changing perspective. *Sem. Cell Biol.* 4, 87-92 (1993).
110. Schell, J.S. The commercialization of biotechnology and the flow of new knowledge. In: *"The Emerging Global Food System"*, G.E. Gaull and R.A. Goldberg (Eds.), Proceedings of the "1991 Nutrasweet CERES Conference" 30.10. - 1.11.1991, John Wiley & Sons, New York, pp. 33-35 (1993).
111. Schell, J.S. Plant biotechnology: A powerful tool to use plant resources and to improve the environmental impact of agriculture. AvH-Magazin Nr. 61, Juli 1993, pp. 9-16 (1993).
112. Schell, J. Gesellschaftliche Erstarrung versus Fortschrittsfähigkeit - Beispiel Gentechnologie. In: *"Erstarrende Gesellschaft in bewegten Zeiten - Gründe, Folgen, Öffnungschancen"*. (Vortrag beim 1. Jahreskolloquium der Alfred Herrhausen Gesellschaft für Internationalen Dialog, Frankfurt/M., 11./ 12.6.1993). Schäffer-Poeschel Verlag, Stuttgart, pp. 179-193 (1993).
113. Schell, J. Gentechnologie kann Agrarprobleme lösen. *Agrarische Rundschau* Gentechnik I, Nr. 6, 1-2 (1993).

114. Schell, J. Plant Biotechnology: State of the Art in Developed Countries and Relevant Safety Considerations. In: *"Proceedings African Regional Conference for International Cooperation on Safety in Biotechnology"* 11-14 October 1993, Harare, Zimbabwe, P.J. van der Meer, P. Schenkelaars, B. Visser, E. Zwangobani (Eds.), pp. 23-29 (1993).
115. Schell, J. Die Auswirkung der molekularen Pflanzenzüchtung auf Landwirtschaft und Umwelt. In: *"Bio- und Gentechnologie - Perspektiven in der Region Köln"* Dokumentation des Forums vom 2.12.1992 im Congreß-Centrum West, KölnMesse, Herausgeber: Stadt Köln - der Oberstadtdirektor - Amt für Stadtentwicklungsplanung, Industrie- u. Handelskammer zu Köln, Moeker Merkur Druck GmbH., Köln, pp. 54-61 (1993).
116. Schell, J., Koncz, C., Spena, A., Palme, K., and Walden, R. Genes involved in the control of growth and differentiation in plants. *Gene* 135, 245-249 (1993).
117. Schell, J., and Logemann, J. Die Wichtigkeit molekularer Pflanzenzüchtung für eine gleichbleibend anbauintensive Landwirtschaft. In: *"Jahrbuch 1992"* Deutsche Akademie der Naturforscher Leopoldina, B. Parthier (Ed.), Leopoldina Reihe 3, Jahrgang 38, Druck-Zuck GmbH., Halle, pp. 181-187 (1993).
118. Schell, J., Palme, K., Reiss, B., and Walden, R. Recent advances in the search for genes involved in the mechanisms of action of auxin and cytokinins. *J. Plant Res.* 3, 221-227 (1993).
119. Schlichter, U., Sohn, A., Peerenbaum, E., Schell, J., and Steinbiss, H.-H. Molecular analysis of the capsid protein gene of a German isolate of barley mild mosaic virus. *Plant Cell Rep.* 12, 237-240 (1993).
120. Schmidt, J., Röhrig, H., John, M., Wieneke, U., Stacey, G., Koncz, C., and Schell, J. Alteration of plant growth and development by *Rhizobium nodA* and *nodB* genes involved in the synthesis of oligosaccharide signal molecules. *Plant J.* 4, 651-658 (1993).
121. Schmülling, T., Fladung, M., Grossmann, K., and Schell, J. Hormonal content and sensitivity of transgenic tobacco and potato plants expressing single *rol* genes of *Agrobacterium rhizogenes* T-DNA. *Plant J.* 3, 371-382 (1993).
122. Schmülling, T., Röhrig, H., Pilz, S., Walden, R., and Schell, J. Restoration of fertility by antisense RNA in genetically engineered male sterile tobacco plants. *Mol. Gen. Genet.* 237, 385-394 (1993).
123. Schmülling, T., and Schell, J. Transgenic tobacco plants regenerated from leaf disks can be periclinal chimeras. *Plant Mol. Biol.* 21, 705-708 (1993).
124. Töpfer, R., Maas, C., Höricke-Grandpierre, C., Schell, J., and Steinbiss, H.-H. Expression vectors for high-level gene expression in dicotyledonous and monocotyledonous plants. *Meth. in Enzymol.* 217, 66-78 (1993).
125. Walden, R., Czaja, I., Schmülling, T., and Schell, J. *Rol* genes alter hormonal requirements for protoplast growth and modify the expression of an auxin responsive promoter. *Plant Cell Rep.* 12, 551-554 (1993).

126. Bako, L., Nuotio, S., Dudits, D., Schell, J., and Koncz, C. Rnapii: A specific target for the cell cycle kinase complex. In: *"Plant Promoters and Transcription Factors"*, (Results and Problems in Cell Differentiation 20), L. Nover (Ed.), Springer-Verlag, Berlin, Heidelberg, pp. 25-64 (1994).
127. Becker, J., Siegert, H., Logemann, J., and Schell, J. Begleitende Sicherheitsforschung zur Freisetzung genetisch veränderter Petunien. In: *"Biologische Sicherheit"*, Band 3, Ch. Buchholz (Ed.) Projektträger Biologie, Energie, Ökologie, Forschungszentrum Jülich GmbH., Werka-Druck GmbH., Linnich, pp. 563-578 (1994).
128. Campos, N., Schell, J., and Palme, K. *In vitro* uptake and processing of maize auxin-binding proteins by ER-derived microsomes. *Plant Cell Physiol.* 35, 153-161 (1994).
129. Dehio, C., and Schell, J. Identification of plant genetic loci involved in a posttranscriptional mechanism for meiotically reversible transgene silencing. *Proc. Natl. Acad. Sci. USA* 91, 5538-5542 (1994).
130. Hewelt, A., Prinsen, E., Schell, J., van Onckelen, H., and Schmülling, T. Promoter tagging with a promoterless *ipt* gene leads to cytokinin induced phenotypic variability in transgenic tobacco plants: implications of gene dosage effects. *Plant J.* 6, 879-891 (1994).
131. Jasper, F., Koncz, C., Schell, J., and Steinbiss, H.-H. *Agrobacterium* T-strand production *in vitro*: Sequence-specific cleavage and 5' protection of single-stranded DNA templates by purified VirD2 protein. *Proc. Natl. Acad. Sci. USA* 91, 694-698 (1994).
132. Koncz, C., Martini, N., Szbados, L., Hrouda, M., Bachmair, A., and Schell, J. Specialized vectors for gene tagging and expression studies. In: *"Plant Molecular Biology Manual"*, S. Gelvin and B. Schilperoort (Eds.), Kluwer Academic Publishers, Dordrecht, Vol. B2, pp. 1-22 (1994).
133. Koncz, C., Nemeth, K., Redei, G.P., and Schell, J. Homology recognition during T-DNA integration into plant genome. In: *"Homologous Recombination in Plants"*, J. Paszkowski (Ed.), Kluwer Academic Publishers, Dordrecht, pp. 167-189 (1994).
134. Magrelli, A., Langenkemper, K., Dehio, C., Schell, J., and Spena, A. Splicing of the *rolA* transcript of *Agrobacterium rhizogenes* in *Arabidopsis*. *Science* 266, 1986-1988 (1994).
135. Reiss, B., Koncz, C., Moore, I., and Schell, J. A family of binary gene vectors with low inter-transformant variation. *Plant Physiol. (Life Sci. Adv.)* 13, 143-149 (1994).
136. Röhrig, H., Schmidt, J., Wieneke, U., Kondorosi, E., Barlier, I., Schell, J., and John, M. Biosynthesis of lipooligosaccharide nodulation factors: *Rhizobium* NodA protein is involved in N-acylation of the chitooligosaccharide backbone. *Proc. Natl. Acad. Sci. USA* 91, 3122-3126 (1994).

137. Röhrig, H., Schmidt, J., Wieneke, U., Kondorosi, E., Barlier, I., Schell, J., and John, M. Role of NodA and NodB proteins in nodulation factor synthesis. In: Proceedings of *"1st European Nitrogen Fixation Conference", Szeged, 28.8.-2.9.1994 and "Workshop on Safe Application of Genetically Modified Microorganisms in the Environment", Szeged, 2.-3.9.1994*, G.B. Kiss, G. Endre (Eds.), Oficina Press, Szeged, pp. 76-80 (1994).
138. Schell, J. Some Aspects of the EU System of Research Funding. In: *"European Research Structures - Changes and Challenges"* Institutional Aspects of European Research Policy, Ringberg Castle, Tegernsee, October 1993. Max-Planck-Gesellschaft E1/94, U. Opolka, L. Yakes-Schröder (Eds.), Wendelstein-Druck, Rosenheim, pp. 42-49 (1994).
139. Schell, J. Closing lecture - ISPMB Congress. *Plant Mol. Biol.* 26, 1695-1699 (1994).
140. Schell, J. Gentechnik in Deutschland: Ein Verzicht wäre unverantwortlich. *Chemie Heute*, Ausgabe 1994/95 (Das Wissenschaftsmagazin des Fonds der Chemischen Industrie), p. 6-11 (1994).
141. Schulte, W., Schell, J., and Töpfer, R. A gene encoding acetyl-coenzyme A carboxylase from *Brassica napus. Plant Physiol.* 106, 793-794 (1994).
142. Sohn, A., Schenk, P., Signoret, P.A., Schmitz, G., Schell, J., and Steinbiss, H.-H. Sequence analysis of the 3'-terminal half of RNA 1 of wheat spindle streak mosaic virus. *Arch. Virol.* 135, 279-292 (1994).
143. Voetz, M., Klein, B., Schell, J., and Töpfer, R. Three different cDNAs encoding acyl carrier proteins from *Cuphea lanceolata. Plant Physiol.* 106, 785-786 (1994).
144. Wada, M., Klein, C., Schell, J., and Reiss, B. A functional assay for *Taq* DNA polymerase in PCR. *BioTechniques* 16, 26-30 (1994).
145. Walden, R., Fritze, K., Hayashi, H., Miklashevichs, E., Harling, H., and Schell, J. Activation tagging: a means of isolating genes implicated as playing a role in plant growth and development. *Plant Mol. Biol.* 26, 1521-1528 (1994).
146. Walden, R., Hayashi, H., Fritze, K., and Schell, J. Using T-DNA tagging to search for genes involved in the mechanism of phytohormone action. In: *"Molecular Botany: Signals and the Environment"* (Biochemical Society Symposium No. 60), D.J. Bowles, P.M. Gilmartin, J.P. Knox, G.G. Lunt (Eds.), Portland Press, London, pp. 199-206 (1994).
147. Walden, R., and Schell, J. Activation T-DNA tagging - a silver bullet approach to isolating plant genes. *Agro Food Industry Hi-Tech* 5 (No. 6), 9-12 (1994).
148. Zettl, R., Schell, J., and Palme, K. Photoaffinity labeling of *Arabidopsis thaliana* plasma membrane vesicles by 5-azido-[7-^{3}H]-indole-3-acetic acid: Identification of a glutathione-S-transferase. *Proc. Natl. Acad. Sci. USA* 91, 689-693 (1994).

149. Beachy, R., Schell-Frederick, E., and Schell, J. La biotechnologie végétale au service de la santé. *Diogène* 172, 97-110 (1995).
150. Beachy, R., Schell-Frederick, E., and Schell, J. Plant biotechnology in the service of human health. *Diogenes* 172, Volume 43/4, 93-104 (1995).
151. Hausmann, L., Schell, J., and Töpfer, R. Cloning of a cDNA coding for a glycerol-3-phosphate dehydrogenase from *Cuphea lanceolata*. In: *"Plant Lipid Metabolism"*, J.-C. Kader, and P. Mazliak (Eds.), Kluwer Academic, Dordrecht, pp. 534-536 (1995).
152. Heyraud-Nitschke, F., Schumacher, S., Laufs, J., Schaefer, S., Schell, J., and Gronenborn, B. Determination of the origin cleavage and joining domain of geminivirus Rep proteins. *Nucleic Acids Res.* 23, 910-916 (1995).
153. Jach, G., Görnhardt, B., Mundy, J., Logemann, J., Pinsdore, E., Leah, R., Schell, J., and Maas, C. Enhanced quantitative resistance against fungal disease by combinatorial expression of different barley antifungal proteins in transgenic tobacco. *Plant J.* 8, 97-109 (1995).
154. John, M., Röhrig, H., Schmidt, J., and Schell, J. Molecular biology in the land of nodules. *Biofutur* 142, 19-21 (1995).
155. Kleczkowski, K., and Schell, J. Phytohormone conjugates: nature and function. *Crit. Rev. Plant Sci.* 14, 283-298 (1995).
156. Laufs, J., Traut, W., Heyraud, F., Matzeit, V., Rogers, S.G., Schell, J., and Gronenborn, B. *In vitro* cleavage and joining at the viral origin of replication by the replication initiator protein of tomato yellow leaf curl virus. *Proc. Natl. Acad. Sci. USA* 92, 3879-3883 (1995).
157. Maas, C., Reichel, C., Schell, J. and Steinbiss, H.-H. Preparation and transformation of monocot protoplasts. In: *"Methods in Cell Biology - Plant Cell Biology"*, Vol. 50, D.W. Galbraith, H. Bohnert, and D.P. Bourque (Eds.), Academic Press, Orlando, pp. 383-399 (1995).
158. Maas, C., and Schell, J. Gentechnologie und Umweltschutz in der Landwirtschaft. In: *"Bildung und Wissenschaft"* II/95, Inter Nationes e.V., Bonn (Ed.), Druckpartner Moser Druck + Verlag GmbH., Rheinbach, pp. 18-19 (1995).
159. Martini, N., Schell, J., and Töpfer, R. Expression of acyl-[ACP] thioesterase in *Cuphea lanceolata* and transgenic rapeseed. In: *"Plant Lipid Metabolism"*, J.-C. Kader, and P. Mazliak (Eds.), Kluwer Academic, Dordrecht, pp. 495-498 (1995).
160. Martini, N., Schell, J., and Töpfer, R. Expression of medium-chain acyl-[ACP] thioesterases in transgenic rapeseed. In: *"Rapeseed Today and Tomorrow"* Vol. 2 (Proceedings of the 9th International Rapeseed Congress, 4.-7. July 1995, Cambridge/U.K.), Organizing Committee GCIRC (Eds.), The Dorset Press, Dorchester, pp. 461-463 (1995).
161. Moore, I., Schell, J., and Palme, K. Subclass-specific sequence motifs identified in Rab GTPases. *Trends Biochem. Soc.* 20, 10-12 (1995).

162. Reichel, C., Feltkamp, D., Walden, R., Steinbiss, H.-H., Schell, J., and Rosahl, S. Inefficient expression of the DNA-binding domain of GAL4 in transgenic plants. *Plant Cell Rep.* 14, 773-776 (1995).
163. Schell, J. Plant biotechnology: State of the art in developed countries and relevant safety considerations. In: *"Pan-European conference on the potential long-term ecological impact of genetically modified organisms"*, Environmental encounters, No. 20, (Proceedings PEC, 24.-26.11.1993, Strasbourg), Council of Europe (Ed.), Council of Europe Press, Netherlands, pp. 25-34 (1995).
164. Schell, J. Progress in plant sciences is our best hope to achieve an economically rewarding, sustainable and environmentally stable agriculture. *Plant Tissue Culture and Biotechnology* 1, 10-12 (1995).
165. Schell, J. Statement zur Gentechnologie. In: *"Gentechnologie - Ethik und wissenschaftlicher Fortschritt"* (Internationales Forum der Universität zu Köln gemeinsam mit der Stadt Köln, 23.6.1994, Historisches Rathaus Köln), Band 4, P. Mittelstaedt (Ed.), Bouvier Verlag, Bonn, pp. 31-33 u. 65-66 (1995).
166. Schell, J. Molecular breeding as well as chemical and biological protection methods have an important role to play to make integrated pest management a practical and desirable reality. *Agro-Food-Industry Hi-Tech* Supplement, Brighton Crop Protection 1995, 4-5 (1995).
167. Schell, J. Crop biotechnology - a necessity for an environmentally friendly and sustainable agriculture. In: Proceedings, Part I, *"Ninth Forum for Applied Biotechnology"* Faculty of Agricultural and Applied Biological Sciences, Gent, 27.-29.9.1995. Med. Fac. Landbouww. Univ. Gent, 60(4a), 1513-1514 (1995).
168. Schell, J., Palme, K., and Walden, R. Molecular approaches to the study of the mechanism of action of auxins. In: *"Plant Hormones - Physiology Biochemistry and Molecular Biology"*, P.J. Davies (Ed.), Kluwer Academic, Dordrecht, pp. 354-371 (1995).
169. Schultze, M., Staehelin, C., Röhrig, H., John, M., Schmidt, J., Kondorosi, E., Schell, J., and Kondorosi, A. *In vitro* sulfotransferase activity of *Rhizobium meliloti* NodH protein: Lipochitooligosaccharide nodulation signals are sulfated after synthesis of the core structure. *Proc. Natl. Acad. Sci. USA* 92, 2706-2709 (1995).
170. Suter-Crazzolara, C., Brzobohaty, B., Gazdova, B., Schell, J., and Reiss, B. T-DNA integration in a new family of repetitive elements of *Nicotiana tabacum. J. Mol. Evol.* 41, 498-504 (1995).
171. Töpfer, R., Hausmann, L., Voetz, M., Walek, J., Schell, J. and Martini, N. Genes for the improvement of the storage oil content of *Brassica napus.* In: *"Rapeseed Today and Tomorrow"*, Vol. 2 (Proceedings of the 9th International Rapeseed Congress, 4.-7. July 1995, Cambridge/U.K.), Organising Committee GCIRC (Eds.), The Dorset Press, Dorchester, pp. 479-481 (1995).

172. Töpfer, R., Martini, N., and Schell, J. Modification of plant lipid synthesis. *Science* 268, 681-686 (1995).

173. Walden, R., Harling, H., Miklashevichs, E., Lubenow, H., and Schell, J. Activation T-DNA tagging: gene isolation and molecular dissection of complex biological pathways. In: *"Induced Mutations and Molecular Techniques for Crop Improvement"* (Proceedings of an International Symposium on the Use of Induced Mutations and Molecular Techniques for Crop Improvement, 19-23 June 1995, Vienna), International Atomic Energy Agency, Vienna (Eds.), IAEA, Austria, pp. 425-431 (1995).

174. Grevelding, C., Suter-Crazzolara, C., von Menges, A., Kemper, E., Masterson, R., Schell, J., and Reiss, B. Characterisation of a new allele of *pale cress* and its role in greening in *Arabidopsis thaliana. Mol. Gen. Genet.* 251, 532-541 (1996).

175. Keller, M., Sneh, B., Strizhov, N., Prudovsky, E., Regev, A., Koncz, C., Schell, J., and Zilberstein, A. Digestion of δ-endotoxin by gut proteases may explain reduced sensitivity of advanced instar larvae of *Spodoptera littoralis* to CryIC. *Insect Biochem. Mol. Biol.* 26, 365-373 (1996).

176. Kleinmann, J.-W., Kleinow, T., Engelhardt, K., Philipp, C., Wegener, D., Schell, J., and Schreier, P. Characterization of two class II chitinase genes from peanut and expression studies in transgenic tobacco plants. *Plant Mol. Biol.* 30, 351-358 (1996).

177. Regev, A., Keller, M., Strizhov, N., Sneh, B., Prudovsky, E., Chet, I., Ginzberg, I., Koncz-Kalman, Z., Koncz, C., Schell, J., and Zilberstein, A. Synergistic activity of a *Bacillus thuringiensis* δ-endotoxin and a bacterial endochitinase against *Spodoptera littoralis* larvae. *Appl. Environ. Microbiol.* 62, 3581-3586 (1996).

178. Reichel, C., Maas, C., Schultze, S., Schell, J., and Steinbiss, H.-H. Cooperative binding to nucleic acids by barley yellow mosaic bymovirus coat protein and characterization of a nucleic acid-binding domain. *Virology* 77, 587-592 (1996).

179. Reichel, C., Mathur, J., Eckes, P., Langenkemper, K., Koncz, C., Schell, J., Reiss, B., and Maas, C. Enhanced green fluorescence by the expression of an *Aequorea victoria* green fluorescent protein mutant in mono- and dicotyledonous plant cells. *Proc. Natl. Acad. Sci. USA* 93, 5888-5893 (1996).

180. Reinemer, P., Prade, L., Hof, P., Neuefeind, T., Huber, R., Zettl, R., Palme, K., Schell, J., Koelln, I., Bartunik, H.D., and Bieseler, B. Three-dimensional structure of glutathione S-transferase from *Arabidopsis thaliana* at 2.2 Å resolution: Structural characterization of herbicide-conjugating plant glutathione S-transferases and a novel active site architecture. *J. Mol. Biol.* 255, 289-309 (1996).

181. Reiss, B., Klemm, M., Kosak, H., and Schell, J. RecA protein stimulates homologous recombination in plants. *Proc. Natl. Acad. Sci. USA* 93, 3094-3098 (1996).
182. Schell, J. Prof. G. Melchers celebrates his ninetieth (90th) birthday! *Mol. Gen. Genet.* 250, 135-136 (1996).
183. Schell, J. Transgenic plants and study of the regeneration of whole plants from individual cells. In: *"Das Ganze und seine Teile"* Europäisches Forum Alpbach 1995, H. Pfusterschmid-Hardtenstein (Ed.), Ibera Verlag, Wien, pp. 515-516 (1996).
184. Schell, J. Biotechnologies et agriculture: les nouveaux defis. Cahiers de la Fondation des Treilles, N°7/Année 1995, 167-186 (1996).
185. Strizhov, N., Keller, M., Koncz-Kalman, Z., Regev, A., Sneh, B., Schell, J., Koncz, C., and Zilberstein, A. Mapping of the entomocidal fragment of *Spodoptera*-specific *Bacillus thuringiensis* toxin CryIC. *Mol. Gen. Genet.* 253, 11-19 (1996).
186. Strizhov, N., Keller, M., Mathur, J., Koncz-Kalman, Z., BOSCH, D., Prudovsky, E., Schell, J., Sneh, B., Koncz, C., and Zilberstein, A. A synthetic *cryIC* gene, encoding a *Bacillus thuringiensis* δ-endotoxin, confers *Spodoptera* resistance in alfalfa and tobacco. *Proc. Natl. Acad. Sci. USA* 93, 15012-15017 (1996).
187. Szekeres, M., Nemeth, K., Koncz-Kalman, Z., Mathur, J., Kauschmann, A., Altmann, T., Redei, G.P., Nagy, F., Schell, J., and Koncz, C. Brassinosteroids rescue the deficiency of CYP 90, a cytochrome P450, controlling cell elongation and de-etiolation in *Arabidopsis. Cell* 85, 171-182 (1996).
188. Van de Samde, K., Pawlowski, K., Czaja, I., Wieneke, U., Schell, J., Schmidt, J., Walden, R., Matvineko, M., Wellink, J., Van Kammen, A., Franssen, H., and Bisseling, T. Modification of phytohormone response by a peptide encoded by *ENOD40* of legumes and a nonlegume. *Science* 273, 370-373 (1996).
189. Walden, R., Maas, C., Martini, N., and Schell, J. Transgenic plants. *European Review* 4, 393-414 (1996).
190. Gadella, Jr., T.W.J., VerebJr., G., Hardi, A.-E., Röhrig, H., Schmidt, J., John, M., Schell, J., and Bisseling, T. Microspectroscopic imaging of nodulation factor-binding sites on living *Vicia sativa* roots using a novel bioactive fluorescent nodulation factor. *Biophys. J.* 72, 1986-1996 (1997).
191. Jach, G., Pinsdore, E., and Schell, J. Direct isolation of poly A^+ RNA from transgenic tobacco leaves with Oligotex. *Qiagen News* 1, 20-21 (1997).
192. John, M., Röhrig, H., Schmidt, J., Walden, R., and Schell, J. Cell signalling by oligosaccharides. *Trends Plant Sci.* 2, 111-115 (1997).
193. Kendall, H.W., Beachy, R., Eisner, T., Gould, F., Herdt, R., Raven, P.H., Schell, J.S., and Swaminathan, M.S. (Eds.) *"Bioengineering of Crops"* Report of the World Bank Panel on Transgenic Crops. Environmentally and Socially

Sustainable Development Studies and Monographs Series 23, Worldbank, Washington (1997).

194. Maas, C., Simpson, C.G., Eckes, P., Schikler, H., Brown, J.W.S., Reiss, B., Salchert, K., Chet, I., Schell, J., and Reichel, C. Expression of intron modified *NPT II* genes in monocotyledonous and dicotyledonous plant cells. *Molecular Breeding* 3, 15-28 (1997).
195. Moore, I., Dieffenthal, T., Zarsky, V., Schell, J., and Palme, K. A homolog of the mammalian GTPase Rab2 is present in *Arabidopsis* and is expressed predominantly in pollen grains and seedlings. *Proc. Natl. Acad. Sci. USA* 94, 762-767 (1997).
196. Reindl, A., Schöffl, F., Schell, J., Koncz, C., and Bako, L. Phosphorylation by a cyclin-dependent kinase modulates DNA binding of the Arabidopsis heat-shock transcription factor HSF1 *in vitro*. *Plant Physiol.* 115, 93-100 (1997).
197. Reiss, B., Kosak, H., Klemm, M., and Schell, J. Targeting of a functional *Escherichia coli* RecA protein in the nucleus of plant cells. *Mol. Gen. Genet.* 253, 695-702 (1997).
198. Röhrig, H., Schmidt, J., Wieneke, U., Walden, R., Schell, J., and John, M. *N*-Acyl galactosamine inhibition of lipo-chitooligosaccharide action. In: *"Biological Fixation of Nitrogen for Ecology and Sustainable Agriculture"* (NATO ASI Series, Vol. G 39), A. Legocki, H. Bothe, A. Pühler (Eds.), Springer Verlag, Berlin, pp. 59-62 (1997).
199. Schell, J. Cotton carrying the recombinant insect poison Bt toxin: no case to doubt the benefits of plant biotechnology. (Commentary) *Current Opinion Biotech.* 8, 235-236 (1997).
200. Schell, J. How might biotechnology contribute to the improvement of the environment? In: *"How Can Biotechnology Benefit the Environment?"* Report of A European Federation of Biotechnology (EFB) Task Group on Public Perceptions of Biotechnology and The Green Alliance (Workshop, 13.01.97, National Museum, London), p. 12 (1997).
201. Schell, J. Plant biotechnology and its future impact on science, agriculture and the environment. In: *"Biotechnology, Patents and Morality"*, S. Sterckx (Ed.), Ashgate Publishing Ltd., Aldershot/UK., pp. 66-71 (1997).
202. Schell, J. Biotechnik in der Pflanzenzucht. *EUmagazin* 11, 15-19 (1997).
203. Schell, J. Les biotechnologies végétales contribueront à rendre l'agriculture intensive à la fois productive et plus respectueuse de l'environnement. *Biofutur* Spécial 172, 62-63 (1997).
204. Schell, J. Umsetzungsdefizite zwischen Wissenschaft und Praxis. (Tagungsbericht VDL-Forum 1997 "Biotechnolgie-Standort Deutschland - Bis zum Jahr 2000 die Nr. 1 in Europa?") *VDL-Journal* 10, 10 (1997).
205. Schulte, W., Töpfer, R., STRACKE, R., Schell, J., and Martini, N. Multi-functional acetyl-CoA carboxylase from *Brassica napus* is encoded by a multi-gene

family: indication for plastidic localization of at least one isoform. *Proc. Natl. Acad. Sci. USA* 94, 3465-3470 (1997).

206. Slabaugh, M.B., Huestis, G.M., Leonard, J., Holloway, J.L., Rosato, C., Hongtrakul, V., Martini, N., Toepfer, R., Voetz, M., Schell, J., and Knapp, S.J. Sequence-based genetic markers for genes and gene families: single-strand conformational polymorphisms for the fatty acid synthesis genes of *Cuphea. Theor. Appl. Genet.* 94, 400-408 (1997).
207. Strizhov, N., Abraham, E., Ökresz, L., Blickling, S., Zilberstein, A., Schell, J., Koncz, C., and Szbados, L. Differential expression of two *P5CS* genes controlling proline accumulation during salt-stress requires ABA and is regulated by *ABA1*, *ABI1* and *AXR2* in *Arabidopsis. Plant J.* 12, 557-569 (1997).
208. Walden, R., Reiss, B., Koncz, C., and Schell, J. The impact of Ti-plasmid-derived gene vectors on the study of the mechanism of action of phytohormones. *Annu. Rev. Phytopathol.* 35, 45-66 (1997).
209. First, N.L., Schell, J., and Vasil, I.K. Prospects and Limitations of Agricultural Biotechnologies: An Update. In: *"Agricultural Biotechnology"*, A. Altman (Ed.), Marcel Dekker, New York, pp. 743-748 (1998).
210. Ginzberg, I., Stein, H., Kapulnik, Y., Szbados, L., Strizhov, N., Schell, J., Koncz, C., and Zilberstein, A. Isolation and characterization of two different cDNAs of Δ^1-pyrroline-5-carboxylate synthase in alfalfa, transcriptionally induced upon salt stress. *Plant Mol. Biol.* 38, 755-764 (1998).
211. Mathur, J., Molnar, G., Fujioka, S., Takatsuto, S., Sakurai, A., Yokota, T., Adam, G., Voigt, B., Nagy, F., Maas, C., Schell, J., Koncz, C., and Szekeres, M. Transcription of the *Arabidopsis CPD* gene, encoding a steroidogenic cytochrome P450, is negatively controlled by brassinosteroids. *Plant J.* 14, 593-602 (1998).
212. Mathur, J., Szbados, L., Schaefer, S., Grunenberg, B., Lossow, A., Jonas-Straube, E., Schell, J., Koncz, C., and Koncz-Kalman, Z. Gene identification with sequenced T-DNA tags generated by transformation of *Arabidopsis* cell suspension. *Plant J.* 13, 707-716 (1998).
213. Moore, I., Gälweiler, L., Grosskopf, D., Schell, J., and Palme, K. A transcription activation system for regulated gene expression in transgenic plants. *Proc. Natl. Acad. Sci. USA* 95, 376-381 (1998).
214. Nemeth, K., Salchert, K., Putnoky, P., Bhalerao, R., Koncz-Kalman, Z., Stankovic-Stangeland, B., Bako, L., Mathur, J., Ökresz, L., Stabel, S., Geigenberger, P., STITT, M., Redei, G.P., Schell, J., and Koncz, C. Pleiotropic control of glucose and hormone responses by PRL1, a nuclear WP protein, in *Arabidopsis. Genes & Development* 12, 3059-3073 (1998).
215. Ökresz, L., Mathe, C., Horvath, E., Schell, J., Koncz, C., and Szbados, L. T-DNA trapping of a cryptic promoter identifies an ortholog of highly

conserved *SNZ* growth arrest response genes in *Arabidopsis. Plant Science* 138, 217-228 (1998).

216. Salamini, F., Reiss, B., Koncz, C., and Schell, J.St. Pflanzenbiotechnologie, die Zukunft für Forschung und Anwendung. *Bayerischer Monatsspiegel* 4, 61-64 (1998).

217. Schell, J. Contribution of plant breeding for making biodiesel rapeseed a realistic proposition. In: *"Plant Oils as Fuels - Present State of Science and Future Developments"* (Proceedings of the Symposium, Potsdam, 16.-18.02.1997), N. Martini, J. Schell (Eds.), Springer, Heidelberg, p. 197-198 (1998).

218. Schell, J. La biotechnologie végétale est nécessaire. *ENA* 285, 27-28 (1998).

219. Schell, J. Plant biotechnology is essential for an economically rewarding and environmentally sustainable agriculture. In: *"Economic Botany in the Tropics"*, Second Edition, S.L. Kochhar (Ed.), Rajiv Beri for Macmillan India Ltd., New Delhi, pp. 510-515 (1998).

220. Schell, J. St. Using transformation to study the mechanism of action of plant hormones. In: *"Bulletin de la Classe des Sciences"*, Académie Royale de Belgique, 6e série, Tome IX, 7-12/1998, Imprimerie Communications s.a., Louvain-la-Neuve, pp. 361-367 (1998).

221. Schell, J., and Tibiracio, A.F. Introduction. In: *"Workshop on Novel Approaches to Study Plant Growth Factors"*, Instituto Juan March Series 87, Ediciones Peninsular, Madrid, pp. 9-10 (1998).

222. Schell, J. St., and Weinand, J. Hat die europäische Agrarpolitik eine Zukunft ohne Pflanzenbiotechnologie? In: *"Max-Planck-Gesellschaft - Jahrbuch 1998"*, S. Deutschmann MPG München (Ed.), Verlag Vandenhoeck & Ruprecht, Göttingen, pp. 73-85 (1998).

223. Schell-Frederick, E., and Schell, J. Plant biology in the service of mankind and his environment. *Agro-Food-Industry Hi-Tech* 9, No. 4, 19-21 (1998).

224. Umeda, M., Bhalerao, R.P., Schell, J., Uchimuya, H., and Koncz, C. A distinct cyclin-dependent kinase-activating kinase of *Arabidopsis thaliana. Proc. Natl. Acad. Sci. USA* 95, 5021-5026 (1998).

225. Weinand, J., und Schell, J. Pflanzen-Biotechnologie kann den negativen Einfluß der Landwirtschaft auf die Umwelt reduzieren. In: *"Moderne Ansätze in der Waldökologie"*, (Berichte des Forschungszentrums Waldökosysteme, Reihe A, Bd. 154), Forschungszentrum Waldökosysteme der Universität Göttingen (Ed.), Selbstverlag, pp. 7-27 (1998).

226. Bhalerao, R.P., Salchert, K., Bako, L., Ökresz, L., Szbados, L., Muranaka, T., Muchida, Y., Schell, J., and Koncz, C. Regulatory interaction of PRL1 WD protein with *Arabidopsis* SNF1-like protein kinases. *Proc. Natl. Acad. Sci. USA* 96, 5322-5327 (1999).

227. Schell, J. Grundlagenforschung und Anwendungsbezug in der Pflanzenzüchtung: Ein "grüner Faden" am Beispiel des Max-Planck-Instituts für Züchtungsforschung. In: "Wirtschaft und Wissenschaft - eine Allianz mit Zukunft in Deutschland?", Ringberg-Symposium Oktober 1998, Max Planck Forum 1/99, U. Opolka (Ed.), Format-Druck, Rosenheim, pp. 283-293 (1999).

228. Schell, J., Bisseling, T., Dülz, M., Franssen, H., Fritze, K., John, M., Kleinow, T., Lessnick, A., Miklashevichs, E., Pawlowski, K., Röhrig, H., Van de Samde, K., Schmidt, J., Steinbiss, H.-H. and Stoll, M. Re-evaluation of phytohormone-independent division of tobacco protoplast-derived cells. *Plant J.* 17, 461-466 (1999).

229. Weinand, J., und Schell, J. Vom Laborphänomen zum Wirtschaftsfaktor. *Mais* 27, 86 (1999).

230. Ferrando, A., Farras, R., Jasik, J., Schell, J., and Koncz, C. Intron-tagged epitope: a tool for facile detection and purification of proteins expressed in *Agrobacterium*-transformed plant cells. *Plant J.* 22, 553-560 (2000).

231. Reiss, B., Schubert, I., Köpchen, K., Wendeler, E., Schell, J., and Puchta, H. RecA stimulates sister chromatid exchange and the fidelity of double-strand break repair, but not gene targeting, in plants transformed by *Agrobacterium. Proc. Natl. Acad. Sci USA* 97, 3358-3363 (2000).

232. Schell, J. Science and agriculture in the 21st century. In: *"Phytosfere '99"* Highlights in European Plant Biotechnology Research and Technology Transfer. (Proceedings of the Second European Conference on Plant Biotechnology, Rome/Italy, 7-9 June 1999). G.E. de Vries and K. Metzlaff (Eds.), Elsevier, Amsterdam, pp. 17-25 (2000).

233. Schell, J., and Koncz, C. The Ti-plasmid and plant molecular biology. In: *"Discoveries in Plant Biology"*, Vol. III, S.-H. Kung and S.-F. Yang (Eds.), World Scientific, Singapore, pp. 393-409 (2000).

234. Schell-Frederick, E., und Schell, J.S. Neuartige Lebensmittel (Novel Food) und Pharmazeutika. *Deutsches Ärzteblatt*, Jg. 97, Heft 28-29, Seiten A1971-1976 (2000).

235. Reiss B, Schubert I, Köpchen K, Wendeler E, Schell J, Puchta H. RecA stimulates sister chromatid exchange and the fidelity of double-strand break repair, but not gene targeting, in plants transformed by Agrobacterium. *Proc Natl Acad Sci USA* 97(7): 3358-63 (2000).

236. Ferrando A, Farràs R, Jásik J, Schell J, Koncz C. Intron-tagged epitope: a tool for facile detection and purification of proteins expressed in Agrobacterium-transformed plant cells. *Plant J.* 22(6): 553-60 (2000).

237. Arras, R., Ferrando, A., Jasik, J., Kleinow, T., Ökresz, L., Tibiracio, A., Salchert, K., del Pozo, C. Schell, J., and Koncz, C. SKP1-SnRK protein kinase interactions mediate proteasomal binding of a plant SCF ubiquitin ligase. *The EMBO Journal*, Vol. 20 N° 11, pp. 2742-2756 (2001).

238. Miklashevichs, E., Röhrig, H., Schell, J. and Schmidt, J. Perception and Signal Transduction of Rhizobial NOD Factors. *Critical Reviews in Plant Sciences*, 20(4):373-394 (2001).

239. Jach, G., Binot, E. Frings, S., Luxa, K. and Schell, J. Use of red fluorescent protein from *Discosoma* sp. (dsRED) as a reporter for plant gene expression. *The Plant Journal* 28(4), 483-491 (2001).

240. Jach G, Binot E, Frings S, Luxa K, Schell J. Use of red fluorescent protein from Discosoma sp. (dsRED) as a reporter for plant gene expression. *Plant J.* 28(4): 483-91 (2001).

241. Ferrando A, Koncz-Kálmán Z, Farràs R, Tiburcio A, Schell J, Koncz C. Detection of in vivo protein interactions between Snf1-related kinase subunits with intron-tagged epitope-labelling in plants cells. *Nucleic Acids Res.* 29(17): 3685-93 (2001).

242. Farrás R, Ferrando A, Jásik J, Kleinow T, Okrész L, Tiburcio A, Salchert K, del Pozo C, Schell J, Koncz C. SKP1-SnRK protein kinase interactions mediate proteasomal binding of a plant SCF ubiquitin ligase. *EMBO J.* 20(11): 2742-56 (2001).

243. Ríos G, Lossow A, Hertel B, Breuer F, Schaefer S, Broich M, Kleinow T, Jásik J, Winter J, Ferrando A, Farrás R, Panicot M, Henriques R, Mariaux JB, Oberschall A, Molnár G, Berendzen K, Shukla V, Lafos M, Koncz Z, Rédei GP, Schell J, Koncz C. Rapid identification of Arabidopsis insertion mutants by non-radioactive detection of T-DNA tagged genes. *Plant J.* 32(2):243-53 (2002).

244. Henriques R, Jásik J, Klein M, Martinoia E, Feller U, Schell J, Pais MS, Koncz C. Knock-out of Arabidopsis metal transporter gene IRT1 results in iron deficiency accompanied by cell differentiation defects. *Plant Mol Biol.* 50(4-5):587-97 (2002).

245. Markmann-Mulisch U, Hadi MZ, Koepchen K, Alonso JC, Russo VE, Schell J, Reiss B. The organization of Physcomitrella patens RAD51 genes is unique among eukaryotic organisms. *Proc Natl Acad Sci USA* 99(5): 2959-64 (2002).

246. Rohrig H, Schmidt J, Miklashevichs E, Schell J, John M. Soybean ENOD40 encodes two peptides that bind to sucrose synthase. *Proc Natl Acad Sci USA* 99(4):1915-20 (2002).

247. Bakó L, Umeda M, Tiburcio AF, Schell J, Koncz C. The VirD2 pilot protein of Agrobacterium-transferred DNA interacts with the TATA box-binding protein and a nuclear protein kinase in plants. *Proc Natl Acad Sci USA* 100(17):10108-13 (2003).

248. Röhrig, H., Schmidt, J., Miklashevichs, E., Schell, J. and John. M. Soybean *ENOD40* encodes two peptides which bind to sucrose synthase. *PNAS* (in press)

Editor's Choice

Transformation of Plant Science in Our Time—the Contribution of Jozef S. Schell (1935–2003)

Editor's note: Plant Physiology does not ordinarily publish obituaries, but is making an exception in this issue to honor Jeff Schell for his enormous contribution to plant science.

The ability to introduce genes into plant species has revolutionized fundamental research and allowed for the fastest development of new varieties in the history of commercial agriculture, as all readers must know. This technology probably represents the most significant breakthrough in plant breeding for the 20th century. The relatively small number of pioneers of this new technology in plant science have been honored by the scientific community, governments, awarding foundations, and scientific academies. With the recent passing away of Jozef (Jeff) Schell (April 2003) , it is appropriate to recall his contribution. He played an enormous role in both the discovery underlying Agrobacterium-mediated transformation, and the vision for how this technology could and should change opportunities for humankind forever.

Jozef S. Schell (1935–2003)

Why was Jeff's influence so noteworthy? There are many reasons. He was an exceptional scientist: charismatic, physically strong, visionary, adventurous, and politically astute. He was an advocate for governments and industries, and he had an immense capacity for hard work. His vision ranged from a comprehension of the intricacies of the wonderful evolutionary process of Agrobacterium T-DNA transfer, to a world where many plants used by people could and would be improved as never before. As the subject of plant genetic engineering matured, he believed, and continuously emphasized to all, that plant biotechnology was essential for our crowded world in order to give everyone a reasonable standard of living and a sustainable environment.

Many research scientists who employ Agrobacterium transformation routinely, and obviously the millions who eat transgenic food or grow transgenic cotton, do not necessarily know how the gene-transfer discoveries were originally made and harnessed. Although many reviews have been written on the subject, it is not widely appreciated that in the 1960s and 1970s Jeff Schell and his colleague Marc Van Montagu in Gent, Belgium, worked as microbiologists, as bacterial geneticists, and not as plant scientists. However, they, like a few others, became intrigued by the long-established evidence that some soil bacteria provoked formation of cancerous tissues on certain plant species. Their papers of the day focused on bacterial genetics, with the discovery, isolation, and mapping of plasmids and recognition of different kinds of agrobacterial plasmids. They described the production of different metabolites, notably nopaline and octapine, by the tumors, and eventually the transfer of T-DNA. Their fundamental research was driven by questions in bacterial genetics and the evolution of an extraordinary plant-microbe interaction; one in which plant cells were induced to proliferate, make, and excrete a metabolite that the particular Agrobacterial strain was exquisitely capable of using as a nitrogen and carbon source and that allowed it to compete more effectively with other microbes. Plant genetic engineering did not emerge as a topic until it was proven that bacterial T-DNA was incorporated into plant chromosomes. Even when this was proven, the job of genetically dissecting the functions of the T-DNA, learning how to isolate the Ti plasmid, learning how to clone the DNA, to create deletions, to recombine DNA into plasmids in vivo, to introduce novel genes into T-DNA on the Ti plasmid, and to shuttle DNA from *E. coli* to Agrobacterium were the stuff only bacterial geneticists could work out. Almost no plant scientist had these skills at the time. Jeff and Marc were the leaders in this new field because they had created the opportunity, were skilled in the thinking and the required techniques, and most importantly, had the foresight to know what was there to be discovered.

Jeff's first publications in the late 1950s and early 1960s were concerned with carbon metabolism in

www.plantphysiol.org/cgi/doi/10.1104/pp.900084.

bacteria. These studies were followed by ones on modifications of phages and restrictions by their hosts. His first paper, coauthored with Marc Van Montagu, on Agrobacterium-related DNA was published in 1972 and focused on Agrobacterium phages. With Rob Schilperoort, he described in 1973 the detection of Agrobacterium DNA in sterile crown gall tissue cultures. In 1974, his group showed that a large plasmid of Agrobacterium was essential for crown gall induction, and this was followed by a series of papers on the plasmids. The isolation of the Ti plasmid was described in 1976, and this breakthrough opened up many new opportunities to study its DNA. By 1977, several groups including those of Schell and Van Montagu in Gent, Mary-Dell Chilton in Eugene Nester's laboratory in Seattle, and Schilperoort's lab in Leiden were actively testing the idea that the Ti plasmid of Agrobacterium might provide a means of inserting genes into plants. In 1978, Jeff and his colleagues (including Marc Van Montagu) published 18 papers covering the origins of crown gall by Ti plasmid transfer and the manipulation of the Ti plasmid by various bacterial genetic tricks, including transposon mutagenesis, cointegration, and transfections to reveal more about the plasmid, the genes it contained, and the requirements for crown gall induction. In 1980 the Gent group published 20 papers, many being symposium articles, to get the Ti–plant cell transformation message out to receptive scientists. By the time the seminal papers were published in 1983, showing the expression of novel chimeric genes introduced into plant cells using a Ti plasmid-derived vector, Jeff and his colleagues had published some 150 papers. Then followed many details of the Ti plasmid genes, their role in oncogenesis, and the behavior of these genes when transferred into plants. Some 23 papers were published in 1984 describing the development of useful vectors and the transfer of several more foreign genes into plants. During this time he also studied the Ri plasmid, the equivalent of the Ti plasmid, in *Agrobacterium rhizogenes* and its function after transfer to plants cells. Once the subject was truly well established, the concept of plant genetic engineering became a topic of worldwide debate, and the plant scientific community embraced these technologies to literally transform the whole field of plant science. During this period, Jeff and many colleagues, including students and postdocs, and collaborators in other laboratories published a large number of papers (another 449 until his death in 2003) on Ti plasmid biology, transformation vectors, plant gene discovery especially in Arabidopsis, promoter analyses, hormone biology, Rhizobium genetics, and plant genetic engineering.

The enormous number of papers that carried his name, first from Gent and then from his very large group in the Max Planck Institute for Plant Breeding, Köln, Germany, are a witness to the huge contribution he and his colleagues made to the development and propagation of plant transformation and molecular genetics. However, the expansion of a technology and its adoption into another area of science worldwide takes not only skilled experiments and provocative publications. It takes teaching of a vision to scientists, research funding agencies, industries and governments, and the training of a new generation of scientists. Jeff did all of this, unceasingly, for twenty years. He had a major impact on students, established scientists, plant breeders, industries, funding agencies, governments, as well as scientific societies and the general public. During the 1980s and 1990s, he sustained a punishing schedule of lectures around the world. In the 1980s he was sought by nearly every plant molecular biology conference, and he wanted to attend them all to learn, to teach, to inspire, and to open up the practice of plant molecular genetics.

If he gave much to plant science everywhere by his teaching, he gave much to European science by accepting in 1978 the Directorship of the Max Planck Institute for Plant Breeding Research in Köln, Germany. This appointment gave Jeff the opportunity to establish a very large team of scientists, to develop the new approaches to plant science, and to train students and postdocs. The position also gave him enhanced authority to talk to European governmental agencies and the European Union (EU). In the early years of plant genetic engineering, he also advised Monsanto, the company that was to become the foremost crop genetic engineering multinational. He also forged a long-term, mutually beneficial, relationship between the Max Planck Institute and Bayer, a company located close to the Max Planck Institute in Germany.

Jeff was a true European. The EU started to offer substantial funds for research in the 1980s, and Jeff and colleagues at the Max Planck Institute in Köln saw the value and necessity for this pan-European resource to spearhead the research that could help address the substantial agricultural, economic, and environmental problems of the future. He taught and lobbied EU scientific leaders about the new plant science. To improve the funding and management of EU plant science programs, the Max Planck Institute and the John Innes Centre (Norwich, England) –the two largest centers for plant molecular genetics in Europe –formed a legal entity under Jeff's chairmanship. Much was changed during the 1990s via this organization and Jeff's tireless and influential leadership. He saw the needs of European and global societies and was not afraid to articulate the opportunities, needs, and requirements for change to anyone.

While being a Director of a team of well over 100 scientists, an international leader, and an advocate for European science and plant molecular biology for agriculture and the environment in general, Jeff found time to do many of the ordinary things that senior scientists do write and review grant proposals and papers and mentor students and postdocs in his laboratory. He was a senior editor of The Plant Journal from 1990 to 1998. He also held the great distinction of being the Chairman of the European Molecular Biology Organisation (EMBO) Council from 1990 to 1995. EMBO is an organization much revered in Europe and via his service to EMBO, Jeff gave much to molecular biology as a whole. Of course, he sat on many boards and committees around the world to give wisdom and guidance to those who sought to further science and plant molecular biology in particular.

Plant science is not renowned for having individuals honored by science and society, but Jeff collected many honors. He was elected to be a Foreign Associate of the US National Academy of Sciences (1985), of the Indian National Science Academy (1998), of the Royal Swedish Academy (1989), and of the Hungarian Academy of Sciences (1993). He won many key prizes, including the Mendel-Medaille of the Deutsche Akademie der Naturforscher Leopoldina (1985), the Otto Bayer-Preis of the Otto-Bayer-Stiftung (1985), Prix Alexandre de Humboldt (1985), the Rank Prize for Nutrition (1987), the IBM Europe Science and Technology Prize (1987), the Wolf Prize in Agriculture (1990), Prix Charles Leopold Mayer of the Academie des Sciences, Paris (1990), the Japan Prize for Biotechnology in Agriculture Sciences by the Science and Technology Foundation of Japan (1998), Premiere Grande Medaille d'Or de l'Academie des Sciences, Paris (1997), the Australia Prize of the Australian Academy of Science (1990), Prix Charles Leopold Mayer of the Academy des Sciences, Paris (1990), the Hansen Gold Medal of the Emil Christian Hansen Foundation, Denmark (1991), and the Wilhelm-Exner-Medaille of Vienna (1995). Several universities awarded him an honorary doctorate degree, including the Hebrew University, Israel (1994), Tel Aviv University, Israel (1997), University of East Anglia, Norwich, United Kingdom (1997), and University Louis Pasteur, Strasbourg (1992). One of his most treasured appointments was of Professeur Honoraire, College de France, Paris (1998).

One cannot look at plant science journals today and their back issues and not recognize that plant science has been transformed since the early 1980s following the activities of those few who believed that crown galls held some valuable biological secrets. Jeff Schell was very much at the helm as he went around the world giving such transforming talks, stimulating so many experiments, writing so many papers, informing politicians and industries, and taking the time here and there to confront "the greens" and skeptics. Two titles of his papers in recent years describe his vision: "Progress in plant science is our best hope to achieve an economically rewarding, sustainable and environmentally stable agriculture" (1995) and "Crop Biotechnology - a necessity for an environmentally friendly and sustainable agriculture" (1995). Let us hope that as we go forward, his vision will be realized.

LITERATURE CITED

Schell J (1995) Progress in plant science is our best hope to achieve an economically rewarding, sustainable and environmentally stable agriculture. Plant Tissue Culture and Biotechnology **1:** 10–12

Schell J (1995) Crop Biotechnology - a necessity for an environmentally friendly and sustainable agriculture. Proceedings Part 1. Ninth Forum for Applied Biotechnology University of Gent. **60:** 1513–1514

Richard Flavell
Chief Scientific Officer
Ceres-inc.
Malibu, California 90265

Shang Fa Yang

Professor of Vegetable Crops
UC Davis, California, USA

1932–2007

1991 — for his remarkable contributions to the understanding of the mechanism of biosynthesis, mode of action and applications of the plant hormone, Ethylene.

CURRICULUM VITAE

Education

B.S.	National Taiwan University	(in Agricultural Chemistry)	1956
M.S.	National Taiwan University	(in Agricultural Chemistry)	1958
Ph.D.	Utah State University	(in Plant Biochemistry)	1962

Postdoctoral Work

University of California Davis	(with Prof. P.K. Stumpt)	1962-63
New York University Medical School	(with Prof. B.N. LaDu)	1963-64
University of California, San Diego	(with Prof. AA. Benson)	1964-65

Research and Professional Experience

Assistant Biochemist (1966-1969), Associate Biochemist (1969-1974) to Professor and Biochemist (1974 to date), Chairman (1989-1990), Department of Vegetable Crops, University of California, Davis.

Visiting Professor: University of Konstanz, Germany, 1974; National Taiwan University, Taipei, 1983; University of Cambridge, England, 1983; Nagoya University, Japan, 1988-1989

Scientific Societies

American Society for Biochemistry and Molecular Biology, American Society of Plant Physiologists, American Society for Horticultural Science, Phytochemical Society of North America, International Plant Growth Substances Association.

Research Activities

Biosynthesis and biochemical action of ethylene in relation to plant senescence; postharvest biochemistry of fruits and vegetables; interaction of ethylene and other plant hormones on plant growth; biochemistry of sulfite and sulfur amino acids.

Honors

American Institute of Biological Sciences - Campbell Award, 1969
J.S. Guggenheim Fellow, 1982
International Plant Growth Substances Association Research Award, 1985
Member of the National Academy of Sciences, USA, 1990
Wolf Prize in Agriculture, 1991
University of California - Davis Faculty Research Lecturer, 1991
American Society for Horticultural Science Outstanding Research Award, 1992
Member of Academia Sinica, 1992

Professional Service

Editorial Board of Plant Physiology, 1974-1992
Associate Editor of Journal Plant Growth Regulation, 1981-1993
Editorial Board of Plant and Cell Physiology, 1987-1991
Editorial Board of Plant Physiology and Biochemistry, 1988-1993
Editorial Board of Acta Phytophysiologica Sinica, 1988-1993

Source: the Academic Senate of the University of California.

IN MEMORIAM

Shang Fa Yang, Emeritus Professor at the University of California, Davis, passed away suddenly and unexpectedly on February 12, 2007 at the age of 74. As the discoverer of the pathway for the biosynthesis of ethylene and the namesake of the Yang Cycle for the regeneration of methionine that initiates this pathway, he leaves a legacy of notable contributions to plant biochemistry and to the University of California.

Shang Fa Yang was born in 1932 in Taiwan where he received his B.S. and M.S. degrees in Agricultural Chemistry from the National University in the late 1950's. He received a scholarship to do graduate work at Utah State University and received his Ph.D. there in 1962 in Plant Biochemistry. He then came to the University of California, Davis, where he did postdoctoral work with Dr. Paul K. Stumpf on higher plants lipid metabolism. Shang Fa was eager to see the East Coast and obtained a fellowship at New York University Medical School where he met his wife, Eleanor, who was studying accounting at NYU. One year later, he returned to California and to plant biochemistry as a postdoctoral scientist at Scripps Institute of Oceanography in La Jolla, CA.

In 1966, Shang Fa was hired as an assistant biochemist in the Department of Vegetable Crops at the University of California, Davis. A strong advocate for his hiring was Harlan K. Pratt, a pioneering researcher in ethylene physiology with whom he initially shared a lab in the newly constructed Mann Laboratory. Shang Fa's early studies on ethylene were assisted by the homemade gas chromatograph that Pratt had cobbled together. Although large and cumbersome, that instrument could measure the parts-per-million concentrations of ethylene produced by plants and remained in use for over 35 years. Armed with this tool and his broad knowledge of chemistry and biochemistry, Shang Fa set out to explore plant ethylene biology.

Since 1934, when ethylene was conclusively shown to be produced by ripening fruit, considerable effort had been expended to discover how the volatile hormone was made in plants. The modern search for the ethylene biosynthetic pathway began in 1965 when it was observed that plants converted methionine to ethylene. This discovery led many scientists around the world on a quest to identify the subsequent steps in the pathway. Shang Fa's first paper on ethylene in 1966 (one of more than 200 publications in his career) explored the intricacies of an in vitro model system for the generation of ethylene from methionine, and his group made many important contributions during this period. In 1977, Shang Fa and his Ph.D. student, Douglas Adams, showed that methionine was converted to S-adenosylmethionine (SAM) and that SAM was a precursor of ethylene. Doug Adams also discovered that when tissues were kept under low oxygen conditions, a treatment known to suppress ethylene biosynthesis and thereby promote postharvest storage of fruits and vegetables, a metabolic intermediate accumulated. The pace quickened and a real race ensued among various labs to identify the intermediate between SAM and ethylene. This intensive effort culminated in 1979 when Adams and Yang identified 1-aminocyclopropane-1-carboxylic acid (ACC) as the final precursor of ethylene.

Assays for ACC were quickly developed and physiological studies into the regulation of ethylene biosynthesis accelerated. For example, Shang Fa's group demonstrated that under low oxygen conditions, such as root flooding, ACC could accumulate and be transported in the xylem to the shoot and subsequently converted to ethylene, serving as a translocatable signal of root stress to the shoots. His group also discovered that ACC could be conjugated to malonate, resulting in an alternative pool of ACC in plant tissues. It had been noted that methionine pools in plants are too low to sustain the observed rates of ethylene synthesis, but Shang Fa and his students resolved this controversy by demonstrating that the methylthio group released from SAM during the synthesis of ACC is recycled to replenish methionine levels. The reactions of this recycling pathway were subsequently christened the Yang Cycle in plant biochemistry texts. As the tools became available for cloning and characterizing the genes responsible for the steps in ethylene biosynthesis, Shang Fa contributed to many studies of the regulation of those genes in fruit ripening, plant growth, wounding and stress responses. He wrote numerous reviews and book chapters that defined ethylene biosynthesis and its role in plant biology for a generation of students and researchers.

In all his work, Shang Fa continually linked his discoveries to practical applications in postharvest biology and plant growth regulation. He used what he knew about physiology to learn more about ethylene biosynthesis, and he applied what he learned about ethylene biosynthesis to contribute to improvements in postharvest storage conditions. He was known for his clarity of thought and the ability to identify and design critical experimental tests of hypotheses. Shang Fa always maintained an open mind and was willing to challenge accepted ideas, even his own, when they proved untenable in the face of experimental data.

Shang Fa had an uncommon faith in humanity and urged his students to always expect the best of people. The coupling of an affable nature and a genuine concern for his students and colleagues enabled Shang Fa to assemble a powerful and effective research group that shared his vision and strove to match his intensity. He also developed an extensive and international network of friends and colleagues. Despite his many honors, he remained humble and always willing to share credit for the many discoveries coming out of his lab or to acknowledge the priority of other groups.

Shang Fa figured prominently at many national and international research conferences over the years and served on the editorial boards of leading journals and as a member of several learned societies. He won many awards and honors, including the Campbell Award of the American Institute of Biological Sciences in 1969, a Guggenheim Fellowship in 1982, the International Plant Growth Substances Association Research Award in 1985, and the Outstanding Researcher Award from the American Society of Horticultural Science in 1992. Shang Fa was named the UC Davis Faculty Research Lecturer in 1992. In 1990 and 1992, he was elected to the National Academy of Sciences, USA and to the Academia Sinica, Taiwan, respectively. In 1991, he received the prestigious international Wolf Prize in Agriculture.

After taking early retirement from University of California in 1994, Shang Fa served as Professor in the Department of Biology at Hong Kong University of Science and Technology from 1994 to 1997, where he established an active research group, and as a Distinguished Research Fellow and the Director of the Institute of Botany at Academia Sinica, Taipei, Taiwan. From 1996 to 1999, he was Vice President of the Academia Sinica and directed its numerous research institutes.

Shang Fa is survived by his wife Eleanor and two sons, Albert and Bryant, who have pursued careers in engineering and chemistry, respectively. While future plant biologists will know of Shang Fa through the Yang Cycle and his many other contributions to our field, students and colleagues who were fortunate enough to know him personally will also remember his humor, his humanity, and his sparkling intellect.

Kent J. Bradford
Alan B. Bennett
John M. Labavitch
Mikal E. Saltveit

List of Publications

1. Yang, S.F. and H.-K. Ho. 1958. Biochemical studies on post-ripening of banana. J. Chinese Chem. Soc. 5: 71-85.
2. Yang. S.F. and G.W. Miller. 1963. Biochemical studies on the effect of fluoride on higher plants. 1. Metabolism of carbohydrates, organic acids and amino acids. Biochem. J. 88: 505-509.
3. Yang, S.F. and G.W. Miller. 1963. Biochemical studies on the effect of fluoride on higher plants. 2. The effect of fluoride on sucrose-synthesizing enzymes from higher plants. Biochem. J. 88: 509-516.
4. Yang, S.F. and G.W. Miller. 1963. Biochemical studies on the effect of fluoride on higher plants. 3. The effect of fluoride on dark carbon dioxide fixation. Biochem. J. 88: 517-522.
5. Yang, S.F. and P.K. Stumpf. 1965. Fat metabolism in higher plants. XXI. Biosynthesis of fatty acids by avocado mesocarp enzyme systems. Biochim. Biophys. Acta 98: 19-26.
6. Yang, S.F. and P.K. Stumpf. 1965. Fat metabolism in higher plants. XXll. Enzymic synthesis of 14-hydroxyl-11-eicosenoic acid by particulate preparations of avocado mesocarp. Biochim. Biophys. Acta 98: 27-35.

7. Yang, S.F., H.S. Ku, and H.K. Pratt. 1966. Ethylene production from methionine as mediated by flavin mononucleotide and light. Biochem. Biophys. Res. Commun. 24: 739-743.
8. Yang, S.F., S. Freer, and A.A. Benson. 1967. Transphosphatidylation by phospholipase D. J. Biol. Chem. 242: 477-484.
9. Ku, H.S., S.F. Yang, and H.K. Pratt. 1967. Enzymic evolution of ethylene from methional by a pea seedling extract. Arch. Biochem. Biophys. 118: 756-758.
10. Welkie, G.W., S.F. Yang, and G.W. Miller. 1967. Metabolite changes induced by cucumber mosaic virus in resistant and susceptible strains of cowpea. Phytopathology 57: 472-475.
11. Rappaport, L., A. Hsu, R. Thompson, and S.F. Yang. 1967. Fate of ^{14}C-gibberellin A3 in plant tissues. Annals of the New York Academy of Sciences 144: 211-218.
12. Yang, S.F., H.S. Ku, and H.K. Pratt. 1967. Photochemical production of ethylene from methionine and its analogues in the presence of flavin mononucleotide. J. Biol. Chem. 242: 5274-5280.
13. Yang, S.F. 1967. Biosynthesis of ethylene. Ethylene formation from methional by horseradish peroxidase. Arch. Biochem. Biophys. 122: 481-487.
14. Yang, S.F. 1968. Biosynthesis of ethylene. *In* P. Wightman, and G. Setterfield (eds.), Biochemistry and Physiology of Plant Growth Substances. The Runge Press. Ltd., Ottawa. pp. 1217-1288.
15. Yang, S.F. 1969. Phospholipase D from savoy cabbage. *In* S.P. Colowick, and N.O. Kaplan (eds.), Methods in Enzymology. Academic Press. New York. pp. 208-211.
16. Ku, H.S., S.F. Yang, and H.K. Pratt. 1969. Ethylene formation from *alpha*-keto-*gamma*-methylthiobutryrate by tomato fruit extracts. Phytochemistry 8: 567-573.
17. Baur, A. and S.F. Yang. 1969. Ethylene production from propanal. Plant Physiol. 44: 189-192.
18. Yang, S.F. 1969. Further studies on ethylene formation from *alpha*-keto-*gamma*-methylthiobutyric acid or *beta*-methylthiopropionaldehyde by peroxidase in the presence of sulfite and oxygen. J. Biol. Chem. 244: 4360-4365.
19. Yang, S.F. 1969. Ethylene evolution from 2-chloroethylphosphonic acid. Plant Physiol. 44: 1203-1204.
20. Baur, A.H. and S.F. Yang. 1969. Precursors of ethylene. Plant Physiol. 44: 1347-1349.
21. Yang, S.F. and A.H. Baur. 1969. Pathways of ethylene biosynthesis. Qualitas Plantarum Materiae Vegetabiles 19: 201-220.
22. Yang, S.F. 1969. A simple method for assay of ^{14}C-formic acid. Anal. Biochem. 32: 519-521.

23. Ku, H.S., S.F. Yang, and H.K. Pratt. 1970. Inactivity of apoperoxidase in indoleacetic acid oxidation and in ethylene formation. Plant Physiol. 45: 358-359.
24. Ku, H.S., S.F. Yang, and H.K. Pratt. 1970. Ethylene production and perioxidase activity during tomato ripening. Plant Cell Physiol. 11: 241-246.
25. Yang, S.F. 1970. Photosensitized conversion of ethionine to ethylene by flavin mononucleotide. Photochem. Photobiol. 12: 419-422.
26. Yang, S.F. 1970. Sulfoxide formation from methionine or its sulfide analogs during aerobic oxidation of sulfite. Biochemistry 9: 5008-5014.
27. Hyodo, H. and S.F. Yang. 1971. Ethylene-enhanced synthesis of phenylalanine ammonia-lyase in pea seedlings. Plant Physiol. 47: 765-770.
28. Baur, A.H., S.F. Yang, and H.K. Pratt. 1971. Ethylene biosynthesis in fruit tissues. Plant Physiol. 47: 696-699.
29. Hyodo, H. and S.F. Yang. 1971. Ethylene-enhanced formation of cinnamic acid 4-hydroxylase in excised pea epicotyl tissue. Arch. Biochem. Biophys. 143: 338-339.
30. Yamaguchi, M., C.W. Chu, and S.F. Yang. 1971. The fate of ^{14}C (2-chloroethyl) phosphonic acid in summer squash, cucumber, and tomato. J. Amer. Soc. Hort. Sci. 96: 606-609.
31. Yang, S.F. and A.H. Baur. 1972. Biosynthesis of ethylene in fruit tissues. *In* D.J. Carr (ed.), Plant Growth Substances 1970, Springer-Verlag, Heidelberg, pp. 510-517.
32. Baur, A.H. and S.F. Yang. 1972. Formation of ethionine from homocysteine and of S-methylmethionine in apple tissue. Phytochemistry 11: 2503-2502.
33. Baur, A.H. and S.F. Yang. 1972. Methionine metabolism in apple tissue in relation to ethylene biosynthesis. Phytochemistry 11: 3207-3214.
34. Yang, S.F. and M.A. Saleh. 1973. Destruction of indole-3-acetic acid during the aerobic oxidation of sulfite. Phytochemistry 12: 1463-1466.
35. Lau, O-L. and S.F. Yang. 1973. Mechanism of a synergistic effect of kinetin on auxin-induced ethylene production: suppression of auxin conjugation. Plant Physiol. 51: 1011-1014.
36. Horng, A.J. and S.F. Yang. 1973. Peroxidase-catalyzed oxidation of indole-3-acetaldehyde to 4-hydroxyquinoline in the presence of bisulfite ion: elimination of pyrrole ring C2 as formic acid. Biochim. Biophys. Acta 321: 456-460.
37. Yang, S.F. 1973. Destruction of tryptophan during the aerobic oxidation of sulfite ions. Environ. Res. 6: 395-402.
38. Chou, T.W. and S.F. Yang. 1973. The biogenesis of ethylene in *Penicillium digitatum.* Arch. Biochem. Biophys. 157: 73-82.
39. Hyodo, H. and S.F. Yang. 1974. The effect of ethylene on the development of phenylalanine ammonia-lyase in potato tuber disks. Zeitschrift fur Pflanzenphysiologie 71: 76-79.

40. Yang, S.F. 1974. The biochemistry of ethylene: biogenesis and metabolism. *In* V.C. Runeckles (ed.), Recent Advances in Phytochemistry, Vol. 7. The Chemistry and Biochemistry of Plant Hormones. Academic Press, New York, pp. 131-164.
41. Lau, O-L. and S.F. Yang. 1974. Synergistic effect of calcium and kinetin on ethylene production by the mungbean hypocotyl. Planta 118: 1-6.
42. Mukerji, S.K. and S.F. Yang. 1974. Phosphoenolpyruvate carboxylase from spinach leaf tissue. Plant Physiol. 53: 829-834.
43. Lau, O-L., S.F. Yang, and K-H. Yung. 1974. Suppression of IAA conjugation by kinetin in relation to a synergistic effect of kinetin on IAA-induced ethylene production. *In* Plant Growth Substances 1973. Hirokawa Publishing Company, Tokyo, pp. 364-393.
44. Lau, O-L., D.P. Mure, and S.F. Yang. 1974. Effect of 2,4-dinitrophenol on auxin-induced ethylene production and auxin conjugation by mung bean tissue. Plant Physiol. 54: 182-185.
45. Yang, S.F. 1974. Ethylene biosynthesis in fruit tissues. *In* Facteurs et Regulation de la Maturation des Fruits. Centre National de la Recherche Scientifique, No. 238, Paris, pp. 171-175.
46. Mure, D.P. and S.F. Yang. 1975. Inhibition of *in vivo* conversion of methionine to ethylene by L-canaline and 2,4-dinitrophenol. Plant Physiol. 55: 79-82.
47. Bowen, J.R. and S.F. Yang. 1975. Photosensitized deamination of sulfur-amino acids by flavin mononucleotide. Photochem. Photobiol. 21: 201-203.
48. Lau, O-L. and S.F. Yang. 1975. Interaction of kinetin and calcium in relation to their effect on stimulation of ethylene production. Plant Physiol. 55: 738-740.
49. Horng, A.J. and S.F. Yang. 1975. Aerobic oxidation of indole-3-acetic acid with bisulfite. Phytochemistry 14: 1425-1428.
50. Mure, D.P. and S.F. Yang. 1975. Conversion of 5-methylthioadenosine to methionine by apple tissue. Phytochemistry 14: 1291-1292.
51. Lau, O-L. and S.F. Yang. 1976. Stimulation of ethylene production in mung bean hypocotyls by cupric ion, calcium ion, and kinetin. Plant Physiol. 57: 88-92.
52. Lau, O-L. and S.F. Yang. 1976. Inhibition of ethylene production by cobaltous ion. Plant Physiol. 58: 114-117.
53. Lau, O-L., W.W. John, and S.F. Yang. 1977. Effect of different cytokinins on ethylene production by mung bean hypocotyls in the presence of indole-3-acetic acid or calcium ion. Physiol. Plant. 39: 1-3.
54. Peiser, G.D. and S.F. Yang. 1977. Chlorophyll destruction by the bisulfite-oxygen system. Plant Physiol. 60: 277-281.
55. Adams, D.O. and S.F. Yang. 1977. Methionine metabolism in apple tissue. Implication of S-adenosylmethionine as an intermediate in the conversion of methionine to ethylene. Plant Physiol. 60: 893-896.

56. Peiser, G.D. and S.F. Yang. 1978. Chlorophyll destruction in the presence of bisulfite and linonenic acid hydroperoxide. Phytochemistry 17: 79-84.
57. Lau, O-L., W.W. John, and S.F. Yang. 1978. Inactivity of oxidation products of indole-3-acetic acid on ethylene production in mung bean hypocotyls. Plant Physiol. 61: 68-71.
58. Hyodo, H., H. Kuroda, and S.F. Yang. 1978. Induction of phenylalanine ammonia-lyase and increase in phenolics in lettuce leaves in relation to the development of russet spotting caused by ethylene. Plant Physiol. 62: 31-35.
59. Yang, S.F. and H.K. Pratt. 1978. The physiology of ethylene in wounded plant tissue. *In* G. Kahl (ed.), Biochemistry of Wounded Plant Tissues. De Gruyter Co., Berlin, pp. 595-622.
60. Yu, Y-B., D.O. Adams, and S.F. Yang. 1979. Regulation of auxin-induced ethylene production in mung bean hypocotyls. Plant Physiol. 63: 589-590.
61. Peiser, G.D. and S.F. Yang. 1979. Ethylene and ethane production from sulfur dioxide-injured plants. Plant Physiol. 63: 142-145.
62. Yu, Y-B. and S.F. Yang. 1979. Auxin-induced ethylene production and its inhibition by aminoethoxy-vinylglycine and cobalt ion. Plant Physiol. 64: 1074-1077.
63. Lizada, M.C.C. and S.F. Yang. 1979. A simple and sensitive assay for l-aminocyclopropane-l-carboxylic acid. Anal. Biochem. 100: 140-145.
64. Adams, D.O. and S.F. Yang. 1979. Ethylene biosynthesis: Identification of 1-aminocyclopropane-1-carboxylic acid as an intermediate in the conversion of methionine to ethylene. Proc. Natl. Acad. Sci. USA 76: 170-174.
65. Peiser, G.D. and S.F. Yang. 1979. Sulfite-mediated destruction of *beta*-carotene. J. Agr. Food Chem. 27: 446-449.
66. Cameron, A.C., C.A. L. Fenton, Y. Yu, D.O. Adams, and S.F. Yang. 1979. Increased production of ethylene by plant tissues treated with l-aminocyclopropane-l-carboxylic acid. HortScience 14: 178-180.
67. Yu, Y-B., D.O. Adams, and S.F. Yang. 1979. 1-aminocyclopropane-carboxylate synthase, a key enzyme in ethylene biosynthesis. Arch. Biochem. Biophys. 198: 280-286.
68. Bradford, K.J. and S.F. Yang. 1980. Xylem transport of 1-amino-cyclopropane-1-carboxylic acid, an ethylene precursor, in waterlogged tomato plants. Plant Physiol. 65: 322-326.
69. Bradford, K.J. and S.F. Yang. 1980. Stress-induced ethylene production in the ethylene-requiring tomato mutant *diageotropica*. Plant Physiol. 65: 327-330.
70. Hoffman, N.E. and S.F. Yang. 1980. Changes of 1-aminocyclopropane-1-carboxylic acid content in ripening fruits in relation to their ethylene production rates. J. Amer. Soc. Hort. Sci. 105: 492-495.
71. Yu, Y-B. and S.F. Yang. 1980. Biosynthesis of wound ethylene. Plant Physiol. 66: 281-285.

72. Yu, Y-B., D.O. Adams, and S.F. Yang. 1980. Inhibition of ethylene production by 2,4-dinitrophenol and high temperature. Plant Physiol. 66: 286-290.
73. Yang, S.F. and D.O. Adams. 1980. Biosynthesis of ethylene. *In* P. K. Stumpf (ed.), The Biochemistry of Plants. Vol. IV Lipids: Structure and Function. Academic Press, New York, pp. 163-175.
74. Yang, S.F., D.O. Adams, C. Lizada, Y. Yu, K.J. Bradford, A.C. Cameron, and N.E. Hoffman. 1980. Mechanism and regulation of ethylene biosynthesis. *In* F. Skoog (ed.), Plant Growth Substances 1979. Springer-Verlag. Berlin. pp. 219-229.
75. Fujino, D.W., M.S. Reid, and S.F. Yang. 1980. Effects of aminooxyacetic acid on postharvest characteristics of carnation. Acta Horticulturae 113: 59-64.
76. Yang, S.F. 1980. Regulation of ethylene biosynthesis. HortScience 15: 238-243.
77. Bufler, G., Y. Mor, M.S. Reid, and S.F. Yang. 1980. Changes in 1-aminocyclopropane-1-carboxylic-acid content of cut carnation flowers in relation to their senescence. Planta 150: 439-442.
78. Yu, Y-B, S.F. Yang, J. Corse, J.A. Kuhnle, and S-S. Hua. 1980. Structures of cytokinins influence synergistic production of ethylene. Phytochemistry 20: 1191-1195.
79. Lizada, M.C.C. and S.F. Yang. 1981. Sulfite-induced lipid peroxidation. Lipids 15: 189-194.
80. Bradford, K.J. and S.F. Yang. 1981. Physiological responses of plants to waterlogging. HortScience 16: 25-30.
81. Apelbaum, A. and S.F. Yang. 1981. Biosynthesis of stress ethylene induced by water deficit. Plant Physiol. 68: 594-596.
82. Adams, D.O. and S.F. Yang. 1981. Ethylene the gaseous plant hormone: mechanism and regulation of biosynthesis. Trends Biochem. Sci. 6: 161-164.
83. Yang, S.F. 1981. Biosynthesis of ethylene and its regulation. *In* J. Friend (ed.), Recent Advances in the Biochemistry of Fruits and Vegetables. Academic Press. London. pp. 98-106.
84. Hoffman, N.E., S.F. Yang, and T. McKeon. 1982. Identification of 1-(malonylamino)cyclopropane-1-carboxylic acid as a major conjugate of 1-aminocyclopropane-1-carboxylic acid, an ethylene precursor in higher plants. Biochem. Biophys. Res. Commun. 104: 765-770.
85. Yung, K.H., S.F. Yang, and F. Schlenk. 1982. Methionine synthesis from 5-methylthioribose in apple tissue. Biochem. Biophys. Res. Commun. 104: 771-777.
86. Hoffman, N.E. and S.F. Yang. 1982. Enhancement of wound-induced ethylene synthesis by ethylene in preclimacteric cantaloupe. Plant Physiol. 69: 317-322.

87. Riov, J. and S.F. Yang. 1982. Autoinhibition of ethylene production in citrus peel discs: suppression of 1-aminocyc1opropane-1-carboxylic acid synthesis. Plant Physiol. 69: 687-690.
88. Yang, S.F., N.E. Hoffman, T. McKeon, J. Riov, C.H. Kao, and K.H. Yung. 1982. Mechanism and regulation of ethylene biosynthesis. *In* P.P. Wareing (ed.), Plant Growth Substances, Academic Press, London, pp. 239-248.
89. Peiser, G.D., M.C.C. Lizada, and S.F. Yang. 1982. Sulfite-induced lipid peroxidation in chloroplasts as determined by ethane production. Plant Physiol. 70: 994-998.
90. Riov, J. and S.F. Yang. 1982. Effect of cycloheximide on ethylene production in intact and excised citrus fruit tissues. J. Plant Growth Regul. 1: 95-104.
91. Cameron, A.C. and S.F. Yang. 1982. A simple method for the determination of resistance to gas diffusion in plant organs. Plant Physiol. 70: 21-23.
92. Riov, J. and S.F. Yang. 1982. Effects of exogenous ethylene on ethylene production in citrus leaf tissue. Plant Physiol. 70: 136-141.
93. Riov, J. and S.F. Yang. 1982. Stimulation of ethylene production in citrus leaf discs by mannitol. Plant Physiol. 70: 142-146.
94. Hoffman, N.E., S.F. Yang, A. Ichihara, and S. Sakamura. 1982. Stereospecific conversion of 1-aminocyclopropane-carboxylic acid to ethylene by plant tissue: conversion of stereoisomers of 1-amino-2-ethylenecyclopropane-carboxylic acid to 1-butene. Plant Physiol. 70: 195-199.
95. Peiser, G.D., M.C.C. Lizada, and S.F. Yang. 1982. Dark metabolism of carbon dioxide and lettuce leaf discs. Plant Physiol. 70: 397-400.
96. Kao, C.H. and S.F. Yang. 1982. Light inhibition of the conversion of 1-aminocyclopropane-1-carboxylic acid to ethylene in leaves is mediated through carbon dioxide. Planta 155: 261-266.
97. McKeon, T.A., N.E. Hoffman, and S.F. Yang. 1982. The effect of plant hormone pretreatments on ethylene production and synthesis of 1-aminocyclopropane-1-carboxylic acid in water-stressed wheat leaves. Planta 155: 437-443.
98. Bradford, K.J., T.C. Hsiao, and S.F. Yang. 1982. Inhibition of ethylene synthesis in tomato plants subjected to anaerobic root stress. Plant Physiol. 70: 1503-1507.
99. Yang, S.F. 1983. Mechanism and regulation of ethylene biosynthesis. *In* T. Akazawa, T. Asahi and H. Imaseki (eds.), The New Frontiers in Plant Biochemistry, Japan Scientific Societies Press, Tokyo, pp. 133-151.
100. Hoffman, N.E., Y. Liu, and S.F. Yang. 1983. Changes in 1-(malonylamino) cyc1opropane-1-carboxylic acid content in wilted wheat leaves in relation to their ethylene production rates and 1-aminocyclopropane-1-carboxylic acid content. Planta 157: 518-523.
101. Hoffman, N.E., J-R. Fu, and S.F. Yang. 1983. Identification and metabolism of 1-(malonyl-amino)cyclopropane-1-carboxylic acid in germinating peanut seeds. Plant Physiol. 71: 197-199.

102. Liu, Y., N.E. Hoffman, and S.F. Yang. 1983. Relationship between the malonylation of 1-aminocyclopropane-1-carboxylic acid and D-amino acids in mung-bean hypocotyls. Planta 158: 437-441.
103. Yang, S.F. 1983. Regulation of ethylene biosynthesis in relation to plant senescence. Bulletin Plant Growth Regulator 11: 9-11.
104. Aharoni, N. and S.F. Yang. 1983. Auxin-induced ethylene production as related to auxin metabolism in leaf discs of tobacco and sugar beet. Plant Physiol. 73: 598-604.
105. Fu, J.R. and S.F. Yang. 1983. Release of heat pretreatment-induced dormancy in lettuce seeds by ethylene or cytokinin in relation to the production of ethylene and the synthesis of 1-aminocyclopropane-1-carboxylic acid during germination. J. Plant Growth Regul. 2: 185-192.
106. McKeon, T.A. and S.F. Yang. 1983. A comparison of the conversion of 1-amino-2-ethylcyclopropane-1-carboxylic acid stereoisomers to 1-butene by pea epicotyls and by a cell free system. Planta 160: 84-87.
107. Kao, C.H. and S.F. Yang. 1983. Role of ethylene in the senescence of detached rice leaves. Plant Physiol. 73: 881-885.
108. Sisler, E.C. and S.F. Yang. 1984. Ethylene, the gaseous plant hormone. Bioscience 34: 234-238.
109. Beyer, E.M., P.W. Morgan, and S.F. Yang. 1984. Ethylene. *In* M.B. Wilkins (ed.), Advanced Plant Physiology. Pitman, London, pp. 111-126.
110. Yang, S.F. and N.E. Hoffman. 1984. Ethylene biosynthesis and its regulation in higher plants. Ann. Rev. Plant Physiol. 35: 155-189.
111. Peiser, G.D., T. Wang, N.E. Hoffman, S.F. Yang, H. Liu, and C.T. Walsh. 1984. Formation of cyanide from carbon 1 of 1-aminocyclopropane-1-carboxylic acid during its conversion to ethylene. Proc. Natl. Acad. Sci. USA 81: 3059-3063.
112. Liu, Y., L. Su, and S.F. Yang. 1984. Metabolism of 1-aminoisobutyric acid in mungbean hypocotyls in relation to metabolism of 1-aminocyclopropane-1-carboxylic acid. Planta 161: 439-443.
113. Lau, O.L., Y. Liu, and S.F. Yang. 1984. Influence of storage atmospheres and procedures on 1-aminocyclopropane-1-carboxylic acid concentration in relation to flesh ripeness in 'Golden Delicious' apple. HortScience 19: 425-426.
114. Su, L., T. McKeon, D. Grierson, M. Cantwell, and S.F. Yang. 1984. Development of 1-aminocyclo-propane-1-carboxylate synthase and polygalacturonase activities during the maturation and ripening of tomato fruits in relation to their ethylene production rates. HortScience 19: 576-578.
115. Yang, S.F. 1984. The formation of ethylene from 1-aminocyclopropane-1-carboxylic acid. *In* R.Y. Fuchs, and E. Chalutz (eds.), Ethylene: Biochemical, Physiological and Applied Aspects. Martinus Nijhoff/Dr. W. Junk Publishers, The Netherlands, pp. 1-10.

116. Yang, S.F. 1984. Reactions of oxidation intermediates of sulphite species with some cellular components of plants. Food Chem. 15: 113-124.
117. Sisler, E.C. and S.F. Yang. 1984. Anti-ethylene effects of cis-2-butene and cyclic olefins. Phytochemistry 23: 2765-2768.
118. Liu, Y., L. Su, and S.F. Yang. 1984. Stereoselectivity of 1-aminocyclopropanecarboxylate malonyl-transferase toward stereioisomers of 1-amino-2-ethylcyclopropanecarboxylic acid. Arch. Biochem. Biophys. 235: 319-325.
119. Liu, Y., N.E. Hoffman, and S.F. Yang. 1985: Promotion by ethylene of the capacity to convert 1-aminocyclopropane-1-carboxylic acid to ethylene in preclimacteric tomato and cantaloupe fruits. Plant Physiol. 77: 407-411.
120. Yang, S.F. 1985. Biosynthesis and action of ethylene. HortScience 20: 41-45.
121. Peiser, G. and S.F. Yang. 1985. Metabolism of sulfur dioxide in 'Thompson Seedless' grape berries. J. Amer. Soc. Hort. Sci. 110: 224-226.
122. Yang, S.F., Y. Liu, L. Su, G.D. Peiser, N.E. Hoffman, and T. McKeon. 1985. Metabolism of 1-aminocyclopropane-1-carboxylic acid. *In* G.A. Tucker and J.A. Roberts (eds.), Ethylene in Plant Development. Buttersworth, London, pp. 9-13.
123. Su, L., Y. Liu, and S.F. Yang. 1985. Relationship between l-aminocyclopropanecarboxylate malonyltransferase and D-amino acid malonyltransferase. Phytochemistry 24: 1141-1145.
124. Liu, Y., L. Su, and S.F. Yang. 1985. Ethylene promotes the capability to malonylate 1-aminocyclopropane-1-carboxylic acid and D-amino acids in preclimacteric tomato fruits. Plant Physiol. 77: 891-895.
125. Liu. Y., N.E. Hoffman, and S.F. Yang. 1985. Ethylene-promoted malonylation of 1-aminocyclopropane-1-carboxylic acid participates in autoinhibition of ethylene synthesis in grapefruit flavedo discs. Planta 164: 565-568.
126. Yang. S.F. 1985. Biosynthesis of ethylene. *In* D.P. Randall (ed.), Current Topics in Plant Biochemistry and Physiology. Volume 4. Univ. of Missouri-Columbia. pp. 126-138.
127. Peiser, G. and S.F. Yang. 1985. Biochemical and physiological effects of SO_2 on nonphotosynthetic processes in plants. *In* W.E. Winner, H.A. Mooney and R.A. Goldstein (eds.), Sulfur Dioxide and Vegetation. Stanford University Press, Stanford, California, pp. 148-161.
128. Yip, W-K. and S.F. Yang. 1986. Effect of thidiazuron, a cytokinin active urea derivative, in cytokinin-dependent ethylene production systems. Plant Physiol. 80: 515-519.
129. Lau, O.L., Y. Liu, and S.F. Yang. 1986. Effects of fruit detachment on ethylene biosynthesis and loss of flesh firmness, skin color and starch in ripening 'Golden Delicious' apples. J. Amer. Soc. Hort. Sci. 111: 731-734.

130. Sisler, E.C., M.S. Reid, and S.F. Yang. 1986. Effect of antagonists of ethylene action on binding of ethylene in cut carnations. Plant Growth Regul. 4: 213-218.
131. Nieder, M., W-K. Yip, and S.F. Yang. 1986. Interferences and specificity of the 1-aminocyclopropane-1-carboxylic acid assay with the hypochlorite reagent. Plant Physiol. 81: 156-160.
132. Yang, S.F., Y. Liu, and O.L. Lau. 1986. Regulation of ethylene biosynthesis in ripening apple fruits. Acta Horticulturae 197: 711-710.
133. Jiao, X-Z., S. Philosoph-Hadas, L-Y. Su, and S.F. Yang. 1986. The conversion of 1-(malonylamino)cyclopropane-1-carboxylic acid to l-aminocyclopropane-1-carboxylic acid in plant tissues. Plant Physiol. 81: 637-641.
134. Tong, C.B., J.M. Labavitch, and S.F. Yang. 1986. The induction of ethylene production from pear cell culture by cell wall fragments. Plant Physiol. 81: 929-930.
135. Philosoph-Hadas, S., N. Aharoni, and S.F. Yang. 1986. Carbon dioxide enhances the development of the ethylene forming enzyme in tobacco leaf discs. Plant Physiol. 82: 925-929.
136. Yang. S.F. 1986. Regulation of plant growth by ethylene and related regulators. *In* Plant Growth Regulators in Agriculture. Food and Fertilizer Technology Center, Taipei, pp. 210-219.
137. Miyazaki, J.H. and S.F. Yang. 1987. The methionine salvage pathway in relation to ethylene and polyamine biosynthesis. Physiol. Plant. 69: 366-370.
138. Wang, T-T. and S.F. Yang. 1987. The physiological role of lipoxygenase in ethylene formation from 1-aminocyclopropane-1-carboxylic acid in oat leaves. Planta 170: 190-196.
139. Miyazaki, J.H. and S.F. Yang. 1987. Metabolism of 5-methylthioribose to methionine. Plant Physiol. 84: 277-281.
140. Yang. S.F. 1987. The role of ethylene synthesis in fruit ripening. *In* W.W. Thomson, E.A. Nothnagel and R.C. Huffaker (eds.), Plant Senescence, Its Biochemistry and Physiology. Amer. Soc. Plant Physiologists. pp. 156-166.
141. Saltveit, M.E., Jr. and S.F. Yang. 1987. Methods in ethylene analysis. *In* A Crozeir (ed.), Principles and Practice of Plant Hormone Analysis. Academic Press, London, pp. 367-401.
142. Yang. S.F. 1987. Regulation of biosynthesis and action of ethylene. Acta Hortic. 201: 53-59.
143. McKeon, T.A. and S.F. Yang. 1987. Biosynthesis and metabolism of ethylene. *In* P.I. Davis (ed.), Plant Hormones and Their Role in Plant Growth and Development. Martinus Nijhoff, pp. 94-112.
144. Miyazaki, J.H. and S.F. Yang. 1987. Inhibition of the methionine cycle enzymes. Phytochemistry 26: 2655-2660.

145. Jiao, X-Z., W-K. Yip, and S.F. Yang. 1987. The effect of light and phytochrome on 1-aminocyclopropane-1-carboxylic acid metabolism in etiolated wheat seedling leaves. Plant Physiol. 85: 643-647.
146. Tong, C.B.S. and S.F. Yang. 1987. Chilling-induced ethylene production by beans and peas. J. Plant Growth Regul. 6: 201-208.
147. Sisler, E.C. and S.F. Yang. 1987. Ethylene. *In* D.W. Newman and K.G. Wilson (eds.), Model in Plant Physiology/Biochemistry/Technology. CRC Press, Vol II, Chapter 20, pp. 99-102.
148. Adams, D.O. and S.F. Yang. 1987. 1-aminocyclopropane-1-carboxylate synthase. *In* W.B. Jakoby and O.W. Griffith (eds.), Methods in Enzymology-Sulfur and Sulfur Amino Acids. Academic Press, pp. 426-429.
149. Yang, S.F. 1987. The biosynthesis and metabolism of 1-aminocyclopropane-1-carboxylic acid in relation to ethylene production. *In* K. Schreiber, H.R. Schute and G. Sembdner (eds.), Conjugated Plant Hormones-Structure, Metabolism and Function. Institute fur Biochemie der Pflanzen der Akademie der Wissenschaften der DDR, East Germany, pp. 92-101.
150. Yip, W-K. and S.F. Yang. 1988. Cyanide metabolism in relation to ethylene production in plant tissues. Plant Physiol. 88: 473-476.
151. Yip, W-K., X-Z. Jiao, and S.F. Yang. 1988. Dependence of *in vivo* ethylene production rate on 1-aminocyclopropane-1-carboxylic acid content and oxygen concentrations. Plant Physiol. 88: 553-558.
152. Satoh, S. and S.F. Yang. 1988. S-adenosylmethionine-dependent inactivation and radiolabeling of 1-aminocyclopropane-1-carboxylate synthase isolated from tomato fruits. Plant Physiol. 88: 109-114.
153. Fujino, D.W., D.W. Burger, S.F. Yang, and K.J. Bradford. 1988. Characterization of an ethylene overproducing mutant of tomato *(Lycopersicon esculentum)* Mill. cultivar VFN8. Plant Physiol. 88: 774-779.
154. Satoh, S. and S.F. Yang. 1989. Specificity of S-adenosyl-L-methionine in the inactivation and the labeling of 1-aminocyclopropane-1-carboxylate synthase isolated from tomato fruits. Arch. Biochem. Biophys. 271: 107-112.
155. Riov, J. and S.F. Yang. 1989. Enhancement of adventitious root formation in mung bean cuttings by 3,5-dihalo-4-hydroxybenzoic acids and 2,4-dinitrophenol. Plant Growth Regul. 8: 277-281.
156. Riov, J. and S.F. Yang. 1989. Ethylene and auxin-ethylene interaction in adventitious root formation in mung bean *(Vigna radiata)* cuttings. J. Plant Growth Regul. 8: 131-141.
157. Wien, H.C., A.D. Turner, and S.F. Yang. 1989. Hormonal basis for low light intensity-induced flower bud abscission of pepper. J. Amer. Soc. Hort. Sci. 114: 981-985.
158. Satoh, S. and S.F. Yang. 1989. Inactivation of 1-aminocyclopropane-1-carboxylate synthase by L-vinylglycine as related to the mechanism based

inactivation of the enzyme by S-adenosyl-L-methionine. Plant Physiol. 91: 1036-1039.

159. Quinn, J.M. and S.F. Yang. 1989. Ethylene-like activity of isocyanides. Plant Physiol. 91: 669-673.
160. Yang, S.F. 1989. Metabolism of 1-aminocyclopropane-1-carboxylic acid in relation to ethylene biosynthesis. *In* J.E. Poulton, J.T. Romeo, and E.E. Conn (eds.), Recent Advances in Phytochemistry, Vol. 23, Plant Nitrogen Metabolism, Plenum Press, New York, pp. 263-287.
161. Satoh, S. and S.F. Yang. 1989. S-Adenosylmethionine-dependent inactivation of l-aminocyclopropane-1-carboxylate synthase isolated from tomato fruits. *In* H. Clijsters (ed.), Biochemical and Physiological Aspects of Ethylene Production in Lower and Higher Plants, Kliewer Academic Publishers, pp. 41-48.
162. Lafuente, M.T., M. Cantwell, S.F. Yang, and V. Rubatzky. 1989. Isocoumarin content of carrots as influenced by ethylene concentration, storage temperature and stress conditions. Acta Horticulturae 258: 523-534
163. Hammer, P.E., S.F. Yang, M.S. Reid, and J.J. Marois. 1990. Postharvest control of *Botrytis cinerea* infections on cut roses using fungistatic storage atmospheres. J. Amer. Soc. Hort. Sci. 115: 102-107.
164. Yang, S.F., W-K. Yip, S. Satoh, J.H. Miyazaki, X. Jiao, Y. Liu, L-Y. Su, and G.D. Peiser. 1990. Metabolic aspects of ethylene biosynthesis. *In* R.P. Pharis and S. Rood (eds.), Plant Growth Substance 1988, Springer-Verlag, Heidelberg, pp. 291-299.
165. Riov, J., E. Dagan, R. Goren, and S.F. Yang. 1990. Characterization of abscisic acid-induced ethylene production in citrus leaf and tomato fruit tissues. Plant Physiol. 92: 48-53.
166. Yip, W-K., J-G. Dong, J.W. Kenny, G.A. Thompson, and S.F. Yang. 1990. Characterization and sequencing of the active site of 1-aminocyclopropane-1-carboxylate synthase. Proc. Natl. Acad. Sci. USA 87: 7930-7934.
167. Yang, S.F., W-K. Yip, and J-G. Dong. 1990. Mechanism and regulation of ethylene biosynthesis. *In* H.E. Flores, R.N. Arteca and J.C. Shannon (eds.), Polyamines and Ethylene: Biochemistry, Physiology and Interactions, Amer. Soc. of Plant Physiologists, pp. 24-35.
168. Yip, W-K., J-G. Dong, and S.F. Yang. 1991. Purification and characterization of 1-aminocyclopropane-carboxylate synthase from apple fruits. Plant Physiol. 95: 251-257.
169. Dong, J-G., W-K. Yip, and S.F. Yang. 1991. Monoclonal antibodies against apple 1-aminocyclopropane-1-carboxylate synthase. Plant Cell Physiol. 32: 25-31.
170. Satoh, S. and S.F. Yang. 1991. Recent progress in research of 1-aminocyclopropane-1-carboxylate synthase, the key enzyme in ethylene biosynthesis. J. Pesticide Sci. 16: 101-108.

171. Dong, J-G., W.T. Kim, W.K. Yip, G.A. Thompson, L. Li, A.B. Bennett, and S.F. Yang. 1991. Cloning of a cDNA encoding 1-aminocyclopropane-1-carboxylate synthase and expression of its mRNA in ripening apple fruit. Planta 185: 38-45.
172. Yang, S.F., W-K. Yip, J-G. Dong, and S. Satoh. 1991. Characterization of 1-aminocyclopropane-1-carboxylate (ACC) synthase, a key enzyme in ethylene biosynthesis: active site peptide sequencing. *In* T. Fukui, H. Kagamiyama, K. Soda, and H. Wada (eds.), Enzymes Dependent on Pyridoxal Phosphate and other Carbonyl Compounds. Pergamon Press, Oxford.
173. Yip, W-K., T. Moore, and S.F. Yang. 1992. Differential accumulation of transcripts for four tomato 1-aminocyclopropane-1-carboxylate synthase homologs under various conditions. Proc. Natl. Acad. Sci. USA 89: 2475-2479.
174. Kim, W.T., A. Silverstone, W-K. Yip, J-G. Dong, and S.F. Yang. 1992. Induction of 1-aminocyclopropane-1-carboxylate synthase mRNA by auxin in mung bean hypocotyls and cultured apple shoots. Plant Physiol. 98: 465-471.
175. Dong, J-G., D. Olson, A. Silverstone, and S.F. Yang. 1992. Sequence of a cDNA coding for a 1-aminocyclopropane-1-carboxylate oxidase homolog from apple fruit. Plant Physiol. 98: 1530-1531.
176. Dong, J-G., J.C. Fernández-Maculet, and S.F. Yang. 1992. Purification and characterization of 1-aminocyclopropane-1-carboxylate oxidase from apple fruit. Proc. Natl. Acad. Sci. USA 89: 9789-9793.
177. Fernández-Maculet, J.C. and S.F. Yang. 1992. Extraction and partial characterization of the ethylene-forming enzyme from apple fruit. Plant Physiol. 99: 751-754.
178. Kim, W.T. and S.F. Yang. 1992. Turnover of 1-aminocyclopropane-1-carboxylic acid synthase protein in wounded tomato fruit tissue. Plant Physiol. 100: 1126-1131.
179. Yip, W.-K. and S.F. Yang. 1993. Ethylene metabolism. *In* P.J. Lea (ed.), Methods in Plant Biochemistry, Vol. 9, Enzymes of Secondary Metabolism. Academic Press, London, pp. 431-446.
180. Yang, S.F. and J.G. Dong. 1993. Recent progress in research of ethylene biosynthesis. Bot. Bull. Acad. Sin. 34: 89-101.
181. Yang, S.F., J.G. Dong, J.C. Fernández-Maculet, and D.C. Olson. 1993. Apple ACC oxidase: purification and characterization of the enzyme and cloning of its cDNA. *In* J.C. Pech et al. (eds.), Cellular and Molecular Aspects of the Plant Hormone Ethylene, Kluwer Academic Publishers, Netherlands, pp. 59-64.
182. Fernández-Maculet, J.C., J.G. Dong, and S.F. Yang. 1993. Activation of 1-aminocyclopropane-1-carboxylate oxidase by carbon dioxide. Biochem. Biophys. Res. Comm. 193: 1168-1173.

183. Kim, W.T. and S.F. Yang. 1994. Structure and expression of cDNAs encoding l-aminocyclopropane-1-carboxylate oxidase homologs isolated from excised mung bean hypocotyls. Planta 194: 223-229.
184. White, M.F., J. Vasquez, S.F. Yang, and J.F. Kirsch. 1994. Expression of apple l-aminocyclopropane-l-carboxylate synthase in *Escherichia coli:* kinetic characterization of wild-type and active-site mutant forms. Proc. Natl. Acad. Sci. USA 91: 12428-12432.
185. Moussatos, V.V., S.F. Yang, B. Ward, and D.G. Gilchrist. 1994. AAL-toxin induced physiological changes in *Lycopersicon esculentum* Mill: roles for ethylene and pyrimidine intermediates in necrosis. Physiol. Mol. Plant Pathol. 44: 455-468.
186. Hyodo, H. and S.F. Yang. 1994. Ethylene. *In* N. Takahashi and Y. Masuda (eds.), Plant Hormone Handbook, Vol. 2, Baihukan, Tokyo, pp. 161-201 (in Japanese).
187. Oetiker, J.H. and S.F. Yang. 1995. The role of ethylene in fruit ripening. Acta Horticulturae 398: 167-178.
188. McKeon, T.A., J.C. Fernández-Maculet, and S.F. Yang. 1995. Biosynthesis and metabolism of ethylene. *In* P.J. Davies, (ed.), Plant Hormones-Physiology, Biochemistry and Molecular Biology, Kluwer Academic Publishers, Netherlands, pp. 118-139.
189. Olson, D.C., J.H. Oetiker, and S.F. Yang. 1995. Analysis of *LE-ACS3,* a 1-aminocyclopropane-1 carboxylic acid synthase gene expressed during flooding in the roots of tomato plants. J. Biol. Chem. 270: 14056-14061.
190. Mizutani, F., J.G. Dong, and S.F. Yang. 1995. Effect of pH on CO_2-activated 1-aminocyclopropane-1 carboxylate oxidase activity from apple fruit. Phytochemistry 39: 751-755.
191. Liu, Y., A.L. Silverstone, Y.M. Wu, and S.F. Yang. 1995. Formation of N-malonyl-L-tryptophan in water-stressed tomato leaves. Phytochemistry 40: 691-697.
192. Wu, Y.M., A.L. Silverstone, Y. Liu, and S.F. Yang. 1995. Partial purification and characterization of tryptophan N-malonyltransferase from tomato leaves. Phytochemistry 40: 699-704.
193. Li, N., S. Huxtabel, S.F. Yang, and S.D. Kung. 1996. Effects of N-terminal deletions on 1-aminocyclopropane-1-carboxylate synthase activity. FEBS Letters. 378: 286-290.
194. Lafuente, M.T., G. Lopez-Galvez, M. Cantwell, and S.F. Yang. 1996. Factors influencing ethylene-induced isocoumarin formation and increased respiration in carrots. J. Amer. Soc. Hort. Sci. 121: 537-542.
195. Li, N., X.N. Jiang, G.P. Cai, and S.F. Yang. 1996. A novel bifunctional fusion enzyme catalyzing ethylene synthesis via 1-aminocyclopropane-1-carboxylic acid. J. Biol. Chem. 271: 25738-25714.

196. Shaw, J.P., Y.S. Chou, R.C. Chang, and S.F. Yang. 1996. Characterization of the ferrous ion binding sites of apple 1-aminocyclopropane-1-carboxylate oxidase by site-directed mutagenesis. Biochem. Biophys. Res. Comm. 225: 697-700.
197. Kim, W.T., A. Campbell, T. Moriguchi, H.C. Yi, and S.F. Yang. 1997. Auxin induces three genes encoding 1-aminocyclopropane-1-carboxylate synthase in mung bean hypocotyls. J. Plant Physiol. 150: 77-84.
198. Kim, J.H., W.T. Kim, B.G. Kang, and S.F. Yang. 1997. Induction of 1-aminocyclopropane-1-carboxylate oxidase mRNA by ethylene in mung bean hypocotyls: Involvement of both protein phosphorylation and dephosphorylation in ethylene signaling. Plant J. 11: 399-405.
199. Lopez-Gomez. R., A. Campbell, J.G. Dong, S.F. Yang, and M.A. Gomez-Lim. 1997. Isolation and analysis of a genomic banana ACC oxidase clone and its expression. Plant Sci. 123: 123-131.
200. Oetiker. J.H., D.C. Olson, O.Y. Shiu, and S.P. Yang. 1997. Differential induction of seven 1-aminocyclopropane-1-carboxylate synthase genes by elicitor in suspension cultures of tomato *(Lycopersicon esculentum).* Plant Mol. Biol. 34: 275-289.
201. Charng. Y.Y., Y. Liu, J.G. Dong, and S.F. Yang. 1997. On 1-aminocyclopropane-1- carboxylic acid (ACC) oxidase: Degradation of α-aminoisobutyric acid and structure-function studies on the CO_2 binding site. *In* A.K. Kanellis, C. Chang, H. Kende, and D. Grierson (eds.), Biology and Biotechnology of the Plant Hormone Ethylene, Kluwer Academic Publishers, Dordrecht, pp. 23-29.
202. Charng, Y.Y., C.W. Sun, S.L. Yan, S.J. Chou, Y.R. Chen, and S.F. Yang. 1997. cDNA sequence of a putative ethylene receptor from carnation petals (Accession No. AFOI6250). (Plant Gene Register PGR 97-144) Plant Physiol. 115: 863.
203. Kung, S.D. and S.F. Yang (eds.). 1998. Discoveries in Plant Biology. Vol. I, World Scientific, Singapore, 371 pp.
204. Saltveit, M.E., S.F. Yang, and W.T. Kim. 1998. History of the discovery of ethylene as a plant growth substance. *In* S.D. Kung and S.F. Yang (eds.), Discoveries in Plant Biology, World Scientific Publishing Company, Singapore, pp. 47-70.
205. Peiser, G. and S.F. Yang. 1998. Evidence for 1-(malonylamino) cyclopropane-1-carboxylic acid being the major conjugate of 1-aminocyclopropane-1-carboxylic acid in tomato fruit. Plant Physiol. 116: 1527-1532.
206. Shiu, O.Y., J.H. Oetiker, W.K. Yip, and S.F. Yang. 1998. Promoter of *LE-ACS7,* an early flooding-induced 1-aminocyclopropane-1-carboxylate synthase gene of the tomato, is tagged by a *Sol3* transposon. Proc. Natl. Acad. Sci. USA 95: 10334-10339.

207. Lin, C.T., K.Y. Fu, M.T. Lin, S.F. Yang, and J.P. Shaw. 1998. Cloning and characterization of a cDNA for 1-aminocyclopropane-1-carboxylate synthase from papaya fruit (Accession No. Y11357) (PGR98-046). Plant Physiol. 116: 1193.

208. Chen, H.H., Y.Y. Charng, S.F. Yang, and J.P. Shaw. 1998. Molecular cloning and sequencing of a broccoli cDNA (Accession No. AF047476) encoding an ETR-type ethylene receptor. (PGR98-088). Plant Physiol. 117: 717.

209. Chen, H.H., Y.Y. Charng. S.F. Yang, and J.P. Shaw. 1998. Isolation and characterization of a broccoli cDNA (Accession No. AF047477) encoding and ERS-type ethylene receptor. (PGR98-123). Plant Physiol. 117: 1126.

210. Yang, S.F. and Y.Y. Charng. 1998. The gaseous plant hormone ethylene: from biosynthesis and action to biotechnology. *In* C.H. Chou and K.T. Shao (eds.), "Plenary Lecture, Frontiers in Biology: The Challenges of Biodiversity, Biotechnology and Sustainable Agriculture", Academia Sinica, Taipei, pp. 25-34.

211. Yip, W.K. and S.F. Yang. 1998. Ethylene biosynthesis in relation to cyanide metabolism. Bot. Bull. Acad. Sin. 39: 1-7.

212. Yang, S.F. and J.H. Oetiker. 1998. Molecular biology of ethylene biosynthesis and its application in horticulture. J. Japan. Soc. Hort. Sci. 67: 1209-1214.

213. Kung. S. D. and S. F. Yang (eds.).1998. Discoveries in Plant Biology. Vol. II. World Scientific, Singapore, 380 pp.

214. Yang, X.X., H.W. Choi, S.F. Yang, and N. Li. 1999. A UV-light activated cinnamic acid isomer regulates plant growth and gravitropism via an ethylene receptor-independent pathway. Aust. J. Plant Physiol. 26: 325-335.

215. Wong WS, W. Ning, P.L. Xu, S.D. Kung, S.F. Yang, and Li N. 1999. Identification of two chilling-regulated 1-aminocyclopropane-1-carboxylate synthase genes from citrus (*Citrus sinensis* Osbeck) fruit. Plant Mol. Biol. 41 (5): 587-600.

216. Zhou H.Q., H.W. Wang, K. Zhu, S.F. Sui, P.L. Xu, S.F. Yang, and Li N. 1999. The multiple roles of conserved arginine 286 of 1-aminocyclopropane-1-carboxylate synthase. Coenzyme binding, substrate binding, and beyond. Plant Physiol. 121: 913-919

217. Ge, L., J.Z. Liu, W.S. Wong, W.L.W. Hsiao, K. Chong, Z.K. Xu, S.F. Yang, S.D. Kung, and Li, N. 2000. Identification of a novel multiple environmental factor-responsive 1-aminocyclopropane-1-carboxylate synthase gene, NT-ACS2, from tobacco. Plant Cell and Environ. 23: 1169-1182.

218. Charng, Y., S.J. Chou, W.T. Jiaang, S.T. Chen, and S.F. Yang. 2001. The catalytic mechanism of 1-aminocyclopropane-1-carboxylic acid oxidase. Arch. Biochem. Biophys. 385: 179-185.

219. Wang, N.N., S.F. Yang, and Y.Y. Charng. 2001. Differential expression of 1-aminocyclopropane-1-carboxylate synthase genes during orchid flower senescence induced by the protein phosphatase inhibitor okadaic acid. Plant Physiol. 126: 253-260.

220. Chen, L.F.O., J.Y. Hwang, Y.Y. Charng, C.W. Sun, and S.F. Yang. 2001. Transformation of broccoli (*Brassica oleracea* var. italica) with isopentenyltransferase gene via *Agrobacterium tumefaciens* for post-harvest yellowing retardation. Molecular Breeding 7: 243-257.
221. Peng, H.P., C.S. Chan, M.C. Shih, and S.F. Yang. 2001. Signaling events in the hypoxic induction of alcohol dehydrogenase gene in Arabidopsis. Plant Physiol. 126: 742-749.
222. Chung, M.C., S.J. Chou, L.Y. Kuang, Y.Y. Charng, and S.F. Yang. 2002. Subcellular localization of 1-aminocyclopropane-1-carboxylic acid oxidase in apple fruit. Plant and Cell Physiol. 43: 549-554.
223. Yang, C.Y., F.H. Chu, Y.T. Wang, Y.T. Chen, S.F. Yang, and J.F. Shaw. 2003. Novel broccoli 1-aminocyclopropane-1-carboxylate oxidase gene (*BoACO3*) associated with the late stage of postharvest floret senescence. J. Agric. Food Chem. 51: 2569-2575.
224. Chen, Y.T., Y.R. Lee, C.Y. Yang, Y.T. Wang, S.F. Yang, and J.F. Shaw. 2003. A novel papaya ACC oxidase gene (*CpACO2*) associated with late stage fruit ripening and leaf senescence. Plant Science 164: 531-540
225. Grichko, V, M. Serek, C.B. Watkins, and S.F. Yang. 2006. Father of 1-MCP - Edward C. Sisler. Biotechnol. Advances 24 (4): 355-356.
226. Huang, L.C., U.L. Lai, S.F. Yang, M.J. Chu, C.I. Kuo, M.F. Tsai, C.W. Sun. 2007. Delayed flower senescence of *Petunia hybrida* plants transformed with antisense broccoli ACC synthase and ACC oxidase genes. Postharvest Biol. and Technol. 46 (1): 47-53

+ Models
PSL-7810; No of Pages 6
ARTICLE IN PRESS

Available online at www.sciencedirect.com
ScienceDirect
ELSEVIER
Plant Science xxx (2008) xxx–xxx
PLANT SCIENCE
www.elsevier.com/locate/plantsci

Shang Fa Yang: Pioneer in plant ethylene biochemistry

Kent J. Bradford *

Department of Plant Sciences, One Shields Avenue, University of California, Davis, CA 95616-8780, USA

Received 19 December 2007; received in revised form 3 January 2008; accepted 4 January 2008

Abstract

Shang Fa Yang was born in Taiwan in 1932. After receiving his B.S. and M.S. degrees in Agricultural Chemistry from the National Taiwan University, he came to the United States in 1958 to pursue a Ph.D. degree at Utah State University. Following three postdoctoral years, he was hired at the University of California, Davis, in 1966 where he was a biochemist and professor for 28 years. After retiring early from UC Davis, he subsequently established a research program at the Hong Kong University of Science and Technology and served as Vice President of Academia Sinica in Taiwan. Yang's research achievements included discovering the biochemical pathway for the synthesis of ethylene by identifying the key steps by which *S*-adenosylmethionine (SAM) is converted into 1-aminocyclopropane-1-carboxylic acid (ACC) and subsequently into ethylene. Yang further demonstrated how the methylthio and ribose moieties from SAM were recycled back into methionine in order to sustain high rates of ethylene synthesis, as in ripening fruits. This recycling pathway is now known as the Yang Cycle. Yang also contributed to the isolation, characterization and cloning of ACC synthase and ACC oxidase, the two enzymes in the ethylene biosynthetic pathway, and to the elucidation of their structure and reaction mechanisms. He made important contributions to auxin, cytokinin, cyanide and sulfur metabolism in plants as well. His work formed the basis for subsequent research that has established ethylene as the most thoroughly characterized of the hormonal biosynthesis and signaling pathways in plants.

Keywords: Shang Fa Yang; Ethylene; ACC; Yang Cycle

1. Introduction

Shang Fa Yang (Fig. 1) was born in Taiwan in 1932. He was the youngest child of a large family (five elder brothers and six elder sisters), and his father was a businessman involved in the production maltose from sugarcane. He attended the National Taiwan University and received his B.S. and M.S. degrees in Agricultural Chemistry in 1956 and 1958, respectively. A scholarship brought him to Utah State University to do graduate work with G.W. Miller on the effects of fluoride on plant biochemistry and metabolism [1]. Yang received his Ph.D. in 1962 in Plant Biochemistry and then accepted a postdoctoral position with Paul K. Stumpf at the University of California, Davis, where he worked on lipid biosynthesis in avocado fruits [2]. He then accepted a postdoctoral fellowship with B.N. LaDu at New York University Medical School in 1963–1964, where he met his future wife Eleanor Liu who was studying accounting there. He subsequently returned to California and worked as a postdoctoral researcher with Andrew A. Benson at Scripps Institute of Oceanography in La Jolla during 1964–1965.

Yang was hired in 1966 as an assistant biochemist in the Department of Vegetable Crops at the University of California, Davis, to study the postharvest biochemistry of fruits and vegetables. California ships fresh produce long distances to markets in the eastern US and internationally, making postharvest storage and physiology of fresh produce of considerable importance to the state. He initially shared a laboratory with Harlan K. Pratt, a pioneering researcher in ethylene physiology, in the newly constructed Mann Laboratory, named after Louis Mann. This laboratory was built specifically for postharvest biology studies and was well equipped with controlled temperature chambers and gas flow equipment. In addition, Pratt had custom-built one of the first gas chromatographs that could measure ethylene in the parts-per-million concentrations that are produced by plants. The gaseous compound was found in the early decades of the 20th century to hasten the ripening of fruits and to cause growth

* Tel.: +1 530 752 6087; fax: +1 530 754 7222.
E-mail address: kjbradford@ucdavis.edu.

0168-9452/$ – see front matter
doi:10.1016/j.plantsci.2008.01.005

Please cite this article in press as: K.J. Bradford, Shang Fa Yang: Pioneer in plant ethylene biochemistry, Plant Sci. (2008), doi:10.1016/j.plantsci.2008.01.005

Fig. 1. Portrait of professor Shang Fa Yang.

distortions in growing plants, generally associated with leaking gas mains. However, research primarily at the Boyce Thompson Institute in the 1930s showed that plants produce ethylene and that it is broadly involved in regulating plant growth and development, including seed germination, root and shoot growth, responses to environmental stresses, flowering, fruit ripening and senescence or death of plant tissues and organs [3]. However, little was known about the pathway of ethylene biosynthesis in plants, and armed with Pratt's gas chromatograph and his broad knowledge of chemistry and biochemistry, Yang set out to explore plant ethylene biology.

2. Ethylene biosynthesis

Ethylene is a simple gaseous compound containing two carbon and four hydrogen atoms and one double bond (C_2H_4). At the time that Yang entered the field, Morris Lieberman and others had evidence that amino acids, particularly methionine, might be a precursor for ethylene production [4,5]. Various *in vitro* systems were being used to convert methionine and other potential precursors into ethylene, and Yang contributed significantly to these studies, using his knowledge of chemistry to explore different reaction mechanisms [6]. He subsequently utilized both *in vitro* and *in vivo* approaches to explore potential intermediates of and mechanisms for the conversion of methionine to ethylene [7,8]. He also studied the mechanism of formation of ethylene from 2-chloroethylphosphonic acid in plant tissues [9]. This compound, under the generic name of ethephon, enabled the commercial application of ethylene for agricultural purposes, as it is taken up by plants and converted into ethylene. It has been widely used as a

Please cite this article in press as: K.J. Bradford, Shang Fa Yang: Pioneer in plant ethylene biochemistry, Plant Sci. (2008), doi:10.1016/j.plantsci.2008.01.005

+ Models
PSL-7810; No of Pages 6
ARTICLE IN PRESS

fruit-ripening agent, to loosen fruits for harvest and for defoliation of cotton before harvest [3].

Continuing studies on the biogenesis of ethylene showed that methionine was converted to *S*-adenosylmethionine (SAM) and that SAM was a precursor of ethylene. When apple tissues were supplied with ^{14}C-SAM under anaerobic conditions that prevented ethylene formation, a labeled compound accumulated in tissues [10]. This discovery stimulated active competition among various groups to identify the unknown intermediate between SAM and ethylene that culminated in the identification by Adams and Yang of 1-aminocyclopropane-1-carboxylic acid (ACC) as the final *in vivo* precursor of ethylene (Fig. 2; [11]). Yang's group quickly developed a sensitive assay for ACC via its chemical conversion to ethylene [12], which facilitated wide-ranging studies by his group on the regulation of ACC and ethylene biosynthesis in plant growth, fruit ripening, and stress responses [13–15].

The ethylene biosynthetic pathway requires only two specific steps, the conversion of SAM to ACC and of ACC to ethylene (Fig. 2), and attention quickly turned to the identification of the enzymes responsible for them. Yang's group and that of Hans Kende at Michigan State University soon reported the properties of 1-aminocyclopropane carboxylate synthase, the enzyme responsible for the conversion of SAM to ACC [16–18]. The subsequent cloning of ACC synthase, in which the Yang, Kende and A. Theologis laboratories were involved, is described in detail by Kende [19]. Isolation of the ethylene-forming enzyme, or ACC oxidase, was a more difficult task, as the enzyme appeared to be membrane-associated and activity was lost upon cellular disruption and fractionation. The gene coding for the enzyme was eventually cloned by D. Grierson's group based upon expression of a ripening-related cDNA and demonstration that its suppression blocked ethylene synthesis [19–21]. Yang contributed to the characterization of the multiple alleles in the ACC synthase and ACC oxidase families and to structure-function studies of the reaction mechanisms of the two enzymes [22–29].

3. ACC metabolism and the Yang Cycle

Ethylene synthesis exhibits multiple points of regulation, including transcriptional control of multiple ACC synthase and ACC oxidase genes and the activation state and/or stability of the enzymes themselves [30]. For example, light, carbon dioxide, oxygen, and water stress all influence the conversion of ACC to ethylene [31]. The conversion of ACC to ethylene is also stereospecific, as demonstrated by the differential conversion of stereoisomers of 1-amino-2-ethylcyclopropane carboxylic acid to 1-butene [32]. This provided an important test for the enzymatic conversion of ACC to ethylene versus non-specific chemical conversion [33] that was important in the subsequent isolation and cloning of ACC oxidase [34,35]. In addition to being converted to ethylene, Yang's group determined that ACC could be malonylated and that this pool of conjugated ACC was largely unavailable for conversion to ethylene (Fig. 2; [36]).

An early dilemma in ethylene biosynthesis was that methionine pools were generally too low in plant tissues to sustain the observed rates of ethylene synthesis. Some had suggested that following ACC formation, the methylthio group from methionine would be attached to an existing homocysteine molecule to form a new methionine molecule, thus recycling only the methylthio group. However, Yang's and Lieberman's groups demonstrated that after ACC is released from SAM, the methylthio group and the ribose moiety from the remaining methylthioadenosine (MTA) are both recycled to replenish methionine levels and sustain ethylene biosynthesis [37–39]. This cycle has been christened the Yang Cycle in plant biochemistry texts (Fig. 2; [40]).

4. Other activities and honors

While most widely known for his work on ethylene biosynthesis and action, Yang also maintained active research programs in other areas of plant growth and metabolism, including on auxin metabolism and action [41–43], on cytokinin action [44,45], on the biological effects of sulfite and sulfur dioxide [46,47], and on cyanide generation and metabolism in plants [48].

A theme throughout all of Yang's work is a linkage to practical applications in postharvest biology and plant growth regulation. He applied his knowledge of ethylene biosynthesis to contribute to improvements in postharvest storage conditions [49–52]. Yang figured prominently at many national and international research conferences and served on the editorial boards of leading journals and as a member of several learned societies. He won many awards and honors, including a Guggenheim Fellowship in 1982, the International Plant Growth Substances Association Research Award in 1985, and the Outstanding Researcher Award from the American Society for Horticultural Science in 1992. In 1992, he was named the UC Davis Faculty Research Lecturer, the highest honor given by that institution for excellence in research. He was elected to the US National Academy of Sciences in 1990 and to the Academia Sinica, Taiwan, in 1992. In 1991, Yang received the prestigious international Wolf Prize in Agriculture, which many consider to be the "Nobel Prize" for agricultural research.

After a distinguished career as professor and biochemist at the University of California, Davis, Yang took early retirement in 1994 to accept a Distinguished Professorship at the Hong Kong University of Science and Technology, where he established an active plant research group in the Department of Biology. In 1995, he was recognized as a Distinguished Research Fellow by the Institute of Botany of the Academia Sinica in Taipei, Taiwan, and returned to Taiwan in 1996 to serve as Vice President of Academia Sinica from 1996 to 1999. In this position he directed its numerous research institutes, including the establishment of a new Institute of Agricultural Biotechnology (subsequently renamed the Agricultural Biotechnology Research Center). In both Hong Kong and Taiwan, Yang played important leadership roles in advancing plant biology and agricultural biotechnology. Following his tenure at

Please cite this article in press as: K.J. Bradford, Shang Fa Yang: Pioneer in plant ethylene biochemistry, Plant Sci. (2008), doi:10.1016/j.plantsci.2008.01.005

+ Models
PSL-7810; No of Pages 6
ARTICLE IN PRESS

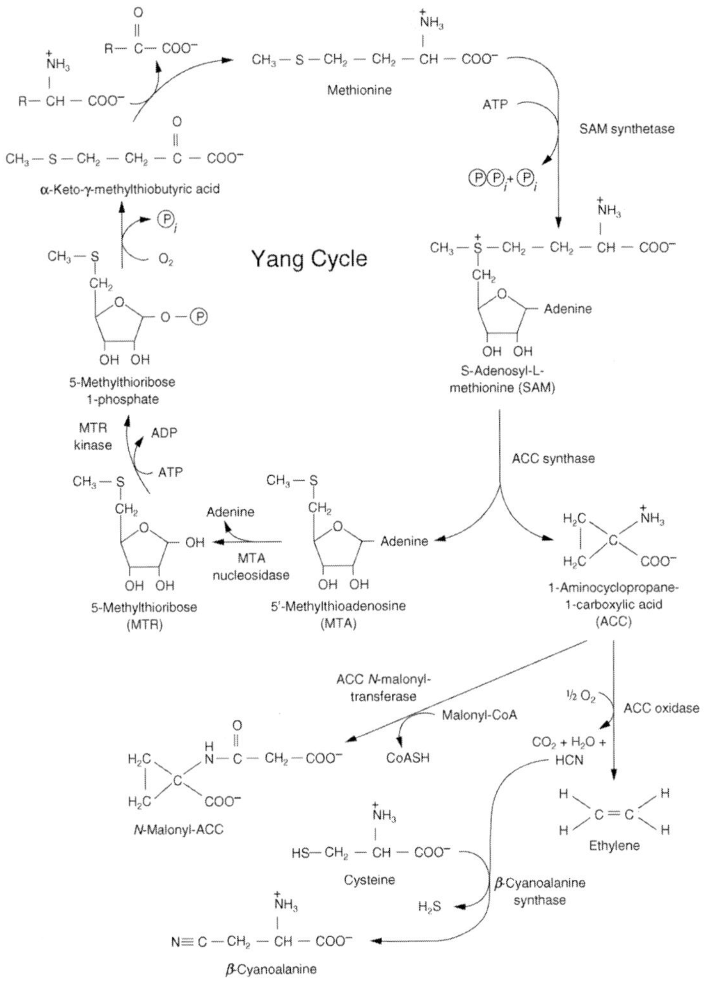

Fig. 2. The Yang Cycle and formation of ethylene and other products from ACC. See text for details.

Academia Sinica, Yang retired to Davis, California in 1999, although he continued to actively publish additional scientific work as recently as 2007.

Shang Fa Yang passed away suddenly and unexpectedly from complications of pneumonia on February 12, 2007 at the age of 74 years. As was his wish, the Shang Fa and Eleanor Yang Scholarly Exchange Endowment was established in October 2007 to support exchange of scholars in the agricultural, biological and chemical sciences between UC Davis and Academia Sinica, Taiwan, two institutions that Yang served with distinction.

5. Concluding remarks

Shang Fa Yang's pioneering studies in the biosynthesis and action of ethylene have ensured his legacy in the history of plant biology. Future plant biologists will know of him through his discovery of ACC and the biosynthetic pathway for ethylene biosynthesis, for the Yang Cycle and for his many other contributions to our field that are described in more than 225 journal articles and book chapters that he published during his career. He was known for his clarity of thought and his ability to identify and design critical experimental tests of hypotheses.

Please cite this article in press as: K.J. Bradford, Shang Fa Yang: Pioneer in plant ethylene biochemistry, Plant Sci. (2008), doi:10.1016/j.plantsci.2008.01.005

Yang maintained an open mind and was willing to challenge accepted ideas, even his own, when they proved untenable in the face of experimental data. As a mentor, Yang was demanding and rigorous yet positive, encouraging and supportive. Many graduate students and postdoctoral associates benefitted greatly from both his mentorship and his subsequent support for their careers. Students and colleagues will remember and miss his humor, his humanity, and his sparkling intellect. This special issue is a fitting tribute to Shang Fa Yang's pioneering contributions to our understanding of the role of ethylene in plant biology.

Acknowledgements

The author is grateful to Eleanor, Albert and Bryant Yang for sharing personal information and thanks Yee-Yung Charng, Neil E. Hoffman, John H. Miyazaki, Galen D. Peiser, Mikal E. Saltveit and Wing Kin Yip for contributing additional information.

References

[1] S.F. Yang, G.W. Miller, Biochemical studies on the effect of fluoride on higher plants. 1. Metabolism of carbohydrates, organic acids and amino acids, Biochem. J. 88 (1963) 505–509.
[2] S.F. Yang, P.K. Stumpf, Fat metabolism in higher plants. XXI. Biosynthesis of fatty acids by avocado mesocarp enzyme systems, Biochim. Biophys. Acta 98 (1965) 19–26.
[3] F.B. Abeles, P.W. Morgan, M.E. Saltveit, Ethylene in Plant Biology, second ed., Academic Press, San Diego, 1992.
[4] M. Lieberman, Biosynthesis and action of ethylene, Annu. Rev. Plant Physiol. 30 (1979) 533–591.
[5] M. Lieberman, A.T. Kunishi, L.W. Mapson, D.A. Wardale, Ethylene production from methionine, Biochem. J. 97 (1965) 449–459.
[6] S.F. Yang, H.S. Ku, H.K. Pratt, Ethylene production from methionine as mediated by flavin mononucleotide and light, Biochem. Biophys. Res. Commun. 24 (1966) 739–743.
[7] A.H. Baur, S.F. Yang, Precursors of ethylene, Plant Physiol. 44 (1969) 1347–1349.
[8] A.H. Baur, S.F. Yang, Methionine metabolism in apple tissue in relation to ethylene biosynthesis, Phytochemistry 11 (1972) 3207–3214.
[9] S.F. Yang, Ethylene evolution from 2-chloroethylphosphonic acid, Plant Physiol. 44 (1969) 1203–1204.
[10] D.O. Adams, S.F. Yang, Methionine metabolism in apple tissue—implication of *S*-adenosylmethionine as an intermediate in conversion of methionine to ethylene, Plant Physiol. 60 (1977) 892–896.
[11] D.O. Adams, S.F. Yang, Ethylene biosynthesis: identification of l-aminocyclopropane-l-carboxylic acid as an intermediate in the conversion of methionine to ethylene, Proc. Natl. Acad. Sci. U.S.A. 76 (1979) 170–174.
[12] M.C.C. Lizada, S.F. Yang, Simple sensitive assay for 1-aminocyclopropane-1-carboxylic acid, Anal. Biochem. 100 (1979) 140–145.
[13] N.E. Hoffman, S.F. Yang, Changes of 1-aminocyclopropane-1-carboxylic acid content in ripening fruits in relation to their ethylene production rates, J. Am. Soc. Hort. Sci. 105 (1980) 492–495.
[14] Y.B. Yu, S.F. Yang, Biosynthesis of wound ethylene, Plant Physiol. 66 (1980) 281–285.
[15] K.J. Bradford, S.F. Yang, Xylem transport of 1-aminocyclopropane-1-carboxylic acid, an ethylene precursor, in waterlogged tomato plants, Plant Physiol. 65 (1980) 322–326.
[16] Y.B. Yu, D.O. Adams, S.F. Yang, 1-Aminocyclopropanecarboxylate synthase, a key enzyme in ethylene biosynthesis, Arch. Biochem. Biophys. 198 (1979) 280–286.
[17] T. Boller, R.C. Herner, H. Kende, Assay for and enzymatic formation of an ethylene precursor, 1-aminocyclopropane-1-carboxylic acid, Planta 145 (1979) 293–303.
[18] H. Kende, Enzymes of ethylene biosynthesis, Plant Physiol. 91 (1989) 1–4.
[19] H. Kende, Ethylene biosynthesis, Annu. Rev. Plant Physiol. Plant Mol. Biol. 44 (1993) 283–307.
[20] C.J.S. Smith, A. Slater, D. Grierson, Rapid appearance of an mRNA correlated with ethylene synthesis encoding a protein of molecular weight 35000, Planta 168 (1986) 94–100.
[21] A.J. Hamilton, G.W. Lycett, D. Grierson, Antisense gene that inhibits synthesis of the hormone ethylene in transgenic plants, Nature 346 (1990) 284–287.
[22] J.G. Dong, W.T. Kim, W.K. Yip, G.A. Thompson, L.M. Li, A.B. Bennett, S.F. Yang, Cloning of a cDNA encoding 1-aminocyclopropane-1-carboxylate synthase and expression of its messenger RNA in ripening apple fruit, Planta 185 (1991) 38–45.
[23] W.T. Kim, S.F. Yang, Structure and expression of cDNAs encoding 1-aminocyclopropane-1-carboxylate oxidase homologs isolated from excised mung bean hypocotyls, Planta 194 (1994) 223–229.
[24] W.-K. Yip, T. Moore, S.F. Yang, Differential accumulation of transcripts for four tomato 1-aminocyc1opropane-1-carboxylate synthase homologs under various conditions, Proc. Natl. Acad. Sci. U.S.A. 89 (1992) 2475–2479.
[25] D.C. Olson, J.H. Oetiker, S.F. Yang, Analysis of LeACS3, a 1-aminocyclopropane-1-carboxylic acid synthase gene expressed during flooding in the roots of tomato plants, J. Biol. Chem. 270 (1995) 14056–14061.
[26] Y. Charng, S.J. Chou, W.T. Jiaang, S.T. Chen, S.F. Yang, The catalytic mechanism of 1-aminocyclopropane-1-carboxylic acid oxidase, Arch. Biochem. Biophys. 385 (2001) 179–185.
[27] N. Li, S. Huxtable, S.F. Yang, S.D. Kung, Effects of N-terminal deletions on 1-aminocyclopropane-1-carboxylate synthase activity, FEBS Lett. 378 (1996) 286–290.
[28] J.F. Shaw, Y.S. Chou, R.C. Chang, S.F. Yang, Characterization of the ferrous ion binding sites of apple 1-aminocyclopropane-1-carboxylate oxidase by site-directed mutagenesis, Biochem. Biophys. Res. Commun. 225 (1996) 697–700.
[29] H.Q. Zhou, H.W. Wang, K. Zhu, S.F. Sui, P.L. Xu, S.F. Yang, N. Li, The multiple roles of conserved arginine 286 of 1-aminocyclopropane-1-carboxylate synthase. Coenzyme binding, substrate binding, and beyond, Plant Physiol. 121 (1999) 913–919.
[30] C.T. Argueso, M. Hansen, J.J. Kieber, Regulation of ethylene biosynthesis, J. Plant Growth Reg. 26 (2007) 92–105.
[31] S.F. Yang, N.E. Hoffman, Ethylene biosynthesis and its regulation in higher plants, Annu. Rev. Plant Physiol. Plant Mol. Biol. 35 (1984) 155–189.
[32] N.E. Hoffman, S.F. Yang, A. Ichihara, S. Sakamura, Stereospecific conversion of 1-aminocyclopropanecarboxylic acid to ethylene by plant tissues—conversion of stereoisomers of 1-amino-2-ethylcyclopropanecarboxylic acid to 1-butene, Plant Physiol. 70 (1982) 195–199.
[33] T.A. McKeon, S.F. Yang, A comparison of the conversion of 1-amino-2-ethylcyclopropane-1-carboxylic acid stereoisomers to 1-butene by pea epicotyls and by a cell-free system, Planta 160 (1984) 84–87.
[34] J.G. Dong, D. Olson, A. Silverstone, S.F. Yang, Sequence of a cDNA coding for a 1-aminocyclopropane-1-carboxylate oxidase homolog from apple fruit, Plant Physiol. 98 (1992) 1530–1531.
[35] A.J. Hamilton, M. Bouzayen, D. Grierson, Identification of a tomato gene for the ethylene-forming enzyme by expression in yeast, Proc. Natl. Acad. Sci. U.S.A. 88 (1991) 7434–7437.
[36] N.E. Hoffman, S.F. Yang, T. McKeon, Identification of 1-(malonylamino)cyclopropane-1-carboxylic acid as a major conjugate of 1-aminocyclopropane-1-carboxylic acid, an ethylene precursor in higher plants, Biochem. Biophys. Res. Commun. 104 (1982) 765–770.
[37] J.H. Miyazaki, S.F. Yang, Metabolism of 5-methylthioribose to methionine, Plant Physiol. 84 (1987) 277–281.
[38] J.H. Miyazaki, S.F. Yang, The methionine salvage pathway in relation to ethylene and polyamine biosynthesis, Physiol. Plant 69 (1987) 366–370.

[39] S.Y. Wang, D.O. Adams, M. Lieberman, Recycling of 5′-methylthioadenosine ribose carbon into methionine in tomato tissue in relation to ethylene production, Plant Physiol. 70 (1982) 117–121.
[40] A. Crozier, Y. Kamiya, G. Bishop, T. Tokota, Biosynthesis of hormones and elicitor molecules, in: B. Buchanan, W. Gruissem, R. Jones (Eds.), Biochemistry & Molecular Biology of Plants, American Society of Plant Biologists, Rockville, MD, 2000, pp. 850–929.
[41] N. Aharoni, S.F. Yang, Auxin-induced ethylene production as related to auxin metabolism in leaf disks of tobacco and sugar beet, Plant Physiol. 73 (1983) 598–604.
[42] Y.B. Yu, D.O. Adams, S.F. Yang, Regulation of auxin-induced ethylene production in mung bean hypocotyls—role of 1-aminocyclopropane-1-carboxylic acid, Plant Physiol. 63 (1979) 589–590.
[43] Y.B. Yu, S.F. Yang, Auxin-induced ethylene production and its inhibition by aminoethoxyvinylglycine and cobalt ion, Plant Physiol. 64 (1979) 1074–1077.
[44] L.F.O. Chen, J.Y. Hwang, Y.Y. Charng, C.W. Sun, S.F. Yang, Transformation of broccoli (*Brassica oleracea* var. *italica*) with isopentenyltransferase gene via *Agrobacterium tumefaciens* for post-harvest yellowing retardation, Mol. Breed. 7 (2001) 243–257.
[45] O. Lau, S.F. Yang, Mechanism of a synergistic effect of kinetin on auxin-induced ethylene production—suppression of auxin conjugation, Plant Physiol. 51 (1973) 1011–1014.
[46] G.D. Peiser, S.F. Yang, Ethylene and ethane production from sulfur dioxide injured plants, Plant Physiol. 63 (1979) 142–145.
[47] G.D. Peiser, M.C.C. Lizada, S.F. Yang, Sulfite-induced lipid peroxidation in chloroplasts as determined by ethane production, Plant Physiol. 70 (1982) 994–998.
[48] G.D. Peiser, T.T. Wang, N.E. Hoffman, S.F. Yang, H.W. Liu, C.T. Walsh, Formation of cyanide from carbon 1 of 1-aminocyclopropane-1-carboxylic acid during its conversion to ethylene, Proc. Natl. Acad. Sci. U.S.A. 81 (1984) 3059–3063.
[49] H. Hyodo, H. Kuroda, S.F. Yang, Induction of phenylalanine ammonia lyase and increase in phenolics in lettuce leaves in relation to development of russet spotting caused by ethylene, Plant Physiol. 62 (1978) 31–35.
[50] S.F. Yang, J.H. Oetiker, Molecular biology of ethylene biosynthesis and its application in horticulture, J. Jpn. Soc. Hort. Sci. 67 (1998) 1209–1214.
[51] L.C. Huang, U.L. Lai, S.F. Yang, M.J. Chu, C.I. Kuo, M.F. Tsai, C.W. Sun, Delayed flower senescence of *Petunia hybrida* plants transformed with antisense broccoli ACC synthase and ACC oxidase genes, Postharvest Biol. Technol. 46 (2007) 47–53.
[52] C.Y. Yang, F.H. Chu, Y.T. Wang, Y.T. Chen, S.F. Yang, J.F. Shaw, Novel broccoli 1-aminocyclopropane-1-carboxylate oxidase gene (BoACO3) associated with the late stage of postharvest floret senescence, J. Agric. Food Chem. 51 (2003) 2569–2575.

Please cite this article in press as: K.J. Bradford, Shang Fa Yang: Pioneer in plant ethylene biochemistry, Plant Sci. (2008), doi:10.1016/j.plantsci.2008.01.005

JOHN E. CASIDA

University of California
Berkeley, California, USA

1993 — for his pioneering studies on the mode of action of insecticides, design of safer pesticides and contributions to the understanding of nerve and muscle function in insects.

AUTOBIOGRAPHICAL SKETCH AND SELECTED PUBLICATIONS

Preface

John E. Casida was a major contributor to the Golden Age of insecticide research. This was an era of glory. DDT, invented by Nobel Laureate Paul Müller, greatly increased crop yields and brought devastating insect-transmitted human diseases under control. Many other insecticides quickly followed to eliminate practically all pest insect problems. It was also a time, as pointed out by Rachel Carson in her book "Silent Spring", when broadscale application of toxic and persistent pesticides damaged natural ecosystems with an impact on life in general. Industries pursued more effective insecticides and governments regulated their safe use. Fundamental knowledge was needed from academic research. This was the origin of the fields of Insect Toxicology and Pesticide Chemistry and Toxicology.

Early Life and Education

John was born in 1929 in Tempe, Arizona. During this recession period his family moved annually from Arizona to Arkansas, Missouri and Wisconsin as teaching and research positions became available for his father Lester Earl Casida (Professor of Reproductive Physiology, University of Wisconsin, Madison). He has an older brother (Lester Earl Casida Jr., Professor of Bacteriology, Pennsylvania State University) and younger sister (Betty Ruth Damerau, violinist, Denver Symphony Orchestra). He attended grade and high schools and University in Madison. John was allowed great freedom to explore, living close to a large arboretum and setting up a basement science laboratory. He participated in all aspects of the Boy Scouts program and high school science and photography clubs. At University he was captain of the fencing team and medalist in saber. Four summers were spent in forestry nursery and vegetable research stations and in programs for seed certification and insect rearing and toxicity testing. Graduate school was interrupted after one year by service in the Korean War as a First Lieutenant in the United States Air Force specializing in Medical Entomology at Camp Detrick (Frederick, Maryland).

Marriage, Family and Travels

John met Katherine (Kati) Faustine Monson, a sculptor and painter, through a mutual friend that was a technician in John's laboratory and said to John he had to meet this fantastic girl. Soon after they met, their busy lives took them in different directions — Kati teaching art on the east coast and John doing research in Wisconsin. Then the events were replayed when another technician in John's laboratory, also a friend of Kati's, quite independently got them back together. The wedding in 1956 was followed by a honeymoon country-hopping through Central America with laboratory and field work in Costa Rica and New Years Eve celebrated in barely pre-Castro Havana. Shared experiences in art, science, travel and pre-Columbian cultures from that time were developed continually thereafter. Two sons, Mark Earl (Professor of Theoretical Chemistry, University of Grenoble, France) and Eric Gerhard (Chief Technical Officer, Excelics Semiconductor, Sunnyvale, California), grew up in Berkeley.

Academic Career

John graduated from the University of Wisconsin at Madison with a B.S. degree in Entomology in 1951, an M.S. in Biochemistry in 1952 and a Ph.D. jointly in Biochemistry, Entomology and Plant Physiology in 1954. He began research in entomology and on insecticides in high school and as an undergraduate in college by field evaluation of botanicals and isolation of their insecticidal ingredients. His Ph.D. thesis was on the metabolic activation of systemic organophosphorus insecticides. Dr. Casida was appointed Assistant Professor of Entomology (Insect Toxicology) at the University of Wisconsin at Madison in 1954 and was promoted to Associate Professor in 1957 and Full Professor in 1961. In 1964 he moved to the University of California at Berkeley as Professor of Entomology and Director of the Environmental Chemistry and Toxicology Laboratory, positions which he has held since then with the additional title of Professor of Toxicology in 1998. While at Wisconsin, Dr. Casida was appointed Haight Traveling Fellow in 1958-1959 for research in enzymology at the University of Stockholm in Sweden and in pesticide metabolism and toxicology at Bayer in Germany, CIBA-Geigy in Switzerland, and Fisons Pest Control in England. John received a Guggenheim Foundation Fellowship in 1970-1971 for research in analytical chemistry at the University of Stockholm, agricultural chemistry at Kyoto University, and insect physiology at Cambridge University. In 1986-1987 Dr. Casida conducted research at the Institute of Molecular Biology and Biotechnology of the University of Crete, Greece and at the Laboratory of Cellular and Molecular Neurobiology, CNRS, Gif-sur-Yvette, France. John was appointed to the newly-created William Mauriece Hoskins Chair in Chemical and Molecular Entomology in 1996 and as Professor of the Graduate School in 2008: these are his current positions on the Berkeley faculty.

Organophosphorus Toxicants

Research in the 1950s and 1960s on organophosphorus and methylcarbamate insecticides, such as schradan and carbaryl, laid the background for safety evaluations by synthesis, structure-activity and metabolism studies on perhaps 50 candidate or commercial anticholinergic pesticides. During this period John discovered the insecticide and anthelmintic butonate, the first specifically-designed selective proinsecticide based on species differences in activation and detoxifying enzymes. Studies with tri-*o*-cresyl phosphate (the casual agent for 40,000 cases of human peripheral neuropathy) identified a saligenin cyclic phosphate as the highly potent activated metabolite and established later that the biochemical lesion involved inhibition of a lysophosphatidylcholine hydrolase designated neuropathy target esterase. Kynurenine formamidase was shown to be the target for organophosphate and methylcarbamate insecticides that are extremely potent in producing embryonic abnormalities (teratogenic effects) in avian embryos. Later studies at Berkeley established major enzymes involved in organophosphate toxicant detoxification and multiple effects on the cannabinoid system and lipid biochemistry and signaling. These are just a few of many sensitive serine hydrolase targets shown to have toxicological significance.

Botanical Insecticides and Synergists

Investigations on natural product insecticides were initiated at Wisconsin by radiosynthesis of [^{14}C]rotenone and [^{14}C]pyrethrin and transferred to California for characterization of their metabolic and photodegradative mechanisms. [^{14}C]Rotenone was then used with NADH-Q oxidoreductase in the first successful radioligand study of the reversible binding of an insecticide to an enzyme. Photoaffinity labeling was employed later to define the specific binding protein for rotenone and a multitude of other respiratory inhibitors. In a major program at Berkeley, fundamental research on the chemistry, stereochemistry, structure-activity relationships, metabolism, photodecomposition and mode of action of pyrethrum flower constituents and on synthetic pyrethroids ultimately had a major influence on pyrethroid discovery, development and safety evaluations. John Casida made seminal observations and important contributions at critical stages in the evolution of the modern synthetic pyrethroids. These resulted from Casida's fundamental research on the photochemistry, metabolism and mode of action of the pyrethrins and more stable analogs conducted by his California team alone or in collaboration with Michael Elliott (Wolf Laureate in Agriculture 1989) from the Rothamsted Experimental Station in England during research periods at Berkeley. Investigation with specifically-prepared [^{14}C]methylenedioxyphenyl compounds demonstrated that important insecticide synergists such as piperonyl butoxide act as both substrates and inhibitors for cytochrome P450-dependent microsomal oxidases, thereby blocking detoxification of pyrethroids (and other insecticides), prolonging their persistence and increasing their potency and cost effectiveness. This research on natural products and synergists led to the following awards: Medal of the VIIth International Congress of Plant Protection in 1970; the American Chemical Society Burdick and Jackson International Award for Research in Pesticide Chemistry in 1970 and the Spencer Award for Research in Agricultural and Food Chemistry in 1978.

Research on natural product insecticides continued through the 1970s and 1980s on isobutylamides (from black pepper), on veratridine from *Sabadilla* (leading to enhanced potency and selectivity) and on nikkomycin (radiosynthesis and inhibition of chitin synthetase). In related research, diflubenzuron and its analogs were shown to block the final polymerization step in chitin synthesis but not chitin synthetase itself. The principal active ingredient of the insecticide ryania (the ground stemwood of *Ryania speciosa*) was identified as 9,21-dehydroryanodine, leading to the radiosynthesis of [9,21-^{3}H]ryanodine and showing that this new radioligand probe acted at the calcium release channel of muscle sarcoplasmic reticulum. To find that on hydrolysis of ryanodine to ryanodol (also a natural product) it became more selective and acted at a different site was surprising. Subsequent work by others established the ryanodine receptor at another site as a major target for selective insecticide action. The toxic principal of blister beetles and purported aphrodisiac cantharidin (the active ingredient of Spanish fly) was prepared as a radioligand and the binding protein isolated and identified as protein phosphatase 2A, thereby explaining its pharmacological and toxicological properties.

GABAergic Insecticides

A major advance in insecticide toxicology came from a chance observation of very simple and highly toxic bicyclophosphates and related bicycloortho-carboxylates with a totally unexpected mode of action, followed by fundamental research on neurobiology. Radioligands were prepared (now available commercially as [^{35}S]TBPS, [^{3}H]TBOB and [^{3}H]EBOB) and receptor binding studies combined with physiological assays established that the new toxicants are noncompetitive blockers of the γ-aminobutyric acid (GABA)-gated chloride channel. The binding site of TBPS and the bicycloorthocarboxylates was discovered to be the same as, or closely coupled to, that of the polychlorocycloalkane insecticides (such as dieldrin, lindane and toxaphene) so finally establishing their mode of action after 3 billion pounds had been used. As further evidence for this mode of action, toxaphene was examined and shown to be a mixture of >188 isomeric polychlorobornanes which were fractionated and the principal toxic hepta- and octachlorobornanes identified. The trace isomers most toxic to insects, fish and mammals are the same as those most active *in vitro* at the receptor. A rational approach to totally new classes of insecticides then led to 1,4-disubstituted-2,6,7-trioxabicyclo[2.2.2]octanes and to 2,5-disubstituted-1,3-dithianes, non-halogenated and non-phosphorus compounds of unusually simple structure, some of which are more active than many commercial insecticides. These new heterocyclic insecticides were optimized for high potency, selective toxicity, and use as photoaffinity ligands and affinity columns. Studies with [^{3}H]EBOB in *Drosophila* established that resistance to nine classes of insecticides is due to a single amino acid change in the chloride channel. Demonstration that the noncompetitive antagonist site is highly expressed and ultra sensitive in a human β_3 subunit $GABA_A$ receptor homopentamer ultimately allowed a model for the precise binding site of multiple classes of natural and synthetic toxicants blocking the chloride channel. The arylpyrazole insecticide fipronil acting at the GABA receptor undergoes a totally unexpected yet facile photodesulfinylation reaction to become much more toxic and persistent. α-Thujone, the principal active ingredient of the aperitif absinthe, was shown to act as a noncompetitive antagonist of the GABA receptor solving a century-old mechanistic puzzle.

Neonicotinoid Insecticides

Nicotine was used for centuries to control insect pests with marginal effectiveness and considerable health risk. The discovery of synthetic alternatives, the neonicotinoids, allowed optimization for potency, safety, and photostability, ultimately leading to the neonicotinoids, currently the third most important class of insecticides. [^{3}H]Neonicotinoids prepared as radioligands showed the unique sensitivity of the nicotinic acetylcholine receptor of insects compared with mammals.

Related chemistry served to isolate pure native *Drosophila* receptor by affinity chromatography and photoaffinity label the binding subunit. The acetylcholine binding protein, used as a surrogate for the nicotinic receptor, was photoaffinity labeled with azidonicotinoids and azidoneonicotinoids and the derivatives identified by mass spectrometry to precisely define the unique binding sites conferring sensitivity differences between mammalian $\alpha_4\beta_2$ and insect subclasses of receptors. Comparative metabolism studies on all commercial neonicotinoids established their ease of biodegradation and the formation of many bioactive metabolites acting on the nicotinic receptor or altering the synchronization of plant growth and response to stress.

Herbicide Research

Herbicide research established that: (a) the metabolic activation of important thiocarbamates involves previously-unrecognized thiocarbamate sulfoxides; (b) the action of dichloroacetamides as safeners lowering the toxicity of thiocarbamates to maize but not to weeds involves inducing the synthesis of glutathione and glutathione *S*-transferase that rapidly detoxify the metabolically-activated herbicide; (c) major thiocarbamate herbicides and the fungicide benomyl are potent acetaldehyde dehydrogenase inhibitors and may therefore sensitize agricultural workers to ethanol intoxification; (d) the mutagenic and carcinogenic activity of chloroallyl thiocarbamates and dibromochloropropane are attributable to 2-haloacrolein metabolites; (e) the stereospecific metabolic conversion of phenoxyaryloxypropionates to their coenzyme A esters greatly enhances their potency as inhibitors of acetyl-coenzyme A carboxylase; (f) the inhibition of protein phosphatase 2A in plants by endothal accounts for its phytotoxicity, as with cantharidin and its effects in mammals; (g) a ^{3}H-labeled *N*-aryltetrahydrophthalimide developed as a radioligand effectively quantitates the protoporphyrinogen IX oxidase site which is very similar in plants and mammals.

Recognition of Lifetime Achievements

The research described above is detailed in a portion of the 737 publications and 31 patents by Dr. Casida and his coworkers. His contributions also involved the training of scientists from throughout the world in pesticide chemistry and toxicology. Sixty-three Ph.D. students have taken their degrees in entomology, molecular toxicology, comparative biochemistry, or agricultural and environmental chemistry under his supervision and he has worked with 160 postdoctoral fellows or visiting scholars, many of whom are now leading scientists in universities, government and industry concerned with pesticide design, development and safety. The National Institutes of Health Program Project “Mode of Action and Metabolism of Organic Toxicants” and related grants with Professor Casida as the Principal

Investigator, first in Madison and then at Berkeley, supported one of the longest continuous major academic programs on pesticide science in the United States. Dr. Casida was a member of the United States National Advisory Environmental Health Sciences Council and currently serves as a consultant in the United States and abroad on agrochemical research. He is a major contributor to international congresses and meetings and had joint research programs in Israel and Hungary and served as advisor to the Ministers of Agriculture and Irrigation in Egypt.

The lifetime contributions of Professor John Casida have been recognized by the Jeffrey Lectureship at the University of New South Wales, Australia in 1983, the Messenger Lectureship at Cornell University in Ithaca in 1985, the Yabuta Lectureship in Fukuoka and the Botyu-Kagaku Lectureship in Kyoto in 1987, the Sterling B. Hendricks Memorial Lectureship of the United States Department of Agriculture in 1992, the Third World Academy of Sciences Lectureship in Sciences, University of Buenos Aires in Argentina, 1997, and the University of California at Berkeley Faculty Research Lectureship (the highest distinction the Berkeley campus can bestow in honoring research accomplishments) in 1998. He received the Distinguished Service Award for research by the United States Department of Agriculture in 1988, the J.E. Bussart Award and the title Fellow by the Entomological Society of America in 1989, the Kôrô-sho Prize of the Pesticide Science Society of Japan in 1995 and was named Honorary Member of the Society of Toxicology in 1996 and the Pesticide Science Society of Japan in 2005. Professor Casida was elected as a member of the United States National Academy of Sciences in 1991, as a foreign member of the Royal Society (United Kingdom) in 1998, as a member of the European Academy of Sciences in 2004, and received the Wolf Foundation Prize in Agriculture in 1993.

Education/Training

University of Wisconsin, Madison: B.S. 1951 Entomology, M.S. 1952 Biochemistry, Ph.D. 1954 Entomology/Biochemistry

Professional Positions

1954-63	Assistant, Associate and Full Professor of Entomology, University of Wisconsin, Madison
1958-59	Visiting Researcher, University of Stockholm and Several European Laboratories
1970-71	Visiting Researcher, Cambridge University, University of Stockholm, and Kyoto University
1986-87	Visiting Researcher, University of Crete (Greece) and Centre National de Recherche Scientifique (Gif-sur-Yvette, France)

1964-present	Professor of Entomology and Toxicology and Director, Environmental Chemistry and Toxicology Laboratory, Dept. of Environmental Science, Policy and Management, University of California, Berkeley
1996-present	William Muriece Hoskins Chair in Chemical and Molecular Entomology, University of California, Berkeley

Award(s)

1958	Haight Fellow
1970	Guggenheim Fellow
1970	International Award for Research in Pesticide Chemistry, American Chemical Society
1978	Spencer Award for Research in Agricultural & Food Chemistry, American Chemical Society
1988	Distinguished Service Award for Research, USDA
1989	J.E. Bussart Award and Fellow of the Entomological Society of America
1991	National Academy of Sciences (U.S.A.), Member
1992	Sterling B. Hendricks Memorial Lectureship, Agricultural Research Service, USDA & American Chemical Society
1993	Wolf Foundation Prize in Agriculture
1994	Founders Award, Society of Environmental Toxicology and Chemistry
1995	Koro-Sho Prize, Pesticide Science Society of Japan
1997	Honorary Member, Society of Toxicology
1997	Honorary Doctor Degree and Third World Academy of Sciences Lectureship, University of Buenos Aires, Argentina
1998	Faculty Research Lecturer, U.C. Berkeley
1998	Royal Society of the United Kingdom, Fellow (FRS) and Foreign Member
2004	European Academy of Sciences, Member
2005	Honorary Member, Pesticide Science Society of Japan

USA and International Patents with Chemical Abstracts Citations

P1. 1957. Stahmann, M.A., T.C. Allen, J.E. Casida and R.K. Chapman. Organophosphoramide N-oxide pesticide product and process for making it. U.S. Pat. 2,814,636. Nov. 26, 1957.

P2. 1959. Casida, J.E. and B.W. Arthur. Process of preparing dimethyl-2,2,2-trichloro-1-n-butyryloxyethylphos-phonate. U.S. Pat. 2,911,435. Nov. 3, 1959.

P3. 1960. Casida, J.E. and B.W. Arthur. Dimethyl-2,2,2-trichloro-1-acetyloxyethyl phosphonate. U.S. Pat. 2,927,880. Mar. 8, 1960.

P4. 1960. Casida, J.E. and B.W. Arthur. Dimethyl-2,2,2-trichloro-1-n-butyryloxyethyl-phosphonate. U.S. Pat. 2,927,881. Mar. 8, 1960.

P5. 1962. Esenther, G.R., T.C. Allen, J.E. Casida and R.D. Shenefelt. Decayed wood extract as termite attractant. U.S. Pat. 3,070,495. Dec. 25, 1962.

P6. 1975. Tilles, H. and J.E. Casida. Method of making carbonoyl sulfoxide derivatives. U.S. Pat. 3,896,169. July 22, 1975.

P7. 1975. Tilles, H. and J.E. Casida. Carbamoyl sulfoxide derivatives. U.S. Pat. 3,928,436. Dec. 23, 1975.

P8. 1985. Palmer, C.J. and J.E. Casida. Pesticides comprising 1,4-bis-substituted 2,6,7-trioxabicyclo[2.2.2]octanes. PCT Int. Appl., 34 pp. CA103(23): 191474k

P9. 1986. Dureja, P., J.E. Casida and L.O. Ruzo. Additives for improved pesticide photostability. U.S. Pat. 4,622,315. Nov. 11, 1986, 6 pp.

P10. 1987. Casida, J.E., C.J. Palmer, J.P. Larkin and I.H. Smith. Trioxabicyclo [2.2.2]octanes as insecticides and acaricides. Eur. Pat. Appl., 65 pp. CA106(21): 176403e

P11. 1987. Casida, J.E., C.J. Palmer, J.P. Larkin and I.H. Smith. Preparation of heterobicycloalkane pesticidal compounds. Eur. Pat. Appl., 52 pp. CA107(1): 2705k

P12. 1988. Palmer, C.J. and J.E. Casida. 1,4-bis-Substituted-2,6,7-trioxabicyclo (2,2,2)-octanes having ethynyl substituted phenyl group. U.S. Pat. 4,772,624. Sept. 20, 1988, 8 pp.

P13. 1988. Casida, J.E., C.J. Palmer, J.P. Larkin and I.H. Smith. Preparation of 1-aryl-2,6,7-trioxabicyclo[2.2.2]octanes as pesticides. Eur. Pat. Appl., 25 pp. CA109(25): 231038d

P14. 1988. Casida, J.E., M. Elliott and D.A. Pulman. Preparation and testing of 4-(tertiary butyl)-1,3-dithianes and oxidized derivatives as insecticides, acaricides, and anthelmintics. Eur. Pat. Appl., 85 pp. CA110(21): 192834j

P15. 1988. Palmer, C.J. and J.E. Casida. 1,4-Bis-substituted 2,6,7-trioxabicyclo [2.2.2]octanes having an ethynyl-substituted phenyl group, and their pesticidal compositions. Eur. Pat. appl., 8 pp. CA110(23): 212833p

P16. 1988. Casida, J.E., M. Elliott and D.A. Pulman. Preparation of dithianes as pesticides and parasiticides. Eur. Pat. Appl., 44 pp. CA110(25): 231651u

P17. 1989. Palmer, C.J. and J.E. Casida. New class of pesticides comprising 1,4-bis-substituted-2,6,7-trioxabicyclo(2,2,2)octanes. U.S. Pat. 4,876,274. Oct. 24, 1989, 12 pp.

P18. 1990. Casida, J.E., C.J. Palmer, J.P. Larkin and I.H. Smith. Pesticidal compounds. U.S. Pat. 4,942,173. July 17, 1990, 10 pp.

P19. 1990. Casida, J.E., C.J. Palmer, J.P. Larkin and I.H. Smith. Pesticidal compounds. U.S. Pat. 4,965,257. Oct. 23, 1990, 19 pp.

P20. 1990. Casida, J.E. and M. Elliott. Preparation of dithiaspiroundecane- and dithiaspiro-undecadiene derivatives as pesticidal compounds. Eur. Pat. Appl., 24 pp. CA113(23): 211963f

P21. 1990. Casida, J.E., M. Elliott and D. Parkin. Preparation of dithianes as pesticides and drugs. Eur. Pat. Appl., 30 pp. CA114(5): 42773g

P22. 1990. Larkin, J.P., J.B. Weston, I.H. Smith, C.J. Palmer and J.E. Casida. Preparation of 2,6,7-trioxabicyclo[2.2.2]octanes and analogs as pesticides. Eur. Pat. Appl., 42 pp. CA114(5): 42795r

P23. 1990. Larkin, J.P., D. Parkin, C.J. Palmer and J.E. Casida. Preparation of 1,3-dithianes as pesticides. Eur. Pat. Appl., 45 pp. CA115(1): 8817k

P24. 1991. Casida, J.E. and M. Elliott. Dialkyl-substituted dithianes and pesticidal compositions. U.S. Pat. 4,985,411. Jan. 15, 1991, 10 pp.

P25. 1991. Casida, J.E., C.J. Palmer, J.P. Larkin and I.H. Smith. Pesticidal compounds. U.S. Pat. 5,057,508. Oct. 15, 1991, 19 pp.

P26. 1991. Palmer, C.J., J.E. Casida and J.P. Larkin. Alkynylphenyl-substituted 1,3-oxathione compounds with pesticidal activity. U.S. Pat. 5,061,726. Oct. 29, 1991, 14 pp.

P27. 1992. Casida, J.E., C.J. Palmer, J.P. Larkin and I.H. Smith. Pesticidal compounds. U.S. Pat. 5,157,029. Oct. 20, 1992.

P28. 1993. Larkin, J.P., J.B. Weston, I.H. Smith, C.J. Palmer and J.E. Casida. Pesticidal compounds. U.S. Pat. 5,204,333. April 20, 1993, 17 pp.

P29. 1994. Palmer, C. and J.E. Casida. Trithiabicyclo(2.2.2)octanes. U.S. Pat. 5,371,075. December 6, 1994, 19 pp.

P30. 1996. Casida, J.E., M. Elliott and D.A. Pulman. Heterocyclic pesticidal compounds. U.S. Pat. 5,502,073. March 26, 1996, 31 pp.

P31. 2001. Latli, B. and J.E. Casida. Substituted pyridines as modulators of the mammalian neuronal nicotinic acetylcholine receptor. U.S. Patent 6,303,638, October 16, 2001, 13 pp.

Selected Publications

(*Reprints included)

Casida, J.E. and T.C. Allen. 1951. A laboratory method for evaluating the phytotoxicity or phytostimulation of insecticides. Science **113**: 553-555.

*[1]Casida, J.E., T.C. Allen and M.A. Stahmann. 1953. Enzymatic and chemical oxidation of dimethylphosphoramides to biologically active dimethylphosphoramide oxides. Nature **172**: 243-245.

Casida, J.E. 1955. Isomeric substituted-vinyl phosphates as systemic insecticides. Science **122**: 587-598.

*Casida, J.E. 1958. Phosphorus-32 pentasulfide: preparation by isotopic exchange and conversion to thiophosphoryl-32 chloride and phosphorus-32 trichloride. Acta Chem. Scand. **12**: 1691-1692.

*[1]Casida, J.E., M. Eto and R.L. Baron. 1961. Biological activity of a tris-*o*-cresyl phosphate metabolite. Nature **191**: 1396-1397.

Dorough, H.W., N.C. Leeling and J.E. Casida. 1963. Nonhydrolytic pathway in metabolism of *N*-methylcarbamate insecticides. Science **140**: 170-171.

Roger, J-C., H. Chambers and J.E. Casida. 1964. Nicotinic acid analogs: effects on response of chick embryos and hens to organophosphate toxicants. Science **144**: 539-548.

Casida, J.E., J.L. Engel, E.G. Essac, F.X. Kamienski and S. Kuwatsuka. 1966. Methylene-C^{14}-dioxyphenyl compounds: metabolism in relation to their synergistic action. Science **153**: 1130-1133.

Fukami, J.-I., I. Yamamoto and J.E.Casida. 1967. Metabolism of rotenone *in vitro* by tissue homogenates from mammals and insects. Science **155**: 713-716.

Tsukamoto, M. and J.E. Casida. 1967. Metabolism of methylcarbamate insecticides by the $NADPH_2$-requiring enzyme system from houseflies. Nature **213**: 49-51.

Ivie, G.W. and J.E. Casida. 1970. Enhancement of photoalteration of cyclodiene insecticide chemical residues by rotenone. Science **167**: 1620-1622.

Casida, J.E. 1970. Mixed-function oxidase involvement in the biochemistry of insecticide synergists. J. Agric. Food Chem. **18**: 753-772.

*[1]Casida, J.E., E.C. Kimmel, M. Elliott and N.F. Janes. 1971. Oxidative metabolism of pyrethrins in mammals. Nature **230**: 326-327.

Maddrell, S.H.P. and J.E. Casida. 1971. Mechanism of insecticide-induced diuresis in *Rhodnius*. Nature **213**: 55-56.

Abernathy, C.O. and J.E. Casida. 1973. Pyrethroid insecticides: esterase cleavage in relation to selective toxicity. Science **179**: 1235-1236.

Bellet. E.M. and J.E. Casida. 1973. Bicyclic phosphorus esters: high toxicity without cholinesterase inhibition. Science **182**: 1135-1136.

*[2]Casida, J.E., R.L. Holmstead, S. Khalifa, J.R. Knox, T. Ohsawa, R.J. Palmer and R.Y. Wong. 1974. Toxaphene insecticide: a complex biodegradable mixture. Science **183**: 520-521.

*[2]Lay, M.-M., J.P. Hubbell and J.E. Casida. 1975. Dichloroacetamide antidotes for thiocarbamate herbicides: mode of action. Science **189**: 287-289.

*[2]Schuphan, I., J.D. Rosen and J.E. Casida. 1979. Novel activation mechanism for the promuta-genic herbicide diallate. Science **205**: 1013-1015.

Wing, K.D., A.H. Glickman and J.E. Casida. 1983. Oxidative bioactivation of *S*-alkyl phosphorothiolate pesticides: stereospecificity of profenofos insecticide activation. Science **219**: 63-65.

Lawrence, L.J. and J.E. Casida. 1984. Interactions of lindane, toxaphene and cyclodienes with brain-specific *t*-butylbicyclophosphorothionate receptor. Life Sci. **35**: 171-178.

Palmer, C.J. and J.E. Casida. 1985. 1,4-Disubstituted 2,6,7-trioxabicyclo [2.2.2]octanes: a new class of insecticides. J. Agric. Food Chem. **33**: 976-980.

Waterhouse, A.L., I.N. Pessah, A.O. Francini and J.E. Casida. 1987. Structural aspects of ryanodine action and selectivity. J. Med. Chem. **30**: 710-716.

Casida, J.E. 1990. Pesticide mode of action: evidence for and implications of a finite number of biochemical targets. Pp. 11-22 in Pesticides and Alternatives: Chemical and Biological Approaches to Pest Control. (Casida, J.E., Ed.), Elsevier Science Publ., New York.

Deng, Y., C.J. Palmer and J.E. Casida. 1991. Housefly brain γ-aminobutyric acid-gated chloride channel: target for multiple classes of insecticides. Pestic. Biochem. Physiol. **41**: 60-65.

Elliott, M., D.A. Pulman, J.P. Larkin and J.E. Casida. 1992. Insecticidal 1,3-dithianes. J. Agric. Food Chem. **40**: 147-151.

*Li, Y.-M. and J.E. Casida. 1992. Cantharidin-binding protein: identification as protein phosphatase 2A. Proc. Natl. Acad. Sci. USA **89**: 11867-11870.

Liu. M.-Y. and J.E. Casida. 1993. High affinity binding of [^{3}H]imidacloprid in the insect acetylcholine receptor. Pestic. Biochem. Physiol. **46**: 40-46.

Tomizawa, M., B. Latli and J.E. Casida. 1996. Novel neonicotinoid-agarose affinity column for *Drosophila* and *Musca* nicotinic acetylcholine receptors. J. Neurochem. **67**: 1669-1676.

Hainzl, D. and J.E. Casida. 1996. Fipronil insecticide: novel photochemical desulfinylation with retention of neurotoxicity. Proc. Natl. Acad. Sci. USA **93**: 12764-12767.

Casida, J.E. and G.B. Quistad. 1998. Golden age of insecticide research: past, present or future? Annu. Rev. Entomol. **43**: 1-16.

Fang, N. and J.E. Casida. 1998. Anticancer action of cubé insecticide: correlation for rotenoid constituents between inhibition of NADH:ubiquinone oxidoreductase and induced ornithine decarboxylase activities. Proc. Natl. Acad. Sci. USA **95**: 3380-3384.

Schuler, F., T. Yano, S. Di Bernardo, T. Yagi, V. Yankovskaya, T.P. Singer and J.E. Casida. 1999. NADH-quinone oxidoreductase:PSST subunit couples electron transfer from iron-sulfur cluster N2 toquinone. Proc. Natl. Acad. Sci. USA **96**: 4149-4153.

*[3]Höld, K.M., N.S. Sirisoma, T. Ikeda, T. Narahashi and J.E. Casida. 2000. α-Thujone (the active component of absinthe): γ-Aminobutyric acid type A receptor modulation and metabolic detoxification. Proc. Natl. Acad. Sci. USA **97**: 3826-3831.

Ratra, G.S., S.G. Kamita and J.E. Casida. 2001. Role of human $GABA_A$ receptor β3 subunit in insecticide toxicity. Toxicol. Appl. Pharmacol. **172**: 233-240.

Winrow, C.J., M.L. Hemming, D.M. Allen, G.B. Quistad, J.E Casida and C. Barlow. 2003. Loss of neuropathy target esterase in mice links organophosphate exposure to hyperactivity. Nat. Genet. **33**: 477-485.

Quistad, G.B, C. Barlow, C.J. Winrow, S.E. Sparks and J.E Casida. 2003. Evidence that mouse brain neuropathy target esterase is a lysophospholipase. Proc. Natl. Acad. Sci. USA **100**: 7983-7987.

Nomura, D.K., D. Leung, K. Chiang, G.B. Quistad, B.F. Cravatt and J.E. Casida. 2005. A brain detoxifying enzyme for organophosphorus nerve poisons. Proc. Natl. Acad. Sci. USA **102**: 6195-6200.

*[4]Chen, L., K.A. Durkin, and J.E. Casida. 2006. Structural model for γ-aminobutyric acid receptor noncompetitive antagonist binding: widely-diverse structures fit the same site. Proc. Natl. Acad. Sci. U.S.A. **103**: 5185-5190.

*[5]Tomizawa, M., T.T. Talley, D. Maltby, K.A. Durkin, K.F. Medzihradszky, A.L. Burlingame, P. Taylor and J.E. Casida. 2007. Mapping the elusive neonicotinoid binding site. Proc. Nat. Acad. Sci. U.S.A. **104**: 9075-9080.

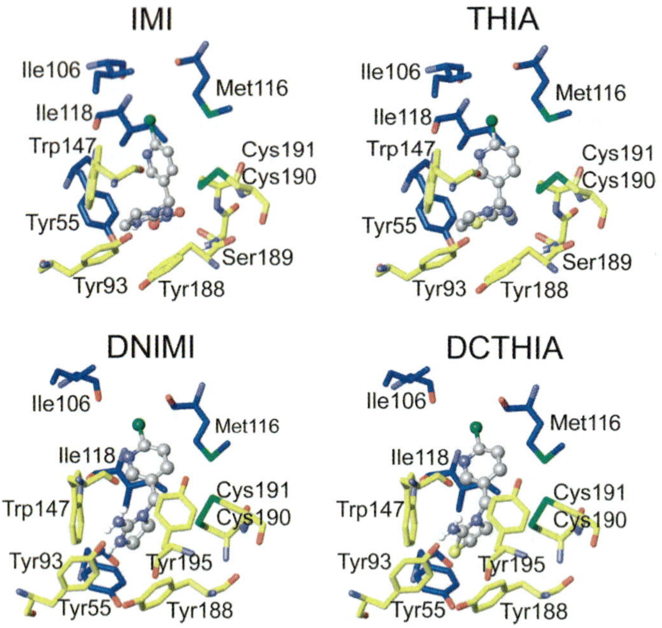

Proposed Interactions of Insecticides and Analogs

Above: at the agonist site of the acetylcholine binding protein (Fig. 4 of Tomizawa et al., 2007).

Below: at the noncompetitive antagonist site of the $GABA_A$ receptor β_3 homopentamer (Fig. 6 of Chen et al., 2006).

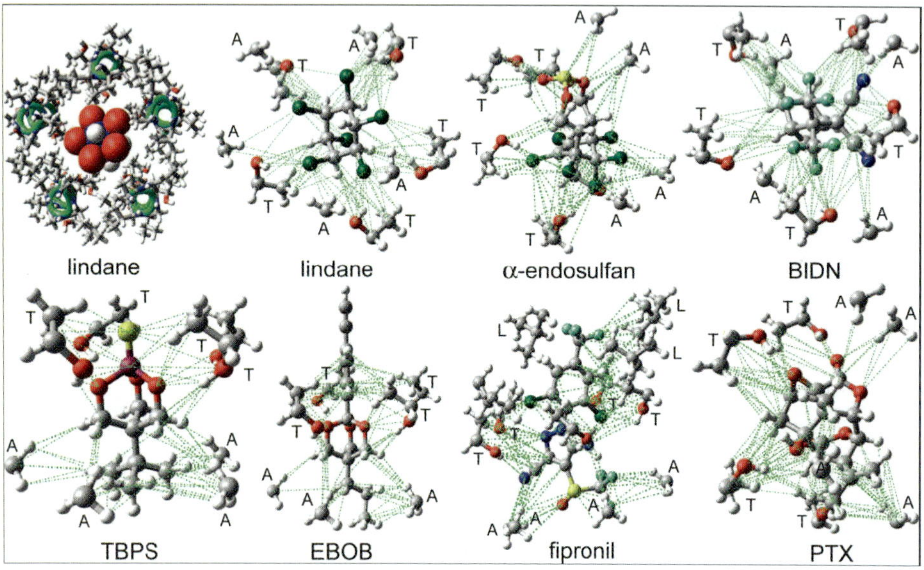

No. 4371 August 8, 1953 NATURE 243

Melbourne; the Manchester and District Branch were responsible for a one-day conference on textile physics; and the Scottish Branch arranged a course of eight lectures at the University of Glasgow on the development of concepts of physical science and a course of seven lectures at the University of Edinburgh on semiconductors. The Education Group held a joint meeting with the Acoustics Group of the Physical Society and a conference on school and university examinations in physics. The summer meeting of the Electronics Group took place at the Clarendon Laboratory, Oxford. The X-ray Analysis Group celebrated the fortieth anniversary of the discovery of X-ray diffraction by a two-day conference in London followed by a dinner at which Prof. M. von Laue was the principal guest, and the Industrial Spectroscopy Group was responsible for the Third International Spectroscopy Colloquium held at High Leigh, Hoddesdon, at which many scientists from overseas were present.

Further repairs and improvements were carried out to the Institute's House at 47 Belgrave Square, London; the Institute is again indebted to Mr. R. S. Whipple for another generous donation, this time of £500 to the Furnishing Fund.

CALIFORNIA INSTITUTE OF TECHNOLOGY

ANNUAL REPORT FOR 1951 52

THE annual report of the president of the California Institute of Technology for the year ended June 30, 1952 (pp. v+110; from the Institute, Pasadena, California, 1953), reflects the current concern in the United States over the shortage of scientific and engineering man-power, though the policy of limiting the freshmen to 180 each year has kept the Institute freer of fluctuations than most other institutions, and there is evidence of a slight increase in enrolments generally which should improve the supply in a few years time. Concern is also expressed lest higher education should depend upon any one source of financial support, particularly upon Federal funds; but the financial report which, together with those of the secretary, the deans and administrative officers, and of the several Divisions, is appended to the president's report, shows that of income and expenditure on organized research totalling nearly 11·5 million dollars, 9·6 millions came from the Federal Government for the Jet Propulsion Laboratory, the South Californian Co-operative Wind Tunnel, and the Vista Project, while a further 1·3 millions was received and spent on other research contracts. Gifts for current operations, however, exceeded 1·2 million dollars, and the president points out that, except for the few projects in such fields as nuclear physics and aeronautics, the educational and research work of the Institute is still mainly supported by private funds.

Among the features of the research during the year is the new technique developed in the Division of Biology by Prof. R. Dulbecco for studying those viruses which attack animal tissue, as compared with those which attack only plants or bacteria. The new synchrotron which has been under construction during the past two years has produced electrons with an energy of 525 MeV., and this beam causes the creation of π-mesons. Active study during the past year of the new unstable sub-atomic particles which occur in the cosmic rays has led to the identification of at least five different types of such particles. In the Division of Chemistry and Chemical Engineering the structure of complex molecules is being determined by the electron-diffraction method and by infra-red spectroscopy. Evidence was obtained during the year that a polypeptide helix without 3·6 amino-acid residues per turn is present not only in some synthetic polypeptides but also in proteins like hair, finger-nail, skin, hæmoglobin and serum albumin. The sequence of amino-acids in proteins continued to be studied by chromatographic analysis, and the emphasis in work in chemical engineering is being gradually changed from the study of the equilibrium thermodynamic properties of fluids at high pressures to that of non-equilibrium phenomena, such as thermal and material transfer, under quiescent and convective conditions.

In the Division of Civil, Electrical and Mechanical Engineering and Aeronautics, application of the Institute's electric analogue computor has led, by means of new electric circuit analogies, to the development of a rapid and general method for the static stress analysis of complex structures. A new research programme on the basic insulation characteristics of rotating machine insulation was started, and a new technique developed for use in the high-speed water tunnel permits the determination of the dynamic force coefficients of submerged bodies. Advances were made in two long-range programmes concerned with turbulence and compressible fluid flow, and in the Division of the Geological Sciences, where earthquake records have been studied for more than twenty years, the recent Tehachapi earthquake and its many after-shocks were fully recorded on a dozen instruments in the Pasadena Laboratory and in several field stations. The shock also focused fresh attention on building design for seismically active regions, and the Division of Engineering is engaged in designing structures having maximum resistance and in predicting the behaviour of a given structure.

ENZYMATIC AND CHEMICAL OXIDATION OF DIMETHYL-PHOSPHORAMIDES TO BIOLOGICALLY ACTIVE DIMETHYL-PHOSPHORAMIDE OXIDES

By J. E. CASIDA, T. C. ALLEN and M. A. STAHMANN

Departments of Biochemistry and Entomology, University of Wisconsin, Madison

SEVERAL dimethylphosphoramides have been investigated as insecticides, and the group as a whole shows great promise in protecting plants against insect pests by chemotherapeutic action. One of these chemicals, octamethylpyrophosphoramide or 'Schradan', has been developed commercially as a systemic insecticide. These alkylamino phosphates differ from the related toxic alkyl phosphates in that they are generally poor *in vitro* inhibitors of cholinesterase, but nevertheless animals exposed to 2–50 mgm./kgm. show typical symptoms of acetylcholine accumulation[1-4]. Furthermore, these alkylamino phosphates produce a differential *in vivo* inhibition of the peripheral cholinesterase[1,5]. They are relatively ineffective inhibitors of chymotrypsin[6].

244 NATURE August 8, 1953 Vol. 172

ENHANCEMENT IN INHIBITORY ACTIVITY OF DIMETHYLPHOSPHORAMIDES BY LIVER SLICES AND PERMANGANATE

Chemicals (*a*)	Rel. hydrolysis, rate (*b*)	Relative effectiveness as chymotrypsin inhibitor (*c*)	Change in effectiveness (*d*) after reaction with: liver slices (*e*)	neutral MnO_4^- (*f*)
(1) 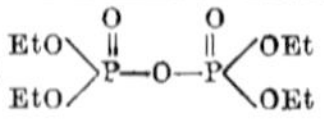	5·0	4·7	− − −	− −
(2) 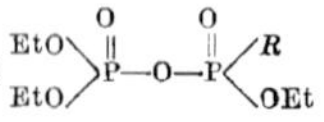	2·9	3·5	− −	−
(3) 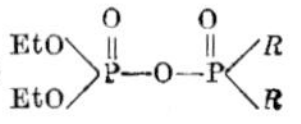	2·8	2·8	−	+
(4) 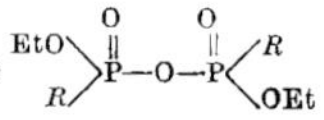	2·7	2·4	−	+
(5) 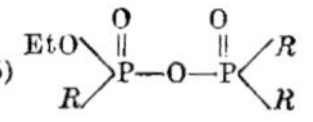	1·8	<0·1	+ + +	+ + + +
(6) 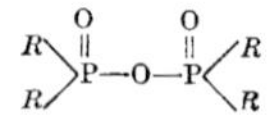	0	<0·1	+ + + +	+ + + +
(7) 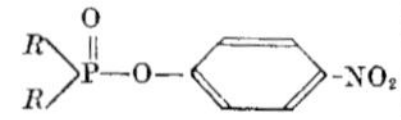	—	0·5	+ +	+ + +
(8) 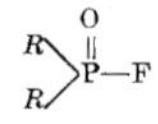	—	<0·1	+ +	+ + +

R = —N(CH$_3$)$_2$ EtO = —OC$_2$H$_5$

(*a*) Sources of chemicals: Victor Chemical Co., 1, 7; Bayer–I. G. Farbenindustrie, 2, 5; Monsanto Chemical Co., 3, 4, 6; and Dr. F. I. Edwards, Bur. Ent. and Plant Quar. U.S. Dep. Agric., 8.

(*b*) Measured manometrically at 37° C. and *p*H 7·4; half-life of first chemical 2·7 hr.; other chemicals expressed as the logarithm of the relative hydrolysis rate based on 5·0 for No. 1.

(*c*) Based on negative logarithm of molar concentration required to effect 50 per cent inhibition of esterase activity of 20 γ crystalline chymotrypsin per ml. in *M*/50 phosphate buffer at *p*H 6·25 with 16 hr. combination time at 4° C. prior to assay.

(*d*) Change in inhibitor potency indicated as follows:

− − − = < 1 per cent of original activity
− − = 1–10 ,, ,, ,,
− = 10–80 ,, ,, ,,
+ = 2–10 fold increase in activity
+ + = 10–100 ,, ,, ,,
+ + + = 100–1,000 ,, ,, ,,
+ + + + = 1,000–1,000,000 fold increase in activity

(*e*) 5 ml. *M*/20 phosphoramide incubated with 2·0 gm. fresh rat liver slices in Krebs–Ringer phosphate buffer at *p*H 6·5 for 3 hr. at 37° C. in 100 per cent oxygen atmosphere, then extracted with 10 ml. of chloroform and successive dilutions of the chloroform extract assayed for inhibitor present.

(*f*) 10 ml. *M*/20 phosphoramide incubated with *M*/20 MnO$_4^-$ in *M*/10 phosphate buffer at *p*H 6·5 until MnO$_4^-$ reduced, and then extracted with 10 ml. of chloroform and successive dilutions of the chloroform extract assayed for inhibitor present. (Compound 1 did not reduce the MnO$_4^-$.)

The dimethyl amide group stabilizes the anhydride bond and makes it very resistant to alkaline hydrolysis[7].

'Schradan' is converted *in vitro* by liver slices[1,2,8], or *in vivo* in mammals or plants[1,9,10] to unstable anticholinesterase agent(s)[1], which are effective inhibitors of chymotrypsin[11]. Permanganate oxidation also changed this alkali-stable pyrophosphate to an unstable material capable of inhibiting the enzymatic activity of chymotrypsin[11] and cholinesterase[4]. The active product from this permanganate oxidation was shown to be a compound that phosphorylates the enzymatically active site of chymotrypsin to add one phosphate residue per molecule of chymotrypsin[11]. This chemical oxidation likewise activated *bis*-(dimethylamino)-*p*-nitrophenyl phosphate and *bis*-(dimethylamino)-fluorophosphine oxide[11]. Liver slices were effective in activating the latter compound[4]. The nature of these active substance(s) has not been previously demonstrated.

Our present work is concerned with the chemical nature of the active substances formed biologically from phosphoramides and their relation to the chemical oxidation products. The specificity of this activation is shown in the accompanying table. Permanganate oxidation or incubation with rat liver slices enhanced the acylating reactivity of many dimethylphosphoramides as measured by chymotrypsin inhibition. The initial activity of these compounds as chymotrypsin inhibitors paralleled their relative rates of hydrolysis. A similar correlation has been made with the stability of the anhydride link in dialkyl phosphates and their cholinesterase inhibiting power *in vitro*[12] and in their effectiveness as contact insecticides. The dimethylphosphoramides shown in the table were exposed to neutral permanganate or liver slices in aqueous buffered solutions and the products then extracted into chloroform. Serial dilutions of this chloroform extract were afterwards assayed for chymotrypsin inhibitory activity. As is shown in the last two columns of the table, this activity generally increased by about the same degree with liver slices or permanganate. With octamethylpyrophosphoramide, compound 6, this increase was about one-million fold. Those phosphoramides which showed no increase in activity were the less-stable compounds where the products might have been destroyed prior to partitioning or were poorly extracted into the chloroform. The products of the various reactions were also tested for *in vitro* inhibition of the true cholinesterase of rat brain. The results showed an increased cholinesterase inhibitory activity which generally paralleled the inhibition observed with chymotrypsin.

The nature of the activation of these phosphoramides was further investigated with octamethylpyrophosphoramide or 'Schradan'. The active constituent from its metabolism by rat liver was purified by counter-current distribution utilizing enzymatic assays for determining inhibitor concentration. The active enzyme inhibitor was separated from the bulk of the unmetabolized phosphoramide with three different pairs of solvent systems. The anti-cholinesterase agent produced in plants or by chemical oxidation possessed the same distribution characteristics and activity against cholinesterase and chymotrypsin as that from the rat liver system, indicating the same active product in all cases.

The mode of formation of the active agent, its decomposition to formaldehyde, distribution behaviour, phosphorylating reactivity and biological

No. 4371 August 8, 1953 NATURE 245

specificity indicate that the activation process is an oxidation and strongly suggest an oxidation of a nitrogen atom of the phosphoramide. Further evidence for this was obtained from the infra-red absorption of 'Schradan' following oxidation. Several new absorption bands appeared at almost the identical positions as new bands which are formed when tertiary alkyl amines are oxidized to the amine oxide. Like the amine oxides, the new 'Schradan' oxide obtained biologically or with peracetic acid or permanganate was decomposed on heating in acid to an aldehyde and monomethyl amine. The yield of formaldehyde from permanganate or biologically oxidized 'Schradan' was directly proportional to the cholinesterase inhibition. This oxide of 'Schradan' showed weak basic properties, for it was less readily extracted from acid solutions than from neutral or alkaline solutions by chloroform. The oxidation increased the reactivity of the pyrophosphate anhydride bond to form a reactive phosphorylating agent which was readily hydrolysed by either acid or alkali. This increased chemical reactivity would appear to be due to the new positive centre formed by the oxidation of the phosphoramide nitrogen.

A variety of biological tissues were effective in oxidizing 'Schradan'. In addition to the higher plants, yeast and bacteria also produced the active oxidation product. Many different insects treated with 'Schradan' similarly oxidized the absorbed material. The tissues found most active in this conversion were the gastric cæcæ of the cockroach. The toxicity of 'Schradan' to insects and mammals appeared in all cases to be due to the enzymatic oxidation *in vivo* to yield this highly effective anti-cholinesterase. The widespread distribution of enzyme systems capable of oxidizing 'Schradan' suggests that such systems must have a rather important biological role and that naturally occurring phosphoramides might also be enzymatically oxidized.

Thus, many dimethylphosphoramides are susceptible to enzymatic and chemical oxidation and thus form a new functional group for which we propose the new name 'phosphoramide oxide'. The phosphoramide oxides of phosphoric acid anhydrides are reactive acylating agents and are particularly effective inhibitors of the physiologically important cholinesterase enzymes.

A more complete account of these studies will appear elsewhere.

Publication of this communication is approved by the director of the Wisconsin Agricultural Experiment Station. The investigation was supported in part by the Research Committee of the Graduate School from funds supplied by the Wisconsin Alumni Research Foundation. [Jan. 23.

[1] DuBois, K. P., Doull, J., and Coon, J. M., *J. Pharmacol. and Exp. Therap.*, **99**, 376 (1950).

[2] Gardiner, J. E., and Kilby, B. A., *Biochem. J.*, **51**, 78 (1952).

[3] Holmstedt, B., *Acta Physiol. Scand.*, **25** (Suppl. 90), 44 (1951).

[4] Aldridge, W. N., and Barnes, J. M., *Nature*, **169**, 345 (1952).

[5] DuBois, K. P., and Doull, J., *Fed. Proc.*, **9**, 268 (1950).

[6] Jansen, E. F., Curl, A. L., and Balls, A. K., *J. Biol. Chem.*, **190**, 557 (1951).

[7] Heath, D. F., and Casapieri, P., *Trans. Farad. Soc.*, **47**, 1093 (1951).

[8] Gardiner, J. E., and Kilby, B. A., *Biochem. J.*, **46**, xxxii (1950).

[9] Hall, S. A., Stohlmann, III, J. Wm., and Schechter, M. S., *Anal. Chem.*, **23**, 1866 (1951).

[10] Casida, J. E., Chapman, R. K., and Allen, T. C., *J. Econ. Ent.*, **45**, 568 (1952).

[11] Casida, J. E., Allen, T. C., and Stahmann, M. A., *J. Amer. Chem. Soc.*, **74**, 5548 (1952).

[12] Aldridge, W. N., and Davison, A. N., *Biochem. J.*, **51**, 62 (1952).

INTRA-HELIX S—S LINKED STRUCTURES FOR INSULIN

By Dr. U. W. ARNDT and Dr. D. P. RILEY

Royal Institution, Albemarle Street, London, W.1

WE have briefly mentioned elsewhere[1] that we have considered some feasible structures for the insulin molecule by combining the amino-acid sequences of Sanger and his co-workers[2] with the α-helix hypothesis of Pauling, Corey and Branson[3] and arrived at a limited number of solutions. As this work was facilitated by the use of the Hartley–Robinson 'Courtauld' atomic models[4] and as we knew that Dr. Conmar Robinson was himself working on the insulin problem, we delayed publishing a fuller account. Dr. Robinson has now published[5] his results, which differ very considerably from our conclusions.

We wish to say, first, that to regard the α-helix as a perfect and immutable structure is to take a much too rigid view, and one that is not required by its chemistry. While, in a synthetic polypeptide composed of a single type of monomer, nearly complete regularity is to be expected, this is no longer so in a natural protein constituted of many different amino-acid residues. Variations in both the angle and translation of the helix are permissible and may be necessitated by the different *R*-groups. Such variations were allowed at the outset by Pauling and Corey, and the characteristic constants of the α-helix are known to vary in the case of the synthetic polypeptides. Our suggestion is that, in the natural proteins, the variations can occur within one and the same chain. In the work to be described here, therefore, we have not hesitated to include apparently distorted structures because of this possibility of distributing the strain over several residues.

The basis of our present argument is that an intra-helix S—S link exists, and that the insulin molecule consists of two types of polypeptide chains, $\alpha_1(LH,C_\beta 2)$ and $\alpha_2(RH,C_\beta 1)$, both, of course, containing only L-C_α atoms, as previously described[1a]. These are identified with Sanger's *A* and *B* fractions: $A \equiv \alpha_1, B \equiv \alpha_2$. There is some experimental evidence for this view. The X-ray scattering curve for powdered native insulin appears to fit better a mixture of the two possible variants of the α-helix than either separately, while that for the separated *B*-fraction corresponds more closely to the calculated α_2 intensity function alone. We have not yet obtained any data for the *A*-fraction; but the purpose of this communication is not so much to discuss our experimental X-ray results as to propose possible hypothetical structures for insulin deduced from consideration of the spatial properties of the α_1- and α_2-helices. Nevertheless, it may be that the scattering method will have provided the essential clue to the structure of this molecule.

We differ from Dr. Robinson in allowing the formation of an intra-helix S—S link between the seventh and eleventh residues of the *A*-chain. It thereby becomes possible to consider molecules containing fewer inter-helix S—S bridges, and this permits the employment of both types of α-helix. Previously[1a], we stated that the presence of proline imposed the α_2-configuration on any chain in which it occurred. This is only partly true, and we should like here to make a fuller statement.

ACTA CHEMICA SCANDINAVICA **12** (1958) 1691—1692

Phosphorus-32 Pentasulfide: Preparation by Isotopic Exchange and Conversion to Thiophosphoryl-32 Chloride and Phosphorus-32 Trichloride

JOHN E. CASIDA*

Department of Entomology, University of Wisconsin, Madison 6, Wisconsin, USA

Certain triesters of phosphoric and thiophosphoric acids are important as pesticides and potential chemical warfare agents. Adequate studies on the metabolism and mode of action of these organophosphorus compounds require their synthesis from high specific activity radiolabeled inorganic intermediates such as phosphorus pentasulfide (P_2S_5), thiophosphoryl chloride ($PSCl_3$), phosphorus trichloride (PCl_3), and phosphoryl chloride ($POCl_3$).

$^{32}POCl_3$ can be readily prepared in high specific activity from the reaction of $H_3{}^{32}PO_4$ with phosphorus pentachloride (PCl_5)[1]. Neutron irradiation is commonly employed to label certain of the other inorganic phosphorus intermediates [2-5]. This method is often inconvenient and the specific activity of the organophosphorus compounds prepared from these irradiated intermediates is usually not greater than 5 mC/g. $^{32}PSCl_3$ can also be prepared from $H_3{}^{32}PO_4$ *via* $^{32}POCl_3$ and $^{32}PCl_3$[6,7] or by an inefficient isotope exchange reaction [8].

Various procedures were tried to achieve isotope exchange between $H_3{}^{32}PO_4$ or $Na_2H^{32}PO_4$ and red phosphorus, $PSCl_3$ or P_2S_5. The products from each attempt were reacted with ethanol [9] or converted to other known organic derivatives [4] and then separated by paper and ion-exchange chromatography [10] to determine the degree of exchange. After correction for the radioactivity remaining on the glassware, the only significant exchange occurred with the sulfur-containing compounds. The percentage exchange was low with $PSCl_3$ but almost complete with P_2S_5.

For preparation of $^{32}P_2S_5$, carrier-free $H_3{}^{32}PO_4$ or $Na_2H^{32}PO_4$ was evaporated in a reaction flask. The acid was preferable when the residue after evaporation was high and the salt when the residue was no more than a few milligrams. Red phosphorus and sulfur were then added in a 2 to 5 molar ratio with a combined weight of 0.20 to 2.0 g. While gassing the flask with dry carbon dioxide, the mixture was heated in one spot with a weak flame to initiate the vigorous reaction [11] that liquified the mixture and formed P_2S_5. When the mixture was then boiled for 10 min the exchange reaction was over 98 % complete. Similar exchange (95–98 %) resulted by slowly melting P_2S_5 in a reaction flask and then boiling for 10 min while gassing with carbon dioxide, and by heating the P_2S_5 for 2 h at 300–500°C in sealed ampules. The specific activity of the $^{32}P_2S_5$ thusly formed remained constant through repeated recrystallizations from carbon disulfide. When this $H_3{}^{32}PO_4$–P_2S_5 exchange reaction was run in the presence of varying amounts of non-labeled H_3PO_4, the percentage of the radioactivity recovered as $H_3{}^{32}PO_4$ closely approximated the percentage of the total phosphorus in the reaction mixture contributed by the H_3PO_4.

$^{32}PSCl_3$ was prepared in 73 % yield from $^{32}P_2S_5$ by reaction with 3 molar equivalents of PCl_5 [12,13] in the presence of anhydrous aluminium chloride comparable to 1.5 times the weight of the sulfur. The reaction was carried out in a closed all-glass system for 1.5 h at 160°C. $^{32}PCl_3$ was then formed in 91 % yield by reaction of the $^{32}PSCl_3$ with 1.1 molar equivalents of triphenyl phosphine at 130–140°C for 2 h according to a known type reaction [14] for obtaining sulfur migration from $PSCl_3$.

$^{32}P_2S_5$ prepared by this isotope exchange reaction has been used in the preparation of 15 different organic phosphate and phosphorothioate insecticides. Yields of the chromatographically-pure insecticides varied from 20–90 % based both on total phosphorus and radioactivity. The specific

* Current address: Institute of Organic Chemistry and Biochemistry, University of Stockholm, Stockholm, Sweden.

activities ranged from 2 to 340 mC/g. A more detailed report on the conversion of the $^{32}P_2S_5$ to various labeled insecticides and the characterization of the radioactive products will be published elsewhere.

Approved for publication by the Director of the Wisconsin Agricultural Experiment Station. This investigation was supported in part by a research contract with the *U. S. Atomic Energy Commission* (Contract No. AT (11—1)-64, Project No. 14). The author gratefully acknowledges the kind advice of Drs. G. Schrader and R. Mühlmann of *Farbenfabriken Bayer*, Elberfeld, Germany, and Dr. C. Simon of *J. R. Geigy*, Basle, Switzerland, and the skilled technical assistance of Mr. T. K. Knoke and Mr. K. Bürgin.

1. Kalinsky, J. L. and Weinstein, A. *J. Am. Chem. Soc.* **76** (1954) 5882.
2. Hein, R. E. and McFarland, R. H. *J. Am. Chem. Soc.* **74** (1952) 1856.
3. Louloudes, S. J., Kaplanis, J. N. and Roan, C. C. *J. Org. Chem.* **21** (1956) 685.
4. Mühlmann, R. and Schrader, G. *Z. Naturforsch.* **12b** (1957) 196.
5. Casida, J. E. *Unpublished results.*
6. Murray, D. H. and Spinks, J. W. *Can. J. Chem.* **30** (1952) 497.
7. Vigne, J. P., Tabau, R. L. and Fondarai, J. *Bull. soc. chim. France* **23** (1956) 459.
8. Vigne, J. P. and Tabau, R. L. *Bull. soc. pharm. Marseille* **5** (1956) 321.
9. Kosolapoff, G. M. *Organophosphorus Compounds,* John Wiley and Sons, New York 1950, pp. 180, 211, 217 and 236.
10. Plapp, F. W. and Casida, J. E. *Anal. Chem.* **30** (1958) *In press.*
11. Schrader, G. *Personal communication.*
12. Thorpe, T. E. *Chem. News* and *J. Phys. Sci.* **24** (1871) 135.
13. Lockau, V. S. and Lüdicke, M. *Z. Naturforsch.* **7b** (1952) 389.
14. Gottlieb, H. B. *J. Am. Chem. Soc.* **54** (1932) 748.

Received September 19, 1958.

1396 NATURE September 30, 1961 Vol. 191

work is under way with a number of Japanese and American varieties of soybeans to determine the effects of variety and location of growth on protein composition.

W. J. Wolf
G. E. Babcock
A. K. Smith

Northern Utilization Research and Development Division,
Agricultural Research Service,
U.S. Department of Agriculture,
Peoria, Illinois.

1 Markley, K. S., ed., *Soybeans and Soybean Products*, **1** (Interscience Publishers, New York, 1950).
2 Cartter, J. L., and Hopper, T. H., *U.S. Dept. Agric. Tech. Bull.* 787, 66 (1942).
3 Csonka, F. A., and Jones, D. B., *J. Agric. Res.*, **46**, 51 (1933).
4 Krober, O. A., *J. Agric. Food Chem.*, **4**, 254 (1956).
5 Van Etten, C. H., Hubbard, J. E., Mallan, J. M., Smith, A. K., and Blessin, C. W., *J. Agric. Food Chem.*, **7**, 129 (1959).
6 Belikov, I. F., and Tyuleneva, N. P., *Masoloboino-Zhirovaya Prom.*, **25**, No. 10, 19 (1959).
7 Wolf, W. J., and Briggs, D. R., *Arch. Biochem. Biophys.*, **85**, 186 (1959).
8 Naismith, W. E. F., *Biochim. Biophys. Acta*, **16**, 203 (1955).
9 Smith, A. K., Belter, P. A., and Johnsen, V. L., *J. Amer. Oil Chemists' Soc.*, **29**, 309 (1952).
10 Bressani, R., and Mertz, E. T., *Cereal Chem.*, **35**, 227 (1958).
11 Jones, R. W., Taylor, N. W., and Senti, F. R., *Arch. Biochem. Biophys.*, **84**, 363 (1959).
12 Jensen, R., *Proc. European Brewery Convention Congr.*, Nice, 1953, 74.

Biological Activity of a Tri-o-Cresyl Phosphate Metabolite

Tri-*o*-cresyl phosphate (TOCP) is metabolized *in vitro* and *in vivo* to form potent esterase inhibitors[1–5]. The nature and biological activity of the metabolites were investigated.

Rats administered radiophosphorus-labelled TOCP were found to excrete radiolabelled di-aryl phosphates, mono-aryl phosphates and phosphoric acid in the urine. Three anti-esterase metabolites were present in the liver, intestine and fæces. Incubation of TOCP with rat liver microsomes fortified with reduced diphosphopyridine nucleotide also yielded three esterase inhibitors of similar chromatographic characteristics to those produced *in vivo*. The major esterase-inhibiting metabolite formed *in vivo* was isolated on silicic acid–'Celite' columns with benzene–ether mixtures. Examination of the infra-red spectrum and hydrolysis products suggested structure I for this metabolite. Reaction of *o*-cresylphosphoryl dichloride with *o*-hydroxybenzyl alcohol yielded 2-(*o*-cresyl)-4H-1,3,2-benzodioxaphosphoran-2-one, b.p. 159–161° C./0·09–0·1 mm. mercury, $n_D^{24.5°} = 1·5584$. This compound was identical with the major metabolite in respect to infra-red spectrum, chromatographic characteristics, and anti-esterase activity. One of the other esterase inhibitors appeared to be the *o*-hydroxymethyl derivative of I. The mechanism of esterase inhibition was investigated by reaction of chymotrypsin (*Xtr*) with I and a phenyl phosphonate analogue, II. Both appeared to phosphorylate by opening the cyclic phosphate structure at the P—O— aryl bond to yield III.

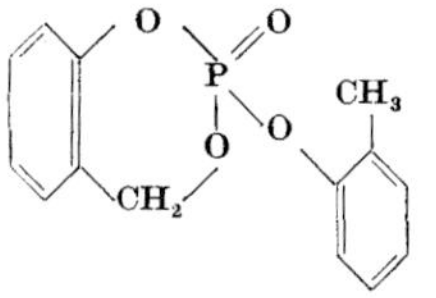

I

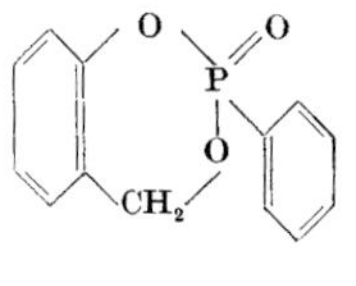

II

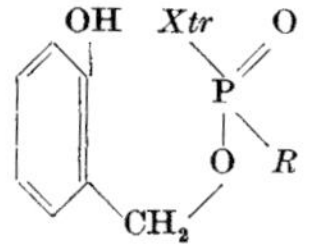

III

An ataxia and demyelination is induced in hens and certain other species by TOCP[6–9]. A similar ataxia in hens resulted from subcutaneous administration of 4–8 mgm./kgm. I, whereas 240 mgm./kgm. II failed to yield ataxia. Preliminary studies indicate a demyelination with I similar to that induced by TOCP. The high activity of I relative to that of TOCP in producing neurotoxicity in hens suggests that the anti-esterase metabolites rather than TOCP *per se* yield the ataxia. TOCP and other ataxia-producing phosphates are known to effect prolonged *in vivo* inhibition of the pseudocholinesterases of hen brain, spinal cord and sciatic nerve[10–13]. The function of this pseudocholinesterase which hydrolyses carboxylic esters of choline is unknown. Experiments with hens established that it also hydrolyses carboxylic esters of thiamine, that is, *O*-acetyl and *O*-propionyl thiamine, an observation consistent with previous reports on the substrate specificity of cholinesterases[14]. The demyelination in hens produced by TOCP is identical with that with certain other organophosphates, and similar lesions appear in hens on a thiamine-deficient diet[8]. Thiamine has been noted as a neuroactive material in peripheral nerves[15] and as a constituent of myelin in peripheral nerves of rabbits[16]. Inhibition of the pseudocholinesterase might result in an accumulation of carboxylic esters of thiamine to impair localized functions of the nerve requiring free thiamine or other thiamine derivatives.

In accordance with this hypothesis, thiamine and other compounds were tested extensively for their effect on the ataxia in hens from TOCP and/or I. Compounds tested alone or in conjunction with thiamine included *O*-acetyl thiamine, tocopherol and cortisone and their acetates, pilocarpine, saponin, and several aldoxime-type cholinesterase reactivators. With varying dosages and multiple administration of these materials before and after the organophosphate, no relief of the neurotoxic symptoms was evident.

TOCP potentiates the toxicity to mammals of malathion (*O,O*-dimethyl S-(1,2-dicarbethoxyethyl) phosphorodithioate)[4,17]. The intraperitoneal LD_{50} for mice was 150 mgm./kgm. for each of I and II. Simultaneous injection of malathion (LD_{50} alone of 1,500 mgm./kgm.) in a 10 : 1 ratio with the cyclic phosphates yielded LD_{50} values of about 1/45th those of either component alone. I and II were similar in this respect, and both yielded marked *in vivo* inhibition of the esterases of plasma and liver hydrolysing the carbethoxy groupings of malathion. The activity of these two compounds in malathion potentiation was greater than that of 120 other di-aryl and tri-aryl phosphates studied which lack the cyclic phosphate structure. The pattern of esterase inhibition in mice resulting from TOCP administration[4] also appeared with I, although the latter acted much more quickly and at a much lower dose.

Metabolic activation occurs with many tri-aryl phosphates containing *o*-methyl, *o*-ethyl, *o*-*n*-propyl and *p*-ethyl groupings[4,5]. *In vivo* hydroxylation of the *o*-methyl and subsequent cyclization to anti-

No. 4796 September 30, 1961 NATURE 1397

esterase metabolites appears to account for the biological activity of TOCP and certain other tri-aryl phosphates containing *o*-methyl groupings. A similar mechanism involving α-hydroxylation and cyclization might explain the conversion of *o*-ethyl and *o*-*n*-propyl derivatives to anti-esterases. A different mechanism must be involved with tri-*p*-ethylphenyl phosphate.

Publication of this communication is approved by the Director of the Wisconsin Agricultural Experiment Station. The investigation was supported in part by the Research Committee of the Graduate School from funds supplied by the Wisconsin Alumni Research Foundation, and by grants from the U.S. Public Health Service, National Institutes of Health (contract *RG*-5304), the U.S. Army Chemical Research and Development Laboratories (grant No. *DA-CML*-18-108-61-*G*-6), and the U.S. Atomic Energy Commission (contract No. *AT* (11-1)-64, Project No. 14).

JOHN E. CASIDA
MORIFUSA ETO *
RONALD L. BARON

Department of Entomology,
University of Wisconsin,
Madison 6, Wisconsin.

* Permanent address: Department of Agricultural Chemistry, Kyushu University, Fukuoka, Japan.

1 Aldridge, W. N., *Biochem. J.*, **56**, 185 (1954).
2 Myers, D. K., Rebel, J. B. J., Veeger, C., Kemp, A., and Simons, E. G. L., *Nature*, **176**, 260 (1955).
3 Davison, A. N., *Biochem. J.*, **61**, 203 (1955).
4 Casida, J. E., *J. Biochem. Pharmacol.*, **5**, 332 (1961).
5 Aldridge, W. N., and Barnes, J. M., *J. Biochem. Pharmacol.*, **6**, 177 (1961).
6 Smith, M. I., and Lillie, R. D., *Arch. Neurol. Psychiat.* (Chicago), **26**, 976 (1931).
7 Smith, M. I., Engel, E. W., and Stohlman, E. F., *Nat. Inst. Hlth. Bull.*, **160**, 1 (1932).
8 Barnes, J. M., and Denz, F. A., *J. Path. Bact.*, **65**, 597 (1953).
9 Geoffroy, H., Slomic, A., Benebadji, M., and Pascal, P., *World Neurology*, **1**, 294 (1960).
10 Earl, C. J., and Thompson, R. H. S., *Brit. J. Pharmacol.*, **7**, 261, 685 (1952).
11 Davison, A. N., *Brit. J. Pharmacol.*, **8**, 212 (1953).
12 Thompson, R. H. S., *Chem. and Indust.*, 749 (1954).
13 Witter, R. F., and Gaines, T. B., *Fed. Proc.*, **19**, 275 (1960).
14 Augustinsson, K.-B., *Acta Physiol. Scand.*, **15**, Supp. 52, 1 (1948).
15 Muralt, A. von, *Vitamins and Hormones*, **5**, 93 (1947).
16 Sjöstrand, F., *Nature*, **157**, 698 (1946).
17 Murphy, S. D., Anderson, R. L., and DuBois, K. P., *Proc. Soc. Exp. Biol.*, **100**, 483 (1959).

Isolation of Sterol Esters from Human Fæces

THE importance of obtaining more precise information regarding the sterol excretion pattern in human beings has recently become apparent and has been emphasized by Aylward[1]. The normal clinical laboratory techniques for stools, namely extraction of dried material or wet extraction coupled with saponification make impossible the isolation of any sterol esters which may be present. This difficulty in isolating (as distinct from estimating) sterol esters was indeed common to all lipid extracts until the methods developed by Börgstrom[2], and others, were applied by Fillerup and Mead[3] to extracts from animal tissues. Using silicic acid, they were able for the first time to separate cholesterol esters from triglycerides. We have applied a modification of this technique to the lipids of human fæces and have been able to isolate cholesterol ester fractions; and, using gas-liquid chromatography, have examined the fatty acids present in the esters.

Because the amount of sterol esters in human fæces is very small, it was necessary to extract large quantities of material. Accordingly, collections from five volunteers were arranged for periods of 6–9 days, and the extraction procedure was commenced immediately on the receipt of each sample. Preliminary experiments indicated that fæcal pigments interfered with the chromatographic separations, and it was therefore decided to extract the stools first with 0·9 per cent saline, a procedure which removes some of the bacterial lipids[4]. The solid obtained by centrifugation was then extracted successively by ethanol, ether and petroleum ether; precautions were taken to minimize oxidation by carrying out the procedures, wherever possible, under oxygen-free nitrogen.

Table 1. COMPONENTS OF PETROLEUM ETHER-SOLUBLE FRACTION FROM HUMAN FÆCES, CHROMATOGRAPHED ON SILICIC ACID (ref. 5)

Fraction	Case 1	Case 2	Case 3	Case 4	Case 5
Hydrocarbons and pigments	8·8	5·2	10·8	Nil	6·8
Sterol esters	3·5	7·1	2·0	1·0	13·0
Triglycerides	2·1	7·1	2·0	0·6	3·4
Sterols	77·1	76·7	65·4	93·6	62·0
Diglycerides, monoglycerides, phospholipids and pigment	8·5	3·9	19·8	4·8	14·7

In order to increase the proportion of sterol ester in the petroleum ether fraction to be chromatographed, the free fatty acids were first removed by washing with 1 per cent sodium carbonate solution; the extract remaining was dissolved in the minimum amount of hexane and fractionated on a column of Mallinckrodt silicic acid, the ratio of adsorbent: extract being kept at about 100 : 1·5. Pre-treatment of the silicic acid and the elution system followed Barron and Hanahan[5]. Positive oxygen-free nitrogen pressure was maintained, 25 ml. fractions were collected, and elution of the sterol esters was judged to be complete when a negative Liebermann–Burchard test was given. The solvent from the fractions was removed at 40° C. by means of a vacuum oven, the fractions were weighed, and the residues were dissolved in petroleum ether and combined according to the peaks shown when the weighed residues were graphed.

Neither free fatty acid nor free sterol was present in the sterol ester fraction, and the weight of the fraction agreed with the estimations of sterol ester made on the original petroleum ether-soluble fraction by a colorimetric method. Table 1 shows the amount of sterol esters isolated by this method, calculated as a percentage of the total material recovered.

The acids obtained by saponification of the sterol esters from extracts 1, 2 and 5 were methylated and submitted to gas-liquid chromatography, diethylene glycol succinate, ethylene glycol adipate, silicone oil, and dimethylchlorosilane being the stationary phases. The fatty acids present in the extract in the unesterified form, and those from the sterol ester fraction, were markedly similar in type and distribution and ranged from C12–C20 +, saturated and unsaturated.

We thank Mr. J. Piercy and Drs. Raymond Greene and D. S. Rideout of the New End Hospital for the facilities provided, and the Nuffield Foundation for a research grant (to P. W.).

FRANCIS AYLWARD
PAMELA WILLS

Department of Chemistry and Food Technology,
Borough Polytechnic,
London, S.E.1.

1 Aylward, F., *Lancet*, ii, 852 (1958).
2 Börgstrom, B., *Acta Physiol. Scand.*, **25**, 101 (1952).
3 Fillerup, D., and Mead, J., *Proc. Soc. Exp. Biol. Med.*, **83**, 574 (1953).
4 Hartman, L., Shorland, F. B., and Cleveley, B., *Biochem. J.*, **69**, 1 (1958).
5 Barron, E. J., and Hanahan, D. J., *J. Biol. Chem.*, **231**, 493 (1958).

the number of isolates per sample for water samples being 8.2 for Lake Michigan and 11.3 for all other tributaries.

As for the metabolic activities of those isolates, there was a general tendency for the number of TDE forming microorganisms to be constant; there were more DDNS forming microbes in the water and silt samples from the tributaries than in the isolates from Lake Michigan; in comparison with the silt samples from Lake Michigan, those from the tributaries were more capable of converting DDT to DDE.

In an attempt to determine the metabolic routes involved in producing DDNS, an important metabolic product of DDT, ^{14}C-TDE was incubated with the active isolates. The results indicated that DDNS formation from TDE in those microbes was the chief metabolic pathway. It is likely, therefore, that the microorganisms first dechlorinate DDT to form TDE and then TDE is dechlorinated in the same way that DDT is dechlorinated to form TDE. In fact, in many isolates the spots corresponding to TDE were also found to accompany the spot representing DDNS. Both DDNS and TDE are highly acricidal compounds. Their toxic effects on aquatic fauna and flora need to be studied and the residues of both metabolites in aquatic environments should be monitored.

This work was supported in part by a grant from the Department of Natural Resources, State of Wisconsin.

F. MATSUMURA
K. C. PATIL
G. M. BOUSH

Department of Entomology,
University of Wisconsin,
Madison, Wisconsin 53706

Received July 29, 1970.

1 Johnson, B. T., Goodman, R. N., and Boldberg, H. S., *Science*, **157**, 560 (1967).
2 Kallman, B. J., and Andrews, A. K., *Science*, **141**, 1050 (1963).
3 Mendel, J., and Walton, M., *Science*, **151**, 1527 (1966).
4 Wedermeyer, G., *Science*, **152**, 647 (1966).
5 Guenz, W. D., and Beard, W. E., *Science*, **154**, 893 (1967).
6 Ko, W. H., and Lockwood, J. L., *Canad. J. Microbiol.*, **14**, 1069 (1968).
7 Patil, K. C., Matsumura, F., and Boush, G. M., *Appl. Microbiol.*, **19**, 879 (1970).
8 Miskus, R. P., Blair, D. P., and Casida, J. E., *J. Agric. Food Chem.*, **13**, 481 (1965).
9 Finley, R. B., and Pillmore, R. E., *Amer. Inst. Biol. Sci. Bull.*, **13**, 41 (1963).
10 Matsumura, F., and Boush, G. M., *Science*, **156**, 959 (1967).
11 Fred, E. B., and Waksman, S. A., *Laboratory Manual of General Microbiology* (McGraw-Hill, New York, 1928).

Oxidative Metabolism of Pyrethrins in Mammals

THE natural pyrethrins present in pyrethrum flowers (*Chrysanthemum cinerariaefolium*) have attracted attention for more than a century because they are powerful insecticides yet have very low toxicity to mammals[1,2]. Concern about the persistence of some insecticides in mammals and in the environment has stimulated a consideration of the degradability of widely used compounds, both old and new. The two principal insecticidal constituents of pyrethrum, pyrethrins I and II (Fig. 1), do not persist in the environment because they are unstable when exposed to light and air[3]. There is some information about the metabolism of pyrethrin I and related compounds in houseflies[4,5], but little is known of the fate of pyrethrum constituents in mammals[6,7] even though man is often exposed to pyrethrins from household aerosols and other sources. We therefore examined the degradation of pyrethrin I and of pyrethrin II in rats fed separately with large amounts of pure tritium-labelled pyrethrins I and II, prepared with high specific activities by a new procedure involving a direct exchange reaction on (+)-

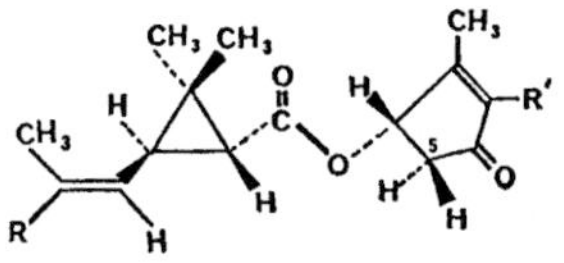

Designation	R	R′
	Pyrethrins	
Pyrethrin I	CH_3	$-CH_2-CH\overset{c}{=}CH-CH=CH_2$
Pyrethrin II	CO_2CH_3	$-CH_2-CH\overset{c}{=}CH-CH=CH_2$
Pyrethrins met. *A*	CO_2H	$-CH_2-CH\overset{c}{=}CH-CHOH-CH_2OH$
Pyrethrins met. *B*	CO_2H	$-CH_2-CHOH-CH\overset{t}{=}CH-CH_2OH$
Pyrethrins met. *C*	CO_2H	$-CH_2-CH\overset{c}{=}CH-CH(CH_2OH)-O-$conj.
Pyrethrins met. *D*	CO_2H	$-CH_2-CH\overset{c}{=}CH-CH=CH_2$
	Allethrin	
Allethrin	CH_3	$-CH_2CH=CH_2$
Allethrin met. *a*	CO_2H	$-CH_2-CHOH-CH_2OH$
Allethrin met. *b*	CO_2H	$-CHOH-CH=CH_2$

Fig. 1 Structures of pyrethrin I, pyrethrin II, allethrin and of compounds found as their metabolites in rats.

pyrethrolone[8]. The metabolites were isolated in milligram quantities from urine; nuclear magnetic resonance (NMR) and mass spectrometry (MS) showed that each of the principal metabolites retains the cyclopropane ester linkage and is formed by modification of both the acid and alcohol moieties. The oxidation of the acid side chain is similar in type to that encountered previously in other systems[4,5], but the establishment of attack on the alcohol part is new and involves the unsaturated side chain as detailed below.

Four samples of labelled pyrethrins, stereochemically and radiochemically pure, were prepared especially for this study: pyrethrins I and II each labelled with ^{3}H in the methyl group and in the 5-methylene group of the cyclopentenolone ring[8]; pyrethrin I labelled with ^{14}C in the carboxyl group of the acid[9]; pyrethrin II labelled with ^{14}C in the methyl of the methoxycarbonyl group. Important features of the labelled materials were their stereochemical purity and very high specific activity ($\sim$0.5 Ci/mmol) which facilitated detection and estimation by scintillation counting and autoradiography. Appropriate mixtures of ^{3}H-labelled and unlabelled pyrethrins[8,10–12] in dimethyl sulphoxide were given by stomach tube to male rats weighing 180–210 g. The low acute toxicity of the pyrethrins was evident, for the rats survived total doses of 450–2,000 mg/kg pyrethrin I and $>$2,900 mg/kg pyrethrin II administered in 2–6 portions within 12–54 h; acute oral LD_{50} values for single doses were 260–420 and $>$600 mg/kg, for pyrethrins I and II, respectively.

Metabolites in the urine of rats receiving 1.2 g of pyrethrin I or 2.4 g of pyrethrin II were isolated by adding ammonium sulphate and ethanol followed by extraction with ether[13]. The ether–ethanol soluble products were separated and purified without structural change by chromatography, initially on silicic acid–'Celite' columns developed with hexane–ether then acetone–methanol mixtures; after methylation (diazomethane) the methyl esters of the metabolites were rechromatographed (^{3}H monitoring) and obtained in high purity as estimated by ultraviolet visualization and by autoradiography respectively, on thin-layer chromatograms [silica gel F_{254}; chloroform–methanol (39 : 1), benzene saturated with formic acid–ether (10 : 3) or ether–benzene–methanol–formic acid (173 : 12 : 11 : 4)]. Both the high specific activities of the products and the NMR spectra indicated that the fractions were substantially pure. The structures of individual metabolites were then assigned by NMR spectroscopy[14] with spectrum accumulation and usually by MS as well, after trimethylsilylation of hydroxyl groups.

NATURE VOL. 230 APRIL 2 1971

The principal metabolite of the pyrethrins excreted in rat urine has structure *A* (Fig. 1). The NMR spectrum indicated that the structure was modified only at R and R′. R was CO_2H because the metabolite after methylation contained the *trans*-methoxycarbonyl group [$R=CO_2CH_3$] as the only alteration on the acid part of the molecule whereas R′ was *cis*-4,5-dihydroxypent-2-en-1-yl based on both NMR and MS. Another metabolite, *B*, is similar but both NMR and MS showed that its alcohol side chain, R′, is *trans*-2,5-dihydroxypent-3-en-1-yl (Fig. 1). Both metabolites *A* and *B* are formed in rats from pyrethrin I and from pyrethrin II by oxidation of the *trans*-methyl group and hydrolysis of the methoxycarbonyl group, respectively; each pyrethrin is further attacked on the pentadienyl side chain, possibly to give initially a 4,5-epoxide from which the two diols are derived. The structure of a third metabolite, *C*, of both pyrethrins is indicated by NMR spectroscopy to be a conjugate of metabolite *A*, with the 4-hydroxy group of the diol side chain esterified with an unidentified aromatic acid. The identified metabolites from pyrethrin II lack the methoxycarbonyl group; ^{14}C from $^{14}CH_3O$-labelled pyrethrin II did indeed appear largely (53%) as $^{14}CO_2$ expired by treated animals. Thus, although the CO_2CH_3 group is hydrolysed quickly, the cyclopropane carboxylic ester group is cleaved only to a minor extent, if at all, because the labelled metabolites detected in the urine were the same whether the rats were fed pyrethrin I labelled with ^3H-alcohol or ^{14}C-acid. Other more polar metabolites in the urine remain to be characterized.

Each metabolite found in the urine was also present in the faeces. Moreover, the faeces but not the urine contained some unmetabolized compound, more of pyrethrin I than of pyrethrin II.

From 64% to 71% of the ^3H in the administered dose of both compounds was recovered in the excreta within 100 h and was distributed almost equally between the urine and faeces. The excreted compounds and their content of ^3H relative to the administered radioactivity were: 14–21% as metabolite *A*, 3.3–4.4% as metabolite *B*, 3.9–6.2% as metabolite *C*, and 4–18% as unmetabolized compounds, with the remainder as unidentified compounds, mostly of greater polarity. About the same proportions of metabolites were obtained whatever the dose in the range of 0.1–400 mg/kg, so the oxidative mechanisms in the mammal seem not to be overloaded even at the higher dose.

Oxidations of the types found here for the pyrethrins are usually initiated by the microsomal mixed-function oxidase (MFO) system fortified with NADPH[4,5,15,16]. Pyrethrin I is readily metabolized in the MFO system derived from housefly abdomens at the *trans*-methyl group on the acid side chain, yielding metabolite *D*[4,5], but here no significant amounts of this product were detected in the excreta of rats treated with pyrethrins I and II. Since even small amounts of rat liver homogenate in the presence or absence of NADPH hydrolysed pyrethrin II to metabolite *D*, this is possibly a transient intermediate in the *in vivo* metabolism by rats, which then produce more extensive degradation.

With allethrin (Fig. 1) two metabolites (*a* and *b*) were identified. As for pyrethrin I, the sites of attack were the *trans*-methyl on the side chain of the acid and the double bond of the alcohol side chain, but with allethrin there was also hydroxylation at the methylene group of the allyl side chain and hydrolysis of the central ester linkage because chrysanthemum dicarboxylic acid and allethrolone were present in the urine.

The ease of oxidative metabolism of pyrethrins and allethrin possibly contributes to, or accounts for, their low toxicity to mammals.

This work was aided by a US Public Health Service grant, an Atomic Energy Commission contract, and grants from the Kenya Pyrethrum Co., the Pyrethrum Marketing Board, S. C. Johnson and Son and the Rockefeller Foundation. For advice and assistance we thank L. Lykken, J. L. Engel and L. C. Gaughan of the Berkeley laboratory. A large part of this work was done while M. Elliott was at Berkeley on leave from Rothamsted.

JOHN E. CASIDA
ELLA C. KIMMEL

Division of Entomology,
University of California,
Berkeley 94720

MICHAEL ELLIOTT
NORMAN F. JANES

Department of Insecticides and Fungicides,
Rothamsted Experimental Station,
Harpenden, Hertfordshire

Received July 31; revised October 8, 1970.

1 Crombie, L., and Elliott, M., *Fortschr. Chem. Org. Naturstoffe*, **19**, 120 (1961).
2 Metcalf, R. L., *Organic Insecticides, Their Chemistry and Mode of Action* (Interscience, New York, 1955).
3 Chen, Y-L., and Casida, J. E., *J. Agric. Food Chem.*, **17**, 208 (1969).
4 Yamamoto, I., and Casida, J. E., *J. Econ. Entomol.*, **59**, 1542 (1966).
5 Yamamoto, I., Kimmel, E. C., and Casida, J. E., *J. Agric. Food Chem.*, **17**, 1227 (1969).
6 Audiffren, M., *J. Pharmacie et Chimie*, **19**, 535 (1934).
7 Menzie, C. M., *Metabolism of Pesticides;* Bur. Sport Fisheries and Wildlife, Special Scientific Report—Wildlife No. 127 (Washington).
8 Elliott, M., Kimmel, E. C., and Casida, J. E., *Pyrethrum Post*, **10**, (2), 3 (1969).
9 Yamamoto, I., and Casida, J. E., *Agric. Biol. Chem. (Tokyo)*, **32**, 1382 (1968).
10 Elliott, M., *J. Chem. Soc.*, 5225 (1964).
11 Maciver, D. R., *Pyrethrum Post*, **9** (4), 41 (1968).
12 Elliott, M., and Janes, N. F., *Chem. and Ind.*, 270 (1969).
13 Slade, M., and Casida, J. E., *J. Agric. Food Chem.*, **18**, 467 (1970).
14 Bramwell, A. F., Crombie, L., Hemesley, P., Pattenden, G., Elliott, M., and Janes, N. F., *Tetrahedron*, **25**, 1727 (1969).
15 *Enzymatic Oxidation of Toxicants* (edit. by Hodgson, E.) North Carolina State University at Raleigh, 1968).
16 Parke, D. V., *The Biochemistry of Foreign Compounds* (Pergamon, London, 1968).

Quasi-Energy—a New Bioenergetic Quantity

IN a mathematical study of the concept of the energetically optimal body weight[1] of an animal, a new bioenergetic quantity appeared which we called "quasi-energy"[2]. It is defined as the total heat produced metabolically by the body, *Q*, multiplied by the specific heat production, *q*, the heat production per unit weight of the body. Thus, if *F* is the quasi-energy and $Q=qw$, *w* being the body weight,

$$F=wq^2 \qquad (1)$$

This quantity is of interest because its variation with body weight resembles a standard potential energy curve in that there are "potential energy wells" and "potential energy barriers" and because data from experiments with shrews have suggested a connexion between specific points on the quasi-energy curve and crucial stages in the life cycle of the animal. The wells and barriers, for example, seem to coincide with the survival period in winter and the reproductive period in spring[3]. Here, by considering the dimensions of quasi-energy and the structure of the equations which define it, we clarify some of its physical characteristics and the way in which it is related to physical energy.

The dimensions of *F* can be expressed in SI units using equation (1): if $q=Q/w$

$$[F]=[Q]^2[w]^{-1} \qquad (2)$$

and the dimensions are $cal^2\ s^{-2}\ N^{-1}$, where N is the newton, the SI force unit defined as $1\ kg\ m\ s^{-1}$. More light can be shed on

References and Notes

1. D. L. Graf and J. R. Goldsmith, *J. Geol.* **64**, 173 (1956); J. R. Goldsmith and H. C. Heard, *ibid.* **69**, 45 (1961).
2. I use "protodolomite" to refer to materials which deviate from ideal dolomite composition, or are imperfectly ordered, or both, but which show unambiguous evidence for a high degree of cation order.
3. Bravais-Miller indices are for the hexagonal unit cell, with cell parameters $a \sim 5$ Å and $c \sim 16$ Å.
4. D. L. Graf, *Am. Mineral.* **46**, 1283 (1961).
5. A. M. Gaines, *Geol. Soc. Am. Spec. Pap. 121* (1968), p. 106.
6. E. D. Glover and R. F. Sippel, *Geochim. Cosmochim. Acta* **31**, 603 (1967).
7. P. K. Weyl, *EPR Publ. 428* (Shell Development Co., Houston, 1965).
8. The majority of the experimental work was performed at the University of Chicago and was supported by grant 3204-C2 from the Petroleum Research Fund, administered by the American Chemical Society.

19 September 1973; revised 14 November 1973 ■

Toxaphene Insecticide: A Complex Biodegradable Mixture

Abstract. *Adsorption and gas-liquid chromatography separate toxaphene into at least 175 polychlorinated 10-carbon compounds including Cl_6, Cl_7, Cl_8, Cl_9, and Cl_{10} derivatives. One toxic component is 2,2,5-*endo,6-*exo,8,9,10-heptachlorobornane. Rats metabolically dechlorinate toxaphene, removing about half of the chlorine from the technical insecticide and from each of seven subfractions of varying composition and toxicity.*

Within the past 25 years 1 billion (10^9) pounds of toxaphene have been applied to crops and livestock for pest insect control. Its use continues at the rate of about 40 million pounds per year, in large part combined with methyl parathion for treatment of cotton. The toxaphene–methyl parathion combination has replaced the toxaphene–DDT combination employed until recently. Despite the fact that toxaphene is used in larger amounts in the United States than any other chlorinated hydrocarbon insecticide, there is insufficient information on several aspects of its chemistry, persistence, and environmental fate (*1*).

Toxaphene is produced by chlorination of camphene to about 67 to 69 percent chlorine by weight, yielding a reproducible but very complex mixture of compounds with an overall average elemental composition of $C_{10}H_{10}Cl_8$ (*2*). No individual component, toxic or otherwise, has previously been isolated in pure form. Despite the difficulties in evaluating toxaphene persistence, it is known from observations on its duration of effectiveness in insect control and from analyses of its residues by a variety of methods that many of the toxaphene constituents degrade under several different environmental conditions more rapidly than certain other chlorinated hydrocarbon insecticides, including DDT (*1*). Increased restrictions on the use of chlorinated insecticide chemicals and improved methodology for working with complex mixtures make it necessary and possible to define the nature of the toxaphene components and their metabolic fate in mammals. We have taken steps in this direction.

Analysis of chlorinated insecticides and other chlorine-containing environmental pollutants normally involves the use of gas-liquid chromatography (GLC) with an electron-capture detector. Technical toxaphene appears to contain 25 to 30 components when analyzed on appropriate GLC columns (*3*); however, many of the peaks detected are due to multiple components that do not separate on GLC. Fractionation of toxaphene on silica gel by thin-layer chromatography (TLC) with pentane as the developer or by column adsorption chromatography with hexane for elution resolves the toxaphene components on a different basis than GLC; the components elute from the GLC column in the general order of increasing degree of chlorination whereas this is not the case for elution from the adsorption column. Examination of the toxaphene fractions from silica gel column chromatography by combination gas-liquid chromatography–mass spectroscopy (MS) techniques (*4*) reveals a complex mixture of at least 175 C_{10} polychloro derivatives made up of $C_{10}H_8Cl_{10}$, $C_{10}H_{18-n}Cl_n$, and $C_{10}H_{16-n}Cl_n$ derivatives where the chlorine number (n) is 6, 7, 8, or 9. It appears likely that the majority of $C_{10}H_{18-n}Cl_n$ compounds are polychlorobornanes since one heptachlorobornane (**1**)

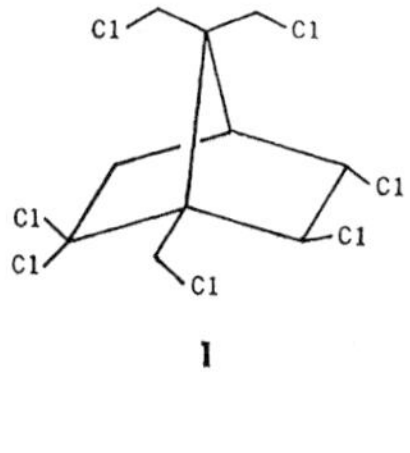

1

has been identified, as described below, and 2-*exo*-10-dichlorobornane is a major intermediate (*5*) in the chlorination of camphene. The $C_{10}H_{16-n}Cl_n$ derivatives are likely to be polychlorobornenes or polychlorotricyclenes or both.

A procedure was devised for isolating individual toxaphene components in pure, crystalline form. It involves separation on a partition column with β-methoxypropionitrile and heptane and then on the silica gel–hexane adsorption column, followed by a repetition of these two steps in sequence, and then preparative GLC and further purification by either sublimation or crystallization (*6*). With suitable monitoring, this sequence of chromatographic steps should permit isolation of any individual component provided it is stable under the chromatographic conditions employed.

By using intraperitoneal acute toxicity in the mouse as the monitoring criterion, two toxic crystalline compounds were isolated, one a $C_{10}H_{11}Cl_7$ and the other a $C_{10}H_{10}Cl_8$ component. These crystalline materials are, respectively, 6 times and 14 times more toxic to mice than technical toxaphene and 2 times and 4 times more toxic to houseflies treated topically. The $C_{10}H_{11}Cl_7$ component has been characterized by x-ray crystallography and by MS and nuclear magnetic resonance studies as 2,2,5-*endo*,6-*exo*,8,9,10-heptachlorobornane (**1**). Crystallization of the racemate from a mixture of hexane and acetone (5 : 1) appears to lead to the separation of the two enantiomers; however, these enantiomers could not be differentiated by the x-ray study. The $C_{10}H_{10}Cl_8$ component has not yet been obtained in crystalline form appropriate for x-ray structure determination. The $C_{10}H_{11}Cl_7$ and $C_{10}H_{10}Cl_8$ toxic components each constitute 2 to 6 percent of technical toxaphene, based on a combination of TLC and preparative GLC analyses, but they are in relatively large amounts compared with many other components. While it is already evident from our study that other toxic materials are present in toxaphene, the two components isolated to date appear to contribute significantly to the mammalian toxicity of commercial toxaphene.

The availability of preparations labeled with ^{36}Cl and ^{14}C (*7*) made it possible to carry out initial studies on the metabolic fate of toxaphene in mammals. Toxaphene labeled with ^{36}Cl was administered orally to rats at about

Source: From *Science* Vol. 183, No. 124, pp. 520-521 (1974). Reprinted with permission from AAAS.

14 mg/kg to determine the extent of dechlorination and rate of elimination of chlorinated compounds from the body over a 14-day period. For analysis, the ^{36}Cl ion was converted to phenylmercuric [^{36}Cl]chloride (*8*), which was recrystallized to a constant specific activity. Comparable studies with [^{14}C]-toxaphene yielded no metabolites labeled with ^{14}C in the recrystallized phenylmercuric chloride fraction, so metabolites of toxaphene other than chloride ion do not interfere in this method of isotope dilution analysis. Most of the [^{36}Cl]toxaphene is metabolized before excretion, less than 0.7 percent of the ^{36}Cl appearing in urine and less than 3 percent in feces as unmetabolized toxaphene. About 5 percent of the administered ^{36}Cl appears in urine and 21 to 24 percent in feces as partially dechlorinated metabolites of toxaphene. The remainder of the excreted ^{36}Cl, accounting for 44 to 57 percent of the administered [^{36}Cl]toxaphene dose, is chloride ion in the urine. For comparison, 90 percent of the ^{36}Cl from labeled NaCl administered to rats is excreted as chloride ion in the urine.

To get an idea of the variation in rate and the degree of metabolism of toxaphene components, we chromatographed [^{36}Cl]toxaphene on the silica gel–hexane column and combined the fractions in the order of elution so that seven subfractions were obtained, each containing one-seventh of the total ^{36}Cl content of [^{36}Cl]toxaphene. These subfractions do not differ greatly in their selectivity for poisoning mice and houseflies, but the toxicity of the subfractions to these organisms varies from one-sixth to three times that of technical toxaphene. The seven subfractions and toxaphene itself are dechlorinated to similar extents and eliminated at similar rates by rats despite a 16-fold toxicity difference for mice and a 10-fold difference for houseflies between the least toxic and the most toxic subfractions.

The results reported here suggest that many of the toxaphene components contain the same or similar biodegradable groupings that undergo in vivo dechlorination in mammals, with about half of the C-Cl bonds on an average being metabolically labile. This does not mean that half of the C-Cl bonds in each component are cleaved since the studies to date involve mixtures of many components, some of which may undergo a small degree of dechlorination while others are more extensively or even completely dechlorinated and fragmented. Two sets of structural features are therefore of interest relative to the individual toxaphene components. An appropriate steric relationship between certain chlorines present in only a few constituents may determine the neurotoxic potency. Biodegradable sites such as chloromethyl and other groupings are probably present in most if not all of the toxaphene components. The composition of technical toxaphene and the structure, metabolic fate, and environmental persistence of the toxaphene components should be defined more completely by the procedures described above.

JOHN E. CASIDA*
ROY L. HOLMSTEAD
SAFY KHALIFA
JOHN R. KNOX†, TOMIHIKO OHSAWA
Division of Entomology and Parasitology, University of California, Berkeley 94720
KENNETH J. PALMER
ROSALIND Y. WONG
Western Regional Research Center, U.S. Department of Agriculture, Albany, California 94710

References and Notes

1. G. E. Guyer, P. L. Adkisson, K. DuBois, C. Menzie, H. P. Nicholson, G. Zweig, C. L. Dunn, *Toxaphene Status Report* (Environmental Protection Agency, Washington, D.C., 1971).
2. G. A. Buntin, U.S. Patent No. 2,565,471 (1951).
3. For GLC we used columns of 3 percent SE-30 on Gas-Chrom Q isothermally at 170°C, or 3 percent Dexil 300 on Varaport-30 with a temperature program between 180° and 270°C.
4. A Finnigan 1015D GLC-MS system with a chemical ionization source was used, with methane at 1 torr as the reagent gas.
5. B. H. Jennings and G. B. Herschbach, *J. Org. Chem.* **30**, 3902 (1965); H. G. Richey, Jr., J. E. Grant, T. J. Garbacik, D. L. Dull, *ibid.*, p. 3909.
6. S. Khalifa, T. R. Mon, J. L. Engel, J. E. Casida, in preparation.
7. The preparations were provided by Hercules Incorporated, Wilmington, Delaware.
8. R. Belcher, J. R. Majer, J. A. Rodriguez-Vazquez, W. I. Stephen, P. C. Uden, *Anal. Chim. Acta* **57**, 73 (1971). The original procedure as described for GLC analysis was modified in our study for radioassay by isotope dilution.
9. Supported in part by PHS grant PO1 ES00049 and grants from the Rockefeller Foundation and Hercules Incorporated. Grateful acknowledgment is made to J. L. Engel and H. T. Gordon of the Division of Entomology and Parasitology for performing many of the animal studies and for suggestions on partition column chromatography, respectively. Thanks are also due to T. R. Mon and R. E. Lundin of the Western Regional Research Center for help in the preparative GLC and nuclear magnetic resonance studies, respectively. A preliminary communication of these results was presented at the American Chemical Society meeting, Chicago, August 1973.
* Send reprint requests to J.E.C.
† Present address: Department of Organic Chemistry, University of Western Australia, Nedlands 6009.

28 September 1973; revised 29 October 1973 ■

Phosphate Absorption Capacity and Acclimation Potential in Plants along a Latitudinal Gradient

Abstract. *The capacity for phosphate absorption by marsh plants is negatively correlated with the soil temperature of the habitat of origin. Species and races from thermally fluctuating environments achieve greater compensatory changes in the phosphate absorption rate through temperature acclimation than their counterparts from more stable environments.*

Related organisms living at very different temperatures may exhibit similar metabolic rates, as a result of both evolutionary modifications and acclimation (a compensatory change in the rate in response to a change in growth temperature). These forms of temperature compensation have been well-documented in animals (*1*) but have only recently received attention by botanists. Evolutionary temperature compensation of plant processes usually (*2, 3*) but not always (*4, 5*) occurs. Acclimation of photosynthesis and respiration generally compensates for seasonal temperature changes, but the extent of this compensation is not consistently correlated with the extent of temperature fluctuation in the habitats from which the species derive (*5, 6*), although this correlation was found in one comprehensive study (*3*). The results of the study presented here indicate that the rate of phosphate absorption is finely attuned to the local thermal regime and that both evolutionary and acclimatory forms of temperature compensation consistently occur in this process.

Species and races of marsh plants were selected for study from habitats along latitudinal gradients of temperature and thermal stability, ranging from the cold, thermally stable soils of the arctic tundra (Barrow, Alaska; latitude, 71°18′N) to the warm, thermally fluctuating soils of a desert oasis (Thousand Palms, California; latitude, 33° 49′N). The soil thermal regimes in this report are characterized by July mean soil temperatures; details of the thermal regimes and temperature measurements emission from both planets is related to the gyrofrequency, a rather interesting relationship exists. The frequency at peak intensities is 0.3 Mhz for Earth and 8 Mhz for Jupiter (*13*), and the observed high-frequency cutoff is about 1.7 Mhz for Earth and 40 Mhz for Jupiter (*11*). The frequencies at peak intensity are thus in a ratio of 27 to 1, and the maximum frequencies are in a ratio of 24 to 1. If the magnetic field strengths of the two planets scale like the gyrofrequencies, the Jovian polar field derived by this method would be 15 to 18 gauss. The Pioneer 11 measurements (*14*) indicate a value of 14 to 23 gauss for the Jovian north pole. With Brown's (*2*) spectrum of the Saturn emissions shown in Fig. 3, we would, by the same logic, infer a polar surface field of 2 gauss for that planet. For a dipole field, this would correspond to 1 gauss at the equator, which is the field strength estimated by Scarf (*15*), using various lines of evidence. This prediction will undergo a crucial test in 1979 with the Pioneer 11 flyby of Saturn.

These similarities are circumstantial and speculative at this point, but we feel, nevertheless, that there may well be a pattern in these radio emissions. The Mariner Jupiter-Saturn missions to be launched in 1977 will carry sophisticated radio astronomy experiments as well as magnetic field and particle experiments capable of determining many of the parameters at Jupiter and Saturn. Thus, in the next 5 or 6 years the reality of our proposed pattern will become known.

M. L. Kaiser
R. G. Stone
Radio Astronomy Branch, Laboratory for Extraterrestrial Physics, Goddard Space Flight Center, Greenbelt, Maryland 20771

References

1. J. W. Warwick, *Radio Astronomical and Satellite Studies of the Atmosphere*, J. Aarons, Ed. (North-Holland, Amsterdam, 1963), pp. 400–424.
2. L. W. Brown, *Astrophys. J.* **198**, L89 (1975).
3. E. A. Benediktov, G. G. Getmantsev, N. A. Mityakov, V. O. Rapoport, A. F. Tarasov, *Kosm. Issled.* **6**, 946 (1968).
4. N. Dunckel, B. Ficklin, L. Rorden, R. A. Helliwell, *J. Geophys. Res.* **75**, 1854 (1970).
5. R. G. Stone, *Space Sci. Rev.* **14**, 534 (1973).
6. D. A. Gurnett, *J. Geophys. Res.* **79**, 4227 (1974).
7. L. W. Brown, *Astrophys. J.* **180**, 359 (1973).
8. J. W. Warwick, *Particles and Fields Near Jupiter* (NASA Publication CR-1685, Government Printing Office, Washington, D.C., 1970).
9. J. Fainberg, L. G. Evans, R. G. Stone, *Science* **178**, 743 (1972).
10. T. R. Hartz, in *The Radiating Atmosphere*, B. M. McCormac, Ed. (Reidel, Dordrecht, Netherlands, 1971), pp. 225–238.
11. For a thorough review, see T. D. Carr and S. Gulkis, *Annu. Rev. Astron. Astrophys.* **7**, 577 (1969).
12. G. M. Gruber and C. Way-Jones, *Nat. Phys. Sci.* **237**, 137 (1972).
13. M. D. Desch and T. D. Carr, *Astrophys. J.* **194**, L57 (1974).
14. M. H. Acuña and N. F. Ness, *J. Geophys. Res.*, in press; E. J. Smith, L. Davis, Jr., D. E. Jones, P. J. Coleman, Jr., D. S. Colburn, P. Dyal, C. P. Sonett, *Science* **188**, 451 (1975).
15. F. L. Scarf, *Cosmic Electrodyn.* **3**, 437 (1973).

30 January 1975; revised 7 April 1975

Dichloroacetamide Antidotes for Thiocarbamate Herbicides: Mode of Action

Abstract. *Thiocarbamate sulfoxides formed on metabolic sulfoxidation of thiocarbamate herbicides in plants and mammals are effective carbamoylating agents for glutathione and other tissue thiols. Dichloroacetamides that protect corn from thiocarbamate herbicide injury induce more rapid detoxification of the thiocarbamate sulfoxides by increasing their rate of carbamoylation of glutathione through elevation of the root glutathione level and glutathione* S*-transferase activity.*

Two recent advances provide new chemical probes, thiocarbamate sulfoxides and dichloroacetamides, useful in elucidating the mode of action of thiocarbamates, one of the most important classes of herbicide chemicals. Biological oxidation to form thiocarbamate sulfoxides probably constitutes the first step in a chain of events leading to inhibition of plant growth (*1*). In thiocarbamate-susceptible corn varieties, this chain of events is apparently disrupted by dichloroacetamide "antidotes," which are effective adjuvants in preventing injury due to thiocarbamates (*2*). Three biochemical observations are also relevant. The metabolism of fatty acids is altered by EPTC (*S*-ethyl *N,N*-dipropylthiocarbamate) in some plant species (*3*), suggesting an interference with coenzyme A (CoASH)–mediated reactions. Thiocarbamate sulfoxides are cleaved by glutathione (GSH) *S*-transferase enzymes of mouse liver (*1*). The much greater tolerance of corn than of oat seedlings to thiocarbamates and their sulfoxides (*1*) extends to atrazine herbicide, a chemical which is metabolized by a GSH *S*-transferase of corn but not oat seedlings (*4*).

We now establish that thiocarbamate sulfoxides in vitro readily carbamoylate CoASH and GSH, important enzyme cofactors, and that the antidotes act in corn to elevate the GSH and GSH *S*-transferase levels, resulting in rapid detoxification of the thiocarbamate sulfoxides. The following scheme illustrates a portion of these reactions and relationships.

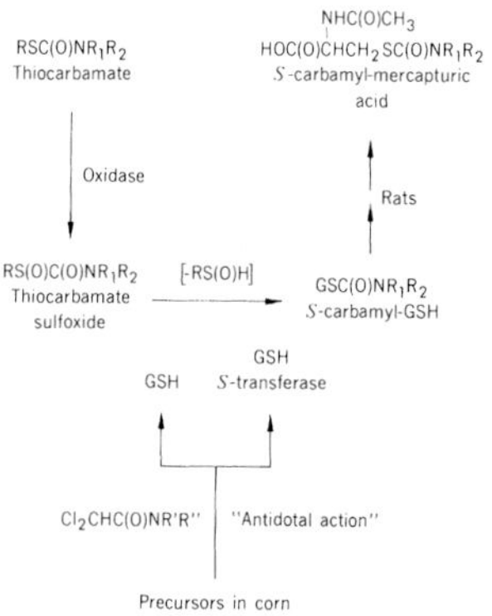

The studies were made with two thiocarbamates (EPTC and butylate), their sulfoxide derivatives prepared by peracid oxidation (*1*), and two dichloroacetamides (R-25788 and R-29148). These two thiocarbamates are used commercially in combination with the antidote R-25788: Eradicane®, which is EPTC plus antidote, and Sutan +®, which is butylate plus antidote (*5*).

EPTC	$CH_3CH_2SC(O)N(CH_2CH_2CH_3)_2$
Butylate	$CH_3CH_2SC(O)N[CH_2CH(CH_3)_2]_2$
R-25788	$Cl_2CHC(O)N(CH_2CH{=}CH_2)_2$
R-29148	$Cl_2CHC(O)N{-}CH_2$ / $(CH_3)_2COCHCH_3$ (ring)

Several initial observations served to focus attention on tissue thiols. The thiocarbamates are converted in mammals and plants to the corresponding sulfoxides, which do not accumulate since they are further metabolized. Thus, EPTC sulfoxide appears as a transient metabolite in the liver of mice 10 minutes after intraperitoneal administration of EPTC at 1.0 mmole/kg (*1*). Further, EPTC sulfoxide is detected in extracts of roots from oat seedlings exposed for 24 hours to [^{14}C]EPTC solutions and of leaves from corn seedlings 24 hours after injection of [^{14}C]EPTC into the stem. The transient nature of the thiocarbamate sulfoxides suggests that they react with tissue constituents such as thiols. This was verified by the finding that CoASH, GSH, and *N*-acetylcysteine are converted to *S*-carbamyl derivatives on reaction with equimolar amounts of either EPTC sulfoxide or butylate sulfoxide in an aqueous medium at *p*H 7.4 (*6*). The alkylsulfenic acids released from the thiocarbamate sulfoxides on carbamoylation of thiols are quite unstable in aqueous medium at physiological *p*H in the presence or absence of biological material, giving predominately the corresponding alkylsulfonic acids (*7*). The thiocarbamate sulfones are even more effective carbamoylating agents than the corresponding sulfoxides, but this observation is probably not relevant to the mode of action of thiocarbamate herbicides since the sulfones are much less effective as herbicides (*1*) and they are not detected as in vivo metabolites of the thiocarbamates in mammals (*1*) or plants.

Source: From *Science* Vol. 189, No. 4199, pp. 287-289 (1975). Reprinted with permission from AAAS.

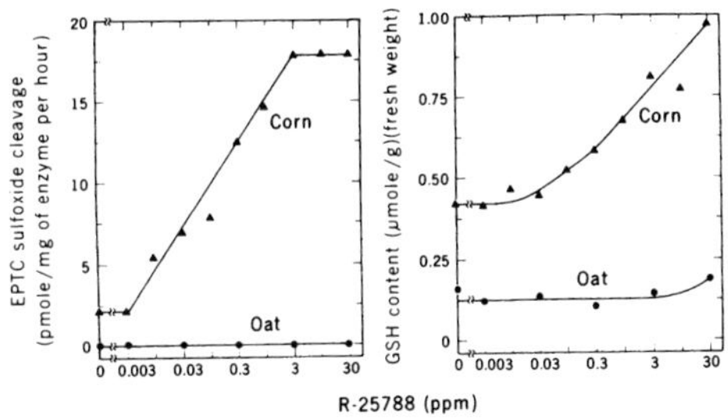

Fig. 1. Effect of exposure for 24 hours of corn and oat seedlings to various concentrations of R-25788 on the root glutathione *S*-transferase or on EPTC sulfoxide cleavage activity, and on the root glutathione content.

The complete inactivity of the thiocarbamates themselves as carbamoylating agents under physiological conditions leads to the hypothesis that the thiocarbamates are oxidized to the corresponding sulfoxides, which then serve as carbamoylating agents for important tissue thiols, such as CoASH and GSH, with or without enzymatic mediation. Whereas the relevance to reactions in vivo of the carbamoylation of CoASH or of other components of CoASH-mediated systems remains to be evaluated, the importance of GSH carbamoylation is firmly established by observations in both mammals and plants, as discussed below.

Male mice and rats were used to evaluate the importance of GSH in mediating the fate in mammals of thiocarbamate herbicide chemicals. The concentration of GSH in the liver (*8*) 3 hours after intraperitoneal administration of EPTC or EPTC sulfoxide (1.5 mmole/kg) to mice is reduced by 26 and 49 percent, respectively, relative to comparable control animals. The reduced GSH content probably results from carbamoylation of the GSH, mediated by GSH *S*-transferase enzymes (*1*), either by the administered sulfoxide or that formed on in vivo sulfoxidation of the thiocarbamate. Conjugates of GSH are normally cleaved and acetylated to form the corresponding mercapturic acids prior to excretion by mammals (*9*). Analysis of the urine (*10*) collected within 24 hours after administration of either EPTC, EPTC sulfoxide, or butylate (0.6 to 1.0 mmole/kg) to rats revealed the corresponding *S*-(*N*,*N*-dialkylcarbamyl)-mercapturic acids in the indicated amounts relative to the administered thiocarbamate or thiocarbamate sulfoxide dose: 5 to 8 percent after oral or intraperitoneal administration of EPTC; 12 percent from intraperitoneal administration of EPTC sulfoxide; 1 percent after oral or intraperitoneal administration of butylate. It appears, therefore, that one pathway for thiocarbamate herbicide metabolism in mammals involves sulfoxidation, carbamoylation of GSH, and degradation of the GSH conjugate to form the corresponding mercapturic acid.

One of several mechanisms by which an additive or antidote might influence the potency of a herbicide involves an alteration in the rate of herbicide detoxification. Accordingly, the thiocarbamate sulfoxide, GSH, GSH *S*-transferase system was examined in the roots of corn and oat seedlings exposed to dichloroacetamides at various concentrations. Germinated corn seedlings (DeKalb XL-66H) with an average root length of 3 to 4 cm and oat seedlings (Curt) with an average root length of 2 to 3 cm were placed on filter paper in petri dishes containing the test solutions so that the roots were partially immersed in the aqueous medium. After 24 hours exposure to selected concentrations of the dichloroacetamides, the roots were cut off, washed with distilled water, and analyzed for GSH content (*8*) and for activity of the root homogenates fortified with GSH in cleavage of EPTC sulfoxide and butylate sulfoxide (*11*).

Dichloroacetamides R-25788 and R-29148 have two concentration-dependent effects on corn seedlings that may be important in reducing corn injury from thiocarbamates. As illustrated in Fig. 1 with R-25788 and EPTC sulfoxide, this antidote elevates the GSH level twofold and the GSH *S*-transferase activity ninefold (*12*). The threshold level of antidote for increased GSH *S*-transferase activity is extremely low, below 0.01 part per minute (ppm) or $5 \times 10^{-8}M$, and a plateau is reached at 3 ppm; beyond this limit no further stimulation occurs. Similar relations are found with the R-29148–EPTC sulfoxide and R-25788–butylate sulfoxide combinations. The activity of the enzyme system in metabolizing the sulfoxides is dependent on GSH fortification. The responsible enzyme is likely to be a GSH *S*-transferase since the principal product of the enzyme reaction with butylate sulfoxide is identified as *S*-(*N*,*N*-diisobutylcarbamyl)-GSH (*13*) and this product is formed in greater amount with enzyme from plants treated with R-25788 than from plants that were not treated. Thus, the antidotes acting in corn increase both the GSH *S*-transferase and GSH, its essential cofactor, which probably results in increased detoxification of the sulfoxides by carbamoylation of the GSH. Oat seedlings also respond to R-25788 with increased GSH concentrations, particularly with antidote concentrations above the 30 ppm shown in Fig. 1, but their activity for metabolizing EPTC sulfoxide is not elevated. However, their level of detoxifying enzyme and cofactor is very low, relative to corn, in any case. Since EPTC sulfoxide is eight times more potent than EPTC in inhibiting oat root elongation (*1*) and antidote R-25788 is not active with oats with either EPTC (*2*) or its sulfoxide, it appears that the level of GSH *S*-transferase activity contributes to the species specificity noted between oat and corn seedlings.

The new chemical probes, thiocarbamate sulfoxides and dichloroacetamide antidotes, used here in investigating thiocarbamate herbicide mode of action may provide a new and useful variable in studies on GSH synthesis and functions. They may also prove useful as general biochemical reagents acting in a manner similar to that in which they serve so importantly in the control of noxious weeds.

MING-MUH LAY, JAMES P. HUBBELL
JOHN E. CASIDA
Division of Entomology and Parasitology, University of California, Berkeley 94720

References and Notes

1. J. E. Casida, R. A. Gray, H. Tilles, *Science* **184**, 573 (1974); J. E. Casida, E. C. Kimmel, M.-M. Lay, H. Ohkawa, J. E. Rodebush, R. A. Gray, C. K. Tseng, H. Tilles, *Environ. Quality Safety*, in press; J. E. Casida, E. C. Kimmel, H. Ohkawa, R. Ohkawa, *Pest. Biochem. Physiol.* **5**, 1 (1975).
2. F. Y. Chang, J. D. Bandeen, G. R. Stephenson, *Can. J. Plant Sci.* **52**, 707 (1972); *Weed Res.* **13**, 399 (1973); F. Y. Chang, G. R. Stephenson, J. D. Bandeen, *Weed Sci.* **21**, 292 (1973); *J. Agr. Food Chem.* **22**, 245 (1974); F. M. Pallos, M. E. Brokke, R. A. Gray, D. R. Arneklev, paper presented at the meeting of the Division of Pesticide Chemistry, American Chemical Society, Chicago, Ill., August 1973; Stauffer Chemical Co., Belgian Patent 782,120 (1972); Belgian Patent 806,038 (1974).
3. W. A. Gentner, *Weeds* **14**, 27 (1966); J. L. Harwood and P. K. Stumpf, *Arch. Biochem. Biophys.* **142**, 281 (1971); P. E. Kolattukudy and L. Brown, *Plant Physiol.* **53**, 903 (1974); G. G. Still, D. G. Davis, G. L. Zander, *ibid.* **46**, 307 (1970); R. E. Wilkinson, *ibid.* **53**, 269 (1974); ——— and W. S. Hardcastle, *Weed Sci.* **18**, 125 (1970); ———, A. E. Smith, paper presented at the meeting of the Weed Science Society of America, Las Vegas, Nev., February 1974.
4. D. S. Frear and H. R. Swanson, *Phytochemistry* **9**, 2123 (1970); ———, F. S. Tanaka, *Recent Adv. Phytochem.* **5**, 225 (1972).
5. *Herbicide Handbook* (Weed Science Society of America, Champaign, Ill., ed. 3, 1974).
6. *S*-(*N*,*N*-Dialkylcarbamyl) derivatives isolated by thin-layer chromatography (TLC) and fractional crystallization (GSH and mercapturic acid derivatives only) were identified by nuclear magnetic resonance (NMR) spectroscopy and functional group tests with appropriate chromogenic agents.
7. Alkylsulfonic acids were identified by NMR and infrared spectroscopy.
8. GSH was determined by homogenizing the samples of liver and plant roots in 5 percent trichloroacetic acid, removing the protein precipitate by centrifugation, neutralizing the supernatant with NaOH, and then reacting the supernatant with 5,5′-dithiobis(2-nitrobenzoic acid) [P. C. Jocelyn, *Biochemistry of the SH Group* (Academic Press, New York, 1972); M. Koivusalo and L. Uotila, *Anal. Biochem.* **59**, 34 (1974)].
9. L. F. Chasseaud, *Drug Metab. Rev.* **2**, 185 (1973); J. L. Wood, in *Metabolic Conjugation and Meta-*

bolic Hydrolysis, W. H. Fishman, Ed. (Academic Press, New York, 1970), vol. 2, p. 261.

10. The methanol-soluble portion of the residue from urine after extraction with ether and lyophilization to dryness was treated with diazomethane and then subjected to gas-liquid chromatography (GLC) analysis on a 3.8 percent OV-101 or 3.0 percent OV-17 column at 150° to 250°C with quantitation of the methylated mercapturic acids using a flame ionization detector. The structural identity of each excreted mercapturic acid was verified by methylation and GLC-mass spectroscopy (chemical ionization) comparison with authentic standards.
11. The standard plant root detoxification assay involved addition of [*S*-alkyl-^{14}C]EPTC sulfoxide (1 nmole) or [*N*-alkyl-^{14}C]butylate sulfoxide (0.5 nmole) to the 17,300*g* supernatant of root homogenates (2.5 to 320 mg of fresh tissue weight equivalent) and GSH (10 μmole) in phosphate buffer (0.1*M*, *p*H 6.8, 1.1 ml). After 2 hours incubation at 25°C, the unmetabolized [^{14}C]EPTC sulfoxide or [^{14}C]butylate sulfoxide was recovered by extraction with chloroform and subjected to liquid scintillation assay. Enzyme activity is expressed as picomoles of thiocarbamate sulfoxide cleaved per milligram of enzyme (fresh tissue weight equivalent) per hour.
12. In a separate but similar study in which were assayed the GSH *S*-transferase activity with EPTC sulfoxide at a high substrate level (1 μmole), corn root enzyme (5 to 40 mg of fresh tissue weight equivalent) prepared with insoluble polyvinylpyrrolidone (*4*) and 40 minutes of incubation, the same relationship was noted of increasing enzyme activity with increasing R-25788 level. In this case, the enzyme activity was elevated two- to fourfold at high antidote levels as compared with no antidote treatment.
13. *S*-(*N*,*N*-[^{14}C]Diisobutylcarbamyl)-GSH formed as a metabolite of [^{14}C]butylate sulfoxide (*11*) was identified by cochromatography with an authentic unlabeled standard (*6*) on two-dimensional TLC with 0.5-mm silica gel F_{254} chromatoplates developed in the first direction with pyridine, isopropanol, water, glacial acetic acid system (100 : 20 : 20 : 1) and in the second direction with the same system in the ratio of 50 : 55 : 45 : 2.
14. Supported in part by PHS grant PO1 ES00049 and PHS fellowship 1 F22 ES00681 (the latter to J.P.H.). R. L. Holmstead of this laboratory assisted in the GLC—mass spectroscopy studies. We thank R. A. Gray and F. M. Pallos of Stauffer Chemical Co. for samples of [^{14}C]EPTC, [^{14}C]-butylate, R-25788, and R-29148. The corn and oat seeds were provided by C. W. Crum of Dekalb Agricultural Research, Inc., DeKalb, Ill., and C. O. Qualset of the University of California, Davis, respectively.

5 January 1975; revised 25 February 1975

Nectar: Its Production and Functions in Trumpet Creeper

Abstract. *Studies of the trumpet creeper,* Campsis radicans *(L.) Seem. (Bignoniaceae), reveal five distinct nectary systems, a phenomenon never before reported among temperate zone plants. Ant activity, centered around the four extrafloral systems, clearly demonstrates the ant-guard symbiosis usually associated only with tropical or subtropical species. Floral nectar, an attractant for hummingbird and bumblebee pollinators, differs chemically from the ant-attracting nectar produced extraflorally.*

The production of nectar by plants and the attraction of certain insects to nectar have been of long-standing fascination to scientists (*1*). Much work has been channeled into the study of floral nectaries and the pollinators associated with them (*2*). Investigations of extrafloral nectaries have been limited until recently. These structures are located on the outer floral (*3*) and vegetative parts, including petioles (*4*), sheaths (*5*), and leaf margins (*6*). A result of this recent work has been a broadening of the field of animal-plant interactions. In recent years many ant-plant associations have been recognized in the tropics and subtropics, ranging from casual temporary alliances to mutualistic symbioses in which the participants are dependent on each other for survival. An example of the latter is the bull's horn acacia, *Acacia cornigera* L., and the acacia ant, *Pseudomyrmex ferruginea* F. Smith (*7*). The ants inhabit the enlarged hollowed stipular thorns and feed on a balanced diet of nectar from petiolar nectaries and Beltian bodies, modified leaf tips rich in protein. In addition to removing encroaching plants, the resident ants drive off invading insects by biting and stinging.

The trumpet creeper, *Campsis radicans*, one of the few temperate representatives of the tropical and subtropical family Bignoniaceae, is a common woody vine of the eastern and midwestern United States. Throughout its range it is host to several genera of ants. The relationship is not obligatory, although varying degrees of protection may be offered by different ant species in return for the extrafloral nectar produced by the plant. Trumpet creeper is possibly unique among temperate species, and certainly among a small number of all nectariferous plants, in possessing five nectary systems. The four extrafloral systems, located on the petiole (Fig. 1, a, b, and d), calyx (Fig. 1, e and f), corolla, and fruit (Fig. 1, h and i), are each visited regularly by ants. This report is perhaps the first documented case of nectaries occurring on developing fruits of any species. The ovarian (floral) nectary (Fig. 1g) aids in attracting the primary and secondary pollinators, hummingbirds and bumblebees, respectively. The extrafloral systems are also present on the Old World *Campsis grandiflora* as well as its hybrid *Campsis* × *tagliabuana* (*C. radicans* × *C. grandiflora*).

The aggressive climbing habit of the trumpet creeper has made it a familiar sight along hedgerows and fences of the Midwest. The flowers are five-parted and bilaterally symmetrical, each with a short tubular calyx and a flaring, tubular, showy, bright orange corolla, borne in dense terminal corymbs of 12 to 35 flowers (n = 21.6) (Fig. 1, c and j). Leaves are opposite, pinnately compound, each with 7 to 11 leaflets. The hybrid closely resembles its American parent in gross morphological features, the most conspicuous differences being the looser inflorescences and the larger flowers. Flowering, which occurs in both taxa from early July until early September, follows the "cornucopia" pattern as described by Gentry (*8*). Fruiting continues into late September, averaging 2.6 fruits per inflorescence.

Material was studied during the summers of 1972 and 1974 from native populations in southern Illinois, and naturalized populations were examined in southeastern New York. The hybrid was grown and observed in the nursery of the Cary Arboretum. Living material was fixed and stored in FAA (formalin, acetic acid, and ethyl alcohol), dehydrated in a tertiary butyl alcohol series, and embedded in TissuePrep. Serial sections were made at 10 μm, stained with safranin, and counterstained with fast green or with Delafield's hematoxylin. In order to determine the carbohydrate components, we collected nectar at different times during the day throughout the growing season, using microcapillary spotting tubes. Analysis was accomplished by thin-layer chromatography (*9*).

The individual extrafloral nectaries are minute and may be easily overlooked by the casual observer. The nectaries are generally circular in outline with a well-defined structure of a cup cavity with a base, surrounded by a wall or rim. Data concerning location, number, and size of the different nectaries and nectar composition are given in Table 1.

The petiolar nectaries are the first to secrete. The nectaries on the youngest three or four pairs of petioles on each branch

Table 1. Nectary systems of the trumpet creeper. Abbreviations: S, sucrose; G, glucose; F, fructose.

Location	Average number	Average height (mm)	Average diameter (mm)	Ratio of sugars		
				S	G	F
Extrafloral						
Petiole—adaxial surface	15.6/petiole	0.14	0.27	1	1	1
Calyx—abaxial lobes	20/flower	0.17	0.26	3	2	1
Corolla—abaxial lobes	25/flower	0.14	0.23	3	2	1
Fruit—scattered	200/fruit	0.15	0.19	1	1	1
Floral						
Ovarian—base of style	1/flower	1.73	3.98	0	1	1

Novel Activation Mechanism for the Promutagenic Herbicide Diallate

Abstract. *The potent bacterial mutagen 2-chloroacrolein is formed from the carcinogenic herbicide* S-2,3-*dichloroallyl diisopropylthiocarbamate (diallate) on incubation with hepatic microsomal monooxygenases or on reaction with* m-*chloroperbenzoic acid. A proposed activation mechanism for this promutagen involves sulfoxidation followed by [2,3] sigmatropic rearrangement and 1,2-elimination reactions. A portion of the highly reactive intermediate, diallate sulfoxide (proximate mutagen), is attacked by glutathione in a reaction which competes with its transformation to the ultimate mutagen, 2-chloroacrolein.*

Thiocarbamate herbicides, important chemicals used in producing corn, rice, sugar beets, vegetables, and other crops, are of two types. Most of them are *S*-alkyl or *S*-benzyl compounds, but there are also three commercial *S*-chloroallyl thiocarbamates: diallate or Avadex (*cis*, *trans* mixture) $[(CH_3)_2CH]_2NC(O)SCH_2C(Cl)=CHCl$; triallate or Avadex BW, $[(CH_3)_2CH]_2NC(O)SCH_2C(Cl)=CCl_2$; and CDEC or sulfallate or Vegadex, $(C_2H_5)_2NC(S)SCH_2C(Cl)=CH_2$. Both types of thiocarbamates appear to have similar effects in inhibiting weed growth (*1*), yet there are marked differences between the two groups in other properties, possibly related to the *S*-chloroallyl moiety. The *S*-alkyl and *S*-benzyl thiocarbamates are proherbicides, undergoing metabolic sulfoxidation to form moderately stable sulfoxides that readily carbamoylate thiol compounds (*2*). Sulfoxide metabolites have not been reported for diallate, triallate, and CDEC. The *S*-chloroallyl thiocarbamates, after metabolic activation, are bacterial mutagens, an unfavorable property not shown by the *S*-alkyl and *S*-benzyl thiocarbamates (*3*). Diallate, the most potent promutagen, received the greatest attention in our investigations.

The metabolic conditions under which the promutagenic *S*-chloroallyl thiocarbamates are activated are those likely to yield sulfoxide derivatives. By oxidizing with peracid at temperatures below 0°C, we were able to prepare diallate and triallate sulfoxides and examine their degradation chemistry (Fig. 1). The *S*-3-chloroallyl thiocarbamate sulfoxides undergo a spontaneous [2,3] sigmatropic rearrangement, followed by a 1,2-elimination reaction (*4*). This sulfoxidation, rearrangement, and elimination sequence converts the promutagen diallate to its ultimate mutagen, 2-chloroacrolein.

The *S*-chloroallyl thiocarbamate sulfoxides were prepared by reacting *cis*-diallate, *trans*-diallate, and triallate with equimolar amounts of *m*-chloroperbenzoic acid in methylene chloride at −15°C. They were isolated as crystalline compounds by working at temperatures below 0°C (*5*). These sulfoxides completely decompose at higher temperatures—for example, on thin-layer chromatography (TLC) at 20°C and on holding in chloroform for 10 minutes (*cis*-diallate sulfoxide) or 1 to 2 hours (*trans*-diallate sulfoxide and triallate sulfoxide) at 40°C. Monitoring the degradation rates and identifying the products by nuclear magnetic resonance (NMR) spectroscopy and other methods (*4*) provided an explanation for the instability of the *S*-chloroallyl thiocarbamate sulfoxides in contrast to the analogous *S*-alkyl and *S*-benzyl sulfoxides. *cis*-Diallate sulfoxide, for example, undergoes a rapid and quantitative conversion to 2-chloroacrolein and the carbamoylsulfenyl chloride (Fig. 1). The *trans* isomer gives the same compounds but more slowly. The products originate from a [2,3] sigmatropic rearrangement to a *S*-*O*-allylsulfenate ester followed immediately by a 1,2-elimination reaction (Fig. 1). Triallate sulfoxide degrades by the same mechanism, yielding 2-chloroacrylyl chloride instead of 2-chloroacrolein. The overall sulfoxidation, rearrangement, elimination reaction sequence proceeds much faster with *cis*-diallate than with *trans*-diallate or triallate.

The applicability of these chemical observations to metabolic conditions was examined in vitro and in vivo with mice and rats, using *cis*- and *trans*-diallate labeled with radiocarbon at the carbonyl position in one preparation and at the allyl group in another preparation. Liver microsomes extensively metabolize the diallate isomers, but only when fortified with reduced nicotinamide adenine dinucleotide phosphate (NADPH). Although this suggests sulfoxidation of *cis*- and *trans*-diallate, the sulfoxides were not detected. However, the sulfoxides are very unstable, as pointed out above, and probably react completely within a few seconds or minutes at physiological temperatures. The possible presence of the

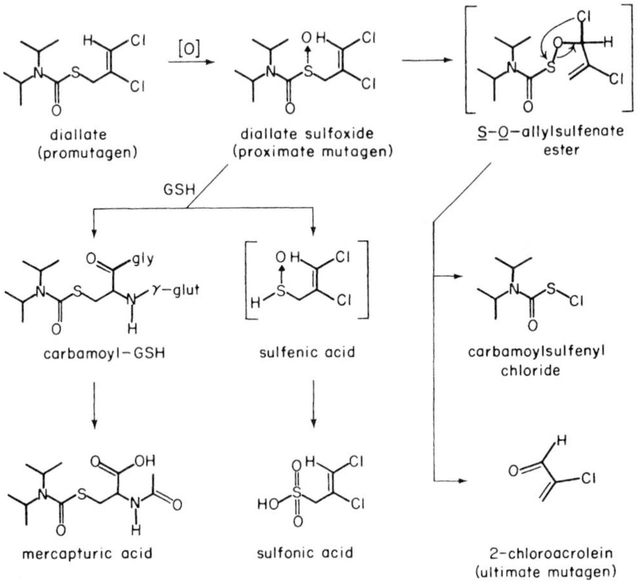

Fig. 1. Metabolic activation of the promutagen *cis*-diallate through *cis*-diallate sulfoxide to the mutagen 2-chloroacrolein. *cis*-Diallate sulfoxide is detoxified by reaction with GSH, yielding as major urinary metabolites the mercapturic and sulfonic acids. Brackets indicate intermediates or compounds probably present but not identified. The same reactions occur with *trans*-diallate, in which the positions of the terminal hydrogen and chlorine are exchanged.

0036-8075/79/0907-1013$00.50/0

Source: From *Science* Vol. 205, No. 4410, pp. 1013-1015 (1979). Reprinted with permission from AAAS.

sulfoxides as transitory intermediates in biological systems was therefore evaluated by identifying other metabolites that would form preferentially or only through the sulfoxides (Fig. 1). In studies with *carbonyl*-^{14}C-labeled *cis*- and *trans*-diallate in mouse microsomal systems (*6*), there was 49 to 67 percent $^{14}CO_2$ without added glutathione (GSH) but only 31 to 40 percent $^{14}CO_2$ when GSH was added; the carbamoyl-GSH derivative accounted for < 1 percent of the radiocarbon in the former case and 28 to 30 percent in the latter case. The diallate isomers do not react with GSH in aqueous acetone, but NMR and TLC studies show that their sulfoxides quickly carbamoylate this important tissue thiol. In the absence of GSH the enzymatically formed [*carbonyl*-^{14}C]diallate sulfoxide undergoes hydrolysis or intramolecular rearrangement reactions, liberating $^{14}CO_2$, whereas with GSH a portion of this intermediate sulfoxide is converted to the relatively stable carbamoyl-GSH derivative. Rats orally given the carbonyl-labeled preparation expired 20 percent $^{14}CO_2$, and their urine contained 62 percent mercapturic acid conjugate (Fig. 1) and 9 percent other metabolites originating from the carbamoyl-GSH derivative (the cysteine and mercaptoacetic acid conjugates). The fate of the chloroallyl portion of the molecule was examined in studies with [*allyl*-^{14}C]diallate. Dichloroallylsulfonic acid, probably formed through the sulfoxide (Fig. 1), was the major metabolite in both mice and rats and their microsome-NADPH systems. Of special importance was the identification of 2-[^{14}C]chloroacrolein in the mouse microsomal oxygenase system (*7*). Analyses by NMR revealed that this aldehyde is not liberated on direct reaction of diallate sulfoxide with GSH in aqueous acetone. These findings establish that under biological conditions diallate sulfoxide undergoes either the major detoxifying GSH conjugation or the minor, competing [2,3] sigmatropic rearrangement reaction followed by 1,2-elimination to liberate a toxicologically significant metabolite.

Mutagenesis studies (Fig. 2) combined with the chemical and biological observations discussed above suggest that the promutagens *cis*- and *trans*-diallate are converted through their sulfoxides (proximate mutagens) to 2-chloroacrolein, the ultimate mutagen (Fig. 1). Diallate is not mutagenic (*8*) unless the microsomal activation system is added, in which case the *trans* isomer gives 25 and the *cis* isomer 40 revertants per nanomole. *cis*-Diallate sulfoxide yields 113 revertants per nanomole, the same value obtained with 2-chloroacrolein (Fig. 2) (*9*); importantly, this is direct mutagenic activity, not dependent on the S9 mix. The greater potency of *cis*- than of *trans*-diallate may be due to more rapid formation of 2-chloroacrolein in the former case.

The fate and mutagenic or herbicidal activity of the diallate isomers appear to depend in large part on formation of the sulfoxides and on their subsequent reactions, which in turn are closely related to tissue GSH levels. The rearrangement and elimination reaction sequence to liberate 2-chloroacrolein occurs only with the portion of the sulfoxide that is not detoxified by reaction with GSH. Herbicidal activity of the diallate isomers may be due to the sulfoxides acting as carbamoylating agents for critical tissue thiols (*2*) or to liberation of 2-chloroacrolein, which is similar in potency to the diallate isomers themselves in inhibiting root growth (*10*).

Triallate sulfoxide, tested immediately after preparation by peracid oxidation, is a potent mutagen (without S9 activation), whereas its decomposition product, 2 chloroacrylyl chloride, is much less active. Since triallate sulfoxide is unstable, it is probably the proximate mutagen. It may be necessary for the chloroacrylyl chloride, which is easily hydrolyzed to the nonmutagenic 2-chloroacrylic acid, to be liberated within the bacteria in order to express its mutagenic activity. Activation of CDEC could involve either initial oxidation at one of the sulfur atoms or hydroxylation at the methylene group adjacent to the sulfur (*11*), the latter reaction leading to spontaneous release of the mutagen, 2-chloroacrolein.

Diallate is carcinogenic in mice (*12*) and CDEC in mice and rats (*13*) but no reports are available on tests with triallate. The findings concerning active mutagenic derivatives from oxidation of *S*-3-chloroallyl thiocarbamate herbicides may have relevance to the mechanism of their carcinogenic activity.

INGOLF SCHUPHAN*
JOSEPH D. ROSEN†
JOHN E. CASIDA
Pesticide Chemistry and Toxicology Laboratory, Department of Entomological Sciences, University of California, Berkeley 94720

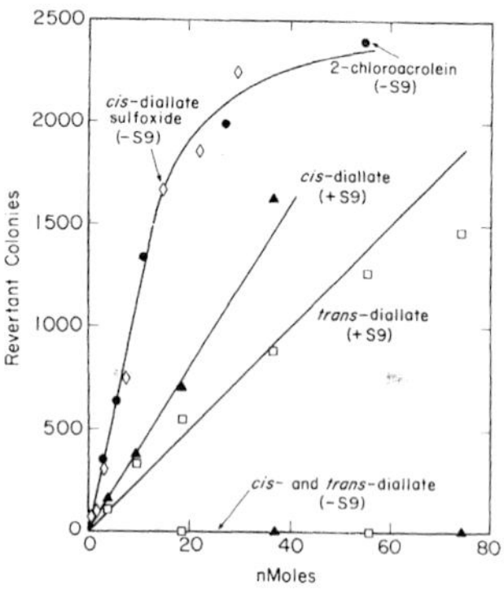

Fig. 2. Mutagenic activity of *cis*- and *trans*-diallate and some of their metabolites, assayed with *Salmonella typhimurium* strain TA 100 sensitive to base-pair substitution mutagens (*14*). The number of spontaneous revertant colonies (~ 120) has been subtracted from the revertant values plotted against amount of mutagen or mixture. Where indicated, S9 mix (containing 20 μl of the 9000*g* supernatant fraction of Arochlor-induced rat liver plus cofactors for xenobiotic metabolism) was incorporated into the top agar. The dose-response relationship is no longer linear at the higher concentrations of 2-chloroacrolein because of accumulation of lethal mutations in the bacteria. 2-Chloroacrolein (− S9) and diallate (+ S9) at 10 to 30 nmole also give significant numbers of revertant colonies with *S. typhimurium* strain TA 98 for detection of frameshift mutagens.

References and Notes

1. *Herbicide Handbook* (Weed Science Society of America, Champaign, Ill., ed. 3, 1974); R. E. Wilkinson, in *Chemistry and Action of Herbicide Antidotes*, F. M. Pallos and J. E. Casida, Eds. (Academic Press, New York, 1978), pp. 85–108.
2. J. E. Casida, R. A. Gray, H. Tilles, *Science* **184**, 573 (1974); J. E. Casida, E. C. Kimmel, M. Lay, H. Ohkawa, J. E. Rodebush, R. A. Gray, C. K. Tseng, H. Tilles, *Environ. Qual. Saf.* **3** (Suppl.), 675 (1975); J. E. Casida, E. C. Kimmel, H. Ohkawa, R. Ohkawa, *Pestic. Biochem. Physiol.* **5**, 1 (1975); J. P. Hubbell and J. E. Casida, *J. Agric. Food Chem.* **25**, 404 (1977).
3. F. De Lorenzo, N. Staiano, L. Silengo, R. Cortese, *Cancer Res.* **38**, 13 (1978); H. C. Sikka and P. Florczyk, *J. Agric. Food Chem.* **26**, 146 (1978).
4. I. Schuphan and J. E. Casida, *Tetrahedron Lett.* (1979), p. 841.
5. The reaction mixture in methylene chloride at 0° to 5°C was washed with cold 5 percent aqueous sodium carbonate, dried over sodium sulfate, and the solvent evaporated below 0°C by using a vigorous stream of nitrogen.
6. Y. S. Chen, I. Schuphan, J. E. Casida, *J. Agric. Food Chem.* **27**, 709 (1979).
7. 2-[^{14}C]Chloroacrolein in the gas phase was trapped in a solution of 2,4-dinitrophenylhydrazine and the resulting hydrazone derivative was isolated, identified by two-dimensional TLC in five different solvent systems, and quantitated by liquid scintillation counting. The yield was < 0.1 percent without NADPH and 1.6 percent with NADPH; that is, the microsomal monooxygenase cofactor is necessary for 2-chloroacrolein formation.
8. Diallate and triallate are not mutagenic even at 135 and 305 nmole per assay plate, respectively (*3*).
9. The mutagenic activity of 2-chloroacrolein on *Salmonella typhimurium* strain TA 100 tested directly or on breakdown of diallate sulfoxide is similar to that of the carcinogen benzo[*a*]pyrene (*14*).
10. 2-Chloroacrolein is approximately equivalent in potency to *cis*- and *trans*-diallate in inhibiting root elongation of oat seedlings. Sulfoxides of the diallate isomers are much less effective, possibly because they decompose on entering the plant without forming 2-chloroacrolein.
11. For an analogous reaction, see Y. S. Chen and J. E. Casida, *J. Agric. Food Chem.* **26**, 263 (1978).

12. J. R. M. Innes *et al.*, *J. Natl. Cancer Inst.* **42**, 1101 (1969).
13. "Bioassay of sulfallate for possible carcinogenicity," *Natl. Cancer Inst. Carcinogenesis Tech. Rep. Ser. 115, DHEW Publ. (NIH) 78-1370* (1978).
14. B. N. Ames, J. McCann, E. Yamasaki, *Mutat. Res.* **31**, 347 (1975).
15. Supported in part by NIH grant 5 P01 ES00049. For advice and assistance we thank B. N. Ames of the Department of Biochemistry (University of California, Berkeley) and Y. -L. Chen and R. L. Holmstead of the Pesticide Chemistry and Toxicology Laboratory.
* Permanent address: Institute of Pesticide Research, Federal Biological Research Center, D-1000 Berlin 33, Federal Republic of Germany.
† Permanent address: Department of Food Science, Cook College, Rutgers University, New Brunswick, N.J. 08903.

12 February 1979; revised 20 April 1979

Visual Resolution and Receptive Field Size: Examination of Two Kinds of Cat Retinal Ganglion Cell

Abstract. *Intraocular recordings from brisk-sustained and brisk-transient ganglion cells in the cat's retina revealed a systematic increase in center size and decrease in spatial cut-off frequency with increasing distance from the area centralis. At any one eccentricity sizes of the centers of sustained and transient cells did not overlap, and the variation in cut-off frequency for each class was constrained to about one-half octave.*

Spatial inhomogeneity within the visual system is reflected in the variation in spatial resolution across the visual field. Thus visual acuity is highest in the fovea and declines progressively as stimulation is moved towards the periphery (*1*). This reduction in spatial resolution is attributed to several factors associated with increasing eccentricity, among them a decline in receptor and ganglion cell density (*2, 3*), a decrease in cortical magnification (*4*), and an increase in receptive field size (*5-7*). We performed experiments to show how both spatial resolution and receptive field size of selected classes of cat retinal ganglion cells vary as a function of eccentricity, and in addition to determine the local variability of these two measures.

The retinal ganglion-cell mosaic is made up of many types of ganglion cells (*6, 7*), which likely subserve different visual functions. It is therefore not sufficient to describe general trends in receptive field properties of ganglion cells taken as a whole; rather, we need knowledge of receptive field properties within a single class as a function of retinal position. This kind of information is most readily obtained through the use of intraocular recording techniques that allow the position of the microelectrode to be varied systematically and accurately. Moreover, it is essential to collect data from as many units as possible within a single retina, for pooling data from several experiments inflates the observed variability.

This report deals only with on-center units with brisk properties (*7*). In all, 180 ganglion cells (138 brisk-sustained units and 42 brisk-transient units) were studied in four cats. In all but seven of these units, we measured, as a function of retinal eccentricity (i) the size of the receptive field center (bar stimuli) and (ii) the spatial cut-off frequency (spatial frequency beyond which there is no significant modulated response) (high-contrast, square-wave gratings). We have specifically avoided examining the unmodulated response observed in brisk-transient units by using fine gratings (*6*).

Experimental methods are essentially the same as those used by Cleland and Levick (*7*). Anesthesia was induced in adult cats by ventilation of 3 to 4 percent halothane in a 2:1 gas mixture of nitrous oxide and carbogen and was maintained during preparatory surgery with 1 to 1.5 percent halothane. To reduce eye movements, the left vagosympathetic trunk was cut and the animal paralyzed by continuous intravenous infusion of an isotonic solution (4 ml/hour) of Flaxedil (5 mg per kilogram of body weight per hour), *d*-tubocurarine (0.4 mg/kg-hour), and glucose. End-tidal P_{CO_2} was monitored and kept at 4 percent by adjusting the stroke volume of the respirator. Bipolar stimulating electrodes (platinum iridium wire in glass) were stereotaxically placed in the left and right optic tracts. The left eye was secured to a micromanipulator by suturing a flap of conjunctiva to a metal ring encircling the globe. The sclera was punctured with a sealed hypodermic needle which allowed the insertion of a tungsten microelectrode (*8*) into the posterior chamber and the precise positioning of the micro-

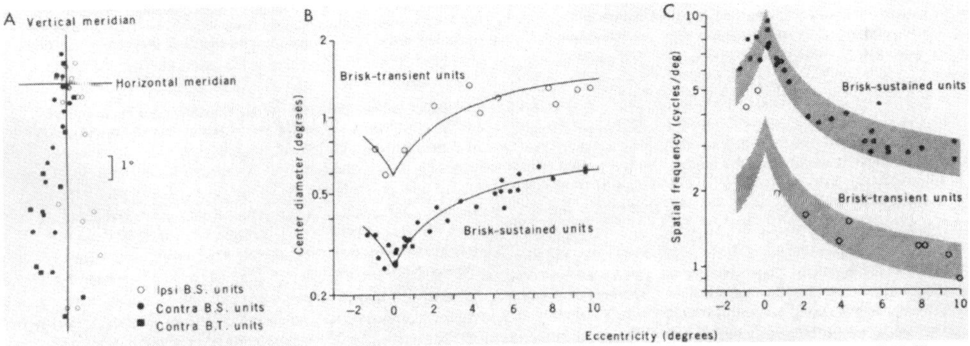

Fig. 1. Results from 42 on-center cells recorded in a single experiment (cat 52). (A) Receptive field positions within the visual field. The two meridians intersect at the area centralis. (B) Center size as a function of eccentricity (angular distance from the area centralis). Peristimulus-time histograms describing the results for a bar (luminance, 25 cd/m², background, 3 cd/m²) moving slowly across the receptive field. Center size was taken as the width of the peak of the histogram at the level of the maintained firing. (C) Spatial cut-off frequency as a function of eccentricity. The shaded area represents a one-quarter octave above and below a hyperbolic curve (reflected) fitted to the sustained units. This area is moved vertically to fit the transient units. Cut-off frequency was estimated from peristimulus-time histograms describing the results of square-wave gratings (mean luminance, 150 cd/m²; contrast, 0.84) moving across the receptive field. Drift frequency, 4 cycles per second.

 0036-8075/79/0907-1015$00.50/0

α-Thujone (the active component of absinthe): γ-Aminobutyric acid type A receptor modulation and metabolic detoxification

Karin M. Höld*, Nilantha S. Sirisoma*, Tomoko Ikeda†, Toshio Narahashi†, and John E. Casida*‡

*Environmental Chemistry and Toxicology Laboratory, Department of Environmental Science, Policy and Management, 114 Wellman Hall, University of California, Berkeley, CA 94720-3112; and †Department of Molecular Pharmacology and Biological Chemistry, Northwestern University Medical School, Chicago, IL 60611-3008

Contributed by John E. Casida, January 31, 2000

α-Thujone is the toxic agent in absinthe, a liqueur popular in the 19th and early 20th centuries that has adverse health effects. It is also the active ingredient of wormwood oil and some other herbal medicines and is reported to have antinociceptive, insecticidal, and anthelmintic activity. This study elucidates the mechanism of α-thujone neurotoxicity and identifies its major metabolites and their role in the poisoning process. Four observations establish that α-thujone is a modulator of the γ-aminobutyric acid (GABA) type A receptor. First, the poisoning signs (and their alleviation by diazepam and phenobarbital) in mice are similar to those of the classical antagonist picrotoxinin. Second, a strain of *Drosophila* specifically resistant to chloride channel blockers is also tolerant to α-thujone. Third, α-thujone is a competitive inhibitor of [^{3}H]ethynylbicycloorthobenzoate binding to mouse brain membranes. Most definitively, GABA-induced peak currents in rat dorsal root ganglion neurons are suppressed by α-thujone with complete reversal after washout. α-Thujone is quickly metabolized *in vitro* by mouse liver microsomes with NADPH (cytochrome P450) forming 7-hydroxy-α-thujone as the major product plus five minor ones (4-hydroxy-α-thujone, 4-hydroxy-β-thujone, two other hydroxythujones, and 7,8-dehydro-α-thujone), several of which also are detected in the brain of mice treated i.p. with α-thujone. The major 7-hydroxy metabolite attains much higher brain levels than α-thujone but is less toxic to mice and *Drosophila* and less potent in the binding assay. The other metabolites assayed are also detoxification products. Thus, α-thujone in absinthe and herbal medicines is a rapid-acting and readily detoxified modulator of the GABA-gated chloride channel.

Absinthe was a popular emerald-green liqueur in the 19th and early 20th centuries. It was commonly imbibed by artists and writers including Vincent van Gogh, Henri de Toulouse-Lautrec, and Charles Baudelaire, often inducing fits and hallucinations and sometimes contributing to psychoses and suicides (1–5). Absinthe became an epidemic health problem and was banned in many countries early in the 20th century, but its use continues legally or illicitly even now (6, 7). The toxic properties of absinthe are attributable to wormwood oil used in making the beverage. Wormwood oil is in itself a prevalent herbal medicine for treating loss of appetite, dyspeptic disorders, and liver and gallbladder complaints (8, 9).

α-Thujone (Fig. 1) generally is considered to be the principal active ingredient of wormwood oil and toxic principle in absinthe (2). The content of β-thujone often exceeds that of α-thujone depending on the plant source, but the β-diastereomer (Fig. 1) is generally of lower toxicity. α-Thujone also is reported to have antinociceptive activity in mice (10). This monoterpenoid occurs in many plants, including *Artemesia* species, sage, and the Thuja tree (4). Extracts of wormwood were used to control gastrointestinal worms with records back to ancient Egyptian times (4). *Artemesia absinthium* and wormwood oil have insecticidal properties (11), and α-thujone was one of the two most toxic monoterpenoids tested against western corn rootworm larvae (12). Public mistrust of synthetic pharmaceuticals and pesticides has led to the increasing popularity of herbal medicines and botanical insecticides even though they have not been subjected to the same rigorous tests of safety and evaluation of toxicological mechanisms (13–15).

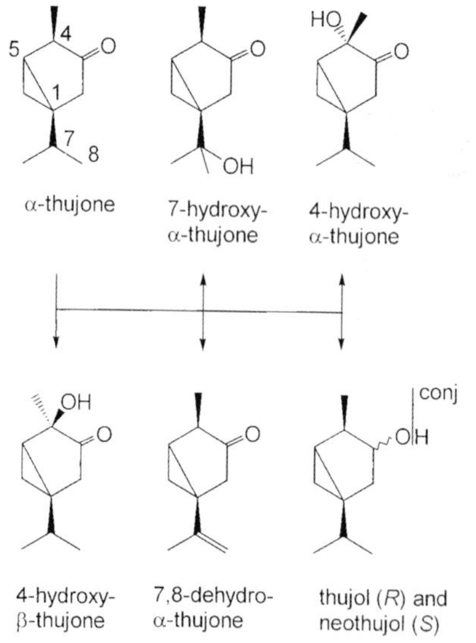

Fig. 1. Structures of α-thujone (1*S*, 4*R*, 5*R*-thujone) and its metabolites in the mouse liver microsomal P450 system, the brain of treated mice, and the urine of treated rabbits. The hydroxythujones and dehydro-α-thujone are observed in the mouse liver microsomal P450 system and in brain whereas thujol and neothujol are identified in the rabbit liver cytosolic ketone reductase system and in urine as conjugates. The major metabolite in mouse brain and the P450 system is 7-hydroxy-α-thujone. β-Thujone is the 1*S*, 4*S*, 5*R* diastereomer (structure not shown).

Abbreviations: EBOB, ethynylbicycloorthobenzoate or 4′-ethynyl-4-*n*-propylbicycloorthobenzoate; GABA, γ-aminobutyric acid; $GABA_A$ receptor, type A GABA receptor; LC_{50}, median lethal concentration.

‡To whom reprint requests should be addressed. E-mail: ectl@nature.berkeley.edu.

The publication costs of this article were defrayed in part by page charge payment. This article must therefore be hereby marked "*advertisement*" in accordance with 18 U.S.C. §1734 solely to indicate this fact.

Article published online before print: *Proc. Natl. Acad. Sci. USA*, 10.1073/pnas.070043297. Article and publication date are at www.pnas.org/cgi/doi/10.1073/pnas.070043297

The toxic effects of α-thujone in mammals are well established but the mode of neurotoxic action is poorly understood. It is porphyrogenic, possibly thereby contributing to the absinthe-induced illness of Vincent van Gogh (5, 16). α-Thujone is neurotoxic in rats (17), and ingestion of wormwood oil containing α-thujone recently resulted in human poisoning (18). The hypothesis that α-thujone activates the CB_1 cannabinoid receptor, based on the structural similarity of thujone enol with tetrahydrocannabinol (19), was not supported experimentally (20). The convulsant action led to multiple speculations on mechanisms, one of which was antagonism of the γ-aminobutyric acid (GABA) receptor system (20), a proposal that was not explored further. α- and β-Thujone are reduced in rabbits from the ketones to the corresponding alcohols (thujol and neothujol) (21) of unknown toxicity but no other metabolites are identified.

The goals of this study are to define the mechanism of neurotoxicity of α-thujone and identify its major metabolites (Fig. 1) and their role in the poisoning process. Emphasis is placed on the hypothesis that the convulsant action is caused by modulating the GABA-gated chloride channel.

Materials and Methods

Chemicals. Sources were: α-thujone (≈99% purity) from Fluka; wormwood oil (3.2% α- and 35% β-thujone) from Lhasa Karnak (Berkeley, CA) and absinthe with 0.4 ppm α-thujone, 5 ppm β-thujone, and 50% (vol/vol) ethanol labeled Herring Absenta (Zaragoza, Spain) with concentrations based on analyses in this laboratory; picrotoxinin, diazepam, and sodium phenobarbital from Sigma; dieldrin and α-endosulfan from Chem Service (West Chester, PA); [^{3}H]ethynylbicycloorthobenzoate ([^{3}H]EBOB) (38 Ci/mmol) from NEN. Although not detailed here, 7-hydroxy-α-thujone, 4-hydroxy-α-thujone, 4-hydroxy-β-thujone, 7,8-dehydro-α-thujone, and a thujol/neothujol mixture were synthesized as standards for comparison with metabolites.

Toxicity to Mice. Male albino Swiss–Webster mice (22–28 g) were treated i.p. with the test compound by using propylene glycol (2 μl/g body weight) as the carrier vehicle. Prophylactic i.p. treatments also were examined for their effect on α-thujone toxicity (100 mg/kg) individually with ethanol (0.5 or 1.0 g/kg as 20% and 40% solutions in saline, 20 min pretreatment), diazepam (1 mg/kg, 15 min pretreatment), or phenobarbital (15 mg/kg, 15 min pretreatment).

Toxicity to *Drosophila*. Fruit flies (*Drosophila melanogaster*) were used in two types of assays: comparing two strains known to be different in sensitivity to insecticidal chloride channel blockers and comparing α-thujone and its metabolites for toxicity to the susceptible strain. The median lethal concentration (LC_{50}) was determined for α-thujone and dieldrin with two strains of *Drosophila*: a dieldrin-resistant *Rdl*$^{MD\text{-}RR}$ strain (22, 23) (obtained from the Bloomington *Drosophila* Stock Center at Indiana University, Bloomington) and the *Canton-S*, wild-type sensitive (*S*) strain. The test chamber was a glass tube (12 × 75 mm) containing a filter paper strip (Whatman no. 1, 8 × 65 mm). Five adult flies were placed in the tube, which then was closed with a single layer of parafilm. A solution of α-thujone or dieldrin in propylene glycol (5 μl) was injected with a 10-μl syringe through the parafilm onto the filter paper after which the tube was covered with a second piece of parafilm. Mortality was recorded after 8 h at 25°C as flies that could not move. The experiment was repeated four times to prepare dosage mortality curves for calculation of resistance ratios (LC_{50} *Rdl*/LC_{50} *S*).

Effect on [^{3}H]EBOB Binding in Mouse Brain Membranes. Mouse brain membranes were prepared and depleted of GABA as described (24). For inhibitor potency assays, the membranes (200 μg protein) were incubated with the test compound (added in DMSO, final concentration 1%) and [^{3}H]EBOB (0.7 nM) in 1.0 ml of 10 mM sodium phosphate, pH 7.5 buffer containing 200 mM sodium chloride at 37°C for 70 min (25). Scatchard analyses were performed with no inhibitor and with 5 and 25 μM α-thujone by using [^{3}H]EBOB at 0.08–26 nM. The inhibitory potency also was compared for ethanol and absinthe (based on ethanol content) with that for ethanol containing 5 μM α-thujone. The incubated mixtures were filtered through GF/C glass fiber filters, then rinsed twice with 5 ml of ice-cold 0.9% sodium chloride, by using a cell harvester. Specific binding was considered to be the difference between total binding and nonspecific binding determined in the presence of 5 μM α-endosulfan {a potent GABA type A ($GABA_A$) receptor antagonist and specific inhibitor of [^{3}H]EBOB binding}.

Effect on GABA-Induced Whole-Cell Currents. Rat dorsal root ganglion neurons were prepared and cultured as described (26). Currents were induced by 10-msec pulses of 300 μM GABA and recorded by using the whole-cell patch clamp technique. The GABA-induced inward current of this preparation was carried by chloride ions through open chloride channels (27). Each cell was tested for the degree of suppression caused by bath application of α-thujone to determine the concentration for 50% inhibition (IC_{50}).

GC-MS Identification and Analysis of α-Thujone and Metabolites. Standard analytical methods of GC-MS and derivatization of alcohol and ketone functionalities were applied to α-thujone and its metabolites. Analyses used the DB-5 fused silica gel capillary column (30 m, 0.25 mm i.d., 0.25 μm film thickness) (J&W Scientific, Folsom, CA). The initial column temperature of 80°C was programmed to 200°C at the rate of 5°C/min, followed by an increase at 20°C/min to 300°C where it was maintained for 2 min. The carrier gas and reagent gas were helium and methane, respectively. Temperatures of the injection port and detector were 250°C and 280°C, respectively. The mass spectrometer was operated in the positive chemical ionization mode. One microliter was injected splitless onto the column. For quantitation, the GC-MS was operated in the selected ion monitoring (SIM) mode, measuring m/z 135 for α-thujone and m/z 151 for the hydroxythujones, dehydrothujone, and (*S*)-(−)-carvone (internal standard). The concentration of each analyte was determined from least-squares equations generated from peak-area ratios of α-thujone, 7-hydroxy-α-thujone, and the internal standard. Identification of α-thujone and metabolites involved comparison with standards by cochromatography and MS fragmentation patterns as parent compounds and two derivatives. Trimethylsilyl ethers were formed on reaction of alcohols with *N*-methyl-*N*-trimethylsilyltrifluoroacetamide and methyloximes on coupling ketones with methoxyamine. These derivatization procedures and MS fragmentation patterns also allowed assignment of some metabolites as hydroxythujones without specifying the position of hydroxylation.

Enzymatic Metabolism. Rabbit or mouse liver cytosol (1 mg protein) or washed mouse liver microsomes (1 mg protein) and NADPH (or other cofactor, 1 mM final concentration) were incubated with α-thujone (30 μg, 0.2 μM final concentration) in 100 mM phosphate, pH 7.4 buffer (1 ml) for 1 h at 37°C. For analysis the internal standard *S* carvone (0.05 μg) was added in ethanol (10 μl), and the mixture was saturated with sodium chloride and extracted with ethyl acetate (3 ml) for 30 min by gentle rocking. The organic extract, recovered by centrifugation at 900 *g*, was almost completely evaporated (but never to dryness) under a stream of nitrogen at room temperature and reconstituted in ethyl acetate (50 μl) for GC-MS analysis. Recovery values by this procedure for α-thujone and the major metabolite were >60% with no degradation during GC.

APPLIED BIOLOGICAL SCIENCES

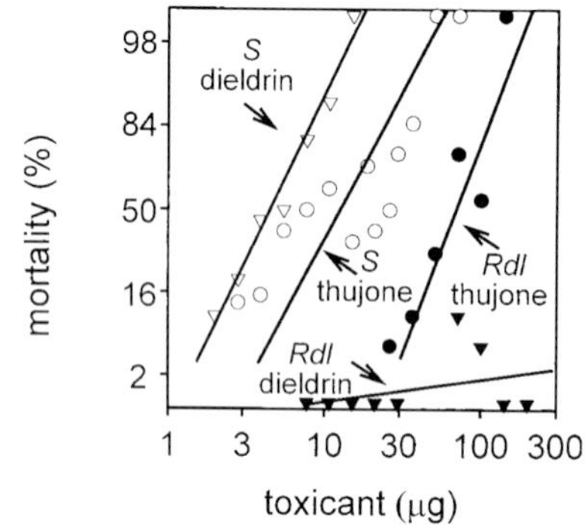

Fig. 2. *Drosophila* of the dieldrin-resistant (*Rdl*) strain are also resistant to α-thujone. The susceptible (*S*) strain is *Canton S*. Concentration is shown on a logarithmic scale and mortality on a probit scale.

Analysis of Brain. Mice were treated i.p. with α-thujone. At appropriate times thereafter the animals were killed and whole brains were removed for analysis. They were rinsed and homogenized in 10 ml of 100 mM phosphate, pH 7.4 buffer. The internal standard was added as above. The mixtures were centrifuged at 1,500 × *g* for 10 min. The pellet was resuspended in 2 ml of phosphate buffer, sonicated for 1 min, and centrifuged, and the supernatant fractions were combined. The samples were extracted with ethyl acetate (6 ml) and analyzed as described in *Enzymatic Metabolism*.

Results

α-Thujone Is a Convulsant. The i.p. LD_{50} of α-thujone in mice is about 45 mg/kg, generally with 0% and 100% mortality at 30 and 60 mg/kg, respectively. Mice at the higher dose undergo a tonic convulsion leading to death within 1 min whereas at 30–45 mg/kg they exhibit tail-raising within the first 2 min, followed by flexion of the trunk and clonic activity of the forelimbs, progressing to generalized and protracted tonic/clonic convulsions that ultimately result in death or recovery. Intraperitoneal administration of diazepam or phenobarbital 15 min before α-thujone at 100 mg/kg results in almost all of the mice surviving this otherwise lethal dose. Ethanol i.p. pretreatment at 1 g/kg (but not at 0.5 g/kg) also protects against the lethal effects of α-thujone at 100 mg/kg.

α-Thujone Cross-Resistance in *Drosophila* Strain Resistant to Dieldrin. Flies of the *Rdl* strain (>55-fold resistant to dieldrin; LC_{50} >275 μg/tube for *Rdl* versus 5 μg/tube for *S*) are 5-fold resistant to α-thujone (LC_{50} 65 μg/tube for *Rdl* versus 12 μg/tube for *S*) (Fig. 2). This finding establishes moderately high insecticidal activity for α-thujone and cross-resistance in the dieldrin-resistant strain.

α-Thujone Inhibition of [^{3}H]EBOB Binding. The IC_{50} of α-thujone for [^{3}H]EBOB binding in mouse brain membranes is 13 ± 4 μM (Fig. 3*A*). The binding of α-thujone is competitive with that of [^{3}H]EBOB based on Scatchard analysis (Fig. 3*B*). For comparison, other IC_{50} values are 29 ± 8 μM for β-thujone, 37 ± 8 μM for wormwood oil (calculated as molecular weight of thujone), and 0.6 ± 0.1 μM for picrotoxinin (inhibition curves not shown).

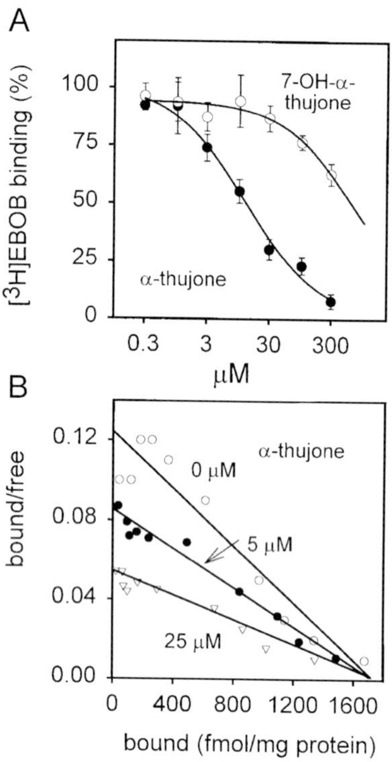

Fig. 3. α-Thujone and 7-hydroxy-α-thujone inhibit [^{3}H]EBOB binding to mouse brain membranes. (*A*) IC_{50} determination for α-thujone and 7-hydroxy-α-thujone (mean ± SEM, $n = 4$). (*B*) Scatchard plots as average of duplicate measurements for [^{3}H]EBOB alone (K_d 2.8 nM and B_{max} 1,700 fmol/mg protein) and with α-thujone at 5 μM (K_d 4.1 and B_{max} 1,700) and 25 μM (K_d 7.2 and B_{max} 1,700).

α-Thujone Modulation of the $GABA_A$ Receptor-Chloride Channel. The currents induced by 300 μM GABA are suppressed with 30 μM bath-applied α-thujone and there is full reversal on washing with α-thujone-free solution (Fig. 4 *A* and *B*). The IC_{50} for α-thujone is 21 μM in suppressing the GABA-induced currents (Fig. 4*C*).

Absinthe, Ethanol, and Ethanol Containing α-Thujone as Inhibitors of [^{3}H]EBOB Binding. The inhibitory effects on [^{3}H]EBOB binding were compared for absinthe, ethanol, and ethanol containing α-thujone to help understand their independent and combined actions on the chloride channel. The IC_{50} for absinthe (based on ethanol content) is 263 ± 47 mM and for ethanol is significantly higher at 370 ± 4 mM (Fig. 5*A*). There is no significant interaction between the effects of ethanol and α-thujone (Fig. 5*B*), i.e., α-thujone (5 μM) inhibition is 20–30% independent of ethanol concentration up to 300 mM.

Metabolism of α-Thujone by Liver Enzymes. Incubation of α-thujone with rabbit (but not mouse) liver cytosol gives thujol and neothujol, identified by GC-MS comparison with authentic

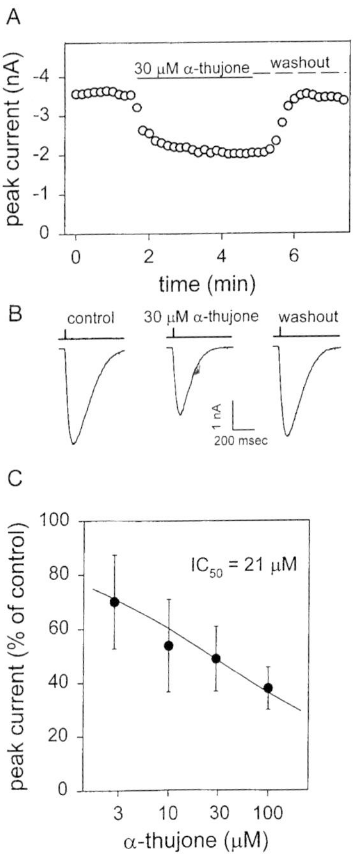

Fig. 4. Suppression of GABA-induced peak currents by bath application of α-thujone. Currents were induced by 300 μM GABA (10 msec) pulses. The peak amplitude of current decreased with 30 μM α-thujone and recovered after washing with α-thujone-free solution. (*A*) Time course of 30 μM α-thujone-induced changes in peak current amplitude. (*B*) Representative current records. (*C*) Concentration-response relationship (mean ± SD, n = 4–5).

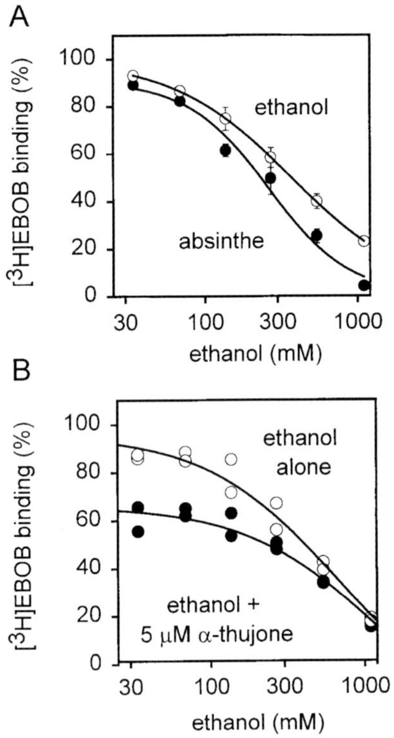

Fig. 5. Absinthe, ethanol, and ethanol containing α-thujone inhibit [^{3}H]EBOB binding to mouse brain membranes. (*A*) Comparison of an absinthe preparation (based on ethanol content) with ethanol (average of duplicate measurements or mean ± SD, n = 6). (*B*) Comparison of ethanol with ethanol containing 5 μM α-thujone (average of duplicate measurements).

standards *per se* and by forming trimethylsilyl (but not methyloxime) derivatives. This enzymatic reduction depends on NADPH but occurs in small yield. Metabolism in mouse liver microsomes is a much more facile reaction and gives no thujol or neothujol but instead different products. α-Thujone is stable on incubation with mouse liver microsomes alone but is almost completely metabolized when NADPH (but not NADP, NADH, or NAD) also is added. Six NADPH-dependent microsomal metabolites are evident by GC-MS, each at higher retention time than the parent α-thujone (Fig. 6). The first-eluting metabolite is identical in GC and MS features to synthetic 7,8-dehydro-α-thujone. The next five metabolites each are converted to trimethylsilyl and methyloxime derivatives, indicating the presence of both an alcohol substituent and a ketone functionality. Synthesis of various hydroxythujones and their comparison with the metabolites (directly, and as trimethylsilyl ethers and methyloximes) identifies the major product as 7-hydroxy-α-thujone and two minor metabolites as the diastereomers of 4-hydroxythujone.

Metabolites in the Brain of α-Thujone-Treated Mice. The brain contains α-thujone, dehydro-α-thujone, and four hydroxythujones (7-hydroxy-α major plus 4-hydroxy-α, 4-hydroxy-β, and one other) also observed in the liver P450 system (Fig. 6). Identifications are based on retention times and MS fragmentation patterns both direct and as trimethylsilyl and methyloxime derivatives. The brain levels of α-thujone and 7-hydroxy-α-thujone are dose- and time-dependent after i.p. injection of α-thujone (Fig. 7). Importantly, α-thujone appears at much lower levels and is less persistent than 7-hydroxy-α-thujone. At severely toxic α-thujone doses (40–60 mg/kg) the levels in brain

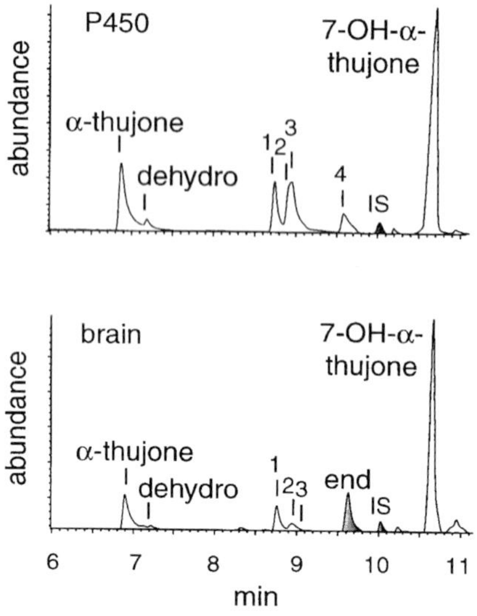

Fig. 6. Representative GC-MS- selected ion monitoring chromatograms for α-thujone and metabolites extracted from the mouse liver microsome-NADPH (P450) system and the brain of α-thujone-treated mice (50 mg/kg, i.p., 10 min after treatment). The major metabolite is 7-hydroxy-α-thujone. Four minor hydroxythujone metabolites are as follows: 1) 4-hydroxy-α; 3) 4-hydroxy-β; 2 and 4) others. Dehydro refers to 7,8-dehydro-α-thujone. Shaded peaks not derived from α-thujone are an endogenous substance (end) and the internal standard (IS). All thujone-derived metabolites fall within the chromatographic region shown.

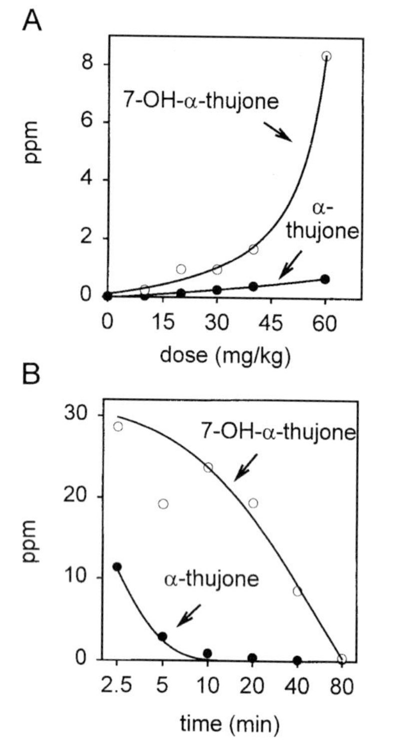

Fig. 7. Brain levels of α-thujone and 7-hydroxy-α-thujone as a function of dose and time in mice treated i.p. with α-thujone. Average of determinations on two mice except for a single determination at 5 min. (*A*) Dose studies at 30 min after treatment. (*B*) Time studies at 50 mg/kg.

at 30 min are 0.3–1.0 ppm for α-thujone and 1.5–8.4 ppm for 7-hydroxy-α-thujone (Fig. 7*A*) with much higher levels (11 and 29 ppm for α-thujone and 7-hydroxy-α-thujone, respectively) at 2.5 min (Fig. 7*B*) when the poisoning signs are most intense. The minor hydroxythujone metabolites are detectable only up to 20 min after the 50 mg/kg α-thujone dose.

Biological Activities of Metabolites. Synthetic standards of the metabolites shown in Fig. 1 except the 4-hydroxy-α-thujone diastereomers were compared with α-thujone for potency as toxicants to mice and *Drosophila* and inhibitors of [^{3}H]EBOB binding. The discriminating levels used were 50 mg/kg i.p. for mice and 50 μg/tube for the *S* strain of *Drosophila*. With mice, α-thujone is lethal, whereas 7-hydroxy-α-thujone, dehydro-α-thujone, and thujol/neothujol are not lethal. With *Drosophila*, α-thujone gives complete mortality, dehydro-α-thujone gives 70% mortality, and 7-hydroxy-α-thujone and thujol/neothujol give about 30% mortality. In the [^{3}H]EBOB binding assay, 7-hydroxy-α-thujone gives an IC_{50} value of 730 ± 265 μM versus 13 ± 4 μM for α-thujone (Fig. 3*A*), whereas the value for dehydro-α-thujone is 149 ± 10 μM (inhibition curve not shown).

Discussion

This study establishes that α-thujone modulates the $GABA_A$ receptor based on four observations. Comparison with picrotoxinin, the classical $GABA_A$ receptor antagonist, revealed similar poisoning signs and in both cases alleviation of the toxicity by diazepam, phenobarbital, and ethanol (28, 29). *Drosophila* with a single point mutation in the *Rdl* GABA receptor subunit of Ala^{302} to Ser conferring resistance to dieldrin (22, 23) is also resistant to α-thujone, albeit to a lesser degree. α-Thujone is a competitive inhibitor of [^{3}H]EBOB binding, i.e., of the noncompetitive blocker site of the GABA-gated chloride channel (25). Most importantly, electrophysiological studies establish that in dorsal root ganglion neurons α-thujone is a reversible modulator of the $GABA_A$ receptor.

Absinthe and wormwood oil contain not only α-thujone as their purported active ingredient but also many other candidate toxicants, including β-thujone and ethanol in the case of absinthe. β-Thujone is less toxic than α-thujone to mice (10) and *Drosophila* and in addition is 2.3-fold less potent in the [^{3}H]EBOB assay (this investigation). Ethanol also enhances neuronal $GABA_A$ receptor function (30) and therefore might suppress the blocking action of α-thujone in absinthe. However, ethanol does not alter the inhibitory action of α-thujone on [^{3}H]EBOB binding. The α- and β-thujone content of the absinthe sample examined here (0.4 and 5 ppm or 2.6 and 33 μM, respectively) may be a contributing factor in the somewhat greater potency of absinthe (based on ethanol content) than of ethanol *per se* in the [^{3}H]EBOB assay. However, the 10 ppm (66 μM) upper limit of the European Commission (6) and particularly the 260 ppm (1710 μM) thujone content of old absinthe (6) would give a detectable to major inhibitory effect beyond that of the ethanol content. Current low levels of α- and β-thujone in absinthe are of much less toxicological concern than the ethanol content (6).

α-Thujone as other monoterpenes is easily metabolized. The single report on metabolism identifies thujol and neothujol probably as conjugates in the urine of thujone-treated rabbits (21). We find enzymatic reduction (possibly by a cytosolic ketone reductase) (31) of α-thujone to thujol and neothujol in low yield by rabbit but not mouse liver cytosol with NADPH. The mouse liver microsomal P450 system rapidly converts α-thujone to 7-hydroxy-α-thujone (major), the diastereomers of 4-hydroxythujone (minor), and other hydroxythujones (minor). Interestingly, the major sites of P450 hydroxylation at the 4- and 7-positions are those involving intermediate tertiary radicals that are more stable than secondary and primary radicals. Dehydro-α-thujone also is observed and may arise from dehydration of the 7-hydroxy compound as a biological reaction because this possible conversion is not an artifact during the extraction and analysis procedure. The various hydroxythujones probably are not the terminal metabolites because they are expected to undergo conjugation and excretion. However, the presence of hydroxythujones in the brain suggests their potential importance in the neurotoxicity.

Metabolic detoxification is a dominant feature of α-thujone neurotoxicity in mice. There are two principal candidate toxicants, α-thujone and its 7-hydroxy metabolite. The 7-hydroxy compound is present in brain at much higher levels than the parent α-thujone, suggesting possible conversion *in situ*, but this oxidation was not observed on incubation of α-thujone with brain microsomes and NADPH. α-Thujone compared with 7-hydroxy-α-thujone is 56-fold more potent in the [^{3}H]EBOB binding assay and much more toxic to mice and houseflies. It appears that all of the metabolites studied here are detoxification products, i.e., less toxic than α-thujone. However, the level in brain of 7-hydroxy-α-thujone is several-fold greater than that of α-thujone (e.g., 29 and 11 ppm, respectively, at the time of severe poising signs), suggesting that either one or both may contribute to the toxic manifestations.

This study establishes that α-thujone acts at the noncompetitive blocker site of the $GABA_A$ receptor and is rapidly detoxified, thereby providing a reasonable explanation for some of the actions of absinthe other than those caused by ethanol, and allowing more meaningful evaluation of risks involved in the continued use of herbal medicines containing α-thujone.

Helpful comments were provided by Wilfred Arnold (University of Kansas Medical Center, Kansas City) and Jeffrey Bloomquist (Virginia Polytechnic Institute, Blacksburg). We thank our former coworker Neil Jacobson for initiating our interest in α-thujone action and our laboratory colleagues Gary Quistad and Susan Sparks for help and advice. This project was supported by National Institute of Environmental Health Sciences Grant R01 ES08419 to J.E.C. and National Institute of Neurological Disorders and Stroke Grant R01 NS14143 to T.N.

APPLIED BIOLOGICAL SCIENCES

1. Vogt, D. D. & Montagne, M. (1982) *Int. J. Addict.* **17,** 1015–1029.
2. Arnold, W. N. (1988) *J. Am. Med. Assoc.* **260,** 3042–3044.
3. Conrad, B., III (1988) *Absinthe: History in a Bottle* (Chronicle Books, San Francisco).
4. Arnold, W. N. (1989) *Sci. Am.* **6,** 112–117.
5. Arnold, W. N. (1992) *Vincent van Gogh: Chemicals, Crises, and Creativity* (Birkhaüser, Boston, MA), pp. 101–137.
6. Strang, J., Arnold, W. N. & Peters, T. (1999) *Br. Med. J.* **319,** 1590–1592.
7. Reese, K. M. (1999) *Chem. Eng. News* **77** (12), 56.
8. Gruenwald, J., Brendler, T. & Jaenicke, C., eds. (1998) *PDR for Herbal Medicines* (Medical Economics, Montvale, NJ), pp. 1181–1183.
9. Blumenthal, M., ed. (1998) *The Complete German Commission E Monographs: Therapeutic Guide to Herbal Medicines* (American Botanical Council, Austin, TX), pp. 664–665.
10. Rice, K. C. & Wilson, R. S. (1976) *J. Med. Chem.* **19,** 1054–1057.
11. Grainge, M. & Ahmed, S. (1988) *Handbook of Plants with Pest-Control Properties* (Wiley, New York), pp. 35–36.
12. Lee, S., Tsao, R., Peterson, C. & Coats, J. R. (1997) *J. Econ. Entomol.* **90,** 883–892.
13. Matthews, H. B., Lucier, G. W. & Fisher, K. D. (1999) *Environ. Health Perspect.* **107,** 773–778.
14. Anonymous (1999) *Environ. Health Perspect.* **107,** A604–A605.
15. Coats, J. R. (1994) *Annu. Rev. Entomol.* **39,** 489–515.
16. Bonkovsky, H. L., Cable, E. E., Cable, J. W., Donohue, S. E., White, E. C., Greene, Y. J., Lambrecht, R. W., Srivastava, K. K. & Arnold, W. N. (1992) *Biochem. Pharmacol.* **43,** 2359–2368.
17. Millet, Y., Jouglard, J., Steinmetz, M. D., Tognetti, P., Joanny, P. & Arditti, J. (1981) *Clin. Toxicol.* **18,** 1485–1498.
18. Weisbord, S. D., Soule, J. B. & Kimmel, P. L. (1997) *N. Engl. J. Med.* **337,** 825–827.
19. del Castillo, J., Anderson M. & Rubottom, G. M. (1975) *Nature (London)* **253,** 365–366.
20. Meschler, J. P. & Howlett, A. C. (1999) *Pharmacol. Biochem. Behav.* **62,** 473–480.
21. Ishida, T., Toyota, M. & Asakawa, Y. (1989) *Xenobiotica* **19,** 843–855.
22. ffrench-Constant, R. H. & Roush, R. T. (1991) *Genet. Res.* **57,** 17–21.
23. ffrench-Constant, R. H., Steichen, J. C., Rocheleau, K., Aronstein, K. & Rousch, R. T. (1993) *Proc. Natl. Acad. Sci. USA* **90,** 1957–1961.
24. Cole, L. M., Lawrence, L. J. & Casida, J. E. (1984) *Life Sci.* **35,** 1755–1762.
25. Cole, L. M. & Casida, J. E. (1992) *Pestic. Biochem. Physiol.* **44,** 1–8.
26. Ikeda, T., Ozoe, Y., Okuyama, E., Nagata, K., Honda, H., Shono, T. & Narahashi, T. (1999) *Br. J. Pharmacol.* **127,** 1567–1576.
27. Ogata, N., Vogel, S. M. & Narahashi, T. (1988) *FASEB J.* **2,** 2895–2900.
28. Enna, S. J. & Bowery, N. G., eds. (1997) *The GABA Receptors* (Humana, Totowa, NJ), 2nd Ed.
29. Kulkarni, S. K., Mehta, A. K. & Ticku, M. K. (1990) *Life Sci.* **46,** 481–487.
30. Mihic, S. J. (1999) *Neurochem. Int.* **35,** 115–123.
31. Felsted, R. L. & Bachur, N. R. (1980) in *Enzymatic Basis of Detoxification*. ed. Jakoby, W. B. (Academic, Orlando, FL), Vol. 1, pp. 281–293.

Structural model for γ-aminobutyric acid receptor noncompetitive antagonist binding: Widely diverse structures fit the same site

Ligong Chen*, Kathleen A. Durkin†, and John E. Casida*‡

*Environmental Chemistry and Toxicology Laboratory, Department of Environmental Science, Policy, and Management and †Molecular Graphics Facility, College of Chemistry, University of California, Berkeley, CA 94720

Contributed by John E. Casida, January 13, 2006

Several major insecticides, including α-endosulfan, lindane, and fipronil, and the botanical picrotoxinin are noncompetitive antagonists (NCAs) for the GABA receptor. We showed earlier that human β_3 homopentameric $GABA_A$ receptor recognizes all of the important GABAergic insecticides and reproduces the high insecticide sensitivity and structure-activity relationships of the native insect receptor. Despite large structural diversity, the NCAs are proposed to fit a single binding site in the chloride channel lumen lined by five transmembrane 2 segments. This hypothesis is examined with the β_3 homopentamer by mutagenesis, pore structure studies, NCA binding, and molecular modeling. The 15 amino acids in the cytoplasmic half of the pore were mutated to cysteine, serine, or other residue for 22 mutants overall. Localization of A-1′C, A2′C, T6′C, and L9′C (index numbers for the transmembrane 2 region) in the channel lumen was established by disulfide cross-linking. Binding of two NCA radioligands [^{3}H]1-(4-ethynylphenyl)-4-*n*-propyl-2,6,7-trioxabicyclo[2.2.2]octane and [^{3}H] 3,3-bis-trifluoromethyl-bicyclo[2,2,1]heptane-2,2-dicarbonitrile was dramatically reduced with 8 of the 15 mutated positions, focusing attention on A2′, T6′, and L9′ as proposed binding sites, consistent with earlier mutagenesis studies. The cytoplasmic half of the β3 homopentamer pore was modeled as an α-helix. The six NCAs listed above plus t-butylbicyclophosphorothionate fit the 2′ to 9′ pore region forming hydrogen bonds with the T6′ hydroxyl and hydrophobic interactions with A2′, T6′, and L9′ alkyl substituents, thereby blocking the channel. Thus, widely diverse NCA structures fit the same GABA receptor β subunit site with important implications for insecticide cross-resistance and selective toxicity between insects and mammals.

β3 homopentamer | transmembrane 2 | insecticide | disulfide trapping | receptor model

Pest insect control in the past 60 years was achieved, in part, by application of >3 billion (3×10^9) pounds of polychlorocycloalkane insecticides, including cyclodienes (e.g., α-endosulfan and dieldrin), lindane and its isomers, and others, which are now highly restricted or banned except for endosulfan and some uses of lindane (1–3). One of the replacement compounds is the phenylpyrazole fipronil. All of these insecticides and the botanical picrotoxinin (PTX) have widely diverse chemical structures but appear to act at the same nerve target. It is therefore important to understand how these compounds work in mammals and insects, or how they do not work when resistant insect strains appear.

The GABA-gated chloride channel is the target for the insecticides and toxicants referred to above based on radioligand binding and electrophysiology studies (3–10). Important radioligands in these developments are [^{3}H]dihydroPTX (4, 11), [^{35}S]t-butylbicyclophosphorothionate (TBPS) (5, 12), [^{3}H]1-(4-ethynylphenyl)-4-*n*-propyl-2,6,7-trioxabicyclo[2.2.2]octane (EBOB) (6), and [^{3}H]3,3-bis-trifluoromethyl-bicyclo[2,2,1]heptane-2,2-dicarbonitrile (BIDN) (8) (Fig. 1*A*). All of these compounds act in mammals and insects as noncompetitive antagonists (NCAs) to block chloride flux so the target is referred to as the GABA receptor NCA-binding site. Vertebrate GABA receptors consist of α, β, γ, ρ, and other subunits in various combinations, for example, $\alpha_1\beta_2\gamma_2$ as a heteropentamer and ρ_1 as a homopentamer (13–15). The molecular localization of the NCA site defined here (Fig. 1*B*) was first indicated by mutagenesis studies (16) as A2′ (17–20), T6′ (21, 22), and L9′ (23, 24) in the cytoplasmic half of the transmembrane 2 domain of the channel (Fig. 2). *Drosophila* resistant to dieldrin (RDL) have a mutation conferring GABA receptor insensitivity identified as A2′S (17). The NCA target of the $GABA_A$ receptor requires a β subunit, and a β_3 homopentamer is sufficient for binding (9, 26). Importantly, the β_3 subunit from human brain, when expressed in insect Sf9 cells, assembles to form a receptor sensitive to all of the important GABAergic insecticides (9) and, surprisingly, reproduces the insecticide sensitivity and structure-activity relationships of the native insect receptor (27). Studies of the GABA receptor NCA site are therefore simplified by using this highly sensitive β_3 homopentamer, an approach verified by showing here that Cys and Ser or Phe mutations in β_3 at each of the 2′, 6′, and 9′ positions greatly reduce or destroy NCA radioligand binding.

This study tested the hypothesis that insecticides and convulsants of many chemical types act at the same GABA receptor site in the same way to initiate insecticidal action and mammalian toxicity. The goal was to characterize the GABA receptor–NCA interaction by using the human $GABA_A$ receptor recombinant β_3 homopentamer as a model. The first step was to prepare Cys and other mutations to scan the cytoplasmic half of M2 and the flanking region (−4′ to 10′), overall 22 mutants involving 15 positions. The mutants were used to identify Cys residues undergoing disulfide cross-linking as a guide to channel pore structure (28). Next, [^{3}H]EBOB and [^{3}H]BIDN were used to identify positions where mutation altered binding (6, 8). Finally, modeling of the NCA-binding domain (29, 30) was applied to the β_3 homopentamer to determine whether the wide diversity of NCAs could fit the same site.

PHARMACOLOGY

Results

Mutagenesis and Protein Expression. The transfection efficiency of each recombinant baculovirus was examined by PCR analysis. The nonrecombinant virus would give one 839-bp band of its polyhedrin region and the recombinant virus incorporating the 1,425-bp β_3 cDNA would appear at 2.3 kb. Each extracted recombinant virus gave only one 2.3-kb band (Fig. 3*A*), indicating a recombination efficiency for the target gene of nearly 100% for all mutants and the WT. Further, all PCR products from

Conflict of interest statement: No conflicts declared.

Abbreviations: BIDN, 3,3-bis-trifluoromethyl-bicyclo[2,2,1]heptane-2,2-dicarbonitrile; Cu:phen, copper:phenanthroline; EBOB, 1-(4-ethynylphenyl)-4-*n*-propyl-2,6,7-trioxabicyclo[2.2.2]octane; M2, transmembrane 2; NCA, noncompetitive antagonist; PTX, picrotoxinin; RDL, resistant to dieldrin; TBPS, t-butylbicyclophosphorothionate.

‡To whom correspondence should be addressed: E-mail: ectl@nature.berkeley.edu.

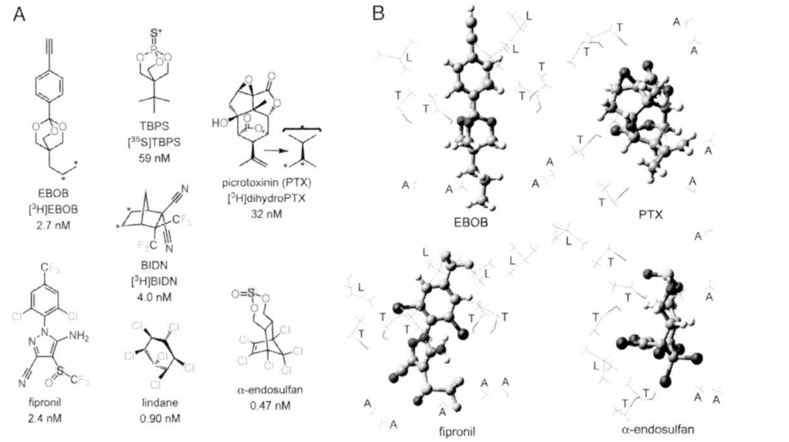

Fig. 1. Structure-activity relationships of seven GABA receptor noncompetitive antagonists. (*A*) Structures of three important insecticides (lindane, fipronil, and α-endosulfan) and four radioligands (asterisk designates labeling position). The high potencies of each compound with the β_3 homopentamer are indicated by the 2.7 nM K_d for [^{3}H]EBOB on direct binding and 0.47–59 nM IC_{50} values for the other compounds in displacing [^{3}H]EBOB binding (9). (*B*) Models of four antagonists positioned as in *A* showing their proposed β_3 homopentamer M2 binding sites in the channel lumen. A, L, and T refer to the side chains of the interacting 2′, 6′, and 9′ residues, respectively.

virus extraction were sequenced and confirmed as the right mutations.

The expression levels of the WT and mutant β_3 subunits were determined by Western blotting. The monoclonal anti-β-chain antibody recognized a very specific band at ≈55 kDa with similar intensity for membrane extracts of the WT and each mutant (Fig. 3*B*). Equal protein transfer levels were determined by Ponceau S staining. As exceptions, two mutants (L3′C and L3′F) were not expressed.

Disulfide Cross-Linking Profiles. Oxidation of the Cys mutants with copper sulfate:1,10-phenanthroline (Cu:phen) resulted in four cases of a molecular mass increase from 55 kDa to ≈130 kDa for the monomer and dimer, respectively, as detected by SDS/PAGE and immunoblotting (Fig. 4). Cys substituents at −1′, 2′ (weak), 6′, and 9′ formed disulfide-linked dimers in the presence but not in the absence of Cu:phen. Only trace amounts of the −1′ and 9′ monomers are left with Cu:phen indicating more extensive reaction possibly due to higher flexibility at these positions. Dimers were not detected under the same conditions for Cys substituents at 0′, 1′, 4′, 5′, 7′, 8′, and 10′, although in some cases there were apparent losses in receptor levels on oxidation. Disulfides at −1′, 2′, and 9′ were completely reversed with DTT, but the one at 6′ was only partially reversed.

	M2: -4' -1' 2' 6' 9'
α1	SVPARTVFGVTTVLT
β1	ASAARVALGITTVLT
β2	ASAARVALGITTVLT
β3	ASA**A**RV**A**LGI**TT**V**L**T
γ2	AVPARTSFGVTTVLT
ρ1	AVPARVPLGITTVLT
ρ2	AVPARVSLGIMTVLT
RDL WT	ATPARV<u>A</u>LGVTTVLT
RDL mutant	ATPARV<u>S</u>LGVTTVLT

Fig. 2. Alignment of the cytoplasmic half of the M2 and flanking sequences of various GABA receptor subunits. The species are human or rat for α, β, and γ, rat for ρ, and *Drosophila* for WT and RDL. Index numbers for positioning in M2 (25) are shown at the top. The β_3 homopentamer region studied here is shown in a box with the channel lumen residues defined in the present investigation by disulfide cross-linking in bold type (−1′, 2′, 6′, and 9′). The resistance-associated RDL mutation (A2′S) in *Drosophila* (17) is underlined.

Effect of Site-Specific Mutations on [^{3}H]EBOB and [^{3}H]BIDN Binding. Membranes (100 μg of protein) from the WT were assayed with [^{3}H]EBOB (1 nM) or [^{3}H]BIDN (2.5 nM) by using incubations for 90 min at 25°C. Specific binding (n = 10) was 2,458 ± 250 dpm for [^{3}H]EBOB and 1,253 ± 100 dpm for [^{3}H]BIDN with nonspecific binding of 496 ± 45 and 225 ± 16 dpm, respectively, i.e., 83–85% specific relative to total binding. Using the same

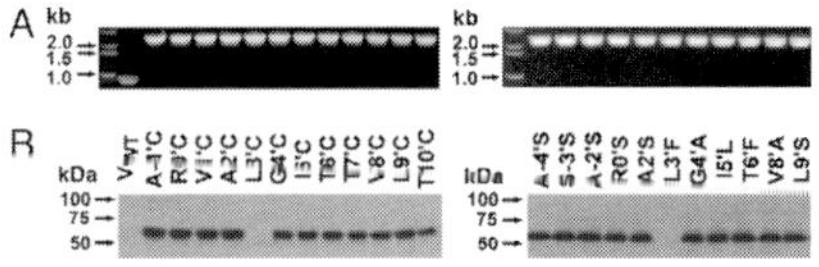

Fig. 3. Baculovirus transfection efficiencies and protein expression levels of WT (S-3′S) and mutant β_3 subunits. (*A*) PCR analysis of recombinant efficiency. (*B*) SDS/PAGE-Western blotting analysis of protein expression level. V_{WT} refers to membrane transfected with WT baculovirus. Samples were treated with 10 mM DTT in sample buffer.

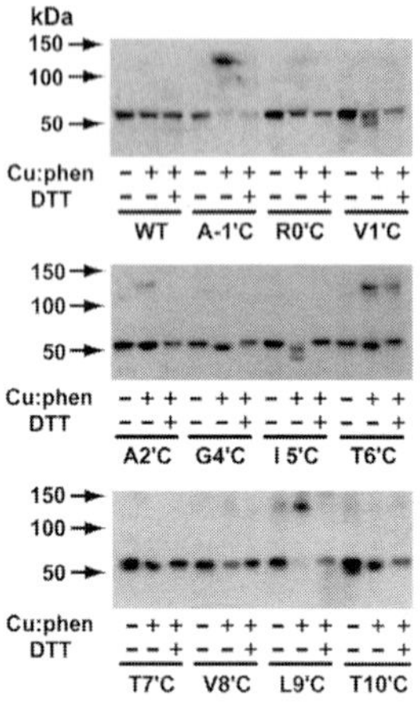

Fig. 4. Disulfide cross-linking profiles. Samples are control without Cu:phen or DTT (−/−), oxidized with Cu:phen but not treated with DTT (+/−), or oxidized with Cu:phen then reduced with 10 mM DTT (+/+). Reactions were terminated with 10 mM *N*-ethylmaleimide before SDS/PAGE-Western blotting analysis.

conditions and amounts of receptors, the mutants were then compared to the WT for both [^{3}H]EBOB and [^{3}H]BIDN binding. The binding activities of A-4′S, T7′C and T10′C were similar to the WT, whereas A-2′S and V1′C gave reduced binding (Fig. 5). All of the rest gave little or no specific binding. It was indeed surprising to find that the low binding for mutants involves the whole segment from A2′ to I5′, in addition to the expected T6′ to L9′, with the two exceptions of L3′ not expressed and T7′ normal. More generally, mutations in the lowest region of the channel have no (−4′ or −3′) or little (−2′) influence on activity, whereas those in the region of −1′ to 10′, except 1′, 7′, and 10′, drastically reduce [^{3}H]EBOB and [^{3}H]BIDN binding. This reduction is not due to interference from oxidation of the Cys moiety because (*i*) DTT did not restore the activity of the eight low-binding mutants (data not shown), and (*ii*) the findings are essentially the same with Ser (0′, 2′, and 9′), Leu (5′), Phe (6′), and Ala (4′ and 8′) as well as for the corresponding Cys mutants. It is assumed that the mutants, which do not bind NCAs, form functional channels that are correctly assembled on

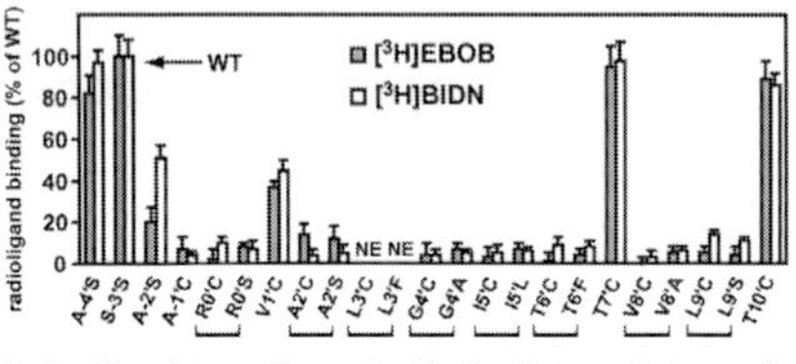

Fig. 5. Effect of site-specific mutations (Cys, Ser, Ala, Leu, or Phe) on specific binding of [^{3}H]EBOB and [^{3}H]BIDN. NE, not expressed. Data are percent of WT (S-3′S) ± SD.

the cell surface because, on the Western blot, they all have a protein of similar size and presumably maturely glycosylated. Most importantly, on an overall basis, the results are essentially the same with [^{3}H]EBOB and [^{3}H]BIDN.

Two methanethiosulfonate (MTS) sulfhydryl-modification reagents provided further information on the NCA site by comparing their effect on [^{3}H]EBOB binding for the Cys active mutants 1′, 7′, and 10′ compared with the WT. With both sulfhydryl reagents, there was a site-dependent effect on [^{3}H]EBOB binding with little inhibition for the T7′C mutant, moderate for T10′C, and almost complete for V1′C.

Structural Model for NCA Binding. Fig. 6 *Upper Left* shows a model of the channel lumen from the 2′ to 9′ positions with five β_3 α-helices and lindane docked into the putative binding site, which it clearly fills to block the pore. Similar models of the six other NCAs also show filling of the pore space.

Attention was focused on A2′, T6′, and L9′, because these residues are in the channel lumen (based on disulfide trapping) and mutations (Cys versus Ser or Phe in each case) at these sites greatly reduce or abolish binding. The interacting sites are shown in Figs. 1*B* and 6. Docking of EBOB positions the A2′ methyls interacting with the normal-propyl and two *O*-methylenes, two T6′ hydroxyls interacting with the oxygens (H---O distance ≈3.1Å), T6′ methyls binding to the phenyl moiety, and, at a slightly longer range, a L9′ methyl also interacting with the ethynyl substituent (evident in Fig. 1*B* but not Fig. 6). TBPS has numerous favorable A2′ interactions with the tertiary-butyl moiety, and the T6′ methyls and hydroxyls interact with the sulfur and cage oxygens. PTX has A2′ methyl interactions with the isopropenyl methyl and methylene and three T6′ hydroxyl hydrogen bonding interactions to three PTX oxygens. BIDN has multiple contact points with A2′ methyls and T6′ methyls and hydroxyls. A cyano nitrogen and a fluorine each form hydrogen bonds to a T6′ hydroxyl. Lindane bridges A2′ methyls and T6′ hydroxyls and methyls, each interacting with multiple chlorines. α-Endosulfan and fipronil have multiple interaction sites and types, with A2′ methyls and T6′ methyls and hydroxyls for both compounds reinforced by L9′ side chains for fipronil. More complete depictions of the β_3 homopentamer model and the docked ligands are given in supporting information, which is published on the PNAS web site.

Discussion

Mutagenesis and Expression. The cytoplasmic half of the M2 region contains 11 amino acids (0′ to 10′), and this number is extended to 15 (−4′ to 10′) with the flanking region of interest. Site-specific mutagenesis introduced Cys at 12 sites (A-1′C to T10′C), Ser at five sites (A-4′S, A-2′S, R0′S, A2′S, and L9′S), and Phe at two sites (L3′F and T6′F). In addition, three mutations were introduced with little change in polarity, i.e., G4′A, I5′L and V8′A. The 3′-position was an exception because L3′C and L3′F did not show detectable expression by Western blotting either in the β_3 homopentamer studied here or the $\alpha_1\beta_3$ heteropentamer (data not shown).

Pore-Lining Residues. The position of pore-lining residues was determined by disulfide cross-linking, cysteine accessibility, and molecular modeling. Cys sulfhydryl substituents in the pore lining can be oxidized to disulfides resulting in dimerization. Disulfide trapping for Cys mutants in the present study places the sulfhydryl substituents of −1′, 2′, 6′, and 9′ within the channel lumen; disulfide trapping of A-1′C, A2′C, and L9′C was not established before. The tight protein packing in the 2′ position (31) may account for the weak dimer formation by limiting the required flexibility and close proximity for disulfide bond formation. Disulfides are not formed with 0′, 1′, 4′, 5′, 7′, 8′, and 10′, indicating they are probably not in the pore or have low

PNAS PNAS PNAS

PHARMACOLOGY

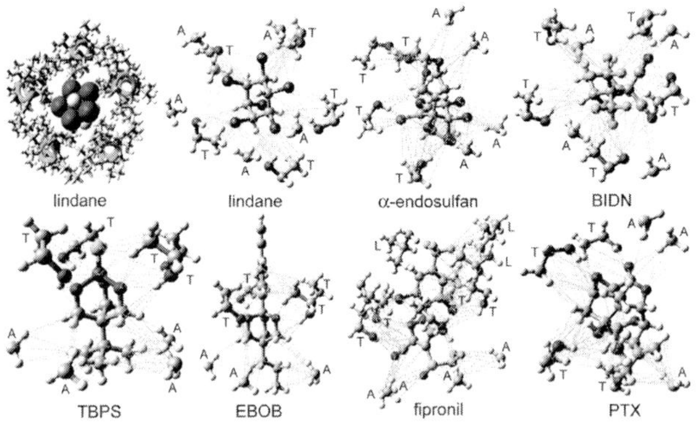

Fig. 6. Proposed interactions of seven noncompetitive antagonists at the same GABA$_A$ receptor β_3 homopentamer binding site. (*Upper Left*) lindane (space fill, red and blue for partial negative and positive charges of chlorine and carbon, respectively) binds to the GABA$_A$ receptor (five β_3 α-helices shown in green) to block the channel pore shown as the 2′ to 9′ positions viewed from the top into the pore. Remaining panels: seven ligands (see Fig. 1*A*) docked at their optimized positions with the perspective chosen for ease of viewing. A, L, and T refer to the side chains of the interacting 2′, 6′, and 9′ residues, respectively. van der Waals contacts are illustrated in green (see text for discussion of hydrogen bonding). The space filling aspects of all of the ligands are most readily evident in supporting information.

mobility/flexibility. For T6′, similar findings are obtained with the α_1T6′Cβ_1T6′C receptor but only in the presence of GABA (28), suggesting that the β_3 homopentamer of the present investigation assumes the spontaneous open state (32). Homology of the GABA receptor β_3 homopentamer with the nicotinic acetylcholine receptor (33) indicates the narrowest gating region of the pore is between 9′ and 14′, suggesting the positioning of L9′C in the pore (24, 31). In the β_1 subunit of the $\alpha_1\beta_1\gamma_2$ receptor, A2′C, T6′C, T7′C (slow reaction rate), V8′C, L9′C, and T10′C are all accessible to a sulfhydryl-modification reagent depending on the state of the channel (31). Methanethiosulfonate reagents in the present β_3 homopentamer study show that V1′C is transiently available in the channel lumen in contrast to T7′C and T10′C, which are not readily accessible. In addition, reaction with the cationic methanethiosulfonate reagent suggests that the anion-selective filter may be below V1′C. Molecular modeling of the β_3 homopentamer as an α-helix (Fig. 6) places −1′, 2′, 6′, and 9′, but not 0′, 1′, 3′, 4′, 5′, 7′, 8′, or 10′, in the channel pore (see supporting information), consistent with the other approaches.

Sites for NCA Interactions. The interacting residues are considered to be A2′ (or more generally the A2′-I5′ hydrophobic pocket) and T6′ (the highly conserved and most important structural determinant) with a supplemental role for L9′. A biophysical calculation model focused on PTX interactions with A2′ and T6′ of the ρ_1 receptor (29). The present study uses site-specific mutations in the β_3 homopentamer to determine the importance of 10 other amino acid residues in NCA binding, i.e., the whole cytoplasmic half of the M2 region. A-4′, S-3′, and A-2′ are apparently outside of the binding site. β_3 homopentamer mutants A-1′C, R0′C, and R0′S block binding, perhaps because of proximity to A2′. Sulfhydryl modification at V1′C impedes [^{3}H]EBOB binding (this study), possibly by overlapping the sensitive A2′ position. Further, for 2′, the low sensitivity of the *Drosophila* RDL homomeric receptor to [^{3}H]EBOB with A2′S (or A2′G) (34) suggests this site for binding with confirmation here from A2′C and A2′S mutants in the β_3 homopentamer. In addition, with V2′C at the α_1 subunit of the $\alpha_1\beta_1\gamma_2$ receptor, PTX protects against sulfhydryl derivatization (18), and a sulfhydryl-reactive fipronil analog [-C(O)CH$_2$Br replaces -S(O)CF$_3$] serves as an irreversible blocker (19). The involvement of 3′, directly or by influencing the neighboring A2′, is shown by L3′F at β_3 of the $\alpha_1\beta_3$ receptor almost abolishing TBPS and PTX binding (20). The structurally critical apolar pocket in the β_3 homopentamer appears to involve A2′, L3′, G4′, and I5′, i.e., a tightly packed and completely hydrophobic region that may play a role in stabilizing the helical structure (31, 35). Although G4′ is on the backside of the helix, the side chains introduced with the G4′C and G4′A mutants appear to perturb the tightly packed 2′-5′ region of the channel lumen to disturb NCA binding. The T6′C and T6′F mutations in the β_3 homopentamer abolish NCA sensitivity, and introducing T6′F in β_2 (or α_1 or γ_2) of $\alpha_1\beta_2\gamma_2$ greatly reduces PTX sensitivity (21). Mutagenesis of the 6′ position of ρ1 and ρ2 receptors from rats showed this site to be important in PTX sensitivity (22). T7′C and V8′C fall outside the pore and, therefore, are not expected to be important binding sites, yet the 8′ mutants block binding, perhaps, by changing the shape of the pore. For L9′, where a mutation can potentially perturb the gating kinetics (24), the L9′C and L9′S mutations for β_3 abolish NCA binding here and L9′S reduces PTX sensitivity in each subunit of $\alpha_1\beta_2\gamma_2$ (24). Finally, with ρ1, several mutations at L9′ also reduce PTX sensitivity (23). Lying outside the pore, T10′C does not affect [^{3}H]EBOB or [^{3}H]BIDN binding.

Widely Diverse NCA Structures Fit the Same Site. The RDL A2′S mutation confers cross-resistance of insects to all classes of commercial NCA insecticides (10, 17), and this cross-resistance also applies to the highly potent model compounds EBOB and BIDN. The effect of all mutations is essentially the same with [^{3}H]EBOB and [^{3}H]BIDN, indicating that they both have the same binding site. More generally, an extremely wide diversity of chemical types, each with configurational specificity, appears to act the same way as GABA receptor NCAs (3, 9, 36, 37). Figs. 1*B* and 6 illustrate how they, in fact, may all fit the same site by showing the proposed interactions of seven NCAs with the β_3 homopentameric receptor.

Favorable hydrophobic interactions are observed for the A2′ methyls with the alkyl substituents of EBOB, TBPS, PTX, and BIDN, the trifluoromethylsulfinyl and pyrazole cyano of fipronil, the hexachlorocyclopentenyl moiety of endosulfan, and the hexachlorocyclohexane isomer lindane. The T6′ methyls interact with the ethynylphenyl and trifluoromethylphenyl substituents of EBOB and fipronil, respectively, the trifluoromethyls and cyanos of BIDN, and the exocyclic oxygen of α-endosulfan. The T6′ hydroxyl substituent hydrogen bonds (H-X distance <3.5 Å) to multiple electronegative sites, i.e., the trioxabicyclooctane oxygens of EBOB and TBPS; the exocyclic oxygen of endosulfan; the epoxy, hydroxyl, and lactone exocyclic oxygens of PTX; the pyrazole, amino, and cyano nitrogens of fipronil; and a cyano nitrogen and fluorine of BIDN. In lindane, four chlorines are <3.5 Å from T6′ hydroxyl hydrogens. On calculating the relative energies of the bound ligands by using MAESTRO/MACROMODEL (Schrödinger LLC, Portland, OR), the most potent γ isomer lindane binds in a more stable configuration than the less active α, β, and δ isomer(s) by >30 kJ/mol and the more active α-endosulfan versus the less potent β-endosulfan by 15 kJ/mol. The L9′ side chains associate with the phenyl group of EBOB and fipronil, the ethynyl of EBOB and the aryl trifluoromethyl and chloro substituents of fipronil, enhancing the potency of these long or extended molecules. The more compact NCAs, including TBPS, lindane, and BIDN, require only A2′ and T6′ for fit lengthwise or lying across the pore, and this positioning probably also applies to α-endosulfan and PTX. These docking proposals are consistent with current structure-activity relationships and may help in further ligand optimization.

The NCAs are chloride channel blockers, i.e., their potency in binding to the NCA site is proportional to their effectiveness in inhibiting chloride flux (38, 39). In the proposed binding site model, the NCAs fill up and actually block the pore, although they also may act allosterically by changing the channel conformation. The internuclear distance across the channel pore is on the order of 8.5 Å, which is the same as or only slightly longer than the distance across multiple types of NCAs (6–8 Å).

NCA Potency and Selectivity Conferred by Subunit Specificity. The β_3 homopentamer has higher NCA sensitivity than other vertebrate GABA receptors and any replacement subunits of those tested reduce ligand affinity (9). The β_3 homopentamer can form a spontaneously opening ion channel (32), potentially facilitating ligand binding. GABA and other agonist modulators affect NCA binding with native and α_1 subunit-containing receptors but not with the β_3 homopentamer (9, 12, 40). As with related ligand-gated ion channels the NCA potency profile varies with subunit composition. Selectivity is conferred by these additional subunits as evident by comparing native receptors with $\alpha_1\beta_2\gamma_2$ heteropentameric and β_3 homopentameric recombinant receptors (9, 27). NCAs with excellent fit for the β_3 homopentamer model may show less favorable docking in the heteropentameric native receptors associated with subunit variation at the 2′ position.

Concluding Remarks. The human GABA$_A$ receptor recombinant β_3 homopentamer retains the NCA site in its most sensitive form, equal to the insect site. Both the β_3 homopentamer pore and principal radioligand [^{3}H]EBOB are symmetrical, thereby greatly facilitating receptor modeling and ligand positioning. Ligands of widely diverse structures approach similar potency when optimized. The effect of mutations is the same for [^{3}H]EBOB and [^{3}H]BIDN binding and possibly for the other NCAs as well. A model for the GABA$_A$ receptor M2 region applied to the β_3 homopentamer brings these observations together to propose structural aspects of the NCA site. Further test of this proposal requires direct rather than indirect structural analysis of the homopentameric and heteropentameric GABA receptors.

Materials and Methods

Site-Directed Mutagenesis. cDNA encoding the human GABA$_A$ receptor β_3 subunit inserted in the pVL1392 baculovirus transfer vector was described in ref. 9. Point mutations were introduced with the QuikChange Site-Directed Mutagenesis kit (Stratagene). Mutagenic oligonucleotides were prepared by Operon (Huntsville, AL). All mutations were confirmed by double-strand DNA sequencing (DNA Sequencing Facility, University of California, Berkeley).

Cell Culture and Protein Expression. Insect Sf9 cells (serum-free adapted, derived from ovaries of *Spodoptera frugiperda*) were maintained by described methods in refs. 9 and 41. Recombinant baculoviruses were constructed by using a Bacfectin-mediated transfection kit (BD Biosciences Clontech). The Invitrogen protocol was used for PCR analysis of recombinant virus. All PCR products were recycled with GelQuick Gel Extraction Kit (Qiagen, Valencia, CA) and then were sequenced as described above. Log phase Sf9 cells were infected with recombinant baculovirus at a multiplicity of infection of 5–8. Cells were harvested at 65 h after infection. They were pelleted at 1,500 × *g* for 5 min and washed once with PBS (155 mM NaCl/3.0 mM NaH_2PO_4/1.0 mM K_2HPO_4, pH 7.4). Cell pellets were stored at −80°C until ready to use.

Membrane Preparation. The pelleted cells were resuspended in PBS and homogenized in a glass tube with a motor-driven Teflon pestle (9, 41). Cellular debris was removed by centrifugation at 500 × *g* for 10 min at 4°C. The supernatant was centrifuged at 100,000 × *g* for 40 min at 4°C, and the resulting pellet was resuspended in PBS and stored at −80°C. Protein concentration was determined with the detergent-compatible Lowry assay (Bio-Rad).

Western Blotting. Membrane preparations were mixed with Laemmli sample buffer (1.5% SDS/5% glycerol/65 mM Tris·HCl, pH 6.8, with or without 10 mM DTT). After boiling at 100°C for 5 min, samples were analyzed by SDS/PAGE (10% acrylamide) by using a Mini-PROTEAN II apparatus (Bio-Rad). Proteins were transferred onto poly(vinylidene difluoride) membranes for 2 h at 100 V and 4°C by using the Transblot apparatus (Bio-Rad). The membranes were blocked in Tris-buffered saline (Bio-Rad) containing 2% nonfat dry milk with 0.5% Tween 20 for 1 h at room temperature and incubated with the mouse anti-GABA$_A$ receptor, β-chain monoclonal antibody (Chemicon International, Temecula, CA), at a dilution of 1:1,000, also for 1 h at room temperature. After three 5-min washings in TBS with 0.5% Tween 20, the blots were incubated with anti-mouse horseradish peroxidase-linked secondary antibodies (Santa Cruz Biotechnology) at a dilution of 1:2,000 for 1 h at room temperature. After extensive washing, immunoreactivity was detected by chemiluminescence kit (PerkinElmer). Finally, the transferred protein was visualized by incubation in Ponceau S solution (Bio-Rad).

Disulfide Cross-Linking and Sulfhydryl Modification. For disulfide cross-linking, the membrane preparation (100 μg of protein) in PBS (100 μl) was oxidized with Cu:phen (100 μM:400 μM) (28, 42) for 5 min at 25°C. The reaction was terminated by adding 10 mM *N*-ethylmaleimide and 1 mM EDTA (final concentrations). After 3 min, the membranes were recovered by centrifugation (20,000 × *g* for 15 min at 4°C), resuspended in PBS, mixed with sample buffer with or without 10 mM DTT, and subjected to SDS/PAGE (10% acrylamide) and Western blot analysis. For sulfhydryl modification, two methanethiosulfonate reagents were used, 2-(trimethylammonium)ethyl methanethiosulfonate bromide and sodium (2-sulfonatoethyl)methanethiosulfonate, under described conditions (18, 43) with analysis for their effect on [^{3}H]EBOB binding.

[^{3}H]EBOB and [^{3}H]BIDN Binding. Assay mixtures contained 1 nM [^{3}H]EBOB (48 Ci/mmol; 1 Ci = 37 GBq) (PerkinElmer) (6, 9) or 2.5 nM [^{3}H]BIDN (50 Ci/mmol) (8) and the recombinant expressed receptor (100 μg protein) in PBS (500 μl final volume) (6). After incubation for 90 min at 25°C, the samples were filtered through GF/B filters (presoaked in 0.2% polyethyleneimine for 3 h) and rinsed three times with ice-cold saline (0.9% NaCl). Nonspecific binding was determined in the presence of 1 μM α-endosulfan for [^{3}H]EBOB or 5 μM unlabeled BIDN for [^{3}H]BIDN by using 5 μl dimethyl sulfoxide to add the displacing agent immediately before incubation. Each experiment was repeated three or more times with duplicate samples. The binding activity of mutants was expressed as percent (mean ± SD) of that for the WT. Supplemental binding studies were made by using DTT preincubation (10 mM for 5 min at 25°C) in deoxygenated PBS continuously bubbling with argon to rule out any spontaneous disulfide formation.

Modeling Receptor–Ligand Interactions. Modeling started from the $\alpha_1\beta_2\gamma_2$ GABA$_A$ receptor based on the homologous nicotinic acetylcholine receptor and acetylcholine binding protein (30). This α-helical structure was reconstructed here as the β_3 homopentamer, i.e., the two β_2 subunits were directly replaced by β_3, because they have the same M2 sequence, then the two α_1 subunits and one γ_2 subunit were replaced with β_3, by using the original α_1 and γ_2 backbone atom positions as a guide, to make the homopentameric model. The cytoplasmic side of the M2 region and adjacent residues are considered, i.e., A-4′ to T10′ (Fig. 2), with particular attention to A2′ to L9′.

All modeling was done with MAESTRO 6.5 (Schrödinger LLC). Macromodel atom types were used to assign partial charges (44). van der Waals contacts were defined as C = (distance between atomic centers)/(radius 1st atom + radius 2nd atom) where good, bad, and ugly contacts are defined as C = 1.3, 0.89, and 0.75 Å, respectively. The antagonists were manually docked into the putative binding site to maximize good contacts, and then the ligand geometry and location were allowed to optimize relative to the β3 homopentamer, which was itself constrained. In this optimization, all settings were left at the default values except a water model (generalized Born/surface area) was used instead of a gas phase model. In each case, sufficient optimization steps were performed as necessary to ensure that the convergence criteria were met.

We thank our department colleagues Jung-Chi Liao, Motohiro Tomizawa, Gary Quistad, Daniel Nomura, and Shannon Liang for helpful advice and Myles Akabas, Gerald Brooks, Richard Olsen, and David Weiss for important suggestions. This work was supported by the William Mureice Hoskins Chair in Chemical and Molecular Entomology (to J.E.C.) and National Science Foundation Grant CHE-0233882 (to K.A.D.). Molecular modeling was performed on workstations purchased with National Science Foundation support matched with an equipment donation from Dell.

1. Casida, J. E. & Quistad, G. B. (1998) *Annu. Rev. Entomol.* **43,** 1–16.
2. Tomlin, C. D. S., ed. (2003) *The Pesticide Manual* (Brit. Crop Prot. Counc., Alton, Hampshire, U.K.), 13th Ed., pp. 1344.
3. Brooks, G. T. (2001) in *Handbook of Pesticide Toxicology: Agents*, ed. Krieger, R. I. (Academic, New York), 2nd Ed., Vol. 2, pp. 1131–1156.
4. Matsumura, F. & Ghiasuddin, S. M. (1983) *J. Environ. Sci. Health B* **18,** 1–14.
5. Lawrence, L. J. & Casida, J. E. (1984) *Life Sci.* **35,** 171–178.
6. Cole, L. M. & Casida, J. E. (1992) *Pestic. Biochem. Physiol.* **44,** 1–8.
7. Casida, J. E. (1993) *Arch. Insect Biochem. Physiol.* **22,** 13–23.
8. Rauh, J. J., Benner, E., Schnee, M. E., Cordova, D., Holyoke, C. W., Howard, M. H., Bai, D., Buckingham, S. D., Hutton, M. L., Hamon, A., *et al.* (1997) *Brit. J. Pharmacol.* **121,** 1496–1505.
9. Ratra, G. S., Kamita, S. G. & Casida, J. E. (2001) *Toxicol. Appl. Pharmacol.* **172,** 233–240.
10. Bloomquist, J. R. (2001) in *Biochemical Sites of Insecticide Action and Resistance*, ed. Ishaaya, I. (Springer, Berlin), pp. 17–41.
11. Ticku, M. K., Ban, M. & Olsen, R. W. (1978) *Mol. Pharmacol.* **14,** 391–402.
12. Squires, R. F., Casida, J. E., Richardson, M. & Saederup, E. (1983) *Mol. Pharmacol.* **23,** 326–336.
13. Barnard, E. A., Skolnick, P., Olsen, R. W., Mohler, H., Sieghart, W., Biggio, G., Braestrup, C., Bateson, A. N. & Langer, S. Z. (1998) *Pharmacol. Rev.* **50,** 291–313.
14. Enna, S. J. & Bowery, N. G., eds. (1997) *The GABA Receptors* (Humana, Totowa, NJ), 2nd Ed.
15. Martin, D. L. & Olsen, R. W., eds. (2000) *GABA in the Nervous System: the View at Fifty Years* (Lippincott, Philadelphia).
16. Chang, Y. & Weiss, D. S. (2000) in *GABA in the Nervous System: the View at Fifty Years*, eds. Martin, D. L. & Olsen, R. W. (Lippincott, Philadelphia), pp. 127–139.
17. ffrench-Constant, R. H., Anthony, N., Aronstein, K., Rocheleau, T. & Stilwell, G. (2000) *Annu. Rev. Entomol.* **48,** 449–466.
18. Xu, M., Covey, D. F. & Akabas, M. H. (1995) *Biophys. J.* **69,** 1858–1867.
19. Perret, P., Sarda, X., Wolff, M., Wu, T.-T., Bushey, D. & Goeldner, M. (1999) *J. Biol. Chem.* **274,** 25350–25354.
20. Buhr, A., Wagner, C., Fuchs, K., Sieghart, W. & Sigel, E. (2001) *J. Biol. Chem.* **276,** 7775–7781.
21. Gurley, D., Amin, J., Ross, P. C., Weiss, D. S. & White, G. (1995) *Recept. Channels* **3,** 13–20.
22. Zhang, D., Pan, Z.-H., Zhang, X., Brideau, A. D. & Lipton, S. A. (1995) *Proc. Natl. Acad. Sci. USA* **92,** 11756–11760.
23. Chang, Y. & Weiss, D. S. (1998) *Mol. Pharmacol.* **53,** 511–523.
24. Chang, Y. & Weiss, D. S. (1999) *Biophys. J.* **77,** 2542–2551.
25. Miller, C. (1989) *Neuron* **2,** 1195–1205.
26. Slany, A., Zezula, J., Tretter, V. & Sieghart, W. (1995) *Mol. Pharmacol.* **48,** 385–391.
27. Ratra, G. S. & Casida, J. E. (2001) *Toxicol. Lett.* **122,** 215–222.
28. Horenstein, J., Wagner, D. A., Czajkowski, C. & Akabas, M. H. (2001) *Nature Neurosci.* **4,** 477–485.
29. Zhorov, B. S. & Bregestovski, P. D. (2000) *Biophys. J.* **78,** 1786–1803.
30. O'Mara, M., Cromer, B., Parker, M. & Chung, S.-H. (2005) *Biophys. J.* **88,** 3286–3299.
31. Goren, E. N., Reeves, D. C. & Akabas, M. H. (2004) *J. Biol. Chem.* **279,** 11198–11205.
32. Wooltorton, J. R. A., Moss, S. J. & Smart, T. G. (1997) *Eur. J. Neurosci.* **9,** 2225–2235.
33. Miyazawa, A., Fujiyoshi, Y. & Unwin, N. (2003) *Nature* **423,** 949–955.
34. Cole, L. M., Roush, R. T. & Casida, J. E. (1995) *Life Sci.* **56,** 757–765.
35. Popot, J.-L. & Engelman, D. M. (2000) *Annu. Rev. Biochem.* **69,** 881–922.
36. Calder, J. A., Wyatt, J. A., Frenkel, D. A. & Casida, J. E. (1993) *J. Comput. Aided Mol. Des.* **7,** 45–60.
37. Ozoe, Y. & Akamatsu, M. (2001) *Pest Manag. Sci.* **57,** 923–931.
38. Obata, T., Yamamura, H. I., Malatynska, E., Ikeda, M., Laird, H., Palmer, C. J. & Casida, J. E. (1988) *J. Pharmacol. Exp. Ther.* **244,** 802–806.
39. Huang, J. & Casida, J. E. (1996) *J. Pharmacol. Exp. Ther.* **279,** 1191–1196.
40. Korpi, E. R., Gründer, G. & Lüddens, H. (2002) *Prog. Neurobiol.* **67,** 113–159.
41. Elster, L., Schousboe, A. & Olsen, R. W. (2000) *J. Neurosci. Res.* **61,** 193–205.
42. Yu, H., Kono, M., McKee, T. D. & Oprian, D. D. (1995) *Biochemistry* **34,** 14963–14969.
43. Slotboom, D. J., Konings, W. N. & Lolkema, J. S. (2001) *J. Biol. Chem.* **276,** 10775–10781.
44. Mohamadi, F., Richards, N. G. J., Guida, W. C., Liskamp, R., Lipton, M., Caufield, C., Chang, G., Hendrickson, T. & Still, W. C. (1990) *J. Comput. Chem.*, **11,** 440–467.

Mapping the elusive neonicotinoid binding site

Motohiro Tomizawa*, Todd T. Talley[†], David Maltby[‡], Kathleen A. Durkin[§], Katalin F. Medzihradszky[‡], Alma L. Burlingame[‡], Palmer Taylor[†], and John E. Casida*[¶]

*Environmental Chemistry and Toxicology Laboratory, Department of Environmental Science, Policy, and Management, University of California, Berkeley, CA 94720-3112; [†]Department of Pharmacology, Skaggs School of Pharmacy and Pharmaceutical Sciences, University of California at San Diego, La Jolla, CA 92093-0650; [‡]Mass Spectrometry Facility, University of California, San Francisco, CA 94143-0446; and [§]Molecular Graphics and Computation Facility, College of Chemistry, University of California, Berkeley, CA 94720-1460

Contributed by John E. Casida, April 10, 2007 (sent for review March 19, 2007)

Two types of structurally similar nicotinic agonists have very different biological and physicochemical properties. Neonicotinoids, important insecticides including imidacloprid and thiacloprid, are nonprotonated and selective for insects and their nicotinic receptors, whereas nicotinoids such as nicotine and epibatidine are cationic and selective for mammalian systems. We discovered that a mollusk acetylcholine binding protein (AChBP), as a structural surrogate for the extracellular ligand-binding domain of the nicotinic receptor, is similarly sensitive to neonicotinoids and nicotinoids. It therefore seemed possible that the proposed very different interactions of the neonicotinoids and nicotinoids might be examined with a single AChBP by using optimized azidochloropyridinyl photoaffinity probes. Two azidoneonicotinoids with a nitro or cyano group were compared with the corresponding desnitro or descyano azidonicotinoids. The four photoactivated nitrene probes modified AChBP with up to one agonist for each subunit based on analysis of the intact derivatized protein. Identical modification sites were observed by collision-induced dissociation analysis for the neonicotinoids and nicotinoids with similar labeling frequency of Tyr-195 of loop C and Met-116 of loop E at the subunit interface. The nitro- or cyano-substituted guanidine/amidine planes of the neonicotinoids provide a unique electronic conjugation system to interact with loop C Tyr-188. The neonicotinoid nitro oxygen and cyano nitrogen contact loop C Cys-190/Ser-189, whereas the cationic head of the corresponding nicotinoids is inverted for hydrogen-bonding and cation-π contact with Trp-147 and Tyr-93. These structural models based on AChBP directly map the elusive neonicotinoid binding site and further describe the molecular determinants of agonists on nicotinic receptors.

acetylcholine binding protein | imidacloprid | neonicotinoid insecticides | nicotinic receptor | photoaffinity labeling

Selective toxicity is critical for insecticide use, combining outstanding effectiveness for pests with safety for humans and wildlife. Neonicotinoids, exemplified by the major imidacloprid (IMI), thiacloprid (THIA), and acetamiprid (ACE) (Figs. 1 and 2), are the most important new class of insecticides of the past three decades and are increasingly replacing the organophosphates and methylcarbamates. The favorable selectivity of the neonicotinoids occurs largely at the target level, which is the agonist binding site of the nicotinic acetylcholine (ACh) receptor (nAChR). Nicotine, the namesake of the nAChR, has been used for pest control since the 17th century despite the risk for people and limited insecticidal efficacy. In contrast, the new and nicotine-like neonicotinoids have low activity on mammalian relative to insect nAChRs, providing a mechanistic basis for their safety (1, 2).

Primary insights into the structure and function of agonist-gated ion channel complexes came from electron microscopy analysis of the *Torpedo* nAChR (3) and x-ray crystallography of ACh binding proteins (AChBPs), soluble homologues of the nAChR extracellular drug-binding domain (4–7). Nicotine and its more potent analog epibatidine (EPI) (Fig. 2) (8) played a major role in structural characterization of the agonist-binding domain. Protonation of these agonists at physiological pH is important for binding site interactions at the superfamily of neurotransmitter-gated ion channels (5, 7, 9–11). Neonicotinoids, on the other hand, are uncharged with an electronegative nitro or cyano pharmacophore (12, 13). The first step in understanding neonicotinoid binding sites was identification of the molecular determinants for target site resistance and site-directed mutagenesis (14–16), but such studies do not provide a complete description of the binding site.

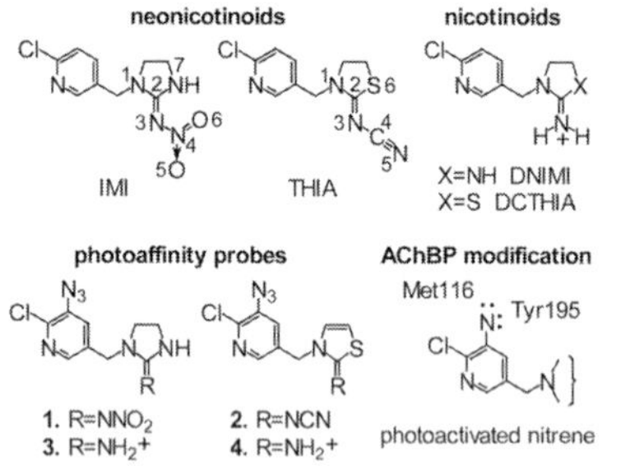

Fig. 1. Structures of neonicotinoids, nicotinoids, and photoaffinity probes. (*Upper*) Neonicotinoid insecticides (IMI and THIA) and the corresponding desnitro and descyano nicotinoid analogs (DNIMI and DCTHIA). (*Lower*) 5-Azido-6-chloropyridin-3-ylmethyl photoaffinity probes (**1–4**) and the photoactivated nitrene that derivatizes the AChBP agonist-binding domain.

We report here that AChBP from the saltwater mollusk *Aplysia californica* surprisingly shows high sensitivity to the neonicotinoids. Photoaffinity labeling combined with MS analysis identifies the sites for specific probe incorporation by using photoreactive neonicotinoid and nicotinoid agonists (probes **1–4** in Fig. 1), defining the structure of the elusive neonicotinoid binding site in comparison with that of the well understood nicotinoid site.

Results

Pharmacological Profiles. The affinities of eight neonicotinoids and four nicotinoids for the *Aplysia* AChBP were evaluated as

Author contributions: M.T. and J.E.C. designed research; M.T., T.T.T., and D.M. performed research; M.T., T.T.T., D.M., K.A.D., K.F.M., A.L.B., P.T., and J.E.C. analyzed data; M.T. and J.E.C. wrote the paper; and K.A.D. provided computational analysis.

The authors declare no conflict of interest.

Abbreviations: ACE, acetamiprid; ACh, acetylcholine; AChBP, ACh binding protein; CID, collision-induced dissociation; DCTHIA, descyanothiacloprid; DNIMI, desnitroimidacloprid; EPI, epibatidine; IMI, imidacloprid; nAChR, nicotinic acetylcholine receptor; THIA, thiacloprid.

[¶]To whom correspondence should be addressed. E-mail: ectl@nature.berkeley.edu.

This article contains supporting information online at www.pnas.org/cgi/content/full/0703309104/DC1.

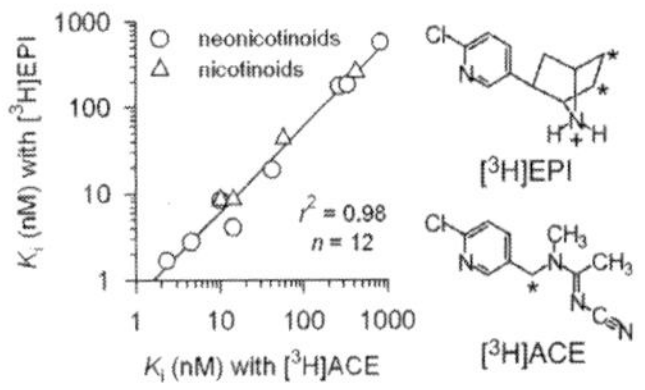

Fig. 2. Correlation plot of K_i values of neonicotinoids and nicotinoids for [^{3}H]ACE and [^{3}H]EPI binding sites in *Aplysia* AChBP. Asterisks indicate positions of tritium labeling for the two radioligands.

competition with the neonicotinoid radioligand [^{3}H]ACE and the nicotinoid probe [^{3}H]EPI (Table 1). Five neonicotinoids had high affinities with K_i values of 2–41 nM [probe **2**, olefin analog of THIA (THIAolf), THIA, nitromethylene analog of IMI (IMI-$CHNO_2$), and ACE], whereas three others showed moderate affinity with K_i values of 180–808 nM (IMI, nitenpyram, and clothianidin). The descyano and desnitro analogs of THIA and IMI [descyanothiacloprid (DCTHIA) and desnitroimidacloprid (DNIMI), respectively] are categorized as nicotinoids because they are iminium cations and act like EPI and nicotine (1, 2). DCTHIA is almost equal to EPI in potency and DNIMI has a six to seven times higher affinity than nicotine. The effect of introducing an azido group at the 5-position was examined with a reference ligand probe **2**, and it gave similar binding affinity to its parent compound THIAolf, consistent with our earlier findings with the nAChR (17). Most interestingly, an excellent correlation (r^2 = 0.98) was evident for the potencies of the 12 ligands in competing for the [^{3}H]ACE and [^{3}H]EPI binding site(s) (Fig. 2).

[^{3}H]5 Photoaffinity Labeling. The azido substituent gave very effective neonicotinoid photoaffinity probes, which, in principle, bind to the specific site and then the reactive nitrene intermediate, generated by photoirradiation, reacts covalently with AChBP (Fig. 1). [^{3}H]**5**, as an easily labeled representative compound, was used for visualization and quantification of the photoderivatization. [^{3}H]**5** photoreacted with 8.4 ± 1.1 pmol sites of AChBP (76 pmol sites) with nonspecific labeling of 0.27 ± 0.04 pmol determined in the presence of IMI (specific labeling 97%) (Fig. 3), occupying 11% of the total AChBP. IMI concentration-dependently inhibited specific incorporation of [^{3}H]**5** into AChBP (38 pmol sites) with 50% protection at ≈500 pmol. Edman sequencing then identified two peptides Gln-184–Lys-203 and Pro-98–Arg-122 with similar levels of radioactivity.

Table 1. Pharmacological profiles of [^{3}H]ACE and [^{3}H]EPI binding sites in *Aplysia* AChBP

Ligand*	K_i ± SEM (nM, n = 3 or 4)† [^{3}H]ACE	[^{3}H]EPI
Neonicotinoids		
Probe **2**	2.3 ± 0.2	1.7 ± 0.1
THIAolf	4.5 ± 0.5	2.8 ± 0.2
THIA	9.9 ± 1.7	8.4 ± 2.8
IMI-$CHNO_2$	14 ± 2	4.1 ± 1.1
ACE	41 ± 3‡	19 ± 3
IMI	267 ± 4	180 ± 40
Nitenpyram	335 ± 21	186 ± 20
Clothianidin	808 ± 74	574 ± 18
Nicotinoids		
DCTHIA	10 ± 1	8.5 ± 1.8
(±)-EPI	14 ± 2	8.6 ± 2.0‡
DNIMI	56 ± 3	44 ± 8
(−)-Nicotine	415 ± 27	260 ± 6

*Chemical structures are given in Fig. 1 and supporting information (SI) Fig. 6.
†Averages for Hill coefficients in [^{3}H]ACE and [^{3}H]EPI bindings are 1.03 and 1.02, respectively.
‡Dissociation constants determined by direct radioligand saturation experiments.

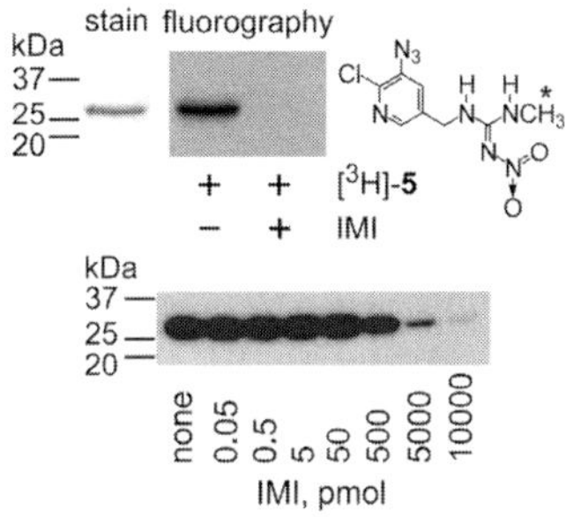

Fig. 3. [^{3}H]5 photoaffinity labeling. (*Upper*) (*Right*) The structure of [^{3}H]**5** is indicated with an asterisk for the position of tritium label. (*Left*) Fluorography of [^{3}H]**5**-derivatized AChBP with Coomassie blue-stained protein. AChBP (76 pmol sites) was photoaffinity-labeled by [^{3}H]**5** (92 pmol) without (−) and with IMI (10,000 pmol) (+). (*Lower*) Protection of photoaffinity labeling by unlabeled IMI is shown. AChBP (38 pmol sites) was reacted with [^{3}H]**5** (46 pmol).

MS Analysis of Photoaffinity-Labeled AChBP. Liquid chromatography/MS of the labeling mixture for each photoaffinity probe revealed unmodified protein and covalently derivatized AChBP based on the expected mass increase (Table 2). The relative intensities of the deconvoluted components established at least 15% single and negligible double labeling in all experiments (SI Fig. 7, for example, data with **1**-nitrene). The calculated mass for the intact, FLAG-tagged [numbered as (−8)DYKDDDDKL(0)] recombinant protein is 25,944 Da. From the mass measured for the unmodified protein and the collision-induced dissociation (CID) data acquired from the tryptic digests we concluded that a mixture of sequences with ragged termini were present. Tyr(-7) to Arg-215 features the closest mass to the measured value: 25,324 Da (−0.1% deviation).

To identify the site(s) of modification the proteins were digested with trypsin, after disulfide-bridge reduction and carbamidomethylation of the free sulfhydryls, and then analyzed by

Table 2. Mass measurements of *Aplysia* AChBP subunit (intact protein) labeled with neonicotinoid and nicotinoid photoaffinity probes

Nitrene*	Molecular mass, Da Unmodified	Modified	Increase
1	25,352	25,620 (25,887)†	268 (535)†
2	25,351	25,615 (25,879)†	264 (528)†
3	25,351	25,575	224
4	25,351	25,590	239

*Structures of azido precursors and photoactivated nitrenes are shown in Fig. 1.
†Mass in parentheses indicates the negligible double modification.

Table 3. Tryptic peptides labeled by nitrene identified by FTICR

Nitrene	*m/z*, observed (z)	Molecular mass, Da*	
		Measured	Theoretical
	Gln-184 to Lys-203 with Tyr-195 modified		
1	906.0771 (3)	2,715.2078	2,715.2091
2	678.5490 (4)	2,710.1647	2,710.1646
3	668.5632 (4)	2,670.2215	2,670.2239
4	672.3003 (4)	2,685.1699	2,685.1694
	Thr-80 to Arg-122 with Met-116 modified		
1	1,224.3602 (4)	4,893.4096	4,893.4063
2	1,223.0987 (4)	4,888.3636	4,888.3621
3	970.6924 (4)	4,848.4229	4,848.4213
4	973.6812 (5)	4,863.3669	4,863.3668

*ΔM are 0.0 to 0.8 ppm.

liquid chromatography/MS/MS. Two-step searches were performed with the CID data, the first with very strict search parameters establishing the identity of the protein and the purity of the preparation. The second search to identify modification(s) was performed allowing an addition of up to 300 Da on any amino acid in the sequence. The labeled peptides were identified by the predicted mass shifts corresponding to the different labeling molecules (Table 3). MS/MS (CID) data provided the final proof, both confirming the identity of the labeled peptides and pinpointing the modification site. Assignments were based on the presence of the properly shifted fragment ions in the CID spectra of the labeled peptides. This evidence identified Tyr-195 and Met-116 in all cases as the labeled sites (SI Figs. 8 and 9 showing the CID data for the **1**-nitrene-labeled peptides as an example). It was observed that some modified fragments may undergo a gas-phase rearrangement reaction and lose the label (SI Fig. 8), a process that could hamper the site assignment, but, fortunately, even peptides bearing the less stable labels yielded sufficient information to identify the modification site.

Structural Models for Neonicotinoid and Nicotinoid Binding Interactions. Structural models for interactions of the neonicotinoids and nicotinoids with the agonist binding domain were established by combining (*i*) the photoaffinity labeling results for precisely positioning the chloropyridinyl moiety with (*ii*) the crystal structure of *Aplysia* AChBP defining the localization and geometry of all relevant nearby residues (Fig. 4). IMI and THIA are calculated to dock in the same orientation into the binding pocket with energies of −7.42 and −7.17 kcal/mol, respectively. The nearby amino acids are: Tyr-93 (loop A); Trp-147 (loop B); Tyr-188, Ser-189, Cys-190, and Tyr-195 (loop C); Tyr-55 (loop D); and Ile-106, Met-116, and Ile-118 (loop E). The chlorine atom of IMI and THIA interacts with the loop E segment and particularly makes van der Waals contacts to the backbone carbonyl oxygen atoms of Ile-106 (4.2 and 4.3 Å, respectively) and Met-116 (4.8 and 4.3 Å, respectively). The pyridine nitrogen of IMI or THIA is expected to undergo hydrogen-bonding with the backbone carbonyl oxygen atoms of Ile-118 (4.3 and 4.2 Å, respectively) and Trp-147 (3.7 and 3.2 Å, respectively) via water bridge(s). Tyr-195 and Met-116 are located at spatially suitable positions to be modified by the photoactivated nitrene of the probes (distance between nitrene and hydroxyl oxygen of Tyr-

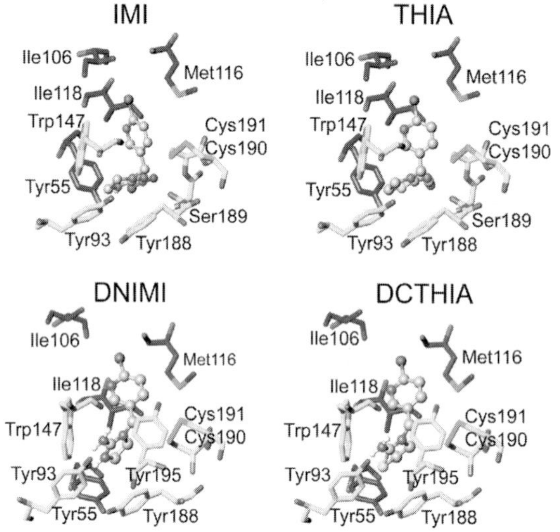

Fig. 4. Structural models for neonicotinoid (*Upper*, IMI and THIA) and nicotinoid (*Lower*, DNIMI and DCTHIA) binding site interactions based on photoaffinity labeling and the crystal structure of EPI–AChBP complex (Protein Data Bank ID code 2BYQ) (7). View angle is from an apical and radial location. Residues in yellow are from the (+)-face and in blue are from the (−)-face. Loop C (shown as a portion Tyr-188 to Cys-191) flexibility (7) may yield multiple similar conformations induced by neonicotinoid binding. Tyr-195 is not displayed for the IMI and THIA models because it would visually obscure their nitro and cyano substituents, respectively, but the positioning is evident from the lower two images (see also SI Fig. 10).

195 or sulfur of Met-116 is presumably < 3 or 4 Å). N1, C2, and S6 (Fig. 1) of the imidazolidine and thiazolidine rings are surrounded by Trp-147 and Tyr-188, and N7 (Fig. 1) of the imidazolidine approaches Tyr-55. The distance between the hydroxyl oxygen of Tyr-55 and the hydrogen on N7 of IMI is 2.1 Å. The guanidine/amidine plane of the two neonicotinoids interacts with Tyr-188 at 3- to 4-Å distance in a slipped-stack geometry attributable to a stabilizing nonbonded interaction analogous to π-stacking. The imine moiety (C2 and N3) faces Tyr-188 and Cys-190. O5 of the nitro group of IMI interacts with loop C Cys-190, whereas O6 is directed toward loop D Tyr-55. The important electronegative tip oxygen O5 (corresponding to that of the active nitroso analogs) (13) or cyano nitrogen N5 makes a hydrogen bond with the backbone NH of Cys-190 (2.2 and 2.7 Å, respectively). IMI nitro O6 also interacts with the hydroxyl group of Tyr-55 (2.2 Å). Further hydrogen-bonding and/or hydrophobic contacts can be predicted between the IMI/THIA tip atom and Cys-190 and Ser-189 within 3 or 4 Å and also a solvent bridge is theoretically possible with the nitro oxygen or cyano nitrogen atoms.

The chloropyridinyl moiety of the nicotinoids is buried in the agonist binding pocket in fundamentally the same direction as that of the neonicotinoids (Fig. 4). The calculated binding energies for DNIMI and DCTHIA are −8.22 and −8.24 kcal/mol, respectively. The chlorine atom of DNIMI and DCTHIA contacts the backbone carbonyl oxygen atoms of Ile-106 (3.2 and 3.5 Å, respectively) and Met-116 (3.2 and 3.1 Å, respectively). The pyridine nitrogen atoms are directed to the backbone carbonyl oxygen atoms of Ile-118 (4.4 and 4.5 Å, respectively) and Trp-147 (3.5 and 3.5 Å, respectively) for solvent bridge(s). In contrast to the neonicotinoids, the iminium end of DNIMI and DCTHIA interacts with Trp-147, i.e., hydrogen-bonding with the backbone carbonyl oxygen (2.0 and 2.5 Å, respectively) and also with the hydroxyl oxygen atom of loop A Tyr-93 (2.2 and 2.1 Å, respectively). Cation-π interactions of DNIMI and DCTHIA to the center of the six-membered aromatic residues are expected for Trp-147 (3.8 and 4.0 Å, respectively), Tyr-93 (4.7 and 4.5 Å), Tyr-195 (both 4.8 Å), and Tyr-188 (5.5 and 5.1 Å) but not for Tyr-55 (7.1 and 7.2 Å).

Discussion

Some Common and Unique Molecular Aspects of Neonicotinoids and Nicotinoids. This study considers two azidoneonicotinoids with a nitro or cyano group and the corresponding azidonicotinoids or iminium analogs. The azido substituent presumably does not greatly alter the fit because it does not appreciably change the potency (17). The neonicotinoids are nonprotonated and the nicotinoids predominantly protonated at physiological pH. The neonicotinoids have an electronegative nitro or cyano tip versus the nicotinoid positively charged end (ammonium or iminium cation). In contrast to the nicotinoids, the neonicotinoids are coplanar between the nitro- or cyano-substituted heterocyclic guanidine/amidine planes (13, 18) providing electronic conjugation, which facilitates partial negative charge (δ^-) flow toward the tip. However, the equivalent region of positive charge to balance the partial negative charge on the tip is not localized on any specific atom but rather is dispersed into the whole heterocyclic moiety (13, 19).

Same Binding Site Pocket. The *Aplysia* AChBP is a homopentamer with five ligand binding sites equally sensitive to neonicotinoids and nicotinoids. The pharmacological profile determined with the neonicotinoid radioligand [^{3}H]ACE is identical to that with the nicotinoid [^{3}H]EPI, suggesting that neonicotinoids and nicotinoids bind to the agonist site without cooperativity in the same manner. Both neonicotinoid and nicotinoid probes specifically bind with an incorporation ratio of one photoactivated (nitrene) probe molecule in each subunit interfacial binding domain based on analysis of the intact derivatized protein. In all cases, the sites for specific derivatization are both Tyr-195 and Met-116 identified by CID. Tyr-195 is located on the loop C segment of the principal or (+)-face subunit of AChBP, whereas Met-116 is on loop E of the complementary or (−)-face subunit and these residues are 3D neighbors to each other. Photoderivatization occurs in similar frequency with Tyr-195 and Met-116, indicating their proximity and reactivity. These labeling results further define the chloropyridinyl position of the neonicotinoids and nicotinoids under physiologically relevant conditions, and it is identical to that of the crystal structure of EPI-bound AChBP (7). This similarity extends to the loop C conformational rearrangement involved in capping the binding pocket for both neonicotinoids and nicotinoids.

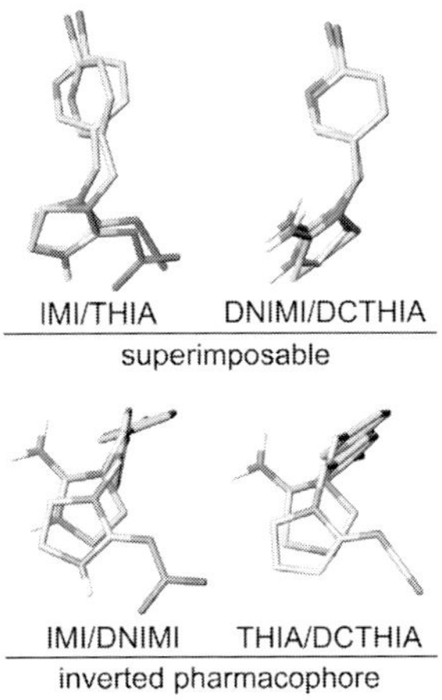

Fig. 5. Conformational comparison of neonicotinoids and nicotinoids as observed in the AChBP binding pocket. The relevant amino acid residues shown in Fig. 4 are not exhibited here. Two neonicotinoids and two nicotinoids are superimposable but the neonicotinoid and nicotinoid pharmacophores are inverted relative to each other.

Unique Binding Site Interactions. The chloropyridinyl moiety is positioned in the same place for the neonicotinoids and nicotinoids and serves as the fulcrum for the rest of the molecules that are superimposable for IMI with THIA and for DNIMI with DCTHIA as observed in the AChBP binding pocket (Fig. 5). The neonicotinoid nitroguanidine or cyanoamidine moiety interacts with loop C Tyr-188 and the nitro oxygen or cyano nitrogen contacts Cys-190/Ser-189. Interestingly, Ser-189 of AChBP corresponds to Glu in the vertebrate neuronal $\alpha4$ and $\alpha7$ nAChRs and faces inside the binding pocket based on the $\alpha4\beta2$ nAChR model (SI Fig. 11), whereas Val, Thr, Pro, Ala, or Ser takes its place in the insect α subunits (20). The Glu negatively charged carboxyl residue possibly repels the electronegative nitro or cyano pharmacophore of the neonicotinoids in the vertebrate system. When Gln-57 on the AChBP (−)-face is substituted by the corresponding Arg/Lys of the insect nAChR β subunit (16), the Arg/Lys is predicted to undergo hydrogen networks with the loop C Cys-190–191 on the (+)-face or α subunit in the

Perry L. Adkisson

Texas A&M University
College Station, Texas, USA

1994/5 — for his contributions to the development and implementation of environmentally beneficial integrated pest management systems for the protection of agricultural crops.

CURRICULUM VITAE

Date and Place of Birth: March 11, 1929, Hickman, Arkansas

EDUCATION:

1946 - Graduate, Armorel High School, Armorel, Arkansas
1950 - B.S. (Agric.), University of Arkansas, Fayetteville, Arkansas
1954 - M.S. (Agron.), University of Arkansas, Fayetteville, Arkansas
1956 - Ph.D. (Ento.), Kansas State University, Manhattan, Kansas
1963 - USPHS Senior Postdoctoral Fellow, Harvard University

EMPLOYMENT BACKGROUND:

1995- - Chancellor Emeritus and Distinguished Professor Emeritus, Texas A&M, College Station, Texas
1991-95 - Regents Professor of Entomology, Texas A&M, College Station, Texas
1991-93 - Regents Professor of Entomology, Texas A&M University, College Station, Texas; and Executive Director, George Bush Presidential Library Center
1986-90 - Chancellor, The Texas A&M University System, College Station, Texas

1983-86 - Deputy Chancellor, The Texas A&M University System, College Station, Texas
1980-83 - Deputy Chancellor for Agriculture, The Texas A&M University System, College Station, Texas
1978-80 - Vice President for Agriculture and Renewable Resources, Texas A&M University, College Station, Texas
1979-95 - Distinguished Professor of Entomology, Department of Entomology, Texas A&M University, College Station, Texas
1967-78 - Head, Department of Entomology, Texas A&M University, College Station, Texas
1963-67 - Professor, Department of Entomology, Texas A&M University, College Station, Texas
1958-63 - Associate Professor, Department of Entomology, Texas A&M University, College Station, Texas
1956-58 - Assistant Professor, Department of Entomology, University of Missouri, Columbia, Missouri
1950 - Instructor (Veterans Program), Vocational Agriculture, Blytheville High School, Blytheville, Arkansas

Military Experience:

1950-53 - Enlisted Man, U.S. Army Medical Corps
1953-79 - Senior Scientist, U.S. Public Health Services, Commissioned Reserve Corps

Honors:

1965 - Faculty Distinguished Achievement Award for Research, Texas A&M University
1967 - J. Everett Bussart Award for Outstanding Research in Economic Entomology, Entomological Society of America
1967 - Award for Outstanding Service, Plains Cotton Growers, Inc., Lubbock, Texas
1977 - Outstanding Entomologist Award, Central Texas Chapter, American Registry of Professional Entomologists
1978 - Man of the Year in Service to Texas Agriculture, Awarded by Progressive Farmer Magazine
1979 - Elected to National Academy of Sciences
1979 - Adventurer in Science Award, International Plant Protection Congress, Washington, D.C.
1979 - Award for Distinguished Service to Professional Entomology, American Registry of Professional Entomologists

1980 - Award for Service in Development of the Integrated Pest Management Programs Benefiting Agricultural Producers of Texas, presented by the Texas Pest Management Association and Producers in Texas
1980 - Distinguished Service in Agriculture Award, Kansas State University, Manhattan, Kansas
1980 - Alexander von Humboldt Award for the most outstanding contribution to agriculture during the past five years for research and developmental work on Integrated Pest Management (IPM), Alexander von Humboldt Foundation, New York
1981 - Award for Distinguished Service to Agriculture, Texas A&M Chapter, Gamma Sigma Delta (Honor Society of Agriculture)
1982 - Distinguished Scientist of the Year Award, Texas Academy of Sciences
1982 - Named to the Presidents Commission on the National Medal of Science (Presidential Appointment, May 1982 for 3-Year Term)
1984 - Named Fellow of the Entomology Society of America at Annual meeting, San Antonio, Texas, December 1984
1985 - Named to the National Science Board (Presidential Appointment, June 1985 for 5-Year Term). Reappointed 1991
1985 - Founders Memorial Lecturer, Entomological Society of America
1986 - Named Fellow of the American Academy of Arts and Sciences at Annual Meeting, Cambridge, Massachusetts, May 1986
1986 - Honored by the Committee on Agriculture, U.S. House of Representatives, for "service to the nation's agriculture"
1987 - Distinguished Service Award, American Institute of Biological Science
1988 - National 4-H Club Distinguished Alumni Award
1990 - Outstanding Alumnus Award, College of Agriculture and Home Economics, University of Arkansas
1990 - Distinguished Alumni Award, University of Arkansas Alumni Association
1990 - William Henry Hatch Memorial Lecturer, Division of Agriculture, National Association of State Universities and Land Grant Colleges
1992 - Honorary Member, International Congress of Entomology
1992 - Distinguished Service Award, American Agricultural Editors Association
1994/95 - Wolf Prize in Agriculture
1995 - "Award of Distinction" in recognition of outstanding services in plant protection. XIII International Congress of Plant Protection. The Hague, The Netherlands. July 2-7, 1995
1997 - Honorary Doctorate of Science, University of Arkansas, May 10, 1997
1997 - World Food Prize
1999 - Named by Progressive Farmer as one of 24 Scientists having the greatest impact on American agriculture in the 20th Century

2003 - Named as one of the 60 individuals making the most outstanding contributions to agriculture during the past 60 years. The Inter-American Institute for Cooperation on Agriculture, Washington, DC

Significant Grants:

National Science Foundation:
Photoperiodic Control of Insect Diapause, 1961-1966

Cotton Incorporated:
Systems Approach to Cotton Insect Control, 1970-1973
Development of Juvenile Hormone as a Third Generation Insecticide, 1970-1973

U.S. Department of Agriculture:
Seasonal Biology of the Boll Weevil in the Texas High Plains, 1965-1968
Potential Attractants for Use in Controlling the Tobacco Budworm, 1966-1968
Improvement of Methods for the Total Population Suppression of the Boll Weevil, 1971-1974

Rockefeller Foundation:
Development of Insect-Resistant Cotton Varieties to Minimize Insecticide Treatments for Control of Bollworms (Heliothis spp.), 1971-1976

International Biological Program/National Science Foundation-Environmental Protection Agency Project:
The Principles, Strategies and Tactics of Pest Population Regulation and Control in Major Crop Ecosystems: Director, Cotton Project; Member, Project Executive Committee, 1970-1978

Environmental Protection Agency/USDA Project:
Development of Comprehensive, Unified, Economically and Environmentally Sound Systems of Integrated Pest Management for Major Crops, 1979-1985

Memberships:

National Academy of Sciences
American Academy of Arts and Sciences
American Association for the Advancement of Science
Entomological Society of America (President, 1974)
International Organization for Biological Control
Southwestern Entomological Society
Sigma Xi
American Registry of Professional Entomologists (President, 1977)
Phi Kappa Phi
Gamma Sigma Delta

PROFESSIONAL ACTIVITIES:

Member, Plains Cotton Growers, Inc., Technical Advisory Committee on Boll Weevil Control, 1968-72
Member, UN/FAO Panel of Experts on Integrated Control, 1968-82
Chairman, Committee on Memorial Lectures, Entomological Society of America (ESA), 1969-71
Chairman, Scientific Advisory Committee to Governor of Texas on Uses of Agricultural Chemicals, 1970-72
Chairman, Texas Pesticide Advisory Committee, 1970-78
Consultant, US/EPA Office of Water Programs, 1971-72
Southern Agricultural Experiment Station Directors' Representative to the Southern Regional Pest Management Working Group, 1971-72
Consultant on Pesticide Use, US/EPA Hazardous Advisory Committee, 1971-72
Member, Governing Board, Entomological Society of America (ESA), 1971-75
Adviser on Pesticides to Texas Legislature, 1972
Member, Texas Structural Pest Control Board, 1972-78
Member, NAS Committee on Biologies of Pest Species (Chairman, 1975-79) 1973-79
President, Entomological Society of America, (ESA), 1974
Member, Executive Committee, N AS Environmental Sciences Board Study Group on Problems of Pest Control, and Member, Cotton Study Team of this Project, 1974-76
Member, Editorial Board, Annual Review of Entomology, 1974-78
Chairman, Governing Council, American Registry of Professional Entomologists, 1975-76
Vice-Chairman, NAS Committee on Acquisition and Use of Scientific and Technical Information in Regulatory Decision-Making, 1976-77
Member, NAS Committee on World Food and Nutrition, 1976-77
President, American Registry of Professional Entomologists, 1977
Member, NRC Committee for the International Union of Biological Sciences, 1979-1985
Member, NAS Class Ill Membership Committee, 1980-83
Member, NAS ad hoc Committee on Relationships between Universities and the U.S. Government 198 1-83

A BRIEF DESCRIPTION OF THE SCIENTIFIC ACHIEVEMENTS

The many contributions of Perry Adkisson to science and society in developing integrated pest management (IPM) systems for the protection of major crops are widely recognized. He utilized the knowledge gained in basic and field research to develop IPM systems that have protected crop yields in a sustainable manner while greatly reducing the use of chemical insecticides. The decreased use of insecticides has lessened the exposure of farm workers to these toxic chemicals and has decreased pesticide pollution of the environment.

Among Adkisson's most important contributions was the development of one of the first semi-synthetic diets for the laboratory rearing of phytophagous insects. He and E. S. Vanderzant developed the Vanderzant-Adkisson Wheat Germ Diet that is widely used for rearing plant-feeding caterpillars in the laboratory under controlled conditions. This diet has greatly facilitated basic studies in insect physiology, developmental biology, toxicology, ecology and genetics. Research in these areas previous to the diet had been very difficult to do when the insects had to be reared on plants or plant parts.

In research using insects reared on artificial diets, Adkisson and his students made many important contributions, elucidating the role of photoperiod, temperature and diet in controlling growth and development of the most important insect pests of cotton in the United States. Especially important was their work demonstrating the role of photoperiod and temperature in controlling the onset and termination of diapause in certain of these pests. Among the most important of these studies was research conducted with Prof. C. M. Williams (Harvard University) which showed that the photoperiod has its primary action on certain receptors in the insect brain by regulating the flow of hormones that control insect growth and development. Adkisson later showed that the onset of diapause in field populations of pest insects whose diapause is photoperiodically controlled can be predicted with considerable precision.

The fundamental knowledge gained in these studies was utilized by Adkisson and co-workers to develop improved IPM systems for cotton. These were designed to make more effective use of cultural and phytosanitation measures for suppressing pests. When these practices were combined with the limited use of insecticides, maximum suppression of the primary pests could be obtained while preserving the natural enemies of key secondary pests.

These programs were implemented on millions of acres of cotton and proved effective in increasing yields. They also saved farmers millions of dollars by greatly reducing the amounts of insecticides used on the crop. The reduced use of insecticides prevented the release of hundreds of tons of these chemicals that otherwise would have entered the environment.

Based on the early success of IPM, Adkisson, Carl Huffaker (co-recipient of the 1994-95 Wolf Prize in Agriculture) and several colleagues organized, gained funding for, and managed the first large national IPM research program in the United States. This project involved more than 250 scientists from seventeen universities who focused their efforts on developing new and improved IPM systems for managing the arthropod pests of six major crops. This project, known as the Huffaker project, was successfully completed in the late 1970's. It was succeeded by an even larger project organized and managed by Adkisson and R. E. Frisbie (Texas). This project was designed to build on the successes achieved by the Huffaker project.

These two projects were successful in developing more effective, environmentally friendly systems of pest management for the target crops. The new 1PM systems were widely implemented by U.S. farmers providing satisfactory pest suppression and greatly reducing the amounts of pesticides applied to U.S. croplands.

The success of these two projects reshaped the philosophies of crop protection specialists across the world. It directed their efforts towards ecologically-based, systems-oriented approaches to pest management and away from the unilateral use of chemical pesticides. 1PM now is accepted internationally as the best approach for protecting crops from insect pests.

Many people were involved in developing 1PM for the protection of crops. Of these, Perry Adkisson as been one of the most influential in developing the concepts and principles of IPM, demonstrating the economic and environmental benefits of 1PM in experimental plots and farmers' fields, and popularizing it on a global basis.

LIST OF PUBLICATIONS

1. Adkisson, Perry L. 1954. The influence of hybridity and boll load upon the incidence of verticillium wilt of cotton. Masters Thesis, University of Arkansas. Fayetteville, AK.
2. Adkisson, Perry L. 1956. The relative susceptibility of the various development stages and instars of the rice weevil, *Sitophilus oryza* (I.) to certain fumigants. Ph.D. Dissertation, Kansas State University. Manhattan, KS.
3. Adkisson, Perry L. 1957. The relative susceptibility of the life history stages of the rice weevil to certain fumigants. J. Econ. Entomol. 50(6):761-764.
4. Adkisson, Perry L. 1957. Influence of irrigation and fertilizer on populations of three species of minds attacking cotton. FAO Plant Protect. Bull. 6(3):33-36.
5. Adkisson, Perry L. 1957. Cotton insect research in Southeast Missouri in 1956. 1956 Progress Report, SEMO Exp. Fields, Mo. Agr. Ext. Sta. Ann. Rpt. 1936:29-36.
6. Adkisson, P. L., S. Kyd and G. W. Thomas. 1957. Cotton insect controls for Missouri. Mo. Agr. Exp. Sta. Ext. Cire. 680.
7. Adkisson, Perry L. 1958. Cotton insect research, 1957. Mo. Agr. Exp. Sta. Bull. 700:4-11.
8. Adkisson, Perry L. 1958. An inexpensive unit for converting a vacuum cleaner to an insect aspirator. J. Kans. Entomol. Soc. 31(4):284-285.
9. Adkisson, Perry L. 1958. Field tests of materials for control of the cotton leafworm. J. Econ. Entomol. 51(2):259.

10. Adkisson, Perry L. 1958. Seed treatment of cotton with systemic insecticides alone and in combination with a fungicidal treatment. J. Econ. Entomol. 51(5):697-700.
11. Adkisson, Perry L. 1958. The influence of fertilizer applications on populations of *Heliothis zea* (Boddie), and certain insect predators. J. Econ. Entomol. 51(6):757-759.
12. Adkisson, P. L., S. Kyd and G. W. Thomas. 1958. 1958 Cotton insect control: Recommendations for Missouri. Mo. Agr. Exp. Sta. Ext. Circ. 685.
13. Adkisson, P. L., L. H. Wilkes and S. P. Johnson. 1958. Chemical, cultural and mechanical control of the pink bollworm. Tex. Agr. Exp. Sta. Bull. 920. 16 p.
14. Wilkes, L. H., P. L. Adkisson and B. J. Cochran. 1959. Effect of spray nozzle types on cotton insect control. Tex. Agr. Exp. Sta. PR 2078.
15. Adkisson, P. L., L. H. Wilkes and B. J. Cochran. 1959. Relative efficiencies of certain spray nozzles for cotton insect control. J. Econ. Entomol. 52(5):985-991.
16. Wilkes, L. H., P. L. Adkisson and B. J. Cochran. 1959. Stalk shredder tests for pink bollworm control, 1958. Tex. Agr. Exp. Sta. PR 2095.
17. Adkisson, Perry L. 1959. The effect of various humidity levels on hatchability of pink bollworm eggs. J. Kans. Entomol. Soc. 32(4): 189-190.
18. Adkisson, P. L., L. H. Wilkes and B. J. Cochran. 1960. Stalk shredding and plowing as measures for controlling the pink bollworm, *Pectinophora gossypiella*. J. Econ. Entomol. 53(3):436-439.
19. Adkjsson, P. L., E. S. Vanderzant, D. L. Bull and W. E. Allison. 1960. A wheat germ medium for rearing the pink bollworm. J. Econ. Entomol. 53(5):759-762.
20. Adkisson, P. L., D. L. Bull and W. B. Allison. 1960. A comparison of certain artificial diets for laboratory cultures of the pink bollworm. J. Econ. Entomol. 53(5):791-793.
21. Bull, D. L. and P. L. Adkisson. 1960. Certain factors influencing diapause in the pink bollworm, *Pectinophora gossypiella*. J. Econ. Entomol. 53(5):793-798.
22. Adkisson, Perry L. 1960. The effect of pink bollworm infestations on cotton produced under high moisture conditions. Tex. Agr. Exp. Sta. PR 2156.
23. Adkisson, Perry L. 1960. The influence of fertilization and irrigation practices on field populations of certain insects of the USA. Proc. XI Intl. Congress Entomol. (Vienna). Band III. (Symposium): 141.
24. Adkisson, P. L. and J. C. Gaines. 1960. Pink bollworm control as related to the total cotton insect control program of Central Texas. Tex. Agr. Exp. Sta. Misc. Publ. 444. 7 p.
25. Reed, D. K. and P. L. Adkisson. 1961. Short-day cotton stocks as possible source of host plant resistance to the pink bollworm. J. Econ. Entomol. 54(3):484-486.

26. Cochran, B. J., L. H. Wilkes and P. L. Adkisson. 1961. Primary tillage practices for pink bollworm control. Tex. Agr. Exp. Sta. PR 2194.
27. Graham, H. M., O. T. Robertson, P. L. Adkisson and L. H. Wilkes. 1961. Further tests of the effectiveness of cotton stalk shredders for controlling the pink bollworm. J. Econ. Entomol. 54(5):1057-1058.
28. Wilkes, L. H., P. L. Adkisson and B. J. Cochran. 1961. Use of an air-carrier sprayer for cotton insect control. Tex. Agr. Exp. Sta. PR 2205.
29. Adkisson, Perry L. 1961. Fecundity and longevity of adult female pink bollworm reared from natural and synthetic diets. J. Econ. Entomol. 54(6):1224-1227.
30. Adkisson, Perry L. 1961. Effect of larval diet on the seasonal occurrence of diapause in the pink bollworm. J. Econ. Entomol. 54(6):7107-71 12.
31. Bull, D. L. and P. L. Adkisson. 1962. Fat content of the larval diet as a factor influencing diapause and growth-rate of the pink bollworm. Ann. Entomol. Soc. Amer. 53(3):499-502.
32. Adkisson, P. L., R. L. Hanna and C. F. Bailey. 1962. Cotton yield and quality losses resulting from various size populations of bollworms. Tex. Agr. Exp. Sta. PR 2235.
33. Wellso, S. G. and P. L. Adkisson. 1962. The morphology of the reproductive system of the female pink bollworm moth, *Pectinophora gossypiella* (Saund.). S. Kansas Entomol. Soc. 35(2):233-235.
34. Adkisson, P. L. and S. G. Wellso. 1962. Effect of DDT poisoning on the longevity and fecundity of the pink bollworm. J. Econ. Entomol. 55(6):842-845.
35. Adkisson, Perry L. 1962. Timing defoliants and desiccants to reduce populations of the pink bollworm in diapause. J. Econ. Entomol. 55(6): 949-951.
36. Wilkes, L. H., B. J. Cochran and P. L. Adkisson. 1962. Further studies of the effectiveness of certain spray nozzle types and sizes and arrangements for controlling boll weevils and bollworms. Tex. Agr. Exp. Sta. PR 2264.
37. Adkisson, P. L., S. Hacskaylo, R. L. Hanna and S. K. Walker. 1962. Flowering rate, boll maturity and yield of cotton treated with various insecticides. Tex. Agr. Exp. Sta. PR 2245.
38. Graham, H. M., L. C. Fife, 0. T. Robertson and P. L. Adkisson. 1962. Pink bollworm population dynamics. In A Summary of Recent Research Basic to the Cultural Control of the Pink Bollworm. Tex. Agr. Exp. Sta. Misc. Publ. 579. pp. 5-10.
39. Lukefahr, M. S., L. C. Pife and P. L. Adkisson. 1962. Pink bollworm diapause studies. In A Summary of Recent Research Basic to the Cultural Control of the Pink Bollworm. Tex. Agr. Exp. Sta. Misc. Publ. 579. pp. 11-13.

40. Adkisson, P. L., O. T. Robertson and L. C. Fife. 1962. Planting date as a factor involved in pink bollworm control. In A Summary of Recent Research Basic to the Cultural Control of the Pink Bollworm. Tex. Agr. Exp. Sta. Misc. Publ. 579. pp. 16-20.
41. Adkisson, P. L., L. H. Wilkes, B. J. Cochran and R. L. Hanna. 1962. Spray nozzle arrangements, types and rates of application for cotton insect control. Tex. Agr. Exp. Sta. Misc. Publ. 595. 12 p.
42. Adkisson, P. L., C. F. Bailey and G. A, Niles. 1962. Cotton stocks screened for resistance to the pink bollworm, 1960-61. Tex. Agr. Exp. Sta. Misc. Publ. 606.
43. Adkisson, P. L., R. A. Bell and S. G. Wellso. 1963. Environmental factors controlling the induction of diapause in the pink bollworm, *Pectinophora gossypiella* (Saunders). J. Insect Physiol. 9:299-310.
44. Bull, D. L. and P. L. Adkisson. 1963. Absorption and metabolism of C 14-labeled DDT by DDT-susceptible and DDT-resistant pink bollworm adults. J. Econ. Entomol. 56(5):641-643.
45. Adkisson, P. L., J. R. Brazzel and J. C. Gaines. 1963. Yield and quality losses resulting from pink bollworm damage to cotton. Tex. Agr. Exp. Sta. Misc. Publ. 632.
46. Adkisson, Perry L. 1963. Timee measurement in the photoperiodic induction of diapause in the pink bollworm. Tex. Agr. Exp. Sta. PR 2274. 4 p.
47. Adkisson, Perry L. 1963. Insect control, a continuing challenge in the Southwest. Cotton Gin and Oil Mill Press. April 13. pp. 9, 10, 34.
48. Adkisson, Perry L. 1963. Time measurement in the photoperiodic control of diapause in an insect. Proc. XVI Intl. Congress of Zoology (Washington, D.C.) 2:51.
49. Henry, P. and P. L. Adkisson. 1963. Relative numbers of bollworms and tobacco bollworms in cotton of Central Texas in 1963. Tex. Agr. Exp. Sta. PR 2289.
50. Adkisson, Perry L. 1963. Relative susceptibility of various geographical races of the pink bollworm to certain insecticides. Tex. Agr. Exp. Sta. PR 2300.
51. Weliso, S. G. and P. L. Adkisson. 1964. Photoperiod and moisture as factors involved in the termination of diapause in the pink bollworm, *Pectinophora gossypielki.* Ann. Entomol. Soc. Amer. 57(2):170-173.
52. Adkisson, P. L., C. F. Bailey and R. L. Hanna. 1964. Effect of the bollworm, *Heliothis zea,* on yield and quality of cotton. J. Econ. Entomol. 57(4): 448-450.
53. Adkisson, P. L., R. L. Hanna and C. F. Bailey. 1964. Estimates of the number of Heliothis larvae per acre in cotton and their relation to the fruiting cycle and yield of the host. J. Econ. Entomol. 57(5):657-663.

54. Adkisson, Perry L. 1964. Comparative effectiveness of several insecticides for controlling bollworms and tobacco budworms. Tex. Agr. Exp. Sta. Misc. Publ. 709, 13 p.
55. Williams, C. M. and P. L. Adkisson. 1964. Photoperiodic control of pupal diapause in the silkworm, *Antheraea pernyi.* Science, 144:569.
56. Cole, C. L. and P. L. Adkisson. 1964, Daily rhythm in the susceptibility of an insect to a toxic agent. Science, 144:1148-1149.
57. Bell, R. A. and P. L. Adkisson. 1964. Photoperiodic reversibility of diapause induction in an insect. Science, 144:1149-1151.
58. Adkisson, Perry L. 1964. Action of the photoperiod in controlling insect diapause. Amer. Nat. XCVIII:357-374.
59. Williams, C. M. and P. L. Adkisson. 1964. Physiology of insect diapause. XIV. An endocrine mechanism for the photoperiodic control of pupal diapause in the oak silkworm, Antheraea pernyi. Biol. Bull. 127:511-525.
60. Williams, C. M., P. L. Adkisson and C. Walcott. 1965. Physiology of insect diapause. X V. T he transmission of photoperiod signals to the brain of the oak silkworm, *Antheraea pernyi.* Biol. Bull. 128:497-501.
61. Cole, C. L. and P. L. Adkisson. 1965. A circadian rhythm in the susceptibility of an insect to an insecticide. Circadian Clocks, J. Aschoff, Ed. North-Holland Publ. Co., Amsterdam. pp. 309-313.
62. Adkisson, Perry L. 1965. Light-dark reactions involved in insect diapause. In Circadian Clocks, J. Aschoff, Ed. North-Holland Publ. Co., Amsterdam. pp. 344-350.
63. Sterling, W. L., S. G. Wellso, P. L. Adkisson and H. W. Dorough. 1965. A cottonseed-meal diet for laboratory cultures of the boll weevil. J. Econ. Entomol. 5 8(5):867-869.
64. Henry, P. and P. L. Adkisson. 1965. Seasonal abundance of bollworms and tobacco budworms on a typical cotton plantation. Tex. Agr. Exp. Sta. Misc. Publ. 767. 6 p.
65. Adkisson, P. L. and S. J. Nemec. 1965. Efficiency of certain insecticides for killing bollworms and tobacco budworms. Tex. Agr. Exp. Sta. PR 2357. 11 p.
66. Adkisson, Perry L. 1965. Present status of insecticide resistance in certain geographical populations of bollworms in Texas. Tex. Agr. Exp. Sta. PR 2358. 5 p.
67. Adkisson, Perry L. 1965. Biological clocks and insect photoperiodism, Proc. 17th Annual Cotton Improvement Conf. (Atlanta, GA):99-104.
68. Adkisson, P. L., J. W. Davis, W. L. Owen and D. R. Rummel. 1965. Evaluation of the 1964 diapause boll weevil control program on the High Plains of Texas. Tex. Agr. Exp. Sta. Dept. Entomol. Tech. Rpt. No. 1. 59 p.

69. Chauthani, A. R. and P. L. Adkisson, 1965. Comparison of two artificial diets for rearing *Heliothis zea* larvae for insecticide tests. J. Econ. Entomol. 58:1163-1164.
70. Adkisson, P. L., D. R. Rummel and W. L. Sterling. 1965. A two-phased control program for reducing diapause boll weevil populations on the High Plains of Texas, 1965. Tex. Agr. Exp. Sta. Dept. of Entomol. Tech. Rpt. No. 2. 6 p.
71. Haima, R. L., J. K. Walker, Jr., and P. L. Adkisson. 1965. Results of field and laboratory studies on the control of cotton insects, 1965. Tex. Agr. Exp. Sta. Dept. of Entomol. Tech.Rpt. No. 3. 24 p.
72. Keimedy, D., P. L. Adkisson, S. R. Aldrich, D. L. Dahlsten, J. E. Davies, G. E. Day, G. Gotsch, J. B. Krier, M. C. Latham, M. D. Meselson, W. W. Murdoch, K. Nair, C. E. Palm, V. W. Ruttan and R. A. Young. 1965. Pest Control: An assessment of present and alternative technologies. Vol. I. Contemporary pest control practices and prospects. NAS, Washington, D.C. 506 p.
73. Chauthani, A. R. and P. L. Adkisson. 1966. Effects of sublethal doses of certain insecticides on eggs, larvae and adults of two species of *Heliothis.* J. Econ. Entomol. 59(5): 1070-1074.
74. Sterling, W. L. and P. L. Adkisson. 1966. An artificial diet for laboratory cultures of boll weevil larvae and adults. J. Econ. Entomol. 59(5):1074-1077.
75. Barry, B. D. and P. L. Adkisson. 1966. Certain aspects of the genetic factors involved in the larval diapause of the pink bollworm. Ann. Entomol. Soc. Amer. 59(1):122-125.
76. Adkisson, Perry L. 1966. Action of light in controlling insect growth and development. Proc. Conf. on Electromagnetic Radiation in Agriculture (Roanoke, VA, 1965): 30-33.
77. Sterling, W. L. and P. L. Adkisson. 1966. Differences in the diapause response of boll weevils from the High Plains and Central Texas and the significance of this phenomenon in revising present fall insecticidal control programs. Tex. Agr. Exp. Sta. Bull. 1047. 7 p.
78. Adkisson, P. L. and S. J. Nemec. 1966. Comparative effectiveness of certain insecticides for killing bollworms and tobacco budworms. Tex. Agr. Exp. Sta. Bull. 1048. 4p.
79. Adkisson, P. L., R. A. Bell, E. C. Berry, C. L. Cole and S. J. Nemec. 1966. Low volume sprays of Guthion and malathion for reducing diapausing boll weevil populations. Tex. Agr. Exp. Sta. PR 2405. 6 p.
80. Adkisson, Perry L. 1966. Present status of the diapause boll weevil control program in the High Plains of Texas. Proc. 15th Ann. Tex. Aviation Conf., Texas A&M Univ. (College Station, TX 1966). pp. Bl-B7.
81. Adkisson, P. L. and S. J. Nemec. 1966. Laboratory and field evaluation of low volume insecticidal sprays for cotton insect control. Proc. 15th Ann. Tex. Aviation Conf., Texas A&M Univ. (College Station, TX 1966). p. Jl-J6.

82. Adkisson, Perry L. 1966. Internal clocks and insect diapause. Science, 154:234-241.
83. Nemec, S. J. and P. L. Adkisson. 1966. Comparative effectiveness of low volume concentrate and water emulsion sprays of certain insecticides for cotton insect control. Tex. Agr. Exp. Sta. Misc. Publ. 813. 6 p.
84. Wellso, S. G. and P. L. Adkisson. 1966. A long-day short-day effect in the photoperiodic control of the pupal diapause of the bollworm, *Heliothis zea* (Boddie). J. Insect Physiol. 12:1455-1465.
85. Adkisson, Perry L. 1966. Diapause boll weevil control on the High Plains of Texas. Agr. Chem. 21(5):56, 57, 60, 62.
86. Adkjsson, P. L., D. R. Rummel, W. L. Sterling and W. L. Owen, Jr. 1966. Diapause boll weevil control: A comparison of two methods. Tei. Agr. Exp. Sta. Bull. 1054. 11 p.
87. Ankersmit, G. W. and P. L. Adkisson. 1967. Photoperiodjc responses of certain geographical strains of *Pectinophora gossypiella* (Lepidoptera). J. Insect Physiol. 13:553-564.
88. Adkisson, P. L. and S. J. Nemec. 1967. Effectiveness of certain organophosphorous insecticides against chlorinated hydrocarbon resistant bollworm and tobacco budworm larvae. J. Econ. Entomol. 60:268-270.
89. Adkisson, Perry L. 1967. Development of resistance by the tobacco budworm to mixtures of toxaphene or Strobane plus DDT. J. Econ. Entomol. 60(3):788-79l.
90. Adkisson, P. L. and S. J. Nemec. 1967. Insecticides for controlling the bollworm, tobacco budworm and boll weevil. Tex. Agr. Exp. Sta. Misc. PubI. 837. 7 p.
91. Nemec, S. J. and P. L. Adkisson. 1967. Effects of adding molasses to certain insecticidal sprays for bollworm control. Tex. Agr. Exp. Sta. PR 2436. 4 p.
92. Hanna, R. L., J. K. Walker, Jr., P. L. Adkisson and S. J. Nemec 1967. Results of field and laboratory studies on the control of cotton insects, 1966. Tex. Agr. Exp. Sta. Dept. Entomol. Tech. Rpt. No. 6. 55 p.
93. Sterling, W. L., P. L. Adkisson, and D. R. Rummel, 1967. Effects of the diapause boll weevil control program on the size and rate of increase of boll weevil populations in the High Plains of Texas. Tex. Agr. Exp. Sta. Dept. Entomol. Tech. Rpt. No. 7. 11 p.
94. Rummel, D. R., W. L. Sterling and P. L. Adkisson. 1967. Evaluation of the 1966 diapause boll weevil control program on the High Plains of Texas. Tex. Agr. Exp. Sta. Dept. Entomol. Tech. Rpt. No. 8. 11 p.
95. Adkisson, Perry L. 1968. Development of resistance by the tobacco budworm to endrin and carbaryl. J. Econ. Entomol. 61(l):37-40.
96. Adkisson, Perry L. 1968. Problems and progress in controlling diapausing boll weevils. Proc. Beltwide Cotton Prod. and Mech. Conf. (Hot Springs, Ark.). pp. 18-20.

124. Adkisson, Perry L. 1972. Use of cultural practices in insect pest management. *In* Implementing Practical Pest Management Strategies. Proc. Natl. Ext. Pest Management Workshop. Purdue Univ. (Lafayette, IN - March 14-16, 1972). pp. 37-50.

125. Adkisson, P. L. and R. Haney. 1972. Cut cotton insecticide costs with integrated control. Tex. Agr. Exp. Sta. Progress. 18(4):4-7.

126. Adkisson, Perry L. 1973. The principles, strategies and tactics of pest control in cotton. In Insects: Studies in Population Management. Ed., P. W. Gier. Ecol. Soc. Austr. Mem. 1:274-282.

127. Cole, C. L., P. L. Adkisson and R. E. Fye. 1973. Seasonal abundance of *Heliothis larvae* on cotton in Presidio, Texas, area. J. Econ. Entomol. 66(2):524-526.

128. Adkisson, Perry L. 1973. Educational, extension and legislative requirements for integrated control. Proc. FAO Conf. Ecol. in Relation to Plant Pest Control. (Rome, Italy - Dec. 11-14, 1972). PP. 299-304.

129. Nemec, S. J. and P. L. Adkisson. 1973. Organophosphate resistance levels in tobacco budworm and bollworm populations in Texas. *In* Investigations of Chemicals for Control of Cotton Insects in Texas. Tex. Agr. Exp. Sta. Dept. Entomol. Tech. Rpt. 73-2. pp. 18-25.

130. Adkisson, P. L. and R. F. Smith. 1973. Pest management in perspective. Proc. Natl. Ext. Pest Management Workshop. (Baton Rouge, LA - March 12-13, 1973).

131. Sterling, W. L. and P. L. Adkisson. 1974. Seasonal incidence of diapause and reproduction in boll weevils inhabiting the High and Rolling Plains of Texas. Tex. Agr. Exp. Sta., Misc. Publ. 1145. 9 p.

132. Adkisson, Perry L. 1974. Opportunities for professional entomologists in private practice. Bull. Entomol. Soc. Amer. 20:277-278.

133. Adkisson, Perry L. 1974. Southern pine beetle research in the Texas Agricultural Experiment Station. Proc. Southern Pine Beetle Symp., Tex. Agr. Exp. Sta. and A.S.F.S. Southern Forest Exp. Sta. (March 1974). p. 6.

134. Smith, R. F., C. B. Huffaker, P. L. Adkisson and L. D. Newsom. 1974. Progress achieved in the implementation of integrated control projects in the USA and tropical countries. EPPO Bull. 43(3):221-239.

135. Adkisson, Perry L. 1975. Entomology, the important profession. Bull. Entomol. Soc. Amer. 21(7):3-4.

136. Rummel, D. R., D. G. Bottrell, P. L. Adkisson and R. C. McIntyre. 1975. An appraisal of a 10-year effort to prevent the westward spread of the boll weevil. Bull. Entomol. Soc. Amer. 21:6-11.

137. Reynolds, H. T., P. L. Adkisson and R. F. Smith. 1975. Cotton Insect Pest Management. j11 Introduction to Insect Pest Management. W. H. Luckmann and R. L. Metcalf, Eds. John Wiley and Sons, NY. pp. 379-443.

138. Beck, S. D., P. L. Adkisson, D. C. Erwin, L. W. Jones, W. W. Murdoch, R. B. Norgaard, J. S. Palaxico, P. W. Santlemann, C. A. Shoemaker, A. D. Tarlock, and B. W. Waddle. 1975. Pest Control: An Assessment of Present and Alternative Technologies. Vols. I-UI. Natl. Acad. Sci., Washington, D.C.
139. Stern, V. M., P. L. Adkisson, O. G. Beingolea and G. A. Viktorov. 1976. Cultural Controls. Chapter In Theory and Practices of Biological Control. C. F. Huffaker. Ed. Academic Press. NY. pp. 593-613.
140. Adkisson, Perry L. 1976. Integrated pest management. Proc. US/USSR Symposium on the Integrated Control of the Insect, Weed and Disease Pests of Cotton, Sorghum and Deciduous Fruits. Tex. Agr. Exp. Sta. Misc. Publ. 1275. pp. 1-3.
141. Maxwell, F. G. and P. L. Adkisson. 1976. Host plant resistance to cotton insects. Proc. US/USSR Symposium on the Integrated Control of the Insect, Weed and Disease Pests of Cotton, Sorghum and Deciduous Fruits'. Tex. Agr. Exp. Sta. Misc. Publ. 1276. pp. 143-151.
142. Adkisson, P. L. and J. W. Smith, Jr. 1976. Economic aspects of insect pest management of peanuts in the southwestern United States. Proc. Int. Inst. Trop. Agr. Ibadan, Nigeria~ (Nov. 11-13, 1976).
143. Adkisson, P. L. and J. K. Walker, Jr. 1976. Economic benefits of growing short-season cottons in areas infested with bolt weevils and bollworms. Proc. USSR/US Symp. on Crop Loss Appraisal. Yalta, USSR. (July 26-29, 1976).
144. Adkisson, Perry L. 1977. Cultural control as part of an integrated pest management system for cotton. Proc. Arab Republic Egypt/UC/AID Pesticide Management Seminar-Workshop, Univ. of Alexandria, Egypt. (March 5-10, 1977). pp. 30-44.
145. Bottrell, D. L. and P. L. Adkisson. 1977. Cotton insect pest management. Ann. Rev. Entomol. Vol. 22. pp. 451-481.
146. Adkisson, Perry L. 1977. Alternatives to the unilateral use of insecticides for insect pest control in certain field crops. Proc. Symp. Ecol. and Agr. Prod., Univ. Tenn. (Knoxville, TN - July 1973). pp. 129-141.
147. Guyer, G. T., P. L. Adkisson, H. C. Chiang, G. W. Fox, C. B. Huffaker, F. G. Maxwell, R. L. Metcalf, H. T. Reynolds, W. L. Roeloffs, B. Swartz, D. E. Weidhaas and H. L. Beemer, Jr. 1977. Insect Control in The People's Republic of China. Natl. Acad. Sci. CSCPRC Rpt. 2. 218 p.
148. Phillips, J. R., P. L. Adkisson and A. P. Gutierrez. 1977. Pest management systems applicable to cotton insect pests. *In* Principles, Strategies and Tactics of Integrated Pest Control. C. B. Huffaker, Ed. John Wiley and Sons, NY.
149. Meola, R. W. and P. L. Adkisson. 1977. Release of prothoraciacotropic hormone and potentiation of developmental ability in the bollworm, *Heliothis zea* Boddie. J. Insect Physiol. 23:683-688.

150. Eden, W. G., P. L. Adkisson (and 14 others). 1977. Analytical studies for the U.S. Environmental Protection Agency. Vol. VII. Pesticide Decision Making. Natl. Acad. Sci. (Washington, D.C.). 100 p.
151. Smith, R. F., J. H. Perkins, P. L. Adkisson, B. E. Day, J. B. Siddall and H. D. Thurston. 1977. Pest control. In Vol. I. Supporting Papers: World Food and Nutrition Study. Natl. Acad. Sci. (Washington, D.C.). pp. 76-138.
152. Meola, R. W. and P. L. Adkisson. 1978. Endocrine aspects of diapause regulation in *Heliothis zea*. Folin Mexicana 39-40:56-57.
153. Sterling, W. L. and P. L. Adkisson. 1978. Population dynamics of the boll weevil inhabiting the High and Rolling Plains of Texas. Environ. Entomol. 7(3):439-444.
154. Adkisson, Perry L. 1978. Systems approach to integrated pest management. Proc. UC/AID-ICA Conf. on Pest Management and the Environment (Bogota, Colombia - Feb. 19-24, 1978).
155. Adkisson, Perry L. 1978. Selective use of insecticides in integrated pest management. Proc. UC/AID-ICA Conf. on Pest Management and the Environment (Bogota, Colombia - Feb. 19-24, 1978).
156. Adkisson, P. L., R. B. Frisbie and R. D. Lacewell. 1978. The need for cost/benefit analyses in the development of integrated pest management programs. Proc. Generale de Central Integrado de Plagas y Enfermedade con Enfasis en Maiz and Saya. Natl. Agr. Univ. de Peru. (Lima, Peru - April 17-May 26, 1978). Vol. I.
157. Adkisson, P. L. and R. F. Smith. 1978. Measurement of crop losses caused by insects. Proc. Generale de Central Integrado de Plagas y Enfermedade con Enfasis en Maiz and Saya. Natl. Agr. Univ. de Peru (Lima, Peru - April 17-May 26, 1978). Vol. I.
158. Adkisson, Perry L. 1978. Development of the systems approach to integrated pest management. Proc. Generale de Central Jntegrado de Plagas y Enfermedade con Enfasis en Maiz and Saya. Nati. Agr. Univ. de Peru (Lima, Peru - April 17-May 26, 1978). Vol. II.
159. Dean, H. A. and P. L. Adkisson. 1978. Imported parasites destroy rhodesgrass scale. Tex. Agr. Prog. 24(2):25-26.
160. Smith, R. F. and P. L. Adkisson. 1979. Expanded horizons of integrated pest control in crop production. Proc. Opening Session and Plenary Session Symposium IX Intl. Congress Plant Prot, (Washington, D.C. - Aug. 5-11, 1979). pp. 29-30.
161. Adkisson, P. L., R. B. Frisbie, J. K. Walker and W. L. Sterling. 1980. Integrated control of the insect pest of cotton. Proc. Entretiens Ecologiques de Dijon. Dijon, France.
162. Adkisson, Perry L. 1980. An ecological approach to the integrated control of the insect pests of sorghum. Proc. Intl. Workshop on Sorghum

Shoot-fly. Intl. Cent. Ins. Physiol. and Ecol. (Nairobi, Kenya, May 5-8, 1980). pp. 12-16.

163. Adkisson, Perry L. 1980. Organizational constraints affecting research and development in integrated pest management. Proc. Intl. Org. Biol. Control on Conference on Future Trends of Integrated Pest Management. (Bellagio, Italy - May 30-June 4, 1980).

164. Adkisson, Perry L. 1980. Desarrollo e implementacion de programs de control de plagas. Proc. CICP/OAS/AID Curso de Control Integrado de Plagas. (Tapachula, Mexico, Sept. 2-14, 1980). 20 p.

165. Adkisson, Perry L. 1980. The essential role of pest control in human health and food production. Proc. Seminar and Workshop on Pest arid Pesticide Management in the Caribbean. (Bridgetown, Barbados - Nov. 3-7, 1980). 8 p.

166. Adkisson, P. L. and V. A. Dyck. 1980. Resistant varieties in pest management systems. In Breeding Plants Resistant to Insects. F. Maxwell and P. Jennings, Eds. John Wiley and Sons, NY. pp. 233-251.

167. Phillips, J. R., A. P. Gutierrez and P. L. Adkisson. 1980. General accomplishments toward better insect control in cotton. In New Technology of Pest Control. Chap. 5. John Wiley and Sons, NY. pp. 123-153.

168. Adkisson, Perry L. 1981. The systems approach to livestock production. Proc. Symp. Systems Approach to Animal Health and Production. (Univ. of Kentucky, Lexington, KY - March 31-April 2, 1981). pp. 18-30.

169. Adkisson, P. L., R. E. Frisbie, J. U. Thomas and G. M. McWhorter. 1981, Organization and implementation of an integrated pest management system. Southwestern Entomol. 6(4):279-287.

170. Cole, C. L. and P. L. Adkisson. 1981. Life history and fecundity of the boll weevil reared in constant and variable temperature regimens. Southwestern Entomol. 6(4):298-302.

171. Adkisson, Perry L. 1981. The approaching crisis in sustaining high-yielding agricultural systems. Plant Disease. 65:940-942.

172. Adkisson, P. L., G. A. Niles, J. K. Walker, L. S. Bird and H. B. Scott. 1982. Controlling cotton's insect pests: A new system. Science, 2 16:19-22.

173. Adkisson, Perry L. 1982. Science and the individual. Texas Journal of Science. 34(2):101-103.

174. Reynolds, H. T., P. L. Adkisson, R. F. Smith and R. E. Frisbie. 1982. Cotton insect pest management. In Introduction to Insect Pest Management (2nd Edition). R. L. Metcalf and W. H. Luckmann, Eds. John Wiley and Sons, NY. pp. 375-441.

175. Cole, C. L. and P. L. Adkisson. 1982. Effects of constant and variable temperature regimens on the survival and rate of increase of the boll weevil. Southwestern Entomol. 7(1):50-55.

176. Marshall, B., P. L. Adkisson (and 23 others). 1983. Strengthening the government-university partnership in science. Report of the ad hoc Committee on Government-University Relationships in Support of Science. Natl. Acad. Sci., Washington, D.C. 234 p.
177. Adkisson, Perry L. 1983. A perspective on pest management. Proc. Third Intl. Working Conf. on Stored-Product Entomol. (Kansas State Univ., Manhattan, KS - Oct. 23-27, 1983).
178. Cole, C. L. and P. L. Adkisson. 1983. The occurrence of diapause in a High Plains and Central Texas strain of the boll weevil. Southwestern Entomol. 8(4):315-319.
179. Adkisson, Perry L. 1984. Biological control as a component of integrated control. Proc. Joint Symp. on the Biol. Control of Insects. (Beijing Univ., Beijing, China - Sept. 26-28, 1982). Academia Sinica Press, The People's Republic of China. pp. 1-8.
180. Croft, B. A., P. L. Adkisson, R. W. Sutherst, and G. A. Simmons. 1984. Applications of ecology for better pest control. Chapter 25 *In* Ecological Entomology. John Wiley and Sons, NY. pp. 763-795.
181. Adkisson, P. L. and R. E. Frisbie. 1984. Increasing food supplies through integrated pest management with special reference to North Africa. Proc. Symp. On Integrated Pest Management and Rationalization of Pesticide Use in the Arab Countries. Algiers, September 16-20. pp. 24-25.
182. Frisbie, R. E. and P.L. Adkisson. 1985. 1PM: Definitions and current status in U.S. agriculture. In Biological Control in Agricultural Integrated Pest Management Systems. M. A. Hoy and D. C. Herzog, Eds. pp. 41-51.
183. Frisbie, R. E. and P. L. Adkisson. 1985. Integrated pest management is at hand. BioScience. 35(2):69.
184. Frisbie, R. E. and P. L. Adkisson, (Eds). 1986. Integrated pest management on major agricultural systems. Tex. Agric. Expt. Sta. MP-1616. 743 p.
185. Adkisson, P. L., R. B. Frisbie, J. G. Thomas and G. M. McWhorter. 1986. Impact of integrated pest management on several crops of the United States. *In* Integrated Pest Management on Major Agricultural Systems. Tex. Agr. Exp. Sta. Misc. Publ. 1616:663-672.
186. Adkisson, Perry L. 1986. Integrated Pest Management. Bull. Ent. Soc. Amer. 32(3): 136-141.
187. Adkisson, Perry L. 1988. The Value of Integrated Pest Management to Crop Production. Proc. Intl. Conf. on Dryland Farming: Challenges in Dryland Agriculture — A Global Perspective. Amarillo/Bushland, TX August 15-19. pp. 905-906.
188. Adkisson, Perry L. 1989. Foreword in 1PM Systems and cotton production. John Wiley and Sons, New York. pp. vii-viii.

189. Adkisson, Perry L. 1 990. Integrated Pest Management: Past and Future. Proc. Entomol. Soc. of Amer. Cent. Natl. Symp. Annual Mtg. San Antonio, TX December 12-13. pp 3-4.

190. Adkjsson, Perry L. 1990. Warning: Eating May be Harmful to Your Health. William Henry Hatch Memorial Lecture, National Association of State Universities and Land Grant Colleges. Kansas City, Missouri. Nov. 12. 9 pp.

191. Adkisson, Perry L. 1991. Integrated Pest Management: Past and Future. In Entomology Serving Society: Emerging Technologies and Challenges - S. B. Vinson and R. L. Metcalf, Eds. Entomological Society of America, Lanham, MD. pp. 320-322.

A Wheat Germ Medium for Rearing the Pink Bollworm[1]

P. L. Adkisson,[2] Erma S. Vanderzant,[3] D. L. Bull[4] and W. E. Allison[4]

Abstract

Composition of an artificial medium for laboratory cultures of the pink bollworm, *Pectinophora gossypiella* (Saund.), is given. The medium has wheat germ meal as the primary constituent but also contains vitamin-free casein, sucrose, Wesson's salts, choline chloride, agar, sodium alginate, vitamins, and water. Growth and development on the pink bollworm on this diet approximated that reported for the insect in the field. The larval developmental period averaged 16.3 days; prepupae 2.2 days; and pupae 8.8 days. Pupae produced on this medium were slightly smaller than those collected from cotton bolls. Oviposition records indicated that females reared on the wheat germ medium would produce more eggs than field-collected moths. Records for several thousand insects indicated that an average of 81.5% of the larvae originally placed on the diet would grow into adults.

There has been considerable research directed toward the development of chemically defined artificial media suitable for laboratory-rearing of the pink bollworm, *Pectinophora gossypiella* (Saund.). Beckman *et al.* (1953) reported that the pink bollworm could be reared on a medium composed mainly of a modified chick ration having egg albumin as the protein constituent. However, development of the insect was slow and the number of adults produced was small with regard to the number of eggs originally placed on the medium. Vanderzant & Reiser (1956a) developed a rearing medium based on albumin on which apparently normal pink bollworms could be produced. The developmental period and pupal weights approximated those reported for pink bollworms in the field. The results of this important research developed many of the principles on which later work was based. Vanderzant *et al.* (1956) reported a method for mass-rearing of the pink bollworm on sprouted peas under aseptic conditions. However, this technique had certain disadvantages since the preparation of the seed was time consuming and great care had to be exercised to avoid contamination. Research by Beck & Stauffer (1950) which led to a purified casein medium for the European corn borer, *Pyrausta nubilalis* (Hbn.) provided the basis for the development by Vanderzant & Reiser (1956b) of a similar type purified casein medium on which the pink bollworm could be successfully reared. The casein medium not only provided a method for future work pertaining to the dietary requirements of the pink bollworm, but it also proved valuable in the development of a rearing medium for laboratory cultures of the boll weevil, *Anthonomus grandis* Boh. (Vanderzant & Davich 1958). The casein diet also provided the basal medium for a study of the rôle of dietary fatty acids in the development of the pink bollworm (Vanderzant *et al.* 1957). It was concluded from this work that the inclusion of corn oil in the medium would meet the insect's requirement for the essential fatty acid, linoleic acid. In subsequent experiments, the amino acid requirements of the pink bollworm were determined (Vanderzant 1957, 1958).

Table 1.—Composition of two wheat germ diets compared with the casein diet for rearing the pink bollworm.

Constituents	Grams		
	Casein	Wheat Germ 1	Wheat Germ 2
Casein, vitamin free	5.00	3.0	3.5
Cysteine hydrochloride	0.10		
Glycine	0.15		
Wheat germ		3.0	3.0
Sucrose	5.00	5.0	3.5
Salts, Wesson's	1.20	1.0	1.0
Cholesterol	0.05		
Corn Oil	0.25		
α-Tocopherol	0.01		
Choline chloride	0.10	0.1	0.1
Cellulose	4.00		
Agar	3.00	2.0	2.5
Sodium alginate	0.50		0.5
Vitamin mixture[a]	1.0 ml.	1.0 ml.	1.0 ml.
Water	80.0 ml.	80.0 ml.	85.0 ml.

[a] The vitamin mixture used for the casein and wheat germ 1 media did not contain inositol.

Results of the investigations reviewed above led to the development of the wheat germ medium. This latter medium contains fewer ingredients and requires less rigid rearing conditions than the casein medium. Synthesis and preliminary testing of the medium under aseptic conditions was performed by Vanderzant.

Composition of Medium.—The composition of the wheat germ medium is relatively simple when compared with the casein diet developed earlier by Vanderzant & Reiser (1956b). This is well illustrated by the list of constituents cited for each medium in table 1. The major constituent is wheat germ which contains sterols, essential fatty acids, and tocopherols. In addition, it also supplies protein, carbohydrates, vitamins, and minerals.

During preliminary testing, two wheat germ media were studied. Cold-rolled wheat germ was used in all tests except one in which heat-treated wheat germ (commonly

[1] Technical contribution No. 3476. Texas Agricultural Experiment Station in cooperation with the Entomology Research Division, Agricultural Research Service, U. S. Department of Agriculture. Accepted for publication March 28, 1960.

[2] Associate Professor, Department of Entomology, Texas Agricultural Experiment Station.
[3] Biochemist, Entomology Research Division, A.R.S., U.S.D.A.
[4] Research Assistants, Texas Agricultural Experiment Station.

Table 2.—A comparison of pupal weights of pink bollworm reared on wheat germ and casein media with pupa collected from cotton bolls.

	Average Weight in Milligrams				
Source	No. of Females	Pupal Weight	No. of Males	Pupal Weight	Average Weight
Wheat germ 1[a]	337	23.0	191	18.0	20.5
Wheat germ 2[a]	124	22.4	60	18.5	20.5
Wheat germ 2[b]	75	21.7	75	16.1	19.5
Casein[a]	110	21.3	85	16.2	18.8
Field[c]	50	—	—	—	23.9

[a] Insects reared aseptically.
[b] Insects reared nonaseptically on diet containing the inhibitor solution.
[c] Unsexed.

sold as cereal) was tested. Although the latter had a lower lipide content, results were similar to those obtained when the cold-rolled wheat germ was used. The differences in the two wheat germ media are indicated in table 1. With the exception of the pupal weights presented in table 2, all records were taken of insects reared on wheat germ 2 medium.

The vitamin mixture was similar to that developed in earlier work (Vanderzant 1957) except the mixture used for the wheat germ medium included inositol since the latter is needed by the boll weevil which is also reared in this laboratory (Vanderzant 1959). Composition of the vitamin solution is presented in table 3.

During preliminary testing, all the insects were reared under aseptic conditions. In subsequent tests, a solution containing chemicals which inhibited the growth of contaminating microorganisms was added to the media. The inhibitor solution contained sorbic acid, methyl paraben (methyl parahydroxybenzoate) and butyl paraben (butyl parahydroxybenzoate) dissolved in 95% ethyl alcohol (Clark 1958). The inhibitors were added to the diet at a concentration equal to 0.2% by weight of each. The inhibitor solution was usually made up in quantities of 100 ml. or more. The proper concentration of the solution was pipetted in each batch of medium as it was being mixed. The use of the inhibitors plus clean equipment and careful technique allowed large numbers of insects to be reared nonaseptically with little contamination.

Preparation of the Media.—Dry ingredients for the medium were weighed separately on a torsion balance. Agar was dissolved in the required amount of cold water and heated to the boiling point. The remainder of the dry ingredients were placed in a Waring Blendor and the boiling water-agar mixture was then added. Other liquid ingredients were pipetted into the mixture and allowed to blend for several minutes. After blending, the hot mixture was poured into plastic "squeeze" bottles. This type bottle was used for dispensing media to the rearing vials. The spout of the plastic bottle was inserted into clean 2-dram shell vials and the media dispensed into them. The vials were filled to about one-third of their volume with medium.

The techniques outlined by Vanderzant & Reiser (1956a, 1956b) were used for rearing larvae aseptically. The vials containing media were plugged with nonabsorbent cotton and autoclaved. After the vials had cooled, one newly hatched larva was placed in each. The larvae were incubated from eggs which had been collected and treated with a sterilizing solution according to the technique of Vanderzant & Reiser (1956a). The vials were kept in an incubator in which a temperature of approximately 85° F. and a relative humidity of 75% to 85% was maintained. The incubator maintained a light cycle of 12 hours of light followed by 12 hours of dark.

Techniques for rearing larvae on the media containing the inhibitor solution were somewhat different from that outlined above. The vials were either heat sterilized or autoclaved before the medium was added to them. After the hot medium was added, the vials were covered with paper and left at room temperature until all free moisture on the sides of the vials had evaporated. Larvae were obtained from eggs washed only in distilled water. The larvae were placed singly in the vial by means of a soft camel's-hair brush. During this process the brush was dipped into an antiseptic solution containing 3000 p. p. m. "Roccal" (benzalkonium chloride) at frequent intervals. After the larvae were transferred, each vial was plugged with clean nonabsorbent cotton and held under continuous light in an incubator at approximately 85° F. and 80% to 90% relative humidity.

Duration of the Developmental Period.—Daily observations of a number of pink bollworms reared on the wheat germ diet containing inhibitors were made to determine the average duration of each developmental stage. These data are reported in table 4. Each vial was numbered so that records could be maintained for each larva and pupa during the entire developmental period. Final records include only insects that pupated within a 60-day test period. Individuals in vials that became contaminated with microorganisms were discarded and were not included in the records. The insects were reared under the conditions previously described for nonaseptic rearing.

Records indicated that larval development required an average of 16.3 days for completion with a range of 13 to 26 days. The mode indicates that the greatest number of individuals completed larval development in 18 days. Only four insects out of the 75 observed required more than 19 days to complete larval development. These data indicate the rapid and uniform rate of development of pink bollworm larvae that may be expected when the insect is reared on wheat germ medium. The duration of the active feeding period on the wheat germ medium is comparable to that reported on bolls and shorter than that reported on squares (Fenton & Owen 1953).

The prepupal stage required an average of 2.2 days and the pupal stage 8.8 days. These figures are about the same

Table 3.—Composition of the vitmain mixture incorporated in the wheat germ media.

Constituent	Mg. per 100 Grams of Medium
Niacinamide	1.00
Calcium pantothenate	1.00
Thiamine (HCl)	0.25
Riboflavin	0.50
Pyridoxine (HCl)	0.25
Folic acid	0.25
Biotin	0.02
Vitamin B_{12}	0.002
Inositol	20.00

Table 4.—Duration of the immature life history stages of the pink bollworm when reared on a wheat germ diet as compared with those reared on cotton bolls and squares.[a]

Stage	Number of Insects	Diet	Days Average	Days Mode	Days Range
Larval	75	Wheat germ[b]	16.3	18	13–26
Prepupal	75	Wheat germ[b]	2.2	2	1– 5
Pupal	75	Wheat germ[b]	8.8	8	5–11
Larval	958	Squares	10.1		6–18
Prepupal	958	Squares	3.8		1–26
Pupal	958	Squares	8.1		6–24
Larval	878	Bolls	16.5		11–25
Prepupal	878	Bolls	3.3		1–26
Pupal	878	Bolls	8.6		6–24

[a] Life history records from bolls and squares were taken from data reported by Fenton & Owen (1953).
[b] Insects reared nonaseptically on diets containing the inhibitor solution.

as those reported for boll- and square-fed pink bollworms by the above authors.

PUPAL WEIGHTS.—Pupal weights for pink bollworms reared aseptically and nonaseptically on wheat germ media are compared with those obtained from cotton bolls grown in the field. Pupae were weighed in milligrams on a Roller-Smith balance. Weights of the pupae are given in table 2. The average weights of pink bollworm pupae reared aseptically on wheat germ media 1 and 2 indicated little difference between the media. The wheat germ medium produced pupae slightly larger than the casein medium and slightly smaller than recorded for pupae recovered from cotton bolls.

Pupal weights obtained from a large number of insects reared for four consecutive generations indicate no deleterious effects from continuous rearing on wheat germ media (table 5).

OVIPOSITION RECORDS.—One pair of moths from the wheat germ medium was confined without food or water on cotton terminals in small glass enclosed cages. Fresh terminals were placed in the cage every 3 days. The terminals, after removal from the cages, were closely examined under a binocular microscope and the number of eggs counted. Moths produced from larvae collected from field-grown cotton bolls were used for comparison. Oviposition records are reported in table 6. Females produced on the wheat germ media had a higher average rate of oviposition than the moths collected from the field, the average number being 62.4 eggs and 32.2 eggs per female, respectively. Published data indicated that considerably greater numbers of eggs could be expected if the moths were furnished water or a sugar solution (Fenton & Owen 1953, Lukefahr & Griffin 1956).

Table 5.—Average pupal weights of four continuous generations of pink bollworms reared on a wheat germ medium containing inhibitor solution.

	Pupal Weights in Milligrams				
Generation	No. of Males	Pupal Weight	No. of Females	Pupal Weight	Population Average
1	238	17.8	479	19.8	19.1
2	243	20.1	596	21.2	20.9
3	237	20.7	430	21.7	21.3
4	150	20.4	418	21.9	21.5

Table 6.—Oviposition records of female pink bollworms reared on a wheat germ medium compared with moths collected from the field.

Source	No. of Females	Eggs per Female	Range
Wheat germ	48	62.4	9–168
Field	19	32.2	2–91

Table 7.—Yields of mature pink bollworms as a percentage of the total number of larvae originally placed on a wheat germ medium containing inhibitor solution.

Batch Number	No. of Vials Infested	No. of Insects Recovered	Per Cent Yield
1	1,161	877	75.5
2	1,292	1,096	84.8
3	993	845	85.1
4	432	344	79.6
Total	3,878	3,162	81.5

YIELD AND CONTAMINATION.—An important consideration in using any artificial diet is the number of mature insects produced in proportion to the number of eggs or larvae originally placed on it. The data presented in table 7 for insects produced on four different batches of wheat germ media indicate that approximately 75% to 85% of the larvae can be expected to live and become adults on any given batch of medium. The overall average yield of adults for the media was 81.5%. Larval death on the media was attributed to several factors, *viz.*, inherent weakness in certain larvae, time elapsed between incubation and transfer to media, larval entanglement in the cotton plugs, larval injury during transfer from incubation chamber to media, larval escape from rearing container, contamination due to poor technique, and mite infestation in certain vials which produced contamination. From experience with other media, the yield of adults reared on the wheat germ medium is considered excellent.

Contamination of the rearing medium by microorganism will lower the yield and impair its usefulness as a laboratory tool. Contamination was usually traceable to poor technique or mite infestations. Good sanitation practices were sufficient to hold contamination at low levels when the antimicrobial agents were included in the diet.

REARING OF OTHER INSECTS.—Some measure of success in rearing certain other lepidopterous insects on the wheat germ media has been attained. No attempt has been made to modify the diet in order to make it a more efficient rearing medium for these species. Preliminary results indicate that the diet has possibilities for the laboratory production of a number of insects other than the pink bollworm.

REFERENCES CITED

Beck, S. D., and J. F. Stauffer. 1950. An aseptic method for rearing European corn borer larvae. Jour. Econ. Ent. 43: 4–6.

Beckman, H. F., S. M. Bruckart, and R. Reiser. 1953. Laboratory culture of the pink bollworm on chemically defined media. Jour. Econ. Ent. 46(4): 627–30.

Clark, E. W. 1958. Personal communication. Ent. Res. Div., Agric. Res. Serv., U.S.D.A., Brownsville, Texas.

Fenton, F. A., and W. L. Owen, Jr. **1953.** The pink bollworm of cotton in Texas. Texas Agric. Expt. Sta. Misc. Publ. **100.**

Lukefahr, M., and J. A. Griffin. **1956.** The effects of food on the longevity and fecundity of pink bollworm moths. Jour. Econ. Ent. **49**(6): 876–7.

Vanderzant, Erma S. 1957. Growth and reproduction of the pink bollworm on an amino acid medium. Jour. Econ. Ent. **50**(2): 219–21.

Vanderzant, Erma S. 1958. The amino acid requirements of the pink bollworm. Jour. Econ. Ent. **51**(3): 309–11.

Vanderzant, Erma S. 1959. Inositol: An indispensable dietary requirement for the boll weevil. Jour. Econ. Ent. **52**(5): 1018–9.

Vanderzant, Erma S., and T. B. Davich. **1958.** Laboratory rearing of the boll weevil: A satisfactory larval diet and oviposition studies. Jour. Econ. Ent. **51**(3): 288–91.

Vanderzant, Erma S., D. Kerur, and R. Reiser. **1957.** The role of dietary fatty acids in the development of the pink bollworm. Jour. Econ. Ent. **50**(5): 606–8.

Vanderzant, Erma S., and R. Reiser. **1956a.** Aseptic rearing of the pink bollworm on synthetic media. Jour. Econ. Ent. **49**(1): 7–10.

Vanderzant, Erma S., and R. Reiser. **1956b.** Studies of the nutrition of the pink bollworm using purified casein media. Jour. Econ. Ent. **49**(4): 454–8.

Vanderzant, Erma S., R. Reiser, and E. E. Ivy. **1956.** Methods for the mass rearing of the pink bollworm. Jour. Econ. Ent. **49**(4): 559–60.

Reprinted from the
JOURNAL OF ECONOMIC ENTOMOLOGY
Vol. 53, No. 5, October, 1960
pp. 759–762

Reprinted from BIOLOGICAL BULLETIN, Vol. 127, No. 3 511–525, December 1964
Printed in U. S. A.

PHYSIOLOGY OF INSECT DIAPAUSE. XIV. AN ENDOCRINE MECHANISM FOR THE PHOTOPERIODIC CONTROL OF PUPAL DIAPAUSE IN THE OAK SILKWORM, ANTHERAEA PERNYI

CARROLL M. WILLIAMS [1] AND PERRY L. ADKISSON [2]

The Biological Laboratories, Harvard University, Cambridge 38, Massachusetts

During the final ten days of larval life, the Pernyi silkworm envelops itself in a stout-walled cocoon within which it pupates. Development may stop right there as the pupa begins a prolonged period of pupal diapause which persists until the following spring. Alternatively, the newly formed pupa may develop into an adult moth without any delay. The moth is then committed to the reproduction of a further generation of pupae which can begin to diapause before the first killing frost. If winter arrives before the larvae can pupate, the insect will experience what is little short of "ecological suicide."

Like so many plants and animals, the Pernyi silkworm minimizes these ecological dangers by monitoring seasonal signals of utmost precision, namely, the lengths of the night and day. For *A. pernyi* the phenomenon is well documented in the detailed studies of the Japanese investigator, Tanaka (1950a, 1950b, 1950c; 1951a, 1951b; see Lees, 1955, for English summary). Thus, when Pernyi larvae are reared under day-lengths longer than 14 hours, they develop without any pupal diapause; at temperate latitudes, photoperiods of this sort are peculiar to late spring and early summer when the season is propitious for a second brood. By contrast, larvae reared under day-lengths shorter than 14 hours (as in late summer and autumn) transform into diapausing pupae.

Little is known about the physiological basis of the photoperiod response. In principle, the minimal mechanism must include, not only a photoreceptor, but also a clockwork-computer which counts the hours of darkness and daylight. Until recently, all that was known was Tanaka's finding that the larval ocelli are not involved in the reception of photoperiod.

But the induction of diapause is only half the story. Of equal significance is the termination of diapause—its timing and synchronization with the seasons.

In a related species, the Cecropia silkworm, the termination of pupal diapause is known to be controlled primarily by environmental temperature. By an apparently direct action on the brain itself, environmental temperature conditions the secretion of a hormone prerequisite for the termination of diapause and the initiation of adult development (Williams, 1956).

This same temperature-sensitive system has generally been presumed for other diapausing pupae including *A. pernyi*. Indeed, as pointed out in de Wilde's

[1] This study was supported, in part, by a grant from the National Science Foundation.

[2] Special Postdoctoral Fellow of the U. S. Public Health Service during the term of the study. Permanent address: Department of Entomology, Texas A & M University, College Station, Texas.

(1962) recent review, insect pupae (unlike all other stages in the life history) are thought to be insensitive to photoperiod.

Some reservations on this point are provoked by a more detailed examination of *A. pernyi.* Thus, when the cocoon is cut open, the pupa is always facing upward in the chamber. Moreover, on inspecting the pupa, itself, one cannot fail to be impressed by the unpigmented, transparent cuticle overlying the brain (Fig. 1). Even if the rest of the pupa is jet-black, the facial cuticle is always pigment-free. By moistening it with alcohol, one can look inside and see the underlying brain. A similar "facial window" is found in all pupae of the genus *Antheraea,* including the American species, *A. polyphemus.*

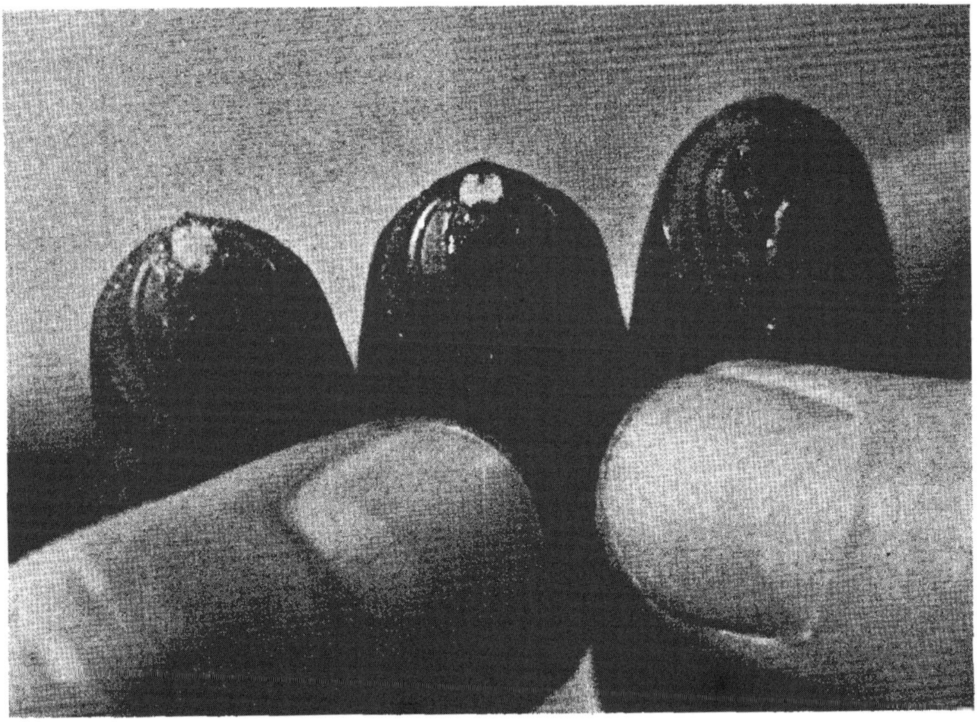

FIGURE 1. Two pupae of *A. pernyi* are here photographed alongside a Cecropia pupa. Pernyi routinely shows a zone of transparent cuticle overlying the pupal brain; the Cecropia pupa does not.

Is it possible that the window has something to do with the transmission of light? This prospect seems to have been reported only by the Russian investigator, Shakhbazov (1961).

Impressed by the facial window, one of us tested, some 16 years ago, the influence of illumination on diapausing pupae of *A. polyphemus.* Groups of pupae were removed from cocoons, subdivided into two lots, and placed at 25° C. in continuous light and darkness, respectively. Illumination had no detectable effects on the rate of termination of diapause.

Subsequently we have learned that continuous light or darkness are both aperiodic signals which, in this sense, are inappropriate tests for photoperiodism. The experiment has now been repeated in a proper manner by making use of specific photoperiods. Certain of the results have been summarized previously (Williams, 1963; Williams and Adkisson, 1964).

Materials and Methods

1. *Experimental animals*

All experiments were performed on pupae of *A. pernyi.* The cocoons were the diapausing first-brood harvested in late July; on August 27 they were shipped from Japan in a series of opaque cardboard boxes. They arrived at Harvard University on September 30.

One hundred pupae were removed from cocoons and inspected under the dissecting microscope to confirm the persistence of diapause. The rest of the cocoons were then subdivided into two lots. One group was spread on tables in a 25 ± 0.5° C. room programmed for a daily illumination of 8 hours; the other group was stored in opaque boxes at 2–3° C.

A few experiments were performed on a subsequent shipment consisting of diapausing cocoons of the second generation harvested in October. This material was stored at 2–3° C.

2. *Photoperiod studies*

Chilled and unchilled cocoons were exposed for 16 weeks to precise regimens of photoperiod. Some pupae were removed from their cocoons at the outset of the experiment; the vast majority were not.

After the insects had been placed at the selected photoperiods, they were inspected at least twice weekly, the number of emerged moths being recorded. Pupae not enclosed in cocoons were examined under the dissecting microscope to ascertain whether adult development had begun. Needless to say, all observations (with the exception of the individuals maintained under continuous darkness) were performed during the photophase of the daily cycle. Chambers maintained in continuous darkness were opened in a dark room and inspected under red light.

3. *Adult development*

The initiation of adult development is signaled by the detachment and retraction of the epidermis from the pupal cuticle. The "zero day of development" was recognized by the initiation of retraction of the epidermis on the ventral side of the tip of the abdomen where the cuticle overlying the genital disc is usually palely pigmented. The overlying cuticle was moistened with 70% ethanol and viewed under a dissecting microscope. In order to eliminate surface reflections, observations of this type are facilitated if a Polaroid filter is placed in front of the microscope lamp and a second "crossed" filter is positioned under the objective lens (Harvey and Williams, 1958).

In pupae of pale pigmentation, the initiation of epidermal retraction was also visible in the pupal wings at the onset of adult development. But in the jet-black pupae which are sometimes encountered, observation of wing retraction was possible only when a zone of the superficial pigment layer was scraped away with a scalpel. A time-table for the adult development of Pernyi will be described in the section on Results.

4. *Experimental chambers*

Two types of chambers were employed. The first consisted of six B.O.D. incubators modified so that each of the three shelves was illuminated from directly overhead by a 15-watt fluorescent lamp (Sylvania cool white, F15T8). The average intensity of illumination was approximately 175 foot-candles (1883 lux).

The second type of chamber was an adaptation of that described by Dutky *et al.* (1962). Five-gallon tin-cans with tight-fitting lids were employed. A 4-watt fluorescent lamp (General Electric cool white F4T5) was installed in the lid of each can to yield an average internal illumination of 110 foot-candles (1184 lux). An electrically driven exhaust fan was attached to the lid and an air-intake placed near the bottom; the intake and exit ports were fitted with coiled lengths of radiator hose to prevent any leakage of light. The assembled chambers were placed in a constant-temperature room at 25 ± 0.5° C. and a relative humidity of 60%. It may be noted that these simple chambers were in every way as satisfactory as the expensive B.O.D. incubators.

The daily cycle of illumination was programmed for each chamber by a Model No. 8001 "Tork" time-clock. In the case of the B.O.D. incubators, the timer's double-throw switch was used to turn on a 50-watt heater (positioned in the rear of the lower shelf) whenever the fluorescent lamps were turned off. This arrangement minimized the temperature fluctuations occasioned by the operation of the lamps.

In all cases (except, of course, the chambers maintained in continuous light or dark) the chambers were programmed for 24-hour daily cycles of light and darkness. For convenience, we shall identify each photoperiod in terms of the duration of the daily light-cycle or "photophase." By so doing, we automatically define the duration of dark-cycle or "scotophase." This terminology seems most straight-forward despite the fact that the length of the night is probably more crucial than the length of the day (Adkisson, 1964).

5. *Photoperiod gradients*

In order to expose opposite ends of individual pupae to different photoperiods, two simple mechanisms were utilized. The first of these consisted of a block of wood in which a series of circular holes, 15 mm. and 17 mm. in diameter, was drilled. A second board was screwed to the bottom of the first to seal the lower ends of the holes. All surfaces were painted with a non-reflecting black paint.

Pupae were selected of appropriate diameters to slip snugly into the holes by friction-fit. Half faced upwards in the holes; half downwards. The entire assembly was placed in a 25° C. incubator programmed for an 8-hour photophase. In this manner one end of each individual was exposed to the 8-hour photophase while the opposite end was maintained in continuous darkness.

In later experiments the set-up was modified to permit one to establish specific photophases on both upper and lower surfaces. For this purpose six rows of

rected to the similar effects of prolonged exposure to continuous light or darkness. Finally, we can state the surprising finding that the response to photoperiod was the same for naked pupae and for those which remained in cocoons.

B. *Chilled pupae*

The preceding experiment was repeated in greater detail, using previously chilled pupae. These individuals were stored at 2–3° C. from the first week of October; for this series of experiments they were removed from storage between November 30 and January 7 and used immediately. Groups of 50 cocoons were placed at ten different photoperiods at 25° C.

The results, recorded in Figure 3, were essentially the same as observed for unchilled pupae, the only major difference being an accelerated response to the regimens which terminated diapause. With these additional data, we now see that the most effective stimulus for the termination of diapause is provided by a 17-hour photophase. And, here again, the 12-hour photophase was most effective in preventing the termination of diapause.

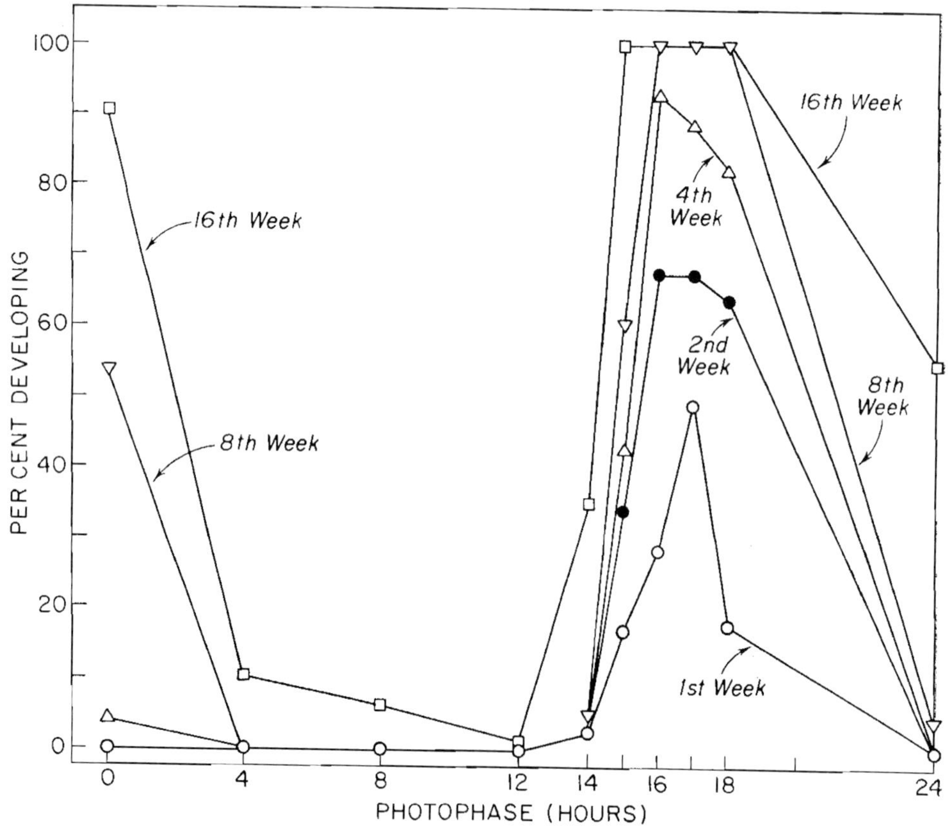

FIGURE 3. The effects of photoperiod on the termination of diapause by previously chilled pupae of *A. pernyi* at 25° C. Diapause is most persistent in the presence of a daily photophase of 12 hours; it is most promptly terminated by a photophase of 17 hours.

3. *The "fine structure" of the photoperiod response*

As is amply demonstrated in Figures 2 and 3, an abrupt transition between "short-" and "long-day" conditions occurs at or near a photophase of 14 hours. This finding was examined in further detail by exposing two groups of 50 unchilled pupae to daily photophases of 13.50 and 13.75 hours, respectively. At the end of eight weeks, 2% of the former group and 22% of the latter group had initiated adult development. This difference signals a discrimination between photophases differing by only 15 minutes.

4. *Effects of preliminary exposure to an 8-hour photophase*

In section 2B of the Results, we observed that the response to photoperiod was accelerated when pupae were first aged at low temperature. A series of experi-

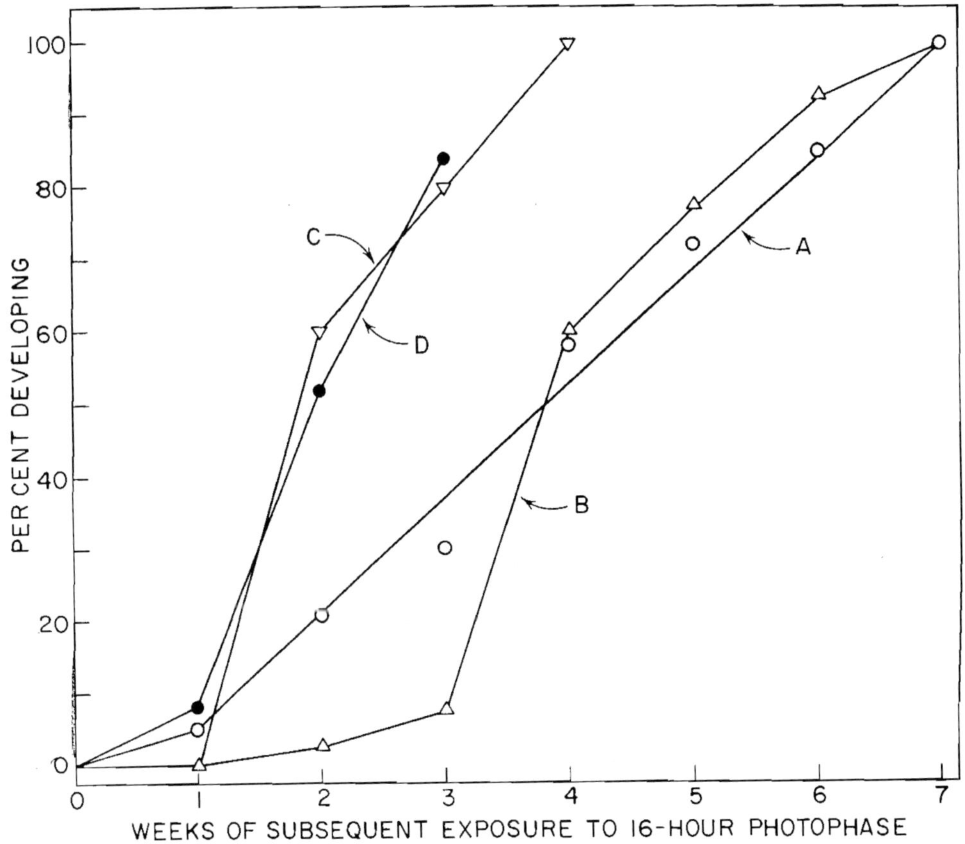

FIGURE 4. Curve A describes the termination of diapause as a function of time; 40 cocoons were incubated at 25° C. in the presence of a long day of 16 hours. In parallel experiments, recorded as Curves B, C, and D, groups of 40 cocoons were first given preliminary exposure at 25° C. to a short-day regimen for 4, 17, and 22 weeks, respectively. The curves describe the termination of diapause after return to the long-day regimen. As noted in Curve B, the preliminary 4-week exposure to the inhibitory photoperiod slowed down the subsequent response. This effect was replaced by a stimulation when the preliminary aging was prolonged to 17 weeks (Curve C) or 22 weeks (Curve D).

ments was designed to test whether the same accelerated response could be induced by aging the pupae at 25° C. in the presence of an inhibitory photophase of 8 hours. To this end, groups of 40 cocoons were first exposed to the 8-hour photophase for 0, 4, 17, and 22 weeks, respectively, and then placed at a 16-hour photophase to induce development.

As summarized (Curve B of Figure 4), the preliminary 4-week exposure to the inhibitory photoperiod slowed down the subsequent response. But when the preliminary exposure was extended to 17 (Curve C) and 22 weeks (Curve D), the inhibition was overcome. When compared to the controls (Curve A), these pupae now showed an accelerated response to the 16-hour photophase reminiscent of that seen after preliminary aging at low temperatures.

5. *Photoperiod gradients*

As described in the section on Methods, 80 unchilled pupae were exposed at 25° C. to an inhibitory 8-hour photophase at one end and a stimulatory photophase of 16 hours at the opposite end. When the experiment was terminated after seven weeks, the results were as follows (see Table II):

TABLE II

Effect of photoperiod gradients on diapausing pupae at 25°C.

Daily photophase (hrs.)		Number of animals	Cumulative % developing after (weeks)						
Head	Abdomen		1	2	3	4	5	6	7
16	8	40	0	2.5	7.5	60.0	77.5	92.5	100
8	16	40	0	0	0	0	0	0	0

All of the animals initiated development when the head-end was exposed to a daily photophase of 16 hours. None initiated development when the head-end was exposed to the inhibitory photophase of 8 hours.

This shows that the reception of the long-day stimulus occurs at the head-end of the pupa; it also shows that exposure of the abdomen to short-day conditions is ineffective in canceling-out a long-day exposure at the head-end.

6. *Transplantation of photosensitivity*

By previously described techniques (Williams, 1946, 1959), the brains were removed from 26 chilled pupae. Each brain was then reimplanted under a plastic window at the tip of the abdomen. The pupae, now with brains in their hind-ends, were positioned in the early version of the "gradient board" and the latter was placed in a 25° C. incubator programmed for an 8-hour photophase. In this manner each individual was exposed at one end to an inhibitory 8-hour photophase while the opposite end was maintained in continuous darkness. During the two-month terms of the experiment, the results were as follows:

Of the 14 individuals whose brainless anterior ends were exposed to the inhibitory 8-hour photophase, 71% initiated adult development. Of the 12 individuals whose brain-containing abdomens were exposed to the inhibitory photophase, none initiated development.

So, by the transplantation of the brain to the tip of the abdomen, the sensitivity to photoperiod was likewise transplanted.

7. *The role of the brain in the photoperiodic response*

A. *Brain removal prior to the initiation of adult development*

Brains were removed from 28 diapausing pupae and the brainless individuals were then stored at 25° C. in the presence of a 17-hour photophase. Despite exposure to this most favorable photoperiod, none of the brainless pupae underwent any development (Table III). Most individuals survived for at least six months and finally died without any trace of adult development.

TABLE III

Effects of brain removal before and after the visible initiation of adult development

Day of adult development	Number of animals	Number of moths formed	% forming moths
Prior to zero	28	0	0
0	15	3	20
1	23	6	26
2	20	17	85

This experiment demonstrates that the brain is indispensable for the initiation of adult development, and that even the most favorable photoperiod becomes completely ineffective in the absence of the brain.

B. *Brain removal after the initiation of development*

A similar group of 58 diapausing pupae was removed from cocoons and exposed to a 17-hour photophase in order to provoke the initiation of adult development. By twice-daily examinations, the zero day of adult development was identified for each individual. Brains were removed on either the zero, first, or second day of adult development and all pupae were then returned to the 17-hour photophase at 25° C. The results are summarized in Table III.

Brain removal on the zero day of adult development completely arrested the further development of 80% of individuals. The same operation performed 24 hours later blocked the further development of 74%. An additional delay of 24 hours (until the "second day of adult development") blocked the development of only 15%.

This experiment shows that in most individuals the brain completes its endocrine function about 60 hours after the visible initiation of adult development.

C. *Effects of photoperiod after the initiation of adult development*

Diapausing second-brood material was used in this experiment. One hundred previously chilled pupae were removed from cocoons, placed at 25° C., and exposed to a 16-hour photophase. On the zero day of adult development, half of the group was returned to the stimulatory 16-hour photophase; the other half was transferred to an inhibitory 8-hour photophase. Both groups emerged as adult moths after an average of 17 days.

Manifestly, photoperiod loses all its influence on the pace of adult development after the latter has actually begun. The formation of the moth then proceeds at a rate dictated by environmental temperature and without any further reference to photoperiod.

Discussion

1. *The induction and termination of diapause in A. pernyi*

As mentioned in the introduction, the detailed studies of Tanaka (1950a, 1950b, 1950c; 1951a, 1951b) have already demonstrated that the induction of diapause in the Pernyi silkworm is controlled by photoperiod. Tanaka's data on the induction of diapause by photophases within the physiological range of 8 to 18 hours are summarized as the hatched line in Figure 5. For comparison, we record as the unbroken line our data for the photoperiodic control of the termination of pupal diapause in previously chilled Pernyi. It is clear that those photoperiods which are effective in inducing diapause are also effective in stabilizing diapause once the latter has begun. Moreover, the photoperiods which are effective in preventing the onset of diapause are precisely the same as those which provoke the termination of diapause.

The fit of the two sets of data is remarkable if one considers that the investigations were performed independently fifteen years apart. In this connection we may note the strategic position of the photophase of 14 hours as the transition between "short-day" and "long-day" conditions.

The obvious inference is that the same photoperiodic mechanism which controls the induction of diapause is retained by the pupa to control the termination of diapause.

2. *The role of the brain in the induction of diapause*

In the Cecropia silkworm the brain is known to play a key role in the induction of pupal diapause. The dormant condition is, in fact, a syndrome of endocrine deficiency due to a failure of the brain to secrete a hormone prerequisite for the initiation of adult development (Williams, 1946, 1952, 1956). This failure is attributable, in turn, to an inactivation of the brain during the prepupal period (Williams, 1952). By contrast, in non-diapausing pupae of *Actias selene, Actias luna,* and *Antheraea pernyi,* the pupal brain is not turned off but retains its full activity. Consequently, within a few days after pupation, sufficient brain hormone is secreted to cause the initiation of adult development. If the brain is excised before this activation is complete, then the potentially non-diapausing pupae are forced to diapause (Williams, 1952; Shappirio and Williams, 1957).

Manifestly, the decision to diapause or not to diapause is dictated by what happens to the brain's endocrine activity during pupation. And in the case of *A. pernyi,* as we have seen, what happens to the brain's activity is conditioned by the photoperiods experienced during larval life.

It is important to note that a photoperiod which induces diapause does not immediately shut-off the larval brain. If it did so, pupation would be blocked and

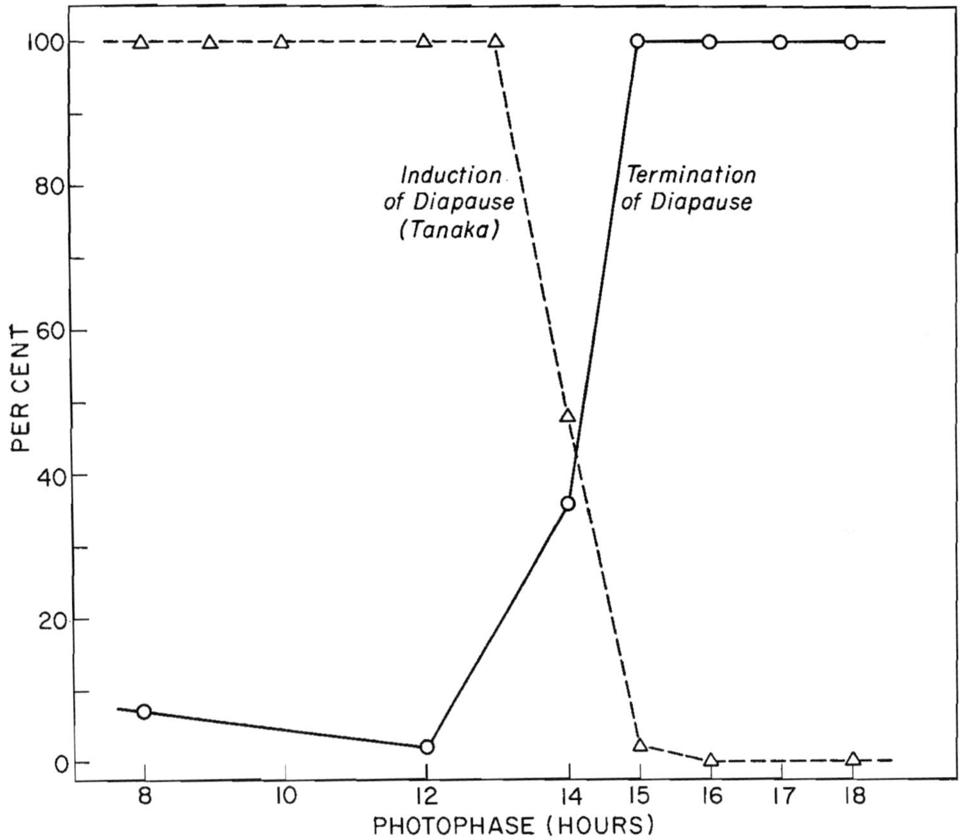

FIGURE 5. The solid line records the effects of photoperiod on the termination of pupal diapause by previously chilled *A. pernyi*. The hatched line shows the influence of photoperiod on the induction of diapause (data of Tanaka). The same short-day conditions which induce pupal diapause are also effective in stabilizing diapause. Moreover, the long-day conditions which prevent pupal diapause are the same as those which cause it to terminate.

one would observe a larval rather than a pupal diapause. The action of photoperiod is to program either the shut-down or the sustained activity of the brain after the latter has secreted sufficient hormone to cause pupation. This state-of-affairs points to some unknown mechanism for the integration and latent storage of daily photoperiod signals accumulated during larval life.

3. *The role of the brain in the termination of diapause by photoperiod*

The results of the present study show that photoperiod is ineffective when it acts on brainless pupae. Even a 17-hour photophase is then incapable of causing the termination of diapause. So, by excising the brain, one effectively removes the vehicle for the photoperiod response.

This conclusion is further affirmed by the experiments in which the brain was removed after the initiation of adult development. During the first 60 hours of adult development, the brain hormone completes its tropic action on the prothoracic glands and the brain, itself, is no longer necessary for the continuation of adult development. By this time, photoperiod has become inconsequential and incapable of influencing the further course of events.

We are therefore persuaded that the photoperiod presides over both the onset and the termination of pupal diapause by controlling the endocrine function of the brain.

4. *Direct or indirect action of photoperiod on the brain?*

In a brief report on the photoperiod response of *A. pernyi,* Shakhbazov (1961) called attention to the transparent facial cuticle which overlies the pupal brain. When this zone was coated with black paint, the pupae behaved as if they were in continuous darkness. Shakhbazov concluded that light is transmitted through both the cocoon and the facial "window" to act on some organ in the pupal head.

In the present study this conclusion has been further documented by exposing individual pupae to photoperiod gradients. Long-day conditions promoted the termination of diapause only when they acted on the head-end. In like manner, short-day conditions were effective in maintaining diapause only when they acted on the head-end. By contrast, exposure of the abdomen to either long- or short-day conditions was without any detectable effects.

The cephalic action of photoperiod has been previously described by Lees (1960) in studies of the aphid, *Megoura.* These remarkable experiments have now been extended and reported in full detail (1964). By pin-pointing tiny beams of light through hollow needles or plastic filaments, Lees was able to show that the photosensitivity of the aphid is confined to the head and that the central region of the head is particularly important as a light pathway.

The present study is in full agreement with Lees' findings and provides the first direct evidence that photoperiod acts on the brain itself. Thus, when the brain was excised from the head and reimplanted into the tip of the abdomen, the entire mechanism of the photoperiod response was thereby transplanted to the hind-end. So, in the case of *A. pernyi* the evidence is little short of conclusive that photoperiod acts directly on the brain.

This conclusion is contrary to that which Beck and Alexander (1964a, 1964b) have recently proposed on the basis of their studies of the termination of larval diapause in the European corn-borer, *Ostrinia nubilalis.* In this species, photoperiod is reported to act on certain cells in the mucosa of the anterior region of the hindgut. These cells are said to secrete a brain-stimulating hormone ("proctodone") which under long-day conditions is in phase with a circadian rhythm of brain reactivity; under short-day conditions, proctodone is ineffective because it is secreted out-of-phase with the endogenous brain rhythm. The new hormone is alleged to play a key role in embryonic and postembryonic development, as well as "in non-diapause growth, polymorphism, periodism, and the several forms of diapause" (Beck and Alexander, 1964b).

It is not our present purpose to discuss the new theory in detail. Proctodone has not entered into our calculations for in *A. pernyi,* as we have seen, it is the head-end which is sensitive to photoperiod and, within the head-end, the brain itself. In *A. pernyi,* we have found no trace of the mechanism described by Beck and Alexander.

5. *The endocrine mechanism*

The present study provides the first experimental proof that the photoperiod acts directly on the brain, itself, to control and modulate the secretion of brain hormone. As mentioned above, this conclusion has long been implicit in the pioneering studies of Lees (1955, 1960, 1964) on the photoperiodic responses of aphids.

We shall postpone to a later occasion detailed consideration of how photoperiod acts on the brain to control the secretion of brain hormone. For present purposes, suffice it to say that the minimal brain mechanism presumably includes the following: (1) a pigment for the absorption of appropriate wave-lengths of light; (2) a timing mechanism which counts the hours of darkness (and perhaps also the hours of light); (3) an output from the clockwork-computer to the neurosecretory cells of the brain; and (4) some sort of physiological "needle-valve" for regulating the secretion and translocation of brain hormone.

Summary

1. In the oak silkworm, *Antheraea pernyi,* photoperiod controls not only the onset of pupal diapause (as previously demonstrated by Tanaka) but also the termination of pupal diapause.

2. At 25° C., short-day conditions (4- to 12-hour photophases) strongly inhibit the termination of pupal diapause; maximal inhibition is by a 12-hour photophase.

3. Long-day conditions (15- to 18-hour photophases) promote the termination of diapause; a 17-hour photophase is the most effective.

4. A 14-hour photophase is transitional between short-days which sustain diapause and long-days which terminate diapause.

5. By various experimental maneuvers, sensitivity to photoperiod was localized in the head-end of the pupa. Short-day illumination of the head-end inhibited the termination of diapause even when the hind-end was exposed to long-day conditions. In like manner, long-day illumination of the head-end was fully effective even when the abdomen received short-day illumination.

6. When the brain was removed from the head and implanted into the tip of the abdomen, the sensitivity to photoperiod was thereby shifted to the hind-end.

7. Additional experiments indicated that the brain is the vehicle for the reception and implementation of photoperiod signals. Brainless pupae are insensitive to photoperiod, while normal pupae are sensitive to photoperiod during the period when development is dependent on the secretion of brain hormone. When this period terminates about 60 hours after the initiation of adult development, photoperiod has lost all its effectiveness.

8. It is concluded that photoperiod acts directly on the brain, itself, to modulate the secretion of brain hormone and thereby to control the termination of pupal diapause.

Note Added to Proof

Since this manuscript was submitted, additional studies have demonstrated that diapausing pupae of *Antheraea polyphemus* respond to photoperiod in essentially the same manner as here described for *A. pernyi*. The same appears to be true for diapausing (unchilled) pupae of *Hyalophora cecropia*.

LITERATURE CITED

ADKISSON, P. L., 1964. Action of the photoperiod in controlling insect diapause. *Amer. Nat.* (in press).

BECK, S. D., AND N. ALEXANDER, 1964a. Hormonal activation of the insect brain. *Science,* **143:** 478–479.

BECK, S. D., AND N. ALEXANDER, 1964b. Proctodone, an insect developmental hormone. *Biol. Bull.,* **126:** 185–198.

DUTKY, S. R., M. S. SCHECHTER AND W. R. SULLIVAN, 1962. A lard-can device for experiments in photoperiodism. *J. Econ. Entomol.,* **55:** 575.

HARVEY, W. R., AND C. M. WILLIAMS, 1958. Physiology of insect diapause. XII. The mechanism of carbon monoxide-sensitivity and -insensitivity during the pupal diapause of the Cecropia silkworm. *Biol. Bull.,* **114:** 36–53.

LEES, A. D., 1955. The Physiology of Diapause in Arthropods. Cambridge University Press.

LEES, A. D., 1960. Some aspects of animal photoperiodism. *Cold Spr. Harb. Symp. Quant. Biol.,* **25:** 261–268.

LEES, A. D., 1964. The location of the photoperiodic receptors in the aphid *Megoura viciae* Buckton. *J. Exp. Biol.,* **41:** 119–133.

SHAKHBAZOV, V. G., 1961. The reaction of the length of daylight and the light receptor of the pupa of the Chinese oak silkworm *Antheraea pernyi G. Dok. Akad. Nauk* SSSR, **140,** No. 1: (AIBS) 944–946.

SHAPPIRIO, D. G., AND C. M. WILLIAMS, 1957. The cytochrome system of the Cecropia silkworm. II. Spectrophotometric studies of oxidative enzyme systems in the wing epithelium. *Proc. Royal Soc. London, Ser. B,* **147:** 233–246.

TANAKA, Y., 1950a. Studies on hibernation with special reference to photoperiodicity and breeding of the Chinese Tussar-silkworm. I. *J. Seric. Sci. Japan,* **19:** 358–371. (*In Japanese*).

TANAKA, Y., 1950b. Studies on hibernation with special reference to photoperiodicity and breeding of the Chinese Tussar-silkworm. II. *J. Seric. Sci. Japan,* **19:** 429–446. (*In Japanese*).

TANAKA, Y., 1950c. Studies on hibernation with special reference to photoperiodicity and breeding of the Chinese Tussar-silkworm. III. *J. Seric. Sci. Japan,* **19:** 580–590. (*In Japanese*).

TANAKA, Y., 1951a. Studies on hibernation with special reference to photoperiodicity and breeding of the Chinese Tussar-silkworm. V. *J. Seric. Sci. Japan,* **20:** 132–138. (*In Japanese*).

TANAKA, Y., 1951b. Studies on hibernation with special reference to photoperiodicity and breeding of the Chinese Tussar-silkworm. VI. *J. Seric. Sci. Japan,* **20:** 191–201. (*In Japanese*).

DE WILDE, J., 1962. Photoperiodism in insects and mites. *Ann. Rev. Entomol.,* **7:** 1–26.

WILLIAMS, C. M., 1946. Physiology of insect diapause: the role of the brain in the production and termination of pupal dormancy in the giant silkworm *Platysamia Cecropia. Biol. Bull.,* **90:** 234–243.

WILLIAMS, C. M., 1952. Physiology of insect diapause. IV. The brain and prothoracic glands as an endocrine system in the Cecropia silkworm. *Biol. Bull.,* **103:** 120–138.

WILLIAMS, C. M., 1956. Physiology of insect diapause. X. An endocrine mechanism for the influence of temperature on the diapausing pupa of the Cecropia silkworm. *Biol. Bull.,* **110:** 201–218.

WILLIAMS, C. M., 1959. The juvenile hormone. I. Endocrine activity of the corpora allata of the adult Cecropia silkworm. *Biol. Bull.,* **116:** 323–338.

WILLIAMS, C. M., 1963. Control of pupal diapause by the direct action of light on the insect brain. *Science,* **140:** 386.

WILLIAMS, C. M., AND P. L. ADKISSON, 1964. Photoperiodic control of pupal diapause in the silkworm, *Antheraea pernyi. Science,* **144:** 569.

Reprint Series
2 April 1982, Volume 216, pp. 19–22

SCIENCE

Controlling Cotton's Insect Pests: A New System

Perry L. Adkisson, George A. Niles, J. Knox Walker, Luther S. Bird, and Helen B. Scott

Source: From *Science* Vol. 216, No. 4541, pp. 19-22 (1982). Reprinted with permission from AAAS.

Controlling Cotton's Insect Pests: A New System

Perry L. Adkisson, George A. Niles, J. Knox Walker
Luther S. Bird, Helen B. Scott

Almost 50 percent of all insecticides applied to crops in the United States are applied to cotton. As a result, most major pest insects of cotton have developed resistance to one or more of these insecticides. Some pests, such as the tobacco budworm [*Heliothis virescens* (Fabricius)] and spider mites (*Tetranychus* species) are now resistant to most of the insecticides registered for use on cotton in the United States. This is rapidly depleting the arsenal of effective insecticides for use on cotton.

An Insecticide-Induced Disaster

The decline in cotton production in northeastern Mexico and southern Texas in the late 1960's and early 1970's is an excellent example of what can happen when insect pests become so resistant to insecticides that control fails. In northeastern Mexico the area of land planted in cotton declined from more than 700,000 acres during the 1960's to less than 1000 acres in 1970 (*1*). Little cotton is grown in the region today. In the Texas Gulf Coast and lower Rio Grande Valley the area planted in cotton declined from 166,000 and 320,000 acres, respectively, in 1968 to 55,000 and 103,000 acres in 1975 (*2*). These reductions occurred because the tobacco budworm developed resistance to all registered insecticides. Although growers treated fields many times, the budworm inflicted such damage that cotton was not profitable to grow.

To understand how this situation developed, it is necessary to review the evolution of insecticide use on the crop in southern Texas, beginning in the 1930's. During this period the boll weevil (*Anthonomus grandis* Boheman), the pink bollworm [*Pectinophora gossypiella* (Saunders)], and the cotton fleahopper [*Pseudatomoscelis seriatus* (Reuters)] were the key pests of the crop. (A key pest occurs annually in a crop and must be controlled to achieve a profitable yield.) The boll weevil was controlled by calcium arsenate dust and the fleahopper by sulfur dust. Although these insecticides permitted profitable production of the crop, substantial yield losses occurred. Early planting of and early destruction of crop residues controlled the pink bollworm.

The advantage of calcium arsenate and sulfur was that they did not kill a great percentage of the insect enemies (parasites and predators) of cotton pests. As a result, outbreaks of two major secondary pests, the tobacco budworm and the bollworm [*Heliothis zea* (Boddie)], occurred only sporadically. (A secondary pest is one that attains crop-damaging numbers only when its natural enemies are decimated.)

Shortly after World War II the new synthetic chlorinated hydrocarbon insecticides, such as toxaphene, DDT, benzene hexachloride, endrin, and dieldrin became available for use on cotton. These had a spectacular effect on cotton production, since they provided almost complete control of the pest insects at an economical cost. Now cotton could be protected throughout the growing season, and 10 to 20 insecticide applications per growing season were common. Varieties were developed whose fruiting periods were more indeterminate, and irrigation and fertilizer were increased so that fruiting could be maintained longer. These practices resulted in major increases in yield.

On the surface, pest insect control in

Summary. Cotton is more heavily treated with insecticides than any other crop in the United States. In southern Texas, this heavy treatment resulted in insecticide-resistant strains of major pests which almost destroyed the industry in the late 1960's and early 1970's. An integrated insect control program based on new short-season cotton varieties and traditional cultural practices has restored production in the area. The new system has been widely implemented because it produces greater net returns by reducing the use of insecticides, fertilizer, and irrigation.

The authors are all at Texas A&M University, College Station 77843. P. L. Adkisson is deputy chancellor for agriculture, G. A. Niles is professor of agronomy, J. K. Walker is professor of entomology, L. S. Bird is professor of plant pathology, and H. B. Scott is associate editor in agricultural communications.

0036-8075/82/0402-0019$01.00/0

southern Texas was without problems. However, the organic insecticides applied to control the boll weevil also heavily decimated the insect enemies of the bollworm, and this pest became more common in cotton fields. The problem was solved by adding 1 to 2 pounds of DDT to the amount of boll weevil poisons used per acre.

In the mid-1950's the boll weevil developed resistance to the chlorinated hydrocarbon insecticides (*3*, *4*). Cotton producers solved this problem by switching to the organophosphorus insecticide methyl parathion. DDT was added to control the bollworm and tobacco budworm, since methyl parathion was highly lethal to their insect enemies but not effective against the worms at the dosage used. A mixture commonly used was 0.5 pound of methyl parathion and 1 to 2 pounds of DDT and toxaphene per acre.

By 1960 the bollworm and tobacco budworm had become difficult to control with DDT. Dosages were increased and treatments were applied at more frequent intervals. By 1965 these pests could no longer be controlled with DDT, endrin, toxaphene and DDT, Strobane and DDT, or the carbamate insecticides (*5–7*) (Table 1).

The problem of resistance by the bollworm and tobacco budworm to the above insecticides was partially resolved by increasing the dosage of methyl parathion to 1 to 2 pounds per acre and reducing the interval between treatments from 4 to 5 days to 2 to 3 days. The control produced was not as complete as that previously obtained with the DDT and methyl parathion mixture, but the treatment was deadly to the boll weevil, almost eliminating it from the cotton fields. The mixture was so effective against the boll weevil that producers no longer thought of it as the primary pest. The bollworm and budworm, the two secondary pests of the calcium arsenate days, had become the major pests of cotton in southern Texas (*8*).

By 1968 the tobacco budworm had developed resistance to methyl parathion (Fig. 1) and could no longer be controlled effectively with any available insecticide (*1*, *8*). Even though many fields were treated 15 to 20 times with all conceivable combinations of insecticides, severe damage was inflicted to cotton across the region. Many farmers suffered almost total losses, and plowed their cotton fields under without a harvest (*8*). The total acreage planted in cotton began its precipitous decline.

The emergence of a strain of tobacco budworm resistant to all insecticides rendered obsolete much of the technology then available for producing cotton. New pest management strategies involving far more limited use of insecticides had to be devised.

Table 1. Increase in the resistance of the bollworm and tobacco budworm to certain organochlorine and carbamate insecticides in southern Texas between 1960 and 1965 (*19*). Values for 1960 and 1961 are from Brazzel (*5*).

Compound	Median lethal dose (milligrams per gram of larva)			
	Bollworm		Tobacco budworm	
	1960	1965	1961	1965
DDT	0.03	1000+	0.13	16.51
Endrin	0.01	0.13	0.06	12.94
Carbaryl	0.12	0.54	0.30	54.57
Strobane and DDT	0.05	1.04	0.73	11.12
Toxaphene and DDT	0.04	0.46	0.47	3.52

Table 2. Per acre comparison of cotton production under different pest management systems in Frio County, Texas, in 1974 (*16*).

Item	Unit	Production technique			
			Cooperating producer		
		Typical*	Before change-over	Short season (40-inch gaps)	Short season (26-inch gaps)
Input					
Fertilizer	Pounds	120	178	72	72
Irrigation	Inches	20	18	12	12
Pesticides	Pounds	9.6	16.9	6.6	6.6
Total energy	Kilocalories × 10^3	3624	3624	2445	2445
Cost	Dollars	278	326	281	279
Cost	Cents per pound	47.60	42.56	33.84	26.90
Production					
Yield	Pounds	500	625	649	765
Gross†	Dollars	340	435	452	532
Net†	Dollars	62	109	170	252

*Based on data in (*17*). †Based on prices of $0.60 per pound for lint and $120 per ton for seed.

The Basis for the New System: Initial Efforts to Reduce Pesticide Use

The boll weevil and cotton fleahopper were the key pests responsible for the problem in the 1960's. The bollworm and tobacco budworm attained damaging numbers only when their natural enemies were killed with insecticides. The solution seemed obvious. Ways had to be devised to control the weevil and fleahopper without inducing outbreaks of the bollworm and budworm.

Research had shown (*9*, *10*) that the boll weevil might be controlled with least disruption to its insect enemies by a combination of measures applied during the harvest season and aimed at reducing the number of adults surviving the winter. These measures included (i) early planting, (ii) use of desiccants and defoliants to terminate the crop and cause shedding of fruit suitable for weevil food and reproduction, (iii) treating the cotton field once or twice with insecticides during the harvest period to kill as many diapausing weevils as possible, (iv) harvesting the crop rapidly, (v) destroying the stalks (this was already being enforced to control the pink bollworm), and (vi) plowing the residue under immediately thereafter. The number of diapausing boll weevils can be so reduced by these practices that damaging outbreaks do not occur during the subsequent season (*9*, *10*). Also, if farmers avoid insecticidal treatments during the early flowering of cotton, bollworms and tobacco budworms can often be controlled by their natural enemies.

The cotton fleahopper is most damaging at the time cotton begins to form squares (unopened flowers). Fortunately, several insecticides, when used at low dosages, destroy enough fleahoppers to allow cotton plants to fruit but do not kill great numbers of the fleahoppers' insect enemies. It also is possible to kill overwintering boll weevils during this period, before they can reproduce, with low dosages of these insecticides.

Reducing irrigation and fertilization to induce early maturation of the cotton crop is also effective, since damage inflicted by the boll weevil, bollworm, and tobacco budworm becomes greater as the season progresses. In addition, one can allow the pest population to grow somewhat larger before initiating insecticide applications, the rationale being that it is better to lose a little cotton to these pests than risk it all by killing their insect enemies.

A program entailing all these measures was implemented in the lower Rio Grande Valley in the fall of 1968. During the 1969 growing season, boll weevils did not multiply to damaging levels, insecticide treatments for this pest were not needed, and outbreaks of bollworms and budworms were averted. The number of insecticide applications was reduced more than 50 percent; many producers did not treat their fields at all. Yields were the second best in 20 years (*8*).

But in 1970 the number of cotton fleahoppers in the area was again extremely high. Cotton farmers had to use insecticides repeatedly, thereby inducing a severe outbreak of budworms and bollworms. Although some fields were treated 20 times or more, crop losses were severe and widespread—and farmers were losing confidence in their ability to produce cotton at a profit (*8*).

Development of the New System

Boll weevils are most vulnerable to insecticides during the harvest season and in the spring, before oviposition occurs. Diapause occurs during late summer and early fall in response to shorter days, cooler temperatures, and maturation of the cotton plant. The adults that emerge during this period are nonreproductive. They feed for several days on cotton fruit and then leave the fields for nearby woods or brushy areas, where they overwinter in leaf litter.

In southern Texas, the overwintering weevils reenter the cotton fields soon after the plants emerge. Reproduction begins as soon as the squares are big enough to support the feeding of the larvae. Generally, the first generation is too small to affect yield. Losses are inflicted by the large second and subsequent generations that develop if not controlled. The overwintering adults should be killed before they can reproduce by treating the cotton one to three times with insecticides before the squares are one-third their full size.

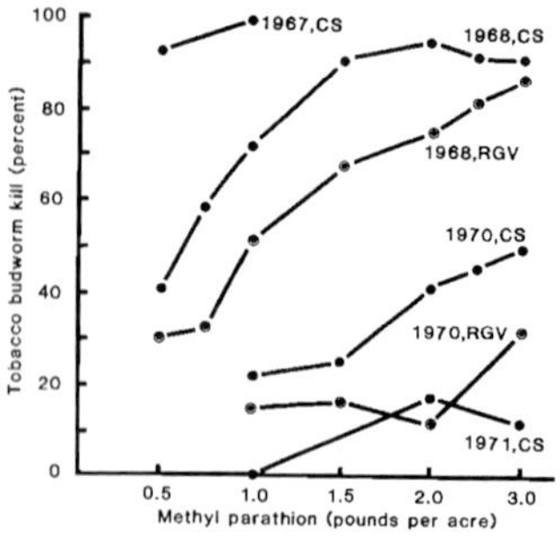

Fig. 1. Increase in resistance of the tobacco budworm to methyl parathion in the lower Rio Grande Valley (*RGV*) and near College Station (*CS*), Texas, between 1967 and 1971.

Knowledge of the boll weevil's life cycle guided the development of a more effective system of cotton pest management in the early 1970's. It was recognized that the system should include three basic components:

1) An areawide control program during the harvest season, combining prompt destruction of stalks and insecticidal treatment of harvested cotton fields to kill as many diapausing weevils as possible [earlier research (*11, 12*) has shown that if stalks are shredded and plowed under by mid-September, weevil numbers are so reduced that they cause negligible yield losses during the subsequent season].

2) Control of overwintered adults in the spring, before they can reproduce, by insecticides applied to cotton before the squares are of sufficient size to support development of the first-generation larvae (these treatments, which also control the cotton fleahopper, should be limited and timed to have the least impact on the insect enemies of the bollworm and tobacco budworm).

3) Cultivation of a rapid-fruiting, short-season cotton variety capable of setting a normal yield of bolls 12 days or older during the first 20 to 30 days of flowering (once a boll reaches 12 days of age, the carpel is so thick that it is safe from weevil attack) (*13*). Such a variety may be harvested in late July and August, before most of the weevils enter diapause.

The rationale for this "short-season" approach to cotton production in Texas is based on the work of Walker and Niles (*14*), who determined the relation between flowering rate in short- and long-season cotton and boll weevil damage (Fig. 2). The short-season variety can produce a much greater percentage of the bolls that produce the final yield during the first generation than an indeterminate variety. The first generation will be small if insecticides are used against the parents early in the spring. The full-season variety, because it fruits later and more slowly, has to set bolls during the period when weevils are most numerous if normal yields are to be attained. If the strategy of controlling overwintering boll weevils and growing a short-season variety is successful, there usually is no need for insecticide treatment of later generations of weevils or of other pests.

Fortuitously, one of us (L.S.B.) had initiated a program in 1963 to develop

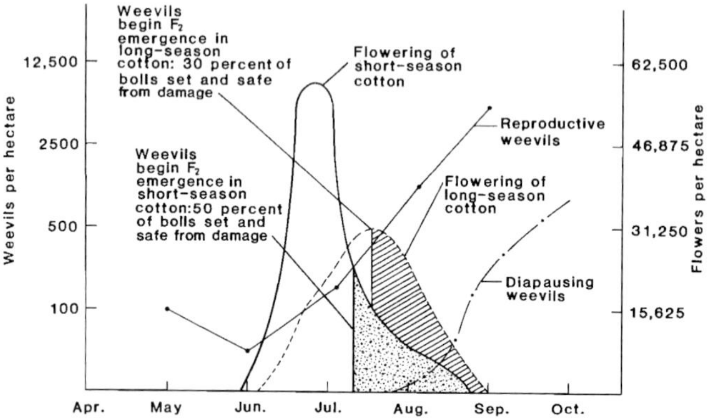

Fig. 2. Fruiting rates of short- and long-season cotton plotted against the boll weevil life cycle. The short-season variety sets bolls early and at a rapid rate, so that 50 percent or more of the yield is set before the second generation (F_2) boll weevils emerge. Only 30 percent of the bolls of the long-season variety are set during this period. Also, the short-season variety matures before many weevils enter diapause, while the long-season variety allows a substantial number to diapause.

cotton varieties whose seeds and seedlings tolerate cold and resist disease, especially blight caused by the bacterium *Xanthomonas malvacearum* (E. F. Smith) Dows. The first strains were tested from 1968 to 1970 and were released in 1973 as varieties TAMCOT SP-21 and TAMCOT SP-37 (*15*). These two varieties met the requirements of the integrated insect control system. Their cold tolerance and disease resistance permitted planting earlier in the season than conventional varieties. They fruited rapidly, retained a higher percentage of early fruit, and gave a normal yield 20 to 30 days earlier than conventional varieties. Moreover, they also were found to have a low level of resistance to the boll weevil.

Since then, varieties TAMCOT SP-21S, TAMCOT SP-37H, and CAMD-E have been released to growers (*15*). These have moderate levels of cold tolerance, resistance to seedling pathogens, and immunity to bacterial blight. In addition, they are moderately resistant to the cotton fleahopper. CAMD-E has moderate resistance to the boll weevil, bollworm, and tobacco budworm. TAMCOT SP-37H has low resistance to the boll weevil. TAMCOT SP-21S, because of its smooth leaf surface, is less heavily infested by the bollworm and tobacco budworm than varieties with hirsute leaves. These varieties have been widely accepted by Texas cotton growers and have greatly improved the efficacy of the new system.

Economic Benefits

One of the most striking demonstrations of the economic benefits of the new system was conducted in 1974 on a private farm in Frio County, Texas (Table 2). The typical cotton farmer in this county spent $278 to grow 1 acre of a conventional variety of cotton under irrigation in an area heavily infested with boll weevils (*16*). With a yield of one bale per acre, the net return was $62 per acre. The cooperating producer, one of the county's best cotton farmers, had been using more fertilizer and insecticide than the typical producer. His costs were greater ($326 per acre), but so were his yields, and the net return was $109 per acre. This farmer's land was used to grow a short-season variety (TAMCOT SP-37), and the amount of fertilizer, irrigation water, and insecticide was reduced 80, 50, and 60 percent, respectively. When the new variety was grown in rows that were spaced 40 inches apart (conventional spacing), the yield was increased over that of the conventional variety, the cost was $281 per acre, and the net return was $170 per acre. When the short-season variety was planted in rows 26 inches apart, the yield was greatly increased, the cost was the same, and the profit increased to $252 per acre. These results caught the attention of cotton farmers in the area and greatly accelerated their changing to the new system (*16*).

Using the new system, cotton producers have not been able to equal the magnitude of the increases in profit achieved in carefully managed demonstration plots, but they have done well. In coastal Texas, cotton yields have increased from 226 to 459 pounds of lint per acre since 1975. The average net return to the cotton producer has increased from $62 to $170 per acre (*18*). In the lower Rio Grande Valley, mean net return to producers of dryland cotton has increased $31 per acre. Insecticide use has decreased from 12.3 pounds of actual toxicant per acre to 1.5 pounds (*17*).

Most cotton farmers in Texas now practice some form of the new system. Insecticide use on cotton has been reduced from a high of 19.3 million pounds in 1964 to 2.3 million in 1976 (*18*). The system has made Texas cotton growers much more competitive economically and has led to a great resurgence of the industry in the state. Texas now produces about half of the nation's cotton.

References and Notes

1. P. L. Adkisson, in *Conn. Agric. Exp. Stn. New Haven Bull.* **708**, 155 (1969).
2. *Texas Cotton Statistics for 1975* (Texas Department of Agriculture, Austin, 1976).
3. J. S. Roussel and D. F. Clower, *La. Agric. Exp. Stn. Circ. 41* (1955).
4. J. R. Brazzel, *Tex. Agric. Exp. Stn. Prog. Rep. 2171* (1961a).
5. ———, *J. Econ. Entomol.* **56**, 561 (1963); *ibid.* **57**, 455 (1964).
6. F. A. Harris, J. B. Graves, S. J. Nemec, S. B. Vinson, D. A. Wolfenbarger, *South. Coop. Ser. Bull.* **169**, 17 (1972).
7. P. L. Adkisson, *Tex. Agric. Exp. Stn. Misc. Publ. 709* (1964); ——— and S. J. Nemec, *Tex. Agric. Exp. Stn. Bull. 1048* (1966); S. J. Nemec and P. L. Adkisson, *Tex. Agric. Exp. Stn. Prog. Rep. 2674* (1969).
8. P. L. Adkisson, paper presented at the Tall Timbers Annual Conference on Ecological Animal Control by Habitat Management, Tallahassee, Fla., 1972.
9. J. R. Brazzel, *Tex. Agric. Exp. Stn. Misc. Publ. 511* (1961b).
10. P. L. Adkisson, D. R. Rummel, W. L. Sterling, W. L. Owen, Jr., *Tex. Agric. Exp. Stn. Bull. 1054* (1966).
11. F. W. Mally, *Report on the Boll Weevil* (Agricultural and Mechanical College of Texas, College Station, 1902).
12. W. D. Hunter, *U.S. Dep. Agric. Farmers' Bull. 500* (1912).
13. R. D. Parker, J. K. Walker, G. A. Niles, J. R. Mulkey, *Tex. Agric. Exp. Stn. Bull. 1315* (1980).
14. J. K. Walker and G. A. Niles, *Tex. Agric. Exp. Stn. Bull. 1109* (1971).
15. L. S. Bird, H. D. Smith, R. W. Hoermann, D. B. McCombs, W. E. Sears, C. W. Horne, *Tex. Agric. Exp. Stn. Prog. Rep. 2666* (1969); L. S. Bird, *Crop Sci.* **16**, 84 (1976); *ibid.* **19**, 410 (1979).
16. M. J. Sprott, R. D. Lacewell, G. A. Niles, J. K. Walker, J. R. Gannaway, *Tex. Agric. Exp. Stn. Misc. Publ. 1250C* (1976).
17. G. S. Collins, R. D. Lacewell, J. W. Norman, *South. J. Agric. Econ.* **11**, 79 (1979); N. P. Clarke, *Texas Agriculture in the 1980's: The Critical Decade* (Texas Agricultural Experiment Station, College Station, 1980).
18. Office of Technology Assessment, *Pest Management Strategies: Present and Future Pest Management Strategies in the Control of Sorghum and Cotton Pests in Texas* (Government Printing Office, Washington, D.C., 1979), vol 2.
19. P. L. Adkisson, paper presented at the FAO/IAEA research coordination meeting on the behavior and ecology of the *Heliothis* complex, Monterrey, Mexico, April 1975.

20. Supported in part by grants from the Rockefeller Foundation, the National Science Foundation, the Environmental Protection Agency, and Cotton, Inc.

Carl B. Huffaker

University of California
Berkeley, California, USA

1914–1995

1994/5 — for his contribution to the development and implementation of environmentally beneficial integrated pest management systems for the protection of agricultural crops.

Professor Carl B. Huffaker performed basic research on crop pests, and subsequently broadened their work to develop biologically-based systems for pest control. Both were major contributors to the development and implementation of integrated pest management, or IPM, systems for crop production. Huffaker successfully demonstrated the efficacy of biological control agents for the control of weeds and insects.

Huffaker was one of the leaders in large-scale programs that demonstrated the effectiveness of IPM systems for crop protection, and that led to the establishment of IPM as an environmentally beneficial and economically favorable system for crop production that is today used throughout the world.

CURRICULUM VITAE

General

Born September 30, 1914, Monticello, Kentucky (USA)
Attended public schools in Monticello, Kentucky, 1920-1933

Training and Degrees

1935-38 - University of Tennessee, Knoxville, B.A.
1939 - University of Tennessee, Knoxville, M.S.
1940 - North Carolina State University, Raleigh
1940-42 - Ohio State University, Columbus, Ph.D.

Positions

1940-41	Teaching Assistant in Zoology, Ohio State University
1941	U.S. Department of Agriculture, Bureau of Entomology (summer)
1941-43	Assistant Entomologist, University of Delaware
1943-45	Associate Entomologist, then Entomologist, Health and Sanitation Division, Institute of Inter-American Affairs (Colombia, Haiti and Dominican Republic)
1946-1984	Assistant Entomologist, then Associate Entomologist, then Entomologist and Professor of Entomology, University of California, Berkeley
1970-1983	Director, International Center for Integrated and Biological Control, University of California, Berkeley and Riverside
1985-	Emeritus Professor, Division of Biological Control, University of California, Berkeley

Honors and Activities

Phi Kappa Phi
Sigma Xi
Guggenheim Fellow
AAAS Fellow
Fellow Franklin Institute
C. W. Woodworth Award
Louis E. Levy Medal of the Franklin Institute
The Journal Premium Award of the Franklin Institute
Rockefeller Foundation Resident-Scholar to Bellagjo Study and Conference Center (Italy)
Honorary Fellow, Royal Entomological Society of London
President, International Organization for Biological Control (1972-1976, 1978-1980)
President, Entomological Society of America (1981)
Member, National Academy of Sciences since 1982
Recipient, Berkeley Citation, University of California, Berkeley

American Association Advancement of Science
Entomological Society of America
Ecological Society of America
Pacific Coast Entomological Society
Pacific Coast Entomological Society
Society Population Ecology
Entomological Society of Canada

Secretary (1958) and Chairman (1959), Biological Control Section, Entomological Society of America

Member, entomological delegation to USSR, 1959 (Department of State auspices)

U. S. Public Health Service, National Institutes of Health grants 1967-1973 in population ecology

National Science Foundation grants 1967-1981

Subcommittee member, U.S. National Committee of the International Biological Programme from its inception in 1966 and member of International Working Group for Biological Control

World coordinator of IBP program on biological control of spider mites, 1966-1970

Member of ad hoc committee to organize a Western Hemisphere Section of the International Organization for Biological Control, then Vice-president of that body

Member of Subcommittee on Weeds, Agricultural Board, National Academy of Sciences, National Research Council, 1966-1968

Director of nationwide, 18-university Integrated Pest Management Project (NSF, EPA, USDA support) 1972-1981

President, International organization for Biological Control, 1972-1977

Member of the board, Division of Environmental Biology, International Union of Biological Sciences, 1973-1977

Member, EPA Office of Pesticides Ad Hoc Committee on Pest Management, 1973-1975

Visited USSR as a participant in U.S./USSR Integrated Pest Management Program, under the Environmental Agreement, was among scientists invited to meet with Russian scientists and to advise on plans for future exchanges 1974

Member, Insect Study Group to visit People's Republic of China under the auspices of the Committee for Scholarly Communication with the People's Republic of China, NAS, 1975

Consultant, United Nations Environmental Programme (1975)

Member, Pesticide Advisory Committee, California State Department of Food and Agriculture, 1976

Participant, UC/AID Pest Management Program, Huancayo, Peru, November 1979

been one of the leading advocates of integrating biological control, along with other selective alternatives to the wholesale use of toxic pesticides, in a holistic system now known as IPM, and has been instrumental in implementing this approach in California in his capacity as Director of the Statewide International Center for Biological and Integrated Control (1970-83). His unrelenting efforts, and the increasing realization that unilateral chemical control will not solve pest problems, eventually led to the endorsement of IPM by government agencies in the USA and elsewhere. The turning point was the initiation of a huge nation-wide project, known as the "Huffaker Project", which he directed during 1972-78. Engaging some 300 scientists from 18 universities and funded jointly by the NSF, EPA and USDA, it was aimed at promoting the development of integrated pest management programs in major agro-ecosystems including soybeans, cotton, alfalfa, pome and stone fruits, citrus and coniferous forests. Summarized in his book, "New Technology of Pest Control" (1980), this unprecedented project had a tremendous impact on the philosophy and practice of plant protection world-wide, and the resulting reduction in pesticide use has already affected environmental quality in many parts of the world.

Professor Huffaker continues to be highly active in both theoretical and practical research, and his seminal publications will continue to influence the world's agricultural and scientific communities for generations to come.

LIST OF SIGNIFICANT PUBLICATIONS

1956. (with C.E. Kennett). Experimental studies on predation. (1) Predation and cyclamen mite populations on strawberries in California. Hilgardia 26: 191-222.

1957. Fundamentals of biological control of weeds. Hilgardia 27: 101-157.

1958. Experimental studies on predation. II. Dispersion factors and predator- prey oscillations. Hilgardia 27: 343-383.

1959. Biological control of weeds with insects. Ann. Rev. Entomol. 4: 251-276.

1964. (with P.S. Messenger). Population ecology - historical development, pp. 45-73, and: The concept and significance of natural control, pp. 74-117, In: P. DeBach, ed., Biological Control of Insect Pests and Weeds. Chapman and Hall, London and Reinhold, New York.

1966. (with C.E. Kennett). Biological control of *Parlatoria oleae* (Colvèe) through the compensatory action of two introduced parasites. Hilgardia 37: 283-355.

1969. (with M. van de Vrie and J.A. McMurtry). The ecology of tetranychid mites and their natural control. Ann. Rev. Entomol. 14: 125-174.

1969. (with M.P. Hassell). Regulatory processes and population cyclicity in laboratory populations of *Anagasta kuhniella* (Zeller) (Lepidoptera: Phycitidae). III. The development of population models. Res. Popul. Ecol. 11: 186-210.

1971. Editor. Biological Control. Plenum Press, New. York, 511 pp.
1976. Editor (with P.S. Messenger). Theory and Practice of Biological Control. Academic Press, New York, 788 pp.
1980. Editor. New Technology of Pest Control. Wiley, New York, 500 pp.
1983. (with D. Rosen). An overview of desired attributes of effective biological control agents, with particular emphasis on mites, pp. 2-11. In: M.A. Hoy, G.L. Cunningham and L. Knutson, ed., Biological Control of Pests by Mites. Univ. Calif. Div. Agric. Publ. 3304.
1984. Editor (with R.L. Rabb). Ecological Entomology. Wiley, New York, 844 pp.
1990. (with A.P. Gutierrez). Natural enemies and prey population regulation, pp. 183-195, and: Evaluation of efficiency of natural enemies in biological control, pp. 473-495, In: O. Rosen, ed., Armored Scale Insects: Their Biology, Natural Enemies and Control. Elsevier, Amsterdam.

Morris Schnitzer
Centre for Land and Biological Resources Research
Agriculture Canada, Ottawa, Ontario, Canada

1995/6 — for the pioneering contributions to our understanding of the chemistry of soil organic matter and its application to agriculture.

CURRICULUM VITAE

Born February 4, 1922 in Bochum, Germany
Immigrated to Canada in May, 1947

Degree Received

B. Sc. Agr. (Soil Chemistry), 1951 - McGill University, Montreal, Canada.
M. Sc. Soil Chemistry, 1952 - McGill University, Montreal, Canada.
Ph.D. Soil Chemistry, 1955 - McGill University, Montreal, Canada.
Postdoctorate studies in organic chemistry, 1961-62, Imperial College of Science and Technology, London, England, with Prof. D. H. R. Barton, Nobel Laureate.

Professional Positions Held

Research and Development Chemist, Aluminum Company of Canada, Arvida, Quebec, 1955-1956.
Research Scientist, Agriculture Canada, 1956-1991.
Retired as Principal Research Scientist in 1991. He is currently Emeritus Distinguished Research Scientist, Agriculture Canada.

Professional Publications

a) Books Written

Schnitzer, M. and S.U. Khan. 1972. Humic Substances in the Environment. Marcel Dekker, New York. 327 Pages.

b) Books edited

- Schnitzer, M. and S.U. Khan. 1978. Soil Organic Matter. Elsevier, Amsterdam, 319 pages.
- Huang, P.M. and M. Schnitzer. 1984. Interaction of Soil Minerals with Natural Organics and Microbes. Soil Science Society of America, Madison, Wisconsin, 606 pages.

c) Scientific papers

355 refereed scientific papers.

Honors and Awards Received

B. Sc. First Class Honours with Distinction (1951)
MacDonald College Alumni Scholarship (1952)
Agricultural Institute of Canada Fellowship (1952)
Fellow, Canadian Society of Soil Science (1971)
Fellow, Soil Science Society of America (1977)
Fellow, American Society of Agronomy (1977)
Honorary Member, International Humic Substances Society (1986)
Fellow of the Royal Society of Canada (1991)
Soil Science Award (1984) of the Soil Science Society of America
Soil Science Distinguished Service Award of the Soil Science Society of America (1995)
1995/96 Co-winner Wolf Prize in Agriculture

Editorial Work

Served on the editorial boards of the Canadian Journal of Soil Science, Soil Science, Geoderma, and Plant and Soil.

Contribution to Societies, Committees, Symposia, etc.

- Chairman, Commission II (Soil Chemistry) of the International Society of soil Science (1978-1982).
- Founding member of the International Humic Substances Society (1982).
- Associate editor of the Canadian Journal of Soil Science (1975-1978).
- Served on the editorial boards of Soil Science, Geoderma, Agrochimica, and Plant and Soil.

- Lectured and participated in scientific meeting and symposia on soil organic matter and humic substances in Canada, the U.S.A., Argentina, India, Israel, France Italy, West Germany, Austria, The Netherlands, Switzerland, the West Indies, and the U.S.S.R.
- Acted as the advisor to CIDA, the international joint committees of Agriculture Canada and the USA department of Agriculture.
- From 1978-1986 was responsible for the design and execution of research programs in soil chemistry and soil biology in the Chemistry and Biology Research Institute of Agriculture Canada in Ottawa.
- Adjunct Professor of Soil Science, University of Guelph (1986). Frequent lecturer at MacDonald College of McGill University.

BRIEF DESCRIPTION OF SCIENTIFIC ACHIEVEMENTS

The main objective of my research was to better understand the chemistry and reactions of humic substances, the major components of soil organic matter in agricultural soils. Over the years, I have developed a variety of novel chemical, physical and spectroscopic methods, and applied these to the analysis of humic acids, fulvic acids and humins, the major humic substances. I have also pioneered the use of sophisticated methods such as ^{13}C NMR spectrometry, gas chromatography/mass spectrometry, and pyrolysis gas chromatography/mass spectrometry in investigations on the chemical structures of humic materials. Many of the methods, which I developed, are now widely employed in soil and environmental science laboratories all over the world. I also used a number of colloid-chemical methods, electron microscopy and X-ray analysis to shed light on shapes, sizes, molecular weights and other molecular properties of humic substances. On the chemical side, I did numerous oxidative degradations of humic materials. Probably, my most important contribution to soil science was the development of model structures for humic acid, soil organic matter and a whole soil. These model structures are in harmony with the chemical, colloid-chemical, ^{13}C NMR, X-Ray and mass spectrometric data on humic substances, which my co-workers and I have obtained over many years. Other research was concerned with metal- and clay-humic interactions and with the identification of N-heterocyclics in soils.

SCIENTIFIC PUBLICATIONS

1. Schnitzer M. and W.A. Delong. A note on the podzolization process. Sci. Agr. 32: 680-681, 1952.
2. Schnitzer M. and W.A. Delong. Note on the reaction of 2,2'-dipyridyl with iron in the presence of organic matter. Cand. J. Agr. Sci. 34: 324-325, 1954.

3. Schnitzer M. and W.A. Delong. Note on relative capabilities of solutions obtained from forest vegetation for mobilization of iron. Can. J. Agr. Sci. 34: 542-543, 1954.
4. Delong, W.A. and M. Schnitzer. Investigations on the mobilization and transport of iron in the forested soils: I. the capacities of leaf extracts and leachates to react with iron. Soil. Sci. Soc. Am. Proc. 19: 360-363, 1955.
5. Schnitzer, M. and W.A. Delong. Investigations on the mobilization and transport of iron in forested soils: II. The nature of the reaction of leaf extracts and leachates with iron. Soil Sci. Soc. Am. Proc. 19: 363-368, 1955.
6. Schnitzer, M. and J.R. Wright. Note on the extraction of organic matter from the B horizon of a podzol soil. Can. J. Agr. Sci. 36: 511-512, 1956.
7. Schnitzer, M. and J.R. Wright. Extraction of organic matter from podzol by means of dilute inorganic acids. Can. J. Soil Sci. 37: 89-95, 1957.
8. Schnitzer, M. Mobilization of iron in podzol soils by aqueous leaf extracts. Chem. & Ind. 1594-1595, 1957.
9. Wright, J.R., M. Schntzer and R. Levick. Some characterizations of the organic matter extracted by dilute inorganic acids from a podzolic B horizon. Can. J. Soil Sci. 38: 49-53, 1958.
10. Schnitzer, M., J.R. Wright and J.G. Desjardins. A comparison of the effectiveness of various extractants for organic matter from two horizons of a podzol profile. Can. J. Soil Sci. 38: 49-53, 1958.
11. Hoffman, I., M. Schnitzer and J.R. Wright. Thermogravimetry of soils. Chem. & Ind. 261, 1958.
12. Wright, J.R. and M. Schnitzer. Alkaline permanganate oxidation of the organic matter of the A_0 and B_{21} horizons of a podzol. Can. J. Soil Sci. 39: 44-53, 1959.
13. Schnitzer, M., J.R. Wright and I. Hoffman. Use of the thermobalance in the analysis of soils and clays. Anal. Chem. 31: 440-444, 1959.
14. Schnitzer, M., D.A. Shearer and J.R. Wright. A study in the infrared of high molecular weight organic matter extracted by various reagents from a podzolic B horizon. Soil Sci. 87: 252-257, 1959.
15. Schnitzer, M. Interaction of iron with rain-fall leachates. J. Soil Sci. 10: 300-308, 1959.
16. Wright, J.R. and M. Schnitzer. Oxygen-containing functional groups in the organic matter of a podzol soil. Nature 184: 1462-1463, 1959.
17. Wright, J.R., I. Hoffman and M. Schnitzer. Application of thermogravimetry to the analysis of carbonates occurring in soils. I. Analysis of pure carbonates and naturally occurring limestones. J. Sci. Food Agr. 11: 163-167, 1960.
18. Hoffman, I., M. Schnitzer and J.R. Wright. Application of thermogravimetry to the analysis of carbonates occurring in soils. II. Analysis of carbonates in soils. J. Sci. Food Agr. 11: 167-172, 1960.

19. Schnitzer, M. and J.R. Wright. Nitric acid oxidation of the organic matter of a podzol. Soil Sci. Soc. Proc. 24: 273-276, 1960.
20. Schnitzer, M. and J.R. Wright. Studies on the oxidation of the organic matter of the A_O and B_h horizons of a podzol. Proc. 7th Intl. Congr. of Soil Sci. II: 112-119, 1960.
21. Wright, J.R. and M. Schnitzer. Functional groups in the organic matter of the A_O and B_h horizons of a podzol. Proc. 7th Intl. Congr. of Soil Sci. II: 120-127, 1960.
22. Wright, J.R. and M. Schnitzer. An estimate of the aromaticity of the organic matter of a Podzol soil. Nature 190: 703-704, 1961.
23. Schnitzer, M. and I. Hoffman. Thermogravimetry of the organic matter of a Podzol soil. Chem. & Ind. 1397-1398, 1961.
24. Schnitzer, M. and J.G. Desjardins. Cryoscopic molecular weight apparatus. Chemist-Analyst 50: 117, 1961.
25. Turner, R.C. and M. Schnitzer. Thermogravimetry of the organic matter of a Podzol. Soil Sci. 93: 225-232, 1962.
26. Schnitzer, M., J.R. Wright and I. Hoffman. High temperature thermogravimetry of chloride and sulphates. Anal. Chim. Acta. 26: 371-377, 1962.
27. Schnitzer, M. and J.G. Desjardins. Molecular and equivalent weights of the organic matter of a Podzol. Soil. Soc. Am. Proc. 26: 362-365, 1962.
28. Barton, D.H.R. and M. Schnitzer. A new experimental approach to the humic acid problem. Nature 198: 417-418, 1963.
29. Wright, J.R. and M. Schnitzer. Metallo-organic interactions associated with podolization. Soil Sci. Soc. Am. Proc. 27: 171-176, 1963.
30. Schnitzer, M. and S.I.M. Skinner. Organo-metallic reactions in soils. 1: Soil Sci. 96: 86-93, 1963.
31. Schnitzer, M. and S.I.M. Skinner. Organo-metallic reactions in soils. 2. Soil Sci. 96: 181-186, 1963.
32. Hoffman, I. and M. Schnitzer. Thermogravimetry: A. valuable analytical tool, Chemistry in Canada 16: 30-32, 1964.
33. Schnitzer, M., R.C. Turner and I. Hoffman. A thermogravimetric study of organic matter of representative Canadian podzol soils. Can. J. Soil Sci. 44: 7-13, 1964.
34. Schnitzer, M. and U.C. Gupta. Some chemical characteristics of the organic matter extracted from the O and B_2 horizon of the grey wooded soil. Soil Sci. Soc. Am. Proc. 28: 374-377, 1964.
35. Schnitzer, M. and I. Hoffman. Pyrolysis of soil organic matter. Soil Sci. Soc. Am. Proc. 28: 520-525, 1964.
36. Schnitzer, M. and S.I.M. Skinner. Organometallic reactions in soil. 3: Soil Sci. 98: 197-203, 1964.

37. Schnitzer, M. and J.G. Desjardins. Gas chromatographic separation of benzene carboxylic acids. J. Gas Chromatography 2: 270-271, 1964
38. Schnitzer, M. and J.G. Desjardins. Further investigations on the alkaline permanganate oxidation of organic matter extracted from a podzole Bh horizon. Can. J. Soil Sci. 44: 272-278, 1964.
39. Schnitzer, M. and U.C. Gupta. Determination of acidity in soil organic matter. Soil Sci. Soc. Am. Proc. 29: 274-277, 1965.
40. Schnitzer, M. and S.I.M. Skinner. Organometallic reactions in soil. 4: Soil Sci. 99: 278-284, 1965.
41. Schnitzer, M. and I. Hoffman. Thermogravimetry of soil humic compounds. Geochim. et Cosmochim. Acta. 29: 859-870, 1965.
42. Schnitzer, M. and S.I.M. Skinner. The carbonyl group in a soil organic matter preparation. Soil Sci. Soc. Am. Proc. 29: 400-405, 1965.
43. Schnitzer, M. Contribution of organic matter to the cation exchange capacity of soils. Nature 207: 667-668, 1965.
44. Schnitzer, M. and J.G. Desjardins. Carboxyl and hydroxyl groups in some organic soils. Can. J. Soil. Sci. 45: 257-264, 1965.
45. Schnitzer, M. and S.I.M. Skinner. A polarographic method for the determination of carbonyl groups in soil humic compounds. Soil Sci., 101: 120-124, 1965.
46. Hoffman I. and M. Schnitzer. Thermogravimetric studies on soil humic compounds. "Thermal Analysis 1965", pages 62-63. McMillan & Co. Ltd., London.
47. Schnitzer, M. The application of infrared spectroscopy to investigations on soil humic compounds. Can. Spectroscopy 10: 121-127, 1965.
48. Schnitzer, M. and I. Hoffman. A thermogravimetric approach to the classification of organic soils. Soil Sci. Soc. Am. Proc. 30: 63-66, 1966.
49. Levescue, M. and M. Schnitzer. Effects of NaOH concentration on the extraction of organic matter and or major inorganic constituents from a soil. Can. J. Soil. Sci., 46: 7-12, 1966.
50. Schnitzer, M. and P.A. Poapst. Effects of soil humic compounds on root initiation. Nature 213: 598-599, 1967.
51. Levesque, M. and M. Schnitzer. Organometallic intereaction in soils 6. Preparation and properties of fulvic acid-metal-phosphates. Soil Sci. 103: 183-190, 1966.
52. Schnitzer, M. and S.I.M. Skinner. Organometallic interactions in soils. 5. Stability constants of Cu-, Fe-, and Zn-fulvic acid complexes. Soil Sci. 102: 361-365, 1966.
53. Schnitzer M. and K. Kodama. Effect of pH on adsorption of a soil humic compound by montmorillonite. Science 153: 70-71, 1966.

54. Schnitzer, M. and J.G. Desjardins. Oxygen containing functional groups in organic soils and their relation to the degree of humification as determined by solubility in sodium pyrophosphate solution. Can. J. Soil Sci. 46: 237-243, 1966.
55. Schnitzer, M. and I. Hoffman. Thermogravimetric analysis of the salts and metal complexes of a soil fulvic acid. Geochim. et Cosmochim. Acta. 31: 7-15, 1967.
56. Hansen E.H. and M. Schnitzer. The alkaline permanganate oxidation of Danish illuvial organic matter. Soil Sci. Soc. Am. Proc. 30: 745-748, 1966.
57. Schnitzer, M. and S.I.M. Skinner. Organometallic interactions in soil: 7. Stability constants of Pb-, Ni-, Mn-, Co-, and Mg-fulvic acid complexes. Soil Sci. 103: 247-252, 1967.
58. Hansen, E.H. and M. Schnitzer. The nitric acid oxidation of Danish illuvial organic matter. Soil Sci. Soc. Am. Proc. 31: 79-85, 1967.
59. Levesque, M. and M. Schnitzer. The extraction of organic matter by base and chelating resin from soils. Can. J. Soil Sci. 47: 76-78, 1967.
60. McKeague, J.A., M. Schnitzer and P.K. Heringa. Properties of an iron pan podzol from Newfoundland. Can. J. Soil Sci. 47: 23-32, 1967.
61. Kodama, H. and M. Schnitzer. X-ray studies of fulvic acid, a soil humic compound. Fuel 46: 87-94, 1967.
62. Morris C.F. and M. Schnitzer. Rapid determination of carbon in soil organic matter extracts by dry combustion. Can. J. Soil Sci. 47: 143-144, 1967.
63. Sowden F.J. and M. Schnitzer. The nitrogen distribution in illuvial organic matter. Can. J. Soil Sci. 47: 111-116, 1967.
64. Schnitzer M. and H. Kodama. Reaction between a podzol fulvic acid and Na-montmorillonite. Soil Sci. Soc. Am. Proc. 31: 632-636, 1967.
65. Schnitzer, M. Humic-fulvic acid relationships in organic soils and humification. Can. J. Soil Sci. 47: 245-250, 1967.
66. Schnitzer, M. and I. Hoffman. Significance of cellulose components in the thermogravimetry of organic soils. Soil Sci. Sco. Am. Proc. 31: 708-709, 1967.
67. Schnitzer, M. Reactions between organic matter and inorganic soil constituents. Proc. 9th Intl. Congress of Soil Sci. I: 635-644, 1968.
68. Hansen, H. and M. Schnitzer. Oxidative degradation of Danish illuvial organic matter in "Soil Chemistry and Fertility", pages 87-92, 1967.
69. Schnitzer, M. and S.I.M. Skinner. Alkali vs. acid extraction of soil organic matter. Soil Sci. 105: 392-396, 1968.
70. Kodama, H. and M. Schnitzer. Effect of interlayer cations on the adsorption of a soil humic compound by montmorillonite. Soil Sci. 106: 73-74, 1968.

71. Hoffman, I. and M. Schnitzer. A thermogravimetric investigation on the effect of solid oxidizing agents on the ignition of carbonaceous materials. Fuel 47: 119-123, 1968.
72. Schnitzer, M. Reaction between fulvic acid, a soil humic compound, and inorganic soil constituents. Soil Sci. Soc. Am. Proc. 33: 75-81, 1969.
73. Schnitzer, M. and S.I.M. Skinner. Gel filtration of fulvic acid, a soil humic compound. In "Isotopes and Radiation in Soil Organic Matter Studies", IAEA, Vienna, 41-56, 1968.
74. Hansen E.H. and M. Schnitzer. Zinc dust distillation of soil humic compounds. Fuel 48: 41-46, 1969.
75. Hansen E.H. and M. Schnitzer. Zinc dust distillation and fusion of a soil humic and fulvic acid. Soil Sci. Soc. Am. Proc. 33: 75-81, 1969.
76. Schnitzer, M. and J.G. Desjardins. Chemical characteristics of natural soil leachates from a humic podzol. Can. J. Soil Sci. 49: 151-158, 1969.
77. Kodama, H. and M. Schnitzer. Thermal analysis of a fulvic acid-montmorillonite complex. Proc. of the 1969 Clay Conference, Tokyo 1: 765-774, 1969.
78. Schnitzer, M. and H. Kodama. Reactions between fulvic acid and, a soil humic compound, and montmorillonite. Israel J. Chem. 7: 141-147, 1969.
79. Poapst, P.A., C. Guenier and M. Schnitzer. Effect of a soil fulvic acid on stem alongation in peas. Soil and Plant 32: 367-372, 1970.
80. Hansen, E.H. and M. Schnitzer. Molecular weight measurements of polycarboxylic acids by vapour pressure osmometry. Anal. Chem. Acta. 46: 247-254, 1969.
81. Schnitzer, M. and S.I.M. Skinner. Free radicals in soil humic compounds. Soil Sci. 108: 383-390, 1969.
82. Schnitzer, M. Characterization of humic substances by spectroscopy. Chapter 4 in "Soil Biochemistry", Volume II, edited by A.D. McLaren and J.J. Skujins, Marcel Dekker, New York, pp. 60-93, 1971.
83. Kodama, H. and M. Schnitzer. Kinetics and mechanisms of the thermal decomposition of fulvic acid. Soil Sci. 109: 265-271, 1970.
84. Schnitzer, M. Metal organic mater interactions in soils and waters. Chapter 13 in "Origin, Distribution, Transport and Fate of Organic Compounds in Aquatic Environments", Marcel Dekker, New York, pp. 297-313, 1971.
85. Schnitzer, M. and E.H. Hansen. Organo-metallic interactions in soils. 8: an evaluation of methods for the determination of stability constants of metal fulvic acid complexes. Soil Sci. 109: 333-340, 1970.
86. Schintzer, M. and J.G. Desjardins. The alkaline permanganate oxidation of methylated and of unmethylated fulvic acid. Soil. Sci. Soc. Am. Proc. 34: 77-79, 1970.

87. Schnitzer, M. Characteristics of organic mater extracted from Podzol B horizons. Can. J. Soil Sci. 50: 199-204, 1970.
88. Ogner, G. and M. Schnitzer. The occurrence of alkanes in fulvic acid, a soil humic fraction. Geochim. Cosmochim. Acta. 34: 921-928, 1970.
89. Schnitzer, M. and G. Ogner. The occurrence of fatty acids fulvic acid, a soil humic fraction. Israel J. Chem. 8: 505-512, 1970.
90. Ogner, G. and M. Schnitzer. Humic substances: Fulvic acid-dialkyl phthalate complexes and their role in pollution. Science 170: 317-318, 1970.
91. Gamble, D., M. Schnitzer and I. Hoffman. Cu^{+2}-fulvic acid chelation equilibrium in 0.1 M KCl at 25°C. Can. J. Chem. 48: 3197-3204, 1970.
92. Poapst, P.A. and M. Schnitzer. Fulvic acid and adventitious root formation. Soil Bio. Biochem. 3: 215-219, 1971.
93. Ogner, G. and M. Schnitzer. Chemistry of fulvic acid, a soil humic fraction, and its relation to lignin. Can. J. Chem. 49: 1053-1063, 1971.
94. Khan, S.U. and M. Schnitzer. Sephadex gel filtration of fulvic acid: The identification of major components in two low molecular weight fractions. Soil Sci. 112: 231-238, 1971.
95. Khan, S.U. and M. Schnitzer. Further investigation on the chemistry of fulvic acid, a soil humic fraction. Can. J. Chem. 49: 2302-2309, 1971.
96. Matsuda, K. and M. Schnitzer. Reactions between fulvic acid, a soil humic acid material, and dialkyl phthalates. Bull. Environmental Contamination and Toxicology 6: 200-204, 1971.
97. Kodama, H. and M. Schnitzer. Evidence for interlamellar adsorption of organic mater by clay in a podzol soil. Can. J. Soil Sci. 51: 509-512, 1971.
98. Schnitzer, M. and H. Kodama. Differential thermal analysis of metal-fulvic acid salts and complexes. Geoderma 7: 93-103, 1972.
99. Riffaldi, R. and M. Schnitzer. Electron spin resonance spectrometry of humic substances. Soil Sci. Soc. Am. Proc. 36: 301-307, 1972.
100. Khan, S.U. and M. Schnitzer. The permanganate oxidation of humic acid, fulvic acids and humins extracted from the Ah horizons of a black Chernozem, a black solod and a black solonetz soil. Can. J. Soil Sci. 52: 43-51, 1972.
101. Khan, S.U. and M. Schnitzer. Permanganate oxidation of humic acids extracted from a grey wooded soil under different cropping systems and fertilizer treatments. Geoderma 7: 113-120, 1972.
102. Matsuda, K. and M. Schnitzer. Oxidation of humic acids from acid soils. Soil Sci. 114: 185-193, 1972.
103. Khan, S.U. and M. Schnitzer. The retention of hydrophobic organic compounds by humic acid. Geochim. et Cosmochim. Acta. 36: 645-654, 1972.
104. Khan, S.U. and M. Schnitzer. The potassium permanganate oxidation of methylated and unmethylated humic acids extracted from solonetz, solod and chernozem Ah horizons. Israel J. Chem. 667-677, 1972.

105. Riffaldi, R. and M. Schnitzer. Effects of diverse experimental conditions on ESR spectra of humic substances. Geoderma 8: 1-10, 1972.
106. Schnitzer, M. and R. Riffaldi. The determination f quinine groups in humic substances. Soil Sci. Soc. Am. Proc. 36: 772-777, 1972.
107. Ortiz de Serra, Maria and M. Schnitzer. The extraction of humic acid by alkali and chelating resin. Can. J. Soil. Sci. 52: 365-374, 1972.
108. Ortiz de Serra, M.I. and M. Schnitzer. The chemistry of humic and fulvic acids extracted from Argentine Soils. I. Analytical characteristics. Soil Biol. Biochem. 5: 281-286. 1972.
109. Ortiz de Serra, M.I. and M. Schnitzer. The chemistry of humic and fulvic acids extracted from Argentine Soils. II. Permanganate oxidation of methylated humic and fulvic acids. Soil Biol. Biochem. 5: 287-296, 1972.
110. Riffaldi, R. and M. Schnitzer. Effects of 6 N HCl hydrolysis on the analytical characteristics and chemical structure of humic acids. Soil Sci. 115: 349-356, 1972.
111. Ortiz, M. I., F.J. Sowden, and M. Schnitzer. The distribution of nitrogen in fungal humic acids. Can. J. Soil Sci. 53: 125-127, 1973.
112. Schnitzer, M. and M.I. Ortiz de Serra and K. Evarson. The chemistry of fungal humic acid-like polymers and of soil humic acids. Soil Sci. Soc. Am. Proc. 37: 229-236, 1973.
113. Schnitzer, M. chemical, spectroscopic and thermal methods for the classification and characterization of humic substances. A chapter in "Humic Compounds in Terrestrial and Aquatic Environments". Centre for Agricultural Publishing and Documentation, Wageningen, pp 293-310. 1975.
114. Gamble, D. and M. Schnitzer. The chemistry of fulvic acid and its reactions with metal ions. Chapter 9 in "Metal-Organic Interactions in Water", edt by S. Singer, Ann Arbor Science Place, Ann Arbor, Michigan, pp 265-302, 1973.
115. Schnitzer, M. the methylation of humic substances. Soil Sci. 117: 94-102, 1974.
116. Schnitzer, M. and M.I. Ortiz de Serra. The sodium-amalgam reduction of soil and fungal humic substances. Geoderma 9: 119-128, 1974.
117. Volk, G.B. and M. Schnitzer. Chemical and spectroscopic methods for assessing subsidence in Florida histosols. Soil Sci. Soc. Am. Proc. 37: 886-1973.
118. Schnitzer, M. and M.I. Ortiz de Serra. The chemical degradation of a humic acid. Can. J. Chem. 51: 1554-1566, 1973.
119. Kodama, H. and M. Schnitzer. Dissolution of chlorite minerals by fulvic acid. Can. J. Soil Sci. 53: 240-243, 1973.
120. Schnitzer, M. Alkaline cupric oxide oxidation of a methylated fulvic acid. Soil Biol. Biochem. 6:1-6, 1974.

121. Schnitzer, M. Investigation on the chemical structure of humic substance by gas chromatography/mass spectrometry. Transactions 10th International Congr. Soil Sci., Moscow II: 294-301, 1974.
122. Schnitzer, M. and S.I.M. Skinner. The low temperature oxidation of humic substances. Can. J. Chem. 52:1072-1080, 1974.
123. Kodama, H. and M. Schnitzer. Further investigations on fulvic acid-Cu^{2}-mont-morillonite interactions. Clays and Clay Minerals 22: 107-110, 1974.
124. Kodam, H. and M. Schnitzer. Adsorption of fulvic acid by non-expanding clay minerals. Transactions 10th Intern. Congr. Soil Sci., Moscow II: 51-66, 1974.
125. Neyroud, J.A. and M. Schnitzer. The alkaline hydrolysis of humic substances. Geoderma 13:171-188, 1975.
126. Neyroud, J.A. and M. Schnitzer. The exhaustive alkaline cupric oxide oxidation of humic acid and fluvic acid. Soil Sci. Soc. Am. Proc. 38: 907-913, 1974.
127. Neyroud, J.A. and M. Schnitzer. The chemistry of high-molecular weight fulvic acid fractions. Can. J. Chem. 52: 4123-4132, 1974.
128. Schnitzer, M. and J.A. Neyroud. Alkanes and fatty acids in humic substances. Fuel 54:17-19, 1975.
129. Schnitzer, M., F.J. Sowden and K.C. Ivarson. Reactions of humic acids with amino acids. Soil Boil. Biochem. 6: 401-407, 1974.
130. Skinner, S.I.M. and M. Schnitzer. Rapid identification by gas chromatography mass spectrometry-computer of organic compounds resulting from the degradation of humic substances. Anal. Chim. Acta. 75:207-211, 1975.
131. Griffith, S.M. and M. Schnitzer. The isolation and characterization of metal-organic complexes from tropical volcanic soils. Soil Sci. 120: 126-131. 1975.
132. Neyroud, J.A. and M. Schnitzer. The mild degradation of humic substances. Agrochimica 19: 116-126, 1975.
133. Schnitzer, M. and H. Kodama. An electron microscopic examination of fulvic acid. Geoderma 13: 279-287, 1975.
134. Schnitzer, M. and E. Vendette. Chemistry of humic substances extracted from an Arctic soil. Can. J. Soil Sci. 55: 99-103, 1975.
135. Griffith, S.M. and Schnitzer. Analytical characteristics of humic acid and fulvic acids extracted from tropical volcanic soils. Soil Sci. Soc. Am. Proc. 39: 861-867, 1975.
136. Griffith, S.M. and M. Schnitzer. Oxidative degradation of humic acid and fulvic acids extracted from tropical volcanic soils. Can. J. Soil Sci. 55: 251-267, 1975.
137. Schnitzer, M. and S.M. Griffith. Novel method for estimating hydrogen bonded CO_2H groups in humic substances. Can. J. Soil Sci. 55: 491-493, 1975.

138. Schnitzer, M. and J.A. Neyroud. Further investigations on the chemistry of fungal humic acids. Soil Biol. Biochem. 7: 365-371, 1975.
139. Sowden, F.J., S.M. Griffith and M. Schnitzer. The distribution of nitrogen in some highly organic tropical volcanic soils. Soil Biol. Biochem. 8: 55-60, 1975.
140. Schnitzer, M. The chemistry of humic substances. Chapter 7 in "Environmental Biogeochemistry", I, edited by J.O. Nriagu, Ann Arbor Science Publishers, Ann Arbor, Michigan, pp 89-107, 1976.
141. Griffith, S.M. and M. Schnitzer. The isolation and characterization of metal-organic complexes from tropical volcanic soils. Chapter 9 in "Environmental Biogeochemistry", I, edited by J.O. Nriagu, Ann Arbor Science Publishers, Ann Arbor, Michigan, pp 117-130, 1976.
142. Griffith, S.M. and M. Schnitzer. The alkaline cupric oxide oxidation of humic and fulvic acids extracted from tropical volcanic soils. Soil Sci. 122: 191-201, 1976.
143. Griffith, S.M., F.J. Sowden and M. Schnitzer. The alkaline hydrolysis of acid resistant nitrogen-containing soil and humic acid residues. Soil Biol. Biochem. 8: 529-531, 1976.
144. Schnitzer, M. and H. Kodama. Reactions of minerals with soil humic substances. Chapter 21 in "Minerals and their roles in the soil environment", edited by J.B. Dixon and S.B. Weed, American Society of Agronomy, Madison, Wisconsin, pp. 741-772, 1977.
145. Schnitzer, M. and H. Kodama. The dissolution of micas by fulvic acid. Geoderma 15: 381-391, 1976.
146. Chen, Y. and M. Schnitzer. Scanning electron microscopy of humic substances and their metal- and clay-complexes. Soil Sci. Soc. Am. Proc. 40: 682-686, 1976.
147. Chen, Y. and M. Schnitzer. Water adsorption on humic substances. Can. J. Soil Sci. 56: 521-524, 1976.
148. Chen, Y., A. Banin and M. Schnitzer. Use of the scanning electron microscope for structural studies on soils and soil components. Scanning Electron Microscopy III, ITT Research Institute, Chicago, Ill. Pp. 425-432, 1976.
149. Senesi, N. and M. Schnitzer. Effects of pH, reaction time, chemical reduction and irradiation on ESR spectra of fulvic acid. Soil Sci. 123: 224-234, 1977.
150. Chen, Y., N. Senesi and M. Schnitzer. Information provided on humic substances by E_4/E_6 ratios. Soil Sci. Soc. Am J. 41: 352-358, 1977.
151. Senesi, N., Y. Chen and M. Schnitzer. The electron and γ-irradiation of humic substances. Fuel 56: 171-176, 1977.
152. Senesi, N., Y. Chen and M. Schnitzer. Hyperfine splitting in ESR spectra of fulvic acid. Soil Biol. Biochem. 9: 371-372, 1977.

153. Senesi, N., S.M. Griffith and M. Schnitzer. Binding of Fe^{+3} by humic materials. Geochim. Cosmochim. Acta. 41: 969-976. 1977.
154. Schnitzer, M. recent findings on the characterization of humic substances extracted from soils from widely different climatic zones. Soil Organic Matter Studies. II International Atomic Energy Agency, Vienna, pp. 117-132, 1977.
155. Neyroud, J.A. and M. Schnitzer. Sur la structure des acides humiques at fulviques du sol. Soil Organic Matter Studies II. International Atomic Energy Agency, Vienna, pp. 157-169, 1977.
156. Senesi, N., Y. Chen and M. Schnitzer. Aggregation-dispersion phenomena in humic substances. Soil Organic Matter Studies II. International Atomic Energy Agency, Vienna, pp. 143-155, 1977.
157. Senesi, N., Y. Chen and M. Schnitzer. The role of free radicals in the oxidation and reduction of fulvic acid. Soil Biol. Biochem. 397-403, 1977.
158. Gamble, D.S., M. Schnitzer and D.S. Skinner. Mn(II)-fulvic acid complexing equilibrium measurements by electron spin resonance spectrometry. Can. J. Soil Sci. 57: 47-53, 1977.
159. Griffith, S.M. and M. Schnitzer. Organic compounds formed by the hydrogen peroxide oxidation of soils. Can. J. Soil Sci. 57: 223-231, 1977.
160. Sowden, F.J., Y. Chen and M. Schnitzer. The nitrogen distribution in soils formed under widely differing climatic conditions. Geochim. et Cosmochim. Acta. 41: 1524-1526, 1977.
161. Chen, Y., F.J. Sowden and M. Schnitzer. Nitrogen in Mediterranean soils. Agrochimica. 21: 7-14, 1977.
162. Kodama, H. and M. Schnitzer. Effect of fulvic acid on the crystallization of Fe(III) oxides. Geoderma 19: 279-291, 1977.
163. Chen, Y., N. Senesi and M. Schnitzer. The chemical degradation of humic and fulvic acids extracted from Mediterranean soils. J. Soil Sci. 29: 350-359, 1978.
164. Chen, Y. and M. Schnitzer. The surface tension of soil humic substances. Soil Sci. 125: 7-15, 1978.
165. Chen, Y., N. Senesi and M. Schnitzer. Chemical and physical characteristics of humic anf fulvic acids extracted from the soils of Mediterranean region. Geoderma 20: 87-104, 1978.
166. Whitby, L.M. and M. Schnitzer. Humic and fulvic acids in sediments and soils of agricultural watersheds. Can. J. Soil Sci. 58: 161-168, 1978.
167. Schnitzer, M. Reactions of humic substances with the minerals in the soil environment. Chapter 51 in Environmental Biogeochemistry and Geomicrobiology. Vol II: 639-647, 1978.
168. Senesi, M. and M. Schnitzer. Free radicals in humic substances. Chapter 36 in Environmental Biogeochemistry and geomicrobiology. Vol II: 467-481, 1978.

169. Chen, Y., S.U. Khan and M. Schnitzer. UV irradiation of dilute fulvic acid solutions. Soil Sci. Soc. Am. J. 42: 292-296, 1978.
170. Mathur, S.P. and M. Schnitzer. A chemical and spectroscopic characterization of some synthetic analogues of humic acids. Soil Sci. Soc. Am. J. 42: 591-596, 1978.
171. Schnitzer, M. Some observations on the chemistry of humic substances. Agrochimica XXII: 216-226, 1978.
172. Schnitzer, M. The chemistry of humic substances. Intern. Congress of soil Sci., Edmonton, Alberta, Canada, I: 51-52, 1978.
173. Schnitzer, M. and M. Levesque. A novel approach to assessing the degree of humification of peats. Soil Sci. 127: 140-145, 1979.
174. Iverson, K.C. and M. Schnitzer. The biodegradability of the "unknown" soil nitrogen. Can. J. Soil Sci. 59: 277-286, 1979.
175. Schnitzer, M. Effect of low pH on the chemical structure and reactions of humic substances. Pages 202-222 in Effects of acid precipitation on terrestrial ecosystems. Edt. by T.C. Hutchinson and M. Havas. Plenum Press, New York, 1979.
176. Schnitzer, M. the chemistry and reactions of humic substances. Pages 807-819 in Ecology and Coal Resource Development (M.K. Wali ed.) Pergamon Press, New York. Vol. II, 1979.
177. Cortez, J. and M. Schnitzer. Nucleic acid bases in soils and their association with organic and inorganic soil constituents. Can. J. Soil Sci. 59: 277-286, 1979.
178. Cortez, J. and M. Schnitzer. Purines and pyrimidines in soils and humic substances. Soil Sci. Soc. Am. J. 43: 958-961, 1979.
179. Ghosh, K. and M. Schnitzer. Macromolecular structure of humic substances. Soil Sci. 129: 266-276, 1980.
180. Ghosh, K. and M. Schnitzer. Some recent advances in the chemistry and reactions of humic substances. Indian J. Chem. LVI: 1090-1093, 1979.
181. Ghosh, K. and M. Schnitzer. UV and visible absorption spectroscopic investigations in relation to macromolecular characteristics of humic substances. J. Soil Sci. 30: 735-745, 1979.
182. Kerndorff, H. and M. Schnitzer. Humic and fulvic acids as indicators of soil and water pollution. J. Water, Air and Soil Pollution 12: 319-329, 1979.
183. Biederbeck, V.O., C.A. Campbell, K. E. Bowren, M. Schnitzer and R.N. McIver. Effect of burning cereal straw on soil properties and grain yields in Saskatchewan. Soil Sci. Soc. Am. J. 44: 103-111, 1980.
184. Schnitzer, M., H. Kodama and K.C. Iverson. Interaction of "protein-like" materials in fulvic acid with montmorillonite, Zeitschr. Pflanzenernahr. Bodenk. 143: 334-343, 1980.

185. Kodama, H. and M. Schnitzer. Effect of fulvic acid on crystallization of aluminum hydroxides. Geoderma 24: 195-205, 1980.
186. Ghosh, K. and M. Schnitzer. Fluorescence excitation spectra of humic substances. Can. J. Soil Sci. 60: 373-379, 1980.
187. Schnitzer, M. Reaction of humic substances with metals and minerals in soils. International Symposium on "Migrations organo-minerales dans les sols temperes". Centre de Pedolologie Biologique, Nancy, France. Pp.229-234, 1981.
188. Griffith, S.M., J. Silver and M. Schnitzer. Hydrazine derivatives at Fe^{3+} site in humic materials. Geoderma 23: 299-302, 1981.
189. Schnitzer, M. and D.A. Hindle. Effect of peracetic acid oxidation on N-containing components of humic materials. Can. J. Soil Sci. 60: 541-548, 1980.
190. Schnitzer, M. and H. Kerndorff. Effects of pollution on humic substances. J. Environ. Sci. Health B 15: 431-456, 1980.
191. Cortez, J. and M. Schnitzer. Reactions of nucleic acid bases with inorganic soil constituents. Soil Biol. Biochem. 13: 173-178, 1981.
192. Ghosh, K. and M. Schnitzer. Fluroscence excitation spectra and viscosity behaviour of a fulvic acid and its copper and iron complexes. Soil Sci. Soc. Am. J. 45: 25-29, 1981.
193. Ghosh, K. and M. Schnitzer. Effects of pH and natural electrolyte concentration on free radical in humic substances. Soil Sci. Soc. Am. J. 44: 975-978, 1981
194. Kerndorff, H. and M. Schnitzer. Sorption of metal on humic acid. Geochim. Cosmochim. Acta. 44: 1701-1708, 1981.
195. Schnitzer, M. and H. Kerndorff. Reactions of fulvic acid with metal ions. J. Water, Air and Soil Pollution 15: 97-108, 1981.
196. Schnitzer, M. and D.A. Hindle. Effects of different methods of acid hydrolysis on the nitrogen distributions in two soils. Plant and Soil 60: 237-243, 1981.
197. Stevenson, I.L. and M. Schnitzer. Transmission electron microscopy of extracted fulvic and humic acids. Soil Sci. 133: 179-185, 1982.
198. Schnitzer, M., L.E. Lowe, J.F. Dormaar and Y. Martel. A procedure for the characterization of soil organic mater. Can. J. Soil Sci. 61: 517-519, 1981.
199. Rauthan, B.S. and M. Schnitzer. Effects of a soil fulvic acid on the growth and nutrient content of cucumber (*Cucumis sativus*) plants. Plant and Soil 63: 491-495, 1981.
200. Hatcher, P., M. Schnitzer, L.W. Dennis and G.E. Maciel. Aromaticity of humic substances in soils. Soil Sci. Soc. Am. J. 45: 1089-1094, 1981.
201. Ivarson, K.C., M. Schnitzer and J. Cortez. The biodegradability of nucleic acid bases absorbed on inorganic and organic soil components. Plant and Soil 64: 343-353, 1982.

202. Schnitzer, M. Extraction, Fractionation and characterization of soil organic matter. A chapter in "Methods of Soil Analysis", Part II, 581-594, Madison, Wisconsin, 1982.
203. Schnitzer, M. and K. Ghosh. Characteristics of water-soluble fulvic acid-copper and fulvic acid-iron complexes. Soil Sci. 134: 354-363, 1982.
204. Ghosh, K. and M. Schnitzer. A scanning electron microscopic study of effects of adding neutral electrolytes to solutions of humic substances. Geoderma 28: 53-56, 1982.
205. Schnitzer, M. Recent advances in humic acid research. Proc. Int. Peat Symp. Bemidji, Minnesota, 17-43, 1982.
206. Spiteller, M. and M. Schnitzer. A comparison of the structural characteristics of pollymaleic acid and a soil fulvic acid. J. Soil Sci. 3: 525-537, 1983.
207. Schnitzer, M. and K.C. Ivarson. Different forms of nitrogen in particle size fractions separated from two soils. Plant and Soil 69: 383-389, 1982.
208. Schnitzer, M. Quo vadis soil organic matter research. Transactions of the 12th Intern. Congress of Soil Sci. 5: 67-78, 1982.
209. Gamble, D.A., M. Schnitzer, H. Kerndorff and C.H. Langford. Multiple metal ion exchange equilibria with humic acid. Geochim. et Cosmochim. Acta. 47: 1311-1323, 1983.
210. Schnitzer, M., P.R. Marshall and D.A. Hindle. The isolation of soil humic and fulvic acid components rich in "unknown" nitrogen. Can. J. Soil Sci. 63: 425-433, 1983.
211. Stevenson, I.L. and M. Schnitzer. Energy-dispersive X-ray microanalysis of saturated fulvic acid-iron and –copper complexes. Soil Sci. 138: 123-126, 1984.
212. Schnitzer, M. and C.M. Preston. Effect of acid hydrolysis on the ^{13}C NMR spectra of humic substances. Plant and Soil 75: 201-211, 1983.
213. Kodama, H., M. Schnitzer and M. Jaakkimainen. Chlorite and biotite weathering by fulvic acid solutions in closed and open systems. Can. J. Soil Sci. 63: 619-629, 1983.
214. Preston, C.M. and M. Schnitzer. Effect of chemical modifications and the extractions on the ^{13}C NMR spectra of humic materials. Soil Sci. Soc. Am. J. 48: 305-311, 1984.
215. Schnitzer, M., M. Barr and R. Hartenstein. Kinetics and characteristics of humic acids produces from simple phenols. Soil Biol. Biochem. 16: 371-376, 1984.
216. Calderoni, G. and M. Schnitzer. Nitrogen distribution as a function of radiocarbon age in paleosol humic acids. Org. Geochem. 5: 203-209, 1984.
217. Chan, Y.K., P.R. Marshall and M. Schnitzer. Proteolytic hydrolysis of humic acid by pronase, in Plantetary Ecology, edt. By D.E. Caldwell, J.A. Brierley and C.L. Brierley, Van Nostrand Reinhard. 562-570, 1985.

218. Schnitzer, M. the nature of nitrogen in humic substances. Chapter 12 in Humic Substances in Soil, Sediment and Water, edt. By G.R. Aiken, D.M. McKnight R.L. Wershaw and P.A. MacCarthy, Wiley Interscience, 303-325, 1985.
219. Schnitzer, M. Reactions of humic substances with metals and minerals. Chapter 24 in Mineral Explorations: Biological Systems and Organic Matter, edt. By D. Carlisle, W.L. Berry, I.R. Kaplan and J. W. Waterson, Rubey Volume V. Prentice-Hall, Englewood Cliffs, N.J., 408-427, 1986.
220. Schnitzer, M. Soil organic matter: its role in the environment. Chapter 10 in Environmental Geochemistry, Short Course Handbook, Volume 10, edt. By M.E. Fleet, Mineralogical Association of Canada, Toronto, Ont., 237-267, 1984.
221. Calderoni, G. and M. Schnitzer. Effects of age on the chemical structure of paleosol humic acids and fulvic acids. Geochim. et Cosmochim. Acta 48: 2045-2051, 1984.
222. Belluomini, G., M. Branca, G. Calderoni and M. Schnitzer. Distribution and geochemical significance of amino acids and amino sugars in a clay suite of the poleocine-pleistocene age from Italy. Org. Geochem. 9: 127-133, 1986.
223. Schnitzer, M. and G. Calderoni. Some chemical characteristics of paleosol humic acids. Chemical Geology. 53: 175-184, 1985.
224. Miles, N.M., C.R. de Kimpe and M. Schnitzer. Effect of soil organic matter oxidation on the collapse of vermiculite. Can J. Soil Sci. 65: 593-597, 1985.
225. Schnitzer, M. and Y.K. Chan. Structural characteristics of a fungal melanin and a soil humic acid. Soil Sci. Soc. Am. J. 50: 67-71, 1986.
226. Schnitzer, M. Water retention by humic substances. Chapter 9 in Peat and Water, edt. by C. Fuchsman. Elsevier Applied Science Publishers, London, 159-176, 1986.
227. Schnitzer, M. Binding of substances by soil mineral colloids. Chapter 4 in Interactions of Soil Minerals with Natural Organics and Microbes, edt. by P.M. Huang and M. Schnitzer, Soil Science Society of America, Madison Wisconsin, 77-101, 1986.
228. Schnitzer, M. and C.M. Preston. The analysis of humic acids by solution and solid state ^{13}C NMR. Soil Sci. Soc. Am. J. 50: 326-331, 1986.
229. Schnitzer, M., D.A. Hindle and M. Meglic. Supercritical-gas extraction of alkanes and alkanoic acids from soils and humic materials. Soil Sci. Soc. Am. J. 50: 913-919, 1986.
230. Schnitzer, M. and Spiteller. The chemistry of "unknown" soil nitrogen. Transactions XIIIth International Congress of Soil Science (Hamburg) II: 473-474, 1986.

231. Biederbeck, V.O., C.A. Campbell and M. Schnitzer. Effects of wheat rotation and fertilization on microorganisms and biochemical properties of a brown loam in Saskatchewan. Transactions XIIIth International Congress of Soil Science (Hamburg) II: 552-553, 1986.

232. MacCarthy, P., R.L. Malcom, M.H.B. Hayes, R.S. Swift, M. Schnitzer and W.L. Campbell. Establishment of a collection of standard humic substances. Transactions XIIIth International Congress of Soil Science (Hamburg) II: 378-379, 1986.

233. Campbell, C.A., M. Schnitzer, J.W.B. Stewart, V.O. Biederbeck and F. Selles. Effects of manure and P fertilizers on properties of a black charnozem in southern Saskatchewan. Can. J. Soil Sci. 66: 601-614, 1986.

234. Arshad, M.A. and M. Schnitzer. The chemistry of a termite fungus comb. Plant and Soil 98: 247-256, 1987.

235. Arshad, M.A., M. Schnitzer and C. Preston. Characteristics of a termite fungus comb studied by chemical and spectroscopic methods. Soil Biol. Biochem. 19: 227-230, 1987.

236. Schnitzer, M. and C.M. Preston. The supercritical gas extraction of a soil with solvents of increasing polarities. Soil Sci. Soc. Am. J. 51: 639-646, 1987.

237. Catroux, G. and M. Schnitzer. Chemical, spectroscopic and biological characteristics of the organic matter in particle size fractions separated from an Aquoll. Soil Sci. Soc. Am. J. 51: 1200-1207, 1987.

238. Preston, C.M. and M. Schnitzer. ^{13}C NMR of humic substances: pH and solvent effects. J. Soil Sci. 38: 667-678, 1987.

239. Arshad, M.A. and M. Schnitzer. Characteristics of the organic matter in a slightly and in a severely crusted soil. Z Pflanzenernachr. Bodenk. 150: 412-416, 1987.

240. Byler, D.M., W.V. Gerasimowicz, H. Susi and M. Schnitzer. FTIR spectra of soil constituents: Fulvic acid and fulvic acid complex with ferric ions. Applied Spectroscopy 41: 1428-1430, 1987.

241. Baldock, J.A., M. Schnitzer and B.D. Kay. Influence of selected cropping treatments and aggregate sizes on soil carbohydrates. Can. J. Soil Sci. 67: 489-499, 1987.

242. Schnitzer, M, J.A. Ripmeester and H. Kodama. Characterization of the organic matter associated with a soil clay. Soil Sci. 145: 448-454, 1988.

243. Schnitzer, M. Selected methods for the characterization of soil humic substances. Humic substances in Soil and Crop Sciences; selected Readings. Special publication of the Soil Sci. Soc. Am. Madison, Wisconsin, 65-89, 1990.

244. Campbell, C.A., V.O. Biederbeck, M. Schnitzer, F. Selles and R.P. Zentner. Effect of 6 years of zero tillage and N fertilizer management on changes in

soil quality of an Orthic Brown Chernozem in southwestern Saskatchewan. Soil Tillage Research. 14: 39-52, 1989.

245. Hatcher, P., M. Schnitzer, A. Vassalo and M.A. Wilson. The chemical structure of three highly aromatic humic acids as determined by dipolar dephasing NMR studies. Geochim. Cosmochim. Acta. 53: 125-130, 1989.

246. Kodama, H., M. Schnitzer and E. Murad. An investigation of iron (III) fulvic acid complexes by Mossbauer spectroscopy and chemical methods. Soil Sci. Soc. Am. J. 52: 994-998, 1988.

247. Preston, C.M., M. Schnitzer. And J.A. Ripmeester. Carbon-13 CPMAS NMR and chemical investigation of HCl/HF deashing of humin from a mineral soil. Soil Sci. Soc. Am. J. 53: 1442-1447, 1989.

248. Arshad, M.S., J.A. Ripmeester and M. Schnitzer. Attempts to improve solid state ^{13}C NMR spectra of whole mineral soils. Can. J. Soil Sci. 68: 593-602, 1988.

249. Arshad, M.A., M. Schnitzer and C.M. Preston. Characterization of humic acids from termite mounds and surrounding soils in different ecological regions of Kenya. Geoderma 42: 213-225, 1988.

250. Schnitzer, M. and H.-R. Schulten. Pyrolysis-soft ionization mass spectrometry of aliphatics extracted from a soil clay and humic substances. The Science of the Total Environment. 81/82: 19-30, 1989.

251. Schnitzer, M. and P. Schuppli. A method of the extraction of organic matter from soil with 0.5 M NaOH and 0.1 M $Na_4P_2O_7$ solutions. Can. J. Soil Sci. 69: 253-262, 1989.

252. Schnitzer, M. and P. Schuppli. Method for the sequential extraction of organic matter from soils and soil fractions. Soil Sci. Soc. Am. J. 53: 1418-1428, 1989.

253. Arshad, M.A. and M. Schnitzer. Chemical characteristics of humic acids from five soils in Kenya. Z. Pflanzenernaehr. Bodenk. 152: 11-16, 1989.

254. Griffith, S.M. and M. Schnitzer. Oxidative degradation of soil humic substances II, edited by M.H.B. Hayes, P. MacCarthy, R.L. Malcom and R.S. Swift, John Wiley, New York, pp. 69-98, 1989.

255. Chen, Y. and M. Schnitzer. Molecular sizes and shapes of humic substances by electron microscopy. Chapter 22 in Humic Substances II, edited by M.H.B. Hayes, P. MacCarthy, R R.L. Malcom and R.S. Swift, John Wiley, New York, pp. 621-638, 1989.

256. Rosell, A., A.E. Andriulo, M. Schnitzer, M.B. Crespo and A.M. Miglierina. Humic acid properties of an Argiudoll soil under two tillage systems. Sci. Tot. Environ. 81/82: 391-400, 1989.

257. Saharinen, M. and M. Schnitzer. Nitrogen in a Finnish agricultural soil and its humic acid compared to two Canadian soils. Sci. Tot. Environ. 81/82: 459-463, 1989.

258. Dinel, H., M. Schnitzer and G. Mehuys. Soil lipids: origin, nature, content, decomposition and effect on soil physical properties. Soil Biochem. 6: 397-429, 1990.
259. Schulten, H.-R. and M. Schnitzer. Aliphatics in soil organic matter associated with clays. Soil Sci. Soc. Am. J. 54: 98-105, 1990.
260. Schnitzer, M. Soil organic matter and soil quality. Agriculture Canada, 1991.
261. De Kempe, C.R. and M. Schnitzer. Low-temperature ashing of humic and fulvic acids. Soil Sci. Soc. Am. J. 54: 399-403, 1990.
262. Schnitzer, M., C. Tarnocia, P. Schuppli and H.-R. Schulten. Nature of the organic matter in tertiary Palescols in the Canadian Arctic. Soil Sci. 149: 257-267, 1990.
263. Schnitzer, M. C. Tarnocia, P. Schuppli, R. Hempfling, R. Muller and H.-R. Schulten. Paleoenvironmental organic indicators in Eocene paleosols from Arctic Canada. Fresenius Z. Anal. Chem. 337: 1882-1884, 1990.
264. Schnitzer, M., H.-R. Schulten, P. Schuppli and D.A. Angers. Extraction of organic matter from soils with water at high pressure and temperatures Soil Sci. Soc. Am. J. 55: 102-108, 1991.
265. Schnitzer, M. Soil organic matter — the next 75 years. Soil Sci. 149: 257-267, 1991.
266. Mathur S.P., M. Schnitzer and P. Schuppli. The distribution of nitrogen in peat-based composts of manure slurries and fisheries wastes. Biol. Agr. Hort. 7: 153-163, 1990.
267. Campbell, C.A., M. Schnitzer, G.P. Lafond, R.V. Zentner and J.E. Knipel. Effects of crop rotations and management practices on soil- and amino- N in a Thin Black Chernozem. Soil Sci. Soc. Am. J. 55: 739-745, 1991.
268. Arshad, M.A., M. Schnitzer, D.A. Angers and J.A. Ripmeester. Effects of till vs no-till on the quality of soil organic matter. Soil Biol. Biochem. 22: 595-599, 1990.
269. Schnitzer, M., H. Kodama and J.A. Ripmeester. Determination of aromaticity of humic substances by X-ray diffraction analysis. Soil Sci. Soc. Am. J. 55: 745-750, 1991.
270. Schnitzer, M. Significance of soil organic matter in soil formation, transport processes in soils and in the formation of soil structures. Barichte uber Landwirtschaft. 206: 63-81, 1992.
271. Schulten, H.-R, B. Plage and M. Schnitzer. A chemical structure for humic substances. Naturwissenschaften 78: 311-312, 1991.
272. Schulten, H.-R and M. Schnitzer. A contribution to solving the puzzle of the chemical structure of humic substances: pyrolysis soft-ionization mass spectrometry. Sci. Tot. Environ. 117/118: 27-39, 1992.
273. Schulten, H.-R and M. Schnitzer. Structural studies of Bainsville humic acids by Curie-point pyrolysis gas chromatography/mass spectrometry. Soil Sci. 153: 205-224, 1992.

274. Schulten, H.-R and M. Schnitzer. Supercritical carbon dioxide extraction of long-chain aliphatics from two soils. Soil Sci. Soc. Am. J. 55: 1603-1611, 1991.
275. Schnitzer, M. and H. Kodama. Interaction between organic and inorganic components in particle size fractions separated from four soils. Soil Sci. Soc. Am. J. 56: 1099-1105, 1992.
276. Schnitzer, M. and H.-R. Schulten. The analysis of soil organic matter by pyrolysis-field ionization mass spectrometry. Soil Sci. Soc. Am. J. 56: 1811-1817, 1992.
277. Campbell, C.A., S.A. Brandt, V.O. Biederbeck, R.P. Zentner and M. Schnitzer. Effect of crop rotations and rotation phase on characteristics of soil organic matter in a dark brown chernozemic soil. Can. J. Soil Sci. 72: 403-416, 1992.
278. Campbell, C.A., K.A. Bowren, M. Schnitzer, R.P. Zentner and L. Townley-Smith. Effect of crop rotations and fertilization on soil organic matter and some biochemical properties of a thick black chernozem. Can. J. Soil Sci. 71: 377-387, 1991.
279. Khan, S.U., M. Schnitzer and H.-R. Schulten. Fate of deltamethrin after nine years of incubation in an organic soil under laboratory conditions. J. Agr. Food Chem. 41: 1143-1151, 1993.
280. Schulten, H.-R. and M. Schnitzer. Temperature-resolved in-source pyrolysis-soft ionization mass spectrometry of soil humic acids. Org. Geochem. 20: 17-25, 1993.
281. Schulten, H.-R. and M. Schnitzer. A state of the art structural concept for humic substances. Naturwissenschaften 80: 29-30, 1993.
282. Sorge, C., M. Schnitzer and H.-R. Schulten. In-source pyrolysis-field ionization mass spectrometry and Currie-point pyrolysis — gas chromatography/mass spectrometry of amino acids in humic substances and soils. Biol. Fertl. Soils 16: 100-110, 1993.
283. Mathur, S.P., G. Owen, H. Dinel, M. Schnitzer and J. Dugan. Determination of compost biomaturity: 1. Literature review. Biol. Agr. Hort. 10: 65-85, 1993.
284. Mathur, S.P., G. Owen, H. Dinel and M. Schnitzer. Determination of compost biomaturity: 2. optical density of water extracts of composts as a reflection of their maturity. Biol. Agr. Hort. 10: 87-108, 1993.
285. Schnitzer, M. H. Dinel, S.P. Mathur, H.-R. Schulten and G. Owen. Determination of compost biomaturity: 3. Evaluation of a colorometric test by ^{13}C NMR spectroscopy and pyrolysi-field ionization mass spectrometry. Biol. Agr. Hort. 10: 109-123, 1993.
286. Monreal, C.M., M.Gregorich, M. Schnitzer, and D.W. Anderson. The quality of soil organic matter as characterized by CP/MAS ^{13}C NMR and Py-FIMS. Chapter 16 in Environmental Impact of Soil Component Interactions, edt. by P.M. Huang, J. Berthelin, J.-M. Bollag, W.B. McGill and A.L. Page. Pp. 207-215, Lewis Publishers Boca Raton, 1995.

287. Preston, C.M., R. Hempfling, H.-R. Schulten, M. Schnitzer, J.A. Torfymow and D.E. Axelson. Characterization of organic matter in the soil profile in a young Douglas-fir plantation. Plant and Soil 158: 69-82, 1994.
288. Schnitzer, M. Organic-inorganic interactions in soil and their effects in soil quality. Chapter 1 in Environmental Impact of Soil Component Interactions, edt. by P.M. Huang, J. Berthelin, J.-M. Bollag, W.B. McGill and A.L. Page. pp. 3-19, Lewis Publishers Boca Raton, 1995.
289. Schnitzer, M. A chemical structure for humic acid. Chemical, ^{13}C NMR, colloid-chemical and electron microscopic evidence in Humic Substances in the Global Environment and Implications on Human Health, edt. by N. Senesi and T.M. Miano. pp. 57-59, Elsevier, 1994.
290. Schnitzer, M., H. Kodama and H.-R. Schulten. Mineral effects on the pyrolysis-field ionization mass spectrometry of fulvic acid. Soil Sci. Soc. Am. J. 58: 1100-1107, 1994.
291. Monreal, C.M., M. Schnitzer, H.-R. Schulten, C.A. Campbell and D.W. Anderson. Soil organic structures in macro and microaggregates of a cultivated brown chernozem. Soil Biol. Biochem. 27: 845-853, 1995.
292. Righi, D., H. Dinel, H.-R. Schulten and M. Schnitzer. Characterization of clay organic matter complexes resistant to oxidation by peroxides. Europ. J. Soil Sci. 46: 423-429, 1995.
293. Sorge, C., M. Schnitzer, P. Leinweber, and H.-R. Schulten. Molecular chemical characterization of organic matter in whole soil and particle-size fractions of a spodosol by pyrolysis-field ionization mass spectrometry. Soil Sci. 158: 189-203, 1994.
294. Schnitzer, M. Research on prodzolization at MacDonald College from 147 to 1959. Can. J. Soil Sci. 75: 155-159, 1995.
295. Schulten, H.-R., C.M. Monreal and M. Schnitzer. Effect of long-term cultivation on the chemical structure of soil organic matter. Naturwissenschaften 82: 42-44, 1995.
296. Schulten, H.-R., C. Sorge and M. Schnitzer. Structural studies on soil nitrogen by Currie-point pyrolysis-gas chromatography/mass spectrometry with nitrogen selective detection. Biol. Fert. Soil 20: 174-184, 1995.
297. Liang, B.C., E.G. Gregorich and M. Schnitzer and R.P. Voroney. Carbon mineralization in soils of different textures as affected by water soluble organic carbon extracted from composted dairy manure. Biol. Fert. Soil 21: 10-16, 1996.
298. Van Bochove, E., D. Conillard, M. Schnitzer and H.-R. Schulten. Pyrolysis-field ionization mass spectrometry of the four phases of cow manure composting. Soil Sci. Soc. Am. J. 60: 1781-1786, 1996.
299. Schnitzer, M. and H.-R. Schulten. Analysis of organic matter in soil extracts and whole soils by pyrolysis mass spectrometry. Adv. Agronomy. 55: 167-217, 1995.

300. Gregorich, E.G., M. Schnitzer, C.M. Monreal and H.-R. Schulten. Management effects and transformation of plant residues into soil organic matter: plant tissues, isolated soil fractions and whole soils. Soil. Sci. 161: 693, 1996.
301. Liang, B.C., E.G. Gregorich and M. Schnitzer. Mineral nitrogen accumulations in soils as affected by water soluble organic carbon extracted from composted dairy manure. Commun. Soil Sci. Plant Anal. 26: 2711-2723, 1995.
302. Dinel, H., M. Schnitzer and S.U. Dumontet. Compost maturity: chemical characteristics of extractable lipids. Compost Sci. Util. 4: 16-25, 1996.
303. Schulten, H.-R. and M. Schnitzer. Three-dimensional models for humic acids and soil organic matter. Naturwissenschaften 82: 487-498, 1995.
304. Liang, B.C., E.G. Gregorich, M. Schnitzer and H.-R. Schulten. Chemical and spectroscopic characterization of water extracts of stockpiled and composted dairy manures and their adsorption by soils. Soil Sci. Sco. Am. J. 60: 1758-1763, 1996.
305. Schnitzer, M. and H.-R. Schulten. New ideas on the chemical make-up of soil organic matter in Future Prospects of Soil Chemistry SSSA Special publication No: 155, pp 153-177, Madison, Wisconsin 1998.
306. Dinel, H., M. Schnitzer and S. Dumontet. Compost maturity: extractable lipids as indicators of organic matter stability. Compost Sci. Util. 4: 6-12, 1996.
307. Monreal, C.M., H. Dinel, M. Schnitzer, D.S. Gamble and V.O. Biederbeck. Impact of carbon sequestration on indicators of soil quality as influenced by management in sustainable agriculture in Soil Process and the Carbon Cycle (ed. By R. Lal, J. Kimble and B.A. Stewart). Adv. in Soil. Sci. 435-458, 1997.
308. Celi, L. M. Schnitzer and M. Negre. Determination of carboxyl groups in humic acids by wet chemical, Fourier-transform infrared spectroscopic, and ^{13}C NMR spectroscopic methods: A comparative study. Soil Sci. 162-189: 190-197, 1997.
309. Schulten H.-R. and M. Schnitzer. Chemical model structures for soil organic matter and soils. Soil Sci. 162: 115-130, 1997.
310. Celi, L., M. Gennari, S.U. Khan and M. Schnitzer. Extractable and non-extractable (bound) residues of scifluorfen in an organic soil. Arg. Foof Chem. 45: 36773680, 1997.
311. Celi, L., M. Gennari, S.U. Khan and M. Schnitzer. Mechanism of acifluorfen interaction with humic acid. Soil Sci. Soc. Am. J. 61: 1659-1665, 1997.
312. Schulten, H.-R., C. Sorge-Lewin and M. Schnitzer. The structure of "unknown" soil nitrogen investigated by analytical pyrolysis. Biol. Fert. Soils 24: 249-254, 1997.
313. Schulten H.-R. and M. Schnitzer. The chemistry of soil nitrogen: a review. Biol. Fert. Soils 26: 1-15, 1998.
314. Schulten H.-R., P. Leinweber and M. Schnitzer. Analytical pyrolysis and computer modeling of humic and soil particles, in Environmental Particles, Vol. 4, IUPAC, 281-324, 1998.

315. Liang, B.C., A.F. MacKenzie, M. Schnitzer, C.M. Monreal, P.R. Voroney. Management induced chemical change in labile organic matter under continuous corn in eastern Canadian soils. Biol. Fert. Soils 26: 88-94, 1998.
316. Pare, T., H. Dinel, M. Schnitzer and S. Dumontet. Transformation of carbon and nitrogen during composting of animal manure and shredded paper. Bio. Fert. Soils 26: 173-178, 1998.
317. Dinel, H., C.M. Monreal and M. Schnitzer. Extractable lipids and organic matter status as influenced by tillage in two catenas. Geoderma 86: 279-293, 1998.
318. Dinel, H. M. Schnitzer and H.-R. Schulten. Chemical and spectroscopic characteristics of colloidal fractions separated from liquid pig manures. Soil Sci. 163: 665-673, 1998.
319. Arshad, C., M. Schnitzer and H.-R. Schulten. Effect of high magnetic field on the chemical composition of soil organic matter. World Congr. Soil Sci., France, Symp. 7, No. 1128: 1-4, 1998.
320. Liang, B.C., E.G. Gregorich, A.F. MacKenzie, M. Schnitzer, P.R. Voroney C.M. Monreal and R.P. Beyaert. Reaction and turn-over of corn residue carbon in eastern Canadian soils. Soil Sci. Soc. Am. J. 62: 1361-1366, 1998.
321. Pare, T., H. Dinel and M. Schnitzer. Extractability of trace metals during co-composting of biosolids and municipal solid wastes. Biol. Fert. Soils 29: 31-37, 1999.
322. Dinel, H., M. Schnitzer, T. Pare, L. Lemee, A. Ambles, E. Topp and N. Pelzer. Effects of direct land application of calcitic lime, and lime- and cement kiln dust-sensitized biosolids on the chemical and spectroscopic characterization of soil lipids. Soil Sci. 164: 322-330, 1999.
323. Dinel, H., T. Pare, M. Schnitzer and N. Pelzer. Direct land application of cement kiln dust and lime-sensitized biosolids: extractability of trace metals and organic matter quality. Geoderma 96: 307-320, 2000.
324. Schnitzer, M. A lifetime perspective on the chemistry of soil organic matter. Adv. Agron. 68: 1-58, 2000.
325. Dinel, H., T. Pare and M. Schnitzer. Carbon and nitrogen mineralization in soil amended with non-tabletized and tebletized poultry manure. Can. J. Soil Sci. 80: 271-276, 2000.
326. Schnitzer, M. The in-situ analysis of organic matter in solids. Can. J. Soil. Sci. 81: 249-254, 2001.
327. Schnitzer, M., H. Dinel, H.-R. Schulten, T. Pare and S. Lafond. Humification of duck farm wastes in Humic Substances, Versatile Components of Plants, Soils and Waters (edt. by E.A. Ghabbour and G. Davies), pp 21-35, Royal Society of Chemistry, Cambridge, UK, 2000.
328. Dumontet, S., H. Dinel, M. Schnitzer, T. Pare and A. Scopa. Composting organic residues: trace metals and microbial pathogens. Can. J. Soil. Sci. 81: 357-367, 2001.

329. Dinel, H., M. Schnitzer, T. Pare, L. Lemee, A. Ambles and S. Lafond. Changes in lipids and sterols during composting. J. Environ. Sci. Health B 36: 651-665, 2001.
330. Tudoret, M.-J., M. Schnitzer and H. Dinel. Some aspects of organic and analytical chemistry of lignite-like materials: A literature review. Agriculture and Agri-Food Canada, 2000.
331. Saharinen, M., H. Dinel and M. Schnitzer. Effects of humic substances on plants and soil properties. Agriculture and Agri-Food Canada, 2000.
332. Lafond, S., T. Pare, H. Dinel, M. Schnitzer, J. Chambers and A. Jaonich. Composting duck excreta-enriched wood shavings: C and N transformations and bacterial pathogen reductions. J. Environ. Sci. Health B 35: 651-665, 2000.
333. Chen, Y., J. Katan, A. Gamliel, T. Aviad and M. Schnitzer. Involvement of soluble organic matter in increased plant growth in solarized soils. Biol. Fert. Soils 32: 28-34, 2000.
334. Schnitzer, M. the chemistry of nitrogen in soils (in E. Keinan and I. Schechter eds.) chemistry for the 21st century. pp 117-128, Wiley-VCH, Weinheim, 2001.
335. Ozdoba, D.M., J. C. Blyth, K.F. Eugler, H. Dinel and M. Schnitzer. Leonadite and humified organic matter in Humic Substances: Structures, Models and Functions (edt. by E.A. Ghabour and G. Davies), pp 309-313, Royal Society of Chemistry, Cambridge, UK, 2001.
336. Schnitzer, M., H. Dinel, T. Pare, H.-R. Schulten and D. Ozdoba. Some chemical and spectroscopic characteristics of six organic ores, in Humic Substances: Structures, Models and Functions (edt. by E.A. Ghabour and G. Davies), pp 315-328, Royal Society of Chemistry, Cambridge, UK, 2001.
337. Dinel, H., M. Schnitzer, H.-R. Schulten, D. Ozdoba and T. Marche. Interaction by principal component analysis of pyrolysis-field ionization mass spectra of lignite ores, Humic Substances: Structures, Models and Functions (edt. by E.A. Ghabour and G. Davies), pp 315-328, Royal Society of Chemistry, Cambridge, UK, 2001.
338. Pare, T., M. Saharinen, M.J. Tucloret, H. Dinel, M. Schnitzer and D. Ozdoba. Response of alfalfa to calcium lignite fertilizer, in Humic Substances: Structures, Models and Functions (edt. by E.A. Ghabour and G. Davies), pp 315-328, Royal Society of Chemistry, Cambridge, UK, 2001.
339. Dinel, H., M. Schnitzer, T. Pare, L. Lemee, A. Ambles, and S. Lafond. Changes in lipids and sterols during composting. J. Environ. Sci. Health B 36(5): 651-665, 2001.
340. McArrthur, D.F.E., H.-R. Schulten, M. Schnitzer, L.M. Kozak and P.M. Huang. Impact of long term cultivation on the status of organic matter in selected Canadian Prerie Soils. Proceedings of the 17th World Congress of Soil Science, Bangkok, 2002.

341. Jokic, A. H.-R. Schulten, M. Schnitzer and P. M. Huang. Maillard Reaction catalyzed by δ-MnO_2: a significant mechanism in formation of "unknown nitrogen" in soils. Proceedings of the 17th World Congress of Soil Science, Bangkok, 2002.
342. Marche, T., M. Schnitzer, H. Dinel, T. Pare, P. Champagne, H.-R. Schulten and G. Facey. Chemical changes during composting of a paper mill sludge-hardwood-sawdust mixture. Geoderma 116: 345-356, 2003.
343. Dinel, H., M. Schnitzer, M. Saharinen, F. Meloche, T. Pare, S. Dumontet, L. Lemee and A. Ambles. Extractable soil lipids and microbial activity as affected by Bt and non Bt maize grown on a silty clay loam soil. J. Environ. Sci. Health B 38: 211-219, 2003.
344. Schnitzer, M. The best teacher I ever had in A.C. Michalos editor, The Althouse Press, The University of Western Ontario, 240-241, 2003.
345. Dinel, H., T. Marche, M. Schnitzer, T. Pare and P. Champagne. Co-composting of paper mill sludge and hardwood sawdust under two types of in-vessel processes. J. Environ. Sci. Health B 39: 139-151, 2004.
346. Jokic, A., H.-R. Schulten, J.N. Cutler, M. Schnitzer and P.M. Huang. A significant abiotic pathway for the formation of unknown" nitrogen in nature. Geophysical Research Letters 31: 1-4, 2004.
347. Schnitzer, M., Organic Matter. Encyclopedia of Soil Sci. Elsevier, pp 85-93, 2005.
348. Jokic, A., H.-R. Schulten, J.N. Cutler, M. Schnitzer and P.M. Huang. Catalysis of the Maillard reaction by δ-MnO_2: A significant abiotic sorptive condensation pathway for the formation of refractory N-containing biomacromolecules in nature, in Soil Abiotic and Biotic Interactions and Impact on the Ecosystem and Human Welfare (edt. by P.M. Huang, A. Violante, J.-M. Bollag and P. Vityakon), pp. 127-152, Science Publishers Inc. Enfield (NH), USA, 2005.
349. Malhi, S.S., Z.H. Wang, M. Schnitzer, C.M. Monreal and J.T. Harapiak. Nitrogen fertilization effects on quality of organic matter in a grassland soil. Nutrient Cycling in Agroecosystems 73: 191-199, 2005.
350. Schnitzer, M., D.E.F. McArthur, H.-R. Schulten, L.M. Kozak and P.M. Huang. Long term cultivation effects on the quantity and quality of organic matter in selected Canadian prairie soils. Geoderma 130: 141-156, 2006.
351. Schnitzer, M.I., C.M. Monreal, G.A.Facey, P.B. Fransham. The conversion of chicken manure to biooil by fast pyrolysis I. Analysis of bio-oils by FTIR and NMR spectroscopy. J. Environ. Sci. Health B 42: 71-77, 2007.
352. Schnitzer, M., C.M. Monreal, G. Jandl, P. Leinweber and P.B. Fransham. The conversion of chicken manure to bio-oil by fast pyrolysis II. Analysis of chicken manure, bio-oils and char by curie-point pyrolysis-gas chromatography/mass spectrometry. J. Environ. Sci. Health B, 42: 79-95, 2007.

353. Schnitzer, M., C.M. Monreal and G. Jandl. The conversion of chicken manure to bio-oil by fast pyrolysis III. Analysis of chicken manure, bio-oils and char by Py-FIMS and Py-FDMS. J. Environ. Sci. Health B, 43, 81-95, 2008.
354. Munoz, C., C.M. Monreal, M. Schnitzer and E. Zagal. Influence of Acacia Caven (mol) coverage on carbon distribution in soil organic carbon fractions in a Mediterranean climate region. Geoderma, 144: 352-360, 2008.
355. Das, D.D., M. Schnitzer, C.M. Monreal and P. Mayer. Acid-base separation of bio-oil derived by fast pyrolysis of chicken manure. Bioresource Technology (In press) 2008.

A REVIEW OF RESEARCH WORK

I was born into a Jewish family in Bochum, Germany, and survived the Holocaust. I immigrated to Canada in May, 1947and enrolled at McGill University in September of the same year. My goal was to study soil chemistry. I obtained my B.Sc. Agr. in 1951, M.Sc. in 1952, and Ph.D in 1955. From 1954 to 1956 I worked as a Research and Development Chemist for the Aluminum Company of Canada (Alcan) in Arvida, Quebec. My task was to develop analytical methods for the analysis of trace metals in aluminum alloys. This work in industry gave me a strong background in analytical chemistry, which underlies all of chemistry, and prepared me for the research that I did later. In 1956, I joined the Research Branch of Agriculture Canada in Ottawa. My first research involved the development and improvement of methods for the extraction of organic matter from soils and sediments and the separation of the extracts into humic and fulvic acids. The organic matter remaining with the soil was humin. Following extractions, each of the three humic fractions was de-ashed and dried so that materials low, or free of ash available for further investigations. Methods of characterization, which we used in late 1950's were elemental analysis (C,H, N, O, S), functional group analysis (CO_2H, alcoholic OH, phenolic OH, carbonyl, quinone, ester, and ether groups), infrared spectrophotometry, and molecular weight measurements. Several of these methods were developed in our laboratory and are still used in many soil laboratories all over the world. My co-workers and I also began studies on the oxidative degradation of humic and fulvic acids by alkaline permanganate solutions and nitric acids followed by the identification of oxidation products in order to gain an insight into the chemical structures of these materials.

From 1961 to 1962 I spent a sabbatical in the organic chemistry department of the Imperial College of Science and Technology in London England working under the guidance of Sir Derek Barton, Nobel Laureate in Organic Chemistry. I did my research on a Spodosol fulvic acid, which I had extracted and purified in Canada. With the help of Prof. Barton, I developed a new approach to solving structural problems in fulvic acid. What we did was to exhaustively methylate the oxygen containing functional groups in fulvic acid so that 50% of it became soluble in benzene. The benzene extract was then separated in alumina into several fractions, which differed in molecular weights, functional groups and spectroscopic properties.

After my return to Canada, my co-workers and I started a long-term investigation on the oxidative degradation of humic acids, fulvic acids, humins as well as whole soils from widely differing locations. We greatly improved the methods of identification of oxidation products by including a gas chromatography-mass spectrometry system. The major oxidation products were aliphatic carboxylic, phenolic and benzene carboxylic acids. From the structural point of view, these studies showed that (a) isolated aromatic rings are important units in all humic substances; (b) aliphatic structures are linking aromatic rings to form aliphatic-aromatic networks; and (c) the structures contains voids of various dimensions that can trap organics and inorganics.

The reductive degradation of humic substances produced methyl-substituted polycyclics ranging from naphthalene to perylene. In other researches in the 1960's, my co-workers and I investigated the chemical nature of metal and clay humic interactions. These interactions include the formation of water soluble simple mixed legand complexes sorption and desorption, dissolution of minerals, adsorption on external clay surfaces and in expanding clay interlayer at pH below 5.0.

During 1970's, my co-workers and I examined the colloid-chemical properties of humic materials. Surface pressure, surface tension, and viscosity measurements showed that under the conditions prevailing in most agricultural soils, humic substances behave like flexible linear polyelectrolyte. This view is also supported by studies with the transmission electron microscope, which showed that in dilute aqueous solutions, humic substances formed flat, elongated, multi-branched filaments, which, at higher humic concentrations coalesced to sheet-like structures perforated by voids of varying dimensions.

Near the end of the 1970's, as ^{13}C NMR spectrometers became commercially available, we focused our attention on developing suitable conditions for the analysis of humic substances by liquid- and solid-state ^{13}C NMR. Until the arrival of ^{13}C NMR spectrometers, most soil chemists thought that the chemical structure of humic substances was predominantly aromatic. ^{13}C NMR, however, demonstrated that concentrations of aliphatic structures in humic substances were often as high and, sometimes, even higher than those aromatic structures. Analyses of humic substances by ^{13}C NMR allowed us to determine the following types of carbons: C in alkanes and fatty acids, in $-OCH_3$ groups, in carbohydrates, in aromatic compounds, phenolic OH groups and in CO_2H groups, esters and amides. From the ratio of aromatic C to total C (aliphatic + aromatic C), it is possible to calculate the aromaticity. Thus, ^{13}C NMR is a powerful fast, and relatively simple method for the analysis of humic, and fulvic acids in either the liquid or solid state, and also for humins and whole soils in the solid state. Valuable information on the chemical structure of humic substances can be obtained by combining ^{13}C NMR with chemical methods. In this manner, effects on the chemical structure of humic substances of different extractants, methylations, hydrolysis, oxidation and reduction can be evaluated.

Electron spin resonance measurement of humic substances showed that these materials are rich in free radicals (unpaired electrons). These free radicals participate in inorganic-organic and organic-organic interactions. From the electron spin resonance spectra it appears that the prominent free radicals in humic substances are semiquinones or substituted semiquinones. We found two types of free radicals in humic substances: (a) permanent ones with long lifetimes; and (b) transient free radicals with relatively short lives (several hours). Transient free radicals in humic materials can be generated by chemical reduction, irradiation or increase in pH. Permanent free radicals appear to stabilize the complex chemical substances of humic materials. Electron spin resonance spectroscopy also provides important information on the co-ordination and symmetry of paramagnetic metal ions complexed by humic substances.

During the 1990's, I collaborated with H.-R. Schulten of the Fresenius Technical University in Wiesbaden, Germany, on developing methods for the mass spectrometric analysis of humic substances. We found that pyrolysis-field ionization mass spectrometry (Py-FIMS) and Curie point-gas chromatography-mass spectrometry (Cp-GC/MS) were suitable for this purpose. The components of humic substances, which could be identified by these methods, were: carbohydrates, phenols, lignin monomers, lignin dimers, n-fatty acids, unsaturated fatty acids, n-alkanes, alkenes, n-alkyl monoesters, n-alkyl diesters, n-alkyl benzenes, naphthalenes, phenanthrenes and N- and S-containing compounds. The methods could be applied to humic acids, fulvic acids, humins, and whole soils. It was no longer necessary to extract the humic substances and de-ashing them prior to chemical analyses.

One important application of Py-FIMS and Cp-GC/MS was in structural studies on humic substances. By both methods we found that benzene, substituted benzenes, and C1 to C13 *n*-alkyl benzenes were the major products. From these data we concluded that n-alkylbenzenes were the main structural features in humic substances and that the aromatic rings were linked covalently by aliphatic chains, which were broken during pyrolysis. We then developed an averaged two dimensional model humic acid structure based on n-alkylbenzenes with oxygen containing functional groups (CO_2H, OH, C=O) on the outside. From the two dimensional structure a three-dimensional structure was derived with the aid of computational chemistry. Throughout our structural work, we saw to it that our model humic acid structure was in harmony with chemical, colloid-chemical, electron microscopic, ^{13}C NMR, X-ray and mass spectrometric data, which my co-workers and I had obtained over a period of many years as well as with exhaustive consultations of the voluminous literature. One of the main features of our model humic acid structure is that it is perforated by voids of various dimensions, which can trap or retain both inorganic and organic molecules such as clays, pesticides, herbicides, carbohydrates and peptides. Another significant point is that we do not consider carbohydrates and proteinaceous material components of the humic acid structure. We believe that the latter are either retained by the voids or adsorbed on the humic acid surfaces. The reason for this is that these compounds are rapidly removed by acid hydrolysis, leaving the humic acid structure intact. Our model structure contains relatively large numbers of CO_2H and OH groups so that there are many opportunities for interacting with metal ions, metal oxides and minerals. There are also many possibilities for the formation of both inter and intra-molecular H-bonds. So far we have developed model structures for humic acids, soil organic matter and a whole soil. As new information on these materials becomes available, it may be necessary to modify and refine the model structures. We hope that our research on model soil organic matter structures will be beneficial to the development of a sustainable agriculture and to the protection of the environment.

To obtain information on the "unknown" N in soils and humic substances, my co-workers and I used Py-GC/MS, with the gas chromatograph being equipped with a N-selective detector. This means that only N-containing compounds were transferred to the mass spectrometer for identification. Employing this approach, we identified over 100 N-containing compounds, many of which were N-heterocyclics. The data included pyrroles, pyrrolidines, imidazoles, pyridines, pyrazines, indoles, quinolines, benzothiazoles, and pyrimidines. In another invention with scientists at the University of Saskatchewan we studied the catalysis of the Maillard reaction by δ-MnO_2 and found that N-heterocyclics could be formed abiotically in this manner.

After the year 2000, I worked only half time and joined a group of scientists at Agriculture and Agri-Food Canada here in Ottawa, who worked on composting of all kinds of organic residues. My job was to elucidate the chemical reactions, which were controlling the decomposition of the organics. One of the interesting conclusions of this research was that the biological and chemical reactions during composting focused mainly on the degradation of the aliphatic structures while the aromatic structures were preserved. In recent years I have worked on the chemistry of bio-oils produced by the fast pyrolysis of chicken manure. The bio-oils produced in this manner contained at least 500 organic compounds. So far we have identified by ^{13}C NMR and by mass spectrometric methods about 350 of these compounds.

My lifetime research was concerned with the chemistry and reactions of soil humic substances and with many N-containing compounds in agricultural soils. I spent close to 50 years on this research. During that time I had many co-workers. These were 30 postdoctoral fellows, from 15 different countries, 10 technicians, and 20 scientists from Canada and outside Canada. Each person who worked with me became a co-author of the paper reporting the research, which he or she did with me (see list of publications). I am most grateful to each of them for their contributions. I learned as much from them as they learned from me.

Finally, let us remember that the soil is our most important resource. It is the presence of organic matter, by acting as a habitat for the microbes, fungi, small animals, etc., which distinguishes the soil from a mass of rock particles and allows it to become a living system. Our ability to produce sufficient food for our expanding population and, at the same time, protect the environment, demands a comprehensive understanding of the physical, chemical, and biological properties of the soil and soil organic matter. I hope that the research, which my colleagues and I have done, will contribute to attaining these objectives.

Frank J. Stevenson

University of Illinois, Urbana, Illinois, USA

1995/6 — for his pioneering contributions to our understanding of the chemistry of soil organic matter and its application to agriculture.

CURRICULUM VITAE

Professor Emeritus, University of Illinois.

1949	Bachelor of Science degree, Brigham Young University.
1953	Ph.D. degree in Soil Science, the Ohio State University.
1953-1990	Research and teaching in soil chemistry, Department of Agronomy, University of Illinois.

Recipient of numerous awards for outstanding research accomplishments, including Fellow of the American Association for the Advancement of Science, Paul A Funk Recognition Award, American Soil Science Society Research Award, Bouyoucos Soil Science Career Award, and the prestigious Wolf Prize in Agriculture (shared with Dr. Morris Schnitzer of Ottawa, Canada).

Over the years, extensive research was conducted on the forms of nitrogen in soils, sediments, and sedimentary rocks; geochemistry of nitrogen; chemical nature of soil organic substances; mineralization-immobilization turnover of fertilizer nitrogen in soils; and reactions of humic substances with micronutrient cations (i.e., copper and zinc) and toxic heavy metals (i.e., lead and cadmium).

DESCRIPTION OF SCIENTITIC ACHIEVEMENTS

Initial research at the University of Illinois dealt with a study of nitrogen compounds in soil, which are of importance to soil fertility in that they serve as sources of nitrogen for plant growth. Modem chromatographic procedures were applied for isolating and identifying nitrogenous biochemicals, notably amino acids and amino sugars. Soils were contained several amino acids not commonly found in plant and animal proteins. Distribution patterns for both amino acids and amino sugars varied with soil type and changes were found to be brought about through cultivation. For amino sugars, distribution patterns varied with depth in the soil profile and were affected by soil management practices.

The work with amino sugars was expanded to an examination of these constituents in the Rhizobia (bacteria responsible for the fixation of atmospheric nitrogen in leguminous plants). Differences were observed in the kinds and amounts of amino sugars in the various Rhizobium species, from which it was postulated that chromatography patterns for amino sugars might serve as a "finger-print" technique for their classification.

An interesting aspect of the soil studies was that some of the nitrogen (particularly in the subsoil) occurred as ammonium held within the lattice structures of clay minerals (i.e., as "fixed ammonium"). Narrowing of the carbon/nitrogen (C/N) with increasing depth in the soil profile was demonstrated to be due to an increase in the percentage of the soil nitrogen as clay-bound ammonium rather than to enrichment with nitrogenous organics, as had commonly been thought.

Attention at this time was directed to nitrogenous constituents in sedimentary rocks and marine sediments, where much of the nitrogen was also found to exist as clay-bound ammonium. Examination of depth samples from the Experimental Mohole (a project of the US National Science Foundation to drill a hole through the earth's crust and into the mantle) showed that, with these sediments, the disappearance of amino acids during digenesis was partly abiotic (i.e., due to chemical processes). With deep-sea sediments from the Argentine Basin, amino acid levels were correlated with climatic changes during Quaternary times, which was attributed to reduced microbial activity during the colder periods, with enhanced preservation of organic substances.

Rocks and potassium-bearing primary silicate minerals (e.g., feldspars and micas) contain small amount of nitrogen and it was found that this nitrogen occurred as ammonium in bound forms. Previous estimates for the total amount of nitrogen in primary rocks of the earth's crust and mantle were found to be high and a revised estimate was made for the geochemical distribution of nitrogen in the earth.

Other research dealt with chemical and biological transformations of fertilizer nitrogen in soils, the physical and chemical properties of humic substances (dark colored components of soils), and reactions of micronutrient cations (e.g., copper and zinc) and toxic heavy metals (i.e., lead and cadmium) with soil organic substances. For the latter, a continuous distribution model was proposed for the calculations of stability constants of metal-humate complexes.

Regarding the chemical properties of humic substances, infrared studies on humic and fulvic acids and their methylated/acetylated derivatives showed that the acidity of the different reactive groups overlapped; accordingly, methods for the determining of functional groups based on acidity are not specific and values obtained there from cannot be used as an absolute measure for functional group content.

In cooperation with others, research was carried out on the mineralization-immobilization turnover of biologically and chemically fixed nitrogen in soil, using the stable isotope nitrogen-15 as a tracer. This research was of importance in that as much as one-third of the nitrogen applied to temperate zone soils as fertilizer is retained in organic forms after the growing season and that an equal amount is lost from the soil-plant system in as yet unexplained ways. Findings of the research have been of use in modeling fertilizer nitrogen transformations in soil, thereby leading to improvements in the management of fertilizer nitrogen for increasing crop yields while at the same time protecting the environment by reducing movement of nitrates into water supplies.

LIST OF SELECTED PUBLICATIONS

Books Written:

F. J. Stevenson, 1982, *Humus Chemistry: Genesis, Composition, Reactions.* John Wiley & Sons, NY (Translated 1994 into Chinese by Dr. Xia Rong-ji and published by the Beijing Agricultural University Press).

F. J. Stevenson. 1986. *Cycles of Soil: C, N, P, S, Micronutrients.* John Wiley & Sons.

F. J. Stevenson. 1994. *Humus Chemistry.* 2nd. Ed., John Wiley & Sons, NY.

F. J. Stevenson and M. A. Cole. 1999. Cycles of Soil: 2nd. Ed., John Wiley & Sons.

Books Edited:

Eliott, L. F. and F. J. Stevenson (Co-Editor). 1976. *Soils for Management of Organic Wastes and Waste Waters.* Amer. Soc. Agron., Madison, WI, 650 pp.

F. J. Stevenson (Editor). 1982. *Nitrogen in Agricultural Soils.* Amer. Soc. Agron., pp. 940.

Book Chapters:

F. J. Stevenson. 1964. Soil Nitrogen. In: *Fertilizer Nitrogen: Its Chemistry and Technology.* Amer. Chem. Soc. Monog. 161, Reinhold, pp. 18-39.

F. J. Stevenson. 1965. Origin and distribution of nitrogen in soil. In: *Soil Nitrogen.* Amer. Soc. Agron., Madison, WI, pp. 1-42.

F. E. Broadbent and F. J. Stevenson.1966. Organic matter interactions. In: *Agricultural Anhydrous Ammonia*: Amer. Soc. Agron., Madison, WI, pp. 169-187.

F. J. Stevenson. 1967. Organic acids. In: *Soil Bioch.* Marcel Dekker, NY, pp. 119-146.

F. J. Stevenson. 1969. Nitrogen origin and distribution. In: *McGraw Hill Yearbook Science and Technology for 1969*, pp. 313-315.

F. J. Stevenson and J. H. A. Butler. 1969. Chemistry of humic acids and related pigments. In: *Organic Geochemistry: Methods and Results.* Springer-Verlag, NY, pp. 534-557.

F. J. Stevenson and G. H. Wagner, 1970, Chemistry of nitrogen in soils. In: *Agricultural Practices and Water Quality.* Iowa State Univ. Press, Ames, IA, pp. 125- 141.

F. J. Stevenson and M. S. Ardakani. 1972. Organic matter interactions involving micronutrients in soil. In: *Micronutrients in Agriculture.* Amer. Soc. Agron., pp. 79-114.

F. J. Stevenson and A. Fitch. 1981. Reactions with soil organic matter, In: *Copper in Soils and Plants.* Academic Press, NY, pp. 69-95.

F. J. Stevenson. 1982. Origin and distribution of nitrogen in soil. In: *Nitrogen in Agricultural Soils.* Amer. Soc. Agron., Madison, WI, 1-42. Also: Organic forms,. pp. 67-122.

F. J. Stevenson. 1983. Trace metal-organic matter interactions in geologic environments. In: *The Significance of Trace Elements in Solving Petrogenetic Problems and Controversies.* Theophrastus Publications, Athens, pp. 671-691.

A. Fitch and F. J. Stevenson. 1983. Stability constants of metal-organic matter complexes. *Ibid.*, pp. 645-669.

F. J. Stevenson. 1985. Geochemistry of soil humic substances. In: *Organic Geochemistry*: Vol. 1, Wiley, pp. 13-52.

F. J. Stevenson. 1985. Nitrogen transformations in soil: A perspective. In: *Nitrogen and the Environment.* Nuclear Institute for Agriculture, Faisaisbad, Pakistan, pp. 7-26.

F. J. Stevenson and K. A. Kelley. 1985. Stabilization, chemical characteristics,.and availability of immobilized nitrogen in soil. *Ibid*, pp. 239-259.

Y. Chen and F. J. Stevenson. 1986. Soil organic matter interactions with trace elements. In: *The Role of Organic Matter in Modern Agr.* Martinus Nijhoff, The Netherlands, pp. 73-116.

F. J. Stevenson and A. Fitch. 1986. Chemistry of complexation of metal ions by soil solution organics. In: *Interaction of Soil Minerals with Natural Organics and Microbes.* Amer. Soc. Agron., pp. 29-58.

F. J. Stevenson. 1989. Reductive degradation of humic substances. In: *Organic Geochemistry*: Vol. 2, Wiley, pp. 122-142.

F. J. Stevenson and X-T. He. 1989. Nitrogen in humic substances as related to soil fertility. In: *Advances in Humic Substances Research Related to Soil and Crop Sciences.* Soil Sci. Soc. Amer. Special Pubi., pp. 91-110.

F. J. Stevenson and G. F. Vance. 1989. Naturally occurring aluminum-organic complexes. In: *The Environmental Chemistry of Aluminum.* CRC Press, Boca Raton, FL, pp. 117-146.

F. J. Stevenson and E. T. Elliott. 1990. Methodologies for assessing organic matter quality. In: *Dynamics of Organic Matter in Tropical Soils.* Univ. Hawaii Press, 173-199

F. J. Stevenson. 1991. Organic matter-micronutrient reactions in soils. In: *Micronutrients in Agriculture.* 2nd Ed., Amer. Soc. Agron., pp. 145-186.

G. F. Vance, F. J. Sikora, and F. J. Stevenson. 1995. Aluminum-organic matter complexes. In: *The Environmental Chemistry of Aluminum.* CRC Press, Boca Raton, FL.

K. A. Kelley and F. J. Stevenson, 1996. Organic forms of Soil N. In: *Humic Substances in Agricultural Ecosystems.* Elsevier, Chapter 10.

ENCYCLOPEDIA ARTICLES & SPECIAL PAPERS:

Among these papers were a number of articles for the Encyclopedia of Earth Sciences, the Encyclopedia of Geochemistry & Environmental Sciences, and Methods of Soil Analysis, and Developments in Soil Science Research.

RESEARCH ARTICLES:

Over 100 articles in a wide variety of research journals. These papers are cited in reviews and book chapters, as noted above.

Neal L. First

University of Wisconsin
Madison, Wisconsin, USA

1996/7 — for his pioneering research in the reproductive biology of livestock.

Professor Neal L. First has made pioneering contributions to animal genetics by the development of systems of bovine embryo cloning, gene transfer, and in-vitro production of livestock embryos. His research has resulted in major advances in the application of biotechnology to reproduction in farm animals, eliminating the need for brood cows in beef cattle breeding.

Nuclear transfer of genetic material is being developed to clone farm animal embryos. His ground-breaking research in this field has led to significant discoveries in reproductive biology. Prof. First's laboratory pioneered the application of this technique to the genetic selection of livestock. He and his colleagues defined the conditions leading to optimal in-vitro fertilization and high frequency embryo development. They developed the defined growth media that are widely used to culture bovine embryos. The procedures developed in Prof. First's laboratory have led to major discoveries throughout the world concerning embryo development in domestic animals.

The new knowledge provided by Prof. First's research is having a major impact on the production of genetically advanced strains of livestock throughout the world. Advances include improved reproduction, growth, lactation, and disease resistance.

in vitro fertilize in vitro matured oocytes and the first to define many of the conditions for efficient IVF such as temperature and timing.

III. In embryo culture his laboratory was the first to characterize the in vitro block to bovine embryo development, to describe the timing of bovine embryonic cell cycles and to describe the time of initiation of embryonic transcription, and the mechanisms initiating bovine embryonic transcription. His students were the first to show that bovine embryos could be cultured past the period of blocked development by co- culture with oviduct epithelial cells or their conditioned media. His laboratory then developed a defined medium CR1[aa] that is widely used to culture bovine embryos.

Together these discoveries led to an efficient system for producing embryos in vitro that is being used to eliminate the brood cow in beef cattle breeding, to propagate valuable animals and to make possible more efficient nuclear and gene transfer. An explosion of information concerning early development and differentiation in domestic animals has resulted in laboratories around the world, stemming directly from the systems developed by Dr. First's laboratory. Dr. First's laboratory was also original in developing cloning of cattle embryos and pigs. It was the first to produce offspring from cultured bovine embryonic stem cells. These two technologies when combined with in vitro production of embryos and homologous DNA recombination provide the basis for effective gene transfer or gene deletion in cattle. The enormous impact on research and commercial animal production will long be felt due to the work of Dr. First and his laboratory for developing systems necessary to the in vitro production of bovine embryos.

SELECTED PUBLICATIONS

Schoff, P.K. and N.L. First. 1995. Manipulation of bovine sperm metabolism and motility using anoxia and phosphodiesterase inhibitors. Cell Motility and the Cytoskeleton 31:140-146.

Jones, J.M. and N.L. First. 1995. Expression of cell cycle control protein cdc25 in cleavage stage bovine embryos. Zygote. 3: May, 1995.

Hageman, L.J., Hillery, F.L., Leibfried-Rutledge, M.L. and First, N.L. 1994. Activation of murme oocytes with Ca2-i- ionophore and cycloheximide. J. Exper. Zool. 271:57-61.

Susko-Parrish, J., Leibfried-Rutledge, M.L., Northey, D., Schutzkus, V. and First, N.L. 1994. Inhibition of protein kinases after an induced calcium transient causes transition of bovine oocytes to embryonic cycles without meiotic completion. Developmental Biology. 166: 729-739.

Rosenkrans, C.F. and First, N.L. 1994. Effect of free amino acids and vitamins on cleavage and developmental rate of bovine zygotes in vitro. J. Anim. Sci. 72: 434-437.

First, N.L., Sims, M.M., Park, S.P., and Kent-First, M.J. 1994. Systems for production of calves from cultured bovine embryonic cells. Reprod. Fertil. and Dev. 6: 553-562.

Sims, M.M. and First, N.L. 1994. Production of calves by transfer of nuclei from cultured inner cell mass cells. Proc. Natl. Acad. Sci. *USA* 90: 6143-6147.

Navara, C.S., First, N.L. and G. Schatten. 1994. Microtubule organization in the cow during fertilization, polyspermy, paerthenogenesis and nuclear transfer: The role of the sperm aster. Dev. Biol. 162:29-40.

Sims, M.M., and First, N.L. 1994. Production of calves by transfer of nuclei from cultured inner cell mass cells. Proc. Natl. Acad. Sci. *USA* 90:6143-6147.

First, N.L., and Leibfried-Rutledge M.L. 1993. Nuclear Transfer in Mammals. lit "In vitro fertilization and Embryo Transfer in Primates" (D.P. Wolf, Stouffer R.L., and Brenner, R.M., Eds.), Springer-Verlag New York, Inc. 3 17-330.

Fulka Jr., J., Leibfried-Rutledge M.L, and First, N.L. 1993. Control of germinal vesicle breakdown in bovine x murine hybrid oocytes. Reprod. Nutr. Dev. 33:411-417.

Kim, T., M.L. Leibfried-Rutledge and First, N.L. 1993. Gene transfer in bovine oocytes using replication-defective retroviral vectors packaged with Gibbon ape leukemia virus envelopes. Molec. Reprod. Dev. 35: 105-113.

Kim, T., M.L. Leibfried-Rutledge and First, N.L. 1993. Gene transfer in bovine oocytes using replication-defective retroviral vectors packaged with Gibbon ape leukemia virus envelopes. Molec. Reprod. Dev. 35: 105-113.

Prather, R.S. and First, N.L. 1993. Cell-to-cell coupling in early-stage bovine embryo: a preliminary report. Theriogenology 39:561-567.

Takahashi, Y. and First, N.L. 1993. In vitro fertilization of bovine oocytes in the presence of theophylline. Anim. Reprod. Sci. 34:1-8.

Takahashi, Y. and First, N.L. 1993. In vitro culture of bovine one-cell embryos fertilized in vitro using synthetic oviduct fluid medium supplemented with fetal calf serum. Anim. Reprod. Sci. 3 1:33-47.

Rosenkrans, Jr., C.F., G.Q. Zeng, G.T. McNamara, P.K. Schoff and First, N.L. 1993. Development of bovine embryos in vitro as affected by energy substrates. Biol. Reprod. 49:459-462.

Rosenkrans, Jr., CF., and First, N.L. 1993. Effect of free amino acids and vitamins on cleavage and developmental rate of bovine zygotes in vitro. J. Anim. Sci. 72:434-437.

Yom, H.C., R.D. Bremel and First, N.L. 1993. Mouse mammary tumor virus promoter directs high-level expression of bovine S1 casein in the milk of transgenic heterozygous and homozygous mice. Anim. Biotech. 4(1):89-107.

Hagemann, L.J. and First, N.L. 1992. Embryonic cytoplasmic extracts rescue murine androgenones to the blastocyst stage. Development 114(4): 997-1001.

Takahashi, Y. and First, N.L. 1993. In vitro development of bovine one-cell embryos: influence of glucose, lactate, pyruvate, amino acids and vitamins. Theriogenology 39:56 1-567.

Eyestone, W.H., J.M. Jones and First, N.L. 1991. Oviduct-conditioned medium for the culture of early bovine embryos to the blastocyst stage. J. Reprod. Fertil. 92:59-64.

Eyestone, W.H. and First, N.L. 1991. Characterization of developmental arrest in early bovine embryos cultured in vitro. Theriogenology 35:613-624.

Hagen, D.R., R.S. Prather, M.M. Sims and First, N.L. 1991. Development of one-cell porcine embryos to the blastocyst stage in simple media. J. Anim. Sci. 69:1147-1150.

First, N.L. and R.S. Prather. 1991. Genomic potential in mammals. Differentiation 48:1-8.

Prather, R.S., M.M. Sims and First, N.L.. 1991. Culture of porcine embryos from one and two cells to blastocysts in sheep oviducts. Theriogenology 35:1147-1152.

Fulka, J., Jr., M.L. Leibfried-Rutledge and First, N.L. 1991. Effect of 6-dimethylaminopurine upon germinal vesicle breakdown of bovine oocytes. Mol. Reprod. Dev. 29:379-384.

Barnes, F.L. and First, N.L. 1991. Embryonic transcription in in vitro cultured bovine embryos. Molec. Reprod. Dev. 29:117-123.

Fulka, J., Jr., M.L. Leibfried-Rutledge and First, N.L. 1991. Effect of 6-dimethylaminopurine upon germinal vesicle breakdown of bovine oocytes. Mol. Reprod. Dev. 29:379-384.

Barnes, F.L. and First, N.L. 1991. Embryonic transcription in in vitro cultured bovine embryos. Molec. Reprod. Dev. 29:117-123.

Hagen, D.R., R.S. Prather and First, N.L. 1991. Response of porcine oocytes to electrical and chemical activation during maturation in vitro. Mol. Reprod. Dev. 28:70-73.

Saeki, K., M. Hoshi, M.L. Leibfried-Rutledge and First, N.L. 1991. In vitro fertilization and development of bovine oocytes matured in serum-free medium. Biol. Reprod. 44:256-260.

Patents

Co-culture of embryos with other cell types
CRlaa embryos cultural system
Nuclear transfer process

Patents Pending

A. Bovine embryonic stem cell derivation
B. Production of bovine parthenotes

Ilan Chet

Faculty of Agriculture
The Hebrew University of Jerusalem
Rehovot, Israel

1998 — for his contributions to the environmentally safe development of world agriculture through innovative approaches in plant breeding and bio-control.

Prof. Ilan Chet is internationally recognized as a pioneer in the field of biological control of plant pathogens which cause major crop losses. This field is of primary importance in the protection of crops as an alternative to environmental contamination by hazardous pesticides. He has pioneered the application of fungal physiology and the use of micro-organisms for biological control of plant diseases. His research combines in-depth knowledge of the physiology and the biochemistry of the fungi at the cellular and molecular levels, seeking explanations for pathogenic processes.

Professor Chet's ability to effectively link fundamental and applied aspects of such diverse fields as plant pathology, soil microbiology and biotechnology led to the development of novel biological control agents which are now successfully used as commercial products. Professor Chet has also trained and mentored a new generation of plant biologists and thus contributed to the wide-spread application of new biological approaches in agriculture.

CURRICULUM VITAE

Personal

Date and place of birth: April 12, 1939, Haifa, Israel.
Nationality: Israeli.
Marital status: Married, with five children.

University Degrees

1962 - B.Sc. with honors. The Hebrew University of Jerusalem, Faculty of Agriculture, Rehovot.
1964 - M.Sc. with honors. The Hebrew University of Jerusalem.
1968 - Ph.D. The Hebrew University of Jerusalem.
Thesis: The structure and behaviour of the fungus *Sclerotium rolfsii.*

Prizes and Awards

1968 - Fulbright Foundation Postdoctoral Travel Grant
1979 - Recipient of Kadma Prize for Plant Protection
1982 - Recipient of the A.Z. Cohen Prize for Plant Protection
1982 - Recipient of the Gottingen University "Heinrich-Christian Burchadt Medal" for distinguished research in biological control
1989 - Recipient of Ulitzky Prize for excellence in soil microbiology
1990 - Recipient of the Rothschild Prize for Agricultural Research
1991 - Honorary doctorate of the Faculty of Natural Sciences, Lund University, Sweden
1991 - Fellow of the American Phytopathological Society
1992 - The Francis Ariowitsch Chair for Agricultural Biotechnology
1994 - Recipient of the Max-Planck Research Award
1996 - Recipient of the International Union of Microbial Societies Arima Prize for Applied Microbiology
1996 - Recipient of the Israel Prize for Agricultural Research
1998 - Recipient of the Wolf Prize for Agriculture
1998 - Elected member of the Israel National Academy of Sciences and Humanities
2001 - Recipient of the Officer's Cross of the Order of Merit of the Federal Republic of Germany award
2003 - Recipient the EMET Prize in Agriculture Awarded by the A.M.N. Foundation
2004 - Elected Member of the European Academy of Sciences and Arts
2004 - Awarded the "lifetime achievement award" by the Trichoderma Society
2004 - Noted by ISI among the most cited scientsts in the world
2005 - Nominated Institute Professor at the Weizmann Institute of Science
2005 - "Paul Harris Fellow" of the Rotary
2006 - Honorary doctorate (Ph.D. honoris causa), Haifa University, Israel
2007 - Honorary associate of Tel Hai College
2007 - Honorary associate of Hadassa College
2008 - Honorary Doctorate (Ph.D. honoris causa), University of Naples, Italy

Activities within the Hebrew University

1965 - Assistant in Microbiology, Faculty of Agriculture, Rehovot
1967 - Instructor in Microbiology

1969	- Lecturer in Microbiology, Department of Plant Pathology and Microbiology, Faculty of Agriculture
1970-72	- Visiting Lecturer, The Ben-Gurion University of the Negev, Beer Sheva, Israel
1972	- Senior Lecturer in Microbiology
1972-74	- Chairman, The Biochemistry and Physiology Interdepartmental Teaching Programme, Faculty of Agriculture
1973-74	- Head, The Interdepartmental Equipment Unit and Equipment Committee
1975	- Associate Professor of Microbiology
1976-77	- Member, The Faculty Teaching Committee
1976	- Member, The Cell & Molecular Biology Sub-Committee for Research Students
1977-79	- Head of the Sub-Committee for Agriculture and Applied Biology and a member of the Hebrew University Committee for Research Students in Natural Sciences
1977-78	- Chairman, The Plant Protection Interdepartmental Teaching Programme, Faculty of Agriculture
1978	- Elected as a Member of the Senate of the Hebrew University
1978	- Professor of Microbiology
1980-81	- Chairman of the Research Committee
1981-82	- Member, The Development Committee
1981-82	- Member, The Board of Directors of the Authority of Research and Development of the Hebrew University
1981-83	- Head, Department of Plant Pathology and Microbiology, Faculty of Agriculture
1981-83	- Chairman of the Committee for Plant Protection
1981-82	- Member of the Appointment Committee of the Hebrew University
1983-86	- Director of the "Otto Warburg Center of Biotechnology in Agriculture", Rehovot, Israel
1984-86	- Member, The University Steering Committee of Biotechnology
1985-86	- Member, The University Committee for Basic Equipment
1986-89	- Dean of the Faculty of Agriculture, Rehovot
1989-90	- Chairman of the Hebrew University Committee of Life Sciences
1990-1992	- Director of the Otto Warburg Center for Biotechnology in Agriculture
1990-1992	- Member of the Executive Committee of the Hebrew University
1990 2001	- Member of the Board of Managers of the Hebrew University of Jerusalem
1990-1992	- Member, University Committee for Academic Policy
1992-2001	- Chairman, The Authority for Research and Development of the Hebrew University

1992-2001 - Vice President for Research and Development of the Hebrew University of Jerusalem
1992-2001 - Member of the Board, Yissum R&D Company of the Hebrew University of Jerusalem
1992-1994 - Member of the Board, Scientific Incubators, Co., Jerusalem
1997-2001 - Member, Advisory Board of Batsheva de Rothschild Foundation
1998-2001 - Chairman, The National Committee for Strategic Infrastructure

Activities within the Weizmann Institute

2001-2005 - Professor in the Department of Biological Chemistry
2001-2006 - President of the Weizmann Instituute
2005-2006 - Professor in the Department of Plant Sciences
2001-2006 - Member of the Board of Yeda Co.

Academic Activities and Honors Abroad

1968 Research Associate in McArdle Laboratory for Cancer Research, Medical Center, The University of Wisconsin, Madison, Wisconsin, USA
1970 Research Fellow in the Laboratory of Applied Microbiology, Harvard University, Cambridge, Massachusetts, USA
1972 Visiting Scientist in Gottingen University, West Germany
1975 Visiting Professor, Laboratory of Applied Microbiology, Division of Applied Sciences, Harvard University, Cambridge, MA, USA
1978 "Richard Merton Professor", Gottingen University, West Germany
1979-80 Visiting Professor, Department of Botany and Plant Pathology, Colorado State University, Fort Collins, CO, USA
1980 Visiting Professor, Harvard University, Cambridge, MA, USA
1984 "DFG Visiting Professor", University of Gottingen, FRG
1986 Visiting Professor, Auburn University and Cornell University, USA
1986 Visiting Professor, Lund University, Lund, Sweden
1989-90 Senior Visiting Scientist, E.I. DuPont de Nemours & Co. Agricultural Experimental Station, Agricultural Products Department, Biotechnology Division, Wilmington, Delaware, USA
1991 Senior Visiting Scientist, E.I. DuPont de Nemours & Co.
1992 Visiting Professor, Cape Town University, South Africa
1994-1995 Visiting Professor, Rutgers University, New Jersey, USA
1995-2006 Visiting Professor, Rutgers University, New Jersey, USA (in summers)

National and International Activities

1972-1974 Member of the Board of Directors and a Treasurer of the Israeli Phytopathological Society

1977	Convenor, The Symposium on Sclerotial Morphogenesis, in the Second in the Second International Mycological Congress, Tampa, USA
1977-78	Member of the Board of Directors of the Israeli Phytopathological Society
1978-79	Head, the Supervising Committee of the Tel-Hai College
1979-82	Member, The Technical Advisory Committee of the United States-Israel Agricultural Research and Development Fund (BARD)
1981-1986	Member of the Editorial Board of the International Journal "Phytoparasitica"
1982	Chairman of the Organizing Committee of the "France-Israel" Symposium on Plant Resistance
1982	Coordinator of the "Gottingen - Jerusalem Symposium" on Soil Plant Relationship
1983	Invited Lecturer, Conference of the French Phytopathological Society on Biological Control, Bordeaux, France
1983	Invited Lecturer, 4th International Congress of Plant Pathology, Melbourne, Australia
1983	Delegate member, the Mycology Division Council of the International Union of Microbiological Societies (IUMS)
1984-1994	Member of the UN Panel for Applied Microbiology and Biotechnology (ICRO)
1984-88	Member, The Special Projects Committee of the International Society for Plant Pathology
1984-1986	Chairman, The Committee of National Council of Research and Development for Agriculture and Biotechnology
1985-1992	Member of the National Committee for Biotechnology
1986-1990	Member of BARD Fellowship Committee
1986-1992	Member of the Commission on Biotechnology of the International Union of Pure and Applied Chemistry (IUPAC)
1987	Keynote Address Speaker of the 25th Anniversary of the Phytopathological Society, South Africa
1987	Invited Lecturer to Meeting of the American Phytopathological Society, Cincinnati, Ohio
1987-1989	Member, The Scientific Advisory Committee of the Applied Science Institute, Ben Gurion University of the Negev
1988	Invited speaker to the Finland-Israel Symposium on Symbiosis, Shoresh, Israel
1988	Member, Organizing Committee, and invited lecturer to the Bat-Sheva Seminar on Host-Fungus Interaction, Jerusalem
1988	Invited speaker to the Niedersachsen Conference - Frontiers in Biology,Chemistry and Physics - Braunschweig, W. Germany

1988	Invited speaker to the VII International Congress on the Global Impact of Microbiology (GIAM), Hong Kong
1988	Invited speaker to the International Congress of Plant Pathology, Kyoto, Japan
1988	Invited speaker to the Annual Meeting of the British Society of Plant Pathology, Reading, England
1989	Invited speaker to the UCLA Symposium on Molecular and Ecological Aspects of Biological Control, Frisco, Colorado
1989-1993	Chairman, The Special Projects Committee of the International Society for Plant Pathology (ISPP)
1989-1994	Member of the Scientific Advisory Committee of the Institute for Cereal Crops Improvement. Tel Aviv University
1989	Invited speaker to the Biotechnology Workshop, Rutgers University, New Jersey, USA
1989	Invited speaker to the Beltsville Symposium on the Rhizosphere and Plant Growth, Beltsville, USA
1990	Invited speaker to the Fourth International Mycological Congress, Regensburg, F.R.G.
1990	Invited speaker to the German Botanical Society Meeting, Regensburg, F.R.G.
1990	Invited speaker to the Workshop on Biological Control of Postharvest Diseases, Sheperdstown, West Virginia, USA
1990-	Member, International Editorial Board, Crop Protection
1991	Invited speaker to the XII International Plant Protection Congress, Rio de Janeiro, Brazil
1991	Invited speaker to the International Symposium on Application of Biotechnology to tree culture protection and utilization. Columbus, Ohio, USA
1991-	Advisory Editor, Wiley Series in Ecological and Applied Microbiology
1992	Invited overview speaker to the International Seminar on "Impact of Biotechnology on Agriculture and Food in Developing Countries", Madras, India
1992	Teaching Course, University of Torino, Italy
1992	Invited speaker, The Center of Bioengineering, Academy of Sciences of Russia
1992-1994	Member, The Committee for Reduction of Pesticides, Ministry of Agriculture
1993-1999	Member, The Scientific Committee of "Med-Campus" of the EEC
1993	Invited speaker, International Conference on Agricultural and Environmental Biotechnology, Torino, Italy

1993	Invited speaker, The British Society of Phytopathology Meeting, Rothamstead, England
1994	Keynote speaker, International Conference on Host-Pathogen Interactions, GIF Conference, Greifswald, Germany
1994	Invited lecturer, Summer Course in Biotechnology to Scandanavian Countries, Uppsala, Sweden
1994	Invited speaker in "The Year of Louis Pasteur Conference" Organized by Institut Pasteur, Microbes, Environment and Biotechnology, May 1995, Papeete, Tahiti
1994	Member of Organizing Committee and speaker in the Katzir Conference on Plant Molecular Biology, Biotechnology and Environment, Koln, Germany
1994	Member of Organizing Committee and speaker in the Conference on "International Cooperation for development of biotechnology", Jerusalem, Israel
1995	Member of Organizing Committee of "IUMS Congress of Microbiology and Mycology" and Chairman of Mycology Program Committee, Jerusalem, Israel
1995	Member of Organizing Committee of the International Mycological Congress (IMC-6), Jerusalem, Israel
1995-	Member of Editorial Board, The European Journal of Plant Pathology
1996-2000	Member, German Israel Coooperation Council for High and Environmental Technology
1996-1997	Member, National Committee of Biotechnology
1997-2003	Chairman of the Rothschild Prize Committee
1997-	Chairman of "Mifal Hapayis" Committee for scientific research Prizes
1997	Invited speaker in the Symposium on Molecular Aspects of Biological Control, Delemont, Switzerland
1997	Invited speaker in the Congress on Techniques and Developments for the Food Industry, Agriculture and the Environment, Perugia, Italy
1998	Invited speaker in the International Symposium on Fungal Control of Pests, Weeds and Diseases, Univ. of Southampton, U.K.
1998	Co-organizer of the BARD Workshop on Soilborne Plant Pathogens, Ramat Rachel, Israel
1998	Co-organizer of the Warburg Symposium on Antibiosis, Göttingen, Germany
1998	Invited speaker in the International Congress of Plant Pathology, Edinburgh, UK
1998	Chairman and invited speaker in the 6th International Mycological Congress, Jerusalem, Israel

1998-2002	Member, External Advisory Group (EAG) to the European Union Research Program
1999	Invited speaker in conference on Application of Biological Control. Alghero, Sardinia, Italy
2000	Invited speaker in the International Conference on Managing Natural Resources, New Delhi, India
þ2000	Invited speaker, European Symposium on Soil Inoculants. Amsterdam, Holland
2000	Keynote Speaker, Third International Symposium on Rhizoctonia, ISR 2000. Taichung, Taiwan
2000-2003	Member, The Advisory Panel of NATO Science Program
2001	Invited speaker and a chairman, in the 3th International Symposium on Chitin Enzymology. May, Senigallia, Italy
2001	Invited speaker in the International workshop on Agricultural Biotechnology. Algehro, Sardinia
2001	Invited lecturer in the New Horizon in Biotechnology conference. Trivandrum, India
2002	Keynote speaker in the workshop "Biotechnology in the service of Agriculture. Warsaw, Poland
2002	Chairman and invited speaker of the session of transgenic plants in FEBS Conference. Istanbul, Turkey
2002	Invited speaker in VIII International Fungal Biology Conference. Guanojuato, Mexico
2003	Invited plenary speaker in Bioacademy 2003 Conference, Tel-Aviv, Israel
2003	Chairman, Ag-Biotech session in Israel Biotech Conference. Tel-Aviv, Israel
2003	Invited participant and speaker in the annual meeting of the World Economic Forum. Davos, Switzerland
2003	Invited participant of "Fortune Brainstorming", Aspen, Co., USA
2003	Invited plenary speaker in "World Knowledge Forum", Seoul, Korea
2003	Keynote speaker on the Beneficial Relations between Universities and Industry. Niedersachsen Symposium, Gottingen, Germany
2004	Elected Member of the Perspective Committee of Research Center Julich, Germany (nomination of the Minister of Science)
2005	Member of the International Board of the International Society of Microbial Ecology
2005	Member of the Advisory Board of the Institute of Life Sciences, Hyderabod, India
2005-	Member of the International Board of the International Society of Microbial Ecology

2006	Chairman, Session on Induced Resistance in the International Trichoderma Workshop, Vienna Austria
2006	Invited speaker, the Max Planck Conference on Innovative Research, Munchen Germany
2006	Member of the International Advisory Committee of the International Society of Microbial Ecology
2006	Chair of Session, 9th International Workshop on Trichderman, Vienna Austria
2006	Invited speaker in the International Conference on Identification and Implementation of Research Topics, Ringberg Castle, Germany
2006	Invited speaker in the International Conference on Medical Plants and Functional food, Tashkent, Uzbekistan
	Keynote speaker in the International Symposium on Acquired Plant Resistance, Budapest, Hungary
2006	Advisor to Evogene Co., Rehovot, Israel
2007	Board member of Makhteshim-Agan Chemical Company
2007	Board member of Fertiseed
2007	Chairman of the Environment and Health Fund, Jerusalem
2007-	Vice President Ilanit-Bilogical Societies
2007-	Chairman of the Health & Environmental Fund
2007-	Chairman of the National Committee for Knowledge Centers
2007	Invited Speaker "Subramanian memorial lecture", Potnugar University, India
2008	Invited Speaker Plant Micro-Interactions Congress, Sorento, Italy
2008	Invited Speaker in Plant Microbe Interaction conference, Salamanca, Spain
2008	Chairman of the National Committee for Stem Cells Centers
2008	President of Ilanit – Biological Societies
2008	Invited Speaker in the Tricoderma conference, Costa Rica
2008	Chairman of session on Biological Control in the International Congress of Plant Protection, Torino, Italy
2008	Invited speaker in Plant-Microbe Conference, Salamanca, Spain

Membership in Scientific Societies and Organizations

Israel Society for Microbiology
Israel Phytopathological Society
Israel Biochemical Society
Israel Society of Cell Biology
American Phytopathological Society
American Society of Microbiology
International Organization for Biotechnology and Bioengineering (IOBB)

International Society of Plant Pathology (ISPP)
International Mycological Association (IMA)
British Mycological Society
International Society for Molecular Plant-Microbe Interactions (IS-MPMI)
European Chitin Society

Grants

Since 1973, Professor Chet has received research grants from the following:

Agencies and foundations:

- Israel Ministry of Agriculture
- Deutsche Forschungsgemeinschaft (DFG)
- National Council for Research and Development (NCRD)
- Federal German Ministry for Science and Technology (BMFT)
- Ministry of Niedersachsen for Science, Germany
- Binational Agricultural Research and Development Fund (BARD)
- The Wolfson Foundation, England
- Agency for International Development (AID)
- Kay Foundation for Biotechnology
- The Szold Foundation
- GIFRID Foundation, Germany
- German-Israeli Agricultural Research and Development (GIARA)
- German-Israeli Foundation (GIF)
- Volkswagen Foundation (VW)
- Dutch-Israel Agricultural Foundation (DIAF)
- Charles Wolfson Charitable Trust
- The Chais Family Charitable Fund
- European Union (E.U)
- Belgian Trilateral Governmental Fund
- Federal German Ministry for Education and Science (BMBF)
- Horowitz Association

Industry:

- Teva Pharmaceutical Industry
- Galil Advanced Technologies
- Bio Technology General Ltd. (BTG)
- Gadot Petrochemical Industries
- Biotechnology Application (BA)
- Makhteshim
- Migal, Israel
- FRM Agricultural Sciences Partnership

Yissum, Israel
Ecogen, USA
Ieves Rocher, France
Israel Biotechnology Research (IBR)
Haifa Chemicals
Mycontrol
Biotechnological Integrated Technologies BIT, (Italy)
Algatech

Patents

1. Henis, Y. and I. Chet. 1971. Method for drug detection. Israel Patent 38,233.
2. Mitchell, R. and I. Chet. 1977. Magnetic separation method. U.S. Patent 4,00-197.
3. Sivan, A., I. Chet and Y. Elad. 1985. Fungicidal compositions and method for using them. Israel Patent 68124.
4. Sivan, A. and I. Chet. 1985. Antifungal compositions containing Trichoderma active against Fusarium. Israel Patent 69368.
5. Chet, I., A. Sivan, and Y. Elad. 1985. Novel isolates of Trichoderma fungicidal compositions containing said isolates and use thereof. European Patent 133878.
6. Shoseyov, O., B.A. Bravdo, R. Ikan and I. Chet. 1987. Production and utilization of a specific endo-beta-glucosidase for food, wine and perfume industries. Israel Patent No. 82980-2, 1987, U.S. Patent Pending, 1988.
7. Chet, I. and Sivan, A. 1988. Antifungal compositions containing Trichoderma active against Fusarium. US Patent 4,784,021.
8. Oppenheim, A., I. Chet and R. Shapira. 1988. A system for excretion of foreign protein, enzymes or polypeptides from E. coli into the culture media, especially for the production and purification of the enzyme chitinase. Israel Patent Pending 85408/3.
9. Eyal, Z., I. Chet and M. Fleishman. 1988. Method of controlling foliar diseases caused by fungal pathogens with fungicidal bacteria and novel pure cultures of fungicidal bacteria. Israel Patent Pending 87322.
10. Spiegel, I., I. Chet and E. Cohn. 1988. Method and composition for combating soil nematodes comprising chitinolytic bacteria and novel pure cultures of certain such bacteria. Israel Patent 87388.
11. Spiegel, I., I. Chet, Cohen, E. and Galper, S. 1989. Nematocidal microorganisms, their isolation and their use as bionematocides. European Patent 89114460.2.
12. Eyal, Z., I. Chet, M. Fleishman and E. Levy. 1989. Method of controlling foliar diseases caused by fungal pathogens with fungicidal bacteria and novel pure cultures of fungicidal bacteria. European Patent No. 89114108.7.
13. Chet, I., A. Sivan and Y. Elad. 1990. A novel isolate of Trichoderma fungicidal compositions containing said isolate and use thereof. U.S. Patent 4,915,944.

14. Ordentlich, A., Z. Weisman , H. Gottlieb and I. Chet. 1991. Antimicrobial compound and compositions. Israel Patent No. 097928.
15. Elad, Y., Zimand, G. and Chet, I. 1991. Novel isolate of Trichoderma harzianum, T-39, fungicidal compositions containing said isolate and use against B. cinerea and S. scherotiorum. European Patent No.466.133.
16. Spiegel, I., Chet, I., Cohn, E. and Galper, S. 1992. Nematicidal strain of Pseudomonas and its use as a biocontrol agent. U.S. Patent No. 5,089,263.
17. Elad, Y. and Chet, I. 1993. Novel isolate of Trichodrema harzianum T-39, fungicidal compositions containing said isolate and use against B. cinerea and S. sclerotiorum.. New Zealand Patent No. 238925.
18. Logemann, J., Jach, G., Gornhardt, B., Mundy, J., Schell, J. Eckes, P. and Chet, I. 1993. Transgenic, pathogens-resistant organisms. German Patent No. P 42 34 131.0.
19. Logemann, J., Jach, G., Gornhardt, B., Mundy, J., Schell, J. Eckes, P. and Chet, I. 1993. Transgenic, pathogens-resistant organisms. European Patent No.93116011.3.
20. Logemann, J., Jach, G., Gornhardt, B., Mundy, J., Schell, J., Eckes, P. and Chet, I. 1994. Transgenic, pathogens-resistant organisms. Canadian Patent 2,108,112.
21. Elad, Y., Zimand, G. and Chet, I. 1993. Isolate of Trichoderma fungicidal compositions containing said isolate and use against B. cinerea and S. sclerotiorum. US Patent No. 5,238,690.
22. Elad, Y., Zimand, G. and Chet, I . 1993. Isolate of Trichoderma harzianum
23. I-952 fungicidal compositions containing said isolate and use against B. cinerea and S. sclerotiorum. US Patent No. 5,266,316.
24. Elad, Y., Zimand, G. and Chet, I. 1994. Novel isolate of Trichoderma harzianum, T-39, fungicidal compositions containing said isolate and use against B. cinerea and S. scherotiorum. Australian Patent No. 649.050.
25. Herrera-Estrella, A. and Chet, I. 1995. Metodo para la obtencion de cepas mejoradas de Trichoderma spp. en su capacidad como agentes de control biologico, y la cepa obtenida por el mismo. 1995. Mexican Patent No. 949442.
26. Nussinovitch, A., Chet, I., Gershon, Z. and Abramsky, M. 1995. Pesticide against soil pests. Israel Patent No. 116008.
27. Elad, Y., Zimand, G. and Chet, I. 1996. Fungicidal compositions containing isolate I-952 of *Trichoderma harzianum* T-39 and their use against *B. cinerea* and *S. sclerotiorum.* Israel Patent No. [11][21] 95066.
28. Shoseyov, O., Chet, I., Bravdo, B-A. and Ikan, R. 1997. Flavor and fragrance enhancing enzymes. European Patent Application EP/0307071 B1.
29. Chet, I., Salt, D., Blaylock, M and Raskin, I. 1998. Microbial isolates promote phytoremediation. United States Patent Number 5,809,693. Date of Patent: Sept. 22, 1998.

30. Herrera-Estrella, A.H., Flores-Martinez, A. and Chet. I. 1999. Method for obtaining strains of Trichoderma spp. United States Serial Number 08/849,345. Filed 07/25/97.
31. Mirelman, D., Abramski, M., Chet, I., Miron, T., Rabinkov, A. and Wilchek, M. 2001. Method for Soil Sterilization from Pathogens. PXT Patent Application No. PCT/US02/01204; Publication No. WO 02/056683.
32. Chet, I. and Viterbo,.A. Fungal Chitinases, Polynucleotide Sequences Encoding Promoters of Same and uses thereof. Patent Pending No. PCT/IL02/00351. Filed 5.5.2002.
33. Chet, I. and Viterbo,.A. Fungal Chitinases, Polynucleotide Sequences Encoding Promoters of Same and uses thereof. Israel Patent Application No. 158732.
34. Chet, I. and Viterbo,.A. Fungal Chitinases, Polynucleotide Sequences Encoding Promoters of Same and uses thereof. US Patent Application No. 10/475.853.
35. Chet, I. and Viterbo,.A. Fungal Chitinases, Polynucleotide Sequences Encoding Promoters of Same and uses thereof. European Patent Application No. 02722649.7.
36. Chet, I. and Viterbo,.A. Fungal Chitinases, Polynucleotide Sequences Encoding Promoters of Same and uses thereof. Indian Patent Application No.1898/ CHENP/2003.
37. Harman, G.E., Khadge, B.R., Lorito, M., Seaman, A.J., Bergstrom, G.C. and Chet, I., Control of Plant Pathogens using *Trichoderma* Spp. United States Patent Pending 19603/4190.
38. Viterbo A. and Chet I. 2005. Antifungal activity of a novel endochitinas gene (chit 36) Bo/288.421, USA.

LIST OF PUBLICATIONS

I. Books

1. Chet, I. 1967. The structure and behaviour of the fungus Sclerotium rolfsii. Ph.D. Thesis. Supervised by Prof. Y. Henis.
2. Huttermann, A. and I. Chet (eds.). 1986. Soil-Plant Interaction. Gottingen Bodenkundische Berichte, FRG, pp. 269.
3. Chet, I. (ed.). 1987. Innovative Approaches to Plant Disease Control. John Wiley & Sons, N.Y. pp. 372.
4. Chet, I. (ed.) 1993. Biotechnology in Plant Disease Control. John Wiley & Sons, N.Y. pp. 373.

II. Publications in Peer-Reviewed Scientific Journals

5. Henis, Y., I. Chet and Z. Avizohar-Hershenzon. 1965. Nutritional and mechanical factors involved in mycelial growth and production of sclerotia

by *Sclerotium rolfsii* in artificial media and amended soil. Phytopathology 55:87-91.

6. Chet, I., Y. Henis and R. Mitchell. 1966. The morphogenic effect of sulfur containing amino acids, glutathione and iodoacetate on *Sclerotium rolfsii.* J. Gen. Microbiol. 45:541-546.
7. Henis, Y. and I. Chet. 1967. Mode of action of ammonia on *Sclerotium rolfsii.* Phytopathology 57:425-427.
8. Chet, I., Y. Henis and R. Mitchell. 1967. Chemical composition of hyphal and sclerotial walls of *Sclerotium rolfsii* Sacc. Can. J. Microbiol. 13:137-141.
9. Henis, Y. and I. Chet. 1968. The effect of nitrogenous amendments on the germinability of sclerotia of *Sclerotium rolfsii* and their accompanying microflora. Phytopathology 58:209-211.
10. Chet, I. and Y. Henis. 1968. The control mechanism of sclerotial formation in *Sclerotium rolfsii* Sacc. J. Gen. Microbiol. 54:231-236.
11. Chet, I. and Y. Henis. 1968. X-ray diffraction analysis of hyphal and sclerotial walls of *Sclerotium rolfsii.* Can. J. Microbiol. 14:815-816.
12. Henis, Y. and I. Chet. 1968. Development biology of sclerotia of Sclerotium rolfsii. Can. J. Bot. 46:947.
13. Henis, Y. and I. Chet. 1968. Experimental soil microbiology. The Hebrew University, Faculty of Agriculture, Rehovot (in Hebrew), pp. 18.
14. Chet, I., Y. Henis and N. Kislev. 1969. Ultrastructure of sclerotia and hyphae of *Sclerotium rolfsii* Sacc. J. Gen. Microbiol. 57:143-147.
15. Chet, I. and Y. Henis 1969. Effect of catechol and disodium EDTA on melanin content of hyphal and sclerotial walls of *Sclerotium rolfsii* Sacc. and the role of melanin in the susceptibility of these walls to (1+3) glucanase and chitinase. Soil Biol. Biochem. 1:131-138.
16. Chet, I. 1969. The role of sclerotial rind in the germinability of sclerotia of *Sclerotium rolfsii.* Can. J. Bot. 47:593-595.
17. Chet, I. and H.P. Rusch. 1969. Induction of spherule formation in *Physarum polycephalum* by polyols. J. Bacteriol. 100:673-678.
18. Melera, P.W., I. Chet and H.P. Rusch. 1970. Electrophoretic characterization of RNA from *Physarum polycephalum.* Biochim. Biophys. Acta 209:569-572.
19. Chet, I. and H.P. Rusch. 1970. RNA differences between sporulating and growing microplasmodia of *Physarum polycephalum* as revealed by sedimentation pattern and DNA-RNA hybridization. Biochim. Biophys. Acta 209:559-568.
20. Chet, I. and H.P. Rusch. 1970. Differences between hybridizable RNA during growth and differentiation of *Physarum polycephalum.* Biochim. Biophys. Acta 213:478-483.
21. Chet, I. and H.P. Rusch. 1970. RNA and protein synthesis during germination of spherules of *Physarum polycephalum.* Biochim. Biophys. Acta 224:620-622.

22. Chet, I.S. Fogel and R. Mitchell. 1971. Chemical detection of microbial prey by bacterial predators. J. Bacteriol. 106:863-867.
23. Huttermann, A. and I. Chet. 1971. Activity of some enzymes in *Physarum polycephalum* III. During spherulation (differentiation) induced by mannitol. Arch. Microbiol. 78:180-192.
24. Mitchell, R., S. Fogel and I. Chet. 1972. Bacterial chemoreception: an important ecological phenomenon inhibited by hydrocarbons. Water Research 6:1137-1140.
25. Okon, Y., I. Chet and Y. Henis. 1972. Lactose-induced synchronous sclerotium formation in *Sclerotium rolfsii* and its inhibition by ethanol. J. Gen. Microbiol. 71:465-470.
26. Assa, Y., B. Gestetner, I. Chet and Y. Henis. 1972. Fungistatic activity of lucerne saponins and digitonin as related to sterols. Life Sciences 11:637-647.
27. Chet, I., N. Retig and Y. Henis. 1972. Changes in total soluble proteins and in some enzymes during morphogenesis of *Sclerotium rolfsii.* J. Gen. Microbiol. 72:451-456.
28. Chet, I. and Y. Henis. 1972. The response of two types of *Sclerotium rolfsii* to factors affecting sclerotium formation. J. Gen. Microbiol. 73:483-486.
29. Okon, Y., I. Chet and Y. Henis. 1973. Effect of lactose, ethanol and cycloheximide on the translocation pattern of radioactive compounds and on sclerotial formation in *Sclerotium rolfsii.* J. Gen. Microbiol. 74:251-258.
30. Chet, I., N. Retig and Y. Henis. 1973. Changes in ribonucleases during differentiation (spherulation) of Physarum polycephalum. Biochim. Biophys. Acta 294:343-347.
31. Kislev, N. and I. Chet. 1973. Scanning electron microscopy of sporulating cultures of the myxomycete Physarum polycephalum. Tissue and Cell 5:349-357.
32. Chet, I., Y. Henis and R. Mitchell. 1973. The effect of biogenic amines and cannabinoids on bacterial chemotaxis. J. Bacteriol. 115:1215-1218.
33. Tirosh, R., A. Oplatka and I. Chet. 1973. Motility in a "cell sap" of the slime mold Physarum polycephalum. FEBS Letters 34:40-42.
34. Chet, I., Y. Zilberstein and Y. Henis. 1973. Chemotaxis of Pseudomonas lachrymans to plant extracts and to water droplets collected from the leaf surfaces of resistant and susceptible plants. Physiol. Plant Pathol. 3:473-479.
35. Henis, Y., Y. Okon and I. Chet. 1973. The relationship between early hyphal branching and formation of sclerotia in *Sclerotium rolfsii.* J. Gen. Microbiol. 79:147-150.
36. Chet, I. and N. Kislev. 1973. Scanning electron microscopy of spherules of *Physarum polycephalum.* Tissue and Cell 5:545-551.

37. Okon, Y., I. Chet, N. Kislev and Y. Henis. 1974. Effect of lactose on soluble glucan production and on the ultrastructure of *Sclerotium rolfsii* Sacc. grown in submerged culture. J. Gen. Microbiol. 81:145-149.
38. Kislev, N. and I. Chet. 1974. Scanning electron microscopy of freeze-fractured sclerotia of *Physarum polycephalum.* Tissue and Cell 6:209-214.
39. Retig, N. and I. Chet. 1974. Catechol-induced resistance of tomato plants to Fusarium wilt. Physiol. Plant pathol. 4:469-475.
40. Assa, Y., I. Chet, B. Gestetner, R. Govrin, Y. Birk and A. Bondi. 1975. The effect of alfalfa saponins on growth and lysis of *Physarum polycephalum.* Arch. Microbiol. 103:77-81.
41. Mitchell, R. and I. Chet. 1975. Bacterial attack of corals in polluted seawater. Microbiol. Ecol. 2:227-233.
42. Okon, Y., I. Chet and Y. Henis. 1975. The effect of glucose and lactose on β-D-galactosidase activity and formation of sclerotia in *Sclerotium rolfsii.* Can. J. Microbiol. 21:1123-1126.
43. Lisker, N., J. Katan, I. Chet and Y. Henis. 1975. Release of cell-bound polygalacturonase and cellulase from mycelium of *Rhizoctonia solani.* Can. J. Microbiol. 21:521-526.
44. Chet, I. 1975. Ultrastructural basis of sclerotial survival in soil. Microbiol. Ecol. 2:194-200.
45. Spiegel, S., Y. Henis, I. Chet and G. Messer. 1975. The effect of heat shock on differentiation of germinating conidia of *Trichoderma viride.* Can. J. Bot. 53:2274-2281.
46. Chet, I., P. Asketh and R. Mitchell. 1975. Repulsion of bacteria from marine surfaces. Appl. Microbiol. 30:1043-1045.
47. Chet, I. and R. Mitchell. 1976. Petroleum hydrocarbons inhibit decomposition of organic matter in seawater. Nature 261:308-309.
48. Kritzman, G., Y. Okon, I. Chet and Y. Henis. 1976. L-threonine metabolism in *Sclerotium rolfsii* and it relationship to morphpgenesis. Isr. J. Med. Sci. 12:696-697.
49. Kritzman, G., Y. Okon, I. Chet and Y. Henis. 1976. Metabolism of L-threonine and its relationship to sclerotium formation in *Sclerotium rolfsii.* J. Gen. Microbiol. 95:78-86.
50. Chet, I. and R. Mitchell. 1976. An enrichment technique for isolation of marine chemotactic bacteria. Microbiol. Ecol. 3:75-78.
51. Chet, I. and R. Mitchell. 1976. The relationship between chemical structure of attractants and chemotaxis by a marine bacterium. Can. J. Microbiol. 22:1206-1208.
52. Segal, L. A., I. Chet and Y. Henis. 1977. A simple quantitative assay for bacterial motility. J. Gen. Microbiol. 98:329-337.
53. Chet, I. and A. Huttermann. 1977. Germination inhibitor in the slime mold *Physarum polycephalum.* FEMS Microbiol. Letts. 1:149-152.

54. Chet, I., D. Timar and Y. Henis. 1977. Physiological and ultrastructural changes occurring during germination of sclerotia of *Sclerotium rolfsii.* Can. J. Bot. 55:1137-1142.
55. Kritzman, G., I. Chet and Y. Henis. 1977. Effect of carbon dioxide on growth and carbohydrate metabolism in *Sclerotium rolfsii.* J. Gen. Microbiol. 100:167-175.
56. Kritzman, G., I. Chet and Y. Henis. 1977. The relationship between rhythmic hyphal growth and circadian formation of sclerotia in *Sclerotium rolfsii.* Sacc. Can. J. Microbiol. 23:959-963.
57. Chet, I. and A. Huttermann. 1977. Melanin biosynthesis during differentiation of *Physarum polycephalum.* Biochem. Biophys. Acta 499:148-155.
58. Chet, I., A. Naveh and Y. Henis. 1977. Chemotaxis of *Physarum polycephalum* towards carbohydrates, amino acids and nucleotides. J. Gen. Microbiol. 102:145-148.
59. Chet, I., A. Naveh and Y. Henis. 1977. Chemotaxis and movement of *Physarum polycephalum* and its response to some neurotransmitters and psychomimetic compounds. J. Mechanochem. and Cell Motility 4:177-187.
60. Chet, I., D. Havkin and J. Katan. 1977. The role of catechol in inhibition of Fusarium wilt. Phytopathol. Z. 91:60-66.
61. Kritzman, G., I. Chet and Y. Henis. 1977. The role of oxalic acid in the pathogenic behaviour of *Sclerotium rolfsii* Sacc. Exper. Mycol. 1:280-285.
62. Huttermann, A., Y. Henis and I. Chet. 1978. Okologische Modellversuche in der Petrischale. Biologie Uns. Z. 8:57-60.
63. Kritzman, G., I. Chet and Y. Henis. 1978. Localization of β-(1,3)-glucanase in mycelium of *Sclerotium rolfsii.* J. Bacteriol. 134:470-475.
64. Kritzman, G., I. Chet, Y. Henis and A. Huttermann. 1978. The use of the brightener "calcofluor white M2R new" in the study of fungal growth. Isr. J. Bot. 27:138-146.
65. Sjoblad, R.D., I. Chet and R. Mitchell. 1978. Chemoreception in the green alga *Danaliella tertiolecta.* Curr. Microbiol. 1:305-308.
66. Hadar, Y., I. Chet and Y. Henis. 1979. Biological control of *Rhizoctonia solani* damping-off with wheat bran culture of *Trichoderma harzianum.* Phytopathology 69:64-68.
67. Netzer, D., G. Kritzman and I. Chet. 1979. β-(1-3) glucanase and quantity of the fungus in relation to Fusarium wilt in resistant and susceptible near isogenic lines of musk melon. Physiol. Plant Pathol. 14:47-55.
68. Grinstein, A., Y. Elad, J. Katan and I. Chet. 1979. Control of *Sclerotium rolfsii* by means of a herbicide and *Trichoderma harzianum.* Plant Dis. Reprt. 63:823-826.
69. Ruben, D.M., Z.R. Frank and I. Chet. 1979. Factors affecting behaviour on developmental synchrony of germinating oospores of *Pythium aphanidermatum.* Phytopathology 70:54-59.

70. Bitton, G., D.A. Chuckran, I. Chet and R. Mitchell. 1979. Resistance of bacterial chemotaxis to blockage in petroleum waters. Marine Pollution Bull. 10:48-49.
71. Sjoblad, R.D., I. Chet and R. Mitchell. 1979. A quantitative assay for algal chemotaxis. Appl. Environ. Microbiol. 36:847-850.
72. Kritzman, G., I. Chet and Y. Henis. 1979. Isolation of extracellular polysaccharides from *Sclerotium rolfsii.* Can. J. Bot. 57:1855-1859.
73. Huttermann, A., M. Gebauer and I. Chet. 1979. Studies on isoenzyme pattern during differentiation (spherulation) of *Physarum polycephalum.* Arch. Microbiol. 120:113-123.
74. Zweck, S., A. Huttermann and I. Chet. 1978. A convenient method for preparing inocula of homogenized mycelium. Exper. Mycol. 2:377-378.
75. Chet, I. and A. Huttermann. 1980. Chemical composition of hyphal walls of *Fomes annosus.* Eur. J. Forest Pathol. 10:65-70.
76. Elad, Y., I. Chet and J. Katan. 1980. *Trichoderma harzianum*: A biocontrol agent effective against Sclerotium rolfsii and Rhizoctonia solani. Phytopathology 70:119-121.
77. Chet, I. and R. Baker. 1980. Induction of suppressiveness to *Rhizoctonia solani* in soil. Phytopathology 70:994-998.
78. Stzejnberg, A., Z. Madar and I. Chet. 1980. Induction and quantification of sclerotia in *Rosellinia necatrix.* Phytopathology 70:525-527.
79. Elad, Y., J. Katan and I. Chet. 1980. Physical, biological and chemical control integrated for soilborne diseases in potatoes. Phytopathology 70:418-422.
80. Okon, Y., L. Cakmakci, I. Nur and I. Chet. 1980. Aerotaxis and chemotaxis of *Azospirillum brasilense.* Microbial. Ecol. 6:277-220.
81. Paster, N. and I. Chet. 1980. Effect of environmental factors on growth and sclerotium formation in *Aspergillus ochraceus.* Can. J. Bot. 58:1844-1850.
82. Kritzman, G., D. Gilan and I. Chet. 1980. Germination inhibitor in *Botrytis allii* spores. Phytoparasitica 8:73-76.
83. Kritzman, G. and I. Chet. 1980. The role of phenols in pathogenicity of *Botrytis allii.* Phytoparasitica 8:27-37.
84. Harman, G.E., I. Chet and R. Baker. 1980. *Trichoderma hamatum* effects on seed and seedling disease induced in radish and pea by *Pythium* spp. or *Rhizoctonia solani.* Phytopathology 70:1167-1172.
85. Hadar, Y., Y. Henis and I. Chet. 1981. The potential for the formation of sclerotia in submerged mycelium of *Sclerotium rolfsii.* J. Gen. Microbiol. 122:137-141.
86. Chet, I. and R. Baker. 1981. Isolation and biocontrol potential of *Trichoderma hamatum* from soil naturally suppressive to *Rhizoctonia solani.* Phytopathology 71:286-290.

87. Chet, I., G.E. Harman and R. Baker. 1981. *Trichoderma hamatum:* its hyphal interactions with *Rhizoctonia solani* and *Pythium* spp. Microbiol. Ecol. 7:29-38.
88. Elad, Y., I. Chet and Y. Henis. 1981. A selective medium for improving quantitative isolation of *Trichoderma* spp. from soil. Phytoparasitica 9:59-67.
89. Elad, Y., Y. Hadar, E. Hadar, I. Chet and Y. Henis. 1981. Biological control of *Rhizoctonia solani* by *Trichoderma harzianum* in carnation. Plant Disease 65:675-677.
90. Kritzman, G., I. Chet and D. Gilan. 1981. Spore germination and penetration of *Botrytis allii* into *Allium cepa* host. Botanical Gazette 142:151-155.
91. Harman, G.E., I. Chet and R. Baker. 1981. Factors affecting efficacy of *Trichoderma hamatum* applied to seeds as a biocontrol agent. Phytopathology 71:569-572.
92. Ruben, D.M., Z.R. Frank, I. Chet and D. Sklan. 1981. Changes in chemical composition of oospores of *Pythium aphanidermatum* during biomodal germination. Can. J. Microbiol. 27:536-543.
93. Elad, Y., I. Chet and Y. Henis. 1981. Biological control of *Rhizoctonia solani* in strawberry fields by *Trichoderma harzianum.* Plant and Soil 60:245-254.
94. Huttermann, A., W.E. Kuhl and I. Chet. 1981. The effect of L-threonine on oxalic acid synthesis by *Fomes annosus.* Eur. J. Forest Pathol. 10:339-344.
95. Haars, A., I. Chet and A. Huttermann. 1981. Effect of phenolic compounds and tannin on growth and laccase activity of *Fomes annosus.* Eur. J. Forest Pathol. 11:67-76.
96. Gerson, U. and I. Chet. 1981. Are allochthonous and autochthonous soil microorganisms r- and k-selected? Rev. Ecol. Biol. Sol. 18:285-289.
97. Platt, M.W., I. Chet and Y. Henis. 1981. Lignocellulose degradation during growth of the mushroom *Pleurotus* sp. Eur. J. Appl. Microbiol. and Biotech. 13:194-195.
98. Platt, M., Y. Bashan, I. Chet and Y. Henis. 1981. Two media for the rapid growth of *Pleurotus* sp. Mushroom Newsletter for Tropics 1:2-5.
99. Elad, Y., Hadar, Y., Hadar, E., I. Chet and Y. Henis, 1981. Biological control of *Rhizoctonia solani* by *Trichoderma harzianum* in carnation. Plant Dis. 65:675-677.
100. Paster, N. and I. Chet. 1982. Influence of controlled atmospheres on the formation and ultrastructure of *Aspergillus ochraceus* sclerotia. Trans. Brit. Mycol. Soc. 78:315-322.
101. Elad, Y., I. Hadar, I. Chet and Y. Henis. 1982. Prevention with *Trichoderma harzianum* Rifai aggr. of reinfestation of *Trichoderma harzianum* and *Rhizoctonia solani* Kuhn of soil fumigated with methyl bromide and improvement of disease control in tomatoes and peanuts. Crop Protection 1:199-211.

102. Hadar, E., Y. Elad, Y. Hadar and I. Chet. 1982. Build-up and decline of *Rhizoctonia solani* inoculum under field conditions. Plant and Soil 65:303-307.

103. Chet, I. and A. Huttermann. 1982. De novo synthesis of polyphenol oxidase (laccase) during formation of sclerotia in *Sclerotium rolfsii.* FEMS Microbiol. Letts. 14:211-215.

104. Elad, Y., I. Chet and Y. Henis. 1982. Degradation of plant pathogenic fungi by *Trichoderma harzianum.* Can. J. Microbiol. 28:719-725.

105. Hadar, Y., Y. Elad, I. Chet and Y. Henis. 1982. Induction of macroscopic strand formation in *Sclerotium rolfsii* by *Trichoderma harzianum.* Isr. J. Bot. 30:156-164.

106. Elad, Y., A. Kalfon and I. Chet. 1982. Control of *Rhizoctonia solani* in cotton by seed coating with Trichoderma spores. Plant and Soil 66:279-281.

107. Elad, Y., I. Chet, P. Boyle and Y. Henis. 1983. Parasitism of Trichoderma spp.on *Rhizoctonia solani* and *Sclerotium rolfsii* — SEM studies and fluorescence microscopy. Phytopathology 73:85-88.

108. Chet, I., Y. Elad, A. Kalfon, Y. Hadar and J. Katan. 1982. Integrated control of soilborne and bulb-borne pathogens in iris. Phytoparasitica 10:229-236.

109. Elad, Y. and I. Chet. 1983. Improved selective media for isolation of Trichoderma or Fusarium spp. Phytoparasitica 11:55-58.

110. Chet, I. and Y. Elad. 1982. Prevention of plant infection by biological means. La Selection des Plantes, Bordeaux, France, INRA Publ. 195-204.

111. Hadar, Y., M. Pines, I. Chet and Y. Henis. 1983. The regulation of sclerotium initiation in *Sclerotium rolfsii* by glucose and cyclic AMP. Can. J. Microbiol. 28:21-26.

112. Elad, Y., Z. Sadovski and I. Chet. 1983. Detection of mycoparasitism by infra-red photomicrography. Microbial Ecol. 9:185-187.

113. Zilberstein, Y., I. Chet and Y. Henis. 1983. Effect of atmosphere on germination of microsclerotia of *Verticillium dahliae* Kleb. Isr. J. Bot. 32:33-36.

114. Lisker, N., N. Paster and I. Chet. 1983. Ochratoxin production by *Aspergillus ochraceus* as affected by methionine and structurally related compounds. Can. J. Microbiol. 29:536-540.

115. Paster, N., N. Lisker and I. Chet. 1983. Ochratoxin A production by *Aspergillus ochraceus* Wilhelm grown under controlled atmospheres. Appl. Environ. Microbiol. 45:1136-1139.

116. Stzejnberg, A., H. Azaizia and I. Chet. 1983. The possible role of phenolic compounds in resistance of horticultural crops to *Dematophora necatrix* Hartig. Phytopathology Z. 107:318-326.

117. Kapulnik, E., A.K. Matoo, E. Chalutz and I. Chet. 1983. The relationship between the production of ethylene by certain fungi, their enzyme

constitution and their ability to accumulate intracellular phosphate. Z. fur Pflanzenphysiologie 109:347-354.

118. Paster, N. and I. Chet. 1983. The possible role of some enzymes in sclerotia formation in *Aspergillus ochraceus*. Can. J. Microbiol. 29:718-723.
119. Edelstein, L., Y. Hadar, I. Chet, Y. Henis and L.A. Segal. 1983. A model for fungal colony growth applied to *Sclerotium rolfsii*. J. Gen. Microbiol. 129:1873-1881.
120. Platt, M.W., Y. Hadar, Y. Henis and I. Chet. 1983. Increased degradation of straw by *Pleurotus ostreatus* sp. 'florida'. Eur. J. Appl. Microbiol. Biotech. 17:1240-142.
121. Zilberstein, Y., I. Chet and Y. Henis. 1983. The influence of microsclerotial source of *Verticillium dahliae* on inoculum quality. Trans. Brit. Mycol. Soc. 81:613-617.
122. Elad, Y., R. Barak and I. Chet. 1983. Possible role of lectins in mycoparasitism. J. Bacteriol. 154:1431-1435.
123. Platt, M.W., I. Chet and Y. Henis. 1983. Growth of *Pleurotus ostreatus* on cotton straw. Mushroom J. 120:425-427.
124. Platt, M.W., J. Trojanowski, I. Chet and A. Huttermann. 1983. Differences in degradation of specifically C-labelled lignin model compounds within the Pleurotus group. Microbiol. Lett. 23:19-21.
125. Ruppel, E.G., R. Baker, G.E. Harman, J.P. Hubbard, R.J. Hecker and I. Chet. 1983. Field tests of *Trichoderma harzianum* as a biocontrol agent of seedling disease in several crops and *Rhizoctonia* root rot of sugar beet. Crop Protection 2:399-408.
126. Elad, Y., R. Barak, I. Chet and Y. Henis. 1983. Ultrastructural studies of the interaction between *Trichoderma* spp. and plant pathogenic fungi. Phytopath. Z. 107:168-175.
127. Elad, Y., Y. Hadar and I. Chet. 1983. The potential of *Trichoderma harzianum* as a biocontrol agent under field conditions. Colloq. INRA 18:305-310.
128. Chet, I. and Y. Elad. 1983. Mechanism of mycoparasitism. Colloques INRA 18:35-40.
129. Chalutz, E., E. Kapulnik and I. Chet. 1983. Fermentative production of ethylene by *Penicillium dicitatum* from citrus fruit peels. Eur. J. Microbiol. Biotechnol. 18:293-297.
130. Elad, Y., R. Barak and I. Chet. 1984. Parasitism of *Sclerotium rolfsii* sclerotia by *Trichoderma harzianum*. Soil Biol. Biochem. 16:381-386.
131. Lifshitiz, R., M. Tabachnik, J. Katan and I. Chet. 1984. The effect of sublethal heating on sclerotia of *Sclerotium rolfsii*. Can. J. Microbiol. 29:1607-1610.
132. Hadar, Y., A. Banin and I. Chet. 1984. Increased inhibitory effect of cation as a clay complex on fungi. Water, Air and Soil Pollution 22:441-446.

133. Sivan, A., Y. Elad and I. Chet. 1984. Biological control of *Pythium aphanidermatum* by a new isolate of *Trichoderma harzianum.* Phytopathology 74:498-501.
134. Shapira, R., D. Sklan, Y. Henis and I. Chet. 1984. Changes in fatty acids during morphogenesis in *Sclerotium rolfsii.* J. Gen. Microbiol. 130:1183-1191.
135. Platt, M.W., Y. Hadar and I. Chet. 1984. Fungal activities involved in lignocellulose degradation by *Pleurotus.* Appl. Microbiol. and Biotechnol. 20:150-154.
136. Baker, R., Y. Elad and I. Chet. 1984. The controlled experiment in the scientific method with special emphasis on biological control. Phytopathology 74:1019-1021.
137. Strashnov, Y., Y. Elad, A. Sivan, Y. Rudich and I. Chet. 1985. Control of *Rhizoctonia solani* fruit rot of tomatoes by *Trichoderma harzianum* Rifai. Crop Protection 4:359-364.
138. Strashnov, Y., Y. Elad, A. Sivan and I. Chet. 1985. Integrated control of *Rhizoctonia solani* Kuhn on bean and carrot seedlings by methyl bromide and *Trichoderma harzianum* Rifai Aggr. Plant Pathol. 34:146-151.
139. Barak, R., Y. Elad, D. Mirelman and I. Chet. 1985. Lectins: a possible basis for specific recognition in *Trichoderma-Sclerotium rolfsii* interaction. Phytopathology 75:458-462.
140. Sadeh, R., G. Kritzman and I. Chet. 1985. The ultrastructure of resting and germinating sclerotia of Botrytis alli. Bot. Gaz. 146:302-307.
141. Platt, M.W., Y. Hadar and I. Chet. 1985. The decolorization of the polymeric dye poly blue (Poly vinalamine sulfonate anthroquinone) by lignin degrading fungi. Appl. Microbiol. and Biotechnol. 21:394-396.
142. Spiegel, Y. and I. Chet. 1985. Chitin synthetase inhibitors and their potential to control the root-knot nematode, *Meloidogyne javanica.* Nematologica 31: 480-482.
143. Zamir, D. and I. Chet. 1985. Application of enzyme electrophoresis for identification of isolates in *Trichoderma harzianum.* Can. J. Microbiol. 31:578-580.
144. Chet, I., J. Trojanowsky and A. Huttermann. 1985. Decolorization of the dyepoly-B-411 and its correlation with lignin degradation by fungi. Microbiol. Letters 29:37-43.
145. Barak, R., A. Maoz and I. Chet. 1985. Antigenic differences among several *Trichoderma* isolates. Can. J. Microbiol. 31:810-816.
146. Paster, N., N. Lisker and I. Chet. 1985. Accumulation of ochratoxin A in sclerotia of *Aspergillus ochraceus.* Can. J. Bot. 63:661-662.
147. Lisker, N., N. Paster and I. Chet. 1985. Effects of cysteine and structurally related compounds on ochratoxin production by *Aspergillus ochraceus.* Can. J. Microbiol. 31:973-977.

148. Freeman, S., A. Sztejnberg and I. Chet. 1985. Evaluation of *Trichoderma* as a biocontrol agent for *Rosellinia necatrix*. Plant Soil 94:163-170.
149. Chang, Y., Y. A. Chang, R. Baker, O. Kleifeld and I. Chet. 1985. Increased growth of plants induced by the biological control agent *Trichoderma harzianum*. Plant Disease 70:145-148.
150. Nadel-Schiffmann, N., A. Michaely, G. Zauberman and I. Chet 1985. Physiological changes occurring in picked climacteric fruit infected with different pathogenic fungi. Phytopath. Z. 113:277-284.
151. Zeidan, O., Y. Elad, Y. Hadar and I. Chet. 1986. Integrating onion in crop rotation as a means of controlling *Sclerotium rolfsii*. Plant Disease 70:426-428.
152. Sivan, A. and I. Chet. 1986. Biological control of *Fusarium* spp. in cotton wheat and musk melon by *Trichoderma harzianum*. Phytopathology Z. 116:39-47.
153. Barak, R. and I. Chet. 1986. Determination by fluorescein diacetate of fungal viability during mycoparasitism. Soil Biol. Biochem. 18:315-319.
154. Barak, R., Y. Elad and I. Chet. 1986. The properties of L-fucose binding agglutinin associated with the cell wall of *Rhizoctonia solani*. Arch. Microbiol. 144:346-349.
155. Spiegel, Y., E. Cohn and I. Chet. 1986. Use of chitin for controlling plant parasitic nematodes. I. Direct effects on nematode reproduction and plant performance. Plant Soil 95:87-95.
156. Elad, Y., Y. Zvieli and I. Chet. 1986. Biological control of *Macrophomina phaseolina* (Tassi) Goid by *Trichoderma harzianum*. Crop Protection 5:288-292.
157. Elad, Y., Z. Sadowsky and I. Chet. 1986. Scanning electron microscopical observations of early stages of interaction between *Trichoderma harzianum* and *Rhizoctonia solani*. Trans. Brit. Mycol. Soc. 88:259-263.
158. Cohen, E., I. Elster and I. Chet. 1986. Properties and inhibition of *Sclerotium rolfsii* chitin synthetase. Pesticide Science 17:175-182.
159. Sivan, A. and I. Chet. 1986. Possible mechanism for biological control of Fusarium spp. by *Trichoderma harzianum*. British Crop Protection Conference on Pests and Diseases, pp. 865-872.
160. Worrall, J.J., I. Chet and A. Huttermann. 1986. Association of rhizomorph formation with laccase activity in *Armillaria* spp. J. Gen. Microbiol. 132:2527-2533.
161. Shapira, R., M. Pines, I. Chet and Y. Henis. 1986. Regulation of glycogen synthetase in *Sclerotium rolfsii*. J. Gen. Microbiol. 132:765-770.
162. Rodriguez-Kabana, R., G. Morgan-Jones and I. Chet. 1987. Biological control of nematodes: soil amendments and microbial antagonists. Plant and Soil 98:237-247.

193. Spiegel, Y., E. Cohn, S. Galper, E. Sharon, and I. Chet. 1991. Evaluation of a newly isolated bacterium, *Pseudomonas chitinolytica* sp. nov. for controlling the root-knot nematode *Meloidogyne javanica*. Biocontrol Science and Technol. 1:115-125.

194. Inbar, J. and I. Chet. 1991. Evidence that chitinase produced by *Aeromonas caviae* is involved in the biological control of soil plant pathogens. Soil Biology and Biochemistry, 23:973-978.

195. Broglie, K., I. Chet, M. Holliday, R. Cressman, P. Biddle, S. Knowlton, C.J. Mauvals, and R. Broglie. 1991. Transgenic plants with enhanced resistance to the fungal pathogen *Rhizoctonia solani*. Science 254:1194-1197.

196. Pe'er, S., Z. Barak, O. Yarden and I. Chet. 1991. Stability of *Trichoderma harzianum amdS* transformants in soil and rhizosphere. Soil Biol. Biochem. 23:1043-1046.

197. Ordentlich, A., Q. Migheli and I. Chet. 1991. Biological control activity of three *Trichoderma* isolates against Fusarium wilts of cotton and muskmelon and lack of correlation with their lytic enzymes. J. Phytopathology 133:177-186.

198. Ordentlich, A., Z. Weisman, H.E. Gottlieb, M. Cojocaru and I. Chet. 1992. Inhibitory furanone produced by the biocontrol agent *Trichoderma harzianum*. Phytochemistry 31:485-486.

199. Inbar, J. and I. Chet. 1992. Biomimics of fungal cell-cell recognition by the use of lectin-coated nylon fibers. J. Bacteriol. 174:1055-1059.

200. Jach, G., S. Logemann, G. Wolf, A. Oppenheim, I. Chet, J. Schell and J. Logemann. 1992. Expression of a bacterial chitinase leads to improved resistance of transgenic tobacco plants against fungal infection. Biopractice 1:33-36.

201. Kleifeld, O. and I. Chet. 1992. *Trichoderma* — Plant interaction and its effect on increased growth response. Plant and Soil 144:267-272.

202. Levy, E., Z. Eyal, I Chet and A. Hochman. 1992. Resistance mechanisms of *Septoria tritici* to antifungal products of *Pseudomonas*. Physiol. Mol. Plant Pathol. 40:163-171.

203. Paster, N., A. Pushinsky, M. Menasherov and I. Chet. 1992. Inhibitory effect of *Aspergillus ochraceus* and *Aspergillus flavus*, and on Aflatoxin formation. J. Sci. Food Agric. 58:589-591.

204. Amir, R., D. Levanon, Y. Hadar and I. Chet. 1992. Formation of sclerotia by *Morchella esculenta*: relationship between media composition and turgor potential in the mycelium. Mycol. Res. 96:943-948.

205. Benhamou, N., K. Broglie, R. Broglie and I. Chet. 1992. Antifungal effect of bean endochitinase on *Rhizoctonia solani*: ultrastructural changes and cytochemical aspects of chitin breakdown. Can. J. Microbiol. 39:318-328.

206. Oppenheim, A.B. and I. Chet. 1992. Biological control of soilborne plant pathogenic fungi by cloned chitinase. Trends Biotechnol. 10:392-394.
207. Amir, R., D. Levanon Y. Hadar and I. Chet. 1993. Morphology and physiology of *Morchella esculenta* during sclerotial formation. Mycological Res. 97:683-689.
208. Elad, Y., G. Zimand, Y. Zaqs, S. Zuriel and I. Chet. 1993. Use of *Trichodrma harzianum* in combination or alternation with fungicides to control cucumber grey mould (*Botrytis cinerea)* under commercial greenhouse conditions. Plant Pathol. 42:324-332.
209. Ocka, Y., I. Chet and Y. Speigel. 1993. Control of the root-knot nematode *Meloidogyne javanica* by *Bacillus cereus.* Biocontrol Sci. Technol. 3:115-126.
210. Sivan, A. and I. Chet. 1993. Integrated control of fusarium crown and root rot of tomato with *Trichoderma harzianum* in combination with methyl bromide or soil solarization. Crop Protection 12:380-386.
211. Fridlender, M., J. Inbar and I. Chet. 1993. Biological control of soilborne plant pathogens by a β-1,3-glucanase producing *Pseudomonas cepacia.* Soil Biol. Biochem. 25: 1211-1221.
212. Sitrit, Y., Z. Barak, Y. Kapulnik, A. Oppenheim and I. Chet. 1993. Expression of *Serratia marcescens* chitinase gene in *Rhizobium meliloti* during symbiosis on alfalfa roots. Mol. Plant Microbe Inter. 6:293-298.
213. Haran, S., H. Schickler, S. Pe'er, S. Logemann, A. Oppenheim and I. Chet. 1993. Increased constitutive chitinase activity in transformed *Trichoderma harzianum.* Biol. Control 3:101-108.
214. Zimand, G., L. Valinsky, S. Manulis, Y. Elad and I. Chet. 1993. DNA fingerprinting of a *Trichoderma* isolate by RAPD procedure. IOBD Bull. 16:173-176.
215. Benhamou, N. and I. Chet. 1993. Hypal interactions between *Trichoderma harzianum* and *Rhizoctonia solani*: ultrastructure and gold cytochemistry of the mycoparasitic process. Phytopathology 83:1062-1071.
216. Benhamou, N., K. Broglie, I. Chet and R. Broglie. 1993. Cytology of infection of 35S-bean chitinase transgenic canola plants by *Rhizoctonia solani*: cytochemical aspects of chitin breakdown *in vivo.* The Plant Journal 4:295-305.
217. Inbar, J. and I. Chet. 1994. A newly isolated lectin from the plant pathogenic fungus *Sclerotium rolfsii*: purification, characterization and role in mycoparasitism. Microbiology 140:651-657.
218. Zimand, G., L. Valinsky, Y. Elad, I. Chet and S. Manulis. 1994. Use of the RAPD procedure for the identification of *Trichoderma* strains. Mycol. Res. 98: 531-534.

219. Amir, R., D. Levanon, Y. Hadar and I. Chet. 1994. The role of source sink relationships in translocation during sclerotia formation by *Morchella esculenta* Mycological Res. 98:1409-1414.
220. Inbar, J., Abramsky, M., Cohen, D. and I. Chet. 1994. Plant growth enhancement and disease control by *Trichoderma harzianum* grown under commercial conditions. Eur. J. Plant Pathol. 100:337-346.
221. Koby, S., H. Schickler, I. Chet and A.B. Oppenheim. 1994. The chitinase encoding Tn-7-based *Chi*A gene endows *Pseudomonas fluorescens* with the capacity to control plant pathogens in soil. Gene 147:81-83.
222. Perrakis, A., I. Tews, Z. Dauter, A.B. Oppenheim, I. Chet, K.S. Wilson and C.E. Vorgias. 1994. Crystal structure of a bacterial chitinase at 2.3 Å resolution. Structure 2:1169-1180.
223. Zimand, G., Y. Elad, G. Kritzman, and I. Chet. 1994. Control of *Botyris cinerea* by *Trichoderma harzianum* (T39): Does the biocontrol agent produce inhibitory substances? Phytoparasitica 22:150-151.
224. Haran, S., H. Schickler, A. Oppenheim and I. Chet. 1995. New components of the chitinolytic system of *Trichoderma harzianum*. Mycol. Research. 99:441-446.
225. Amir, R., D. Levanon, Y. Hadar and I. Chet. 1995. Factors affecting translocation and sclerotia formation by *Morchella esculenta*. Exp. Mycol. 19: 61-70.
226. Amir, R., E. Steudle, D. Levanon, Y. Hadar and I. Chet. 1995. Turgor changes in *Morchella esculenta* during translocation and sclerotia formation. Exp. Mycol. 19: 129-136.
227. Sitrit, Y., Vorgias, C.E., I. Chet and A.B. Oppenheim. 1995. Cloning and primary structure of the *chi*A gene from *Aeromonas caviae*. J. Bacteriol. 177:4187-4189.
228. Salt, D.E., M. Blaylock, N.P.B.A. Kumar, V. Dushenkov, B.D. Ensley, I. Chet, and I. Raskin. 1995. Phytoremediation: A novel strategy for the removal of toxic metals from the environment using plants. Nature Biotechnology 13:468-474.
229. Chernin, L., Z. Ismailov, S. Haran and I. Chet. 1995. Chitinolytic *Enterobacter agglomerans* antagonistic to fungal plant pathogens. Appl Environ. Microbiol. 61:1720-1726.
230. Inbar, J. and I. Chet. 1995. The role of recognition in the induction of specific chitinases during mycoparasitism by *Trichoderma harzianum*. Microbiology 141:2823-2829.
231. Pleban, S., F. Ingel and I. Chet. 1995. Control of *Rhizoctonia solani* and *Sclerotium rolfsii* in the greenhouse using endophytic *Bacillus* spp. Eur. J. Plant Pathol. 101:665-672.

232. Spiegel, Y., I. Chet, I. Mor, O. Kleifeld, E. Sharon and M. Bar-Eyal. 1995. Trichoderma Reparation as biological control of the root-knot nematode *Meloidogyne javanics*: Evaluation of the nematocidal activity and preliminary studies of the mechanism. ARO publication, series E, 1995m no. 1210 (appeared in Hebrew in Hassadeh 75:67-72).

233. Benhamou, N. and Ilan Chet. 1996. Parasitism of sclerotia of *Sclerotium rolfsii* by *Trichoderma harzianum:* ultrastructural and cytochemical aspects of the interaction. Phytopathology 86:405-416.

234. Chernin, L., A. Brandis, Z. Ismailov and I. Chet. 1996. Pyrrolnitrin production by an *Enterobacter agglomerans* strain with a broad spectrum of antagonistic activity towards fungal and bacterial phytopathogens. Current Microbiol. 32:208-212.

235. Inbar, J., A. Menendez, and I. Chet. 1996. Hyphal interactions between *Trichoderma harzianum* and *Sclerotinia sclerotiorum* and its role in biological control. Soil Biol. & Biochem. 28:757-763.

236. Zimand, G., Y. Elad and I. Chet. 1996. Effect of *Trichoderma harzianum* on *Botrytis cinerea* pathogenicity. Phytopathology 86:1255-1262.

237. Rousseau, A., N. Benhamou, I. Chet and Y. Piché, 1996. Mycoparasitism of the extramatrical phase of *Glomus intraradices by Trichoderma harzianum.* Phytopathology 86, 434-443.

238. Haran, S., H. Schickler, A. Oppenheim and I. Chet. 1996. Differential expression of *Trichoderma harzianum* chitinases during mycoparasitism. Phytopathology 86:980-985.

239. Regev, A., M. Keller, N. Strizhov, B. Sneh, E. Prudovsky, I. Chet, I. Ginzberg, Z. Koncz-Kalman, C. Koncz, J. Schell, and A. Zilberstein. 1996. Synergistic activity of a *Bacillus thuringiensis* δ-endotoxin and a bacterial endochitinase against *Spodoptera littoralis* larvae. App. Environ. Microbiol. 62:3581-3586.

240. Chernin, L.S., L. de la Fuente, V. Sobolev, S. Haran, C.E. Vorgias, A.B. Oppenheim and I. Chet. 1997. Molecular cloning, structural analysis, and expression in *Escherichia coli* of a chitinase gene from *Enterobacter agglomerans.* Appl. Environ. Microbiol. 63:834-839.

241. Maas, C., C.G. Simpson, P. Eckes, H. Schickler, J.W.S. Brown, B. Reiss, K. Salchert, I. Chet, J. Schell and C. Reichel. 1997. Expression of intron modified *NPT II* genes in monocotyledonous and dicotyledonous plant cells. Mol. Breeding 3:15-28.

242. Benhamou, N. and I. Chet. 1997. Cellular and molecular mechanisms involved in the interaction between *Trichoderma harzianum* and *Pythium ultimum.* Appl. Environ. Microbiol. 63:2095-2099.

243. Flores, A., I. Chet and A. Herrera-Estrella. 1997. Improved biocontrol activity of *Trichoderma harzianum* by overexpression of the proteinase-encoding gene *prb*1. Current Genetics 31:30-37.

244. Pleban, S., L. Chernin and I. Chet. 1997. Chitinolytic activity of endophytic strain of *Bacillus cereus.* Lett. Appl. Microbiol. 25:284-288.
245. Sela, S., H. Schickler, I. Chet and Y. Spiegel. 1997. Purification and characterization of a *Bacillus* cereus collagenolytic/proteolytic enzyme and its effect on *Meloidogyne javanica* cuticular proteins. Eur. J. Plant Path. 104:59-67.
246. Gomez, I, I. Chet and A. Herrera-Estrella. 1997. Genetic diversity and vegetative compatibility among *Triochoderma harzianum* isolates. Mol. Gen. Genetics 256:127-135.
247. Schickler, H. and I. Chet. 1997. Heterologous chitinase gene expression to improve plant defense against phytopathogenic fungi. J. Industrial Microbiol. & biotechnol. 19:196-201.
248. Oka, Y., I. Chet and Y. Spiegel. 1997. An immunoreactive protein to wheat-germ agglutinin antibody is induced in oat roots following invasion of the cereal cyst nematode *Heterodera avenae*, and by Jasmonate. Mol. Plant-Microbe Interactions 10:961-969.
249. Oka, Y., I. Chet, More, M. and Y. Spiegel. 1997. A fungal parasite of *Meloidogyne javanica* eggs: Evaluation of its use to control the root-knot nematode. Biocontrol Science and Technol. 7:489-497.
250. Oka, Y., I. Chet and Y. Spiegel. 1997. Are pathogenesis-related poteins induced by Meloidogyne *javanica* or *Heterodera avenae* invasion? J. Nematol. 29:501-508.
251. Oka, Y., I. Chet and Y. Spiegel. 1997. Accumulation of lectins in cereal roots invaded by the cereal cyst nematode *Heterodera venae.* Physiol. Mol. Olant Pathol. 51:333-345.
252. Zoubenko, O., F. Uckun, Y. Hur, I. Chet and N. Tumer. 1997. Plant resistance to fungal infection induced by nontoxic pokeweed antiviral protein mutants. Nature Biotechnol. 15(10):992-996.
253. Oka, Y., I. Chet, Mor, M. and Y. Spiegel. 1997. A fungal parasite of *Meloidogyne javanica* eggs: Evaluation of its use to control the root-knot nematode. Biocontrol Science and Technol. 7:489-497.
254. Danai, O., Olenik, I., Hadar, Y., I. Chet and D. Levanon. 1998. The role of light in the morphogenesis of *Pleurotus ostreatus.* Int. J. Mushroom Sci. 2:31-37.
255. Oka, Y., I. Chet and Y. Spiegel. 1997. Accumulation of lectins in cereal roots invaded by the cereal cyst nematode *Heterodera avenae.* Physiol. Mol. Plant Pathol. 51:333-345.
256. Dickman, M.B. and I. Chet. 1998. Biodegradation of oxalic acid: A potential new approach to biological control. Soil Biol. Biochem. 30:1195-1197.
257. Schickler, H., B-C. Danin-Gehali, S. Haran and I. Chet. 1998. Electro-phoretic characterization of chitinases as a tool for the identification of *Trichoderma harzianum* strains. Mycol. Res. 102:373-377.

258. Sela, S., H. Schickler, I. Chet and Y. Spiegel. 1998. Putrification and characterization of *Bacillus cereus* collagenolytic/proteolytic enzymes and its effect on *Meloidogyne juvanica* cuticular proteins. Eur. J. Plant Pathol. 104: 59-67.
259. Schickler, H., Haran, S., A. Oppenheim and I. Chet. 1998. Induction of the *Trichoderma* harzianum chitinolytic system is triggered by the chitin monomer, *N*-acetylglucosamine. Mycol. Res. 102:1224-1226.
260. Chernin, L.S., M.K. Winson, J.M. Thompson, S. Haran, B.W. Bycroft, I. Chet, P. Williams and G.S.A.B. Stewart. 1998. Chitinolytic activity in *Chromobacterium violaceum:* Substrate analysis and regulation by quorum sensing. J. Bacteriol. 180:4435-4441.
261. Cortes, C., A. Gutierrez, V. Olmedo, J. Inbar, I. Chet and A. Herrera-Estrella. 1998. The expression of genes involved in parasitism by *Trichoderma harzianum* is triggered by a diffusable factor. Mol. Gen. Gen. 260:218-225.
262. Yedidia I., N. Benhamou, and I. Chet. 1999. Induction of defense responses in cucumber plants (*Cucumis sativus* L.) by the biocontrol agent *Trichoderma harzianum.* Appl. Environ. Microbiol. 65:1061-1070.
263. Carsolio C., N. Benhamou, Haran, S., C. Cortés, A. Gutiérrez, I, Chet and A. Herrera-Estrella. 1999. Role of the *Trichoderma harzianum* endochitinase gene, *ech42,* in mycoparasitism. Appl. Environ. Microbiol. 65:929-935.
264. Salt, D.E., N. Benhamou, M. Leszczyniecka, I. Raskin and I. Chet. 1999. Rhizobacterial enhanced Cadmium A accumulation in plants. Int. J. Phytoremed. 1:67-79.
265. Cohen-Kupiec, R., K.E. Broglie, D. Friesem, R.M. Broglie and I. Chet. 1999. Molecular characterization of a novel β-1,3 exo-glucanase related to mycoparasitism of *Trichoderma harzianum.* Gene 226:147-154.
266. Omero, C., J. Inbar, V. Rocha-Ramirez, A. Herrera-Estrella, I. Chet and B.A. Horwitz. 1999. G protein activators and cAMP promote mycoparasitic behaviour in *Trichoderma harzianum.* Mycol. Res. 103:1637-1642.
267. Shaul, O., S. Galili, H. Volpin, I. Ginzberg, Y. Elad, I. Chet and Y. Kapulnik. 1999. Mycorrhiza-induced changes in disease severity and PR protein expression in tobacco leaves. Mol. Plant Microbe Interaction 12:1000-1007.
268. Ramot, O., R. Cohen-Kupiec and I. Chet. 1999. Regulation of β-1,3-glucanase by carbon starvation in the mycoparasite *Trichoderma harzianum.* Mycol. Res. 104:415-420.
269. Oka, Y., H. Koltai, M. Bar-Eyal, M. Mor, E. Sharon, I. Chet and Y. Spiegel. 2000. New strategies for the control of plant-parasitic nematodes. Pest Manag. Sci. 56:983-988.
270. Yedidia I., N. Benhamou, Y. Kapulnik and I. Chet. 2000. Induction and accumulation of PR proteins activity during early stages of roots colonization by the mycoparasite *Trichoderma harzianum* strain T-203. Plant Physiol. Biochem. 38:863-873.

271. Huttermann, A., A.S. Hamza, I. Chet, A. Majcherczyk, T. Fouad, R. Badr, R. Cohen, L. Persky and Y. Hadar. 2000. Recycling of agricultural wastes by white-rot fungi for the production of fodder for ruminants. Agro-Food-Industry Hi-Tech. Nov./Dec. 29-32.
272. Omero, C., B.A.Horwitz and I. Chet. 2001. A convenient in vivo detection system for extracellular N-acetylglucosaminidases using a fluorogenic substrate in the biocontrol agent *Trichoderma harzianum.* J. Microbiol. Methods 43:165-169.
273. Guenoune, D., S. Galili, D.A. Phillips, H. Volpin, I. Chet, Y Okon and Y. Kapulnik. 2001. The defence response elicited by the pathogen *Rhizoctonia solani* is suppressed by colonization of the AM-fungus *Glomus intraradices.* Plant Science 160:925-932.
274. Chernin, L., Ismailov, Z. nd I. Chet. 2001. Genetic manipulations of bacterial enzymes and antibiotics to enhance biocontrol of phytopathogens, pp. 3-4. Proc. Achievements of Biotechnology for the Future of Mankind, Samarkand.
275. Sharon, E., M. Bar-Eyal, I. Chet, A. Herrera-Estrella, O. Kliefeld and Y. Spiegel. 2001. Biocontrol of the root-knot nematode *Meloidogyne javanica* by *Trichoderma harzianum.* Phytopathology 91:687-693.
276. Viterbo, A., S. Haran, D. Friesem, O. Ramot and I. Chet. 2001. Antifungal activity of a novel endochitinase gene (chit36) from *Trichoderma harzianum* Rifai TM. FEMS Microbiol. Lett. 200:169-174.
277. Yedidia, I., A.K. Srivastva, Y. Kapulnik and I. Chet. 2001. Effect of *Trichoderma harzianum* on microelement concentrations and increased growth of cucumber plants. Plant and Soil 235: 235-242.
278. Rocha-Ramirez, V., C. Omero, I. Chet, B.A. Horwitz,and A. Herrera-Estrella 2002. A *Trichoderma atroviride* G protein *a* subunit gene, *tga1*, involved in mycoparasitic *coiling* and conidiation. Eukaryotic Cell 1:594-605.
279. Shaul-Keinan O., V. Gadkar, I. Ginzberg, J.M. Granzweig, I. Chet, Y. Elad, S. Wininger, E. Belausov, Y. Eshed, N. Atzmon, Y. Ben-Tal and Y Kapulnik. 2002. Hormone concentrations in tobacco roots change during arbuscular mycorrhizal colonization with *Glomus intraradices.* New Phytologist 154:501-507.
280. Zohar-Perez C., E. Ritte L. Chernin I. Chet and A. Nussinovitch. 2002. Preservation of chitinolytic *Pantoae agglomerans* in a viable form by cellular dried alginate-based carriers. Biotechnol. Prog. 18:1133-1140.
281. Viterbo, A., O. Ramot, L. Chernin and I. Chet. 2002. Significance of lytic enzymes from Trichoderma spp. in Antonie van Loeuwenhoek 81:549-556.
282. Kamensky, M., M. Ovadis, I. Chet, and L. Chernin. 2003. Soil-borne strain ICI4 of *Serratia plymuthica* with multiple mechanisms of antifungal activity provides biocontrol of *Botrytis cinerea* and *Sclerotinia sclerotiorum* diseases. Soil Biol. Biochem. 35:323-331.

283. Viterbo, A., O. Monteroi, O. Ramot, D. Friesem, E. Monte, A. Llobell and I. Chet. 2003. Expression regulation of the endochitinase chit36 from *Trichoderma asperellum*. Current Genetics 45:205-213.

284. Zohar-Perez, C., L. Chernin, I. Chet and A. Nussinovitch. 2003. Structure of dried cellular alignate matrix containing fillers provides extra protection for microorganisms against UVC. Radiation Res. 160:198-204.

285. Yedidia, I., M. Shoresh, Z. Kerem, N. Benhamou, Y. Kapulnik, and I. Chet. 2003. Concomitant induction of systemic resistance to *Pseudomonas syringae* pv. *lachrymans* in cucumber by *Trichoderma asperellum* (T-203) and accumulation of phytoalexins. Appl. Environ. Microbiol. 69:7343-7353.

286. Zohar-Perez, C., I. Chet, and A. Nussinovitch. 2004. Irregular Surface-Textural Features of Dried Alginate-Filler Beads. Food hydrocolloids 18:249-258.

287. Ramot, O., A. Viterbo, D. Friesem, A. Oppenheim and I. Chet. 2004 Regulation of two homodimer hexosaminidases in the mycoparasitic fungus *Trichoderma asperellum* by glucosamine. Current Genetics 45:205-213.

288. Ovadis, M., X. Liu, S. Gavriel, Z. Usmailov, I. Chet, and L. Chernin. 2004 The Global Regulator Genes from Biocontrol *Serratia plymuthica* IC1270: Cloning, Sequencing, and Functional Studies. J. Bacteriol. 186:4986-4993.

289. Zohar-Perez C., I. Chet and A. Nussinovitch. 2004. Unexpected distribution of immobilized microorganisms within alginate beads. Biotechnology Bioengineering 88:671-674.

290. Viterbo A., M. Harel and I. Chet. 2004. Isolation of two aspartyl proteases from *Trichoderma asperellum* expressed during colonization of cucumber roots. FEMS Microbiol. Lett. 238: 151-158.

291. Shoresh M., I. Yedidia and I. Chet. 2005. Involvement of jasmonic acid/ ethylene signalling pathway in the systemic resistance induced in cucumber by *Trichoderma asperellum*. Phytopathology 95:76-84.

292. Viterbo A., M. Harel, B.A. Horwitz, I. Chet, and P.K. Mukherjee. 2005. *Trichoderma* MAP-kinase signaling is involved in induction of plant resistance. Appl. & Environ. Microbiol. 71:6241-6246.

293. Tadmor Z., S. Sachs, I. Chet, Y. Spiegel, I. Ravina, G. Manor, and S. Yannai. 2005. An Innovative Approach: The Use of Di-Nitrogen Tetroxide for Soil Fumingation. Biosystem Engineering 91:413-419.

294. Khmel I.A., M.I. Ovadis, A.V. Mayatskaya, A.M. Veselovskii, I.A. Bass, V.A.Lipasova, A. Bolshoy, I. Chet and L.S. Chernin. 2005. Activity of *Serratia plymuthica* IC1270 gene *chi*A promoter region in *Eschrichia coli* mutants deficient in global regulators of transcription. J. Basic Microbiol. 45: 426-439.

295. Zohar-Perez, C., I. Chet and A. Nussinovitch. 2005. Mutual Relationships Between Soils and Biological Carrier Systems. Biotechnology Bioengineering 92:54-60.

296. Meziane, H., S. Gavriel, Z. Ismailov, I. Chet, L. Chernin and M. Höfte. 2006. Control of green and blue mould on orange fruit by *Serratia plymuthica* strains IC14 and IC1270 and putative modes of action. Postharvest Biology and Technology 39:125-133.
297. Viterbo A. and I. Chet. 2006. *TasHyd1,* a new hydrophobin gene from the biocontrol agent *Trichoderma apenellum,* is involved in plant root colonization. Molecular plant pathology (in press).
298. Shoresh M., A. Gal-On, D. Leibman, and I. Chet. 2006. Charecterization of a MAPK gene from cucumber required for *Trichoderma*-conferred plant resistance. Plant Physiol. (in press).
299. Sharon, E, I. Chet, A. Viterbo, M. Bar-Eyal, H. Nagan, G.J. Samuels and Y. Spiegel. 2007. Parasitism of Trichoderma on Meloidogynejavanica and role of the gelatinous matrix. Eur. Plant Pathol. (in press).
300. Viterbo, A., E. Wiest, Y. Brotman, I. Chet and C. Kenerley. 2007. The 18mer peptaibols from *Trichoderma virens* Elicit plant defense responses. Molecular Plant Pathol. 8:737-746.
301. Makovitzki, A., A. Viterbo, Y. Brotman, I. Chet and Y. Shai. 2007. Inhibition of fungal and bacterialplant pathogens *in vitro* and *in vivo* with ultrashot cationic lipopeptide. Appl. Environ. Microbiol. 73:6629-6636.

III. Chapters and Reviews

302. Chet, I., Y. Assa and B. Gestetner. 1972. The effect of lucerne saponins on *Physarum polycephalum.* First European Conference of Physarum, Gottingen, W. Germany.
303. Chet, I., N. Retig and Y. Henis. 1972. The relationship between RNA and ribonucleases during mannitol-induced spherulation of *Physarum polycephalum.* First European Conference of Physarum, Gottingen, W. Germany.
304. Chet, I. 1973. Changes in ribonucleic acid during differentiation of *Physarum polycephalum.* In: A. Huttermann (ed.), "*Physarum polycephalum*, Object of Research in Cell Biology". Ber. Deutsch Lot. Ges. 86:77-92. Gustav Fisher Verlag, Stuttgart.
305. Henis, Y. and I. Chet. 1975. Microbiological control of plant pathogens. Advances in Applied Microbiology 19:85-111.
306. Chet, I. and Y. Henis. 1975. Sclerotial morphogenesis in fungi. Annual Review of Phytopathology 13:169-192.
307. Chet, I. 1976. Melanin and polyphenol oxidase inhibitors, pp. 87-88. In: "Biodeterioration of Tropical Wood, Chemical Basis for Natural Resistance". Naval Research Lab., Washington DC.
308. Chet, I. and R. Mitchell. 1976. Ecological aspects of microbial chemotactic behaviour. Ann. Rev. Microbiol. 30:221-239.

309. Chet, I. and R. Mitchell. 1976. Control of marine fouling by chemical repellents. In: S.M. Sharpley and A.M. Kaplan (eds.) "International Biodegradation Symposium". Applied Science Publishers Ltd., England, 3:515-521.

310. Chet, I. and A. Huttermann. 1976. Melanogenesis during sporulation of *Physarum polycephalum.* Third European Physarum Workshop, Bern, Switzerland, p. 13.

311. Chet, I. 1977. Sclerotial morphogenesis in *Sclerotium rolfsii.* pp. 93. Second International Mycological Congress. Tampa, U.S.A.

312. Mitchell, R. and I. Chet. 1978. Indirect ecological effect of pollution. In: R. Mitchell (ed.), Water Pollution Microbiology, Vol. 2:177-179, Wiley Interscience.

313. Chet, I., Y. Hadar, Y. Elad, J. Katan and Y. Henis. 1978. Biological control of soilborne plant pathogens by *Trichoderma harzianum*, pp. 585-591. In: B. Schippers and W. Gam (eds.), "Soil-Borne Plant Pathogens", Academic Press.

314. Chet, I. 1980. The relationships among soil microorganisms and their potential in biological control of soil borne plant pathogens. In A. Zuckerman (ed.) "Dynamic Aspects of Host Parasitic Relationships". Keter Publishing House, Jerusalem Ltd., 3:145-154.

315. Huttermann, A., A. Haars, S. Fleming, P. Cwielong, I. Chet and L. Dimitri. 1981. Biological and biotechnological control of *Fomes annosus*, pp.6-15. In: K. Mauka (ed.), "Root and Butt Rot in Scotch Pine Stands".

316. Baker, R. and I. Chet. 1982. Induction of suppressiveness, p. 88. In: R.W. Schneider (ed.), "Suppressive Soils and Plant Disease". American Phytopathology Society, St. Paul, MN.

317. Huttermann, A., A. Haars, S. Flemming, P. Cwielong, I. Chet and L. Dimitri. 1982. Biochemical and Biotechnological control of *Fomes annosus.* In: K. Manka (ed.), "Root and Butt Rots in Scotch Pine Stands". Polish Forestry Institute, Warsdawa.

318. Elad, Y., A. Sivan, O. Zeidan, O. Kleifeld and I. Chet. 1983. Biological and integrated control of soil pathogens as a means for reduction of fungicides. In: H. I. Shuval (ed.), "Development in Ecology and Environmental Quality". Balaban International Science Services, Vol. 2:587-594.

319. Chet, I. and A. Huttermann. 1984. Biochemical manipulation of phytopathogenic fungi, pp. 248-256. In: Proceedings of the Sixth International Conference on Root and But Rots of Forest Trees, IUFRO.

320. Chet, I. and Y. Henis. 1985. Trichoderma as a biocontrol agent against soilborne root pathogens, 110-112. In: L.A. Parker, K.S. Moore, P.T.W. Wang, and J.F. Kollmorgen (eds.), "Ecology and Management of Soilborne Plant Pathogens". American Phytopathological Society Press, St. Paul, MN, USA.

321. Chet, I., E. Cohen and I. Elster. 1985. The role of chitinase and chitin synthetase inhibitors in controlling plant pathogenic fungi, pp. 237-240. In: R.A.A. Muzzarelli, C. Jeuniaux and G.W. Gooday (eds.), "Chitin in Nature and Technology, Plenum Publishing Corp., N.Y.

322. Chet, I. and Y. Elad. 1985. Biological and integrated control of soilborne plant pathogens: mechanisms and application. Proceedings of Gottingen Symposium.

323. Nordbring-Hertz, B. and I. Chet. 1986. Fungal lectins and agglutinins, pp. 393-408. In: D. Mirelman (ed.), "Microbial Lectins and Agglutinins". John Wiley & Sons, N.Y.

324. Sivan, A. and I. Chet. 1986. Trichoderma harzianum: an effective biocontrol agent of Fusarium spp. In: V. Jansen, A. Kjoller and L.H. Sorenson (eds.). "Microbial Communities in Soil". Elsevier Applied Science Publishers, London, New York.

325. Chet, I. 1986. Saprophyte-parasite interaction of Trichoderma in the rhizosphere, pp. 489-492. In: F. Megusar and M. Gantar (eds.) Proc. IV ISME, "Perspectives in Microbial Ecology".

326. Shoseyov, O., B.A. Bravdo, R. Ikan and I. Chet. 1987. Monoterpene glycoside hydrolysis by a beta-glucosidase from immobilized *Aspergillus niger*, pp. 63-71. In: A. Scienza and G. Versini (eds.), Proc. Int. Symp. The Aromatic Substances in Grapes and Wines.

327. Chet, I. (ed).1987. Trichoderma-application, mode of action and potential use as a biocontrol agent of soilborne plant pathogenic fungi, pp. 137-160. In: I. Chet (ed.), "Innovative Approaches to Plant Disease Control", John Wiley & Sons, NY.

328. Chet, I. 1990. Biological control of soilborne pathogens with fungal antagonists in combination with soil treatments. pp. 15-25. In: D. Hornby, R.J. Cook, Y. Henis, W.H. Ko, A.D. Rovira, B. Schippers and P.R. Scott (eds.), "Biological Control of Soilborne Pathogens". CAB Publishing House, N.Y.

329. Chet, I. 1989. Mycoparasitism — recognition, physiology and ecology, pp. 723-733. In: R. Baker and P. Dunn (eds.), "New Directions in Biological Control", Vol. 112, Alan R. Liss Inc., New York.

330. Levy, E., Z. Eyal, S. Carmely, Y. Kashman and I. Chet. 1989 Suppression of Septoria tritici blotch and leaf rust on wheat seedling leaves by antibiotic producing Pseudomonas. In: Septoria diseases of cereals. Proceedings of the 3rd International Septoria Workshop. Zurich-Reckenhold, Switzerland.

331. Sivan, A. and I. Chet. 1992. Microbial control of plant diseases, pp. 335-354. In: R. Mitchell (ed.), "New Concepts in Environmental Microbiology", Wiley-Liss, New York.

332. Oppenheim, A.B., Hirsch, Y., Koby, S. and I. Chet. 1992. Protein secretion in biotechnology, pp. 69-76. In: M.D. White, S. Revveny and A. Sahafferman

(eds.) "Biologicals from Recombinant Microorganisms and Animal Cells Production and Recovery", VCH Weinheim/Derfield Beaver and Balaban.

333. Broglie, R., Broglie, K., Roby, D. and I. Chet. 1992. The production of transgenic plants with enhanced resistance to microbial pathogens, 265-276, In: S.D. Kung and R. Wu (eds.), Transgenic Plants, Vol. 1: Engineering and Utilization, Academic Press, New York (373 p).

334. Broglie, R., Broglie, K., N. Benhamou and I. Chet. 1993. The molecular role of lytic enzymes in plant resistance to fungal pathogens, pp. 139-156. In: I. Chet (ed., Biotechnology in Plant Disease Control, John Wiley , New York (373 p).

335. Chet, I., Barak, Z. and Oppenheim, A. 1993. Genetic engineering of microorganisms for improved biocontrol activity, pp. 211-235. In: I. Chet (ed.), Biotechnology in Plant Disease Control, John Wiley, New York (373 p).

336. Oppenheim, A.B. and I. Chet. 1992. Biological control of soilborne plant pathogenic fungi by cloned chitinase. Trends in Biotechnology 10:392-394.

337. Chet, I. and Y. Hadar. 1992. Future directions and potential of agricultural biotechnology, pp. 11-20. Proceedings of the International Seminar on Impacts of Biotechnology in Agriculture and Food in Developing Countries, Madras, India. ICSU, COSTED and ANBS.

338. Zimand, G., L. Valinsky, Y. Elad, I. Chet and S. Manulis. 1993. DNA fingerprinting of a Trichoderma isolate by the RAPD procedure. In: N.J. Fokkema, J. Köhl and Y. Elad (eds.), Biological Control of Foliar and Post-Harvest Diseases. Proc. Workshop, Wageningen, The Netherlands. IOBC/WPRS Bulletin, Bulletin OILB/SROP Vol. 16:173-176.

339. Chet, I., H. Schickler, S. Haran and A.B. Oppenheim. 1993. Cloned chitinases and their role in biological control of plant pathogenic fungi, pp. 375-392 In: R. Muzzarelli (ed.), Chitin Enzymology , Eur. Chitin Soc. Ancona.

340. Perrakis, A., K.S. Wilson, I. Chet, A.B. Oppenheim and C.E. Vorgias. 1993. Phylogenetic relationships of chitinases. pp. 217-232. In: R. Muzarelli (ed.), Chitin Enzymology. Eur. Chitin Soc., Ancona.

341. Chet, I., K. Broglie, R. Shapira and R. Broglie. 1994. Biotechnological approaches for plant disease control. In: J.M. Slavicek (ed.), The Application of Biotechnology to Tree Culture, Protection and Utilization. Proceedings of the Beltsville Symposium, 1993.

342. Ikan, R., V. Weinstein, Y. Milner, B. Bravdo, O. Shoseyov, D. Segal, A. Altman and I. Chet. 1993. Natural glycosides as potential odorants and flavorants. Eds. D. Palevitch and E. Putievsky, Proceedings of International Symposium on Medicinal and Aromatic Plants, Tiberias, March 22-25, pp. 17-28.

343. Elad, Y. and I. Chet. 1995. Practical approaches for biocontrol agents implementation. pp. 323-338. In: R. Reuveni (ed.), Novel Approaches to Integrated Pest Management. Lewis Publishers, Inc., and CRC Press, USA.

344. Chet, I., J. Inbar, S. Haran, S. Pleban and H. Schickler. 1995. Antagonistic microorganisms in biological control. Proceedings of the "Year of Louis Pasteur" Symposium, Microbes, Enviornment, Biotechnology, Tahiti, May 1995. pp.
345. Chet, I. and Inbar, J. 1994. Biological control of fungal pathogens. Applied Biochemistry and Biotechnology. 47:37-43.
346. Shaul, O., Y. Elad, S. Gallili, H. Volpin, Y. Kapulnik and I. Chet. 1995. Peroxidase and pathogenicity-related proteins in plant tissues infected by *Botrytis cinerea.* Aspects of Appl. Biol., Physiological Responses of Plants to Pathogens, 42:1-8.
347. Jiazhi, W., S. Pleban, L. Chernin and I. Chet. 1996. Biological control of fungal phytopathogenes by soil-borne and endophytic bacteria, pp. 116-124. In: T. Wenhua, R.J. Cook and A. Rovira (eds.) Advances in Biological Control of Plant Diseases. China Agricultural University Press, Beijing.
348. Haran, S., H. Schickler and I. Chet. 1996. Molecular mechanisms of lytic enzymes involved in the biocontrol activity of *Trichoderma harzianum.* Microbiology (review article) 142:2321-2331.
349. Inbar, J. and I. Chet. 1996. The role of lectins in recognition and adhesion of the mycoparasitic fungus Trichoderma spp. to its host, pp. 229-231. In: I. Kahane and I. Ofek (eds.) Towards Anti-Adhesion Therapy for Microbial Diseases. Advances in Experimental Medicine and Biology, Vol. 408. Plenum Press.
350. Inbar, J. and I. Chet. 1997. Lectins and Biocontrol, pp. 1-20. In: G.G. Stewart, and I. Russell (eds), Critical Reviews in Biotechnology, Vol. 17(1). CRC Press, Boca Raton.
351. Chet, I. and J. Inbar. 1997. Biological Control: Fungi, pp. 65-80. In: T. Anke (ed). Fungal Biotechnology. Chapman & Hall, New York.
352. Chet, I., J. Inbar and Y. Hadar. 1997. II Fungal Antagonists and Mycoparasites. In: Wicklow/Soderstrom (eds.) The Mycota IV: Environmental and Microbial Relationships, pp. 165-184.
353. S. Haran, H. Schickler, A. Oppenheim and I. Chet 1997. Characterization and *in vivo* detection of chitinase activity, pp. 255-260. In: R.A.A. Muzzarelli and M.G. Peter (eds.), Chitin Handbook. Atec, Grottammare, Italy.
354. Herrera-Estrella and I. Chet. 1998. Biocontrol of bacteria and phytopathogenic fungi, pp. 263-282. In: A. Altman (ed.) Agricultural Biotechnology, Marcel Dekker, New York, Basel, Hong Kong.
355. Spiegel, Y. and I. Chet. 1998. Evaluation of Trichoderma spp as a biocontrol agent against soilborne fungi and plant-parasitic nematodes in Israel. Integrated Pest Management Review. 3:1-7.
356. Chet, I. 1998. Cloning chitinases for plant protection: hope or hazard for the environment, pp. 25-30. In: R.A.A.Muzzarelli (ed.), Chitin Enzymology. Plenum Press.

357. Chet, I., N. Benhamou and S. Haran. 1998. Mycoparasitism and lytic enzymes. In: G. Harman and C. Kubicek (eds.) Trichoderma and Gliocladium. Vol. 2(7): 153-172.
358. Cohen-Kupiec, R. and I. Chet. 1998. The molecular biology of chitin digestion. Curr. Opinion in Biotechnol. 9:270-277.
359. Cohen-Kupiec, R., D. Friesem and I. Chet. 1998. Improving biocontrol activity of T*richoderma harzianum* by molecular biology techniques. In: B.K. Duffy, U. Rosenberger and G. Defago (eds.) Biological Control of Fungi and Bacterial Plant Pathogens IOBC Bulletin. Vol. 21(9):233-239.
360. Herrera-Estrella, A. and I. Chet. 1999. Chitinases in biological control, pp. 171-184. In: P. Jolles and R.A.A. Muzzarelli (eds.). Chitin and Chitinases (EXS). Birkhauser Publ. Ltd., Switzerland.
361. Chernin, L., Z. Ismailov, I. Khmel, N. Lemanova, A. Perebityuk and I. Chet. 1999. Development of new a pproaches for the biological control of crop diseases: A joint Israel-CIS venture, pp. 65-74. In: O.L. Figovsky (ed.) Scientific Israel-Technological Advantages. Vol. 1, No. 3. Scientific Cooperation Israel-the CIS and the Baltic States.
362. Chernin, L. and I. Chet. 2000. Biological control: An environmentally friendly technology for sustainable agriculture, p. 72. Proc. Int. Conf. On Managing Natural Resources for Sustainable Agricultural Production in the 21st Century, New Delhi. Cambridge Printing Works, New Delhi.
363. Rymon, D. and I. Chet. 2000. Potencial de la biotecnologia en el agro israeli, pp. 40-43. In: Economia y Negocios: Guia de Agrotecnologia en Israel. Edicion 2000.
364. Chernin, L. and I. Chet. 2001. Microbial enzymes in the biological control of plant pathogens and pests, pp. 171-225. In: R. G. Burns and R. P. Dick (eds), Enzymes in the Environment. Marcel Dekker, New York, USA.
365. Chet, I. and L. Chernin. 2001. Biocontrol, microbial agents in soil, pp. 450-465. In: Encyclopedia of Environmental Microbiology. John Wiley & Sons, Inc.
366. Viterbo, A., O. Ramot and I. Chet. 2001. Cloning of chitinases as a means for biocontrol and improvement of plant protection, pp. 41-45. In: R.A.A. Muzzarelli (ed.), Chitin Enzymology. Atec, Italy.
367. Zhou, L., M. Ovadis, Z. Ismailov, I. Chet and L. Chernin. 2001. Chitinolitic activity of *Enterobacter agglomerans* enhanced by Pseudomonas' regulatory genes or exogenous quorum-sensing signals, pp. 241-247. In: R.A.A. Muzzarelli, (ed.), Chitin Enzymology. Atec, Italy.
368. Chernin, L., S. Klein, I. Khmel, M. Veselova, A. Metlitskaya, I.A. Bass, A. Mayatskaya, M. Kholmeckaya, E. Lobanok, X. Liu and I. Chet. 2004. Intercellular communication in the environment and biocontrol activity of plant-associated bacteria producing N-acyl homoserine lactone signal molecules. "Scientific Israel – Technological Advantages", Vol 6 (1-2):19-30.

369. Klein, S., M. Veselova, A. Mayatskaya, I. Khmel, I. Chet and L. Chernin. 2004. Biocontrol activity of phenazine-producing rhizobacterium *Pseudomonas chlororaphis* 449, pp. 145-150. In: R. Sikora, S. Gowen, R. Hauschild, and S. Kiewnick (eds.), Multitrophic Integrations in Soil. IOBC wprs Bulletin, 27.
370. Gavriel, S., Z. Ismailov, M. Ovadis, I. Chet and L. Chernin. 2004). Comparative study of the role of chitinase and pyrrolnitrin for biocontrol activity of *Serratia plymuthica* IC1270, pp. 347-350. In: The IOBC/WPRS Bull 27(8), Working Group Meeting on "Management of plant diseases and arthropod pests by BCAs and their integration in agricultural systems", S. Michele all'Adige, Trentino, Italy.
371. Harman, G.E., C.R. Howell, A. Viterbo, I. Chet and M. Lorito. 2004. Trichoderma species — opportunistic, avirulent plant symbionts. Nature Microbiol. Rev. 2:43-56.
372. Viterbo, A., J. Inbar, Y. Hadar and I. Chet. 2006. Plant disease biocontrol and induced resistance via fungal mycoparasites. In: D.T. Wicklow and B. Söderström (eds.), The Mycota IV. 2nd ed: Environmental and Microbial Relationships. Springer-Verlag Berlin, Germany.
373. Viterbo, A., J. Inbar, Y. Hadar and I. Chet. 2007. 8 Plant disease biocontrol and induced resistance via fungal microparasites. In: C.B. Kubicek and I.S. Druzhinina (eds.), Environmental and Microbial Relationships, 2nd ed., Springer Verlag Berlin Heidelberg, Germany.

MOLECULAR PLANT PATHOLOGY (2007) **8**(6), 737–746 DOI: 10.1111/J.1364-3703.2007.00430.X

The 18mer peptaibols from *Trichoderma virens* elicit plant defence responses

ADA VITERBO[1]*, ARIC WIEST[2]†, YARIV BROTMAN[1], ILAN CHET[1] AND CHARLES KENERLEY[2]

[1]*Department of Plant Science, The Weizmann Institute of Science, 76100, Rehovot, Israel*
[2]*Department of Plant Pathology and Microbiology, Texas A&M University, College Station, TX 77843, USA*

SUMMARY

Peptaibols, the products of non-ribosomal peptide synthetases (NRPS), are linear peptide antibiotics produced by *Trichoderma* and other fungal genera. *Trichoderma virens* strain Gv29-8, a well-known biocontrol agent and inducer of plant defence responses, produces three lengths of peptaibols, 11, 14 and 18 residues long, with several isoforms of each. Disruption of the NRPS gene, *tex1*, encoded by a 62.8-kb uninterrupted open reading frame, results in the loss of production of all forms of 18-residue peptaibols. *Tex1* is expressed during all *Trichoderma* developmental stages (germinating conidia, sporulating and non-sporulating mycelia) examined on solid media. Expression analysis by reverse transcriptase PCR shows that in Gv29-8 wild-type the abundance of *tex1* transcript is greater during co-cultivation with cucumber seedling roots than when grown alone. Cucumber plants co-cultivated with *T. virens* strains disrupted in *tex1* show a significantly reduced systemic resistance response against the leaf pathogen *Pseudomonas syringae* pv. *lachrymans*, and reduced ability to produce phenolic compounds with inhibitory activity to the bacteria as compared with plants grown in the presence of wild-type. Two synthetic 18-amino-acid peptaibol isoforms (TvBI and TvBII) from Gv29-8 when applied to cucumber seedlings through the transpiration stream can alone induce systemic protection to the leaf pathogenic bacteria, induce antimicrobial compounds in cucumber cotyledons and up-regulate hydroxyperoxide lyase (*hpl*), phenylalanine ammonia lyase (*pal1*) and peroxidase (*prx*) gene expression. These data strongly suggest that the 18mer peptaibols are critical in the chemical communication between *Trichoderma* and plants as triggers of non-cultivar-specific defence responses.

**Correspondence*: Tel.: +972 8 9344466; Fax: +972 8 9344112;
E-mail: ada.viterbo@weizmann.ac.il.
†*Present address*: Fungal Genetics Stock Center-School of Biological Sciences, SBS 404, University of Missouri, Kansas City, MO 64110, USA.

INTRODUCTION

Peptaibols are a class of linear, short-chain-length (≤ 20 residues) peptides of fungal origin. Typical features are the presence of α-amino isobutyric acid moieties, acetylation of the N-terminus of the peptide chain and reduction of the C-terminus to an amino alcohol (Grigoriev *et al.*, 2003). In a previous study we have shown that *tex1*, a non-ribosomal peptide synthetase (NRPS), is responsible for peptaibol production in *Trichoderma virens* (Wiest *et al.*, 2002). NRPSs are large proteins capable of enzymatic assembly of small peptides, characterized by an ordered modular arrangement. Each module consists of approximately 1100-amino-acid residues (Marahiel *et al.*, 1997) and catalyses the incorporation of a single amino acid residue. Modular order parallels the order of the residues in the product (Marahiel *et al.*, 1997). A single module of an NRPS, at minimum, consists of three domains: adenylation (A-domain), thiolation (T-Domain) and condensation (C-domain). The coding sequence of *tex1* represents the largest known open reading frame (ORF), 62.8kb, and contains modules for the incorporation of 18 amino acids as well as motifs for the acetylation of the N-terminus and the reduction of the C-terminus. Even though a peptaibol database (http://public-1.cryst.bbk.ac.uk/peptaibol/home.shtml) provides documentation of over 300 subfamilies of these unique compounds, their role in biological systems has only been partially demonstrated for a few individual compounds (Chug and Wallace, 2001; Rebuffat *et al.*, 2000; Schirmbock *et al.*, 1994). The *in vitro* antibacterial activity of several peptaibols has been shown (Szekeres *et al.*, 2005), suggesting a role for these compounds as competitive inhibitors of other microbes in soil or the rhizosphere. The antibiotic functions of peptaibols arise from their membrane insertion and pore-forming abilities (Duclohier and Wroblewski, 2001). There are a few reports indicating that peptaibols may also represent a novel class of plant elicitors. Exogenous application of the 20-residue peptaibol alamethicin, produced by *T. viride*, has been shown to induce defence responses in *Phaseolus lunatus* (lima bean) (Engelberth *et al.*, 2000) and *Arabidopsis thaliana* (Chen *et al.*, 2003). In lima bean these responses include biosynthesis of volatile compounds and salicylate biosynthesis.

A. thaliana showed increased production of methyl salicylate. Furthermore, Chrysospermin, a 19-residue peptaibol from *Apiocrea chrosospermin*, afforded *Nicotiana tabacum* protection from tobacco mosaic virus infection (Kim *et al.*, 2000).

As a ubiquitous soil inhabitant and rhizosphere competent fungus, *T. virens* has been used successfully as a biological control agent for the management of plant pathogens (Fravel, 2005; Lumsden *et al.*, 1996). Several mechanisms of biocontrol have been proposed for *T. virens*, including competition, mycoparasitism and the induction of plant defence responses due to colonization of plant root intercellular spaces (Baek *et al.*, 1999; Harman *et al.*, 2004; Howell, 2003; Pozo *et al.*, 2004). The induction of both local and systemic resistance in cotton has been demonstrated by the application of the proteinaceous elicitor SM1 (Djonovic *et al.*, 2006). Considering the large number of metabolites secreted and the intimate contact with the root epidermis by *T. virens*, as well as the expanding list of diverse elicitors produced by fungi, peptaibols may be part of a signalling cascade resulting in greater colonization by *T. virens* or plant resistance induction.

Several types of peptaibols are produced by *T. virens* strain GV29-8, including 11-, 14- and 18-amino-acid residue isoforms. Our initial *tex1* disruptant strains (Gv234 and Gv223) were the result of multiple integration events, and we have reported that they produced no peptaibol isoforms (Wiest *et al.*, 2002). In order to generate single homologous integration disruptants, the fungal transformation was repeated. In the present study, analysis of peptaibols from the new disruptants revealed that the 11- and 14-residue isoforms were still produced, while all forms of the 18-residue peptaibol were absent. These results, similar to the findings of Wei *et al.* (2005), suggest that the NRPS TEX1 is only responsible for the production of 18-residue isomers.

To explore the significance of peptaibol production in *T. virens*, we examined the expression patterns of *tex1* in an axenic plant system along with several developmental and nutrient conditions. We also took advantage of the new *tex1* knockouts and synthetic 18-residue peptaibols to test whether these peptides alone are necessary or sufficient to induce plant defence responses *in vivo*. Our data strongly suggest that in addition to their antibacterial activity, peptaibols also have a role as signalling molecules during the development of symbiotic interactions between *Trichoderma* and plants.

RESULTS

tex1 expression analysis by RT-PCR

RT-PCR was used to investigate expression of *tex1*. Northern analysis was not attempted because the probability of isolating intact *tex1* mRNA was low as RNA is prone to shearing during extraction procedures. Using this indirect technique, *tex1* transcript was detected differentially. Intensity of RT-PCR products was used as a preliminary indicator of *tex1* expression following amplification with primer pairs for module 1 (AW1 and AW2). *tex1* was expressed at all developmental stages for at least one of the conditions on solid media (Fig. 1A). Transcript

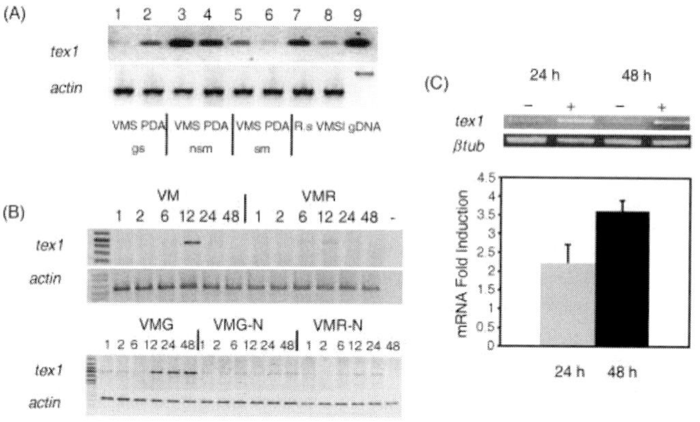

Fig. 1 (A) *Tex1* expression during developmental stages of Gv29-8. Conditions used: germinating spores (gs) for 12 h on solid VMS (lane1) or PDA (lane 2); 2-day-old non-sporulating mycelia (nsm) on solid VMS (lane 3) or PDA (lane 4); 5-day-old sporulating mycelia (sm) from solid VMS (lane 5) or PDA (lane 6); mycelia of Gv29-8 indirectly confrontating a culture of *R. solani* (lane 7); 5-day-old VMS liquid culture of Gv29-8 (lane 8); genomic DNA (lane 9). (B) *Tex1* expression in liquid media differing in carbon and ammonium source. Conditions used: VM (Vogel's minimal with no carbon), VMR (VM + 0.5% *R. solani* cell walls), VMG (VM with 1.5% glucose), VMG-N (no ammonium), VMR-N (0.5% *R. solani* cell walls without ammonium). (C) *Tex1* expression under cucumber co-cultivation (+) and without (–) at 24 h and 48 h. Mycelia were pooled for RNA extraction from three different hydroponic culture boxes each containing 25 seedlings. Graph illustrates quantification of mRNA fold induction. The values obtained for each band were normalized to β-tubulin. The results are averages of three independent experiments and the error bars indicate standard deviations.

level was higher among conidia germinating on the complex medium PDA than the minimal medium VMS. However, in both types of mycelia (sporulating and non-sporulating), *tex1* was expressed at higher levels in VMS than in PDA. There appears to be transient induction in VM liquid culture at 12 h and consistent expression in VMG at 12, 24 and 48 h (Fig. 1B). In the absence of a nitrogen source, the levels appear faint, indicating a partial dependence on nitrogen. Simulated mycoparasitism towards *Rhizoctonia solani* did not induce *tex1* expression either on solid or in liquid cultures. *tex1* transcript was detected at a higher level in PGM with cucumber than PGM alone at both 24 and 48 h (Fig. 1C).

tex1 Disruption and confirmation

We previously reported the detection of three lengths of peptaibol isoforms in Gv29-8, but not in two other *tex1* disruptants strains, Gv223 and Gv234 (Wiest *et al.*, 2002). These two original disruption strains were generated using hygromycin resistance as a selectable marker. These strains contained multiple copies of the disruption vector and had reduced growth compared with Gv29-8. New disruptant strains were generated using complementation of arginine auxotrophy as a selectable marker. The colossal size of *tex1* precluded elimination of the entire gene from the genome; therefore, *arg2* was inserted between modules 10 and 11 to interrupt the reading frame. This region was chosen as it was the furthest upstream module characterized at the time.

tex1 disruptants were generated by replacing a 3-kb *Nde*I fragment of module 10 with the *arg2* gene using disruption vector pSKO2 (Fig. 2). Of 85 stable transformants, 23 were screened by Southern analysis using *Nco*I- and *Xho*I-digested genomic DNA. A homologous double crossover event was identified by a 3.0-kb *Nco*I fragment and a 3.8-kb *Xho*I fragment. In the wild-type strain or transformants containing ectopic copies of the disruption vector, bands of 12.5 kb (*Nco*I) and 8.5 kb (*Xho*I) were detected. Three transformants, GvL1, GvC3 and GvC12, contained single-copy homologous integrations of the disruption vector (Fig. 3). Several ectopic integrants were also detected (data not shown) but were not analysed further. The three transformants had growth rates, colony morphology and sporulation rates comparable with the wild-type (data not shown).

Two of the homologous single-integrant disruptants were further tested by RT-PCR. RNA isolated from mycelia of Gv29-8, GvL1 and GvC12 was used as template for cDNA synthesis. Transcripts of *tex1* were detected by amplification of cDNA fragments using primer pairs located upstream (P1: AW1 and AW2 in module 1) and downstream (P2: RT1 and RT2 in module 14; P3: AW3 and AW4 in module 18) of the disruption event (Fig. 4A). Strains GvL1 and GvC12 yielded PCR products upstream of the disruption, but not downstream, indicating that *tex1* was not fully transcribed in these strains (Fig. 4B). In Gv29-8, transcripts of *tex1* were detected using all primer pairs. No products were amplified in the negative control reactions. To ensure cDNA preparations were free of genomic DNA contamination, amplification of actin, including genomic introns, was used as a control. All cDNA preparations yielded a single intronless 576-bp actin PCR product while the genomic DNA control yielded a 940-bp PCR product.

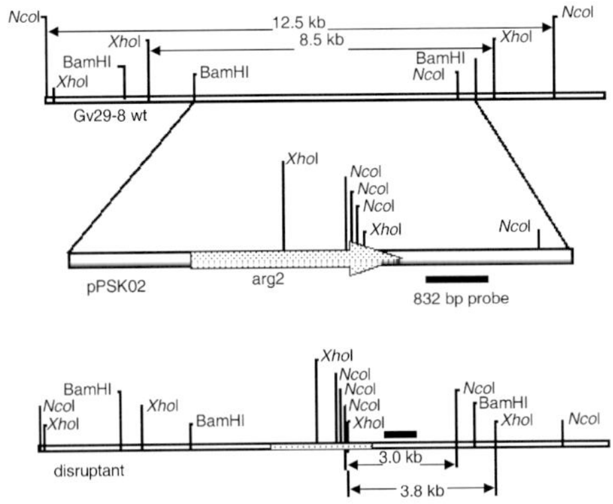

Fig. 2 *Tex1* disruption vector pPSKO2. The disruption vector cassette is shown relative to the wild-type region. A homologous double crossover disruption event yields a 3.0-kb *Nco*I fragment and a 3.8-kb *Xho*I fragment when probed with the 832-bp region shown.

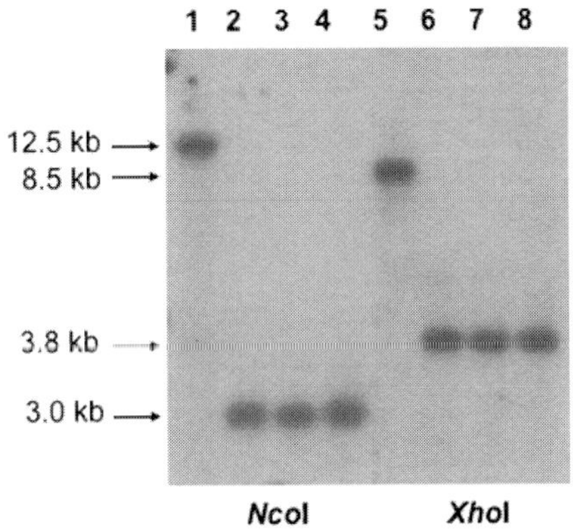

Fig. 3 Confirmation of *tex1* disruptants. Strains Gv29-8, GvL1, GvC3 and GvC12 were digested with either *Nco*I (lanes 1–4) or *Xho*I (lanes 5–8) and probed with the 832-bp probe indicated in Fig. 2. Wild-type bands, shown in lanes 1 and 5, are 12.5 kb (*Nco*I) and 8.5 kb (*Xho*I). The presence of 3.0-kb *Nco*I (lanes 2–4) and 3.8-kb *Xho*I (lanes 6–8) fragments confirms homologous disruption of *tex1* in GvL1, GvC3 and GvC12. DNA fragment sizes are indicated by arrows.

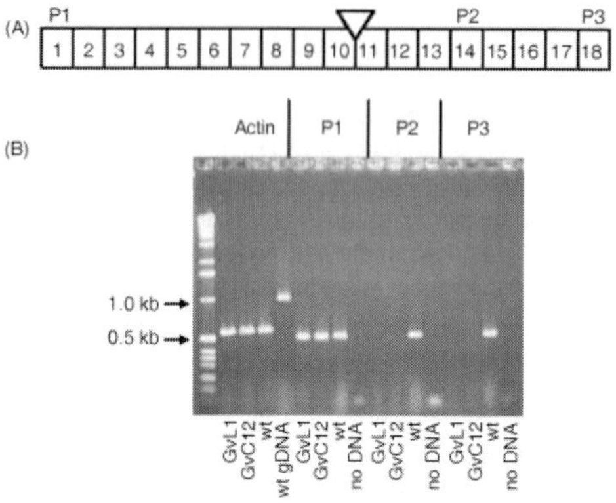

Fig. 4 (A) Schematic representation of RT-PCR primers: RT-PCR primer pairs 1 (P1), 2 (P2) and 3 (P3) shown in relation to *tex1* modules. The triangle between modules 10 and 11 indicates the location of the disruption event. (B) RT-PCR analysis of GvL1, GvC12 and Gv29-8 (lanes 1, 2 and 3, respectively, under each heading). Lane 4 under actin is positive control and under primer headings, lane 4 is negative control.

Peptaibol detection

Using MALDI-TOF (matrix-assisted laser desorption/ionization-time of flight mass spectrometry) on lyophilized fungal tissue harvested from cultures grown for 9 days on VMS without shaking, we detected only the 11- and 14-residue peptaibol forms in the *tex1* disruptant strains. The wild-type produced all three isoforms (Fig. 5). Peptaibols were detected from tissue co-cultivated with and without cucumber (data not shown)

Induced resistance towards the leaf pathogen *P. syringae* pv. *lachrymans*

Two *tex1* mutants (GvL1 and GvC12) were tested in a plant interaction assay as described in Yedidia *et al.* (1993). No significant differences could be detected between *tex1* disruptants and wild-type in the ability to colonize roots (Fig. 6A). However, the protective effect conferred towards the leaf pathogen *Pseudomonas syringae* pv. *lachrymans* (*Psl*) was significantly reduced in cucumber seedlings colonized by *tex1* mutants (Fig. 6B). The aglyconic phenols fraction extracted from such plants had an almost undetectable antimicrobial activity towards *Psl* bacteria (Fig. 6C). Clear zones of bacterial inhibition were seen with increasing amounts of these fractions prepared from cultures of Gv29-8 (Fig. 6C).

RT-PCR analysis shows that priming of *hpl*, *pal1* and *prx* expression in leaves of cucumber seedlings inoculated with GvL1 or GvC12 mutants and challenged with the pathogen *Psl* is clearly less pronounced than in seedlings inoculated with Gv29-8 wild-type (Fig. 6D).

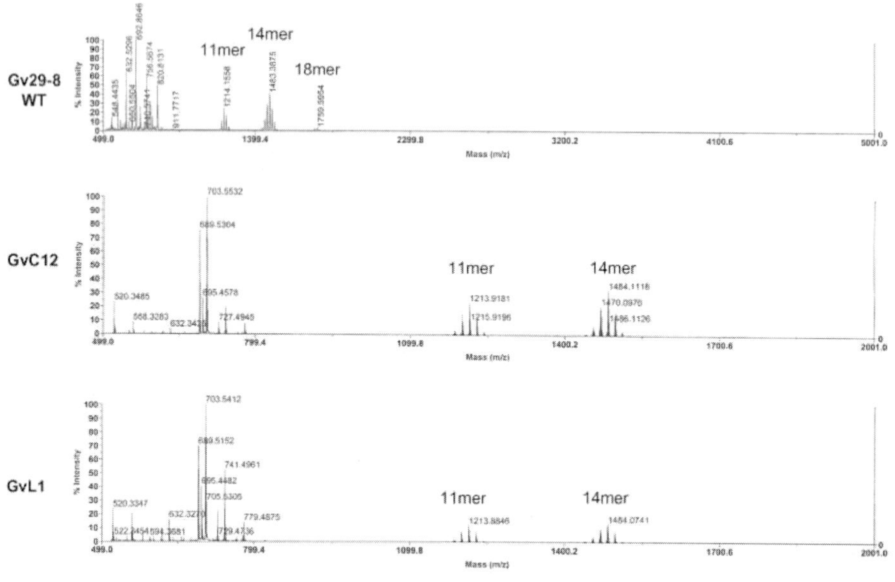

Fig. 5 MALDI-TOF of lyophilized fungal tissue harvested from 9-day-old stationary cultures grown in VMS. Strains GvL1 and GvC12 are disrupted in *tex1*.

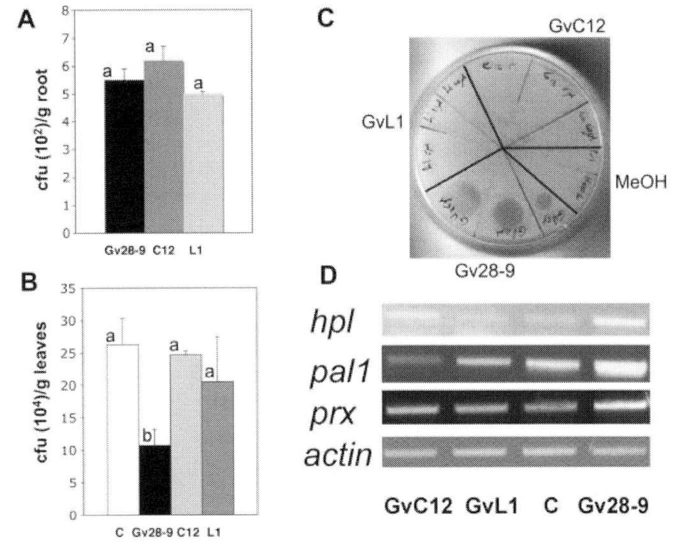

Fig. 6 (A) *T. virens* colony forming units per gram of root tissue for the wild-type (Gv29-8) and two *tex1* disruptants, GvL1 and GvC12. (B) Effect of *Trichoderma* root inoculation on multiplication of *Psl* in challenged cotyledons, 96 h after infection. (C) Bioassay comparing the antimicrobial activity of 5–15 μL of the aglycone fraction obtained 48 h post-challenge with *Psl*, by acid hydrolysis of crude phenolics extract of cucumber leaves from seedlings inoculated with Gv18-9 and deletion mutants L1 and C12, 48 h prior to challenge with *Psl*. (D) RT-PCR analysis of systemic up-regulation of *pal1*, *hpl* and *prx* genes by Gv29-8, GvL1 and GvC12. Cucumber seedlings pretreated with the different *Trichoderma* isolates were infected after 48 h with *Psl* and RNA was extracted from cotyledons 48 h post bacterial challenge. Control (C) are seedlings challenged with *Psl* and not pre-inoculated with *Trichoderma*. Three seedlings were pooled for each extraction. Gel is representative of one out of three independent extractions.

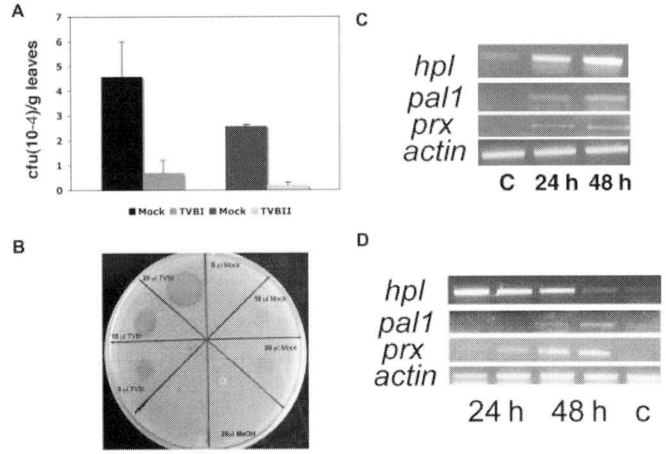

Fig. 7 Plant elicitation responses by synthetic 18mer peptaibols TVBI and TVBII. (A) *Psl* multiplication in cotyledons of cucumber seedlings treated with 9.6 nmol of TvBI or TvBII (and mock) 48 h prior to bacterial challenge. The results are the average of three independent experiments. (B) Bioassay comparing the antimicrobial activity of the aglycone fraction obtained by acid hydrolysis of crude phenolics extract of cucumber cotyledons from seedlings treated for 48 h with 9.6 nmol of TvBI (and mock), 48 h post challenge with *Psl*. (C) RT-PCR analysis of *hpl*, *pal1* and *prx* expression in cotyledons 24 h and 48 h after application of TvBII (9.6 nmol) applied through the transpiration stream. Three seedlings were pooled for each extraction, and three independent extractions were performed per time point. (D) RT-PCR analysis of *hpl*, *pal1* and *prx* expression in cotyledons 24 h and 48 h after injection with TvBI (lanes 1 and 3) and TVBII (lanes 2 and 4) (3 nmol). As a control cotyledons were infiltrated with 0.1% methanol. The gel is representative of one experiment out of five independent injections.

Synthetic TvBI and TvBII peptides when applied through the transpiration stream (9.6 nmol) for 48 h prior to the bacteria challenge could mimic the protective effect of wild-type Gv29-8 (Fig. 7A). The phenol aglyconic extract from such plants also presented growth-inhibitory activity towards the *Psl* bacteria at a much higher extent than the extract obtained from plants challenged with the bacteria without prior incubation with the two 18mer peptaibols (Fig. 7B). When synthetic TvBI or TvBII peptides (5, 10, 20 nmol) were incubated overnight with cultures of *Psl* or *P. syringae* DC3000, no growth inhibition (OD_{600}) could be detected (data not shown).

Induction of defence-related genes expression by synthetic 18mer peptaibols TvBI and TvBII

Induction of *hpl*, *pal1* and *prx* genes by synthetic TvBI and TvBII was followed in cucumber cotyledons that were directly injected with a peptide solution (3 nmol) or that were harvested from cucumber seedlings after the peptide solution (9.6 nmol) was applied through the transpiration stream. In both experimental systems a clear up-regulation of these genes could be detected by RT-PCR analysis even after 24 h (Fig. 7C–D). Expression of β-glucanase and chitinase1 genes was not affected (data not shown). No phytotoxic symptoms were detected in infiltrated cucumber leaves or treated seedlings at these concentrations.

DISCUSSION

Peptaibols are the largest known products produced by peptide synthetases and over 300 have currently been identified (http://www.cryst.bbk.ac.uk/peptaibol/home.shtml) (Chugh and Wallace, 2001). Because of the similar structure of most peptaibols and the evolutionary relatedness of the producing fungi, it is likely that all peptaibols are produced by NRPS enzymes encoded by genes similar to *tex1*. Our study shows that disruptants of *tex1* still produce 11- and 14-residue peptaibols. This suggests that there is more than one NRPS responsible for peptaibol production in *T. virens*. In a recent study (Wei *et al.*, 2005) the *tex1* homologue was examined in a commercially relevant *T. virens* strain, G20. As our MALDI-TOF data of the strains disrupted in *tex1* indicated, this gene is only responsible for the production of an 18mer peptaibol, and not 11mer and 14mer peptaibols. Two additional NRPS adenylate domains were identified in *T. virens*, and sequence comparisons suggest that these might be part of a separate peptaibol synthesis NRPS gene. It is likely that there are two additional NRPS genes, one producing the 11-residue isoforms and one producing the 14-residue isoforms. The predicted sizes of these genes are approximately 38 and 49 kb, respectively. Indeed, the genomic sequence of *Trichoderma reesei* reveals two large NRPSs of 14 and 18 modules of approximately 51 and 69.3kb, respectively. (http://genome.jgi-psf.org/Trire2/Trire2.home.html).

Tex1 expression is generally detectable in all the fungal developmental stages we examined, being down-regulated by low nitrogen levels in liquid cultures and up-regulated during plant–root interaction. In a previous study (Schirmbck *et al.*, 1994) it was shown that *Botrytis cinerea* cell walls trigger the production of both cell wall hydrolytic enzymes and peptaibols in *T. harzianum*. In the present study, simulated mycoparasitism of *T. virens* with *R. solani* cell walls or mycelium did not show *tex1* gene induction, suggesting that the products of the NRPS may not be directly involved in this specific interaction. All three *tex1* disruptants were capable of contacting and coiling around the hyphae of *R. solani*, indicating that the peptaibol 18mer isoforms are not required for the initial stages of mycoparasitism (data not shown). Both synthetic 18mer peptaibols had weak antifungal activity at 200 μg/mL when tested against *R. solani* on PDA plates (data not shown).

Engelberth *et al.* (2000) showed that emission of ethylene, jasmonic acid and volatile compounds related to the octadecanoid signalling pathway are inducible responses to treatment of lima bean plants with alamecithin, an ion channel-forming peptaibol from *T. viride*. Interestingly, these compounds and pathways are the same as those involved in the *Trichoderma* plant-induced resistance (Shoresh *et al.*, 2005; Yedidia *et al.*, 2003). These observations led us to verify the direct involvement of peptaibols in the elicitation of *Trichoderma*–plant-induced resistance.

tex1 mutants still retain the 11- and 14-amino-acids peptaibols, but these strains do not provide full systemic protection as the wild-type strain, indicating that the 18-amino-acid peptaibols have a dominant effect in initiating the very first steps of the signalling cascade of the plant defence response, most probably due to their membrane channel-forming properties (Engelberth *et al.*, 2000). Changes in membrane potential are the initial responses in many signalling pathways (Ehrhardt *et al.*, 1992).

Yedidia *et al.* (2003) demonstrated that the inhibition of *Psl* proliferation in cucumber leaves by application of *T. asperellum* to the roots correlated with the accumulation of antimicrobial aglyconic phenolic compounds. Concomitantly, expression of *pal1* and *hpl*, genes related to the phytoalexin synthesis pathway, was increased in roots and cotyledons when plants were treated with *T. asperellum*. Exogenous application of synthetic peptides of TvBI and TvBII confirmed that these isoforms are effective inducers of defence responses in cucumber against *Psl*. The defence response involves the production of antimicrobial compounds originating through the signalling cascade of *pal1* and *hpl* pathways (Fig. 7). Jabs *et al.* (1997) have already suggested that the induction of phytoalexin biosynthesis in plants might be directly linked to ion fluxes and subsequent intracellular signalling. The protective effect of the 18-residue peptaibols against the leaf pathogen is mediated by the induction of the plant disease defence system. At comparable and even higher concentrations of the peptides, no inhibitory effect on growth of the pathogen could be seen after 24 h of direct incubation of bacteria with the peptides (data not shown). As *Psl* are Gram-negative bacteria, the lipopolysaccharides of their outer membrane form a strong diffusion barrier against hydrophobic molecules as peptaibols. This effect has been shown for alamethicin (Duclohier and Wroblewski, 2001) and trichokonins from *T. koningii* (Song *et al.*, 2006). As the *tex1* transformants are capable of producing the 11 and 14mer peptaibols, the interesting quandary is the inability of these two peptaibols to induce plant defence responses. There is evidence for major differences in the relative membrane activities of peptaibols depending on the length of the peptide chain and amino acid composition (Grigoriev *et al.*, 2003). Efficiency of membrane pore formation by peptaibols containing > 17 amino acids was much greater than for smaller peptaibols. Peptaibols of 11 residues appear to lack the ability to span membranes, greatly decreasing their membrane activity (Shenkarev *et al.*, 2004).

Therefore, it has been suggested that the capacity of channel formation decreases in parallel with the decreasing number of the constituting amino acids (Berg *et al.*, 2003).

In addition, it is possible that the genes coding for the 11- and 14-amino-acid peptaibols are less affected by the interaction with the plant roots and more involved as component of mycoparasitism.

In the communication zone between plant and *Trichoderma* there are many molecules that can act as PAMPs/elicitors that can affect different signalling pathways, in turn affecting different classes of pathogens. In *T. virens* we have already shown that the small hydrophobin-like protein SM1 can induce expression of defence responses in cotton and maize (Djonovic *et al.*, 2006). Here we present genetic and chemical evidence for the first time that also 18mer peptaibols in *T. virens* are potent elicitors of defence responses in cucumber against the leaf pathogen *P. syringae* pv. *lachrymans* and can be assumed as major players in the 'antigenic' potential of *Trichoderma virens* to trigger non-cultivar-specific defence responses.

EXPERIMENTAL PROCEDURES

Strains, culture conditions and fermentation

The wild-type strain (WT) of *Trichoderma virens*, Gv29-8, was isolated from a cotton field in College Station, TX. Strain Tv10.4, an arginine auxotroph, was created by a point mutation in the small subunit of a carbamoyl phosphate synthetase (*arg2*) after treatment with 4-nitroquinoline-1-oxide (NQO) (Baek and Kenerley, 1998). Vogel's minimal medium (Vogel, 1956) was used with different carbon sources: VMS (1.5% sucrose), VMR (0.5% *Rhizoctonia solani* cell wall), VMG (glucose 1.5%). Ammonium was omitted from treatments labelled VM-N, VMG-N and VMR-N. Stock cultures were stored at –80 °C as conidial suspensions in 30% glycerol.

tex1 Disruption vectors and Southern analysis

The *tex1* disruptants were generated by replacing a 3-kb *Nde*I fragment coding for module 10 with the *arg2* gene. The disruption vector pPSK2 contained 1.65- and 2.4-kb fragments from cosmid CMK40B7 flanking the selectable *arg2* gene. This plasmid was transformed as described (Thomas and Kenerley, 1989) into the arginine auxotrophic strain Tv10.4 with selection for arginine prototrophy. DNA was extracted as described (Xu *et al.*, 1996) from 23 putative transformants, digested with either *Nco*I or *Xho*I, and screened by Southern blot analysis to identify strains containing the disruption. A 832-bp probe was amplified from pPSK2 using primers PS9348 (5′-TCAACCATCGGCCACCCTGTC-3′) and PS3-11 (5′-GTGACAAGATGGATGGC-3′). Based on *tex1* sequence data, a positive disruption event was expected to yield a 3.0-kb band (vs. a 12.5-kb band in the wild-type) for the *Nco*I digest and 3.8-kb band (vs. a 8.5-kb band in the wild-type) for the *Xho*I digest.

Growth and sporulation assays

Transformants and the WT strains were assayed for radial growth and colony morphology. Agar plugs from actively expanding colonies were placed in the centre of VMS, PDA or WA (water agar) plates. Production of aerial hyphae and colony colour and morphology were visually inspected after 7 days and hyphal extension recorded at 24 and 48 h at 27 °C. Conidial production among the various strains was assayed by inoculating PDA with conidial suspensions (1×10^6 mL^{-1}). After 10 days of incubation, three agar plugs (5 mm diameter) were removed from each of five repetitions and placed in 10 mL of sterile water containing 0.1% triton. Conidial concentrations were determined by haemacytometer. Surface area of growth for each day was determined by photographing fungal growth and importing the images into ImageJ software (http://rsb.info.nih.gov/) for calculations. The treatments contained four repetitions and each experiment was repeated at least twice. Data analysis were performed by analysis of variance and Fisher's PLSD test ($P < 0.05$).

Detection of peptaibols by MALDI-TOF

Mycelia harvested from 9-day-old stationary cultures grown in liquid VMS were filtered, flash frozen in liquid nitrogen and lyophilized overnight. Dried cells were dissolved in water/ethanol/acetonitrile (1 : 1 : 1). All samples were analysed in a saturated alpha-CHCA matrix solubilized in 30% acetonitrile, 0.07% TFA. A mixture of 1 μL matrix and 1 μL sample was prepared directly on the plate. The samples were analysed using a MALDI VOYAGER ELITE time-of-flight mass spectrometer with a nitrogen laser giving a 337-nm output. The ions were accelerated with a voltage of 20 kV. Measurements were performed in the delayed extraction mode, allowing the determination of monoisotopic mass values. The mass spectrometer was used in the positive ion detection and reflector mode (Texas A&M University, College Station, TX, and Anagnostec Luckenwalde, Germany).

Peptide synthesis

Peptides TvBI (Ac-UGAVUQUAUSLUPLUUQV-OH) and TVBII (Ac-UGALUQUAUSLUPLUUQV-OH) (Wiest *et al.*, 2002) were synthesized and purchased from EZBiolab Inc., IN, USA. The peptides were dissolved in methanol to give a 6 mM stock solution (10 mg/mL). This solution was diluted in water to different concentrations as detailed in the different experiments.

Cucumis sativus seedling growth conditions

Cucumis sativus seeds (Kfir, Gedera Seeds, Israel) were surface disinfested in 2% NaOCl for 2 min and thoroughly washed with sterile distilled water. Growth containers with plant growth medium (PGM) (Yedidia *et al.*, 2003) were aerated through 0.45-μm pore-size filters. Plants were grown in a controlled environment: 26 °C, 80% relative humidity, light at 300 $\mu E/m^2/s$, and a circadian cycle of 14 h of light and 10 h of darkness. *T. virens* was inoculated as described (Yedidia *et al.*, 2003). Briefly, germinated *T. virens* conidia were inoculated into the PGM of 7-day-old seedlings to a final concentration of approximately 10^5 germinated conidia/mL.

Root colonization assay

Root colonization assays were performed according to Viterbo *et al.* (2005). Briefly, roots were detached 48 h post-inoculation and extensively washed in water. After sterilization in 1% NaOCl for 2 min, the roots were washed with sterile distilled water, weighed and homogenized using an ULTRA-TURRAX apparatus (Janke & Kunkel, Staufen, Germany) in 20 mL water for 1 min. Serial dilutions were assayed for CFU on *Trichoderma*-selective medium (Elad *et al.*, 1981) at 30 °C.

Phenolic extraction from cucumber cotyledons

Phenolic extraction was performed 48 h post-challenge inoculation with *P. syringae* pv. *lachrimans* according to Yedidia *et al.* (1993). Briefly, fresh foliar material was ground to a fine powder in liquid N_2 and extracted in 80% acidified methanol [10 g (fresh weight) per 100 mL). The mixture was maintained for 24 h in the dark under nitrogen. The extract was filtered through glass fibre and concentrated with a rotoevaporator. The aqueous residue was partitioned against hexane and ethyl acetate and subjected to acid hydrolysis. The hydrolysate was cooled and partitioned against ethyl acetate. This fraction was dried, and resuspended in absolute methanol [2.5 g (fresh weight)/mL].

Microbial bioassay

Cells of *P. syringae* pv. *lachrymans* were cultured in trypticase soy broth. Crude phenolic extracts were further concentrated by Speed-Vac and adjusted to 100 μL with absolute methanol. Different amounts of the concentrated samples were pipetted on to trypticase soy agar plates (TSA) and dried. Bacterial suspensions (200 μL) were mixed into 3 mL of soft TSA and overlaid on the dried plates. Antimicrobial activity of the extract was assayed 48 h after bacterial application and appeared as clear lytic circles on the plates (Yedidia *et al.*, 2003).

Elicitation of plant defence responses by synthetic TvBI and TVBII

Cucumber seeds were germinated and cultivated as described in the previous section. After 6 days, seedlings with fully expanded cotyledons were transferred to small vials containing 400 μL of a solution of 9.6 nmol of a methanol stock of TvBI or TvBII diluted in water or mock (methanol/water 4 : 1000). The vials were then placed in a sterile polycarbonate culture box (Djonovic *et al.*, 2006). After 48 h the seedlings were inoculated with 10 μL of bacterial suspension as described in Yedidia *et al.* (2003). Water was added after the first 24 h to the small vials as needed. Cotyledons were sampled at different times for RNA extraction, bacterial counting or phenolic extraction. Synthetic peptaibols at different concentrations (3–9 nmol) were also directly injected into cucumber leaves or cotyledon tissues through a needless syringe on the abaxial surface.

RNA extraction from *T. virens*

Transcriptional analysis of *tex1* during developmental stages of *T. virens* was performed as described by Djonovic *et al.* (2006). Developmental stages included: germinating spores (GS), non-sporulating mycelia (nSM) and sporulating mycelia (SM) grown on solid VMS or PDA; mycelia grown in liquid VMS; and mycelia of *T. virens* indirectly confronting *R. solani* (Cortes *et al.*, 1998). For the GS stage, conidia from 5-day-old cultures of PDA were spread on to cellophane overlaid on VMS or PDA and incubated for 12 h at 27 °C in the dark. The germinated spores were then scraped off the cellophane, and frozen in liquid nitrogen for RNA extraction. An agar plug from an actively growing culture was placed on cellophane overlaid on a VMS or PDA plate to obtain the nSM stage. After 2 days' growth in the dark, the mycelia were harvested. For the SM stage, mycelia were collected with a scalpel from 5-day-old cultures on VMS or PDA. Mycelium was harvested from liquid VMS following inoculation with conidia (final concentration 10^6 spores/mL) and incubation for 5 days at room temperature. Indirect confrontation with *R. solani* was conducted by placing an agar plug of *R. solani* on to VMS, followed by two cellophane discs and then an agar plug of *T. virens*. The mycelia of *T. virens* were allowed to grow for approximately 3 days on the cellophane and then harvested from areas that overlapped with the hyphae of *R. solani*. Assays to determine the expression of *tex1* under various nutritional conditions were performed by inoculating liquid VMS with conidia (final concentration 10^6 spores/mL), and incubating for 24 h at 25 °C while shaking. Germlings were then harvested, washed with sterile distilled water, and an equal fresh weight transferred to other liquid media (VM, VMS, VM-N, VMG, VMG-N, VMR, VMR-N and PGM). Mycelia were filtered on miracloth, ground in liquid nitrogen and RNA extracted (Jones *et al.*, 1985) and quantified by UV spectroscopy.

RNA was also extracted from *T. virens* mycelium wrapped around the roots as described in Viterbo *et al.* (2005). Control RNA for this experiment was extracted from cultures grown in PGM amended with 0.05% glucose.

tex1 RT-PCR analysis: primer design and reaction conditions

RNA was treated with DNase and 2.5 µg was used as template for reverse transcription using random hexamers as per the manufacturer's protocol (First-Strand cDNA Synthesis Kit, Amersham, Piscataway, NJ).

Primers were designed to amplify three regions of *tex1* (Fig. 3a). Primer pairs AW1 (5′-TGCTTCCTGGGAGGCTACTA-3′) and AW2 (5′-TATGGAAAGGCATCGTCCTC-3′) were used to amplify a 496-bp fragment from module 1, which is upstream of the insertion site of the disruption vector pPSKO2. The two pairs amplify regions downstream of the disruption event. Primer pairs RT1 (5′-GGGCTATGGCTCTCCTCTCT-3′) and RT2 (5′-AAGACCAGGCATTCTGCTCTA-3′) were used to amplify a 498-bp fragment from module 14; primer pairs AW3 (5′-CGAATGGGTCTAAGCGATGT-3′) and AW4 (5′-TGGTAACGTCAGCGACAGAG-3′) were used to amplify a 501-bp fragment from module 18. Primers from module 13, tex1GvF (5′-GGCAAGCCCAAGGGGATG-3′) and tex1GvR (5′-TATAGACTCTCTTGTTG-3′), were used in the RT-PCR from cucumber roots.

PCR was performed in 25-µL volumes containing 1 µL cDNA, 10 pmol of each primer, 1 unit Taq polymerase (Invitrogen) at 1 cycle of 94 °C for 5 min; 30 cycles of 94 °C for 30 s, 60 °C for 30 s and 72 °C for 20 s; followed by 1 cycle of 72 °C for 7 min. Positive controls, containing Gv29-8 genomic DNA, and negative controls with no DNA were included in all experiments.

RNA isolation from plant tissue and RT-PCR analysis of plant defence genes expression

Total RNA was extracted from cotyledons using the EZ-RNA total RNA isolation kit (Biological Industries Co., Beit-Haemek, Israel). Reverse transcription analysis was performed as in Viterbo *et al.* (2005). Primers for the amplification of *hpl* and *pal1* were as described in Yedidia *et al.* (2003) and for *prx* as in Shoresh *et al.* (2005).

ACKNOWLEDGEMENTS

This work was supported by 'The Dr Alexander and Myrna Strelinger Endowment Fund' and a United States–Israel Binational Agricultural Research and Development Fund (BARD) (US-3507-04) and a grant from the US National Science Foundation (IOB0445650).

REFERENCES

Baek, J.M., Howell, C.R. and Kenerley, C.M. (1999) The role of an extracellular chitinase from *Trichoderma virens* Gv29-8 in the biocontrol of *Rhizoctonia solani*. *Curr. Genet.* **35**, 41–50.

Baek, J.M. and Kenerley, C.M. (1998) The arg2 gene of *Trichoderma virens*: cloning and development of a homologous transformation system. *Fungal Genet. Biol.* **23**, 34–44.

Berg, A., Grigoriev, P.A., Degenkolb, T., Neuhof, T., Hartl, A., Schlegel, B. and Grafe, U. (2003) Isolation, structure elucidation and biological activities of trichofumins A, B, C and D, new 11 and 13mer peptaibols from *Trichoderma* sp. HKI 0276W. *J. Pept. Sci.* **9**, 810–816.

Chen, F., D'Auria, J.C., Tholl, D., Ross, J.R., Gershenzon, J., Noel, J.P. and Pichersky, E. (2003) An *Arabidopsis thaliana* gene for methylsalicylate biosynthesis, identified by a biochemical genomics approach, has a role in defense. *Plant J.* **36**, 577–588.

Chugh, J.K. and Wallace, B.A. (2001) Peptaibols: models for ion channels. *Biochem. Soc. Trans.* **4**, 565–570.

Cortes, C., Gutierrez, A., Olmedo, V., Inbar, J., Chet, I. and Herrera-Estrella, A. (1998) The expression of genes involved in parasitism by *Trichoderma harzianum* is triggered by a diffusible factor. *Mol. Gen. Genet.* **260**, 218–225.

Djonovic, S., Pozo, M.J., Dangott, L.J., Howell, C.R. and Kenerley, C.M. (2006) Sm1, a proteinaceous elicitor secreted by the biocontrol fungus *Trichoderma virens* induces plant defense responses and systemic resistance. *Mol. Plant–Microbe Interact.* **19**, 838–853.

Duclohier, H. and Wroblewski, H. (2001) Voltage-dependent pore formation and antimicrobial activity by alamethicin and analogues. *J. Membrane Biol.* **184**, 1–12.

Ehrhardt, D.W., Atkinson, E.M. and Long, S.R. (1992) Depolarization of alfalfa root hair membrane potential by *Rhizobium meliloti* Nod factors. *Science*, **256**, 998–1000.

Elad, Y., Chet, I. and Henis, Y. (1981) A selective medium for improving quantitative isolation of *Trichoderma* spp. from soil. *Phytoparasitica*, **9**, 59–67.

Engelberth, J., Koch, T., Schuler, G., Bachmann, N., Rechtenbach, J. and Boland, W. (2000) Ion channel-forming alamethicin is a potent elicitor of volatile biosynthesis and tendril coiling. Cross talk between jasmonate and salicylate signaling in lima bean. *Plant Physiol.* **125**, 369–377.

Fravel, D.R. (2005) Commercialization and implementation of biocontrol. *Annu. Rev. Phytopathol.* **43**, 337–359.

Grigoriev, P.A., Schlegel, B., Kronen, M., Berg, A., Härtl, A. and Gräfe, U. (2003) Differences in membrane pore formation by peptaibols. *J. Pept. Sci.* **9**, 763–768.

Harman, G.E., Howell, C.R., Viterbo, A., Chet, I. and Lorito, M. (2004) *Trichoderma* species—opportunistic, avirulent plant symbionts. *Nat. Rev. Microbiol.* **2**, 43–56.

Howell, C.R. (2003) Mechanisms employed by *Trichoderma* species in the biological control of plant diseases. *Plant Dis.* **87**, 4–10.

Jabs, T., Tschope, M., Colling, C., Hahlbrock, K. and Scheel, D. (1997) Elicitor-stimulated ion fluxes and O_2- from the oxidative burst are essential components in triggering defense gene activation and phytoalexin synthesis in parsley. *Proc. Natl Acad. Sci.* **94**, 4800–4805.

Jones, J.D.G., Dunsmuir, P. and Bedbrook, J. (1985) High level of expression of introduced chimeric genes in regenerated transformed plants. *The EMBO. J.* **4**, 2411–2418.

Kim, Y.H., Yeo, W.H., Kim, Y.S., Chae, S.Y. and Kim, K.S. (2000) Antiviral activity of antibiotic peptaibols, chrysospemins B and D, produced by *Apiocrea* sp. 14T against TMV infection. *J. Microbiol. Biotechnol.* **10**, 522–528.

Lumsden, R.D., Walter, J.F. and Baker, C.P. (1996) Development of *Gliocladium virens* for damping-off disease control. *Can. J. Plant Pathol.* **18**, 463–468;.

Marahiel, M.A., Stachelhaus, T. and Mootz, H.D. (1997) Modular peptide synthetases involved in nonribosomal peptide synthesis. *Chem. Rev.* **97**, 2651–2674.

Pozo, M.J., Baek, J.M., Garcia, J.M. and Kenerley, C.M. (2004) Functional analysis of *tvsp1*, a serine protease-encoding gene in the biocontrol agent *Trichoderma virens*. *Fungal Genet. Biol.* **41**, 336–348.

Rebuffat, S., Goulard, C., Hlimi, S. and Bodo, B. (2000) Two unprecedented natural Aib-peptides with the (Xaa-Yaa-Aib-Pro) motif and an unusual C-terminus: structures, membrane-modifying and antibacterial properties of pseudokonins KL III and KL VI from the fungus *Trichoderma pseudokoningii*. *J. Pept. Sci.* **10**, 519–533.

Schirmböck, M., Lorito, M., Wang, Y.L., Hayes, C.K., Arisan-Atac, I., Scala, F., Harman, G.E. and Kubicek, C.P. (1994) Parallel formation and synergism of hydrolytic enzymes and peptaibol antibiotics, molecular mechanisms involved in the antagonistic action of *Trichoderma harzianum* against phytopathogenic fungi. *Appl. Environ. Microbiol.* **60**, 4364–4370.

Shenkarev, Z.O., Balashova, T.A., Yakimenko, Z.A., Ovchinnikova, T.V. and Arseniev, A.S. (2004) Peptaibol zervamicin IIB structure and dynamics refinement from transhydrogen bond J couplings. *Biophys. J.* **86**, 3687–3699.

Shoresh, M., Yedidia, I. and Chet, I. (2005) Involvement of jasmonic acid/ethylene signaling pathway in the systemic resistance induced in cucumber by *Trichoderma asperellum* T203. *Phytopathology*, **95**, 76–84.

Song, X.Y., Shen, Q.T.X.T., Chen, X.L., Sun, C.Y. and Zhang, Y.Z. (2006) Broad-spectrum antimicrobial activity and high stability of trichokonins from *Trichoderma koningii* SMF2 against plant pathogens. *FEMS Microbiol. Lett.* **260**, 119–125.

Szekeres, A., Leitgeb, B., Kredics, L., Antal, Z., Hatvani, L., Manczinger, L. and Vagvolgyi, C. (2005) Peptaibols and related peptaibiotics of *Trichoderma*. A review. *Acta Microbiol. Immunol. Hung.* **52**, 137–168.

Thomas, M.D. and Kenerley, C.M. (1989) Transformation of the mycoparasite *Gliocladium*. *Curr. Genet.* **15**, 415–420.

Viterbo, A., Harel, M., Horwitz, B.A., Chet, I. and Mukherjee, P.K. (2005) *Trichoderma* MAP-kinase signaling is involved in induction of plant systemic resistance. *Appl. Environ. Microbiol.* **71**, 6241–6246.

Vogel, H.J. (1956) A convenient growth medium for *Neurospora* (Medium N). *Microbiol. Genet. Bull.* **13**, 42–43.

Wei, X., Yang, F. and Straney, D.C. (2005) Multiple non-ribosomal peptide synthetase genes determine peptaibol synthesis in *Trichoderma virens*. *Can. J. Microbiol.* **51**, 423–429.

Wiest, A., Grzegorski, D., Xu, B.W., Goulard, C., Rebuffat, S., Ebbole, D.J., Bodo, B. and Kenerley, C. (2002) Identification of peptaibols from *Trichoderma virens* and cloning of a peptaibol synthetase. *J. Biol. Chem.* **277**, 20862–20868.

Xu, B.W., Wild, J.R. and Kenerley, C.M. (1996) Enhanced expression of a bacterial gene for pesticide degradation in a common soil fungus. *J. Ferment. Bioeng.* **81**, 473–481.

Yedidia, I., Shoresh, M., Kerem, Z., Benhamou, N., Kapulnik, Y. and Chet, I. (2003) Concomitant induction of systemic resistance to *Pseudomonas syringae* pv. *lachrymans* in cucumber by *Trichoderma asperellum* (T-203) and accumulation of phytoalexins. *Appl. Environ. Microbiol.* **69**, 7343–7353.

8 Plant Disease Biocontrol and Induced Resistance via Fungal Mycoparasites

A. Viterbo[1], J. Inbar[1], Y. Hadar[2], I. Chet[1]

CONTENTS

I. Introduction

Due to the adverse environmental effects of pesticides that create health hazards for human and other nontarget organisms, including the pests' natural enemies, these chemicals have been the object of substantial criticism in recent years. The development of safer, environmentally feasible control alternatives has therefore become a top priority. In this context, biological control is becoming an urgently needed component of agriculture.

[1] Department of Plant Sciences, Weizmann Institute of Science, Rehovot 76100, Israel
[2] Department of Plant Pathology and Microbiology, The Hebrew University of Jerusalem, Faculty of Agriculture, Rehovot 76100, Israel

Biological control of plant pathogens is defined as the use of biological processes to lower inoculum density of the pathogen, with the aim of reducing its disease-producing activities (Baker and Cook 1974).

Biological control may be achieved by both direct and indirect strategies. Indirect strategies include the use of organic soil amendments that enhance the activity of indigenous microbial antagonists against a specific pathogen. Another indirect approach, cross-protection, involves the stimulation of plant self-defense mechanisms against a particular pathogen by prior inoculation of the plant rhizosphere with a nonvirulent strain or other nonpathogenic rhyzo-competent bacteria or fungi. Successful protection resulting in induced resistance has been documented for viruses, bacterial pathogens, and fungi (van Loon et al. 1998; Harman et al. 2004; Haas and Defago 2005).

The direct approach involves the introduction of specific microbial antagonists into soil or plant material (Cook and Baker 1983). These antagonists have to proliferate and establish themselves in the appropriate ecological niche in order to be active against the pathogen. Antagonists are microorganisms with the potential to interfere with the growth and/or survival of plant pathogens, and thereby contribute to biological control. Antagonistic interactions among microorganisms in nature include parasitism or lysis, antibiosis, and competition. These microbial interactions serve as the basic mechanisms via which biocontrol agents operate. An elucidation of the mechanisms involved in biocontrol activity is considered to be one of the key factors in developing useful biocontrol agents. Of the numerous biocontrol agents examined, only a few have been subjected to a thorough analysis of the mechanisms involved in the suppression of the pathogen. In this chapter, we shall use examples to briefly demonstrate the different mechanisms, and will concentrate on mycoparasitism.

Environmental and Microbial Relationships, 2nd Edition
The Mycota IV
C. P. Kubicek and I. S. Druzhinina (Eds.)

A. Antibiosis

Handelsman and Parke (1989) restricted the definition of antibiosis to those interactions that involve a low-molecular-weight diffusible compound, or an antibiotic produced by a microorganism that inhibits the growth of another microorganism. This definition excluded proteins or enzymes that kill the target organism. Baker and Griffin (1995) extended the scope of the definition to "inhibition or destruction of an organism by the metabolic production of another." This definition includes small toxic molecules, volatiles, and lytic enzymes.

The production of inhibitory metabolites by fungal biocontrol agents has been reported in the literature over the last five decades (Bryan and McGowan 1945; Dennis and Webster 1971a, b; Ghisalberti and Sivasithamparam 1991). *Gliocladium virens* is a common example of the role of antibiotics in biological control by fungal antagonists. Gliovirin is a diketopiperazine antibiotic that appears to kill the fungus *Pythium ultimum* by causing coagulation of its protoplasm.

Howell and Stipanovic (1983) obtained gliovirin-deficient mutants of *G. virens* via ultraviolet mutagenesis. These mutants failed to protect cotton seedlings from *P. ultimum* damping-off when applied to the seeds, whereas the normal parent strain protected the seedlings. Moreover, a gliovirin-overproducing mutant provided control similar to that of the wild type, although it exhibited a much lower growth rate. A combination of *G. virens* treatment of cotton seed with reduced levels of the fungicide metalaxyl provided diseased suppression equal to that of a full fungicide treatment (Howell 1991). An extensive review on antibiosis and production of *Trichoderma* secondary metabolites is provided in Howell (1998). Baker and Griffin (1995) concluded that the impact of antibiosis in biological control is uncertain. Even in cases where antifungal metabolite production by an agent reduces disease, other mechanisms may also be operating. Synergism among various lytic enzymes, and between enzymes and antibiotics has been shown to be very critical for the activity of many biocontrol agents (reviewed by Woo et al. 2002).

B. Competition

Many plant pathogens require exogenous nutrients to successfully germinate, penetrate, and infect host tissue (Baker and Griffin 1995). Garrett (1965) concluded that the most common cause of death in a microorganism is starvation. Therefore, competition for limiting nutritional factors, mainly carbon, nitrogen, and iron, may result in biological control of plant pathogens.

Research over the years has concentrated on competition by bacterial biocontrol agents, mainly for iron (Fe). However, fungal antagonists have received very little attention. Sivan and Chet (1989) found that a strain of *T. harzianum* (T-35) that controls *Fusarium* spp. on various crops may operate via competition for nutrients and rhizosphere colonization.

The potential of microorganisms, which are applied as a seed treatment, to proliferate and establish along the developing root system has been named rhizosphere competence (Ahmad and Baker 1987). When T-35 conidia were applied to soil enriched with chlamydospores of *F. oxysporum* f. sp. *melonis* and f. sp. *vasinfectum*, and amended with low levels of glucose and asparagine, the ability of the chlamydospores to germinate was reduced. This inhibitory effect could be reversed by adding an excess of glucose and asparagine or of seedling exudates to the soil. After its application as a seed treatment, this strain effectively colonized the rhizosphere of melon and cotton, and prevented colonization of these roots by *F. oxysporum*. Thus, competition for carbon and nitrogen in the rhizosphere, as well as rhizosphere competence itself may be involved in the biocontrol of *F. oxysporum* by *T. harzianum* strain T-35 (Sivan and Chet 1989).

C. Mycoparasitism

Mycoparasitism is defined as a direct attack on a fungal thallus, followed by utilization of its nutrients by the parasite. The term hyperparasitism is sometimes used to describe a fungus that is parasitic on another parasitic pathogenic fungus. Barnett and Binder (1973) divided mycoparasitism into: (1) necrotrophic (destructive) parasitism, in which the relationships result in death and destruction of one or more components of the host thallus, and (2) biotrophic (balanced) parasitism, in which the development of the parasite is favored by a living, rather than a dead host structure.

Necrotrophic mycoparasites tend to be more aggressive, have a broad host range extending to wide taxonomic groups, and are relatively unspecialized in their mode of parasitism. The antagonistic activity of necrotrophic mycoparasites is attributed to the production of antibiotics, toxins, or hydrolytic enzymes in proportions that cause the death and destruction of their host. Biotrophic mycoparasites, on the other hand, tend to have a more restricted host range and produce specialized structures to adsorb nutrients from their host (Manocha 1990).

The parasitic relationships between fungi and their significance in biological control are the subjects of this chapter. Both types of mycoparasitism are described and discussed in the scope of their contribution to biological control. We will concentrate on the morphological, biochemical and molecular aspects of mycoparasitism in relation to biological control. The ecological aspects of this phenomenon are discussed in Jeffries (1997).

II. Mycoparasites as Biocontrol Agents

Many comprehensive reviews on mycoparasitism in biological control in general have been published over recent years (Chet 1990; Deacon 1991; Elad and Chet 1995; Benitez et al. 2004). Due to their nature, only a few examples of biotrophic mycoparasites as biocontrol agents exist (Ayers and Adams 1981; Sztejnberg et al. 1989; Adams 1990). Necrotrophic mycoparasites, being more common, saprophytic in nature, and less specialized in their mode of action, are easier to study. As a result, the majority of the mycoparasites used as biocontrol agents in greenhouse or field trials to date have been necrotrophs. In this chapter, we will emphasize those examples in which the research that has been carried out is both applied and basic in nature.

A. Biotrophic Mycoparasites

1. *Sporidesmium sclerotivorum*

Sporidesmium sclerotivorum is a dermatiacesus hyphomycete that was isolated from field soil by Uecker et al. (1980). In nature, the fungus has been found to be an obligate parasite on sclerotia of *Sclerotinia sclerotiorum*, *S. minor*, *S. trifoliorum*, *Sclerotium cepivorum*, and *Botrytis cinerea* (Ayers and Adams 1981). In response to chemicals released by the host's sclerotia, macroconidia of *S. sclerotivorum* in the soil germinate and the germ tubes infect the sclerotia. When volatile compounds secreted by sclerotia of *Sclerotinia minor*, *Sclerotinia sclerotiorum*, and *Sclerotium rolfsii* were tested to determine if they could stimulate germination of conidia of *S. sclerotivorum*, none of the chemicals alone or a combination of all chemicals induced germination (Fravel et al. 2002). The hyphae penetrate the intercellular matrix of the conidia, which is composed mainly of β-glucans (Ayers et al. 1981). The production and activity of haustoria by the mycoparasite stimulate the host sclerotia to increase their glucanase, and probably other enzyme activities, resulting in the degradation of glucan into available glucose (Bullock et al. 1986). The mycoparasite establishes itself in the sclerotia, where its mycelium grows out into the surrounding soil to infect additional sclerotia and to produce new macroconidia (Adams et al. 1984). The interaction between *S. sclerotivorum* and *Sclerotinia minor* depends on both the host and parasite density (Adams 1986). The infection process is favored by soil pH, water potential and temperature (20–22 °C) (Adams and Ayers 1980).

Under field conditions, a single application of an *S. sclerotivorum* preparation at a concentration of 10^2 or 10^3 macroconidia g^{-1} soil caused a 75–95% reduction in the number of sclerotia of *S. minor* per plot. Control of lettuce drop caused by *S. minor* in these plots varied from 40–83% in four consecutive lettuce crops (Adams and Ayers 1982). These results were significant but not economically important (Adams 1990). Adams (1990) concluded that "one of the biggest obstacles to practical biological control is the large quantity of the agent necessary to achieve biological control when applied directly to soil in the field." He therefore suggested two alternatives: (1) to add sclerotia of *S. minor* or a nonpathogenic *Sclerotinia* that is also a host of *S. sclerotivorum* that are already infected by the mycoparasite; (2) to apply a low dosage of the mycoparasite preparation to a diseased crop, and then immediately incorporate the treated crop into the soil. This latter procedure ensures that a high percentage of the mycoparasites will be present in the soil in close contact with the sclerotia of the pathogen. Although the author assumes that these alternatives are easier and more practical, neither has been explored to any significant extent (Adams 1990). Field studies were later conducted (Del Rio et al. 2002) to evaluate the effectiveness of *S. sclerotivorum* to control *Sclerotinia* stem rot of soybean. Experimental plots were infested with *S. sclerotivorum* macroconidia at a rate of 0, 2, or 20 spores per cm^2. Two years later, the disease was completely suppressed in all plots. *S. sclerotivorum* was retrieved from all infested plots at all locations 2 years

pathogenic fungus *S. sclerotiorum*. In Canada, Huang (1977) found that sclerotia of *S. sclerotiorum* in roots and stems of sunflower, at the end of the season, became infected with the parasite *C. minitans*. This infection actually provided natural biological control of this pathogen in the field. Applying *C. minitans* to the seed furrow in field trials, in soil naturally or artificially infested with *S. sclerotiorum*, produced 42–78% disease control of sunflower wilts over 2 successive years (Huang 1980). *C. minitans* is a destructive parasite that kills both hyphae and sclerotia of *S. sclerotiorum*. By using scanning electron microscopy, it was shown that hyphae of *C. minitans* grow intracellularly in the infected sclerotia (Phillips and Price 1983; Tu 1984). Phillips and Price (1983), based on transmission electron microscopic studies, concluded that penetration of the rind cells of *S. sclerotiorum* sclerotia by *C. minitans* is due to physical pressure, rather than enzymatic lysis of the cell wall. In a later study, Huang and Kokko (1987) found, by transmission electron microscopy, that there was destruction and disintegration of the sclerotial tissues, caused by penetration of the parasitic hyphae. Evidence from cell-wall etching at the penetration site suggests that chemical activity is indeed required for hyphae of *C. minitans* to penetrate the thick, melanized rind walls. The medullary tissue infected by *C. minitans* showed signs of plasmolysis, aggregation and vacuolization of the cytoplasm, and dissolution of the cell walls. The authors concluded that cell wall-lysing enzymes, responsible for the degradation of *S. sclerotiorum* hyphae, may also play a significant role in the dissolution and degradation of the sclerotial rind wall at the penetration site and other affected areas (Huang and Kokko 1987). Glucanase, chitinase, cellulase, and xylanase enzyme activities were recently reported in sclerotia-containing cultures of *C. minitans* (Kaur et al. 2005).

Infection of *S. sclerotiorum* hyphae by the hyperparasite *C. minitans* has been reported by several workers (Huang and Hoes 1976; Tu 1984). However, researchers are not in complete agreement on the mode of hyperparasitism. Using light microscopy, Huang and Hoes (1976) observed that hyphal tips of *C. minitans* invade hyphae of *S. sclerotiorum* by direct penetration, without forming any special structure. Host cytoplasm disintegrates and cell walls collapse as a result of infection. Microconidia and intrahyphal hyphae were produced by *S. sclerotiorum* in infected colonies.

Production of appressoria by *C. minitans* when it comes into contact with the undamaged hyphae of *S. sclerotiorum* in dual culture on potato dextrose agar (PDA) was observed by Tu (1984). He stated that hyphal penetration by the hyperparasite sometimes occurs without the formation of appressoria, but only on damaged host cells.

Huang and Kokko (1988), using scanning electron microscopy, confirmed previous reports from light microscopic studies that hyphal tips of *C. minitans* invade the host hyphae by direct penetration, without developing appressoria, and that indentation of the host cell wall at the point of penetration is often evident. No functional distinction between main branch and side branch hyphae of the hyperparasite was found, and tips of either type of hypha are capable of invading host hyphae by direct penetration.

4. *Gliocladium* and *Trichoderma* spp.

The morphological borders between *Trichoderma* and *Gliocladium* are blurred. Therefore, in recent years molecular methods have been applied as an aid to resolving the taxonomy and systematic of *Trichoderma* and *Gliocladium*. *Gliocladium virens* is now generally recognized as belonging to the genus *Trichoderma* (Gams and Bisset 1998).

Particular attention has been paid to species identification of the genus *Hypocrea*/*Trichoderma* that has proved problematic when traditional methods are used. An update on the taxonomy and phylogeny of the 88 taxa (which occur as 14 holomorphs, 49 teleomorphs and 25 anamorphs in nature) of *Hypocrea*/*Trichoderma*, confirmed by a combination of morphological, physiological and genetic approaches, is presented in Druzhinina and Kubicek (2005).

Several species of *Gliocladium* have been reported to be hyperparasites of many fungi. The biology, ecology, and potential of this genus for biological control of plant pathogens have been extensively reviewed in a comprehensive treatise by Papavizas (1985). Huang (1978) reported that *G. catenulatum* parasitizes *S. sclerotiorum* and *Fusarium* spp. It kills the host by direct hyphal contact, causing the affected cells to collapse or disintegrate. Pseudoappressoria are formed by the hyperparasite, but hyphae derived from these do not penetrate the host cell walls. Vegetative hyphae of all species tested, and macroconidia of *Fusarium* spp. are susceptible to this hyperparasite, but chlamydospores of *Fusarium equiseti* are resistant.

Phillips (1986) studied aspects of the biology of *G. virens* and its parasitism of sclerotia of *S. sclerotiorum* in soil. *G. virens* parasitized and decayed sclerotia of *S. sclerotiorum*, *S. minor*, *Botrytis cinerea*, *Sclerotium rolfsii*, and *Macrophomina phaseolina* on laboratory media, and caused a reduction in the survival of sclerotia of *S. sclerotiorum* in soil. However, parasitism of the mycelium was not detected.

A strain of *G. virens* isolated from the parasitized hyphae of *R. solani* by Howell (1982) significantly suppressed damping-off in cotton seedlings by this pathogen and by *Pythium ultimum*. Treatment with *G. virens* more than doubled the number of surviving cotton seedlings grown in soil infested with either pathogen. *G. virens* parasitized *R. solani* by coiling around, and penetrating the hyphae. *P. ultimum* was not parasitized by *G. virens*, but was strongly inhibited by antibiosis. Treatment of soil infested with propagules of *R. solani* or *P. ultimum* with *G. virens* resulted in a 63% reduction in the number of viable *R. solani* sclerotia after 3 weeks of incubation, whereas oospores of *P. ultimum* were unaffected. Strains of *G. virens* were separated into two distinct groups, P and Q, on the basis of secondary metabolite production in vitro (Howell et al. 1993).

Gliovirin was very inhibitory to *P. ultimum*, but exhibited no activity against *R. solani*, and strains that produced it (P group) were more effective seed-treatment biocontrol agents of disease incited by *P. ultimum*. Conversely, gliotoxin was more active against *R. solani* than against *P. ultimum*, and strains that produced it (Q group) were more effective seed treatments for controlling disease incited by *R. solani*. Based on these results, the authors suggested that it may be necessary to treat seeds with a combination of strains in order to broaden the disease control spectrum.

Howell (1987) isolated mutants of *G. virens*, obtained by irradiation with ultraviolet light, that showed no mycoparasitic activity. The selected mutants retained the same antibiotic complement as the parent strains. Peat moss-Czapek's broth cultures of parent and mutant strains were similarly effective as biocontrol agents of cotton seedling disease induced by *R. solani*, and as antagonists of *R. solani* sclerotia in soil. In the light of these results, Howell (1987) concluded that mycoparasitism is not a major mechanism in the biological control of *R. solani*-incited seedling disease by *G. virens*.

In addition, Pachenari and Dix (1980) concluded that *G. virens* need not make intimate contact with *Botrytis allili* to cause severe internal disorganization of host cells, coagulation of cytoplasm, vacuolation, and loss of contents from organelles. Cultures of *B. allili* parasitized by *G. roseum* contained considerable β-(1-3)-glucanase and chitinase, and the cytoplasm coagulated without physical contact. *G. virens* isolate G1-21 was grown on various solid and liquid media: wheat bran and peanut hull meal (PHM), as well as spent glucose tartrate broth (GTB), Czapek-Dox broth (CDB), and potato dextrose broth (PDB) (Lewis et al. 1991). Aqueous extracts of these media caused leakage of carbohydrates and electrolytes from hyphae of the soil-borne plant pathogen *R. solani*, and its mycelial weight was reduced. Size fractionation experiments indicated that it was a combination of factors associated with *G. virens*, rather than a single one, which induced this phenomenon. Gliotoxin was detected in culture filtrates from *G. virens* grown on bran and PHM media. Gliotoxin preparations induced leakage of carbohydrates and electrolytes from *R. solani*, and caused a concomitant reduction in mycelial weight, which suggests the action of a leakage factor (Lewis et al. 1991). The authors speculated that hydrolytic enzymes such as β-1-3-glucanase, β-1-4-glucanase, chitinase, and protease, shown to be produced by isolates of *G. virens* (Roberts and Lumsden 1990), have the potential to act on *R. solani* cell walls and membranes. The role of extracellular chitinase in the biocontrol activity of *Trichoderma virens* was later examined using genetically manipulated strains of this fungus. The *T. virens* strains in which the chitinase gene (*cht42*) was disrupted (KO) or constitutively overexpressed (COE) were constructed through genetic transformation. Biocontrol activity of the KO and COE strains were significantly decreased and enhanced, respectively against cotton seedling disease incited by *Rhizoctonia solani* when compared with the wild-type strain (Baek et al. 1999).

More than 60 years ago, Weindling (1932) was the first to demonstrate the mycoparasitic nature of fungi from the genus *Trichoderma*. He suggested their potential use as biocontrol agents of plant pathogenic fungi. However, the first report on a biological control experiment using *Trichoderma* spp. under natural field conditions came 40 years later, by Wells et al. (1972) who used *T. harzianum* grown on an autoclaved mixture of ryegrass seeds and soil to control *Sclerotium rolfsii* Sace. Since then, more *Trichoderma* isolates have been obtained from natural habitats, and used in biocontrol trials against several soil-borne plant pathogenic fungi under both greenhouse and field conditions (Chet 1990; Harman and Lumsden 1990; Harman 2006).

A seed treatment was developed by Harman et al. (1980) to reduce the amount of *Trichoderma* added to the soil to control soil-borne plant pathogenic fungi. *T. hamatum* conidia applied in the laboratory, to seeds of pea and radish as a Methocel slurry, provided protection to seeds and seedlings from *Pythium* spp. and *R. solani*, respectively, almost as effectively as fungicide seed treatment. Establishment of the mycoparasite and long-term action were demonstrated, as the propagules of *T. hamatum* increased approximately 100-fold in soils planted with treated seeds. Population densities of *R. solani* and *Pythium* spp. were lower in soils containing *T. hamatum* than in soils lacking this antagonist. Replanting these soils once, or even twice with untreated seeds yielded lower disease incidence than in soils originally planted with untreated seeds. Addition of chitin or *R. solani* cell walls to the coating of seeds previously treated with a conidial suspension increased both the ability of *T. hamatum* to protect the seeds against *Pythium* spp. or *R. solani*, and the population density of *Trichoderma* in the soil. *T. hamatum* with chitin, but without *R. solani* cell walls, effectively reduced damping-off caused by *Pythium* spp., compared to seed treatment containing only *T. hamatum* (Harman et al. 1980). Sivan et al. (1984) applied a peat-bran mixture (1:1 v/v) preparation of *T. harzianum* (isolate 315) to either soil or rooting mixture, and efficiently controlled damping-off induced by *Pythium aphanidermatum* in pea, cucumber, tomato, pepper, and gypsophila. Several isolates of *T. harzianum* and *T. hamatum* were found to antagonize and control *Macrophomina phaseolina* in beans and melon (Elad et al. 1986). Isolates of *T harzianum* and *T. hamatum* antagonized and controlled *Rosellinia necatrix* in almond seedlings (Freeman et al. 1986). Sztejnberg et al. (1987) combined sublethal soil heating with an application of *T. harzianum* to yield better control of *R. necatrix* than that achieved by either treatment alone.

Sivan and Chet (1986) isolated a new *Trichoderma harzianum* isolate (T-35) from the rhizosphere of cotton plants grown in fields infested with *Fusarium*. In a further study, the isolate was tested in biological control trials over two successive growing seasons against *Fusarium* crown rot of tomato in fields naturally infested with *F. oxysporum* f. sp. *radici lycopersici* (Sivan et al. 1987). *T. harzianum* was applied as a seed coating or as a wheat branpeat (1:1, v/v) preparation introduced into the tomato rooting mixture. *Trichoderma*-treated transplants were better protected against *Fusarium* crown rot than untreated controls when planted in MB-fumigated or nonfumigated infested fields. The total yield of tomatoes in the *T. harzianum*-treated plots increased as much as 26.2% over the controls. Integrated control of *Verticillium dahliae* in potato by *T. harzianum* and the fungicide Captan was reported by Ordentlich et al. (1990).

C. Induced Systemic Resistance by *Trichoderma* spp.

Some *Trichoderma* rhizosphere-competent strains colonize entire root surfaces with morphological features reminiscent of those seen during mycoparasitism (Yedidia et al. 1999). Penetration of the root tissue is usually limited to the first or second layers of cells, and occurs only in the intercellular spaces. *Trichoderma* strains capable of establishing such interaction induce metabolic changes in plants that increase resistance to a wide range of plant-pathogenic microorganisms and viruses (Harman et al. 2004; Fig. 8.1). This response seems to be broadly effective for many plants, which indicates that there is little or no plant specificity.

At least three classes of substances that elicit plant defense responses have been identified. These elicitors include proteins, peptides, and low-molecular-weight compounds (Harman et al. 2004; Viterbo et al. 2004). The systemic response in plants occurs through the jasmonic

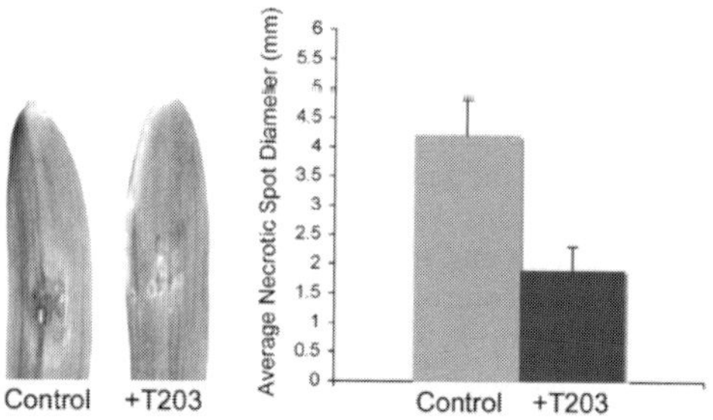

Fig. 8.1. Induced resistance toward the leaf pathogen *Cochliobolus heterostrophus* in maize. Seedling roots were infected with germinated *Trichoderma* spores (10^5 ml^{-1}) 48 h prior to pathogen leaf infection (800 spores). The symptoms were recorded 72 h after infection

acid/ethylene signaling pathway in a way similar to the rhizobacteria-induced systemic resistance (van Loon et al. 1998; Shoresh et al. 2005). Several studies have shown that root colonization by *Trichoderma* strains results in massive changes in plant gene expression patterns and metabolome. Changes in plant metabolism lead to the accumulation of antimicrobial compounds. In cucumber, root colonization by *T. asperellum* strain T-203 causes an increase in phenolic glucoside levels in leaves, which are strongly inhibitory to a range of bacteria and fungi (Yedidia et al. 2003). The protection afforded by the biocontrol agent is associated with the accumulation of mRNA of two defense genes: the phenylpropanoid pathway gene phenylalanine ammonia lyase (*PAL*) and the lipoxygenase pathway gene hydroxyperoxide lyase (*HPL*) (Yedidia et al. 2003). Increased levels of other defense-related plant enzymes, such as peroxidases, chitinases, and β-1,3-glucanases, have been recorded in *Trichoderma*-treated cucumber seedlings upon pathogen challenge (Shoresh et al. 2005). This potentiation in the gene expression enables *Trichoderma*-treated plants to be more resistant to subsequent pathogen infection. The MAPK signal transduction pathways, both of the plant and of *Trichoderma*, are important for the induction of systemic resistance (Viterbo et al. 2005; Shoresh et al. 2006).

D. Mycoparasitism in Suppressive Environments (Soils and Composts)

Suppression of soil-borne plant pathogens occurs in environments such as field soils, and soils amended with organic matter or compost as the organic component in media for container-grown plants. A review on this topic for field soils was recently published by Stone et al. (2004).

Pathogen suppressiveness has been defined by Cook and Baker (1983) as "soils in which the pathogen does not establish or persist, establishes but causes little or no damage, or establishes and causes disease for a while but thereafter the disease is less important, although the pathogen may persist in the soil."

Baker and Cook (1974) divided suppression mechanisms into two broad categories defined as general and specific. General suppression is a result of total microbial activity. In contrast, specific suppression applies when bacteria or fungi, individually or as a group, are responsible for the suppression effect. Mycoparasitism, the focus of this chapter, is a major mechanism of specific suppression. For example, Chet and Baker (1981) reported on a soil which was suppressive to *R. solani* of carnation near Bogota, Colombia. This soil contained high levels of organic matter (35%), was highly acidic (pH 5.1), and its main microbiological component was the antagonistic fungus *T. hamatum* at a population density of 8×10^5 propagules g^{-1}. The level of this mycoparasite in mineral-conducive soil was four orders of magnitude lower.

In another, related study, Henis et al. (1978) showed the effect of successive plantings on the development of suppression. A soil sown with radish every week became suppressive to *R. solani* by the fourth sowing, and was even more suppressive by the fifth and subsequent sowings. The population of *T. harzianum*, antagonistic to *R. solani*, increased with successive sowings of radish (Liu and Baker 1980), possibly in response to increases in the amount of *R. solani* in the soil resulting from its parasitism of the radish seedlings. The addition of *T. harzianum* spores to a conducive soil at the same density as that found in the suppressive soil caused the conducive soil to become suppressive. *Trichoderma* spp. were also reported to be responsible for the suppression of the take-all disease caused by *Gaeumannomyces graminis*. Low pH conditions were found to be favorable to *Trichoderma*, and to enhance suppression (Simon and Sivasithamparam 1990).

A practical approach to utilizing suppression in agriculture is the use of suppressive composts, mainly in container media. Composting is the breakdown of organic waste material by a succession of mixed populations of microorganisms in a thermophilic aerobic environment. The final product is compost or humus, which is the stabilized organic matter populated by microorganisms capable of suppressing soil-borne plant pathogens. Disease-suppressive effects of composts have been investigated intensively over the past two decades, and were recently reviewed by Noble and Coventry (2005) and Zinati (2005). Compost of a wide variety of waste materials (hardwood or pine bark, municipal sludge, grape marc, or cattle manure) is an economically and ecologically sound alternative to pesticides.

The mechanisms of suppression in composts do not differ substantially from those described for soils, and can be either general or specific. Physiological profiling, and the use of DNA-based techniques such as denaturing gradient gel electrophoresis (DGGE) may lead to an improved understanding of the changes in microbial communities associated with disease control resulting from compost amendment of soil, sand, or peat. Nelson et al. (1983) identified specific strains of four *Trichoderma* spp. and isolates of *Gliocladium virens* as the most effective fungal hyperparasites of *R. solani* present in bark compost. A few of the 230 other fungal species also showed activity, but most were ineffective. Kwok et al. (1987) described synergistic interactions between *T. hamatum* and *Flavobacterium balustinum*. Several other bacterial strains, including *Enterobacter*, *Pseudomonas*, and *Xanthomonas* spp., also interacted with the *Trichoderma* isolate in suppression of *Rhizoctonia* damping-off (Kwok et al. 1987). Composted grape marc was effective in suppressing disease caused by *S. rolfsii* in beans and chickpeas (Gorodecki and Hadar 1990). Hadar and Gorodecki (1991) placed sclerotia of S. *rolfsii* on composted grape mare to isolate hyperparasites of this pathogen. Viability of sclerotia decreased from 100% to less than 10% within 40 h. It remained close to 100% for sclerotia placed on a conducive peat mix. *Penicillium* spp. and *Fusarium* spp. were observed by scanning electron microscopy to colonize the sclerotia. *Trichoderma* populations in the grape mare compost were at very low levels (10^2 cfu $g-^1$ dry weight). The hyperparasites present in this compost are therefore quite different from those isolated from tree bark compost, where *Trichoderma* and *Gliocladium* isolates predominate.

In conclusion, suppression of soil-borne plant pathogens in field soil or container media is brought about by antagonistic microorganisms. Such systems could be a source for mycoparasities to be used in biocontrol, or to be incorporated into integrated disease control programs. The inoculation of composts with biological control agents may improve the efficacy and reliability of disease control obtained.

III. Hyphal Interactions in Mycoparasitism

A. Biotrophs

Piptocephalis virginiana is a haustorial biotrophic mycoparasite that parasitizes fungi belonging to the order Mucorales exclusively (Manocha 1981). Attachment of a biotrophic mycoparasite to its host surface is considered to be an essential prerequisite step for further penetration of the host by the parasite (Manocha and Chen 1990). *P. virginiana* attaches to the surface of both the compatible *Choanephora cucurbitarum* and *Mortierella pusilla*, and the incompatible *Phascolomyces articulosus* hosts, but not to the surface of the nonhost *Mortierella candelabrum* (Manocha 1985; Manocha et al. 1986). Comparative research was performed by Manocha and his coworkers in an attempt to unravel the molecular basis for specificity and recognition in this system. Cytological and biochemical investigations were carried out to study the structure and chemical composition of cell walls of host and nonhost species (Manocha 1981, 1987). The germ tubes of the biotrophic mycoparasite *P. virginiana* were found to attach to the cell-wall surface of the host, but not to that of the nonhost (Manocha 1985; Manocha et al. 1986). This attachment could be specifically inhibited by chitobiose and chitotriose. The authors therefore suggested a possible involvement of carbohydrate-binding proteins in the specificity of this interaction. A comparison of protein and glycoprotein profiles of cell-wall extracts revealed marked differences between host and nonhost species. Two high-molecular-weight glycoproteins were observed only in the extract of host cell walls, being absent in that of the nonhost (Manocha 1985; Manocha et al. 1986). Further isolation and characterization of the host cell surface proteins revealed that attachment and appressorium formation by the parasite germ tubes could be inhibited by treating host cell-wall fragments with 0.1 M NaOH or pronase E. Furthermore, the two purified glycoproteins were able to agglutinate both nongerminated and germinated spores of the mycoparasite. Arabinose, glucose, and N-acetylglucosamine could totally inhibit this agglutination. These glycoproteins were suggested to be two subunits of a carbohydrate-binding agglutinin present on the host cell surface, and to be involved in agglutination and attachment of the mycoparasite germ tubes (Manocha and Chen 1991).

Using fluorescein isothiocyanate-labeled lectin-binding techniques, Manocha et al. (1990) were able to show differences in the distribution pattern of glycosyl residues at the level of the cell wall between fungi that are hosts and those that are nonhosts of the mycoparasite *P. virginiana*, and at the protoplast level between compatible and incompatible hosts.

The cell walls of the compatible hosts (*C. cucurbitarum* and *M. pusilla*) and the incompatible host (*P. articulosus*), as well as that of the mycoparasite itself, contain glucose and N-acetylglucosamine. In the nonhost (*M. candelabrum*), however, other sugars such as fucose, N-acetylgalactosamine, and galactose could also be detected. These latter sugars could be detected on both the host and the parasite surface after mild treatment with proteinase or when grown in liquid medium. The researchers speculated that the failure of the mycoparasite to attach to the host cells after proteinase treatment or in liquid culture may be due to the appearance of galactose and galactosamine at the host cell surface. The idea that N-acetylglucosamine and glucose may be involved in the attachment of *P. virginiana* to its host cell surface was supported by the observation that pretreatment of the mycoparasite germ tubes with N-acetylguclosamine or glucose inhibited their attachment to the host cells. In addition, the germ tubes attached to agarose beads coated with glucose or with N-acetylglucosamine, but not with N-acetylgalactosamine (Manocha et al. 1990).

The protoplast surfaces of compatible hosts contained all of the above-listed sugars, and these protoplasts could attach to the germ tube of the mycoparasite. Only lectins specific for N-acetylglucosamine and glucose were bound at the protoplast surface of the incompatible host; these protoplasts did not attach to the mycoparasite germ tubes. Indications were found for different factors being responsible for attachment and for appressorium formation, as pretreatment of the mycoparasite with glucose and N-acetylglucosamine inhibited its attachment to the host cell surface, but had no obvious effect on appressorium formation. On the other hand, appressorium formation was inhibited by heat treatment of host cell-wall fragments that still permitted attachment (Manocha et al. 1990). The authors therefore suggested a model for the recognition between *P. virginiana* and its host fungi that operates at two levels at least: the cell wall, and the protoplast surface. At the cell-wall level, the attachment probably involves carbohydrate-binding agglutinins that recognize specific sugar residues on the host but not on the nonhost cell wall. After the initial recognition and attachment, at the protoplast level, the parasite distinguishes compatible from incompatible hosts. The mechanism of this distinction is not clear. Yet, it seems that protoplast membrane sugars are not a major factor in recognition at this level (Manocha et al. 1990). Immunofluorescence microscopy was used to detect, in the mycoparasite *P. virginiana*, the presence of a complementary glycoprotein that binds specifically to the host cell surface glycoproteins. This technique revealed surface localization of the protein on the germ tubes of *P. virginiana*. Fluorescence was also observed at the surface of the germinated spores and hyphae of the host *M. pusilla*, after treatment with complementary protein from *P. virginiana*, and with primary antibody prepared against the complementary protein (Manocha et al. 1997).

B. Necrotrophs

As early as 1932, Weindling reported the coiling of *Trichoderma* spp. hyphae around hyphae of other fungi. These strains were later shown to actually be a species of *Gliocladium* (Webster and Lomas 1964). Dennis and Webster (1971a, b, c) published an extensive report on the antagonistic properties of species groups of *Trichoderma*. The hyphal interaction between *Trichoderma* and plant pathogenic fungi was first comprehensively studied in their work. Since then, numerous studies on the hyphal interaction and coiling phenomenon of *Trichoderma* around its host hyphae have been carried out with the use of light and electron microscopy (Chet et al. 1981; Elad et al. 1983a; Baker 1987; Inbar and Chet 1992; Omero et al. 1999; Rocha-Ramirez et al. 2002; Fig. 8.2).

The destructive mode of parasitism in *Trichoderma* appears to be a process consisting of several consecutive events initiated by attraction and directed growth of *Trichoderma* toward its host, probably by chemotropism. Positive chemotropism was found in *Trichoderma* (Chet et al. 1981), as it could detect its host from a distance and begin to branch in an atypical way. These branches grew toward the pathogenic host fungi. Similar behavior was also found in *Pythium nunn* (Lifshitz et al. 1984a), *P. oligandrum* (Lewis et al. 1989), and in *Gliocladium* spp. (Huang 1978). This event is presumably a response of the antagonist to the chemical gradient of an attractant coming from the host. However, no specific stimuli other than amino acids and simple sugars have thus far been detected (R. Barak and I. Chet, unpublished data). Hence, the specificity of the phenomenon is not clear. Apparently, it is not an essential step for mycoparasitism, although it may hold some advantage for

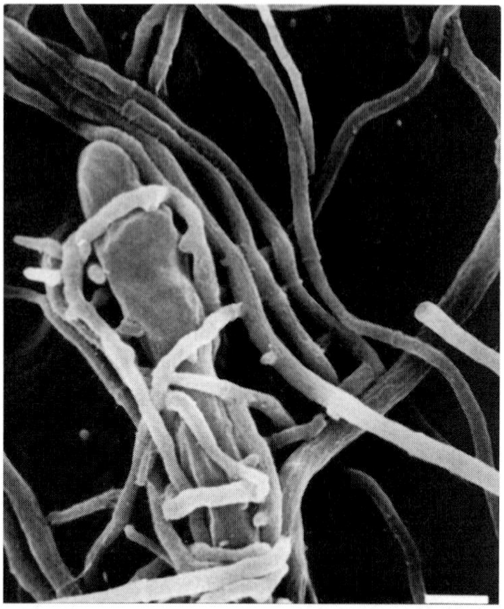

Fig. 8.2. A mycoparasitic relationship. Scanning electron micrograph of *T. harzianum* hyphae coiling around those of the plant pathogenic fungus *S. sclerotiorum*. *Bar* 10 μm

the antagonist. Subsequently, contact is made, and in some cases, *Trichoderma* coils around or grows along the host hyphae and forms hook-like structures, presumably appressoria, which probably aid in penetrating the host hyphal cell wall (Chet et al. 1981; Elad et al. 1983b). The coiling phenomenon and appressoria formation have been reported for other mycoparasites as well (Tu 1984; Lifshitz et al. 1984a). However, Deacon (1976) concluded that in the case of *P. oligandrum*, coiling of the antagonist around its host hyphae indicates temporary host resistance, rather than susceptibility. Nevertheless, in *Trichoderma*, this reaction was found to be rather specific, and *Trichoderma* attacks only a few fungi. Moreover, Dennis and Webster (1971c), using plastic threads of a diameter similar to that of *P. ultimum* hyphae, concluded that the coiling of the *Trichoderma* is not merely a thigmotropic response. The *Trichoderma* hyphae never coiled around the threads, but rather grew over or followed them in a straight course. This led to the idea that there is a molecular basis for the specificity. However, despite the fact that first observations and reports of this phenomenon were published decades ago, we are only now on the verge of being able to understand it. Reviews dealing with cellular interactions in fungi (Tunlid et al. 1992; Manocha and Sahai 1993), and the specificity of attachment of fungal parasites to their hosts (Manocha and Chen 1990) have been published. The physiology and biochemistry of biotrophic mycoparasitism in particular have been extensively reviewed by Manocha (1990).

Attachment and "recognition" between the mycoparasite and its host appears to be essential, and a crucial stage for successful continuation of the process. Lectins are sugar-binding proteins or glycoproteins of nonimmune origin that agglutinate cells and/or precipitate glycoconjugates (Goldstein et al. 1980). First discovered in plants and later in other organisms, they are involved in interactions between the cell surface and its extracellular environment (Barondes 1981). Indeed, lectins were found to be produced by some soil-borne plant pathogenic fungi such as *R. solani* and *S. rolfsii* (Barak et al. 1985, 1986), and by different members of the Sclerotiniaceae (Kellens et al. 1992).

Therefore, a role for lectins in the recognition and specificity of attachment between *Trichoderma* and its host fungi was suggested. However, no conclusive evidence to support this hypothesis was available at the time. In an attempt to test this hypothesis, Inbar and Chet (1992) used a novel approach based on the binding of lectins to a surface of nylon fibers. This biomimetic system imitates the host hyphae, and enables an examination of the role of lectins in mycoparasitism. Inert nylon fibers were chemically activated to enable the covalent binding of the lectins (Inbar and Chet 1992).

Concanavalin A, a plant lectin that is similar to the lectin of *S. rolfsii* (LSR) in its carbohydrate specificity (cf. they are both specific to D-glucose and D-mannose), was used first to establish the system. The *Trichoderma* recognized the LSR-treated fibers as a host, and attached and coiled around them in a pattern similar to that seen with real host hyphae (Inbar and Chet 1992; Fig. 8.3). In contrast, in the untreated control, no interaction could be observed – the *Trichoderma* grew uninterruptedly over and along the fibers, exactly as outlined by Dennis and Webster (1971c). These findings provided the first direct evidence for the role of lectins in mycoparasitism. The researchers were able to show that inert nylon fibers coated with fungal lectins mimic the real host hyphae, and can stimulate the parasite to coil around them.

A novel lectin was isolated and purified from the culture filtrate of the soil-borne plant pathogenic fungus *S. rolfsii* (Inbar and Chet 1994). Agglutination of *E. coli* cells by the purified lectin

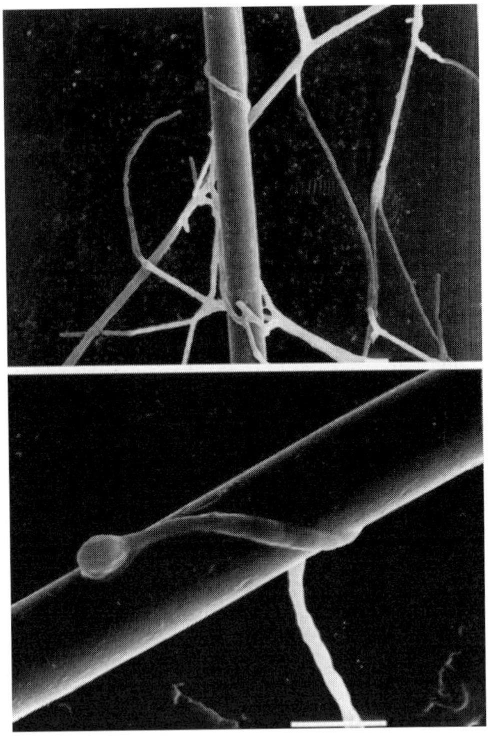

Fig. 8.3. Biomimetic systems for simulating the interaction between *Trichoderma* and plant pathogenic fungi. **a** Scanning electron micrograph of *T. harzianum* hyphae coiling around inert nylon fibers coated with a surface lectin from the plant pathogenic fungus *S. rolfsii*. Bar 10 μm. **b** Appressorium formation by *T. harzianum* grown on nylon fibers coated with the *S. rolfsii* lectin. *Bar* 10 μm

could be inhibited by the glycoproteins mucin and asialomucin. Proteases, as well as β-1,3-glucanase, were found to be totally destructive to the agglutination activity, indicating that both protein and β-1,3-glucan are necessary for agglutination. Using the biomimetic system, it was apparent that the presence of the purified agglutinin of the surface of the fibers significantly induces mycoparasitic behavior in *T. harzianum*, compared with the untreated ones or with those treated with nonagglutinating extracellular proteins from *S. rolfsii* (Inbar and Chet 1994).

It was later demonstrated that induction of chitinolytic enzymes in *Trichoderma* is elicited by the recognition signal (i.e., lectin–carbohydrate interactions). It was postulated that recognition is the first step in a cascade of antagonistic events triggering the parasitic response in *Trichoderma* (Inbar and Chet 1995).

The same biomimetic system was used to test the involvement of signal transduction pathways in the induction of coils in *T. harzianum* (Omero et al. 1999). Two activators of G protein-mediated signal transduction induced coiling of hyphae around nylon fibers. The peptide toxin mastoparan increased coiling more than twofold in comparison with controls. The activator fluoroaluminate (A1F4) had a similar effect, whereas aluminum ions alone were ineffective; cAMP increased coiling about threefold. Although the two G-protein activators, mastoparan and fluoroaluminate, have very different modes of action, they share the Ga subunit as a target. Based on these results, it was proposed that a signal for mycoparasitic behavior from the host cell surface is transduced by heterotrimeric G protein(s) and mediated by cAMP.

Rocha-Ramirez et al. (2002) isolated a *T. atroviride* G-protein alpha-subunit (Ga) gene (*tga1*). Transgenic lines overexpressing *tga1* showed a delayed sporulation and coiled at a higher frequency, compared to the wild type. Likewise, transgenic lines that expressed an activated mutant protein with no GTPase activity did not sporulate and coiled at a higher frequency. Lines that expressed an antisense version of the gene were hypersporulating and coiled at a much lower frequency in the biomimetic assay. The loss of *tga1* in these mutants correlated with the loss of GTPase activity stimulated by the peptide toxin Mas-7. The application of Mas-7 to growing mycelia raises intracellular cAMP levels, suggesting that *tga1* can activate adenylyl cyclase. In contrast, cAMP levels and cAMP-dependent protein kinase activity drop when diffusible host signals are encountered and the mycoparasitism-related genes *ech42* and *prb1* are highly expressed. These results demonstrated that the product of the *tga1* gene is involved in both coiling and conidiation.

Penetration and degradation of the host cell wall under the coiling and interaction sites are evident by visual observation, fluorescent indicators, and enzymatic studies. Using scanning electron microscopy, lysed sites and penetration holes were found in hyphae of *R. solani* and *S. rolfsii* following removal of *Trichoderma* spp. hyphae (Elad et al. 1983b). The cell walls of Basidiomycota and Ascomycota contain chitin and laminarin (P glucan) but no cellulose. Oomycota contain β-glucans and cellulose and relatively small amounts of chitin (<1.5%). Therefore, to penetrate the host cell wall, mycoparasites should have a system of hydrolytic enzymes that can degrade these components. Enzymatic degradation of fungal cell walls occurs mainly via the excretion of the extracellular enzymes β-1-3-glucanase and chitinase. Indeed, high β-1-3-glucanase and chitinase activities were detected in dual cultures when *T. harzianum* parasitized *S. rolfsii*, contrasting with the low levels found with either fungus alone. Cycloheximide prevented antagonism, and enzymatic activity was diminished (Elad et al. 1983b). Using gold cytochemistry, *T. harzianum* hyphae were shown to coil around and penetrate cells of *R. solani*, causing extensive damage such as cell-wall alteration, plasma membrane retraction, and cytoplasm aggregation (Benhamou and Chet 1993).

The involvement and importance of lytic enzymes, mainly chitinase, in the biological control of plant pathogens by both fungi and rhizobacterial agents (Ordentlich et al. 1988; Inbar and Chet 1991; Sahai and Manocha 1993; Viterbo et al. 2002), as well as their involvement in the defense of plants against pathogenic infection (Boller 1985; Broglie et al. 1991) is well documented.

IV. Molecular Aspects and Genetic Engineering in Mycoparasitism

Trichoderma is one the most frequently used biocontrol agents in agriculture. The role of lytic enzymes in its mycoparasitic activity has recently been largely reviewed (Viterbo et al. 2002; Benitez et al. 2004). The sensing of the host in the *Trichoderma* mycoparasitic interaction and gene activation has been the subject of extensive studies in the last few years (Inbar and Chet 1995; Zeilinger et al. 1999; Brunner et al. 2003). The pattern of induction of different cell wall-degrading enzymes differs from one *Trichoderma* strain to another. It is believed that *Trichoderma* secrete exochitinases constitutively at low levels. When chitinases degrade fungal cell walls, they release oligomers that induce other chitinases, and attack begins.

In the last decade, the significance of several newly isolated lytic enzymes has been demonstrated by overexpression and deletion of the respective genes (Pozo et al. 2004; Hoell et al. 2005). Molecular approaches and genetic engineering techniques have been applied to gain a better and more basic understanding of the system, as well as to develop superior and improved strains of biocontrol agents with enhanced activity (Mendoza et al. 2003; Brunner et al. 2005). The chitinase gene *chiA*, encoding one of the chitinases from *Serratia marcescens*, a well-known biocontrol agent, was isolated and cloned into *E. coli* (Shapira et al. 1989). *E. coli* transformed by the *chiA* gene, under the oLpL operator and promoter of bacteriophage, expressed and excreted the corresponding protein into the growth medium. Almost pure *S. marcescens* chitinase from *E. coli* or whole viable cells were used in greenhouse experiments against *S. rolfsii* in beans, and *R. solani* in cotton. Using the chitinase preparation in the irrigation water effectively reduced the number of diseased plants. Whole viable cells of transformed *E. coli* were also effective in inhibiting *S. rolfsii*, but to a lesser degree. The genetically engineered *E. coli*, a nonsoil bacterium, served here as a model system to demonstrate the role of chitinase in controlling a chitin-containing plant pathogen.

It is suggested that the introduction of such engineered genes into soil bacteria will increase control efficiency by combining high expression of a gene coding for a lytic enzyme with rhizosphere competence. Southern blot analysis of the *chi*A gene cloned from *S. marcescens* showed homology to one of the *Trichoderma* chitinase genes. Based on this, the *chiA* gene was used as a probe to isolate a chitinase gene from a cDNA library prepared from *T. harzianum* (T-35) grown on chitin (Chet et al. 1993). The chitinase gene from T. *harzianum* (T-35) was cloned in a Bluescript plasmid under the lac promoter. When the transformed *E. coli* was plated on LB+0.2% chitin plates and induced by 1 mM IPTG, the bacteria showed chitinolytic activity. In greenhouse experiments, irrigation of bean seedlings with 10^7 cfu g^{-1} soil day^{-1} of *E. coli* XLIBlue, transformed with the *Trichoderma* chitinase gene induced by 1 mM IPTG, resulted in significant biocontrol activity. Suppression of the disease caused by *S. rolfsii* was obvious. The treated plants exhibited a better growth rate than untreated controls. After 18 days, the growth rate of the plants irrigated with the transformed bacteria was similar to that of uninfected plants (Chet et al. 1993).

In an attempt to increase its effectiveness, *T. harzianum* protoplasts were cotransformed using two plasmids: pSL3chiAII, containing a bacterial chitinase gene from *S. marcescens* under the control of a constitutive viral promotor, and p35SR2, a marker for selection after transformation, encoding for acetamidase. Two transformants showed increased constitutive chitinase activity (specific activity 11 and 5 times higher than the recipient; Fig. 8.4), and excreted a protein of ca. 58 kDa, the expected size of S. *marcescens* chitinase, when grown on synthetic medium. Antagonistic activity of the transformants was significantly higher than that of the wild-type *T. harzianum*, as evaluated by testing their ability to overgrow the plant pathogen *S. rolfsii* in dual culture (Haran et al. 1993). The major advantage of such genetic manipulations is the ability to isolate genes from one strain and introduce them into other varieties of fungi, bacteria, or plants. This enhances the potency of biocontrol agents and makes a single strain consistently effective against more than one plant pathogenic fungus, without the hazardous effects of chemical pesticides.

This approach was taken by Broglie et al. (1991) who, in a pioneering work, produced seedlings constitutively expressing a bean chitinase gene under the control of the cauliflower mosaic virus 35S promoter. The timing of the natural host defense mechanism was modified to produce fungus-resistant plants with increased ability to survive in soil infested with the fungal pathogen *R. solani*, delaying the development of disease symptoms.

Since then, genetic manipulations of valuable crop plants with one or more cell wall-degrading enzymes from mycoparasitic fungi have been considered a potent tool for improving plant resistance

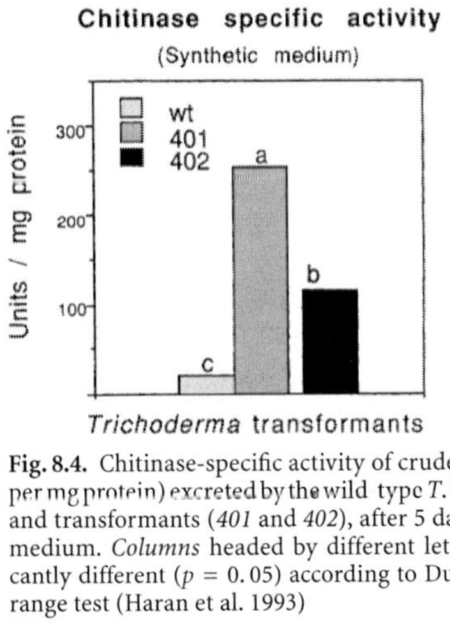

Fig. 8.4. Chitinase-specific activity of crude enzyme (units per mg protein) excreted by the wild type *T. harzianum* (*wt*) and transformants (*401* and *402*), after 5 days on synthetic medium. *Columns* headed by different letters are significantly different ($p = 0.05$) according to Duncan's multiple range test (Haran et al. 1993)

to fungal pathogens. Transgenic apple plants expressing *T. atroviride* endochitinase and exochitinase, singly or in combination, were produced and screened for resistance to *Venturia inaequalis*, the causal agent of apple scab (Bolar et al. 2001). Plants expressing both enzymes at the same time were more resistant, demonstrating for the first time in planta synergism between the two enzymes. Constitutive expression of *Trichoderma* endochinase can be exploited to enhance resistance to fungal pathogens in important forest tree species. The *ech42* from *T. harzianum* was introduced into forest trees, black spruce (*Picea mariana*) and hybrid poplar (*Populus nigra* x *Populus maximowiczii*), by *Agrobacterium*-mediated transformation. In vitro assays demonstrated that the transgenic poplars had increased resistance to the leaf rust pathogen *Melampsora medusae*. Seedlings of transgenic spruce lines showed an increased resistance to the spruce root pathogen *Cylindrocladium floridanum* (Noël et al. 2006).

V. Conclusions

Mycoparasitism is a quite common, and yet exciting phenomenon. It appears to play an important role in biological control, even though it should be pointed out that mycoparasitism is only one specific aspect in the whole complex system of biological control of plant diseases. For example, the direct effects of root-colonizing *Trichoderma* spp. on plants are at least as important as the direct effects on pathogens – or perhaps more so. These fungi have profound impacts on plant growth and development, and they also induce resistance to a variety of classes of plant pathogens.

Mycoparasitism is a complex process that includes the following steps: (1) chemotrophic growth of the antagonist toward the host; (2) recognition of the host by the mycoparasite; (3) attachment; (4) excretion of extracellular enzymes; and (5) lysis and exploitation of the host.

Mycoparasitism occurs under appropriate ecological conditions. The population and activity of the mycoparasite can be increased by relatively specific substances, such as chitin. The gene coding for chitinase is only one example of genes with mycoparasitic activity. Other potential genes are those coding for β-1,3-glucanase, protease, and lipase.

Engineering various chitinases together with other genes that may act as antifungal agents may lead to better protection of plants against pathogenic fungi. It may therefore be possible to improve mycoparasitism, and to enhance the plants resistance response by integrating cloned chitinase with different lytic enzymes and other available antifungal polypeptides. By understanding the mode of action of biocontrol mycoparasitic fungi, we should be able to manipulate the fungal agent, the plant, and their interactions to achieve more effective and safer plant resistance to various biotic and abiotic stresses.

Acknowledgements. This work was supported by *The Dr. Alexander and Myrna Strelinger Endowment Fund, The Judith and Avraham Goldwasser* Foundation, and a United States-Israel Binational Agricultural Research and Development Fund (BARD) (US-3507-04).

References

Abo-Foul S, Raskin VI, Sztejnberg A, Marder JB (1996) Disruption of chlorophyll organization and function in powdery mildew-diseased cucumber leaves and its control by the hyperparasite *Ampelomyces quisqualis*. Phytopathology 86:195–199

Adams PB (1986) Effect of soil temperature, moisture, and depth on survival and activity of *Sclerotinia minor, Sclerotium cepivorum* and *Sporidesmium sclerotivorum*. Plant Disease 71:170

Adams PB (1990) The potential of mycoparasites for biological control of plant diseases. Annu Rev Phytopathol 28:59–72

Adams PB, Ayers WA (1980) Factors affecting parasitic activity of *Sporidesmium sclerotivorum* on sclerotia of *Sclerotinia minor* in soil. Phytopathology 70:366–368

Adams PB, Ayers WA (1982) Biological control of *Sclerotinia* lettuce drop in the field by *Sporidesmium sclerotivorum*. Phytopathology 72:485–488

Adams PB, Marois JJ, Ayers WA (1984) Population dynamics of the mycoparasite, *Sporidesmium sclerotivorum*, and its host, *Sclerotinia minor* in soil. Soil Biol Biochem 16:627–633

Ahmad JS, Baker R (1987) Rhizosphere competence of *Trichoderma harzianum*. Phytopathology 77:182–189

Ayers WA, Adams PB (1981) Mycoparasitism and its application to biological control of plant diseases. In: Papavizas GC (ed) Biological control in crop protection, vol 15. Allanheld, Osmun Totowa, NJ, pp 91–105

Ayers WA, Barnett EA, Adams PB (1981) Germination of macroconidia and growth of *Sporidesmium sclerotivorum* in vitro. Can J Microbiol 27:664–669

Baek JM, Howell CR, Kenerley CM (1999) The role of an extracellular chitinase from Trichoderma virens Gv29-8 in the biocontrol of *Rhizoctonia solani*. Curr Genet 35:41–50

Baker R (1987) Mycoparasitism: ecology and physiology. Can J Plant Pathol 9:370–379

Baker KF, Cook RJ (1974) Biological control of plant pathogens. The American Phytopathological Society, St Paul, MN

Baker R, Griffin GJ (1995) Molecular strategies for biological control of fungal plant pathogens. In: Reuveni R (ed) Novel approaches to integrated pest management. Lewis, CRC Press, Boca Raton, FL, pp 153–182

Barak R, Elad Y, Mirelman D, Chet I (1985) Lectins: a possible basis for specific recognition in *Trichoderma-Sclerotium rolfsii* interaction. Phytopathology 75:458–462

Barak R, Elad Y, Chet I (1986) The properties of L-fucose binding agglutinin associated with the cell wall of *Rhizoctonia solani*. Arch Microbiol 144:346–349

Barnett HL, Binder FL (1973) The fungal host-parasite relationship. Annu Rev Phytopathol 11:273–292

Barondes SH (1981) Lectins: their multiple endogenous cellular functions. Annu Rev Biochem 50:207–231

Benhamou N, Chet I (1993) Hyphal interactions between *Trichoderma harzianum* and *Rhizoctonia solani*: ultrastructure and cytochemistry of the antagonistic process. Phytopathology 83:1062–1071

Benitez T, Rincon AM, Limon MC, Codon AC (2004) Biocontrol mechanisms of *Trichoderma* strains. Int Microbiol 7:249–60

Bolar JP, Norelli JL, Harman GE, Brown SK, Aldwinckle HS (2001) Synergistic activity of endochitinase and exochitinase from *Trichoderma atroviride* (*T. harzianum*) against the pathogenic fungus (*Venturia inaequalis*) in transgenic apple plants. Transgenic Res 10:533–43

Boller T (1985) Induction of hydrolases as a defence reaction against pathogens. In: Key JL, Kosuge T (eds) Cellular and molecular biology of plant stress. Alan R Liss, New York, pp 247–262

Boosalis MG (1956) Effect of soil temperature and green-manure amendment of unsterilized soil on parasitism of *Rhizoctonia solani* by *Penicillium vermiculatum* and *Trichoderma* spp. Phytopathology 46:473–478

Broglie K, Chet I, Holliday M, Cressman R, Biddle P, Knowlton S, Mauvais CJ, Broglie R (1991) Transgenic plants with enhanced resistance to the fungal pathogen *Rhizoctonia solani*. Science 254:1194–1197

Brunner K, Peterbauer CK, Mach RL, Lorito M, Zeilinger S, Kubicek CP (2003) The Nag1 N-acetylglucosaminidase of *Trichoderma atroviride* is essential for chitinase induction by chitin and of major relevance to biocontrol. Curr Genet 43:289–95

Brunner K, Zeilinger S, Ciliento R, Woo SL, Lorito M, Kubicek CP, Mach RL (2005) Improvement of the fungal biocontrol agent *Trichoderma atroviride* to enhance both antagonism and induction of plant systemic disease resistance. Appl Environ Microbiol 71:3959–3965

Bryan PW, McGowan JC (1945) Viridin: a highly fungistatic substance produced by *Trichoderma viride*. Nature 151:144–145

Bullock S, Adams PB, Willetts HJ, Ayers WA (1986) Production of haustoria by *Sporidesmiicm sclerotivorurn* in sclerotia of *Sclerotinia minor*. Phytopathology 76:101–103

Chet I (1990) Mycoparasitism – recognition, physiology and ecology. In: Baker R, Dunn P (eds) New directions in biological control: alternatives for suppressing agricultural pests and diseases. Alan R Liss, New York, pp 725–733

Chet I, Baker R (1981) Isolation and biocontrol potential of *Trichoderma hamatum* from soil naturally suppressive to *Rhizoctonia solani*. Phytopathology 71:286–290

Chet I, Harman GE, Baker R (1981) *Trichoderma hamatum*: its hyphal interactions with *Rhizoctonia solani* and *Pythium* spp. Microbial Ecol 7:29–38
Chet I, Barak Z, Oppenheim A (1993) Genetic engineering of microorganisms for improved biocontrol activity. In: Chet I (ed) Biotechnology in plant disease control. Wiley-Liss, New York, pp 211–235
Cook RJ, Baker KF (1983) The nature and practice of biological control of plant pathogens. American Phytopathological Society, St Paul, MN
Deacon JW (1976) Studies on *Pythium oligandrum*, an aggressive parasite of other fungi. Trans Br Mycol Soc 66:383–391
Deacon JW (1991) Significance of ecology in the development of biocontrol agents against soil-borne plant pathogens. Biocontrol Sci Technol 1:5–20
Del Rio LE, Martinson CA, Yang XB (2002) Biological control of *Sclerotinia* stem rot of soybean with *Sporidesmium sclerotivorum*. Plant Disease 86:999–1004
Dennis C, Webster J (1971a) Antagonistic properties of species-groups of *Trichoderma*. I. Production of non volatile antibiotics. Trans Br Mycol Soc 57:25–39
Dennis C, Webster J (1971b) Antagonistic properties of species-groups of *Trichoderma*. II. Production of volatile antibiotics. Trans Br Mycol Soc 57:41–48
Dennis C, Webster J (1971c) Antagonistic properties of species-groups of *Trichoderma*. III. Hyphal interaction. Trans Br Mycol Soc 57:363–369
Druzhinina I, Kubicek CP (2005) Species concepts and biodiversity in *Trichoderma* and *Hypocrea*: from aggregate species to species clusters? J Zhejiang Univ Sci B 6:100–112
Duo-Chuan LI, Chen S, Jing LU (2005) Purification and partial characterization of two chitinases from the mycoparasitic fungus *Talaromyces flavus*. Mycopathologia. 159:223–229
Elad Y, Chet I (1995) Practical approaches for biocontrol agents implementation. In: Reuveni R (ed) Novel approaches to integrated pest management. Lewis, CRC Press, Boca Raton, FL, pp 323–338
Elad Y, Barak R, Chet I, Henis Y (1983a) Ultrastructural studies of the interaction between *Trichoderma* spp. and plant pathogenic fungi. Phytopathol Z 107:168–175
Elad Y, Chet I, Boyle P, Henis Y (1983b) Parasitism of *Trichoderma* spp. on *Rhizoctonia solani* and *Sclerotium* scanning electron microscopy and fluorescence microscopy. Phytopathology 73:85–88
Elad Y, Lifshitz R, Baker R (1985) Enzymatic activity of the mycoparasite *Pythium nunn* during interaction with host and non-host fungi. Physiol Plant Pathol 27:131
Elad Y, Zvieli Y, Chet I (1986) Biological control of *Macrophomina phaseolina* (Tassi) by *Trichoderma harzianum*. Crop Protect 5:288–292
Fahima T, Henis Y (1990) Interaction between pathogen, host and biocontrol agent: multiplication of *Trichoderma hamatum* and *Talaromyces flavus* roots of diseased and healthy hosts. In: Hornby D (ed) Biological control of soil-borne plant pathogens. CABI, Wallingford, pp 165–180
Fahima T, Madi L, Henis Y (1992) Ultrastructure and germinability of *Verticillium* microsclerotia parasitized by *Talaromyces flavus* on agar medium and in treated soil. Biocontrol Sci Technol 2:69–78
Fravel DR, Keinath AP (1991) Biocontrol of soilborne plant pathogens with fungi. In: Keister DL, Cregan PB (eds) The rhizosphere and plant growth. Kluwer, Dordrecht, pp 237–243
Fravel DR, Connick WJ Jr, Grimm CC, Lloyd SW (2002) Volatile compounds emitted by sclerotia of *Sclerotinia minor*, *Sclerotinia sclerotiorum*, and *Sclerotium rolfsii*. J Agric Food Chem 50:3761–3764
Freeman S, Sztejnberg A, Chet I (1986) Evaluation of *Trichoderma* as a biocontrol agent for *Rosellinia necatrix*. Plant Soil 94:163–170
Gams WX, Bisset JX (1998) Morphology and identification of *Trichoderma*. In: Kubicek CP, Harman GE (eds) *Trichoderma* and *Gliocladium*. Taylor and Francis, Padstow, pp 3–25
Garrett SD (1965) Toward biological control of soil-borne plant pathogens. In: Baker KF, Synder WC (eds) Ecology of soil-borne plant pathogens. University of California Press, Berkeley, CA, pp 4–17
Ghisalberti EL, Sivasithamparam K (1991) Antifungal antibiotics produced by *Trichoderma* spp. Soil Biol 23:1011–1020
Goldstein IJ, Hughes RC, Monsigny M, Osawa T, Sharon N (1980) What should be called a lectin? Nature 285:66
Gorodecki B, Hadar Y (1990) Suppression of *Rhizoctonia solani* and *Sclerotium rolfsii* in container media containing composted separated cattle manure and composted grape marc. Crop Protect 9:271
Haas D, Defago G (2005) Biological control of soil-borne pathogens by fluorescent pseudomonads. Nature Rev Microbiol 3(4):307–319
Hadar Y, Gorodecki B (1991) Suppression of germination of sclerotia of *Sclerotium rolfsii* in compost. Soil Biol Biochem 23:303–306
Handelsman J, Parke JL (1989) Mechanism in biocontrol of soilborne plant pathogens. In: Kosuge T, Nester EW (eds) Plant microbe interactions, vol 3. McGraw-Hill, New York, pp 27–61
Haran S, Schickler H, Pe'er S, Logemann S, Oppenheim A, Chet I (1993) Increased constitutive chitinase activity in transformed *Trichoderma harzianum*. Biol Control 3:101–108
Harman GE (2006) Overview of mechanisms and uses of *Trichoderma* spp. Phytopathology 96:190–194
Harman GE, Lumsden RD (1990) Biological disease control. In: Lynch JM (ed) The rhizosphere. Wiley, New York, pp 259–280
Harman GE, Chet I, Baker R (1980) *Trichoderma hamatum* effects on seed and seedling disease induced in radish and pea by *Pythium* spp. or *Rhizoctonia solani*. Phytopathology 70:1167–1172
Harman GE, Howell CR, Viterbo A, Chet I, Lorito M (2004) *Trichoderma* species–opportunistic, avirulent plant symbionts. Nature Rev Microbiol 2(1):43–56
Hashioka Y, Nakai Y (1980) Ultrastructure of pycnidial development and mycoparasitism of *Ampelomyces quisqualis* parasitic on *Erysiphales*. Trans Mycol Soc Jpn 21:329 338
Henis Y, Ghaffar A, Baker R (1978) Integrated control of *Rhizoctonia solani* damping-off of radish: effect of successive plantings, PCNB, and *Trichoderma harzianum* on pathogen and disease. Phytopathology 68:900–907

Hoell IA, Klemsdal SS, Vaaje-Kolstad G, Horn SJ, Eijsink VG (2005) Overexpression and characterization of a novel chitinase from *Trichoderma atroviride* strain P1. Biochim Biophys Acta 15:180–190
Howell CR (1982) Effect of *Gliocladium virens* on *Pythium ultimum*, *Rhizoctonia solani* and damping-off of cotton seedling. Phytopathology 72:496–498
Howell CR (1987) Relevance of mycoparasitism in the biological control of *Rhizoctonia solani* by *Gliocladium virens*. Phytopathology 77:992–994
Howell CR (1991) Biological control of *Pythium* damping-off of cotton with seed coating preparation of *Gliocladium virens*. Phytopathology 81:738–741
Howell CR (1998) The role of antibiosis in biocontrol. In: Harman GE, Kubicek CP(eds) *Trichoderma* and *Gliocladium*, vol 2. Taylor and Francis, Padstow, pp 173–184
Howell CR, Stipanovic RD (1983) Gliovirin, a new antibiotic from *Gliocladium virens* and its role in the biological control of *Pythium ultimum*. Can J Microbiol 29:321–324
Howell CR, Stipanovic RD, Lumsden RD (1993) Antibiotic production by strains of *Gliocladium virens* and its relation to the biocontrol of cotton seedling diseases. Biocontrol Sci Technol 3:435–441
Huang HC (1977) Importance of *Coniothyrium minitans* in survival of sclerotia of *Sclerotinia sclerotiorum* in wilted sunflower. Can J Bot 55:289–295
Huang HC (1978) *Gliocladium catenulatum:* hyperparasite of *Sclerotinia sclerotiorum* and *Fusarium* species. Can J Bot 56:2243–2246
Huang HC (1980) Control of *Sclerotinia* wilt of sunflower by hyperparasites. Can J Plant Pathol 2:26–32
Huang HC, Hoes JA (1976) Penetration and infection of *Sclerotinia sclerotiorum* by *Coniothyrium minitans*. Can J Bot 54:406–410
Huang HC, Kokko EG (1987) Ultrastructure of hyperparasitism of *Coniothyrium minitans* on sclerotia of *Sclerotinia sclerotiorum*. Can J Bot 65:2483–2489
Huang HC, Kokko EG (1988) Penetration of hyphae of *Sclerotinia sclerotiorum* by *Coniothyrium minitans* without the formation of appressoria. J Phytopathol 123:133–139
Inbar J, Chet I (1991) Evidence that chitinase produced by *Aeromonas caviae* is involved in the biological control of soil plant pathogens by this bacterium. Soil Biol Biochem 23:973–978
Inbar J, Chet I (1992) Biomimics of fungal cell-cell recognition by use of lectin-coated nylon fibers. J Bacteriol 174:1055–1059
Inbar J, Chet I (1994) A newly isolated lectin from the plant pathogenic fungus *Sclerotium rolfsii*: purification, characterization and its role in mycoparasitism. Microbiology 140:651–657
Inbar J, Chet I (1995) The role of recognition in the induction of specific chitinases during mycoparasitism by *Trichoderma harzianum*. Microbiology 141:2823–2829
Jeffries P (1997) Mycoparasitism. In: Wicklow DT, Söderström B (eds) The Mycota vol IV, 1st edn. Environmental and microbial relationships. Springer, Berlin Heidelberg New York, pp 149–164
Kaur J, Munshi GD, Singh RS, Koch E (2005) Effect of carbon source on production of lytic enzymes by the sclerotial parasites *Trichoderma atroviride* and *Coniothyrium minitans*. J Phytopathol 153:274–279
Kellens JTC, Goldstein IJ, Peumans WJ (1992) Lectins in different members of the Sclerotiniaceae. Mycol Res 96:495–502
Kritzman G, Chet I, Henis Y, Huttermann A (1978) The use of the brightener Calcofluor White M2R New in the study of fungal growth. Isr J Bot 27:138–146
Kwok OCH, Fahy PC, Hoitink HAJ, Kuter FA (1987) Interaction between bacteria and *Trichoderma hamatum* in suppression of *Rhizoctonia* damping-off in bark compost media. Phytopathology 77:1206–1212
Lewis K, Whipps JM, Cooke RC (1989) Mechanisms of biological disease control with special reference to the case study of *Pythium oligandrum* as an antagonist. In: Whipps JM, Lumsden RD (eds) Biotechnology of fungi for improving plant growth. Cambridge University Press, Cambridge, pp 191–217
Lewis JA, Roberts DP, Hollenbeck MD (1991) Induction of cytoplasmic leakage from *Rhizoctionia solani* hyphae by *Gliocladium virens* and partial characterization of a leakage factor. Biocontrol Sci Technol 1:21–29
Lifshitz R, Dupler M, Elad Y, Baker R (1984a) Hyphal interactions between *Pythium nunn* and several soil fungi. Can J Microbiol 30:1482–1487
Lifshitz R, Stanghellini ME, Baker R (1984b) A new species of *Pythium* isolated from soil in Colorado. Mycotaxon 20:373–379
Liu SY, Baker R (1980) Mechanism of biological control in soil suppressive to *Rhizoctonia solani*. Phytopathology 70:404–412
Manocha MS (1981) Host specificity and mechanism of resistance in a mycoparasitic system. Physiol Plant Pathol 18:257–267
Manocha MS (1985) Specificity of mycoparasite attachment to the host cell surface. Can J Bot 63:772–778
Manocha MS (1987) Cellular and molecular aspects of fungal host-mycoparasite interaction. J Plant Diseases Protect 94:431–444
Manocha MS (1990) Cell-cell interaction in fungi. J Plant Diseases Protect 97:655–669
Manocha MS, Chen Y (1990) Specificity of attachment of fungal parasites to their hosts. Can J Microbiol 36:69–76
Manocha MS, Chen Y (1991) Isolation and partial characterization of host cell surface agglutinin and its role in attachment of a biotrophic mycoparasite. Can J Microbiol 37:377–383
Manocha MS, Sahai AS (1993) Mechanisms of recognition in necrotrophic and biotrophic mycoparasites. Can J Microbiol 39:269–275
Manocha MS, Balasubramanian R, Enskat (1986) Attachment of a mycoparasite with host but not with nonhost *Mortierella* species. In: Bailey J (ed) Biology and molecular biology of plant pathogen interactions. Springer, Berlin Heidelberg New York, pp 59–69
Manocha MS, Chen Y, Rao N (1990) Involvement of cells surface sugars in recognition attachment and appressorium formation by a mycoparasite. Can J Microbiol 36:771–778
Manocha MS, Xiong D, Govindsamy V (1997) Isolation and partial characterization of a complementary protein from the mycoparasite *Piptocephalis virginiana* that specifically binds to two glycoproteins at the host cell surface. Can J Microbiol 43:625–632

McLaren DL, Huang HC, Rimmer SR (1986) Hyperparasitism of *Sclerotinia sclerotiorum* by *Talaromyces flavus*. Can J Plant Pathol 8:43–48

Mendoza-Mendoza A, Pozo MJ, Grzegorski D, Martinez P, Garcia JM, Olmedo-Monfil V, Cortes C, Kenerley C, Herrera-Estrella A (2003) Enhanced biocontrol activity of *Trichoderma* through inactivation of a mitogen-activated protein kinase. Proc Natl Acad Sci USA 23:15965–15970

-Nelson EB, Kuter GA, Hoitink HAJ (1983) Effect of fungal antagonists and compost age on suppression of *Rhizoctonia* damping-off in container media amended with composted hardwood bark. Phytopathology 73:1457–1462

Noble R, Coventry E (2005) Suppression of soil-borne plant diseases with composts: a review. Biocontrol Sci Technol 15:3–20

Noël A, Levasseur C, Le VQ, Séguin A (2006) Enhanced resistance to fungal pathogens in forest trees by genetic transformation of black spruce and hybrid poplar with a *Trichoderma harzianum* endochitinase gene. Physiol Mol Plant Pathol 67:92–99

Omero C, Inbar J, Rocha-Ramirez V, Herrera-Estrella A, Chet I, Horwitz BA (1999) G protein activators and cAMP promote mycoparasitic behaviour in *Trichoderma harzianum*. Mycol Res 103:1637–1642

-Ordentlich A, Elad Y, Chet I (1988) The role of chitinase of *Serratia marcescens* in biocontrol of *Sclerotium rolfsii*. Phytopathology 78:84–88

Ordentlich A, Nachmias A, Chet I (1990) Integrated control of *Verticillium dahliae* in potato by *Trichoderma harziccnum* and captan. Crop Protect 9:363–366

Pachenari A, Dix NJ (1980) Production of toxins and wall degrading enzymes by *Gliocladium roseum*. Trans Br Mycol Soc 74:561–566

Papavizas GC (1985) *Trichoderma* and *Gliocladium:* biology, ecology and the potential for biocontrol. Annu Rev Phytopathol 23:23–54

Phillips AJL (1986) Factors affecting the parasitic activity of *Gliocladium virens* on the sclerotia of *Sclerotinia sclerotiorum* and a note on its host range. J Phytopathol 116:212–220

Phillips AJL, Price K (1983) Structural aspects of the parasitism of sclerotia of *Sclerotinia sclerotiorum* (Lib.) de Bary by *Coniothyrium minitans* Campb. Phytopathol Z 107:193–203

Pozo MJ, Baek JM, Garcia JM, Kenerley CM (2004) Functional analysis of *tvsp1*, a serine protease-encoding gene in the biocontrol agent *Trichoderma virens*. Fungal Genet Biol 41:336–348

Roberts DP, Lumsden RD (1990) Effect of extracellular metabolites from *Gliocladium virens* on germination of sporangia and mycelial growth of *Pythium ultimum*. Phytopathology 80:461–465

Rocha-Ramirez V, Omero C, Chet I, Horwitz BA, Herrera-Estrella A (2002) *Trichoderma atroviride* G-protein alpha-subunit gene tga1 is involved in mycoparasitic coiling and conidiation. Eukaryot Cell 1:594–605

Sahai AS, Manocha MS (1993) Chitinases of fungi and plants: their involvement in morphogenesis and host-parasite interaction. FEMS Microbiol Rev 11:317–338

Shapira R, Ordentlich A, Chet I, Oppenheim AB (1989) Control of plant diseases by chitinase expressed from cloned DNA in *Escherichia* coli. Phytopathology 79:1246–1249

Shoresh M, Yedidia I, Chet I (2005) Involvement of jasmonic acid/ethylene signaling pathway in the systemic resistance induced in cucumber by *Trichoderma asperellum* T203. Phytopathology 95:76–84

Shoresh M, Gal-On A, Leibman D, Chet I (2006) Characterization of a mitogen-activated protein kinase gene from cucumber required for trichoderma-conferred plant resistance. Plant Physiol 142:1169–1179

Simon A, Sivasithamparam K (1990) Effect of crop rotation, nitrogenous fertilizer and lime on biological suppression of the take-all fungus. In: Hornby D (ed) Biological control of soil-borne plant pathogens. CABI, Wallingford, pp 215–226

Sivan A, Chet I (1986) Biological control of *Fusarium* spp. in cotton, wheat and muskmelon by *Trichoderma harzianum*. Phytopathol Z 116:39–47

Sivan A, Chet I (1989) The possible role of competition between *Trichoderma harzianum* and *Fusarium oxysporum* on rhizosphere colonization. Phytopathology 79:198–203

Sivan A, Elad Y, Chet I (1984) Biological control effects of a new isolate of *Trichoderma harzianum* on *Pythium aphanidermatum*. Phytopathology 74:498

Sivan A, Ucko O, Chet I (1987) Biological control of *Fusarium* crown rot of tomato by *Trichoderma harzianum* under field conditions. Plant Disease 71:587–592

Stone AG, Scheurell SJ, Darby HM (2004) Suppression of soilborne diseases in field agricultural systems: organic matter management, cover cropping and other cultural practices. In: Magdoff F, Weil RR (eds) Soil organic matter in sustainable agriculture. CRC Press LLC, Boca Raton, FL, pp 131–177

Sullivan RF, White JF (2000) *Phoma glomerata* as a mycoparasite of powdery mildew. Appl Environ Microbiol 66:425–427

Sundheim L, Krekling T (1982) Host-parasite relationships of the hyperparasite *Ampelomcyes quisqualis* and its powdery mildew host *Sphaerotheca fuliginea*. Phytopathol Z 104:202–210

Sztejnberg A, Abo-Foul S (1990) The hyperparasite *Ampelomyces quisqualis* increases yield and photosynthesis of powdery mildew-infected cucumber and zucchini. In: Proc APS/CPS Annu Meet, Grand Rapids, MI, August 1990

Sztejnberg A, Freeman S, Chet I, Katan J (1987) Control of *Rosellinia necatrix* in soil and apple orchard by solarization and *Trichoderma harzianum*. Plant Disease 71:365–369

Sztejnberg A, Galper S, Mazar S, Lisker N (1989) *Ampelomyces quisqualis* for biological and integrated control of powdery mildews in Israel. J Phytopathol 124:285–295

Sztejnberg A, Galper S, Lisker N (1990) Conditions for pycnidial production and spore formation by *Ampelomyces quisqualis*. Can J Microbiol 36:193–198

Tu JC (1984) Mycoparasitism by *Coniothyrium minitans* and its effects on sclerotia germination. Phytopathol Z 109:261–268

Tunlid A, Jansson HB, Nordbring-Hertz B (1992) Fungal attachment to nematodes. Mycol Res 96:401–412

1974-1986	Professor, Department of Plant Science, University of Manitoba
1986-	Professor and Senior Scholar, Department of Plant Science, University of Manitoba
Since 1987	Professor Emeritus

MAJOR HONORS AND AWARDS (only most prominent ones):

Royal Bank Award	1975
Fellow, Agricultural Institute of Canada	1975
Honorary Life Member, Canadian Seed Growers Association	1976
Queen's Jubilee Medal	1977
Grindley Medal	1978
H.R. MacMillan Laureate in Agriculture	1980
Agronomy Merit Award	1980
CSP Foods Canola Award	1981
Manitoba Inst. of Agrologists Distinguished Agrologist Award	1981
Canadian Barley & Oilseeds Conference Award	1982
Honorary Life Member, Manitoba Institute of Agrologists	1984
Officer of the Order of Canada	1985
Group Consultant of International de Recherche sur le Colya (GCIRC) International Award for Research in Rapeseed	1987
GCIRC International Award for Research in Rapeseed	1987
McANSH Award	1989

VARIETIES RELEASED:

Soybeans
- Portage 1964
- Altona 1966

Rape
- Tanka 1963
- Target 1966
- Turret 1970
- Tower 1974
- Regent 1977
- Reston 1982
- Pivot 1985

Turnip rape
- Polar 1970

MEMBERSHIP IN COMMITTEES AND ASSOCIATIONS:

Research Committee, Canola Council of Canada
Canadian Committee on Fats and Oils
Canadian Barley and Oil Seeds Conference
Sub-committee on Oilseeds and Special Crops, Canada Committee on Grain Breeding
Manitoba Stock Seed Distribution Committee, Manitoba Branch, Canadian Seed Growers Association
Oilseeds and Special Crops Committee, Manitoba Agronomists Conference
Manitoba Agricultural Services Committee
Manitoba Institute of Agrologists
Agricultural Institute of Canada
Swedish Seed Association
American Contract Bridge League

DESCRIPTION OF THE SCIENTIFIC ACHIEVEMENTS

Professor Baldur R. Stefansson has made major contributions to the development of canola over a period of more than 30 years. It is mainly due to his influence and vision that oilseed rape has been transformed from a marginal edible oilseed covering 2.7 million hectares in 1950 to a leading oilseed crop in the world today covering in excess of 27 million hectares.

Prof. Stefansson has been one of the first to apply modern chemical analytical tools to monitor the composition of fatty acids of the canola oil, and thus to rapidly recognize and follow the development of new fatty acid compositions. This has eventually resulted in canola being one of the most nutritious oils for food use today. Canola is among the first crops presently undergoing a second revolution due to the advent of biotechnology. This second revolution will further advance the use of canola for the benefit of mankind.

Baldur Stefansson has been referred to as the father of canola. When he began his career as a plant breeder in the early 1950's, oilseed rape world production was approximately 2.8 million metric tons. It was a crop used primarily as an industrial oil, with limited application as an edible oil in the western world. It became clear, in the mid-1950's, that there were anti-nutritional components in both the oil and the meal that gave cause for concern when used for animal consumption. The presence in diets of high amounts of erucic acid, a C22 monounsaturated fatty acid, was associated with deposition of fat in the heart, skeletal muscle, and adrenal glands of rodents, impairing the animal's growth. Furthermore, the high levels of glucosinolates in rapeseed meal had goitrogenic effects when used as animal feed.

In spite of the obstacles to improving the quality of the seed, Stefansson recognized rapeseed's potential as an edible oilseed crop in temperate climates. He knew that the chemical composition of the oil required modification, if the crop was to be widely accepted. He had the wisdom and foresight to realize that, in addition to his skills as a breeder, he required the collaboration of a chemist in order to make use of the newly developed technique, gas chromatography, to permit analyses of the fatty acid composition of rapeseed oil.

Stefansson undertook a survey of rape accessions from many parts of the world, looking for variability in the content of erucic acid in the oil fraction and quickly established that variability existed. After surveying about 4000 lines, he discovered a forage rape, *Liho*, that had wide variability in its erucic acid content. Through breeding and selection using *Liho* as a parent, he and his colleague, Dr. K. Downey, demonstrated that erucic acid could be essentially eliminated from rape oil (Stefansson *et al.*, 1961).

The value of an oilseed crop is considerably enhanced if the meal has superior nutritional value. Stefansson recognized this and embarked on a program to reduce both erucic acid and glucosinates: At the same time, he selected for increased oil content and increased meal protein content, a first for oilseed breeders. As a result of this, his early varieties (Tanka, Target and Turret) were widely accepted because they were high yielding with an above average oil and protein content. In 1974 and 1977, Dr. Stefansson released Tower and Regent, two highly successful cultivars that had less than 5 percent erucic acid and less than 3 mg/g glucosinolates in the meal. These were the first 'double zero' rape cultivars worldwide. This dramatic modification eventually led to the name, 'Canola', to describe cultivars with low erucic acid and low glucosinolate content. Dr. Stefansson was directly involved in the licensing and release of seven oilseed rape cultivars and developed breeding lines that were the parents of a new generation of canola that had reduced linolenic acid content, providing greater stability and improved nutrition. He also selected lines that were extremely high in erucic acid, to produce an industrial oil useful in nylon manufacturing and in marine lubrication.

Dr. Stefansson's scientific contributions can not be measured in the form of publications as one might measure a laboratory scientist's efforts. He is more to be measured by what he has left mankind — a tailor-made oilseed crop with a world production of 27.2 million metric tons in 1993-94, making it the world's third most important oilseed crop. It is a major oilseed crop in Canada, Europe, India, and China and its cultivation is rapidly spreading throughout the world. Its human nutritional value is judged amongst the best of any plant oil and its meal is valued as a high protein supplement for animal feed. Because of its ease of genetic transformation, it as become a favorite host for incorporating foreign genes. It is safe to say that much of the interest in the crop today can be traced to the vision and pioneering work of Baldur Stefansson.

LIST OF PUBLICATIONS

Stefaneson, B.R. 1959. A method for estimating the percentage of hybrids in seedlots of first generation advance sunflowers by means of a seedling characteristic. Can. J. Plant Sci. 39: 71-74.

Howell, R.W., C.J. Wargel, C.A. Brim, E.E. Hartwig, J.W. Lambert, I.R. Torpson, B.R. Stefansson, J.K. Park, W.E. Seigler and B.K. Webb. 1960. Response of soybeans to seed treatment with gibberellin under simulated commercial conditions. Agron. J. 52: 144-146.

Stefansson, B.R., F.W. Hougen and R.K. Downey. 1961. Note on the isolation of rape plants with seed oil free from erucic acid. Can. J. Plant Sci. 41: 218-219.

Stefansson, B.R. and F.W. Hougen. 1964. Selection of rape plants (*Brassica napus*) with seed oil practically free from erucic acid. Can. J. Plant Sci. 44: 359-364.

Kondra, Z.P. and B.R. Stefansson. 1965. Inheritance of erucic and eicosenoic acid content of rapeseed oil (*Brassica napus*). Can. J. Genetics and Cytol. 7: 505-510.

Gross, A.T.H. and B.R. Stefansson. 1966. Effect of planting date on protein oil and fatty acid content of rapeseed and turnip rape. Can. J. Plant Sci. 46: 389-395.

Stefansson, B.R. 1966. Registration of Portage soybeans. Crop. Sci. 6: 612.

Stefansson, B.R. 1966. Altona, a new variety of soybeans. Can. J. Plant Sci. 46: 693.

Stefansson, B.R. 1966. Target, a new variety of summer rape. Can. J. Plant Sci. 46: 694.

Stefaneson, B.R. 1968. Registration of Altona soybeans. Crop Sci. 8: 777.

Stefansson, B.R. 1969. Registration of Target rape. Crop. Sci. 9: 394-395.

Stefansson, B.R. and A.K. Storgaard. 1969. Correlations involving oil and fatty acids in rapeseed. Can. J. Plant Sci. 49: 573-580.

Kondra, Z.P. and B.R. Stefanseon. 1970. A maternal effect on the fatty acid composition of rapeseed oil (*Brassica napus*). Can. J. Plant Sci. 50: 345-346.

Kondra, Z.P. and B.R. Stefansson. 1970. Inheritance of the major glucosinolates of rapeseed (*Brassica napus*) meal. Can. J. Plant Sci. 50: 643-647.

Stefansson, B.R. 1971. Registration of Polar summer turnip rape. Crop Sci. 11: 134.

Stefanason, B.F. 1971. Registration of Turret summer rape. Crop Sci. 11: 135.

Fowler, D.B. and B.R. Stefansson. 1972. Effects of the mutagenic agent ethyl methanesulfonate on the M_1 generation of rape (*Brassica napus*). Can. J. Plant Sci. 52: 53-62.

Stefansson, B.R. and Z.P. Kondra. 1974. Tower Summer Rape. Can. J. Plant Sci. 54: 343-344.

Fowler, D.B. and B.R. Stefansson. 1975. Ethyl-methanesulfonate-induced mutations in rape (*Brassica napus*). Can. J. Plant Sci. 55: 817-821.

Grami, Bahram and B.R. Stefansson. 1977. Gene action for protein and oil content in summer rape. Can. J. Plant Sci. 57: 625-631.

Grami, Bahram, RJ. Baker and B.R. Stefansson. 1977. Genetics of protein and oil content in summer rape: Heritability, number of effective factors, and correlations. Can. J. Plant Sci. 57: 937-943.

Grami, Bahram and B.R. Stefansson. 1977. Paternal and maternal effects on protein and oil content in summer rape. Can. J. Plant Sci. 57: 945-949.

Sernyk, J.L. and B.R. Stefansson. 1982. White flower color in rape (*Brassica napus* L.) associated with a radish (*Raphanus sativus* L.) chromosome. Can. J. Genetics and Cytology 24:729-734.

Sernyk, J.L. and B.R. Stefanesson. 1983. Heterosis in summer rape (*Brassica napus* L.). Can. J. Plant Science 63:407-413.

Fan, Z., S.R. Rimmer and B. R. Stefansson. 1983. Inheritance of resistance to *Albugo candida* in rape (*Brassica napus* L.). Can. J. Genetics and Cytology. 25:420-424.

Fan, Z., W. Tai and B.R. Stefansson. 1984. Male sterility in *Brassica napus* L. associated with an extra chromosome. Can. J. of Genetics and Cytology. 27:467-471.

Fan, Z. and B.R. Stefansson. 1986. Influence of temperature on sterility of two cytoplasmic male sterility systems in rape (*Brassica napus* L.) Can. J. Plant Sci. 66: 221-227.

Fan, Z., B.R. Stefansson and J.L. Sernyk. 1986. Maintainers and restorers of three male-sterility-inducing cytoplasms in rape (*Brassica napus* L.). Can. J. Plant Sci. 66: 229-234.

Li, S., Y. Qian, Z. Wu and B.R. Stefansson. 1988. Genetic male sterility in rape (*Brassica napus* L.) conditioned by interaction of genes at two loci. Can. J. Plant Sci. 68: 1115-1118.

Contributions to Books

Stefansson, B.R. 1973. Oilseeds: production, marketing and processing. In Grains and Oilseeds, section D-11, International Grains Institute, Winnipeg, Manitoba. pp. 1-29.

Stefansson, B.R. 1974. Rapeseed Breeding in Western Canada. In The Story of Rapeseed in Western Canada, edited by D.A. McLeod, Saskatchewan Wheat Pool. pp. 22-25.

Stefansson, B.R. 1974. Rapeseed. In Principles and Practices of Commercial Farming, Faculty of Agriculture, University of Manitoba, Winnipeg, Manitoba. pp. 162-165.

Stefansson, B.R. et al. 1975. Breeding Rapeseed and Mustard Crops. In Oilseed and Pulse Crops in Western Canada, Western Co-operative Fertilizers Ltd., Calgary, Alberta. pp. 165-183.

Stefansson, B.R. 1975. Oilseeds: Production. In Grains and Oilseeds, Handling, Marketing, Processing. International Grains Institute, Winnipeg, Manitoba. pp. 607-616.

Stefansson, B.R. 1976. Tower rapeseed, a major step in the development of high quality rapeseed. In Fats and Oils in Canada, Department of Industry, Trade and Commerce, Ottawa. pp. 2-7.

Hougen, F.W. and B.R. Stefansson. 1982. Rapeseed. Chap. 6, in Advances in Cereal Science and Technology. Vol. V. Edited by Y. Pomeranz. Amer. Assoc. of Cereal Chemists, Inc., St. Paul, Minnesota. pp. 261-289.

Stefansson, B.R. 1983. The development of improved rapeseed cultivars. Chap. 6, in High and Low Erucic Acid Rapeseed Oils, edited by J.K.G. Kramer, F.D. Sauer and W.J. Pigden. Academic Press Canada, Toronto. pp. 143-159.

Stefansson, B.R. 1983. Oilseeds. In The Canadian Encyclopedia, Hurtig Pub. Ltd., Edmonton, Alberta. pp. 1314-1315.

Conference Proceedings

B.R. Stefaneson and F.W. Hougen. Breeding for Low Erucic Acid Content in Rape. Proc. Canadian Barley and Oil Seeds Conf., Winnipeg, February 1960. p. 27.

B.R. Stefanseon. Reply to Question re Soybeans. Proc. Man. Agronomists' Ann. Conf., 1967. p. 26.

Stefansson, B.R. Expected Changes in Rapeseed Varieties in the Future. Proc. Man. Agronomists' Ann. Conf., 1969. pp. 23-24.

Stefansson, B.R. Plant Breeding and New Varieties of Rapeseed. Proc. 2nd Ann. Meeting, Rapeseed Assoc. of Canada, Saskatoon, Mar. 1969. pp. 44-46.

Stefansson, B.R. Fatty Acid Composition of Fats and Oils. Proc. Man. Agronomists' Ann. Conf., 1970. pp. 49-51.

Stefansson, B.R. Influence of Temperature and Soil on Seed Composition. Proc. Int. Conf. on the Science, Technology and Marketing of Rapeseed and Rapeseed Products. Ste. Adele, Quebec, Sept. 20-23, 1970. pp. 86-91.

Stefansson, B.R. Rapeseed Varieties Present and Future. Proc. Man. Agronomists' Ann. Conf., 1971. pp. 59-61.

Stefansson, B.R. Progress Report on "Double Zero" Rapeseed Varieties. Proc. Canadian Barley and Oil Seeds Conf., Winnipeg, Feb. 1972. p. 103.

Stefansson, B.R. What's New in Oilseed Varieties – Rapeseed. Proc. Canadian Barley and Oil Seeds Conf., Winnipeg February 1974. pp. 90-93.

Stefansson, B.R. Rapeseed Varieties. Proc. 7th Ann. Meeting, Rapeseed Assoc. of Canada, Calgary, March 1974. pp. 46-47.

Stefansson, B.R. Rapeseed for 1975. Proc. Man. Agronomists' Ann. Conf. 1974. pp. 74-76.

Stefansson, B.R. Summer Rape (Argentine Type) Varieties Proc. 9th Ann. Meet. Rapeseed Assoc. of Canada, Winnipeg, March 1976. pp. 37-39.

Stefansson, B.R. Canadian Rapeseed — The Best in the World and Getting Better. Proc. Manitoba Agronomists' Ann. Conf., 1978. pp. 39-43.

Stefansson, B.R. Selection for Oil and Protein in Rapeseed Proc. 5th Int. Rapeseed Conf. Vol. 1, Malmo, Sweden, June 12-16, 1978. pp. 113-114.

Stefansson, B.R. Rapeseed Varieties of the Future. Proc. Canadian Barley and Oil Seeds Conf., Winnipeg, Feb. 1980. pp. 75-80.

__________ Rapeseed in China. Report on Canola Technical Seminars People's Republic of China, Nov. 9-22, 1980. pp. 14-17.

__________ Canola/Rapeseed Variety Development. Proc. 14th Ann. Conf., Canola Council of Canada, Vancouver, Mar. 1981. pp. 47-50.

__________ Report on Canada/Japan Canola Consultation Tokyo, Japan, Nov 7-11, 1983.

Gurdev S. Khush

International Rice Research Institute
Makati City, Philippines

2000 — for his extraordinary contribution to theoretical research in plant genetics, evolution and breeding especially of rice, with regard to food production and alleviation of hunger.

CURRICULUM VITAE

Date and Place of Birth: August 22, 1935, Rurkee, Punjab, India
Nationality: Indian
Business Address: Adjunct Professor
Dept. of Plant Sciences
University of California, Davis, CA 95616

Education

(1) Matriculation Punjab University 1951 (topped the list of successful candidates of the school in first division)
(2) B.Sc. (Agri) Punjab Agricultural University, Ludhiana, Punjab, India 1955 (First division)
(3) Ph.D. (Genetics) University of California, Davis, Davis, California 1960

Professional Positions Held

(1) Research Assistant, University of California, Davis, June 1957 to July 1960
(2) Assistant Geneticist, University of California, Davis, August 1960 to July 1967

(3) Plant Breeder, International Rice Research Institute (IRRI), August 1967 to June 1972
(4) Plant Breeder and Head, Plant Breeding Dept., (IRRI), July 1972 to December 1985
(5) Principal Plant Breeder and Head, Division of Plant Breeding, Genetics and Biochemistry, (IRRI), January 1986 to February 2002
(6) Adjunct Professor University of California, Davis 2002-to date

National Awards

(1) Best all round student award of Punjab Agricultural University, 1955
(2) Achievement Award by the Bureau of Plant Industry, Manila, Philippines, 1983
(3) K. Ramiah Medal, National Academy of Agricultural Sciences, 1997
(4) Medal for the "Cause of Agriculture and Rural Development" of Vietnam, 1998
(5) International Agricultural Cooperation Prize, Ministry of Agriculture, Government of China, 1999
(6) Friendship Award, Government of China, 1999
(7) B.P. Pal Memorial Award, Indian Science Association, 2000
(8) Padma Shri Award by President of India, 2000
(9) Gold Medal, Ministry of Agriculture, Islamic Republic of Iran, 2000
(10) Gopalan Oration Award and Gold Medal 2000
(11) International Cooperation Award, Government of Egypt, 2001
(12) Citation by President, Government of Philippines, 2002
(13) Plaque of Appreciation by the President of Indonesia, 2002
(14) Amrik Singh Cheema Award by Young Farmer's association of Punjab, 2006
(15) Swaminathan Award for leadership in agriculture by TAAS, 2006
(16) Golden Sickle Award, Government of Thailand, 2007
(17) NCCPB Genetics and Plant Breeding Award, 2007
(18) CV Raman Medal, Indian National Science Academy, 2007

International Awards

(1) Borlaug Award for Achievements in Plant Breeding, 1977
(2) Japan Prize, Science and Technology Foundation of Japan, 1987
(3) International Agronomy Award, American Society of Agronomy, USA, 1989
(4) Emil M. Mrak International Award, University of California, Davis, USA, 1990
(5) World Food Prize, World Food Prize Foundation, USA, 1996
(6) Rank Prize, Rank Prize Foundation, United Kingdom, 1998
(7) Wolf Prize in Agriculture, Wolf Prize Foundation, Israel, 2000

(8) International Scientific and Technological Cooperation Award, Government of China, 2001
(9) International Khwarizmi Award, Science and Technology Organisation of Iran, 2004

Membership in Professional Societies

(1) Genetic Society of America
(2) American Society of Agronomy
(3) Crop Science Society of America
(4) Crop Science Society of the Philippines
(5) Indian Society of Genetics and Plant Breeding
(6) Society for the Advancement of Breeding Researches in Asia and Oceania (SABRAO)
(7) President, SABRAO (1997-2001)

Honorary Degrees

(1) Doctor of Science (Honoris Causa), Punjab Agricultural University, India 1987
(2) Doctor of Science (Honoris Causa), Tamil Nadu Agricultural University, India 1995
(3) Doctor of Science (Honoris Causa), C.S. Azad University of Agriculture and Technology, India 1995
(4) Doctor of Science (Honoris Causa), G.B. Pant University of Agriculture and Technology, India 1996
(5) Doctor of Science (Honoris Causa), De Montfort University, Leicester, England 1998
(6) Doctor of Science (Honoris Causa), Assam Agricultural University, India, 2000
(7) Doctor of Science (Honoris Causa), Cambridge University, United Kingdom, 2000
(8) Doctor of Science (Honoris Causa) N.D. University of Agriculture and Technology, India 2003
(9) Doctor of Science (Honoris Causa) Ohio State University, USA, 2006
(10) Doctor of Science (Honoris Causa) Guru Nanak Dev University, India, 2007

Fellowship of Scientific Societies

(1) Honorary Fellow Award, Crop Science Society of the Philippines, 1986
(2) American Society of Agronomy Fellows Award, 1987
(3) Indian Society of Genetics and Plant Breeding, Fellows Award, 1991

(4) Crop Science Society of America, Fellows Award, 1996
(5) Honorary Fellow Award, Bangladesh Agronomy Society, 1999

Fellowship of Scientific Academies

(1) New York Academy of Sciences, 1975
(2) Indian National Science Academy, 1977
(3) Foreign Associate, U.S. National Academy of Sciences, 1989
(4) Third World Academy of Sciences, 1990
(5) Indian Academy of Sciences, 1991
(6) National Academy of Agricultural Sciences (India) 1991
(7) Royal Society of London, 1995
(8) Russian Academy of Agricultural Sciences, 1997
(9) World Innovation Foundation, 2000
(10) Chinese Academy of Sciences, 2002
(11) National Academy of Sciences, India 2003

Membership in Editorial Boards of Scientific Journals

(1) Journal of Theoretical and Applied Genetics (TAG) (1980-1994)
(2) Plant Breeding Abstracts (1988-1994)
(3) Breeding Science (2000-2006)
(4) Genetics and Plant Breeding (1990-2000)
(5) Crop Improvement (1985-)
(6) Oryza (1987-)
(7) Journal of Genetics (1985-2007)
(8) Rice Genetics Newsletter (Founding Editor)
(9) Journal of Crop Production (1995-)
(10) Scientia Agricultura Sinica (2000-)

Buildings Named in Honor of G. S. Khush

(1) G.S. Khush Genetic Engineering Laboratories, Agricultural Biotechnology Research Institute of Iran
(2) G.S. Khush Biotechnology Laboratories, Punjab Agricultural University, Ludhiana, India
(3) Khush Hall, International Rice Research Institute, Los Banos Philippines

Major Research Contributions

(1) Studied the cytotaxomic relationships of cultivated rye and wild species of genus Secale.

(2) Carried out cytogenetic explorations of tomato genome through induced deficiencies and primary, secondary, tertiary, telo and compensating trisomics.
(3) Associated linkage groups of rice with respective chromosomes through primary trisomic analysis and determined the orientation of linkage and position of centromeres using secondary trisomics and telotrisomics of rice.
(4) Identified numerous genes for disease and insect resistance in rice.
(5) Led the rice breeding program of IRRI for 35 years. More than 320 breeding lines developed at IRRI have been released as varieties by national rice improvement programs throughout the world. Numerous other lines have been used in the hybridization programs. IRRI developed breeding materials and their progenies are now planted to 60% of the world rice land. One of the IRRI varieties, IR36, was planted to 11 million hectares of rice land in 1980s. No other variety of rice or any other food crop has been planted that widely before. World rice production more than doubled from 257 million tons in 1966 to 600 million tons in 2000. Increased production from these varieties feeds 1 billion more rice consumers annually.

DESCRIPTION OF SCIENTIFIC CONTRIBUTIONS

The 1960's was a decade of despair with regard to the world's ability to cope with the food-population balance in the developing countries. The cultivated land frontier was closing in most Asian countries, while population growth rates were accelerating, owing to rapidly declining mortality rates resulting from advancements in modern medicine and health care. International organizations and concerned professionals were busy organizing seminars and conferences to raise awareness regarding ensuing food crisis and to mobilize global resources to tackle the problem on emergency basis. In a well known book entitled "Times of Famine" published in 1967, the Paddock brothers predicted "Ten years from now, parts of the underdeveloped world will be suffering from famines. In 15 years, the famines will be catastrophic, and revolution and social turmoil and economic upheavals will sweep across areas of Asia, Africa and Latin America". Thanks to the life long contributions of Dr. Borlaug and his colleagues in Mexico and Dr. Khush and his colleagues in Philippines in developing improved varieties of wheat and rice respectively, large scale famines and social and economic upheavals were averted.

More than 330 improved varieties of rice developed under the leadership of Dr. Khush have been released in rice growing countries of Asia, Africa and Latin America. IRRI bred varieties have also been used as parents in national rice breeding programs. It is estimated that 60% of world's rice land is now planted to IRRI bred rice varieties or their progenies. The widescale adoption of these varieties has impacted food production, food security and environmental sustainability.

Impact on food production

The gradual replacement of traditional varieties of rice by improved ones, together with associated improvement in farm management practices, has had a dramatic effect on the growth of rice output. Farmers harvest 5-7 tons of unmilled rice from improved varieties compared with 1-3 tons from traditional varieties. Since 1966, when the first high-yielding variety of rice was released, the area planted to rice increased marginally from 126 million hectares to 152 million hectares (20%), whereas average rice yield has increased from 2.1 tons to 4.0 tons per hectare (95%). The total rice production has increased from 257 million tons in 1966 to 620 million tons in 2006 (140%). Most of the major rice producing countries in Asia where 92% of the world's rice is produced became self sufficient and several have exportable surplus.

Impact on food security

In many rice growing countries, the growth in rice production has outstripped the rise in population, leading to a substantial increase in cereal consumption and caloric intake per capita. During 1965-1990, the daily caloric supply in relation to requirement improved from 81% to 120% in Indonesia, from 86% to 110% in China, from 82 to 99% in the Philippines and from 89 to 94% in India.

The increase in per capita availability of rice and a decline in the cost of production per ton of output contributed to a decline in the real price of rice in the international and domestic markets. The unit cost of production is about 20-30% lower for high yielding varieties than for traditional varieties of rice and the price of rice adjusted for inflation is 40% lower than in the mid 1960s. The decline in food prices has benefited the urban poor and rural landless, who are not directly involved in food production but who spend more than one half their income on food.

Impact on landless workers

The diffusion of high yielding varieties has also contributed to a growth in income for rural landless workers. High yielding varieties require more labor per unit of land, because of increased intensive care in agricultural operations and the harvesting of a larger output. The labor requirement has also increased because of the higher intensity of cropping, which has been made possible by the reduction in crop growth time. As farm incomes increase, better-off farm households substitute leisure for family labor and hire more landless workers to do the job. The marketing of a larger volume of produce and an increased demand for non farm goods and services, resulting from larger farm incomes, has generated additional employment in rural trade, transport, and construction activities. The economic miracle underway in many Asian countries was triggered by the growth in agricultural income and its equitable distribution, which helped expand the domestic market for non farm goods and services.

Impact on environmental sustainability

In sharp contrast to developed countries, where more of the environmental problems have been urban and industrial, the critical environmental problems in most of the low income, developing countries are still rural, agricultural and poverty related. More than half the world's very poor live on lands that are environmentally fragile and rely on natural resources over which they have little control. Land hungry farmers resort to cultivating unsuitable areas, such as erosion prone hillsides, semi-arid areas where soil degradation is rapid and the tropical forests, where crop yields on cleared land drop sharply after just a few years.

The widespread adoption of high-yielding varieties has helped most Asian countries meet their growing food needs from productive lands, and has reduced the pressure to open up more fragile lands for agriculture. Had 1961 yields still prevailed today, three times more land in India and two times more land in China would be needed to equal 2000 rice production. If the Asian countries attempted to produce a 2000 harvest at the yield levels of 1960s, most of the forests, woodlands and range lands would disappear and mountainsides would be eroded, with disastrous consequences for the upper water shed and productive lowlands, the extinction of wildlife habitats and the destruction of biodiversity. To produce 2000 world rice production of 600 million tons at the yield levels of 1965, 130 million hectares more land would be required.

One of the major achievements of IRRI's rice improvement program has been the development of varieties with multiple resistance to diseases and insects. Most of the high yielding varieties are resistant to major diseases and insects and can be grown without insecticide applications. The insecticide use on rice in Asia has been drastically reduced. The reduced needs for application of pesticides has facilitated the adoption of integrated pest management practices. Reduced pesticide use has resulted in (i) enhanced environmental quality, (ii) improved health of farming communities, (iii) made safer food available, and (iv) protected useful fauna and flora.

An independent assessment carried out by The Royal Society of London concluded that forests and agricultural land act as "carbon sinks" that reduce greenhouse gases. Improved varieties of rice help reduce the atmospheric carbon dioxide in two ways. Firstly, because they are more productive they utilize more carbon dioxide as compared to traditional varieties. Secondly, their adoption has led to conservation of forests which are excellent carbon sinks.

Contributions to rice genetics

Understanding of the basic genetics of the crop species is crucial for its improvement. The status of rice genetics was much behind that of other important food crops such as maize, wheat, barley and tomatoes. The linkage map of rice was poorly understood and there was no agreement on chromosome numbering system and gene symbolization. Dr. Khush established a complete series of primary trisomics of rice and utilized these for associating linkage groups with cytologically identifiable chromosomes. He then organized an international symposium on rice genetics in 1985 which was attended by more than 200 rice geneticists. Discussions among participants led to the birth of Rice Genetics Cooperative for international collaboration in rice genetics (RGC). Dr. Khush served as the secretary of RGC from its inception in 1985 till 2005. RGC established the rules for gene symbolization and chromosome numbering system. It also publishes a Rice Genetics Newsletter which was edited by Dr. G.S. Khush and a Japanese colleague till 2000. Dr. Khush also organized three more International Rice Genetics Symposia in 1990, 1995 and 2000.

In cooperation with colleagues at Cornell University, Dr. Khush established the first molecular genetic map of rice. The molecular map of rice now comprises more than 4000 markers and is the best known of that of any crop species. He and his colleagues have tagged many genes of economic importance with molecular markers and this has permitted the application of molecular marker aided selection in rice breeding.

Dr. Khush's research on the genetics of host-plant interactions has led to the identification of useful genes for disease and insect resistance, which have been utilized for developing disease and insect resistant varieties at IRRI and by national program scientists. Dr. Khush is a world renowned geneticist and author of a widely used text "Cytogenetics of Aneuploids" published by Academic Press, New York. Another book on "Host Plant Resistance" to Insects which he co-authored with Dr. N. Panda is a standard reference on the subject. He has written 24 review articles, 84 book chapters, 180 research papers in refereed journals and reviewed 10 books.

Dr. Khush has trained numerous rice breeders. He supervised 20 post-doctoral fellows, served as major professor of 16 Ph.D. degree candidates, 25 M.Sc. students and 60 non degree trainees. He has visited 60 rice growing countries to observe rice research and production systems and served as consultant to 15 national rice improvement programs.

Dr. Khush has served on the editorial boards of Journal of Theoretical and Applied Genetics (TAG), Plant Breeding Abstracts, Journal of Genetics, Journal of Genetics and Breeding, Crop Improvement, Journal of Crop Production, Oryza and is the founder editor of Rice Genetics Newsletter.

Dr. Khush has served as a scientific advisor to International Foundation of Science, Sweden (1990-2000), The Third World Academy of Science, Italy (1995-2002), member of Scientific Advisory Committee of the Rockefeller Foundation's International Program on Rice Biotechnology (1985-2000), Member Technical Advisory Group of Agricultural Biotechnology for Sustainable Productivity at Michigan State University (1989-1995) and has been serving as a member of Scientific Advisory Committee, Department of Biotechnology, Government of India since

Dr. Khush has received several International Awards such as Borlaug Award by Coromandal Fertilizers (1977), New Delhi, Japan Prize from Science and Technology Foundation of Japan (1987), International Agronomy Award (1989), Emil M. Marak International Award, University of California, Davis (1990), World Food Prize (1996), Rank Prize in Nutrition (1998), Wolf Prize in Agriculture (2000), Padma Shri Award from President of India (2000), Scientific and Technological Collaboration Award from Government of China and Golden Sickle Award From Government of Thailand.

Dr. Khush has also been honored with Doctor of Science (Honoris Causa) degrees from ten universities including University of Cambridge, England and Ohio State University.

Dr. Khush has been elected to some of the World's most prestigious academies such as Indian National Science Academy, Indian Academy of Sciences, National Academy of Agricultural Sciences (India), The Third World Academy of Sciences, Foreign Associate U.S. National Academy of Sciences and The Royal Society of London (FRS).

LIST OF PUBLICATIONS (from 1961-2007)

Books

1973 **Khush GS** Cytogenetics of Aneuploids. Academic Press New York 302 pp.

1984 **Khush GS** Plant breeding lecture (In Chinese) CNRRI, Hangzhou, China 98 pp.

1995 **Khush GS** (With N. Panda). Host Plant resistance to insects. CAB, International 431 pp.

Books Edited by GS Khush

1986 Proceedings of the First International Rice Genetics Symposium, Manila, Philippines. International Rice Research Institute, Los Banos. 932 pp.

1991 Rice Genetics 11. Proceedings of the Second International Rice Genetics Symposium. Manila, Philippines. Int Rice Research Institute, Los Banos 844 pp.

1991 Rice Biotechnology (with G.H. Toenniessen). Wallingford, UK. 320 pp.

1992 Nodulation and Nitrogen Fixation in Rice (With J. Bennett). Manila Philippines Int. Rice Research Institute, Los Banos 136 pp.

1994 Apomixis: Exploiting Hybrid Vigor in Rice. Int. Rice Research Institute, Los Banos, Philippines. 76 pp.

1996 Rice Genetics 111. Proceedings of Third International Rice Genetics Symposium, Manila Philippines, Int. Rice Rice Research Institute, Los Banos 1011 pp.

2000 Aromatic Rices (with R.K. Singh and U.S. Singh). Enfield (USA) and Plymouth (UK), 292 pp.

2001 Rice Genetics 1V (D.S. Brar and B. Hardy). Proceedings of Fourth International Rice Genetics Symposium, Manila, Philippines. Int. Rice Research Institute, Los Banos. 488 pp.

Referred Scientific Publications and Review Articles

1. 1961. Stebbins, G. L., and G. S. Khush. Variation in the organization of the guard cell complex in the leaf epidermis of monocotyledons and its bearing on their phylogeny. Am. J. Botany 48:51-59.
2. 1961. Khush, G. S., and G. L. Stebbins. Cytogenetic and evolutionary studies in *Secale. I.* Some new data on the ancestry of *S. cereale.* Am. J. Botany 48:723-750.
3. 1961. Rick, C. M., and G. S. Khush. X-ray induced deficiencies of chromosome 11 in the tomato. Genetics 46:1389-1393.
4. 1962. Khush, G. S. Cytogenetic and evolutionary studies in *Secale.* II. Interrelationships of the wild species. Evolution 16:848-896.
5. 1963. Khush, G. S. Cytogenetic and evolutionary studies in *Secale.* III. Cytogenetics of weedy ryes and origin of cultivated rye. Econ. Botany 17:60-71.
6. 1963. Khush, G. S. Cytogenetic and evolutionary studies in *Secale.* IV. Secale vavilovii and its biosystematic status. Zeitschrift fur Pflanzenzuchtung. 50:34-43.
7. 1963. Khush, G. S. and C. M. Rick. Meiosis in hybrids between *Lycopersicon esculantum* and *Solanum pennellii.* Genetica 33:167-183.
8. 1964. Rick, C. M., W. H. Dempsey, and G. S. Khush. Further studies on the primary trisomics of the tomato. Canad. J. Genet. and Cytol. 6:93-108.
9. 1964. Khush, G. S., C. M. Rick, and R. W. Robinson. Genetic activity in a heterochromatic chromosome segment of the tomato. Science 145:1432-1434.
10. 1965. Rick, C. M. and G. S. Khush, Chromosome engineering in *Lycopersicon.* Hered. 20 (Supl.):8-20.
11. 1966. Khush, G. S., and C. M. Rick. The origin, identification, and cytogenetic behavior of tomato monosomics. Chromosoma 18:407-420.

12. 1967. Khush, G. S., and C. M. Rick. Novel compensating trisomics of the tomato: Cytogenetics, monosomic analysis, and other applications. Genetics 56:297-307.
13. 1967. Khush, G. S. and C. M. Rick. Studies on the linkage map of chromosome 4 of the tomato and on the transmission of induced deficiencies. Genetica 38:74-94.
14. 1967. Khush, G. S., and C. M. Rick. Tomato tertiary trisomics: origin, identification, morphology, and use in determining position of centromeres and arm location of markers. Canad. J. Genet. Cytol. 9:610-631.
15. 1967. Khush, G. S., and C. M. Rick. Haplo-triplo-disomics of the tomato:origin, cytogeneticsand utilization as a source of secondary trisomics. Biol. Zbl. 86:257-265.
16. 1968. Khush, G. S., C. M. Rick. Tomato telotrisomics: origin, identification, and use in linkage mapping. Cytologia 33:137-148.
17. 1968. Khush, G. S., and C. M. Rick. Cytogenetic analysis of tomato genome by means of induced deficiencies. Chromosoma 23:452-484.
18. 1968. Reeves, A. F., G. S. Khush, and C. M. Rick. 1968. Segregation and Recombination of Trisomics: A reconsideration. Canad. J. Genet. Cytol. 10:937-940.
19. 1969. Khush, G. S., and C. M. Rick. Tomato secondary trisomics: Origin, identification, morphology and use in cytogenetic analysis of the genome. Heredity 24:129-146.
20. 1969. Suneson, C. A., K. O. Rachie, and G. S. Khush. A dynamic population of weedy rye. Crop Science 9:121-124.
21. 1969. Beachell, H. M., and G. S. Khush. Objectives of the IRRI breeding program. SABRAO Newsletter 1:69-80.
22. 1969. Rick, C. M., and G. S. Khush. Cytogenetic exploration in the tomato genome. pp. 145-168 In R. Bogart, ed. Genetic Lectures. Oregon State University Press.
23. 1970. Khush, G. S. Breeding rice for resistance to grassy stunt. Paper presented at the 13th Session of the International Rice Commission Working Party on Rice Production and Protection, Teheran, Iran, December 1970.
24. 1971. Khush, G. S. Rice Breeding for disease and insect resistance at IRRI. Oryza 8:111-119.
25. 1971. Khush, G. S., E. Torres, and R. Aquino. Exploiting wild germplasm of Oryza for improving cultivated rices. In Crop Science Society of the Philippines. Proceedings of the second annual scientific meeting, May 4-6, 1971. College, Laguna, Philippines. Pages 311-320.
26. 1972. Khush, G. S., and H. M. Beachell. Breeding for disease and insect resistance at IRRI. In International Rice Research Institute, Rice Breeding. Los Banos, Philippines. pp. 309-322.

27. 1972. Murty, V. V. S., and G. S. Khush. Studies on the inheritance of resistance to bacterial leaf blight in rice varieties. In International Rice Research Institute. Rice Breeding. Los Banos, Philippines. pp. 301-307.
28. 1972. Beachell, H. M., G. S. Khush, and R. C. Aquino. IRRI's international breeding program. In International Rice Research Institute. Rice Breeding, Los Banos, Philippines. pp. 89-106.
29. 1972. Beachell, H. M., G. S. Khush, and B. O. Juliano. Breeding for high protein content in rice. In International Rice Research Institute. Rice Breeding. Los Banos, Philippines. pp. 419-430.
30. 1972. Beachell, H. M. and G. S. Khush. Rice varietal improvement in Asia. In International Conference on Tropical and Subtropical Agriculture. Conference Papers, American Society of Agricultural Engineers. Pages 156-160.
31. 1973. Khush, G. S. Cytogenetics of Aneuploids. Academic Press, New York, 301 pp.
32. 1973. Murty, V. V. S., G. S. Khush, and N. F. Jensen. Inheritance of resistance to bacterial *Xanthomonas oryzae* (Uyeda et Ishiyama) Dowson in rice. Allelic relationships of resistance genes in donor varieties. Jap. Journ. Breed. 23:325-328.
33. 1974. Khush, G. S., W. R. Coffman and T. T. Chang. Current status of the varietal development program at IRRI. Paper presented during the Annual Rice Research Review Meeting, WARDA, July 15-20, 1974, Monrovia, Liberia.
34. 1974. Martinez, C. R. and G. S. Khush. Sources and inheritance of resistance to brown planthopper in some breeding lines of rice *Oryza saliva L.* Crop Science 14:264-267.
35. 1974. Khush, G. S. and K. C. Ling. Inheritance of resistance to grassy stunt virus and its vector in rice. Jour. Hered. 65:135-137.
36. 1975. Pathak, M. D., and G. S. Khush. Control of upland rice insects through varietal resistance. In International Rice Research Institute. Major Research in Upland Rice. Los Banos, Philippines. pp. 117-125.
37. 1975. Ou, S. H., K. C. Ling, H. E. Kauffman, and G. S. Khush. Diseases of upland rice and their control through varietal resistance. In International Rice Research Institute. Major Research in Upland Rice. Los Banos, Philippines. pp. 126-135.
38. 1975. Khush, G. S. Rice. In Handbook of Genetics, R. C. King Edit. Plenum Press, New York. Vol. 2, 31-58 pp.
39. 1976. Librojo, V., H. E. Kauffman, and G. S. Khush. Genetic analysis of bacterial blight resistance in four varieties of rice. SABRAO J. 8:105-110.
40. 1976. Khush, G. S. Breeding for resistance in rice. In The Genetic Basis of Epidemics in Agriculture. Annals the New York Academy of Sciences, 287:296-308.

41. 1976. Siwi, B. H. and G. S. Khush. New genes for resistance to the green leafhopper in rice. Crop Science 17:17-20.
42. 1976. Lakshminarayana, A., and G. S. Khush. New genes for resistance to brown planthopper in rice. Crop Science 17:96-100.
43. 1977. Khush, G. S., K. C. Ling, R. C. Aquino and V. M. Aguiero. 1977. Breeding for resistance to grassy stunt in rice. In Proceedings 3rd Int. Congr. SABRAO, Canberra, Australia. Plant Breeding Papers l:4(b):3-9.
44. 1977. Khush, G. S., and W. R. Co man. Genetic Evaluation and Utilization (GEU) Program — The Rice Improvement Program of the International Rice Research Institute. Theor. Appl. Genet. 51:97-110.
45. 1977. Khush, G. S. Disease and insect resistance in rice. Adv. Agron. 29:265-341.
46. 1977. Olufowote, J. O., G. S. Khush, and H. E. Kauffman. Genetics of bacterial blight resistance in rice. Phytopathology 67:772-775.
47. 1977. Petpisit, V., G. S. Khush, and H. E. Kau man. Inheritance of resistance to bacterial blight in rice. Crop Science 17:551-554.
48. 1977. Sujadi, S. and G. S. Khush. Studies on linkage relations of genes controlling disease and insect resistance and nature of endosperm in rice. Euphytica 26:337-342.
49. 1977. Sidhu, G. S., and G. S. Khush. Dominance reversal of a bacterial blight resistance gene in some varieties of rice. Phytopathology 67:461-463.
50. 1978. Khush, G. S. Breeding methods and procedures employed at IRRI for developing rice germplasm with multiple resistance to diseases and insects. In International Symposium on Methods of Crop Breeding. Tropical Agr. Res. Center, Japan. Series II:69-76.
51. 1978. Siwi, B. H., G. S. Sidhu, and G. S. Khush. Genetic analysis of rice variety Ptb 8 for resistance to green leafhopper and brown planthopper. Contr. Centr. Res. Inst. Agric. Bogor. 47:7-10.
52. 1978. Sidhu, G. S., G. S. Khush and T. W. Mew. Genetic analysis of bacterial blight resistance in seventy-four varieties of rice, Oryza sativa L. Theor. Appl. Genet. 53:105-111.
53. 1978. Sidhu, G. S. and G. S. Khush. Genetic analysis of brown planthopper resistance in twenty varieties of rice, *Oryza saliva L.* Theor. Appl. Genet. 53:199-204.
54. 1979. Ikehashi, H. and G. S. Khush. Breeding for blast resistance in IRRI. In International Rice Research Institute. Rice Blast Workshop. Los Banos, Philippines. pp. 69-80.
55. 1979. Khush, G. S. Genetics of and breeding for resistance to brown planthopper. In International Rice Research Institute. Brown Planthopper: A Threat to Rice Production in Asia. Los Banos, Laguna, Philippines. pp. 321-332.

56. 1979. Pathak, M. D., and G. S. Khush. Studies on varietal resistance to the brown planthopper at IRRIIn International Rice Research Institute. Brown Planthopper: A Threat to Rice Production in Asia. Los Banos, Philippines. pp. 285-302.
57. 1979. Gallun, R. L., and G. S. Khush. Genetic factors affecting expression and stability of resistance. In F.G. Maxwell and P.R. Jennings, eds. Breeding Plants Resistant to Insects. John Wiley & Sons, New York. pp. 63-85.
58. 1979. Khush, G. S. Breeding for multiple disease and insect resistance in rice. In M. K. Harris, ed. Biology and Breeding for resistance to Arthropods and Pathogens in Agricultural Plants. Texas A&M University, College Station, Texas. pp. 341-354.
59. 1979. Khush, G. S., C. M. Paule, and N. M. dela Cruz. Rice grain quality evaluation and improvement at IRRI. In. International Rice Research Institute. Proc. of Workshop on Chemical Aspects of Rice Grain Quality. Los Banos, Laguna, Philippines. pp. 21-31.
60. 1979. Ikehashi, H., and G. S. Khush. Methodology of assessing appearance of the rice grain including chalkiness and whiteness. In International Rice Research Institute. Proc. of Workshop on Chemical Aspects of Rice Grain Quality. Los Banos, Laguna, Philippines. pp. 223-229.
61. 1979. Crill, J. P. and G. S. Khush. Effective and stable control of rice blast with monogenic resistance. Food and Fertilizer Technology Center, Taipei City, Taiwan. Extension Bulletin No. 128.
62. 1979 . Sidhu, G. S. and G. S. Khush. Linkage relationships of some genes for disease and insect resistance and semidwarf stature in rice. Euphytica 28:233-237.
63. 1979. Sidhu, G. S., G. S. Khush and F. G. Medrano. A dominant gene for resistance to whitebacked planthopper in rice and its association with some other traits. Euphytica 28:227-232.
64. 1980. Khush, G. S. Genetics and breeding of photoperiod sensitive rice varieties. In Proceedings of the Photoperiod-Sensitive Transplanted Rice. Oct. 1977. Bangladesh Rice Research Institute, Joydebpur. pp. 37-44.
65. 1980. Sidhu, G. S., G. S. Khush and T. W. Mew. Genetic analysis of bacterial blight resistance in seventy cultivars of rice, Oryza sativa L. from Indonesia. Crop Improvement 6:19-25.
66. 1981. Khush, G. S. and R. C. Chaudhary. Role of resistant varieties in integrated pest management of rice. Food and Fertilizer Technology Center, Taipei City, Taiwan. Extension Bulletin No. 162.
67. 1981. Khush, G. S. Breeding rice for multiple disease and insect resistance. In Rice Improvement in China and other Asian Countries. International Rice Research Institute, Los Banos, Philippines. pp. 220-237.

68. 1981. Virmani, S. S., R. C. Chaudhary and G. S. Khush. Current outlook on hybrid rice. Oryzae 18:67-84.
69. 1981. Mew, T. W. and G. S. Khush. Breeding for bacterial blight resistance in rice. In Proc. 5th Int. Conf. Pathol. Bac. Cali, Colombia. pp. 504-510.
70. 1981. Horino, 0., T. W. Mew, G. S. Khush and A. Ezuka. Comparison of two differential systems for distinguishing pathogenic groups of *Xanthomonas campestris pv. oryzae.* Ann. Phytopath. Soc. Jpn 147:1-14.
71. 1981. Angeles, E., R., G. S. Khush and E. A. Heinrichs. New genes for resistance to whitebacked planthopper in rice, *Oryza sativa L.* Crop Sci. 21:47-50.
72. 1981. Chaudhary, R. C., S. S. Virmani and G. S. Khush. Patterns of pollen abortion in some cytoplasmic male sterile lines of rice. Oryza 18:140-142.
73. 1981. Hernandez, J. E. and G. S. Khush. Genetics of resistance to whitebacked planthopper in some rice varieties. Oryza 18:44-50.
74. 1982. Nair, R. V., T. M. Masajo, and G. S. Khush. Genetic analysis of resistance to whitebacked planthopper in twenty two varieties of rice, *Oryza saliva L.* Theor. Appl. Genet. 61:19-22.
75. 1982. Saini, S. S., G. S. Khush, and E. A. Heinrichs. Genetic analysis of resistance to whitebacked planthopper, *Sogalella furcifera* (Horvath), in some rice varieties. Crop Protec. 1:289-297.
76. 1982. Virmani, S. S., R. C. Aquino, and G. S. Khush. Heterosis breeding in rice *(Oryza saliva L.).* Theor. Appl. Genet. 63:373-380.
77. 1983. Singh, R. J., G. S. Khush and T. W. Mew. A new gene for resistance to bacterial blight in rice, *Oryza saliva L.* Crop Sci. 23:558-560.
78. 1983. Yoshimura, A., T. W. Mew, G. S. Khush and T. Omura. Inheritance of resistance to bacterial blight in rice cultivar CAS 209. Phytopathology 73:1409-1412.
79. 1983. Sivasubramanian, V. and G. S. Khush. Genetic analysis of resistance to bacterial blight in 27 varieties of rice. Oryza 20:151-155.
80. 1983. Young, J. B., S. S. Virmani and G. S. Khush. Cytogenic relationship among cytoplasmic-genetic male sterile, maintainer and restorer lines of rice. Philip. J. Crop Sci. 8(3):119-124.
81. 1983. Zapata, F. J., G. S. Khush, J. P. Crill, M. H. Heu, R. O. Romero, L. B. Torrizo and M. Alejar. Rice Anther Culture at IRRI. In Cell and Tissue Culture Techniques for Cereal Crop Improvement. Science Press Beijing, China and International Rice Research Institute, Manila, Philippines. pp. 27-46.
82. 1984. Harahap, Z., P. Soetjipto and G. S. Khush. Strategies for varietal improvement in tidal swamp rice. In Workshop on Research Priorities in Tidal Swamp Rice. International Rice Research Institute, Los Banos, Philippines. Pages 175-181.

83. 1984. Khush, G. S. IRRI breeding program and its worldwide impact on increasing rice production. In J.P. Gustafson, ed. Gene Manipulation in Plant Improvement. Plenum Press, New York. pp. 61-94.
84. 1984. Khush, G. S. and B. O. Juliano. Rice varietal improvement for protein content at the International Rice Research Institute. In Cereal Grain Protein Improvement. International Atomic Energy Agency, Vienna. pp. 199-202.
85. 1984. Khush, G. S. Terminology for rice growing environments. In Terminology for Rice Growing Environments. International Rice Research Institute, Los Banos, Philippines. pp. 5-10.
86. 1984. Khush, G. S. Plant Breeding Lectures (in Chinese) China National Rice Research Institute, Hangzhou 98 pp.
87. 1984. Chaudhary, R. C., G. S. Khush and E. A. Heinrichs. Varietal resistance to rice stemborers in Asia. Insect Sci. and Its Application. 5:447-463.
88. 1984. Khush, G. S., R. J. Singh, S. C. Sur and A. L. Librojo. Primary trisomics of rice:origin, morphology, cytology and use in linkage mapping. Genetics 107:141-163.
89. 1984. Khush, G. S. Breeding rice for resistance to insects. Protection Ecology 7:147-165.
90. 1984. Avesi, G. M. and G. S. Khush. Genetic analysis for resistance to the green leafhopper, *Nephotettix virescens* (Distant), in some cultivars of rice, *Oryza saliva L.* Crop Protection 3:41-51.
91. 1984. Yoshimura, A., T. W. Mew, G. S. Khush and T. Omura. Genetics of bacterial blight resistance in a breeding line of rice. Phytopathology 74:773-777.
92. 1985. Khush, G. S. and S. S. Virmani. Breeding rice for disease resistance. In G. E. Russell, ed. Progress in Plant Breeding. Butterworths, London. pp. 239-279.
93. 1985. Yoshimura, A., T. Omura, T. W. Mew and G. S. Khush. Genetic behavior of resistance to bacterial blight in differential rice cultivars in the Philippines. Bull. Inst. Trop. Agr. Kyushu Univ., Japan 8:1-54.
94. 1985. Miah, M.A.A., E. D. Earle, and G. S. Khush. Inheritance of callus formation ability in anther culture of rice, *Oryza saliva L.* Theor. Appl. Genet. 70:113-226.
95. 1985. Mohanty, H. K. and G. S. Khush. Diallel analysis of submergence tolerance in rice, *Oryza saliva L.* Theor. Appl. Genet. 70:467-473.
96. 1985. Wu, C. F. and G. S. Khush. A new dominant gene for resistance to whitebacked planthopper in rice. Crop Sci. 25:505-509.
97. 1985. Kaw, R. N. and G. S. Khush. Heterosis in traits related to low temperature tolerance in rice. Phil. J. Crop Sci. 10(2)93-105.
98. 1985. Kaw, R. N. and G. S. Khush. Heterosis in traits related to low temperature tolerance in rice. Phil. J. Crop. Sci. 10:93-105.

99. 1985. Khush, G. S. Rice varietal improvement. International Rice Commission Newsletter. XXXIV(2):111-125.
100. 1985. Khush, G. S., A. N. M. Rezaul Karim and E. R. Angeles. Genetics of resistance of rice cultivar ARC 10550 to Bangladesh brown planthopper biotype. Jour. Genet. 64:121-125.
101. 1986. Khush, G. S. and Kumar, I. Genetic selection for improved nutritional quality of fibrous crop residues of cereal crops — a plant breeders viewpoint. In R. M. Dixon, Ed. Ruminant feeding systems utilizing fibrous agricultural residues. Canberra, Australia. pp. 25-32.
102. 1986. Khush, G. S., Application of chromomere engineering in crop improvement. In Li Zhensheng and M. S. Swaminathan, Eds. Proceedings of the First International Symposium on Chromosome Engineering in Plants. Man, China. pp. 162-171.
103. 1986. Brar, D. S. and G. S. Khush. Wide hybridization and chromosomal manipulation in cereals. In D.A. Evans, W. R. Sharp and P. V. Ammirato, eds. Handbook of Plant Cell Culture. Vol. 4. MacMillan Publ. Co., New York. pp. 221-263.
104. 1986. Akbar, M., G. S. Khush and D. HilleRisLambers. Genetics of salt tolerance in rice. In Rice Genetics. International Rice Research Institute, P.O. Box 933, Manila, Philippines. pp. 399-410.
105. 1986. Angeles, E. R., G. S. Khush and E. A. Heinrichs. Inheritance of resistance to planthoppers and leafhoppers in rice. In Rice Genetics. International Rice Research Institute, P.O. Box 933, Manila, Philippines. pp. 537-550.
106. 1986. Chaudhary, B. P., P. S. Srivastava, M. N. Shrivastava and G. S. Khush. Inheritance of resistance to gall midge in some cultivars of rice. In Rice Genetics. International Rice Research Institute, P. O. Box, Manila, Philippines. pp. 523-528.
107. 1986. Jena, K. K. and G. S. Khush. Production of monosomic alien addition lines of *Oryza sativa* having a single chromosome of O. *officinalis*. In Rice Genetics. International Rice Research Institute, P.O. Box 933, Manila, Philippines. pp. 199-209.
108. 1986. Kaw, R. N. and G. S. Khush. Combining ability for low temperature tolerance in rice. In Rice Genetics. International Rice Research Institute, P.O. Box 933, Manila, Philippines. pp. 593-612.
109. 1986. Khush, G. S. and R. J. Singh. Relationships between linkage groups and cytologically identifiable chromosomes of rice. In Rice Genetics, International Rice Research Institute, P.O. Box 933, Manila, Philippines. pp. 239-248.
110. 1986. Kumar, I., and G. S. Khush. Genetics of amylose content in rice *(Oryza sativa L.)* Jour. Genet. 65:1-11.
111. 1986. Librojo, A. L. and G. S. Khush. Chromosomal location of some mutant genes through the use of primary trisomics of rice. In Rice Genetics.

International Rice Research Institute, P.O. Box 933, Manila, Philippines. pp. 249-255.

112. 1986. Shahi, B. B. and G. S. Khush. Genetic analysis of cold tolerance in rice. In Rice Genetics. International Rice Research Institute, P.O. Box 933, Manila, Philippines. pp. 429-436.
113. 1986. Kumar, I. and G. S. Khush. Gene dosage effects of amylose content in rice endosperm. Jpn. J. Genet. 61:559-568.
114. 1987. Khush, G. S. Development of rice varieties suitable for double cropping. In Tropical Agriculture Research Center, Ministry of Agriculture, Forestry and Fisheries, Japan. Tropical Agriculture Research Series No. 20. pp. 235-246.
115. 1987. Khush, G. S. Rice breeding: past, present and future. J. Genet. 66:195-216.
116. 1987. Kumar, I., G. S. Khush and B. O. Juliano. Genetic analysis of waxy locus in rice *(Oryza sativa L.)* Theor. Appl. Genet. 73:481-488.
117. 1987. Ruangsook, B. and G. S. Khush. Genetic analysis of resistance to green leafhopper, *Nephotettix virescens* (Distant) in some selected rice varieties. Crop Protection 6:244-249.
118. 1987. Kumar, I., and G. S. Khush. Genetic analysis of different amylose levels in rice. Crop Sci. 27:1167-1172.
119. 1987. Velusamy, R., E. A. Heinrichs and G. S. Khush. Genetics of field resistance to brown planthopper. Crop Sci. 27:199-200.
120. 1988. Kumar, I. and G. S. Khush. Inheritance of amylose content in rice *(Oryza saliva L.)* Euphytica 38:261-269.
121. 1988. Kumar, I. and G. S. Khush. Genetic analysis of different amylose levels in rice. Crop Sci. 27:1167-1172.
122. 1988. Wu, K. S., J. C. Glaszmann and G. S. Khush. Chromosomal locations of ten isozyme loci in rice *(Oryza sativa L.)* through trisomic analysis. Biochemical Genetics 26:303-320.
123. 1988. Ranjhan, S., J. C. Glaszmann, D. A. Ramirez and G. S. Khush. Chromosomal localization of four isozyme loci by trisomic analysis. Theor. Appl. Genet. 75:541-545.
124. 1988. Kabir, M. A. and G. S. Khush. Genetic analysis of resistance to brown planthopper in rice *(Oryza saliva L.)*. Plant Breeding 100:54-58.
125. 1988. Ghani, M. U., and G. S. Khush. A new gene for resistance to green leafhopper, *Nephotettix virescens* (Distant) in rice. J. Genet. 67:151-159.
126. 1988. Glaszmann, J. C., B. G. de los Reyes and G. S. Khush. Electrophoretic variation of isozymes in plumules of rice *(Oryza saliva L.)* - a key to the identification of 76 alleles at 24 loci. International Rice Research Institute, P. O. Box 933, Manila, Philippines. IRRI Research Paper Series No. 134.

127. 1988. McCouch, S. R., G. Kochert, Z. H. Yu, Z. Y. Wang, G. S. Khush, W. R. Coman and S. D. Tanksley. Molecular mapping of rice chromosomes. Theor. Appl. Genet. 76:815-829.
128. 1988. Saha, A.K. and G. S. Khush. Genetic analysis for resistance to four Philippine races of bacterial blight in three cultivars of rice *(Oryza saliva L.).* SABRAO Jour. 20:83-90.
129. 1988. Khush, G. S., B. O. Juliano and D. B. Roxas. Genetic selection for improved nutritional quality of rice straw — A plant breeder's viewpoint. In J. D. Reed et al. (Eds). Plant breeding and the nutritive value of crop residues. International Livestock Center for Africa, P. O. Box 5689, Addis Ababa, Ethiopia. pp. 261-281.
130. 1988. Khush, G. S. and S. S. Virmani. Plant breeding strategies for biological control of plant diseases. In K. G. Mukerjee and K. L. Garg, Eds. Biocontrol of Plant Diseases. Vol. II. CRC Press Inc., Boca Raton, Florida. pp. 37-42.
131. 1988. Chang, T.T., D.V. Seshu and G.S. Khush. The rice seed exchange and evaluation program of IRRI. In Rice Seed Health. International Rice Research Institute, P.O. Box 933, Manila, Philippines. pp. 21-32.
132. 1988. Khush, G.S., I.Kumar and S.S. Virmani. Grain quality of hybrid rice. In Hybrid Rice. International Rice Research Institute, P.O. Box 933, Manila, Philippines. pp. 201-215.
133. 1988. Mew, T.W., F.M. Wang, J.T. Wu, KR Lin and G.S. Khush. Disease and insect resistance in hybrid rice. In Hybrid Rice. International Rice Research Institute, P.O. Box 933, Manila, Philippines. pp. 189-200.
134. 1988. Khush, G.S. and D.S. Brar. Wide hybridization in plant breeding. In Plant Breeding and Genetic Engineering. A.H. Zakri Ed. SABRAO, Malaysia. pp. 141-188.
135. 1989. Khush, G.S. Innovative approaches to rice improvement. In Rice Farming Systems, New Directions. International Rice Research Institute, P.O. Box 933, Manila, Philippines. pp. 71-84.
136. 1898. Khush, G.S. Multiple disease and insect resistance for increased yield stability in rice. In Progress in Irrigated Rice Research. International Rice Research Institute, P.O. Box 933, Manila, Philippines. pp. 79-92.
137. 1989. Khush, G.S., D.J. Mackill and G.S. Sidhu. Breeding rice for resistance to bacterial blight. In Bacterial Blight of Rice. International Rice Research Institute, P.O. Box 933, Manila, Philippines. pp. 207-217.
138. 1989. Ogawa, T. and G.S. Khush. Major genes for resistance to bacterial blight in rice. In Bacterial Blight of Rice. International Rice Research Institute, P.O. Box 933, Manila, Philippines. pp. 177-192.
139. 1989. Sahu, R.K. and G.S. Khush. Inheritance of resistance to bacterial in seven cultivars of rice. Plant Disease 73:688-690.

140. 1989. Sahu, R.K. and G.S. Khush. Genetics of resistance to four races of *Xanthomonas campestris pv. oryzae* in some rice cultivars. Plant Breed. 102:232-236.
141. 1989. Jena, K.K. and G. S. Khush. Monosomic alien addition lines of rice: production, morphology, cytology and breeding behavior. Genome 32:449-455.
142. 1989. Dela Cruz, Normita, I. Kumar, R.P. Kaushik and G.S. Khush. Effect of temperature during grain development on stability of cooking quality components in rice. Japan. J. Breed. 39:299-306.
143. 1989. Khush, G.S. and K.K. Jena. Biosystematic status of *Oryza indandamanica* Ellis. In Proceedings of the 6ᵗʰ International Congr. SABRAO, Tsukuba, Japan. pp. 179-198.
144. 1989. Abdullah, R., J.A. Thompson, G.S. Khush, R.P. Kaushik and E.C. Cocking. Protoclonal variation in the seed progeny of plants regenerated from rice protoplasts. Plant Science 65:97-101.
145. 1989. Tang, S.X., G. S. Khush and B.O. Juliano. Diallele analysis of gel consistency in rice *(Oryza sativa L.)*. Sabrao Journal 21:135-142.
146. 1989. Tang, S.X., G.S. Khush and B.O. Juliano. Variation and correlation of four cooking and eating quality indices of rices. Philipp. J. Crop Sci. 14:45-49.
147. 1990. Khush, G.S. Rice breeding - accomplishments and challenges. Plant Breeding Abstracts 60:461-467.
148. 1990. Khush, G.S., S.S. Virmani and D.S. Brar. Expectations of plant breeders from tissue culture. In S.S. Bhojwani (Ed.) Plant Tissue Culture: Applications and Limitations. Elsevier Amsterdam - Oxford - New York - Tokyo. pp. 412-423.
149. 1990. Ogawa, T., T. Yamamoto, G.S. Khush and T.W. Mew. Genetics of resistance in rice cultivars, Chugoku 45 and Java 14 to Philippine and Japanese races of bacterial blight pathogen. Japan. J. Breed. 40:77-90.
150. 1990. Ogawa, T., T. Yamamoto, G.S. Khush and T.W. Mew. Genetics of resistance in rice cultivars, Zenith and Cempo Selak to Philippine and Japanese races of bacterial blight pathogen. Japan. J. Breed. 40:183-192.
151. 1990. Ogawa, T., T. Yamamoto, G.S. Khush and T.W. Mew. Genetics of resistance in rice cultivar Sateng to Philippine and Japanese races of bacterial blight pathogen. Japan. J. Breed. 40:329-338.
152. 1990. Ogawa, T., T. Yamamoto, G.S. Khush and T.W. Mew. Genetics of resistance in rice cultivar, IR20 and Semora Mangga to Philippine and Japanese races of bacterial blight pathogen. Japan. J. Breeding 40:435-447.
153. 1990. Jena, K.K. and G.S. Khush. Introgression of genes from *Oryza officinalis* Well ex Watt to cultivated rice, *O. saliva L.* Theor. Appl. Genet. 80:737-745.

154. 1990. Glaszmann, J.C., R.N. Kaw and G.S. Khush. Genetic divergence among cold tolerant rices *(Oryza saliva L.)*. Euphytica 45:95-104.

155. 11990. Juliano, B.O., C.M. Perez, R. Kaushik and G.S. Khush. Grain properties of 1R36-based starch mutants. Starch 42:256-260.

156. 1990. Ishii, T., O. Panaud, D. S. Brar and G. S. Khush. Use of non radioactive digoxigenin-labelled DNA probes for RFLP analysis in rice. Plant Molecular Biology Reporter 8:167-171.

157. 1990. Chaudhary, R.C. and G.S. Khush. Breeding rice varieties for resistance against Chilo spp. of stem borers in Asia and Africa. Insect Sci. Applic. 11:659-669.

158. 1990. Mir, G.N. and G.S. Khush. Genetic analysis of some rice cultivars for resistance to bacterial blight. Indian J. Mycol. Plant Pathol. 20(2):130-134.

159. 1991. Khush, G.S. and D.S. Brar. Genetics of resistance to insects in crop plants. Adv. in Agron. 45:223-274.

160. 1991. Khush, G. S. and T. Kinoshita. Rice karyotype, marker genes, and linkage groups. In G.S. Khush and G.H. Toenniessen (Eds) Rice Biotechnology. C.A.B. International Wallingford, England and International Rice Research Institute, P.O. Box 933, Manila, Philippines. pp. 83-108.

161. 1991. Khush, G.S. and R.J. Singh. Chromosome architecture and aneuploids in rice. In P.K. Gupta and T. Tsuchiya, Eds. Chromosome Engineering in Plants: Genetics, Breeding, Evolution Part A. Elsevier, Amsterdam. pp. 577-598.

162. 1991. Ogawa, T., G.A. Bustos, R.E. Tabien, G.O. Romero, N. Endo and G.S. Khush. Grouping of rice cultivars based on reaction pattern to Philippine races of bacterial blight pathogen *(Xanthomonas campestris* pv. *oryzae)*. Japan J. Breed. 41:109-119.

163. 1991. Ogawa, T., T. Yamamoto, G.S. Khush and T.W. Mew. Resistance and its inheritance to bacterial blight of 1R8 rice cultivar group. Japan. J. Breed. 41:211-222.

164. 1991. Endo, N., G.A. Bustos, T. Ogawa and G.S. Khush. Rice cultivar groups in Myanmar based on reaction to bacterial blight. Japan. J. Breed. 41:289-300.

165. 1991. Ogawa, T., T. Yamamoto, G. S. Khush and T.W. Mew. Breeding of near isogenic lines of rice with single genes for resistance to bacterial blight pathogen *(Xanthomonas campestris* pv. *oryzae)*. Japan. J. Breed. 41:523-529.

166. 1991. Tang, S.X., G.S. Khush and B.O. Juliano. Genetics of gel consistency in rice. J. Genet. 70:69-78.

167. 1991. Kaushik, R.P. and G. S. Khush. Genetic analysis of endosperm mutants in rice, *Oryza sativa L.* Theor. Appl. Genet. 83:146-152.

168. 1991. Kaushik, R.P. and G.S. Khush. Endosperm mutants in rice: gene expression in Japonica and indica backgrounds. Cereal Chemistry 68:487-491.

169. 1991. Brar, D.S., B.G. delos Reyes, O. Panaud, A. Sanchez, and G.S. Khush. Genetic mapping in rice using isozymes and RFLP markers. In Rice Genetics II. International Rice Research Institute, P.O. Box 933, Manila, Philippines. pp. 137-145.
170. 1991. McCouch, S.R., G. S. Khush and S.D. Tanksley. Tagging genes for disease and insect resistance via linkage to RFLP markers. In Rice Genetics II. International Rice Research Institute, P.O. Box 933, Manila, Philippines. pp. 443-449.
171. 1991. Mir, G.N. and G. S. Khush. Genetics of resistance to bacterial blight in rice *(Oryza sativa L.).* Indian J. Genet. 51:72-78.
172. 1992. Khush, G.S. Selecting rice for simply inherited resistances. In H.S. Stalker and J.P. Murphy, Eds. Plant Breeding in 1990's. C.A.B. International Wallingford, England. pp. 303-323.
173. 1992. Khush, G.S. and G.H. Toenniessen. Biotechnology and pest management in 2000 AD. In A. Aziz, S.A. Kadir and H.S. Barlow, Eds. Pest Management and the Environment in 2000. C.A.B. International, Wallingford, England. pp. 348-359.
174. 1992. Khush, G. S. and D. S. Brar. Overcoming barriers in hybridization. In G. Kallo and J.B. Chaudhury, Eds. Distant hybridization of crop plants. Theor. Appl. Genet. Monograph Ser. 16:47-61.
175. 1992. Bonman, J.M., G.S. Khush, and R.J. Nelson. Breeding rice for resistance to pests. Ann. Rev. Phytopath. 30:485-506.
176. 1992. Pooni, H.S., Ish Kumar and G.S. Khush. A comprehensive model for disomically inherited metrical traits expressed in triploid tissues. Heredity 69:166-174.
177. 1992. Taura, S., T. Ogawa, R.E. Tabien, G.S. Khush, A. Yoshimura and T. Omura. Resistance gene of rice cultivar, Taichung Native I to Philippine races of bacterial blight pathogen. Jap. J. Breed. 42:195-202.
178. 1992. Endo, N., T. Ogawa and G.S. Khush. Genetic analysis of Myanmar rice cultivars for resistance to bacterial blight. Jap. J. Breed. 42:341-352.
179. 1992. Jena, K.K., G.S. Khush and G. Kochert. RFLP analysis of rice *(Oryza saliva L.)* introgression lines. Theor. Appl. Genet. 84:608-616.
180. 1993. Brar, D.S. and G.S. Khush. Application of biotechnology in integrated pest management. J. Insect Sci. 6:7-14.
181. 1993. Khush, G.S. Biotechnology in Agriculture. In M. Nazim and N. Polunin, eds. Environmental challenges: From Stockholm to Rio and beyond. Published by Energy and Environment Society of Pakistan and Foundation for Environmental Conservation, Geneva, Switzerland. pp. 117-148.
182. 1993. Khush, G.S. Varietal needs for different environments and breeding strategies. In K. Muralidharan and E.A. Siddiq, Eds. New Frontiers in Rice Research, Directorate of Rice Research, Hyderabad, India. pp. 68-75.

183. 1993. Khush, G.S., D.S. Brar, F.J. Zapata, R. Nelson, S. McCouch and D.G. Bottrell. In B.C. Imrie and J.B. Hacker, Eds. Focused Plant Improvement: Towards Responsible and Sustainable Agriculture. Proceedings Tenth Australian Plant Breeding Conference. Published by Conference Organizing Committee, Canberra, Australia. Biotechnology for rice improvement Vol. I. pp. 258-279.
184. 1993. Khush, G. S. Breeding rice for sustainable agricultural systems. In D.R. Buxton, R. Shibles, R.A. Forsberg, B.L. Blad, K.H. Asay, G.M. Paulson, R.F. Wilson, Eds. International Crop Science I. Crop Science Society of America, Madison, Wisconsin, USA. pp. 189-199.
185. 1993. Bennett, J, and G. S. Khush. In McGraw-Hill Yearbook of Science and Technology. McGraw-Hill Inc., New York. pp. 321-324.
186. 1993. Pooni, H.S., I. Kumar and G.S. Khush. Genetical control of amylose content in a diallel set of rice crosses. Heredity 71:603-613.
187. 1994. Abenes, M.L.P., R.E. Tabien, S.R. McCouch, R. Ikeda, P. Ronald, G.S. Khush and N. Huang. Orientation and integration of the classical and molecular map of chromosome I 1 in rice. Euphytica 76:81-87.
188. 1994. Fukui, K., N. Ohmido and G.S. Khush. Variability in rDNA loci in the genus *Oryza* detected through fluorescence in situ hybridization. Theor. Appl. Genet. 87:893-899.
189. 1994. Ishii, T., D. S. Brar, D. S. Multani and G. S. Khush. Molecular tagging of genes for brown planthopper resistance and earliness introgressed from *Oryza australiensis* into cultivated rice, O. *saliva.* Genome 37:217-221.
190. 1994. Jena, K.K., G.S. Khush and G. Kochert. Comparative RFLP mapping of a wild rice, *Oryza officinalis* and cultivated rice, O. *saliva.* Genome 37:382-389.
191. 1994. Multani, D. S., K.K. Jena, D. S. Brar, B. G. delos Reyes, E.R. Angeles, and G. S. Khush. Development of monosomic alien addition lines and introgression of genes from *Oryza australiensis* Domin. to cultivated rice, O. *saliva L.* Theor. Appl. Genet. 88:102-109.
192. 1994. Pooni, H. S., Ish Kumar and G. S. Khush. A general method of detecting additive dominance, and epistatic variation for metrical traits. V. Triple test cross analysis of disomically inherited traits expressed in triploid tissues. Heredity 72:563-569.
193. 1994. Sanchez, A.C. and G.S. Khush. Chromosomal location of some marker genes in rice using primary trisomics. J. Hered. 85:297-300.
194. 1994. Virmani, S.S., G.S. Khush and P.L. Pingali. Hybrid rice for tropics: Potentials, research priorities and policy issues. In R. S. Paroda and Mangala Rai, Eds. Hybrid research and development needs in major cereals in the Asia-Pacific region. Food and Agriculture Organization of the United Nations, Regional Office for Asia and Pacific, Bangkok. pp. 61-86.

195. 1994. Khush, G.S. and R.C. Aquino. Breeding tropical japonicas for hybrid rice production. In Virmani S.S., Ed. Hybrid rice technology: New developments and future prospects. International Rice Research Institute, P.O. Box 933, Manila 1099, Philippines. pp. 33-36.
196. 1994. Khush, G.S., D.S. Brar, F.J. Zapata, R. Nelson, S. McCouch, and D.G. Bottrell. Rice biotechnology at IRRI. In Towards enhanced and sustainable agricultural productivity in the 2000's: Breeding research and biotechnology. Taichung District Agricultural Improvement Station and SABRAO. Pages 387-414.
197. 1994. Khush, G.S. Increasing the genetic yield potential of rice: prospects and approaches. International Rice Commission Newsletter 43:1-8.
198. 1994. Brar, D. S. and G. S. Khush. Cell and tissue culture for plant improvement. In A.S. Basra, ed. Mechanisms of plant growth and improved productivity. Modern Approaches, Marcel Dekker, Inc., New York. pp. 229-278.
199. 1994. Khush, G. S. Rice Genetics and Breeding. Encyclopedia of Agricultural Science 3:607-616.
200. 1995. Panda, N. and G. S. Khush. Host Plant Resistance to Insects. C.A.B. International, Wallingford, England and International Rice Research Institute, P.O. Box 933, Manila, Philippines. 431 pp.
201. 1995. Khush, G. S. Modern varieties - their real contribution to food supplies and equity. Geo Journal 335(3):275-284.
202. 1995. Khush, G.S. Breaking the yield barrier of rice. Geo Journal 35(3):329-332.
203. 1995. Alejar M, F.J. Zapata, D. Senadhira, G.S. Khush and S.K. Datta SK. Utilization of anther culture as a breeding tool in rice improvement. In M. Terzi et al., eds. Current issues in plant molecular and cellular biology. Kluwer Academic Publishers, Netherlands. Pages 137-143.
204. 1995. Zheng, K., N. Huang, J. Bennett, and G.S. Khush. PCR-based marker-assisted selection in rice breeding, International Rice Reserach Institute, Manila Philippines. IRRI Discussion Paper Series No 12.
205. 1995. Brar, D. S. and G. S. Khush. Wide hybridization for enhancing resistance to biotic and abiotic stresses in rainfed lowland rice. In Fragile lives in fragile ecosystems. Proc. International Rice Research Conference. IRRI, P.O. Box 933, Manila, Philippines. Pages 901-910.
206. 1995. Hittalmani, S, T. Mew, G.S.Khush and N. Huang. DNA marker-assisted breeding for disease resistance in rice. In Fragiles lives in fragile ecosystems. Proc. International Rice Research Conference, IRRI, P.O. Box 933, Manila, Philippines. pp. 947-961.
207. 1995. Katiyar, S.K., Y. Tan, Y. Zhang, B. Huang, Y. Xu, L. Zhao, N. Huang, G.S. Khush and J. Bennett. Molecular tagging of gall midge resistance genes

in rice. In Fragile lives in fragile ecosystems. Proc. of International Rice Research Conference. IRRI, P.O. Box 933, Manila, Philippines. pp. 935-948.

208. 1995. Khush, G. S., S. S. Virmani and P.K. Aggarwal. Expectations of plant breeders from crop physiology and modelling. In Applications of systems approaches in plant breeding. SARP Research Proceedings, AB-DLO, TPE-WAU. P.K. Aggarwal, R.B. Mathews, M.J. Kropff and H. H. van Laar (eds). Wageningen and IRRI, P.O. Box 933, Manila, Philippines. pp. 13-22.
209. 1995. Khush, G.S., V.T. Xuan, N.V. Luat, B.C. Buu, D.T. Tuan and V.T. Hoang. Vietnam-IRRI collaboration in rice varietal improvement. In Vietnam and IRRI: A partnership in rice research. G.L. Denning and V.T. Xuan, eds. IRRI and Ministry of Agriculture and Food Industry, Hanoi, Vietnam. pp. 55-60.
210. 1995. Luu, N.T. D.T. Tuan, D.S. Brar, B.G. delos Reyes and and G.S. Khush. Classification of traditional rice germplasm from Vietnam based on isozyme pattern. In Vietnam and IRRI: A partnership in rice research. G L. Denning and V.T. Xuan, eds. IRRI and Ministry of Agriculture and Food Industry, Hanoi, Vietnam. pp. 81-83.
211. 1995. Khush, G. S. Biotechnology approaches to rice improvement. In Induced mutations and molecular techniques for crop improvement. International Atomic Energy Agency, Vienna, Austria. pp. 455-475.
212. 1995. Bharaj T. S., S. S. Virmani and G. S. Khush. Chromosomal location of fertility restoring genes for 'wild abortive' cytoplasmic male sterility using primary trisomics in rice. Euphytica 83:169-173.
213. 1995. Dalmacio R, D. S. Brar, T. Ishii, L. S. Sitch, S. S. Virmani and G. S. Khush. Identification and transfer of a new cytoplasmic male sterility source from *Oryza perennis* into indica rice *(O. saliva).* Euphytica 82:221-225.
214. 1995. Ghareyazie B, N. Huang, G. Second, J. Bennett and G. S. Khush. Classification of rice germplasm. I. Analysis using ALP and PCR-based RFLP. Theor. Appl. Genet. 91:218-227.
215. 1995. Ishii T, D.S. Brar, G. Second, K. Tsunewaki and G.S. Khush. Nuclear genome differentiation in Asian cultivated rice as revealed by RFLP analysis. Jap. J. Genet. 70:643-652.
216. 1995. Kabir, A.M., M. Aktaruzzaman and G. S. Khush. Allelic relationships of dominant genes for resistance to brown planthopper in five rice varieties. Sabrao Journal 27:31-33.
217. 1996. Khush, G.S. and S.S. Virmani. Haploids in Plant Breeding. In S. M. Jain, S. K. Sopory & R. E. Veilleux, eds. In Vitro Haploid Production in Higher Plants. Kluwer Academic Publisher. Pages 9-31.
218. 1996. Khush, G.S. Prospects of and approaches to increasing the genetic yield potential of rice. In R.E. Evenson, R.W. Herdt and M. Hossain, eds. Rice Research in Asia: Progress and Priorities. Pages 59-71.

219. 1996. Kohli, A., B. Ghareyazie, H. S. Kim, G. S. Khush and J. Bennett. Cytosine methylation implicated in silencing of B-glucuronidase genes in transgenic rice. In G.S. Khush, ed. Rice Genetics III. International Rice Research Institute, P.O. Box 933, Manila, Philippines. pp. 825-828.
220. 1996. Brar, D. S., R. Dalmacio, R. Elloran, R. Aggarwal, ER. Angeles and G. S. Khush. Gene transfer and molecular characterization of introgression grom wild *Oryza* species into rice. In G. S. Khush, ed. Rice Genetics III. International Rice Research Institute, P.O. Box 933, Manila, Philippines. pp. 477-486.
221. 1996. Khush, G. S., K. Singh, T. Ishii, A. Parco, N. Huang, D. S. Brar and D. S. Multani. Centromere mapping and orientation of the cytological, classical, and molecular linkage maps of rice. In G.S. Khush, ed. Rice Genetics III. International Rice Research Institute, P.O. Box 933, Manila, Philippines. pp. 57-75.
222. 1996. Robeniol, J.A., S.V. Constantino, A.P. Resurreccion, C.P. Villareal, B. Ghareyazie, B.R. Lu, S.K. Katiyar, N. Huang, C.A. Menguito, E.R. Angeles, H.Y. Fu, S. Reddy, W. Park, S.R. McCouch, G.S. Khush and J. Bennett. Sequence-tagged sites and low-cost DNA marker technology for rice. In G.S. Khush, ed. Rice Genetics III. International Rice Research Institute, P.O. Box 933, Manila, Philippines. pp. 293-306.
223. 1996. Khush, G.S. Biotechnology for rice improvement. In R. Ishii and T. Horie, eds. Crop Research in Asia: Achievements and Perspectives. Proceedings of 2nd Asian Crop Science Congress. "Towards Improvement of Food Production under Steep Population Increase and Global Environment Change" Crop Science Society of Japan, Tokyo. pp. 320-325.
224. 1996. Khush, G. S. Genetic improvement of rice for weed management. In R. Naylor, ed. Herbicide in Asian rice: transition in weed management. Palo Alto (California): Institute for International Studies, Stanford University, and Manila (Philippines): International Rice Research Institute. pp. 201-207.
225. 1996. Khush, G.S. and S. Peng. Breaking the yield frontier of rice. In M.P. Reynolds, S. Rajaram and A. McNab, eds. Increasing the yield potential in wheat: Breaking the barriers. CIMMYT D.F. Mexico. pp. 36-51.
226. 1996. Datta, S.K., K. Datta, L. Torrizo, M.F. Alam, J. Tu, Y. Fan, I. Altosaar, J. Bentur, and G. S. Khush. Production of efficient Bt rice and application. In Proceedings of 2nd Pacific Rim Conference on Biotechnology of *Bacillus thuringiensis* and its impact on the enviroment. The Pacific Rim Bt Conference, Chiang Mai, Thailand. Pages 401-411.
227. 1996. Singh, K., T. Ishii, A. Parco, N. Huang, D. S. Brar and G. S. Khush. Centromere mapping and orientation of the molecular linkage map of rice *(Oryza sativa L.)*. Proc. Natl. Acad. Sci. USA 93:6163-6168.

228. 1996. Singh, K., D.S. Multani and G.S. Khush. Secondary trisomics and telotrisomics of rice: Origin, characterization and use in determining the orientation of chromosome map. Genetics 143:517-529.

229. 1996. Tan, Y.J., Y. Zhang, B.C. Huang, Y.K. Xu, L.X. Zhao, S.K. Katiyar, N. Huang, G.S. Khush, and J. Bennett. Analysis of resistance genetics and mapping of resistance gene in a rice variety Duokang 1 to gall midge. Acta. Phytophylacica Sinica 23:315-320.

230. 1996. Zhang, G., E.R. Angeles, M.L.P. Abenes, G.S. Khush and N. Huang. RAPD and RFLP mapping of bacterial blight resistance gene *xa-13* in rice. Theor. Appl. Genet. 93:65-70.

231. 1996. Huang, N. , B. Courtois, G. S. Khush, H. Lin, G. Wang, P. Wu and K. Zheng. Association of quantitative trait loci for plant height with major dwarfing genes in rice. Heredity 74:130-137.

232. 1996. Mukhopadhay, K., N.K. Garrison, D.M. Hinton, C.W. Bacon, G.S. Khush, H.D. Peck and N. Datta. Identification and characterization of bacterial endophytes of rice. Mycopathologia 134:151-159.

233. 1997. Khush, G.S. Origin, dispersal, cultivation and variation of rice. Plant Molecular Biology 35:25-34.

234. 1997. Brar, D. S. and G. S. Khush. Alien introgression in rice. Plant Molecular Biology 35:35-47.

235. 1997. Bennett, J., M.B. Cohen, S. K. Katiyar, B. Ghareyazie and G. S. Khush. Enhancing insect resistance in rice through biotechnology. In N. Carozzi and M. Koziel, eds. Advances in insect control: The role of transgenic plants. Francis and Taylor, London. pp. 75-93.

236. 1997. Brar, D.S. and G.S. Khush. Wide hybridization for rice improvement: Alien gene transfer and molecular characterization of introgression. In M.P. Jones, M. Dingkuhn, D.E. Johnson and S.O. Fagade, eds. Interspecific hybridization: Progress and Prospects. WARDA, 01 BP 2551, Bouake, Cote d'Ivoire. pp. 21-29.

237. 1997. Aggarwal, P.K., M.J. Kropff, P.S. Teng and G.S. Khush. The challenge of integrating systems approach and limitations. In M.J. Kropff, P.S. Teng, P.K. Aggarwal, J. Bouma, B.A.M. Bouman, J.W. Jones, H.H. Van Laar, eds. Application of Systems Approaches at Field Level Vol. 2. Kluwer Academic Publishers, Netherlands. pp. 1-23.

238. 1997. Datta, K., L. Torrizo, N. Oliva, M.F. Alam, C. Wu, E. Abrigo, A. Vasquez, J. Tu, C. Quimio, M. Alejar, Z. Nicola, G.S. Khush and S.K. Datta. Production of transgenic rice by protoplast, biolistic and Agrobacterium systems. In Proceedings, Fifth International Symposium on Rice Molecular Biology. Yi Hsien Pub. Co. Taipei, Taiwan. Pages 159-167.

239. 1997. Endo, N., T. Ogawa and G. S. Khush. Isozyme classification of Myanmar rice cultivars resistant to bacterial blight. Breed. Sci. 47:27-32.

263. 1999. Peng, S., R.C. Laza, R.M. Visperas, A.L. Sanico, K.G. Cassman and G.S. Khush. Grain yield of rice cultivars and lines developed in the Philippines since 1966. Crop Science 40:307-314.
264. 1999. Peng S., K. G. Cassman, S. S. Virmani, J. Sheehy and G. S. Khush. Yield potential trends of tropical rice since the release of IR8 and the challenge of increasing rice yield potential. Crop Science 39:1552-1559.
265. 1999. Jalodar N.B., N.W. Blackhall, T.P.W. Hartman, D.S. Brar, G.S. Khush, M.R. Davey, E.C. Cocking, and J.B. Power. Intergeneric somatic hybrids of rice *[Oryza saliva L. (+) Porleresia coarctala* (Roxb.) Tateoka]. Theor. Appl. Genet. 99:570-577.
266. 2000. Fischer, K.S., J. Barton, G.S. Khush, H. Leung and R. Cantrell. Collaborations in rice. Science 290:279-280.
267. 2000. Khush, G. S. New plant type of rice for increasing the yield potential of rice. In J.S. Nanda, ed. Rice breeding and genetics. Science Publishers, Inc., P.O. Box 699, Enfield, New Hampshire 03748, USA. Pages 99-108.
268. 2000. Jena K.K. and G.S. Khush. Exploitation of alien species in rice improvement — opportunities, achievements and future challenges. In J.S. Nanda, ed. Rice breeding and genetics. Science Publishers, Inc., P.O. Box 699, Enfield, New Hampshire 03748, USA. Pages 271-285.
269. 2000. Singh, R.J. and G. S. Khush. Cytogenetics of rice. In J. S. Nanda, ed. Rice breeding and genetics. Science Publishers Inc. P.O. Box 699, Enfield, New Hampshire 03748, USA. Pages 287-311.
270. 2000. Dela Cruz N, and G. S. Khush. Rice grain quality evaluation procedures. In R.K. Singh and G. S. Khush, eds. Aromatic Rices. Science Publishers, Inc. Enfield U.S.A. and Plymouth, U.K. Pages 15-28.
271. 2000. Khush, G.S. Taxonomy and origin of rice. In R.K. Singh, U.S. Singh, G.S. Khush, eds. Aromatic Rices. Science Publishers, Inc. Enfield, U.S.A. and Plymouth, U.K. Pages 5-13.
272. 2000. Hossain M, J. Bennett, S.K. Datta, H. Leung and G. S. Khush. Biotechnology research in rice for Asia: Priorities, focus and directions. In M. Qaim, A.F. Krattiger and J. von Braun, eds. Agricultural Biotechnology in Developing Countries: Towards Optimizing the Benefits for the Poor. Kluwer Academic Publishers. Pages 99-120.
273. 2000. Khush G.S. and H. Leung. Plant genome research and breeding strategies for sustainable food production in 21st century. In L. Esaki, ed. New frontiers of science and technology. Proceedings of the International Conference on Science Frontier Tsukuba 999, November 17-19, 1999 at Tsukuba Center, Japan. Universal Academy Press Inc., Tokyo, Japan. Pages 15-27.
274. 2000. Khush G.S. Strategies for increasing the yield potential of rice. In Sheehy JE, Mitchell PL and Hardy B (eds). Redesigning rice photosynthesis to increase yield. International Rice Research Institute, Manila, Philippines. Pages 207-211.

275. 2000. Tu J., K. Datta, G.S. Khush, Q. Zhang, and S.K. Datta. Field performance of *Xa21* transgenic indica rice *(Oryza saliva L.),* IR72. Theor. Appl. Genet. 101:15-20.
276. 2000. Tu J., G. Zhang, K. Datta, C. Xu, Y. He, Q. Zhang, G. S. Khush and S.K. Datta. Field performance of transgenic elite commercial hybrid rice expressing *Bacillus thuringiensis* S-endotoxin. Nature Biotechnology 18:1101-1104.
277. 2000. Tsunematsu, H., M.J. Yanoria, L.A. Ebron, N. Hayashi, 1. Endo, H. Kato, T. Imbe and G. S. Khush. Development of monogenic lines of rice for blast resistance. Breeding Science 50:229-234.
278. 2000. Sripongpangkul, K., G.B. T. Posa, D. Senadhira, D. S. Brar, N. Huang, G. S. Khush, and Z. Li. Genes/QTLs affecting flood tolerance in rice. Theor. Appl. Genet. 101:1074-1081.
279. 2000. Angeles, E.R. and G.S. Khush. Genetics of resistance to green leafhopper in five cultivars of rice, *Oryza saliva L.* SABRAO Journal 32:1-4.
280. 2000. Sanchez, A.C., D.S. Brar, N. Huang, Z. Li and G.S. Khush. Sequence tagged site marker-assisted selection for three bacterial blight resistance genes in rice. Crop Sci. 40:792-797.
281. 2000. Zhongchao, Y, J. Chen, L. Zeng, M. Goh, H. Leung, G.S. Khush, and G.L. Wang. Characterizing rice lesion mimic mutants and identifying mutant with broad spectrum resistance to rice blast and bacterial blight. MPMI 13:869-876.
282. 2000. Angeles, E.R. and G.S. Khush. Genetic analysis of resistance to green leafhopper, *Nephotettix virescens* (Distant) in three varieties of rice. Plant Breeding 119:446-448.
283. 2000. Sanchez, A.C. and G.S. Khush. Chromosomal localization of five mutant genes in rice, Oryza sativa L. using primary trisomics. Plant Breeding 119-84-86.
284. 2000. Imbe, T., H. Tsunematsu, H. Kato and G. S. Khush. Genetic analysis of blast resistance in IR varieties and resistant breeding. In D. Tharreau et al., eds. Advances in Rice Blast Research. Kluwer Academic Publishers. Netherlands. Pages 1-8.
285. 2000. Sheehy, J., P. Mitchell, J. Dionora, T. Tsukaguchi, S.B. Peng and G.S. Khush. Unlocking the yield barrier in rice through a nitrogen-led improvement in the radiation conversion factor. Plant Prod. Sci. 3(4): 372-374.
286. 2001. Khush, G. S. Green revolution: the way forward. Nature Reviews Genetics 2: 815-822.
287. 2001. Khush G. S. Challenges for meeting the global food and nutrient needs in the new millennium. Proc. Nutrition Society 60:15-26.
288. 2001. Khush G.S. Super rice for increasing the genetic yield potential. In Reeves, E.C.R. ed. Encyclopedia of Genetics. Fitzroy Dearborn Publishers, London, Chicago. Pages 663-667.

289. 2001. Khush G.S. Rice improvement at IRRI: An example of International Collaboration, Breeding Research 3:281-286.

290. 2001. Khush G.S and D.S. Brar Rice genetics from Mendel to functional genomics. In Rice Genetics IV Science Publishers, G. S. Khush, D. S. Brar, B. Hardy, eds. New Delhi, and International Rice Research Institute, Los Banos, Laguna. Pages -25.

291. 2001. Bennett J, X. Bi, A. Kathiresan, and G.S. Khush. Molecular tools for achieving synthetic apomixis in rice. In Rice Genetics IV. Science Publishers, Khush G.S., D.S. Brar and B. Hardy, eds. New Delhi, and International Rice Research Insitute, Manila. pp 377-401.

292. 2001. Kihupi, A.N., E.R. Angeles and G.S. Khush. Genetic analysis of resistance to bacterial blight, *Xanthomonas oryzae pv. Oryzae* in rice, *Oryza sativa* L. Euphytica 117:39-46.

293. 2001. Lou, L.J., Z.K. Li, H.W. Mei, Q.Y. Shu, R. Tabien, D.B. Zhong, C.S. Ying, J.W. Stansel, G.S. Khush and A.H. Paterson. Overdominant epistatic loci are the primary genetic basis of inbreeding depression and heterosis in rice. II. Grain yield components. Genetics 158:1755-1771.

294. 2001. Li, L.K., L.J. Luo, H.W. Mei, D.L. Wang, Q.Y. Shu, R. Tabien, D.B. Zhong, C.S. Ying, J.W. Stansel, G.S. Khush and A.H. Paterson. Overdominant epistatic loci are the primary genetic basis of inbreeding depression and heterosis in rice. I. Biomas and grain yield. Genetics 158:1737-1753.

295. 2001. Li, L.K., A.C. Sanchez, E.R. Angeles, S. Singh, J. Domingo, N. Huang and G.S. Khush. Are dominant and recessive plant disease resistance genes similar?: A case study of rice R genes and *Xanthomonas oryzae pv. Oryzae* races. Genetics 159:757-765.

296. 2001. Singh, S., J. Sidhu, N. Huang, Y. Vikal, Z. Li, D. S. Brar, H. S. Dhaliwal and G. S. Khush. Pyramiding three bacterial blight resistance genes *(xa5, xa13* and *Xa21)* using marker-assisted selection into indica rice cultivar PR106. Theor. Appl. Genet. 102-1011 - 1015.

297. 2002. Hittalmani, S., H. E. Shashidhar, P. G. Bagli, N. Huang, J. S. Sidhu, V. P. Singh, and G. S. Khush. Molecular mapping of quantitative trait loci for plant growth, yield and yield related traits across three diverse locations in doubled haploid rice population. Euphytica 125: 207-214.

298. 2002. Jauhar, P.P. and G. S. Khush. Importance of biotechnology in global food security. Page 107-128. In R. Lal, D. Hanson, N. Uphoff and S. Slack eds. Food security and environmental quality in the developing world. Lewis Publishers.

299. 2003. Bennett, J. and G. S. Khush. A strategy for enhancing salt tolerance in crops through molecular breeding. J. Crop Breed. 7: 1-30.

300. 2003. Brar, D. S. and G. S. Khush. Utilization of wild species of genus Oryza in rice improvement. Pages 283-309 In monograph on genus Oryza.

J.S. Nanda and S.D. Sharma eds. Science Publishers Inc. Enfield (NH) USA, Plymouth, USA.

301. 2003. Khush, G.S., D.S. Brar, P.S. Virka, S.X. Tang, S.S. Malik, G.A. Bastos, Y.T. Lee, R. McNally, L.N. Trinh. Y. Jiang and M.A.M. Shata. Classifying rice germplasm by isozyme polymorphism and origin of cultivated rice. Discussion paper series No 46. Los Banos (Philippines): International Rice Research Institute. 279 P.
302. 2003. Kobayashi, S. Y. Fukua, S. Morita, T. Sato, M. Osaki. G.S. Khush Quantitative trait loci affecting flag leaf development in rice (Oryza SativaL.) Breeding Science 53: 255-262.
303. 2003. Kobayashi, S., Y. Fukula, T. Sato, M. Osaki and G. S. Khush. Molecular Marker dissection of rice (*Oryza Sativa*) plant architecture under temperate and tropical climates. Theor. Appl. Genet. 107-1350-1356.
304. 2003. Lee, K. S., S. Rasabandith, E. R. Angeles and G. S. Khush. Inheritance of resistance to bacterial blight of rice. Phytopathology 93: 147-152.
305. 2003. Multani, D. S., G. S. Khush, B. G. delos Reyes, D. S. Brar. Alian gene introgression and development of monosomic alien addition lines from *Oryza latifolia* Desv. to rice, Oryza Sativa L. Theor. Appl. Genet. 107: 395-405.
306. 2003. Peng, S. and G. S. Khush. Four decades of breeding for varietal improvement of irrigated lowland rice in the International Rice Research Institute. Plant Production Science 6-157-164.
307. 2004. Ebron, L. A., Y. Fukuta, T. Imbe, H. Kato, J.M.T. Yanoria, H. Tsunematsu, G. S. Khush and M. Yokoo. Estimation of genes for blast resistance in elite indica type rice (*Oryza sativa* L). Varieties bred at the international Rice Research Institute. Breeding Science. 53: 381-387.
308. 2004. Kabayashi, S., Y. Fukuta, T. Yagif T. Sato, M. Osaki, G. S. Khush. Identification and characterization of quantitative trait loci affecting spikelet number per panicle in rice (*Oryza sativa* L) Field Crops Research 89: 253-262.
309. 2004. Khush, G. S., E. Angeles, P. S. Virk and D. S. Brar. Breeding rice for resistance to tungro virus at IRRI. SABRAO Journal 36: 101-106.
310. 2004. Nematzadeh, Gh. A., N. Huang and G. S. Khush. Mapping the gene for aroma in rice (*Oryza sativa* L.) by bulk segregation analysis via RAPD markers. J. Agric. Sci. Technol. 6:129-137.
311. 2004. Peng, S., J. Huang, J. E. Sheehy, R. C. Laza, R. M. Visperas, X. Zhong, G. S. Centano, G.S. Khush and K. G. Cassman. Rice yields decline with higher night temperature from global warming. Proc. Natl. Acad. Sciences (USA) 101(27): 9971-9975.
312. 2005. Ebron, L. A., Y. Fukuta, T. Imbe, H. Kato, M. J. T. Yanoria, H. Tsunematsu, G. S. Khush, N. Kobayashi and M. Yokoo. Identification of blast resistance

genes in elite indica type varieties of rice (*Oryza sativa* L). SABRAO Journal 37: 19-31.

313. 2005. Khush, G. S. Green Revolution: challenges ahead. Pages 37-51. In R. Tuberosa, R. L. Phillips and M. Gale (eds.) In the wake of the double helix. From green revolution to the gene revolution. University of Bologna, Italy.
314. 2005. Khush, G. S. Taxonomy, ecology and agronomy of rice cultivation vis-a-vis genetics engineering of rice. Pages 26-37. In V.L. Chopra, S. Shantaram and R. P. Sharma eds. Biosafetu of transgenic rice. National Academy of Agricultural Sciences, New Delhi.
315. 2005. Khush, G. S. What it will take to feed 5.0 billion rice consumers in 2030? Plant Molecular Biology 59: 1-6.
316. 2005. Khush, G. S. and P.S. Virk IR varieties and their impact. Los Banos (Philippines): International Rice Research Institute. 163 P.
317. 2005. Wu, J. L., C. Wu, C. Lei, M. Baraoidan, A. Bordeos, M. R.S. Madamba, M. Ramos-Pamplona, R. Mauleon, A. Portugal, V. J. Ulat, R. Brukiewich, G. Wang, J. Leach, G. S. Khush and Hei Leung. Chemical- and irradiation-induced mutants of indica rice IR64 for forward and reverse genetics. Plant Molecular Biology. 59: 85-97.
318. 2006. Baisakh, N., S,. Rehana, M. Rai, N. Oliva, J. Tan, D. J. MacKill, G. S. Khush, K. Datta and S.K. Datta·. Marker free transgenic (MFT) near-isogenic introgression lines (NILS) of golden indica rice (Cv. IR64) with accumulation of provitamin A in the endosperm tissue. Plant Biotechnology Journal 4: 467-475.
319. 2006. Brar, D. S. and G. S. Khush. Cytogenetic manipulation and germplasm enhancement of rice (*Oryza sativa* L). Pages 115-158. *In* R. J. Singh and P.P. Jauhar eds. Genetic resources, chromosome engineering and crop improvement. Taylor and Francis, Boca Raton, London, New York.
320. 2006. Kobayashi, S., D. Araki, M. Osaki, G. S. Khush, Y. Fukuta· Localization, validation and characterization of plant type QTLs on chromosome 4 and 6 in rice (*Oryza sativa* L) Field Crops Research. 96: 106-112.
321. 2006. Subudhi, P. K., T. Sasaki, G. S. Khush. Rice. Pages 1-78 *In* Kole C. (ed) Genome mapping and molecular breeding in plants. Springer: Berlin, Heidelberg, New York.
322. 2007. Khush, G.S., E.R. Angeles, and D.S. Brar. Genetics analysis of resistance to green leafhopper, *Nephotettix virescens* (Distant), in IR rice varieties. SABRAO Journal 39(2):79-88.

Roger N. Beachy

Danforth Plant Science Center
St. Louis, Missouri, USA

2001 — for the use of recombinant DNA technology, to revolutionize plant and animal sciences, paving the way for applications to neighboring fields.

Roger Beachy, Ph.D., founding president of the Donald Danforth Plant Science Center, a non-profit private research institute in St. Louis, MO, earned his Ph.D. in Plant Pathology at Michigan State Univ.; he conducted post-doctoral positions at the Univ. of Arizona and at Cornell Univ., NY. He previously held academic positions at Washington Univ., St. Louis, and The Scripps Research Institute, La Jolla, CA, where he was co-founder of the International Laboratory for Tropical Agricultural Biotechnology. He is a member of the U.S. National Academy of Sciences and a Fellow of the American Academy of Microbiology, and the American Assoc. for the Advancement of Science. He has received numerous awards including the Wolf Prize in Agriculture, the D. Robert Hoagland Award from the Society of Plant Biologists and Ruth Allen Award from the American Phytopathological Society. Beachy serves as Chair-Elect of the AAAS Section on Agriculture, Food and Renewable Resources and is President of the International Assoc. of Plant Biotechnology, among other activities. He is recognized for his work in molecular virology and gene expression in plants, and for pioneering research in developing transgenic plants that are resistant to virus infection. His research includes: studies of mechanisms of transgenic virus resistance including, in rice and sweet potato; characterizing functional activities of transcription factors; and developing a chemical gene switching system for use in plants. Beachy has been a vocal advocate for safe applications of biotechnology to enhance efficiency of agriculture and to reduce reliance on agricultural chemicals.

CURRICULUM VITAE

President	**Professor**
Donald Danforth Plant Science Center 975 North Warson Road St. Louis, Missouri	Department of Biology Washington University St. Louis, Missouri

Education:

1966 **B.A.** - Biology, Goshen College, Goshen, Indiana
1973 **Ph.D.** - Plant Pathology, Michigan State University (advisor H. H. Murakishi), East Lansing, Michigan

Post-doctoral:

1973	Department of Agricultural Biochemistry, University of Arizona, Tucson, Arizona
1973-1976	Department of Plant Pathology (advisor M. Zaitlin), Cornell University, Ithaca, New York
1976-1978	U.S.D.A. Nutrition Laboratory, (advisor J. F. Thompson), Cornell University, Ithaca, New York

Professional Experience:

Jan. 1999 to present	President and Director, Donald Danforth Plant Science Center, 975 North Warson Road, St. Louis, Missouri
Jan. 1999 to present	Professor, Department of Biology, Washington University, St. Louis, Missouri
June 1991 to Jan. 1999	Professor and Scripps Family Chair in Cell Biology, Head, Division of Plant Biology, Co-Director, Inter-national Laboratory for Tropical Agricultural Bio-technology (ILTAB), The Scripps Research Institute, La Jolla, California
1992 to 2005	Adjunct Professor, Department of Biology, Peking University, Beijing, China
1997 to 1999	Adjunct Professor, Department of Plant Pathology, University of California, Riverside, California
Jan. 1986 to June 1991	Professor of Biology, Director of the Center for Plant Science and Biotechnology, Washington University, St. Louis, Missouri

July 1983 to Dec. 1985	Associate Professor of Biology, Washington University, St. Louis, Missouri
Sept. 1978 to July 1983	Assistant Professor of Biology, Washington University, St. Louis, Missouri
July 1976 to Sept. 1978	Research Associate, U.S. Plant, Soil and Nutrition Laboratory, Cornell University, Ithaca, New York
Jan. 1973 to July 1976	Post-doctoral Research Associate and NIH Fellow, Department of Plant Pathology, Cornell University, Ithaca, New York

PROFESSIONAL SOCIETY AFFILIATION:

National Academy of Sciences, American Society for Plant Physiology, American Phytopathological Society, American Association for the Advancement of Science, Sigma Xi, American Society for Virology, International Society of Plant Molecular Biology, American Society for Biochemistry & Molecular Biology, American Society of Plant Biologists, American Society for Cell Biology.

HONORS:

Betty Klepper Crop Science Society of America Endowed Lecture Award, 2006
Wolf Prize in Agriculture, 2001
Dennis Robert Hoagland Award, American Society of Plant Biologists, 2000
Fellow, St. Louis Academy of Science, 2000
Fellow, American Academy of Microbiology, 2000
Honorary Doctor of Science, Michigan State University, 2001
Culture for Service Award, Goshen College, 2001
Scientist of the Year, R&D Magazine, 1999
Member, National Academy of Sciences, U.S.A., 1997
Commonwealth Award for Science and Industry, Bank of Delaware, 1991
Ruth Allen Award, American Phytopathological Society, 1990
Fellow, American Association for the Advancement of Science, 1989

LOCAL RECOGNITION:

Agri-Business Leader of the Year, St. Louis Agri-Business Club, 2004
Honoree, Huntington's Disease Society of America-St. Louis Chapter, 2002
Entrepreneur of the Year, Ernst & Young, 2001
Wm. D. Phillips Technology Advancement Award, St. Louis County Economic Council, 2000
Headliner of the Year Award, San Diego Press Club, 1995

OTHER PROFESSIONAL ACTIVITIES:

Current:

Member, *St. Louis Post-Dispatch* Community Advisory Board, 2007-2008
Member, *Tropical Plant Biology* Advisory Board, 2007-2008
Chair, AAAS Section on Agriculture, Food, and Renewable Resources, 2008-2009
President, International Association for Plant Biotechnology (IAPB), 2006-2010
Member, AAAS Annual Meeting Scientific Program Committee, 2006-2009
Editor, Proceedings of the National Academy of Sciences Editorial Board, 2003-2008
Member, National Research Council Governing Board of NAS, 2006-2009
Member, Missouri Biotechnology Association (MOBIO) Board, 2006-present
Chairman, Governor's Advisory Council for Plant Biotechnology for the State of Missouri, 2005-present
Member, Malaysia's International Advisory Panel (IAP) on Biotechnology Development, 2005-present
Member, Board on Agriculture and Natural Resources, National Research Council, National Academy of Sciences, 2005-2009
Member, FertiSeeds, Ltd.; Science Advisory Board; Ness Ziona, Israel, 2004-2006

Past:

Member, American Society for Gravitational and Space Biology, 2004-present
Councilor, National Academy of Sciences, 2003-2006
Member, National Academy of Sciences Council Committee on Budget and Internal Affairs, 2003-2006
Member, UC-Davis Office of Research External Research Advisory Board, 2003-2006
Member, Editorial Board, Current Opinion in Plant Biology, 2002-present
Member, Portland State University Project Advisory Group for "Public Goods and University-Industry Relationships in Agricultural Biotechnology", 2002-2006
Member, International Crops Research Institute for the Semi-Arid Tropics (ICRISAT); Governing Board of Directors; Pantancheru, India, 2001-2007
Member, DOE Biomass Research and Development Technical Advisory Committee, 2003-2005
Member, NASA's Biological and Physical Research Maximization and Prioritization (REMAP) Task Force, 2002
Member, National Academy of Sciences/National Research Council Opportunities in Agriculture Committee, 2001-2002
Member, Burrill and Company; Board of Advisors; San Francisco, CA, 2000-2007
Member, German American Academic Council, National Academy of Sciences, 1999-2000

Advisor, World Agricultural Forum, 1998-present
NAS/NRC Commission on Life Sciences, Board on Biotechnology, Committee on Biobased Industrial Products, 1998-2001
Councilor, American Society for Virology, 1996-1999
Councilor for Plant Virology, American Society for Virology, 1996-1999
Advisor, South Korean Association for Plant Molecular Biology and Biotechnology, 1995
Editor, Journal of Virology, 1995-2000
Editor, Plant Molecular Biology, supplement: Molecular Breeding, 1994-1998
Member, Scientific Advisory Board - Bio/Technology, 1994-1997
Advisor, International Working Group on Tropical Virology, Indian Society of Virology, 1993-present
Member, County of San Diego Science Advisory Board, 1992-1999
Member, National Committee, International Union of Biological Sciences (IUBS), 1991-1997
Board of Directors, IDEALS, 1990-1996
Associate Editor, Journal of Cell Biology, 1989-1992
NAS/NRC Commission on Life Sciences, Board on Biology, Committee on Introduction of Genetically Modified Microorganisms and Plants into the Environment, 1989-1990
Editorial Board, Annual Review of Plant Physiology, 1988-1990
U.S.D.A., Working Group on National Biological Impact Assessment Program, 1988
Editor, Plant Cell Reports, 1987-1995
U.S.D.A., Workshop for Recombinant DNA Release Guidelines, 1986
NSF, Metabolic Biology Panel, 1984-1987
DOE-BER Review Panel, 1983
NIH Post-doctoral Review Panel, 1982-1985
NIH Postdoctoral Fellow, Division of Allergy and Infectious Diseases, 1975-1976
NAS/NRC Commission on Life Sciences, Board on Agriculture and Natural Resources, Committee on Pest and Pathogen Control through Management of Biological Control Agents
NRC Committee on Biological Control of Diseases

Local/Regional Professional Activities:

Member, Board of Trustees, St. Louis Symphony; St. Louis, MO, 2005-present
Board of Advisors, World Agricultural Forum; St. Louis, MO, 1999-present
Center for Emerging Technologies; Board of Directors, St. Louis, MO, 1999-2006
Member, Board of Directors, NIDUS Center for Scientific Enterprise; St. Louis, MO, 1999-present

Member, Board of Trustees, The Academy of Science of St. Louis; St. Louis, MO, 2000-present
Member, Board of Trustees, St. Louis Science Center; St. Louis, MO, 2003-present
Member, Science Advisory Board Chlorogen, Inc.; St. Louis, MO, 2003-2007
St. Louis Regional Chamber & Growth Association (RCGA), St. Louis, MO 2002-present
Member, Technology Board, Midwest Bancshares; St. Louis, MO, 2001-present
The World Affairs Council of St. Louis; St. Louis, MO, 2000-2005

PATENTS PENDING OR ISSUED:

1. U.S. Serial #06/917,027 - Protection of Plants Against Viral Infection; Capsid Protein: ISSUED
2. U.S. Serial #07/789,738; Patent #5,824,857 - Plant Promoter: ISSUED
3. U.S. Serial #08/231,209 - Protection of Plants Against Viral Infection Using a Movement Protein
4. U.S. Serial #08/192,477 - Use of TMV to Produce Peptides: ISSUED
5. U.S. Serial #08/678,559 - Method for Using *Tobacco mosaic virus* to Overproduce Peptides and Proteins
6. U.S. Serial #08/192,152 - A Cassette to Accumulate Multiple Proteins Through Synthesis of a Self-Processing Polypeptide
7. Int'l. Patent Appl. No. PCT/US97/10376 - Cassava Vein Mosaic Virus Promoters and Uses Thereof
8. U.S. Serial #09/622,500 - Int'l. Patent Appl. No. PCT/US99/04716 - Resistance in Plants to Infection by ssDNA Virus Using Inoviridae Virus ssDNA-Binding Protein, Compositions and Methods of Use: ISSUED
9. PCT/US01/50748: RF2a and RF2b Transcription Factors
10. U.S. Serial #10/888,613. Methods and compositions for regulating gene expression in plant cells

NAMED LECTURES:-

Storer Life Sciences Endowment Lecture Series, University of California-Davis, 2008
National Academy of Sciences Lectureship Series, University of Minnesota, 2007
The Horning Endowment Lecture, Oregon State University, 2007
President's Circle, NAS, Washington, D.C., 2006
Sackler Colloquium, Organizer, Washington, D.C., 2006
Betty Klepper Endowed Lecturer, Crop Science Society of America, Washington, DC, 2006
Stadler Genetics Symposium, Columbia, MO 2006
Boyer Symposium: Marine and Terrestrial Molecular Bioscience-New Frontiers, Philadelphia, PA, 2005

A. von Humboldt Symposium, The Biotechnology Revolution: Scientific and Societal Aspects, Berlin/Potsdam, Germany, 2003
USDA Beltsville Area Distinguished Lecture Series, 2002
Colorado State University Thorton-Massa Lecture Series, 2002
National African Summit, Washington, D.C., 2000
Milton Zaitlin Lecture, Cornell University, 1999
Dayton Lecture, University of Illinois at Champaign-Urbana, 1999
Varner Lecture: Washington University, 1998
College of Agricultural Sciences' Distinguished Lecture Series, Pennsylvania State University, 1995
Kosuge Memorial Lectureship, University of California at Davis, 1994
Loomis Lecture Series; Iowa State University, 1993
J.C Walker Lecture Series; University of Wisconsin at Madison, 1993
Nelson Lecture Series; Montana State University, 1992

Selected other Invited Lectures: 2000-2008

EAGLES (European Action on Global Life Sciences), Alexandria, Egypt, 2008
Science and Technology in Society Forum, Kyoto, Japan, 2007
Japan External Trade Organization Biotechnology Open Symposium, Tokyo, Japan, 2007
Plant Sciences Symposium University of California–Davis, 2007
Genomics in Agricultural Research, Purdue University, West Lafayette, Indiana, 2007
Genomics in Business International Conference, Amsterdam, The Netherlands, 2007
Chemical Heritage Foundation, Philadelphia, PA, 2007
American Society for Microbiology, Washington, DC, 2007
American Chemical Society Symposium, Chicago, 2007
Ohio Valley Affiliates for Life Sciences, Cincinnati, Ohio, 2007
100th Anniversary University of California – Riverside Citrus Research Center & Agricultural Experiment Station, Riverside, 2007
Ohio Plant Biotechnology Consortium Retreat, Wooster, Ohio, 2006
NCPGR Symposium on "Shaping the Future of Food and Nutritional Security Through Genomics," New Delhi, India, 2006
13th Wold Congress "Food Is Life" Symposium, Nantes, France, 2006
Hungarian Academy of Sciences, Budapest, Hungary, 2006
Agricultural Biotechnology Research for Public Goods and Private Goods, Keynote Address, Jurys Hotel, Washington, DC, 2006
Inter-departmental Plant Science Symposium, Texas A&M University, 2006
Centennial Celebration and World Presidential Forum, China Agricultural University, Beijing, China, 2005
ICV Symposium at the IUMS 2005 Conference, San Francisco, CA, 2005

Gordon Research Conference, Tilton, NH, 2005
Boyer Symposium, Newark, DE, 2005
NABC 17th Annual Meeting, Nashville, TN, 2005
BIO 2005, Philadelphia, PA, 2005
CGIAR Trustee meeting (IARC), Beltsville, MD, 2005
Plant Made Pharmaceuticals Conference, Montreal, Canada, 2005
ABSP II, New Delhi, India, 2005
JWG on Agricultural Biotechnology, New Delhi, India, 2004
ABIC Meeting, Cologne, Germany, 2004
Institute of Food Technology, Las Vegas, NV, 2004
RedBIO/FAO Latin American Meeting on Agricultural Biotechnology, Punta Cana, Dominican Republic, 2004
2004 North American International Trade Corridor Partnership, Kansas City, MO, 2004
2004 Environmental Biology Symposium, Taipei, Taiwan, 2004
Bibliotecha Alexandria Forum on Biotechnology and Sustainable Development, Alexandria, Egypt, 2004
American Association for the Advancement of Science, Seattle, WA, 2004
University of Guelph, Guelph, Ontario, Canada, 2004
Missouri Summit on Agroterroism , Columbia, MO, 2003
22nd Annual Meeting of the American Society for Virology, Sacramento, CA, 2003
BIO 2003 Annual Convention, Washington, DC, 2003
USDA Ministerial Conference on Agricultural Science and Technology, Sacramento, CA, 2003
43rd Midwest Regional Developmental Biology Meeting, Kansas City, MO, 2003
"In the Wake of the Double Helix: From the Green Revolution to the Gene Revolution", International Congress, Bologna, Italy, 2003
BioTechnica America, Monterrey, CA, 2003
DuPont Pioneer Conference on Chemical Switches, Chestertown, MD, 2003
BioVision 2003, The World Life Sciences Forum, Lyon, France, 2003
Tree Fruit Technology Roadmap, Leesburg, VA, 2003
Wabash College, Wabash, IN, 2003
Gordon Research Conference, Los Angeles, CA, 2003
Biotechnology in Colorado and Around the World, Colorado Biotech Association, 2003
BioMalaysia 2002, Keynote Address, Kuala Lumpur, Malaysia, 2002
American Society for Virology, Lexington, KY, 2002
American Chemical Society Sixth Annual Green Chemistry & Engineering Conference, Washington, DC, 2002
BIO2002 International Biotechnology Convention, Toronto, Canada, 2002
DuPont Seminar Series, Wilmington, DE, 2002

Regional Commerce and Growth Association's Missouri Life Sciences Summit, Lake of the Ozarks, MO, 2002
University of California-Irvine 15th Annual Chief Executive Roundtable Retreat, 2002
University of Minnesota Plant Disease & Pest Management Symposium, 2002
University of California-Irvine Virology Seminar, 2002
North Carolina State University Emerging Issues Forum, 2002
International Rice Genome Meeting, Forum X, Japan, 2002
Michigan State University Plant Breeding & Genetics Symposium, 2001
American Farm Bureau Annual Meeting, Orlando, FL, 2001
University of Minnesota Conference Series, Minneapolis, MN, 2001
Seminar Series for University of California at Davis, 2001
Knowledge Millennium II: Biotechnology – The New World, New Delhi, India, 2001
Forum Agrosante, Paris, France, 2001
National Academy of Sciences 2001 Symposium, Washington, DC, 2001
World Agricultural Forum World Congress, St. Louis, MO, 2001
2001 Congress on In Vitro Biology, St. Louis, MO, 2001
International Seminar on Future of Wheat and Wheat Research, Paris, France, 2001
2001 Cold Spring Harbor Laboratory Summer Course, Cold Spring Harbor, NY, 2001
Global Consortium of Higher Education & Research for Agriculture, San Francisco, CA, 2001
Society for Industrial Microbiology Annual Meeting, St. Louis, MO, 2001
Seminar on Agricultural Biotechnology at Academia Sinica in Taipei, Taiwan, 2001
American Chemical Society, Chicago, IL, 2001
Pharmacia/Monsanto Annual Symposium, 2001
Keynote Speaker, Iowa State University Symposium, 2001
National Association of State University and Land Grant Colleges, 2001
National Association of Conservation Districts Symposium, Colorado Springs, CO, 2000
Princeton University, Department of Molecular Biology, Princeton, NJ, 2000
Bio 2000, Boston, MA, 2000
U.S.-India High Level Science and Technology Dialogue, 2000
Missouri Venture Forum, St. Louis, MO2000
Michigan State University, East Lansing, MI, 2000
John Innes Centre, Norwich, United Kingdom, 2000
Goshen College, Goshen, IN, 2000
Max-Planck-Institut, Cologne, Germany, Symposium, 2000
Great Plains Cereals Biotechnology Symposium, Manhattan, KS, 2000

American Dietetic Association Food and Nutrition Conference and Exhibition, Denver, CO, 2000

Buenos Aires Plant Biology Lectures, Buenos Aires, Argentina 2000

35th Midwest Regional American Chemical Society Symposium, St. Louis, MO, 2000

The Commercial Club of Chicago, Chicago, IL, 2000

Kasetsart University, Nakhon Pathom, Thailand, 2000

LIST OF PUBLICATIONS

Invited Articles, Books - 1974 to present:

1. Zaitlin, M. and R.N. Beachy. 1974. Protoplasts and separated cells: Some new vistas for plant virology. In: "Proc. of the 3rd Intern. Congress for Plant Tissues and Cell Culture." Leicester, pp. 265-286.
2. Zaitlin, M. R.N. Beachy, G. Bruening, C.P. Romaine, and R. Scalla. 1976. Translation of tobacco mosaic virus RNA. In "Animal Virology," eds. D. Baltimore, A. Huang, and C.F. Fox, Academic Press, pp. 567-581.
3. Bruening, G., R.N. Beachy, and M. Zaitlin. 1979. Replication of RNA plant viruses. In "Molecular Biology of Plants," eds. I. Rubenstein, R.L. Phillips, C.E. Green, and B.G. Gengenbach, Academic Press pp. 241-272.
4. Beachy, R.N., J.F. Thompson, and J.T. Madison. 1979. Isolation and characterization of messenger RNAs that code for the subunits of soybean seed protein. In "The Plant Seed: Development, Preservation, and Germination," eds., Rubenstein, Green, Phillips and Gengenbach, Academic Press, New York, pp. 67-84.
5. Beachy, R.N., K.A. Barton, J.F. Thompson, and N.P. Jarvis. 1980. The mRNAs that code for soybean seed proteins. In "Genome Organization and Expression in Plants," C. Leaver, ed., Plenum Press, N.Y., pp. 273-281.
6. Walbot, V., R.N. Beachy, and M.C. Yao. 1980. Molecular techniques applied to polyploids. In "Polyploidy: Biological Relevance," W.H. Lewis, ed., Plenum Publ. Corp., N.Y., pp. 529-535.
7. Beachy, R.N. Molecular aspects of seed storage protein biosynthesis. 1981. In "Critical Reviews in Food Science and Nutrition", CRC Press, Vol. 16(2): 187-198.
8. Beachy, R.N., J.J. Doyle, B.F. Ladin, and M.A. Schuler. 1983. Structure and expression of genes encoding the soybean 7S seed storage proteins. In "Structure and Function of the Plant Genome", Plenum Press, N.Y.
9. Beachy, R.N., J. Bryant, J.J. Doyle, K. Kitamura, and B.F. Ladin. 1983. Molecular characterization of a soybean variety lacking a subunit of the 7S seed storage protein. ed. R.B. Goldberg. *Plant Mol Biol* A.R. Liss, Inc., N.Y. pp. 413-422.

10. Doyle, J.J., R.N. Beachy, and W.H. Lewis. 1984. Evolution of rDNA in *Claytonia* polyploid complexes. In: "Plant Biosystematics," ed. W.G. Grant, Academic Press, N.Y., pp. 321-342.
11. Beachy, R.N. 1984. Toward an understanding of gene expression in plants. In: "Proceedings of the Stadler Genetics Symposium," ed. P. Gustafson, Plenum Press, Inc., pp. 605-626.
12. Beachy, R.N. and R. Fraley. 1985. Potentials for applications of genetic engineering technology to soybeans. In "New Protein Foods," eds. A.M. Altschul and H. Wilke, Academic Press, N.Y., pp. 89-105.
13. Beachy, R.N., P. Abel, M.J. Oliver, B. De, R.T. Fraley, S.G. Rogers, and R.B. Horsch. 1985. Potential for applying genetic transformation to studies of viral pathogenesis and cross-protection. In: "Biotechnology in Plant Sciences: Relevance to Agriculture in the 1980's", eds. M. Zaitlin, P. Day, and A. Hollaender, Academic Press, pp. 265-276.
14. Beachy, R.N., P.P. Abel, R.S. Nelson, S.G. Rogers, and R.T. Fraley. 1987. Transgenic plants that express the coat protein gene of TMV are resistant to infection by TMV. In: "Molecular Strategies for Crop Protection," eds. C.S. Arntzen and C.A. Ryan. A.R. Liss, Inc., N.Y., pp. 205-213.
15. Bray, E.A., and R.N. Beachy. February, 1986. A modulation by abscisic acid of genes encoding bð-ðconglycinin in developing soybean cotyledons. In: "Proceedings of UCLA Symposium on Plant Growth Regulators," Lake Tahoe, CA.
16. Jaworski, E.G., R.T. Fraley, S.G. Rogers, R.B. Horsch, R.N. Beachy, and N.-H. Chua. 1987. Genetic transformation of plants. In: "Protein Engineering: Applications in Science, Medicine, and Industry," eds. M. Inouye and R. Sarma. Academic Press, Inc., pp. 383-391.
17. Beachy, R.N., D.M. Stark, C.M. Deom, M.J. Oliver, and R.T. Fraley. 1987. Expression of sequences of *tobacco mosaic virus* in transgenic plants and their role in disease resistance. In: "Tailoring Genes for Crop Improvement," eds. G. Bruening, J. Harada, T. Kosuge and J. Hallaender, Plenum Publ. Corp., N.Y., pp. 169-180.
18. Beachy, R.N., S.G. Rogers and R.T. Fraley. 1987. Genetic transformation to confer resistance to plant virus disease. In: "Genetic Engineering," Vol. 9, eds. J. Setlow, Plenum Press, N.Y., pp. 229-247.
19. Beachy, R.N., P. Powell Abel, R.S. Nelson, J.C. Register III, N. Tumer, and R.T. Fraley. 1987. Genetic engineering of plants for protection against virus diseases. In: "Plant Resistance to Viruses," Ciba Foundation Symposium, eds. D. Evered and S. Harnett, Wiley and Sons, pp. 151-169.
20. Beachy, R.N. 1988. Virus cross-protection in transgenic plants. In: "Plant Gene Research. Temporal and Spatial Regulation of Plant Genes," eds. D.P.S. Verma and R.B. Goldberg, Springer-Verlag, N.Y., pp. 313-327.

46. Beachy, R.N., J.H. Fitchen, and M.B. Hein. 1996. Use of plant viruses for delivery of vaccine epitopes. In: "Engineering Plants for Commercial Products and Applications," eds. Glenn B. Collins and Robert J. Shepherd, Annals of the New York Academy of Sciences, New York 792:43-50.
47. Beachy, R.N., J.H. Fitchen. 1996. Pathogen-derived resistance to plant viruses and plant viruses as vaccines. In: "Microbe Hunters – Then and Now," eds. H. Koprowski and M. Oldstone, Medi-Ed Press, Bloomington. pp. 295-305.
48. Beachy, R.N. and C.M. Fauquet. 1996. La cooperation scientifique nord-sud en biotechnologie vegetale: l'ILTAB un modele parmi d'autres. In "Les Sciences Hors D'Occident Au XXE Siecle," 7:277-283. ORSTOM Editions, Paris.
49. Beachy, R.N. 1997. Mechanisms and applications of pathogen-derived resistance in transgenic plants. In Current Opinion in Biotechnology. 8:215-220.
50. Voloudakis, A.E., Y. Yin and R. N. Beachy. 1999. Recombinant protein expression in plants. In: "Gene Expression Systems: Using Nature for the Art of Expression. Academic Press. pp. 429-461.
51. Beachy, R.N., J. Bennetzen, B. Chassy, M. Chrispeels, J. Chory, J. Ecker, J. Noel, S. Kay, C. Dean, C. Lamb, J. Jones, C. Santerre, J. Schroeder, J. Umen, M. Yanofsky, S. Wessler, Y. Zhao, and W. Parrott. 2002. Letter to the Editor: Divergent perspectives on GM food. In "Nature Biotechnology," 20:1-2.
52. Beachy, R.N. 2003. IP Policies and Serving the Public. Science 299:473.
53. Frommer, W.B. and R. N. Beachy. 2003. Editorial overview. Plant biotechnology: A future for plant biotechnology? Naturally! In "*Current Opinion in Plant Biology*," 6:147-149.
54. Beachy, R.N. 2003. Foreward. In *Encyclopedia of Plant and Crop Science*; 1st Ed; Goodman, R.M., Ed.; Taylor & Francis Group, LLC: New York, NY.
55. Yadav, J. S., R. N. Beachy, and C. M. Fauquet. 2005. Control of plant virus diseases. In *Encyclopedia of Plant and Crop Science*; 1st Ed; Goodman, R.M., Ed.; Taylor & Francis Group, LLC: New York, NY. pp. 1-6. DOI: 10.1081./E-EPCS-120019914.
56. Berg, R. Howard, R.N. Beachy. 2006. Fluorescent Protein Applications in Plants. In *Fluorescent Proteins,* Second Edition, Volume 87 of *Methods in Cell Biology*. Elsevier Inc. In print.

Journal Articles ~ 1970 to present

1. Murakishi, H.H., J.X. Hartmann, L.E. Pelcher, and R.N. Beachy. 1970. Improved inoculation of cultured plant cells resulting in high virus titer and crystal formation. *Virology* 41:365-367.
2. Murakishi, H.H., J.X. Hartmann, R.N. Beachy, and L.E. Pelcher. 1971. Growth curve and yield of *tobacco mosaic virus* in tobacco callus cells. *Virology* 43:62-68.

3. Beachy, R.N. and H.H. Murakishi. 1971. Local lesion formation in tobacco tissue culture. *Phytopathology* 61:877-878.
4. Beachy, R.N. and H.H. Murakishi. 1973. Effect of cycloheximide on TMV synthesis in callus from hypersensitive tobacco. *Virology* 55:320-328.
5. Zaitlin, M. and R.N. Beachy. 1974. The use of protoplasts and separated cells in plant virus research. *Adv Virus Res* 19:1-37.
6. Beachy, R.N. and M. Zaitlin. 1975. Replication of *tobacco mosaic virus* VI. Replicative intermediate and TMV-RNA- related RNAs associated with polyribosomes. *Virology* 63:84-97.
7. Beachy, R.N. and H.H. Murakishi. 1976. Changes in soluble proteins in callus cells of hypersensitive tobacco inoculated with *tobacco mosaic virus. In Vitro* 12:517-520.
8. Bruening, G., R.N. Beachy, R. Scalla, and M. Zaitlin. 1976. In vitro and in vivo translation of the ribonucleic acids of a cowpea strain of *tobacco mosaic virus. Virology* 71:498-517.
9. Beachy, R.N., M. Zaitlin, G. Bruening, and H.W. Israel. 1976. A genetic map for the cowpea strain of TMV. *Virology* 73:498-507.
10. Zaitlin, M., G. Bruening, R.N. Beachy, and R. Scalla. 1976. The cowpea strain of TMV is a pseudo-multicomponent virus. *Ann Microbiol* 127A:37-38.
11. Beachy, R.N. and M. Zaitlin. 1977. Characterization and *in vitro* translation of the RNAs from less-than-full-length, virus-related, nucleoprotein rods present in *tobacco mosaic virus* preparations. *Virology* 81:160-169.
12. Zaitlin, M., R.N. Beachy, and G. Bruening. 1977. Lack of molecular hybridization between RNAs of two strains of TMV: a reconsideration of the criteria for strain relationships. *Virology* 8:237-241.
13. Beachy, R.N., J.F. Thompson, and J.T. Madison. 1978. Isolation of polyribosomes and messenger RNA active in *in vitro* synthesis of soybean seed proteins. *Plant Physiol* 61:139-144.
14. Beachy, R.N., K. Barton, J.F. Thompson and J.T. Madison. 1980. *In vitro* synthesis of the α and α' subunits of the 7S storage protein (conglycinin) of soybean seeds. *Plant Physiol* 65:990-994.
15. Beachy, R.N., N.P. Jarvis, and K.A. Barton. 1981. Biosynthesis of subunits of the soybean 7S storage protein. *J Mol Appl Genet* 1:19-27.
16. Meinke, D.W., J. Chen, and R.N. Beachy. 1981. Expression of storage protein genes during soybean seed development. *Planta* 153:130-139.
17. Bernal-Lugo, I., R.N. Beachy, and J.E. Varner. 1981. The response of barley aleurone layers to gibberellic acid includes the transcription of new sequences. *Biochim Biophys Acta* 102:617-623.
18. Barton, K.A., J.F. Thompson, J.T. Madison, R. Rosenthal, N.P. Jarvis, and R.N. Beachy. The biosynthesis and processing of high molecular weight precursors of soybean glycinin subunits. *J Biol Chem* 257:6089-6095.

19. Schuler, M.A., B.F. Ladin, J.C. Polacco, G. Freyer, and R.N. Beachy. 1982. Structural sequences are conserved in the genes coding for the a, a', and a-subunits of the soybean 7S seed storage protein. *Nucleic Acids Res* 10:8245-8261.
20. Schuler, M.A., E.S. Schmidt, and R.N. Beachy. 1982. Closely related families of genes code for the α and α'-subunits of the soybean 7S storage proteins. *Nucleic Acids Res* 10:8225-8244.
21. Foard, D.E., P.A. Gutag, B.F. Ladin, R.N. Beachy, and B.A. Larkins. 1982. In vitro synthesis of Bowman-Birk and related soybean protease inhibitors. *Plant Mol Biol* 1:227-243.
22. Schuler, M.A., J.J. Doyle and R.N. Beachy. 1983. Nucleotide homologies between the glycosylated seed storage proteins of *Glycine max* and *Phaseolus vulgaris. Plant Mol Biol* 2:119-127.
23. Ladin, B.F., J.J. Doyle and R.N. Beachy. 1984. Molecular characterization of a deletion mutation affecting the α'-subunit of β-conglycinin of soybean. *J Mol App Genet* 2:372-380.
24. Vance, V.S. and R.N. Beachy. 1984. Detection of genomic-length soybean mosaic virus RNA on polyribisomes of infected soybean leaves. *Virology* 132:271-281.
25. Vance, V.S. and R.N. Beachy. 1984. Cloning of soybean mosaic virus RNA: Detection of a subgenomic mRNA in infected leaves. *Virology* 138:26-36.
26. Doyle, J.J., B.F. Ladin, and R.N. Beachy. 1985. Antigenic relationship of legume seed proteins to the 7S seed storage protein of soybean. *Biochem Syst Ecol* 13:123-132.
27. Doyle, J.J. and R.N. Beachy. 1985. Ribosomal gene variation in soybean (Glycine) and its relatives. *Theor Appl Genet* 70: 369-376.
28. Shattuck-Eidens, D.M. and R.N. Beachy. 1985. Degradation of β-conglycinin in early stages of soybean embryogenesis. *Plant Physiol* 78:895-898.
29. Bray, B.A. and R.N. Beachy. 1985. Regulation by ABA of β-conglycinin expression in cultured developing soybean cotyledons. *Plant Physiol* 79:746-750.
30. Beachy, R.N., Z.-L. Chen, R.B. Horsch, S.G. Rogers, N.J. Hoffmann, and R.T. Fraley. 1985. Accumulation and assembly of soybean β-conglycinin in seeds of transformed petunia plants. *EMBO J* 4:3047-3053.
31. Reinero, A., and R.N. Beachy. 1986. Association of TMV coat protein with thylakoid membrane in virus infected leaves. *Plant Mol Biol* 6:291-301.
32. Doyle, J.J., Schuler, M.A., J. Slightom, W.D. Goddette, V.E. Zenger, and R.N. Beachy. 1986. The glycosylated seed storage proteins of *Glycine max* and *Phaseolus vulgaris*: structural homologies of genes and proteins. *J Biol Chem* 261:9228-9238.

33. Powell-Abel, P.A., R.S. Nelson, B. De, N. Hoffmann, S.G. Rogers, R.T. Fraley, and R.N. Beachy. 1986. Delay of disease development in transgenic plants that express the tobacco mosaic virus coat protein gene. *Science* 232:738-743.
34. Oliver, M.J., C.M. Deom, B.K. De, and R.N. Beachy. 1986. In vitro transcription and translation of cloned cDNA's encoding the 30 kD protein gene of TMV. *Virology* 155:277-283.
35. Chen, Z.L., M.A. Schuler, and R.N. Beachy. 1986. Functional analysis of regulatory elements in a plant embryo-specific gene. *Proc Natl Acad Sci USA* 83:8560-8564.
36. Ladin, B.F., M. L. Tierney, D.W. Meinke, P. Hosángadi, M. Vieth, and R.N. Beachy. 1987. Developmental regulation of β-conglycinin in soybean axes and cotyledons. *Plant Physiol* 84:35-41.
37. Nelson, R.E., P. Powell Abel, and R.N. Beachy. 1987. Lesions and virus accumulation in inoculated transgenic tobacco plants expressing the coat protein gene of *tobacco mosaic virus*. *Virology* 158:126-132.
38. Tumer, N.E., K.M. O'Connell, R.S. Nelson, P.R. Sanders, and R.N. Beachy. 1987. Expression of alfalfa mosaic virus coat protein gene confers cross-protection in transgenic tobacco and tomato plants. *EMBO J* 6:1181-1188.
39. Bray, E.A., S. Naito, N.-S. Pan, E. Anderson, P. Dubé, and R.N. Beachy. 1987. Expression of the β-subunit of β-conglycinin in seeds of transgenic plants. *Planta* 172:364-370.
40. Tierney, M.L., E.A. Bray, R.D. Allen, R.E. Droug, Y. Ma, J. Slighton, and R.N. Beachy. 1987. Isolation and characterization of a genomic clone encoding the β-subunit of β-conglycinin. *Planta* 172:356-363.
41. Della-Cioppa, G., G.M. Kishore, R.N. Beachy, and R.T. Fraley. 1987. Protein trafficking in plant cells. *Plant Physiol* 84:965-968.
42. Deom, C.M., M.J. Oliver, and R.N. Beachy. 1987. The 30-kilodalton gene product of *tobacco mosaic virus* potentiates virus movement. *Science* 237:389-394.
43. Lawton, M.A., M.L. Tierney, I. Nakamura, E. Anderson, Y. Komeda, P. Dubé, N. Hoffmann, R.T. Fraley, and R.N. Beachy. 1987. Expression of a soybean β-conglycinin gene under the control of the cauliflower mosaic virus 35S and 19S promoters in transformed petunia tissues. *Plant Mol Biol* 9:315-324.
44. Nelson, R.S., S.M. McCormick, X. Delannay, P. Dubé, J. Layton, E.J. Anderson, M. Kaniewska, R.K. Proksch, R.B. Horsch, S.G. Rogers, R.T. Fraley, and R.N. Beachy. 1988. Virus tolerance, plant growth, and field performance of transgenic tomato plants expressing coat protein from *tobacco mosaic virus*. *Bio/technology* 6:403-409.

45. Chen, Z.-L., N.-S. Pan, and R.N. Beachy. 1988. Identification of a developmentally regulated enhancer element in a soybean embryo-specific gene. *EMBO J* 7:297-302.
46. Naito, S., P.H. Dubé, and R.N. Beachy. 1988. Differential expression of developmentally regulated genes in transgenic plants. *Plant Mol Biol* 11:109-123.
47. Register III, J.C., and R.N. Beachy. 1988. Resistance to TMV in transgenic plants results from interference with an early event in infection. *Virology* 166:524-532.
48. Reinero, A., and R.N. Beachy. 1989. Inhibition of Photosystem II and accumulation of viral coat protein in thylakoids of leaves infected by *tobacco mosaic virus. Plant Physiol* 89:111-116.
49. Chen, Z.-L., S. Naito, I. Nakamura, and R.N. Beachy. 1989. Regulated expression of genes encoding soybean β-conglycinins in transgenic plants. *Dev Genet* 10:112-122.
50. Eggenberger, A.L., D.M. Stark, and R.N. Beachy. 1989. The nucleotide sequence of a soybean mosaic virus coat protein-coding region and its expression in *Escherichia col, Agrobacterium tumefaciens* and tobacco callus. *J Gen Virol* 70:1853-1860.
51. Carr, J.P., R.N. Beachy, and D.F. Klessig. 1989. Are the PR1 proteins of tobacco involved in genetically engineered resistance to TMV? *Virology* 169:470-473.
52. Anderson, E.J., D.M. Stark, R.S. Nelson, N.E. Tumer, and R.N. Beachy. 1989. Transgenic plants that express the coat protein genes of TMV or AlMV interfere with disease development of some non-related viruses. *Phytopathology* 12:1284-1290.
53. Hodgson, R.A.J., R.N. Beachy, and H.B. Pakrasi. 1989. Selective inhibition of photosystem II in spinach by *tobacco mosaic virus*: An effect of the viral coat protein. *FEBS Lett.* 245(1,2):267-270.
54. Powell, P.A., D.M. Stark, P. Sanders, and R.N. Beachy. 1989. Protection against *tobacco mosaic virus* in transgenic plants that express TMV antisense RNA. *Proc Natl Acad Sci USA* 86:6949-6952.
55. Allen, R.D., F. Bernier, P.A. Lessard, and R.N. Beachy. 1989. Nuclear factors interact with a soybean β-conglycinin enhancer. *Plant Cell* 1:623-631.
56. Stark, D.M., and R.N. Beachy. 1989. Protection against potyvirus infection in transgenic plants: Evidence for broad-spectrum resistance. *Bio/technology* 7:1257-1262.
57. Nejidat, A., and R.N. Beachy. 1989. Decreased Levels of TMV Coat Protein in Transgenic Tobacco Plants at Elevated Temperatures Reduce Resistance to TMV Infection. *Virology* 173:531-538.

58. Osbourn, J.K., J.W. Watts, R.N. Beachy, and T.M.A. Wilson. 1989. Disassembly of pseudovirus particles is inhibited in electroporated protoplasts from transgenic tobaccos expressing *tobacco mosaic virus* coat protein. Evidence that virus replication is also inhibited at a later stage of infection. *Virology* 172:370-373.
59. Register III, J.C., and R.N. Beachy. 1989. A transient protoplast assay for capsid protein-mediated protection: Effect of capsid protein aggregation state on protection against *tobacco mosaic virus*. *Virology* 173:656-663.
60. Wolf, S., C.M. Deom, R.N. Beachy, and W.J. Lucas. 1989. Movement protein of *tobacco mosaic virus* modified plasmodesmatal size exclusion limit. *Science* 246:377-379.
61. Powell, P.A., P.R. Sanders, N. Tumer, and R.N. Beachy. 1990. Protection against *tobacco mosaic virus* infection in transgenic plants requires accumulation of capsid protein rather than coat protein RNA sequences. *Virology* 175:124-130.
62. Deom, C.M., K.R. Schubert, S. Wolf, C.A. Holt, W.J. Lucas, and R.N. Beachy. 1990. Molecular characterization and biological function of the movement protein of *tobacco mosaic virus* in transgenic plants. *Proc Natl Acad Sci USA* 87:3284-3288.
63. Wisniewski, L.A., P.A. Powell, R.S. Nelson, and R.N Beachy. 1990. Local and systemic movement of *tobacco mosaic virus* (TMV) in tobacco plants that express the TMV coat protein gene. *Plant Cell* 2:559-567.
64. Franche, C., D. Bogusz, C. Fauquet, C. Schöpke, C. Jobe, J. Pease, and R.N. Beachy. 1990. A la frontière de la balistique et de la biologie végéale: le canon à particules. *ORSTOM Actualités*, 29:8-10.
65. Nejidat, A., and R.N. Beachy. 1990. Transgenic tobacco plants expressing a coat protein gene of *tobacco mosaic virus* are resistant to some other tobamoviruses. *Mol Plant Microbe Interact* 3:247-251.
66. Sebastiani, F.L., L.B. Farrell, M.A. Schuler, and R.N. Beachy. 1990. Complete sequence of cDNA of a-subunit of soybean β-conglycinin. *Plant Mol Biol* 15:197-201.
67. Nejidat, A., W.G. Clark, and R.N. Beachy. 1990. Engineered resistance against plant virus diseases. *Physiol Plant* 80:662-668.
68. Beachy, R.N., S. Loesch-Fries, and N.E. Tumer. 1990. Coat protein-mediated resistance against virus infection. *Annu Rev Phytopathol* 28:451-474.
69. Rochester, D.E., W. Kositratana, and R.N. Beachy. 1990. Systemic movement and symptom production following agroinoculation with a single DNA of tomato yellow leaf curl virus (Thailand) geminivirus. *Virology* 178:520-526.
70. Nejidat, A., F. Cellier, C.A. Holt, R. Gafny, A.L. Eggenberger, and R.N. Beachy. 1990. Transfer of the movement protein gene between two tobamoviruses: Influence on local lesion development. *Virology* 180:318-326.

71. Clark, W.G., J.C. Register, III, A Nejidat, D.A. Eichholtz, P.R. Sanders, R.T. Fraley, and R.N. Beachy. 1990. Tissue-specific expression of the TMV coat protein in transgenic tobacco plants affects the level of coat protein-mediated virus protection. *Virology* 179:640-647.
72. Holt, C.A., R.A.J. Hodgson, F.. Coker, R.N. Beachy, and R.S. Nelson. 1990. Characterization of the masked strain of *tobacco mosaic virus*: Identification of the region responsible for symptom attenuation by analysis of an infectious cDNA clone. *Mol Plant Microbe Interact* 3(6):17-423.
73. Farrell L.B., and R.N. Beachy. 1990. Manipulation of β-glucuronidase for use as a reporter in vacuolar targeting studies. *Plant Mol Biol* 15:821-825.
74. Deom, C.M., S. Wolf, C.A. Holt, W.J. Lucas, and R.N Beachy. 1991. Altered function of the *tobacco mosaic virus* movement protein in a hypersensitive host. *Virology* 180:251-256.
75. Atkins, D., R. Hull, B. Wells, K. Roberts, P. Moore, and R.N. Beachy. 1991. The TMV 30K movement protein in transgenic tobacco plants is localized to plasmodesmata. *J Gen Virol* 72:209-211.
76. Wu, X., R.N. Beachy, T.M.A. Wilson, and J G. Shaw. 1990. Inhibition of uncoating of *tobacco mosaic virus* particles in protoplasts from transgenic tobacco plants that express the viral coat protein gene. *Virology* 179:893-895.
77. Lessard, P.A., R.D. Allen, F. Bernier, J.D. Crispino, T. Fujiwara, and R.N. Beachy. 1991. Multiple nuclear factors interact with upstream sequences of differentially regulated β-conglycinin genes. *Plant Mol Biol* 16:397-413.
78. Jones, M., K. Gough, I. Dasgupta, B.L. Subba Rao, J. Cliffe, R. Qu, P. Shen, M. Kaniewska, M. Blakebrough, J.W. Davies, R.N. Beachy, and R. Hull. 1991. Rice tungro disease is caused by an RNA and a DNA virus. *J Gen Virol* 72:757-761.
79. Holt, C.A., and R.N. Beachy. 1991. *In vivo* complementation of infectious transcripts from mutant *tobacco mosaic virus* cDNAs in transgenic plants. *Virology* 181:109-117.
80. Berna, A., R. Gafny, S. Wolf, W.J. Lucas, C.A. Holt and R.N. Beachy. 1991. The TMV movement protein: Role of the C-terminal 73 amino acids in subcellular localization and function. *Virology* 182:682-689.
81. Franche, C., D. Bogusz, C. Schöpke, C. Fauquet, and R.N. Beachy. 1991. Transient gene expression in cassava using high-velocity microprojectiles. *Plant Mol Biol* 17:493-498.
82. Sebastiani, F.L., L.B. Farrell, M. Vasquez, and R.N. Beachy. 1991. Conserved amino acid sequences among plant proteins sorted to protein bodies and plant vacuoles: can they play a role in protein sorting?. *Eur J Biochem* 199:441-450.

83. Wolf, S., C.M. Deom, R.N. Beachy, and W.J. Lucas. 1991. Plasmodesmatal function is probed using transgenic tobacco plants that express a virus movement protein. *Plant Cell* 3:593-604.
84. Niblett, C.L., L.A. Calvert, T.L. Kendall, D.M. Stark, K.R. Zagula, C.E. Smith, R.N. Beachy, and S.A. Lommel. 1991. cDNA Cloning and Nucleotide Sequence of the Wheat Streak Mosaic Virus Capsid Protein Gene. *J Gen Virol* 72:499-504.
85. Qu, R., M. Bhattacharyya, G.S. Laco, A. de Kochko, B.L. Subba Rao, M.B. Kaniewska, J.S. Elmer, D.E. Rochester, C.E. Smith, and R.N. Beachy. 1991. The complete sequence of rice tungro bacilliform virus DNA — comparison with commelina yellow mottle virus and caulimoviruses. *Virology* 185:354-364.
86. Jain, R.K., N.M. McKern, S.A. Tolin, J.H. Hill, O.W. Barnett, M. Tosic, R.E. Ford, R.N. Beachy, M.H. Yu, C.W. Ward, and D.D. Shukla. 1992. Confirmation that fourteen potyvirus isolates from soybean are strains of one virus by comparing coat protein peptide profiles. *Phytopathology.* 82:294-299.
87. Gafny, R., M. Lapidot, A. Berna, C. Holt, C.M. Deom, and R.N. Beachy. 1992. Effects of terminal deletion mutations on function of the movement protein of *tobacco mosaic virus. Virology* 187:499-507.
88. Fujiwara, T., P.A. Lessard, and R.N. Beachy. 1992. Seed-specific repression of GUS activity in tobacco plants by antisense RNA. *Plant Mol Biol* 20:1059-1069.
89. Perl, M., R. Gafney, and R.N. Beachy. 1992. Phosphodiesterase activities in transgenic tobacco plants associated with the movement protein of *tobacco mosaic virus. Theor Appl Genet* 84:730-734.
90. Deom, C.M., M. Lapidot, and R.N. Beachy. 1992. Plant virus movement proteins. *Cell* 69:221-224.
91. Ding, B., J.S. Haudenshield, R.J. Hull, S. Wolf, R.N. Beachy, and W.J. Lucas. 1992. Secondary plasmodesmata are specific sites of localization of the *tobacco mosaic virus* movement protein in transgenic tobacco plants. *Plant Cell* 4:915-928.
92. Moore, P.J., C.A. Fenczik, C.M. Deom, and R.N. Beachy. 1992. Developmental changes in the structure of plasmodesmata in transgenic tobacco expressing the movement proteins of *tobacco mosaic virus. Protoplasma* 170:115-127.
93. Huttner, S.L., C. Arntzen, R.N. Beachy, G. Bruening, L. DeFrancesco, E. Nester, C. Qualset, and A. Vidaver. 1992. Revising oversight of genetically modified plants. *Bio/technology* 10(9):967-971.
94. Fujiwara, T., and R.N. Beachy. 1993. Expression of soybean seed storage protein genes in transgenic plants: Their effects on expression of a neighboring gene and position dependency. *Plant Cell Physiol* 34(1):13-20.

95. Fujiwara, T., P.A. Lessard, and R.N. Beachy. 1993. Inactivation of the nopaline synthase gene by double transformation: reactivation by segregation of the induced DNA. *Plant Cell Rep* 12:133-138.
96. Shen, P., M. Kaniewska, C. Smith, and R.N. Beachy. 1993. Nucleotide sequence and genomic organization of rice tungro spherical virus. *Virology* 193:621-630.
97. Li, L., R. Qu, A deKochko, C. Fauquet, and R.N. Beachy. 1993. An improved rice transformation system using the biolistic approach. *Plant Cell Rep* 12:250-255.
98. Padgett, H.S., and R.N. Beachy. 1993. Nucleotide sequence and mutational analysis of an infectious cDNA clone of Ob, a tobamovirus that overcomes the N resistance gene of *Nicotiana*. *Plant Cell* 5:577-586.
99. Lucas, W.J., A. Olesinski, R.J. Hull, J.S. Haudenshield, C.M. Deom, R.N. Beachy, and S. Wolf. 1993. Influence of the *tobacco mosaic virus* 30-kDa movement protein on carbon metabolism and photosynthate partitioning in transgenic tobacco plants. *Planta* 190:88-96.
100. Nelson, R.S., G. Li, R.A.J. Hodgson, R.N. Beachy, and M. H. Shintaku. 1993. Impeded phloem-dependent accumulation of the masked strain of *tobacco mosaic virus*. *Mol Plant Microbe Interact* 6(1):45-54.
101. Reimann-Philipp, U. and R.N. Beachy. 1993. Coat protein-mediated resistance in transgenic tobacco expressing the TMV coat protein from tissue specific promoters. *Mol Plant Microbe Interact* 6: 323-330.
102. Mathews, H., C. Schöpke, R. Carcamo, P. Chavarriaga, C. Fauquet, and R.N. Beachy. 1993. Improvement of somatic embryogenesis and plant recovery in cassava. *Plant Cell Rep* 12:328-333.
103. Bhattacharyya-Pakrasi, M., J. Peng, J.S. Elmer, G. Laco, P. Shen, M.B. Kaniewska, H. Konowicz, F. Wen, T.K. Hodges, and R.N. Beachy. 1993. Specificity of a promoter from the rice tungro bacilliform virus for expression in phloem tissues. *Plant J* 4:71-79.
104. Lessard, P.A., R.D. Allen, T. Fujiwara, and R.N. Beachy. 1993. Upstream regulatory sequences from two β-conglycinin genes. *Plant Mol Biol* 22: 873-885.
105. Nakamura, I., P.H. Dube, and R.N. Beachy. 1993. Accumulation of the products of β-conglycinin a'-subunit gene constitutively expressed in seeds and non-seed tissues of the transgenic petunia plants. *Plant Cell Physiol* 34(6):865-872.
106. Lapidot, M., R. Gafny, B. Ding, S. Wolf, W.J. Lucas, and R.N. Beachy. 1993. A dysfunctional movement protein of *tobacco mosaic virus* that partially modifies the plasmodesmata and limits virus spread in transgenic plants. *Plant J* 4(6): 959-970.

107. Reimann-Philipp, U., and R.N. Beachy. 1993. The mechanism(s) of coat protein-mediated resistance against *tobacco mosaic virus. Semin Virol* 4(6): 349-356.
108. Fujiwara, T., and R.N. Beachy. 1993. Tissue-specific and temporal regulation of a β-conglycinin gene: roles of the RY repeat and other *cis*-acting elements. *Plant Mol Biol* 24:261-272.
109. Rance, I., W. Tian, H. Mathews, A. de Kochko, R.N. Beachy, and C.M. Fauquet. 1994. Partial desiccation of mature embryo-derived calli, a simple treatment that dramatically enhances the regeneration ability of *Indica* rice. *Plant Cell Rep* 13:647-651.
110. Ngon a Yassi, M. C. Ritzenthaler, C. Brugidou, C. Fauquet, and R.N. Beachy. 1994. Nucleotide sequence and genome characterization of rice yellow mottle virus RNA. *J Gen Virol* 75:249-257.
111. Marcos, J.F., and R.N. Beachy. 1994. *In vitro* characterization of a cassette to accumulate multiple proteins through synthesis of a self-processing polypeptide. *Plant Mol Biol* 24:495-503.
112. Rochester, D.E., J.J. DePaulo, C.M. Fauquet, and R.N. Beachy. 1994. Complete nucleotide sequence of the geminivirus, tomato yellow leaf curl virus (Thailand isolate). *J Gen Virol* 75: 477-485.
113. Wenzhong, T., I. Rance, E. Sivamani, C. Fauquet, and R.N. Beachy. 1994. Improvement of plant regeneration frequency *in vitro* in indica rice. *Chin J Genet* 21(3):1-9.
114. Laco, G., and R.N. Beachy. 1994. Rice tungro bacilliform virus encodes reverse transcriptase, DNA polymerase, and ribonuclease H activities. *Proc Natl Acad Sci USA* 91:2654-2658.
115. De Kochko, A., R. Qu, P. Marmey, I. Rance, C. Fauquet, and R.N. Beachy. 1994. Protein slot blotting; an easy, rapid and reliable technique to identify the expression of a protein in transgenic plants. *Plant Mol Biol Rep* 12(3):274-278.
116. Deom, C.M., Z.H. Xian, R.N. Beachy, and A.K. Weissinger. 1994. Influence of heterologous tobamovirus movement protein and chimeric-movement protein genes on cell-to-cell and long-distance movement. *Virology* 205:198-209.
117. Rance, I.M., W. Tian, H. Mathews, A. deKochko, R.N. Beachy, and C. Fauquet. 1994. Partial desiccation of mature embryo-derived calli, a simple treatment that dramatically enhances the regeneration ability of indica rice. *Plant Cell Rep* 13:647-651.
118. Cooper, B., M. Lapidot, J.A. Heick, J.A. Dodds, and R.N. Beachy. 1995. A defective movement protein of TMV in transgenic plants confers resistance to multiple viruses whereas the functional analog increases susceptibility. *Virology* 206:307-313.

119. Brugidou, C., C. Holt, M. Ngon a Yassi, S. Zhang, R.N. Beachy, and C. Fauquet. 1995. Synthesis of an infectious full-length cDNA clone of rice yellow mottle virus and mutagenesis of the coat protein. *Virology* 206:108-115.
120. Padidam, M., R.N. Beachy, and C.M. Fauquet. 1995. Sequence comparison can be used to identify and classify geminiviruses. *J Gen Virol* 76:249-263.
121. Padidam, M., R.N. Beachy, and C.M. Fauquet 1995. Tomato leaf curl geminivirus from India has abipartite genome and coat protein is not essential for infectivity. *J Gen Virol* 76:25-35.
122. Yin, Y. and R.N. Beachy. 1995. The regulatory regions of the rice tungro bacilliform virus promoter and interacting nuclear factors in rice (*Oryza sativa L.*). *Plant J* 7(6):969-980.
123. Laco, G.S., S.B.H. Kent, and R.N. Beachy. 1995. Analysis of the proteolytic processing and activation of the rice tungro bacilliform virus reverse transcriptase. *Virology* 208:207-214.
124. Zahalak, M., R.N. Beachy, and T. Thiel. 1995. Expression of the movement protein of tobacco mosaic virus in the cyanobacterium *Anabaena sp.* strain PCC 7120. *Mol Plant Microbe Interact* 8(2):192-199.
125. Clark, W.G., J.H. Fitchen, and R.N. Beachy. 1995. Studies of coat protein-mediated resistance to TMV: I. The PM2 assembly defective mutant confers resistance to TMV. *Virology* 208:485-491.
126. Arce-Johnson, P., T.W. Kahn, U. Reimann-Philipp, R.R. Bustamante, and R.N. Beachy. 1995. The amount of movement protein produced in transgenic plants influences the establishment, local movement, and systemic spread of infection by movement protein-deficient *tobacco mosaic virus. Mol Plant Microbe Interact* (3):415-423.
127. Clark, W. Gregg, J. Fitchen, A. Nejidat, C. M. Deom, and R.N. Beachy. 1995. Studies of coat protein-mediated resistance to TMV: II. Challenge by a mutant with altered virion surface does not overcome resistance conferred by TMV CP. *J Gen Virol* 76(10):2613-2617.
128. Fitchen, J., R.N. Beachy, and M.B. Hein. 1995. Plant virus expressing hybrid coat protein and added murine epitope elicits autoantibody response. *Vaccine* 13(12):1051-1057.
129. Heinlein, M., B.L. Epel, H.S. Padgett, and R.N. Beachy. 1995. Interaction of tobamovirus movement proteins with the plant cytoskeleton. *Science* 270:1983-1985.
130. Fenczik, C.A., H.S. Padgett, C.A. Holt, S.J. Casper, and R.N. Beachy. 1995. Mutational analysis of the movement protein of odontoglossum ringspot virus to identify a host-range determinant. *Mol Plant Microbe Interact* 8(5):666-673.
131. Bryant, J.L., P. Hosangadi, and R.N. Beachy. 1995. The β-conglycinins of soybean remain assembled in a 7S conformation while undergoing proteolytic

cleavage during germination and early seedling growth. *Plant Cell Physiol* 36:1059-1065.

132. Marcos, J.F., R.N. Beachy, R.A. Houghten, S.E. Blondelle, and E. Perez-Paya. 1995. Inhibition of a plant virus infection by analogs of melittin. *Proc Natl Acad Sci USA* 92:12466-12469.
133. Sivamani, E., P. Shen, N. Opalka, R.N. Beachy, and C.M. Fauquet. 1995. Selection of large quantities of embryogenic subcultured calli from *Indica* rice seeds for production of fertile transgenic plants using the biolistic method. *Plant Cell Rep* 15:322-327.
134. Cooper, Bret, I. Schmitz, A.L.N. Rao, R.N. Beachy, and J.A. Dodds. 1996. Cell-to-cell transport of movement-defective cucumber mosaic and *tobacco mosaic viruses* in transgenic plants expressing heterologous movement protein genes. *Virology* 216:208-213.
135. Schopke, C., N. Taylor, R. Carcamo, K.K. N'Da, P. Marmey, G.G. Henshaw, R.N. Beachy, and C.M. Fauquet. 1996. Regeneration of transgenic cassava plants (*Manihot esculenta* Crantz) from microbombarded embryogenic suspension cultures. *Nat Biotechnol* 14:731-735.
136. Epel, B.L., H.S. Padgett, M. Heinlein, and R.N. Beachy. 1996. Plant virus movement protein dynamics probed with a gfp-protein fusion *Gene* 173(1):75-79.
137. Aleman, M.-E., J.F. Marcos, C. Brugidou, R.N. Beachy, and C.M. Fauquet. 1996. The complete nucleotide sequence of yam mosaic virus (Ivory Coast isolate) genomic RNA. *Arch Virol* 141:1259-1278.
138. Verdaguer, B., A. de Kochko, R.N. Beachy, and C.M. Fauquet. 1996. Isolation and expression in transgenic plants of the cassava vein mosaic virus (CVMV) promoter. *Plant Mol Biol* 31:1129-1139.
139. Zhang, S., L. Chen, R. Qu, P. Marmey, R.N. Beachy, and C.M. Fauquet. 1996. Regeneration of fertile transgenic *Indica* (Group 1) rice plants following microprojectile-transformation of embryogenic suspension culture cells. *Plant Cell Rep* 15:465-469.
140. Padidam, M., R.N. Beachy, and C.M Fauquet. 1996. The role of AV2 ("precoat") and coat protein in viral replication and movement in tomato leaf curl geminivirus. *Virology* 224:390-404.
141. Padgett, H.S., B.L. Epel, M.H. Heinlein, Y. Watanabe, and R.N. Beachy. 1996. Distribution of tobamovirus movement protein in infected cells and implications for cell-to-cell spread of infection. *Plant J* 10:1079-1088.
142. Qu, R., A. de Kochko, L. Zhang, P. Marmey, L. Li, W. Tian. S. Zhang, C.M. Fauquet, and R.N. Beachy. 1996. Analysis of a large number of independent transgenic rice plants produced by the biolistic method. *In vitro Cell Dev Biol Plant* 32:233-240.

143. Beachy, R.N., H.S. Padgett, T. Kahn, M. Bendahmane, J.H. Fitchen, M. Heinlein, Y. Watanabe, and B.L. Epel. 1996. Cellular and molecular mechanisms of coat protein and movement protein mediated resistance against TMV. *Molec Plant Microbe Interact* 259-264.

144. Kahn, T.W., R.N. Beachy, and M. M. Falk. 1997. Cell-free expression of a GFP fusion protein allows quantitation *in vitro* and *in vivo. Curr Biol* 7:R207-R208.

145. Beachy, R.N. 1997. The 2',5'A antiviral system in plants: A dose of mammalian medicine. *Nat Biotechnol* 14:1538-39.

146. Arce-Johnson, P., U. Reimann-Philipp, H.S. Padgett, R. Rivera-Bustamante, and R.N. Beachy. 1997. Requirement of the movement protein for long distance spread of *tobacco mosaic virus* in grafted plants. *Mol Plant Microbe Interact* 10:691-699.

147. Marcos, J.F. and R.N. Beachy. 1997. Transgenic accumulation of two plant virus proteins on a single self-processing polypeptide. *J Gen Virol* 78:1771-1778.

148. Aleman-Verdaguer, M.-E., C. Goudou-Urbino, J. Dubern, R.N. Beachy, and C.M. Fauquet. 1997. Analysis of the sequence diversity of the P1, HC, P3, Nib and CP genomic regions of several yam mosaic potyvirus isolates: implications for the intraspecies molecular diversity of potyviruses. *J Gen Virol* 78:1253-1264.

149. Padgett, H.S., Y. Watanabe, and R.N. Beachy. 1997. Identification of the tobamovirus protein that activates the N gene-mediated hypersensitive response. *Mol Plant Microbe Interact* 10:709-715.

150. Yin, Y., Q. Zhu, S. Dai, C. Lamb, and R.N. Beachy. 1997. RF2a, a bZIP transcriptional activator of the phloem-specific rice tungro bacilliform virus promoter, functions in vascular development. *EMBO J* 16:5247-5259.

151. Yin, Y., L. Chen, and R.N. Beachy. 1997. Promoter elements required for phloem-specific gene expression from the RTBV promoter in rice. *Plant J* 12:1179-1188.

152. Bendahmane, M., J.H. Fitchen, G. Zhang, and R.N. Beachy. 1997. Studies of coat protein mediated resistance to TMV: Correlation between assembly of mutant coat proteins and resistance. *J Virol* 71:7942-7950.

153. Ceriani, M.F., J.F. Marcos, H.E. Hopp, and R.N. Beachy. 1997. Simultaneous accumulation of multiple viral coat proteins from a TEV-NIa based expression vector. *Plant Mol Biol* 36:239-248.

154. Oparka, K.J., D.A.M. Prior, S. Santa Cruz, H.S. Padgett, and R.N. Beachy. 1997. Gating of epidermal plasmodesmata is restricted to the leading edge of expanding infection sites of *tobacco mosaic virus. Plant J* 12:781-789.

155. Ares, X., G. Calamante, S. Cabral, P. Hemenway, R.N. Beachy and A. Mentaberry. 1998. Transgenic plants expressing PVX ORF2 protein (p24) are resistant to TMV and Ob tobamoviruses. *J Virol* 72:24-31.

156. Kahn, T.W., M. Lapidot, M. Heinlein, C. Reichel, B. Cooper, R. Gafny, and R.N. Beachy. 1998. Domains of the TMV movement protein involved in subcellular localization. *Plant J* 15(1):15-25.
157. Heinlein, M., M. Wood, T. Thiel, and R.N. Beachy. 1998. Targeting and modification of prokaryotic cell-cell junctions by *tobacco mosaic virus* cell-to-cell movement protein. *Plant J* 14(3): 345-351.
158. Heinlein, M., H.S. Padgett, S. Gens, B. Pickard, S.J. Casper, B.L. Epel, and R.N. Beachy. 1998. Changing patterns of localization of TMV movement protein and replicase to endoplasmic reticulum and microtubules during infection. *Plant Cell* 10:1107-1120.
159. Lewis, J.K., M. Bendahamane, T.J. Smith, R.N. Beachy, and G. Siuzdak. 1998. Identification of viral mutants by mass spectrometry. *Proc Natl Acad Sci USA* 95:8596-8601.
160. Opalka, N., C. Brugidou, C. Bonneau, M. Nicole, R.N. Beachy, M. Yeager, and C.M. Fauquet. 1998. Movement of rice yellow mottle virus between xylem cells through pit membranes. *Proc Natl Acad Sci USA* 95: 3323-3328.
161. Verdaguer, B., A. de Kochko, C. Fux, R.N. Beachy, and C.M. Fauquet. 1998. Functional organization of the cassava vein mosaic virus (CsVMV) promoter. *Plant Mol Biol* 37:1055-1067.
162. Bonneau, C., C. Brugidou, L. Chen, R.N. Beachy, and C. Fauquet. 1998. Expression of the rice yellow mottle virus P1 protein *in vitro* and *in vivo* and its involvement in virus spread. *Virology* 244:79-86.
163. Chen, L., S. Zhang, R.N. Beachy, and C.M. Fauquet. 1998. Establishing a protocol for consistent large-scale production of transgenic rice plants. *Plant Cell Rep* 18:25-31.
164. De Kochko, A., B. Verdaguer, N. Taylor, R. Carcamo, R.N. Beachy, and C.M. Fauquet. 1998. Cassava vein mosaic virus (CsVMV), type species for a new genus of plant double stranded DNA viruses? *Arch Virol* 43:945-962.
165. Franck-Duchenne, M., Y. Wang, S.B. Tahar and R.N. Beachy. 1998. *In vitro* stem elongation of sweet pepper in media containing 24-epi brassinolide. *Plant Cell Tissue Organ Cult* 53.79-84.
166. Gonzalez, A.E., C. Schopke, N.J. Taylor, R.N. Beachy, and C.M. Fauquet. 1998. Regeneration of transgenic cassava plants *(Manihot esculenta* Crantz) through *Agrobacterium*-mediated transformation of embryogenic suspension cultures. *Plant Cell Rep* 17:827-831.
167. Reichel, C. and R.N. Beachy. 1998. *Tobacco mosaic virus* infection induces severe morphological changes of the endoplasmic reticulum. *Proc Natl Acad Sci USA* 95:11169-11174.
168. Chen, L., P. Marmey, N. Taylor, J.P. Brizard, C. Espinoza, P. D'Cruz, H. Huet, S. Zhang, A.deKochko, R.N. Beachy, and C.M. Fauquet. 1998. Expression to inheritance of multiple transgenes in rice plants. *Nat Biotechnol* 16:1060-1064.

195. Beachy, R. N. and M. Heinlein. 2000. Role of P30 in replication and spread of TMV. *Traffic* 1:540-544.
196. Qu, C., L. Liljas, N. Opalka, C. Brugidou, M. Yeager, R. N. Beachy, C. M. Fauquet, J.E. Johnson, and T. Lin. 2000. 3D domain swapping of a molecular switch for quasi-equivalent symmetry modulates the stability of an icosahedral virus. *Structure* 8:1095-1103.
197. Dai, S., P. Zheng , P. Marmey , S. Zhang, W. Tian, S. Chen , R. N. Beachy, and C. M. Fauquet. 2001. Comparative analysis of transgenic rice plants obtained by *Agrobacterium*-mediated transformation and particle bombardment. *Mol Breed* 7: 25-33.
198. Petruccelli, S., S. Dai, R. Carcamo, Y. Yin, S. Chen, and R. N. Beachy. 2001. Transcription factor RF2a alters expression of the rice tungro bacilliform virus promoter in transgenic tobacco plants. *Proc Natl Acad Sci USA* 98:7635-7640.
199. Kotlizky, G., A. Katz, J. van der Laak, V. Boyko, M. Lapidot, R.N. Beachy, M. Heinlein, and B. Epel. 2001. A dysfunctional movement protein of *tobacco mosaic virus* interferes with targeting of wild-type movement protein to microtubules. *Mol Plant Microbe Interact* 14:895-904.
200. Dai, S., R. Carcamo, Z. Zhang, S. Chen, and R. N. Beachy. 2001. Bacterial cytosine deaminase gene used as a conditional negative selection marker in transgenic rice plants. *Plant Cell Rep* 20:738-743.
201. Bendahmane, M., J. Szecsi, I. Chen, R.H. Berg, and R.N. Beachy. 2002. Characterization of mutant tobacco mosaic virus coat protein that interferes with virus cell-to-cell movement. *Proc Natl Acad Sci USA* 99:3645-3650.
202. Zhu, Q., M. Ordiz, T. Dabi. R.N. Beachy, and C. Lamb. 2002. Rice TATA binding protein interacts functionally with transcription factor IIB and the RF2a bZIP transcriptional activator in an enhanced plant *in vitro* transcription system. *Plant Cell* 14: 795-803.
203. Ordiz, M.I., C.F. Barbas III, and R. N. Beachy. 2002. Regulation of transgene expression in plants with polydactyl zinc finger transcription factors. *Proc Natl Acad Sci USA* 99:13290-13295. DOI 10.1073/pnas.202471899.
204. Beachy, R.N. and M.I. Ordiz. 2002. The fingers of gene regulation in plants. *ISB News Report:* December 2002:5-6.
205. Stege, J.T., X. Guan, T. Ho, R.N. Beachy, and C.F. Barbas III. 2002. Controlling gene expression in plants using synthetic zinc finger transcription factors. *Plant J* 32:1077-1086. DOI 10.1046/j.1365-313X.2007.01492.X
206. Brugidou, C. N. Opalka, M. Yeager, R.N. Beachy, and C.M. Fauquet. 2002. Stability of *Rice Yellow Mottle Virus* and cellular compartmentalization during the infection process in *Oryza sativa* (L.). *Virology* 297:98-108.
207. Atkinson, R.C., R.N. Beachy, G. Conway, F.A. Cordova, M.A. Fox, K.A. Holbrook, D.F. Klessig, R. L. McCormick, P.M. McPherson, H.R. Rawlings IIII, R. Rapson,

L.N. Vanderhoef, J.D. Wiley, and C.E. Young. 2003. Public Sector Collaboration for Agricultural IP Management. *Science* 301:174-175.

208. Dai, S., S. Petruccelli, M.I. Ordiz, Z. Zhang, S. Chen, and R.N. Beachy. 2003. Functional analysis of RF2a, a rice transcription factor. *J Biol Chem* 278:36396-36402.
209. Whitelaw, C.A., W.B. Barbazuk, G. Pertea, A.P. Chan, F. Cheung, Y. Lee, L. Zheng, S. van Heeringen, S. Karamycheva, J.L. Bennetzen, P. SanMiguel, N. Lakey, J. Bedell, Y. Yuan, M.A. Budiman, A. Resnick, S. Van Aken, T. Utterback, S. Riedmuller, M. Williams, T. Feldblyum, K. Schubert, R. Beachy, C. M. Fraser, J. Quackenbush. 2003. Enrichment of gene-coding sequences in maize by genome filtration. *Science* 302:2118-2120.
210. Dai, S., Z. Zhang, S. Chen, R.N. Beachy. 2004. RF2b, a rice bZIP transcription activator, interacts with RF2a and is involved in symptom development of rice tungro disease. *Proc Natl Acad Sci USA* 101:687-692.
211. Asurmendi, S., H. Berg, J. Koo, and R.N. Beachy. 2004. Coat protein regulates formation of replication complexes during TMV infection. *Proc Natl Acad Sci USA* 101:1415-1420.
212. Koo, J.C., S. Asurmendi, J. Bick, T. Woodford-Thomas, and R.N. Beachy. 2004. Ecdysone agonist-inducible expression of a coat protein gene from Tobacco Mosaic Virus confers viral resistance in transgenic *Arabidopsis. Plant J* 37:439-448.
213. Brill, L.M., S. Dechongkit, B. DeLaBarre, J. Stroebel, R.N. Beachy, and M. Yeager. 2004. Dimerization of recombinant TMV movement protein. *J Virol* 78:3372-3377.
214. Voloudakis, A.E., C.A. Malpica, M-E. Aleman-Verdaguer, D.M. Stark, C.M. Fauquet, and R.N. Beachy. 2004. Structural characterization of Tobacco Etch Virus coat protein mutants. *Arch Virol* 149:699-712.
215. Hossain, T., I. Rosenberg, J. Selhub, G. Kishore, R.N. Beachy, and K. Schubert. 2004. Enhancement of folates in plants through metabolic engineering. *Proc Natl Acad Sci USA* 101:5158-5163.
216. Kawakami, S., Y. Watanabe, R.N. Beachy. 2004. Tobacco mosaic virus infection spreads cell to cell as intact replication complexes. *Proc Natl Acad Sci USA* 101:6291-6296.
217. Beachy, R.N. 2004. Continuing the Effort. *AgBioForum* 7(1&2):1-1.
218. Marmey, P., A. Rojas-Mendoza, A. de Kochko, R.N. Beachy, and C.M. Fauquet. 2005. Characterization of the protease domain of rice tungro bacilliform virus responsible for the processing of the capsid protein from the polyprotein. *Virol J* 2:33-45.
219. Voloudakis, A.E., M-E. Aleman-Verdaguer, H.S. Padgett, and R.N. Beachy. 2005. Characterization of resistance in transgenic *Nicotiana benthamiana* encoding N-terminal deletion and assembly mutants of the tobacco etch potyvirus coat protein. *Arch Virol* 150:2567-2582.

220. Dai, S., Z. Zhang, J. Bick, and R. N. Beachy. 2006. Essential role of box II *cis* element and cognate host factors in regulating the promoter of rice tungro bacilliform virus. *J Gen Virol* 87:715-722. DOI 10.1099/vir.0.81488-0.
221. Bazzini, A., S. Asurmendi, H. E. Hopp, and R. N. Beachy. 2006. TMV and PVX coat proteins confer heterologous interference to PVX and TMV infection, respectively. *J Gen Virol* 87:1005-1012. DOI 10.1009/vir.0.81396-0.
222. Petruccelli S., M. Otegui, F. Lareu, O. Tran Dinh, A.-C. Fitchette, A. Circosta, M. Rumbo, M. Bardor, R. Carcamo, V. Gomord, and R.N. Beachy. 2006. A KDEL-tagged monoclonal antibody is efficiently retained in the endoplasmic reticulum in leaves, but is both partially secreted and sorted to protein storage vacuoles in seeds. *Plant Biotechnol J* 4:511-517. DOI: 10.1111/j.1467-7652.2006.00200.x
223. Kouassi, N.K., L. Chen, C. Siré, M. Bangratz-Reyser, R.N. Beachy, C.M. Fauquet, and C. Brugidou. 2006. Expression of rice yellow mottle virus coat protein enhances virus infection in transgenic plants. *Arch Virol.* 151: 2111-2122. DOI 10.1007/s00705-006-0802-3.
224. Fujiki, M., S. Kawakami, R.W. Kim, and R.N. Beachy. 2006. Domains of *Tobacco mosaic virus* movement protein essential for its membrane association. *J Gen Virol* 87:2699-2707.
225. Bazzini, A., H.E. Hopp, R.N. Beachy, and S. Asurmendi. 2006. Posttranscriptional gene silencing does not play a significant role in PVX coat protein mediated resistance. *Phytopathology*, Vol. 96, No. 11. 1175-1178.
226. Liu, Y., S. Dai, R.N. Beachy. 2007. Role of the C-terminal domains of rice bZIP proteins RF2a and RF2b in regulating transcription. Biochem J. 405: 243-249. DOI 10.1042/BJ2006/375.
227. Bazzini, A., HE. Hopp, R.N. Beachy and Sebastian Asuremendi. 2007. Infection and co-accumulation of Tobacco Mosaic Virus proteins alter microRNA accmumulation, correlating with symptom on plant development. PNAS. 104 (29) 12157-12162. DOI 10.1073/pnas 0705/14104.
228. Asurmendi, S., R.H. Berg, T.J. Smith, M. Bendahmane, and R.N. Beachy. 2007. Aggregation of TMV CP plays a role in CP functions and in coat-protein mediated resistance. Virology. 366: 98-106. DOI 10.1016/j.viol.2007.03.014.
229. Bendahmane, M., I.Chen, S. Asurmendi, A. Bazzini, J. Szecsi, and R.N. Beachy. 2007. Coat protein-mediated resistance to TMV infection of Nicotiana tabacum involves multiple modes of interference by coat protein. Virology. 366: 107-116.
230. Berg, H. and R.N. Beachy. 2008. Fluorescent Protein Applications in Plants. Methods in Cell Biology, 85: 153-177. DOI: 10.1016/S0091-679X(08)85008-X.

RESEARCH PAPERS

VIRUS TOLERANCE, PLANT GROWTH, AND FIELD PERFORMANCE OF TRANSGENIC TOMATO PLANTS EXPRESSING COAT PROTEIN FROM TOBACCO MOSAIC VIRUS

Richard S. Nelson, Sheila M. McCormick*†, Xavier Delannay*, Philip Dube, Jeanne Layton*, Edwin J. Anderson, Maria Kaniewska, Robert K. Proksch*, Robert B. Horsch*, Stephen G. Rogers*, Robert T. Fraley*, Roger N. Beachy

Plant Biology Program, Department of Biology, Washington University, St. Louis, MO 63130. *Monsanto Company, 700 Chesterfield Village Pkway, St. Louis, MO 63198. †Present address: USDA-ARS, Plant Gene Expression Center, 800 Buchanan St., Albany, CA 94710.

Two transgenic tomato plants that express the coat protein (CP) of the common (U_1) strain of tobacco mosaic virus (TMV) were produced from cultivar VF36 using gene transfer techniques. CP-expressing plants were partially resistant to infection and symptom development caused by TMV and tomato mosaic virus (ToMV) strains L, 2, or 2^2. Strains 2 and 2^2 normally overcome the natural resistance genes present in many commercial tomato cultivars. In the field, no more than 5% of the CP-expressing plants inoculated with TMV exhibited visual systemic disease symptoms by fruit harvest compared with 99% of the VF36 plants. Lack of visual symptoms was associated with lack of virus accumulation in the CP-expressing plants. In terms of agronomic traits, leaf and stem dry weight accumulation in greenhouse-grown uninoculated CP-expressing (line 306) and nonexpressing plants were essentially equal. In field analyses, tomato fruit yields of the VF36 plants decreased 26–35% due to virus infection, whereas yields of the CP-expressing plants were unaffected. Yields from one CP-expressing line were equal to that of the uninoculated VF36 plants suggesting that expression of the CP gene does not intrinsically cause yield depression. The results from these growth chamber, greenhouse and field experiments indicate the potential for use of genetically engineered protection in agriculture.

Virus infection can decrease yields by 20% or more in tomato plants grown under production conditions[1]. Although tobacco mosaic virus (TMV) can and does infect tomatoes, most of the yield loss is attributed to infection by strains of tomato mosaic virus (ToMV), a tobamovirus closely related to TMV[2]. Improved hygiene and the use of natural cross-protection (infection of plants with a mild strain of virus to protect against infection by more virulent strains) has helped to control severe outbreaks of virus infection. However, the use of virus-resistant tomato cultivars has provided the best control against ToMV, where a gene-for-gene relationship exists for resistance[2–5]. There are two loci for resistance to ToMV in tomato, Tm-1 and Tm-2; the Tm-2 locus has two alleles, Tm-2 and Tm-2^2. Tomato lines carrying no genetic resistance develop symptoms when inoculated with strain O virus. The resistance genes Tm-1, Tm-2 and Tm-2^2 can be overcome by tomato mosaic virus strains 1, 2, and 2^2, respectively. Most cultivars used in commercial tomato production carry Tm-2 or Tm-2^2 resistance. Although the Tm-2^2 resistance has proven durable to date, the appearance of more aggressive strains of 2^2 that reduce the usefulness of the Tm-2^2 locus is possible.

Recently it was shown that the expression of the coat protein (CP) of TMV[6] and of alfalfa mosaic virus (AlMV)[7–9] in transgenic plants results in protection of those plants against virus infection by TMV or AlMV, respectively. This "engineered" protection mimics cross-protection in that it is less effective at high levels of inoculum, largely overcome by inoculation with viral RNA, and exhibits some degree of strain specificity. An important manifestation of genetic cross-protection is that the inoculated leaves show fewer chlorotic or necrotic lesions compared with control plants[7–10]. There is also reduced rate of systemic spread of the virus if infection occurs[10]. Evidence to be published shortly indicates that CP derived resistance has been achieved for potato virus X and cucumber mosaic virus[11–12].

In this paper we report that expression of the TMV coat protein gene in transgenic tomato plants results in the absence or decrease in disease symptoms in these plants after inoculation with TMV or ToMV, including those ToMV strains that overcome the natural resistance genes in tomato. We also report that protection is effective in

The above article is reprinted with permission from Macmillan Publishers Ltd., *Bio/Technology*, Vol. 6, pp. 403-409 (April 1988).

greenhouse and field experiments and that CP expression does not affect the agronomic characteristics of the CP-expressing tomato lines in the absence of virus inoculation.

RESULTS

Presence and expression of the coat protein gene in tomato. Two plants (306 and 329) were regenerated from independent transformation events (see Experimental Protocol), self-fertilized, and the progeny analyzed for the presence of the TMV coat protein (CP) gene by Southern blot analysis. Line 306 plants (R1 generation) and a progeny line obtained by self-fertilization of line 306 (line 306-98, R3 generation), have one gene insert per haploid genome as determined by copy number reconstruction and border fragment analyses (data not shown). Line 329 plants (R1 generation) contained at least two inserts at different sites as determined by border fragment analysis (data not shown).

Plant lines 306 and 306-98 (R1 and R3 generation, respectively) were analyzed for CP mRNA accumulation by northern blot analysis (Fig. 1). The hybridization signal from the 0.9–1.0 kb RNA homologous to the CP gene in line 306-98 was approximately twice that observed for the RNA from line 306. Whereas transgenic tobacco lines that harbor the pTM319 plasmid accumulate a 2.1 kb RNA as well as the 0.9 kb RNA (Figure 1, lane 4), the larger RNA molecule was not observed in the tomato plants analyzed (Fig. 1).

Coat protein concentration in the youngest fully-expanded leaves from plant lines 306 and 329 (R1 generation) were approximately 0.05% of the total leaf protein extracted from plants grown under greenhouse conditions (data not shown). Although CP concentrations were similar, segregation ratios were dissimilar between lines 306 and 329, in which the CP segregated at 3:1 (expression:nonexpression) and 15:1 or greater ratios, respectively (line 306: 229 plants analyzed, 169 plants expressing; line 329: 29 plants analyzed, 28 plants expressing). These segregation ratios indicate one and two or more active loci in lines 306 and 329, respectively, and support the gene copy number analyses described above. Subsequent generations of 306-98 (R3 and R4 generations) were confirmed to be homozygous by CP analyses (data not shown).

Virus inoculation and symptom development in CP-expressing and control plants. Seedling progeny of line 306 that express the CP-gene [CP(+)] and that did not express the CP-gene [CP(−)], of transgenic plants harboring genes other than the CP gene, and of the nontransformed parental line (VF36) were inoculated with increasing concentrations of U_1-TMV (Table 1). Although increasing the virus concentration caused more rapid appearance of systemic disease symptoms in the control plants, the CP(+) plants showed no such increase. Protection was also observed against a highly virulent strain of TMV (PV230, ATCC designation) and to a lesser extent, against a strain of tomato mosaic virus (strain L, Table 1).

Experiments were also conducted to compare disease development in a CP(+) transgenic line (306), nontransgenic line (cv. VF 36) and near isogenic lines (cv. Craigella) that carried determinants for genetic resistance (Tm-1, Tm-2 or Tm-2^2). Strain 2 not only caused rapid symptom development in VF36 plants that carry no resistance genes against ToMV, but also in plants carrying the Tm-2 resistance gene. CP(+) plants inoculated with ToMV 2 exhibited either a delay in symptom development or escaped disease. Similar results were obtained after ToMV 2^2 inoculation except that plants carrying the Tm-2^2 resistance were overcome and plants carrying Tm-2 resistance were resistant. These data indicate that CP(+) plants protected fully or partially against all the TMV and ToMV strains tested. The other resistance genes (Tm-1, Tm-2, and Tm-2^2) may confer more complete resistance against most strains, but are highly susceptible to specifically adapted strains.

In addition to visual observations, the accumulation of TMV (PV230) in inoculated and systemically infected leaves was determined by protein immunodot blot analysis 14 days post inoculation (data not shown). Both inoculated and systemic leaves of all VF36 plants contained high levels of TMV. Two of the five CP(+) plants escaped virus infection, while three of the five became infected and accumulated low levels of virus in the inoculated leaves. Only one CP(+) plants became systematically infected. None of the resistant lines (Tm-1, Tm-2, and Tm-2^2) accumulated any virus and PV230 can be considered a type O strain. Thus, the correlation between visual observations (Table 2) and virus accumulation in the upper leaves was complete.

Plant growth of CP-expressing and control plants under greenhouse conditions. CP(+) plants from line 306 (R1 generation) and homozygous line 306-98 (R2 generation) were analyzed for leaf and stem dry weight accumulation through 88 days after planting in the greenhouse (Fig. 2). Leaf and stem dry weight accumulations in line 306 and 306-98 plants were, at minimum, equal to those in VF36 plants. It appears, therefore, that expression of

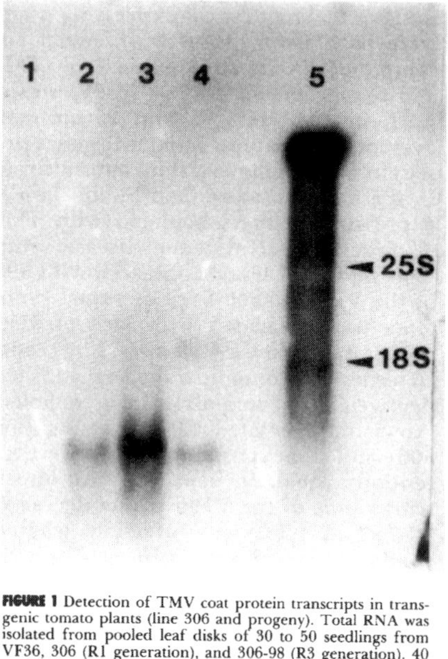

FIGURE 1 Detection of TMV coat protein transcripts in transgenic tomato plants (line 306 and progeny). Total RNA was isolated from pooled leaf disks of 30 to 50 seedlings from VF36, 306 (R1 generation), and 306-98 (R3 generation). 40 μg RNA per plant type was fractionated on an agarose gel containing formaldehyde. After transfer to nitrocellulose the RNA was probed with a nick-translated TMV CP fragment. Lanes 1, 2, 3 and 4 contain RNA from VF36, 306, 306-98, and a tobacco plant (lines 3404, 20 μg total RNA) expressing the TMV CP, respectively. Lane 5 contains full length U_1-TMV RNA (6.4 kb). The position to which the 25S and 18S (2 kb) plant ribosomal RNAs migrated is noted.

the CP gene in the transgenic plants (these plants did not express nopaline) does not alter the growth and development of these plants. By 88 days after planting (DAP) we observed that the VF36 plants had inadvertently become infected with TMV and thus terminated the formal experiment.

CP expression and virus resistance under field conditions. After receiving USDA-APHIS approval, the CP-expressing and nonexpressing plants (cv. VF36) were grown in the field to determine their agronomic characteristics. CP(+) plants from lines 306-98 (R4 generation) and 329-3 (R2 generation) were analyzed for CP expression under field conditions prior to inoculation with virus (Fig. 3). Although field expression levels were slightly lower than that observed under growth chamber or greenhouse conditions (0.02% vs. 0.05% of the total extracted leaf protein, respectively) we do not consider the difference significant.

Plants of each line were inoculated with different concentrations of the U_1 strain of TMV at 39 and 54 DAP. In each case CP(+) plants expressed a delay or, in the majority, an absence of systemic disease symptoms compared with VF36 plants (Fig. 4a and b). By the beginning of fruit harvest for both inoculation dates only 3 of 96 of the 306-98 plants and 0 of 96 of the 329-3 plants bore disease symptoms compared with 95 of 96 VF36 plants. The visual observations were substantiated by quantitating virus accumulation in systemically infected leaves of these plants (Table 3). CP accumulated in all plants showing visual symptoms. CP(+) plants not showing symptoms did not accumulate TMV with the exception of two 306-98 plants that had weak visual symptoms and a small but measurable accumulation of virus. These experiments clearly demonstrate a high level of resistance to TMV strain U_1 under field conditions.

Fruit yield under field conditions. Mature fruit appeared on uninoculated VF36 and CP(+) plants at similar times in these field trials. Tomato yields from virus-inoculated CP-expressing plants were unchanged compared with the uninoculated CP-expressing plants, while

TABLE 1 Protection against systemic symptom development in coat protein (CP) expressing transgenic plants in the presence of increasing virus inoculum concentrations. CP-expressing (line 306, R1 generation) and various control plants not expressing CP were inoculated with the U_1 or PV230 (ATCC designation) strains of TMV or with the L strain of ToMV and observed for symptom development in growth chambers (U_1 inoculations) or in the greenhouse (PV230 and L inoculations).

Plant Type[a]	Virus	Virus Inoculum Concentration (μg/ml)	Days Post Inoculation					
			5	6	7	8	9	30
			Percent Plants Showing Systemic Symptoms					
+CP(7)[b]	U_1	0.5	0	0	0	0	0	0
−CP(6)			17	17	17	17	50	83
−CP,ssu(9)			11	22	56	67	67	100
+CP(9)	U_1	2.0	0	0	0	0	0	22
−CP(6)			0	17	67	83	100	100
−CP,ssu(7)			0	29	43	86	100	100
+CP(10)	U_1	5.0	0	0	0	0	0	0
−CP(4)			0	50	75	75	100	100
−CP,ssu(8)			0	0	88	100	100	100
+CP(11)	U_1	20.0	0	0	0	0	0	9
−CP(4)			25	50	75	75	75	100
−CP,ssu(8)			25	38	88	100	100	100
VF36(4)			50	75	75	100	100	100
			Days Post Inoculation					
			7	8	9	10	13	29
+CP(10)	PV230	2.0	0	0	0	0	0	0
−CP(6)			17	67	83	100	100	100
−CP,ssu(5)			40	60	60	60	60	80
+CP(13)	PV230	20.0	0	0	0	8	23	54
−CP(3)			0	33	100	100	100	100
−CP,ssu(6)			17	50	100	100	100	100
			Days Post Inoculation					
			8	9	10	12	20	29
+CP(12)	L	2.0	0	0	0	0	33	58
−CP(4)			0	0	0	50	100	100
−CP,ssu(6)			17	17	67	83	83	100
+CP(13)	L	20.0	0	0	0	0	46	62
−CP(3)			33	66	66	100	100	100
−CP,ssu(6)			33	50	100	100	100	100

[a] +CP, Segregating progeny expressing CP
−CP, Segregating progeny not expressing CP
−CP,ssu, Progeny of plants harboring the neomycin phosphotransferase gene driven by the promoter from the small subunit (ssu) of ribulose bisphosphate carboxylase.
VF36, Nontransformed parental line.
[b] Numbers in parentheses equal the sample size for the particular treatment.

yields from virus-inoculated VF36 plants were substantially lower than yields of the uninoculated VF36 plants (Table 4). The major effect of virus infection on the VF36 plants was on number of fruits/plant rather than on weight/fruit. Tomato yields from 306-98 plants, regardless of virus inoculation, were equal to the yield of uninoculated VF36 plants. Thus the expression of the CP gene has not affected fruit production in this genetic background under these conditions. Yields of 329-3 were depressed presumably due to as yet undefined effects of plant transformation and/or regeneration.

DISCUSSION

Two transgenic tomato plant lines (306 and 329) that express a chimeric gene encoding the coat protein (CP) of the common strain of TMV were produced. Line 306 contained and expressed a single gene, while 329 expressed two or more genes. Expression levels for CP were determined to be approximately 0.05% or less of the extractable leaf protein under our experimental conditions (data not shown; Fig. 3). Similar levels were determined for leaves of transgenic tobacco plants expressing the CP gene[6,10]. Expression of the TMV-CP gene was stable through all tomato generations produced to date (R2 generation for 329, R4 generation for 306). Müller et al.[13] have shown that homozygous transgenic lines of *Nicotiana tabacum*, produced using a binary vector, allowed for 0.06% or fewer revertants to occur after backcross to the parental line. They concluded that meiotic instability in their plants was not greater than the spontaneous mutation rate of other plant genes and therefore acceptable for commercialization. Other workers[14] have noted some genetic instability after backcrossing transgenic *L. esculentum* × *L. pennellii* plants with *L. esculentum*. The instability, however, may be due to the interspecies cross or more likely to the complicated arrangement of the T-DNA itself in the particular transformant.

In addition to the high degree of protection against TMV (U_1 and PV230), protection was observed against three strains of ToMV (L, 2, and 2^2) in the CP(+) tomato plants (Tables 1 and 2). Although the protection provided by the CP of the U_1 strain of TMV against ToMV strains was somewhat less than that provided against the TMV strains, it is novel in that previous reports using classical cross-protection suggested that TMV could not protect against ToMV[15,16]. Our data demonstrate that it may be useful to introduce CP-expression into other resistant lines to add breadth to existing resistance genes. In addition, it will be interesting to determine if expression of a ToMV-CP will increase the protection against these strains in transgenic plants.

Analysis of virus accumulation within CP(+) plants after inoculation with TMV (PV230) showed that these plants accumulate less virus in the inoculated and systemic leaves compared with VF36 plants, and that many plants escaped infection (data not shown). Thus, there is blockage in infection and/or replication of virus in the inoculated leaves. Previous reports with both TMV and AlMV CP-expressing tobacco plants[7–10] and AlMV expressing tomato plants[7] indicate a similar blockage in the inoculated leaves. Thus, there may be a commonality in the protection mechanism across several plant species expressing unrelated virus CP genes.

Expressing the TMV CP gene did not affect vegetative traits since the presence and expression of this gene did not alter seed germination percentages (data not shown), the rate of stem elongation (data not shown), leaf or stem dry matter accumulation, or dry matter partitioning between leaf and stem (Fig. 4), in line 306 and its progeny. In addition, reproductive traits such as the number of flowers/plant (data not shown), fruit maturation date (data not shown), and fruit yield (Table 4) were also unaffected. Future experiments will address processing and fruit quality questions as well as the expression level of the CP gene in transgenic fruit. Expression of the CP gene in the tomato fruit of the transgenic plant is not novel since it has been known for many years that tomato plants infected with TMV contains CP-enshrouded virus in their fruit[15,17].

In terms of pathogen protection, both of the CP-expressing lines, 306 and 329, displayed nearly complete protection against TMV infection in the field (Fig. 4, Table 3), and the yields of tomato fruits from these plants were equal to those of uninoculated CP-expressing plants.

TABLE 2 Comparison of symptom development in CP-expressing, non-expressing, and genetic resistant lines after virus inoculation under greenhouse conditions. VF36, nontransformed parental cultivar; 306, CP-expressing line (R1 generation); Tm-1, Tm-2, and Tm-2^2, genetic resistant lines of the cv. *Craigella*. Seedlings were inoculated with ToMV strains (2 or 2^2) or a TMV strain (ATCC designation PV230). Numbers in parentheses indicate the sample size. PV230 inoculum concentration was 2 μg/ml while those for ToMV-2 and 2^2 were approximately 3 times the infectivity of PV230 as determined by comparison of lesion numbers on local lesion host plants (*Nicotiana tabacum* cv. Xanthi 'nc').

		Days Post Inoculation						
Plant Line	**Virus Strain**	**4**	**5**	**6**	**7**	**8**	**11**	**14**
		Percent Plants Showing Systemic Symptoms						
VF36 (5)		80	100	100	100	100	100	100
306 (5)	ToMV-2	0	0	0	20	20	60	60
Tm-1 (5)		0	0	0	0	0	0	0
Tm-2 (5)		40	60	100	100	100	100	100
Tm-2^2 (5)		0	0	0	0	0	0	0
VF36 (5)		0	0	20	100	100	100	100
306 (5)		0	0	0	0	0	20	20
Tm-1 (5)	ToMV-2^2	0	0	0	0	0	0	0
Tm-2 (5)		0	0	0	0	0	0	0
Tm-2^2 (5)		0	40	40	40	40	60	60
VF36 (5)		60	80	100	100	100	100	100
306 (5)	TMV-PV230	0	0	0	0	0	0	20
Tm-1 (5)		0	0	0	0	0	0	0
Tm-2 (5)		0	0	0	0	0	0	0
Tm-2^2 (5)		0	0	0	0	0	0	0

In contrast, fruit yields of VF36 plants infected with TMV were 26–35% lower than the uninoculated VF36 plants (Table 4). It is apparent that, as has been found in natural cross-protection experiments[16,18], the genetically engineered cross-protection has the potential for maintaining yields even under the severe infection pressures in these experiments. The protection effect would be considerably greater if the challenge virus would have been a more severe strain such as PV230 which causes marked chlorosis and stunting in tomatoes.

The results of these growth chamber, greenhouse and field experiments indicate that expression of the CP gene of U_1-TMV provides a high level of protection against several different strains of TMV and ToMV. As depicted in Figure 4 and Table 3 not only do most plants escape disease symptom development, but less virus accumulates in the transgenic CP(+) plants than in the non-expressing CP(−) controls. Recently evidence has been presented[19,20] that expression of viral satellite RNAs in transgenic plants suppresses replication and/or symptom expression caused by infection of the helper virus. While this approach to plant protection may be a second method for protecting plants against some virus diseases, it suffers from two problems: (1) satellite RNAs have been identified in a very limited number of virus situations; and (2) in some situations disease symptoms resulting from infection by helper viruses alone can be dramatically aggravated by the presence of satellite RNA as in the case of tomato lethal necrosis[21]. The breadth of application of the CP derived protection and its relative safety, as well as efficacy, therefore appear to have advantages over that provided by satellite RNAs.

TABLE 3 Accumulation of U_1-TMV after inoculation of CP-expressing or nontransformed plants under field conditions. Inoculations were carried out as described under Materials and Methods on CP-expressing plants from lines 306-98 (R4 generation) and 329-3 (R2-generation) and nontransformed plants from the parental cv. VF36. Leaves from plants inoculated at 44 days after planting (DAP, early inoculation, 10 μg/ml U_1 virus) were harvested at 49 days post inoculation (DPI) while leaves from plants inoculated at 59 DAP (late inoculation, 40 μg/ml U_1 virus) were harvested at 39 DPI. One subterminal leaflet from each of 3 branches per plant were harvested and each sample was analyzed for U_1-TMV CP by ELISA. Sampled leaves were greater than 5 cm in length with the next younger leaf (if present) less than 5 cm in length. For VF36 plants each value represents the mean ± S.D. for 4 replicates with each replicate containing values from 4 plants. For 306-98 and 329-3 plants, values are given in ranges with the number of plants within the range in parentheses following the value. Values are in ng CP per 500 ng protein. The limit of detection was 0.25 ng per 500 ng of protein in the assay.

	Time of Virus Inoculation	
Plant Line	Early	Late
VF36	46±9	47±16
306-98	6(1)	15(1)
	0.25-1(3)	0.25-1(3)
	<0.25(12)	<0.25(12)
329-3	0.25-1(7)	0.25-1(3)
	<0.25(9)	<0.25(13)

TABLE 4 Fruit weight (Mg/ha) of CP-expressing or non-expressing field grown plants under various virus inoculum regimes. Inoculations were carried out on plants as described in Table 3 and Figure 4. VF36, nontransformed parental cultivar; 306-98, R4 generation of a CP-expressing line; 329-3, R2 generation of a CP-expressing line. Harvest was begun at 98 DAP and continued through 128 DAP. Fruit was harvested at the physiological stage of color break and fruit number and fresh weight determined. Twelve plants per replicate were harvested. There were four replicates per plant line-virus treatment, except for the uninoculated VF36 fruit yield which was composed of three replicates. Each value represents the mean ± S.E. for the replicates. The experimental design was a split plot and statistical analysis was by analysis of variance.

	Time of Virus Inoculation		
Plant Line	Uninoculated	Early	Late
VF36	84.0±6.4	62.0±3.4*	54.6±2.3*
306-98	88.3±2.0	85.8±6.2	77.6±5.5
329-3	29.9±2.2⁺	33.3±2.1	34.1±2.9

*Significantly different from uninoculated treatment ($p<0.05$).
⁺Significantly different from VF36 uninoculated treatment ($p<0.05$).

EXPERIMENTAL PROTOCOL

Sources of plants and virus strains. *Lycopersicon esculentum* cv. VF36 seed was obtained from the Tomato Genetics Stock Center, UC, Davis. Near isogenic lines (cv. *Craigella*) carrying the genetic resistances Tm-1, Tm-2, or Tm-2^2 were obtained from the Glasshouse Crops Research Institute, Sussex, England. Tobacco mosaic virus strains U_1 and PV230 were obtained from Milton

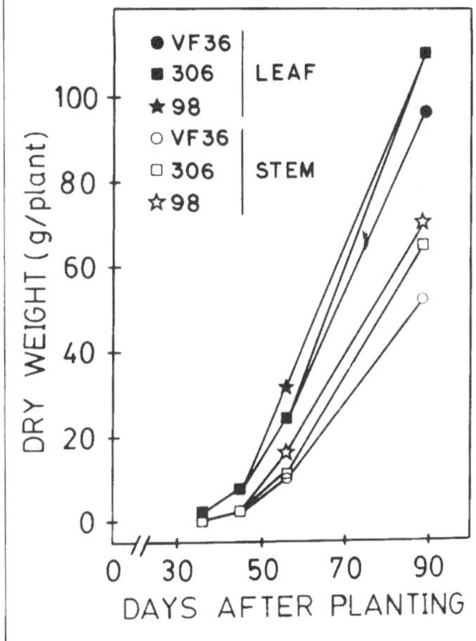

FIGURE 2 Analysis of leaf and stem dry weight accumulation in transgenic and nontransformed plants under greenhouse conditions. The experimental design was a randomized complete block with three replicates for nontransformed cv. VF36 and CP-expressing plant line 306 (R1 generation) and two replicates for the CP-expressing homozygous line 306-98 (98, R2 generation). Each replicate contained 3–4 plants. No statistically significant decrease in dry matter accumulation was observed for plants expressing the CP gene compared with VF36. Statistical analysis was by F test followed by an lsd determination at the 0.05 level with a correction made for the missing replicate for line 306-98. The only statistically significant observation was a decrease in stem weight of VF36 compared with the CP-expressing plants at 88 days after planting.

Zaitlin (Cornell University, Ithaca, NY) and the American Type Culture Collection, respectively. Tomato mosaic virus strains 2 and 2^2 were obtained from A. Th. B. Rast (Institute of Phytopathological Research, Wageningen, the Netherlands). Tomato mosaic virus strain L was obtained from M. Zaitlin.

Plant transformation. VF36 tomato leaf pieces or cotyledons were transformed with *A. tumafaciens* harboring pTM319[6] as described by McCormick et al.[22] and transgenic plants were regenerated. Progeny (R1 generation) of the self-fertilized primary transformant, plant number 306, were scored for kanamycin resistance by the leaf callus assay[22], and for the presence of TMV coat protein (CP) by immunoblot analysis[6]. Although the nopaline synthase gene was carried on pTM319, the parent and the progeny of plant 306 were silent for nopaline expression. There was a 100% correlation in the progeny between kanamycin resistance and CP expression (40 plants analyzed). Homozygotes in line 306 were identified by self-fertilizing R1 generation plants and analyzing progeny for kanamycin resistance. Progeny of a second self-fertilized primary transformant, plant number 329, were scored for nopaline synthase activity[23] and TMV coat protein. Nopaline and CP expression segregated together in the 19 R1 progeny of plant 329 analyzed.

Isolation of plant RNA and northern blot analysis. Total RNA was isolated from pooled leaf disks from 30 to 50 seedlings from VF36, 306, or 306-98 as described by Powell Abel et al.[6] The RNA was fractionated by electrophoresis to nitrocellulose[6]. The

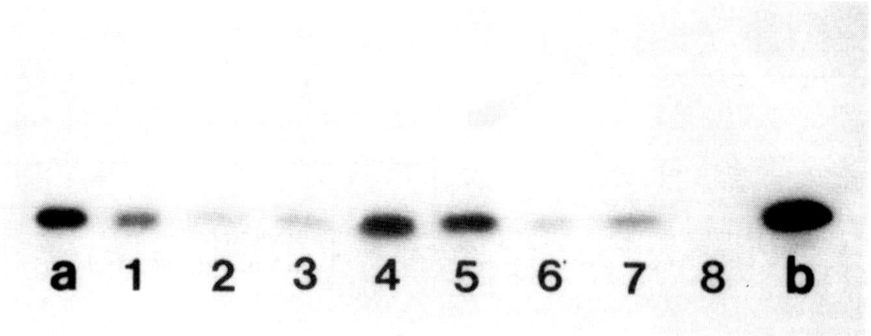

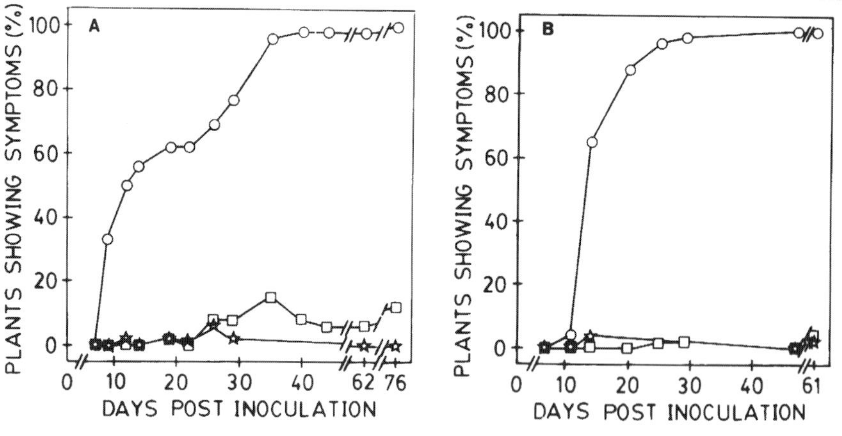

FIGURE 3 Accumulation of coat protein (CP) expression in transgenic plants under field conditions. Leaf samples from single plants were taken eight days after transplanting from the greenhouse to the field. 40 μg of protein extracted from one expanded leaflet of either VF36, 306-98 (R4 generation), or 329-3 (R2 generation) plants were analyzed by SDS-PAGE and immunoblot. Lanes 1, 4, and 5 contain samples from 306-98 plants; lanes 2, 3, 6, and 7 contain samples from 329-3 plants; lane 8 contains a sample from a greenhouse grown VF36 plant; and lane a and b contain, respectively, 7.5 ng and 30 ng of U_1 virus standard.

FIGURE 4 Systemic symptom development in CP-expressing and nontransformed plants under field conditions after virus inoculation. Panel A shows disease development in nontransformed plants of cv. VF36 (○), and CP-expressing plants of line 306-98 (R4 generation, □) and 329-3 (R2 generation, ☆) inoculated with 10 μg/ml of strain U_1 TMV on terminal leaflets of three successive leaves. The youngest inoculated leaf was greater than 3 cm in length. Plants were inoculated at 44 days after planting (eight days after transplanting to the field) and observations were made on 48 plants per plant line. Panel B shows results from plants inoculated with 40 μg/ml of strain U_1 TMV on terminal leaflets of two successive leaves. The youngest inoculated leaf was greater than 5 cm in length. Plants were inoculated at 59 days after planting (23 days after transplanting) and observations were made on 48 plants per plant line.

blot was probed with a ^{32}P-labeled DNA fragment containing only the CP sequence.

Isolation of plant protein and immunological analyses. Protein was extracted from leaves of plants from lines 306, 329, and 306-98 and analyzed by immunoblot reactions for the presence or absence of CP (+CP and −CP, respectively) in the transgenic plants as described by Powell Abel et al.[6] In experiments in which virus accumulation after inoculation was determined, analyses were by dot blot immunoassay[10] or enzyme linked immunosorbant assay (double sandwich ELISA)[24].

Growth, inoculation, observation and sampling of plants. Seeds were germinated in a greenhouse under natural light conditions, supplemented in winter months with light (~75 $\mu E.m^{-2}.sec^{-1}$) to produce 14 h days. Seedlings were transplanted into 4 in. pots approximately 7 days after planting. Approximately 7 days following transplanting, leaf tissue was taken from the 1st and/or 2nd leaf above the cotyledons, frozen in liquid N_2, and later analyzed for CP expression. In addition, fresh leaf tissue was analyzed for kanamycin resistance (lines 306 and 306-98) or the presence of nopaline (line 329). In growth chamber and greenhouse experiments involving virus inoculations, approximately 21 days after planting, two unsampled terminal leaflets were dusted with carborundum (330 grit, Fisher Scientific) and inoculated with purified virus diluted to designated concentrations with inoculation buffer (20 mM sodium phosphate, pH 7.2, 1 mM EDTA). After inoculation, the leaflets were rinsed with water and the plants placed either in growth chambers under previously reported conditions[10] or left in the greenhouse. Individual plants were scored daily for systemic disease symptom development (chlorotic areas on the leaves above the inoculated leaves leading to mosaic symptoms in later developing leaves). To determine virus accumulation, leaf tissue was sampled from both inoculated and systemic leaves, frozen in liquid N_2, and then ground in extraction buffer (see previous section) in ratios of 1:2 or 2:1 (ml buffer:g fresh wt. of tissue), respectively. In the greenhouse experiment to determine leaf and stem dry weight accumulation, plants were transplanted at 8 days after planting (DAP) into 4 in. pots, and then at 36 DAP into 3 gallon pots and allowed to grow under greenhouse conditions. Leaf tissue was collected at various dates, and consisted of leaf lamina, midrib, and petioles, while stem tissue consisted of main plus side stems from the cotyledon node upward. After sampling, tissue was dried in paper bags at 80°C for 48 h in forced air ovens and then weighed. The experimental design was a randomized complete block design with three replicates per plant genotype and one missing replicate for line 306-98. Field experiments were initiated by germinating seed (VF36; 306-98, R4 generation; 329-3, R2 generation) in transplanting trays. The seedlings were grown in a greenhouse followed by transplanting to the field at 36 DAP. The field test was located in Jersey County, Illinois, on a Muscatine silt loam soil. Permission to conduct this first field test of transgenic tomatoes was granted by the Biological Assessment Support Staff, Plant Protection and Quarantine within the Animal and Plant Health Inspection Service, United States Department of Agriculture on June 1, 1987 (application #000038)[25]. The experimental design was a randomized split plot design with virus treatments assigned to the whole plots and genotypes assigned to the subplots. There were four replications. Each subplot consisted of a single row ~3.6 m in length containing 12 plants 31 cm apart. Distance between rows was 1.5 m within the main plots. The main plots were separated by a 6 m border. Plants were not staked and irrigation was provided as needed. The virus treatments included an early inoculation of plants at 44 DAP with 10 μg/ml of the U_1 strain of TMV on three leaves per plant, a late inoculation of plants at 59 DAP with 40 μg/ml of U_1-TMV on two leaves per plant, and uninoculated plants. Virus inoculations were executed by abrading leaves with a sponge soaked in a suspension of virus containing 1% carborundum. In addition to visual observations, virus accumulation in plants was quantitated immunologically. Fruit harvest began at 98 DAP and continued through 128 DAP. Fruit was harvested on a continuous basis after reaching the color break stage. Fruit fresh weight, number of fruits, and date of maturation were determined from this data.

Acknowledgments

The authors thank W. Gregg Clark for supplying purified DNA fragments used as probes and Curtis Holt for providing conjugated antibody. In addition, we thank Michael G. Dyer and Donna L. Droste for germination and maintenance of tomato plants at Washington University, Bill Schuler, Judy Erbschloe and Alan Lloyd for growth and maintenance of plants at Monsanto and Brad LaVallee and Jesse Hart for field assistance. Matt Harline is thanked for technical assistance. Thanks go to all the researchers at Washington University and Monsanto who helped from time to time on the field experiment to make it a success. We also thank the Monsanto Registration and Regulatory Affairs groups for their efforts in obtaining USDA approval for the field test. Nancy Burkhart is thanked for her usual expert preparation of the manuscript. This research was supported by a grant from the Monsanto Company.

Received 14 December 1987; accepted 3 February 1988.

References

1. Broadbent, L. 1976. Epidemiology and control of tomato mosaic virus. Annu. Rev. Phytopathol. **14**:75–96.
2. Brunt, A. A. 1986. Tomato Mosaic Virus, p. 181–204. *In*: The Plant Viruses: The Rod-shaped Plant Viruses. (M. H. V. Van Regenmortel and Heinz Fraenkel-Conrat (Eds.). Plenum Press, New York.
3. Hall, T. J. 1980. Resistance at the Tm-2 locus in the tomato to tomato mosaic virus. Euphytica **29**:189–197.
4. Ponz, F. and Bruening, G. 1986. Mechanisms of resistance to plant viruses. Annu. Rev. Phytopathol. **24**:355–381.
5. Van Loon, L. C. 1987. Disease induction by plant viruses. Adv. Virus Res. **33**:205–255.
6. Powell Abel, P., Nelson, R. S., De, B., Hoffman, N., Rogers, S. G., Fraley, R. T., and Beachy, R.N. 1986. Delay of disease development in transgenic plants that express the tobacco mosaic virus coat protein gene. Science **232**:738–743.
7. Tumer, N. E., O'Connell, K. M., Nelson, R. S., Sanders, P. R., Beachy, R. N., Fraley, R. T., and Shah, D. M. 1987. Expression of alfalfa mosaic virus coat protein gene confers cross-protection in transgenic tobacco and tomato plants. EMBO J. **6**:1181–1188.
8. Loesch-Fries, L. S., Merlo, D., Zinnen, T., Burhop, L., Hill, K., Krahn, K., Jarvis, N., Nelson, S., and Halk, E. 1987. Expression of alfalfa mosaic virus RNA 4 in transgenic plants confers virus resistance. EMBO J. **6**:1845–1851.
9. Van Dun, M. P., Bol, J. F., and Van Vloten-Doting, L. 1987. Expression of alfalfa mosaic virus and tobacco rattle virus coat protein genes in transgenic tobacco plants. Virology **159**:299–305.
10. Nelson, R. S., Powell Abel, P., and Beachy, R. N. 1987. Lesions and virus accumulation in inoculated transgenic tobacco plants expressing the coat protein gene of tobacco mosaic virus. Virology **158**:126–132.
11. Hemenway, C., Fang, R.-X., Kaniewski, W., Chua, N-H., and Tumer, N. E. 1988. Analysis of the mechanism of protection in transgenic plants expressing potato virus X protein or its antisense RNA. EMBO J., in press.
12. Cuozzo, M., O'Connell, K. M., Kaniewski, W., Fang, R.-X., Chua, N-H., and Tumer, N. E. 1988. Viral protection in transgenic plants expressing the cucumber mosaic virus coat protein or its antisense RNA. Bio/Technology, in press.
13. Muller, A. J., Mendel, R. R., Schiemann, J., Simoens, C., and Inzé, D. 1987. High meiotic stability of a foreign gene introduced into tobacco by *Agrobacterium*-mediated transformation. Mol. Gen. Genet. **207**:171–175.
14. Chyi, Y. S., Jorgensen, R. A., Goldstein, D., Tanksley, S. D., and Loaiza-Figueroa, F. 1986. Locations and stability of *Agrobacterium*-mediated T-DNA insertions in the *Lycopersicon* genome. Mol. Gen. Genet. **204**:64–69.
15. Broadbent, L. 1964. The epidemiology of tomato mosaic VII. The effect of TMV on tomato fruit yield and quality under glass. Ann. Appl. Biol. **54**:209–224.
16. Rast, A. Th. B. 1975. Variability of tobacco mosaic virus in relation to control of tomato mosaic in glasshouse tomato crops by resistance breeding and cross protection. Verslagen Van Landbouradkundige Des Rizksland Agricultural Research Reports **834**:1–76.
17. Boyle, J. S. and Wharton, D. C. 1957. The experimental reproduction of tomato internal browning by inoculation with strains of tobacco mosaic virus. Phytopathology **47**:199–207.
18. Ahoonmanesh, A. and Shalla, T. A. 1981. Feasibility of cross-protection for control of tomato mosaic virus in fresh market field-grown tomatoes. Plant Disease **65**:56–58.
19. Harrison, B. D., Mayo, M. A., and Baulcombe, D. C. 1987. Virus resistance in transgenic plants that express cucumber mosaic virus satellite RNA. Nature **328**:799–802.
20. Gerlach, W. L., Llewellyn, D., and Haseloff, J. 1987. Construction of a plant disease resistance gene from the satellite RNA of tobacco ringspot virus. Nature **328**:802–805.
21. Kaper, J. M. and Waterworth, H. E. 1977. Cucumber mosaic virus associated RNA 5: Causal agent for tomato necrosis. Science **196**:429–431.
22. McCormick, S., Niedermeyer, J., Fry, J., Barnason, A., Horsch, R., and Fraley, R. 1986. Leaf disc transformation of cultivated tomato (*L. esculentum*) using *Agrobacterium tumefaciens*. Plant Cell Reports 5:81–84.
23. Rogers, S. G., Horsch, R. B., and Fraley, R. T. 1986. Gene transfer in plants: production of transformed plants using Ti plasmid vectors. Meth. in Enzymol. **118**:627–640.
24. Clark, M. F. and Bar-Joseph, M. 1984. Enzyme immunosorbent assays in plant virology. Methods in Virology **7**:51–85.
25. Anonymous. 1987. Genetically engineered tomato plant: Determination of plant pest risk status. Federal Register **52**:29403–29404.

James E. Womack

Texas A&M University
College Station, Texas, USA

2001 — for the use of recombinant DNA technology, to revolutionize plant and animal sciences, paving the way for applications to neighboring fields.

CURRICULUM VITAE

Present Position and Address:

Distinguished Professor of Veterinary Pathobiology and Genetics
Director, Center for Animal Biotechnology and Genomics
Department of Veterinary Pathobiology
College of Veterinary Medicine and Biomedical Sciences
Texas A&M University
College Station, Texas 77843-4467

Education:

Degree	*Conferring Institute*	*Field*	*Year*
B.S.	Abilene Christian College	Mathematics Ed	1964
Ph.D.	Oregon State University	Genetics	1968

Professional Experience and Academic Appointments:

1968-1971	Assistant Professor, Abilene Christian College
1971-1973	Associate Professor, Abilene Christian College
1973-1975	Visiting Scientist, The Jackson Laboratory
1975-1977	Staff Scientist, The Jackson Laboratory
1977-1983	Associate Professor, Texas A&M University
1983-Present	Professor, Veterinary Pathology, Texas A&M University
1987-Present	W.P. Luse Professor, Texas A&M University
1989-1996	Director, Center for Animal Genetics, Institute of Biosciences & Technology
1990-1993	Interim Asst. Department Head, Veterinary Pathobiology, Texas A&M University
2001-Present	Distinguished Professor, Texas A&M University
2001-Present	Director, Center for Animal Biotechnology and Genomics, Texas A&M University

Awards and Honors:

1983	Alumni Citation Award, Abilene Christian University
1986	Beecham Award for Research Excellence
1987	Faculty Distinguished Achievement Award for Research, Texas A&M University
1990	Carrington Award for Research in Cell Biology
1990	McMaster Fellow, CSIRO, Australia
1993	Faculty of Genetics Research Award, Texas A&M University
1993	CIBA Prize for Research in Animal Health, Basel, Switzerland
1996	Outstanding Texas Geneticist, Texas Genetics Society
1997	Honor Award, U.S.D.A.
1999	Fellow, American Association for the Advancement of Science
1999	Elected to National Academy of Sciences, USA
2001	Wolf Prize (Agriculture), Israel
2001	Distinguished Professor, Texas A&M University
2003	Charter Member, Texas Academy of Science, Engineering and Medicine
2006	Distinguished Service Award, Texas Genetics Society
2006	Outstanding Alumnus of the Year, Abilene Christian University
2007	Dean's Impact Award, CVM, Texas A&M University
2008	Bush Excellence Award for Faculty in International Research

PROFESSIONAL ORGANIZATIONS:

American Genetic Association, Member -	1981-
President -	1985
Executive Vice President -	1996-
Genetics Society of America -	1968-
American Association for the Advancement of Science -	1969-
Fellow -	1999-
The American Society for Human Genetics -	1972-
Texas Genetics Society -	1978
Board of Directors -	1982-1987
President -	1989
International Society for Animal Genetics -	1988-
Chairman, Committee on Gene Mapping -	1988-1996
President -	2000-2006
Human Genome Organization (HUGO) -	1993-
Steering Committee, NIH Bovine Genome Sequencing Project -	2004-
Technical Committee, NIH Bovine Genome Sequencing Project -	2004-

EDITORIAL BOARDS:

Genomics, -	1987-1995
The Journal of Heredity -	1987-
Biochemical Genetics -	1987-
Animal Biotechnology -	1989-
Mammalian Genome -	1990-
Genome Research -	1995-2004
Animal Genetics -	1991-1999

GRADUATE STUDENTS (TEXAS A&M):

Name	*Degree*	*Institution*	*Completion Dates*
Frank H. Deal	M.S. GENE	TAMU	1981
Cheryl A. Coates	M.S. GENE	TAMU	1983
Gregory K. Fontenot	M.S. GENE	TAMU	1985
Linda R. Adkison	Ph.D. GENE	TAMU	1986
James C. McAvin	M.S. GENE	TAMU	1986
Lori Peterson	M.S. GENE	TAMU	1986
Donna Rich	M.S. GENE	TAMU	1986
Marla E. Foreman	Ph.D. GENE	TAMU	1988
Laura K. Faber	M.S. GENE	TAMU	1988
Robert Livingston	M.S. GENE	TAMU	1988
Yvonne D. Moll	M.S. GENE	TAMU	1988
Ingrid Doodeheefver Lyles	M.S. GENE	TAMU	1989
David W. Threadgill	Ph.D. GENE	TAMU	1989
Susan M. Zneimer	Ph.D. GENE	TAMU	1989
Kelly McRae Butts	M.S. GENE	TAMU	1990
Deborah S. Threadgill	Ph.D. GENE	TAMU	1990
Tammy Tobin-Jansen	Ph.D. GENE	TAMU	1990
Nan Zhang	Ph.D. GENE	TAMU	1991
Donna L. Day	M.S. GENE	TAMU	1992
Allan B. Dietz	Ph.D. GENE	TAMU	1992
Lei Li	Ph.D. MBCH	TAMU	1993
Holly L. Neibergs	Ph.D. GENE	TAMU	1993
Anne M. Ryan	Ph.D. GENE	TAMU	1993
Jutta Weiss	Ph.D. MBCH	TAMU	1993
Peter Skirpstunas	M.S. GENE	TAMU	1994
John Miggins	M.S. TOXI	TAMU	1996
Qing Gao	Ph.D. GENE	TAMU	1996
Penny Riggs	Ph.D. GENE	TAMU	1996

Yaping Yang	Ph.D. GENE	TAMU	1996
Elisabete Amaral	Ph.D. GENE	UNESP Brazil	1996
Kathryn M. Meurs	Ph.D. GENE	TAMU	1997
Thomas Brodigan	M.S. MBCH	TAMU	1998
Caird Rexroad III	Ph.D. GENE	TAMU	1998
Monica Amarante	Ph.D. GENE	UNESP Brazil	1998
Leonardo Sena	Ph.D. GENE	UFDP Brazil	2003
Scott Dindot	Ph.D. GENE	TAMU	2002
Stephanie McKay	M.S. VTMI	TAMU	2002
Stephen White	Ph.D. GENE	TAMU	2003
Kristen Taylor	Ph.D. GENE	TAMU	2004
Ralph Callicott	Ph.D. GENE	TAMU	current
Erin Gillenwaters	Ph.D. GENE	TAMU	current
Brittany Jones	Ph.D. GENE	TAMU	current
Catherine Busch	M.S. GENE	TAMU	current

POSTDOCTORAL FELLOWS:

Name	*Ph.D. Institution*	*Dates*
Eric Hallerman	The Hebrew University	1986-1987
Duke Rogers	University of California, Berkeley	1987-1989
Daniel Gallagher	Texas A&M University	1989-1993
Claude Schelling	Univ. of Berne, Switzlerland	1993-1995
Jorg Schlapher	Univ. of Berne, Switzerland	1995-1996
Prapti Mahyuddin	Institute for Animal Production, Indonesia	1998-2001
Tom Goldammer	Univ. of Rostock, Germany	1999-2001
Edward Cargill	Texas A&M University	2005-2006
Christopher Seabury	Texas A&M University	2006-

MY UNIQUE CAREER PATH TO AGRICULTURAL RESEARCH: AN AUTOBIOGRAPHICAL SKETCH

James E. Womack, Distinguished Professor
Texas A&M University, College Station, TX USA, April, 2008

I was honored to share the 2001 Wolf Prize in Agriculture with Roger Beechy for our respective work in animal and plant genomics. My award reads "*for the use of recombinant DNA technology to revolutionize animal sciences, paving the way for applications to neighboring fields.*" I'll try to recount my career path to this highest honor in agricultural research which was anything but a straight line. In fact, the winding path of my academic training and research career is marked by relatively few tracks pointing in the direction of livestock genomics. There are subtle hoofprints, however, which are largely overshadowed by chance encounters with a variety of scientific mentors, colleagues, friends and family who unknowingly shaped my path.

I was born in Anson, Texas in 1941 and didn't range far in the first 23 years of my life except for a couple of years when my dad was stationed in Michigan before the end of WWII. My parents purchased a small stock farm near Dad's homeplace after the war and built a modest house while Dad taught at a country school and completed his education on the G.I. bill at Abilene Christian College. My mother had only a high school education but could have and should have been an engineer. She could design, construct or repair anything from a baby dress to a cattle barn. Dad was offered a job with the Hawley Independent School District in the fall of 1946, the year I started first grade. He taught all the high school English courses, coached all the boys and girls athletic teams, and drove the school bus that traveled a 50 mile route to pick up about 30 students which when added to the locals produced a student body of just under 100 students spread over 12 grades. Less than a month into the school year, the Superintendent resigned and Dad was tapped for that job, a position he held until his retirement 33 years later. Over those years, he never gave up coaching basketball and retired with a national record for career wins and was inducted into the Texas High School Basketball Hall of Fame in 2005. My two sisters and brother also became teachers and needless to say, we were all exposed to basketball. In addition to basketball, I was exposed to farm life. Our homestead was several hundred acres and we raised beef cattle and horses, partly to supplement Dad's teaching salary but mostly, I think for the educational experience it provided the children. I made my spending money from a few steers we pasture finished each year and by breaking and selling raw horses we bought each summer from a dealer in Mexico. I enjoyed 4H and FFA and assumed I would someday have a small ranch of my own. I finished all 12 grades of public education at Hawley, graduating with a senior class of 16 students. Despite being less than six feet tall, I had a number of basketball scholarship offers, including a couple from major universities. Still a homeboy, I chose to follow Dad's footsteps at nearby Abilene Christian College.

I entered college with the intent of becoming a high school math and science teacher and basketball coach. My main interest, however, was playing basketball. I had a good career that actually spanned 5 years, thanks to a "redshirt" sophomore year. Although not a prolific scorer as in high school, I started two years, one year as team captain and played in two NCAA postseason tournaments. My closest friends to this day are some of my basketball teammates. I met Raby Jean Beakley during my freshman year but we didn't start dating until a couple of years later. She came from a ranching family and a small school and we were seated next to each other in one class by a professor who knew us both and thought we should get to know each other. He was right. She finished her teaching degree in 1962 and taught for a year before we married in June of 1963. I still didn't have my degree and had another year of basketball ahead of me. I had taken a number of education courses along with chemistry, physics, and math and was still planning a career in teaching and coaching. I had never taken a biology course. At some point it struck me that most public school teachers couldn't afford the luxury of my life long dream of living in the country, owning some land, and raising horses and cattle. Raby and I spent a lot of time on a ranch owned by her dad and with a dentist cousin who also had a nice cattle ranch. I came to the conclusion that my quickest route to land and cattle was through dentistry, not through teaching. The requirements for dental school included courses in embryology and comparative anatomy both of which had lab requirements and one other biology course. I chose genetics as my other biology course since it was the only course offered in the fall that didn't include a laboratory. I was already pushing my limit with labs that conflicted with basketball practice. All three courses required permission of the instructor since I had none of the prerequisite biology courses. Probably the biggest single moment in shaping my scientific career was my request for permission to take introductory genetics. Dr. Jim Throneberry was a cynical sort who apparently had some bad experiences with athletes in his classes. He let me know that I was welcome to give it a try but he was pretty sure I couldn't pass his course. I took his expectation of my failure as a personal challenge and immediately went to the bookstore and purchased the textbook, an introductory genetics text by Irwin Herskowitz. By the first day of class, I had read much of the text and was fascinated by the supplemental material I had discovered at the back of the book. The Supplement consisted of part of Mendel's 1867 letter to Nageli, and the Nobel Prize lectures of Morgan (1934), Muller (1946), Beadle (1958), Tatum (1958), Kornberg (1959) and Lederberg (1958). Although dental school was still my goal, I was convinced that genetics would provide most of the scientific excitement of the last half of the century.

After the first exam, Jim Throneberry became a friend and mentor and he along with Drs. Tommy McCord (organic chemistry) and Alvie Davis (biochemistry) began to try to convince me that dental school would be a waste of my talents and interest. Nonetheless, I applied to the Baylor College of Dentistry, and received my letter of acceptance for the fall class of 1964. I was still on track to graduate with a degree in Math Education, but now with a good background in biology and still a strong interest in genetics. I had broken my foot in the first basketball game of the season and had not worked my way back into the starting lineup until near the end of the season. Raby was already establishing herself as an outstanding elementary school teacher and loved her job at Dyess Elementary. It is hard to believe now but Texas schools had just come to grips with integration and the Dyess school was the first in Abilene to open their doors to both black and white students. Raby's love for children lessened the challenges of integration for both her and her students. As with every first or second grade she has taught, she still gets appreciative phone calls and emails from her former students. It was difficult for her to resign her job in Abilene but she applied, was interviewed, and quickly hired for the Fall of 1964 in Richardson, near Dallas and the Baylor College of Dentistry. Three days in August of 1964 marked another sharp turn in my career path. I was invited to work in the office of a dentist friend who suspected my interest in dentistry might be waning if in fact, it had ever been genuine. After the first day, I came home excited about all the new things I'd seen and learned. After the second day, I told Raby it was pretty much a repeat of the day before. After the third day, I told her I couldn't do that for the rest of my life. Her reaction was totally unselfish and left me no room to ever doubt that she loved me. She said "we'll do what makes you happy." I couldn't shake my interest in genetics. I talked to Dr. Throneberry who suggested I teach biology labs at ACC in the fall while applying to graduate programs in genetics. Raby resigned the job in Richardson and was quickly welcomed back in Abilene. We forfeited our $100 deposit at Baylor and I settled into an academic career, understanding fully that we would probably never have our little ranch with cows and horses.

I enjoyed teaching general biology labs and began to envision a career as a professor of genetics. In retrospect, I know I would have found a research career even if I had gone to dental school but now my interest was graduate school and genetics. Without the benefit of the internet, selection of a graduate school in 1964 was by word of mouth and Peterson's Guide. My love for the outdoors, including hunting and fishing, pulled me toward the Pacific Northwest and eventually to Oregon State University where Dr. Ralph Bogart, an Animal Scientist, directed a large, well funded, interdisciplinary Genetics Institute. I accepted an offer from Abilene Christian for support of $200 per month with the agreement that I would return to their faculty upon receiving my PhD. This was a nice supplement to the NASA fellowship Ralph provided. We arrived in Corvallis in August of 1965 pulling all our belongings in a small covered trailer I had made over the summer. Raby was pregnant with our first child and I was excited about the breadth of course offerings in genetics at the University. My mother had been diagnosed with cancer just a few months before we left making the decision to move so far away a difficult one. My family assisted us with the move and enjoyed some time in Oregon while helping us move into our new apartment. Watching them drive away brought tears to our eyes. Our daughter Wendy was due in December but didn't arrive until January of 1966. We tried to induce labor with weekend trips following streams up the rough logging roads or climbing over the massive timbers drifted up on the beautiful Oregon beaches. But like everything else she has done in her life, Wendy arrived when she chose, not at our convenience. Raby's dad had passed away in 1964 and her mom spent quite a bit of time with us over the next three years, providing greatly appreciated day care as Raby was able to teach between the birth of our two children. I spent a lot of time in the graduate student cubicles in the basement of Withycombe Hall with people like Jerry Reeves, Paul Humes, Carol Nix, Prentice Schilling, and Joe Templeton, who with his wife Jamie had moved up with us from Abilene Christian. I had never doubted the quality of my education from a small public school and a small college in West Texas and found my courses in genetics and radiation biophysics challenging but not overwhelming and totally delightful. Ralph was a quantitative geneticist by training but had incorporated considerable physiological genetics into his research program. When he suggested I might find a PhD research project in the wealth of data he was collecting from selection experiments in cattle, I distinctly recall telling him I didn't really want to study cattle genetics or animal breeding. I was more interested in laboratory animals and biomedical applications. I had done some reading on mutagenesis and we agreed on a radiation mutagenesis project in mice with emphasis on quantitative traits. I measured selection response for body weight and fertility traits in irradiated populations and published two papers from my dissertation research. More importantly, I learned about mammalian mutagenesis and became fascinated with mouse genetics. Ralph taught me a lot about mentoring although I'm sure I never practiced it like he did. He had an enormous garden and orchard that was planted with poor starving graduate students in mind. Of course, we spent some of our weekend time tilling the soil. He and Fran became adopted grandparents to his students' children. He guided us into our postdoctoral work and he was disappointed that I had made a commitment to Abilene Christian. He thought I should look for a postdoc in a medical school or research institute but he accepted my decision and encouraged me to grow and develop as a teacher.

Our second child was already overdue when I got my PhD degree in May of 1968. Our good friends from church, Gail and Chuck Woosley, escorted Raby to the graduation ceremony and had parked a car near Gill Coliseum for Raby if she needed a quick exit. She didn't. Louis and Janice Stone, who along with Pat and Carolyn Agnew, and Larry and Ella Rogers, have been lifelong friends from basketball days at ACC, had come to Corvallis to help us move. Raby, Wendy, the baby, and her mother would fly back to Texas, and the Stones would alternate with me in driving the U-Haul truck (we had outgrown the home made trailer) and our car. We waited and waited and waited. As before, the doctor had to induce labor and this was not an easy one for Raby....or Jimmy who was born with some ugly forceps bruises which fortunately disappeared before he and his mother, sister and grandmother arrived back in Texas. I would arrive a few days later with the Stones, our belongings and two black lab puppies.

The summer of 1968 was both a happy and sad one, but memorable. Thanks to a gift of stocks from Raby's parents, and her teaching salary from the beginning of our marriage, we had managed to buy 160 acres of farm/pasture land near Hawley, only 12 miles from where I would begin work in the fall and only 6 miles from my parents. Mom was very sick and I am thankful for the time she had with our two children before her death that fall. They made her smile. The property had an old and very small house in which we were determined to live. My grandpa Hollums, a carpenter by trade, helped me add a room and refurbish the house that summer, an experience I treasure. We built what we thought was a rattlesnake proof fence around the back yard but were proved wrong on several occasions. Despite the snakes, the sandstorms, the red mud, and constant concern about a half mile of river and a large pond, it was a nice place to raise young children. They had wonderful pets including a young skunk which is a story in itself. Raby was quickly employed by the Abilene schools and I began teaching genetics, cell biology and zoology in the fall of 1968 at what was now Abilene Christian University. I even bought that small herd of cattle I had long coveted although the ranching venture was a financial failure, thanks to cattle thieves who knew we left the property unoccupied for most of the daylight hours.

My most vivid memories of my first teaching experience include the high quality of the undergraduate students at ACU. The Biology department offered a Master's degree and I took on a few grad students, including Loren Skow who is now a colleague on the Veterinary faculty at Texas A&M. Dr. Clark Stevens was a wonderful department head and we were blessed with teachers like Mike Kemp, Kenneth Williams, and Roy Shake. Although I enjoyed teaching, my graduate school experience had kindled a burning desire for research that wouldn't go away. I wanted to know more about mutagenesis in mammals and wrote a proposal to the National Foundation March of Dimes to develop a biochemical genetic assay of mutation in mice using allozymes as an endpoint. I was familiar with mutagenesis studies of Tom Roderick and Earl Green at the Jackson Laboratory and had contacted Tom who agreed to be a collaborator in my proposal. If funded, I would come to Bar Harbor to study in the summers and continue my teaching on a nine month basis with ACU. The work was funded and we made our first of several trips to Bar Harbor in the summer of 1970. Although we didn't discover any radiation induce allozyme variants in one summer of starch gels, we developed the concept into a germinal mutation assay the NIEHS would eventually use on a grand scale in their chemical mutagenesis program. I also developed a lasting professional and personal friendship with Tom Roderick and a number of other scientists at The Jackson Laboratory such as Murriel Davisson, Eva Eicher and Ben Taylor, who along with visitors like John Hutton and Verne Chapman had begun to use allozyme variants as markers for populating the mouse linkage map. Enzyme markers were especially exciting to me because the human gene map was in its infancy and was being developed through somatic cell genetics and biochemical markers. By mapping the orthologous gene products in mice we had the first opportunity to do focused comparative gene mapping in mammals, albeit we were doing linkage mapping in mice while people like Frank Ruddle and Tom Shows were building synteny maps in humans. I was able to continue my work in comparative mapping during the academic semesters at ACU despite a heavy teaching load, thanks to bright young undergraduates like Mark Sharp. Mark and I demonstrated conservation of autosomal linkage in rats and mice, a novel concept in the pre-genomics era. While I still enjoyed teaching, I found myself wanting more hours in the laboratory and more opportunity to compare the genomes of humans, mice and other mammals.

Our second summer in Bar Harbor was probably our best summer ever. Raby and I had been offered and had accepted the role of Supervisors of The Jackson Laboratory's summer research program for college students. We were house parents to a terrific group of university students who were placed in various labs for their first real research experience. They included Don Stern who was to become a New York attorney and later Chairman of the Jackson Laboratory's Board of Governing Trustees, Clair Francomono M.D. who became Chief of the Human Genetics Branch of NHGRI before joining the Johns Hopkins medical faculty, Mike Lynes, Professor of Molecular and Cellular Medicine at the University of Connecticut and a dozen other equally talented and lively young people with whom I have sadly lost contact. Tom and I continued our collaboration which by now was directed more toward comparative gene mapping using allozyme variation between inbred strains and recently derived laboratory strains of *M.m.molossinus* and *M.m.castaneous*. Thanks to Tom I now had two summers of the postdoctoral research experience I had missed by returning directly to ACU after my PhD.

I still loved ACU as I do today. I loved the faculty and administration. The students were still bright and stimulating. And I was able to piece together some small grants and continue research on a small scale, largely with undergraduate student help during the academic year and with summer trips to beautiful Mount Desert Island and The Jackson Laboratory. Unfortunately, the research that had captured my interest required large numbers of mice and a research infrastructure that didn't exist at ACU or any other university focused primarily on undergraduate education. When Tom Roderick was offered a two-year appointment in Washington as a scientific program director for the Atomic Energy Commission, he made me an offer I couldn't refuse. He asked me to oversee his laboratory in his absence. I requested and received a leave of absence from ACU and in the summer of 1973 we moved to Bar Harbor expecting to be there for a two year stint as a Visiting Scientist. One phone call to one of her references and Raby had a teaching job. The children loved Mount Desert Island and we adapted to the winters. For the first time in my life I was fully engaged in genetics research.

Tom had great people in his laboratory and Norm Hawes and Muriel Davisson ran the cytogenetics component of the lab without missing a beat. My oversight of

allow me to have graduate students in mouse genetics and as much or as little classroom teaching as I wanted. My mission was also to help build broader programs in comparative genetics in the College, branching into genetics of domestic species and better use of clinical cases as models of human disease. I continued to do comparative mapping and although I received another March of Dimes grant and a NIH grant for mapping the Ah receptor in mice, I found funding for mouse genetics much more difficult to come by than it had been at JAX. I kept up with the comparative mapping community and particularly Steve O'Brien with whom I had frequent conversations about whether the comparison of mouse and human chromosomes were sufficient to tell the story of mammalian chromosome evolution. His background with murine retroviruses and his position at the National Cancer Institute made the domestic cat genome and it's associated leukemia virus an obvious target for comparative mapping in his laboratory. By the late 1970's he had made a hybrid somatic cell panel for the cat and had demonstrated extensive conservation of synteny with the human genome clearly showing the mouse to be the odd man out in genome evolution. It was late one evening at Steve and Diane's house in Bethesda when I bounced my idea to explore a genome from another species off Steve. My veterinary college appointment suggested the dog might be a logical choice because of the morphological and behavioral diversity between breeds and its clinical presentation of a large number of genetic diseases. Although it would be mapping another carnivore, the large number of chromosomes in the dog suggested that its genome would be very different from the cat. Steve got another beer from the refrigerator and turned up the volume on Willie and Wayland and the Boys. Another reasonable project I suggested was to map the genome of domestic cattle. Its genome would represent an unmapped mammalian order and a map of the cattle genome should be of interest to the USDA and other agricultural agencies. "Let me get this straight, Hoss", Steve said, "you're in Texas, right? Texas is cows and cowboys, not dogs." After about the sixth cycle of "Blue Eyes Crying in the Rain" I went to bed thinking about agricultural research for the first time since telling Doc Bogart I wasn't interested in cattle genetics for a dissertation project.

I grossly overestimated the interest of the USDA and other agricultural agencies in funding gene mapping. Interestingly, my inquiries to different institutes at the NIH about developing and mapping bovine genetic models of human disease prompted variants of the reply "too bad you're not working with dogs." Fortunately for me, Neville Clarke, Director of the Texas Agricultural Experiment Station, was willing to listen to my rationale for mapping cattle genes. He provided state funds to construct a hybrid cell panel and to begin genotyping with biochemical markers. With Cindy Sharp making the hybrids and Yvonne Moll doing the genotyping we began constructing the map while I beat the bushes looking for funding to expand the map into a really meaningful tool, both for comparative genome analysis and for the eventual improvement of animal health and productivity. Although they had not embraced gene mapping, the USDA and animal breeders were interested in the concept of genetic engineering in the early 1980's and became especially excited when Palmiter and Brinster produced transgenic mice expressing and responding to introduced growth hormone genes. I did everything within my power to tie gene mapping to the genetic engineering bandwagon. Dewey Kraemer led a strong embryo physiology group at A&M and he appreciated the eventual role of genome mapping in the success of genetic engineering of farm animals. My first external funding for the cattle map came from the Hillcrest Foundation of Dallas after a fascinating meeting between Dewey, myself and Mr. W. C. Caruth of the Hillcrest Board of Trustees. Mr. Caruth was intrigued by a popular article he had read predicting that genetic engineering could produce wooly pigs, transgenic pigs producing wool in addition to meat. We didn't promise Mr. Caruth a wooly pig but we promised significant advances in the genetic mapping and genome manipulation of cattle in exchange for a generous $100,000 grant from the Foundation that Dewey and I would divide between our laboratories. The cattle map was off the ground and we published our first cow map in Steve's "Genetic Maps" in 1984 and the first cow-human comparative map in the Journal of Heredity in 1986. The cattle genome proved to be more conserved with human than the mouse genome but not to the extent of the cat-human comparison.

I came to know Morris Soller of the Hebrew University through my appreciation for his far-sighted projections of "genome genetics" in livestock and the use of gene markers in selection for quantitative traits. He was clearly the theoretical leader of the genomic era of animal breeding. We teamed up to write a BARD grant in 1985, proposing to significantly expand the cattle somatic cell map using DNA markers and also to build a RFLP map for linkage analysis of quantitative traits. We got our grant although RFLPs were soon to be replaced by microsatellites as markers of choice for linkage mapping. We got another BARD grant in 1990 and I have treasured our collaboration and friendship. Several small grants from the USDA to my laboratory were forthcoming in the 1980's, mostly through Animal Health and with ties to genes for disease resistance. My first competitive grant from the USDA purely for genome mapping in cattle came in 1986. I also received some support from the Kleberg Foundation. I spent much of the 1980's in the dual role of trying to get my own mapping efforts funded and in trying to convince funding agencies of the value of genome maps for livestock species. The idea of a human genome initiative was gaining traction in the early 80's and we were beginning to imagine having access to the complete sequence of the human genome and several model organism genomes including the mouse. With the human genome sequence projected to cost $3.5 billion, it was difficult to imagine that the genomes of livestock species would ever be sequenced. My pitch was for funding of comparative gene mapping which would identify regions of conservation between cattle (or pigs or sheep) and humans and mice on a fine scale. The mouse and human sequence could then be reasonably well exploited to find important livestock genes. This idea complemented proposals by Soller, Jacques Beckmann and others to develop markers for building linkage maps and mapping quantitative trait loci (QTL). I always assumed that finding the genes underlying QTL would come from mining the sequenced human and mouse genomes at the addresses suggested by the comparative livestock maps.

The concept of an international bovine genome project was discussed in a workshop setting for the first time to my knowledge at the opening of the Internatioinal Trypanotolerance Center in the Gambia in March of 1987 followed by a meeting of scientists from the European Economic Community in Brussels in December of 1987. Mapping the bovine genome received considerable attention at the XXI Conference on Animal Blood Groups and Biochemical Polymorphisms in Torino in July of 1988. I was an invited speaker at that conference and remember it well, having popped a couple of stitches from my triple bypass surgery two months earlier. This group was to soon become the International Society for Animal Genetics (ISAG) and would play a major role in animal gene mapping over the next 20 years through its workshop structure at biennial meetings. I had the privilege of serving as President of ISAG a few years later. I also had a wonderful opportunity visit Australia as a McMaster Fellow with Jay Hetzel's CSIRO group in Rockhampton in 1990, doing a little lab work but mostly helping campaign for funding of their mapping program and I participated in the organization of the Canadian animal mapping program in a workshop in Ottawa later that year. These and many other international experiences were important to me professionally but more important to me personally for the wonderful friends Raby and I discovered among animal gene mappers.

Several events in the early 1990's were important to the US effort in bovine genomics (the term "genomics" was coined by Tom Roderick in 1986 for the title of the journal Ruddle and McKusick proposed to accompany the human genome initiative). Charlie Arntzen had taken the position of Director of the Texas Agricultural Experiment Station and the Institute of Biosciences and Technology at Texas A&M. He continued the support of my mapping laboratory that Neville had begun and also supported and helped organize a Banbury Conference on Mapping the Genomes of Agriculturally Important Animals at Cold Spring Harbor in the summer of 1990. This conference brought representatives of the human genome project (Jim Watson, Victor McKusick, David Housman, and Ray White) together with some of the animal mapping community (Michel Georges, Jay Hetzel, Brian Kirkpatrick, Harris Lewin, Joan Lunney, Larry Schook, Loren Skow, Morris Soller, Alan Teale and myself) in the presence of administrators from the USDA and several of the land grant universities. The conference was followed in the fall by the first Allerton Conference organized by Lewin, Schook and others at the University of Illinois. These two workshops led to the organization of a USDA National Animal

Genome Research Progam in 1993 for which I have served as Cattle Coordinator for the last 15 years.

Most of the ideas and essentially all of the data credited to me over the years have come from the excellent students, postdocs and technical staff who have blessed my laboratory. More than 40 graduate students have completed their degrees with me at Texas A&M and I can't really name one without naming them all (which I've done in my c.v). These students and staff developed the hybrid somatic cell panel into a valuable resource for comparative mapping, beginning with biochemical markers and evolving through Southern blotting and PCR genotyping of DNA markers, always with an eye toward the comparative maps of homologous genes in humans and mice. Dan Gallagher, an exceptional postdoctoral fellow, brought the cytogenetics expertise to the lab to assign syntenic groups to chromosomes, not an easy chore with 29 acrocentric autosomes in the bovine karyotype. Dan played a critical role in standardizing the karyotype of domestic cattle which became the prototype for chromosome identification in other artiodactyls. The construction of radiation hybrid panels allowed us to order markers in the cattle genome and define conservation and disruption of gene order within conserved syntenic groups. These panels, especially when genotyped with ESTs and BAC end sequences in our collaboration with Harris Lewin's laboratory, became valuable tools for fine scale mapping, for mining gene loci underlying traits of economic and biological interest, and ultimately for assembly of the cattle genome sequence. The citation for my election to the National Academy of Sciences USA in 1999 reads "*Womack launched the discipline of livestock genomics with his initial map of the bovine genome and the demonstration of extensive chromosomal conservation in cattle, humans, and mice. His international leadership resulted in a comprehensive bovine linkage map and the first genetic mapping of economically important traits in cattle.*" It is obvious to me, if not to everyone familiar with my work, that my students and collaborators are equally deserving of this and other recognitions I have received. Perhaps I am considered a leader in the discipline because I got an early start and couldn't work fast enough to get out of the way. A highlight of my career was in April 2007 when a group of my laboratory alumni organized a symposium in my honor entitled Advances in Applied and Comparative Genetics. The two-day scientific program was comprised entirely of presentations from my former students, postdocs and collaborators.

Early in this decade, it became obvious that the genomes of human and selected model organisms would be sequenced ahead of schedule and under budget by the National Human Genome Research Institute. A call went out for white paper proposals for sequencing of additional genomes that could aid in informing human genetics. I was able to organize a National Academy of Science workshop in early 2002 to discuss the status of domestic animal genomics and to help prioritize different species for genome sequencing. Shortly thereafter, Richard Gibbs and

George Weinstock of the Baylor College of Medicine Human Genome Center, Steve Kappes of the USDA, Larry Schook of the University of Illinois, and Loren Skow and myself of Texas A&M wrote a white paper proposal for sequencing the bovine genome. In September of that year the cattle and dog genomes were approved with high priority for sequencing. Our work had only begun, however. The cost for 7-8x coverage of a 3000 megabase genome had come down over the years to about $50 million. While appreciating the comparative information the bovine genome could provide medical genetics, the NIH was well aware that there were, or should be, significant agricultural interest in sequencing this genome. They agreed to fund half the total cost and, in all fairness, asked us to come up with the other $25 M. Despite pledges from the USDA, the dairy and beef industries, and international agricultural agencies, the funding came together very slowly. As of March, 2003 we had only $15 M committed. In fact, we were within a week of losing our high priority status when Texas Governor Rick Perry committed the final $10 M. This would have never happened without the efforts of a wonderful lady from South Texas. Mrs.Anne Armstrong, former U.S. Ambassador to Great Britain and a member of the Board of Regents at Texas A&M, learned about the bovine sequencing project. She and her husband, Tobin, owned and operated the Armstrong Ranch in Armstrong, Texas and were influential in the beef cattle industry as well as in State and National politics. She recognized the potential of genome sequencing to the improvement of beef cattle health and productivity and used her influence to persuade the Governor's office to insure its initiation. I suppose the bovine genome would have ultimately been sequenced without the help of Anne Armstrong, but it would have certainly been delayed by several years.

I am planning to reap some of the benefits of the cattle genome sequence for a few years before I retire. I have aggressively avoided administrative responsibilities at Texas A&M and have maintained a laboratory totally with external funding. Jan Elliot, Elaine Owens and Mary Jewell are still the nucleus of my research team. We continue to train graduate students and postdocs and are focusing our efforts on finding diversity in the bovine genome that translates to differential susceptibility to pathogens. Raby and I plan to retire to our little ranch in Robertson County where we currently relax (and work) on weekends and have the occasional barbecue with family, friends and guests. We enjoy our little cattle herd (with one bison) and I've (gently) broken and trained a young filly who has become more of a pasture pet than a working horse. I can imagine being very happy spending my retirement years with the animals on our ranch and I'm sure I'll have questions about the composition of their genomes, the same kind of questions that have excited me about going to work every day for the past 40 years. In the early days of human gene mapping, Frank Ruddle responded to the question of "why map genes" with the reply " because mapping genes is good for you." I kept that quote on my office wall for a number of years and can now summarize my own career with the statement "mapping genes has not only been good for me, mapping genes has been good to me."

James E. (Jim) Womack. Commencement Address. Texas A&M University, August 11, 2006

Biographical Statement for Introduction: James E. (Jim) Womack was born in Anson, TX in 1941 and attended Hawley, TX public schools, grades 1-12. He enrolled in Abilene Christian College in 1959 on a basketball scholarship and captained two NCAA Div II tournament teams before graduating in 1964. Although he received his B.S. degree in Math Education, he committed late in his degree program to advanced study in Genetics, a discipline basking in recent discoveries of DNA structure and function. He entered the graduate program in Genetics at Oregon State University in 1965 and received his PhD degree in 1968. His research in radiation mutagenesis in mice provided the foundation for a career in mammalian genetics that continues today. Returning to Abilene Christian as a member of the Biology faculty from1968-73, Womack taught Genetics and Cell Biology to undergraduates, many of whom matriculated to Texas A&M as graduate and professional students and at least eight who eventually became members of the A&M faculty. The opportunity to pursue advanced research in mammalian genetics drew him to a position of Staff Scientist at The Jackson Laboratory in Bar Harbor, Maine in 1973.

Womack was recruited by the College of Veterinary Medicine at Texas A&M in 1977 where he has advanced through the academic ranks to an endowed professorship in 1987 and to Distinguished Professor of Veterinary Pathobiology in 2001. He is a member of the *Faculty of Genetics* and holds a joint appointment in the *Department of Medical Biochemistry and Genetics.* He is currently Director of the Center for Animal Biotechnology and Genomics and teaches an undergraduate course in Mammalian Genetics. Womack shifted his research emphasis from laboratory animals to domestic animals in the early 1980's and has been an international leader in advancing genomic research in cattle and other livestock species. He has published, along with his 32 graduate students and post-docs and a large number of collaborators, more than 300 peer reviewed articles in scientific journals. He serves as Coordinator for the USDA Cattle Genome Research Program, is the current President of the *International Society for Animal Genetics*, and is past President of both the *American Genetics Association* and the *Texas Genetics Society.* Womack received the Faculty Distinguished Achievement Award in Research from the Association of Former Students in 1987 and was the recipient of the 1994 CIBA Prize for research in animal health. He was elected to the National Academy of Sciences USA in 1999, becoming the fifth member of the Texas A&M faculty to receive this honor. In 2001 Womack was recipient of the Wolf Prize, the world's most prestigious award for research in agriculture "for the use of recombinant DNA technology to revolutionize animal sciences, paving the way for applications in neighboring fields."

Professor Womack lives in College Station with his wife of 43 years, Raby (Beakley) Womack and has two children, Wendy Hill of Jemez Springs, N.M. and James M. Womack of Austin.

COMMENCEMENT ADDRESS:

President Gates, Regent Jones, Provost Prior, Dean Watson Mr. Taylor, General Van Alstyne, and others on the stage, I am deeply honored to be here among you and to address this 2006 graduating class of Fightin' Texas Aggies.

Howdy Ags! Congratulations on a job well done. And congratulations to your families who have supported you in every step of this long and difficult journey.

Graduates, I'm here today as a representative of your faculty to thank you for the privilege of working with you these last four..(or five)…(or six) years. I'm here to wish you well in the many endeavors that tomorrow holds. And I'm here to give you one last lecture before you leave. Ags, along with your family, we the faculty have invested our lives in you, we care about your future, and we love you. So, bear with me for a few minutes.

A new world awaits you outside the doors of this arena. A world full of excitement and opportunities, and challenges. A world my generation inherited, and molded and shaped for you, improving it in some areas and really messing it up in others.

I sat where you are in the early 1960s, not at this great university that adopted me 29 years ago, but at a fine little school in West Texas. I had a wonderful undergraduate experience. I went through college on a basketball scholarship (I was taller then). I played in a couple of NCAA tournaments. I found and married my life-long partner. I could solve mathematical equations with the best of them and I had read and absorbed just about everything that had been written at that time about a wonderful new molecule called DNA. But what I got that day was a diploma, my education was only beginning.

My diploma did not prepare me to deal with a war being fought half-way around the world that was taking the lives of some of my classmates and splitting our great country down the middle. We haven't improved the world much for you in 40 years, have we?

But we've done some good things, especially in science and technology. We got you into space, we connected you to the world through the internet, we gave you Dolly the sheep and we sequenced your genome. You will undoubtedly figure out what to do with these gifts, then continue to advance technology in ways I cannot even imagine. My new world of the 60's and your new world today are not that different, otherwise I wouldn't be so presumptuous as to offer you a few words of advice.

So, what can I say to help you succeed in a world that is both ever changing and never changing. I'm certainly not here to tell you how to become rich or famous, and if I were it wouldn't be from first hand experience. However, if you can look up from what you are doing 30 years from today and say "I have contributed some little something to making this world a better place, I have great friends and family who love me, and I wouldn't trade with anyone what I do for a living", you are a success.

That's my definition of success, now let me give you three keys to finding it. And let me tell you from the beginning, these aren't original ideas.

1. **Work Hard.** (Have you heard that before?) Don't allow yourself to fail at anything for lack of effort. Don't be ashamed to be an overachiever, in fact, strive for it. All of you are talented or you wouldn't be here, some of you are blessed with exceptional talents. For you gifted Aggies who managed to get your diploma despite spending more of the last 6 years in the Dixie Chicken than in Sterling Evans, I say "shame on you". But now, it's time to get with it. If you don't push your talents to their limits, then a step beyond, you will look back on your life with serious regrets. Be an overachiever in your job..and that begins with finding work that you enjoy doing. If you love your job you will go the extra mile.

My admonition to work hard extends beyond your job, however. A successful marriage or any worthwhile relationship requires hard work. And nothing requires more effort than parenthood. Hard work does not guarantee success in any endeavor, but less than your best effort is surely an invitation for failure. We would all do well to heed the words of King Solomon who wrote in the book of Ecclesiastes "Whatever you find to do, do it with all your might."

2. **Live Smart.** Don't ever, ever, ever stop learning. Learn from your mistakes, you can turn your wounds into wisdom. Choose your friends wisely. Surround yourself with the kind of people you would like to emulate. You are Aggies, you are going to be bosses…don't be afraid to hire people who are smarter than you. I've never taken a graduate student into my laboratory who wasn't smarter than me. They haven't made me rich or famous, but I wouldn't trade my job of teaching and research at this university for any in the world. (There went any hope I had for getting a raise this year.) Select your life-long mate very carefully. Marry up, as the saying goes, and I'm not talking about money. It worked for me. The person you will learn the most from the rest of your life is probably the person you wake up with every morning. And besides that, your spouse will contribute the DNA that makes your children smarter than you… and they will be…if you don't believe me, ask them when they are 16.

Elbert Hubbard, American essayist from some100 yrs ago said "The recipe for perpetual ignorance is to be satisfied with your opinion and content with your knowledge."

Stay informed, and weigh conflicting information wisely. Keep up with the technical and scientific advances of the world, whether you are trained in science or not. Your new world is a technical one and you are going to be a juror in the court of public opinion on many issues outside the discipline named in your diploma.

You will decide whether stem cell research is allowed to proceed with its life saving potential.

You will determine whether starving children in developing countries will eat food that has been genetically modified to grow in harsh environments.

You will make political decisions to stop environmental deterioration and restrain global warming.

Don't make your decisions on the basis of my obvious biases. Read and learn, and always weigh the motivation behind the source of your information.

And above all, don't let other people do your thinking for you. I encourage loyalty to your religion, to your country and to this great university. But blind loyalty is wasted loyalty. Participate in shaping the future of the institutions you hold dear, but be an informed participant.

3. Be a team player. Regardless of how hard you work, or how smart you become, your success and happiness is going to boil down to relationships. The world has become far too complex to go it alone. Globalization is a reality. You may spend your working hours alone with a computer, but on the other end of that broad band or satellite beam are real people. Whether in our professional lives or in our personal lives, our relationships with people are the keys to our success. Of course, developing business and personal relationships begins with our choices of associates. But it doesn't end there, it requires putting the success of the team ahead of personal gratification, and that's an Aggie tradition. In the end, team play translates to unselfishness. It requires taking our work seriously without taking ourselves seriously. There is no room for an overbearing ego in a team endeavor, or in any successful relationship. Team play is a big part of your heritage as Aggies, take it with you.

So, my three keys to success for you today are pretty simple, Work Hard, Work Smart, and Work Together. It's perfectly legitimate in our academic world to use the ideas of other people, so long as we give proper credit. I told you earlier I am an old jock and most of my personal philosophy has at one time or another been posted on a locker room wall. Some of you may recognize my keys to success as the motto of our outstanding basketball coach Billy Gillispie, who demands, and I emphasize demands, that his Aggies **"Play hard, Play smart, and Play together"**. I recommend Coach Gillispie's motto as a sound rule for a successful life. We've seen it work miracles here on this floor, I believe it will work for you outside the doors of this arena.

Gig em' Ags, and God Bless You.

LIST OF PUBLICATIONS

(Selected Publications from a total of 331 in Refereed Journals):

Womack, J.E., Yan, D.L.S., and Potier, M.: Gene for neuraminidase activity on mouse chromosome 17 near *H-2:* pleiotropic effects on multiple hydrolases. *Science* 212:63-65, 1981.

O'Brien, S.J., Moore, J.L., Martin, M.S., and Womack, J.E.: Evidence for the horizontal acquisition of murine AKR virogenes by recent horizontal infection of the germ line. *J. Exptl. Med.* 155:1120-1132, 1982.

Womack, J.E. and Moll, Y.D.: A gene map of the cow: Conservation of linkage with mouse and man. *J. Hered.* 77:2-7, 1986.

O'Brien, S.J., Seuanez, H.M., and Womack, J.E.: Mammalian genome organization: an evolutionary view. *Ann. Rev. Genet.* 22:223-251, 1988.

Threadgill, D.S., Krause, J.P., Krawetz, S.A., and Womack, J.E.: Evidence for the evolutionary origin of human chromosome 21 from comparative gene mapping in the cow and mouse. *Proc. Natl. Acad. Sci. USA.* 88:154-158, 1991.

Gallagher, D.S. and Womack, J.E.: Chromosome conservation in the bovidae. *J. Hered.* 83:287-298, 1992.

O'Brien, S.J., Womack, J.E., Lyons, L.A., Moore, K.J., Jenkins, N.A., and Copeland, N.G.: Anchored reference loci for comparative genome mapping in mammals. *Nature Genetics* 3, 103-112, 1993.

Georges, M., Dietz, A.B., Mishra, A., Nielsen, D., Sargeant, L., Sorensen, A., Steele, M.R., Zhao, X., Leipold, H., and Womack, J.E.: Microsatellite mapping of the gene causing weaver disease in cattle will allow the study of an associated quantitative trait locus. *Proc. Natl. Acad. Sci. USA* 90:1058-1062, 1993.

Georges, M., Drinkwater, R., King, T., Mishra, A., Moore, S.S., Nielsen, D., Sargeant, L.S., Sorensen, A., Steele, M.R., Zhao, X., Womack, J.E. and Hetzel, J: Microsatellite mapping of a gene affecting horn development in *Bos taurus. Nature Genetics* 4:206-210, 1993.

Barendse, W., Armitage, S.M., Kossarek, L.M., Shalom, A., Kirkpatrick, B.W., Ryan, A.M., Clayton, D., Li, L., Neibergs, H.L., Zhang, N., Grosse, W.M., Weiss, J., Creighton, P., McCarthy, F., Ron, M., Teale, A.J., Fries, R., McGraw, R.A., Moore, S.S., Georges, M., Soller, M., Womack, J.E., and Hetzel, D.J.S.: A genetic linkage map of the bovine genome. *Nature Genetics* 6:227-235, 1994.

Womack, J.E. and Kata, Srinivas R.: Bovine genome mapping: evolutionary inference and the power of comparative genomics. *Current Opinion in Genetics & Development* 5:725-733, 1995.

Lyons, L.A., Laughlin, T.F., Copeland, N.G., Jenkins, N.A., Womack, J.E. and O'Brien, J.O.: Comparative anchor tagged sequences (CATS) for integrative mapping of mammalian genomes. *Nature Genetics* 15:47-56, 1997.

Womack, J.E., Johnson, J.S., Owens, E.K., Rexroad, C.E. III, Schläpfer, J., and Yang, Y.-P.: A whole-genome radiation hybrid panel for bovine gene mapping. *Mammalian Genome* 8:854-856, 1997.

Yang, Y.-P. and Womack, J.E.: Parallel radiation hybrid mapping: A powerful tool for high-resolution genomic comparison. *Genome Research* 8:731-736, 1998.

O'Brien, S.J., Menotti-Raymond, M., Murphy, W.J., Nash, W.G., Wienberg, J., Stanyon, R., Copeland, Neal G., Jenkins, N.A., Womack, J.E., and Graves, J.A.M.: The promise of comparative genomics in mammals. *Science* 286:458-481, 1999.

Winter, A., Kramer, W., Werner, F.A., Kollers, S., Kata, S., Durstewitz, G., Buitkamp, J., Womack, J.E., Thaller, G., and Fries, R.: Association of a lysine-232/alanine polymorphism in a bovine gene encoding acyl-CoA:diacylglycerol acyltransferase (DGAT1) with variation at a quantitative trait locus for milk fat content. *Proc. Natl. Acad. Sci. USA* 99-9300-9305, 2000.

Ozawa, A., Band, M.R., Larson, J.H., Donovan, J., Green, C.A., Womack, J.E., and Lewin, H.A.: Comparative organization of cattle chromosome 5 revealed by comparative mapping by annotation and sequence similarity and radiation hybrid mapping. *Proc. Natl. Acad. Sci. USA* 97:4150-4155, 2000.

Goldammer, T., Kata, S.R., Brunner, R.M., Dorroch, U., Sanftleben, H., Schwerin, M., and Womack, J.E.: A comparative radiation hybrid map of bovine chromosome 18 and homologous chromosomes in human and mice. *Proc. Natl. Acad. Sci. USA* 99:2106-2111, 2002.

Takeda, H., Takami, M., Oguni, T., Tsuji, T., Yoneda, K., Sato, H., Ihara, N., Itoh, T., Kata, S.R., Mishina, Y., Womack, J.E., Moritomo, Y., Sugimoto, Y. and Kunieda, T.: Positional cloning of the gene LIMBIN responsible for bovine chondrodysplastic dwarfism. *Proc. Natl. Acad. Sci. USA* 99:10549-10554, 2002.

Winter, A., Krämer, W., Werner, F.A., Kollers, S., Kata, S., Durstewitz, G., Buitkamp, J., Womack, J.E., Thaller, G., and Fries, R.: Association of a lysine-232/alanine polymorphism in a bovine gene encoding acyl-CoA:diacylglycerol acyltransferase (*DGAT1*) with variation at a quantitative trait locus for milk fat content. *Proc. Natl. Acad. Sci. USA* 99:9300-9305, 2002.

White, S.N., Taylor, K.H., Abbey, C.A., and Gill, C.A., and Womack, J.E.: Haploytype variation in bovine Toll-like receptor 4 and computational predication of a positively selected ligand-binding domain. *Proc. Natl. Acad. Sci. USA* 100:10364-10369, 2003.

Larkin, D.M., der Wind, A.E., Rebeiz, M., Schweitzer, P.A., Bachman, S., Green, C., Wright, C.L., Campos, E.J., Benson, L.D., Edwards, J., Liu, L., Osoegawa, K., Womack, J.E., de Jong, P., and Harris, L.A.: A cattle-human comparative map built with cattle BAC-ends and human genome sequence. *Genome Research* 13:1966-1973, 2003.

Murphy, W.J,. Larkin D.M., Everts-van der Wind A., Bourque G., Tesler G., Auvil L., Beever J.E., Chowdhary B.P., Galibert F., Gatzke L., Hitte C., Meyers S.N., Milan D., Ostrander E.A., Pape G., Parker G.H., Raudseep T., Rogatcheva M.B., Schook

L.B., Skow L.C., Welge M., Womack J.E., O'Brien S.J., Pevzner P.A., and Lewin H.A.: Dynamics of mammalian chromosome evolution inferred from multispecies comparative maps. *Science* 309, 613-617, 2005.

Everts-van der Wind , A., Larkin, D.M., Green, C.A., Elliott, J.S., Olmstead, C., Chiu, R., Schein, J.E., Marra, MA, Womack, J.E., Lewin, H.A.: A high-resolution whole-genome cattle-human comparative map reveals details of mammalian chromosome evolution. *Proc Natl Acad Sci USA*; 102:18526-18531, 2006.

Sugimoto, M., Fujikawa A., Womack, J.E., Sugimoto Y.: Evidence that bovine forebrain embryonic zinc finger-like gene influences immune response associated with mastitis resistance. *Proc Natl Acad Sci USA* 103:6454-6459, 2006.

33 Book Chapters Including the Following:

Womack, J.E.: Biochemical loci: mouse. In: P.L. Altman and D.D. Katz, eds., *Inbred and Genetically Defined Strains of Laboratory Animals, Part I.* Federation of American Societies for Experimental Biology, Bethesda, pp. 96-100, 1979.

Womack, J.E.: Genetically controlled biochemical variants: rat. In P.L. Altman and D.D. Katz, eds., *Inbred and Genetically Defined Strains of Laboratory Animals, Part I.* Federation of American Societies for Experimental Biology, Bethesda. p. 305, 1979.

Womack, J.E.: Linkage of mammalian isozyme loci: a comparative approach. In: M.C. Rattazzi, J.G. Scandalios, and G.S. Whitt, eds., *Isozymes: Current Topics in Biological and Medical Research*, Vol. 6., Alan R. Liss, Inc., New York, pp. 207-246, 1982.

Womack, J.E.: Post-translational modification of enzymes: processing genes. In: M.C. Rattazzi, J.G. Scandalios, and G.S. Whitt, eds., *Isozymes: Current Topics in Biological and Medical Research*, Vol. 7., Alan R. Liss, Inc., New York, pp. 175-186, 1983.

O'Brien, S.J., Seuanez, H.N. and Womack, J.E.: On the Evolution of Genome Organization in Mammals. In: *Molecular Evolutionary Genetics*, Chapter 8, Plenum Publishing Corp., New York, pp. 519-589, 1985.

Womack, J.E.: Gene mapping in the cow. In: R.A. McFeely, ed., Domestic Animal Cytogenetics, Advances. *Veterinary Science and Comparative Medicine*, Vol. 34. Academic Press, Orlando. pp. 251-271, 1990.

Womack, J.E.: Strategies and Technologies for Comparative Gene Mapping. In L.B. Schook, H.A. Lewin, and D.G. McLaren, eds., *Gene Mapping: Strategies, Techniques and Applications*, Marcel Dekker, Inc., New York, pp. 3-19, 1991.

Womack, J.E.: Genome Analysis in Farm Animals. In: J.P. Gustafson and R.B. Flavell, ed., *Genomes of Plants and Animals*: 21st Stadler Genetics Symposium. Plenum Press, New York, pp. 127-137, 1996.

Ryan, A.M. and Womack, J.E.: A molecular genetic approach to improved animal health. *The effect of interferon genotype on the severity of experimental bovine herpesvirus-1 infection.* Veterinary Clinics of North America: Food Animal Practice. Vol. 13 pp. 401-409, 1997.

Womack, J.E.: Mapping Animal Genomes. In: *Advances in Veterinary Medicine, Academic Press,* Vol. 40, pp. 157-189, 1997.

Serov, O.L., Chowdhary, B., Womack, J.E, and Marshall Graves, J.A: Comparative gene mapping, chromosome painting and the reconstruction of the ancestral mammalian karyotype. In: A. Ruvinsky and J. Marshall Graves, eds., *Mammalian Genomics,* pp. 349-392, 2005.

174 Invited Presentations, Symposia, Colloquia and Named Lectures Including the Following:

Keynote Address: Abilene Christian University Science Symposium - The new genetics: practical and ethical considerations, Abilene, TX, 1983.

Invited Seminar: "Somatic Cell and Molecular Approaches to Gene Mapping in Cattle." International Trypanotolerance Centra, Banjul, The Gambia, March, 1987.

Invited Symposium Presentation: "Organization of Animal Genomes: The Merger of Cytogenetics and Molecular Biology." American Dairy Science Association and 5th North American Symposium on Cytogenetics of Cell Biology of Domestic Animals. Columbia, MO, June, 1987.

Invited Speaker: "Strategies for the Development of a Bovine Gene Map." XXI International Conference on Animal Blood Groups and Biochemical Polymorphisms. Torino, Italy, July 1988.

Invited Presentation: "Current Status of the Bovine Gene Map." Banbury Conference on Mapping the Genomes of Agriculturally Important Animals, Cold Spring Harbor Laboratory, Cold Spring Harbor, NY, 1990.

Invited Lecture: "Goals for Mapping Domestic Animal Genomes." Allerton Conference on Mapping Domestic Animal Genomes: Needs and Opportunities, University of Illinois, Champaign-Urbana, IL, 1990.

Invited Speaker: "Gene Mapping in Livestock." Agriculture Canada Workshop on Animal Gene Resources, Ottawa, Canada, November, 1990.

Invited Presentation: Status of the Bovine Gene Map Workshop on Bovine Genome: Mapping and Trypanotolerance. Nairobi, Africa, April 1992.

Invited Speaker: "Genomic Research to Advance Animal Production." AAAS Annual Meeting. Chicago, IL, February, 1992.

Invited Lecture: A. L. Bortree Lecture, sponsored by the Department of Veterinary Science, College of Agriculture, Penn State University, State College, PA, February 1993.

Invited Lecture: Gene Mapping Programs and their Application to Improved Animal Production. VII World Conference on Animal Production in Edmonton, Alberta, Canada, June 1993.

Invited Speaker: "Comparative mapping in Eutherian Mammals." Symposium on Genome Organization and Evolution in Higher Mammals. XVII International Congress of Genetics, Birmingham, England, August 1993.

Invited Lecture: "Application of the Bovine Gene Map to Identifying Economic Trait Loci." Miles, Inc. Distinguished Lecture Series, Washington State University, Pullman, WA, March/April, 1994.

Symposium Speaker: XXIV Meeting of International Society for Animal Genetics "New Developments in Genome Mapping." Prague, Czech Republic, July/August 1994.

Keynote Speaker: "Comparative Genomics and Animal Improvement." Colloquium on the Future of Animal Genetics, U.C. Davis, March, 1995.

Invited Speaker: "Integrating the Syntenic, Physical and Linkage Maps." USDA Beltsville Symposium XX, Biotechnology's Role in the Genetic Improvement of Farm Animals, Beltsville, MD, May, 1995.

Invited Speaker: "Genome Analysis in Farm Animals." 21st Stadler Genetics Symposium. Columbia, MO, May, 1995.

Invited Speaker: "Current Status of Comparative Maps — Livestock." HUGO Comparative Genome Organization Workshop. Fraser Island, Qld. Australia, December, 1995.

Invited Speaker: "From the Human Genome to Animal Physiology." Banbury Conference, Genomics to Physiology and Beyond: How Do We Get There? Cold Spring Harbor, NY, February, 1997.

Invited Symposium Speaker: "Evolutionary Implications of Genome Mapping in Farm Animals." 9th International Congress on Genes, Gene Families and Isozymes, San Antonio, April, 1997.

Invited Plenary Lecture: "High-Resolution Comparative Mapping of the Bovine Genome." ISAG 2000, Minneapolis, MN, July, 2000.

Invited Lecture: "Animal Genomes and Animal Genomics: What Do We Know and Where Do We Go?" Animal Genomics 2000 Sympoisum. North Carolina State University, Raleigh, NC, August 2000.

College of Medicine Convocation Address, Texas A&M University, College Station, TX, October 2001.

Invited Lecture: "The Genomics Revolution: Marching into the Millennium with the Secrets of Life." University Distinguished Lectures Series, College Station, TX, December 2001.

Invited Speaker: "Genomes in Animal Agriculture." Annual Meeting of the AAAS, Boston, MA, February, 2002.

Invited Presentation: "Comparative Genomics." Horizons in Livestock Sciences Conference, Queenland, Australia, May, 2003.

Invited Speaker: "Application of Comparative Gene Maps to Finding Important Genes in Livestock." Korean Conference on Innovation Science and Technology. Jeju, Korea, November 2003.

Verne Chapman Memorial: "Barnyard Genomics: Cattle and Other Domestic Animals." 20th International Mammalian Genome Conference, Charleston, S.C., Nov. 2006.

Invited Presentation: The Bovine Genome Project." US-Australia National Academies Workshop on Animal Genomics. Irvine, CA, July 2007.

Dr. James Womack named ACU's Outstanding Alumnus of the Year

By Tamara Thompson

On a shelf in Dr. James Womack's modest office at Texas A&M University is a gray book, published in 1962. Its edges are worn and its 543 pages are yellowed now, but if not for that book and a skeptical Abilene Christian University biology professor, the world of genetics might be very different today.

Womack ('63), a former Hawley (Texas) High School basketball star who at 5'10" made the all-state team twice and averaged 30 points a game, now chuckles as he recalls how his career path took a sharp turn.

Planning to teach and coach, Womack entered ACU in 1959 on a basketball scholarship, lettered four years, and captained the 1962-63 Wildcat team. By his junior year, he decided he wanted to be a dentist instead, but lacked the biology prerequisites. He visited Dr. James Throneberry, who taught Genetics, the only course that didn't interfere with basketball workouts.

"He gave me a challenge that I took very personally," Womack says. "He said in the first place, normally athletes didn't pass his class, and second, that without the prerequisites, I didn't have a chance. But if I wanted to fail his course, I was welcome to sign up. With a challenge like that, I went to the bookstore and bought the book, and to make a long story short, I just read the textbook before the first class."

The textbook – which he has kept – fascinated Womack, especially a supplement containing Nobel Prize-winning papers on the structure of DNA. "This was the early 1960s and everything was just breaking in genetics. By the time the course was over, I had decided I didn't want to go to dental school," Womack says.

Even so, he was accepted into Baylor College of Dentistry, and the summer after his graduation, he and his new wife, Raby (Beakley '62), prepared to go. In mid-August, an Abilene dentist invited Womack to come help in his office. "About the third or fourth day of actually working with him, I came home and told Raby, 'I can't do this the rest of my life,' " Womack says, laughing. "I've got to pursue the genetic stuff."

So he stayed at ACU, teaching labs. "Throneberry became my big advocate and helped me get in grad school and kind of pushed me on my way to becoming a geneticist," says Womack.

But not just any geneticist.

Now distinguished professor and director of the Center for Animal Biotechnology and Genomics at Texas A&M, Womack has been recognized nationally and internationally for his pioneering work in the cattle genome. Most notably, he received the 2001 prestigious Wolf Prize in agriculture, which many deem equivalent to the Nobel Prize since it does not honor agriculturists.

For the past 20 years, Womack, who received his Ph.D. at Oregon State University, has focused his research interests on mapping the bovine genome, with an emphasis on how the genome relates to that of other mammals, especially humans and mice.

Womack has led the research in sequencing the bovine genome, on target to be complete in early 2007. Although the actual sequencing is being done at Baylor College of Medicine's Human Genome Center, it is the culmination of his work with the bovine genome and believed to be his finest accomplishment.

"In the last year, seeing the sequence published has been very rewarding to me," Womack says. "I had no idea when we started building maps of the cattle genome that we would ever, in my lifetime, see the complete sequence." He credits advances in technology for making the sequencing happen faster than anticipated.

"The practical applications of my research are mostly in disease resistance. I'm interested in why individuals, whether they be cattle or humans, ... respond differently to pathogens, whether they be bacterial, viral, parasites, whatever," Womack says, explaining that when the genetic causes for these differences are determined, proper treatment will follow.

"The practical goal of my research is to be able to understand these individual differences, so that whether it be veterinary medicine or human medicine, we can use genomic information to better treat infectious diseases," Womack adds.

Citing Womack's work, the Wolf Prize selection panel noted, "Dr. Womack has provided the foundation for the development of marker-assisted selections in cattle, comparing the bovine and human genomes within the bovine genetic map ... A whole new generation of scientists will use techniques provided by Womack to clone the genes affecting economically important traits in mammals."

As a result of his work, Womack believes in a decade or so, "we're going to have genetic markers in cattle, for example, so that we can identify with a little DNA test which cattle will be susceptible to a particular disease, a particular parasite, which one is more likely to be infected with mastitis ... That's probably the biggest goal of my research, to be able to offer some avenues to genetically select animals that don't require the antibiotics, animals that don't require the pesticides in use today."

In addition to the Wolf Prize, Womack received the 1996 Outstanding Texas Geneticist Award from the Texas Genetic Society, the CIBA Prize for Research in Animal Health (awarded from Switzerland), the Carrington Award for Research in Cell Biology, the Beecham Award for Research Excellence, and two Distinguished ACU Alumni Citations. In 1999, he was named to the prestigious National Academy of Sciences. He has lectured in 16 countries, published more than 300 peer-reviewed articles in scientific journals, and served as president of the International Society for Animal Genetics, the American Genetics Association, and the Texas Genetics Society.

"He is known as the bovine genome leader in the whole world. Everyone looks to him ... and respects him for that," says Dr. Joe Templeton ('64), professor of genetics and pathobiology at Texas A&M and a former ACU and Oregon State classmate. "Scientifically his credentials are sparkling, but he's still a Christian, no matter what award he's received or what he's recognized for."

As a scientist and a Christian, Womack finds discussions on creation inevitably arise. He likes to make a distinction between how and why.

"I believe that our presence here as humans is by God's design, and I believe we're different from other animals," he says. "If that's what intelligent design means, I'm 100 percent in that camp. But what I hear are people saying that intelligent design is a science – in our schools we should be teaching biology as intelligent design – and I don't go that far. I think biology has to be taught as biology, and math as math, and physics as physics ... So I hold intelligent design as a theology very strongly, but I don't believe it's a substitute for hard science ... I don't think any of our sciences tell us why we're here – but I don't think we can learn anything about how we're here by either a creationist or an intelligent design approach to biology."

"Because we had the same undergraduate institution and were both at A&M, we had a lot of discussions about where science and religion fit in," says Caird (Eugene '93) Rexroad, one of two ACU graduates to do doctoral studies under Womack at Texas A&M, and now a molecular biologist with the USDA in Virginia. "One thing about Jim is he never wanted to and doesn't like to push his philosophies on people. He has a way of asking the right questions to get somebody to think and consider all options."

Despite Womack's success, "he's never forgotten where he came from, and things that were important to him before he was successful still are," says Templeton.

Church and family are two of those things. Womack and his wife have been members of the A&M Church of Christ since moving to College Station in 1977, and he has served as a deacon for 20 years. The couple lives in College Station and maintains a 100-acre farm near Wheelock with 30 cows, one bison and two horses. He and Raby, a retired second-grade teacher, built a log cabin there they plan to expand and eventually move into when he retires. It will be close to their daughter, Wendy (Womack '90) Faltys, a nurse in Franklin, and their son, James Michael Womack ('90), a business owner in Austin.

Throughout his years at A&M, Womack has not only influenced the future of genetics, but also the lives of future geneticists. More than 30 graduate students such as Rexroad did their doctoral studies under Womack's tutelage and have gone on to successful careers in genetics.

In late April, his former graduate students will hold a symposium at A&M sharing how Womack's influence, in and out of the lab, benefited their careers. It will be a time to celebrate, once again, the spark ignited in a young man by a book and a challenge.

Reprinted by permission from Abilene Christian University's ACU Today Magazine, Winter 2007 issue — photo by Bill Albrecht/Abilene Christian University.

Fuller W. Bazer

Texas Agricultural Experiment Station
Texas A&M University, College Station, Texas, USA

2002/3 — for discoveries of Interferon-t and other pregnancy-associated proteins, which clarified the biological mystery of signaling between embryo and mother to maintain pregnancy, with profound effects on the efficiency of animal production systems, as well as human health and well-being.

Among his key discoveries, Bazer isolated a uterine protein called uteroferrin and identified that it is a hematopoietic growth factor that influences the survival of the neonate and may be useful in treating diseases, such as leukemia and osteoporosis.

Bazer determined that estrogen in pigs and interferon-t in ruminant species, are the signals for pregnancy maintenance. The ability of interferon-t to suppress transcription of the estrogen receptor gene, provides a model for potential treatment of estrogen-dependent tumors.

F.W. Bazer exemplifies how devotion to basic research in agriculture can lead to practical outcomes that impact both animal production, as well as human health and well-being.

KEY SCIENTIFIC PAPERS

1975 Bazer FW. Uterine protein secretions. J Anim Sci. 41:1376-1382.

1977 Bazer FW, Thatcher WW. Theory of maternal recognition of pregnancy in swine based on estrogen controlled endocrine versus exocrine secretion of prostaglandin F_2 by the uterine endometrium. Theriogenology 14:397-401.

1979 Bazer FW, Roberts RM, Basha SMM, Zavy MT, Caton D, Barron DH. Study of ovine uterine secretions obtained from unilaterally pregnant ewes. J Anim Sci 49:1522-1527.

1986 Bazer FW, Vallet JL, Roberts RM, Sharp DC, Thatcher WW. Role of conceptus secretory products in the establishment of pregnancy. J Reprod Fertil 76:841-850.

1989 Vallet JL, Bazer FW. The effect of ovine trophoblast protein-1, oestrogen and progesterone on oxytocin-induced phosphatidylinositol turnover in endometrium of sheep. J Reprod Fertil 87:755-761.

1990 Mirando MA, Henry JP, Beers S, Pontzer CH, Torres BA, Johnson HM, Bazer FW. Onset of secretion of protein with antiviral activity in pig conceptuses. J Reprod Fertil 88:197-203.

1991 Bazer FW, Worthington-White D, Fliss MFV, Gross S. Uteroferrin: A progesterone-induced hematopoietic growth factor of uterine origin. J Exp Hematol 1991; 19:910-915.

1994 Harney JP, Smith LC, Simmen RCM, Fliss AE, Bazer FW. Retinol-binding protein immunolocalization of protein and abundance of messenger ribonucleic acid in conceptus and maternal tissues during pregnancy in pigs. Biol Reprod 50:1126-1135.

1996 Spencer TE, Bazer FW. Ovine interferon tau suppresses transcription of the estrogen receptor and oxytocin receptor genes in the ovine endometrium. Endocrinology 137:1144-1147.

1996 VanHeeke G, Ott TL, Strauss A, Ammaturo D, Bazer FW. High yield expression and secretion of the pregnancy recognition hormone ovine-τ interferon by *Pichia pastoris*. J Interferon Res 16:119-126.

1997 Laurenz JC, Hadjisavas M, Chovanic GW, Bazer FW. Myelosuppression in the pig (*Sus scrofa*): Uteroferrin reduces the myelosuppressive effects of 5-flourouracil in young pigs. Comp Biochem Physiol 116A:369-377.

1998 Spencer TE, Ott TL, Bazer FW. Expression of interferon regulatory factors one and two in the ovine endometrium: Effects of pregnancy and ovine interferon tau. Biol Reprod 58:1154-1162.

1998 Tuo W, Brown WC, Rogers E, Zhu D, Lin G, Smith R, Bazer FW. Trophoblast IFN-τ differentially induces lymphopenia and neutropenia in lambs. J Interferon Cytokine Res 18:731-737.

1999 Johnson GA, Spencer TE, Burghardt RC, Bazer FW. 1999. Ovine osteopontin: I. Cloning and expression of mRNA in the uterus during the peri-implantation period. Biol Reprod 61: 884-891.

1999 Johnson GA, Burghardt RC, Spencer TE, Newton GR, Ott TL, Bazer FW. 1999. Ovine osteopontin: II. Osteopontin and $\alpha v\beta 3$ integrin expression in the uterus and conceptus during the peri-implantation period. Biol Reprod 61:892-899.

2000 Chen C, Spencer TE, Bazer FW. 2000. Expression of hepatocyte growth factor and its receptor *c-met* in the ovine uterus. Biol Reprod. 62:1844-1850.

2000 Chen C, Spencer TE, Bazer FW. 2000. Fibroblast growth factor-10: A stromal mediator of epithelial function in the uterus and conceptus. Biol Reprod. 63:959-966.

2000 Ka H, Spencer TE, Johnson GA, Bazer FW. 2000. Keratinocyte growth factor: Expression by endometrial epithelia of the porcine uterus. Biol Reprod. 62:1772-1778.

2001 Ka H, Jaeger LA, Johnson GA, Spencer TE, Bazer FW. Keratinocyte growth factor in up-regulated in the porcine uterine endometrium and functions in trophectoderm cell proliferation and differentiation. *Endocrinology* 2001; 142:2303-2310.

2001 Choi Y, Johnson GA, Burghardt RC, Berghman LR, Joyce MM, Taylor KM, Stewart MD, Bazer FW, Spencer TE.Interferon regulatory factor-two restricts expression of interferon-stimulated genes to the endometrial stroma and glandular epithelium of the ovine uterus. Biol Reprod. 2001; 65:1038-1049.

2002 Gray CA, Burghardt RC, Johnson GA, Bazer FW, Spencer TE. Evidence that an absence of endometrial gland secretions in uterine gland knockout (UGKO) ewes compromises conceptus survival and elongation. Reproduction 2002; 124:289-300.

2005 Fleming JGW, Spencer TE, Safe SH, Bazer FW. The ovine uterine oxytocin receptor gene: Regulation of expression by estradiol and role of interferon tau signaling for establishment of pregnancy in ruminants. Endocrinology 2005; 147(2):899-911.

2005 Song G, Spencer TE, Bazer FW. Cathepsins in the ovine uterus: regulation by pregnancy, progesterone and interferon tau. Endocrinology 2005; 146:4825-4833.

2007 Ka H, Al-Ramadan S, Johnson GA, Jaeger LA, Burghardt RC, Spencer TE, Bazer FW. Regulation of Fibroblast Growth Factor-7 Expression by Progesterone and Estradiol-17β and Their Receptors in the Porcine Endometrium. Biol Reprod 2007: 172-180.

the dosage of estradiol described for the nonpregnant gilts. The gilts were hysterectomized on day 40 of gestation. Treatment did not significantly affect conception rate, embryonic survival, litter size or dry weight of the embryos. However, gilts receiving the higher levels of progesterone had heavier empty uterine weights ($P<.01$), greater allantoic fluid volume ($P<.01$) and longer placentae ($P<.05$).

In general, neither the passive immunization experiments nor the progesterone therapy experiment suggested a primary effect of treatment on the fetus. Rather, the effects observed were on placental development. Therefore, the primary role of porcine uterine protein secretions may be related to placental development.

Steroid Binding Proteins. Blastokinin has been shown to bind progesterone (Urzua *et al.*, 1970; Arthur *et al.*, 1972) and estradiol (Arthur *et al.*, 1972). The blastocoelic fluid of rabbit blastocysts is known to contain blastokinin at least from day 4 to 6.75 of pregnancy (Kirchner, 1969; Hamana and Hafez, 1970; Petzholdt, 1974). It is also known that progesterone is present in blastocoelic fluid of unattached blastocysts (days 5 and 6 p.c., $\bar{x}$ = 21.2 ng/ml) and attached blastocysts (days 7, 8 and 9 p.c., $\bar{x}$ = 5.9 ng/ml) and it has been suggested that uterine secretions may convey the progesterone to the blastocoelic fluid (Seamark and Lutwak-Mann, 1972). Furthermore, Kirchner (1972) has demonstrated the presence of channels in the zona pellucia through which blastokinin diffused into the blastocyst. Taken collectively, these data suggest that the principal role of blastokinin may be as a progesterone-binding protein for delivery of that steroid to the blastocyst where the progesterone may: (1) serve to regulate metabolic activity of the blastocyst directly or (2) be converted to another steroid, e.g., estradiol which may affect metabolic activity of the blastocyst, uterine blood flow, water movement to the implantation site and/or other aspects of implantation and blastocyst development which are not readily apparent.

Support for the first possibility is based on data which indicate that day 5 rabbit blastocysts grew best and took up significantly greater amounts of uridine and amino acids in medium containing progesterone and uterine protein secretions than they did in medium containing uterine protein secretions alone or progesterone alone (El Banna and Daniel, 1972).

The second possibility is based partly on recent evidence which suggests that rat (Dickmann and Dey, 1974) and mouse (Dey and Dickmann, 1974) embryos have weak Δ^5-3 β-hydroxysteroid dehydrogenase (3 β-HSD) activity as early as the 8-cell stage and strong 3 β-HSD activity by the blastocyst stage. These authors suggest that the presence of this enzyme is indicative of steroidogenic activity. More convincing data have been presented by Perry *et al.* (1973) who found that pig blastocysts could convert tritium labeled androstenedione, dehydroepiandrosterone and progesterone to estradiol and estrone.

In the pig, there is evidence that progesterone is taken up by the pregnant uterus, i.e., there is a positive arterial-venous difference while estrogen, primarily estrone, is produced by the placenta of the pregnant uterus (Knight *et al.*, 1974c). Furthermore, ovariectomized gilts treated with 3.3 mg progesterone/kg/day had significantly ($P<.01$) greater allantoic fluid volume than those receiving only 1.1 mg progesterone/kg/day (94.4 *vs* 177.4 ml) according to Knight *et al.* (1974a). The movement of water into the pregnant uterus (allantoic fluid) is an estrogen related phenomenon (Alexander and Williams, 1966).

Eiler and Nalbandov (1973) and F. W. Bazer and W. W. Thatcher (*unpublished results*) have found progesterone in uterine washings from pigs. Total uterine protein recovered from gilts on day 15 of the estrous cycle has been incubated with tritium-labeled progesterone, estrone, estradiol and prostaglandin, (F. W. Bazer and W. W. Thatcher, *unpublished data*). Progesterone binding, after addition of dextran coated charcoal to each fraction to remove any free steroid, was associated with Sephadex G-200 gel filtration uterine protein Fractions I and V. There was no evidence of estradiol, estrone or prostaglandin binding to the uterine protein secretions. Progesterone binding to allantoic fluid protein of the cow, mare and sow have also been found (W. W. Thatcher, D. C. Sharp and F. W. Bazer, *unpublished data*). The available data support the hypothesis that blastokinin and certain proteins in the genital tract secretions of the gilt, mare and cow may serve to deliver progesterone or other steroids to the blastocyst where they may be sequestered, as such, or converted to a different steroid(s) e.g., estrogens, which may act to enhance the process of implantation and continued blastocyst development.

Leucine Aminopeptidase. Shomberg (1967, 1969) reported that porcine granulosa cells, cultured as monolayers *in vitro*, were destroyed by a macromolecule in pig uterine flushings obtained between days 13 and 18 of the estrous cycle. This cytolytic factor is believed to be leucine aminopeptidase (D. W. Shomberg, *personal communication*). F. W. Bazer, R. M. Roberts and N. J. Baldwin (*unpublished data*) have found leucine aminopeptidase (LAP) activity in pig uterine secretions to be associated with progesterone and to increase in gilts ovariectomized on day 4 and treated with progesterone from days 4 through 60 after onset of estrus. This protein is also present in allantoic fluid collected between days 35 and 100 of gestation. As with the purple protein, LAP was first detectable in allantoic fluid after placental areolae formation. LAP has also been demonstrated in rabbit uterine flushings and its presence appears to be controlled by progesterone (Beier, 1974). The specific role of this protein in the reproductive tract is not known; however, its generally accepted function is to hydrolyze L-peptides by splitting off N-terminal residues with a free α-amino group (Mahler and Cordes, 1966).

Cathepsins and Lysozyme. Cathepsins and lysozyme are intracellular proteolytic enzymes. Cathepsins A, B_1, C and D and lysozyme have been detected in uterine flushings of ovariectomized progesterone-treated gilts and allantoic fluid from pregnant gilts (W. W. Pollard, N. J. Baldwin, R. M. Roberts and F. W. Bazer, *unpublished data*). As with the other enzymes, their role in the female reproductive tract is unknown.

CONCLUSION

It is known that proteins, in general, may serve as enzymes, carrier molecules for steroids, vitamins, minerals and etc., and possible regulators of genetic activity. The data presented in this paper indicate that proteins present in the uterine lumen play similar roles. We are now challenged with the task of determining the biological significance of these proteins in the reproductive process.

LITERATURE CITED

Adams, C. E. 1958. Egg development in the rabbit: the influence of post-coital ligation of the uterine tube and of ovariectomy. J. Endocrinol. 16:283.

Adams, C. E. 1973. Asynchronous egg transfer in the rabbit. J. Reprod. Fertil. 35:613.

Alberts, H. J. and M. Neves e Castro. 1961. The protein components of rat uterine fluid. An analysis of its antigens by immunoelectrophoresis and Ouchterlony gel diffusion technic. Fertil. and Steril. 12:142.

Alden, R. H. 1942. Aspects of the egg-ovary-oviduct relationship in the albino rat. II. Egg development within the oviduct. J. Exp. Zool. 90:171.

Alexander, G. and D. Williams. 1966. Progesterone and placental development in sheep. J. Endocrinol. 34:241.

Arthur, A. T. and J. C. Daniel, Jr. 1972. Progesterone regulation of blastokinin production and maintenance of rabbit blastocysts transferred into uteri of castrate recipients. Fertil. and Steril. 23:115.

Arthur, A. T., B. D. Cowan and J. C. Daniel, Jr. 1972. Steroid binding to blastokinin. Fertil. and Steril. 23:85.

Bazer, Fuller W., T. T. Chen, J. W. Knight, D. Schlosnagle, Norma J. Baldwin and R. M. Roberts. 1975. Presence of a progesterone-induced, uterine specific, acid phosphatase in allantoic fluid of gilts. J. Anim. Sci. 41:1112.

Beier, H. M. 1968. Uteroglobin: a hormone-sensitive endometrial protein involed in blastocyst development. Biochim. Biophs. Acta 160:289.

Beier, H. M., W. Kühnel and G. Petry. 1971. Uterine secretion proteins as extrinsic factors in preimplantation development. Adv. Biosci. 6:165.

Beier, H. M., U. Mootz and W. Kühnel. 1972. Asynchronous egg transfer during delayed secretion in the rabbit. VII Int. Congress of Animal Reproduction, Munich. 3:1891.

Beier, H. M. 1974. Oviducal and uterine fluids. J. Reprod. Fertil. 37:221.

Brambell, C. E. 1933. Allantochorionic differentiations of the pig studied morphologically and histochemically. Amer. J. Anat. 52:397.

Chen, T. T. and Fuller W. Bazer. 1973. Effect of antiserum to porcine fraction IV uterine protein on the conceptus. J. Anim. Sci. 37:304. (Abstr.)

Chen, T. T., Fuller W. Bazer, J. J. Cetorelli, William Pollard and R. Michael Roberts. 1973. Purification and properties of a progesterone-induced basic glycoprotein from the uterine fluids of pigs. J. Biol. Chem. 248:8560.

Chen, T. T., Fuller W. Bazer, B. M. Gebhardt and R. M. Roberts. 1975. Synthesis and movement of porcine purple uterine protein. J. Anim. Sci. 40:170. (Abstr.).

Daniel, J. C. 1972. Preliminary attempts to terminate pregnancy by immunological attach on uterine protein. Experientia. 28:700.

Dey, S. K. and Z. Dickman. 1974. Δ^5-3β-hydroxysteroid dehydrogenase activity in mouse morula and blastocysts. Proc. 7th Ann. Meeting Soc. Study Reprod., Ottawa. pp. 149–150. (Abstr.).

Dickman, Z. and S. K. Dey. 1974. Steroidogenesis in the preimplantation rat embryo and its possible influence on morula-blastocyst transformation and implantation. J. Reprod. Fertil. 37:91.

Eiler, H. and A. V. Nalbandov. 1973. Estrogen and progesterone content in the lumen of the reproductive tract at different stages of the estrous cycle in gilts and vaginal flux in gilts and women. Proc. 6th Ann. Meet. Soc. Study Reprod., Athens. p. 111–112. (Abstr.).

El-Banna, A. A. and J. C. Daniel, Jr. 1972. Stimulation of rabbit blastocysts *in vitro* by progesterone and uterine proteins in combination. Fertil and Steril. 23:101.

Hamana, K. and E. S. E. Hafez. 1970. Disc electrophoretic patterns of uteroglobin and serum proteins in rabbit blastocoelic fluid. J. Reprod. Fertil. 21:555.

Hansel, W. and S. E. Echternkamp. 1972. Control of ovarian function in domestic animals. Amer. Zool. 12:225.

Johnson, M. H. 1972. The protein composition of secretions from pregnant and pseudopregnant rabbit uteri with and without copper intrauterine device. Fertil. and Steril. 21:123.

Junge, J. M. and R. J. Blandau. 1958. Studies on the electrophoretic properties of cornual fluids of rats in heat. Fertil. and Steril. 9:353. Kane, M. T. and

Kane, M. T. and R. H. Foote. 1970a. Culture of two- and four-cell rabbit embryos to the blastocyst stage in serum and serum extracts. Biol. Reprod. 2:245.

Kane, M. T. and R. H. Foote. 1970b. Culture of 2- and 4-cell rabbit embryos to the expanding blastocyst stage in synthetic media. Proc. Soc. Exp. Biol. Med. 133:921.

Kane, M. T. and R. H. Foote. 1970c. Fractionated serum dialysate and synthetic media for culturing 2- and 4-cell rabbit embryos. Biol. Reprod. 2:356.

Kirby, D. R. S. 1962. The influence of the uterine environment on the development of mouse eggs. J. Embryol. Exp. Morphol. 10:496.

Kirchner, C. 1969. Untersuchungen an uterusspezifischen glykoproteinen wahrend der fruhen Graviditat des Kaninchens *Oryctolagus cuniculus*. Wilh. Roux' Arch. Entwicklungsmech. Organ. 164:97.

Kirchner, C. 1972. Immune histologic studies on the synthesis of a uterine-specific protein in the rabbit and its passage through the blastocyst coverings. Fertil. and Steril. 23:131.

Knight, J. W., Fuller W. Bazer and H. D. Wallace. 1973a. Effects of superovulation and unilateral ovariectomy-hysterectomy on porcine uterine protein secretions. J. Anim. Sci. 36:61.

Knight, J. W., Fuller W. Bazer and H. D. Wallace. 1973b. Hormonal regulation of procine uterine protein secretions. J. Anim. Sci. 36:546.

Knight, J. W., Fuller W. Bazer and H. D. Wallace. 1974a. Effect of progesterone induced increase in uterine secretory activity on development of the porcine conceptus. J. Anim. Sci. 39:743.

Knight, J. W., Fuller W. Bazer, H. D. Wallace and C. J. Wilcox. 1974b. Dose-response relationships between exogenous progesterone and estradiol and procine uterine protein secretions. J. Anim. Sci. 39:747.

Knight, J. W., Fuller W. Bazer, W. W. Thatcher and C. J. Wilcox. 1974c. Steroid levels in gilts during gestation. J. Anim. Sci. 39:215. (Abstr.).

Krishnan, R. S. and J. C. Daniel, Jr. 1967. "Blastokinin:" Inducer and regulator of blastocyst development in the rabbit uterus. Science 158:490.

Mahler, H. R. and E. H. Cordes. 1966. Biological Chemistry. Harper and Row, New York.

Murray, F. A., Fuller W. Bazer, J. W. Rundell, C. K. Vincent, H. D. Wallace and A. C. Warnick. 1971. Development failure of swine embryos restricted to the oviducal environment. J. Reprod. Fertil. 24:445.

Murray, F. A., Jr., Fuller W. Bazer, H. D. Wallace and A. C. Warnick. 1972. Quantitative and qualitative variation in the secretion of protein by the procine uterus during the estrous cycle. Biol. Reprod. 7:314.

Orsini, M. W. and A. McLaren. 1967. Loss of the zona pellucida in mice and the effect of tubal ligation and ovariectomy. J. Reprod. Fertil. 13:485.

Perry, J. S. and I. W. Rowlands. 1962. Early pregnancy in the pig. J. Reprod. Fertil. 4:175.

Perry, J. S., R. B. Heap and E. C. Amoroso. 1973. Steroid hormone production by pig blastocysts. Nature. 245:45.

Petzholdt, U. 1974. Micro-disc electrophoresis of soluble proteins in rabbit blastocysts. J. Embryol. Exp. Morph. 31:479.

Pincus, G. and R. E. Kirsch. 1936. The sterility in rabbits produced by injection of oestrone and related compounds. Amer. J. Physiol. 115:219.

Pope, C. E. and B. N. Day. 1972. Development of pig embryos following restriction to the ampullar portion of the oviduct. J. Reprod. Fertil. 31:135.

Ringler, I. 1961. The composition of rat uterine luminal fluid. Fed. Proc. 15:281.

Schlosnagle, David C., Fuller W. Bazer, John C. M. Tsibris and R. Michael Roberts. 1974. An iron-containing phosphatase induced by progesterone in the uterine fluids of pigs. J. Biol. Chem. 235:7574.

Schomberg, D. W. 1967. A demonstration *in vitro* of luteolytic activity in pig uterine flushings. J. Endocrinol. 38:359.

Schomberg, D. W. 1969. The concept of a uterine luteolytic hormone, p. 383. *In* K. W. McKerns (Ed.) The Gonads. Appleton-Century-Crofts, New York.

Seamark, R. F. and Cecilia Lutwak-Mann. 1972. Progestins in rabbit blastocysts. J. Reprod. Fertil. 29:147.

Squire, G. D., Fuller W. Bazer and F. A. Murray. 1972. Electrophoretic patterns of procine uterine protein secretions during the estrous cycle. Biol. Reprod. 7:321.

Stevens, K. R., H. D. Hafs and A. G. Hunter. 1964. Immunochemical and electrophoretic properties of oestrous rabbit uterine fluid proteins obtained by uterine ligation. J. Reprod. Fertil. 8:319.

Urzua, M. A., R. Stambaugh, G. Flickinger and L. Mastroianni, Jr. 1970. Uterine and oviduct fluid protein patterns in the rabbit before and after ovulation. Fertil. and Steril. 21:860.

Westman, A., E. Jorpes and G. Widstram. 1931. Untersuchungen uber den Schleimhautzyklus in der Tuba Uterina, seine hormonale regulierung und die Bedeutung des Tubensekrets fur die Vitalitat der befruchteten Eier. Acta Obstet. et Gynecol. Scand. 11:279.

Whitten, W. K. and J. D. Biggers. 1968. Complete development *in vitro* of the pre-implantation stages of the mouse in a chemically defined medium. J. Reprod. Fertil. 17:399.

Wintenberger-Torres, S. 1956. Les rapports entre l'oef en segmentation et le tractus maternel chez la brebis. Third Int. Congr. Animal Reprod., Cambridge 1:62.

BIOLOGY OF REPRODUCTION **61**, 884–891 (1999)

Ovine Osteopontin: I. Cloning and Expression of Messenger Ribonucleic Acid in the Uterus During the Periimplantation Period[1]

Greg A. Johnson,[3,5] Thomas E. Spencer,[3,5] Robert C. Burghardt,[4,5] and Fuller W. Bazer[2,3,5]

Departments of Animal Science[3] and Veterinary Anatomy & Public Health,[4] Center for Animal Biotechnology and Genomics,[5] Institute of Biosciences and Technology, Texas A&M University System Health Science Center, College Station, Texas 77843-2471

ABSTRACT

Trophoblast-derived interferon tau (IFNτ) acts on the endometrium to increase secretion of several proteins during the pregnancy recognition period in ruminants. One of these is a 70-kDa acidic protein that has not been identified. Our hypothesis was that the 70-kDa acidic protein is osteopontin (OPN). OPN is an acidic glycoprotein that fragments upon freezing and thawing or treatment with proteases including thrombin. OPN contains a Gly-Arg-Gly-Asp-Ser (GRGDS) sequence that binds to cell surface integrins to promote cell-cell attachment and cell spreading. Using antisera to recombinant human OPN, both 70-kDa and 45-kDa proteins were identified in uterine flushings from pregnant ewes by Western blotting. A clone containing the entire ovine OPN cDNA coding sequence was isolated by screening a Day 15 pregnant ovine endometrial cDNA library with a partial ovine OPN cDNA. In pregnant ewes, steady-state levels of OPN endometrial mRNA increased ($P < 0.01$) after Day 17. In both cyclic and pregnant ewes, in situ hybridization analysis showed that OPN mRNA was localized on unidentified immune cells within the stratum compactum of the endometrium. In pregnant ewes, OPN mRNA was also expressed by the glandular epithelium. Results suggest that progesterone and/or IFNτ induce expression and secretion of OPN by uterine glands during the periimplantation period and that OPN may induce adhesion between luminal epithelium and trophectoderm to facilitate superficial implantation.

INTRODUCTION

Ruminants are spontaneous ovulators that undergo uterine-dependent estrous cycles until a viable conceptus produces interferon tau (IFNτ), the signal for maternal recognition of pregnancy. Ovine IFNτ is secreted by the conceptus trophectoderm between Days 10 and 21 of gestation and blocks transcription of genes for ER and OTR in sheep uterine epithelium [1]. These events attenuate pulsatile release of PGF, prevent regression of the CL, and ensure maintenance of a progestational environment that is required for endometrial support of superficial implantation, placentation, and fetal/placental development.

In addition to signaling pregnancy recognition, IFNτ increases secretion of several endometrial proteins. These proteins are hypothesized to support conceptus development during the periimplantation period [2]. Specifically, they may act as nutrients, growth factors, immunomodulatory proteins for protection against conceptus allograft rejection, protease inhibitors, enzymes, or molecules that promote adhesion between conceptus trophectoderm and uterine LE. The secretion of at least 11 endometrial proteins is known to be up-regulated by IFNτ [3], including 1) β_2 microglobulin [4], 2) ubiquitin cross-reactive protein [5], 3) granulocyte chemotactic protein-2 [6], 4) Mx [7], and 5) 2′,5′-oligoadenylate synthetase [8]. Another unidentified, secreted 70-kDa protein has a pI of approximately 4 [3].

Osteopontin (OPN), also known as the early T-cell activation-1 (Eta-1) cytokine, is a 70-kDa, acidic (pI 4–5) glycoprotein [9, 10]. Upon freezing and thawing or treatment with proteases, the 70-kDa protein gives rise to 45-kDa and 24-kDa fragments [11]. OPN has been localized to the LE of both mouse and human uterine endometrium [12, 13] and is secreted by a number of tissues [13–15]. OPN/Eta-1 binds sheep red blood cell (SRBC) glycophorin and inhibits the ability of SRBC to interact with mouse T-lymphocytes that express surface receptors specific for SRBC glycophorin [11]. This rosette inhibition activity has been used as an indicator of a viable conceptus [16] and is attributed to early pregnancy factor (EPF) [17, 18].

Our hypothesis was that the 70-kDa IFNτ-regulated protein was OPN. Therefore, the objectives were to determine 1) if OPN was secreted into the uterine lumen of pregnant ewes, 2) clone a full-length ovine OPN cDNA, and 3) determine temporal and spatial alterations in OPN mRNA expression during the ovine estrous cycle and early pregnancy.

MATERIALS AND METHODS

Animals

All experimental and surgical procedures involving animals were approved by the Agricultural Animal Care and Use Committee, Texas A&M University (Animal Use Protocols 7-286 and AG-239AG).

Mature western-range ewes of primarily Rambouillet breeding were observed daily for estrous behavior in the presence of vasectomized rams. After experiencing at least two estrous cycles of normal duration (16–18 days), ewes were assigned randomly on Day 0 (estrus/mating) to cyclic or pregnant status. Ewes assigned to pregnant status were mated to intact rams three times at 12-h intervals beginning at estrus. Fifty-two ewes were ovariohysterectomized (n = 4 ewes/day) on Day 1, 3, 5, 7, 9, 11, 13, or 15 of the estrous cycle or Day 11, 13, 15, 17, or 19 of gestation. At hysterectomy, uteri (excluding those from Day 19) were flushed with 0.9% NaCl, and flushes were frozen at −80°C. Pregnancy was verified by recovery of an apparently normal conceptus in uterine flushes. Several sections (1–1.5 cm) of uterine wall from the middle of each uterine horn were fixed in fresh 4% paraformaldehyde in PBS (pH 7.2). After 24 h, fixed tissues were changed to 70% ethanol for 24 h and then embedded in Paraplast-Plus (Oxford Labware, St. Louis, MO). The remaining endometrium was physically

Accepted May 24, 1999.
Received February 19, 1999.
[1]Research supported by USDA-NRICGP 98-35203-6337 to F.W.B. and R.C.B.
[2]Correspondence: Fuller W. Bazer, Department of Animal Science and Center for Animal Biotechnology and Genomics, 442D Kleberg Center, Texas A&M University, College Station, TX 77843-2471. FAX: 409 862 2662; e-mail: fbazer@cvm.tamu.edu

separated by dissection from myometrium, frozen in liquid nitrogen, and stored at −80°C.

Western Blot Analysis

Uterine flushings from cyclic and pregnant ewes were dialyzed against 3 liters of cold dialysis buffer (10 mM Tris, pH 8.2) in Spectrapor 3 dialysis membrane (cutoff, M_r 3500; Spectrum, Houston, TX) for 24 h with stirring at 4°C. Dialysis buffer was changed three times. Samples were then concentrated by snap-freezing in liquid nitrogen followed by lyophilization. Proteins were then resuspended in 10 mM Tris (pH 8.2). Endometrium was thawed and immediately homogenized in extraction buffer (10 mM Tris [pH 7], 1 mM EDTA, 1 mM dithiothreitol [DTT], 100 μg/ml PMSF) at a ratio of 1 g tissue per 5 ml buffer. Homogenates were sonicated for 30 sec with a Mini Ultrasonic Cell Disrupter (Sonics & Materials, Inc., Danbury, CT) and clarified by centrifugation (10 000 × *g* for 15 min at 4°C). Concentrations of protein in dialyzed uterine flushings and endometrial extracts were determined using a Bradford protein assay (Bio-Rad Laboratories, Hercules, CA) with BSA as the standard. Proteins in uterine flushings (8 μg) and uterine extracts (120 μg) were denatured in Laemmli buffer, separated on 10% (total monomer) 1D or 2D-SDS-PAGE and transferred to nitrocellulose [19]. Blots were blocked overnight in TBST (20 mM Tris [pH 7.5], 137 mM NaCl, 0.05% Tween-20) containing 5% dried milk. Blots were washed three times for 5 min each in TBST and then incubated with a cocktail containing rabbit anti-human OPN IgG (LF-123 and LF-124; 5 μg/ml) [20], or normal rabbit serum (5 μg/ml) in TBST containing 2% dried milk while rocking overnight at 4°C. Blots were then washed three times for 10 min each in TBST with goat anti-rabbit IgG-horseradish peroxidase conjugate (1:15 000 dilution of 1 mg/ml stock; KPL, Bethesda, MD) for 1 h at room temperature while rocking. Blots were washed three times for 10 min each in TBST, and immunoreactive proteins were detected using enhanced chemiluminescence (Amersham Life Sciences, Arlington Heights, Rochester, NY).

Cloning of Partial Ovine OPN cDNA

The bovine OPN sequence [21] was used to derive a forward primer beginning at base 225 (5′-TGATGATAA-CAGCCAGGACGA-3′) and a reverse primer beginning at base 500 (5′-GTGAAGTCCTCCTCTGTGGC-3′). A 291-base pair (bp) OPN cDNA fragment was then amplified by polymerase chain reaction (PCR) from a Day 15 pregnant ovine endometrial cDNA library (Lambda ZAP II library; Stratagene Cloning Systems, La Jolla, CA) as described previously [22, 23]. The 291-bp OPN cDNA was then cloned into the pCR-II vector (InVitrogen, Carlsbad, CA) and sequenced.

cDNA Library Screening

A Day 15 pregnant ovine endometrial Lambda Zap II cDNA library (Stratagene) was screened with the partial ovine OPN cDNA using standard methods [22]. Positive plaques from the initial library screen were subsequently rescreened three times until plaque pure. One clone (10.2.1) containing the entire ovine OPN cDNA coding region was sequenced in both directions using forward and reversed primers as well as internal primers.

Northern Blot Analysis

Total cellular RNA was isolated from cyclic and pregnant endometrial samples using the Trizol reagent (Gibco-BRL, Grand Island, NY). The quantity of RNA was assessed spectrophotometrically, and integrity of RNA was examined by gel electrophoresis in a 1% denaturing agarose gel [22]. Total RNA (20 μg) was loaded onto a 1.2% agarose gel, electrophoresed, and transferred to a 0.2-μm nylon membrane as previously described [23]. The Northern blot was then hybridized with a radiolabeled antisense cRNA probe generated from linearized ovine OPN (10.2.1) plasmid template, washed and visualized by autoradiography for 16 h at −80°C (X-OMAT AR Film; Kodak, Rochester, NY) as previously described [23]. The antisense OPN cRNA probe was made against the OPN 10.2.1 plasmid template in pCR-II, linearized by restriction with *Not*I, using T7 RNA polymerase, [α-^{32}P]UTP (Amersham), and the Riboprobe Gemini kit (Promega, Madison, WI).

Slot Blot Hybridization Analysis

Steady-state levels of OPN mRNA were measured in cyclic and pregnant endometrial samples using slot blot hybridization analysis. For each ewe, denatured total cellular RNA (20 μg) was hybridized with radiolabeled antisense cRNA probes generated by in vitro transcription with [α-^{32}P]UTP (Amersham) as described above. Plasmid templates for ovine OPN (10.2.1) and 18S rRNA (pT718S; Ambion, Austin, TX) were used. The radioactivity in each slot was quantitated using an Instant Imager (Packard Instruments, Meridan, CT) and expressed as total counts.

In Situ Hybridization Analysis

The OPN mRNA was localized in uterine tissue sections by in situ hybridization analysis. Uterine tissue sections were deparaffinized in xylene and then rehydrated to water through a graded series of alcohol. Tissue sections were postfixed in 4% paraformaldehyde in PBS and then digested with Proteinase K (20 μg/ml) in PK digestion buffer (50 mM Tris, 5 mM EDTA, pH 8) for 8 min at 37°C. Sections were then refixed for 5 min in 4% paraformaldehyde, rinsed twice for 5 min each in PBS, dehydrated through a graded series of alcohol, and then dried at room temperature for 30 min. Sections were hybridized with radiolabeled antisense or sense cRNA probes generated from a linearized ovine OPN (10.2.1) plasmid template using in vitro transcription with [α-^{35}S]UTP (specific activity: 3000 Ci/mmol; Amersham). Antisense and sense cRNA probes were made from the OPN 10.2.1 plasmid restricted with *Xho*I and T3 RNA polymerase for sense and restricted with *Not*I and T7 RNA polymerase for antisense. Radiolabeled cRNA probe (5 × 10^6 cpm/slide) was denatured in 75 μl hybridization buffer (50% formamide, 0.3 M NaCl, 20 mM Tris-HCL [pH 8], 5 mM EDTA [pH 8], 10 mM sodium phosphate [pH 8], single-strength Denhardt's, 10% dextran sulfate, 0.5 mg/ml yeast RNA, 100 mM DTT) at 70°C for 10 min. Hybridization solution was applied to the middle of each slide and a coverslip placed gently on top. Slides were then incubated in a humidified chamber containing 50% formamide/5-strength SSC and hybridized overnight at 55°C. Coverslips were removed by placing slides in 5-strength SSC/10 mM βME for 30 min at 55°C. Sections were then washed as follows: 50% formamide/double-strength SSC/50 mM βME for 20 min at 65°C; single-strength TEN (0.05 M NaCl/10 mM Tris [pH 8]/5 M EDTA) for 10 min at room temperature; and then three times in single-strength TEN for 10 min at 37°C. Sections were then digested with DNase-free RNase (10 μg/ml) in single-strength TEN for 30 min at 37°C to remove nonspecifically bound probe and washed as follows: single-strength TEN for 30 min at 37°C; 50% formamide/double-strength SSC/50 mM βME for 20 min at 65°C; double-strength SSC for 15 min at room temperature; 0.1-strength SSC for 12 min at room temperature; 70% ethanol containing 0.3 M ammonium acetate for 5 min at room temperature; 95% ethanol containing 0.03 M ammonium acetate for 1 min at room temperature; twice in 100% ethanol; and three times in single-strength TEN for 10 min at 37°C. Liquid film emulsion autoradiography was performed using Kodak NTB-2 liquid photographic emulsion [22]. Slides were stored at 4°C for 5 days, developed in Kodak D-19 developer, counterstained with Harris' modified hematoxylin in acetic acid (Fisher, Fairlawn, NJ), dehydrated through a graded series of alcohol to xylene, coverslipped, and evaluated by both brightfield and darkfield microscopy with a Zeiss Photomicroscope III (Carl Zeiss Inc., Thornwood, NY).

Stainsall Detection of OPN Protein

Endometrium was thawed and homogenized, and the concentration of protein was determined as described for Western blotting. Protein (120 μg) in endometrial extracts was denatured in Laemmli buffer and separated in 10% gels by two-dimensional (2D) SDS-PAGE. After electrophoresis, gels were fixed (10% acetic acid:45% methanol:45% H_2O) for 1 h at room temperature, and then soaked in 50% methanol: 50% H_2O for 4 h with two changes of this solution to remove acetic acid. Phosphoproteins were detected by staining overnight at room temperature (in the dark) with Stainsall [24]: 5 mg Stainsall (1-ethyl-2-[3-(1-ethylnaphtho[1,2-D]thiazolin-2-ylidene)-2-methylpropenyl]naphtho-[1,2-D]thiazolium bromide; Sigma, St. Louis, MO) in 5 ml formamide, 25 ml isopropanol, 0.5 ml 3 M Tris-HCl (pH 8.8), and 69.5 ml H_2O. Gels were then soaked in 40% methanol/60% H_2O, placed on a gel light-box, and photographed through transmitted white light.

Statistical Analysis

Data were subjected to least-squares analysis of variance (ANOVA) using the general Linear Models (GLM) procedures of the Statistical Analysis System (Users guide 1990, Ver. 6; SAS Institute, Cary, NC). Slot blot hybridization data (total counts) were adjusted for differences in sample loading using the 18S rRNA data as a covariate [25]. All tests of statistical significance were performed using the appropriate error terms according to the expectation of mean squares. Data are presented as least-square means (LSM) with standard errors (SE).

RESULTS

Western blot analyses of ovine uterine flushings from Day 15 cyclic and pregnant ewes indicated the presence of 70-kDa and 45-kDa immunoreactive proteins (Fig. 1). The amount of immunoreactive 70-kDa and 45-kDa proteins was greater in uterine flushings from pregnant as compared to cyclic ewes.

Hybridization and screening of the ovine endometrial cDNA library with the 291-bp ovine OPN cDNA fragment resulted in isolation of 20 positive plaques. Several plaques were purified and cDNA inserts recovered by self excision.

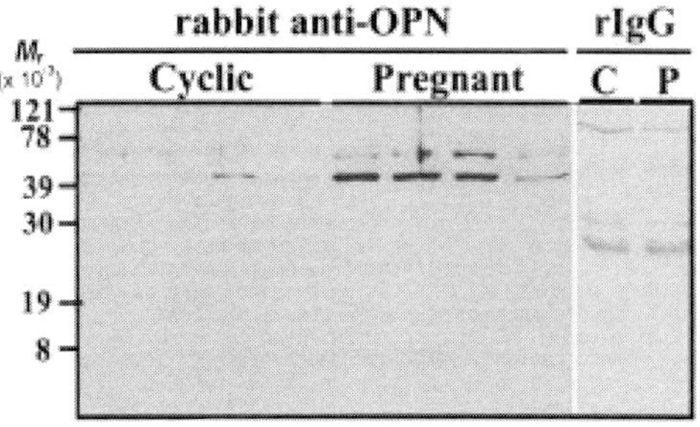

FIG. 1. Detection of OPN in ovine uterine flushings from Day 15 of the estrous cycle and pregnancy using Western blotting. Each lane represents proteins in uterine flushings from different ewes. Immunoreactive proteins were detected using either polyclonal rabbit anti-human OPN IgG or rabbit immunoglobulin (rIgG). Positions of prestained molecular weight standards are indicated on the left. Immunoreactive OPN was greater in uterine flushings from pregnant compared to cyclic ewes.

One clone, found to be large enough to potentially contain the entire coding sequence, was sequenced in both directions and the nucleotide sequence for ovine OPN cDNA (clone 10.2.1) is shown in Figure 2. The inferred amino acid sequence predicts that ovine OPN is a single chain protein of 278 amino acids with a hydrophobic leader sequence of 16 amino acids characteristic of secreted proteins. Ovine OPN contains 1) a potential calcium phosphate apatite binding region of consecutive Asp residues beginning at amino acid 86, 2) a cell-attachment Gly-Arg-Gly-Asp-Ser (GRGDS) sequence (residues 151–155), 3) a thrombin KS cleavage site (residues 161 and 162), and 4) two glutamines that are recognized substrates for transglutaminase in other species.

Nucleotide and amino acid similarities of ovine OPN with bovine, rat, mouse, and porcine OPN are summarized in Table 1. Sequences that are highly conserved in all of these species include 1) four of the 9 or 10 residues in the poly-Asp region, 2) the GRGDS, 3) 15 serines that include an SSEEK sequence (residues 26–30), 4) three threonines (residues 131, 140, and 145), 5) an NES sequence (residues 79–81), and 6) glutamines at positions 50 and 52. A feature of the ovine OPN sequence, shared with bovine OPN only, is deletion of 22 amino acids that would otherwise be inserted between residues 196 and 197. Also, ovine and bovine OPN have a KS, rather than RS putative thrombin cleavage site.

The OPN cDNA detected a ~1-kilobase (kb) mRNA in Northern blot analysis of ovine endometrial total RNA (Fig. 3A). The amount of OPN mRNA increased in pregnant ewes. Steady-state levels (Fig. 3B) did not increase in cyclic ewes but did increase in pregnant ewes after Day 17 ($P < 0.05$).

In situ hybridization analysis of cyclic and pregnant uteri revealed two distinct patterns of OPN mRNA expression (Fig. 4). During the estrous cycle and early pregnancy, expression of transcript was observed in a small percentage of cells scattered throughout the endometrium. The cells that expressed OPN mRNA became concentrated in the stratum compactum immediately beneath the LE on Days 1, 5, and 7 of the estrous cycle. This expression was intermittent, and extended along only a small percentage of the total circumference of the uterine wall. In pregnant ewes, OPN mRNA was present in the GE, where expression was

```
1     ggcacgaggagcactgcatcagcatcacaggggactggactcttctctgc
51    tgctgcagaccaagaaaaaaatcatgaccATGAGAATTGCAGTGATTTGC
                                    M  R  I  A  V  I  C    7
101   TTCTGCCTCTTGGGCATTGCCTCCGCCCTTCCAGTTAAACCGACCAGTTC
       F  C  L  L  G  I  A  S  A  L  P  V  K  P  T  S  S 24
151   TGGCAGCTCTGAGGAAAAGCAGCTTAACAACAAATACCCAGATGCTGTAG
        G  S  S  E  E  K  Q  L  N  N  K  Y  P  D  A  V   40
201   CCACATGGCTAAAGCCTGACCCATCTCAGAAGCAGACTTTCCTAGAACCC
      A  T  W  L  K  P  D  P  S  Q  K  Q  T  F  L  E  P  57
251   CAGAATTCTGTGTCCTCTGAGGAAACTGATGACAACAAACAAAATACCCT
       Q  N  S  V  S  S  E  E  T  D  D  N  K  Q  N  T  L 74
301   CCCGAGTAAGTCCAATGAAAGCCCTGAGCAAACAGACGATCTAGATGATG
        P  S  K  S  N  E  S  P  E  Q  T  D  D  L  D  D   90
351   ATGATGAAAACAGCCAGGAAGTCAACTCTGATGACTCCGACGATGCTGAA
      D  D  E  N  S  Q  E  V  N  S  D  D  S  D  D  A  E  107
401   ACCCCTGATGACTCTGACCATTCCAACGAGTCTCACCATTCTGATGAATC
       T  P  D  D  S  D  H  S  N  E  S  H  H  S  D  E  S 124
451   TGATGAAGCTGATTTTCCCACTGACATTCCAACAATCGCAGTTTTCACTC
        D  E  A  D  F  P  T  D  I  P  T  I  A  V  F  T   140
501   CACCTTTCCCTACGGAAAGCACAAATGATGGCCGAGGTGATAGTGTGGCT
      P  P  F  P  T  E  S  T  N  D  G  R  G  D  S  V  A  157
551   TATGGACTGAAGTCAAAATCTAAGAAGTTCCGCCGATCTAACGTTGAGAG
       Y  G  L  K  S  K  S  K  K  F  R  R  S  N  V  E  S 174
601   TCCAGATGCCACAGAGGAGGACTTCACATCACACATAGAGAGTGAGGAGA
        P  D  A  T  E  E  D  F  T  S  H  I  E  S  E  E   190
651   TGCATGACGCACCTAAGAAGACGAGTCAGCTGACTGACCACAGCGAAGAA
      M  H  D  A  P  K  K  T  S  Q  L  T  D  H  S  E  E  207
701   ACCAACAGTGACGAGCTTCCCAAAGAACTCACGCCAAAGGCCAAGGAGGA
       T  N  S  D  E  L  P  K  E  L  T  P  K  A  K  E  E 224
751   AAGCAAGCATTCCAATCGGATCGAGAGTCAGGAAAATTCCAAACTCAGCC
        S  K  H  S  N  R  I  E  S  Q  E  N  S  K  L  S   240
801   AAGAATTCCATAGCCTTGAAGACAAGCTAGACCTAGATCATAAGAGTGAA
      Q  E  F  H  S  L  E  D  K  L  D  L  D  H  K  S  E  257
851   GAAGACAAGCGCCTGAAAATCCGCATTTCTCATGAATTAGATAGTGTCTC
       E  D  K  R  L  K  I  R  I  S  H  E  L  D  S  V  S 274
901   TTCTGAGGTCAACTGAaaggagaaaatcaagttcttactttgcttttagt
        S  E  V  N  *                                    278
951   aaaaagaaaaggatatgttaaagcagggtgggagacaatatgaaatgcgt
1001  atttctcagcttagttggtgaatgtatatgtgtgtagatctggaaacaga
1051  tcatgtttttgatcattagtttagtgtgtggcttcatggtaacaccctta
1101  taaactaaaagcttcagggtttagtctatgttctttccacataaaaaatg
1151  caaaccatcacagcattttaatgtttgctaccctttttatgaatagaaatt
1201  catgtagaagcaaacaaaatactgttacacactttttaagaaggaatataa
1251  aatttcacgtcactatgatcttttgtttttttaaattagtatatattttgt
1301  tgtgattatttttttctgttgtgaataaatctttttatcttgaatgtaaaaa
1351  aaaaaaaaaaaaccggacacgattattgtcagcaatgttaagagacatcc
1401  tgcatgtggggcaggaggtgtcttgttctagccaggaacgaagacaggag
1451  ttgtggaaaagatgtccacagggcagtttcctcgcagcctgcatggagtc
1501  ccagcagatggcgcagtcatcgttattgacagccagctcctctggagttg
1551  ccactgcaaacctggcttccatgttcccgaccattcgcaagtagttcttg
1601  tgccggcggatgcgacgctgcacttcatgaaacaggtaacgcagctgcat
1651  gaagatgaccaagctggccatggacaaccagatgttgccaaacaacaaca
1701  tgtgaatatggtgcatgaggtccaaggacaagagagtgagctccatgacg
1751  aatctgtgtaatagacatacgtcccctttccttcccatgtcccttcatgg
1801  ttgaggtcccagaggtgaattacatatcgtaaaatcacatgagcagtcct
1851  cactgtcacaagaagagactctgcagccatgaaagccaacgtgtgcatcc
1901  catgggtgtagcccgtgacgcaacagacgactgctaattccacagcagga
1951  aagcagcatggcgatcagaagggacagaactcggccgtggctgctcatgg
2001  gggtggtggagaaaaagaaagatattcaaaccggtccttgcagagtgaac
2051  catcaggtgcagaaataccagtccagcaaaccagaggcaccacatgacca
2101  cttcttccactgtttggacattcagcaccccgaaaatgaaaatgaacttg
2151  tagaaaataaaattccaaaacttgtctttgaggtgctgtctttcgctcac
2201  tcgaagagggccaaacacaatacactggattagcttagccaccaacatca
2251  aaatgcagcaagcagtattcactagaacccacacgaacaggctgtccgag
2301  agcaggtactgggccacgtcgcgggcccggggtcccccggcgctcggccg
2351  gacgagcgcggccgggtctggctgcaccctcgt
```

FIG. 2. Nucleotide and inferred amino acid sequences of ovine OPN (GenBank accession number AF152416). Nucleotides are numbered to the left, and amino acids are numbered to the right. The hydrophobic leader sequence, GRGDS (sequence for cell attachment), KS (sequence for thrombin cleavage), polyaspartic acid domain, substrates for transglutaminase (glutamines), and polyadenylation site are shaded.

first detected in some ewes on Day 13 of pregnancy. Expression of OPN mRNA by GE increased markedly in all ewes between Days 15 and 19 of pregnancy. However, the temporal rise in glandular OPN mRNA varied between ewes. Day 19 conceptuses did not express OPN mRNA (Fig. 4).

TABLE 1. Nucleotide and amino acid similarities of ovine OPN with bovine, rat, mouse, and porcine OPN.

Source	Sequence similarities (%)		GenBank accession No.
	Nucleotide	Amino acid	
Bovine	95.2	91.7	M66236
Human	67.3	57.0	J04765
Rat	61.5	51.4	AB001382
Mouse	53.6	53.2	J04806
Porcine	73.1	63.8	X16575

Western blot and Stainsall analysis of endometrial proteins from Day 17 pregnant ewes separated by 2D-PAGE detected a 70-kDa protein (Fig. 5). Western blots were performed using affinity purified rabbit polyclonal antibody raised against recombinant human OPN (rhOPN), whereas the Stainsall reagent stains phosphoproteins and is established as the method of choice for detection of OPN [24, 26, 27]. The stained protein exhibited a range (~5–6) of isoelectric species (Fig. 5).

DISCUSSION

This is the first report of OPN expression and secretion by the ovine uterus. Prior to these studies, it was hypothesized that OPN was the IFNτ-regulated acidic 70-kDa se-

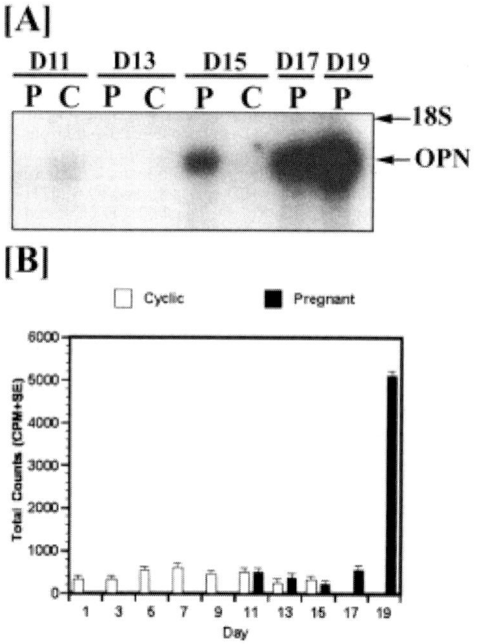

FIG. 3. Detection of OPN mRNA in ovine endometrium. **A**) Northern blot analyses of OPN mRNA (20 μg/lane) in ovine endometrium from cyclic (C) and pregnant (P) ewes. Endometrial OPN mRNA (~1 kb) increased during early pregnancy. **B**) Steady-state levels of OPN mRNA in ovine endometrium during the estrous cycle and pregnancy. There was a day × pregnancy status interaction ($P < 0.05$) and, within pregnant ewes, there was an effect of Day ($P < 0.05$).

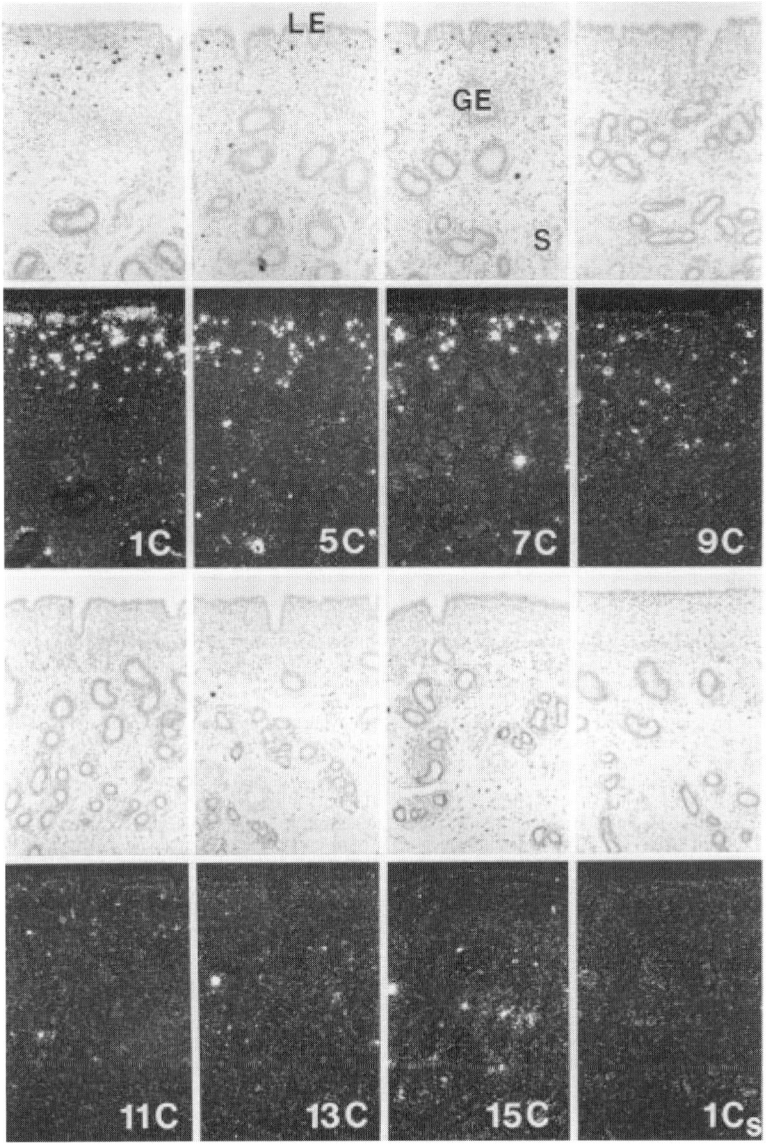

FIG. 4. In situ hybridization analysis of OPN mRNA in ovine endometrium during the estrous cycle (this page, C panels) and pregnancy (opposite page, P panels). **C panels**) Corresponding brightfield and darkfield images of endometrium in cycling animals. Note OPN transcript in scattered cells immediately beneath LE on Days 1, 5, and 7 of the estrous cycle (1C, 5C, 7C) and the decline in number of these cells later in the cycle. A representative section from Day 1 hybridized with radiolabeled sense cRNA probe ($1C_S$) serves as a negative control. **P panels**) Brightfield and darkfield images of endometrium in pregnant animals. Note that OPN transcript appears in GE and increases during pregnancy but is absent in the Day 19 conceptus (19P, TR). A representative section from Day 19 of pregnancy hybridized with radiolabeled sense cRNA probe ($19P_S$) serves as a negative control. All sections represent examples of maximal OPN mRNA expression for each day observed. LE, Luminal epithelium; GE, glandular epithelium; S, stroma; TR, trophectoderm. ×60.

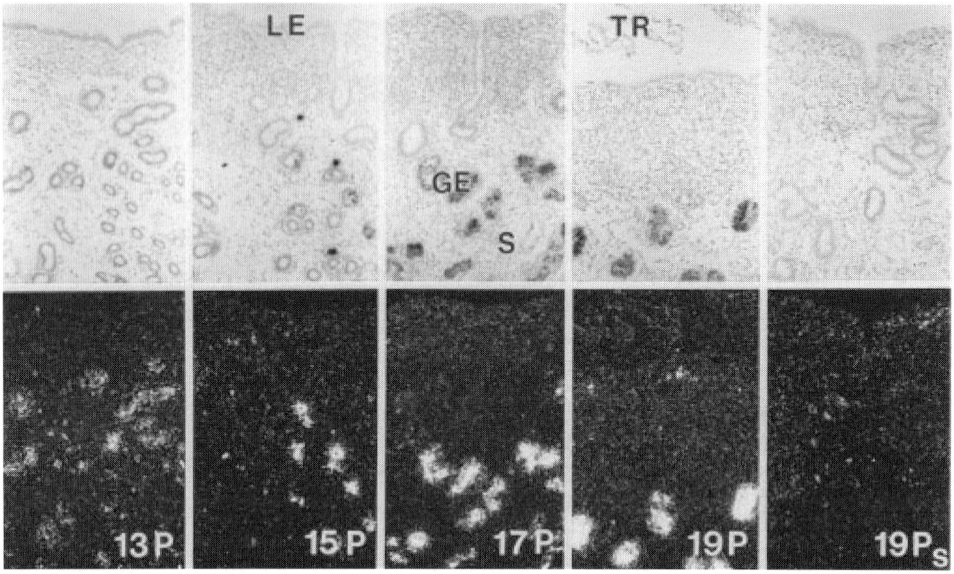

FIG. 4. Continued.

cretory protein. This hypothesis was reinforced by the presence of increased OPN protein in uterine flushings from pregnant ewes. However, when endometrial protein extracts from Day 17 pregnant ewes were examined by 2D-PAGE, the 70-kDa OPN protein exhibited a number of isoelectric variants ranging in pI from 5–6. Because the IFNτ-regulated 70-kDa protein has a single pI of 4, identification of this protein as OPN is questionable.

The cDNA sequence for ovine OPN shares high homology with those for other mammalian species [28–31] and is most similar to bovine OPN (91.7% amino acid sequence similarity) [21]. Features of OPN are that it 1) is a glycosylated phosphoprotein, 2) contains an RGD peptide integrin binding sequence, 3) is secreted, and 4) is susceptible to cleavage by proteases [9]. The inferred OPN amino acid sequence is consistent with these properties. Ovine OPN has a 16 amino acid hydrophobic signal peptide, GRGDS cell attachment and KS thrombin cleavage sites, 41 serines of which 13 phosphoserines are found in the same location in all species, and three threonine *O*-glycosylation sites as well as a potential *N*-glycosylation (NES) site.

Consistent with the presence of OPN protein in uterine flushings, steady-state levels of endometrial OPN mRNA increased during pregnancy. In situ hybridization analysis indicated an increase in OPN mRNA in the GE of some pregnant ewes as early as Day 13, and all ewes by Day 19. In cyclic and pregnant ovine uteri, OPN mRNA was localized to different cell types. In cyclic ewes, OPN mRNA was detected in scattered immune cells that represent a small percentage of total cells in the endometrium. Secretion of OPN by these cells may be responsible for the presence of the 70- and 45-kDa OPN forms in uterine flushings from cyclic ewes. In pregnant ewes, there was significant induction of OPN gene transcript in GE. This increase of GE expression is likely responsible for increased OPN protein in uterine flushings from pregnant ewes.

Pregnancy-specific expression of OPN mRNA in ovine uterine epithelial cells is not surprising. In humans, OPN mRNA has been localized to luminal epithelial cells of the gastrointestinal tract, gall bladder, pancreatic ducts, distal tubules of the kidney, lung, thyroid, GE of the breast, mucinous epithelium of the endocervix, and fallopian tubes [13]. The secretory phase glands of both nonpregnant and pregnant human uteri express OPN [13, 32]. In mice, the

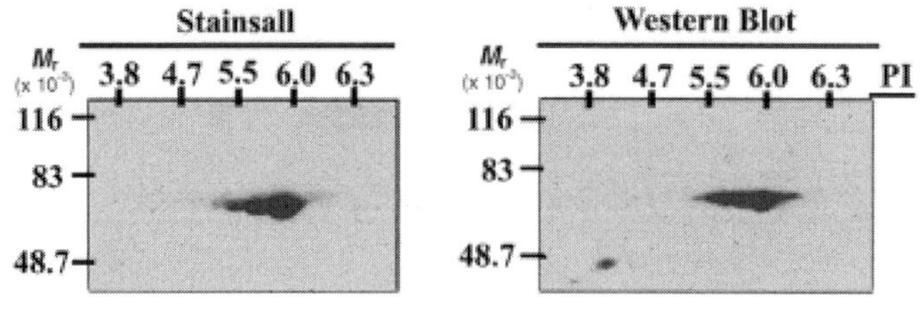

FIG. 5. Detection of OPN (10% 2D-PAGE) in Day 17 pregnant ovine endometrial extracts (120 μg total protein) with Stainsall and Western blotting. Positions of prestained molecular weight standards are indicated on the left. The isoelectric variants are most likely the result of variable degrees of posttranslational modification of a single protein. OPN extracted from rat bone matrix is also heterogenous with respect to phosphorylation [45]. OPN was the only endometrial protein that stained with Stainsall on these gels.

uterine epithelium, adjacent to deciduoma elicited in pseudopregnant females, expresses OPN, while the epithelium in the contralateral horn does not [12]. Results of the present study indicate that OPN mRNA is not present in LE or the conceptus but restricted to the endometrial glands of Day 13–19 pregnant ewes.

The role of OPN during the estrous cycle is unknown. The OPN-positive cells in uterine stroma are probably immune cells because 1) OPN/Eta-1 has been cloned and sequenced from activated T helper lymphocytes [33] and may be the most abundant protein secreted by various T lymphocyte populations [11, 34, 35]; 2) OPN/Eta-1 is expressed by monocytes and macrophages after tissue injury [36]; 3) osteoclasts, which deposit OPN/Eta-1 into lacunae of resorbing bone, are specialized tissue macrophages [37]; and 4) OPN/Eta-1 has been detected in activated $CD8^-$ $CD4^-$ natural killer cells, including granulated metrial gland cells [39]. The immunological role of OPN/Eta-1 may be significant since it binds the CD44 receptor on lymphocytes and monocytes to induce chemotaxis of these cells out of the bloodstream and into sites of inflammation [10]. The OPN/Eta-1 gives rise to 24- and 45-kDa proteins that bind antigen and suppress T-helper cells, respectively [11]. And, the 45-kDa fragment increases immunoglobulin production by B lymphocytes [39].

During early pregnancy, OPN may act as an adhesion molecule. It binds primarily to $\alpha_v\beta_3$ integrin heterodimer on tissues via its GRGDS sequence to promote cell-cell attachment, cell spreading, and cell-extracellular matrix communication [40, 41]. In women [42], expression of $\alpha_v\beta_3$ is unique to the "implantation window" and increases in response to progesterone in uteri of baboons [43]. Temporal and spatial alterations in expression of extracellular matrix proteins (ECMs) and integrins by the uterus and conceptuses of pigs during the periimplantation period have also been reported [44]. Thus, our working hypothesis is that modulation of expression of OPN by progesterone and/or sheep trophoblast interferons may induce expression and secretion of OPN by uterine epithelium. This OPN then binds $\alpha_v\beta_3$ integrin heterodimer expressed by trophectoderm and/or uterus to 1) stimulate changes in morphology of trophectoderm and extra-embryonic endoderm that result in elongation of the conceptus and 2) induce adhesion between luminal epithelium and trophectoderm essential for attachment and superficial implantation. Experiments designed to determine if OPN expression is regulated in vivo by progesterone and/or IFNτ are in progress.

ACKNOWLEDGMENTS

The authors thank Dr. Marian F. Young for the rabbit polyclonal antibodies to recombinant human OPN (LF-123 and LF-124); Dr. Mary C. Farach-Carson of the University of Delaware Biological Sciences Department for advice concerning Stainsall detection of OPN; and Dr. Shawn W. Ramsey and Mr. Todd Taylor of the Texas A&M University Sheep & Goat Center for care and management of ewes. Photomicrographs were prepared using facilities in the College of Veterinary Medicine Image Analyses Laboratory, which is supported, in part, by NIH Grant P30 ES09106.

REFERENCES

1. Spencer TE, Bazer FW. Ovine interferon tau suppresses transcription of the estrogen receptor and oxytocin receptor genes in the ovine endometrium. Endocrinology 1996; 137:1144–1147.
2. Bazer FW. Establishment of pregnancy in sheep and pigs. Reprod Fertil Dev 1989; 1:237–242.
3. Vallet JL, Bazer FW, Roberts RM. The effect of ovine trophoblast protein-one on endometrial protein secretion, and cyclic nucleotides. Biol Reprod 1987; 37:1307–1316.
4. Vallet JL, Barker PJ, Lamming GE, Skinner N, Huskisson NS. A low molecular weight endometrial secretory protein which is increased by ovine trophoblast protein-1 is β_2 microglobulin-like protein. J Endocrinol 1991; 130:R1-R4.
5. Austin KJ, Ward SK, Teixeira MG, Dean VC, Moore DW, Hansen TR. Ubiquitin cross-reactive protein is released by the bovine uterus in response to interferon during early pregnancy. Biol Reprod 1996; 54:600–606.
6. Teixeira MG, Austin KJ, Perry DJ, Dooly VD, Johnson GA, Francis BR, Hansen TR. Bovine granulocyte chemotactic protein-2 is secreted by the endometrium in response to interferon-tau (IFN-τ). Endocrine 1997; 6:31–37.
7. Ott TL, Spencer TE, Lin JY, Kim H-T, Gerami B, Bartol FF, Wiley AA, Bazer FW. Effects of the estrous cycle and early pregnancy on uterine expression of Mx protein in sheep (*Ovis aries*). Biol Reprod 1998; 59:784–789.
8. Short EC, Geisert RD, Helmer SD, Zavy MT, Fulton RW. Expression of antiviral activity and induction of 2′,5′-oligoadenylate synthetase by conceptus secretory proteins enriched in bovine trophoblast protein-1. Biol Reprod 1991; 44:261–268.
9. Butler WT, Ridall AL, McKee MD. Osteopontin. In: Bilezikian JP, Raisz LG, Rodan GA (eds), Principals of Bone Biology. New York: Academic Press, Inc.; 1996: 167–181.
10. Weber GF, Cantor H. The immunology of Eta-1/osteopontin. Cytokine Growth Factor Rev 1996; 7:241–248.
11. Fresno M, McVay-Boudreau L, Nabel G, Cantor H. Antigen-specific T-lymphocyte clones. II. Purification and biological characterization of an antigen-specific suppressive protein synthesized by cloned T-cells. J Exp Med 1981; 153:1260–1274.
12. Nomura S, Wills AJ, Edwards JK, Heath JK, Hogan BLM. Developmental expression of 2ar (osteopontin) and SPARC (osteonectin) RNA as revealed by in situ hybridization. J Cell Biol 1988; 106:441–450.
13. Brown LF, Berse B, Van DeWater L, Papadopoulos-Sergiose A, Peruzzi CA, Manseau EJ, Dvorak HF, Senger DR. Expression and distribution of osteopontin in human tissues: widespread association with luminal epithelial surfaces. Mol Biol Cell 1992; 3:1169–1180.
14. Senger DR, Perruzzi CA, Papadopoulos A, Tenen DG. Purification of a human milk protein closely similar to tumor secreted phosphoproteins and osteopontin. Biochim Biophys Acta 1989; 996:43–48.
15. Khori K, Nomura S, Kitamura Y, Nagata T, Yoshioka K, Iguchi M, Yamate T, Umekawa T, Suzuki Y, Shinohara H. Structure and expression of the mRNA encoding urinary stone protein (osteopontin). J Biol Chem 1993; 268:15180–15184.
16. Mesrogli M, Schneider J, Maas DHA. Early pregnancy factor as a marker for the earliest stages of pregnancy sera. J Biol Chem 1988; 264:2266–2271.
17. Morton H, High V, Clunie GJA. Immunosuppression detected in pregnant mice by rosette inhibitor test. Nature 1974; 249:459–460.
18. Clark IQ, Orosco C, Cock IE, Clark FM. The 'early pregnancy factor' revisited: the effect of ammonium sulfate on the capacity pf pregnant mouse sera to express activity in the rosette inhibitor assay. J Reprod Fertil 1994; 100:279–289.
19. Weise DW, Newton GR, Emish GC. The effects of day of the estrous cycle or pregnancy on protein secretion by caprine endometrial tissue. Biol Reprod 1993; 49: 522–527.
20. Fisher LW, Stubbs III JT, Young MF. Antisera and cDNA probes to human and certain animal model bone matrix noncollagenous proteins. Acta Orthop Scand 1995; 66:61–65.
21. Kerr JM, Fisher LW, Termine JP, Young MF. Cloning and RNA distribution of bovine osteopontin. Gene 1992; 108:227–234.
22. Ausebel FM, Brent R, Kinston RE, Moore DD, Seidmen JG, Smith HA, Struhl KA. Current Protocols in Molecular Biology. New York: John Wiley and Sons; 1992.
23. Spencer TE, Ing NA, Ott TL, Mayes JS, Becker WC, Watson GH, Mirando MA, Bazer FW. Intrauterine injection of ovine interferon-tau (oIFN-τ) alters oestrogen receptor and oxytocin receptor expression in the endometrium of cyclic ewes. J Mol Endocrinol 1995; 15:203–220.
24. Termine JD, Wientroub S, Fisher LW. Methods for the isolation and testing of skeletal tissue matrix proteins. In: Dickson GR (ed), Methods of Calcified Tissue Preparation. Elsevier, Amsterdam; 1984:547–563.
25. Steele RGD, Torrie JH. Principles and Procedures of Statistics. New York: McGraw-Hill; 1980.
26. Van Dijk S, D'Errico JA, Somerman MJ, Farach-Carson MC, Butler WT. Evidence that a non-RGD domain in rat osteopontin is involved in cell attachment. J Bone Miner Res 1993; 8:1499–1505.

27. Devoll RE, Pinero GJ, Appelbaum ER, Dul E, Troncoso P, Butler WT, Farach-Carson MC. Improved immunohistochemical staining of osteopontin (OPN) in paraffin-embedded archival bone specimens following antigen retrieval: anti-human OPN antibody recognizes multiple molecular forms. Calcif Tissue Int 1977; 60:380–386.
28. Craig AM, Denhardt DT. The murine gene encoding secreted phosphoprotein 1 (osteopontin), promotor structure, activity, induction in vivo by estrogen and progesterone. Gene 1991; 100:163–171.
29. Oldberg A, Franzen A, Heinegard D. Cloning and sequence analysis of rat bone sialoprotein (osteopontin) cDNA reveals an Arg-Gly-Asp cell binding sequence. Proc Natl Acad Sci USA 1986; 83:8819–8823.
30. Kiefer MC, Bauer DM, Barr PJ. The cDNA and derived amino acid sequence of human osteopontin. Nucleic Acids Res 1989; 17:3306.
31. Wrana JL, Zhang Q, Sodek J. Full length cDNA sequence of porcine secreted phosphoprotein-I (SPP-1, osteopontin). Nucleic Acids Res 1989; 17:10119.
32. Young MF, Kerr JM, Termine JD, Weever UM, Wang MG, McBride OW, Fisher LW. cDNA cloning, mRNA distribution and heterogeneity, chromosomal location, and RFLP analysis of human osteopontin (OPN). Genomics 1990; 7:491–502.
33. Patarca R, Freeman GJ, Singh RP. Structural and functional studies of the early T-lymphocyte activation 1 (Eta-1) gene. Definition of a novel T cell-dependent response associated with genetic resistance to bacterial infection. J Exp Med 1989; 170:145–161.
34. Fresno M, McVay-Boudreau L, Cantor H. Antigen-specific T-lymphocyte clones. J Exp Med 1982; 155:981–993.
35. Cone RE, Rosenstein RW, Janeway CA. Affinity-purified antigen-specific products produced by T-cells share epitopes recognized by heterologous antisera against several different antigen-specific products from T-cells. Cell Immunol 1983; 82:232–245.
36. Murray CE, Giachelli CM, Schwartz SM, Vracko R. Macrophages express osteopontin during repair of myocardial necrosis. Am J Pathol 1994; 145:1450–1462.
37. Appelbaum RA, Connor JR, James IE, Rykaczewski EL. Human osteoclasts, not osteoblasts, deposit osteopontin into resorption surfaces: an in vitro and in vivo study of remodeling bone. J Bone Miner Res 1995; 10:1666–1680.
38. Pollack SB, Linnemeyer PA, Gill S. Induction of osteopontin mRNA expression during activation of murine NK cells. J Leukocyte Biol 1994; 55:398–400.
39. Fresno M, DerSimonian H, Nabel G, Cantor H. Proteins synthesized by inducer T-cells: evidence for a mitogenic peptide shared by inducer molecules that stimulate different cell types. Cell 1982; 30:707–713.
40. Miyauchi A, Alvarez J, Greenfield EM, Teti A, Grano M, Colucci S, Zambonin-Zallone A, Ross FP, Teitelbaum SL, Cheresh D, Hruska KA. Recognition of osteopontin and related peptides by an $\alpha_v\beta_3$ integrin stimulates immediate cell signals in osteoclasts. J Biol Chem 1991; 266:20369–20374.
41. Ross FP, Chapper J, Alvarez JI, Sander D, Butler WT, Farach-Carson MC, Mintz KA, Robey PG, Teitelbaum SL, Cheresh DA. Interactions between the bone matrix proteins osteopontin and bone sialoprotein and the osteoclast integrin $\alpha_v\beta_3$ potentiate bone resorption. J Biol Chem 1993; 263:19433–19436.
42. Lessey BA, Lei Y, Castelbaum AJ, Yowell CW, Buck CA, Sun J. Further characterization of endometrial integrins during the menstrual cycle and pregnancy. Fertil Steril 1994; 62:497–506.
43. Fazleabas AT, Bell SC, Fleming S, Sun J, Lessey BA. Distribution of integrins and the extracellular matrix proteins in the baboon endometrium during the menstrual cycle and early pregnancy. Biol Reprod 1997; 56:348–356.
44. Bowen JA, Bazer FW, Burghardt RC. Spatial and temporal analysis of integrin and muc-1 expression in porcine uterine epithelium and trophectoderm in vitro. Biol Reprod 1997; 56:409–415.
45. Safran JP, Butler WT, Farach-Carson MC. Modulation of osteopontin post-translational state by 1,25-$(OH)_2$-vitamine D_3. J Biol Chem 1998; 273:29935–29941.

0013-7227/01/$03.00/0
Endocrinology

Vol. 142, No. 6
Printed in U.S.A.

Keratinocyte Growth Factor Is Up-Regulated by Estrogen in the Porcine Uterine Endometrium and Functions in Trophectoderm Cell Proliferation and Differentiation*

HAKHYUN KA, LAURIE A. JAEGER, GREG A. JOHNSON†, THOMAS E. SPENCER, AND FULLER W. BAZER

Center for Animal Biotechnology and Genomics (H.K., L.A.J., G.A.J., T.E.S., F.W.B.) and Departments of Animal Science (H.K., G.A.J., T.E.S., F.W.B.) and Veterinary Anatomy and Public Health (L.A.J.), Texas A&M University, College Station, Texas 77843-2471

ABSTRACT

Keratinocyte growth factor (KGF) is expressed by uterine endometrial epithelial cells during the estrous cycle and during pregnancy in pigs, whereas KGF receptor is expressed in conceptus trophectoderm and endometrial epithelia. In particular, KGF expression in the endometrium is highest on day 12 of pregnancy. This corresponds to the period of maternal recognition of pregnancy in pigs, which is signaled by large amounts of estrogen secreted by conceptus trophectoderm acting on the endometrium. Our hypothesis is that estrogens of conceptus origin stimulate endometrial epithelial KGF expression, and, in turn, secreted KGF stimulates proliferation and differentiation of conceptus trophectoderm. To determine the factors affecting KGF expression in the uterus, endometrial explants from gilts on day 9 of the estrous cycle were cultured in the presence of 17β-estradiol, catechol estrogens, or progesterone. 17β-Estradiol stimulated the expression of KGF ($P < 0.05$), whereas catechol estrogens had no effect ($P > 0.05$). Between days 9 and 15 of pregnancy, proliferating cell nuclear antigen was abundant in conceptuses, but was barely detectable in uterine endometrial epithelia. To determine the effects of KGF on conceptus trophectoderm, porcine trophectoderm (pTr) cells were treated with recombinant rat KGF (rKGF). rKGF increased the proliferation of pTr cells ($P < 0.01$) as measured by [^{3}H]thymidine incorporation. rKGF elicited phosphorylation of KGF receptor and activated the mitogen-activated protein kinase (ERK1/2) cascade in pTr cells. pTr cell differentiation was affected by rKGF, because it increased expression of urokinase-type plasminogen activator, a marker for differentiation in pTr cells. Collectively, these results indicate that estrogen, the pregnancy recognition signal from the conceptus in pigs, increases uterine epithelial KGF expression, and, in turn, KGF stimulates the proliferation and differentiation of conceptus trophectoderm. (*Endocrinology* **142:** 2303–2310, 2001)

KERATINOCYTE GROWTH factor/fibroblast growth factor-7 (KGF/FGF-7), is a paracrine mediator of epithelial-mesenchymal interactions in various organs, including those of the reproductive tract (1, 2). The receptor for KGF (KGFR), which is also called FGF receptor 2IIIb (FGFR2IIIb), is an alternative splice variant of the *bek* gene product and is present only on epithelial cells. KGF is expressed in ovary, uterus, prostate, and mammary gland, and uterine expression of KGF has been identified in several species, including humans, primates, rodents, and sheep (2, 3). In these species, mesenchymal expression of KGF and epithelial expression of KGFR suggested that KGF is a stromal cell-derived paracrine mediator of epithelial-mesenchymal interactions in the uterus. In the pig, however, KGF is expressed in the uterine endometrial epithelia and secreted into the uterine lumen, whereas the KGFR is expressed in both endometrial epithelia and conceptus trophectoderm (4). These findings suggest that in the pig, which is the only species possessing a true epitheliochorial type of placentation, KGF may play a role in paracrine epithelial-epithelial interactions between conceptus and uterus during early pregnancy (4).

In the pig, KGF expression in the endometrium is highest on day 12 of pregnancy during the period of maternal recognition of pregnancy (4). In male and female reproductive organs, KGF gene expression in rodents is up-regulated by steroid hormones such as androgens, estrogen, and progesterone (P_4). The KGF gene has an androgen response element in the promoter region (5), and testosterone increased the expression of KGF in prostate and seminal vesicles (6, 7). In primate endometrium, P_4 increases KGF expression during the luteal phase, suggesting that KGF is a mediator of P_4 action on epithelial cells or a progestamedin (8). It has also been suggested that KGF is induced by estrogen in the female genital tract of mice during neonatal development (9). 17β-Estradiol (E_2) treatment *in vivo* increases KGF messenger RNA (mRNA) and protein expression in the mouse mammary gland (10). In the porcine uterus, estrogens are secreted by conceptus trophectoderm beginning on days 11 and 12 of pregnancy and are the signal for maternal recognition of pregnancy (11). Progesterone levels also increase up to 30 ng/ml in plasma during early pregnancy (12). Therefore,

Received January 8, 2001.
Address all correspondence and requests for reprints to: Dr. Fuller W. Bazer, Department of Animal Science and Center for Animal Biotechnology and Genomics, 442D Kleberg Center, Texas A&M University, College Station, Texas 77843-2471. E-mail: fbazer@cvm.tamu.edu.
* This work was supported by USDA Grant NRICGP 2000–02290 (to F.W.B. and L.A.J.) and NIH Grant P30-ES-09106 (to Image Analysis Laboratory of College of Veterinary Medicine).
† Present address: Department of Animal and Veterinary Science, University of Idaho, Moscow, Idaho 83844-2330.

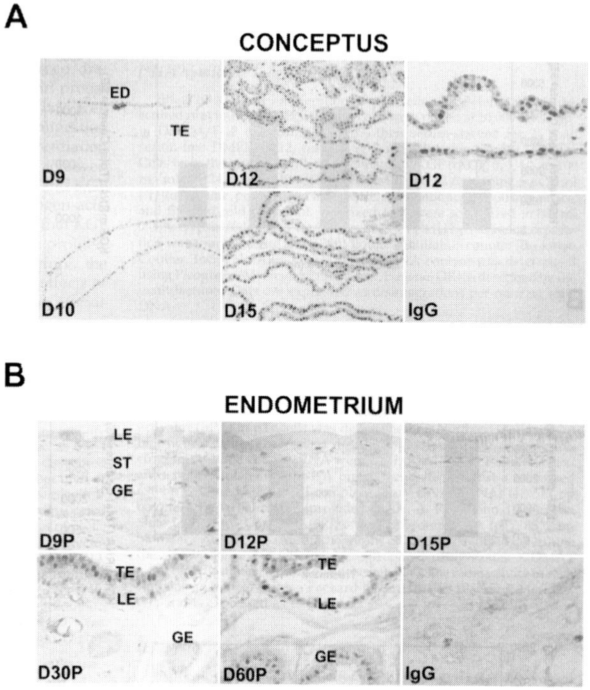

FIG. 3. Representative photomicrographs illustrating the distribution of immunoreactive PCNA protein in the conceptus (A) and the uterine endometrium (B) on various days (D) of pregnancy. Nuclear staining was not observed when irrelevant mouse IgG was substituted for primary antibodies (IgG). ED, Embryonic disc; TE, trophectoderm; LE, luminal epithelium; GE, glandular epithelium; ST, stroma; Magnification, ×70 for conceptus and ×170 for endometrium. A panel representing D12 conceptus on the *top right* (A) is shown at higher magnification (×170).

extracellular signal-regulated kinases 1 and 2 (ERK1/2), was assessed by immunoprecipitation and Western blot analysis. Briefly, monolayer cultures of pTr cells were grown to 75% confluence on 75-cm^2 tissue culture flasks and then incubated in serum-free DMEM/F-12 containing 0.1% BSA for 48 h. Whole cell extracts were prepared as previously described (22). The protein concentration of the lysate supernatant was determined by Bradford assay (Bio-Rad Laboratories, Inc., Burlingame, CA) using BSA as the standard and 1 mg of each extract used for immunoprecipitation. Five micrograms of KGFR antibody (Santa Cruz Biotechnology, Inc., catalogue no. sc-122) or normal rabbit IgG were added to each extract, and bound proteins were purified using protein A/G plus agarose as described previously (22). Immunoprecipitated proteins were separated by SDS-PAGE and analyzed by Western blotting (as described below) with antibody to phosphotyrosine (Santa Cruz Biotechnology, Inc.; catalogue no. sc-7020) diluted 1:100 in 5% BSA-TBST (Tris-buffered saline/0.1% Tween-20).

In similar experiments pTr cells were serum-starved and treated with rKGF as described above. In addition to KGF treatment, some cells were pretreated with 0 or 50 μM of the MAPK/ERK kinase 1 (MEK1) inhibitor, PD98059 (New England Biolab, Beverly, MA; catalogue no. 9900L) for 1 h, then treated with 0 or 10 ng/ml rKGF for 60 min. Twenty micrograms of whole cell extract protein from each sample were separated by SDS-PAGE and transferred to nitrocellulose as described previously (21). Blots were blocked for 1 h at 4 C with either 5% BSA-TBST for phospho-specific antibodies or 5% nonfat milk-TBST for other antibodies, and then incubated with primary antibodies overnight at 4 C. Monoclonal antibodies to phospho-ERK1/2 (pERK1/2; 1:400 dilution in 5% BSA-TBST; catalogue no. sc-7383) and phospho-tyrosine (1:100 dilution in 5% BSA-TBST; catalogue no. sc-7020), and goat polyclonal antibody to ERK1/2 (1:400 dilution in 2% milk-TBST; catalogue no. sc-94-g) were obtained from Santa Cruz Biotechnology, Inc. Blots were then incubated with rabbit antigoat IgG or goat antimouse IgG conjugated to peroxidase (Kirkegaard & Perry Laboratories, Bethesda, MD) for 1 h at room temperature, and immunoreactive proteins were detected using enhanced chemiluminescence (Amersham Pharmacia Biotech, Arlington Heights, NY) according to the manufacturer's recommendations.

Effect of KGF on pTr cell differentiation

To determine whether KGF affects functional cell differentiation, pTr cells were treated with increasing doses of rKGF (0, 1, 10, and 100 ng/ml) in serum-free DMEM/F-12 for 24 h. Expression of uPA was used as a marker for pTr cell differentiation by Northern and slot blot analyses.

Statistical analysis

All quantitative data were subjected to least squares ANOVA using the general linear models procedures of the Statistical Analysis System (SAS Institute, Inc., Cary, NC) (23). Data from dose-response studies on KGF and uPA expression were analyzed by least squares regression analysis. Slot blot data (total counts) were analyzed using the 18S ribosomal RNA data as a covariate to correct for differences in sample loading. Preplanned contrasts (control *vs.* E_2 plus ICI, E_2 *vs.* E_2 plus ICI, control *vs.* catechol estrogens; control *vs.* E_2; P_4 *vs.* E_2 plus P_4) were used to test for effects of treatments in slot blot analyses. All tests of statistical significance were performed using the appropriate error terms according to the expectation of mean squares. Data are presented as least squares means with SEs.

Results

Effects of estrogens on KGF expression

KGF expression was increased (quadratic, $P < 0.05$) by E_2 (Fig. 1A). The increase in KGF expression by E_2 was blocked by addition of the ER antagonist, ICI (Fig. 1B; E_2 *vs.* E_2 plus ICI, $P < 0.05$). Catechol estrogens, 2OH-E_2 and 4OH-E_2, did not affect KGF expression in endometrium (Fig. 1C; control *vs.* 2OH-E_2 and control *vs.* 4OH-E_2, $P > 0.05$).

Effects of P_4 on KGF expression

Increasing doses of P_4 decreased (quadratic, $P = 0.07$) KGF expression in the uterine endometrium (Fig. 2A). However, E_2 increased ($P < 0.01$) KGF expression in both the absence and presence of P_4 (Fig. 2B).

PCNA protein expression

Immunoreactive PCNA protein was detected in nuclei of cells in the embryonic disc and in the trophectoderm (TE) near the embryonic disc of day 9 conceptuses (Fig. 3A). All TE cells of conceptuses recovered between days 10 and 15 of pregnancy expressed PCNA. In contrast, PCNA was not detected in endometrial luminal epithelium (LE) or glandular epithelium (GE) between days 9 and 12 of the estrous cycle or pregnancy. Although PCNA was barely detectable on day 15 of the estrous cycle and pregnancy, it was strongly detectable in LE and TE from days 30 and 60 of pregnancy, and was abundant in GE from day 60 of pregnancy (Fig. 3B).

Characterization of pTr cells

The epithelial phenotype of pTr cells (Fig. 4A) was confirmed based on a cobblestone morphology (a) and by positive cytokeratin (b) and negative vimentin (c) immunostaining compared with the IgG control (d). KGFR, expressed in porcine trophectoderm and endometrial epithelial cells (4), and IFNδ, expressed by porcine trophectoderm (24), were detected in pTr cells by RT-PCR (Fig. 4, B and C).

Effect of KGF on pTr cell proliferation

DNA synthesis by pTr cells, as measured by [^{3}H]thymidine incorporation, increased (quadratic, $P < 0.01$) in response to treatment with increasing doses of rKGF (Fig. 5A).

Phosphorylation of KGFR and MAPK by KGF

Treatment of pTr cells with rKGF resulted in phosphorylation of KGFR and ERK1/2 within 10 min, and phosphorylation of ERK1/2 was maintained up to 4 h (Fig. 5, B and C). Pretreatment of cells with PD98059, a MEK1 inhibitor, blocked the ability of rKGF to induce phosphorylation of ERK1/2 (Fig. 5D).

Effect of KGF on pTr cell differentiation

The specificity of the bovine uPA probe was confirmed by Northern blot analysis of total RNA from pTr cells, and a single transcript of about 2.5 kb was detected (Fig. 6A). Treatment of pTr cells with increasing doses of rKGF increased (quadratic, $P < 0.01$) the expression of uPA (Fig. 6B).

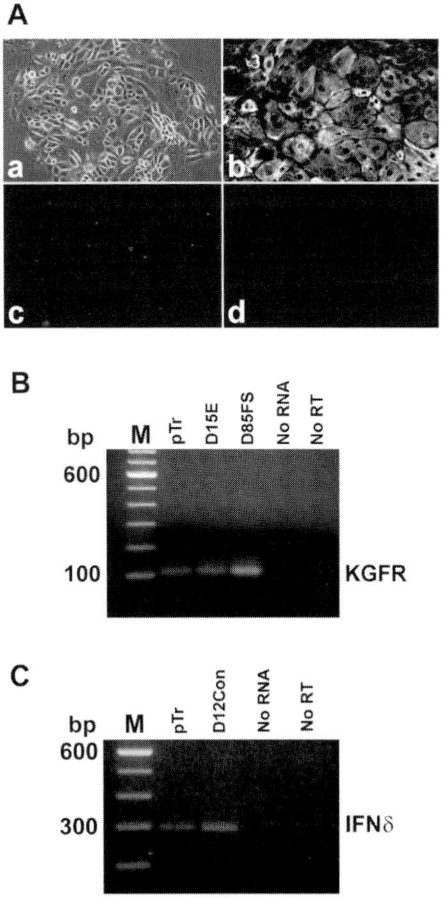

FIG. 4. Characterization of pTr cells. A, Phase contrast microscopy of pTr cells (a). Immunofluorescence analysis of expression of cytokeratin (b), vimentin (c), and the negative control for immunofluorescence (d). RT-PCR analysis of KGFR (B) and IFNδ (C) expression by pTr cells is shown when primers specific for 104-bp KGFR and 296-bp IFNδ were used for PCR amplification. Note that pTr cells express KGFR and IFNδ, which are porcine trophectoderm markers. D15E, Day 15 endometrium; D85FS, day 85 fetal skin; No RNA, no total RNA; No RT, no reverse transcriptase; D12Conc, day 12 conceptus.

Discussion

Synthesis and secretion of estrogens by conceptuses have been reported for many species, including pigs, ruminants, and horses (25). However, the ability to synthesize estrogens is especially significant for pig conceptuses, because these

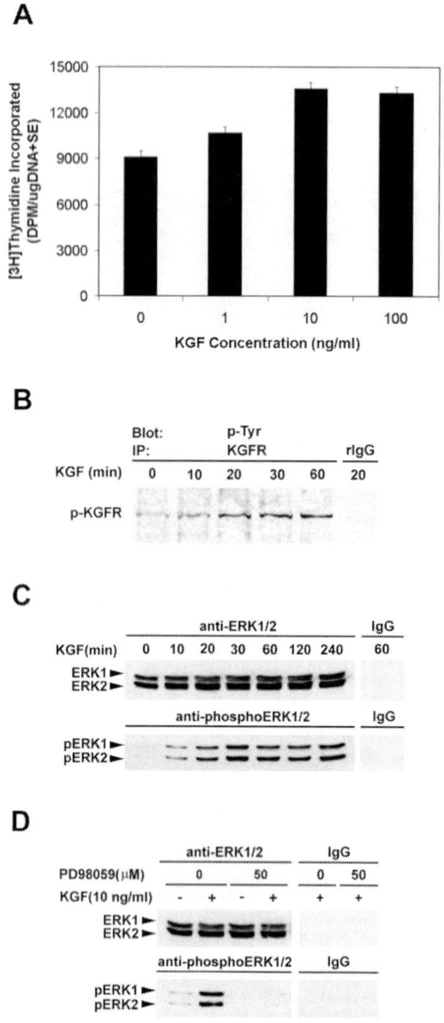

FIG. 5. Effect of KGF on pTr cell proliferation, KGFR activation, and the MAPK signaling pathway. A, Effect of KGF on [^{3}H]thymidine incorporation in pTr cells. rKGF increased [^{3}H]thymidine incorporation in a dose dependent manner in pTr cells (quadratic, $P < 0.01$). B, Activation of KGFR phosphorylation by KGF treatment. pTr cells were treated with rKGF (10 ng/ml) for 0, 10, 20, 30, and 60 min, and 1 mg of each cell lysate was immunoprecipitated with 5 μg KGFR antibody. A duplicate 30 min point was immunoprecipitated with 5 μg normal rabbit IgG. Immunoprecipitated proteins were analyzed by Western blotting using anti-phosphotyrosine. C, Activation of ERK1/2 phosphorylation by KGF treatment between 0 and 20 min. pTr cells were treated with rKGF (10 ng/ml) for 0, 10, 20, 30, 60, 120, or 240 min, and 20 μg of each cell lysate were separated by SDS-PAGE and analyzed for ERK1/2 and phosphorylated ERK1/2 (pERK1/2) by Western blotting. D, Inhibition of KGF effect on ERK1/2 phosphorylation by PD98059, a MEK1 inhibitor, is demonstrated. pTr cells were pre-treated with PD98059 (0 or 50 μM) for 60 min and treated with rKGF (0 or10 ng/ml) for 60 min, and 20 μg of each cell lysate was separated by SDS-PAGE and analyzed for ERK1/2 and pERK1/2 by Western blotting.

estrogens are responsible for maternal recognition of pregnancy as well as stimulation of uterine secretory activity (11, 25, 26). The results of the present study indicate that KGF expression is up-regulated by E_2 and are consistent with recent reports that estrogen increases KGF expression (9, 10). In mouse mammary gland, E_2 treatment *in vivo* increased the expression of KGF mRNA and protein in both peripubertal and mature mice (10).

Catechol estrogens, 2OH-E_2 and 4OH-E_2, are secreted by porcine conceptuses on days 12 and 13 (27), but did not affect KGF expression in porcine endometrium in the present study. This may be due to the low affinity of catechol estrogens for the ER compared with E_2 (28, 29) or perhaps the inherent instability of catechol estrogens. It should be noted that various forms of estrogens are present in the porcine uterine lumen during pregnancy, such as estrone, estradiol, estriol, catechol estrogens, and their sulfated forms (27, 30, 31). Levels of estradiol in the uterine lumen of pregnant pigs (~4.3 nM) are significantly higher than those in nonpregnant pigs (~0.26 nM) between days 9 and 15, and E_2 predominates over all other estrogens in the uterine lumen (30). Therefore, it is likely that E_2 is the predominant estrogen that acts via ERα to increase endometrial KGF expression in the pregnant pig uterus. It remains to be determined whether ERβ is expressed in the porcine uterine endometrium.

E_2 probably up-regulates KGF mRNA expression in endometrial epithelial cells through direct interaction with ER present in these cells. ERα protein is localized to LE and GE cells of the porcine endometrium (32), and up-regulation of KGF expression by E_2 is inhibited by the ER antagonist, ICI 182,780. Indeed, the promoter region of the human KGF gene contains an estrogen response element (33). ERα expression in endometrial epithelial cells during early pregnancy is unique in the pig uterus compared with other species, such as sheep and primates, in which the endometrial epithelial cells do not express ERα during the P_4-dominant period (16, 34). In pigs, ERα staining in LE and GE is readily detectable from days 5–12 of the estrous cycle and pregnancy, then decreases, but remains detectable until day 15 of the estrous cycle and pregnancy. ERα is absent in stromal cells between days 5 and 15 of the estrous cycle and pregnancy (32).

P_4 is the major hormone responsible for the establishment and maintenance of pregnancy. During diestrus and early pregnancy, production of P_4 by the corpus luteum begins to increase (12) and is associated with significant increases in uterine secretory activity (13). P_4 treatment in the primate uterus increases KGF expression (8). However, the results of the present study indicated that P_4 decreased KGF expression in porcine endometrium. The mechanism of the P_4-

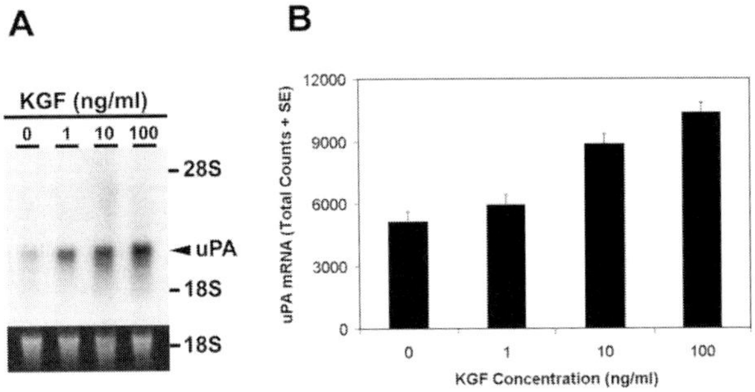

FIG. 6. Effect of KGF on expression of uPA, a marker for pTr cell differentiation. Northern (A) and slot blot (B) analyses of uPA expression. A single transcript of uPA (~2.5 kb) was detected (A), and KGF increased uPA expression (B) in a dose-dependent manner in pTr cells (quadratic, $P < 0.01$).

mediated decrease in KGF expression is not known. Interestingly, E_2 increased KGF expression in the presence of P_4, a situation to which the endometrial epithelium would be exposed *in vivo* during pregnancy. As P_4 down-regulates endometrial ERα in ovine uterus (16, 35, 36), the decreased expression of KGF in endometrial explants treated with P_4 alone and E_2 plus P_4 may be the result of a P_4-mediated decrease in ER and an attenuated response to E_2. We cannot rule out the effects of any residual levels of estradiol on decreased KGF expression by P_4 in this system. It is also possible that P_4 inhibits KGF expression through a direct mechanism or indirectly by modulation of other unknown factors which then down-regulate KGF expression.

KGF affects epithelial cell proliferation in various tissues (37). In the present study KGF increased DNA synthesis of pTr cells, which are of epithelial cell origin and express KGFR, and IFNδ, a porcine trophectoderm cell-specific marker (24). Porcine conceptuses undergo dramatic morphological changes during early pregnancy (13, 14). Between days 10 and 12 of pregnancy there is a rapid transition from spherical (9–10 mm in diameter) to tubular (10–50 mm in length) and elongated filamentous forms (>100 mm long) (14). In the present study trophectoderm cell proliferation was detected between days 9 and 15 of conceptus development as reported previously (14). Thus, *in vitro* results of the present study suggest that KGF of endometrial epithelial origin may increase the proliferation of conceptus trophectoderm during the periimplantation phase of development. This hypothesis is supported by results from PCNA staining of conceptuses *in vivo* between days 9–15 of pregnancy, indicating high intensity PCNA staining from day 10 of pregnancy and thereafter. The PCNA staining in the conceptuses is coordinate with increasing KGF expression in the endometrium between days 10 and 15 of pregnancy (4). The lack of detection of PCNA in the LE and GE during early pregnancy suggests that epithelial cells do not proliferate to any great degree at this time. Therefore, although KGF is present in the uterine lumen, and KGFR is expressed in LE and GE, it is unlikely that KGF affects epithelial cell proliferation in the endometrium during early pregnancy. However, it is possible that KGF affects differentiation of endometrial epithelial cells.

In addition to effecting proliferation, KGF was found to alter conceptus trophectoderm cell differentiation in this study. In various cell types, uPA, aromatase, surfactant protein A and D, syndecan-1, and Na^+/K^+-adenosine triphosphatase are increased by KGF (1, 37, 38). In particular, KGF increases uPA expression and activity in human uterine exocervical epithelial cells (39) and keratinocytes (40). In this study KGF increased uPA expression in pTr cells. Differentiating pig conceptuses produce uPA from trophoblast (41), and pig conceptus trophectoderm produces uPA in a biphasic manner between days 10 and 12 and between days 14 and 16 of early pregnancy (42), coinciding with estrogen production by conceptuses. Therefore, increased expression of uPA by KGF in pTr cells suggests that KGF expression increases within endometrial LE in response to estrogen, is secreted into the uterine lumen, and may be an important regulator of uPA secretion by conceptus trophectoderm.

Like most receptor tyrosine kinase-activating growth factors, KGFR signals through the MAPK pathway (15). In the present study phosphorylation of KGFR and ERK1/2, members of the MAPK family, was detected in KGF-treated pTr cells, suggesting that effects on trophectoderm proliferation and differentiation were mediated by interaction of KGF with the KGFR. Among the several FGFR isoforms, KGF recognizes only KGFR with high affinity and biological activity (43), precluding the possibility that ERK1/2 is activated by other members of the FGFR family. The precise mechanisms of intracellular KGF signaling for proliferation and differentiation in pTr cells remain to be determined.

In summary, the results of the present study indicate that E_2, a pregnancy recognition signal in pigs, increases KGF expression in the uterine endometrium, and that KGF increases the proliferation of conceptus trophectoderm and stimulates the expression of uPA, a marker for differentiation. Thus, KGF of endometrial origin affects both the growth and differentiation of trophectoderm during this crucial phase of conceptus development in pigs.

Acknowledgments

The authors thank Dr. Robert C. Burghardt for assistance with immunofluorescence image capture and use of microscopy and imaging facilities in the Image Analysis Laboratory of College of Veterinary Medicine. We also thank Dr. Wallace L. McKeehan for providing recombinant rat KGF.

References

1. **Rubin JS, Bottaro DP, Chedid M, Miki T, Ron D, Cheon HG, Taylor WG, Fortney E, Sakata H, Finch PW, LaRochelle WJ** 1995 Keratinocyte growth factor. Cell Biol Int 19:399–411
2. **Cooke PS, Buchanan DL, Kurita T, Lubahn DB, Cunha GR** 1998 Stromal-epithelial cell communication in the female reproductive tract. In: Bazer FW (ed) The Endocrinology of Pregnancy, chapt 17. Humana Press, Totowa, pp 491–506
3. **Chen C, Spencer TE, Bazer FW** 2000 Fibroblast growth factor-10: a stromal mediator of epithelial function in the ovine uterus. Biol Reprod 63:959–966
4. **Ka H, Spencer TE, Johnson GA, Bazer FW** 2000 Keratinocyte growth factor: expression by endometrial epithelia of the porcine uterus. Biol Reprod 62:1772–1778
5. **Fasciana C, van der Made AC, Faber PW, Trapman J** 1996 Androgen regulation of the rat keratinocyte growth factor (KGF/FGF7) promoter. Biochem Biophys Res Commun 220:858–863
6. **Yan G, Fukabori Y, Nikolaropoulos S, Wang F, McKeehan WL** 1992 Heparin-binding keratinocyte growth factor is a candidate stromal to epithelial cell andromedin. Mol Endocrinol 6:2123–2128
7. **Alarid ET, Rubin JS, Young P, Chedid M, Ron D, Aaronson SA, Cunha GR** 1994 Keratinocyte growth factor functions in epithelial induction during seminal vesicle development. Proc Natl Acad Sci USA 91:1074–1078
8. **Koji T, Chedid M, Rubin JS, Slayden OD, Csaky KG, Aaronson SA, Brenner RM** 1994 Progesterone-dependent expression of keratinocyte growth factor mRNA in stromal cells of the primate endometrium: keratinocyte growth factor as a progestomedin. J Cell Biol 125:393–401
9. **Hom YK, Young P, Thomson AA, Cunha GR** 1998 Keratinocyte growth factor injected into female mouse neonates stimulates uterine and vaginal epithelial growth. Endocrinology 139:3772–3779
10. **Pedchenko VK, Imagawa W** 2000 Estrogen treatment in vivo increases keratinocyte growth factor expression in the mammary gland. J Endocrinol 165:39–49
11. **Bazer FW, Ott TL, Spencer TE** 1998 Endocrinology of the transition from recurring estrous cycles to establishment of pregnancy in subprimate mammals. In: Bazer FW (ed) The Endocrinology of Pregnancy, chapt 1. Humana Press, Totowa, pp 1–34
12. **Guthrie HD, Henricks DM, Handlin DL** 1972 Plasma estrogen, progesterone, and luteinizing hormone prior to estrus and during pregnancy in pigs. Endocrinology 91:675–679
13. **Roberts RM, Bazer FW** 1988 The functions of uterine secretions. J Reprod Fertil 82:875–892
14. **Geisert RD, Brookbank JW, Roberts RM, Bazer FW** 1982 Establishment of pregnancy in the pig. II. Cellular remodeling of the porcine blastocyst during elongation on day 12 of pregnancy. Biol Reprod 27:941–955
15. **Szebenyi G, Fallon JF** 1999 Fibroblast growth factors as multifunctional signaling factors. Int Rev Cytol 185:45–106
16. **Spencer TE, Bazer FW** 1995 Temporal and spatial alterations in uterine estrogen receptor and progesterone receptor gene expression during the estrous cycle and early pregnancy in the ewe. Biol Reprod 53:1527–1543
17. **Johnson GA, Spencer TE, Hansen TR, Austin KJ, Burghardt RC, Bazer FW** 1999 Expression of the interferon tau inducible ubiquitin cross-reactive protein in the ovine uterus. Biol Reprod 61:312–318
18. **Corps AN, Brigstock DR, Littlewood CJ, Brown KD** 1990 Receptors for epidermal growth factor and insulin-like growth factor I on preimplantation trophectoderm of the pig. Development 110:221–227
19. **Gupta A, Dekaney CM, Bazer FW, Madrigal MM, Jaeger LA** 1998 β Transforming growth factors (TGFβ) at the porcine conceptus-maternal interface. II. Uterine TGFβ bioactivity and expression of immunoreactive TGFβs (TGFβ1, TGFβ2, and TGFβ3) and their receptors (type I and type II). Biol Reprod 59:911–917
20. **Taylor KM, Gray CA, Joyce MM, Stewart MD, Bazer FW, Spencer TE** 2000 Neonatal ovine uterine development involves alterations in expression of receptors for estrogen, progesterone, and prolactin. Biol Reprod 63:1192–1204
21. **Johnson GA, Burghardt RC, Newton GR, Bazer FW, Spencer TE** 1999 Development and characterization of immortalized ovine endometrial cell lines. Biol Reprod 61:1324–1330
22. **Stewart MD, Johnson GA, Vyhlidal CA, Burghardt RC, Safe SH, Yu-Lee LY, Bazer FW, Spencer TE** 2000 Interferon tau activates multiple STAT proteins and has complex effects on interferon-responsive gene transcription in endometrial epithelial cells. Endocrinology 142:98–107
23. **SAS Institute** 1990 SAS User's Guide: Statistics, version 6. SAS Institute, Cary
24. **Lefevre F, Guillomot M, D'Andrea S, Battegay S, La Bonnardiere C** 1998 Interferon-δ: the first member of a novel type I interferon family. Biochimie 80:779–788
25. **Heap RB, Flint AFP, Staple LD** 1983 Endocrinology of trophoblast in farm animals. In: Loke YW, Whyte A (eds) Biology of Trophoblast. Elsevier, New York, pp 353–400
26. **Geisert RD, Zavy MT, Wettemann RP, Biggers BG** 1987 Length of pseudopregnancy and pattern of uterine protein release as influenced by time and duration of oestrogen administration in the pig. J Reprod Fertil 79:163–172
27. **Mondschein JS, Hersey RM, Dey SK, Davis DL, Weisz J** 1985 Catechol estrogen formation by pig blastocysts during the preimplantation period: biochemical characterization of estrogen-2/4-hydroxylase and correlation with aromatase activity. Endocrinology 117:2339–2346
28. **Martucci C, Fishman J** 1976 Uterine estrogen receptor binding of catechol estrogens and of estetrol (1,3,5(10)-estratriene-3,15α,16[alpha,17beta]-tetrol). Steroids 27:325–333
29. **Kuiper GG, Carlsson B, Grandien K, Enmark E, Haggblad J, Nilsson S, Gustafsson JA** 1997 Comparison of the ligand binding specificity and transcript tissue distribution of estrogen receptors α and β. Endocrinology 138:863–870
30. **Stone BA, Seamark RF** 1985 Steroid hormones in uterine washings and in plasma of gilts between days 9 and 15 after oestrus and between days 9 and 15 after coitus. J Reprod Fertil 75:209–221
31. **Geisert RD, Renegar RH, Thatcher WW, Roberts RM, Bazer FW** 1982 Establishment of pregnancy in the pig. I. Interrelationships between preimplantation development of the pig blastocyst and uterine endometrial secretions. Biol Reprod 27:925–939
32. **Geisert RD, Brenner RM, Moffatt RJ, Harney JP, Yellin T, Bazer FW** 1993 Changes in oestrogen receptor protein, mRNA expression and localization in the endometrium of cyclic and pregnant gilts. Reprod Fertil Dev 5:247–260
33. **Finch PW, Lengel C, Chedid M** 1995 Cloning and characterization of the promoter region of the human keratinocyte growth factor gene. J Biol Chem 270:11230–11237
34. **Brenner RM, West NB, McClellan MC** 1990 Estrogen and progestin receptors in the reproductive tract of male and female primates. Biol Reprod 42:11–19
35. **Spencer TE, Mirando MA, Watson GH, Ott TL, Bazer FW** 1996 Effects of interferon-τ and progesterone on estrogen-stimulated expression of receptors for estrogen, progesterone and oxytocin in the endometrium of ovariectomized ewes. Reprod Fertil Dev 8:843–853
36. **Spencer TE, Gray A, Johnson GA, Taylor KM, Gertler A, Gootwine E, Ott TL, Bazer FW** 1999 Effects of recombinant ovine interferon tau, placental lactogen, and growth hormone on the ovine uterus. Biol Reprod 61:1409–1418
37. **Werner S** 1998 Keratinocyte growth factor: a unique player in epithelial repair processes. Cytokine Growth Factor Rev 9:153–165
38. **Maatta A, Jaakkola P, Jalkanen M** 1999 Extracellular matrix-dependent activation of syndecan-1 expression in keratinocyte growth factor-treated keratinocytes. J Biol Chem 274:9891–9898
39. **Zheng J, Siren V, Vaheri A** 1996 Keratinocyte growth factor enhances urokinase-type plasminogen activator activity in HPV16 DNA-immortalized human uterine exocervical epithelial cells. Eur J Cell Biol 69:128–134
40. **Tsuboi R, Sato C, Kurita Y, Ron D, Rubin JS, Ogawa H** 1993 Keratinocyte growth factor (FGF-7) stimulates migration and plasminogen activator activity of normal human keratinocytes. J Invest Dermatol 101:49–53
41. **Strickland S, Reich E, Sherman MI** 1976 Plasminogen activator in early embryogenesis: enzyme production by trophoblast and parietal endoderm. Cell 9:231–240
42. **Fazleabas AT, Geisert RD, Bazer FW, Roberts RM** 1983 Relationship between release of plasminogen activator and estrogen by blastocysts and secretion of plasmin inhibitor by uterine endometrium in the pregnant pig. Biol Reprod 29:225–238
43. **Ornitz DM, Xu J, Colvin JS, McEwen DG, MacArthur CA, Coulier F, Gao G, Goldfarb M** 1996 Receptor specificity of the fibroblast growth factor family. J Biol Chem 271:15292–15297

Reproduction (2002) **124**, 289–300

Research

Evidence that absence of endometrial gland secretions in uterine gland knockout ewes compromises conceptus survival and elongation

C. A. Gray[1], R. C. Burghardt[2], G. A. Johnson[1]*, F. W. Bazer[1] and T. E. Spencer[1†]

[1]*Center for Animal Biotechnology and Genomics and Department of Animal Science, Texas A&M University, College Station, Texas 77843-2471, USA; and* [2]*Department of Veterinary Anatomy and Public Health, Texas A&M University, College Station, Texas 77843-4458, USA*

Endometrial glands are necessary for conceptus implantation and growth. In the ovine uterine gland knockout (UGKO) model, blastocysts hatch normally but fail to survive or elongate. This peri-implantation defect in UGKO ewes may be due to the absence of endometrial glands or, alternatively, to the lack of certain epithelial adhesion molecules or the inability of the endometrium to respond to signals from the conceptus. Two studies were performed to examine these hypotheses. In study one, normal ($n = 8$) and UGKO ($n = 12$) ewes were mated at oestrus (day 0) with intact rams and their uteri were flushed 14 days after oestrus. Normal ewes ($n = 4$) were also flushed on 14 days after oestrus. Uterine flushes from bred normal ewes contained filamentous conceptuses ($n = 7$ of 8), whereas those from UGKO ewes contained no conceptus ($n = 5$ of 12), a growth-retarded, tubular conceptus ($n = 6$ of 12), or a fragmented, filamentous conceptus ($n = 1$ of 12). In all groups, expression of mucin 1 and integrin α_v, α_5, β_3 and β_5 was localized at the apical surface of the endometrial luminal epithelium with no detectable differences between normal and UGKO ewes. Uterine flushes from pregnant ewes, but not cyclic or UGKO ewes, contained abundant immunoreactive interferon τ and the cell adhesion proteins, osteopontin and glycosylation-dependent cell adhesion molecule one. In study two, UGKO ewes were fitted with uterine catheters 5 days after oestrus, infused with recombinant ovine interferon τ or control proteins from 11 to 15 days after oestrus, and underwent hysterectomy 16 days after oestrus. Expression of several interferon τ-stimulated genes (*ISG17*, *STAT1*, *STAT2* and *IRF-1*) was increased in the endometrium from interferon τ-infused UGKO ewes. These results support the hypothesis that the defects in conceptus elongation and survival in UGKO ewes are due to the absence of endometrial glands and their secretions rather than to alterations in expression of anti-adhesive or adhesive molecules on the endometrial luminal epithelium or to the responsiveness of the endometrium to the conceptus pregnancy recognition signal.

Introduction

Uterine glands are present in the endometrium of all mammalian uteri and develop after birth in many species (for reviews, see Bartol *et al.*, 1999; Gray *et al.*, 2001a). Endometrial glands synthesize, secrete or transport substances that are necessary for conceptus (that is, the embryo–fetus and placental membranes) survival and growth (for reviews, see Bazer, 1975; Roberts and Bazer 1988; Simmen and Simmen, 1990; Fazleabas *et al.*, 1994; Kane *et al.*, 1997; Carson *et al.*, 2000). A role for the endometrial glandular epithelium (GE) has been demonstrated in rodents, in which the absence of glandular secretions, such as leukaemia inhibitory factor and calcitonin, compromised embryo survival by disrupting establishment of uterine receptivity and embryo development (Stewart *et al.*, 1992; Zhu *et al.*, 1998). Similarly, peri-implantation pregnancy loss was observed in the uterine gland knockout (UGKO) ewe model (Gray *et al.*, 2000a, 2001b,c).

The UGKO ewe is a novel model in which endometrial gland morphogenesis has been epigenetically ablated by inappropriate exposure of neonatal ewes to a 19-norprogestin from birth to 8 weeks of age (Spencer *et al.*, 1999a; Gray *et al.*, 2000a). The progestin specifically ablates development of the glands within the endometrium without altering development of the uterine myometrium or

*Present address: Department of Animal and Veterinary Science, Agricultural Biotechnology Building, University of Idaho, Moscow, Idaho 83844-2330, USA
†Correspondence
Email: tspencer@tamu.edu

1470-1626/2002

other Müllerian duct-derived female reproductive tract structures (Gray *et al.*, 2000b, 2001b). Despite repeated matings with fertile rams, adult UGKO ewes do not establish pregnancy (Gray *et al.*, 2000a, 2001b,c). Transfer of hatched blastocysts recovered from superovulated normal donor ewes into the uteri of UGKO ewes failed to ameliorate the pregnancy defect (Gray *et al.*, 2001c). Normal hatched blastocysts were found in the uterine flushings of mated UGKO ewes 6 and 9 days, but not 14 days, after mating (Gray *et al.*, 2001b,c). Uterine flushings taken 14 days after mating from mated UGKO ewes contained either no conceptus or a severely growth-retarded conceptus that had failed to elongate from a tubular to filamentous form (Gray *et al.*, 2001c). Therefore, UGKO ewes exhibit a peri-implantation pregnancy defect, the timing of which correlates with most of the embryo loss that occurs during pregnancy in livestock and humans (Bazer, 1975; Kane *et al.*, 1997).

Implantation in ruminants is a highly co-ordinated process that involves apposition, attachment, and adhesion of the conceptus trophectoderm to the endometrial luminal epithelium (LE) (Guillomot *et al.*, 1981; Guillomot, 1995). In sheep, the peri-implantation period is marked by rapid elongation of the conceptus from a tubular to filamentous form on days 12–13 of gestation and the production of large amounts of interferon τ (IFN-τ), a type I interferon that is the signal for maternal recognition of pregnancy (Bazer *et al.*, 1997). Elongation of the conceptus is critical for developmentally regulated production of IFN-τ (Farin *et al.*, 1989) and requires the uterus, as hatched blastocyts fail to elongate *in vitro* unless transferred to the uterus (Heyman *et al.*, 1984). Apposition of conceptus trophectoderm and endometrial LE is initiated on day 14, adhesion occurs on day 15, and firm attachment on days 16–18 of gestation (Guillomot *et al.*, 1981). Adhesion of conceptus trophectoderm to the LE is temporally regulated by non-adhesive and adhesive factors on the apical surface of the endometrial LE (Burghardt *et al.*, 1997, in press; Johnson *et al.*, 2001a). It is hypothesized that, initially, non-adhesive factors, such as mucin 1 (Muc-1), sterically impair interactions between adhesive glycoproteins expressed on the apical surfaces of conceptus trophectoderm and LE by means of their extensive glycosylation and extended extracellular structure (Carson *et al.*, 2000; Johnson *et al.*, 2001a). In sheep, immunoreactive Muc-1 expression on the LE decreases progressively between day 9 and day 17 of early pregnancy, presumably to unmask adhesive glycoproteins on the LE for interaction with the trophectoderm (Johnson *et al.*, 2001a). Integrins are thought to be the dominant glycoproteins that regulate trophectoderm adhesion. During the peri-implantation period in ewes, integrin subunits α_v, α_4, α_5, β_1, β_3 and β_5 were constitutively expressed on the conceptus trophectoderm as well as the apical surface of the endometrial LE (Johnson *et al.*, 2001a). During the peri-implantation period, in addition to constitutive expression of integrins, two molecules involved in cell adhesion, osteopontin (Johnson *et al.*, 1999a,b, 2001a) and glycosylation-dependent cell adhesion molecule 1 (GlyCAM-1) (Spencer *et al.*, 1999b), are secreted by the endometrial GE and are thought to bind to trophectoderm and LE to stimulate elongation, adhesion and attachment of the ovine conceptus (Johnson *et al.*, 2001a).

The available evidence supports the hypothesis that the failure of peri-implantation conceptus survival and elongation in UGKO ewes is due to an absence of endometrial glands and, by default, their secretory products. Alternatively, the pregnancy defect in UGKO ewes could be attributed to deficient expression of adhesion molecules on the LE or inability of the endometrium to respond to conceptus signals such as IFN-τ. Two studies were conducted to test these hypotheses by determining differences in normal and UGKO ewes on: (1) distribution of Muc-1 and integrin subunit expression in the uterine LE 14 days after mating; (2) abundance of osteopontin and GlyCAM-1 in the uterine lumen 14 days after mating; and (3) endometrial expression of IFN-τ-stimulated genes in response to intrauterine administration of recombinant ovine IFN-τ.

Materials and Methods

Animals

Experimental and surgical procedures complied with the Guide for Care and Use of Agriculture Animals and were approved by the Institutional Agricultural Animal Care and Use Committee of the Texas A&M University System Agricultural Experiment Station (Animal Use Protocol 7-286).

UGKO ewes were produced as described by Spencer *et al.* (1999a) and Gray *et al.* (2000a) by implanting crossbred Rambouillet ewe lambs with a single Synchromate B® (Sanofi, Overland Park, KS) implant within 12 h of birth and every 2 weeks thereafter for a total of 8 weeks. Implants were inserted s.c. into the periscapular area and released approximately 6 mg norgestomet (17α-acetoxy-11β-methyl-19-norpreg-4-ene-3,20-dione), a potent synthetic 19-norprogestin, over 14 days (Bartol *et al.*, 1988). Normal control ewes did not receive implants.

Study one

Adult UGKO ($n = 18$) and normal ewes ($n = 12$) were given two i.m. injections (at 07:00 and 17:00 h) of 10 mg prostaglandin $F_{2\alpha}$ (Lutalyse, Upjohn, Kalamazoo, MI) 9 days apart to synchronize oestrus. Ewes were monitored each day for oestrous behaviour using vasectomized rams. All UGKO and some normal control ewes ($n = 8$) were mated at oestrus (day 0) and at 12 and 24 h after oestrus with intact rams of proven fertility. The remaining normal control ewes ($n = 4$) were assigned to cyclic status, and oestrus was determined using vasectomized rams.

Fourteen days after oestrus or mating, all ewes were subjected to mid-ventral laparotomy, and their uterine lumina were flushed with 20 ml sterile saline. Uterine flushes were analysed under a dissecting microscope to recover conceptuses, if any, and determine their morphology. Conceptuses were fixed in 4% paraformaldehyde in PBS (pH 7.2). Uterine flushes were clarified by centrifugation (2000 ***g*** for 30 min at 4°C), and aliquots were snap-frozen in liquid nitrogen and stored at –80°C. Several sections (1–1.5 cm) from the middle of each uterine horn were snap–frozen in Tissue-Tek OCT compound (Miles, Oneonta, NY). Several sections from the middle region of each uterine horn were also fixed in 4% paraformaldehyde in PBS (pH 7.2). After 24 h, fixed tissues were changed to 70% ethanol for 24 h and then dehydrated and embedded in Paraplast-Plus (Oxford Labware, St Louis, MO). The endometrium was physically dissected from myometrium from the remainder of each uterine horn ipsilateral to the ovary bearing the corpus luteum and then snap–frozen in liquid nitrogen and stored at –80°C for RNA extraction.

Study two

Fourteen days after mating, another group of UGKO ewes (*n* = 8) was subjected to mid-ventral laparotomy, their uterine lumina flushed with 20 ml sterile saline, and their uterine horns fitted with catheters as described by Spencer *et al.* (1995). Ewes were given two i.m. injections (07:00 and 17:00 h) of 10 mg prostaglandin $F_{2\alpha}$ (Lutalyse, Upjohn, Kalamazoo, MI) to synchronize oestrus, and were monitored twice a day for oestrous behaviour using a vasectomized ram, and then assigned randomly to receive daily intra-uterine injections (at 07:00 and 19:00 h each day) of control proteins (6 mg serum proteins per day) or recombinant ovine IFN-τ (roIFN-τ; 2×10^7 antiviral units per day) from day 11 to day 15 after mating. Recombinant oIFN-τ was produced from a synthetic gene construct in *Pichia pastoris* and purified as described by Van Heeke *et al.* (1996). Preparation of control proteins and roIFN-τ was performed as described by Spencer *et al.* (1995). Blood samples were collected on days 11–15 after mating, via jugular venepuncture into Vacutainer evacuated blood collection tubes with sodium heparin (Becton-Dickinson, Franklin Lakes, NJ), and plasma was stored at –20°C. All ewes were hysterectomized 16 days after mating. At hysterectomy, portions (approximately 1.0 cm) from the middle region of each uterine horn were fixed in fresh 4% paraformaldehyde in PBS for 24 h and embedded in Paraplast-Plus (Oxford Labware). From the remainder of each uterine horn, endometrium was dissected from myometrium, frozen separately in liquid nitrogen, and stored at –80°C.

Histological analyses

Conceptus tissues were sectioned (5 μm) and stained with haematoxylin and eosin as described by Gray *et al.* (2000a). Uteri were sectioned (5 μm) and stained with Masson's trichrome stain. This procedure stains nuclei black, cytoplasm and muscle fibres red, and extracellular matrix (ECM) components blue. For this stain, uterine sections were deparaffinized in CitraSolv (Fisher Scientific; Fairlawn, NJ) and rehydrated through a graded alcohol series to distilled water. Tissues were then incubated for 1 h at 55°C in Bouin's solution (71% (v/v) picric acid, 24% formaldehyde (40%), and 5% (v/v) glacial acetic acid) and rinsed in water. Slides were incubated sequentially at room temperature for 5 min each in Weigert's iron haematoxylin (50% (v/v) ethanol (95%), 4% (v/v) ferric chloride (29% aqueous), 1% (v/v) hydrochloric acid, and 1% (w/v) haematoxylin), biebrich scarlet-acid fuchsin solution (90% (v/v) biebrich scarlet (1% aqueous), 9% (v/v) acid fuchsin (1% aqueous), and 1% (v/v) glacial acetic acid), phosphomolybdic–phosphotungstic acid solution (2.5% (v/v) phosphomolybdic acid, 2.5% (w/v) phosphotungstic acid), aniline blue solution (2.5% (w/v) aniline blue, 2% (v/v) glacial acetic acid), and then in 1% glacial acetic acid (v/v) for 5 min. Slides were then dehydrated through alcohol to xylene, and coverslips fixed with Permount (Fisher Scientific, Fair Lawn, NJ). Photomicrographs of stained tissues were captured using a Zeiss Axioplan2 photomicroscope (New York, NY) fitted with a Hamamatsu chilled 3CCD colour camera (Hamamatsu, Hamamatsu City).

Western blot analyses

Uterine flushes (2 ml) from day 14 cyclic, pregnant and UGKO ewes were concentrated using Centricon-3 columns (Amicon, Beverly, MA). Protein content was determined using a Bradford protein assay (Bio-Rad, Hercules, CA) with bovine serum albumin (BSA) as the standard. Uterine flush proteins (30 μg) were denatured, separated by SDS-PAGE using 12% acrylamide gels, and transferred to nitrocellulose membranes as described by Spencer *et al.* (1999b). Membranes were blocked for 1 h at room temperature with 5% (w/v) milk–TBST (20 mmol Tris l^{-1} (pH 7.5), 137 mmol NaCl l^{-1}, 0.05% (v/v) Tween 20) and then incubated with mouse anti-ovine IFN-τ (HL129; 4 μg ml^{-1}) (Swann *et al.*, 1999), rabbit anti-human osteopontin (LF-123 and LF-124; 1 : 2500 each) (Johnson *et al.*, 1999a), or rabbit anti-rat GlyCAM-1 (CAM02; 1 μg ml^{-1}; kindly provided by S. D. Rosen (University of California) (Singer and Rosen, 1996) in 5% milk-TBST as described by Spencer *et al.* (1999b). Negative control blots were performed in which primary antibody was replaced by mouse IgG (IFN-τ), rabbit IgG (GlyCAM-1) or normal rabbit serum (osteopontin) at the same concentration used for the respective primary antibodies. After an overnight incubation at 4°C, membranes were washed for 30 min with TBST and then incubated with either goat anti-mouse or goat anti-rabbit IgG–horseradish peroxidase-conjugated secondary antibody (KPL, Bethesda, MD) for 1 h at room temperature. Membranes were again washed with TBST for 30 min before detection by chemiluminescence using a Super Signal West Pico kit (Pierce; Rockford, IL) and Kodak X-OMAT AR film.

Immunocytochemical analyses

Antibodies used for immunocytochemistry included rabbit anti-Muc-1 (generously provided by D. Carson, University of Delaware); rabbit anti-α_v (AB1930), α_5 (AB1928), β_3 (AB1932), and β_5 (AB1926) from Chemicon (Temecula, CA); normal rabbit IgG (15006) from Sigma (St Louis, MO); and fluorescein-conjugated goat anti-rabbit IgG (65-611) from Zymed (San Francisco, CA).

Proteins were localized in frozen uterine tissue sections (8–10 μm) by immunofluorescence staining as described by Johnson *et al.* (2001a). Briefly, frozen uterine tissues were sectioned (8 μm) with a Hacker-Bright OTF cryostat (Hacker Instruments, Fairfield, NJ) and mounted on Superfrost/Plus microscope slides (Fisher Scientific, Pittsburgh, PA). Frozen sections were fixed in –20°C methanol for 10 min, permeabilized with 0.3% (v/v) Tween-20 in 0.02 M PBS, and then blocked in antibody dilution buffer (two parts 0.02 mol PBS l^{-1}, 1.0% BSA, 0.3% Tween-20 (pH 8.0) and one part glycerol) containing 5% normal goat serum for 1 h at room temperature. Immunoreactive protein was then detected using a fluorescein-conjugated secondary antibody for 1 h at room temperature. Slides were overlaid with Prolong antifade mounting reagent and then coverslips were added (Molecular Probes, Eugene, OR). Representative fluorescence images of cross-sections for each antibody and for each ewe were recorded digitally.

Progesterone radioimmunoassay

Concentrations of progesterone in plasma samples were determined using an Active Progesterone Radioimmunoassay kit (Diagnostic Systems Laboratories, Webster, TX) as described by Gray *et al.* (2000a). Assay sensitivity was 0.1 ng ml^{-1}, and the intra-assay coefficient of variation was 5%.

RNA isolation and analyses

Total cellular RNA was isolated from frozen endometrium using the Trizol reagent (Gibco-BRL, Grand Island, NY) and analysed for concentration and quality. Denatured total cellular RNA (20 μg) for each ewe was analysed by slot blot hybridization using radiolabelled antisense cRNA probes generated by transcription *in vitro* with [α-^{32}P] UTP (Amersham Pharmacia Biotech, Piscataway, NJ) as described by Spencer *et al.* (1995). Plasmid templates containing cDNAs for ovine IFN-stimulated gene 17 (ISG17; kindly provided by T. Hansen, University of Wyoming), human signal transducer and activator of transcription (STAT) 1, human STAT2, ovine IFN regulatory factor one (IRF-1), and 18S rRNA (pT718S; Ambion, Austin, TX) were used to produce radiolabelled cRNA probes as described by Choi *et al.* (2001). Hybridization signals were detected by exposing washed slot blots to a phosphoimager screen and visualized using a Typhoon 8600 Variable Mode Imager (Molecular Dynamics, Sunnyvale, CA).

Statistical analyses

All quantitative data were subjected to least-squares analysis of variance (LS-ANOVA) using General Linear Models procedures of the Statistical Analysis System (SAS, 1990). Analyses of steady-state amounts of endometrial mRNA determined by slot blot hybridization included the 18S rRNA as a covariate in LS-ANOVA to correct for differences in sample loading. In all analyses, error terms used in tests of significance were identified according to the expectation of the mean squares for error (Steele and Torrie, 1980). Data are presented as least-squares means with overall SE.

Results

Study one: conceptus morphology in uterine flushes

One to three filamentous conceptuses were present in the uterine flushes of most normal mated control ewes ($n = 7$ of 8) mated with an intact fertile ram. The uterine flushes of UGKO ewes contained either no conceptus ($n = 5$ of 12) or degenerating conceptuses ($n = 7$ of 12). The conceptus morphology in UGKO uterine flushes was classified as fragmented filamentous ($n = 1$) or degenerating tubular ($n = 6$).

Study one: histoarchitecture of the uterus

Uteri of all control ewes exhibited extensive, normal gland development in the intercaruncular endometrium (Fig. 1a–d). As expected, none of the UGKO ewes displayed uterine gland development characteristic of controls (Fig. 1e,f). In UGKO ewes, uterine glands were most often either absent or sporadically distributed at very low density. In uteri from normal day 14 cyclic and pregnant ewes, caruncular areas of the endometria consisted of densely packed stromal cells and showed decreased staining for ECM components as compared with the intercaruncular areas of the endometrium. The stroma of the intercaruncular endometrium in ruminants can be divided into the stratum compactum beneath the LE, consisting of densely packed stroma, and the stratum spongiosum nearer the myometrium, consisting of less dense stroma. The stratum spongiosum of the intercaruncular endometrium of normal ewes contained many endometrial glands and showed pronounced ECM compared with the stroma in the stratum compactum zone and caruncular area. In comparison, endometrium of uteri from UGKO ewes was most similar to that of the caruncular area of the uteri of control ewes, because it lacked characteristic intercaruncular endometrium. Furthermore, the lumen of the uterus was greater in width with the caruncular areas not protruding as far into the uterine lumen, as observed in the uteri of normal ewes. The UGKO endometrium consisted primarily of dense stromal cells with very little detectable ECM except in the deep stroma on the border of the myometrium. This area in which ECM staining was detected corresponds to the area in which large numbers of blood vessels are present.

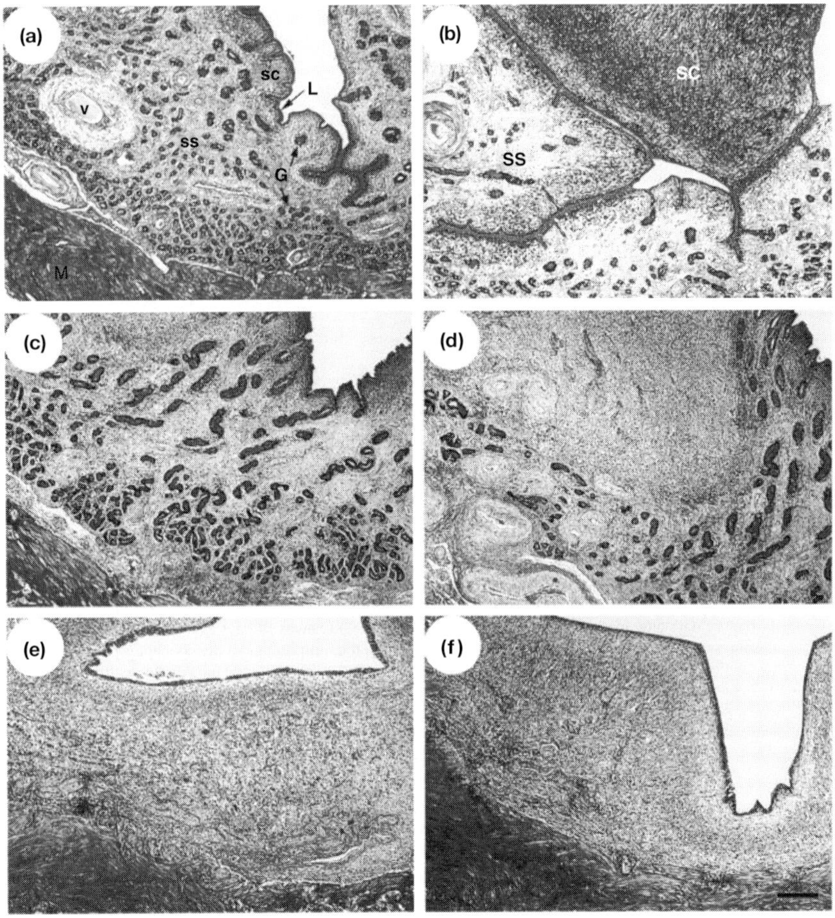

Fig. 1. Representative photomicrographs of the uterus from cyclic (a,b), pregnant (c,d) and uterine gland knockout (UGKO) ewes (e,f) 14 days after mating, using a Masson's trichrome stain. This procedure stains nuclei black, cytoplasm and muscle fibres red, and extracellular matrix (ECM) components blue. The left column (a,c) from normal ewes depicts intercaruncular endometrial regions and the right column (b,d) depicts caruncular endometrial areas. No distinct caruncular or intercaruncular regions are distinguishable in uteri from UGKO ewes (e,f). L: luminal epithelium; G: glandular epithelium; SC: stratum compactum; SS: stratum spongiosum; M: myometrium; V: blood vessel. Scale bar represents 200 μm.

Myometrial histoarchitecture did not appear different between uteri from UGKO and control ewes.

Study one: IFN-τ, osteopontin and GlyCAM-1 in uterine flushes

Filamentous conceptuses were present in most uterine flushes from normal ewes 14 days after mating (Fig. 2a). In contrast, growth-retarded conceptuses were observed in the uterine flush of some UGKO ewes. A representative degenerating tubular conceptus found in the uterine flush of an UGKO ewe is shown (Fig. 2c).

Immunoreactive IFN-τ was absent in uterine flushes obtained from cyclic ewes (Fig. 2b, lanes 1–4), but detected in uterine flushes obtained from pregnant ewes (Fig. 2b, lanes 5–8). The amount of IFN-τ present in the uterine

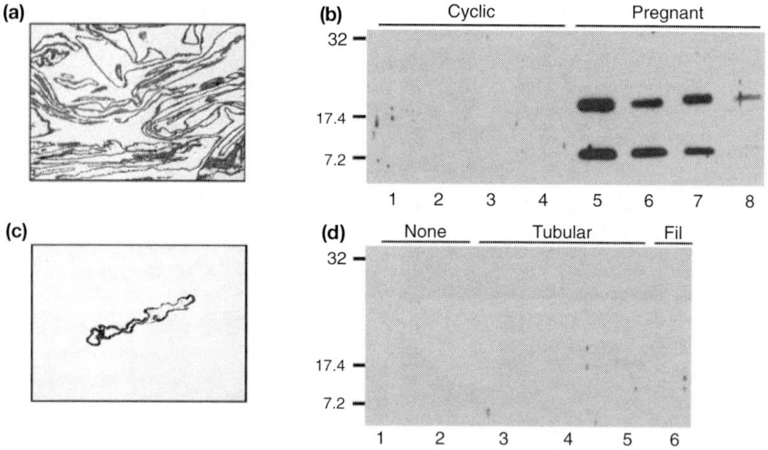

Fig. 2. Conceptus morphology and uterine flush content of interferon τ (IFN-τ) from cyclic, pregnant and uterine gland knockout (UGKO) ewes. Photomicrographs (a,c) are haematoxylin and eosin-stained conceptuses that were present in the uterine flushes. (a) Elongating filamentous conceptus representative of conceptuses found in the uterine flushes from normal bred ewes 14 days after mating. (c) Growth-retarded, or tubular, conceptus detected in the uterine flush of an UGKO ewe 14 days after mating. Photomicrographs are shown at the same magnification (× 164). (b,d) Western blot analyses of IFN-τ in concentrated uterine flushes obtained from (b) normal ewes on day 14 of oestrus (Cyclic, lanes 1–4) or 14 days after mating (Pregnant, lanes 5–8) and (d) UGKO ewes 14 days after mating. (d) Uterine flushings from UGKO ewes were loaded depending upon the morphology of the conceptus present in the uterine flush. No conceptus (lanes 1–2), a growth-retarded or degenerating tubular conceptus (lanes 3–5), or a fragmented filamentous conceptus (Fil; lane 6) was detected in the uterine flushes from UGKO ewes. Each lane in the blots represents proteins from a single flush from an individual ewe (30 μg per lane). Immunoreactive protein was detected using mouse monoclonal antibody directed against ovine IFN-τ. Positions of pre-stained molecular weight standards (× 10^{-3}) are indicated.

flushes corresponded to the number of conceptuses present in the uterine flush. The uterine flush from the ewe represented in lane 5 contained three filamentous conceptuses, whereas those in lanes 6 and 7 were from ewes with two conceptuses, and lane 8 represents a flush that contained a single conceptus. Western blot analyses of uterine flushes that contained a conceptus from UGKO ewes is presented (Fig. 2d). Immunoreactive IFN-τ was not detected in uterine flushes from UGKO ewes containing no conceptus (lanes 1 and 2), degenerating tubular conceptuses (lanes 3 and 4), or a fragmenting filamentous conceptus (lane 6). Immunoreactive IFN-τ was within detection limits in the uterine flush of one UGKO ewe that contained a degenerating, tubular conceptus (lane 5).

Immunoreactive osteopontin was absent in uterine flushings from cyclic ewes (Fig. 3a, lanes 1–4), but present in uterine flushings from pregnant ewes (Fig. 3a, lanes 5–8). Several molecular mass forms (70, 45, and 25 kDa) of immunoreactive osteopontin were detected in the uterine flushings from pregnant ewes. The 45 and 25 kDa forms of osteopontin are cleavage products of the 70 kDa form (Johnson *et al.*, 1999a). The 70 kDa form of osteopontin was present in only one uterine flush (Fig. 3a, lane 5), which contained three filamentous conceptuses. Immunoreactive osteopontin was not detectable in any uterine flush obtained from UGKO ewes, regardless of the presence of a degenerating conceptus (Fig. 3b, lanes 1–6).

Immunoreactive GlyCAM-1 was detectable in two of the uterine flushes obtained from cyclic ewes (Fig. 4a, lanes 2 and 4), but abundant in all uterine flushes obtained from pregnant ewes (Fig. 4a, lanes 5–8). Small amounts of GlyCAM-1 were detected in uterine flushings from three UGKO ewes (Fig. 4b, lanes 1, 3 and 5). However, the amount of GlyCAM-1 in uterine flushings from UGKO ewes was not abundant and appeared similar to that in normal cyclic ewes.

Study one: expression of Muc-1 and integrin subunits in the endometrium

Immunoreactive Muc-1 was restricted to the apical surface of uterine LE and GE (Fig. 5). Apical staining of

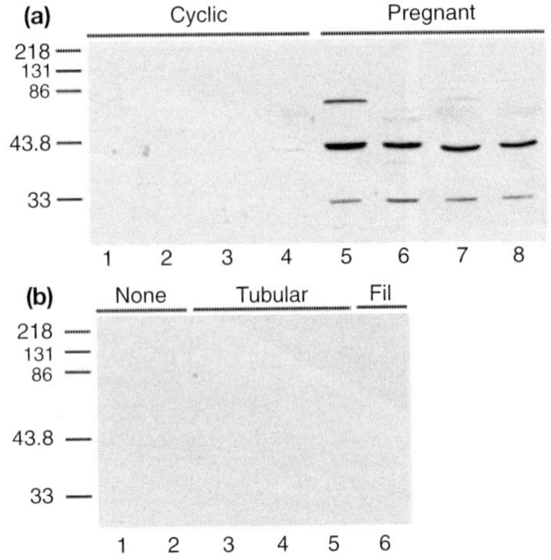

Fig. 3. Western blot analyses of osteopontin in concentrated uterine flushes obtained from (a) normal ewes on day 14 of oestrus (Cyclic, lanes 1–4) or 14 days after mating (Pregnant, lanes 5–8) and (b) uterine gland knockout (UGKO) ewes 14 days after mating. Flushes from UGKO ewes were loaded depending upon the morphology of the conceptus present in the uterine flush. No conceptus (lanes 1–2), a growth-retarded or degenerating tubular conceptus (lanes 3–5), or a fragile filamentous conceptus (Fil; lane 6) was detected in the uterine flushes from UGKO ewes. Each lane in the blots represents a single flush from an individual ewe (30 µg per lane). Immunoreactive protein was detected using a cocktail of polyclonal rabbit anti-human osteopontin IgG (LF-123 and LF-124). Positions of pre-stained molecular weight standards ($\times 10^{-3}$) are indicated.

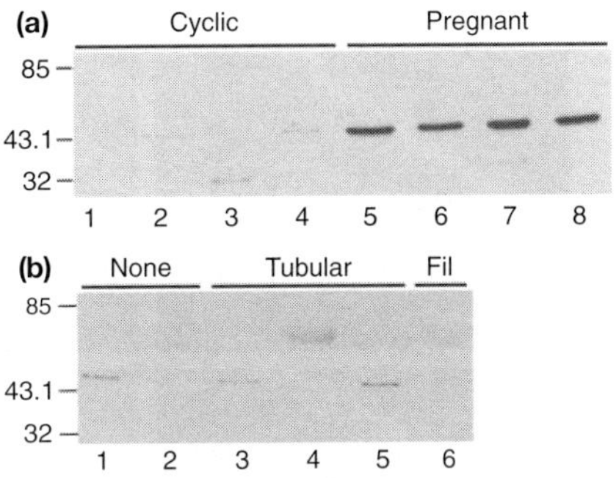

Fig. 4. Western blot analyses of glycosylation-dependent cell adhesion molecule 1 (GlyCAM-1) in concentrated uterine flushes obtained on from (a) normal ewes on day 14 of oestrus (Cyclic, lanes 1–4) or 14 days after mating (Pregnant, lanes 5–8) and (b) uterine gland knockout (UGKO) ewes. Flushes from UGKO ewes were loaded depending upon the morphology of the conceptus present in the uterine flush. No conceptus (lanes 1–2), a growth-retarded or degenerating tubular conceptus (lanes 3–5), or a fragmenting filamentous conceptus (Fil; lane 6) was detected in the uterine flushes from UGKO ewes. Each lane in the blots represents a single flush from an individual ewe (30 µg per lane). Immunoreactive protein was detected using rabbit anti-rat GlyCAM-1 (CAM02) IgG. Positions of pre-stained molecular weight standards ($\times 10^{-3}$) are indicated.

Muc-1 was detected on the LE of all uteri and was not appreciably different in comparisons among uteri from cyclic, pregnant and UGKO ewes. Immunoreactive α_v, α_4, α_5, β_3 and β_5 integrin subunits were detected at the apical surface of uterine LE and GE of normal cyclic and pregnant ewes (Fig. 5) and were not affected by pregnancy status in normal ewes. Overall, the relative intensity of immunoreactive integrin subunit expression was not different in the LE of the endometrium of UGKO ewes compared with that for normal cyclic and pregnant ewes.

Study two: peripheral plasma concentrations of progesterone

Concentrations of progesterone in plasma obtained each day during the intra-uterine infusion period were not affected by treatment (day × treatment, $P = 0.628$). Plasma progesterone concentrations remained constant throughout the uterine infusion of either control proteins or roIFN-τ.

Study two: steady-state concentrations of mRNAs encoding interferon-stimulated gene in the endometrium

Results of the slot blot hybridization analyses to determine steady state levels of endometrial mRNAs are summarized (Table 1). Infusion of roIFN-τ into the uterus of UGKO ewes increased endometrial concentrations of ISG17 ($P < 0.001$), STAT1 ($P < 0.001$), STAT2 ($P < 0.001$) and IRF-1 ($P < 0.003$) as compared with endometrium from ewes infused with control proteins.

Discussion

The results of the present study extend previous reports describing both gross and histological differences between uteri from normal and UGKO ewes, including differences in the stromal organization of the endometria of UGKO ewes (Gray *et al.*, 2000b, 2001b). The present results are the first to indicate that uteri of adult UGKO ewes, which lack an intercaruncular area consisting entirely of compact stroma, also have major differences in expression of ECM components as assessed by Masson's trichrome staining.

Gray *et al.* (2001c) found that normal conceptuses were present in the uterine lumen of UGKO ewes on days 6 and 9, but not on day 14, after mating. Similarly, in the present study, uterine flushings of UGKO ewes contained either no

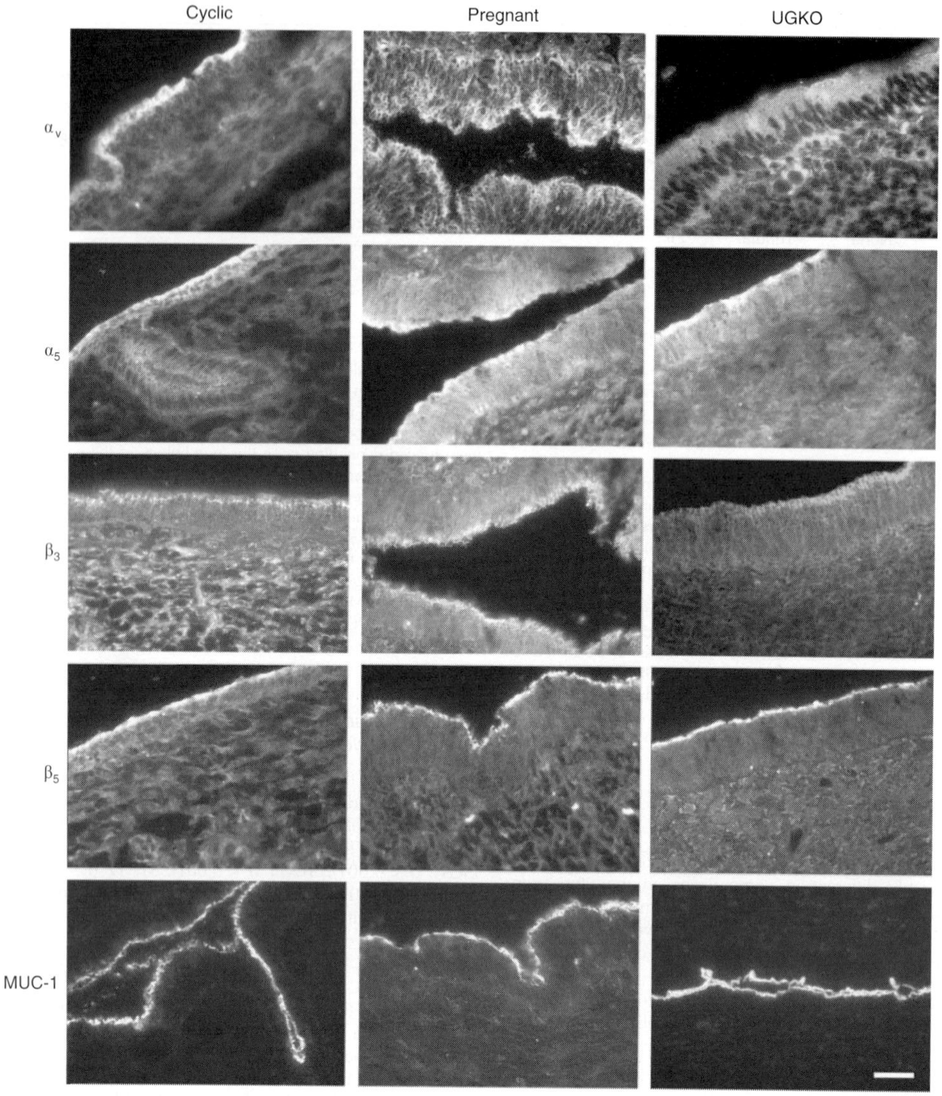

Fig. 5. Immunofluorescent localization of integrins α_v, α_5, β_3, β_5 and MUC-1 in the endometrium of normal ewes on day 14 of oestrus (Cyclic; left column) or 14 days after mating (Pregnant; middle column), or uterine gland knockout (UGKO) ewes after mating (UGKO; right column). Frozen cross-sections of the uterine wall from control and UGKO ewes were fixed in methanol and blocked in 5% normal goat serum. Sections were incubated overnight at 4°C with 20 μg ml^{-1} rabbit IgG against α_v, α_5, β_3, β_5 or MUC-1, and then incubated with fluorescein-conjugated goat anti-rabbit IgG. Photomicrographs were taken using a fluorescein filter. Scale bar represents 200 μm.

Table 1. Steady state concentrations of endometrial mRNA encoding interferon (IFN)-stimulated gene 17 (ISG17), signal transducer and activator of transcription 1 (STAT1), STAT2, and IFN regulatory factor 1 (IRF-1)

Intrauterine protein treatment	Endometrial expression of mRNA[a]			
	ISG17	STAT1	STAT2	IRF-1
Control	218 053	7 728 198	443 682	602 494
roIFN-τ	3 086 827	31 034 609	2 668 866	1 059 946
Overall SE	102 246	669 813	31 321	41 152
P value[b]	0.001	0.001	0.001	0.003
Fold increase	14	4	6	2

[a]Data are presented as relative light units with SE.
[b]One-way ANOVA (Control versus IFN- τ).

conceptus or a growth-retarded conceptus on day 14 after mating. Taken together, the evidence indicates that endometrial glands and, by default, their secretions are not required for development of the conceptus to the hatched blastocyst state, but are crucial for blastocyst elongation. Normal conceptuses by day 14 after mating have developed from a tubular to an elongated filamentous form, secreted large amounts of IFN-τ, and transiently contacted the maternal epithelium in preparation for implantation (Guillomot, 1995).

In the present study, uterine flushes from UGKO ewes on day 14 after mating had either no detectable or very low amounts of IFN-τ associated with severely growth retarded conceptuses. Secretion of IFN-τ by the trophectoderm serves as a marker for conceptus health and state of development. IFN-τ is secreted by the ovine conceptus between days 8 and 21 after mating, with maximum amounts secreted on days 14–16 as the conceptus develops from a tubular to a filamentous form (Ashworth and Bazer, 1989; Farin *et al.*, 1989). In the present study, the uterine flush of UGKO ewes did not contain the high concentrations of immunoreactive IFN-τ characteristic of uterine flushings from normal ewes on day 14 after mating. The lack of appreciable IFN-τ in the uterine flush of UGKO ewes is correlated directly with the lack of a conceptus or the presence of a growth-retarded tubular conceptus in the uteri of UGKO ewes.

In ewes, IFN-τ acts on the endometrium to induce or increase the expression of several inteferon-stimulated genes. IFN-τ-stimulated genes include ISG17/ubiquitin cross-reactive protein (Johnson *et al.*, 1999b, 2000), 2′,5′-oligoadenylate synthetase (OAS) (Mirando *et al.*, 1991; Johnson *et al.*, 2001b; Stewart *et al.*, 2001a), STATs 1 and 2 (Choi *et al.*, 2001; Stewart *et al.*, 2001a,b), and IRF-1 and -9 (Choi *et al.*, 2001; Stewart *et al.*, 2002). Despite retarded conceptus development and absence of IFN-τ production by UGKO conceptuses, findings from the present study indicate that the endometrium of UGKO ewes is responsive to IFN-τ in terms of induction of or increases in expression of IFN-τ-stimulated genes. The ability of IFN-τ to stimulate increases in IFN-τ-stimulated genes in the UGKO uterus is in agreement with previous studies in normal cyclic ewes infused with IFN-τ (Spencer *et al.*, 1999c; Johnson *et al.*, 2000; Choi *et al.*, 2001). Taken together, available results indicate that the endometrium of UGKO ewes can respond properly to the pregnancy recognition signal produced by the ovine conceptus, but is unable to produce the proper secretions to stimulate conceptus development.

The process of superficial implantation in sheep involves a sequence of events, including removal of anti-adhesion molecules on the LE, expression of receptors on the LE and trophectoderm, and secretion of adhesive factors by the endometrial LE and GE (Bowen and Burghardt 2000; Johnson *et al.*, 2001a; Burghardt *et al.*, in press). Muc-1 is a heavily glycosylated mucin thought to project above the apical surface of LE cells to sterically block cell–cell and cell–ECM adhesion (Wesseling *et al.*, 1995) and, thereby, trophectoderm access to the uterine LE. In both humans and rodents, the expression pattern of glycoproteins, such as MUC-1, on the uterine LE may control the accessibility of integrin receptors to their ligands and provide a barrier to invasiveness (Carson *et al.*, 2000). The implantation adhesion cascade in rodents is initiated after downregulation of MUC-1 (Carson *et al.*, 1998). In sheep, MUC-1 is also expressed on the apical surface of the LE and is down-regulated during implantation to allow interactions of integrins and their ligands (Johnson *et al.*, 2001a). Integrins comprise a family of intrinsic membrane proteins of non-covalently linked α and β subunit heterodimers. Integrins serve as receptors and bind ECM and other ligands to aid in cellular adhesion, reorganization of cytoskeletal molecules or signal transduction (Miyamoto *et al.*, 1995, Burghardt *et al.*, 1997, in press). In sheep, integrin expression on the apical surface of the LE is constitutive (Johnson *et al.*, 2001a). The ovine endometrial LE and trophectoderm express integrin subunits α_v, α_4, α_5, β_1, β_3 and β_5 (Johnson *et al.*, 2001a). The present study examined expression of the glycoprotein MUC-1 and integrin subunits α_v, α_5, β_3 and β_5 by the endometrial epithelium of normal and UGKO ewes 14 days after mating. There were no differences between normal and UGKO ewes in the patterns of Muc-1 and integrin subunit expression on the endometrial LE. Previous studies have demonstrated similar results for temporal expression of steroid hormone receptors (Gray *et al.*, 2000a) and LE-specific genes (Gray *et al.*, 2001c) in the endometrium of UGKO compared with normal ewes. Taken together, the results of these studies support the contention that the endometrial LE of UGKO ewes is not abnormal in phenotype and is not the underlying cause of conceptus growth retardation and mortality.

In several species, conceptus implantation involves molecules that are secreted by the endometrial glands during pregnancy. In rodents, the endometrial glands secrete leucocyte inhibitory factor and calcitonin, which act on the uterine LE or conceptus to promote conceptus development, establishment of uterine receptivity and implantation (Carson *et al.*, 2000). Available evidence from the ovine UGKO model supports the contention that the endometrial glands also secrete molecules that support conceptus survival and development. The endometrial glands of the pregnant ovine uterus synthesize and secrete osteopontin, which is hypothesized to play a role in conceptus attachment, adhesion and elongation during peri-implantation (Johnson *et al.*, 1999a,b, 2001a). Osteopontin is a 70 kDa acidic glycoprotein component of the ECM (Butler *et al.*, 1996) that gives rise to 25 and 45 kDa fragments upon treatment with proteases and freezing or thawing (Weber and Cantor, 1996). Osteopontin binds to cell surface integrins present on both trophectoderm and LE, via its RGD sequence present on the 70 and 45 kDa forms, to promote cell–cell attachment during the period of implantation (Johnson *et al.*, 1999b). Osteopontin engages a number of receptors, including the integrins $\alpha_v(\beta_1$, β_3 or $\beta_5)$ and $(\alpha_4$, α_5, α_8 or $\alpha_9)\beta_1$ (Denhardt *et al.*, 2001). In the present study, osteopontin was found to be absent in uterine flushings from cyclic and UGKO ewes and present in all three forms in uterine flushings from pregnant ewes 14 days after mating. This pattern of osteopontin expression in normal ewes is similar to that reported by Johnson *et al.* (1999a). If osteopontin is essential for normal conceptus elongation and implantation in sheep, then its absence in the uterine milieu of UGKO ewes would be detrimental to conceptus survival and development, despite the presence of the integrin receptors for osteopontin on the endometrial LE. The process of conceptus elongation and implantation probably requires the interaction of ECM ligands and integrin receptors on the trophectoderm and endometrial LE (Johnson *et al.*, 1999b, 2001a; Carson *et al.*, 2000; Burghardt *et al.*, in press). Results from the present study concur with this hypothesis and provide new insight into the physiology involved in ECM-mediated conceptus–uterine interactions involved in conceptus elongation and implantation in ruminants.

Another molecule that may be required for conceptus attachment during the peri-implantation period in the ovine uterus is GlyCAM-1 (Spencer *et al.*, 1999b). GlyCAM-1 is a sulphated member of the mucin family of glycoproteins that is normally expressed by the endothelium of lymph nodes (Lasky *et al.*, 1992). This glycoprotein serves as a ligand for L-selectin which, in turn, activates integrins and promotes fibronectin adhesion to aid in extravasation of lymphocytes in lymph nodes (Rosen, 1993; Hwang *et al.*, 1996; Giblin *et al.*, 1997). In pregnant ewes, GlyCAM-1 in the LE and GE was low on days 11 and 13, increased on day 15, and was abundant on days 17 and 19 after mating (Spencer *et al.*, 1999b). In pregnant ewes, the relative amount of immunoreactive GlyCAM-1 in uterine flushings was low on days 11 and 13, but high on days 15 and 17 after mating. Similar to findings for osteopontin, the abundance of GlyCAM-1 was low or undetectable in uterine flushings from cyclic and UGKO ewes, but was abundant in uterine flushings from pregnant ewes 14 days after mating. Thus, the absence of GlyCAM-1 in the uterine environment of UGKO ewes could also impede conceptus elongation and implantation. It is likely that several other molecules are also lacking in the UGKO uterus, resulting in conceptus mortality and growth retardation during the peri-implantation period. The lack of a closed uterine lumen with discreet protruding caruncles may also contribute to defects in conceptus survival and elongation observed in UGKO ewes. However, available results support the hypothesis that defects in peri-implantation conceptus survival and growth in UGKO ewes are not the result of alterations in the endometrial expression of either anti-adhesion molecules or integrin receptors, but are rather the result of an absence of endometrial glands and their secretions, including molecules involved in cell–cell adhesions such as osteopontin and GlyCAM-1. A functional genomics and proteomics approach will be necessary to determine which uterine factors are absent in UGKO ewes as compared with normal ewes during early pregnancy.

These studies were supported by Grants 1998-35203-6322 and 2001-35203-10700 from the NRI Competitive Grants Program/CSREES/USDA and, in part, by P30 ES09106. The authors appreciate the gift of reagents by T. Hansen (University of Wyoming) and D. Carson (University of Delaware).

References

Ashworth CJ and Bazer FW (1989) Changes in ovine conceptus and endometrial function following asynchronous embryo transfer or administration of progesterone *Biology of Reproduction* **40** 425–433

Bartol FF, Wiley AA, Coleman DA, Wolfe DF and Riddel MG (1988) Ovine uterine morphogenesis: effects of age and progestin administration and withdrawal on neonatal endometrial development and DNA synthesis *Journal of Animal Science* **66** 3000–3009

Bartol FF, Wiley AA, Floyd JG, Ott TL, Bazer FW, Gray CA and Spencer TE (1999) Uterine differentiation as a foundation for subsequent fertility *Journal of Reproduction and Fertility Supplement* **54** 285–300

Bazer FW (1975) Uterine protein secretions: relationship to development of the conceptus *Journal of Animal Science* **41** 1376–1382

Bazer FW, Spencer TE and Ott TL (1997) Interferon tau: a novel pregnancy recognition signal *American Journal of Reproductive Immunology* **37** 412–420

Bowen JA and Burghardt RC (2000) Cellular mechanisms of implantation in domestic farm animals *Cell and Developmental Biology* **11** 93–104

Burghardt RC, Bowen JA, Newton GR and Bazer FW (1997) Extracellular matrix and the implantation cascade in pigs *Journal of Reproductive Fertility Supplement* **52** 151–164

Burghardt RC, Johnson GA, Jaeger LA, Ka H, Garlow JE, Spencer TE and Bazer FW Integrins and extracellular matrix proteins at the maternal–fetal interface in domestic animals *Cells, Tissues and Organs* (in press)

Butler WT, Ridall AL and McKee MD (1996) Osteopontin. In *Principals of Bone Biology* pp 167–181 Eds JP Bilezikian *et al.* Academic Press, New York

Carson DD, DeSouza MM, Kardon R, Zhou X, Lagow E and Julian J (1998) Mucin expression and function in the female reproductive tract *Human Reproduction Update* **4** 459–464

Carson DD, Bagchi I, Dey SK, Enders AC, Fazleabas AT, Lessey BA and Yoshinaga K (2000) Embryo implantation *Developmental Biology* **223** 217–237

Choi Y, Johnson GA, Burghardt RC, Berghman LR, Joyce MM, Taylor KM, Stewart MD, Bazer FW and Spencer TE (2001) Interferon regulatory factor two restricts expression of interferon- stimulated genes to the endometrial stroma and glandular epithelium of the ovine uterus *Biology of Reproduction* **65** 1038–1049

Denhardt DT, Giachelli CM and Rittling SR (2001) Role of osteopontin in cellular signaling and toxicant injury *Annual Review of Pharmacological Toxicology* **41** 723–749

Farin CE, Imakawa K and Roberts RM (1989) *In situ* localization of mRNA for the interferon, ovine trophoblast protein 1, during early embryonic development of the sheep *Molecular Endocrinology* **3** 1099–1107

Fazleabas AT, Hild-Petito S and Varhaage HG (1994) Secretory proteins and growth factors of the baboon (*Papio anubis*) uterus: potential roles in pregnancy *Cell Biology International* **18** 1145–1153

Giblin PA, Hwang ST, Katsumo TR and Rosen SD (1997) Ligation of L-selectin on T lymphocytes activates β1 integrins and promotes adhesion to fibronectin *Journal of Immunology* **159** 3498–3507

Gray CA, Bartol FF, Taylor KM, Wiley AA, Ramsey WS, Ott TL, Bazer FW and Spencer TE (2000a) The ovine uterine gland knockout model: effects of gland ablation on the estrous cycle *Biology of Reproduction* **62** 448–456

Gray CA, Taylor KM, Bazer FW and Spencer TE (2000b) Mechanisms regulating norgestomet inhibition of endometrial gland morphogenesis in the neonatal ovine uterus *Molecular Reproductive Development* **57** 67–78

Gray CA, Bartol FF, Tarleton BJ, Wiley AA, Johnson GA, Bazer FW and Spencer TE (2001a) Developmental biology of uterine glands *Biology of Reproduction* **65** 1311–1323

Gray CA, Bazer FW and Spencer TE (2001b) Effects of neonatal progestin exposure on female reproductive tract structure and function in the adult ewe *Biology of Reproduction* **64** 797–804

Gray CA, Taylor KM, Ramsey WS, Hill JR, Bazer FW, Bartol FF and Spencer TE (2001c) Endometrial glands are required for pre-implantation conceptus elongation and survival *Biology of Reproduction* **64** 1608–1613

Guillomot M (1995) Cellular interactions during implantation in domestic ruminants *Journal of Reproduction and Fertility Supplement* **49** 39–51

Guillomot M, Fléchon J-E and Wintenberger-Torres S (1981) Conceptus attachment in the ewe: an ultrastructural study *Placenta* **2** 169–182

Heyman Y, Camous S, Fevre J, Meziou W and Martal J (1984) Maintenance of the corpus luteum after uterine transfer of trophoblastic vesicles to cyclic cows and ewes *Journal of Reproduction and Fertility* **70** 533–550

Hwang ST, Singer MS, Giblin PA, Yednock TA, Bacon KB, Simon SI and Rosen SD (1996) GlyCAM-1, a physiologic ligand for L-selectin, activates β2 integrins on naive peripheral lymphocytes *Experimental Medicine* **184** 1343–1348

Johnson GA, Burghardt RC, Spencer TE, Newton GR, Ott TL and Bazer FW (1999a) Ovine osteopontin II: osteopontin and αvβ3 integrin expression in the uterus and conceptus during the periimplantation period *Biology of Reproduction* **61** 892–899

Johnson GA, Spencer TE, Hansen TR, Austin KJ, Burghardt RC and Bazer FW (1999b) Expression of the interferon tau inducible ubiquitin cross-reactive protein in the ovine uterus *Biology of Reproduction* **61** 312–318

Johnson GA, Spencer TE, Burghardt RC, Joyce MM and Bazer FW (2000) Interferon-tau and progesterone regulate ubiquitin cross-reactive protein expression in the ovine uterus *Biology of Reproduction* **62** 622–627

Johnson GA, Bazer FW, Jaeger LA, Ka H, Garlow JE, Pfarrer C, Spencer TE and Burghardt RC (2001a) Muc-1, integrin and osteopontin expression during the implantation cascade in sheep *Biology of Reproduction* **65** 820–828

Johnson GA, Stewart MD, Gray CA, Choi Y, Burghardt RC, Yu-Lee LY, Bazer FW and Spencer TE (2001b) Effects of the estrous cycle, pregnancy, and interferon tau on 2′,5′-oligoadenylate synthetase expression in the ovine uterus *Biology of Reproduction* **64** 1392–1399

Kane MT, Morgan PM and Coonan C (1997) Peptide growth factors and preimplantation development *Human Reproduction Update* **3** 137–157

Lasky LA, Singer M, Dowbenko D, Imai Y, Henzel W, Grimley C, Finnie C, Gillett N, Watson S and Rosen S (1992) An endothelial ligand for L-selectin is a novel mucin-like molecule *Cell* **69** 927–938

Mirando MA, Short EC, Jr, Geisert RD, Vallet JL and Bazer FW (1991) Stimulation of 2′,5′-oligoadenylate synthetase activity in sheep endometrium during pregnancy, by intrauterine infusion of ovine trophoblast protein-1, and by intramuscular administration of recombinant bovine interferon- alpha I1 *Journal of Reproduction and Fertility* **93** 599–607

Miyamoto S, Teramoto H, Coso OA, Gutkind JS, Burbelo PD, Akiyama SK and Yamada KM (1995) Integrin function: molecular hierarchies of cytoskeletal and signaling molecules *Journal of Cell Biology* **131** 791–805

Roberts RM and Bazer FW (1988) The function of uterine secretions *Journal of Reproduction and Fertility* **82** 875–892

Rosen SD (1993) L-Selectin and its biological ligands *Histochemistry* **100** 185–191

SAS (1990) *SAS User's Guide: Statistics* Version 6. Statistical Analysis System Institute, Cary, NC

Simmen RCM and Simmen FA (1990) Regulation of uterine and conceptus secretory activity in the pig *Journal of Reproduction and Fertility Supplement* **40** 279–292

Singer MS and Rosen SD (1996) Purification and quantification of L-selectin-reactive GlyCAM-1 from mouse serum *Journal of Immunoligical Methods* **196** 153–161

Spencer TE, Becker WC, George P, Mirando MA, Ogle TF and Bazer FW (1995) Ovine interferon-tau inhibits estrogen receptor up-regulation and estrogen-induced luteolysis in cyclic ewes *Endocrinology* **136** 4932–4944

Spencer TE, Stagg AG, Joyce MM, Jenster G, Wood CG, Bazer FW, Wiley AA and Bartol FF (1999a) Discovery and characterization of endometrial epithelial messenger ribonucleic acids using the ovine uterine gland knockout model *Endocrinology* **140** 4070–4080

Spencer TE, Bartol FF, Bazer FW, Johnson GA and Joyce MM (1999b) Identification and characterization of glycosylation-dependent cell adhesion molecule 1-like protein expression in the ovine uterus *Biology of Reproduction* **60** 241–250

Spencer TE, Gray CA, Ott TL, Johnson GA, Ramsey WS and Bazer FW (1999c) Differential effects of intrauterine and subcutaneous administration of recombinant ovine interferon tau on endometrial gene expression of cyclic ewes *Biology of Reproduction* **61** 464–470

Steele RGD and Torrie JH (1980) *Principles and Procedures of Statistics* McGraw–Hill, New York

Stewart CL, Kaspar P, Brunet LJ, Bhatt H, Gadi I, Köntgen F and Abbondanzo SJ (1992) Blastocyst implantation depends on maternal expression of leukaemia inhibitory factor *Nature* **359** 76–79

Stewart DM, Johnson GA, Vyhlidal CA, Burghardt RC, Safe SH, Yu-Lee LY, Bazer FW and Spencer TE (2001a) Interferon-tau activates multiple signal transducer and activator of transcription proteins and has complex effects on interferon-responsive gene transcription in ovine endometrial epithelial cells *Endocrinology* **142** 98–107

Stewart DM, Johnson GA, Bazer FW and Spencer TE (2001b) Interferon-tau (IFNt) regulation of IFN-stimulated gene expression in cell lines lacking specific IFN-signaling components *Endocrinology* **142** 1786–1794

Stewart MD, Choi Y, Johnson GA, Yu-Lee Ly LY, Bazer FW and Spencer TE (2002) Roles of Stat1, Stat2, and interferon regulatory factor-9 (IRF-9) in interferon tau regulation of IRF-1 *Biology of Reproduction* **66** 393–400

Swann SL, Bazer FW, Villarete LH, Chung A and Pontzer CH (1999) Functional characterization of monoclonal antibodies to interferon-tau *Hybridoma* **18** 399–405

Van Heeke G, Ott TL, Strauss A, Ammaturo D and Bazer FW (1996) High yield expression and secretion of the ovine pregnancy recognition hormone interferon-tau by *Pichia pastoris. Journal of Interferon and Cytokine Research* **16** 119–126

Weber GF and Cantor H (1996) The immunology of Eta-1/osteopontin *Cytokines and Growth Factor Review* **7** 241–248

Wesseling J, Van der Valk SW, Vos HL, Sonnenberg A and Hilkens J (1995) Episialin (MUC-1) overexpression inhibits integrin mediated cell adhesion to extracellular matrix components *Journal of Cell Biology* **129** 255–265

Zhu LJ, Bagchi MK and Bagchi IC (1998) Attenuation of calcitonin gene expression in pregnant rat uterus leads to a block in embryonic implantation *Endocrinology* **139** 330–339

Received 11 February 2002.
First decision 27 March 2002.
Accepted 11 April 2002.

0013-7227/05/$15.00/0
Printed in U.S.A.

Endocrinology 146(11):4825–4833

doi: 10.1210/en.2005-0768

Cathepsins in the Ovine Uterus: Regulation by Pregnancy, Progesterone, and Interferon Tau

Gwonhwa Song, Thomas E. Spencer, and Fuller W. Bazer

Center for Animal Biotechnology and Genomics and Department of Animal Science, Texas A&M University, College Station, Texas 77843

Cathepsins (CTS) are peptidases that have biological roles in degrading extracellular matrix, catabolism of intracellular proteins, and processing of prohormones. Expression of CTSB, CTSD, CTSH, CTSK, CTSL, CTSS, and CTSZ genes was detected in the endometria of cyclic and early pregnant ewes with distinct temporal and spatial expression patterns. In the d 18 and 20 conceptus, expression of CTSB, CTSD, CTSL, and CTSZ mRNA was detected in the trophectoderm. Of particular note, CTSL mRNA was the most abundant CTS mRNA in the ovine endometrium and detected only in the luminal epithelium and superficial glandular epithelium of cyclic and pregnant ewes. CTSL mRNA increased 8-fold between d 10 and 18 in endometria of pregnant ewes, whereas it declined between d 14 and 16 in cyclic ewes. CTSL protein was also detected in conceptus trophectoderm, and pro-CTSL was detected in uterine flushings from ewes between d 12 and 16 of pregnancy. In ovariectomized and catheterized ewes, CTSL mRNA in the endometrium was increased by progesterone and intrauterine injections of ovine interferon (IFN)τ. Other endometrial CTS genes were also regulated by progesterone alone (CTSB, CTSK, CTSS, and CTSZ) or progesterone and IFNτ (CTSH, CTSK, CTSS, and CTSZ). These results indicate that CTS of endometrial and conceptus origin may regulate endometrial remodeling and conceptus implantation, endometrial CTS genes are regulated by ovarian and placental hormones, and CTSL is a novel IFNτ-stimulated gene expressed only in luminal epithelium and superficial glandular epithelium of the endometrium. (*Endocrinology* 146: 4825–4833, 2005)

CATHEPSINS (CTS) ARE a family of lysosomal proteinases active in an acidic environment (1). They have the ability to degrade extracellular matrix (ECM) molecules, including collagens, laminin, fibronectin, and proteoglycans and are also involved in the catabolism of intracellular proteins and prohormone processing. A member of the cysteine proteinase family, CTSB, can activate matrix metalloproteinases (MMPs) and urokinase type plasminogen activator (2), and the closely related CTSL can cleave prourokinase type plasminogen activator into the active form (2). On the other hand, inactive precursors of these CTS can be activated by MMPs (1). In humans, CTSB, CTSH, CTSK, CTSL, and CTSS are expressed in the proliferative and secretory phase endometria and appear to be required for normal uterine development and function as well as menstruation (3). Available evidence supports the concept that a variety of proteases as well as their specific inhibitors regulate trophoblast invasion in many species (*e.g.* mouse, rat, cat, pig, and human) during conceptus implantation (3–10). Specifically, these studies implicate CTS in regulation of uterine receptivity for implantation and trophoblast invasion in a number of mammals (see Refs. 11–14 for review).

Regulation of CTS expression in the ovine uterus and conceptus has not been reported. Trophoblast invasion in ruminants (sheep, cattle, goats) is limited to fusion of migrating binucleate cells with uterine epithelium, but considerable tissue remodeling and angiogenesis occur within the endometrium at implantation, which is associated with the cysteine and serine proteases and production of MMPs by the endometrium and conceptus (12, 14). Endometrial function during this period of pregnancy appears to be primarily regulated by progesterone from the corpus luteum and hormones from the conceptus, including interferon (IFN)τ (15, 16). IFNτ is the signal for maternal recognition of pregnancy in ruminants and is produced between d 10 and 21–25 of pregnancy in sheep by the mononuclear trophoblast cells of the conceptus (17, 18). In sheep, IFNτ acts in a paracrine manner on endometrial epithelia to inhibit transcription of the estrogen receptor-α and oxytocin receptor genes (17, 19), thereby preventing endometrial release of luteolytic pulses of prostaglandin F2α (20). The antiluteolytic actions of IFNτ are required for maintenance of a functional corpus luteum and secretion of progesterone, the essential hormone of pregnancy (20). IFNτ also induces or stimulates expression of a number of genes in the endometrium that are hypothesized to play important biological roles in conceptus implantation (21). This study determined effects of the estrous cycle, pregnancy, progesterone, and IFNτ on expression of selected CTS genes in the ovine endometrium. Results indicated that a number of CTS genes are expressed in the endometrium and conceptus during early pregnancy and regulated by progesterone and/or IFNτ. In particular, CTSL was found to be novel gene stimulated by progesterone and IFNτ only in endometrial luminal (LE) and superficial ductal glandular epithelia (sGE).

First Published Online August 11, 2005

Abbreviations: C, Cyclic; CTS, cathepsin; CX, control; ECM, extracellular matrix; IFN, interferon; LE, luminal epithelium; MMP, matrix metalloproteinase; P, progesterone; PR, progesterone receptor; PX, pregnant; ro, recombinant ovine; sGE, superficial glandular epithelium.

***Endocrinology* is published monthly by The Endocrine Society (http://www.endo-society.org), the foremost professional society serving the endocrine community.**

Materials and Methods

Animals

Mature crossbred Suffolk ewes (*Ovis aries*) were observed daily for estrus in the presence of vasectomized rams and used in experiments only after they had exhibited at least two estrous cycles of normal duration (16–18 d). All experimental and surgical procedures were in compliance with the Guide for the Care and Use of Agriculture Animals and approved by the University Laboratory Animal Care and Use Committee of Texas A&M University.

Experimental design

Study 1. At estrus (d 0), ewes were mated to either an intact or vasectomized ram as described previously (22) and then hysterectomized (n = 5 ewes/d) on d 10, 12, 14, or 16 of the estrous cycle or d 10, 12, 14, 16, 18, or 20 of pregnancy. Pregnancy was confirmed on d 10–16 after mating by the presence of a morphologically normal conceptus(es) in the uterus. At hysterectomy, several sections (~0.5 cm) from the midportion of each uterine horn ipsilateral to the corpus luteum were fixed in fresh 4% paraformaldehyde in PBS (pH 7.2). After 24 h, fixed tissues were changed to 70% ethanol for 24 h and then dehydrated and embedded in Paraplast-Plus (Oxford Labware, St. Louis, MO). Several sections (1–1.5 cm) from the middle of each uterine horn were embedded in Tissue-Tek OCT compound (Miles, Oneonta, NY), frozen in liquid nitrogen vapor, and stored at −80 C. The remaining endometrium was physically dissected from myometrium, frozen in liquid nitrogen, and stored at −80 C for subsequent RNA or protein extraction. In monovulatory pregnant (PX) ewes, uterine tissue samples were marked as either contralateral or ipsilateral to the ovary bearing the corpus luteum. No tissues from the contralateral uterine horn were used for study. Uterine flushes were clarified by centrifugation (3000 × *g* for 30 min at 4 C) and frozen at −80 C for Western blot analysis.

Study 2. Cyclic (C) ewes (n = 20) were checked daily for estrus and then ovariectomized and fitted with indwelling uterine catheters on d 5 as described previously (23). Ewes were then assigned randomly (n = 5 per treatment) to receive daily im injections of progesterone and/or a progesterone receptor (PR) antagonist (ZK 136,317; Schering AG, Berlin, Germany) and intrauterine infusions of control serum proteins and/or recombinant ovine IFNτ protein as follows: 1) 50 mg progesterone (P, d 5–16) and 200 μg control (CX) serum proteins (d 11–16) (P+CX); 2) P and 75 mg ZK 136,317 (d 11–16) and CX proteins (P+ZK+CX); 3) P and IFNτ (2×10^7 antiviral units, d 11–16) (P+IFN); or 4) P and ZK and IFNτ (P+ZK+IFN). Steroids were administered daily in corn oil vehicle. Both uterine horns of each ewe received twice-daily injections of either CX proteins (50 μg/horn per injection) or IFNτ (5×10^6 antiviral units/horn per injection). Recombinant ovine IFNτ was produced in *Pichia pastoris* and purified as described previously (24). Proteins were prepared for intrauterine injection as described previously (23). This regimen of progesterone and recombinant ovine (ro)IFNτ mimics the effects of progesterone and the conceptus on endometrial expression of hormone receptors and IFNτ-stimulated genes during early pregnancy in ewes (25–28). All ewes were hysterectomized on d 17, and the uteri and endometria processed as described in study 1.

RNA isolation

Total cellular RNA was isolated from frozen ipsilateral endometrium (studies 1 and 2) using Trizol reagent (Life Technologies, Inc.-BRL, Bethesda, MD) according to manufacturer's recommendations. The quantity and quality of total RNA was determined by spectrometry and denaturing agarose gel electrophoresis, respectively.

Cloning of partial cDNAs for ovine CTSB, CTSK, CTSL, CTSH, CTSS, CTSD, and CTSZ

Partial cDNAs for ovine CTSB, CTSD, CTSK, CTSL, CTSH, CTSS, and CTSZ mRNAs were amplified by RT-PCR using total RNA from endometrial tissues from d 16–18 of pregnancy using specific primers (Table 1). PCR amplification was conducted as follows for ovine CTSB, CTSK, CTSL, CTSH, CTSS, CTSD and CTSZ : 1) 95 C for 5 min; 2) 95 C for 45 sec; 59.1 C (for CTSB and CTSH), 56.5 C (for CTSD, CTSK, CTSL, and CTSZ), or 64.5 C (for CTSS) for 1 min; and 72 C for 1 min for 35 cycles; and 3) 72 C for 10 min. Partial cDNAs of the correct size were cloned into pCRII using a T/A cloning kit (Invitrogen) and their sequences verified by sequencing.

TABLE 1. Sequences of primers used for RT-PCR and cloning

Gene	Sequence (5′–3′): forward and reverse	GenBank accession no.	Product size (bp)
CTSB	GCAAAACACCACTTGGAAGG AGGAACTGCATCCAAAATGC	L06075	581
CTSD	ACCTTCGACATCCACTACGG GTAGCTCTCGCACCTCATCC	AF164143	520
CTSH	GGTCAGAGCCTCAGAATTGC CATCGTTCAGTGTGATGTTGG	NM_004390	420
CTSK	GGGTCTCACGGTTCTACTGC CAGTCCACCAGGTTCTGAGG	NM_000396	504
CTSL	AATGGAGGAGAGCAGTGTGG CCTTCATAAGGGCCTTCTCC	X91755	579
CTSS	CCTGGAAGCACAAGTGAAGC GAATGGCTCGCGTCTATACC	X62001	330
CTSZ	GGCCTCATGAGTACCTGTCC TTGCCATTATGCCACAGC	NM_001336	501

Slot blot hybridization analyses

Steady-state levels of mRNA in ovine endometria were assessed by slot blot hybridization as described previously (28, 29). Radiolabeled antisense and sense cRNA probes were generated by *in vitro* transcription using linearized plasmid template, RNA polymerases, and [α-^{32}P]uridine 5-triphosphate. Denatured total endometrial RNA (20 μg) from each ewe in studies 1 and 2 was hybridized with radiolabeled cRNA probes. To correct for variation in total RNA loading, a duplicate RNA slot membrane was hybridized with radiolabeled antisense 18S cRNA (pT718S; Ambion, Austin, TX). After washing, the blots were digested with ribonuclease A and radioactivity associated with slots quantified using a Typhoon 8600 MultiImager (Molecular Dynamics, Piscataway, NJ). Data are expressed as relative units.

In situ hybridization analyses

Location of mRNA expression in sections (5 μm) of the ovine uterus was determined by radioactive *in situ* hybridization analysis as described previously (28, 29). Radiolabeled antisense and sense cRNA probes were generated by *in vitro* transcription using linearized plasmid template, RNA polymerases, and [α-^{35}S]uridine 5-triphosphate. Deparaffinized, rehydrated, and deproteinated uterine tissue sections were hybridized with radiolabeled antisense or sense cRNA probes. After hybridization, washing, and ribonuclease A digestion, slides were dipped in NTB-2 liquid photographic emulsion (Kodak, Rochester, NY) and exposed at 4 C for 2 wk. Slides were developed in Kodak D-19 developer, counterstained with Gill's hematoxylin (Fisher Scientific, Fairlawn, NJ), and then dehydrated through a graded series of alcohol to xylene. Coverslips were then affixed with Permount (Fisher Scientific). Images of representative fields were recorded under bright-field or dark-field illumination using an Eclipse 1000 photomicroscope (Nikon Instruments Inc., Lewisville, TX) fitted with a Nikon DXM1200 digital camera.

Immunohistochemistry

Immunocytochemical localization of immunoreactive CTSL protein in the ovine uterus was performed as described previously (22) in uterine tissue cross-sections from studies 1 and 2 using rabbit antihuman CTSL polyclonal antibody (catalog no. 3192-100; BioVision, Mountain View, CA) at a final concentration of 1 μg/ml. Antigen retrieval was performed by using boiling citrate buffer as described previously (30). Negative controls included substitution of the primary antibody with nonimmune rabbit IgG (Sigma Chemical Co., St. Louis, MO) at the same final concentration.

Song *et al.* • Progesterone and IFNτ Regulate Endometrial CTSL Endocrinology, November 2005, 146(11):4825–4833 **4827**

Western blot analyses

Protein concentrations of uterine flushes were determined using the Bradford protein assay (Bio-Rad Laboratories, Hercules, CA) with BSA as the standard. Proteins were denatured and separated by 12% SDS-PAGE and Western blot analysis conducted as described previously (22) by using enhanced chemiluminescence (SuperSignal West Pico, Pierce, Rockford, IL) and X-OMAT AR x-ray film (Kodak) according to the manufacturer's recommendations. Immunoreactive CTSL protein was detected using rabbit antihuman CTSL polyclonal antibody (catalog no. 3192-100; BioVision) at 0.5 μg/ml.

Statistical analyses

Data from slot blot hybridization analyses were subjected to least squares ANOVA using the general linear models procedures of the Statistical Analysis System (Cary, NC). Slot blot hybridization data were corrected for differences in sample loading using the 18S rRNA data as a covariate. Data from study One were analyzed for effects of day, pregnancy status (C or PX), and their interaction. Effects of day were determined by least squares regression analysis. Data from study 2 were analyzed using preplanned orthogonal contrasts (P+CX *vs.* P+IFN, P+CX *vs.* P+ZK+CX, and P+IFN *vs.* P+ZK+IFN). Data are presented as least squares means with overall SE values.

Results

Effects of estrous cycle and pregnancy on expression of CTS mRNAs in ovine endometrium (study 1)

Steady-state levels of ovine *CTSB, CTSD, CTSH, CTSK, CTSL, CTSS,* and *CTSZ* mRNAs in endometria from C and PX ewes were determined by slot blot hybridization analyses (Fig. 1).

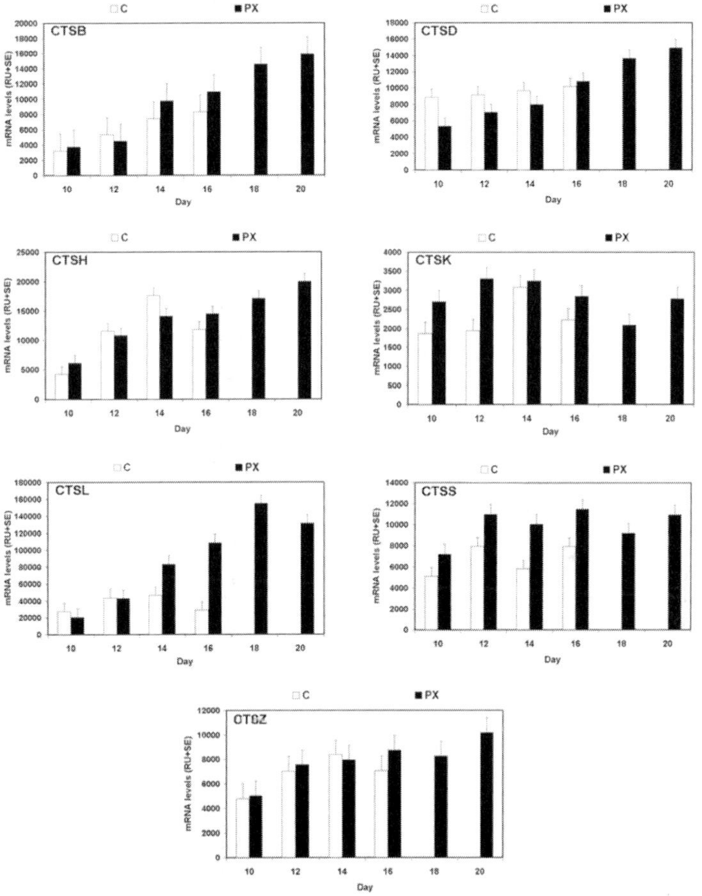

FIG. 1. Steady-state levels of *CTSB, CTSD, CTSH, CTSK, CTSL, CTSS,* and *CTSZ* mRNAs in endometria from C and PX ewes as determined by slot blot analysis. See text for description of effects of day of the C or PX on mRNA levels in the endometrium.

Expression of *CTSB* mRNA was lowest on d 10 and increased to d 16 or 20 in C and PX ewes, respectively (linear effect of day, $P < 0.01$). Endometrial levels of *CTSD* mRNA did not change in C ewes but increased from d 10–20 in PX ewes (linear effect of day, $P < 0.01$). *CTSH* mRNA levels increased from d 10–14 in C ewes and from d 10–20 in PX ewes (linear effect of day, $P < 0.01$). In contrast, *CTSK* mRNA did not change ($P > 0.10$) in endometria of C and PX ewes. *CTSL* mRNA was affected ($P < 0.05$) by day, status, and their interaction. In C ewes, *CTSL* mRNA increased from d 10–14 and then declined to d 16 (quadratic effect of day, $P < 0.05$). In PX ewes, *CTSL* mRNA increased about 8-fold between d 10 and 18 (linear effect of day, $P < 0.01$). Furthermore, *CTSL* mRNA levels in the endometrium were greater on d 14 and 16 in PX than C ewes (day × status, $P < 0.05$). Endometrial *CTSS* and *CTSZ* mRNA levels were not affected by pregnancy status or day or their interaction ($P > 0.10$).

In situ hybridization analyses determined the location of *CTS* gene expression in endometria. In C and PX ewes, *CTSB* mRNA was detected in the endometrial LE, ductal sGE, stratum compactum stroma and cells distributed throughout the stroma that appeared to be immune cells based on their morphology (Fig. 2). Abundant *CTSB* mRNA was detected in the trophectoderm of the conceptus. *CTSD* mRNA was expressed at low levels in the endometrial LE and sGE; however, abundant *CTSD* mRNA was detected in the trophectoderm of the conceptus. *CTSH* mRNA was expressed at moderate levels in the endometrial LE and GE, particularly on d 18 and 20 in PX ewes. In C and PX ewes, *CTSK* mRNA was expressed at moderate levels in the endometrial LE and stroma as well as cells within the stroma that appeared to be immune cells based on their morphology and location.

CTSL mRNA was the most abundant CTS genes expressed in the endometrium, and it was detected only in endometrial LE and sGE (Fig. 3). Furthermore, *CTSL* mRNA was expressed by conceptus trophectoderm on d 18 and 20 of PX. *CTSS* mRNA was detected at low levels in the endometrial LE and cells within the stroma that appeared to be immune cells based on their morphology and distribution. The number of *CTSS* mRNA-positive immune-like cells increased between d 14 and 16 of pregnancy. *CTSZ* mRNA was detected at low levels specifically in the endometrial LE and sGE as well as conceptus trophectoderm on d 18 and 20 of pregnancy. No differences in expression of *CTS* mRNAs in the LE or stroma of the intercaruncular endometria were found when compared with the caruncular endometria in the uterus of either C or PX ewes (data not shown).

Collectively, results of slot blot and *in situ* hybridization analyses indicated that *CTSL* mRNA was the most abundant CTS gene expressed in the endometrium and the only CTS

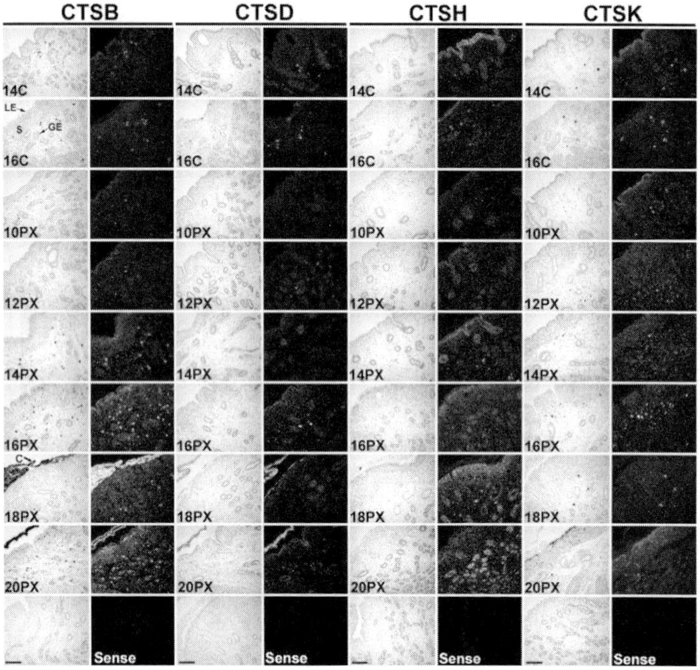

FIG. 2. *In situ* hybridization analyses of *CTSB*, *CTSD*, *CTSH*, and *CTSK* mRNAs in uteri of C and PX ewes. Cross-sections of the uterine wall from C and PX ewes were hybridized with radiolabeled antisense or sense ovine CTS cRNA probes. C, Conceptus; S, stroma. *Scale bar*, 10 μm.

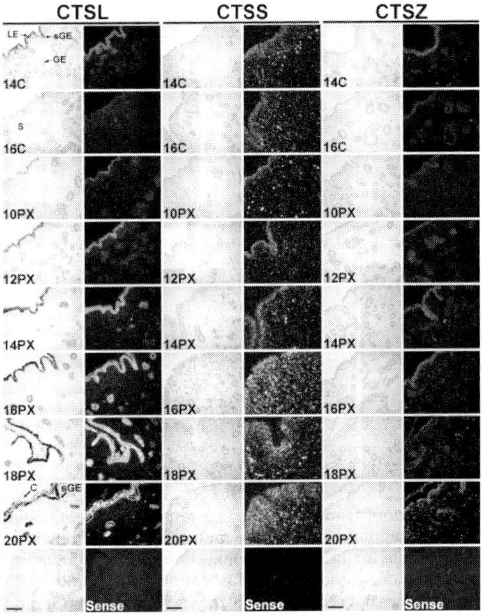

FIG. 3. *In situ* hybridization analyses of *CTSL*, *CTSS*, and *CTSZ* mRNAs in uteri of C and PX ewes. Cross-sections of the uterine wall from C and PX ewes were hybridized with radiolabeled antisense or sense ovine CTS cRNA probes. C, Conceptus; S, stroma. *Scale bar*, 10 μm.

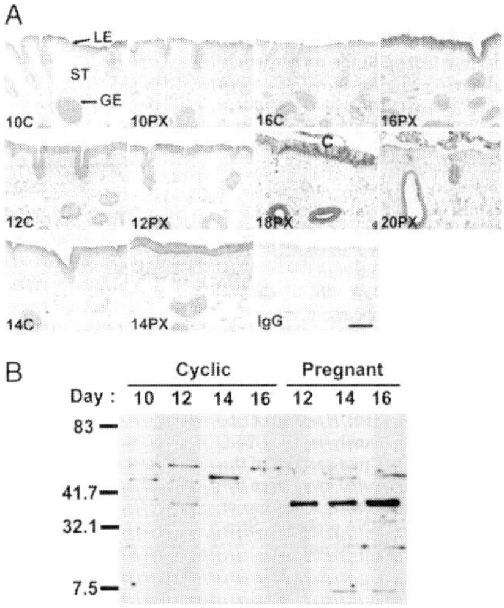

FIG. 4. CTSL protein in the endometrium and uterine flush of C and PX ewes from study 1. A, Immunoreactive CTSL protein was localized in sections of the uterus using a rabbit antihuman CTSL polyclonal antibody. For the IgG control, normal rabbit IgG was substituted for the primary antibody. Sections were not counterstained. C, Conceptus; ST, stroma. *Scale bar*, 10 μm. B, Representative Western blot analysis of CTSL in uterine flushings. Proteins in uterine flushings were analyzed by 12% SDS-PAGE (10 μg/lane), and immunoreactive protein was detected by Western blot analysis using rabbit antihuman CTSL polyclonal antibody that detects both the proenzyme and the mature forms of CTSL.

in the endometrium that appeared to be regulated by progesterone and a product of the conceptus. Therefore, CTSL protein was studied in the uterus.

CTSL protein in the endometrium and uterine lumen (study 1)

Consistent with *in situ* hybridization analyses, immunoreactive CTSL protein was observed predominantly in the LE and sGE in the endometrium of C and PX ewes (Fig. 4A). In PX ewes, the amount of immunoreactive CTSL protein increased from d 10–16 and was observed predominantly near the apical surface of the LE. Less immunoreactive CTSL protein was detected in the stroma and conceptus trophectoderm.

Western blot analyses detected abundant levels of the 38- to 40-kDa form of pro-CTSL in the uterine flushings from PX but not C ewes (Fig. 4B). Furthermore, the cleaved and active forms of CTSL, made up of 21- and 5-kDa subunits, were also detected at very low abundance in uterine flushings from PX ewes.

Effects of progesterone and IFNτ on endometrial CTS expression (study 2)

To determine whether P and IFNτ regulated CTS gene expression in the endometrium, a study was conducted as described in *Materials and Methods* (Fig. 5A). As illustrated in Fig. 5B, treatment with P increased *CTSL* mRNA in the endometrium (P+CX *vs.* P+ZK+CX, $P < 0.001$), which was further stimulated by about 3-fold in ewes receiving intrauterine administration of roIFNτ (P+CX *vs.* P+IFN, $P < 0.01$), but roIFNτ did not stimulate *CTSL* mRNA in ewes receiving the ZK antiprogestin (P+IFN *vs.* P+ZK+IFN, $P > 0.10$). *In situ* hybridization analyses revealed that *CTSL* mRNA was expressed abundantly only in the endometrial LE and sGE of ewes treated with P (P+CX and P+IFN) (Fig. 5C).

Endometrial *CTSB* mRNA was stimulated by P (P+CX *vs.* P+ZK+CX, $P < 0.02$) but decreased by roIFNτ in ewes receiving P (P+CX *vs.* P+IFN, $P < 0.01$), whereas roIFNτ increased *CTSB* mRNA in ewes receiving P and ZK (P+ZK+CX *vs.* P+ZK+IFN, $P < 0.04$) (Fig. 6). Expression of *CTSD* mRNA was not affected ($P > 0.10$) by steroid or intrauterine roIFNτ treatment. Endometrial *CTSH* mRNA was increased by IFNτ (P+CX *vs.* P+IFN, $P < 0.001$) but not affected by other treatments ($P > 0.10$). *CTSK* mRNA was decreased by P (P+CX *vs.* P+ZK+CX, $P < 0.02$) but increased by roIFNτ in ewes receiving P (P+CX *vs.* P+IFN, $P < 0.01$) or P+ZK (P+ZK+CX *vs.* P+ZK+IFN, $P < 0.001$). *CTSS* mRNA was also stimulated by P (P+CX *vs.* P+ZK+CX, $P < 0.02$). In ewes receiving P only, roIFNτ decreased *CTSS* mRNA in the endometrium (P+CX *vs.* P+IFN, $P = 0.06$). *CTSZ* mRNA was slightly stimulated by P (P+CX *vs.* P+ZK+CX, $P < 0.05$) and increased by roIFNτ in ewes re-

endometrial LE and sGE of ovariectomized ewes treated with P for 12 d, but this expression was prevented by administration of the PR antagonist ZK 136,317. Continuous exposure of the sheep uterus to P for 8–10 d down-regulates PR expression in endometrial LE and sGE but not stroma or myometrium (25). PRs are present in the endometrial epithelia of P+ZK-treated sheep (42) because PR antagonists prevent the inhibitory effects of P on the PR gene expression. Consequently, P modulation of *CTSL* mRNA may be attributed, at least in part, to down-regulation of PR by P that occurs in LE and sGE between d 10 and 12 of the cycle and pregnancy (15, 41). Thus, PR loss in endometrial epithelia may reprogram these cells, allowing them to increase expression of genes associated with implantation (15, 16). Alternatively, P may act on PR-positive stromal cells to induce them to express growth factors or changes in the ECM that regulate expression of selected epithelial genes (15).

In addition to regulation by P, the present studies indicate that *CTSH*, *CTSK*, *CTSL*, and *CTSZ* are regulated by IFNτ. IFNτ is the pregnancy recognition hormone in sheep that acts on the endometrium to prevent development of the luteolytic mechanism, thereby maintaining the corpus luteum and production of P (16). Of particular note, *CTSL* is a novel gene stimulated by IFNτ in endometrial LE and sGE as expression between d 10 and 18 of early pregnancy and parallels the increase in production of IFNτ by the elongating conceptus, which is maximal on d 16 (43). In study 2, intrauterine administration of roIFNτ increased *CTSL* mRNA but only in P-treated ewes. One hypothesis is that IFNτ can stimulate transcription of the *CTSL* gene only in the absence of repression by liganded PR. Alternatively, the PR-positive stroma may produce a progestamedin that is also required for LE and sGE to respond to IFNτ (16). The signaling pathway whereby IFNτ regulates transcription of the *CTSL* gene is not known, but it clearly does not involve the classical Janus kinase-signal transducer and activator of transcription signaling pathway (15, 19, 26, 29). To date, *WNT7A* and *LGALS15* (galectin-15) are the only other genes identified in endometrial LE and sGE that are induced or stimulated by IFNτ, respectively (26, 44). Thus, the diverse actions of IFNτ on the endometrium include repression of genes, including ERα, to abrogate activation of the luteolytic mechanism as well as stimulation of genes that are critical to implantation, placentation, and conceptus growth and development (15). Knowledge of mechanisms whereby IFNτ stimulates *CTSL* gene expression in the endometrial LE and sGE is expected to unravel a nonclassical signaling pathway for type I IFNs. Future studies will focus on the role of CTSL, other CTS family members, and their inhibitors in endometrial remodeling and conceptus implantation and placentation.

Acknowledgments

Received June 24, 2005. Accepted August 2, 2005.

Address all correspondence and requests for reprints to: Fuller W. Bazer, Center for Animal Biotechnology and Genomics, 442 Kleberg Center, 2471 TAMU, Texas A&M University, College Station, Texas 77843-2471. E-mail: fbazer@cvm.tamu.edu.

References

1. **Kirschke H, Barrett AJ, Rawlings ND** 1998 Lysosomal cysteine proteases. Oxford, UK: Oxford University Press
2. **Schmitt M, Goretzki L, Janicke F, Calvete J, Eulitz M, Kobayashi H, Chucholowski N, Graeff H** 1991 Biological and clinical relevance of the urokinase-type plasminogen activator (uPA) in breast cancer. Biomed Biochim Acta 50:731–741
3. **Jokimaa V, Oksjoki S, Kujari H, Vuorio E, Anttila L** 2001 Expression patterns of cathepsins B, H, K, L and S in the human endometrium. Mol Hum Reprod 7:73–78
4. **Afonso S, Romagnano L, Babiarz B** 1997 The expression and function of cystatin C and cathepsin B and cathepsin L during mouse embryo implantation and placentation. Development 124:3415–3425
5. **Elangovan S, Moulton BC** 1980 Blastocyst implantation in the rat and the immunohistochemical distribution and rate of synthesis of uterine lysosomal cathepsin D. Biol Reprod 23:663–668
6. **Li W, Jaffe RC, Verhage HG** 1992 Immunocytochemical localization and messenger ribonucleic acid levels of a progesterone-dependent endometrial secretory protein (cathepsin L) in the pregnant cat uterus. Biol Reprod 47:21–28
7. **Li WG, Jaffe RC, Fazleabas AT, Verhage HG** 1991 Progesterone-dependent cathepsin L proteolytic activity in cat uterine flushings. Biol Reprod 44:625–631
8. **Verhage HG, Boomsma RA, Mavrogianis PA, Li W, Fazleabas AT, Jaffe RC** 1989 Immunological characterization and immunocytochemical localization of a progesterone-dependent cat endometrial secretory protein. Biol Reprod 41: 347–354
9. **Geisert RD, Blair RM, Pratt T, Zavy MT** 1997 Characterization and proteolytic activity of a cathepsin L-like polypeptide in endometrium and uterine flushings of cycling, pregnant and steroid-treated ovariectomized gilts. Reprod Fertil Dev 9:395–402
10. **Roberts RM, Bazer FW, Baldwin N, Pollard WE** 1976 Progesterone induction of lysozyme and peptidase activities in the porcine uterus. Arch Biochem Biophys 177:499–507
11. **Salamonsen LA, Woolley DE** 1996 Matrix metalloproteinases in normal menstruation. Hum Reprod 11(Suppl 2):124–133
12. **Salamonsen LA** 1999 Role of proteases in implantation. Rev Reprod 4:11–22
13. **Curry Jr TE, Osteen KG** 2003 The matrix metalloproteinase system: changes, regulation, and impact throughout the ovarian and uterine reproductive cycle. Endocr Rev 24:428–465
14. **Salamonsen LA, Nagase H, Woolley DE** 1995 Matrix metalloproteinases and their tissue inhibitors at the ovine trophoblast-uterine interface. J Reprod Fertil 49:29–37
15. **Spencer TE, Johnson GA, Burghardt RC, Bazer FW** 2004 Progesterone and placental hormone actions on the uterus: insights from domestic animals. Biol Reprod 71:2–10
16. **Spencer TE, Johnson GA, Bazer FW, Burghardt RC** 2004 Implantation mechanisms: insights from the sheep. Reproduction 128:657–668
17. **Spencer TE, Ott TL, Bazer FW** 1996 τ-Interferon: pregnancy recognition signal in ruminants. Proc Soc Exp Biol Med 213:215–229
18. **Spencer TE, Bazer FW** 2002 Biology of progesterone action during pregnancy recognition and maintenance of pregnancy. Front Biosci 7:d1879–d1998
19. **Fleming JA, Choi Y, Johnson GA, Spencer TE, Bazer FW** 2001 Cloning of the ovine estrogen receptor-α promoter and functional regulation by ovine interferon-τ. Endocrinology 142:2879–2887
20. **Spencer TE, Burghardt RC, Johnson GA, Bazer FW** 2004 Conceptus signals for establishment and maintenance of pregnancy. Anim Reprod Sci 82–83: 537–550
21. **Hansen TR, Austin KJ, Perry DJ, Pru JK, Teixeira MG, Johnson GA** 1999 Mechanism of action of interferon-τ in the uterus during early pregnancy. J Reprod Fertil Suppl 54:329–339
22. **Spencer TE, Bartol FF, Bazer FW, Johnson GA, Joyce MM** 1999 Identification and characterization of glycosylation-dependent cell adhesion molecule 1-like protein expression in the ovine uterus. Biol Reprod 60:241–250
23. **Spencer TE, Gray A, Johnson GA, Taylor KM, Gertler A, Gootwine E, Ott TL, Bazer FW** 1999 Effects of recombinant ovine interferon τ, placental lactogen, and growth hormone on the ovine uterus. Biol Reprod 61:1409–1418
24. **Van Heeke G, Ott TL, Strauss A, Ammaturo D, Bazer FW** 1996 High yield expression and secretion of the ovine pregnancy recognition hormone interferon-τ by *Pichia pastoris*. J Interferon Cytokine Res 16:119–126
25. **Spencer TE, Becker WC, George P, Mirando MA, Ogle TF, Bazer FW** 1995 Ovine interferon-τ regulates expression of endometrial receptors for estrogen and oxytocin but not progesterone. Biol Reprod 53:732–745
26. **Kim S, Choi Y, Bazer FW, Spencer TE** 2003 Identification of genes in the ovine endometrium regulated by interferon τ independent of signal transducer and activator of transcription 1. Endocrinology 144:5203–5214
27. **Johnson GA, Stewart MD, Gray CA, Choi Y, Burghardt RC, Yu-Lee LY, Bazer FW, Spencer TE** 2001 Effects of the estrous cycle, pregnancy, and interferon τ on 2′,5′-oligoadenylate synthetase expression in the ovine uterus. Biol Reprod 64:1392–1399
28. **Spencer TE, Stagg AG, Ott TL, Johnson GA, Ramsey WS, Bazer FW** 1999 Differential effects of intrauterine and subcutaneous administration of recom-

binant ovine interferon τ on the endometrium of cyclic ewes. Biol Reprod 61:464–470

29. **Choi Y, Johnson GA, Burghardt RC, Berghman LR, Joyce MM, Taylor KM, Stewart MD, Bazer FW, Spencer TE** 2001 Interferon regulatory factor-two restricts expression of interferon-stimulated genes to the endometrial stroma and glandular epithelium of the ovine uterus. Biol Reprod 65:1038–1049
30. **Taylor KM, Gray CA, Joyce MM, Stewart MD, Bazer FW, Spencer TE** 2000 Neonatal ovine uterine development involves alterations in expression of receptors for estrogen, progesterone, and prolactin. Biol Reprod 63:1192–1204
31. **Green JA, Xie S, Quan X, Bao B, Gan X, Mathialagan N, Beckers J-F, Roberts RM** 2000 Pregnancy-associated bovine and ovine glycoproteins exhibit spatially and temporally distinct expression patterns during pregnancy. Biol Reprod 62:1624–1631
32. **Peltier MR, Hansen PJ** 2001 Immunoregulatory activity, biochemistry, and phylogeny of ovine uterine serpin. Am J Reprod Immunol 45:266–272
33. **Johnson GA, Burghardt RC, Bazer FW, Spencer TE** 2003 Osteopontin: roles in implantation and placentation. Biol Reprod 69:1458–1471
34. **Johnson GA, Burghardt RC, Joyce MM, Spencer TE, Bazer FW, Gray CA, Pfarrer C** 2003 Osteopontin is synthesized by uterine glands and a 45-kDa cleavage fragment is localized at the uterine-placental interface throughout ovine pregnancy. Biol Reprod 69:92–98
35. **Wimsatt WA** 1950 Hew histological observations on the placenta of the sheep. Am J Anat 87:391–436
36. **Wimsatt WA** 1951 Observations of the morphogenesis, cytochemistry and significance of the binucleate giant cells of the placenta of ruminants. J Anat 89:233–282
37. **Nomura T, Katunuma N** 2005 Involvement of cathepsins in the invasion, metastasis and proliferation of cancer cells. J Med Invest 52:1–9
38. **Babiarz BS, Romagnano LC, Kurilla GM** 1992 Interaction of mouse ectoplacental cone trophoblast and uterine decidua *in vitro*. In Vitro Cell Dev Biol 28A:500–508
39. **Cheon Y-P, DeMayo FJ, Bagchi MK, Bagchi IC** 2004 Induction of cytotoxic T-lymphocyte antigen-2β, a cysteine protease inhibitor in decidua: a potential regulator of embryo implantation. J Biol Chem 279:10357–10363
40. **Hamilton RT, Bruns KA, Delgado MA, Shim JK, Fang Y, Denhardt DT, Nilsen-Hamilton M** 1991 Developmental expression of cathepsin L and c-rasHa in the mouse placenta. Mol Reprod Dev 30:285–292
41. **Spencer TE, Bazer FW** 1995 Temporal and spatial alterations in uterine estrogen receptor and progesterone receptor gene expression during the estrous cycle and early pregnancy in the ewe. Biol Reprod 53:1527–1543
42. **Johnson GA, Spencer TE, Burghardt RC, Taylor KM, Gray CA, Bazer FW** 2000 Progesterone modulation of osteopontin gene expression in the ovine uterus. Biol Reprod 62:1315–1321
43. **Farin CE, Imakawa K, Roberts RM** 1989 *In situ* localization of mRNA for the interferon, ovine trophoblast protein-1, during early embryonic development of the sheep. Mol Endocrinol 3:1099–1107
44. **Gray CA, Adelson DL, Bazer FW, Burghardt RC, Meeusen EN, Spencer TE** 2004 Discovery and characterization of an epithelial-specific galactin in the endometrium that forms crystals in the trophectoderm. Proc Natl Acad Sci USA 101:7982–7987

***Endocrinology* is published monthly by The Endocrine Society (http://www.endo-society.org), the foremost professional society serving the endocrine community.**

0013-7227/06/$15.00/0
Printed in U.S.A.

Endocrinology 147(2):899–911

doi: 10.1210/en.2005-1120

Estrogen Regulates Transcription of the Ovine Oxytocin Receptor Gene through GC-Rich SP1 Promoter Elements

JoAnn G. W. Fleming, Thomas E. Spencer, Stephen H. Safe, and Fuller W. Bazer

Center for Animal Biotechnology and Genomics (J.G.W.F., T.E.S., F.W.B.) and Departments of Animal Science (T.E.S., F.W.B.) and Veterinary Physiology and Pharmacology (S.H.S., F.W.B.), Texas A&M University, College Station, Texas 77843

Establishment of pregnancy in ruminants results from paracrine signaling by interferon τ (IFNT) from the conceptus to uterine endometrial luminal epithelia (LE) that prevents release of luteolytic prostaglandin $F_{2\alpha}$ pulses. In cyclic and pregnant ewes, progesterone down-regulates progesterone receptor (PGR) gene expression in LE. In cyclic ewes, loss of PGR allows for increases in estrogen receptor α (ESR1) and then oxytocin receptor (OXTR) gene expression followed by oxytocin-induced prostaglandin $F_{2\alpha}$ pulses. In pregnant ewes, IFNT inhibits transcription of the ESR1 gene, which presumably inhibits OXTR gene transcription. Alternatively, IFNT may directly inhibit OXTR gene transcription. The 5′ promoter/enhancer region of the ovine OXTR gene was cloned and found to contain predicted binding sites for activator protein 1, SP1, and PGR, but not for ESR1. Deletion analysis showed that the basal promoter activity was dependent on the region from −144 to −4 bp that contained only SP1 sites. IFNT did not affect activity of the OXTR promoter. In cells transfected with ESR1, E2, and ICI 182,780 increased promoter activity due to GC-rich SP1 binding sites at positions −104 and −64. Mutation analyses showed that the proximal SP1 sites mediated ESR1 action as well as basal activity of the promoter. In response to progesterone, progesterone receptor B also increased OXTR promoter activity. SP1 protein was constitutively expressed and abundant in the LE of the ovine uterus. These results support the hypothesis that the antiluteolytic effects of IFNT are mediated by direct inhibition or silencing of ESR1 gene transcription, thereby precluding ESR1/SP1 from stimulating OXTR gene transcription. (*Endocrinology* 147: 899–911, 2006)

MATERNAL RECOGNITION OF pregnancy is the physiological process whereby the conceptus signals its presence to the maternal system and prolongs lifespan of the corpus luteum (CL) (1). Sheep experience uterine-dependent estrous cycles until establishment of pregnancy (2). The estrous cycle is dependent on the uterus, because it releases prostaglandin $F_{2\alpha}$ (PGF) in a pulsatile manner to induce luteolysis during late diestrus. The luteolytic pulses of PGF are produced by the endometrial luminal epithelium (LE) and superficial ductal glandular epithelium (sGE) and are generated by oxytocin binding to oxytocin receptors (OXTR) on those epithelia (3–6). Endometrial epithelial expression of OXTR is regulated primarily by receptors for progesterone [progesterone receptor (PGR)] and estrogen [estrogen receptor α (ESR1)] (7–9). During proestrus and estrus (d −3 to 0), estrogen from ovarian follicles increases ESR1, PGR, and OXTR expression. During diestrus (d 3–15), progesterone levels increase and act via PGR to "block" expression of ESR1 and OXTR in LE and glandular epithelium (GE) between d 5 and 11 after onset of estrus (10–12). However, continuous exposure of the uterus to progesterone down-regulates expression of PGR in LE/sGE after d 11 and GE after d 13, allowing for rapid increases in expression of ESR1 on d 12–13 and then OXTR on d 14 in those epithelia (9, 10, 13, 14). ESR1, presumably activated by estrogen from ovarian follicles or possibly growth factors from the stroma, stimulates transcription of the OXTR gene in the ovine endometrium (15–18). Oxytocin, secreted from as early as d 9 from posterior pituitary and/or the CL, binds to OXTR to induce pulsatile release of luteolytic PGF between d 14–16 (3). In response to four to five luteolytic pulses of PGF over a 25-h period, the CL undergoes functional and structural regression. The systemic loss in progesterone between d 15–16 allows the ewe to return to estrus, complete the 17-d estrous cycle, and experience another opportunity to mate and establish pregnancy.

Interferon τ (IFNT), the pregnancy recognition signal in ruminants, is expressed by conceptus trophectoderm between d 10 and 21 of gestation (19) and acts in a paracrine antiluteolytic manner on the endometrial LE and sGE to inhibit *OXTR* gene expression (10, 13, 20). IFNT does not inhibit basal production of PGF, which is higher in pregnant than cyclic ewes, and the conceptus and IFNT do not decrease expression of *PTGS2* [prostaglandin-endoperoxide synthase 2 (prostaglandin G/H synthase and cyclooxygenase)], the rate-limiting enzyme in prostaglandin production, in endometrial epithelia of pregnant ewes (5, 21). The antiluteolytic effects of IFNT to inhibit *OXTR* gene expression in the endometrial LE and sGE have been hypothesized to involve: 1) direct stabilization of epithelial PGR expression that, in turn, extends progesterone inhibition of ESR1 and

First Published Online October 27, 2005

Abbreviations: AP-1, Activator protein 1; CL, corpus luteum; E3, 17β-estradiol; ERE, ESR1 response element; ESR1, estrogen receptor α; FBS, fetal bovine serum; GAS, γ activation site; GE, glandular epithelium; IFNT, interferon τ; IRF, interferon regulatory factor; IRFE, interferon regulatory factor element; ISRE, interferon-stimulated response element; LE, luminal epithelium; OXTR, oxytocin receptor; PGF, prostaglandin $F_{2\alpha}$; PGR, progesterone receptor; PGR-B, progesterone receptor B form; PRE, progesterone response element; sGE, superficial ductal glandular epithelium.

Endocrinology **is published monthly by The Endocrine Society (http://www.endo-society.org), the foremost professional society serving the endocrine community.**

OXTR expression; 2) direct inhibition or silencing of *ESR1* gene transcription that, in turn, indirectly inhibits *OXTR* gene expression; and 3) direct inhibition of *OXTR* gene transcription. IFNT does not inhibit PGR loss in the endometrial epithelia (12, 16), but rather appears to indirectly inhibit *OXTR* gene transcription by silencing *ESR1* gene transcription in the endometrial LE (15, 16, 22). Indeed, the 2.7-kb 5′ promoter/enhancer region of the ovine *ESR1* gene was found to contain four interferon regulatory factor elements (IRFEs) and one interferon-stimulated response element (ISRE) that were functional in binding interferon regulatory factor (IRF)2 (23). IRF2 is a transcriptional repressor that is expressed specifically in the endometrial LE and sGE of the ovine uterus and increases in expression from d 10–16 of pregnant but not cyclic ewes (24). In transfection assays, IFNT inhibited transcriptional activity of the ovine *ESR1* promoter, and analyses of sequential 5′-deletion mutants of the ovine *ESR1* promoter indicated that the effects of IFNT may be mediated by IRFEs as well as other elements (23). Physiological evidence indicates that estrogen stimulates *OXTR* gene expression and that IFNT does not directly inhibit *OXTR* gene expression in the sheep uterus (15, 25). Direct effects of estrogen and IFNT on the ovine *OXTR* gene have not been studied previously, because the gene had not been cloned. The bovine *OXTR* gene promoter region was cloned and found to contain an IRFE, ESR1 response element (ERE) half sites, and SP1 sites, and could be transactivated by estrogen if cells were cotransfected with ESR1 and steroid receptor coactivator 1 (26). Curiously, IRF2 overexpression increased activity of the bovine *OXTR* promoter, but a direct effect of IFNT on promoter activity was not reported. In all mammals, estrogen is considered a key regulator of *OXTR* gene expression; however, all of the studied *OXTR* genes lack a complete ERE, suggesting estrogen induction of *OXTR* gene transcription may be indirect rather than due to direct ESR1/ERE interactions (27).

Our working hypothesis is that estrogen acts indirectly to regulate *OXTR* gene expression in the endometrium of the ovine uterus and that the antiluteolytic effects of IFNT are manifest on silencing or inhibition of *ESR1* gene transcription in LE and sGE subsequent to PGR down-regulation, which then precludes the ability of estrogen or perhaps growth factors to activate ESR1 and stimulate *OXTR* gene expression. These actions of IFNT prevent formation of OXTR and abrogate uterine release of luteolytic pulses of PGF (4, 8, 23, 28). The objective of the present studies was to determine if estrogen and/or IFNT regulate *OXTR* promoter activity. The promoter region of the ovine *OXTR* gene was cloned, sequenced, and analyzed for functional *cis*-elements mediating effects of steroid hormones and IFNT. Relevant findings are that IFNT does not affect *OXTR* promoter activity nor inhibit estrogen induction of *OXTR* promoter activity. Two GC-rich SP1 binding sites, located within 140 bp of the translational start site, regulated basal activity of the promoter as well as hormonal responsiveness to 17β-estradiol (E2), ICI 182,780, and progesterone. Thus, estrogen regulation of ovine *OXTR* promoter activity involves ESR1/SP1 interactions.

Materials and Methods

Cells and reagents

The 2fTGH (parental) and U3A (STAT1-deficient 2fTGH) immortalized cells (29) were maintained in DMEM-F12 medium (Sigma-Aldrich Corp., St. Louis, MO) supplemented with 5% fetal bovine serum (FBS) (Hyclone, Logan, UT) and penicillin/streptomycin sulfate/amphotericin B solution (Invitrogen, Carlsbad, CA). Recombinant ovine IFNT (10^8 antiviral units/mg) was prepared and assayed as described previously (30). Restriction endonucleases, T4 DNA ligase, T4 DNA kinase, and recombinant human SP1 protein were purchased from Promega (Madison, WI). Vent *Taq* polymerase (New England Biolabs, Beverly, MA), AmpliTaq polymerase (Applied Biosystems, Foster City, CA), and ExTaq polymerase (Takara, Kyoto, Japan) were used. R5020 was purchased from PerkinElmer Life Sciences (Boston, MA), and E2 was from Sigma-Aldrich Corp. Plasmid DNAs were purified by the alkaline lysis method according to the manufacturer's instructions (Qiagen, Valencia, CA).

Cloning of the 5′ upstream region and luciferase constructs

A 273-bp DNA fragment of the 5′ end of the coding region of the ovine *OXTR* gene was PCR amplified in 25-μl reactions containing 40 ng ovine genomic DNA (a gift from Dr. C. Gill, Texas A&M University, College Station, TX), PCR Optimized Buffer N (Invitrogen), 125 μM dNTPs, 2.5 μM forward primer (5′-GGAGGCGGTCAACGGGAG-3′), 2.5 μM reverse primer (5′-CGTGATGTCCCACAGAAGC-3′), and 1 U AmpliTaq polymerase using an Eppendorf Mastercycler thermocycler with conditions of: 1) 94 C for 5 min; 2) 94 C for 30 sec, 53 C for 30 sec, and 72 C for 30 sec for 35 cycles; and 3) 72 C for 7 min. The product was cloned into pCRII Dual using the TA Cloning kit (Invitrogen). The cloned 273-bp *OXTR* fragment was excised with *Eco*RI, random prime labeled with [α-^{32}P]dCTP, and used to screen an ovine λ genomic library cloned in λBluestar (Novagen, Madison, WI), which was kindly provided by Dr. J. C. DeMartini (Colorado State University, Fort Collins, CO). A recombinant pBluestar phagemid DNA, containing approximately 9.8 kb of ovine genomic DNA, was excised from plaque-purified isolates by Cre-Lox excision according to the manufacturer's instructions. Physical mapping, Southern blotting with the 273-bp *OXTR* probe, and sequencing with T7 and T3 primers and several OXTR-specific primers were done to identify the *OXTR* coding and upstream sequences in the clone. The 5′ terminus of the coding sequence and the contiguous upstream sequences were sequenced on both strands by the dideoxy chain termination method. The *OXTR* upstream region was immediately adjacent to the *Not*I site in the vector multiple cloning site. The upstream region was excised as an 826-bp fragment using the vector *Not*I and a *Not*I site at −4 relative to the ATG (+1) of the *OXTR* coding sequences and cloned in pCRII Dual in both orientations as determined by physical mapping of the unique *Bgl*II site at −144. The OXTR(−830/−4) upstream region then was directionally subcloned into the luciferase vector pGL3Basic (Promega). Computer-assisted prediction of transcription factor binding sites in the promoter sequence was performed with TESS (Transcription Element Search System) using TRANSFAC version 4.0 (30) and MatInspector using Matrix Family Library Version 2.4 (Genomatix, Munich, Germany) (31).

Truncations of the upstream region at approximately 100-bp intervals were made by PCR amplification in 50-μl reactions containing 1 ng pCRII-OXTR(−830/−4), 25 mM TAPS (*N*-[Tris(hydroxymethyl)-methyl]-3-aminopropane-sulfonic acid, pH 9.3 at room temperature), 50 mM KCl, 2 mM $MgCl_2$, 200 μM dNTPs, 0.2 μM each forward primer (Table 1), 0.2 μM T7 primer as reverse primer, and 1.25 U ExTaq in a Perkin-Elmer 9700 thermocycler using conditions of: 1) 94 C for 2 min; 2) 94 C for 1 min, 50 C for 1 min, 72 C for 1 min for 30 cycles; and 3) 72 C for 10 min. The truncated PCR products were cloned into pCRII Dual, physically mapped to determine orientation, and directionally subcloned into the pGL3Basic vector. The OXTR(−144/−4) LUC truncated clone was made by digestion of OXTR(−220/−4) LUC plasmid DNA with *Bgl*II, reinserting the *Bgl*II (−144/−4) fragment into the luciferase vector, and physically mapping the construct to confirm proper orientation. Point mutations were introduced into the SP1 sites at −104 and −64 of the OXTR(−220/−4) truncated pCRII subclone by PCR-directed mutagenesis using the primers listed in Table 1 and Vent *Taq* polymerase and ExTaq (31). A OXTR(−220/−4) pCRII clone containing an introduced mutation at the −64 SP1 site was used as the template for the construction of OXTR−220 constructs with mutated SP1 sites at both

TABLE 1. Sequences of primers for cloning and gel shift assays

Primer sequence (5′→3′)	Primer name/genotype
Truncations	
CTGAGAACAGAGGTTAGG	OXTR-711 forward
GGATGGTCTTGAGTAAGG	OXTR-584 forward
GGAGTAACTAGTTGGGAG	OXTR-469 forward
GCTGATTTCCGGTGTCTC	OXTR-340 forward
GAGGCATACATGAGTCG	OXTR-220 forward
ACGCACGGTCTTCTCACG	OXTR-206 forward
TAATACGACTCACTATAGGG	T7 reverse
EMSA oligonucleotides	
GCTCG***CCCCGCCCC***GATCGAAT	SP1 consensus
TGCAGCTCCACGCGCACCTCGC	Nonspecific competitor
TCGGGCCCCGCCCCGCGCGCCG	wt OTR-64
GGAGCCCCCACCCCGCCCCCAGGCACG	wt OTR-104
TCGGGCCCAGACCCGCGCGCCG	mut OXTR-64
GGAGCCCCCACCCAGACCCCAGGCACG	mut1 OXTR-104
GGAGCCCCAAACCCGCCCCCAGGCACG	mut2 OXTR-104
GGAGCCCCAAACCAGACCCCAGGCACG	mut4 OXTR-104
PCR mutagenesis	
ACCGGTACCGAGCTCATATGAGGCATACATGAGTCG	5′ terminus (ovine)
CGTTGTAAAACGACGGCCAG	M13(−20) reverse
GCGGCCGCCGCGGTGTGCCCCGGCCTTCG	3′ terminus (ovine)
GGACTGCGCGTCGGGCCCAGACCCGCGCGCCGCACC	OXTR mut −64
GCAGGGGAATCCAGGAGCCCCCACCCAGACCCCAGGCACG	mut1 OXTR −104
GCAGGGGAATCCAGGAGCCCCAAACCCGCCCCCAGGCACG	mut2 OXTR −104
GCAGGGGAATCCAGGAGCCCCAACCCCGCCCCCAGGCACG	mut3 OXTR −104
GGAGCCCCAAACCAGACCCCAGGCACG	mut4 OXTR −104

The complement of the consensus SP1 binding site is shown in *bold italics*. The sequences of the SP1 binding sites in the consensus oligonucleotide and the predicted wild-type (wt) SP1 sites at −104 and −64 in the ovine OXTR promoter are *double-underlined*. The OXTR-104 SP1 site contains overlapping putative nonconsensus and consensus SP1 binding sites, respectively, as discussed in the text and illustrated in Fig. 1. Bases in the ovine OXTR −104 and −64 SP1 sites that were mutated by C→A transversions are underscored by a *single line*. The resulting mutations are listed as "mut" if they changed C residues immediately adjacent to the central A or G residues in a putative SP1 site.

−104 and −64. Mutated DNAs were confirmed by sequencing, and fragments containing only the desired point mutations were directionally subcloned into the luciferase vector as described above.

Transient transfections and luciferase assays

Cells were subcultured into 12-well plates (67–75% confluent) and transiently transfected as described previously (23) with the following modifications. Luciferase constructs (500 ng/well) were cotransfected with either pEF1-Myc-His-LacZ (500 ng/well; Invitrogen) or expression plasmids for ovine IRF1, ovine IRF2, human ESR1wt, human ESR1IIc, or human progesterone receptor B (PGR-B) (500 ng/well except where noted). The ovine IRF1 and IRF2 mammalian overexpression plasmids have been described previously (24). The ESR1wt (wild-type human ESR1) and ESR1IIc (a DNA binding domain mutant construct) overexpression plasmids have been described previously (32, 33). The PGR-B plasmid, pSV40-hPGR-B that overexpresses the B form of human PGR, was kindly provided by Dr. M.-j. Tsai (Baylor College of Medicine, Houston, TX). Transfected cells were grown in 10% FBS for approximately 14–16 h before treatment for 24 h in serum-free medium. Phenol red-free DMEM-F12 medium and dextran-coated charcoal-stripped FBS were substituted in experiments when testing for effects of steroids. Steroid agonists and antagonists were dissolved in 100% ethanol. For the control, cells were treated with the same volume of 100% ethanol alone. Luciferase and protein assays were done as described previously (23, 24). Transfection assays were repeated a minimum of three times.

EMSA

Double-stranded oligonucleotide primers listed in Table 1 were end-labeled to high specific activity with [γ-^{32}P]ATP and T4 DNA kinase by standard methods. Purified SP1 protein (54 ng/reaction) was incubated in 10 μl reactions containing 4% glycerol, 1 mM $MgCl_2$, 0.5 mM EDTA, 50 mM NaCl, 10 mM Tris-HCl (pH 7.5), 0.5 mM DTT, 0.05 mg/ml poly(dI-dC) at room temperature for 15 min. A 100-fold excess of unlabeled specific or nonspecific competitor oligonucleotides was included in some reactions. The nonspecific competitor was a 22-nt oligonucleotide with the same base composition as the SP1 consensus oligonucleotide to minimize charge differences, but having a different sequence (Table 1). Radiolabeled probes (10 fmol per reaction) were added to reactions that were incubated at room temperature for an additional 15 min. Binding reactions were electrophoresed in 5% native polyacrylamide gels in 1× TBE. Imaging of dried gels was done with a Molecular Dynamics Typhoon phosphorimager.

Immunohistochemistry

Mature crossbred Suffolk ewes (*Ovis aries*) were observed daily for estrus in the presence of vasectomized rams and used in experiments only after they had exhibited at least two estrous cycles of normal duration (16–18 d). All experimental and surgical procedures were in compliance with the Guide for the Care and Use of Agriculture Animals and approved by the University Laboratory Animal Care and Use Committee of Texas A&M University.

At estrus (d 0), ewes were mated to either an intact or vasectomized ram as described previously (34) and then hysterectomized (n = 5 ewes per day) on either d 10, 12, 14, or 16 of the estrous cycle or d 10, 12, 14, 16, 18, or 20 of pregnancy. Pregnancy was confirmed on d 10–16 after mating by the presence of a morphologically normal conceptus(es) in the uterus. At hysterectomy, several sections (~0.5 cm) from the mid-portion of each uterine horn ipsilateral to the CL were fixed in fresh 4% paraformaldehyde in PBS (pH 7.2). After 24 h, fixed tissues were changed to 70% ethanol for 24 h and then dehydrated and embedded in Paraplast-Plus (Oxford Labware, St. Louis, MO). In monovulatory pregnant ewes, uterine tissue samples were marked as either contralateral or ipsilateral to the ovary bearing the CL. No tissues from the contralateral uterine horn were used for study.

Immunoreactive SP1 proteins were localized in cross-sections (5 μm) of the uterus using a rabbit polyclonal antibody to human SP1 (catalog SC-59; Santa Cruz Biotechnology Inc., Santa Cruz, CA) at a final working concentration of 0.1 μg/ml and a VectaStain Rabbit IgG Elite ABC kit (Vector Laboratories, Burlingame, CA) using methods described previously (35). Antigen retrieval using boiling citrate buffer was performed as described previously (35, 36). The chromagen used for peroxidase localization was 3,3′-diaminobenzidine tetrahydrochloride from Sigma Chemical Co. (St. Louis, MO). Negative controls were performed in which the primary antibody was substituted with the same concentration of purified normal rabbit IgG from Sigma Chemical Co. Multiple tissue sections from each ewe were processed as sets within an experiment. Sections were not counterstained before affixing coverslips. Representative photomicrographs of uterine tissues were taken using a Nikon Eclipse 1000 photomicroscope (Nikon Instruments Inc., Lewisville, TX) fitted with a Nikon DXM1200 digital camera. Digital images were captured and assembled using Adobe Photoshop (Adobe Systems, Seattle, WA).

Statistical analyses

The effects of steroid hormones, ICI 182,780, or IFNT on the activity of promoter reporter constructs in transient transfection assays were analyzed by least squares ANOVA using the General Linear Models procedure of the Statistical Analysis System (Cary, NC). A P value of 0.10 or less was considered statistically significant. Unless denoted, data are reported as mean with SD. Effects of treatment were determined using orthogonal or nonorthogonal contrasts using the PDIFF option of General Linear Models.

Results

Isolation and sequence analysis of the OXTR 5′ upstream region

A phagemid DNA containing approximately 9.8 kbp of ovine genomic DNA was isolated from a λ ovine genomic library by conventional screening. Southern blotting, physical mapping, and sequence analysis with several *OXTR*-specific primers revealed that the clone contained a 1059-bp region that included the 5′ end of the coding sequence of the ovine *OXTR* gene contiguous to 830 nt of upstream sequence. The 1059-bp clone was sequenced in both directions and deposited into GenBank (accession AY163261). Sequence comparisons of the ovine upstream region with the previously characterized *Bos taurus OXTR* promoter region and adjacent coding regions (GenBank AF100633) indicated that the two sequences were 91% identical. Although the coding sequence displayed high homology with the *OXTR* cDNA from other species (human, dog, pig, gorilla, rat, mouse, chicken), the promoter region had very little or no homology to those of other species except for bovine.

Bioinformatic analyses found that the ovine *OXTR* promoter region contained multiple putative binding sites for the transcription factors SP1 and activator protein 1 (AP-1), which regulates target genes by steroid hormone receptors (see Refs. 37 and 38 for review). Putative nonconsensus SP1 sites were found at −748, −543, and −444 relative to the translational start site (ATG +1) in addition to consensus sites at −104 and −64 (Fig. 1). The putative −64 SP1 binding site is identical to the consensus SP1 binding motif (5′-CCCCGCCCC-3′, sense strand). The −104 putative binding site (5′-CCCCACCCCGCCCC-3′, sense strand) also contains a consensus binding motif that is immediately adjacent to and overlaps a sequence that resembles a nonconsensus SP1 binding site (5′-CCCCACCCC-3′) (Fig. 1). Nonconsensus SP1 sites having the sequence 5′-CCCCACCCC-3′ have been reported for the long terminal repeats of the human endogenous retrovirus H family of human retrovirus-like elements (39). Alternatively, part of the sequence of the OXTR −104 SP1 site may contain a CACCC box at the 5′ end. The CACCC elements bind SP1 and related Sp/Kruppel-like factor transcription factors (40) and are involved in regulation of both basal (41) and induced levels of transcription in other systems (42). A consensus AP-1 binding site at −680 and nonconsensus AP-1 sites at −574, −329, and −221 were identified, which also mediate hormone actions (38). No full progesterone response elements (PREs) were found, but several putative PRE half sites were detected at −639, −486, −477, and −187. No full EREs or ERE half sites were found by computer-assisted analysis of the ovine OXTR upstream region. Furthermore, elements mediating effects of type I or II IFNs or STATs, including ISRE, IRFE, or γ activation sites (GAS) (43), were also not found by bioinformatic analyses.

IFNT does not regulate OXTR promoter activity

To determine whether IFNT regulated activity of the *OXTR* promoter, transient transfection assays were conducted in 2fTGH parental and U3A (STAT1 null) cells, which are responsive to IFNT (44–46) in a STAT-dependent (2fTGH) and STAT1-independent (U3A) manner. These cells have been extensively used by our laboratory as a model for the endometrial stroma and LE, respectively (44–46). Cells were transfected with the full-length OXTR(−830/−4) and sequential 5′-deletion reporter constructs and treated with different doses of recombinant ovine IFNT (10^3–10^6 antiviral units). No dose-dependent effects of IFNT on promoter activity were detected ($P > 0.10$) with any of the *OXTR* promoter-reporter constructs in either 2fTGH or U3A cells (data not shown). Cells were also cotransfected with the OXTR(−830/−4) promoter-reporter construct and either

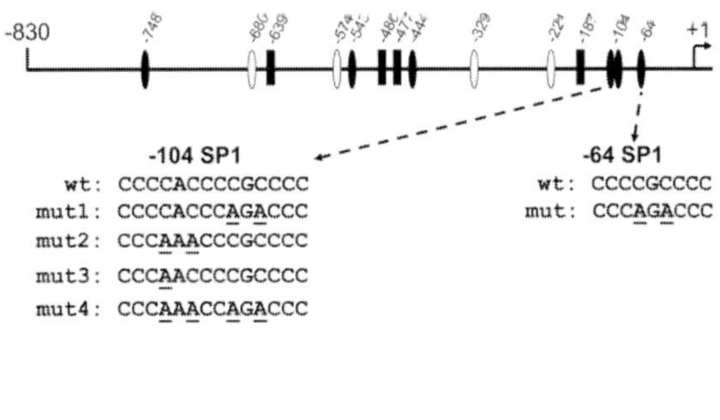

FIG. 1. Schematic representation of putative transcription factor binding sites in the promoter of the ovine *OXTR* gene. The location of mutations made in the SP1 sites of the minimal promoter are also denoted below the schematic. The sequences of the OXTR wt −104 and −64 SP1 sites are shown together with mutations created by C→A transversions (*underlined*). The −104 SP1 site contains overlapping putative nonconsensus and consensus SP1 binding sites 5′ (CCCCACCCC) and 3′ (CCCCGCCCC).

Endocrinology, February 2006, 147(2):899–911 **903**

pEF1-Myc/His-LacZ, ovine IRF1, or ovine IRF2 overexpression vectors. Relative to the LacZ control, overexpression of ovine IRF1 or ovine IRF2 did not affect ($P > 0.10$) activity of the *OXTR* promoter constructs in either cell type (data not shown). These results are consistent with the lack of IFN-responsive elements (GAS, ISRE, or IRFE) in the ovine *OXTR* promoter.

In 2fTGH cells transfected with the OXTR(−830/−4) construct and ESR1wt, E2 (10^{-8} M) increased ($P < 0.01$) luciferase activity, and this induction was not affected ($P > 0.10$) by IFNT treatment (Fig. 2). Treatment with IFNT lowered ($P < 0.05$) the basal activity of the ovine *OXTR* promoter; however, the fold induction by E2 was the same in untreated and IFNT-treated cells. Similar results were obtained using U3A cells (data not shown). The basal activity of several promoter-reporter constructs appears to be nonspecifically affected by IFNT, presumably due to effects on cell metabolism due to induction of an antiviral state (data not shown). These results indicate that IFNT does not inhibit E2-induced transactivation in cells transfected with *OXTR* promoter-derived constructs.

Estrogen and antiestrogen stimulate OXTR promoter activity

In 2fTGH and U3A cells, basal activity of the *OXTR* promoter changed with length of the promoter-reporter construct (Figs. 3 and 4). The full-length OXTR(−830/−4)-LUC construct had considerably lower basal activity compared ($P < 0.01$) to all other 5′-deletion constructs. Basal activity of the OXTR(−340/−4) construct was approximately 1.5-fold greater ($P < 0.05$) than the OXTR(−220/−4) construct containing the minimal promoter.

2fTGH cells were cotransfected with *OXTR* promoter-reporter constructs and ESR1wt (Fig. 3A) or ESR1IIc (Fig. 4B), a DNA binding domain mutant derived from ESR1wt, and treated with vehicle as a control or 10^{-8} M E2 for 24 h. In cells transfected with ESR1wt, E2 treatment increased ($P < 0.05$) activity of the OXTR(−830/−4) promoter-LUC construct as well as all 5′ deletions (Fig. 3A). Similarly, E2 treatment increased ($P < 0.01$) activity of all *OXTR* promoter-reporter

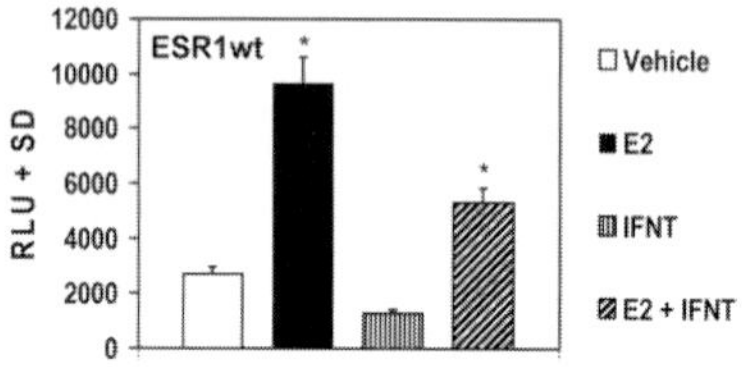

FIG. 2. Effects of E2 and IFNT in 2fTGH cells. Cells were cotransfected with constructs containing the full-length OXTR (−830/−4) promoter insert and ESR1wt, treated with vehicle, E2 (10^{-8} M), recombinant ovine IFNT (10^4 antiviral units), or E2 and IFNT for 24 h, and luciferase activity was determined as described in *Materials and Methods*. Significant induction ($P < 0.10$) is indicated with an *asterisk*, and results are expressed as mean relative light units (RLU) with SD. Four replicate determinations for each treatment group were conducted in each experiment. A representative experiment of three independent experiments with similar results is presented. Similar results were obtained using U3A cells (data not shown).

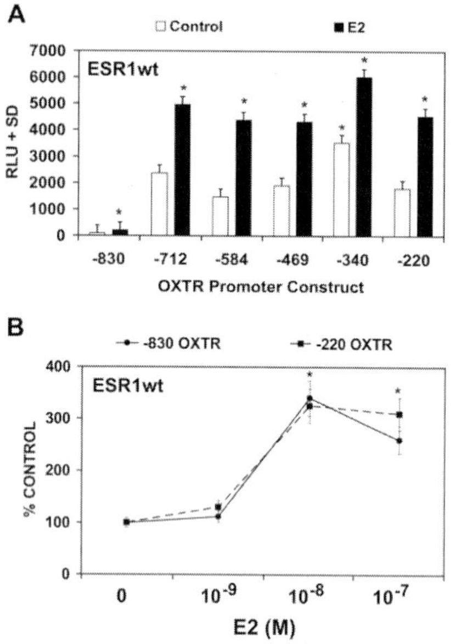

FIG. 3. Comparative activation of OXTR constructs by wild-type human ESR1 in 2fTGH cells. A, Cells were cotransfected with constructs containing the *OXTR* promoter fragments and ESR1wt and treated with vehicle as a control or E2 (10^{-8} M) for 24 h, and luciferase activity was determined as described in *Materials and Methods*. Significant induction ($P < 0.10$) is indicated with an *asterisk*, and results are expressed as mean relative light units (RLU) with SD. Four replicate determinations for each treatment group were conducted in each experiment. A representative experiment of three independent experiments with similar results is presented. B, Cells were cotransfected with constructs containing the OXTR promoter fragments and ESR1wt, treated with vehicle as a control (0) or increasing amounts of E2 for 24 h, and luciferase activity was determined as described in *Materials and Methods*. Significant induction ($P < 0.10$) is indicated with an *asterisk*, and results are expressed as mean percentage of vehicle control with SD for three independent experiments. Four replicate determinations for each treatment group were conducted in each experiment. Similar results were obtained using U3A cells (data not shown).

constructs in cells cotransfected with ESR1IIc (Fig. 4A). The fold increase in *OXTR* promoter activity was consistently greater in cells transfected with ESR1IIc than ESR1wt. Dose response studies found that E2 at a concentration of greater than 10^{-9} M increased activity of the full-length OXTR(−830/−4) promoter as well as the minimal OXTR(−220/−4) promoter in cells cotransfected with ESR1wt (Fig. 3B) or ESR1IIc (Fig. 4B). These results indicate that *cis* elements in the minimal promoter (−220/−4) are sufficient to mediate ligand-activated ESR1 effects. Furthermore, ESR1 induction of the *OXTR* promoter does not require DNA binding, suggesting protein-protein interactions through another transcription factor. Estrogen induces activity of many genes with GC-rich promoters in breast cancer cells transfected with ESR1 or

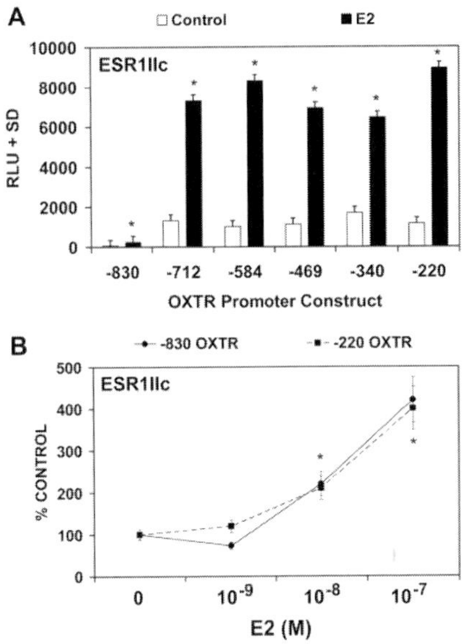

FIG. 4. Comparative activation of OXTR constructs by a variant human ESR1IIc in 2fTGH cells. A, Cells were cotransfected with constructs containing the OXTR promoter fragments and the DNA-binding domain mutant ESR1IIc, treated with vehicle as a control or E2 (10^{-8} M) for 24 h, and luciferase activity was determined as described in *Materials and Methods*. Significant induction ($P < 0.10$) is indicated with an *asterisk*, and results are expressed as mean relative light units (RLU) with SD. Four replicate determinations for each treatment group were conducted in each experiment. A representative experiment of three independent experiments with similar results is presented. B, Cells were cotransfected with constructs containing the OXTR promoter fragments and ESR1IIc and treated with vehicle as a control (0) or increasing amounts of E2 for 24 h, and luciferase activity was determined as described in *Materials and Methods*. Significant induction ($P < 0.10$) is indicated with an *asterisk*, and results are expressed as mean percentage of vehicle control with SD for three independent experiments. Four replicate determinations for each treatment group were conducted in each experiment.

ESR1IIc, even though the genes have no EREs in their promoters (see Refs. 37 and 47).

ICI 182,780 is a pure antiestrogen that inhibits liganded ESR1 from transactivating estrogen-responsive genes containing an ERE(s) (48). However, ICI can exhibit partial agonist activity in some cells and induce ESR1 activation of some estrogen-responsive genes containing consensus GC-rich motifs in the promoter (49). To examine the effects of E2 and ICI 182,780 on ovine *OXTR* promoter activity, 2fTGH and U3A cells were cotransfected with ESR1wt or ESR1IIc and OXTR(−830/−4)-LUC, OXTR(−220/−4)-LUC, 3xERE-TATA-LUC, or 3xSP1-LUC reporters. Cells were then treated for 24 h with E2 (10^{-8} M) or ICI 182,780 (10^{-6} M). In 2fTGH cells, E2 stimulated ($P < 0.01$) activity of the *OXTR* reporters in cells transfected with ESR1wt or ESR1IIc (Fig. 5). E2 also induced ($P < 0.01$) activity of 3xERE-TATA-LUC in cells cotransfected with ESR1wt but not ESR1IIc, consistent with the requirement for DNA bound ESR1. ICI stimulated ($P < 0.01$) *OXTR* promoter activity in cells cotransfected with ESR1wt and ESR1IIc. As expected, ICI did not affect activity of the 3xERE-TATA-LUC in 2fTGH cells cotransfected with ESR1IIc. Both E2 and ICI stimulated ($P < 0.01$) activity of the 3xSP1-LUC reporter in 2fTGH cells transfected with either ESR1wt or ESR1IIc. These results are consistent with E2 and ICI activation of ovine *OXTR* promoter activity through an ESR1/SP1-mediated pathway as previously observed for several hormone-responsive genes with GC-rich promoters (see Refs. 37 and 47).

Minimal OXTR promoter contains functional SP1 elements

Deletion analysis showed that the basal promoter activity and minimal hormone-responsive region was between −220 to −4 bp in the *OXTR* promoter. This region contains an AP-1 site, PRE half site, and three SP1 sites (Fig. 1). The functionality of the SP1 elements was determined by EMSA using recombinant human SP1 protein (Fig. 6). As illustrated in Fig. 6A, the predicted SP1 binding site at −64 in the *OXTR* promoter bound SP1 protein (lane 6). As a positive control, SP1 protein-DNA binding was observed with an oligonucleotide containing a consensus SP1 binding site (lane 2). SP1 binding to the −64 OXTR oligonucleotide was specific, because it was competitively decreased by incubation with a 100-fold excess of unlabeled SP1 consensus oligonucleotide (lane 5) or the unlabeled −64 OXTR oligonucleotide itself (lane 3). Mutation of two critical C residues in the −64 OXTR oligonucleotide rendered it unable to bind SP1 protein (lane 7) or compete for binding of SP1 protein to the wild-type −64 OXTR oligonucleotide (lane 4). In other experiments, 100-fold excess of unlabeled wt −64 OXTR oligonucleotide prevented binding of SP1 protein to the consensus SP1 oligonucleotide, whereas the mutant −64 OXTR oligonucleotide was unable to compete (data not shown). These results demonstrate that the ovine *OXTR* gene promoter contains a high affinity SP1 binding site at position −64.

As illustrated in Fig. 6B, the GC-rich sites at −104 in the ovine OXTR promoter also bound SP1 protein (lane 2) in EMSAs. Incubation with a 100-fold excess of the wild-type −104 OXTR oligonucleotide (lane 3) or consensus SP1 oligonucleotide (lane 4), but not a nonspecific oligonucleotide (lane 5), competitively decreased SP1 protein binding to the radiolabeled −104 OXTR oligonucleotide. The sequence of the −104 SP1 binding site in the OXTR promoter contains a 5′ nonconsensus GC-rich SP1 binding site adjacent to a 3′ consensus site (Fig. 1). Mutation of the 3′ consensus GC-rich site prevented SP1 protein binding (mut1) (lane 6), whereas mutation of the 5′ nonconsensus binding site in the −104 OXTR oligonucleotide (mut2) did not affect SP1 binding (lane 7). Furthermore, SP1 binding was not observed when both the 5′ and 3′ predicted SP1 elements were mutated together (mut4) (lane 8). Collectively, EMSA results indicate that two GC-rich sites in the ovine *OXTR* promoter at −64 and −104 bp bind SP1 protein.

To determine whether the GC-rich sites at −104 and −64 are critical for responsiveness to ESR1, these sites in the −220 OXTR-LUC reporter construct were mutated (see Fig. 1). Due to the complex structure of the −104 SP1 element, mutation

Endocrinology, February 2006, 147(2):899–911 **905**

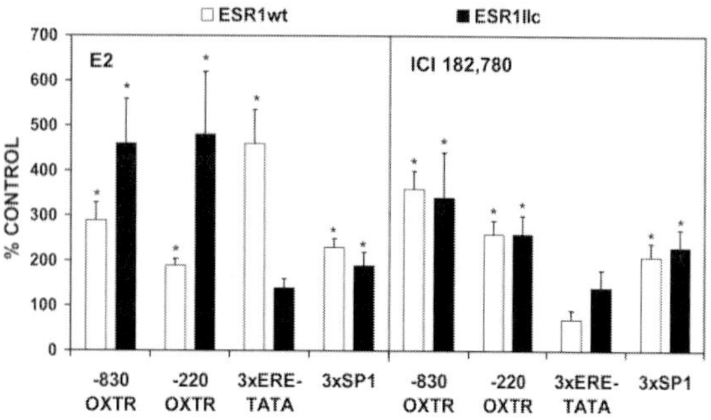

FIG. 5. Comparative activation of OXTR, consensus ERE, and consensus SP1 constructs by wild-type ESR1 and a variant human ESR1IIc in 2fTGH cells. Cells were cotransfected with constructs containing the OXTR(−830/−4) promoter, OXTR(−220/−4) promoter, 3xERE, or 3xSP1 fragments and the wild-type ESR1 or DNA-binding domain mutant ESR1IIc, treated with vehicle as a control and E2 (10^{-8} M) or ICI 182,780 (10^{-6} M) for 24 h, and luciferase activity was determined as described in *Materials and Methods*. Significant induction ($P < 0.10$) is indicated with an *asterisk*, and results are expressed as mean percentage of vehicle control with SD. Four replicate determinations for each treatment group were conducted in each experiment. A representative experiment of three independent experiments with similar results is presented.

of either the 5′ nonconsensus SP1 site or the 3′ consensus SP1 element was performed, because it could alter the integrity of the overlapping site and affect the ability of the −104 sequence to bind SP1. Therefore, in addition to mutations of the −104 site tested in initial gel shift analyses, some C→A transversions that mutated either the consensus or nonconsensus portion of the site, without altering the other overlapping moiety, were also introduced. The oligonucleotides used to create the mutations were tested initially for their ability to bind SP1 protein by EMSA (Fig. 7). Mutation of the 5′ nonconsensus site did not affect SP1 binding regardless of whether the overlapping 3′ consensus portion remained wild-type (mut3; lane 5) or was slightly altered (mut2; lane 4). Mutation of the 3′ consensus sequence alone (mut1, lane 3) substantially reduced SP1 binding compared with the wild-type −104 OXTR oligonucleotide (lane 2), but did not eliminate binding entirely. Mutation of both 5′ and 3′ sites (mut4) completely prevented SP1 binding (lane 6). In competition experiments (Fig. 7B), excess unlabeled wt −104 oligonucleotide and the mut1, mut2, and mut3 oligonucleotides reduced or prevented SP1 binding to the radiolabeled wt −104 oligonucleotide (lanes 6, 7, and 8). As expected, competition with the fully mutated mut4 oligonucleotide did not affect SP1 binding to the radiolabeled −104 OXTR wt oligonucleotide (lane 9). Collectively, these results indicate that SP1 binding to the −104 OXTR sequence depends primarily on the 3′ consensus GC-rich element. However, the 5′ nonconsensus GC-rich element appears to weakly bind SP1 protein, suggesting that the two overlapping motifs at −104 may act cooperatively to bind SP1.

Transient transfection assays were then conducted to determine the relative contributions of the −104 and −64 SP1 binding sites on hormonal responsiveness of the ovine *OXTR* promoter (Fig. 8). Wild-type or mutant OXTR(−220/−4) promoter-reporter constructs were cotransfected with ESR1IIc into 2fTGH cells and left unstimulated or treated with E2 (10^{-8} M). Mutation of either the 5′ nonconsensus site (constructs 2 and 3), the 3′ consensus site (construct 4), or both sites (construct 5) at −104 reduced ($P < 0.05$) basal luciferase expression in unstimulated cells by 25–40% compared with the wild-type reporter (construct 1) depending on the location of the mutation. Mutation of the −64 SP1 site reduced ($P < 0.01$) basal luciferase expression to very low levels, regardless of whether the −104 site was wt (construct 6), double mutated (constructs 7 and 9), or point mutated (construct 8). Indeed, basal luciferase levels of the −64 SP1 mutants were about 10% of wild-type parental levels. Mutation of the 5′ consensus SP1 site or both the nonconsensus and consensus sites at −104 and the −64 SP1 site (construct 10) further reduced ($P < 0.10$) basal activity to about 2% of activity observed for the wt construct.

Estradiol-induced activity, in all cells transfected with the minimal OXTR(−220/−4) promoter, was lower ($P < 0.10$) in constructs containing a mutation in the −104 and −64 SP1 binding sites (Fig. 8). Mutation of both sites at position −104 further reduced stimulation by estradiol, and this was consistent with the EMSA data suggesting that both the left and right moieties are involved in binding SP1 protein. However, mutation of the −64 SP1 site decreased ($P < 0.10$) stimulation by estradiol, regardless of whether the −104 SP1 site was wt or mutated. Overall, these results suggest that both the −104 and −64 SP1 sites in the proximal *OXTR* promoter regulates basal levels of *OXTR* gene transcription, and that both SP1 binding sites, especially the consensus sequences, regulate E2 stimulation of *OXTR* gene expression.

Progesterone stimulates OXTR promoter activity

During metestrus and estrus in cyclic ewes, PGR return to the OXTR-expressing endometrial epithelia as systemic progesterone declines to undetectable levels (9, 10, 12–14). The prevailing theory is that PGR do not directly inhibit *OXTR* gene expression, but rather the progesterone block to OXTR formation is indirect via progesterone inhibition of ESR1 expression (4, 8, 28). To determine whether the *OXTR* promoter is regulated by PGR, 2fTGH cells were cotransfected with human PGR-B and OXTR promoter-LUC constructs and then treated with R5020 (10^{-8} M), a nonmetabolizable form of progesterone (Fig. 9). As observed previously, basal activity of the OXTR promoter constructs increased ($P < 0.05$) by deletion of the region from −830 to −711. Treatment with R5020 stimulated ($P < 0.05$) the activity of all OXTR promoter-reporter constructs. Next, the same experiment was repeated with the reporter constructs and PGR-B, but cells were treated with a range of R5020 doses. Dose-dependent effects of R5020 (10^{-14}–10^{-5} M) on activity of the −830 and −220 OXTR promoter constructs were observed in 2fTGH cells cotransfected with PGR-B, with concentrations

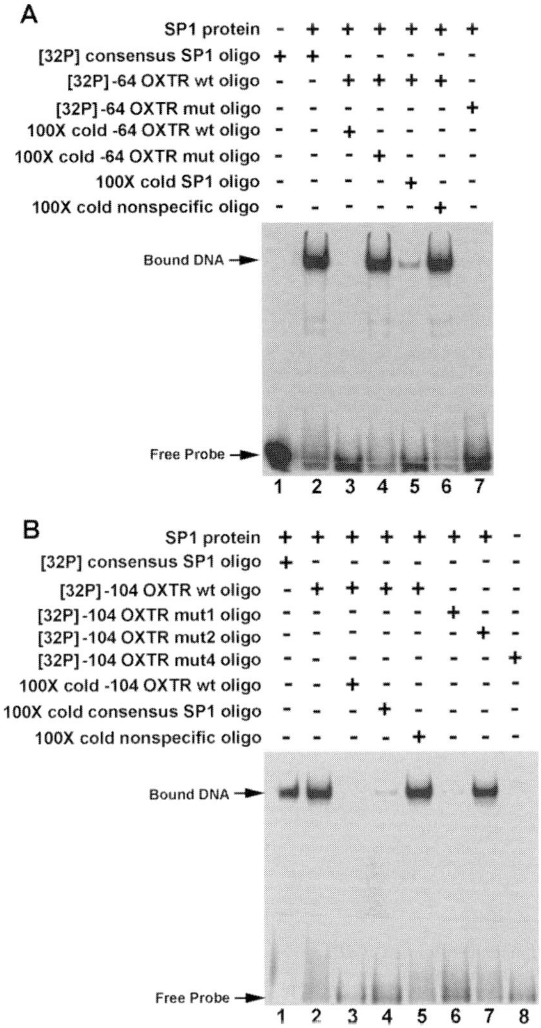

FIG. 6. Gel EMSAs. A, SP1 binding to −64 region of the ovine OXTR promoter. Gel mobility shift assays were conducted by incubating recombinant human SP1 protein with radiolabeled [^{32}P] −64 OXTR wild-type (wt) or mutant (mut) oligonucleotides (see Fig. 1) and various unlabeled oligonucleotides for competition as described in *Materials and Methods*. B, SP1 binding to −104 region of the ovine OXTR promoter. Gel mobility shift assays were conducted by incubating recombinant human SP1 protein with radiolabeled [^{32}P] −104 OXTR wild-type (wt) or mutant (mut) oligonucleotides (see Fig. 1) and various unlabeled oligonucleotides for competition as described in *Materials and Methods*. Results in these assays are from duplicate experiments.

as low as 10^{-10} M R5020 stimulated transactivation in cells transfected with OXTR constructs (data not shown).

Deletion of the AP-1 and PRE half site does not affect E2 or R5020 stimulation of OXTR promoter activity

The minimal OXTR(−220/−4) promoter contains an AP-1 site at −220 and a PRE half site at position −187. To determine whether these sites are involved in hormone responsiveness, the sites were sequentially removed by 5′ deletion. The *OXTR* promoter constructs were then transfected into 2fTGH cells along with ESR1IIc or PGR-B and treated with E2 (10^{-8} M) or R5020 (10^{-8} M) for 24 h, respectively (Fig. 10). Deletion of the AP-1 site and PRE half site did not affect ($P >$ 0.10) the stimulatory effects of E2 and R5020 on *OXTR* promoter activity. These results indicate that the SP1 binding sites at −104 and −64 are the only *cis* elements required for basal activity of the *OXTR* promoter and its responsiveness to stimulation by ligand-activated ESR1 and PGR-B.

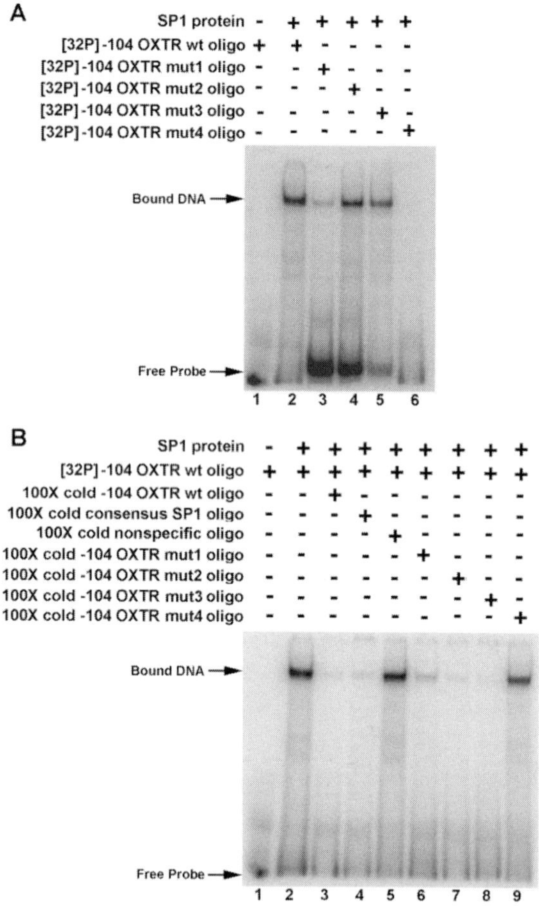

FIG. 7. Gel EMSAs. A, SP1 binding to SP1 binding site mutants in the −104 region of the ovine *OXTR* promoter. Gel mobility shift assays were conducted by incubating recombinant human SP1 protein with radiolabeled [^{32}P] −104 OXTR wild-type (wt) or mutant (mut) oligonucleotides (see Fig. 1) as described in *Materials and Methods*. B, Gel mobility shift assays were conducted by incubating recombinant human SP1 protein with radiolabeled [^{32}P] −104 OXTR wild-type (wt) oligonucleotide (see Fig. 1) and various unlabeled wt and mutant (mut) oligonucleotides for competition as described in *Materials and Methods*. Results in these assays are from duplicate experiments.

Immunolocalization of SP1 protein in the endometrium of cyclic and pregnant ewes

To confirm that SP1 was expressed in the endometrium, SP1 protein was studied in the uterus of cyclic and early pregnant ewes by immunohistochemistry (Fig. 11). Immunoreactive SP1 protein was observed in all endometrial cell types of uteri from ewes on d 10–16 of the cycle and d 10–20 of pregnancy. Although all endometrial cell types expressed SP1 protein, it was most abundant in the endometrial LE and GE. No differences in SP1 protein abundance were detected between the endometrial cells of cyclic and pregnant ewes or between LE of intercaruncular and caruncular areas of the endometrium (data not shown). In pregnant ewes, SP1 protein was also detected in conceptus trophectoderm.

Discussion

The gene for OXTR exists as a single copy in all studied species, and the promoter region of the *OXTR* genes of the human (50), rat (51, 52), bovine (53, 54), mouse (55), and vole (56) are available. Comparative studies of the promoter regions of homologous genes from several species often reveal conserved and important regulatory elements. However, the promoter regions of the ovine and bovine *OXTR* genes are not well conserved with other species, despite significant sequence conservation found in the coding sequence. The human *OXTR* gene promoter can be regulated by proinflammatory cytokines, such as IL-1 and IL-6 via activation of the binding elements for nuclear factor-IL-6 or STAT3 (57). Serum increases activity of the human *OXTR* promoter through

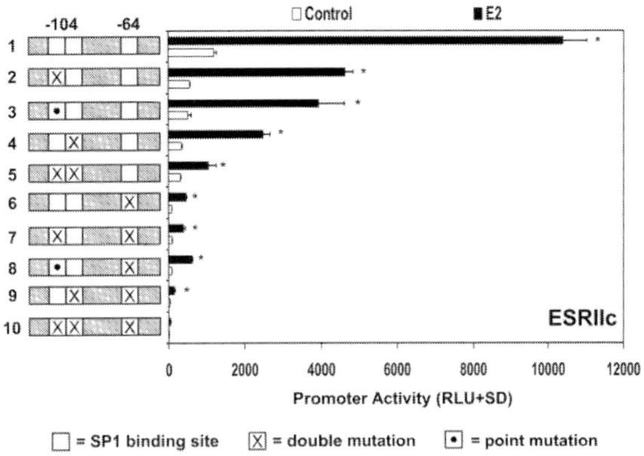

FIG. 8. Comparative activation of *OXTR* constructs by a variant human ESR1IIc in 2fTGH cells. Cells were cotransfected with constructs containing the wild-type or mutant *OXTR* promoter fragments from the hormone-responsive −220 to −4 region and the DNA-binding domain mutant ESR1IIc, treated with vehicle as a control or E2 (10^{-8} M) for 24 h, and luciferase activity was determined as described in *Materials and Methods*. Significant induction ($P < 0.10$) is indicated with an *asterisk*, and results are expressed as mean relative light units (RLU) with SD. Four replicate determinations for each treatment group were conducted in each experiment. A representative experiment of three independent experiments with similar results is presented.

a protein kinase C-dependent pathway that can be synergistically augmented by dexamethasone (58). In rabbit amnion cells, *in vitro* treatment with forskolin and/or cortisol activates *OXTR* gene transcription (59). Protein kinase A, protein kinase C, and nerve growth factor-dependent *OXTR* up-regulation is also observed in MCF-7 and SK-N-SH cells transfected with the rat *OXTR* gene promoter (60). In the human, basal promoter activity is conferred by the 85-bp flanking the 5′ end of the human *OXTR* gene. In this sequence, there is a consensus ETS binding sequence. GABP (GA binding protein transcription factor), which binds to the ETS element, cooperates with AP-1 (FOS/JUN) to activate human *OXTR* promoter transcription in Hs578T cells (61). However, elements that regulate activity of the human and rat *OXTR* promoter/enhancer regions are not present in the ovine and bovine *OXTR* promoters, suggesting that the identified pathways regulating *OXTR* expression in other species may not be the same for ruminants.

In every studied mammal, estrogen is considered a key

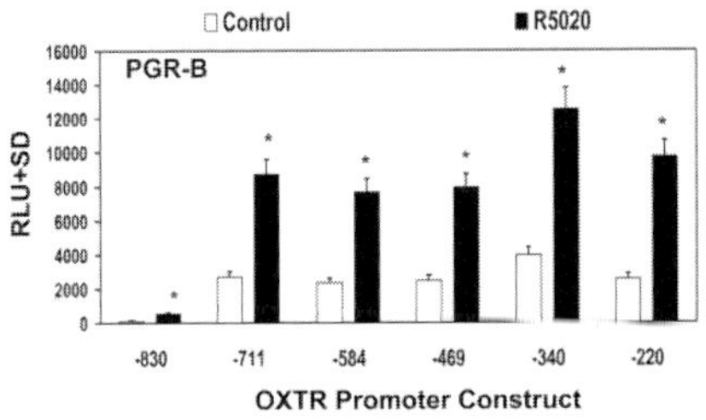

FIG. 9. Comparative activation of *OXTR* constructs by wild-type human PGR-B in 2fTGH cells. Cells were cotransfected with constructs containing the *OXTR* promoter fragments and PGR-B, treated with vehicle as a control or the nonmetabolizable R5020 progestin (10^{-8} M) for 24 h, and luciferase activity was determined as described in *Materials and Methods*. Significant induction ($P < 0.10$) is indicated with an *asterisk*, and results are expressed as mean relative light units (RLU) with SD. Four replicate determinations for each treatment group were conducted in each experiment. A representative experiment of three independent experiments with similar results is presented.

regulator of *OXTR* gene expression (27). Similarly in sheep, *in vivo* administration of estrogen increases *OXTR* mRNA and the number of oxytocin binding sites in the endometrium (15, 16, 22, 62). The classical pathway for estrogen regulation of target genes involves ligand-activated ESR1/ERE interactions or ERE half sites (37, 47). In the rat *OXTR* gene, a palindromic ERE was identified about 4 kb upstream of the transcriptional start site (52). This sequence was hormone-responsive in transfection studies in human MCF-7 breast cancer cells, but only when the 3.3-kb fragment between the ERE and the basal promoter of the rat *OXTR* gene was truncated. In the wild-type construct, this ERE was not activated by estrogen. However, the human *OXTR* gene does not have a palindromic ERE in the area comparable to that of the rat gene (27). Therefore, estrogen induction of ovine *OXTR* gene transcription may be indirect rather than due to direct ESR1/ERE interactions.

The present study found that an estrogen agonist (E2) and antagonist (ICI 182,780) transactivated the ovine *OXTR* promoter in cells transfected with ESR1. Deletion and mutation analyses found that induction of ovine *OXTR* promoter activity by E2-activated ESR1 is dependent on GC-rich SP1 binding sites at −104 and −64 in the minimal promoter, and basal activity of this promoter is also regulated by the same GC-rich SP1 elements adjacent to the translational start site of the *OXTR* gene. One caveat of the present studies is that human fibrosarcoma cells were used for transfection assays instead of an ovine endometrial epithelial cell line the endogenous expresses the *OXTR* gene. Indeed, the GC-rich motifs in several gene promoters are *cis*-acting elements for an increasing number of ligand-activated nuclear and orphan receptors that interact with SP1 and related proteins (see Refs. 37 and 47). In the present study, SP1 transcription factor was constitutively expressed in nearly all cell types of the endometrium from both cyclic and pregnant ewes and was particularly abundant in LE and sGE. Therefore, the primary determinant of *OXTR* gene regulation in the endometrial epithelia is expression of the ESR1 gene. Administration of estrogen on d 11 or 12 postestrus induces structural and functional luteolysis in cyclic ewes with the following

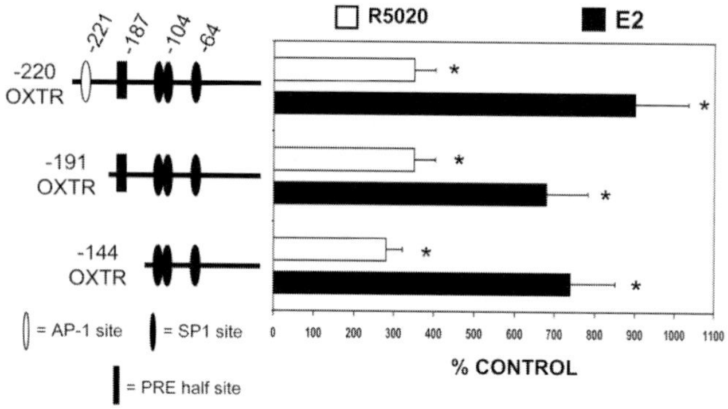

FIG. 10. Comparative activation of *OXTR* constructs by ESR1IIc or PGR-B in 2fTGH cells. Cells were cotransfected with constructs containing the *OXTR* promoter fragments and ESR1IIc or PGR-B, treated with vehicle as a control (0) and E2 (10^{-8} M) with ESR1IIc or R5020 progestin (10^{-8} M) with PGR-B for 24 h, and luciferase activity was determined as described in *Materials and Methods*. Significant induction ($P < 0.10$) is indicated with an *asterisk*, and results are expressed as mean percentage of vehicle control with SD for three independent experiments. Four replicate determinations for each treatment group were conducted in each experiment.

temporal events: 1) increase ESR1 mRNA and protein in endometrial epithelia between 12–24 h; 2) moderate increase in endometrial OXTR density between 12–36 h; 3) large increase in OXTR density between 36–48 h; 4) a decline in concentrations of progesterone in plasma after 36 h; and 5) decrease in CL weight by 48 h (15, 17). Similar temporospatial alterations in *ESR1* and *OXTR* gene expression occur in the endometrium of cyclic ewes between d 12–16 postestrus (9, 10, 12). Although exogenous estrogen can clearly elicit development of the endometrial luteolytic mechanism and induce *OXTR* promoter activity, the precise physiological role of estrogen in the luteolytic mechanism *in vivo* during a natural estrous cycle has not been adequately investigated. Estrogen may not be needed to increase *OXTR* gene expression, because ESR1 can be activated in a ligand-dependent manner by estrogens or a ligand-independent manner by growth factors (63), such as IGF-I and IGF-II that are abundant in the ovine endometrial stroma (64). Furthermore, ICI 182,780, a pure antiestrogen, will not be useful for *in vivo* investigations, because it activated ESR1 and induced *OXTR* promoter activity in the present study. Similarly, the antiestrogens 4′-hydroxytamoxifen (4-OHT) and ICI 182,780 activate reporter gene activity in cells transfected with constructs containing GC-rich promoters with SP1 and SP3 binding sites (65). In addition to the *OXTR*, many other endometrial genes that are regulated by hormone and orphan receptors are likely to involve SP1 and related transcription factors.

Development of the endometrial luteolytic mechanism and uterine release of luteolytic pulses of PGF occur in the presence of high circulating levels of progesterone. However, the endometrial epithelia are not responsive to receptor-dependent actions of progesterone, because they are PGR-negative between d 11–15 of the estrous cycle. As progesterone levels decrease and estrogen levels increase during proestrus (d 15–17), PGR protein returns to the endometrial epithelia of the ovine uterus (10, 12). In the present study, PGR also increased activity of the ovine *OXTR* promoter through GC-rich SP1 sites rather than the AP-1 site or PRE half sites also present in the promoter of the *OXTR* gene. Progestin-dependent regulation of *OXTR* through GC-rich sites is not unprecedented becausePGR/SP1 activation of glycodelin A and CDKN1A (cyclin-dependent kinase inhibitor 1A or p21) in endometrial and breast cancer cells, respectively, is also due to specific GC-rich SP1 binding sites in their promoters (66, 67). Indeed, uterine *OXTR* mRNA and protein increases in the endometrial epithelia and stroma after d 15 and is maximal at estrus, which is coincident with *PGR* mRNA and protein as well as SP1 expression in the same cells (10, 12, 13). Therefore, the progesterone block to *OXTR* formation must be due to inhibition of *ESR1* gene transcription via direct or indirect actions that does not involve SP1.

In the present study, IFNT did not affect *OXTR* promoter activity or E2 induction of *OXTR* promoter activity. These

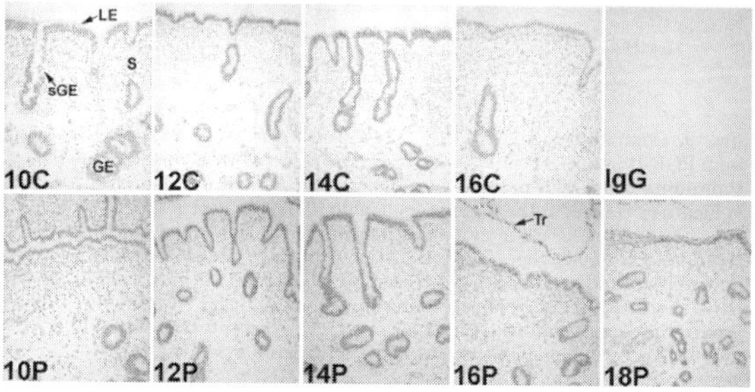

FIG. 11. SP1 protein in the endometrium of cyclic and pregnant ewes. Immunoreactive SP1 protein was localized in sections of the uterus using a rabbit antihuman SP1 polyclonal antibody. For the IgG control, normal rabbit IgG was substituted for the primary antibody. Sections were not counterstained. S, stroma; Tr, conceptus trophectoderm. All photomicrographs are shown at ×150 magnification.

results were consistent with the lack of classical interferon regulatory elements (ISRE, IRFE, or GAS) in the ovine *OXTR* promoter and the lack of IRF1 and IRF2 effects on *OXTR* promoter activity in transient transfection assays. Collectively, results of the present and previous studies continue to support our theory that the antiluteolytic effects of IFNT from the conceptus are to inhibit or silence ESR1 gene transcription, which in turn precludes ESR1 stimulation of *OXTR* gene transcription and, therefore, oxytocin-induced release of luteolytic pulses of PGF by uterine epithelia. Future experiments are needed to address how progesterone down-regulates PGR gene expression, what factors critically regulate PGR inhibition of *ESR1* gene transcription in the endometrial epithelia of the uterus, and if SP1 and related transcription factors regulate other endometrial genes in the ovine uterus.

Acknowledgments

We thank Haijun Gao for assistance with immunohistochemical analysis of SP1 protein and other members of our laboratories who contributed to the research presented in this paper.

Received September 1, 2005. Accepted October 19, 2005.

Address all correspondence and requests for reprints to: Fuller W. Bazer, Center for Animal Biotechnology and Genomics, 442 Kleberg Center, 2471 TAMU, Texas A&M University, College Station, Texas 77843-2471. E-mail: fbazer@cvm.tamu.edu.

Financial support for these studies was obtained from National Institutes of Health R01 Grant HD32534 and National Institute on Environmental Health Sciences 5 P30 ES09106-03.

References

1. **Bazer FW, Thatcher WW, Hansen PJ, Mirando MA, Ott TL, Plante C** 1991 Physiological mechanisms of pregnancy recognition in ruminants. J Reprod Fertil Suppl 43:39–47
2. **Bazer FW** 1992 Mediators of maternal recognition of pregnancy in mammals. Proc Soc Exp Biol Med 199:373–384
3. **McCracken JA, Custer EE, Lamsa JC** 1999 Luteolysis: a neuroendocrine-mediated event. Physiol Rev 79:263–323
4. **Spencer TE, Bazer FW** 2002 Biology of progesterone action during pregnancy recognition and maintenance of pregnancy. Front Biosci 7:d1879–d1898
5. **Charpigny G, Reinaud P, Tamby JP, Creminon C, Martal J, Maclouf J, Guillomot M** 1997 Expression of cyclooxygenase-1 and -2 in ovine endometrium during the estrous cycle and early pregnancy. Endocrinology 138:2163–2171
6. **Gray C, Bartol FF, Taylor KM, Wiley AA, Ramsey WS, Ott TL, Bazer FW, Spencer TE** 2000 Ovine uterine gland knock-out model: effects of gland ablation on the estrous cycle. Biol Reprod 62:448–456
7. **Spencer TE, Ott TL, Bazer FW** 1996 τ-Interferon: pregnancy recognition signal in ruminants. Proc Soc Exp Biol Med 213:215–229
8. **Spencer TE, Bazer FW** 2004 Conceptus signals for establishment and maintenance of pregnancy. Reprod Biol Endocrinol 2:49
9. **Ayad VJ, Matthews EL, Wathes DC, Parkinson TJ, Wild ML** 1991 Autoradiographic localization of oxytocin receptors in the endometrium during the oestrous cycle of the ewe. J Endocrinol 130:199–206
10. **Wathes DC, Hamon M** 1993 Localization of oestradiol, progesterone and oxytocin receptors in the uterus during the oestrous cycle and early pregnancy of the ewe. J Endocrinol 138:479–492
11. **McCracken J, Schramm W, Okulicz WC** 1984 Hormone receptor control of pulsatile secretion of PGF-2α from the ovine uterus during luteolysis and its abrogation in early pregnancy. Anim Reprod Sci 7:31–55
12. **Spencer TE, Bazer FW** 1995 Temporal and spatial alterations in uterine estrogen receptor and progesterone receptor gene expression during the estrous cycle and early pregnancy in the ewe. Biol Reprod 53:1527–1543
13. **Stevenson KR, Riley PR, Stewart HJ, Flint AP, Wathes DC** 1994 Localization of oxytocin receptor mRNA in the ovine uterus during the oestrous cycle and early pregnancy. J Mol Endocrinol 12:93–105
14. **Wathes DC, Mann GE, Payne JH, Riley PR, Stevenson KR, Lamming GE** 1996 Regulation of oxytocin, oestradiol and progesterone receptor concentrations in different uterine regions by oestradiol, progesterone and oxytocin in ovariectomized ewes. J Endocrinol 151:375–393
15. **Spencer TE, Becker WC, George P, Mirando MA, Ogle TF, Bazer FW** 1995 Ovine interferon-τ inhibits estrogen receptor up-regulation and estrogen-induced luteolysis in cyclic ewes. Endocrinology 136:4932–4944
16. **Spencer TE, Becker WC, George P, Mirando MA, Ogle TF, Bazer FW** 1995 Ovine interferon-τ regulates expression of endometrial receptors for estrogen and oxytocin but not progesterone. Biol Reprod 53:732–745
17. **Hixon JE, Flint AP** 1987 Effects of a luteolytic dose of oestradiol benzoate on uterine oxytocin receptor concentrations, phosphoinositide turnover and prostaglandin F-2α secretion in sheep. J Reprod Fertil 79:457–467
18. **Vallet JL, Lamming GE, Batten M** 1990 Control of endometrial oxytocin receptor and uterine response to oxytocin by progesterone and oestradiol in the ewe. J Reprod Fertil 90:625–634
19. **Roberts RM, Ealy AD, Alexenko AP, Han CS, Ezashi T** 1999 Trophoblast interferons. Placenta 20:259–264
20. **Wathes DC, Lamming GE** 1995 The oxytocin receptor, luteolysis and the maintenance of pregnancy. J Reprod Fertil Suppl 49:53–67
21. **Kim S, Choi Y, Spencer TE, Bazer FW** 2003 Effects of the estrous cycle, pregnancy and interferon τ on expression of cyclooxygenase two (COX-2) in ovine endometrium. Reprod Biol Endocrinol 1:58
22. **Spencer TE, Bazer FW** 1996 Ovine interferon τ suppresses transcription of the estrogen receptor and oxytocin receptor genes in the ovine endometrium. Endocrinology 137:1144–1147
23. **Fleming JA, Choi Y, Johnson GA, Spencer TE, Bazer FW** 2001 Cloning of the ovine estrogen receptor-α promoter and functional regulation by ovine interferon-τ. Endocrinology 142:2879–2887
24. **Choi Y, Johnson GA, Burghardt RC, Berghman LR, Joyce MM, Taylor KM, Stewart MD, Bazer FW, Spencer TE** 2001 Interferon regulatory factor-two restricts expression of interferon-stimulated genes to the endometrial stroma and glandular epithelium of the ovine uterus. Biol Reprod 65:1038–1049
25. **Vallet JL, Bazer FW** 1989 Effect of ovine trophoblast protein-1, oestrogen and progesterone on oxytocin-induced phosphatidylinositol turnover in endometrium of sheep. J Reprod Fertil 87:755–761
26. **Telgmann R, Bathgate RA, Jaeger S, Tillmann G, Ivell R** 2003 Transcriptional regulation of the bovine oxytocin receptor gene. Biol Reprod 68:1015–1026
27. **Kimura T, Saji F, Nishimori K, Ogita K, Nakamura H, Koyama M, Murata Y** 2003 Molecular regulation of the oxytocin receptor in peripheral organs. J Mol Endocrinol 30:109–115
28. **Spencer TE, Johnson GA, Burghardt RC, Bazer FW** 2004 Progesterone and placental hormone actions on the uterus: insights from domestic animals. Biol Reprod 71:2–10
29. **Pellegrini S, John J, Shearer M, Kerr IM, Stark GR** 1989 Use of a selectable marker regulated by α interferon to obtain mutations in the signaling pathway. Mol Cell Biol 9:4605–4612
30. **Van Heeke G, Ott TL, Strauss A, Ammaturo D, Bazer FW** 1996 High yield expression and secretion of the ovine pregnancy recognition hormone interferon-τ by *Pichia pastoris*. J Interferon Cytokine Res 16:119–126
31. **Ausebel FM, Brent R, Kinston RE, Moore DD, Seidmen JG, Smith HA, Struhl KA** 1998 Current protocols in molecular biology. New York: John Wiley and Sons
32. **Stoner M, Wang F, Wormke M, Nguyen T, Samudio I, Vyhlidal C, Marme D, Finkenzeller G, Safe S** 2000 Inhibition of vascular endothelial growth factor expression in HEC1A endometrial cancer cells through interactions of estrogen receptor α and Sp3 proteins. J Biol Chem 275:22769–22779
33. **Sun G, Porter W, Safe S** 1998 Estrogen-induced retinoic acid receptor α1 gene expression: role of estrogen receptor-Sp1 complex. Mol Endocrinol 12:882–890
34. **Spencer TE, Bartol FF, Bazer FW, Johnson GA, Joyce MM** 1999 Identification and characterization of glycosylation-dependent cell adhesion molecule 1-like protein expression in the ovine uterus. Biol Reprod 60:241–250
35. **Taylor KM, Gray CA, Joyce MM, Stewart MD, Bazer FW, Spencer TE** 2000 Neonatal ovine uterine development involves alterations in expression of receptors for estrogen, progesterone, and prolactin. Biol Reprod 63:1192–1204
36. **Gray CA, Taylor KM, Bazer FW, Spencer TE** 2000 Mechanisms regulating norgestomet inhibition of endometrial gland morphogenesis in the neonatal ovine uterus. Mol Reprod Dev 57:67–78
37. **Safe S, Kim K** 2004 Nuclear receptor-mediated transactivation through interaction with Sp proteins. Prog Nucleic Acids Res Mol Biol 77:1–36
38. **Kushner PJ, Agard DA, Greene GL, Scanlan TS, Shiau AK, Uht RM, Webb P** 2000 Estrogen receptor pathways to AP-1. J Steroid Biochem Mol Biol 74:311–317
39. **Sjottem E, Anderssen S, Johansen T** 1996 The promoter activity of long terminal repeats of the HERV-H family of human retrovirus-like elements is critically dependent on Sp1 family proteins interacting with a GC/GT box located immediately 3′ to the TATA box. J Virol 70:188–198
40. **Black AR, Black JD, Azizkhan-Clifford J** 2001 Sp1 and kruppel-like factor family of transcription factors in cell growth regulation and cancer. J Cell Physiol 188:143–160
41. **Wang X, Talamantez JL, Adamo ML** 1998 A CACCC box in the proximal exon 2 promoter of the rat insulin-like growth factor I gene is required for basal promoter activity. Endocrinology 139:1054–1066
42. **Marini MG, Asunis I, Porcu L, Salgo MG, Loi MG, Bruchietti A, Cao A, Moi P** 2004 The distal β-globin CACCC box is required for maximal stimulation of the β-globin gene by EKLF. Br J Haematol 127:114–117

43. **Stark GR, Kerr IM, Williams BR, Silverman RH, Schreiber RD** 1998 How cells respond to interferons. Annu Rev Biochem 67:227–264
44. **Stewart MD, Johnson GA, Bazer FW, Spencer TE** 2001 Interferon-τ (IFNτ) regulation of IFN-stimulated gene expression in cell lines lacking specific IFN-signaling components. Endocrinology 142:1786–1794
45. **Stewart MD, Choi Y, Johnson GA, Yu-Lee LY, Bazer FW, Spencer TE** 2002 Roles of Stat1, Stat2, and interferon regulatory factor-9 (IRF-9) in interferon τ regulation of IRF-1. Biol Reprod 66:393–400
46. **Kim S, Choi Y, Bazer FW, Spencer TE** 2003 Identification of genes in the ovine endometrium regulated by interferon τ independent of signal transducer and activator of transcription 1. Endocrinology 144:5203–5214
47. **Safe S** 2001 Transcriptional activation of genes by 17 β-estradiol through estrogen receptor-Sp1 interactions. Vitam Horm 62:231–252
48. **Frasor J, Stossi F, Danes JM, Komm B, Lyttle CR, Katzenellenbogen BS** 2004 Selective estrogen receptor modulators: discrimination of agonistic versus antagonistic activities by gene expression profiling in breast cancer cells. Cancer Res 64:1522–1533
49. **Kim K, Thu N, Saville B, Safe S** 2003 Domains of estrogen receptor α (ERα) required for ERα/Sp1-mediated activation of GC-rich promoters by estrogens and antiestrogens in breast cancer cells. Mol Endocrinol 17:804–817
50. **Inoue T, Kimura T, Azuma C, Inazawa J, Takemura M, Kikuchi T, Kubota Y, Ogita K, Saji F** 1994 Structural organization of the human oxytocin receptor gene. J Biol Chem 269:32451–32456
51. **Rozen F, Russo C, Banville D, Zingg HH** 1995 Structure, characterization, and expression of the rat oxytocin receptor gene. Proc Natl Acad Sci USA 92:200–204
52. **Bale TL, Dorsa DM** 1997 Cloning, novel promoter sequence, and estrogen regulation of a rat oxytocin receptor gene. Endocrinology 138:1151–1158
53. **Ivell R, Bathgate R, Rust W, Balvers M, Morley S** 1995 Structure and organization of the bovine oxytocin receptor gene. Adv Exp Med Biol 395:295–300
54. **Bathgate R, Rust W, Balvers M, Hartung S, Morley S, Ivell R** 1995 Structure and expression of the bovine oxytocin receptor gene. DNA Cell Biol 14:1037–1048
55. **Kubota Y, Kimura T, Hashimoto K, Tokugawa Y, Nobunaga K, Azuma C, Saji F, Murata Y** 1996 Structure and expression of the mouse oxytocin receptor gene. Mol Cell Endocrinol 124:25–32
56. **Young LJ, Waymire KG, Nilsen R, Macgregor GR, Wang Z, Insel TR** 1997 The 5′ flanking region of the monogamous prairie vole oxytocin receptor gene directs tissue-specific expression in transgenic mice. Ann NY Acad Sci 807: 514–517
57. **Schmid B, Wong S, Mitchell BF** 2001 Transcriptional regulation of oxytocin receptor by interleukin-1β and interleukin-6. Endocrinology 142:1380–1385
58. **Copland JA, Jeng YJ, Strakova Z, Ives KL, Hellmich MR, Soloff MS** 1999 Demonstration of functional oxytocin receptors in human breast Hs578T cells and their up-regulation through a protein kinase C-dependent pathway. Endocrinology 140:2258–2267
59. **Jeng YJ, Lolait SJ, Soloff MS** 1998 Induction of oxytocin receptor gene expression in rabbit amnion cells. Endocrinology 139:3449–3455
60. **Bale TL, Dorsa DM** 1998 NGF, cyclic AMP, and phorbol esters regulate oxytocin receptor gene transcription in SK-N-SH and MCF7 cells. Brain Res Mol Brain Res 53:130–137
61. **Hoare S, Copland JA, Wood TG, Jeng YJ, Izban MG, Soloff MS** 1999 Identification of a GABP α/β binding site involved in the induction of oxytocin receptor gene expression in human breast cells, potentiation by c-Fos/c-Jun. Endocrinology 140:2268–2279
62. **Spencer TE, Mirando MA, Mayes JS, Watson GH, Ott TL, Bazer FW** 1996 Effects of interferon-τ and progesterone on oestrogen-stimulated expression of receptors for oestrogen, progesterone and oxytocin in the endometrium of ovariectomized ewes. Reprod Fertil Dev 8:843–853
63. **Smith CL** 1998 Cross-talk between peptide growth factor and estrogen receptor signaling pathways. Biol Reprod 58:627–632
64. **Stevenson KR, Gilmour RS, Wathes DC** 1994 Localization of insulin-like growth factor-I (IGF-I) and -II messenger ribonucleic acid and type 1 IGF receptors in the ovine uterus during the estrous cycle and early pregnancy. Endocrinology 134:1655–1664
65. **Saville B, Wormke M, Wang F, Nguyen T, Enmark E, Kuiper G, Gustafsson JA, Safe S** 2000 Ligand-, cell-, and estrogen receptor subtype (α/β)-dependent activation at GC-rich (Sp1) promoter elements. J Biol Chem 275:5379–5387
66. **Gao J, Mazella J, Seppala M, Tseng L** 2001 Ligand activated hPR modulates the glycodelin promoter activity through the Sp1 sites in human endometrial adenocarcinoma cells. Mol Cell Endocrinol 176:97–102
67. **Owen GI, Richer JK, Tung L, Takimoto G, Horwitz KB** 1998 Progesterone regulates transcription of the p21(WAF1) cyclin-dependent kinase inhibitor gene through Sp1 and CBP/p300. J Biol Chem 273:10696–10701

Endocrinology is published monthly by The Endocrine Society (http://www.endo-society.org), the foremost professional society serving the endocrine community.

BIOLOGY OF REPRODUCTION **77**, 172–180 (2007)
Published online before print 28 March 2007.
DOI 10.1095/biolreprod.106.056309

Regulation of Expression of Fibroblast Growth Factor 7 in the Pig Uterus by Progesterone and Estradiol[1]

Hakhyun Ka,[3,4,5,7] Saeed Al-Ramadan,[3,4,6] David W. Erikson,[4,6] Greg A. Johnson,[4,6] Robert C. Burghardt,[4,6] Thomas E. Spencer,[4,5] Laurie A. Jaeger,[4,6] and Fuller W. Bazer[2,4,5]

Center for Animal Biotechnology and Genomics,[4] Departments of Animal Science,[5] and Veterinary Integrative Biosciences,[6] Texas A&M University, College Station, Texas 77843
Department of Biological Resources and Technology,[7] Yonsei University, Wonju 220-710, South Korea

ABSTRACT

Fibroblast growth factor 7 (FGF7) stimulates cell proliferation, differentiation, migration and angiogenesis. The consensus is that FGF7, expressed by mesenchymal cells, binds FGF receptor 2IIIb (FGFR2) on epithelia, thereby mediating epithelial-mesenchymal interactions. The pig uterus is unique in that FGF7 is expressed by the luminal epithelium (LE) and FGFR2 is expressed by the LE, glandular epithelium (GE), and trophectoderm to effect proliferation and differentiated cell functions during conceptus development and implantation. *FGF7* expression by the uterine LE of pigs increases between Days 9 and 12 of the estrus cycle and pregnancy, as circulating concentrations of progesterone increase, progesterone receptors (PGR) in the uterine epithelia decrease, and the conceptuses secrete estradiol-17beta (E_2), for pregnancy recognition. Furthermore, E_2 increases the expression of *FGF7* in pig uterine explants. The present study investigates the relationships between progesterone, E_2, and their receptors and the expression of FGF7 in the pig uterus in vivo. Pigs were ovariectomized on Day 4 of the estrus cycle and injected i.m. daily from Day 4 to Day 12 with either corn oil (CO), progesterone (P4), P4 and ZK317,316 (PZK), E_2, P4 and E_2 (PE), or P4 and ZK and E_2 (PZKE). All gilts (n = 5/treatment) were hysterectomized on Day 12. The results suggest that: 1) P4 is permissive to FGF7 expression by down-regulating PGR in LE; 2) P4 stimulates PGR-positive uterine stromal cells to release an unidentified progestamedin that induces FGF7 expression by LE; 3) E_2 and P4 can induce FGF7 when PGR are rendered nonfunctional by ZK; and 4) E_2 from conceptuses interacts via estrogen receptor alpha, but not estrogen receptor beta in LE to induce maximal expression of FGF7 in LE on Day 12 of pregnancy in pigs.

endometrium, estradiol receptor, progesterone receptor

INTRODUCTION

Fibroblast growth factor 7 (FGF7), also called keratinocyte growth factor, is a member of the FGF superfamily, which has been reported to stimulate cell proliferation, differentiation, migration, and vascular angiogenesis [1]. The preponderance of evidence indicates that FGF7 is expressed in cells of mesenchymal (stromal) origin. However, FGF7 receptors (FGF receptor 2IIIb, FGFR2) are present only on epithelial cells [2]. Therefore, the prevailing opinion is that mesenchymal-derived FGF7 binds receptors on epithelia to mediate epithelial-mesenchymal interactions in various organs, including the reproductive tract [2, 3]. In sharp contrast to this consensus opinion, the results of our studies have demonstrated that *FGF7* mRNA is expressed in the endometrial epithelia of pigs, that *FGFR2* mRNA is present in both the endometrial epithelia and conceptus trophectoderm, and that FGF7 stimulates the trophectoderm, but not endometrial epithelial cells, to undergo proliferation and differentiation, which suggest that FGF7 is a paracrine mediator of interactions between the uterus and conceptus [4–6].

Several factors, including cytokines, growth factors, and hormones, are known to affect FGF7 expression in various tissues. Interleukin (IL) 1, IL6, platelet-derived growth factor-β, and transforming growth factor-α increase *FGF7* mRNA expression [7, 8]. Parrott and coworkers have shown that FGF7 and hepatocyte growth factor (HGF) stimulate FGF7 expression in ovarian surface epithelia via a positive autocrine feedback mechanism [9]. Steroid hormones also regulate FGF7 expression in reproductive organs. In the uterine endometria of rhesus monkeys, progesterone (P4) increases stromal expression of FGF7, which may mediate P4-induced increases in epithelial cell proliferation and spiral artery development [10]. In mice, estrogen increases mammary gland expression of FGF7 and may play a role in gland development [11]. Androgens also stimulate FGF7 expression in stromal cells of the prostate [12]. It is significant that the promoter region of the *FGF7* gene contains various regulatory factor binding sites, including steroid response elements for estrogen and glucocorticoids in humans [13, 14] and for androgens in rats [15].

Northern blot and in situ hybridization analyses of *FGF7* in the pig endometrium indicate expression during the estrus cycle and peri-implantation period. *FGF7* expression was first detected in LE between Days 9 and 12 of both the estrus cycle and pregnancy, peaked on Day 12 of pregnancy, and remained high through Day 20 [4]. Since the increase in FGF7 expression on Day 12 of pregnancy was significantly higher than that on Day 12 of the estrus cycle, subsequent studies using an endometrial explant culture system were performed to examine the hormonal and cytokine regulation of FGF7. In these studies, estradiol-17β (E_2), but not P4, increased the level of *FGF7* mRNA [5]. However, the peri-implantation period of pregnancy is highly complex and cannot be duplicated in in vitro culture systems. In pigs, the uterine environment is influenced by the overlapping events of E_2 release by

[1]Supported by USDA-NRICGP 2000–02290 (F.W.B. and L.A.J.) and NIH P30 ES09106 (G.A.J., R.C.B., T.E.S., F.W.B.).
[2]Correspondence: Fuller W. Bazer, Department of Animal Science and Center for Animal Biotechnology and Genomics, 442D Kleberg Center, Texas A&M University, College Station, TX 77843-2471.
FAX: 979 845 9938; e-mail: fbazer@cvm.tamu.edu
[3]These authors contributed equally to this work.

Received: 7 August 2006.
First decision: 17 August 2006.
Accepted: 22 March 2007.

ISSN: 0006-3363. http://www.biolreprod.org

conceptuses for maternal recognition of pregnancy and by extended exposure to P4, which is the hormone of pregnancy, to mediate developmental changes in the uterus that are conducive to successful implantation and placentation [16, 17]. It is known that estrogen receptor α (ESR1) is present in uterine endometrial epithelial cells between Days 12 and 15 of the estrus cycle and pregnancy in the pig, whereas P4 receptors (PGR) are absent from epithelial cells [18, 19]. Given that E_2 is secreted by pig conceptuses into the uterine lumen [16, 17], it is likely that FGF7 expression in the LE is up-regulated by E_2 via ESR1 that is present in the endometrial epithelia. However, since P4 is the dominant hormone during diestrus of the estrus cycle and during pregnancy and since FGF7 is expressed by LE between Day 12 and Day 15 of both the estrus cycle and pregnancy, it is possible that P4 is required as a permissive hormone to allow E_2 to stimulate FGF7 expression by endometrial LE. Therefore, we hypothesized that P4 is required as a permissive hormone for E_2-induced FGF7 expression in the LE of pig endometrium, and we conducted a study to investigate the relationship between P4 and E_2 and their receptors on FGF7 expression in vivo.

MATERIALS AND METHODS

Animals and Tissue Collection

All experimental and surgical procedures complied with the Guide for the Care and Use of Laboratory Animals and were approved by the Institutional Agricultural Animal Care and Use Committee of Texas A&M University. In all, 26 sexually mature gilts of similar age, weight, and genetic background were ovariectomized on Day 4 after the onset of estrus (Day 0) and were assigned randomly to be treated daily at 0700 h from Day 4 through Day 12 as follows: 1) 200 mg P4 in corn oil vehicle (CO; Sigma Chemical Co., St. Louis, MO) (5 gilts); 2) P4 plus 75 mg ZK317,316 (ZK, which was generously provided by Dr. Kristoff Chwalisz, Schering AG, Berlin, Germany) (5 gilts); 3) P4 plus E_2 (100 μg; Sigma) (5 gilts); 4) P4 plus ZK plus E_2 (5 gilts); 5) estradiol benzoate (E_2, 5 mg in 5 ml of CO/day) (3 gilts); and 6) CO alone (3 gilts). The doses of hormones used are those that have been shown to maintain pregnancy in ovariectomized gilts [20, 21] or to induce pseudopregnancy in pigs [22].

All the gilts were hysterectomized on Day 12, and uterine flushings were obtained by introducing and recovering 20 ml of sterile saline per uterine horn at hysterectomy. The flushings were clarified by centrifugation (3000 × *g* for 10 min at 4°C), aliquoted, and frozen at −80°C until analyzed. Several tissue sections (~0.5 cm) from the middle of each uterine horn were fixed in 4% paraformaldehyde in PBS (pH 7.2) and embedded in Paraplast-Plus (Oxford Laboratory, St. Louis, MO). The remaining endometrium was physically dissected from the myometrium, frozen in liquid nitrogen, and stored at −80 C for RNA extraction.

Northern Blot Hybridization

Total cellular RNA was isolated from endometrial tissues using Trizol reagent (Invitrogen, Carlsbad, CA) according to manufacturer's recommendations. Total endometrial RNA (20 μg) was loaded onto a 1.2% (wt/vol) agarose gel, electrophoresed, and transferred to a 0.2-μm pore size nylon membrane, as described previously [23]. The blot was then hybridized with [^{32}P]-radiolabeled antisense cRNA probes generated from a linearized porcine *FGF7* partial cDNA [4]. Autoradiographs of the Northern blots were prepared using Kodak X-OMAT x-ray film (Eastman Kodak Co., Rochester, NY).

Slot Blot Hybridization

The steady-state levels of *FGF7* were assessed by slot blot hybridization, as described previously [4]. Denatured total endometrial RNA (20 μg) from each pig was analyzed using a [^{32}P]-radiolabeled antisense pig cRNA probe [4]. To correct for variations in total RNA loading, a duplicate RNA slot membrane was hybridized with radiolabeled antisense 18S rRNA (pTRI 18S; Ambion, Austin, TX). Following washing, nonspecific hybridization was removed by RNase A digestion. The radioactivity associated with each slot was quantitated by electronic autoradiography using an Instant Imager (Packard Instrument Company, Meridian, CT) and is expressed as total counts.

RT-PCR Analysis

Total endometrial cDNAs were produced using the SuperScript III First-Strand Synthesis System for RT-PCR (Invitrogen) with 1 μl (100 ng) of total RNA. PCR was performed with Platinum *Taq* DNA polymerase (Invitrogen) in 50-μl reactions that contained 2 μl of cDNA, 5 μl of 10× PCR buffer, 1 μl of 10 mM dNTP mixture, 1.5 μl of 50 mM $MgCl_2$, 0.2 μl of Platinum *Taq* DNA polymerase, and 1 μl of either 10 μM ERS1 (5′-AGGGAAGCTCCTA-TTTGCTCC-3′) or 10 μM ERβ (5′-GCTTCGTGGAGCTCAGCCTG-3′) sense primer and 1 μl of either 10 μM ERS1 (5′-CGGTGGATGTGGTCC-TTCTCT-3′) or 10 μM ERβ (5′-AGGATCATGGCCTTGACACAGA3′) antisense primer [24, 25]. Amplification was performed as previously described [20], with initial denaturation at 94°C for 10 min, followed by 35 cycles of denaturation at 94°C for 30 sec, annealing at 58°C for 30 sec, extension at 72°C for 2 min, and a final extension step at 72°C for 5 min. The PCR products (5 μl) were analyzed on a 2% (wt/vol) agarose gel. The resulting gels were scanned using the BioRad ChemiDoc XRS scanner and analyzed with the Quantity One software (Bio-Rad).

In Situ Hybridization

The location of *FGF7* mRNA in uterine tissue sections was determined by in situ hybridization analysis, as described previously [26]. Briefly, deparaffinized, rehydrated, and deproteinated uterine tissue sections (~5 μm) were hybridized with radiolabeled antisense or sense porcine *FGF7* cRNA probes that were generated from linearized plasmid templates by in vitro transcription with [^{35}S]-UTP (Perkin Elmer Life Sciences). The plasmid templates were partial cDNAs for porcine *FGF7* [4]. After hybridization, washing, and digestion with RNase A, autoradiography was performed using Kodak NTB-2 liquid photographic emulsion (Eastman Kodak). The slides were exposed at 4°C for 2 wk, developed in Kodak D-19 developer, counterstained with Harris modified hematoxylin (Fisher Scientific, Fairlawn, NJ), dehydrated through a graded series of alcohol to Citrisolv (Decon Laboratories Inc., King of Prussia, PA), and protected with coverslips.

Western Blot Analysis

FGF7 was purified from 1 mg of total protein from each uterine flushing by incubating with 200 μl heparin-beaded agarose (Sigma) at 4°C overnight and washing three times with Hanks balanced salt solution (Sigma) and centrifugation at 10 000 × *g* for 5 min. Proteins that bound to the heparin-beaded agarose were denatured in Laemmli buffer and separated by SDS-PAGE, transferred to nitrocellulose, and blocked with 5% (wt/vol) nonfat milk-TBST (Tris-buffered saline with 0.1% Tween-20), as described previously [4]. Blots were incubated with a goat polyclonal antibody against human FGF7 synthetic peptide (2 μg/ml; Santa Cruz Biotechnology, Santa Cruz, CA) or normal goat IgG (Sigma) as the primary antibody in 2% (wt/vol) milk-TBST overnight at 4°C, rinsed for 30 min at room temperature with TBST, incubated with peroxidase-conjugated rabbit anti-goat IgG (1:10 000; Zymed Laboratories Inc., San Francisco, CA) as the secondary antibody for 1 h at room temperature, and then rinsed again for 30 min at room temperature with TBST. Immunoreactive proteins were detected using enhanced chemiluminescence (Amersham/Pharmacia, Arlington Heights, NY) according to the manufacturer's recommendations.

Immunohistochemistry

The expression of PGR and ESR1 proteins was evaluated in paraformaldehyde-fixed, paraffin-embedded, uterine tissue cross-sections (4 μm) using procedures described previously [23]. Briefly, boiling citrate buffer was used for retrieval of PGR, and pronase E (0.5 mg/ml in PBS) was used to retrieve the ESR1 antigen. Proteins were detected with mouse anti-human PGR IgG (2 μg/ml 2C5 PGR; Zymed Labs) and rat anti-human-ESR1 IgG (2 μg/ml H222; Abbott Laboratories, Chicago, IL) and visualized with the Super ABC Mouse/Rat IgG Kit (Biomeda, Foster City, CA). For negative controls, primary antibodies were substituted with mouse (for PGR) or rat (for ESR1) IgG at the same concentrations as the primary antibodies.

Photomicrography

Digital photomicrographs of representative bright-field (immunohistochemistry) or bright-field and dark-field illumination (in situ) images were evaluated with a Zeiss Axioplan2 microscope (Carl Zeiss, Thornwood, NY) fitted with an Axiocam HR digital camera. Digital images for immunohistochemistry and in situ hybridization were recorded using the Axiovision 4.3 software. All

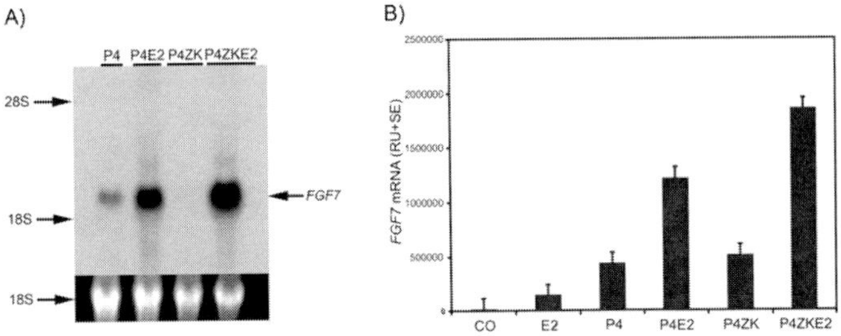

FIG. 1. Detection of *FGF7* mRNA in the endometrium. **A**) Northern blot analysis of *FGF7* mRNA (20 μg) in the endometria of ovariectomized pigs treated with P4, P4 + E_2 (P4E2), P4 + ZK (P4ZK), and P4ZKE2. The positions of the *28S* (4.7-kb) and *18S* (1.8-kb) rRNAs are indicated. A single transcript for *FGF7* (~2.4 kb) is detected. **B**) Steady-state levels of *FGF7* mRNAs in CO-, E_2-, P4-, P4E2-, P4ZK-, and P4ZKE2-treated pig endometria, expressed as least square means of the relative units of counts per minute with overall SEM, normalized for differences in sample loading using *18s* rRNA and representing 20 μg of total endometrial mRNA per sample.

immunohistochemistry and in situ hybridization figures were assembled using Adobe Photoshop 8.0 (Adobe Systems Inc., San Jose, CA).

Statistical Analysis

All quantitative data were subjected to least-squares ANOVA using the general linear models procedures of the Statistical Analysis System (SAS Institute, Cary, NC). Slot blot hybridization data were corrected for differences in sample loading using the 18S rRNA data as a covariate. Orthogonal contrasts were used to determine the effects of treatment. Preplanned contrasts were CO vs. E_2, E_2 vs. P4, P4 vs. P4 + E_2, P4 + ZK vs. P4 + ZK + E_2, and P4 vs. P4 + ZK. All tests of significance were performed using the appropriate error terms according to the expectation of the mean squares for error. A P value equal to or less than 0.05 was considered statistically significant. Data are presented as least-square means (LSM) with standard errors (SEM).

RESULTS

Steady-State Levels of FGF7 *mRNA in the Pig Endometrium and of FGF7 Protein in the Uterine Lumen*

The *FGF7* cRNA detected a single transcript of ~2.4 kb in Northern blot analysis of pig endometrial total RNA (Fig. 1A). The steady-state levels of endometrial *FGF7* mRNA were not different between pigs that received CO or E_2 ($P > 0.10$), or between pigs that received P4 or P4ZK ($P > 0.10$). However *FGF7* mRNA increased in pigs injected with P4 compared to those injected with E_2 ($P < 0.05$). The combination of P4 and E_2 increased endometrial *FGF7* mRNA compared to P4 alone ($P < 0.01$), and the addition of E_2 (PZKE2) to the P4ZK treatment increased the level of *FGF7* mRNA significantly ($P > 0.01$). The results of the Western blotting analysis to detect FGF7 protein purified from uterine flushings using heparin-coated agarose beads are shown in Figure 2. An immunoreactive FGF7 protein of 17-kDa was detected in the flushings from pigs treated with P4 and E_2, and two molecular mass variants of the FGF7 protein, 17 kDa and 25 kDa, were detected in the flushings from pigs treated with the combination of P4, E_2, and ZK, which indicates the secretion of FGF7 into the uterine lumen of pigs (Fig. 2).

Relationships Between FGF7 *mRNA, PGR, and ESR1 in Pig Endometria*

Previous studies have reported the presence of *ESR2* mRNA and ESR2 protein in the pig uterus [27–29]. A specific PCR product was detected in the pig endometrium using primers for *ESR2*. Sequence analysis indicated that the product was pig *ESR2*. However, although ESR1 was readily detected by RT-PCR, ESR2 was barely detected above background in total endometrium from pigs from all the treatment groups (Fig. 3). In all the endometrial samples, the level of ESR1 was ~8-fold higher than that of ESR2. In addition, *ESR2* was not detected above the level of the sense cRNA probe or irrelevant IgG background using in situ hybridization or immunohistochemistry, respectively (data not shown). In situ hybridization analysis of pig endometrium localized *FGF7* mRNA to the LE of ovariectomized pigs treated with P4, P4 and E_2 or P4, E_2, and ZK. However, *FGF7* mRNA was not detected in the endometrial LE of pigs treated with CO vehicle, P4 and ZK or E_2 alone (Fig. 4). Immunohistochemistry for PGR revealed expression in the endometrial LE of pigs treated with CO vehicle or E_2, whereas PGR expression was absent in the LE of pigs treated with P4 alone or P4 in combination with ZK and/or E_2. In contrast, PGR was expressed in the endometrial stromal cells of pigs in all the treatment groups (Fig. 4). Immunohistochemistry for ESR1 revealed expression in the LE, stromal cells, and GE of all pigs in all treatment groups. In the following section, the relationships between treatment, PGR and ESR1 status, and *FGF7* mRNA expression are summarized by treatment group (Fig. 4).

Ovariectomized pigs injected with corn oil. In the absence of ovarian steroid hormones following ovariectomy, the absence of P4 and E_2 resulted in default expression of both PGR and ESR1 in the endometrial LE, stromal cells, and GE (Fig. 4). The presence of PGR in the LE is considered to preclude the expression of *FGF7* mRNA in CO-treated pigs.

Ovariectomized pigs injected with progesterone. Continuous 9-day exposure of pigs to P4 resulted in the down-regulation of PGR in endometrial LE but not in stromal cells, and did not affect the expression of ESR1 in LE, GE or stromal cells (Fig. 4). The combined effects of P4-induced down-regulation of PGR in LE and the maintenance of PGR in stromal cells is considered to enable P4 to interact with PGR in stromal cells to induce the synthesis and secretion of a putative progestamedin that induces *FGF7* in the LE.

Ovariectomized pigs treated with progesterone and ZK137,316. It appears that 75 mg ZK daily does not recapitulate the previous results obtained for sheep [30] to block completely the effects of P4 on the pig endometrium. PGR in the LE and stromal cells are differentially sensitive to the effects of ZK. Although ZK did not inhibit P4-induced down-regulation of PGR in endometrial LE in 5/5 pigs, ZK bound PGR in the endometrial stromal cells, thereby rendering the stromal cells unresponsive to P4 (Fig. 4). Therefore, it seems likely that a progestamedin was not synthesized or released by stromal cells in response to P4 so as to mediate the induction of *FGF7* mRNA in the LE of the 3/5 pigs that received both P4 and ZK.

Ovariectomized pigs treated with estradiol benzoate. The absence of P4 to down-regulate PGR and the effect of E_2 in inducing PGR resulted in the expression of PGR in the LE, which is considered to have prevented the induction of *FGF7* mRNA in the LE by E_2 (Fig. 4).

Ovariectomized pigs treated with progesterone and E_2. Treatment of pigs with both P4 and E_2 allowed P4 to down-regulate the PGR in the LE and allowed E_2 to induce *FGF7* mRNA expression in the endometrial LE (Fig. 3). Furthermore, the interaction of P4 with PGR in stromal cells probably allows the induction of a progestamedin that acts on the LE to induce the expression of *FGF7* mRNA. Alternatively, the expression of *FGF7* mRNA may be induced in this treatment group through a combination of these two mechanisms.

Ovariectomized pigs treated with progesterone, ZK, and E_2. The action of ZK on PGR in stromal cells inhibits the production of the putative progestamedin that is necessary for the induction of *FGF7* expression in the LE (Fig. 4). Therefore, in the absence of functional PGR, E_2 action via ESR1 is considered to be sufficient for the induction of *FGF7* mRNA expression in pig endometrial LE.

DISCUSSION

The results of the present study of pig endometria indicate that: 1) P4 is permissive for FGF7 expression through its action in down-regulating PGR in LE; 2) P4 stimulates PGR-positive uterine stromal cells to release an unidentified progestamedin, which induces FGF7 expression by the LE; 3) the combined effects of E_2 and P4 or E_2, P4, and ZK can induce FGF7 due to the effect of P4 in down-regulating PGR or the effect of ZK in blocking PGR function and allowing E_2 to act on PGR-negative LE; and 4) E_2 from conceptuses interacts via ESR1 in LE to induce maximal expression of FGF7 in PGR-negative LE on Day 12 of pregnancy in pigs. The group that received both P4 and E_2 is the most physiologically relevant, with P4 from the CL being permissive for the actions of E_2 from pig conceptuses during the peri-implantation period. A model that summarizes the present working hypothesis for hormonal regulation and the role of uterine FGF7 during early pregnancy in pigs is presented in Figure 5. The mechanism responsible for the lack of expression of FGF7 by endometrial GE is not known.

P4, which is the hormone of pregnancy in all mammals, is critical for the control of the temporal and spatial (cell-specific) changes in gene expression within the uterus that ensure synchrony between uterine and conceptus (embryo/fetus and associated membranes) development [31]. Indeed, treatment with exogenous P4 significantly alters the expression of a number of genes in rodent, primate, and sheep uteri, as measured in microarray analyses [32–34]. Although similar studies have not been performed in pigs, P4 increases the expression of calbindin-D9k [35], vascular endothelial growth factor [36], FGF2 and two of its receptors, FGFR1 and FGFR2 [37], and the $\alpha 4$, $\alpha 5$, and $\beta 1$ integrin receptor subunits [38], as well as suppressing the expression of MUC1 [38]. Importantly, P4 increases the expression levels of various uterine secretory proteins, components of the histotroph, which is hypothesized to support conceptus development in pigs [39–41]. Previous studies with endometrial explant cultures and in vivo steroid replacement experiments have failed to demonstrate FGF7

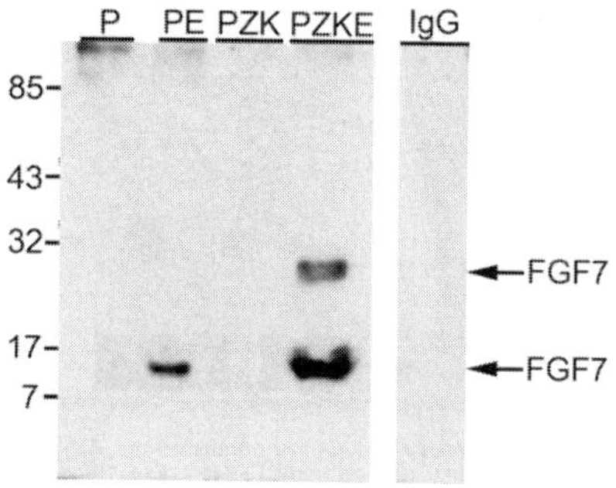

FIG. 2. Western blot analysis of FGF7 in porcine uterine luminal flushings. The positions of the prestained molecular mass standards are indicated. Immunoreactive FGF7 is detected (arrows) in the uterine flushings from ovariectomized pigs treated with P4 + E_2 (P4E2) or P4 + ZK + E_2(P4ZKE2).

regulation by P4 in the pig uterus, since the investigators assumed that increases in FGF7 expression during the estrus cycle and pregnancy were most likely influenced by P4 [5, 42]. The results of the present study clearly indicate that FGF7 is induced in the uterine LE of ovariectomized pigs treated with P4, and that the expression of FGF7 is blocked by ZK, a PGR antagonist.

The permissive effects of P4 on FGF7 expression are mediated by PGR in the pig endometrium [43, 44]. In most mammalian uteri, PGR are expressed in endometrial epithelia and stromal cells during the early to midluteal phase, allowing direct regulation of genes by P4. However, in sheep, the presence of PGR in LE precludes the expression of most P4-regulated genes in the LE until continuous exposure of the endometrium to P4 down-regulates PGR expression exclusively in the LE and GE [31]. This paradigm of loss of PGR from uterine epithelia prior to implantation appears to be common among mammals [45] and predicates that the endocrine effects of ovarian P4 on endometrial epithelia during the peri-implantation period are mediated indirectly by either P4-induced paracrine-acting factors (progestamedins) produced by PGR-positive stromal cells or by the induction of factors in the

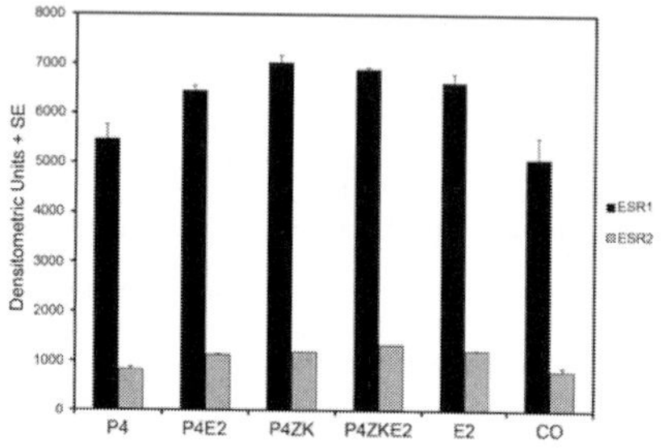

FIG. 3. RT-PCR analysis of *ERS1* and *ESR2* mRNAs in the endometria of pigs treated with P4, P4 + E_2 (P4E2), P4 + ZK (P4ZK), and P4ZKE2. Densitometric analyses of *ESR1* and *ESR2* transcripts from the endometria of pigs in different treatment groups following PCR amplification reveal that *ESR1* mRNA is maintained at higher levels than *ESR1* mRNA in all the treatment groups.

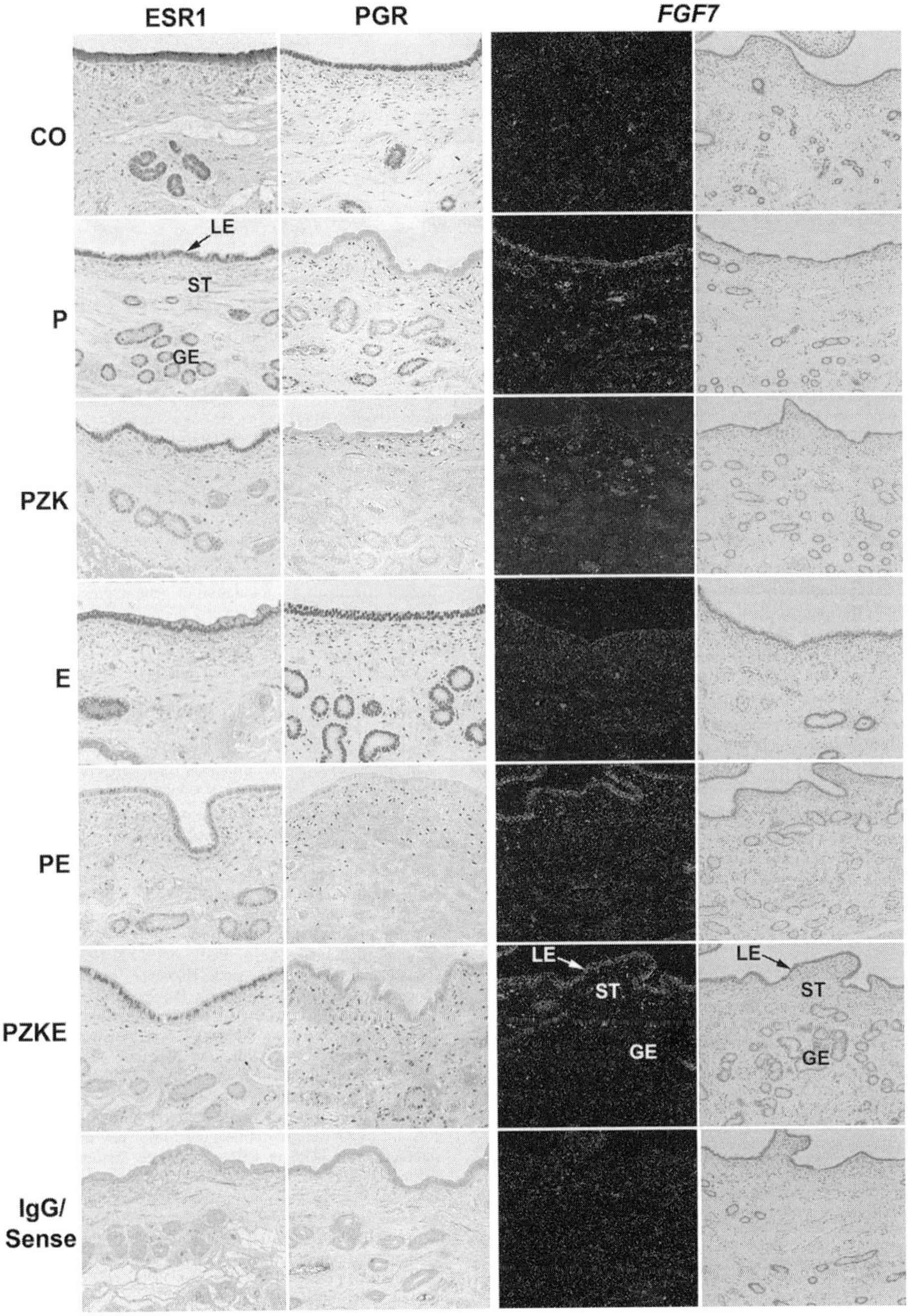
ESR1
PGR
FGF7
CO
P
PZK
E
PE
PZKE
IgG/
Sense
LE
ST
GE

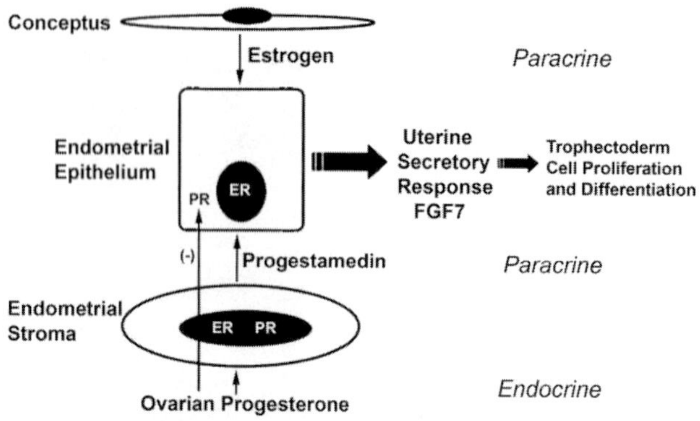

FIG. 5. Schematic illustration of the proposed hormonal regulation of FGF7 expression in the porcine pregnant uterine endometrium. During early pregnancy, continuous exposure of the uterine epithelia to progesterone down-regulates PGR, thereby eliminating PGR-dependent inhibition of expression of most P4-regulated genes, e.g., those in the LE. Therefore, the endocrine effects of P4 in inducing the expression of FGF7 in endometrial LE and its secretion into the uterine lumen are mediated by a paracrine-acting factor(s) (progestamedin) produced by the PGR-positive stromal cells. Furthermore, in PGR-negative epithelial cells, estrogen produced by pig conceptuses binds to ESR1 in the LE to induce FGF7 expression. The combined endocrine/paracrine effects of ovarian P4 and conceptus estrogens are likely responsible for the high levels of *FGF7* mRNA expression in the endometria and of FGF7 protein in the uterine lumen on Day 12 of pregnancy. During the estrus cycle, FGF7 expression increases in the LE during the P4-dominated luteal phase, whereas maximal levels of FGF7 are attained on Day 12 of pregnancy after the PGR are down-regulated and when LE is stimulated by high levels of estrogen released by pig conceptuses for pregnancy recognition and P4 can stimulate the secretion of a progestamedin(s) from uterine stromal cells. It is hypothesized that secreted FGF7 acts on the conceptus to stimulate the proliferation and differentiation of the trophectoderm.

LE that simultaneously down-regulate PGR and either allow or stimulate the expression of endometrial genes [30, 46]. In pigs, the expression of PGR in endometrial LE and GE is down-regulated by Day 10 of the estrus cycle and pregnancy, whereas the expression of PGR is maintained in stromal cells and the myometrium [19]. Removal of PGR from LE correlates with loss of MUC1 and expression of secreted phosphoprotein 1 (SPP1; also known as osteopontin) on the apical surface of the LE, which exposes integrins to extracellular matrix proteins for trophoblast attachment to the uterus [38, 47]. This is also the period during which the endometrium releases many cytokines and growth factors into the uterine lumen of the pig to support conceptus development and trophoblast elongation [48]. Although the loss of PGR from the endometrial epithelia of pigs is well established [19], the present results are the first to

FIG. 4. Interrelationships between *FGF7* mRNA, PGR protein, and estrogen receptor α protein in pig endometria. First column: nuclear localization of ESR1 protein in the luminal epithelia (LE) of corn oil (CO) and E_2-treated (E2) pigs, whereas PGR is present in stromal cells (ST) of the endometria from pigs in all the treatment groups. A section from an E_2-treated pig stained with nonimmune mouse IgG (IgG) serves as a negative control. The width of each field is 540 μm. Second column: nuclear immunostaining for PGR in the LE, GE, and ST of endometria from pigs in all treatment groups. A section from an estradiol valerate-treated pig stained with nonimmune rat IgG serves as a negative control. The width of each field is 540 μm. Third and fourth columns: in situ hybridization analysis of *FGF7* mRNA expression in pig endometria. The left and right panels represent corresponding bright-field and dark-field images, respectively, of endometria from pigs in each treatment group. A representative section from a P4-treated pig hybridized with a radiolabeled sense cRNA probe (Sense) serves as a negative control. Note that *FGF7* mRNA is detectable only in the LE, and that the hybridization signal is evident only in the endometria from pigs treated with P4, P4 + E_2 (PE2), and P4 + ZK + E_2 (PZKE2). The width of each field is 690 μm.

dissect effectively P4 regulation of gene expression in the LE during the estrus cycle and early pregnancy.

Three conclusions can made based on the results of the present study. First, similar to sheep [23], P4 negatively autoregulates PGR in the LE, but not PGR in the endometrial stromal cells of pigs. The expression of PGR in the LE and stromal cells was detected in pigs that received CO but it was down-regulated in the LE of all pigs treated with P4. Second, similar to the results for P4-regulated genes in the endometrial epithelia of sheep [23, 30], the presence of PGR in pig LE precludes the induction of *FGF7*. All pigs with PGR in the LE, i.e., those treated with CO alone or E_2 alone, failed to express *FGF7* in LE. Third, the combined effects of P4-induced down-regulation of PGR in LE and P4 interaction with PGR in stromal cells result in the expression of a progestamedin(s) from stromal cells that acts on LE to induce *FGF7*. Pigs treated with P4 exhibited down-regulation of PGR in LE concomitant with induction of FGF7. Furthermore, FGF7 expression requires functional stromal PGR, since ZK treatment ablated P4-induced FGF7 expression. However, in the absence of functional PGR (treatment with P4, E_2, and ZK), E_2 alone interacts with ESR1 to induce FGF7 expression.

The mechanisms by which P4 both down-regulates PGR and up-regulates the expression of other genes within the uterine LE of pigs are not understood. Induction of stromal-derived progestamedins is one explanation for these phenomena, although an attractive alternative hypothesis has been put forward by Geisert and colleagues [46]. They propose that P4 interacts with PGR in LE to stimulate factors that activate nuclear factor kappa B (NF-κB), which then functions to inhibit PGR expression and activate transcription of genes that are believed to be involved in implantation [46]. The results of the present study are consistent with both of these theories of endometrial gene regulation by P4.

Estrogens secreted by pig conceptuses on Day 12 of gestation comprise the maternal recognition signal that switches the secretion of endometrial prostaglandin F2α from the endocrine to exocrine direction to prevent CL regression [16]. In addition, conceptus estrogens modulate uterine gene expression to support uterine secretions and the controlled inflammatory-like events that characterize changes in conceptus morphology and uterine remodeling for implantation in pigs [49]. Indeed, secreted SPP1 is induced by estrogen in LE, and it initially localizes to the LE in close proximity to the Day-12 implanting conceptuses [47], whereas conceptus secretion of estrogens correlates with conceptus secretion of IL1β and may modulate uterine responses to this cytokine [50]. The importance of estrogen for the early survival of pig conceptuses is underscored by pregnancy loss in response to premature exposure of the pregnant uterus to estrogen. Administration of estrogen on Days 9 and 10 of pig pregnancy is associated with altered expression of SPP1 and cyclooxygenase 1 in LE [46, 51] and degeneration of conceptuses by Day 15 [52]. In pigs, ESR1 is readily detectable in LE from Day 5 through Day 12 of the estrus cycle and pregnancy, then decreases, but remains detectable until Day 15 [18]. The presence of ESR1 in LE provides a mechanism by which estrogens secreted by the elongating pig conceptus can stimulate the changes in uterine function that are necessary for the maintenance of pregnancy [18]. Previous reports, using endometrial explant cultures and in vivo steroid replacement experiments, have strongly suggested that the induction of FGF7 in uterine LE during early pregnancy in pigs is stimulated primarily by conceptus estrogens [5, 42]. The results of the present study support these previous reports and further indicate that FGF7 is induced in the uterine LE of ovariectomized pigs injected with E_2, but only when P4 has down-regulated PGR in the LE. The uterine LE of pigs treated with E_2 alone does not express *FGF7* mRNA, probably due to the lack of P4 to down-regulate PGR in the LE, whereas pigs treated with a combination of E_2 and P4 exhibit *FGF7* expression. Furthermore, functional stromal PGR, and therefore progestamedins, are not required for E_2 induction of *FGF7* because the addition of ZK did not alter the effects of the combination of P4 and E_2.

There is a discrepancy between the results of the present study and those in the previous reports with respect to the fining that estrogen in the absence of P4 can increase endometrial expression of FGF7 [5, 37]. Ka et al. [5] used endometrial explant cultures from Day 9 of the estrus cycle, at which time-point the PGR are significantly reduced compared to Days 0–5 [19, 53]. It is likely that the PGR were already decreased to levels that were insufficient to prevent the induction of *FGF7*. Although the study conducted by Wollenhaupt et al. [42] employed ovariectomized, steroid-replaced pigs and detected a significant increase in endometrial *FGF7* mRNA in response to estradiol benzoate compared to P4 treatment, the levels of expression of *FGF7* mRNA overlapped between gilts that were treated with vehicle and estradiol benzoate. Furthermore, the effects on FGF7 expression of estradiol benzoate alone did not differ from the combined effects of P4 and estradiol benzoate, which did not differ from the results of treatment with either vehicle or P4 alone. Interestingly, increases in FGF7 protein were detected primarily in the vascular smooth muscle cells and endothelium, whereas weak expression in LE was not affected by the different treatments [42]. In contrast, the present study utilized ZK, which is a PGR antagonist, to dissect the hormonal regulation of FGF7, and in situ hybridization analyses to understand cell-specific changes in *FGF7* expression. The results clearly indicate the complex and overlapping regulation of FGF7 in endometrial LE by P4 and E_2.

The FGF7 synthesized and secreted by uterine epithelial cells into the uterine lumen can stimulate the proliferation and differentiation of conceptus trophectoderm [4, 5] by influencing DNA synthesis and through motility, differentiation, cytoprotective, and antiapoptotic effects on cells [54]. Western blotting detected an immunoreactive FGF7 protein of about 25 kDa only in the uterine flushings of pigs treated with E_2. Interestingly, a 17-kDa FGF7 protein was also detected in uterine flushings from pigs treated with E_2; this protein may be a cleavage fragment of the native 25-kDa FGF7 protein. These results indicate that E_2 regulates the secretion of FGF7 from LE cells. It is known that estrogen (of conceptus origin or administered exogenously) induces pseudopregnancy, modulates the redirection of prostaglandin F2α from primarily endocrine secretion into the underlying vasculature to exocrine secretion into the uterine lumen [16], and increases the secretory activities of endometrial epithelia directly or by increasing uterine expression of prolactin receptors on uterine epithelia [55]. Therefore, we hypothesize that FGF7 secretion is regulated by conceptus estrogens during pregnancy and by ovarian estrogens during the estrus cycle. The detailed cellular mechanism by which estrogen affects epithelial FGF7 secretion remains to be determined.

In summary, the uterine environment of early pregnancy in pigs is complex and influenced by the overlapping actions of conceptus estrogens for pregnancy recognition and the permissive effects of P4 acting on uterine epithelia to down-regulate PGR and/or via PGR expressed by uterine stromal cells to stimulate the expression of a progestamedin(s) that mediates the developmental changes in uterine functions necessary for the establishment and maintenance of pregnancy. The dynamic regulation of FGF7 by estrogen and P4 reflects this complexity, and strongly suggests important roles for FGF7 in stimulating the proliferation and differentiation of the pig conceptus during elongation and implantation. The results of the present study provide new insights into the intricate interplay between the actions of estrogen and P4 to modify gene expression in the peri-implantation pig uterus. A similar interplay between pregnancy recognition signals and progesterone has been reported for sheep [56, 57]. However, the magnitude and extent of the influence that conceptus estrogens impart on the early pregnant uterus are remarkable and unique to the pig. Clearly, further studies are warranted to dissect further the endocrine and paracrine regulation of gene expression in the uterus of pregnant pigs.

REFERENCES

1. Szebenyi G, Fallon JF. Fibroblast growth factors as multifunctional signaling factors. Int Rev Cytol 1999; 185:45–106.
2. Rubin JS, Bottaro DP, Chedid M, Miki T, Ron D, Cheon HG, Taylor WG, Fortney E, Sakata H, Finch PW, LaRochelle WJ. Keratinocyte growth factor. Cell Biol Int 1995; 19:399–411.
3. Cooke PS, Buchanan DL, Kurita T, Lubahn DB, Cunha GR. Stromal-epithelial cell communication in the female reproductive tract. In: Bazer FW (ed.), The Endocrinology of Pregnancy. Totowa: Humana Press; 1998:491–506.
4. Ka H, Spencer TE, Johnson GA, Bazer FW. Keratinocyte growth factor: expression by endometrial epithelia of the porcine uterus. Biol Reprod 2000; 62:1772–1778.
5. Ka H, Jaeger LA, Johnson GA, Spencer TE, Bazer FW. Keratinocyte growth factor is up-regulated by estrogen in the porcine uterine endometrium and functions in trophectoderm cell proliferation and differentiation. Endocrinology 2001; 142:2303–2310.
6. Ka H, Bazer FW. Effects of keratinocyte growth factor on the uterine endometrial epithelial cells in pigs. Asian-Aust J Anim Sci 2005; 18: 1708–1714.
7. Chedid M, Rubin JS, Csaky KG, Aaronson SA. Regulation of keratinocyte growth factor gene expression by interleukin 1. J Biol Chem 1994; 269: 10753–10757.
8. Li D-Q, Tseng SCG. Differential regulation of keratinocyte growth factor and hepatocyte growth factor/scatter factor by different cytokines in human corneal and limbal fibroblasts. J Cell Physiol 1997; 172:361–372.

9. Parrott JA, Mosher R, Kim G, Skinner MK. Autocrine interactions of keratinocyte growth factor, hepatocyte growth factor, and kit-ligand in the regulation of normal ovarian surface epithelial cells. Endocrinology 2000; 141:2532–2539.
10. Koji T, Chedid M, Rubin JS, Slayden OD, Csaky KG, Aaronson SA, Brenner RM. Progesterone-dependent expression of keratinocyte growth factor mRNA in stromal cells of the primate endometrium: keratinocyte growth factor as a progestomedin. J Cell Biol 1994; 125:393–401.
11. Pedchenko VK, Imagawa W. Estrogen treatment in vivo increases keratinocyte growth factor expression in the mammary gland. J Endocrinol 2000; 165:39–49.
12. Yan G, Fukabori Y, Nikolaropoulos S, Wang F, McKeehan WL. Heparin-binding keratinocyte growth factor is a candidate stromal-to-epithelial-cell andromedin. Mol Endocrinol 1992; 6:2123–2128.
13. Finch PW, Lengel C, Chedid M. Cloning and characterization of the promoter region of the human keratinocyte growth factor gene. J Biol Chem 1995; 270:11230–11237.
14. Zhou J, Finch PW. Identification of a novel transcriptional regulatory element within the promoter region of the keratinocyte growth factor gene that mediates inducibility to cyclic AMP. Biochim Biophys Acta 1999; 1446:71–81.
15. Fasciana C, van der Made AC, Faber PW, Trapman J. Androgen regulation of the rat keratinocyte growth factor (KGF/FGF7) promoter. Biochem Biophys Res Commun 1996; 220:858–863
16. Bazer FW, Thatcher WW. Theory of maternal recognition of pregnancy in swine based on estrogen controlled endocrine versus exocrine secretion of prostaglandin F2alpha by the uterine endometrium. Prostaglandins 1977; 14:397400
17. Bazer FW, Ott TL, Spencer TE. Endocrinology of the transition from recurring estrous cycles to establishment of pregnancy in subprimate mammals. In: Bazer FW (ed.), The Endocrinology of Pregnancy. Totowa: Humana Press; 1998:1–34.
18. Geisert RD, Brenner RM, Moffatt RJ, Harney JP, Yellin T, Bazer FW. Changes in oestrogen receptor protein, mRNA expression and localization in the endometrium of cyclic and pregnant gilts. Reprod Fertil Dev 1993; 5:247–260.
19. Geisert RD, Pratt TN, Bazer FW, Mayes JS, Watson GH. Immunocytochemical localization and changes in endometrial progestin receptor protein during the porcine oestrous cycle and early pregnancy. Reprod Fertil Dev 1994; 6:749–760.
20. Knight J, Bazer FW, Wallace HD. Effect of progesterone induced increase in uterine secretory activity on development of the porcine conceptus. J Anim Sci 1974; 39:743–746.
21. Knight JW, Bazer FW, Wallace HD, Wilcox CJ. Dose response relationships between exogenous progesterone and estradiol and porcine uterine protein secretions. J Anim Sci 1974; 39:747–751
22. Frank M, Bazer FW, Thatcher WW, Wilcox CJ. A study of prostaglandin F_2 as the luteolysin in swine: III. Effects of estradiol valerate on prostaglandin F, progestins, estrone and estradiol concentrations in the utero-ovarian vein of non-pregnant gilts. Prostaglandins 1977; 14:1183–1196
23. Spencer TE, Bazer FW. Temporal and spatial alterations in uterine estrogen receptor and progesterone receptor gene expression during the estrous cycle and early pregnancy in the ewe. Biol Reprod 1995; 53(6): 1527–1543.
24. Kalbe CM, Mau M, Wooenhaupt, Rehfeldt C. Evidence for estrogen receptor alpha and beta expression in skeletal muscle of pigs. Histomen Cell Biol 2007; 127:95–107.
25. Pfaffle MW, Lange IG, Daxenberger A, Meyer HH. Tissue-specific expression pattern of estrogen receptors (ER): quantification of ER alpha and ER beta mRNA with real-time RT-PCR. APMIS 2001;109:345–355.
26. Johnson GA, Spencer TE, Hansen TR, Austin KJ, Burghardt RC, Bazer FW. Expression of the interferon tau inducible ubiquitin cross-reactive protein in the ovine uterus. Biol Reprod 1999; 61:312–318.
27. Kowalski AA, Graddy LG, Vale-Cruz DS, Choi I, Katzenellenbogen BS, Simmen FA, Simmen RC. Molecular cloning of porcine estrogen receptor-beta complementary DNAs and developmental expression in periimplantaion embryos. Biol Reprod 2002; 66:760–769.
28. Cardenas H, Pope WF. Attentuation of estrogenic effects by dihydrotestosterone in the pig uterus is associated with downregulation of the estrogen receptors. Biol Reprod 2004; 70:297–302.
29. Cadenas H, Pope WF. Estrogen receptors in the uterus and ovarian follicles of gilts treated with dihydrotestosterone. Domest Anim Endocrinol 2005; 29:523–533.
30. Johnson GA, Spencer TE, Burghardt RC, Taylor KM, Gray CA, Bazer FW. Progesterone modulation of osteopontin gene expression in the ovine uterus. Biol Reprod 2000; 62:1315–1321.
31. Spencer TE, Johnson GA, Burghardt RC, Bazer FW. Progesterone and placental hormone actions on the uterus: insights from domestic animals. Biol Reprod 2004; 71:2–10.
32. Ace CI, Okulicz WC. Microarray profiling of progesterone-regulated endometrial genes during the rhesus monkey secretory phase. Reprod Biol Endocrinol 2004; 2:54.
33. Jeong JW, Lee KY, Kwak I, White LD, Hilsenbeck SG, Lydon JP, DeMayo FJ. Identification of murine uterine genes regulated in a ligand-dependent manner by the progesterone receptor. Endocrinology 2005; 146:3490–4505.
34. Gray CA, Abbey CA, Beremand PD, Choi Y, Farmer JL, Adelson DL, Thomas TL, Bazer FW, Spencer TE. Identification of endometrial genes regulated by early pregnancy, progesterone, and interferon tau in the ovine uterus. Biol Reprod 2006; 74:383–394.
35. Yun SM, Choi KC, Kim IH, An BS, Lee GS, Hong EJ, Oh GT, Jeung EB. Dominant expression of porcine Calbindin-D9k in the uterus during a luteal phase. Mol Reprod Dev 2004; 67:251–256.
36. Welter H, Wollenhaupt K, Tiemann U, Einspanier R. Regulation of the VEGF-system in the endometrium during steroid-replacement and early pregnancy of pigs. Exp Clin Endocrinol 2003; 111:33–40.
37. Welter H, Wollenhaupt K, Einspanier R. Developmental and hormonal regulated gene expression of fibroblast growth factor 2 (FGF-2) and its receptors in porcine endometrium. Steroid Biochem Mol Biol 2004; 88: 295–304.
38. Bowen JA, Bazer FW, Burghardt RC. Spatial and temporal analyses of integrin and Muc-1 expression in porcine uterine epithelium and trophectoderm in vivo. Biol Reprod 1996; 55:1098–1106.
39. Roberts RM, Bazer FW. The functions of uterine secretions. J Reprod Fertil 1988; 82:875–892.
40. Simmen RC, Simmen FA, Hofig A, Farmer SJ, Bazer FW. Hormonal regulation of insulin-like growth factor gene expression in pig uterus. Endocrinology 1990; 127:2166–2174.
41. Harney JP, Ott TL, Geisert RD, Bazer FW. Retinol-binding protein gene expression in cyclic and pregnant endometrium of pigs, sheep, and cattle. Biol Reprod 1993; 49:1066–1073.
42. Wollenhaupt K, Welter H, Brüssow K-P, Einspanier R. Regulation of endometrial fibroblast growth factor 7 (FGF-7) and its receptor FGFR2IIIb in gilts after sex steroid replacements, and during the estrous cycle and early gestation. J Reprod Dev 2005; 51:509–519.
43. Kastner P, Krust A, Turcotte B, Stropp U, Tora L, Gronemeyer H, Chambon P. Two distinct estrogen-regulated promoters generate transcripts encoding the two functionally different human progesterone receptor forms A and B. EMBO J 1990; 9:1603–14.
44. Cheon YP, Li Q, Xu X, DeMayo FJ, Bagchi IC, Bagchi MK. A genomic approach to identify novel progesterone receptor regulated pathways in the uterus during implantation. Mol Endocrinol 2002; 16:2853–2871.
45. Carson DD, Bagchi I, Dey SK, Enders AC, Fazleabas AT, Lessey BA, Yoshinaga K. Embryo implantation. Dev Biol 2000; 223:217–37.
46. Geisert RD, Ross JW, Ashworth MD, White FJ, Johnson GA, Hu J, DeSilva U. Maternal recognition of pregnancy signal or endocrine disruptor: the two faces of oestrogen during establishment of pregnancy in the pig. In: Kraeling R, Ashworth C (eds.), Control of Pig Reproduction VII. Nottingham: Nottingham University Press; 2005:131–145.
47. White FJ, Ross JW, Joyce MM, Geisert RD, Burghardt RJ, Johnson GA. Steroid regulation of cell specific secreted phosphoprotein 1 (osteopontin) expression in the pregnant porcine uterus. Biol Reprod 2005; 73:1294–1301.
48. Geisert RD, Yelich JV. Regulation of conceptus development and attachment. J Reprod Fertil 1997; 52:133–149.
49. Geisert RD, Thatcher WW, Roberts RM, Bazer FW. Establishment of pregnancy in the pig: III. Endometrial secretory response to estradiol valerate administered on day 11 of the estrous cycle. Biol Reprod 1982; 27:957–968.
50. Ross JW, Malayer JR, Ritchey JW, Geisert RD. Involvement of the interleukin-1β system in porcine trophoblastic elongation and at the fetal-maternal interface during peri-implantation development. Biol Reprod 2003; 69:1251–1259.
51. Ashworth MD, Ross JW, Hu J, White FJ, Stein DR, DeSilva U, Johnson GA, Spencer TE, Geisert RD. Porcine endometrial prostaglandin synthase expression during the estrous cycle, early pregnancy and following endocrine disruption of pregnancy. Biol Reprod 2006; 74:1007–1015.
52. Morgan GL, Geisert RC, Zavy MT, Fazleabas AT. Development and survival of porcine blastocysts on day 16 of pregnancy following oestrogen administration on day 9 or days 9 and 10. J Reprod Fertil 1987; 80:33–41.
53. Sukjumlong S, Dalin AM, Sahlin L, Persson E. Immunohistochemical studies on the progesterone receptor (PR) in the sow uterus during the

oestrous cycle and in inseminated sows at oestrus and early pregnancy. Reproduction 2005; 129:349–359.
54. Finch PW, Rubin JS, Miki T, Ron D, Aaronson SA. Human KGF is FGF-related with properties of a paracrine effector of epithelial cell growth. Science 1989; 245:752–755.
55. Young KH, Kraeling RR, Bazer FW. Effects of prolactin on conceptus survival and uterine secretory activity in pigs. J Reprod Fertil 1989; 86: 713–722.
56. Song G, Spencer TE, Bazer FW. Cathepsins in the ovine uterus: regulation by pregnancy, progesterone, and interferon tau. Endocrinol 2005; 146: 4825–4833.
57. Song G, Spencer TE, Bazer FW. Progesterone and interferon-tau regulate cyctatin C in the endometrium. Endocrinol 2006; 147:3478–3483.

R. Michael Roberts

Department of Animal Sciences
University of Missouri, Columbia, Missouri, USA

2002/3 — for discoveries of Interferon-t and other pregnancy-associated proteins, which clarified the biological mystery of signaling between embryo and mother to maintain pregnancy, with profound effects on the efficiency of animal production systems, as well as human health and well-being.

CURRICULUM VITAE

Date and Place of Birth: October 23, 1940; Ilkley, Yorkshire, England

EDUCATION:

Bradford Grammar School, England	- 1951-59
B.A. (Botany) Oxford University, England	- 1962
D. Phil. Oxford University, England	- 1965

APPOINTMENTS:

- Research Associate, State University of New York at Buffalo NY, 1965-67.
- Assistant Professor of Biology, State University of New York at Buffalo, NY, 1967-68.
- Senior Research Fellow, UK Atomic Energy Authority, Radiochemical Centre, Amersham, England, 1968-69.
- Assistant Professor of Biochemistry, University of Florida, January 1970-September 1972.

- Associate Professor of Biochemistry, University of Florida, September 1972-June 1976.
- Professor of Biochemistry, University of Florida, July 1976-October 31, 1985.
- Acting Chairman, Department of Biochemistry and Molecular Biology, University of Florida, September 1978-September 1979.
- Professor of Animal Sciences and Biochemistry, University of Missouri-Columbia, November 1, 1985-present.
- J. Fred McKenzie Distinguished Professor in Reproductive Biology, University of Missouri-Columbia, July 1988-1996.
- Chair, Veterinary Pathobiology, University of Missouri-Columbia 1995-1998.
- Curators' Professor, University of Missouri-Columbia 1996-present.
- Chief Scientist, National Research Initiative (NRICGP/CSREES/USDA) 1998-2000. (50% Appointment).
- Director, Life Sciences Center, University of Missouri-Columbia, January 1, 2004-2005.

Awards and Honors:

- State Scholar, Oxford University, England, 1959-62.
- Christopher Welch Scholar, Oxford University, England, 1962-65.
- Senior Research Fellowship, United Kingdom Atomic Energy Authority, 1968-69.
- Research Career Development Award from the U.S. Public Health Service, 1972-1977.
- Fellowship, World Health Organization, Cambridge, England, Spring 1977.
- Senior Fellowship, North Atlantic Treaty Organization, University of Manchester, England. Summer 1977.
- First Prize, Electrofocus/78 Competition sponsored by LKB Instruments, Inc., for Advancement of Knowledge Applications & Techniques of Electrofocusing; Atlanta, Georgia.
- Sigma Chi, Distinguished Faculty Award, University of Florida, April 1984.
- Basic Science Research Prize, College of Medicine, University of Florida, May 1984.
- Vice-Chairman, Gordon Conference on Mammalian Genital Tract, Plymouth State College, July 6-11, 1986.
- Chairman, Gordon Conference on Reproductive Tract Biology, Brewster Academy, July 4-8, 1988.
- Research Award, Society for Study of Reproduction, July, 1990.
- NIH MERIT Award, 1990-2000.
- 1990 Sydney A. Asdell Lecturer, Cornell University.
- United States Department of Agriculture Distinguished Scientist, 1992.
- Amoroso Lecturer, Society for the Study of Fertility, 1994.
- Milstein Award, International Society for Interferon and Cytokine Research, 1995.

- 1st Recipient, Research Award, College of Agri., Food & Natural Resources, University of Missouri-Columbia, 1995.
- Member, National Academy of Sciences, elected 1996.
- The Alexander vom Humboldt Award for Agriculture, 1996.
- President's Award for Research and Creativity, University of Missouri system, 1997.
- Foreign Specialist, National Institute of Animal Industry, Japan, 1998.
- Docteur Honoris Causa (Honorary Doctorate) from the University of Liege, Belgium, 1998.
- Nalbandov Memorial Lecturer, University of Illinois, 2000.
- Wolf Prize for Agriculture 2003; For the discoveries of interferon-tau and other pregnancy-associated proteins, which clarified the biological mystery of signaling between the embryo and mother to maintain pregnancy.
- Larry Ewing Memorial Lecturer, John Hopkins University School of Public Health (2003).
- University of Missouri Faculty/Alumni Award 2005.
- Scientific American Top 50 List (Dec) 2005 Acknowledged for accomplishment in research and technological leadership.
- Patent Recipient Awardee-University of Missouri-Columbia Technology Transfer Showcase 2006.
- Carl G. Hartman Award-Recognition of a research career and scholarly activities in the field of reproductive biology 2006.
- International Federation Placenta Association 2007 Senior Investigator Award.

Society Membership:

American Society of Biological Chemists
Society for Study of Reproduction
American Society for Advancement of Science
International Society for Interferon and Cytokine Research
American Society for Cell Biology
Endocrine Society

AUTOBIOGRAPHY

I was born in my parents' home on October 23, 1940, in Menston, a village in West Yorkshire in the postal district of Ilkley, a few miles north of the industrial cities of Bradford and Leeds. My father was a wool sorter in Bradford, while my mother had been employed until my birth in a shirt factory. My one brother, Glyn, was also a war baby, born in February 1944. I was enrolled in Menston Primary School from the autumn of 1945 until the summer of 1951 and then in Bradford (Boys) Grammar School until 1959. The latter was a "Direct Grant Grammar

School", comprised of fee-paying as well as scholarship pupils, and had high academic standards, with a quite ruthless preoccupation with placing its students into the Universities of Oxford and Cambridge. My father badly wanted his eldest son to develop a career that provided more job security, better pay, shorter working hours, and less dirt than he himself had experienced as a textile worker. His preference was for me to become a dentist, although even he acknowledged later that such a decision would have been a disaster for an individual so acutely lacking fine motor skills, patience, and an ability to visualize anything in three dimensions. Nevertheless, at the age of 13, I was funneled (quite happily I should add) into an academic track that provided increasing emphasis on science, particularly mathematics, physics and chemistry. My interest in biology was fostered by my dad's knowledge of birds and local plants, family walks, and having the freedom to roam anywhere in the local countryside. My mother, who had left school at 13, was an enthusiastic co-pupil. Unfortunately neither of them was much help when it came to algebra, physics, Latin and French.

I realized in my early teens that I enjoyed science, particularly chemistry. My main chemistry teacher at Bradford Grammar School, a Mr. J. B. Bentley, was a great influence on my schoolboy academics. He was a remarkable man, Oxford trained, a WW II veteran, broadly read, authoritative yet wonderfully encouraging in his restrained English fashion. I found the formal course work of biology less challenging and less interesting than chemistry and physics, but was always fascinated by what I read outside the classroom. At 17, I passed my "A Level" national examinations sufficiently well to be awarded a State Scholarship, which would pay my way at any university that cared to accept me. I remained at Bradford Grammar for an extra year to prepare for the entrance examinations to Oxford and was offered a place at Lincoln College in December 1958, leaving me free to concentrate on rugby, cricket, and other pleasures for the rest of the academic year.

My scholarship paid tuition and provided full living expenses for three years at Oxford. Its annual value was about the equivalent of what my father earned in a year. I chose to major in Biochemistry, which at that time was really chemistry with some biochemistry tagged on. I remember Hans Krebs (after he had won the Nobel Prize, but before he became Sir Hans) wandering through our lab class not quite knowing what to say to us students and everyone feeling slightly awkward. I was particularly unsettled during my first term in residence, uncomfortable in the Oxford environment where everyone seemed smarter and more accomplished than I. I also wasn't enjoying inorganic chemistry and toyed with the idea of switching to medicine but, for reasons that I still don't quite understand, decided I wanted to work with plants and to major in botany. At that time, the Professor and Head of Department was the cytogeneticist, Cyril Darlington, a controversial, figure because of his views on eugenics and his inability to be politically correct on just about any subject, but who ensured that classical botany was never much emphasized. Instead students were exposed to extensive amounts of genetics and plant biochemistry. Even though my decision to become a plant scientist appeared radically unsound to my father, I thoroughly enjoyed my remaining years as an undergraduate, and graduated with an "Upper Second" honors degree in June of 1962.

There are two events during my undergraduate life that are worth remarking upon, since they have had a profound effect on my future. The first was that I married Susan Elizabeth Davies, a technician in the department of Botany, in January 1961. We had our first child, Mark Glyn, in the late summer of that year. My college was silently aghast, and the addition of a family certainly strained what financial resources we had available, particularly as it was impossible, even in the 1960s, for a young mother to find child care and to continue to work. Our flat was also pretty primitive and certainly cheap by Oxford standards; it had an outside lavatory, a bath tub in the kitchen, and no central heating. In the record, arctic-like winter of 1963, when Susan was pregnant with our second child, Samantha Clare, everything seemed to freeze, including the lavatory. Despite this unpropitious start and impoverished state, I was extremely happy and certainly a more motivated individual than before. Our marriage has survived for 47 years.

The second noteworthy occurrence was that in March of my final undergraduate year I was awarded a prestigious Christopher Welch Scholarship in biology, which would pay me a stipend of £500 (now £13,500) a year to pursue a D.Phil. degree at Oxford. It was an honor to accept a scholarship previously held by such scientists as John Eccles, J. Z. Young, Peter Medawar, Anne McLaren, Alec Jeffreys, and John Gurdon, among others. The examination for this award was competitive and involved written and oral sections, which required that the applicant map out a research problem they proposed to tackle. I developed a theoretical approach to studying the biosynthesis of cell wall polysaccharides in roots through the incorporation of radioactive precursors and then using autoradiography to follow patterns of deposition along files of cells from their stem cell origins in the meristem through the main zone of elongation further along the root. Oddly, this theoretical scheme worked almost perfectly when I came to perform real experiments during my graduate training.

My nominal supervisor during my graduate student days was a plant biochemist, Vernon Butt, a nice man but who was largely absent from the laboratory during my three years of research as he tried to recover from the premature death of his wife and to bring up two young children. I worked more closely with Lionel Clowes, who discovered the quiescent centre within root meristems, and Barrie Juniper, a microscopist, with whom I published my first paper on the production of mucus-like, lubricating polysaccharides by the maize root cap (Juniper and Roberts 1966). My thesis eventually led to the publication of six papers, but getting Vernon to even read them (I was in the U.S.A. by then) was like pulling teeth. The gambit that eventually worked was to submit and publish a paper without his authorship (Roberts 1967). I had no trouble after that with getting him to edit and co-author the subsequent manuscripts.

In the summer of 1964, I attended the International Botanical Congress in Edinburgh and had the opportunity to meet several plant scientists from North America, who were recruiting postdoctoral fellows. The post-Sputnik, U.S. science expansion was in full swing. Although Buffalo New York was not the place of my dreams, I eventually decided to join Frank Loewus there in the Department of Biology. America seemed such a potential adventure, and it was. I had never travelled by air before; I didn't possess a driver's license, and fondly believed that we could get about on bicycles, just as we had in Oxford. That first winter in Buffalo, we had 135 inches of snow, but at least we were warmer than we had been in our frigid flat on Divinity Road, Oxford. I drove a VW bug illegally (you need an automobile to survive a Buffalo winter) for six months before taking my test and became an expert on ice.

My U.S. education began in September 1965. Frank Loewus was best known for his work in ascorbic acid biosynthesis, but had recently discovered the inositol pathway in plants in which myo-inositol is oxidatively cleaved to D-glucuronic acid. I was able to demonstrate that myo-inositol, via this cleavage reaction, is a precursor of the uronic acids and pentoses incorporated into cells wall pectins and hemicelluloses (Roberts, Deshusses et al. 1968). I also had the opportunity to study the formation of apiose, an unusual sugar found in the walls of the pond weed *Lemna gibba*, and of phytic acid, the polyphosphorylated form of inositol (Roberts, Shah et al. 1967). Frank taught me an awful lot, particularly some carbohydrate chemistry and the use of rigorous methods for identifying metabolites (with no help from mass spectrometry). The lack of course work associated with an Oxford D.Phil. and the narrow scope of the prior undergraduate curriculum allows a motivated student enormous license to follow what interests him, but the down side can be a profound ignorance of all sorts of other useful stuff, a fact I soon began to appreciate. Another feature of the Loewus laboratory was that it required scrupulously annotated note books, another valuable training lesson. I thoroughly enjoyed my time in Frank's laboratory and owe him a great deal about how to run and how not to run a research group. In my third postdoctoral year, I was also appointed as a temporary Assistant Professor and given the opportunity to lecture and run a laboratory course, all useful experience for the future, although a low grade in 1967 could condemn a student to a vacation in Vietnam.

The late 1960s were heady times. As a green card recipient (Brits were a favored nationality in those days), I was obliged to register for the draft myself, but my advanced age (24) and two children meant I was unlikely ever to be called up for war duty. In any case, we were only 20 minutes from the Canadian border. The Vietnam War was not yet at its peak and popular when we first arrived, but out of hand and facing massive protests by the time we left in 1968. That year the U.S. experienced the deaths of Robert Kennedy and Martin Luther King, riots in central cities, including the predominantly black areas of Buffalo, the Johnson resignation, and the extraordinary Democratic Convention in Chicago, which we watched from our hotel room in New York, just before embarking on the great ocean liner, "France", for Southampton at the end of August. I look back fondly to those years in the U.S.A. and the freedom of being a postdoc. Our family were also liberated by the possibilities of traveling where we wanted by automobile, camping where we chose, and the seductive opportunities of having little extra money in our pockets.

I returned to England in August 1968 as a Fellow of the Atomic Energy Authority to work at the Radiochemical Center in Amersham, then a government owned, commercial enterprise that sold radiochemicals. It was an odd 16 months. I had to sign the "Official Secrets Act" although, in retrospect, that seems no more bizarre than signing a Loyalty Oath for my subsequent job at the University of Florida. I was given a lot of freedom to pursue topics in which I was interested, learned a lot of enzymology, and was able to make use of the almost limitless supplies of labeled compounds available for metabolic tracer studies. At the same time, fixed work hours, the limited availability of equipment, and the constant risk of radioactive contamination were irksome. My mail was opened, and, I assume, read, but I benefitted from being surrounded by excellent professionals and a knowledge that I had a job for life should I have wanted such security, which I didn't. After having a series of job interviews in botany departments at U.K. universities I also decided that I didn't desire that sort of job either, and I seized upon the opportunity to interview for a position in the Department of Biochemistry at the University of Florida in Gainesville, then something of a backwater tucked away in North Florida. I was offered that position, and the family moved to Gainesville in January 1970.

At Florida, I was hired as a plant biochemist but my laboratory was located in the Medical School. I continued to concentrate on plant cell wall formation and on determining whether plants could utilize glucosamine as a precursor of glycoproteins (Roberts, Connor et al. 1972). I wrote my first NSF grant application that first month of 1970 with nothing much in the way of preliminary results, but by late summer was funded. How the world has changed! I also began to gravitate towards animal biochemistry through collaborations. I teamed up with Owen Rennert, a pediatrician, to determine whether there was anything abnormal about the glycosaminoglycans secreted by cultured dermal fibroblasts from patients with cystic fibrosis (there wasn't) (Welch and Roberts 1975), but on the basis of a few ideas wrote a successful application for a Career Development Award to NIH, a grant that paid my salary for five years from 1972 through 1977. Also, through a peculiar series of circumstances, one of which was being the faculty advisor for the University of Florida cricket team, I became acquainted with the reproductive laboratory had no experience with the required technology. Moreover, those were days before the appearance of "kits" and other aids to neophytes such as me who wanted to become "molecular biologists". To accomplish the cloning quickly, I established a collaboration with Keith Marotti at the UpJohn Co. in Kalamazoo, Michigan. An energetic postdoctoral fellow, Kaz Imakawa, was dispatched to Keith's laboratory to establish cDNA expression libraries in lambda phage with RNA extracted from ovine embryos and to screen these libraries with an antiserum prepared against the purified oTP-1. The first sequencing data emerged in late 1986 and immediately revealed that oTP-1 was allied to the type I interferons (IFN), proteins known for their abilities to protect cells from virus. That phone call with Keith when he described the results was one of those "wow moments" that happen infrequently but sufficiently often to make research so rewarding an experience. Despite the clear signal that we were dealing with an IFN, however, it was clear that the resemblance to human, mouse, and bovine IFNα and IFNβ, the best characterized kinds of Type I IFN at the time, was relatively distant (~60 % and 30 % amino acid sequence identities, respectively). Moreover, the predicted primary structure of oTP-1 indicated that it possessed an extra sequence of six amino acids on its carboxyl tail, which was absent on IFNα, which are generally 166 residues in length. On the other hand, oTP1 did bear a closer resemblance (~70 % identity) to a newly discovered Type I IFN, eventually named IFNω. These data were published in late 1987 in Nature (Imakawa, Anthony et al. 1987), but not before the manuscript had been misplaced for several weeks. Soon after this publication, bTP-1, the bovine equivalent of oTP-1 was also cloned (Imakawa, Hansen et al. 1989). Both the native proteins and their recombinant forms possessed an antiviral potency similar to that reported for other Type I IFN. On the basis of this information, I chose to name the oTP-1 and bTP-1 proteins interferon-tau (IFNτ), although it required several years for the official nomenclature committee to affirm this designation, largely because there was concern that this new group was part of the established IFNω family with which it shared structural characteristics (Allen and Diaz 1994).

A question immediately arose was whether relatives of the IFNτ serve as signaling molecules for maternal recognition of pregnancy in other mammals and especially the human. We were excited to find that antiviral activity was released by murine and porcine embryos (Cross, Farin et al. 1990). The nature of the former activity still remains mysterious, whereas that of the pig is due to the secretion of a mixture of IFN-gamma (Lefevre, Martinat-Botte et al. 1990), a cytokine structurally unrelated to Type I IFN, and IFN-delta, a novel Type I IFN, most closely related to IFNβ (Lefevre and Boulay 1993). No antiviral activity could be detected in the medium in which human blastocysts had been cultured (Roberts, Cross et al. 1991). Ultimately, my group went on to show what we had long anticipated, namely that the gene family encoding the IFNτ (*IFNT*) was of recent evolutionary origin (Leaman and Roberts 1992). We showed that it had diverged from the IFNω gene (*IFNW*) lineage approximately 36 million years ago (Roberts, Liu et al. 1997). As a result of this recent origin, the *IFNT* were confined to a single sub-order, the Ruminantia, a grouping that contains among others, cattle and sheep and their relatives, antelope, deer, and giraffe. The *IFNT* are absent from rodents, primates, and even other taxa within the Artiodactyl order, such as swine. The ruminant ungulates are an especially successful group of mammals with a geographic spread from equatorial regions, e.g. wildebeest, impala, to the arctic, e.g. musk oxen, reindeer, and elk, and from wetland, e.g. water buffalo, to desert, e.g. addax. Many, such as the American bison, once roamed in huge numbers and have only become diminished as a result of recent human activity. The emergence of the *IFNT* correlated in time with the origin of the Ruminantia and with the evolution of the highly efficient, non-invasive placentation that characterizes this taxon. I like to believe that the recruitment of this interferon for reproductive purposes has had a great deal to do with the triumphant rise of the magnificent animals that occupy this sub-order of mammals.

A second and more persistent question has been whether the IFNτ possess some unique features that especially equip them for their role in maternal recognition of pregnancy. In our studies at Missouri, we found that the IFNτ have an antiviral activity comparable or superior to other type I IFN (Alexenko, Leaman et al. 1997), that they bind to the same receptor as other Type I IFN (Li and Roberts 1994), and that this receptor is especially concentrated on the surface epithelium of the uterus targeted by IFNτ (Rosenfeld, Han et al. 2002). My group also showed that the ability of different forms of IFNτ to cause a pseudopregnancy in sheep is strongly correlated with their potency as antiviral agents (Niswender, Li et al. 1997). The IFNτ activate the same signal transduction system as other Type I IFN and regulate most, if not all the same genes (Chen, Antoniou et al. 2007). Conversely, other Type 1 IFN can extend estrous cycle length in ewes provided they are supplied in amounts that would evoke an equivalent antiviral response to the IFNτ (Ealy, Green et al. 1998). As a result of such observations, I had begun, by the late 1990s, to favor the alternative hypothesis, namely that the proteins had not evolved unusual properties but instead their unique role depended upon their being produced in the right place, at the right time, and in sufficient quantities to trigger maternal responses that would allow the pregnancy to be maintained. Accordingly the attention of the laboratory became focused not so much on the proteins as on *IFNT* gene expression itself and specifically on the control elements of these genes. It was already known that *IFNT* expression was unlike that of the other known Type I IFN. The genes were not upregulated in response to viruses and other pathogens, but instead were expressed constitutively as part of a program associated with conceptus development (Roberts, Cross et al. 1992). This expression is confined to a single layer of cells known as trophectoderm, the precursor tissue of the fetal days a year of sunshine. Certainly, Lake of the Ozarks wasn't much compensation, but I felt energized by the change and relished having full control over the animal studies. Susan and I enjoyed the courtesy of Midwest. Getting your car fixed or calling a plumber were somehow pleasanter events than they had been. The university was smaller and more accessible than that of the University of Florida, and Columbia was a delightful place in which to re-settle. I was treated well; all agreements with the administration were kept and generally exceeded. I had dependable support from my Dean, Roger Mitchell, equivalent to an extra grant, which has been maintained to this date. That source of funding allowed me to try new directions and take risks that would have been much more difficult with just federal grant support. No, there have been no regrets, even though I never imagined I could spend so long in one place. I began to receive more recognition for my work, particularly for the research on interferon-tau. In 1989, Fuller Bazer and I shared the Research Award from the Society for Study of Reproduction, the professional society I still love best. I continued to maintain at least two grants from NIH, and in 1992 received one of the best presents of my career, a Merit Award from the institute at NIH, NIHCD, which has supported me for over 30 years. That Merit Award was not only an honor, it provided an additional stable source of funding for the next ten years. The research on interferon was also recognized outside the reproductive biology community when I was awarded the 1995 Milstein Award by International Society for Interferon and Cytokine Research. I suppose that 1996 has to be my banner year because I was elected to the U.S. National Academy of Sciences, a huge thrill not just for me but for my laboratory, and later in the year received the Alexander Von Humboldt Award for Agriculture.

The research that most likely gained me NAS membership and provided the main basis for my selection as the 2002/2003 Wolf Awardee for Agriculture was undoubtedly that on interferon-tau, work that can be considered important for several reasons. First, it described a Type 1 IFN functioning in a constitutive biological process unrelated to pathogenesis. Second, the find provided the first example of a cytokine involved in trophoblast signaling, plus an example of how a defined trophoblast product might intervene in maternal immune responses. Third, the *IFNT* would appear to be the most recent mammalian Type 1 gene family to have arisen and provide a clear illustration of how a control region of a gene can be commandeered and then refined to provide a radically changed pattern of expression. Most importantly, IFNτ illustrated why cattle, sheep, and goats are able to recognize that they are pregnant even before the conceptus has attached to the uterine wall and why many pregnancies fail at this stage of development. I was immensely proud to receive the Wolf Prize for this and other work on farm species with my long-time collaborator, Fuller Bazer, and to accept the prize simultaneously with three other friends, Oliver Smithies, Mario Capecchi, and Ralph Brinster, who shared the Prize for Medicine.

References

Alexenko, A. P., D. W. Leaman, et al. (1997). "The antiproliferative and antiviral activities of IFN-tau variants in human cells." *J Interferon Cytokine Res* **17**(12): 769-79.

Allen, G. and M. O. Diaz (1994). "Nomenclature of the human interferon proteins." *J Interferon Res* **14**(4): 223-6.

Antanaitis, B. C., P. Aisen, et al. (1980). "The novel "g' = 1.74" EPR spectrum of pink and purple uteroferrin." *J Biol Chem* **255**(23): 11204-9.

Bartol, F. F., R. M. Roberts, et al. (1985). "Characterization of proteins produced in vitro by periattachment bovine conceptuses." *Biol Reprod* **32**(3): 681-93.

Baumbach, G. A., P. T. Saunders, et al. (1984). "Uteroferrin has N-asparagine-linked high-mannose-type oligosaccharides that contain mannose 6-phosphate." *Proc Natl Acad Sci U S A* **81**(10): 2985-9.

Buhi, W. C., C. A. Ducsay, et al. (1982). "Iron transfer between the purple phosphatase uteroferrin and transferrin and its possible role in iron metabolism of the fetal pig." *J Biol Chem* **257**(4): 1712-23.

Chen, T. T., F. W. Bazer, et al. (1973). "Purification and properties of a progesterone-induced basic glycoprotein from the uterine fluids of pigs." *J Biol Chem* **248**(24): 8560-6.

Chen, Y., E. Antoniou, et al. (2007). "A microarray analysis for genes regulated by interferon-{tau} in ovine luminal epithelial cells." *Reproduction* **134**(1): 123-35.

Couso, R., L. Lang, et al. (1986). "Phosphorylation of the oligosaccharide of uteroferrin by UDP-GlcNAc:glycoprotein N-acetylglucosamine-1-phosphotransferases from rat liver, Acanthamoeba castellani, and Dictyostelium discoideum requires alpha 1,2-linked mannose residues." *J Biol Chem* **261**(14): 6326-31.

Cross, J. C., C. E. Farin, et al. (1990). "Characterization of the antiviral activity constitutively produced by murine conceptuses: absence of placental mRNAs for interferon alpha and beta." *Mol Reprod Dev* **26**(2): 122-8.

Das, P., T. Ezashi, et al. (2008). "Combinatorial Roles of Protein Kinase A, Ets2, and 3',5'-Cyclic-Adenosine Monophosphate Response Element-Binding Protein-Binding Protein/p300 in the Transcriptional Control of Interferon-{tau} Expression in a Trophoblast Cell Line." *Mol Endocrinol* **22**(2): 331-43.

Das, P., T. Ezashi, et al. (2008). "Effects of FGF2 and oxygen in the BMP 4-driven differentiation of trophoblast from human embryonic stem cells." *Stem Cell Research* **1**(1): 61-74.

Ealy, A., J. Green, et al. (1998). "Differences in biological activities among recombinant variants of ovine interferon-tau." *Biology of Reproduction* **58**(566-573).

Ezashi, T., P. Das, et al. (2008). "The Role of Homeobox Protein Distal-Less 3 and Its Interaction with ETS2 in Regulating Bovine Interferon-Tau Gene Expression-Synergistic Transcriptional Activation with ETS2." *Biol Reprod.*

Ezashi, T., P. Das, et al. (2005). "Low O2 tensions and the prevention of differentiation of hES cells." *Proc Natl Acad Sci U S A* **102**(13): 4783-8.

Ezashi, T., A. D. Ealy, et al. (1998). "Control of interferon-tau gene expression by Ets-2." *Proc Natl Acad Sci U S A* **95**(14): 7882-7.

Ezashi, T. and R. M. Roberts (2004). "Regulation of interferon-tau (IFN-tau) gene promoters by growth factors that target the Ets-2 composite enhancer: a possible model for maternal control of IFN-tau production by the conceptus during early pregnancy." *Endocrinology* **145**(10): 4452-60.

Gaber, B. P., J. P. Sheridan, et al. (1979). "Resonance Raman Scattering from Uteroferrin, the Purple Glycopprotein of the Porcine Uterus." *J. Biol. Chem.* **254**: 8340-8342.

Godkin, J. D., F. W. Bazer, et al. (1982). "Purification and properties of a major, low molecular weight protein released by the trophoblast of sheep blastocysts at day 13-21." *J Reprod Fertil* **65**(1): 141-50.

Green, J. A., T. E. Parks, et al. (2005). "The establishment of an ELISA for the detection of pregnancy-associated glycoproteins (PAGs) in the serum of pregnant cows and heifers." *Theriogenol.* **63**: 1481-1503.

Green, J. A., S. Xie, et al. (2000). "Pregnancy-associated bovine and ovine glycoproteins exhibit spatially and temporally distinct expression patterns during pregnancy." *Biol Reprod* **62**(6): 1624-31.

Hunt, D. F., J. R. Yates, et al. (1987). "Sequence homology in the metalloproteins: purple acid phosphatase from beef spleen and uteroferrin from porcine uterus." *Biochem. & Biophys. Res. Comm.* **144**: 1154-1160.

Imakawa, K., R. V. Anthony, et al. (1987). "Interferon-like sequence of ovine trophoblast protein secreted by embryonic trophectoderm." *Nature* **330**(6146): 377-9.

Imakawa, K., T. R. Hansen, et al. (1989). "Molecular cloning and characterization of cDNA's corresponding to bovine trophoblast protein-1. A comparison with ovine trophoblast protein-1 and bovine interferon-*I." *Molec. Endocrinol* **3**: 127-139.

Juniper, B. E. and R. M. Roberts (1966). "Polysaccharide Synthesis and the Fine Structure of Root Cells." *J. Roy. Microscop.* **85**: 63-72.

Ketcham, C. M., G. A. Baumbach, et al. (1985). "The type 5, Acid Phosphatase from Spleen of Humans with Hairy Cell Leukemia." *J. Biol. Chem.* **260**: 5768-5776.

Lang, L., M. Reitman, et al. (1984). "Recognition of a Protein-dependent Determinant allows Specific Phosphorylation of Oligosaccharides present on Lysosomal Enzymes." *J. Biol. Chem.* **259**: 14663-14671.

Larson, M. A., K. Kimura, et al. (2001). "Sexual dimorphism among bovine embryos in their ability to make the transition to expanded blastocyst and in the expression of the signaling molecule IFN-tau." *Proc Natl Acad Sci U S A* 98(17): 9677-82.

Leaman, D. W. and R. M. Roberts (1992). "Genes for the trophoblast interferons in sheep, goat, and musk ox and distribution of related genes among mammals." *J Interferon Res* 12(1): 1-11.

Lefevre, F. and V. Boulay (1993). "A novel and atypical type one interferon gene expressed by trophoblast during early pregnancy." *J Biol Chem* 268(26): 19760-8.

Lefevre, F., F. Martinat-Botte, et al. (1990). "Interferon-gamma gene and protein are spontaneously expressed by the porcine trophectoderm early in gestation." *Eur J Immunol* 20(11): 2485-90.

Li, J. and R. M. Roberts (1994). "Interferon-tau and interferon-alpha interact with the same receptors in bovine endometrium. Use of a readily iodinatable form of recombinant interferon-tau for binding studies." *J Biol Chem* 269(18): 13544-50.

Malathy, P. V., K. Imakawa, et al. (1990). "Molecular cloning of the uteroferrin-associated protein, a major progesterone-induced serpin secreted by the porcine uterus, and the expression of its mRNA during pregnancy." *Mol. Endocr.* 4: 428-440.

Meyer, M. D., P. J. Hansen, et al. (1995). "Extension of corpus luteum lifespan and reduction of uterine secretion of prostaglandin F2 alpha of cows in response to recombinant interferon- tau." *J Dairy Sci* 78(9): 1921-31.

Meyer, M. D., P. J. Hansen, et al. (1995). "Effect of bovine interferon-tau on body temperature and plasma progesterone concentrations in cyclic dairy cows." *J Dairy Sci* 78(7): 1470-6.

Niswender, K. D., J. Li, et al. (1997). "Effect of variants of interferon-tau with mutations near the carboxyl terminus on luteal life span in sheep." *Biol Reprod* 56(1): 214-20.

Renegar, R. H., F. W. Bazer, et al. (1982). "Placental Transport and Distribution of Uteroferrin in the Fetal Pig." *Biol. Reprod.* 27: 1247-1260.

Roberts, R., J. Cross, et al. (1991). Trophoblast Interferons. In: *Growth Factors in Reproduction.* D. W. Schomberg, Serono Symposia, USA, Springer-Verlag, New York: 235-245.

Roberts, R. M. (1967). "Incorporation of ^{14}C-labeled D-glucuronate and D-galactose into Segments of the Root Tip of Corn." *Phytochemistry* 6: 525-533.

Roberts, R. M. and F. W. Bazer (1988). "The functions of uterine secretions." *J Reprod Fertil* 82(2): 875-92.

Roberts, R. M., A. B. Connor, et al. (1972). "The Formation of Glycoproteins in Tissues of Higher Plants. Specific Labeling Using (1-^{14}C) glucosamine." *Biochem. J.* 125: 999-1008.

Roberts, R. M., J. C. Cross, et al. (1992). "Interferons as hormones of pregnancy." *Endocr Rev* **13**(3): 432-52.

Roberts, R. M., J. Deshusses, et al. (1968). "Inositol Metabolism in Plants. V. Conversion of Myo-inositol to Uronic Acid and Pentose Units of Acidic Polysaccharides in Root-tips of *Zea mays*." *Plant Physiol.* **43**: 979-989.

Roberts, R. M., L. Liu, et al. (1997). "New and atypical families of type I interferons in mammals: comparative functions, structures, and evolutionary relationships." *Prog Nucleic Acid Res Mol Biol* **56**: 287-325.

Roberts, R. M., T. J. Raub, et al. (1986). "Role of uteroferrin in transplacental iron transport in the pig." *Fed Proc* **45**(10): 2513-8.

Roberts, R. M., R. H. Shah, et al. (1967). "Inositol Metabolism in Plants. IV. Biosynthesis of Apiose in *Lemna* and *Petroselinum*." *Plant Physiol.* **42**: 659-666.

Rosenfeld, C. S., K. M. Grimm, et al. (2003). "Striking variation in the sex ratio of pups born to mice according to whether maternal diet is high in fat or carbohydrate." *Proc Natl Acad Sci U S A* **100**(8): 4628-32.

Rosenfeld, C. S., C. S. Han, et al. (2002). "Expression of interferon receptor subunits, IFNAR1 and IFNAR2, in the ovine uterus." *Biol Reprod* **67**(3): 847-53.

Schalue-Francis, T. K., P. W. Farin, et al. (1991). "Effect of injected bovine interferon-alpha I1 on estrous cycle length and pregnancy success in sheep." *J Reprod Fertil* **91**(1): 347-56.

Schlosnagle, D. C., E. G. Sander, et al. (1976). "The Requirement of an Essential Thiol Group and Ferric Iron for the Activity of the Progesterone-Induced, Porcine Uterine, Purple Phosphatase." *J. Biol. Chem.* **251**: 4680-4685.

Stallings-Mann, M. L., M. C. Burke, et al. (1994). "Purification, characterization, and cDNA cloning of a Kunitz-type proteinase inhibitor secreted by the porcine uterus." *J. Biol. Chem.* **269**(39): 24090-24094.

Stallings-Mann, M. L., W. E. Trout, et al. (1993). "Porcine uterine retinol-binding proteins are identical gene products to the serum retinol-binding protein." *Biol Reprod* **48**(5): 998-1005.

Welch, D. W. and R. M. Roberts (1975). "Complex Saccharide Metabolism in Cystic Fibrosis Fibroblasts." *Pediatrics Res.* **9**: 698-702.

Xie, S., J. Geen, et al. (1997). "The diversity and evolutionary relationships of the pregnancy-associated glycoproteins, an aspartic proteinase subfamily consisting of many trophoblast-expressed genes." *Proc. Natl. Acad, Sci, USA* **94**: 12809-12816.

Xie, S. C., B. G. Low, et al. (1991). "Identification of the major pregnancy-specific antigens of cattle and sheep as inactive members of the aspartic proteinase family." *Proc Natl Acad Sci U S A* **88**(22): 10247-51.

REFEREED PUBLICATIONS

1. Juniper, B.E. and Roberts, R.M. (1966) Polysaccharide Synthesis and the Fine Structure of Root Cells. *J. Roy. Microscop.* 85, 63-72.
2. Roberts, R.M. and Loewus, F. (1966) Inositol Metabolism in Plants III. Conversion of Myo-inositol-2-^{3}H to Cell Wall Polysaccharides in Sycamore (*Acer pseudoplatanus L.*) Cell Culture. *Plant Physiol.* 41, 1489-1498.
3. Roberts, R.M. (1967) Incorporation of ^{14}C-labeled D-glucuronate and D-galactose into Segments of the Root Tip of Corn. *Phytochemistry* 6, 525-533.
4. Roberts, R.M., Shah, R., and Loewus, F. (1967) Conversion of Myo-inositol-2-^{14}C to Labeled 4-0-methyl-glucuronic Acid in the Cell Wall of Maize Root Tips. *Arch. Biochem. Biophys.* 119, 590-593.
5. Roberts, R.M., Shah, R.H. and Loewus, F. (1967) Inositol Metabolism in Plants. IV. Biosynthesis of Apiose in *Lemna* and *Petroselinum*. *Plant Physiol.* 42, 659-666.
6. Roberts, R.M. and Butt, V.S. (1967) Rates of Incorporation of Label from D-[^{14}C] Glucose into Different Tissues of the Root-tip of Corn. *New Phytologist* 66, 563-567.
7. Roberts, R.M. and Butt, V.S. (1967) Patterns of Cellulose Synthesis in Maize Root-tips. *Exper. Cell Res.* 46, 495-510.
8. Roberts, R.M., Shah, R., Golebiewski, A. and Loewus, F. (1967) Incorporation of Methanol into Pectic Substances. *Plant Physiol.* 42, 1737-1742.
9. Roberts, R.M. and Butt, V.S. (1968) Patterns of Incorporation of Pentose and Uronic Acid into Cell Walls of Maize Root-tips. *Exp. Cell Res.* 51, 519-530.
10. Roberts, R.M., Deshusses, J. and Loewus, F. (1968) Inositol Metabolism in Plants. V. Conversion of Myo-inositol to Uronic Acid and Pentose Units of Acidic Polysaccharides in Root-tips of *Zea mays*. *Plant Physiol.* 43, 979-989.
11. Roberts, R.M., Deshusses, J. and Loewus, F. (1968) Inositol Metabolism in Plants. Metabolism of Myo-inositol Metabolism in Plants. VI. Conversion of Myo-inositol to Phytic Acid in *Wolffiella floridana*. *Plant Physiol.* 43, 1710-1716.
12. Roberts, R.M. (1968) The Metabolism of L-fucose by Sycamore (*Acer pseudoplatanus L.*) Cell Cultures. *Arch. Biochem. Biophys.* 128, 818-820.
13. Roberts, R.M. and Butt, V.S. (1969) Patterns of Incorporation of D-galactose into Cell Wall Polysaccharide of Growing Maize Roots. *Planta* 84, 250-262.
14. Roberts, R.M. and Tovey, K.C. (1969) Trehalase Activity in *Selaginella martensii*. *Arch. Biochem. Biophys.* 133, 408-412.
15. Roberts, R.M. (1970) The Incorporation of D-glucosamine-^{14}C into Root Tissues of Higher Plants. *Plant Physiol.* 45, 263-267.

16. Roberts, R.M. and Tovey, K.C. (1970) A Critical Study of the DEAE paper Disc Technique Used in Radiochemical Enzyme Assays. *Anal. Biochem.* 34, 582-590.
17. Tovey, K.C. and Roberts, R.M. (1970) An Artifact in the Chromatography of Sugar Nucleotides Using Solvents Containing Ammonium Acetate. *J. Chromatog.* 47, 287-290.
18. Tovey, K.C. and Roberts, R.M. (1970) Studies on the Pyrophosphorylases of Wheat. *Plant Physiol.* 46, 406-411.
19. Roberts, R.M., Butt, V.S. (1970) Incorporation of ^{14}C-L-arabinose into Polysaccharides of Maize Root-tips. *Planta (Berl.)* 94, 175-183.
20. Roberts, R.M., Heishman, A. and Wicklin, C. (1971) Growth Inhibition and Metabolite Pool Levels in Plant Tissues Fed D-glucosamine and D-galactose. *Plant Physiol.* 48, 36-42.
21. Roberts, R.M. (1971) The Metabolism of D-mannose-^{14}C to Polysaccharide in Corn Roots. Specific Labeling of L-galactose, D-mannose and L-fucose. *Arch. Biochem. Biophys.* 145, 685-692.
22. Roberts, R.M. (1971) The Formation of UDP-glucuronic Acid in Plants. UDP-glucuronic Acid Pyrophosphorylase from Barley Seedling. *J. Biol. Chem.* 246, 4995-5002.
23. Kirby, E.G. and Roberts, R.M. (1971) The Localized Incorporation of ^{3}H-L-fucose into Cell Wall Polysaccharides of Cap and Epidermis of Corn Roots. *Planta* 99, 211-221.
24. Roberts, R.M., Connor, A.B. and Cetorelli, J.J. (1972) The Formation of Glycoproteins in Tissues of Higher Plants. Specific Labeling Using (1-^{14}C) glucosamine. *Biochem. J.* 125, 999-1008.
25. Roberts, R.M., Cetorelli, J.J., Kirby, E.G. and Ericson, M. (1972) Location of Glycoproteins That Contain Glucosamine in Plant Tissues. *Plant Physiol.* 50, 531-535.
26. Baig, M.M. and Roberts, R.M. (1973) Comparative Studies on the Carbohydrate-containing Membrane Components of Normal and Adenosine 3′:5′-cyclic Monophosphate-treated CHO cells. *Biochem. J.* 134, 329-339.
27. Roberts, R.M., Walker, A. and Cetorelli, J.J. (1973) Fucose-containing Glycopeptides from the Cell Surface of Chinese Hamster Ovary Cells Grown in the Presence or Absence of Cyclic AMP. *Nature New Biol.* 244, 86-89.
28. Chen, T.T., Bazer, F.W., Cetorelli, J.J., Pollard, W.E. and Roberts, R.M. (1973) Purification and Properties of a Progesterone-induced Basic Glycoprotein from the Uterine Fluids of Pigs. *J. Biol. Chem.* 248, 8560-8566.
29. Roberts, R.M. and Harrer, E. (1973) Determination of L-galactose in Polysaccharide Material. *Phytochemistry* 12, 2679-2682.
30. Roberts, R.M. and Loewus, F. (1973) The conversion of D-Glucose-6-^{14}C to Cell Wall Polysaccharide Material in *Zea mays* in Presence of High Endogenous Levels of Myoinositol. *Plant Physiol.* 52, 646-650.

31. Roberts, R.M. and Yuan, B. O-C. (1974) Chemical Modification of the Plasma Membrane Polypeptides of Cultured Mammalian Cells as an Aid to Studying Protein Turnover. *Biochemistry* 13, 4846-4855.
32. Schlosnagle, D.C., Bazer, F.W., Tsibris, J.C.M. and Roberts, R.M. (1974) An Iron-containing Phosphatase Induced by Progesterone in the Uterine Fluids of Pigs. *J. Biol. Chem.* 249, 7574-7579.
33. Roberts, R.M. and Pollard, W.E. (1975) The Incorporation of D-glucosamine into Glycolipids and Glycoproteins of Membrane Preparation from *Phaseolus aureus* hypocotyls. *Plant Physiol.* 55, 431-436.
34. Baig, M.M., Cetorelli, J.J. and Roberts, R.M. (1975) Plasma Membrane Proteins and Glycoproteins from Human Fibroblasts Derived from Normal Individuals and Patients with Cystic Fibrosis. *J. Pediatrics* 86, 72-76.
35. Welch, D.W. and Roberts, R.M. (1975) Complex Saccharide Metabolism in Cystic Fibrosis Fibroblasts. *Pediatrics Res.* 9, 698-702.
36. Chen, T.T., Bazer, F.W., Gebhardt, G. and Roberts, R.M. (1975) Uterine Secretions in Mammals: Synthesis and Placental Transport of a Purple Acid Phosphatase in Pigs. *Biol. Reprod.* 13, 304-313.
37. Bazer, F.W., Chen, T.T., Knight, J.S., Schlosnagle, D.C. and Roberts, R.M. (1975) Presence of a Progesterone-Induced, Uterine Specific Acid Phosphatase in Allantoic Fluid of Gilts. *J. Anim. Sci.* 41, 1112-1119.
38. Roberts, R.M. and Yuan, B. O-C (1975) Radiolabeling of Mammalian Cells in Tissue Culture Using Acetic Anhydride. Potential Value for Studying the Dynamics of Protein Turnover in Living Cells. *Arch. Biochem. Biophys.* 171, 226-233.
39. Roberts, R.M. and Yuan, B.O-C. (1975) Turnover of Plasma Membrane Polypeptides in Non-Proliferating Cultures of CHO Cells and Human Skin Fibroblasts. *Arch. Biochem. Biophys.* 171, 234-244.
40. Stein, G.S., Roberts, R.M., Davis, J.L., Stein, J.L., Thrall, C.L., VanVeen, J. and Welch, D.W. (1975) Are Glycoproteins and Glycosaminoglycans Components of the Eukaryotic Genome? *Nature* 258, 639-641.
41. Hall, J.L. and Roberts, R.M. (1975) Biochemical Characteristics of Membrane Fractions Isolated from Maize (*Zea mays*) Roots. *Ann. Bot.* 39, 983-993.
42. Roberts, R.M. and Bazer, F.W. (1976) Phosphoprotein Phosphatase Activity of the Progesterone-Induced Purple Glycoprotein of the Porcine Uterus. *Biochem. Biophys. Res. Commun.* 68, 450-455.
43. VanVeen, J., Roberts, R.M. and Noonan, K.D. (1976) An Analysis of Lectin-initiated Cell Agglutinability in a Series of CHO Sub-clones which Respond Morphologically to growth in Dibutyryl Cyclic AMP. *J. Cell Biol.* 70, 204-216.
44. Schlosnagle, D.C., Sander, E.G., Bazer, F.W. and Roberts, R.M. (1976) The Requirement of an Essential Thiol Group and Ferric Iron for the Activity of the Progesterone-Induced, Porcine Uterine, Purple Phosphatase. *J. Biol. Chem.* 251, 4680-4685.

45. VanVeen, J., Noonan, K.D. and Roberts, R.M. (1976) A Correlation between Membrane Glycopeptide Composition and Changes in Agglutinability Induced by Dibutyryl 3′:5′ Cyclic AMP in Chinese Hamster Ovary Cell. *Exptl. Cell Res.* 130, 405-413.
46. Roberts, R.M., Bazer, F.W., Baldwin, N. and Pollard, W.E. (1976) Progesterone Induction of Lysozyme and Peptidase Activities in the Porcine Uterus. *Arch. Biochem. Biophys.* 177, 499-507.
47. Basha, S.M.M. and Roberts, R.M. (1977) A Simple Colorimetric Method for the Determination of Tryptophan. *Anal. Biochem.* 77, 378-386.
48. Basha, S.M.M., Horst, M.N., Bazer, F.W. and Roberts, R.M. (1978) Purification and Properties of Aminopeptidases from Reproductive Fluids of Pigs. *Arch. Biochem. Biophys.* 185, 174-184.
49. Basha, S.M.M., Bazer, F.W. and Roberts, R.M. (1979) The Secretion of a Uterine Specific, Purple Phosphatase by Cultured Explants of Porcine Endometrium Dependency upon the State of Pregnancy of the Donor Animal. *Biol. Reprod.* 20, 431-441.
50. Horst, M.N. and Roberts, R.M. (1979) Analysis of Polypeptide Turnover Rates in Chinese Hamster Ovary Cell Plasma Membranes Using Two-Dimensional Electrophoresis. *J. Biol. Chem.* 254, 5000-5007.
51. Horst, M.N., Baumbach, G. and Roberts, R.M. (1979) Two-Dimensional Electrophoretic Analysis of Concanvalin-A Binding Components in the Plasma Membranes of Chinese Hamster Fibroblasts. *FEBS Lett.* 100, 385-388.
52. Gaber, B.P., Sheridan, J.P., Bazer, F.W. and Roberts, R.M. (1979) Resonance Raman Scattering from Uteroferrin, the Purple Glycoprotein of the Porcine Uterus. *J. Biol. Chem.* 254, 8340-8342.
53. Bazer, F.W., Roberts, R.M., Basha, S.M.M., Zavy, M.T., Caton, D. and Barron, D. (1980) Method for Obtaining Ovine Uterine Secretions from Unilaterally Pregnant Ewes. *J. Anim. Sci.* 49, 1522-1527.
54. Basha, S.M.M., Bazer, F.W. and Roberts, R.M. (1980) Effect of Conceptus on Quantitative and Qualitative Aspects of Porcine Uterine Secretions. *J. Reprod. Fert.* 60, 41-48.
55. Basha, S.M.M., Bazer, F.W., Geisert, R.D. and Roberts, R.M. (1980) Progesterone-induced Uterine Secretions in Pigs. Recovery from Pseudopregnant and Unilaterally Pregnant Gilts. *J. Anim. Sci.* 50, 113-123.
56. Horst, M.N., Basha, S.M.M. Baumbach, G.A., Mansfield, E. and Roberts, R.M. (1980) Alkaline Urea Solubilization, Two-dimensional Electrophoresis and Lectin Staining of Mammalian Cell Plasma Membrane and Plant Seed Proteins. *Anal. Biochem.* 102, 399-408.
57. Horst, M.N., Baumbach, G.A. and Roberts, R.M. (1980) Isolation of a Domain of the Plasma Membrane in CHO Cells which Contains Iodinatable Acidic Glycoproteins of High Molecular Weight. *Biochim. Biophys. Acta* 600, 48-61.

58. Mullins, D.E., Bazer, F.W. and Roberts, R.M. (1980) Secretion of a Progesterone-induced Inhibitor of Plasminogen Activator by the Porcine Uterus. *Cell* 20, 865-872.
59. Mullins, D.E., Horst, M.N., Bazer, F.W. and Roberts, R.M. (1980) Isolation and Characterization of a Plasma Membrane Fraction Derived from the Luminal Surface of the Pig Uterus During the Estrous Cycle and Early Pregnancy. *Biol. Reprod.* 22, 1181-1192.
60. Antanaitis, B.C., Aisen, P., Lillenthal, H.R., Roberts, R.M. and Bazer, F.W. (1980) the Novel "g′=1.74" EPR Spectrum of Pink and Purple Uteroferrin. *J. Biol. Chem.* 255, 11204-11209.
61. Basha, S.M.M. and Roberts, R.M. (1981) The Glycoproteins of Plant Seeds. Analysis by Two-Dimensional Polyacrylamide Gel Electrophoresis and by their Lectin Binding Properties. *Plant Physiol* 67, 936-942.
62. Adams, K.L., Bazer, F.W. and Roberts, R.M. (1981) Progesterone-induced Secretion of a Retinol-Binding Protein in the Pig Uterus. *J. Reprod. Fert.* 62, 39-47.
63. Mancarella, D.A., Basha, S.M.M., Mullins, D.E., Mansfield, E., Bazer, F.W. and Roberts, R.M. (1981) Properties of a Membrane-associated L-Leucine-naphthylamidase (Leucine Aminopeptidase) from the Porcine Uterus. *Biol. Reprod.* 24, 879-888.
64. Bartol, F.F., Thatcher, W.W., Bazer, F.W., Kimball, F.A., Chenault, J.R., Wilcox, C.J. and Roberts, R.M. (1981) Effects of the Estrous Cycle and Early Pregnancy on Bovine Uterine, Luteal and Follicular Responses. *Biol. Reprod.* 25, 759-776.
65. Godkin, J.D., Bazer, F.W., Moffatt, J., Sessions, F. and Roberts, R.M. (1982) Purification and Properties of a Major, Low Molecular Weight Protein Released by the Trophoblast of Sheep Blastocysts at Day 13-21. *J. Reprod. Fert.* 65, 141-150.
66. Fitzgerald, L.A., Baumbach, G.A., Horst, M.N., Noonan, K.D. and Roberts, R.M. (1982) The Purification and Immunocytochemical Localization of the Major, Iodinatable, Cell Surface Glycoproteins of CHO Cells. *J. Cell Sci.* 52, 405-424.
67. Zavy, M.T., Sharp, D.C., Bazer, F.W., Fazleabas, A., Sessions, F. and Roberts, R.M. (1982) Identification of Stage-Specific and Hormonally-Induced Polypeptides in the Uterine Protein Secretions of the Mare During the Oestrous Cycle and Pregnancy. *J. Reprod. Fert.* 64, 199-207.
68. Buhi, W.C., Gray, W.J., Mansfield, E.A., Chun, P.W., Ducsay, C.A., Bazer, F.W. and Roberts, R.M. (1981) Iron Content and Molecular Weight of Uteroferrin and a Comparison of Its Iron and Copper Forms. *Biochim. Biophys. Acta* 601, 32-38.

69. Denny, J.B. and Roberts, R.M. (1982) A New Immunoreactive Probe for the Isolation and Analysis of Plasma Membrane Polypeptides. Synthesis and Properties of Isethionyl 3-(N-2,4-Dinitrophenyl)-aminopropioimidate. *J. Biol. Chem.* 257, 2460-2468.
70. Buhi, W.C., Ducsay, C.A., Bazer, F.W. and Roberts, R.M. (1982) Iron Transfer Between the Purple Phosphatase Uteroferrin and Transferrin and Its Possible Role in Iron Metabolism of the Fetal Pig. *J. Biol. Chem.* 257, 1712-1721.
71. Fazleabas, A.T., Bazer, F.W. and Roberts, R.M. (1982) Purification and Properties of a Progesterone-Induced, Plasmin/trypsin Inhibitor from the Uterine Secretions of Pigs and Its Immunocytochemical Localization in the Pregnant Uterus. *J. Biol. Chem.* 257, 6886-6897.
72. Ducsay, C.A., Buhi, W.C., Bazer, F.W. and Roberts, R.M. (1982) Role of Uteroferrin in Iron Transport and Macromolecular Uptake by Allantoic Epithelium of the Porcine Conceptus. *Biol. Reprod.* 26, 729-743.
73. McDowell, K., Sharp, D.C., Zavy, M.T., Fazleabas, A., Roberts, R.M. (1982) and Bazer, F.W., Partial Characterization of the Equine Uteroferrin-like Protein. *J. Reprod. Fert. Suppl.* 32, 329-334.
74. Masters, R.A., Roberts, R.M., Lewis, G.S., Thatcher, W.W., Bazer, F.W. and Godkin, J.D. (1982) High Molecular Weight Glycoproteins Released by Expanding Pre-attachment Sheep, Pig and Cow Blastocysts in Culture. *J. Reprod. Fert.* 66, 571-583.
75. Godkin, J.D., Bazer, F.W., Geisert, R.D., Lewis, G.S. and Roberts, R.M. (1982) Synthesis and Release of Polypeptides by Pig Conceptuses During the Period of Blastocyst Elongation and Attachment. *Biol. Reprod.* 27, 977-987.
76. Geisert, R.D., Renegar, R.H., Thatcher, W.W., Roberts, R.M. and Bazer, F.W. (1982) Establishment of Pregnancy in the Pig: I. Interrelationships Between Preimplantation Development of the Pig Blastocyst and Uterine Endometrial Secretions. *Biol. Reprod.* 27, 925-939.
77. Geisert, R.D., Brookbank, J.W., Roberts, R.M. and Bazer, F.W. (1982) Establishment of Pregnancy in the Pig: II. Cellular Remodeling of the Porcine Blastocyst During Elongation on Day 12 of Pregnancy. *Biol. Reprod.* 27, 941-955.
78. Geisert, R.D., Thatcher, W.W., Roberts, R.M. and Bazer, F.W. (1982) Establishment of Pregnancy in the Pig: III. Endometrial Secretory Response to Estradiol Valerate Administered on Day 11 of the Estrous Cycle. *Biol. Reprod.* 27, 957-965.
79. Zavy, M.T., Clark, W.R., Sharp, D.C., Roberts, R.M. and Bazer, F.W. (1982) Comparison of Glucose, Fructose, Ascorbic Acid and Glucose phosphate Isomerase Enzymatic Activity in Uterine Flushings from Nonpregnant and Pregnant Gilts and Pony Mares. *Biol. Reprod.* 27, 1147-1158.
80. Renegar, R.H., Bazer, F.W. and Roberts, R.M. (1982) Placental Transport and Distribution of Uteroferrin in the Fetal Pig. *Biol. Reprod.* 27, 1247-1260.

81. Fazleabas, A.T., Geisert, R.D., Bazer, F.W. and Roberts, R.M. (1983) The Relationship Between Release of Plasminogen Activator and Estrogen by Blastocysts and Secretion of Plasmin Inhibitor by Uterine Endometrium in the Pregnant Pig. *Biol. Reprod.* 29, 225-238.
82. Buhi, W.C., Ducsay, C.A., Bartol, F.F., Bazer, F.W. and Roberts, R.M. (1983) A Function of the Allantoic Sac in the Metabolism of Uteroferrin and Maternal Iron by the Fetal Pig. *Placenta* 4, 455-470.
83. Godkin, J.D., Bazer, F.W. and Roberts, R.M. (1984) Ovine Trophoblast Protein 1 (oTP-1), an Early Secreted Blastocyst Protein, Binds Specifically to Uterine Endometrium and Affects Protein Synthesis. *Endocrinol.* 114, 120-130.
84. Fazleabas, A.T., Mead, R.A., Rourke, R.A. and Roberts, R.M. (1984) Presence of an Inhibitor of Plasminogen Activator in Uterine Fluid of the Western Spotted Skunk During Delayed Implantation. *Biol. Reprod.* 30, 311-322.
85. Ducsay, C.A., Buhi, W.C., Bazer, F.W., Roberts, R.M. and Combs, G.E. (1984) Role of Uteroferrin in Placental Iron Transport. II. Effect of Maternal Iron Treatment on Fetal Iron, Uteroferrin Content and Neonatal Hemoglobin. *J. Anim. Sci.* 59, 1303-1308.
86. Fitzgerald, L.A., Denny, J.B., Baumbach, G., Ketcham, C.M. and Roberts, R.M. (1984) The Effect of Altered Oligosaccharide Structure on the Cell Surface Number, Distribution and Turnover of the High Molecular Weight Acidic Glycoprotein of CHO Cells. *J. Cell Sci.* 67, 1-23.
87. Godkin, J.D., Bazer, F.W., Thatcher, W.W. and Roberts, R.M. (1984) Proteins Released by Cultured Day 15-16 Conceptuses Prolong Luteal Maintenance When Introduced into the Lumen of Cyclic Ewes. *J. Reprod. Fert.* 71, 57-64.
88. Baumbach, G.A., Saunders, P.T.K., Bazer, F.W. and Roberts, R.M. (1984) Uteroferrin has N-asparagine linked, High Mannose-type Oligosaccharides Which Contain Mannose-6-phosphate. *Proc. Natl. Acad. Sci. USA.* 81, 2985-2989.
89. Segerson, E.C., Moffatt, R.J., Bazer, F.W. and Roberts, R.M. (1984) Suppression of Phytohemagglutinin- stimulated Lymphocyet Blastogenesis by Ovine Uterine Milk Protein. *Biol. Reprod.* 30, 1175-1186.
90. Zavy, M.T., Roberts, R.M. and Bazer, F.W. (1984) Acid Phosphatase and Leucine Aminopeptidase Activity in the Uterine Flushings of Nonpregnant and Pregnant Gilts. *J. Reprod. Fert.* 72, 503-507.
91. Lang, L., Reitman, M., Tang, J., Roberts, R.M. and Kornfeld, S. (1984) Recognition of a Protein-dependent Determinant allows Specific Phosphorylation of Oligosaccharides present on Lysosomal Enzymes. *J. Biol. Chem.* 259, 14663-14671.
92. Fazleabas, A.T., Bazer, F.W., Hansen, P.J., Geisert, R.D. and Roberts, R.M. (1985) Differential Patterns of Secretory Protein Synthesis within the Pig Uterine Endometrium in Response to Progesterone. *Endocrinol.* 116, 240-245.

93. Ketcham, C.M., Baumbach, G.A., Bazer, F.W. and Roberts, R.M. (1985) The type 5, Acid Phosphatase from Spleen of Humans with Hairy Cell Leukemia. *J. Biol. Chem.* 260, 5768-5776.
94. Saunders, P.T.K., Renegar, R.H., Raub, T.J., Baumbach, G.A., Atkinson, P.H., Bazer, F.W. and Roberts, R.M. (1985) The Carbohydrate Structure of Porcine Uteroferrin and the Role of the High Mannose Chains in Promoting Uptake by the Reticuloendothelial Cells of the Fetal Liver. *J. Biol. Chem.* 260, 3658-3665.
95. Raub, T.J., Bazer, F.W. and Roberts, R.M. (1985) Localization of the Iron Transport Glycoprotein, Uteroferrin, in the Porcine Endometrium and Placenta by Using Immunocolloidal Gold. *Anat. Embryol.* 171, 253-258.
96. Bartol, F.F., Roberts, R.M., Godkin, J.D., Bazer, F.W, Lewis, G.S. and Thatcher, W.W. (1985) Characterization of Proteins produced *in vitro* by Preattachment Bovine Conceptuses. *Biol. Reprod.* 32, 681-693.
97. Hansen, P.J., Bazer, F.W. and Roberts, R.M. (1985) Appearance of β-Hexosaminidase and other Lysosomal-like Enzymes in the Uterine Lumen of Gilts, Ewes and Mares in response to Progesterone. *J. Reprod. Fert.* 73, 411-424.
98. Godkin, J.D., Bazer, F.W. and Roberts, R.M. (1985) Protein Production by Cultures Established from Day 14-16 Sheep and Pig Conceptuses. *J. Reprod. Fert.* 74, 377-382.
99. Hansen, P.J., Anthony, R.V., Bazer, F.W., Baumbach, G.A. and Roberts, R.M. (1985) In Vitro synthesis and secretion of ovine trophoblast protein-1 during the period of maternal recognition of pregnancy. *Endocrinol.* 117, 1424-1430.
100. Bartol, F., Roberts, R.M., Bazer, F.W. and Thatcher, W.W. (1985) Characterization of proteins produced *in vitro* by bovine endometrial explants. *Biol. Reprod.* 33, 745-759.
101. Ducsay, C.A., Buhi, W.C., Bazer, F.W., and Roberts, R.M. (1986) Role of uteroferrin in placental iron transport in swine: relationship between uteroferrin levels and iron deposition in the conceptus during gestation. *J. Anim. Sci.* 62, 706-716.
102. Raub, T.J., Denny, J.B., and Roberts, R.M. (1986) Cell surface glycoproteins of CHO cells. I. internalization and rapid recycling. *Exptl Cell Res.* 165, 73-91.
103. Raub, T.J., and Roberts, R.M. (1986) Cell surface glycoproteins of CHO cells. II. surface distribution and pathway of internalization. *Exptl. Cell Res.* 165, 92-107.
104. Baumbach, G.A., Ketcham, C.M., Richardson, D.E. Bazer, F.W. and Roberts R.M. (1986) Isolation and characterization of a high molecular weight stable pink form of uteroferrin from uterine secretions and allantoic fluid of pigs. *J. Biol. Chem.* 261, 12869-12878.

105. Fincher, K.B., Bazer, F.W., Hansen, P.J., Thatcher, W.W., and Roberts, R.M. (1986) Proteins secreted by the sheep conceptus suppress induction of uterine prostaglandin $F_2\alpha$ release by oestradiol and oxytocin. *J. Reprod. Fert.* 76, 425-433.
106. Knickerbocker, J.J., Thatcher, W.W., Bazer, F.W., Drost, M., Barron, D.H., Fincher, K.B. and Roberts, R.M. (1986) Proteins secreted by day 16 to day 18 bovine conceptuses extend corpus luteum functions in cows. *J. Reprod. Fert.* 77, 381-391.
107. Hansen, P.J., Ing, N.H., Moffatt, R.J., Baumbach, G.A., Saunders, P.T.K., Bazer, F.W. and Roberts, R.M. (1987) Biochemical characterization and biosynthesis of the uterine milk proteins of the pregnant sheep uterus. *Biol. Reprod.* 36, 405-418.
108. Moffatt, R.J., Bazer, F.W., Hansen, P.J., Chun, P.W., and Roberts, R.M. (1987) Purification, secretion and immunocytochemical localization of the uterine milk proteins, major progesterone-induced proteins in uterine secretions of the sheep. *Biol. Reprod.* 36, 419-430.
109. Couso, R., Lang, L., Roberts, R.M. and Kornfeld, S. (1986) Phosphorylation of the oligosaccharide of uteroferrin by UDP-GlcNAc:glycoprotein N-acetylglucosamine-1-phosphotransferases from rat liver. *Acanthamoeba castellani*, and *Dictyostelium discoideum* requires α1,2-linked mannose residues. *J. Biol. Chem.* 261, 6326-6331.
110. Knickerbocker, J.J., Thatcher, W.W., Bazer, F.W., Barron, D.H., and Roberts, R.M. (1986) Inhibition of uterine prostaglandin $F_2\alpha$ production by bovine conceptus secretory proteins. *Prostaglandins* 31, 777-793.
111. Bazer, F.W., Vallet, J.L. Roberts, R.M., Sharp, D.C. and Thatcher, W.W. (1986) Role of conceptus secretory products in establishment of pregnancy. *J. Reprod. Fert.* 76, 841-850.
112. Roberts, R.M., Baumbach, G.A., Saunders, P.T.K., Raub, T.J., Renegar, R.H. and Bazer, F.W. (1986) Possible function of carbohydrate on glycoproteins secreted into the uterus of pigs during pregnancy. *Mol. Cell. Biochem.* 72, 67-79.
113. Young, K.H., Bazer, F.W., Simpkins, J.W. and Roberts, R.M. (1987) Effects of Early Pregnancy and Acute 17β-estradiol Administration on Porcine Uterine Secretion, Cyclic Nucleotides and Catecholamines. *Endocrinol.* 120, 254-263.
114. Helmer, S.D., Hansen, P.J., Anthony, R.V., Thatcher, W.W., Bazer, F.W. and Roberts, R.M. (1987) Identification of bovine trophoblast protein-1, a secretory protein immunologically related to ovine trophoblast protein-1. *J. Reprod. Fert.* 79, 83-91.
115. Murray, M.K., Segerson, E.C., Hansen, P.J., Bazer, F.W. and Roberts, R.M. (1987) Suppression of lymphocyte activation by a high molecular weight glycoprotein released from preimplantation ovine and porcine conceptuses. *Amer. J. Reprod. Immunol. and Microbiol.* 14, 38-44.

116. Hunt, D.F., Yates, J.R., Shabanowitz, J., Zhu, N.-Z., Zirino, T., Averill, B.A., Daurat-Larraque, S.T., Shewale, J.G., Roberts, R.M. and Brew, K. (1987) Sequence homology in the metalloproteins: purple acid phosphatase from beef spleen and uteroferrin from porcine uterus. *Biochem. & Biophys. Res. Comm.* 144, 1154-1160.

117. Moffatt, R.J., Bazer, F.W., Roberts, R.M. and Thatcher, W.W. (1987) Secretory function of the ovine uterus: effects of gestation and steroid replacement therapy. *J. Anim. Sci.* 65, 1400-1410.

118. Imakawa, K. Anthony, R.V., Kazemi, M., Marotti, K.R., Polites, H.G., and Roberts, R.M. (1987) Interferon-like sequence of ovine trophoblast protein secreted by embryonic trophectoderm. *Nature (London)* 330, 377-379.

119. Vallet, J.L., Bazer, F.W. and Roberts, R.M. (1987) The effect of ovine trophoblast protein-one on endometrial protein secretion and cyclic nucleotides. *Biol. Reprod.* 37, 1307-1316.

120. Roberts, R.M., Murray M.K., Burke M.G., and Bazer F.W. (1987) Hormonal control and function of secretory proteins. *Adv. Exp. Biol.* 230, 137-50.

121. A. Simmen, R.C.M., Baumbach, G.A. and Roberts, R.M. (1988) Molecular cloning and temporal expression during pregnancy of the messenger RNA encoding uteroferrin, a progesterone-induced secretory protein. *Mol. Endocrinol.* 2, 253-262.

a.B. Lottenberg, R., Sjak-Shie, N., Fazleabas, A.T. and Roberts, R.M. (1988) Aprotinin inhibits urokinase but not tissue type plasminogen activator. *Thrombosis Res.* 49, 549-556.

122. Kazemi, M., Malathy, P.V., Keisler, D.H. and Roberts, R.M. (1988) Ovine trophoblast protein-1 and bovine trophoblast protein-1 are present as specific components of uterine flushings of pregnant ewes and cows. *Biol. Reprod.* 39, 457-463.

123. Hansen, T.R., Imakawa, K., Polites, H.G., Marotti, K.R., Anthony, R.V. and Roberts, R.M. (1988) Interferon RNA of embryonic origin is expressed transiently during early pregnancy in the ewe. *J. Biol. Chem.* 263, 12801-12804.

124. Anthony, R.V., Helmer, S.D., Sharif, S.F., Roberts, R.M., Hansen, P.J., Thatcher, W.W. and Bazer, F.W. (1988) Synthesis and processing of ovine trophoblast protein-1 and bovine trophoblast protein-1, conceptus secretory proteins involved in the maternal recognition of pregnancy. *Endocrinol.* 123, 1274-1280.

125. Wettemann, R.P., Bazer, F.W., Thatcher, W.W., Caton, D. and Roberts, R.M. (1988) Conceptus development, uterine response, blood gases and endocrine function of gilts exposed to increased ambient temperature during early pregnancy. *Theriogenol.* 30, 57-72.

126. Ketcham, C.M., Roberts, R.M., Simmen, R.C.M. and Nick, H.S. (1989) Molecular cloning of the type 5, iron-containing, tartrate resistant acid phosphatase from human placenta. *J. Biol. Chem.* 264, 557-563.
127. Imakawa, K., Hansen, T.R., Malathy, P.V., Anthony, R.V., Polites, H.G., Marotti, K.R. and Roberts, R.M. (1989) Molecular cloning and characterization of cDNA's corresponding to bovine trophoblast protein-1. A comparison with ovine trophoblast protein-1 and bovine interferon-α_{II}. *Mol. Endocrinol.* 3, 127-139.
128. Allen, S.H., Nuttleman, P.R., Ketcham, C.M. and Roberts, R.M. (1989) Purification and characterization of human bone tartrate-resistant acid phosphatase. *J. Bone & Mineral Res.* 4, 47-55.
129. Roberts, R.M., Imakawa, K., Niwano, Y., Kazemi, M., Malathy, P.V., Hansen, T.R., Glass, A.A. and Kronenberg, L.H. (1989) Interferon production by the preimplantation sheep embryo. *J. Interferon Res.* 9, 175-187.
130. Sharif, S.F., Francis, H., Keisler, D.H. and Roberts, R.M. (1989) Correlation between the release of ovine trophoblast protein-1 by the conceptus and the production of polypeptides by the maternal endometrium of ewes. *J. Reprod. Fert.* 85, 471-476.
131. Hansen, T.R., Kazemi, M., Keisler, D.H., Malathy, P.V., Imakawa, K. and Roberts, R.M. (1989) Complex binding of the embryonic interferon, ovine trophoblast protein-1, to endometrial receptors. *J. Interferon Res.* 9, 215-225.
132. Ing, N.H. and Roberts, R.M. (1989) The major progesterone-modulated proteins secreted into the sheep uterus are members of the serpin family of serine protease inhibitors. *J. Biol. Chem.* 264, 3372-3379.
133. Murray, M.K., Malathy, P.V., Bazer, F.W. and Roberts, R.M. (1989) Structural relationship, biosynthesis and immunocytochemical localization of the uteroferrin-associated basic glycoproteins. *J. Biol. Chem.* 264, 4143-4150.
134. Roberts, R.M. (1989) Minireview: Conceptus interferons and maternal recognition of pregnancy. *Biol. Reprod.* 40, 449-452.
135. Allen, B.S., Ketcham, C.M., Roberts, R.M., Nick, H.S. and Ostrer, H. (1989) Localization of the Human Type 5, tartrate-resistant acid phosphatase gene by in situ hybridization. *Genomics* 4, 597-600.
136. Cross, J.C. and Roberts, R.M. (1989) Porcine conceptuses secrete an interferon during the preattachment period of early pregnancy. *Biol. Reprod.* 40, 1109-1118.
137. Farin, C.E., Imakawa, K. and Roberts, R.M. (1989) *In situ* localization of mRNA for the interferon, ovine trophoblast protein-1, during early embryonic development of the sheep. *Mol. Endocr.* 3, 1099-1107.
138. Simmen, R.C.M., Srivivas, V. and Roberts, R.M. (1989) cDNA sequence, gene organization and progesterone induction of messenger RNA for uteroferrin, a porcine uterine iron transport protein. *DNA* 8, 543-554.

139. Roberts, R.M. (1989) Minireview: A Novel Group of Interferons Associated with the Early Ovine and Bovine Embryo. *J. Interferon Res.* 9, 373-378.
140. Niwano, Y., Hansen, T.R., Kazemi, M., Malathy, P-V., Johnson, H.D., Roberts, R.M. and Imakawa, K. (1989) Suppression of T-lymphocyte blastogenesis by ovine trophoblast protein-1 and human interferon-α may be independent of interleukin-2 production. *Amer. J. Reprod. Immunol.* 20, 21-26.
141. Ing, N.H., Francis, H., McDonnell, J.J., Amann, J.F. and Roberts, R.M. (1989) Progesterone induction of the uterine milk proteins: major secretory proteins of sheep endometrium. *Biol. Reprod.* 41, 643-654.
142. Malathy, P.V., Imakawa, K., Simmen, R.C.M. and Roberts, R.M. (1990) Molecular cloning of the uteroferrin-associated protein, a major progesterone-induced serpin secreted by the porcine uterus, and the expression of its mRNA during pregnancy. *Mol. Endocr.* 4, 428-440.
143. Roberts, R.M., Malathy, P.V., Hansen, T.R., Farin, C.E. and Imakawa, K. (1990) Bovine conceptus products involved in pregnancy recognition. *J. Anim. Sci.* 68, 28-38.
144. Nephew, K.P., McClure, K.E., Day, M.L., Xie, S., Roberts, R.M. and Pope, W.F. (1990) Effects of intramuscular administration of recombinant bovine interferon-α_I1 during the period of maternal recognition of pregnancy. *J. Anim. Sci.* 68, 2766-2770.
145. Roberts, R.M., Schalue-Francis, T., Francis, H. and Keisler, D. (1990) Maternal recognition of pregnancy and embryonic loss. *Theriogenol.* 33, 175-183.
146. Gignac, S.M., Buschle, M., Roberts, R.M., Pettit, G.R., Hoffbrand, A.V. and Drexler, H.G. (1990) Differential expression of TRAP isoenzyme in B-CLL cells treated with different inducers. *Leukemia & Lymphoma* 3, 19-29.
147. Farin, C.E., Imakawa, K., Hansen, T.R., McDonnell, J.J., Murphy, C.N., Farin, P.W. and Roberts, R.M. (1990) Expression of trophoblastic interferon genes in sheep and cattle. *Biol. Reprod.*, 43, 210-218.
148. McDowell, K.J., Sharp, D.C., Fazleabas, A.T. and Roberts, R.M. (1990) Two-dimensional polyacrylamide gel electrophoresis of proteins synthesized and released by conceptuses and endometria from pony mares. *J. Reprod. Fert.*, 89, 107-115.
149. Clawitter, J., Trout, W.E., Burke, M.G., Araghi, S. and Roberts, R.M. (1990) A novel family of progesterone-induced, retinol-binding proteins from uterine secretions of the pig. *J. Biol. Chem.* 265, 3248-3255.
150. Nuttleman, P.R. and Roberts, R.M. (1990) Transfer of iron from uteroferrin (purple acid phosphatase) to transferrin related to acid phosphatase activity. *J. Biol. Chem.* 265, 12192-12199.
151. Kazemi, M., Amann, J.F., Keisler, D.H., Ing, N.H., Roberts, R.M., Morgan, G. and Wooding, F.B.P. (1990) A progesterone-modulated, low molecular weight

protein from the uterus of the sheep is associated with crystalline inclusion bodies in uterine epithelium and embryonic trophectoderm. *Biol. Reprod.* 43, 80-96.

152. Cross, J.C., Farin, C.E., Sharif, S.F. and Roberts, R.M. (1990) Character-ization of the antiviral activity constitutively produced by murine conceptuses: absence of placental mRNAs for interferons alpha and beta. *Mol. Reprod. Dev.* 26, 122-128.
153. Klemann, S.W., Li, J., Imakawa, K., Cross, J.C., Francis, H. and Roberts, R.M. (1990) The production, purification, and bioactivity of recombinant bovine trophoblast protein-1 (bovine trophoblast interferon). *Mol. Endocr.* 4, 1506-1514 .
154. Raub, T.J., Koroly, M.J. and Roberts, R.M. (1990) Endocytosis of wheat germ agglutinin-binding sites from the cell surface into a tubular endosomal network. *J. Cell. Physiol.* 143, 1-12.
155. Raub, T.J., Koroly, M.J. and Roberts, R.M. (1990) Rapid endocytosis and recycling of wheat germ agglutinin-binding sites on CHO cells: Evidence for two compartments in a nondegradative pathway. *J. Cell. Physiol.* 144, 52-61.
156. Klemann, S.W., Imakawa, K. and Roberts, R.M. (1990) Sequence variability among ovine trophoblast interferon cDNA. *Nucl. Acids Res.* 18, 6724.
157. Hansen, T.R., Leaman, D.W., Cross, J.C., Mathialagan, N., Bixby, J.A. and Roberts, R.M. (1991) The genes for the trophoblast interferons and the related interferon-II possess distinct 5′-promoter and 3′-flanking sequences. *J. Biol. Chem.* 266, 3060-3067.
158. Cross, J.C. and Roberts, R.M. (1991) Constitutive and trophoblast-specific expression of a class of bovine interferon genes. *Proc. Natl. Acad. Sci. USA.* 88, 3817-3821.
159. Schalue-Francis, T.K., Farin, P.W., Cross, J.C., Keisler, D. and Roberts, R.M. (1991) Effect of injected bovine interferon-α_I1 on oestrous cycle length and pregnancy success in sheep. *J. Reprod. Fert.* 91, 347-356.
160. Wooding, F.B.P., Morgan, G. and Roberts, R.M. (1991) Quantitative immunogold ultracryomicrotome studies of the distribution of peri-implantation proteins in the sheep. *J. Cell & Tissue Res.* 265, 83-93.
161. Farin, C.E., Cross, J.C., Tindle, N.A., Murphy, C.N., Farin, P.W. and Roberts, R.M. (1991) Induction of trophoblastic interferon expression in ovine blastocysts after treatment with double-stranded RNA. *J. Interferon Res.* 11, 151-157.
162. Hansen, T.R., Cross, J.C., Farin, C.E., Imakawa, K. and Roberts, R.M. (1991) Slowed transcription and rapid messenger RNA turnover contribute to a decline in synthesis of ovine trophoblast protein during *in vitro* culture. *Biol. Reprod.* 45, 94-100.

163. Baumbach, G.A., Saunders, P.T.K., Ketcham, C.M., Bazer, F.W. and Roberts, R.M. (1991) Uteroferrin contains complex and high mannose-type oligosaccharides when synthesized *in vitro. Mol. Cell. Biochem.* 105, 107-117.

164. Xie, S., Low, B.G., Nagel, R.J., Kramer, K.K., Anthony, R.V., Zoli, A.P., Beckers, J-F. and Roberts, R.M. (1991) Identification of the major pregnancy-specific antigens of cattle and sheep as inactive members of the aspartic proteinase family. *Proc. Natl. Acad. Sci. USA.* 88, 10247-10251.

165. Trout, W.E., McDonnell, J.J., Kramer, K.K., Baumbach, G.A. and Roberts, R.M. (1991) The retinol-binding protein of the expanding pig blastocyst: Molecular cloning and expression in trophectoderm and embryonic disc. *Mol. Endocr.* 5, 1533-1540.

166. Francis, H., Keisler, D.H. and Roberts, R.M. (1991) Induction of the synthesis of the pregnancy-specific protein p70 in the endometrium by intramuscular injection of recombinant bovine interferon-α_I1 in nonpregnant ewes. *J. Reprod. Fert.* 93, 367-374.

167. Leaman, D.W. and Roberts, R.M. (1992) Genes for the trophoblast interferons in sheep, goat and musk ox and distribution of related genes among mammals. *J. Interferon Res.* 12, 1-11.

168. Trout, W.E., Hall, J.A., Stallings-Mann M.L., Galvin, J.M., Anthony, R.V. and Roberts, R.M. (1992) Steroid regulation of the synthesis and secretion of retinol-binding protein by the uterus of the pig. *Endocrinol.* 130, 2557-2564.

169. Garverick, H.A., Moser, M.T., Keisler, D.H., Hamilton, S.A., Roberts, R.M. and Smith, M.F. (1992) Luteal function after intrauterine infusion of recombinant bovine interferon-α_I1 into postpartum beef cows expected to have short or normal luteal phases. *J. Reprod. Fert.* 94, 319-325.

170. Hernandez-Ledezma, J.J., Sikes, J.D., Murphy, C.N., Watson, A.J., Schultz, G.A. and Roberts, R.M. (1992) Expression of bovine trophoblast interferon in conceptuses derived by *in vitro* techniques. *Biol. Reprod.* 47, 374-380.

171. Mathialagan, N., Bixby, J.A. and Roberts, R.M. (1992) Expression of interleukin-6 in porcine, ovine and bovine preimplantation conceptuses. *Mol. Reprod. & Dev* 32, 324-330.

172. Skopets, B., Li, J., Thatcher, W.W., Roberts, R.M. and Hansen, P.J. (1992) Inhibition of lymphocyte proliferation by bovine trophoblast protein-1 (Type I trophoblast interferon) and bovine interferon-α_I1. *Vet. Immunol. Immunopath.* 34, 81-96.

173. A. Leaman, D.W., Cross, J.C. and Roberts, R.M. (1992) Genes for the trophoblast interferons and their distribution among mammals. *Reprod. Fertil. Dev.* 4, 349-353.

174. B. Roberts, R. M., Cross, J. C., and Douglas, W. L. (1992) Interferons as Hormones of Pregnancy. *Endocrine Rev.* 13, 432-452.

175. Ling, P. and Roberts, R.M. (1993) Uteroferrin and intracellular tartrate-resistant acid phosphatases are the products of the same gene. *J. Biol. Chem.* 268, 6896-6902.

176. Stallings-Mann, M.L., Trout, W.E. and Roberts, R.M. (1993) Porcine uterine retinol-binding proteins are identical gene products to the serum retinol-binding protein. *Biol. Reprod.* 48, 998-1005.

177. Hernandez-Ledezma, J.J., Villanueva, C., Sikes, J.D. and Roberts, R.M. (1993) Effects of CZB *versus* Medium 199 and of conditioning culture media with either bovine oviductal epithelial cells or Buffalo rat liver cells on the development of bovine zygotes derived by *in vitro* maturation-*in vitro* fertilization procedures. *Theriogenol.* 39, 1267-1277.

178. Hernandez-Ledezma, J.J., Mathialagan, N., Villanueva, C., Sikes, J.D. and Roberts, R.M. (1993) Expression of bovine trophoblast interferons by *in vitro*-derived blastocysts is correlated with their morphological quality and stage of development. *Mol. Reprod. Dev.* 36, 1-6.

179. Juengel, J.L., Imakawa, K., Farin, P.W., Youngquist, R.S., Roberts, R.M., Smith, M.F. and Garverick, H.A. (1993) Detection of mRNA for inhibin α- and β_A-subunits in bovine ovarian tissues and the effect of *in vivo* administration of GnRH. *Dom. Anim. Endocrinol.* 10, 207-218.

180. Ling, P. and Roberts, R.M. (1993) Overexpression of uteroferrin, a lysosomal acid phosphatase found in porcine uterine secretions, results in its high rate of secretion from transfected fibroblasts. *Biol. Reprod.* 49, 1317-1327.

181. Kramer, K.K., Duffy, J.Y., Klemann, S.W., Bixby, J.A., Low, B.G., Pope, W.F. and Roberts, R.M. (1994) Selective cloning of cDNA for secretory proteins of early embryos. *J. Biol. Chem.* 269, 7255-7261.

182. Leaman, D.W., Cross, J.C. and Roberts, R.M. (1994) Multiple regulatory elements are required to direct trophoblast interferon gene expression in choriocarcinoma cells and trophectoderm. *Mol. Endocr.* 8, 456-468.

183. Li, J. and Roberts, R.M. (1994) Interferon-τ and interferon-α interact with the same receptors in bovine endometrium. *J. Biol. Chem.* 269, 13544-13550.

184. Stallings-Mann, M.L., Burke, M.G., Trout, W.E. and Roberts, R.M. (1994) Purification, characterization, and cDNA cloning of a Kunitz-type proteinase inhibitor secreted by the porcine uterus. *J. Biol. Chem.* 269, 24090-24094.

185. Li, J. and Roberts, R.M. (1994) Structure-function relationships in the interferon-τ (IFN-τ): Changes in receptor binding and in antiviral and antiproliferative activities resulting from site-directed mutagenesis performed near the carboxyl terminus. *J. Biol. Chem.* 269, 24826-24833.

186. Xie, S., Low, B.G., Nagel, R.J., Beckers, J-F. and Roberts, R.M. (1994) A novel glycoprotein of the aspartic proteinase gene family expressed in bovine placental trophectoderm. *Biol. Reprod.* 51, 1145-1153.
187. Szafranska, B., Xie, X., Green, J. and Roberts, R.M. (1995) Porcine pregnancy-associated glycoproteins: new members of the aspartic proteinase gene family expressed in trophectoderm. *Biol. Reprod.* 53, 21-28.
188. Kubisch, H.M., Larson, M.A., Funahashi, H., Day, B.N. and Roberts, R.M. (1995) Pronuclear visibility, development and transgene expression in IVM-IVF-derived porcine embryos. *Theriogenol.* 44, 391-401.
189. Kubisch, H.M., Hernandez-Ledezma, J.J., Larson, M.A., Sikes, J.D. and Roberts, R.M. (1995) Expression of two transgenes in *in vitro* matured and fertilized bovine zygotes after DNA microinjection. *J. Reprod. Fertil.* 104, 133-139.
190. Xie, S., Green, J., Beckers, J.F. and Roberts, R.M. (1995) The gene encoding bovine pregnancy-associated glycoprotein-1, an inactive member of the aspartic proteinase family. *Gene* 159, 193-197.
191. Li, J., Alexenko, A.P. and Roberts, R.M. (1995) A bifunctional vector suitable for both site-directed mutagenesis and recombinant expression of interferon-τ in *Escherichia coli. Prot. Express. Purif.* 6, 401-407.
192. Meyer, M.D., Hansen, P.J., Thatcher, W.W., Drost, M. and Roberts, R.M. (1995) Effect of bovine interferon-τ on body temperature and plasma progesterone concentrations in cyclic dairy cows. *J. Dairy Sci.* 78, 1470-1476.
193. Meyer, M.D., Hansen, P.J., Thatcher, W.W., Drost, M., Badinga, L., Roberts, R.M., Li, J., Ott, T.L. and Bazer, F.W. (1995) Extension of corpus luteum lifespan and reduction of uterine secretion of prostaglandin $F_{2\alpha}$ of cows in response to recombinant interferon-τ. *J. Dairy Sci.* 78, 1921-1931.
194. Senda, T., Saitoh, S-I., Mitsui, Y., Li, J. and Roberts, R.M. (1995) A three-dimensional model of interferon-τ. *J. Interferon & Cytokine Res.* 15, 1053-1060.
195. Smith, G.W., Gentry, P.C., Roberts, R.M. and Smith, M.F. (1996) Ontogeny and regulation of luteinizing hormone receptor messenger ribonucleic acid within the ovine corpus luteum. *Biol. Reprod.* 54, 76-83.
196. Xie, S., Nagel, R.J., Green, J., Beckers, J-F. and Roberts, R.M. (1996) Trophoblast-specific processing and phosphorylation of pregnancy-associated glycoprotein-1 in day 15 to 25 sheep placenta. *Biol. Reprod.* 54, 122-129.
197. Smith, G.W., Gentry, P.C., Long, D.K., Smith, M.F. and Roberts, R.M. (1996) Differential gene expression within the ovine corpus luteum: identification of secreted protein acidic and rich in cysteine as a major secretory product of small but not large luteal cells. *Endocrinol.* 137, 755-762.
198. Liu, L., Leaman, D.W., Bixby, J.A. and Roberts, R.M. (1996) A type I ovine interferon with limited similarity to IFN-α, IFN-ω and IFN-τ: gene structure,

biological properties and unusual species specificity. *Biochim. Biophys. Acta* 1294, 55-62.

199. Liu, L. and Roberts, R.M. (1996) Silencing of the gene for the β subunit of human chorionic gonadotropin by the embryonic transcription factor Oct-3/4. *J. Biol. Chem.* 271, 16683-16689.
200. Liu, L., Leaman, D.W. and Roberts, R.M. (1996) The interferon-τ genes of the giraffe, a nonbovid species. *J. Interferon & Cytokine Res.* 16, 949-951.
201. 199a. Roberts, R.M. (1996) Interferon-t and Pregnancy. J interferon & cytokine Res 16, 271-273.
202. Guruprasad, K., Blundell, T.L., Xie, S., Green, J., Szafranska, B., Nagel, R.J., McDowell, K., Baker, B.B. and Roberts, R.M. (1996) Comparative modeling and analysis of amino substitutions suggests that the family of pregnancy-associated glycoproteins includes both active and inactive aspartic proteinases. *Protein Eng.* 9, 849-856.
203. Kubisch, H.M., Larson, M.A., Eichen, P.A., Wilson, J.M. and Roberts, R.M. (1997) Adenovirus-mediated gene transfer by perivitelline microinjection of mouse, rat, and cow embryos. *Biol. Reprod.* 56, 119-124.
204. Niswender, K.D., Li, J., Powell, M.R., Loos, K.R., Roberts, R.M., Keisler, D.H. and Smith, M.F. (1997) Effects of variants of interferon-τ with mutations near the carboxyl terminus on luteal life span in sheep. *Biol. Reprod.* 56, 214-220.
205. Duffy, J.Y., Stallings-Mann, M.L., Trout, W.E., Roberts, R.M. (1997) Expression of a Plasmin/Trypsin Kunitz Inhibitor by Pig Trophoblast. *Mol. Reprod. Dev.* 46, 443-449.
206. Smith, G.W., Gentry, P.C., Bao, B., Long, D.K., Roberts, R.M., and Smith, M.F. (1997) Control of extracellular matrix remodeling within ovarian tissues: localization and regulation of gene expression of plasminogen activator inhibitor type-1 within the ovine corpus luteum. *J. Reprod. & Fert.* 110, 107-114.
207. Liu, L., Leaman, D., Villalta, M., Roberts, R.M. (1997) Silencing of the Gene for the α-Subunit of Human Chorionic Gonadotropin by the Embryonic Transcription Factor Oct-3/4. *Mol. Endocrinol.* 11, 1651-1658.
208. Han, C.-S., Mathialagan, N., Klemann, S., and Roberts, R.M. (1997) Molecular Cloning of Ovine and Bovine Type I Interferon Receptor Subunits from Uteri, and Endometrial Expression of Messenger Ribonucleic Acid for Ovine Receptors During the Estrous Cycle and Pregnancy. *Endocrinol.* 138, 4757-4767.
209. Xie, S., Green, J., Bao, B., Beckers, J-F, Valdez, K., Hakami, L., Roberts, R.M. (1997) Multiple Pregnancy-Associated Glycoproteins Are Secreted by Day 100 Ovine Placental Tissue. *Biol. Reprod.* 57, 1384-1393.

210. Xie, Sancai. Green. J., Bixby, J.B., Szafranska, B., DeMartini, J.C., Hecht, S., Roberts, R.M. (1997) The diversity and evolutionary relationships of the pregnancy-associated glycoproteins, an aspartic proteinase subfamily consisting of many trophoblast-expressed genes. *Proc. Natl. Acad. Sci. USA* 94, 12809-12816.

211. Alexenko, A.P., Leaman, D.W., Li, J., and Roberts, R.M. (1997) The antiproliferative and antiviral properties of IFN-τ variants in human cells. *J. Interfer. Cytokine Res.* 17, 769-779.

212. Ealy, A. D., Green, A. G., Alexenko, A., Keisler, D., and Roberts, R.M. (1998) Different ovine interferon-tau genes are not expressed identically and their protein products display different activities. *Biol. Reprod.* 58, 566-573.

213. Kubisch, H. Michael, Larson, M. and Roberts, R.M. (1998) Relationship between age of blastocyst formation and interferon-τ secretion by in vitro-derived bovine embryos. *Mol. Reprod. Dev.* 49, 254-260.

214. Ealy, A.D., Alexenko, A.P., Keisler, D.H., and Roberts, R.M. (1998) Loss of the signature six carboxyl amino acid tail from ovine interferon-tau does not affect biological activity. *Biol. Reprod.* 58, 1463-1468.

215. Ezashi, T., Ealy, A.D., Ostrowski, M.C., and Roberts, R.M. (1998) Control of interferon-τ gene expression by Ets-2. *Proc. Natl. Acad. Sci. USA.* 95, 7882-7887.

216. Roberts, R.M., Liu, L., Guo, Q., Leaman, D., Bixby, J. (1998) The Evolution of the Type I Interferons. *J. Interferon & Cytokine Res.* 18, 805-816.

217. Green, J., Xie, S., Szafranska., B, Gan, X., Newman, A., McDowell, K., and Roberts, R.M. (1999) Identification of a New Aspartic Proteinase expressed by the outer chorionic cell layer of the equine placenta. *Biol. Reprod.* 60, 1069-1077.

218. Roberts, R.M., Ealy, A.D., Alexenko, A.P., Han, C.-S., and Ezashi, T. (1999) Trophoblast Interferons. *Placenta* 20, 259-264.

219. Winkelman, G.L., Roberts, R.M., Peterson, A.J., Alexenko, A.P., and Ealy, A.D. (1999) Identification of the Expressed Forms of Ovine Interferon-Tau in the Periimplantation conceptus: Sequence Relationships and Comparative Biological Activities. *Biol. Reprod.* 61, 1592-1600.

220. Alexenko, A.P., Ealy, A.D., and Roberts, R.M. (1999) The Cross-Species Antiviral Activities of Different IFN-τ Subtypes on Bovine, Murine and Human Cells: Contradictory Evidence for Therapeutic Potential. *J. Interferon & Cytokine Res.* 19, 1335-1341.

221. Green, J.A., Xie, S., Quan, X., Bao, B., Gan, X., Mathialagan, N., Beckers, J.-F., and Roberts, R.M. (2000) Pregnancy-Associated Bovine and Ovine Glycoproteins Exhibit Spatially and Temporally Distinct Expression Patterns During Pregnancy. *Biol. Reprod.* 62, 1624-1631.

222. Hughes, A.L., Green, J.A., Garbayo J.M., and Roberts, R.M. (2000) Adaptive diversification within a large family of recently duplicated, placentally expressed genes. *Proc. Natl. Acad. Sci. USA.* 97, 3319-3323.
223. Hughes, A.L., and Roberts R.M. (2000) Independent origin of IFN-α and IFN-β in birds and mammals. *J. Interferon Cytokine Res.* 20, 737-740.
224. Garbayo, J.A., Green, J.A., Manikkam, M., Beckers, J.F., Kiesling, D.O., Ealy, A.D., Roberts, R.M. (2000) Caprine pregnancy-associated glycoproteins (PAG): Their cloning, expression and evolutionary relationship to other PAG. *Mol. Reprod. Dev.* 57, 311-322.
225. Alexenko, A.P., Ealy, A.D., Bixby, J.A., & Roberts, R.M. (2000) A classification for the interferon-τ. *J. Interferon & Cytokine Res.* 20, 817-822.
226. Do, H.J., Kim, J.H., Abeydeera, L.R., Han, Y.M., Green, J., Roberts, R.M., Matteri, R.L., Day, B.N., and Prather, R.S. (2001) Expression of pregnancy-associated glycoprotein 1 and 2 genes in in vitro-, in vivo- and parthenogenetically-derived preimplantation pig embryos. *Zygote* 9, 245-250. *PMID 11508744*
227. Kubisch, H.M., Larson, M.A., Ealy, A.D., Murphy, C.N., Roberts R.M. (2001) Genetic and environmental determinants of interferon-τ secretion by in vivo- and in vitro-derived bovine blastocysts. *Anim. Reprod. Sci.* 66, 1-13. *PMID 11343838*
228. Kubisch, H.M., Larson, M.A., Kiesling, D.O., Roberts, R.M. (2001) Control of interferon-τ secretion by in vitro-derived bovine blastocysts during extended culture and outgrowth formation. *Mol. Reprod. Dev.* 58, 390-397.
229. Han C. -S., Chen Y., Ezashi, T., Roberts R.M. (2001) Antiviral activities of the soluble extracellular domains of Type 1 Interferon Receptors. *Proc. Natl. Acad. Sci. USA.* 98, 6138-6143. *PMID 11344274*
230. Ealy, A.D., Larson S.F., Liu L., Alexenko A.P., Winkleman G.L., Kubisch H.M., Bixby J.A., Roberts R.M. (2001) Polymorphic forms of expressed bovine interferon-tau genes: Relative transcript abundance during early placental development, promoter sequences of genes and biological activity of protein products. *Endocrinol.* 142, 2906-2915. *PMID 11416010*
231. Chen, X., Rosenfeld C.S., Roberts R.M., Green J.A. (2001) An Aspartic Proteinase expressed in the Yolk Sac and neonatal stomach of the mouse. *Biol. Reprod.* 65, 1092-1101. *PMID 11566730*
232. Szafranska, B., Miura R., Ghosh D., Ezashi T., Xie S., Roberts R.M., Green J.A. (2001) The Gene for Porcine Pregnancy-associated glycoprotein 2 (poPAG2): Its structural Organization and Analysis of its Promoter. *Mol. Reprod. Dev.* 60, 137-146. *PMID 11553911*
233. Larson, M.A., Kimura K., Kubisch H.M., Roberts R.M. (2001) Sexual dimorphism among bovine embryos in their ability to make the transition to expanded blastocyst and in the expression of the signaling molecule IFN-τ. *Proc. Natl. Acad. Sci. USA.* 98, 9677-9682. *PMID 11481449*

234. Ezashi T., Ghosh D., Roberts R.M. (2001) Repression of Ets-2 induced transactivation of the interferon-tau promoter by Oct-4. *Mol. Cell. Biol.* 21, 7883-7891. *PMID 11689681*

235. Rosenfeld, C.S., Han C.S., Alexenko A.P., Spencer T.E., Roberts R.M. (2002) The Expression of interferon receptor subunits, IFNAR1 and IFNAR2, in the Ovine Uterus. *Biol Reprod.* 6, 847-853. *PMID 12193393*

236. Ghosh, D., Ezashi T., Ostrowski M.C., Roberts R.M. (2003) A central role for Ets-2 in the transcriptional regulation and cyclic AMP-responsiveness of the human chorionic gonadotropin-beta subunit gene. Mol. Endocrinol, 17, 11-26. *PMID 12511603*

237. MacLean II, J.A., Chakrabarty, A., Xie, S., Bixby, J.A., Roberts, R.M. and Green J.A. (2003) A family of Kunitz proteins from trophoblast: Expression of the trophoblast Kunitz domain proteins (TKDP) in cattle and sheep. Mol. Reprod. Dev. 65, 30-40. *PMID 12658631*

238. Roberts, R.M., Ezashi, T., Rosenfeld, C.S., Ealy, A.D., and Kubisch, H.M. (2003) Evolution of the Interferon-τ genes and their promoters, and maternal-trophoblast interactions in control of their expression. Reproduction, Suppl 61, 239-251.*PMID 14635939*

239. Rosenfeld, C.S., Grimm K., Livingston K., Brokman A., Lamberson W.E., Roberts R.M. (2003) Striking variation in the sex ratio of pups born to mice according to whether maternal diet is high in fat or carbohydrate. Proc. Natl. Acad. Sci. USA 100, 4628-4632. *PMID 12672968*

240. Hughes, A.L., Green J.A., Piontkivska H., Roberts R.M. (2003) Aspartic proteinase phylogeny and the origin of pregnancy-associated glycoproteins. Mol. Biol. Evol. 20, 1940-45. *PMID 12949149*

241. Green, J.A, Parks, T.E., Avalle, P.A., Teluga, B.P., McLain, A.L., Peterson, A.J., McMillan W., Mathialagan, N., Hook, R.R., Xie, S., Roberts, R.M. (2005) The establishment of an ELISA for the detection of preganancy-associated glycoproteins (PAG'S) in the serum of pregnant cow and heifers. *Theriogenology.* 63, 1481-503. *PMID 15725453*

242. Kimura, K., Spate, L.D. Green, M.P.and Roberts, R.M. (2004) Effects of Oxidative Stress and Inhibitors of the Pentose Phosphate Pathway on Sexually Dimorphic Production of IFN-τ by Bovine Blastocysts. Mol. Reprod. Dev. 68, 88-95. *PMID 15039952*

243. Kimura, K., Spate, L.D., Green, M.P., Murphy, C.N., Seidel Jr , G.E., Roberts, R.M. (2004) Sexual Dimorphism in Interferon-tau Production by *In vivo*-Derived Bovine Embryos. Mol. Reprod. Dev. 67, 193-199. *PMID 14694435*

244. Roberts, R.M, Ezashi, T., Das, P. (2004) Trophoblast Gene Expression: transcription factors in the specification of early trophoblast. Reprod. Biol. & Endocrinology. 2, 47. *PMID 15236655*

245. Ezashi, T., Roberts, R.M. (2004) Regulation of Interferon-t (IFN-t) Gene Promoters by Growth Factors that
246. Target the Ets-2 Composite Enhancer: A Possible Model for Maternal Control of IFN-t Production by the Conceptus during Early Pregnancy. *Endocrinology* 145, 4452-60. *PMID 15217985*
247. Maclean II, J.A., Roberts, R.M. and Green, J.A. (2004) Atypical Kunitz-Type Serine Proteinase Inhibitors Produced by the Ruminant Placenta. *Biology of Reproduction* 71, 455-463. *PMID 15070828*
248. Wang, S. and Roberts, R.M. (2004) Interaction of Stress-Activated Protein Kinase-Interacting Protein-1 with the Interferon Receptor Subunit IFNAR2 in Uterine Endometrium. *Endocrinology* 145, 5820-5831. *PMID 155345682*
249. Roberts RM. (2004). Impact of Maternal Diet on Reproductive Outcome: Foreword. *Biol Reprod.* 71, 1045.
250. Rosenfeld, C. S. and Roberts, R.M. (2004). "Maternal Diet and Other Factors Affecting Offspring Sex Ratio: A Review." *Biol Reprod* 71, 1063-1070. *PMID 15229140*
251. Wang, S. and Roberts, R.M. (2005) The evolution of the Sin1 gene product, a little known protein implicated in stress responses and type I interferon signaling in vertebrates. *BMC Evol Biol.* 7;5(1):13 *PMID 15698473*
252. Ghosh D, Sachdev, S., Hannink M, Roberts, R.M. (2005) Coordinate regulation of basal and cAMP-activated expression of hCG{alpha} by Ets-2 and CREB. *Mol Endocrinol.* 1049-1066. *PMID 15637148*
253. Wooding, F.B.P., Roberts, R.M. and Green, J.A. (2005) Light and electron microscope immunocytochemical studies of the distribution of pregnancy associated glycoproteins (PAGs) throughout pregnancy in the cow: possible functional implications. *Placenta.* Nov. 26(10):807-27. *PMID 16226131*
254. Ezashi T., Das P., Roberts R.M. (2005) Low O2 tensions and the prevention of differentiation of human embryonic stem cells. *PNAS* 102, 4783-4788. *PMID 15772165*
255. Roberts, R.M. (2005) Embryo culture conditions: What embryos like best. *Endocrinology* 146, 2140-2141. *PMID 15833923*
256. Green, M.P., L.D. Spate, Bixby, J.A.. Ealy, A.D Roberts, R.M. (2005) A Comparison of the Anti-Luteolytic Activities of Recombinant Ovine Interferon-Alpha and –Tau in Sheep. *Biol Reprod* 73, 1087-1093 *PMID 16079305*
257. Kimura, K, Spate, L.D., Green, M.P., Roberts, R.M. (2005) Effects of D-glucose concentration, D-fructose, and inhibitors of enzymes of the pentose phosphate pathway on the development and sex ration of bovine blastocysts. *Molecular Reproduction and Development.* 72:2, 201-207. *PMID 15968626*
258. Chen Y Green J.A. Antoniou, E., Ealy A.E. Mathialagan N.P., Walker A.M., Avalle M.P., Rosenfeld C.S., Hearne, L., Roberts R.M. (2006) Effect of Interferon tau Administration on Endometrium of Nonpregnant Ewes: A Comparison with Pregnant Ewes. Endocrinology, 147:2127-2137. *PMID 16469802*

259. Chakrabarty, A, Green J.A., Roberts R.M. (2006) Origin and evolution of the TKDP gene family. *GENE*, 373. 35-43 *PMID 16567063*

260. Chakrabarty A., MacLean J, Roberts R.M. Hughes, A Green, J.A. (2006) Rapid Evolution of the trophoblast Kunitz domain proteins (TKDPs) – A multi-gene family expressed in the trophoblast of ruminants. J. Mol. Evol. 63: 274-282 *PMID 16830095*

261. Whyte, J J.,Roberts R.M., Rosenfeld, C.S. (2007) Fluorescent in situ hybridization for sex chromosome determination before and after fertilization in mice. Theriogenology, 67:1022-1031. *PMID 17215034*

262. Roberts, R.M., Yong, H.J., Smith, S. (2006) What Drives the Formation of Trophectoderm During Early Embryonic Development? J of Reprod & Dev, 52, S87-S97. *[link to article]*

263. Ezashi, T., Das. P., Roberts R.M. (2006) Regulation of IFN-tau gene expression by uterine factors: interaction between the conceptus and maternal environment during early pregnancy in cattle and sheep. J of Reprod & Dev, 52, S99-S109. *[link to article]*

264. Green, JA, Roberts RM. (2006) Establishment of an ELISA for the detection of native bovine pregnancy-associated glycoprotein secreted by trophoblast binucleate cells. Methods Mol Med. 122: 321-30. *PMID 16511990*

265. Chen, Y., Antoniou, E., Liu, Z., Hearne, L., Roberts, RM (2007) A Microarray analysis for genes regulated by interferon-tau in Ovine luminal epithelial cells, Reproduction 134. 123-135. *PMID 17641094*

266. Chakrabarty, A. and Roberts R.M. (2007) Ets-2 and C/EBP-beta are important mediators of ovine trophoblast Kunitz domain protein-1 gene expression in trophoblast. BMC Molecular Biology. 8:14 *PMID 17326832*

267. Alexenko, A., Mao, J., Ellersieck, M., Davis, A., Whyte, J.J., Rosenfeld, R.S., Roberts R.M. (2007) The contrasting effects of ad libitum and restricted feeding of a diet very high in saturated fats on sex ratio and metabolic hormones in mice. Biol Reprod. 77(4): 599-604. *PMID 17522073*

268. Roberts, RM (2007) Interferon-Tau, a Type 1 Interferon involved in maternal recognition of pregnancy. Cytokine and Growth Factors Review. 18 (5-6) 403-8 *PMID 17662642*

269. Fountain, E.D., Mao, J., Whyte, J.J., Mueller, K.E., Ellersieck, M.R., Will, M.R., Roberts, R.M., MacDonald, R., Rosenfeld, C.S. (2008) Effects of diets enriched in omega-3 and omega-6 polyunsaturated fatty acids on offspring sex-ratio and maternal behavior in mice. Biol of Reprod. 78, 211-217. *PMID 17928632*

270. Das P, Ezashi, T., Gupta, R, Roberts RM. (2008) Combinatorial roles of protein kinase A, Ets2 and CBP/p300 in the transcriptional control of interferon-tau expression in a trophoblast cell line. Mol. Endocrinol. 22:331-43 *PMID 17975022*

271. Schulz, L., Ezashi, T., Das, P., Westfall, SD, Livingston, KA Roberts RM. (2008) Human embryonic stem cells as models for trophoblast differentiation. Placenta 29 Suppl A, Trophoblast Research. 22:S10-S16 *Science Direct*
272. Roberts, RM, Chen, Y., Ezashi, T & Walker A. (2008) Interferons and the Maternal-Conceptus Dialog in Mammals Seminars in Cell and Developmental Biology Semin Cell Dev Biol. 19:170-7. *PMID 198032074*
273. Das, P., Ezashi, T., Schulz, LC., Westfall, SD., Livingston, KA., Roberts, RM. (2008) Effects of FGF2 and Oxygen in BMP4-Driven differentiation of trophoblast from human embryonic stem cells. Stem Cell Research. 1:1 61-74. *Science Direct 18735061*
274. Ezashi, T., Das, P., Gupta R., Walker, A., Roberts, R.M. (2008) The Role of Homeobox Protein Distal-less3 and its interaction with Ets2 in Regulating Interferon-Tau Gene Expression-Synergistic transcriptional activation with Ets2. Biol Reprod. March 5, 2008 DOI:10.1095
275. Westfall, SD., Sachdev, S., Das, P., Hearne, L., Hannink, M., Roberts, RM., Ezashi, T. Identification of Oxygen Sensitive Transcriptional Programs in Human Embryonic Stem Cell. Stem Cell & Development (accepted for Publication 1/08)
276. Ghosh, D, Srivastava, G, Xu, D, Roberts, RM. A link between SIN1 (MAPKAP1) and Poly(rC) Binding protein-2 (PCBP2) in counteracting environmental stress. (accepted for Publication 2/08)

277. CHAPTERS IN BOOKS AND INVITED REVIEWS
278. Roberts, R.M. and Cetorelli, J.J. (1973) UDP-D-glucuronic Acid Pyrophosphorylase and the Formation of UDP-D-glucuronic Acid in Plants. In: *Biogenesis of Plant Cell Wall Polysaccharides* (F. Loewus, ed.). Academic Press, New York, p. 49.
279. Noonan, K.D., VanVeen, J. and Roberts, R.M. (1977) The Role of the Plasma Membrane and Intracellular Microskeletal Elements in Determining the Lectin Agglutinin-ability of the two CHO sub-clones. In: *Cell Shape and Surface Architecture* (Revel, J.P, Hennig, U. and Fox, C.F. eds.). Alan R. Liss, Inc., New York. pp 521-530.
280. Bazer, F.W., Sharp, D.C. and Roberts, R.M. (1978) Collection and Analysis of Female Genital Tract Secretions. In: *Methods in Mammalian Embryology.* (Daniel, J.C. ed.). Academic Press. pp. 503-528.
281. Bazer, F.W. Roberts, R.M. and Thatcher, W.W. (1979) Action of Hormones on the Uterus and Effect on Conceptus Development. XIII Biennial Symposium on Animal Reproduction. *J. Anim. Sci. 49, Supplement II*, pp. 35-45.
282. Roberts, R.M. and Bazer, F.W. (1980) The Properties, Function and Hormonal Control of Uteroferrin, the Purple Protein of the Pig Uterus. In: *Steroid Induced Proteins* (Beato, M., ed.). Elsevier-North Holland., pp. 133-149.

283. Horst, M.N. and Roberts, R.M. (1979) Characterization of Plasma Membrane Polypeptides from Chinese Hamster Ovary Cells by Two-dimensional Electrophoresis. In: *Isofocus 78* (Haglund, Westerfield and Ball, eds.). Elsevier-North Holland, pp. 99-118.

284. Bazer, F.W., Sharp, D., Thatcher, W.W. and Roberts, R.M. (1980) Comparative Approach to Mechanism in the Maintenance of Early Pregnancy. In: *Reproductive Processes and Contraception.* (McKerns, K.W. ed.). Plenum Press, N.Y. pp 581-617.

285. Mullins, D.E., Bazer, F.W. and Roberts, R.M. (1980) The Pig Uterus Secretes a Progesterone-induced Inhibitor of Plasminogen Activator. In: *Cellular and Molecular Aspects of Implantation* (Glasser, S.R. and Bullock, D.W. eds.). Academic Press, NY

286. Stein, G.S., Roberts, R.M., Stein, J.L. and Davis, J.L. (1981) Nuclear Glycoproteins and Glycosaminoglycans. In: *The Cell Nucleus, IX Nuclear Particles, Part B* (Busch, H. ed.). pp. 341-357.

287. Denny, J.B. and Roberts, R.M. (1981) Turnover of Plasma Membrane Polypeptides in Transformed and Normal Cells Using $^{125}I/^{131}I$ Double Label Techniques. In: *Cancer Cell Organelles: Methodological Surveys in Biochemistry, Vol. 10* (Reid, E. and Morre, D.J. eds.). Ellis-Horwood Lts. Chichester. pp. 388-395.

288. Baumbach, G.A., Horst, M.N., Olympio, M.A., Aldrich, H.C. and Roberts, R.M. (1981) Isolation and Characterization of Plasma Membranes from Cultured Cells. In: *Cancer Cell Organelles: Methodological Surveys in Biochemistry, Vol. 10* (Reid, E. and Morre, D.J. eds.). Ellis-Horwood Ltd., Chichester, pp. 351-370.

289. Bazer, F.W., Roberts, R.M., Sharp, D.C. and Thatcher, W.W. (1981) Uterine Proteins Synthesized During the Progestative Period and Pregnancy. In: *Uterus et Fecondite.* (Boury-Heyler, C.I., Mauleon, P. and Rochet, Y., eds.). Masson Publishing Co., Paris, pp 17-32.

290. Bazer, F.W., Geisert, R.D., Thatcher, W.W. and Roberts, R.M. (1982) The Establishment and Maintenance of Pregnancy. In: *Control of Reproduction in Pig.* (Cole, D.I.A. and Foxcroft, G.R. eds.). Butterworth Co., London, pp. 227-252.

291. Roberts, R.M., Baumbach, G.A., Buhi, W.C., Denny, J.B., Fitzgerald, L.A., Babelyn, S.F. and Horst, M.N. (1983) Analysis of Membrane Polypeptides by Two-dimensional Polyacrylamide Gel Electrophoresis. In: *Receptor Biochemistry and Methodology. Vol. III. Molecular and Chemical Characterization of Membrane Receptors.* (Ventner, J.C. and Harrison, L. eds Alan Liss, Co. NY, .) pp. 61-113.

292. Bazer, F.W. and Roberts, R.M. (1983) Biochemical Aspects of Conceptus-Endometrial Interactions. In: *New Frontiers in Mammalian Reproduction and Development.* (Stone, W.H. ed.). *J. Exp. Zool.* 228, 373-383.

293. Roberts, R.M., Bazer, F.W. and Thatcher, W.W. (1984) Biochemical Interactions Between Blastocyst and Endometrium in the Large Domestic Animals. J. Biosciences, 6 (Suppl. 2), 63-74. (Indian Academy of Sciences). *Proceedings of a Symposium on The Blastocyst* held at the Indian Institute of Sciences, Bangalore, India, Dec. 1983.
294. Roberts, R.M. and Bazer, F.W. (1984) Uteroferrin, a Protein in Search of a Function. *Bioessays. Vol. 1*, pp. 8-11.
295. Thatcher, W.W., Bartol, F.F., Knickerbocker, J.J., Curl, J.S., Wolfenson, D., Bazer, F.W. and Roberts, R.M. (1984) Maternal recognition of pregnancy in cattle. *J. Dairy Sci.* 67, 2797-2811.
296. Thatcher, W.W., Knickerbocker, J.J., Bartol, F.F., Bazer, F.W., Roberts, R.M. and Drost, M. (1984) Maternal recognition of pregnancy in relation to the survival of transferred embryos: Endocrine aspects. *Theriogenol.* 23, 129-143.
297. Roberts, R.M. Godkin, J.D., Bazer, F.W., Fincher, K.B., Thatcher, W.W., Knickerbocker, J.J. and Bartol, F.F. (1985) Antiluteolysins produced by mammalian conceptuses. In: *Implantation of the Human Embryo* (R.G. Edwards, J. Purdy and S. Steptoe eds.). Academic Press, NY pp. 253-282.
298. Bazer, F.W., Roberts, R.M., Thatcher, W.W. and Sharp, D.C. (1985) Mechanisms Related to Establishment of Pregnancy. In: *Early Pregnancy Factors.* (Ellendorff, F. and Koch, E. eds.) Perinatology Press. pp. 13-23.
299. Thatcher, W.W., Knickerbocker, J.J., Helmer, S.D., Hansen, P.J., Bartol, F.F., Roberts, R.M. and Bazer, F.W. (1985) Characteristics of Conceptus Secretory Proteins and Their Effect on $PGF_2\alpha$ Secretion by the Maternal Endometrium in Cattle. In: *Early Pregnancy Factors* (Ellendorff, F. and Koch, E. eds.) Perinatology Press. pp. 25-27.
300. Thatcher, W.W., Bazer, F.W., Sharp, D.C. and Roberts, R.M. (1986) Interrelationships Between Uterus and Conceptus to Maintain Corpus Luteum Function in Early Pregnancy: Sheep, Cattle, Pigs and Horses. In: *Uterine-Conceptus Control of Corpus Luteum, J. Anim. Sci.* 62 , 25-46.
301. Roberts, R.M., Raub, T.J. and Bazer, F.W. (1986) Role of uteroferrin in transplacental iron transport in the pig. *Fed. Proc. of the Federation of American Societies for Experimental Biology* 45, 2513-2518.
302. Roberts, R.M. and Bazer, F.W. (1988) The function of uterine secretions. *J. Reprod. Fertil.* 82, 875-892.
303. Roberts, R.M., Murray, M.K., Burke, M.G., Ketcham, C.M. and Bazer, F.W. (1987) Hormonal Control and Function of Secretory Proteins. In: *Cell and Molecular Biology of the Uterus* (W. W. Leavitt, ed.) Plenum Publishing Corporation, pp.137-150.
304. Bazer, F.W., Vallet, J.L., Ashworth, C.J., Anthony, R.V. and Roberts, R.M. (1987) The role of ovine conceptus secretory proteins in the establishment of pregnancy. In: *Cell and Molecular Biology of the Uterus* (Wendell W. Leavitt, ed.) Plenum Publishing Corporation, pp 221-234.

305. Roberts, R.M., Kronenberg, L.A., Malathy, P.V. and Imakawa, K. (1988) Embryonic interferons and maternal recognition of pregnancy. In: *The Biology of the Interferon System 1988* (Y. Kawade and S. Kobayashi, eds.), Kodansha Scientific Press, Ltd., Tokyo, pp 315-321.
306. Imakawa, K. and Roberts R.M. (1989) Interferons and maternal recognition of pregnancy. In: *Development of Preimplantation Embryos and Their Environment* (K. Yoshinaga and T. Mori, eds.), Alan R. Liss, Inc., New York NY, pp 347-358.
307. Roberts, R.M., Farin, C.E. and Imakawa, K. (1989) Embryonic mediators of maternal recognition of pregnancy. In: *Blastocyst Implantation* (K. Yoshinaga ed.), Serono Symposia USA, Adams Publishing Group, Ltd., Boston, MA. pp 25-30.
308. Roberts, R.M., Cross, J.C., Farin, C.E., Hansen, T.R., Klemann, S.W. and Imakawa, K. (1990) Interferons at the placental interface. In: *Proc. 2nd Symp. on the Genetic Engineering of Animals* (W. Hansel and B.J. Weir, eds.) *J. Reprod. Fertil.* 41, 63-74.
309. Roberts, R.M., Farin, C.E., Cross, J.C. (1990) Trophoblast proteins and maternal recognition of pregnancy. *Oxford Reviews of Reproductive Biology* (S.R. Milligan, ed.), Oxford Univ. Press, Oxford. Vol. 12, pp 147-180.
310. Roberts, R.M., Cross, J.C., Farin, C.E., Kramer, K., Schalue-Francis, F. and Hansen, T.R. (1990) Trophoblast interferons as Antiluteolysins. In: *Establishing a Successful Human Pregnancy* (R.G. Edwards, ed.), Serono Symposia Publications, Raven Press, New York. Vol. 66, pp 257-268.
311. Roberts, R.M. (1991) A role for interferons in early pregnancy. *Bioessays.* 13, pp 121-126.
312. Roberts, R.M., Klemann, S.W., Leaman, D.W., Bixby, J.A., Cross, J.C., Farin, C.E., Imakawa, K. and Hansen, T.R. (1991) The polypeptides and genes for ovine and bovine trophoblast protein-1. *J. Reprod. Fert, Suppl. 43*, pp 3-12.
313. Roberts, R.M., Cross, J.C., Farin, C.E., Farin, P.W., Kramer, K.K., Francis, H., Librach, C. and Fisher, S.J. (1991) Trophoblast Interferons. In: *Growth Factors in Reproduction*, David W. Schomberg (ed.), Serono Symposia, USA, Springer-Verlag, New York, pp 235-245.
314. Roberts, R.M. (1991) Embryonic loss and conceptus interferon production. In: J.F. Strauss III and C.R. Lyttle (Eds.), *Uterine and Embryonic Factors in Early Pregnancy*, Plenum Press, New York, pp 21-31.
315. Roberts, R.M., Cross, J.C. and Leaman, D.W. (1991) Unique features of the trophoblast interferons. *Pharmac. Ther.* 51, 329-345.
316. Roberts, R.M., Anthony, R.V., Thatcher, W.W., Hansen, P.J. and Bazer, F.W. (1991) Conceptus proteins as Antiluteolysins. In: *Development of Pre-Implantation Embryos* (G.S. Toteja, R.S. Sharma, B.K. Singh, S. Mokkapati, B.N. Saxena, eds.), Indian Council of Medical Research, New Delhi, pp 17-27.

317. Roberts, R.M., Leaman, D.W. and Cross, J.C. (1992) Minireview: Role of interferons in maternal recognition of pregnancy in ruminants. *Proc. Soc. Exp. Biol. Med.* 200, 7-18.
318. Roberts, R.M., Cross, J.C. and Leaman, D.W. (1992) Interferons as Hormones of Pregnancy. *Endocrine Rev.* 13, 432-452.
319. Roberts, R.M., Trout, W.E., Mathialagan, N., Stallings-Mann, M. and Ling, P. (1993) Uterine secretory activity and embryo Development. In: *Preimplantation Embryo Development* (B.D. Bavister, ed.), Serono Symposia, Springer-Verlag, New York, Chap. 17, pp 229-243.
320. Roberts, R.M., Leaman, D.W., Hernandez-Ledezma, J.J. and Cosby, N.C. (1993) Trophoblast interferons: expression during development and gene organization. In: M.J. Soares, S. Handwerger, and F. Talamantes (Eds), *Trophoblast Cells: Pathways for Maternal-embryonic Communication*, Springer-Verlag, New York, NY, pp 206-221.
321. Roberts, R.M., Xie, S. and Trout, W.E. (1993) Embryo-uterine interactions in pigs during week 2 of pregnancy. In: G.R. Foxcroft, M.G. Hunter, C. Doberska (eds.), *Control of Pig Reproduction IV*, Proceedings of the Forth International Conference on Pig Reproduction, Univ. of Missouri-Columbia, May 1993. *J. Reprod. Fertil.* 48, 171-186.
322. Roberts, R.M. and Anthony, R.V. (1994) Molecular biology of trophectoderm and placental hormones. In: J.K. Findlay (ed.), *Molecular Biology of the Female Reproductive System*, Chapter 14. Academic Press, San Diego CA, pp 395-440.
323. Roberts, R.M., Kramer, K.K., Xie, S., Duffy, J., Nagel, R.J. and Wooding, F.B.P. (1994) Identification of secretory proteins released by preimplantation embryos of the sheep. In: L. Mastroianni, Jr., H.J.T. Coelingh Bennink, S. Suzuki and H.M. Vener (eds.), *Gamete and Embryo Quality*, Proceedings of the Ivth Organon, Round Table Conference, Thessaloniki, Greece, 24-25 June, 1993. The Parthenon Publishing Group, Pearl River, NY, Chapter 14, pp 209-223.
324. Mathialagan, N. and Roberts, R.M. (1994) Review Article: A role for cytokines in early pregnancy. *Ind. J. Physiol. Pharmacol.* 38, 143-162.
325. {Leaman, 1992 #3257}
326. Mathialagan, N. and Roberts, R.M. (1995) Type I interferons and their role in maternal recognition of pregnancy and fecundity in domestic ruminant species. In: *Cytokines in Animal Health and Disease* (M.J. Myers and M.P. Murtaugh, eds.), Marcel Dekker, Inc., New York, pp 215-234.
327. Roberts, R.M., Xie, S., Nagel, R.J., Low, B., Green, J. and Beckers, J-F. (1995) Glycoproteins of the aspartyl proteinase gene family secreted by the developing placenta. In: *Aspartic Proteinases: Structure, Function, Biology and Biomedical Implications*, K. Takahashi (ed.), Plenum Press, New York. *Adv. Exp. Med. Biol.* 362, 231-240.

328. Roberts, R.M., Mathialagan, N., Duffy, J.Y. and Smith, G. (1995) Regulation and Regulatory Role of Proteinase Inhibitors. *Critical Reviews in Eukaryotic Gene Expression.* 5, 385-436.
329. Roberts, R.M., Mathialagan, N., Duffy, J.Y., Stallings-Mann, M.L. and Trout, W.E. (1995) Proteinase inhibitors at the trophoblast-uterine interface: roles in implantation or immunomodulation. In: S.K. Dey (ed.) *Molecular and Cellular Aspects of Periimplantation Processes*, Serono Symposia USA. Springer-Verlag, New York, pp 253-267.
330. Roberts, R.M., Xie, S. and Mathialagan, N. (1996) Maternal recognition of pregnancy. *Biol. Reprod.* 54, 294-302.
331. Roberts, R.M. (1996) Interferon-τ and Pregnancy. Milstein Award Lecture, Annual Meeting of Soc. Interferon & Cytokine Res, Nov. 6, 1995, Baltimore, MD. *J. Interferon & Cytokine Res.*16, 271-273.
332. Roberts, R.M. and Liu, L. (1996) Interferon-ω. In: B.B. Aggarwal and J.U. Gutterman (eds.) *Human Cytokines, Handbook for Basic and Clinical Research*, Vol. II, Chapter 9. Blackwell Sciences, Cambridge, MA, pp 168-177.
333. Roberts, R.M. (1996) Interferons as hormones of pregnancy. In: J. Sengupta, D. Ghosh (eds.) *Cellular and Molecular Signaling in Reproduction*, New Age International Publishers, New Delhi, pp 187-194.
334. Roberts, R.M., Liu, L., Alexenko, A. (1997) New and atypical families of Type I interferons in mammals: comparative functions, structures and evolutionary relationships. *Prog. Nucleic Acid Res. & Mol. Biol.* 56, 287-326.
335. Green, Jonathan A., Xie, S., and Roberts, R.M. (1998) Pepsin-related molecules secreted by trophoblast. *Reviews of Reproduction* 3, 62-69.
336. Green, Jonathan, Xie, S., Gan, X., and Roberts, R.M. (1998) An aspartic proteinase expressed in the equine placenta. *Aspartic Proteinases* MNG James (ed). Plenum Press, New York. pp 162-167.
337. Roberts, R.M., Xie, S., and Green, J. (1998) Pregnancy-associated glycoproteins. *Handbook of Proteolytic Enzymes* F. Woessner, N. Rawlings, A. Barrett (eds). Academic Press, San Diego, California. pp 911-914.
338. Green, J., Xie, S., and Roberts, R.M. (1998) Horse placental aspartic proteinase. *Handbook of Proteolytic Enzymes* F. Woessner, N. Rawlings, A. Barrett (eds). Academic Press, San Diego, California. pp 914-915.
339. Liu, L., Leaman, D., and Roberts, R.M. (1999) Possible Role of the Transcription Factor Oct-3/4 in Control of Human Chorionic Gonadotropin Expression. *Embryo Implantation: Molecular, Cellular and Clinical Aspects* DD Carson (ed). Springer-Verlag, New York, New York. pp 261-270.
340. Ezashi, T., and Roberts, R.M. (2000) Yeast One-Hybrid System: A Genetic System to Identify DNA-Binding Proteins. *Yeast Hybrid Technologies* L. Zhu, G. J. Hannon (eds). Eaton Publishing, Natick, Massachutes, pp 177-195.
341. Roberts, R.M. (2001) The Place of Farm Species in the New Genomics World of Reproductive Biology. *Biol. Reprod.* 64, 409-417.

342. Roberts, R.M. (2001) Making the Case for Productivity. *Agbioforum 3*, Special Issue: Understanding the Differences in US-EU Biotech Regulation. N. Kalaitzandonakes (ed). www.agbioforum.org.
343. Rosenfeld, C.S., Roberts, R.M., Lubahn, D.B. (2001) Estrogen Receptor- and Aromatase Deficient Mice Provide Insight into the Roles of Estrogen within the Ovary and Uterus. *Mol. Reprod. Dev.* 59, 336-346.
344. Roberts, R.M. (2001) National Bovine Genomics Projects: Present Status, Future Directions, and why they are important. Factors Affecting Calf Crop: Biotechnology of Reproduction, (Eds: M.J. Fields, J.V. Yelich, R.S. Sand) CRC Press, Boca Raton, FL. pp 165-175.
345. Rosenfeld, C.S., Wagner, J.S., Roberts, R.M., Lubahn, D.B. (2001) Intraovarian actions of Oestrogen. *J. Reprod. Fertil.* 122, 215-226.
346. Roberts, R.M. (2004) Interferon-τ. Encyclopedia of Endocrine Diseases. L. Martini (ed). Elsevier Academic Press. London.
347. Green, M.P. and Roberts, R.M. (2004) Pregnancy Recognition and Signaling. *Encyclopedia of Animal Science.* (eds: W.G. Pond & A.W. Bell) Marcel Dekker, N.Y.
348. Green, J.A., and Roberts R.M. (2004) Pepsin F. In Handbook of Proteolytic Enzymes, Second Edition (A.J. Barrett, N.D. Rawlings, and J.F. Woessner, Jr., eds) Elsevier Academic Press, London.
349. Roberts R.M. , Xie, S., Green, J.A. (2004) Pregnancy-associated glycoproteins. In Handbook of Proteolytic Enzymes, Second Edition (A.J. Barrett, N.D. Rawlings, and J.F. Woessner, Jr., eds) Elsevier Academic Press, London.
350. Jonathan A. Green and R. Michael Roberts (2006). Establishment of an ELISA for the Detection of Native Bovine Pregnancy-Associated Glycoproteins Secreted by Trophoblast Binucleate Cells. In Soares, M., and Hunt J.S. Placenta and Trophoblast: Methods and Protocols, Vol II. Humana Press, Totowa, NJ.
351. Roberts R., Ezashi, T., Westfall, S., Sachev, S., Hannink, M. (2006) Epigenetics and Embryonic Stem Cells. In *Epigenetics: Principles, Protocols and Practice* (ed. JL Xue) China: SSMPF Publishing, pp 355-372
352. Roberts, RM. (2007) Genetically modified organisms for Agricultural Food Production. In labeling genetically modified food: The philosophical and legal debate. Edited by Paul Weirich, Oxford University Press. pp 10-16.

353. BOOKS
354. Longman's Publishers, England, *Plant Cell Structure and Metabolism.* This is under joint authorship with J. Hall and T. Flowers. (First edition, 1974) Second edition (1982) This book has also been translated into Japanese, Polish and Italian.
355. Yokendo Ltd Publishers, Japan, *Cloned Animal and Placentation.* Eds: R. Michael Roberts, R. Yanagimachi, T. Kariya, K. Hashizume. 2000.

356. The National Academies Press, Washington DC, *Animal Biotechnology Science-Based Concerns.* Coauthor, National Research Council of the National Academies. 2002.
357. The National Academies Press, Washington, D.C., *Animal Care & Management at the National Zoo.* Chair and Co-author, National Research Council of the National Academies, 2004.

358. PATENTS
359. "Use of interferons of the alpha family to enhance fertility in mammals" (with P.J. Hansen, W.W. Thatcher and K. Imakawa co-inventors). U.S. Patent No Filed 3/04/88.
360. "Compositions and Methods for early pregnancy diagnosis" U.S. Patent No. 6,869,770, March 22 2005.
361. "Early Pregnancy Diagnosis Using PAGS (pregnancy-associated glycoproteins) Mexican Patent 233668, January 10, 2006.

362. SEMINAR AND SYMPOSIUM PRESENTATIONS SINCE 1979
363. Roswell Park Memorial Hospital, Buffalo, Department of Cell Biology. 7/3/79. Plasma membrane domains of CHO cells.
364. State University of New York at Buffalo, Department of Cell & Molecular Biology. 7/5/79. Hormonal control properties and function of uteroferrin, an iron-containing phosphatase involved in transplacental transport of iron.
365. University of Geneva, Switzerland, Department of Biochemistry. 9/21/79. Uteroferrin, a acid phosphatase involved in transplacental transport of iron.
366. Swiss Institute for Experimental Cancer Research, Lausanne, Switzerland. 9/24/79. Dynamics and turnover of CHO plasma membrane polypeptides.
367. Regensburg, W. Germany. Department of Anatomy. 7/29/79. Hormonal control properties and function of uteroferrin, an iron-containing phosphatase involved in transplacental transport of iron.
368. Marburg, W. Germany. International Symposium on Progesterone-induced Proteins. 7/27/79. The properties, function and hormonal control of synthesis of uteroferrin, the purple protein of the pig uterus.
369. Annual Meeting of the "Iron Club". Massachusetts Institute of Technology. 12/18/79. Uteroferrin.
370. Massachusetts General Hospital, Department of Gynecology. 12/19/79. Hormonal Control properties and function of uteroferrin, an iron-containing phosphatase involved in transplacental transport of iron.
371. Florida State University, Department of Gynecology. 12/19/79. Hormonal Control properties and function of uteroferrin, an iron-containing phosphatase involved in transplacental transport of iron.

372. University of Miami, Department of Biochemistry. 3/7/80. Hormonal Control properties and function of uteroferrin, an iron-containing phosphatase involved in transplacental transport of iron.
373. Albert Einstein College of Medicine, N.Y., Departments of Biophysics and Hematology 13/12/80. Hormonal Control properties and function of uteroferrin, an iron-containing phosphatase involved in transplacental transport of iron.
374. Downstate Medical Center, Brooklyn, N.Y., Department of Cell Biology. 3/13/80. Membrane dynamics of CHO Cells.
375. New York University Medical School, N.Y., Department of Cell Biology. 7/14/81. Hormonal Control properties and function of uteroferrin, an iron-containing phosphatase involved in transplacental transport of iron.
376. Gordon Conference (Mammalian Genital Tract), Colby-Sawyer College. 7/7/80. Secretory activity of the trophoblast and endometrium in the sheep and the pig.
377. University of Hanover, W. Germany. 7/15/80. The role of uterine secretions in the development of the pig blastocyst.
378. National Institute for Dairying Shinfield, Reading, England. 8/6/80. Secretion of the early blastocyst.
379. Institute for Animal Physiology, Babraham, Cambridge, England. 8/7/80. Secretory activity of the trophoblast and endometrium in the sheep and cow.
380. International Conference on Cancer Cell Organelles, University of Surrey, Guildford, England. 8/29/80. Isolation and characterization of plasma membranes from cultured cancer cells.
381. University of Arkansas Medical Center, Little Rock, Arkansas. 1/27/81. Plasma membrane dynamics in CHO cells.
382. The Rockefeller University, NY. 1/29/81. Plasma membrane dynamics in CHO cells.
383. Institute of Animal Physiology, Babraham, Cambridge, England. 3/28/82. The function of uteroferrin.
384. Medical College of South Carolina, Charleston. 5/7/82. Uteroferrin, a progesterone-induced, iron transport protein.
385. Gordon Research Conference, Mammalian Genital Tract, Wolfsborough, NH. 7/6/82. Interactions between blastocyst and endometrium.
386. U.S. Department of Agriculture, Beltsville, MD. 12/3/82. Transplacental iron transport in pigs.
387. Iron Club Meeting, NY. 12/15/82. Uteroferrin.
388. Washington University, St. Louis, MO, Division of Oncology. 4/1/83. Uteroferrin, an iron-containing lysosomal-like acid phosphatase.
389. North Carolina State University, Departments of Biochemistry & Physiology. 4/13/83. Uteroferrin, a progesterone-induced iron-containing acid phosphatase.

390. Bangalore, India. Invited participant in the Indo-US Workshop on Blastocyst Research. 12/6/83-12/10/83. Subject: Two-dimensional polyacrylamide gel electrophoresis of secreted proteins, 2) Immunocytochemical localization of tissue antigens.
391. Bangalore, India. Symposium on Blastocyst Research 12/12/83-12/16/84. Biochemical interactions between the blastocyst and endometrium.
392. Bangalore, India. 12/16/83. Uteroferrin, an iron-containing lysosomal-like acid phosphatase. Department of Biochemistry, Indian Institute of Science.
393. Symposium on Reproductive Biology of Domestic Animals, honoring Professor Donald H. Barron. Institute of Food and Agricultural Sciences, University of Florida. 4/5/84. Uteroferrin: The Purple Phosphatase of the Pig Uterus.
394. Gordon Research Conference, Mammalian Genital Tract, Plymouth, NH. 7/9/84. Chairman of session Extracellular microenvironments.
395. Upjohn Co., Kalamazoo, MI. 10/8/84. Conceptus protein secretion and maternal recognition of pregnancy.
396. Department of Anatomy, Texas Tech University, Lubbock, TX. 11/6/84. Protein synthesis and secretion by the early trophoblast and its role in maternal recognition of pregnancy.
397. Department of Cell Biology, Baylor College of Medicine, Houston, TX 11/7/84. Rapid surface membrane shuttling and membrane turnover in CHO cells.
398. Second Bourn Hall Meeting, Cambridge, England on Implantation of the Human Embryo. 11/20/84. Embryonic Antiluteolysins.
399. Monsanto Chemical Co., Agricultural Division, St. Louis, MO. 12/4/84. Proteins secreted into the uterine tract of the large domestic farm animals during early pregnancy.
400. Visiting Scientist, Department of Physiology, University of Kansas Medical Center, Kansas City (1/14/85-1/15/85). The role of trophoblast secretory products in maternal recognition of pregnancy (1/15/85).
401. Department of Animal Science, University of Missouri, Columbia, MO. 1/16/85. The role of trophoblast secretory products in maternal recognition of pregnancy in the sheep and cow.
402. Department of Biochemistry, University of Missouri, Columbia, MO. 3/5/85. Uteroferrin: A progesterone-induced acid phosphatase with a role in iron metabolism and transport.
403. Federation Meetings, Anaheim, CA (4/22/85). Role of uteroferrin in transplacental iron transport in the pig. (Theme symposium: Placental Transport).
404. NIH, Environmental Health Sciences, Research Triangle Park, NC 5/3/85. Role of trophoblast secreted proteins in maternal recognition of pregnancy.

405. American Society of Zoologists, Baltimore, MD (12/30/85) Possible functions of carbohydrates on glycoproteins secreted into the uterus of pigs during pregnancy (Symposium: The Significance of Protein Glycosylation in Molecular and Cellular Recognition).
406. University of Wisconsin, Madison, WI. (2/17/86). Production of antiluteolysins by sheep and cow trophectoderm.
407. National Institute of Health and Family Welfare, Indian Council of Medical Research, New Delhi (4/5/86). Conceptus Proteins as antiluteolysins. In the Indo-US Workshop on Development of pre-implantation embryos.
408. Upjohn Company, Kalamazoo, Michigan (10/1/86). Uteroferrin and other iron-containing acid phosphatases.
409. College of Agriculture, University of Nebraska (10/17/86). Biotechnology and the Animal Sciences.
410. School of Medicine Colloquium, University of Missouri (11/2/86). Polypeptides involved in maternal recognition of pregnancy: identification, characterization and molecular cloning.
411. Brymar Seminar, University of Missouri (11/13/86). Technology and breeding.
412. Synergen Company, Boulder, Colorado 12/4/86. Protein secretion by early embryos of cow and ewe: paracrine hormones involved in maternal recognition of pregnancy.
413. Department of Physiology, Colorado State University, Ft. Collins, Colorado 12/5/86. Polypeptides involved in maternal recognition of pregnancy in cow and ewe.
414. Symposium on Cell Biology of the Uterus, National Institutes of Health, Bethesda, Maryland 12/12/86. Hormonal control and function of secretory proteins.
415. Department of Cell Biology, Baylor College of Medicine, Houston, TX 2/26/87. Purification, function and molecular cloning of proteins involved in maternal recognition of pregnancy in the sheep and cow.
416. Seminar presented to Biotechnology Group at Oklahoma State University, 4/9/87. The purification, function and molecular cloning of proteins involved in maternal recognition of pregnancy in the ewe and cow.
417. Department of Animal Sciences, Oklahoma State University, 4/10/87. Biotechnology in the Animal Sciences.
418. Department of OB/GYN, Albert Einstein Medical Center, Philadelphia, 5/18/87. Maternal recognition of pregnancy in Clinical Applications of Reproductive Immunology.
419. Department of OB/GYN, Albert Einstein Medical Center, Philadelphia, 5/18/87. Proteins secreted by the trophoblast: factors influencing the outcome of pregnancy in Clinical Applications of Reproductive Immunology.

420. Society for Study of Fertility, York, England 7/08/87. Functions of uterine secretions in early embryonic development. In the Meeting Symposium: The milieux of the egg and early embryo.
421. Blastocyst-Implantation Discussion Group, University of Illinois 7/19/87. Ovine trophoblast protein-1, a polypeptide implicated in mediating maternal recognition of pregnancy in sheep, is an interferon of the alpha class.
422. Department of Biochemistry and Molecular Biology, Gainesville, FL. 9/24/87. Purification, properties and molecular cloning of ovine trophoblast protein-1, an unusual interferon of the alpha class.
423. Biochemistry-Biophysics Program, Washington State University, Pullman, WA. 11/03/87. Characterization and molecular cloning of a novel interferon from sheep trophoblast.
424. Department of Biochemistry, University of Kansas Medical School, Kansas City, KA. 12/04/87. Interferons and maternal recognition of pregnancy.
425. Abbott Laboratories (Diagnostic Division), Abbott Park, IL. 12/10/87. The characterization of acid phosphatase derived from human placenta, spleen and bone.
426. AAAS Annual Meeting 1988, Boston, MA. Symposium on "Frontiers of Reproductive Biology" 2/13/88. Interferons and Maternal Recognition of Pregnancy.
427. Indo-US Symposium on "Factors influencing the establishment of pregnancy," All-India Institute of Medical Sciences, New Delhi, 2/17/88. Interferons secreted by the trophoblast are involved in maternal recognition of pregnancy in sheep and cattle.
428. Indian Institute of Immunology, New Delhi, 2/17/88. Interferons and maternal recognition of pregnancy.
429. Vanderbilt University, Reproductive Biology Forum 4/5/88. Characterization and molecular cloning of a novel interferon from sheep trophoblast.
430. Lexington Hormone Research Symposium, University of Kentucky 5/20/88. The possible role of interferons in maternal recognition of pregnancy.
431. Current Trends in Reproductive Biology Research, Tufts University School of Veterinary Sciences 7/01/88. The role of interferons in early pregnancy in farm animals.
432. Gordon Conference, Reproductive Tract Biology, 7/08/88, Brewster Academy. Possible function of interferons produced by the trophoblast in maternal recognition of pregnancy.
433. Institute of Medicine/National Acad. Sciences Conference on Basic Science Foundation of Medically Assisted Conception, 8/23/88, Newporter Resort, Irvine CA. Uterine Receptivity, Maternal Recognition of Pregnancy and Early Embryonic Loss.

434. Department of Physiology and Biophysics, University of Illinois, Urbana, 10/14/88. Embryonic interferons and maternal recognition of pregnancy.
435. 1988 Annual Meeting of the International Society for Interferon Research and ISIR-JSIR Meeting on Interferon and Cytokines, Kyoto, Japan, 11/18/88. Embryonic interferons and maternal recognition of pregnancy.
436. Molecular Biology Group, University of Nebraska-Lincoln, 12/07/88. A novel role for interferons in pregnancy.
437. UCLA Colloquium, Embryo Development and Paracrine Relationships, Taos N.M., 2/03/89. Role of uterine secretory activity in support of the preimplantation embryo.
438. CIBA-GEIGY Corp., Biotechnology, Basel, Switzerland 3/29/89. Role of interferons in early pregnancy.
439. Third International Conference on Pig Reproduction, University of Nottingham 4/13/89. Chairman, Session on Early Pregnancy.
440. Royal Postgraduate Medical School, Hammersmith Hospital, London, 4/15/89. Iron-containing acid phosphatases: Structure and possible function.
441. Second Symposium on the Genetic Engineering of Animals, Cornell University, 6/25/89 - 6/28/89. Interferons at the placental interface.
442. The 1989 Biennial Animal Reproduction Symposium, Lexington, Kentucky 7/31/89. Conceptus products involved in pregnancy recognition.
443. Cell Biology Symposium, 22nd Annual Meeting of the Society for Study of Reproduction, Columbia, Missouri 8/06/89. Trophoblast interferons and maternal recognition of pregnancy.
444. Patrick Steptoe Memorial Symposium: Establishment of Successful Human Pregnancy 9/22/89. Blastocyst uterine interactions.
445. Department of Physiology, The University of Arizona 10/20/89. Structure, Function and Control of Expression of Embryonic Interferon.
446. The Johns Hopkins University, Department of Population Dynamics, 12/05/89. Structure, Function and Control of Expression of Embryonic Interferons.
447. Third International Ruminant Reproduction Symposium, Nice, France, 3/25/90- 3/28/90. The polypeptides and genes for ovine and bovine trophoblast protein-1.
448. Serono Symposium on Growth Factors in Reproduction, Savannah, Georgia, 4/01/90- 4/04/90. Trophoblast Interferons.
449. Tufts University, Boston. Progress in Cell, Molecular and Developmental Biology, 4/10/90. Embryonic Interferons: molecular cloning, structure and role in maternal recognition of pregnancy.
450. The Procter and Gamble Company, Corporate Research Division, Cincinnati, OH, 4/19/90. The biological and molecular properties of tartrate-resistant acid phosphatase.

451. University of Pennsylvania Medical Center, Division of Reproductive Biology, 5/22/90. Trophoblast Interferons.
452. AgTechnology '90 Conference, St. Louis MO, 9/17/90. Interferons as Potential Enhancers of Fertility in Domestic Animals.
453. Workshop on Uterine and Embryonic Factors in Early Pregnancy, Rockefeller Foundation, Bellagio, Italy, 10/08/90. Interferon-like factors.
454. 23rd Annual Postgraduate Course: Course VII, Periconceptional Molecular Biology and Growth Factors, The American Fertility Society, Washington DC, 10/13/90. Maternal Recognition of Pregnancy and Use of polymerase chain reaction to study genes and gene products.
455. Beijing (China) Workshop on Reproductive Biology, sponsored by the Rockefeller Foundation, 10/19-10/21/90. Several lectures on implantation and maternal recognition of pregnancy.
456. Plenary Speaker, Beijing International Symposium on Reproductive Biology, Beijing, China, 10/23/90. Trophoblast Interferons.
457. Plenary Speaker, 1990 Annual Meeting of the International Society for Interferon Research, San Francisco, 11/16/90. Unique Features of the Trophoblast Interferons.
458. The 2nd Sydney A. Asdell Lecture - 1990, Cornell University, Ithaca NY. 11/28/90. Embryonic Interferons: Discovery, Distribution, Molecular Cloning and Function in Early Pregnancy.
459. NIH Workshop on Molecular Approaches in Reproductive Endocrinology - February 1991, Parklawn Bldg. Conference Center. Trophoblast Interferons.
460. Lincoln University, Canterbury, New Zealand, March 1, 1991. Trophoblast Interferons.
461. Prince Henry's Institute of Medical Research, Melbourne, Australia - March 13, 1991. Trophoblast interferons: a unique class of type 1 interferons with regard to structure and expression.
462. Prince Henry's Institute of Medical Research, Melbourne, Australia - March 13, 1991. Do the trophoblast interferons have a unique function in pregnancy?
463. Monash University, Melbourne, Australia, March 15, 1991. Trophoblast Interferons.
464. Serono Symposium: Frontiers in Reproductive Endocrinology - April 23, 1991, Washington DC. Trophoblast Interferons and Maternal Recognition of Pregnancy.
465. Serono Symposium: Preimplantation Embryo Development - August 18, 1991, Newton, MA. Uterine Secretory Activity and Embryo Development.
466. Washington University, St. Louis, Missouri, September 3, 1991. Gene expression in trophectoderm of preimplantation embryos: trophoblast interferons and aspartyl proteinases.

467. Washington University, St. Louis, Missouri, September 4, 1991. Role of uterine secretions in early development of the embryo.
468. University of Minnesota, St. Paul, Minnesota, November 20, 1991. Trophoblast interferons: structure, distribution, control of expression and function during pregnancy.
469. University of Illinois, Champaign-Urbana, May 11, 1992. Interferons: Novel roles in the endocrinology of pregnancy.
470. Serono Symposium *Trophoblast Cells: Pathways for Maternal-Embryonic Communications,* Las Vegas, August 9, 1992. Interferons: Functions in pregnancy, gene distribution among mammals and trophectoderm-restricted expression.
471. The Procter & Gamble Company, Cincinnati OH, October 6, 1992. Tartrate-resistant Acid Phosphatases.
472. Genzyme Corporation, Framingham MA, October 26, 1992. Interferons as Hormones of Pregnancy.
473. University of Missouri-Kansas City, December 10, 1992. Interferons as Hormones of Pregnancy.
474. Triangle Conference on Reproductive Biology (Conception), Research Triangle Park NC, January 9, 1992. Factors produced by the early conceptus involved in maternal recognition of pregnancy.
475. Department of Physiology, All Indian Institute of Medical Sciences, New Delhi, February 4, 1993. Trophoblast Interferons.
476. Workshop on Molecular Biology Techniques in Reproduction Research, Bangalore, India, February 11, 1993. Farming of Embryos: *In Vitro* Fertilization to produce large numbers of blastocysts for experimental use.
477. Discussion Meeting on Recent Trends in Molecular Aspects of Reproductive Biology, Indian Institute of Science, Bangalore, India, February 15, 1993. Trophoblast Interferons.
478. Department of Physiology, Colorado State University, Ft. Collins, May 11, 1993. Cloning and characterization of secretory products of pre-implantation ovine conceptuses.
479. Fourth International Conference on Pig Reproduction, Columbia, Missouri, May 25, 1993. Embryo-uterine interactions in the pig.
480. IVth Organon Round Table Conference, Thessaloniki, Greece, June 24-25, 1993. Identification of secretory proteins released by preimplantation embryos.
481. VIIIth World Congress on *In Vitro* Fertilization and Alternate Assisted Reproduction, Kyoto Japan, September 12-15, 1993. Interferons and other factors released by the perimplantation conceptus.
482. The 5th International Conference on Aspartic Proteinases, Gifu, Japan, September 19-24, 1993. Glycoproteins of the aspartyl proteinase gene family secreted by the developing placenta.

510. Monsanto Corporation, Chesterfield, Missouri, September 26, 1997. Pregnancy Associated Glycoproteins.
511. Serono Symposium, Embryo Implantation: Molecular, Cellular and Clinical Aspects, Newport Beach, California, October 6, 1997. Silencing of hCG Alpha and hCG Beta Gene Expression.
512. 18th Annual Meeting of the American Society of Reproductive Immunology. Chicago, IL, May 10, 1998. Interferon-tau: basis of trophoblast-specific expression and evolutionary origins.
513. Departments of Physiology and Ob-Gyn, University of Illinois School of Medicine, Chicago, May 12, 1998. Interferons and maternal recognition of pregnancy.
514. Symposium: Maternal Recognition of Pregnancy: Gainesville and Beyond (In memory of Dr. R. Jeffrey Moffatt), Texas A&M University, College Station, Texas, August 12, 1998. Conceptus interferons: an evolutionary perspective.
515. USDA/ARS Annual Reproductive Group. Beltsville, MD, October 16, 1998. Pregnancy-associated glycoproteins: Multigene family of aspartic proteinases expressed in placenta.
516. USDA/ARS; Beltsville, MD. October 16, 1998. Development of an ideal pregnancy test in cattle.
517. Visiting Professor, National Institute of Animal Industry. Presented seminars on IFN-τ and pregnancy associated-glycoproteins at four locations in Japan. Japan. October 25-November 6, 1998.
518. Development of the ideal pregnancy test for cattle, University of Liege, Belgium. November 25, 1998.
519. Keynote Speaker: International Workshop on Embryogenesis and Implantation. National Institute of Animal Industry, Japan. Hawaii, HI, February 1-4, 1999. Genomes of farm animal species: A forgotten side of agricultural research and one vital to understanding reproductive processes.
520. Life Sciences Consortium Colloquium. Pennsylvania State University. March 2, 1999. Rapid Evolutionary Changes in the Mechanisms for Maternal Recognition of Pregnancy in Placental Mammals.
521. S.W. Medical School, Dallas Texas. May 7,1999. Pregnancy-Associated Glycoproteins: An Extremely Large, Rapidly Evolving Family of Peptide-Binding Molecules Expressed in Placenta.
522. Dept. of Animal Science Seminar. Michigan State University. May 20, 1999. Variability and Evolution in Mechanisms for Maternal Recognition of Pregnancy.
523. VIIIth International Aspartic Proteinase Conference. Funibal, Madeira, Portugal. September 7-12, 1999. Pregnancy-associated glycoproteins (PAGs): a large family of aspartic proteinases expressed in placenta of Artiodactyls.

524. Department of Ob/Gyn, University of Pennsylvania, Philadelphia. March 29, 2000. Rapid Functional Evolution of Genes Expressed Specifically in Placenta.
525. Departments of Animal Sciences and Biochemistry. University of Wyoming, Laramie. March 31, 2000. Evolution of Interferon-tau gene function.
526. Life Sciences Symposium. University of Wyoming. April 1, 2000. Post Genomic Biology in Agriculture.
527. 49th Beef Cattle Short Course. Florida Cattle Short Course 2000. Gainesville, Florida. May 3-5, 2000. National Genome Projects: Present Status, Future Directions, Why They are Important.
528. Symposium on US-EU Policy Issues in Animal Production. University of Missouri-Columbia. May 22, 2000. U.S. Perspective.
529. Keynote Speaker, 33rd Annual Meeting of the Society for the Study of Reproduction. Madison, WI. July 15, 2000. The Place of farm species in the New Genomics World of Reproductive Biology.
530. A Nalbandov Memorial Lecture. University of Illinois, Urbana, November 29, 2000. Future of Reproductive Sciences in the Post Genomic World.
531. AAI Symposium Comparative View of Fetal-Maternal Immunological Interactions. FASEB Annual Meeting, Orlando FL, April 2, 2001. Expression of Interferons by Trophoblast in Ungulates.
532. Department of Physiology, University KUMUS Medical School, Kansas City, Missouri. April 15, 2001. The Evolution of a Reproductive Function among Type I Interferons.
533. Jackson Laboratory, Bar Harbor, MN. June 13, 2001. Evolution of mechanisms for Maternal Recognition of Pregnancy.
534. International Society for Interferon and Cytokine Research (ISICR) annual meeting, Cleveland, OH. October 7-11, 2001. Effect of three distinct IFN-tau variants on progression and severity of experimental allergic encephalomyelitis (EAE) in mice.
535. Reproductive Forum, Texas A&M University, College Station, TX. November 2, 2001. Developmental Control of Interferon-tau Expression.
536. Department of Animal Production/Animal Physiology, University of Nottingham, Nottingham, England. November 13, 2001. Developmental control of Interferon-Tau Expression.
537. Monash University, Melbourne, Australia. February 19, 2002. Evolution of the interferon system for a role in maternal recognition of pregnancy.
538. Department of Obstetrics & Gynecology, Adelaide University, Australia. February 22, 2002. Evolution of the interferon system for a role in maternal recognition of pregnancy.
539. Symposium : Sperm Storage, Sorting and Sexual Dimorphism. Mid West meeting of the American Society of Animal Sciences, Des Moines IA, March

19, 2002. Sexual dimorphism among blastocysts may provide for sex ratio adjustment in the bovine.

540. Department of Biomedical Sciences, Southwest Missouri State University, Springfield, MO. April 11, 2002. Interferons in Disease and Pregnancy.
541. Department of Veterinary Medicine, University of Pennsylvania, Philadelphia, PA. April 16, 2002. The Evolution of Mechanisms of Maternal Recognition of Pregnancy.
542. Mammalian Gametogenesis & Embryogenesis Conference, Connecticut College, New London CT. June 30-July 5, 2002. Mechanisms of Genetic Quality Control: The in utero environment and its influence on sex ratio.
543. Sixth International Symposium on Reproduction in Domestic Ruminants, Crieff, Scotland, UK. August 14-17, 2002. Ovary-uterine-embryo interactions.
544. Department of Physiology, University of Cambridge, England, UK. November 15, 2002. Does maternal diet influence the sex of offspring born?
545. Department Cell Biology, University of Southampton, England, UK. November 19, 2002. Does maternal diet influence the sex of offspring born?
546. Koret School of Veterinary Medicine, Tel Aviv, Israel. May 12, 2003. The relationship between diet and sex of offspring.
547. Annual Meeting of the Israel Fertility Association, Tel Aviv, Israel. May 14, 2003. Pregnancy recognition signals and uterine biology.
548. Havemeyer Foundation Workshop, Embryonic and Fetal Nutrition, Ravello, Italy. May 15-18, 2003. Can maternal nutrition influence the sex of offspring?
549. Larry Ewing Memorial Lecture, John Hopkins University, Baltimore MD. October 8-9, 2003. Sexual dimorphism during early embryonic development and the effects of diet on sex of offspring.
550. Keynote Speaker, Kansas IDeA Biomedical Focus Group Symposium, University of Kansas Medical Center, Kansas City KS. November 1, 2003. The Culture of Collaboration.
551. Reproductive Biology Seminar Series, University of Illinois. May 5, 2004. How the diet of the mother affects the sex of her offspring.
552. Missouri. Agriculture Leaders Forum with USDA Undersecretary Bill Hawks, University of Missouri-Columbia. October 12, 2004
553. Advantages of Agriculturally Important Domestic Species as Biomedical Models Workshop, Michigan State University. October 29-31 2004. Are the Domestic Farm Species Redundant as Models in Biomedical Research? Does Mighty Mouse Rule Supreme?
554. International Symposium on Stem Cells: Premises and Promises for Research and Therapeutics. September 18-21, 2005 Mumbai India. Hypoxia and prevention of human embryonic stem cell differentiation
555. Saturday Morning Science Lecture Jan 28, 2006. University of Missouri. How Diet Influences Sex of Offsprings.

556. Life Sciences and Society Symposium- Invited Speaker March 8, 2006 University of Missouri. Stem Cells: The Upside and Downside of Stem Cell Science.
557. Saturday Morning Science Lecture Sept 16, 2006. Stem Cells: Technology Promise and Concern
558. Center for Reproductive Science Workshop, UCSF. October 12, 2006. Embryonic Stem Cells as Models for Trophoblast Differentiation under High and Low Oxygen Conditions.
559. William Woods University Invited Speaker October 24, 2006. Stem Cells: Technology Promise and Concern
560. University of Connecticut Invited Speaker. November 15, 2006. Lineage Commitment to Trophectoderm.
561. 33rd Annual Conference of the Int'l Embryo Transfer Society. Kyoto, Japan. January 10, 2007 Lineage Specification in Mouse.
562. Second Okayama Symposium in Reproductive Biology. Okayama Japan. Jan 11, 2007. IFNT Regulated Genes.
563. Keystone Symposia Reproduction: Advances and Challenges (C1-2007) Santa Fe, New Mexico Feb 24, 2007. Maternal Diet and Reproductive Outcome.
564. 2nd SGI Endometrial Satellite Symposium Reno, Nevada. March 13, 2007 Implantation and Interferon Response.
565. International Federation of Placenta Associations Meeting "Placenta-Platform for Life. Kingston, Ontario Canada. August 17-21, 2007. Human Embryonic Stem Cells as Models for Trophoblast Differentiation.
566. Seminar presentation (Invited Speaker). November 27, 2007. UC Santa Barbara. "Human Embryonic Stem Cells as Models of Placental Cell Development."
567. Seminar presentation (Invited Speaker). March 7, 2008. University of Florida-Gainesville. "Can a mother's diet influence the sex of her offspring?"
568. Seminar presentation (Invited Speaker) April 15, 2008. University of Santa Barbara "Interferon-tau: function, control of expression and evolutionary origin".

GRADUATE STUDENTS

1. Jacques Van Veen (Ph.D.): 1972-1975. "Cell Surface Properties in Relation to Growth and Form in Chinese Hamster Ovary Cells in Culture."
2. Daniel W. Welch (Ph.D./M.D.): 1971-1976. "Complex Carbohydrate Metabolism in Skin Fibroblast Cultures from Patients with Cystic Fibrosis of the Pancreas."
3. William Buhi (Ph.D.): 1975-1981. The Role of Uteroferrin Iron Transport of the Pig Conceptus.

4. Deborra Mullins (Ph.D.): 1977-1979.
   Title of Thesis: "A Study of Steroid Hormone-induced Changes in Epithelial Cells of Reproductive Tissue."
5. Kathy Adams (M.S.): 1977-1979. "Retinol Binding Protein in Uterine Flushings of Pigs."
6. George A. Baumbach (Ph.D.) 1980-1984.
   Gene Expression in the Porcine Uterus: Biosynthesis and Secretion of Progesterone-induced Proteins.
7. John B. Denny (Ph.D.) 1980-1982. Dynamics of CHO Cell Surface Glycoproteins.
8. Catherine Ketcham (Ph.D.) 1984-1988. Molecular cloning of the human placental acid phosphatases.
9. Nancy Ing (Ph.D.) 1982-1988 (Joint Ph.D.-D.V.M. Candidate) Uterine milk proteins of the sheep: Biosynthesis, secretion and molecular cloning.
10. Michael Burke (Ph.D.) 1984-1989. Uterine plasmin inhibitor of the pig.
11. Peter Nuttleman (M.S.) 1986-1989. Transfer of uteroferrin iron to fetal transferrin.
12. Janet Clawitter (M.S.) 1986-1989. Retinol-binding proteins from pig uterine secretions.
13. Jay Cross (Ph.D.) 1987-1991. Trophoblast interferons.
14. Ping-Ling (Ph.D.) 1987-1992. Iron-containing acid phosphatases.
15. Robert Nagel (M.S.) 1991-1993. Ovine pregnancy-associated glycoproteins.
16. Kyle Kramer (Ph.D.) 1987-1993. Ovine conceptus secretory proteins.
17. Douglas Leaman (Ph.D.) 1989-1993. Molecular biology of the trophoblast interferons.
18. Junzhi Li (Ph.D.) 1988-1994. Structure-function studies on interferon-tau.
19. Melody Stallings (Ph.D.) 1991-1994. Progesterone-induced expression of secretory proteins in the porcine uterus.
20. Suzanne Clark (M.S.) 1991-1994. Recombinant human tartrate-resistant acid phosphatase.
21. Limin Liu (Ph.D.) 1990-1996. Evolution and function of trophoblast and omega interferon genes.
22. Jonathan Green (Ph.D.) 1992-present. Structure, function and evolution of pregnancy-associated glycoproteins.
23. Jodie Duffy (Ph.D.) 1992-1996. Secretory products of pre-implantation ovine trophoblast.
24. George Smith (Ph.D.) 1993-1996. Differential gene expression by two luteal cell populations.
25. Melissa Larson (Ph.D.) 1993-2000. Time of transgene integration in livestock species.

26. Chun Sheng Han (Ph.D.) 1994-1999. Type I interferon receptors.
27. James MacLean (Ph.D.) 1994-2000. Kunitz proteinases as immunoprotective agents.
28. Xinsheng Gan (M.S.) 1996-1998. Pregnancy-associated glycoproteins of non-ruminant species.
29. Xiaodi Chen (Ph.D.) 1996-2000. An aspartic proteinase expressed in placenta and neonatal stomach of the mouse.
30. Phillip Wang (Ph.D.) 1997-2004. Interactions of cytoplasmic components with the intracellular domain of IFNAR2.
31. Anindita Chakrabarty (Ph.D.) 1998-2004. The evolution and function of the repeat regions of the trophoblast-domain Kunitz inhibitors.
32. Yizhen Chen (Ph.D.) 1999-2005. Responses to interferon-τ in mammalian tissues and cells.
33. Angela Walker (M.S.: joint DVM candidate) 2000-present. Comparison of glycosylated and non-glycosylated interferon-τ.
34. Padmalaya Das (Ph.D.), 2001-2006. Transcription factors controlling interferon-tau expression.
35. Rangan Gupta (Ph.D.), 2005-present
36. Steven Daniel Smith (Ph.D.) 2005-present

Partial direction of thesis work of David Schlosnagle (Ph.D. in Chemistry, 1974), Thomas Chen (Ph.D. in Animal Science, 1974), M.T. Zavy (Ph.D. in Animal Science 1980), C.A. Ducsay (Ph.D. in Animal Science, 1981), R.D. Geisert (Ph.D. in Animal Science 1982), R.H. Renegar (Ph.D. in Animal Science, 1982), F.F. Bartol (Ph.D. in Animal Science, 1982), and J. Moffatt (Ph.D. in Animal Science, 1983), K. McDowell (Ph.D. Animal Science, 1986), J. Vallet (Ph.D. Animal Science, 1988).

Michael D. Olympio (1978-1979) was awarded 2nd Prize for Research (Watson Clinic Award) in the Medical Student Research Competition in 1979.

John Denny won the 1st prize in the Graduate Research Competition in the College of Medicine in 1981.

Rosalind Masters won the 1st prize for Research (Watson Clinic Award) in the Medical Student Research Competition in 1982. Rosalind Masters won the 1st prize for a written paper in the AOA competition for Medical Student Research in 1982.

George Baumbach won 2nd prize in the Graduate Research Competition in the College of Medicine, University of Florida, 1983.

Catherine Ketcham won 1st prize in the Graduate Research Competition, College of Medicine, University of Florida, 1987.

Jay Cross won second place in the "Young Investigator" competition at the SSR Meeting in Seattle, 1988.

Kyle Kramer won the first place prize in the biological sciences interactions portion of the Ninth Annual Research and Creative Activities Forum at the University of Missouri in 1992.

Doug Leaman won 1st Place in the "Young Investigator" competition at the SSR Meeting in Raleigh NC, 1992.

George Smith won the Superior Graduate Achievement Award for 1993-94 from the University of Missouri Graduate Association and the Graduate School.

Limin Liu won 1st Place in the "Young Investigator" competition at the SSR Meeting in Ann Arbor, MI, 1994.

Shushong (Phillip) Wang won 1st Place in the "Young Investigator" poster competition at the SSR Meeting in Cincinnati 2004.

Steven D. Tanksley

Cornell University, Ithaca, New York, USA

2004 — for innovative development of hybrid rice and discovery of the genetic basis of heterosis in this important food staple.

CURRICULUM VITAE

Education

Colorado State University	B.S.	1976	Agronomy
University of California, Davis	Research Assistant	1976-79	Genetics
University of California, Davis	Ph.D.	1979	Genetics
University of California, Davis	Postdoctoral Fellow	1979-81	Genetics

Professional Experience

1994-Present	Liberty Hyde Bailey Professor of Plant Breeding, Professor of Plant Biology, Cornell University
1991-1994	Professor, Department of Plant Breeding & Biometry, Cornell University
1985-1991	Associate Professor, Department of Plant Breeding & Biometry, Cornell University
1981-1985	Assistant Professor, Department of Horticulture and Plant Genetic Engineering Laboratory, New Mexico State University, Las Cruces

Honors/Activities

Editor, Genetics 1988-92
Editor, Molecular Breeding 1996-97
Appointed to Liberty Hyde Bailey Professorship, 1994
Elected to the National Academy of Sciences, USA, 1995
Award for Outstanding Contributions to Biology, University of California, Davis, 1989
Outstanding Undergraduate Mentor Award, Cornell University, 1989
Outstanding Research Award (Gamma Sigma Delta), 1996
Chair, Cornell Genomics Initiative, 1997-2005
Alexander von Humboldt Foundation Award, 1998
Martin Gibbs Medal, American Society of Plant Physiology, 1999
Wolf Foundation Prize in Agriculture, 2004
Kumho Award in Plant Molecular Biology and Biotechnology Korea, 2005
Appointed as an Einstein Professor of the Chinese Academy of Sciences, 2006
Appointed as Honorary Professor Wuhan Botanical Garden (China), 2006
Rank Prize Award, 2008

SCIENTIFIC ACHIEVEMENTS

Steve Tanksley's research has resulted in major contributions in our understanding of plant genome organization and in concepts and infrastructure that improve dramatically the efficiency of plant breeding programs. He published the first molecular map of tomato in 1992. This proof-of-concept map was followed up with work providing some of the first clear experimental evidence of how molecular maps and markers could be applied to agriculture. Tanksley's early work was extended to molecular mapping of other major crop species, including the Solanaceous crops, potato and pepper, and cereal crops. The markers developed in Tanksley's lab were distributed widely to researchers throughout the world and facilitated the flow of information and scientific collaboration. His work to produce the molecular map of rice became a cornerstone of the Rockefeller Foundation's International Program on Rice Biotechnology and was pivotal in the development of the USDA's National Research Initiative in Plant Genomes.

In 1988, Tanksley demonstrated for the first time that quantitatively inherited traits spanning an entire genome could be dissected into their corresponding Mendelian factors (quantitative trait loci, QTLs) using a comprehensive RFLP linkage map. This led to a cascade of experiments by other researchers aimed at detecting and mapping QTLs in a wide array of other organisms. Within Tanksley's own group, QTL analysis in rice led to the discovery of the genetic basis of heterosis in this key food staple. More recently, Tanksley has used the power of QTL analysis to explore ways of efficiently utilizing wild and unadapted germplasm resources in plant improvement. Working in rice and tomato, he and his colleagues are applying the molecular maps and markers they developed to identify rate-limiting genes associated with crop performance. Recently they discovered genes from a wild ancestor of rice that can boost the yields of domesticated rice by 15-18%. This approach provides an effective mechanism for expanding the gene pool of cultivated crop varieties.

The ability to clone genes of agronomic importance and to determine their function is a major objective of biology today. Tanksley's group used the tomato molecular map to accomplish the first map-based cloning of a resistance gene in any crop species in 1993. They isolated the *Pto* gene for resistance to bacterial speck and demonstrated that a protein kinase played a critical role in this plant-microbe interaction. This work has become one of the classic studies in disease resistance gene isolation and contributed major new insights leading to an understanding of the kinds of signal transduction pathways that are active in host-pathogen recognition.

Tanksley's group successfully extended map-based cloning to QTLs of agronomic significance where nothing is known of the target gene other than the phenotype it confers. The QTL *Fw2.2* is responsible for 30% of the difference in tomato fruit size between the domesticated tomato and their wild relatives. *Fw2.2* was cloned using a multi-tiered strategy including high resolution mapping with RFLPs and RAPDs on near isogenic lines, physical mapping using yeast artificial chromosomes and cosmic clones, and finally complementation tests to confirm the identity of the gene. This seminal research set the stage for a better understanding of both evolution under natural selection and crop improvement under human selection. Furthermore, this outstanding achievement in resolving the genetic basis of complex phenotypes has opened the way for similar studies in other organisms.

Another major contribution was the development of comparative maps that clarified evolutionary relationships among genera and provided an essential framework for exploring questions of gene homoeology in plants. Tanksley showed that the same RFLP probes could be used to map different members of plant families. Tanksley's first comparative map (tomato-potato) was published in 1988. At the level of macro-synteny, these species appeared to be differentiated by only 5 major translocations. Following the investigation of the Solanaceae family, Tanksley's group demonstrated the extent of macro-synteny in the distantly related sub-families of the grasses, notably rice and maize. Today synteny is accepted as the rule throughout crop plant genetics and the concept pervades plant genetics research.

Tanksley's current work demonstrates a new marker-assisted breeding strategy that is having a direct impact on plant improvement. His group uses maps and markers to identify new yield-enhancing genes/QTLs from wild species and transfer them to elite cultivars of tomato and rice. Based on an "Advanced Backcross" strategy, Tanksley aims to demonstrate an effective new concept in breeding that is based on the ability of molecular marker analysis to identify favourable genes or QTLs from almost any cross and efficiently introgress them into existing high-yielding cultivars. By looking for novel alleles in wild and unadapted germplasm, this approach broadens the gene pool of cultivated crop species. The power of advanced backcross has been demonstrated in tomato and in rice, with yield increases of 15-20%. When a yield enhancement of 18% was documented in an wild-cultivated BC2 rice testcross generation compared to one of the most widely planted hybrids in China, rice breeders took an immediate interest. This was the first example of a molecular strategy for plant improvement that does not simply try to overlay marker based selection on a traditional breeding framework. The impact of this work is likely to be felt most profoundly in international circles where staple food production is a pressing concern and yield per acre is an important target in the face of increasing population pressure. The work is also impacting on the private sector where novel quality traits are of interest and reservoirs of wild species are now a proven source of genes.

Steve Tanksley has promoted the genomic revolution in biology across a broad front. His group developed algorithms that facilitated the analysis of the complete Arabidopsis genome sequence and led to the discovery that at least 60% of this small genome is involved in segmental duplications. Tanksley was the first to demonstrate the micro-syntenic relationships between the genomes of tomato and arabidopsis. He also created a large publicly available EST database in tomato (Solanaceae Genome Network http://www.sgn.cornell.edu/) and used this to elucidate structural detail of the tomato genome. The same dataset has been used to unite the sequenced maps of diverse plant species. His group screened the tomato ESTs against the arabidopsis genomic sequence and reported the identification of a set of 1000 genes (referred to as a conserved ortholog set, or COS markers) which have remained relatively stable in sequence since the early radiation of dicotyledonous plants. These COS markers also are conserved in the genomes of other plant families and can be used for comparative mapping studies between highly divergent genomes. Tanksley's most recent work has focused on the genetic dissection of adaptive traits in tomato.

In summary, Tanksley has provided many of the tools and generated much of the fundamental knowledge and concepts that are the basis of plant genetics and crop improvement today.

LIST OF PUBLICATIONS

1. Chen K, Cong B, Wing R, Vrebalov J, Tanksley S (2007) Regulatory Mutation in a Cell Elongation Transcription Factor Responsible for the Evolution of Autogamy in Cultivated Tomatoes. Science 318, 643.

2. McCarthy, A., Biget, L., Lin, C., Petiard, V., Tanksley, S., and McCarthy, J. Cloning (2007) Expression, crystallization and preliminary X-ray analysis of the XMT and DXMT N-methyltransferases from Coffea canephora (robusta). Acta Crystallographica F63, 304-307.
3. Tanksley S, Fulton T (2007) Dissecting quantitative trait variation—examples from the tomato. Euphytica 145: 365-370.
4. Hobolth A, R. Nielsen, Y. Wang, F. Wu, and S. D. Tanksley. (2006) CpG + CpNpG analysis of protein-coding sequences from tomato. Mol. Biol. Evol. 2006 23:1318-1323.
5. Hinniger C, Caillet V, Michoux F, Amor M, Tanksley S, Lin C, McCarthy J (2006) Isolation and Characterization of cDNA Encoding Three Dehydrins Expressed During Coffea canephora (Robusta) Grain Development. Annals of Botany 97:755-765.
6. Carlson J, Leebens-Mack J, Wall P, Zahn L, Mueller L, Landherr L, Hu Y, Ilut D, Arrington J, Choirean S, Becker A, Field D, Tanksley S, Ma H, dePamphilis C (2006) EST database for early flower development in California poppy (Eschscholzia californica Cham., Papaveraceae) tags over 6000 genes from a basal eudicot. Plant Molecular Biology 62 351-369.
7. Wang Y, R. S. van der Hoeven, R. Nielsen, L. A. Mueller and S. D. Tanksley (2006) Characteristics of the tomato nuclear genome as determined by sequencing undermethylated EcoRI digested fragments. Theor Appl Genet 112: 72-84.
8. Wang Y, Tang X,Cheng Z, Mueller M, Giovannoni J, Tanksley S (2006) Euchromatin and Pericentromeric Heterochromatin: Comparative Composition in the Tomato GenomeGenetics 2006 172: 2529-2540.
9. Wu F, Mueller L, Crouzillat D, Petiard V(2006) Bioinformatics and Phylogenetics to Identify Large Sets of Single Copy, Orthologous Genes (COSII) for Comparative, Evolutionary and Systematic Studies: A Test Case in the Euasterid Plant Clade Genetics; 174: 1407-142.
10. Cong B, Tanksley SD (2006) FW2.2 and cell cycle control in developing tomato fruit: a possible example of gene co-option in the evolution of a novel organ. Plant Mole Biol 62:867-880.
11. Barrero L.S., B. Cong, F. Wu, and S.D. Tanksley 2006 Developmental characterization of the *fasciated* locus and mapping of candidate genes involved in the control of floral meristem size and carpel number in tomato. Genome 49:991-1006.
12. Frary A, Xu Y, Liu J, Mitchell Sh, Tedeschi E, Tanksley SD (2005) Development of a set of PCR-based anchor markers encompassing the tomato genome and evaluation of their usefulness for genetics and breeding experiments. Theor Appl Genet 111:291.

13. Albert, V.A., D.E. Soltis, J. Carlson, W.G. Farmerie, K. Wall, D.C. Ilut, T. M. Solow, L.A. Mueller, L.L. Landherr, Y. Hu, M. Buzgo, S. Kim, M-J. Yoo, M.W. Frohlich, R. Perl-Treves, S. Schlarbaum, B. Bliss, X. Zhang, S. Tanksley, D.G. Oppenheimer, P.S. Soltis, H. Ma, C.W. dePamphilis, and J.H. Leebens-Mack. (2005) Floral gene resources from basal angiosperms for comparative genomics research. BMC Plant Biology 5:5.
14. Yogeeswaran K, Frary A, York T, Amenta A, Lesser A, Nasrallah J, Tanksley S, Nasrallah M (2005)Comparative genome analyses of Arabidopsis spp.: Inferring chromosomal rearrangement events in the evolutionary history of A. thaliana. Genome Res., Apr 2005; 15: 505-515.
15. York T, Durrett R, Tanksley S, Nielsen R (2005) Genetical Research. 85 Bayesian and maximum likelihood estimation of genetic maps 85:159-168.
16. Lin,-Chenwei, Mueller L, McCarthy J, Crouzillat D, Petiard V, Tanksley S. 2005 Coffee and tomato share common gene repertoires as revealed by deep sequencing of seed and cherry transcripts. Theoretical-and-Applied-Genetics. 112(1): 114-130.
17. Barrero L.S., and S. D. Tanksley 2004 Evaluating the Genetic Basis of Multiple-locule Fruit in a Broad Cross Section of Tomato Cultivars. Theor Appl Genet 109: 669-679.
18. Frary, Amy, Fritz, Lisa A., and S. D. Tanksley 2004 A Comparative Study of the Genetic Bases of Natural Variation in Morphology of Tomato Leaves, Sepals, and Petals. Theor Appl Genet 109: 523-533.
19. van der Knaap, E., Sanyal, A. Jackson, S.A. S.D. Tanksley 2004 High-Resolution Fine-Mapping and FISH analysis of *sun* a Locus Controlling Tomato Fruit Shape, Reveals a Region of the Tomato Genome Prone to DNA Rearrangements. Genetics 168(4):2127-40.
20. Chen, Kai-Yi, Tanksley, S. D. High Resolution Mapping and Functional Analysis of *se2.1*: a Major Stigma Exsertion QTL Associated with the Evolution from Allogamy to Autogamy in the Genus *Lycopersicon.* Genetics 168, 1563-1573.
21. Yates H, Frary A, Doganlar S, Frampton A, Eannetta N, Uhlig J, Tanksley S (2004) Comparative fine mapping of fruit quality QTLs on chromosome 4 introgressions derived from two wild tomato species Euphytica 135:283-296.
22. Frary, A., T. M. Fulton, D. Zamir, S. D. Tanksley 2004 Advanced Backcross QTL Analysis of a Lycopersicon Esculentum x L. Pennellii Cross and Identification of Possible Orthologs in the Solanaceae. Theor Appl Genet 108, no. 3: 485.
23. Tanksley, S. D. (2004) The Genetic, Developmental and Molecular Bases of Fruit Size and Shape Variation in Tomato. Plant Cell Special Issue 16: S181-S189.
24. Ronning et al 2003 Comparative analyses of potato expressed sequence Tag libraries. Plant Phys 131:419-429.

25. Van der Knapp E, Tanksley S D. 2003 The Making of a Bell Pepper-Shaped Tomato Fruit: Identification of Loci Controlling Fruit Morphology In Yellow Stuffer Tomato. Theor Appl Genet 107:139-147.
26. Cong B, Liu J, Tanksley SD. 2002. Natural Alleles at a Tomato Fruit Size Quantitative Trait locus Differ by Heterochronic Regulatory Mutations. Proc Natl Acad Sci 99:13606-13611.
27. Fulton, T. M., R. van der Hoeven, N.T. Eannetta, S.D. Tanksley. 2002. Identification, Analysis and Utilization of Conserved Ortholog Set (COS) Markers for Comparative Genomics in Higher Plants. Plant Cell 14:1457-1467.
28. Fulton, T. M., P. Bucheli, E. Voirol, J. Lopez, V. Petiard, and S. D. Tanksley. 2002. Quantitative Trait Loci (QTL) Affecting Sugars, Organic Acids and Other Biochemical Properties Possibly Contributing to Flavor, Identified in Four Advanced Backcross Populations of tomato. Euphytica 127:163-177.
29. Doganlar, S., A. Frary, M-C. Daunay, R. N. Lester, and S. D. Tanksley. 2002. A Comparative Genetic Linkage Map of Eggplant (*Solanum melongena*) and its Implications for Genome Evolution in the Solanaceae Genetics 161: 1697-1711.
30. Doganlar, S., A. Frary, M-C. Daunay, R.N. Lester, and S. D. Tanksley. 2002. Conservation of Gene Function in the Solanaceae as Revealed by Comparative Mapping of Domestication Traits in Eggplant Genetics 161: 1713-1726.
31. Durrett R T, Chen K-Y, Tanksley S 2002. A Simple Formula Useful for Positional Cloning. Genetics 160: 353-355.
32. Giovannoni J et al. Genetic Control of Fruit Quality and Prospects for Nutrient Modification. HortScience. 37. P. 453-456.
33. Giovannoni J. 2002. Molecular Biology of Fruit Maturation and Ripening. Annual. Review Plant Physiol. Plant. Mol. Biol. Vol. 52: 725-749.
34. Liu, Jiping, J. Van Eck, B. Cong, and S. D. Tanksley . 2002. A New Class of Regulatiry Genes Underlying the Cause of Pear-Shaped Tomato Fruit PNAS. Vol. 99: 13302-13306.
35. Payton P et al. Creation of a baseline gene expression profile for tomato fruit ripening: characterization of ripening, ethylene signaling, and light signaling genes. 53, No. 377. 2023-2030.
36. Van der Hoeven, R., C. Ronning, J. Giovannoni, G. Martin, S.D. Tanksley. 2002. Deductions about the number, organization and evolution of genes in the tomato genome based on analysis of a large EST collection and selective genomic sequencing. Plant Cell 14(7):1441-56.
37. Ku H-M, Liu J, Doganlar S, Tanksley S 2001. Exploitation of Arabidopsis-Tomato Synteny to Construct a High-Resolution Map of the *ovate* Locus in Tomato. Genome 44:470-475.

38. Lippman, Z, Tanksley 2001. Dissecting the genetic pathway to extreme fruit size in tomato using a cross between the small-fruited wild species *L. pimpinellifolium* and *L. esculentum*, var. 'Giant Heirloom'. Genetics 158:413-422.
39. Monforte A, Freidman E, Zamir D, Tanksley S 2001. Comparison of a set of allelic QTL-NILs for chromosome 4 of tomato: Deductions about natural variation and implications for germplasm utilization. Theor Appl Genet 102:572-590.
40. Van der Knapp E, Tanksley S 2001. Identification and characterization of a novel locus controlling early fruit development in tomato. Theor Appl Genet 103:353-358.
41. Brommonschenkel S, Frary A, Frary A, Tanksley S 2000. The broad spectrum tospovirus resistance gene *Sw-5* of tomato is a homolog of the root knot nematode resistance gene *Mi.* 13(10):1130-8.
42. Doganlar S, Tanksley S, Mutschler M 2000. Identification and molecular mapping of loci controlling fruit ripening time in tomato. Theor Appl Genet 100(2):249-255.
43. Doganlar S, Frary A, Tanksley S 2000. The genetic basis of seed weight variation: tomato as a model system. Theor Appl Genet 100(8):1267-1273.
44. Frary A, Nesbitt TC, Frary A, Grandillo S, van der Knaap E, Cong B, Liu J, Meller J, Elber R, Alpert K, Tanksley S 2000. Cloning and Transgenic Expression of *fw2.2*: a Quantitative Trait Locus Key to the Evolution of Tomato Fruit. Science 289:85-87.
45. Fulton T M, Grandillo S, Beck-Bunn T, Fridman E, Frampton A, Lopez J, Petiard V, Uhli J, Zamir D, Tanksley S 2000. Advanced backcross QTL analysis of a *Lycopersicon esculentum* x *L. parviflorum* cross Theor Appl Genet 100:1025-1042.
46. Ku H-M, Vision T, Liu J, Tanksley S 2000. Comparing Sequenced Segments of the Tomato and Arabidopsis Genomes: Large-Scale Duplication Followed by Selective Gene Loss Creates a Network of Synteny. PNAS 97: 9121-9126.
47. Ku H, Grandillo S, Tanksley S 2000. fs8.1, a major QTL, sets the pattern of tomato carpel shape well before anthesis. Theor Appl Genet 101:873-878.
48. Monforte A, Tanksley S 2000. Development of a set of near isogenic and backcross recombinant inbred lines containing most of the *Lycopersicon hirsutum* genome in a *L. esculentum* genetic background: A tool for gene mapping and gene discovery. Genome 43:803-813.
49. Monforte A, Tanksley S 2000. Fine mapping of a quantitative triat locus (QTL) from Lycopersicon hirsutum chromosome 1 affecting fruit characteristics and agronomic traits: breaking linkage among QTLs affecting different traits and dissection of heterosis for yield. Theor Appl Genet 100:471-479.

50. Vision T, Brown D, Tanksley S 2000. The origins of genomic duplications in *Arabidopsis.* Science 290:2114-2117.
51. Vision T, Brown D, Shmoys D, Durret R, Tanksley S 2000. Selective Mapping: A Strategy for Optimizing the Construction of High-Density Linkage Maps. Genetics 155:407-429.
52. Grandillo S, Ku H-M, Tanksley S (1999) Identifying loci responsible for natural variation in fruit size and shape in tomato. Theor Appl Genet 99:978-987.
53. Hamilton CM, Frary A, Xu Y, Tanksley S, Zhang H-B (1999) Construction of tomato genomic DNA libraries in a binary-BAC (BIBAC) vector. The Plant Journal 18:223-229.
54. Haanstra JPW, Wye C, Verbakel H, Meijer-Dekens F, van den Berg P, Odinot P, van Heusden AW, Tanksley D, Lindhout P, Peleman J. (1999) An integrated high-density RFLP-AFLP map of tomato based on two Lcopersicon esculentum x L. pennellii F2 populations. Theor Appl Genet 99:254-271.
55. Ku H-M, Doganlar S, Chen K-Y, Tanksley S (1999) The genetic basis of pear-shaped tomato fruit. Theor Appl Genet 9:844-850.
56. Monforte A, Zamir D, Tanksley S (2000) Comparison of a set of allelic QTL-NILs for chromosome 4 of tomato: Deductions about natural variation and implications for germplasm utilization (in press).
57. Monforte A, Tanksley S (2000) Fine mapping of a quantitative triat locus (QTL) from Lycopersicon hirsutum chromosome 1 affecting fruit characteristics and agronomic traits: breaking linkage among QTLs affecting different traits and dissection of heterosis for yield. Theor Appl Genet 100:471-479.
58. Doganlar S, Tanksley S, Mutschler M (2000)) Identification and molecular mapping of loci controlling fruit ripening time in tomato. Theor Appl Genet 100(2):249-255.
59. Doganlar S, Frary A, Tanksley S (2000) The genetic basis of seed weight variation: tomato as a model system. Theor Appl Genet 100(8): 1267-1273.
60. Brommonschenkel S, Frary A, Frary A, Tanksley S (2000) The broad spectrum tospovirus resistance gene *Sw-5* of tomato is a homolog of the root knot nematode resistance gene *Mi.* 13:1030.
61. Bucheli, Peter, Elisabeth Voirol, Rosa de la Torre, Joaquin Lopez, Andre Rytz, Steven Tanksley and Vincent Petiard. 1999. Definition of non-volatile markers for flavor of tomato (*Lycopersicon esculentum* Mill.) as tools in selection and breeding. J. Food Agric. Chem. 47 (2) 659-664.
62. Bernacchi, D., T. Beck-Bunn, Y. Eshed, J. Lopez, V. Petiard, J. Uhlig, D. Zamir and S. Tanksley. 1998. Advanced backcross QTL analysis of tomato. I. Identification of QTL for traits of agronomic importance from *Lycopersicon hirsutum.* Theor Appl Genet 97:381-397.

63. Bernacchi, D., T. Beck-Bunn, d. Emmatty, Y. Eshed, S. Inai, J. Lopez, V. Petiard, H. Sayama, J. Uhlig, D. Zamir and S. Tanksley. 1998. Advanced backcross QTL analysis of tomato. II: Evaluation of near-isogenic lines carrying single donor introgressions for desirable wild QTL-alleles derived from *Lycopersicon hirsutum* and *L. pimpinellifolium*. Theor Appl Genet 97:170-180.
64. Doganlar, Sami, Jeff Dodson, Brad Gabor, Teresa Beck-Bunn, Curt Crossman and Steven D. Tanksley. 1998. Molecular mapping of the *py-1* gene for resistance to corky root rot (*Pyrenochaeta lycopersici*) in tomato. Theor Appl Genet 97:784-788.
65. Grandillo, Silvana, Daniel Zamir and Steven D. Tanksley. 1999. Genetic improvement of processing tomatoes: A 20 years perspective. Euphytica 110:85-97.
66. Diwan. N, R. Fluhr, Y. Eshed, S.D. Tanksley and D. Zamir. 1998. Mapping of Ve in tomato: A gene conferring resistance to the broad spectrum pathogen, Verticillium dahliae race 1. Theor Appl Genet 98(2):315-319.
67. Tanksley, Steven D., Dario Bernacchi, Teresa Beck-Bunn, Davy Emmatty, Yuval Eshed, Shuji Inai, Joachin Lopez, Vincent Petiard, Haruki Sayama, John Uhlig, Daniel Zamir. 1998. Yield and quality evaluations on a pair of processing tomato lines nearly isogenic for the Tm2a Gene for resistance to the tobacco mosaic virus. Euphytica 99:77-83.
68. Bernacchi and Steven D. Tanksley. 1997. An interspecific backcross of Lycopersicon esculentum x L. hirsutum: Linkage analysis and a QTL study of sexual compatibility factors and floral traits. Genetics 147:861-877.
69. Brommonschenkel, S.H. and S.D. Tanksley. 1997. Map-based cloning of the tomato genomic region that spans the Sw-5 tospovirus resistance gene in tomato. Mol Gen Genet 256:121-126.
70. Chunwongse, J., S. Doganlar, C. Crossman, J. Jiang, and S.D. Tanksley. 1997. High resolution genetic map of the Lv resistance locus in tomato. Theor Appl Genet 95:220-223.
71. Doganlar, Sami, Anne Frary, and Steven D. Tanksley. 1997. Production of interspecific F1 hybrids, BC1, BC2, and BC3 populations between Lycopersicon esculentum and two accessions of Lycopersicon peruvianum carrying new root-knot nematode resistance genes. Euphytica 95(2) 203-207.
72. Fulton, Theresa M., James C. Nelson, Steven D. Tanksley. 1997. Introgression and DNA marker analysis of Lycopersicon peruvianum, a wild relative of the cultivated tomato, into Lycopersicon esculentum, followed through three successive backcross generations. Theor Appl Genet 95:895-902.
73. Fulton, Theresa M., Teresa Beck-Bunn, Davy Emmatty, Yuval Eshed, Joaquin Lopez, John Uhlig, Daniel Zamir, and Steven D. Tanksley. 1997. QTL analysis of an advanced backcross of Lycopersicon peruvianum to the cultivated tomato and comparisons with QTLs found in other wild species. Theor Appl Genet 95:881-894.

74. Tanksley, Steven D. and Susan R. McCouch. 1997. Seed banks and molecular maps: unlocking genetic potential from the wild. Science 277:1063-1066.
75. Xiao, J., J. Li, S. Grandillo, S.N. Ahn, L.P. Yuan, S.D. Tanksley and S.R. McCouch. Identificatoin of trait-improving QTL alleles from a wild rice relative, *O. rufipogon* Genetics 150:899-909.
76. Alpert, Kevin B. and Steven D. Tanksley. 1996. High-resolution mapping and isolation of a yeast artificial chromosome contig containing fw 2.2: A major fruit weight quantitative locus in tomato. Proc. Natl. Acad. Sci. 93:15503-15507.
77. Autrique, Enrique, Miloudi M. Nachit, Philippe Monneveux, Steven D. Tanksley, and Mark E. Sorrels. 1996. Genetic diversity in durum wheat based on RFLPs, morphophysiological traits, and coefficient of parentage. Crop Sci. 36:735-742.
78. Broun, Pierre and Steven D. Tanksley. 1996. Characterization and genetic mapping of simple repeat sequences in the tomato genome. Mol. Gen. Genet 150:39-49.
79. Frary, Amy, Gernot G. Presting and Steven D. Tanksley. 1996. Molecular mapping of the centromeres of tomato chromosomes 7 and 9. Mol. Gen. Genet. 250:295-304.
80. Ganal, M.W. and S.D. Tanksley. 1996. Recombination around the Tm2a and Mi resistance genes in different crosses of Lycopersicon peruvianum. Theor. Appl. Genet. 92:101-108.
81. Grandillo, S. and S.D. Tanksley. 1996. Genetic analysis of RFLPs, GATA microsatellites and RAPDs in a cross between L. esculentum and L. Pimpinellifolium. Theor. Appl. Genet. 92:957-965.
82. Grandillo, S. and S.D. Tanksley. 1996. QTL analysis of horticultural traits differentiating the cultivated tomato from the closely related species Lycopersicon pimpinellifolium. Theor. Appl. Genet. 92:935-951.
83. Grandillo, Silvana, Hsin-Mei Ku and Steven D. Tanksley. 1996. Characterization of fs8.1, a major QTL influencing fruit shape in tomato. Molecular Breeding 2:251-160.
84. Hamilton, Carol M., Ane Frary, Candice Lewis, and Steven D. Tanksley. 1996. Stable transfer of intact high molecular weight DNA into plant chromosomes. Proc. Natl. Acad. Sci. 93:9975-9979.
85. Pillen, K., M.W. Ganal, and S.D. Tanksley. 1996. Construction of a high-resolution genetic map and YAC-contigs int he tomato Tm-2a region. Theor. Appl. Genet. 93:228-233.
86. Pillen, Klaus, Kevin B. Alpert, James J. Giovannoni, Martin W. Ganal, and Steve D. Tanksley. 1996. Rapid and reliable screening of a tomato YAC library exclusively based on PCR. Plant Mol. Biology Reporter 14:58-67.

87. Tanksley, S.D. and J.C. Nelson. 1996. Advanced backcross QTL analysis: a metod for the simultaneous discovery and transfer of valuable QTLs from unadapted germplasm into elite breeding lines. Theor. Appl. Genet. 92:191-203.
88. Tanksley, S.D., S. Grandillo, T.M. Fulton, D. Zmir, Y. Eshed, V. Petiard, J. Lopez, and T. Beck-Bunn. 1996. Advanced backcross QTL analysis in a cross between an elite processing line of tomato and its wild relative L. pimpinellifolium. Theor. Appl. Genet. 92:213-224.
89. Xiao, Jinhua, Jiming Li, Longping Yuan and Steve D. Tanksley. 1996. Dominance is the major genetic basis of heterosis in rice as revealed by QTL analysis using molecular markers. Genetics 140:745-754.
90. Xiao, Jinhua, Silvana Grandillo, Sang Nag Aho, Susan R. McCouch, Steven D. Tanksley, Jiming Li, and Longping Yuan. 1996. Genes from wild rice improve yield. Scientific Correspondence - Nature 384:223-224.
91. Xiao, J., J. Li, L.P. Yuan, S.R. McCouch and S.D. Tanksley. 1996. Genetic diversity and its relationship to hybrid performance and heterosis in rice as revealed by PCR-based markers. Theor. Appl. Genet. 92:637-643.
92. Yencho, G. Craig, Meredith W. Bonierbale, Ward M. Tingey, Robert L. Plaisted, and Steven D. Tanksley. 1996. Molecular markers locate genes for resistance to the Colorado potato beetle, Leptinotarsa decemlineata, in hybrid Solanum tuberosum x S. berthaultii potato progenies. Entomologia Experimentalis et Applicata 81: 141-154.
93. Yu, Z. H., D. J. Mackill, J. M. Bonman, S. R. McCouch, E. Guiderdoni, J. L. Notteghem, and S. D. Tanksley. 1996. Molecular mapping of genes for resistance to rice blast (Pyricularia grisea Sacc.). Theor Appl Gene 93:859-863.
94. Azanza, F., D. Kim, S.D. Tanksley, J.A. Juvik. 1995. Genes from Lycopersicon chmielewskii affecting tomato quality during fruit ripening. Theor. App. Genet. 91:495-504.
95. Dixon, M.S., D.A. Jones, K. Hatzixanthis, M.W. Ganal, S.D. Tanksley, and J.D.G. Jones. 1995. High-resolution mapping of the physical location of the tomato Cf-2 gene.
96. Ganal, Martin W., R. Simon, S. Brommonschenkel, M. Arndt, M.S. Phillips, S. D. Tanksley, and A. Kumar. 1995. Genetic mapping of a wide spectrum nematode resistance gene (Hero) against Globodera rostochiensis in tomato. Molecular Plant-Microbe Interactions 6:886-891.
97. Giovannoni, James J., Erick N. Noensie, Diane M. Ruezinsky, Xianghuai Lu, Samantha L. Tracy, Martin W. Ganal, Gregory B. Martin, Klaus Pillen, Kevin Alpert, Steven D. Tanksley. 1995. Molecular genetic analysis of the ripening-inhibitor and non-ripening loci of tomato: a first step in genetic map-based cloning of fruit ripening genes. Mol Gen Genet 248:195-206.

98. O'Donoughue, L.S., S.F. Kianian, P.J. Rayapati, G.A. Penner, M.E. Sorrells, S.D. Tanksley, R.L. Phillips, H.W. Rines, M. Lee, G. Fedak, S.J. Molnar, D. Hoffman, C.A. Salas, B. Wu, E. Autrique, and A. Van Deynze. 1995. A molecular linkage map of cultivated oat. Geneome 38:368-380.
99. Roder, Marion S., Mark E. Sorrells, and Steven D. Tanklsey. 1995. Pulsed-field gel analysis of 5S and satellite DNA in barley. Genome 38:153-157.
100. Roder, Marion, Jens Plaschke, Susanne U. Konig, Andreas Borner, Mark E. Sorrells, Steven D. Tanksley, and Martin W. Ganal. 1995. Abundance, variability and chromosomal location of microsatellites in wheat. Mol. Gen. Genet. 246:327-333.
101. Tanksley, Steven D., Martin W. Ganal and Gregory B. Martin. 1995. Chromosome landing: a paradigm for map-based gene cloning in plants with large genomes. Trends in Genetics 11:63-68.
102. Yen, Hsiao-Ching, Sanghyeob Lee, Steven D. Tanksley, Michael B. Lanahan, Harry J. Klee, and James J. Giovannoni. 1995. The tomato never-ripe locus regulates ethylene-inducible gene expression and is linked to a homolog of the arabidopsis ETR1 Gene. Plant Physiol. 107:1343-1353.
103. Yu Z.H., S.R. McCouch, T. Kinoshita, S. Sato, and S.D. Tanksley 1995. Association of Morphological and RFLP Markers in Rice (*Oryza sativa* L.). Genome 38: 38:566-574.
104. Arumuganathan, K., Gregory B. Martin, Hakan Telenius, Steven D. Tanksley, and Elizabeth D. Earle. 1994. Chromosome 2-specific DNA clones from flow-sorted chromosomes of tomato. Mol. Gen. Genet. 242:551-558.
105. Autrique, Enrique, Ravi P. Singh, Steven D. Tanksley, and Mark E. Sorrells. 1994. Molecular markers for four leaf rust resistance genes introgressed into wheat from wild relatives. Crop Science Genome 38:75-83.
106. Azanza, F., T.E. Young, D. Kim, S.D. Tanksley, and J.A. Juvik. 1994. Characterization of the effects of introgressed segments of chromosome 7 and 10 from Lycopersicon chmielewskill on tomato soluble solids, pH, and yield. Theor. Appl. Genet. 87:965-972.
107. Bonierbale, M.W., R.L. Plaisted, O. Pineda, and S.D. Tanksley. 1994. QTL analysis of trichome-mediated insect resistance in potato. Theor. Appl. Genet. 87:973-987.
108. Causse, M.A., T.M. Fulton, Y.G. Cho, S.N. Ahn, J. Chunwongse, K. Wu, J. Xiao, Z. Yu, P.C. Ronald, S.E. Harrington, G. Second, S.R. McCouch, and S.D. Tanksley. 1994. Saturated molecular map of the rice genome based on an interspecific backcross population. Genetics 138:1251-1274.
109. Chunwongse, J., T.B. Bunn, C. Crossman, J. Jiang, and S.D. Tanksley. 1994. Chromosomal localization and molecular marker tagging of the powdery mildew resistance gene (Lv) in tomato. Theor. Appl. Genet. 89:76-79.

110. Ma, Z.Q., M.E. Sorrells, and S.D. Tanksley. 1994. RFLP markers linked to powdery mildew resistance genes Pm1, Pm2, Pm3 and Pm4 in wheat. Genome. 37:871-875.
111. Martin, Gregory B., Anne Frary, Tiyun Wu, Sergio Brommonschenkel, Julapark Chunwongse, Elizabeth D. Earle, and Steven D. Tanksley. 1994. A member of the tomato Pto gene family confers sensitivity to fenthion resulting in rapid cell death. The Plant Cell 6:1543-1552.
112. O'Donoughue, L.S., E. Souza, S.D. Tanksley, and M.E. Sorrells. 1994. Relationships among North American oat cultivars based on restriction fragment length polymorphism. Crop Sci. 34:1251-1258.
113. Ori, N., Paran, I., Aviv, D., Eshed, Y., Tanksley, S., Zamir, D., and Fluhr, R. 1994. A genomic serarch for the gene conferring resistance to Fusarium wilt in tomato. Euphytica 79:201-204.
114. Paul, E., M. Goto, and S.D. Tanksley. 1994. SolGenes - a Solanaceae database. Euphytica 79:181-186.
115. Wing, Rod A., Hong-Bin Zhang, and Steven D. Tanksley. 1994. Map-based cloning in crop plants. Tomato as a model system: I. Genetic and phsyical mapping of jointless. Mol. Gen. Genet. 242:681-688.
116. Wing, Rod A., Vipin K. Rastogi, Hong-bin Zhang, Andrew H. Paterson, and Steven D. Tanksley. 1994. An improved method of plant megabase DNA isolation in agarose microbeads suitable for physical mapping and YAC cloning. The Plant Journal. 4:893-898.
117. Ahn, S. and S.D. Tanksley. 1993. Comparative linkage maps of rice and maize genomes. Proc. Natl. Acad. Sci. USA 90:7980-7984.
118. Ahn, S., J.A. Anderson, M.E. Sorrells, and S.D. Tanksley. 1993. Homoelogous relationships of rice, wheat and maize chromosomes. Mol. Gen. Genet. 241:483-490.
119. Anderson, J.A., G.A. Churchill, J.E. Autrique, S.D. Tanksley, and M.E. Sorrells. 1993. Optimizing parental selection for genetic linkage maps. Genome 31:181-186.
120. Anderson, J.A., M.E. Sorrells, and S.D. Tanksley. 1993. Detection of QTLs affecting pre-harvest sprouting resistance in wheat by RFLPs. Crop Sci. 33:453-459.
121. Anderson, James A., Mark E. Sorrells, and Steven D. Tanksley. 1993. RFLP analysis of genomic regions associated with resistance to preharvest sprouting in wheat. Crop Sci. 33:453-459.
122. Chunwongse, J., G.B. Martin, and S.D. Tanksley. 1993. Pre-germination genotypic screening using PCR amplification of half-seeds. Theor. Appl. Genet. 86:694-698.
123. Da Silva, Jorge A.G., Mark E. Sorrells, William L. Burnquist, and Steven D. Tanksley. 1993. RFLP linkage map and genome analysis of Saccharum spontaneum. Genome 36:782-791.

124. deVicente, M.C. and S.D. Tanksley. 1993. QTL analysis of transgressive segregation in an interspecific tomato cross. Genetics 134:585-596.
125. Mackill, D.J., M.A. Salam, Z.Y. Wang, and S.D. Tanksley. 1993. A major photoperiod sensitivity gene tagged with RFLP and isozyme markers in rice. Theor. Appl. Genet. 85:536-540.
126. Martin, Gregory B., Sergio H. Brommonschenkel, Julapark Chunwongse, Anne Frary, Martin W. Ganal, Rody Spivey, Tiyun Wu, Elizabeth D. Earle, and Steven D. Tanksley. 1993. Map-based cloning of a protein kinase gene conferring disease resistance in tomato. Science 262:1432-1436.
127. Pineda, O., M.W. Bonierbale, R.L. Plaisted, B.B. Brodie, and S.D. Tanksley. 1993. Identification of RFLP markers linked to the H1 gene conferring resistance to the potato cyst nematode Globodera rostochiensis. Genome 36:152-156.
128. Roder, Marion S., Nora L.V. Lapitan, Mark E. Sorrells, and Steven D. Tanksley. 1993. Genetic and physical mapping of barley telomeres. Mol. Gen. Genet. 238-303.
129. Tanksley, Steven D. 1993. Mapping polygenes. Annu. Rev. Genet. 27:205-233.
130. Tanksley, S.D., T.M. Fulton, and S.R. McCouch. 1993. Linkage map of rice (Oryza sativa) (2N=24) in O'Brien, S.J. (ed.). Genetic Maps 6.61-6.79.
131. Wu, Kun-Sheng and Steven D. Tanksley. 1993. Genetic and physical mapping of telomeres and macrosatellites of rice. Plant Molecular Biology 22:861-872.
132. Ahn, S. N., C. N. Bollich, and S.D. Tanksley. 1992. RFLP tagging of a gene for aroma in rice. Theor. Appl. Genet. 84:825-828.
133. Anderson, J.A., Y. Ogihara, M.E. Sorrells, and S.D. Tanksley. 1992. Developmnt of a chromosomal arm map for wheat based on RFLP markers. Theor. Appl. Genet. 83:1035-1043.
134. Broun, Pierre, Martin W. Ganal, and Steven D. Tanksley. 1992. Telomeric arrays display high levels of heritable polymorphism among closely related plant varieties. Proc. Natl. Acad. Sci. USA 89:1354-1357.
135. Goffreda, J.C., W.B. Burnquist, S.C. Beer,S.D. Tanksley, and M.E. Sorrells. 1992. Application of molecular markers to assess genetic relationships among accessions of wild oat, Avena sterilis. Theor. Appl. Genet. 5:146-151.
136. Ma, Z.-Q., B.S. Gill, M.E. Sorrells, and S.D. Tanksley. 1992. RFLP markers linked to two Hessian fly-resistant genes in wheat (Triticum tauschii (coss.) Schmal. Theor. Appl. Genet. 85:750-754.
137. Martin, Gregory B., de Vicente, M. Carmen, and Steven D. Tanksley. 1992. High resolution linkage analysis and physical characterization of the Pto bacterial resistance locus in tomato. MPMI.
138. Martin, Gregory B., Martin W. Ganal, and Steven D. Tanksley. 1992. Construction of a yeast artificial chromosome library of tomato and identification of cloned segments linked to two disease resistance loci. Mol. Gen. Genet. 233:25-32.

139. O'Donoughue, L., Z. Wang, M. Roder, M. Leggett, M.E. Sorrells, and S.D. Tanksley. 1992. An RFLP based linkage map of oats based on a cross between to diploid taxa (Avena atlantica x A. hirtula). Genome 35:765-771.
140. Prince, J.P., F. Loaiza-Figueroa, and S.D. Tanksley. 1992. Restriction fragment length polymorphism and genetic distance among Mexican accessions of Capsicum. Genome 35:726-732.
141. Prince, James P. and Steven D. Tanksley. 1992. Restriction fragment length polymorphisms in plant breeding and genetics. Proc. Roy. Soc. Edinburgh 99B:23-29.
142. Roder, M.S., M.E. Sorrells, and S.D. Tanksley. 1992. 5S ribosomal gene clusters in wheat: pulsed field gel elecrophoresis reveals a high degree of polymorphism. Mol. Gen. Genet. 232:215-220.
143. Ronald, P.C., B. Albano, R. Tabien, L. Abenes, K. Wu, S. McCouch, and S.D. Tanksley. 1992. Genetic and physical analysis of the rice bacterial blight disease resistance locus, Xa21. Mol. Gen. Genet. 236:ll3-120.
144. Sorrells, M.E., J.A. Anderson, Y. Ogihara, and S.D. Tanksley. 1992. Development and application of a chromosomal arm map for wheat based on RFLP markers. p. 3-17 in: Gill,B.S., W.J. Raupp, and H. Corke (eds.) Progress in genome mapping of wheat and rlated species: Proceedings of the 2nd public workshop of the International Triticease Mapping Initiative, Manhattan, Kansas, 1991. Report No. 10. Genetic Resource Conservation Program, Davis, CA.
145. Tanksley, S., M. Causse, T. Fulton, N. Ahn, Z. Wang, K. Wu, J. Xiao, P. Ronald, Z. Yu, G. Second, and S. McCouch. 1992. A high density molecular map of the rice genome. Rice Genetics Newsletter 9:111-115.
146. Tanksley, S.D., M.W. Ganal, J.P. Prince, M.C. deVicente, M.W. Bonierbale, P. Broun, T.M. Fulton, J.J. Giovanonni, S. Grandillo, G.B. Martin, R. Messeguer, J.C. Miller, L. Miller, A.H. Paterson, O. Pineda, M.S. Roder, R.A. Wing, W. Wu, and N.D. Young. 1992. High density molecular linkage maps of the tomato and potato genomes. Genetics 132:1141-1160.
147. Wang, Z. Y., G. Second, and S. D. Tanksley. 1992. Polymorphism and phylogenetic relationships among species in the genus Oryza as determined by analysis of nuclear RFLPs. Theor. Appl. Genet. 83:565-581.
148. Wu, K.K., W. Burnquist, M.E. Sorrells, T.L. Tew, P.H. Moore, and S.D. Tanksley. 1992. The detection and estimation of linkage in polyploids using single-dose restriction fragments. Theor. Appl. Genet. 83:294-300.
149. Xiao, J., T. Fulton, S. McCouch, S. Tanksley, N. Kishimoto, R. Chsawa, Y. Ukai, and A. Saito. 1992. Progress in integration of the molecular maps of rice. Rice Genet. Newsletter 9:124-128.
150. de Vicente, M.C. and S.D. Tanksley. 1991. Genome-wide reduction in recombination of backcross progeny derived from male versus female gametes in an interspecific cross of tomato. Theor. Appl. Genet. 83:173-178.

151. Ganal, Martin W., Nora L. V. Lapitan and Steven D. Tanksley. 1991. Macrostructure of the tomato telomeres. The Plant Cell 3:87-94.
152. Ganal, Martin W., Meredith W. Bonierbale, Marion S. Roeder, William D. Park and Steven D. Tanksley. 1991. Genetic and physical mapping of the patatin genes in potato and tomato. Mol. Gen. Genet. 225:501-509.
153. Giovannoni, James J., Rod A. Wing, Martin W. Ganal, and Steven D. Tanksley. 1991. Isolation of molecular markers from specific chromosomal intervals using DNA pools from existing mapping populations. Nucleic Acids Research 19:6553-6558.
154. Heun, M., A.E. Kennedy, J.A. Anderson, N.L.V. Lapitan, M.E. Sorrells, and S.D. Tanksley. 1991. Construction of a restriction fragment length polymorphism map for barley (Hordeum vulgare). Genome 34:437-447.
155. Lapitan, Nora L.V., Martin W. Ganal, and Steven D. Tanksley. 1991. Organization of the 5S ribosomal RNA genes in the genome of tomato. Genome 34:509-514.
156. Martin, Gregory, John G. K. Williams and Steven D. Tanksley. 1991. Rapid identification of markers linked to a Pseudomonas resistance gene in tomato by using random primers and near-isogenic lines. Proc. Natl. Acad. Sci. USA 88:2336-2340.
157. McCouch, S.R., M.L. Abenes, R. Angeles, G.S. Khush, and S.D. Tanksley. 1991. Molecular tagging of a recessive gene, Xa-5, for resistance to bacterial blight of rice. Rice Genet. Newsl. 8:143-145.
158. Messeguer, R. M. Ganal, M.C. de Vicente, N.D. Young, H. Bolkan, and S.D. Tanksley. 1991. High resolution RFLP map around the root knot nematode resistance gene (Mi) in tomato. Theor. Appl. Genet. 82:529-536.
159. Messeguer, Ramon, Martin W. Ganal, John C. Steffens and Steven D. Tanksley. 1991. Characterization of the level, target sites and inheritance of cytosine methylation in tomato nuclear DNA. Plant Molecular Biology 16:753-770.
160. Paterson, A.H., S.D. Tanksley, and M.E. Sorrells. 1991. DNA markers in plant improvement. Adv. Agron. 46:40-90.
161. Paterson, Andrew H., Susan Damon, John D. Hewitt, Daniel Zamir, Haim D. Rabinowitch, Stephen E. Lincoln, Eric S. Lander, and Steven D. Tanksley. 1991. Mendelian factors underlying quantitative traits in tomato: comparison across species, generations, and environments. Genetics 127:181-197.
162. Yu, Z. H., D. J. Mackill, J. M. Bonman and S. D. Tanksley. 1991. Tagging genes for blast resistance in rice via linkage to RFLP markers. Theor. Appl. Genet. 81:471-476.
163. Ganal, M. W., G. B. Martin, R. Messeguer and S. D. Tanksley. 1990. Application of RFLPs, physical mapping and large DNA technologies to the cloning of important genes from crop plants. AgBiotech News and Information 2:835-840.

164. Miller, J. C. and S. D. Tanksley. 1990. RFLP analysis of phylogenetic relationships and genetic variation in the genus Lycopersicon. Theor. Appl. Genet. 80:437-448.
165. Miller, J. C. and S. D. Tanksley. 1990. Effect of different restriction enzymes, probe source, and probe length on detecting restriction fragment length polymorphism in tomato. Theor. Appl. Genet. 80: 385-389.
166. Paterson, Andrew H., Joseph W. DeVerna, Brenda Lanini and Steven D. Tanksley. 1990. Fine mapping of quantitative trait loci using selected overlapping recombinant chromosomes, in an interspecies cross of tomato. Genetics 124:735-742.
167. Tai, Thomas H. and Steven D. Tanksley. 1990. A rapid and inexpensive method for isolation of total DNA from dehydrated plant tissue. Plant Molecular Biology Report 8:297-303.
168. Arumuganathan, K., J. P. Slattery, S. D. Tanksley, and E. D. Earle. 1991. Preparation and flow cytometric analysis of metaphase chromosomes of tomato. Theor. Appl. Genet. 82:101-111.
169. Ganal, Martin W. and Steven D. Tanksley. 1989. Analysis of tomato DNA by pulsed field gel electrophoresis. Plant Mol. Biol. Rep. 7:17-27.
170. Ganal, Martin W., Nevin D. Young and Steven D. Tanksley. 1989. Pulsed field gel electrophoresis and physical mapping of large DNA fragments in the Tm-2a region of chromosome 9 in tomato. Mole. Gen. Genet. 215:395-400.
171. Lapitan, Nora L.V., Martin W. Ganal, and Steven D. Tanksley. 1989. Somatic chromosome karyotype of tomato based on in situ hybridization of the TGRI satellite repeat. Genome 32:992-998.
172. Roder, M., K. Arumuganathan, N.L. Lapitan, S.D. Tanksley, and E.D. Earle. 1989. A rapid method for large scale isolation of tomato metaphase chromosomes. In:G. Galling (ed.), Proceedings of the Braunschweig Symposium on Applied Plant Molecular Biology. pp. 137-140.
173. Tanksley, S. D., N. D. Young, A. H. Paterson, and M. W. Bonierbale. 1989. RFLP mapping in plant breeding: New tools for an old science. Bio/Technology 7:257-264.
174. Wang, Z. Y. and S. D. Tanksley. 1989. Restriction fragment length polymorphism in Oryza sativa L. Genome 32:1113-1118.
175. Young, N.D. and S. D. Tanksley. 1989. RFLP analysis of the size of chromosomal segments retained around the Tm-2 locus of tomato during backcross breeding. Theor. Appl. Genet. 77:353-359.
176. Young, Nevin D. and Steven D. Tanksley. 1989. Restriction fragment polymorphism maps and the concept of graphical genotypes. Theor. Appl. Genet. 77:95-101.
177. Bernatzky, R., E. Pichersky, V. S. Malik, and S. D. Tanksley. 1988. CRI — A dispersed repeated element associated with the CAB-1 locus in tomato. Plant Mole. Biol. 10:423-434.

178. Bonierbale, Meredith, Robert L. Plaisted, and Steven D. Tanksley. 1988. RFLP maps of potato and tomato based on a common set of clones reveal modes of chromosomal evolution. Genetics 120:1095-1103.

179. Ganal, Martin W., Nora L. V. Lapitan, and Steven D. Tanksley. 1988. A molecular and cytogenetic survey of major repeated DNA sequences in tomato (Lycopersicon esculentum). Mole. Gen. Genet. 213:262-268.

180. Lapitan, Nora L. V., Martin W. Ganal, and Steven D. Tanksley. 1988. Generation of a tomato somatic chromosome karyotype based on in situ hybridization of the TGR1 satellite repeat.

181. Loaiza-Figueroa, Fernando and Steven D. Tanksley. 1988. Genetics of a second locus determining pungency in chili peppers (Capsicum). J. Hered.

182. McCouch, S. R., G. Kochert, Z. Yu, Z. Wang, G. S. Khush, W. R. Coffman, and S. D. Tanksley. 1988. Molecular mapping of rice chromosomes. Theor. Appl. Genet. 76:815-829.

183. Paterson, Andrew H., Eric S. Lander, John D. Hewitt, Susan Peterson,Stephen E. Lincoln, and Steven D.Tanksley. 1988. Resolution of quantitative traits into Mendelian factors by using a complete RFLP linkage map. Nature 335:721-726.

184. Tanksley, S. D. and J. Hewitt. 1988. Use of molecular markers in breeding for soluble solids content in tomato — a re-examination. Theor. Appl. Genet. 75:811-823.

185. Tanksley, Steven D., Joyce Miller, Andrew Paterson, and Robert Bernatzky. 1988. Molecular mapping of plant chromosomes. In: Chromosome Structure and Function (J. Gustafson and R. Appels, eds.). Plenum. pp. 157-172.

186. Tanksley, Steven D., Robert Bernatzky, Nora L. Lapitan, and James P. Prince. 1988. Conservation of gene repertoire but not gene order in pepper and tomato. Proc. Natl. Acad. Sci. 85:6419-6423.

187. Young, Nevin D., Daniel Zamir, Martin W. Ganal, and Steven D. Tanksley. 1988. Use of isogenic lines and simultaneous probing to identify DNA markers tightly linked to the Tm-2a gene in tomato. Genetics 120:579-585.

188. Tanksley, S. D. and E. Pichersky. 1988. Organization and evolution of sequences in the plant nuclear genome. In: L. D. Gottlieb and S. K. Jain (eds.). Plant Evolutionary Biology. Chapman & Hall.

189. Zamir, D. and S. D. Tanksley. 1988. Tomato genome is comprised largely of fast-evolving low copy-number sequences. Mole. Gen. Genet. 213:254-261.

190. Pichersky, E., N. E. Hoffman, R. Bernatzky, B. Piechulla, S. D. Tanksley, and A. R. Cashmore. 1987. Molecular characterization and genetic mapping of DNA sequences encoding the Type 1 chlorophyla/b-binding polypeptide of photosystem 1 in Lycopersicon esculentum (tomato). Plant Molecular Biology 9:205-216.

191. Pichersky, E., N. E. Hoffman, V. S. Malik, R. Bernatzky, S. D. Tanksley, L. Szabo, and A. R. Cashmore. 1987. The tomato Cab-4 and Cab-5 genes encode a second type of CAB polypeptides localized in Photosystem II. Plant Molecular Biology 9:109-120.
192. Tanksley, S. D. 1987. Nuclear genome organization in tomato and related diploid species. American Naturalist 130:S46-S56.
193. Tanksley, S. D. and D. Zamir. 1987. Double tagging of a male-sterile gene in tomato using a morphological and enzymatic marker gene. Hort. Sci. 23:387-388.
194. Tanksley, S. D., M. A. Mutschler and C. M. Rick. 1987. Linkage map of the tomato. In:
195. Bernatzky, R. and S. D. Tanksley. 1986. Methods for detection of single or low copy sequences in tomato on Southern blots. Plant Mole. Biol. Reptr. 4:37-41.
196. Bernatzky, R. and S. D. Tanksley. 1986. Towards a saturated linkage map in tomato based on isozymes and cDNA clones. Genetics 112:887-898.
197. Bernatzky, R., and S. D. Tanksley. 1986. Genetics of actin-related sequences in tomato. Theor. Appl. Genet. 72:3l4-321.
198. Bernatzky, R.. and S. D. Tanksley. 1986. Majority of random cDNA clones correspond to single loci in the tomato genome. Mole. Gen. Genet. 203:8-14.
199. Chyi, Y. S., R. A. Jorgensen, D. Goldstein, S. D. Tanksley, and F. Loaiza. 1986. Locations and stability of Agrobacterium-mediated T-DNA insertions in the Lycopersicon genome. Mol. Gen. Genet. 204:64-69.
200. Pichersky, E., R. Bernatzky, S. D. Tanksley, and A. P. Cashmore. 1986. Evidence for selection as a mechanism in the concerted evolution of Lycopersicon esculentu (tomato) RBCS genes. Proc. Natl. Acad. Sci. 83:3880-3884.
201. Vallejos, C. E., S. D. Tanksley, and R. Bernatzky.1986. Localization in the tomato genome of DNA restriction fragments containing sequences coding for rDNA (45s). ribulose bisphosphate carboxylase and chlorophyll a/b binding protein. Genetics 112:93-105.
202. Young, N. D., J. C. Miller, and S. D. Tanksley. 1987. Rapid chromosomal assignment of multiple genomic clones in tomato using primary trisomics. Nucleic Acids Research 115:9339-9348.
203. Bernatzky, R. and S. D. Tanksley. 1985. Chromosomal locations of nuclear sequences homologous to random cDNA clones in tomato. Plant genetics, UCLA symposium on molecular and cellular biology.
204. Peffley, E. B., J. N. Corgan, K. E. Horak, and S. D. Tanksley. 1985. Electrophoretic analysis of Allium alien addition lines. Theor. Appl. Genet. 71:176-184.
205. Pichersky, E., R. Bernatzky, S. D. Tanksley, R. W. Breidenbach, A. P. Kausch, and A. R. Cashmore. 1985. Molecular characterization and genetic mapping

of two clusters of genes encoding chlorophyll a/b binding proteins in Lycopersicon esculentum (tomato). Gene 40:247-258.

206. Tanksley, S. D. and F. Loaiza. 1985. Gametophytic self-incompatibility is controlled by a single locus on chromosome 1 in tomato. Proc. Natl. Acad. Sci. 82:5093-5096.
207. Tanksley, S. D. and G. Kuehn. 1985. Genetics, subcelllular localization and molecular characterization of 6-phosphogluconate dehydrogenase isozymes in tomato. Biochem. Genet. 23:442-454.
208. Tanksley, S. D., C. M. Rick, and C. E. Vallejos. 1984. Tight linkage between a nuclear male-sterile locus and an enzyme marker in tomato. Theor. Appl. Genet. 68:109-113.
209. Tanksley, S. D. 1984. Linkage relationships and chromosomal locations of enzyme-coding genes in pepper, Capsicum annuum. Chromosoma 89:352.
210. Tanksley, S. D. 1984. High rates of cross-pollination in chile pepper. HortScience 19:580.
211. Tanksley, S. D., and J. Iglesias-Olivas. 1984. Inheritance and transfer of multiple-flower character from Capsicum chinense into Capsicum annuum. Euphytica 33:769-777.
212. Tanksley, S. D. 1983. Molecular markers in plant breeding. Plant Molecular Biology Reporter 1: 3-8.
213. Vallejos, C. E., and S. D. Tanksley. 1983. Segregation of isozyme markers and cold tolerance to an interspecific backcross of tomato. Theor. Appl. Genet. 66:241-247.
214. Zamir, D., S. D. Tanksley, and R. A. Jones. 1982. Haploid selection for low temperature tolerance of tomato pollen. Genetics 101:129-137.
215. Arus, P., S. D. Tanksley, T. J. Orton, and R. A. Jones. 1981. Electrophoretic variation as a tool for determining seed puriity and for breeding hybrid varieties of Brassica oleracea. Euphytica 31:
216. Rick, C. M., and S. D. Tanksley. 1981. Genetic variation in Solanum pennellii: comparisons with two other sympatric tomato species. Plant Syst. Evol. 139:11-45.
217. Tanksley, S. D. and R. A. Jones. 1981. Application of alcohol dehydrogenase allozymes in testing the genetic purity of F1 hybrids of tomato. HortScience 16:179-181.
218. Tanksley, S. D., D. Zamir, and C. M. Rick. 1981. Evidence for extensive overlap of sporophytic and gametophytic gene expression in Lycopersicon esculentum. Science 313:453-455.
219. Tanksley, S. D., H. Medina Filho, and C. M. Rick. 1981. Using naturally occurring enzyme variation to detect and map genes controlling quantitative traits in an inter-specific backcross of tomato. Heredity 49:11-25.

220. Tanksley, S.D. and R. A. Jones. 1981. Effect of 0_2 stress on tomato ADH's: description of a second ADH coding gene. Biochem. Genet. 19:397.
221. Tanksley, S.D., H. Medina Filho, and C. M. Rick. 1981. The effect of isozyme selection on an interspecific backcross of tomato. Theor. Appl. Genetics 60:291-296.
222. Zamir, D., S. D. Tanksley, and R. A. Jones. 1981. Low temperature effect on selective fertilization in pollen mixtures of wild and cultivated tomato species. Theor. Appl. Genetics 59:235-238.
223. Zamir, D., S. D. Tanksley, and R. A. Jones. 1981. The genetic origin of plantlets derived from male sterile tomato anthers. Plant Science Letters 21:223-227.
224. Tanksley S, Rick C (1980) Isozymic gene linkage map of the tomato: Applications in genetics and breeding 58:161.
225. Tanksley S, Rick C (1980) Genetics of esterases in species of Lycopersicon. Theor Appl Genet 56:209.
226. Tanksley S (1979) Linkage, chromosomal association, and expression of Adh-I and Pgm-2 in tomato. Biochemical Genetics 17:1159.
227. Rick, C. M., Tanksley, S. D., and Fobes, J. F. (1979). A pseudoduplication in Lycopersicon pimpinellifolium. Proc. Natl. Acad. Sci. 763435–3439.
228. Rick C, Fobes J, Tanksley (1979) Evolution of mating systems in Lycopersicon hirsutum as deduced from genetic variation in electrophoretic and morphological characters. Plant Systematics and Evolution 132:279.

Books and Monographs

Isozymes in Plant Genetics and Breeding 1983. S. D. Tanksley, T. J. Orton (editors). Elsevier Scientific Publishing Company, Amsterdam.

Longping Yuan

China National Hybrid Rice Research and Development Center Mapoling, Hunan Province, China

2004 — for innovative development of hybrid rice and discovery of the genetic basis of heterosis in this important food staple.

CURRICULUM VITAE

Marital Status: Married
Date & place of birth: Sept. 7, 1930, Peking
Nationality: Chinese
Position/Title: Director General, China National Hybrid Rice Research & Development Center
Honorary President of Hunan Academy of Agriculture Science
Degrees: Honorary Ph. D, Chinese University of Hong Kong, 2001
B.Sc. Agronomy, South-west Agricultural University, in 1953
Scientific Field: Hybrid Rice Breeding
Address (Institution): China National Hybrid Rice Research & Development Center, Changsha, Hunan 410125, P.R. China

Profession & Appointments:

1. Foreign associate of the National Academy of Sciences (NAS)
2. Member of the Chinese Academy of Engineering (CAE)
3. Senior Plant Breeder, Hunan Academy of Agricultural Sciences
4. Professor of Hunan Agricultural University, Qinghua University, Wuhan University

MEMBERSHIP OF ACADEMIES AND SOCIETIES:

1. Vice Chairman of China's Society of Crop Science
2. Chairman of Hunan Agronomy Society
3. Deputy Director of Hunan Provincial Natural Science Foundation
4. Vice Chairman of National Expert & Consultant Group for Field Crop Production of China Ministry of Agriculture
5. Vice Chairman of Science & Technology Association of Hunan Province

AWARDS & HONOURS:

1. Outstanding Achievement Award, ASPA, 2005
2. World Food Prize, World Food Prize Foundation, Des Moines, USA, 2004
3. Wolf Prize, Wolf Foundation, Jerusalem, Israel, 2004
4. Golden Sickle Award, Bangkok, Thailand, 2004
5. Roman Magsaysay Award, Roman Magsaysay Award Foundation, Manila, Philippines, 2001
6. Supreme State Sci. & Tech. Award, given by Chinese Government, Beijing, China, 2000
7. New Name of Minor Planet (Yuan Longping=1996 SDI), named by Minor Planet Center, 1999
8. Outstanding Scientist, given by Ministry of Science and Technology and Ministry of Personnel, P.R. China, 1999
9. The Fukui International Koshihikari Rice Prize, Japan, 1998
10. Distinguished Pioneer Scientist in Heterosis, "The Genetics and Exploitation of Heterosis in Crop International Symposium", Mexico, 1997
11. NIKKEI Asian Prize Award, Tokyo, Japan, 1996
12. World Food Security and Sustainability Medal awarded by FAO, Quebec, Canada, 1995
13. Ho Leung Ho Lee Foundation Prize, Beijing, China, 1994
14. Alan Shawn Feinstein World Hunger Award for research and education, Brown University, U.S.A. 1993
15. Meritorious Scientist, Hunan Provincial Government, Changsha, China, 1992
16. The Rank Prize for Agronomy and Nutrition, London, England, 1988
17. The UNESCO Science Prize in the recognition of the outstanding contribution in the field of science and technology for development, Paris, France, 1987
18. The WIPO Gold Medal for the outstanding inventor, 1985
19. The first Special-class National Invention Prize, Beijing, China, 1981

BIOGRAPHY

For his breakthrough achievement in developing the genetic materials and technologies essential for breeding high-yielding hybrid rice varieties, Prof. Yuan

Longping won the World Food Prize in 2004 – the United Nations' Food and Agricultural Organization's International Year of Rice.

Born in Peking in 1930, Yuan Longping graduated from Southwest Agricultural College in China in 1953, and then taught crop genetics and breeding at Hunan Anjiang Agricultural School. He began research there in hybrid rice development in 1964 and was transferred to the Hunan Academy of Agricultural Sciences in 1971 to serve as a research professor.

Professor Yuan is widely acknowledged for his discovery of the genetic basis of heterosis in rice – a phenomenon in which the progeny of two distinctly different parents grow faster, yield more, and resist stress better than either parent. In developing his three-line system of hybrid rice, Professor Yuan and his team soon produced a commercial hybrid rice variety called Nan-you No. 2, which was released in 1974. With yields 20 percent higher than previous varieties, Professor Yuan's new crop immediately began to improve food availability in China.

In the subsequent three decades, planting of this new crop has spread so widely that now about 60 percent of China's rice production area is planted in hybrid rice with a 20 percent higher yield over conventional rice varieties. This translates into food to nourish approximately 70 million more people per year in China alone.

Beyond this exceptional accomplishment, Professor Yuan has built an additional legacy of combating food shortages and hunger. He has developed a new technique for increasing hybrid seed yields through out-pollination, an improved two-line system of hybrid rice, and other strategies to further improve hybrid rice. His colleagues have carried on his work with his collaboration and supervision to develop new strains of "super hybrid rice" that produce almost 10 tons per hectare. With higher yields, farmers have increased rice production while simultaneously shifting millions of hectares out of rice and into alternatives such as fish, vegetables, fruits, and other food and fiber crops, giving more balanced diets and a higher standard of living to rural Chinese families. Additionally, based on the new production technique Prof. Yuan developed in 1975 to obtain higher amounts of Nan-you No. 2 seed, China was able to establish its own hybrid seed industry, which today provides additional revenue and training opportunities to thousands of farmers.

The impact of Prof. Yuan's ingenuity has been felt beyond China's rice industry. Researchers and producers of other crops in China have successfully used the two-line system for rice to explore similar systems for hybrid sorghum and rapeseed with increased yields. He has also played a key role in developing hybrid rice throughout Asia and to Africa and the Americas. Since 1980, he has trained thousands of scientists and researchers from over 25 countries and has served as a chief consultant to the FAO. Farmers in more than ten other countries besides China, including the United States, have thus benefited from his work, gaining access to a technology they may otherwise never have enjoyed.

In addition to the 2004 World Food Prize, Prof. Yuan's honors and awards include China's State Supreme Science and Technology Award, the 2001 Magsaysay Award, the UN FAO Medal of Honor for Food Security, and the 2004 Wolf Prize in Agriculture. Prof. Yuan currently directs China's National Hybrid Rice Research and Development Center and is a member or leader of several national committees, conferences, and foundations that support agriculture, science, and technology in China.

Professor Yuan Longping's pioneering research has helped transform China from food deficiency to food security within three decades. His accomplishments and clear vision helped create a more abundant food supply and, through food security, a more stable world. Professor Yuan's distinguished life's work has caused many to call him the "Father of Hybrid Rice," while his continuing research offers even more promise for world food security and adequate nutrition for the world's poor.

LIST OF PUBLICATIONS

1. A preliminary report on male sterility in rice, Oriza sativa L. 1966, Scientia Sinica Bulletin (in Chinese with English abstract).
2. Hybrid rice breeding—practice and theory, 1977, Scientia Agricultura Sinica (in Chinese).
3. Key techniques for hybrid rice seed production, 1977, Genetics and Breeding (in Chinese).
4. Hybrid rice breeding in China, 1980, Selected papers from the 1979 international Rice research Conference, IRRI.
5. A Concise Course in Hybrid Rice, 1985, published by Hunan Science and Technology Publishing House (A Chinese-English bilingual textbook).
6. Current Status of Hybrid Rice Research and Development. International Symposium on Hybrid Rice, Changsha, Hunan, China, 6-10 Oct., 1986.
7. Organization of a Hybrid Rice Breeding Program, Hybrid Rice Proceedings of International Symposium, 1986, Changsha, China.
8. The strategies for hybrid rice breeding, 1987, Hybrid Rice (in Chinese).
9. The Potential Vse of Apomixis in Crop Improvement, Hybrid Rice, 1988(4):1-3.
10. Breeding for Wide Compatibility Rice Line, Hybrid Rice, 1989(2):35-38.
11. Rice Line 84-15 as an Apomictic Material Needs Scientific Evidence. Hybrid Rice, 1989(4):3-4.
12. New Progresses on Rice Apomixis Research, Hybrid Rice, 1990(1):29-32.
13. Progress of two-line system hybrid rice breeding, 1990, Scientia Agricultura Sinica.
14. Technical Strategy of Breeding for TGMS Lines. Hybrid Rice, 1992(1):1-4.
15. Purfication of T(P)GMS Lines of Rice and Their Production of Foundation Seeds. Hybrid Rice. 1994(6):1-3.

16. Technology of hybrid rice production, a manual published by FAO, 1995 (in English).
17. Studies on Two-Line System Hybrid Rice. Hunan Agr. Sci., 1995(6):4-5.
18. Progress of Two-line System Hybrid Rice Breeding Sci. Agr. Sinica, 1990(3):1-6.
19. Prospects of Rice Yield Potential from the View Point of Plant Breeding, Hybrid Rice, 1996(4):1-2.
20. Hybrid Rice Breeding in China. Advances in Hybrid Rice Technology: Proceedings of the 3rd International Symposium on Hybrid Rice, 1996, India.
21. Breeding Strategies for Development of Intersubspecific Hybrid Rice. Hybrid Rice, 1996(2):1-3.
22. Genes from Wild Rice Improve Yield. Nature. 1996, Nov. 21, 384(6606): 223-224.
23. A Strategy for Developing Wide Spectrum Compatibility Rice Line. Sci. Agr. Sinica, 1997, 30(4):1-8.
24. Explaiting Crop Heterosis by Two-line System Hybrid: Current, Status and Future Prospects. Proc. Int. Symp. On Two-line System Heterosis Breed in Crops. 1997, Changsha, Hunan, China.
25. Current Status, Objectives and Prospects of Research and Development of Two-Line Hybrid Rice. Research of Agricultural Modernization, 1997(1): 1-3 Preliminary Experiences on SRI for Growing Super Hybrid Rice.
26. Hybrid Rice Breeding for Super High Yield. Hybrid Rice. 1997(6):1-6.
27. Hybrid Rice Development and Use: Innovative Approach and Challenges. The 19th Session of International Rice Commission, 1998, Cairo, Egypt.
28. Hybrid Rice Breeding for Super High Yield XVIIIth Int. Congr. Genet., 1998, Beijing, China.
29. Retrospect, Current Status and Outlook of Hybrid Rice Breeding. China Rice, 1999(4)3-6.
30. Hybrid Rice — Genetics, Breeding and Seed Production. Plant Breeding Review, Vol. 17. 1999.
31. The System of Rice Intensification (SRI). Hybrid Rice, 2001(4):1-3.
32. Preliminary Experiences on SRI for Growing Super Hybrid Rice. SRI International Conference, 2002, Sanya, China.
33. Hybrid Rice Technology for Food Security in the world. The 4th International Hybrid Rice Symposium, 2002, Hanoi, Vietnam.
34. The Second Generation of Hybrid Rice in China, The Commission's 20th Session, Bangkok, Thailand, July 23-26, 2002.
35. Recent Progress in Breeding Super Hybrid Rice in China. International Rice Congress 2002, Beijing, China.
36. L.P. Yuan. Hybridriceology, Published by *China Agriculture Press*, 2002, Beijing.
37. J. Yu, S. Hu, J. Wang, G. K. S. Wong, S. Li, B. Liu, Y. Deng, J. Sun, J. Tang, Y. Chen, X. Huang, W. Lin, C. Ye, W. Tong, L. Cong, J. Geng, Y. Han, L. Li, W. Li, G. Hu, X. Huang, W. Li, J. Li, Z. Liu, L. Li, J. Liu, Q. Qi, J. Liu, L. Li,

T. Li, X. Wang, H. Lu, T. Wu, M. Zhu, P. Ni, H. Han, W. Dong, X. Ren, X. Feng, P. Cui, X. Li, H. Wang, X. Xu, W. Zhai, Z. Xu, J. Zhang, S. He, J. Zhang, J. Xu, K. Zhang, X. Zheng, J. Dong, W. Zeng, L. Tao, J. Ye, J. Tan, X. Ren, X. Chen, J. He, D. Liu, W. Tian, C. Tian, H. Xia, Q. Bao, G. Li, H. Gao, T. Cao, J. Wang, W. Zhao, P. Li, W. Chen, X. Wang, Y. Zhang, J. Hu, J. Wang, S. Liu, J. Yang, G. Zhang, Y. Xiong, Z. Li, L. Mao, C. Zhou, Z. Zhu, R. Chen, B. Hao, W. Zheng, S. Chen, W. Guo, G. Li, S. Liu, M. Tao, J. Wang, L. Zhu, Longping Yuan and H. Yang. A draft sequence of the rice genome (*Oryza sativa* L. ssp. indica). *Science,* 2002(296): 79-92.

38. Progress in Breeding Super Hybrid Rice in China. International Rice Forum, 2004, Manila, Philippines.
39. J. Li, J. Xiao, S. Grandillo, L. Jiang, Y. Wan, Q. Deng, Longping Yuan and S. R. McCouch. QTL detection for rice grain quality traits using an interspecific backcross population derived from cultivated Asian (*O. sativa* L.) and African (*O. glaberrima* S.) rice. *Genome,* 2004(47): 697-704.
40. J. Yu, J. Wang, W. Lin, S. Li, H. Li, J. Zhou, P. Ni, W. Dong, S. Hu, C. Zeng, J. Zhang, Y. Zhang, R. Li, Z. Xu, S. Li, X. Li, H. Zheng, L. Cong, L. Lin, J. Yin, J. Geng, G. Li, J. Shi, J. Liu, H. Lv, J. Li, J. Wang, X. Wang, D. Li, D. Lium X. Zhang, Z. Ji, W. Zhao, Y. Sun, Z. Zhang, J. Bao, Y. Han, L. Dong, J. Ji, P. Chen, S. Wu, J. Liu, Y. Xiao, D. Bu, J. Tan, L. Yang, C. Ye, J/ Zhang, J. Xu, Y. Zhou, Y. Yu, B. Zhang, S. Zhuang, H. Wei, B. Liu, M. Lei, H. Yu, Y. Li, H. Xu, S. Wei, X. He, L. Fang, Z. Zhang, Y. Zhang, X. Huang, Z. Su, W. Tong, J. Li, Z. Tong, S. Li, J. Ye, L. Wang, L. Fang, T. Lei, C. Chen, H. Chen, Z. Xu, H. Li, H. Huang, F. Zhang, H. Xu, N. Li, C. Zhao, S. Li, L. Dong, Y. Huang, L. Li, Y. Xi, Q. Qi, W. Li, B. Zhang, W. Hu, Y. Zhang, X. Tian, Y. Jiao, X. Liang, J. Jin, L. Gao, W. Zheng, B. Hao, S. Liu, W. Wang, Longpiing Yuan, M. Cao, J. McDermott, R. Samudrala, J. Wang, G. K. S. Wong and H. Yang. The genomes of Oryza sativa: A history of duplications. *PloS Biology,* 2005(3): 266-81.

CONTRIBUTIONS TO THE DEVELOPMENT OF HYBRID RICE

1. ***Development of practicable three-line system hybrid rice.*** Prof. Yuan started his research on hybrid rice in 1964. After nine year's hard work, he and his assistants in 1973 successfully developed the genetic tools essential for hybrid rice breeding, namely, the cytoplasmic male sterile line (A Line), cytoplasmic male sterility maintainer line (B line) and cytoplasmic male sterility restorer line (R line). The first commercialized hybrid rice variety, Nan-you No. 2, with good heterosis and high yield potential was also developed by him in 1974. Hybrid rice was released for commercial production in 1976. In recent years hybrid rice is planted to around 15 million ha each year and it has been proven

that hybrid rice outyields the modern improved inbred varieties by around 20% for many years.

2. ***Development of hybrid rice seed production technology.*** Previously, it seemed impossible to obtain high seed yield because of the thought that the self-pollinating characteristic in rice would restrict its outcrossing rate. Prof. Yuan and his assistants, however, developed the hybrid seed production technique in 1975. On average the hybrid seed yield is around 2.7 tons per ha and the ratio of field areas between hybrid seed production and commercial production of hybrid rice is 1 to 120-150 nowadays.

3. ***Development of strategies for further improvement of hybrid rice.*** Prof. Yuan is not content with what he has achieved. He has always been pursuing to evolve new approaches to enhance the heterosis level and to simplify the methodology in hybrid rice technology. He proposed an overall strategy for further improvement of hybrid rice, i.e. the "three developmental phases of hybrid rice breeding":

   To enhance the heterosis level: intervarietal heterosis (phase I) → intersubspecific heterosis (phase II) → distant heterosis (phase III).

   To simplify the methodology: three line or CMS system (phase I) → two line or P(T)GMS system (phase II) → one line or apomixis system (phase III).

   Each of the phases would mark a new breakthrough in rice breeding and result in a great increase in yield if it is attained.

4. ***Development of practicable two-line system hybrid rice.*** Two-line system has two advantages over the three-line system, namely, simpler procedure of seed production and higher chance of developing better heterotic hybrids. Under Prof. Yuan's wise leadership and technical guidance, his research team succeeded in developing a whole package of two-line system hybrid rice technology in 1995. The planting area under two-line hybrid rice reached 3.5 million ha recently in China and the newly released two-line hybrids outyielded the three-line check hybrids by about 10%.

5. ***Developing super-high yielding hybrid rice and exploring new approaches to further increase yield potential of rice.*** Prof. Yuan is an imaginative scientist. He realized early on that there is always a yield stagnation after a breakthrough in rice breeding and dedicated to further improve the yield potential of hybrid rice. Recently, he proposed a unique morphological model and a strategy of utilizing inter-subspecific (indica/japonica) heterosis for developing super-high yielding hybrid rice, i.e. super hybrid rice. Several inter-subspecific hybrids developed according to his model yielded 12 t/ha at multi-location yield trials. The average yield of the super hybrid rice is 8.5 t/ha on very large scale commercial production (2 million ha) in recent years.

6. ***Extending hybrid rice technology outside China.*** Development of hybrid rice to benefit the people worldwide is one of Prof. Yuan's lifelong noble wishes. He concerns not only China's food problems but also the world's hungry. It is he who shares his valuable knowledge, experience, idea and breeding materials especially CMS lines with institutes including IRRI outside China to help develop hybrid rice in other countries. Since 1980, he and his colleagues have trained more than 600 scientists and technicians from 25 countries of Asia, Africa and South America on hybrid rice technology at China National Hybrid Rice Research and Development Center and outside China such as in Indonesia and Bangladesh. He was invited as a chief consultant on hybrid rice project outside China by FAO. In 1998, FAO published a book titled "Technology of Hybrid Rice Production" written by Prof. Yuan, which has been distributed to a lot of countries. Prof. Yuan and his assistants have served in India, Vietnam, Bangladesh, Myanmar and USA more than 30 times for consultancy. Currently, hybrid rice is commercialized in 9 countries including India, Vietnam, Philippines, Bangladesh, Pakistan, Myanmar, Indonesia, Egypt and USA. Besides, many other countries are actively exploring this technology to enhance rice yield.
7. Due to his great contribution to the development of hybrid rice in the world, he is honored with the name of "Father of hybrid rice" worldwide.

PROF. YUAN'S IMPACT ON THE WORLD FOOD SECURITY

1. ***Greatly increasing the food production in China.*** It has been proven practically for 32 years that hybrid rice cultivars outyield the modern semi-dwarf pure line counterparts by more than 20%. In recent years, hybrid rice covered about 15 million ha – constituting half of China's rice cultivation area while contributing around 57% of the total rice production in China. The average yield is 7.2 tons/ha for hybrid rice against 5.8 tons/ha for conventional pure line rice. The grains increased by hybrid rice can feed 70 million more people each year. It is obvious that hybrid rice has played a very significant role in achieving food self-sufficiency since 1980s in China, which has the largest population in the world.
2. ***Creating an effective approach for increasing rice production outside China.*** The whole package of hybrid rice technology developed mainly by Prof. Yuan now is available to most rice growing countries. Generally speaking, rice hybrids can outyield improved pure line rice varieties by around 20%. For example, in Vietnam the area under hybrid rice was more than 65000 ha in 2005. The average yield of rice hybrids is 6.3 t/ha while the yield of inbred varieties is 4.5 t/ha. Due to planting hybrid rice on large area for years, Vietnam becomes the second largest rice export country in Asia. In the Philippines, both the

present president Arroyo and the former president Estrada who witnessed the remarkable yield advantage in farmers' fields strongly encourage the nation's farmers to plant hybrid rice for increasing rice yield and farmers' income and ensuring the food security of the country. Because of the success of developing hybrid rice outside China in recent years, the important role of hybrid rice technology in enhancing grain yield was emphasized at the 19th session of the International Rice Commission (held in Cairo, 7-9 September 1998). Also at the session, it was suggested that FAO should take active measures to speed up the development of hybrid rice in the world.

3. ***Offering a feasible approach to enhance rice yield in the world this century.*** The successful development of two-line hybrid rice and the ongoing research on developing super hybrid rice offer a bright opportunity to achieve food sufficiency in the new century in China and other developing countries as well. In 2010, the population in China will reach 1.36 billion and 50 million ton more food is needed. The situation may be even worse in some other Asian countries. Because it is impossible to expand the arable land, to raise the yield per unit area is the only solution. Developing two-line system hybrid rice, especially super hybrid rice may play a more important role in raising rice yield to fill the food gap in this century. If 50% of the world paddy field were covered by rice hybrids and 2 tons grains per ha were increased, it means 400-500 million more people can be fed. Therefore, the government as well as non-governmental organizations in and outside China pay great attention to the super hybrid rice program.

4. ***Offering a new way for heterosis utilization in other crops.*** Encouraged and inspired by the success in developing two-line system hybrid rice, this new approach is being explored in some other crops, especially those in which the three line system has not yet been available. To date, much progress has been made in the development of two-line system hybrids of sorghum, rape seed, etc. For example, a two-line system hybrid of sorghum outyielded the three-line check hybrid variety by more than 10% in yield trial in 1998 in China and this was reconfirmed from 1999 to 2001 on large scale commercial production.

5. ***Directing the hybrid rice research for the future.*** His "three developmental phases of hybrid rice breeding" strategies offer the future direction of hybrid rice research and are actually followed by most hybrid rice breeders. Especially, the newly proposed model of plant type and approaches for developing super hybrid rice are highly appreciated and followed by many scientists in and outside China.

6. ***Providing socio-economic benefits.*** Before the release of hybrid rice, the planting area under rice was about 35 million ha in China and even so, the production could not meet the need of rice in the country. After the exploitation of hybrid rice technology, as well as improved cultivation practices, the land for rice production reduced year by year due to the increased yield per unit area. In recent years the rice area is reduced to about 30 million ha and 5 million ha of rice land has been shifted to the production of some other important or more profitable alternatives, such as fish, vegetables, fruits, cotton, rape seed, etc., not only contributing a lot to the sufficient and balance food supply but also helping enhance living standard of the people in China. In addition, a good number of jobs has been created because of hybrid seed business, esp. in the rural area a lot of woman labors are involved in hybrid seed production.

Kexue Tongbao, Volume 17, No. 7 - 15 April 1966

BOTANY

A PRELIMINARY REPORT ON MALE STERILITY IN RICE, *Oryza satbra* L

L. P. YUAN
Anjiang Agricultural School, Hunan

Through panicle to panicle inspection at heading stage, six male-sterile rice plants were found in the paddy fields where several varieties were grown from 1964 to 1965 in Hunan Province. The frequency of occurrence of such male-sterile rice plants under natural conditions was estimated to he 0.13 per cent. According to the characteristics of their anthers and pollen grains, these male-sterile plants can he distinguished into three different types:

(1) *Pollen-free type.* Two plants of this type were found to be completely male-sterile. Their anthers are smaller and thinner than those of the normal plant as shown in Fig. 1. They are colourless, remain unbroken after flowering, and contain no pollen grains.

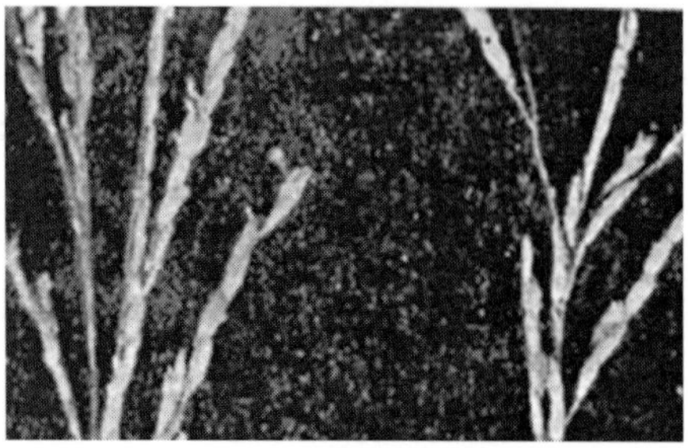

Fig. 1. Rice anthers 2 hours after flowering. Left – normal panicle; right – male-sterile panicle (pollen-free type).

(2) *Pollen abortive type.* Two plants of this kind were found to be completely male-sterile. Their pale-yellow anthers, smaller than normal ones, fail to dehisce after flowering, but contain poorly developed pollen grains. These pollen grains are not only smaller in size, but also irregular in form. They show no blue colour reaction when treated with KI solution.

(3) *Partially male-sterile type.* Of this type two plants were found. On one plant the anthers in most of the florets fail to dehisce after flowering, but all the anthers contain normal pollen grains. On the other plant the anthers in most of the florets contain poorly developed pollen grains, and remain unbroken after flowering.

No significant difference was found between the above-mentioned male-sterile plants and their normal check plants in respect of other morphological characters including the pistil. The seeds produced by the pollen-free male-sterile plants through artificial crosses, have not been sown so far. The panicles arising from the stubbles of these plants remain to be completely male-sterile.

The two original, partially male-sterile plants produced partially male-sterile progenies and male-fertile progenies.

Completely male-sterile, partially male-sterile, and male-fertile plants were obtained from the progenies of the two pollen-abortive male-sterile plants through open pollination. The F_1 completely male-sterile plants were crossed artificially with fifteen varieties, the average rate of seed setting being 74.1 per cent. Three crosses had a higher rate of seed setting above 90 per cent. This indicates that the pistils of these male-sterile plants function as usual preliminary data suggest that the male sterility from this source might be used for breeding male-sterile lines.

THE STRATEGIES FOR HYBRID RICE BREEDING

Yuan Longping
Hunan Hybrid Rice Research Center (1987)
(In Chinese - English Abstract)

The exploitation of hybrid rice has played a key role in increasing grain yield in China. However, the heterosis employed now is inter-vatietal and the method being used belongs to the three line system. From a strategic point of view, the development of hybrid rice breeding can be divided into three phases according to the heterosis level and breeding methodology.

1. In terms of methodology: three line or CMS (cytoplasmic male sterile) system two line or TGMS (thermo sensitive genic male sterile) and PGMS (photoperiod-sensitive genic male sterile) system → one line or apomixis system.
2. In terms of heterosis level: inter-varietal heterosis inter-subspecific heterosis → distant heterosis.

PROGRESS IN BREEDING OF SUPER HYBRID RICE

Yuan Longping
(China National Hybrid Rice R&D Center)

Introduction

The success in the development of hybrid rice in China is a great breakthrough in rice breeding and a technological innovation in rice production, being an effective approach to markedly increasing rice yield on a large scale.

China is the first country in the world to exploit heterosis in rice commercially. Research on breeding hybrid rice was initiated in 1964. The genetic tools (viz., male sterile, maintainer and restorer line) essential to develop F_1 hybrid were successfully developed in 1973. The hybrid combinations with good heterosis were selected in 1974, and seed production techniques were fundamentally completed in 1975. In 1976, hybrid rice was released to farmers. Since then, the acreage of hybrid rice has increased rapidly each year. At present, there are about 29 million hectares under rice in China of which nearly 60% is by hybrid rice.

It has been proven for many years that hybrid rice varieties have more than 20 percent yield advantage over improved inbred varieties. In recent years the nationwide average yield of hybrid rice is 7.1 t/ha, about 1.4 t/ha higher than that of improved inbred varieties which is 5.7 t/ha. (The worldwide average yield is 3.9 t/ha, the yield in Japan is 6.3 t/ha, and in India is 3 t/ha.) These figures indicate that hybrid rice has been playing a critical role in solving the food problem of China.

Rice is the main crop food feeding about 60% of China's population and still has great yield potential especially in hybrid rice to be tapped. In order to meet food requirement for all Chinese people in the 21st century, a super rice breeding program was set up by China Ministry of Agriculture in 1996. The yield targets for hybrid rice are listed below (Table 1).

Table 1. Yield standards of super rice in China

Phase	Hybrid rice[a]			% increase
	First cropping	Second cropping	Single cropping	
1996 level	7.50	7.50	8.25	0
Phase I (1996-2000)	9.75	9.75	10.50	More than 20%
Phase II (2001-2005)	11.25	11.25	12.00	More than 40%

[a]In t ha^{-1} at 2 locations with 6.7 ha each in 2 consecutive years.

With morphological improvement plus the use of intersubspecific (indica/japonica) heterosis, several pioneer two-line super hybrid rice varieties had been developed and met the Phase I yield standard of single cropping rice by 2000. There were more than 20 demonstration locations with 6.7 ha or 67 ha each, where their average yield was more than 10.5 t ha^{-1} in 2000. A combination, P64S/9311, was released for commercial production in 2001. Since then the area under this hybrid has expanded very fast and reached 2 million ha in recent years (Fig. 1). Its average yield is 8.4 t ha^{-1} while that of the nationwide rice was 6.3 t ha^{-1}. Another combination, P64S/E32, had created a record yield of 17.1 t ha^{-1} in an experimental plot (720m^2) in 1999 (Fig. 2).

Fig. 1. P64S/9311 was released for commercial production in 2001.

Fig. 2. P64S/E32 created a record yield of 17.1t ha^{-1} in an experimental plot in 1999.

Recently the efforts are focused on breeding Phase II super hybrid rice and good progress has been made. Some promising combinations with 13 t ha^{-1} of yield potential have been developed. Among them, the best one is P88S/0293, which yielded over 12 t ha^{-1} at four 6.7-ha or 67-ha locations in 2003 and twelve 6.7-ha or 67-ha locations in 2004. It means the yield target of Phase II is attained one year ahead of the plan.

Technical approaches

Crop improvement practices have indicated, up to now, that there are only two effective ways to increase the yield potential of crops through plant breeding, *i.e.*, morphological improvement and the use of heterosis. However, the potential is very limited when using morphological improvement alone and heterosis breeding will produce undesirable results if it is not combined with morphological improvement. Any other breeding approaches and methods, including high technology such as genetic engineering, must be incorporated into good morphological characters and strong heterosis; otherwise, there will be no actual contributions to a yield increase. On the other hand, the further development of plant breeding for a super yield target must rely on progress in biotechnology.

Morphological improvement

A good plant type is the foundation for super high yield. Since Dr. Donald proposed the concept of ideotype in 1968, many rice breeders pay great attention to this idea and propose various models for super high-yielding rice. Among them, the well-known one is the "new plant type" proposed by Dr. Khush at IRRI. Its main features are: (1) large panicles, with 250 spikelets per panicle; (2) fewer tillers, 3-4 productive tillers per plant; and (3) a short and sturdy culm. Whether these models can realize super high yield or not has yet to be proved.

Based on our studies, especially inspired by the striking characteristics of the high-yielding combination P64S/E32, which has made a record yield of 17.1 t ha^{-1}, we have found that the super high-yielding rice variety has the following morphological features:

1. Tall erect-leaf canopy

The upper three leaf blades should be long, erect, narrow, V-shaped, and thick (Fig. 3). Long and erect leaves have a larger leaf area, can accept light on both sides, and will not shade each other. Therefore, light is used more efficiently and air ventilation is also better within such a canopy. Narrow leaves occupy a relatively small space and thus allow a higher effective leaf area index. A V-shape makes the leaf blade stiffer so that it is not prone to be droopy. Thick leaves have a higher photosynthetic function and are not easily senescent. These morphological features signify a large source of assimilates that are essential to super high yield.

Fig. 3. The upper three leaf blades should be long, erect, narrow, V-shape and thick.

2. Lower panicle position

The tip of the panicle is only 60-70 cm above the ground during the ripening stage. Because the plant's center of gravity is quite low, this architecture enables the plant to be highly resistant to lodging. Lodging resistance is also one of the essential characters required for breeding a super high-yielding rice variety.

3. Bigger panicle size

Grain weight per panicle is around 5 g and the number of panicles is about 300 m^{-2}. Theoretically, the yield potential is 15 t ha^{-1} in this case.

Grain yield = biomass × harvest index. Nowadays, the harvest index (HI) is very high (above 0.5). A further raising of the rice yield ceiling should rely on increasing biomass because further improvement of the HI is limited. From the viewpoint of morphology, to increase plant height is an effective and feasible way to increase biomass. However, this approach will cause lodging. To solve this problem, many breeders are trying to make the stem thicker and sturdier, but this approach usually results in a decrease in HI. Therefore, it is difficult to obtain a super high yield in this way. The plant model of a taller canopy can combine the advantages of a higher biomass, higher HI, and higher resistance to lodging.

Raising the level of heterosis

According to our studies, heterosis level in rice has the following general trend: indica/japonica > indica/javanica > japonica/javanica > indica/indica > japonica/japonica. Indica/japonica hybrids possess a very large sink and rich source, the yield potential of which is 30% higher than that of intervarietal indica hybrids being used commercially (Figs. 4 to 6). Therefore, efforts have been focused on using indica/japonica heterosis to develop super hybrid rice. However, many problems exist in indica/japonica hybrids, especially their very low seed set, which must be solved to practically use their heterosis (Table 2.). Making use of wide

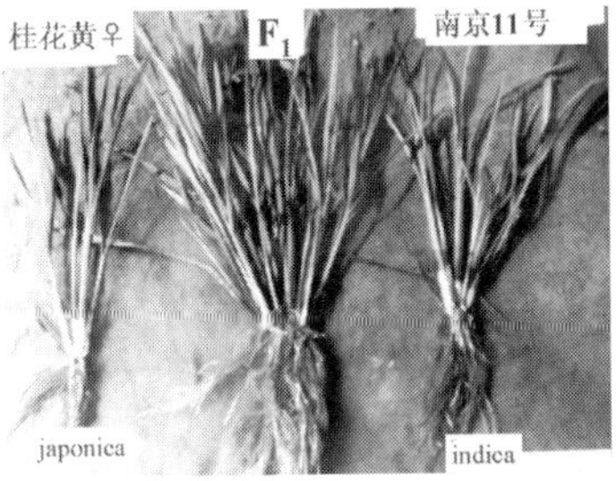

Fig. 4. Hybrid vigour in plant.

Fig. 5. Hybrid vigour in root.

Fig. 6. Hybrid vigour in panicle.

Table 2. Yield potential of an indica/japonica hybrid

Combination	Plant height (cm)	Number of spikelets/ panicle	Number of spikelets/plant	Seed setting rate %	Actual yield (kg/ha)
Chengte 232 (japonica) × 26 Zhaizao (indica)	120	269.4	1779.4	54.0	8250
Weiyou 35 (indica/indica)	89	102.6	800.3	92.9	8625
Increase %	34.8	162.8	122.4	-41.9	-4.3

compatibility (WC) genes and selecting intermediate-type lines instead of typical indica or japonica lines as parents, several intersubspecific hybrid varieties with stronger heterosis and normal seed set have been successfully developed.

By means of biotechnology

This is another important approach for developing super hybrid rice. So far, three encouraging results have been obtained in this research area.

The use of favorable genes from wild rice: Based on molecular analysis and field experiments, two yield-enhancing QTLs from wild rice (*Oryza rufipogon* L.) were identified. Each of the QTLs contributed to a yield advantage of 18% over the high-yielding check hybrid Weiyou 64 (one of the most elite hybrids). By means of molecular marker-assisted backcrosses and field selection, an excellent R line (Q611) carrying one of these QTLs was developed. Its hybrid, J23A/Q611, outyielded the check hybrid by 35% in a replicated trial for the second cropping in 2001. Its yield in farmers' fields planted as the second cropping is around 10 t ha^{-1}, about 2 t ha^{-1} higher than that of the CK.

Using genomic DNA from barnyard grass (*Echinochloa crus-galli*) to create a new resource of rice: The total DNA of barnyard grass was introduced into a restoring line (R207) by the spike-stalk injection method and variants occurred in the D_1. From these variants, new elite stable R lines containing some DNA fragments of barnyard grass have been developed (Figs. 7 and 8). The most outstanding one is RB207, and its agronomic characters such as number of spikelets per panicle and grain weight are much better than those of the original R207. Particularly, its hybrid, GD S/RB207, has a good plant type and very strong heterosis (Fig. 9). The estimated yield was more than 15 t ha^{-1} in our experimental plot in 2002. Its yield potential at different ecologic locations is now being evaluated. Preliminary results indicated that this hybrid performed very well in higher elevation regions. At a mountainous location (600m above sea level), its average yield was 13 t ha^{-1} in a large scale (6.5ha) demonstration (Fig. 10).

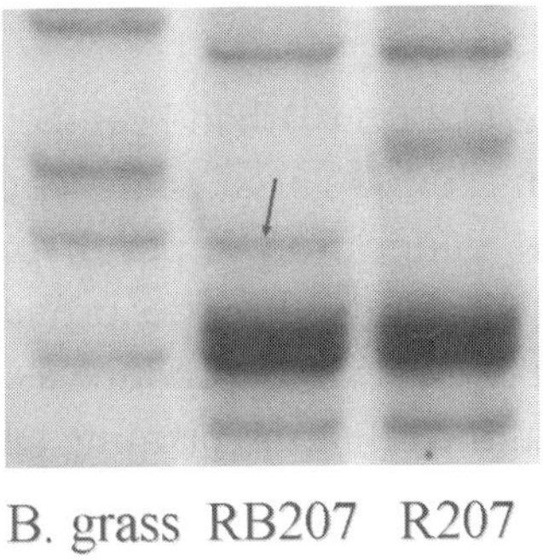

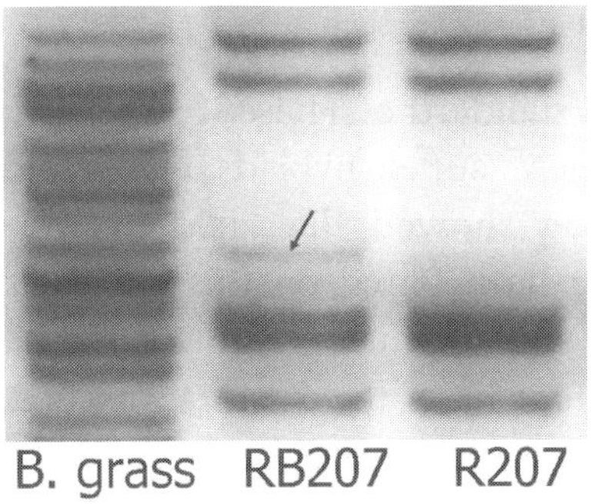

Figs. 7 and 8. DNA fragments from B. grass.

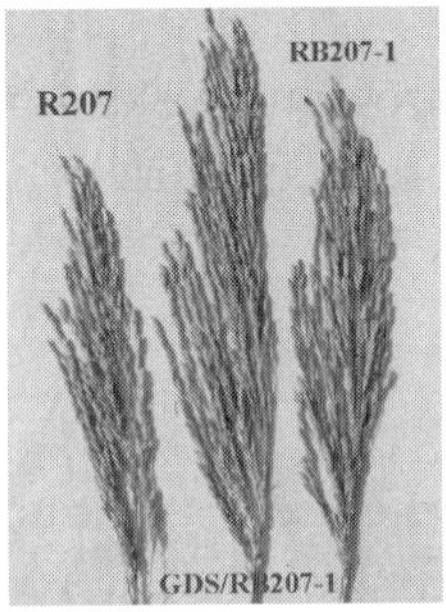

Fig. 9. Strong heterosis of GDS/RB207.

Fig. 10. GDS/RB207 on a large scale demonstration.

C_4 genes from maize have been cloned and successfully transferred into rice plant. Preliminary test showed that the photosynthetic efficiency of the leaves of some rice plants containing C_4 genes was 15-30% higher than that of the CK. By using this transgenic plant as donor to introduce C_4 genes into super hybrid rice is underway. It is expected that the yield potential of C_4 super hybrid rice can be further increased by a big margin theoretically.

Based on the above progress, the Phase III super hybrid rice program is proposed, in which the yield target is 13.5 t ha^{-1} and could be fulfilled by 2010 (Fig. 11).

Fig. 11. "Water-fall rice".

Prospects

The yield standard of Phase II super hybrid rice (12 t ha^{-1}) had been achieved by 2004. These super hybrids were released in 2006. Large scale commercial production showed the yield of the second generation super hybrids was 10-10.5 t ha^{-1}. Based on this achievement, a "planting three to produce four" project is initiated, i.e. planting three ha of super hybrids to produce as much as growing four ha of the existing varieties. It is planned that the planting area under such super hybrids would be extended to 4 million ha in five years, in which an equivalent of 5.3 million ha of existing rice varieties would be produced. Therefore, implementation of this project is not only very important for food security but also can save 1.3 million ha of land for other more profitable uses, thus to help farmers alleviate poverty and become rich.

By reaching the target of Phase III super hybrid rice program, we can increase a yield of 3 t ha^{-1}, and will produce 30 million t of more rice yearly when it is commercialized up to 10 million ha.

Conclusion

Chinese people can not only meet their food demand by themselves, but also help other developing countries to solve their food shortage problems.

THE RICE GENOME

A Draft Sequence of the Rice Genome (*Oryza sativa* L. ssp. *indica*)

Jun Yu,[1,2,3,4*] Songnian Hu,[1*] Jun Wang,[1,2,5*] Gane Ka-Shu Wong,[1,2,4*] Songgang Li,[1,5] Bin Liu,[1] Yajun Deng,[1,6] Li Dai,[1] Yan Zhou,[2,7] Xiuqing Zhang,[1,3] Mengliang Cao,[8] Jing Liu,[2] Jiandong Sun,[1] Jiabin Tang,[1,3] Yanjiong Chen,[1,6] Xiaobing Huang,[1] Wei Lin,[2] Chen Ye,[1] Wei Tong,[1] Lijuan Cong,[1] Jianing Geng,[1] Yujun Han,[1] Lin Li,[1] Wei Li,[1,9] Guangqiang Hu,[1] Xiangang Huang,[1] Wenjie Li,[1] Jian Li,[1] Zhanwei Liu,[1] Long Li,[1] Jianping Liu,[1] Qiuhui Qi,[1] Jinsong Liu,[1] Li Li,[1] Tao Li,[1] Xuegang Wang,[1] Hong Lu,[1] Tingting Wu,[1] Miao Zhu,[1] Peixiang Ni,[1] Hua Han,[1] Wei Dong,[1,3] Xiaoyu Ren,[1] Xiaoli Feng,[1,3] Peng Cui,[1] Xianran Li,[1] Hao Wang,[1] Xin Xu,[1] Wenxue Zhai,[3] Zhao Xu,[1] Jinsong Zhang,[3] Sijie He,[3] Jianguo Zhang,[1] Jichen Xu,[3] Kunlin Zhang,[1,5] Xianwu Zheng,[3] Jianhai Dong,[2] Wanyong Zeng,[3] Lin Tao,[2] Jia Ye,[2] Jun Tan,[2] Xide Ren,[1] Xuewei Chen,[3] Jun He,[2] Daofeng Liu,[3] Wei Tian,[2,6] Chaoguang Tian,[1] Hongai Xia,[1] Qiyu Bao,[1] Gang Li,[1] Hui Gao,[1] Ting Cao,[1] Juan Wang,[1] Wenming Zhao,[1] Ping Li,[3] Wei Chen,[1] Xudong Wang,[3] Yong Zhang,[1,5] Jianfei Hu,[1,5] Jing Wang,[1,5] Song Liu,[1] Jian Yang,[1] Guangyu Zhang,[1] Yuqing Xiong,[1] Zhijie Li,[1] Long Mao,[3] Chengshu Zhou,[8] Zhen Zhu,[3] Runsheng Chen,[1,9] Bailin Hao,[2,10] Weimou Zheng,[1,10] Shouyi Chen,[3] Wei Guo,[11] Guojie Li,[12] Siqi Liu,[1,2] Ming Tao,[1,2] Jian Wang,[1,2] Lihuang Zhu,[3†] Longping Yuan,[8†] Huanming Yang[1,2,3†]

We have produced a draft sequence of the rice genome for the most widely cultivated subspecies in China, *Oryza sativa* L. ssp. *indica*, by whole-genome shotgun sequencing. The genome was 466 megabases in size, with an estimated 46,022 to 55,615 genes. Functional coverage in the assembled sequences was 92.0%. About 42.2% of the genome was in exact 20-nucleotide oligomer repeats, and most of the transposons were in the intergenic regions between genes. Although 80.6% of predicted *Arabidopsis thaliana* genes had a homolog in rice, only 49.4% of predicted rice genes had a homolog in *A. thaliana*. The large proportion of rice genes with no recognizable homologs is due to a gradient in the GC content of rice coding sequences.

Rice is the most important crop for human consumption, providing staple food for more than half the world's population. The euchromatic portion of the rice genome is estimated to be 430 Mb in size (*1–3*), which is the smallest of the cereal crops. It is 3.7 times larger than that of *A. thaliana* (*4–6*), and 6.7 times smaller than that of the human (*7*, *8*). The well-established protocols for high-efficiency genetic transformation, widespread availability of high-density genetic and physical maps (*9*, *10*), and high degrees of synteny among cereal genomes (*11–15*) combine to make rice a unique organism for studying the physiology, developmental biology, genetics, and evolution of plants. The International Rice Genome Sequencing Project (IRGSP) (*16*) has already delivered a substantial amount of sequence for the *japonica* (*Nipponbare*) subspecies, in bacterial artificial chromosome (BAC) and P1-derived artificial chromosome (PAC)–sized contigs. Working independently, Monsanto and Syngenta (*17*, *18*) established proprietary working drafts for *japonica*, in April 2000 and February 2001, respectively. The Monsanto sequence has been used to assist in the efforts of the IRGSP.

We are releasing a draft genome sequence for rice from *93-11* (*19*), which is a cultivar of *Oryza sativa* L. ssp. *indica*, the major rice subspecies grown in China and many other Asia-Pacific regions. It is the paternal cultivar of a super-hybrid rice, *Liang-You-Pei-Jiu* (*LYP9*), which has 20 to 30% more yield per hectare than the other rice crops in cultivation (*20*). The maternal cultivar of *LYP9* is *Pei-Ai 64s* (*PA64s*), which has a major background of *indica* and a minor background of *japonica* and *javanica*, two other commonly cultivated subspecies. We have also produced a low-coverage draft sequence for *PA64s*. A preliminary assembly and analysis on a subset of this sequence was published in the *Chinese Science Bulletin* (*21*). Our discussion will focus largely on the genome landscape of rice, how it differs from that of the other sequenced plant, *A. thaliana*, and how both plant genomes differ from that of the human. We will show that rice genes exhibit a gradient in GC content, codon usage, and amino acid usage. This compositional gradient reflects a unique phenomenon in the evolutionary history of rice, and perhaps all monocot plants, but not eudicot plants. As a result, about one-half of the predicted rice genes have no obvious homolog in *A. thaliana*, whereas the other half is almost a replica of the *A. thaliana* gene set.

The entire rice genome sequence can be downloaded from our Web site at http://btn.genomics.org.cn/rice. Following our announcement of the rice genome sequence at the annual Plant, Animal and Microbe Genomes (PAG X) conference, in San Diego, during the ensuing period from 14 January to 2 March 2002, this sequence was downloaded 556 times, and the BLAST search facilities were used 7008 times by 343 individuals. This sequence has also been deposited at the DNA Data Bank of Japan/European Molecular Biology Laboratory/GenBank under the project accession number AAAA00000000. The version described in this paper is AAAA01000000.

Experimental design. The rice genome project at the Beijing Genomics Institute has been designed in two stages. This is a report on stage I, the primary objective of which was to generate a draft sequence of rice at ~4× coverage for *93-11*. A similar amount of data will eventually be generated for

[1]Beijing Genomics Institute/Center of Genomics and Bioinformatics, Chinese Academy of Sciences, Beijing 101300, China. [2]Hangzhou Genomics Institute–Institute of Bioinformatics of Zhejiang University–Key Laboratory of Bioinformatics of Zhejiang Province, Hangzhou 310007, China. [3]Institute of Genetics, Chinese Academy of Sciences, Beijing 100101, China. [4]University of Washington Genome Center, Department of Medicine, Seattle, WA 98195, USA. [5]College of Life Sciences, Peking University, Beijing 100871, China. [6]Medical College, Xi'an Jiaotong University, Xi'an 710061, China. [7]Fudan University, Shanghai 200433, China. [8]National Hybrid Rice R&D Center, Changsha 410125, China. [9]Laboratory of Bioinformatics, Institute of Biophysics, Chinese Academy of Sciences, Beijing 100101, China. [10]Institute of Theoretical Physics, Chinese Academy of Sciences, Beijing 100080, China. [11]Digital China Ltd., Beijing 100080, China. [12]Institute of Computing Technology, Chinese Academy of Sciences, Beijing 100080, China.

*These authors contributed equally to this work.
†To whom correspondence should be addressed. E-mail: hyang@genomics.org.cn.

Source: From *Science* Vol. 296, No. 5565, pp. 79-92 (2002). Reprinted with permission from AAAS.

THE RICE GENOME

PA64s, but at present there is only enough data to estimate polymorphism rates between rice cultivars. The sequence reads were acquired on high-throughput capillary machines (MegaBACE 1000, 10 to 11 runs per machine per day). Concurrent with the data acquisition, we developed a software package (*22*) to identify and mask repetitive sequences and to correctly assemble these sequence reads into contigs and scaffolds, even though cereal genomes contain far more repetitive sequence than many other genomes (*23*, *24*). We generated 87,842 expressed sequence tags (ESTs), against our ultimate goal of 1,000,000 ESTs, to provide confirmatory evidence for gene identification, and for gene expression analysis. Comparing the *93-11* contig assemblies with the public data, we generated a set of polymorphic markers for genetic analysis. In stage II of the project, our objective will be to obtain a high-quality sequence, fully integrated with the physical/genetic maps, and with complete gene annotations.

We used a "whole-genome shotgun" approach, as successfully applied to *Drosophila melanogaster* (*25*) and *Homo sapiens* (*8*). Our data are complementary to those of the IRGSP, which is sequencing *Nipponbare*, a cultivar of the subspecies *japonica*, with a "clone-by-clone" approach. If we assume a euchromatic rice genome size of 430 Mb, and a Phred Q20 (*26*, *27*) read length of 500 base pairs (bp), then 1× coverage would be equivalent to 0.86 million sequence reads, or 1 million reads after the typical success rate of 80 to 85% is factored in. Shotgun libraries were constructed with a variety of methods for clone-insert preparation (*28*–*30*), to minimize the likelihood of systematic biases in genome representation. A total of 55 plasmid libraries were constructed for *93-11* and *PA64s*, with a 2-kb nominal clone-insert size. Overall, we prepared 2.75 million plasmid DNA samples (*31*, *32*). Sequencing was performed on both ends of the inserts. By the 21 October 2001 freeze, there were 4.62 million successful reads, indicating an 84% success rate. The average Q20 read length was 546 bp.

Assembling the draft. Genomic studies of grasses, especially the cereal crops, have indicated that the intergenic regions between genes are inhabited by clusters of nested retrotransposons (*23*, *24*, *33*), which compose almost half of the rice genome, and substantially larger fractions of other crop plants like *Zea mays* (maize) and *Triticum aestivum* (wheat). Our sequence assembler software was designed to handle highly repetitive genomes without having to first characterize the repeats in any traditional biological sense. The focus was on contiguity at the scaffold level, instead of complete assembly across all the repeats. However, error probabilities would be computed for every base that was successfully assembled.

A typical assembly, based on our software RePS (Repeat-masked Phrap with Scaffolding) (*22*), is shown in Fig. 1. We began by computing the number of times that any 20-bp sequence (20-nucleotide oligomer, 20-mer) appeared in the data set. Those 20-mers that appeared more often than a fixed threshold were flagged as mathematically defined repeats (MDRs). RePS made no effort to identify biologically defined repeats (BDRs), because if a 20-mer was repeated in the MDR sense, it would complicate the sequence assembly, regardless of its biological context (e.g., microsatellites, transposable elements, multigene families, recently duplicated chromosomal segments, or pseudogenes). Instead, it masked the MDRs, so that they were invisible to the sequence assembler Phrap (*34*). This reduced the computational load by many orders of magnitude, while minimizing the likelihood of making a false join. However, it also introduced another class of gaps, repeat masked gaps (RMGs), distinct from the Lander-Waterman gaps (LWGs) that are usually encountered in sequencing. In a RMG, the gap sequence is actually in the data set, but it was not usable because it was made invisible to Phrap by the masking. In a LWG, the gap sequence is missing, as a result of sampling statistics (*35*). Some of the RMGs could be closed with the clone-end pairing information, assuming that both clone ends were not fully masked. After repeat-gap closure, and regardless of the nature of the remaining gaps, RePS was used to analyze the clone-end pairing information to construct scaffolds—nonoverlapping contigs linked together in the correct order and orientation. LWGs were usually small, easy to close by polymerase chain reaction. Gaps larger than a few kb were usually RMGs due to the nested retrotransposons in the

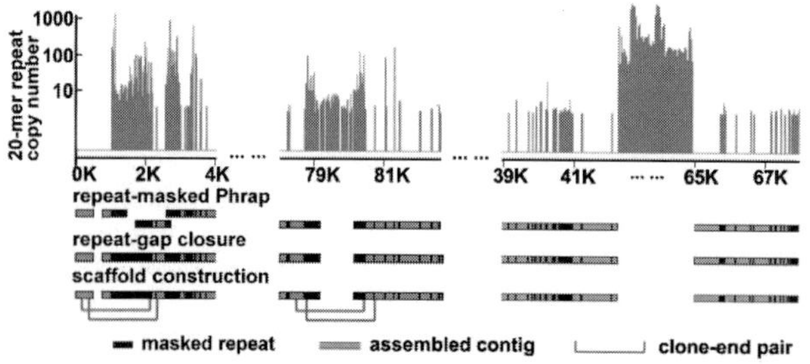

Fig. 1. Typical RePS assembly, with *93-11* (*indica*) contigs aligned to finished BAC sequences from *GLA* (*indica*) (GenBank accession numbers AL442007 and AL512542). Exact 20-mer repeats are indicated by the blue histogram bars, with bar heights proportional to estimated copy number in *93-11* (*indica*). Three stages are shown: repeat-masked Phrap, repeat-gap closure, and scaffold construction. First, we mask exact the 20-mer repeats and use Phrap to assemble the data on the basis of the unique sequence. Second, we use the clone-end pairing information to close smaller repeat masked gaps (RMGs) ignored by Phrap because of the masking. However, larger RMGs and gaps due to sampling statistics, Lander-Waterman gaps (LWGs), cannot be so closed. Third, we use the clone-end pairing information to construct scaffolds—sets of nonoverlapping contigs linked together in the correct order and orientation. A LWG at 0.5 kb is scaffolded over. RMGs at 1.5 and 2.5 kb are closed, and another at 80 kb is scaffolded over. The RMG between 42 and 65 kb is too large to scaffold across given a clone-insert size of 2 kb.

Table 1. Sequence assembly statistics for *93-11* (*indica*). The Q20 read lengths refer to the usable part of the sequence with error probabilities less than 10^{-2}. Masking 20-mer repeats eliminated 42.2% of the sequence by length. Some reads were partially masked, but 18.7% of reads were fully masked. The N50 contig or scaffold sizes define that size above which 50% of the assembly was found. To estimate the assembled-equivalent size of the unused reads, we divided total Q20 lengths by the 4.2× depth of reads in the assembled contigs. This resulted in an assembled-equivalent size of 104 Mb, of which 78 Mb was fully masked reads. The total genome size was thus estimated to be 466 Mb.

Basic shotgun data	
Total genome size (Mb)	466
Number of reads	3,565,386
Q20 read lengths (bp)	546
Shotgun coverage	4.2
Exact 20-nt oligomer repeats	
Length of fraction masked	42.2%
No. of fully masked reads	18.7%
Sequence assembly	
Total contig size (Mb)	361
N50 contig size (kb)	6.69
Total scaffold size (Mb)	362
N50 scaffold size (kb)	11.76
Unassembled data	
Fully masked reads (Mb)	78
All other reads (Mb)	26

intergenic regions between genes. Whether these gaps should be closed or not remains to be resolved.

Shotgun data for *93-11* (Table 1) and *PA64s* were assembled separately, to allow for large differences in their genome sequences. In *93-11*, there were 3.57 million sequence reads after removal of the ones containing mitochondrial, chloroplastic, and bacterial sequence. Our RePS assembly yielded 127,550 contigs with an N50 size (i.e., the size above which 50% of the total assembly is found) of 6.69 kb. The total contig length was 361 Mb. These contigs were linked into 103,044 scaffolds with an N50 size of 11.76 kb, or a 1.8-fold increase over the initial contigs. The total scaffold length was 362 Mb. In contrast, for the *PA64s* data set, we had only 1.05 million sequence reads. With such low coverage, the N50 contig and scaffold sizes were much smaller, at 1.88 and 1.97 kb, respectively. These statistics differ slightly from those reported in the *Chinese Science Bulletin* (*21*), because of improvements in the RePS software. Remaining gaps between scaffolds are probably larger than the clone-insert size of 2 kb; otherwise, we would have been able to bridge them. We cannot provide a gap size distribution, but in the rice BACs that have been sequenced, repeat cluster sizes up to 25 kb have been observed.

The total contig and scaffold lengths fall far short of the previously estimated euchromatic genome size of 430 Mb. Where is the missing DNA? In the initial phase of the RePS assembly, 42.2% of the sequence was identified as a MDR and masked. A total of 18.7% of all the reads were fully masked and not immediately usable. Even though some were later incorporated into the assembly, with the clone-end pairing information, a large number of fully masked reads, and some partially masked reads, remained unused. To estimate the effective-assembled size of the unused reads, we defined an empirical coverage based on the depth of reads in the assembled contigs, 4.2×. The effective-assembled size for the unused fully masked and partially masked reads was thus estimated as 78 and 26 Mb, respectively, resulting in a total genome size of 466 Mb. That this is larger than the previous estimates is reasonable, given that whole-genome shotgun data inevitably contain some amount of heterochromatin DNA.

Quality assessments. We assumed that any large cluster of MDRs was an intergenic region and that we could safely avoid having to assemble across such a region. If so, then most of the "functional sequence" that encodes genes, and their immediate regulatory elements, should lie in our 361 Mb of assembled contigs. To confirm that this was indeed the case, we gathered all the publicly available sequence-tagged sites (STSs) and full-length cDNA sequences, as well as our own ESTs, and searched for them in our assembled contigs, using BLAST (*36*). Fortunately, a dense physical map of STS markers had already been established (*37*) for *japonica*. A total of 2845 markers were analyzed, and on the basis of sequence identity, 91.5% of their total length could be found in our contigs. Similarly, 24,776 UniGene clusters were assembled from 87,842 ESTs for *93-11*, and 93.8% of their total length could be found in our contigs. Finally, 907 nonredundant cDNA sequences were extracted from GenBank release 125 (15 August 2001), and 90.8% of their total length could be found in our contigs. Averaged across these three data sets, the functional coverage was 92.0%.

The quality metrics that matter for gene identification are (i) contiguity on the length scale of a gene, (ii) single-base error probability, and (iii) contig assembly accuracy on the length scale of a gene. As will be detailed in a later section, the mean gene size for rice is about 4.5 kb. Considering that our N50 scaffold size is only 11.76 kb, larger scaffolds would reduce the number of genes that are split across scaffolds, and this is a key objective in stage II of the project. The number most often cited is the single-base error probability, which the International Human Genome Sequencing Consortium (*7*) determined should be 10^{-4} or better, based on a human polymorphisms rate of 10^{-3}. Actually, as is detailed in a later section, rice polymorphism rates are closer to 10^{-2}, so an error rate of 10^{-4} is better than needed. On the basis of Phrap estimates (*26, 27, 34*), 94.2, 90.8, and 83.5% of the *93-11* sequence had an error rate of better than 10^{-2}, 10^{-3}, and 10^{-4}, respectively. However, most of the problematic bases were at the ends of the contigs. When we restricted this calculation to contigs greater than 3 kb and ignored bases within 500 bp of the ends, 97.3, 96.1, and 92.5% of the *93-11* sequence had an error rate of better than 10^{-2}, 10^{-3}, and 10^{-4}, respectively. It is important to bear these error rates in mind when comparing two sequences to estimate polymorphism rates.

Assembly accuracy is an often overlooked but nevertheless important quality metric. When the sequence reads are joined together in the wrong order or orientation, some of the exons will be arranged in the wrong order or orientation. This will confuse any gene-annotation program. For example, a 2-kb segment that is flanked by a pair of inverted repeats might be assembled in the wrong orientation. Comparison of independently assembled BACs would not necessarily detect the mistake, because the problem is due to sequence content, not data quality, and the same mistake could be made in both BACs. Comparison with existing physical or genetic maps validate assembly accuracy on the Mb length scale, but that is much larger than the size of most genes. Clone-end pairing information does validate a contig assembly on the kb length scale of the genes. However, when the clone ends are also used to assemble the sequence, they do not qualify as an independent confirmation. To address this problem, we aligned cDNA sequences (i.e., experimentally derived transcripts) with the genome sequence.

We removed obvious redundancies by eliminating any cDNA that was more than 90% contained inside another. Transposon sequences identified by RepeatMasker (*38*), generally in the 3′-untranslated region, were trimmed off to minimize the number of ambiguous hits. Alignments were allowed to span multiple contigs. Within any one contig, a putative misassembly was flagged whenever an exon was missing from the middle of the chain, in the wrong order, or in the wrong orientation. Missing splice sites resulting from minor sequencing errors, and partial alignments resulting from missing sequences at the end of a contig, were not counted. All putative misassemblies were validated by visual inspection, to ensure that no better alignments could be found. If in the end, the best alignment remained problematic, we concluded that there must have been a misassembly. One might think that lower quality cDNA sequences would contribute to the problematic alignments, and that this procedure would only set an upper bound on the number of misassemblies. However, we doubt that this is a serious problem. Substitutional errors might be common in cDNA sequences, but they would not trigger our detection algorithm. Only exon-sized rearrangements, especially those that change the order and orientation, would do so, but such rearrangements are rare in cDNA sequences.

We benchmarked our misassembly detection procedure on two of the most recently completed model organism genomes: *A. thaliana*, which is of finished quality (*4*), and *Drosophila melanogaster*, from the Celera 13× whole-genome-shotgun sequence (*25*). For *A. thaliana*, we detected problems in 0.2% of 4804 genes, and for *D. melanogaster*, we detected problems in 1.1% of 1889 genes. For *93-11* contigs, we detected problems in 1.1% of 907 genes, which was comparable to the *D. melanogaster* data.

Compositional gradients. The rice genome has compositional properties that differentiate it from the other sequenced plant genome, *A. thaliana*, and introduce unique difficulties for genome analysis. Here, we show data on exon, intron, and gene sequences derived from alignment of cDNAs with genomic sequence. Indeed, for Figs. 3 through 7, all of the gene models were derived from cDNA alignments, not gene-prediction programs. GenBank release 125 (15 August 2001) was used for the *A. thaliana* figures, and for the rice cDNAs. The rice genome sequence was our *93-11* assembly. The human cDNA sequence was downloaded on 2 March 2001 from NCBI-RefSeq ftp://ncbi.nlm.nih.gov/refseq/H_sapiens and the human genome sequence was downloaded on 27 February 2001 from ftp://ncbi.nlm.nih.gov/genomes/H_sapiens, immediately after the initial annotation papers.

Genomic, exon, and intron GC contents. The average genomic GC content for prokaryotes and eukaryotes varies widely. It ranges from less than 22% in the human malaria parasite, *Plasmodium falciparum*, to more than 68% in the large amplicon of *Halobactrium* sp. NRC1 (*39*). Local heterogeneity in GC content can be enormous, ranging from 26 to 65% in the human genome alone. In contrast, AG content (purine) is homogeneous (*40–43*), fluctuating by just a few percent about a mean of 50%. Compositional heterogeneity has been debated for more than 30 years (*44–47*). Discussions have focused on the characterization of the human genome as a mosaic of GC-rich and AT-rich "isochores," which are observed in warm-blooded vertebrates, but not in cold-blooded vertebrates. More recently, an elevated GC content in the *Gramineae* (grass) genomes was reported, extending perhaps to all monocot genomes, but not to eudicot genomes (*48*). It is not known whether or how this phenomenon is related to isochores.

Major differences between sequence content in *A. thaliana*, rice, and human are observable even at the simplest level, from distributions of genomic GC content. Traditionally, GC content was computed on a large window size, typically in the 100s of kilobases, to mimic the original Cs_2SO_4 density gradient experiments (*49*, *50*). We have found that smaller windows are more informative, because when these windows are larger than a typical gene size, they obscure differences between intergenic DNA and genes. We used a 500-bp window size, to obtain a smaller size than that of most plant genes (Fig. 2). As previously reported (*51*), the *A. thaliana* distribution displayed a "shoulder" on the AT-rich side, which could be attributed to the sizable fraction of the genome that was in intergenic DNA. The primary peak at 0.382 was nearly identical to the 0.388 GC content of the average *A. thaliana* gene. In contrast, no shoulder was observed in rice. However, a "tail" was apparent on the GC-rich side. The human distribution also displayed no shoulder, but a minor tail might have been present. To analyze these features, we plotted GC content distributions for exons and introns (Fig. 3). Rice exons exhibited a GC-rich tail, but rice introns did not, indicating that the GC-rich tail in the rice genomic distribution was primarily due to the exons.

Variation in GC content within genes. The key question is whether the increase in exon GC content was due to many genes with a few GC-rich exons or to a subset of GC-rich genes. Equivalently, was most of the variation in exon GC content within genes or between genes? After the GC contents of individual exons and introns were plotted as a function of genomic length (i.e., the sum of exon and intron lengths), it was apparent that most of the variation was within genes (Fig. 4). Contrary to the expectation that, in the human genome, large genes are on average more AT-rich than small genes, we found that at least one exon of exceptionally high GC content could be found in almost every rice gene, including the largest ones. Moreover, when the GC content of the protein-coding regions was plotted as a function of position along the direction of transcription, starting

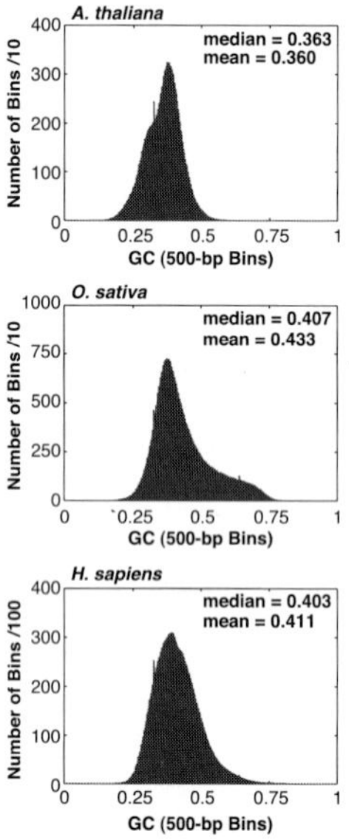

Fig. 2. Distributions for genomic GC content in *A. thaliana*, *O. sativa*, and *H. sapiens*, computed over a bin size of 500 bp. Note that for bins/10 = 100, the number of bins with that GC content is 1000.

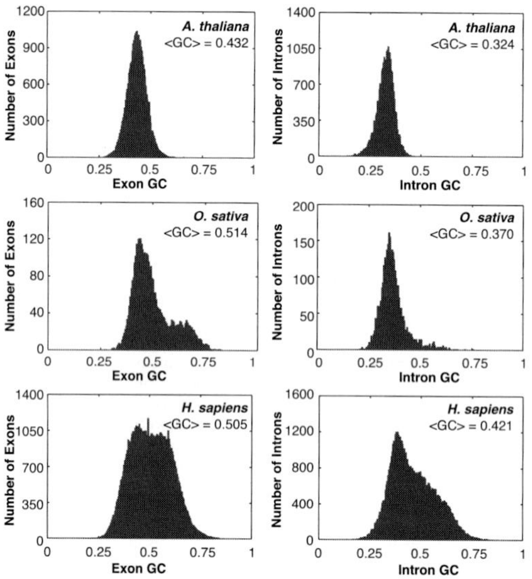

Fig. 3. GC content distribution for exons and introns in *A. thaliana*, *O. sativa*, and *H. sapiens*. All exon and intron sequences were derived from cDNA-to-genomic alignments. Mean GC content is computed on a length-weighted basis as $\langle GC \rangle = \sum_i L_i \cdot GC_i / \sum_i L_i$, where GC_i and L_i are the GC content and length for the *i*th segment (exon or intron).

THE RICE GENOME

from the 5′ end, we observed a negative gradient in the GC content of rice genes (Fig. 5). Typically, the 5′ end was up to 25% richer in GC content than the 3′ end. These gradients would extend to about 1 kb from the 5′ end, before finally petering out. The magnitudes of these gradients varied. A few genes had zero gradients, but almost no genes had positive gradients. In contrast, for *A. thaliana*, no comparable gradients were observed. Examining hundreds of best available homologs (i.e., possible orthologs), we found that the GC content of rice genes was equal to or exceeded that of their *A. thaliana* counterparts, at all positions along the coding region.

A more detailed analysis of this compositional gradient will be presented elsewhere (*52*). The important point is that, not only is there a gradient in GC content, but there are also gradients in the patterns of codon and amino acid usage. The former is a novel challenge for the ab initio gene-prediction programs that rely on codon-usage statistics, and the latter makes it more difficult to do protein homology searches across the monocot-eudicot divide.

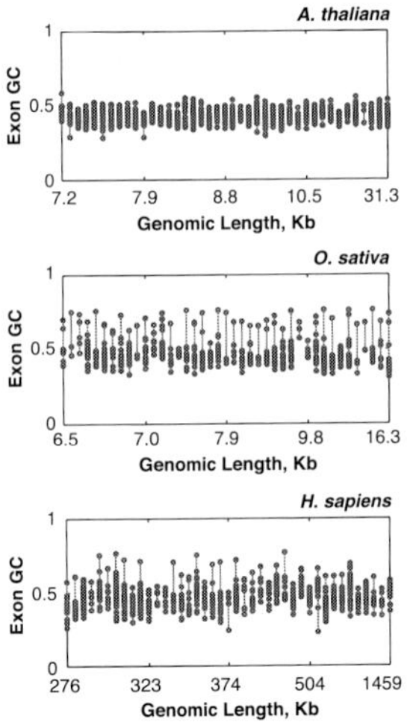

Fig. 4. GC content for individual exons as a function of their gene size, in *A. thaliana*, *O. sativa*, and *H. sapiens*. All exon and intron sequences were derived from cDNA to genomic alignments. Each data point is a single exon. Exons for the same gene are plotted at the same abscissa and connected by a vertical line. The genes are sorted by size, where gene size is defined as the sum of exon and intron lengths. To make the figure legible, we use constant spacing between genes, thus resulting in nonuniform abscissa labels. We show only the 41 largest genes for which the entire cDNA could be aligned to genomic sequence. Given the draft nature of the rice genome, some of the largest rice genes had to be omitted.

Genic and intergenic DNA. We examined exon and intron distributions for every plant, vertebrate, and invertebrate organism with more than a hundred or so genes in GenBank, either by cDNA alignments or by parsing annotations. Numerical summaries (Web Supplement 1) are available on *Science* Online at www.sciencemag.org/cgi/content/full/296/5565/79/DC1. Exon sizes are narrowly constrained, but intron sizes can be highly variable within and between organisms. Intron-size distributions tend to be bimodal, weakly (most organisms) or strongly (human). There is always a sharp "spike" at some organism-specific minimum size, which is about 90 bp for plants and vertebrates (Fig. 6). There is also a broad "hump" due to the larger introns. The magnitude of this hump is highly variable between organisms and can be difficult to ascertain precisely, because of the systematic biases against the complete sequencing of large genes.

Although the existence of this acquisition bias is known, the magnitude of its effect on our perception of intron and gene sizes is not well appreciated. For example, in the initial annotation of the human genome, the reported mean gene size of 27 kb turned out to be an underestimate by a factor of 3 (*53*). To correct for this bias, one need only realize that the bias against complete sequencing of large genes is equivalent to the bias against production of large genomic contigs. Thus, the correction can be made by restricting the computation of the mean gene size to cDNA alignments in contigs above a minimum size, and extrapolating to the limit of infinite contigs. For the human genome, the extrapolated mean gene size was 72 kb. In *A. thaliana*, there was no appreciable bias, because most of the contigs were already much larger than the genes, which had a mean size of 2.4 kb. This was larger than the published gene size of 2.0 kb, but only because we included UTRs, whereas the published numbers did not. In rice, there was a small acquisition bias, given the draft nature of our sequence. Nevertheless, the extrapolated mean gene size was only 4.5 kb, much smaller than in the human, consistent with the relatively small hump in the rice intron-size distribution.

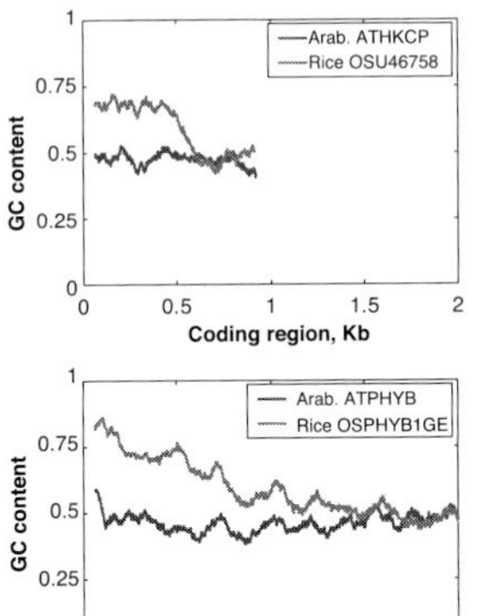

Fig. 5. GC content for homologous genes in *A. thaliana* and *O. sativa* as a function of gene position from the 5′ to 3′ end, computed on a sliding 129-bp window (equal to the median exon size in rice). Only the coding region is shown. GenBank locus identifiers are specified in the legend. The smaller gene is "potassium channel beta subunit," and the larger gene is "phytochrome B."

A preliminary gene count can be estimated from the mean gene size, for comparison against the number of genes identified by the gene-prediction programs. Our estimated 4.5-kb mean gene size for rice is similar to the maximal gene density of one per 4 to 5 kb, based on analyses of syntenic loci across many plant species (*54*). Assuming that the rice intergenic fraction is equal to the 42.2% of the sequence that was in MDRs, and taking 466 Mb as the genome size, the estimated number of rice genes is 59,855. One could also include CpG islands in the gene count, although not every CpG island is associated with a gene, so that this number can at best be considered an upper bound (*55*). Including both assembled contigs and unused reads, 138,485 CpG islands were identified by the standard algorithm (*56*). Either way, rice almost certainly has more genes than *A. thaliana* (*4*), which has only 25,498. It might even have more genes than the human (*7*, *8*), which has 30,000 to 40,000, although the actual gene count remains controversial. The idea that plants might have more genes than humans is not new, as it was predicted before our analysis (*57*, *58*).

Where did the transposons end up? The significance of these size distributions is that a prominent "hump" in the intron-size distribution, as observed for the human, is evidence of extensive transposon activity in the evolution of intron size. RepeatMasker (*38*) identified at least one transposon, and often many more than one, in almost every human intron larger than 1 kb. It rarely found a transposon in smaller introns less than 1 kb, not only in the human, but in every organism analyzed. The negative result might have been due to the incomplete status of the transposon database on which RepeatMasker relies. To support this claim, we introduce an argument that does not rely on knowledge of the sequences of all the extant transposons.

The main assumption was that any transposon should have inserted into the genome many times before becoming inactive. Despite subsequent degradation of these transposon sequences, portions should remain in many different places throughout the genome. RePS computed the copy number for every 20-mer sequence in the genome, indicating how many times each occurred in the genome. We could therefore determine the copy number required to account for all of a particular sequence data set. These data sets would include all exons, introns, and known transposons (Fig. 7). For the exons and introns, we used cDNA-to-genomic alignments. For transposons, we used RepBase 6.6 (*59*), a database of consensus sequences for every known family or subfamily of transposons. In plants, exons and introns were fully accounted for by 20-mers with copy numbers of less than 10. Transposons required much higher copy numbers of 10 to 10^2 in *A. thaliana* and 10^2 to 10^3 in rice. One could legitimately ask if the absence of large MDR clusters in our rice assemblies was a confounding factor in the intron analysis. We therefore performed an analysis on introns from finished BAC sequences and found no detectable differences. Strikingly, in the human, extremely large copy numbers of 10^4 to 10^5 were required to fully account for the introns, as was observed with the transposons. Human exons, however, were found at the same low copy numbers as in plants.

The copy number analysis shows that few plant transposons are in the introns, and by definition, plant transposons must be located in the intergenic regions between genes. Conversely, analyses of gene size show that most human transposons are in the introns (*53*). We believe that this dichotomy in where the transposons ended up reflects a fundamental difference in plant and vertebrate genomes. The dichotomy is not due to any lack of transposons in plants, because plant genomes contain many transposons. At least 24.9% of the rice genome was identifiably of transposon origins, based on a weighted average of assembled contigs and unused reads, but the correct percentage is likely to be much higher, because the transposon databases on which RepeatMasker relied were incomplete. *A. thaliana*, being a more compact genome, had a reported transposon fraction of 10%, although we suspect that this too is an underestimate.

Repetitive sequences. We deal with three classes of repeats: simple repeats [e.g., $(CAG)_n$], complex repeats (i.e., transposable elements or TEs), and mathematically defined repeats (MDRs). Here, we focus on the first two classes, which we called biologically defined repeats (BDRs). As with intron and gene sizes, acquisition biases must be factored in, so that we do not introduce additional discrepancies among the published studies. For example, a survey of 73,000 sequence-tagged connectors, totaling 48 Mb of sequence from *japonica* (*60*), found that 63% of identified TEs were retrotransposons (e.g., *copia* and *gypsy*). However, a survey of 910 kb of rice genomic sequence (*61*) found that 18.6% of identified TEs were retrotransposons. Most of the remainder were MITEs,

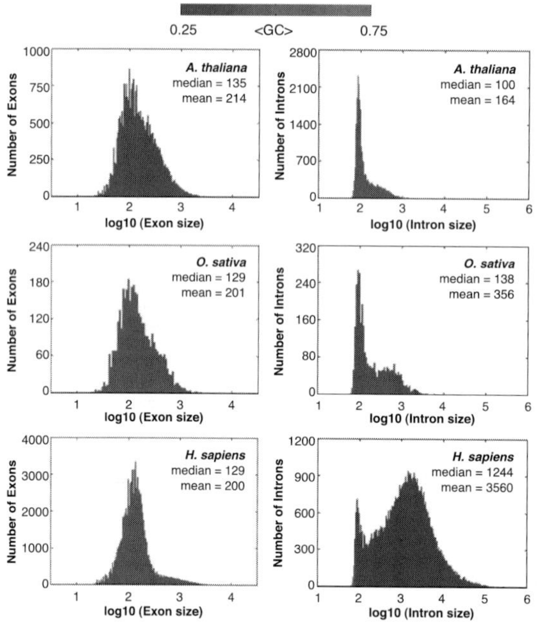

Fig. 6. Exon- and intron-size distributions for *A. thaliana*, *O. sativa*, and *H. sapiens*, with color indicating averaged GC content for exons or introns at that size range. All exon and intron sequences were derived from cDNA-to-genomic alignments.

Table 2. Simple repeats. Shown are tandem repeats with periods 1 to 4 (mono-, di-, tri-, and tetranucleotide) and the totality of repeats with all periods. The index *n* is the number of periodic units. For example, AGTTAGTT is a tetranucleotide of $n = 2$. We compute mean GC contents of the observed repeats in each category. Repeat content is then given as a percentage by length, normalized with respect to the data set (assembled contigs, fully masked reads, or cDNAs).

	93-11 assembled contigs				*93-11* fully masked reads				Full-length cDNAs			
	n = 6–11		n > 11		n = 6–11		n > 11		n = 6–11		n > 11	
	% GC	% of data set	% GC	% of data set	% GC	% of data set	% GC	% of data set	% GC	% of data set	% GC	% of data set
Mononucleotides	7.63	1.7847	27.65	0.0680	20.34	0.6953	21.02	0.0154	24.08	0.6303	1.31	0.7125
Dinucleotides	35.77	0.0904	13.08	0.0847	41.86	0.0553	4.38	0.0294	46.85	0.0573	31.11	0.0394
Trinucleotides	71.79	0.0454	10.08	0.0106	67.20	0.0098	20.81	0.0012	83.05	0.1335	66.67	0.0043
Tetranucleotides	28.77	0.0072	24.90	0.0032	37.35	0.0020	31.90	0.0010	50.00	0.0018	0.00	0.0000
All periods		**1.9277**		**0.1665**		**0.7624**		**0.0469**		**0.8229**		**0.7561**

THE RICE GENOME

or miniature inverted-repeat TEs, which accounted for 71.6% of identified TEs. Similarly large discrepancies were encountered in analyses of microsatellite distributions using mixed data from BAC ends, ESTs, and finished BAC/PACs (*62*).

For the *93-11* sequence, it is particularly important that we analyze those sequence reads that were not assembled into contigs. Tables 2 and 3 thus summarize repeat contents in the two largest components: 361 Mb of assembled contigs and 78 Mb of unused fully masked reads. Weighted averages for the entire rice genome were also computed. For comparison, we show repeat content in 907 nonredundant full-length cDNAs from GenBank release 125 (15 August 2001). Absolute numbers are not listed because, with so much of the genome in unassembled reads, and with so many of the transposons nested inside some other transposon, accurate counts were not feasible. Results are listed as a fraction of total sequence length.

Simple sequence repeats (SSRs). SSRs are particularly useful for developing genetic markers. They are believed to vary through DNA replication slippage (*63–65*), and are related to genetic instability (*66*). In Table 2, we describe SSR content for two sectors, n = 6 to 11 units and $n > 11$ units, to emphasize that the number of SSRs dropped substantially after 11 units. The SSR content for *93-11* was 1.7% of the genome, lower than in the human, where it was 3% (*7*). The overwhelming majority of rice SSRs were mononucleotides, primarily $(A)_n$ or $(T)_n$, and with n = 6 to 11. In contrast, for the human, the greatest contributions came from dinucleotides. Notably, trinucleotides with n = 6 to 11 were a barometer of gene content. The basic effect was captured by the ratio of trinucleotide to dinucleotide content, which was 2.33, 0.50, and 0.18 in cDNAs, assembled contigs, and fully masked reads, respectively. As required for a barometer, these numbers are well correlated with presumed gene content. In addition, the GC content of these trinucleotides was high, consistent with the high GC content of many rice exons.

Complex sequence repeats (TEs). Transposons identified by RepeatMasker (*38*) were assigned into three classes. Class I repeats are retrotransposons, primarily *Ty1/Copia*-like and *Ty3/Gypsy*-like. Class II repeats are DNA transposons, including *Ac/Ds*, *En/Spm*, *Mariner*-like, and *Mutator* elements. Class III repeats are a previously unknown type of short DNA transposons called MITEs (*67*, *68*). The two common examples are *Stowaway* and *Tourist*. Recently, an active family of *tourist-like* MITEs was identified in maize (*69*). Programs like RepeatMasker identify sequences that share at least 50% identity with a known TE. Because TEs are under no selective constraints after they insert in a genome, they tend to diverge from their ancestral sequence, and become unrecognizable over a time scale of a hundred million years (*70*). Identifiable repeat content is thus a function of TE age and completeness of the TE databases. The numbers listed in Table 3 must therefore be considered underestimates.

Fully masked reads were composed of 59% identifiable TEs. Assembled contigs were only 16%. Of these TEs, the amount in class I and class III repeats was 97 and 1%, respectively, for fully masked reads, but 42 and 40% for assembled contigs. This extremely biased distribution is notable, because class I repeats reportedly inhabit the intergenic regions (*23*, *24*), and class III repeats are found near, although not necessarily in, the genes (*71*). Thus, we had 92.0% functional coverage despite having only 361 Mb in assembled contigs, in a genome of total size 466 Mb. The reason class I repeats failed to assemble is apparent when one examines their mean size. Class III repeats were usually smaller than 671 bp, but class I repeats were as large as 7 kb. Our ability to close repeat-masked-gaps, or RMGs, was limited by the clone-insert sizes. For this assembly, the clone-insert sizes were only 2 kb, although we plan to use larger sizes for the next stage of the rice genome project.

Finally, the TEs in rice cDNAs constituted only 1% of the sequence, which is much lower than the 4% that was reported for human genes (*72*). Gene-associated TEs, in human and other vertebrates, have been proposed to play crucial roles in creating new genes (*73*) and in changing the regulatory circuitry to promote evolution in the host genome (*74*).

Rice gene annotations. Gradients in GC content and codon usage for rice genes create special problems in the gene-annotation process (*52*). Because rice genes have different compositional properties at their 5′ and 3′ ends, it is difficult to train a program to perform well under all circumstances. Some ab initio gene-prediction programs can use different codon-usage statistics for different genes, on the basis of regional GC content, but none use different codon-usage statistics at different positions along the same gene. Unless the gradient is explicitly modeled, or perhaps, codon-usage statistics are abandoned altogether, performance will be subject to the vagaries of the training process. With this in mind, we set out to survey all of the programs trained for rice: FGeneSH (*75*), GeneMark (*76*), GenScan (*77*), GlimmerM

Table 3. Complex repeats. Transposons identified by RepeatMasker are assigned to three classes. Each class has a number of families (e.g., *tourist*-like MITEs), and each family has a number of different subfamilies. The number of subfamilies is listed, as well as their total and mean size. Repeat content for each family is given as a percentage by length, normalized with respect to the data set (assembled contigs, fully masked reads, or cDNAs) or with respect to all identified transposons.

		Number	Total (bp)	Mean (bp)	*93-11* assembled contigs		*93-11* fully masked reads		Full-length cDNAs	
					% of data set	% of repeats	% of data set	% of repeats	% of data set	% of repeats
Class I	LINEs	5	18,997	3,799	1.1905	7.43	0.1318	0.22	0.0257	2.51
	SINEs	7	1,254	179	0.0888	0.55	0.0047	0.01	0.0268	2.61
	gypsy-like	19	105,614	5,559	3.7285	23.28	41.6894	70.35	0.1238	12.07
	copia-like	5	35,151	7,030	1.7175	10.72	15.8506	26.75	0.0869	8.47
	Subtotal				**6.7254**	**41.99**	**57.6766**	**97.33**	**0.2631**	**25.65**
Class II	*Ac/Ds* TEs	3	1,567	522	0.1099	0.69	0.0145	0.02	0.0000	0.00
	En/Spm TEs	3	5,558	1,853	0.2590	1.62	0.2770	0.47	0.0000	0.00
	MULEs	22	25,800	1,173	2.4500	15.30	0.6378	1.08	0.1807	17.62
	Subtotal				**2.8190**	**17.60**	**0.9293**	**1.57**	**0.1807**	**17.62**
Class III	Stowaway-like	70	16,112	230	2.2370	13.97	0.1247	0.21	0.1910	18.62
	tourist-like	77	19,933	259	3.7405	23.35	0.3228	0.54	0.3451	33.65
	Unknown MITEs	2	1,341	671	0.4950	3.09	0.2080	0.35	0.0458	4.46
	Subtotal				**6.4725**	**40.41**	**0.6556**	**1.11**	**0.5818**	**56.73**
	Grand Total	**213**	**231,327**	**1,086**	**16.0169**	**100.00**	**59.2615**	**100.00**	**1.0255**	**100.00**

(*78*), and RiceHMM (*79*). Strictly speaking, GenScan was trained for maize, another monocot with GC content gradients.

Assessment of gene-prediction programs. All the gene-prediction programs were pretrained by the authors and tested against our cDNA-to-genomic alignments. These comparisons may favor the program that was trained on the largest and most recent data set, but that information was not available to us. Performance was measured at the base pair and the exon levels, and then plotted as a function of position from 5′ to 3′ end (Fig. 8). Sensitivity is the probability that the actual coding region is correctly predicted (1 minus false-negative rate). Specificity is the probability that the predicted coding region corresponds to the actual coding region (1 minus false-positive rate). Some programs, including GenScan, had sensitivities that were extremely dependent on position, although this was not the case when we applied these performance metrics to human genes. This suggests that the compositional gradients were indeed a source of error. That GenScan would be affected is significant because, in the most recent comparative analysis of human genes (*80*), two of the most successful programs were FGeneS (a variant of FGeneSH) and GenScan. For rice, however, FGeneSH is the most successful program. It is not obvious why, although the documentation states that FGeneSH places more weight on signal terms (e.g., splice sites, start and stop codons) than on content terms (i.e., codon usage).

Submitting our *93-11* assembly to the FGeneSH Web site returned 75,659 predictions. However, only 53,398 were complete, in the sense that initial and terminal exons were both present; 7489 had only an initial exon, 11,367 had only a terminal exon, and 3405 had neither. When we include predictions without both an initial and terminal exon as only half a gene, we obtain an upper bound of 64,529 genes. Without correcting for sensitivity or specificity, the estimated gene count is 53,398 to 64,529. This is similar to the 59,855 genes that we predicted from considerations of gene size and repeat content. How good are these predictions? We have reservations about the absolute value of the performance metrics, because FGeneSH was probably trained on a gene set with considerable overlap to our reference cDNAs. These metrics may not tell us how well FGeneSH performs for rice genes with substantially different compositional properties. However, their relative values should be interpretable. Namely, base-level specificities were better than base-level sensitivities, indicating that false-negatives are more likely to be a problem than false-positives. The program is more likely to miss an exon fragment than to label something part of an exon by mistake. Sensitivities and specificities were much worse at the exon level, implying that that the exon-intron boundaries are not precisely defined, even when the presence of a gene is correctly detected.

Two pieces of evidence qualify our level of confidence in the gene predictions. First, if the sensitivity is really as good as suggested, then we ought to be able to find most of the ESTs in the predicted gene set. We thus performed a comparison against the 24,776 UniGene clusters assembled from our 87,842 ESTs. The result was that only 77.3% of these clusters could be found in the FGeneSH predictions. Second, the mean size of the predicted coding regions in rice was only 328 residues, or 73.5% of the predicted coding regions in *A. thaliana*, which averaged 446 residues. This was the case even though we restricted the mean to complete genes with initial and terminal exons. Although it is possible that rice genes are intrinsically smaller than *A. thaliana* genes, we believe that this discrepancy reflects a deeper problem that is related to the compositional gradients, as will be explained below.

Functional classification of rice genes.

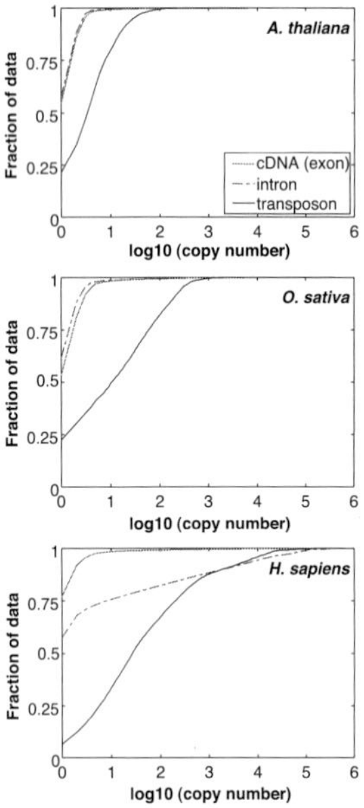

Fig. 7. Cumulative copy numbers for exons, introns, and known transposons in *A. thaliana*, *O. sativa*, and *H. sapiens*. We determined the copy number of 20-mers in each genome, and then mapped these 20-mers back to exons, introns, and known transposons for each genome. All exon and intron sequences were derived from cDNA-to-genomic alignments. The analyzed transposons were the consensus sequences for the known families or subfamilies of transposons. We show here the fraction of each data set that is in 20-mers up to the indicated copy numbers.

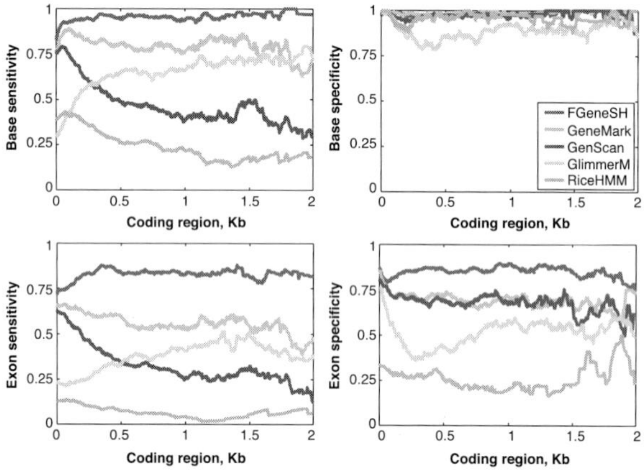

Fig. 8. Performance metrics for ab initio gene-prediction programs, as a function of gene position from 5′ to 3′ end, when compared against cDNA-to-genomic alignments at the same loci. Sensitivity is the probability that the coding region is correctly predicted (1 minus false-negative rate). Specificity is the probability that the predicted coding region is real (1 minus false-positive rate). At the exon level, both splice sites must be correctly predicted for an exon to be counted as correct.

Although 25,426 genes have been identified in *A. thaliana*, fewer than 10% have been documented experimentally (*81*). Consequently, functional classification of plant genes must rely heavily on homology, coupled with a few nonhomology-based methods, such as phylogenetic profiling, correlated gene expression, and conserved gene orders. Only 27.3 and 36.3% of *A. thaliana* genes have been classified by InterPro (*82*) and Gene Ontology Consortium (*83*), respectively. To establish functional classifications for rice genes, we performed protein-to-protein sequence comparisons against *A. thaliana* annotations, and adopted classifications from the best match to *A. thaliana*. We considered only those 53,398 predictions from FGeneSH with initial and terminal exons. When multiple hits were found, we selected the one with the longest extent of homology. We required that at least 25% of the protein length be matched. This is a low-threshold setting, but as we will explain below, it was necessary. In total, 15.9 and 20.4% of rice gene predictions were classified by InterPro and Gene Ontology Consortium, respectively. As a percentage of classified genes, the predicted gene sets for rice and *A. thaliana* are similarly distributed among different functional categories (Fig. 9). We depict Gene Ontology Consortium because more genes were classified. Tables of predicted rice genes and their functional classifications (Web supplement 2), as well as InterPro figures (Web supplement 3), are available on *Science* Online at www.sciencemag.org/cgi/content/full/296/5565/79/DC1.

We advise extreme caution in interpreting minor differences in functional classification between the predicted gene sets for rice and *A. thaliana*. With such a large fraction of the genes unclassified, intrinsic uncertainties in any classification scheme are amplified into artifactual differences. For example, the largest difference for InterPro was in signal transduction genes, but no notable difference was observed for Gene Ontology Consortium. Furthermore, focusing on small differences that had a high likelihood of being artifactual would distract from the major difference between rice and *A. thaliana*, which as we will show next, lies almost entirely in those genes with no functional classification.

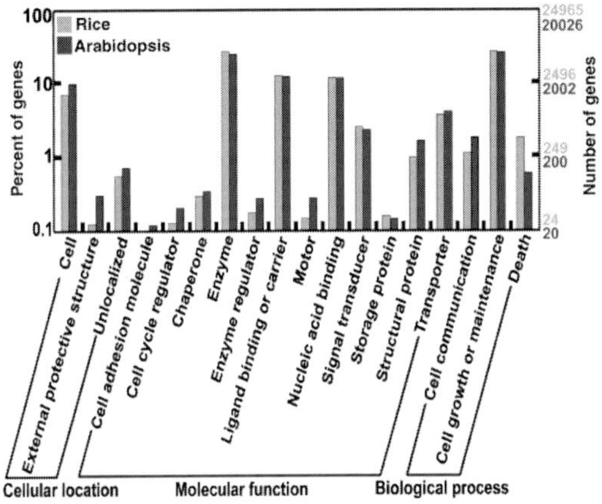

Fig. 9. Functional classification of rice genes, according to Gene Ontology Consortium, and assigned by homology to categorized *A. thaliana* genes. In this ontology, "biological process," "cellular location," and "molecular function" are treated as independent attributes. Only 36.3% of the 25,426 predicted genes for *A. thaliana* are classified. For rice, only 20.4% of the 53,398 complete predictions, with both initial and terminal exons, could be classified.

Table 4. Polymorphism rates relative to *93-11* (*indica*). Comparisons were made to finished BAC sequences from *GLA* (*indica*) and *Nipponbare* (*japonica*), as well as to *PA64s* contigs. Rates were computed for repeated and unique regions, in single-base substitutions (SNPs) and insertion-deletions (InDels). The numbers given for "unalignable" are a gross underestimate because RePS assemblies omit many of the fully masked reads that correspond to the unalignable regions of Fig. 14.

	Nipponbare (*japonica*)	*PA64s*	*GLA* (*indica*)
SNPs in repeated sequence (%)	0.88	0.68	0.65
InDels in repeated sequence (%)	0.33	0.45	0.27
SNPs in unique sequence (%)	0.50	0.35	0.50
InDels in unique sequence (%)	0.14	0.16	0.15
Repeated sequence fraction (%)	24.1	25.5	22.8
Unique sequence fraction (%)	74.8	74.3	74.1
Parts unalignable by BLAST (%)	1.1	0.3	3.1

***A. thaliana* comparisons.** In general, there are two ways to compare gene sets: through colinearity and homology. Colinearity of plant genomes has been studied extensively (*84, 85*). For analyses done within a plant family, high degrees of colinearity have been consistently observed. Across the monocot-eudicot divide, with rice and *A. thaliana* as representative species, observed degrees of colinearity have been considerably lower (*86–89*). For example, an analysis of a 340-kb segment on rice chromosome 2 identified 56 putative genes (*88*). Homologs for 22 (39%) of them were identified in *A. thaliana*, but were distributed among 5 chromosomal segments, with several small-scale inversions. Another study of 126 rice BACs, totaling 20 Mb of sequence and with 3011 putative genes, identified homologs in *A. thaliana* for 1747 (58%) of these genes (*89*). Typically, each 150-kb BAC mapped to three or more chromosomes. Notwithstanding the absence of colinearity, the finding that only half of the rice genes had a homolog in *A. thaliana* was unexpected. Although these analyses were based on predicted genes, which have not yet been confirmed, we do not believe that this was why so few rice genes had a homolog in *A. thaliana*, because a similar analysis was done with 27,294 unique ESTs from *Z. mays* (maize), and only 62% of the open reading frames had a homolog in *A. thaliana* (*90*).

We focus exclusively on homology, rather than orthology, because extensive gene duplications in *A. thaliana* (*4, 91*) and rice make strict one-to-one pairing relations, the classic definition for orthology (*92*), difficult to determine. A mere 35% of *A. thaliana* genes are unique and 37.4% belong to gene families with more than five members. Segmental duplications larger than 100 kb in size constitute 58% of the genome, and 17% of the genes are arranged in tandem arrays. In comparisons of rice with *A. thaliana*, and vice versa, we sought to compute the degree of homology in each direction, and the extent to which gene duplications in *A. thaliana* are replicated in rice when decomposed by functional classification. Even this modest objective was not easy to accomplish, because of unexpected complications introduced by the compositional gradients in rice.

Homology between monocots-eudicots. The complete set of 25,426 annotated *A. thaliana* genes was downloaded from the *Arabidopsis* Information Resource Web site (*93*) on 29 November 2001. As a control, 1441 proteins were downloaded from SwissProt (*94*) on the same day. The rice genes were restricted to the 53,398 predictions from FGeneSH with initial and terminal exons. We compared protein sequence to all six reading frames of the genome sequence by means of TblastN (*36*). Therefore, if the homology search failed, it would not be due to a gene being missing from the annotation of the target genome. The expectation value cutoff was set to 10^{-7}. This was not a sensitive parameter, as most hits were either very good or very bad. What mattered was the "coverage rule." We projected every hit back to the protein query, and unless a minimum fraction of the protein was covered, none of the hits were accepted. The hits had to occur in the same order in both the query and the target, and they all had to be in the same orientation. When a homolog spanned more than one scaffold, the coverage rule was imposed on each scaffold. From this rule, we estimated the number of homologs per gene, the extent of the homology, and the percentage amino acid identity (*95*).

The asymmetry in the monocot-eudicot analysis was striking (Fig. 10). About 80.6% of *A. thaliana* genes had a homolog in rice. The mean extent of homology was 80.1% of the protein length, and there was 60.0% amino acid identity. If instead of the full set of annotated genes, we had used SwissProt genes, 94.9% of the genes would have had a homolog, across 86.7% of the protein length and at 72.9% amino acid identity. Presumably, there were more homologs in the SwissProt data because they were more biased toward highly conserved proteins. In contrast, only 49.4% of predicted rice genes had a homolog in *A. thaliana*. The mean extent of homology was 77.8% of the protein length, and there was 57.8% amino acid identity. For brevity, predicted rice genes with a homolog in *A. thaliana* are called WH genes, and those with no homologs are called NH genes. We identified two distinct problems in this analysis, both attributable to the compositional gradients in rice. One was the poor quality of the FGeneSH predictions for NH genes, and the other was related to the probability of identifying a TblastN hit even with a perfect gene annotation. We did use ESTs to confirm that NH genes were not false predictions, but first, we will discuss what we believe to be the true problems.

We had previously observed that rice gene predictions were only 73.5% the size of *A. thaliana* gene predictions. This discrepancy is not due to the WH half of the rice genes. It is due to the NH half, which was on average 49.4% smaller than the WH half (Fig. 11). To analyze the problem, we randomly sampled 3000 WH genes and 3000 NH genes, and applied the analyses of Fig. 3 to Fig. 7. In general, WH genes resembled the "gold standard" based on alignment of cDNA to genomic sequence. NH genes exhibited a number of striking differences. First, the decreased coding region size was clearly due to a decrease in the number of exons, not to a decrease in the size of the exons. The GC-rich tail in NH gene exon distribution was twice as large as normal (Fig. 11), suggesting that NH genes had more pronounced GC content gradients than either WH genes or those cDNAs retrieved from GenBank. It is plausible that FGeneSH performance would have faltered on NH genes, because NH genes did not resemble those genes on which FGeneSH was presumably trained. NH genes also had twice as many introns as normal in the 200- to 2000-bp range (Fig. 11). This would be consistent with some of these missing exons being combined with their flanking introns. The preponderance of anomalous subminimal introns would be consistent with exon fragments being mistakenly called introns. However, NH genes could not be transposon sequences, because a 20-mer analysis confirmed that their constituent sequences were found in the genome at low copy numbers, much like WH and cDNA-derived genes.

Although we do not entirely ascribe the small size of NH genes to a failure by FGeneSH to detect exons, it is likely that more exons were missed than for WH genes. Thus, it would be more difficult to identify a homolog for these genes in *A. thaliana*. However, even for experimentally derived gene

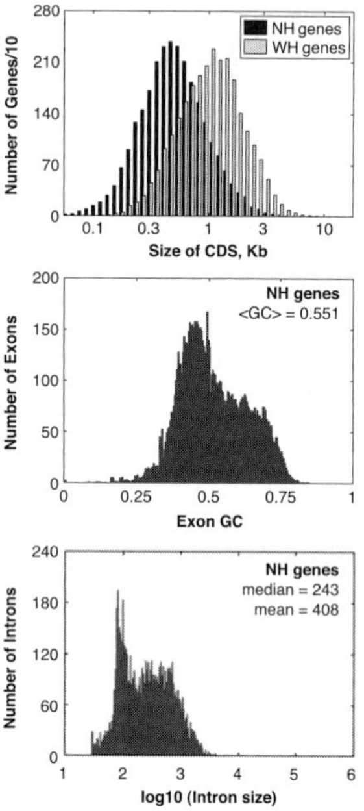

Fig. 11. Size distribution of predicted rice genes with a homolog (WH), and with no homolog (NH), in *A. thaliana*, plus exon GC content and intron size for a random sampling of 3000 NH genes. Gene size refers to the size of the predicted coding region.

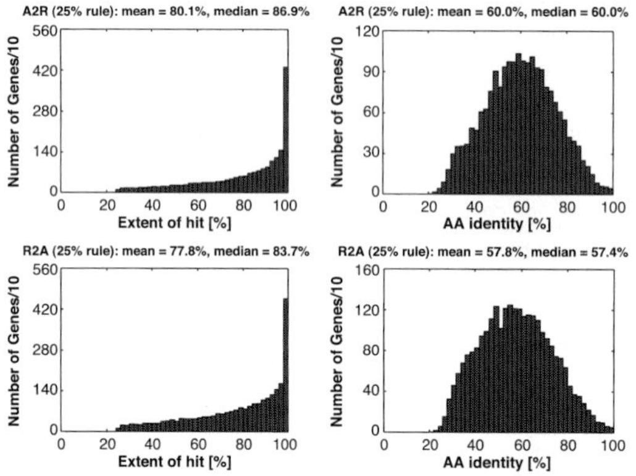

Fig. 10. Distributions in extent of homology and maximum amino acid identity, for *Arabidopsis*-to-rice and rice-to-*Arabidopsis* comparisons. These values are based on a comparison of predicted protein sequence against all six reading frames of the target genome sequence.

sequences, like cDNAs, the probability of identifying a TblastN hit, as a function of the position, dropped precipitously near the 5′ end of the genes (*52*). Far from the end, the probability was about 90%, but within the first few hundred bases near the 5′ end, the probability dropped to less than 50%. This was another consequence of the compositional gradients in rice. The magnitude of the effect was unexpected. We had thought that selective constraints on coding sequences would have limited the number of amino acid changes, despite pressure from rising GC content. However, this was not the case. Homology searches were more likely to fail with the smaller NH genes because the problematic region was a larger fraction of their total length, and our "coverage rule" required that the TblastN hits cover a minimum fraction of the coding region. Even in the *Arabidopsis*-to-rice analysis, where the gene predictions were more reliable, 83.2%, 80.6%, 69.5%, and 48.5% of *A. thaliana* genes had a homolog in rice, for coverage rules of 0, 25, 50, and 75%. We had to use a relatively low coverage rule of 25%. Given the typical protein and protein domain sizes of 446 and 100 residues (*96–98*), respectively, this was equivalent to one protein domain.

Alternatively, what if the problem were due to scaffold size? Half of the NH genes were identified in a scaffold that was smaller than 7.1 kb. However, as a function of scaffold size, predicted coding regions for NH genes were almost always the same size. NH genes found in scaffolds greater than 7.1 kb were only 7% larger than those found in scaffolds less than 7.1 kb. Scaffold size could not have been responsible for the small size of the NH genes. Perhaps NH genes are not real genes at all. Are they even expressed? Looking back at our EST confirmation analysis, we found that 42.9% of WH genes were confirmed by a UniGene cluster, compared with 15.4% of NH genes. Assuming that all WH genes are real, this would imply that (15.4/42.9) × 100% = 35.9% of NH genes are real. However, if we adjust for their being 49.4% smaller than normal, attributing this size deficit to missed exons, then 72.7% of NH genes are real. Certainly, not every NH gene is real, but many are. To be conservative, we can adjust our gene count estimates by a factor of (0.494 + 0.727 × 0.506), resulting in a revised gene count of 46,022 to 55,615.

Considering the relatively recent divergence between monocots and eudicots, 145 to 206 million years ago, it is surprising to find so many genes in rice with no homolog in *A. thaliana*. Even more intriguing, this absence of homology for NH genes extended to other sequenced organisms, including *D. melanogaster*, *Caenorhabditis elegans*, *Saccharomyces cerevisiae*, and *Schizosaccharomyces pombe*. Although WH genes had a 30.5% probability of being homologous to at least one gene in one of these organisms, NH genes had a 2.4% probability. Hence, the major difference between rice and *A. thaliana* gene sets lies in that half of the predicted rice gene set with essentially no homologs in any organism, and whose functions are largely unclassifiable.

Duplication between monocots-eudicots. Having established the major difference between the gene sets for rice and *A. thaliana*, we now consider the similarity. We had reported that 80.6% of the predicted *A. thaliana* genes, and 94.9% of the SwissProt genes, had a homolog in rice. The actual number is likely to be even higher, because the gradients kept us from identifying potential homologs for smaller genes. We know that, within *A. thaliana*, the genes are highly duplicated. Are these genes duplicated in the same manner when mapped to rice? As a proxy for the number of gene homologs within and between genomes, we used the "hits per gene," as defined in the notes (*95*). Considering that, in the *Arabidopsis*-to-rice comparison, we used a low coverage rule of 25% to compensate for the gradients, it was inevitable that we would experience more difficulty than usual in distinguishing between duplicated domains and duplicated genes. Thus, the number of hits per gene is an overestimate of the number of gene homologs.

Comparing *Arabidopsis*-to-*Arabidopsis* (A2A), the mean and median hits per gene were 38.2 and 6.0, similar to the mean and median of 33.4 and 5.0 that we observed comparing *Arabidopsis*-to-rice (A2R) (Fig. 12). That the A2R numbers would be slightly smaller makes sense, given the 145 to 206 million years of divergence. We further note that the means were large only because of a few outliers, some with up to 1000 hits. The identity of these outliers included protein kinase, cytochrome P450, putative disease resistance, and many "unknown" genes. It is difficult to draw any conclusions about the last category, but the others are highly duplicated gene families, which confirms that these outliers were not computational artifacts. The maximum amino acid identity was independent of the number of hits, but the minimum amino acid identity decreased with the number of hits, which would be consistent with an increasing occurrence of hits to ever larger families of related but divergent genes. Although the number of hits was dependent on the functional classification, it was similarly distributed among the different functional categories for A2R and A2A (Fig. 13). Therefore, not only was it possible to identify a homolog in rice for almost every *A. thaliana* gene, but the patterns of gene duplication in one were largely replicated in the other.

The most parsimonious explanation is that the rice gene set is essentially a "superset" of the *A. thaliana* gene set. However, we are unable to say how many of these additional genes that are unique to rice are functionally novel, or merely unrecognizable, because of gradients in rice amino acid usage. It does seem unlikely that so many novel genes would arise within only 145 to 206 million years, and therefore, we suspect that a massive duplication event (or a series of duplication events) occurred, after which many of the rice genes were rendered unrecognizable

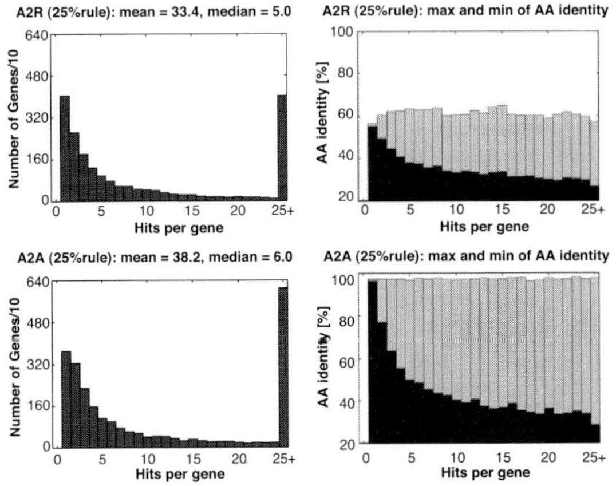

Fig. 12. Distributions in number of hits per gene and maximum-versus-minimum amino acid identity, for *Arabidopsis*-to-rice and *Arabidopsis*-to-*Arabidopsis* comparisons. "Hits per gene" is a proxy for the number of gene homologs, between and within genomes.

by compositional gradients. Some may have been inactivated, and now exist only as pseudogenes. However, until we can compensate for the confounding effects of compositional gradients, we cannot explore the extent to which rice (*99*) and many other plants, including *A. thaliana*, are hybrid (*100*) or allopolyploid (*101*, *102*) in origin.

Rice polymorphisms. Differences between subspecies or cultivars of rice must be described at two levels, gross and nucleotide. At the gross level, we found kilobase-sized regions of high similarity interspersed with kilobase-sized regions of no similarity. One such example is shown in Fig. 14, which was based on a comparison of two overlapping BACs from *indica* and *japonica*. Every unalignable region coincided with a cluster of MDRs, traceable to length differences of 0.7 to 25 kb between the two source sequences, distributed in almost equal proportions between insertions and deletions. To the extent that BDRs could be identified, in roughly half of the unalignable regions, they belong to the class of nested retrotransposons that inhabit the intergenic regions between genes. This is another confirmation of the observation that genome sizes change rapidly in grasses (*103*). On the basis of the available 259 kb of overlapping finished BAC sequences, from *Nipponbare* (*japonica*) and *GLA* (*indica*), all on rice chromosome 4, we would estimate that 16% of the *indica* and *japonica* genome is unalignable by this definition.

At the nucleotide level, excluding the unalignable regions, we define polymorphism rates for repeated and unique sequence, partitioned in single-base substitutions (single-nucleotide polymorphisms, SNPs) and insertion-deletion polymorphisms (InDels). By repeated sequence, we mean MDRs. Three different comparisons are shown in Table 4. Two are based on the alignment of *93-11* contigs to finished BAC sequences from *Nipponbare* (*japonica*) and *GLA* (*indica*), totaling 11.8 and 0.9 Mb, respectively. The other is a comparison of *93-11* and *PA64s* contigs. One might question the accuracy of a polymorphism rate based on rough draft sequence, particularly the low-coverage *PA64s* sequence. However, as we noted in our "quality assessments" section, most of the errors are in the small contigs and at the ends of the contigs. Thus, we restricted this analysis to contigs larger than 3 kb, with 500 bp trimmed off both ends. Overall, there was twice as much variation in the repeated regions as in the unique regions. Substitution rates were two to three times as large as InDel rates. Remarkably, there was very little difference among the three pairwise comparisons. For *93-11* to *PA64s*, averaged over repeated and unique regions, the SNP and InDel rates were 0.43 and 0.23%, respectively. Combining the SNP and InDel rates, we obtained an overall rate of 0.67%. Although the numbers are not exactly comparable, the measured polymorphism rate in maize was 0.96% (*104*).

SNPs are useful in genetic mapping (*105*), and are either directly applicable to phenotypes or indirectly applicable through linkage and association studies. Polymorphisms in the unique regions are particularly useful because, unlike those in the repeated regions, they are more reliably genotyped with existing high-throughput technologies, which always involve some sort of hybridization step. We expect that genome-wide SNP mapping in plants (*106*) will become more popular as new technologies become available, especially as some are customized for plants (*107*).

Concluding remarks. In the initial annotation of the human genome (*7*, *8*), alternative splicing was proposed as a method by which protein diversity could be generated from the surprisingly small number of genes that were identified. The idea that there is extensive alternative splicing in human genes has been supported by analyses of EST data (*108*–*112*). Alternative splicing is often associated with the exon recognition model (*113*) of pre-mRNA splicing. Exon recognition is facilitated by exonic splicing enhancers—short, degenerate sequences located in the exons that are recognized by a multitude of RNA binding factors (*114*, *115*). Because it is the exons that are recognized by the splicing machinery, the intron sequence content is less critical, and transposon insertions into the intron are more readily tolerated. Thus, the preponderance of large transposons-filled introns in the human genome is consistent with extensive alternative splicing.

The presence of relatively few transposons inside plant introns suggests that exon recognition is not a common process for plant genes. Indeed, exonic splicing enhancers have yet to be identified in plants (*116*). The corollary is that there should be relatively little alternative splicing in plant genes. Analysis of the EST data confirms that *A. thaliana* has substantially less alternative splicing than vertebrates or invertebrates (*117*). However,

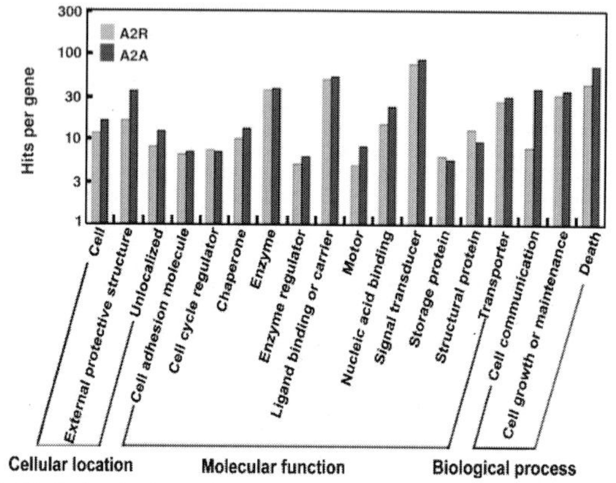

Fig. 13. Distributions in the number of hits per gene, sorted according to Gene Ontology Consortium, for *Arabidopsis*-to-rice and *Arabidopsis*-to-*Arabidopsis* comparisons. This figure shows only the 36.3% of predicted *A. thaliana* genes that are classified.

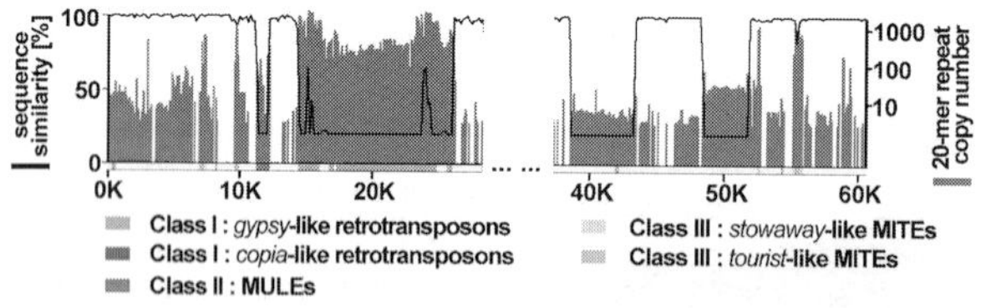

Fig. 14. Comparison of *GLA* (*indica*) and *Nipponbare* (*japonica*) BAC sequences (GenBank accession number AL442110 and AL606449, respectively). Exact 20-mer repeats are indicated by blue histogram bars, with bar heights proportional to copy number in *93-11* (*indica*). Sequence similarity is almost 100%, or unalignable and set to 20% by BLAST. Every unalignable region coincides with a cluster of MDRs, but RepeatMasker fails to identify a BDR in half of these regions.

protein diversity must be generated for the organism to evolve. Our analysis has demonstrated extensive gene duplications in rice and *A. thaliana*, which are highly correlated with each other when decomposed by functional classification. The conclusion is that protein diversity in plants is generated primarily through gene duplications, whereas in vertebrates, it is generated through gene duplications and alternative splicing. This would explain why rice has so many genes. However, as a method of generating protein diversity, gene duplications come at the cost of an increase in transcriptional noise (*118*). Perhaps, at some level of complexity, alternative splicing becomes preferred.

Looking to the future, we intend to improve our draft sequence by adding more reads from large-insert clones, filling any gaps that are likely to contain genes, and integrating the resultant sequence with existing physical and genetic maps. The large-insert clones are necessary to correctly assemble across the large repeat clusters that are sprinkled throughout the rice genome. Until then, the BAC-end sequences (*119*) may not be useful because they are too large to bridge adjacent contigs, and instead skip intervening contigs, resulting in a morass of interleaving scaffolds. One should also be wary of large-scale differences between *indica* and *japonica*. In any event, the final assembly will be made freely available to the research community. We will then apply the experiences gained from the rice genome project to other agriculturally important crops, including *Z. mays* (maize) and *T. aestivum* (wheat).

References and Notes

1. T. Sasaki, B. Burr, *Curr. Opin. Plant Biol.* **3**, 138 (2000).
2. N. A. Eckardt, *Plant Cell* **12**, 2011 (2000).
3. K. Arumuganathan, E. D. Earle, *Plant Mol. Biol. Rep.* **9**, 208 (1991).
4. The *Arabidopsis* Genome Initiative, *Nature* **408**, 796 (2000).
5. B. Martienssen, W. R. McCombie, *Cell* **10**, 571 (2001).
6. M. Bevan *et al.*, *Curr. Opin. Plant Biol.* **4**, 105 (2001).
7. International Human Genome Sequencing Consortium, *Nature* **409**, 860 (2001).
8. J. C. Venter *et al.*, *Science* **291**, 1304 (2001).
9. Q. Tao *et al.*, *Cell Res.* **4**, 127 (1994).
10. Y. Umehara, A. Miyazaki, H. Tanoue, *Mol. Breed.* **1**, 79 (1995).
11. M. Bevan, G. Murphy, *Trends Genet.* **15**, 211 (1999).
12. G. L. Wang *et al.*, *Plant J.* **7**, 525 (1995).
13. M. D. Gale, K. M. Devos, *Proc. Natl. Acad. Sci. U.S.A.* **95**, 1971 (1998).
14. J. Messing, V. Llaca, *Proc. Natl. Acad. Sci. U.S.A.* **95**, 2017 (1998)
15. S. Goff, *Curr. Opin. Plant Biol.* **2**, 86 (1999).
16. Major Web sites for rice genome data: http://rgp.dna.affrc.go.jp; http://www.genome.clemson.edu; http://ars-genome.cornell.edu; http://www.tigr.org/tdb/e2k1/osa1/BACmapping/description.shtml.
17. R. J. Davenport, *Science* **291**, 807 (2001).
18. D. Dickson, D. Cyranoski, *Nature* **409**, 551 (2001).
19. Z. Y. Dai, B. H. Zhao, X. J. Liu (in Chinese), *Jiangsu Agric. Sci.* **4**, 13 (1997).
20. L. P. Yuan (in Chinese), *Hybrid Rice* **1**, 1 (1997).
21. J. Yu *et al.*, *Chin. Sci. Bull.* **46**, 1937 (2001).
22. J. Wang *et al.*, *Genome Res.*, in press. The software can be obtained by e-mailing the authors at reps@genomics.org.cn.
23. J. L. Bennetzen, *Plant Cell* **12**, 1021 (2000).
24. A. Kumar, J. L. Bennetzen, *Annu. Rev. Genet.* **33**, 479 (1999).
25. M. D. Adams *et al.*, *Science* **287**, 2185 (2000).
26. B. Ewing, L. Hillier, M. C. Wendl, P. Green, *Genome Res.* **8**, 175 (1998).
27. B. Ewing, P. Green, *Genome Res.* **8**, 186 (1998).
28. For the plasmid shotgun libraries, a DNA isolation protocol was modified from Sambrook and Russell (*29*). Fresh leaves at the seeding stage were ground in liquid nitrogen before complete lysis (*30*). Purified high–molecular weight genomic DNA was sonicated and sized on agarose gels, selecting for fragments of size 1.5 to 3.0 kb. QIAEX Gel Extraction Kit (QIAGEN) was used to purify DNA from the gel slices. Genomic fragments were ligated to Sma I–linearized pUC18 plasmids and transformed into DH10B-competent cells by electroporation.
29. J. Sambrook, J. D. Russell, *Molecular Cloning*, (Cold Spring Harbor Laboratory Press, Cold Spring Harbor, NY, ed. 3, 2001).
30. S. Hatano, J. Yamaguchi, A. Hirai, *Plant Sci.* **83**, 55 (1992).
31. Single colonies were grown in 96-deep-well plates, and plasmid DNA was prepared by alkaline lysis (*32*). Quality of DNA and insert sizes were examined by agarose gel electrophoresis. Purified plasmid DNA (200 ng; Amersham Pharmacia Biotech, Beijing) was used for the sequencing reactions. DNA sequencing was done with MegaBACE 1000 capillary sequencers (Amersham Pharmacia Biotech, Beijing). Machine parameters were adjusted for high output (10 to 11 runs a day on average).
32. H. C. Birnboim, *Methods Enzymol.* **100**, 243 (1983).
33. N. Jiang, S. R. Wessler, *Plant Cell* **13**, 2553 (2001).
34. P. Green, http://www.phrap.org.
35. E. S. Lander, M. S. Waterman, *Genomics* **2**, 231 (1988).
36. S. F. Altschul, W. Gish, *Methods Enzymol.* **266**, 460 (1996).
37. Sources for STS, STR, restriction fragment length polymorphism sequences: http://ars-genome.cornell.edu/; http://www.ncbi.nlm.nih.gov/; http://rgp.dna.affrc.go.jp/publicdata/geneticmap2000/index.html.
38. A. F. Smit, P. Green, http://ftp.genome.washington.edu/RM/RepeatMasker.html.
39. S. P. Kennedy, W. V. Ng, S. L. Salzberg, L. Hood, S. DasSarma, *Genome Res.* **11**, 1641 (2001).
40. E. Chargaff, *Experientia* **6**, 201 (1950).
41. R. Rolfe, M. Meselson, *Proc. Natl. Acad. Sci. U.S.A.* **45**, 1039 (1959).
42. N. Sueoka, J. Marmur, P. Doty, *Nature* **183**, 1427 (1959).
43. D. R. Forsdyke, J. R. Mortimer, *Gene* **261**, 127 (2000).
44. G. Bernardi, *Gene* **259**, 31 (2000).
45. A. Eyre-Walker, L. D. Hurst, *Nature Rev. Genet.* **2**, 540 (2001).
46. C. Gautier, *Curr. Opin. Genet. Dev.* **10**, 656 (2000).
47. S. Karlin, A. M. Campbell, J. Mrazek, *Annu. Rev. Genet.* **32**, 185 (1998)
48. N. Carels, G. Bernardi, *Genetics* **154**, 1819 (2000).
49. J. Filipski, J. P. Thiery, G. Bernardi, *J. Mol. Biol.* **80**, 177 (1973).
50. G. Bernardi *et al.*, *Science* **228**, 953 (1985).
51. G. K. S Wong, D. A. Passey, Y. Z. Huang, Z. Yang, J. Yu, *Genome Res.* **10**, 1672 (2000).
52. G. K. S. Wong *et al.*, *Genome Res.*, in press.
53. G. K. S. Wong, D. A. Passey, J. Yu, *Genome Res.* **11**, 1672 (2001).
54. C. Feuillet, B. Keller, *Proc. Natl. Acad. Sci. U.S.A.* **96**, 8265 (1999)
55. I. Ashikawa, *Plant J.* **26**, 617 (2001).
56. F. Larsen, G. Gundersen, L. Lopez, H. Prydz, *Genomics* **13**, 1095 (1992).
57. J. Messing, *Trends Genet.* **6**, 196 (2001).
58. R. Sánchez-Fernández, P. A. Rea, T. G. E. Davies, J. O. D. Coleman, *Trends Plant Sci.* **6**, 348 (2001).
59. J. Jurka, *Trends Genet.* **16**, 418 (2000); http://www.girinst.org/index.html.
60. L. Mao *et al.*, *Genome Res.* **10**, 982 (2000).
61. K. Turcotte, S. Srinivasan, T. Bureau, *Plant J.* **25**, 169 (2001).
62. S. Temnykh *et al*,. *Genome Res.* **11**, 1441 (2001).
63. R. I. Richards, G. R. Sutherland, *Cell* **70**, 709 (1992).
64. ———, *Nature Genet.* **6**, 114 (1994).
65. C. Schlötterer, D. Tautz, *Nucleic Acids Res.* **20**, 211 (1992).
66. D. Tautz, M. Trick, G. Dover, *Nature* **322**, 652 (1986).
67. S. A. Surzycki, W. R. Belknap, *Proc. Natl. Acad. Sci. U.S.A.* **97**, 245 (2000).
68. R. Tarchini, P. Biddle, R. Wineland, S. Tingey, A. Rafalski, *Plant Cell* **12**, 381 (2000).
69. X. Zhang *et al.*, *Proc. Natl. Acad. Sci. U.S.A.* **98**, 12572 (2001).
70. A. F. Smit, *Curr. Opin. Genet. Dev.* **6**, 743 (1996).
71. T. E. Bureau, P. C. Ronald, S. R. Wessler, *Proc. Natl. Acad. Sci. U.S.A.* **93**, 8524 (1996).
72. A. Nekrutenko, W. H. Li, *Trends Genet.* **17**, 619 (2001).
73. M. Long, *Curr. Opin. Genet. Dev.* **11**, 673 (2001).
74. F. Jacob, *Science* **196**, 1141 (1977).
75. A. A. Salamov, V. V. Solovyev, *Genome Res.* **10**, 516 (2000); http://www.softberry.com/berry.phtml?topic=gfind.
76. A. V. Lukashin, M. Borodovsky, *Nucleic Acids Res.* **26**, 1107 (1998). http://opal.biology.gatech.edu/GeneMark/eukhmm.cgi?org=O.sativa.
77. C. Burge, S. Karlin, *J. Mol. Biol.* **268**, 78 (1997); http://genes.mit.edu/GENSCAN.html.
78. A. L. Delcher, D. Harmon, S. Kasif, O. White, S. L. Salzberg, *Nucleic Acids Res.* **27**, 4636 (1999); http://www.tigr.org/softlab/glimmerm.
79. K. Sakata *et al.*, *Abstracts of 4th Annual Conference on Computational Genomics* (2000), p. 31; http://rgp.dna.affrc.go.jp/RiceHMM.
80. S. Rogic, A. K. Mackworth, F. B. Ouellette, *Genome Res.* **11**, 817 (2001).
81. P. Breyne, M. Zabeau, *Curr. Opin. Plant Biol.* **4**, 42 (2001).
82. R. Apweiler *et al.*, *Nucleic Acids Res.* **29**, 44 (2001).
83. The Gene Ontology Consortium, *Nature Genet.* **25**, 25 (2000).
84. M. D. Gale, K. M. Devos, *Science* **282**, 656 (1998).
85. R. Schmidt, *Curr. Opin. Plant Biol.* **3**, 97 (2000).
86. K. M. Devos, J. Beales, Y. Nagamura, T. Sasaki, *Genome Res.* **9**, 825 (1999).
87. A. van Dodeweerd *et al.*, *Genome* **42**, 887 (1999).
88. K. Mayer *et al.*, *Genome Res.* **11**, 1167 (2001).
89. H. Liu, R. Sachidanandam, L. Stein, *Genome Res.* **11**, 2020 (2001).
90. V. Brendel, S. Kurtz, V. Walbot, *Genome Biol.* **3**, reviews 1005.1 (2002).
91. M. Devan *et al.*, *Curr. Opin. Plant Biol.* **4**, 105 (2001).
92. W. Fitch, *Syst. Zool.* **19**, 99 (1970).
93. *Arabidopsis* genome annotations: http://www.arabidopsis.org.
94. E. Gasteiger, E. Jung, A. Bairoch, *Curr. Iss. Mol. Biol.* **3**, 47 (2001); http://www.expasy.com.
95. For each protein query, we created an array with one element for each amino acid position. Blast_hits() recorded the number of times that each position was covered by a TblastN hit. Each hit had associated with it a score for the percentage of identically matched amino acids. AA_identity() recorded the maximum and minimum score at each position, across all TblastN hits. "Extent of hit," quoted as a percentage of the protein length, is the number of nonzero elements in Blast_hits(). "AA identity" and "hits per gene" are the median values of AA_identity() and Blast_hits(), computed over positions with one or more hits. We used the median, instead of the mean, to minimize the likelihood of counting a highly duplicated domain when the entire protein is not duplicated.
96. S. A. Islam, J. Luo, M. J. Sternberg, *Protein Eng.* **8**, 513 (1995).
97. R. Sowdhamini, S. D. Rufino, T. L. Blundell, *Fold Des.* **1**, 209 (1996)
98. S. J. Wheelan, A. Marchler-Bauer, S. H. Bryant. *Bioinformatics* **16**, 613 (2000).
99. S. Ge, T. Sang, B. R. Lu, D. Y. Hong, *Proc. Natl. Acad. Sci. U.S.A.* **96**, 14400 (1999).
100. L. H. Rieseberg, *Annu. Rev. Ecol. Syst.* **28**, 359 (1997).
101. J. Masterson, *Science* **264**, 421 (1994).
102. L. Comai, *Plant Mol. Biol.* **43**, 387 (2000).
103. C. M. Vicient, M. J. Jaaskelainen, R. Kalendar, A. H. Schulman, *Plant Physiol.* **125**, 1283 (2001).
104. M. I. Tenaillon *et al.*, *Proc. Natl. Acad. Sci. U.S.A.* **98**, 9161 (2001).
105. L. Kruglyak, *Nature Genet.* **17**, 21 (1997).

THE RICE GENOME

106. R. J. Cho *et al.*, *Nature Genet.* **23**, 203 (1999).
107. E. Drenkard *et al.*, *Plant Physiol.* **124**, 1483 (2000).
108. J. Hanke *et al.*, *Trends Genet.* **15**, 389 (1999).
109. A. A. Mironov, J. W. Fickett, M. S. Gelfand, *Genome Res.* **9**, 1288 (1999).
110. L. Croft *et al.*, *Nature Genet.* **24**, 340 (2000).
111. D. Brett *et al.*, *FEBS Lett.* **474**, 83 (2000).
112. B. Modrek, A. Resch, C. Grasso, C. Lee, *Nucleic Acids Res.* **29**, 2850 (2001).
113. S. M. Berget, *J. Biol. Chem.* **270**, 2411 (1995).
114. B. J. Blencowe, *Trends Biochem. Sci.* **25**, 106 (2000).
115. M. L. Hastings, A. R. Krainer, *Curr. Opin. Cell Biol.* **13**, 302 (2001).
116. Z. J. Lorkovic, D. A. Wieczorek Kirk, M. H. Lambermon, W. Filipowicz, *Trends Plant Sci.* **5**, 160 (2000).
117. D. Brett *et al.*, *Nature Genet.* **30**, 29 (2002).
118. A. P. Bird, *Trends Genet.* **11**, 94 (1995).
119. Source for BAC-end sequences: http://www.genome.clemson.edu/projects/rice/fpc.
120. We are indebted to faculty and staff at the Beijing Genomics Institute, whose names were not listed, but who also contributed to the team effort (www.genomics.org.cn). We are indebted to our scientific advisors, M. V. Olson, L. Bolund, R. Waterston, E. Lander, and M.-C. King, for their long-term support. We are grateful to R. Wu and C. Herlache for editorial assistance on the manuscript. We thank Amersham Pharmacia Biotech (China) Ltd., SUN Microsystems (China) Inc., and Dawning Computer Corp. for their support and service. This project was jointly sponsored by the Chinese Academy of Science, the Commission for Economy Planning, the Ministry of Science and Technology, the Zhejiang Provincial Government, the Hangzhou Municipal Government, the Beijing Municipal Government, and the National Natural Science Foundation of China. The analysis was supported in part by the National Institute of Environmental Health Sciences (grant 1 RO1 ES09909).

14 November 2001; accepted 20 February 2002

Michel A. J. Georges

University of Liège, Liège, Belgium

2006/2007 — for groundbreaking discoveries in genetics and genomics, laying the foundations for improvements in crop and livestock breeding, and sparking important advances in plant and animal sciences.

CURRICULUM VITAE

Date of birth: July 18th, 1959
Nationality: Belgian
Father of three lovely children (Mathieu, Anouk, Delphine)

Education:

Doctor in Veterinary Medicine, University of Liège, 1983.
Master of Science, Molecular Biology, Free University of Brussels, 1985.
Agrégation de l'Enseignement Supérieur, University of Liège, 1991.

Professional experience:

1983-1988: Assistant Professor, Genetics (Prof. R. Hanset), Faculty of Veterinary Medicine, University of Liège.

1989-1993: Senior Scientists, then Director of Research, Genmark Inc., Salt Lake City, Utah.

MAPPING, FINE-MAPPING AND POSITIONAL CLONING OF QUANTITATIVE TRAIT LOCI (QTL)
(*Annual Reviews of Genomics and Human Genetics* 8: 131-162, 2007)

As previously mentioned, most medically and agronomically important traits are thought to be under the influence of multiple genes mapping to QTL, and the "hunt" for these QTL is one of the major objectives of modern genetics. For more than 15 years now, Michel Georges has been at the forefront in the development and application of molecular and statistic methods towards that goal.

He played a key role in the realization of the first whole genome QTL scan performed in a mammal that lead to the identification of QTL influencing blood pressure in the SHRSP rat (*Nature* 353: 521-529, 1991). He performed the first whole genome QTL scan in livestock in order to identify QTL influencing milk yield and composition in cattle. This let to the location of multiple QTL and a seminal paper in the field (*Genetics* 139: 907-920, 1995). After this pioneering work, his team performed multiple whole genome scans in cattle, pigs and sheep, which all lead to the identification of multiple QTL influencing a broad range of agronomically important phenotypes (f.i. *Mammalian Genome* 9: 540-544, 1998; *Animal Genetics* 29: 107-115, 1998; *J Dairy Sci* 82: 2514-2516, 1999; *J Dairy Sci* 83: 795-806, 2000; *Genet Selec Evol* 34: 371-388, 2002).

The mapping resolution achieved with conventional QTL mapping techniques is typically in the 20-30 centimorgan (cM) range, therefore insufficient for positional cloning or effective marker assisted selection. To overcome these limitations, the laboratory of Michel Georges has devised original identity-by-descent (IBD) and linkage disequilibrium (LD) methods for the fine-mapping of QTL. Application of these novel approaches has allowed them to position QTL in chromosome intervals as small as 250 Kb (f.i. *Proc. Natl. Acad. Sci. USA* 96: 9252-9257, 1999; *Genetics* 161: 275-287, 2002; *Genetics* 163: 253-266, 2003; *Genetics* 165: 277-285, 2003).

Achieving this high-resolution mapping of QTL ultimately allowed the team of Michel Georges to identify the causal genes and mutations for two QTL in cattle and one QTL in porcine. The first two were shown to correspond to structural mutations in the diacylglycerol acyl transferase 1 (*DGAT1*) and growth hormone receptor (*GHR*) genes respectively, while the latter one was shown — in collaboration with Dr. Leif Andersson from Uppsala — to correspond to a regulatory mutation in the porcine insulin-like growth factor 2 (*IGF2*) gene (*Genome Research* 12: 222-231, 2002; *Genetics* 163: 253-266, 2003; *Nature* 425: 832-836, 2003; *Proc. Natl. Acad. Sci. USA* 101: 2398-2403, 2004) (Fig. 2).

It is noteworthy that despite considerable investments committed by the scientific community towards the identification of QTL in human, rodent models, plants and animals in recent years, only very few successes have been achieved to date (see f.i. *Science* 298: 2345, 2002 for a review). The successes of Michel Georges' team are therefore rightfully considered quite remarkable.

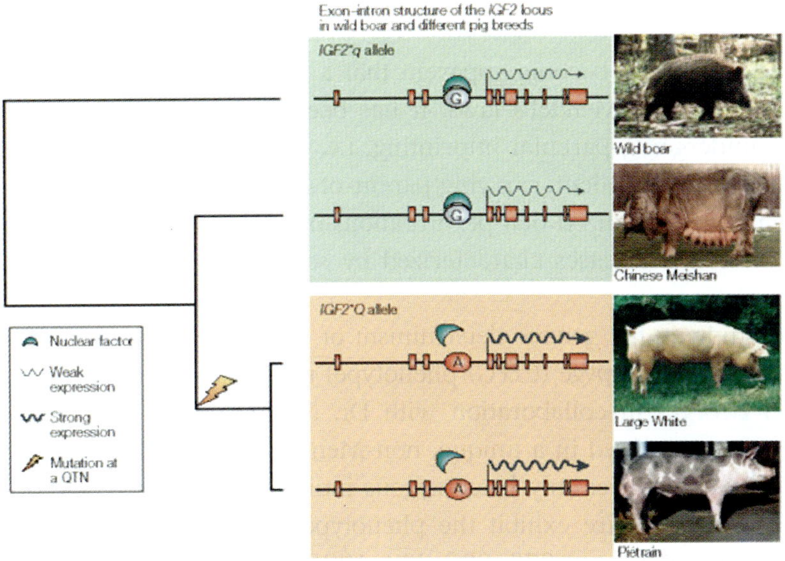

Figure 2: The *IGF2* quantitative trait locus in pigs. Insulin-like growth factor 2 (*IGF2*) was first identified as a paternally expressed quantitative trait locus (QTL) in INTERCROSSES between the European wild boar and Large White domestic pigs; and between Large White and Piétrain pigs. In the wild-boar intercross, the QTL allele from the domestic pig was associated with high muscularity, less backfat and a larger heart. Sequence analysis showed that the *IGF2* haplotypes in the Swedish Large White and Piétrain pigs were identical by descent, whereas the Belgian Large White and wild-boar haplotypes were similar, which indicated the presence of two alleles that are denoted *Q* and *q* for high and low muscle growth, respectively. Another intriguing observation was the large sequence divergence (~1%) between the two haplotypes. This led to the suspicion that the two haplotypes might have an Asian and European origin, in line with previous finding that some European breeds, including the Large White, are hybrids of Asian and European domestic pigs. Sequence analysis of *IGF2* haplotypes segregating in an intercross between Chinese Meishan and Large White pigs confirmed this hypothesis. The Meishan allele, which was functionally equivalent to *IGF2*q*, was nearly identical to *IGF2*Q* at the sequence level. These data provided conclusive evidence that the causative mutation for *IGF2*Q* was a G-to-A substitution at nucleotide 3,072 in intron 3. *IGF2* was identified as a positional candidate gene, but the quantitative trait nucleotide (QTN) was identified by pure genetics: linkage analysis to deduce QTL genotypes combined with an analysis of the minimum shared haplotype. Functional studies showed a plausible mechanism for the QTL effect (see figure). First, the mutation does not affect the imprinting or METHYLATION STATUS of the QTN region and the region is undermethylated in skeletal muscle. Second, the wild-type sequence binds a nuclear factor and this interaction is abrogated by the mutation and by methylation. Third, transfection analysis indicated that the wild-type sequence functions as a silencer element, whereas the mutant sequence is a significantly weaker silencer. Finally, expression analysis showed an approximately threefold upregulation of *IGF2* expression in postnatal skeletal and cardiac muscle but not in prenatal muscle or in liver. The result is consistent with phenotypic data showing that *IGF2*Q* are associated with high muscle growth and a larger heart, but has no effect on birth weight or the size of the liver. The *IGF2* QTL is truly adaptive from a pig production point of view as it does not affect birth weight but supports muscle growth after birth. The photographs of the wild boar, Meishan, Piétrain and Large White pigs were provided by B. Kristiansson, Quality Genetics AB, J.-M. Beduin and the Roslin Institute, respectively. (from *Nature Reviews Genetics* 5: 202-212, 2004)

Polar overdominance at the ovine callipyge locus and other imprinted QTL

In recent years, it has become apparent that a growing number of genes do not operate according to Mendel's laws. It has been shown in particular that some genes are undergoing parental imprinting, i.e. paternal and maternal alleles are functionally non-equivalent, carrying parent-of-origin specific epigenetic marks. It is also increasingly realized that perturbations of such imprinted genes underlie a series of inherited diseases characterized by so-called "parent-of-origin" effects in man.

While studying the genetic determinism of a muscular hypertrophy in sheep (referred to as the callipyge (CLPG) phenotype) using molecular markers, the team of Michel Georges in collaboration with Dr. N. Cockett demonstrated that this phenotype is transmitted in a unique, non-Mendelian way which was referred to as polar overdominance: only heterozygous individuals having received the CLPG mutation form their sire exhibit the phenotype (*Proc. Natl. Acad. Sci. USA* **91**: 3019-3023, 1994; *Science* **273**: 236-238, 1996). The gene has been mapped to a 400 Kb chromosome interval (e.g. *Mammalian Genome* **12**: 141-149, 2001; *Mammalian Genome* **12**: 183-185, 2001). Complete sequencing, annotation and functional analysis of 400 kilobases within this interval lead to the identification of six novel genes which were shown to undergo parental imprinting and be preferentially expressed in skeletal muscle (*Genome Research* **11**: 850-862, 2001; submitted). It was subsequently shown that the CLPG mutation enhances the expression of four of these genes without altering their imprinting status (*Nature Genetics* **27**: 367-369, 2001). The CLPG mutation was shown by others and by Michel Georges's team to be a single base pair substitution in a highly conserved dodecamer motif thought to act as a skeletal muscle specific silencer element (*Genome Research* **12**: 1496-1506, 2002; *Genetics* **163**: 453-456, 2003). The fact that only the padumnal heterozygotes express the phenotype ("polar overdominance") was then shown to result from the trans-inhibition of the genes expressed from the paternal allele by the genes expressed from the maternal allele, and the consequent ectopic expression of DLK1 exclusively in skeletal muscle of callipyge animals (*Trends in Genetics* **19**: 248-252, 2003; *Current Biology* **14**: 1858-1862, 2004). Michel Georges' team has recently shown that this trans-inhibition involves RNA interference and micro-RNA genes expressed exclusively from the maternal allele (*Current Biology* **15**: 743-749, 2005) (Fig. 3).

Polar overdominance at the ovine callipyge locus is becoming a classical topic in the emerging epigenetic field, as testified by the fact that Michel Georges was invited to give a plenary lecture at last years 69th Cold Spring Harbor Symposium on Quantitative Biology devoted to "Epigenetics".

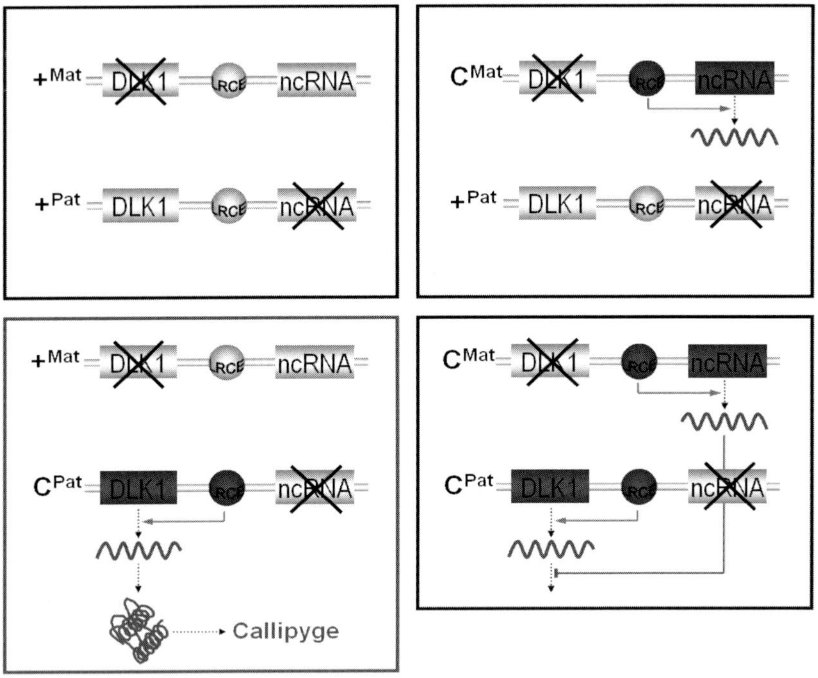

Figure 3: Working model for polar overdominance, involving (i) a paternally expressed growth promoter (DLK1), (ii) a maternally expressed *trans*-acting (red arrow) growth repressor thought to correspond to one or more non-coding RNA genes (*ncRNA*), (iii) a *cis*-acting (green arrow) LRCE affected by the *CLPG* mutation. The crosses correspond to the parental imprinting effects. The functional status in postnatal skeletal muscle is illustrated for the four *CLPG* genotypes: active elements are shown in red, inactive elements in white. Only the $+^{Mat}/C^{Pat}$ individuals express the callipyge phenotype (boxed in red) due to ectopic expression of the DLK1 protein. (from *Current Biology* **14**: 1858-1862, 2004)

The callipyge phenotype was the first example in livestock of a phenotype subject to parental imprinting. Subsequently, the teams of Michel Georges and Leif Andersson demonstrated that one of the QTL they had identified in the pig as having a major effect on muscle mass and fat deposition was in fact subject to parental imprinting as well (*Nature Genetics* 21: 155-156, 1999). This QTL was subsequently shown to be due to a regulatory mutation in the porcine IGF2 gene (see above).

These results have spurred efforts amongst the scientific community to identify other such imprinted QTL. There is growing evidence that parental imprinting may play a much more important role in the determinism of complex traits in livestock than originally anticipated. This may have major consequences for future breeding strategies as conventional selection methods can't cope effectively with imprinted genes.

Polymorphic miRNA-target interactions

Using the same positional cloning approach, Michel Georges' team has recently identified a QTL with major effect on muscular development in the sheep. The QTL was fine-mapped to a region shown to harbour the myostatin gene known from previous work to cause double-muscling in cattle. Detailed analysis of the ovine myostatin gene identified a A to G transition in the 3'UTR revealing an illegitimate target site for miRNAs that are highly expressed in skeletal muscle. miRNAs are a recently discovered class of small non-coding RNA genes that play a key "epigenetic" role during developmental by causing either the degradation or the translational inhibition of target mRNAs. A very large proportion of genes is now thought to be under the control of batteries of miRNAs. It was demonstrated that the identified A to G transition indeed leads to the translational inhibition of the myostatin transcripts and thus to muscular hypertrophy (Fig. 4) (*Nature Genetics* 38: 813-818, 2006).

Most importantly, however, subsequent bioinformatic analyses of SNPs databases for human and mice, demonstrated that mutations creating or destroying putative miRNA target sites are abundant and might thus be important effectors of phenotypic variation. This information has been compiled in a publicly accessible database (www.patrocles.org). This discovery opens a new field of investigation not only for

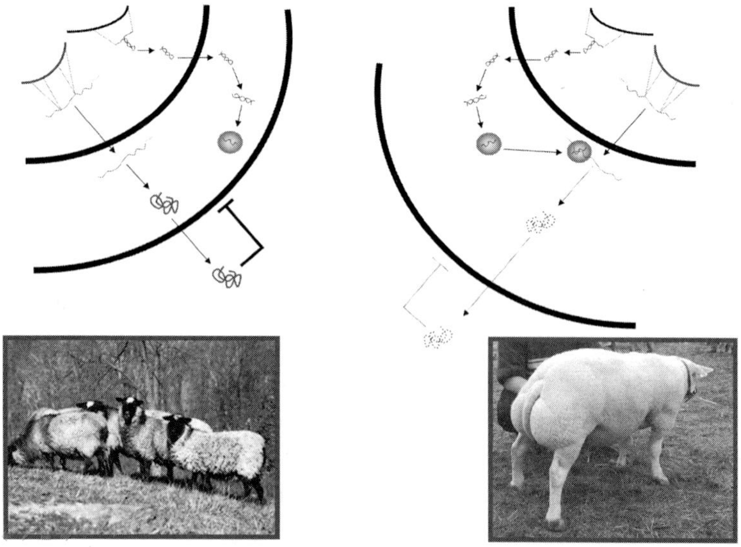

Figure 4: Schematic representation of the effect of the *g+6723G-A* mutation in the 3'UTR of the ovine *GDF8* gene. In wild-type animals (such as the Romanov sheep shown in the left picture) the *GDF8* transcripts (in blue) in the sarcoplasm don't interact with microRNAs *miR-1* and *miR-206* despite their high level of expression in skeletal muscle. Hence GDF8 protein is produced at regular levels, controlling muscle growth by retro-inhibition. In mutant animals (such as the Texel sheep shown in the right picture), *GDF8* transcripts (in red) become targets for *miR-1* and *miR-206* as a result of the *g+6723G-A* mutation. This leads to translational inhibition of the *GDF8* transcripts, reduced GDF8 protein levels, relaxation of the retro-inhibition and hence enhanced muscle growth. (from *71st CSH Symposium on Quantitative Biology* Vol. 69, pp 477-483, 2006)

Michel Georges' team but for the scientific community in general (*Current Opinion in Genetics & Development* 17: 1-11, 2007).

Genome wide association (GWA) analysis of Crohn's disease (CD)

Michel Georges' team has recently moved to the new, multidisciplinary GIGA genome centre on the premises of the Faculty of Medicine of the University of Liège. Teaming up with clinicians (particularly Dr. Edouard Louis), they have applied their expertise in the molecular dissection of complex traits to the genetic analysis of common human diseases including Inflammatory Bowel Disease (IBD). They were amongst the first teams in the world and certainly the first team in Belgium to perform a large scale, genome wide association study using recently emerged high throughput SNP genotyping technology. This work — let in collaboration with the Centre National de Genotypage in Paris (Prof. Mark Lathrop) as well as four other Belgian universities (KUL, RUG, UCL and ULB) — resulted in the identification of a novel locus (5p13.1) contributing to inherited susceptibility to CD. The mining of transcriptome data in conjunction with SNP data let to the identification of a prostanoid receptor (PTGER4) as causal gene thereby opening novel therapeutic avenues (*PloS Genetics*: e58, 2007) (Fig. 5).

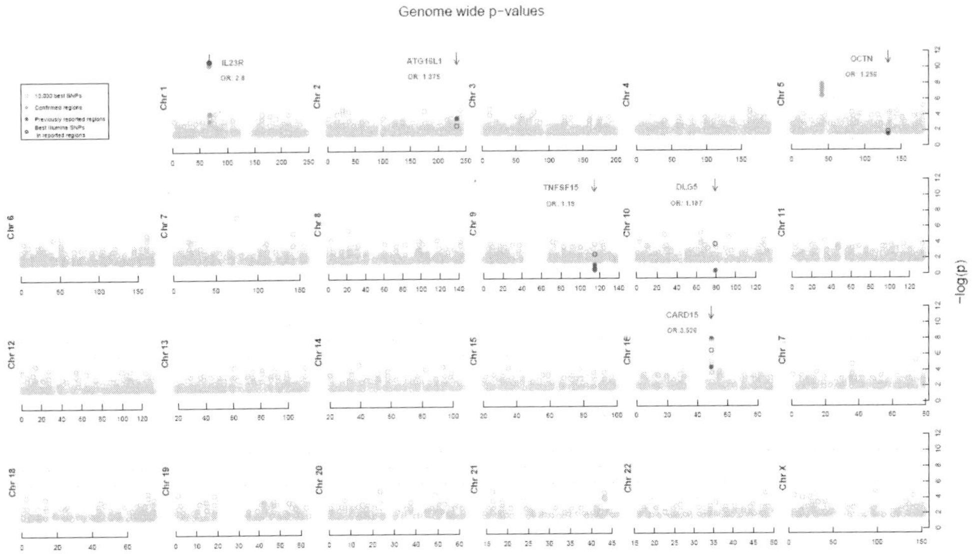

Figure 5: Results of the whole genome association for Crohn Disease. P-values (-log(p)) for the 10,000 best SNPs out of 311,882 are shown (gray circles). The position of previously described susceptibility loci are marked by red arrows. The p-values obtained in our cohorts with the reportedly associated SNPs/mutations are shown by the red dots, and the corresponding odds ratios (OR) indicated. The p-values obtained with SNPs included in the Illumina panel at ≤ 50 Kb from these SNPs/mutations are marked by black circles. Two novel susceptibility loci identified in this study are marked by green dots and correspond respectively to the *IL23R* gene on chromosome 1 and the *PTGER4* gene on chromosome 5. (from *PLoS Genetics* e58, 2007)

These result "have placed Belgium on the map" in the extremely competitive "GWA" field. The laboratory of Michel Georges is presently very actively involved in a large scale meta-analysis in collaboration with the NIDDK and Broad.

Institute in the US, and the Welcome Trust Case Control Consortium in the UK. The results of this unique study, involving ~10,000 CD patients, will be presented at the 2008 American Society of Human Genetics meeting and should result in a major publication in the near future.

Michel Georges' team continues to study the molecular mechanisms involving PTGER4-dependent CD susceptibility as well as the contribution of Copy Number Variation to IBD susceptibility.

Mendelian traits, transgenic livestock and rodent models of susceptibility to infection

In addition to these most salient achievements, Michel Georges' team (i) has been very active over the years in the development and application of bioinformatic, statistical and molecular methods for the positional identification of QTL (e.g. *Science* **235**: 683-684, 1987; *Genomics* **18**: 270-276, 1993; *Genome Research* **6**: 580-589, 1996; *Genetics* **149**: 1547-1555, 1998; *Proc. Natl. Acad. Sci. USA* **96**: 9252-9257, 1999; *Genetics* **161**: 275-287, 2002; *Genetics* **163**: 253-266, 2003; *Genetics* **165**: 277-285, 2003), (ii) has mapped several genes underlying Mendelian traits including the gene causing progressive degenerative myeloencephalopathy called "Weaver" disease (*Proc. Natl. Acad. Sci. USA* **90**: 1058-1062, 1993), the Polled gene determining the presence or absence of horns (*Nature Genetics* **4**: 206-210, 1993), the roan gene involved in a developmental anomaly of the female genital tract called White Heifer disease (*Mammalian Genome* **7**:138-142, 1996), and the gene causing syndactyly in cattle (*Genome Research* **6**: 580-589, 1996). Michel Georges' team has recently developed genome-wide SNP arrays for the bovine and demonstrated their extreme efficacy to map and subsequently clone genes underlying recessive defects. In a few months they have mapped and cloned the genes causing two types of "Congenital Muscular Dystonia" (CMD) as well as "Harlequin Ichthyosis" and mapped the genes causing "Crooked Tailed Syndrome" and "Renal Lipofuscinosis" in cattle (*Charlier et al.*, in preparation). These findings have resulted in the development of diagnostic tests which have been extensively used in the field, eliminating CMD from the Belgian Blue Cattle breed in a matter of months.

Michel Georges' team has also been active in designing strategies to utilize gene information in breeding programs whether by marker assisted selection (e.g. *Livestock Production Science* **54**: 229-250, 1998) or by means of genetic engineering and transgenics (e.g. *Genesis* **35**: 227-238, 2003; *Proc. Natl. Acad. Sci. USA* **102**: 6413-6418, 2005).

More recently they have expanded their activities in rodent models to study the genetic determinism of resistance to infectious agents in collaboration with Dr. Desmecht. Results of genome-wide QTL scans for resistance to Mouse Pneumonia Virus (a mouse model of RSV dependent bronchiolitis in children) and Sendai Virus are being prepared for publication.

Team work

The previously described scientific achievements are obviously the result of intense team work (Fig. 6). Michel Georges is presently heading a group of approximately 30, comprising 1/3 senior scientists and post-doctoral fellows, 1/3 PhD students and 1/3 laboratory technicians. The team comprises people originating from at least six countries with backgrounds ranging from computer engineering to veterinarians via biologists, chemists, physicists, and agronomists. Special credit is expressed towards the group leaders, especially Carole Charlier and Wouter Coppieters.

The team benefits from an annual research budget of the order of 2.5 million Euros per year of which 65% is provided by Belgian and European public funding organizations and 35% by private industry from Belgium, the Netherlands, New Zealand and Canada.

Figure 6: Unit of Animal Genomics 2002. (i) Dimitri Pirottin, Cécile Libioulle, Latifa Karim, Anne Cornet, Fabienne Marcq, Luc Grobet, Alain Empain, Patricia Simon, Bernard Grisart, (ii) Carole Charlier, Fabrice Moreau, Valérie Marot, Karin Segers, Xavier Hubin, Fabienne Davin, Catherine Collette, Erica Davis, Wouter Coppiteres, Maria Smit, (iii) Michel Georges, Minh Nguyen, Benoît Brouwers, Laurence Moreau, Myriam Mni, Francesca Baraldi, Nadine Cambisano, (iv) Bernadette Marcq, Jong-Jo Kim, Paulette Berzi.

This important corporate contribution is perceived as beneficial as it has raised the professional standards of the team, has provided critical mass, has opened avenues for the utilization of the results in the field, even if it has imposed constraints on sharing results with the scientific community. Members of Michel Georges' team are inventors on more than five patents issued over the last five years.

PhDs and postdoctoral fellows

Since he has been holding a professorship at the University of Liège (1994 – present), Michel Georges has been promoter of seven completed PhD theses (Carole Charlier, Luc Grobet (hese d'aggrégation) Frederic Farnir, Stéphane Berghmans, Karin Segers, Dimitri Pirottin, Erica Davis), while five more are in progress. Carole Charlier is now a permanent research fellow of the Belgian National Science Foundation (FNRS); Luc Grobet and Frederic Farnir are professors in respectively Embryology and Statistics at the University of Liège; Stéphane Berghmans has completed a postdoctoral fellowship at the Dana Farber Cancer Institute in Boston (Dr. Zon); Karin Segers is senior scientists in the department of Human Genetics at the University of Liège; Erica Davis is a postdoctoral fellowship at the Johns Hopkins University School of Medicine (Dr. Katsanis); and Dimitri Pirotin is completing research in the laboratory of Michel Georges.

During the same period, the laboratory of Michel Georges has hosted 24 postdoctoral fellows from nine nationalities (Belgium, France, Japan, Korea, Spain, Switzerland, United Kingdom, United States Vietnam). Moreover, many scientists from around the world have visited the Georges' laboratory for short stays.

INTERNATIONAL RECOGNITION

Invited lectures

Over the last 15 years, Michel Georges has been invited on average ~5 times per year to speak at international conferences, including the meetings of the International Society of Animal Genetics (1992, 1994, 2004), the World Congress on Genetics Applied to Livestock Production (1990, 1994, 1998, 2002, 2006), the Plant and Animal Genome Conference (2002, 2003, 2004, 2006, 2007), the International Congress of Genetics (1993, 2003, 2006, 2008), HUGO's Human Genome Meetings (1999, 2001, 2007), Gordon Conferences (1993, 1999, 2003), Keystone Meetings (2004, 2007, 2x2008) and Banbury Conferences (1990, 1995).

The invited lectures that he is most proud of, however, are the three plenary lectures that he was invited to give respectively at the 68th, 69th and 71st Cold Spring Harbour Symposia on Quantitative Biology devoted respectively to "The Genome of Homo Sapiens" (2003), "Epigenetics" (2004), and "Regulatory RNAs"

(2006) as well as for the Cold Spring Harbour Meeting on "The Biology of Genomes" in 2007. In these, his contributions were respectively on the positional identification of QTN in livestock in 2003, on the molecular dissection of callipyge in 2004, on polymorphic miRNA-target interactions in 2006, and on the identification of a novel risk locus for Crohn disease in 2007.

Michel Georges has been invited on several occasions to give courses including at the University of Guelph (Canada), at the University of Leon (Spain), at "Internordic Courses" in Uppsala (Sweden) and Lillehammer (Norway), at the University of Helsinki (Finland), by INRA, at a Wellcome Trust Advanced Course and at the Summer Institute in Statistical Genetics (NCSU and University of Washington, USA).

Editorial boards and peer review

Michel Georges is editor-in-chief of *Genetics Selection Evolution*, and associate editor of *PloS Genetics.*

He is a member of the editorial board of *Mammalian Genome, Genetical Research, Genome Research* and *Physiological Genomics.*

He has been solicited for peer review by numerous journals including *Animal Genetics, Animal Biotechnology, Annales de Médecine Vétérinaire, Genetics, Genetical Research Genetics Selection Evolution, Genome Research, Genomics, Human Molecular Genetics, Journal of Experimental Biology, Journal of Heredity, Livestock Production Science, Mammalian Genome, Nature, Nature Genetics, Nature Reviews Genetics, PNAS, Science, Theoretical and Applied Genetic* and *Trends in Genetics.*

Miscellaneous

Michel Georges is or has been a member (i) of the Steering Committee for the ESF Scientific Program on Integrated Approaches for Functional Genomics, (ii) of the NFWO commission on "Medical cell biology and genetics", (iii) of the Scientific Advisory Board of the Centre National de Génotypage (CNG) – Evry, France, (iv) of the Scientific Advisory Board of the Institut National de la Recherche Agronomique (INRA)- France, (v) of the Scientific Advisory Board of the Consortium National de Recherche en Génomique (CNRG) – France, and (vi) of the RIKEN SNP Research Center Advisory Council (SRAC) – Japan.

His laboratory is part of two European Networks of Excellence: (i) The Epigenome, devoted to the study of epigenetic phenomena (http://www. epigenome-noe.net/) and (ii) Eadgene, devoted to the study of host-pathogen interactions in livestock (http://lotus5.vitamib.com/hnb/eadgene/eadgene.nsf/web/frame?openform).

He is one of the original founders of the GIGA/CBIG project at the University of Liège that has culminated with the creation of a new genome centre regrouping over 175 scientists in the fall of 2006.

The work of Michel Georges' team has been covered by review articles in *Science*, *Scientific American* (e.g. W.W. Gibbs The unseen genome: beyond DNA. *Scientific American* 12/01/2003) *La Recherche* and *Nature Reviews Genetics.*

Michel Georges has recently organized the European Institute in Statistical Genetics at the University of Liège, in collaboration with Professor Bruce Weir from the University of Washington in Seattle. It has attracted 223 students from 27 countries which have followed a total of 450 modules.

(http://www.biostat.washington.edu/sisg07/)

In 2007, Michel Georges was awarded the Wolf prize for agriculture, which is arguably the most prestigious prize in this field.

MAJOR PUBLICATIONS

Articles in international journals with peer review:

[x]: impact factor of journal, year of publication

|Corresponding author|: mentioned when M.G. not last author despite being corresponding author.

1. **GEORGES, M.**; LEQUARRE, A.S.; HANSET, R.; VASSART, G. (1987). Genetic variation of the bovine thyroglobulin locus studied at the DNA level. *Animal Genetics* 18:41-50. [1.4]
2. **GEORGES, M.**; COCHAUX, P. LEQUARRÉ, A.S.; YOUNG, M.W.; VASSART, G. (1987). DNA fingerprinting in man using a mouse probe related to part of the Drosophila "Per" gene. *Nucleic Acids Research* 15:7193. [6.6]
3. **GEORGES, M.**; SWILLENS, S.; BOUQUET, Y.; LEQUARRÉ, A.S.; HANSET, R. (1987) Genetic linkage between the bovine plasma protease inhibitor 2 (Pi-2) and S blood group loci. *Animal Genetics* 18:311-316. [1.4]
4. VASSART, G.; **GEORGES, M.**; MONSIEUR, R.; BROCAS, H.; LEQUARRE, A.S.; CHRISTOPHE, D. (1987). A sequence in M13 phage detects hypervariable minisatellites in human and animal DNA. *Science* 235: 683-684. [24.68]
5. **GEORGES, M.**; LEQUARRÉ, A.S.; CASTELLI, M.; HANSET, R.; VASSART, G. (1988). DNA fingerprinting in domestic animals using four different minisatellite probes. *Cytogenet Cell Genet* 47:127-131. [1.2]
6. **GEORGES, M.**; HILBERT, P.; LEQUARRÉ, A.; LECLERC, V.; HANSET, R.; VASSART, G. (1988). Use of DNA bar codes to resolve a canine paternity dispute. *JAVMA* 193: 9-12. [1.4]
7. FARBER, C.M.; **GEORGES, M.**; DEBOCK, G.; VERHEST, A.; SIMON, P.; VERSCHRAEGEN-SPAE, M. (1989). Demonstration of spontaneous XX/XY chimerism by DNA fingerprinting. *Hum Genet* 182:1-3. [4]

8. FRIES, R.; BECKMANN, J.S.; **GEORGES, M.**; SOLLER, M.; WOMACK, J. (1989). The bovine gene map. *Animal Genetics* 20:3-29. [1.4]
9. HILBERT, P.; MARCOTTE, A.; SCHWERS, A.; HANSET, R.; VASSART, G.; **GEORGES, M.** (1989). Analysis of genetic variation in the Belgian Blue Cattle Breed using DNA sequence polymorphism at the growth hormone, low density lipoprotein receptor, -subunit of glycoprotein hormones and thyroglobulin loci. *Animal Genetics* 20:383-394. [1.4]
10. **GEORGES, M.**; LATHROP, M.; HILBERT, P.; MARCOTTE, A.; SCHWERS, A.; SWILLENS, S.; VASSART, G.; HANSET, R. (1990). On the use of DNA fingerprints for linkage studies in cattle. *Genomics* 6: 461-474. [3.5]
11. PERRET, J.; SHIA, Y.; FRIES, R.; VASSART, G.; **GEORGES, M.** (1990). A polymorphic satellite sequence maps to the pericentric region of the bovine Y chromosome. *Genomics* 6:482-490. [3.5]
12. **GEORGES, M.**; LATHROP, M.; BOUQUET, Y.; HILBERT, P.; MARCOTTE, A.; SCHWERS, A.; ROUPAIN, J.; VASSART, G.; HANSET, R. (1990) Linkage relationships among 19 genetic markers in cattle. Evidence for linkage between two pairs of blood group systems: B-Z and S-F/V respectively. *Animal Genetics* 21:95-105. [1.4]
13. STREYDIO, C.; SWILLENS, S.; **GEORGES, M.**; SZPIRER, C.; VASSART G. (1990). Structure, evolution and chromosomal localisation of the human pregnancy-specific ß1 glycoprotein gene family. *Genomics* 6:579-592. [3.5]
14. JULIER, C.; DE GOUYON, B.; **GEORGES, M.**; GUENET, J.L.; AVNER, P.; LATHROP, G.M. (1990). Minisatellite linkage maps in the mouse by cross-hybridization with human VNTR probes. *Proc. Natl. Acad. Sci. USA* 87: 4585-4589. [10.3]
15. **GEORGES, M.**; MASSEY, J.M. (1991) Velogenetics or the synergistic use of Marker Assisted Selection and Germ-Line Manipulation. *Theriogenology* 35(1): 151-156. [1.8]
16. MOORE, S.S.; SARGEANT, L.; KING, T.J.; MATTICK, J.S.; **GEORGES, M.**; HETZEL, D.J.S. (1991). The conservation of dinucleotide microsatellites among mammalian genomes allows the use of heterologous PCR primer pairs in closely related species. *Genomics* 10: 654-660. [3.5]
17. STEELE, M.; **GEORGES, M.** (1991) Generation of bovine multisite haplotypes using random cosmids. *Genomics* 10: 889-904. [3.5]
18. **GEORGES, M.**; GUNAWARDANA, A.; THREADGILL, D.; LATHROP, M.; OLSAKER, I.; MISHRA, A.; SARGEANT, L.; STEELE, M.; TERRY, Ch.; ZHAO. X.; HOLM, T.; FRIES, R.; WOMACK, J. (1991). Characterization of a set of variable number of tandem repeat markers conserved in Bovidae. *Genomics* 11: 24-32. [3.5]
19. HILBERT, P.; LINDPAINTER, K.; SERIKAWA, T.; SOUBRIER, F.; CARTWRIGHT, P.; DUBAY, C.; JULIER, C.; TAKAHASI, S.; VINCENT, M.; BECKMANN, J.; GANTEN, D.; **GEORGES, M.** & LATHROP, M. (1991). Chromosomal mapping

of two genetic loci associated with hereditary hypertension in the rat. *Nature* 353:521-529. [31]

20. DIETZ, A.B.; **GEORGES, M.**; THREADGILL, D.W.; WOMACK, J.E.; SCHULER, L.A. (1992). Somatic cell mapping, polymorphism and linkage analysis of bovine prolactin related proteins and placental lactogen. *Genomics* 14: 137-143. [3.5]
21. MACKINNON, M.J.; **GEORGES, M.** (1992). The effects of selection on linkage analysis for quantitative traits. *Genetics* 132: 1177-1185. [4.9]
22. DELHAISE, F.; ZHAO, X.; DESSY, F.; **GEORGES, M.** (1993). Quantitative estimation of chimerism in mice using microsatellite markers. *Molecular Reproduction and Development* 34: 127-132. [2.5]
23. **GEORGES, M.**; LATHROP, M.; DIETZ, A.B.; LEFORT, A.; LIBERT, F.; MISHRA, A.; NIELSEN, D.; SARGEANT, L.S.; STEELE, M.R.; ZHAO, X.; LEIPOLD, H.; WOMACK, J.E. (1993). Microsatellite mapping of the gene causing weaver disease in cattle will allow the study of an associated QTL. *Proc. Natl. Acad. Sci. USA* 90: 1058-1062. [10.3]
24. **GEORGES, M.**; DRINKWATER, R.; LEFORT, A.; LIBERT, F.; KING, T.; MISHRA, A.; NIELSEN, D.; SARGEANT, L.S.; SORENSEN, A.; STEELE, M.R.; ZHAO, X.; WOMACK, J.E.; HETZEL, J. (1993). Microsatellite mapping of a gene affecting horn development in Bos taurus. *Nature Genetics* 4: 206-210. [38.8]
25. LIBERT, F.; LEFORT, A.; OKIMOTO, R.; **GEORGES, M.** (1993). Construction of a bovine genomic library of large yeast artificial chromosome clones. *Genomics* 18:270-276. [3.5]
26. NEIBERGS, H.L.; GALLAGHER, D.S.; **GEORGES, M.**; WOMACK, J.E. (1993). Physical mapping of Inhibin beta-A in domestic cattle. *Mammalian Genome* 4: 328-332. [2.7]
27. BARENDSE, W.; ARMITAGE, S.M.; KIRKPATRICK, B.W.; MOORE, S.S.; **GEORGES, M.**; WOMACK, J.E.; HETZEL, J. (1993). A genetic map of index markers on bovine chromosome 21. *Genomics* 18: 598-601. [3.5]
28. BARENDSE, W.; ARMITAGE, S.M.; RYAN, A.M.; MOORE, S.S.; CLAYTON, D.; **GEORGES, M.**; WOMACK, J.E.; HETZEL, J. (1993). A genetic map of index markers on bovine chromosome 1. *Genomics* 18: 602-608. [3.5]
29. BARENDSE, W.; ARMITAGE, S.M.; KOSSAREK, L.M.; KIRKPATRICK, B.W.; RYAN, A.M.; SHALOM, A.; CLAYTON, D.; LI, L.; NEIBERGS, H.L.; ZHANG, N.; GROSSE, W.M.; WEISS, J.; CREIGHTON, P.; MCCARTHY, F.; RON, M.; TEALE, A.J.; FRIES, R.; MCGRAW, R.A.; MOORE, S.S.; SOLLER, M.; **GEORGES, M.**; WOMACK, J.E.; HETZEL, D.J.S. (1994). A genetic map of the bovine genome shows unexpected chromosomal rearrangements between humans and cattle. *Nature Genetics* 6:227-235. [38.8]

30. COCKETT, N.E.; JACKSON, S.P.; SHAY, T.D.; NIELSEN, D.; GREEN, R.D.; **GEORGES, M.** (1994). Chromosomal localization of the callipyge gene in sheep (Ovis aries) using bovine DNA markers. *Proc. Natl. Acad. Sci. USA* 91: 3019-3023. [10.3]
31. **GEORGES, M.**; NIELSEN, D.; MACKINNON, M.; MISHRA, A.; OKIMOTO, R.; PASQUINO, A.T.; SARGEANT, L.S.; SORENSEN, A.; STEELE, M.R.; ZHAO, X.; WOMACK, J.E.; HOESCHELE, I. (1995). Mapping quantitative trait loci controlling milk production by exploiting progeny testing. *Genetics* 139: 907-920. [4.9]
32. CHARLIER, C.; COPPIETERS, W.; FARNIR, F.; GROBET, L.; LEROY, P.; MICHAUX, C.; MNI, M.; SCHWERS, A.; VANMANSHOVEN, P.; HANSET, R. & **GEORGES, M.** (1995) The mh gene causing double-muscling in cattle maps to bovine chromosome 2. *Mammalian Genome* 6: 788-792. [2.7]
33. CHARLIER, C.; DENYS, B.; BELANCHE, J.I.; COPPIETERS, W.; GROBET, L.; MNI, M.; WOMACK, J.; HANSET, R. & **GEORGES, M.** (1996). Microsatellite mapping of a major determinant of *White Heifer Disease*: the bovine *roan* locus. *Mammalian Genome* 7:138-142. [2.7]
34. CHARLIER, C.; FARNIR, F.; BERZI, P.; VANMANSHOVEN, P.; BROUWERS, B.; **GEORGES, M.** (1996). IBD mapping of recessive traits in livestock: application to map the bovine syndactyly locus to chromosome 15. *Genome Research* 6:580-589. [9.6]
35. **GEORGES, M.** & ANDERSSON, L. (1996). Livestock genomics comes of age. *Genome Research* 6:907-921. [9.6]
36. COCKETT, N.; JACKSON, S.; SHAW, T.; FARNIR, F.; SNOWDER, G.; NIELSEN, D.; **GEORGES, M.** (1996). Polar overdominance at the ovine callipyge locus. *Science* 273:236-238. [24.7]
37. SAMSON, M.; LIBERT, F.; DORANZ, B.J.; RUCKER, J.; LIESNARD, C.; FARBER, C-M.; SARAGOSTI, S.; LAPOUREMOULIE, C.; COGNAUX, J.; FORCEILLE, C.; MUYLDERMANS, G.; VERHOFSTEDE, C.; BURTONBOY, G.; **GEORGES, M.**; IMAI, T.; RANA, S.; YI, Y.; SMYTH, R.J.; COLLMAN, R.G.; DOMS, R.W.; VASSART, G.; PARMENTIER, M. (1996). Resistance to HIV-1 infection in caucasian individuals bearing mutant alleles of the CCR-5 chemokine receptor gene. *Nature* 382: 722-725. [31]
38. CAI, L.; SCHALKWIJK L.C.; STEHLI, A.S.; ZEE, R.Y.L.; HAAF, T.; **GEORGES, M.**; LEHRACH, H.; LINDPAINTER, K. (1996). Construction and characterization of a 10-genome equivalent yeast artificial chromosome library for the laboratory rat, Rattus norvegicus. *Genomics* 38: 385-392. [3.5]
39. MILINKOVITCH, M.C.; LEDUC, R.G.; ADACHI, J.; FARNIR, F.; **GEORGES, M.**; HASEGAWA, M. (1996). Effects of character weighting and species sampling on phylogeny reconstruction: a case study based on DNA sequence data in cetaceans. *Genetics* 144:1817-1833. [4.9]

40. BARENDSE, W.; VAIMAN, D.; KEMP, S.J.; SUGIMOTO, Y.; ARMITAGE, S.M.; WILLIAMS, J.L.; SUN, H.S.; EGGEN, A.; AGABA, M.; ALEYASIN, S.A.; BAND, M.; BISHOP, M.D.; BUITKAMP, J.; BYRNE, K.; COLLINS, F.; COOPER, L.; COPPETTIERS, W.; DENYS, B.; DRINKWATER, R.D.; EASTERDAY, K.; ELDUQUE, C.; ENNIS, S.; ERHARDT, G.; FERRETTI, L.; FLAVIN, N.; GAO, Q.; **GEORGES, M.**; GURUNG, R.; HARLIZIUS, B.; HAWKINS, G.; HETZEL, J.; HIRANO, T.; HULME, D.; JORGENSEN, C.; KESSLER, M.; KIRKPATRICK, B.W.; KONFORTOV, B.; KOSTIA, S.; KUHN, C.; LENSTRA, J.A.; LEVEZIEL, H.; LEWIN, H.A.; LEYHE, B.; LIL, L.; BURRIEL, I.M.; MCGRAW, R.A.; MILLER, J.R.; MOODY, D.E.; MOORE, S.S.; NAKANE, S.; NIJMAN, I.J.; OLSAKER, I.; POMP, D.; RANDO, A.; RON, M.; SHALOM, A.; TEALE, A.J.; THIEVEN, U.; URQUHART, B.G.D.; VAGE, D.I.; VANDEWEGHE, A.; VARVIO, S.; VELMALA, R.; VIKKI, J.; WEIKARD, R.; WOODSIDE, C.; WOMACK, J.E.; ZANOTTI, M.; ZARAGOZA, P. (1997) A medium-density genetic linkage map of the bovine genome. *Mammalian Genome* 8: 21-28. [2.7]
41. DUNNER, S.; CHARLIER, C.; FARNIR, F.; BROUWERS, B.; CANON, J.; **GEORGES, M.** Towards interbreed IBD fine mapping of the *mh* locus: double-muscling in the *Asturiana de los Valles* breed involves the same locus as in the *Belgian Blue* cattle breed.(1997) *Mammalian Genome* 8: 430-435. [2.7]
42. **GEORGES, M.** (1997). From QTL mapping to QTL cloning: mice at the rescue. *Genome Research* 7: 665-666. [9.6]
43. GROBET, L.; ROYO MARTIN, L.J.; PONCELET, D.; PIROTTIN, D.; BROUWERS, B.; RIQUET, J.; SCHOEBERLEIN, A.; DUNNER, S.; MENISSIER, F.; MASSABANDA, J.; FRIES, R.; HANSET, R.; **GEORGES, M.** (1997) A deletion in the myostatin gene causes double-muscling in cattle. *Nature Genetics* 17:71-74. [38.8]
44. GROBET, L.; PONCELET, D.; ROYO MARTIN, L.J.; BROUWERS, B.; PIROTTIN, D.; MICHAUX, CH.; MENISSI ER , F.; ZANOTTI, M.; DUNNER, S.; **GEORGES, M.** (1998). Molecular definition of an allelic series of mutations disrupting the myostatin function and causing double-muscling in cattle. *Mammalian Genome* 9: 210-213. [2.7]
45. COPPIETERS, W.; MACKINNON, M.; **GEORGES, M.** (1998). A note on the effect of selection on linkage analysis for quantitative traits. *Journal of Heredity*, 89:193-195. [1.7]
46. MACKINNON, M.; **GEORGES, M.** (1998) A bottom-up approach towards marker assisted selection. *Livestock Production Science* 54:229-250. [1]
47. COPPIETERS, W.; RIQUET, J.; ARRANZ, J.-J.; BERZI, P.; CAMBISANO, N.; GRISART, B.; KARIM, L.; MARCQ, F.; SIMON, P.; VANMANSHOVEN, P.; WAGENAAR, D.; **GEORGES, M.** (1998) A QTL with major effect on milk yield and composition maps to bovine chromosome 14. *Mammalian Genome* 9: 540-544. [2.7]

48. COPPIETERS, W.; KVASZ, A.; ARRANZ, J.J.; GRISART, B.; FARNIR, F.; MACKINNON, M. & **GEORGES, M.** (1998). A rank-based non parametric method to map QTL in outbred half-sib pedigrees: application to milk production in a grand-daughter design. *Genetics* 149:1547-1555. [4.9]
49. ARRANZ, J.-J.; COPPIETERS, W.; BERZI, P.; CAMBISANO, N.; GRISART, B.; KARIM, L.; MARCQ, F.; RIQUET, J.; SIMON, P.; VANMANSHOVEN, P.; WAGENAAR, D.; **GEORGES, M.** (1998) A QTL affecting milk yield and composition maps to bovine chromosome 20: a confirmation. *Animal Genetics* 29: 107-115. [1.4]
50. PIROTTIN, D.; PONCELET , D.; GROBET, L.; ROYO, L.J.; BROUWERS, B.; MASABANDA, J.; TAKEDA, H.; FRIES, R.; SUGIMOTO, K.; WOMACK, J.; DUNNER, S.; **GEORGES, M.** (1999). High resolution human-bovine comparative mapping based on a closed YAC contig spanning the bovine *mh* locus. *Mammalian Genome* 10: 289-293. [2.7]
51. NEZER, C.; MOREAU, L.; BROUWERS, B.; COPPIETERS, W.; DETILLEUX, J.; HANSET, R.; KARIM, L.; KVASZ, A.; LEROY, P.; **GEORGES, M.** (1999) An imprinted QTL with major effect on muscle mass and fat deposition maps to the IgfII locus in pigs. *Nature Genetics* 21: 155-156. [38.8]
52. COCKETT, N.E.; SHAY, T.; BEEVER, J.E.; NIELSEN, D.M.; ALBRETSEN, J.; **GEORGES, M.**; PETERSON, K.; STEPHENS, A.; VERNON, W.; TIMOFEEVSKAIA, O.; SOUTH, S.; MORK, J.; MACIULIS, A.; HOLYOAK, G.R.; BUNCH, T.D. (1999). Localization of the locus causing Spider Lamb Syndrome to the distal end of ovine chromosome 6. *Mammalian Genome* 10: 35-38. [2.7]
53. LIEN, S.; COCKETT, N.E.; KLUNGLAND, H.; ARNHEIM, N.; **GEORGES, M.**; GOMEZ-RAYA, L. (1999). High-resolution gametic map of the sheep callipyge region: linkage heterogeneity among rams detected by sperm typing. *Animal Genetics* 30: 42-46. [1.4]
54. RIQUET, J.; COPPIETERS, W.; CAMBISANO, N.; ARRANZ, J.-J.; BERZI, P.; DAVIS, S.; GRISART, B.; FARNIR, F.; KARIM, L.; MNI, M.; SIMON, P.; TAYLOR, J.; VANMANSHOVEN, P.; WAGENAAR, D.; WOMACK, J.E.; **GEORGES, M.** (1999). Identity-by-descent fine-mapping of QTL in outbred populations: application to milk production in dairy cattle. *Proc. Natl. Acad. Sci. USA* 96: 9252-9257. [9.04]
55. COPPIETERS, W.; KVASZ, A.; ARRANZ, J.-J.; GRISART, B.; RIQUET, J.; FARNIR, F.; **GEORGES, M.** (1999). The great-grand-daughter design: a simple strategy to increase the power of a grand-daughter design. *Genetical Research* 74: 189-199. [2.5]
56. SPELMAN, R.J.; HUISMAN, A.E.; SINGIREDDY, S.R.; COPPIETERS, W.; ARRANZ, J.J.; **GEORGES, M.**; GARRICK, D.J. (1999). Quantitative Trait Loci analysis on 17 nonproduction traits in the New Zealand dairy population. *J Dairy Sci* 82: 2514-2516. [2.1]

57. FARNIR, F.; COPPIETERS, W.; ARRANZ, J.-J.; BERZI, P.; CAMBISANO, N.; GRISART, B.; KARIM, L.; MARCQ, F.; MOREAU, L.; MNI, M.; NEZER, C.; SIMON, P.; VANMANSHOVEN, P.; WAGENAAR, D.; **GEORGES, M.** (2000). Extensive genome-wide linkage disequilibrium in cattle. *Genome Research* 10: 220-227. [9.6]
58. IKONEN, T.; BOVENHUIS, H.; OJALA, M.; UOTTINEN, O.; **GEORGES M.** (2001). Associations between casein haplotypes and first lactation milk production traits in Finnish Ayrshire cows. *J Dairy Sci.* 84(2):507-14. [2.1]
59. KARIM, L.; COPPIETERS, W.; GROBET, L.; VALENTINI, A.; **GEORGES, M.** (2001). Convenient genotyping of six myostatin mutations causing double-muscling in cattle using a multiplex oligonucleotide ligation assay. *Animal Genetics* 31: 396-399. [1.4]
60. SEGERS, K.; VAIMAN, D.; BERGHMANS, S.; SHAY, T.; BEEVER, J.; COCKETT, N.; **GEORGES, M.** & CHARLIER, C. (2001). Construction and characterization of an ovine BAC contig spanning the callipyge locus. *Animal Genetics* 31: 352-359. [1.4] |corresponding author|
61. SHAY, T.; BERGHMANS, S.; SEGERS, K.; MEYERS, S.; WOMACK., J.; BEEVER, J.; **GEORGES, M.**; CHARLIER, C. & COCKETT, N. E. (2001). Fine-mapping and construction of a bovine contig spanning a 4.6 centimorgan interval containing the CLPG locus. *Mammalian Genome* 12: 141-149. [2.7] |corresponding author|
62. BERGHMANS, S.; SEGERS, K.; SHAY, T.; **GEORGES, M.**; COCKETT, N. E. & CHARLIER, C. (2001). Breakpoint mapping positions the callipyge gene within a 400 kilobase chromosome segment containing the Dlk1 and Gtl-2 genes. *Mammalian Genome* 12: 183-185. [2.7] |corresponding author|
63. CHARLIER, C.; SEGERS, K.; KARIM, L.; SHAY, T.; GYAPAY, G.; COCKETT, N.; **GEORGES, M.** (2001). The callipyge (CLPG) mutation enhances the expression of the coregulated *DLK1, GTL2, PEG11* and *MEG8* genes in *cis* without affecting their imprinting status. *Nature Genetics* 27:367-369. [26.5]
64. CHARLIER, C.; SEGERS, K.; WAGENAAR, D.; KARIM, L.; BERGHMANS, S.; JAILLON, O.; SHAY, T.; WEISSENBACH, J.; COCKETT, N.; GYAPAY, G.; **GEORGES, M.** (2001). Human - ovine comparative sequencing of a 250 kilobase imprinted domain encompassing the callipyge (*clpg*) gene and identification of six imprinted transcripts: *DLK1, DAT, GTL2, PEG11, antiPEG11* and *MEG8. Genome Research* 11:850-862. [9.6]
65. PAILHOUX, E.; VIGIER, B.; VAIMAN, D.; SCHIBLER, L.; VAIMAN, A.; CRIBIU, E.; NEZER, C.; **GEORGES, M.**; SUNDSTROM, J.; PELLINIEMI, L.J.; FELLOUS, M.; COTINOT, C. (2001) Contribution of domestic animals to the identification of new genes involved in sex determination. *J. Exp. Zoology* 290: 700-708. [1.8]

66. PAULSEN, M.; TAKADA, S.; YOUNGSON, N.A.; BENCHAIB, M.; CHARLIER, C.; SEGERS, K.; **GEORGES, M.**; FERGUSON-SMITH, A. (2001) Detailed sequence analysis of the imprinted *Dlk1-Gtl2* locus in three mammalian species identifies highly conserved genomic elements and a domain structure different from the *Igf2-H19* region. *Genome Research* 11: 2085-2094. [9.6]
67. BIDWELL, C.A.; SHAY, T.L.; **GEORGES, M.**; BEEVER, J.E.; BERGHMANS, S.; COCKETT, N.E. (2001) Differential expression of the GTL2 gene within the callipyge region of ovinechromosome 18. *Animal Genetics* 32:248-256. [1.4]
68. GRISART, B.; COPPIETERS, W.; FARNIR, F.; KARIM, L.; FORD, C.; CAMBISANO, N.; MNI, M.; REID, S.; SPELMAN, R.; **GEORGES, M.** & SNELL, R. (2002). Positional candidate cloning of a QTL in dairy cattle: Identification of a missense mutation in the bovine DGAT gene with major effect on milk yield and composition. *Genome Research* 12: 222-231. [9.6] |corresponding author|
69. FARNIR, F.; GRISART, B.; COPPIETERS, W.; RIQUET, J.; BERZI, P.; CAMBISANO, N.; KARIM, L.; MNI, M.; MOISIO, S.; SIMON, P.; WAGENAAR, D.; VILKKI, J.; **GEORGES, M.** (2002). Simultaneous mining of linkage and linkage disequilibrium to fine-map QTL in outbred half-sib pedigrees: revisiting the location of a QTL with major effect on milk production on bovine chromosome 14. *Genetics* 161: 275-287. [4.9]
70. NEZER, C.; MOREAU, L.; WAGENAAR; D.; **GEORGES, M.** (2002). Results of a whole genome scan targeting QTL for growth and carcass characteristics in a Piétrain X Large White intercross. *Genetics, Selection, Evolution* 34: 371-388. [1.3]
71. AMARGER, V.; NGUYEN, M.; VAN LARE, A.S.; NEZER, C.; **GEORGES, M.**; ANDERSSON, L. (2002). Comparative sequence analysis of the INS-IGF2-H19 gene cluster in pigs. *Mammalian Genome* 13: 388-398. [2.7]
72. KIM, J.J.; **GEORGES, M.** (2002). Evaluation of a new fine-mapping method exploiting linkage disequilibrium: a case study analysing a QTL with major effect on milk composition on bovine chromosome 14. *Asian-Aust. J. Anim. Sci.* 15: 1250-1256. [0.4]
73. SMIT, M.; SEGERS, K.; SHAY, T.; BARALDI, F.; GYAPAY, G.; SNOWDER, G.; **GEORGES, M.**; COCKETT, N.; CHARLIER, C. (2003). Mosaicism of Solid Gold supports the causality of a non-coding A to G transition in the determinism of the callipyge phenotype. *Genetics* 163: 453-456. [4.9] |corresponding author|
74. BLOTT, S.; KIM, J.-J.; MOISIO, S.; SCHMIDT-KÜNTZEL, A.; CORNET, A.; BERZI, P.; CAMBISANO, N.; FORD, C.; GRISART, B.; JOHNSON, D.; KARIM, L.; SIMON, P.; SNELL, R.; SPELMAN, R.; WONG, J.; VILKKI, J.; **GEORGES, M.**; FARNIR, F.; COPPIETERS, W. (2003). Molecular dissection of a QTL: a phenylalanine to tyrosine substitution in the transmembrane domain of the

bovine growth hormone receptor is associated with a major effect on milk yield and composition. *Genetics* 163: 253-266. [4.93] |corresponding author|

75. DUNNER, S.; MIRANDA, M.E.; AMIGUES, Y.; CANON, J.; **GEORGES, M.**; HANSET, R.; WILLIAMS, J.; MENISSIER, F. (2003) Haplotype diversity of the myostatin gene among beef cattle breeds. *Genetics Selection Evolution* 35: 103-118. [1.3]
76. GROBET, L.; PIROTTIN, D.; FARNIR, F.; PONCELET, D.; ROYO, L.J.; BROUWERS, B.; CHRISTIANS, E.; DESMECHT, D.; COIGNOUL, F.; KAHN, R.; **GEORGES, M.** (2003). Modulating skeletal muscle mass by post-natal, muscle-specific inactivation of the myostatin gene. *Genesis* 35: 227-238. [2.9]
77. **GEORGES, M.**; CHARLIER, C.; COCKETT, N. (2003). Polar overdominance at the ovine callipyge locus supports trans interaction between the products of reciprocally imprinted genes. *Trends in Genetics* 19:248-252. [12]
78. NEZER, C.; COLLETTE, C.; MOREAU, L.; BROUWERS, B.; KIM, J.J.; GIUFFRA, E.; BUYS, N.; ANDERSSON, L.; **GEORGES, M.** (2003). Haplotype sharing refines the location of an imprinted QTL with major effect on muscle mass to a 90 Kb chromosome segment containing the porcine *IGF2* gene. *Genetics* 165: 277-285. [4.9]
79. VAN LAERE, A.-S.; NGUYEN, M.; BRAUNSCHWEIG, M.; NEZER, C.; COLLETTE, C.; MOREAU, L.; ARCHIBALD, A.L.; HALEY, C.S.; BUYS, N.; TALLY, M.; ANDERSSON, G.; **GEORGES, M.** & ANDERSSON, L. (2003). Positional identification of a regulatory mutation in *IGF2* causing a major QTL effect on muscle growth in the pig. *Nature* 425: 832-836. [31] |co-corresponding author|
80. WERNER, F.A.O.; DURSTEWITZ, G.; HABERMANN, F.A.; THALLER, G.; KRAMER, W.; KOLLERS, S.; BUITKAMP, J.; **GEORGES, M.**; BREM, G.; MOSNER, J.; FRIES, R. (2004) Detection and characterization of SNPs useful for identity control and parentage testing in major European dairy breeds. *Animal Genetics* 35: 44-48. [1.4]
81. GRISART B., FARNIR F., KARIM L., CAMBISANO N., KIM J.-J., KVASZ A., MNI M., SIMON P., FRERE J.-M., COPPIETERS W., **GEORGES M.** (2004). Genetic and functional demonstration of the causality of the *DGAT1 K232A* mutation in the determinism of the BTA14 QTL affecting milk yield and composition. *Proc. Natl. Acad. Sci. USA* 101: 2398-2403. [10.3]
82. ANDERSSON, L.; **GEORGES, M.** (2004) Domestic animal genomics: deciphering the genetics of complex traits. *Nature Reviews Genetics* 5: 202-212. [25.7]
83. LAVILLE, E.; BOUIX, J.; SAYD, T.; BIBE, B.; ELSEN, J.M.; LARZUL, C.; EYCHENNE, F.; MARCQ, F.; **GEORGES M.** (2004) Effects of a quantitative trait locus for muscle hypertrophy from Belgian Texel sheep on carcass conformation and muscularity. *J. Anim. Sci* 82: 3128-3137. [1.6]

84. DAVIS, E.; HARKEN JENSEN, C.; SCHRODER, H.D.; SHAY, T.; KLIEM, A.; COCKETT, N.; **GEORGES, M.**; CHARLIER, C. (2004). Ectopic expression of DLK1 protein in skeletal muscle of padumnal heterozygotes causes the callipyge phenotype. *Current Biology* 14: 1858-1862. [11.9] |corresponding author|
85. PIROTTIN, D.; GROBET, L.; ADAMANTIDIS, A.; FARNIR, F.; HERENS, C; SCHRODER, H.D.; **GEORGES, M.** (2005). Transgenic engineering of a male-specific muscular hypertrophy. *Proc. Natl. Acad. Sci. USA* 102: 6413-6418 [10.3]
86. DAVIS, E.; CAIMENT, F.; TORDOIR, X.; CAVAILLÉ, J.; FERGUSON-SMITH, A.; COCKETT, N.; **GEORGES, M.**; CHARLIER, C. (2005). RNAi-mediated allelic trans-interaction at the imprinted callipyge locus. *Current Biology* 15: 743-749 |corresponding author| [11.9]
87. SMIT, M.A., TORDOIR, X.; GYAPAY, G.; COCKETT, N.; **GEORGES, M.**; CHARLIER, C. (2005). BEGAIN: a novel imprinted gene that generates paternally expressed transcripts in a tissue- and promotor-specific manner. *Mammalian Genome* 16: 801-814. |corresponding author| [2.4]
88. QUAN, X.; LAES, J.-F.; STIEBER, D.; RIVIÈRE, M.; RUSSO, J.; WEDEKIND, D.; COPPIETERS, W.; FARNIR, F.; **GEORGES, M.**; SZPIRER, J.; SZPIRER, C. (2006) Genetic identification of distinct loci controlling mammary tumor multiplicity, latency and aggressiveness in the rat. *Mammalian Genome* 17: 310-321. [2.4]
89. SANDOR, C.; FARNIR, F.; MEUWISSEN, T.; COPPIETERS, W.; **GEORGES, M.** (2006). Linkage d.isequilibrium on the bovine X chromosome: characterization -and use in QTL mapping. *Genetics* 173: 1777-1786. [4.3]
90. HARMEGNIES, N.; FARNIR, F.; DAVIN, F.; GEERTS, I.; BUYS, N.; **GEORGES, M.**; COPPIETERS, W. (2006). Measuring the extent of linkage disequilibrium in commercial pig populations. *Animal Genetics* 37: 225-231. |corresponding author| [2.4]
91. TAKEDA, H.; CAIMENT, F.; SMIT, M.; HIARD, S.; TORDOIR, X.; COCKETT, N.; **GEORGES, M.** & CHARLIER,5 C. (2006). The callipyge mutation enhances bidirectional long-range DLK1-GTL2 intergenic transcription in cis. *Proc. Natl. Acad. Sci. USA* 103: 8119-8124. |corresponding author| [10.2]
92. VIITALA, S., SZYDA, J., BLOTT, S., SCHULMAN, N., LIDAUER, M., MAKI-TANILA, A., **GEORGES, M.**, VILKKI, J. (2006) The role of the bovine growth hormone receptor and prolactin receptor genes in milk, fat and protein production in finnish ayrshire dairy cattle. *Genetics* 173:2151-2164. [4.3]
93. MORREEL, K., GOEMINNE, G., STORME, V., STERCK, L., RALPH, J., COPPIETERS, W., BREYNE, P., STEENACKERS, M., **GEORGES, M.**, MESSENS, E., BOERJAN, W. (2006) Genetical metabolomics of flavonoid biosynthesis in Populus: a case study. *Plant J.* 47:224-37. [7.0]

94. CLOP, A.; MARCQ, F.; TAKEDA, H.; PIROTTIN, D.; TORDOIR, X.; BIBÉ, B.; BOUIX, J.; CAIMENT, F.; ELSEN, J.M.; EYCHENNE, F.; LARZUL, C.; LAVILLE, E.; MEISH, F.; MILENKOVIC, D.; TOBIN, J.; CHARLIER, C.; **GEORGES, M.** (2006). A mutation creating a potential illegitimate miRNA target site in the myostatin gene affects muscularity in sheep. *Nature Genetics* 38:813-818. [25.8]
95. HARMEGNIES, N.; DAVIN, F.; DE SMET, S.; BUYS, N.; **GEORGES, M.**; COPPIETERS, W. (2006). Results of a whole genome QTL scan for growth, carcass composition, and meat quality in a porcine four-way cross. *Animal Genetics* 37: 543-553 |corresponding author| [2.4]
96. WUNDERLICH, K.R.; ABBEY, C.A.; CLAYTON, D.R.; SONG, Y.; SCHEIN, J.E.; **GEORGES, M.**; COPPIETERS, W.; ADELSON, D.L.; TAYLOR, J.F.; DAVIS, S.L.; GILL, C.A. (2006) A 2.5-Mb contig constructed from Angus, Longhorn and horned Hereford DNA spanning the polled interval on bovine chromosome 1. *Anim Genet.* 37:592-594. [2.4]
97. HAMELIN, M.; SAYD, T.; CHAMBON, C.; BOUIX, J.; BIBE, B.; MILENKOVIC, D.; LEVEZIEL, H.; **GEORGES, M.**; CLOP, A.; MARINOVA, P.; LAVILLE, E. (2006) Proteomic analysis of ovine muscle hypertrophy. *J Anim Sci.* 84:3266-3276. [1.6]
98. LIBIOULLE, C.; LOUIS, E.; HANSOUL, S.; SANDOR, C.; FARNIR, F.; FRANCHIMONT, D.; DE WIT, O.; DE VOS, M.; VERMEIRE, S.; DEMARCHE, B.; GUT, I.; HEATH, S.; MNI, M.. ZELENIKA, D.; BELAICHE, J.; RUTGEERTS, P.; LATHROP, M.; **GEORGES, M.** (2007) A novel susceptibility locus for Crohn's disease identified by whole genome association maps to a gene desert on chromosome 5p13.1 and modulates the level of expression of the prostaglandin receptor EP4. *PloS Genetics* 4:e58 [7.7]
99. **GEORGES, M.**; COPPIETERS, W.; CHARLIER, C. (2007) Polymorphic miRNA-mediated gene regulation: contribution to phenotypic variation and disease. *Current Opinion in Genetics & Development* 17: 1-11 [10]
100. **GEORGES, M.** (2007) Mapping, fine-mapping and cloning QTL in domestic animals. *Annual Review of Genomics and Human Genetics* 8: 131-162. [10.8]
101. CHARLIER, C.; COPPIETERS, W.; AGERHOLM, J.S.; CAMBISANO, N.; CARTA, E.; DESMECHT, D.; DIVE,M.; FASQUELLE, C.; FRENNET, J.C.; HANSET, R.; HUBIN, X.; JORGENSEN, C.; KARIM, L.; KENT, M.; HARVEY, K.; PEARCE, B.R.; ROLLIN, F.; SIMON, P.; TAMA, N.; NIE, H.; VANDEPUTTE, S.; LIEN, S.; LONGERI, M.; FREDHOLM, M.; HARVEY, R.J.; **GEORGES, M.** (2007) Highly effective SNP-based association mapping and management of recessive defects in livestock. Nature Genetics, in the press. [25.8]

BOOK CHAPTERS

1. **GEORGES, M.** (1991). Hypervariable Minisatellites and their use in animal breeding. In: Gene Mapping: strategies, techniques and applications, pp. 89-112. Ed. by L.B. Schook, Lewin and McLaren. Marcel Dekker, New York.
2. **GEORGES, M.** (1991). Perspectives for marker assisted selection and velogenetics in animal breeding. in Animal Applications of Research in Mammalian Development, pp. 285-325. Current Communications in Cell & Molecular Biology 4. Ed. Pedersen, McLaren and First. Cold Spring Harbor Laboratory Press.
3. LATHROP, M.; CARTWRIGHT, P.; WRIGHT, S.; NAKAMURA, Y.; **GEORGES, M.** (1991). Data analysis for linkage studies. In: Gene Mapping: strategies, techniques and applications, pp. 177-198. Ed. by L.B. Schook, Lewin and McLaren. Marcel Dekker, New York.
4. **GEORGES, M.** (1997). Recent progress in mammalian genomics and its implication for the selection of candidate transgenes in livestock species. In: Transgenic animals: generation and use. Pp.519-524. Eds. Houdebine. Harwood Academic Publisher.
5. **GEORGES, M.** (1997) Case history in animal improvement: Mapping complex traits in ruminants. Pp. 229-240. In Molecular analysis of complex traits. Ed. A. Patterson, CRC Press.
6. **GEORGES, M.** (1998). Mapping genes underlying production traits in livestock. Pp. 77-101. In Animal Breeding: Technology for the 21st century. Ed. J. Clark. Harwood Academic Publishers.
7. **GEORGES, M.**; ANDERSSON, L. (2004). Positional identification of structural and regulatory quantitative trait nucleotides in livestock. In "The Genome of Homo Sapiens". Cold Spring Harbor Symposia on Quantitative Biology, Vol. 68, pp 179-187. Edited by Bruce Stillman and David Stewart, Cold Spring Harbor Laboratory.
8. **GEORGES, M.**; CHARLIER, C.; SMIT, M.; DAVIS, E.; SHAY, T.; TORDOIR, X.; TAKEDA, H.; CAIMAN, F.; COCKETT, N. (2005). Toward molecular understanding of polar overdominance at the ovine callipyge locus. In "Epigenetics" Cold Spring Harbor Symposium on Quantitative Biology, Vol. 69, pp 477-483. Edited by Bruce Stillman and David Stewart, Cold Spring Harbor Laboratory.
9. **GEORGES, M.**, CLOP, A.; MARCQ, F.; TAKEDA, H.; PIROTTIN, D.; TORDOIR, X.; HIARD, S.; BIBÉ, B.; BOUIX, J.; CAIMENT, F.; ELSEN, J.M.; EYCHENNE, F.; LARZUL, C.; LAVILLE, E.; MEISH, F.; MILENKOVIC, D.; TOBIN, J.; CHARLIER, C. (2006). Polymorphic miRNA-target interactions. In "Regulatory RNAs" Cold Spring Harbor Symposium on Quantitative Biology, Vol. 71, pp .... Edited by Bruce Stillman and David Stewart, Cold Spring Harbor Laboratory.

10. **GEORGES, M.**; CLOP, A.; MARCQ, F.; TAKEDA, H.; PIROTTIN, D.; HIARD, S.; TORDOIR, X.; CAIMENT, F.; MEISH, F.; BIBE, B.; BOUIX, J.; ELSEN, J.M.; EYCHENNE, F.; LAVILLE, E.; LARZUL, C.; MILENKOVIC, D.; TOBIN, J.; CHARLIER, C. (2006) Polymorphic Polymorphic MicroRNA-Target Interactions: A Novel Source of Phenotypic Variation. *Cold Spring Harb Symp Quant Biol.* 71:343-350.

CITATION ANALYSIS

Total impact factor: 819
Average impact factor per publication: 8.4
Total number of citations: 7,457
Average number of citations per publication: 51.08
h-index: 41 (41 publications with more than 40 citations)
16 publications with more than 100 citations.

Published items per year

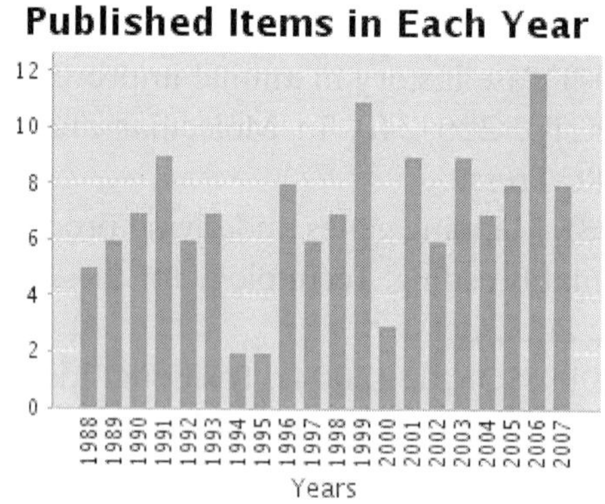

Citations per year

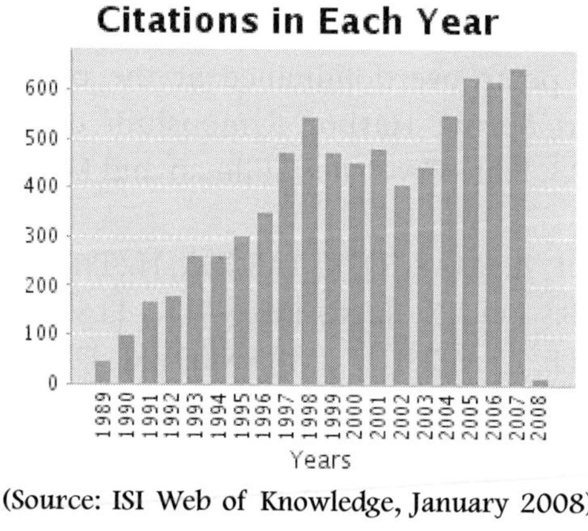

(Source: ISI Web of Knowledge, January 2008)

Selected publications

COCKETT, N.; JACKSON, S.; SHAW, T.; FARNIR, F.; SNOWDER, G.; NIELSEN, D.; **GEORGES, M.** (1996). Polar overdominance at the ovine callipyge locus. *Science* 273:236-238.

GROBET, L.; ROYO MARTIN, L.J.; PONCELET, D.; PIROTTIN, D.; BROUWERS, B.; RIQUET, J.; SCHOEBERLEIN, A.; DUNNER, S.; MENISSIER, F.; MASSABANDA, J.; FRIES, R.; HANSET, R.; **GEORGES, M.** (1997) A deletion in the myostatin gene causes double-muscling in cattle. *Nature Genetics* 17:71-74.

CHARLIER, C.; SEGERS, K.; KARIM, L.; SHAY, T.; GYAPAY, G.; COCKETT, N.; **GEORGES, M.** (2001). The callipyge (CLPG) mutation enhances the expression of the coregulated *DLK1, GTL2, PEG11* and *MEG8* genes in *cis* without affecting their imprinting status. *Nature Genetics* 27:367-369.

CLOP, A..; MARCQ, F.; TAKEDA, H.; PIROTTIN, D.; TORDOIR, X.; BIBÉ, B.; BOUIX, J.; CAIMENT, F.; ELSEN, J.M.; EYCHENNE, F.; LARZUL, C.; LAVILLE, E.; MEISH, F.; MILENKOVIC, D.; TOBIN, J.; CHARLIER, C.; **GEORGES, M.** (2006). A mutation creating a potential illegitimate miRNA target site in the myostatin gene affects muscularity in sheep. *Nature Genetics* 38:813-818.

LIBIOULLE, C.; LOUIS, E.; HANSOUL, S.; SANDOR, C.; FARNIR, F.; FRANCHIMONT, D.; DE WIT, O.; DE VOS, M.; VERMEIRE, S.; DEMARCHE, B.; GUT, I.; HEATH, S.; MNI, M.; ZELENIKA, D.; BELAICHE, J.; RUTGEERTS, P.; LATHROP, M.; **GEORGES, M.** (2007) A novel susceptibility locus for Crohn's disease identified by whole genome association maps to a gene desert on chromosome 5p13.1 and modulates the level of expression of the prostaglandin receptor EP4. *PloS Genetics* 4:e58

CHARLIER, C.; COPPIETERS, W.; AGERHOLM, J.S.; CAMBISANO, N.; CARTA, E.; DESMECHT, D.; DIVE, M.; FASQUELLE, C.; FRENNET, J.C.; HANSET, R.; HUBIN, X.; JORGENSEN, C.; KARIM, L.; KENT, M.; HARVEY, K.; PEARCE, B.R.; ROLLIN, F.; SIMON, P.; TAMA, N.; NIE, H.; VANDEPUTTE, S.; LIEN, S.; LONGERI, M.; FREDHOLM, M.; HARVEY, R.J.; **GEORGES, M.** (2007) Highly effective SNP-based association mapping and management of recessive defects in livestock. Nature Genetics, in the press.

Ronald L. Phillips

University of Minnesota
St. Paul, Minnesota, USA

2006/7 — for groundbreaking discoveries in genetics and genomics, laying the foundations for improvements in crop and livestock breeding, and sparking important advances in plant and animal sciences.

Rank: Regents Professor, University of Minnesota

CURRICULUM VITAE

Date and place of birth: January 1, 1940, Huntington County, Indiana, USA

EDUCATIONAL HISTORY:

1961 B.S. Purdue University – Major Crop Science.
1963 M.S. Purdue University – Major Plant Breeding and Genetics.
1966 Ph.D. University of Minnesota – Major Genetics-Cytogenetics and Plant Breeding Specialties.
1967 Postdoctorate, Cornell University – Major Genetics.

PROFESSIONAL POSITIONS:

1966-67 NIH Trainee, Cornell University.
1967-68 Research Associate, University of Minnesota.
1968-72 Assistant Professor, University of Minnesota.
1972-76 Associate Professor, University of Minnesota.
1976-93 Professor, University of Minnesota.

1979 (5 mos.)	Program Director, USDA Competitive Research Grants Office, Washington, DC.
1981 (1 mo.)	Visiting Professor, Italian National Research Council and Experimental Institute for Cereal Crops.
1984 (3 wks.)	Visiting Professor, University of Guelph, Ontario, Canada.
1986 (1 mo.)	Visiting Professor, People's Republic of China.
1990 (3 wks.)	Foreign Specialist, Japan.
1991-1994	Director, Plant Molecular Genetics Inst., University of Minnesota.
1996-1998	Chief Scientist, USDA-National Research Initiative Competitive Grants Program.
1993-present	Regents Professor, University of Minnesota.
1999-present	McKnight Presidential Chair in Genomics, University of Minnesota.
2000-2005	Director, Ctr. for Microbial and Plant Genomics, University of Minnesota.

SELECTED HIGHLIGHTS

Grants Management:

- USDA Grants Manager for the Genetic Mechanisms Program (1979) – 5 months in Washington, D.C.
- USDA Chief Scientist in charge of the National Research Initiative (1996-98) – 50% time in Washington, D.C.
- Grants panels for USDA (1978-80), NSF (1976, 2002), DOE (1979, 1982), USAID (Chair, 2002)
- Interagency Working Group (OSTP) on Plant Genome, Chair (1997-98). The Working Group's Report was prepared in response to Senator Kit Bond's request for a plan for Congress. The report resulted in establishment of the NSF Plant Genome Research Program.
- Advisory Committee for University of Arizona NSF Chromatin Functional Genomics Grant (2003-2005)
- Advisory Committee for Purdue University NSF Comparative Genomics Grant (2004-present)
- Scientific Advisory Board, Chair, USDA RiceCAP grant (2004-present)

Recent NSF Grants Received:

- NSF Plant Genome Research Program 1998-2001 ($1.8m), 2002-2007 ($3.1m): RL Phillips and HW Rines, Co-PIs.

Research Center Advisories:

- Palm Oil Research Institute of Malaysia Program Advisory Committee (1993-2001)

- Scientific Liaison Officer for USAID to IRRI (1999-2003)
- Donald Danforth Plant Science Center Scientific Advisory Board (2000-present)
- Noble Foundation Non-resident Fellow (2002-2006)
- International Rice Research Institute (IRRI) Board of Trustees – Program Chair (2004-present)

Honors and Awards:

- Fellow: American Association for the Advancement of Science (1980), Crop Science Society of America (1985), American Society of Agronomy (1985)
- Crop Science Research Award – Crop Science Society of America (1988)
- National Academy of Sciences (elected 1991)
- Purdue University Distinguished Agriculture Alumni Award (1993)
- Regents Professor, University of Minnesota (1993-present)
- Dekalb Genetics Crop Science Distinguished Career Award (1997)
- Honorary doctorate – Purdue University (2000)
- McKnight Presidential Chair in Genomics (2000-present)
- Wolf Prize in Agriculture, co-recipient (2007)

Special Offices:

- American Association for the Advancement of Science: Section "O" Chair (1989-1990), Program Committee (2003-2005); Task Force on Women in Agricultural Sciences (2006-present)
- Crop Science Society of America, President (1999-2000)
- National Academy of Sciences: Section Chair (2001-2003), Class Membership Committee (1994-95, 2007)
- University of Minnesota Plant Molecular Genetics Institute, Co-founder and Director (1991-94); Microbial and Plant Genomics Institute, Co-founder and Director (2000-2005)
- International Crop Science Society, Vice President (2004-present)
- World Food Prize, Committee and Youth Institute Faculty (2004-present)
- Council of Scientific Society Presidents, Chair (2006)

Student Training:

- Advised 61 M.S. and Ph.D. student theses
- Advised 20 postdoctoral scientists
- Teaching: Graduate course in Plant Cytogenetics (1970-present)
- Cold Spring Harbor summer course in Plant Molecular and Developmental Biology, Guest Instructor (1984, 1985, 1990, 1991)
- President's Distinguished Faculty Mentoring Program (1988-1994, 2002-2005)
- Project Ag-Grad (African graduate student program), Founder and Co-director (1986-present)

- Unique Educational Opportunities Grant, University of Minnesota, Chair (1986-present)
- Burnham/Chang Scholarship for high school students, University of Minnesota, Chair (1993-present)
- Eisenhower Program High School Biology Teachers Course, Guest Lecturer (2000-present)
- Multicultural Undergraduate Research Program (2001)
- Golden Opportunity Scholars Institute, Crop Science Society of America, Mentor (2006-present), Committee Member (2007-present)
- University of Minnesota/IRRI graduate shuttle program (2006-present)

Symposium Organization:

- National Academy of Sciences Colloquium on "Protecting our food supply: The value of plant genome initiatives", Irvine, CA, Co-organizer (1997)
- University of Bologna, Italy, International Conference: "From the Green Revolution to the Gene Revolution", Co-organizer (2002-2005)
- University of Minnesota, National Symposium: "Intellectual Property for the Public Good", Organizer (2003-2004)

Publications:

- 142 Refereed Journal Articles
- 72 Chapters, or edited books
- 343 Abstracts

BRIEF BIOGRAPHY

Hallmarks of Dr. Phillips' career involve innovative research in plant biology and genetics leading to technology used in genetic engineering of crops, genomics, trait selection, student education, and national/international service.

Dr. Phillips is Regents Professor and McKnight Presidential Chair in Genomics, University of Minnesota. He earned B.S. and M.S. degrees from Purdue University and a Ph.D. from the University of Minnesota; his postdoctoral training was at Cornell University. Dr. Phillips has advised over 60 graduate theses and taught a course in cytogenetics as a faculty member in the Department of Agronomy and Plant Genetics for over 40 years. Dr. Phillips was the co-recipient of the prestigious Wolf Prize in Agriculture in Israel in 2007 for "ground breaking research in service of mankind." In 1991, he was elected a member of the National Academy of Sciences. Currently, he serves on the Board of Trustees of the premier International Rice Research Institute in the Philippines and on the Scientific Advisory Board of the Donald Danforth Plant Science Center. Other awards include an honorary doctorate from Purdue University, Fellow of AAAS (American Association for the Advancement of Science), ASA (American Society of Agronomy), and CSSA (Crop Science Society of America), the Purdue University Agriculture Distinguished Alumni Award, the Dekalb Genetics Crop Science Distinguished Career Award, and the Crop Science Society of America Research Award. Dr. Phillips served as Chief Scientist of the US Department of Agriculture (1996-1998) in charge of the National Research Initiative Competitive Grants Program and chaired the Interagency Working Group that wrote the plan for the NSF Plant Genome Research Initiative of the U.S. He served as President of the Crop Science Society of America in 2000 and Chair of the Council of Scientific Society Presidents in 2006.

GENERAL DESCRIPTION OF ACCOMPLISHMENTS

Throughout his career, Dr. Phillips has coupled the techniques of plant genetics and molecular biology to enhance the understanding of basic biology of cereal crops and to improve these species by innovative methods. His research program at the University of Minnesota was one of the early programs in modern plant biotechnology related to agriculture. He is a founding member and former Director of both the Plant Molecular Genetics Institute and the Microbial and Plant Genomics Institute of the University of Minnesota. He has served on numerous editorial boards, edited six books, and published over 140 refereed journal articles, 70 chapters, and 340 abstracts. Dr. Phillips has been invited to teach, present the results of his research, and/or serve in an advisory capacity at numerous university, governmental, and industrial institutions in the U.S. and abroad.

Dr. Phillips conducts research and teaching in plant genetics applied to plant improvement with an attempt to bridge basic and applied aspects. The research objectives have been to develop and apply molecular genetic information to the improvement of important crop traits, to develop and evaluate somatic cell genetic systems for manipulating crop species, to develop and use genetic and molecular biological selection procedures, and to develop high-throughput genomic mapping procedures. As Regents Professor and member of the U.S. National Academy of Sciences, he participates in addressing University-wide, national, and international issues.

Specific Research Accomplishments:

Modern agricultural biotechnology reflects Regents Professor Phillips' significant contributions in research, education, and national/international public service. His applications of molecular biology toward understanding plant genomes started in 1971 with the chromosomal location of important protein machinery genes (ribosomal DNA) to the Nucleolus Organizer Region in maize using DNA/RNA hybridization.[10] This was the first localization of ribosomal RNA genes to chromosome in plants. The high estimated number of copies of these genes per cell was an unexpected result at the time.

His laboratory, along with postdoctoral scientist Ed Green, were the first to regenerate complete corn plants from cells in tissue culture,[18] thus providing in 1975 a key technology needed for the development of genetically engineered cereals, a technology directly applied for corn and indirectly for other cereals. The use of genetically modified corn with insect resistance, herbicide tolerance and other traits has expanded tremendously around the world. The inbred line used in the first successful regeneration of corn plants from cells is still widely used over thirty years later. In addition, the corn cell line most widely used today in academia and industry for genetic engineering was developed in Dr. Phillips' laboratory by Armstrong in the mid 1980s.

Green and Phillips developed a laboratory selection system for identifying cells and/or plants high in essential amino acids,[17] and lines have been obtained high in threonine and methionine. The high methionine trait discovered in the Phillips lab has been transferred to elite lines of maize and recently released to the public.

Variation induced during the tissue culture process was extensively documented by Dr. Phillips and his students showing the occurrence of mutations, chromosome breakage, DNA methylation alterations, and activation of transposable elements.[44, 46, 47, 48, 49, 51, 54, 56, 62, 69, 156, 157, 163, 164, 167, 168, 186, 189, 194]

Events in the corn kernel endosperm were well described in the Phillips lab especially with Kowles showing a dramatic but normal developmental increase in DNA levels per cell.[35, 36, 42, 55, 59, 81, 89, 105] Some cells were shown to contain nearly 200X the level of DNA in the basic chromosome complement. Because the endosperm is 85% of the kernel, the increase in DNA leading to larger cells must play a major role in corn yields.[81]

A major gene influencing flowering date in corn was identified by use of molecular genetic markers.[64] After considerable follow up in collaboration with Tubersosa, Salvi, and DuPont/Pioneer scientists, the gene was isolated by map-based cloning and subsequently shown, unexpectedly, to be a non-coding sequence 70kb upstream of a flowering gene.[139] This gene caused a 10-day difference in flowering in the material analyzed. Genetically engineered plants with this gene showed expected changes in flowering time.

Dr. Phillips and his students have played a central role in the production of useful molecular genetic marker maps for oats and wild rice and identified chromosomal regions responsible for several important developmental, agronomic, and nutritional traits.[79, 97, 109, 110, 111, 115, 117, 125, 135, 156, 176, 179, 193, 203, 213]

Phillips has presented widely-discussed ideas on how to interpret progress in plant improvement programs with a restricted genetic base, emphasizing *de novo* variation. The paper co-authored with Rasmusson[85] has been the topic of discussion in several courses.

With colleagues, he has edited several books that advance the field of plant genetics, including one on *DNA-based Markers in Plants* co-edited with Vasil[36, 178, 201] and, he has served on several prestigious editorial boards, including the *Proceedings of the National Academy of Sciences* and *Advances in Agronomy*.

Together with Rines and several students and postdoctoral scientists, the Phillips lab developed oat lines with individual corn chromosomes added (Oat-Maize Addition Lines, or OMAs) that allow rapid and efficient mapping of corn sequences to chromosome.[83, 84, 114] All 10 corn chromosomes have been individually transferred to oat by standard wide-hybridization followed by embryo rescue techniques.[113, 187] In addition to these lines being used extensively for mapping, they have been used for chromosome sorting, centromere studies, mapping of gene families, identification of chimeric BACs (Bacterial Artificial Chromosomes), chromosome pairing observations, evolutionary studies, and study of C4 versus C3 photosynthesis. The OMAs were subsequently irradiated with gamma rays to break apart the corn chromosome to form "Radiation Hybrids."[120, 207, 209] These lines (approximately 600) allow a gene sequence to be located to a sub-region of the chromosome.[129]

Recent isolation of a DNA sequence associated with corn centromeres[96] has formed the basis of developing a maize artificial chromosome by Ananiev and colleagues at DuPont/Pioneer that should be useful for introducing multiple genes in genetically engineered corn.

Together with Jacobs and Diez-Gonzalez, transgenic lines of corn have been produced that possess genes coding for a toxin against the pathogenic *E. coli* O157:H7[136] which is the cause of many recalls of hamburger, spinach, and other foods. The concept is that this corn feed will eliminate the pathogenic *E. coli* strain from the intestinal tract of the cattle thus preventing contamination of food.[130]

More recently the Phillips lab has been determining the genetic basis of a high-oil corn that has an embryo which is 31% of the kernel and contains 50% oil to give an overall oil level of 20%.

Professor Phillips' national influence on plant science is notable. He chaired the 1998 report to Congress from the President's Office of Science and Technology Policy that established the U.S. National Plant Genome Initiative of the NSF. He has been a major advocate for science in agriculture as Chief Scientist for the U.S. Department of Agriculture in charge of the National Research Initiative, as a member of the National Academy of Sciences serving as Chair of his section, on a National Research Council transgenic plants study commission, as Director of the Center for Microbial and Plant Genomics (now the Microbial and Plant Genomics Institute) with 85 faculty members and a new microbial and plant genomics building (which he played a major role in securing), as scientific advisor to many research institutes including the Donald Danforth Plant Science Center, the Noble

Foundation, and the USDA-ARS University of California-Berkeley Plant Gene Expression Center, as well as a member and Chair of the Board of Directors of the Council of Scientific Society Presidents.

International agricultural programs have extensively enlisted Dr. Phillips' advice. He serves as Program Chair on the Board of Trustees for the International Rice Research Institute, Vice President of the International Crop Science Society, and served for 10 years on the Biology Program Advisory Committee for the Palm Oil Research Institute of Malaysia. Dr. Phillips has been a co-organizer of several international conferences including two in Japan and one recently in Italy (http://www.doublehelix.too.it) and has often been invited to present papers at international meetings.

Part of Dr. Phillips' renown is due to his outstanding teaching and training of more than 50 excellent graduate students and 20 postdoctoral scientists in the field of plant genetics. Dr. Phillips has been asked to participate in many forums around the U.S. and world due to his many research contributions and clear and informative manner of presentation.

Related Educational Contributions:

In the Heritage Room of the University of Minnesota McNamara Gateway Center, Regents Professor Phillips was recently pictured in a greenhouse discussing a genomics experiment with students. A quote from him is included in the display: "To discover something no one has ever known in the history of the world is a thrill to behold, and to see it through the eyes of one of your students is a thrill unsurpassed". This statement sums up the genuine enjoyment that Professor Phillips experiences from interaction with students. In advising graduate students, he holds the philosophy that it is important to give them his very best ideas – a trait said to have been the reason for the success of R.A. Emerson at Cornell University who in the pioneering days of plant genetics had two Nobel Prize winners among a group of five students, all of whom made significant contributions to the discipline.

Professor Phillips views everything that he does as education, whether it is formal teaching, advising research centers here or abroad, presenting seminars at other universities or at national/international meetings, or publishing research results (141 refereed publications, 72 chapters in books, and 343 abstracts). As reflected in one of his two commencement speeches to University of Minnesota Graduate School graduates – the one entitled "Like Fine China: Easy to Break but Difficult to Mend" – he used the "fine china" metaphor to talk about the importance of integrity in professional and personal life. He has guest-lectured in courses in the University of Minnesota Carlson School of Management on biotechnology and its ethical, social, economic and applied implications, as well as in the Law School Joint Degree Program and the Humphrey Institute on issues of world food production and the role of technology.

Professor Phillips' teaching experience began while he was still an undergraduate. He served as teaching assistant to his undergraduate advisor, employer, and Master's advisor, Professor Wayne Keim, whose name is inscribed on the Wall of Great Teachers at Purdue University, and who provided much of the teaching philosophy that Professor Phillips uses daily. With his clear and concise lectures, carefully crafted exams, and positive approach to grading, Professor Keim provided a great role model. Professor Phillips also benefited as a teaching assistant from several other professors during his undergraduate and graduate career. Professor Charles Burnham at the University of Minnesota served as his Ph.D. advisor and Adrian Srb at Cornell as his postdoctoral advisor. Both of these mentors authored widely-used genetics textbooks and each provided unique insights about the teaching profession.

Teaching a graduate course in cytogenetics (the science combining observations on chromosome behavior from cell to cell and generation to generation with the correlated inheritance patterns) for over thirty-five years, Professor Phillips has always viewed this activity as useful for keeping up-to-date in the field. Students credit this course with providing an especially useful integration of their various courses in genetics. One of the books used in the course was compiled by Professors Phillips and Burnham and includes reprints of landmark papers, together with explanations by the two of them elucidating the importance of the papers to the development of the field.[143] And, he, along with Vasil edited a prominent book in the field, "DNA-Based Markers in Plants."[178] Professor Phillips was invited to teach condensed versions of his course in both Italy and China. In China, a 111-page book was developed from his lectures[158] and distributed around the country.

Another course, "Plant Biotechniques," was developed by Professor Phillips and taught by postdoctoral researchers in molecular biology. This was considered sufficiently innovative that a publication ensued in the Journal of Natural Resources, Life Sciences, and Education.[82] The postdoctoral scientists presented the techniques, underlying theories, and applications of the latest protocols used in their laboratories for which they had hands-on experience. The students appreciated learning about the techniques being used in campus laboratories, and the postdoctoral research associates benefited from having additional teaching experience before entering into the work force.

In 2002, the organizers of a symposium on careers at the American Society of Agronomy (ASA) meetings invited Professor Phillips and three others to talk to undergraduates about their professional experiences. Professor Phillips discussed the differences between his actual career path in comparison to what his career expectations were when he was a student. Some of the points included his surprise at how much writing was involved, the amount of international travel, and how much of his work was socially controversial even though based on solid science. At a previous ASA meeting, he had been asked to review the advances in crop science during the previous decade (1990s); the talk was videotaped and distributed in both English and Spanish. At the 50th anniversary Crop Science Society of America meeting in 2005, he was asked to present a major plenary talk on the development of the science of genetics within the crop science context. His paper was entitled "Genetic Tools from Nature and the Nature of Genetic Tools."[212] In 2007, Professor Phillips was invited to be a centennial speaker in the 100th anniversaries of the Department of Crop Science at Oregon State University and the Department of Agronomy at Purdue. In addition to the book on cytogenetics used in his class, Professor Phillips has co-edited four symposia books[144, 145, 146, 147] and two editions of the book on "DNA-Based Markers in Plants."[178, 201] He also has been a sub-editor for an encyclopedia series on agricultural science.[206]

Professor Phillips has assisted in the organization of several national and international meetings. One was an international congress held at the University of Bologna in Italy: "In the wake of the double helix; from the green revolution to the gene revolution," which was attended by 250 scientists from 35 countries. Eight papers from the meeting were published in the Proceedings of the National Academy of Sciences, including one from the Phillips lab.[209] A book and CD of review papers have been published from the meeting.[210] In conjunction with that meeting, the U.S. Department of State asked Professor Phillips to present two seminars at different universities and a press conference on biotechnology in Italy, a country not generally sympathetic toward genetically modified crops. He also was invited to present a talk entitled "Are GMOs Safe?" at a meeting of AAAS (American Association for the Advancement of Science). He recently served on the Program Committee of AAAS, which orchestrates the conference program for 10,000 attendees (including 1,000 journalists). The 2007 home page for the AAAS Annual Meeting featured a picture of Professor Phillips along with quotes about the meeting and a biographical sketch.

One of the recent symposia organized and chaired by Professor Phillips was the University of Minnesota President's 21st Century Interdisciplinary Conference on "Intellectual Property Rights for the Public Good: Obligations of U.S. Universities to Developing Countries," co-sponsored by the Center for Microbial and Plant Genomics (now the Microbial & Plant Genomics Institute) and the Consortium on Law and Values in Health, Environment & the Life Sciences. The first issue of a new journal entitled Minnesota Journal of Law, Science and Technology (edited by students and faculty) was devoted to papers from this meeting, including an introductory paper by Professor Phillips.[132] He now serves on the Editorial Advisory Board of the Minnesota Journal of Law, Science & Technology. The suggestion for a "humanitarian use clause" to be included in university intellectual property policies was a major outcome of the meeting. Such a clause would facilitate the transfer of technology to the developing world. Professor Phillips and three colleagues in the U of M Law School wrote a summary of the symposium published in The Scientist (April 29, 2004).

Serving as an international advisor has allowed Professor Phillips to impart his knowledge of science and influence goals. Among his numerous experiences, he served on the Program Advisory Committee of the Palm Oil Research Institute of Malaysia for ten years, chairing it for the last four; he was the Scientific Liaison Officer for the U. S. Agency for International Development to the International Rice Research Institute (IRRI) in the Philippines and currently is a member and Program Chair of the IRRI Board of Trustees; he helped forge the IRRI/CIMMYT alliance research program on information technology, intensive cereal production, and the computer knowledge bank; he also consulted and helped establish a new applied molecular genetics activity and laboratory at the Center for Maize and Wheat Improvement (CIMMYT) in Mexico. In 2006, Professor Phillips was invited by the National Academy of Sciences to join a delegation to India to establish a tripartite agricultural biotechnology program with the US, India, and Pakistan.

In 1983, the Council of Bishops of the United Methodist Church called on the various state organizations to form programs important for alleviating hunger at the local, national and international levels. Professor Phillips was appointed to the task force to develop a plan. He suggested that the Council could provide support for students from developing countries to undertake graduate study in production agriculture at the University of Minnesota, with the expectation that after obtaining a M.S. and/or Ph.D. the students would return to their home countries and teach and work on behalf of food production for the rest of their careers. The idea was adopted and Professor Phillips became the Director of a program called Project AgGrad. In 2006, Professor Phillips led the development of a MOU (Memorandum of Understanding) between the University of Minnesota and the International Rice Research Institute (IRRI) that offers graduate students short-term exposures to the world of international agriculture. The first issue (2008) of the new journal, RICE, has a lead article on this opportunity for students.

Professor Phillips played a major role in founding two interdisciplinary centers at the University of Minnesota; the Plant Molecular Genetics Institute in 1980 and the Center for Microbial and Plant Genomics (CMPG) in 2000. These two units have now been merged into the single Microbial and Plant Genomics Institute. He served as Director of the CMPG for five years. He developed two seminar series that have proved very valuable to faculty, students and staff in the field of genomics. The first, entitled the "National Academy of Sciences Lectureship" seminar series, has brought members of the National Academy of Sciences to campus to present seminars on their latest research, to share their professional experiences related to ethics and careers in science with the graduate students and postdoctoral researchers, as well as meet with members of the press. The second is the "Leading Edge Technologies" seminar series, and sponsors visits of researchers who are developing new and powerful technologies of interest to University of Minnesota scientists.

He also proposed and chaired the development of a grant proposal at the request of the University of Minnesota Foundation, which helped to raise $20 million from the Cargill Foundation and the State of Minnesota for the Cargill Building for Microbial and Plant Genomics and $15 million from the McKnight Foundation for funding 15 McKnight Presidential Endowed Chairs.

Several University of Minnesota committees which Professor Phillips has chaired have had significant impact. The report for the reorganization of biological sciences at the U of M (commonly known as the "Phillips report") played a major role in streamlining the organization of biology and bringing together scientists and students with common interests. He also chaired a task force of Regents Professors and McKnight Distinguished Professors, at the request of the Provost and Executive Vice President and the Faculty Consultative Committee, charged with describing the value of the University of Minnesota without the usual emphasis on numbers (students, faculty, patents, grants, or economic impact). This report has become known as the "Lighthouse Report.

Finally, Professor Phillips is continually asked to give talks about his research, and frequently speaks about the science and issues associated with genomics and biotechnology. In total, he has given at least 140 invited talks to Minnesota groups, 100 talks in other states, plus 30 to companies, and over 40 international presentations.

Professor Phillips' supportiveness of students is noteworthy. He not only administers the University of Minnesota Burnham/Chang fund for the support of high school students, but he also established a fund to provide unique educational experiences for graduate students in his department, such as short-term research training opportunities in other laboratories. And in 2007, he donated the $50,000 Wolf Prize he was awarded to establish graduate fellowships at the University of Minnesota.

Professor Phillips has advised 34 Ph.D., 22 M.S., and 3 Senior Theses. He also has served for 35 years as the chair of the H.K. Hayes Memorial Lecture and Graduate Student Award. The graduate student recipient is chosen based on excellence in research, scholarship, and leadership.

Professor Phillips' achievements have been recognized in many ways, including his appointment as a Regents Professor and the first McKnight Presidential Chair at the University of Minnesota, his receipt of an honorary doctorate from Purdue University, his election as a member of the National Academy of Sciences, and his elections as Fellow in the Crop Science Society of America, the American Society of Agronomy, and the American Association for the Advancement of Sciences, and his award of the Wolf Prize in Agriculture.

List of Graduate Student Advisees:

Kowles, R.V., Professor - St. Mary's University, Winona, MN
Stout, J.T., Senior Scientist (d) - Upjohn Company, Kalamazoo, MI

Givens, J.F., Retired - Millipore Company/Stock Broker, Lexington, KY
Springer, W.D., Senior Manager - Malt-O-Meal, Northfield, MN
McMullen, M.S., Professor- North Dakota State Univ., Fargo, ND
Kurvink, K.D., Professor - Bethlehem Univ., Bethlehem, PA
Murphy, T.C., Senior Scientist - Dow Agrosciences, Indianapolis, IN
Buescher, P.J., Landscape Architect - Bourgenois, IL
Albertsen, M.C., Senior Scientist - Pioneer Hi-Bred Intl., Johnston, IA
McCoy, T.C., Dean - Montana State Univ., Bozeman, MT
Thompson, S.A., Global Leader - Dow Agrosciences, Indianapolis, IN
Benzion, G., Patent Examiner - U.S. Patent & Trademark Office, Washington, D.C.
Lee, M., Professor - Iowa State Univ., Ames, IA
West, D.P., Scientist - Brown Seed Co., Prescott, WI
Block, L.P., Business - Steamboat Springs, CO
Rhoades, C.A., Homemaker - Former scientist, Calgene, Palo Alto, CA
Johnson, S.S., Self-employed, Plant Breeder
Tuberosa, R., Professor - University of Bologna, Italy
Armstrong, C.L., Senior Scientist - Monsanto Co., St. Louis, MO
Peschke, V.M., Science Writer - Self-employed science writer; Former Sr. Scientist, Monsanto
Bullock, W.P., Senior Scientist - Garst Seed Co., IA
Benner, M.S., Professor - Ridder Univ., PA
Jellen, E.N., Professor - Brigham Young Univ., Provo, UT
Muenchrath, D.A., Asst. Professor (d) - Iowa State Univ., Ames, IA
Enomoto, S., Research Assoc. - Univ. of Minnesota, St. Paul, MN
Kaeppler, S.M., Professor - Univ. of Wisconsin, Madison, WI
Kim, T.S., Senior Scientist - RDA, Suewon, Korea
Schweizer, L., Sr. Res. Investigator - Oncology Dept, Bristol Myers Squibb, Princeton, NJ
Rooney, W.L., Professor - Texas A&M Univ., College Sta., TX
Graham, G.I., Senior Scientist - Pioneer Hi-Bred Intl., Johnston, IA
Nunez, V.M., Scientist - Garst Seed Co., Slater, IA
Parentoni, S.N., Scientist - Embrapa, Sete Lagos, Brazil
Krone, T.L., Senior Scientist - Pioneer Hi-Bred Inc., Johnston, IA
Milach, S.C.K.M., Senior Scientist - Pioneer Hi-Bred Inc., Brazil
Al-Saady, N.A., Professor - University of Oman
Olhoft, P.M., Scientist - CIBA-GEIGY, Triangle Park, NC
Vladutu, C.I., Scientist - Romania
Riera-Lizarazu, O., Assoc. Prof. - Oregon State Univ. Corvallis, OR
MacLaughlin, J.E., Scientist - Rutgers Univ., New Jersey
Robert, V.J-M., Postdoc - North Dakota State Univ., Fargo, ND
Olsen, M.S., Scientist - Monsanto Co., Northfield, MN
Start, M.A., Scientist - Syngenta, Northfield, MN

Springer, N.A., Asst. Prof. - Univ. of Minnesota, St. Paul, MN
Rasmussen, R.D., Scientist - STA Laboratories, Longmont, CO
Imle, P.T., Co-Owner - Pine Lake Wild Rice, Inc., Grand Rapids, MN
Baumgarten, A.M., Scientist - Pioneer Hi-Bred Intl., Johnston, IA
Odland, W.E., Scientist - Pioneer Hi-Bred Intl., Mankato, MN
Walch, M.D., Scientist - Pioneer Hi-Bred Intl., Wilmer, MN
Kianian, P.M.A., Teacher - Moorhead State University, Moorhead, MN
Jacobs, J.L., Scientist - Monsanto Co., St. Louis, MO
Cabral, C., Current Ph.D. student from Brazil
Garcia, N., Current M.S. student from Philippines
Kahler, A., Current Ph.D. student from South Dakota

List of Postdoctoral Scientists:

Green, C.E., V.P. (r) - Sembios, Woodland, CA
Mascia, P., V.P. - Ceres, Inc., Malibu, CA
McMullen, M., Scientist - USDA-ARS/University of Missouri
Yerk-Davis, G., Assoc. Prof. - University of Missouri-Columbia, MO
Grombacher, A., Scientist - Pioneer Hi-Bred Intl., Canada
Wu, B.-C., Res. Assoc. - Boston University Medical School Human Genome Center
Chen, Z., Professor - University of Texas, Austin, TX
Fox, S., Scientist - AgCanada/Univ. of Saskatchewan, SK
Kianian, S., Assoc. Prof. - North Dakota State Univ., Fargo, ND
Ni, W., Scientist - Dow Agrosciences, Indianapolis, IN
Ananiev, E. , Scientist (d) - Pioneer Hi-Bred Intl., Johnston, IA
Kennard, W., Scientist - Monsanto, Ankeny, IA
Vales, I., Asst. Prof. - Oregon State Univ., Corvallis, OR
Riera-Lizarazu, O., Assoc. Prof. - Oregon State Univ., Corvallis, OR
Kynast, R., Curator - Jodrell Laboratory, Royal Botanic Gardens, Kew, London, UK
Jackson, S., Professor - Purdue Univ., W. Lafayette, IN
Portyanko, V., Postdoc - Univ. of Minnesota, St. Paul, MN
Okagaki, R., Postdoc - Univ. of Minnesota, St. Paul, MN
Baumgartner, A., Scientist - Pioneer Hi-Bred Intl., Johnston, IA

JOURNAL ARTICLES AND OTHER REFEREED PUBLICATIONS

1. Phillips, R.L. 1965. Determination of chromosome numbers in six Lotus species. Crop Sci. 5:370.
2. Phillips, R.L. 1967. The association of linkage group V with chromosome 2 in *Neurospora crassa.* J. Hered. 58:263-265.
3. Phillips, R.L. and A.M. Srb. 1967. A new white ascospore mutant of *Neurospora crassa.* Can. J. Genet. Cytol. 9:766-775.

4. Phillips, R.L. and W.F. Keim. 1968. Seed pod dehiscence in *Lotus* and interspecific hybridization involving *L. corniculatus* L. Crop Sci. 8:18-21.
5. Phillips, R.L. 1968. Cyanogenesis in *Lotus* species. Crop Sci. 8:123-124.
6. Phillips, R.L. 1968. Cytogenetic location of the gene for the purple plant color in maize. J. Hered. 59:56.
7. Phillips, R.L. 1969. Recombination in *Zea mays* L. I. Location of genes and interchanges in chromsomes 5, 6 and 7. Genetics 61:107-116.
8. Phillips, R.L. 1969. Recombination in *Zea mays* L. II. Cytogenetic studies of recombination in reciprocal crosses. Genetics 61:117-127.
9. Phillips, R.L., C.R. Burnham, and E.B. Patterson. 1971. Advantages of chromosomal interchanges that generate haplo-viable deficiency-duplications. Crop Sci. 11:525-528.
10. Phillips, R.L., R.A. Kleese, and S.S. Wang. 1971. The nucleolus organizer region of maize (*Zea mays* L.): Chromosomal site of DNA complementary to ribosomal RNA. Chromsoma (Berl.) 36:79-88.
11. Kleese, R.A. and R.L. Phillips. 1972. Electrophoretic mutants as useful markers for chromosome aberrations. Genetics 72:537-540.
12. Burnham, C.R., J.T. Stout, W.H. Weinheimer, R.V. Kowles, and R.L. Phillips. 1972. Chromosome pairing in maize. Genetics 71:111-126.
13. Burnham, C.R., and R.L. Phillips. 1972. Linkage Groups: Plants. Part IV, Corn. Contribution to Biology Data Book Second Edition, Vol. 1, pp. 85-88. Fed. Amer. Socs. Exptl. Biol.
14. Stout, J.T., and R.L. Phillips. 1973. Two independently inherited electrophoretic variants of the lysine-rich histones of maize. Proc. Nat. Acad. Sci. U.S.A. 70:3043-3047.
15. Phillips, R.L., D.F. Weber, R.A. Kleese, and S.S. Wang. 1974. The nucleolus organizer region of maize (*Zea mays* L.): Tests for ribosomal gene compensation or magnification. Genetics 77:285-297.
16. Green, C.E., R.L. Phillips, and R.A. Kleese. 1974. Tissue cultures of maize (*Zea mays* L.): Initiation, maintenance, and organic growth factors. Crop Sci. 14:54-58.
17. Green, C.E., and R.L. Phillips. 1974. Potential selection system for mutants with increased lysine, threonine, and methionine in cereal crops. Crop Sci. 14:827-831.
18. Green, C.E., and R.L. Phillips. 1975. Plant regeneration from tissue culture of maize. Crop Sci. 15:417-421.
19. Givens, J.F., and R.L. Phillips. 1976. The nucleolus organizer region of maize (*Zea mays* L.): Ribosomal RNA gene distribution and nucleolar interactions. Chromosoma (Berl.) 57:103-117.
20. Liang, G.H., A.S. Wang, and R.L. Phillips. 1977. Control of ribosomal RNA gene multiplicity in wheat. Can J. Genet. Cytol. 19:425-435.

21. Phillips, R.L. 1977. Genetic engineering for crop improvement. 32nd Ann. Corn and Sorghum Res. Conf. pp. 6-20.
22. Phillips, R.L., A.S. Wang, I. Rubenstein, and W.D. Park. 1979. Hybridization of ribosomal RNA to maize chromosomes. Maydica 24:7-21.
23. Phillips, R.L., E.D. Garber, O.L. Miller, Jr., and H.H. Kramer. 1979. C. R. Burnham. Maydica 24:3-6.
24. Albertsen, M.C., and R.L. Phillips. 1981. Developmental cytology of 13 genetic male sterile loci in maize. Can. J. Genet. Cytol. 23:195-208.
25. Phillips, R.L., P.R. Morris, F. Wold, and B.G. Gengenbach. 1981. Seedling screening for lysine-plus-threonine resistant maize. Crop Sci. 21:601-607.
26. Mascia, P.H., I. Rubenstein, R.L. Phillips, A.S. Wang, and Lu Zhen Xiang. 1981. Localization of the 5S rRNA genes and evidence for diversity in the 5S rDNA region of maize. Gene 15:7-20.
27. Burnham, C.R., R.L. Phillips, and M.C. Albertsen. 1981. Inheritance of male sterility in flax involving nuclear-cytoplasmic interaction, including methods of testing for cytoplas-mic restoration. Crop Sci. 21:659-663.
28. McCoy, T.J., R.L. Phillips, and H.W. Rines. 1982. Cytogenetic analysis of plants regenerated from oat (*Avena sativa*) tissue cultures; high frequency of partial chromosome loss. Can J. Genet. Cytol. 24:37-50.
29. McMullen, M.S., R.L. Phillips, and D.D. Stuthman. 1982. Meiotic irregularities in *Avena sativa* L./*A. sterilis* L. hybrids and breeding implications. Crop Sci. 22:890-897.
30. McCoy, T.J., and R.L. Phillips. 1982. Chromosome stability in maize (*Zea mays*) tissue cultures and sectoring in some regenerated plants. Can. J. Genet. Cytol. 24:559-565.
31. Phillips, R.L., A.S. Wang, and R.V. Kowles. 1983. Molecular and developmental cytogenetics of gene multiplicity in maize. Stadler Symp. 15:105-118.
32. Buescher, P.J., R.L. Phillips, and R. Brambl. 1984. Ribosomal RNA contents of maize genotypes with different ribosomal gene numbers. Biochem. Genet. 22:923-930.
33. Luby, J.J., D.D. Stuthman, and R.L. Phillips. 1985. Micronuclei frequency and character coherence in *Avena sativa* L./*A. fatua* crosses. Theor. Appl. Genet. 69:367-373.
34. Phillips, R.L., and B.A. McClure. 1985. Elevated proteinbound methionine in seeds of a lysine-plus-threonine resistant maize line. Cereal Chem. 62:213-218.
35. Kowles, R.V., and R.L. Phillips. 1985. DNA amplification patterns in maize endosperm nuclei during kernel development. Proc. Natl. Acad. Sci. (USA) 82:7010-7014.
36. Phillips, R.L., R.V. Kowles, M.D. McMullen, S. Enomoto, and I. Rubenstein. 1985. Developmentally-timed changes in maize endosperm DNA. UCLA Symp. Plant Genetics Alan R. Liss, Inc., pp. 739-754.

37. McMullen, M.D., B. Hunter, R.L. Phillips, and I. Rubenstein. 1986. The structure of the maize ribosomal DNA spacer region. Nucl. Acids Res. 14:4953-4968.
38. Rhodes, C.A., R.L. Phillips, and C.E. Green. 1986. Cytogenetic stability of aneuploid maize tissue cultures. Can. J. Genet. Cytol. 28:374-384.
39. Rhodes, C.A., C.E. Green, and R.L. Phillips. 1986. Factors affecting tissue culture initiation from maize tassels. Plant Sci. 46:225-232.
40. Tuberosa, R., and R.L. Phillips. 1986. Isolation of methotrexatetolerant cell lines of corn. Maydica 31:215-225.
41. Wang, A.S., R.L. Phillips, and C.C. Mi. 1986. Cell cycle parameters and accumulation of metaphase cells in maize suspension cultures. Plant Sci. 46:53-61.
42. Kowles, R.V., M.D. McMullen, and R.L. Phillips. 1986. Gene expression in developing kernels. First Ann. Penn. State Symp. in Plant Physiol., University Park, Penn.
43. Tuberosa, R., and R.L. Phillips. 1986. Selection and partial characterization of corn (*Zea mays* L.) callus lines tolerant to methotrexate. Genetica Agraria 40:479.
44. Lee, M., and R.L. Phillips. 1987. Genomic rearrangements in maize induced by tissue culture. Genome 29:122-128.
45. Johnson, S.S., R.L. Phillips, and H.W. Rines. 1987. Meiotic behavior in progeny of tissue culture regenerated oat plants (*Avena sativa* L.) carrying neartelocentric chromosomes. Genome 29:431-438.
46. Johnson, S.S., R.L. Phillips, and H.W. Rines. 1987. Possible role of heterochromatin in chromosome breakage induced by tissue culture in oats (*Avena sativa* L.). Genome 29:439-446.
47. Peschke, V.M., R.L. Phillips, and B.G. Gengenbach. 1987. Discovery of transposable element activity among progeny of tissue culturederived maize plants. Science 238:804-807.
48. Lee, M. and R.L. Phillips. 1987. Genetic variants in progeny of regenerated maize plants. Genome 29:834-838.
49. Armstrong, C.L. and R.L. Phillips. 1988. Genetic and cytogenetic variation in plants regenerated from organogenic and friable, embryogenic tissue cultures of maize. Crop Sci. 28:363-369.
50. Lee, M., J.L. Geadelmann, and R.L. Phillips. 1988. Agronomic evaluation of inbred lines derived from tissue cultures of maize. Theor. Appl. Genet. 75:841-849.
51. Benzion, G., and R.L. Phillips. 1988. Cytogenetic stability of maize tissue cultures: A cell line pedigree analysis. Genome 30:318-325.
52. Stadler, J., R. Phillips, and M. Leonard. 1989. Mitotic blocking agents for suspension cultures of maize 'Black Mexican Sweet' cell lines. Genome 32:475-478.

53. Benner, M.S., R.L. Phillips, J.A. Kirihara, and J.W. Messing. 1989. Genetic analysis of methionine-rich storage protein accumulation in maize. Theor. Appl. Genet. 78:761-767.
54. Phillips, R.L. 1989. Somaclonal and gametoclonal variation. Genome 31:1119-1120.
55. Kowles, R.V., F. Srienc, and R.L. Phillips. 1990. Endoreduplication of nuclear DNA in the developing maize endosperm. Devel. Genet. 11:125-132.
56. Peschke, V.M. and R.L. Phillips. 1991. Activation of the maize transposable element *Suppressor-mutator* (*Spm*) in tissue culture. Theor. Appl. Genet. 81:90-97.
57. Peschke, V.M., R.L. Phillips, and B.G. Gengenbach. 1991. Genetic and molecular analysis of tissue culture-derived *Ac* elements. Theor. Appl. Genet. 82:121-129.
58. Sullivan, T.D., L.I. Strelow, C.A. Illingworth, R.L. Phillips, and O.E. Nelson, Jr. 1991. Analysis of maize *Brittle-1* alleles and a defective *Suppressor-Mutator*-induced mutable allele. Plant Cell 3:1337-1348.
59. Kowles, R.V., M.D. McMullen, G. Yerk, R.L. Phillips, S. Kraemer, and F. Srienc. 1992. Endosperm mitotic activity and endoreduplication in maize affected by defective kernel mutations. Genome 35:68-77.
60. Phillips, R.L. 1993. Plant genetics: out with the old, in with the new? Am. J. Clinical Nutrition 58 (Suppl.):2595-2635.
61. Phillips, R.L. 1993. Cytogenetic manipulation of polymitotic (*po*). Maydica 38:85-92.
62. Kaeppler, S.M., and R.L. Phillips. 1993. DNA methylation and tissue culture-induced variation in plants. In Vitro Cell. Dev. Biol. 29P:125-130.
63. Stapleton, A.E. and R.L. Phillips. 1993. A fertile field: Maize genetics '93. Plant Cell 5:723-727.
64. Phillips, R.L., T.S. Kim, S.M. Kaeppler, S.N. Parentoni, D.L. Shaver, R.E. Stucker, and S.J. Openshaw. 1993. Genetic dissection of maturity using RFLPs. Proc. 17th Annual Corn & Sorghum Industry Res. Conf. pp. 135-150.
65. Muenchrath, D.A. and R.L. Phillips. 1993. Relationship of maize seedling response to lysine-plus-threonine medium and whole kernel amino acid profile. Crop Sci. 33:1095-1099.
66. Kaeppler, S.M., R.L. Phillips, and T.S. Kim. 1993. Use of near-isogenic lines derived by backcrossing or selfing to map qualitative traits. Theor. Appl. Genet. 87:233-237.
67. Jellen, E.N., R.L. Phillips, and H.W. Rines. 1993. C-banded karyotypes and polymorphisms in hexaploid oat accessions (*Avena* spp.) using Wrights' stain. Genome 36:1129-1137.
68. Jellen, E.N., W.L. Rooney, R.L. Phillips, and H.W. Rines. 1993. Characterization of the hexaploid oat *Avena byzantina* cv. Kanota monosomic series using C-banding and RFLPs. Genome 36:962-970.

69. Kaeppler, S.M. and R.L. Phillips. 1993. Tissue culture-induced DNA methylation variation in maize. Proc. Natl. Acad. Sci. USA 90:8773-8776.
70. Kim, T.S., R.L. Phillips, and M.Y. Eun. 1993. Identification of genomic regions controlling maturity in maize (*Zea mays* L.). Crop Production and Improvement Technology in Asia. pp. 379-397. Korean Soc. Crop Sci.
71. Jellen, E.N., R.L. Phillips, and H.W. Rines. 1994. Chromosomal localization and polymorphisms of ribosomal DNA in oat (*Avena* spp.). Genome 37:23-32.
72. Phillips, R.L. and E.N. Jellen. 1994. Videotaped lectures in a graduate cytogenetics course. J. Nat. Res. Life Sci. Educ. 23:16-18.
73. Phillips, R.L., S.M. Kaeppler, and P. Olhoft. 1994. Genetic instability of plant tissue cultures: Breakdown of normal controls. Proc. Natl. Acad. Sci. USA 91:5222-5226.
74. Young, N.D. and R.L. Phillips. 1994. Cloning plant genes known only by phenotype. Plant Cell 6:1193-1195.
75. Phillips, R.L., H.W. Rines, E.N. Jellen, W.L. Rooney, B.C. Wu, S. Kianian, and O. Riera-Lizarazu. 1994. Oat genome analysis via molecular markers and oat x corn crosses. U.S.-Japan Joint Seminar, Kansas State University.
76. Rooney, W.L., H.W. Rines, and R.L. Phillips. 1994. Identification of RFLP markers linked to crown rust resistance genes Pc91 and Pc92 in oat. Crop Sci. 34:940-944.
77. Rooney, W.L., E.N. Jellen, R.L. Phillips, H.W. Rines, and S.F. Kianian. 1994. Identification of homoeologous chromosomes in hexaploid oat (*A. byzantina* cv Kanota) using monosomics and RFLP analysis. Theor. Appl. Genet. 89:329-335.
78. Jellen, E.N., R.L. Phillips, H.W. Rines, and W.L. Rooney. 1995. Molecular genetic identification of *Avena* chromosomes related to the group 1 chromosomes of the Triticeae. Genome 38:185-189.
79. O'Donoughue, L.S., S.F. Kianian, P.J. Rayapati, G.A. Penner, M.E. Sorrells, S.D. Tanksley, R.L. Phillips, H.W. Rines, M. Lee, G. Fedak, S.J. Molnar, D. Hoffman, C.A. Salas, B. Wu, E. Autrique, and A. Van Deynze. 1995. A molecular linkage map of cultivated oat. Genome 38:368-380.
80. Phillips, R.L., M.A. Matzke, and K. Oono. 1995. Treasure your exceptions. Plant Cell 7:1522-1527.
81. Schweizer, L., G.L. Yerk-Davis, R.L. Phillips, F. Srienc, and R.J. Jones. 1995. Dynamics of maize endosperm development and DNA endoreduplication. Proc. Natl. Acad. Sci. USA 92:7070-7074.
82. Kianian, S.F., D.C. Bittel, R.L. Phillips, and M. Gallo-Meagher. 1996. Postodoctoral research associates as an educational resource: opportunities and constraints. J. Nat. Resources, Life Sciences & Education 25:144-147.
83. Riera-Lizarazu, O., H.W. Rines, and R.L. Phillips. 1996. Cytological and molecular characterization of oat x maize partial hybrids. Theor. Appl. Genet. 93:123-135.

84. Ananiev, E.V., O. Riera-Lizarazu, H.W. Rines, and R.L. Phillips. 1997. Oat-maize chromosome addition lines: A new system for mapping the maize genome. Proc. Natl. Acad. Sci. USA 94:3524-3529.
85. Rasmusson, D.C. and R.L. Phillips. 1997. Plant breeding progress and genetic diversity from *de novo* variation and elevated epistasis. Crop Sci 37:303-310.
86. Martin, M., R. Phillips, and G. Gardner. 1997. Conducting and administering biotechnology research at a land-grant university. J. Agric. and Human Values.
87. Phillips, R.L. and J. Suresh. 1997. Cytogenetic basis for fertile plants in a maize nuclear male sterility system. Maydica 42:127-131.
88. Kianian, S.F., B-C. Wu, S.L. Fox, H.W. Rines, and R.L. Phillips. 1997. Aneuploid marker assignment in hexaploid oat with the C genome as a reference for determining remnant homoeology. Genome 40:386-396.
89. Kowles, R.V., G.L. Yerk, K.M. Haas, and R.L. Phillips. 1997. Maternal effects influencing DNA endoreduplication in developing endosperm of *Zea mays.* Genome 40:798-805.
90. Milach, S.C.K., H.W. Rines, and R.L. Phillips. 1997. Molecular genetic mapping of dwarfing genes in oat. Theor. Appl. Genet. 95:783-790.
91. Jellen, E.N., H.W. Rines, S.L. Fox, D.W. Davis. R.L. Phillips, and B.S. Gill. 1997. Characterization of "Sun II" oat monosomics through C-banding and identification of eight new "Sun II" monosomics. Theor. Appl. Genet. 95:1190-1195.
92. Milach, S.C.K., H.W. Rines, R.L. Phillips, D.D. Stuthman, and J. Morikawa. 1998. Inheritance of a new dwarfing gene in oat. Crop Sci. 38:356-360.
93. Phillips, R.L., and M. Freeling, 1998. Plant genomics and our food supply: An introduction. Proc. Natl. Acad. Sci. USA 95: 1969-1970.
94. Editor of 15 papers in 1998 Proc. Natl. Acad. Sci. USA 95: 1971-2032: National Academy of Sciences Colloquium on "Protecting our Food Supply: The Value of Plant Genome Initiatives."
95. Chen, Z.J., R.L. Phillips, and H.W. Rines, 1998. Maize DNA enrichment by representational difference analysis. Theor. Appl. Genet. 97:337-344.
96. Ananiev, E.V., R.L. Phillips, and H.W. Rines, 1998. Chromosome-specific molecular organization of maize (*Zea mays L.*) centromeric regions. Proc. Natl. Acad. Sci. USA 95:13073-13078.
97. Kianian, S.F., M.A. Egli, R.L. Phillips, H.W. Rines, D.A. Somers, B.G. Gengenbach, F.H. Webster, S.M. Livingston, S. Groh, L.O. O'Donoughue, M.E. Sorrells, D.M. Wesenberg, D.D. Stuthman, and R.G. Fulcher, 1999. Association of a major groat oil content QTL and an acetyl-CoA carboxylase gene in oat. Theor. Appl. Genet. 98:884-894.
98. Ananiev, E.V., R.L. Phillips, and H.W. Rines, 1998. Complex structure of knob DNA on maize chromosome 9: Retrotransposon invasion into heterochromatin. Genetics 149: 2025-2037.

99. Ananiev, E.V., R.L. Phillips, and H.W. Rines, 1998. A knob-associated tandem repeat in maize capable of forming fold-back DNA segments: Are chromosome knobs megatransposons? Proc. Natl. Acad. Sci. USA 95: 10785-10790.
100. Kennard, W., R. Phillips, R. Porter, and A. Grombacher. 1999. A comparative map of wild rice (*Zigania palustris* L. 2n=2x=30). Theor. Appl. Genet. 99:793-799.
101. DeKoeyer, D.L., R.L. Phillips, and D.D. Stuthman. 1999. Changes in genetic diversity during seven cycles of recurrent selection for grain yield in oat, *Avena sativa* L. Plant Breeding 118:37-43.
102. Vladutu, C., J. McLaughlin, and R.L. Phillips. 1999. Fine mapping and characterization of linked quantitative trait loci involved in the transition of the maize apical meristem from vegetative to generative structures. Genetics 153:993-1007.
103. Bass, H.W., O. Riera-Lizarazu, E.V. Ananiev, S.J. Bordoli, H.W. Rines, R.L. Phillips, J.W. Sedat, D.A. Agard, and W.Z. Cande. 2000. Evidence for the coincident initiation of homolog pairing and synapsis during the telomere-clustering (bouquet) stage of meiotic prophase. J. Cell Sci. 113:1033-1042.
104. Riera-Lizarazu, O., M.I. Vales, E.V. Ananiev, H.W. Rines, and R.L. Phillips. 2000. Production and characterization of maize-chromosome 9 radiation hybrids derived from an oat-maize addition line. Genetics 156:327-339.
105. Engelen-Eigles, G., R.J. Jones, and R.L. Phillips. 2000. DNA endore-duplication in maize endosperm cells: The effect of exposure to short-term high temperature. Plant, Cell and Environment 23:657-663.
106. Cao, X., N.M. Springer, M.G. Muszynski, R.L. Phillips, S. Kaeppler, and S.E. Jacobsen. 2000. Conserved plant genes with similarity to mammalian *de novo* DNA methyltransferases. Proc. Natl. Acad. Sci. USA 97:4979-4984.
107. Ananiev, E.V., R.L. Phillips, and H.W. Rines. 2000. Complex structure of knobs and centromeric regions in maize chromosomes. Tsitol Genet. 34:11-15.
108. Muehlbauer, G.J., O. Riera-Lizarazu, R.G. Kynast, D. Martin, R.L. Phillips, and H.W. Rines. 2000. A maize-chromosome 3 addition line of oat exhibits expression of the maize homeobox gene *liguleless3* and alterations of cell fates. Genome 43:1055-1064.
109. Kennard, W., R. Phillips, R. Porter, and A. Grombacher. 2000. A comparative map of wild rice (*Zizania palustris* L. 2n=2x=30). Theor. Appl. Genet. 101:677-684.
110. Kianian, S.F., R.L. Phillips, H.W. Rines, R.G. Fulcher, F.H. Webster, and D.D. Stuthman. 2000. Quantitative trait loci influencing β glucan content in oat (*Avena sativa*, 2n=6x=42). Theor. Appl. Genet. 101:1039-1048.
111. Groh, S., A. Zacharias, S.F. Kianian, G.A. Penner, J. Chong, H.W. Rines, and R.L. Phillips. 2001. Comparative AFLP mapping in two hexaploid oat populations. Theor. Appl. Genet. 102:876-884.

112. Li, L.J., K. Aramuganathan, H.W. Rines, R.L. Phillips, O. Riera-Lizarazu, D. Sandhu, Y. Zhou, and K.S. Gill. 2001. Flow cytometric sorting of maize chromosome 9 from an oat-maize chromosome addition line. Theor. Appl. Genet. 102:658-663.
113. Kynast, R.G., O. Riera-Lizarazu, M.I. Vales, R.J. Okagaki, S.B. Maquieira, G. Chen, E.V. Ananiev, W.E. Odland, C.D. Russell, A.O. Stec, S.M. Livingston, H.A. Zaia, H.W. Rines, and R.L. Phillips. 2001. A complete set of maize individual chromosome additions to the oat genome. Plant Physiol. 125: 1216-1227.
114. Okagaki, R.J., R.G. Kynast, S.M. Livingston, C.D. Russell, H.W. Rines, and R.L. Phillips. 2001. Mapping maize sequences to chromosomes using oat-maize chromosome addition materials. Plant Physiol. 125:1228-1235.
115. Groh, S., S.F. Kianian, R.L. Phillips, H.W. Rines, D.D. Stuthman, D.M. Wesenberg, and R.G. Fulcher. 2001. Analysis of factors influencing milling yield and their association to other traits by QTL analysis in two hexaploid oat populations. Theor. Appl. Genet. 103:9-18.
116. Jackson, S.A., P. Zhang, W.P. Chen, R.L. Phillips, B. Friebe, S. Muthukrishnan, and B.S. Gill. 2001. High-resolution structural analysis of biolistic transgene integration into the genome of wheat. Theor. Appl. Genet. 103:56-62.
117. Fox, S.L., E.N. Jellen, S.F. Kianian, H.W. Rines, and R.L. Phillips. 2001. Assignment of RFLP linkage groups to chromosomes using monosomic F_1 analysis in hexaploid oat. Theor. Appl. Genet. 102:320-326.
118. Engelen-Eigles, G., R.J. Jones, and R.L. Phillips. 2001. DNA endore-duplication in maize endosperm cells is reduced by high temperature during the mitotic phase. Crop Sci. 41:1114-1121.
119. Salvi, S., R. Tuberosa, and R.L. Phillips. 2001. Development of PCR-based assays for allelic discrimination in maize by using the 5'-nuclease procedure. Molec. Breeding 8:169-176.
120. Kynast, R.G., R.J. Okagaki, H.W. Rines, and R.L. Phillips. 2002. Maize individualized chromosome and derived radiation hybrid lines and their use in functional genomics. Funct. Integr. Genomics 2:60-69.
121. Ananiev, E.V., M.I. Vales, R.L. Phillips, and H.W. Rines. 2002. Isolation of A/D and C genome specific dispersed and clustered repetitive DNA sequences from *Avena sativa*. Genome 45:431-441.
122. Salvi, S., R. Tuberosa, E. Chiapparino, M. Maccaferri, S. Veillet, L. van Beuningen, P. Isaac, K. Edwards, and R.L. Phillips. 2002. Toward positional cloning of *vgt1*, a QTL controlling the transition from vegetative to the reproductive phase in maize. Plant Molec. Biol. 48:601-613.
123. Milach, S.C.K., H.W. Rines, and R.L. Phillips. 2002. Plant height components and gibberellic acid response of oat dwarf lines. Crop Sci. 42:1147-1154.

124. Chen, G., H. Rines, K. Leonard, D. Stuthman, R. Phillips, and S. Ding. 2002. Identification of QTLs for horizontal resistance to crown rust in oat. J. Zhejiang Univ. (Agri. & Life Sci.) 27:151-155.
125. Wright, C.P., N.A. Tinker, S.F. Kianian, M.E. Sorrells, L.S. O'Donoughue, D.L. Hoffman, S. Groh, G.J. Scoles, C.D. Li, F.H. Webster, R.L. Phillips, H.W. Rines, S.M. Livingston, K.C. Armstrong, G. Fedak, and S.J. Molnar. 2003. A molecular map in Kanota x Ogle hexaploid oat (Avena sp.) enhanced by additional markers and a robust framework. Genome 46: 28-47.
126. Hass, B.L., J.C. Pires, R. Porter, R.L. Phillips, and S.A. Jackson. 2003. Comparative genetics at the gene and chromosome levels between rice (*Oryza sativa*) and wild rice (*Zizania palustris*). Theor. Appl. Genet. 107:773-782.
127. Salvi, S., M. Morgante, K. Fengler, R. Meeley, E. Ananiev, S. Svitashev, E. Bruggemann, X. Niu, B. Li, S.V. Tingey, et al. 2003. Proc. of 57th Corn and Sorghum Conference.
128. Okagaki, R.J., and R.L. Phillips. 2004. Maize DNA-sequencing strategies and genome organization. Genome Biol. Vol. 5, Issue 5, Article 223, pp. 1-3.
129. Kynast, R.G., R.J. Okagaki, M.W. Galatowitsch, S.R. Granath, M.S. Jacobs, A.O. Stec, H.W. Rines, and R.L. Phillips. 2004. Dissecting the maize genome by using chromosome addition and radiation hybrid lines. Proc. Natl. Acad. Sci. USA 101: 9921-9926. (cover picture).
130. Schamberger, Gerry P., Ronald L. Phillips, Jennifer L. Jacobs, and Francisco Diez-Gonzalez. 2004. Reduction of *Escherichia coli* O157:H7 populations in cattle by addition of colicin E7-producing *E. coli* to feed. Appl. Environ. Microbiol. 70:6053-6060.
131. Vales, M.I., O. Riera-Lizarazu, H.W. Rines, and R.L. Phillips. 2004. Transmission of maize chromosome 9 rearrangements in oat-maize radiation hybrids. Genome 47:1202-1210.
132. Phillips, R.L. 2004. Intellectual property rights for the public good: Obligations of U.S. universities to developing countries. MN J. Law, Sci. & Tech. 6:177-185.
133. Buhinicek, I., I. Pejic, J. Suresh, A. Vragolovic, and R. L. Phillips. 2004. Genetic divergence of elite maize inbred lines comparing to Illinois high oil source. Austrian J. Agric. Res. 55:29-35.
134. Dinu, I., R.J. Hayes, R. Kynast, R.L. Phillips and C.A. Thill. 2005. Novel inter-series hybrids in *Solanum*, section *Petota*. Theor. Appl. Genet., 110:403-415.
135. Portyanko, V.A., G. Chen, H.W. Rines, R.L. Phillips, K.J. Leonard, G.E. Ochochi, and D.D. Stuthman. 2005. Quantitative trait loci for partial resistance to crown rust, *Puccinia coronata*, in cultivated oat, *Avena sativa* L. Theor. Appl. Genet. 111:313-324.
136. Jacobs, J.L., F. Diez Gonzalez, M.D. Stern, and R.L. Phillips. 2005. Detection of transgenic maize Cry1Ab protein subjected to ruminal digestion. J. Am. Feed Sci. 14:655-664.

137. Odland, W., A. Baumgarten, and R. Phillips. 2006. Ancestral rice blocks define multiple related regions in the maize genome. The Plant Genome 1:S41-S48 (Supplement to Crop Sci. 46).
138. Baumgarten, A.M., J. Suresh, G. May, and R.L. Phillips. 2007. Mapping QTLs contributing to *Ustilago maydis* resistance in specific plant tissues of maize. Theor. Appl. Genet. 114:1229-1238.
139. Salvi, S., G. Sponza, M. Morgante, D. Tomes, X. Niu, K.A. Fengler, R. Meeley, E.V. Ananiev, S. Svitashev, E. Bruggemann, B. Li, C.F. Hainey, S. Radovic, G. Zaina, J.-A. Rafalski, S.V. Tingey, G.H. Miso, R.L. Phillips, and R. Tuberosa. 2007. Conserved noncoding genomic sequences associated with a flowering-time quantitative trait locus in maize. Proc. Natl. Acad. Sci. USA 104:11376-11381.
140. Okagaki, R.J., M.S. Jacobs, A.O. Stec, R.G. Kynast, E. Buescher, H.W. Rines, M.I. Vales, O. Riera-Lizarazu, M. Schneerman, G. Doyle, K.L. Friedman, R.W. Staub, D.F. Weber, T.L. Kamps, I.F.E. Amarillo, C.D. Chase, and R.L. Phillips. 2008. Maize centromere mapping: A comparison of physical and genetic strategies. J. Hered. 92:85-93.
141. Makarevitch, I., R.L. Phillips, and N.M. Springer. 2008. Profiling expression changes caused by a segmental aneuploid in maize. BMC Genomics 9:7.
142. Phillips, R.L., N. Magor, D. Shires, H. Leung, S. McCouch, and D. McIntosh. 2008. Student Opportunity: short-term exposure to international agriculture. Rice 1: (in press).

Book chapters, reviews, or monographs:

143. Phillips, R.L., and C.R. Burnham, Eds. *Cytogenetics*. In *Benchmark Papers in Genetics Series*. Dowden, Hutchinson, and Ross, Stroudsburg, PA 489 pp. (1977).
144. Rubenstein, I., R.L. Phillips, C.E. Green, and R.J. Desnick, Eds. *Molecular Genetic Modification of Eucaryotes*. Academic Press. 171 pp. (1977).
145. Rubenstein, I., R.L. Phillips, C.E. Green, and B.G. Gengenbach, Eds. *Molecular Biology of Plants*. Academic Press. 343 pp. (1979)
146. Rubenstein, I., R.L. Phillips, C.E. Green, and B.G. Gengenbach, Eds. *The Plant Seed: Development, Preservation, and Germination*. Academic Press. 266 pp. (1979).
147. Rubenstein, I., B.G. Gengenbach, R.L. Phillips, and C.E. Green, Eds. *Genetic Improvement of Crops: Emergent Techniques*. University of Minnesota Press. 232 pp. (1980).
148. Phillips, R.L. Molecular cytogenetics of the nucleolus organizer region. *In* Maize Breeding and Genetics, Chapter 43, John Wiley & Sons, NY (1978).
149. Phillips, R.L. Plant Cytogenetics. *In* Staining Procedures, 4th Edition, Chapter 8. Williams and Wilkins Co., Baltimore. (pp. 341-359). 1981.

150. Phillips, R.L. Pollen and pollen tubes. *In* Staining Procedures, 4th Edition, Chapter 9. Williams and Wilkins Co., Baltimore. (pp. 361-366). 1981.
151. Phillips, R.L., and A.S. Wang. 1982. *In situ* hybridization with maize meiotic cells. *In* Maize for Biological Research. W.F. Sheridan, Ed. Plant Molec. Biol. Assoc., Charlottesville, VA.
152. Bashaw, E.L., J. James, J.R. Laughnan, and R.L. Phillips. Chromosomal and cytoplasmic manipulations: Discussion panel. *In* Plant Breeding II. K. J. Frey, Ed. The Iowa State University Press, Ames, IA pp. 138-146.
153. Phillips, R.L. 1983. Genetic engineering of plants: some perspectives on the conference, the present, and the future. *In* Genetic Engineering of Plants, C. Meredith and T. Kosuge, Eds., Plenum Press.
154. Phillips, R.L., and A.S. Wang. 1984. Chromosome analysis. *In* Cell Culture and Somatic Cell Genetics of Plants. Vol. I. Laboratory Techniques Chapter 78, pp. 712-727.
155. Wang, A.S., and R.L. Phillips. 1984. Synchronization of suspension culture cells. *In* Cell Culture and Somatic Cell Genetics of Plants. Vol. I. Laboratory Techniques, Chapter 21, pp. 175-181.
156. Rines, H.W., S.S. Johnson, and R.L. Phillips. 1986. Tissue culture induced variation in oats. *In* 2nd Intl. Oats Conf., D.A. Lawes and H. Thomas (eds.) Aberystwyth, Wales, July 15-18, 1985. Martinas Nijhoff Publishers, Dordrecht, Netherlands, pp. 29-33.
157. Benzion, G., R.L. Phillips, and H.W. Rines. 1986. Case histories of genetic variability *in vitro*: oats and maize. *In* Plant Regeneration and Genetic Variability, Vol. III. Cell Culture and Somatic Cell Genetics of Plants series. I. K. Vasil, Ed. Academic Press, New York, Chapter 22, pp. 435-448.
158. Phillips, R.L. 1987. Lectures in molecular cytogenetics. Published by The Beijing Agricultural University, People's Republic of China. 111 pp.
159. Phillips, R.L., M.D. McMullen, S. Enomoto, and I. Rubenstein. 1987. Ribosomal DNA in maize. p. 201-214. *In* J.P. Gustafson and R. Appels (eds.) Chromosome Structure and Function: Impact of New Concepts. 18th Stadler Genetics Symp. Columbia, MO. Plenum Press. New York.
160. Kowles, R.V., and R.L. Phillips. 1988. Endosperm development in maize. Intl. Rev. Cytl. 112:97-136.
161. Phillips, R.L. 1988. Maize genetic engineering update. Proc. Corn Refiners Assoc. September 19-22, 1988. St. Louis, MO. pp. 1-8.
162. Phillips, R.L., D.A. Somers, and K.A. Hibberd. 1988. Cell/tissue culture and *in vitro* manipulation. *In* Corn and Corn Improvement 3rd Ed., (Ed. G.F. Sprague and J.W. Dudley). Amer. Soc. Agron., Madison, WI. p. 345-387.
163. Lee, M.D., and R.L. Phillips. 1988. Chromosomal basis of somaclonal variation. Ann. Rev. Plant Physiol. Plant Mol. Biol. 39:413-437.

164. Phillips, R.L., and V.M. Peschke. 1988. Discovery of *Ac* activity among progeny of tissue culture-derived maize plants. *In* Plant Transposable Elements, Ed. O. Nelson, Plenum Press, New York. p. 305-315.
165. Somers, D.A., R.L. Phillips, and H.W. Rines. 1988. Corn and oat tissue cultures and genetic variation in regenerated plants. *In* Cell and Tissue Culture in Field Crop Improvement. J. Bay-Petersen and T.M. Chu (eds.). FFTC Symp., Tsukuba, Japan, Food and Fertilizer Tech. Center. Book Series No. 38, Taipei, Taiwan. pp 57-66.
166. Rines, H.W., R.L. Phillips, W.P. Bullock, and E.N. Jellen. 1989. Biotechnology applications in oat breeding in the U.S. p. 100-107. *In*: B. Mattsson and R. Lyhagen (ed.) Proc. 3rd Int'l. Oat Conf., Lund, Sweden, 4-8 July 1988. Svälof AB, Sweden.
167. Phillips, R.L. 1990. Genomic reorganization induced by plant tissue culture. *In* R.S. Sangwan and B.S. Sangwan-Norreel (eds.) The Impact of Biotechnology in Agriculture. Proc. Intl. Conf.: The meeting point between fundamental and applied *in vitro* culture research, pp. 237-246. Amiens, France. Kluwer Academic Publ., Dordrecht.
168. Phillips, R.L., S.M. Kaeppler, and V.M. Peschke. 1990. Do we understand somaclonal variation? *In*: H.J.J. Nijkamp, L.H.W. Van Der Plas, and J. Van Aartrijk (eds.). Progress in Plant Cellular and Molecular Biology. Proc. VIIth Int. Congr. Plant Tissue and Cell Culture, pp. 131-141. Amsterdam, The Netherlands. Kluwer Academic Publ., Dordrecht.
169. McMullen, M.D., R.L. Phillips, and I. Rubenstein. 1991. Molecular analysis of the nucleolus organizer region in maize. *In* Chromosome Engineering in Plants: Genetics, Breeding, and Evolution. Part A. Eds. P.K. Gupta and Tsuchiya, T. Elsevier Science Publ., Amsterdam, pp. 561-576.
170. Phillips, R.L. 1991. Prospects of cell and molecular biology/breeding interface in crop improvement. *In*: Proc. 1991 PORIM International Palm Oil Conf., 9-14 Sept. 1991, Kuala Lumpur, Malaysia.
171. Phillips, R.L., D.J. Plunkett, and S.M. Kaeppler. 1992. Novel approaches to the induction of genetic variation and plant breeding implications. *In* Stalker, H.T., and J.P. Murphy (eds.) Plant Breeding in the 1990's. North Carolina State University Symposium, pp. 389-408. CAB Intl Press, Oxon, England.
172. Davis, D.W., O. Riera-Lizarazu, H.W. Rines, and R.L. Phillips. 1992. Characterization of wide-cross derived oat haploids and their progeny. Proc. 4th Intl. Oat Conf., Adelaide, South Australia, pp. 109-111.
173. Rines, H.W., R.L. Phillips, E.N. Jellen, W.L. Rooney, S.F. Kianian, and B.-C. Wu. 1992. Toward an integrated chromosomal/molecular genetic map in oat. Proc. 4th Intl. Oat Conf., Adelaide, South Australia, pp. 97-100.
174. Peschke, V.M., and R.L. Phillips. 1992. Genetic implications of somaclonal variation in plants. Adv. Genet. 30:41-75.

175. Kowles, R.V., G.L. Yerk, F. Srienc, and R.L. Phillips. 1992. Maize endosperm tissue as an endoreduplication system. *In*: Setlow, J.K. (ed.) Genetic Engineering, Principles and Methods. Vol. 14, pp. 65-88. Plenum Press, New York.
176. Rines, H.W., R.L. Phillips, and D.A. Somers. 1992. Applications of tissue culture to oat improvement. *In* Marshall, H.G., and M.E. Sorrells (eds.) Oat Science and Technology. pp. 777-791. Am. Soc. Agron., Crop Sci. Soc. Am. Madison, WI.
177. Phillips, R.L., and S.A. Eberhart. 1993. Novel methodology in plant breeding-discussion. International Crop Science I. eds. D.R. Buxton, R. Shibles, R.A. Forsberg, B.L. Blad, K.H. Asay, G.M. Paulsen, and R.F. Wilson. Crop Sci. Soc. Am., Madison, pp. 647-648.
178. Phillips, R.L., and I.K. Vasil (eds). 1994. DNA-based markers in plants. Kluwer Academic Publishers. Dordrecht, The Netherlands 384 pp.
179. O'Donoughue, L.S., J.P. Rayapati, S.F. Kianian, M.E. Sorrells, S.D. Tanksley, M. Lee, H.W. Rines, and R.L. Phillips. 1994. Development of RFLP-based linkage maps in diploid and hexaploid oat (*Avena* sp.). *In*: Phillips, R.L., and I. K. Vasil (eds.) DNA-Based Markers in Plants. Kluwer Academic Publishers, Dordrecht, The Netherlands pp. 359-374.
180. Freeling, M., V. Walbot (eds.) 1994. The Maize Handbook. Ch. III Genetics Protocols. R. Phillips and J. Birchler (eds.), Ch V Cell Culture Protocols. R. Phillips (ed.). Springer Verlag, New York.
181. Kowles, R.V., G.L. Yerk, L. Schweizer, F. Srienc, and R.L. Phillips. 1994. Flow cytometry for endosperm nuclear DNA. *In* The Maize Handbook. M. Freeling and V. Walbot (eds). Springer Verlag, New York pp. 400-406.
182. Phillips, R.L. 1994. Classification of pollen abortion in the field. *In* The Maize Handbook. M. Freeling and V. Walbot (eds.). Springer Verlag, New York pp. 297-298.
183. Phillips, R.L. 1994. Chromosomal translocations involving the nucleolus organizer region or satellite of chromosome 6. *In* The Maize Handbook. M. Freeling and V. Walbot (eds.). Springer Verlag, New York pp. 342-345.
184. Livingston, S.M., and R.L. Phillips. 1994. *In situ* hybridization of DNA and RNA probes to maize chromosomes. *In* The Maize Handbook. M. Freeling and V. Walbot (eds.). Springer Verlag, New York pp. 504-508.
185. Kowles, R.V., G.L. Yerk, F. Srienc, and R.L. Phillips. 1994. Absorption cytophotometry of nuclear DNA. *In* The Maize Handbook. M. Freeling and V. Walbot (eds.). Springer Verlag, New York pp. 396-400.
186. Phillips, R.L., and P. Olhoft. 1995. Genomic reprogramming by tissue culture. *In* Modification of gene expression and non-Mendelian inheritance. K. Oono and F. Takaiwa (eds.). Natl. Inst. Agrobiol. Resources, Tsukuba, Japan, pp. 1-14.

187. Rines, H.W., O. Riera-Lizarazu, and R.L. Phillips. 1995. Disomic maize chromosome-addition oat plants derived from oat x maize crosses. *In* Modification of gene expression and non-Mendelian inheritance. K. Oono and F. Takaiwa (eds.). Natl. Inst. Agrobiol. Resources, Tsukuba, Japan, pp. 235-251.

188. Phillips, R.L., H.W. Rines, E.N. Jellen, W.L. Rooney, B.C. Wu, S. Kianian, and O. Riera-Lizarazu. 1995. Oat genome anlaysis via molecular markers and oat x corn crosses. *In* Classical and molecular cytogenetic analysis. Proc. U.S. Japan Joint Symp. (W.J. Raupp and B.S. Gill, eds). Kansas Ag Exp Stn Rep. 95:352-D, Manhattan, pp. 49-57.

189. Olhoft, P. and R.L. Phillips. 1995. Genetic and epigenetic changes induced by maize tissue culture. *In* Use of induced mutations and molecular techniques for crop improvement. FAO/IAEA Intl. Symp., Vienna, Austria, June 19-23.

190. Kaeppler, S.M., R.L. Phillips, and P. Olhoft. 1996. Molecular basis of heritable tissue culture-induced variation in plants. *In* Somaclonal variation and induced mutation in crop improvement. Jain, S.M., B.S. Ahloowalia, and D.S. Brar, eds. Kluwer Academic Publishers, The Netherlands, Dordrecht.

191. Rines, H.W., O. Riera-Lizarazu, S.B. Maguieira, and R.L. Phillips. 1996. Wide crosses for haploids. *In* V Intl. Oat Conf. and VII Intl. Barley Genetics Symp. G. Scoles and B. Rossnagal, eds. pp. 207-212.

192. Rines, H.W., O. Riera-Lizarazu, V.M. Nunez, D.W. Davis, and R.L. Phillips. 1997. Oat haploids from anther culture and from wide hybridizations. *In In vitro* haploid production of haploids in higher plants. S.M. Jain, S.K. Sopory, and R.E. Veillaux (eds.). Kluwer Acad. Publ., Dordrecht. Vol. 4:205-221.

193. Kianian, S., M. Egli, R. Phillips, H. Rines, D. Somers, B. Gengenbach, F. Webster, S. Livingston, S. Groh, L. O'Donoughue, M. Sorrells, D. Wesenberg, D. Stuthman, and G. Fulcher. 1998. Association of acetyl-CoA carboxylase and groat oil content in oat. In: J. Sanchez, E. Cerdo-Olmedo, and E. Martinez-Force (eds.), Adv. Plant Lipid Res., pp. 629-632.

194. Olhoft, P.M. and R.L. Phillips. 1999. Genetic and epigenetic instability in tissue culture and regenerated progenies. *In* Plant Response to Environmental Stresses: From phytohormones to genomic reorganization. Ed. H.R. Lerner, M. Dekker Inc., New York, pp. 111-148.

195. Phillips, R.L. 1999. Unconventional sources of genetic diversity: *de novo* variation and elevated epistasis. In: Plant Breeding in the Turn of the Millennium. Federal University of Vicosa, Brazil, March 2-3, pp. 103-131.

196. Phillips, R.L. 1999. Research needs in heterosis. In: Genetics and exploitation of heterosis in crops. pp. 501-508. Ed. J.G. Coors and S. Pandey. Amer. Soc. Agron., Madison, WI.

197. Phillips, R.L. 2000. Synteny among cereal genomes and its utilization in agriculture. In: Challenging New Frontiers of Scientific and Technical Importance Symp. Tsukuba, Japan.
198. Phillips, R.L. 2000. Biotechnology and agriculture in today's world: A brief comment on their roles for improving human nutrition. In: Improving Human Nutrition Through Agriculture. Symp. International Food Policy Research Institute and International Rice Research Institute, Philippines.
199. Imle, P.T., R.L. Phillips, and R.A. Porter. 2000. Molecular genetics of wild rice. In: Wild Rice Research and Management Conference Proc.
200. Olsen, M.S. and R.L. Phillips. 2001. Molecular genetic improvement of protein quality in maize. Encyl. Life Support Syst-UNESCO.
201. Phillips, R.L., and I.K. Vasil (eds.) 2001. DNA-based Markers in Plants. 2nd Ed., Kluwer Academic Publishers, Dordrecht, The Netherlands, 512 pp.
202. Phillips, R.L., and I.K. Vasil. 2001. Introduction: Molecular marker maps of major crop species. In: DNA-based markers in plants. Eds. R.L. Phillips and I.K. Vasil. Kluwer Academic Press, The Netherlands, pp. 167-168.
203. Kianian, S.F., S.L. Fox, S. Groh, N. Tinker, L.S. O'Donoughue, P.J. Rayapati, R.P. Wise, M. Lee, M.E. Sorrells, G. Fedak, S.J. Molnar, H.W. Rines, and R.L. Phillips. 2001. Molecular marker linkage maps in diploid and hexaploid oat (Avena sp.). In: DNA-based markers in plants. Eds. R.L. Phillips and I.K. Vasil. Kluwer Academic Press, The Netherlands, pp. 443-462.
204. Riera-Lizarazu, O., M.I. Vales, and R.L. Phillips. 2001. A compilation of molecular genetic maps of cultivated plants. In: DNA-based markers in plants. Eds. R.L. Phillips and I.K. Vasil. Kluwer Academic Press, The Netherlands, pp. 463-497.
205. NRC Committee on Environmental Impacts associated with Commercialization of Transgenic Plants. 2002. Environmental effects of transgenic plants: the scope and adequacy of regulation. National Academy Press. Washington, DC, 320 pp.
206. Okagaki, R.J., R.G. Kynast, H.W. Rines, and R.L. Phillips. 2004. Radiation hybrid mapping. In: Encyclopedia of Plant & Crop Science. Ed., R.M. Goodman, Marcel Dekker, Inc., New York, pp. 1074-1077.
207. Phillips, R.L., H.W. Rines, R.J. Okagaki, R.G., Kynast, R. Donahue, and W.E. Odland. 2003. Genetic and physical mapping of the maize genome through radiation hybrids. Proc. Intl. Rice Research Conference 2003. Gurdev Khush Symposium. In: Rice Science: Innovations and Impact for Livelihood. Eds. T.W. Mew, D.S. Brar, S. Peng, D. Dawe, and B. Hardy, pp. 77-90.
208. Tuberosa, R., and R.L. Phillips. 2003. Progress in map-based cloning of *Vgt1*, a QTL controlling flowering time in maize. 57th Corn and Sorghum Seed Research Conf., Dec., 2002.

209. Rines, H.W., R.L. Phillips, R.G. Kynast, R.J. Okagaki, W.E. Odland, A.O. Stec, M.S. Jacobs, and S.R. Granath. 2005. Maize chromosome additions and radiation hybrids in oat and their use in dissecting the maize genome. In: In the wake of the double helix: from the green revolution to the gene revolution. Eds. Tuberosa, R., R.L. Phillips, and M. Gale, Avenue Media, Bologna, Italy, pp. 427-442.
210. Tuberosa, R., R.L. Phillips, and M. Gale. (Eds.) 2005. In the wake of the double helix: From the green revolution to the gene revolution. Avenue Media, Bologna, Italy. 772 pp. CD produced in 2006.
211. Rines, H.W., S.J. Molnar, N.A. Tinker, and R.L. Phillips. 2006. Oat. In: Kole, C. (ed.). Genome Mapping and Molecular Breeding in Plants: Cereals and Millets Vol. 1. Springer, Inc., NY, USA. pp. 211-242.
212. Phillips, R.L. 2006. Genetic tools from nature and the nature of genetic tools. In: CSSA Golden Anniversary Symposium. Ed. C. Stuber. Crop Sci. 46:2245-2252.
213. Phillips, R.L., W.E. Odland, and A.L. Kahler. 2007. Rice as a reference genome and more. In: 5th Intl. Rice Genetics Symp., Eds. D.S. Brar, D. Mackill, and B. Hardy. pp. 3-15.
214. Phillips, R.L., and H.W. Rines. 2008. Genetic analyses with oat maize addition and radiation hybrid lines. In: The Maize Handbook: Domestication, Genetics, and Genome, Eds. J.L. Bennetzen and S.C. Hake, Springer-Life Sciences, New York, NY (in press).

Potential Selection System for Mutants with Increased Lysine, Threonine, and Methionine in Cereal Crops[1]

C. E. Green and R. L. Phillips[2]

ABSTRACT

Simultaneous addition of lysine and threonine inhibits growth in maize (*Zea mays* L.) callus cultures and embryo cultures from several cereals. Methionine or homoserine reverses lysine-threonine inhibition. Isolation of feedback inhibition resistance mutants is proposed for several cereals by selecting callus or seedlings which grow in the presence of lysine-threonine. Resulting mutants may be overproducers of lysine, threonine, or methionine.

Additional index words: **Protein quality, Tissue culture, Embryo culture.**

THE grain of most crops is deficient in one or more of the amino acids; tryptophan, lysine, threonine, and methioine which are essential for human or monogastric animal nutrition (13). A continuing goal of geneticists and plant breeders is the identification of genes and the selection of varieties which supply increased levels of these amino acids. Well known examples of success are the single gene mutations in maize (*Zea mays* L.) and barley (*Hordeum vulgare* L.), which alter the quantity of specific endosperm proteins (12, 15). In higher plants (as in bacteria) lysine, threonine, and methionine are synthesized in a branched pathway from aspartate (Fig. 1) (7, 20). The first and third enzymes in this pathway, aspartokinase (EC 2.7.2.4) and homoserine dehydrogenase, (EC 1.1.1.3) respectively, are known to be regulated by endproduct feedback inhibition. Lysine or the combination of lysine and threonine are known to inhibit the activity of aspartokinase (4, 23, 24), while threonine inhibits the activity of homoserine dehydrogenase (3). Feedback inhibition, is a regulatory mechanism by which organisms, including plants, efficiently regulate the synthesis of cellular metabolites. Regulatory mutants, resistant to feedback inhibition or enzyme repression, have been well defined in prokaryotes (1, 21), in lower eukaryotes (17, 18, 19), and recently tryptophan and methionine overproducers have been isolated from plant cell cultures (5, 22). Feedback inhibition mutants result in overproduction of the pathway endproduct(s) due to altered regulatory sites which do not allow normal inhibition of the enzyme. In crop plants these mutants might provide a mechanism to increase the total synthesis of nutritionally limiting amino acids. Recent reports discuss additional literature, potential uses, and limitations of regulatory mutants in higher plants (14, 18).

In the present work we report *in vivo* regulation of the lysine, threonine, and methionine biosynthetic pathway in several crop plants and discuss this regulation and its feasibility as the basis of a selection system for potential overproducer mutants of these amino acids. We have incorporated brief materials and methods information into the figure legends.

RESULTS AND DISCUSSION

The initiation and maintenance of somatic callus cultures derived from excised mature embryos of maize single cross Oh51A × Os420 have been described previously (11). Figure 2-A shows the typical response of these cultures to 1 m*M* L-lysine, L-threonine, L-

[1] Contribution from Department of Agronomy and Plant Genetics, University of Minnesota, St. Paul, MN 55101. Paper no. 8616, Scientific Journal Series, Minnesota Agricultural Experiment Station. Received Aug. 1, 1974.

[2] Assistant and Associate Professor, respectively.

methionine, and L-homoserine. The growth rate of the maize callus cultures was stimulated by lysine; inhibited by threonine and methionine; and unaffected by homoserine. Although much reduced as compared to the lysine growth rate, the lysine-methionine combination had a growth rate similar to the control. The threonine-methionine combination inhibited the growth rate of the cultures but did not exceed the inhibition of threonine or methionine individually. In contrast to the lysine stimulation and the partial threonine inhibition the simultaneous addition of 1 m*M* lysine and 1 m*M* threonine to the cultures completely prevented callus growth.

One potential interpretation of this result was that lysine and threonine cooperatively inhibited aspartokinase which prevented the normal lysine, threonine, and methionine synthesis. Since the exogenous lysine and threonine in the culture medium provided these amino acids, growth inhibition could have resulted from the inability to synthesize methionine. If true, the addition of methionine to the lysine-threonine medium should allow growth of the cultures.

An additional test would be the replacement of methionine in the lysine-threonine culture medium with homoserine. Homoserine, a metabolic intermediate of methionine biosynthesis, can be converted into methionine in several enzymatic steps (Fig. 1). Figure 2-A shows the complete reversal of 1 m*M* lysine-1 m*M* threonine inhibition by 0.1 m*M* methionine or 0.4 m*M* homoserine. Although these observations are not direct evidence of a concerted lysine-threonine feedback inhibition on aspartokinase of maize, they are identical to *in vivo* responses observed with Common duckweed (*Lemna minor*, L.) (24), common marchantia (*Marchantia polymorpha*, L.) (8, 9), and rice (*Oryza sativa*, L.) callus (10).

An alternative explanation that the synergistic action of lysine-threonine was due to combined inhibition of aspartokinase and homoserine dehydrogenase can not be excluded. This is not supported, however, by the failure of the methionine-threonine combination to reverse the partial inhibition caused by threonine or methionine individually in maize callus (Fig

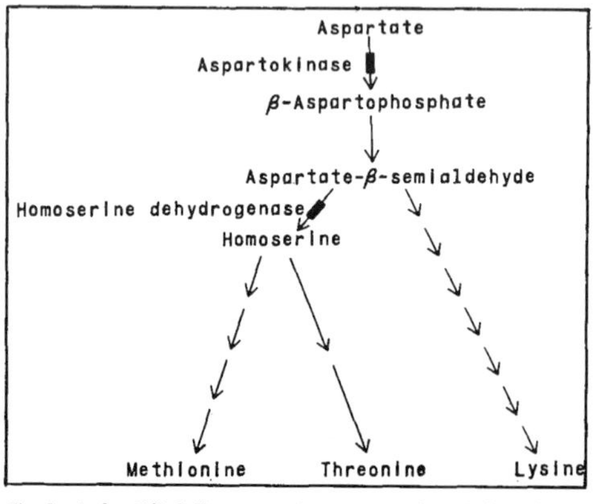

Fig. 1. A simplified diagrammatic representation of the L-lysine, L-threonine, and L-methionine biosynthetic pathway in higher plants.

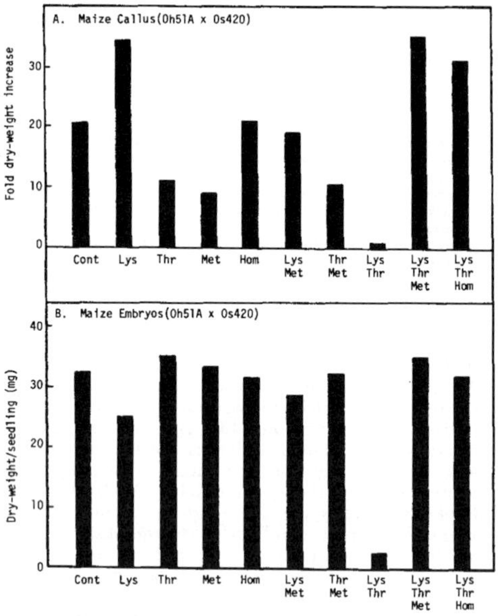

Fig. 2. Growth of Oh51A × Os420 maize callus (A) and embryos (B) in the presence of L-lysine (Lys), L-threonine (Thr), L-methionine (Met), and L-homoserine (Hom).

Method: Callus was grown for 30 days at 28C on modified Linsmaier and Skoog medium with 2 mg/liter 2,4-D (11). The embryos were sterilized and excised as previously described (11) and grown on callus medium without 2,4-D for 6 days at 28C. A 16 hour photoperiod with an intensity of 1200 lux was provided by fluorescent lights. Cont designates untreated controls. Amino acid concentrations were as follows: (A) 1mM for each in all treatments except the Lys-Thr-Met and Lys-Thr-Hom combinations which contained 0.1 mM Met and 0.4 mM Hom respectively; (B) 2.5 mM for each in all treatments except the Lys-Thr-Met and Lys-Thr-Hom combinations which contained 0.25 mM Met and 1.0 mM Hom, respectively. All material was dried at 70C for 48 hours for dry weight determination. Fold dry weight increase is the ratio of the final dry weight to the initial dry weight and was calculated from the mean of 16 replications/treatment.

2-A). In addition, as will be discussed later, methionine or threonine did not inhibit the growth of maize seedlings (Fig. 2-B).

This lysine-threonine inhibition has immediate application in the isolation of feedback resistant mutants in tissue cultures of crop plants. However, current cell culture limitations such as the inability to routinely differentiate plants from desired cultures in some crops may hamper mutant recovery and utilization. Logistically feasible selection procedures applicable to intact plants would provide an alternate recovery method for potential overproducers of specific metabolites, such as lysine, threonine, or methionine.

Embryos, excised from mature corn kernels, grew into vigorous plants when cultured on maize callus medium which lacked 2,4-dichlorophenoxyacetic acid (2,4-D). Figure 2-B shows the response of single cross Oh51A × Os420 embryos to 2.5 m*M* lysine, threonine, methionine, and homoserine. In contrast to the callus results, lysine slightly inhibited embryo growth. The other amino acid treatments permitted normal or near normal growth except the 2.5 m*M* lysine-2.5m*M* threonine combination which severely inhibited growth (Fig. 2-B). This inhibition was reversed by 0.25 m*M* methionine or 1.0 m*M* homoserine. It was evident that maize tissue cultures and embryos responded similarly to concerted lysine-threonine inhibition and to reversal by methionine or homoserine. Further work with inbred W23 embryos indicated that 1 m*M* lysine-1m*M* threonine concentrations were sufficient to cause concerted growth inhibition (Fig. 3-B) and that growth could be restored in inhibited embryos by transfer to rescue medium containing lysine, threonine, and methionine within 6 days. In addition, concentrations of methionine as low as 0.02 m*M* effectively reversed 1 m*M* lysine-1 m*M* threonine inhibition.

Embryos from other cereal crops were cultured to test the occurrence of this regulatory mechanism in other species. These embryos produced well developed plants on the maize callus medium (Fig. 3). Barley, wheat (*Triticum aestivum* L.), and rye (*Secale cereale* L.) embryos showed concerted inhibition when grown on culture medium with 1 m*M* lysine-1 m*M* threonine (Fig. 3-C_2, D_2, and E_2). This inhibition was reversed by incorporating 0.1 m*M* methionine (Fig. 3-C_3, D_3, and E_3) or 0.4 m*M* homoserine (Fig. 3-C_4, D_4, and E_4) into the lysine-threonine medium. The complete inhibition observed in sorghum (*Sorghum vulgare* Pers) and rice embryos at 1 m*M* lysine-1 m*M* threonine was only partially reversed by 0.1 m*M* methionine or 0.4 m*M* homoserine. For these embryos these concentrations of lysine and threonine may have caused secondary inhibitory effects because other methionine or homoserine concentrations did not improve the reversal. However, when the lysine-threonine concentrations were reduced to 0.5 m*M*, sorghum and rice embryos were still effectively inhibited (Fig. 3-F_2, G_2) and then were more fully reversed by 0.5 m*M* methionine (Fig. 3-F_3, G_3) or 0.2 m*M* homoserine (Fig. 3-F_4, G_4).

Oats (*Avena sativa* L.) was an exception to the concerted lysine-threonine inhibition observed in the other plant species. At 1 m*M* lysine-1 m*M* threonine, embryo growth was inhibited about 50% (Fig. 3-H_2) and reversed by 0.1 m*M* methionine and 0.4 m*M* homoserine (Fig. 3-H_3 and H_4). Inhibition significantly increased with 2.5 m*M* lysine-2.5 m*M* threonine but

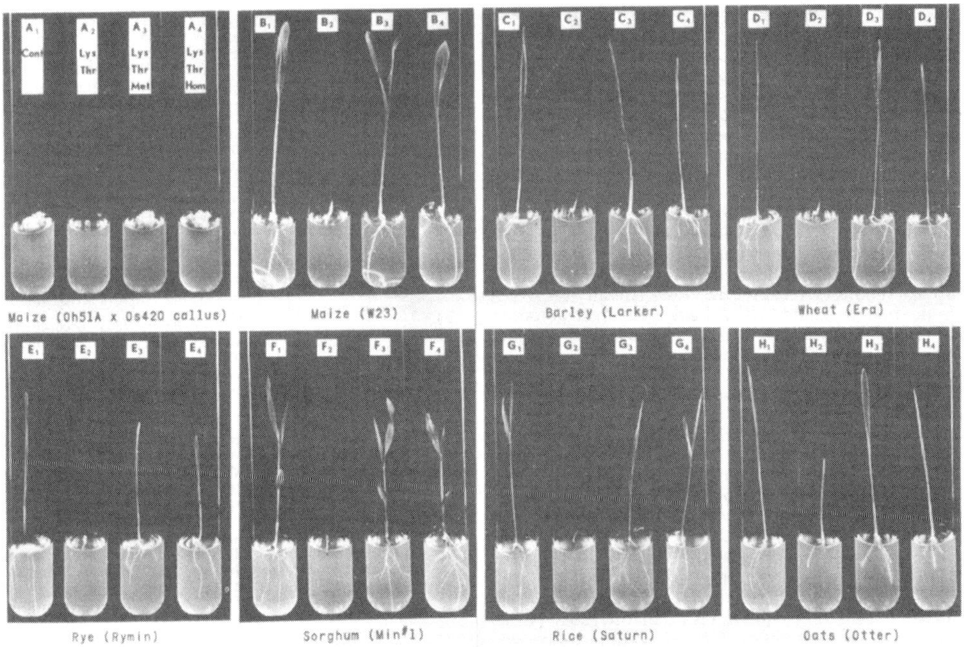

Fig. 3. Growth of embryos from various cereal crops in the presence of L-lysine (Lys), L-threonine (Thr), L-methionine (Met) and L-homoserine (Hom).

Method. Culture medium and procedures were as described in Fig. 1 except that embryos other than maize were sterilized by dusting dry kernels (oat and rice were dehulled first) with Captan (N-[(trichloromethyl)thio]-4-cyclohexene-1,2-dicarboxyimide). Embryos were removed and cultured after germination for 16 to 24 hours in sterile petri dishes in a 10 mg/liter chloramphenicol solution. Treatments are as follows: A_1-H_1 are untreated controls; A_2-E_2 and H_2 contain 1 mM Lys and 1 mM Thr; F_2 and G_2 contain 0.5 mM Lys and 0.5 mM Thr; A_3-E_3 and H_3 contain 1 mM Lys, 1 mM Thr, and 0.1 mM Met; F_3 and G_3 contain 0.5 mM Lys, 0.5 mM Thr, and 0.05 mM Met; A_4-E_4 and H_4 contain 1 mM Lys, 1 mM Thr, and 0.4 mM Hom; F_4 and G_4 contain 0.5 mM Lys, 0.5 mM Thr, and 0.2 mM Hom.

was only partially reversed by 0.1 to 1.0 m*M* methionine or 0.5 to 2.5 m*M* homoserine. These results have been observed in 11 oat varieties and may reflect a different or less sensitive feedback regulation in the lysine, threonine, and methionine biosynthetic pathway in oats. Nonetheless, the concerted lysine-threonine inhibition and its specific reversal in six of seven plant species examined was surprising based on the diversity of feedback regulations in this pathway among different species of bacteria (16).

Lysine-threonine inhibition provides a potential selection system in several crops for mutants which could overproduce one or more of the amino acids lysine, threonine, or methionine. Such mutants should have altered feedback inhibition controls which are less sensitive to lysine-threonine inhibition. The desired mutants whether isolated from tissue culture, cultured embryos, or germinating whole kernels would be detected among the normally inhibited popnlation by their growth in the presence of lysine and threonine. Thus, within a large population only those individuals which grow would be selected. This selection procedure can be easily applied to the large cell populations available in tissne cultures.

In addition, our analysis suggests that isolation of overproducer mutants in whole plauts by screening diverse or mutagenized populations should be more promising than previously reported (2). 1) As well as inhibiting maize callus and embryos lysine-threonine also inhibits root growth in germinating maize kernels. A whole kernel screening procedure would facilitate handling large populations. 2) Based on experience in lower eukaryotes and plant tissue culture, feedback inhibition resistant mutants should behave as dominants and therefore can be isolated in the first generation in diploids as heterozygotes (18). This would significantly reduce the time and population size required for mutant isolation. In addition, permease mutants, which would grow in this selection system because of impaired lysine and/or threonine uptake should be recessive and not generally recovered in the first generation in diploids (14). 3) A mutation at either the lysine or threonine binding site would result in resistance to lysine-threonine inhibition because they individually do not cause significant inhibition in seedlings. Therefore, two available sites should improve the mutation frequency, but the presence of secondary feedback inhibition sites may require selection of double mutants. 4) Mutation frequency has been shown to be sufficiently high in ethyl methauesulfonate-treated maize kernels to permit recovery of significant numbers of enzymatic point mutations (6).

On the other hand, there are certain questions that are currently unresolved. For example, will feedback resistance mutants derived from this selection system in callus cells or seedlings result in increased lysine, threonine, or methionine in the grain? Also, will these mutants allow healthy plant development and will sufficient essential amino acids accumulate to be nutritionally significant? The resolution of these and other questions must await the isolation and analysis of feedback inhibition resistance mutants. Work is currently in progress to select lysine-threonine-resistance mutants in cell cultures and seedlings and to isolate and study the regulatory properties of key enzymes in this pathway, like aspartokinase and homoserine dehydrogenase.

REFERENCES

1. Beckwith, J., and D. Zipser. 1970. The lactose operon. Cold Spring Harbor Laboratory, New York. 437 p.
2. Brock, R. D., E. A. Friederich, and J. Langridge. 1972. The modification of amino acid composition of higher plants by mutation and selection. p. 329-338. *In* Proceedings of the FAO/IAEA/GSF meeting on nuclear techniques for seed protein improvement. IAEA, Vienna.
3. Bryan, J. K. 1968. Studies on the catalytic and regnlatory properties of homoserine dehydrogenase of *Zea mays* roots. Biochim. Biophys. Acta. 171:205-216.
4. Bryan, P. A., R. D. Cawley, C. E. Brunner, and J. K. Bryan. 1970. Isolation and characterization of a lysine-sensitive aspartokinase from a multicellular plant. Biochem. Biophys. Res. Comm. 41:1211-1217.
5. Carlson, P. S. 1973. Methionine snlfoximine-resistant mutants of tobacco. Science 180:1366-1368.
6. Chonrey, P. S., and D. Schwartz. 1971. Ethyl methane snlfonate-induced mutations of the Sh_1 proteiu in maize. Mutation Res. 12:151-157.
7. Dougall, D. K., and M. M. Fulton. 1967. Biosynthesis of protein amino acids in plant tissue cultures. IV. Isotope competition experiments using glucose-U-14C and potential intermediates. Plant Physiol. 42:941-945.
8. Dunham, V. L., and J. K. Bryan. 1969. Synergistic effects of metabolically related amino acids on the growth of a mnlticellular plant. Plant Physiol. 44:1601-1608.
9. ———, and ———. 1971. Synergistic effects of metabolically related amino acids on the growth of a multicellular plant. II. Studies of ^{14}C-amino acid incorporation. Plant Physiol. 47:91-97.
10. Furuhashi, K., and M. Yatazawa. 1970. Methionine-lysine threonine-isoleucine interrelationships in the amino acid nutrition of rice callns tissue. Plant & Cell Physiol. 11: 569-578.
11. Green, C. E., R. L. Phillips, and R. A. Kleese. 1974. Tissue enltures of maize (*Zea mays* L.): Initiation, maintenance, and organic growth factors. Crop Sci. 14:54-58.
12. Mnnek, L. 1970. Increasing the nutritional valne in cereal protein; Basic research on the hily character. p. 319-330. *In* Proceedings of the FAO/IAEA meeting on improving plant protein by nuclear techniques. IAEA, Vienna.
13. Nelson, O. E. 1969. Genetic modification of protein quality in plants. Advan. Agron. 21:171-194.
14. ———. 1973. Breeding for specific amino acids. p. 303-311. *In* A. M. Srb (ed.) Genes, enzymes, and populations. Plenum Press, New York.
15. ———, E. T. Mertz, and L. S. Bates. 1965. Second mutant gene affecting the amino acid pattern of maize endosperm proteins. Science 150:1469-1470.
16. Paulus, H., and E. Gray. 1967. Multivalent feedback inhibition of aspartokinase iu *Bacillus polymyxa*. J. Biol. Chem. 242:4980-4986.
17. Polacco, J., and S. R. Gross. 1973. Genetic regulatory mechanism in the fungi. p. 71-83. *In* A. M. Srb (ed.) Genes, enzymes, and populations. Plenum Press, New York.
18. Rasse-Messenguy, F., and G. R. Fink. 1973. Feedback-resistant mntants in histidine biosynthesis in yeast. p. 85-95. *In* A. M. Srb (ed.) Geues, enzymes, and populations. Plenum Press, New York.
19. Sloger, M., and L. D. Owens. 1974. Control of free methionine production in wild type and ethionine-resistant mntants of *Chlorella sorokiniana*. Plant Physiol. 53:469-473.
20. Umbarger, E., and B. D. Davis. 1962. Pathway of amino acid biosynthesis. p. 167-251. *In* I. C. Gunsalus and R. Y. Stanier (eds.) The bacteria, Vol. III. Academic Press Iuc., New York.
21. ———. 1968. Regulation of amino acid metabolism. Annu. Rev. Biochem. 38:323-370.
22. Widholm, J. M. 1972. Anthanilate synthetase from 5-methyltryptophan-susceptible and resistant cultnred *Daucus carota* cells. Biochim. Biophys. Acta 279:48-57.
23. Wong, K. F., and D. T. Dennis. 1973. Aspartokinase from wheat germ. Isolation, characterization, and regulation. Plant Physiol. 51:322-326.
24. ———, and ———. 1973. Aspartokinase iu *Lemna minor* L. Plant Physiol. 51:327-331.

Plant Regeneration from Tissue Cultures of Maize[1]

C. E. Green and R. L. Phillips[2]

ABSTRACT

Effective utilization of cell and tissue culture methods in *Zea mays* research requires cultures capable of plant regeneration. These differentiated plants would provide a direct link with conventional genetic and breeding procedures.

Maize callus from embryo scutellar tissues was initiated and maintained on MS medium inorganic components, Straus medium vitamins and amino acids, 20 g sucrose and 8 g agar per liter, and 2 mg 2,4-dichlorophenoxyacetic acid (2,4-D)/liter. Callus has been maintained by subculture every 21 to 28 days and has remained capable of differentiation for 9 months. Regeneration of complete plants was accomplished by subculture of callus to 0.25 mg 2,4-D/liter for 30 days followed by transfer to 2,4-D-free culture medium. At 0.25 mg 2,4-D/liter numerous curled and wrinkled leaves developed. Approximately 200 complete plants have been differentiated. after transfer to the 2,4-D-free medium. Root tip cells from five plants indicated that each had 20 chromosomes. After transplantation to soil, 10 to 15% of the plants survived and grew normally. The optimum embryo age for scutellar callus initiation was 18 days post-pollination. Hormone combinations such as 1 mg 2,4-D, 4 mg α-naphthaleneacetic acid (NAA), and 0.05 mg 6-(γ,γ-dimethyl allylamino)-purine (2iP)/liter may increase the efficiency of scutellar callus initiation.

***Additional index words:* *Zea mays* L., Embryo culture, Scutellum, Hormones, Differentiation.**

THE ability to develop tissue and cell cultures from cereal plants has increased rapidly in recent years. Examples include: somatic callus initiation and plant regeneration in sorghum [*Sorghum bicolor* (L.) Moench] (11), wheat (*Triticum monococcum* L.) (3, 17), rice (*Oryza sativa* L.) (21), and sugarcane (*Saccharum officinarcm* L.) (7); anther culture and haploid plant formation in rice (14), wheat (*Triticum aestivum* L.) (15, 16), and barley (*Hordeum vulgare* L.) (1); and protoplast formation and culture in sugarcane (9).

In maize (*Zea mays* L.), endosperm cultures have been available for many years (20). They have been used for biochemical studies (5, 18, 22) and for protoplast formation and culture (12). More recently, maize somatic callus cultures have been initiated from seedling shoot sections (10), immature inflourescences (8), and mature embryos (6). Plants were not regenerated in these somatic or endosperm callus cultures.

Current cell culture methods have fostered new genetical and biochemical techniques in plants. These techniques coupled with the well-documented genetics and physiology of maize could offer many research opportunities as efficient tissue and cell culture procedures continue to be developed. One of the most critical needs has been to establish callus cultures from which complete plants could be regenerated. This would provide a direct link between cell culture research and conventional genetic and breeding procedures. This paper describes parameters for initiating and maintaining callus cultures from immature maize embryo scutellum and the subsequent regeneration of complete plants.

MATERIALS AND METHODS

Primary genotypes used in this study included the field-corn inbred A118 (*r—r*; colorless aleurone, embryo scutellum and plumule, and dilute sun red plant color) and the homozygous R-navajo (*R—nj;* pigmented aleurone crown, embryo scutellum and plumule, and seedling) stocks. Field-corn inbreds A619, A632, B9A, and W64A were obtained from J. Geadelmann, Dep. of Agron. and Plant Genet., U. of Minn.

Ears were removed from field or greenhouse-grown plants at 14 through 24 days after pollination. Within 30 minutes after removal from the plant, they were broken into 5 to 8 cm segments. They were then sterilized for 20 min by submerging in a solution of 1 g Al lab detergent (A & L Laboratories, Minneapolis, Minn.)/100 ml 2.5% sodium hypochlorite. The ear segments were then transferred through three sterile deionized-distilled water rinses. The isolation of embryos was usually begun immediately; however, sterile ears could be stored without apparent damage to the embryos for at least 12 hours at 4 C.

Immature embryos were isolated from the ear by cutting off the kernel crown, removing the endosperm with a narrow spatula, and then transferring the embryo, located at the kernel base, onto culture medium. Embryos were placed on the solid culture medium with the rounded scutellar side exposed and the flat plumule-radicle axis side in contact with the medium.

Callus was initiated and maintained on MS culture medium which contained the inorganic components of Murashige and Skoog medium (13); 7.7 mg glycine/liter, 1.98 g L-asparagine/liter, 1.3 mg niacin/liter, 0.25 mg thiamine-HCl/liter, 0.25 mg pyridoxine-HCl/liter, and 0.25 mg Ca pantothenate/liter from Straus medium (4, 19); 20 g sucrose/liter; and 8 g agar/liter. Varying concentrations of the plant hormones, 2,4-dichlorophenoxyacetic acid (2,4-D), p-chlorophenoxyacetic acid (p-CPA), α-naphthaleneacetic acid (NAA), and 6-(γ,γ-dimethylallylamino)-purine (2iP) were used. The medium was adjusted to pH 6.0 with 0.4 *N* NaOH and autoclaved for 20 min at 15 psi.

The embryos and subsequent callus cultures were incubated at 28 to 30 C with a 16/8 hour photoperiod from cool-white fluorescent lights with an intensity of 2,000 lux. Callus cultures were maintained by subculturing every 21 to 28 days to MS containing 2 mg 2,4-D/liter.

Root tips from differentiated plants were prefixed at 4 C in saturated 8-hydroxyquinoline for 4 hours, transferred to fresh Farmer's solution (3 parts 95% ethanol:1 part glacial acetic acid), fixed for 24 hours at room temperature, and stored at —10 C in 70% ethanol. Chromosome counts were made on metaphase cells from squashed root tips stained with propionocarmine and observed by phase-contrast microscopy.

RESULTS AND DISCUSSION

Culture Initiation

Initial success in establishing totipotent callus cultures from maize was achieved by aseptically isolating 20-day post-pollination embryos from the cross A118 × *R-njR-nj*. With the plumule-radical axis side placed in contact with MS medium containing 1,

[1] Contribution from Dep. of Agron. and Plant Genet., U. of Minn., St. Paul, MN 55108. Paper no. 8960, scientific journal series, Minn. Agric. Exp. Stn. Received Jan 6. 1975.

[2] Assistant and associate professors, respectively.

2, 4, or 6 mg 2,4-D/liter, embryos were incubated at 28 to 30 C. This orientation allowed germination, but suppressed subsequent embryo growth and promoted callus formation from the scutellum. In comparison, when the rounded, scutellar embryo surface was placed in contact with the medium and the plumule-radical axis was oriented upwards, rapid germination and extensive shoot and root growth occurred, even in the presence of 4 to 6 mg 2,4-D/liter. No scutellar callus was formed, although slow growing compact yellow callus frequently developed from roots and/or the first node of the shoot after 4 weeks incubation. These calli were generally similar to the somatic cultures previously reported (6), which were easily maintained and differentiated numerous roots, but not shoots.

Culture initiation began with enlargement of the entire exposed scutellar surface, which resulted in a 6 to 8 mm diameter dome-shaped scutellum within a few days. After 10 days, the scutellum surface had developed a bumpy and irregular shape and by 14 days, nodular white or pale-yellow callus was visible in localized regions. These primary cultures developed 1 to 2 cm callus masses after 30 days, which frequently exhibited early organizational events; such as localized chlorophyll development and the formation of occassional light-green 1 to 2 mm long leaves from the chlorophyll regions.

Table 1. Ability of immature embryos from two maize lines to form scutellar callus cultures capable of plant regeneration.

Genotype	Source	Embryos isolated	Callus clones at 2	Clones differentiating leaves at 0.25
			— mg 2,4-D/liter —	
A188 × R-njR-nj	Field, 1973	7*	5	3 (3)‡
	Field, 1974	220†	174	77 (8)
A188	Greenhouse, 1974	106†	91	29 (15)
	Field, 1974	260†	141	55 (5)

* 20 day post-pollination embryos. † 18 day post-pollination embryos. ‡ ( ) represents the number of totipotent callus clones tested on MS medium with 0 mg 2,4-D/liter for plant differentiation. All tested clones differentiated complete plants.

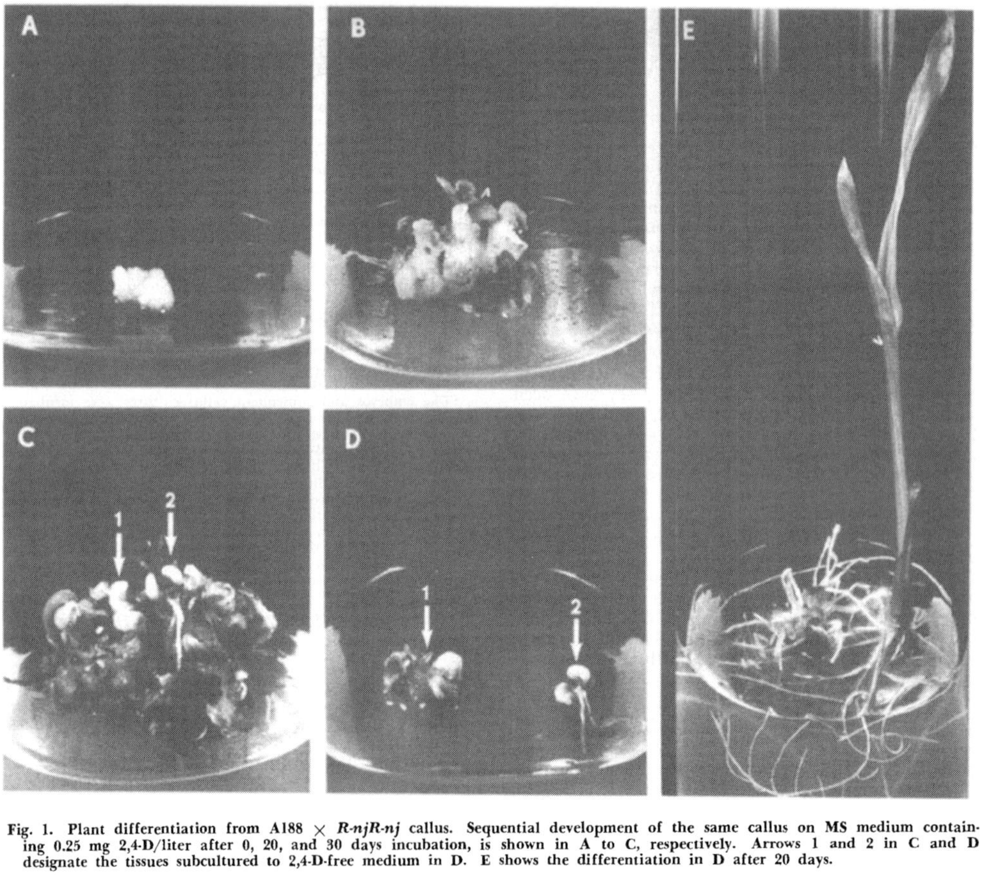

Fig. 1. Plant differentiation from A188 × *R-njR-nj* callus. Sequential development of the same callus on MS medium containing 0.25 mg 2,4-D/liter after 0, 20, and 30 days incubation, is shown in A to C, respectively. Arrows 1 and 2 in C and D designate the tissues subcultured to 2,4-D-free medium in D. E shows the differentiation in D after 20 days.

Totipotent callus cultures were most successfully initiated from the scutellum on MS medium containing 2 mg 2,4-D/liter. Medium containing 4 mg 2,4-D/liter was less effective. MS containing 1 mg 2,4-D/liter allowed increased embryo shoot and root growth and limited callus formation from the scutellum. Embryo shoot and root growth was minimal and the scutellum produced slow growing callus on medium containing 6 mg 2,4-D/liter. No plants were differentiated from these cultures. During subculture, the callus was easily separated from scutellar tissue, which had not grown, and the remainder of the embryo. The original A188 × *R-njR-nj* cultures have been maintained for 19 months on MS medium with 2 mg 2,4-D/liter by subculturing every 21 to 28 days. Localized areas of chlorophyll and small leaves have continued to form, especially near the end of each culture period. This organization was apparently associated with the gradual depletion of 2,4-D in the culture medium.

Totipotent scutellar callus has been initiated from four independent plant populations. These were A188 × *R-njR-nj* crosses and sibbed or selfed A188 (Table 1). Among the populations, callus initiation on MS medium containing 2 mg 2,4-D/liter ranged from 54 to 85% of the embryos tested. These callus clones were maintained by subculture and their totipotency tested by subsequent subculture to MS medium containing 0.25 mg 2,4-D/liter. After 30 days incubation, the clones differentiating leaves varied from 21 to 43% of the isolated embryos.

The plumule and radicle tissues of embryos had a characteristic fate during scutellar callus initiation which was related to embryo age and development at isolation. The shoot and root primordia in A188 × *R-njR-nj* or A188 embryos isolated from ears after 20 days development formed 2 to 3 cm plumules and 0.5 cm radicles during the initial 10 days of incubation on MS medium containing 2 mg 2,4-D/liter. These tissues did not elongate further during subsequent incubation and were usually necrotic after 30 days. No plumule or radicle elongation occurred during incubation of 14-day embryos while 5 to 8 cm plumules and 0.5 cm radicles developed during incubation of 24-day embryos.

Differentiation

The differentiation and growth of plants from A188 × *R-njR-nj* scutellar callus is illustrated in Fig. 1. The visibly unorganized 0.5 cm diameter callus in Fig. 1-A was obtained from MS medium containing 2 mg 2,4-D/liter and transferred to MS containing 0.25 mg 2,4-D/liter. By 10 days, the rapid growth of this callus was accompanied by localized chlorophyll development and the formation of small, 1 to 2 mm leaves. The growth and development at 20 days (Fig. 1-B) included many light-green leaves and white compact structures that resembled the organized scutellum of the original embryos. By 30 days, many curled and wrinkled leaves, additional white scutellar-like structures (Fig. 1-C, arrows 1 and 2) and short roots had formed.

The differentiation of complete seedlings was accomplished by transferring cultures with many small leaf structures from MS containing 0.25 mg 2,4-D/liter to 2,4-D-free medium (Fig. 2-B). In order to closely observe seedling development, the organized culture in Fig. 1-C was subdivided into 12 small fragments which were inoculated onto 2,4-D-free MS medium. Two of these (Fig. 1-D, arrows 1 and 2) cor-

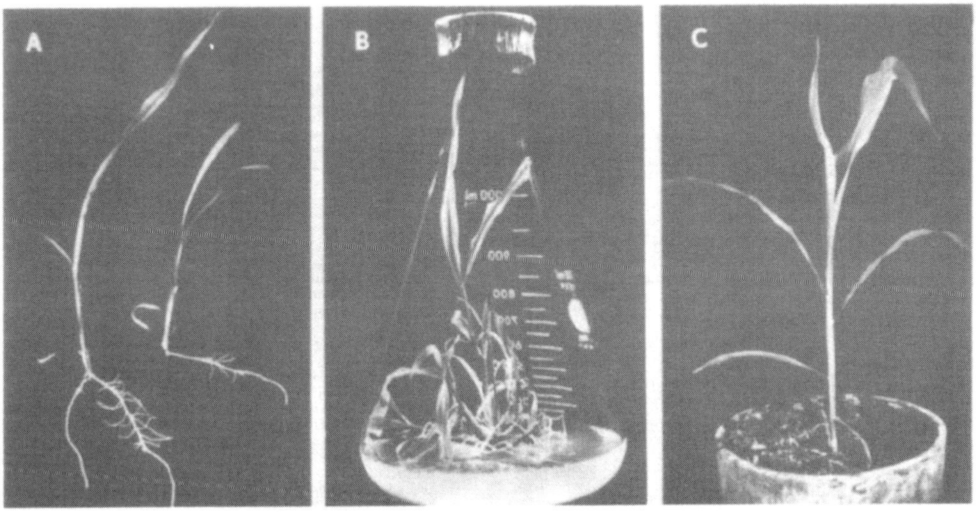

Fig. 2. Differentiated A188 × *R-njR-nj* plants. A typical root-shoot development; B, multiple plant differentiation at 0 mg 2,4-D/liter; C, five-leaf-stage differentiated plant established in soil.

Table 2. Influence of embryo age and genotype on scutellar callus formation.

Inbred	Post-pollination embryo age, days 14	16	18	20	22	24
	— embryos forming differentiating cultures*, % —					
A188	18	15	21	14	9	0
A619	0	5	6	5	0	0
A632	0	0	0	0	0	0
B9A†	0	5	8	0	0	0
W64A	0	0	11	0	0	0

* Each value represents the percent of 60 embryos. † Subsequently, these B9A cultures died during the fourth month of growth (fourth subculture).

respond to the tissues indicated by the arrows in Fig. 1-C. The plant shown in Fig. 1-E was one of seven which differentiated during 20 days incubation. This plant emerged between the lobes of the white scutellar-like tissue (Fig. 1-D, arrow 2) and had attached roots. Typical shoot-root associations and development in the differentiated plants is shown in Fig. 2-A. To date, approximately 200 seedlings have been differentiated from A188 × *R-njR-nj* callus, and the original cultures remain totipotent after 19 months. Cultures, which did not form plants on 2,4-D-free medium, frequently developed extensive roots as shown by the tissues on the left in Fig. 1-E. Recently, these results have been independently confirmed by the initiation of totipotent cultures and subsequent plant differentiation from the maize line Alhexo Single Kernel Cycle IV in another laboratory (2). In addition, primary cultures obtained from inbred W64A scutellum also have been observed to regenerate maize plants. These cultures, however, were not successfully maintained by subculture (23).

When two-leaf-stage differentiated A188 × *R-njR-nj* or A188 plants were transplanted to steam sterilized soil 10 to 15% of the plants became established and grew (Fig. 2-C). This limited success was apparently due to inadequate root development in the culture medium. Roots were frequently thin and total root mass was small in relation to shoot tissues. Differentiated plants grown in soil had wide leaves, ears, and tassels. Pollen examined from one plant was 90 to 95% fertile. Although many abnormal leaf shapes were observed during plant differentiation, no albino, straited, or chimeral leaves or plants were observed. Chromosome counts in root tip cells from five differentiated plants indicated that each had 20 chromosomes. Meiotic tissue has not yet been examined cytologically nor have crosses been attempted to recover progeny from differentiated plants.

Preliminary results indicated that p-CPA at 2 and 4 mg/liter was less effective than 2,4-D for scutellar callus initiation from 18-day A188 × *R-njR-nj* embryos. Although callus formation with p-CPA was less prolific, it became more highly organized with numerous small leaves present after 30 days. Plant differtiation from these cultivars was vigorous on MS medium with 0 mg p-CPA/liter. When NAA and 2iP were combined at 2, 4, or 8 mg/liter and 0.05 mg/liter respectively, scutellar callus initiation was very limited. Promising results were obtained, however, from the hormone combination 1 mg 2,4-D/liter, 4 mg NAA/liter, and 0.05 mg 2iP/liter wherein 25 of 32 embryos developed callus which has been maintained through three subcultures to date. Leaves developed rapidly in 18 of these callus clones when subcultured to MS containing 1 mg NAA/liter and 0.05 mg 2iP/liter. These preliminary results suggest that differentiating cultures can be obtained with hormones other than 2,4-D and that hormone combinations may increase the efficiency of culture initiation.

Influence of Embryo Age and Genotype

The influence of embryo age on the formation of callus and plant differentiation was studied by isolating embryos from the field grown inbreds; A188, A619, A632, B9A, and W64A (Table 2), which had been sib or self-pollinated. Embryos were isolated from ears harvested at 2-day intervals from 14 through 24 days after pollination. Although embryo size varied slightly between genotypes, typical embryo lengths were as follows: 1 mm at 14 days, 3 mm at 18 days, and 5 mm at 22 days. Sixty embryos from each genotypes at each age were explanted onto MS medium with 2 mg 2,4-D/liter and incubated for 30 days. Scutellar calli were subcultured to MS medium with 2 mg 2,4-D/liter for 21 days and then to medium with 0.25 mg 2,4-D/liter to detect cultures capable of plant differentiation. Variability existed between genotypes in ability to initiate differentiating cultures. For example, A188 formed differentiating cultures from embryos at all ages except 24 days, while none were recovered from A632. However, the optimal embryo age for totipotent culture initiation was 18 days for A188, A619, B9A, and W64A. Maize embryos are approaching maturity in embryological development and size at 22 days after fertilization. The poor response of older embryos may reflect a requirement for rapidly growing scutellar tissue for culture initiation.

REFERENCES

1. Clapman, D. 1973. Haploid *Hordeum* plants from anthers *in vitro*. Z. Pflanzenzuecht. 69:142-155.
2. Freeling, M. Personal communication.
3. Gamborg, O. L., and D. E. Eveleigh. 1968. Culture methods and detection of glucanases in suspension cultures of wheat and barley. Can. J. Biochem. 46:417-421.
4. Graebe, J. E., and G. D. Novelli. 1966. A practical method for large scale plant tissue culture. Exp. Cell Res. 41:509-520.
5. ———, and ———. 1966. Amino acid incorporation in a cell-free system submerged tissue cultures of *Zea mays* L. Exp. Cell Res. 41:521-534.
6. Green, C. E., R. L. Phillips, and R. A. Kleese. 1974. Tissue cultures of maize (*Zea mays* L.): initiation, maintenance, and organic growth factors. Crop Sci. 14:54-58.
7. Heinz, D. J., and G. W. P. Mee. 1969. Plant differentiation from callus tissue of *Saccharum* species. Crop Sci. 9:346-348.
8. Linsmaier-Bedner, E. M., and T. W. Bedner. 1972. Light and hormonal control of root formation in *Zea mays* callus cultures. Dev. Growth & Differ. 14:165-174.
9. Maretzki, A., and L. G. Nickell. 1973. Formation of protoplasts from sugarcane cell suspensions and the regeneration of cell cultures from protoplasts. p. 51-56. *In* Protoplasts et fusion de cellules somatique vegetales. Collog. Int. C. N. R. S. No. 212 Inst. Natl. Res. Agron. (Paris).
10. Mascarenhas, A. F., B. M. Sayagover, and V. Jagannathan. 1965. Studies on the growth of callus cultures of *Zea mays* L. p. 283-291. *In* C. V. Ramakrishnan (ed.) Tissue culture. Dr. W. Junk, Publisher The Hague.
11. Masteller, V. J., and D. J. Holden. 1970. The growth of and organ formation from callus tissue of sorghum. Plant Physiol. 45:362-364.

12. Motoyashi, F. 1971. Protoplasts isolated from callus cells of maize endosperm. Exp. Cell Res. 68:452-456.
13. Murashige, T., and F. Skoog. 1962. A revised medium for rapid growth and bioassays with tobacco tissue cultures. Physiol. Plant. 15:473-497.
14. Niizeki, H., and K. Oono. 1968. Induction of haploid rice plant from anther culture. Japan. Acad. Proc. 44:554-557.
15. Ouyang, T. -w, H. Hu, C. -c. Chuang, and C. -c. Tseng. 1973. Induction of pollen plants from anthers of *Triticum aestivum* L. cultured *in vitro*. Sci. Sin. 16:79-95.
16. Picard, E., and J. deBuyser. 1973. Obtaining haploid plants of *Triticum aestivum* L. by the culture of anthers *in vitro*. C. R. Acad. Sci. Ser. D. 277:1463-1466.
17. Shimada, T., T. Sasakuma, and K. Tsunewaki. 1969. *In vitro* culture of wheat tissues. I. Callus formation, organ redifferentiation and single cell culture. Can. J. Genet. Cytol. 11:294-304.
18. Straus, J. 1958. Spontaneous changes in corn endosperm tissue cultures. Science 128:537-538.
19. ———. 1960. Maize endosperm tissue grown *in vitro*. III. Development of a synthetic medium. Am. J. Bot. 47:641-647.
20. Straus, J., and C. D. LaRue. 1954. Maize endosperm tissue grown *in vitro*. I. Cultural requirements. Am. J. Bot. 41: 687-694.
21. Tamura, S. 1968. Shoot formation in calli orignated from rice embryo. Japan. Acad., Proc. 44:544-548.
22. Tsai, S. 1970. Carbohydrates and amylase enzymes of maize endosperm tissue cultures. Ph.D. thesis., Pa. State U. U. Microfilms, Ann Arbor, Mich. (diss. abstr. 71-16:680).
23. ——— and R. Creech. Personal communication.

Proc. Natl. Acad. Sci. USA
Vol. 82, pp. 7010–7014, October 1985
Genetics

DNA amplification patterns in maize endosperm nuclei during kernel development

(chromatin/polyploidy/polytenization/replication/cytophotometry)

RICHARD V. KOWLES* AND RONALD L. PHILLIPS

Department of Agronomy and Plant Genetics, University of Minnesota, St. Paul, MN 55108

Communicated by John D. Axtell, July 1, 1985

ABSTRACT Increased DNA levels in centrally located endosperm nuclei are shown to be related to endosperm development in *Zea mays*. Mitotic activity sharply decreases in endosperm cells 10–12 days after pollination. At this time nuclear size and DNA content per nucleus (where C = haploid content) sharply increase until peak levels are reached at about 14–18 days after pollination. Mean DNA content per endosperm nucleus in strain A188 was shown by Feulgen cytophotometry to increase to about 90C by this peak stage, with the pattern being remarkably consistent over four consecutive growing seasons. Some individual nuclei achieved levels of >200C. Most other strains compared during one growing season averaged even higher peak levels of DNA per nucleus than did A188. Individual nuclei in those strains reached levels as high as 690C. A decrease in DNA level was observed in older endosperms with most strains. Endosperm mutant strains did not show a significant reduction in DNA. Opaque-2 mutants in several backgrounds achieved higher levels of DNA per nucleus. DNA levels from F_1 endosperms did not indicate heterosis. Regardless of differences in DNA content, the pattern of DNA increasing as development proceeds followed by a DNA decrease was observed for most strains. Cytological studies reveal much variation in chromatin strandedness, a maximum of three nucleoli, a maximum of three nucleolar organizer regions, and ≈30 diffuse chromatin masses in older endosperm tissue. A form of DNA amplification, perhaps polytenization, appears to be occurring during endosperm development.

The developmental biology of maize (*Zea mays* L.) is important because of the prominence of maize in plant genetics research and its importance as a major agricultural food crop. Understanding the development of the endosperm of maize becomes especially important because this tissue makes up 85–90% of the mature kernel dry weight. Endosperm tissue obviously serves a critical role in overall kernel development.

The initial ploidy level of the endosperm is 3X, because two polar nuclei of the central cell of the embryo sac fuse with one of the two sperm nuclei of the pollen grain. Within several hours, the initial triploid nucleus undergoes rapid and synchronous divisions that ensue for several days. Both nuclear divisions and cytokineses eventually cease in the more central regions of the tissue, whereas these activities persist for the longest time only in the peripheral endosperm regions. The outermost layer of cells, the aleurone, cytologically behaves like a meristem providing additional cells to the interior region of the kernel. The nuclei of the nonperipheral cells continue to increase in size during the early period of endosperm development. Aside from the morphological aspects of kernel development reported by Lampe (1), Randolph (2), and Kiesselbach (3), the genetic and molecular behaviors of maize endosperm during the process of development have not been fully described. Nuclei of the central endosperm regions have been characterized with regard to nuclear size, chromosome knobs, nucleoli, and chromatin strandedness (4). DNA increases were suggested based primarily upon observed increases in nuclear size. Similar observations have been made for maize endosperm cells grown under *in vitro* conditions (5). Reports based only upon nuclear volume indicated increased DNA contents ranging from 6C to 384C (6, 7). One report based upon Feulgen cytophotometry showed maximum DNA levels of 24C in kernels of young ears (8). None of these investigations quantified DNA levels in relation to endosperm development.

MATERIALS AND METHODS

Stocks and Sample Collections. Inbred A188 plants were grown during the 1980–1983 growing seasons at the University of Minnesota, St. Paul. Other strains were grown during the 1983 season. Kernels were collected at 1- to 3-day intervals following self- or cross-pollinations and were immediately fixed in 95% ethanol/glacial acetic acid, 3:1.

Cytological Observations. Squashes were made with propionic carmine stain. Mean nuclear volume was calculated from measurements made with a Zeiss digital image analyzer system (MOP-3). Mitotic index was calculated as the fraction of the total number of cells in mitosis.

Autoradiography. Three kernels from ears at 8, 12, or 16 days after pollination (ap) were placed in 1.5 ml of sterile modified kernel medium (9). After moderate shaking at 28°C for 2 hr, 20 μl of [^{3}H]thymidine was added at a concentration of 1 mCi/ml (specific activity, 80.1 Ci/mmol; 1 Ci = 37 GBq). Incubation was continued for 16–20 hr on a shaker at 28°C. Kernels were fixed in 95% ethanol/glacial acetic acid, 3:1. Centrally located endosperm tissue was squashed on slides in a drop of 45% acetic acid. Cover glasses were removed by the dry ice method. The preparations were then subjected to a 70%, 50%, and 30% ethanol series and 0.3 M NaCl/30 mM sodium citrate, pH 7.5, for 5 min each. Slides were dipped into Kodak NTB-2 liquid emulsion diluted 2:1 with distilled water at 42°C, dried for 2 hr, placed into an opaque slide box with drierite, and exposed at 4°C for 7 days. Development was accomplished with D-19 developer for 1 min, stop bath for 30 sec, Kodak fixer for 2 min, and washing for 5 min. The slides were placed into 5% Giemsa for 6 min, briefly rinsed in 0.01 M phosphate buffer (pH 6.9), air-dried, and mounted in Euparal.

Cytophotometry. Slides were prepared with endosperm tissue fixed in 95% ethanol/glacial acetic acid, 3:1. Tissue was taken from only the central endosperm regions for those

The publication costs of this article were defrayed in part by page charge payment. This article must therefore be hereby marked "*advertisement*" in accordance with 18 U.S.C. §1734 solely to indicate this fact.

Abbreviation: ap, after pollination.

*Present address: Biology Department, St. Mary's College, Winona, MN 55987.

samples large enough to facilitate such tissue removal—that is, 10 days ap and later. This tissue was placed in a drop of 45% acetic acid and squashed on slides that also contained fixed *Gallus* (chicken) erythrocytes as an internal reference (10). Cover glasses were removed by the dry ice method, which also removed much of the starch from around the nuclei. Slides were immediately subjected to the Feulgen reaction series and made permanent according to the following schedule: 5 min each in 70%, 50%, and 30% ethanol and distilled water; hydrolysis in 5 M HCl for 20 min at room temperature; cold distilled water for 5 min; Schiff's reagent for 1 hr; bleach (distilled water/1 M HCl/10% potassium metabisulfite, 18:1:1) for 10 min; and 5 min each in distilled water, 30%, 50%, 70%, 95%, and 100% ethanol. Slides were then mounted in Euparal.

A two-wavelength microspectrophotometric method was employed (11, 12) with a Zeiss Universal microscope with an MP 101 photometer. From spectral curves, 550 nm and 488 nm were selected as the two appropriate wavelengths. Percent transmission at each of the two wavelengths was obtained for both endosperm nuclei and chicken erythrocyte nuclei on the same slide and then was converted into arbitrary DNA units (13). A mean DNA content per chicken erythrocyte of 5.33 pg was determined with the use of an interference microscope at the U.S. Department of Agriculture Forest Service, North Central Forest Experiment Station (Rhinelander, Wisconsin). This value contrasts somewhat with other reports that vary between 2.5 and 3.97 pg (10; 14–16). Calculations of DNA (pg) per endosperm nucleus were based upon arbitrary DNA units measured for endosperm nuclei and chicken erythrocyte nuclei and the known amount of DNA per chicken erythrocyte.

These values were converted to DNA C levels (where C = haploid content) by dividing the pg of DNA per maize nucleus by 5.035 pg, the amount of DNA determined to be in a haploid maize nucleus (17). This DNA value was confirmed by measuring root-tip anaphase figures (2N) compared with the chicken erythrocytes.

RESULTS AND DISCUSSION

Rapid development of endosperm tissue occurs in the kernel following pollination (Fig. 1*A*). Particular note can be made of the changes in overall endosperm size that occur between 8 and 12 days ap; this is due to an increase in cell number and, to some extent, cell and nuclear size. Thereafter, increases in cell and nuclear size rather than cell number are the basis for changes in the central endosperm region. Microscopic studies of centrally located endosperm tissue reveal extremely large nuclei, especially later than 12 days ap. An indication of the extent to which these nuclei can develop is given in Fig. 1*B*.

The mitotic index for the central endosperm tissue peaked 8–10 days ap and then sharply decreased. Only 1 of 413 nuclei located in the central region was found to be undergoing mitosis 14 days ap or later. Nonmitotic endosperm nuclei of these later developmental stages were shown by [^{3}H]thymidine autoradiography to be actively synthesizing DNA. Silver grains were observed directly over nuclear chromatin at stages later than 12 days ap. Some of the [^{3}H]thymidine incorporation could be the result of DNA repair.

In A188, the strain most intensively studied, DNA level per nucleus and nuclear size surge 8–10 days ap and peak between 14 and 18 days ap. DNA content per nucleus reached a maximum mean level of 90–100C. Nuclear volume and DNA content means for each ap date are significantly correlated ($r = 0.92$; $P < 0.001$). DNA content of interphase nuclei measured before the rapid increase averaged between 4 and 5.6C. Some of these nuclei were presumably in the S or G_2 phase of the nuclear cycle, at least partially accounting

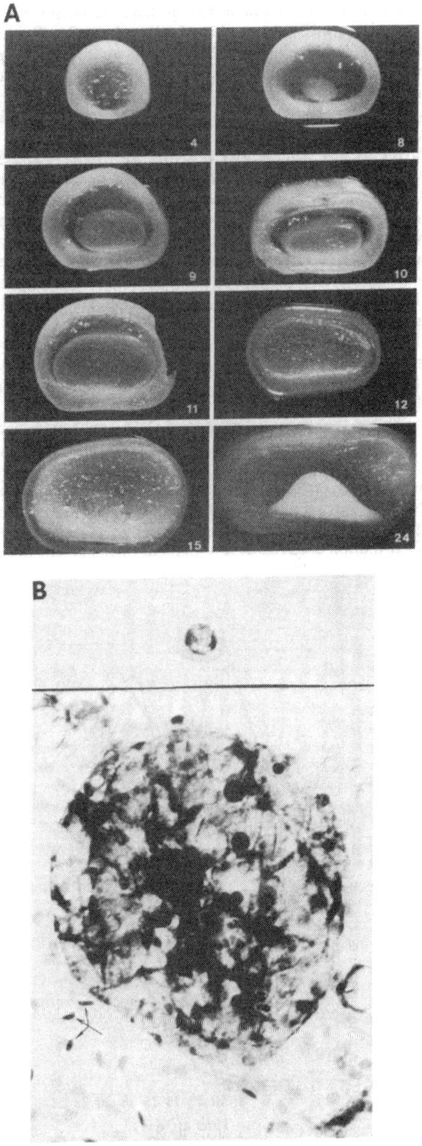

FIG. 1. (*A*) Development of the endosperm as illustrated by kernel cross sections. Four days ap, the endosperm is the small structure at the kernel base. The bulk of the kernel consists of nucellar tissue with surrounding carpels; 8 days ap, the endosperm is easily discerned as the body in the lower part of the kernel; 9–12 days ap, the endosperm has compressed the nucellus toward the outer edge of the kernel; 15 days ap, the volume of the kernel is mostly endosperm; 24 days ap, the embryo is additionally observed within the endosperm mass. (*B*) A maize root-tip nucleus (top) compared with a 22-day ap endosperm nucleus (bottom) against a background of chicken erythrocyte nuclei (arrows). (Nuclear stain, propionic carmine; original magnification, ×200.)

for the mean DNA content being >3C from the outset. We are not sure that a sufficient number of cells were in S or G_2 phase to account for the relatively high C levels at the early stages. A possibility is that some cells may have already commenced DNA replication in the absence of cytokinesis. Fig. 2*A* illustrates the timing relationships among mitotic index, nuclear volume, and DNA per nucleus for inbred A188.

Mean nuclear volume was 5.3×10^4 μm^3 18 days ap (1981 data); this constitutes a 20-fold increase in mean nuclear volume compared to 4 days ap. Regardless of different environmental conditions from year to year, the pattern of DNA increases per nucleus over four consecutive growing seasons showed remarkable similarity (Fig. 2*B*). The period of ≈10–12 days ap (beginning 8 days ap in one case) appears to be a critical time in endosperm development marked by drastic changes in DNA levels.

The rate of DNA amplification during the 10- to 16-day ap period of endosperm development is noteworthy. In the A188 strain, the mean increase in DNA per nucleus during this 6-day period was about 480 pg (1981 data). Based on the assumption that repeated rounds of DNA replication lead to an exponential increase, each round of replication is accomplished in ≈22 hr over this time period. This means that the last round of replication during this period requires linear DNA synthesis at a rate of 916.4 μm/sec per nucleus, which approximates 2.75×10^6 base pairs per second per nucleus.

Distribution patterns of DNA content per nucleus in A188

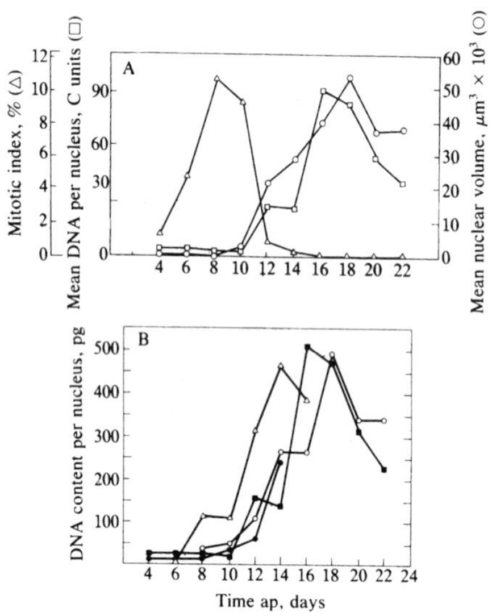

FIG. 2. (*A*) Mitotic index (△), DNA content (□), and nuclear volume (○) for 4-day through 22-day ap maize (A188) endosperm. Mitotic index is based on 352–970 cells per ap date. DNA content and nuclear volume are based on 20–47 nuclei per ap date. (*B*) DNA content per nucleus from centrally located A188 endosperm tissue over four consecutive growing seasons: 1980 (●), 1981 (■), 1982 (○), 1983 (△). The 1980 measurements were only made through 14 days ap, and the DNA (pg) in that year was based upon reference nuclei (*Gallus* erythrocytes) prepared on separate slides. All other calculations were made by using *Gallus* erythrocytes prepared on the same slide (internal reference). An earlier increase and peak in DNA content occurred in 1983 than in the other three growing seasons, all of which began a rapid DNA increase at about 10 days ap.

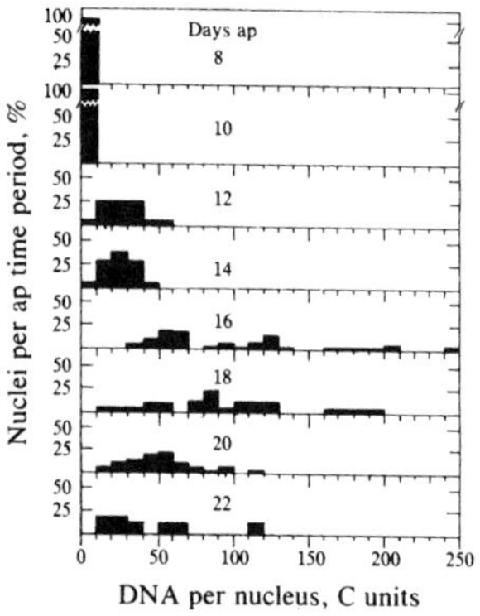

FIG. 3. Distribution patterns of DNA content of centrally located A188 nuclei over the 8-day to 22-day ap period. Note that some of the nuclei contain >200C amounts of DNA at 16 days ap.

for the ap periods are displayed in Fig. 3. Distribution patterns of nuclear size are very similar to those of DNA content. Some nuclei at 16 and 18 days ap reach DNA levels of >200C. The distributions also show that such large nuclei are not found at the earlier dates ap, all of which minimizes the possibility that the DNA content pattern could be due to erroneous sampling.

Such DNA increases raise the question of whether the larger nuclei result from nuclear fusion, polyploidy (>30 chromosomes), polytenization (endoreduplication), under-replication, or preferential gene amplification. From squashes and sections of endosperm tissue, Duncan and Ross (4)

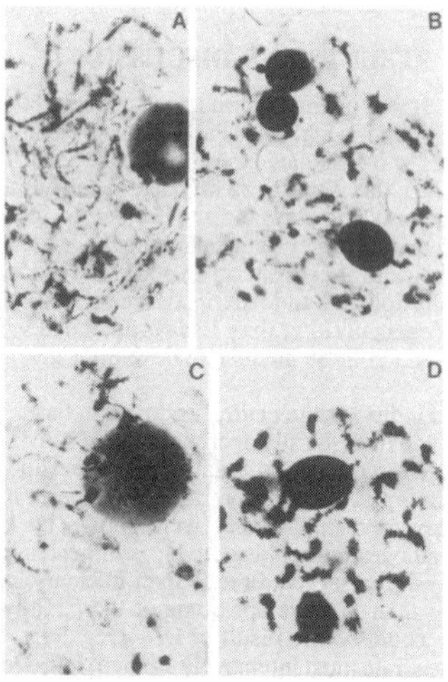

FIG. 4. Cytological aspects of endosperm nuclei. (*A*) An A188 endosperm nucleus at 22 days ap depicting various degrees of chromatin strandedness. (*B*) Three prominent nucleoli in an endosperm nucleus. (*C*) One large nucleolus showing three distinct nucleolar organizer regions (arrows). (*D*) An 18-day ap endosperm nucleus from A188 with ≈30 diffuse chromosome-like bodies. (Nuclear stain, propionic carmine; original magnifications: *A*–*C*, ×500; *D*, ×200.)

observed chromatin multistrandedness, never more than three nucleoli, a constant knob number, and ≈30 chromosome-like bodies in older tissue. They concluded that the endosperm cells were basically triploid and that the vast increases in nuclear size were due to endomitotic events. Tschermak-Woess and Enzenberg-Kunz (7) reported occasional smaller nucleoli, called secondary nucleoli, alongside of the main large nucleolus. Still, their photomicrographs show only three deeply stained nucleolar organizer regions per nucleus.

Our observations also show chromatin multistrandedness in nuclei of 12-day ap and older endosperm (Fig. 4*A*). Much variation exists in chromatin strandedness—that is, from areas of single strands to areas of many strands with chromomeres sometimes appearing in a side-by-side orientation. Individual chromosomes cannot be followed throughout their length after mitotic activity ceases at about 14 days ap because of the lack of condensation. A variety of chemical pretreatments over temperature gradients did not successfully individualize the chromosomes. In addition, we always observed one, two, or a maximum of three fully developed nucleoli per endosperm nucleus (Fig. 4*B*). Three nucleolar organizer regions were always observed in nuclei with only one large nucleolus (Fig. 4*C*). In older tissue, some nuclei were observed with ≈30 diffuse chromosome entities consistent with a 3X chromosome number in maize (Fig. 4*D*). These chromosome-like structures, or chromatin masses, are much wider than normal maize chromosomes. *In situ* hybridization of maize endosperm nuclei with labeled rRNA resulted in only three silver grain clusters over nucleoli (18).

The cytological investigations indicate that polyploidy (beyond triploidy) is unlikely as the main condition of these particular endosperm cells. Punnett (6) reported some hexaploid cells in 8-day ap tissue from greenhouse-grown plants. No hexaploid cells were reported in the 10- to 14-day ap time period. The excellent cytological techniques of Lin (19, 20) revealed 95% of the nuclei to be triploid in 6- to 10-day ap tissue. The other 5% consisted mostly of 6X, 9X, and an infrequent number of 12X. Lin provided evidence that endosperm tissue is not uniformly triploid and proposed nuclear fusion as a possible explanation. We have also observed occasional nuclei that appeared to be polyploid. Although a polyploid condition beyond triploidy may exist in a small number of nuclei, especially in younger tissue and in the more peripheral regions, the overall cytological evidence has led us to discount conventional polyploidy as the main explanation for the high DNA levels in the central regions of the endosperm. We suggest that, at the minimum, some form of polytenization occurs in maize endosperm nuclei during development. Polytene chromosomes have been found in suspensor cells, cotyledons, antipodals, synergids, and other tissues in a number of plant species (21, 22).

Mean DNA content per nucleus during endosperm development is presented in Table 1 for 17 different strains, including endosperm mutants in various backgrounds and F_1 crosses. In all strains, the mean DNA content per nucleus

Table 1. Mean DNA per nucleus (C levels) during endosperm development for 17 maize strains

Time ap, days	Number of nuclei	C level
	A188	
6	20	6 ± 0.3
8	41	22 ± 2.0
10	42	22 ± 2.3
12	41	62 ± 5.1
14	30	93 ± 8.7
16	29	68 ± 7.0
18	43	67 ± 8.9
20	17	49 ± 7.9
	B37	
8	33	23 ± 2.9
12	47	37 ± 4.8
14	45	66 ± 7.8
16	45	95 ± 8.0
19	43	141 ± 17.5
	L289	
9	18	30 ± 5.8
13	45	94 ± 8.9
15	45	155 ± 16.7
17	45	80 ± 6.9
	Wilbur's Knobless Flint	
8	44	12 ± 1.0
12	32	78 ± 10.8
15	52	137 ± 12.6
17	46	147 ± 11.4
19	42	125 ± 11.0
	Zapalote Chico	
8	40	18 ± 2.5
12	40	65 ± 8.9
14	50	47 ± 4.8
16	37	55 ± 4.8
19	42	112 ± 10.2
	A188 opaque-2	
9	32	8 ± 0.9
13	25	114 ± 23.6

Time ap, days	Number of nuclei	C level
	A188 waxy	
7	22	8 ± 1.3
9	21	3 ± 0.2
11	23	12 ± 1.4
13	23	55 ± 10.9
	B37 waxy	
11	21	54 ± 6.5
13	33	22 ± 2.1
16	42	59 ± 5.5
17	40	123 ± 13.3
19	48	134 ± 13.9
	B37 floury-2	
9	32	13 ± 1.2
11	33	19 ± 2.3
13	41	93 ± 12.3
15	20	106 ± 25.5
17	40	161 ± 13.6
19	39	162 ± 16.1
	B37 opaque-2	
9	15	40 ± 7.6
11	32	43 ± 4.3
13	35	57 ± 6.2
15	34	72 ± 10.2
17	45	178 ± 19.1
19	44	168 ± 15.0
20	45	114 ± 10.6
	B37 opaque-2 floury-2	
8	24	6 ± 1.0
10	43	30 ± 2.8
12	69	26 ± 1.6
14	42	58 ± 5.0
	B37 sugary-2	
9	23	15 ± 2.0
11	25	45 ± 6.8
19	28	80 ± 9.1

Time ap, days	Number of nuclei	C level
	A619 opaque-2	
9	22	6 ± 0.4
11	26	42 ± 5.7
13	27	33 ± 3.0
18	22	153 ± 24.6
20	30	103 ± 14.4
	A188♀ × B37♂ (F_1)	
8	33	32 ± 4.3
10	41	57 ± 7.0
14	22	35 ± 8.5
15	35	55 ± 6.7
	B37♀ × A188♂ (F_1)	
8	23	21 ± 2.4
10	23	23 ± 3.4
12	41	67 ± 8.0
14	31	57 ± 9.0
16	43	78 ± 8.9
17	42	57 ± 5.4
	Zapalote♀ × Wilbur's Flint♂ (F_1)	
9	41	30 ± 2.7
12	21	66 ± 8.5
15	23	120 ± 18.7
18	32	87 ± 7.6
	Wilbur's Flint♀ × Zapalote♂ (F_1)	
10	42	34 ± 3.0
12	24	57 ± 7.9

C levels are expressed as means ± SEM. All strains and crosses have highly significant differences in DNA levels among days ap.

varies over the different ap periods in a highly significant manner. These highly significant deviations exist in spite of the tremendous variation in DNA content among the nuclei within each ap date (see Fig. 3). The mean DNA content per nucleus is drastically increased in the central tissue of the endosperm during development.

Opaque-2 in a B37 background possessed the highest peak for mean DNA content per nucleus, reaching 178C. This is almost twice as much DNA per nucleus as that attained by A188, the line initially investigated. A188 may prove to be one of the lower DNA strains. Opaque-2 in the A619 background also showed a relatively large amount of DNA per nucleus, peaking at 153C. The floury-2 mutant in the B37 background reached a value of 162C. Both opaque-2 and floury-2 in B37 reached higher mean DNA contents per nucleus than the B37 inbred without an endosperm mutation, although B37 may not have reached the peak at the dates sampled. The highest value for an individual nucleus was 690C observed in opaque-2 with an A188 background; opaque-2 in the B37 background had a nucleus with a 635C DNA level. With the possible exception of B37 sugary-2, endosperm mutations do not appear to dramatically alter the mean DNA per endosperm nucleus over the course of development. The B37 sugary-2 strain must be investigated further, especially since sugary-2 reduces seed size (23).

Four seasons of data for A188 indicate that the mean DNA level decreases somewhat after achieving the peak at about 14–18 days of endosperm development. Several of the strains reported here also appear to decrease in mean DNA levels at later stages of endosperm development—namely, L289, Wilbur's Knobless Flint, Zapalote × Wilbur's Flint, B37 × A188, B37 opaque-2, and A619 opaque-2. This decrease could be due to DNA degradation rather than preferential loss of larger nuclei or sampling error. The other strains may not have been sampled late enough to have reached the peak. The overall patterns, however, are consistent in demonstrating an initial low DNA level and increasing levels during development at least to a point around 14–18 days ap.

Values of maximum mean DNA content per nucleus were statistically compared for various groups of interest. An F test of all of the strains indicated highly significant differences in maximum DNA values. The five strains A188, B37, L289, Wilbur's Knobless Flint, and Zapalote Chico reveal a rather large range and highly significant differences in maximum DNA levels. B37 and A619 converted to opaque-2 achieve high mean DNA levels of 178C and 152C, respectively. The A188 opaque-2 strain was only sampled through 13 days ap. Nonetheless, the 114C for opaque-2 in A188 at 13 days ap surpassed all maximum mean DNA levels observed in the A188 strain without an endosperm mutation over four consecutive growing seasons. The endosperm DNA level resulting from the F_1 cross of Zapalote Chico and Wilbur's Knobless Flint was very similar to the lower value of the two parents—namely, Zapalote Chico (Table 1). Other F_1 crosses were made between the B37 and A188 strains. From the reciprocal crosses, the resultant endosperm possessed lower DNA levels per nucleus than the parental strains. In these few preliminary F_1 crosses, at least, heterosis does not appear to exist relative to DNA content per endosperm nucleus.

Foremost among the questions prompted by these results is that relating to the purpose of the extensive DNA amplification. Certain genes such as the rRNA genes, for which multiplicity is deemed necessary for highly active synthesis, are already found to number between 5000 and 12,000 per 2C nucleus (24) without additional amplification.

The maize endosperm is extremely heterogeneous within a single kernel relative to cell size, nuclear DNA content, and, to some extent, morphological appearance. How this heterogeneity relates to gene expression during endosperm development is another question. The repeatable nature of the DNA increase indicates that it is fundamental to kernel development. Also, since kernel development is fundamental to yield *per se*, understanding the plasticity of the maize nuclear genome during development may lead to useful applications in genetics and breeding research.

This is paper 13,407, Scientific Journal Series, Minnesota Agricultural Experiment Station (St. Paul, MN). Additional support was provided by grants from the National Science Foundation (PCM-82155757), Pioneer HiBred International, Inc. (Des Moines, IA), and the McKnight Foundation.

1. Lampe, L. (1931) *Bot. Gaz. (Chicago)* **151,** 337–377.
2. Randolph, L. F. (1936) *J. Agric. Res. (Lahore)* **53,** 881–916.
3. Kiesselbach, T. A. (1949) *Nebr. Agric. Exp. Stn. Res. Bull.* **161,** 1–96.
4. Duncan, R. E. & Ross, J. G. (1950) *J. Hered.* **41,** 259–268.
5. Straus, J. (1954) *Am. J. Bot.* **41,** 833–839.
6. Punnett, H. H. (1953) *J. Hered.* **44,** 257–259.
7. Tschermak-Woess, E. & Enzenberg-Kunz, U. (1965) *Planta* **64,** 149–169.
8. Swift, H. (1950) *Genetics* **36,** 643–654.
9. Gengenbach, B. G. (1977) *Planta* **134,** 91–93.
10. Dhillon, S. S., Berlyn, G. P. & Miksche, J. P. (1977) *Am. J. Bot.* **64,** 117–121.
11. Mendelsohn, M. L. (1958) *J. Biophys. Biochem. Cytol.* **4,** 407–414.
12. Berlyn, G. P. & Cecich, R. A. (1976) in *Modern Methods in Forest Genetics*, ed. Miksche, J. P. (Iowa State University Press, Ames, IA), pp. 1–18.
13. Mendelsohn, M. L. (1958) *J. Biophys. Biochem. Cytol.* **4,** 415–424.
14. LeGal, M. F., Hallet, J. N. & Rey, L. (1984) *Protoplasma* **120,** 113–122.
15. Dhillon, S. S. & Miksche, J. P. (1983) *Histochem. J.* **15,** 21–37.
16. Resslar, P. M., Stucky, J. M. & Miksche, J. P. (1981) *Am. J. Bot.* **68,** 149–153.
17. Phillips, R. L., Weber, D. F., Kleese, R. A. & Wang, S. S. (1974) *Genetics* **77,** 285–297.
18. Phillips, R. L., Wang, A. S. & Kowles, R. V. (1983) *Stadler Genet. Symp.* **15,** 105–118.
19. Lin, B.-Y. (1975) Dissertation (University of Wisconsin, Madison).
20. Lin, B.-Y. (1977) *J. Hered.* **68,** 143–149.
21. Nagl, W. (1981) *Int. Rev. Cytol.* **73,** 21–53.
22. Nagl, W. (1982) in *Cell Growth*, ed. Nicolini, C. (Plenum, New York), pp. 619–651.
23. Glover, D. V., Crane, P. L., Misra, P. S. & Mertz, E. T. (1975) in *High Quality Protein Maize*, ed. Bauman, L. F., Mertz, E. T., Carballo, A. & Sprague, E. W. (Dowden, Hutchinson & Ross, Stroudsburg, PA), pp. 228–240.
24. Phillips, R. L. (1978) in *Maize Breeding and Genetics*, ed. Walden, D. B. (Wiley-Interscience, New York), pp. 711–741.

Proc. Natl. Acad. Sci. USA
Vol. 94, pp. 3524–3529, April 1997
Agricultural Sciences

Oat–maize chromosome addition lines: A new system for mapping the maize genome

(corn/chromosome-specific/cosmid library/repetitive DNA/cloning)

EVGUENI V. ANANIEV*, OSCAR RIERA-LIZARAZU*, HOWARD W. RINES*†, AND RONALD L. PHILLIPS*‡

*Department of Agronomy and Plant Genetics and Plant Molecular Genetics Institute, University of Minnesota, 1991 Upper Buford Circle, St. Paul, MN 55108; and †Plant Sciences Research Unit, U.S. Department of Agriculture–Agricultural Research Service, St. Paul, MN 55108

Contributed by Ronald L. Phillips, February 10, 1997

ABSTRACT **Novel plants with individual maize chromosomes added to a complete oat genome have been recovered via embryo rescue from oat (*Avena sativa* L., $2n = 6x = 42$) × maize (*Zea mays* L., $2n = 20$) crosses. An oat–maize disomic addition line possessing 21 pairs of oat chromosomes and one maize chromosome 9 pair was used to construct a cosmid library. A multiprobe (mixture of labeled fragments used as a probe) of highly repetitive maize-specific sequences was used to selectively isolate cosmid clones containing maize genomic DNA. Hybridization of individual maize cosmid clones or their subcloned fragments to maize and oat genomic DNA revealed that most high, middle, or low copy number DNA sequences are maize-specific. Such DNA markers allow the identification of maize genomic DNA in an oat genomic background. Chimeric cosmid clones were not found; apparently, significant exchanges of genetic material had not occurred between the maize-addition chromosome and the oat genome in these novel plants or in the cloning process. About 95% of clones selected at random from a maize genomic cosmid library could be detected by the multiprobe. The ability to selectively detect maize sequences in an oat background enables us to consider oat as a host for the cloning of specific maize chromosomes or maize chromosome segments. Introgressing maize chromosome segments into the oat genome via irradiation should allow the construction of a library of overlapping fragments for each maize chromosome to be used for developing a physical map of the maize genome.**

Chromosome addition lines of different plant species (1–5) have been generated to introgress valuable genes from wild or cultivated relatives into host plant species. Alien chromosome additions have been used for gene mapping (2, 6, 7) and serve as an enriched source of markers for positional cloning and constructing physical maps of specific chromosomes. The discovery of maize-chromosome retention in oat "haploids" after oat × maize crosses and the recovery of stable maize chromosome-addition oat lines (8, 9) should allow the development of a system of chromosome analysis similar to that available in mammalian hybrid-cell systems (10–12). Such a system may be used for gene assignment, isolation of chromosome-specific probes (13), flow sorting (14) and microdissection of chromosomes (15), development of chromosome-specific "paints" of fluorochrome-labeled DNA fragments (16–18), physical mapping, and selective isolation and mapping of cDNAs of a particular chromosome (19, 20).

In this paper we describe an approach for isolating clones containing large-fragment maize DNA of a single-chromosome origin from the oat–maize chromosome addition lines that will assist in the construction of physical maps for maize chromosomes. The approach is based on cloning genomic DNA of an oat–maize chromosome-addition line in an appropriate vector and subsequent use of maize-specific dispersed repetitive DNA sequences as detection probes to isolate clones carrying maize genomic DNA. Such an approach was successfully applied to the isolation of human-specific DNA fragments in cosmid libraries constructed from DNA of human-rodent hybrid cell lines carrying individual human chromosomes. In those experiments the probes included Alu-repeats (21), a Cot1 DNA fraction (22), or labeled total human DNA (23).

Plant repetitive sequences that are apparently species-specific were first isolated from rye (24, 25) and later found in many other species (26). Comparative studies revealed a strong correlation between the proportion of species-specific repeated families in a genome and phylogenetic relationships (27, 28). About 90–95% of all randomly tested genomic repeated sequences from barley were detected in wheat and rye (27). According to DNA reassociation studies, less-related species such as maize and wheat have less than 10% nucleotide sequences in common (28). The success of isolating maize DNA from oat–maize chromosome-addition lines depends on how many maize-specific (relative to oat) high-copy-number dispersed nucleotide sequences are found in the maize genome. Genome analysis in grasses, including maize and oat, reveals that a major portion of genomic DNA consists of families of repetitive sequences dispersed throughout the genome (26, 29, 30). According to DNA–DNA renaturation data in maize (28), unique sequences of average length 2,100 bp are interspersed with mid-repetitive sequences. Direct analysis of sequences adjacent to several maize genes reveals that these genes are flanked by highly repetitive DNA sequences (31). Thus, maize large DNA inserts cloned in an appropriate vector will likely carry some kind of dispersed maize-specific nucleotide sequence. A cosmid vector was chosen for this project because of the high cloning efficiency and relatively large insertion size. Because the oat genome is about 11,300 Mb, an added maize chromosome with median size of about 250 Mb would constitute about 2–4% of the total nuclear DNA of an oat–maize chromosome addition line.

We found that the major part of the maize genome consists of nucleotide sequences that do not cross-hybridize to oat genomic sequences under standard hybridization conditions. This predominant nonhomology between oat and maize genomic sequences means that many maize genomic sequences can be directly used for Southern blot hybridization on DNA from oat–maize chromosome addition lines to selectively detect sequences of maize origin. A mixture of highly repetitive dispersed DNA sequences of maize was used as a maize-specific multiprobe to screen a cosmid library of the chromosome addition line. A group of maize-specific cosmids with

0027-8424/97/943524-6$2.00/0
PNAS is available online at **http://www.pnas.org**.

Abbreviation: RFLP, restriction fragment length polymorphism; Mb, megabase(s).
‡To whom reprint requests should be addressed.

inserts from only maize chromosome 9 was isolated and characterized. Based on the large difference in nucleotide sequence composition between oat and maize and the means to efficiently isolate and identify maize DNA fragments from a chromosome addition line, we propose oat as an effective host for the cloning of maize DNA segments to construct physical maps for maize chromosomes.

MATERIALS AND METHODS

Maize and Oat Strains. Oat–maize disomic addition lines for maize chromosomes 2, 3, 4, 7, and 9 were derived from plants recovered following sexual crosses of oat by maize (8, 9). In these crosses about one-third of the recovered plants contained a haploid set of 21 oat chromosomes plus one or more maize chromosomes as a result of incomplete maize-chromosome elimination during early embryo development. Partial self-fertility due to the production of unreduced gametes by these plants yielded disomic maize-chromosome-addition oat plants ($2n = 42 + 2$). The presence of maize chromosomes was verified cytologically for each plant in the current study. The chromosome addition lines, the maize parent lines Seneca 60 and A188, and the oat parent lines Starter-1 and Sun II were used for DNA extraction.

Isolation and Analysis of DNA. Leaves of 2- to 4-week-old seedlings grown in a growth chamber were used for nuclei isolation in pH 9.5 buffer according to the protocol of Liu and Whitter (32). High molecular weight DNA was purified by phenol extraction from nuclei after lysis of the nuclei suspension in an equal volume of the same buffer supplemented with 2% sarkosyl. After two phenol extractions, DNA was precipitated with two volumes of ethanol in the presence of 0.3 M NaOAc, dissolved in TE buffer (10 mM Tris·Cl/1 mM EDTA, pH 8.0), treated with RNase (50 mg/ml), and extracted with phenol-chloroform.

*Sau*3A genomic DNA fragments were cloned into the *Bam*HI site of dephosphorylated plasmid vector pBlueScript (pBS) II KS (Stratagene). *Eco*RI subfragments from cosmid clones were cloned into the *Eco*RI site of the dephosphorylated pBS KS vector according to standard procedures (33). Insertions were amplified with the help of forward and reverse primers (Stratagene) and purified by agarose gel electrophoresis.

Gel-blot analysis of plant and cosmid DNA was carried out as described by Sambrook *et al.* (33) with several modifications (34). DNA fragments and total plant DNA were labeled by random primer extension (35).

Seventeen clones carrying maize repetitive nucleotide sequences (30) were kindly provided by J. Bennetzen (Purdue University, West Lafayette, IN). A clone containing the maize 185-bp knob repeat (36) was provided by W. Peacock (Commonwealth Scientific and Industrial Research Organization, Canberra, Australia).

Cosmid Library Construction and Screening. Cosmid library construction and screening were done using the cosmid vector SuperCos 1 (Stratagene) and packaging extract Giga-Pack II (Stratagene) with protocols provided by the manufacturer. Total nuclear DNA was partially digested with *Sau*3A, dephosphorylated, and ligated to the cosmid vector. Ligation products were packaged and the library propagated in *Escherichia coli* XL1-Blue MR. Cosmid libraries were constructed for genomic DNAs of the parental maize (Seneca 60) and oat (Starter-1) lines, as well as for the oat–maize chromosome 9 addition line.

The size of the insertion in a cosmid clone was determined as the sum of *Eco*RI subfragments after fractionation in gels with different concentrations of agarose from 0.6% up to 1.5% to achieve satisfactory resolution of long and short DNA subfragments.

RESULTS

Cloning of Maize-Specific Repetitive Sequences. When labeled maize total genomic DNA was used as a probe in blot hybridization, little cross-hybridization to oat genomic DNA occurred under standard conditions (65°C in 6× SSC) in comparison with strong hybridization to maize genomic DNA (data not shown). This hybridization pattern indicated that a significant portion of the repeated nucleotide sequences of maize and oat are not shared. To isolate maize repeated DNA sequences specific to maize relative to oat (hereafter referred to simply as "maize-specific" sequences) a maize plasmid genomic library was constructed with fragments from a complete *Sau*3A digest and screened with labeled total maize DNA. Clones were isolated that gave a high signal with total labeled maize DNA as the probe. Labeled oat genomic DNA hybridized only to a small extent to the same set of clones on a replica filter. Purified plasmids with *Sau*3A insertions, together with a group of maize repetitive DNA sequences in plasmids (see *Materials and Methods*), were cut with appropriate restriction enzymes to release insertions. They were screened by blot hybridization with labeled maize and oat DNA. Clones showing a high signal similar to the signal of an 18S–26S rDNA sequence were considered as highly repetitive. From this screening, 15 plasmids from the *Sau*3A maize genomic library, 6 Zpr clones (from Bennetzen), and the one 185-bp knob repeat clone were selected as highly repetitive maize-specific nucleotide sequences. Insertions from these 22 selected plasmids were combined to form a composite multiprobe. This multiprobe revealed very strong hybridization to maize genomic DNA on a Southern blot and no detectable hybridization to oat genomic DNA (Fig. 1). The same multiprobe was used to screen cosmid partial libraries of genomic DNA of maize and oat (about 10,000 clones each). Almost all colonies (90–95%) in the maize cosmid library revealed strong or moderate signals. At the same time, only a few colonies, all with relatively weak signals, were detected in the oat cosmid library. Taken together, these results indicate that this composite multiprobe is highly specific for maize DNA and is suitable for detecting maize DNA fragments in a cosmid library made from an oat–maize chromosome-addition line.

Isolation of Maize-Specific Cosmids from an Oat–Maize Addition-Line Cosmid Library. The multiprobe was used to screen a cosmid library made from an oat–maize chromosome addition line carrying maize chromosome 9. The screening and rescreening of about 5,000 clones led to the isolation of 29 hybridization-positive individual colonies. The resulting clones constituted a maize chromosome 9 library. The cloned maize DNA fragments in the selected 29 clones had a median size of about 39 kb. Together they comprised more than 1 Mb of maize genomic DNA. *Eco*RI restriction enzyme digestion enabled cutting out the cloned DNA fragments from the vector and identification of the number and size of *Eco*RI subfragments in the original insertions (Fig. 2*a*). Fractionation of the samples on gels of various agarose concentrations from 0.6 to 1.5% enabled detection of doublets and additional small subfragments below 0.5 kb not detected in the 0.8% agarose gel blots shown.

Blot hybridization of labeled oat genomic DNA to purified DNA recovered from each cosmid clone cut by *Eco*RI did not reveal strong cross-hybridization to any of the *Eco*RI subfragments (data not shown). Only after substantial overexposure were weak signals revealed and then only for a portion of the DNA fragments. At the same time, most (>90%) of the *Eco*RI fragments revealed strong, medium, or weak hybridization to labeled maize genomic DNA (Fig. 2*b*). The difference in hybridization to oat vs. maize genomic DNA indicates that the selected cloned DNA fragments likely originated from the maize chromosome, with no evidence that any were chimeric in origin.

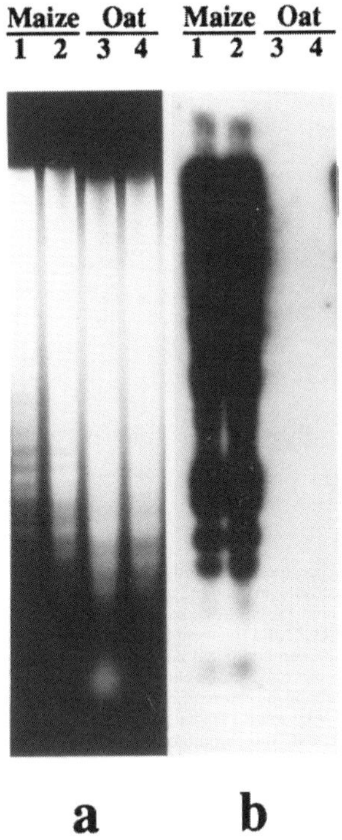

FIG. 1. Blot hybridization of a labeled multiprobe, composed of 22 maize-specific repetitive DNA sequences, to genomic DNA digests of 2 maize and 2 oat varieties. Strong signal is seen over lanes with maize DNA but not those with oat DNA. (*a*) EtdBr-stained 0.8% agarose gel after separation of DNA samples cut with *Eco*RI. (*b*) Autoradiogram after hybridization to the P-32 labeled multiprobe.

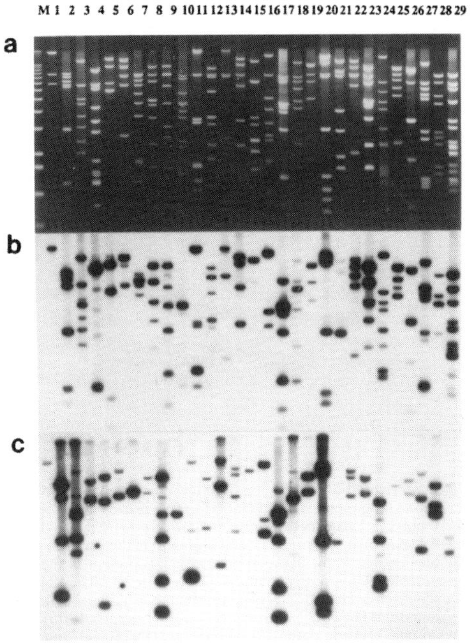

FIG. 2. Blot panel of 29 maize-specific cosmid clones isolated from a cosmid library of an oat–maize chromosome 9 addition line. (*a*) Cosmids (1–29) are cut by *Eco*RI restriction enzyme and size-fractionated in an 0.85% agarose gel stained with EtdBr (M = molecular weight marker, 1-kb ladder). (*b*) Labeled, total genomic maize DNA as a probe shows strong, medium, or weak hybridization to almost 90% of all *Eco*RI subfragments present in the cosmid clones. (*c*) Labeled multiprobe of highly repeated maize DNA sequences shows hybridization to from one to five *Eco*RI fragments in each lane.

The *Eco*RI digests of the set of selected cosmid clones were hybridized with the maize-specific repeated-sequence multiprobe that was used to screen the cosmid library (Fig. 2*c*). The multiprobe highlighted about 100 of the more than 200 *Eco*RI fragments identified in this set of cosmid clones. On average about three fragments per cosmid were homologous to various repeated DNA sequences present in the multiprobe. At the same time a portion of the *Eco*RI fragments did not hybridize to the multiprobe nor to total oat DNA but did hybridize to labeled total maize DNA. These fragments may be added to the collection of maize-specific repeated DNA sequences to make the multiprobe an even more efficient screening tool.

Maize-Specific Clones Originate from Maize Chromosome 9. *Eco*RI subfragments that showed no readily detectable hybridization to maize genomic DNA as the probe were identified in 12 of the 29 cosmid clones. Presumably they are low copy number or unique nucleotide sequences. Altogether they comprise 11% of the *Eco*RI fragments or about 6.5% of the total amount of maize DNA cloned in this set of cosmid clones. Twenty-one of them were recloned in plasmids and, as illustrated for five fragments in Fig. 3, used as probes on blot panels of DNA samples from addition lines for maize chromosomes 2, 3, 4, 7, and 9. Two parental maize and two parental oat lines were also on this blot panel (Fig. 3). Nine of these 21 *Eco*RI fragments revealed one or several bands with DNA specific for chromosome 9 of maize and did not hybridize to the DNA of the other four maize chromosomes tested (Fig. 3 *a*–*c*). Seven of these nine fragments revealed restriction fragment length polymorphism (RFLP) in the two maize stocks, A188 and Seneca 60 (Fig. 3 *a* and *b*). Another 11 *Eco*RI fragments produced multiple bands with DNA of all maize chromosomes tested, which is characteristic for low-copy-number families of dispersed repeated sequences (10–100 copies) (Fig. 3 *d* and *e*). Eight of these sequences produced a chromosome-specific pattern of hybridization and potentially could be used for chromosome identification (Fig. 3 *d* and *e*).

Several *Eco*RI fragments belonging to medium and highly repetitive classes of nucleotide sequences from different cosmid clones of maize chromosome 9 were used as probes. These revealed a complex pattern of hybridization for all maize chromosomes on the panel of chromosome-addition lines (Fig. 4*b*). Labeled highly repetitive DNA from cosmid 1 (Fig. 4*c*), as well as the full set of 29 cosmids (data not shown), gave strong hybridization to all maize chromosomes and did not reveal any chromosome 9-specific repetitive nucleotide sequences. Only the 185-bp knob-repeat revealed apparent chromosome specificity (Fig. 4*d*). Many copies of 185-bp repeats are located on chromosome 9, and a small fraction is present on chromosome 4 in these materials. Several weak bands are seen with DNA of maize chromosomes 2 and 3. This result is compatible with reports that most 185-bp knob DNA sequences are organized in the form of clusters on different chromosomes and relative knob size can differ greatly among chromosomes in different maize lines (31).

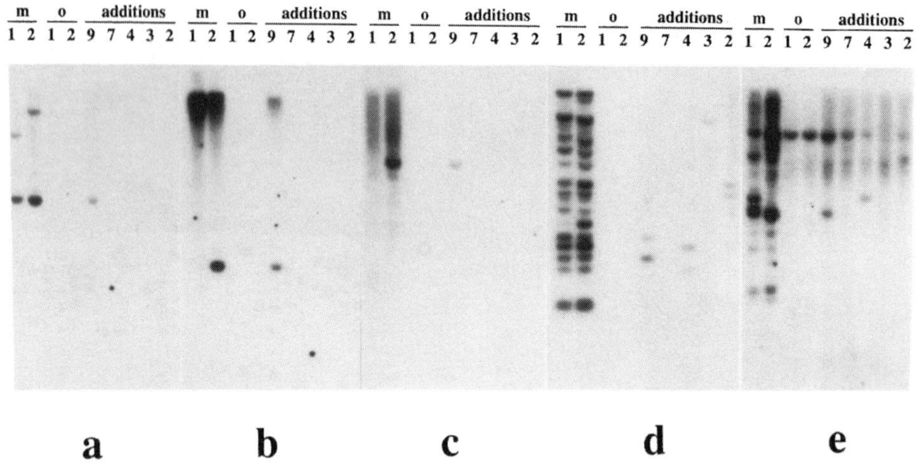

FIG. 3. Hybridization of unique and low-copy-number DNA sequences isolated from maize-specific cosmids shown in Fig. 2 to a blot panel of oat–maize chromosome addition lines carrying maize chromosomes 9, 7, 4, 3, and 2, respectively; m1 and m2 are maize stocks A188 and Seneca 60; o1 and o2 are oat stocks Sun II and Starter-1. (*a*) The 2.5-kb *Eco*RI fragment from cosmid 15 shows one band on the chromosome 9 addition line. An additional polymorphic band is present in the parental stocks of maize. (*b*) The 1.8-kb *Eco*RI fragment from cosmid 28 detects two bands. One band, polymorphic between m1 and m2, is present in the chromosome 9 addition line. An additional nonpolymorphic band is present on chromosome 9 and in both parental stocks. (*c*) The 2.1-kb *Eco*RI fragment from cosmid 10 shows one band on chromosome 9. (*d*) The 1.4-kb *Eco*RI fragment from cosmid 20 detects about 20 bands in parental maize stocks and several bands among the chromosome addition lines. The band pattern is chromosome-specific. No cross-hybridization occurred to oat DNA in *a–d*. (e) The 2.9-kb *Eco*RI fragment from cosmid 6 detects several polymorphic bands in parental maize stocks. One nonpolymorphic band is seen in maize, oat, and chromosomes 9, 7, 4, and 2 addition lines. Additional bands are seen in other chromosome addition lines.

Most Cloned Maize Nucleotide Sequences Are Maize-Specific. Seven out of 11 unique or low-copy-number sequences showed no cross-hybridization to oat genomic DNA (Fig. 3 *a–d*). This result indicates that a significant portion (>50%) of low-copy-number maize sequences have diverged enough to show no cross-hybridization to oat genomic DNA.

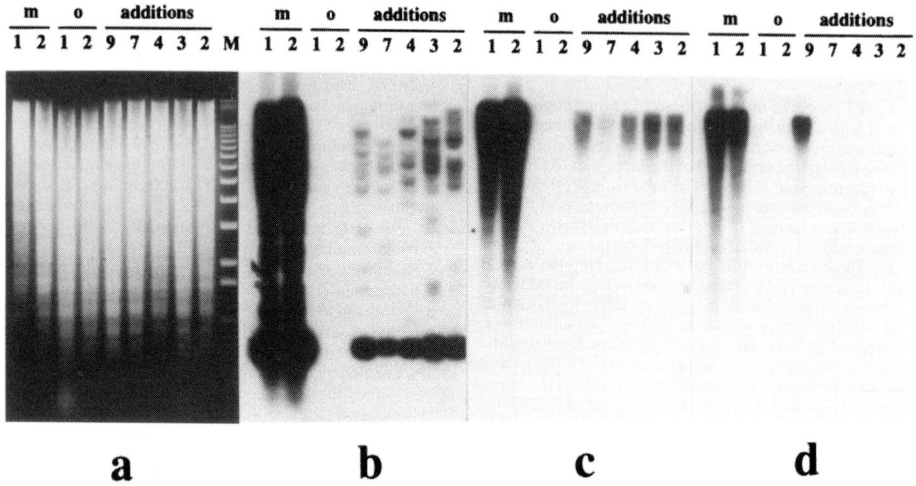

FIG. 4. Hybridization of medium and highly repetitive cloned maize DNA sequences to a blot panel of oat–maize chromosome addition lines carrying maize chromosomes 9, 7, 4, 3, and 2, respectively; m1 and m2 are maize stocks A188 and Seneca 60; o1 and o2 are oat stocks Sun II and Starter-1. (*a*) EtdBr-stained 0.8% agarose gel; lane M is a 1-kb molecular weight marker ladder. (*b*) The 0.7-kb *Eco*RI fragment of cosmid 10 shows multiple bands on different maize chromosomes as well as one common band for all chromosomes. (*c*) Cosmid 1, with a 40-kb insertion, shows strong hybridization signal over maize (lines m1 and m2) and over chromosome addition lines 9, 7, 4, 3, and 2; no hybridization is seen over oat DNA (lines o1 and o2). (*d*) With the 185-bp knob repeat, most of the hybridization signal is located on chromosome 9. No hybridization to oat DNA is detected.

Eleven of 14 (80%) middle repetitive sequences (class 10–100 copies) are also maize-specific. Two among the remaining three showed only faint cross-hybridization to oat genomic DNA. Highly repetitive sequences were preselected in this study, and the process of selection showed that most highly repetitive maize sequences are also maize-specific (Fig. 4 *b–d*).

DISCUSSION

Maize Species-Specific Sequences. The cross between oat and maize, two species from different subfamilies of the Graminae, represents the widest combination of crop species to date yielding stable, fertile partial hybrids (37). However, little is known about the comparative structure and composition of the two genomes. Highly conserved sequences such as rRNA genes or tubulin cDNAs cross-hybridize to corresponding genes in the maize and oat genomes. Unique maize RFLP probes also have been used to detect sequences in the oat genome (9). A series of cDNA probes has shown 71% conservation of linkage associations between maize and oat according to Van Deynze *et al.* (38); however, almost 50% of maize *Pst*I unique genomic probes do not hybridize to oat genomic DNA under standard conditions (39).

We report herein that the proportion of maize and oat nucleotide sequences that cross-hybridize to each other is low under standard hybridization conditions (65°C, 6× SCC). Among the set of low-copy-number sequences isolated and tested, more than 50% appeared to be maize specific. The proportion of maize-specific sequences among the middle or highly repetitive maize elements was about 80–95% and among those that do hybridize to oat, most hybridize only weakly; therefore, even total genomic DNA of maize can serve as an efficient probe to identify maize-specific cosmids in an oat genomic background. But total maize DNA is less reliable for this use than the described multiprobe because it gives more false positive clones during primary screening of the cosmid library. The oat–maize chromosome addition lines are especially attractive as a source for isolation of region-specific maize DNA fragments. It is possible to identify even small portions of a maize chromosome in an oat genome by probing with total maize genomic DNA, a multiprobe of maize repeated nucleotide sequences, or many other types of cloned maize sequences. In the case of the highly conserved nucleotide sequences, which are common in both maize and oat genomes, RFLPs may be used to differentially identify maize DNA in the oat genome.

Efficiency of Recovering Maize-Specific Cosmid Clones. The efficiency of identifying maize-specific cosmid clones depends on copy number and the pattern of distribution of repeated sequences chosen to comprise the multiprobe. Studies of genome structure in maize (28, 40), especially analysis of repeated DNA sequences around several maize genes (30, 31), show that unique sequences with a median size of 2.1 kb are surrounded by a long, complex array of repeated sequences that belong to many different families, many of which appear to be of retrotransposon origin (31).

Our multiprobe highlighted *Eco*RI subfragments in each of the 29 cosmid clones of chromosome 9 analyzed, detecting about 100 of the total of more than 200 fragments. Each clone contained one or more of the repeated sequences present in the multiprobe. The same group of *Eco*RI fragments were classified as highly repetitive nucleotide sequences because they showed strong hybridization to labeled total maize genomic DNA. In the same set of clones at least 50 additional *Eco*RI fragments showed a strong hybridization signal to maize genomic DNA and may be classified as additional highly repetitive nucleotide sequences. These additional sequences could be included in our collection of maize-specific repeated nucleotide sequences to increase the efficiency of the multiprobe used in searching for maize-specific cosmids.

The number and diversity of repeats differ among clones. Some of the cosmids consist entirely of several different repeated sequences; others have only one copy of the tested repeats. Regions of maize chromosomal DNA consisting only of low-copy-number sequences with a composite length exceeding 40 kb, the size of the DNA insertion in most cosmids, would not be detected in a cosmid library by screening with the multiprobe.

Because the 29 analyzed clones of maize chromosome 9 were selected with the help of the multiprobe, they may carry more repeated DNA sequences than a random set of clones. Therefore, a control experiment was conducted to analyze the distribution of repetitive sequences in a set of 38 cosmid clones randomly chosen from a cosmid library constructed from DNA of the parental maize line Seneca 60. Only 2 of 38 maize cosmid clones revealed cross-hybridization to oat genomic DNA, and they were found to be composed of rDNA sequences. Thirty-four of the remaining 36 cosmid clones carried maize-specific highly repetitive sequences detected by the multiprobe (data not shown). This experiment indicates that the probability of recovering maize-specific cosmid clones in a genomic library of maize chromosome 9 with the help of the primary multiprobe is around 95%.

Oat as a Host for Cloning Large Maize Chromosomal DNA Fragments. Impressive achievements have been made in cloning large (>1 Mb) DNA fragments in yeast artificial chromosomes (41). However, the difficulties encountered make this method of cloning a eukaryotic genome expensive and problematic for wide use. Radiation hybrids (derivative cell lines from irradiated somatic cell hybrids) are an attractive alternative for cloning subfragments of a chromosome. Irradiation to produce chromosome breakage and segregation of genetic material has been applied to a number of mammalian somatic cell hybrids containing individual alien chromosomes (10). A disadvantage of this approach in some cases is that a host genome and an alien chromosome have too many in-common nucleotide sequences, thus making difficult the direct analysis, identification, and isolation of the alien chromosome fragments.

The results of the research reported here allow us to propose the use of oat as a host for cloning maize genomic DNA. The substantial differences in the repetitive DNA composition of oat and maize genomes and the high level of genetic stability of the maize-addition chromosomes make it possible to apply conventional molecular methods to the study of maize genetic material directly in the oat genome. This approach would be modeled after that used in mammalian systems (10). Radiation would be used to induce translocations of maize segments into the oat genome, from which maize DNA could subsequently be isolated. In comparison with other possible maize-cloning systems using phages, bacteria, or yeast, the advantage of this approach is that all maize chromosome fragments will derive from a known chromosome. The high proportion of maize sequences that may serve as maize-specific DNA probes in this system allows the identification of almost any maize chromosome segment in oat or any cloned maize DNA fragment in a library. The current availability of an extensive collection of mapped DNA markers for all maize chromosomes (42) may allow us to identify the boundaries of cloned subfragments and to generate a collection of overlapping subfragments representing the whole chromosome. This approach would allow the cloning in oat of subfragments of maize chromosomes of a wide size range. Chromosomal subfragments 5–30 Mb in size may be considered optimal because they can be aligned relative to the genetic map and, if cloned in bacterial artificial chromosomes or even in cosmid vectors, may be arranged in contigs to generate detailed physical maps for each subfragment. Thus, 10–60 subchromosomal fragments may comprise a collection of overlapping large DNA fragments of one particular maize chromosome. Taking into account the high stability of maize genetic material in the oat background, oat lines with maize DNA represent a continuous, renewable source of large segments of maize chromosomal DNA.

This is a joint publication of the University of Minnesota and U.S. Department of Agriculture–Agricultural Research Service and is Paper No. 22,475 in the Scientific Journal Series, Minnesota Agricultural Experiment Station. Mention of a trademark or proprietary product does not constitute a guarantee or warranty by the University of Minnesota or the USDA-ARS and does not imply approval over other products that also may be suitable. This research was funded in part by the Plant Molecular Genetics Institute, University of Minnesota, and The Quaker Oats Company.

1. Islam, A. K. M. R., Shepherd, R. & Sparrow, D. H. B. (1981) *Heredity* **46,** 161–174.
2. Riley, R. & Law, C. N. (1984) *Stadler Genet. Symp.* **16,** 301–322.
3. Thomas, H., Leggett, J. M. & Jones, I. T. (1975) *Euphytica* **24,** 717–724.
4. Jena, K. K. & Khush, G. S. (1989) *Genome* **32,** 449–455.
5. Jiang, J, Morris, K. L. & Gill, B. S. (1994) *Chromosome Res.* **2,** 3–13.
6. Appels, R. & Moran, L. B. (1984) *Stadler Genet. Symp.* **16,** 529–558.
7. Kanazin, V., Ananiev, E. V. & Blake, T. (1993) *Genome* **36,** 1023–1028.
8. Rines, H. W., Riera-Lizarazu, O. & Phillips, R. L. (1995) in *Modification of Gene Expression and Non-Mendelian Inheritance*, eds. Oono, K. & Takiwa, F. (National Institute of Agrobiological Resources, Tsukuba, Japan), pp. 235–251.
9. Riera-Lizarazu, O., Rines, H. W. & Phillips, R. L. (1996) *Theor. Appl. Genet.* **93,** 123–135.
10. Cox, D. R., Burmeister, M., Price, E. R., Kim, S. & Myers, R. M. (1990) *Science* **250,** 245–250.
11. Reichester, M. & Parsons, B. (1976) *J. Cell. Physiol.* **88,** 167–179.
12. Olson, M. V. (1993) *Proc. Natl. Acad. Sci. USA* **90,** 4338–4344.
13. Clarke, B., Stancombe, P., Money, T., Foote, T. & Moore, G. (1992) *Nucleic Acids Res.* **20,** 1289–1292.
14. Wang, M. L., Leitch, A. R., Schwarzacher, T., Heslop-Harrison, J. S. & Moore, G. (1992) *Nucleic Acids Res.* **20,** 1897–1901.
15. Jung, C., Claussen, U., Horsthemke, B., Fisher, F. & Herrmann, R. G. (1992) *Plant Mol. Biol.* **20,** 503–511.
16. Carter, N. P. (1994) in *Cell Biology: A Laboratory Handbook*, ed. Celis, J. E. (Academic, San Diego), pp. 442–449.
17. Vega, M., Abbo, S., Feldman, M. & Levy, A. A. (1994) *Proc. Natl. Acad. Sci. USA* **91,** 12041–12045.
18. Fuchs, J., Houben, A., Brandes, A. & Shubert, I. (1996) *Chromosoma* **104,** 315–320.
19. Lovett, M., Kere, J. & Hinton, L. M. (1991) *Proc. Natl. Acad. Sci. USA* **88,** 9628–9632.
20. Parimo, S., Patanjali, S. R., Shukla, H., Chaplin, D. D. & Weisman, S. M. (1991) *Proc. Natl. Acad. Sci. USA* **88,** 9623–9627.
21. Monaco, A. P., Lam, V. M. S., Zehetner, G., Lenon, G. G., Douglas, C., Nizetic, D., Goodfellow, P. N. & Lehrach, H. (1991) *Nucleic Acids Res.* **19,** 3315–3318.
22. Graw, S. L., Buckler, A. J., Britt, D. E., Jackson, C. L., Taruscio, D., Baldini, A., Ward, D. C. & Housman, D. E. (1992) *Somatic Cell Mol. Genet.* **18,** 269–284.
23. Pritchard, C. A., Casher, D., Uglum, E., Cox, D. R. & Myers, R. M. (1989) *Genomics* **4,** 408–418.
24. Bedbrook, J. R., O'Dell, M. & Flavell, R. B. (1980) *Nature (London)* **228,** 133–320.
25. Appels, R., Reddy, P., McIntyre, C. L., Moran, L. B., Frankel, O. H. & Clarke, B. C. (1989) *Genome* **31,** 122–133.
26. Flavell, R. B. (1986) *Philos. Trans. R. Soc. Lond.* **312,** 227–242.
27. Ananiev, E. V. (1992) in *Barley Genetics, Biochemistry, Molecular Biology and Biotechnology*, ed. Shewry, P. R. (CAB International, Wallingford, U.K.), pp. 135–150.
28. Hake, S. & Walbot, V. (1980) *Chromosoma* **79,** 251–270.
29. Rimpau, I., Smith, D. B. & Flavell, R. B. (1980) *Heredity* **44,** 131–149.
30. Bennetzen, J. L., Schrick, K., Springer, P. S., Brown, W. E. & SanMiguel, P. (1994) *Genome* **37,** 565–576.
31. San Miguel, P., Tikhonov A., Jin, Y. K., Motchulskaia, N., Zakharov, D., Melake-Berhan, A., Springer, P., Edwards, K., Lee, M., Avramova, Z. & Bennetzen, J. L. (1996) *Science* **274,** 765–768.
32. Liu, Y. & Whitter, R. F. (1994) *Nucleic Acids Res.* **22,** 2168–2169.
33. Sambrook, J., Fritsch, E. F. & Maniatis, T. (1989) in *Molecular Cloning: A Laboratory Manual*, (Cold Spring Harbor Laboratory, Cold Spring Harbor, NY), pp. 1.60–1.62.
34. Helentjaris, T. (1995) in *UMC Maize RFLP Procedures Manual, UMC Maize RFLP Laboratory* (University of Missouri-Columbia, Columbia, MD), pp. 11–12.
35. Feinberg, A. P. & Vogelstein, B. (1984) *Anal. Biochem.* **137,** 266–267.
36. Peacock, W. J., Dennis, E. S., Rhoades, M. & Pryor, A. J. (1981) *Proc. Natl. Acad. Sci. USA* **78,** 4490–4494.
37. Rines, H. W. & Dahleen, L. S. (1990) *Crop Sci.* **30,** 1073–1078.
38. Van Deynze, A. E., Nelson, J. C., O'Donoughue, L. S., Ahn, S. N., Siripoonwiwat, W., Harrington, S. E., Yglesias, E. S., Braga, D. P., McCouch, S. R. & Sorrells, M. E. (1995) *Mol. Gen. Genet.* **249,** 349–356.
39. Yu, G. X., Bush, A. L. & Wise, R. P. (1996) *Genome* **39,** 155–164.
40. Gupta, M., Sheperd, N. S., Bertman, I. & Saedler, H. (1984) *EMBO J.* **3,** 133–139.
41. Haldi, M., Perrot, V., Saumier, M., Desai, T., Cohen, D., Cherif, D., Ward, D. & Lander, E. S. (1995) *Genomics* **24,** 478–484.
42. Gardiner, J. M., Coe, E. H., Melia-Hancock, S., Hoisington, D. A. & Chao, S. (1993) *Genetics* **134,** 917–930.

Dissecting the maize genome by using chromosome addition and radiation hybrid lines

Ralf G. Kynast*, Ron J. Okagaki*, Mark W. Galatowitsch*, Shannon R. Granath*, Morrison S. Jacobs*, Adrian O. Stec*, Howard W. Rines*†, and Ronald L. Phillips*‡

*Department of Agronomy and Plant Genetics and Center for Microbial and Plant Genomics, University of Minnesota, 1991 Upper Buford Circle, St. Paul, MN 55108; and †Plant Science Research Unit, Department of Agriculture–Agricultural Research Service, St. Paul, MN 55108

Contributed by Ronald L. Phillips, May 13, 2004

We have developed from crosses of oat (*Avena sativa* L.) and maize (*Zea mays* L.) 50 fertile lines that are disomic additions of individual maize chromosomes 1–9 and chromosome 10 as a short-arm telosome. The whole chromosome 10 addition is available only in haploid oat background. Most of the maize chromosome disomic addition lines have regular transmission; however, chromosome 5 showed diminished paternal transmission, and chromosome 10 is transmitted to offspring only as a short-arm telosome. To further dissect the maize genome, we irradiated monosomic additions with γ rays and recovered radiation hybrid (RH) lines providing low- to medium-resolution mapping for most of the maize chromosomes. For maize chromosome 1, mapping 45 simple-sequence repeat markers delineated 10 groups of RH plants reflecting different chromosome breaks. The present chromosome 1 RH panel dissects this chromosome into eight physical segments defined by the 10 groups of RH lines. Genomic *in situ* hybridization revealed the physical size of a distal region, which is represented by six of the eight physical segments, as being ≈20% of the length of the short arm, representing ≈one-third of the genetic chromosome 1 map. The distal ≈20% of the physical length of the long arm of maize chromosome 1 is represented by a single group of RH lines that spans >23% of the total genetic map. These oat–maize RH lines provide valuable tools for physical mapping of the complex highly duplicated maize genome and for unique studies of interspecific gene interactions.

Plants with one chromosome (monosomic) or one pair of homologous chromosomes (disomic) of an alien donor species added to the entire recipient species chromosome complement serve to dissect the donor genome into individual chromosome entities and separate them from their own genome remnant. The transfer liberates the added chromosome (pair) from the interactive gene expression network of the donor genome and puts the chromosome's genes into the environment of the host genome. This new structural and functional situation can create novel orthologous and nonhomologous gene-to-gene interactions and, hence, helps to answer fundamental questions about gene expression control, inheritance, and syntenic correspondence among different plant species, especially those with large genomes, including maize, with a 1C content of ≈2.7 billion base pairs [Plant DNA C-Values Database (Release 2.0, January 2003), M. D. Bennett and I. J. Leitch, http://rbgkew.org.uk/cval/homepage.html] and a subgenome structure reflecting ancient tetraploidy (1).

By crossing maize to oat, (oat × maize)F_1 proembryos were generated, of which 5–10% could be rescued *in vitro*. Molecular and cytological analyses showed retention of one or more maize chromosomes in addition to the haploid oat genome in 34% of the F_1 plants (2–7). Because haploid oat frequently develops unreduced gametes (8), subsequent self-fertilization of (oat × maize)F_1 plants with one maize chromosome added to the haploid oat genome ($n = 3x + 1 = 22$) can produce F_2 offspring with one homologous maize chromosome pair added to the doubled haploid (hexaploid) oat genome ($2n = 6x + 2 = 44$) among other euploid and aneuploid types (9).

A complete series of oat–maize chromosome addition lines (10) has enabled markers and genes to be physically allocated to maize chromosomes without a need for detectable polymorphisms. These unique plant materials confirmed interchromosomal duplicate loci on a large scale, one of the obstacles to whole-genome sequencing. Numerous locus duplications complicate the reassembly of a set of shotgun DNA sequences. Oat–maize chromosome addition lines, however, physically separate these interchromosomal maize orthologs and paralogs from each other and make them accessible to mapping, sequencing, and cloning, even in cases where the duplicated loci of interest carry genes with monomorphic allelic sequences.

A second obstacle that impedes sequencing of the complete maize genome by today's technology is the repetitive nature of ≈85% of the maize DNA. Thus, sequencing strategies are being tested that accomplish the targeted sequencing of less repetitive gene-rich regions (11, 12). These strategies must involve technologies that are capable of arranging those gene islands along the chromosomes and bridging long gaps between contigs. Generating random breaks in the maize chromosome in an identified monosomic oat–maize chromosome addition line and maintaining diminutive maize chromosomes or pieces translocated into oat can provide DNA panels of radiation hybrid (RH) lines, which allow for a presence vs. absence test of markers without the need for polymorphisms (13). With sufficient resolution that is determined by the number and distribution of breaks along the maize chromosomes, RH lines can contribute to placing contigs in the correct order. A panel of RH lines for maize chromosome 9 has demonstrated the efficient mapping of molecular markers (14).

This report summarizes the status for the oat–maize chromosome addition line production and characterization, including the irregular transmission behavior of maize chromosomes 5 and 10 in oat, the last two maize chromosomes recovered as fertile oat–maize addition lines. We also illustrate the development of oat–maize RH lines from addition plants for maize chromosome 1. We show the use of these RH lines for physical mapping and relating genetic map distances to physical chromosome segment sizes.

Materials and Methods

Plant Material. Plants of oat (*Avena sativa* L.) cultivars GAF-Park, Kanota, Preakness, Starter, Stout, Sun II, and the MN-hybrid (MN97201-1 × MN841801-1) were grown and crossed by maize

This report was presented at the International Congress "In the Wake of the Double Helix: From the Green Revolution to the Gene Revolution," held May 27–31, 2003, at the University of Bologna, Bologna, Italy. The scientific organizers were Roberto Tuberosa, University of Bologna, Bologna, Italy; Ronald L. Phillips, University of Minnesota, St. Paul, MN; and Mike Gale, John Innes Center, Norwich, United Kingdom. The Congress web site (www.doublehelix.too.it) reports the list of sponsors and the abstracts.

Abbreviations: BC, backcross; GISH, genomic *in situ* hybridization; RH, radiation hybrid; SSR marker, simple-sequence repeat marker.

‡To whom correspondence should be addressed. E-mail: phill005@umn.edu.

PLANT BIOLOGY

(*Zea mays* L.) lines Seneca 60, A188, A619, B73, Mo17, the (A188 × W64A)F_1 hybrid, and a line carrying the allele *bz1-mum9*. More than 200 putative F_1-hybrid plants indicating successful (oat × maize) hybridization were recovered and analyzed for vigor, seed set, maize chromosome retention in various tissues, and maize chromosome transmission to F_2 offspring (5, 10). F_2 plants were tested by PCR-based markers, test crosses, genomic *in situ* hybridization (GISH), and chromosome counting for presence, stability, and transmission of maize chromosomes added to the oat genome (15). F_2 plants were propagated for production of F_3 and subsequent generations. Backcross (BC_1) plants monosomic for the maize chromosome addition were produced by crossing disomic oat–maize chromosome addition plants back to their corresponding parental oat lines. Batches of 150–300 BC_1 seeds were irradiated with γ rays from a ^{137}Cs source at different intensities (20–50 krad) to induce as many maize chromosome breaks as possible without seriously damaging the vigor of the seeds or resulting seedlings. BC_1F_2 offspring with transmitted maize chromosome deficiencies or oat–maize chromosome translocations (RH plants) were selected by a PCR assay with primers specific for Grande 1 (16) and CentA (17) and used for physically mapping molecular markers to the particular maize chromosome segments.

Genomic DNA Extraction. For a limited number of PCRs (75 or fewer) from single plants, DNA was extracted by the use of the REDExtract-N-Amp Plant PCR kit (Sigma). For larger numbers of PCRs (>75) from single plants, DNA was extracted by using either the DNeasy Plant Mini kit (Qiagen, Valencia, CA) or the acetyltrimethylammonium bromide procedure (18). For labeling and use as probe in GISH experiments, genomic maize DNA was extracted from leaf cell nuclei and purified through a CsCl gradient (19).

PCR. PCRs were accomplished by the use of the REDExtract-N-Amp Plant PCR kit, according to the vendor's recommendations. F_1 plantlets and seedlings of the consecutive generations were screened for the presence of maize sequences by using maize-specific primers for the long terminal repeat of the highly dispersed retrotransposon Grande 1 (16) and for the highly centromere-specific retrotransposon-like repeat CentA (17). Individual maize chromosomes or chromosome segments in F_1 plantlets, addition lines, and RH seedlings were identified by using maize-specific primers for simple-sequence repeat (SSR) markers (10) that were selected from the maize genetics and genomics database (www.maizegdb.org). BC_1F_2 plants (putative RH plants) were tested for presence vs. absence of maize chromosome segments by a PCR assay with 45 SSR markers specific for maize chromosome 1 (*p-umc1354* to *p-umc2244* spanning 1,120 map units, according to the IBM2 map). Markers were selected from the maize genetics and genomics database mentioned above.

Cytology. Root tips (1.5–2 cm) of oat, maize, and oat–maize chromosome addition and RH lines were pretreated, fixed, and stored as described in ref. 10. Root tips for chromosome counting were prepared as described in ref. 20. Meristem cells were squashed in 2% wt/vol Aceto-Orcein (Carolina Biological Supply). Root tips for GISH were prepared as described in ref. 10. Further steps of RNase treatment, postfixation, and *in situ* hybridization were as described earlier in ref. 21 except that total genomic maize DNA was labeled by the use of the ULYSIS Alexa Fluor 488 nucleic acid labeling kit (Molecular Probes) and probed on slides without using unlabeled competitor DNA. Hybridization was carried out in 40% formamide in 1.5× SSC (225 mM NaCl/22.5 mM trisodium citrate, pH 7.0) at 37°C. Posthybridization stringency washes were carried out in 40% formamide in 1.5× SSC at 42°C. Chromosomes were counterstained with propidium iodide. Signals were visualized and captured by using an Axioskop microscope equipped for epifluorescence (Zeiss) and a Magnafire charge-coupled device camera (Optronics International, Chelmsford, MA).

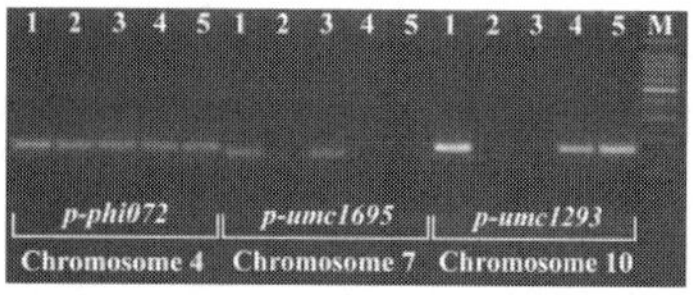

Fig. 1. PCR products from DNA of the F_1 (Starter × B73) plant F_1-5133-1 when chromosome-specific SSR markers are used; electrophoresis in a 3.5% agarose gel shows the different elimination of maize chromosomes in individual tillers. Lanes 1, young plantlet; lanes 2, tiller F_1-5133-1/a; lanes 3, tiller F_1-5133-1/b; lanes 4, tiller F_1-5133-1/c; lanes 5, tiller F_1-5133-1/d; and lane M, standard 100-bp ladder.

Results and Discussion

Maize Chromosome Elimination in (Oat × Maize)F_1 Hybrids. In oat × maize crosses, maize chromosomes are occasionally retained (2). This situation is distinct in that the maize genome is completely eliminated in hybridizations between maize and wheat or barley. There is only one report of a maize chromosome being retained in wheat; however, the maize chromosome was not transmitted to offspring (23). The timing of elimination differs as well. The maize chromosome elimination process in oat sometimes extends over longer periods of time compared with that in F_1 hybrids generated from wheat × maize and barley × maize crosses (24). In 70% of (wheat × maize)F_1 embryos, one or more maize chromosomes were eliminated at the first mitosis. By the eight-cell stage, the embryos had lost all maize chromosomes (24). Maize chromosome elimination from (oat × maize)F_1 embryos starts at an early stage in embryogenesis as well (4). However, as an example of the extended time of maize chromosome elimination from oat, in the (oat × maize)F_1 plant F_1-5133-1 with maize chromosomes 4, 7, and 10 all detected at a young plant stage, a consecutive elimination of individual maize chromosomes was detected by the PCR analysis of genomic DNA extracted from tissues of flag leaves of different tillers from the same plant shortly after meiosis. The first tiller retained only maize chromosome 4, thus eliminated chromosomes 7 and 10. The second tiller eliminated chromosome 10, thus retained chromosomes 4 and 7. The third and fourth tillers eliminated chromosome 7, thus retained chromosome 4 and chromosome 10 as a short-arm telosome (Fig. 1). We observed further instances where maize chromosomes were lost in later growth stages, particularly from plantlets with two or more originally retained maize chromosomes in their complements (results not shown).

(Oat × Maize)F_1 Hybrids and F_2 Offspring in Different Genetic Backgrounds. Our initial work was conducted with the maize chromosome donor Seneca 60 and the oat recipient Starter. In the last 2 years we have tested several other combinations of maize and oat lines. A total of 201 (oat × maize)F_1 plants have been generated from various maize and oat backgrounds, of which 68 F_1 hybrids retained one or more maize chromosome(s) in their complements. All 10 maize chromosomes could be recovered, with each occurring at different frequencies as single additions and in combination with other maize chromosomes. No obvious preferential combination for two or more specific maize chromosomes was detected in multiple additions in haploid oats. The frequency of recovery of a particular maize chromosome, the

Table 1. Fertile oat–maize chromosome addition lines

Added maize chromosome	Oat host	Maize donor	Addition	No. of lines
1	Starter	Seneca 60	Disomic	1
2	Starter	Seneca 60	Disomic	9
2	Starter	*bz1-mum9*	Disomic	1
2	Sun II	Seneca 60	Disomic	2
3	Sun II	Seneca 60	Disomic	1
3	Preakness	Seneca 60	Disomic	1
4	Starter	Seneca 60	Disomic	6
4	Starter	A188	Disomic	2
4	Starter	B73	Disomic	1
5	Starter	Seneca 60	Disomic	1
5	MN-hybrid	Seneca 60	Disomic	1
5	Sun II	Seneca 60	Disomic	1
6	Starter	Seneca 60	Disomic	1
6	MN-hybrid	Seneca 60	Disomic	1
6	Sun II	Seneca 60	Disomic	1
7	Starter	Seneca 60	Disomic	4
7	GAF-Park	Seneca 60	Disomic	1
8	Starter	Seneca 60	Disomic	1
8	Starter	*bz1-mum9*	Disomic	1
8	GAF-Park	Seneca 60	Monosomic	1
9	Starter	Seneca 60	Disomic	7
9	GAF-Park	Seneca 60	Disomic	1
9	Sun II	Seneca 60	Disomic	1
10S	Sun II	Seneca 60	Ditelosomic	1
1 + 9	Starter	Seneca 60	Double disomic	1
4 + 6	Starter	Seneca 60	Double disomic	1

This table lists all available addition lines and is an update of an earlier list that involved fewer lines (15).

fertility of that plant, and the stability of a maize chromosome appeared to be primarily dependent on the particular maize chromosome interacting with the oat background. Furthermore, the ability to recover a particular maize chromosome in F_1 hybrids was not correlated with the ability to produce a fertile addition line for that chromosome. For instance, 20 F_1 hybrids with maize chromosome 5 were produced, making the chromosome 5 the most frequently recovered chromosome, either as a single chromosome addition or in combination with other maize chromosomes; chromosome 2 addition plants were the next most frequent with 15 F_1 plants produced. Yet, only one of the chromosome 5 plants was fertile and transmitted the chromosome 5 to offspring resulting in the fertile disomic addition OMAd5.59 in Sun II oat background. This finding can be contrasted with chromosome 4, which has been recovered as an addition line nine times, with three of these addition plants being fertile and transmitting the chromosome 4 to offspring. With respect to transmission of maize chromosomes, fertile disomic addition lines fall into different categories. Addition lines carrying chromosomes 2, 3, 4, 6, and 9 exhibited little or no problems with transmission (nearly 100% maternal and paternal transmission rate). Lines carrying chromosomes 1 and 7 initially had poor transmission of the maize chromosome, but after several generations they show good transmission of the maize chromosome (>80% transmission rate), possibly due to selective breeding for stable diploid offspring. Chromosomes 1, 5, and 8 additions have fertility problems even after several generations of selection, and chromosome 10 additions have transmitted only short-arm derivatives to offspring. Eleven disomic additions for different maize chromosomes, including 2, 4, 5, 6, 7, 8, and 9 in different oat backgrounds, have been added to the previously reported set (15), making a total of 50 fertile addition lines (Table 1).

Table 2. Seed set from fertile oat–maize chromosome 10S additions

	No. of F_2 seeds				
F_1-plant panicle	Total	Tested	Maize-positive	Disomic	Monosomic
F_1-0289-1/a	51	10	9	3	6
F_1-0289-1/b	59	10	9	5	4
F_1-0289-1/c	48	10	0	0	0
F_1-0289-1/d	31	10	10	8	2
F_1-5133-1/a	6	6	6	3	3
F_1-5133-1/b	0	—	—	—	—
F_1-5133-1/c	1	1	1	0	1*
F_1-5133-1/d	2	2	2	1†	1*

*Monosomic for maize chromosome 4 and monosomic for maize chromosome 10S derivative.
†Disomic for maize chromosome 4.

Irregular Maize Chromosome 5 Transmission in Oat. Disomic maize chromosome 5 additions recovered earlier in two different oat backgrounds (OMAd5.09 in Starter and OMAd5.17 in MN-hybrid) show significantly diminished paternal transmission, whereas maternal transmission of the maize chromosome 5 is only moderately reduced. Crossing a OMAd5.09 plant as male back to Starter produced four monosomic maize additions among 58 BC_1 offspring, giving a paternal transmission rate of 6.9% (4 of 58). In 30 F_4 offspring of a disomic addition from the line OMAd5.09, only 2 plants were disomic additions, which corresponds to a minimal paternal transmission of 6.7% (2 of 30). Twenty-one plants were monosomic additions, which indicates a probable maternal transmission of 76.7% (23 of 30), although one or two of the monosomic additions could be from paternal transmission. In a further experiment analyzing 25 F_4 offspring of a disomic addition from the line OMAd5.17, only two plants were disomic additions, indicating a paternal transmission of 8% (2 of 25). Seventeen plants were monosomic additions, which indicates a probable maternal transmission of 76% (19 of 25). Taking data for both oat backgrounds together, the irregular low paternal transmission of the maize chromosome 5 (6.7–8%) clearly accounts for the low frequency of disomic addition offspring from disomic chromosome 5 addition plants.

Maize Chromosome 10 Transmission. Chromosome 10 is the smallest chromosome, with 190 Mbp in the Seneca 60 complement. Yet, it was the most frequently eliminated chromosome in (oat × maize)F_1 hybrids, indicating a low tolerance for its presence in oat (4, 10). The first recovered monosomic addition of chromosome 10 occurred in a haploid of GAF-Park oat background (10). Since that time, the plant has been vegetatively propagated by tiller cloning under short-day conditions. Periodically, tiller clones have been moved to a long-day regime to induce flowering for self-pollination or backcrossing with GAF-Park pollen. After screening thousands of spikelets over a period of ≈3 years, we recently recovered one seed. This one seed, however, did not possess maize chromatin; thus, a maize chromosome addition offspring has not been produced by this plant.

In this new series of oat × maize crosses, we recovered 11 plantlets that retained maize chromosome 10 as single or multiple chromosome additions, 9 plantlets in Starter oat background, and 2 plantlets in Sun II background. One chromosome 10-positive F_1 plant (F_1-0289-1, Seneca 60 × Sun II) set 189 F_2 seeds in its first four panicles (Table 2). The PCR assay with Grande 1 showed that 28 F_2 plants originating from three panicles were maize-positive, and 2 F_2 plants were maize-negative. GISH analyses revealed 16 disomic and 12 monosomic addition plants. All 10 F_2 plants originating from the fourth panicle were maize-negative. In all 28 maize-positive plants, only

PLANT BIOLOGY

chromosome 10 short-arm-specific SSR markers (*p-phi041*, *p-phi117*, and *p-umc1293*) were present; none of the long-arm-specific markers (*p-umc1249*, *p-umc1196*, *p-umc1176*, and *p-umc1084*) tested was detected (22). We, therefore, assume that the chromosome transmitted to offspring in every case is a short-arm telocentric derivative of chromosome 10 (22). The derived line disomic for the chromosome 10 short-arm telocentric was labeled OMAdt10S.20 and seeds (F_3 offspring) have already been distributed (Table 1).

The F_1 hybrid F_1-5133-1 originated from crosses of Starter oat with maize B73. This hybrid possessed the three maize chromosomes 4, 7, and 10 in addition to the haploid oat complement at a young growth stage. In DNA samples from both the third and fourth tillers (F_1-5133-1/c and F_1-5133-1/d), SSR markers were present for chromosome 4 and chromosome 10. Although all three short-arm-specific markers for chromosome 10 were present in the two DNA samples, neither sample showed evidence for long-arm-specific markers. Therefore, we assume that chromosome 10, which was accompanied by maize chromosome 4, also was a telocentric short-arm derivative of chromosome 10. PCR analysis of the three F_2 offspring from the panicles of the third and fourth tiller of plant F_1-5133-1 (Table 2) showed that only two F_2 plants had the short-arm-specific SSR markers for chromosome 10, and none had any of the long-arm-specific SSR markers for chromosome 10. All three F_2 plants had the chromosome 4-specific SSR markers. All six F_2 offspring from the F_1-5133-1/a panicle were positive for B73 chromosome 4, three as monosomic and three as disomic additions. The tiller F_1-5133-1/b did not set seed.

The generation of a fertile disomic telocentric addition for chromosome 10 (OMAdt10S.20) is a major breakthrough in our efforts to develop a complete series of fertile oat–maize addition lines (22). However, this observation raises the question of why does only the short arm of an added maize chromosome 10 transmit in oat. Does the long arm possess a gene that prevents transmission in this alien background? High sterility occurs in the highly stable whole chromosome 10 addition in GAF-Park oat and the two independent events of short-arm derivatives of chromosome 10 in Sun II and Starter oat, where the long-arm telocentrics could not be established. The situation appears similar to the difficulties of generating a disomic euplasmic addition line for Betzes barley chromosome 1H and for its long-arm telosome 1HL in Chinese Spring wheat (25). The difficulties in wheat (26–28) appear to be caused by the interaction of the gene *Shw* (sterility in hybrid with wheat) with the wheat background causing sterility. However, the sterility was alleviated by the simultaneous addition of monosomic or disomic chromosome 6H to the 1H addition (29). Perhaps we could select (oat × maize)F_1 hybrids for the simultaneous additions of other chromosomes with chromosome 10 to possibly allow fertility and transmission of the whole chromosome 10. In addition, it may be feasible to use additional maize genotypes that possess a different allele of the presumed gene on chromosome arm 10L responsible for the sterility. In the corresponding wheat–barley addition situation, chromosome 1H of the closely related wild barley (*Hordeum vulgare* L. subsp. *spontaneum*) was added to wheat without causing a severe effect of sterility (30).

Oat–Maize Chromosome 1 RHs. Monosomic oat–maize chromosome 1 addition seeds, the foundation for the development of oat–maize chromosome 1 RH lines, were treated with γ rays at two levels. These levels were 180 BC_1 seeds treated with 40 krad and 120 BC_1 seeds treated with 35 krad. A total of 46 maize-positive plants, as indicated by the presence of the markers Grande 1 and/or CentA, were recovered from the 40-krad treatments. The 35-krad treatment generated 54 maize-positive plants. These 100 BC_1 plants were allowed to self-pollinate. Of these, 91 panicles produced 340 BC_1F_2 offspring that tested negative and 171 BC_1F_2 offspring that tested positive for maize chromatin in their genomes, indicating successful transmission of maize segments. It is notable that after the γ radiation treatment of 300 monosomic addition seeds, only 100 plants retained their maize chromosomes or a diminutive maize chromosome derivative. This finding indicates that a majority of the breaks generated maize fragments that were eliminated from somatic tissues. Earlier results showed a certain level of somatic instability for whole chromosome 1 addition plants resulting in chromosome loss (7, 10).

Table 3. Groups of RH lines from independent chromosome mutation events that define the same breakpoints on chromosome 1

Treatment, krad	No. of BC_1 plants	No. of BC_1F_2 plants with similar breaks	No. of independent events*	Panel group
40	2	3	2	1
35	3	3	3	1
40	1	1	1	2
40	1	3	1	3
35	1	2	1	4
35	1	1	1	5
40	1	1	1	6
35	2	2	2	7
35	2	2	2	8
40	2	20	2	9
35	4	5	4	9
35	1	1	1	10

*Each irradiated BC_1 plant producing offspring with a rearranged added maize chromosome either as a deletion or a translocation with an oat chromosome represents at least one independent mutation event. Chromosomal BC_1-plant mutants, however, are not necessarily based only on one single rearrangement event because of potentially different mutations in different embryo cells during irradiation. Thus, the resulting chimeric nature of the radiated BC_1 plants among their different tillers can produce and transmit more then one maize chromosome derivative to the corresponding BC_1F_2-offspring plants.

A set of 45 SSR markers distributed along maize chromosome 1 was used to determine by a presence vs. absence test for each marker approximate points of maize chromosome breakage in the 171 BC_1F_2 plants. All 45 SSR markers were present in 98 BC_1F_2 plants, which represent 50 families. These plants were considered as possessing either a whole maize chromosome without a break or a reciprocal oat–maize translocation. These plants will be self-pollinated, and offspring of those with reciprocal translocations will be selected for the segregating translocated chromosomes. Nine BC_1F_2 plants, forming five families, showed complex rearrangements, including interstitial deletions and multiple translocations with oat. Forty-four BC_1F_2 plants constituted 21 families, each representing one likely independent (chromosome rearrangement) event (Table 3). These 44 RH plants were placed into 10 panel groups, with plants within a group resulting from similar maize chromosome breaks based on marker analysis (Table 3 and Fig. 2). Fig. 2 illustrates the definition of eight segments by seven breaks in selected RH lines for maize chromosome 1 representing the 10 groups (Table 3). Plants with only one break in their maize chromosome, and thus possessing only one deficiency or one oat–maize translocation, are shown in the first panel (Fig. 2). The markers shown in the left column are the first and last marker present or absent and frame the breakpoints. The points define six segments on the short arm (*p-umc1354* to *p-umc168*), one large segment spanning the centromere region (*p-umc1626* to *p-mmc0041*), and one additional segment on the long arm (*p-bnlg1720* to *p-umc2244*).

The apparently single breakpoint in the long arm of chromo-

SSR Marker	Map Coordinates According IBM2 Neighbors 1 Map	Genetic Bins	OMAd1.07	Group 1	Group 2	Group 4	Group 5	Group 6	Group 7	Group 9	Group 3	Group 8	Group 10	RH Segments
p-umc1354	0.00	1.00	+	-	-	-	-	-	-	-	+	+	+	1
p-umc1071	85.20	1.01	+	-	-	-	-	-	-	-	+	+	+	
p-umc1727	97.97		+	+	-	-	-	-	-	-	+	+	+	2
p-umc1397	226.40	1.03	+	+	-	-	-	-	-	-	+	+	+	
p-umc1479	257.40		+	+	+	-	-	-	-	-	-	+	+	3
p-umc1514	301.37		+	+	+	-	-	-	-	-	-	+	+	
p-bnlg2238	326.70	1.04	+	+	+	+	-	-	-	-	-	+	+	4
p-umc2217	358.40		+	+	+	+	+	-	-	-	-	+	+	5
p-bnlg1811	386.40		+	+	+	+	+	-	-	-	-	+	+	
p-umc2025	417.00	1.05	+	+	+	+	+	+	-	-	-	+	+	6
p-umc1689	445.10		+	+	+	+	+	+	-	-	-	+	+	
p-umc1626	464.08		+	+	+	+	+	+	+	-	-	-	+	7
p-mmc0041	787.49	1.08	+	+	+	+	+	+	+	-	-	-	+	
p-bnlg1720	858.39	1.09-1.10	+	+	+	+	+	+	+	+	-	-	-	8
p-umc2244	1120.30	1.12	+	+	+	+	+	+	+	+	-	-	-	

Fig. 2. Panel of the first RH lines for maize chromosome 1. Shown are the 15 SSR markers that frame the seven breakpoints, hence define the RH segments between *p-umc1354* (most distal on the short arm) and *p-umc2244* (most distal on the long arm) markers representing a genetic distance of more than 1,120 map units according to the IBM2 map.

some 1 as defined by groups 9 and 10 is represented by seven independent breakage events. This break is marked proximally by the SSR marker *p-mmc0041* and distally by the SSR marker *p-bnlg1720* (Fig. 2). The portion (7 of 21) of lines that break at the same point or very similar points indicates a preferential breakage and/or transmission of the chromosome segments to offspring. On the short arm, five independent events (5 of 21) demonstrate a common break in the range that is marked by the SSRs *p-umc1071* and *p-umc1727* defining segment 1. All other breakpoints are defined by one or two events. Most striking is that we did not observe a centric break resulting in either a telocentric maize chromosome or a centric maize–oat translocation. This situation left intact a large area spanning the centromere and the proximal regions on both arms, segment 7 (Fig. 2), and it is in strong contrast to earlier results from the production of maize chromosome 9 RHs (12) that showed that the level of chromosome 9 breakage across the chromosome was relatively constant.

The physical sizes of the segments differ remarkably as shown by GISH experiments (Fig. 3). In the line 1.07.3-001.3-03 (a sibling from group 7), the maize chromosome shows a primary constriction that defines the deficient short arm to ≈80% of its regular WT metaphase length. This result would mean that the missing element (20%) represents a genetic size of at least 445 map units separated into 6 distinct segments by 8 of 10 groups. On the other hand, GISH of line 1.07.2-007.3-04 (a sibling from group 9) shows the distal maize chromosome fragment translocated to an oat chromosome. The fragment length corresponds to ≈20% of the long-arm WT length in metaphase and visualizes segment 8. Even considering that the definition of the single segment by two markers varies over a considerable genetic distance, the segment 7 spans approximately the proximal 80% of the short and the proximal 80% of the long arm of the genetic map of maize chromosome 1. The line 1.07.1-020.3-01 (sibling of group 3) shows by GISH analysis a fragment of ≈15% of the WT short-arm length translocated to an oat chromosome. This fragment visualizes the length of the segments 1 and 2 together marked by SSRs *p-umc1397* and *p-umc1479*.

Summary

The current set of disomic oat–maize addition lines involves all maize chromosomes in different oat backgrounds with the exception of chromosome 10. The maize chromosome 10 addition progeny has only the short arm; a fertile disomic telocentric addition line is available. The whole chromosome 10 added to haploid GAF-Park oat does allow the availability of DNA. Although not fertile, and therefore not capable of producing disomic addition offspring, we continue to maintain the original plant vegetatively by tiller cloning under short-day growing conditions. The leaves show remarkable somatic stability for the added maize chromosome over a period of >3 years. The plant serves as a source for chromosome 10 genomic DNA and RNA.

The complete series of DNAs made from each maize chromosome addition has been used as a powerful tool to allocate genes and markers to chromosome. Ananiev *et al.* (31) used the oat–maize chromosome 9 addition line as the DNA source to construct a chromosome-specific cosmid library allowing the isolation of maize-specific repetitive DNA families. The low level of cross-hybridization under standard conditions between oat and maize genomic DNA makes it possible to screen libraries for maize species-specific sequences (31).

Oat–maize addition lines are ideal for mapping gene families and markers that have more than one copy on different chromosomes likely because of the duplicative nature of maize. For example, Okagaki *et al.* (32) mapped 350 ESTs and sequence tagged sites to chromosomes by a presence vs. absence test and

Fig. 3. GISH of metaphase chromosomes from root tips of three RH plants of the maize chromosome 1 panel. (*A*) Plant BC_1F_2, 1.07.3-001.3-3 (sibling of group 7), arrow points to the deficient short arm of maize chromosome 1; the chromosome lost ≈20% of its short arm. *p-umc1626* is the most distal present marker tested (see also Fig. 2). The yellow-painted chromosome visualizes the segments 7 and 8 representing the genetic distance of 656–675 map units (*B*) Plant BC_1F_2 1.07.2-007.3-4 (sibling of group 9), arrow points to the translocation fragment visualizing the RH segment 8. The translocation fragment accounts for ≈20% of the long-arm length representing the genetic distance of 261–332 map units (*C*) Plant BC_1F_2 1.07.1-020.3-1 (sibling of group 3), arrow points to the translocation fragment visualizing the RH segments 1 and 2 accounting for ≈15% of the short-arm length representing the distance of 226–257 map units.

demonstrated the usefulness of the complete addition line set. However, the true power of the addition lines as a tool for maize genomics and genetics may be that no marker polymorphism is required for large-scale mapping. The value of the plant material for gene expression studies has already been shown in an analysis of interchromosomal interaction with respect to expression of the maize gene *liguleless 3* on chromosome 3 (33) or the reduced susceptibility against the fungal pathogen *Puccinia coronata* f. sp. *avenae* in the oat–maize chromosome 5 addition lines (unpublished data).

With the development of RH lines from the oat–maize additions, markers can be placed to chromosome regions. Visualization by GISH of the rearranged maize chromosome fragments together with marker data helps to relate physical sizes to genetic distances along the chromosome arms. Besides the use of addition and radiation hybrid lines for mapping purposes, the extensive dissection of the maize genome provides powerful material for the targeted cloning of chromosome-specific DNA and to study chromosome-specific structures and their behavior in an alien background.

This is a joint contribution of the Minnesota Agricultural Experiment Station and the Department of Agriculture–Agricultural Research Service. This material is based upon work supported by the National Science Foundation Grant 011134.

1. Ahn, S. & Tanksley, S. D. (1993) *Proc. Natl. Acad. Sci. USA* **90,** 7980–7984.
2. Rines, H. W. & Dahleen, L. S. (1990) *Crop Sci.* **30,** 1073–1078.
3. Machan, F., Nesvadba, Z. & Ohnoutkova, L. (1995) *Genet. Slechteni (Prague)* **31,** 1–10.
4. Riera-Lizarazu, O., Rines, H. W. & Phillips, R. L. (1996) *Theor. Appl. Genet.* **93,** 123–135.
5. Rines, H. W., Riera-Lizarazu, O., Maquieira, S. B. & Phillips, R. L. (1996) in *Proceedings of the Fifth International Oat Conference and Eighth International Barley Genetic Symposium*, eds. Scoles, G. & Rossnagel, B. (University Extension Press, Saskatoon, SK, Canada), pp. 207–212.
6. Rines, H. W., Riera-Lizarazu, O., Nunez, V. M., Davis, D. M. & Phillips, R. L. (1997) in *In Vitro Production of Haploids in Higher Plants 4*, eds. Jain, S. M., Sopory, S. K. & Veilleux, R. E. (Kluwer, Dordrecht, The Netherlands), pp. 205–221.
7. Kynast, R. G., Okagaki, R. J., Odland, W. E., Russell, C. D., Livingston, S. M., Rines, H. W. & Phillips, R. L. (2000) *Maize Genetics Cooperation Newsletter* **74,** 60–61.
8. Davis, D. W. (1992) M.S. thesis (University of Minnesota, St. Paul).
9. Rines, H. W., Riera-Lizarazu, O. & Phillips, R. L. (1995) in *Modification of Gene Expression and Non-Mendelian Inheritance*, eds. Oono, K. & Takaiwa, F. (National Institute of Agrobiological Resources, Tsukuba, Japan), pp. 235–251.
10. Kynast, R. G., Riera-Lizarazu, O., Vales, M. I., Okagaki, R. J., Maquieira, S. B., Chen, G., Ananiev, E. V., Odland, W. E., Russell, C. D., Stec, A. O., *et al.* (2001) *Plant Physiol.* **125,** 1216–1227.
11. Palmer, L. E., Rabinowicz, P. D., O'Shaughnessy, A. L., Balija, V. S., Nascimento, L. U., Dike, S., de la Bastide, M., Martienssen, R. A. & McCombie, W. R. (2003) *Science* **302,** 2115–2117.
12. Whitelaw, C. A., Barbazuk, W. B., Pertea, G., Chan, A. P., Cheung, F., Lee, Y., Zheng, L., van Heeringen, S., Karamycheva, S., Bennetzen, J. L., *et al.* (2003) *Science* **302,** 2118–2120.
13. Cox, D. R., Burmeister, M., Price, E. R., Kim, S. & Myers, R. M. (1990) *Science* **250,** 245–250.
14. Riera-Lizarazu, O., Vales, M. I., Ananiev, E. V., Rines, H. W. & Phillips, R. L. (2000) *Genetics* **156,** 327–339.
15. Kynast, R. G., Okagaki, R. J., Rines, H. W. & Phillips, R. L. (2002) *Funct. Integr. Genomics* **2,** 60–69.
16. Monfort, A., Vicient, C. M., Raz, R., Puigdomenech, P. & Martinez-Izquierdo, J. A. (1995) *DNA Res.* **2,** 255–261.
17. Ananiev, E. V., Phillips, R. L. & Rines, H. W. (1998) *Proc. Natl. Acad. Sci. USA* **95,** 13073–13078.
18. Saghai-Maroof, M. A., Soliman, K. M., Jorgensen, R. A. & Allard, R. W. (1984) *Proc. Natl. Acad. Sci. USA* **81,** 8014–8018.
19. Sambrook, J., Fritsch, E. F. & Maniatis, T. (1989) *Molecular Cloning: A Laboratory Manual* (Cold Spring Harbor Lab. Press, Plainview, NY), 2nd Ed., pp. 1.42–1.46.
20. Feulgen, R. & Rössenbeck, M. (1924) *Z. Physiol. Chem.* **135,** 203–248.
21. Pickering, R. A., Hill, A. M. & Kynast, R. G. (1997) *Genome* **40,** 195–200.
22. Kynast, R. G., Okagaki, R. J., Granath, S. R., Rines, H. W. & Phillips, R. L. (2003) *Maize Genetics Cooperation Newsletter* **77,** 62–63.
23. Comeau, A., Nadeau, P., Plourde, A., Simard. R., Maes, O., Kelly, S., Harper, L., Lettre, J., Landry, B. & St-Pierre, C.-A. (1992) *Plant Sci. (Shannon, Irel.)* **81,** 117–125.
24. Laurie, D. A. & Bennett, M. D. (1989) *Genome* **32,** 953–961.
25. Islam, A. K. M. R., Shepherd, K. W. & Sparrow, D. H. B. (1981) *Heredity* **46,** 161–174.
26. Taketa, S. & Takeda, K. (1997) *Genes Genet. Syst.* **72,** 101–106.
27. Taketa, S. & Takeda, K. (1998) *Proc. Int. Wheat Genet. Symp. 9th* **3,** 72–74.
28. Taketa, S., Choda, M., Ohashi, R., Ichii, M. & Takeda, K. (2002) *Genome* **45,** 617–625.
29. Islam, A. K. M. R. & Shepherd, K. W. (2000) *Euphytica* **111,** 145–149.
30. Taketa, S. & Takeda, K. (2001) *Breed. Sci.* **51,** 199–206.
31. Ananiev, E. V., Riera-Lizarazu, O., Rines, H. W. & Phillips, R. L. (1997) *Proc. Natl. Acad. Sci. USA* **94,** 3524–3529.
32. Okagaki, R. J., Kynast, R. G., Livingston, S. M., Russell, C. D., Rines, H. W. & Phillips, R. L. (2001) *Plant Physiol.* **125,** 1228–1235.
33. Muehlbauer, G. J., Riera-Lizerazu, O., Kynast, R. G., Martin, D., Phillips, R. L. & Rines, H. W. (2000) *Genome* **43,** 1055–1064.

John Anthony Pickett, CBE, DSc, FRS

Rothamsted Research, Harpenden
Hertfordshire, United Kingdom

2008 — for remarkable discoveries of mechanisms governing plant-insect and plant-plant interactions. Their scientific contributions on chemical ecology have fostered the development of integrated pest management and significantly advanced agricultural sustainability.

For many years, Professor John A. Pickett has contributed significantly to the field of chemical ecology and agriculture worldwide, and has achieved an international reputation equaled by few. He has opened up new areas of research with novel approaches and then applied basic research results by developing innovative control strategies for agricultural pests and weeds. By integrating techniques from molecular biology, electrophysiology, biochemistry and behavior, he has positioned himself at the cutting-edge of the field. As a result, Pickett has become a world leader in the development of sustainable, environmentally-sound methods of insect pest management, based on use of semio-chemicals to manipulate the behavior of pests and their natural enemies.

CURRICULUM VITAE:

Place/date of birth: Leicester, United Kingdom, 21 April 1945
Marital status: Married, two children
Professional title and affiliation:
Scientific Director, Rothamsted Centre for Sustainable Pest and Disease Management, and Head, Department of Biological Chemistry, Rothamsted Research
Special Professor, School of Biology, University of Nottingham
School: King Edward VII Grammar School, Coalville (1956-1963)

Qualifications:

B.Sc., Honours chemistry, 2(i), 1967, University of Surrey (1963-1967)
Ph.D., Organic chemistry, 1971, University of Surrey, Compounds from dinitriles and hydrazine, Professor J.A. Elvidge, 1967-1970
Chartered Chemist, 1975
D.Sc., University of Nottingham, 1993
Chartered Scientist, 2004

Honours/Awards:

Special Professor, University of Nottingham, 1991
The Rank Prize, Nutrition and Crop Husbandry, 1995
Fellow of the Royal Society, 1996
Member of the Deutsche Akademie der Naturforscher Leopoldina, 2001
International Society of Chemical Ecology Medal, 2002
CBE, for Services to Biological Chemistry, 2004
Honorary Life Membership, the Association of Applied Biologists, 2004
Foreign Member of the Royal Swedish Academy of Agriculture and Forestry, 2005
Wolf Foundation Prize in Agriculture, 2008

Professional affiliations:

Fellow of the Royal Society of Chemistry, 1982
Fellow of the Royal Entomological Society, 1989
Member of the American Chemical Society, 1993
Member of the Royal Institution, 1997

Academic Positions:

External examiner, MSc Course, Pest Management, Imperial College at Silwood Park, 1992-1995
Chairman, Advisory Committee, School of Applied Chemistry, University of North London, 1993-1995
Honorary Member, Academic Staff, University of Reading, 1995-
External Examiner, Environmental Science, University of Sussex, 1997-2000

Career:

1970-1972 - Postdoctoral fellowship, University of Manchester Institute of Science and Technology, Synthesis and photochemistry of perfluoroalkylpyridazines, Professor R.N. Haszeldine, FRS.

1972-1976 - Senior scientist, Chemistry Department, Brewing Research Foundation, Redhill, Surrey. Chemical studies on flavour active components of hops and malt.

1976-1983 - Principal Scientific Officer, Department of Insecticides and Fungicides, Rothamsted Experimental Station: Leading and coordinating studies on semiochemical aspects of insect chemical ecology.

1984-present - Head, Band 2, Individual Merit, of the Biological Chemistry Division (formerly of the Department of Insecticides and Fungicides, 1984-1993, then of the Biological and Ecological Chemistry Department, 1993-2001), Rothamsted (Individual Merit Promotion to Grade Band 2, 1993).

DISTINGUISHED LECTURES:

- Distinguished Lecture in Life Sciences, 1991, Boyce Thompson Institute for Plant Research at Cornell University
- Alfred M. Boyce Lecture, University of California, Riverside, 1993
- "Insects and Chemical Signals: A Volatile Situation", Royal Institution Discourse, November 1996
- "Keynotes in Natural Resources" lecture, Swedish University of Agricultural Sciences, Uppsala, 2 March 1998
- The Woolhouse Lecture, The Society for Experimental Biology Annual Meeting, University of York, 26 March 1998
- The 1999 Barrington Memorial Lecture, University of Nottingham, 29 April 1999
- The 2000 Cameron-Gifford Lecture, University of Newcastle, 9 February 2000
- The Andersonian Chemical Society Centenary Lecture, University of Strathclyde, 5 April 2006
- 17th Annual H.R. MacCarthy Pest Management Lecture, University of British Columbia, Canada, "Exploiting induced and constitutive plant stress signalling in crop protection", 10th October 2007
- The Royal Society Croonian Lecture, 3rd June 2008

POSTS OF DISTINCTION:

- Editorial Board, Journal of Chemical Ecology, 1991-
- Councillor, International Society of Chemical Ecology, 1991-2000
- Visiting Group IPO-DLO, Netherlands, 1993
- Vice-President, the International Society of Chemical Ecology, 1994
- President, International Society of Chemical Ecology, 1995
- Expert, committee for evaluation of INCO-Copernicus, European Commission DGXII, 1996
- Scientific Adviser, International Foundation for Science, Stockholm, Sweden, 1996-
- Member of Royal Society Conference Grants Committee, 1997-1999
- Member of Royal Society Sectional Committee 9, 1997-2000

- Chairman, Organiser, Royal Society Discussion Meeting, April 1998; co-editor of associated volume "Insecticide Resistance: from Mechanisms to Management", CABI publishing, 1999
- Chairman, Organiser, Biotechnology in Agriculture, IBC UK Conference, London, 6-7 July, 1998
- Guest Editor, The Biochemist, Insect Supersense, August 1998
- Vice-Chairman, Executive Committee, Ninth International Congress of Pesticide Chemistry, London, 2-7 August 1998
- Proposer, Chairman, Novartis Foundation Symposium 223, Insect-Plant Interactions and Induced Plant Defence, 13-15 October 1998
- Presenter of expert evidence to the House of Lords Select Committee on the European Communities, Organic Farming and the European Union, May 1999
- Member of Peer Review Committee on Environmental Biology and Chemical Ecology in the Boyce Thompson Institute for Plant Research at Cornell University, October 1999
- Chairman, Royal Society Working Group on the Future of Sites of Special Scientific Interest (SSSIs), 2000-1
- Member of Royal Society Council, 2000-2002
- Chairman, Royal Society Copus Grants Panel, 2002-2004
- Member, Plant Sciences Advisory Board, University of Reading, 2002-
- Member of review panel of the Swiss National Science Foundation, National Centre of Competence in Research (NCCR), Plant Survival in Natural and Agricultural Ecosystems, 2002-
- Member of the Royal Society Working Group monitoring the Planetary Protection activities associated with the Beagle 2 probe to Mars, 2002-2003
- Chairman of Scientific Advisory Board for the Max-Planck-Gesellschaft, Institute of Chemical Ecology, Jena, Germany, 2007- (and Member from 2003)
- Member of the BBSRC Appointments Board, 2004-2007
- Chairman of the International Centre of Insect Physiology and Ecology (ICIPE) Governing Council, Nairobi, 2007- (and Member from 2005)
- Editorial Board, International Journal of Tropical Insect Science, 2006-
- Member of the Royal Society's Dorothy Hodgkin Fellowship Selection Panel, 2006-2008
- Chairman of the Royal Society Working Group on Development for Biofuels, 2006-2007
- Chairman of the Royal Society Sectional Committee 9, 2007-2009
- Member of Governing Body of Uppingham School as representative for the Royal Society, 2007-
- Member of the Royal Society's Soirée Committee, 2008-2010
- Editorial Board, Phytochemistry, 2008-

SCIENTIFIC INTERESTS:

Head of Department: The objectives of the Department are to make crop protection involving natural and synthetic chemical safer and more effective. The Department is responsible for important discoveries, including the synthetic pyrethroids, and has a prominent world position in research areas such as chemical ecology and evolutionary response of insects to insecticidal stress.

Personal research: The chemical ecology of interactions between insects, and some other animals, and between insects and plants. This specifically involves the chemical characterisation of molecular structures that influence the development and behaviour (semiochemicals) of insects and other organisms. Pickett was the first to identify aphid, mosquito and sandfly pheromones. Research extends to the biochemistry and molecular biology of secondary plant metabolites that act as semiochemicals and the mechanisms by which they are employed by insects. The long term objectives are to develop pheromones and other semiochemicals for new methods of pest control. This is exemplified particularly by his recent work in Africa. His recent studies have turned more to devising novel ways of controlling vectors of pathogens attacking the human population and farm animals.

LIST OF PUBLICATIONS:

1. J.A. Elvidge and J.A. Pickett (1972) Heterocyclic imines and amines. Part 13. 3,6-Dihydrazinopyridazine and the nature of the reaction between 3,6-dimethoxypyridazine and hydrazine. *Journal of the Chemical Society Perkin I*, 1483-1488.
2. J.A. Elvidge and J.A. Pickett (1972) Heterocyclic imines and amines. Part 14. Products from 2,5-diiminopyrrolidine (succinimidine) and hydrazine. *Journal of the Chemical Society Perkin I*, 2346-2351.
3. M.G. Barlow, R.N. Haszeldine and J.A. Pickett (1978) Heterocyclic polyfluoro-compounds. Part 26. Synthesis of 3,6-bistrifluoromethyl pyridazines and -di-hydropyridazines. *Journal of the Chemical Society Perkin I*, 378-380.
4. J.A. Pickett (1973) Use of micro-bead anion-exchange resin for direct estimation of flavour nucleotides in complex solutions. *Journal of Chromatography* 81, 156-159.
5. J.A. Pickett (1974) Estimation of nucleotides in beers and their effect on flavour. *Journal of the Institute of Brewing* 80, 42-47.
6. D.R.J. Laws and J.A. Pickett (1974) Brewing, malting and allied processes. *Reports on the Progress of Applied Chemistry* 59, 345-357.
7. J.A. Pickett, J. Coates and F.R. Sharpe (1975) Distortion of essential oil composition during isolation by steam distillation. *Chemistry and Industry*, 571-572.

8. J.A. Pickett, J. Coates and F.R. Sharpe (1975) Improvement of hop aroma in beer. *Proceedings of the European Brewery Convention, 15th Congress, Nice*, 123-140.
9. D.R.J. Laws and J.A. Pickett (1975) Brewing, malting and allied processes. *Reports on The Progress of Applied Chemistry* **60**, 451-463.
10. J.A. Pickett (1976) Studies on flavour-active sulphur components of hops and beer. *Proceedings of the Analytical Division of the Chemical Society*, 215-217.
11. J.A. Pickett, J. Coates and F.R. Sharpe (1976) Procedure for non-destructive concentration of flavour-active components of beer. *Journal of the Institute of Brewing* **82**, 228-233.
12. J.A. Pickett, J. Coates and F.R. Sharpe (1976) Chemical characterisation of differences between ales and lagers. *Journal of the Institute of Brewing* **82**, 233-238.
13. J.A. Pickett, T.L. Peppard and F.R. Sharpe (1976) Effect of 'sulphuring' on hop oil composition. *Journal of the Institute of Brewing* **82**, 288-289.
14. J.A. Pickett, F.R. Sharpe and T.L. Peppard (1976) Stability of essential oil of hops. *Journal of the Institute of Brewing* **82**, 330-333.
15. J.A. Pickett and F.R. Sharpe (1976) Effect of reduction in hop oil content on rate of deterioration of alpha-acid in hops. *Journal of the Institute of Brewing* **82**, 333.
16. D.R.J. Laws, N.A. Bath and J.A. Pickett (1977) Production of solvent-free isomerized extracts. *Journal of the American Society of Brewing Chemists* **35**, 187-191.
17. D.R.J. Laws, N.A. Bath, J.A. Pickett, C.S. Ennis and A.G. Wheldon (1977) Preparation of hop extracts without using organic solvents. *Journal of the Institute of Brewing* **83**, 39-40.
18. J.A. Pickett, T.L. Peppard and F.R. Sharpe (1977) Recent developments in low temperature steam distillation of hop oil. *Journal of the Institute of Brewing* **83**, 302-304.
19. J.A. Pickett, F.R. Sharpe and T.L. Peppard (1977) Aerial oxidation of humulene. *Chemistry and Industry*, 30-31.
20. D.R.J. Laws, T.L. Peppard, F.R. Sharpe and J.A. Pickett (1978) Recent developments in imparting hop character to beer. *Journal of the American Society of Brewing Chemists* **36**, 69-72.
21. J.A. Pickett (1979) Behaviour controlling chemicals. *Education in Chemistry* **16**, 44-47.
22. A.W. Ferguson, J.B. Free, J.A. Pickett and M. Winder (1979) Techniques used for studying honeybee pheromones involved in clustering, and experiments on the effect of Nasonov and queen pheromones. *Physiological Entomology* **4**, 339-344.

23. D.C. Griffiths and J.A. Pickett (1980) A potential application of aphid alarm pheromones. *Entomologia Experimentalis et Applicata* 27, 199-201.
24. J.A. Pickett and D.C. Griffiths (1980) Composition of aphid alarm pheromones. *Journal of Chemical Ecology* 6, 349-360.
25. J.A. Pickett, I.H. Williams, A.P. Martin and M.C. Smith (1980) The Nasonov pheromone of the honey bee, *Apis mellifera* L. (Hymenoptera: Apidae). Part I. Chemical characterisation. *Journal of Chemical Ecology* 6, 425-434.
26. J.A. Pickett and J.W. Stephenson (1980) Plant volatiles and components influencing behaviour of the field slug, *Deroceras reticulatum* (Mu"ll). *Journal of Chemical Ecology* 6, 435-444.
27. J.B. Free, A.W. Ferguson and J.A. Pickett (1981) Evaluation of the various components of the Nasonov pheromone used by clustering honeybees. *Physiological Entomology* 6, 263-268.
28. I.H. Williams, J.A. Pickett and A.P. Martin (1981) The Nasonov pheromone of the honey bee, *Apis mellifera* L. (Hymenoptera: Apidae). Part II. Bioassay of the components using foragers. *Journal of Chemical Ecology* 7, 225-237.
29. J.A. Pickett, I.H. Williams, M.C. Smith and A.P. Martin (1981) The Nasonov pheromone of the honey bee, *Apis mellifera* L. (Hymenoptera: Apidae). Part III. Regulation of pheromone composition and production. *Journal of Chemical Ecology* 7, 543-554.
30. J.B. Free, J.A. Pickett, A.W. Ferguson and M.C. Smith (1981) Synthetic pheromones to attract honeybee (*Apis mellifera*) swarms. *Journal of Agricultural Science Cambridge* 97, 427-431.
31. I.H. Williams, J.A. Pickett and A.P. Martin (1981) Attraction of honeybees to flowering plants by using synthetic Nasonov pheromone. *Entomologia Experimentalis et Applicata* 30, 199-201.
32. J.A. Pickett, I.H. Williams and A.P. Martin (1982) (*Z*)-11-Eicosen-1-ol, an important new pheromonal component from the sting of the honey bee, *Apis mellifera* L. (Hymenoptera: Apidae). *Journal of Chemical Ecology* 8, 163-175.
33. B.R. Laurence and J.A. Pickett (1982) *Erythro*-6-Acetoxy-5-hexadecanolide, the major component of a mosquito oviposition attractant pheromone. *Journal of the Chemical Society Chemical Communications*, 59-60.
34. R.W. Gibson, A.D. Rice, J.A. Pickett, M.C. Smith and R.M. Sawicki (1982) The effects of the repellents dodecenoic acid and polygodial on the acquisition of non-, semi- and persistent plant viruses by the aphid *Myzus persicae*. *Annals of Applied Biology* 100, 55-59.
35. I.H. Williams, J.A. Pickett and A.P. Martin (1982) Nasonov pheromone of the honeybee, *Apis mellifera* L. (Hymenoptera: Apidae). Part IV. Comparative electroantennogram responses. *Journal of Chemical Ecology* 8, 567-574.

36. J.A. Pickett, G.W. Dawson, R.W. Gibson, D.C. Griffiths, A.D. Rice, R.M. Sawicki, M.C. Smith and C.M. Woodcock (1982) Controlling aphid behaviour. *Les Colloques de l'INRA* 7, 243-252.
37. J.B. Free, A.W. Ferguson, J.A. Pickett and I.H. Williams (1982) Use of unpurified Nasonov pheromone components to attract clustering honeybees. *Journal of Apicultural Research* 21, 26-29.
38. J.B. Free, I.H. Williams, J.A. Pickett, A.W. Ferguson and A.P. Martin (1982) Attractiveness of (*Z*)-11-eicosen-1-ol to foraging honeybees. *Journal of Apicultural Research* 21, 151-156.
39. D.C. Griffiths, J.A. Pickett and C.M. Woodcock (1982) Behaviour of alatae of *Myzus persicae* (Sulzer) (Hemiptera: Aphididae) on chemically treated surfaces after tethered flight. *Bulletin of Entomological Research* 72, 687-693.
40. G.W. Dawson, D.C. Griffiths, J.A. Pickett, M.C. Smith and C.M. Woodcock (1982) Improved preparation of (*E*)-β-farnesene and its activity with economically important aphids. *Journal of Chemical Ecology* 8, 1111-1117.
41. G.W. Dawson, R.W. Gibson, D.C. Griffiths, J.A. Pickett, A.D. Rice and C.M. Woodcock (1982) Aphid alarm pheromone derivatives affecting settling and the transmission of plant viruses. *Journal of Chemical Ecology* 8, 1377-1388.
42. G.G. Briggs, G.W. Dawson, R.W. Gibson, D.C. Griffiths, J.A. Pickett, A.D. Rice, M.F. Stribley and C.M. Woodcock (1983) Compounds derived from the aphid pheromone that interfere with colonization and virus transmission by aphids. *Aspects of Applied Biology* 2, 1983, *Pests, Diseases, Weeds and Weed Beet in Sugar Beet*, 41-43.
43. J.A. Pickett, G.W. Dawson, D.C. Griffiths, Liu X., E.D.M. Macaulay, I.H. Williams and C.M. Woodcock (1983) Stabilizing pheromones for field use: propheromones. *Proceedings of the 10th International Congress of Plant Protection* 1, 271.
44. D.C. Griffiths, G.W. Dawson, J.A. Pickett and C.M. Woodcock (1983) Uses of the aphid alarm pheromone and derivatives. *Proceedings of the 10th International Congress of Plant Protection* 1, 272.
45. G.G. Briggs, G.W. Dawson, R.W. Gibson, D.C. Griffiths, J.A. Pickett, A.D. Rice, M.F. Stribley and C.M. Woodcock (1983) Compounds derived from the aphid alarm pheromone of potential use in controlling colonization and virus transmission by aphids. *Proceedings of the Fifth International Congress of Pesticide Chemistry (IUPAC) Kyoto, 1982* 2, 117-122.
46. J.B. Free, A.W. Ferguson and J.A. Pickett (1983) Effect of the components of the Nasonov pheromone on its release by honeybees at the hive entrance. *Journal of Apicultural Research* 22, 155-157.
47. R.W. Gibson, J.A. Pickett, G.W. Dawson, A.D. Rice and C. Venables (1983) Pyrethroid insecticides and aphid repellents as means of controlling potato

virus Y. *Proceedings of the 10th Anniversary Conference of the International Potato Center, Peru, 1982*, 87-88.

48. R.W. Gibson and J.A. Pickett (1983) Wild potato repels aphids by release of aphid alarm pheromone. *Nature* **302**, 608-609.
49. G.W. Dawson, D.C. Griffiths, J.A. Pickett and C.M. Woodcock (1983) Decreased response to alarm pheromone by insecticide resistant aphids. *Naturwissenschaften* **70**, 254-255.
50. J.B. Free, A.W. Ferguson and J.A. Pickett (1983) A synthetic pheromone lure to induce worker honeybees (*Apis mellifera* L.) to consume water and artificial forage. *Journal of Apicultural Research* **22**, 224-228.
51. J.A. Pickett, G.W. Dawson, D.C. Griffiths, Liu Xun, E.D.M. Macaulay and C.M. Woodcock (1984) Propheromones: an approach to the slow release of pheromones. *Pesticide Science* **15**, 261-264.
52. R.W. Gibson, J.A. Pickett, G.W. Dawson, A.D. Rice and M.F. Stribley (1984) Effects of aphid alarm pheromone derivatives and related compounds on non- and semi-persistent plant virus transmission by *Myzus persicae*. *Annals of Applied Biology* **104**, 203-209.
53. Liu X., E.D.M. Macaulay and J.A. Pickett (1984) Propheromones that release pheromonal carbonyl compounds in light. *Journal of Chemical Ecology* **10**, 809-822.
54. J.A. Pickett (1984) Prospects for new chemical approaches to insect control. *Chemistry and Industry*, 657-660.
55. M.M. Blight, J.A. Pickett, M.C. Smith and L.J. Wadhams (1984) An aggregation pheromone of *Sitona lineatus*. *Naturwissenschaften* **71**, S.480.
56. J.B. Free, J.A. Pickett, A.W. Ferguson, J.R. Simpkins and C. Williams (1984) Honeybee Nasonov pheromone lure. *Bee World* **65**, 175-181.
57. G.C. Scott, J.A. Pickett, M.C. Smith, C.M. Woodcock, P.G.W. Harris, R.P. Hammon and H.D. Koetecha (1984) Seed treatments for controlling slugs in winter wheat. *Proceedings of the British Crop Protection Conference: Pests and Diseases*, 133-138.
58. J.A. Pickett, G.W. Dawson, J.B. Free, D.C. Griffiths, W. Powell, I.H. Williams and C.M. Woodcock (1984) Pheromones in the management of beneficial insects. *Proceedings of the British Crop Protection Conference: Pests and Diseases*, 247-254.
59. G.W. Dawson, D.C. Griffiths, J.A. Pickett, M.C. Smith and C.M. Woodcock (1984) Natural inhibition of the aphid alarm pheromone. *Entomologia Experimentalis et Applicata* **36**, 197-199.
60. B.R. Laurence, K. Mori, T. Otsuka, J.A. Pickett and L.J. Wadhams (1985) Absolute configuration of mosquito oviposition attractant pheromone, 6-acetoxy-5-hexadecanolide. *Journal of Chemical Ecology* **11**, 643-648.

61. B.R. Laurence and J.A. Pickett (1985) An oviposition attractant pheromone in *Culex quinquefasciatus* Say (Diptera: Culicidae). *Bulletin of Entomological Research* 75, 283-290.
62. J.A. Pickett (1985) Production of behaviour-controlling chemicals by crop plants. *Philosophical Transactions of the Royal Society of London* 310, 235-239.
63. J.B. Free, J.A. Pickett, A.W. Ferguson, J.R. Simpkins and M.C. Smith (1985) Repelling foraging honeybees with alarm pheromones. *Journal of Agricultural Science, Cambridge* 105, 255-260.
64. C. Wall, J.A. Pickett, D.G. Garthwaite and N. Morris (1985) A female sex pheromone in the pea midge, *Contarinia pisi. Entomologia Experimentalis et Applicata* 39, 11-14.
65. Liu Xun, Z-n. Zhang, Kong Jie, J.A. Pickett, Pan Yongcheng, Xie Yige and Gu Jiechen (1985) Field attractant activity of the synthetic sex pheromone of diamondback moth, *Plutella xylostella. Acta Ecologica Sinica* 5, 249-256.
66. J.A. Pickett (1986) Biotechnology in the service of pest control. *Proceedings of National Agricultural Conference, 6 March 1986.*
67. G.G. Briggs, G.R. Cayley, G.W. Dawson, D.C. Griffiths, E.D.M. Macaulay, J.A. Pickett, M.M. Pile, L.J. Wadhams and C.M. Woodcock (1986) Some fluorine-containing pheromone analogues. *Pesticide Science* 17, 441-448.
68. I.K. Kigatiira, J.W.L. Beament, J.B. Free and J.A. Pickett (1986) Using synthetic pheromone lures to attract honeybee colonies in Kenya. *Journal of Apicultural Research* 25, 85-86.
69. E.D.M. Macaulay, G.W. Dawson, Liu Xun and J.A. Pickett (1986) Field performance of synthetic diamondback moth sex pheromones. *Aspects of Applied Biology* 12, 105-116.
70. J.A. Pickett and D.C. Griffiths (1986) Electrostatic sprayers for behaviour-controlling chemicals. *AFRC Science Sprays & Sprayers,* 18-19.
71. G.W. Dawson, D.C. Griffiths, J.A. Pickett, L.J. Wadhams and C.M. Woodcock (1986) Plant compounds that synergise activity of the aphid alarm pheromone. *Proceedings of the British Crop Protection Conference – Pests and Diseases,* 829-834.
72. G.W. Dawson, D.C. Griffiths, A. Hassanali, J.A. Pickett, R.T. Plumb, B.J. Pye, L.E. Smart and C.M. Woodcock (1986) Antifeedants: a new concept for control of barley yellow dwarf virus in winter cereals. *Proceedings of the British Crop Protection Conference – Pests and Diseases,* 1001-1008.
73. J.A. Pickett (1986) Honey bee pheromones: some recent developments in controlling honey bee behaviour. *The Gooding Memorial Lecture 1985,* pp. 1-11. (Published by the Central Association of Bee–Keepers).
74. G.W. Dawson, D.C. Griffiths, N.F. Janes, A. Mudd, J.A. Pickett, L.J. Wadhams and C.M. Woodcock (1987) Identification of an aphid sex pheromone. *Nature* 325, 614-616.

75. J.A. Pickett, G.W. Dawson, D.C. Griffiths, A. Hassanali, L.A. Merritt, A. Mudd, M.C. Smith, L.J. Wadhams, C.M. Woodcock and Z-n. Zhang (1987) Development of plant-derived antifeedants for crop protection. In: *Pesticide Science and Biotechnology*, pp. 125-128. Editors R. Greenhalgh and T.R. Roberts. (Blackwell Scientific Publications).
76. D.C. Griffiths and J.A. Pickett (1987) Novel chemicals and their formulation for aphid control. *Proceedings of the 14th International Symposium on Controlled Release of Bioactive Materials* **14**, 243-244.
77. B. Mauchamp and J.A. Pickett (1987) Juvenile hormone-like activity of (*E*)-β-farnesene derivatives. *Agronomie* **7**, 523-529.
78. J.A. Pickett, G.W. Dawson, B.R. Laurence, M.M. Pile, M.C. Smith and L.J. Wadhams (1986) Development of the oviposition attractant pheromone for control of *Culex* sp. mosquitoes. *Abstracts 6th International Congress Pesticide Chemistry, IUPAC, Ottawa, 1986*, 2C-07.
79. J.A. Pickett, G.R. Cayley, G.W. Dawson, D.C. Griffiths, S.H. Hockland, B. Marples, R.T. Plumb and C.M. Woodcock (1986) Use of the alarm pheromone and derivatives against aphid-mediated damage. *Abstracts 6th International Congress Pesticide Chemistry, IUPAC, Ottawa, 1986*, 2C-08.
80. 79a. S.H. Hockland, G.W. Dawson, D.C. Griffiths, B. Marples, J.A. Pickett and C.M. Woodcock (1986) The use of aphid alarm pheromone (*E*)-β-farnesene to increase effectiveness of the entomophilic fungus *Verticillium lecanii* in controlling aphids on chrysanthemums under glass. In: *Fundamental and Applied Aspects of Invertebrate Pathology*, p. 252. Editors R.A. Samson, J.M. Vlak and R. Peters. (The Netherlands: Society of Invertebrate Pathology).
81. Liu Xun, Kong Jie, Z-n. Zhang, J.A. Pickett, Pan Yongcheng and Meng Xiaoyun (1987) Field attraction of the photosensitive propheromone to the diamondback moth, *Plutella xylostella* (L). *Sinozoologia* **5**, 15-19.
82. G.W. Dawson, D.C. Griffiths, J.A. Pickett, L.J. Wadhams and C.M. Woodcock (1987) Plant derived synergists of alarm pheromone from turnip aphid, *Lipaphis (Hyadaphis) erysimi* (Homoptera, Aphididae). *Journal of Chemical Ecology* **13**, 1663-1671.
83. G.W. Dawson, D.C. Griffiths, J.A. Pickett, R.T. Plumb, C.M. Woodcock and Z-n. Zhang (1988) Structure/activity studies on aphid alarm pheromone derivatives and their field use against transmission of barley yellow dwarf virus. *Pesticide Science* **22**, 17-30.
84. J.A. Pickett (1988) Integrating use of beneficial organisms with chemical crop protection. *Philosophical Transactions of the Royal Society of London* B **318**, 203-211.
85. J.A. Pickett (1988) Chemical pest control – the new philosophy. *Chemistry in Britain* **24**, 137-142.
86. J.A. Pickett (1988) The future of semiochemicals in pest control. *Aspects of Applied Biology* **17**, 397-406.

87. W.A. Otieno, T.O. Onyango, M.M. Pile, B.R. Laurence, G.W. Dawson, L.J. Wadhams and J.A. Pickett (1988) A field trial of the synthetic oviposition pheromone with *Culex quinquefasciatus* Say (Diptera: Culicidae) in Kenya. *Bulletin of Entomological Research* 78, 463-470.
88. G.W. Dawson, D.C. Griffiths, L.A. Merritt, A. Mudd, J.A. Pickett, L.J. Wadhams and C.M. Woodcock (1988) The sex pheromone of the greenbug, *Schizaphis graminum. Entomologia Experimentalis et Applicata* 48, 91-93.
89. D.C. Griffiths, A. Hassanali, L.A. Merritt, A. Mudd, J.A. Pickett, S.J. Shah, L.E. Smart, L.J. Wadhams and C.M. Woodcock (1988) Highly active antifeedants against coleopteran pests. *Proceedings of the Brighton Crop Protection Conference – Pests and Diseases*, 1041-1046.
90. Y. Asakawa, G.W. Dawson, D.C. Griffiths, J-Y. Lallemand, S.V. Ley, K. Mori, A. Mudd, M. Pezechk–Leclaire, J.A. Pickett, H. Watanabe, C.M. Woodcock and Z-n. Zhang (1988) Activity of drimane antifeedants and related compounds against aphids, and comparative biological effects and chemical reactivity of (–)– and (+)– polygodial. *Journal of Chemical Ecology* 14, 1845-1855.
91. G.W. Dawson, N.F. Janes, A. Mudd, J.A. Pickett, A.M.Z. Slawin, L.J. Wadhams and D.J. Williams (1989) The aphid sex pheromone. *Pure and Applied Chemistry* 61, 555-558.
92. Z-n. Zhang, Liu Xun and J.A. Pickett (1988) Several aphid alarm pheromone analogues possessing biological activity. *Acta Entomologica Sinica* 31, 435-438.
93. I.F. Henderson, G.G. Briggs, N.P. Coward, G.W. Dawson, J.A. Pickett, J.I. Bullock and L.F. Larkworthy (1989) A new group of molluscicidal compounds. *1989 BCPC Monograph No. 41 Slugs and Snails in World Agriculture*, 289-294.
94. W.J. Airey, I.F. Henderson, J.A. Pickett, G.C. Scott, J.W. Stephenson and C.M. Woodcock (1989) Novel chemical approaches to mollusc control. *1989 BCPC Monograph No. 41 Slugs and Snails in World Agriculture*, 301-307.
95. G.W. Dawson, J.A. Pickett and L.J. Wadhams (1989) S344 Case Study 2. Pheromones. In: *Organic chemistry: a synthesis approach*, pp. 1-24. The Open University.
96. J. Polonsky, S.C. Bhatnagar, D.C. Griffiths, J.A. Pickett and C.M. Woodcock (1989) Activity of quassinoids as antifeedants against aphids. *Journal of Chemical Ecology* 15, 993-998.
97. G.W. Dawson, D.L. Hallahan, A. Mudd, M.M. Patel, J.A. Pickett, L.J. Wadhams and R.M. Wallsgrove (1989) Secondary plant metabolites as targets for genetic modification of crop plants for pest resistance. *Pesticide Science* 27, 191-201.
98. D.C. Griffiths, J.A. Pickett, L.E. Smart and C.M. Woodcock (1989) Use of insect antifeedants against aphid vectors of plant virus disease. *Pesticide Science* 27, 269-276.

99. G.W. Dawson, B.R. Laurence, J.A. Pickett, M.M. Pile and L.J. Wadhams (1989) A note on the mosquito oviposition pheromone. *Pesticide Science* 27, 277-280.
100. M.M. Blight, J.A. Pickett, L.J. Wadhams and C.M. Woodcock (1989) Antennal responses of *Ceutorhynchus assimilis* and *Psylliodes chrysocephala* to volatiles from oilseed rape. *Aspects of Applied Biology* 23, 329-334.
101. J.A. Pickett (1989) Semiochemicals for aphid control. *Journal of Biological Education* 23, 180-186.
102. J.A. Pickett, L.J. Wadhams and C.M. Woodcock (1989) Chemical ecology and pest management: some recent insights. *Insect Science and its Application* 10, 741-750.
103. J.A. Pickett (1989) Towards zero pesticide residues: the biomanagement of pests and diseases of oilseed rape. *AFRC Institute of Arable Crops Research Report for 1989*, pp. 79-82.
104. G.W. Dawson, A. Mudd, J.A. Pickett, M.M. Pile and L.J. Wadhams (1990) Convenient synthesis of mosquito oviposition pheromone and a highly fluorinated analog retaining biological activity. *Journal of Chemical Ecology* 16, 1779-1789.
105. P.I. Sopp, A. Palmer and J.A. Pickett (1990) The effect of a plant-derived antifeedant on *Tetranychus urticae* and *Phytoseiulus persimilis*: "A first look". *SROP/WPRS Bulletin XIII/5 (1990)*, 198-201.
106. G.W. Dawson, D.C. Griffiths, L.A. Merritt, A. Mudd, J.A. Pickett, L.J. Wadhams and C.M. Woodcock (1990) Aphid semiochemicals - a review, and recent advances on the sex pheromone. *Journal of Chemical Ecology* 16, 3019-3030.
107. J. Hardie, M. Holyoak, J. Nicholas, S.F. Nottingham, J.A. Pickett, L.J. Wadhams and C.M. Woodcock (1990) Aphid sex pheromone components: age-dependent release by females and species-specific male response. *Chemoecology* 1, 63-68.
108. J.A. Pickett (1990) Gas chromatography-mass spectrometry in insect pheromone identification: three extreme case histories. In: *Chromatography and Isolation of Insect Hormones and Pheromones*, pp. 299-309. Editors A.R. McCaffery and I.D. Wilson. (Plenum Press).
109. C.A.M. Campbell, G.W. Dawson, D.C. Griffiths, J. Pettersson, J.A. Pickett, L.J. Wadhams and C.M. Woodcock (1990) Sex attractant pheromone of damson-hop aphid *Phorodon humuli* (Homoptera, Aphididae). *Journal of Chemical Ecology* 16, 3455-3465.
110. D.C. Griffiths, S.P. Maniar, L.A. Merritt, A. Mudd, J.A. Pickett, B.J. Pye, L.E. Smart and L.J. Wadhams (1991) Laboratory evaluation of pest management strategies combining antifeedants with insect growth regulator insecticides. *Crop Protection* 10, 145-151.

111. A. Cork, D.R. Hall, R.J. Hodges and J.A. Pickett (1991) Identification of major component of male-produced aggregation pheromone of larger grain borer, *Prostephanus truncatus* (Horn) (Coleoptera: Bostrichidae). *Journal of Chemical Ecology* 17, 789-803.
112. S.F. Nottingham, J. Hardie, G.W. Dawson, A.J. Hick, J.A. Pickett, L.J. Wadhams and C.M. Woodcock (1991) Behavioral and electrophysiological responses of aphids to host and nonhost plant volatiles. *Journal of Chemical Ecology* 17, 1231-1242.
113. J.A. Pickett (1991) Novel approaches to crop protection. *Proceedings of the Symposium on Poverty Alleviation through Chemistry for Improved Food Production, Colombo, Sri Lanka, 17 May, 1991*, 57-62.
114. J.A. Pickett (1991) Lower terpenoids as natural insect control agents. In: *Ecological Chemistry and Biochemistry of Plant Terpenoids*, pp. 297-313. Editors J.B. Harborne and F.A. Tomas-Barberan. (Clarendon Press, Oxford).
115. J.A. Pickett (1991) Pheromones: will their promise in insect pest control ever be achieved? *Bulletin of Entomological Research* 81, 229-232.
116. J.A. Pickett, L.J. Wadhams and C.M. Woodcock (1991) New approaches to the development of semiochemicals for insect control. *Proceedings of the Conference on Insect Chemical Ecology, Tábor 1990*, 333-345. (Academia Prague and SPB Acad. Publ., The Hague).
117. J.A. Pickett (1992) Even safer insecticides? *Chemistry and Industry* 1, 25-26.
118. M.M. Blight, G.W. Dawson, J.A. Pickett and L.J. Wadhams (1991) The identification and biological activity of the aggregation pheromone of *Sitona lineatus. Aspects of Applied Biology* 27, 137-142.
119. J.A. Pickett, L.J. Wadhams, C.M. Woodcock and J. Hardie (1992) The chemical ecology of aphids. *Annual Review of Entomology* 37, 67-90.
120. R. Garraway, L.D. Leake, M.G. Ford and J.A. Pickett (1991) The action of a range of volatile compounds on a tentacular preparation of the field slug, *Deroceras reticulatum* (Müll.). *Pesticide Science* 33, 240-242.
121. F. Rodriguez, D.L. Hallahan, J.A. Pickett and F. Camps (1992) Characterization of the Δ^{11}-palmitoyl-CoA-desaturase from *Spodoptera littoralis* (Lepidoptera: Noctuidae). *Insect Biochem. Molec. Biol.* 22, 143-148.
122. J.A. Pickett (1992). Pest control - novel approaches based on plant secondary metabolism. *Proceedings V Congresso Nazionale della Società Italiana di Fitochimica, Ischia, Napoli, 30 May-2 June, 1990.* 11 pp.
123. R. Garraway, L.D. Leake, M.G. Ford and J.A. Pickett (1992) The development of a chemoreceptive neurophysiological assay for the field slug, *Deroceras reticulatum* (Müll.). *Pesticide Science* 34, 97-98.
124. D.L. Hallahan, J.A. Pickett, L.J. Wadhams, R.M. Wallsgrove and C.M. Woodcock (1992) Potential of secondary metabolites in genetic engineering of crops for resistance. In: *Plant Genetic Manipulation for Crop Protection,*

pp. 215-248. Editors A.M.R. Gatehouse, V.A. Hilder and D. Boulter. (C.A.B. International, Wallingford).

125. J.A. Pickett, L.J. Wadhams and C.M. Woodcock (1992) Molecular determinants of pheromone activity. In: *Neurotox '91. Molecular Basis of Drug and Pesticide Action*, pp. 339-348. Editor I.R. Duce. (Elsevier Applied Science, London and New York).

126. J.A. Pickett (1992) Potential of novel chemical approaches for overcoming insecticide resistance. In: *Proceedings SCI Symposium "Resistance '91: Achievements and Developments in Combating Pesticide Resistance", July 1991, Harpenden, U.K.*, pp. 354-365. Editors I. Denholm, A.L. Devonshire and D.W. Hollomon. (Elsevier Applied Science, London and New York).

127. J.A. Pickett, W. Powell, L.J. Wadhams, C.M. Woodcock and A.F. Wright (1991) Biochemical interactions between plant-herbivore-parasitoid. *Proceedings 4th European Workshop on Insect Parasitoids, April 3-5 1991, Perugia, Italy*, pp. 1-14. Editor F. Bin. (REDIA, Vol. LXXIV, n.3, Appendice).

128. J.A. Pickett and C.M. Woodcock (1992) The future of chemical pest control. *Korean Journal of Applied Entomology* 31, 304-313.

129. J.A. Pickett, B.J. Pye, L.J. Wadhams, C.M. Woodcock and C.A.M. Campbell (1992) Potential applications of semiochemicals in aphid control. *1992 BCPC Monograph No. 51 Insect Pheromones and Other Behaviour-Modifying Chemicals: Application and Regulation*, 29-33.

130. J.A. Pickett and C.M. Woodcock (1992) Semiochemicals: molecular determinants of activity and biosynthesis. In: *Insect Molecular Science. Proceedings 16th Symposium of the Royal Entomological Society London, Imperial College London, September 1991*, pp. 141-149. Editors J.M. Crampton and P. Eggleston. (Academic Press, London).

131. J.A. Guldemond, A.F.G. Dixon, J.A. Pickett, L.J.Wadhams and C.M. Woodcock (1992) The role of host-plant odour and sex pheromones in mate recognition in the aphid *Cryptomyzus*. *Proceedings 8th International Symposium on Insect-Plant Relationships, Dordrecht*, pp. 119-121. Editors S.B.J. Menken, J.H. Visser and P. Harrewijn. (Kluwer Academic Publishers).

132. M.H. Pham-Delègue, M.M. Blight, M. Le Métayer, F. Marion-Poll, A.L. Picard, J.A. Pickett, L.J. Wadhams and C.M. Woodcock (1992) Plant chemicals involved in honeybee-rapeseed relationships: behavioural, electrophysiological and chemical studies. *Proceedings 8th International Symposium on Insect-Plant Relationships, Dordrecht*, pp. 129-130. Editors S.B.J. Menken, J.H. Visser and P. Harrewijn. (Kluwer Academic Publishers).

133. G. Powell, J. Hardie and J.A. Pickett (1992) Effects of the plant-derived antifeedant polygodial on aphid host selection behaviour. *Proceedings 8th International Symposium on Insect-Plant Relationships, Dordrecht*, pp. 181-182. Editors S.B.J. Menken, J.H. Visser and P. Harrewijn. (Kluwer Academic Publishers).

134. M.M. Blight, A.J. Hick, J.A. Pickett, L.E. Smart, L.J. Wadhams and C.M. Woodcock (1992) Volatile plant metabolites involved in host-plant recognition by the cabbage seed weevil, *Ceutorhynchus assimilis. Proceedings 8th International Symposium on Insect-Plant Relationships, Dordrecht*, pp. 105-106. Editors S.B.J. Menken, J.H. Visser and P. Harrewijn. (Kluwer Academic Publishers).
135. J.A. Pickett (1993) Honey bee pheromones: some recent developments in controlling honey bee behaviour. In: *Keeping Bees*, pp. 163-171. Compiled by J.B. Free. (Republication of *The Gooding Memorial Lecture 1985*). (The Central Association of Bee–Keepers, Ilford).
136. J.A. Pickett and C.M. Woodcock (1993) Chemical ecology of plants and insects: helping crops to help themselves. *Interdisciplinary Science Reviews* **18**, 68-72.
137. A.J. Mordue (Luntz), A. Blackwell, B.S. Hansson, L.J. Wadhams and J.A. Pickett (1992) Behavioural and electrophysiological evaluation of oviposition attractants for *Culex quinquefasciatus* Say (Diptera: Culicidae). *Experientia* **48**, 1109-1111.
138. J.M. Cottrell, I.F. Henderson, J.A. Pickett and D.J. Wright (1993) Evidence for glycosaminoglycans as a major component of trail mucus from the terrestrial slug, *Arion ater* L. *Comp. Biochem. Physiol.* **104B**, 455-468.
139. J. Hardie, S.F. Nottingham, G.W. Dawson, R. Harrington, J.A. Pickett and L.J. Wadhams (1992) Attraction of field-flying aphid males to synthetic sex pheromone. *Chemoecology* 3, 113-117.
140. C.A.M. Campbell, J. Pettersson, J.A. Pickett, L.J. Wadhams and C.M. Woodcock (1993) Spring migration of damson-hop aphid, *Phorodon humuli* (Homoptera, Aphididae), and summer host plant-derived semiochemicals released on feeding. *Journal of Chemical Ecology* **19**, 1569-1576.
141. G.W. Robertson, D.W. Griffiths, J.A.T. Woodford, A.N.E. Birch, J.A. Pickett and L.J. Wadhams (1993) A comparison of the flower volatiles from hawthorn and four raspberry cultivars. *Phytochemistry* **33**, 1047-1053.
142. G. Powell, J. Hardie and J.A. Pickett (1993) Effects of the antifeedant polygodial on plant penetration by aphids, assessed by video and electrical recording. *Entomol. exp. appl.* **68**, 193-200.
143. J.A. Pickett, L.J. Wadhams and C.M. Woodcock (1993) New approaches for semiochemicals in insect control. In: *New Frontiers in Rice Research*, pp. 235-240. Editors K. Muralidharan and E.A. Siddiq. (Directorate of Rice Research, Hyderabad).
144. J.A. Guldemond, A.F.G. Dixon, J.A. Pickett, L.J. Wadhams and C.M. Woodcock (1993) Specificity of sex pheromones, the role of host plant odour in the olfactory attraction of males, and mate recognition in the aphid *Cryptomyzus*. *Physiological Entomology* **18**, 137-143.

145. J.A. Pickett, L.J. Wadhams and C.M. Woodcock (1993) Exploiting behaviourally active phytochemicals in crop protection. In: *Phytochemistry and Agriculture*, pp. 62-75. Editors T.A. van Beek and H. Breteler. (Clarendon Press, Oxford).
146. G.W. Dawson, K.J. Doughty, A.J. Hick, J.A. Pickett, B.J. Pye, L.E. Smart and L.J. Wadhams (1993) Chemical precursors for studying the effects of glucosinolate catabolites on diseases and pests of oilseed rape (*Brassica napus*) or related plants. *Pesticide Science* **39**, 271-278.
147. G.W. Dawson, A.J. Hick, R.N. Bennett, A. Donald, J.A. Pickett and R.M. Wallsgrove (1993) Synthesis of glucosinolate precursors and investigations into the biosynthesis of phenylalkyl- and methylthioalkylglucosinolates. *The Journal of Biological Chemistry* **268**, 27154-27159.
148. A. Blackwell, A.J. Mordue (Luntz), B.S. Hansson, L.J. Wadhams and J.A. Pickett (1993) A behavioural and electrophysiological study of oviposition cues for *Culex quinquefasciatus. Physiological Entomology* **18**, 343-348.
149. A.J. Mordue (Luntz), A. Blackwell, B.S. Hansson, L.J. Wadhams and J.A. Pickett (1993) Oviposition attractants for *Culex quinquefasciatus. IOBC/WPRS Bulletin* **16**, 335-340.
150. J.A. Pickett (1993) Pheromones and other semiochemicals in the control of aphids. *IOBC/WPRS Bulletin* **16**, 81.
151. L.E. Smart, M.M. Blight, J.A. Pickett and B.J. Pye (1994) Development of field strategies incorporating semiochemicals for the control of the pea and bean weevil, *Sitona lineatus* L. *Crop Protection* **13**, 127-135.
152. J.A. Pickett (1994) From the laboratory to the field: the major problem in developing pheromones and other semiochemicals for pest management. *1994 BCPC Monograph no. 59: Comparing glasshouse & field pesticide performance II*, 317-318.
153. C. Höller, S.G. Micha, S. Schulz, W. Francke and J.A. Pickett (1994) Enemy-induced dispersal in a parasitic wasp. *Experientia* **50**, 182-185.
154. D.L. Hallahan, S-M. C. Lau, P.A. Harder, D.W.M. Smiley, G.W. Dawson, J.A. Pickett, R.E. Christoffersen and D.P. O'Keefe (1994) Cytochrome *P*-450-catalysed monoterpenoid oxidation in catmint (*Nepeta racemosa*) and avocado (*Persea americana*); evidence for related enzymes with different activities. *Biochimica et Biophysica Acta* **1201**, 94-100.
155. E.C. Cocking, N.W. Blackhall, D.S. Brar, M.R. Davey, G.S. Khush, J.K. Ladha, J.A. Pickett, J.B. Power and P.R. Shewry (1994) Biotechnological approaches to rice genetic improvement. *Proceedings, Food Security in Asia, The Royal Society, London, 1st November, 1994*, pp. 23-26.
156. J.A. Pickett, L.J. Wadhams and C.M. Woodcock (1994) Attempts to control aphid pests by integrated use of semiochemicals. *Brighton Crop Protection Conference – Pests and Diseases -1994*, 1239-1246.

157. J. Pettersson, J.A. Pickett, B.J. Pye, A. Quiroz, L.E. Smart, L.J. Wadhams and C.M. Woodcock (1994) Winter host component reduces colonization by bird-cherry-oat aphid, *Rhopalosiphum padi* (L.) (Homoptera, Aphididae), and other aphids in cereal fields. *Journal of Chemical Ecology* 20, 2565-2574.
158. J. Hardie, R. Isaacs, J.A. Pickett, L.J. Wadhams and C.M. Woodcock (1994) Methyl salicylate and (-)-(1*R*,5*S*)-myrtenal are plant-derived repellents for black bean aphid, *Aphis fabae* Scop. (Homoptera: Aphididae). *Journal of Chemical Ecology* 20, 2847-2855.
159. K.J. Doughty, G.A. Kiddle, B.J. Pye, R.M. Wallsgrove and J.A. Pickett (1995) Selective induction of glucosinolates in oilseed rape leaves by methyl jasmonate. *Phytochemistry* 38, 347-350.
160. G. Powell, J. Hardie and J.A. Pickett (1995) Responses of *Myzus persicae* to the repellent polygodial in choice and no-choice video assays with young and mature leaf tissue. *Entomologia Experimentalis et Applicata* 74, 91-94.
161. D.L. Hallahan, J.M. West, R.M. Wallsgrove, D.W.M. Smiley, G.W. Dawson, J.A. Pickett and J.G.C. Hamilton (1995) Purification and characterization of an acyclic monoterpene primary alcohol:$NADP^+$ oxidoreductase from catmint (*Nepeta racemosa*). *Archives of Biochemistry and Biophysics* 318, 105-112.
162. J.A. Pickett, T.M. Butt, K.J. Doughty, R.M. Wallsgrove and I.H. Williams (1995) Minimising pesticide input in oilseed rape by exploiting natural regulatory processes. *Proceedings of the Ninth International Rapeseed Congress, Rapeseed Today and Tomorrow, 4-7 July, 1995, Cambridge, U.K.*, Volume 2, F1, pp. 565-571. (Dorset Press, Dorchester).
163. M.M. Blight, C.H. Bock, K.J. Doughty, J.K. Fieldsend and J.A. Pickett (1995) Release of isothiocyanates from *Brassica rapa* seedlings during infection by *Alternaria brassicae. Proceedings of the Ninth International Rapeseed Congress, Rapeseed Today and Tomorrow, 4-7 July, 1995, Cambridge, U.K.*, Volume 2, F13, pp. 604-606. (Dorset Press, Dorchester).
164. M.M. Blight, A.J. Hick, J.A. Pickett, L.E. Smart, L.J. Wadhams and C.M. Woodcock (1995) Oilseed rape volatiles cueing host-plant recognition by the seed weevil, *Ceutorhynchus assimilis*: chemical, electrophysiological and behavioural studies. *Proceedings of the Ninth International Rapeseed Congress, Rapeseed Today and Tomorrow, 4-7 July, 1995, Cambridge, U.K.*, Volume 3, I21, pp. 1031-1033. (Dorset Press, Dorchester).
165. M.M. Blight, J.A. Pickett, J. Ryan, L.J. Wadhams and C.M. Woodcock (1995) Recognition of oilseed rape volatiles by pollen beetles, *Meligethes spp.*: electrophysiological and chemical studies. *Proceedings of the Ninth International Rapeseed Congress, Rapeseed Today and Tomorrow, 4-7 July, 1995, Cambridge, U.K.*, Volume 3, I25, pp. 1043-1045. (Dorset Press, Dorchester).

166. A. Hick, G. Dawson and J. Pickett (1995) Chemical studies on glucosinolate biosynthesis. In: *Abstracts International Conference on Biological Challenges for Organic Chemistry, St. Andrews, 10-13 July, 1995,* The Royal Society of Chemistry, P.54.

167. D. Smiley, G. Dawson, J. Pickett and D. Knight (1995) Chemical studies on cyclopentanoid biosynthesis. In: *Abstracts International Conference on Biological Challenges for Organic Chemistry, St. Andrews, 10-13 July, 1995,* The Royal Society of Chemistry, P.55.

168. R. Lilley, J. Hardie, L.A. Merritt, J.A. Pickett, L.J. Wadhams and C.M. Woodcock (1995) The sex pheromone of the grain aphid, *Sitobion avenae* (Fab.) (Homoptera, Aphididae). *Chemoecology* **5/6,1**, 43-46.

169. J.A. Pickett, L.J. Wadhams and C.M. Woodcock (1995) Exploiting chemical ecology for sustainable pest control. *1995 BCPC Symposium Proceedings No. 63: Integrated Crop Protection: Towards Sustainability?,* 353-362.

170. G. Powell, J. Hardie and J.A. Pickett (1995) Behavioural evidence for detection of the repellent polygodial by aphid antennal tip sensilla. *Physiological Entomology* **20**, 141-146.

171. J.A. Pickett, L.J. Wadhams and C.M. Woodcock (1995) Non-host interactions in insect chemical ecology. In: *Proceedings of the 1st International Conference on Insects: Chemical, Physiological and Environmental Aspects, September 26-29, 1994, Ladek Zdroj, Poland,* pp. 126-133. Editors D. Konopinska, G. Goldsworthy, R.J. Nachman, J. Nawrot, I. Orchard, G. Rosinski and W. Sobótka. (University of Wroclaw).

172. M.M. Blight, J.A. Pickett, L.J. Wadhams and C.M. Woodcock (1995) Antennal perception of oilseed rape, *Brassica napus* (Brassicaceae), volatiles by the cabbage seed weevil *Ceutorhynchus assimilis* (Coleoptera, Curculionidae). *Journal of Chemical Ecology* **21**, 1649-1664.

173. J.A. Pickett (1996) Gas chromatography-mass spectrometry in insect pheromone identification: three extreme case histories. In: *Techniques in Plant-Insect Interaction and Biopesticides. Proceedings IFS Workshop in Chemical Ecology, Santiago, 24-30 September, 1995,* 25-37. Editor H.M. Niemeyer. (International Foundation for Science, Stockholm).

174. G. Powell, J. Hardie and J.A. Pickett (1996) Effects of the repellent polygodial on stylet penetration behaviour and non-persistent transmission of plant viruses by aphids. *Journal of Applied Entomology* **120**, 241-243.

175. J.A. Pickett, E. Bartlet, J. Buxton, L.J. Wadhams and C.M. Woodcock (1996) Chemical ecology of adult vine weevil. *Mitteilungen aus der Biologischen Bundesanstalt für Land- und Forstwirtschaft,* H. 316, 41-45. (Berlin-Dahlem).

176. G.W. Dawson, J.A. Pickett and D.W.M. Smiley (1996) The aphid sex pheromone cyclopentanoids: synthesis in the elucidation of structure and biosynthetic pathways. *Bioorganic and Medicinal Chemistry* **4**, 351-361.

177. C.A. Clarke, A. Cronin, W. Francke, P. Philipp, J.A. Pickett, L.J. Wadhams and C.M. Woodcock (1996) Mating attempts between the Scarlet Tiger Moth, *Callimorpha dominula* L., and the Cinnabar Moth, *Tyria jacobaeae* L. (Lepidoptera: Arctiidae), involve a common sex pheromone composition. *Experientia* 52, 636-638.

178. J.A. Pickett and C.M. Woodcock (1996) The role of mosquito olfaction in oviposition site location and in the avoidance of unsuitable hosts. In: *Olfaction in Mosquito-Host Interactions. CIBA Foundation Symposium No. 200*, 109-123. Editor G. Cardew. (John Wiley & Sons Ltd., Chichester).

179. K.J. Doughty, M.M. Blight, C.H. Bock, J.K. Fieldsend and J.A. Pickett (1996) Release of alkenyl isothiocyanates and other volatiles from *Brassica rapa* seedlings during infection by *Alternaria brassicae. Phytochemistry* 43, 371-374.

180. J.S. Waterhouse, J. Ke, J.A. Pickett and P.J. Weldon (1996) Volatile components in dorsal gland secretions of the collared peccary, *Tayassu tajacu* (Tayassuidae, Mammalia). *Journal of Chemical Ecology* 22, 1307-1314.

181. C.J. Dodds, M.G. Ford, I.F. Henderson, L.D. Leake, A.P. Martin, J.A. Pickett, L.J. Wadhams and P. Watson (1996) Slug chemical ecology: electrophysiological and behavioural studies. In: *Slug and Snail Pests in Agriculture, British Crop Protection Council Symposium Proceedings No. 66*, 73-81. BCPC, Farnham.

182. J.G.C. Hamilton, G.W. Dawson and J.A. Pickett (1996) 9-Methylgermacrene-B; proposed structure for novel homosesquiterpene from the sex pheromone glands of *Lutzomyia longipalpis* (Diptera: Psychodidae) from Lapinha, Brazil. *Journal of Chemical Ecology* 22, 1477-1491.

183. J.G.C. Hamilton, G.W. Dawson and J.A. Pickett (1996) 3-Methyl-α-himachalene; proposed structure for the novel homosesquiterpene sex pheromone of *Lutzomyia longipalpis* (Diptera: Psychodidae) from Jacobina, Brazil. *Journal of Chemical Ecology* 22, 2331-2340.

184. P.M. Lösel, M. Lindemann, J. Scherkenbeck, J. Maier, B. Engelhard, C.A.M. Campbell, J. Hardie, J.A. Pickett, L.J. Wadhams, A. Elbert and G. Thielking (1996) The potential of semiochemicals for control of *Phorodon humuli* (Homoptera: Aphididae). *Pesticide Science* 48, 293-303.

185. P.M. Lösel, M. Lindemann, J. Scherkenbeck, C.A.M. Campbell, J. Hardie, J.A. Pickett and L.J. Wadhams (1996) Effect of primary-host kairomones on the attractiveness of the hop-aphid sex pheromone to *Phorodon humuli* males and gynoparae. *Entomologia Experimentalis et Applicata* 80, 79-82.

186. L.E. Smart, J.A. Pickett and W. Powell (1997) "Push-Pull" strategies for pest control. *Grain Legumes No. 15*, 14-15.

187. J.A. Pickett, L.J. Wadhams and C.M. Woodcock (1997) First steps in the use of aphid sex pheromones. In: *Insect Pheromone Research: New Directions*, pp. 439-444. Editors R.T. Cardé and A.K. Minks. (Chapman and Hall, New York).

188. J.E. Fuentes-Contreras, W. Powell, L.J. Wadhams, J.A. Pickett and H.M. Niemeyer (1996) Influence of wheat and oat cultivars on the development of the cereal aphid parasitoid *Aphidius rhopalosiphi* and the generalist aphid parasitoid *Ephedrus plagiator*. *Annals of Applied Biology* **128**, 181-187.
189. A.J. Hick, J.A. Pickett, D.W.M. Smiley, L.J. Wadhams and C.M. Woodcock (1997) Higher plants as a clean source of semiochemicals and genes for their biotechnological production. In: *Phytochemical Diversity: A Source of New Industrial Products*, pp. 220-236. Editors S. Wrigley, M. Hayes, R. Thomas and E. Chrystal. The Royal Society of Chemistry, Cambridge.
190. Z.R. Khan, K. Ampong-Nyarko, P. Chiliswa, A. Hassanali, S. Kimani, W. Lwande, W.A. Overholt, J.A. Pickett, L.E. Smart, L.J. Wadhams and C.M. Woodcock (1997) Intercropping increases parasitism of pests. *Nature* **388**, 631-632.
191. M.M. Blight, M. Le Métayer, M-H. Pham-Delègue, J.A. Pickett, F. Marion-Poll and L.J. Wadhams (1997) Identification of floral volatiles involved in recognition of oilseed rape flowers, *Brassica napus* by honeybees, *Apis mellifera*. *Journal of Chemical Ecology* **23**, 1715-1727.
192. B.J. Gabrys, H.J. Gadomski, Z. Klukowski, J.A. Pickett, G.T. Sobota, L.J. Wadhams and C.M. Woodcock (1997) Sex pheromone of cabbage aphid *Brevicoryne brassicae*: identification and field trapping of male aphids and parasitoids. *Journal of Chemical Ecology* **23**, 1881-1890.
193. G. Powell, J. Hardie and J.A. Pickett (1997) Laboratory evaluation of antifeedant compounds for inhibiting settling by cereal aphids. *Entomologia Experimentalis et Applicata* **84**, 189-193.
194. J.A. Pickett, L.J. Wadhams and C.M. Woodcock (1997) Developing sustainable pest control from chemical ecology. *Agriculture, Ecosystems and Environment* **64**, 149-156.
195. A. Guerrero, J. Feixas, J. Pajares, L.J. Wadhams, J.A. Pickett and C.M. Woodcock (1997) Semiochemically induced inhibition of behaviour of *Tomicus destruens* (Woll.) (Coleoptera: Scolytidae). *Naturwissenschaften* **84**, 155-157.
196. J. Hardie, L. Peace, J.A. Pickett, D.W.M. Smiley, J.R. Storer and L.J. Wadhams (1997) Sex pheromone stereochemistry and purity affect field catches of male aphids. *Journal of Chemical Ecology* **23**, 2547-2554.
197. A. Quiroz, J. Pettersson, J.A. Pickett, L.J. Wadhams and H.M. Niemeyer (1997) Semiochemicals mediating spacing behavior of bird cherry-oat aphid, *Rhopalosiphum padi* feeding on cereals. *Journal of Chemical Ecology* **23**, 2599-2607.
198. K.S. Boo, I.B. Chung, K.S. Han, J.A. Pickett and L.J. Wadhams (1998) Response of the lacewing *Chrysopa cognata* to pheromones of its aphid prey. *Journal of Chemical Ecology* **24**, 631-643.

199. D.L. Hallahan, J.M. West, D.W.M. Smiley and J.A. Pickett (1998) Nepetalactol oxidoreductase in trichomes of the catmint *Nepeta racemosa. Phytochemistry* **48**, 421-427.
200. M.C. Luszniak and J.A. Pickett (1998) Self-defence for plants. *Chemistry in Britain* **34**, 29-32.
201. S. Al Abassi, M.A. Birkett, J. Pettersson, J.A. Pickett and C.M. Woodcock (1998) Ladybird beetle odour identified and found to be responsible for attraction between adults. *Cellular and Molecular Life Sciences* **54**, 876-879.
202. Y. Du, G.M. Poppy, W. Powell, J.A. Pickett, L.J. Wadhams and C.M. Woodcock (1998) Identification of semiochemicals released during aphid feeding that attract parasitoid *Aphidius ervi. Journal of Chemical Ecology* **24**, 1355-1368.
203. M. Subchev, A. Harizanov, W. Francke, S. Franke, E. Plass, A. Reckziegel, F. Schröder, J.A. Pickett, L.J. Wadhams and C.M. Woodcock (1998) Sex pheromone of female vine bud moth, *Theresimima ampellophaga* comprises (2*R*)-butyl (7*Z*)-tetradecenoate. *Journal of Chemical Ecology* **24**, 1141-1151.
204. N.G. Agelopoulos and J.A. Pickett (1998) Headspace analysis in chemical ecology: effects of different sampling methods on ratios of volatile compounds present in headspace samples. *Journal of Chemical Ecology* **24**, 1161-1172.
205. G. Powell, J. Hardie and J.A. Pickett (1998) The effects of antifeedant compounds and mineral oil on stylet penetration and transmission of potato virus Y by *Myzus persicae* (Sulz.) (Hom., Aphididae). *Journal of Applied Entomology* **122**, 331-333.
206. J.A. Pickett, L.J. Wadhams and C.M. Woodcock. (1998) Insect supersense: mate and host location by insects as model systems for exploiting olfactory interactions. *The Biochemist*, August 1998, 8-13.
207. J.A. Pickett (1998) Insects and chemical signals: a volatile situation. *Proceedings of the Royal Institution of Great Britain* **69**, 243-257. (Oxford University Press).
208. J.A. Pickett (1998) Pest semiochemicals in arable crop protection. *Pesticide Science* **54**, 290-291.
209. J.G.C. Hamilton, A.M. Hooper, K. Mori, J.A. Pickett and S. Sano (1999) 3-Methyl-α-himachalene is confirmed, and the relative stereochemistry defined, by synthesis as the sex pheromone of the sandfly *Lutzomyia longipalpis* from Jacobina, Brazil. *Chemical Communications*, 355-356.
210. A.J. Hick, M.C. Luszniak and J.A. Pickett (1999) Volatile isoprenoids that control insect behaviour and development. *Natural Product Reports* **16**, 39-54.
211. R.T. Glinwood, D.W.M. Smiley, J. Hardie, J.A. Pickett, W. Powell, L.J. Wadhams and C.M. Woodcock (1999) Aphid sex pheromones: manipulation of beneficial insects for aphid population control. *Pesticide Science* **55**, 208-209.

212. N. Agelopoulos, M.A. Birkett, A.J. Hick, A.M. Hooper, J.A. Pickett, E.M. Pow, L.E. Smart, D.W.M. Smiley, L.J. Wadhams and C.M. Woodcock (1999) Exploiting semiochemicals in insect control. *Pesticide Science* 55, 225-235.
213. A.M. Hooper, J.A. Bennison, M.C. Luszniak, J.A. Pickett, E.M. Pow and L.J. Wadhams (1999) *Verbena x hybrida* flower volatiles attractive to Western flower thrips, *Frankliniella occidentalis. Pesticide Science* 55, 660-662.
214. E.M. Pow, A.M. Hooper, M.C. Luszniak, J.A. Pickett and L.J. Wadhams (1998) Novel strategies for improving biological control of western flower thrips on protected ornamentals - attraction of western flower thrips to verbena plants. *Proceedings British Crop Protection Conference - Pests and Diseases, 1998*, 417-422. BCPC, Farnham.
215. M.A. Birkett, B.P.S. Khambay, J.A. Pickett, L.J. Wadhams and C.M. Woodcock (1999) Strategies for developing natural products as crop protection agents employing neurotoxicological and other neurophysiological modes of action. In: *Progress in Neuropharmacology and Neurotoxicology of Pesticides and Drugs.* SCI Special Publication No. 232, pp. 184-192. Editor D.J. Beadle (Royal Society of Chemistry, Cambridge).
216. J.A. Pickett, K. Chamberlain, G.M. Poppy and C.M. Woodcock (1999) Exploiting insect responses in identifying plant signals. In: *Insect-Plant Interactions and Induced Plant Defence. Novartis Foundation Symposium 223.* Editor J. Goode. pp. 253-265. (John Wiley & Sons Ltd., Chichester).
217. N.G. Agelopoulos, A.M. Hooper, S.P. Maniar, J.A. Pickett and L.J. Wadhams (1999) A novel approach for isolation of volatile chemicals released by individual leaves of a plant *in situ. Journal of Chemical Ecology* 25, 1411-1425.
218. T.O. Olagbemiro, M.A. Birkett, A.J. Mordue (Luntz) and J.A. Pickett (1999) Production of (5*R*,6*S*)-6-acetoxy-5-hexadecanolide, the mosquito oviposition pheromone, from the seed oil of the summer cypress plant, *Kochia scoparia* (Chenopodiaceae). *Journal of Agricultural and Food Chemistry* 47, 3411-3415.
219. J.A. Pickett, D.W.M. Smiley and C.M. Woodcock (1999) Secondary Metabolites in Plant-Insect Interactions: Dynamic Systems of Induced and Adaptive Responses. In: *Advances in Botanical Research* 30, pp. 91-115. (Academic Press, London).
220. A.M. Hooper, J.G.C. Hamilton, K. Mori, J.A. Pickett and S. Sano (1999) Volatile isoprenoids that control insect behaviour. *Chem. Listy, Symposia*, 93 S21-S23.
221. J. Hardie, J.A. Pickett, E.M. Pow and D.W.M. Smiley (1999) Aphids. In: *Pheromones of Non-Lepidopteran Insects Associated with Agricultural Plants*, pp. 227-250. Editors J. Hardie and A.K. Minks. (CAB International).

222. G. Powell, S.P. Maniar, J.A. Pickett and J. Hardie (1999) Aphid responses to non-host epicuticular lipids. *Entomologia Experimentalis et Applicata* **91**, 115-123.
223. E.M. Pow, J.A. Bennison, M.A. Birkett, M.C. Luszniak, M. Manjunatha, J.A. Pickett, I.S. Segers, L.J. Wadhams, L.R. Wardlow and C.M. Woodcock (1999) Behavioural responses of western flower thrips (*Frankliniella occidentalis*) to host plant volatiles. *Proceedings Sixth International Symposium on Thysanoptera, Antalya, Turkey, April 27-May 1, 1998*, pp. 121-128.
224. L.E.G. Mboera, K.Y. Mdira, F.M. Salum, W. Takken and J.A. Pickett (1999) Influence of synthetic oviposition pheromone and volatiles from soakage pits and grass infusions upon oviposition site-selection of *Culex* mosquitoes in Tanzania. *Journal of Chemical Ecology* **25**, 1855-1865.
225. M. Manjunatha, J.A. Pickett, L.J. Wadhams and F. Nazzi (1998) Response of western flower thrips, *Frankliniella occidentalis* and its predator *Amblyseius cucumeris* to chrysanthemum volatiles in olfactometer and greenhouse trials. *Insect Science and its Application* **18**, 139-144.
226. L.M.C. Rebouças, M. do S.B. Caraciolo, A.E.G. Sant'Ana, J.A. Pickett, L.J. Wadhams and E.M. Pow (1999) Composição química da glândula abdominal da fêmea da mariposa *Castnia licus* (Drury) (Lepidoptera: Castniidae): possíveis feromônios e precursores. *Química Nova* **22**, 645-648.
227. J.G.C. Hamilton, A.M. Hooper, H.C. Ibbotson, S. Kurosawa, K. Mori, S. Muto and J.A. Pickett (1999) 9-Methylgermacrene-B is confirmed as the sex pheromone of the sandfly *Lutzomyia longipalpis* from Lapinha, Brazil, and the absolute stereochemistry defined as *S*. *Chemical Communications*, 2335-2336.
228. P.A. Shah, J.A. Pickett and J.D. Vandenberg (1999) Responses of Russian Wheat Aphid (Homoptera: Aphididae) to Aphid Alarm Pheromone. *Environmental Entomology* **28**, 983-985.
229. K. Chamberlain, J.A. Pickett and C.M. Woodcock (2000) Plant signalling and induced defence in insect attack. *Molecular Plant Pathology* **1**, 67-72.
230. N.G. Agelopoulos, K. Chamberlain and J.A. Pickett (2000) Factors affecting volatile emissions of intact potato plants, *Solanum tuberosum*: variability of quantities and stability of ratios. *Journal of Chemical Ecology* **26**, 497-511.
231. K.S. Boo, M.Y. Choi, I.B. Chung, V.F. Eastop, J.A. Pickett, L.J. Wadhams and C.M. Woodcock (2000) Sex pheromone of the peach aphid, *Tuberocephalus momonis*, and optimal blends for trapping males and females in the field. *Journal of Chemical Ecology* **26**, 601-609.
232. W. Francke, E. Plass, N. Zimmermann, H. Tietgen, T. Tolasch, S. Franke, M. Subchev, T. Toshova, J.A. Pickett, L.J. Wadhams and C.M. Woodcock (2000) Major sex pheromone component of female herald moth *Scoliopteryx libatrix* is the novel branched alkene (6*Z*,13)-methylheneicosene. *Journal of Chemical Ecology* **26**, 1135-1149.

233. M.A. Birkett, C.A.M. Campbell, K. Chamberlain, E. Guerrieri, A.J. Hick, J.L. Martin, M. Matthes, J.A. Napier, J. Pettersson, J.A. Pickett, G.M. Poppy, E.M. Pow, B.J. Pye, L.E. Smart, G.H. Wadhams, L.J. Wadhams and C.M. Woodcock (2000) New roles for *cis*-jasmone as an insect semiochemical and in plant defense. *Proceedings of the National Academy of Sciences USA* **97**, 9329-9334.
234. L.E.G. Mboera, W. Takken, K.Y. Mdira and J.A. Pickett (2000) Sampling gravid *Culex quinquefasciatus* (Diptera: Culicidae) in Tanzania with traps baited with synthetic oviposition pheromone and grass infusions. *Journal of Medical Entomology* **37**, 172-176.
235. L.E.G. Mboera, W. Takken, K.Y. Mdira, G.J. Chuwa and J.A. Pickett (2000) Oviposition and behavioral responses of *Culex quinquefasciatus* to skatole and synthetic oviposition pheromone in Tanzania. *Journal of Chemical Ecology* **26**, 1193-1203.
236. S. Al Abassi, M.A. Birkett, J. Pettersson, J.A. Pickett, L.J. Wadhams and C.M. Woodcock (2000) Response of the seven-spot ladybird to an aphid alarm pheromone and an alarm pheromone inhibitor is mediated by paired olfactory cells. *Journal of Chemical Ecology* **26**, 1765-1771.
237. G.J. Devine, A. Ingvarsdóttir, W. Mordue, A.W. Pike, J.A. Pickett, I. Duce and A.J. Mordue (Luntz) (2000) Salmon lice, *Lepeophtheirus salmonis*, exhibit specific chemotactic responses to semiochemicals originating from the salmonid, *Salmo salar*. *Journal of Chemical Ecology* **26**, 1833-1847.
238. E.N. Barata, J.A. Pickett, L.J. Wadhams, C.M. Woodcock and H. Mustaparta (2000) Identification of host and nonhost semiochemicals of eucalyptus woodborer *Phoracantha semipunctata* by gas chromatography-electronantennography. *Journal of Chemical Ecology* **26**, 1877-1895.
239. Z.R. Khan, J.A. Pickett, J. van den Berg, L.J. Wadhams and C.M. Woodcock (2000) Exploiting chemical ecology and species diversity: stem borer and striga control for maize and sorghum in Africa. *Pest Management Science* **56**, 957-962.
240. L.M. Field, J.A. Pickett and L.J. Wadhams (2000) Molecular studies in insect olfaction. *Insect Molecular Biology* **9**, 545-551.
241. H. Whitney, O. Sayanova, M.J. Lewis, J. Pickett and J.A. Napier (2000) Isolation of two putative acyl-acyl carrier protein desaturase enzymes from *Kochia scoparia*. *Biochemical Society Transactions* **28**, 623-624.
242. S. Al Abassi, M.A. Birkett, J. Pettersson, J.A. Pickett, L.J. Wadhams and C.M. Woodcock (2001) Response of the ladybird parasitoid *Dinocampus coccinellae* to toxic alkaloids from the seven-spot ladybird, *Coccinella septempunctata*. *Journal of Chemical Ecology* **27**, 33-43.
243. J.A. Pickett and G.M. Poppy (2001) Switching on plant genes by external chemical signals. *Trends in Plant Science* **6**, 137-139.

244. C.H. Hansen, U. Wittstock, C.E. Olsen, A.J. Hick, J.A. Pickett and B.A. Halkier (2001) Cytochrome P450 CYP79F1 from Arabidopsis catalyzes the conversion of dihomomethionine and trihomomethionine to the corresponding aldoximes in the biosynthesis of aliphatic glucosinolates. *The Journal of Biological Chemistry* **276**, 11078-11085.
245. C.M. Hartfield, C.A.M. Campbell, J. Hardie, J.A. Pickett and L.J. Wadhams (2001) Pheromone traps for the dissemination of an entomopathogen by the damson-hop aphid *Phorodon humuli*. *Biocontrol Science and Technology* **11**, 401-410.
246. B. Donato, S.J. Brooks, J.A. Pickett and J. Hardie (2001) *Peyerimhoffina gracilis* (Schneider, 1851) (Neur.: Chrysopidae): a green lacewing new to Britain. *Entomologist's Record* **113**, 131-135.
247. M.A. Birkett, K. Chamberlain, A.M. Hooper and J.A. Pickett (2001) Does allelopathy offer real promise for practical weed management and for explaining rhizosphere interactions involving higher plants? *Plant and Soil* **232**, 31-39.
248. K. Chamberlain, E. Guerrieri, F. Pennachio, J. Pettersson, J.A. Pickett, G.M. Poppy, W. Powell, L.J. Wadhams and C.M. Woodcock (2001) Can aphid-induced plant signals be transmitted aerially and through the rhizosphere? *Biochemical Systematics and Ecology* **29**, 1063-1074.
249. C. Costantini, M.A. Birkett, G. Gibson, J. Ziesmann, N'F. Sagnon, H.A. Mohammed, M. Coluzzi and J.A. Pickett (2001) Electroantennogram and behavioural responses of the malaria vector *Anopheles gambiae* to human-specific sweat components. *Medical and Veterinary Entomology* **15**, 259-266.
250. Z.R. Khan, A. Hassanali, T.M. Khamis, J.A. Pickett and L.J. Wadhams (2001) Mechanisms of *Striga hermonthica* suppression by *Desmodium* spp. *Proceedings of the BCPC Conference – Weeds 2001*, Brighton, UK, 12-15 November 2001, 895-900.
251. J.S. Waterhouse, M. Hudson, J.A. Pickett and P.J. Weldon (2001) Volatile components in dorsal gland secretions of the white-lipped peccary, *Tayassu pecari*, from Bolivia. *Journal of Chemical Ecology* **27**, 2459-2469.
252. E.N. Barata, H. Mustaparta, J.A. Pickett, L.J. Wadhams and J. Araujo (2002) Encoding of host and non-host plant odours by receptor neurones in the eucalyptus woodborer, *Phoracantha semipunctata* (Coleoptera: Cerambycidae). *Journal of Comparative Physiology A* **188**, 121-133.
253. J.A. Pickett and T.O. Olagbemiro (2002) Repellents. In *Encyclopedia of Pest Management*, pp. 696-697. (Marcel Dekker).
254. Z.R. Khan, J.A. Pickett, L. Wadhams and F. Muyekho (2001) Habitat management strategies for the control of cereal stemborers and striga in maize in Kenya. *Insect Science and Its Application.* **21**, 375-380.

255. A. Ingvarsdóttir, M.A. Birkett, I. Duce, R.L. Genna, W. Mordue, J.A. Pickett, L.J. Wadhams and A.J. Mordue (Luntz) (2002) Semiochemical strategies for sea louse control: host location cues. *Pest Management Science* **58**, 537-545.
256. A.M. Hooper, B. Donato, C.M. Woodcock, J.H. Park, R.L. Paul, K.S. Boo, J. Hardie and J.A. Pickett (2002) Characterization of (1*R*,4*S*,4a*R*,7*S*,7a*R*)-dihydronepetalactol as a semiochemical for lacewings, including *Chrysopa* spp. and *Peyerimhoffina gracilis*. *Journal of Chemical Ecology* **28**, 849-864.
257. I. Prosser, A.L. Phillips, S. Gittings, M.J. Lewis, A.M. Hooper, J.A. Pickett and M.H. Beale (2002) (+)-(10*R*)-Germacrene A synthase from goldenrod, *Solidago canadensis*;. cDNA isolation, bacterial expression and functional analysis. *Phytochemistry* **60**, 691-702.
258. Z.R. Khan, A. Hassanali, W. Overholt, T.M. Khamis, A.M. Hooper, J.A. Pickett, L.J. Wadhams and C.M. Woodcock (2002) Control of witchweed *Striga hermonthica* by intercropping with *Desmodium* spp., and the mechanism defined as allelopathic. *Journal of Chemical Ecology* **28**, 1871-1885.
259. A. Ingvarsdóttir, M.A. Birkett, I. Duce, W. Mordue, J.A. Pickett, L.J. Wadhams and A.J. Mordue (Luntz) (2002) Role of semiochemicals in mate location by parasitic sea louse, *Lepeophtheirus salmonis*. *Journal of Chemical Ecology* **28**, 2107-2117.
260. W. Powell and J.A. Pickett (2003) Manipulation of parasitoids for aphid pest management: progress and prospects. *Pest Management Science* **59**, 149-155.
261. J.A. Pickett, H.B. Rasmussen, C.M. Woodcock, M. Matthes and J.A. Napier (2003) Plant stress signalling: understanding and exploiting plant-plant interactions. *Biochemical Society Transactions* **31**, 123-127.
262. M.A. Birkett and J.A. Pickett (2003) Aphid sex pheromones: from discovery to commercial production. (Molecules of Interest Review). *Phytochemistry* **62**, 651-656.
263. G. Jones, C.A.M. Campbell, J. Hardie, J.A. Pickett, B.J. Pye and L.J. Wadhams (2003) Integrated management of two-spotted spider mite *Tetranychus urticae* on hops using hop β-acids as an antifeedant together with the predatory mite *Phytoseiulus persimilis*. *Biocontrol Science and Technology* **13**, 241-252.
264. S. Chen, E. Glawischnig, K. Jørgensen, P. Naur, B. Jørgensen, C-E. Olsen, C.H. Hansen, H. Rasmussen, J.A. Pickett and B.A. Halkier (2003) CYP79F1 and CYP79F2 have distinct functions in the biosynthesis of aliphatic glucosinolates in Arabidopsis. *The Plant Journal* **33**, 923-937.
265. R. Glinwood, J. Pettersson, E. Ahmed, V. Ninkovic, M. Birkett and J. Pickett (2003) Change in acceptability of barley plants to aphids after exposure to allelochemicals from couch-grass (*Elytrigia repens*). *Journal of Chemical Ecology* **29**, 261-274.

266. K.S. Boo, S.S. Kang, J.H. Park, J.A. Pickett and L.J. Wadhams (2003) Field trapping of *Chrysopa cognata* (Neuroptera: Chrysopidae) with aphid sex pheromone components in Korea. *Journal of Asia-Pacific Entomology* **6**, 29-36.

267. T.J. Bruce, J.A. Pickett and L.E. Smart (2003) *Cis*-jasmone switches on plant defence against insects. *Pesticide Outlook* **14**, 96-98.

268. M.A. Birkett, K. Chamberlain, E. Guerrieri, J.A. Pickett, L.J. Wadhams and T. Yasuda (2003) Volatiles from whitefly-infested plants elicit a host-locating response in the parasitoid, *Encarsia formosa. Journal of Chemical Ecology* **29**, 1589-1600.

269. T.J.A. Bruce, J.L. Martin, J.A. Pickett, B.J. Pye, L.E. Smart and L.J. Wadhams (2003) cis-Jasmone treatment induces resistance in wheat plants against the grain aphid, Sitobion avenae (Fabricius) (Homoptera: Aphididae). Pest Management Science 59, 1031-1036.

270. M.K. Tsanuo, A. Hassanali, A.M. Hooper, Z.R. Khan, F. Kaberia, J.A. Pickett and L.J. Wadhams (2003) Isoflavanones from the allelopathic aqueous root exudate of *Desmodium uncinatum. Phytochemistry* 64, 265-273.

271. M. Matthes, J.A. Napier, J.A. Pickett & C.M. Woodcock (2003) New chemical signals in plant protection against herbivores and weeds. *Proceedings BCPC International Congress – Crop Science & Technology 2003, 10-12 November, SECC Glasgow,* 9C-4, 1227-1236.

272. H.M. Whitney, L.V. Michaelson, O. Sayanova, J.A. Pickett and J.A. Napier (2003) Functional characterisation of two cytochrome b_5-fusion desaturases from *Anemone leveillei*: the unexpected identification of a fatty acid Δ^6-desaturase. *Planta* **217**, 983-992.

273. C.A.M. Campbell, F.J. Cook, J.A. Pickett, T.W. Pope, L.J. Wadhams and C.M. Woodcock (2003) Responses of the aphids *Phorodon humuli* and *Rhopalosiphum padi* to sex pheromone stereochemistry in the field. *Journal of Chemical Ecology* 29, 2225-2234.

274. C.H. Hansen, L. Du, P. Naur, C.E. Olsen, K.B. Axelsen, A.J. Hick, J.A. Pickett and B.A. Halkier (2001) CYP83B1 is the oxime-metabolizing enzyme in the glucosinolate pathway in *Arabidopsis. The Journal of Biological Chemistry* 276, 24790-24796.

275. L.S. Gohole, W.A. Overholt, Z.R. Khan, J.A. Pickett and L.E.M. Vet (2003) Effects of molasses grass, *Melinis minutiflora* volatiles on the foraging behavior of the cereal stemborer parasitoid, *Cotesia sesamiae. Journal of Chemical Ecology* **29**, 731-745.

276. M. Subchev, A. Mircheva, J. Pickett, L. Wadhams, C. Woodcock, A. dos Santos, S. Franke and W. Francke (2003) Sex pheromone of the leaf-miner *Phyllonorycter platani:* (*Z*10)-tetradecenyl acetate. *Journal of Chemical Ecology* **29**, 2391-2396. (Originally published online July 18, 2003, Rapid

Communications, pp. RC93-98 (*http://www.kluweronline.com/issn/0098-0331*).

277. J-J. Zhou, G-A. Zhang, W. Huang, M.A. Birkett, L.M. Field, J.A. Pickett and P. Pelosi (2004) Revisiting the odorant-binding protein LUSH of *Drosophila melanogaster*: evidence for odour recognition and discrimination. *FEBS Letters* 558, 23-26.

278. J-J. Zhou, W. Huang, G-A. Zhang, J.A. Pickett and L.M. Field (2004) "Plus-C" odorant-binding protein genes in two *Drosophila* species and the malaria mosquito *Anopheles gambiae*. *Gene* 327, 117-129.

279. M.A. Birkett, C.J. Dodds, I.F. Henderson, L.D. Leake, J.A. Pickett, M.J. Selby and P. Watson (2004) Antifeedant compounds from three species of Apiaceae active against the field slug, *Deroceras reticulatum* (Müller). *Journal of Chemical Ecology* 30, 563-576.

280. A.M. Hooper and J.A. Pickett (2004) Semiochemistry. In *Encyclopaedia of Supramolecular Chemistry*, 1270-1277, Marcel Dekker, New York.

281. S.H. Goldansaz, S. Dewhirst, M.A. Birkett, A.M. Hooper, D.W.M. Smiley, J.A. Pickett, L. Wadhams and J.N. McNeil (2004) Identification of two sex pheromone components of the potato aphid, *Macrosiphum euphorbiae* (Thomas). *Journal of Chemical Ecology* 30, 819-834.

282. T.O. Olagbemiro, M.A. Birkett, A.J. Mordue (Luntz) and J.A. Pickett (2004) Laboratory and field responses of the mosquito, *Culex quinquefasciatus*, to plant-derived *Culex* spp. oviposition pheromone and the oviposition cue skatole. *Journal of Chemical Ecology* 30, 965-976.

283. E. Bartlet, M.M. Blight, J.A. Pickett, L.E. Smart, G. Turner and C.M. Woodcock (2004) Orientation and feeding responses of the pollen beetle, *Meligethes aeneus*, to candytuft, *Iberis amara*. *Journal of Chemical Ecology* 30, 913-925.

284. Y. Nakashima, M.A. Birkett, B.J. Pye, J.A. Pickett and W. Powell (2004) The role of semiochemicals in the avoidance of the seven-spot ladybird, *Coccinella septempunctata*, by the aphid parasitoid, *Aphidius ervi*. *Journal of Chemical Ecology* 30, 1103-1116.

285. K.L. Falk, C. Vogel, S. Textor, S. Bartram, A. Hick, J.A. Pickett and J. Gershenzon (2004) Glucosinolate biosynthesis: demonstration and characterization of the condensing enzyme of the chain elongation cycle in *Eruca sativa*. *Phytochemistry* 65, 1073-1084.

286. J.A. Pickett. (2004) New opportunities in neuroscience, but a great danger that some may be lost. In: *Neurotox '03: Neurotoxicological Targets from Functional Genomics and Proteomics*, pp. 1-10. Editors D.J. Beadle, I.R. Mellor and P.N.R. Usherwood. (Society of Chemical Industry, London).

287. K.-M.V. Jensen, J.B. Jespersen, M.A Birkett, J.A. Pickett, G. Thomas, L.J. Wadhams and C.M. Woodcock (2004) Variation in the load of the horn fly,

Haematobia irritans, in cattle herds is determined by the presence or absence of individual heifers. *Medical and Veterinary Entomology* **18**, 275-280.

288. S.M. Kimani, S.C. Chhabra, W. Lwande, Z.R. Khan, A. Hassanali and J.A. Pickett (2000) Airborne volatiles from *Melinis minutiflora* P. Beauv., a non-host plant of the spotted stem borer. *Journal of Essential Oil Research* **12**, 221-224.
289. Z.R. Khan and J.A. Pickett (2004) The 'push-pull' strategy for stemborer management: a case study in exploiting biodiversity and chemical ecology. In: *Ecological Engineering for Pest Management: Advances in Habitat Manipulation for Arthropods*, pp. 155-164. Editors G. Gurr, S. Wratten and M. Altieri. (CSIRO Publishing, Australia).
290. S. Textor, S. Bartram, J. Kroymann, K.L. Falk, A. Hick, J.A. Pickett and J. Gershenzon (2004) Biosynthesis of methionine-derived glucosinolates in *Arabidopsis thaliana*: recombinant expression and characterization of methylthioalkylmalate synthase, the condensing enzyme of the chain-elongation cycle. *Planta* **218**, 1026-1035.
291. J.D. Blande, J.A. Pickett and G.M. Poppy (2004) Attack rate and success of the parasitoid *Diaeretiella rapae* on specialist and generalist feeding aphids. *Journal of Chemical Ecology* **30**, 1781-1795.
292. H.M. Whitney, O. Sayanova, J.A. Pickett and J.A. Napier (2004) Isolation and expression pattern of two putative acyl-ACP desaturase cDNAs from *Bassia scoparia*. *Journal of Experimental Botany* **55**, 787-789.
293. M.A. Birkett, N. Agelopoulos, K.-M.V. Jensen, J.B. Jespersen, J.A. Pickett, H.J. Prijs, G. Thomas, J.J. Trapman, L.J. Wadhams and C.M. Woodcock (2004) The role of volatile semiochemicals in mediating host location and selection by nuisance and disease-transmitting cattle flies. *Medical and Veterinary Entomology* **18**, 313-322.
294. A.M. Hooper, E. Napper and J.A. Pickett (2004) Alarm pheromones of insects. In *Encyclopedia of Entomology, Vol. 3*, 50-59. Editor John L. Capinera. Kluwer Academic Publishers.
295. Z-X. Li, J.A. Pickett, L.M. Field and J-J. Zhou (2005) Identification and expression of odorant-binding proteins of the malaria-carrying mosquitoes *Anopheles gambiae* and *Anopheles arabiensis*. *Archives of Insect Biochemistry and Physiology* **58**, 175-189.
296. J. Zhu, J.J. Obrycki, S.A. Ochieng, T.C. Baker, J.A. Pickett and D. Smiley (2005) Attraction of two lacewing species to volatiles produced by host plants and aphid prey. *Naturwissenschaften* **92**, 277-281.
297. S. Gouinguené, J.A. Pickett, L.J. Wadhams, M.A. Birkett and T.C.J. Turlings (2005) Antennal electrophysiological responses of three parasitic wasps to caterpillar-induced volatiles from maize (*Zea mays mays*), cotton (*Gossypium herbaceum*), and cowpea (*Vigna unguiculata*). *Journal of Chemical Ecology* **31**, 1023-1038.

298. N. Ngumbi, A.J. Ngi-Song, E.N.M. Njagi, R. Torto, L.J. Wadhams, M.A. Birkett, J.A. Pickett, W.A. Overholt and B. Torto (2005) Responses of the stem borer larval endoparasitoid *Cotesia flavipes* (Hymenoptera: Braconidae) to plant derived synomones: laboratory and field cage experiments. *Biocontrol Science and Technology* **15**, 271-279.

299. M.D. Soler Cruz, M.C. Vega Robles, J.B. Jespersen, O. Kilpinen, M. Birkett, S. Dewhirst and J. Pickett (2005) Scanning electron microscopy of foreleg tarsal sense organs of the poultry red mite, *Dermanyssus gallinae* (DeGeer) (Acari: Dermanyssidae). *Micron* **36**, 415-421.

300. S.P. Jacobs, A.P. Liggins, J-J. Zhou, J.A. Pickett, X. Jin and L.M. Field (2005) *OS-D*-like genes and their expression in aphids (Hemiptera: Aphididae). *Insect Molecular Biology* **14**, 423-432.

301. J. Pettersson, V. Ninkovic, R. Glinwood, M.A. Birkett and J.A Pickett (2005) Foraging in a complex environment – semiochemicals support searching behaviour of the seven spot ladybird. *European Journal of Entomology* **102**, 365-370.

302. J.A. Pickett, M.A. Birkett, T.J.A. Bruce, K. Chamberlain,R. Gordon-Weeks, M.C. Matthes, C.B. Moraes, J.A. Napier,L.E. Smart, L.J. Wadhams and C.M. Woodcock (2005) *cis*-Jasmone as an allelopathic agent through plant defence induction.pp 122-127 In *Proceedings of the 4th World Congress on Allelopathy, 21-26 August 2005, Eds JDI Harper M An, H Wu JH Kent, Charles Sturt University, Wagga Wagga, NSW, Australia. International Allelopathy Society.*

303. J. Pettersson, V. Ninkovic, R. Glinwood, S. Al Abassi, M. Birkett, J. Pickett and L. Wadhams (2005) Foraging behaviour of *Coccinella septempunctata* (L.): volatiles and allelobiosis. *Proceedings International Symposium on Biological Control of Aphids and Coccids, Tsuruoka, Japan, September 25-29, 2005*, 187-192.

304. J.A. Pickett, T.J.A. Bruce, K. Chamberlain, A. Hassanali, Z.R. Khan, M.C. Matthes, J.A. Napier, L.E. Smart, L.J. Wadhams and C.M. Woodcock (2006) Plant volatiles yielding new ways to exploit plant defence. In: *Chemical ecology: from gene to ecosystem*, pp. 161-173. Editors M. Dicke and W. Takken. (Springer, Netherlands).

305. R.J.E. Bailey, M.A. Birkett, A. Ingvarsdóttir, A.J. Mordue (Luntz), W. Mordue, B. O'Shea, J.A. Pickett and L.J. Wadhams (2006) The role of semiochemicals in host location and non-host avoidance by salmon louse (*Lepeophtheirus salmonis*) copepodids. *Canadian Journal of Fisheries and Aquatic Sciences* **63**, 448-456.

306. A. Couty, H. van Emden, J.N. Perry, J. Hardie, J.A. Pickett and L.J. Wadhams (2006) The roles of olfaction and vision in host-plant finding by the diamondback moth, *Plutella xylostella. Physiological Entomology* **31**, 134-145.

307. J-J.Zhou, Y. Kan, J. Antoniw, J.A. Pickett and L.M. Field (2006) Genome and EST analyses and expression of a gene family with putative functions in insect chemoreception. *Chemical Senses* (advance access published March 31, 2006), 1-13.
308. Z.R. Khan, A. Hassanali and J.A. Pickett (2006) Managing polycropping to enhance soil system productivity: a case study from Africa. *In* Biological Approaches to Sustainable Soil Systems, ed. N. Uphoff, CRC Press, pp. 575-586.
309. C.A.O. Midega, Z.R. Khan, J. van den Berg, C.K.P.O. Ogol, J.A. Pickett and L.J. Wadhams (2006) Maize stemborer predator activity under 'push-pull' system and Bt-maize: a potential component in managing Bt resistance. *International Journal of Pest Management* 52, 1-10.
310. K. Chamberlain, Z.R. Khan, J.A. Pickett, T. Toshova and L.J. Wadhams (2006) Diel periodicity in the production of green leaf volatiles by wild and cultivated host plants of stemborer moths, *Chilo partellus* and *Busseola fusca*. *Journal of Chemical Ecology* 32, 565-577.
311. M.H. Beale, M.A. Birkett, T.J. Bruce, K. Chamberlain, L.M. Field, A.K. Huttly, J.L. Martin, R. Parker, A.L. Phillips, J.A. Pickett, I.M. Prosser, P.R. Shewry, L.E. Smart, L.J. Wadhams, C.M. Woodcock and Y. Zhang (2006) Aphid alarm pheromone produced by transgenic plants affects aphid and parasitoid behaviour. *Proceedings of the National Academy of Sciences USA* 103, 10509-10513.
312. A.M. Hooper, J-B. Farcet, N.P. Mulholland and J.A. Pickett (2006) Synthesis of 9-methylgermacrene B, racemate of the sex pheromone of *Lutzomyia longipalpis* (Lapinha), from the renewable resource, *Geranium macrorrhizum* essential oil. *Green Chemistry* 8, 513-515.
313. Z.R. Khan, J.A. Pickett, L.J. Wadhams, A. Hassanali and C.A.O. Midega (2006) Combined control of *Striga hermonthica* and stemborers by maize-*Desmodium* spp. intercrops. *Crop Protection* 25, 989-995.
314. T. Kalule, Z.R. Khan, G. Bigirwa, J. Alupo, S. Okanya, J.A. Pickett and L.J. Wadhams (2006) Farmers' perceptions of importance, control practices and alternative hosts of maize stemborers in Uganda. *International Journal of Tropical Insect Science* 26, 71-77.
315. I.M. Prosser, R.J. Adams, M.H. Beale, N.D. Hawkins, A.L. Phillips, J.A. Pickett and L.M. Field (2006) Cloning and functional characterisation of a *cis*-muuroladiene synthase from black peppermint (*Mentha x piperita*) and direct evidence for a chemotype unable to synthesise farnesene. *Phytochemistry* 67, 1564-1571.
316. K. Schönrogge, M.G. Gardner, G.W. Elmes, E.K.V. Napper, D.J. Simcox, J.C. Wardlaw, J. Breen, B. Barr, J.J. Knapp, J.A. Pickett and J.A. Thomas (2006) Host propagation permits extreme local adaptation in a social parasite of ants. *Ecology Letters* 9, 1032-1040.

317. Birkett, M.A., Chamberlain, K., Khan, Z.R., Pickett, J.A., Toshova, T., Wadhams, L.J. and Woodcock, C.M. (2006) Electrophysiological responses of the lepidopterous stemborers *Chilo partellus* and *Busseola fusca* to volatiles from wild and cultivated host plants. *Journal of Chemical Ecology* 32, 2475-2487.
318. Z.R. Khan, C.A.O. Midega, A. Hassanali, J.A. Pickett, L.J. Wadhams and A. Wanjoya (2006) Management of witchweed, *Striga hermonthica,* and stemborers in sorghum, *Sorghum bicolor,* through intercropping with greenleaf desmodium, *Desmodium intortum. International Journal of Pest Management* 52, 297-302.
319. S.M. Cook, Z.R. Khan and J.A. Pickett (2007) The use of push-pull strategies in integrated pest management. *Annual Review of Entomology* 52, 375-400.
320. J.A. Pickett, M.A. Birkett, M.C. Blassioli Moraes, T.J.A. Bruce, K. Chamberlain, R. Gordon-Weeks, M.C. Matthes, J.A. Napier, L.E. Smart, L.J. Wadhams and C.M. Woodcock (2007) *cis*-Jasmone as allelopathic agent in inducing plant defence. *Allelopathy Journal* 19, 109-118.
321. A.C. Skelton, M.A. Birkett, J.A. Pickett and M.M. Cameron (2007) Olfactory responses of medically and economically important mites (Acari: Epidermoptidae and Acaridae) to volatile chemicals. *Journal of Medical Entomology* 44, 367-371.
322. Z.R. Khan, C.A.O. Midega, A. Hassanali, J.A. Pickett and L.J. Wadhams (2007) Assessment of different legumes for the control of *Striga hermonthica* in maize and sorghum. *Crop Science* 47, 728-734.
323. J.D. Blande, J.A. Pickett and G.M. Poppy (2007) A comparison of semiochemically mediated interactions involving specialist and generalist *Brassica*-feeding aphids and the braconid parasitoid *Diaeretiella rapae. Journal of Chemical Ecology* 33, 767-779.
324. H.M. Mohamed, Z.R. Khan, J.M. Mueke, A. Hassanali, E. Kairu and J.A. Pickett (2007) Behaviour and biology of *Chilo partellus* (Swinhoe) on *Striga hermonthica* (Del.) Benth. infested and uninfested maize plants. *Crop Protection* 26, 998-1005.
325. Z.R. Khan, C.A.O. Midega, L.J. Wadhams, J.A. Pickett and A. Mumuni (2007) Evaluation of Napier grass (*Pennisetum purpureum*) varieties for use as trap plants for the management of African stemborer (*Busseola fusca*) in a push-pull strategy. *Entomologia Experimentalis et Applicata* 124, 201-211.
326. E. Kazana, T.W. Pope, L. Tibbles, M. Bridges, J.A. Pickett, A.M. Bones, G. Powell and J.T. Rossiter (2007) The cabbage aphid: a walking mustard oil bomb. *Proceedings of the Royal Society B* 274, 2271-2277.
327. T.J.A. Bruce and J.A. Pickett (2007) Plant defence signalling induced by biotic attacks. *Current Opinion in Plant Biology* 10, 387-392.

328. A.M. Hooper, S. Dufour, S. Willaert, S. Pouvreau and J.A. Pickett (2007) Synthesis of (2*S*,7*S*)-dibutyroxynonane, the sex pheromone of the orange wheat blossom midge, *Sitodiplosis mosellana* (Géhin) (Diptera: Cecidomyiidae), by diastereoselective silicon-tethered ring-closing metathesis. *Tetrahedron Letters* 48, 5991-5994.
329. J.A. Pickett, Z.R. Khan, A. Hassanali and A.M. Hooper (2007) Chemicals involved in post-germination inhibition of *Striga* by *Desmodium*: opportunities for utilizing the associated allelopathic traits. In: *Integrating New Technologies for* Striga *Control: Towards Ending the Witch-hunt*, pp. 61-70. Editors G. Ejeta and J. Gressel. (World Scientific, Singapore, New Jersey, London).
330. Z.R. Khan, C.A.O. Midega, A. Hassanali and J.A. Pickett (2007) Field developments on *Striga* control by *Desmodium* intercrops in a "push-pull" strategy. In: *Integrating New Technologies for* Striga *Control: Towards Ending the Witch-hunt*, pp. 241-252. Editors G. Ejeta and J. Gressel. (World Scientific, Singapore, New Jersey, London).
331. J.A. Pickett and R.T. Glinwood (2007) Chemical Ecology. In: *Aphids as Crop Pests*, pp. 235-260. Editors H.F. van Emden and R.H. Harrington. (CABI, Wallingford).
332. S.M. Guchu, A. Yenesew, M.K. Tsanuo, N.K. Gikonyo, J.A. Pickett, A.M. Hooper and A. Hassanali (2007) *C*-methylated and *C*-prenylated isoflavonoids from root extract of *Desmodium uncinatum*. *Phytochemistry* 68, 646-651.
333. T.J.A. Bruce, M.C. Matthes, J.A. Napier and J.A. Pickett (2007) Stressful "memories" of plants: evidence and possible mechanisms. *Plant Science* 173, 603-608.
334. J.A. Pickett, M.A. Birkett, T.J.A. Bruce, K. Chamberlain, R. Gordon-Weeks, M.C. Matthes, J.A. Napier, L.E. Smart and C.M. Woodcock (2007) Developments in aspects of ecological phytochemistry: the role of *cis*-jasmone in inducible defence systems in plants. *Phytochemistry* 68, 2937-2945. *(PSE 50 Special Issue)*.
335. A. Hassanali, H. Herren, Z.R. Khan, J.A. Pickett and C.M. Woodcock (2008) Integrated pest management: the push-pull approach for controlling insect pests and weeds of cereals, and its potential for other agricultural systems including animal husbandry. Integrated Pest Management. *Philosophical Transactions of the Royal Society London B (Theme Issue: Sustainable Agriculture I)* 363, 611-621.
336. Z.R. Khan, C.A.O. Midega, D.M. Amudavi, A. Hassanali and J.A. Pickett (2008) On-farm evaluation of the 'push-pull' technology for the control of stemborers and striga weed on maize in western Kenya. *Field Crops Research* 106, 224-233.
337. Z.R. Khan, D.M. Amudavi, C.A.O. Midega, J.M. Wanyama and J.A. Pickett (2008) Farmers' perceptions of a 'push-pull' technology for control of cereal stemborers and *Striga* weed in western Kenya. *Crop Protection* 27, 976-987.

338. C.A.O. Midega, Z.R. Khan, J. van den Berg, C.K.P.O. Ogol, A.S. Dippenaar-Schoeman, J.A. Pickett and L.J. Wadhams (2008) Response of ground-dwelling arthropods to a 'push-pull' habitat management system: spiders as an indicator group. *Journal of Applied Entomology* 132, 248-254.
339. K. Schönrogge, E.K.V. Napper, M.A. Birkett, C.M. Woodcock, J.A. Pickett, L.J. Wadhams and J.A. Thomas (2008) Host recognition by the specialist hoverfly *Microdon mutabilis*, a social parasite of the ant *Formica lemani*. *Journal of Chemical Ecology* 34, 168-178. (DOI 10.1007/s10886-007-9417-8).
340. J.G. Logan, M.A. Birkett, S.J. Clark, S. Powers, N.J. Seal, L.J. Wadhams, A.J. Mordue (Luntz) and J.A. Pickett (2008) Identification of human-derived volatile chemicals that interfere with attraction of *Aedes aegypti* mosquitoes. *Journal of Chemical Ecology* 34, 308-322.
341. T.J.A. Bruce, M.C. Matthes, K. Chamberlain, C.M. Woodcock, A. Mohib, B. Webster. L.E. Smart, M.A. Birkett, J.A. Pickett and J.A. Napier (2008) *cis*-Jasmone induces *Arabidopsis* genes that affect the chemical ecology of multitrophic interactions with aphids and their parasitoids. *Proceedings of the National Academy of Sciences USA* 105, 4553-4558.
342. Z.R. Khan, D.G. James, C.A.O. Midega and J.A. Pickett (2008) Chemical ecology and conservation biological control. *Biological Control* 45, 210-224. (doi:10.1016/j.biocontrol.2007.11.009).
343. Z.R. Khan, C.A.O. Midega, E.M. Njuguna, D.M.Amudavi, J.W. Wanyama and J.A. Pickett (2008) Economic performance of the 'push-pull' technology for stemborer and Striga control in smallholder farming systems in western Kenya. Crop Protection 27, 1084-1097. (doi:10.1016/j.cropro.2008.01.005).

Popular Articles, 1999 onwards

J.A. Pickett, D.W.M. Smiley and L.J. Wadhams. Food safety: insects to the rescue. *ACTIN MAFF LINK Newsletter*. February, 1999.

J.A. Pickett. *Striga* control by intercropping with *Desmodium* species. 2000. In: Haustorium. (Parasitic Plants Newsletter. Official Organ of the International Parasitic Seed Plant Research Group). Number 37, August 2000.

M.J. Uddin, N.J. Holliday, P.A. MacKay, W. Powell and J.A. Pickett (2001) IPM of insect pests in seed alfalfa. *Forage Seed News* 8, 37-42.

Sex pheromone of pea aphids – an integrated pest management component as it attracts the natural enemies. Joint Annual Meeting 2000 of the Entomological Societies of America, Canada and Quebec.

J.A. Pickett (2002) Stem borer and striga control for maize and sorghum in Africa. *Tropical Agriculture Association Newsletter, June 2002*, 25-26.

J.A. Pickett *et al.* (2002). "Growing the Future". Insect Pests and their Control. Report of a workshop held at Rothamsted Research 6-8 March 2002.

J.A. Pickett *et al.* (2002) Closing discussion: the way forward. Salmon Farming: towards an Integrated pest Management strategy for sea lice. SCI meeting at Department of Zoology, University of Aberdeen, Scotland, 18-19 June, 2001. *Pest Management Science* 58, 622-629.

Z.R. Khan, F.N. Muyekho, E. Njuguna, J.A. Pickett, L.J. Wadhams, N. Dibogo, A. Ndiege, G. Genga and C. Lusweti (2005) A primer on planting and managing 'push-pull' fields for stemborer and striga weed control in maize. A step-by-step guide for farmers. Editor A. Ng'eny-Mengech. ICIPE Science Press, 48 pp.

J.A. Pickett (2006) Book review: Insect-plant biology,second edition. Schoonhoven, L.M., van Loon, Joop J.A., Dicke, M. (eds.). Oxford University Press, pp. 421 + xvii. ISBN: 0-19-852594-8 (paperback, £34.95)/0-19-852594-X (hardback, £75.00). *Phytochemistry* 67, 1970.

The Quiet Revolution: Push-Pull Technology and the African Farmer. Gatsby Occasional Paper. The Gatsby Charitable Foundation, April 2005. 26 pp.

J.A. Pickett (2007) Pheromones and other scents for the alleviation, worldwide, of many pest problems in plant, animal and human health. *Transactions of the Leicester Literary & Philosophical Society* 101, 29-32.

D. Amudavi, Z. Khan and J. Pickett (2007) Enhancing the Push-Pull strategy. *Leisa Magazine*, 23.4 December 2007, 8-10.

Patents

D.R.J. Laws and J.A. Pickett (1975) Preparation of hop oil. Brit. Pat. No. 1501098 and 9 foreign equivalents.

D.R.J. Laws, N.A. Bath, C.S. Ennis, J.A. Pickett and A.G. Wheldon (1977) Preparation of iso-α-acids. Brit. Pat. No. 1576729, and 11 foreign equivalents.

J.B. Free, J.A. Pickett, A.W. Ferguson and M.C. Smith (1981) Bee pheromone. Brit. Pat. No. 8106906, 8206254.

G.W. Dawson, D.C. Griffiths and J.A. Pickett (1981) Improvements in or relating to pheromones. Brit. Pat. No. 2106503 and various foreign equivalents.

B.R. Laurence and J.A. Pickett (1981) Improvements in or relating to pheromones. Brit. Pat. No. 2111481 and various foreign equivalents.

E.D.M. Macaulay and J.A. Pickett (1983) Improvements in or relating to behaviour modifying chemicals. Brit. Pat. No. 2140421 and various foreign equivalents.

G.G. Briggs, G.R. Cayley, B.R. Laurence and J.A. Pickett (1985) Improvements in or relating to pheromones. Brit. Pat. No. 8504483, 8604222.

A.J. Mordue (Luntz), J.A. Pickett and M.A. Birkett (23 November 2005) Semiochemicals for sea lice monitoring and control. UK patent no. GB 2388544.

James H. Tumlinson
Pennsylvania State University
University Park, Pennsylvania, USA

2008 — for remarkable discoveries of mechanisms governing plant-insect and plant-plant interactions. His scientific contributions on chemical ecology have fostered the development of integrated pest management and significantly advanced agricultural sustainability.

CURRICULUM VITAE

Ralph O. Mumma Professor and Director of the Center for Chemical Ecology, Department of Entomology, The Pennsylvania State University, 111 Chemical Ecology Laboratory, University Park, PA 16802.

Date of Birth: February 28, 1938

EDUCATION:

B.S.	Virginia Military Institute, Lexington, Virginia	1960	Chemistry
M.S.	Mississippi State University, State College, Mississippi	1966	Organic Chemistry *Entomology minor*
Ph.D.	Mississippi State University, State College, Mississippi	1969	Organic Chemistry *Biochemistry minor*

PROFESSIONAL EXPERIENCE:

1964-1969 Chemist, USDA-ARS, Boll Weevil Research Laboratory, State College, MS.

1969-1970	Postdoctoral Research Associate with R. M. Silverstein, New York State College of Forestry, Department of Chemistry, Syracuse, NY.
1970-1972	Research Chemist, USDA-ARS, Insect Attractants, Behavior and Basic Biology Research Laboratory, Gainesville, FL.
1970-1975	Adjunct Assistant Professor; University of Florida, Institute of Food and Agricultural Sciences (IFAS), Department of Entomology and Nematology, Gainesville, FL.
1975-1982	Adjunct Associate Professor and member of Doctoral Faculty; University of Florida, IFAS, Department of Entomology and Nematology, and Department of Chemistry, Gainesville, FL.
1982-Present	Adjunct Professor and member of Doctoral Faculty; University of Florida, IFAS, Departments of Entomology and Nematology, and Chemistry, Gainesville, FL.
1972-2003	Research Leader, USDA, ARS; Center for Medical, Agricultural, and Veterinary Entomology (formerly, Insect Attractants, Behavior and Basic Biology Research Laboratory), Gainesville, FL.
2003-Present	Ralph O. Mumma Professor of Entomology, Department of Entomology, The Pennsylvania State University, University Park, PA.
2006-Present	Director, Center for Chemical Ecology, The Pennsylvania State University, University Park, PA.

Field of Specialization:

Insect chemical communication and chemical ecology: defining chemical communication systems, including pheromones and other semiochemicals that mediate insect-insect and plant-insect interactions; biosynthesis of pheromones and plant chemical signals; insect behavior, including learning, mediated by semiochemicals. Emphasis is on developing fundamental knowledge and principles that can be applied in environmentally safe, ecologically sound, sustainable pest management programs.

Honors:

1975	U.S. Department of Agriculture, Superior Service Award, (Boll Weevil Pheromone Development Group) "In recognition of the discovery of and for pioneering the development of the pheromone of the boll weevil as a technique for detection, survey, suppression, or elimination of the pest."
1983	U.S. Department of Agriculture, Superior Service Award, "For outstanding service in the isolation, identification, and synthesis of pheromones of a number of major pest insects and providing science and industry with chemicals for insect research and control."

1984 U.S. Department of Agriculture, Agricultural Research Service, Distinguished Research Scientist of the Year.

1986 Burdick and Jackson International Award for Research in Pesticide Chemistry, presented by Agrochemicals Division of the American Chemical Society.

1990 J.E. Bussart Memorial Award from the Entomological Society of America for research accomplishments in the area of insect semiochemicals and associated behavior.

1991 Florida Entomological Society Annual Research Award (Jointly with W. J. Lewis and T. C. J. Turlings).

1994 LeTourneau Memorial Lecture, University of Idaho.

1995 USDA, The Secretary of Agriculture's Award for Personal and Professional Excellence "For Pioneering Research on Insect Pheromones that Provided the Basis for Control of Major Insect Pests, Including the Boll Weevil, Thereby Reducing Environmental Contamination by Pesticides".

1996 Elected a Fellow of the Entomological Society of America.

1997 Elected to the National Academy of Sciences.

1998 Distinguished Lecturer in Life Sciences, Boyce Thompson Institute for Plant Research, Cornell University, Ithaca, NY.

1998 Alfred M. Boyce Lecturer, University of California at Riverside.

1998 Inducted into ARS Hall of Fame.

1999 Strickland Lecturer, University of Alberta, Edmonton.

1999 Joseph LeConte Lecturer, Georgia Southern University, Statesboro.

2000 Recognition Award in Insect Physiology, Biochemistry and Toxicology, from the Entomological Society of America.

2002 Kenneth A. Spencer Award for Outstanding Achievement in Agricultural and Food Chemistry.

2002 ISI Essential Science Indicators; listed Tumlinson's publications in the top 1% in terms of total citations earned in the field of Environment/Ecology.

2003 Presidential Rank Award as a Meritorious Senior Professional in USDA, ARS.

2003 Jean-Marie Delwart Foundation International Prize (with Dr. W.J. Lewis) for chemical communication.

2005 Silver Medal Award of the International Society of Chemical Ecology.

2008 Wolf Prize in Agriculture.

Other Professional Activities:

- Member of the editorial boards of the Journal of Chemical Ecology and of Biological Control: Theory and Application in Pest Management.
- Vice President, International Society of Chemical Ecology, 1997.

- President, International Society of Chemical Ecology, 1998.
- Member, Board on Agriculture and Natural Resources, National Research Council, National Academy of Sciences, April, 2001-2003.
- Served as member of review panel and moderated a session for the meeting: "Malaria control: a reconsideration of the role of DDT", Board on Global health and Board on Agriculture and Natural Resources, The National Academies, Washington, DC, July 21-22, 2004.

Graduate and Post-Doctoral Students:

Masters Students

Basilios E. Mazomenos 1978 Greece
Thomas M. Dykstra 1994 U.S.
Mary Donohue 2000 U.S.

Doctoral Students

Peter E. A. Teal 1981 Canada
Fred J. Eller 1990 U.S.
Ted C. J. Turlings 1991 The Netherlands
Ursula Rose 1997 Germany
Yasmin Cardoza 2002 Honduras
Ezra Schwartzberg, current U.S.
Emily Hohlfeld, current U.S.
Elizabeth Bosak, current U.S.
Tracy Conklin, current U.S.
Christy Harris, current U.S.
Sean Halloran, current U.S.

Postdoctoral

Hajime Sugie Japan
Tatsugi Chuman Japan
Stuart Krasnoff U.S.
Hans Alborn Sweden
Philip McCall England
Nianbai Fang China
John Loughrin U.S.
Yoav Gazit Israel
Paul Pare U.S.
Eric Schmelz U.S.
Consuelo De Moraes Brazil
Alonzo Suaso Honduras
Jurgen Engleberth Germany
Cameron Lait Canada
Naoki Mori Japan
Baldwyn Torto Ghana
Juan Huang China
Katalin Boeroeczky Hungary
Irmgard Seidl-Adams Germany
Naoko Yoshinaga Japan

BRIEF SUMMARY OF RESEARCH CAREER

(Numbers in superscript refer to the list of publications that follow.)

My research career, which began in 1964, has been devoted to the study of chemical ecology, in particular the chemistry of insect communication and behavior. All of my investigations have been interdisciplinary efforts. I was Research leader for the Chemistry Research Unit at the Center for Medical, Agricultural and Veterinary Entomology (formerly Insect Attractants, Behavior and Basic Biology

Research Laboratory) from 1972 until 2003. With my colleagues and students I identified and synthesized numerous pheromones of various types, kairomones, biologically active plant constituents, and insect herbivore-produced elicitors of plant defenses for over 40 insect species spanning 13 families and 5 orders. Most of my investigations dealt with insects of considerable economic importance in one or more areas of the world and resulted in several novel discoveries that were the first of their kind and subsequently led to similar discoveries by others. In the last 20 years my research has focused on the chemical ecology of tritrophic insect-plant interactions. My colleagues and I have taken an active role in transferring the technology developed through basic research into practical tools for insect pest management and have had several successes, notably the boll weevil and Japanese beetle pheromones and lures for the Mediterranean fruit fly, among others. I have given over 150 lectures on my research throughout the United States and in several foreign countries. Some examples of my research accomplishments follow.

For my Ph.D. dissertation I isolated, identified, and synthesized the four terpenoid components of the boll weevil pheromone[9]. This pheromone was applied as one of the three primary tools in an integrated program that resulted in the successful eradication of the boll weevil from the U.S., which had significant economic and environmental impacts. For example, in Georgia, cotton acreage increased from 115,000 in 1983 (pre-eradication) to 1.5 million in 1995, yield from 112,000 bales to 2 million, and net crop revenues from $187 to over $451 per acre. In the same period, the average number of insecticide treatments decreased from 14.4 to 5.4 per acre.

As a Postdoctoral Associate in the laboratory of Prof. R. M. Silverstein, at the New York State College of Forestry, I identified the sex pheromone of the Indian meal moth[12] and the trail pheromone of the leaf cutting ant, *Atta texana*[13].

In 1970 I accepted a position as a Research Chemist at the USDA, ARS, Insect Attractants, Behavior and Basic Biology Laboratory in Gainesville, Florida, where my colleagues and I isolated, identified, synthesized, and developed lures and trapping systems for monitoring of over 20 species of pest insects. One of the most notable examples was the sex pheromone of the Japanese beetle[46]. This was the first time that an enantiomer of an optically active pheromone was clearly demonstrated to be a potent inhibitor of attraction to the pheromone. Collaboration with colleagues and industry scientists resulted in development of a trapping system used around airports and other vulnerable locations throughout the world to guard against introduction of this pest, and over 1.5 million lures per year have been sold in the U.S. for several years.

In a collaboration spanning more than 20 years, Dr. W. J. Lewis and I, with our students and colleagues, conducted an in depth investigation of the chemical ecology and behavioral mechanisms that enable insect parasitoids to locate their herbivorous insect hosts. The goal of this research was to develop more effective methods for biological control of agricultural pests. We discovered that plants actively produce volatile organic compounds, in response to feeding damage by Lepidopterous larvae, that attract parasitic wasps, the natural enemies of the herbivores[164]; further, that female parasitoids can learn the chemical cues that guide them to their hosts by associating odors with a host recognition signal produced by their hosts, and that learning plays a major role in the parasitoids' host foraging behavior[127]. It enables them to exploit the vast array and dynamic nature of the chemical cues produced by the feeding of their hosts on different plant species and other factors in the environment. In further investigations we discovered that plants respond systemically[181,215], as well as locally, to herbivore damage, that the release of plant volatiles follows a diurnal rhythm with maximum emission during the day[196,240], when the natural enemies are foraging, that many of the volatile organic compounds released by the plants in response to herbivore damage are synthesized *de novo*[217,222], rather than being released from storage, and that plants release different volatile blends in response to damage by different species of herbivores[226]. In the latter case, parasitoid wasps that specialize on a single species of herbivore can distinguish volatiles induced by hosts from those induced by non-hosts feeding on the same plant. These results clearly illustrate the active, dynamic ability of plants to defend themselves against herbivory, by releasing chemical signals that recruit natural enemies of their attackers.

Equally intriguing was the discovery that plant defensive reactions are triggered by a substance or substances in the oral secretions or regurgitant of the feeding caterpillars[218]. This discovery led to the identification and synthesis of the first herbivore-produced, herbivore-specific elicitor of plant volatiles from the oral secretion of beet armyworm larvae, and the demonstration that this elicitor is produced in the insects from precursors obtained from the plants on which they feed[228]. The elicitor, *N*-(17-hydroxylinolenoyl)-L-glutamine, was named volicitin. The similarity of the structure of this insect-produced elicitor with the structures of the components of the octadecanoid signaling pathway in plants suggests the possibility of developing methods to enhance and manipulate the chemical defenses of crops to increase their resistance to insect pests. Subsequently, volicitin and/or analogous compounds were discovered in several different species of Lepidopteran larvae as well as in other insects. The fact that these compounds are synthesized by the herbivores[250] even though they induce plant defensive reactions against the insects indicates that they play a critical role in the digestion and or metabolism of the insect herbivores. Research aimed at elucidating the roles of the elicitors in insect metabolism as well as the mechanism by which they induce plant defenses is ongoing. Recently the Tumlinson laboratory identified another class of insect-produced elicitors (named caeliferins)[269] of plant defenses from the regurgitant of grasshoppers. It is now evident that plants can detect and respond to a broad array of compounds produced by their attackers.

Another feature of this interesting and complex interplay between insects and plants is that plants can detect and react to chemical signals from neighboring plants. Chemical signals from a plant damaged by insect herbivores alert neighbors to prime their defenses so as to respond more strongly to subsequent attack than if they had not been forewarned. Thus, corn seedlings previously exposed to "green leaf" volatiles from damaged corn seedlings emitted at least twice the quantity of volatile terpenes when attacked by caterpillars as seedlings that had not been exposed to the "alarm" signals[256]. This ability of plants to prime their defenses to respond more strongly when attacked, but only if attacked, has since been reported in other plant species.

In 2003, I accepted the Ralph O. Mumma endowed chair in chemical ecology, in the Department of Entomology at The Pennsylvania State University, where my students and I are continuing to investigate the chemical ecology of plant-insect interactions. In 2006, I became the director of the Center for Chemical Ecology at Penn State. This center promotes the collaboration of 24 faculty and their students from seven departments and two colleges across the university in interdisciplinary investigations of the chemical ecology of numerous biological systems.

LIST OF PUBLICATIONS

1. Minyard, J. P., Tumlinson, J. H., Hedin, P. A., and Thompson, A. C. Constituents of the cotton bud. Terpene hydrocarbons. J. Agr. Food Chem. 13:599-602. 1965.
2. Minyard, J. P., Tumlinson, J. H., Thompson, A. C., and Hedin, P. A. Constituents of the cotton bud. Sesquiterpene hydrocarbons. J. Agr. Food Chem. 14:332-336. 1966.
3. Minyard, J. P., Tumlinson, J. H., Thompson, A. C., and Hedin, P. A. Constituents of the cotton bud. The carbonyl compounds. J. Agr. Food Chem. 15:517-524. 1967.
4. Tumlinson, J. H., Minyard, J. P., Hedin, P. A., and Thompson, A. C. Reaction chromatography. I. Gas-liquid/thin-layer chromatographic derivatization technique for the identification of carbonyl compounds. J. Chromatogr. 29:80-87. 1967.
5. Minyard, J. P., Tumlinson, J. H., Thompson, A. C., and Hedin, P. A. II. Gas-liquid/thin-layer chromatographic derivatization technique for the identification of alcohols. J. Chromatogr. 29:88-93. 1967.
6. Tumlinson, J. H., Hardee, D. D., Minyard, J. P., Thompson, A. C., Gast, R. T., and Hedin, P. A. Boll weevil sex attractant: Isolation studies. J. Econ. Entomol. 61:470-474. 1968.
7. Cross, W. H., Hardee, D. D., Nichols, F., Mitchell, H. C., Mitchell, E. B., Huddleston, P. M., and Tumlinson, J. H. Attraction of female boll weevils to

traps baited with males or extracts of males. J. Econ. Entomol. 62:154-161. 1969.

8. Tumlinson, J. H. Isolation, identification, and partial synthesis of boll weevil sex attractant. A dissertation submitted to the faculty of Mississippi State University in partial fulfillment of the requirements for the Degree of Doctor of Philosophy in the Department of Chemistry. 1969.
9. Tumlinson, J. H., Hardee, D. D., Gueldner, R. C., Thompson, A. C., Hedin, P. A., and Minyard, J. P. Sex pheromones produced by the male boll weevil: Isolation, identification, and synthesis. Science 166:1010-1012. 1969.
10. Tumlinson, J. H., Gueldner, R. C., Hardee, D. D., Thompson, A. C., Hedin, P. A., and Minyard, J. P. The boll weevil sex attractant. pp. 41-59. In Morton, B. (Ed.) Chemicals Controlling Insect Behavior, Academic Press, New York. 1970. (Book Chapter)
11. Tumlinson, J. H., Gueldner, R. C., Hardee, D. D., Thompson, A. C., Hedin, P. A., and Minyard, J. P. Identification and synthesis of the four compounds comprising the boll weevil sex attractant. J. Org. Chem. 36:2616-2621. 1971.
12. Brady, U. E., Tumlinson, J. H., Brownlee, R. G., and Silverstein, R. M. Sex stimulant and attractant in the Indian meal moth and in the almond moth. Science 171:802-804. 1971.
13. Tumlinson, J. H., Silverstein, R. M., Moser, J. C., Brownlee, R. G., and Ruth, J. M. Identification of the trail pheromone of a leaf-cutting ant, *Atta texana*. Nature 234:348-349. 1971.
14. Hardee, D. D., McKibben, G. H., Gueldner, R. C., Mitchell, E. B., Tumlinson, J. H., and Cross, W. H. Boll weevils in nature respond to grandlure, a synthetic pheromone. J. Econ. Entomol. 65:97-100. 1972.
15. Tumlinson, J. H., Moser, J. C., Silverstein, R. M., Brownlee, R. G., and Ruth, J. M. A volatile trail pheromone of the leaf-cutting ant, *Atta texana*. J. Insect Physiol. 18:809-814. 1972.
16. Tumlinson, J. H., Mitchell, E. R., Browner, S. M., Mayer, M. S., Green, N., Hines, R., and Lindquist, D. A. *Cis*-7-Dodecen-1-ol, a potent inhibitor of the cabbage looper sex pheromone. Environ. Entomol. 1(3):354-358. 1972.
17. Tumlinson, J. H., Mitchell, E. R., Browner, S. M., and Lindquist, D. A. A sex pheromone for the soybean looper. Environ. Entomol. 1(4):466-468. 1972.
18. Mitchell, E. R. and Tumlinson, J. H. An attractant for males of *Spodoptera dolichos* (Lepidoptera: Noctuidae). Ann. Entomol. Soc. Am. 66(4):917-918. 1973.
19. Sower, L. L., Vick, K. W., and Tumlinson, J. H. (*Z,E*)-9,12- Tetradecadien-1-ol: A chemical released by female *Plodia interpunctella* that inhibits the sex pheromone response of male *Cadra cautella*. Environ. Entomol. 3(1):120-122. 1974.

20. Yonce, C. E., Tumlinson, J. H., Gentry, C. R., and Mitchell, E. R. Extraction and field bioassay of the sex pheromone of the lesser peachtree borer. Environ. Entomol. 3(3):569-570. 1974.
21. Tumlinson, J. H., Yonce, C. E., Doolittle, R. E., Heath, R. R., Gentry, C. R., and Mitchell, E. R. Sex pheromones and reproductive isolation of the lesser peachtree borer and the peachtree borer. Science 185:614-616. 1974.
22. Tumlinson, J. H., Heath, R. R., and Doolittle, R. E. Application of chemical ionization mass spectrometry of epoxides to the determination of olefin position in aliphatic chains. Anal. Chem. 46:1309-1312. 1974.
23. Hendricks, D. E. and Tumlinson, J. H. A field cage bioassay system for testing candidate sex pheromones of the tobacco budworm. Ann. Entomol. Soc. Am. 67(4):547-552. 1974.
24. Mitchell, E. R., Tumlinson, J. H., Copeland, W. W., Hines, R. W., and Brennan, M. M. Tobacco budworm: Production, collection, and use of natural pheromone in field traps. Environ. Entomol. 3(4):711-713. 1974.
25. McLaughlin, J. R., Mitchell, E. R., Chambers, D. L., and Tumlinson, J. H. Perception of *Z*-7-dodecen-1-ol and modification of the sex pheromone response of male loopers. Environ. Entomol. 3(4):677-680. 1974.
26. Gueldner, R. C., Tumlinson, J. H., III, Hardee, D. D., Hedin, P. A., Thompson, A. C., and Minyard, J. P. Boll Weevil Sex Attractant. U. S. Patent Office #3,813,443, May 28, 1974. (Patent)
27. Tumlinson, J. H., III, Mitchell, E. R., Browner, S. M., Mayer, M. S., Green, N., Hines, R. W., and Lindquist, D. A. Method for Disrupting Pheromone Communication with *Cis*-7-Dodecen-1-ol. United States Patent #3,832, 461, Aug. 27, 1974. (Patent)
28. Tumlinson, J. H., Hendricks, D. E., Mitchell, E. R., Doolittle, R. E., and Brennan, M. M. Isolation, identification, and synthesis of the sex pheromone of the tobacco budworm. J. Chem. Ecol. 1(2):203-214. 1975.
29. Nielsen, D. G., Purrington, F. F., Tumlinson, J. H., Doolittle, R. E., and Yonce, C. E. Response of male clearwing moths to caged virgin females, female extracts, and synthetic sex attractants. Environ. Entomol. 4(3):451-454. 1975.
30. Heath, R. R., Tumlinson, J. H., Doolittle, R. E., and Proveaux, A. T. Silver nitrate-high pressure liquid chromatography of geometrical isomers. J. Chrom. Sci. 13:380-382. 1975.
31. McLaughlin, J. R., Mitchell, E. R., and Tumlinson, J. H. Evaluation of some formulations for dispensing insect pheromones in field and orchard crops. Proc. Int. Controlled Release Pesticide Symposium, Sept. 8-10, Wright State University, Dayton, Ohio, pp 209-215. 1975. (Proceedings)
32. McLaughlin, John R., Doolittle, R. E., Gentry, C. R., Mitchell, E. R., and Tumlinson, J. H. Response to pheromone traps and disruption of pheromone communication in the lesser peachtree borer and the peachtree borer (Lepidoptera: Sesiidae). J. Chem. Ecol. 2(1):73-81. 1976.

33. Tumlinson, J. H., and Heath, R. R. Structure elucidation of insect pheromones by microanalytical methods. J. Chem. Ecol. 2(1):87-99. 1976.
34. Tumlinson, J. H., Mitchell, E. R., and Chambers, D. L. Manipulating complexes of insect pests with various combinations of behavior-modifying chemicals, pp. 53-66. In M. Beroza, Ed., Advances in Chemistry, ACS Symp. Ser. No. 23, Pest Management with Insect Sex Attractants and Other Behavior-Controlling Chemicals. 1976. (Book Chapter)
35. Yonce, C. E., Gentry, C. R., Tumlinson, J. H., Doolittle, R. E., and Nielsen, D. G. Lesser peachtree borer: Influence of trap height, substrates, concentration, and trap design on capture of male moths with females and with a synthetic pheromone. Environ. Entomol. 5(3):417-420. 1976.
36. Yaginuma, K., Kumakura, M., Tamaki, Y., Yushima, T., and Tumlinson, J. H. Sex attractant for the cherry tree borer, *Synanthedon hector* Butler (Lepidoptera: Sesiidae). Appl. Ent. Zool. 11(3):266-268. 1976.
37. Karandinos, M. G., Tumlinson, J. H., and Eichlin, T. D. Field evidence of synergism and inhibition in the Sesiidae sex pheromone system. J. Chem. Ecol. 3(1):57-64. 1977.
38. Heath, R. R., Tumlinson, J. H., Doolittle, R. E., and Duncan, J. H. Analytical and preparative separation of geometrical isomers by high efficiency silver nitrate liquid chromatography. J. Chromatogr. Sci. 15:10-13. 1977.
39. Haniotakis, G. E., Mazomenos, B. E., and Tumlinson, J. H. A sex attractant of the olive fruit fly, *Dacus oleae* and its biological activity under laboratory and field conditions. Ent. Exp. Appl. 21:81-87. 1977.
40. Yonce, C. E., Gentry, C. R., Tumlinson, J. H., Doolittle, R. E., Mitchell, E. R., and McLaughlin, J. R. Seasonal distribution of the lesser peachtree borer in Central Georgia as monitored by pupal skin counts and pheromone trapping techniques. Environ. Entomol. 6(2):203-206. 1977.
41. Greany, P. D., Tumlinson, J. H., Chambers, D. L., and Boush, G. M. Chemically mediated host finding by *Biosteres (Opius) longicaudatus*, a parasitoid of tephritid fruit fly larvae. J. Chem. Ecol. 3(2):189-195. 1977.
42. McLaughlin, John R., Tumlinson, J. H., and Sharp, J. L. Absence of synergism in the response of Florida lesser peachtree borer males to synthetic sex pheromone. Fla. Entomol. 60(1):27-30. 1977.
43. Gentry, C. R., Yonce, C. E., Tumlinson, J. H., and Blythe, J. L. Capture of male lesser peachtree borers and peachtree borers at pheromone sources in the field. J. Ga. Entomol. Soc. 12(2):145-150. 1977.
44. Holloway, R. L., Childers, S., Tumlinson, J. H., and Yonce, C. E. Trapping the lesser peachtree borer with (*E*,*Z*)3,13-octadecadien-1-ol acetate. J. Ga. Entomol. Soc. 12:365-368. 1977.
45. Tamaki, Y., Yushima, T., Oda, M., Kida, K., Kitamura, K., Yabuki, S., and Tumlinson, J. H. Attractiveness of 3,13-octadecadienyl acetates for males of clearwing moths. Japanese J. Appl. Ent. Zool. 21(2):106-107. 1977.

46. Tumlinson, J. H., Klein, M. G., Doolittle, R. E., Ladd, T. L., and Proveaux, A. T. Identification of the female Japanese beetle sex pheromone: Inhibition of male response by an enantiomer. Science 197:789-792. 1977.
47. Tumlinson, J. H. Animal Communication by Pheromones. H. H. Shorey (ed.). Academic Press, New York, 1976. (Book Review) Pan-Pacific Entomol. 53(3):180. 1977.
48. Barry, M. W., Nielsen, D. G., Purrington, F. F., and Tumlinson, J. H. Attractivity of pheromone blends to male peachtree borer, *Synanthedon exitiosa*. Environ. Entomol. 7(1):13. 1978.
49. Nielsen, D. G., Purrington, F. F., Campbell, R. L., Wilmot, T. R., Capizzi, J., and Tumlinson, J. H. Sex attractants for Sequoia pitch moth and strawberry crown moth. Environ. Entomol. 7(4):544-546. 1978.
50. Sharp, J. L., McLaughlin, J. R., James, J., and Tumlinson, J. H. Seasonal abundance of *Synanthedon pictipes* and *S. exitiosa* in North Central Florida. Environ. Entomol. 7(4):589-591. 1978.
51. Voerman, S., Minks, A. K., Vanwetswinkel, G., and Tumlinson, J. H. Attractivity of 3,13-octadecadien-1-ol acetates to the male clearwing moth *Synanthedon myopaeformis* (Borkhausen) (Lepidoptera, Sesiidae). Ent. Exp. & Appl. 23:301-304. 1978.
52. Mitchell, E. R., Tumlinson, J. H., and Baumhover, A. H. *Heliothis virescens*: Attraction of males to blends of (*Z*)-9-tetradecen-1-ol formate and (*Z*)-9-tetradecenal. J. Chem. Ecol. 4(6):709-716. 1978.
53. Heath, R. R., Proveaux, A. T., and Tumlinson, J. H. A simple terminator for high efficiency liquid chromatography columns. J. HRC & CC 1(6):317-319. 1978.
54. Sharp, J. L., McLaughlin, J. R., James, J., Eichlin, T. D., and Tumlinson, J. H. Seasonal occurrence of male Sesiidae in North Central Florida determined with pheromone trapping methods. Fla. Entomol. 61(4):245-250. 1978.
55. Sales, F. M., Tumlinson, J. H., McLaughlin, J. R., and Sailer, R. I. Comportamento do parasitoide. *Trissolcus basalis* (Wollaston) em resposta a queromonios produzidos pelio hospedeiro, *Nezara viridula* (L.). Fitossanidade, Fortaleza 2(3):88. 1978.
56. Sales, F. M., McLaughlin, J. R., Sailer, R. I., and Tumlinson, J. H. Temporal analysis of the ovipositional behavior of the female egg parasitoid, *Trissolcus basalis* (Wollaston). Fitossanidade, Fortaleza 2(3):80-83. 1978.
57. Tumlinson, J. H. Recent discoveries in insect pheromone chemistry, pp. 315-322. In Geissbuhler, H. (Ed.), Advances In Pesticide Science. Pergamon Press, Oxford & NY. 1979. (Book Chapter)
58. Tumlinson, J. H. The need for biological information in developing strategies for applying semiochemicals, pp. 301-311. In Ritter, F. J. (Ed.), Chemical Ecology: Odour Communication in Animals. Elsevier/North-Holland Biomedical Press, The Netherlands. 1979.

59. Gentry, C. R., Yonce, C. E., Blythe, J. L., and Tumlinson, J. H. Lesser peachtree borer: Recovery of marked native males in pheromone baited traps. Environ. Entomol. 8(2):218-220. 1979.
60. Heath, R. R., McLaughlin, J. R., Tumlinson, J. H., Ashley, T. R., and Doolittle, R. E. Identification of the white peach scale sex pheromone, an illustration of micro techniques. J. Chem. Ecol. 5(6):941-953. 1979.
61. Heath, R. R., Jordan, J. R., Sonnet, P. E., and Tumlinson, J. H. Potential for the separation of insect pheromones by gas chromatography on columns coated with cholesteryl cinnamate, a liquid-crystal phase. J. HRC & CC 2:712-714. 1979.
62. Tumlinson, J. H. The chemistry of Sesiidae pheromones, pp. 1-10. In Pheromones of the Sesiidae (formerly Aegeriidae). USDA/SEA AAR-NE-6, 83 pp. 1979. (Proceedings)
63. Tumlinson, J. H., Klein, M. G., Doolittle, R. E., and Ladd, Jr., T. L. Sex Pheromone Produced by the Female Japanese Beetle: Specificity of Male Response of Enantiomers. U.S. Patent #4,179,446. 10/18/79. (Patent)
64. Chow, Y. S., Mayer, M. S., and Tumlinson, J. H. Electroantennogram response of *Plodia interpunctella* to its sex pheromone and wing gland extracts. Bull. Inst. Zool. 19(1):27-31. 1980.
65. Heath, R. R., Burnsed, G. E., Tumlinson, J. H., and Doolittle, R. E. Separation of a series of positional and geometrical isomers of olefinic aliphatic primary alcohols and acetates by capillary gas chromatography. J. Chromatogr. 189:199-208. 1980.
66. Doolittle, R. E., Tumlinson, J. H., Proveaux, A. T., and Heath, R. R. Synthesis of the sex pheromone of the Japanese beetle. J. Chem. Ecol. 6(2):473-485. 1980.
67. Heath, R. R., Doolittle, R. E., Sonnet, P. E., and Tumlinson, J. H. Sex pheromone of the white peach scale: Highly stereoselective synthesis of the stereoisomers of pentagonol propionate. J. Org. Chem. 45(14): 2910-2912. 1980.
68. Cross, J. H., Tumlinson, J. H., Heath R. R., and Burnett, D. E. Apparatus and procedure for measuring release rates from formulations of lepidopteran semiochemicals. J. Chem. Ecol. 6(4):759-770. 1980.
69. Cross, J. H., Mitchell, E. R., Tumlinson, J. H., and Burnett, D. E. Selection of a polyethylene tubing formulation of (*Z*)-9-tetradecen-1-ol formate and its use in disrupting pheromone communication in *Heliothis zea* (Boddie). J. Chem. Ecol. 6(4):771-779. 1980.
70. Vander Meer, R. K., Glancey, B. M., Lofgren, C. S., Glover, A., Tumlinson, J. H., and Rocca, J. The poison sac of red imported fire ant queens: Source of a pheromone attractant. Ann. Entomol. Soc. Am. 73(5): 609-612. 1980.
71. Tumlinson, J. H., Mitchell, E. R., and Sonnet, P. E. Sex pheromone components of the beet armyworm, *Spodoptera exigua.* J. Environ. Sci. Health, A16(2):189-200. 1981.

72. Klein, M. G., Tumlinson, J. H., Ladd, Jr., T. L., and Doolittle, R. E. Japanese beetle (Coleoptera: Scarabaeidae): Response to synthetic sex attractant plus phenethyl propionate: Eugenol. J. Chem. Ecol. 7(1):1-7. 1981.
73. Tumlinson, J. H. Chemistry of the queen pheromone of the red imported fire ant. Proc. Tall Timbers Conf. on Ecological Animal Control by Habitat Management No. 7, pp. 155-161. 1981. (Proceedings)
74. Teal, P. E. A., McLaughlin, J. R., and Tumlinson, J. H. Analysis of the reproductive behavior of *Heliothis virescens* (F.) under laboratory conditions. Ann. Entomol. Soc. Am. 74(3):324-330. 1981.
75. Teal, P. E. A., Heath, R. R., Tumlinson, J. H., and McLaughlin, J. R. Identification of a sex pheromone of *Heliothis subflexa* (Gn.) (Lepidoptera: Noctuidae) and field trapping studies using different blends of components. J. Chem. Ecol. 7(6):1011-22. 1981.
76. Tumlinson, J. H., Teal, P. E. A., and Heath, R. R. Chemical ethology: A holistic approach to the study of insect pheromones. Proc. lst Japan/USA Symp. on IPM, Tsukuba, Japan, Sept. 29-30, 1981, pp. 19-31. 1981. (Proceedings)
77. Ladd, T. L., Klein, M. G., and Tumlinson, J. H. Phenethyl propionate + eugenol + geraniol (3:7:3) and japonilure: A highly effective joint lure for Japanese beetles. J. Econ. Entomol. 74(6):665-667. 1981.
78. Johnson, D. W., Mitchell, E. R., Tumlinson, J. H., and Allen, G. E. Velvetbean caterpillar: Response of males to virgin females and pheromone in the laboratory and field. Fla. Entomol. 64(4):529-533. 1981.
79. Prokopy, R. J., Averill, A. L., Bardinelli, C. M., Bowdan, E. S., Cooley, S. S., Crnjar, R. M., Dundulis, E. A., Roitberg, C. A., Spatcher, P. J., Tumlinson, J. H., and Weeks, B. L. Site of production of an oviposition-deterring pheromone component in *Rhagoletis pomonella* flies. J. Insect Physiol. 28(1):1-10. 1982.
80. Solomon, J. D., Oliveria, F. L., Tumlinson, J. H., and Doolittle, R. E. Occurrence of clearwing borers (Sesiidae) in west central Mississippi. J. Ga. Entomol. Soc. 17(1):4-12. 1982.
81. Guss, P. L., Tumlinson, J. H., Sonnet, P. E., and Proveaux, A. T. Identification of a female-produced sex pheromone of the western corn rootworm. J. Chem. Ecol. 8(2):545-556. 1982. (Book Chapter)
82. Tumlinson, J. H., Heath, R. R., and Teal, P. E. A. Analysis of chemical communications systems of Lepidoptera, pp. 1-25. In Leonhardt, B. A., and Beroza, M. (Eds.), Insect Pheromone Technology: Chemistry and Applications. ACS Symp. Series No. 190. 1982.
83. Lewis, W. J., Nordlund, D. A., Gueldner, R. C., Teal, P. E. A., and Tumlinson, J. H. Kairomones and their use for management of entomophagous insects. XIII. Kairomonal activity for *Trichogramma* spp. of abdominal tips, excretion, and a synthetic sex pheromone blend of *Heliothis zea* (Boddie) moths. J. Chem. Ecol. 8(10):1323-1331. 1982.

84. Tumlinson, J. H., and Teal, P. E. A. The sophisticated language of insect chemical communication. J. Ga. Entomol. Soc. 17(4)2nd Supplement:11-23. 1982. (Proceedings)
85. Teal, P. E. A., McLaughlin, J. R., Tumlinson, J. H., and Rush, R. R. A sex attractant for *Platysenta videns* (Gn.) similar to the sex pheromone of *Heliothis virescens* (F.). Fla. Entomol. 65(4):583-584. 1982.
86. Tumlinson, J. H. The chemical basis for communication between the sexes in *Heliothis virescens* and other insect species. Proceedings Les Mediateurs Chimiques, 16-20 Nov. 1981, Versailles, France, pp. 193-201. 1982. (Proceedings)
87. Mitchell, E. R., Sugie, H., and Tumlinson, J. H. *Spodoptera exigua*: Capture of feral males in traps baited with blends of pheromone components. J. Chem. Ecol. 9(1):95-104. 1983.
88. Lofgren, C. S., Glancey, B. M., Glover, A., Rocca, J., and Tumlinson, J. H. Behavior of workers of *Solenopsis invicta* (Hymenoptera: Formicidae) to the queen recognition pheromone: Laboratory studies with an olfactometer and surrogate queens. Ann. Entomol. Soc. Am. 76:(1)44-50. 1983.
89. Teal, P. E. A., Carlysle, T. C., and Tumlinson, J. H. Epidermal glands in terminal abdominal segments of female Heliothis virescens (F.) (Lepidoptera: Noctuidae). Ann. Entomol. Soc. Am. 76(2):242-247. 1983.
90. Rocca, J. R., Tumlinson, J. H., Glancey, B. M., and Lofgren, C. S. The queen recognition pheromone of *Solenopsis invicta*, preparation of (*E*)-6- (1-pentenyl)-2H-pyran-2-one. Tetrahedron Lett. 24(18):1889-1892. 1983.
91. Rocca, J. R., Tumlinson, J. H., Glancey, B. M., and Lofgren, C. S. Synthesis and stereochemistry of tetrahydro-3,5-dimethyl-6-(1-methyl- butyl)-2H-pyran-2-one, a component of the queen recognition pheromone of *Solenopsis invicta*. Tetrahedron Lett. 24(18):1893-1896. 1983.
92. Glancey, B. M., Lofgren, C. S., Rocca, J. R., and Tumlinson, J. H. Behavior of disrupted colonies of *Solenopsis invicta* towards queens and pheromone-treated surrogate queens placed outside the nest. Sociobiology. 7:283-288. 1983.
93. Mitchell, E. R., Sugie, H., and Tumlinson, J. H. Rubber septa as dispensers for the fall armyworm sex attractant. J. Environ. Sci. Health, A18(3):463-470. 1983.
94. Guss, P. S., Carney, R. L., Sonnet, P. E., and Tumlinson, J. H. Stereospecific sex attractant for *Diabrotica cristata* (Harris) (Coleoptera: Chrysomelidae). Environ. Entomol. 12(4):1296-1297. 1983.
95. Heath, R. R., Tumlinson, J. H., Leppla, N. C., McLaughlin, J. R., Dueben, B., Dundulis, E., and Guy, R. H. Identification of a sex pheromone produced by female velvetbean caterpillar moth. J. Chem. Ecol. 9(5):645-656. 1983.

96. Guss, P. L., Tumlinson, J. H., Sonnet, P. E., and McLaughlin, J. R. Identification of a female-produced sex pheromone from the southern corn rootworm, *Diabrotica undecimpunctata howardi* Barber. J. Chem. Ecol. 9(9):1363-1375. 1983.
97. Teal, P. E. A., Tumlinson, J. H., McLaughlin, J. R., Heath, R. R., and Rush, R. A. (Z)-11-Hexadecen-1-ol: A behavioral modifying chemical present in the pheromone gland of female *Heliothis zea* (Lepidoptera: Noctuidae). J. Can. Ent. 116:777-779. 1984.
98. Guss, P. L., Sonnet, P. E., Carney, R. L., Branson, T. F., and Tumlinson, J. H. Response of *Diabrotica virgifera, D. v. zeae*, and *D. porracea* to stereoisomers of 8-methyl-2-decyl propanoate. J. Chem. Ecol. 10(7):1123-1131. 1984.
99. Heath, R. R., and Tumlinson, J. H. Techniques for purifying, analyzing, and identifying pheromones, pp. 287-322. In Hummel, H. E. and Miller, T. A. (Eds.) Techniques in Pheromone Research, Springer-Verlag, New York, NY. 1984. (Book Chapter)
100. Guss, Paul L., Tumlinson, James H., III, Sonnet, Philip E., and McLaughlin, John R. Synthetic pheromone 10-methyl-2-tridecanone and its use in controlling the southern corn rootworm and related diabroticites. P.C. 6918, U.S. Patent No. 4,474,991, October 2, 1984. Also: P.C. N6931, U.S. Patent No. 4,565,695, January 21, 1986. (Patent)
101. Glancey, B. M., Rocca, J., Lofgren, C. S., and Tumlinson, J. Field tests with synthetic components of the queen recognition pheromone of the red imported fire ant, *Solenopsis invicta* Buren. Sociobiology 9(1):19-30. 1984.
102. Guss, P. L., Sonnet, P. E., Carney, R. L., Tumlinson, J. H., and Wilkin, P. J. Response of northern corn rootworm, *Diabrotica barberi* Smith and Lawrence, to stereoisomers of 8-methyl-2-decyl propanoate. J. Chem. Ecol. 11(1):21-26. 1985.
103. Tumlinson, J. H. The beetles: pheromonal chemists par excellence, pp. 367-380. In Paul A. Hedin (ed.), Bioregulators for Pest Control. ACS Symposium Series 276. ACS, Washington, DC. 1985. (Book Chapter)
104. Teal, P. E. A., Mitchell, E. R., Tumlinson, J. H., Heath, R. R., and Sugie, H. Identification of volatile sex pheromone components released by the southern armyworm, *Spodoptera eridania* (Cramer). J. Chem. Ecol. 11(6):717-725. 1985.
105. Snow, J. Wendell, Eichlin, T. D., and Tumlinson, J. H. Seasonal captures of clearwing moths (Sesiidae) in traps baited with various formulations of 3,13-octadecadienyl acetate and alcohol. J. Agric. Entomol. 2(1):73-84. 1985.
106. Doolittle, R. E., Tumlinson, J. H., and Proveaux, A. T. Determination of double bond position in conjugated dienes by chemical ionization mass spectrometry with isobutane. Anal. Chem. 57(8):1625-1630. 1985.

107. Mitchell, Everett R., J. H. Tumlinson, and Jeremy N. McNeil. Field evaluation of commercial pheromone formulations and traps using a more effective sex pheromone blend for the fall armyworm (Lepidoptera: Noctuidae). J. Econ. Entomol. 78:1364-1369. 1985.
108. Teal, P. E. A., Tumlinson, J. H., and Heath, R. R. Chemical and behavioral analyses of volatile sex pheromone components released by calling *Heliothis virescens* (F.) females (Lepidoptera: Noctuidae). J. Chem. Ecol. 12(1):107-126. 1986.
109. Tumlinson, J. H., Teal, P. E. A., Mitchell, E. R., and Heath, R. R. Sex pheromones of economically important North American *Spodoptera* species. Int. Conf. on Integr. Plant Prot., Budapest, Hungary, July 4-9, 1983. Proc. Int. Conf. on Integr. Plant Prot. Vol. 4:140-146. 1986. (Proceedings)
110. Teal, P. E. A., and Tumlinson, J. H. Terminal steps in pheromone bio-synthesis by *Heliothis virescens* and *H. zea*. J. Chem. Ecol. 12(2):353-366. 1986.
111. Heath, R. R., Coffelt, J. A., Sonnet, P. E., Proshold, F. I., Dueben, B. D., and Tumlinson, J. H. Identification of a sex pheromone produced by female sweetpotato weevil. J. Chem. Ecol. 12(6):1489-1503. 1986.
112. Krysan, J. L., Wilkin, P. H., Tumlinson, J. H., Sonnet, P. E., Carney, R. L., and Guss, P. L. Responses of *Diabrotica lemniscata* and *D. longicornis* (Coleoptera: Chrysomelidae) to stereoisomers of 8-methyl-2- decanol propanoate and studies on the pheromone of *D. longicornis.* Ann. Entomol. Soc. Am. 79(4):742-746. 1986.
113. Tumlinson, J. H., Mitchell, E. R., Teal, P. E. A., Heath, R. R., and Mengelkoch, L. J. Sex pheromone of the fall armyworm, *Spodoptera frugiperda* (J. E. Smith): Identification of components critical to attraction in the field. J. Chem. Ecol. 12(9):1909-1926. 1986.
114. Schneiderman, A. M., Hildebrand, J. G., Brennan, M. M., and Tumlinson, J. H. Trans-sexually grafted antennae alter pheromone-directed behavior in a moth. Nature 323:801-803. 1986.
115. Heath, R. R., and Tumlinson, J. H. Correlation of retention times on a liquid crystal capillary column with reported vapor pressures and half-lives of compounds used in pheromone formulations. J. Chem. Ecol. 12(11):2081-2088. 1986.
116. Heath, R. R., Teal, P. E. A., Tumlinson, J. H., and Mengelkoch, L. J. Prediction of release ratios of multicomponent pheromones from rubber septa. J. Chem. Ecol. 12(12):2133-2143. 1986.
117. Szöcs, G., Tóth, M., Bestmann, H. J., Vostrowsky, O., Heath, R. R., and Tumlinson, J. H. Polyenic hydrocarbons as sex attractants for geometrids and amatids (Lepidoptera) found by field screening in Hungary. Z. Naturforsch 42c:165-168. 1987.

118. Chuman, T., Guss, P. L., Doolittle, R. E., McLaughlin, J. R., Krysan, J. L., Schalk, J. M. and Tumlinson, J. H. Identification of a female-produced sex pheromone from banded cucumber beetle, *Diabrotica balteata* LeConte (Coleoptera: Chrysomelidae). J. Chem. Ecol. 13(7):1601-1616. 1987.
119. Teal, P. E. A., and Tumlinson, J. H. Induced changes in sex pheromone biosynthesis of *Heliothis* moths (Lepidoptera: Noctuidae), pp. 79-94. In Law, J. (ed). Molecular Entomology. Alan R. Liss, Inc., New York. 1987. (Book Chapter)
120. Teal, P. E. A. and Tumlinson, J. H. The role of alcohols in pheromone biosynthesis by two noctuid moths that use acetate pheromone components. Archiv. Insect Biochem. & Physiol. 4:261-269. 1987.
121. Stowe, M., Tumlinson, J. H., and Heath, R. R. Chemical Mimicry: Bolas spiders emit components of moth prey species sex pheromone. Science. 236:964-967. 1987.
122. Chuman, T., Landolt, P. J., Heath, R. R., and Tumlinson, J. H. Isolation, identification and synthesis of a male-produced sex pheromone of the papaya fruit fly, *Toxotrypana curvicauda* Gerstaecker (Diptera: Tephritidae). J. Chem. Ecol. 13(9):1979-1992. 1987.
123. Tumlinson, J. H. and Teal, P. E. A. Relationship of structure and function to biochemistry in insect pheromone systems. pp. 3-26. In Prestwich, G. D. and Blomquist, G. J. (eds), Pheromone Biochemistry, Academic Press. 1987. (Book Chapter)
124. Sonnet, P. E., Guss, P. L., Tumlinson, J. H., McGovern, T. P. and Cunningham, R. T. Asymmetric synthesis of selected insect pheromones, pp. 388-400. In Synthesis and Chemistry of Agrochemicals, ACS Symposium Series 355, Washington, D. C. 1987. (Book Chapter)
125. Noldus, L. P. J. J., Lewis, W. J., Tumlinson, J. H., and van Lenteren, J. C. Olfactometer and wind tunnel experiments on the role of sex pheromones of noctuid moths in the foraging behaviour of *Trichogramma* spp. Proc. of 2nd Int. Sym. *Trichogramma* and Other Egg Parasites. Guangzhou, China, Nov. 10-15, 1986. Ed. INRA, Paris, (Les Colloques de l'INRA no. 43), pp. 223-238. 1988. (Proceedings)
126. Eller, F. J., Tumlinson, J. H. and Lewis, W. J. Beneficial arthropod behavior mediated by airborne semiochemicals. II. Evidence for the use of volatile attractants by the parasitoid *Microplitis croceipes* (Cresson) (Hymenoptera: Braconidae). J. Chem. Ecol. 14(2):425-434. 1988.
127. Lewis, W. J. and Tumlinson, J. H. Host detection by chemically mediated associative learning in a parasitic wasp. Nature. 331:257-259. 1988.
128. Tumlinson, J. H. Insect pheromone systems. Comments Agric. and Food Chemistry. 1(3):115-146. 1988.

129. Drost, Y. C., Lewis, W. J., and Tumlinson, J. H. Beneficial arthropod behavior mediated by airborne semiochemicals. V. Influence of rearing method, host plant, and adult experience on host-searching behavior of *Microplitis croceipes* (Cresson), a larval parasitoid of *Heliothis*. J. Chem. Ecol. 14:1607-1616. 1988.
130. Heath, R. R., Coffelt, J. A., Proshold, F. I., Sonnet, P. E., and Tumlinson, J. H., III. (Z)-3-Dodecen-1-ol (E)-2-butenoate and its use in monitoring the sweetpotato weevil. U. S. Patent No. 4,732,756, March 22, 1988. (Patent)
131. Guss, P. L., Tumlinson, J. H., III, Sonnet, P. E., and Proveaux, A. T. Synthetic pheromone 8-methyl-2-decanol propanoate. U. S. Patent No. 4,732,524, March 29, 1988. (Patent)
132. Herard, F., Keller, M. A., Lewis, W. J., and Tumlinson, J. H. Beneficial arthropod behavior mediated by airborne semiochemicals. III. Influence of age and experience on flight chamber responses of *Microplitis demolitor* Wilkinson. J. Chem. Ecol. 14:1583-1596. 1988.
133. Herard, F., Keller, M. A., Lewis, W. J., and Tumlinson, J. H. Beneficial arthropod behavior mediated by airborne semiochemicals. IV. Influence of host diet on host-oriented flight chamber responses of *Microplitis demolitor* Wilkinson. J. Chem. Ecol. 14:1597-1606. 1988.
134. Landolt, P. J., Heath, R. R., Agee, H. R., Tumlinson, J. H., and Calkins, C. O. A sex pheromone-based trapping system for the papaya fruit fly, *Toxotrypana curvicauda* Gerstaecker (Diptera: Tephritidae). J. Econ. Entomol. 81:1163-1169. 1988.
135. Mitchell, E. R., Heath, R. R., Tumlinson, J. H. WORT: Wind-oriented trap for evaluating several pheromone blends simultaneously. J. Econ. Entomol. 81:966-969. 1988.
136. Eller, F. J., Tumlinson, J. H., and Lewis, W. J. Beneficial arthropod behavior mediated by airborne semiochemicals: Source of volatiles mediating the host-location flight behavior of *Microplitis croceipes* (Cresson) (Hymenoptera: Braconidae), a Parasitoid of *Heliothis zea* (Boddie) (Lepidoptera: Noctuidae). Environ. Entomol. 17:745-753. 1988.
137. Eller, F. J., Boucias, D. G. and Tumlinson, J. H. Interactions between *Microplitis croceipes* (Cresson) (Hymenoptera: Braconidae) and a nuclear polyhedrosis virus of *Heliothis zea* (Boddie) (Lepidoptera: Noctuidae). Environ. Entomol. 17(6):977-982. 1988.
138. Tumlinson, J. H. Contemporary frontiers in insect semiochemical research. J. Chem. Ecol. 14(11):2109-2130. 1988.
139. Teal, P. E. A. and Tumlinson, J. H. Properties of cuticular oxidases used for sex pheromone biosynthesis by *Heliothis zea*. J. Chem. Ecol. 14(11):2131-2145. 1988.

140. Chuman, T., Sivinski, J., Heath, R. R., Calkins, C. O., and Tumlinson, J. H., Battiste, M. A., Wydra, R. L., Strekowski, L., and Nation, J. L. Suspensolide, a new macrolide component of male Caribbean fruit fly (*Anastrepha suspensa* (Loew)) volatiles. Tetrahedron Lett. 29(50):6561-6564. 1988.
141. Battiste, M. A., Rocca, J. R., Wydra, R. L., Tumlinson, J. H., and Chuman, T. Total synthesis and structure proof of (3E,8E)-Suspensolide. Tetrahedron Lett. 29(50):6565-6568. 1988.
142. Prokopy, R. J., Powers, P. J., Heath, R. R., Dueben, B. D., and Tumlinson, J. H. Comparative laboratory methods for assaying behavioral responses of *Rhagoletis pomonella* flies to host marking pheromone. J. Appl. Ent. 106:437-443. 1988.
143. Kovalev, B. G., Tumlinson, J. H., Bolgar, T. S., Kolomoets, T. P., Kovalenko, V. M., and Heath. R. R. Isolation and identification of a component of the sex pheromone of *Aegeria apiformis*. Translated from Khimiya Prirodnykh Soedinenii, No. 1, 125-128. 1988.
144. Teal, P. E. A. and Tumlinson, J. H. Isolation, identification and biosynthesis of compounds produced by male hairpencil glands of *Heliothis virescens* (F.) J. Chem. Ecol. 15(1):413-427. 1989.
145. Tumlinson, J. H. Insect chemical communication systems. Pure & Appl. Chem. 61(3):559-562. 1989. (Proceedings)
146. Teal, P. E. A. and Tumlinson, J. H. Neurohormonal induction of pheromone biosynthesis by *Heliothis zea* (Boddie) during the photophase. Can. Entomol. 121:43-46. 1989.
147. Nordlund, D. A., Lewis, W. J., and Tumlinson, J. H. Habitat and host location behavior of *Microplitis croceipes*. Southwestern Ent. Suppl. 12:39-48. 1989. (Proceedings)
148. Teal, P. E. A., Tumlinson, J. H., and Oostendorp. A. Enzyme-catalyzed pheromone synthesis by *Heliothis* moths, pp. 332-343. In Whitaker, J. R. and Sonnet, P. E. (Eds.) Biocatalysis in Agricultural Biotechnology, Am. Chem. Soc. 1989. (Book Chapter)
149. Teal, P. E. A., Tumlinson, J. H. and Oberlander, H. Neural regulation of sex pheromone biosynthesis in *Heliothis* moths. Proc. Natl. Acad. Sci. USA. 86:2488-2492. 1989.
150. Turlings, T. C. J., Tumlinson, J. H., Lewis, W. J. and Vet, L. E. M. Beneficial arthropod behavior mediated by airborne semiochemicals. VIII. Learning of host related odors induced by a brief contact experience with host by-products in *Cotesia marginiventris* (Cresson), a generalist larval parasitoid. J. Insect Behav. 2:217-225. 1989.
151. Landolt, P. J., Tumlinson, J. H., and Brennan, M. M. Attraction of *Amphion floridensis* (Lepidoptera: Sphingidae) to bombykal (E,Z)-10,12-hexadecadienal. Fla. Entomol. 72(2):324-327. 1989.

152. Tumlinson, J. H., Brennan, M. M., Doolittle, R. E., Mitchell, E. R., Brabham, A., Mazomenos, B. E., Baumhover, A., and Jackson, D. M. Identification of a pheromone blend attractive to *Manduca sexta* (L.) males in a wind tunnel. Archiv. Insect Biochem. & Physiol. 10:255-271. 1989.
153. Kaissling, K. E., Hildebrand, J. G., and Tumlinson, J. H. Pheromone receptor cells in the male moth *Manduca sexta*. Arch. Insect Biochem. & Physiol. 10:273-279. 1989.
154. Christensen, T. A., Hildebrand, J. G., Tumlinson, J. H. and Doolittle, R. E. The sex pheromone blend of *Manduca sexta*: Responses of central olfactory interneurons to antennal stimulation in male moths. Arch. Insect. Biochem. & Physiol. 10:281-291. 1989.
155. Tumlinson, J. H. Chemical analysis and identification of insect pheromones, pp. 73-92. In Ridgway, R. L., Silverstein, R. M., and Inscoe, M. N. (Eds.), Behavior-Modifying Chemicals for Insect Management, Applications of Pheromones and Other Attractants. Marcel Dekker, Inc., New York and Basel. 1990. (Book Chapter)
156. Chuman, T., Guss, P. L., Doolittle, R. E., McLaughlin, J. R., and Tumlinson, J. H. 6,12-Dimethylpentadecan-2-one and its use in monitoring and controlling the banded cucumber beetle. U. S. Patent # 4,871,537, October 3, 1989. (Patent)
157. Krysan, J. L., McDonald, I. C., and Tumlinson, J. H. Phenogram based on allozymes and its relationship to classical biosystematics and pheromone structure among eleven diabroticites (Coleoptera: Chrysomelidae). Ann. Entomol. Soc. Am. 82(5):574-581. 1989.
158. Turlings, T. C. J., Scheepmaker, J. W. A., Vet, L. E. M., Tumlinson, J. H., and Lewis, W. J. How contact foraging experiences affect the preferences for host-related odors in the larval parasitoid *Cotesia marginiventris* (Cresson) (Hymenoptera: Braconidae). J. Chem. Ecol. 16:1577-1589. 1990.
159. Doolittle, R. E., Brabham, A., and Tumlinson, J. H. Sex pheromone of *Manduca sexta* (L.): Stereoselective synthesis of (10*E*,12*E*,14*Z*)-10,12,14-hexadecatrienal and isomers. J. Chem. Ecol. 16:1131-1153. 1990.
160. Domek, J. M., Tumlinson, J. H., and Johnson, D. T. Responses of Male Green June Beetles *Continis nitida* (L.) (Coleoptera:Scarabaeidae) to Female Volatiles in a Flight Tunnel. J. Insect Behavior. 3(2):271-276. 1990.
161. Eller, F. J., Tumlinson, J.H., and Lewis, W. J. Intraspecific competition in *Microplitis croceipes* (Cresson) (Hymenoptera:Braconidae), a parasitoid of *Heliothis* species (Lepidoptera:Noctuidae). Ann. Entomol. Soc. Am. 83(3):504-508. 1990.
162. Heath, R. R., Chambers, D. L., Tumlinson, J. H., and Landolt, P. J. Controlled release of Trimedlure Isomer C from a compressed disk and the evaluation of its attractiveness to the Mediterranean Fruit Fly (Diptera:Tephritidae). J. Econ. Entomol. 83:(3):819-822. 1990.

163. Schalk, J. J., McLaughlin, J. R., and Tumlinson, J. H. Field response of feral male banded cucumber beetles to the sex pheromone. 6,12-Dimethylpentadecan-2-one. Fla. Entomol. 1990. 73(2):271-276. 1990.
164. Turlings, T. C. J., Tumlinson, J. H., and Lewis, W. J. Exploitation of Herbivore-Induced Plant Odors by Host-Seeking Parasitic Wasps. Science. 250:1251-1253. 1990.
165. Lewis, W. J., Vet, Louise E. M., Tumlinson, J. H., van Lenteren, J. C., and Papaj, D. R. Variations in parasitoid behavior essential element of a sound biological control theory. Environ. Entomol. 19(5):1183-1193. 1990.
166. Teal, P. E. A., Tumlinson, J. H., and Oberlander, H. Endogenous suppression of pheromone production in virgin female moths. Experientia. 46:1047-1050. 1990.
167. Tumlinson, J. H., Mitchell, E. R., and Yu, H. S. Analysis and field evaluation of the volatile blend emitted by calling virgin females of the beet armyworm moth, *Spodoptera exigua* (Hübner). J. Chem. Ecol. 16(12):3411-3424. 1990.
168. Tumlinson, J. H. and Teal, P.E.A. Endogenous regulation of pheromone biosynthesis in *Heliothis* moths, pp. 243-255. In Hagedorn, H. H., Hildebrand, J. G., Kidwell, M., and Law, J. H. (Eds.), Molecular Insect Science. Plenum Press. New York. 1990. (Book Chapter)
169. Noldus, L.P.J.J., Lewis, W. J., and Tumlinson, J. H. Beneficial Arthropod Behavior Mediated by Airborne Semiochemicals. IX. Differential Response of *Trichogramma pretiosum*, an Egg Parasitoid of *Heliothis zea*, to Various Olfactory Cues. J. Chem. Ecol. 16(12):3531-3544. 1990.
170. Doolittle, R. E., Tumlinson, J. H., Brabham, A., Brennan, M. M., and Mitchell, E. R. Sex Pheromone Blend of the Tobacco Hornworm. Identification and Stereoselective Synthesis, pp. 491-503. In Baker, D.R., Fenyes, J. G., and Moberg, W. K. (Eds.), Synthesis and Chemistry of Agrochemicals II, ACS Symposium Series No. 443. Washington, DC. American Chemical Society. 1991. (Book Chapter)
171. Teal, P. E. A., and Tumlinson, J. H. Sex pheromone production in moths, endogenous regulation of initiation and inhibition, pp. 66-77. In Hedin, P.A., (Ed.), Naturally Occurring Pest Bioregulators. ACS Symposium Series No. 449, Washington, DC. American Chemical Society. 1991. (Book Chapter)
172. Turlings, T. C. J., Tumlinson, J. H., Eller, F., and Lewis, W. J. Larvae-damaged Plants: Source of volatile synomones that guide the parasitoid *Cotesia marginiventris* to the microhabitat of its hosts. Entomol. exp. appl. 58:75-82. 1991.
173. McLaughlin, J. R., Tumlinson, J. H., and Mori, K. Responses of male *Diabrotica balteata* (Coleoptera: Chrysomelidae) to stereoisomers of the sex pheromone 6, 12-dimethylpentadecan-2-one. J. Econ. Entomol. 84(1):99-102. 1991.
174. Turlings, T. C. J. and Tumlinson, J. H. Do parasitoids use herbivore-induced plant chemical defenses to locate hosts? Fla. Entomol. 74(1):42-50. 1991.

175. Lewis, W. J., Tumlinson, J. H., and Krasnoff, S. Chemically Mediated Associative Learning; an Important Function in the Foraging Behavior of *Microplitis croceipes* (Cresson). J. Chem. Ecol. 17(7):1309-1325. 1991.
176. Christensen, T. A., Itagaki, H., Teal, P. E. A., Jasensky, R. D., Tumlinson, J. H. and Hildebrand. J. G. Innervation and neural regulation of the sex pheromone gland in female *Heliothis* moths. Proc. Natl. Acad. Sci. USA. 88:4971-4975. 1991.
177. Heath, R. R., Landolt, P. J., Tumlinson, J. H., Chambers, D. L., Murphy, R., Doolittle, R. E., Dueben, B. D., Sivinski, J. J., and Calkins, C. O. Analysis, synthesis, and formulation of sex pheromone components of the male Mediterranean fruit flies attractive to females. J. Chem. Ecol. 17(9):1925-1940. 1991.
178. Turlings, T. C. J., Tumlinson, J. H., Heath, R. R., Proveaux, A. T., and Doolittle, R. E. Isolation and identification of allelochemicals that attract the larval parasitoid, *Cotesia marginiventris* (Cresson), to the micro-habitat of one of its hosts. J. Chem. Ecol. 17:2235-2251. 1991.
179. Tumlinson, J. H. and Lewis, W. J. Chemically Mediated Foraging Behavior of Parasitoids: New Implications for Biocontrol, pp. 161-178, In: Vinson, S. B., and Metcalf, R. L. (Eds.), Entomology Serving Society: Emerging Technologies and Challenges, Entomological Society of America, Lanham, Maryland. Symposium held Dec. 12-13, 1989, San Antonio, Texas. 1991. (Proceedings)
180. Teal, P. E. A., and Tumlinson, J. H. Lipidic pheromones. In Current Opinion in Structural Biology. Vol. 2, pp. 475-481. 1992. (Book Chapter)
181. Turlings, T. C. J., and Tumlinson, J. H. Systemic release of chemical signals by herbivore-injured corn. Proc. Natl. Acad. Sci. USA. 89:8399-8402. 1992.
182. Fang, N., Tumlinson, J. H., Teal, P. E. A., and Oberlander, H. Fatty acyl pheromone precursors in the sex pheromone gland of female tobacco hornworm moths, *Manduca sexta* (L.). Insect Biochem. Molec. Biol. 22(7):621-631. 1992.
183. Eller, F. J., Tumlinson, J. H., and Lewis, W. J. Effect of host diet and preflight experience on the flight response of *Microplitis croceipes* (Cresson). Physiol. Entomol. 17:235-240. 1992.
184. Tumlinson, J. H., Turlings, T. C. J., and Lewis, W. J. The semiochemical complexes that mediate insect parasitoid foraging. Agric. Zool. Reviews, Vol. 5:221-252. 1992.
185. Christensen, T. A., Lehman, H. K., Teal, P. E. A., Itagaki, H., Tumlinson, J. H., and J. G. Hildebrand. Diel changes in the presence and physiological actions of octopamine in the female sex-pheromone glands of heliothine moths. Insect Biochem. Molec. Biol. 22(8):841-849. 1992.
186. Turlings, T. C. J., McCall, P. J., Alborn, H. T., and Tumlinson, J. H. Systemic releases of volatiles by herbivore-damaged plants: what possible functions? pp. 365-366. In Menken, S. B. J., Visser, J. H., and Harrewijn, P. (Eds.),

Insect-Plant Relationships, Proc. 8th Int. Symp. Dordrecht: Kluwer Acad. Publ., 1992. (Proceedings)

187. Turlings, T. C. J., Wäckers, F., Vet, L. E. M., Lewis, W. J. and Tumlinson, J. H. Learning of host-finding cues by hymenopterous parasitoids, pp. 51-78. In Lewis, A. C. and Papaj, D. R. (Eds.), Insect Learning: Ecological and Evolutionary Perspectives, Chapman and Hall, New York. 1993. (Book Chapter)
188. Tumlinson, J. H., Lewis, W. J., and Vet, L. E. M. How parasitic wasps find their hosts. Scientific American. 268(3):100-106. 1993.
189. Tumlinson, J. H., Turlings, T. C. J., and Lewis, W. J. Semiochemically medicated foraging behavior in beneficial parasitic insects. Arch. Insect Biochem. Physiol. 22(3/4):385-391. 1993.
190. Turlings, T. C. J., McCall, P. J., Alborn, H. T., and Tumlinson, J. H. An elicitor in caterpillar oral secretions that induces corn seedlings to emit volatiles attractive to parasitic wasps. J. Chem. Ecol. 19(3):411-425. 1993.
191. Teal, P. E. A., Oostendorp, A. and Tumlinson, J. H. Induction of pheromone production in females of *Heliothis virescens* (F.) and *H. subflexa* (Gn.) (Lepidoptera: Noctuidae) during the photophase. Can. Entomol. 125: 355-366. 1993.
192. McCall, P. J., Turlings, T. C. J., Lewis, W. J., and Tumlinson, J. H. Role of plant volatiles in host location by the specialist parasitoid *Microplitis croceipes* (Cresson) (Braconidae: Hymenoptera). J. Insect Beh. 6(5):625-639. 1993.
193. Mitchell, E. R., and Tumlinson, J. H. Response of *Spodoptera exigua* and *S. eridania* (Lepidoptera: Noctuidae) males to synthetic pheromone and *S. exigua* females. Fla. Entomol. 77(2):237-247. 1994.
194. Tumlinson, J. H., Mitchell, E. R., Doolittle, R. E. and Jackson, D. M. Field tests of Synthetic Manduca sexta sex pheromone. J. Chem. Ecol. 20(3):579-591. 1994.
195. Abernathy, R. L., Teal, P.E.A., and Tumlinson, J. H. Age and crowding affects the amount of sex pheromone and the oviposition rates of virgin and mated females of *Helicoverpa zea* (Lepidoptera: Noctuidae). Ann. Entomol. Soc. Am. 87(3):350-354. 1994.
196. Loughrin, J. H., Manukian, A., Heath, R. R., Turlings, T. C. J., and Tumlinson, J. H. Diurnal cycle of emission of induced volatile terpenoids by herbivore-injured cotton plants. Proc. Natl. Acad. Sci. USA. 91:11836-11840. 1994.
197. McCall, P. J., Turlings, T. C. J., Loughrin, J. H., Proveaux, A. T., and Tumlinson, J. H. Herbivore-induced volatile emissions from cotton (*Gossypium hirsutum* L.) Seedlings. J. Chem. Ecol. 20(12):3039-3050. 1994.
198. Lewis, W. J., Sheehan, W., and Tumlinson, J. H. Unraveling the story of how parasitoids find their hosts, pp. 671-680. In Rosen, D., Bennett, F. D. and J. L. Capinera (Eds.), Pest Management in the Subtropics: Biological Control — a Florida Perspective, Intercept Limited, Andover. 1994. (Book Chapter)

199. Stowe, M. K., Turlings, T. C., Loughrin, J. H., Lewis, W. J., and Tumlinson, J. H. The chemistry of eavesdropping, alarm, and deceit. Proc. Natl. Acad. Sci. USA. 92:23-28. 1995. (Proceedings)
200. Fang, N., Teal, P. E. A., Doolittle, R. E., and Tumlinson, J. H. Biosynthesis of conjugated olefinic systems in the sex pheromone gland of female tobacco hornworm moths, *Manduca sexta* (L.). Insect Biochem. Molec. Biol. 25(1):39-48. 1995.
201. Abernathy, R. L., Nachman, R. J., Teal, P. E. A., Yamashita, O., and Tumlinson, J. H. Pheromonotropic activity of naturally occurring pyrokinin insect neuropeptides (FxPRLamide) in *Helicoverpa zea*. Peptides. 16(2):215-129. 1995
202. Fang. N., Teal, P. E. A., and Tumlinson, J. H. PBAN regulation of pheromone biosynthesis in female Tobacco Hornworm moths, *Manduca sexta* (L.). Arch. Insect Biochem. Physiol. 29:35-44. 1995.
203. Turlings, T. C. J., Loughrin, J. H., McCall, P. J., Röse, U., Lewis, W. J., and Tumlinson, J. H. How caterpillar-damaged plants protect themselves by attracting parasite wasps. Proc. Nat. Acad. Sci. USA. 92:4169-4174. 1995. (Proceedings)
204. Fang, N., Teal, P. E. A., and Tumlinson, J. H. Characterization of oxidase(s) associated with the sex pheromone gland in *Manduca sexta* (L.) females. Arch. Insect Biochem. & Physiol. 29:243-257. 1995.
205. Loughrin, J. H., Manukian, A., Heath, R. R., and Tumlinson, J. H. Volatiles emitted by different cotton varieties damaged by feeding beet armyworm larvae. J. Chem. Ecol. 21(8):1217-1227. 1995.
206. Berg, B. G., Tumlinson, J. H., and Mustaparta, H. Chemical communication in Heliothine moths. IV. Receptor neuron responses to pheromone compounds and formate analogues in the male tobacco budworm moth *Heliothis virescens*. J. Comp. Physiol. A. 177:527-534. 1995.
207. Fang, N., Teal, P. E. A., and Tumlinson, J. H. Correlation between glycerolipids and pheromone aldehydes in the sex pheromone gland of female tobacco hornworm moths, *Manduca sexta* (L.). Arch. Insect Biochem. & Physiol. 30:321-336. 1995.
208. Alborn, H. T., Lewis, W. J., and Tumlinson, J. H. Host-specific recognition kairomone for the parasitoid *Microplitis croceipes* (CRESSON). J. Chem. Ecol. 21(11):1697-1708. 1995.
209. Teal, P. E. A., Abernathy, R. L., Nachman, R. J., Fang, N., Meredith, J. A., and Tumlinson, J. H. Pheromone biosynthesis activating neuropeptides: Functions and chemistry. Peptides 17(2):337-344. 1996.
210. Tumlinson, J. H., Teal, P. E. A., and Fang, N. The integral role of triacyl glycerols in the biosynthesis of the aldehydic sex pheromones of *Manduca sexta* (L.). Bioorganic and Medicinal Chemistry. 4(3):452-460. 1996.

211. Gazit, Y., Lewis, W. J., and Tumlinson, J. H. Arrestment of *Telenomous remus* (Hymenoptera: Scelionidae) by a kairomone associated with eggs of its host, *Spodoptera frugiperda* (Lepidoptera: Noctuidae). Biol. Control. 6:283-290. 1996.
212. Teal, P. E. A. and Tumlinson, J. H. Effects of interspecific hybridization between *Heliothis virescens* and *H. zea* on the sex pheromone communication system, pp 535-547. In Carde, R. T. and Minks, A. K. (eds.) Insect Pheromone Research: New Directions. Proceedings of the First International Symposium on Insect Pheromones. Chapman & Hall, New York. 1997. (Proceedings) (Book chapter).
213. Paré, P. W. and Tumlinson, J. H. Plant volatile signals in response to herbivore feeding. Fla. Entomol. 79(2):93-103. 1996.
214. Fang, N., Teal, P. E. A., and Tumlinson, J. H. Effects of decapitation and PBAN injection on amounts of triacylglycerols in the sex pheromone gland of *Manduca sexta* (L.). Arch. Insect Biochem. Physiol. 32:249-260. 1996.
215. Röse, U. S. R., Manukian, A., Heath, R. R., and Tumlinson, J. H. Volatile semiochemicals released from undamaged cotton leaves: a systemic response of living plants to caterpillar damage. Plant Physiology. 111:487-495. 1996.
216. Heath, R. R., Tumlinson, J. H. and Landolt, P. J. Interfacing analytical chemistry with IPM - the Florida experience pp. 167-181. In Rosen, D., Bennett, F. D. and Capinera, J. L. (Eds.) Pest Management in the Subtropics: Integrated Pest Management - A Florida Perspective. Intercept, Andover, Hants., U.K. 1996. (Book Chapter)
217. Paré, P. W. and Tumlinson, J. H. Induced synthesis of plant volatiles. Nature. 385:30-31. 1997.
218. Alborn, H. T., Turlings, T. C. J., Jones, T. H., Stenhagen, G., Loughrin, J. H., and Tumlinson, J. H. An elicitor of plant volatiles from beet armyworm oral secretion. Science. 276:945-949. 1997.
219. Stapel, J. O., Cortesero, A. M., De Moraes, C. M., Tumlinson, J. H., and Lewis, W. J. Extrafloral nectar, honeydew, and sucrose effects on searching behavior and efficiency of *Microplitis croccipes* (Hymenoptera: Braconidae) in cotton. Environ. Entomol. 26(3):617-623. 1997.
220. Cortesero, A.M., De Moraes, C. M., Stapel, J. O., Tumlinson, J. H., and Lewis, W. J. Comparisons and contrasts in host foraging strategies of two larval parasitoids with different degrees of host specificity. J. Chem. Ecol. 23(6):1589-1606. 1997.
221. Röse, U. S. R., Alborn, H. T., Makranczy, G., Lewis, W. J., and Tumlinson, J. H. Host recognition by the specialist endoparasitoid *Microplitis croceipes* (Hymenoptera: Braconidae): Role of host- and plant-related volatiles. J. Insect Behav. 10(3):313-330. 1997.
222. Paré, P. W. and Tumlinson, J. H. *De novo* biosynthesis of volatiles induced by insect herbivory in cotton plants. Plant Physiology. 114(1):1161-1167. 1997.

223. Lewis, W. J., van Lenteren, J. C., Phatak, S. C. and Tumlinson, J. H. A total system approach to sustainable pest management. Proc. Natl. Acad. Sci. USA 94:12243-12248. 1997.
224. Paré, P. W. and Tumlinson, J. H. Cotton volatiles synthesized and released distal to the site of insect damage. Phytochemistry. 47(4):521-526. 1997.
225. Röse, U. S. R., Lewis, W. J., and Tumlinson, J. H. Specificity of systemically released cotton volatiles as attractants for specialist and generalist parasitic wasps. J. Chem. Ecol. 24(2):303-319. 1998.
226. De Moraes, C. M., Lewis, W. J., Paré, P. W., Alborn, H. T., and Tumlinson, J. H. Herbivore-infested plants selectively attract parasitoids. Nature. 393:570-573. 1998.
227. Fónagy, A., Teal, P. E. A., Meredith, J. M., Kormendy, C., and Tumlinson, J. H. The purification, isolation and partial identification of a new pheromonotropic peptide from the moth *Mamestra brassicae* (Lepidoptera). Proceedings of the 18th Congress of European Comparative Endocrinology. Annals NY Academy Sciences. 839:488-490. 1998. (Proceedings)
228. Paré, P. W., Alborn, H. T., and Tumlinson, J. H. Concerted biosynthesis of an insect elicitor of plant volatiles. Proc. Natl. Aacd. Sci. USA. 95:13971-13975. 1998.
229. Paré, P. W. and Tumlinson, J. H. Plant volatiles as a defense against insect herbivores. Plant Physiology 121: 325-331. 1999.
230. Paré, P. W., Lewis, J. W., and Tumlinson, J. H. Induced plant volatiles: biochemistry and effects on parasitoids, pp. 167-180. In Agrawal, A.A., Tuzun, S. and Bent, E. (eds.) Induced Plant Defenses Against Pathogens and Herbivores: Biochemistry, Ecology and Agriculture. American Phytopathological Society Press, St. Paul, MN. 1999. (Book Chapter)
231. Landolt, P. J., Tumlinson, J. H., and Alborn, H. T. Attraction of Colorado potato beetle (Coleoptera: Chrysomelidae) to damaged and chemically-induced potato plants. Environ. Entomol. 28(6): 973-978. 1999.
232. Tumlinson, J. H., Paré, P. W., and Lewis, W. J. Plant production of volatile semiochemicals in response to insect derived elicitors, pp. 95-109. In Chadwick, D.J. and Goode, J. (eds.) Insect-plant interactions and induced plant defence. Novartis Foundation Symposium No. 223. John Wiley Journal. 1999.
233. Turlings, T. C. J., Alborn, H. T., Loughrin, J. H., and Tumlinson, J. H. Volicitin, an elicitor of maize volatiles in the oral secretion of *Spodoptera exigua*: its isolation and bio-activity. J. Chem. Ecol. 26(1): 189-202. 2000.
234. Alborn, H. T., Jones, T. H., Stenhagen, G. S., and Tumlinson, J. H. Identification and synthesis of volicitin and related components from beet armyworm oral secretions. J. Chem. Ecol. 26(1): 203-220. 2000.

235. Tumlinson, J. H., Paré, P. W., Alborn, H. T., and Lewis, W. J. Chemically mediated tritrophic plant-insect interactions. In de Wit, P. J. G. M., Bisseling, T., and Stiekema, W. J. (Eds.) Biology of Plant-Microbe Interactions, Vol. 2. Proc. 9th Intl. Congress on Molecular Plant-Microbe Interactions, pp 378-383. 1999.
236. De Moraes, C. M., Lewis, W. J., and Tumlinson, J. H. Examining plant-parasitoid interactions in tritrophic systems. Annals da Sociedade Entomologica do Brasil 29(2): 189-203. 2000.
237. Tumlinson, J. H., Alborn, H. T., Loughrin, J. H., Turlings, T. C. J., and Jones, T. H. Plant Volatile Elicitor from Insects. U. S. Patent # 6,054,483; April 25, 2000. (M-3499); U. S. Patent # 6,207,712 B1; March 27, 2001.
238. Frey, M., Stettner, C., Paré, P. W., Schmelz, E. A., Tumlinson, J. H., and Gierl, A. An herbivore elicitor activates the gene for indole emission in maize. PNAS 97(26): 14801-14806. 2000.
239. Mori, N., Alborn, H. T., Teal, P. E. A., and Tumlinson, J. H. Enzymatic decomposition of elicitors of plant volatiles in *Heliothis virescens* and *Helicoverpa zea.* J. Ins. Phys. 47: 749-757. 2001.
240. De Moraes, C. M., Mescher, M. C., and Tumlinson, J. H. Caterpillar-induced nocturnal plant volatiles repel conspecific females. Nature. 410: 577-580. 2001.
241. Schmelz, E.A., Alborn, H. T., and Tumlinson, J. H. The influence of intact-plant and excised-leaf bioassay designs on volicitin- and jasmonic acid-induced sesquiterpene volatile release in *Zea mays.* Planta. 214: 171-179. 2001.
242. Cardoza, Y. J., Alborn, H. T., and Tumlinson, J. H. *In vivo* volatile emissions from peanut plants induced by simultaneous fungal infection and insect damage. J. Chem. Ecol. 28(1): 161-174. 2002.
243. Engelberth, J., Schmelz, E.A., Alborn, H. T., Cardoza, Y. J., Huang, J. and Tumlinson, J. H. Simultaneous quantification of Jasmonic acid and Sclicylic acid in plants by vapor phase extraction and gas chromatograph-chemical ionization-mass spectrometry. Analytical Biochemistry. 312: 242-250. 2003.
244. Schmelz, E. A., Alborn, H. T., Banchio, E., and Tumlinson, J. H. Quantitative relationships between induced jasmonic acid levels and volatile emission in *Zea mays* during *Spodoptera exigua* herbivory. Planta. 216:665-673. 2003.
245. Schmelz, E. A., Alborn, H. T., and Tumlinson, J. H. Synergistic interactions between volicitin, jasmonic acid and ethylene mediate insect-induced volatile emission in Zea mays. Physiologia Plantarum. 117: 403-412. 2003.
246. Cardoza, Y.J., C.G. Lait, E.A. Schmelz, J. Huang, and J.H. Tumlinson. Fungus-induced biochemical changes in peanut plants and their effect on development of beet armyworm, Spodoptera exigua Hübner (Lepidoptera: Noctuidae), larvae. Environmental Entomology 32: 220-228. 2003.

247. Suazo, A., Torto, B., Teal, P.E.A. and Tumlinson, J. H. Response of the small hive beetle (Aethina tumida Murray) to honey bee (Aphis Mellifera L.) and beehive-produced volatiles. Apidologie. 34: 525-534. 2003.
248. Olson, D. M., Rains, G. C., Meiners, T., Takasu, K., Tertuliano, M., Tumlinson, J. H., Wackers, F. L., Lewis, W. J. Parasitic Wasps Learn and Report Diverse Chemicals with Unique Conditionable Behaviors. Chemical Senses. 28: 545-550. 2003.
249. Aldrich, J. R., Bartelt, R. J., Dickens, J. C., Knight, A. L., Light, D. M., Tumlinson, J. H. Insect chemical ecology research in the United States Department of Agriculture — Agricultural Research Service. Pest Management Science. 59: 777-787. 2003.
250. Lait C.G., H.T. Alborn, P.E.A. Teal and J.H. Tumlinson III. Rapid biosynthesis of N-linolenoyl-L-glutamine, an elicitor of plant volatiles, by membrane associated enzyme(s) in Manduca sexta. Proceedings of the National Academy of Sciences USA 100: 7027-7032. 2003.
251. Cardoza, Y.J., Teal, P. E. A., Tumlinson, J. H. Effect of Peanut Plant Fungal Infection on Oviposition Preference by Spodoptera exigua and on Host-Searching Behavior by Cotesia marginiventris. Environmental Entomology. 32: 970-976. 2003.
252. Huang, J., Cardoza, Y. J., Schmelz, E. A., Raina, R., Engelberth, J., Tumlinson, J. H. Differential volatile emissions and salicylic acid levels from tobacco plants in response to different strains of Pseudomonas syringae. Planta. 217: 767-775. 2003.
253. Alborn, H. T., Brennan, M. M., Tumlinson, J. H. Differential Activity and Degradation of Plant Volatile Elicitors in Regurgitant of Tobacco Hornworm (Manduca sexta) Larvae. Journal of Chemical Ecology. 29: 1357-1372. 2003.
254. Schmelz, E. A., Engelberth, J., Alborn, H. T., O'Donnell, P., Sammons, M., Toshima, H. and Tumlinson, J. H. Simultaneous analysis of phytohormones, phytotoxins, and volatile organic compounds in plants. PNAS, 100: 10552-10557. 2003.
255. Schmelz, E. A., Alborn, H. T., Engelberth, J., Tumlinson, J. H. Nitrogen Deficiency Increases Volicitin-Induced Volatile Emission, Jasmonic Acid Accumulation, and Ethylene Sensitivity in Maize. Plant Physiology. 133: 295-306. 2003.
256. Engelberth, J., Alborn, H. T., Schmelz, E. A., and Tumlinson J. H. Airborne signals prime plants against insect herbivore attack. PNAS, 101: 1781-1785. 2004.
257. Rose, U. S. R. and Tumlinson J. H. Volatiles released from cotton plants in response to *Helicoverpa zea* feeding damage on cotton flower buds. Planta 218: 824-832. 2004.

258. De Moraes, C. M., Schultz, J. C., Mescher, M. C., and Tumlinson, J. H. Induced plant signaling and its implications for environmental sensing. Journal of Toxicology and Environmental Health, Part A, 67:819-834, (2004).
259. Schmelz, E. A., Engelberth, J., Tumlinson, J. H., Block, A., and Alborn, H.T. The use of vapor phase extraction in metabolic profiling of phytohormones and other metabolites. The Plant Journal 39: 790-808 (2004).
260. Tumlinson, J. H. and Lait, C. G. Biosynthesis of fatty acid amide elicitors of plant volatiles by insect herbivores. Arch. Insect Biochem. Physiol. 58: 54-68 (2005)
261. Gomez, S.K., Cox, M.M., Bede, J.B., Inoue, K., Alborn, H. T., Tumlinson, J.H., and Korth, K.L. Lepidopteran herbivory and oral factors induce transcripts encoding novel terpene synthases in *Medicago truncatula.* Archives of Insect Biochemistry and Physiology 58:114-127 (2005).
262. Huang, J., Schmelz, E.A., Alborn, H.T., Engelberth, J., and Tumlinson, J.H. Phytohormones mediate volatile emission during the interaction of compatible and incompatible pathogens: The role of ethylene in Pseudomonas syringae infected tobacco, J. Chem. Ecol. 31: 439-459 (2005).
263. Rose, U. S. R. and Tumlinson J. H. Systemic induction of volatile release in cotton: How specific is the signal to herbivory. Planta 222: 327-335 (2005).
264. Torto, B., Suazo, A., Alborn, H., Tumlinson, J. H., and Teal, P.E.A. Response of the small hive beetle (Aethina tumida) to a blend of chemicals identified from honeybee (Apis mellifera) volatiles. Apidologie 36: 523-532 (2005).
265. Rose, U. S. R., Lewis, W. J. and Tumlinson J. H. Extrafloral nectar from cotton (*Gossypium hirsutum*) as a food source for parasitic wasps. Functional Ecology 20: 67-74 (2006).
266. Cardoza, Y. J. and Tumlinson, J. H. Compatible and Incompatible Xanthomonas Infections Differentially Affect Herbivore-Induced Volatile Emission by Pepper Plants. J. Chem. Ecol. 32: 1755-1768 (2006).
267. Torto, B., Boucias, D.G., Arbogast, R.T., Tumlinson, J.H., and Teal, P.E.A. Multitrophic interaction facilitates parasite–host relationship between an invasive beetle and the honey bee. PNAS 104: 8374-8378 (2007).
268. Engelberth, J., Seidl-Adams, I., Schultz, J.C., and Tumlinson, J.H. Insect elicitors and exposure to green leafy volatiles differentially up-regulate major octadecanoids and transcripts of 12-oxo phytodienoic acid reductases in Zea mays. Molecular Plant-Microbe Interactions 20: 707-716 (2007).
269. Alborn, H.T., Hansen, T.V., Jones, T.H., Bennett, D.C., Tumlinson, J.H., Schmelz, E.A., and Teal, P.E.A. Disulfooxy fatty acids from the American bird grasshopper, *Schistocerca Americana*, elicitors of plant volatiles. PNAS 104: 12976-12981 (2007).

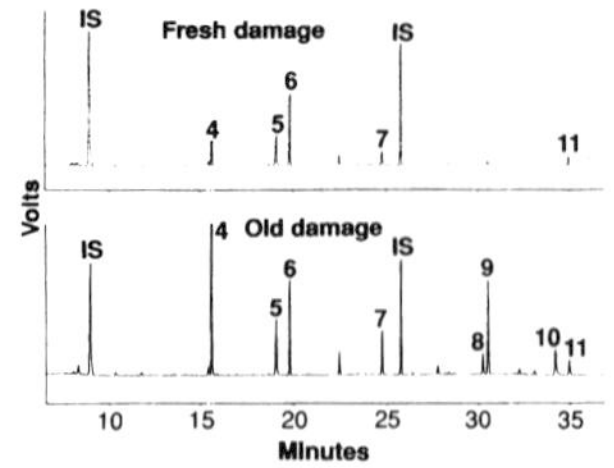

Fig. 2. Chromatographic profiles of collected volatiles from three corn seedlings damaged by BAW caterpillars over 2 hours. The volatiles were collected either just after damage took place (fresh damage) or 16 hours later on the following day (old damage). The collection procedure was the same as described in the legend to Fig. 1. No BAW larvae were present during the collections. Note that compounds 1 through 3 were not detected; these compounds only show up in significant amounts when the caterpillars are actively damaging the plants (*14*). Peak numbers correspond with the numbers and compounds given with Fig. 1. Each internal standard (*n*-octane and *n*-nonyl acetate) represents 1 μg.

fense. When under herbivore attack, however, their flexible defense expressions will allow a fast induced production of carbon-based defensive chemicals (*3*). We suspect that the terpenoids released by corn seedlings serve in a direct defense against herbivores. In addition, their volatility and high turnover rate should make them reliable indicators of the presence of hosts for parasitoids. They would be even more reliable if the induced response is specific for damage inflicted by herbivores. This was tested.

Ten third instar BAW caterpillars were allowed to feed on three corn seedlings for 2 hours (late afternoon). During the same period, six other seedlings were damaged artificially with a razor blade, whereby the damage done by the caterpillars was roughly mimicked. Three of the artificially damaged corn seedlings, as well as three undamaged seedlings, were treated with the regurgitated gut contents of other corn-fed BAW caterpillars (*15*). The following morning, volatiles were collected from all four treatment groups for 2 hours. Striking differences in terpenoid and indole release were found among the collections when analyzed by gas chromatography (Fig. 3). Again, leaves with larval damage released the compounds in relatively large amounts. Seedlings that only underwent artificial damage released far less. The artificially damaged seedlings that were treated with caterpillar regurgitant released the most dominant compounds and in amounts similar to those found for the larval-damaged seedlings. The control seedlings, undamaged and treated with regurgitant, released virtually no detectable amounts of volatiles.

Other seedlings that were subjected to these treatments were tested for attractiveness to experienced (*16*) *C. marginiventris* females in a flight tunnel (*17*). In two-choice tests, the insects strongly preferred the leaves with larval damage over leaves with just artificial damage (Fig. 4). The artificially damaged leaves that had been treated with caterpillar regurgitant were clearly preferred over the leaves with only artificial damage. When given the choice between leaves with larval damage and artificially damaged leaves treated with regurgitant, the females showed no preference (Fig. 4C). Fewer females flew to undamaged leaves treated with regurgitant than to artificially damaged leaves (Fig. 4D). This is in agreement with the observation that artificial damage alone does result in the release of some terpenoids (Fig. 3) and shows that the regurgitant by itself did not elicit attraction. The results strongly indicate that the observed plant response is greatly enhanced by the feeding of BAW caterpillars. It involves not only damage, but requires a factor in the regurgitant (most likely in the saliva) of the caterpillars as well. Whether this factor involves enzymes, microorganisms, or something else has yet to be determined.

We also found that volatiles collected from BAW larvae feeding on corn were attractive to the parasitoid when applied on a piece of paper (*14*). An equivalent synthetic blend containing all 11 compounds was attractive as well (*14*). Response depended on an insect's experience: females that had experienced BAW caterpillars feeding on corn (*16*) responded more to a natural blend than to a synthetic blend. Females that had experienced only a synthetic blend responded equally well to the synthetic and natural blend (*14*).

It is clear that experience plays a major role in the behavior of the parasitoids (*16*), and the preference for the terpenoid-releasing plants may have been learned during the preflight experience. Again, this demonstrates the highly flexible host-searching behavior exhibited by these insects that allows them to learn odors closely associated with their hosts.

Many herbivores have developed variable levels of resistance to plant-produced chemicals, making those chemicals less effective in direct defense. Attraction of the natural enemies of herbivores, however, may result in an additional advantage to the plants, thereby maintaining selection pressures that favor the production of these chemicals in the observed high quantities. Cost-benefit anal-

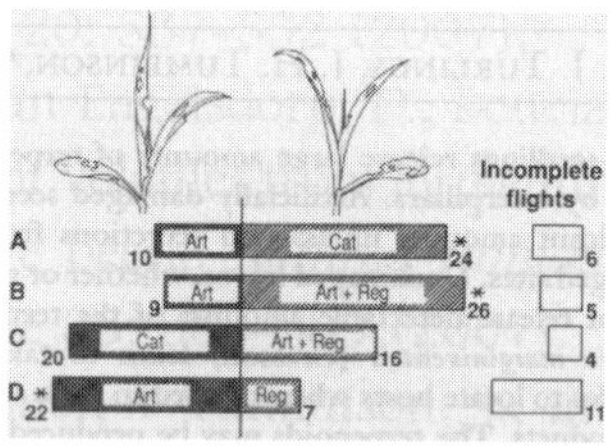

Fig. 4. Responses during two-choice flight tunnel tests (*17*) by experienced *Cotesia marginiventris* females to corn seedlings that underwent various treatments. A day after the leaves were treated, females had the opportunity to choose between the odors released by seedlings with (**A**) artificial damage (Art) or caterpillar damage (Cat); (**B**) artificial damage (Art) or artificial damage treated with regurgitant (Art + Reg); (**C**) caterpillar damage (Cat) or artificial damage treated with regurgitant (Art + Reg); (**D**) artificial damage (Art) or no damage treated with regurgitant (Reg). On five different days eight females were tested to each combination ($n = 40$). The open bars represent the females that did not fly to the odor sources. Total numbers are given with each bar. Asterisks indicate statistically significant preferences for a particular odor (χ^2 test, $P < 0.05$).

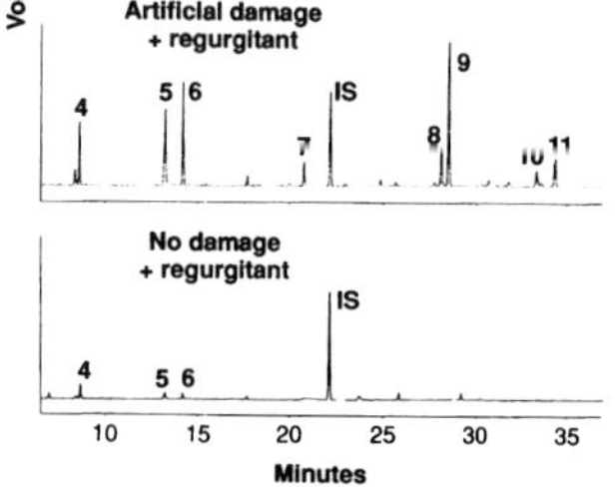

Fig. 3. Chromatographic profiles of corn seedlings with natural caterpillar damage, or with various artificial treatments. The day before collections took place, the seedlings were either damaged by BAW caterpillars for 2 hours (caterpillar damage), artificially damaged with a razor blade during the same period (artificial damage), artificially damaged and treated with caterpillar regurgitant (*15*) (artificial damage + regurgitant), or left undamaged but treated with regurgitant (*15*) (no damage + regurgitant). The collection procedure was the same as described in the legend to Fig. 1. Peak numbers correspond to the numbers and compounds given in Fig. 1.

yses concerning plant defensive strategies should consider how plants can safeguard themselves against severe herbivore injury by attracting predators or parasitoids (*18*). The terpenoids are reliable cues for the parasitoids because they are closely associated with herbivore damage and they are released even during the frequent pauses in eating by the caterpillars (Fig. 2). We do not yet know whether the induced reaction is limited to the damaged sites, or whether it is systemic as has been shown in other studies (*19, 20*).

Our results indicate an active release of chemicals by plants that is exploited by host-searching parasitoids. It is likely that the terpenoids and indole are involved in other types of interactions as well. They may, for example, act as oviposition deterrents for herbivorous insects searching for sites to deposit their eggs or function in communication between plants (*20, 21*). More knowledge about the injury-dependent production of airborne semiochemicals by plants may point to new possibilities for biological control of pest insects.

REFERENCES AND NOTES

1. D. F. Rhoades in *Herbivores, Their Interactions with Secondary Plant Metabolites*, G. Rosenthal and D. Janzen, Eds. (Academic Press, New York, 1979), pp. 4–54.
2. M. Kogan and J. Paxton in *Plant Resistance to Insects*, P. A. Hedin, Ed. (American Chemical Society, Washington, DC, 1983), p. 153.
3. P. D. Coley, J. P. Bryant, F. S. Chapin III, *Science* **230**, 895 (1985).
4. P. W. Price *et al.*, *Annu. Rev. Ecol. Syst.* **11**, **41** (1980); P. W. Price, in *Interactions of Plant Resistance and Parasitoids and Predators of Insects*, D. J. Boethel and R. D. Eikenbary, Eds. (Wiley, New York, 1986), pp. 11–20.
5. H. Nadel and J. J. M. van Alphen, *Entomol. Exp. Appl.* **45**, 181 (1987).
6. M. Dicke and M. W. Sabelis, *Neth. J. Zool.* **38**, **148** (1988); M. Dicke *et al.*, *J. Chem. Ecol.* **16**, 381 (1990).
7. D. A. Nordlund, W. J. Lewis, M. A. Altieri, in *Novel Aspects of Insect-Plant Interactions*, P. Barbosa and D. K. Letourneau, Eds. (Wiley, New York, 1988), pp. 65–90; D. W. Whitman, in *Biologically Active Natural Products: Potential Use in Agriculture*, H. G. Cutler, Ed. (American Chemical Society, Washington, DC, 1988), pp. 386–396.
8. J. J. M. van Alphen and L. E. M. Vet, in *Insect Parasitoids*, J. K. Waage and D. J. Greathead, Eds. (Academic Press, London, 1986), pp. 23–61; W. J. Lewis and J. H. Tumlinson, *Nature* **331**, 257 (1988); L. E. M. Vet and A. W. Groenewold, *J. Chem. Ecol.*, in press.
9. D. R. Papaj and R. J. Prokopy, *Annu. Rev. Entomol.* **34**, 315 (1989); L. E. M. Vet, W. J. Lewis, D. R. Papaj, J. C. van Lenteren, *J. Insect Behav.* **3**, **471** (1990); W. J. Lewis, L. E. M. Vet, J. H. Tumlinson, J. C. van Lenteren, D. R. Papaj, *Environ. Entomol.* **19**, 1113 (1990).
10. T. C. J. Turlings, J. H. Tumlinson, F. J. Eller, W. J. Lewis, *Entomol. Exp. Appl.*, in press.
11. The glass collection chambers, similar to those described by J. H. Cross [*J. Chem. Ecol.* **6**, 781 (1980)], consisted of two parts. Purified, humidified air entered the first part through a 4-cm-long 0.25-inch outer diameter inlet, which connected onto a wider section (6 cm long; 3 cm inner diameter) that contained a glass frit. After passing through the frit the air entered the second part (15 cm long; 3 cm inner diameter) which contained the odor source. The second part tapered into a 4-cm long, 0.25-inch outer diameter, outlet. Both parts had fitting ball joints that were clamped together. Super Q (*12*) traps were connected to the outlet with brass Swagelock fittings containing teflon ferules. Air passed through the chambers at a rate of 300 ml/min. The air was pushed in at the inlet side and pulled at the outlet, such that the pressure inside the system was slightly higher than outside. See T. C. J. Turlings *et al.* (*14*) for details.
12. Super Q, 80/100 (Cat. No. 2735) (Alltech Associates Inc., Deerfield, IL).
13. J. H. Visser, S. van Straten, H. Maarse, *J. Chem. Ecol.* **5**, 13 (1979).
14. T. C. J. Turlings *et al.*, in preparation.
15. The leaves were treated by grabbing a caterpillar with a pair of forceps, pinching the head region with another pair until regurgitation was induced. The regurgitant was then immediately rubbed over a site on a seedling that had just previously been damaged. All artificially damaged sites on three of the seedlings were treated this way. As a control equal amounts of regurgitant were rubbed over the leaves of three seedlings that did not receive any type of damage.
16. Before release in the tunnel (*17*) each test insect was given a 20-s contact experience with a plant-host complex of BAW caterpillars on corn. Experience on host-infested leaves dramatically increases the subsequent responses by *C. marginiventris* females to host-related odors in olfactometric bioassays [T. C. J. Turlings, J. H. Tumlinson, W. J. Lewis, L. E. M Vet, *J. Insect Behav.* **2**, 217 (1988); T. C. J. Turlings, J. W. A. Scheepmaker, L. E. M. Vet, J. H. Tumlinson, W. J. Lewis, *J. Chem. Ecol.* **16**, 1577 (1990)].
17. Responses of individual *C. marginiventris* females to odors were tested in a Plexiglas flight tunnel. Dimensions of the tunnel were 60 cm by 60 cm in cross section and 2.4 m long. Air was pulled through the tunnel at 0.15 m/sec and was exhausted through a 30-cm flexible pipe with a fan. More details on the tunnel are given by F. J. Eller, J. H. Tumlinson, and W. J. Lewis [*Environ. Entomol.* **17**, 745 (1988)]. Seedlings used as odor sources were placed 20 cm apart and 30 cm from the tunnel floor approximately 80 cm upwind from the insect release point. After the females were experienced (*16*), they were released into the tunnel from a glass funnel (*10*). Their responses and choices were recorded. If a female, after three trials, had not flown all the way to a source the flight was considered incomplete.
18. M. W. Sabelis and M. C. M. de Jong, *Oikos* **53**, **247** (1988); M. Dicke and M. W. Sabelis, in *Variation in Growth Rate and Productivity of Higher Plants*, H. Lambers, M. L. Cambridge, H. Konings, T. L. Pons, Eds. (SPB Academic, The Hague, 1989), pp. 341–358.
19. T. R. Green and C. A. Ryan, *Science* **175**, 776 (1972); D. Bowles, *Nature* **343**, 314 (1990).
20. M. Dicke, M. W. Sabelis, J. Takabayashi, J. Bruin, M. A. Posthumus, *J. Chem. Ecol.* **16**, 3091 (1990).
21. I. T. Baldwin and J. C. Schultz, *Science* **221**, 277 (1983); D. F. Rhoades, *Recent Adv. Phytochem.* **19**, 195 (1985); H. J. Zeringue, *Phytochemistry* **26**, 1357 (1987).
22. We thank L. E. M. Vet, H. T. Alborn, M. Dicke, E. A. Bernays, H. J. Brockmann, F. L. Petitt, and P. J. Landolt for critically reading and commenting on the original manuscript. T. J. Rossignol provided technical assistance. Synthetic α-*trans*-bergamotene was generously provided by D. B. McIlwaine. Supported in part by a grant from the International Division of the USDA, OICD, and a Fulbright grant provided by the USIA.

21 May 1990; accepted 21 August 1990

Proc. Natl. Acad. Sci. USA
Vol. 91, pp. 11836–11840, December 1994
Plant Biology

Diurnal cycle of emission of induced volatile terpenoids by herbivore-injured cotton plants

(beet armyworm/plant–insect interaction)

JOHN H. LOUGHRIN*, ARA MANUKIAN, ROBERT R. HEATH, TED C. J. TURLINGS†, AND JAMES H. TUMLINSON‡

Insect Attractants, Behavior and Basic Biology Research Laboratory, U.S. Department of Agriculture, Agricultural Research Service, 1700 Southwest 23rd Drive, Gainesville, FL 32604

Communicated by John H. Law, August 15, 1994

ABSTRACT **Cotton plants attacked by herbivorous insect pests emit relatively large amounts of characteristic volatile terpenoids that have been implicated in the attraction of natural enemies of the herbivores. However, the composition of the blend of volatile terpenes released by the plants varies remarkably throughout the photoperiod. Some components are emitted in at least 10-fold greater quantities during the photophase than during the scotophase, whereas others are released continuously, without conforming to a pattern, during the entire time that the plants are under herbivore attack. The diurnal pattern of emission of volatile terpenoids was determined by collecting and analyzing the volatile compounds emitted by cotton plants subjected to feeding damage by beet armyworm larvae *in situ*. The damage was allowed to proceed for 3 days, and volatile emission was monitored continuously. During early stages of damage high levels of lipoxygenase-derived volatile compounds [e.g., (*Z*)-3-hexenal, (*Z*)-3-hexenyl acetate] and several terpene hydrocarbons [e.g., α-pinene, caryophyllene] were emitted. As damage proceeded, high levels of other terpenes, all acyclic [e.g., (*E*)-β-ocimene, (*E*)-β-farnesene], were emitted in a pronounced diurnal fashion; maximal emissions occurred in the afternoon. These acyclic terpenes followed this diurnal pattern of emission, even after removal of the caterpillars, although emission was in somewhat smaller amounts. In contrast, the emission of cyclic terpenes almost ceased after the caterpillars were removed.**

Plant odors have long been of interest because they attract phytophagous insects. Additionally, a rapidly growing body of evidence has implicated plant odors in the attraction of species that prey on or parasitize herbivorous insect pests (1). In the cases so far reported, plants that were nearly odorless before feeding damage emitted large quantities of volatile compounds in a delayed response to herbivore feeding (2). These induced odors have been shown to be powerful attractants for parasitic Hymenoptera (2) and predatory mites (3). However, this mechanism of self-defense by plants has been explored in only a few species.

In recent studies of both corn, *Zea mays* (2), and cotton, *Gossypium hirsutum* (4), seedlings we found that plants release a significantly greater number of compounds and larger amounts of total volatile compounds after overnight feeding damage by insects than when they have been freshly damaged by the insects. However, there was no previous indication that plants respond to herbivore damage with a diurnal rhythm of volatile compound emission, although a number of investigations have demonstrated the rhythmic nature of volatile compound release from flowers (5, 6). In many cases it appears that the release of volatile compounds by flowers is timed to coincide with the period of greatest activity of their pollinators. For example, Heath *et al.* (7) found that emission of floral odor by night-blooming jessamine peaked in the first 2 hr of the scotophase, coincident with the period of maximum feeding activity of the nocturnal moths that pollinate the jessamine. If plants are signaling to natural enemies of their herbivore attackers, it would seem likely that they would emit volatile compounds in greater quantities during the period when the natural enemies are actively hunting for hosts or prey. Thus, we investigated the release of volatile compounds from beet armyworm-damaged cotton leaves throughout three consecutive photoperiods to detect any diurnal variations in volatile compound emissions that might occur.

MATERIALS AND METHODS

Cotton plants, cv. Delta Pineland 90, were greenhouse grown in 16-cm-diameter pots in top soil/vermiculite/peat moss, 60:20:20, until ≈1.5 mo old. The plants were 25–30 cm tall and had not set flower buds. Beet armyworms (*Spodoptera exigua* Hübner) were reared by the method of King and Leppla (8).

The collection apparatus for volatile compounds used a push/pull technique and has been described in detail by Heath and Manukian (9). Briefly, the apparatus consisted of an 8-cm-diameter by 34-cm-tall glass sleeve that contained the plant. Air entered the top of the sleeve through multiple layers of an activated charcoal-infused fabric and passed down over the plant at a rate of 5 liters·min^{-1}. At the bottom of the sleeve a split plate with a hole in the center closed loosely around the stem of the plant like a guillotine to prevent air from flowing back into the sleeve from outside. The bottom of the sleeve was connected to a base that accommodated eight filter traps, arranged concentrically, for volatile compound collection. During collection, air was pulled through one of the traps at a rate of 1 liter·min^{-1}, and the remaining 4 liters·min^{-1} were vented out the bottom of the apparatus. Each trap contained 50 mg of Super Q adsorbent (Alltech Associates) and was conditioned before use by rinsing with 2 ml of CH_2Cl_2. Timing of the volatile compound collections was accomplished by computer-automated switching valves operated under the control of software developed in this laboratory (9, 10).

Volatile compounds were collected during May and June, 1993, in Gainesville, Florida. In the first series of experiments

The publication costs of this article were defrayed in part by page charge payment. This article must therefore be hereby marked "*advertisement*" in accordance with 18 U.S.C. §1734 solely to indicate this fact.

*Present address: Department of Entomology, University of Kentucky, S225 Agricultural Science Center North, Lexington, KY 40546-0091.
†Present address: Department of Applied Entomology, Division of Phytomedicine, Eidgenössiche Technische Hochschule Zentrum/CLS, Clausiusstrasse 21, CH 8092 Zürich, Switzerland.
‡To whom reprint requests should be addressed.

second-instar caterpillars were isolated at 0800 hr and starved until placed on the plants at 1400 hr. Five larvae were placed on each plant in the collection apparatus and the hole at the bottom of the chamber around the plant stem was loosely plugged with cotton to prevent larval escape. Volatile compound collections were begun at 1500 hr, were of 3-hr duration in each trap, and lasted for 60 hr, during which time caterpillars were allowed to feed continuously. All experiments were replicated five times, and the resulting data were subjected to ANOVA, and means were compared by a least-significant-difference test (11). An equal number of head-space collections from undamaged cotton plants were done for comparison.

Volatile compounds were eluted from the traps with 150 μl of CH_2Cl_2, 600 ng each of *n*-octane and nonyl acetate were added as internal standards, and 2-μl aliquots of all samples were analyzed by gas chromatography (GC) on a 50-m × 0.25 mm (i.d.) fused silica column with a 0.25-μm-thick bonded methyl silicone stationary phase (Quadrex, New Haven, CT). Injections were made in the splitless mode for 30 sec, and the gas chromatograph was operated under the following conditions: injector 220°C, detector 240°C, column oven 50°C for 3 min, then programmed at 5°C·min^{-1} to 190°C, He carrier gas linear flow velocity 21 cm·sec^{-1}. Selected samples were also analyzed on a 50 m × 0.25 mm (i.d.) fused silica column with a 0.25-μm-thick bonded cyanopropyl silicone (CPS-1; Quadrex) stationary phase operated under the following conditions: 30-sec splitless delay after injection, oven temperature 60°C for 1 min and then programmed at 5°C·min to 180°C, injector temperature 220°C, detector temperature 260°C, He carrier gas linear flow velocity 19 cm·sec^{-1}. For mass spectral analyses, 1-μl samples were introduced via a methyl silicone column, operated as before, into a Finnigan-MAT ITS40 (ion trap) mass spectrometer operated in the electron impact mode. Constituents of the plant volatile emission were identified by comparison of mass spectra with spectra in the Environmental Protection Agency–National Institutes of Health data base and spectra obtained of authentic compounds, and by comparison of GC retention times with those of authentic standards on both capillary columns.

RESULTS AND DISCUSSION

Representative chromatograms of head-space volatile compounds from beet armyworm-damaged cotton plants are shown in Fig. 1. Early stages of damage were characterized by high levels of lipoxygenase-derived volatile compounds [e.g., (*Z*)-3-hexenal, (*Z*)-3-hexenyl acetate] and a number of terpene hydrocarbons such as α-pinene, myrcene, and caryophyllene (Fig. 1*A*). As damage progressed, however, a number of other terpenes, which in early stages of damage had been released only in relatively small amounts, started to increase (Fig. 1*B*). Thus, whereas in early stages of damage a mixture of cyclic (e.g., α-pinene, caryophyllene) and acyclic terpenes (myrcene) predominated, later different terpenes, which appear to have been induced by beet armyworm-feeding damage, prevailed. These latter terpenes were all acyclic and were identified as (*E*)-β-ocimene, linalool, (*E*)-4,8-dimethyl-1,3,7-nonatriene, (*E*)-β-farnesene, (*E,E*)-α-farnesene, and (*E,E*)-4,8,12-trimethyl-1,3,7,11-tridecatetraene.

The increase in emission of the inducible terpenes began on the morning after the start of feeding damage and peaked during the afternoon. For example, immediately after the start of feeding during the 1500- to 1800-hr collection period, the cotton plants emitted high levels of α-pinene (>5 μg over the 3-hr collection period) but only small amounts of (*E*)-β-ocimene and (*E*)-4,8-dimethyl-1,3,7-nonatriene (Fig. 2). However, the next day levels of the latter terpenes started to rise rapidly, and in the peak hours of emission (i.e., from 1200 to 1500 hr and from 1500 to 1800 hr) production of (*E*)-β-ocimene and (*E*)-4,8-dimethyl-1,3,7-nonatriene averaged over 20 μg and 13 μg, respectively. These values were >10-fold increases in emission. Similar changes occurred in

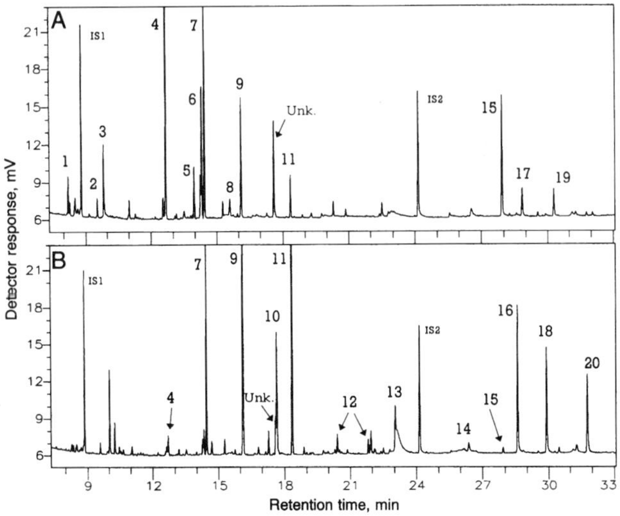

FIG. 1. Typical chromatograms of volatile compounds emitted from 1500 to 1800 hr by cotton plants subjected to beet armyworm-feeding damage. (*A*) First day of feeding damage. (*B*) Third day of feeding damage. Peak identities: 1, (*Z*)-3-hexenal; 2, (*E*)-2-hexenal; 3, (*Z*)-3-hexenyl acetate; 4, α-pinene; 5, β-pinene; 6, myrcene; 7, (*Z*)-3-hexenyl acetate; 8, limonene; 9, (*E*)-β-ocimene; 10, linalool; 11, (*E*)-4,8-dimethyl-1,3,7-nonatriene; 12, isomeric hexenyl butyrates; 13, indole; 14, (*Z*)-jasmone; 15, caryophyllene; 16, (*E*)-β-farnesene; 17, α-humulene; 18, (*E,E*)-α-farnesene; 19, γ-bisabolene; 20, (*E,E*)-4,8,12-trimethyl-1,3,7,11-tridecatetraene. IS1, *n*-octane; IS2, nonyl acetate; Unk., identity unknown.

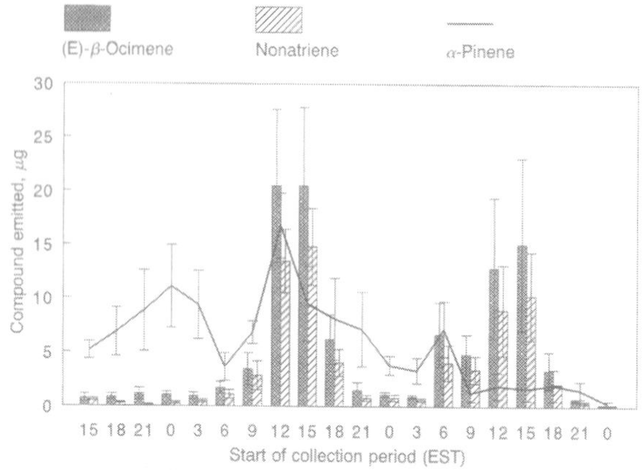

FIG. 2. Emission of α-pinene, (*E*)-β-ocimene, and (*E*)-4,8-dimethyl-1,3,7-nonatriene by cotton plants subjected to beet armyworm-feeding damage over the course of several days. Data represent the mean of five replications ± SE. EST, Eastern standard time.

the emission of the sesquiterpenes (Fig. 3), where in early stages of damage high levels of caryophyllene were emitted, whereas the following day high levels of (*E*)-β-farnesene and (*E,E*)-α-farnesene were released.

The emission of the inducible terpenoids exhibited distinctly diurnal periodicity. As illustrated in Figs. 2 and 3, emissions peaked strongly in the late afternoon on the second and third days with relatively low levels of emission at night. In contrast to the inducible terpenes, the emission of the other terpenes, such as α-pinene and caryophyllene, followed no distinct pattern. Rather, levels of these compounds slowly increased until they peaked on the second day of damage and then slowly declined thereafter.

Relatively low levels of volatile compounds were collected from undamaged plants, and the most abundant components were α-pinene, (*E*)-β-ocimene, (*E*)-4,8-dimethyl-1,3,7-nonatriene, and (*E,E*)-4,8,12-trimethyl-1,3,7,11-tridecatetraene (Fig. 4). A diurnal pattern of periodicity was again noted for the latter three compounds. Although the possible effects of minor damage to the cotton plants from insects present in the greenhouse cannot be excluded, the levels detected from undamaged plants may suggest that insect-feeding damage amplifies an internal rhythm in cotton leaves. Beet armyworm larvae are nearly odorless (2) and thus can be excluded as a source of volatile compounds.

Because cotton plants accumulate large amounts of free monoterpenes and sesquiterpenes in specialized structures (12), the lack of a pattern of emission for compounds such as caryophyllene is probably due to their release by simple breakage of glands as a consequence of caterpillar feeding. Also, lipoxygenase-derived compounds, arising from the hydroperoxidation of fatty acids containing a *cis,cis*-1,4-pentadiene moiety, are known to be released as a result of mechanical damage (13). Therefore, the emission of such compounds should decline rapidly upon the removal of the caterpillars. To test this hypothesis, the caterpillars were removed at 0850 hr on the second day of feeding damage. As shown in Fig. 5, the levels of α-pinene and (*Z*)-3-hexenal rapidly declined after the removal of the caterpillars, whereas the emission of (*E*)-β-ocimene still peaked in the late afternoon, although at lower levels than when the caterpillars were allowed to remain on the plants.

The pronounced changes that occurred in volatile emissions during the three afternoons that the plants were mon-

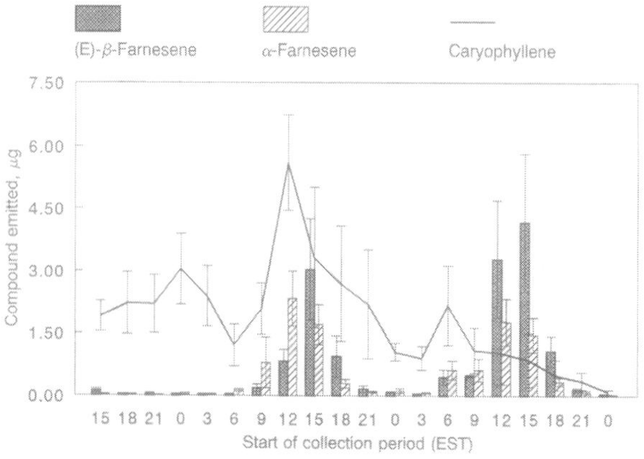

FIG. 3. Emission of caryophyllene, (*E*)-β-farnesene, and (*E,E*)-α-farnesene by cotton plants subjected to beet armyworm-feeding damage over the course of several days. Data represent the mean of five replications ± SE.

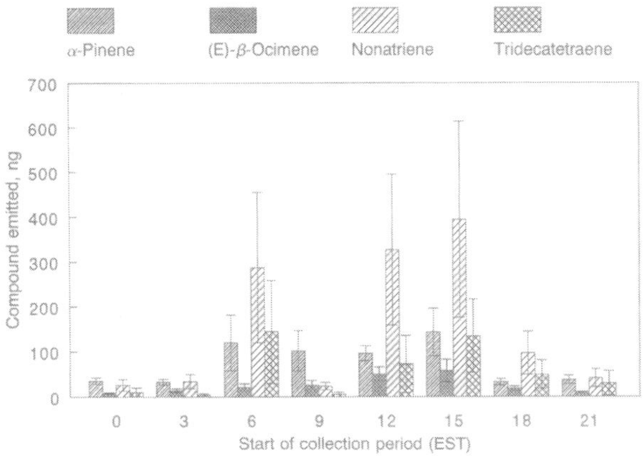

FIG. 4. Emission of α-pinene, (*E*)-β-ocimene, (*E*)-4,8-dimethyl-1,3,7-nonatriene, and (*E,E*)-4,8,12-trimethyl-1,3,7,11-tridecatetraene from undamaged cotton plants. Data represent the mean of five replications ± SE.

itored are summarized in Table 1, along with levels emitted by undamaged plants during the same time period. On the first day of feeding damage, from 1500 to 1800 hr, total production of identified components averaged ≈6000 ng·hr^{-1}. On the second day during the same time period, this production had increased to >35,000 ng·hr^{-1}. The striking increase in the levels of (*Z*)-3-hexenyl acetate that occurred warrants particular mention; during the same sampling period the levels of this compound had increased ≈20-fold from the first to the second day of feeding damage. Unlike the changes observed for the inducible terpenes, however, this compound declined after the second day of damage and did not conspicuously increase again.

The mechanism whereby terpenoid release is induced by the caterpillar-feeding damage is unknown. It is possible that release from glycosidically bound forms is involved (14). The terpene hydrocarbons could then arise via acid-catalyzed dehydration of parent alcohols (15). Alternatively, a group of terpene synthases could be inducible by the feeding damage. Mechanical damage has been shown to induce terpene cyclase activity in *Abies grandis* (16), although in this species the products produced by inducible cyclase activity did not differ greatly from those of constitutive activity.

The release of terpenoids by cotton differs from that found previously for corn (2). Whereas cotton begins to release several terpenoids as soon as damage occurs, corn appears to have no free terpenoids available for immediate release. Terpenoids are released by corn plants only several hours after insect herbivore damage begins. This difference may be a reflection of two distinct strategies that the plants use in defense against herbivores. As suggested by Coley *et al.* (17), fast-growing annuals like corn may initially invest most of their energy in growth rather than defense. In corn, energy appears to be committed to defense only when the plant is under attack. Slower growing perennials like cotton may invest in some constitutive defenses to be braced for and perhaps "discourage" possible herbivore attacks.

In addition to the constitutive terpenoids, our results show that cotton produces inducible terpenoids in response to herbivory. While these may add to the blend of toxic chemicals available to combat herbivores, they also may provide parasitoids and predators with reliable signals pointing out the location of potential hosts or prey. The odors induced by

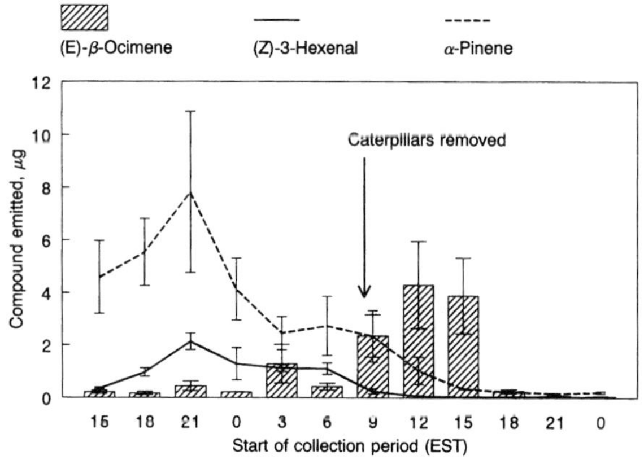

FIG. 5. Emission of (*Z*)-3-hexenal, α-pinene, and (*E*)-β-ocimene with caterpillars removed from the plant at 0850 hr Eastern standard time (EST). Data represent the mean of five replications ± SE.

Table 1. Composition of volatile blends collected between 1500 and 1800 hr from undamaged cotton plants and cotton plants subjected to continual beet armyworm-feeding damage for 3 days

Compound	ng of compound emitted			
	Undamaged plants	Beet armyworm-damaged plants		
		Day 1	Day 2	Day 3
(*Z*)-3-Hexenal	28*	799*†	1,603†	108*
(*E*)-2-Hexenal	21*	140*	618†	135*
(*Z*)-3-Hexenol	15*	610*†	1,515†	48*
(*Z*)-3-Hexenyl acetate	46*	1,325*	27,700†	3,710*
(*Z*)-3-Hexenyl butyrate	ND*	96*	812†	117*
(*E*)-2-Hexenyl butyrate	ND*	9*	458†	485†
(*Z*)-3-Hexenyl 2-methylbutyrate	ND*	6*	568†	200*
(*E*)-2-Hexenyl 2-methylbutyrate	ND*	3*	458†	329†
(*Z*)-Jasmone	ND*	181*†	482†	144*
Indole	ND*	3,330†	8,000†	4,160†
α-Pinene	143*	5,170*†	9,510†	1,680*
β-Pinene	30*	835*	1,520*	288*
Myrcene	77*	1,137*†	4,430†	1,470*†
Limonene	6*	505*†	737†	280*†
(*E*)-β-Ocimene	56*	772*	20,500†	15,100†
Linalool	10*	35*	2,510†	1,640†
(*E*)-4,8-Dimethyl-1,3,7-nonatriene	394*	616*	14,800†	10,200†
(*E,E*)-4,8,12-Trimethyl-1,3,7,11-trideca-tetraene	135*	103*	1,420†	1,530†
Caryophyllene	26*	1,910*†	3,310†	861*†
α-Humulene	11*	467*†	858†	271*†
γ-Bisabolene	10*	526*†	1,140†	270*†
(*E*)-β-Farnesene	5*	121*	3,030†	4,160†
(*E,E*)-α-Farnesene	2*	29*	1,700†	1,460†

ND, not detected.
Data represent the mean of five replications. Means within rows followed by the same symbol (* or †) are not significantly different ($P > 0.05$, least significant difference test).

the caterpillar-feeding damage have a distinctly floral character. In fact, it is remarkable how closely the rhythms of volatile emission noted here parallel those from some flowers (5, 6). In *Nicotiana sylvestris*, for example, while some compounds are emitted in a rhythmic fashion, others show no particular periodicity. It is interesting to note that many parasitic wasps exploit nectar as a food source (18). The few reports of studies of the diurnal patterns of parasitoid foraging in the field (19, 20) and our observations of the foraging behavior of the beet armyworm larval parasitoid *Cotesia marginiventris* in the laboratory indicate that these wasps forage during the middle to latter part of the photophase. Perhaps the induced release of large amounts of certain odors in damaged plants is similar to the periodic release of specific odors by flowers, in that it provides a clear and reliable signal, above the background of other plant odors, that attracts natural enemies of the herbivorous attackers.

We thank A. T. Proveaux for technical assistance, Jessica Lockerman for assistance, and the DeltaPine Corporation for donation of seed.

1. Tumlinson, J. H., Lewis, W. J. & Vet, L. E. M. (1993) *Sci. Am.* **268** (3), 100–106.
2. Turlings, T. C. J., Tumlinson, J. H. & Lewis, W. J. (1990) *Science* **250,** 1251–1253.
3. Dicke, M. & Sabelis, M. W. (1988) *Neth. J. Zool.* **38,** 148–165.
4. McCall, P. J., Turlings, T. C. J., Loughrin, J. H., Proveaux, A. T. & Tumlinson, J. H. (1994) *J. Chem. Ecol.*, in press.
5. Matile, P. & Altenburger, R. (1988) *Planta* **174,** 242–247.
6. Loughrin, J. H., Hamilton-Kemp, T. R., Andersen, R. A. & Hildebrand, D. F. (1991) *Physiol. Plant.* **83,** 492–496.
7. Heath, R. R., Landolt, P. J., Dueben, B. & Lenczewski, B. (1992) *Environ. Entomol.* **21,** 854–859.
8. King, E. G. & Leppla, N. C. (1984) *Advances and Challenges in Insect Rearing* (U.S. Gov. Printing Service, Washington, DC).
9. Heath, R. R. & Manukian, A. (1994) *J. Chem. Ecol.* **20,** 593–608.
10. Manukian, A. & Heath, R. R. (1993) *Sci. Comput. Autom.* **9,** 27–40.
11. Sokal, R. R. & Rohlf, F. J. (1981) *Biometry—The Principles and Practice of Statistics in Biological Research* (Freeman, New York), p. 244.
12. Elzen, G. W., Williams, H. H., Bell, A. A., Stipanovic, R. D. & Vinson, S. B. (1985) *J. Agric. Food Chem.* **33,** 1079–1082.
13. Hatanaka, A., Kajiwara, T. & Sekiya, J. (1987) *Chem. Phys. Lipids* **44,** 341–361.
14. Gäbler, A., Boland, W., Preiss, U. & Simon, H. (1991) *Helv. Chimica Acta* **74,** 1773–1779.
15. Engel, K.-H. & Tressl, R. (1983) *J. Agric. Food Chem.* **31,** 998–1002.
16. Lewinsohn, E., Gijzen, M. & Croteau, R. B. (1992) in *Regulation of Isopentenoid Metabolism*, ACS Symposium Series 497, eds. Nes, W. D., Parish, E. J. & Trzaskos, J. M. (Am. Chem. Soc., Washington, DC), pp. 8–17.
17. Coley, P. D., Bryant, J. P. & Chapin, F. S. (1985) *Science* **230,** 895–899.
18. Leius, K. (1960) *Can. Entomol.* **92,** 369–376.
19. Lewis, W. J., Sparks, A. N., Jones, R. L. & Barras, D. J. (1972) *Environ. Entomol.* **1,** 468–471.
20. Snow, J. W. & Burton, R. L. (1967) *J. Georgia Entomol. Soc.* **2,** 47–52.

An Elicitor of Plant Volatiles from Beet Armyworm Oral Secretion

H. T. Alborn, T. C. J. Turlings,* T. H. Jones, G. Stenhagen, J. H. Loughrin,† J. H. Tumlinson‡

The compound *N*-(17-hydroxylinolenoyl)-L-glutamine (named here volicitin) was isolated from oral secretions of beet armyworm caterpillars. When applied to damaged leaves of corn seedlings, volicitin induces the seedlings to emit volatile compounds that attract parasitic wasps, natural enemies of the caterpillars. Mechanical damage of the leaves, without application of this compound, did not trigger release of the same blend of volatiles. Volicitin is a key component in a chain of chemical signals and biochemical processes that regulate tritrophic interactions among plants, insect herbivores, and natural enemies of the herbivores.

The intriguing defensive reaction of plants, whereby plant volatiles induced by insect herbivore injury attract natural enemies of the herbivores, is triggered when a substance in the oral secretion of the insect herbivores contacts damaged plant tissue. We have isolated, identified, and synthesized *N*-(17-hydroxylinolenoyl)-L-glutamine (Fig. 1), which we have named volicitin, from the oral secretion of beet armyworm (BAW) (*Spodoptera exigua* Hübner) caterpillars. Synthesized and natural volicitin induce corn (*Zea mays* L.) seedlings to release the same blend of volatile terpenoids and indole released when they are damaged by caterpillar feeding (*1*). This blend of volatile compounds attracts females of the parasitic wasp *Cotesia marginiventris*, natural enemies of BAW caterpillars, to the damaged corn plants (*2*).

The similarity of the structure of the insect-produced elicitor with the structures of the precursors of eicosanoids and prostaglandins involved in signaling in insects and other animals (*3*) and the components of the octadecanoid signaling pathway in plants (Fig. 1) (*4*) indicates a link between these two systems. The octadecanoid pathway is involved in induction of biosynthesis and release of volatiles in response to insect herbivore feeding (*5*, *6*).

We collected oral secretions (about 5 μl per caterpillar) by squeezing third to fifth instar BAW caterpillars that had been fed on corn seedlings, causing them to regurgitate (*1*). Biological activity was determined by collection and capillary gas chromatographic (GC) analysis (*7*) of volatiles from corn seedlings treated with oral secretion or subsequent fractions thereof (*8*).

The crude oral secretion was acidified, centrifuged, and filtered to remove proteins and solids (*9*). When the filtered oral secretion was fractionated on a reversed phase solid-phase extraction cartridge (*9*), the total activity of the original crude secretion eluted from the reversed phase cartridge with 50% CH_3CN in H_2O, indicating a molecule of medium polarity.

We further purified the active material from solid-phase extraction by a series of reversed-phase high-performance liquid chromatography (rpHPLC) fractionations using three sets of conditions (*10*). Only one active component, detected by monitoring ultraviolet (UV) absorption at 200 nm and with no absorption above 220 nm, eluted from the final column, and its biological activity was equivalent to that of the original crude oral secretion. The active material was extracted into CH_2Cl_2 from an acidified (pH 3) aqueous solution but not from an aqueous solution at pH 8 (*11*). All biological activity could be extracted from the organic phase back into pH 8 buffer, indicating lipid character and an acidic functional group. A CH_2Cl_2 solution of the active material was also fractionated on a

H. T. Alborn, T. C. J. Turlings, J. H. Loughrin, J. H. Tumlinson, Center for Medical, Agricultural, and Veterinary Entomology, United States Department of Agriculture, Agricultural Research Service, 1700 Southwest 23rd Drive, Gainesville, FL 32608, USA.
T. H. Jones, Department of Chemistry, Virginia Military Institute, Lexington, VA 24450, USA.
G. Stenhagen, Department of Organic Chemistry, Chalmers University of Technology, S-412 96 Göteborg, Sweden.

*Present address: University of Neuchatel, Institute of Zoology, CH-2007 Neuchatel, Switzerland.
†Present address: Community Research Service, Kentucky State University, Frankfort, KY 40601, USA.
‡To whom correspondence should be addressed. E-mail: tumlnsn@nervm.nerdc.ufl.edu

Source: From *Science* Vol. 276, No. 5314, pp. 945-949 (1997). Reprinted with permission from AAAS.

normal-phase diol column (*12*). The active component eluted from this column with MeOH (Me designates methyl). Rechromatography of the active component on the final rpHPLC column indicated it was greater than 99% pure.

Purified active compound was applied to artificially damaged leaves of intact corn seedlings (*2*). The volatiles released were the same as those induced by treatment with BAW oral secretion. Application of only buffer resulted in the release of significantly smaller quantities and different proportions of volatiles (Fig. 2).

The active compound was identified by mass and infrared spectroscopy and by chemical transformations. Fast atom bombardment mass spectrometry (FAB-MS) analysis (*13*) indicated the presence of only one compound with diagnostic peaks at mass-to-charge ratio (m/z) 423.280 $(M + H)^+$ in the positive ion mode and at m/z 421.2733 $(M - H)^-$ in the negative ion mode. The addition of sodium chloride to the FAB matrix resulted in reduced intensity of the m/z 423.280 and the appearance of m/z 445.2628 $(MH + Na)^+$. Thus, the active compound is a weak acid with a molecular weight of 422.274 daltons for the neutral molecule in acid form, and its elemental composition is $C_{23}H_{38}N_2O_5$ (422.278 daltons).

Daughter ions of the sodium salt, m/z 445, obtained by FAB-MS/MS (*13*), included a dominant ion at m/z 427 (445 − 18), whereas daughter ions of m/z 423 gave a strong peak at m/z 405 (423 − 18), both indicating a loss of H_2O. The lower mass region showed a characteristic pattern of peaks at m/z 147, 130, 129, 101, 84, 67, and 56, which is most con-

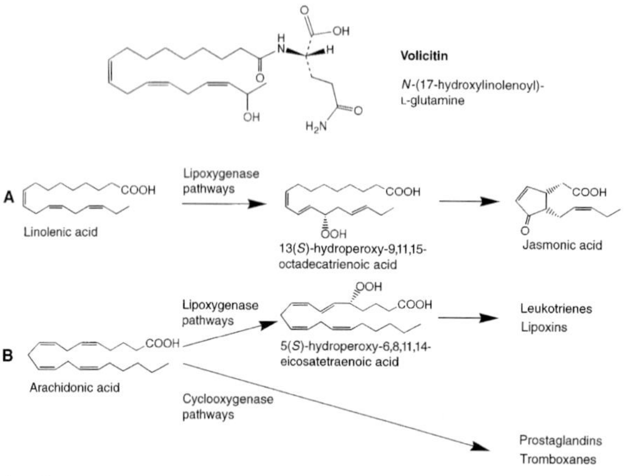

Fig. 1. Structure of volicitin. (**A**) The biosynthetic pathway leading to jasmonic acid in plants and (**B**) the biosynthetic pathways leading to prostaglandins and leukotrienes in animals.

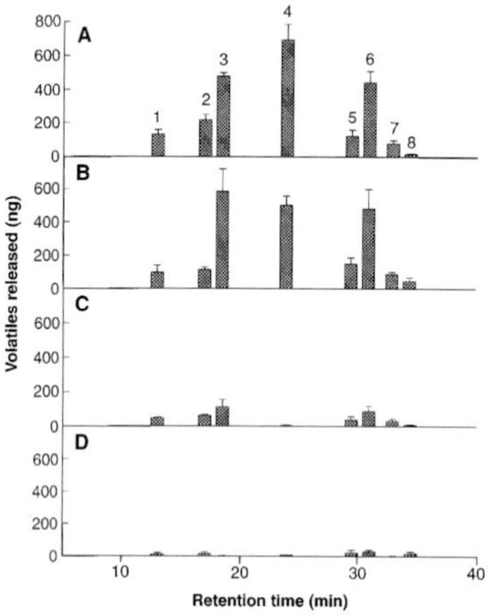

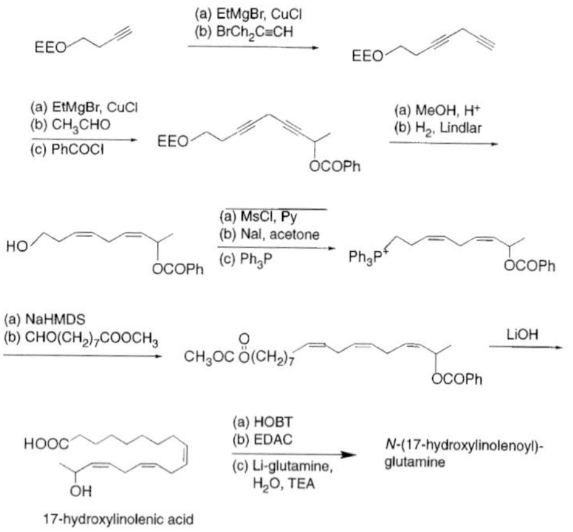

Fig. 2 (**left**). Average amount (nanograms per 2 hours) (n = 4) of volatiles collected from three intact Ioana corn seedlings that had been artificially damaged and treated with (**A**) 15 μl of BAW oral secretion per seedling on the damage sites, (**B**) 15 μl of oral secretion equivalents of pure natural volicitin, (**C**) 15 μl of buffer (*8*), or (**D**) undamaged control plants. At 9:00 p.m. a 1-cm^2 area of the second leaf of three-leaf seedlings was scratched with a clean razor blade and the test solution immediately rubbed over the damaged site. The next morning at 9:00 a.m. the seedlings were cut off above the root, and volatiles were collected and analyzed as described (*7, 8*). Bars with the same retention time in each graph represent the following compounds: 1, hexenyl acetate; 2, linalool; 3, (*E*)-4,8-dimethyl-1,3,7-nonatriene; 4, indole; 5, α-*trans*-bergamotene; 6, (*E*)-β-farnesene; 7, (*E*)-nerolidol; 8, (3*E*,7*E*)-4,8,12-trimethyl-1,3,7,11-tridecatetraene. **Fig. 3** (**right**). Synthesis scheme for volicitin.

sistent with the electron impact (EI) mass spectrum of glutamine (*14*). Subtraction of glutamine, linked by an ester or amide bond, gave $C_{18}H_{30}O_3$ as the elemental composition of the second part of the molecule, which is consistent with a hydroxy C_{18} acid with three double bonds.

When the active sample was lyophilized and treated with MeOH and acetic anhydride (*15*), GC/MS analysis (*16*) of the product revealed two prominent peaks. Chemical ionization (CI)/MS analysis of the first of these peaks with a retention time of 21.05 min revealed a prominent $(M + 1)^+$ ion at *m/z* 144, and EI/MS analysis revealed a molecular ion at *m/z* 143 and diagnostic ions at *m/z* 84 (base peak), 56, and 41, identifying it as the methyl ester of pyroglutamate, which confirmed the presence of glutamine. The CI mass spectrum of the second peak (retention time of 27.55 min) contained a very weak *m/z* 309 $(M + 1)^+$ ion and a predominant ion at *m/z* 291 due to loss of H_2O $(M + 1 - 18)^+$. Loss of MeOH gave an ion at *m/z* 277 $(M + 1 - 32)^+$, and the loss of both H_2O and MeOH gave an ion at *m/z* 259 $(M + 1 - 18 - 32)^+$. The EI spectrum of the same peak showed no molecular ion but a strong peak at *m/z* 290 due to the loss of H_2O $(M - 18)^+$, and a fragmentation pattern of ions characteristic of a straight-chain unsaturated hydrocarbon. These results were consistent with the methyl ester of a hydroxy acid. A smaller peak in the chromatogram had retention characteristics (retention time of 26.09 min) and a mass spectrum consistent with the acetate of the same hydroxy acid methyl ester (*17*).

Fourier transform infrared analysis (*18*) of the hydroxy acid methyl ester peak from GC produced a spectrum with a weak absorption at 3646 cm^{-1}, indicating an alcohol, and absorption bands at 3019, 2935, and 2865 cm^{-1}, typical of an unbranched, nonconjugated unsaturated hydrocarbon chain. The absence of absorption bands in the 2000 to 2500 cm^{-1} and the 960 to 980 cm^{-1} regions eliminated the possibility of an acetylene or trans double bond, respectively. The intensity of the 3019 cm^{-1} peak indicated three cis double bonds. Absorption at 1758 cm^{-1} confirmed the presence of a methyl ester.

Partial reduction (*19*) of the methyl ester of the C_{18} hydroxy acid resulted in a mixture of mono- and di-unsaturated products as identified by GC/MS. Subsequent ozonolysis (*20*) of this mixture and GC/MS analysis produced three diagnostic GC peaks with $(M + 1)^+$ ions at *m/z* 187, 229, and 271, corresponding to $H(CO)(CH_2)_n(CO)OCH_3$ with $n = 7$, 10, and 13, respectively. Methyl linolenate treated in the same way gave identical products. Thus, the olefinic bonds in the chain are located on carbons 9, 12, and 15, and the alcohol group is on either the 17th or 18th carbon.

The methyl ester was saturated by treatment with PdO/H_2 overnight (*21*). GC/MS (EI) analysis of the product showed an *m/z* 299 $(M - 15)^+$ ion and *m/z* 270/271 $(M - 44/M - 43)^+$ ions, indicative of β-cleavage of a secondary alcohol on C17. EI mass spectra of a pyrrolidide derivative (*21*) of the saturated product produced diagnostic ions at *m/z* 309 $(M - 44)^+$ and *m/z* 338 $(M - 15)^+$, confirming the C17 location of the hydroxyl group.

Only an amide bond between glutamine and the acid moiety of the hydroxy acid would be consistent with the results of all analyses. Thus, the active compound isolated from the oral secretion of BAW larvae is *N*-(17-hydroxylinolenoyl) glutamine, which we name volicitin.

Racemic 17-hydroxylinolenic acid was synthesized (Fig. 3) (*22*) and coupled with D- and L-glutamine (*23*). The crude synthetic conjugates purified on HPLC showed retention characteristics identical to the natural product. FAB-MS/MS and GC/MS analyses showed the synthetic and natural products to be identical.

Volicitin synthesized with D- or L-glutamine contained about 15% of the opposing form (*24*), indicating racemization during the step in which the fatty acid was coupled to glutamine. However, the active natural compound consisted exclusively of the L-glutamine form. Enantiomerically pure D- and L-glutamine forms of synthetic volicitin were collected from a chiral column (*24*) for bioassay.

The concentration of natural volicitin was estimated by HPLC detector response to be about 20 pmol per microliter of BAW oral secretion. Bioactivity of the L-glutamine form of the synthetic volicitin was equivalent to that of oral secretion (Fig. 4). The D-glutamine form was not active (Fig. 4). Racemic 17-hydroxylinolenic acid, D-glutamine, and L- glutamine at 300 and 900 pmol per excised corn seedling showed no activity. Thus, the biologically active compound is *N*-(17-hydroxylinolenoyl)-L-glutamine. At this time the configuration about the asymmetric 17th carbon in the fatty acid chain remains unknown.

In plants, the synthesis and release of volatile compounds appear to be induced by jasmonic acid, which is produced from linolenic acid by the octadecanoid signaling pathway (*6*). Jasmonates also stimulate other physiological and defensive processes in plants (*4*, *6*, *25*), and the amino acid conjugates of jasmonic acid are involved in physiological and developmental processes in many plants (*6*, *26*). Therefore, the presence of an elicitor that is an octadecatrienoate conjugated to an amino acid suggests that the elicitor molecule is involved with the octadecanoid pathway in the herbivore-damaged plants.

This elicitor activity is not diet-related and thus does not originate from the plants (*1*), although the fatty acid moiety is probably derived from linolenic acid obtained from the diet. The oral secretion of insects fed an artificial diet or filter paper is as active as that from insects fed on plants (*1*). Volicitin is also related in structure to eicosapentaenoic and arachidonic acids from the fungus *Phytophthora infestans* that elicit the production of fungitoxic sesquiterpenes in potato (*27*).

The octadecanoid signaling pathway in plants is similar in many ways to the eicosanoid pathways in animals that produce prostaglandins and leukotrienes (*4*, *25*). In insects eicosanoids may mediate cellular responses to bacterial infections as well as regulate other physiological functions (*3*).

Both corn and cotton respond to BAW damage and to the oral secretions of BAW applied to damaged leaves by producing and releasing terpenoids and indole (*1*, *2*, *28*–*30*). Although some compounds, such as indole, ocimene, and farnesene, are released by both plants, others are unique to each plant. Both plants respond systemically to BAW oral secretion by releasing induced volatiles from undamaged leaves of injured plants (*29*, *31*). In cotton the induced volatile compounds are

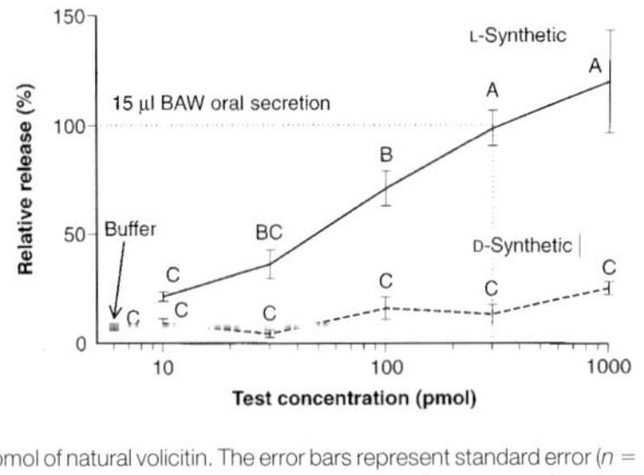

Fig. 4. Relative release of volatiles collected for 2 hours from three LG11 corn seedlings that had been treated with 10, 30, 100, 300, or 1000 pmol per plant of the D-glutamine (D-Synthetic) or the L-glutamine (L-Synthetic) forms of volicitin in 500 µl of buffer (*8*), or with buffer only. The combined amount in nanograms of caryophyllene, α-*trans*-bergamotene, (*E*)-β-farnesene, (*E*)-nerolidol, and (3*E*,7*E*)-4,8,12-trimethyl-1,3,7,11-tridecatetraene was used to calculate the release relative to that of seedlings treated with 15 µl of BAW oral secretion, which is equivalent to ~300 pmol of natural volicitin. The error bars represent standard error ($n = 5$). Data points topped by the same letter do not differ significantly [Tukey test, $P \leq 0.05$ (*33*)].

known to be synthesized de novo (*30*).

Volicitin accounts for the total activity of BAW oral secretion, and boiling the secretion for 30 min did not diminish its activity; thus, there is no evidence for enzymatic activity in eliciting volatiles in this case. In contrast, a β-glucosidase in the saliva of *Pieris brassicae* caterpillars elicits the release of volatile compounds from cabbage leaves (*32*). The sequence of events between the introduction of β-glucosidase and the emission of volatiles is unknown (*5*), but the simple release of a terpenoid or other volatile compound from a glycoside by the direct action of such an enzyme from the attacking herbivore obviously occurs by a different mechanism than the induction of de novo biosynthesis by a small molecule like volicitin. Also, it cannot account for the delayed release of herbivore-induced volatiles (*28*) or the systemic release of induced volatiles (*29*, *31*). Thus, different plant species may use different mechanisms to produce and release volatiles and may respond to different elicitors. Whether closely related insect species, particularly those that feed on the same types of host plants, produce volatile elicitors with the same or very similar structures remains unknown.

REFERENCES AND NOTES

1. T. C. J. Turlings, P. J. McCall, H. T. Alborn, J. H. Tumlinson, *J. Chem. Ecol.* **19**, 411 (1993).
2. T. C. J. Turlings, J. H. Tumlinson, W. J. Lewis, *Science* **250**, 1251 (1990).
3. D. W. Stanley-Samuelson, *J. Insect. Physiol.* **40**, 3 (1994).
4. E. E. Farmer and C. A. Ryan, *Trends Cell Biol.* **2**, 236 (1992).
5. J. Hopke, J. Donath, S. Blechert, W. Boland, *FEBS Lett.* **352**, 146 (1994).
6. T. Krumm, K. Bandemer, W. Boland, *ibid.* **377**, 523 (1995).
7. T. C. J. Turlings, J. H. Tumlinson, R. R. Heath, A. T. Proveaux, R. E Doolittle, *J. Chem. Ecol.* **17**, 2235 (1991).
8. Corn, *Z. mays* L., var. Ioana and LG11 sweet corn, was grown in a greenhouse in a potting soil mixture. Natural light was supplemented with 400-W high-pressure sodium lamps placed 1 m above the trays to maintain a 14-hour light (6 a.m. to 8 p.m.):10-hour dark cycle. Temperature varied from 30° to 35°C during the day and was kept at 20°C at night. BAW oral secretion (45 μl or the equivalent of each sample) was concentrated to dryness and dissolved in 1500 μl of 50 mM Na_2HPO_4 buffer, titrated to pH 8 with 1 M citric acid, which was divided into three 500-μl portions in 1-ml glass vials. At 9 to 11 p.m., a 10- to 14-day-old corn seedling was cut off above the root with a razor blade and transferred to a vial and allowed to draw up the solution for 12 hours in complete darkness. The three seedlings for each treatment were then combined in a volatile collection chamber [15 cm long, 3 cm inner diameter (ID)] under artificial light. Purified, humidified air was drawn for 2 hours through the chamber at 300 ml/min and through a polymer adsorbent (Super Q 80/100 catalog number 2735, Alltech Associates, Deerfield, IL). The adsorbent was removed from the chamber and extracted with 150 μl of CH_2Cl_2 (GC², Burdick and Jackson, Muskegon, MI), then internal standard (600 ng of nonyl acetate in 30 μl of CH_2Cl_2) was added.
9. Crude oral secretion was centrifuged at 16,000*g* for 30 min to eliminate solid material, and the supernatant was then filtered through a 0.22-μm sterilizing membrane (Millex GV, Millipore, Bedford, MA). An equal amount of 50 mM pH 3.3 phosphate buffer was added, and the precipitated proteins were removed by centrifugation as before. A 0.5-ml sample of the supernatant was put on a 6-ml activated, octadecyl solid-phase extraction cartridge (Bakerbond, J. T. Baker, Phillipsburg, NJ) and eluted with 2-ml volumes of H_2O, 50% CH_3CN (Low-UV HPLC grade, Burdick and Jackson) in H_2O, and CH_3CN. All fractions were concentrated to near dryness under vacuum (Speed Vac rotary concentrator, Savant Instruments, Farmingdale, NY) and redissolved in 0.5 ml of 50 mM pH 8 buffer, for bioassay.
10. The active material from solid-phase extraction was fractionated by HPLC [LDC 4100 pump with SM5000 diode array UV detector (LDC Analytical, Riviera Beach, FL)], monitoring wavelengths from 190 to 360 nm. A reversed-phase column (Waters Nova Pac C_{18} 4 μm, 4-mm ID by 150-mm long column, Waters, Milford, MA) was eluted (1 ml/min) with a solvent gradient of 0 to 25% CH_3CN in H_2O in 15 min, followed by an increase to 100% CH_3CN in 15 min. Eluate was collected in 2-ml fractions. All activity was in the 6- to 8-ml fraction, which was further fractionated on a C_{18} reversed-phase column with different selectivity (ODS-AQ S-5 200 Å, 250 mm long, 4.6-mm ID, YMC Company, Kyoto, Japan) with the same solvent gradient. All active material eluted in the 20- to 22-ml fraction. This fraction contained two overlapping peaks, which could be separated on the same column with a solvent gradient of 20 to 60% CH_3CN in H_2O from 0 to 20 min.
11. A 2-ml sample of the active fraction from the final HPLC reversed-phase column (*10*) was concentrated to dryness under vacuum and redissolved in 2 ml of H_2O, and then 100 μl of acetic acid and 2 ml of CH_2Cl_2 were added and the mixture shaken for 5 min. The H_2O and organic phases were separated and concentrated to dryness under vacuum. Bioassay of the fractions redissolved in pH 8 buffer showed all active material to be present in the organic phase, and HPLC analysis (*10*) showed the peak to be present in this fraction.
12. A 0.5-ml sample of the CH_2Cl_2 fraction was put on an activated 3-ml 10SPE diol cartridge (Bakerbond, J. T. Baker), which was eluted with 3 ml of CH_2Cl_2 and then 2 ml of MeOH. The two fractions were concentrated to dryness under vacuum and each redissolved in 0.5 ml of 50 mM pH 8 buffer for bioassay and HPLC analysis.
13. Pure elicitor [40 ng in 4 μl of 50% CH_3CN in H_2O and 1 μl of trifluoroacetic acid (TFA)] was added to a glycerol matrix and analyzed with FAB-MS on a VG Zabspec instrument (VG Analytical, Fison Instruments, Manchester, England). To obtain sodium adducts, we replaced the TFA with 1 μl of a 1 M sodium chloride solution. High-resolution mass measurements were obtained by adding polyethylene glycol (1 μl) with an average mass of 400 daltons (PEG 400) to the glycerol matrix to give reference ions of known mass for calibration of the mass scale. Possible elemental compositions were established allowing a maximum of 30 carbons, 2 to 8 oxygens, and 2, 4, or 6 nitrogens. The mass window for the calculations was limited to an error of 10 milli mass units. Daughter ion spectra were obtained from samples in the same FAB matrix as above with a tandem four-sector mass spectrometer (JEOL HX/HX110A, Tokyo, Japan). The nitrogen collision gas was adjusted to give a 60% reduction in intensity of the mother ion.
14. National Institute of Standards and Technology (Gaithersburg, MD), mass spectral library on CD-ROM, version 1.0, January 1995. Characteristic ions for glutamine are *m/z* 130, 129, 101, 84 (base peak), and 56.
15. Purified active compound equivalent to 100 μl of oral secretion was methanolyzed with MeOH and Ac_2O following the procedures of J. M. L. Mee [*Biomed. Mass Spectrom.* **4**, 178 (1977)]. The sample was concentrated to dryness, dissolved in a mixture of 50 μl of dry MeOH and 50 μl of Ac_2O, and heated to 100°C for 10 min under N_2 in a sealed ampoule. It was concentrated to dryness by a stream of N_2, then 50 μl of CH_2Cl_2 was added and the sample was analyzed by GC/MS.
16. Methanolysis products were analyzed on a Finnigan TSQ 700 mass spectrometer (Finnigan MAT, San Jose, CA) interfaced to an HP 5890 gas chromatograph (Hewlett-Packard, Palo Alto, CA). Injections were made in the splitless mode at 225°C (1 μl containing the equivalent of 2 μl of oral secretion). A polar capillary column (OV351, 25 m long, 0.25-mm ID, Scandinaviska Genetec, Kungsbacka, Sweden) was held at 60°C for 3 min and then increased 10°C per minute to 250°C and held at that temperature for 18 min. The carrier gas was He and the reaction gas was CH_4 for chemical ionization. The ion source temperature was 200°C in both EI and CI mode.
17. The acetate had a shorter retention time (26.06 min) on the polar OV-351 column than the hydroxy acid methyl ester (27.55 min), but a longer retention time (25.54 min) than the hydroxy acid methyl ester (25.38 min) on a nonpolar column (DB1, 25 m long, 0.25-mm ID, J&W Scientific, Folsom, CA). The CI mass spectrum showed ions at *m/z* 351 (weak) (M + 1), *m/z* 309 (weak) (loss of MeOH), *m/z* 291 (base peak) (loss of acetic acid), and *m/z* 259 (loss of both MeOH and acetic acid).
18. The methanolysis products were analyzed on a Hewlett-Packard (Palo Alto, CA) model 5965B Fourier transform infrared spectrometer interfaced to a model 5890 GC, equipped with a DB1 column (25 m long, 0.32-mm ID, J&W Scientific). Injections (2 μl, equivalent of 4 μl of oral secretion) were made in the splitless mode at 225°C. The carrier gas was N_2 and the temperature program the same as in (*16*).
19. The methyl ester of the hydroxy acid was partially saturated following the procedures of A. B. Attygalle, G. N. Jham, A. Svatos, R. T. S. Frighetto, and J. Meinwald [*Tetrahedron Lett.* **36**, 5471 (1995)]. After concentration to dryness the sample was redissolved in 10 μl of ethanol, then 30 μl of 10% hydrazine in ethanol and 30 μl of 0.6% hydrogen peroxide in ethanol was added and the solution heated to 60°C for 1 hour. After the solution cooled to room temperature, 35 μl of 5% HCl in H_2O was added. The solution was extracted with two 40-μl portions of hexane (GC², Burdick and Jackson) and the hexane solution washed with four 50-μl portions of H_2O before being analyzed by GC/MS (*24*).
20. Partially reduced hydroxy acid methyl ester (*19*) was ozonized following the procedures of M. Beroza and B. Bierl [*Anal. Chem.* **38**, 1976 (1966); *ibid.* **39**, 1131 (1967)] and the product analyzed by GC/MS (*16*).
21. The hydroxy acid methyl ester was dissolved in 50 μl of ethyl acetate, a few grains of PdO was added, and saturation was achieved by bubbling H_2 through the solution for 18 hours. A pyrrolidide derivative of the saturated methyl ester was prepared following the procedures of B. Å. Andersson [*Prog. Chem. Fats Other Lipids* **16**, 279 (1978)].
22. Racemic 17-hydroxy linolenic acid was synthesized (Fig. 3) by the condensation of the appropriate triphenylphosphonium ylid, prepared from the alkyl triphenylphosphonium iodide by the silazide method [H. J. Bestman, W. Stransky, O. Vostrowsky, *Chem. Ber.* **109**, 1694 (1976); H. J. Bestman *et al.*, *Liebigs Ann. Chem.* **1987** 417 (1987)], with methyl 9-oxononanoate [F. C. Pennington, W. D. Clemer, W. M. McLamore, V. V. Bogert, I. A. Solomons, *J. Am. Chem. Soc.* **75**, 109 (1953)]. The phosphonium salt was obtained by standard methodology from the ethoxy ethyl ether of 3,6-heptadiyn-1-ol [W. Huang, S. P. Pulaski, G. Meinwald, *J. Org. Chem.* **48**, 2270 (1983)] by sequential condensation with acetaldehyde and benzoyl chloride, followed by deprotection, partial hydrogenation, and conversion of the primary alcohol to the phosphonium iodide.
23. D- or L-glutamine (1.1 to 1.2 equivalents) was converted into its lithium salt by the procedures of R. Kramell, J. Schmidt, G. Schneider, G. Sembdner, and K. Schreiber [*Tetrahedron* **44**, 5791 (1988)] and coupled to the hydroxy acid by procedures modified from J. C. Sheehan,

J. Preston, and P. A. Cruickshank [*J. Am. Chem. Soc.* **87**, 2492 (1965); W. Konig and R. Geiger, *Chem. Ber.* **103**, 788 (1970)]. The coupling was carried out in *N,N*-dimethylformamide (DMF) (Sigma) with addition of 5 to 10% H_2O to increase solubility of the salt and 1.2 equivalents of 1-hydroxybenzotriazole (Sigma) to reduce racemization of glutamine. The product was purified by rpHPLC (*10*) on the YMC column eluted with 25% CH_3CN in 0.4 mM ammonium acetate buffer (Aldrich) at a flow rate of 1.2 ml/min. The retention time of the synthetic and natural elicitor was identical.

24. The D- and L-glutamine forms of synthetic volicitin were separated on a 250-mm, 4.6-mm ID chirobiotic T column (Advanced Separation Technologies, Whippany, NJ) eluted with 10% CH_3CN in 10 mM ammonium acetate buffer, pH 4.5, at a flow rate of 1 ml/min (*10*). The synthetic L-form had a retention time of 5.25 min, identical to that of the natural elicitor, and the D-form had a retention time of 9.03 min.
25. E. E. Farmer and C. A. Ryan, *Plant Cell* **4**, 129 (1992).
26. R. Kramell *et al.*, *J. Plant Growth Regul.* **14**, 29 (1995).
27. K. E. Ricker and R. M. Bostock, *Physiol. Mol. Plant Pathol.* **44**, 65 (1994), and references therein.
28. J. H. Loughrin, A. Manukian, R. R. Heath, T. C. J. Turlings, J. H. Tumlinson, *Proc. Natl. Acad. Sci. U.S.A.* **91**, 11836 (1994).
29. U. S. R. Röse, A. Manukian, R. R. Heath, J. H. Tumlinson, *Plant Physiol.* **111**, 487 (1996).
30. P. W. Pare and J. H. Tumlinson, *Nature* **385**, 30 (1997).
31. T. C. J. Turlings and J. H. Tumlinson, *Proc. Natl. Acad. Sci. U.S.A.* **89**, 8399 (1992).
32. L. Mattiacci, M. Dicke, M. A. Posthumus, *ibid.* **92**, 2036 (1995).
33. Significant differences in relative release rates of volatiles were tested by Tukey's studentized range test after analysis of variance with a significance level of 5% (SYSTAT, Systat Inc., Evanston, IL).
34. This project was funded in part by a grant from the Swedish Natural Science Research Council. We thank H. Karlsson, A. T. Proveaux, and D. Powell for assistance with mass spectrometric analysis, J. Lockerman and S. Sharp for oral secretion collection, and M. Brennan for technical assistance. We also thank J. G. Millar, C. A. Ryan, and G. G. Still for helpful comments on the manuscript.

24 January 1997; accepted 17 March 1997

Proc. Natl. Acad. Sci. USA
Vol. 95, pp. 13971–13975, November 1998
Plant Biology

Concerted biosynthesis of an insect elicitor of plant volatiles

P. W. PARÉ, H. T. ALBORN, AND J. H. TUMLINSON*

U.S. Department of Agriculture–Agricultural Research Services, Center for Medical Agricultural and Veterinary Entomology, 1700 Southwest 23rd Drive, Gainesville, FL 32608

Contributed by J. H. Tumlinson, September 25, 1998

ABSTRACT **A variety of agricultural plant species, including corn, respond to insect herbivore damage by releasing large quantities of volatile compounds and, as a result, become highly attractive to parasitic wasps that attack the herbivores. An elicitor of plant volatiles, *N*-(17-hydroxylinolenoyl)-L-glutamine, named volicitin and isolated from beet armyworm caterpillars, is a key component in plant recognition of damage from insect herbivory. Chemical analysis of the oral secretion from beet armyworms that have fed on ^{13}C-labeled corn seedlings established that the fatty acid portion of volicitin is plant derived whereas the 17-hydroxylation reaction and the conjugation with glutamine are carried out by the caterpillar by using glutamine of insect origin. Ironically, these insect-catalyzed chemical modifications to linolenic acid are critical for the biological activity that triggers the release of plant volatiles, which in turn attract natural enemies of the caterpillar.**

Several studies have shown the active role of herbivore-damaged plants in the attraction of insect predators and parasitoids. Volatile plant compounds released in response to insect feeding serve as a chemical signal for herbivore natural enemies (for a review, see refs. 1 and 2). Recent work suggests that this as well as other plant defense responses are triggered by an active component or components associated with the feeding herbivore that allows the plant to differentiate between general wounding and damage caused by chewing insects. In cotton, induced volatiles that are synthesized in response to wounding are released in greater quantities as a result of caterpillar feeding than with mechanical damage alone (3), and, in tobacco, higher concentrations of the defense-signaling molecule jasmonic acid result from herbivore damage by hornworm caterpillars than from mechanical damage designed to mimic herbivory (4). At the transcriptional level, potato mRNAs involved in plant defense accumulate more rapidly with insect-derived elicitor(s) in contact with the damaged leaves than with mechanical damage alone (5).

Thus far, two elicitors of plant volatiles have been identified and reported from chewing insects. Dicke's group has identified a β-glucosidase, present in the regurgitant of *Pieris brassicae* caterpillars that triggers the same emissions of volatiles in cabbage plants as larvae that feed on the plant (6). Because enzyme activity in the regurgitant is retained when caterpillars are fed on β-glucosidase free diet, enzyme activity does not appear to be plant derived. Volicitin, *N*-(17-hydroxylinolenoyl)-L-glutamine, identified from the oral secretion of beet armyworm caterpillars, induces corn seedlings to synthesize and release volatile chemical signals (7) that attract parasitic wasps. (ref. 8; for review, see refs. 9 and 10). Volicitin has not been found in plant tissues, although the structure of the molecule suggests that it may interact with the octadecanoid signaling pathway in plants (7). The collection of volatiles from plants fed on by different herbivores suggests that each insect species may produce its own signature molecule(s) that allows plants to differentiate among herbivorous attackers (11, 12). The current study was undertaken to determine the biogenetic origin of volicitin. By feeding beet armyworms corn seedlings labeled with $^{13}CO_2$, we obtained chemical evidence that the caterpillars acquire linolenic acid [an essential fatty acid in the diet of Lepidoptera (13)] from plants and that the insects subsequently hydroxylate and conjugate the fatty acid with glutamine. Thus, the modification by the insect of linolenic acid of plant origin provides a distinct chemical cue that allows a plant to differentiate between herbivore damage and other types of wounding that trigger the octadecanoid defense signaling pathway.

MATERIALS AND METHODS

Plant Growth and Labeling Conditions. Corn seeds, *Zea mays* L., variety LG11 sweet corn, were given 4 days to germinate under a moist paper towel in the dark. To reduce the amount of stored ^{12}C-labeled carbon available for the growing seedling, seeds were trimmed by using a razor blade to remove ≈60–70% of the endosperm without cutting into the germinated seedling. Nine excised seedlings were planted in sterilized Metromix 300 potting soil (Scotts-Sierra Horticulture, Marysville, OH). The pot was placed in a 48- × 16-cm (diameter) glass growth chamber, and synthetic premixed air (Cambridge Isotope Laboratories, Cambridge, MA and Airco, Riverton, NJ), which contained 1,800 μM/liter CO_2 (^{13}C 99%), 20.7% oxygen, and a balance of nitrogen was introduced by flushing the chamber at 500 ml min^{-1} for 10 min and then reducing to a flow of 50 ml min^{-1}. Synthetic air passed up over the potted plants contained within the volatile collection apparatus and out one of eight ports from the glass lid through a plastic tube into a water bubbler. One metal halide and two 400-W high-pressure sodium lamps positioned 10 cm above the chamber provided a 16-h light/8-h dark photoperiod; an air-cooled glass panel directly below the lights insulated the plants from the lamp heat source.

Collection of Plant Volatiles. Corn seedling leaves were damaged with a stainless steel wire, fed through one of the top ports of the growth chamber to scrape the leaves against the glass chamber. Plant volatiles then were collected from 1,200–1,500 h by drawing synthetic ^{13}C-labeled air (100 ml min^{-1}) through Super-Q adsorbent traps (Alltech Associates). Compounds were eluted from the adsorbent filters with 150 μl of dichloromethane; 1-μl aliquots were analyzed by capillary GC on a 50-m × 0.25-mm (i.d.) fused silica column with a 0.25-μm-thick bonded methyl silicone stationary phase (Quadrex, New Haven, CT). The column was held at 60°C for 5 min and then was increased 5°C per min to 225°C and was held at that temperature for 30 min. Helium was used as a carrier gas at a linear flow velocity of 18 cm sec^{-1}. To determine the amount of ^{13}C incorporated into each compound, samples were analyzed by a Finnigan-MAT ITS40 (ion trap) mass

The publication costs of this article were defrayed in part by page charge payment. This article must therefore be hereby marked "*advertisement*" in accordance with 18 U.S.C. §1734 solely to indicate this fact.

0027-8424/98/9513971-5$0.00/0
PNAS is available online at www.pnas.org.

*To whom reprint requests should be addressed. e-mail: jtumlinson@gainesville.usda.ufl.edu.

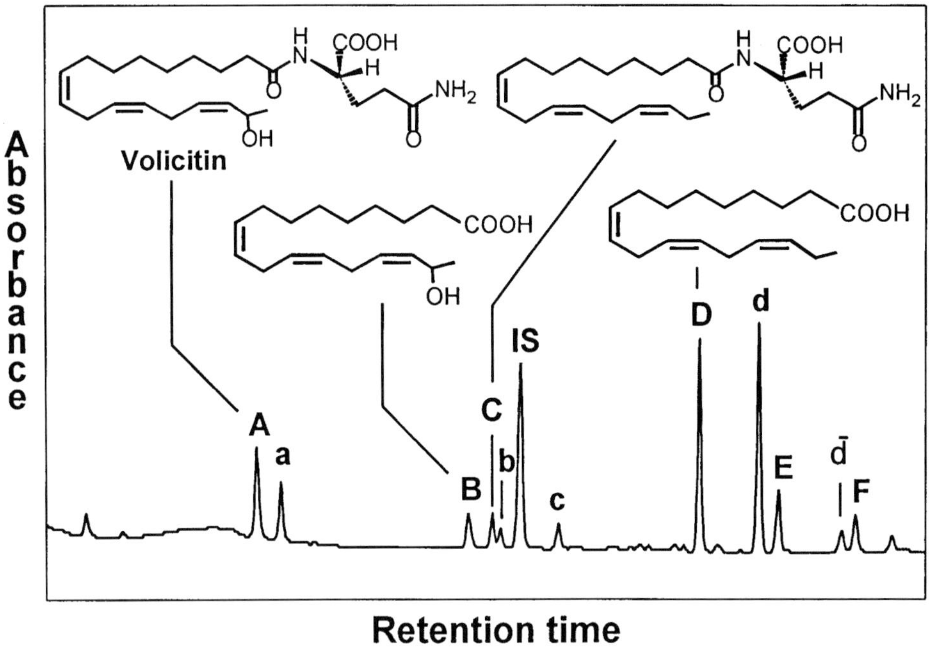

FIG. 1. HPLC profile of volicitin (A) and other fatty acid derivatives detected in the oral secretions of beet armyworms at a wavelength of 200 nm. Compounds and approximate retention times include: (A) *N*-(17-hydroxylinolenoyl)-L-glutamine, 7.5 min; (a) *N*-(17-hydroxylinoleoyl)-L-glutamine, 8.0 min; (B) 17-hydroxy linolenic acid, 11.1 min; (C) *N*-linolenoyl-L-glutamine, 11.9 min; (b) 17-hydroxy linoleic acid, 12 min; (IS) *N*-palmitoleoyl-L-glutamine (internal standard), 12.3 min; (c) *N*-linoleoyl-L-glutamine, 12.6 min; (D) linolenic acid, 15.2 min; (d) linoleic acid, 16.2 min; (E) unknown, 16.6 min; (d̄) oleic acid, 17.1 min; and (F) impurity, 17.8 min.

spectrometer (Finnigan-MAT, San Jose, CA) in the chemical-ionization mode, with isobutane as the reagent gas. Injections were made via a septum-equipped programmable injector held at 40°C for 0.25 min, then programmed at 170°C min^{-1} to 270°C onto a 30-m × 0.25-mm (i.d.) fused silica column with 0.25-μm-thick bonded 5% phenyl methyl silicone (DB-5MS; J & W Scientific, Folsom, CA) held at 40°C for 5 min, then programmed at 5°C min^{-1} to 260°C; He carrier gas linear flow velocity was 19 cm sec^{-1}. Source temperature was adjusted to 120 ± 20°C to optimize the molecular ion abundance. Selected mass ions were quantified via computer software analysis. The fraction of each compound that incorporated ^{13}C was computed on a molecule basis (14).

Plant Chemical Analysis. Lipid and amino acid analysis was based on procedures of Harborne (15). For lipid extraction, frozen plant tissue (0.1 g) was homogenized with chilled isopropanol (5 ml) and was extracted with diethyl ether (5 ml); the supernatant fractions, after centrifugation, were combined and concentrated to dryness. For methanolysis, 2 ml MeOH:HCl (3:1) was added to the extract and was heated for 30 min at 100°C. Hydrolyzed lipids were diluted with H_2O (5 ml), were extracted with CH_2Cl_2 (2 × 4 ml), and were rinsed with saturated $NaHCO_3$ (2 × 4 ml). A concentrated extract was redissolved in a minimum amount of Hex:EtOAc (5:1) and was applied to a 6-ml activated silica gel solid-phase extraction cartridge that was eluted with the same solvent. Fractions containing oleic, linoleic, and linolenic methyl esters initially were identified by comigration with standards on silica gel TLC plates (Merck). Glutamine and glutamic acid were extracted from frozen tissue (0.1 g) with 65% EtOH (5 ml); the supernatant, after centrifugation, was concentrated to dryness and was redissolved in H_2O (400 μl), and the supernatant, after centrifugation, was collected. The aqueous extract was concentrated to dryness and derivatized with 300 μl MeOH:Ac_2O (5:1) by heating for 10 min at 100°C (16). Identification of plant fatty acids, glutamine, and glutamic acid methyl esters were confirmed by comparison with methylated synthetic standards by GC-MS analysis following the same procedures as the analysis of plant volatiles except that samples were injected onto a polar capillary column (OV351, 25-m × 0.25-mm i.d., Quadrex, New Haven, CT), which was held at 60°C for 5 min and then was increased 5°C per min to 230°C and held at that temperature for 30 min.

Caterpillar Regurgitant Collection and Analysis. Beet armyworms were reared on an artificial pinto bean diet following the method of King and Leppla (17). Insects were transferred to feed on corn seedlings at least 48 h before labeling experiments. Regurgitation was induced by holding fourth instar beet armyworm caterpillars with forceps and gently pinching behind the head with a second pair (18). The oral secretion from five larvae was collected, 5 μl of an aqueous *N*-palmitoleoyl-L-glutamine solution (1 $\mu g\, \mu l^{-1}$) was added as an internal standard, and the mixture was centrifuged. The supernatant was fractioned by HPLC [LDC 4100 pump with SM5000 diode array UV detector (LDC Analytical, Rivera Beach, FL)] monitoring wavelength 200 nm on a reversed-phase column (C_{18} ODS-AQ S-5 200 Å, 250 mm long, 4.6-mm i.d., YMC, Kyoto). Components were eluted (1 ml min^{-1}) with a solvent gradient of 40 to 100% CH_3CN, containing 0.8% glacial acetic acid, in H_2O, containing 0.4% glacial acetic acid, over 10 min and then were held at 100% CH_3CN, still

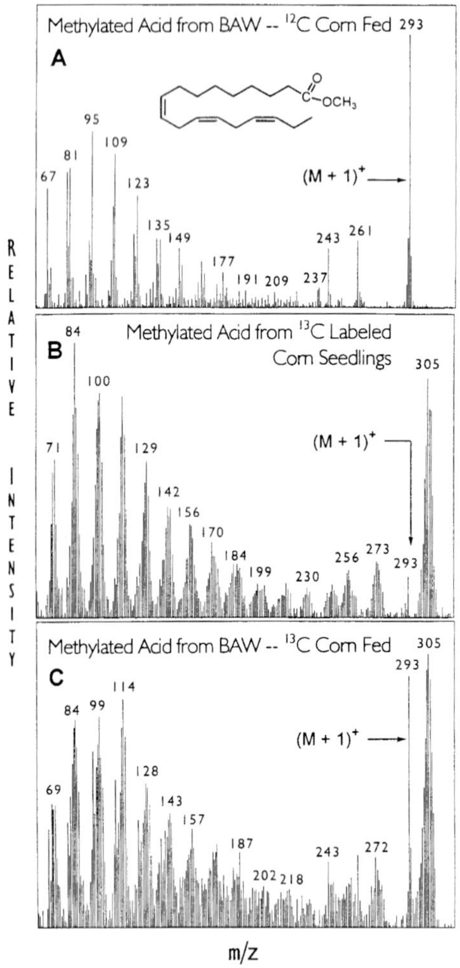

FIG. 2. Chemical ionization mass spectra of methylated linolenic acid from: beet armyworms (BAW) fed on unlabeled corn seedlings for 48 h (*A*), extracted corn seedlings grown for 12 days with $^{13}CO_2$ enrichment (*B*), and beet armyworms fed on unlabeled corn plants and then fed for 6 h on ^{13}C-labeled corn seedlings (C).

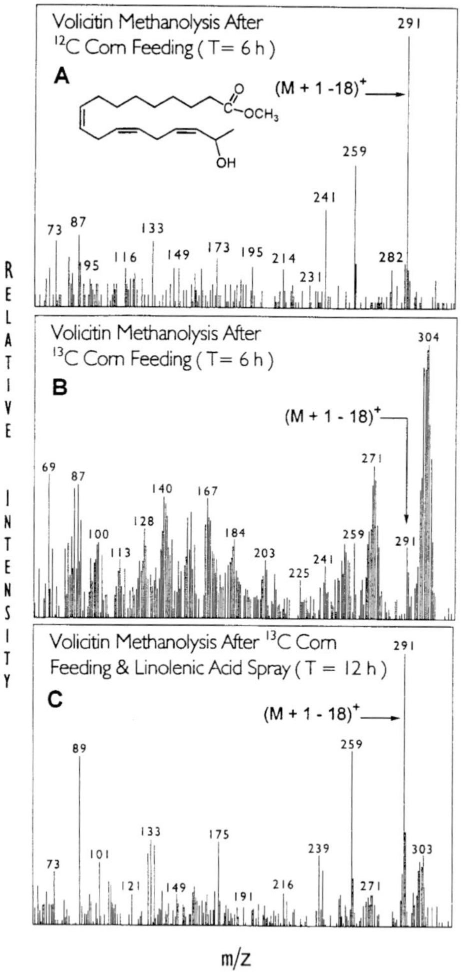

FIG. 3. Chemical ionization mass spectra of methyl 17-hydroxylinolenate prepared by transesterification of volicitin from oral secretion of caterpillars fed 6 h on unlabeled corn seedlings (*A*), 6 h on ^{13}C-labeled corn seedlings (*B*), and 12 h on ^{13}C-labeled seedlings with unlabeled linolenic acid added to the leaves after 6 h (*C*). The 291 ion resulted from loss of water from the molecular ion $(M + 1 - 18)^+$.

containing 0.8% acetic acid, for 10 min. Fractions were methylated with MeOH and Ac_2O as noted above (10).

Unlabeled Linolenic Acid Applications. Linolenic acid (Sigma) in EtOH (1:9) was diluted to a 1% acid solution with H_2O. The solution was applied to labeled corn seedling with an atomize sprayer to cover the corn leaves with a thin mist of unlabeled acid.

RESULTS

To assess rate of synthesis and the source of the chemical components used by beet armyworm (*Spodoptera exigua* Hübner) to assemble volicitin, caterpillars were fed on corn seedlings that were labeled uniformly with ^{13}C. Plants were labeled isotopically by growing seedlings in a closed container into which synthetic premixed air that contained 1,800 μM/liter $^{13}CO_2$ flowed continuously. As a preliminary experiment to estimate the degree of incorporation of ^{13}C into the fatty acids of the plant, leaves were damaged mechanically and the hexenals and hexenols, released as a result of lipoxygenase activity, were collected and analyzed by GC-MS (19). These analyses indicated that linolenic acid in 9-day-old seedlings had incorporated substantial levels of ^{13}C [e.g., 84% enrichment of (*Z*)-3-hexenol]. Seedlings grown for 12 days in an enclosed glass chamber to ensure a high level of ^{13}C labeling were used to feed beet armyworm larvae. Subsequent to the feeding experiment (see below), the remaining portions of the leaves of each plant were extracted and analyzed to determine the

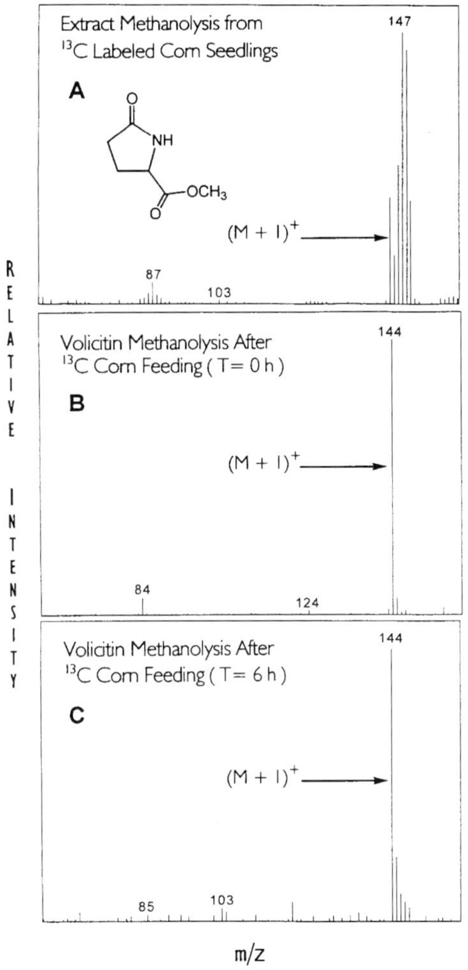

FIG. 4. Chemical ionization mass spectra of methyl pyroglutamate (the product of glutamine methylation) from extracted corn seedlings that were grown for 12 days with $^{13}CO_2$ enrichment (*A*), volicitin from caterpillars that fed on unlabeled corn seedlings for 48 h (*B*), and volicitin from beet armyworms that fed on unlabeled corn plants and then fed 6 h on ^{13}C-labeled corn seedlings (*C*).

percentage of incorporation of the label in linolenic acid and glutamine.

Early fourth instar beet armyworms fed on an artificial diet were transferred to unlabeled corn plants for a minimum of 48 h before being placed on the labeled corn seedlings, to ensure that the ratio of fatty acids in the oral secretions was based on plant constituents and not our rearing diet. Oral secretion was collected by gently squeezing caterpillars after they had fed on labeled seedlings for 6 h, causing them to regurgitate. The oral secretion supernatant from the caterpillars was injected, without further purification, onto a reverse phase HPLC column and was eluted with an acetonitrile-water gradient. Peaks eluting from this column (Fig. 1) were collected, lyophilized, and treated with methanol and acetic anhydride, and the derivatized products were analyzed by GC-MS. Analysis of compounds purified from oral secretions of insects fed on unlabeled and labeled seedlings revealed the consistent presence of nine compounds (Fig. 1). In addition to volicitin, beet armyworm oral secretions contained free 17-hydroxylinolenic acid, *N*-linolenoyl-L-glutamine and free linolenic acid and an analogous series of compounds with a linoleic acid (two double bond) backbone. These compounds have been identified by chemical ionization–MS and fast atom bombardment–MS, and their structures were confirmed by comparisons with synthetic standards (H.T.A., T. H. Jones, G. S. Stenhagen, and J.H.T., unpublished work).

The mass spectral data for the methyl ester of linolenic acid (Fig. 2) as well as the fatty acid portions from the other beet armyworm components demonstrated extensive incorporation of ^{13}C. There are two distinct peak aggregations at m/z 293 and m/z 305 in the spectra of the methyl ester of free linolenic acid from the ^{13}C-labeled corn seedlings and from the caterpillars that have fed on ^{13}C-labeled corn seedlings. The m/z 293 ion $(M+1)^+$ represents linolenic acid molecules devoid of ^{13}C whereas m/z 311 $(M+19)^+$ corresponds to linolenic acid in which all carbon is ^{13}C-labeled. The fraction of the fatty acid portion of each compound that incorporated ^{13}C on a molecular basis with 6 h of feeding on the labeled corn seedlings is as follows: *N*-(17-hydroxylinolenoyl)-L-glutamine, 63%; *N*-(17-hydroxylinoleoyl)-:L-glutamine, 52%; 17-hydroxylinolenic acid, 61%; *N*-linolenoyl-L-glutamine, 83%; *N*-linoleoyl-L-glutamine, 51%; linolenic acid, 80%; and linoleic acid, 57%. In fact, within 6 h, the acids in beet armyworm oral secretions contained a level of ^{13}C-labeling comparable to what is found in the labeled seedlings. For the fatty acids analyzed from the caterpillars, as well as the labeled corn seedlings that they fed on, the pattern of ^{13}C incorporation followed a gaussian distribution with m/z 305 representing slightly more than half of the carbon atoms containing the ^{13}C label. This indicates that the fatty acid portion of volicitin and the other conjugated compounds in the insect oral secretion are obtained directly from the plant.

To determine whether the beet armyworm hydroxylates the fatty acid portion of volicitin or whether the entire 17-hydroxylinolenic acid molecule is plant-derived, a pulse–chase experiment was conducted. Caterpillars that had fed on ^{13}C-labeled corn for 6 h so that the linolenic acid moiety of volicitin was ^{13}C-labeled (Fig. 3*B*) were moved to ^{13}C-labeled seedlings that had been sprayed with unlabeled linolenic acid and were allowed to feed for an additional 6 h. The marked decrease of ^{13}C in the methyl 17-hydroxylinolenoate derived from volicitin (Fig. 3*C*) and free linolenic acid, both collected from the beet armyworm, indicates that hydroxylation of the relatively large amount of free unlabeled linolenic acid sprayed onto the plant leaves is done by the beet armyworm. The linolenic and linoleic acid components from caterpillars that had fed on unsprayed ^{13}C-labeled seedlings for the same total time of 12 h maintained high ^{13}C incorporation levels. To ensure that the decrease in ^{13}C incorporation with application of unlabeled linolenic acid was a specific response for linolenic acid derivatives, linoleic acid was collected from caterpillars that had fed on sprayed seedlings and was found to sustain the high level of ^{13}C labeling. The 17-hydroxylinolenic acid methyl ester and the 17-hydroxylinoleic acid methyl ester were not detected in derivatized plant extracts, although these compounds were identified when plant tissue was spiked with the corresponding synthetic acid before extraction and methylation.

In contrast to the fatty acid portion of volicitin, which rapidly incorporated ^{13}C and showed a similar mass spectral labeling pattern as the linolenic acid from the plant (Fig. 2*B*), the glutamine incorporated little ^{13}C relative to the glutamine from the plant (Fig. 4), indicating that the plant was not catalyzing the coupling of glutamine to the fatty acids. It also appears that labeled glutamine from the plant ingested by the insect was diluted with an unlabeled insect source of glutamine before coupling to form the volicitin molecule. This could be caused by either a large glutamine pool present in the caterpillar's oral secretions or a physically separate glutamine source. Because the plants are ^{13}C-labeled from the time that photosynthesis begins, the possibility that unlabeled glutamine is derived from an isolated source separate from current photosynthate and apart from the major pool of plant glutamine is unlikely.

These biochemical data demonstrate that the plant supplies linolenic acid, which is required for growth and development of beet armyworms, and also provides this fatty acyl chain for the synthesis of volicitin, the modified elicitor of plant volatiles that is central to signaling between plants and natural enemies of the caterpillars that attack them. It is not clear why the caterpillar (or gut symbionts in the caterpillar) adds L-glutamine and the hydroxy group to linolenic acid or whether volicitin plays a role in the metabolism of the herbivore. It is also not clear how volicitin interacts on a biochemical level to induce synthesis of plant volatiles; however, we do know that volicitin provides a rapid, clear, and reliable signal for initiation of synthesis and release of volatile compounds. Thus the plant and the herbivore are inexorably linked through a signaling molecule of dual origin and effect.

We are grateful for technical assistance from P. M. Brennan and A. T. Proveaux. We thank C. A. Ryan, R. B. Croteau, J. Gershenzon, F. C. Schroeder, P. E. A. Teal, and T. C. J. Turlings for their constructive comments concerning the initial manuscript.

1. Tumlinson, J. H., Lewis, W. J. & Vet, L. E. M. (1993) *Sci. Am.* **268,** 100–106.
2. Paré, P. W. & Tumlinson, J. H. (1996) *Florida Entomol.* **79,** 93–103.
3. Paré, P. W. & Tumlinson, J. H (1997) *Nature (London)* **385,** 30–31.
4. McCloud, E. S. & Baldwin, I. T. (1998) *Planta* **203,** 430–435.
5. Korth, K. L & Dixon, R. A. (1997) *Plant Physiol.* **115,** 1299–1305.
6. Mattiacci, L., Dicke, M. & Posthumus, M. A. (1995) *Proc. Natl. Acad. Sci. USA* **92,** 2036–2040.
7. Alborn, H. T., Turlings, T. C. J., Jones, T. H., Stenhagen, G., Loughrin, J. H. & Tumlinson, J. H. (1997) *Science* **276,** 945- 949.
8. Turlings, T. C. J., Tumlinson, J. H. & Lewis, W. J. (1990) *Science* **250,** 1251–1253.
9. Dicke, M. (1994) *J. Plant Physiol.* **143,** 465–472.
10. Takabayashi, J. & Dicke, M. (1996) *Trends Plant Sci.* **1,** 109–113.
11. De Moraes, C. M., Lewis, W. J., Paré, P. W., Alborn, H. T. & Tumlinson, J. H. (1998) *Nature (London)* **393,** 570–572.
12. Takabayashi, J., Dicke, M. & Posthumus, M. A. (1991) *Chemoecology* **2,** 1–6.
13. Chapman, R. F. (1982) *The Insects: Structure and Function* (Harvard Univ. Press, Cambridge, MA).
14. Paré, P. W. & Tumlinson, J. H (1997) *Plant Physiol.* **114,** 1161–1167.
15. Harborne, J. B (1984) *Phytochemical Methods* (Chapman & Hall, London).
16. Mee, J. M. L. (1977) *Biomed. Mass Spectrom.* **4,** 178–182.
17. King, E. G. & Leppla, N. C. (1984) *Advances and Challenges in Insect Rearing.* (U. S. Government Printing Service, Washington, D. C).
18. Turlings, T. C. J., McCall, P. J., Alborn, H. T. & Tumlinson, J. H. (1993) *J. Chem. Ecol.* **19,** 411–425.
19. Vick, B. A. & Zimmerman, D. C. (1987) *Plant Physiol.* **85,** 1073–1078.

Plant Physiology, October 1999, Vol. 121, pp. 325–331, www.plantphysiol.org

Update on Plant-Insect Interactions

Plant Volatiles as a Defense against Insect Herbivores

Paul W. Paré[1] and James H. Tumlinson*

Center for Medical, Agricultural and Veterinary Entomology, United States Department of Agriculture, Agricultural Research Services, Gainesville, Florida 32608

Leaves normally release small quantities of volatile chemicals, but when a plant is damaged by herbivorous insects, many more volatiles are released. The chemical identity of the volatile compounds varies with the plant species and with the herbivorous insect species. These volatiles attract both parasitic and predatory insects that are natural enemies of the herbivores. They may also induce defense responses in neighboring plants. Such chemicals, which function in communication between and among species, as well as those that serve as messengers between members of the same species, are called semiochemicals (from the Greek "semeion," a mark or signal) (Law and Regnier, 1971).

Semiochemicals emitted from a diverse group of plants and insects mediate key processes in the behavior of specific insects. Volatile phytochemicals can serve as airborne semiochemicals, promoting or deterring interactions between plants and insect herbivores. For example, wheat seedlings without herbivore damage attract aphids, whereas odors released from wheat seedlings with a high density of aphids repel other aphids (Quiroz et al., 1997). For swallowtail butterflies, volatiles from host plants enhance the effect of contact stimulants, increasing landing rates and oviposition relative to non-host plants (Feeny et al., 1989).

In addition to the bouquet of compounds that render leaves attractive or disagreeable to herbivores, volatile terpenoids and other compounds emitted from leaves in response to insect damage allow insect parasitoids (such as parasitic wasps) and predators to distinguish between infested and noninfested plants, and thus aid in locating hosts or prey (Fig. 1). These phytodistress signals, which result in an active interaction between herbivore-damaged plants and a third trophic level, have been described for several agro-ecosystems. Examples include lima bean and apple plants, which produce volatiles that attract predatory mites when damaged by spider mites (Takabayashi and Dicke, 1996), and corn and cotton plants, which release volatiles that attract hymenopterous parasitoids that attack larvae of several Lepidoptera species (Tumlinson et al., 1993). In the latter case, a parasitoid female injects her eggs when she stings, and the eggs hatch into wasp larvae inside the caterpillar. Once the caterpillar has been stung, its reproductive cycle is terminated and a new generation of wasps is produced.

In all plants reported thus far, there are notable similarities in the structure of the volatile compounds that are emitted from insect-damaged leaves and from leaves distal to the site of damage. The structural uniformity in the chemical emissions from different plants with insect feeding suggests the activation of a common set of biosynthetic pathways shared by a wide range of plant families, and that the products are detectable to a broad spectrum of insect parasitoids and predators (Fig. 2). The ability of host-seeking insects to recognize and respond to such chemical cues and differentiate them from background odors indicates that insect-damaged plants emit volatile chemicals that are clearly distinguishable from those released in response to other types of damage or those released from undamaged plants. The plant's ability to differentiate between herbivore damage and a general wound response suggests the presence of elicitors associated with insect feeding that are absent from other types of leaf damage.

PLANTS RESPOND TO INSECT FEEDING DAMAGE BY RELEASING GREATER AMOUNTS OF A VARIETY OF VOLATILES

An undamaged plant maintains a baseline level of volatile metabolites that are released from the surface of the leaf and/or from accumulated storage sites in the leaf. These constitutive chemical reserves, which often include monoterpenes, sesquiterpenes, and aromatics, accumulate to high levels in specialized glands or trichomes (Paré and Tumlinson, 1997a). In addition, green-leaf odors consisting of a blend of saturated and unsaturated six-carbon alcohols, aldehydes, and esters are produced by autolytic oxidative breakdown of membrane lipids and are released when leaves are mechanically damaged. This pattern of constitutive compounds has been analyzed in the field for perennials, including beech (Tollsten and Müller, 1996) and ash (Markovic et al., 1996) trees, as well as under greenhouse conditions for many herbaceous annuals, including

[1]Present address: Department of Chemistry and Biochemistry, Texas Tech University, Lubbock, TX 79409.

* Corresponding author; e-mail jtumlinson@gainesville.usda.ufl.edu; fax 352–374–5707.

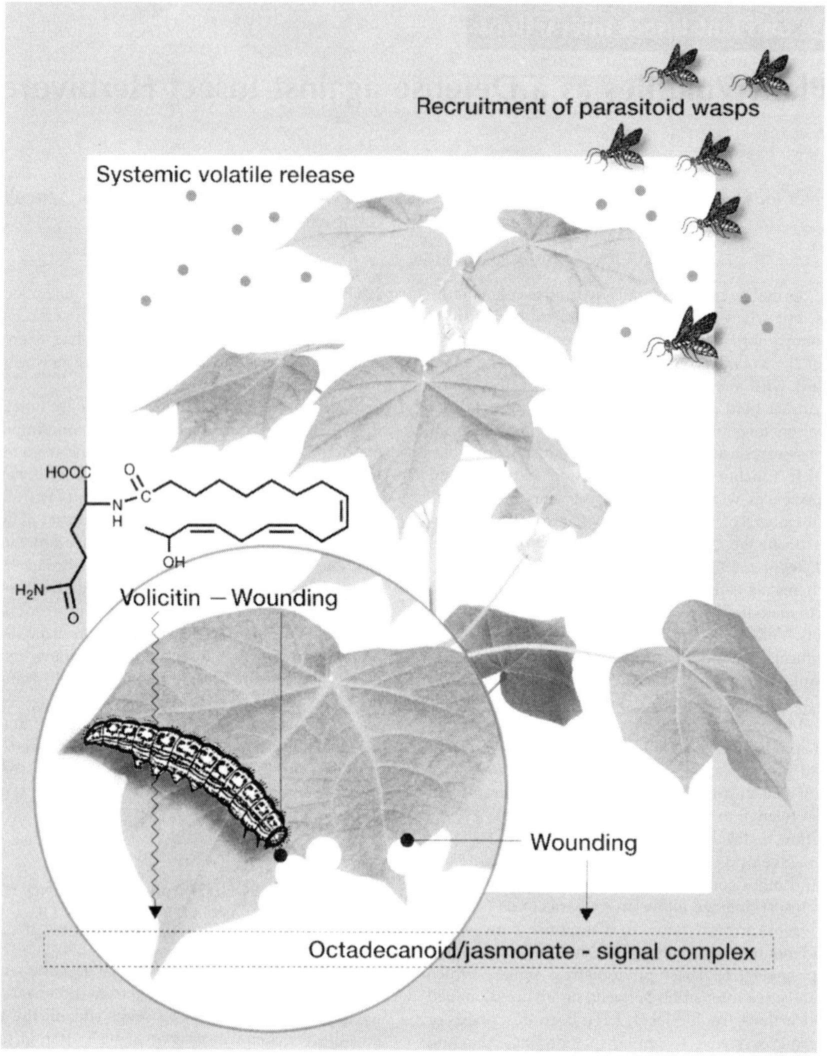

Figure 1. Schematic representation indicating an increase of volatile compounds released by plants in response to insect feeding triggered by an interaction of elicitors such as volicitin in the oral secretions of insect herbivores with damaged plant tissue. Volatile semiochemicals are then used by natural enemies of herbivores such as parasitoid wasps to locate their hosts.

brussels sprouts (Mattiacci et al., 1994) and cucumber (Takabayashi et al., 1994).

Plants respond to insect feeding damage by releasing a variety of volatiles from the damaged site, and the profile of the volatiles emitted is markedly different from those of undamaged or mechanically damaged plants. In cotton, breakage of leaf glands causes stored terpenes to be released in much higher levels, and the emissions of lipoxygenase pathway green-leaf volatiles are also increased. While the release of these metabolites correlates closely with leaf damage from insect feeding (Loughrin et al., 1994), a subset of terpenes, the nitrogen-containing com-

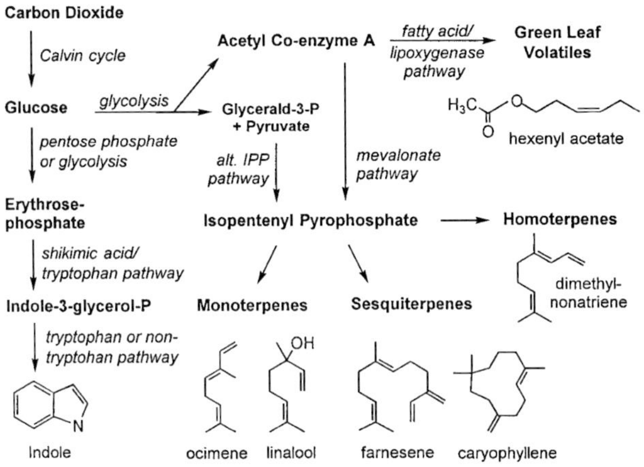

Figure 2. Biosynthetic pathways leading to the release of plant volatiles. Indole, a product of the shikimic acid pathway, is formed from indole-3-glycerol-P either as an intermediate in Trp biosynthesis or by a Trp-independent pathway leading to a family of nitrogen-containing defense compounds (e.g. 2,4-dihydroxy-7-methoxy-1,4-benzoxazin-3-one) (Frey et al., 1997). Sesquiterpenes are synthesized via the isopentenyl pyrophosphate (IPP) intermediate following the classical mevalonate pathway, whereas monoterpenes and diterpenes are synthesized via an alternative IPP pathway with glyceraldehyde-3-P and pyruvate identified as the direct precursors of IPP (Lichtenthaler et al., 1997). The mevalonate pathway is localized in the cytosol and reactions for the non-mevalonate pathway are localized in plastids. The homoterpene (*E*)-4,8-dimethyl-1,3,7-nonatriene and (*E,E*)-4,8,12-trimethyl-1,3,7,11-tridecatetraene are derived from their 15 and 20 carbon precursors, farnesyl- and geranylgeranyl-pyrophosphate, respectively, by a series of enzymatic steps with the overall loss of four carbon units (Donath and Boland, 1994). The green-leaf volatiles are derived from linolenic acid via a 13-hydroperoxylinolenic acid intermediate (Blee, 1998). This oxidized linolenic acid, instead of losing water and committing the molecule down the defense signaling jasmonic acid pathway, is cleaved to form two fragments of 12 and six carbon units (Fig. 3). The variety of green-leaf volatiles are formed from this second pathway by multiple rearrangement steps of the six-carbon (*Z*)-3-hexenal.

pound indole, and hexenyl acetate are also released in much higher levels with insect feeding, but in a diurnal cycle that is decoupled from short-term insect damage. These compounds, linalool and (*E*)-β-ocimene (monoterpenes), (*E,E*)-α-farnesene and (*E*)-β-farnesene (sesquiterpenes), nonatriene and tridecatetraene (homoterpenes), and indole and (*Z*)-3-hexenyl acetate, have an emissions profile more similar to the light cycle, with low emissions at night and high levels during the periods of maximal photosynthesis.

Chemical labeling studies have established that the compounds released in much greater quantities during the day and specifically in response to insect damage are synthesized de novo and are not stored in the plant (Paré and Tumlinson, 1997b). These induced compounds rapidly incorporate a high level of label when plants damaged by feeding caterpillars are held in volatile collection chambers under an atmosphere containing ^{13}C-CO_2. The high incorporation of ^{13}C detected by mass spectral analysis, and the rapid turnover of this label in experiments where short pulses of ^{13}C-CO_2 were used indicate that its production is tightly coupled with photosynthesis. A consistent, several-hour delay between when insect feeding begins and emission of the induced volatiles supports the hypothesis that a series of biochemical reactions, including gene expression, protein assembly, and/or enzyme induction, is required for the synthesis and release of these compounds.

RELEASE OF VOLATILES FROM UNDAMAGED LEAVES OF A DAMAGED PLANT INDICATES A SYSTEMIC SIGNAL

In addition to the release of volatiles at the site of herbivore feeding, analysis of volatile emissions from unharmed leaves of insect-damaged plants has established that there is a systemic response. In both corn (Turlings and Tumlinson, 1992) and cotton (Röse et al., 1996), leaves distal to the site of herbivore feeding showed an increase in the release of volatiles. The chemical blend of volatiles from undamaged cotton leaves differs from the volatiles collected from the entire plant. The products of the lipoxygenase pathway, including the hexenals and hexenols, which are released from freshly cut or damaged tissue, are not detected in the systemically released volatiles, with the exception of (*Z*)-3-hexenyl acetate. One explanation is that these six-carbon compounds can only be released from undamaged leaf tissue when they are converted to the acetate form (Paré and Tumlinson, 1998).

The activation of the lipoxygenase pathway in undamaged leaves suggests a mechanism analogous to that proposed by Farmer and Ryan (1992), wherein a mobile signal such as systemin can transmit information from the damaged site to distal leaves, triggering the lipoxygenase pathway and resulting in a cascade of signals activating several defense responses in plants. Some of the monoterpenes and sesquiterpenes, as well as indole and isomeric hexenyl butyrates and 2-methyl butyrates, are also only released from damaged leaves (Röse et al., 1996). The induced terpenoids that are synthesized de novo in cotton leaves in response to herbivore damage are also released systemically from undamaged leaves of a caterpillar-damaged plant. Chemical labeling experiments established that the systemic volatiles are synthesized at the site of release, suggesting that a mobile chemical messenger is transported from the damage location to distal, undamaged leaves to trigger synthesis and volatile release (Paré and Tumlinson, 1998).

Chemical labeling experiments using herbivore-damaged plants in combination with an analysis of the volatiles released has only been reported for cotton. However, since many of the compounds emitted from corn during the day have also been shown to be induced in cotton, and the quantities released increase with increased light intensity, it can be speculated that these volatiles are also synthesized de novo in corn plants. It is interesting that similar compounds are emitted in response to insect herbivore damage in several agricultural species, including cucumber, apple, lima bean, corn, and cotton (see Table I). Both among individual plants of the same species and between different plant species, whether the blend of volatile compounds is induced through a common signaling pathway or if their emissions are triggered by different signaling mechanisms is not yet known.

THE SYNTHESIS OF VOLATILES HAS A HIGH METABOLIC COST

Terpenes are an important source of olefinic compounds involved in the formation of phytotoxic products. For example, in conifers (Buchbauer et al., 1994) and broadleaf tree species (Monson and Fall, 1989), an array of terpene hydrocarbons are released from plants during times of photosynthesis. These naturally produced isoprenoids are known to form photooxidants and ozone in combination with nitrogen oxides. As a result, increased amounts of terpenes can act as pollutants, increasing the stress to the plant. The metabolic cost of these phytochemical emissions can also be high. In particular, terpenoids are more expensive to manufacture per gram than most other primary and secondary metabolites due to the need for extensive chemical reduction (Gershenzon, 1994). Defensive compound production costs in terms of reproductive success can depend on the level of herbivory. When herbivore levels are low, chemically induced wild-type tobacco plants produce fewer seeds than their noninduced counterparts. With intermediate herbivory, chemically induced plants experience less feeding on the foliage and have a higher fitness level than noninduced, insect-damaged control plants (Baldwin, 1998; Mitchell-Olds et al., 1998). It appears that volatiles need to be judiciously synthesized and safely stored, as increased synthesis can be costly and potentially toxic to the plant. However, decreases in terpene accumulation may make an individual plant more vulnerable to insect pest attacks or temperature stress.

With or without insect feeding, plants usually release a variety of terpenes during periods of high temperature. Although the biological function of terpene production is not fully understood, one proposed explanation for these emissions is that it is a strategy for responding to high temperatures (Mlot, 1995). It has been suggested that fat-soluble hydrocarbons dissolve into the thylakoid membrane and keep the chloroplast from degrading when temperatures exceed the plant's biological optimum. These hydrocarbons evaporate as the temperature rises, so that terpene volatilization cools the chloroplasts. However, since the evaporative cooling of terpenes is relatively small compared with a solvent such as water, this explanation is not universally accepted.

VOLATILES FROM INSECT-DAMAGED PLANTS ATTRACT NATURAL ENEMIES OF THE HERBIVORES

The task for a female parasitoid to locate lepidopteran caterpillar hosts would most often be unproductive if she were simply to rely on visual cues. Unlike insect pollinators seeking out well-marked flower targets, parasitoids are searching for small herbivores that are often well camouflaged and mostly inhabit the undersides of leaves. Therefore, the chances of parasitoids finding hosts by random searching are remote. Both McCall et al. (1993) and Steinberg et al. (1993) have shown by wind tunnel flights and GC analysis the weak allure and low abundance that herbivore odors alone provide for parasitoids. In contrast, the chemicals released from herbivore-damaged plants appear to contain critical chemical information that draws parasitoids to air streams spiked with these plant odors in the laboratory and to damaged plants placed among a group of undamaged neighbors in the field.

Table I. *Diverse plant species with shared volatile terpenes released in response to herbivory*

Plant	(*E*)-β-Ocimene	Linalool	(*E*)-4,8-Dimethyl-1,3,7-Nonatriene	(*E,E*)-α-Farnesene	(*E*)-β-Farnesene	(*E,E*)-4,8,12-Trimethyl-1,3,7,11-Tridecatetraene	Reference
Cucumber	+		+	+		+	Takabayashi et al. (1994)
Apple	+		+	+		+	Takabayashi et al. (1991)
Lima bean	+	+	+		+	+	Takabayashi et al. (1994)
Cotton	+	+	+	+	+	+	Paré and Tumlinson (1997a)
Corn	+	+	+	+	+	+	Turlings et al. (1990)
Tobacco	+	+		+	+		De Moraes et al. (1998)
Potato	+	+		+	+		Bolter et al. (1997)

To examine whether systemically released chemicals alone provide sufficient chemical cues to attract parasitic wasps, herbivore-damaged leaves were removed immediately before flight tests. Wind tunnel experiments showed that systemically released components were detectable at levels sufficient to direct parasitoids to their hosts (Cortesero et al., 1997). In cotton and tobacco field trials using female wasps (*Cardiochiles nigriceps*), the ratio of landings on host (tobacco budworm) damaged versus undamaged plants was high: approximately 95% to 5%, respectively, in systemic or whole-plant volatile emissions (De Moraes et al., 1998). Interestingly, this specialist parasitic wasp, using chemical cues released by the plant, can distinguish plants infested by her host *Heliothis virescens* from those infested by *Helicoverpa zea,* a closely related, non-host herbivore species. In tobacco, cotton, and maize, each plant produces a herbivore-specific blend of volatile components in response to a particular herbivore species feeding on the leaves, and these differences are observable by GC chemical analyses and detectable by parasitic wasps.

PARASITIC WASPS LEARN CHEMICAL CUES ASSOCIATED WITH HOSTS

Although the volatile compounds released by insect herbivore damage are similar among the several plant species studied thus far, the specific blends are quite distinct, varying in both the number of compounds and the actual structures produced. Thus, the task of finding a host is more complicated for the parasitoid when the host feeds on several different plant species. The wasps have overcome this obstacle by developing the ability to learn chemical cues associated with the presence of a host (Lewis and Tumlinson, 1988). The chemicals to which a female wasp is exposed during interactions with her host familiarize her with particular host location cues. A successful host experience increases the wasp's responsiveness to host-associated chemicals. For example, an oviposition experience on the plant-host complex significantly increases the oriented flight and landing responses of females of the aphid parasitoid *Aphidius ervi* relative to those that aren't allowed to sting but that are exposed to undamaged or host-damaged plants (Du et al., 1997). This underscores the importance of the oviposition experience in combination with host-damaged plant cues. Interestingly, female wasps can also learn volatile odors associated with food sources and use them to locate necessary food (Lewis and Takasu, 1990).

ENVIRONMENTAL CONDITIONS MODULATE VOLATILE EMISSIONS

Differences in the amount of volatiles released by individual plants can vary with environmental conditions that influence the plant's physiology. Several species, including corn, cotton, and lima bean, respond to reduced light (due to either lower light intensity or shorter daylength) with a decline in the release of herbivore-induced volatiles. Based on studies with lima bean, water stress also seems to directly affect volatile release (Takabayashi et al., 1994). With less water available for the plant, elevated levels of volatiles are released from infested individuals relative to non-water-stressed controls. Correlating this with insect preference showed that predatory mites selected plants that were infested and water-stressed over those that were infested but not water-stressed (Takabayashi et al., 1994). The addition of high levels of mineral and/or organic nitrogen fertilizers significantly decreased the constitutive volatiles extracted from celery (Van Wassenhove et al., 1990). With volatile analysis and flight studies for plants under different nutritional conditions, the role of these volatiles in attracting wasps to their herbivore hosts may be more clearly assigned.

ENZYMES AND ELICITORS FROM INSECT HERBIVORES TRIGGER VOLATILE RELEASE

Key to the emissions of plant signals for the foraging success of parasitoids are substances in the oral secretion of herbivores. Recent work suggests that volatile emissions and other plant defense responses are potentiated by a component or components associated with the feeding herbivore that allows the plant to differentiate between general wounding and damage due to chewing insects. In cotton, induced volatiles that are synthesized in response to wounding are released in greater quantities as a result of caterpillar feeding than as a result of mechanical damage alone (Paré and Tumlinson, 1997a). In tobacco, higher concentrations of the defense-signaling molecule jasmonic acid result from herbivore damage by hornworm caterpillars than from mechanical damage designed to mimic herbivory (McCloud and Baldwin, 1998). At the transcriptional level, potato mRNAs involved in plant defense accumulate more rapidly with insect-derived elicitor(s) in contact with the damaged leaves than with mechanical damage alone (Korth and Dixon, 1997).

Thus far, two oral secretion products from chewing insects have been identified that augment the release of plant volatiles. A β-glucosidase present in the regurgitant of *Pieris brassicae* caterpillars triggers the same emissions of volatiles in cabbage plants as induced by feeding caterpillars (Mattiacci et al., 1995). Since enzyme activity in the regurgitant is retained when caterpillars are fed a β-glucosidase-free diet, enzyme activity does not appear to be plant derived. Presumably, the enzyme acts to cleave sugars coupled to organic compounds that then become more volatile and are released. In contrast, a low-M_r fatty acid derivative, *N*-(17-hydroxylinolenoyl)-L-Gln (volicitin), has been identified from the oral secretions of beet armyworm caterpillars and induces corn seedlings to release volatile chemical signals (Alborn et al., 1997).

Analysis of volicitin from beet armyworms fed ^{13}C-labeled corn seedlings demonstrated that the caterpillar synthesizes this elicitor by adding a hydroxyl group and Gln to linolenic acid obtained directly from the plant on which the caterpillar feeds (Paré et al., 1998). Thus, although the precursor of volicitin is obtained from plants, the bioactive product has only been found in the caterpillar. This strongly suggests that these molecules play an important yet still unknown role in metabolism or some

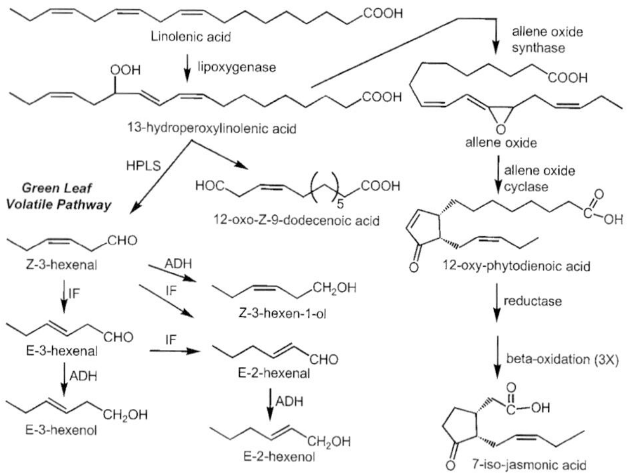

Figure 3. Select intermediates in the metabolic conversion of linolenic acid to jasmonic acid and a series of hexenyl or green-leaf volatiles catalyzed by the enzymes hydroperoxide lyase (HPLS), isomerization factor (IF), and alcohol dehydrogenase (ADH) (Blee, 1998).

other process critical to the life of the herbivorous insects. Although it is known that the plant provides linolenic acid, which is essential for most lepidopteran larvae (Stanley-Samuelson, 1994), it is seemingly detrimental to the insect to chemically convert this fatty acid into an elicitor that triggers plant defense. The full implications of this are not yet understood.

It has been suggested that jasmonic acid, which is produced from linolenic acid by the octadecanoid signaling pathway (see Fig. 3), is a key regulatory component in the transduction sequence that triggers synthesis and release of volatile compounds by plants (Krumm et al., 1995). Jasmonates also stimulate other physiological and defensive processes in plants (Farmer and Ryan, 1992), and the amino acid conjugates of jasmonic acid are involved in physiological and developmental processes in many plants (Kramell et al., 1995). Therefore, the structure of volicitin, an octadecatrienoate conjugated to an amino acid, suggests that the elicitor molecule interacts with the octadecanoid pathway in herbivore-damaged plants.

THE MECHANISMS THAT REGULATE THE SYNTHESIS AND RELEASE OF PLANT VOLATILES ARE POORLY UNDERSTOOD

There is still much to learn about the chemical interactions between plants and insect herbivores that lead to the synthesis and release of volatiles by the plants. Only one herbivore-specific volatile elicitor, volicitin, has been identified, but we know from preliminary investigations of the chemistry and activity of oral secretions of other insect herbivores that other compounds, some analogous in structure to volicitin, are also active. Furthermore, damage of a plant by different herbivore species can induce the release of volatile blends with different proportions of constituents. Thus, distinct responses are induced by elicitors of different structures from different herbivore species. However, we don't know the biochemical mechanisms by which these elicitors trigger biosynthesis and release of plant volatiles. Do they interact with the octadecanoid signaling pathway, and if so, how? Do they regulate the release of linolenic acid, the production of jasmonic acid, or the activation of the oxidative burst, all of which are associated with the wounding of plant tissue? Also, we have no knowledge of the mechanism leading to the systemic release of volatiles. Does the original, herbivore-produced elicitor serve as a mobile messenger, triggering whole-plant volatile synthesis? Or are secondary messengers employed to transmit the signal to sites distal to the site of damage? Furthermore, why do herbivores produce compounds that activate plant chemical defenses? What function, if any, do these compounds serve in herbivore metabolism or defense?

The answers to these and similar questions should lead to the development of more effective methods for the biological control of insect pests with natural enemies. It may also lead to the development of new plant varieties with enhanced chemical defenses or to methods of "inoculating" plants with elicitors to increase their resistance to insect pests.

Received May 20, 1999; accepted June 16, 1999.

LITERATURE CITED

Alborn HT, Turlings TC, Jones TH, Stenhagen G, Loughrin JH, Tumlinson JH (1997) An elicitor of plant volatiles from beet armyworm oral secretion. Science **276:** 945–949

Baldwin IT (1998) Jasmonate-induced responses are costly but benefit plants under attack in native populations. Proc Natl Acad Sci USA **95:** 8113–8118

Blee E (1998) Phytooxylipins and plant defense reactions. Prog Lipid Res **37:** 33–72

Bolter CJ, Dicke M, Van Loon JJ, Visser JH, Posthumus MA (1997) Attraction of Colorado potato beetle to herbivore-damaged plants during herbivory and after its termination. J Chem Ecol **23:** 1003–1023

Buchbauer G, Jirovetz L, Wasicky M, Nikiforov A (1994) Comparative investigation of Douglas fir headspace samples, essential oils, and extracts (needles and twigs) using GC-FID and GC-FTIR-MS. J Agric Food Chem **42:** 2852–2854

Cortesero AM, De Moraes CD, Stapel JO, Tumlinson JH, Lewis WJ (1997) Comparisons and contrasts in host-foraging strategies of two larval parasitoids with different degrees of host specificity. J Chem Ecol **23:** 1589–1606

De Moraes CM, Lewis WJ, Paré PW, Alborn HT, Tumlinson JH (1998) Herbivore-infested plants selectively attract parasitoids. Nature **393:** 570–573

Donath J, Boland W (1994) Biosynthesis of acyclic homoterpenes in higher plants parallels steroid hormone metabolism. J Plant Physiol **143:** 473–478

Du Y, Poppy GM, Powell W, Wadhams LJ (1997) Chemically mediated associative learning in the host foraging behavior of the apid parasitoid *Aphidius ervi* (Hymenoptera: Braconidae). J Insect Behav **10:** 509–522

Farmer EE, Ryan CA (1992) Octadecanoid precursors of jasmonic acid activate the synthesis of wound-inducible proteinase inhibitors. Plant Cell **4:** 129–134

Feeny P, Stadler E, Ahman I, Carter M (1989) Effects of plant odor on oviposition by the black swallowtail butterfly, *Papilio polyxenes* (Lepidoptera: Papilionidae). J Insect Behav **2:** 803–827

Frey M, Chomet P, Glawischnig E, Stettner C, Grun S, Winklmair A, Eisenreich W, Bacher A, Meeley RB, Briggs SP, Simcox K, Gierl A (1997) Analysis of a chemical plant defense mechanism in grasses. Science **277:** 696–699

Gershenzon J (1994) Metabolic costs of terpenoid accumulation in higher plants. J Chem Ecol **20:** 1281–1328

Korth KL, Dixon RA (1997) Evidence for chewing insect-specific molecular events distinct from a general wound response in leaves. Plant Physiol **115:** 1299–1305

Kramell R, Atzorn R, Schneider G (1995) Occurrence and identification of jasmonic acid and its amino conjugates induced by osmotic stress in barley leaf tissue. J Plant Growth Regul **14:** 29–36

Krumm T, Bandemer K, Boland W (1995) Induction of volatile biosynthesis in the lima bean (*Phaseolus lunatus*) by leucine- and isoleucine conjugates of 1-oxo- and 1-hydroxyindan-4-carboxylic acid: evidence for amino acid conjugates of jasmonic acid as intermediates in the octadecanoid signaling pathway. FEBS Lett **377:** 523–529

Law JH, Regnier FE (1971) Pheromones. Annu Rev Biochem **40:** 533–548

Lewis WJ, Takasu K (1990) Use of learned odours by a parasitic wasp in accordance with host and food needs. Nature **348:** 635–636

Lewis WJ, Tumlinson JH (1988) Host detection by chemically mediated associative learning in a parasitic wasp. Nature **331:** 257–259

Lichtenthaler HK, Rohmer M, Schwender J (1997) Two independent biochemical pathways for isopentenyl diphosphate and isoprenoid biosynthesis in higher plants. Physiol Plant **101:** 643–652

Loughrin JH, Manukian A, Heath RR, Turlings TC, Tumlinson JH (1994) Diurnal cycle of emission of induced volatile terpenoids by herbivore-injured cotton plants. Proc Natl Acad Sci USA **91:** 11836–11840

Markovic I, Norris DM, Phillips JK, Webster FX (1996) Volatiles involved in the nonhost rejection of *Fraxinus pennsylvanica* by *Lymantria dispar* larvae. J Agric Food Chem **44:** 929–935

Mattiacci L, Dicke M, Posthumus MA (1994) Induction of parasitoid attracting synomone in brussels sprouts plants by feeding of *Pieris brassicae* larvae: role of mechanical damage and herbivore elicitor. J Chem Ecol **20:** 2229–2247

Mattiacci L, Dicke M, Posthumus MA (1995) β-Glucosidase: an elicitor of herbivore-induced plant odor that attracts host-searching parasitic wasps. Proc Natl Acad Sci USA **92:** 2036–2040

McCall PJ, Turlings TC, Lewis WJ, Tumlinson JH (1993) Role of plant volatiles in host location by the specialist parasitoid *Microplitis croceipes* cresson (Braconidae: Hymenoptera). J Insect Behav **6:** 625–639

McCloud ES, Baldwin IT (1998) Herbivory and caterpillar regurgitants amplify the wound-induced increases in jasmonic acid but not nicotine in *Nicotiana sylvestris*. Planta **203:** 430–435

Mitchell-Olds T, Gershenzon J, Baldwin I, Boland W (1998) Chemical ecology in the molecular era. Trends Plant Sci **3:** 362–365

Mlot C (1995) A clearer view of why plants make haze. Science **268:** 641–642

Monson RK, Fall R (1989) Isoprene emission from aspen leaves. Plant Physiol **90:** 267–274

Paré PW, Alborn HT, Tumlinson JH (1998) Concerted biosynthesis of an insect elicitor of plant volatiles. Proc Natl Acad Sci USA **95:** 13971–13975

Paré PW, Tumlinson JH (1997a) De novo biosynthesis of volatiles induced by insect herbivory in cotton plants. Plant Physiol **114:** 1161–1167

Paré PW, Tumlinson JH (1997b) Induced synthesis of plant volatiles. Nature **38:** 30–31

Paré PW, Tumlinson JH (1998) Cotton volatiles synthesized and released distal to the site of insect damage. Phytochemistry **47:** 521–526

Quiroz A, Pettersson J, Pickett JA, Wadhams LJ, Niemeyer HM (1997) Semiochemicals mediating spacing behavior of bird cherry-oat aphid, *Rhopalosiphum padi,* feeding on cereals. J Chem Ecol **23:** 2599–2607

Röse US, Manukian A, Heath RR, Tumlinson JH (1996) Volatile semiochemicals released from undamaged cotton leaves: a systemic response of living plants to caterpillar damage. Plant Physiol **111:** 487–495

Stanley-Samuelson DW (1994) Prostaglandins and related eicosanoids in insects. Adv Insect Physiol **24:** 115–212

Steinberg S, Dicke M, Vet LE (1993) Relative importance of infochemicals from first and second trophic level in long-range host location by the larval parasitoid *Cotesia glomerata*. J Chem Ecol **19:** 47–59

Takabayashi J, Dicke M (1996) Plant-carnivore mutualism through herbivore-induced carnivore attractants. Trends Plant Sci **1:** 109–113

Takabayashi J, Dicke M, Posthumus MA (1991) Variation in composition of predator-attracting allelochemicals emitted by herbivore-infested plants: relative influence of plant and herbivore. Chemoecology **2:** 1–6

Takabayashi J, Dicke M, Posthumus MA (1994) Volatile herbivore-induced terpenoids in plant-mite interactions: variation caused by biotic and abiotic factors. J Chem Ecol **20:** 1329–1354

Tollsten L, Müller PM (1996) Volatile organic compounds emitted from beech leaves. Phytochemistry **43:** 759–762

Tumlinson JH, Lewis WJ, Vet LEM (1993) How parasitic wasps find their hosts. Sci Am **268:** 100–106

Turlings TC, Tumlinson JH (1992) Systemic release of chemical signals by herbivore-injured corn. Proc Natl Acad Sci USA **89:** 8399–8402

Turlings TC, Tumlinson JH, Lewis WJ (1990) Exploitation of herbivore-induced plant odors by host-seeking parasitic wasps. Science **250:** 1251–1253

Van Wassenhove FA, Dirinck PJ, Schamp NM, Vulsteke GA (1990) Effect of nitrogen fertilizers on celery volatiles. J Agric Food Chem **38:** 220–226

Caterpillar-induced nocturnal plant volatiles repel conspecific females

Consuelo M. De Moraes*, Mark C. Mescher† & James H. Tumlinson*

* *USDA-ARS, CMAVE, PO Box 14565, Gainesville, Florida 32604, USA*
† *Department of Entomology, University of Georgia, Athens, Georgia 30602, USA*

Plants respond to insect herbivory by synthesizing and releasing complex blends of volatile compounds, which provide important host-location cues for insects that are natural enemies of herbivores[1–3]. The effects of these volatile blends on herbivore behaviour have been investigated to only a limited extent[4,5], in part because of the assumption that herbivore-induced volatile emissions occur mainly during the light phase of the photoperiod[6,7]. Because many moths—whose larvae are some of the most important insect herbivores—are nocturnal, herbivore-induced plant volatiles have not hitherto been considered to be temporally available as host-location cues for ovipositing females. Here we present chemical and behavioural assays showing that tobacco plants (*Nicotiana tabacum*) release herbivore-induced volatiles during both night and day. Moreover, several volatile compounds are released exclusively at night and are highly repellent to female moths (*Heliothis virescens*). The demonstration that tobacco plants release temporally different volatile blends and that lepidopteran herbivores use induced plant signals released during the dark phase to choose sites for oviposition adds a new dimension to our understanding of the role of chemical cues in mediating tritrophic interactions.

Feeding by insect herbivores induces plants to release chemical signals that serve as important foraging cues for parasitoids and predators, and thus enhance the plants' defence[1–3,8–10]. Synthesis and release of these chemical signals is an active physiological process triggered by substances in the oral secretion of herbivores[11,12]. The recent discovery that plant volatiles can transmit herbivore-specific information that allows natural enemies to identify particular herbivore species demonstrated that chemically mediated plant–insect interactions are more sophisticated and complex than was previously appreciated[13]. However, the role of chemical signals in plant–herbivore interactions remains largely unexplored. Some researchers have examined the effects of constitutive plant volatiles and herbivore-induced daytime volatiles on conspecific herbivores[4,14–17] including some lepidopterans[18], but the effect of herbivore-induced plant volatiles on moths that are active at night has been neglected. The fact that several major terpene components of herbivore-induced plant volatiles have high emissions during the periods of maximal photosynthesis[6,7] may explain why little attention has been paid to the importance of these volatiles to female moths searching for oviposition sites at night. To our knowledge, this study represents the first demonstration that plants emit herbivore-induced volatile blends that exhibit systematic temporal variation, that some volatile compounds are released exclusively at night, and that female moths exploit these specific night-time signals to avoid oviposition on previously damaged plants.

Gas chromatographic analysis of volatiles collected in two-hour intervals continuously for seven days revealed consistent differences in the composition of volatile blends released by *H. virescens*-infested tobacco plants ($n = 6$) during the light and dark phases of the photoperiod (Fig. 1a). Visual and auditory observations confirmed that larvae fed during both the light and dark phases. Seven major compounds were consistently released during both light and dark phases, but usually in lesser amounts during the dark phase (Fig. 1a). In addition, five compounds ((*Z*)-3-hexenyl butyrate, (*Z*)-3-hexenyl isobutyrate, (*Z*)-3-hexenyl acetate, (*Z*)-3-hexenyl tiglate, and one unidentified compound) were produced only during the dark phase. Others—(*E*)-2-hexenal and three unidentified compounds—were produced in significantly larger amounts during the dark than the light period. Thus, the qualitative and quantitative composition of volatile blends emitted by tobacco plants in response to feeding by *H. virescens* larvae can differ significantly between night and day.

Repetition of our analysis using two other species of lepidopteran larvae ($n = 6$ per species), *Manduca sexta* and *Helicoverpa zea*, provided further evidence of induced volatile release from tobacco plants during the dark period (Fig. 1b). Although the volatile

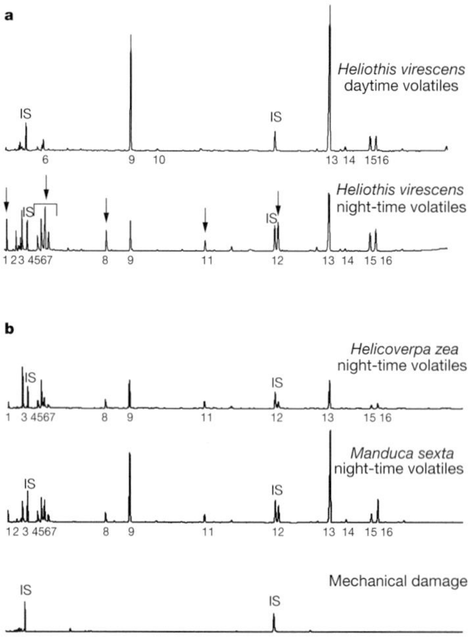

Figure 1 Gas chromatographic analysis of induced plant volatiles. **a**, Diurnal and nocturnal profiles of volatiles released from tobacco plants during a 2-h interval after 48 h of feeding by *H. virescens*. Arrows represent volatiles that are present only (or in significantly larger amounts) in the nocturnal profile. **b**, Nocturnal profiles of volatiles released from tobacco plants during a 2-h interval after 48 h of feeding by *H. virescens*, *H. zea* or *M. sexta* compared with mechanical damage. Represented are: 4, (*E*)-2-hexenal; 5, (*Z*)-3-hexen-1-ol; 8, (*Z*)-3-hexenyl acetate; 9, (*E*)-β-ocimene; 10, linalool; 11, (*Z*)-3-hexenyl butyrate; 12, (*Z*)-3-hexenyl tiglate; 13, β-caryophyllene; 14, α-humulene; 15, (*E,E*)-α-farnesene; 16, unidentified sesquiterpene; compounds 1–3, 6, 7 are unidentified compounds; IS, internal standards (*n*-octane and *n*-nonyl-acetate).

profiles revealed quantitative differences in plant response to the three herbivores, all caterpillar species induced release of the same volatile compounds from tobacco during the night (Fig. 1). To determine whether plants continue nocturnal volatile production in the absence of continuous feeding by the caterpillars, *H. virescens* were removed from the plants at 15:00 after 48 hours of feeding. In this case, plants still emitted volatiles during the dark phase although in smaller amounts.

In behavioural trials, mated *H. virescens* females spent a significantly greater proportion of time (80% of the observed hour) in an area with only undamaged plants than in an area with both damaged and undamaged plants (Fig. 2A, a). Moths exhibited a tendency to fly south when first released; however, if the damaged plants were placed to the south, the moths would often turn and fly in the opposite direction. The same preference was displayed in oviposition. Females selected only non-infested plants for oviposition (Fig. 2A, b; $F_{1,10} = 427$, $P < 0.0001$) and also avoided uninfested plants close to infested plants.

Because the preference of *H. virescens* for uninfested plants might conceivably be explained by visual cues associated with plant damage, we repeated the behavioural assays using synthetic volatile blends. Synthetic blends of the major volatile compounds that were produced in significant amounts and emitted by the plants during the dark period were formulated on rubber septa[19,20] to release volatiles in approximately the same proportions and amounts as released by herbivore-damaged plants. In behavioural assays, *H. virescens* demonstrated a significant preference ($F_{1,10} = 134$, $P < 0.0001$) for untreated tobacco plants over those with septa releasing the synthetic blends (Fig. 2B, b). In control trials, the response to plants with septa containing only solvent (hexane) was statistically indistinguishable ($F_{1,10} = 1.91$, $P > 0.05$) from that to untreated plants (Fig. 2B, a). Thus, female *H. virescens* are able to identify infested plants on the basis of chemical cues consisting of volatile compounds emitted by the plants at night.

Some of the volatiles in the nocturnal volatile profile were also present in the diurnal profile. However, others were released only at night. To determine the importance of volatiles produced at night, relative to those produced during the day, we repeated the behavioural assays using the diurnal blends. Although these daytime volatiles produced avoidance behaviour (Fig. 2B, c; $F_{1,10} = 55$, $P < 0.001$), they were significantly less repellent than the nocturnal volatiles (Fig. 2B, b), indicating that the moths are specifically repelled by the night-time volatile blends. To further examine the extent to which the repellence of the night-time volatiles is due to the presence of compounds produced exclusively at night, the experiment was repeated using a synthetic blend containing only these compounds. The presence of these exclusively nocturnal compounds alone was sufficient to explain the moth repellence effect (Fig. 2B, d; $F_{1,10} = 145$, $P < 0.0001$).

It is well established that induced plant volatiles function as signals between plants and the natural enemies of insect herbivores. The discovery that the herbivores themselves exploit the information present in night-time volatile blends to avoid oviposition on previously damaged plants reveals a new dimension of chemically

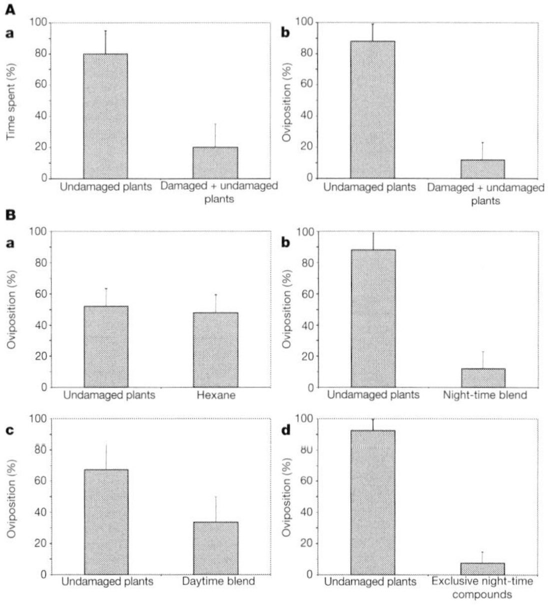

Figure 2 Response of female moths to plant volatiles. **A**, Nocturnal flight responses of three mated *H. virescens* females to tobacco plants. On one side of the cage, only undamaged plants (6) were used. On the other side, two plants that had been infested by ten third-instar *H. virescens* larvae were included. Larvae were removed from the plants before moths were introduced. **a**, Per cent of time during one hour of observation spent on each side of the cage, **b**, per cent of eggs oviposited on each side of the cage. **B**, Oviposition preference by *H. virescens* females for tobacco plants that had been treated with synthetic blend. On one side of the cage only undamaged plants (6) were used, on the other side two plants were treated (synthetic blends on rubber septa) placed among four undamaged and untreated plants. **a**, Solvent (hexane) used as control. **b**, A nocturnal blend compared to undamaged plants. **c**, A diurnal blend compared to undamaged plants. **d**, A blend containing compounds released exclusively at night compared to undamaged plants.

mediated plant–insect interactions and raises a number of issues. The fitness advantages to herbivores of avoiding oviposition on induced plants are obvious, as such plants are likely to host not only larvae that represent potential competitors for the moth's offspring but also potentially a population of natural enemies attracted by the volatile blend[21]. Moreover, the associated induction of direct defence mechanisms means that infested plants are likely to contain chemical toxins and to be of lower nutritional value than uninfested plants[22,23]. It is less clear whether plants benefit significantly from the release of nocturnal volatiles or whether such release is merely a physiological by-product associated with diurnal volatile production. The protection enjoyed by undamaged plants that reside near induced plants is interesting, but it is not certain whether this phenomenon would have been significant in the ancestral environment or represents only an effect of high-density agricultural cultivation. If plants do benefit from advertising their status to herbivores, it raises the question of whether herbivore-induced signals first evolved as parasitoid attractants or as herbivore repellents; alternatively, the dual functions of herbivore-induced plant signals may have evolved simultaneously.

The recognition that plant volatile signals, long known to be important in the mediation of plant–parasitoid interactions, also transmit information to herbivores expands our view of tritrophic systems and is significant with regard to our understanding of the selective pressures governing the evolution of such signals. We are currently exploring whether moths can interpret the herbivore-specific information that these signals convey to parasitoids[13] or rather use these signals only as generalized indicators of insect feeding. If moths can interpret the higher-order information content of these signals, then they may be making sophisticated choices based on the likely presence of particular larval competitors and perhaps even of particular predators and parasitoids. □

Methods

Volatile collection and analysis

All experiments were conducted in an insect-free greenhouse (temperature 29 ± 4 °C). Ten third-instar larvae of either *H. virescens*, *H. zea* or *M. sexta* were allowed to feed continuously on the leaves of an eight-week-old, greenhouse-grown tobacco plant (*Nicotiana tabacum* strain K326) enclosed in a volatile collection chamber[24], beginning the night before collection commenced.

Plant volatiles were collected 24-h a day in 2-h intervals by pulling 1 l of the air passing over the plant (5 l min^{-1}) through Super Q adsorbent (25 mg) traps at the base of the volatile-collection chamber; the remainder of the air vented out of the bottom of the system[24]. Traps were rinsed with 150 μl methylene chloride, 400 ng of *n*-octane and nonyl acetate were added as internal standards, and samples were analysed by gas chromatography and mass spectrometry[13] (electron impact and chemical ionization with isobutane reagent gas). Volatile compounds were identified by comparison of chromatographic retention times and mass spectra with those of commercially available standards analysed on the same instruments. Quantification was based on peak area (flame ionization detector) relative to that of internal standards.

Average volatile released by plants from 11:00 to 13:00, 50 h after initial damage: (ng h^{-1}, s.d.): (*E*)-2-hexenal (190.93, 23.06); (*Z*)-3-hexen-1-ol (62.04, 40.08); (*E*)-β-ocimene (5510.46, 789.72); indole (212.67, 45.03); β-caryophyllene (7481.88, 823.40); α-humulene (221.88, 13.08); (*E,E*)-α-farnesene (1220.48, 152.81), unidentified sesquiterpene (1148.25,165.85).

Average volatile released by plants from 19:00 to 21:00, 58 h after initial damage: (ng h^{-1}, s.d.): (*E*)-2-hexenal (488.44, 36.78); (*Z*)-3-hexen-1-ol (792.70, 160.12); (*Z*)-3-hexenyl acetate (987.99, 97.12); (*E*)-β-ocimene (1610.46/349.32); (*Z*)-3-hexenyl isobutyrate (123.84, 23.76); (*Z*)-3-hexenyl butyrate (488.85, 121.67) indole (212.67, 45.03); (*Z*)-3-hexenyl tiglate (1328.87, 198.76), β-caryophyllene (2452.88, 223.46); α-humulene (83.48, 35.17); (*E,E*)-α-farnesene (888.88, 154.98), unidentified sesquiterpene (960.59, 200.91).

Behavioural assays

Twelve eight-week-old, potted, greenhouse-grown tobacco plants were placed in a screened cage (4 × 2 × 3 m) with six plants on each side of the cage (80 cm between plants). On one side only undamaged plants were used, on the other side two of the plants had been fed on by 10 laboratory-reared third-instar *H. virescens* (two larvae per leaf) for 48 h. Larvae were removed from plants before moth release. Three mated *H. virescens* females (18 females per treatment) were released in the central sector of the cage at dusk. To account for a possible directional preference by the moths, two cages were used so in one the damaged plants were on the north end and in the other they were on the south end of the cage. The moths' behaviour was visually observed for one hour (from approximately 19:00 to 20:00) and the time (min.) spent by moths on each side of the cage was recorded. Egg numbers per plant on each side of the cage were counted at 06:00 the next morning. Similar procedures were used for assays measuring oviposition and behaviour of female moths in response to plants with or adjacent to synthetic blends, but in this case plants on both sides of the cages were undamaged. For these treatments, pots of undamaged plants with volatile-releasing rubber septa on wooden sticks (three per plant) replaced the damaged plants. We evaluated the response of moths to three synthetic volatile blends (nocturnal volatiles, exclusively nocturnal volatiles, and diurnal volatiles). Rubber septa without the synthetic blend were placed on the control side to neutralize any visual preference. Each bioassay was conducted on three days (two repetitions each) to account for day-to-day variation. Preference between the undamaged and damaged plants by the herbivores was subjected to analysis of variance (an arc-sine square root transformation of the percentage data was used; SAS Institute, Cary, North Carolina).

Synthetic blends

Synthetic blends were formulated according to a method developed to predict the release ratio of volatile compounds from rubber septa[19,20]. Calculations of predicted release ratios are based on relative vapour pressures[19] of the components and the original amounts released in the natural blend. Blends were dissolved in hexane and 0.3 ml of a blend solution was pipetted onto a rubber septum. Volatiles released by the blend when formulated on rubber septa were sampled and analysed. The relative amounts were adjusted to correct for slight deviations from the predicted amount. Compounds used to prepare synthetic blends were obtained from Sigma-Aldrich or from Chemical Samples and analysed by gas chromatography and mass spectrometry to determine purity. All synthetic compounds were at least 98% pure except for ocimene where both isomers were present (60% *trans* and 40% *cis*).

Release rates of synthetic blends: daytime blend 1, (*E*)-2-hexenal (200 ng h^{-1}); 2, (*E*)-β-ocimene (5,500 ng h^{-1}); 3, β-caryophyllene (7,300 ng h^{-1}); 4, α-humulene (200 ng h^{-1}); 5, (*E,E*)-α-farnesene (1,180 ng h^{-1}). Night-time blend 1, (*E*)-2-hexenal (500 ng h^{-1}); 2, (*Z*)-3-hexen-1-ol (850 ng h^{-1}); 3, (*Z*)-3-hexenyl acetate (1,000 ng h^{-1}); 4, (*E*)-β-ocimene (1,500 ng h^{-1}); 5, (*Z*)-3-hexenyl isobutyrate (100 ng h^{-1}); 6, (*Z*)-3-hexenyl butyrate (500 ng h^{-1}); 7, (*Z*)-3 hexenyl tiglate (1,200 ng h^{-1}); 8, β-caryophyllene (2,200 ng h^{-1}); 9, (*E,E*)-α-farnesene (900 ng h^{-1}). Exclusive night-time blend 1, (*Z*)-3-hexen-1-ol (850 ng h^{-1}); 2, (*Z*)-3-hexenyl acetate (1,000 ng h^{-1}); 3, (*Z*)-3-hexenyl isobutyrate (100 ng h^{-1}); 4, (*Z*)-3-hexenyl butyrate (500 ng h^{-1}); 5, (*Z*)-3-hexenyl tiglate (1,200 ng h^{-1}).

Received 19 December 2000; accepted 31 January 2001.

1. Dicke, M. Local and systemic production of volatile herbivore-induced terpenoids: their role in plant-carnivore mutualism. *J. Plant Physiol.* **143,** 465–472 (1994).
2. Turlings, T. C. J., Tumlinson, J. H. & Lewis, W. J. Exploitation of herbivore-induced plant odors by host-seeking parasitic wasps. *Science* **250,** 1251–1253 (1990).
3. McCall, P. J., Turlings, T. C. J., Lewis, W. J. & Tumlinson, J. H. Role of plant volatiles in host location by the specialist parasitoid *Microplitis croceipes* Cresson (Braconidae:Hymenoptera). *J. Insect Behav.* **6,** 625–639 (1993).
4. Pallini, A., Jansen, A. & Sabelis, M. Odour-mediated responses of phytophagous mites to conspecific and heterospecific competitors. *Oecologia* **110,** 179–185 (1997).
5. Karban, R. & Baldwin, I. T. *Induced Responses to Herbivory* (Univ. Chicago Press, Chicago, Illinois, 1997).
6. Paré, P. W. & Tumlinson, J. H. Induced synthesis of plant volatiles. *Nature* **385,** 30–31 (1997).
7. Loughrin, J. H., Manukian, A., Heath, R. R. & Tumlinson, J. H. Diurnal cycle of emission of induced volatile terpenoids by herbivore-injured cotton plants. *J. Chem. Ecol.* **21,** 1217–1227 (1994).
8. Turlings, T. C. J. *et al.* How caterpillar-damaged plants protect themselves by attracting parasitic wasps. *Proc. Natl Acad. Sci. USA* **92,** 4169–4174 (1995).
9. Karban, R. Resistance against spider mites in cotton induced by mechanical abrasion. *Entomol. Exp. Appl.* **37,** 137–141 (1985).
10. Dicke, M. *et al.* Isolation and identification of volatile kairomone that affects acarine predator-prey interactions. Involvement of host plant in its production. *J. Chem. Ecol.* **16,** 381–396 (1990).
11. Mattiacci, L., Dicke, M. & Posthumus, M. A. Induction of parasitoid attracting synomone in brussels sprouts plants by feeding of *Pieris brassicae* larvae: Role of mechanical damage and herbivore elicitor. *J. Chem. Ecol.* **20,** 2229–2247 (1994).
12. Alborn, H. T. *et al.* An elicitor of plant volatiles from beet armyworm oral secretion. *Science* **276,** 945–948 (1997).
13. De Moraes, C. M. *et al.* Herbivore-infested plants selectively attract parasitoids. *Nature* **393,** 570–573 (1998).
14. Dicke, M. Volatile spider-mite pheromone and host plant kairomone, involved in spaced-out gregariousness in spider mite *Tetranychus urticae*. *Physiol. Entomol.* **11,** 251–262 (1986).
15. Schultz, S. *et al.* Host plant selection of the Colorado Potato Beetle as influenced by damage induced volatiles of the potato plant. *Naturwissenschaften* **84,** 212–217 (1997).
16. Harari, A. R. Ben-Yakir, D. & Rosen, D. Mechanisms of aggregation behavior in *Maladera matrida* Argaman (Coleoptera: Scarabeidae). *J. Chem. Ecol.* **20,** 361–371 (1994).
17. Landolt, P. J., Tumlinson, J. H. & Alborn, D. H. Attraction of Colorado potato beetle (Coleoptera : Chrysomelidae) to damaged and chemically induced potato plants. *Environ. Entomol.* **28,** 973–978 (1999).
18. Landolt, P. J. Effects of host plant leaf damage on cabbage looper moth attraction and oviposition. *Entomol. Exp. Appl.* **67,** 79–85 (1993).
19. Heath, R. R. & Tumlinson, J. H. Prediction of release ratios of multicomponent pheromones from rubber septa. *J. Chem. Ecol.* **12,** 2081–2088 (1986).
20. Heath, R. R., Teal, P. E. A., Tumlinson, J. H. & Mengelkoch, L. J. Correlation of retention times on liquid crystal capillary column with reported vapor pressures and half-lives of compounds used in pheromone formulations. *J. Chem. Ecol.* **12,** 2133–2143 (1986).
21. Thaler, J. S. Jasmonate-inducible plant defenses cause increased parasitism of herbivores. *Nature* **399,** 686–588 (1999).

22. Agrawal, A. A. Induced responses to herbivory and increased plant performance. *Science* **279,** 1201–1202 (1998).
23. Coley, P. D., Bryant, J. P. & Chapin, F. S. Resource availability and plant antiherbivore defense. *Science* **230,** 895–899 (1985).
24. Heath, R. R. & Manukian, A. An automated-system for use in collecting volatile chemicals released from plants. *J. Chem. Ecol.* **20,** 593–608 (1994).

Acknowledgements

We thank G. W. G. De Moraes for discussions; H. T. Alborn, J. G. Hildebrand, P. J. Landolt, W. J. Lewis, K. G. Ross and J. R. Ruberson for comments on the manuscript; and B. Dueben and M. Sammons for technical assistance.

Correspondence and requests for materials should be addressed to J.H.T. (e-mail: jtumlinson@gainesville.usda.ufl.edu).

Airborne signals prime plants against insect herbivore attack

Juergen Engelberth*, Hans T. Alborn, Eric A. Schmelz, and James H. Tumlinson*†

Center for Medical, Agricultural, and Veterinary Entomology, U.S. Department of Agriculture Agricultural Research Service, 1700 SW 23rd Drive, Gainesville, FL 32608

Contributed by James H. Tumlinson, December 4, 2003

Green leafy volatiles (GLV), six-carbon aldehydes, alcohols, and esters commonly emitted by plants in response to mechanical damage or herbivory, induced intact undamaged corn seedlings to rapidly produce jasmonic acid (JA) and emit sesquiterpenes. More importantly, corn seedlings previously exposed to GLV from neighboring plants produced significantly more JA and volatile sesquiterpenes when mechanically damaged and induced with caterpillar regurgitant than seedlings not exposed to GLV. The use of pure synthetic chemicals revealed that (*Z*)-3-hexenal, (*Z*)-3-hexen-1-ol, and (*Z*)-3-hexenyl acetate have nearly identical priming activity. Caterpillar-induced nocturnal volatiles, which are enriched in GLV, also exhibited a strong priming effect, inducing production of larger amounts of JA and release of greater quantities of volatile organic compounds after caterpillar regurgitant application. In contrast, GLV priming did not affect JA production induced by mechanical wounding alone. Thus, GLV specifically prime neighboring plants against impending herbivory by enhancing inducible chemical defense responses triggered during attack and may play a key role in plant–plant signaling and plant–insect interactions.

Plant defenses against herbivorous insects include both chemical and physical mechanisms that directly affect the performance of the herbivores (1). Additionally, a countermeasure against damaging insect herbivores that may be even more effective is the release of volatile organic compounds (VOC), consisting mainly of products of the shikimic acid-pathway, fatty acid-derived products and terpenes, which attract parasitoids and predators, natural enemies of the actively feeding arthropods (2, 3). Herbivore-induced VOC have also been shown to decrease oviposition rates and increase egg predation on the emitting plant in nature (4, 5). Although insect-induced VOC can serve direct and indirect defense functions, neighboring (or receiver) plants may also perceive and respond to these signals. Pathogen infection of plants often leads to the release of methyl salicylate (6), which serves as a potential signal for the induction of defense-related genes in neighboring plants. Also, after wounding and herbivore damage some plants emit methyl jasmonate (MeJA), which has been shown to effectively turn on defense genes (7). However, not all plants release MeJA, calling into question its role as a general volatile defense signal. Green leafy volatiles (GLV) consist mainly of degradation products derived from C_{18} fatty acids (linolenic and linoleic acid), which, after being transformed to a hydroperoxide by a lipoxygenase, are cleaved into C_{12} and C_6 components by hydroperoxide lyase (HPL). Depending on the C_{18}-substrate, HPL produces either (*Z*)-3-hexenal [(*Z*)-3-HAL)] or hexanal (8). Further processing by alcohol dehydrogenase, acetylation, and isomerization leads to the production of the remaining C_6-components, like (*Z*)-3-hexenol [(*Z*)-3-HOL)], (*Z*)-3-hexenyl acetate [(*Z*)-3-HAC], and the respective E-isomers. The C_{12}-component is processed to traumatin, which has long been hypothesized to play an important role in the wound response of plants (9). GLV are typically released locally by plants immediately after wounding or herbivore damage (8) but can also be induced and released systemically (10). Previous studies indicated that GLV induce certain defense-related genes (11–13). However, although plants treated with (E)-2-hexenal released significantly greater quantities of VOC than control plants, they released significantly less than plants treated with MeJA or damaged by insect herbivores (13). Also, in other studies, treatment of plants with six-carbon aldehydes induced less than the complete set of defense-related genes and resulted in a moderate plant response relative to MeJA at both the physiological and molecular levels (11, 12). Two important questions arose from these findings: Which signaling pathways are involved, and is the immediate but rather moderate activation of plant defense responses the main function of this signaling? To answer these questions, we have started a comprehensive investigation of the effects of naturally released GLV on neighboring plants. We used GLV from wounded plant tissue, caterpillar-induced night-time volatiles enriched in GLV, and pure C_6 compounds to examine their effects on intact corn plants. We discovered not only that GLV stimulate transient jasmonic acid (JA) biosynthesis and VOC release in corn, but also that exposure to GLV primed corn plant defenses to respond more strongly against subsequent attack by herbivorous insects by increasing JA biosynthesis and VOC release.

Materials and Methods

Plant and Insect Material. Corn (*Zea mays* cv. Delprim) was grown as reported (14). Beet armyworm (BAW, *Spodoptera exigua*) eggs were obtained from W. J. Lewis (Insect Biology and Population Management Research Laboratory, U.S. Department of Agriculture, Agricultural Research Service, Tifton, GA) and reared on an artificial diet based on pinto beans (15). Late first and early second instar larvae were selected for the induction of nocturnal volatiles.

Preparation of Crude Regurgitant Elicitor (CRE) from Larvae of BAW. BAW were transferred to feed on corn seedlings at least 48 h before collection of regurgitant. Regurgitation was induced by holding fourth instar BAW caterpillars with forceps and gently pinching behind the head with a second pair. The regurgitant from 40–50 caterpillars was collected, boiled for 5 min to inactivate degrading enzymes (16), centrifuged to remove cell debris and denatured proteins, and the supernatant diluted 1:1 in buffer (50 mM sodium phosphate, pH 8) before use (referred to as CRE).

Chemicals. (*Z*)-3-HAL (50% in triacetin), (*Z*)-3-HOL (98% pure), cis jasmone (85% pure), methyl salicylate, and MeJA were purchased from Sigma–Aldrich. (*Z*)-3-HAC was made from (*Z*)-3-HOL by acetylation with acetyl chloride and purity estimated by GC and GC/MS analysis (95% pure). Dihydro MeJA (Bedoukian Research, Danbury, CT) was converted to dihydro

Abbreviations: JA, jasmonic acid; SA, salicylic acid; GLV, green leafy volatiles; CIV, caterpillar-induced volatiles; VOC, volatile organic compounds; (*Z*)-3-HAL, (*Z*)-3-hexenal; (*Z*)-3-HOL, (*Z*)-3-hexenol; (*Z*)-3-HAC, (*Z*)-3-hexenyl acetate; MeJA, methyl jasmonate; CRE, crude regurgitant elicitor; FW, fresh weight.

*Present address: Department of Entomology, Pennsylvania State University, Pesticide Research Laboratory, University Park, PA 16802.

†To whom correspondence should be addressed. E-mail: jht2@psu.edu.

JA (DhJA) by alkaline hydrolysis. [$^2H_6$]salicylic acid (SA) was purchased from CDN Isotopes (Pointe-Claire, QC, Canada). All solvents used were analytical grade.

Plant Treatments. Two different sets of experiments were conducted, one to measure the direct effect of GLV on JA, SA, and VOC production in plants, and the second to determine the effect of GLV treatment on subsequent plant response to wounding and treatment with caterpillar regurgitant. To measure the short-term production of JA and SA, intact corn plants (receiver plants) were exposed to various volatiles in 6-l Plexiglas cylinders. After 0, 30, 60, and 180 min, the intact receiver plants were removed from the chamber and the leaves frozen in liquid nitrogen for further analysis. For short-term exposure to GLV vapors, 2 g of cut-leaf material from 2- to 3-wk-old corn plants was placed on a small dish and added to the chamber with the receiver corn plant. Control plants were held in the chambers for equal periods of time in ambient air. For induction with synthetic compounds, 20 μg of (*Z*)-3-HAL, (*Z*)-3-HOL, (*Z*)-3-HAC, or cis jasmone (17) (dissolved in dichloromethane, 1 μg/μl), was pipetted onto a cotton ball in the Plexiglas cylinder. Controls consisted of a plant in a chamber with 20 μl of pure dichloromethane or 20 μg of triacetin in dichloromethane [for treatments with (*Z*)-3-HAL)] on a cotton ball. For short-term exposure to caterpillar-induced volatiles (CIV) emitted from neighboring infested plants, 2-wk-old intact corn plants held in 200-ml glass tubes were infested with 20–25 BAW caterpillars (late first to early second instar) in the morning. After 5 h, each glass tube with a caterpillar-infested corn plant (source plant) was connected by Teflon tubing (i.d. 5 mm) to a 6-liter Plexiglas cylinder containing an intact receiver plant. A vacuum was attached to each Plexiglas cylinder, and fresh charcoal-purified air was pulled at ≈200 ml/min over the infested plant and then through the Plexiglas cylinder to allow the entrained nocturnal volatiles to flow over the intact plants. As a control, uninfested corn seedlings were used as source plants.

A second control was performed by using CRE-induced corn plants as a source for VOC. Plants were induced through application of 5 μl of CRE on one wounded site on each leaf (three leaves total) and then transferred to the 200-ml (vol) glass tubes. After 5 h, the tubes were connected to the 6-l Plexiglas cylinders with the receiver plants as described above for CIV.

For overnight exposure to GLV, cut corn leaf material was added to the 200-ml glass cylinder, and GLV was drawn over the receiver plants in 6-liter cylinders at 200 ml/min, as described above. In the controls, the 200-ml cylinders were empty. For induction with synthetic compounds, 20 μg of (*Z*)-3-HAL, (*Z*)-3-HOL, or (*Z*)-3-HAC (dissolved in dichloromethane, 1 μg/μl) was pipetted onto a cotton ball in the Plexiglas cylinder. For controls, 20 μl of dichloromethane was added to the cotton ball. Corn plants were exposed to CIV overnight (15 h) in the same manner described for short-term exposure to CIV (see above).

For induction with CRE, intact corn plants were exposed to GLV, CIV, or the respective synthetic chemical overnight, as described above. After 15 h, plants were removed from Plexiglas cylinders. An area of ≈2 × 10 mm on the third leaf of each plant was scratched with a razor blade and 5 μl of CRE from BAW immediately added to the wounded site. For controls, buffer only was added to the wounded site. Plants were harvested 0, 30, 60, and 180 min after application of CRE, and JA and SA quantified (see below). In a separate experiment, VOC were collected from plants exposed to GLV, CIV, and synthetic compounds overnight and then treated with CRE (see below).

To estimate the amount of each compound in the volatiles to which the corn plants were exposed (Table 1), a Super Q filter-trap (Alltech Associates) was connected to the downstream side of the Plexiglas cylinders containing the treated plants, and air was drawn through the trap at 200 ml/min for various periods of time, depending on the experiment. VOC were eluted from the trap with dichloromethane, nonyl acetate added as an internal standard, and samples were analyzed by GC and GC/MS (3). The identity of each compound was confirmed by comparison of retention times and mass spectra with those of authentic chemicals, and quantities were determined by comparison of peak areas with peak area of internal standard. The average amounts of GLV released from the cut-leaf material and after caterpillar damage are shown in Table 1.

Table 1. Amount of GLV (in ng) released by source plants upstream from receiver plants during periods of treatment

		CIV	GLV	Control
30 min	Z-3-HAL	356 ± 95	4,872 ± 633	ND
	Z-3-HOL	108 ± 53	2,736 ± 546	ND
	Z-3-HAC	595 ± 569	3,720 ± 1,052	ND
Overnight	Z-3-HAL	2,177 ± 470	653 ± 339	44 ± 37
	Z-3-HOL	1,210 ± 128	3,137 ± 245	44 ± 30
	Z-3-HAC	3,540 ± 833	4,102 ± 663	207 ± 184

Data are from caterpillar-infested corn plants (CIV), cut leaf material (GLV), and control plants during the 30-min and overnight incubation period. Data represent mean ± SD ($n = 4$). ND, not detected.

Quantification of JA and SA. Extraction and quantification were performed as described (18, 19). In brief, plant tissues were frozen in liquid N_2, and ≈100 mg of each sample was transferred to 2-ml screw-cap FastPrep tubes (Qbiogene, Carlsbad, CA) containing 1 g of Zirmil beads (1.1 mm; SEPR Ceramic Beads and Powders, Mountainside, NJ). Dihydro JA and [$^2H_6$]SA (100 ng) were added to the 2-ml tubes before sample addition. The samples were mixed with 300 μl of 1-propanol/H_2O/HCl (2:1:0.002) and shaken for 30 s in a FastPrep FP 120 tissue homogenizer (Qbiogene). Dichloromethane (1 ml) was added to each sample, reshaken for 10 s in the homogenizer, and centrifuged at 11,300 × *g* for 30 s. The bottom dichloromethane/1-propanol layer was then transferred to a 4-ml glass screw-cap vial, with care taken to avoid transfer of the upper aqueous layer. The organic phase was evaporated by a constant airstream and 100 μl of diethyl ether/methanol (9:1, vol/vol) added. Carboxylic acids were converted into methyl esters by the addition of 2 μl of a 2.0 M solution of trimethylsilyldiazomethane in hexane. The vials were then capped, vortexed, and allowed to sit at room temperature for 30 min. Excess trimethylsilyldiazomethane was then destroyed by adding an equivalent molar amount of acetic acid to each sample.

Volatile metabolites were separated from the complex mixture by vapor-phase extraction as described in ref. 19. The trapped volatiles were then eluted with 150 μl of dichloromethane and analyzed by chemical ionization-GC/MS (19).

Analysis of Released VOC. Plants exposed to GLV, CIV, and pure synthetic compounds overnight (15 h) were transferred to 200-ml glass cylinders and VOC collected for 1 h without further treatment. In a second experiment, intact corn plants were exposed overnight as described above. After 15 h, plants were removed from the Plexiglas incubation chamber and induced with CRE as described above. Plants were subsequently transferred to the 200-ml glass cylinder, and VOC were collected for 1-h periods beginning 30 min after induction. Between sequential volatile collections from the same plant, there was a 30-min delay, during which the plant was watered. Volatiles were collected by pulling purified air at 200 ml/min over the plants and through Super Q filter traps. Analysis of the trapped VOC was performed as described in ref. 3. The amounts of linalool,

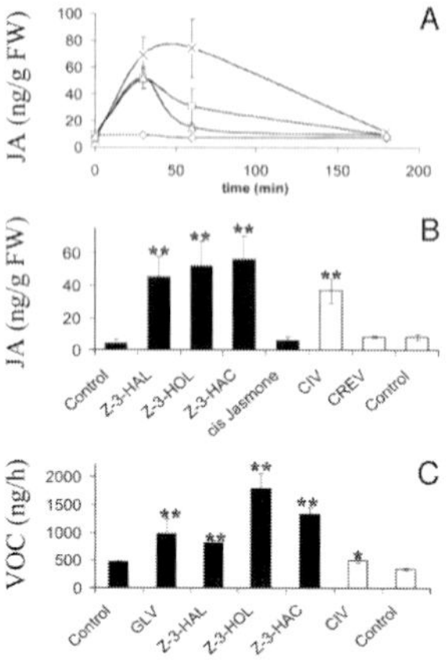

Fig. 1. Effects of GLV, CIV, and pure C_6 compounds on JA production and the release of volatiles in intact corn plants. Error bars represent SD ($n = 4$). Data were analyzed for significance with *t* test (*, $P \leq 0.05$; **, $P \leq 0.01$). (*A*) Induction of JA in corn seedlings by GLV and 20 μg or 1 mg of (*Z*)-3-HAC in a 6-liter Plexiglas container. Corn plants were incubated for 0, 30, 60, and 180 min. Controls were treated the same way, except that no volatile compounds were added to the incubation chambers. ◇, control; △, GLV-treated; □, ≤20 μg of (*Z*)-3-HAC; ×, 1 mg of (*Z*)-3-HAC. Twenty micrograms of (*Z*)-3-HAC corresponds to 30 nM maximum concentration in the gas phase, and 1 mg corresponds to 1.66 μM maximum concentration in the gas phase. Data points are connected by smoothed lines. (*B*) Black bars indicate levels of JA in intact corn plants after 30-min exposure to volatiles from synthetic compounds (*Z*)-3-HAL, (*Z*)-3-HOL, (*Z*)-3-HAC, or cis jasmone (20 μg each). White bars indicate JA levels in corn plants after 30-min exposure to CIV, crude regurgitant elicitor-induced volatiles (CREV), and control. Intact uninfested corn plants were used as a source of volatiles in the controls. (*C*) Induction of VOC after overnight exposure to GLV, synthetic C_6 compounds, CIV, or controls, as described above. The amounts of linalool, 4,8-dimethylnona-1,3,7-triene, β-caryophyllene, bergamotene, and β-farnesene emitted by the plants during the volatile collection period were measured and summed to obtain total volatiles.

4,8-dimethylnona-1,3,7-triene, β-caryophyllene, bergamotene, and β-farnesene emitted by the plants during the volatile collection period were measured and summed to obtain an estimate of total volatiles.

Statistical Analysis. At least three replicates of all experiments were conducted. Data were analyzed for significance with *t* test ($P < 0.05$). Treatments were compared to appropriate controls.

Results

GLV Induce JA Production and Volatile Emission. When we exposed intact hydroponically grown corn seedlings to wound-induced GLV by adding cut-leaf material to the incubation chamber, JA was induced transiently, reaching a maximum [52 ng/g fresh weight (FW)] 30 min after exposure (Fig. 1*A*). This initial burst of JA declined rapidly and reached the baseline levels of untreated control plants after 2–3 h (9 ng/g FW; control 8 ng/g FW). The analysis of the released GLV revealed that predominantly the (*Z*)-3-isomers were released after wounding and caterpillar infestation. When individual seedlings in incubation chambers were exposed to the vapors of (*Z*)-3-HAL, (*Z*)-3-HOL, (*Z*)-3-HAC, or cis jasmone, evaporated from cotton balls in concentrations comparable to those released by cut-leaf material, all C_6 compounds tested induced JA in comparable amounts (45–56 ng/g FW JA, compared to 4 ng/g FW in the control plants; Fig. 1*B*). Higher concentrations of synthetic compounds did not elevate the amount of induced JA significantly above this level but maintained it over a longer period (Fig. 1*A*). Synthetic compounds were still active at concentrations as low as 3 nM in the gas phase, inducing 15–20 ng/g FW of JA compared to 5 ng/g FW in control plants. Levels of endogenous JA did not change in control plants and plants treated with cis jasmone (17) (Fig. 1*B*). SA was not affected by treatment with GLV.

In a second experiment, nocturnal VOCs from plants infested with BAW, consisting predominantly of (*Z*)-3-HAL, (*Z*)-3-HOL, and (*Z*)-3-HAC, were used as a source of naturally released GLV. Importantly, plants exposed to these CIV also exhibited an increase of JA after 30 min (38 ng/g FW; Fig. 1*B*), although these plants received less GLV compared to those plants exposed to cut-leaf material (Table 1). SA was again not affected by this treatment, even with longer treatments of 8 h (data not shown). To examine the relevance of GLV further, CRE-induced volatiles (CREV), which contain only trace amounts of GLV, were used to induce plants. CREV failed to induce JA in receiver plants (Fig. 1*B*).

After overnight treatment of corn plants with either GLV or pure C_6 compounds, we observed the induction of small but significant amounts of VOC compared to control plants (Fig. 1*C*). Overnight exposure of corn plants to CIV also resulted in a slight stimulation of volatile release by receiver plants (Fig. 1*C*).

GLV Pretreatment Enhances JA Production and Volatile Emission in Response to Treatment with Caterpillar Regurgitant. To test the priming hypothesis, corn plants were pretreated with either wound-induced GLV, CIV, or synthetic C_6 compounds overnight (15 h), as described above. Then, CRE was applied to the plants the next day as a mimic of actual caterpillar feeding. We measured CRE-induced JA and the release of induced VOC as indicators of the induction of defense responses. After overnight exposure to GLV, CIV, and synthetic C_6 compounds, the resting levels of JA in these plants were the same as in the control plants (8 ng/g FW). Thirty minutes after induction with CRE, the level of endogenous JA rose in control plants to 96 ng/g FW, whereas in GLV-pretreated plants, 190 ng/g FW were found (Fig. 2*A* and *B*). The level of JA in GLV-pretreated plants remained higher over a period of 3 h but followed the same trend as JA in CRE-induced control plants (Fig. 2*A*). The same effect was observed for pretreatment with pure C_6 compounds and CIV (Fig. 2*B*). Surprisingly, wound-induced JA was not affected by previous exposure of corn plants to GLV (Fig. 2*C*).

The consequences of GLV pretreatment were further demonstrated by their effect on the release of VOC induced by application of CRE. Corn plants exposed overnight to GLV, CIV, and pure C_6 compounds released VOC without further treatment. However, this effect was significantly enhanced by induction with CRE. The GLV-pretreated plants released ≈4 μg of total VOC 4–6 h postinduction compared to 2.5 μg from nonprimed control plants (Fig. 3*A*). The same effect was observed for pretreatment with CIV (Fig. 3*B*) and pure C_6 compounds (Fig. 3*C*). GLV-pretreated plants reached the maximum release rate of unprimed plants earlier and exhibited a higher absolute release rate

Discussion

Our results clearly demonstrate a specific function for GLV in priming the defenses of corn plants against herbivorous insects. Both JA production, which is important to direct and indirect

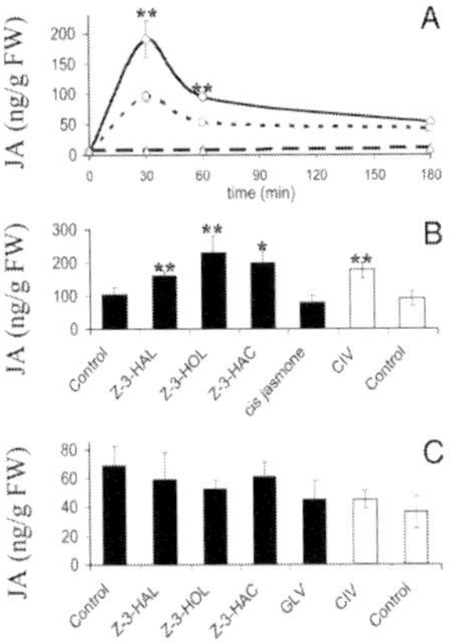

Fig. 2. Effects of pretreatment with GLV and pure chemicals on subsequent CRE- or wound-induced JA. Error bars represent SD ($n = 4$). Data for induction of JA by application of CRE (with or without previous exposure to GLV) were analyzed for significance with t test (*, $P \leq 0.05$; **, $P \leq 0.01$). (*A*) Induction of JA after application of CRE to GLV-pretreated intact corn plants. ◇, GLV plus CRE; □, control plus CRE; △, control. Data points are connected by smoothed lines. (*B*) Effects of overnight exposure to synthetic chemicals on CRE-induced JA in intact corn plants. JA was quantified from leaf tissue 30 min after application of CRE. (*C*) Effects of GLV, CIV, and pure C_6 compounds on wound-induced JA. Corn plants were exposed to synthetic C_6 compounds, GLV, CIV, and the respective control for 15 h. One leaf then was wounded with a razor blade, and JA was quantified after 30 min.

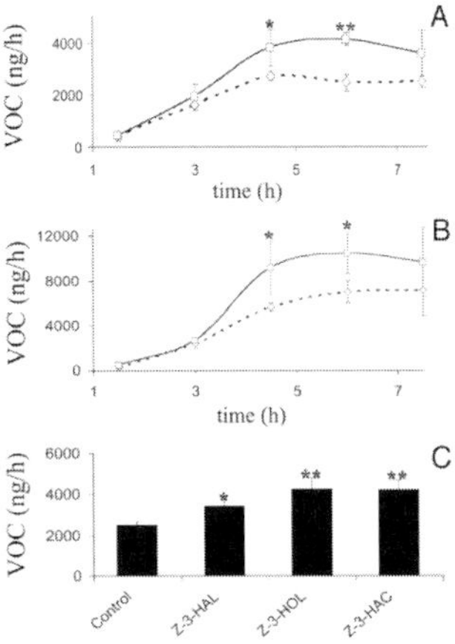

Fig. 3. Effects of pretreatment with GLV, pure chemicals, and CIV on subsequent CRE-induced volatiles. Error bars represent SD ($n = 4$). Data were analyzed for significance with t test (*, $P \leq 0.05$; **, $P \leq 0.01$). After induction with CRE, volatiles were collected during the period of maximum release rate (4–5 h after induction). Note that experiments were done with different batches of plants and CRE and at different times of the year. (*A*) Induction of volatiles by CRE in corn seedlings after overnight exposure to GLV. □, GLV plus CRE; ◇, control plus CRE. Data points are connected by smoothed lines. (*B*) Induction of volatiles by CRE in corn seedlings after overnight exposure to CIV. □, CIV plus CRE; ◇, control plus CRE. Data points are connected by smoothed lines. (*C*) Induction of volatiles by CRE in corn seedlings after overnight exposure to pure C_6 compounds.

defenses, and VOC release in response to simulated herbivore attack were significantly enhanced in plants previously exposed to GLV. Furthermore, only elicitor-induced JA was affected, whereas wound-induced production of JA remained unchanged.

GLV are important components of the VOC blends released by plants as a defense against attacking insect herbivores (2, 3). Moreover, unlike the induced emission of VOC like terpenes, indole, and methyl salicylate, which may not be released for hours after the beginning of herbivore damage, GLV are released almost immediately after wounding and the onset of herbivory and are considered typical wound signals (8).

GLV have been described as inducers of defense-related processes in various plant species; however, the signaling mechanisms involved were not clearly defined (11–13, 20). Furthermore, gene expression analysis indicated that a subset of defense-related genes was induced after exposure to GLV, all of which were related to JA action. JA, also first considered to be a typical wound signal (7), plays an important role in the activation of defensive functions in plants (7, 21, 22). The specific attribute of exogenous GLV to induce JA in corn plants, as shown herein, was the first indication of a possible mechanism linking this highly volatile wound signal to defense responses. The induction of subsets of defense-related genes (11, 12) might reflect the relatively low and transient induction of JA after exposure to GLV (compare Figs. 1*A* and 2*A*).

C_6 compounds are known to induce the release of VOC in tomato plants exposed to physiological concentrations (13), but the response of these plants was lower compared to those exposed to MeJA or actual insect herbivory. This coincides well with our own findings. Corn plants exposed to GLV, CIV, or pure C_6 compounds also released significant amounts of VOC compared to control plants. Comparing the total amounts of released VOC after exposure to GLV with that induced by application of CRE revealed that application of CRE induced four to six times more VOC, although only one leaf was damaged.

GLV may also mediate direct defense against insect pests. Aphids feeding on potato plants depleted in HPL activity (23) exhibited a 2-fold increase in fecundity compared to those feeding on wild-type plants. Interestingly, in this example, the plants did not show any phenotypic differences compared to wild-type plants with regard to wound-induced gene expression, further indicating a specific function of GLV in defense-related processes rather than in the wound response. Together with our own results, these examples demonstrate the capability of GLV to affect plant defense responses against various threats in different plant species. However, it is obvious that all these responses are moderate compared to those after actual insect herbivore damage and would provide only a very low level of protection against attack by herbivorous insects. This led to the question of whether the capability of GLV to induce JA and thus JA-responsive defense related genes is the primary function of these compounds, or whether the rather modest induction of defense-related processes primes the receiver plants against pending attack.

Cost/benefit analyses demonstrate that maintaining a high level of defense without being actually threatened negatively affects plant performance (24, 25) and would make such a mechanism beneficial only under conditions with a high probability of attack. However, priming by GLV provides a different way of responding to the threat of insect herbivory. Defense-related processes are turned on but incompletely compared to actual herbivore damage (11, 12). More importantly, the plant prepares, by a yet-unknown mechanism, to respond more intensely when it is subsequently attacked, as demonstrated herein for the induction of JA and release of VOC. In this way, the plant avoids great biochemical investments, which would affect the general physiology significantly unless actually attacked (24, 25).

A further, equally important aspect of priming is the specificity of the signaling process. As mentioned previously, GLV induce a subset of defense-related genes (11, 12) or cause the release of VOC (13), all processes related to JA. However, in addition to the induction of defense responses, JA is involved in various developmental processes and responses to other environmental factors in plants (21, 22). This raises the question of which JA signaling processes are induced by GLV. In corn, there are at least three different ways of inducing JA, all related to wounding but different in their regulation. GLV induces JA transiently; wounding also results in JA production; application of CRE on wounded sites combines the wound response and an elicitor response, resulting in higher amounts of induced JA. Priming with GLV specifically promoted only the CRE-induced JA production, whereas wound-induced JA or JA induced by a second application of GLV after 15 h (data not shown) was not affected. Furthermore, SA was never affected by any of the treatments with GLV, giving further proof for the specificity of the GLV signal in insect herbivore defense response.

The induction and release of VOC by plants as a response to insect herbivore damage have been demonstrated to be an effective defense strategy. Recruiting parasites and predators of the herbivore (2, 3) as well as repelling female moths, thereby avoiding egg deposition (4), helps the plant to reduce damage. Additionally, parasitization of attacking insect herbivores increased seed production in mature corn plants (26), demonstrating clearly the fitness benefits of this defense measure. It is obvious that the effectiveness of this strategy strongly depends on the timely release of a strong VOC signal. Priming corn plants with GLV results in a faster and more intense release of these VOC when induced with CRE and could give them a competitive advantage over nonprimed plants.

Our results demonstrate that GLV induce defense responses in neighboring plants via induction of JA followed by the release of low levels of typical herbivore-induced VOC. Furthermore, a specific priming of corn plants against subsequent insect herbivore attack is initialized, allowing them to respond more rapidly by enhanced JA production and an increased release of VOC (26, 27). The principle of priming plants against pathogen infections by chemicals that mimic endogenous defense-related signaling compounds is well established (28). A comparable mechanism has not been shown to date for priming against insect herbivore attack.

A further indication for the specificity of this signaling is the inactivity of other VOC in this system. However, other classes of VOC have been demonstrated to be interplant defense signals in other plant species (11, 29–31). The mechanism of priming may benefit receiver plants by reducing investment in defenses until the onset of actual herbivory. Thus, the effect of GLV is far-reaching and influences both directly and indirectly the entire tritrophic complex of plants, insect herbivores, and natural enemies of the herbivores. Future research is now directed toward the underlying molecular mechanism of this process in plants.

We thank the two reviewers for helpful advice that significantly improved the manuscript. This work was supported in part by a grant from the Defense Advanced Research Project Agency.

1. Karban, R. & Baldwin I. T. (1997) *Induced Responses to Herbivory* (Univ. of Chicago Press, Chicago).
2. Takabayashi, J. & Dicke, M. (1996) *Trends Plant Sci.* 109–113.
3. Turlings, T. C. J., Tumlinson, J. H., Heath, R. R., Proveaux, A. T. & Doolittle, R. E. J. (1991) *Chem. Ecol.* **17,** 2235–2251.
4. DeMoraes, C. M., Mescher, M. C. & Tumlinson, J. H. (2001) *Nature* **410,** 577–580.
5. Kessler, A. & Baldwin, I. T. (2001) *Science* **291,** 2141–2144.
6. Shulaev, V., Silverman, P. & Raskin, I. (1997) *Nature* **386,** 718–721.
7. Creelman, R. A. & Mullet, J. E. (1997) *Annu. Rev. Plant Physiol. Plant Mol. Biol.* **48,** 355–381.
8. Hatanaka, A. (1993) *Phytochemistry* **34,** 1201–1218.
9. Vick, B. A. (1993) in *Lipid Metabolism in Plants*, ed. Moore, T. S., Jr. (CRC Press, Boca Raton, FL), pp. 167–191.
10. Paré, P. W. & Tumlinson, J. H. (1997) *Plant Physiol.* **114,** 1161–1167.
11. Bate, N. J. & Rothstein, S. J. (1998) *Plant J.* **16,** 561–569.
12. Arimura, G., Ozawa, R., Nishioka, T., Boland, W., Koch, T., Kühnemann, F. & Takabayashi, J. (2002) *Plant J.* **29,** 87–98.
13. Farag, M. A. & Parc, P. W. (2002) *Phytochemistry* **61,** 545–554.
14. Schmelz, E. A., Alborn, H. T. & Tumlinson, J. H. (2001) *Planta* **214,** 171–179.
15. King, E. G. & Leppla, N. C. (1984) *Advances and Challenges in Insect Rearing* (U.S. Govt. Printing Office, Washington, DC).
16. Mori, N., Alborn, H. T., Teal, P. E. & Tumlinson, J. H. (2001) *J. Insect Physiol.* **47,** 749–757.
17. Birkett, M. A., Campbell, C. A. M., Chamberlain, K., Guerrieri, E., Hick, A. J., Martin, J. L., Matthes, M., Napier, J. A., Pettersson, J., Picket, J. A., *et al.* (2000) *Proc. Natl. Acad. Sci. USA* **97,** 9329–9334.
18. Schmelz, E. A., Engelberth, J., Alborn, H. T., O'Donnel, P., Sammons, M., Toshima, H. & Tumlinson, J. H. (2003) *Proc. Natl. Acad. Sci.* USA **100,** 10552–10557.
19. Engelberth, J., Schmelz, E. A., Alborn, H. T., Cardoza, Y. J., Huang, J. & Tumlinson, J. H. (2003) *Anal. Biochem.* **312,** 242–250.
20. Sivasankar, S., Sheldrick, B. & Rothstein, S. J. (2000) *Plant Physiol.* **122,** 1335–1342.
21. Creelman, R. A. & Mullet, J. E. (1995) *Proc. Natl. Acad. Sci. USA* **92,** 4114–4119.
22. Beale, M. H. & Ward, J. L. (1998) *Nat. Prod. Rep.* **15,** 533–548.
23. Vancanneyt, G., Sanz, C., Farmaki, T., Paneque, M., Ortego, F., Castanera, P. & Sanchez-Serrano, J. J. (2001) *Proc. Natl. Acad. Sci. USA* **98,** 8139–8144.
24. Heil, M. & Baldwin, I. T. (2002) *Trends Plant Sci.* **7,** 61–67.
25. Baldwin, I. T. (1998) *Proc. Natl. Acad. Sci. USA* **95,** 8113–8118.
26. Fritzsche Hoballah, M. E. & Turlings, T. C. J. (2001) *Evol. Ecol. Res.* **3,** 553–565.
27. Fritzsche Hobalah, M. E., Tamo, C. & Turlings, T. C. J. (2002) *J. Chem. Ecol.* **28,** 951–968.
28. Conrath, U., Pieterse, C. M. J. & Mauch-Mani, B. (2002) *Trends Plant Sci.* **7,** 210–216.
29. Arimura, G., Ozawa, R., Shimoda, T., Nishioka, T., Boland, W. & Takabayashi, J. (2000) *Nature* **406,** 512–515.
30. Baldwin, I. T. & Schultz, J. C. (1983) *Science* **221,** 277–279.
31. Tscharnke, T., Thiessen, S., Dolch, R. & Boland, W. (2001) *Biochem. Syst. Ecol.* **29,** 1025–1047.

Disulfooxy fatty acids from the American bird grasshopper *Schistocerca americana*, elicitors of plant volatiles

Hans T. Alborn*[†], Trond V. Hansen*[‡], Tappey H. Jones[§], Derrick C. Bennett*, James H. Tumlinson[†§], Eric A. Schmelz*, and Peter E. A. Teal*

*Center for Medical, Agricultural, and Veterinary Entomology, United States Department of Agriculture, 1600/1700 Southwest 23rd Drive, Gainesville, FL 32608; [§]Department of Chemistry, Virginia Military Institute, Lexington, VA 24450; and [§]Department of Entomology, Center for Chemical Ecology, 111 Chemical Ecology Laboratory, University Park, PA 16802

Contributed by James H. Tumlinson, June 27, 2007 (sent for review April 23, 2007)

A previously unidentified class of compounds has been isolated from the regurgitant of the grasshopper species *Schistocerca americana*. These compounds (named here "caeliferins") are composed of saturated and monounsaturated sulfated α-hydroxy fatty acids in which the ω-carbon is functionalized with either a sulfated hydroxyl or a carboxyl conjugated to glycine via an amide bond. The regurgitant contains a series of these compounds with fatty acid chains of 15–20 carbons and in varying proportions. Of these, the 16-carbon analogs are predominant and are also most active in inducing release of volatile organic compounds when applied to damaged leaves of corn seedlings. Caeliferins are nonlepidopteran elicitors identified in insect herbivores. This adds a category of insect herbivore-produced elicitors of plant responses, providing further evidence of the ability of plants to detect and respond to a broad range of insect herbivore-produced compounds.

American grasshopper | insect herbivory | regurgitant | caeliferin

Insect herbivory provokes direct defenses that impair herbivore performance and induces plants to release volatile organic compounds (VOC) that attract natural enemies of the herbivores (1, 2). The composition of insect-induced VOC varies with different species and even different varieties of plants (3, 4). They are composed of components synthesized by several different biosynthetic pathways, including the lipoxygenase, shikimate, and isoprenoid pathways (5). Induced VOC may be stored in glands, as in cotton, and be released by herbivore damage to the glands (5, 6). However, many plants, such as corn, release VOC soon after herbivore-induced *de novo* biosynthesis. For some plants, such as lima beans, mechanical damage is sufficient to induce the release of substantial quantities of VOC (7, 8). In most described cases, plants damaged by insect herbivores or treated with regurgitant release greater quantities and/or different proportions of VOC than plants receiving only mechanical damage (5, 9–17). Furthermore, specialist parasitoids can distinguish volatiles induced by host larvae from volatiles induced by nonhost larvae (18, 19). Thus, herbivore-produced substances play a critical role in induction of plant defenses and herbivore-specific volatile emissions.

The first identified nonenzymatic insect herbivore-produced elicitor of plant volatiles was volicitin, *N*-(17-hydroxylinolenoyl)-L-glutamine, isolated from the regurgitant of beet armyworm, (20, 21). Subsequently, other analogous fatty acid–amino acid conjugates have been identified from regurgitant of larvae of several species of Lepidoptera (22–24). This class of elicitors typically consists of linolenic, linoleic, and oleic acids obtained from the diet or plant on which the larvae feed (25) or from their 17-hydroxy analogs conjugated with glutamine or glutamic acid. Thus far, volicitin, *N*-linolenoyl-L-glutamine, and *N*-linolenoyl-L-glutamate have been demonstrated to have significant activity in inducing plants to produce and release VOC (21, 22). However, not all plants respond to these elicitors (8, 26). For example, cow pea releases VOC when fed on by fall armyworm larvae, but not when treated with fatty acid–amino acid elicitors. This phenomenon led to the identification of a previously unidentified type of peptide elicitor in fall armyworm regurgitant, derived by proteolysis from the chloroplastic ATP synthase of the plant on which the herbivore feeds (26). Thus, *Spodoptera* larvae produce at least two different types of elicitors, both derived in part from the plant, that share similar activity, i.e., induced release of volatiles, but on different host plants of the herbivores. Salivary enzymes of insect herbivores also may influence plant defense responses (27). Therefore, complex interactions between herbivore-produced enzymes and elicitors are likely responsible for the specificity in the blend of VOC released by plants attacked by Lepidoptera herbivores.

Corn seedlings fed on by the grasshopper *Schistocerca americana* or treated with its regurgitant were previously found to release VOC similar to that induced by caterpillar feeding or regurgitant treatment (28). Here we describe the isolation, identification, and synthesis of a previously unidentified class of sulfated fatty acids found in regurgitant of *S. americana* and their biological activity as elicitors of induced VOC in corn seedlings.

Results

Corn seedlings treated with 5 μl of *S. americana* regurgitant in an excised plant bioassay emitted amounts of VOC equivalent to emissions by seedlings treated with 100 pmol of volicitin or seedlings exposed to grasshopper feeding for 2 h the previous day. The VOC were composed of indole, monoterpenes, sesquiterpenes, and their 11- and 16-carbon homologs. However, no *N*-acyl fatty acid-type elicitors could be detected in the grasshopper regurgitant by HPLC or liquid chromatography (LC)/MS methods developed for analyses of Lepidoptera larvae regurgitant (21) (data not shown).

No loss of regurgitant activity was found after protein precipitation. Activity was measured as the combined amount of four induced terpenoids relative to the amount induced by crude regurgitant. The most active regurgitant fraction eluted from a

Author contributions: H.T.A., J.H.T., E.A.S., and P.E.A.T. designed research; H.T.A., T.V.H., T.H.J., and D.C.B. performed research; H.T.A., T.H.J., D.C.B., and E.A.S. contributed new reagents/analytic tools; H.T.A. analyzed data; and H.T.A., J.H.T., and P.E.A.T. wrote the paper.

The authors declare no conflict of interest.

Abbreviations: amu, atomic mass unit; CI, chemical ionization; EI, electron impact; LC, liquid chromatography; MW, molecular weight; VOC, volatile organic compounds.

[†]To whom correspondence may be addressed. E-mail: hans.alborn@ars.usda.gov or jht2@psu.edu.

[‡]Present address: Department of Pharmaceutical Chemistry, University of Oslo, N-0316 Oslo, Norway.

This article contains supporting information online at www.pnas.org/cgi/content/full/0705947104/DC1.

www.pnas.org/cgi/doi/10.1073/pnas.0705947104

product resulted in three peaks; the mono-acetylated dimethyl ester of the complete compound (minus the sulfate) with a MW of 429, the acetylated methyl ester seen before with a MW of 372 and finally the acetylated methyl ester of glycine (MW 131), giving the final structure of caeliferin B16:0 in Fig. 3. Also, homologs with C_{17} and C_{18} fatty acid chains (with traces of C15 and C19) were found in analogy with caeliferin A16:0. There were also traces of an unsaturated version of caeliferin B16:0 (peak 1–3 in Fig. 4*A*), consequently named caeliferin B16:1 (Fig. 3). Proton NMR analysis of fraction 4 showed the same single proton at a substituted carbon α to a carboxylic acid (δ = 4.45 ppm) (broad) but, as expected, with no indication that an alcohol-substituted ω-carbon was present. However, the two protons at δ = 2.14 ppm confirmed the amide, and finally a singlet with 2 protons at δ = 3.62 ppm confirmed the presence of glycine.

LC/MS made it possible to collect and bioassay the identified caeliferins and several of the related peaks in Fig. 4*A*. However, the LC/MS trace could not be used for quantitative estimates due to different response factors for caeliferin A and B. Therefore the absolute amount of the collected compounds in each fraction was established by derivatization and GC analysis. In the first intact plant bioassay, we used the same regurgitant equivalents (0.5 μl) for each fraction tested, maintaining their natural relative concentrations (as in Fig. 1). The results (Fig. 4*B*) were also very comparable to the earlier results. Caeliferin A16:1 is naturally only a minor component ($\approx$200 pmol/μl) compared with caeliferin A16:0 ($\approx$4 nmol/μl) and B16:0 ($\approx$500 pmol/μl). In a second assay (Fig. 4*C*) we bioassayed the major compounds at the same concentration (100 pmol per seedling). In this experiment caeliferin A16:1 was significantly more active than any other component, including caeliferin A16:0. The latter was now approximately as active as its C_{17} and C_{18} analogs and caeliferin B16:0.

Discussion

We report a new class of insect herbivore-produced plant volatile elicitors with strikingly different chemical structures than those previously described. Yet, the basic moiety of these structures is a fatty acid, such as the fatty acid–amino acid conjugates of the Lepidopteran elicitors, although the most active, as well as the most abundant, compounds from *S. americana* have 16- rather than 18-carbon chains. Preliminary analyses of regurgitant collected from several other species of Orthoptera indicates that the caeliferins may be present in most, if not all grasshoppers, members of the suborder Caelifera, but not in crickets or katydids in suborder Ensifera. This was the basis for naming these new compounds caeliferins. Interestingly, regurgitant of at least some crickets (31) and katydids (unpublished data) contain fatty acid amides, although we have not been able to find even traces of these in grasshopper regurgitant.

In the Lepidoptera, the fatty acid moiety of the elicitors is derived from the diet on which the larvae feed (25), and variation in proportions of dietary fatty acids results in similar changes in proportions of the corresponding elicitors (32). The composition of the regurgitant of laboratory-reared *S. americana* is very consistent and does not change with diet or maturing of the insect (data not shown). However, regurgitant from field-collected insects can contain up to 5-fold more of caeliferin A16:1, as well as more of the glycine-containing caeliferin B16:1 and B16:0 than laboratory-reared insects. Furthermore, the regurgitant composition in wild grasshoppers will change into laboratory-reared proportions within a week of caging in the laboratory, independent of what diet the insects are fed. Whether this is diet-related or due to other factors, such as crowding, is not yet known. The origin, or precursors of the *S. americana* elicitors is not known. On the basis of the natural preference for even carbon chain lengths, we would have expected C_{16} and C_{18} to dominate over C_{17} and, if the origin was plant lipids, then we would expect more C_{18} than C_{16} and also double bonds to be cis. However, in *S. americana* regurgitant, the relative abundance of caeliferin A16:0, A17:0, and A18:0 is $\approx$30:5:1, and the double bond in caeliferin A16:1 is trans. We have not found hydroxylation at any but the 2 and 16 positions. The 2-hydroxylation might explain the unusual lengths in carbon chains as the first step in a one-carbon chain-shortening sequence. Furthermore, the ω-hydroxyl might be the result of enzymatic reduction of an amide precursor (caeliferin B). However, that still leaves us with a unique and unknown precursor. It appears that all caeliferins in all species investigated are totally sulfated in the pH 5 regurgitant, whereas alkaline or stronger acid conditions result in partially or completely desulfated compounds.

The range of biological activity and function of caeliferins in grasshoppers is not fully understood. Sulfated compounds, for example, glycosaminoglycans, and proteins, have been found in insect ovaries, eggs, and fat bodies (33–35). The hydrophilic sulfate groups make caeliferins especially useful in bridging lipids and water, and as such they may function in digestion. Additionally, they may also be an important part of chemical defense, because grasshoppers regurgitate copious amounts of frothy secretions as a defense against attack. Indeed, investigations (36–38) have shown that both compounds from host plants and insect produced water-soluble components of the regurgitant have deterrent activity (36). Although the deterrent effect of caeliferins remains to be tested, it is probable that the emulsifying properties are important in aiding the water-based regurgitant to carry lipophillic toxic compounds.

It has been well demonstrated that induced plant volatiles play a role in attracting natural enemies of Lepidopteran herbivores, but there is no behavioral evidence that we are aware of that indicates this occurs for grasshoppers. Jasmonic acid signaling is a key player in induced plant response to wounding and feeding damage. Recently it has been shown that a sulfotransferase in *Arabidopsis thaliana* inactivates 12-hydroxyjasmonate by transforming it to its corresponding sulfate (39). Caeliferins might interact with this system, but at this stage we don't know the full range of activity of caeliferins on plants and signal cascades.

It has been reported that *S. americana* avoids diet with added monoterpenoids (40). We know, from electroantennogram experiments, that *S. americana* detects most of the volatile compounds released by induced corn seedlings. It is therefore likely that, under natural conditions, the solitary *S. americana* will stop feeding on and even avoid induced plants. Thus, the plants will directly benefit from the release of VOC in response to grasshopper feeding and caeliferins. Most of the time the American grasshopper lives a solitary life, but occasionally they transform into a gregarious stage, for example, when crowded. At that stage, plant volatiles could aid in aggregation. Caging will transform wild, solitary, nymphal as well as adult *S. americana* from solitary to gregarious stage, with the typical change in coloration occurring within a few days. This has so far made it difficult to study changes in behavior, including their response to induced plant volatiles. This physical change occurs in parallel with the change in caeliferin composition, mentioned earlier.

S. americana feeds on a broad range of plants (41). There is now evidence for dramatic variation in the response, or lack of response, to volicitin- and inceptin-types of elicitors from Lepidopteran larvae among plant species (26). It is an intriguing possibility that analogs of caeliferin A and B have different levels of activity on the wide variety of host plants of *S. americana* or even on different varieties of the same species of plants.

Induced release of volatiles is just one of many levels of defense against phytophagous insects, and it would be surprising if the known elicitors do not also trigger nonvolatile defenses. An investigation of the activity of these elicitors on several plant species in inducing volatile emission, as well as increased levels of jasmonic acid and other plant hormones. The discovery of caeliferins provides new biological tools and directions to explore the physiological ecology and interactions of insects and plants.

Materials and Methods

Insect Rearing. *S. americana* were either obtained from a colony maintained at the Department of Entomology and Nematology at the University of Florida or field-collected in Gainesville, FL. The insects were fed lettuce, wheat, or corn leaves and kept in cages in a temperature-controlled greenhouse (25°C during the day/20°C during the night) with natural light, supported by a mixture of high-pressure sodium and metal halide lamps (400W Lucalox; GE, Piscataway, NJ) to provide 14 h of day/10 h of night. To facilitate reproduction and egg laying, cups with moist vermiculite were kept in the cages.

Plants. All plants were grown in 1-gallon containers (Miracle Grow no. 92695 potting soil fertilized with 1 teaspoon of Osmocote 14-14-14; Scotts-Sierra Horticulture, Marysville, OH). Feed corn variety Delprim (Delley Seeds and Plants, Delley, Switzerland) was planted 10 seeds per pot and maintained on a 12-h dark and 12-h light cycle. One week after germination, the number of seedlings per pot were either reduced to six of uniform size for the cut stem bioassay or transferred to hydroponic chambers (18). The light intensity was 305 $\mu mol \cdot m^{-2} \cdot S^{-1}$ of photosynthetically active radiation supplied by a mixture of high-pressure sodium and metal halide lamps (400W GE Lucalox) at the top of the plant canopy with day/night temperatures at 25°C/20°C, respectively. The relative humidity ranged from 60% to 70%. For both bioassays, 10- to 12-day-old seedlings that contained three to four leaves were used for treatments and volatile collections.

Plant Bioassay. For both bioassays, plants were treated ≈2 h into the scotophase and incubated for ≈16–17 h on the same day/night schedule under which they were grown but with reduced daytime light (2,000 lux) before volatile collection.

Excised Plant. At the time of treatment, the corn seedlings were cut off near the base of the stem with a razor blade, and the cut end was immediately immersed in 600 μl of 50 mM Na_2HPO_4 buffer (pH 8) (control) or a test solution in 600 μl of the same buffer, contained in a 1.5-ml vial. Five-microliter regurgitant equivalents (MRE) of the crude regurgitant or subsequent fractions thereof were used to treat each seedling in all bioassays. After incubation, the stem of each plant was recut, wrapped in wet cotton, and placed in a glass volatile collection chamber, and volatiles were collected as described below.

Intact Plant. At the time of treatment the two oldest leaves of individual plants each received two superficial damage sites by scratching the surface (2 mm × 10 mm) with a razor blade approximately equidistant between the base and tip of the leaf (29). Immediately after damage, 5 μl of the same buffer solution as above containing 0.5-μl regurgitant equivalents (MRE) was applied evenly to the four damage sites. The plants remained in their hydroponic solution during incubation and then were removed from their container, their roots were gently wrapped in soft tissue and dipped in hydroponic solution, the seedlings were placed in the volatile collection chambers, and volatiles were collected as described below.

Volatile Collections. Volatile chambers (30 cm long × 4 cm i.d.) were placed under the same type and intensity of light as used in rearing and volatiles, were collected on Super Q 80/100 (catalog no. 2735; Alltech Associates, Deerfield, IL) as described in refs. 22 and 28, and were analyzed by gas chromatography (*Analysis 5* in *SI Text* and SI Fig. 9). Volatiles were collected for 2 h for the cut-stem assays and for 1 h for the intact-plant assays. The response by the plant to each test sample was measured as the combined amount of four induced terpenoids: *E*-β-caryophyllene, *E*-α-bergamotene, α-humulene, and *E*-β-farnesene. Although several more volatile components are released by Delprim cultivar in response to insect herbivory (5), these four components show the strongest induced response. All bioassays also included positive controls (0.5 or 5 μl of crude regurgitant, depending on bioassay) and a negative control (buffer only). All samples were replicated three to four times within each bioassay and each bioassay was repeated three to four times (giving a minimum total of $n = 9$ per treatment). To reduce the variation in response between bioassays, a correlation factor was calculated for each bioassay to give the mean release of induced volatiles for the regurgitant treatments a response of 100. The correlation factor was then used for all samples within that bioassay. Data were analyzed using a Tukey pairwise multiple comparison test.

Collection and Initial Fractionation of Regurgitant. Regurgitant was collected from both wild and laboratory-reared nymphs and adults as described earlier for Lepidoptera caterpillars (22). The regurgitant was diluted 50:50 with acetonitrile, centrifuged at 16,000 × *g* for 15 min to remove precipitated proteins, and then filtered through 0.45- and 0.22-μm sterilizing membranes. The supernatant was concentrated to original volume by a gentle stream of N_2 and stored at −70°C until used. A C_{18} solid-phase extraction column (mega bond elute; Waters, Milford, MA) was activated by eluting with 40 ml of acetonitrile and then with 40 ml of water. Two milliliters of concentrated supernatant was added to the top of the column and separated into three fractions by eluting with 20 ml of water, followed by 20 ml of 50:50 water:acetonitrile and then by 20 ml of acetonitrile with gravity flow. Each fraction was concentrated to the original volume before bioassaying.

HPLC Purification. The active 50% acetonitrile-water fraction was further fractionated by HPLC under acidic conditions (AS3500 Autosampler, Constametric 4100 Pump, and Spectromonitor 3200 variable wavelength detector; Thermo Separation Products, Riviera Beach, FL), while monitoring UV absorption at 220 nm. With the temperature maintained at 60°C, a C_{18} reverse-phase column (ODS-AMQ, S-5 μm, 200 A, 250 × 4.6 mm i.d.; YMC, Kyoto, Japan) was eluted with a solvent gradient (1.0 ml/min) of 10–70% acetonitrile in water for 20 min followed by 70% acetonitrile for 10 min., with both solvents containing 1 mM NH_4Ac:AcOH buffer, pH 4.5 [acetic acid (Aldrich, Milwaukee, WI), acetonitrile UV (B & J Brand High Purity Solvent, Muskegon, MI), and water (Milli-Q UV Plus System; Millipore, Bedford, MA)]. Combined fractions from several repeated separations were evaporated under vacuum to near dryness and desalted on activated C_{18} solid-phase extraction columns (SPE C18; J.T. Baker, Phillipsburg, NJ) that were rinsed with 2 ml of water, followed by 2 ml of 50% acetonitrile in water. The latter fraction was reduced to dryness and redissolved in bioassay buffer (50 mM phosphate buffer, pH 8) before bioassaying.

The most active fractions from the first HPLC separation were combined and further fractionated by a modified solvent gradient and neutral conditions by using the same column and system as above with UV absorption monitored at 210 nm. The column was eluted with a solvent gradient (1.0 ml/min) of 15–60% acetonitrile in water in 20 min followed by 60% acetonitrile for 10 min, with both solvents containing 5 mM NH_4Ac buffer, pH 7.0. Two-minute fractions were collected and desalted as described above before bioassaying.

LC/MS Conditions. A Thermo Finnigan LCQ Deca XP Max was used with electro spray ionization (sheet gas, 30 arbitrary units; sweep gas, 5 arbitrary units; spray voltage, 5.00 kV; capillary temperature 275°C; and capillary voltage, 3.0 V). The Thermo Separations spectra HPLC system consisted of a quaternary pump P4000, autosampler AS 3000, and diode array detector UV6000. The tertiary solvents consisted of (*i*) methanol with 0.05% formic acid, (*ii*) water with 10 mM ammonium formate, and (*iii*) 90% acetonitrile/10% water with 10 mM ammonium formate. With the column temperature maintained at 60°C and a solvent flow of 1.0 ml/min, the C_{18} column (ODS-AMQ, S-5 μm, 200 A, 250 × 4.6 mm i.d.; YMC) was eluted with a solvent composition starting with 4:90:6 (*i:ii:iii*) for 1 min, followed by a gradient to 4:50:30 in 13 min and a fast ramping to 4:0:96 in 2 min and was then kept at that composition for 5 min. UV absorption was monitored at 190–400 nm, and the solvent flow between the UV detector and MS electro spray interface was split 9:1 with a low-volume micro needle valve splitter P450 (Upchurch Scientific, Oak Harbor, WA) making it possible to obtain spectra of eluted compounds and simultaneously collect 90% of the injected material for bioassaying.

GC/MS Conditions. Derivatized samples were analyzed by GC/MS (6,890/5,973 GC/MS; Agilent, Palo Alto, CA) in both EI and isobutane CI mode. Samples were injected in the splitless mode at 220°C or with cold on-column. The methyl silicone column, (HP1, 30 m × 0.25 mm i.d. × 0.1-μm film thickness; Agilent) was kept at 40°C for 1 min and was then temperature programmed at 10°C/min to 260°C. The He carrier gas flow rate was 30 cm/sec (constant flow), and the transfer line temperature was 250°C. The ion source temperature was 220°C in EI mode and 250°C in CI mode.

Derivatizations for GC/MS. For methanolysis and ethanolysis 10 μl of regurgitant supernatant after protein precipitation, or the same amount of regurgitant equivalents of HPLC fraction, was added to a vial and evaporated to dryness with N_2. Two hundred microliters of dry methanol/HCl (10:1) or ethanol/HCl was added, and the sealed vial was heated to 100°C for 30 min. The mixture was allowed to cool to room temperature, 500 μl of methylene chloride was added, and the solution was extracted twice with 500 μl of water. The organic phase was evaporated to dryness with N_2, and 500 μl of fresh methylene chloride was added for GC/MS analyses.

For acetylation, 20 μl of acetyl chloride was added to the above solution, which then was heated to 90°C for 30 min. The solution was then extracted three times with water before GC/MS analyses.

HPLC fractions derivatized for GC/MS analyses (acetylated methanolysis products) were ozonized as described in refs. 20 and 42 and analyzed by GC/MS by using on-column injection.

Double bonds were reduced by gently drying methylene chloride solutions of acetylated methylesters of pure HPLC fractions with N_2 and redissolving in ethyl acetate. A few grains of Pd on carbon were added, H_2 was gently bubbled through the solution for 1 h, and the solution analyzed on GC/MS.

NMR Analyses. ^{1}H NMR spectra were obtained with Varian (Palo Alto, CA) Unity 600 and Bruker (Billerica, MA) Avance 500 instruments by using a Wilmad (Buena, NY) 520 1A microbore tube. Samples were dissolved in D_2O, and ^{1}H NMR chemical shifts are reported with respect to internal tetramethylsilane.

We thank Peggy Brennan, Mary Searle, Spencer Walse (all at the Center for Medical, Agricultural, and Veterinary Entomology at the United States Department of Agriculture) and Jim Rocca (University of Florida, Gainesville, FL) for technical assistance; and Paul Pare (Texas Tech University, Lubbock, TX), Naoki Mori (Kyoto University, Kyoto, Japan), Ted Turlings (University of Neuchatel, Neuchatel, Switzerland), and John Pickett (Rothamsted Experimental Station, Harpenden, U.K.) for helpful comments.

1. Turlings TCJ, Tumlinson JH, Lewis WJ (1990) *Science* 250:1251–1253.
2. Kessler A, Baldwin IT (2001) *Science* 291:2141–2144.
3. Loughrin JH, Manukian A, Heath RR, Tumlinson JH (1995) *J Chem Ecol* 1217–1227.
4. Gouinguene S, Degen T, Turlings TCJ (2001) *Chemoecol* 11:9–16.
5. Paré PW, Tumlinson JH (1997) *Nature* 385:30–31.
6. Rose USR, Manukian A, Heath RR, Tumlinson JH (1996) *Plant Physiol* 111:487–495.
7. Mithofer A, Wanner G, Boland W (2005) *Plant Physiol* 137:1160–1168.
8. Spiteller D, Pohnert G, Boland W (2001) *Tetrahedron Lett* 42:1483–1485.
9. Arimura G, Huber DPW, Bohlman J (2004) *Plant J* 37:603–616.
10. Degenhardt DC, Lincoln DE (2006) *J Chem Ecol* 32:725–743.
11. De Moraes CM, Mescher MC, Tumlinson JH (2001) *Nature* 410:577–580.
12. Litvak ME, Monson RK (1998) *Oecologia* 114:531–540.
13. Roese USR, Tumlinson JH (2005) *Planta* 222:327–335.
14. Rodrigues-Saona C, Crafts-Brandner SJ, Williams L, Pare PW (2002) *J Chem Ecol* 28:1733–1747.
15. Turlings TCJ, Bernasconi M, Bertossa R, Bigler F, Caloz G, Dorn S (1998) *Biolog Cont* 11:122–129.
16. Turlings TCJ, Alborn, HT. Loughrin JH, Tumlinson JH (2000) *J Chem Ecol* 26:189–202.
17. Williams L, Rodriguez-Saona C, Pare PW, Crafts-Brandner SJ (2005) *Arch Biochem Phys* 58:84–96.
18. De Moraes CM, Lewis WJ, Pare PW, Alborn HT, Tumlinson JH (1998) *Nature* 393:570–573.
19. Du YJ, Poppy GM, Powell W, Pickett JA, Wadhams LJ, Woodcock CM (1998) *J Chem Ecol* 24:1355–1368.
20. Alborn HT, Turlings TCJ, Jones TH, Stenhagen G, Loughrin JH, Tumlinson JH (1997) *Science* 276:945–949.
21. Alborn HT, Jones TH, Stenhagen GS, Tumlinson JH (2000) *J Chem Ecol* 26:203–220.
22. Alborn HT, Brennan MM, Tumlinson JHT (2003) *J Chem Ecol* 29:1357–1372.
23. Mori N, Alborn HT, Teal EA, Tumlinson JH (2001) *J InsectPhysiol* 47:749–757.
24. Pohnert G, Jung V, Haukioja E, Lempa K, Boland W (1999) *Tetrahedron* 55:11275–11280.
25. Paré PW, Alborn HT, Tumlinson JH (1998) *Proc Natl Acad Sci USA* 95:13971–13975.
26. Schmelz EA, Carroll MJ, LeClere S, Phipps SM, Meredith J, Chourey PS, Alborn HT, Teal PEA (2006) *Proc Natl Acad Sci USA* 103:8894–8899.
27. Musser RO, Hum-Musser SM, Eichenseer H, Peiffer M, Ervin G, Murphy JB, Felton GW (2002) *Nature* 416:599–600.
28. Turlings TCJ, McCall PJ, Alborn HT, Tumlinson JH (1993) *J Chem Ecol* 19:411–425.
29. Schmelz EA, Alborn HT, Banchio E, Tumlinson JH (2003b) *Planta* 216:665–673.
30. Mee JML, Korth J, Halpern B (1977) *Biomed Mass Spectrom* 4:178–181.
31. Yoshinaga N, Aboshi T, Ishikawa C, Fukui M, Shimoda M, Nishida. R., Lait CG, Tumlinson JHT, Mori N (2007) *J Chem Ecol* 33:1376–1381.
32. De Moraes CM, Mescher MC (2004) *Proc Natl Acad Sci USA* 101:8993–8997.
33. Costa A, Werneck CC, Nasciutti LE, Masuda H, Atella GC, Silva LCF (2001) *Insect Biochem Mol Biol* 31:31–40.
34. Giorgi F, Cecchettini A, Locci MT, Masetti M, Bradley JT (1995) *Develop Biol* 167:379–387.
35. Giorgi F, Falleni A, Cecchettini A, Gremigni V (1998) *Biol Cell* 90:183–197.
36. Ortego F, Evans PH, Bowers WS (1997) *J Chem Ecol* 23:1941–1950.
37. Sword GA (2001) *Oecologia* 128:416–421.
38. Calcagno MP, Avila JL, Rudman I, Otero LD, Alonso-Amelot ME (2004) *Physiol Entomol* 29:123–128.
39. Gidda SK, Miersch O, Levitin J, Wasternack C, Varin L (2003) *J Biol Chem* 278:17895–17900.
40. Bernays EA, Chapman RF (1977) *Ecol Entomol* 2:1–18.
41. Capinera JH (1993) *Pop Ecol* 22:127–133.
42. Beroza and Bierl (1966) *Anal Chem* 38:1976–1986.

W. Joe Lewis

Agricultural Research Service
US Department of Agriculture, Tifton, Georgia, USA

2008 — for remarkable discoveries of mechanisms governing plant-insect and plant-plant interactions. His scientific contributions on chemical ecology have fostered the development of integrated pest management and significantly advanced agricultural sustainability.

CURRICULUM VITAE

Supervisory Research Entomologist (Super Grade), United States Department of Agriculture, Agricultural Research Service (USDA-ARS); Crop Protection and Management Research Laboratory; Tifton, Georgia, 31793, USA (Retired September, 2006)

EDUCATION:

Ph.D.; 1968	Mississippi State University; major, Entomology; minor, Zoology
M.S.; 1965	Mississippi State University; major, Entomology; minor, Genetics
B.S.; 1964	Mississippi State University; major, Entomology; minor, Composite; (Outstanding Senior of The Year Award, Department of Entomology)
	A.A. 1962 Copiah-Lincholn Jr. College; major, Engineering.

RESEARCH EXPERIENCE:

1999-2006	Supervisory Research Entomologist, Crop Protection and Management Research Unit, USDA-ARS, Tifton, GA.

1987-99	Research Entomologist, Insect Biology and Population Management Research Laboratory, USDA-ARS, Tifton, GA.
1984-87	Research Entomologist, Insect Attractants, Behavior and Basic Biology Research Laboratory, USDA-ARS, Gainesville, FL. (Duty location, Tifton, GA)
1971-2006	Adjunct Faculty, Department of Entomology and Nematology, University of Florida.
1969-2006	Adjunct Faculty, Department of Entomology, University of Georgia.
1967-84	Research Entomologist, Southern Grain Insects Research Laboratory, Tifton, GA, USDA-ARS, Tifton, GA.
1965-67	Research Assistant, Mississippi Agricultural Experiment Station, Mississippi State, MS.

Notable Awards and Recognitions:

- Awarded Wolf Prize in Agriculture (with Drs. J. H. Tumlinson and John S. Pickett), 2008, by Wolf Foundation Council, Israel.
- Received Special Congressional Recognition for Outstanding Achievement, Service, and Public Distinction, in recognition of scientific and community contributions, Congressman Sanford D. Bishop, Jr., 2006.
- Granted Super Grade Rank, for Meritorious Service, Secretary of Agriculture, USDA, ARS, 2005.
- Awarded Jean-Marie Delwart Foundation International Prize (with Dr. J. H. Tumlinson) for Chemical Communication, 2003.
- E. T. York Lecturer, Auburn University, Auburn, AL, 2003.
- Awarded Invitational Fellowship for Research, Japan Society for the Promotion of Science, 2003.
- Elected as Fellow, Georgia Entomological Society, 2001.
- Received Outstanding Senior Scientist of Year Award–Nationwide, USDA-ARS, 1999.
- Together with Drs. J. H. Tumlinson, USDA-ARS and T. C. Baker, Iowa State University, was awarded a $4.3 Million grant from DARPA (Defense Advanced Research Programs Agency) to investigate the prospects using the chemical detection and learning capacity of parasitic wasps for deployment in security, agricultural, and human health environments. Breakthrough findings have emerged from the project, and Dr. Lewis is the lead inventor on a patent application on such uses filed through the USDA-ARS., 1998-2002.
- Received a combination of USDA, EPA, Georgia Cotton Commission, and private Foundation funds (over $200,000) to lead an interdisciplinary/multi-agency team of scientists in the pilot demonstration of a total systems approach for sustainable cotton production directly in cooperation with producers in Georgia and the Southeast in the interest of natural resource conservation and sustainable production systems, 1966-1998.

- Appointed by the National Research Council/National Academy of Sciences, Board on Agriculture to serve on a Panel currently charged with reviewing and developing a comprehensive report on the status, barriers, and opportunities for biological/natural control of pests, 1992-1993.
- Presented with the prestigious Founder's Memorial Lecturer Award by the Entomological Society of America, for outstanding research in understanding natural enemies and application in biological control, 1990.
- Achievement Award for Collaborative Research (Co-recipients: J. H. Tumlinson; T. C. J. Turlings), The Florida Entomological Entomological Society, for outstanding research accomplishments related to semiochemicals and parasitoid behavior which resulted in publications appearing simultaneously in Science and Nature. 1991.
- Requested and appointed by Academic Press to organize, launch, and serve, along with Drs. R. Charudatton (University of Florida), Harry Kaya (University of California, Davis), and Charlie Rogers (ARS, Tifton) as Charter Editors of a new Journal, "Biological Control: Theory and Application", the first such journal dedicated to the subject of biological control. The publication has become a well-established international journal, 1991-present.
- Together with Dr. J. H. Tumlinson organized and co-led a USDA-OICD funded program of cooperative research with colleagues in The Netherlands and France (1985-90). Hosted 7 visiting scientists from these countries, generally graduate students or post-doctorates, to conduct research for 1 to 2 years, in his laboratory under his supervision.
- Awarded invitational Visiting Professor assignment, Division of Biological Control, University of California, Berkeley, April 1 - June 30, 1980.
- Granted $350,000 from the Research Development and Pilot Test Funds (1973-77) to pioneer the role of kairomones in the foraging behavior of parasitic insects.
- Together with D. A. Nordlund and R. L. Jones, was contracted by John Wiley and Sons, Inc. to edit a book on *Semiochemicals: Their Role in Pest Control*, the first comprehensive coverage of this subject. The incumbent also authored the introductory and co-authored the summary and conclusion chapters.
- Selected to serve on a National Academy of Science delegation of Biological Control Specialists to visit and exchange Biological Control Technology with the People's Republic of China, 1982.

Selected Invitational Presentations From Recent Years:

"Foraging for Host and Food Resources: Comparisons, Contrasts, and Interacting Influences," in a Symposium on "Critical Issues In Host Selection Behavior," International Congress of Entomology, Florence, Italy, August, 1997. (Invitational).

"Understanding an Fostering Natural Enemies as a Component of the Agroecosystem," XVII Congress of Entomology, Rio de Janeiro, Brazil, August, 1998. (Invitational)

Invited to serve as a Discussant in a Novartis Foundation Symposium on "Insect-Plant Interactions and Induced Plant Defense," London, United Kingdom, October, 1998.

"Sustainable Farm Systems Applied to Your Community," 14th Annual Sustainable Agriculture Conference, Carolina Farm Stewardship Association, Highpoint, North Carolina, November, 1999. **(Keynote Address)**.

"Building Sustainable Communities and Schools," Future Farms 2000 Conference: New Ideas for Family Farms and Rural Communities," Kerr Center for Sustainable Agriculture, Oklahoma City, Oklahoma, February, 2000. **(Keynote Address)**.

"Sustainable Pest Management: Building It In As A Part Of The System," **(Keynote Address)**, and "Agriculture As A Part Of Ecologically-Based Communities," **(After Dinner Address)**, Georgia Organics Association Annual Meeting, Fort Valley, Georgia, March, 2000.

"Mechanisms of Detection of Chemical Signals by Parasitic Wasps," Controlled Biological And Biomimetic Systems and Tissue Based Biosensors Meeting 2000, San Antonio, TX, Sponsored by DARPA/DOD, April, 2000. (Invitational).

"Achieving Ecologically-Based Communities," as part of a symposium on "Economic Development Should Mean More Than Just Making Money," (Invitational) National Community Service Conference 2000. Orlando, FL, June, 2000.

"The Role of Extrafloral Nectar in The Tritrophic Interactions of Plants, Herbivores, and Parasitoids," in a symposium on "The Effect of Plant-Derived Food Supplements on Tri-Trophic Interactions," (Invitational), XXI International Congress of Entomology, Iguassu Falls, Brazil, August, 2000.

"Fostering Ecologically-Based Agriculture as A Component of Sustainable Communities," (Invitational Seminar), Embrapa National Center of Agrobiology Research and Federal Rual University of Rio De Janeiro at Seropedica, Seropedica, Brazil, August, 2000.

"A Whole-System Approach to Pest Management," and "Using Ecological Principles to Help Develop Sustainable Human Communities," (Invitational Seminars), University of Vermont, Burlington, Vermont, October, 2000.

"Sustainable Management of Insect Pests: Building it into the System," Symposium on Integrated Agricultural Systems, (Invitational), 2000 Annual Meeting of the American Society of Agronomy/Crop Science Society of America/Soil Science society of America, Minneapolis, MN, November, 2000.

"Whole-System vs. Reactive Approaches: What's the Big Deal," **(Keynote Presentation)** Workshop on "A Systems Approach to Pest Management in Fresh

Market Vegetables," Sponsored by Northeast Region SARE, Simsbury, CT, December, 2000.

"Ecosystem Component Interactions" and "Human Communities as Living Dynamic Systems," (Invitational) Conference on Systems In Agriculture and Land Management, Sponsored by Holistic Resource Management of Texas, Inc. and Texas Agricultural Extensions Service, Fort Worth, TX March, 2001.

"From Biological Control to Sustainable Agriculture and Rural Communities," Georgia Entomological Society Annual Meeting **(Invited presentation in connection with election as a Fellow of the Georgia Entomological Society)**, April, 2001.

"Ecologically-Based Agriculture and Rural Communities" (Invitational), Professional Consultation on "Economy, Ecology, and Community – Building a Win-Win Agriculture" Sponsored by Leopold Center for Sustainable Agriculture, Ames, IA, July, 2001.

"Ecologically-Based Communities: From School House to Farm House," **(Keynote Address)**, "Integrated Pest Management – Row crop practices that foster the natural strengths of our agro-ecosystems," (Invitational), 13th Annual Southern Sustainable Agricultural Working Group, Chattanooga, TN, January, 2002.

"Biological Control as a Component of Sustainable Agriculture," (Invitational) Virginia Biological Farming Conference, Front Royal VA, February, 2002.

"An Overview of Whole Systems Pest Management," (Invitational) in symposium on "Putting Whole System Pest Management into Practice," 4th IPM Work Conference, Indianapolis, IN, April, 2003.

Series of lectures on "Insect Parasitoids and Sustainable Agriculture," (Invitational) Kyushu University, University of Kyoto, and Saga University, Japan, November, 2003.

"Intrinsic-Based Education as Derived from Ecological Models," (Invitational) Faculty of Education and Instituted of Ecological Economics, University of Vermont, Burlington, VT, October, 2004.

Key Extension, Administrative and Outreach activities include:

- Served as a collaborator with the International Communication Agency in their preparation of an exhibit on "U. S. Advances in Control of Crop Insect Pests," Peking, China, 1979.
- Appointed and served on the SEA-USDA Southern Region Advisory Committee on Biological Control and Pilot Testing, 1973-80.
- Appointed as an ARS key scientist on the Maize Work Team for Exchange of Biological Agents Between the USSR and USA, 1977-81.
- Assigned as leader of behavioral manipulation component of a multi-location pilot test on use of *Trichogramma* for control of *Heliothis* on cotton, 1979-1982.

- Served as Chair of the Advisory Committee on Integrated Pest Management to the Tall Timbers Research Foundation, 1979-1985.
- At the request and under the joint sponsorship of the Intersociety Consortium of Plant Protection, Tall Timbers Research Foundation and Winrock International, co-organized a forum (Federal and State administrators and scientists) to explore various plans and alternative ways of coordinating national efforts on biological control. A plan was proposed by which USDA-SEA, professional societies, and private foundations could work jointly to develop a national research program on biological control. Morrilton, AR, 1980.
- Assigned (along with J. H. Tumlinson) by Office of International Cooperative Development to visit The Netherlands and France, including organizing a multinational workshop in France to develop a U.S.-French-Dutch cooperative research program on Semiochemicals and Biological Control, 1984.
- At the request of the Organizational Committee of International Vedalia Symposium on Biological Control (centennial celebration of Biological Control), organized and moderated a workshop on "The Role of Behavior and Semiochemicals in the Interaction Between Natural Enemies and Their Hosts or Prey," Riverside, CA, 1989.
- Appointed by the Area Director, South Atlantic Area, USDA-ARS to serve as agency representative and facilitator in the ARS Regional Vision Development Conferences held at Beltsville, MD and Atlanta, GA, June and July, 1995, as part of the Agency's in-depth strategic planning effort.
- Appointed by the National Research Council/National Academy of Sciences, Board on Agriculture to serve on a Panel charged with reviewing and developing a comprehensive report on the status, barriers, and opportunities of biological/natural control of pests, 1993-1995. The committee's report, "Ecologically Based Pest Management: New Solutions for a New Century," was published and released in April, 1996.
- At the invitation of Sustainable Agricultural Network: The National Outreach Arm of the USDA Sustainable Agricultural Research and Education (SARE) program, served on a task force to develop a brochure, "*A Whole-Farm Approach to Managing Pests: Naturalize Your Farming System*," (Bulletin, 5/2000) for which the incumbent's work (# 151) served as a guiding basis. He is currently working with them on similar follow-up literature. He also frequently serves on the review panel for the SARE, Southern Region to evaluate proposals for their grants. 1998-2004.
- Advised a study group of Defense Sciences Research Council on discoveries in 'search and find' aspects of tritrophic interactions and applications of this knowledge to technology for identification of buried targets, Herndon, VA, October, 1997, San Diego, CA, August, 1998, Arlington, VA, April, 1999. A $25 million program by DARPA-DOD on Controlled Biological Systems, including a

$4.3 million project by incumbent in cooperation with scientists at Gainesville, FL and Ames, IA on use of parasitic wasps as chemical detectors was guided by these and related discussions.

- At the request of USDA Council on Sustainable Development along with Communities In Schools, Inc., participated in a planning workshop to develop programmatic and funding strategies to better foster sustainable economic development and educational outreach activities for rural communities, Washington, DC, September, 1999.
- Invited to serve on a panel member at a Workshop on Bt Crop Resistance Management, sponsored by EPA and USDA to determine current and future issues on refuge design and deployment, education and compliance, and monitoring and remedial action as related to management of resistance to Bt toxins in transgenic plants and to develop plans for future guidelines and policies on these matters, Memphis, TN, August, 1999.
- Led an interdisciplinary/multi-agency team of scientists (involving soil and water, plant pathology, horticulture, entomology, including scientists, extension specialists, agricultural consultants, NRCS representatives, Georgia Cotton Commission, and grower representatives) to launch a project on a total systems approach for sustainable cotton production directly in cooperation with the producers in Georgia and the Southeast in the interest of natural resource conservation and sustained economic competitiveness, 1998-2002.
- At the request of American Farmland Trust participated as an advisor/discussant in a series of meetings with partners stakeholders, including USDA, EPA, and farmers/producers to discuss programs to advance community-wide/whole systems agriculture, Memphis, TN, August, 1999; Bartowe/Douglas, GA, March, 2000; Washington, DC, April, 2000.
- Invited to serve as a Professional Consultation Discussant on "Economy, Ecology, and Community – Building a Win-Win Agriculture," as part of a panel sponsored by Leopold Center for Sustainable Agriculture, Ames, IA, to develop their future objective and plans, July, 2001.
- Invited and sponsored by the Japan Society for Promotion of Science to visit, lecture, and advise faculty and students for 14 days at Kyushu University, University of Kyoto, Saga University, and other institutions to strengthen programs on insect parasitoids and sustainable agriculture at these institutions, November 2003.

Professional Organizations and Leadership Activities:

Dr. Lewis's has an extension record in service activities involving membership and activities in numerous professional and honor societies including the Entomological Society of America, International Society of Chemical Ecology, Georgia Entomological Society, International Working Group on Entomophagous Insects, American

Association for the Advancement of Science, etc. His services have consisted of numerous Office roles, organizing many symposia/conferences, editorial duties, and a range of professional board activities. Specific examples include:

- Member, Resolutions Committee, Georgia Entomological Society, 1970.
- Member, Paper Evaluation Committee, Georgia Entomological Society, 1970.
- Member, Program Committee, Tifton Sigma Xi Club, 1972-74.
- Member, Committee to Select Future Site, Georgia Entomological Society Meeting, 1975.
- Designated, under the auspices of the Biological Control and Behavioral Sections, to organize and co-moderate, along with Dr. K. S. Hagen, a conference on kairomones and Natural Enemy Behavior at XV International Congress of Entomology, 1976.
- Member of the Biological Control Subcommittee and selected as Group Leader for Behavioral Manipulation of Entomophagous Arthropods: Technical Committee, Regional Research Project S-59, 1973-85.
- Assigned, under auspices of Section C (together with Dr. Richard L. Jones and Donald A. Nordlund), to organize a symposium on "Recent Advances in Biological Control Technology: Interactions of Entomophages and Semiochemicals," National Meeting of the Entomological Society of America, Houston, TX, 1978.
- Member of Committee for Section VI: Biological Component of Plant Protection for IX International Congress of Plant Protection, Washington, DC, and was organizer and moderator of a Congress Symposium on a Behavioral Chemicals: Role and Employment in Plant Protection," 1979.
- Appointed to chair a Task Force Committee to prepare a report on "The Status and Potential Use of Behavioral Chemicals in Pest Management" for the FAO/UNEP Panel of Experts on Integrated Pest Control, 1979.
- Member, Program Committee, Georgia Entomological Society, 1980-81.
- President, Southeastern Biological Control Working Group, 1981-82.
- Member, Examining Board (Biological Control Specialty Area), American Registry of Professional Entomologists, 1985-91.
- Secretary, Subsection Ca, Entomological Society of America, 1985.
- Vice-Chair, Subsection Ca, Entomological Society of America, 1986.
- Chair, Subsection Ca, Entomological Society of America, 1987.
- Member, Special Committee on Professional Affairs/Membership Services, American Registry of Professional Entomologists, 1986-87.
- Member, Awards Committee, Southeastern Branch, Entomological Society of America, 1987-88.
- Requested and appointed by Academic Press to organize, launch, and serve, along with Drs. R. Charudatton (University of Florida), Harry Kaya (University of California, Davis), and Charlie Rogers (ARS, Tifton) as Charter Editor of a new Journal, "Biological Control: Theory and Application," the first such journal

dedicated to the subject of biological control. The journal has been highly successful and is established as major forum for the discipline. He served as Editor, 1989-1998 and on Editorial Board 1998-2005.

- Nominations Committee for Nomination of Officers, Georgia Entomological Society, 1998.

SUMMARY STATEMENT OF PROFESSIONAL ACCOMPLISHMENTS AND ACTIVITIES

Dr. W. Joe Lewis grew up as a sharecropper's son in south Mississippi. He began his career as a Research Entomologist at Mississippi Agricultural Experiment Station, Mississippi State, MS in 1965. Upon completing his B.S. degree in 1964, he continued his graduate studies and received a M.S. in 1965 and Ph.D. in 1968 in Entomology from Mississippi State University. From 1967 until his retirement in 2006, he was a research entomologist in the USDA, ARS, Crop Protection and Management Research Unit, Tifton, GA. – with joint faculty appointments with the University of Georgia and University of Florida. He is Vice-Mayor of the City of Tifton, a role he has served for over ten years with twelve years as an elected member to the City Council – and recently was elected, unapposed, for a new 4-year term. He is married (Wife, Beth) with two children (Son, Alan Lewis and Daughter, Joanne DeLoach) and four grandchildren.

Dr. Lewis is recognized worldwide for major crosscutting impact on the fundamental science of pest management, and on sustainable agriculture and community development practices through research unraveling how ecosystems work. The models for his studies have been behavioral and chemical interactions between plants and insects, but the applications of his findings are of broad significance. This is evidenced by over 200 refereed scientific publications and book chapters, including four papers in prestigious journals of Nature and Science, and three in Proceedings of the National Academy of Sciences, and invitational paper in Scientific American. His work has been highlighted extensively in the popular press, including CNN Science and Technology, BBC/ Discovery Channel, Business Week, Kiplinger Agricultural Letter, New York Times, Organic Gardening, Intervoice, National Public Radio and BBC Wildlife, Fortune Magazine, Atlanta Magazine, and NBC Today Show. Radio Netherlands produced and broadcast, internationally, a half hour documentary, Rural Renaissance, based on Dr. Lewis' ecologically-based rural community studies.

In addition to his recent selection by the Wolf Foundation of Israel for the world renown "Wolf Prize in Agriculture," his numerous honors, awards, and achievements include the prestigious Founder's Memorial Lecturer Award by the Entomological Society of America; Honorary Visiting Professor, University of California, Berkeley; the E. T. York Distinguished Lecturer, Auburn University; the

C. P. Alexander Distinguished Lecturer, University of Massachusetts; Appointed by Academic Press as Charter Editor of the International Journal Biological Control: Theory and Applications; named U.S. Department of Agriculture-Agriculture Research Service Outstanding Scientist of the Year; received Special Congressional Recognition for Outstanding Achievement, Service and Public Distinction; decorated internationally by the Jean-Marie Delwart Foundation Award and Prize (conferred in Brussels, in conjunction with the Royal Academy of Sciences of Belgium); an Invitational Fellowship for Research by the Japan Society for Promotion of Science for a series of lectures in Japan; and by nomination for Israel's Wolf Foundation Prize in Agriculture. In 2005, Dr. Lewis was appointed to Supergrade rank by the Secretary of Agriculture. Also, he has been awarded numerous major grants, including $4.3 million from the Department of Defense to explore the potential utility of trained insects as biological detectors for monitoring for various materials of medical, agricultural, security, and food safety concerns.

Dr. Lewis receives extensive requests for presentations, including numerous keynote addresses. His extensive international influence on scientists and students is evidenced by a continual string of those who have come from worldwide to work in his lab under his guidance and training, including from the countries of Australia, Brazil, Benin, France, Germany, the Netherlands, Japan, Poland, Sweden, Switzerland and from throughout the U.S.

Dr. Lewis has been very active in a wide range of civic activities. In addition to his role as Vice-Mayor and City Council member for Tifton, he serves as a member of the Downtown Development Authority and is very active in Historic Preservation and Downtown redevelopment. He helped develop and serves as a member of the Service Delivery Coordinating committee between the City of Tifton and Tift County through which an extensive set of joint services have been planned and implemented via inter-local governmental agreements. Prior to being elected to the Tifton City Council, he served as a member of the Greater Tifton/Tift County Planning and Zoning Commission for eight years, as Chair for four years. For over twenty years he served as a member and Chair of the Tift County Board of Elections. He has been active in numerous other aspects of community service, including serving as the Tift County representative to the Regional Board on Mental Health, Mental Retardation and Substance Abuse, charged by the State of Georgia with planning and funding services for these needs in a 10-County area. He and his wife are currently very active in approaches for improving education and workforce quality.

DESCRIPTION OF SCIENTIFIC ACHIEVEMENTS

(Key Research Accomplishments and Impact)

A. **Elucidated the involvement and vital role of chemical cues (kairomones) in host finding activities of parasitoids and cooperated with Dr. R. L. Jones and others to isolate and identify key kairomones for *Microplitis croceipes* and**

Trichogramma evanescens. **(1970-73).** *Impact:* Prior insect behavioral chemical research focused predominantly on intraspecific roles of sex pheromones, while the importance and identity of chemical cues in host finding by beneficial insects were unknown. These findings helped spur a major expansion in the scope of research on odorant mediated behavior to one based on a broader appreciation of semiochemicals as an array of cues that mediate inter- and intraspecific interactions among plants, herbivores, and entomophages. (Publications: #16, #17, and #20)

B. **In cooperation with Drs. R. L. Jones, H. R. Gross, and others discovered that kairomones (which, in the case of *Trichogramma*, was the sex pheromone of the host moth) as extracts or synthetics can be used to manipulate and enhance the searching efficacy of released or natural parasitoids in field plot situations (1975-79).** *Impact:* The use of kairomones as a management tool to augment field releases of parasitoids by altering their behavior was unexplored. This research introduced an entirely new concept for use of parasitoids as pest control agents, which had not been previously considered, perhaps because of the long standing assumption of random search by parasitoids. These findings together with the expanded view of the role of semiochemicals sparked, on an international scale, major new research on semiochemicals/natural enemies as documented in a book edited by Dr. Lewis and his colleagues, "*Semiochemicals: Their Role in Pest Control*" — John Wiley & Sons (1981), the first comprehensive coverage of this subject. (Publications: #23, #32, and #36)

C. **With Post-Doctorates, students, and colleagues developed a flight tunnel system effective for studying in-flight responses of *Microplitis croceipes* and other parasitoids to airborne odors, determined primary sources of host plant and host odors eliciting the responses and influence of various preflight experiences on the response, and analyzed the behavioral sequence involved in the flights (1985-88).** *Impact:* Earlier studies by the incumbent had demonstrated the importance of contact kairomones in host trails to the host finding behavior of *M. croceipes*, whereas oriented responses to airborne odors, though strongly implicated, had gone unexplored because of lack of techniques to study such responses under controlled conditions. These breakthrough developments and findings opened the way for systematic studies of parasitoid responses to airborne odors and was a catalyst for major expansions in research on foraging behavior of parasitoids. For example, following Dr. Lewis' lead and based on training in his laboratory, similar research projects were established at numerous other locations, including the University of Massachusetts, University of Maryland, and at locations in The Netherlands, France, Japan, and Australia. Furthermore, based on technology stemming from these techniques gained while working in the incumbent's laboratory, a Doctoral student, Lucas Noldus, launched and is now Director and major owner of Noldus Information Technology,

findings fill major gaps in understanding past failures and insuring future reliability of biological control agents, and based on these reports similar research has been initiated in the USA, Switzerland, France, and Japan. The sophisticated foraging system of parasitoids demonstrated by Dr. Lewis and others is a central premise for Dr. E. F. Knipling's recent book "*Principles of Insect Parasitism Analyzed from New Perspectives: Practical Implications for Regulating Insect Populations by Biological Means*". Furthermore, these findings in combination with 5 & 6 above have received extensive popular press, including Business Week, Kiplinger Agricultural Letter, Organic Gardening, New York Times, BBC Wildlife, CNN Science and Technology, and documentaries by the discovery channel. (Publications: #106, #140, 147, #155, and #158)

H. **With Dr. J. H. Tumlinson and a team of associated cooperators, used cotton production as a model example, to formulate and demonstrate, a fundamental shift to a total system approach to sustainable pest management involving the use of ecosystem management, crop attributes/multitrophic level interactions, and minimal use of therapeutics — the basis for which is largely drawn from the fundamental discoveries of Dr. Lewis and cooperators (1993-99).** *Impact:* Despite major efforts toward alternatives to conventional pesticides, pest management science has continued by-and-large to regress to a continued array of new therapeutic intervention tools and failed to make a mainstream shift to understanding and redesigning cropping systems so that natural and renewable strengths keep pests within acceptable bounds on a sustained basis. On-farm work, was conducted directly with cotton growers, and showed a $60/acre higher net annual profit along with major benefits of improved soil and water health and reduced environmental contamination. These demonstrations catalyzed rapid expansion of the acreage under sustainable practices in Georgia and other areas of the region. For example, sustainable practices in Coffee County, Georgia increased from 200 acres in the early 1990's to over 30,000 acres. These concepts, approaches, and demonstrations are having major impact on overall crop management philosophy and practices at consultant and grower levels, research directions, and funding and policy decisions. For example, the SARE agency has employed their seminal PNAS paper and associated work for nationwide use as a standard in guiding constituents toward systems approaches, major funding decisions and research directions by EPA, ARS, and other institutions are being shaped by the work, and various institutions are using the work as a key focus in educational/training programs. (Publications: #151, #174, #149, #159, #165 and #175)

I. **In cooperation with Dr. J. H. Tumlinson, Post-Doctorates and other colleagues, and with major funding by the Department of Defense, developed and demonstrated methodology for training parasitic wasps to quickly learn any chemical of interest in association with food or hosts and to display a**

distinct, reliable behavior when these chemicals are subsequently detected (1999-2002). *Impact:* This finding and development opens immense prospects for using insects and other invertebrates as biological sensors, in a similar fashion as dogs, to detect and monitor for conditions/materials of concern in a wide array of situations. The use of insects and related invertebrates has many advantages over dogs for this purpose, including the fact that they can be produced and trained much quicker and less costly. (Publications: #181, #185, and #190)

Teaching and Mentorship Achievements

Dr. Lewis has taught and mentored a continuous string of young scientists from throughout the world who are now leading major programs and initiatives of their own as shown below:

Three Doctoral Graduate students conducted their research directly under Dr. Lewis' supervision in his laboratory: G. Morrison - Ph.D., 1985, University of Georgia, Was selected outstanding graduate student by the Southwestern Branch of ESA during his studies; F. Wäckers – Ph.D., Agricultural University, Wageningen, The Netherlands (Currently Professor/Director, Centre for Sustainable Agriculture Lancaster Environment Centre, Lancaster University, United Kingdom); C. De Moraes - Ph.D., 1998, University of Georgia, (Supported by funds from Brazil); Selected as the outstanding Ph.D. student of the Department of Entomology and for the College of Agriculture and Environmental Sciences (E. Broadus Browne Award), Currently on the faculty at Pennsylvania State University. *Also, he has served on the committee of five Doctoral and two Master's Degree graduate students.*

Twelve Post-Doctorates have conducted research studies in Dr. Lewis' laboratory under his guidance: M. Altieri, 1979-1980, Currently a Professor, University of California, Berkeley; M. Keller, 1984-86, Currently an Associate Professor, University of Adelaide, Australia, Waite Campus and Deputy Head of the School of Agriculture Food and Wine; D. W. Whitman, 1986-1988, Currently Professor and Insect Curator, Illinois State University; T. Mueller, 1987-1988, Currently an Entomologist, Collier Enterprises, Immokalee, FL; W. Martin, 1988-1990; Currently an Entomologist with Biosys, Gainesville, FL; W. Sheehan, 1990-1992, Currently employed with an environmental consulting and waste management firm, Athens, GA; K. Takasu, 1990-1994, (Sponsored by Funds from Japan) Currently on the Faculty of Agriculture, Kobe University, Kobe 657, Japan; J. Ruberson, 1992-1994, Currently an Associate Professor, University of Georgia, Tifton, GA; A. Cortesero, 1993-1997, (Sponsored by funds from France) Currently on Faculty of University of Renne, Renne, France; D. Olson, 1998-2000, Currently a Research Entomologist, USDA-ARS, Tifton, GA; J. Tomberlin, 2001-2002, Currently on Faculty at Texas A&M University; M. Tertuliano, 2000-Present.

Fourteen Visiting Scientists have conducted visiting research assignments in Dr. Lewis' laboratory under his guidance: J. Dmoch - 1981, Professor, University of Warsaw, Poland (Supported by funds from Poland); Y. Drost -1984-86, (OICD funds) Visiting Scholar, Groningen University, The Netherlands, eighteen-month visit; O. Zanen - 1985-86, Visiting Scholar (OICD funds), Groningen University, The Netherlands, one-year visit; L. Noldus -1986, Visiting Scholar (OICD funds), Agricultural University, Wageningen, The Netherlands, one-year visit, currently Director and major owner of Noldus Information Technology, Wageningen, The Netherlands, an enterprise that develops computer software and instrumentation for recording and analyzing behavioral studies of animals, humans and machines (currently marketed in 50 countries); F. Hérard -1985-86, Research Entomologist, USDA-ARS, European Parasite Laboratory, Behoust, France, (supported by funds from home laboratory), one-year visit; G. Prevost - 1986-1989, (OICD funds) Visiting Scholar, University of Lyon, Lyon, France, two-year visit; W. van Giessen - 1988, (OICD funds) Visiting Scholar, Agricultural University, Wageningen, The Netherlands, six-month visit; L. Vet -1988, (OICD funds) Professor, Agricultural University, Wageningen, The Netherlands, six-week visit; Janet Strong-Gunderson, 1992, Visiting Scientist, Oak Ridge Associated Universities Laboratory, Oak Ridge, TN, two-month visit; A. Demata, 1993, Visiting Scholar Agricultural University, Wageningen, The Netherlands, six-month visit; O. Stapel, 1993-1997, Visiting Scholar, Agricultural University, Wageningen, The Netherlands. two-year visit with extension; T. Meiners, 1999, Visiting Scientist (DOD funds), Institut fuer Biologie, Angewandte Zoologie/Oekologie der Tiere, Freie Universitaet Berlin Haderslebener, Germany, six-month visit, C. Bonifay, 1999, Visiting Scientist (DOD funds), Swiss Federal Institute of Technology, Zurich, Switzerland three-month visit; M. D'Alessandro, 2000, Visiting Scientist (DOD funds), Swiss Federal Institute of Technology, Zurich, Switzerland, six-month visit.

Extension, Administrative, and Other Outreach Achievements:

In addition to his primary emphasis on research, Dr. Lewis has had major involvement and impact in Extension and Administrative areas. On-farm demonstrations and education by Dr. Lewis and cooperators catalyzed rapid expansion of the acreage under sustainable practices in Georgia and other areas of the region. For example, sustainable practices in Coffee County, Georgia increased from 200 acres in the early 1990's to over 30,000 acres. These concepts, approaches, and demonstrations are having major impact on overall crop management philosophy and practices at consultant and grower levels, research directions, and funding and policy decisions. For example, the SARE agency has employed their seminal PNAS paper and associated work for nationwide use as a standard in guiding constituents toward systems approaches, major funding decisions and research directions by EPA, ARS, and other institutions are being shaped by the work, and various institutions are using the work as a key focus in educational/ training programs.

Also, Dr. Lewis and cooperators formulated principles, derived from natural ecosystems, for guiding sustainable human communities, including publishing a benchmark publication (through the Kerr Center for Sustainable Agriculture and Rural Communities) with an analysis of how modern trends are eroding inherent strengths in local communities and proposed fundamental redirections for achieving ecologically-based human communities, and is working directly with local governments, educational systems, business leaders, and advocacy organizations to implement redirections toward sustainable communities. As Vice–Mayor, of his hometown, Tifton, Georgia, he has led implementation of these principles in successful "smart growth" land-use patterns, greenway development, historic preservation, and local "self-sufficiency" in waste management and fiber optic/ digital technology. Tifton's success is highly recognized at the State level and used through the Georgia Municipal Association and Department of Community Affairs as a model for wise strategies in sustainable community development (He was presented Special Congressional recognition in May, 2000). More recently, with the cooperation of key education organizations, he is underway with the development of pilot programs for schools in South Georgia to connect students with ecological values and the importance of sustainable living practices. A Radio Netherlands documentary, Rural Renaissance, broadcast internationally, recently highlighted his work in Tifton. He is working in cooperation with the Kerr Center, SARE-USDA, and other organizations to implement these sustainable practices on a National and International basis.

LIST OF PUBLICATIONS

1. Vinson, S. B. and Lewis, W.J. A method of host selection by Cardiochiles nigriceps. J. Econ. Entomol. 58: 869-871. 1965.
2. Lewis, W.J. and Brazzel, J.R. Biological relationships between Cardiochiles nigriceps and the Heliothis complex. J. Econ. Entomol. 59: 8320-823. 1966.
3. Lewis, W. J. and Brazzel, J. R. Incidence of parasitic insects on the Bollworm in Mississippi. Mississippi State Univ. 727: 1-7 pp. 1966.
4. Lewis W. J. and Brazzel, J. R.and Vinson, S.B. Heliothis subflexa, a host for Cardiochiles nigriceps. J. Econ. Entomol. 60: 615-616. 1967.
5. Lewis, W. J. Studies of the suitability of certain Heliothis spp. as a host for Cardiochile nigriceps Vierick. Ph.D. Thesis Miss. State Univ. pp. 56. 1968.
6. Lewis, W.J. and Brazzel, J.R. A three-year study of parasites of the bollworm and the tobacco budworm in Mississippi. J. Econ. Entomol. 61: 673-676. 1968.

7. Lewis, W. J. and Vinson, S. B. Egg and larval development of Cardiochiles nigriceps. Annals of the Entomological Society of America 61(3): pp. 561-565. 1968.
8. Lewis, W. J. and Vinson, S. B. Immunological relationships between the parasite Cardiochiles nigriceps Vierick and certain Heliothis species. J. Insect. Physiol. 14: pp. 613-626. 1968.
9. Lewis, W. J. and Redlinger, L. M. Suitability of eggs of the almond moth, Cadra cautella, of various ages for parasitism by Trichogramma evanescens. Ann. Entomol. Soc. Am. 62: 1482-1484. 1969.
10. Vinson, S. B. and Lewis, W. J. Electrophoretic study of protein changes between several developmental stages of three species of Heliothis (Lepidoptera: Noctuidae). Comp. Biochem. Physiol. 28: 215-220. 1969.
11. Lewis, W. J. Life history and anatomy of Microplitis croceipes (Hymenoptera: Braconidae), a parasite of Heliothis spp. (Lepidoptera: Noctuidae). Ann. Entomol. Soc. Am. 63: 67-70. 1970.
12. Lewis, W. J. Study of species and instars of larval Heliothis parasitized by Microplitis croceipes. J. Econ. Entomol. 63: 363-365. 1970.
13. Lewis, W. J. and Burton, R. L. Rearing Microplitis croceipes in the laboratory with Heliothis zea as hosts. J. Econ. Entomol. 63: 656-658. 1970.
14. Snow, J. W., Lewis, W. J., and Sparks, A. N. Mating of normal and sterilized colonized and native corn earworms with emphasis on failure of pairs to separate after copulation. J. Econ. Entomol. 63: 1873-1876. 1970.
15. Jones, R. L. and Lewis, W. J. Physiology of the host parasite relationship between Heliothis zea and Microplitis croceipes. J. Econ. Entomol. 17: 921-927. 1971.
16. Jones, R. L., Lewis, W. J., Bowman, M. C., Beroza, M., and Bierl, B. A. Host-seeking stimulant for parasite of corn earworm: isolation, identification, and synthesis. Science 173: 842-843. 1971.
17. Lewis, W. J. and Jones, R. L. Substance that stimulates host-seeking by Microplitis croceipes (Hymenoptera: Braconidae), a parasite of Heliothis species. Ann. Entomol. Soc. Am. 64: 471-473. 1971.
18. Lewis, W. J. and Snow, J. W. Fecundity, sex ratios, and egg distribution by Microplitis croceipes, a parasite of Heliothis. J. Econ. Entomol 64: 6-8. 1971.
19. Lewis, W. J., Snow, J. W., and Jones, R. L. A pheromone trap for studying populations of Cardiochiles nigriceps, a parasite of Heliothis virescens. J. Econ. Entomol. 64: 1417-1421. 1971.
20. Lewis, W. J., Sparks, A. N., and Redlinger, L. M. Moth odor: A method of host-finding by Trichogramma evanescens. J. Econ. Entomol. 64: 557-558. 1971.
21. Lewis, W. J. and Vinson, S. B. Suitability of certain Heliothis (Lepidoptera: Noctuidae) as hosts for the parasite Cardiochiles nigriceps. Ann. Entomol. Soc. Am. 64: 970-972. 1971.

22. Barras, D. J., Kisner, R. L., Lewis, W. J., and Jones, R. L. Effects of the parasitoid, Microplitis croceipes, on the haemolymph proteins of the corn earworm, Heliothis zea. Comp. Biochem. Physiol. 43: 940-947. 1972.
23. Lewis, W. J., Jones, R. L., and Sparks, A. N. A host-seeking stimulant for the egg parasite Trichogramma evanescens: its source and a demonstration of its laboratory and field activity. Ann. Entomol. Soc. Am. 65: 1087-1089. 1972.
24. Lewis, W. J., Sparks, A. N., Jones, R. L., and Barras, D. J. Efficiency of Cardiochiles nigriceps as a parasite of Heliothis virescens on cotton. Environ. Entomol. 1: 468-471. 1972.
25. Lewis, W. J. and Young, J. R. Parasitism by Trichogramma evanescens of eggs from tepa-sterilized and normal Heliothis zea. J. Econ. Entomol. 65: 705-711. 1972.
26. Snow, J. W., Sparks, A. N., and Lewis, W. J. Seasonal capture of corn earworm adults in light traps near Tifton, Georgia, compared with captures in traps baited with virgin females. J. Georgia Entomol. Soc. 7: 85-89. 1972.
27. Snow, J. W., Young, J. R., Lewis, W. J., and Jones, R. L. Sterilization of adult fall armyworms by gamma irradiation and its effect on competitiveness. J. Econ. Entomol. 65(5): 1431-1433. 1972.
28. Jones, R. L., Lewis, W. J., Beroza, M., Bierl, B. A., and Sparks, A. N. Host-seeking stimulants (kairomones) for the egg parasite, Trichogramma evanescens. Environ. Entomol. 2: 593-596. 1973.
29. Vinson, S. B. and Lewis, W. J. Teratocytes: growth and numbers in the hemocoel of Heliothis virescens attacked by Microplitis croceipes. J. Invertebr. Pathol. 22: 351-355. 1973.
30. Nordlund, D. A., Lewis, W. J., Gross, H. R., Jr., and Harrell, E. A. Description and evaluation of a method for field application of Heliothis zea eggs and kairomones for Trichogramma. Environ. Entomol. 3: 981-984. 1974.
31. Gross, H. R., Jr., Lewis, W. J., Jones, R. L., and Nordlund, D. A. Kairomones and their use for management of entomophagous insects: III. Stimulation of Trichogramma achaeae, T. pretiosum, and Microplitis croceipes with host-siiking stimuli at time of release to improve their efficiency. J. Chem. Ecol. 1: 431-438. 1975.
32. Lewis, W. J., Jones, R. L., Nordlund, D. A., and Sparks, A. N. Kairomones and their use for management of entomophagous insects: I. Evaluation for increasing rates of parasitization by Trichogramma spp. in the field. J. Chem. Ecol. 1: 343-360. 1975.
33. Lewis, W. J., Jones, R. L., Nordlund, D. A., and Gross, H. R., Jr. Kairomones and their use for management of entomophagous insects: II. Mechanism causing increase in rate of parasitization by Trichogramma spp. J. Chem. Ecol. 1: 349-360. 1975.

34. Copeland, E. L., Young, J. R., and Lewis, W. J. Labeling of Trichogramma pretiosum by rearing on eggs from 32P-fed adults of Heliothis zea. Ann. Entomol. Soc. Am. 69: 804-806. 1976.
35. Jones, R. L., Lewis, W. J., Gross, H. R., Jr., and Nordlund, D. A. Use of kairomones to promote action by beneficial insect parasites. ACS Symp. Series No. 23, Pest Management with insect Sex Attractants and Other Behavior Controlling Chemicals. pp. 119-134. 1976.
36. Lewis, W. J., Jones, R. L., Gross, H. R., Jr., and Nordlund, D. A. The role of kairomones and other behavioral chemicals in host finding by parasitic insects. Behav. Biol. 16: 267-289. 1976.
37. Lewis, W. J., Jones, R. L., Nordlund, D. A., and Gross, H. R., Jr. Kairomones and their role in pest management, pp. 7-14. Proceedings 8th Ann For Insect Work Conf, Hamden, CT, USDA For. Serv. Gen. Tech. Rep. NE-27. 20 pp. 1976.
38. Lewis, W. J., Nordlund, D. A., Gross, H. R., Jr., Perkins, W. D., Knipling, E. F., and Voegele, J. Production and performance of Trichogramma reared on eggs of Heliothis zea and other hosts. Environ. Entomol. 5: 449-452. 1976.
39. Nordlund, D. A. and Lewis, W. J. Terminology of chemical releasing stimuli in intraspecific and interspecific interactions. J. Chem. Ecol. 2: 211-220. 1976.
40. Nordlund, D. A., Lewis, W. J., Jones, R. L., and Gross, H. R., Jr. Kairomones and their use for management of entomophagous insects: IV. Effect of kairomones on productivity and longevity of Trichogramma pretiosum Riley (Hymenoptera: Trichogrammatidae). J. Chem. Ecol. 2: 67-72. 1976.
41. Todd, J. W. and Lewis, W. J. Incidence and oviposition patterns of Trichopoda pennipes (F.), a parasite of the southern green stink bug, Nezara viridula (L.). J. Georgia Entomol. Soc. 11: 50-54. 1976.
42. Lewis, W. J., Jones, R. L., Nordlund, D. A., and Gross, H. R., Jr. Kairomones and their use for management of entomophagous insects. Proc. Int. Colloq. C.N.R.S. on Insect Behavior and The Signals Issued from the Trophic Surroundings. 265: 455-469. 1977.
43. Lewis, W. J., Nordlund, D. A., Gross, H. R., Jr., Jones, R. L., and Jones, S. L. Kairomones and their use for management of entomophagous insects: V. Moth scales as a stimulus for predation of Heliothis zea (Boddie) eggs by Chrysopa carnea Stephens larvae. J. Chem. Ecol. 3: 483-487. 1977.
44. Nordlund, D. A., Lewis, W. J., Jones, R. L., Gross, H. R., Jr., and Hagen, K. S. Kairomones and their use for management of entomophagous insects: VI. An examination of the kairomones for the predator Chrysopa carnea Stephens at the oviposition sites of Heliothis zea (Boddie). J. Chem. Ecol. 3: 507-511. 1977.

45. Nordlund, D. A., Lewis, W. J., Todd, J. W., and Chalfant, R. B. Kairomones and their use for management of entomophagous insects: VII. The involvement of various stimuli in the differential response of Trichogramma pretiosum Riley to two suitable hosts. J. Chem. Ecol. 3: 513-518. 1977.
46. Lewis, W. J., Beevers, M., Nordlund, D. A., Gross, H. R., Jr., and Hagen, K. S. Kairomones and their use for management of entomophagous insects: IX. Investigations of various kairomone-treatment patterns for Trichogramma spp. J. Chem. Ecol. 5: 673-680. 1979.
47. Sauls, C. E., Nordlund, D. A., and Lewis, W. J. Kairomones and their use for management of entomophagous insects: VIII. Effect of diet on the kairomonal activity of frass from Heliothis zea (Boddie) larvae for Microplitis croceipes (Cresson). J. Chem. Ecol. 5: 363-369. 1979.
48. Lewis, W. J. and Nordlund, D. A. Employment of parasitoids and predators for fall armyworm control. Fla. Entomol. 63: 433-438. 1980.
49. Lewis, W. J., Hagen, K. S., Roelofs, W. L., and Schoonhoven, L. M. Status and potential use of behavioural chemicals in pest management. FAO Plant Protect. Bull. 28: 121-128. 1980.
50. Morrison, G., Lewis, W. J., and Nordlund, D. A. Spatial differences in Heliothis zea egg density and the intensity of parasitism by Trichogramma spp.: An experimental analysis. Environ. Entomol. 9: 79-85. 1980.
51. Altieri, M. A., Lewis, W. J., Nordlund, D. A., Gueldner, R. C., and Todd, J. W. Chemical interactions between plants and Trichogramma wasps in Georgia soybean fields. Protect. Ecol. 3: 259-263. 1981.
52. Beevers, M., Lewis, W. J., Gross, H. R., Jr., and Nordlund, D. A. Kairomones and their use for management of entomophagous insects: X. Laboratory studies on manipulation of host-finding behavior of Trichogramma pretiosum Riley with a kairomone extracted from Heliothis zea (Boddie) moth scales. J. Chem. Ecol. 7: 635-648. 1981.
53. Gross, H. R., Jr., Harrell, E. A., Lewis, W. J., and Nordlund, D. A. Trichogramma spp.: concurrent ground application of parasitized eggs, supplemental Heliothis zea host eggs, and host-seeking stimuli. J. Econ. Entomol. 74: 227-229. 1981.
54. Gross, H. R., Jr. Employment of kairomones in the management of parasitoids, pp. 137-150. In: D. A. Nordlund, R. L. Jones, and W. J. Lewis (ed.), Semiochemicals: Their Role in Pest Control. Wiley, NY, 306 pp. 1981.
55. Gross, H. R., Jr., Lewis, W. J., and Nordlund, D. A. Trichogramma pretiosum: effect of prerelease parasitization experience on retention in release areas and efficiency. Environ. Entomol. 10: 554-556. 1981.
56. Jackson, R. D. and Lewis, W. J. Summary of significance and employment strategies for semiochemicals, pp. 283-295. In: D. A. Nordlund, R. L. Jones, and W. J. Lewis (ed.), Semiochemicals: Their Role in Pest Control. Wiley, NY, 306 pp. 1981.

57. Lewis, W. J. Semiochemicals: their role with changing approaches to pest control, pp. 3-12. In: Semiochemicals: Their Role in Pest Control. Wiley, NY, 306 pp. 1981.
58. Lewis, W. J., Nordlund, D. A., and Gueldner, R. C. Semiochemicals influencing behavior of entomophages: Roles and strategies for their employment in pest control, pp. 225-242. Le Mediateurs Chimiques, Les Colloques de L'INRA, Paris., 1981.
59. Nordlund, D. A. Semiochemicals: a review of the terminology, pp. 13-28. In: D. A. Nordlund, R. L. Jones, and W. J. Lewis (ed.), Semiochemicals: Their Role in Pest Control. Wiley, NY, 306 pp. 1981.
60. Nordlund, D. A., Jones, R. L., and Lewis, W. J. Semiochemicals: Their Role in Pest Control, Wiley, NY, 1981.
61. Nordlund, D. A., Lewis, W. J., and Gross, H. R., Jr. Elucidation and employment of semiochemicals in the manipulation of entomophagous insects, pp. 463-475. In: E. R. Mitchell (ed.), Management of Insect Pests with Semiochemicals: Concepts and Practice. Plenum, NY, 514 pp. 1981.
62. Morrison, G. and Lewis, W. J. The allocation of searching time by Trichogramma pretiosum in host-containing patches. Ent. Exp. & Appl. 30: 31-39. 1981.
63. Nordlund, D. A., Lewis, W. J., Gross, H. R., Jr., and Beevers, M. Kairomones and their use for management of entomophagous insects. XII. The stimulatory effects of host eggs and the importance of host-egg density to the effective use of kairomones for Trichogramma pretiosum Riley. J. Chem. Ecol. 7: 909-917. 1981.
64. Lewis, W. J., Nordlund, D. A., Gueldner, R. C., Teal, P. E. A., and Tumlinson, J. H. Kairomones and their use for management of entomophagous insects. XIII. Kairomonal activity for Trichogramma spp. of abdominal tips, excretion, and a synthetic sex pheromone blend of Heliothis zea (Boddie) moths. J. Chem. Ecol. 1323-1331. 1982.
65. Greenblatt, J. A. and Lewis, W. J. Chemical environment manipulation for pest insects control. Environmental Management 7: 35-41. 1983.
66. Nordlund, D. A., Lewis, W. J., and Gueldner, R. C. Kairomones and their use for management of entomophagous insects. XIV. Response of Telenomus remus to abdominal tips of Spodoptera frugiperda, (z)-9-tetradecene-1-ol acetate and (z)-9-dodecene-1-ol acetate. J. Chem. Ecol. 9: 695-701. 1983.
67. Altieri, M. A., Martin, P. B., and Lewis, W. J. A quest for ecologically based past management systems. Environmental Management 7: 91-100. 1984. Greany, P. D., Vinson, S. B., and Lewis, W. J. Insect parasitoids: Finding new opportunities for biological control. BioScience 34: 690-696. 1984.
68. Greany, P. D., Vinson, S. B., and Lewis, W. J. Insect parasitoids: Finding new opportunities for biological control. BioScience 34: 690-696. 1984.

69. Gross, H. R., Jr., Lewis, W. J., Beevers, M., and Nordlund, D. A. Trichogramma pretiosum (Hymenoptera: Trichogrammatidae): effects of augmented densities and distributions of Heliothis zea (Lepidoptera: Noctuidae) host eggs and kairomones on field performance. Environ. Entomol. 13: 981-985. 1984.
70. Gueldner, R. C., Nordlund, D. A., Lewis, W. J., Thean, J. E., and Wilson, D. M. Kairomones and their use for management of entomophagous insects. XV. Identification of several acids in scales of Heliothis zea moths and comments on their possible role as kairomones for Trichogramma pretiosum. J. Chem. Ecol. 10: 245-251. 1984.
71. King, E. G., Bouse, L. F., Bull, D. L., Dickerson, W. M., Lewis, W. J., Liapis, P., Lopez, J. D., Morrison, R. K., and Phillips, J. R. Potential for management of Heliothis spp in cotton by augmentative releases of Trichogramma pretiosum, pp. 232-236. Proceedings Beltwide Cotton Prod Res Conf, Memphis, TN, National Cotton Council of Am., 381 pp. 1984.
72. Lewis, W. J. and Nordlund, D. A. Semiochemicals influencing fall armyworm parasitoid behavior: implications for behavioral manipulation. Fla. Entomol. 67: 343-349. 1984.
73. Morrison, G. and Lewis, W. J. Inferring parasitoid searching behaviors from host mortality data: A note of caution. Environ. Entomol. 13: 7-14. 1984.
74. Nordlund, D. A., Chalfant, R. B., and Lewis, W. J. Arthropod populations, yield and damage in monocultures and polycultures of corn, beans and tomatoes. Agric. Ecosys. Environ. 11: 353-367. 1984.
75. Dmoch, J., Lewis, W. J., Martin, P. B., and Nordlund, D. A. Role of host-produced stimuli and learning in host selection behavior of Cotesia (=Apanteles) marginiventris (Cresson). J. Chem. Ecol. 11: 453-463. 1985.
76. Keller, M. A. and Lewis, W. J. Movements by Trichogramma pretiosum (Hymenoptera: Trichogrammatidae) released into cotton. Southwest. Entomol. 8: 99-109. 1985.
77. Keller, M. A., Lewis, W. J., and Stinner, R. E. Biological and practical significance of movement by Trichogramma species: A review. Southwest. Entomol. 8: 138-155. 1985.
78. Lewis, W. J., Gross, H. R., Jr., and Nordlund, D. A. Biological control of bollworm and tobacco budworm in cotton by augmentative releases of Trichogramma — behavioral manipulation of Trichogramma (Hymenoptera: Trichogrammatidae). Southwest. Entomol. Suppl. No. 8: 49-55. 1985.
79. Lewis, W. J. and Nordlund, D. A. Behavior-modifying chemicals to enhance natural enemy effectiveness, pp. 89-101. In: M. A. Hoy and G. A. Herzog (ed.), Biological Control in Agricultural Integrated Pest Management Systems. Academic Press, Orlando, FL, 589 pp. 1985.

80. Nordlund, D. A., Chalfant, R. B., and Lewis, W. J. Response of Trichogramma pretiosum females to extracts from two plants attacked by Heliothis zea. Agric. Ecosys. Environ. 12: 127-133. 1985.
81. Nordlund, D. A. and Lewis, W. J. Response of females of the Braconid parasitoid, Microplitis demolitor to frass of larvae of the noctuids, Heliothis zea and Trichoplusia ni and to 13methylhentriacontane. Entomol. Exp. Appl. 38: 109-112. 1985.
82. Drost, Y. C., Lewis, W. J., Zanen, P. O., and Keller, M. A. Beneficial arthropod behavior mediated by airborne semiochemicals. Part I. Flight behavior and influence of preflight handling of Microplitis croceipes (Cresson). J. Chem. Ecol. 12: 1247-1262. 1986.
83. King, E. G., Bouse, L. F., Bull, D. L., Coleman, R. J., Dickerson, W. A., Lewis, W. J., Lopez, J. D., Morrison, R. K., and Phillips, J. R. Management of Heliothis spp. in cotton by augmentative releases of Trichogramma pretiosum Riley. J. Appl. Entomol. 101: 2-10. 1986.
84. Nordlund, D. A., Chalfant, R. B., and Lewis, W. J. Response of Trichogramma pretiosum females to volatile synomones from tomato plants. J. Entomol. Sci. 20: 372-376. 1986.
85. Nordlund, D. A., Lewis, W. J., Vinson, S. B., and Gross, H. R., Jr. Behavioral manipulation of entomophagous insects, pp. 104-115. In: S. J. Johnson, E. G. King, and J. R. Bradley, Jr. (ed.), Theory and Tactics of Heliothis Population Management: I - Cultural and Biol. Control. Southern Coop. Series Bull. 316. 161 pp. 1986.
86. Nordlund, D. A., Strand, M. R., Lewis, W. J., and Vinson, S. B. Role of kairomones from host accessory gland secretion in host recognition by Telenomus remus and Trichogramma pretiosum, with partial characterization. Entomol. Exp. Appl. 44: 37-43. 1987.
87. Drost, Y. C., Lewis, W. J., and Tumlinson, J. H. Beneficial arthropod behavior mediated by airborne semiochemicals. V. Influence of rearing method, host plant, and adult experience on host-searching behavior of Microplitis croceipes (Cresson), a larval parasitoid of Heliothis. J. Chem. Ecol. 14: 1607-1616. 1988.
88. Eller, F. J., Tumlinson, J. H., and Lewis, W. J. Beneficial arthropod behavior mediated by airborne semiochemicals. II. Olfactometric studies of host location by the parasitoid Microplitis croceipes (Cresson) (Humenoptera: Braconidae). J. Chem. Ecol. 14: 425-434. 1988.
89. Eller, F. J., Tumlinson, J. H., and Lewis, W. J. Beneficial arthropod behavior mediated by airborne semiochemicals: Source of volatiles mediating the host-location flight behavior of Microplitis croceipes (Cresson) (Hymenoptera: Braconidae), a parasitoid of Heliothis zea (Boddie) (Lepidoptera: Noctuidae). Environ. Entomol. 17: 745-753. 1988.

90. Hamm, J. J., Styer, E. L., and Lewis, W. J. A baculovirus pathogenic to the parasitoid Microplitis croceipes (Hymenoptera: Braconidae). J. Invertebr. Pathol. 52: 189-191. 1988.
91. Herard, F, Keller, M. A., and Lewis, W. J. Rearing Microplitis demolitor Wilkinson (Hymenoptera: Braconidae) in the laboratory for use in studies of semiochemical mediated searching behavior. J. Entomol. Sci. 23: 105-111. 1988.
92. Herard, F., Keller, M. A., Lewis, W. J., and Tumlinson, J. H. Beneficial arthropod behavior mediated by airborne semiochemicals. III. Influence of age and experience on flight chamber responses of Microplitis demolitor Wilkinson. J. Chem. Ecol. 14: 1583-1596. 1988.
93. Herard, F., Keller, M. A., Lewis, W. J., and Tumlinson, J. H. Beneficial arthropod behavior mediated by airborne semiochemicals. IV. Influence of host diet on host-oriented flight chamber responses of Microplitis demolitor Wilkinson. J. Chem. Ecol. 14: 1597-1606. 1988.
94. Lewis, W. J., Sonnet, P. E., and Nordlund, D. A. Responses of braconid parasitoids Microplitis croceipes (Cresson) and M. demolitor Wilkinson to stereoisomers of kairomone 13-methylhentriacontane. J. Chem. Ecol. 14: 883-888. 1988.
95. Lewis, W. J. and Tumlinson, J. H. Host detection by chemically mediated associative learning in a parasitic wasp. Nature 331: 257-259. 1988.
96. Noldus, L. P., Lewis, W. J., Tumlinson, J. H., and van Lenteren, J. C. Olfactometer and wind tunnel experiments on the role of sex pheromones of noctuid moths in the foraging behavior of Trichogramma spp., pp. 223-228. IIld International Symposium Guangzhou (China), Nov. 10-15, 1986. INRA Paris, 1988.
97. Nordlund, D. A., Lewis, W. J., and Altieri, M. A. Influence of plant produced allelochemicals on the host/prey selection behavior of entomophagous insects., pp. 65-90. In: P. Barbosa and D. Letourneau (ed.), Novel Aspects of Insect-Plant Interactions. J. Wiley and Sons, 1988.
98. Keller, M. A. and Lewis, W. J. Behavior modifying chemicals to increase the efficacy of predators and parasitoids of Heliothis spp, pp. 449-481. In: E. G. King and R. D. Jackson (ed.), Proceedings, Biological Control of Heliothis: Increasing the Effectiveness of Natural Enemies, Rekha Printers Pvt. Ltd., 1989.
99. Lewis, W. J. and Gross, H. R., Jr. Comparative studies on field performance of Heliothis larval parasitoids, Microplitis croceipes and Cardiochiles nigriceps, at varying densities and under selected host plant conditions. Fla. Entomol. 72: 6-14. 1989.
100. Nordlund, D. A., Lewis, W. J., and Tumlinson, J. H. Habitat and host location behavior of Microplitis croceipes. Southwest. Entomol. 12: 39-48. 1989.

101. Turlings, T. C. J., Tumlinson, J. H., Lewis, W. J., and Vet, L. E. M. Beneficial arthropod behavior mediated by airborne semiochemicals. VII. Learning of host-related odors induced by a brief contact experience with host by-products in Cotesia marginiventris (Cresson). J. Insect Behav. 2: 217-225. 1989.
102. Zanen, P. O., Lewis, W. J., Carde, R. T., and Mullinix, B. G. Beneficial arthropod behavior mediated by airborne semiochemicals. VI. Flight responses of female Microplitis croceipes (Cresson), a braconid endoparasitoid of Heliothis spp., to varying olfactory stimulus conditions created with a turbulant jet. J. Chem. Ecol. 15: 141-168. 1989.
103. Eller, F. J., Tumlinson, J. H., and Lewis, W. J. Intraspecific competition in Microplitis croceipes (Hymenoptera: Braconidae), a parasitoid of Heliothis species (Lepidoptera: Noctuidae). Ann. Entomol. Soc. Am. 83: 504-508. 1990.
104. Hamm, J. J., Styer, E. L., and Lewis, W. J. Comparative virogenesis of filamentous virus and polydnavirus in the female reproductive tract of Cotesia marginiventris (Hymenoptera: Braconidae). J. Invertebr. Pathol. 55: 357-374. 1990.
105. Lewis, W. J. and Martin, W. R., Jr. Semiochemicals for use with parasitoids: Status and future. J. Chem. Ecol. 16: 3067-3089. 1990.
106. Lewis, W. J. and Takasu, K. Use of learned odours by a parasitic wasp in accordance with host and food needs. Nature 348: 635-636. 1990.
107. Lewis, W. J., Vet, L. E. M., Tumlinson, J. H., van Lenteren, J. C., and Papaj, D. R. Variations in parasitoid foraging behavior: essential element of a sound biological control theory. Environ. Entomol. 19: 1183-1193. 1990.
108. Noldus, L. P., Lewis, W. J., and Tumlinson, J. H. Beneficial arthropod behavior mediated by airborne semiochemicals. IX. Differential response of Trichogramma pretiosum, an egg parasitoid of Heliothis zea, to various olfactory cues. J. Chem. Ecol. 16: 3531-3544. 1990.
109. Prevost, G. and Lewis, W. J. Heritable differences in the response of the braconid wasp Microplitis croceipes to volatile allelochemicals. J. Insect Behav. 3: 277-287. 1990.
110. Turlings, T. C. J., Scheepmaker, J. W. A., Vet, L. E. M., Tumlinson, J. H., and Lewis, W. J. How contact foraging experiences affect preferences for host-related odors in the larval parasitoid Cotesia marginiventris (Cresson) (Hymenoptera: Braconidae). J. Chem. Ecol. 16: 1577-1589. 1990.
111. Turlings, T. C. J., Tumlinson, J. H., and Lewis, W. J. Exploitation of herbivore-induced plant odors by host-seeking parasitic wasps. Science 250: 1251-1253. 1990.
112. Vet, L. E. M., Lewis, W. J., Papaj, D. R., and van Lenteren, J. C. A variable-response model for parasitoid foraging behavior. J. Insect Behav. 3: 471-490. 1990.

113. Lewis, W. J., Tumlinson, J. H., and Krasnoff, S. Chemically mediated associative learning: An important function in the foraging behavior of Microplitis croceipes (Cresson). J. Chem. Ecol. 17: 1309-1325. 1991.
114. Noldus, L. P., van Lenteren, J. C., and Lewis, W. J. How Trichogramma parasitoids use moth sex pheromones as kairomones: Orientation behavior in a wind tunnel. Physiol. Entomol. 16: 313-327. 1991.
115. Tumlinson, J. H. and Lewis, W. J. Chemically mediated foraging behavior of parasitoids: New implications for biocontrol, pp. 161-178. In: Entomology Serving Society: Emerging technologies and challenges, Entomological Society of America. Lanham, MD, 1991.
116. Turlings, T. C. J., Tumlinson, J. H., Eller, F. J., and Lewis, W. J. Larval-damaged plants: source of volatile synomones that guide the parasitoid Cotesia marginiventris to the micro-habitat of its host. Entomol. Exp. Appl. 58: 75-82. 1991.
117. Eller, F. J., Tumlinson, J. H., and Lewis, W. J. Effect of host diet and preflight experience on the flight response of Microplitis croceipes (Cresson). Physiol. Entomol. 17: 235-240. 1992.
118. Hamm, J. J., Styer, E. L., and Lewis, W. J. Three viruses found in the braconid parasitoid Microplitis croceipes and their implications in Biological Control Programs. Biol. Control 2: 329-336. 1992.
119. Sheehan, W., Wackers, F. L., and Lewis, W. J. Discrimination of previously searched, host-free sites by Microplitis croceipes (Hymenoptera: Braconidae). J. Insect Behav. 6: 323-331. 1992.
120. Tumlinson, J. H., Turlings, T. C. J., and Lewis, W. J. The semiochemical complexes that mediate insect parasitoid foraging. Agric. Zool. Reviews 5: 221-252. 1992.
121. McCall, P. J., Turlings, T. C. J., Lewis, W. J., and Tumlinson, J. H. Role of plant volatiles in host location by the special parasitoid Microplitis croceipes (Cresson) (Braconidae: Hymenoptera). J. Insect Behav. 6: 625-639. 1993.
122. Prokopy, R. J. and Lewis, W. J. Application of learning to pest management, pp. 308-342. In: D. R. Papaj and A. Lewis (ed.), Insect Learning: Ecological and Evolutionary Perspectives. Chapman and Hall, New York, 398 pp. 1993.
123. Ruberson, J. R., Herzog, G. A., and Lewis, W. J. Parasitism of the beet armyworm, Spodoptera exigua, in south Georgia cotton, 2: 993-997. Proceedings Beltwide Cotton Conference, 1993.
124. Sheehan, W., Wackers, F. L., and Lewis, W. J. Discrimination of previously searched, host-free sites by Microplitis croceipes (Hymenoptera: Braconidae). J. Insect Behav. 6: 323-331. 1993.
125. Takasu, K. and Lewis, W. J. Host-and-food-foraging of the parasitoid Microplitis croceipes: Learning and physiological state effects. Biol. Control 3: 70-74. 1993.

150. Lewis, W. J., Haney, P. B., Reed, R., and Walker, A. A total systems approach for sustainable cotton production in Georgia and southeast: first year results. Proc. Beltwide Cotton Conf. 2: 1129-1134. 1997.
151. Lewis, W. J., van Lenteren, J. C., Phatak, S. C., and Tumlinson, J. H. A total system approach to sustainable pest management. Proc. Natl. Acad. Sci. USA 94: 12243-12248. 1997.
152. Reed, R., Phatak, S. C., Page, A., Haney, P. B., and Lewis, W. J. Conservation tillage in Coffee County cotton. Proc. Beltwide Cotton Conf. 1: 621-622. 1997.
153. Rose, U. S. R., Alborn, H. T., Makranczy, G., Lewis, W. J., and Tumlinson, J. H. Host recognition by the specialist endoparasitoid Microplitis croceipes. J. Insect Behav. 10(3): 313-330. 1997.
154. Ruberson, J. R., Phatak, S. C., and Lewis, W. J. Insect population in a cover crop/strip tillage system. Proc. Beltwide Cotton Conf. 2: 1121-1123. 1997.
155. Stapel, J. O., Cortesero, A. M., De Moraes, C. M., Tumlinson, J. H., and Lewis, W. J. Extrafloral nectar, honeydew, and sucrose effects on searching behavior and efficiency of Microplitis croceipes (Hymenoptera: Braconidae) in cotton. Environ. Entomol. 26(3): 617-623. 1997.
156. Stapel, J. O., Ruberson, J. R., Gross, H. R., Jr., and Lewis, W. J. Progeny allocation by the parasitoid Lespesia archippivora (Diptera: Tachinidae) in larvae of Spodoptera exigua (Lepidoptera: Noctuidae). Environ. Entomol. 26(2): 265-271. 1997.
157. De Moraes, C. M., Lewis, W. J., Pare, P. W., Alborn, H. T., and Tumlinson, J. H. Herbivore-infested plants selectively attract parasitoids. Nature 393: 570-573. 1998.
158. Lewis, W. J., Stapel, J. O., Cortesero, A. M., and Takasu, K. Understanding how parasitoids balance food and host needs: Importance to biological control. Biol. Control 11: 175-183. 1998.
159. Stapel, J. O., Waters, D. J., Ruberson, J. R., and Lewis, W. J. Development and behavior of Spodoptera exigua (Lepidoptera: Noctuidae) larvae in choice tests with food substrates containing toxins of Bacillus thuringiensis. Biol. Control 11: 29-37. 1998.
160. De Moraes, C. M. and Lewis, W. J. Analyses of two parasitoids with convergent foraging strategies. J. Insect Behav. 12: 571-583. 1999.
161. Olson, D. M., , Phatak S. C., and Lewis, W. J. Influence of Nitrogen Levels on Cotton Plant/Insect Interactions in a Conservation Tillage System, In: J. E. Hook (ed.), Proceedings of the 22nd Annual Southern Conservation Tillage Conference for Sustainable Agriculture, July 6-8, 1999. Tifton, GA, GA Agr. Exp. Stn. Special Pub. 95. 1999.
162. Pare', P. W., Lewis, W. J., and Tumlinson, J. H. Induced plant volatiles: Biochemistry and effects on parasitoids, 167-180. In: A. A. Agrawal, S. Tuzun, and E. Bent (ed.), Inducible plant defenses against pathogens and herbivores:

Biochemistry, ecology, and agriculture. American Phytopathological Society Press, St. Paul, MN, 1999.

163. Tumlinson, J. H., Pare', P. W., and Lewis, W. J. Plant production of volatile semiochemicals in response to insect derived elicitors, pp. 95-109. In: D. J. Chadwick and J. Goode (ed.), Insect-plant interactions and induced plant defense. John Wiley Journal, Novartis Foundation Symposium No. 223, 1999.

164. Tumlinson, J. H., Pare', P. W., Alborn, H. T., and Lewis, W. J. Chemically mediated tritrophic plant-insect interactions, In: P. J. G. M. de Wit, T. Bisseling, and W. J. Stiekema (ed.), Biology of Plant-Microbe Interactions Vo. 2. Proc. of the 9th International Congress on Molecular Plant-Microbe Interactions, July 25-30, 1999. Amsterdam, the Netherlands, 1999.

165. Stapel, J. O., Lewis, W. J. Phatak, S. C., Ruberson, J. R. Insect pest management as a component of a sustainable cotton production system. Proc. Beltwide Cotton Conf. 1998. 2: 1107-1111, 1999.

166. Rains, G. C., Meiners, T., Takasu, K., Olson, D. M., Lewis, W. J., Tumlinson, J., and Cardoza, Y. Development of a programmable whole-organism wasp sensor for monitoring crop conditions from volatile chemicals. ASAE Paper No. 003062. 2000.

167. DeMoraes, C.M., W.J. Lewis, P.W. Pare, H.T. Alborn and J.H. Tumlinson Hebivore-infested plants selectively attract parasitoids Nature 393: 570-573. 1999.

168. DeMoraes, C.M., A.M. Cortesero, J.O. Stapel and W.J. Lewis. Intrinsic and extrinsic interactions between two larval parasitoids of Heliothis virescens. Ecol. Entomol. 24: 402-410 1999

169. Wäckers, F.L. and W.J. Lewis. A comparison of color-, shape- and pattern-learning by the hymenopteran parasitoid Microplitis croceipes. J. Comp. Physiol. A. 184: 387-393 1999.

170. Lewis, W. J. and Jay, M. Ecologically-Based Communities; Putting it all together at the local level. 48p The Kerr Center for Sustainable Agriculture, Poteau, OK, 2000.

171. Lewis, W. J. (Lead Inventor). Utilization of invertebrate learning for flexible and sensitive monitoring and identification of chemicals. Invention Disclosure Serial No. 60/195,218: Docket Number 0065.00. 2000.

172. De Moraes, C. M., Lewis, W. J., and Tumlinson, J. H. Examining plant-parasitoid interactions in tritrophic systems. An. Soc. Entomol. Brasil 29(2): 189-203. 2000.

173. Stapel, J. O., Cortesero, A. M., Lewis, W. J. Disruptive Sublethal Effects of Insecticides on Biological Control: Altered Foraging Ability and Life Span of A Parasitoid After Feeding On Extrafloral Nectar of Cotton Treated With Systemic Insecticides. Biological Control 17: 243-249. 2000.

Reprinted from Nature, Vol. 331, No. 6153, pp. 257-259, 21 January 1988

Host detection by chemically mediated associative learning in a parasitic wasp

W. J. Lewis* & J. H. Tumlinson†

Insect Attractants, Behavior and Basic Biology Research Laboratory, Agricultural Research Service, US Department of Agriculture, Gainesville, Florida 32604, USA

Parasitic insects use chemical cues to locate their hosts, and prior experiences can modify their responses to these odours[1-4]. Females of the parasitic wasp *Microplitis croceipes* experienced by contact with host faeces, orient and fly upwind to odours from their hosts, larvae of the moth *Heliothis zea*[5]. We use flight tunnel studies to show that associative learning occurs during encounters with host faeces. When females touch the faeces with their antennae they learn to recognize and subsequently fly to various volatile odours, even novel and otherwise unattractive odours like vanilla, associated with the faeces. They link these volatile odours with a water extractable nonvolatile chemical in the faeces, evidently a host-specific recognition cue. The association of tracking cues with host by-products, without the need for direct contact with the host, is a valuable adaptation for locating cryptic and evasive hosts.

When females of *M. croceipes* are presented with fresh faeces or the water extract of faeces (Fig. 1) on filter paper, they immediately rub the material with their antennae and sometimes probe it with their abdomens. This intense examination continues for several minutes unless interrupted. Subsequently, these females exhibit characteristic host-seeking flight responses to attractants in a wind tunnel. Females presented with the hexane extract (Fig. 1) do not exhibit the sustained antennal rubbing and subsequently do not respond to attractants any better than naive females (Fig. 2). When the hexane and water extracts were tested as attractants in the wind tunnel, the hexane extract was significantly better than other sources including fresh faeces (Fig. 3).

As the water extract showed some attraction in the wind tunnel, we tested whether actual contact with the material is a necessary part of the experience, or whether exposure to certain volatiles simply increases the probability that females will recognize and subsequently respond to those volatiles. Females were exposed for 1 min to volatiles from the water extract on filter paper in closed Petri dishes, but were prevented from contacting the filter paper by a piece of cheese cloth. Of 20 females exposed in this way, none responded to the hexane extract in the wind tunnel, whereas 80% of a similar group of females allowed to contact the water extract on filter paper with only their antennae did respond. Thus, water extractable, non-volatile components of host faeces, contacted directly by *M. croceipes* female antennae, provide a cue causing the females to respond subsequently to volatile attractants.

The most likely explanation of the flight response of the parasitoid to the hexane extract subsequent to antennal rubbing of the water extract was considered to be either one or a combination of sensitization, associative learning and/or priming. As prior exposure to hexane extract did not improve response and actual contact with the water extract was required (indicating a non-volatile material), sensitization seemed an unlikely explanation. We adopted associative learning as our working hypothesis on the assumption that the limited attraction of the water extract (Fig. 3) was a result of contamination with hexane soluble volatiles. Thus, contact with the water extract (or raw faeces) would result in the association of volatile hexane soluble chemicals with a nonvolatile water soluble host-recognition material. This was supported by the fact that exposure to the

* Present address: Insect Biology and Population Management Research Laboratory ARS, USDA Tifton, Georgia 31793, USA.

† To whom correspondence should be addressed.

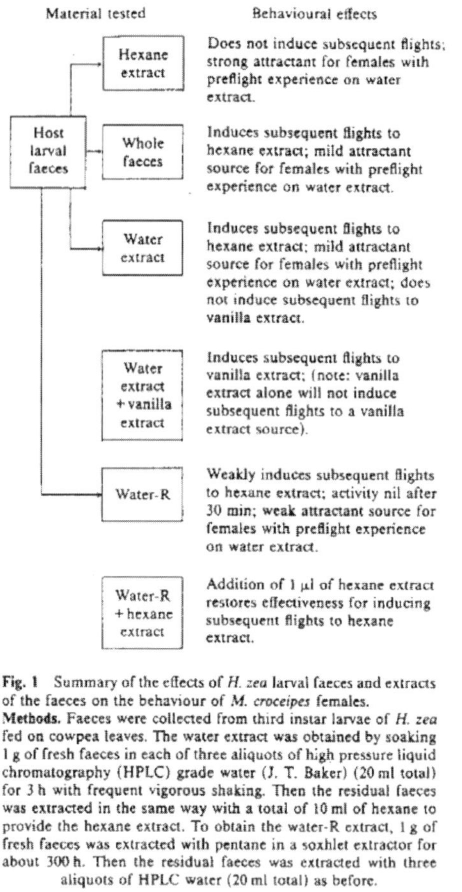

Fig. 1 Summary of the effects of *H. zea* larval faeces and extracts of the faeces on the behaviour of *M. croceipes* females.
Methods. Faeces were collected from third instar larvae of *H. zea* fed on cowpea leaves. The water extract was obtained by soaking 1 g of fresh faeces in each of three aliquots of high pressure liquid chromatography (HPLC) grade water (J. T. Baker) (20 ml total) for 3 h with frequent vigorous shaking. Then the residual faeces was extracted in the same way with a total of 10 ml of hexane to provide the hexane extract. To obtain the water-R extract, 1 g of fresh faeces was extracted with pentane in a soxhlet extractor for about 300 h. Then the residual faeces was extracted with three aliquots of HPLC water (20 ml total) as before.

water extract (water-R) of the residue from faeces exhaustively extracted with pentane resulted in very weak responses, whereas activity could be restored by the addition of hexane extract (Fig. 1).

To test this associative learning hypothesis, we added a novel odour, vanilla extract (Nielsen-Massey) to filter paper together with the water extract. A significant number of females exposed to this combination subsequently were attracted to vanilla odour in a flight tunnel. Naive females or females exposed to the water extract without vanilla did not fly to the vanilla source (Fig. 4). This shows that associative learning does occur. When exposed to vanilla extract alone, females do not rub it with their antennae, and subsequently are not attracted to a vanilla source.

Taken together with the fact that the hexane extract alone does not condition flight responses, these results show that a nonvolatile water extractable component (unconditioned stimulus) acts together with hexane soluble volatile chemicals (conditioned stimuli) to elicit antennal rubbing and the association process. The parasitoid subsequently flies to sources of the volatiles thereby learned. An analogous process has been shown in rats, where novel odours were potentiated when the taste receptors were contacted[6]. To show that the water extractable substance (a kairomone[7]) does not occur in the larvae of *Trichophisia ni*, a non-host species, females were exposed, both with and without the presence of hexane extracts of *H. zea*

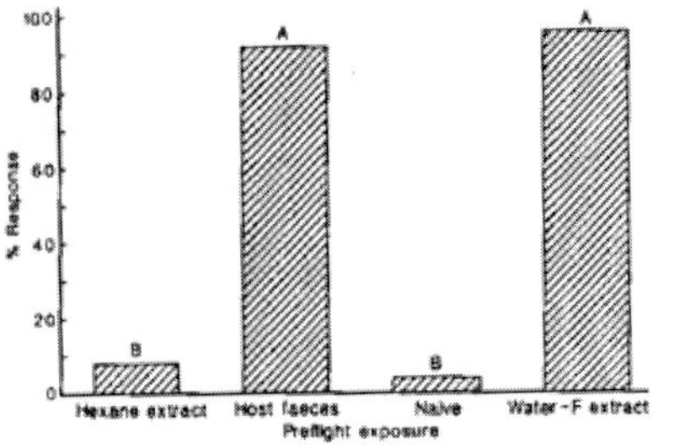

Fig. 2 Flight response of *M. croceipes* females to hexane extract of *H. zea* larval faeces after no experience (naive) or contact with fresh host faeces, with water extract of host faeces (water-F), or with hexane extract of host faeces. Bars capped by different letters are significantly different, $P < 0.01$, Waller-Duncan K-ratio t-test, minimum significant difference = 18.9%, $n = 96$ wasps (four replications, four treatments, six wasps per replication per treatment).
Methods. *M. croceipes* were reared on the larvae of *H. zea* fed on an artificial diet[13]. Females were allowed to mate and then held in 30 × 30 × 20 cm cages until bioassay. The wind tunnel and general bioassay procedure have been described previously[5]. All flight responses were tested at 26–28 °C, at a wind speed of 56 cm s^{-1} and at a light intensity of 2,500 lx. The hexane extract (14 μl = 1.4 mg of faeces) used as the attractant source was pipetted onto a triangular (9.5 × 6.5 cm) piece of Whatman number 1 filter paper suspended at the centre of the upwind end of the tunnel. Females, tested individually, were released from a 4-dram shell vial 80 cm downwind and in the plume from the odour source. The behaviour of the females in the wind tunnel was observed until a complete flight occurred or five minutes elapsed. A complete response involved the host-seeking flight behaviour described by Drost *et al.*[5] and included approach and landing on the odour source. For each observation, a female was given two chances to make a successful flight. Materials to which females were exposed before flight testing—28 μl (= 1.4 mg faeces) of water-F extract, 14 μl hexane extract, or ~2 mg of host faeces—were placed on a 12.5 cm diameter sheet of Whatman number 40 filter paper in a Petri dish. Females were allowed to walk directly from a vial onto the test material and then to investigate the material with their antennae for 45 s.

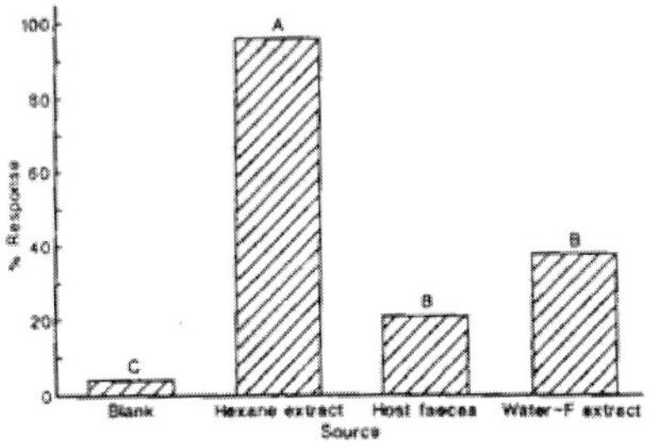

Fig. 3 Flight response of *M. croceipes* females to pure hexane (blank), fresh host faeces, water extract of host faeces (water-F), and hexane extract of host faeces, after contact with water extract of host faeces. Bars capped by different letters are significantly different, $P < 0.01$, Waller-Duncan K-Ratio t-test, minimum significant difference = 17.2%, $n = 96$ wasps (four replications, four treatments, six wasps per replication per treatment).
Methods. See legend to Fig. 2.

(host) faeces, to water extracts of faeces of the larvae of *T. ni* fed on cowpea leaves. These females did not respond to hexane extracts of either *H. zea* or *T. ni* faeces.

Our findings show that the host-seeking behaviour of *M.*

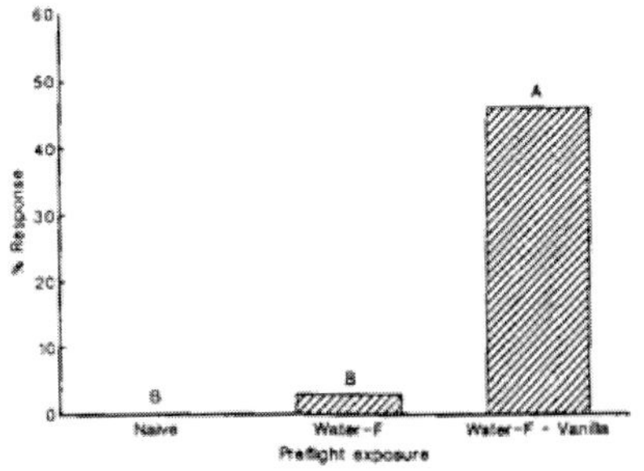

Fig. 4 Flight response of *M. croceipes* females to a source of vanilla extract (Nielsen Massey) in a wind tunnel after no experience (naive), or contact with water-F, or water-F plus vanilla extract. Bars capped by different letters are significantly different, $P < 0.01$, Waller-Duncan K-Ratio t-test, minimum significant difference = 12.4%, $n = 90$ wasps (three replications, three treatments, ten wasps per replication per treatment).
Methods. This experiment was conducted in the same way as previous experiments (Figs 2 and 3). Pure vanilla extract (80 μl) was pipetted onto a piece of Whatman number 1 filter paper and used as the attractant source as described before. Materials used for preflight conditioning of the females were prepared as before. When water-F and vanilla were combined, 28 μl of water-F was first placed on the filter paper and then 1 μl of vanilla extract was added to the centre of the moist spot.

croceipes females is mediated by both a nonvolatile, contact chemical and by volatile chemical cues. The nonvolatile substance is water extractable and distinct from the hexane soluble volatile cues that attract females. The nonvolatile compound is also host specific. The volatile attractants may be plant-derived. The role of the relatively nonvolatile, hexane soluble kairomone, 13-methylhentriacontane, previously identified for this parasitoid[8], is not known. It may facilitate the association process described here by enhancing the antennal rubbing behaviour.

M. croceipes parasitizes *Heliothis* species, which are highly polyphagous. Thus the ability to locate its hosts among other plant feeders in diverse systems of vegetation is crucial. Learned responses to varying trail odours through association with a host recognition kairomone in the host faeces may give *M. croceipes* females an adaptive advantage. Associative learning in connection with host-detection by parasitoid and plant feeding insects has been shown by a number of authors[2–5,9–12]. Contact with the target organism (or a model) and subsequent oviposition, however, generally was necessary. We believe this to be the first demonstration of associative learning in insects where encounter with the target organism is not involved. This phenomenon provides a valuable way of finding hosts or prey species and may be widespread in parasitic and predatory systems as the target organisms (but not their by-products) often elude the searching parasitoid or predator.

Received 31 August; accepted 30 October 1987.

1. Arthur, A. P. in *Semiochemicals, Their Role in Pest Control* (eds Nordlund, D. A., Jones, R. L. & Lewis, W. J.) 97–120 (Wiley, New York, 1981).
2. Vet, L. E. M. *Neth. J. Zool.* 33, 225–248 (1983).
3. Vet, L. E. M. & van Opzeeland, K. *Oecologia* 63, 171–177 (1984).
4. Vet, L. E. M. & van Opzeeland, K. *Neth. J. Zool.* 35, 497–504 (1985).
5. Drost, Y. C., Lewis, W. J., Zanen, P. O. & Keller, M. A. *J. chem. Ecol.* 12, 1247–1262 (1986).
6. Palmetino, C. C., Rusiniak, K. W. & Garcia, J. *Science* 208, 753–755 (1980).
7. Nordlund, D. A. in *Semiochemicals, Their Role in Pest Control* (eds Nordlund, D. A., Jones, R. L. & Lewis, W. J.) 13–28 (Wiley, New York, 1981).
8. Jones, R. L., Lewis, W. J., Bowman, M. C., Beroza, M. & Bierl, B. A. *Science* 173, 842–843 (1971).
9. Papaj, D. R. *Anim. Behav.* 34, 1281–1288 (1986).
10. Papaj, D. R. & Prokopy, R. J. *J. chem. Ecol.* 12, 1125–1143 (1986).
11. Traynier, R. M. M. *Physiol. Entomol.* 9, 465–472 (1984).
12. Vinson, S. B., Barfield, C. S. & Henson, R. D. *Physiol. Entomol.* 2, 157–164 (1977).
13. Lewis, W. J. & Burton, R. L. *J. econ. Entomol.* 63, 656–658 (1970).

Printed in Great Britain — Turnergraphic Limited, Basingstoke, Hampshire

31. Hsu, K. I., Cita, M. B. & Ryan W. B. F. *Init. Rep. DSDP* Leg 42, 943-949 (1977).
32. Dañobeitia, Ju. J., Alonso, B. & Maldonado, A. *Mar. Geol.* (in the press).
33. Long, R. E. & Backhouse, R. W. *Geophys. J. R. astr. Soc.* **44,** 677-688 (1976).
34. Woollard, G. P. *Bur. Gravimetrique Int. Bull. d'Inform.* **36,** 106-110 (1975).
35. LASE study group. *Mar. Petrol Geol.* **3,** 234-242 (1986).
36. White, R. S. *et al. Geophys. J. R. astr. Soc.* **89,** 265-272 (1987).
37. ECORS Pyrenees Team *Nature* **331,** 508-511 (1988).
38. Marti, J., Grachev, A., Mitjavila, J., Aparicio, A. & Roca, E. *Terra Abstr.* **2,** 114 (1990).
39. *International Bathymetric Chart of the Mediterranean* (Intergovernmental Oceanographic Commission).
40. Stoffa, P. L., Diebold, J. B. & Buhl, P. *Geophys. Prosp.* **30,** 25-57 (1982).

ACKNOWLEDGEMENTS. This work was supported by the NSF.

Use of learned odours by a parasitic wasp in accordance with host and food needs

W. J. Lewis & Keiji Takasu*

Insect Biology & Population Management Research Laboratory, US Department of Agriculture, Agricultural Research Service, PO Box 748, Tifton, Georgia 31793, USA
* Visiting scientist with the Department of Entomology, Coastal Plain Experiment Station, University of Georgia, Tifton, Georgia 31793, USA, from the Division of Entomology, National Institute of Agro-Environmental Sciences, Kannondai 3-1-1, Tsukuba, Ibaraki 305, Japan

ADULT parasitic wasps need nectar or some other fragrant food source[1,2], but the effect of food odours on their behaviour has not been investigated to any extent[3]. Females of the parasitic wasp *Microplitis croceipes* learn volatile odours associated with host sites and use them to find hosts more effectively[4,5]. We have now performed flight tunnel experiments that show that this wasp also uses airborne odours to find food sources. Females learn associatively and subsequently fly to volatile odours presented for smell while they are feeding on sugar water. Furthermore, they can learn two novel odours associated with separate host and food resources and then make an accurate choice between these odours on the basis of their relative host and food needs. This ability of parasitic wasps to link different odours with specific resources and then effectively to use them as cues for choice between competing needs is of fundamental and applied importance.

Females of *M. croceipes* fed on wildflower honey subsequently flew up-wind to the odour of honey, whereas naive females and females fed a pure aqueous sucrose solution both responded much more weakly (Fig. 1). To determine whether this oriented response to food was learned, female wasps were provided with either one of two arbitrarily selected odours, vanilla or chocolate, while feeding on sucrose. A droplet of the vanilla extract or chocolate extract was placed near the sucrose solution so that the wasp could smell the odour while simultaneously tasting the sucrose, but was not allowed to taste or otherwise contact the odour. Females so trained showed a clear preference for the respective odour of their training when subsequently provided with a choice of these two odours in the flight chamber (Fig. 2). Females given the smell of vanilla or chocolate by the same procedure, but separately from feeding, did not respond to either odour.

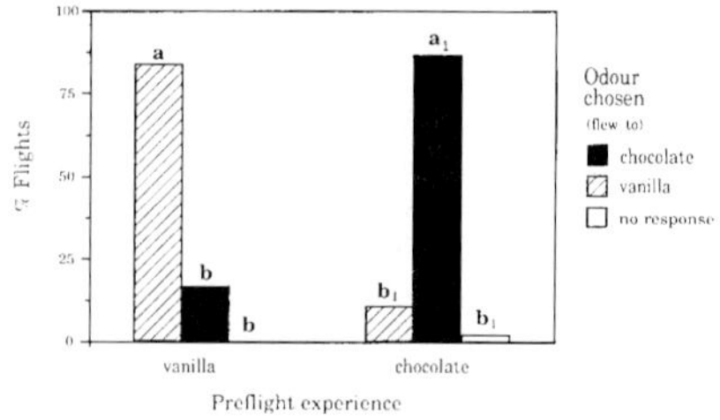

FIG. 2 Flight responses by *M. croceipes* females to vanilla or chocolate extract after smelling one of these odours while feeding on sugar water (preflight experience). Bars within the same treatment group capped by different letters are significantly different, $P<0.01$, Waller-Duncan K-ratio t-test; minimum significant difference, 14.3%, $n=64$ wasps (8 replications, 4 wasps per replication per preflight treatment).
METHODS. Females were reared and handled as described for Fig. 1. The two-day-old females were given a 10-μl droplet of sugar water (20% sucrose) as described for Fig. 1. In addition, however, a 10-μl droplet of either vanilla (Nielsen-Massey) or chocolate (McCormick) extract was placed 5 mm away from the sugar water for simultaneous smelling while feeding (care was taken that the females did not touch the extract). Wasps were permitted feeding/smelling training sessions of 5 s and sessions were repeated 5 times for each female with 2-min intervals between sessions. Twenty minutes after the final training session, females were released in the flight chamber as described in Fig. 1 and their response observed to a choice of vanilla extract (10 μl) or chocolate (10 μl) suspended 8 cm apart at the centre of the up-wind end of the chamber.

In earlier studies, females presented with odours in association with hosts and host frass likewise linked these odours to hosts and flew to the odours in search of host resources[4,5]. We conducted tests to determine whether females could learn and

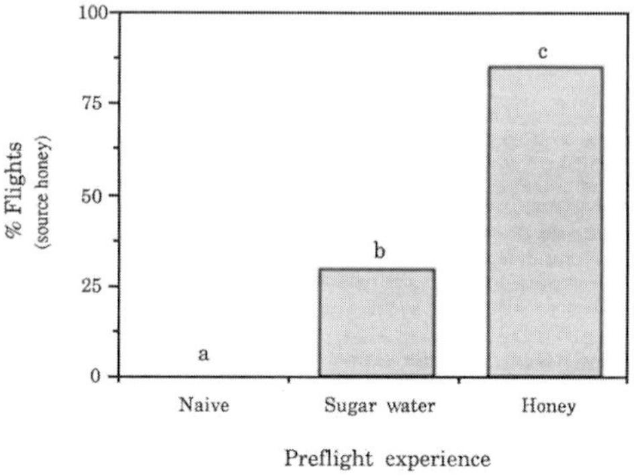

FIG. 1 Flight response of *M. croceipes* females to honey after no experience (naive), or feeding on sugar water or honey. Bars capped by different letters are significantly different, $P<0.01$, Waller-Duncan K-ratio t-test; minimum significant difference, 21.9%, $n=90$ wasps (5 replications, 6 wasps per replication per preflight treatment).
METHODS. *M. croceipes* were reared on larvae of *H. zea* fed an artificial diet. Females were allowed to mate and then held with a water supply in 30 × 30 × 20 cm cages for 2 days after emergence. The 2-day-old females were each exposed to a drop of 20% sucrose water (10 μl) or honey (Powers, wildflower) (50 mg) placed on the bottom of a polystyrene Petri dish (8 × 1.5 cm). Females were allowed to walk directly from a vial onto the test material and then to feed on the material for 5 seconds. This experience was repeated 3 times with 2-min intervals. The wind tunnel has been described[4]. All flight responses were tested at 26–28 °C at a wind speed of 56 cm s^{-1} and at a light intensity of 2,000 lux. As an attractant source 50 mg of honey was pipetted onto a piece of Whatman number 1 filter paper suspended at the centre of the up-wind end of the tunnel. Females, tested individually, were released from a 4-dram shell vial 80 cm downwind and in the plume from the odour source. The behaviour of the females in the wind tunnel was observed until a complete flight occurred or 5 min had elapsed. A complete response involved the host-seeking flight behaviour described by Drost *et al.*[4] and included approach and landing on the odour source. For each observation, a female was given two chances to make a successful flight. Flight tests were conducted 20 min post-experience.

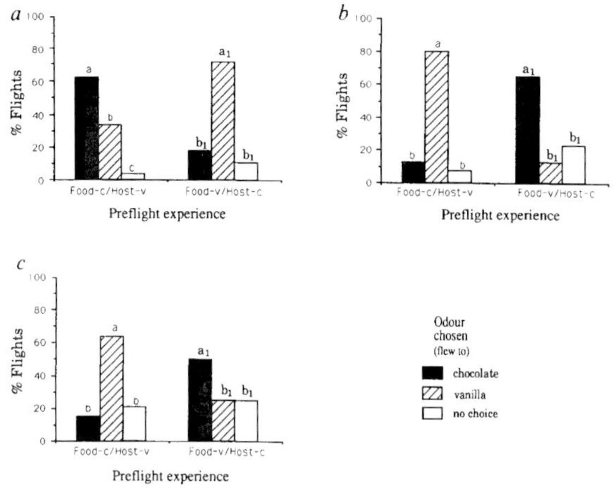

FIG. 3 Flight responses to vanilla or chocolate extract by hungry (*a*), well fed (*b*), or trained hungry then well fed (*c*) females, with preflight training indicated. Food-C/Host-V: females given training experience of chocolate-to-food and vanilla-to-hosts; Food-V/Host-C: females given training experience of vanilla-to-food and chocolate-to-hosts (the order of food-odour and host-odour training was alternated and the results combined). Bars within same tratment group capped by different letters are significantly different, $P<0.01$, Waller-Duncan *K*-ratio *t*-test; minimum significant difference, 10.5% (*a* and *b*) and 16.3% (*c*); $n=62$ wasps for each test group (6 replications, 6 wasps per replication per preflight treatment).

METHODS. Females were reared to 2 days old as described in Fig. 1. Hungry females were deprived of food (except water) until time for training and testing (2 days old). Well fed females were fed 20% sucrose water for 5 min when one day old. Training procedures were the same for hungry and well fed females. The training procedure to link odour (chocolate or vanilla) to food was as described in Fig. 2. The training procedure to link odour to hosts consisted of placing frass (1.4 mg from *H. zea* larvae fed on cowpea seedlings) and a third instar larvae on a filter paper plus 10 μl of either chocolate or vanilla extract. Wasps were allowed to antennate the frass and sting the larva. This training was repeated 3 times for each female with 2-min intervals in between. The order of training for the food-odour and host-odour was alternated evenly. Forty minutes after the last training the wasps' flight responses to chocolate versus vanilla were tested in the flight chamber as described in Fig. 2. The females used in Fig. 3*c* were each provided with a 3-min feeding session on 20% sucrose water 20 min after training and 20 min before testing.

differentially link two separate odours to their food and host foraging modalities and then if they could choose accurately between these learned cues on the basis of their current resource need. The females of one group (50% hungry and 50% well fed) were trained to associate chocolate with food and vanilla with hosts, whereas a second group of females (also 50% hungry and 50% well fed) were trained using these same two odours in reverse order. The training consisted of repeated presentation of the odour in association with either food or host (for details, see figure legends). The flight responses of hungry or well fed females trained in this way (Fig. 3*a* and *b*) were evaluated shortly (40 min) afterwards in a flight chamber by offering them a flight choice between vanilla and chocolate odours. The results in Fig. 3*a* and *b* indicate that the females were able to learn both odours successfully and could learn the combination in either direction (vanilla-to-hosts and chocolate-to-food, or chocolate-to-hosts and vanilla-to-food). Regardless of the direction of training, the hungry females showed a flight preference for the odour learned in connection with food (sucrose) and the well fed females preferred the odour learned as associated with hosts.

A final test was needed to ensure that the flight response was indeed a function of food and host needs, rather than a difference in how well the hungry or well fed wasps learned the odour associations. Therefore a group of hungry females were each trained to one of the host and food odour combinations (50% chocolate-to-food and vanilla-to-hosts; 50% vanilla-to-food and chocolate-to-hosts). Twenty minutes after training, each female was brought to the well fed state by a 3-min feeding session on pure aqueous sucrose solution, then their odour preferences were tested 20 min after the feeding (Fig. 3*c*). This group of well fed wasps, although trained while hungry, showed a preference for the host-associated odour like the well fed females (trained and tested while well fed) shown in Fig. 3*b*, in contrast to the hungry females (trained and tested while hungry) shown in Fig. 3*a*. These results confirm that the wasps readily learn both odour associations and can retrieve this information and respond in accordance with current need. Note that the alternative of training well fed wasps and testing them while hungry was not attempted. Memory decay during the extended period (~2 days) necessary between training and reaching a hungry state for testing makes this option impractical.

Learned responses to odours associated with hosts has been demonstrated for several types of parasitic wasp[6-8], but our results are the first to show that parasitic wasps associatively learn and respond to odours affiliated with their food. Further, this finding reveals how parasitic wasps like *M. croceipes* can locate their food as well as polyphagous hosts under highly variable conditions. The female olfactory system constantly monitors the array of odours encountered during foraging. The associative occurrence of the odours with the different resources (hosts or food) are perceived and the linkages learned. Linked odours become cues for seeking those host or food resources and the rank order of responses to these learned cues can be varied in accordance with the relative levels of need for the different resources.

The finding of these remarkable capabilities by solitary parasitic wasps for processing olfactory information and adapting it for foraging advantages indicates a system that may rival that described for social honeybees[9] and even rats[10]. This discovery increases our insight into invertebrate learning and ethology and should be of value in improving the use of parasititic wasps as biological control agents. □

Received 20 August; accepted 18 October 1990.

1. Leius, K. *Can. Entomol.* **42,** 369–376 (1960).
2. van Emden, H. F. *Entomol. Mon. Mag.* **98,** 265–270 (1963).
3. Vinson, S. B. in *Comprehensive Insect Physiology Biochemistry and Pharmacology* (eds Kerkut, G. A. & Gilbert, L. I.) 417–469 (Pergamon, Oxford, 1985).
4. Drost, Y. C., Lewis, W. J., Zanen, P. O. & Keller, M. A. *J. chem. Ecol.* **12,** 1247–1262 (1986).
5. Lewis, W. J. & Tumlinson, J. H. *Nature* **331,** 257–259 (1988).
6. Arthur, A. P. in *Semiochemicals, Their Role in Pest Control* (eds Nordlund, D. A., Jones, R. L. & Lewis, W. J. (Wiley, New York, 1981).
7. Vet, L. E. M. *Neth. J. Zool.* **33,** 225–248 (1983).
8. Vinson, S. B., Barfield, C. S. & Henson, R. D. *Physiol. Entomol.* **2,** 157–164 (1977).
9. Gould, J. L. in *The Biology of Learning* (eds Marler P. & Terrace, H. S.) 149–180 (Springer, Berlin, 1984).
10. Palmerino, C. C., Rusiniak, K. W. & Garcia, J. *Science* **208,** 753–755 (1980).

Proc. Natl. Acad. Sci. USA
Vol. 94, pp. 12243–12248, November 1997

Perspective

A total system approach to sustainable pest management

*W. J. Lewis**†, *J. C. van Lenteren*‡, *Sharad C. Phatak*§, *and J. H. Tumlinson, III*¶

*Insect Biology and Population Management Research Laboratory, Agricultural Research Service, United States Department of Agriculture, P.O. Box 748, Tifton, GA 31793; ‡Department of Entomology, Agricultural University, P.O. Box 8031, 6700 EH Wageningen, The Netherlands; §Department of Horticulture, Coastal Plain Experiment Station, University of Georgia, P.O. Box 748, Tifton, GA 31793; and ¶United States Department of Agriculture-Agricultural Research Service, Center for Medical, Agricultural, and Veterinary Entomology, P.O. Box 14565, Gainesville, FL 32604

Contributed by J. H. Tumlinson, III, August 13, 1997

ABSTRACT **A fundamental shift to a total system approach for crop protection is urgently needed to resolve escalating economic and environmental consequences of combating agricultural pests. Pest management strategies have long been dominated by quests for "silver bullet" products to control pest outbreaks. However, managing undesired variables in ecosystems is similar to that for other systems, including the human body and social orders. Experience in these fields substantiates the fact that therapeutic interventions into any system are effective only for short term relief because these externalities are soon "neutralized" by countermoves within the system. Long term resolutions can be achieved only by restructuring and managing these systems in ways that maximize the array of "built-in" preventive strengths, with therapeutic tactics serving strictly as backups to these natural regulators. To date, we have failed to incorporate this basic principle into the mainstream of pest management science and continue to regress into a foot race with nature. In this report, we establish why a total system approach is essential as the guiding premise of pest management and provide arguments as to how earlier attempts for change and current mainstream initiatives generally fail to follow this principle. We then draw on emerging knowledge about multitrophic level interactions and other specific findings about management of ecosystems to propose a pivotal redirection of pest management strategies that would honor this principle and, thus, be sustainable. Finally, we discuss the potential immense benefits of such a central shift in pest management philosophy.**

The therapeutic approach of killing pest organisms with toxic chemicals has been the prevailing pest control strategy for over 50 years. Safety problems and ecological disruptions continue to ensue (1), and there are renewed appeals for effective, safe, and economically acceptable alternatives (2). Considerable effort has been directed toward such alternatives, and new technology has been implemented and is still emerging. However, the major trend has been toward the use of modern chemistry and molecular biology to replace traditional pesticides with less hazardous chemicals or nontoxic biologically based products; but these means are still therapeutics. Thus, the classic treadmill effect in pursuit of remediation of the symptoms persists (2) while tolls due to pests grow higher by some estimates. Crop losses due to arthropods, diseases, and weeds, though disputed by some as a valid measure, have increased on a world basis from 34.9% in 1965 (3) to 42.1% in 1988–1990 (4) despite the intensification of pest control.

In this report, we argue that the central weakness in how we think about pest management as a component of agricultural systems has not been addressed. We must go beyond replacing toxic chemicals with more sophisticated, biologically based agents and re-examine the entire paradigm around the therapeutic approach, including how and why those therapeutics are used. Truly satisfactory solutions to pest problems will require a shift to understanding and promoting naturally occurring biological agents and other inherent strengths as components of total agricultural ecosystems and designing our cropping systems so that these natural forces keep the pests within acceptable bounds. Recent discoveries in multitrophic interactions (5) together with renewed emphasis on broader based ecosystem management (6) indicate powerful prospects for this direction. Although we address the subject primarily from a perspective of arthropod pests, similar cases can generally be made for other pests [see Cook *et al.* (7) for background information important to related views for other pests].

Premise of a Revised Approach

The underlying principle of our position is that components of agricultural ecosystems interact, and, through a set of feedback loops, maintain "balance" within functional fluctuating bounds. Furthermore, therapeutic interventions into these systems are met by countermoves that "neutralize" their effectiveness [see Flint and van den Bosch (8) and Cook and Baker (9) for an elegant discussion of this point]. We are taught this basic principle from our earliest training in ecology but often overlook it in practice for various reasons, including our tendency in science to divide things into specialized parts, i.e., to apply a reductionist approach. The basic principle for managing undesired variables in agricultural systems is similar to that for other systems, including the human body and social systems. On the surface, it would seem that an optimal corrective action for an undesired entity is to apply a direct external counter force against it. However, there is a long history of experiences in medicine and social science where such interventionist actions never produce sustainable desired effects. Rather, the attempted solution becomes the problem [See Waltzlawick *et al.* (10) for a discussion of this subject with coverage of underlying mathematical principles.] We find vivid examples to this end in the problems of addiction as a consequence of the use of drugs for treatment of pain or mental distress and black market crime as a repercussion to the use of prohibition as an intended solution for alcoholism. Thus, as a matter of fundamental principle, application of external corrective actions into a system can be effective only for short

The publication costs of this article were defrayed in part by page charge payment. This article must therefore be hereby marked "*advertisement*" in accordance with 18 U.S.C. §1734 solely to indicate this fact.

Abbreviations: IPM, Integrated Pest Management; Bt, *Bacillus thuringiensis*.

†To whom reprint requests should be addressed. e-mail: wjl@tifton.cpes.peachnet.edu.

term relief. Long term, sustainable solutions must be achieved through restructuring the system so that inherent forces that function via feedback mechanisms such as density dependence are added and/or function more effectively.

The foundation for pest management in agricultural systems should be an understanding and shoring up of the full composite of inherent plant defenses, plant mixtures, soil, natural enemies, and other components of the system. These natural "built in" regulators are linked in a web of feedback loops and are renewable and sustainable. The use of pesticides and other "treat-the-symptoms" approaches are unsustainable and should be the last rather than the first line of defense. A pest management strategy should always start with the question "Why is the pest a pest?" and should seek to address underlying weaknesses in ecosystems and/or agronomic practice(s) that have allowed organisms to reach pest status.

Attempts for Change: No Real Change

Throughout the debate on alternative methods for controlling pests, various ideas have been expressed and new approaches have emerged. Three subject areas, Biological Control, Integrated Pest Management (IPM), and Biotechnology, have achieved particular importance in our quest for better pest control strategies.

Biological Control. Biological control has a long history of use in pest management and has gained renewed interest because of problems encountered with the use of pesticides. The term "biological control" has been used, at times, in a broad context to encompass a full spectrum of biological organisms and biologically based products including pheromones, resistant plant varieties, and autocidal techniques such as sterile insects. The historical and more prevalent use of this term is restricted to use of natural enemies to manage populations of pest organisms.

Biological control has been spectacularly successful in many instances, with a number of pest problems permanently resolved by importation and successful establishment of natural enemies. These importation successes have been limited largely to certain types of ecosystems and/or pest situations such as introduced pests in perennial ecosystems. On the other hand, this approach has met with limited success for major pests of row crops or other ephemeral systems. In these situations, the problem is often not the lack of effective natural enemies but management practices and a lack of concerted research on factors that determine the success or failure of importation attempts in the specific agroecosystem setting. Thus, importation programs, to date, are largely a matter of trial and error based on experience of the individual specialists involved.

Conservation of natural enemies received more attention as part of a cultural management approach before the advent of synthetic pesticides. Since that time, this realm of biological control has been neglected. The term "conservation" tends to limit one's vision to a passive approach of acknowledging that natural enemies are valuable and should be harmed no more than necessary. It is important that we develop a more active approach that seeks to understand natural enemies and how they function as a part of the ecosystem and to promote their effectiveness by use of habitat management (landscape ecology) and other cultural management approaches.

Augmentation through propagation and release of natural enemies is an area of biological control that has received much attention in recent years. These efforts include research on *in vitro* and *in vivo* mass rearing technology and on transport and release methodology for area-wide population suppression and for field-to-field therapeutic treatments. Although the development of this technology is valuable, it is an extension of the treat-the-symptoms paradigm. In principle, natural enemies used in these methodologies are biopesticides, and the general approaches differ from conventional pesticidal applications only in the kind of products used. From this "product formulation" perspective and from our existing infrastructure, the major emphasis in augmentation schemes becomes focused on how to produce and transport a large number of natural enemies at a low cost. Less emphasis is placed on how natural enemies function and how we can promote their natural effectiveness.

In keeping with the historical therapeutic-based attitude and existing infrastructure, most concentrated efforts for biological control appear to be directed toward the "rear and release" augmentation, followed by importation and thirdly by conservation. This order of priorities should be reversed. First, we need to understand, promote, and maximize the effectiveness of indigenous populations of natural enemies. Then, based on the knowledge and results of these actions, we should fill any key gaps by importation. Finally, therapeutic propagation and releases should be used as a backup to these programs when necessary.

IPM. Throughout our quest for alternative pest control measures, the IPM concept has by far received the most attention as a comprehensive pest management approach. IPM has had a varied history, has been defined in many ways, and has been implemented under an array of different connotations. The term was first used as "integrated control" by Bartlett (11) and was further elaborated on by Stern *et al.* (12) in reference to the concept of integrating the use of biological and other controls in complementary ways. The term was later broadened to embrace coordinated use of all biological, cultural, and artificial practices (13). Subsequently, under the term "IPM," various authors have advocated the principle of incorporating the full array of pest management practices together with production objectives into a total systems approach. See Flint and van den Bosch (8) for a comprehensive and ecologically based discussion of this concept and the potential benefits of its implementation.

The principles discussed by Flint and van den Bosch (8) are, in our opinion, solid and on target. They make a thorough case for a comprehensive long term pest management program based on knowledge of an ecosystem that weighs economic, environmental, and social consequences of interventions. However, as translated into practice, IPM has been primarily a monitoring program in which thresholds are established and chemicals are used only on an as-needed basis. Much less emphasis has been placed on understanding and promoting inherent strengths within systems to limit pest populations through use of approaches such as landscape ecology. In other words, IPM programs have been operated with pesticide management objectives rather than pest management objectives. We hasten to add that their use has been of major benefit and has greatly reduced the quantity of pesticides used. Furthermore, activities remain underway to refocus IPM toward the achievement of its full objectives (14, 15). However, our point is that, again, the tendency has been to remain centered on a monitor and treat-the-symptoms approach vs. the more fundamental question of "Why is the pest a pest?"

Biotechnology. Although biotechnology is not a pest management approach as such, we include it because it is receiving major emphasis and is being geared to provide a wave of new products for pest management. In fact, many seem to view biotechnology as an innovative means for providing safe and effective tools that will essentially resolve pest management problems. Major technological advances in chemistry, biochemistry, behavior, neurophysiology, molecular genetics, and genetic engineering have resulted in an array of biorational products and materials that are less toxic and hazardous to humans and the environment than conventional pesticides. These products include genetically engineered plants for stronger resistance to pests, plants, and natural enemies with high tolerance to pesticides and sophisticated formulations and delivery methods for biopesticides, semiochemicals, and other new tools. The biorational/biologically based materials provided are potentially valuable advancements that have an

traits such as tissue-specific and damage-induced chemical defenses that work in harmony with natural systems.

Genetic engineering and other such technologies are powerful tools of great value in pest management. But, if their deployment is to be sustainable, they must be used in conjunction with a solid appreciation of multitrophic interactions and in ways that anticipate countermoves within the systems. Otherwise, their effectiveness is prone to neutralization by resistance in the same manner as with pesticides.

Therapeutics. Therapeutics have a valuable role in ecologically based pest management strategies, but they should be viewed as backups rather than as primary lines of defense. Also, therapeutics should be recognized as potentially disruptive and used as unobtrusively as possible. The key principle is that they should be geared toward bringing a pest organism into acceptable bounds with as little ecological disruption as possible. Synthetic products, natural products, and living organisms can be effective as therapeutics, and the fact that a product is natural and/or nontoxic does not necessarily mean it is less disruptive than synthetic products. The important thing is to work as much in harmony as possible with the system's inherent defenses.

A wide array of therapeutic products are available, and more are being developed with modern technology. A vast arsenal of natural products identified from plants, insects, and microorganisms is being synthesized and formulated for use as biopesticides. Semiochemicals such as sex pheromones and natural enemy attractants can be used as baits and lures to disrupt pest activity and promote natural enemy presence. Pathogens, parasitoids, as reared *in vivo* or *in vitro*, are available and are being touted as therapeutic tools. All of these organisms and/or their by-products are important biofriendly alternatives to toxic, broad spectrum, conventional pesticides. Still, our primary pest management tactic should be maximization of built in pest reduction features of an ecosystem. Therapeutic tools should be used as secondary backups. Overreliance on them will return pest management strategies to a treadmill situation (Fig. 1).

Another problem is the tendency to seek therapeutics that give us the quickest effect. Sales of biological insecticides amount to about $110 million annually, and Bt is the main product ($90 million). Generally, microbial organisms work slowly relative to synthetic pesticides. Therefore, industry has as first priority formulation of microbials to obtain faster kill and is less interested in long term pest reduction effects. Thus, the role that microbials could play in orchard and forest pest management, as well as in programs like control of grasshoppers in Sahelian, Africa, is neglected (39).

Retarded development of pests may be more desirable than quick kill in certain situations. For example, Bt products are considered unacceptable for controlling beet armyworms in cotton because of their slow killing action. Yet, some studies indicate that a slow kill may be more preferable when examined from a larger perspective. As indicated above, *C. marginiventris* is a key parasitoid for managing the beet armyworm and interventions should avoid disrupting this natural enemy. Beet armyworm larvae intoxicated by sublethal dosages of MVP (Mycogen, San Diego) (a Bt-derived biopesticide) experience retarded development and feeding and are subject to higher parasitism than nontreated beet armyworm larvae (40). In other words, an effective, nondisruptive way to manage a moderate beet armyworm outbreak may be to retard its development and damage while giving the parasitoids time to work, thereby strengthening the parasitoids' effect during subsequent generations. A similar effect was reported earlier for Bt and a parasitoid of gypsy moths (41). A quick kill may provide more immediate results but destroys a resource for parasitoids and limits their presence with subsequent generations of pests, thus leading to resurgence.

We must remember—our primary objective in pest management is not to eliminate a pest organism but to bring it into acceptable bounds. The role of therapeutics is not to replace natural systems. Rather, their role is to serve as complements while the system is temporarily out of balance. From that perspective, it is clear that interventions that interfere with the restoration of balance are counterproductive. Waage (39) suggests that biopesticides could form the "methadone of IPM," helping agroecosystems to recover from the habit of calendar spraying while we are redesigning and nurturing them to a more self-renewing capacity.

Potential Benefits

The benefits of a total system approach would be immense, directly to farming and indirectly to society. The approach takes into account impacts on our natural resources such as the preservation of flora and fauna, quality and diversity of landscape, and conservation of energy and nonrenewable resources. Long term sociological benefits would also emerge in areas of employment, public health, and well being of persons associated with agriculture (42, 43).

In The Netherlands, prototypes of various multidisciplinary, arable farming systems have been evaluated on a semi-practical scale (44). In 1979, a national experimental farm for the development and comparison of alternative farming systems was set up in Nagele (one of the "polders"). The size of the farm was 72 hectares (almost 300 acres). Among other studies, integrated and conventional farming practices were compared for seed potatoes, dry peas, carrots, onions, sugar beets, and winter wheat. Crop protection and other management practices with the integrated approach followed the basic principles discussed herein.

Over a 15-year period, pesticide use on these integrated farms was reduced over 90% (Fig. 2). They found that pesticides, and fertilizers, can be decreased through implementation of alternative practices based on intensified knowledge of the ecosystem. Artificial fertilizers are replaced by organic manure and effective use of crop residues. Insect, weed, and disease problems are reduced through natural control by the enriched natural enemy fauna, the use of weed-competitive or disease- and pest-resistant varieties (with an emphasis on durable systems for resistance), reduction of nitrogen fertilization, and judicious use of chemical pest control based on careful population sampling and decision thresholds. Results from these demonstration farms have been so encouraging that implementation of integrated farming is being enforced by the

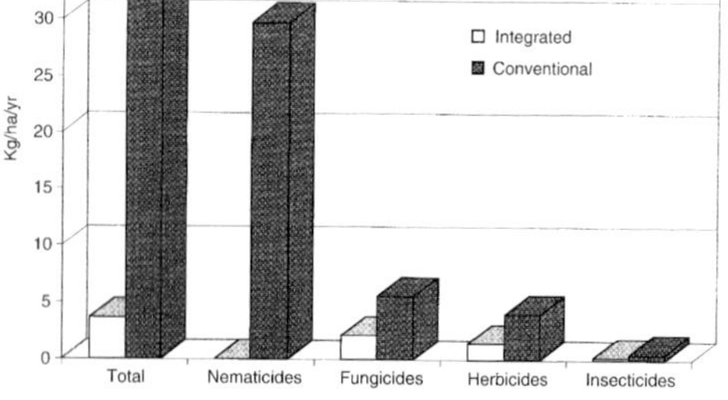

FIG. 2. Average use of pesticides (kilogram active ingredient/hectare/year) in conventional and integrated farming demonstrations in The Netherlands (1986–1990); after Wijnands and Kroonen-Backbier (43). ha, hectare.

Dutch Ministry of Agriculture to reduce environmental pollution and to create a firmer basis for survival of agriculture in the longer term.

Yields were somewhat lower on the demonstration farms but were compensated for by cost reduction through lower pesticide and fertilizer inputs. Thus, the net short term profits of the demonstration farms were equal to those of the conventional farms. We emphasize the short term economic aspect of sustainable farming because immediate profitability figures, along with the environmental concerns, are crucial to adoption of the practices. However, the eventual consequences of conventional farming are so severe, environmentally, socially, and economically, that it is wise to initiate changes even under situations in which short term economic benefits are marginal. Bio-friendly agriculture and good economics, over the long term, clearly go hand in hand.

CONCLUSION

Recent quests for effective, safe, and lasting pest management programs have been targeted primarily toward development of new and better products with which to replace conventional toxic pesticides. We assert that the key weakness with our pest management strategies is not so much the products we use but our central operating philosophy. The use of therapeutic tools, whether biological, chemical, or physical, as the primary means of controlling pests rather than as occasional supplements to natural regulators to bring them into acceptable bounds violates fundamental unifying principles and cannot be sustainable. We must turn more to developing farming practices that are compatible with ecological systems and designing cropping systems that naturally limit the elevation of an organism to pest status. We historically have sold nature short, both in its ability to neutralize the effectiveness of ecologically unsound methods as well as its array of inherent strengths that can be used to keep pest organisms within bounds. If we will but understand and work more in harmony with nature's checks and balances we will be able to enjoy sustainable and profitable pest management strategies, which are beneficial to all participants in the ecosystem, including humans.

We are grateful to Karen Idoine and Drs. Charlie E. Rogers (now deceased), G. T. Fincher, John Garcia, Bryan Beirne, Thomas Eisner, R. James Cook, and Fred Gould for providing important advice and suggestions throughout the drafting and revision of this paper.

1. Wright, L. (1996) *New Yorker*, Jan. 15th, 42–48, 50–53, 55.
2. Benbrook, C. M. (1996) *Pest Management at the Crossroads* (Consumer's Union, Yonkers, NY).
3. Cramer, H. H. (1967) *Bayer Pflanzenschutznachrichten* **20,** 1–523.
4. Oerke, E. C., Dehne, H. W., Shoenbeck, F. & Weber, A. (1994) *Estimated Losses in Major Food and Cash Crops* (Elsevier, Amstedam).
5. Tumlinson, J. H., Lewis, W. J. & Vet, L. E. M. (1993) *Sci. Am.* **268,** 100–106.
6. Altieri, M. (1995) *Agroecology. The Science of Sustainable Agriculture.* (Westview, Boulder, CO).
7. Cook, R. J., Gabriel, C. J., Kelman, A., Tolin, S. & Vidaver, A. K. (1995) *BioScience* **45,** 354–357.
8. Flint, M. L. & van den Bosch, R. (1981) *Introduction to Integrated Pest Management* (Plenum, New York).
9. Cook, R. J. & Baker, K. F. (1983) *The Nature and Practice of Biological Control of Plant Pathogens* (Am. Phytopathol. Soc., St. Paul, MN).
10. Watzlawick, P., Weakland, J. & Fisch, R. (1974) *Change: Principles of Problem Formation and Problem Resolution* (Norton, New York).
11. Barlett, B. R. (1956) *Agric. Chem.* **11,** 42–44.
12. Stern, V. M., Smith, R. F. van den Bosch, R. & Hagen, K. S. (1959) *Hilgardia* **29,** 81–101.
13. van den Bosch, R. & Stern, V. M. (1962) *Annu. Rev. Entomol.* **7,** 367–387.
14. Board on Agriculture, National Research Council, Committee on Pest and Pathogen Control Through Management of Biological Control Agents and Enhanced Natural Cycles and Processes (1996) *Ecologically Based Pest Management: New Solutions for a New Century* (National Academy Press, Washington, DC).
15. van Lenteren, J. C., Benuzzi, M., Nicoli, G. & Maini, S. (1992) in *Biological Control and Integrated Crop Protection: Toward Environmentally Safer Agriculture*, eds. van Lenteren, J. C., Minks, A. K. & de Ponti, O. M. B. (Purdoc, Wageningen, The Netherlands), pp. 77–89.
16. Edwards, C. A., Lal, R., Madden, P., Miller, R. H. & House, G. (1990) *Sustainable Agricultural Systems* (St. Lucie, Delray Beach, FL).
17. Blumberg, A. Y. & Crossley, D. A. (1982) *Agro-Ecosyst.* **8,** 247–253.
18. McPherson, R. M., Smith, J. C. & Allen, W. A. (1982) *Environ. Entomol.* **11,** 685–689.
19. Fleischer, S. J. & Gaylor, M. J. (1987) *Environ. Entomol.* **16,** 379–389.
20. Phatak, S. C. (1993) in *Cover Crops: Cotton Relay Cropping* (2nd Organic Growers Conference, Calif. Inst. Rural Studies, Davis, CA), pp. 269–274.
21. Ruberson, J. R., Herzog, G. A., Lambert, W. R. & Lewis, W. J. (1994) *Florida Entomol.* **77,** 440–453.
22. Bugg, R. L., Wäckers, F. L., Brunson, K. E., Dutcher, J. D. & Phatak, S. C. (1991) *J. Econ. Entomol.* **84,** 408–416.
23. Griffith, D. R., Mannering, J. V. & Box, J. E. (1986) in *No-Tillage and Surface-Tillage Agriculture: The Tillage Revolution,* eds. Sprague, M. A. & Triplett, G. B. (Wiley, New York), pp. 19–57.
24. Altieri, M. A. (1994) *Biodiversity and Pest Management in Agroecosystems,* (Food Products Press, New York).
25. Luna, J. M. & House, G. J. (1990) in *Sustainable Agricultural Systems,* eds. Edwards, C. A., Lal, R., Madden, P., Miller, R. H. & House, G. J. (St. Lucie, Delray Beach, FL), pp. 157–173.
26. Olkowski, W., Daar, S. & Olkowski, H. (1991) *Common Sense Pest Control* (Taunton, Newton, CT).
27. All, J. N. & Musick, G. J. (1986) in *No-Tillage and Surface-Tillage Agriculture: The Tillage Revolution,* eds. Sprague, M. A. & Triplett, G. B. (Wiley, New York), pp. 347–387.
28. Dicke, M. & Sabelis, M. W. (1988) *Neth. J. Zool.* **38,** 148–165.
29. Dicke, M., Sabelis, M. W., Takabayashi, J., Bruin, J. & Posthumus, A. (1990) *J. Chem. Ecol.* **16,** 3091–3118.
30. Turlings, T. C. J., Tumlinson, J. H. & Lewis, W. J. (1990) *Science* **250,** 1251–1253.
31. Turlings, T. C. J. & Tumlinson, J. H. (1992) *Proc. Natl. Acad. Sci. USA* **89,** 8399–8402.
32. Röse, U. S. R., Manukian, A., Heath, R. R. & Tumlinson, J. H. (1996) *Plant Physiol.* **111,** 487–495.
33. Loughrin, J. H., Manukian, A., Heath, R. R. & Tumlinson, J. H. (1995) *J. Chem. Ecol.* **21,** 1217–1227.
34. Lewis, W. J. & Takasu, K. (1990) *Nature (London)* **348,** 635–636.
35. Stapel, J. O., Cortesero, A. M. Demoraes, C. M., Tumlinson, J. H. & Lewis, W. J. (1997) *Environ. Entomol.* **26,** 617–623.
36. Ryan, C. A. & Farmer, E. E. (1991) *Annu. Rev. Plant Physiol. Plant Mol. Bio.* **42,** 651–743.
37. Pearce, G., Ryan, C. A. & Liljegren, D. (1988) *Planta* **175,** 527-531.
38. Gould, F. (1991) *Am. Sci.* **79,** 496–507.
39. Waage, J. K. (1994) *Biodiversity, Business and Biotechnology: Shaping the Future of Biological Control* (Wageningen Agricultural University, Wageningen, The Netherlands).
40. Soares, G. G. Jr., Lewis, W. J., Strong-Gunderson, J. M., Waters, D. J. & Hamm, J. J. (1994) *Proceedings of the Second Canberra Bacillus thuringiensis Meeting* (Commonwealth Scientific and Industrial Research Organization, Canberra, Australia).
41. Weseloh, R. M., Andreadis, T. G., Moore, R. E. B., Anderson, F. F., Dubois, N. R. & Lewis, F. B. (1983) *J. Invert. Pathol.* **42,** 99–103.
42. Vereijken, P., Edwards, C., El Titi, A., Fougeroux, A. & Way., M. (1986) *Proceedings of Workshop on Integrated Farming. Bull. I. O. B. C./W. P. R. S./IX* **2,** 1–34.
43. Wijnands, F. G. & Kroonen-Backbier, B. M. A. (1993) in *Modern Crop Protection: Developments and Perspectives,* ed. Zadoks, J. C. (Wageningen Pers, Wageningen, The Netherlands), pp. 227–234.
44. van Lenteren, J. C. (1997) in *Modern Agriculture and the Environment,* eds. Rosen, D., Tel-or, E. & Chen, Y. (Kluwer, Dordrecht), pp. 17–31.

Published in ATLANTA February 2006, 77-78, 96, 98,100

STING OPERATION

HOW A BUG EXPERT IN TIFTON TRAINS WASPS TO FIGHT IN THE WAR ON TERROR

BY LUKE DITTRICH, ILLUSTRATION BY CELIA JOHNSON

One day, when Joe Lewis was 11 or 12 years old, he snatched one of his father's roosters and smeared it with smut from a wash pot until its neck feathers were smudgy and dark. Joe's grandfather had told him that roosters recognize one another by the coloration of their neck feathers, so Joe had decided to find out what would happen if he messed with that coloration. This particular rooster was the top cock in the local pecking order, the one all the other roosters habitually deferred to in matters of food and females. When the boy returned the besmirched bird to its flock, what happened next was so exciting that he repeated the process many times in the weeks that followed.

"He'd go to take charge and take back his domain," Joe remembers, "and all these other roosters would go, 'Who are you?' That was one of my earliest scientific studies. I could spend all day watching rooster fights!" Joe Lewis is now 62 years old, and this is a story about one of his more recent scientific studies. This is also a story about screaming plants, chocoholic wasps and why the Department of Defense thinks educated insects might protect us from terrorists.

Joe grew up on 4.5 acres of cotton in South Mississippi, a place he now calls "the back end of the world." His father was a share cropper who did not know how to read or write. The family lived close to the soil, as farmers say. A mule pulled the plow. Geese cut the grass. Dogs helped the hunt. Joe plucked the fiber from the bolls. Life on the cotton patch was a complex balance of chores and processes partitioned between nature and man.

I. Joe has learned many things by studying bugs. Strangely enough, the bugs that he studies have also learned many things from Joe.

The cotton had enemies. If you had been there, in those fields, and you had knelt in the dirt, and you had looked closely at the underbellies of the plants, the enemies would have revealed themselves. You might, for example, have seen a small caterpillar the color of faded parchment, cutting at a leaf with serrated jaws, moving as slowly as a minute hand, ingesting healthy plant matter at one end and ejecting useless brown waste out the other. If you had watched long enough, the enemies of the plant's enemies would have revealed themselves too. A tiny wasp, even smaller than the caterpillar, might have lit upon the worm's back and impaled it with its ovipositor.

The two kinds of bugs, worms and wasps, had a common enemy as well. Every afternoon, Joe's father would tie a red handkerchief around his face, haul a sack of dust out into the field and coat his crops with DDT and other poisons, tainting wasp, worm and human alike. When Joe left the cotton patch, he went to Mississippi State University. One of the first books he read there was Rachel Carson's *Silent Spring*, a popular environmentalist work that excoriated chemical pesticides like those Joe's father used. Carson recommended an alternative, something called biological control, which, rather than killing your crop's enemies with artificial dust, would kill them with their own natural enemies. If Joe's father had employed such methods, he might have, for example, dealt with his bollworm problem by sprinkling his cotton plants with wasps, rather than DDT. The wasps would then have done what they do naturally: lay voracious offspring in the bollworms' bellies, stunting their growth and shortening their lives. The concept was not new. Chinese farmers purchased nests of mercenary ants to kill pests in their citrus groves as far back as the third century, and such methods eventually spread around the world, becoming ubiquitous by the 1800s. By the 20th century, however, the cheapness and relative effectiveness of chemical pesticides had almost made biological control obsolete. Joe found Carson's ideas hugely appealing. He particularly liked the idea that they might offer men like his father an alternative to the killing dust. As Joe soon learned, however, biological control was a process as primitive and unrefined as it was ancient. Humans, for all their complexity and smarts, knew very little about effectively harnessing the swarms of insects that surrounded, outnumbered, helped and hindered them. For example, everyone knew wasps killed bollworms, but nobody knew how a tiny wasp could find a tiny bollworm in the middle of a 4.5-acre cotton patch. If we could figure out what was guiding the wasps, maybe we could exploit that knowledge to help the wasps hunt, making them stronger allies. Until we learned such things, though, biological control methods were unlikely to ever replace cheap and dirty chemicals.

So Joe decided to study bugs.

In his Tifton laboratory, Joe Lewis uses wasps' natural attraction to scents and trains the bugs to be drawn to certain odors - chocolate, for instance, or explosives.

II. When he was a boy, back in the back end of the world, he knew only a little about the caterpillars he called bollworms and the tiny black wasps that preyed on them. Now he knows a lot more.

Dr. Joe Lewis is a big man with a broad, handsome face, floppy grey hair and a sort of exuberant politeness that probably serves him well in his role as vice mayor of Tifton, a small city in South Georgia surrounded by thousands of acres of cotton and peanuts. Politics is a part-time thing for Joe. His main job is with the United States Department of Agriculture's Agriculture Research Service, where he works as a behavioral entomologist at the USDA's Crop Protection and Management Research Laboratory in Tifton. When he was a boy, back in the back end of the world, he knew only a little about the caterpillars he called bollworms and the tiny black wasps that preyed on them. Now he knows a lot more. These days, he calls the bollworms *Helicoverpa zea* and the wasps *Microplitis croceipes*, and he writes papers about them with lofty titles like "Herbivore-Infested Plants Selectively Attract Parasitoids."

Today, when he describes a field of cotton, or a field of any other crop, he sounds like he's describing a single living organism. His descriptions are anthropomorphic. The cotton plants are the body, the bollworms are a virus, the wasps are white blood cells, and every part of the field is reliant upon every other part in an intricate and delicate balance. Joe has learned many things by studying bugs. Strangely enough, the bugs that he studies have also learned many things from Joe. Here is one thing Joe has learned by studying bugs: Plants scream.

When a bollworm slashes through the chlorophyllic flesh of a cotton plant, for example, the plant pumps a combination of 12 chemical compounds out through its pores.

Those chemicals are its scream, and only bugs hear it. If a wasp who is hungry for bollworms senses such a distress signal, it reacts, following the signal to its source, attacking the worm, defending the plant. The screams are very specific. The one just described might be translated into English as: "I am a cotton plant, and a bollworm is eating me!" Had the cotton plant been attacked by an aphid, the signal would have been different, and the wasp, who has no use for aphids, would have ignored it. In that case, a passing aphid-starved ladybug might instead respond to the plant's scream and feast.

The most important thing Joe has learned from bugs is that bugs can learn from Joe.

This was a surprise — not just to Joe but to the entire scientific community, which had long thought bugs to be hard-wired creatures that basically did everything by instinct. In the late 1980s, however, Joe and his former Mississippi State University classmate and long-time USDA colleague, Dr. J.H. Tumlinson, discovered that bugs, in fact, have quite malleable minds and are constantly learning and adapting to their changing circumstances. For example, if a female parasitic wasp encounters a bollworm on a stalk of corn, she will lay her egg in the worm and, while doing so, her antennae will energetically sample the cornstalk. When she flies off, the odors of the cornstalk and the satisfaction of egg-laying will be newly linked in her mind. The next time she detects cornstalk odors, she will track them to their source to see if she can find another host.

This is called "associative learning." It is exactly the same process that caused dogs to salivate when Pavlov rang his tuning fork, and it is a process that, before Joe came along, everyone thought was exclusive to mammals and other relatively large-brained creatures.

When Joe learned that bugs learn, he published a paper in the scientific journal *Nature* detailing his findings. The paper described a simple experiment in which Joe demonstrated he could teach a wasp to love vanilla. He did this by exposing a wasp to the smell of vanilla while it examined its naturally beloved bollworm feces. After this exposure, the wasp would become a vanilla-hunting fiend, flying straight toward the scent in a wind tunnel. To make sure this particular species of wasp didn't have some strange innate predisposition to vanilla, Joe repeated the experiment using a different scent. Soon, his trained wasps were flying just as fixedly toward the scent of chocolate.

Within the Department of Defense, there is something called the Defense Advanced Research Projects Agency, or DARPA. It used to be known as ARPA, which is why the Internet, which DARPA invented, used to be known as the ARPANET.

DARPA's mission, in its own words, is to "prevent technological surprise from harming our national security by sponsoring revolutionary, high-payoff research that bridges the gap between fundamental discoveries and their military use." Many DARPA-developed technologies — stealth fighters and global positioning satellites, for instance — have already changed the world. Many other technologies still in development — performance-enhancing exoskeletons, cognition-enhancing neural implants — will probably change the future. An important part of what DARPA does consists of keeping tabs on the latest breakthroughs in every scientific field. If DARPA hears of a new idea or theory that might have some sort of military application, it approaches the scientists behind the idea.

That is why, in the summer of 1999, DARPA approached Joe Lewis and asked whether he and his chocolate and vanilla-loving wasps would like to become defense contractors.

Humans are very bright, but our senses are very dull. Thousands of species of nonhuman creatures see better than us, hear better than us and detect smells better than us. We have always recognized the superior sensitivity of certain animals, and exploited it, utilizing these animals as what organizations like DARPA would call "biological sensors." Dogs are the most familiar example. The first domesticated dogs were basically organic burglar alarms. They could be counted on to bark at an intruder long before we could sense the stranger's approach. Eventually, we employed dogs for more sophisticated detection work. Starting in the seventh century A.D., we began training them to actively track other animals and humans. Starting in the last century, we also trained them to sniff out bombs, bongs and bodies.

Even with the recent invention of computerized "electronic noses" — devices that cost about $10,000 and attempt to replicate biological olfactory systems in silicon-dogs still possess the best readily available chemical-detecting technology on the planet. Our most sophisticated electronic noses are only about a tenth as sensitive as the nose of a trained canine. But although detection dogs are useful and effective, they are also big and expensive. They require years of training, experienced handlers and are impossible to use surreptitiously.

DARPA wanted Joe to investigate whether his trained wasps might, in essence, be smaller, faster, cheaper bloodhounds.

When DARPA first approached him, Joe was a little skeptical. "We knew we could train [wasps] to natural odors like vanilla and chocolate, but we didn't know whether we could train them to these totally unnatural things that had no part in their natural history, like explosives and nerve gases."

"But anyway, we said, 'We'll look into it.'"

"Do you want to smell it?"

Dr. Glen Rains, an agricultural engineer at the University of Georgia who has worked closely with Dr. Lewis in his wasp-training work, holds up a little brown bottle. The label indicates it comes from a chemical manufacturing company called Aldrich and contains something called cadaverine. Cadaverine is just what you might guess it is: a chemical compound released by human bodies when they decay. Believe it or not, it hardly smells at all.

Cadaverine is one of the dozens of compounds that Joe has trained his wasps to seek out. In fact, once he started on his DARPA work, Joe quickly discovered that there doesn't appear to be a single chemical compound that his wasps can't learn to love. Before long, he was churning out TNT-sniffing wasps, corpse-sniffing wasps and, yes, nerve gas–sniffing wasps. Best of all, his wasps appeared to have olfactory systems just as acutely sensitive as dogs and were able to detect quantities of any given chemical compound in ratios as low as four parts to a billion.

Once he'd demonstrated that his wasps could, in fact, become bloodhounds, Joe's next problem was figuring out how to harness them.

The way DARPA saw things was a little simplistic, Joe says. He recalls that his DARPA liaison wondered why they couldn't just release a few thousand bomb-sniffing wasps and follow them until they clouded around the head of, for example, someone with explosives hidden under his clothes. "We said, 'Well, wasps don't do that.' The thing about parasitic insects is, if you put them in a room, or any unnatural environment, they're distracted, and they'll just fly to the light."

What Joe needed was the wasp equivalent of a bloodhound's leash —something that would keep the insects under control and undistracted as they sniffed out whatever it was they were trained to sniff out.

It doesn't look like much.

For something that represents the culmination of more than a decade of science, for the prototype of what is perhaps the first viable insect-based biosensor, for a device that has the potential to detect four molecules of a specific chemical in a billion-molecule soup, the Wasp Hound looks, frankly, like a piece of junk.

It consists of a few bucks worth of white PVC plastic tubing, a $20 fan salvaged from a computer, a $5 USB cable, a $1 light-emitting diode, an $80 Webcam and some wires. The Wasp Hound's slightly superhero name comes courtesy of Dr. Rains, Joe's colleague, who, apart from being an agricultural engineer, is also a comic book fan with an office full of Spiderman memorabilia. One of Dr. Rains' students, Samuel Utley, programmed most of the device's software.

Here's how the Wasp Hound works:

1. Insert five pre-trained wasps.
2. Attach the USB cable to a laptop running a proprietary program called "Visual Cortex."
3. Turn the Wasp Hound on and move it to the item or location you wish to test.

That's it. If the wasps detect the chemical they have been trained to detect, they will cease their aimless milling and cluster around a hole in the bottom of the Wasp Hound. The hole is right above the fan, which sucks air inside. The computer program analyzes the Web cam's overhead view of the clustering wasps. When they cluster thickly enough, it sets off an alarm, indicating the presence of the chemical in question.

With properly trained wasps, the Wasp Hound can detect the merest whiff of TNT in a pile of suitcases. A different batch of wasps might pick up a smidgen of cadaverine emanating from beneath a field of wildflowers. Other wasps might notice a few molecules of aflatoxin, a dangerous toxin, in an otherwise robust crop of peanuts.

The Wasp Hound is about as compact as the best currently existing electronic nose, more effective and only about 2 percent as expensive. Dr. Rains sees a sweet sort of back-tracking in the superiority of wasp-driven chemical sensors over silicon-driven ones.

"With technology, you're always trying to get away from the way things used to be, but as we go further and further, we kind of start going back. Look at the medical field: Lots of doctors are going back to using maggots to help clean wounds! So you know, things kind of come back around. Nature's had millions of years to develop systems that are as efficient as possible."

Will bugs soon be battling bad guys?

Maybe.

DARPA might choose to enlist Joe's trained wasps in the war on terror. In the field, Wasp Hound–like devices could feasibly sniff out caches of explosives or send out warnings at the first hint of nerve gas. Now that DARPA has jump-started his research and gleaned what's possible, Joe's not even sure he'd know if the government decided to pursue future military applications. "They might step in again. We may or may not know. If they want to do something more, they'd probably want to classify it and do all of it in-house."

In the meantime, Joe and Dr. Rains and their respective institutions, the USDA and the University of Georgia, have applied for a patent on the Wasp Hound. They don't plan to produce and market a commercial version of the device themselves but, rather, would like an outside company to lease the patent and take over from there. There are still problems to work out to make the technology practical. The most pressing is how to go about mass-producing trained wasps. The current training process is done one wasp at a time and takes about 20 minutes to get a cartridge-filling quantity of five wasps ready to go. They also have to be retrained every 48 hours or so until the end of their lives, which typically comes about 21 days after they're born. Apart from the problem of mass producing these short-lived worker wasps, there's also the inevitable inertia that always attends the birth of a completely new and unknown sort of product.

"We're doing stuff in an area where there's no industry that currently exists," Joe says. "It's not like developing a new pesticide. There's no infrastructure in place that takes invertebrate organisms and uses them for devices for chemical detection. There's an incredible need in the whole area. But there's a lack of infrastructure."

Joe would just as soon leave the birthing of a new industry and all the associated headaches to someone else. He'd rather get back to doing what he does best: learning about bugs. And now that he's learned just about everything he can about wasps and proven that they can, in theory at least, double for dogs, he's decided to try teaching a new bug some new tricks.

Usually when Joe's father finished dusting the cotton plants with DDT, he'd go back to the house and sprinkle a little pile of the poison on the kitchen windowsill to keep mosquitoes at bay. DDT has since been banned, but as anyone who's ever seen a digital watch's plastic casing melt after an application of a DEET-based repellent knows, we still engage in potentially hazardous chemical warfare against mosquitoes all the time. A lot of people wish there were safer ways to fight them.

That's why, today, if you visit Joe's laboratory, you're more likely to find mosquitoes flying around in his wind tunnel than wasps. Mosquitoes, Joe discovered just last year, have malleable, trainable minds too, though you have to substitute blood for bollworm feces when you're teaching them. Joe envisions several mosquito-devastating scenarios growing out of his work. For example, you could use bait stations to train a small number of mosquitoes to be drawn to a certain scent, then count on the insects' natural "aggression" behavior —their tendency to follow one another when in pursuit of food — to give other nearby mosquitoes a nose for the same scent. By lacing these bait stations with chemicals that sterilize mosquitoes, you could nip an entire community in the bud.

Joe has just begun his work with mosquitoes, and he's hesitant to divulge much more about it, but the basic principles are the same as his work with wasps. "Once you can find out what they're doing and play games with them," he explains, "there are all sorts of ways you can intervene and alter them."

A half-century ago, a boy from the back end of the world grabbed a rooster and smeared its neck with smut. He had begun playing games with the creatures he shared his cotton patch with. He hasn't stopped yet.